SECTION I:
PERIODICAL LITERATURE AND ESSAYS

Sample Entry

HUMAN EXPERIMENTATION/ . . . / EMBRYOS AND FETUSES[1]

Lott, Jason P.; Savulescu, Julian.[2] Towards a global human embryonic stem cell bank.[3] *American Journal of Bioethics*[4] 2007 August[5]; 7(8)[6]: 37-44[7]. 27 refs.[8] NRCBL: 18.5.4; 15.1; 19.5; 14.5; 9.3.1[9]. SC: an[10].

Keywords: *embryonic stem cells; *tissue banks; cell lines; cloning; donors; embryo disposition; ethnic groups; economics; financial support; in vitro fertilization; incentives; informed consent; international aspects; justice; mandatory programs; minority groups; moral policy; nuclear transfer techniques; organ transplantation; policy analysis; racial groups; remuneration; resource allocation; scarcity; social discrimination; standards; stem cell transplantation; transplant recipients;[15] Proposed Keywords: embryo donation; haplotypes; tissue typing[16]

Abstract: An increasingly unbridgeable gap exists between the supply and demand of transplantable organs. Human embryonic stem cell technology could solve the organ shortage problem by restoring diseased or damaged tissue across a range of common conditions. However, such technology faces several largely ignored immunological challenges in delivering cell lines to large populations. We address some of these challenges and argue in favor of encouraging contribution or intentional creation of embryos from which widely immunocompatible stem cell lines could be derived. Further, we argue that current immunological constraints in tissue transplantation demand the creation of a global stem cell bank, which may hold particular promise for minority populations and other sub-groups currently marginalized from organ procurement and allocation systems. Finally, we conclude by offering a number of practical and ethically oriented recommendations for constructing a human embryonic stem cell bank that we hope will help solve the ongoing organ shortage problem.[18]

1. Subject heading: **HUMAN EXPERIMENTATION/ . . . / EMBRYOS AND FETUSES** *[Note: the ellipsis indicates that an intervening phrase of an exceptionally long subject heading has been omitted.]*
2. Author(s):**Lott, Jason P.; Savulescu, Julian.**
3. Title of article: Towards a global human embryonic stem cell bank.
4. Title of journal: *American Journal of Bioethics*
5. Date of publication: 2007 August
6. Volume and issue number (if available): 7(8)
7. Pagination: 37-44
8. References: 27 refs. (optional)
9. NRCBL (all classification numbers): 18.5.4; 15.1; 19.5; 14.5; 9.3.1
10. SC (Subject Captions): an [analytical] (optional)
11. Identifiers: (optional; not present here)
12. Note: additional information (optional; not present here)
13. Conference: (optional; not present here)
14. Comments: information about related publications (optional; not present here)
15. Keywords: *embryonic stem cells; *tissue banks; etc. (optional)
16. Proposed Keywords: embryo donation; haplotypes; tissue typing (optional)
17. Keyword Identifiers: (optional; not present here)
18. Abstract: An increasingly unbridgeable gap exists between the supply and demand of transplantable organs. Human embryonic stem cell technology . . . . (optional)

BIBLIOGRAPHY
OF
BIOETHICS

BIBLIOGRAPHY OF BIOETHICS

Volume 34

Editors

LeRoy Walters

Tamar Joy Kahn

Doris Mueller Goldstein

Associate Editors

Richard M. Anderson

Laura Jane Bishop

Martina Darragh

Harriet Hutson Gray

Lucinda Fitch Huttlinger

Patricia C. Martin

Hannelore S. Ninomiya

Anita Lonnes Nolen

Susan Cartier Poland

Managing Editor

Mara R. Snyder

Editorial Production

Roxie France-Nuriddin

KENNEDY INSTITUTE OF ETHICS
GEORGETOWN UNIVERSITY
Box 571212
WASHINGTON, DC 20057-1212

Publication of the annual *Bibliography of Bioethics* is a project of the Kennedy Institute of Ethics, Georgetown University. This book benefits from acquisition, classification, and database development activities funded by Contract HHSN276200553501C with the National Library of Medicine, and by Grant 5 P41 HG01115 from the National Human Genome Research Institute. It also reflects the contributions of The Anderson Partnership; Max M. and Marjorie B. Kampelman; and the National Endowment for the Humanities.

ISSN 0363-0161
ISBN 978-1-883913-15-1

This volume of the *Bibliography of Bioethics*
is dedicated to the memory of
our dear colleague and friend
Frances Amitay Abramson
January 13, 1933 - August 10, 2007

Associate Editor of the
Bibliography of Bioethics, 1980-2007
Library and Information Services
Kennedy Institute of Ethics

Contents

Staff xiii

Editorial Advisory Board xv

Introduction 3

> The Field of Bioethics
>
> The Scope of the Bibliography
>
> Arrangement of the Bibliography
>
> The Bibliography of Bioethics: History and Current Availability on the
>   World Wide Web
>
> Acknowledgments
>
> Distribution of the Bibliography
>
> International Bioethics Exchange Project

Section I: Periodical Literature and Essays — Subject Entries

> Abortion 13
>
> > *Legal Aspects*
> > *Moral and Religious Aspects*
> > *Social Aspects*
>
> Advance Directives 20
>
> AIDS 22
>
> > *Confidentiality*
> > *Human Experimentation*
> > *Legal Aspects*
>
> Animal Experimentation 26
>
> Artificial Insemination and Surrogate Mothers 30
>
> Assisted Suicide 32
>
> Behavior Control 35
>
> Behavioral Genetics 36
>
> Behavioral Research 36
>
> Bioethics and Medical Ethics 38
>
> > *Commissions*
> > *Education*
> > *History*
> > *Legal Aspects*
> > *Philosophical Aspects*

Religious Aspects

Biomedical Research 61

    Research Ethics and Scientific Misconduct
    Social Control of Science and Technology

Blood Banking, Donation, and Transfusion 77

Capital Punishment 78

Care for Specific Groups 79

    Aged
    Fetuses
    Indigents
    Mentally Disabled
    Minorities
    Minors
    Substance Abusers
    Women

Chimeras and Hybrids 99

Cloning 102

    Legal Aspects

Codes of Ethics 109

Confidentiality 110

Contraception 115

Cryobanking of Sperm, Ova, and Embryos 118

Cultural Pluralism 119

Death and Dying 119

    Attitudes to Death
    Determination of Death
    Terminal Care
        Minors

Drug Industry 127

Electroconvulsive Therapy 135

Enhancement 136

Ethicists and Ethics Committees 137

Eugenics 142

Euthanasia and Allowing to Die 145

    Legal Aspects
    Minors

Philosophical Aspects
Religious Aspects

Gene Therapy 166

Genetic Counseling 168

Genetic Databases and Biobanks 176

Genetic Engineering and Biotechnology 182

Genetic Enhancement 191

Genetic Privacy 193

Genetic Research 197

Genetic Screening 206

   Legal Aspects

   Socioeconomic Aspects

Genetically Modified Organisms and Food 220

Genetics and Genomics 224

Genetics and Human Ancestry 229

Genome Mapping and Sequencing 232

Health Care 233

   Health Care Economics
      Managed Care Programs
   Health Care Quality

Health, Concept of 244

Human Experimentation 245

   Ethics Committees and Policy Guidelines
      Legal Aspects
   Informed Consent
   Regulation
   Special Populations
      Aged and Terminally Ill
      Embryos and Fetuses
         Legal Aspects
         Philosophical and Religious Aspects
      Foreign Nationals
      Mentally Disabled
      Minors
      Prisoners
      Women

In Vitro Fertilization 299

Informed Consent 302

Incompetents
Minors

International Health and Human Rights 313

International Migration of Health Care Professionals 315

Involuntary Commitment 315

Journalism and Publishing 317

Malpractice and Professional Misconduct 320

Medical Education 321

Mental Health, Concept of 325

Mental Health Therapies and Neurosciences 326

Nanotechnology 335

Nursing Ethics and Philosophy 337

Organ and Tissue Transplantation 342
   Allocation
   Donation and Procurement
      Economic Aspects
      Legal Aspects
   Xenotransplantation

Patents 356

Patient Relationships 359

Pharmacogenetics 367

Philosophy of Medicine 368

Professional Ethics 371

Professional Professional Relationship 375

Psychopharmacology 376

Psychotherapy 380

Public Health 381

Quality and Value of Life 383

Reproductive Technologies 389

Resource Allocation 395

Right to Health Care 399

Sex Determination 402

Sexuality 404

Sociology of Medicine 405

Stem Cell Research 406

Telemedicine and Informatics 420

Torture, Genocide, and War Crimes 421

Treatment Refusal 422

    Mentally Ill

Truth Disclosure 423

War and Terrorism 426

Section II: Periodical Literature and Essays — Author Index 431

Section III: Monographs — Subject Entries 641

    Monographs: Contents

Section IV: Monographs — Title Index 691

Staff

Editorial Advisory Board

INTRODUCTION

INTRODUCTION

The Field of Bioethics

Bioethics can be defined as the systematic study of value questions that arise in health care delivery and in biomedicine. Specific bioethical issues that have recently received national and international attention include euthanasia, assisted suicide, new reproductive technologies, cloning, human experimentation, genetic engineering, the neurosciences, abortion, informed consent, acquired immunodeficiency syndrome (AIDS), organ donation and transplantation, and managed care and other concerns in the allocation of health care resources.

As this list of topics suggests, the field of bioethics includes several dimensions. The first is the ethics of the professional patient relationship. Traditionally, the accent has been on the duties of health professionals–duties that, since the time of Hippocrates, have frequently been delineated in codes of professional ethics. In more recent times the rights of patients have also received considerable attention. Research ethics, the study of value problems in biomedical and behavioral research, constitutes a second dimension of bioethics. During the 20th century and the start of the 21st century, as both the volume and visible achievements of such research have increased, new questions have arisen concerning the investigator-subject relationship and the potential social impact of biomedical and behavioral research and technology. In recent years a third dimension of bioethics has emerged–the quest to develop reasonable public policy guidelines for both the delivery of health care and the allocation of health care resources, as well as for the conduct of research.

No single academic discipline is adequate to discuss these various dimensions of bioethics. For this reason bioethics has been, since its inception in the late 1960s, a cross-disciplinary field. The primary participants in the interdisciplinary discussion have been physicians and other health professionals, biologists, psychologists, sociologists, lawyers, historians, and philosophical and religious ethicists.

During the past thirty-three years there has been a rapid growth of academic, professional, and public interest in the field of bioethics. One evidence of this interest is the establishment of numerous research institutes and teaching programs in bioethics, both in the United States and abroad. Professional societies, federal and state legislatures, and the courts have also turned increasing attention to problems in the field. In addition, there has been a veritable explosion of literature on bioethical issues.

The literature of bioethics appears in widely scattered sources and is reported in diverse indexes which employ a bewildering variety of subject headings. This annual *Bibliography* is the product of a unique information retrieval system designed to identify the central issues of bioethics, to develop a subject classification scheme appropriate to the field, and to provide comprehensive, cross-disciplinary coverage of current English-language materials on bioethical topics.

Volume 34 of the *Bibliography* contains one year's worth of the literature garnered by this comprehensive information system. Specifically, it includes selected citations that were acquired in 2007 by two projects at the Kennedy Institute of Ethics: the National Reference Center for Bioethics Literature (NRCBL) and the National Information Resource on Ethics & Human Genetics (NIREHG).

The Table of Contents includes a list of subject headings used to arrange the citations. Most citations are listed once, under their primary subject heading. Classification numbers at the end of each citation represent additional topics covered by the publication. These classification numbers are drawn from the NRCBL's Classification Scheme, reproduced on the inside front cover. The inside back cover offers a key to going from brief citations in Section II to more complete citations in Section I.

BIBLIOGRAPHY OF BIOETHICS
The Scope of the Bibliography

This thirty-fourth volume of the *Bibliography of Bioethics* includes materials which discuss the ethical aspects of the following major topics and subtopics:

BIOETHICS, MEDICAL ETHICS, AND
PROFESSIONAL ETHICS
 Codes of Ethics
 Commissions
 Education
 Ethicists and Ethics Committees
 History
 Nursing Ethics and Philosophy
 Philosophy of Medicine
 Professional Ethics
 Quality and Value of Life
DEATH AND DYING
 Advance Directives
 Assisted Suicide
 Attitudes to Death
 Capital Punishment
 Determination of Death
 Euthanasia and Allowing to Die
 Terminal Care
GENETICS AND GENOMICS
 Behavioral Genetics
 Chimeras and Hybrids
 Eugenics
 Gene Therapy
 Genetic Counseling
 Genetic Databases and Biobanks
 Genetic Engineering and Biotechnology
 Genetic Enhancement
 Genetic Privacy
 Genetic Research
 Genetic Screening
 Genetically Modified Organisms and Food
 Genetics and Human Ancestry
 Genome Mapping and Sequencing
 Patents
 Pharmacogenetics
HEALTH CARE AND PUBLIC HEALTH
 AIDS
 Blood Banking, Donation and Transfusion
 Care for Specific Groups
 Drug Industry
 Health, Concept of
  Mental Health, Concept of
 Health Care
  Health Care Economics

 Health Care Quality
 Organ and Tissue Transplantation
 Public Health
 Resource Allocation
 Right to Health Care
 Sexuality
 Telemedicine and Informatics
MENTAL HEALTH THERAPIES AND
NEUROSCIENCES
 Behavior Control
 Electroconvulsive Therapy
 Involuntary Commitment
 Psychopharmacology
 Psychotherapy
PATIENT RELATIONSHIPS
 Confidentiality
 Informed Consent
 Treatment Refusal
 Truth Disclosure
REPRODUCTION AND REPRODUCTIVE
TECHNOLOGIES
 Abortion
 Artificial Insemination and Surrogate Mothers
 Cloning
 Contraception
 Cryobanking of Sperm, Ova, and Embryos
 In Vitro Fertilization
 Sex Determination
RESEARCH
 Animal Experimentation
 Behavioral Research
 Biomedical Research
 Enhancement
 Human Experimentation
 Nanotechnology
 Research Ethics and Scientific Misconduct
 Social Control of Science and Technology
 Stem Cell Research
SOCIOLOGY OF MEDICINE
 Cultural Pluralism
 Journalism and Publishing
 Malpractice and Professional Misconduct
 Medical Education
 Professional Professional Relationship

INTRODUCTION

WAR AND HUMAN RIGHTS ABUSES
    International Health and Human Rights
    International Migration of Health Care Professionals

Torture, Genocide, and War Crimes
War and Terrorism

This volume of the *Bibliography* cites 5,955 documents (primarily in English) that discuss ethical and related public policy aspects of the topics and subtopics listed above. Documents cited in this volume include journal and newspaper articles, laws, court decisions, monographs, and chapters in books. Most of the documents listed were published since 2004. In the *Periodical Literature and Essays* section, for example, 3,037 of the 5,201 entries were published in 2007; 1,316 in 2006; and 373 in 2005; therefore, 91 per cent of the literature cited in Section I was published from 2005 to 2007.

A cross-disciplinary monitoring system has been devised in an effort to secure documents falling within the subject-matter scope outlined above. Among the reference tools and databases searched for pertinent citations are the following:

AGRICOLA
All England Law Reports (subject index)
ATLA Religion Database
Choice
Cumulative Index to Nursing and Allied Health Literature
    (CINAHL)
Current Contents: Social and Behavioral Sciences
Digital Dissertations and Theses (UMI Proquest)
Dominion Law Reports (subject index)
ERIC
GPO Access
Library Journal
Mental and Physical Disability Law Reporter

Month in Review (GAO reports and other publications)
New Titles in Bioethics
NLM Catalog
PAIS International
Philosopher's Index
POPLINE
PsycInfo
PUBMED
Social Sciences Index
Sociological Abstracts
Specialty Law Digest: Health Care
Tarlton Law Library Legal Bibliography Series
WorldCat

In addition, the *Bibliography* staff directly monitors hundreds of web sites on an ongoing basis as well as 184 journals and newspapers for articles falling within the scope of bioethics. Those preceded by an asterisk (*) have given permission for abstracts to be included in this volume. It is important to note that the journal articles cited in this volume are actually drawn from many more journals than those listed below.

Academic Medicine
*Accountability in Research
Agriculture and Human Values
AIDS and Public Policy Journal
America
*American Journal of Bioethics
*American Journal of Law and Medicine
American Journal of Nursing
American Journal of Psychiatry
*American Journal of Public Health
Annals of Health Law
*Annals of Internal Medicine
APA Newsletter on Philosophy and Medicine
*Archives of Internal Medicine
ATLA: Alternatives to Laboratory Animals
*Bioethics
Bioethics Forum
BMC Medical Ethics [electronic resource]
*BMJ (British Medical Journal)

British Journal of Nursing
*Cambridge Quarterly of Healthcare Ethics
Canadian Medical Association Journal
CCAR Journal
Cerebrum
*Christian Bioethics
Christian Century
Community Genetics
Criminal Justice Ethics
DePaul Journal of Health Care Law
Developing World Bioethics
Dolentium Hominum
Environmental Ethics
Ethical Human Psychology and Psychiatry
Ethical Perspectives
Ethical Theory and Moral Practice
*Ethics
*Ethics and Behavior
Ethics and Intellectual Disability

5

BIBLIOGRAPHY OF BIOETHICS

Ethics and Medicine

Ethics and Medics

*Eubios Journal of Asian and International Bioethics

European Journal of Health Law

First Things: A Monthly Journal of Religion and Public Life

Formosan Journal of Medical Humanities

Free Inquiry

Genetic Testing

Genetics in Medicine

GeneWatch

Georgetown Journal of Legal Ethics

*Hastings Center Report

Health Affairs

Health and Human Rights

*Health Care Analysis

Health Care Ethics USA [online]

Health Law in Canada

Health Law Journal

Health Law Review

Health Matrix (Cleveland)

Health Policy [ISSN 0168-8510]

Health Progress

*HEC Forum

Human Genome News

Human Life Review

Human Reproduction

Human Reproduction and Genetic Ethics

Human Research Report

Humane Health Care [electronic resource]

Hypatia

IDHL: International Digest of Health Legislation [online]

Indian Journal of Medical Ethics

International Journal of Applied Philosophy

*International Journal of Bioethics (Journal International de Bioéthique)

International Journal of Health Services

International Journal of Law and Psychiatry

International Journal of Technology Assessment in Health Care

IRB: Ethics and Human Research

Issues in Law and Medicine

Issues in Science and Technology

*JAMA

Jewish Medical Ethics and Halacha

JONA's Healthcare Law, Ethics, and Regulation

Journal of Advanced Nursing

Journal of Applied Animal Welfare Science

Journal of Applied Philosophy

*Journal of Bioethical Inquiry

Journal of Biolaw and Business

*Journal of Clinical Ethics

Journal of Contemporary Health Law and Policy

Journal of Ethics

Journal of General Internal Medicine

Journal of Genetic Counseling

Journal of Halacha and Contemporary Society

Journal of Health Care Law and Policy

*Journal of Health Politics, Policy and Law

Journal of Information Ethics

Journal of Intellectual Disability Research

Journal of Law and Health

Journal of Law and Religion

*Journal of Law, Medicine and Ethics

Journal of Legal Medicine

*Journal of Medical Ethics

Journal of Medical Genetics

*Journal of Medical Humanities

*Journal of Medicine and Philosophy

Journal of Moral Education

Journal of Nursing Administration

Journal of Nursing Law

Journal of Palliative Care

Journal of Philosophy, Science and Law

Journal of Professional Nursing

Journal of Psychiatry and Law

Journal of Public Health Policy

Journal of Religion and Health

Journal of Religious Ethics

Journal of Social Philosophy

Journal of the American Academy of Psychiatry and the Law

Journal of the American College of Dentists

Journal of the American Geriatrics Society

Judaism

*Kennedy Institute of Ethics Journal

*Lancet

Law and the Human Genome Review (Revista de Derecho y Genoma Humano)

Legal Medical Quarterly

Linacre Quarterly

Literature and Medicine

Medical Ethics & Bioethics (Medicinska Etika & Bioetika)

Medical Humanities

Medical Humanities Review

Medical Law International

Medical Law Review

Medicine and Law

Medicine, Conflict and Survival

Medicine, Health Care and Philosophy

Mental Retardation [ISSN 0047-6765]

*Milbank Quarterly

Minnesota Medicine

Monash Bioethics Review

Nanoethics
*National Catholic Bioethics Quarterly
*Nature
Nature Biotechnology
Nature Medicine
NCEHR Communiqué (National Council on Ethics in Human Research)
New Atlantis
*New England Journal of Medicine
New Genetics and Society
New Scientist
New York Times
Newsweek
Notizie de Politeia
Notre Dame Journal of Law, Ethics and Public Policy
*Nursing Ethics
Omega: Journal of Death and Dying
Online Journal of Issues in Nursing
Origins
Perspectives in Biology and Medicine
Perspectives on the Professions: Ethical & Policy Issues
Pharos
*Philosophy and Public Affairs
Philosophy and Public Policy Quarterly
Politics and the Life Sciences

Professional Ethics Report
Protecting Human Subjects
Psychiatric Services
Public Affairs Quarterly
Romanian Journal of Bioethics (Revista Romana de Bioetica)
*Science
Science and Engineering Ethics
Science as Culture
Science, Technology, and Human Values
Sh'ma
Social Justice Research
Social Philosophy and Policy
*Social Science and Medicine
Social Theory and Practice
Society and Animals
*Theoretical Medicine and Bioethics
Tradition
Update (Loma Linda University Ethics Center)
U.S. News and World Report
Virtual Mentor: Ethics Journal of the American Medical Association [electronic resource]
Washington Post
Women's Health Issues
Yale Journal of Health Policy, Law, and Ethics

All documents cited by the *Bibliography* are in the collection of the NRCBL.

Arrangement of the *Bibliography*

This volume of the *Bibliography of Bioethics* is divided into five parts:

1. Introduction
2. Section I: Periodical Literature and Essays — Subject Entries
3. Section II: Periodical Literature and Essays — Author Index
4. Section III: Monographs — Subject Entries
5. Section IV: Monographs — Title Index.

Sections 2 and 4 constitute the core of the *Bibliography*.

Section 1: Periodical Literature and Essays — Subject Entries

This Section, one of the two main parts of the *Bibliography*, contains usually one entry for each of the documents selected by the bioethics information retrieval system during the preceding year. In Volume 34 of the *Bibliography*, entries for 5,201 documents have been included in the Section. The format of these documents is as follows:

Journal articles	3,853
Essays in books	979
Newspaper articles	223
Reports and grey literature	129
Legal documents	17

Section I is organized under 78 major subject headings, of which 14 are further divided by subheadings. Each subheading is separated from the major subject term by a slash.

BIBLIOGRAPHY OF BIOETHICS

Readers of the *Bibliography* should first scan the alphabetic list of subject headings in the Table of Contents to determine where citations of interest to them are likely to be found.

Section I includes cross references of two types. *See* cross references lead the reader from terms that are not used as subject headings to terms that are used. *See also* cross references suggest additional subject headings where the reader may find citations of related interest.

Citations appear alphabetically by author, with anonymous citations at the beginning of the section, sorted alphabetically by title. Entries with both corporate and personal authors are sorted by the corporate author. As explained below, the citations are accompanied by NRCBL Classification Scheme numbers as well as, in some cases, Subject Captions denoting approach or content. Subject Caption definitions can be found on page footers. Abstracts are again included in this volume. Several optional fields provide additional information: identifiers (such as persons, places, organizations, acronym equivalents), conference information, comments regarding related publications, and general notes.

With the support of a grant from the U.S. National Human Genome Research Institute, bibliographers began indexing the genetics literature early in 2007 using the *Bioethics Thesaurus for Genetics* (see http://bioethics.georgetown.edu/nirehg/index.htm), a controlled vocabulary developed specifically for this project, using the *Bioethics Thesaurus*, originally developed for the BIOETHICSLINE database, as its foundation. The indexed citations are accompanied by keywords, proposed keywords, and keyword identifiers (usually proper nouns) in Section I. In addition to enhancing the representation of the content of each indexed citation, the keywords are used to generate more precise subject categories for the genetics literature. New subject headings appearing in this volume are: CHIMERAS AND HYBRIDS, GENETIC DATABASES AND BIOBANKS, GENETIC ENGINEERING AND BIOTECHNOLOGY, GENETIC ENHANCEMENT, GENETICALLY MODIFIED ORGANISMS AND FOOD, and PHARMACOGENETICS.

Now eighteen data elements may appear in a citation. A sample subject heading and entry for a journal article follow:

> **HUMAN EXPERIMENTATION/ . . . / EMBRYOS AND FETUSES** [1]
> **Lott, Jason P.; Savulescu, Julian.**[2] Towards a global human embryonic stem cell bank.[3] *American Journal of Bioethics*[4] 2007 August[5]; 7(8)[6]: 37-44[7]. NRCBL: 18.5.4; 15.1; 19.5; 14.5; 9.3.1 [8]. SC: an [9].
>> Keywords: *embryonic stem cells; *tissue banks; cell lines; cloning; donors; embryo disposition;[15] Proposed Keywords: embryo donation; haplotypes; tissue typing[16]
>> Abstract: An increasingly unbridgeable gap exists between the supply and demand of transplantable organs. Human embryonic stem cell technology could solve the organ shortage problem by restoring diseased or damaged . . . .[18]

1. Subject heading: **HUMAN EXPERIMENTATION/ . . . / EMBRYOS AND FETUSES** *[Note: the ellipsis indicates that an intervening phrase of an exceptionally long subject heading has been omitted.]*
2. Author(s): **Lott, Jason P.; Savulescu, Julian.**
3. Title of article: Towards a global human embryonic stem cell bank.
4. Title of journal: *American Journal of Bioethics*
5. Date of publication: 2007 August
6. Volume and issue number (if available): 7(8)
7. Pagination: 37-44
8. References: 27 refs. (optional)
9. NRCBL (all classification numbers): 18.5.4; 15.1; 19.5; 14.5; 9.3.1
10. SC (Subject Captions): an [analytical] (optional; not present here)
11. Identifiers: (optional; not present here)
12. Note: additional information (optional; not present here)
13. Conference: (optional; not present here)
14. Comments: information about related publications (optional; not present here)
15. Keywords: *embryonic stem cells; *tissue banks; . . . (optional)

16. Proposed Keywords: embryo donation; haplotypes; tissue typing (optional)
17. Keyword Identifiers: (optional; not present here)
18. Abstract: An increasingly unbridgeable gap . . . . (optional)

The journal article is the most prevalent publication type. The title field may be augmented by terms in square brackets which indicate additional aspects of the document, such as: letter, editorial, and news. The complete NRCBL Classification Scheme can be found on the inside front cover, and the Subject Captions equivalents are on alternating footers in Section I. The inside back cover displays the Subject Heading Key for Section II, leading the reader from the primary, i.e. first, NRCBL number to the corresponding Subject Heading(s) in Section I. Most citations appear only once in this volume.

Section II: Periodical Literature and Essays — Author Index

Citations in the Author Index are followed by the primary NRCBL Classification Number (Subject). Citations that have no personal or corporate author are listed at the end of the Author Index for Periodical Literature and Essays under ANONYMOUS. The two-page SUBJECT HEADING KEY FOR SECTION II appears on the inside back cover; it provides subject heading equivalents in Section I for the subject numbers appearing at the end of each citation in Section II.

Section III: Monographs — Subject Entries

These records have been derived from the annual publication of the NRCBL's *New Titles in Bioethics*, and cite monographs added to the collection in 2007 that cover bioethics and related areas of ethics and applied ethics. The NRCBL Classification Scheme (reproduced in full on the inside front cover) provides the arrangement for these citations. The Monographs section includes 754 records for books, reports, audiovisuals, special issues of journals, and new periodical subscriptions. Only subject headings actually occurring in Volume 34 are included on this list.

The monograph citations are arranged according to the primary subject category of the volume, and then, under subject category, by author, editor, producer, or title. Each citation in the Section usually appears only once. Classification numbers at the end of each citation represent additional bioethics topics covered by the publication. Monograph entries also include acquisition information, especially important for the so-called "gray literature." Monographs in foreign languages are included in the *Bibliography*.

Section IV: Monographs — Title Index

This Section provides a title index to all the entries in the Monographs Section. The title is followed by the subject section and author within which the complete citation can be found.

The *Bibliography of Bioethics*: History and Current Availability on the World Wide Web

Through December 2000, the entries in all of the annual volumes of the *Bibliography of Bioethics* were available online in BIOETHICSLINE, a database produced for the National Library of Medicine (NLM) by the Bioethics Information Retrieval Project at the Kennedy Institute of Ethics, Georgetown University. As of 2001, NLM incorporated its subject-oriented databases–like BIOETHICSLINE– into two large databases, PubMed/MEDLINE for journal articles and related documents, and LOCATOR*plus*/NLM Catalog for books and related documents.

Citations from the *Bibliography of Bioethics* are available on the World Wide Web via the ETHXWeb and GenETHX databases, maintained by NRCBL. Access to these databases, along with searching information, is available through the Web gateway of the Kennedy Institute of Ethics at http://bioethics.georgetown.edu. In addition, a comprehensive NRCBL publication provides advice for database searchers: *Bioethics Searcher's Guide to Online Information Resources*. (See "Distribution" paragraph below for ordering information.)

BIBLIOGRAPHY OF BIOETHICS

Acknowledgments

It is a pleasure to acknowledge the assistance of several people and organizations who played significant roles in the production of this thirty-fourth volume of the *Bibliography of Bioethics*. Although this publication is not a direct product of federal funding, it depends upon critical support from the National Library of Medicine and the National Human Genome Research Institute, both at the National Institutes of Health. We wish to thank, in particular, our NLM Project Officers, Sara Tybaerts, Martha Cohn, and Susan Von Braunsberg; our NLM Contracting Officer, Alex Navas; and Joy Boyer, Program Director, The Ethical, Legal, and Social Implications Program, National Human Genome Research Institute, for their interest and support. Other support is provided by the National Endowment for the Humanities, the Anderson Partnership, Max M. and Marjory B. Kampelman, and many publishers and individuals who contributed copies of books and journal articles to NRCBL.

Patricia Milmoe McCarrick, former reference librarian at NRCBL, continues as a library volunteer. Mark Rosetti, a Georgetown University student, carried out document acquisition and data entry tasks.

Distribution of the *Bibliography of Bioethics* and Related Publications

Inquiries about purchasing Volumes 10-34 of the *Bibliography* or the current edition of *Bioethics Searcher's Guide to Online Information Resources* should be directed to Library Publications, Kennedy Institute of Ethics, Georgetown University, Box 571212, Washington, DC 20057-1212, telephone 202-687-3885 or 888-BIO-ETHX (outside the Washington, DC metropolitan area); fax 202-687-6770, e-mail: bioethics@georgetown.edu.

International Bioethics Exchange Project (IBEP)

IBEP, a project of the Kennedy Institute of Ethics, promotes research and education in bioethics in the developing world by donating multiple volumes of the *Bibliography* to libraries abroad in order to encourage the development of bioethics reference resources in those countries. In turn, IBEP is eager to collect documents about bioethics from the exchange participants. Any books, policy statements, periodicals and other materials about bioethical issues in the participant countries that are donated to IBEP are added to the NRCBL collection and considered for inclusion in the *Bibliography*. This project relies upon the support of donors to underwrite the transport of the volumes to the developing country library.

To date libraries in the following countries have become participants in the project: Argentina, Belarus, Brazil, Burkina Faso, Cameroon, China, Congo, Costa Rica, Croatia, Eritrea, Gabon, Grenada, Israel, Jamaica, Kenya, Korea (South), Liberia, Lithuania, Madagascar, Mexico, Nigeria, Papua New Guinea, Philippines, Poland, Portugal, Romania, Rwanda, Saint Lucia, Sierra Leone, Slovakia, Slovenia, South Africa, Sri Lanka, Sudan, Thailand, Trinidad and Tobago, Turkey, Uzbekistan, Venezuela, Yemen, and Zambia.

Contributions in support of an IBEP library or donations of bioethics books, reprints, audiovisual materials, and other documents should be sent to Lucinda Fitch Huttlinger, Acquisitions Librarian, Kennedy Institute of Ethics, Georgetown University, Box 571212, Washington, DC 20057-1212; Telephone: +202-687-6433; Toll-free telephone: 1-888-BIO-ETHX (U.S. and Canada); FAX +202-687-6770; e-mail: bioethics@georgetown.edu. All donations are reviewed for inclusion in the NRCBL collection as well as for this *Bibliography*.

The staff welcomes suggestions for the improvement of future volumes of the *Bibliography of Bioethics*. Please send all comments to:

> Editors, *Bibliography of Bioethics*
> Kennedy Institute of Ethics, Box 571212
> Georgetown University
> Washington, DC 20057-1212

July 22, 2008

SECTION I:
PERIODICAL LITERATURE AND ESSAYS

SUBJECT ENTRIES

SECTION I: PERIODICAL LITERATURE AND ESSAYS
SUBJECT ENTRIES

ABORTION

Baggot, Paddy Jim. Hard cases do not justify partial birth abortion. *Linacre Quarterly* 2007 August; 74(3): 248-252. NRCBL: 12.1.

Blackmer, Jeff. Clarification of the CMA's position concerning induced abortion [letter]. *CMAJ/JAMC: Canadian Medical Association Journal* 2007 April 24; 176(9): 1310. NRCBL: 12.1; 12.2. Comments: S. Rodgers and J. Downie. Abortion: ensuring access [editorial]. CMAJ/JAMC: Canadian Medical Association Journal 2006 July 4; 175(1): 9.

Cahill, H.; Holland, S. Abortion and other 'beginning of life' issues. *In:* Holland, Stephen, ed. Introducing Nursing Ethics: Themes in Theory and Practice. Salisbury: APS, 2004: 29-48. NRCBL: 12.1; 4.1.3. SC: cs.

Cole, Andrew. Botched abortions kill more than 66,000 women each year [news]. *BMJ: British Medical Journal* 2007 October 27; 335(7625): 845. NRCBL: 12.1; 9.5.5; 9.8; 20.1.

Daleiden, Joseph L. Abortion. *In his:* The Science of Morality: The Individual, Community, and Future Generations. Amherst, NY: Prometheus Books, 1998: 348-372. NRCBL: 12.1; 12.3; 4.4; 12.4.2. SC: an; le.

de Costa, Caroline; de Costa, Naomi. Medical abortion and the law. *University of New South Wales Law Journal* 2006; 29(2): 218-223. NRCBL: 12.1; 9.7. SC: le.

Dyer, Clare. Experts clash over reducing abortion limit [news]. *BMJ: British Medical Journal* 2007 October 20; 335(7624): 789. NRCBL: 12.1; 12.4.1. Identifiers: Great Britain (United Kingdom).

Horst, Jason M. The meaning of "life": the morning-after pill, the question of when life begins, and judicial review. *Texas Journal of Women and the Law* 2007 Spring; 16(2): 205. NRCBL: 12.1; 9.7; 4.4. SC: le.

Kmietowicz, Zosia. Make access to early abortions easier and quicker, say doctors. *BMJ: British Medical Journal* 2007 July 7; 335(7609): 14. NRCBL: 12.1; 12.5.3.

Ledewitz, Bruce. Protecting posterity: economics, abortion politics, and the law. *Conservation Biology* 2006 August; 20(4): 940-941. NRCBL: 12.1; 12.2; 12.4.1. SC: le.

McLachlan, Hugh V.; Swales, J. Kim. Abortion. *In their:* From the Womb to the Tomb: Issues in Medical Ethics. Glasgow, Scotland: Humming Earth, 2007: 1-20. NRCBL: 12.1; 4.4; 12.3. SC: an.

Pesce, Andrew. Abortion laws in Australia: time for consistency? *University of New South Wales Law Journal* 2006; 29(2): 224-226. NRCBL: 12.1; 9.7. SC: le.

Saugstad, Ola Didrik. Non-selective fetal reduction is malpractice. *Journal of Perinatal Medicine* 2006; 34(5): 355-358. NRCBL: 12.1; 14.4; 8.5; 1.3.2.

Snodgrass, Mary Ellen. Abortion. *In her:* Historical Encyclopedia of Nursing. Santa Barbara, CA: ABC-CLIO, 1999: 1-4. NRCBL: 12.1; 4.1.3; 12.5.1.

ABORTION/ LEGAL ASPECTS

Italy: abortion blunder rekindles debate. *New York Times* 2007 August 28; p. A13. NRCBL: 12.4.2. SC: po; le.

Adkins, Jason A. Meet me at the (West Coast) hotel: the Lochner Era and the demise of Roe v. Wade. *Specialty Law Digest: Health Care Law* 2007 July; (339): 9-43. NRCBL: 12.4.1. SC: le.

Annas, George J. The Supreme Court and abortion rights. *New England Journal of Medicine* 2007 May 24; 356(21): 2201-2207. NRCBL: 12.4.1; 12.3; 12.5.3. SC: le. Identifiers: Gonzalez v. Carhart.

Azira bt Tengku Zainuden, Tengku Noor. Partial-birth abortion from the perspective of Malaysian criminal law. *Formosan Journal of Medical Humanities* 2007 July; 8(1-2): 1-13. NRCBL: 12.4.3.

Bodger, Jessica Ansley. Taking the sting out of reporting requirements: reproductive health clinics and the constitutional right to informational privacy. *Duke Law Journal* 2006 November; 56(2): 583-609. NRCBL: 12.4.2; 8.4; 8.3.2; 9.5.5; 10. SC: an; le.

NRCBL: National Reference Center for Bioethics Literature Classification Scheme   See inside front cover for terms.

13

Breen, John M.; Scaperlanda, Michael A. Never get out'a the boat: Stenberg v. Carhart and the future of American law. *Connecticut Law Review* 2006 November; 39(1): 297-323. NRCBL: 12.4.2; 7.1; 12.4.1. SC: an; le.

Caplan, Arthur; Marino, Thomas A. The role of scientists in the beginning-of-life debate: a 25-year retrospective. *Perspectives in Biology and Medicine* 2007 Autumn; 50(4): 603-613. NRCBL: 12.4.2; 1.3.5; 1.3.9; 4.4; 5.1; 15.1; 18.5.4; 2.2. SC: le.

Carranza, María. The therapeutic exception: abortion, sterilization and medical necessity in Costa Rica. *Developing World Bioethics* 2007 August; 7(2): 55-63. NRCBL: 12.4.1; 11.3; 12.5.1. SC: le.

Abstract: Based on the case of Rosa, a nine-year-old girl who was denied a therapeutic abortion, this article analyzes the role played by the social in medical practice. For that purpose, it compares the different application of two similar pieces of legislation in Costa Rica, where both the practice of abortion and sterilization are restricted to the protection of health and life by the Penal Code. As a concept subject to interpretation, a broad conception of medical necessity could enable an ample use of the therapeutic exception and a liberal use of both surgeries. The practice of therapeutic sterilization has been generalized in Costa Rica and has become the legitimate way to distribute contraceptive sterilization. In contrast, therapeutic abortion is very rarely practiced. The analysis carried out proposes that it is the difference in social acceptance of abortion and sterilization that explains the different use that doctors, as gatekeepers of social morality, make of medical necessity.

Center for Reproductive Rights. The world's abortion laws. New York, NY: Center for Reproductive Rights, 2005 April: 4 p. [Online]. Accessed: http://www. reproductiverights.org/pub_fac_abortion_laws.html [2007 April 2]. NRCBL: 12.4.1; 21.1. SC: em; le.

Charo, R. Alta. The partial death of abortion rights. *New England Journal of Medicine* 2007 May 24; 356(21): 2125-2128. NRCBL: 12.4.2; 12.3. SC: le; rv. Identifiers: Gonzalez v. Carhart; United States.

Coghlan, Andy. Pro-choice? Pro-life? No choice. *New Scientist* 2007 October 20-26; 196(2626): 8-9. NRCBL: 12.4.1; 11.1; 12.5.1.

Cook, R.J.; Ortega-Ortiz, A.; Romans, S.; Ross, L.E. Legal abortion for mental health indications. *International Journal of Gynecology and Obstetrics* 2006 November; 95(2): 185-190. NRCBL: 12.4.2; 17.1; 8.1; 7.1.

Davis, Gayle; Davidson, Roger. "A fifth freedom" or "hideous atheistic expediency"? The medical community and abortion law reform in Scotland, c.1960-1975. *Medical History* 2006 January; 50(1): 29-48. NRCBL: 12.4.1; 12.5.1; 2.2; 4.1.2. SC: le. Identifiers: Medical Termination of Pregnancy Bill (1966); Abortion Act (1967).

de Roubaix, M. Ten years hence — has the South African Choice on Termination of Pregnancy Act, Act 92 of 1996, realised its aims? A moral-critical evaluation. *Medicine and Law: The World Association for Medical Law* 2007 March; 26(1): 145-177. NRCBL: 12.4.2; 12.3; 4.1.3; 7.3; 8.3.2. SC: le; an.

Abstract: The South African Choice on Termination of Pregnancy Act (Act 92 of 1996) (CTOP) passed by parliament ten years ago, aims to promote female reproductive autonomy through legitimising free access to abortion up to 20 weeks of gestation. The article critically evaluates CTOP and highlights three societal concerns: the effect of CTOP on the self-esteem of nurses who perform abortion; the effect on general societal morality, and its desirability. CTOP has enjoyed mixed success. On the plus side, it has furthered female reproductive autonomy, has decreased early pregnancy maternal mortality and has advanced non-racialism through equal access to safe abortion. On the minus side, it remains controversial; the majority of the population opposes abortion on request, predominantly based on religiously-informed intuitions on the value of ante-natal life. Officials and managers of public health care facilities are often obstructive, and TOP personnel victimised and socially stigmatised. An unacceptably high rate of unsafe abortion prevails, particularly in rural areas and amongst adolescents, but also in certain urban areas. The prime causes are inadequate public education, attitudinal problems, and lack of psychological support for TOP personnel, the segregation of ante-natal care and abortion services, inadequate training, research, communication and contraceptive services, absence of incentives for TOP personnel and "traditional" gender roles and male power-based domination in reproductive choices. Corrective measures include a goal directed educational programme and initiatives like value clarification workshops which have been effective in changing negative attitudes of participants, and may thus address stigmatisation, improve working conditions of TOP personnel, promote societal tolerance and acceptance, and informed consent. Of particular concern are the questions of informed consent, minors, promotion of counselling and contraceptive services (particularly for adolescents), conscientious objection and the protection of compliant (and non-compliant) personnel.

DeMarco, Donald. Fetal pain: real or relative? *Human Life Review* 2007 Winter; 33(1): 68-71. NRCBL: 12.4.2; 12.4.4; 11.1.

Deutsche Gesellschaft für Humangenetik = German Society of Human Genetics [DGfH]. Committee for Public Relations and Ethical Issues. Statement on the revision of section 218 of the German Penal Code with elimination of the so-called embryopathic indication for terminating pregnancy [position statement]. *Medizinische Genetik* 1995; 7: 360-361 (3 p.) [Online]. Accessed: http://www.medgenetik.de/sonderdruck/en/218_Revision.pdf [2007 February 12]. NRCBL: 12.4.2; 15.2; 9.5.5.

Diniz, Debora. Selective abortion in Brazil: the anencephaly case. *Developing World Bioethics* 2007 August; 7(2): 64-67. NRCBL: 12.4.1; 11.3; 12.5.1. SC: le.

Abstract: This paper discusses the Brazilian Supreme Court ruling on the case of anencephaly. In Brazil, abortion is a crime against the life of a fetus, and selective abortion of non-viable fetuses is prohibited. Following a paradigmatic case discussed by the Brazilian Supreme Court in 2004, the use of abortion was authorized in the case of a fetus with anencephaly. The objective of this paper is to analyze the ethical arguments of the case, in particular the strategy of avoiding the moral status of the fetus, the cornerstone thesis of the Catholic Church.

Drazen, Jeffrey M. Government in medicine [editorial]. *New England Journal of Medicine* 2007 May 24; 356(21): 2195. NRCBL: 12.4.1; 1.3.5. SC: le. Identifiers: Gonzalez v. Carhart.

Dresser, Rebecca. Protecting women from their abortion choices. *Hastings Center Report* 2007 November-December; 37(6): 13-14. NRCBL: 12.4.2; 9.5.5; 17.1. SC: le.

Dyer, Clare. Girl carrying anencephalic fetus is granted right to travel. *BMJ: British Medical Journal* 2007 May 19; 334(7602): 1026. NRCBL: 12.4.2; 20.5.2. Identifiers: Ireland.

Ehlrich, J. Shoshanna. Shifting boundaries: abortion, criminal culpability and the indeterminate legal status of adolescents. *Wisconsin Women's Law Journal* 2003 Spring; 18(1): 77-116. NRCBL: 12.4.2; 1.1; 1.3.5; 9.5.7. SC: le.

European Commission of Human Rights. Brüggemann v. Germany [Date of Decision: 12 July 1977]. European Human Rights Reports 1977; 3: 244-258. NRCBL: 12.4.4; 12.4.3. SC: le.

Abstract: Two German women, Brüggemann and Scheuten, assert that German law on abortion interferes with their right to respect for their private life under the European Convention on Human Rights. Specifically, they allege that they are not free to abort an unwanted pregnancy. In 1975, the German constitutional court voided a 1974 law allowing abortion during the first twelve weeks of pregnancy without any particular reason; thus German law remained as banning any abortion except in the case of medical grounds, i.e. saving the mother's life or health. After that decision, in the next year, 1976, a new law was passed, which continued to hold abortion as a criminal offence, but allowed that in situations of distress, a doctor may perform an abortion with the woman's consent after consultation. The European Commission on Human Rights found that German abortion law did not interfere with a woman's right to respect for her private life, because Article 8(1) of the European Convention on Human Rights "cannot be interpreted as meaning that pregnancy and its termination are, as a principle, solely a matter of the private life of the mother." The Commission reasoned that "pregnancy cannot be said to pertain uniquely to the sphere of private life." Earlier jurisprudence held that "the claim to respect

for private life is automatically reduced to the extent that the individual himself brings his private life into contact with public life or into close connection with other protected interests".

Finnis, John. "A vote decisive for . . . a more restrictive law". *In:* Watt, Helen, ed. Cooperation, Complicity and Conscience: Problems in Healthcare, Science, Law and Public Policy. London: Linacre Centre, 2005: 269-295. NRCBL: 12.4.1; 12.3. SC: le; an.

Finnis, John. Restricting legalised abortion is not intrinsically unjust. *In:* Watt, Helen, ed. Cooperation, Complicity and Conscience: Problems in Healthcare, Science, Law and Public Policy. London: Linacre Centre, 2005: 209-245. NRCBL: 12.4.1; 12.3. SC: an; le.

Flood, Patrick J. Is international law on the side of the unborn child? *National Catholic Bioethics Quarterly* 2007 Spring; 7(1): 73-95. NRCBL: 12.4.2; 21.1; 4.4; 12.4.1. SC: le.

Ford, Mary. The consent model of pregnancy: deadlock undiminished. *McGill Law Journal* 2005; 50: 619-666. NRCBL: 12.4.2; 4.4; 8.3.1; 10. SC: an; le.

Forsythe, Clarke D. A lack of prudence. *Human Life Review* 2007 Fall; 33(4): 15-21. NRCBL: 12.4.2. SC: le. Identifiers: Partial Birth Abortion Act; Gonzales v. Carhart.

Goldberg, Jordan. The Commerce Clause and federal abortion law: why progressives might be tempted to embrace federalism. *Fordham Law Review* 2006 October; 75(1): 301-354. NRCBL: 12.4.1; 1.3.5. SC: le.

Gornall, Jonathan. Where do we draw the line? Numerous attempts have been made to change the rules on abortion since it was legalised 40 years ago. *BMJ: British Medical Journal* 2007 February 10; 334(7588): 285-289. NRCBL: 12.4.1; 12.3; 12.5.1; 21.1. Identifiers: Great Britain (United Kingdom).

Gostin, Lawrence O. Abortion politics: clinical freedom, trust in the judiciary, and the autonomy of women. *JAMA: The Journal of the American Medical Association* 2007 October 3; 298(13): 1562-1564. NRCBL: 12.4.2; 8.1; 9.5.5. SC: le.

Greene, Michael F. The intimidation of American physicians: banning partial-birth abortion. *New England Journal of Medicine* 2007 May 24; 356(21): 2128-2129. NRCBL: 12.4.1; 1.3.5. SC: le. Identifiers: Gonzalez v. Carhart.

Harte, Colin. The opening up of a discussion: a response to John Finnis. *In:* Watt, Helen, ed. Cooperation, Complicity and Conscience: Problems in Healthcare, Science, Law and Public Policy. London: Linacre Centre, 2005: 246-268. NRCBL: 12.4.1; 12.3. SC: an; le.

NRCBL: National Reference Center for Bioethics Literature Classification Scheme   See inside front cover for terms.

15

Hatziavramidis, Katie. Parental involvement law for abortion in the United States and the United Nations conventions on the rights of the child: can international law secure the right to choose for minors? *Texas Journal of Women and the Law* 2007 Spring; 16(2): 185-204. NRCBL: 12.4.2; 8.3.2; 21.1. SC: le.

Hewson, Barbara. Abortion in Poland: a new human rights ruling. *Conscience* 2007 Summer; 28(2): 34-35. NRCBL: 12.4.2; 21.1. SC: le. Identifiers: Tysiac v. Poland.

Huff, Sarah A. The abortion crisis in Peru: finding a woman's right to obtain safe and legal abortions in the convention on the elimination of all forms of discrimination against women. *Boston College International and Comparative Law Review* 2007 Winter; 30(1): 237-248. NRCBL: 12.4.2; 10; 12.4.3. SC: le.

Hunter, Nan D. Justice Blackmun, abortion, and the myth of medical independence. *Brooklyn Law Review* 2006 Fall; 72(1): 147-197. NRCBL: 12.4.2; 1.1; 4.1.2; 9.5.5; 10. SC: le.

Joyce, Theodore; Kaestner, Robert. State reproductive policies and adolescent pregnancy resolution: the case of parental involvement laws. *Journal of Health Economics* 1996 October; 15(5): 579-607. NRCBL: 12.4.1; 8.3.2; 12.4.3. SC: le.

Kamm, F.M. Ronald Dworkin's views on abortion and assisted suicide. *In:* Burley, Justine, ed. Dworkin and His Critics: With Replies by Dworkin. Malden, MA: Blackwell Pub., 2004: 218-240. NRCBL: 12.4.2; 4.4; 20.5.1; 1.1. SC: an. Identifiers: Life's Dominion.

Kolenc, Antony B. Easing abortion's pain: can fetal pain legislation survive the new judicial scrutiny of legislative fact-finding? *Texas Review of Law and Politics* 2005 Fall; 10(1): 171-228. NRCBL: 12.4.2; 9.5.8; 4.4. SC: le.

Lee, Ellie. Young women, pregnancy, and abortion in Britain: a discussion of 'law in practice'. *International Journal of Law, Policy and the Family* 2004 December; 18(3): 283-304. NRCBL: 12.4.2; 8.3.2. SC: em; le.

Maloy, Richard H.W. Will new appointees to the Supreme Court be able to effect an overruling of Roe v. Wade? *Western New England Law Review* 2005; 28(1): 29-55. NRCBL: 12.4.1; 1.3.5.

McAllister, Marc Chase. Human dignity and individual liberty in Germany and the United States as examined through each country's leading abortion cases. *Tulsa Journal of Comparative and International Law* 2004 Spring; 11(2): 491-520. NRCBL: 12.4.2; 12.4.1; 4.4; 21.1.

Miller, Suellen; Billings, Deborah L. Abortion and postabortion care: ethical, legal, and policy issues in developing countries. *Journal of Midwifery and Women's Health* 2005 July-August; 50(4): 341-343. NRCBL: 12.4.1; 12.5.3; 21.1; 12.3; 4.1.3. SC: cs.

Miyazaki, Michiko. The history of abortion-related acts and current issues in Japan. *Medicine and Law: The World Association for Medical Law* 2007 December; 26(4): 791-799. NRCBL: 12.4.2; 15.5; 2.2. SC: le.
    Abstract: In Japan abortion is categorized into two types by law; one is illegal feticide and the other is legal abortion. The present criminal law forbids feticide in principle and the life of a fetus is protected. However, abortion can be practiced under the "Eugenic Protection Act" established in 1948 (currently referred to as the "Maternal Protection Act"), and is readily available in Japan. In this paper, I have traced the historical origins of abortion law and attempted to clarify the problems related to the current laws relating to artificial abortion. As a result, the existence of contradictions between attitudes toward the life of the fetus and that of the mother, women's right to self determination, and women's rights under current legislation has been clarified.

Molinelli, A.; Picchioni, D.M.; Celesti, R. Voluntary interruption of pregnancy in Europe: medico-legal issues and ethical approach to the regulation. *Minerva Ginecologica* 2005 April; 57(2): 217-223. NRCBL: 12.4.1; 12.3. SC: le; em.

Mushaben, Joyce; Giles, Geoffrey; Lennox, Sara. Women, men and unification: gender politics and the abortion struggle since 1989. *In:* Jarausch, K.H., ed. After Unity: Reconfiguring German Identities. Oxford: Berghahn Books, 1997: 137-172. NRCBL: 12.4.2; 12.5.2; 10; 21.1.

O'Dowd, Adrian. Doctors don't need second signature for abortion [news]. *BMJ: British Medical Journal* 2007 October 27; 335(7625): 844. NRCBL: 12.4.3. SC: le.

O'Dowd, Adrian. MPs want to drop second doctor's signature for abortion [news]. *BMJ: British Medical Journal* 2007 November 10; 335(7627): 960. NRCBL: 12.4.1. SC: le. Identifiers: Great Britain (United Kingdom); Members of Parliament.

O'Dowd, Adrian. No evidence backs reduction in abortion time limit, minister says [news]. *BMJ: British Medical Journal* 2007 November 3; 335(7626): 903. NRCBL: 12.4.1. SC: le. Identifiers: Great Britain (United Kingdom).

O'Toole, Leslie C.; Sobel-Read, Kevin B. Pharmacist refusals: a new twist on the debate over individual autonomy. *Gender Medicine* 2006 March; 3(1): 13-17. NRCBL: 12.4.3; 12.3; 8.1; 9.7; 1.1.

Petersen, Kerry. Classifying abortion as a health matter: the case for de-criminalising abortion laws in Australia. *In:* McLean, Sheila A.M., ed. First Do No Harm: Law, Ethics, and Healthcare. Aldershot, England; Burlington, VT: Ashgate, 2006: 353-368. NRCBL: 12.4.2; 12.5.3; 7.1; 8.4; 8.3.2. SC: le.

Pozgar, George D. Issues of procreation. *In his:* Legal Aspects of Health Care Administration. 9th edition. Sudbury,

---

MA: Jones and Bartlett Publishers, 2004: 345-364. NRCBL: 12.4.2; 12.4.3; 11.3; 11.4. SC: le; cs.

Pringle, Helen. Abortion and disability: reforming the law in South Australia. *University of New South Wales Law Journal* 2006; 29(2): 207-217. NRCBL: 12.4.2; 4.4; 4.2; 4.3. SC: le.

Rakowski, Eric. Reverence for life and the limits of state power. *In:* Burley, Justine, ed. Dworkin and His Critics: With Replies by Dworkin. Malden, MA: Blackwell Pub., 2004: 241-263. NRCBL: 12.4.2; 4.4; 20.5.1; 1.1; 12.3. SC: an. Identifiers: Ronald Dworkin; Life's Dominion.

Roden, Gregory J. Unborn persons, incrementalism and the silence of the lambs. *Human Life Review* 2007 Fall; 33(4): 22-32. NRCBL: 12.4.2; 9.5.5; 4.4. SC: le.

Rodgers, Sanda. Abortion denied: bearing the limits of law. *In:* Flood, Colleen M., ed. Just Medicare: What's In, What's Out, How We Decide. Buffalo, NY: University of Toronto Press, 2006: 107-136. NRCBL: 12.4.2; 9.5.5. SC: le. Identifiers: Canada; Badgley Report.

Sanger, Carol. Regulating teenage abortion in the United States: politics and policy. *International Journal of Law, Policy and the Family* 2004 December; 18(3): 305-318. NRCBL: 12.4.2; 8.3.2; 12.4.1.

Sciolino, Elaine. Low turnout undercuts Portugal vote on abortion law. *New York Times* 2007 February 12; p. A3. NRCBL: 12.4.1. SC: po.

Sciolino, Elaine. Portugal to vote on putting end to abortion ban. *New York Times* 2007 February 11; p. A1, A14. NRCBL: 12.4.1.

Sharma, S. Legalization of abortion in Nepal: the way forward. *Kathmandu University Medical Journal* 2004 July-September; 2(3): 177-178. NRCBL: 12.4.1. SC: le.

Sherwin, Susan. Abortion through a feminist lens. *In:* Baylis, Françoise; Downie, Jocelyn; Freedman, Benjamin; Hoffmaster, Barry; Sherwin, Susan, eds. Health Care Ethics in Canada. Toronto: Harcourt Brace Canada, 1995: 441-447. NRCBL: 12.4.1; 12.4.2; 12.3; 10. SC: le.

Spence, Des. A time of change in abortion. *BMJ: British Medical Journal* 2007 December 15; 335(7632): 1266. NRCBL: 12.4.3.

Statham, H.; Solomou, W.; Green, J. Late termination of pregnancy: law, policy and decision making in four English fetal medicine units. *BJOG: An International Journal of Obstetrics and Gynaecology* 2006 December; 113(12): 1402-1411. NRCBL: 12.4.2; 12.5.2.

Sylva, Douglas; Yoshihara, Susan. Rights by stealth: the role of UN human rights treaty bodies in the campaign for an international right to abortion. *National Catholic Bioethics Quarterly* 2007 Spring; 7(1): 97-128. NRCBL: 12.4.1; 12.2; 21.1; 12.4.2; 12.4.4. SC: le.

Tanne, Janice Hopkins. US Congress asked to suspend funding for Planned Parenthood [news]. *BMJ: British Medical Journal* 2007 November 3; 335(7626): 903. NRCBL: 12.4.2; 9.3.1. SC: le.

Tanne, Janice Hopkins. US Supreme Court approves ban on "partial birth abortion". *BMJ: British Medical Journal* 2007 April 28; 334(7599): 866. NRCBL: 12.4.2.

United States. Supreme Court. U.S. Supreme Court partial-birth abortion decision. *Origins* 2007 May 3; 36(46): 749-753. NRCBL: 12.4.4. SC: le.

Vendittelli, F.; Pons, J.C. Elective abortions for minors: impact of the new law in France. *European Journal of Obstetrics, Gynecology, and Reproductive Biology* 2007 January; 130(1): 107-113. NRCBL: 12.4.2; 9.5.7. SC: le.

Villanueva, Tiago. Portugal is ready to decriminalise abortion [news]. *BMJ: British Medical Journal* 2007 February 17; 334(7589): 332. NRCBL: 12.4.1.

Wonkam, Ambroise; Hurst, Samia A. Acceptance of abortion by doctors and medical students in Cameroon [letter]. *Lancet* 2007 June 16-22; 369(9578): 1999. NRCBL: 12.4.3; 12.3; 8.1. SC: em.

ABORTION/ MORAL AND RELIGIOUS ASPECTS

Audi, Robert. Preventing abortion as a test case for the justifiability of violence. *Journal of Ethics* 1997; 1(2): 141-163. NRCBL: 12.3; 4.4; 9.1; 1.1; 21.7. SC: an.

Breger, Marshall. Freedom to choose. For Jewish women, a pro-choice stance on abortion often reflects their political values and culture as much as their views on reproductive freedom. *Moment* 1999 August; 24(4): 28-29. NRCBL: 12.3; 12.5.1; 1.2.

Clark, Tom C. Religion, morality and abortion: a constitutional appraisal. *Loyola University Los Angeles Law Review* 1969; 2: 1-11. NRCBL: 12.3; 12.4.1.

de Roubaix, J.A.M; van Niekerk, A.A. Separation-survivability — the elusive moral cut-off point? *South African Medical Journal = Suid-afrikaanse Tydskrif vir Genees- kunde* 2006 July; 96(7): 623-626. NRCBL: 12.3; 4.4; 12.4.2.

Eberl, Jason T. Issues at the beginning of human life: abortion, embryonic stem cell research, and cloning. *In his:* Thomistic Principles and Bioethics. London; New York: Routledge, 2006: 62-94. 19 fn. NRCBL: 12.3; 1.1; 4.4; 14.5; 18.5.4; 15.1. SC: an.

> Keywords: *abortion; *beginning of life; *cloning; *embryo research; *embryos; *moral status; abortifacients; adult stem cells; embryo disposition; fetal stem cells; in vitro fertilization; moral complicity; natural law; personhood; philosophy; reproductive technologies; Proposed Keywords: blastocysts; Keyword Identifiers: *Thomas Aquinas

---

NRCBL: National Reference Center for Bioethics Literature Classification Scheme   See inside front cover for terms.

17

Galvão, Pedro. Boonin on the future-like-ours argument against abortion. *Bioethics* 2007 July; 21(6): 324-328. NRCBL: 12.3. SC: an.

Abstract: I argue that David Boonin has failed in his attempt to undermine Donald Marquis's future-like-ours argument against abortion. I show that the ethical principle advanced by Boonin in his critique to that argument is unable, contrary to what he claims, to account for the wrongness of infanticide. Then I argue that Boonin's critique misrepresents Marquis's argument. Although there is a way to restate his critique in order to avoid the misrepresentation, the success of such restatement is precluded by the wrongness of infanticide.

Joffe, Carole; Shields, Wayne C. Morality and the abortion provider. *Contraception* 2006 July; 74(1): 1-2. NRCBL: 12.3; 8.1; 7.1.

Kavanaugh, John F. In defense of human life [editorial]. *America* 2007 November 26; 197(17): 8. NRCBL: 12.3; 4.4.

Luna, Florencia. Internal reasons and abortion. *In:* Luna, Florencia; Herissone-Kelly, Peter, ed. Bioethics and Vulnerability: a Latin American View. Amsterdam; New York: Rodopi, 2006: 39-47. NRCBL: 12.3; 1.1.

Lustig, B. Andrew. The church and the world: are there theological resources for a common conversation? *Christian Bioethics* 2007 May-August; (13)2: 225-244. NRCBL: 12.3; 1.2; 7.1. SC: an.

Abstract: Abortion is an especially salient issue for considering the general problematic of religiously based conversation in the public square. It remains deeply divisive, fully thirty-four years after Roe v. Wade. Such divisiveness cannot be interpreted as merely an expression of profound differences between "secular" and "religious" voices, because differences also emerge among Christian denominations, reflecting different sources of moral authority, different accounts of moral discernment, and different judgments about the appropriate relations between law and morality in the context of pluralism. As this paper explores, however, despite those differences, a generally identifiable "Christian" position concerning the moral status of abortion can be distinguished from secular philosophical judgments on the issue, which is important for Christian engagement with public policy debate.

Marquis, Don. Abortion revisited. *In:* Steinbock, Bonnie, ed. The Oxford Handbook of Bioethics. Oxford; New York: Oxford University Press, 2007: 395-415. NRCBL: 12.3; 4.4.

Raghavan, Ramesh. A question of faith. *JAMA: The Journal of the American Medical Association* 2007 April 4; 297(13): 1412. NRCBL: 12.3; 1.2. SC: sc.

Rakhudu, M.A.; Mmelesi, A.M.M.; Myburgh, C.P.H.; Poggenpoel, M. Exploration of the views of traditional healers regarding the termination of pregnancy (TOP) law. *Curationis* 2006 August; 29(3): 56-60. NRCBL: 12.3;

21.7. Identifiers: South Africa. Note: Abstract in Afrikaans.

Rakowski, Eric. Ronald Dworkin, reverence for life, and the limits of state power. *Utilitas* 2001 March; 13(1): 33-64. NRCBL: 12.3; 20.5.1; 1.1. Note: Discussion of Life's Dominion, by Ronald Dworkin.

Reiman, Jeffrey. The pro-life argument from substantial identity and the pro-choice argument from asymmetric value: a reply to Patrick Lee [see correction in Bioethics 2007 September; 21(7): 407]. *Bioethics* 2007 July; 21(6): 329-341. NRCBL: 12.3; 1.1. SC: an.

Abstract: Lee claims that foetuses and adult humans are phases of the same identical substance, and thus have the same moral status because: first, foetuses and adults are the same physical organism, and second, the development from foetus to adult is quantitative and thus not a change of substance. Versus the first argument, I contend that the fact that foetuses and adults are the same physical organism implies only that they are the same thing but not the same substance, much as living adults and their corpses are the same thing (same body) but not the same substance. Against Lee's second argument, I contend that Lee confuses the nature of a process with the nature of its result. A process of quantitative change can produce a change in substance. Lee also fails to show that foetuses are rational and thus have all the essential properties of adults, as required for them to be the same substance. Against the pro-choice argument from asymmetric value (that only the fact that a human has become conscious of its life and begun to count on its continuing can explain human life's asymmetric moral value, i.e. that it is vastly worse to kill a human than not to produce one), Lee claims that foetus's lives are asymmetrically valuable to them before consciousness. This leads to counterintuitive outcomes, and it confuses the goodness of life (a symmetric value that cannot account for why it is worse to kill a human than not produce one) with asymmetric value.

Walter, James J. Theological parameters: Catholic doctrine on abortion in a pluralist society. *In:* Walter, James J.; Shannon, Thomas A., eds. Contemporary Issues in Bioethics: A Catholic Perspective. Lanham, MD: Rowman and Littlefield Publishers, 2005: 145-178. NRCBL: 12.3; 4.4. SC: le.

Wilcockson, Michael. Abortion and infanticide. *In his:* Issues of Life and Death. London: Hodder and Stoughton Educational, 1999: 33-55. NRCBL: 12.3; 4.4; 12.4.1; 20.5.2; 1.2. SC: cs; le. Identifiers: Sanctity of Life (SOL).

Yacoub, Ahmed Abdel Aziz. The prevention and termination of pregnancy. *In his:* The Fiqh of Medicine: Responses in Islamic Jurisprudence to Developments in Medical Science. London, UK: Ta-Ha Publishers Ltd., 2001: 202-232. NRCBL: 12.3; 1.2; 11.1. SC: le.

ABORTION/ SOCIAL ASPECTS

Making abortion legal, safe, and rare [editorial]. *Lancet* 2007 July 28-August 3; 370(9584): 291. NRCBL: 12.5.1; 12.3; 21.1.

Bolks, Sean M.; Evans, Diana; Polinard J.L.; Wrinkle, Robert D. Core beliefs and abortion attitudes: a look at Latinos. *Social Science Quarterly* 2000; 81(1): 253-260. NRCBL: 12.5.2; 9.5.4. SC: em.

Boonstra, Heather D.; Gold, Rachel Benson; Richards, Cory L.; Finer, Lawrence B. Abortion in women's lives. New York: Guttmacher Institute, 2006; 44 p. [Online]. Accessed: http://www.guttmacher.org/pubs/2006/05/04/ AiWL.pdf [2007 April 3]. NRCBL: 12.5.1. SC: em.

Brown, Hannah. Abortion round the world. *BMJ: British Medical Journal* 2007 November 17; 335(7628): 1018-1019. NRCBL: 12.5.2; 12.4.2; 21.1.

Brown, R.W.; Jewell, R.T.; Rous, J.J. Abortion decisions among Hispanic women along the Texas-Mexico border. *Social Science Quarterly* 2000; 81(1): 237-252. NRCBL: 12.5.2; 9.5.4. SC: em.

Echevarria, Laura. Personally opposed, but. *Human Life Review* 2007 Spring; 33(2): 30-38. NRCBL: 12.5.1; 12.5.2; 21.1.

Hashiloni-Dolev, Yael. Abortions on embryopathic grounds: policy and practice in Israel and Germany. *In her:* A Life (Un)Worthy of Living: Reproductive Genetics in Israel and Germany. Dordrecht: Springer, 2007: 83-104. 24 fn. NRCBL: 12.5.1; 15.2. SC: em.
> Keywords: *congenital disorders; *fetal development; *genetic counseling; *selective abortion; attitude of health personnel; cross cultural comparison; eugenics; fetuses; genetic disorders; historical aspects; interviews; Jewish ethics; legal aspects; males; prenatal diagnosis; professional role; public opinion; religion; Roman Catholic ethics; statistics; value of life; viability; women's rights; wrongful life; Keyword Identifiers: *Germany; *Israel; Twentieth Century

Hashiloni-Dolev, Yael. Sex chromosome anomalies (SCAs) in Israel and Germany: assessing "birth defects" and medical risks according to the importance of fertility. *In her:* A Life (Un)Worthy of Living: Reproductive Genetics in Israel and Germany. Dordrecht: Springer, 2007: 105-117. 3 fn. NRCBL: 12.5.1; 15.2. SC: em.
> Keywords: *chromosome abnormalities; *genetic counseling; *selective abortion; attitude of health personnel; childbirth; cross cultural comparison; culture; infertility; prenatal diagnosis; public opinion; reproductive technologies; risk; Keyword Identifiers: *Germany; *Israel; Klinefelter syndrome; Turner syndrome

International Planned Parenthood Federation [IPPF]. Death and denial: unsafe abortion and poverty. London: International Planned Parenthood Federation, 2006 January: 18 p. [Online]. Accessed: http://content.ippf.org/ output/ORG/files/13108.pdf [2007 April 30]. NRCBL: 12.5.1; 9.5.10; 20.1. SC: em.

Kaposy, Chris; Sastre, María Teresa Muñoz; Peccarisi, Céline; Legrain, Elizabeth; Mullet, Etienne; Sorum, Paul. The real-life consequences of being denied access to an abortion. *American Journal of Bioethics* 2007 August; 7(8): 34-36; author reply W3. NRCBL: 12.5.1; 12.5.2; 9.2; 9.5.7. Comments: María Teresa Muñoz Sastre, Céline Peccarisi, Elizabeth Legrain, Etienne Mullet, and Paul Sorum. Acceptability in France of induced abortion for adolescents. American Journal of Bioethics 2007 August; 7(8): 26-32.

Lazarus, Jeffrey V.; Nielsen, Stine; Jakubcionyte, Rita; Kuliesyte, Esmeralda; Liljestrand, Jerker. Factors affecting attitudes towards medical abortion in Lithuania. *European Journal of Contraception and Reproductive Health Care* 2006 September; 11(3): 202-209. NRCBL: 12.5.2.

Mitchell, Ellen M.H.; Trueman, Karen; Gabriel, Mosotho; Bock, Lindsey B. Bickers. Building alliances from ambivalence: evaluation of abortion values clarification workshops with stakeholders in South Africa. *African Journal of Reproductive Health* 2005 December; 9(3): 89-99. NRCBL: 12.5.1; 12.5.2; 7.1. SC: le; em. Identifiers: Choice in Termination of Pregnancy Act (1996).

Morroni, Chelsea; Myer, Landon; Tibazarwa, Kemilembe. Knowledge of the abortion legislation among South African women: a cross-sectional study. *Reproductive Health [electronic]* 2006; 3:7, 5 p. Accessed: http://www. reproductive-health-journal.com/content/pdf/1742-4755- 3-7.pdf [2007 April 19]. NRCBL: 12.5.2; 12.4.1; 9.5.5. SC: le. Identifiers: Choice on Termination of Pregnancy Act 1996.

Nunes, Fred. Abortion: thinking clearly about controversial public policy. *African Journal of Reproductive Health* 2004 December; 8(3): 11-26. NRCBL: 12.5.2; 12.4.2.

Replogle, Jill. Abortion debate heats up in Latin America. *Lancet* 2007 July 28-August 3; 370(9584): 305-306. NRCBL: 12.5.1; 12.3; 21.1.

Sastre, María Teresa Muñoz; Peccarisi, Céline; Legrain, Elizabeth; Mullet, Etienne; Sorum, Paul. Acceptability in France of induced abortion for adolescents. *American Journal of Bioethics* 2007 August; 7(8): 26-32. NRCBL: 12.5.2; 9.5.7. SC: em. Comments: American Journal of Bioethics 2007 August: 34-36.

Sedgh, Gilda; Henshaw, Stanley; Singh, Susheela; Åhman, Elizabeth; Shah, Iqbal H. Induced abortion: estimated rates and trends worldwide. *Lancet* 2007 October 13-19; 370(9595): 1338-1345. NRCBL: 12.5.2. SC: em.
> Abstract: BACKGROUND: Information on incidence of induced abortion is crucial for identifying policy and programmatic needs aimed at reducing unintended pregnancy. Because unsafe abortion is a cause of maternal morbidity and mortality, measures of its incidence are also important for monitoring progress towards Millennium Development Goal 5. We present new worldwide

NRCBL: National Reference Center for Bioethics Literature Classification Scheme   See inside front cover for terms.

19

estimates of abortion rates and trends and discuss their implications for policies and programmes to reduce unintended pregnancy and unsafe abortion and to increase access to safe abortion. METHODS: The worldwide and regional incidences of safe abortions in 2003 were calculated by use of reports from official national reporting systems, nationally representative surveys, and published studies. Unsafe abortion rates in 2003 were estimated from hospital data, surveys, and other published studies. Demographic techniques were applied to estimate numbers of abortions and to calculate rates and ratios for 2003. UN estimates of female populations and livebirths were the source for denominators for rates and ratios, respectively. Regions are defined according to UN classifications. Trends in abortion rates and incidences between 1995 and 2003 are presented. FINDINGS: An estimated 42 million abortions were induced in 2003, compared with 46 million in 1995. The induced abortion rate in 2003 was 29 per 1000 women aged 15-44 years, down from 35 in 1995. Abortion rates were lowest in western Europe (12 per 1000 women). Rates were 17 per 1000 women in northern Europe, 18 per 1000 women in southern Europe, and 21 per 1000 women in northern America (USA and Canada). In 2003, 48% of all abortions worldwide were unsafe, and more than 97% of all unsafe abortions were in developing countries. There were 31 abortions for every 100 livebirths worldwide in 2003, and this ratio was highest in eastern Europe (105 for every 100 livebirths). INTERPRETATION: Overall abortion rates are similar in the developing and developed world, but unsafe abortion is concentrated in developing countries. Ensuring that the need for contraception is met and that all abortions are safe will reduce maternal mortality substantially and protect maternal health.

ADVANCE DIRECTIVES
See also DEATH AND DYING; TREATMENT REFUSAL

Pennsylvania: advance directive; DNR order. *Mental and Physical Disability Law Reporter* 2007 March-April; 31(2): 312-313. NRCBL: 20.5.4; 20.5.1. SC: le.

Allen, Charlotte F. Back off! I'm not dead yet. I don't want a living will. Why should I? [op-ed]. *Washington Post* 2007 October 14; p. B1, B4. NRCBL: 20.5.4.

Baumrucker, Steven J. Durable power of attorney versus the advance directive: who wins, who suffers? *American Journal of Hospice and Palliative Care* 2007 February-March; 24(1): 68-73. NRCBL: 20.5.4; 20.5.1; 8.3.3; 9.4.

Bianchi, Susan B. Living will for a handicapped child [letter]. *Health Affairs* 2007 September-October; 26(5): 1507. NRCBL: 20.5.4; 20.5.2.

Birnbacher, Dieter; Dabrock, Peter; Taupitz, Jochen; Vollmann, Jochen. Wie sollen Ärzte mit Patientenverfügungen umgehen? Ein Vorschlag aus interdisziplinärer Sicht [How should physicians deal with advance directives? A proposal from an interdisciplinary point of view].

Ethik in der Medizin 2007 June; 19(2): 139-147. NRCBL: 20.5.4; 8.1.

Bockenheimer-Lucius, Gisela. Behandlungsbegrenzung durch eine Patientenverfügung - im individuellen Fall auch mit Blick auf neue therapeutische Möglichkeiten! = Limitation of treatment by an advanced directive - in the individual case also with regard to the therapeutic possibilities! *Ethik in der Medizin* 2007 March; 19(1): 5-6. NRCBL: 20.5.4; 20.5.1. SC: le. Identifiers: Austria; Germany.

Cohen, Cynthia B. Philosophical challenges to the use of advance directives. *In:* Khushf, George, ed. Handbook of Bioethics: Taking Stock of the Field From a Philosophical Perspective. Dordrecht; Boston: Kluwer Academic, 2004: 291-314. NRCBL: 20.5.4; 1.1.

Davis, John K. Precedent, autonomy, advance directives, and end-of-life care. *In:* Steinbock, Bonnie, ed. The Oxford Handbook of Bioethics. Oxford; New York: Oxford University Press, 2007: 349-374. NRCBL: 20.5.4; 1.1.

Dyer, Clare. BMA gives advice on withdrawing treatment [news]. *BMJ: British Medical Journal* 2007 April 7; 334(7596): 711. NRCBL: 20.5.4; 8.3.3. SC: le. Identifiers: British Medical Association; Mental Capacity Act 2005.

Dyer, Clare. Patients win right to have their advance decisions honoured by medical staff [news]. *BMJ: British Medical Journal* 2007 October 6; 335(7622): 688-689. NRCBL: 20.5.4. SC: le. Identifiers: Great Britain (United Kingdom); Mental Capacity Act 2005.

Ehrlich, Joseph B. Schiavo: cold justice. Did the courts pursue, in view of the high court's decision in Troxell v. Granville, the correct conclusion to the Schiavo case? *The Journal of Law in Society Wayne State University Law School* 2005 Fall; 7(1): 1-15. NRCBL: 20.5.4; 4.4; 8.3.3; 20.3.3; 20.5.1. SC: le.

Gallagher, Romayne. Does knowledge of ethics and end-of-life issues inform choices in advance care planning scenarios? [letter]. *Journal of the American Geriatrics Society* 2007 October; 55(10): 1695-1696. NRCBL: 20.5.4; 20.5.2. SC: em.

Hoffenberg, Raymond. Advance healthcare directives. *Clinical Medicine* 2006 May-June; 6(3): 231-233. NRCBL: 20.5.4; 21.1; 7.1.

Jones, Sian; Jones, Bridie. Advance directives and implications for emergency departments. *British Journal of Nursing* 2007 February 22-March 7; 16(4): 220-223. NRCBL: 20.5.4; 8.3.3. SC: le. Identifiers: Mental Capacity Act 2005.

Kelk, Constantijn. Previously expressed wishes related to end of life decisions. *In:* Gevers, J.K.M.; Hondius, E.H.; Hubben, J.H., eds. Health Law, Human Rights and the Biomedicine Convention: Essays in Honour of Henriette Roscam Abbing. Leiden; Boston: Martinus Nijhoff Publishers, 2005: 131-145. NRCBL: 20.5.4; 20.7; 20.5.1. SC: le.

Kressel, Laura M.; Chapman, Gretchen B.; Leventhal, Elaine. The influence of default options on the expression of end-of-life treatment preferences in advance directives. *JGIM: Journal of General Internal Medicine* 2007 July; 22(7): 1007-1010. NRCBL: 20.5.4; 9.5.2. SC: em.

Kring, Daria L. The patient self-determination act: has it reached the end of its life? *JONA's Healthcare Law, Ethics, and Regulation* 2007 October-December; 9(4): 125-133. NRCBL: 20.5.4. SC: le.
Abstract: The Patient Self-determination Act requires that patients be informed in writing of their right to accept or refuse treatment and the right to an advance directive. For the past 15 years, hospitals have been providing these written materials, yet advance directives are still not adequately directing end-of-life care. Barriers and facilitators to implementation of this law are discussed, as well as the role of nursing management in meeting its true intent.

Madoff, Ray D. Autonomy and end-of-life decision making: reflections of a lawyer and a daughter. *Buffalo Law Review* 2005 Summer; 53(3): 963-971. NRCBL: 20.5.4; 1.1; 20.3.3. SC: cs; le.

Matesanz, Mateu B. Advance statements: legal and ethical implications. *Nursing Standard* 2006 September 20-26; 21(2): 41-45. NRCBL: 20.5.4; 4.1.3; 9.2.

Michalowski, Sabine. Advance refusals of life-sustaining medical treatment: the relativity of an absolute right. *Modern Law Review* 2005 November; 68(6): 958-982. NRCBL: 20.5.4; 8.3.3; 8.3.4; 1.1; 1.2. SC: le. Identifiers: Great Britain (United Kingdom); Mental Capacity Act of 2005.

Miller, Mark. Making decisions about advance health care directives. *In:* Hamel, Ronald, ed. Making Health Care Decisions: A Catholic Guide. Liguori, MO: Liguori Publications, 2006: 91-108. NRCBL: 20.5.4; 1.2.

Mirarchi, Ferdinando L.; Conti, Lucia. Living wills and DNR: is patient safety compromised? *Human Life Review* 2007 Fall; 33(4): 66-73. NRCBL: 20.5.4; 20.5.1.

Morrison, Wynne. Thoughts on advance directives. *Journal of Palliative Medicine* 2006 April; 9(2): 483-484. NRCBL: 20.5.4.

Nakashima, David Y. Your body, your choice: how mandatory advance health-care directives are necessary to protect your fundamental right to accept or refuse medical treatment. *University of Hawai'i Law Review* 2004 Winter; 27(1): 201-231. NRCBL: 20.5.4; 8.3.4; 8.4; 20.5.1. Identifiers: Patient Self Determination Act; Uniform Health Care Decisions Act.

Nelson, Margaret K. Listening to Anna. *Health Affairs* 2007 May-June; 26(3): 836-840. NRCBL: 20.5.4.

Perkins, Henry S. Controlling death: the false promise of advance directives. *Annals of Internal Medicine* 2007 July 3; 147(1): 51-57. NRCBL: 20.5.4. SC: cs.
Abstract: Advance directives promise patients a say in their future care but actually have had little effect. Many experts blame problems with completion and implementation, but the advance directive concept itself may be fundamentally flawed. Advance directives simply presuppose more control over future care than is realistic. Medical crises cannot be predicted in detail, making most prior instructions difficult to adapt, irrelevant, or even misleading. Furthermore, many proxies either do not know patients' wishes or do not pursue those wishes effectively. Thus, unexpected problems arise often to defeat advance directives, as the case in this paper illustrates. Because advance directives offer only limited benefit, advance care planning should emphasize not the completion of directives but the emotional preparation of patients and families for future crises. The existentialist Albert Camus might suggest that physicians should warn patients and families that momentous, unforeseeable decisions lie ahead. Then, when the crisis hits, physicians should provide guidance; should help make decisions despite the inevitable uncertainties; should share responsibility for those decisions; and, above all, should courageously see patients and families through the fearsome experience of dying.

Rodriguez, Keri L.; Young, Amanda J. Patients' and healthcare providers' understandings of life-sustaining treatment: are perceptions of goals shared or divergent? *Social Science and Medicine* 2006 January; 62(1): 125-133. NRCBL: 20.5.4; 20.5.1; 8.1. SC: cs; em.

Rozzini, Renzo; Trabucchi, Marco. Advance directives and quality of end-of-life care: pros and cons in older people [letter]. *Journal of the American Geriatrics Society* 2007 September; 55(9): 1472. NRCBL: 20.5.4; 9.5.2; 9.8; 20.4.1.

Samanta, Ash; Samanta, Jo. Advance directives, best interests and clinical judgement: shifting sands at the end of life. *Clinical Medicine* 2006 May-June; 6(3): 274-278. NRCBL: 20.5.4; 20.5.1; 9.4; 7.1; 8.1.

Shaw, Jim. POLST: honoring wishes at the end of life. *Health Care Ethics USA* 2007 Winter; 15(1): 8-12. NRCBL: 20.5.4; 20.5.1. Identifiers: Physician Orders for Life-Sustaining Treatment.

Simon, Alfred. Editorial. *Ethik in der Medizin* 2007 June; 19(2): 89-90. NRCBL: 20.5.4. SC: le. Identifiers: Recommendations for dealing with durable powers of attorney and advance directives in medical practice.

Stocking, C.B.; Hougham, G.W.; Danner, D.D.; Patterson, M.B.; Whitehouse, P.J.; Sachs, G.A. Speaking of research advance directives: planning for future research participation. *Neurology* 2006 May 9; 66(9): 1361-1366. NRCBL: 20.5.4; 18.2; 8.3.3.

Sweet, Victoria. Thy will be done. *Health Affairs* 2007 May-June; 26(3): 825-830. NRCBL: 20.5.4.

Szasz, Thomas. Honouring advance decisions: you don't in psychiatry [letter]. *BMJ: British Medical Journal* 2007 November 3; 335(7626): 900. NRCBL: 20.5.4; 8.3.4; 9.5.1; 17.1. Identifiers: Great Britain (United Kingdom). Comments: Clare Dyer. Patients win right to have their advance decisions honoured by medical staff. BMJ: British Medical Journal 2007 October 6; 335(7622): 688-689.

Tice, Martha A. Patient safety: honoring advanced directives. *Home Healthcare Nurse* 2007 February; 25(2): 79-81. NRCBL: 20.5.4; 20.4.1; 4.1.3; 6.

Volandes, Angelo E.; Abbo, Elmer D. Flipping the default: a novel approach to cardiopulmonary resuscitation in end-stage dementia. *Journal of Clinical Ethics* 2007 Summer; 18(2): 122-139. NRCBL: 20.5.4; 9.5.2; 9.5.1; 20.4.1; 8.3.3; 17.1. SC: em.

Willmott, Lindy; White, Ben; Howard, Michelle. Refusing advance refusals: advance directives and life-sustaining medical treatment. *Melbourne University Law Review* 2006 April; 30(1): 211-243. NRCBL: 20.5.4; 8.3.4. SC: le. Identifiers: Australia.

Witherspoon, Gerald S. What I learned from Schiavo. *Hastings Center Report* 2007 November-December; 37(6): 17-20. NRCBL: 20.5.4; 20.3.1; 20.5.1.

ADVISORY COMMITTEES ON BIOETHICS
See BIOETHICS AND MEDICAL ETHICS/ COMMISSIONS

AFRICAN AMERICANS AS RESEARCH SUBJECTS *See* HUMAN EXPERIMENTATION/ SPECIAL POPULATIONS

AGED *See* CARE FOR SPECIFIC GROUPS/ AGED; HUMAN EXPERIMENTATION/ SPECIAL POPULATIONS/ AGED AND TERMINALLY ILL

AGRICULTURE *See* GENETICALLY MODIFIED ORGANISMS AND FOOD

AIDS

Altman, Dennis. Taboos and denial in government responses. *International Affairs* 2006 March; 82(2): 257-268. NRCBL: 9.5.6; 21.1.

Bayer, Ronald. Ethics and public policy: engaging the moral challenges posed by AIDS. *AIDS Patient Care and STDs* 2006 July; 20(7): 456-460. NRCBL: 9.5.6; 1.3.5; 9.1.

Benatar, Solomon R. Achieving gold standards in ethics and human rights in medical practice. *PLoS Medicine* 2005 August; 2(8): e260. NRCBL: 9.5.6; 9.8; 9.2; 7.1; 21.1.

Bennett, Rebecca. Routine antenatal HIV testing and informed consent: an unworkable marriage? *Journal of Medical Ethics* 2007 August; 33(8): 446-448. NRCBL: 9.5.6; 8.3.1; 9.5.5.
Abstract: This paper considers the ethics of routine antenatal HIV testing and the role of informed consent within such a policy in order to decide how we should proceed in this area — a decision that ultimately rests on the relative importance we give to public health goals on the one hand and respect for individual autonomy on the other.

Chan, K.Y.; Reidpath, D.D. Future research on structural and institutional forms of HIV discrimination. *AIDS Care* 2005 July; 17(Supplement 2): S215-S218. NRCBL: 9.5.6; 9.5.4; 21.1. Identifiers: UNAIDS Protocol for the Identification of Discrimination against People Living with HIV; Asia Pacific.

Chan, K.Y.; Reidpath, D.D. Methodological considerations in the measurement of institutional and structural forms of HIV discrimination. *AIDS Care* 2005 July; 17(Supplement 2): S205-S213. NRCBL: 9.5.6; 9.5.4; 21.1. Identifiers: UNAIDS Protocol for the Identification of Discrimination against People Living with HIV.

Christie, Timothy; Asrat, Getnet A.; Jiwani, Bashir; Maddix, Thomas; Montaner, Julio S.G. Exploring disparities between global HIV/AIDS funding and recent tsunami relief efforts: an ethical analysis. *Developing World Bioethics* 2007 April; 7(1): 1-7. NRCBL: 9.5.6; 9.4; 21.1.
Abstract: Objective: To contrast relief efforts for the 26 December 2004 tsunami with current global HIV/AIDS relief efforts and analyse possible reasons for the disparity. Methods: Literature review and ethical analysis. Results: Just over 273,000 people died in the tsunami, resulting in relief efforts of more than US$10 bn, which is sufficient to achieve the United Nation's long-term recovery plan for South East Asia. In contrast, 14 times more people died from HIV/AIDS in 2004, with UNAIDS predicting a US$8 bn funding gap for HIV/ AIDS in developing nations between now and 2007. This disparity raises two important ethical questions. First, what is it that motivates a more empathic response to the victims of the tsunami than to those affected by HIV/ AIDS? Second, is there a morally relevant difference between the two tragedies that justifies the difference in the international response? The principle of justice requires that two cases similarly situated be treated similarly. For the difference in the international response to the tsunami and HIV/AIDS to be justified, the tragedies have to be shown to be dissimilar in some relevant respect. Are the tragedies of the tsunami disaster and the HIV/AIDS pandemic sufficiently different, in relevant respects, to justify the difference in scope of the response by the international community? Conclusion: We detected no morally relevant distinction between the tsunami and the HIV/AIDS pandemic that justifies the disparity. Therefore, we must conclude that the international response to HIV/AIDS violates the fundamental principles of justice and fairness.

Cohen, Jon. Brazil, Thailand override big pharma patents. *Science* 2007 May 11; 316(5826): 816. NRCBL: 9.5.6; 9.7; 9.3.1; 21.1; 5.3.

Coombes, Rebecca. Bad blood. *BMJ: British Medical Journal* 2007 April 28; 334(7599): 879-880. NRCBL: 9.5.6; 8.3.1; 19.4. Identifiers: haemophilia; factor VIII; hepatitis; HIV.

de Zulueta, Paquita; Boulton, Mary. Routine antenatal HIV testing: the responses and perceptions of pregnant women and the viability of informed consent. A qualitative study. *Journal of Medical Ethics* 2007 June; 33(6): 329-336. NRCBL: 9.5.6; 1.3.5; 8.3.1; 9.1; 9.5.5. SC: em.
Abstract: This qualitative cross-sectional survey, undertaken in the antenatal booking clinics of a hospital in central London, explores pregnant women's responses to routine HIV testing, examines their reasons for declining or accepting the test, and assesses how far their responses fulfil standard criteria for informed consent. Of the 32 women interviewed, only 10 participants were prepared for HIV testing at their booking interview. None of the women viewed themselves as being particularly at risk for HIV infection. The minority (n = 6) of the participants who declined testing differed from those who accepted, by interpreting test acceptance as risky behaviour, privileging the negative outcomes of HIV positivity and expressing an inability to cope with these, should they occur. Troublingly, only a minority of women (n = 9) had a broad understanding of the rationale for the test, and none fulfilled the standard criteria for informed consent. This study suggests that, although routine screening combined with professional recommendation may be successful in increasing uptake, this may be at the cost of eroding informed consent. Protecting third parties (notably fetuses) from a preventable disease may outweigh the moral duty of respecting autonomy, enshrined in Western bioethical tradition. Nevertheless, such a policy should be made transparent, debated in the public domain and negotiated with women seeking antenatal care.

Delpierre, Cyrille; Cuzin, Lise; Lert, France. Routine testing to reduce late HIV diagnosis in France. *BMJ: British Medical Journal* 2007 June 30; 334(7608): 1354-1356. NRCBL: 9.5.6; 9.1.

Fielder, Odicie; Altice, Frederick L. Attitudes toward and beliefs about prenatal HIV testing policies and mandatory HIV testing of newborns among drug users. *AIDS and Public Policy Journal* 2005 Fall-Winter; 20(3-4): 74-91. NRCBL: 9.5.6; 9.1; 7.1; 9.5.7; 9.5.9. SC: em.

Gilling-Smith, Carole. Risking parenthood? Serious viral illness, parenting and the welfare of the child. *In:* Shenfield, Françoise; Sureau, Claude, eds. Contemporary Ethical Dilemmas in Assisted Reproduction. Abingdon: Informa Healthcare, 2006: 57-69. NRCBL: 9.5.6; 9.5.1; 9.5.7; 14.4.

Gray, Glenda. When bodies remember [review of Experiences and Politics of AIDS in South Africa, by Didier Fassin]. *New England Journal of Medicine* 2007 October 25; 357(17): 1783-1784. NRCBL: 9.5.6; 21.1.

Hammill, M.; Burgoine, K., Farrell, F.; Hemelaar, J.; Patel, G.; Welchew, D.E.; Jaffe, H.W. Time to move towards opt-out testing for HIV in the UK. *BMJ: British Medical Journal* 2007 June 30; 334(7608): 1352-1354. NRCBL: 9.5.6; 8.3.1; 9.1.

Letamo, Gobopamang. The discriminatory attitudes of health workers against people living with HIV. *PLoS Medicine* 2005 August; 2(8): e261 (0715-0716). NRCBL: 9.5.6; 8.1; 9.2.

Luna, Florencia. To procreate or not to procreate? AIDS and reproductive rights. *In:* Luna, Florencia; Herissone-Kelly, Peter, ed. Bioethics and Vulnerability: a Latin American View. Amsterdam; New York: Rodopi, 2006: 49-62. NRCBL: 9.5.6; 14.1; 15.2; 9.5.8; 9.5.5; 1.1.

Myer, Landon; Moodley, Kaymanthri; Hendricks, Fahad; Cotton, Mark. Healthcare providers' perspectives on discussing HIV status with infected children. *Journal of Tropical Pediatrics* 2006 August; 52(4): 293-295. NRCBL: 9.5.6; 9.5.7; 8.1; 8.2.

Patton, Cindy. Bullets, balance, or both: medicalisation in HIV treatment. *Lancet* 2007 February 24-March 2; 369(9562): 706-707. NRCBL: 9.5.6; 7.3; 7.1; 8.1; 4.1.1; 4.2.

Paxton, Lynn A.; Hope, Tony; Jaffe, Harold W. Pre-exposure prophylaxis for HIV infection: what if it works? *Lancet* 2007 July 7-13; 370(9581): 89-93. NRCBL: 9.5.6; 9.2; 9.7.

Peters, Jeremy W. New Jersey requires HIV test in pregnancy. *New York Times* 2007 December 27; p. B3. NRCBL: 9.5.6; 9.5.5.

Reidpath, D.D.; Brijnath, B.; Chan, K.Y. An Asia Pacific six-country study on HIV-related discrimination: introduction. *AIDS Care* 2005 July; 17(Supplement 2): S117-S127. NRCBL: 9.5.6; 9.5.4; 21.1. SC: em. Identifiers: India; Thailand; Phillipines; China; Vietnam; Indonesia; UNAIDS Protocol for the Identification of Discrimination Against People Living with HIV (2000).

Reidpath, D.D.; Chan, K.Y. HIV/AIDS discrimination in the Asia Pacific [editorial]. *AIDS Care* 2005 July; 17(Supplement 2): S115-S116. NRCBL: 9.5.6; 9.5.4; 21.1.

Reis, Chen; Heisler, Michele; Amowitz, Lynn L.; Moreland, R. Scott; Mafeni, Jerome O.; Anyamele, Chukwuemeka; Iacopino, Vincent. Discriminatory attitudes and practices by health workers toward patients with HIV/AIDS in Nigeria. *PLoS Medicine* 2005 August; 2(8): e246 (0743-0752). NRCBL: 9.5.6; 8.1; 9.2. SC: em.

Rennie, Stuart. Do the ravages of the HIV/AIDS epidemic ethically justify mandatory HIV testing? [letter]. *Developing World Bioethics* 2007 April; 7(1): 48-49.

NRCBL: National Reference Center for Bioethics Literature Classification Scheme			See inside front cover for terms.

23

NRCBL: 9.5.6. Identifiers: Africa. Comments: Peter A. Clark. Mother-to-child transmission of HIV in Botswana: an ethics perspective on mandatory testing. Developing World Bioethics 2006; 6: 1-12.

Rosen, Sydney; Sanne, Ian; Collier, Alizanne; Simon, Jonathon L. Rationing antiretroviral therapy for HIV/AIDS in Africa: choices and consequences. *PLoS Medicine* 2005 November; 2(11): e303: 1098-1104. NRCBL: 9.5.6; 9.2; 21.1; 9.3.1. SC: em.

Schüklenk, Udo; Kleinschmidt, Anita. Rethinking mandatory HIV testing during pregnancy in areas with high HIV prevalence rates: ethical and policy issues. *American Journal of Public Health* 2007 July; 97(7): 1179-1183. NRCBL: 9.5.6; 8.3; 9.1; 9.5.5; 21.1.

Abstract: We analyzed the ethical and policy issues surrounding mandatory HIV testing of pregnant women in areas with high HIV prevalence rates. Through this analysis, we seek to demonstrate that a mandatory approach to testing and treatment has the potential to significantly reduce perinatal transmission of HIV and defend the view that mandatory testing is morally required if a number of conditions can be met. If such programs are to be introduced, continuing medical care, including highly active antiretroviral therapy, must be provided and pregnant women must have reasonable alternatives to compulsory testing and treatment. We propose that a liberal regime entailing abortion rights up to the point of fetal viability would satisfy these requirements. Pilot studies in the high-prevalence region of southern African countries should investigate the feasibility of this approach.

Smith, Charles B.; Battin, Margaret P.; Francis, Leslie P.; Jacobson, Jay A. Should rapid tests for HIV infection now be mandatory during pregnancy? Global differences in scarcity and a dilemma of technological advance. *Developing World Bioethics* 2007 August; 7(2): 86-103. NRCBL: 9.5.6; 9.5.5.

Abstract: Since testing for HIV infection became possible in 1985, testing of pregnant women has been conducted primarily on a voluntary, 'opt-in' basis. Faden, Geller and Powers, Bayer, Wilfert, and McKenna, among others, have suggested that with the development of more reliable testing and more effective therapy to reduce maternal-fetal transmission, testing should become either routine with 'opt-out' provisions or mandatory. We ask, in the light of the new rapid tests for HIV, such as OraQuick, and the development of antiretroviral treatment that can reduce maternal-fetal transmission rates to %, whether that time is now. Illustrating our argument with cases from the United States (US), Kenya, Peru, and an undocumented Mexican worker in the US, we show that when testing is accompanied by assured multi-drug therapy for the mother, the argument for opt-out or mandatory testing for HIV in pregnancy is strong, but that it is problematic where testing is accompanied by adverse events such as spousal abuse or by inadequate intrapartum or follow-up treatment. The difference is not a 'double standard', but reflects the presence of conflicts between the health interests of the mother and the fetus –

conflicts that would be abrogated by the assurance of adequate, continuing multi-drug therapy. In light of these conflicts, where they still occur, careful processes of informed consent are appropriate, rather than opt-out or mandatory testing.

Tarantola, Daniel; Gruskin, Sofia. New guidance on recommended HIV testing and counselling [comment]. *Lancet* 2007 July 21-27; 370(9583): 202-203. NRCBL: 9.5.6; 8.3.1; 9.8.

Wicker, Sabine; Rabenau, Holger; Gottschalk, Rene; Spickhoff, Andreas. Nadelstichverletzung des behandelnden Arztes bei der Untersuchung einer nichteinwilligungsfähigen Patientin - Darf ein HIV-Test durchgeführt werden? [Needlestick injuries incurred while examining a patient incapable of giving consent. May an HIV-test be performed?]. *Ethik in der Medizin* 2007 September; 19(3): 215-220. NRCBL: 9.5.6; 8.3.1; 16.3. SC: cs.

Yang, Y.; Zhang, K.L.; Chan, K.Y.; Reidpath, D.D. Institutional and structural forms of HIV-related discrimination in health care: a study set in Beijing. *AIDS Care* 2005 July; 17(Supplement 2): S129-S140. NRCBL: 9.5.6; 9.5.4. SC: em. Identifiers: China.

AIDS/ CONFIDENTIALITY

Dixon-Mueller, Ruth; Germain, Adrienne. HIV testing: the mutual rights and responsibilities of partners. *Lancet* 2007 December 1-7; 370(9602): 1808-1809. NRCBL: 9.5.6; 8.4; 8.3.4; 9.2; 7.1; 21.1.

Johnson, Sandra. New CDC guidelines for HIV screening: ethical implications for health care providers. *Health Care Ethics USA* 2007 Winter; 15(1): 5-7. NRCBL: 9.5.6; 9.1; 8.4. Identifiers: Centers for Disease Control.

Lifson, Alan R.; Rybicki, Sarah L. Routine opt-out HIV testing [commentary]. *Lancet* 2007 February 17-23; 369(9561): 539-540. NRCBL: 9.5.6; 9.5.1; 8.3.1; 9.3.1; 7.1; 8.4.

Pozgar, George D. Acquired immunodeficiency syndrome. *In his:* Legal Aspects of Health Care Administration. 9th edition. Sudbury, MA: Jones and Bartlett Publishers, 2004: 375-385. NRCBL: 9.5.6; 8.4; 8.5. SC: le; cs.

World Medical Association [WMA]. The World Medical Association statement on HIV/AIDS and the medical profession. *Indian Journal of Medical Ethics* 2007 April-June; 4(2): 84-86. NRCBL: 9.5.6; 8.4; 9.8; 7.1; 7.2; 21.1.

AIDS/ HUMAN EXPERIMENTATION

Ansell, Nicola; Van Blerk, Lorraine. Joining the conspiracy? Negotiating ethics and emotions in researching (around) AIDS in Southern Africa. *Ethics Place and Environment* 2005 March; 8(1): 61-82. NRCBL: 9.5.6; 18.5.9; 7.1; 21.1. SC: cs.

Cleaton-Jones, Peter. The first randomised trial of male circumcision for preventing HIV: what were the ethical issues? *PLoS Medicine* 2005 November; 2(11): e287: 1073-1075. NRCBL: 9.5.6; 18.2; 9.5.1; 10.

Cohen, Jon. Feud over AIDS vaccine trials leads prominent Italian researchers to court. *Science* 2007 August 10; 317(5839): 738-739. NRCBL: 9.5.6; 9.7; 18.5.9. SC: le.

de la Gorgendière, Louise. Rights and wrongs: HIV/ AIDS research in Africa. *Human Organization* 2005 Summer; 64(2): 166-178. NRCBL: 9.5.6; 18.5.9; 14.1; 10; 7.1. SC: em.

Delaney, Martin. AIDS activism and the pharmaceutical industry. *In:* Santoro, Michael A.; Gorrie, Thomas M., eds. Ethics and the Pharmaceutical Industry. Cambridge; New York: Cambridge University Press, 2005: 300-325. NRCBL: 9.5.6; 9.7; 18.5.1; 18.2; 18.6.

Farquhar, Carey; John-Stewart, Grace C.; John, Francis N.; Kabura, Marjory N.; Kiarie, James N. Pediatric HIV type 1 vaccine trial acceptability among mothers in Kenya. *AIDS Research and Human Retroviruses* 2006 June; 22(6): 491-495. NRCBL: 9.5.6; 9.7; 18.5.2.

Garber, Mandy; Hanusa, Barbara H.; Switzer, Galen E.; Mellors, John; Arnold, Robert M. HIV-infected African Americans are willing to participate in HIV treatment trials. *JGIM: Journal of General Internal Medicine* 2007 January; 22(1): 17-42. NRCBL: 9.5.6; 18.5.1. SC: em.

Greif, Karen F.; Merz, Jon F. Protecting the public: the FDA and new AIDS drugs. *In their:* Current Controversies in the Biological Sciences: Case Studies of Policy Challenges from New Technologies. Cambridge, MA: MIT, 2007: 117-147. NRCBL: 9.5.6; 9.7; 9.8; 18.5.1. SC: le.

Luna, Florencia. AIDS, research, and acceptable codes. *In:* Luna, Florencia; Herissone-Kelly, Peter, ed. Bioethics and Vulnerability: a Latin American View. Amsterdam; New York: Rodopi, 2006: 87-109. NRCBL: 9.5.6; 6; 18.2; 18.5.9.

Moodley, Kaymanthri. Microbicide research in developing countries: have we given the ethical concerns due consideration? *BMC Medical Ethics* 2007; 8(10): 7 p. [electronic]. Accessed: http://www.biomedcentral.com/content/pdf/1472-6939-8-10.pdf [2007 December 18]. NRCBL: 9.5.6; 18.5.9; 9.7; 18.5.2; 18.5.3.

Muula, Adamson. Malawi: ethical challenges of HIV and AIDS in Malawi, southern Africa. *In:* Davis, Anne J.; Tschudin, Verena; de Raeve, Louise, eds. Essentials of Teaching and Learning in Nursing Ethics: Perspectives and Methods. New York: Churchill Livingstone Elsevier, 2006: 291-300. NRCBL: 9.5.6; 4.1.3; 7.2; 21.7; 10; 18.5.9.

Pence, Gregory E. Utilitarians vs. Kantians on stopping AIDS. *In his:* The Elements of Bioethics. Boston:

McGraw-Hill, 2007: 81-108. NRCBL: 9.5.6; 1.1; 9.7; 9.3.1; 9.4; 18.3.

Richardson, Henry S. Gradations of researchers' obligation to provide ancillary care for HIV/AIDS in developing countries. *American Journal of Public Health* 2007 November; 97(11): 1956-1961. NRCBL: 9.5.6; 18.5.9.

Slomka, Jacquelyn; McCurdy, Sheryl; Ratliff, Eric A.; Timpson, Sandra; Williams, Mark L. Perceptions of financial payment for research participation among African-American drug users in HIV studies. *JGIM: Journal of General Internal Medicine* 2007 October; 22(10): 1403-1409. NRCBL: 9.5.6; 18.5.1; 9.3.1. SC: em.

Zion, Deborah. Ethics, disease and obligation. *In:* Bennett, Belinda; Tomossy, George F., eds. Globalization and Health: Challenges for Health Law and Bioethics. Dordrecht: Springer, 2006: 47-57. NRCBL: 9.5.6; 9.7; 21.1; 18.5.9.

AIDS/ LEGAL ASPECTS

Cohen, Jon. Feud over AIDS vaccine trials leads prominent Italian researchers to court. *Science* 2007 August 10; 317(5839): 738-739. NRCBL: 9.5.6; 9.7; 18.5.9. SC: le.

Crook, Jamie. Balancing intellectual property protection with the human right to health. *Berkeley Journal of International Law* 2005; 23(3): 524-550. NRCBL: 9.5.6; 1.3.2; 9.7; 21.1. SC: le. Identifiers: TRIPS (Trade-Related Aspects of Intellectual Property Rights); Doha Declaration.

Elamon, J. A situational analysis of HIV/AIDS-related discrimination in Kerala, India. *AIDS Care* 2005 July; 17(Supplement 2): S141-S151. NRCBL: 9.5.6; 9.5.4; 21.1. SC: em; le.

Forman, Lisa. Claiming equity and justice in health: the role of the South African right to health in ensuring access to HIV/AIDS treatment. *In:* Flood, Colleen M., ed. Just Medicare: What's In, What's Out, How We Decide. Buffalo, NY: University of Toronto Press, 2006: 80-104. NRCBL: 9.5.6; 9.2; 9.1; 1.3.5. SC: an; le.

Greif, Karen F.; Merz, Jon F. Protecting the public: the FDA and new AIDS drugs. *In their:* Current Controversies in the Biological Sciences: Case Studies of Policy Challenges from New Technologies. Cambridge, MA: MIT, 2007: 117-147. NRCBL: 9.5.6; 9.7; 9.8; 18.5.1. SC: le.

Halpern, Scott D.; Metkus, Thomas S.; Fuchs, Barry D.; Ward, Nicholas S.; Siegel, Mark D.; Luce, John M.; Curtis, J. Randall. Nonconsented human immunodeficiency virus testing among critically ill patients: intensivists' practices and the influence of state laws. *Archives of Internal Medicine* 2007 November 26; 167(21): 2323-2328. NRCBL: 9.5.6; 8.3.3; 9.1. SC: le.
Abstract: BACKGROUND: Human immunodeficiency virus (HIV) testing can improve care for many critically ill patients, but state laws and institutional policies may

NRCBL: National Reference Center for Bioethics Literature Classification Scheme   See inside front cover for terms.

25

impede such testing when patients cannot provide consent. METHODS: We electronically surveyed all US academic intensivists in 2006 to determine how state laws influence intensivists' decisions to perform nonconsented HIV testing and to assess intensivists' reliance on surrogate markers of HIV infection when unable to obtain HIV tests. We used multivariate logistic regression, clustered by state, to identify factors associated with intensivists' decisions to pursue nonconsented HIV testing. RESULTS: Of 1026 responding intensivists, 765 (74.6%) had encountered decisionally incapacitated patients for whom HIV testing was wanted. Of these intensivists, 168 pursued testing without consent and 476 first obtained surrogate consent to testing. Intensivists who believed nonconsented HIV testing was ethical (odds ratio, 3.8; 95% confidence interval, 2.2-6.5) and those who believed their states allowed nonconsented testing when medically necessary (odds ratio, 2.3; 95% confidence interval, 1.6-3.4) were more likely to pursue nonconsented HIV tests; actual state laws were unrelated to testing practices. Of the intensivists, 72.7% had ordered tests for perceived surrogate markers of HIV infection in lieu of HIV tests; more than 90% believed these tests were sufficiently valid to base clinical decisions on. CONCLUSIONS: Most US intensivists have encountered decisionally incapacitated patients for whom HIV testing may improve care. Intensivists' decisions to pursue nonconsented testing are associated with their personal ethics and often erroneous perceptions of state laws, but not with the laws themselves. Uniform standards enabling nonconsented HIV testing may minimize inappropriate influences on intensivists' decisions and reduce intensivists' reliance on perceived surrogate markers of immunodeficiency.

Khoat, D.V.; Hong, L.D.; An, C.Q.; Ngu, D.; Reidpath, D.D. A situational analysis of HIV/AIDS-related discrimination in Hanoi, Vietnam. *AIDS Care* 2005 July; 17(Supplement 2): S181-S193. NRCBL: 9.5.6; 9.5.4. SC: em; le. Identifiers: UNAIDS Protocol for the Identification of Discrimination Against People Living with HIV.

Merati, T.; Supriyadi; Yuliana, F. The disjunction between policy and practice: HIV discrimination in health care and employment in Indonesia. *AIDS Care* 2005 July; 17(Supplement 2): S175-S179. NRCBL: 9.5.6; 9.5.4. SC: em; le.

Ortega, N.L.; Bicaldo, B.F.; Sobritchea, C.; Tan, M.L. Exploring the realities of HIV/AIDS-related discrimination in Manila, Philippines. *AIDS Care* 2005 July; 17(Supplement 2): S153-S164. NRCBL: 9.5.6; 9.5.4; 21.1. SC: em; le.

Pottker-Fishel, Carrie Gene. Improper bedside manner: why state partner notification laws are ineffective in controlling the proliferation of HIV. *Health Matrix: The Journal of Law-Medicine* 2007 Winter; 17(1): 147-179. NRCBL: 9.5.6; 9.1. SC: le.

Pozgar, George D. Acquired immunodeficiency syndrome. *In his:* Legal Aspects of Health Care Administra-

tion. 9th edition. Sudbury, MA: Jones and Bartlett Publishers, 2004: 375-385. NRCBL: 9.5.6; 8.4; 8.5. SC: le; cs.

Reidpath, D.D.; Chan, K.Y. HIV discrimination: integrating the results from a six-country situational analysis in the Asia Pacific. *AIDS Care* 2005 July; 17(Supplement 2): S195-S204. NRCBL: 9.5.6; 9.5.4; 21.1. SC: em; le. Identifiers: China; India; Indonesia; Phillippines; Thailand; Vietnam.

Sloth-Nielsen, Julia. Of newborns and nubiles: some critical challenges to children's rights in Africa in the era of HIV/AIDS. *In:* Freeman, Michael, ed. Children's Health and Children's Rights. Boston: Martinus Nijhoff Publishers, 2006: 73-85. NRCBL: 9.5.6; 9.5.7; 6; 21.1. SC: le.

Sringernyuang, L.; Thaweesit, S.; Nakapiew, S. A situational analysis of HIV/AIDS-related discrimination in Bangkok, Thailand. *AIDS Care* 2005 July; 17(Supplement 2): S165-S174. NRCBL: 9.5.6; 9.5.4. SC: em; le. Identifiers: UNAIDS Protocol for the Identification of Discrimination against People Living with HIV.

Zetola, Nicola M. Association between rates of HIV testing and elimination of written consents in San Francisco [letter]. *JAMA: The Journal of the American Medical Association* 2007 March 14; 297(10): 1061. NRCBL: 9.5.6; 8.3.1. SC: le.

ALLOCATION *See* ORGAN AND TISSUE TRANSPLANTATION/ ALLOCATION; RESOURCE ALLOCATION

ALLOWING TO DIE *See* EUTHANASIA AND ALLOWING TO DIE

ANIMAL EXPERIMENTATION

NIH to make chimpanzee breeding moratoriam permanent [news]. *ATLA: Alternatives to Laboratory Animals* 2007 June; 35(3): 300. NRCBL: 22.2; 5.3.

Spare the apes [news]. *New Scientist* 2007 September 15-21; 195(2621): 4. NRCBL: 22.2; 21.1.

Allen, Colin; Bekoff, Marc. Animal minds, cognitive ethology, and ethics. *Journal of Ethics* 2007; 11(3): 299-317. NRCBL: 22.1; 1.1; 4.4.

American Association for Laboratory Animal Science. AALAS position statement on the humane care and use of laboratory animals. *Comparative Medicine* 2006 December; 56(6): 534. NRCBL: 22.2; 6.

Armstrong, Susan J.; Botzler, Richard G. Animal experimentation. *In their:* The Animal Ethics Reader. London; New York: Routledge, 2003: 245-307. NRCBL: 22.2; 16.1; 22.3; 21.1.

Baldwin, Ann; Bekoff, Marc. Too stressed to work: scientists must provide lab animals with decent living conditions or accept that their results could be useless [commentary]. *New Scientist* 2007 June 2-8; 194(2606): 24. NRCBL: 22.2.

Balls, Michael; Combes, Robert. Putting replacement first [editorial]. *ATLA: Alternatives to Laboratory Animals* 2007 June; 35(3): 297-298. NRCBL: 22.2.

Bayne, Kathryn A.; Harkness, John E. Welfare of research animals. *In:* Kulakowski, Elliott C.; Chronister, Lynne U., eds. Research Administration and Management. Sudbury, MA: Jones and Bartlett, 2006: 577-581. NRCBL: 22.2; 5.3.

Benatar, David. Unscientific ethics: science and selective ethics. *Hastings Center Report* 2007 January-February; 37(1): 30-32. NRCBL: 22.2; 1.3.9; 1.3.7.

Bergmeier, L. Animal welfare is not just another bureaucratic hoop [letter]. *Nature* 2007 July 19; 448(7151): 251. NRCBL: 22.2; 1.3.7.

Bermúdez, José Luis. Thinking without words: an overview for animal ethics. *Journal of Ethics* 2007; 11(3): 319-335. NRCBL: 22.1; 4.4. SC: an.

Boon, Mieke. Comments on Thompson: research ethics for animal biotechnology. *In:* Korthals, Michiel; Bogers, Robert J., eds. Ethics for Life Scientists. Dordrecht, The Netherlands: Springer, 2004: 121-125. NRCBL: 22.2; 15.1; 1.3.9.

Bruckner, Donald W. Considerations on the morality of meat consumption: hunted-game versus farm-raised animals. *Journal of Social Philosophy* 2007 Summer; 38(2): 311-330. NRCBL: 22.3; 22.1; 1.1. SC: an.

Buck, Victoria. Who will start the 3Rs ball rolling for animal welfare? [letter]. *Nature* 2007 April 19; 446(7138): 856. NRCBL: 22.2; 1.3.7. Comments: Hanno Würbel. Publications should include an animal-welfare section. Nature 2007 March 15; 446(7133): 257.

Burgess-Jackson, Keith. Doing right by our animal companions. *Journal of Ethics* 1998; 2(2): 159-185. NRCBL: 22.1. SC: an.

Canadian Council on Animal Care [CCAC]. Facts and figures: CCAC animal use survey 2002. Ottawa, Canada: Canadian Council on Animal Care, 2002: 13 p. [Online]. Accessed: http://www.ccac.ca/en/Publications/Facts_Figures/pdfs/ aus2002en-all.pdf [2007 April 26]. NRCBL: 22.2. SC: em.

Carruthers, Peter. Invertebrate minds: a challenge for ethical theory. *Journal of Ethics* 2007; 11(3): 275-297. NRCBL: 22.1; 1.1; 4.4. SC: an.

Chandna, Alka; Stephens, Martin L.; Runkle, Deborah; Pippin, John J.; Greek, Ray; Perry, Seth. What have we learned from the use of animals in scientific research? [letters and reply]. *Chronicle of Higher Education* 2007 March 23; 53(29): B17-B18. NRCBL: 22.2; 22.1.

Comments: Seth Perry. Lives in the balance. Chronicle of Higher Education 2007 January 26; 53(21): B14.

Cohen, Andrew I. Contractarianism, other-regarding attitudes, and the moral standing of nonhuman animals. *Journal of Applied Philosophy* 2007 May; 24(2): 188-201. NRCBL: 22.1. SC: an.

Abstract: Contractarianism roots moral standing in an agreement among rational agents in the circumstances of justice. Critics have argued that the theory must exclude nonhuman animals from the protection of justice. I argue that contractarianism can consistently accommodate the notion that nonhuman animals are owed direct moral consideration. They can acquire their moral status indirectly, but their claims to justice can be as stringent as those among able-bodied rational adult humans. Any remaining criticisms of contractarianism likely rest on a disputable moral realism; contractarianism can underwrite the direct moral considerability of nonhuman animals by appealing to a projectivist quasi-realism.

Cohen, Jon. NIH to end chimp breeding for research [news]. *Science* 2007 June 1; 316(5829): 1265. NRCBL: 22.2.

Cohen, Jon. The endangered lab chimp [news]. *Science* 2007 January 26; 315(5811): 450-452. NRCBL: 22.2; 21.1.

Dandie, Geoff. Report of the International Consensus meeting on carbon dioxide euthanasia of laboratory animals. *ANZCCART News (Australian and New Zealand Council for the Care of Animals in Research and Teaching)* 2006; 19(2): 1-7 [Online]. Accessed: http://www.adelaide.edu.au/ANZCCART/news/AN19_2.pdf [2008 February 15]. NRCBL: 22.1; 21.1.

Davey, Gareth; Wu, Zhihui. Attitudes in China toward the use of animals in laboratory research. *ATLA: Alternatives to Laboratory Animals* 2007 June; 35(3): 313-316. NRCBL: 22.2; 7.1. SC: em.

Dennison, Tania; Leach, Matthew. Animal research, ethics and law. *In:* Pullen, Sophie; Gray, Carol, eds. Ethics, Law and the Veterinary Nurse. New York: Elsevier Butterworth Heinemann, 2006: 103-116. NRCBL: 22.2; 22.1; 18.2. SC: le. Identifiers: Great Britain (United Kingdom); Animals (Scientific Procedures) Act 1986.

Douglas, Kate. Just like us. Humans have rights, other animals don't — no matter how human-like they are. *New Scientist* 2007 June 2-8; 194(2606): 46-49. NRCBL: 22.1; 22.2.

Faden, Ruth R.; Duggan, Patrick S.; Karron, Ruth. Who pays to stop a pandemic? [op-ed]. *New York Times* 2007 February 9; p. A19. NRCBL: 22.3; 9.1; 7.1. SC: po.

Fox, Marie. Exposing harm: the erasure of animal bodies in healthcare law. *In:* McLean, Sheila A.M., ed. First Do No Harm: Law, Ethics, and Healthcare. Aldershot, Eng-

NRCBL: National Reference Center for Bioethics Literature Classification Scheme   See inside front cover for terms.

27

land; Burlington, VT: Ashgate, 2006: 543-559. 86 fn. NRCBL: 22.1; 15.1; 19.1; 22.2. SC: an; le.
> Keywords: *animal welfare; *genetically modified animals; *legal aspects; animal experimentation; animal organs; biotechnology; ethical theory; organ transplantation; speciesism; suffering; Proposed Keywords: *harm; laboratory animals; Keyword Identifiers: Great Britain; Nuffield Council on Bioethics

Francescotti, Robert. Animal mind and animal ethics: an introduction. *Journal of Ethics* 2007; 11(3): 239-252. NRCBL: 22.1; 4.4; 1.1.

Hackam, Daniel G. Translating animal research into clinical benefit [editorial]. *BMJ: British Medical Journal* 2007 January 27; 334(7586): 163-164. NRCBL: 22.2; 18.1.

Hauskeller, Michael. The reification of life. *Genomics, Society and Policy* 2007 August; 3(2): 70-81. 27 fn. NRCBL: 22.1; 1.1; 15.1. SC: an.
> Keywords: *animal rights; *animal welfare; *commodification; *genetically modified animals; *moral policy; animal experimentation; biotechnology; genetic patents; Proposed Keywords: laboratory animals

Abstract: What's wrong – fundamentally wrong – with the way animals are treated (…) isn't the pain, the suffering, isn't the deprivation. (…) The fundamental wrong is the system that allows us to view animals as our resources, here for us – to be eaten, or surgically manipulated, or exploited for sport or money.' Tom Regan made this claim 20 years ago. What he maintains is basically that the fundamental wrong is not the suffering we inflict on animals but the way we look at them. What we do to them, what we believe we are allowed to do to them, is dependent on how we perceive or conceptualize them. We not only treat them as resources but prior to this we already think of them as resources, and when we look at them, all we tend to see is resources. In our perception of them they exist not for themselves but 'for us'. But obviously it can only be fundamentally wrong in a moral sense to view them that way if it is wrong in a factual sense, that is, if animals are in fact not 'for us'. But is it wrong?

Hewson, Caroline J. Veterinarians who swear: animal welfare and the veterinary oath. *Canadian Veterinary Journal. La Revue Vétérinaire Canadienne* 2006 August; 47(8): 807-811. NRCBL: 22.1; 4.1.1.

Ito, Shigehiko. Beyond standing: a search for a new solution in animal welfare. *Santa Clara Law Review* 2005-2006; 46(2): 377-418. NRCBL: 22.1; 16.1. SC: le.

King, Lesley; Rowan, Andrew N. The mental health of laboratory animals. *In:* McMillan, Frank, ed. Mental Health and Well-Being in Animals. Ames, IA: Blackwell Pub., 2005: 259-276. NRCBL: 22.2; 22.1; 4.3.

Knight, Andrew. The poor contribution of chimpanzee experiments to biomedical progress. *Journal of Applied Animal Welfare Science* 2007; 10(4): 281-308. NRCBL: 22.2.

Lamey, Andy. Food fight!: Davis versus Regan on the ethics of eating beef. *Journal of Social Philosophy* 2007 Summer; 38(2): 331-348. NRCBL: 22.1; 1.1. Identifiers: Steven Davis; Tom Regan.

Lassen, Jesper; Gjerris, Mickey; Sandoe, Peter. After Dolly — ethical limits to the use of biotechnology on farm animals. *Theriogenology* 2006 March 15; 65(5): 992-1004. 14 refs. NRCBL: 22.3; 14.5; 15.1. SC: em.
> Keywords: *animal cloning; *animal welfare; *biotechnology; *genetically modified animals; agriculture; animal experimentation; animal organs; attitudes; focus groups; food; genetically modified organisms; interviews; public opinion; risk; wedge argument; Proposed Keywords: *domestic animals; laboratory animals; Keyword Identifiers: *Denmark; European Union

Markowitz, Hal; Timmel, Gregory B. Animal well-being and research outcomes. *In:* McMillan, Frank, ed. Mental Health and Well-Being in Animals. Ames, IA: Blackwell Pub., 2005: 277-283. NRCBL: 22.2; 22.1; 4.3.

Montgomery, Charlotte. Research: keeping humans alive. *In her:* Blood Relations: Animals, Humans, and Politics. Toronto: Between the Lines, 2000: 80-127. NRCBL: 22.2. SC: le.

Musch, Timothy I.; Carroll, Robert G.; Lane, Pascale H.; Talman, William T. A broader view of animal research [letter]. *BMJ: British Medical Journal* 2007 February 10; 334(7588): 274. NRCBL: 22.2.

Norcross, Alastair. Animal experimentation. *In:* Steinbock, Bonnie, ed. The Oxford Handbook of Bioethics. Oxford; New York: Oxford University Press, 2007: 648-667. NRCBL: 22.2; 1.1.

Olsson, I. Anna S.; Hansen, Axel K.; Sandøe, Peter. Ethics and refinement in animal research [letter]. *Science* 2007 September 21; 317(5845): 1680. NRCBL: 22.2; 22.1.

Perel, Pablo; Roberts, Ian; Sena, Emily; Wheble, Philipa; Briscoe, Catherine; Sandercock, Peter; Macleod, Malcolm; Mignini, Luciano E.; Jayaram, Predeep; Khan, Khalid S. Comparison of treatment effects between animal experiments and clinical trials: systematic review. *BMJ: British Medical Journal* 2007 January 27; 334(7586): 197-200. NRCBL: 22.2; 18.1. SC: em.

Abstract: OBJECTIVE: To examine concordance between treatment effects in animal experiments and clinical trials. Study design Systematic review. DATA SOURCES: Medline, Embase, SIGLE, NTIS, Science Citation Index, CAB, BIOSIS. STUDY SELECTION: Animal studies for interventions with unambiguous evidence of a treatment effect (benefit or harm) in clinical trials: head injury, antifibrinolytics in haemorrhage, thrombolysis in acute ischaemic stroke, tirilazad in acute ischaemic stroke, antenatal corticosteroids to prevent neonatal respiratory distress syndrome, and bisphosphonates to treat osteoporosis. Review methods Data were extracted on study design, allocation concealment, number of randomised animals, type of model, intervention, and out-

---

come. RESULTS: Corticosteroids did not show any benefit in clinical trials of treatment for head injury but did show a benefit in animal models (pooled odds ratio for adverse functional outcome 0.58, 95% confidence interval 0.41 to 0.83). Antifibrinolytics reduced bleeding in clinical trials but the data were inconclusive in animal models. Thrombolysis improved outcome in patients with ischaemic stroke. In animal models, tissue plasminogen activator reduced infarct volume by 24% (95% confidence interval 20% to 28%) and improved neurobehavioural scores by 23% (17% to 29%). Tirilazad was associated with a worse outcome in patients with ischaemic stroke. In animal models, tirilazad reduced infarct volume by 29% (21% to 37%) and improved neuro- behavioural scores by 48% (29% to 67%). Antenatal corticosteroids reduced respiratory distress and mortality in neonates whereas in animal models respiratory distress was reduced but the effect on mortality was inconclusive (odds ratio 4.2, 95% confidence interval 0.85 to 20.9). Bisphosphonates increased bone mineral density in patients with osteoporosis. In animal models the bisphos- phonate alendronate increased bone mineral density compared with placebo by 11.0% (95% confidence interval 9.2% to 12.9%) in the combined results for the hip region. The corresponding treatment effect in the lumbar spine was 8.5% (5.8% to 11.2%) and in the combined results for the forearms (baboons only) was 1.7% (-1.4% to 4.7%). CONCLUSIONS: Discordance between animal and human studies may be due to bias or to the failure of animal models to mimic clinical disease adequately.

Quigley, Muireann. Non-human primates: the appropriate subjects of biomedical research? *Journal of Medical Ethics* 2007 November; 33(11): 655-658. NRCBL: 22.2. SC: an.

Abstract: Following the publication of the Weatherall report on the use of non-human primates in research, this paper reflects on how to provide appropriate and ethical models for research beneficial to humankind. Two of the main justifications for the use of non-human primates in biomedical research are analysed. These are the "least-harm/greatest-good" argument and the "capacity" argument. This paper argues that these are equally applicable when considering whether humans are appropriate subjects of biomedical research.

Quill, Elizabeth. Congress considers higher fines for mistreating laboratory animals [news]. *Chronicle of Higher Education* 2007 October 26; 54(9): A20-A21. NRCBL: 22.2. SC: le.

Quill, Elizabeth. Researchers call for self-regulation in care of lab animals. *Chronicle of Higher Education* 2007 October 26; 54(9): A20, A22. NRCBL: 22.2; 1.3.7.

Rayasam, Renuka. Cloning around. An FDA ruling could spur growth at a Texas company that has the leg up on duplicating animals. *U.S. News and World Report* 2007 January 8; 142(1): 46-47. NRCBL: 22.3; 14.5; 5.3. SC: po.

Rollin, Bernard. The ethics of referral. *The Canadian Veterinary Journal.* 2006 July; 47(7): 717-718. NRCBL: 22.1; 7.3; 8.1.

Rollin, Bernard E. An ethicist's commentary on characterizing of convenience euthanasia in ethical terms. *Canadian Veterinary Journal. La Revue Vétérinaire Canadienne* 2006 August; 47(8): 742. NRCBL: 22.1; 4.1.1.

Rollin, Bernard E. Animal mind: science, philosophy, and ethics. *Journal of Ethics* 2007; 11(3): 253-274. NRCBL: 22.1; 22.2; 1.3.11; 1.1; 4.4.

Rudnick, Abraham. Other-consciousness and the use of animals as illustrated in medical experiments. *Journal of Applied Philosophy* 2007 May; 24(2): 202-208. NRCBL: 22.2; 1.1. SC: an.

Abstract: Ethicists such as Peter Singer argue that consciousness and self-consciousness are the principal considerations in discussing the use of animals by humans, such as in medical experiments. This paper raises an additional consideration to factor into this ethical discussion. Ethics deal with the intentional impact of subjects on each other. This assumes a meta-representational ability of subjects to represent states of mind of others, which may be termed other-consciousness. The moral weight of other-consciousness is manifest in the notion of responsibility, where humans lacking in other-consciousness (such as individuals with autism) may not be held responsible for their harmful actions towards others. As responsibility implies not only duties but also rights and more generally high moral status, it follows that other-consciousness grants high moral status, other things being equal — recognizing that other factors grant moral status too. Other-consciousness also increases the capacity for suffering, both due to increased freedom (and consequently increased possibility of restriction of freedom) and to increased empathy (with suffering of others). Hence, the more an animal is other-conscious, the more it deserves high moral status and the more it can suffer, other things being equal, and consequently, the less it should be used for human purposes. Further study is required to elucidate to what extent animals used by humans, such as in medical experiments, particularly primates and other highly evolved mammals, are other-conscious.

Schiermeier, Quirin. Primate work faces German veto [news]. *Nature* 2007 April 26; 446(7139): 955. NRCBL: 22.2. Identifiers: Germany

Shapiro, Kenneth. Animal experimentation. *In:* Waldau, Paul; Patton, Kimberly, eds. A Communion of Subjects: Animals in Religion, Science, and Ethics. New York: Columbia University Press, 2006: 533-543. NRCBL: 22.2; 1.2; 1.3.9; 1.1.

Sherwin, C.M. Animal welfare: reporting details is good science [letter]. *Nature* 2007 July 19; 448(7151): 251. NRCBL: 22.2; 1.3.7.

NRCBL: National Reference Center for Bioethics Literature Classification Scheme   See inside front cover for terms.

29

Simmons, Aaron. A critique of Mary Anne Warren's weak animal rights view. *Environmental Ethics* 2007 Fall; 29(3): 267-278. NRCBL: 22.1; 1.1. SC: an.

Spiers, Alexander S.D. Studies in animals should be more like those in humans [letter]. *BMJ: British Medical Journal* 2007 February 10; 334(7588): 274. NRCBL: 22.2.

Stafleu, F.R.; Tramper, R.; Vorstenbosch, J.; Joles, J.A. The ethical acceptability of animal experiments: a proposal for a system to support decisionmaking. *Lab Animal* 1999 July; 33(3): 295-303. NRCBL: 22.2; 5.2.

Sunstein, Cass R. Slaughterhouse jive [review of Introduction to Animal Rights: Your Child or the Dog? by Gary L. Francione]. *New Republic* 2001 January 29; 224(5): 40-45. NRCBL: 22.1; 1.1; 7.1; 22.2. SC: an; le.

Thompson, Paul B. Research ethics for animal biotechnology. *In:* Korthals, Michiel; Bogers, Robert J., eds. Ethics for Life Scientists. Dordrecht, The Netherlands: Springer, 2004: 105-120. NRCBL: 22.2; 1.3.9; 22.3; 15.1; 5.3. SC: an.

Touitou, Yvan.; Smolensky, Michael H.; Portaluppi, Francesco. Ethics, standards, and procedures of animal and human chronobiology research. *Chronobiology International* 2006; 23(6): 1083-1096. NRCBL: 22.1; 18.2; 1.3.7.

Watts, Geoff. Animal testing: is it worth it? *BMJ: British Medical Journal* 2007 January 27; 334(7586): 182-184. NRCBL: 22.2. SC: rv. Identifiers: Great Britain (United Kingdom).

Weatherall, David. Animal research: the debate continues. *Lancet* 2007 April 7-13; 369(9568): 1147-1148. NRCBL: 22.2; 5.3; 4.4; 1.3.7; 7.1; 15.1; 18.5.4.

Weatherall, David; Munn, Helen. Moving the primate debate forward [editorial]. *Science* 2007 April 13; 316(5822): 173. NRCBL: 22.2.

Würbel, Hanno. Publications should include an animal-welfare section [letter]. *Nature* 2007 March 15; 446(7133): 257. NRCBL: 22.2; 1.3.7.

ARTIFICIAL INSEMINATION AND SURROGATE MOTHERS
See also REPRODUCTIVE TECHNOLOGIES

Austria. Bioethikkommission beim Bundeskanzleramt Austria. Bioethics Commission at the Federal Chancellery. Praimplantationsdiagnostik (PID): Bericht der Bioethikkommission beim Bundeskanzleramt [Preimplantation genetic diagnosis [PGD]: report of the Bioethics Commission of the Federal Chancellery]. Vienna, Austria: Bioethikkommission beim Bundeskanzleramt, 2004 July: 71 p. Refs.: p. 68-71. NRCBL: 14.2; 18.5.4; 8.3.1; 12.1; 15.2. Note: Also published in English.

Keywords: *preimplantation diagnosis; *public policy; legal aspects; prenatal diagnosis; embryos; advisory committees; moral status; international aspects; disabled persons; selective abortion; social discrimination; beginning of life; government regulation; motivation; value of life; Keyword Identifiers: *Austria

Austria. Bioethikkommission beim Bundeskanzleramt Austria. Bioethics Commission at the Federal Chancellery. Preimplantation genetic diagnosis (PGD): report of the Bioethics Commission at the Federal Chancellery. Vienna, Austria: Bioethikkommission beim Bundeskanzleramt, 2004 July: 67 p. Refs.: p. 64-67. NRCBL: 14.2; 18.5.4; 8.3.1; 12.1; 15.2. SC: le. Note: Also published in German.

Keywords: *preimplantation diagnosis; *public policy; legal aspects; international aspects; prenatal diagnosis; embryos; moral status; advisory committees; beginning of life; government regulation; motivation; value of life; disabled persons; eugenics; social discrimination

Bahadur, Guido. Till death do us part: to be or not to be . . . a parent after one's death? *In:* Shenfield, Françoise; Sureau, Claude, eds. Contemporary Ethical Dilemmas in Assisted Reproduction. Abingdon: Informa Healthcare, 2006: 29-42. NRCBL: 14.2; 8.3.3; 19.5; 20.1. SC: cs.

Case, Gretchen. The few and the proud. *Atrium* 2007 Summer; 4: 9. NRCBL: 14.2; 19.5; 1.3.2; 9.3.1. Identifiers: commercial egg donation campaign.

Costa, Rosely Gomes. Racial classification regarding semen donor selection in Brazil. *Developing World Bioethics* 2007 August; 7(2): 104-111. NRCBL: 14.2; 9.5.4.

Abstract: Brazil has not yet approved legislation on assisted reproduction. For this reason, clinics, hospitals and semen banks active in the area follow Resolution 1358/92 of the Conselho Federal de Medicina, dated 30 September 1992. In respect to semen donation, the object of this article, the Resolution sets out that gamete donation shall be anonymous, that is, that the donor and recipients (and the children who might subsequently be born) shall not be informed of each other's identity. Thus, since recipients are unaware of the donor's identity, semen banks and the medical teams involved in assisted reproduction become the intermediaries in the process. The objective of this article is to show that, in practice, this represents disrespect for the ethical principles of autonomy, privacy and equality. The article also stresses that the problem is compounded by the racial question. In a country like Brazil, where racial classification is so flexible and goes side by side with racist attitudes, the intermediary role played by semen banks and medical teams is conditioned by their own criteria of racial classification, which are not always the same as those of donors and semen recipients. The data presented in this paper were taken from two semen banks located in the city of São Paulo (Brazil). At the time of my research, they were the only semen banks in the state of São Paulo and supplied semen to the capital (São Paulo city), the state of São Paulo, and to cities in other Brazilian states where semen banks were not available.

Daniels, Ken; Meadows, Letitia. Sharing information with adults conceived as a result of donor insemination. *Human Fertility* 2006 June; 9(2): 93-99. NRCBL: 14.2; 8.4.

Davis, A. Surrogates and outcast mothers: racism and reproductive politics in the nineties. *In:* James, J., ed. The Angela Y. Davis Reader. Malden, MA: Blackwell, 1998: 210-221. NRCBL: 14.2; 14.1; 9.5.4; 10.

Day, Michael. Number of sperm donors rises in UK despite removal of anonymity [news]. *BMJ: British Medical Journal* 2007 May 12; 334(7601): 971. NRCBL: 14.2; 19.5.

De Jonge, Christopher; Barratt, Christopher L. Gamete donation: a question of anonymity. *Fertility and Sterility* 2006 February; 85(2): 500-501. NRCBL: 14.2; 19.5.

Ernst, Erik; Ingerslev, Hans Jakob.; Schou, Ole; Stoltenberg, Meredin. Attitudes among sperm donors in 1992 and 2002: a Danish questionnaire survey. *Acta obstetricia et gynecologica Scandinavica* 2007 March; 86(3): 327-333. NRCBL: 14.2; 1.3.5; 14.1; 8.4. SC: le; em.

Hershberger, Patricia; Klock, Susan C.; Barnes, Randall B. Disclosure decisions among pregnant women who received donor oocytes: a phenomenological study. *Fertility and Sterility* 2007 February; 87(2): 288-296. NRCBL: 14.2; 9.5.5; 8.4; 1.3.1. SC: em.

Legge, M.; Fitzgerald, R.; Frank, N. A retrospective study of New Zealand case law involving assisted reproduction technology and the social recognition of 'new' family. *Human Reproduction* 2007 January; 22(1): 17-25. NRCBL: 14.2; 21.7; 14.1. SC: le; an.

Mavroforou, A.; Koumantakis, E.; Mavrophoros, D.; Michalodimitrakis, E. Medically assisted human reproduction: the Greek view. *Medicine and Law: The World Association for Medical Law* 2007 June; 26(2): 339-347. NRCBL: 14.2. SC: le.

Abstract: Medically assisted human reproduction is a controversial issue that has attracted heated debate over the last two decades. In December 2002 the Greek Parliament passed a law with major social and scientific impact: the Medically Assisted Human Reproduction Act 3089/02. This law establishes the parameters of so-called surrogate motherhood, protects the anonymity of semen donors and sets the legal framework through which a woman's artificial fertilization after her husband's death is allowed. This article aims to discuss the legal ramifications of medically assisted human reproduction and especially the moral and social issues concerning the introduction of surrogate motherhood in Greece.

McGregor, J.; Dreifuss-Netter, F. France and the United States: the legal and ethical differences in assisted reproductive technology (ART). *Medicine and Law: The World Association for Medical Law* 2007 March; 26(1): 117-135. NRCBL: 14.2; 14.4; 14.6. SC: le; an.

Abstract: In this paper, we will look at the development and the current status of the laws in France (and to some extent other EU nations) and the United States. In doing so, many differences will emerge about the views and values of assisted reproductive technology in the two countries. We will try to articulate some of those underlying moral, political, and legal differences.

McLachlan, Hugh V.; Swales, J. Kim. Posthumous insemination and consent. *In their:* From the Womb to the Tomb: Issues in Medical Ethics. Glasgow, Scotland: Humming Earth, 2007: 251-267. NRCBL: 14.2; 8.3.1; 20.1. SC: an; le. Identifiers: Diane Blood.

McLachlan, Hugh V.; Swales, J. Kim. Surrogate motherhood. *In their:* From the Womb to the Tomb: Issues in Medical Ethics. Glasgow, Scotland: Humming Earth, 2007: 85-170. NRCBL: 14.2; 9.3.1. SC: an. Identifiers: Brazier Report; Warnock Report.

McLeod, Carolyn. For dignity or money: feminists on the commodification of women's reproductive labour. *In:* Steinbock, Bonnie, ed. The Oxford Handbook of Bioethics. Oxford; New York: Oxford University Press, 2007: 258-281. NRCBL: 14.2; 4.4; 9.5.5; 9.3.1.

Robertson, John A. Compensation and egg donation for research. *Fertility and Sterility* 2006 December; 86(6): 1573-1575. NRCBL: 14.2; 14.4; 9.3.1; 18.2; 18.5.4. SC: le.

Suzuki, Kohta; Hoshi, Kazuhiko; Minai, Junko; Yanaihara, Takumi; Takeda, Yasuhisa; Yamagata, Zentaro. Analysis of national representative opinion surveys concerning gestational surrogacy in Japan. *European Journal of Obstetrics, Gynecology, and Reproductive Biology* 2006 May 1; 126(1): 39-47. NRCBL: 14.2; 21.7; 9.3.1.

Willmott, Lindy. Surrogacy: ART's forgotten child. *University of New South Wales Law Journal* 2006; 29(2): 227-232. NRCBL: 14.2. SC: le. Identifiers: assisted reproductive technology.

Yee, Samantha; Hitkari, Jason A.; Greenblatt, Ellen M. A follow-up study of women who donated oocytes to known recipient couples for altruistic reasons. *Human Reproduction* 2007 July; 22(7): 2040-2050. NRCBL: 14.2; 9.5.5; 19.5. SC: em.

Zahedi, F.; Larijani, B.; Isikoglu, M.; Senol, Y.; Berkkanoglu, M.; Ozgur, K.; Donmez, L. Considerations of third-party reproduction in Iran [letter and reply]. *Human Reproduction* 2007 March; 22(3): 902-903. NRCBL: 14.2; 1.2. Comments: M. Isikoglu, et al. Public opinion regarding oocyte donation in Turkey. Human Reproduction 2006 January; 21(1): 318-323.

ARTIFICIAL NUTRITION AND HYDRATION
See EUTHANASIA AND ALLOWING TO DIE

NRCBL: National Reference Center for Bioethics Literature Classification Scheme   See inside front cover for terms.

31

ASSISTED REPRODUCTIVE TECHNOLO-GIES *See* REPRODUCTIVE TECHNOLOGIES

ASSISTED SUICIDE
See also EUTHANASIA AND ALLOWING TO DIE

California suicide bill 'implicitly anti-Catholic' [news brief]. *America* 2007 May 28; 196(19): 7. NRCBL: 20.7; 20.5.1; 1.2. SC: le.

American Nurses Association [ANA]. Ethics and human rights position statements: assisted suicide. Silver Spring, MD: American Nurses Association, 1994 December 8: 5 p. [Online]. Accessed: http://www.nursingworld.org/ readroom/position/ethics/etsuic.htm [2007 April 19]. NRCBL: 20.7; 20.5.1; 8.1.

Appel, Jacob M. A suicide right for the mentally ill? A Swiss case opens a new debate. *Hastings Center Report* 2007 May-June; 37(3): 21-23. NRCBL: 20.7; 20.5.1; 8.1; 17.1. SC: le. Identifiers: Switzerland.

Banerjee, Albert; Birenbaum-Carmeli, Daphna. Ordering suicide: media reporting of family assisted suicide in Britain. *Journal of Medical Ethics* 2007 November; 33(11): 639-642. NRCBL: 20.7; 1.3.7; 20.5.1. SC: em.
Abstract: OBJECTIVE: To explore the relationship between the presentation of suffering and support for euthanasia in the British news media. METHOD: Data was retrieved by searching the British newspaper database LexisNexis from 1996 to 2000. Twenty-nine articles covering three cases of family assisted suicide (FAS) were found. Presentations of suffering were analysed employing Heidegger's distinction between technological ordering and poetic revealing. FINDINGS: With few exceptions, the press constructed the complex terrain of FAS as an orderly or orderable performance. This was enabled by containing the contradictions of FAS through a number of journalistic strategies: treating degenerative dying as an aberrant condition, smoothing over botched attempts, locating the object of ethical evaluation in persons, not contexts, abbreviating the decision making process, constructing community consensus and marginalising opposing views. CONCLUSION: The findings of this study support the view that news reporting of FAS is not neutral or inconsequential. In particular, those reports presenting FAS as an orderly, rational performance were biased in favor of technical solutions by way of the legalisation of euthanasia and/or the involvement of medical professionals. In contrast, while news reports sensitive to contradiction did not necessarily oppose euthanasia, they were less inclined to overtly support technical solutions, recognising the importance of a trial to address the complexity of FAS.

Basta, Lofty L. Dying on one's own terms. *American Journal of Geriatric Cardiology* 2006 July-August; 15(4): 250-252. NRCBL: 20.7; 20.5.1; 8.1. SC: le.

Battin, Margaret P.; van der Heide, Agnes; Ganzini, Linda; van der Wal, Gerrit; Onwuteaka-Philipsen, Bregje D. Legal physician-assisted dying in Oregon and the Netherlands: evidence concerning the impact on patients in "vulnerable" groups. *Journal of Medical Ethics* 2007 October; 33(10): 591-597. NRCBL: 20.7; 20.5.1. SC: em; le.
Abstract: Background: Debates over legalisation of physician-assisted suicide (PAS) or euthanasia often warn of a "slippery slope", predicting abuse of people in vulnerable groups. To assess this concern, the authors examined data from Oregon and the Netherlands, the two principal jurisdictions in which physician-assisted dying is legal and data have been collected over a substantial period. Methods: The data from Oregon (where PAS, now called death under the Oregon Death with Dignity Act, is legal) comprised all annual and cumulative Department of Human Services reports 1998–2006 and three independent studies; the data from the Netherlands (where both PAS and euthanasia are now legal) comprised all four government-commissioned nationwide studies of end-of-life decision making (1990, 1995, 2001 and 2005) and specialised studies. Evidence of any disproportionate impact on 10 groups of potentially vulnerable patients was sought. Results: Rates of assisted dying in Oregon and in the Netherlands showed no evidence of heightened risk for the elderly, women, the uninsured (inapplicable in the Netherlands, where all are insured), people with low educational status, the poor, the physically disabled or chronically ill, minors, people with psychiatric illnesses including depression, or racial or ethnic minorities, compared with background populations. The only group with a heightened risk was people with AIDS. While extralegal cases were not the focus of this study, none have been uncovered in Oregon; among extralegal cases in the Netherlands, there was no evidence of higher rates in vulnerable groups. Conclusions: Where assisted dying is already legal, there is no current evidence for the claim that legalised PAS or euthanasia will have disproportionate impact on patients in vulnerable groups. Those who received physician-assisted dying in the jurisdictions studied appeared to enjoy comparative social, economic, educational, professional and other privileges.

Belgium. Ministry of Justice. The Belgian Act on Euthanasia of May 28th, 2002. *Ethical Perspectives* 2002; 9(2-3): 182-188 [Online]. Accessed: http://www. kuleuven.ac.be/cbmer/viewpic.php?LAN=E&TABLE= DOCS&ID=23 [2007 April 17]. NRCBL: 20.7; 20.5.1; 20.5.3. SC: le.

Berghmans, Ron L.P.; Widdershoven, Guy A.M. Physician-assisted dying in the Netherlands. *EACME Newsletter* 2007 July; (17): 3 p. [Online]. Accessed: http://www.eacmeweb.com/newsletter/n17.htm [2007 August 15]. NRCBL: 20.7; 20.5.1; 8.1. SC: em.

Bergner, Daniel. Death in the family: Booth Gardner, a former governor of Washington State who has Parkinson's, is urgently lobbying for a doctor-assisted-suicide law. His son is among those fighting him every step of the way. *New York Times Magazine* 2007 December 2; p. 38-45, 60, 76, 78, 80. NRCBL: 20.7; 20.5.1; 20.5.4; 20.3.3. SC: po; le.

Button, James. Dealing in the desire for death. *Sydney Morning Herald* 2007 February 3; 4 p. [Online]. Accessed: http://www.smh.com.au/news/world/dealing-in-the-desire-for-death/2007/02/02/116 9919531030.html [2008 February 8]. NRCBL: 20.7; 20.5.1; 4.4.

Canada. Supreme Court; Brock, Dan W.; Callahan, Daniel. Euthanasia and assisted suicide. *In:* Baylis, Françoise; Downie, Jocelyn; Freedman, Benjamin; Hoffmaster, Barry; Sherwin, Susan, eds. Health Care Ethics in Canada. Toronto: Harcourt Brace Canada, 1995: 527-571. NRCBL: 20.7; 20.5.1; 1.1. SC: le. Note: Section includes: Introduction — Voluntary active euthanasia / Dan W. Brock — When self-determination runs amok / Daniel Callahan — Sue Rodriguez v. the Attorney General of Canada and the Attorney General of British Columbia / Supreme Court of Canada.

Craig, Alexa; Cronin, Beth; Eward, William; Metz, James; Murray, Logan; Rose, Gail; Suess, Eric; Vergara, Maria E. Attitudes toward physician-assisted suicide among physicians in Vermont. *Journal of Medical Ethics* 2007 July; 33(7): 400-403. NRCBL: 20.7; 20.5.1; 20.3.2. SC: em; le.

Abstract: Background: Legislation on physician-assisted suicide (PAS) is being considered in a number of states since the passage of the Oregon Death With Dignity Act in 1994. Opinion assessment surveys have historically assessed particular subsets of physicians. Objective: To determine variables predictive of physicians' opinions on PAS in a rural state, Vermont, USA. Design: Cross-sectional mailing survey. Participants: 1052 (48% response rate) physicians licensed by the state of Vermont. Results: Of the respondents, 38.2% believed PAS should be legalised, 16.0% believed it should be prohibited and 26.0% believed it should not be legislated. 15.7% were undecided. Males were more likely than females to favour legalisation (42% vs 34%). Physicians who did not care for patients through the end of life were significantly more likely to favour legalisation of PAS than physicians who do care for patients with terminal illness (48% vs 33%). 30% of the respondents had experienced a request for assistance with suicide. Conclusions: Vermont physicians' opinions on the legalisation of PAS is sharply polarised. Patient autonomy was a factor strongly associated with opinions in favour of legalisation, whereas the sanctity of the doctor–patient relationship was strongly associated with opinions in favour of not legislating PAS. Those in favour of making PAS illegal overwhelmingly cited moral and ethical beliefs as factors in their opinion. Although opinions on legalisation appear to be based on firmly held beliefs, approximately half of Vermont physicians who responded to the survey agree that there is a need for more education in palliative care and pain management.

Dieterle, J.M. Physician assisted suicide: a new look at the arguments. *Bioethics* 2007 March; 21(3): 127-139. NRCBL: 20.7; 20.5.1; 8.1; 1.1; 10; 21.1.

Abstract: In this paper, I examine the arguments against physician assisted suicide (PAS). Many of these arguments are consequentialist. Consequentialist arguments rely on empirical claims about the future and thus their strength depends on how likely it is that the predictions will be realized. I discuss these predictions against the backdrop of Oregon's Death with Dignity Act and the practice of PAS in the Netherlands. I then turn to a specific consequentialist argument against PAS – Susan M. Wolf's feminist critique of the practice. Finally, I examine the two most prominent deontological arguments against PAS. Ultimately, I conclude that no anti-PAS argument has merit. Although I do not provide positive arguments for PAS, if none of the arguments against it are strong, we have no reason not to legalize it.

Dworkin, Gerald. Physician-assisted death: the state of the debate. *In:* Steinbock, Bonnie, ed. The Oxford Handbook of Bioethics. Oxford; New York: Oxford University Press, 2007: 375-392. NRCBL: 20.7; 20.5.1.

Dyer, Clare. Dignitas is forced to offer its services from a former factory [news]. *BMJ: British Medical Journal* 2007 December 8; 335(7631): 1176. NRCBL: 20.7; 20.5.1.

Fenner, Dagmar. Ist die Institutionalisierung und Legalisierung der Suizidbeihilfe gefährlich? Eine kritische Analyze der Gegenargumente = Is the institutionalization and legalization of assistance to suicide dangerous? A critical analysis of counterarguments. *Ethik in der Medizin* 2007 September; 19(3): 200-214. NRCBL: 20.7; 20.5.1; 1.1; 8.1. SC: an.

Ferreira, N. Latest legal and social developments in the euthanasia debate: bad moral consciences and political unrest. *Medicine and Law: The World Association for Medical Law* 2007 June; 26(2): 387-407. NRCBL: 20.7; 20.5.1; 21.1. SC: rv.

Abstract: Several events that took place during recent years, such as the French Act on the rights of patients and the end of life, the Terri Schiavo case and Lord Joffe's proposal for an Assisted Dying Bill in the United Kingdom, have triggered the debate on euthanasia more than ever. It is therefore opportune to revisit basic notions related thereto and to make a comparative analysis of the legal regime of euthanasia in several countries in Europe and elsewhere, as well as to try to see how the public awareness of the problem has of late developed. There seems to be a clear trend in many legal systems towards an increasing respect for the patient's right to self-determination. However, we are still looking at a complex social game, where legal and medical terminology are manipulated and euphemisms are invented in order to accommodate bad moral consciences and avoid political unrest.

Fish, Mark. The health professional and the dying patient. *In:* Gunning, Jennifer; Holm, Søren, eds. Ethics, law, and society. Volume 1. Aldershot, Hants, England; Burlington,

NRCBL: National Reference Center for Bioethics Literature Classification Scheme   See inside front cover for terms.

33

VT: Ashgate, 2005: 239-241. NRCBL: 20.7; 20.5.1; 20.3.2.

Ganz, Freda DeKeyser; Musgrave, Catherine F. Israeli critical care nurses' attitudes toward physician-assisted dying. *Heart and Lung* 2006 November-December; 35(6): 412-422. NRCBL: 20.7; 20.5.1; 4.1.3; 20.3.2. SC: em.

Ganzini, Linda; Beer, Tomasz M.; Brouns, Matthew C. Views on physician-assisted suicide among family members of Oregon cancer patients. *Journal of Pain and Symptom Management* 2006 September; 32(3): 230-236. NRCBL: 20.7; 20.5.1; 8.1; 9.5.1.

Gevers, Sjef. Evaluation of the Dutch legislation on euthanasia and assisted suicide. *European Journal of Health Law* 2007 December; 14(4): 369-379. NRCBL: 20.7; 20.5.1. SC: le. Identifiers: Netherlands. Note: Includes a Summary of the Evaluation Report.

Guevin, Benedict M. Extraordinary treatment or suicide? *Ethics and Medics* 2007 May; 32(5): 1-2. NRCBL: 20.7; 20.5.1; 8.1; 1.2. Identifiers: Italy; Piergiorgio Welby.

Hicks, Madelyn Hsiao-Rei. Physician-assisted suicide: a review of the literature concerning practical and clinical implications for UK doctors. *BMC Family Practice* 2006 June 22; 7: 39-55. NRCBL: 20.7; 20.5.1; 8.1. SC: rv.

Kamm, F.M. Ending life. *In:* Rhodes, Rosamond; Francis, Leslie P.; Silvers, Anita, eds. The Blackwell Guide to Medical Ethics. Malden, MA: Blackwell Pub., 2007: 142-161. NRCBL: 20.7; 20.5.1; 20.5.4.

Kapp, Marshall B. The US Supreme Court decision on assisted suicide and the prescription of pain medication: limit the celebration. *Journal of Opioid Management* 2006 March-April; 2(2): 73-74. NRCBL: 20.7; 20.5.1; 1.3.5; 9.7. SC: le.

Ledger, Sylvia Dianne. Euthanasia and assisted suicide: there is an alternative. *Ethics and Medicine: An International Journal of Bioethics* 2007 Summer; 23(2): 81-94. NRCBL: 20.7; 20.5.1; 8.1; 20.4.1; 4.4; 7.2. SC: rv.

Long, Robert Emmet. Dr. Kevorkian and assisted suicide. *In his:* Suicide. New York: H.W. Wilson, 1995: 76-114. NRCBL: 20.7; 20.5.1; 9.7; 8.1; 4.4; 2.2.

Long, Robert Emmet. Kevorkian's critics. *In his:* Suicide. New York: H.W. Wilson, 1995: 115-144. NRCBL: 20.7; 20.5.1; 8.1. SC: an.

Maitra, Robin T.; Harfst, Anja; Bjerre, Lise M.; Kochen, Michael M.; Becker, Annette. Do German general practitioners support euthanasia? Results of a nationwide questionnaire survey. *European Journal of General Practice* 2005 September-December; 11(3-4): 94-100. NRCBL: 20.7; 20.5.1; 8.1. SC: em.

Marker, Rita L. Suicide by any other name. *Human Life Review* 2007 Winter; 33(1): 78-94. NRCBL: 20.7; 20.5.1. SC: le. Identifiers: Oregon.

McAneeley, Lindsay N. Physician assisted suicide: expanding the laboratory to the state of Hawai'i. *University of Hawai'i Law Review* 2006 Winter; 29(1): 269-299. NRCBL: 20.7; 1.1; 8.1; 20.5.1. SC: le.

Novak, David. Physician-assisted suicide. *In his:* The Sanctity of Human Life. Washington, DC: Georgetown University Press, 2007: 111-171. NRCBL: 20.7; 8.1; 20.5.1; 1.2. SC: an; cs.

Pakes, Francis. The legislation of euthanasia and assisted suicide: a tale of two scenarios. *International Journal of the Sociology of Law* 2005 June; 33(2): 71-84. NRCBL: 20.7; 20.5.1; 8.1; 21.1. SC: le.

Pozgar, George D. Health care ethics. *In his:* Legal Aspects of Health Care Administration. 9th edition. Sudbury, MA: Jones and Bartlett Publishers, 2004: 387-422. NRCBL: 20.7; 20.5.1; 19.5; 20.2.1; 20.5.4; 8.3.3. SC: le; cs.

Quill, Timothy E. Legal regulation of physician-assisted death: the latest report cards. *New England Journal of Medicine* 2007 May 10; 356(19): 1911-1913. NRCBL: 20.7; 20.5.1. SC: em; le.

Quill, Timothy E. Physician assisted death in vulnerable populations [editorial]. *BMJ: British Medical Journal* 2007 September 29; 335(7621): 625-626. NRCBL: 20.7; 20.5.1; 9.5.1. Identifiers: Oregon; Netherlands.

Smith, Stephen W. Some realism about end of life: the current prohibition and the euthanasia underground. *American Journal of Law and Medicine* 2007; 33(1): 55-95. NRCBL: 20.7; 20.5.1; 20.3.1; 20.3.2. SC: em.

Snyder, Lois. Bioethics, assisted suicide, and the "right to die". *Annals of Clinical Psychiatry* 2001; 13(1): 13-18. NRCBL: 20.7; 1.1; 7.1; 8.1; 20.5.1. SC: an.

Stephenson, Jeffrey. Assisted dying: a palliative care physician's view. *Clinical Medicine (London, England)* 2006 July-August; 6(4): 374-377. NRCBL: 20.7; 20.5.1; 20.4.1; 7.1; 8.1.

Tonks, Alison. Physician assisted deaths: no "slippery slope" in the Netherlands and Oregon. *BMJ: British Medical Journal* 2007 May 19; 334(7602): 1029. NRCBL: 20.7; 20.5.1; 8.1.

Tucker, Kathryn L. Federalism in the context of assisted dying: time for the laboratory to extend beyond Oregon, to the neighboring state of California. *Willamette Law Review* 2005; 41(5): 863-880. NRCBL: 20.7; 20.5.1; 8.1. SC: le.

Tuffs, Annette. Swiss hospitals admit to allowing assisted suicide on their wards under guidelines [news]. *BMJ: Brit-*

---

ish Medical Journal 2007 November 24; 335(7629): 1064-1065. NRCBL: 20.7; 20.5.1.

Turone, Fabio. Doctor helps Italian patient die [news]. *BMJ: British Medical Journal* 2007 January 6; 334(7583): 9. NRCBL: 20.7; 20.5.1; 8.1.

van der Heide, Agnes; Onwuteaka-Philipsen, Bregje D.; Rurup, Mette L.; Buiting, Hilde M.; van Delden, Johannes J.M.; Hanssen-de Wolf, Johanna E.; Janssen, Anke G.J.M.; Pasman, H. Roeline W.; Rietjens, Judith A.C.; Prins, Cornelis J.M.; Deerenberg, Ingeborg M.; Gevers, Joseph K.M.; van der Maas, Paul J.; van der Wal, Gerrit. End-of-life practices in the Netherlands under the Euthanasia Act. *New England Journal of Medicine* 2007 May 10; 356(19): 1957-1965. NRCBL: 20.7; 20.3.2; 20.5.1. SC: em; le.

Abstract: BACKGROUND: In 2002, an act regulating the ending of life by a physician at the request of a patient with unbearable suffering came into effect in the Netherlands. In 2005, we performed a follow-up study of euthanasia, physician-assisted suicide, and other end-of-life practices. METHODS: We mailed questionnaires to physicians attending 6860 deaths that were identified from death certificates. The response rate was 77.8%. RESULTS: In 2005, of all deaths in the Netherlands, 1.7% were the result of euthanasia and 0.1% were the result of physician-assisted suicide. These percentages were significantly lower than those in 2001, when 2.6% of all deaths resulted from euthanasia and 0.2% from assisted suicide. Of all deaths, 0.4% were the result of the ending of life without an explicit request by the patient. Continuous deep sedation was used in conjunction with possible hastening of death in 7.1% of all deaths in 2005, significantly increased from 5.6% in 2001. In 73.9% of all cases of euthanasia or assisted suicide in 2005, life was ended with the use of neuromuscular relaxants or barbiturates; opioids were used in 16.2% of cases. In 2005, 80.2% of all cases of euthanasia or assisted suicide were reported. Physicians were most likely to report their end-of-life practices if they considered them to be an act of euthanasia or assisted suicide, which was rarely true when opioids were used. CONCLUSIONS: The Dutch Euthanasia Act was followed by a modest decrease in the rates of euthanasia and physician-assisted suicide. The decrease may have resulted from the increased application of other end-of-life care interventions, such as palliative sedation.

Watson, Katie; Quill, Timothy. A conversation with Dr. Timothy Quill [interview]. *Atrium* 2007 Summer; 4: 18-20. NRCBL: 20.7; 20.5.1.

Ziegler, Stephen J. Euthanasia and the administration of neuromuscular blockers without ventilation: should physicians fear prosecution? *Omega* 2006; 53(4): 295-310. NRCBL: 20.7; 20.5.1; 9.7. SC: cs; em; le.

Ziegler, Stephen; Bosshard, Georg. Role of non-governmental organisations in physician assisted suicide. *BMJ: British Medical Journal* 2007 February 10; 334(7588): 295-298. NRCBL: 20.7; 20.5.1; 8.1; 9.1; 21.1.

ATTITUDES TO DEATH *See* DEATH AND DYING/ ATTITUDES TO DEATH

BEHAVIOR CONTROL
See also CARE FOR SPECIFIC GROUPS/ MENTALLY DISABLED; ELECTROCONVULSIVE THERAPY; INVOLUNTARY COMMITMENT; MENTAL HEALTH THERAPIES AND NEUROSCIENCES; PSYCHOPHARMACOLOGY; PSYCHOTHERAPY

The future of mind control. *Economist* 2002 May 25; 363(8274): 11. NRCBL: 17.3; 14.5; 4.4.

Gaber, Tarek A-Z.K. Medico-legal and ethical aspects in the management of wandering patients following brain injury: questionnaire survey. *Disability and Rehabilitation* 2006 November 30; 28(22): 1413-1416. NRCBL: 17.3; 7.1; 8.1. SC: le.

Irving, Kate. Governing the conduct of conduct: are restraints inevitable? *Journal of Advanced Nursing* 2002 November; 40(4): 405-412. NRCBL: 17.3; 9.5.2. Identifiers: Australia.

Paterson, Brodie; Duxbury, Joy. Restraint and the question of validity. *Nursing Ethics* 2007 July; 14(4): 535-545. NRCBL: 17.3.
Abstract: Restraint as an intervention in the management of acute mental distress has a long history that predates the existence of psychiatry. However, it remains a source of controversy with an ongoing debate as to its role. This article critically explores what to date has seemingly been only implicit in the debate surrounding the role of restraint: how should the concept of validity be interpreted when applied to restraint as an intervention? The practice of restraint in mental health is critically examined using two post-positivist constructions of validity, the pragmatic and the psychopolitical, by means of a critical examination of the literature. The current literature provides only weak support for the pragmatic validity of restraint as an intervention and no support to date for its psychopolitical validity. Judgements regarding the validity of any intervention that is coercive must include reference to the psychopolitical dimensions of both practice and policy.

Walther, Guy. Freiheitsentziehende Maßnahmen in Altenpflegeheimen - rechtliche Grundlagen und Alternativen der Pflege = Restraints in long-term care nursing homes for the elderly - legal aspects and alternatives. *Ethik in der Medizin* 2007 December; 19(4): 289-300. NRCBL: 17.3; 17.4; 9.5.2. SC: le.

Wolverson, M. Restrictive physical interventions. *In:* Holland, Stephen, ed. Introducing Nursing Ethics: Themes in Theory and Practice. Salisbury: APS, 2004: 111-130. NRCBL: 17.3; 2.1; 4.4. SC: cs.

NRCBL: National Reference Center for Bioethics Literature Classification Scheme   See inside front cover for terms.

35

BEHAVIORAL GENETICS

Backlar, Patricia. Genes and behavior: will genetic information change the way we see ourselves? *Community Mental Health Journal* 1996 June; 32: 205-209. 18 refs. NRCBL: 15.6; 15.1.
> Keywords: *behavioral genetics; *genetic information; *self concept; clinical genetics; eugenics; genetic determinism; genetic research; mental disorders; right not to know

Berghmans, R.; de Jong, Johan; Tibben, A.; de Wert, G. Genetics of alcoholism: ethical and societal implications. *EACME Newsletter* 2007 July; (17): 2 p. [Online]. Accessed: http://www.eacmeweb.com/newsletter/n17.htm [2007 August 15]. NRCBL: 15.6; 18.5.1; 18.6.

Canli, Turhan. When genes and brains unite: ethical implications of genomic neuroimaging. *In:* Illes, Judy, ed. Neuroethics: Defining the Issues in Theory, Practice, and Policy. New York: Oxford University Press, 2006: 169-183. 53 refs. NRCBL: 15.6; 17.1.
> Keywords: *behavioral genetics; *neurosciences; behavioral research; brain; emotions; employment; genetic privacy; genetic screening; law enforcement; marketing; politics; Proposed Keywords: *brain imaging; personality

de Melo-Martín, Immaculada. When is biology destiny? Biological determinism and social responsibility. *Philosophy of Science* 2003 December; 70(5): 1184-1194. 33 refs. 2 fn. NRCBL: 15.6; 10; 3.1; 15.9. SC: an.
> Keywords: *genetic determinism; *obligations of society; *socioeconomic factors; *values; disadvantaged persons; eugenics; intelligence; justice; men; sex factors; sociobiology; women

Hamer, Ian. Pastor, the gene made me do it! *Concordia Journal* 1997 January; 23: 18-26. NRCBL: 15.6; 1.2; 15.1.

Hil, Richard; Hindmarsh, Richard. Body talk: genetic screening as a device of crime regulation. *In:* Betta, Michela, ed. The Moral, Social, and Commercial Imperatives of Genetic Testing and Screening: the Australian Case. Dordrecht: Springer, 2006: 55-70. 46 fn. NRCBL: 15.6; 1.3.5; 15.5.
> Keywords: *aggression; *behavioral genetics; *genetic screening; *violence; behavior disorders; eugenics; racial groups; social problems; XYY chromosome; Proposed Keywords: *antisocial behavior; *crime; attention deficit disorder with hyperactivity; forensic genetics

Levitt, Mairi; Manson, Neil. My genes made me do it? The implications of behavioural genetics for responsibility and blame. *Health Care Analysis: An International Journal of Health Philosophy and Policy* 2007 March; 15(1): 33-40. 26 refs. 2 fn. NRCBL: 15.6; 1.1; 1.3.5.
> Keywords: *behavioral genetics; *criminal law; *genetic determinism; *violence; adolescents; aggression; behavior control; behavior disorders; behavioral research; dangerousness; DNA fingerprinting; genetic predisposition; genetic research; genetic screening; law enforcement; legal liability; philosophy; socioeconomic factors; Proposed Keywords: *culpability; Keyword Identifiers: Great Britain; United States

Abstract: The idea of individual responsibility for action is central to our conception of what it is to be a person. Behavioural genetic research may seem to call into question the idea of individual responsibility with possible implications for the criminal justice system. These implications will depend on the understandings of the various agencies and professional groups involved in responding to violent and anti-social behaviour, and, the result of negotiations between them over resulting practice. The paper considers two kinds of approaches to the question of responsibility and 'criminal genes' arising from a sociological and philosophical perspective respectively. One is to consider the social context and possible practical implications of research into 'criminal genes' which will later be examined through interviews and discussions with a range of experts including lawyers and social workers. A second and different kind of approach is to ask whether the findings of behavioural genetics ought to have implications for attributions of responsibility. Issues of genetic influence are central to both approaches.

MacKellar, Calum. Ethics and genetics of human behaviour [commentary]. *Ethics and Medicine: An International Journal of Bioethics* 2007 Spring; 23(1): 7-9. 6 fn. NRCBL: 15.6; 4.3.
> Keywords: *behavioral genetics; *genetic research; eugenics; genetic determinism; mental disorders; risks and benefits; social discrimination; stigmatization

McPhate, Gordon. Ensoulment revised in response to genetics, neuroscience and out-of-body experiences. *In:* Deane-Drummond, Celia; Scott, Peter Manley, eds. Future Perfect?: God, Medicine and Human Identity. New York: T and T Clark International, 2006: 100-112. NRCBL: 15.6; 17.1; 1.2; 1.1.

Radulovic, Jelena; Stankovic, Bratislav. Genetic determinants of emotional behavior: legal lessons from genetic models. *DePaul Law Review* 2007 Spring; 56(3): 823-836. NRCBL: 15.6; 15.11; 1.3.8.

Sternberg, Robert J. Not a case of black and white. *New Scientist* 2007 October 27-November 2; 196(2627): 24. NRCBL: 15.6; 15.11.

BEHAVIORAL RESEARCH
See also BIOMEDICAL RESEARCH; HUMAN EXPERIMENTATION

Ashcraft, Mark H.; Krause, Jeremy A. Social and behavioral researchers' experiences with their IRBs. *Ethics and Behavior* 2007; 17(1): 1-17. NRCBL: 18.4; 18.2; 1.3.9; 7.1. SC: em.
Abstract: A national survey on researchers' experiences with their institutional review boards (IRBs) is presented, focused exclusively on social and behavioral researchers. A wide range of experiences is apparent in the data, especially in terms of turnaround time for submitted protocols, incidence of data collection without prior IRB approval, and stated reasons for "going solo." Sixty-two percent felt that the turnaround time they typically experience is "reasonable," and 44% said they had not experi-

enced long delays in obtaining approval. However, 48% of respondents reported either conducting a project without IRB approval or modifying an existing project without IRB approval, with anticipated time for approval being the dominant reason offered for doing so. This adds a new dimension to the widely discussed "national IRB crisis" (e.g., Illinois White Paper, 2005). The article concludes with 2 preliminary recommendations for IRB reform.

Barsky, Lauren E.; Donner, William R. Lessons from the field: human needs often complicate ethical duties in disaster research. Hurricane Katrina investigation. *Protecting Human Subjects* 2007 November (15): 18-19. NRCBL: 18.4; 18.2; 18.5.1.

Broom, Alex. Ethical issues in social research. *Complementary Therapies in Medicine* 2006 June; 14(2): 151-156. NRCBL: 18.4; 4.1.1; 8.4.

Burke, John; Diehl, Dawn; Durosinmi, Brenda; McGinnis, Troy A. The privacy of stigmatized persons. *Journal of Empirical Research on Human Research Ethics* 2007 March; 2(1): 65-67. NRCBL: 18.4; 18.5.1; 8.4; 10; 18.5.5.

Fischer, Henry W. Protecting human subjects from themselves . . . after the disaster. *Protecting Human Subjects* 2007 November (15): 20-21. NRCBL: 18.4; 18.2; 18.5.9; 18.3.

Flagel, David C.; Best, Lisa A.; Hunter, Aren C. Perceptions of stress among students participating in psychology research: a Canadian survey. *Journal of Empirical Research on Human Research Ethics* 2007 September; 2(3): 61-67. NRCBL: 18.4; 18.3; 18.2; 7.2. SC: em. Identifiers: Canada.

Abstract: It has been shown that properly conducted interviews in sensitive clinical contexts are negligibly stressful. The present study sought to extend these results and determine the perception of stress by research participants in nonclinical settings. Students enrolled in first year psychology courses typically have the option to receive class credit for research participation in studies assumed to pose minimal risk to participants. The perceptions of 101 student volunteers were examined to determine if they felt that research participation was stressful and, if so, what components of the process caused their stress. Participants completed a short survey indicating the reasons they served as research participants and the degree to which participation was stressful. They indicated that research participation was a valuable learning experience and the majority felt no stress associated with participation. Stress was reported by some due to concerns about confidentiality and evaluation by others of their personal performance. In addition, the majority of students reported having no knowledge of the ethical review process that preceded their participation. It is suggested that students should be informed of the ethical review process.

Holm, Søren; Bortolotti, Lisa. Large scale surveys for policy formation and research—a study in inconsistency. *Theoretical Medicine and Bioethics* 2007; 28(3): 205-220. NRCBL: 18.4; 9.1; 18.3; 21.1. SC: an. Identifiers: Europe.

Abstract: In this paper we analyse the degree to which a distinction between social science and public health research and other non-research activities can account for differences between a number of large scale social surveys performed at the national and European level. The differences we will focus on are differences in how participation is elicited and how data are used for government, research and other purposes. We will argue that the research / non-research distinction does not account for the identified differences in recruitment or use and that there are no other convincing justifications. We argue that this entails that eliciting participation by coercion or manipulation becomes very difficult to justify.

Jacobson, Nora; Gewurtz, Rebecca; Haydon, Emma. Ethical review of interpretive research: problems and solutions. *IRB: Ethics and Human Research* 2007 September-October; 29(5): 1-8. NRCBL: 18.4; 18.2.

Kilpatrick, Dean G. The ethics of disaster research: a special section. *Journal of Traumatic Stress* 2004 October; 17(5): 361-362. NRCBL: 18.4; 18.2; 18.5.1.

Miller, Arthur G. What can the Milgram obedience experiments tell us about the Holocaust? Generalizing from the social psychology laboratory. *In:* Miller, Arthur G. ed. The Social Psychology of Good and Evil. New York: The Guilford Press, 2004: 193-239. NRCBL: 18.4; 21.4; 2.2. SC: rv.

North, Carol S.; Pfefferbaum, Betty; Tucker, Phebe. Ethical and methodological issues in academic mental health research in populations affected by disasters: the Oklahoma City experience relevant to September 11, 2001. *CNS Spectrums* 2002 August; 7(8): 580-584. NRCBL: 18.4; 4.3; 18.2; 18.5.2; 18.5.6.

Oczak, Malgorzata; Niedzwienska; Agnieszka. Debriefing in deceptive research: a proposed new procedure. *Journal of Empirical Research on Human Research Ethics* 2007 September; 2(3): 49-59. NRCBL: 18.4; 8.2; 18.2. SC: em. Identifiers: Poland.

Abstract: This study examines the effectiveness of a new debriefing procedure designed specifically to address possible negative consequences of participation in deceptive research. The new debriefing includes an extended educational procedure that enables participants to gain insight into relevant deceptive practices and how to recognize and deal effectively with them, and thus end their participation with a positive and beneficial learning experience. The usefulness of the new tool was analyzed in a suggestibility study in which we compared the effects of the standard debriefing and the new procedure in terms of participants' mood, self-esteem, and attitudes toward psychological experiments. The most important result was that at the end of the study subjects who received the new debriefing system expressed more positive mood

NRCBL: National Reference Center for Bioethics Literature Classification Scheme   See inside front cover for terms.

37

and more positive attitudes toward research than those who received the standard debriefing system. The implications of these results for generalizing to other kinds of deception research are discussed.

Øhrstrøm, Peter; Dyhrberg, Johan. Ethical problems inherent in psychological research based on Internet communication as stored information. *Theoretical Medicine and Bioethics* 2007; 28(3): 221-241. NRCBL: 18.4; 1.3.12; 8.4; 18.3.

Abstract: This paper deals with certain ethical problems inherent in psychological research based on Internet communication as stored information. Section 1 contains an analysis of research on Internet debates. In particular, it takes into account a famous example of deception for psychology research purposes. In section 2, the focus is on research on personal data in texts published on the Internet. Section 3 includes an attempt to formulate some ethical principles and guidelines, which should be regarded as fundamental in research on stored information.

Owen, Michael. Ethical review of social and behavioral science research. *In:* Kulakowski, Elliott C.; Chronister, Lynne U., eds. Research Administration and Management. Sudbury, MA: Jones and Bartlett, 2006: 543-556. NRCBL: 18.4; 18.2; 18.3; 18.5.1; 18.5.9.

Ross, Colin A. Ethics of CIA and military contracting by psychiatrists and psychologists. *Ethical Human Psychology and Psychiatry* 2007; 9(1): 25-34. NRCBL: 18.4; 8.2; 1.3.5; 17.3; 17.4; 17.5.

Rutherford, Alexandra. The social control of behavior control: behavior modification, individual rights, and research ethics in America, 1971-1979. *Journal of the History of the Behavioral Sciences* 2006 Summer; 42(3): 203-220. NRCBL: 18.4; 18.6; 2.2. SC: le.

Watts, Geoff. Quite reasonably emotional. *Lancet* 2007 January 13-19; 369(9556): 90-91. NRCBL: 18.4; 1.1.

Weitlauf, Julie C.; Ruzek, Josef I.; Westrup, Darrah A.; Lee, Tina; Keller, Jennifer. Empirically assessing participant perceptions of the research experience in a randomized clinical trial: the women's self-defense project as a case example. *Journal of Empirical Research on Human Research Ethics* 2007 June; 2(2): 11-24. NRCBL: 18.4; 10; 18.3; 18.5.3. SC: em.

Abstract: A growing body of empirical literature has systematically documented the reactions to research participation among participants in trauma focused research. To date, the available data has generally presented an optimistic picture regarding participants' ability to tolerate and even find benefit from their participation. However, this literature has been largely limited to cross-sectional designs. No extant literature has yet examined the perceptions of participants with psychiatric illness who are participating in randomized clinical trials (RCTs) designed to evaluate the efficacy or effectiveness of novel trauma treatments. The authors posit that negative experiences of, or poor reactions to, the research experience in the context of a trauma-focused RCT may elevate the risk of participation. Indeed, negative reactions may threaten to undermine the potential therapeutic gains of participants and promoting early drop out from the trial. Empirically assessing reactions to research participation at the pilot-study phase of a clinical trial can both provide investigators and IRB members alike with empirical evidence of some likely risks of participation. In turn, this information can be used to help shape the design and recruitment methodology of the full-scale trial. Using data from the pilot study of the Women's Self-Defense Project as a case illustration, we provide readers with concrete suggestions for empirically assessing participants' perceptions of risk involved in their participation in behaviorally oriented clinical trials.

Zimbardo, Philip. When good people do evil. *Yale Alumni Magazine* 2007 January-February; 70(3): 40-47. NRCBL: 18.4; 8.2; 18.3; 18.6. Identifiers: Stanley Milgram; Obedience (film).

BIOBANKS *See* GENETIC DATABASES AND BIOBANKS

BIOETHICISTS *See* ETHICISTS AND ETHICS COMMITTEES

BIOETHICS AND MEDICAL ETHICS
See also CODES OF ETHICS; ETHICISTS AND ETHICS COMMITTEES; NURSING ETHICS AND PHILOSOPHY; PROFESSIONAL ETHICS

Alpert, Joseph S. Ethical precepts for cardiologists. *Current Cardiology Reports* 2005 January; 7(1): 1-2. NRCBL: 2.1; 7.3; 18.2.

Andorno, R. Global bioethics at UNESCO: in defence of the Universal Declaration on Bioethics and Human Rights. *Journal of Medical Ethics* 2007 March; 33(3): 150-154. NRCBL: 2.1; 21.1; 2.4.

Abstract: The Universal Declaration on Bioethics and Human Rights adopted by the United Nations Educational, Scientific, and Cultural Organisation (UNESCO) on 19 October 2005 is an important step in the search for global minimum standards in biomedical research and clinical practice. As a member of UNESCO International Bioethics Committee, I participated in the drafting of this document. Drawing on this experience, the principal features of the Declaration are outlined, before responding to two general charges that have been levelled at UNESCO's bioethical activities and at this particular document, are outlined. One criticism is to the effect that UNESCO is exceeding its mandate by drafting such bioethical instruments — in particular, the charge is that it is trespassing on a topic that lies in the responsibility of the World Health Organization. The second criticism is that UNESCO's reliance on international human rights norms is inappropriate.

Bavastro, Paolo. Europäische Initiative gegen Bio-Ethik und deren Folgen. *In:* Neuer-Miebach, Therese; Wunder, Michael, eds. Bio-Ethik und die Zukunft der Medizin.

Bonn: Psychiatrie-Verlag, 1998: 155-158. NRCBL: 2.1; 21.1.

Biller-Andorno, Nikola. Epilogue: cross-cultural discourse in bioethics: it's a small world after all. *In:* Roetz, Heiner, ed. Cross-Cultural Issues in Bioethics: The Example of Human Cloning. New York: Rodopi, 2006: 459-463. NRCBL: 2.1; 21.7.

Blacksher, Erika. Bioethics and politics: a values analysis of the mission of the Center for Practical Bioethics. *American Journal of Bioethics* 2007 October; 7(10): 34-36. NRCBL: 2.1; 21.1. Comments: Myra J. Christopher. "Show me" bioethics and politics. American Journal of Bioethics 2007 October; 7(10): 28-33.

Boyd, Kenneth. Medical ethics: Hippocratic and democratic ideals. *In:* McLean, Sheila A.M., ed. First Do No Harm: Law, Ethics, and Healthcare. Aldershot, England; Burlington, VT: Ashgate, 2006: 29-38. NRCBL: 2.1; 2.2; 6; 20.5.1.

Byrne, Margaret. Against Bioethics, by Jonathan Baron [book review]. *DePaul Journal of Health Care Law* 2007; 10(4): 535-542. NRCBL: 2.1.

Capron, Alexander Morgan. Imagining a new world: using internationalism to overcome the 10/90 gap in bioethics. *Bioethics* 2007 October; 21(8): 409-412. NRCBL: 2.1; 2.2; 21.1.
  Abstract: The IAB Presidential Address was delivered by Alexander Capron to the internationally gathered audience at the Closing Ceremony of the 8th World Congress of Bioethics, Beijing on 9th August 2006.

Cassell, Eric J. Unanswered questions: bioethics and human relationships. *Hastings Center Report* 2007 September-October; 37(5): 20-23. NRCBL: 2.1; 8.1.

Christopher, Myra J. "Show me" bioethics and politics. *American Journal of Bioethics* 2007 October; 7(10): 28-33. NRCBL: 2.1; 21.1. Comments: American Journal of Bioethics 2007 October; 7(10): 34-44.
  Abstract: Missouri, the "Show Me State," has become the epicenter of several important national public policy debates, including abortion rights, the right to choose and refuse medical treatment, and, most recently, early stem cell research. In this environment, the Center for Practical Bioethics (formerly, Midwest Bioethics Center) emerged and grew. The Center's role in these "cultural wars" is not to advocate for a particular position but to provide well researched and objective information, perspective, and advocacy for the ethical justification of policy positions; and to serve as a neutral convener and provider of a public forum for discussion. In this article, the Center's work on early stem cell research is a case study through which to argue that not only the Center, but also the field of bioethics has a critical role in the politics of public health policy.

De Vries, Raymond G.; Turner, Leigh; Orfali, Kristina; Bosk, Charles L. Social science and bioethics: morality

from the ground up. *Clinical Ethics* 2007 March; 2(1): 33-35. NRCBL: 2.1; 7.1.

Dimitrov, Borislav D.; Glutnikova, Zlatka; St. Dimitrova, Bogdana. Education and practice of medical ethics in Bulgaria after political and socio-economic changes in the 90's [commentary]. *Ethics and Medicine: An International Journal of Bioethics* 2007 Spring; 23(1): 11-14. NRCBL: 2.1; 2.3.

Dyer, Clare. GMC guidance on conscience goes too far, says BMA [news]. *BMJ: British Medical Journal* 2007 October 6; 335(7622): 688. NRCBL: 2.1; 11.1; 12.1; 20.5.1. Identifiers: Great Britain (United Kingdom);General Medical Council; British Medical Association.

Emanuel, Ezekiel J. Researching a bioethical question. *In:* Gallin, John I.; Ognibene, Frederick P., eds. Principles and Practice of Clinical Research. 2nd edition. Oxford: Academic, 2007: 27-38. NRCBL: 2.1; 18.1; 18.3; 18.5.1. SC: rv.

English, Veronica; Hamm, Danielle; Harrison, Caroline; Sheather, Julian; Sommerville, Ann. Ethics briefings. *Journal of Medical Ethics* 2007 February; 33(2): 123-124. NRCBL: 2.1; 9.3.1; 9.4; 9.5.1; 12.4.1; 21.1. Identifiers: abortion; conflicts of interests; HPV vaccine; patient non-compliance; medical tourism.

Faber, Berit A. Bioethics in Europe. *In:* Gunning, Jennifer; Holm, Søren, eds. Ethics, law, and society. Volume 1. Aldershot, Hants, England; Burlington, VT: Ashgate, 2005: 41-44. NRCBL: 2.1; 21.1.

Fuchs, Ursel. Bürger gegen Bio-Ethik: Internationale Initiative gegen die geplante Bio-Ethik-Konvention. *In:* Neuer-Miebach, Therese; Wunder, Michael, eds. Bio-Ethik und die Zukunft der Medizin. Bonn: Psychiatrie-Verlag, 1998: 165-166. NRCBL: 2.1; 21.1; 5.1.

Fuchs, Ursel. Experten entscheiden. Unter sich. Und über uns. Die Bio-Ethik-Konvention geht alle an. *In:* Neuer-Miebach, Therese; Wunder, Michael, eds. Bio-Ethik und die Zukunft der Medizin. Bonn: Psychiatrie-Verlag, 1998: 130-138. NRCBL: 2.1; 5.1; 21.1.

Garcia, Jorge L.A. Revisiting African American perspectives on biomedical ethics: distinctiveness and other questions. *In:* Prograis, Lawrence J.; Pellegrino, Edmund D., eds. African American Bioethics: Culture, Race, and Identity. Washington, DC: Georgetown University Press, 2004: 1-23. NRCBL: 2.1; 21.7; 9.5.4. Conference: Symposium on African American Perspectives in Bioethics and Second Annual Conference on Health Disparities, held on September 23-24, 2004, at Georgetown University.

Garner, Samual A. Dear bioethics, the country needs you. *American Journal of Bioethics* 2007 October; 7(10): 38-39. NRCBL: 2.1; 21.1. Comments: Myra J. Christo-

---

NRCBL: National Reference Center for Bioethics Literature Classification Scheme   See inside front cover for terms.

39

pher. "Show me" bioethics and politics. American Journal of Bioethics 2007 October; 7(10): 28-33.

Gbadegesin, Segun. The moral weight of culture in ethics. *In:* Prograis, Lawrence J.; Pellegrino, Edmund D., eds. African American Bioethics: Culture, Race, and Identity. Washington, DC: Georgetown University Press, 2004: 25-45. NRCBL: 2.1; 21.7. Conference: Symposium on African American Perspectives in Bioethics and Second Annual Conference on Health Disparities, held on September 23-24, 2004, at Georgetown University.

Gilam, Lynn. What is bioethics all about? *Monash Bioethics Review* 2000 October; 19(4): 51-54. NRCBL: 2.1; 21.1; 9.1. SC: an. Comments: Tim Smyth. Bioethics, politics and policy development. Monash Bioethics Review 2000 October; 19(4): 38-45.

Glick, Shimon; Jotkowitz, Alan. Compromise and dialogue in bioethical disputes. *American Journal of Bioethics* 2007 October; 7(10): 36-38. NRCBL: 2.1; 21.1. Identifiers: Israel. Comments: Myra J. Christopher. "Show me" bioethics and politics. American Journal of Bioethics 2007 October; 7(10): 28-33.

Gorovitz, Samuel. The centrality of marginalization. *Monash Bioethics Review* 2000 October; 19(4): 49-51. NRCBL: 2.1; 21.1; 9.1. SC: an. Comments: Tim Smyth. Bioethics, politics and policy development. Monash Bioethics Review 2000 October; 19(4): 38-45.

Graumann, Sigrid. Ethik in der Medizin und ihre Aufgaben in der Politik [Ethics in medicine and its role in politics]. *Ethik in der Medizin* 2006 December; 18(4): 359-363. NRCBL: 2.1; 21.1.

Griffith, Ezra E.H. Personal narrative and an African American perspective on medical ethics. *In:* Prograis, Lawrence J.; Pellegrino, Edmund D., eds. African American Bioethics: Culture, Race, and Identity. Washington, DC: Georgetown University Press, 2004: 105-125. NRCBL: 2.1; 9.5.4. Conference: Symposium on African American Perspectives in Bioethics and Second Annual Conference on Health Disparities, held on September 23-24, 2004, at Georgetown University.

Gross, Jed Adam; Moreno, Jonathan D.; Berger, Sam. Gray, not red: the hue of neoconservative bioethics. *American Journal of Bioethics* 2007 October; 7(10): 22-25; author reply W1-W3. NRCBL: 2.1; 5.1; 15.1; 21.1. Comments: Jonathan D. Moreno and Sam Berger. Biotechnology and the new right: neoconservatism's red menace. American Journal of Bioethics 2007 October; 7(10): 7-13.

Hansson, Mats G.; Kihlbom, Ulrik; Tuvemo, Torsten; Olsen, Leif A.; Rodriguez, Alina. Ethics takes time, but not that long. *BMC Medical Ethics [electronic]* 2007; 8:6. 7 p. NRCBL: 2.1; 8.1.
  Abstract: Background: Time and communication are important aspects of the medical consultation. Physician behavior in real-life pediatric consultations in relation to

ethical practice, such as informed consent (provision of information, understanding), respect for integrity and patient autonomy (decision-making), has not been subjected to thorough empirical investigation. Such investigations are important tools in developing sound ethical praxis. Methods: 21 consultations for inguinal hernia were video recorded and observers independently assessed global impressions of provision of information, understanding, respect for integrity, and participation in decision making. The consultations were analyzed for the occurrence of specific physician verbal and nonverbal behaviors and length of time in minutes. Results: All of the consultations took less than 20 minutes, the majority consisting of 10 minutes or less. Despite this narrow time frame, we found strong and consistent association between increasing time and higher ratings on all components of ethical practice: information, ($ß = .43$), understanding ($ß = .52$), respect for integrity ($ß = .60$), and decision making ($ß = .43$). Positive nonverbal behaviors by physicians during the consultation were associated particularly with respect for integrity ($ß = .36$). Positive behaviors by physicians during the physical examination were related to respect for children's integrity. Conclusion: Time was of essence for the ethical encounter. Further, verbal and nonverbal positive behaviors by the physicians also contributed to higher ratings of ethical aspects. These results can help to improve quality of ethical practice in pediatric settings and are of relevance for teaching and policy makers.

Hatfield, Amy J.; Kelley, Shana D. Case study: lessons learned through digitizing the National Commission for the Protection of Human Subjects of Biomedical and Behavioral Research collection. *Journal of the Medical Library Association* 2007 July; 95(3): 267-270. NRCBL: 2.1; 1.3.12.

Hennig, Wolfgang. Bioethics in China: although national guidelines are in place, their implementation remains difficult. *EMBO Reports* 2006 September; 7(9): 850-854. NRCBL: 2.1; 2.2; 2.4; 18.2. SC: rv.

Kettner, Matthias. Medizinethik in den Medien — Befunde und Aufgaben in Theorie und Praxis [Medical ethics in the media — findings and challenges in theory and practice]. *Ethik in der Medizin* 2006 December; 18(4): 353-358. NRCBL: 2.1; 1.3.7.

Khroutski, Konstantin S. BioCosmological approach in world bioethics. *Eubios Journal of Asian and International Bioethics* 2007 November; 17(6): 167-171. NRCBL: 2.1; 21.7.

Lanzerath, Dirk. Die Eigenständigkeit der Bioethik und ihr Verhältnis zur Biopolitik [The autonomy of bioethics and its relationship to biopolitics]. *Ethik in der Medizin* 2006 December; 18(4): 364-368. NRCBL: 2.1; 21.1.

Lindemann, Hilde. Obligations to fellow and future bioethicists: publication. *In:* Eckenwiler, Lisa A.; Cohn, Felicia G., eds. The Ethics of Bioethics: Mapping the

Moral Landscape. Baltimore, MD: Johns Hopkins University Press, 2007: 270-277. NRCBL: 2.1; 1.3.7; 7.3.

Lo, Bernard. Principles in the ethical care of underserved patients. *In:* King, Talmadge E.; Wheeler, Margaret B., eds. Medical Management of Vulnerable and Underserved Patients: Principles, Practice, and Populations. New York: McGraw-Hill Medical Pub. Division, 2007: 47-55. NRCBL: 2.1; 8.1; 9.4; 8.3.3.

Lorenz, Rolf J. Tübinger Initiative gegen die geplante Bio-Ethik-Konvention. *In:* Neuer-Miebach, Therese; Wunder, Michael, eds. Bio-Ethik und die Zukunft der Medizin. Bonn: Psychiatrie-Verlag, 1998: 159-162. NRCBL: 2.1; 21.1.

Macpherson, Cheryl Cox. Global bioethics: did the Universal Declaration on Bioethics and Human Rights miss the boat? *Journal of Medical Ethics* 2007 October; 33(10): 588-590. NRCBL: 2.1; 6; 21.1. SC: an.

Marshall, Mary Faith. ASBH and moral tolerance. *In:* Eckenwiler, Lisa A.; Cohn, Felicia G., eds. The Ethics of Bioethics: Mapping the Moral Landscape. Baltimore, MD: Johns Hopkins University Press, 2007: 134-144. NRCBL: 2.1; 7.3; 21.4; 12.4.2. Identifiers: American Society of Bioethics and Humanities.

Maschke, Karen J. The federalist turn in bioethics? *Hastings Center Report* 2007 November-December; 37(6): 3. NRCBL: 2.1; 1.3.5. Comments: James W. Fossett, Alicia R. Ouellette, Sean Philpott, David Magnus, and Glenn McGee. Federalism and bioethics: states and moral pluralism. Hastings Center Report 2007 November-December; 37(6): 24-35.

McGee, Glenn; Bjarnadóttir, Dyrleif. Abuses of science in medical ethics. *In:* Rhodes, Rosamond; Francis, Leslie P.; Silvers, Anita, eds. The Blackwell Guide to Medical Ethics. Malden, MA: Blackwell Pub., 2007: 289-302. NRCBL: 2.1; 1.3.9.

McNeill, Paul M. Should bioethics play football? *Monash Bioethics Review* 2000 October; 19(4): 46-49. NRCBL: 2.1; 9.1; 21.1. SC: an. Conference: Symposium on the Role of Bioethics and Bioethicists; Sydney, Australia; 2000 July 7; Australian Bioethics Association and the Australian Institute of Health, Law and Ethics at the University of Sydney. Comments: Tim Smyth. Bioethics, politics and policy development. Monash Bioethics Review 2000 October; 19(4): 38-45.

Meilaender, Gilbert. Human dignity and public bioethics. *New Atlantis* 2007 Summer; (17): 33-52. NRCBL: 2.1; 4.4; 5.1.

National Reference Center for Bioethics Literature. News from the National Reference Center for Bioethics Literature (NRCBL) and the National Information Resource on Ethics and Human Genetics (NIREHG). *Ken-nedy Institute of Ethics Journal* 2007 December; 17(4): 399-403. NRCBL: 2.1; 1.3.12; 15.1.

> Keywords: *bioethical issues; *bioethics; *databases; *genetics; *information dissemination; *literature; *terminology; eugenics; genetic patents; human experimentation; professional organizations; publishing; Proposed Keywords: *abstracting and indexing; *information services; publications; Keyword Identifiers: *Bioethics Thesaurus for Genetics; *ETHXWeb; *GenETHX; *National Information Resource on Ethics and Genetics; *National Reference Center for Bioethics Literature; Bioethics Thesaurus; Georgetown University; Kennedy Institute of Ethics

Paul, Jobst. Das bio-ethische Netzwerk. *In:* Neuer-Miebach, Therese; Wunder, Michael, eds. Bio-Ethik und die Zukunft der Medizin. Bonn: Psychiatrie-Verlag, 1998: 60-71. NRCBL: 2.1; 5.1; 21.1.

Prograis, Lawrence J. An African American's internal perspective on biomedical ethics. *In:* Prograis, Lawrence J.; Pellegrino, Edmund D., eds. African American Bioethics: Culture, Race, and Identity. Washington, DC: Georgetown University Press, 2004: 153-158. NRCBL: 2.1; 9.5.4. Conference: Symposium on African American Perspectives in Bioethics and Second Annual Conference on Health Disparities, held on September 23-24, 2004, at Georgetown University.

Salter, Brian; Salter, Charlotte. Bioethics and the global moral economy: the cultural politics of human embryonic stem cell science. *Science, Technology and Human Values* 2007 September; 32(5): 554-581. NRCBL: 2.1; 18.5.4; 15.1; 21.1. SC: an; em.

Saunders, William L. Washington insider [summary]. *National Catholic Bioethics Quarterly* 2007 Winter; 7(4): 661-669. 20 fn. NRCBL: 2.1; 2.4; 12.4.4; 18.5.4; 15.1; 9.7; 11.1. Identifiers: bioethical issues.

> Keywords: *bioethical issues; advisory committees; chimeras; cloning; contraception; embryo research; embryonic stem cells; federal government; government financing; government regulation; legal aspects; politics; preimplantation diagnosis; research support; siblings; tissue typing; value of life; Keyword Identifiers: Great Britain; Human Fertilisation and Embryology Authority; President's Council on Bioethics; United States

Schaub, Diana. Bioethics and "The Public Interest". *New Atlantis* 2007 Winter; 15: 135-140. NRCBL: 2.1; 1.3.7.

Schmidt, Harald. Whose dignity? Resolving ambiguities in the scope of "human dignity" in the Universal Declaration on Bioethics and Human Rights. *Journal of Medical Ethics* 2007 October; 33(10): 578-584. NRCBL: 2.1; 6; 21.1. SC: an.

> Abstract: In October 2005, the United Nations Educational, Scientific and Cultural Organization adopted the Universal Declaration on Bioethics and Human Rights (UDBHR). A concept of central importance in the declaration is that of "human dignity". However, there is lack of clarity about its scope, especially concerning the question of whether prenatal human life has the same dignity and rights as born human beings. This ambiguity has im-

NRCBL: National Reference Center for Bioethics Literature Classification Scheme   See inside front cover for terms.

41

plications for the interpretation of important articles of the declaration, including 2©), 4, 8, 10 and 11. The paper applies relevant provisions of the UDBHR to specific cases, addresses problems of internal consistency and considers attempts at clarifying the scope of "human dignity" by the negotiating parties. An analysis of the important relationship between the UDBHR and the Universal Declaration of Human Rights, to which the UDBHR refers in its title and elsewhere, shows that because of a crucial emphatic asymmetry, a broad reading according to which the UDBHR must be understood to ascribe human rights and dignity to prenatal life is untenable. However, the view that the UDBHR confers human rights and dignity on humans from the moment of birth onwards is robust and defensible. This conclusion is important for a proper understanding of the declaration and its use, as stated in Articles 1(2) and 22, the latter urging states ". . . to give effect to the principles . . . in this declaration". Similarly, it has implications for the use of the declaration in the wider context of bioethics-related law and policy, as well as in academic and other discussions where increasing reference to the UDBHR is likely.

Sharpe, Virginia A. Strategic disclosure requirements and the ethics of bioethics. *In:* Eckenwiler, Lisa A.; Cohn, Felicia G., eds. The Ethics of Bioethics: Mapping the Moral Landscape. Baltimore, MD: Johns Hopkins University Press, 2007: 170-180. NRCBL: 2.1; 5.3; 7.3; 1.3.2; 9.7; 9.3.1.

Smith, Ian A. A new defense of Quinn's Principle of Double Effect. *Journal of Social Philosophy* 2007 Summer; 38(2): 349-364. NRCBL: 2.1. SC: an.

ten Have, H.; Ang, T.W. UNESCO's Global Ethics Observatory. *Journal of Medical Ethics* 2007 January; 33(1): 15-16. NRCBL: 2.1; 1.3.9; 1.3.12; 2.2.
Abstract: The Global Ethics Observatory, launched by the United Nations Educational, Scientific, and Cultural Organization in December 2005, is a system of databases in the ethics of science and technology. It presents data on experts in ethics, on institutions (university departments and centres, commissions, councils and review boards, and societies and associations) and on teaching programmes in ethics. It has a global coverage and will be available in six major languages. Its aim is to facilitate the establishment of ethical infrastructures and international cooperation all around the world.

Toth-Fejel, Tihamer; Dodsworth, Chris; Lahl, Jennifer. Syntactic measures of bias (and a perspective on the essential issue of bioethics). *American Journal of Bioethics* 2007 October; 7(10): 40-42. NRCBL: 2.1; 4.4. Comments: Myra J. Christopher. "Show me" bioethics and politics. American Journal of Bioethics 2007 October; 7(10): 28-33.

Trotter, Griffin. Left bias in academic bioethics. *In:* Eckenwiler, Lisa A.; Cohn, Felicia G., eds. The Ethics of Bioethics: Mapping the Moral Landscape. Baltimore, MD: Johns Hopkins University Press, 2007: 108-117. NRCBL: 2.1; 21.1.

Victoroff, Michael S. Guide to critical care ethics not ready for prime time [review of Critical Care Ethics: A Practice Guide from the ACCM Ethics Committee, edited by Dan R. Thompson and Heidi B. Kummer]. *Managed Care* 2006 July; 15(7): 14-16. NRCBL: 2.1; 9.6. Identifiers: American College of Critical Care Medicine.

Vollmann, Jochen. Ethik in der klinischen Medizin — Bestandsaufnahme und Ausblick [Ethics in clinical medicine — taking stock and prospects]. *Ethik in der Medizin* 2006 December; 18(4): 348-352. NRCBL: 2.1; 9.6.

Wiesemann, Claudia. Die Beziehung der Medizinethik zur Medizingeschichte und Medizinetheorie [The relationship of medical ethics to the history of medicine and medical theory]. *Ethik in der Medizin* 2006 December; 18(4): 337-341. NRCBL: 2.1; 7.1.

Wilkinson, Stephen. Eugenics and the criticism of bioethics [review of Genetic Politics: From Eugenics to Genome, by Ann Kerr and Tom Shakespeare]. *Ethical Theory and Moral Practice* 2007 August; 10(4): 409-418. NRCBL: 2.1; 15.5. SC: an.

Wunder, Michael. Grafenecker Erklärung zur Bio-Ethik. *In:* Neuer-Miebach, Therese; Wunder, Michael, eds. Bio-Ethik und die Zukunft der Medizin. Bonn: Psychiatrie-Verlag, 1998: 182-195. NRCBL: 2.1; 21.1.

BIOETHICS AND MEDICAL ETHICS/ COMMISSIONS
See also ETHICISTS AND ETHICS COMMITTEES

Dissanayake, V.H.W; Lanerolle, R.D.; Mendis, N. Research ethics and ethical review committees in Sri Lanka: a 25 year journey. *Ceylon Medical Journal* 2006 September; 51(3): 110-113. NRCBL: 2.4; 9.6.

Dzur, Albert W.; Levin, Daniel. The primacy of the public: in support of bioethics commissions as deliberative forums. *Kennedy Institute of Ethics Journal* 2007 June; 17(2): 133-142. NRCBL: 2.4; 5.1; 1.3.5; 2.1; 1.1; 7.1. SC: an.
Abstract: In a 2004 article, we argued that bioethics commissions should be assessed in terms of their usefulness as public forums. A 2006 article by Summer Johnson argued that our perspective was not supported by the existing literature on presidential commissions, which had not previously identified commissions as public forums and that we did not properly account for the political functions of commissions as instruments of presidential power. Johnson also argued that there was nothing sufficiently unique about bioethics commissions to make the public forum perspective particularly applicable. We respond by arguing that analysis of commissions' work as public forums fits well within the literature on commissions, especially on their agenda-setting functions, and that the political functions of commissions are often compatible with their functioning as public forums. We also demonstrate how the origins and concerns of bioethics

make public forum analysis particularly applicable to bioethics commissions.

Gianelli, Diane M.; Davis, F. Daniel. News from the President's Council on Bioethics. *Kennedy Institute of Ethics Journal* 2007 December; 17(4): 397-398. NRCBL: 2.4; 20.2.1.

Hanna, Kathi E. A brief history of public debate about reproductive technologies: politics and commissions. *In:* Knowles, Lori P.; Kaebnick, Gregory E., eds. Reprogenetics: Law, Policy, and Ethical Issues. Baltimore: Johns Hopkins University Press, 2007: 197-225. 50 refs., 3 fn. NRCBL: 2.4; 14.1; 15.1; 18.5.4; 2.2.

    Keywords: *advisory committees; *embryo research; *public policy; *reproductive technologies; bioethical issues; cloning; decision making; embryonic stem cells; federal government; fetal research; genetic intervention; historical aspects; human experimentation; organization and administration; policy making; Keyword Identifiers: *United States; Biomedical Ethics Advisory Committee; Ethics Advisory Board; Human Fetal Tissue Transplantation Research Panel; National Bioethics Advisory Commission; National Commission for the Protection of Human Subjects; President's Commission for the Study of Ethical Problems; President's Council on Bioethics; Twentieth Century

Johnson, Summer. A rebuttal to Dzur and Levin: Johnson on the legitimacy and authority of bioethics commissions. *Kennedy Institute of Ethics Journal* 2007 June; 17(2): 143-152. NRCBL: 2.4; 5.1; 1.3.5; 2.1; 1.1; 7.1. SC: an. Comments: Albert W. Dzur and Daniel Levin. The primacy of the public: in support of bioethics commissions as deliberative forums. Kennedy Institute of Ethics Journal 2007 June; 17(2): 133-142.

    Abstract: Bioethics commissions have been critiqued on the basis that they are not sufficiently public or are too reliant upon expertise to have legitimacy or authority in regard to public policy debates. Adequately assessing the legitimacy and authority of commissions requires thinking clearly about the "publics" these commissions serve, the primary tasks of public bioethics, and how those tasks might be performed with a certain kind of ethical expertise and limited authority that makes them legitimate players in public policy debates concerning bioethics.

Kuczewski, Mark G. Democratic ideals and bioethics commissions: the problem of expertise in an egalitarian society. *In:* Eckenwiler, Lisa A.; Cohn, Felicia G., eds. The Ethics of Bioethics: Mapping the Moral Landscape. Baltimore, MD: Johns Hopkins University Press, 2007: 83-94. NRCBL: 2.4; 5.3; 21.1.

Puljak, Livia. Croatia founded a national body for ethics in science. *Science and Engineering Ethics* 2007 June; 13(2): 191-193. NRCBL: 2.4; 1.3.9.

Riley, Margaret Foster; Merrill, Richard A. Regulating reproductive genetics: a review of American bioethics commissions and comparison to the British Human Fertilisation and Embryology Authority. *Columbia Science and Technology Law Review* 2005; 6(1): 1-64. 363 fn. NRCBL: 2.4; 14.1; 14.4; 18.5.4; 15.1; 18.2. SC: le.

    Keywords: *advisory committees; *government regulation; *regulation; *reproductive technologies; bioethical issues; biotechnology; comparative studies; legislation; organization and administration; preimplantation diagnosis; Keyword Identifiers: *Great Britain; *Human Fertilisation and Embryology Authority; *United States; Biomedical Ethics Advisory Commission; Ethics Advisory Board; Human Embryo Research Panel; Human Fertilisation and Embryology Act 1990 (Great Britain); President's Council on Bioethics; President's Commission for the Study of Ethical Problems

Sherlock, Richard. Bioethics in liberal regimes: a review of the President's Council. *Ethics and Medicine: An International Journal of Bioethics* 2007 Fall; 23(3): 169-188. 10 fn. NRCBL: 2.4; 2.2; 2.1; 1.2; 9.1; 14.1; 15.1; 15.2; 9.5.4; 10; 5.1. SC: rv.

    Keywords: *advisory committees; *bioethical issues; *bioethics; biomedical enhancement; biotechnology; cloning; cultural pluralism; democracy; embryo research; embryonic stem cells; eugenics; euthanasia; genetic engineering; historical aspects; human dignity; human experimentation; morality; preimplantation diagnosis; public policy; religion; reproductive technologies; Proposed Keywords: publications; Keyword Identifiers: *President's Council on Bioethics; *United States; National Bioethics Advisory Commission; National Commission for the Protection of Human Subjects and Biomedical and Behavioral Research; President's Commission for the Study of Ethical Problems in Biomedical and Behavioral Research; Twentieth Century; Twenty-First Century

Thomas, Cordelia. Public dialogue and xenotransplantation. *Medicine and Law: The World Association for Medical Law* 2007 December; 26(4): 801-815. NRCBL: 2.4; 19.1; 22.2; 21.7. SC: le.

    Abstract: Toi te Taiao: the Bioethics Council was established in 2002 to enhance New Zealand's understanding of the cultural, ethical and spiritual aspects of biotechnology and ensure that the use of biotechnology has regard for the values held by New Zealanders. In 2005, the Bioethics Council focused on xenotransplantation. A series of dialogue events were held, the public had the opportunity to participate in an online discussion forum and were able to make written submissions. There is worldwide interest in the potential of this biotechnology to cure or alleviate a number of serious health conditions. However, there are concerns about the risks, especially the potential for cross species infection. Such risks have not yet been reliably quantified, but any decision about safety and effectiveness is also about cultural, ethical and spiritual factors. This paper considers some of the outcomes from the dialogue process and the reflections of the Bioethics Council on these. It contrasts the process with that of classic consultation and concludes that, although the process may be more costly and time consuming than the traditional consultative approach, it enables the role of science to be appreciated in its full context, including appreciation of the uncertainties of natural systems and the relevance of cultural, ethical and spiritual human values. It will be suggested that the public are able to interweave ethical concerns with scientific knowledge to engage in meaningful dialogue, resulting in useful recommendations.

NRCBL: National Reference Center for Bioethics Literature Classification Scheme   See inside front cover for terms.

43

United Nations Educational, Scientific and Cultural Organisation [UNESCO]. Division of Ethics of Science and Technology. Guide No.2: bioethics committees at work: procedures and policies. Paris, France: United Nations Educational, Scientific and Cultural Organisation, Division of Ethics of Science and Technology, 2005; 72 p. [Online]. Accessed: http://unesdoc.unesco.org/images/0014/ 001473/147392e.pdf [2007 April 19]. NRCBL: 2.4; 9.6; 18.2.

BIOETHICS AND MEDICAL ETHICS/ EDUCATION
See also MEDICAL EDUCATION

Beckmann, Jan P. Ethik in der Medizin in Aus- und Weiterbildung aus der Sicht der Philosophie [Ethics in medicine in education and continuing education from the philosophical point of view]. *Ethik in der Medizin* 2006 December; 18(4): 369-373. NRCBL: 2.3; 1.1; 7.2.

Benatar, D. Moral theories may have some role in teaching applied ethics. *Journal of Medical Ethics* 2007 November; 33(11): 671-672. NRCBL: 2.3; 1.1. SC: an.
Abstract: In a recent paper, Rob Lawlor argues that moral theories should not be taught in courses on applied ethics. The author contends that Dr Lawlor's arguments overlook at least two important roles that some attention to ethical theories may play in practical ethics courses. The conclusion is not that moral theory must be taught, but rather that there is more to be said for it than Dr Lawlor's arguments reveal.

Childress, James F. Mentoring in bioethics: possibilities and problems. *In:* Eckenwiler, Lisa A.; Cohn, Felicia G., eds. The Ethics of Bioethics: Mapping the Moral Landscape. Baltimore, MD: Johns Hopkins University Press, 2007: 260-269. NRCBL: 2.3; 7.3; 8.4.

Claudot, Frédérique; Alla, François; Ducrocq, Xavier; Coudane, Henry. Teaching ethics in Europe. *Journal of Medical Ethics* 2007 August; 33(8): 491-495. NRCBL: 2.3; 7.2; 21.1. SC: em.
Abstract: AIM: To carry out an appropriate overview and inventory of the teaching of ethics within the European Union Schools of Medicine. METHODS: A questionnaire was sent by email to 45 randomly selected medical schools from each of 23 countries in the European Union in February 2006. RESULTS: 25 schools of medicine from 18 European countries were included (response rate = 56%). In 21 of 25 medical schools, there was at least one ethics module. In 11 of 25 medical schools, the teaching of ethics was transversal. Only one of the responding schools did not teach ethics. The mean time invested in ethics teaching was 44 h during the overall curriculum. CONCLUSIONS: Ethics now has an established place within the medical curriculum throughout the European Union. However, there is a notable disparity in programme characteristics among schools of medicine.

Cowley, Christopher. Why medical ethics should not be taught by philosophers. *Discourse* 2005 Autumn; 5(1): 50-63 [Online]. Accessed: http://prs.heacademy.ac.uk/publications/autumn2005.pdf [2006 December 6]. NRCBL: 2.3; 7.2; 1.1.

Daher, Michel. Current trends in medical ethics education. *Journal Medical Libanais* 2006 July-September; 54(3): 121-123. NRCBL: 2.3.

Derse, Arthur R. The evolution of medical ethics education at the Medical College of Wisconsin. *WMJ: Official publication of the State Medical Society of Wisconsin* 2006 June; 105(4): 18-20. NRCBL: 2.3; 7.2; 2.2.

Kipnis, Kenneth. The expert ethics witness as teacher. *In:* Rasmussen, Lisa, ed. Ethics Expertise: History, Contemporary Perspectives, and Applications. Dordrecht: Springer, 2005: 269-279. NRCBL: 2.3; 2.1; 7.3; 1.3.5. SC: le.

Lawlor, Rob. Moral theories in teaching applied ethics. *Journal of Medical Ethics* 2007 June; 33(6): 370-372. NRCBL: 2.3; 1.1. SC: an.
Abstract: It is argued, in this paper, that moral theories should not be discussed extensively when teaching applied ethics. First, it is argued that, students are either presented with a large amount of information regarding the various subtle distinctions and the nuances of the theory and, as a result, the students simply fail to take it in or, alternatively, the students are presented with a simplified caricature of the theory, in which case the students may understand the information they are given, but what they have understood is of little or no value because it is merely a caricature of a theory. Second, there is a methodological problem with appealing to moral theories to solve particular issues in applied ethics. An analogy with science is appealed to. In physics there is a hope that we could discover a unified theory of everything. But this is, of course, a hugely ambitious project, and much harder than, for example, finding a theory of motion. If the physicist wants to understand motion, he should try to do so directly. We would think he was particularly misguided if he thought that, to answer this question, he first needed to construct a unified theory of everything.

Leget, Carlo; Olthuis, Gert. Compassion as a basis for ethics in medical education. *Journal of Medical Ethics* 2007 October; 33(10): 617-620. NRCBL: 2.3; 7.2; 8.1.
Abstract: The idea that ethics is a matter of personal feeling is a dogma widespread among medical students. Because emotivism is firmly rooted in contemporary culture, the authors think that focusing on personal feeling can be an important point of departure for moral education. In this contribution, they clarify how personal feelings can be a solid basis for moral education by focusing on the analysis of compassion by the French phenomenologist Emmanuel Housset. This leads to three important issues regarding ethics education: (1) the necessity of a continuous attention for and interpretation of the meaning of language, (2) the importance of examining what aspect of "the other" touches one and what it is that evokes the urge to act morally and (3) the need to relate oneself to the community, both to the medical community

and to collectively formulated rules and laws. These issues can have a place in medical education by means of an ethical portfolio that supports students in their moral development. First, keeping a portfolio will improve their expression of the moral dimension of medical practice. Second, the effects of self-knowledge and language mastery will limit the pitfalls of emotivism and ethical subjectivism and will stimulate the inclination to really encounter the other. Third, it will show medical students from the start that their moral responsibility is more than following rules and that they are involved personally.

Liaschenko, Joan. Teaching feminist ethics. *In:* Davis, Anne J.; Tschudin, Verena; de Raeve, Louise, eds. Essentials of Teaching and Learning in Nursing Ethics: Perspectives and Methods. New York: Churchill Livingstone Elsevier, 2006: 203-215. NRCBL: 2.3; 1.1; 10; 4.1.3.

Moodley, Kaymanthri. Teaching medical ethics to undergraduate students in post-apartheid South Africa, 2003-2006. *Journal of Medical Ethics* 2007 November; 33(11): 673-677. NRCBL: 2.3; 7.2; 21.1.

Abstract: The apartheid ideology in South Africa had a pervasive influence on all levels of education including medical undergraduate training. The role of the health sector in human rights abuses during the apartheid era was highlighted in 1997 during the Truth and Reconciliation Commission hearings. The Health Professions Council of South Africa (HPCSA) subsequently realised the importance of medical ethics education and encouraged the introduction of such teaching in all medical schools in the country. Curricular reform at the University of Stellenbosch in 1999 presented an unparalleled opportunity to formally introduce ethics teaching to undergraduate students. This paper outlines the introduction of a medical ethics programme at the Faculty of Health Sciences from 2003 to 2006, with special emphasis on the challenges encountered. It remains one of the most comprehensive undergraduate medical ethics programmes in South Africa. However, there is scope for expanding the curricular time allocated to medical ethics. Integrating the curriculum both horizontally and vertically is imperative. Implementing a core curriculum for all medical schools in South Africa would significantly enhance the goals of medical education in the country.

Oguz, N. Yasemin; Kavas, M. Volkan; Aksu, Murat. Teaching thanatology: a qualitative and quantitative study. *Eubios Journal of Asian and International Bioethics* 2007 November; 17(6): 172-177. NRCBL: 2.3; 20.1; 20.3.2. SC: em. Note: Article contains Appendix 1: Focus Group Sessions Interview Guide and Appendix 2: Response Questions.

Parker, Michael. Deliberation and moral courage: the UK Genethics Club as a case study. *Notizie di Politeia* 2006; 22(81): 78-83. 11 refs. NRCBL: 2.3; 2.1; 15.1; 9.6; 15.2; 15.3.

Keywords: *clinical genetics; *ethicists; *ethics; *morality; *professional role; *teaching methods; case studies; cultural pluralism; genetics; health personnel; interdisciplinary communication; interprofessional relations; researchers; science; Proposed Keywords: character; Keyword Identifiers: *UK Genetics Club; Causing Death and Saving Lives (Glover, Jonathan); Great Britain

Pegoraro, Renzo; Putoto, Giovanni. Findings from a European survey on current bioethics training activities in hospitals. *Medicine, Health Care and Philosophy* 2007 March; 10(1): 91-96. NRCBL: 2.3; 7.2; 21.1. SC: em.

Abstract: While much work has been done on improving undergraduate education in bioethics, particularly in medicine, less has been said about continuing education of health care workers, particularly non-medical and nursing personnel. Hospitals bring together a variety of professional and non-professional groups in the place where clinical dilemmas are daily events, and would seem ideal places to conduct an ongoing bioethics dialogue. Yet evidence that this is being achieved is sparse. The European Hospital (-Based) Bioethics Program (EHBP) brings together both current and aspirant members of the EU as partners in a project that aims to assess the current situation with regard to bioethics education in hospitals, identify shortfalls, and address these. In order to achieve the first objective of the EHBP a survey of the current training activities (focused on activities in hospitals) in clinical bioethics in Europe was carried out. The results are presented in this paper, along with a discussion about the implications for the EHBP to address these issues.

Rabe, Marianne. Ethik in der Pflegeausbildung [Ethics in nursing education]. *Ethik in der Medizin* 2006 December; 18(4): 379-384. NRCBL: 2.3; 4.1.3; 7.2.

Roberts, Laura Weiss; Warner, Teddy D.; Dunn, Laura B.; Brody, Janet L.; Hammond, Katherine A. Green; Roberts, Brian B. Shaping medical students' attitudes toward ethically important aspects of clinical research: results of a randomized, controlled educational intervention. *Ethics and Behavior* 2007; 17(1): 19-50. NRCBL: 2.3; 7.2; 18.1; 1.3.9; 7.1; 18.3. SC: em.

Abstract: The effects of research ethics training on medical students' attitudes about clinical research are examined. A preliminary randomized controlled trial evaluated 2 didactic approaches to ethics training compared to a no-intervention control. The participant-oriented intervention emphasized subjective experiences of research participants (empathy focused). The criteria-oriented intervention emphasized specific ethical criteria for analyzing protocols (analytic focused). Compared to controls, those in the participant-oriented intervention group exhibited greater attunement to research participants' attitudes related to altruism, trust, quality of relationships with researchers, desire for information, hopes about participation and possible therapeutic misconception, importance of consent forms, and deciding quickly about participation. The participant-oriented group also agreed more strongly that seriously ill people are capable of making their own research participation decisions. The criteria-oriented intervention did not affect learners' attitudes about clinical research, ethical duties of investigators, or research participants' decision making. An empathy-focused approach affected medical students' at-

NRCBL: National Reference Center for Bioethics Literature Classification Scheme   See inside front cover for terms.

45

tunement to research volunteer perspectives, preferences, and attributes, but an analytically oriented approach had no influence. These findings underscore the need to further examine the differential effects of empathy-versus analytic-focused approaches to the teaching of ethics.

Turrens, Julio F. Teaching research integrity and bioethics to science undergraduates. *Cell Biology Education* 2005 Winter; 4(4): 330-334. NRCBL: 2.3; 1.3.9. SC: em.

Yalisove, Daniel. From the ivory tower to the trenches: teaching professional ethics to substance abuse counselors. *In:* Kleinig, John; Einstein, Stanley, eds. Ethical Challenges for Intervening in Drug Use: Policy, Research and Treatment Issues. Huntsville, TX: Office of International Criminal Justice; Criminal Justice Center, Sam Houston State University, 2006: 507-527. NRCBL: 2.3; 9.5.9; 17.1; 1.1; 1.3.1.

BIOETHICS AND MEDICAL ETHICS/ HISTORY

Carson, Ronald A. Engaged humanities: moral work in the precincts of medicine. *Perspectives in Biology and Medicine* 2007 Summer; 50(3): 321-333. NRCBL: 2.2; 2.1; 4.1.2; 7.1. SC: rv.

Fowler, Marsha; Tschudin, Verena. Ethics in nursing: an historical perspective. *In:* Davis, Anne J.; Tschudin, Verena; de Raeve, Louise, eds. Essentials of Teaching and Learning in Nursing Ethics: Perspectives and Methods. New York: Churchill Livingstone Elsevier, 2006: 13-25. NRCBL: 2.2; 4.1.3; 7.2.

Jonsen, Albert R. The history of bioethics as a discipline. *In:* Khushf, George, ed. Handbook of Bioethics: Taking Stock of the Field From a Philosophical Perspective. Dordrecht; Boston: Kluwer Academic, 2004: 31-51. NRCBL: 2.2. SC: rv.

Lauritzen, Paul. Daniel Callahan and bioethics. Where the best arguments take him. *Commonweal* 2007 June 1; 134(11): 8-13. NRCBL: 2.2; 1.1; 12.3.

Levine, Carol. Analyzing Pandora's box: the history of bioethics. *In:* Eckenwiler, Lisa A.; Cohn, Felicia G., eds. The Ethics of Bioethics: Mapping the Moral Landscape. Baltimore, MD: Johns Hopkins University Press, 2007: 3-23. NRCBL: 2.2; 2.1. SC: rv.

Maehle, Andreas-Holger. Professional ethics and discipline: the Prussian Medical Courts of Honour, 1888-1920. *Medizinhistorisches Journal* 1999; 34(3-4): 309-338. NRCBL: 2.2; 6; 7.1; 21.1; 7.4. SC: em.

Majumdar, Sisir K. History of evolution of the concept of medical ethics. *Bulletin of the Indian Institute of History of Medicine (Hyderabad)* 2003 January-June; 33(1): 17-31. NRCBL: 2.2; 2.1; 1.1; 1.3.1; 1.2.

McCullough, Laurence B. John Gregory's medical ethics and the reform of medical practice in eighteenth-century Edinburgh. *Journal of the Royal College of Physicians of Edinburgh* 2006 March; 36(1): 86-92. NRCBL: 2.2.

Sass, Hans-Martin. Fritz Jahr's 1927 concept of bioethics. *Kennedy Institute of Ethics Journal* 2007 December; 17(4): 279-295. NRCBL: 2.2; 1.1; 22.1; 16.1; 2.3.

Abstract: In 1927, Fritz Jahr, a Protestant pastor, philosopher, and educator in Halle an der Saale, published an article entitled "Bio-Ethics: A Review of the Ethical Relationships of Humans to Animals and Plants" and proposed a "Bioethical Imperative," extending Kant's moral imperative to all forms of life. Reviewing new physiological knowledge of his times and moral challenges associated with the development of secular and pluralistic societies, Jahr redefines moral obligations towards human and nonhuman forms of life, outlining the concept of bioethics as an academic discipline, principle, and virtue. Although he had no immediate long-lasting influence during politically and morally turbulent times, his argument that new science and technology requires new ethical and philosophical reflection and resolve may contribute toward clarification of terminology and of normative and practical visions of bioethics, including understanding of the geoethical dimensions of bioethics.

Vanderpool, Harold Y. A revisionist look at the fall and rise of medical ethics [review of Disrupted Dialogue: Medical Ethics and the Collapse of Physician-Humanist Communication (1770-1980), by Robert M. Veatch]. *Medical Humanities Review* 2005 Spring-Fall; 19(1-2): 30-34. NRCBL: 2.2; 7.1; 7.2; 6; 7.3.

BIOETHICS AND MEDICAL ETHICS/ LEGAL ASPECTS

Ethics briefings. *Journal of Medical Ethics* 2007 April; 33(4): 247-248. NRCBL: 2.1; 8.3.3; 8.4; 9.5.6; 9.5.7; 10; 11.3; 12.4.1; 14.1; 20.5.1. SC: le. Identifiers: Great Britain (United Kingdom); Portugal.

Boyle, Joseph. The bioethics of global biomedicine: a natural law reflection. *In:* Engelhardt, H. Tristram, ed. Global Bioethics: The Collapse of Consensus. Salem, MA: M&M Scrivener Press, 2006: 300-334. NRCBL: 2.1; 1.1; 21.1. SC: le.

Duttge, Gunnar. Zukunftsperspektiven der Medizinethik — aus Sicht des Rechts [Future prospects of medical ethics — from the legal point of view]. *Ethik in der Medizin* 2006 December; 18(4): 331-336. NRCBL: 2.1. SC: le.

English, Veronica; Hamm, Danielle; Harrison, Caroline; Mussell, Rebecca; Sheather, Julian; Sommerville, Ann. Ethics briefings. *Journal of Medical Ethics* 2007 July; 33(7): 433-434 [see correction Journal of Medical Ethics 2007 October; 33(10): 620]. NRCBL: 2.1; 8.3.2; 8.3.3; 8.4; 9.1; 10; 12.4.1; 17.1; 17.7; 20.5.1. SC: le. Identifiers: euthanasia; mental health law; confidentiality; abortion.

Fossett, James W.; Ouellette, Alicia R.; Philpott, Sean; Magnus, David; McGee, Glenn. Federalism and bioethics: states and moral pluralism. *Hastings Center Report* 2007 November-December; 37(6): 24-35. 45 fn. NRCBL: 2.1; 1.3.5; 18.6; 18.5.4; 15.1; 11.1; 9.7. SC: cs; an; le.
 Keywords: *bioethical issues; *cultural pluralism; *federal government; *government regulation; *political systems; *state government; advisory committees; attitudes; bioethics; constitutional law; contraception; democracy; embryo research; embryonic stem cells; empirical research; ethicists; judicial role; government financing; legal aspects; policy analysis; politics; policy making; public opinion; research support; Supreme Court decisions; Keyword Identifiers: *United States
Abstract: Bioethicists are often interested mostly in national standards and institutions, but state governments have historically overseen a wide range of bioethical issues and share responsibility with the federal government for still others. States ought to have an important role. By allowing for multiple outcomes, the American federal system allows a better fit between public opinion and public policies.

Fry-Revere, Sigrid. Legal trends in bioethics. *Journal of Clinical Ethics* 2007 Spring; 18(1): 72-90. NRCBL: 2.1; 2.5.3; 12.4.4; 18.5.4; 15.1; 19.5; 8.3.5; 8.4; 15.3; 20.4.1; 8.1. SC: le.

Fry-Revere, Sigrid. Legal trends in bioethics. *Journal of Clinical Ethics* 2007 Summer; 18(2): 162-188. NRCBL: 2.1; 1.3.8; 12.4.4; 8.3.5; 19.5; 8.4; 9.5.6. SC: le.

Fry-Revere, Sigrid; Koshy, Sheeba. Legal trends in bioethics. *Journal of Clinical Ethics* 2007 Fall; 18(3): 294-328. NRCBL: 2.1; 14.1; 14.5; 12.4.4; 15.1; 18.5.4; 21.1; 9.7; 8.3.5; 19.5; 20.5.3. SC: le.
 Keywords: *bioethical issues; *legal aspects; *state government; *trends; abortion; access to information; cloning; confidentiality; conscience; embryo research; embryonic stem cells; euthanasia; government regulation; health disparities; immunization; informed consent; international aspects; investigational drugs; mandatory programs; mentally ill persons; Supreme Court decisions; terminally ill; Keyword Identifiers: *United States

Fry-Revere, Sigrid; Koshy, Sheeba; Leppard IV, John. Legal trends in bioethics. *Journal of Clinical Ethics* 2007 Winter; 18(4): 404-424. NRCBL: 2.1; 9.1; 9.7; 12.4.4; 14.1; 21.1; 9.3.2; 19.5; 8.3.1; 20.5.3; 9.5.6. SC: le.

Höfling, Wolfram. Das "Menschenrechtsübereinkommen zur Bio-Medizin" und die Grund- und Menschenrechte. *In:* Neuer-Miebach, Therese; Wunder, Michael, eds. Bio-Ethik und die Zukunft der Medizin. Bonn: Psychiatrie-Verlag, 1998: 72-86. NRCBL: 2.1; 5.3; 21.1. SC: le.

Kipnis, Kenneth. The expert ethics witness as teacher. *In:* Rasmussen, Lisa, ed. Ethics Expertise: History, Contemporary Perspectives, and Applications. Dordrecht: Springer, 2005: 269-279. NRCBL: 2.3; 2.1; 7.3; 1.3.5. SC: le.

Laurie, Graeme. The autonomy of others: reflections on the rise and rise of patient choice in contemporary medical law. *In:* McLean, Sheila A.M., ed. First Do No Harm: Law, Ethics, and Healthcare. Aldershot, England; Burlington, VT: Ashgate, 2006: 131-149. NRCBL: 2.1; 2.2; 20.7; 20.5.1; 8.1. SC: le.

Mauler, Valerie. Improving public health: balancing ethics, culture, and technology. *Georgetown Journal of Legal Ethics* 2007 Summer; 20(3): 817-833. 113 fn. NRCBL: 2.1; 21.7; 18.5.4; 15.1. SC: le.
 Keywords: *bioethics; *cultural pluralism; *embryo research; *embryonic stem cells; *human rights; *international aspects; *legal aspects; non-Western World; public health; regulation; values; Western World; Keyword Identifiers: United States

McPhee, John; Stewart, Cameron. Recent developments. *Journal of Bioethical Inquiry* 2006; 3(3): 125-131. NRCBL: 2.1; 20.7; 20.5.3; 14.6. SC: cs; le.

Petroni, Angelo Maria. Perspectives for freedom of choice in bioethics and health care in Europe. *In:* Engelhardt, H. Tristram, ed. Global Bioethics: The Collapse of Consensus. Salem, MA: M&M Scrivener Press, 2006: 238-270. NRCBL: 2.1; 9.1; 21.1. SC: le. Identifiers: European Union.

Riley, Margaret Foster; Merrill, Richard A. Regulating reproductive genetics: a review of American bioethics commissions and comparison to the British Human Fertilisation and Embryology Authority. *Columbia Science and Technology Law Review* 2005; 6(1): 1-64. 363 fn. NRCBL: 2.4; 14.1; 14.4; 18.5.4; 15.1; 18.2. SC: le.
 Keywords: *advisory committees; *government regulation; *regulation; *reproductive technologies; bioethical issues; biotechnology; comparative studies; legislation; organization and administration; preimplantation diagnosis; Keyword Identifiers: *Great Britain; *Human Fertilisation and Embryology Authority; *United States; Biomedical Ethics Advisory Commission; Ethics Advisory Board; Human Embryo Research Panel; Human Fertilisation and Embryology Act 1990 (Great Britain); President's Council on Bioethics; President's Commission for the Study of Ethical Problems

Smith, George P. Procreational autonomy or theological restraints. *In his:* The Christian Religion and Biotechnology: A Search for Principled Decision-making. Dordrecht: Springer, 2005: 61-84. NRCBL: 2.1; 1.2; 12.3; 14.4. SC: le.

Stevens, M.L. Tina. Intellectual capital and voting booth bioethics: a contemporary historical critique. *In:* Eckenwiler, Lisa A.; Cohn, Felicia G., eds. The Ethics of Bioethics: Mapping the Moral Landscape. Baltimore, MD: Johns Hopkins University Press, 2007: 59-73. NRCBL: 2.1; 5.3; 18.5.4; 15.1. SC: le. Identifiers: California; Proposition 71; stem cell research.

Thomas, Cordelia. Public dialogue and xenotransplantation. *Medicine and Law: The World Association for Medical Law* 2007 December; 26(4): 801-815. NRCBL: 2.4; 19.1; 22.2; 21.7. SC: le.

NRCBL: National Reference Center for Bioethics Literature Classification Scheme   See inside front cover for terms.

47

Abstract: Toi te Taiao: the Bioethics Council was established in 2002 to enhance New Zealand's understanding of the cultural, ethical and spiritual aspects of biotechnology and ensure that the use of biotechnology has regard for the values held by New Zealanders. In 2005, the Bioethics Council focused on xenotransplantation. A series of dialogue events were held, the public had the opportunity to participate in an online discussion forum and were able to make written submissions. There is worldwide interest in the potential of this biotechnology to cure or alleviate a number of serious health conditions. However, there are concerns about the risks, especially the potential for cross species infection. Such risks have not yet been reliably quantified, but any decision about safety and effectiveness is also about cultural, ethical and spiritual factors. This paper considers some of the outcomes from the dialogue process and the reflections of the Bioethics Council on these. It contrasts the process with that of classic consultation and concludes that, although the process may be more costly and time consuming than the traditional consultative approach, it enables the role of science to be appreciated in its full context, including appreciation of the uncertainties of natural systems and the relevance of cultural, ethical and spiritual human values. It will be suggested that the public are able to interweave ethical concerns with scientific knowledge to engage in meaningful dialogue, resulting in useful recommendations.

BIOETHICS AND MEDICAL ETHICS/ PHILOSOPHICAL ASPECTS

Adams, John. Prescribing: the ethical dimension. *Nurse Prescriber [electronic]* 2004 July; 1(7): 3 p. Accessed: http://journals.cambridge.org/action/displayJournal?jid=NPR [2007 May 11]. NRCBL: 2.1; 1.1; 9.7.

Agich, George J. Autonomy as a problem for clinical ethics. *In:* Nys, Thomas; Denier, Yvonne; Vandevelde, Toon, eds. Autonomy and Paternalism: Reflections on the Theory and Practice of Health Care. Leuven; Dudley, MA: Peeters, 2007: 71-91. NRCBL: 2.1; 1.1; 8.1; 9.5.1. SC: an.

Andre, Judith. Learning to listen: second-order moral perception and the work of bioethics. *In:* Eckenwiler, Lisa A.; Cohn, Felicia G., eds. The Ethics of Bioethics: Mapping the Moral Landscape. Baltimore, MD: Johns Hopkins University Press, 2007: 220-228. NRCBL: 2.1; 22.1; 9.5.10; 21.1; 1.1.

Árnason, Vilhjálmur. The global and the local: fruitful tensions in medical ethics. *Ethik in der Medizin* 2006 December; 18(4): 385-389. NRCBL: 2.1; 1.1; 21.1.

Arras, John D. The way we reason now: reflective equilibrium in bioethics. *In:* Steinbock, Bonnie, ed. The Oxford Handbook of Bioethics. Oxford; New York: Oxford University Press, 2007: 46-71. NRCBL: 2.1; 1.1.

Austin, Wendy. The ethics of everyday practice: healthcare environments as moral communities. *ANS: Advances in Nursing Science* 2007 January-March; 30(1): 81-88. NRCBL: 2.1; 4.1.3; 1.1.

Baker, Robert B.; McCullough, Laurence B. The relationship between moral philosophy and medical ethics reconsidered. *Kennedy Institute of Ethics Journal* 2007 September; 17(3): 271-276. NRCBL: 2.1; 1.1; 4.1.2; 1.3.1; 2.2. SC: an.

Abstract: Medical ethics often is treated as applied ethics, that is, the application of moral philosophy to ethical issues in medicine. In an earlier paper, we examined instances of moral philosophy's influence on medical ethics. We found the applied ethics model inadequate and sketched an alternative model. On this model, practitioners seeking to change morality "appropriate" concepts and theory fragments from moral philosophy to valorize and justify their innovations. Goldilocks-like, five commentators tasted our offerings. Some found them too cold, since they had already abandoned applied ethics; others too hot, since they still find the applied ethics model to their taste. We reply that the appropriation model offers an empirically testable account of the historical relationship between moral philosophy and medical ethics that explains why practitioners appropriate concepts and fragments from moral philosophy. In contrast, the now fashionable common morality theory neither explains moral change nor why practitioners turn to moral philosophy.

Baker, Robert; McCullough, Laurence. Medical ethics' appropriation of moral philosophy: the case of the sympathetic and the unsympathetic physician. *Kennedy Institute of Ethics Journal* 2007 March; 17(1): 3-22. NRCBL: 2.1; 1.3.1; 2.2; 1.1; 7.1; 9.5.3; 21.4; 4.1.2; 20.5.1. SC: an. Identifiers: Alfred Hoche; Karl Binding.

Abstract: Philosophy textbooks typically treat bioethics as a form of "applied ethics"—i.e., an attempt to apply a moral theory, like utilitarianism, to controversial ethical issues in biology and medicine. Historians, however, can find virtually no cases in which applied philosophical moral theory influenced ethical practice in biology or medicine. In light of the absence of historical evidence, the authors of this paper advance an alternative model of the historical relationship between philosophical ethics and medical ethics, the appropriation model. They offer two historical case studies to illustrate the ways in which physicians have "appropriated" concepts and theory fragments from philosophers, and demonstrate how appropriated moral philosophy profoundly influenced the way medical morality was conceived and practiced.

Bayertz, Kurt. Struggling for consensus and living without it: the construction of a common European bioethics. *In:* Engelhardt, H. Tristram, ed. Global Bioethics: The Collapse of Consensus. Salem, MA: M&M Scrivener Press, 2006: 207-237. NRCBL: 2.1; 21.1; 1.1.

Beauchamp, Tom L. History and theory in "applied ethics". *Kennedy Institute of Ethics Journal* 2007 March; 17(1): 55-64. NRCBL: 2.1; 1.3.1; 2.2; 1.1; 7.1; 4.1.2. SC: an.

Abstract: Robert Baker and Laurence McCullough argue that the "applied ethics model" is deficient and in need of a replacement model. However, they supply no clear meaning to "applied ethics" and miss most of what is important in the literature on methodology that treats this question. The Baker-McCullough account of medical and applied ethics is a straw man that has had no influence in these fields or in philosophical ethics. The authors are also on shaky historical grounds in dealing with two problems: (1) the historical source of the notion of "practical ethics" and (2) the historical source of and the assimilation of the term "autonomy" into applied philosophy and professional ethics. They mistakenly hold (1) that the expression "practical ethics" was first used in a publication by Thomas Percival and (2) that Kant is the primary historical source of the notion of autonomy as that notion is used in contemporary applied ethics.

Beauchamp, Tom L.; DeGrazia, David. Principles and principlism. *In:* Khushf, George, ed. Handbook of Bioethics: Taking Stock of the Field From a Philosophical Perspective. Dordrecht; Boston: Kluwer Academic, 2004: 55-74. NRCBL: 2.1; 1.1. SC: rv.

Beckmann, Jan P. Ethik in der Medizin in Aus- und Weiterbildung aus der Sicht der Philosophie [Ethics in medicine in education and continuing education from the philosophical point of view]. *Ethik in der Medizin* 2006 December; 18(4): 369-373. NRCBL: 2.3; 1.1; 7.2.

Bedford-Strohm, Heinrich. Justice and long-term care: a theological ethical perspective. *Christian Bioethics* 2007 September-December; (13)3: 269-285. NRCBL: 2.1; 1.2; 1.1; 9.5.1; 9.5.2; 9.5.10.
Abstract: The relevance of justice for the current debate on long-term care is explored on the basis of demographic and economic data, especially in the U.S. and Germany. There is a justice question concerning the quality and availability of long-term care for different groups within society. Mapping the justice debate by discussing the two main opponents, John Rawls and Robert Nozick, the article identifies fundamental assumptions in both theories. An exploration of the biblical concept of the "option for the poor" and its influence on a new "ecumenical social teaching from below" leads to the conclusion that a Christian ethical account of long-term care will argue for a system that guarantees decent care to every citizen. The German model of Soziale Pflegeversicherung is presented as one possible option for putting this ethical guideline into political practice. In a final reflection, the role of religious affiliation for long-term care is discussed by looking at empirical data and by naming seven dimensions of faith-driven long-term care.

Benatar, D. Moral theories may have some role in teaching applied ethics. *Journal of Medical Ethics* 2007 November; 33(11): 671-672. NRCBL: 2.3; 1.1. SC: an.
Abstract: In a recent paper, Rob Lawlor argues that moral theories should not be taught in courses on applied ethics. The author contends that Dr Lawlor's arguments overlook at least two important roles that some attention to ethical theories may play in practical ethics courses. The
conclusion is not that moral theory must be taught, but rather that there is more to be said for it than Dr Lawlor's arguments reveal.

Biller-Andorno, Nikola. The global, the local, and the parochial - a commentary on Vilhjálmur Árnason. *Ethik in der Medizin* 2006 December; 18(4): 390-392. NRCBL: 2.1; 1.1; 21.1.

Boyle, Joseph. Casuistry. *In:* Khushf, George, ed. Handbook of Bioethics: Taking Stock of the Field From a Philosophical Perspective. Dordrecht; Boston: Kluwer Academic, 2004: 75-88. NRCBL: 2.1; 1.1; 1.2.

Boyle, Joseph. The bioethics of global biomedicine: a natural law reflection. *In:* Engelhardt, H. Tristram, ed. Global Bioethics: The Collapse of Consensus. Salem, MA: M&M Scrivener Press, 2006: 300-334. NRCBL: 2.1; 1.1; 21.1. SC: le.

Buchanan, Allen. Social moral epistemology and the role of bioethicists. *In:* Eckenwiler, Lisa A.; Cohn, Felicia G., eds. The Ethics of Bioethics: Mapping the Moral Landscape. Baltimore, MD: Johns Hopkins University Press, 2007: 288-296. NRCBL: 2.1; 1.1; 8.1; 1.3.2.

Capaldi, Nicholas. How philosophy and theology have undermined bioethics. *Christian Bioethics* 2007 January-April; (13)1: 53-66. NRCBL: 2.1; 1.2; 1.1; 7.1.
Abstract: This essay begins by distinguishing among the viewpoints of philosophy, theology, and religion; it then explores how each deals with "sin" in the bioethical context. The conclusions are that the philosophical and theological viewpoints are intellectually defective in that they cripple our ability to deal with normative issues, and are in the end unable to integrate Christian concepts like "sin" successfully into bioethics. Sin is predicated only of beings with free will, though only in Western Christianity must all sins be committed with knowledge and voluntarily. Without the notions of free will, sin, and a narrative of redemption, bioethics remains unable to provide itself with an adequate normative framework. Bioethics, and morality in general, remain a morass precisely because there has been a failure to translate Christian morality into fully secular and scientist terms.

Carson, Ronald A. Engaged humanities: moral work in the precincts of medicine. *Perspectives in Biology and Medicine* 2007 Summer; 50(3): 321-333. NRCBL: 2.2; 2.1; 4.1.2; 7.1. SC: rv.

Chambers, Tod. The virtue of attacking the bioethicist. *In:* Eckenwiler, Lisa A.; Cohn, Felicia G., eds. The Ethics of Bioethics: Mapping the Moral Landscape. Baltimore, MD: Johns Hopkins University Press, 2007: 281-287. NRCBL: 2.1; 1.1; 7.3.

Cherry, Mark J. Preserving the possibility for liberty in health care. *In:* Engelhardt, H. Tristram, ed. Global Bioethics: The Collapse of Consensus. Salem, MA: M&M Scrivener Press, 2006: 95-130. NRCBL: 2.1; 1.1; 9.1; 21.7.

NRCBL: National Reference Center for Bioethics Literature Classification Scheme   See inside front cover for terms.

49

Childress, James F. Methods in bioethics. *In:* Steinbock, Bonnie, ed. The Oxford Handbook of Bioethics. Oxford; New York: Oxford University Press, 2007: 15-45. NRCBL: 2.1; 1.1.

Clouser, K. Danner; Gert, Bernard. Common morality. *In:* Khushf, George, ed. Handbook of Bioethics: Taking Stock of the Field From a Philosophical Perspective. Dordrecht; Boston: Kluwer Academic, 2004: 121-141. NRCBL: 2.1; 1.1. SC: rv.

Cowley, Christopher. Why medical ethics should not be taught by philosophers. *Discourse* 2005 Autumn; 5(1): 50-63 [Online]. Accessed: http://prs.heacademy.ac.uk/publications/autumn2005.pdf [2006 December 6]. NRCBL: 2.3; 7.2; 1.1.

Cutter, Mary Ann G. Expert moral choice in medicine: a study of uncertainty and locality. *In:* Rasmussen, Lisa, ed. Ethics Expertise: History, Contemporary Perspectives, and Applications. Dordrecht: Springer, 2005: 125-137. NRCBL: 2.1; 1.1. SC: an.

Davis, John K. Intuition and the junctures of judgment in decision procedures for clinical ethics. *Theoretical Medicine and Bioethics* 2007; 28(1): 1-30. NRCBL: 2.1; 1.1. SC: an.

Abstract: Moral decision procedures such as principlism or casuistry require intuition at certain junctures, as when a principle seems indeterminate, or principles conflict, or we wonder which paradigm case is most relevantly similar to the instant case. However, intuitions are widely thought to lack epistemic justification, and many ethicists urge that such decision procedures dispense with intuition in favor of forms of reasoning that provide discursive justification. I argue that discursive justification does not eliminate or minimize the need for intuition, or constrain our intuitions. However, this is not a problem, for intuitions can be justified in easy or obvious cases, and decision procedures should be understood as heuristic devices for reaching judgments about harder cases that approximate the justified intuitions we would have about cases under ideal conditions, where hard cases become easy. Similarly, the forms of reasoning which provide discursive justification help decision procedures perform this heuristic function not by avoiding intuition, but by making such heuristics more accurate. Nonetheless, it is possible to demand too much justification; many clinical ethicists lack the time and philosophical training to reach the more elaborate levels of discursive justification. We should keep moral decision procedures simple and user-friendly so that they will provide what justification can be achieved under clinical conditions, rather than trying to maximize our epistemic justification out of an overstated concern about intuition.

Delkeskamp-Hayes, Corinna. Implementing health care rights versus health care cultures: the limits of tolerance, Kant's rationality, and the moral pitfalls of international bioethics standardization. *In:* Engelhardt, H. Tristram, ed. Global Bioethics: The Collapse of Consensus. Salem, MA: M&M Scrivener Press, 2006: 50-94. NRCBL: 2.1; 1.1; 21.1; 21.7.

Delkeskamp-Hayes, Corinna. Resisting the therapeutic reduction: on the significance of sin. *Christian Bioethics* 2007 January-April; (13)1: 105-127. NRCBL: 2.1; 1.2; 4.2; 1.1; 7.1; 21.7.

Abstract: Sin-talk, though politically incorrect, is indispensable. Placing human life under the "hermeneutic of sin" means acknowledging that one ought to aim flawlessly at God, and that one can fail in this endeavor. None of this can be appreciated within the contemporary post-Christian, mindset, which has attempted to reduce religion to morality and culture. In such a secular context, the guilt-feelings connected with the recognition of sin are considered to be harmful; the eternal benefit of a repentance is disregarded. Nevertheless, spirituality appears to have therapeutic benefits. Therefore attempts are made to re-locate within healthcare a religion shorn of its transcendent claims, so as then to harvest the benefits of a spirituality "saved from sin". This reduction of religiosity to its therapeutic function is nourished by a post-modern constructivist construal of religion. This article critically examines the dis-ingenuity marring such recasting, as well as the incoherence of related attempts to reduce transcendence to solidarity, and to re-shape the significance of religious rituals.

Delkeskamp-Hayes, Corinna. Societal consensus and the problem of consent: refocusing the problem of ethics expertise in liberal democracies. *In:* Rasmussen, Lisa, ed. Ethics Expertise: History, Contemporary Perspectives, and Applications. Dordrecht: Springer, 2005: 139-163. NRCBL: 2.1; 1.1; 1.3.5.

Dikova, Rossitza. Introduction [to issue on Philanthropy, Caritas, Diakonia — European approaches to Christians' service in the world]. *Christian Bioethics* 2007 September-December;(13)3: 245-250. NRCBL: 2.1; 1.2; 9.5.1; 9.5.2; 9.5.10; 1.1.

Dzur, Albert W.; Levin, Daniel. The primacy of the public: in support of bioethics commissions as deliberative forums. *Kennedy Institute of Ethics Journal* 2007 June; 17(2): 133-142. NRCBL: 2.4; 5.1; 1.3.5; 2.1; 1.1; 7.1. SC: an.

Abstract: In a 2004 article, we argued that bioethics commissions should be assessed in terms of their usefulness as public forums. A 2006 article by Summer Johnson argued that our perspective was not supported by the existing literature on presidential commissions, which had not previously identified commissions as public forums and that we did not properly account for the political functions of commissions as instruments of presidential power. Johnson also argued that there was nothing sufficiently unique about bioethics commissions to make the public forum perspective particularly applicable. We respond by arguing that analysis of commissions' work as public forums fits well within the literature on commissions, especially on their agenda-setting functions, and that the political functions of commissions are often compatible with their functioning as public forums. We also demonstrate how the origins and concerns of bioethics make public forum analysis particularly applicable to bioethics commissions.

Ebbesen, Mette; Pedersen, Birthe D. Using empirical research to formulate normative ethical principles in biomedicine. *Medicine, Health Care and Philosophy* 2007 March; 10(1): 33-48. NRCBL: 2.1; 1.1. SC: an; em.
Abstract: Bioethical research has tended to focus on theoretical discussion of the principles on which the analysis of ethical issues in biomedicine should be based. But this discussion often seems remote from biomedical practice where researchers and physicians confront ethical problems. On the other hand, published empirical research on the ethical reasoning of health care professionals offer only descriptions of how physicians and nurses actually reason ethically. The question remains whether these descriptions have any normative implications for nurses and physicians? In this article, we illustrate an approach that integrates empirical research into the formulation of normative ethical principles using the moral-philosophical method of Wide Reflective Equilibrium (WRE). The research method discussed in this article was developed in connection with the project 'Bioethics in Theory and Practice'. The purpose of this project is to investigate ethical reasoning in biomedical practice in Denmark empirically. In this article, we take the research method as our point of departure, but we exclusively discuss the theoretical framework of the method, not its empirical results. We argue that the descriptive phenomenological hermeneutical method developed by Lindseth and Norberg (2004) and Pedersen (1999) can be combined with the theory of WRE to arrive at a decision procedure and thus a foundation for the formulation of normative ethical principles. This could provide health care professionals and biomedical researchers with normative principles about how to analyse, reason and act in ethically difficult situations in their practice. We also show how to use existing bioethical principles as inspiration for interpreting the empirical findings of qualitative studies. This may help researchers design their own empirical studies in the field of ethics.

Elliott, Carl. The tyranny of expertise. *In:* Eckenwiler, Lisa A.; Cohn, Felicia G., eds. The Ethics of Bioethics: Mapping the Moral Landscape. Baltimore, MD: Johns Hopkins University Press, 2007: 43-46. NRCBL: 2.1; 4.1.1; 7.3.

Engelhardt, H. Tristram. Bioethics as politics: a critical reassessment. *In:* Eckenwiler, Lisa A.; Cohn, Felicia G., eds. The Ethics of Bioethics: Mapping the Moral Landscape. Baltimore, MD: Johns Hopkins University Press, 2007: 118-133. NRCBL: 2.1; 21.1; 5.3; 1.1.

Engelhardt, H. Tristram. The search for global morality: bioethics, the culture wars, and moral diversity. *In:* Engelhardt, H. Tristram, ed. Global Bioethics: The Collapse of Consensus. Salem, MA: M&M Scrivener Press, 2006: 18-49. NRCBL: 2.1; 1.1; 21.1; 21.7. Identifiers: Francis Fukuyama.

Engelhardt, H. Tristram. Why ecumenism fails: taking theological differences seriously. *Christian Bioethics* 2007 January-April; (13)1: 25-51. NRCBL: 2.1; 1.2; 1.1.
Abstract: Contemporary Christians are separated by foundationally disparate understandings of Christianity itself. Christians do not share one theology, much less a common understanding of the significance of sin, suffering, disease, and death. These foundational disagreements not only stand as impediments to an intellectually defensible ecumenism, but they also form the underpinnings of major disputes in the culture wars, particularly as these are expressed in healthcare. There is not one Christian bioethics of sin, suffering, sickness, and death. In this article, the character of the moral-theological visions separating the various Christianities and thus their bioethics is examined. Particular emphasis is placed on the differences that set contemporary Western theology at odds with the theology of the Christianity of the first millennium. As is shown, the ground for this gulf lies in the divide between traditional and post-traditional views of the appropriate role of philosophy in theology, a difference rooted in disparate understandings of the meaning of church and of the meaning of the logos, the Son of God.

English, Veronica; Mussell, Rebecca; Sheather, Julian; Sommerville, Ann. Autonomy and its limits: what place for the public good? *In:* McLean, Sheila A.M., ed. First Do No Harm: Law, Ethics, and Healthcare. Aldershot, England; Burlington, VT: Ashgate, 2006: 117-130. NRCBL: 2.1; 1.1; 8.4; 15.3; 17.1.

Fan, Ruiping. Bioethics: globalization, communitization, or localization? *In:* Engelhardt, H. Tristram, ed. Global Bioethics: The Collapse of Consensus. Salem, MA: M&M Scrivener Press, 2006: 271-299. NRCBL: 2.1; 1.1; 21.1.

Fisher, Anthony. Cooperation in evil: understanding the issues. *In:* Watt, Helen, ed. Cooperation, Complicity and Conscience: Problems in Healthcare, Science, Law and Public Policy. London: Linacre Centre, 2005: 27-64. NRCBL: 2.1; 1.1; 1.2.

Fowler, Marsha; Tschudin, Verena. Ethics in nursing: an historical perspective. *In:* Davis, Anne J.; Tschudin, Verena; de Raeve, Louise, eds. Essentials of Teaching and Learning in Nursing Ethics: Perspectives and Methods. New York: Churchill Livingstone Elsevier, 2006: 13-25. NRCBL: 2.2; 4.1.3; 7.2.

Fox, Daniel M. Selective appropriation, medical ethics, and health politics: the complementarity of Baker, McCullough, and me. *Kennedy Institute of Ethics Journal* 2007 March; 17(1): 23-30. NRCBL: 2.1; 1.3.1; 2.2; 1.1; 7.1; 4.1.2. SC: an.
Abstract: Baker and McCullough (2007) criticize a 1979 article by this author for insufficiently appreciating how physicians have appropriated ideas from moral philosophy. This rejoinder argues that the two articles are complementary. The 1979 article summarized evidence that leading physicians in the nineteenth and twentieth centuries appropriated ideas from moral philosophy and re-

NRCBL: National Reference Center for Bioethics Literature Classification Scheme   See inside front cover for terms.

51

lated disciplines that reinforced their political goals of self-regulation and dominance of the allocation of resources for health. In retrospect the 1979 article also urged bioethicists to appropriate ideas from other disciplines, including moral philosophy, which would contribute to improving the health of populations.

Frank, Arthur W. Social bioethics and the critique of autonomy. *Health* 2000 July; 4(3): 378-394 [Online]. Accessed: http://hea.sagepub.com/cgi/reprint/4/3/378 [2007 May 10]. NRCBL: 2.1; 1.1; 7.1; 9.4; 10. SC: rv.

Fulford, K.W.M. (Bill); Thornton, Tim; Graham, George. From bioethics to values-based practice. *In their:* Oxford Textbook of Philosophy and Psychiatry. Oxford; New York: Oxford University Press, 2006: 498-538. NRCBL: 2.1; 1.1; 7.2; 6.

Gentry, Glenn. Rawls and religious community: ethical decision making in the public square. *Christian Bioethics* 2007 May-August; (13)2: 171-181. NRCBL: 2.1; 1.2; 7.1; 1.1. SC: an.
Abstract: While most people may initially agree that justice is fairness, as an evangelical Protestant I argue that, for many religious comprehensive doctrines, the Rawlsean model does not possess the resources necessary to sustain tolerance in moral decision making. The weakness of Rawls's model centers on the reasonable priority of convictions that arise from private comprehensive doctrines. To attain a free and pluralistic society, people need resources sufficient to provide reasons to tolerate actions that are otherwise intolerable. In addition to arguing for the deficiency of the Rawlsean political model, I sketch out a preliminary model of ambassadorship that offers religious communities, and in particular Protestant evangelicals, the necessary resources to engage the broader society tolerantly while maintaining their religious convictions. As a citizen of the church and a member of another kingdom, Christians serve as ambassadors to those who are not of the heavenly kingdom. I take this model to be more ambitious than that of a sojourner who lives in the land but is isolated as much as possible from society, while more modest than that of reconstructionists who seek to implement their own sacred law on all others.

Giampietro, Anthony E. Improving the Catholic approach to healthcare? [review of More Humane Medicine: A Liberal Catholic Bioethics, by James F. Drane]. *HEC (Healthcare Ethics Committee) Forum* 2007 September; 19(3): 261-270. NRCBL: 2.1; 1.2; 1.1.

Gormally, Luke. Why not dirty your hands? Or: on the supposed rightness of (sometimes) intentionally cooperating in wrongdoing. *In:* Watt, Helen, ed. Cooperation, Complicity and Conscience: Problems in Healthcare, Science, Law and Public Policy. London: Linacre Centre, 2005: 12-26. NRCBL: 2.1; 1.1; 1.2.

Hanson, Stephen S. Moral acquaintances: Loewy, Wildes, and beyond. *HEC (Healthcare Ethics Committee) Forum* 2007 September; 19(3): 207-225. NRCBL: 2.1; 1.1; 1.2; 4.1.2; 21.7. SC: cs.

Helmchen, Hanfried. Ethics as a focus of controversy in postmodern antagonisms. *In:* Schramme, Thomas; Thome, Johannes, eds. Philosophy and Psychiatry. Berlin; New York: De Gruyter, 2004: 347-351. NRCBL: 2.1; 1.1.

Holland, S. Theories, principles, and types of arguments. *In his:* Introducing Nursing Ethics: Themes in Theory and Practice. Salisbury: APS, 2004: 1-28. NRCBL: 2.1; 1.1.

Holm, S.; Takala, T. High hopes and automatic escalators: a critique of some new arguments in bioethics. *Journal of Medical Ethics* 2007 January; 33(1): 1-4. NRCBL: 2.1; 1.1. SC: an.
Abstract: Two protechnology arguments, the "hopeful principle" and the "automatic escalator", often used in bioethics, are identified and critically analysed in this paper. It is shown that the hopeful principle is closely related to the problematic precautionary principle, and the automatic escalator argument has close affinities to the often criticised empirical slippery slope argument. The hopeful principle is shown to be really hopeless as an argument, and automatic escalator arguments often lead nowhere when critically analysed. These arguments should therefore only be used with great caution.

Hooft, Stan van. Bioethics and caring. *In his:* Caring about Health. Aldershot, England; Burlington, VT: Ashgate, 2006: 27-39. NRCBL: 2.1; 1.1; 1.3.1. SC: an.

Jennings, Bruce. Autonomy. *In:* Steinbock, Bonnie, ed. The Oxford Handbook of Bioethics. Oxford; New York: Oxford University Press, 2007: 72-89. NRCBL: 2.1; 1.1.

Johnson, Summer. A rebuttal to Dzur and Levin: Johnson on the legitimacy and authority of bioethics commissions. *Kennedy Institute of Ethics Journal* 2007 June; 17(2): 143-152. NRCBL: 2.4; 5.1; 1.3.5; 2.1; 1.1; 7.1. SC: an. Comments: Albert W. Dzur and Daniel Levin. The primacy of the public: in support of bioethics commissions as deliberative forums. Kennedy Institute of Ethics Journal 2007 June; 17(2): 133-142.
Abstract: Bioethics commissions have been critiqued on the basis that they are not sufficiently public or are too reliant upon expertise to have legitimacy or authority in regard to public policy debates. Adequately assessing the legitimacy and authority of commissions requires thinking clearly about the "publics" these commissions serve, the primary tasks of public bioethics, and how those tasks might be performed with a certain kind of ethical expertise and limited authority that makes them legitimate players in public policy debates concerning bioethics.

Jonsen, Albert R. How to appropriate appropriately: a comment on Baker and McCullough. *Kennedy Institute of Ethics Journal* 2007 March; 17(1): 43-54. NRCBL: 2.1; 1.3.1; 2.2; 1.1; 7.1; 4.1.2. SC: an.
Abstract: The article by Baker and McCullough in this issue posits that bioethics has generally applied moral theories to practical problems. They propose that, rather than this "application," bioethicists should "appropriate" aspects of ethical theory. This article disagrees that bioethical writing is primary "application." It agrees that "ap-

propriation" is the most suitable approach to bioethical analysis but claims that the description of appropriation provided by Baker and McCullough is inadequate. It must be supplemented by the rhetorical concept of "invention."

Keulartz, Jozef. Comments on Gert: Gert's common morality: old-fashioned or untimely? *In:* Korthals, Michiel; Bogers, Robert J., eds. Ethics for Life Scientists. Dordrecht, The Netherlands: Springer, 2004: 141-145. NRCBL: 2.1; 1.1.

Khushf, George. Introduction: taking stock of bioethics from a philosophical perspective. *In:* Khushf, George, ed. Handbook of Bioethics: Taking Stock of the Field From a Philosophical Perspective. Dordrecht; Boston: Kluwer Academic, 2004: 1-28. NRCBL: 2.1; 1.1.

King, Nancy M.P. The glass house: assessing bioethics. *In:* Eckenwiler, Lisa A.; Cohn, Felicia G., eds. The Ethics of Bioethics: Mapping the Moral Landscape. Baltimore, MD: Johns Hopkins University Press, 2007: 297-309. NRCBL: 2.1; 1.1; 7.3.

Kirklin, D. Minding the gap between logic and intuition: an interpretative approach to ethical analysis. *Journal of Medical Ethics* 2007 July; 33(7): 386-389. NRCBL: 2.1; 1.1; 12.1; 19.5; 20.5.1. SC: an.
Abstract: In an attempt to be rational and objective, and, possibly, to avoid the charge of moral relativism, ethicists seek to categorise and characterise ethical dilemmas. This approach is intended to minimise the effect of the confusing individuality of the context within which ethically challenging problems exist. Despite and I argue partly as a result of this attempt to be rational and objective, even when the logic of the argument is accepted — for example, by healthcare professionals — those same professionals might well respond by stating that the conclusions are unacceptable to them. In this paper, I argue that an interpretative approach to ethical analysis, involving an examination of the ways in which ethical arguments are constructed and shared, can help ethicists to understand the origins of this gap between logic and intuition. I suggest that an argument will be persuasive either if the values underpinning the proposed argument accord with the reader's values and worldview, or if the argument succeeds in persuading the reader to alter these. A failure either to appreciate or to acknowledge those things that give meaning to the lives of all the interested parties will make this objective far harder, if not impossible, to achieve. If, as a consequence, the narratives ethicists use to make their arguments seem to be about people living in different circumstances, and faced with different choices and challenges, from those the readers or listeners consider important or have to face in their own lives, then the argument is unlikely to seem either relevant or applicable to those people. The conclusion offered by the ethicist will be, for that individual, counterintuitive. Abortion, euthanasia and cadaveric organ donation are used as examples to support my argument.

Kuczewski, Mark G. Empirical metaethical absolutism in contemporary Catholic bioethics [review of Contemporary Catholic Health Care Ethics, by David F. Kelly]. *Medical Humanities Review* 2005 Spring-Fall; 19(1-2): 76-80. NRCBL: 2.1; 1.1; 1.2.

LaFollette, Eva; LaFollette, Hugh. Private conscience, public acts [editorial]. *Journal of Medical Ethics* 2007 May; 33(5): 249-254. NRCBL: 2.1; 1.1; 4.1.1.

Lance, Mark Norris; Little, Margaret Olivia. Defending moral particularism. *In:* Dreier, James, ed. Contemporary Debates in Moral Theory. Oxford: Blackwell, 2006: 305-321. NRCBL: 2.1; 1.1. SC: an.

Lantz, Göran. Is health care ethics useful? *In:* Østnor, Lars, ed. Bioetikk og teologi: Rapport fra Nordisk teologisk nettverk for bioetikks workshop I Stockholm 27.-29. September 1996. Oslo: Nord. teol. nettv. for bioetikk, 1996: 101-106. NRCBL: 2.1; 1.1; 4.1.1.

Lauritzen, Paul. Daniel Callahan and bioethics. Where the best arguments take him. *Commonweal* 2007 June 1; 134(11): 8-13. NRCBL: 2.2; 1.1; 12.3.

Lawlor, Rob. Moral theories in teaching applied ethics. *Journal of Medical Ethics* 2007 June; 33(6): 370-372. NRCBL: 2.3; 1.1. SC: an.
Abstract: It is argued, in this paper, that moral theories should not be discussed extensively when teaching applied ethics. First, it is argued that, students are either presented with a large amount of information regarding the various subtle distinctions and the nuances of the theory and, as a result, the students simply fail to take it in or, alternatively, the students are presented with a simplified caricature of the theory, in which case the students may understand the information they are given, but what they have understood is of little or no value because it is merely a caricature of a theory. Second, there is a methodological problem with appealing to moral theories to solve particular issues in applied ethics. An analogy with science is appealed to. In physics there is a hope that we could discover a unified theory of everything. But this is, of course, a hugely ambitious project, and much harder than, for example, finding a theory of motion. If the physicist wants to understand motion, he should try to do so directly. We would think he was particularly misguided if he thought that, to answer this question, he first needed to construct a unified theory of everything.

Liaschenko, Joan. Teaching feminist ethics. *In:* Davis, Anne J.; Tschudin, Verena; de Raeve, Louise, eds. Essentials of Teaching and Learning in Nursing Ethics: Perspectives and Methods. New York: Churchill Livingstone Elsevier, 2006: 203-215. NRCBL: 2.3; 1.1; 10; 4.1.3.

Lindemann, Gesa. Die faktische Kraft des Normativen [The real power of normative ethics]. *Ethik in der Medizin* 2006 December; 18(4): 342-347. NRCBL: 2.1; 1.1; 7.1.

London, Alex John. Clinical equipoise: foundational requirement or fundamental error? *In:* Steinbock, Bonnie,

---

NRCBL: National Reference Center for Bioethics Literature Classification Scheme   See inside front cover for terms.

53

ed. The Oxford Handbook of Bioethics. Oxford; New York: Oxford University Press, 2007: 571-596. NRCBL: 2.1; 4.1.2; 18.1.

Majumdar, Sisir K. History of evolution of the concept of medical ethics. *Bulletin of the Indian Institute of History of Medicine (Hyderabad)* 2003 January-June; 33(1): 17-31. NRCBL: 2.2; 2.1; 1.1; 1.3.1; 1.2.

McCullough, Laurence B. Towards a professional ethics model of clinical ethics. *Journal of Medicine and Philosophy* 2007 January-February; 32(1): 1-6. NRCBL: 2.1; 4.1.2; 8.1; 1.3.1.

McKay, Angela. Publicly accessible intuitions: "neutral reasons" and bioethics. *Christian Bioethics* 2007 May-August; (13)2: 183-197. NRCBL: 2.1; 1.2; 7.1; 1.1; 4.4; 20.7; 20.5.1; 8.1. SC: an.

Abstract: This article examines Leon Kass's contention that a choice for physician-assisted suicide is "undignified." Although Kass is Jewish rather than Christian, he argues for positions that most Christians share, and he argues for these positions without presupposing the truth of specific religious claims. I argue that although Kass has some important intuitions, he too readily assumes that these intuitions will be shared by his audience, and that this assumption diminishes the force of his argument. An examination of the limitations of Kass's argument is helpful insofar as it illustrates the real challenge faced by religious believers who wish to defend their beliefs in the "public forum." For it illustrates that what needs to be made "accessible" is the Judeo-Christian understanding of man and his place in the world. While I do not wish to claim that this task is impossible, I do think that it is far more difficult than most realize. Like all important tasks, however, unless we wrestle with the difficulties it raises, our arguments will strike many as unconvincing.

Moreno, Jonathan D. The triumph of autonomy in bioethics and commercialism in American healthcare. *CQ: Cambridge Quarterly of Healthcare Ethics* 2007 Fall; 16(4): 415-419. NRCBL: 2.1; 2.2; 1.1; 9.3.1; 7.3; 1.3.2. SC: an.

Müller, Denis. The original risk: overtheologizing ethics and undertheologizing sin. *Christian Bioethics* 2007 January-April; (13)1: 7-23. NRCBL: 2.1; 1.2; 1.1.

Abstract: The project of articulating a theological ethics on the basis of liturgical anthropology is bound to fail if the necessary consequence is that one has to quit the forum of critical modern rationality. The risk of Engelhardt's approach is to limit rationality to a narrow vision of reason. Sin is not to be understood as the negation of human holiness, but as the negation of divine holiness. The only way to renew theological ethics is to understand sin as the anthropological and ethical expression of the biblical message of the justification by faith only. Sin is therefore a secondary category, which can only by interpreted in light of the positive manifestation of liberation, justification, and grace. The central issue of Christian ethics is not ritual purity or morality, but experience, confession and recognition of our own injustice in our dealing with God and men.

Myser, Catherine. White normativity in U.S. bioethics: a call and method for more pluralist and democratic standards and policies. *In:* Eckenwiler, Lisa A.; Cohn, Felicia G., eds. The Ethics of Bioethics: Mapping the Moral Landscape. Baltimore, MD: Johns Hopkins University Press, 2007: 241-259. NRCBL: 2.1; 21.7; 1.1; 18.6; 1.3.1.

Neale, Ann. Who really wants health care justice? *Health Progress* 2007 January-February; 88(1): 40-43. NRCBL: 2.1; 1.1; 9.1. SC: em.

Nelson, Hilde Lindemann. Four narrative approaches to bioethics. *In:* Khushf, George, ed. Handbook of Bioethics: Taking Stock of the Field From a Philosophical Perspective. Dordrecht; Boston: Kluwer Academic, 2004: 163-181. NRCBL: 2.1; 1.1.

Nelson, James Lindemann. Trusting bioethicists. *In:* Eckenwiler, Lisa A.; Cohn, Felicia G., eds. The Ethics of Bioethics: Mapping the Moral Landscape. Baltimore, MD: Johns Hopkins University Press, 2007: 47-55. NRCBL: 2.1; 4.1.1; 7.3; 1.1.

Nys, Thomas. A bridge over troubled water: paternalism as the expression of autonomy. *In:* Nys, Thomas; Denier, Yvonne; Vandevelde, Toon, eds. Autonomy and Paternalism: Reflections on the Theory and Practice of Health Care. Leuven; Dudley, MA: Peeters, 2007: 147-165. NRCBL: 2.1; 1.1; 8.1.

Padela, Aasim I. Islamic medical ethics: a primer. *Bioethics* 2007 March; 21(3): 169-178. NRCBL: 2.1; 4.1.2; 1.2. SC: rv.

Abstract: Modern medical practice is becoming increasingly pluralistic and diverse. Hence, cultural competency and awareness are given more focus in physician training seminars and within medical school curricula. A renewed interest in describing the varied ethical constructs of specific populations has taken place within medical literature. This paper aims to provide an overview of Islamic Medical Ethics. Beginning with a definition of Islamic Medical Ethics, the reader will be introduced to the scope of Islamic Medical Ethics literature, from that aimed at developing moral character to writings grounded in Islamic law. In the latter form, there is an attempt to derive an Islamic perspective on bioethical issues such as abortion, gender relations within the patient-doctor relationship, end-of-life care and euthanasia. It is hoped that the insights gained will aid both clinicians and ethicists to better understand the Islamic paradigm of medical ethics and thereby positively affect patient care.

Parker, C. Perspectives on ethics. *Journal of Medical Ethics* 2007 January; 33(1): 21-23. NRCBL: 2.1; 1.1; 8.1; 18.5.4; 21.7.

Abstract: In his recent paper about understanding ethical issues, Boyd suggests that traditional approaches based on principles or people are understood better in terms of perspectives, especially the perspective-based approach of hermeneutics, which he uses for conversation rather than controversy. However, we find that Boyd's undefined contrast between conversation and controversy does not point to any improvement in communication: disputes occur during conversation and controversy may be conducted in gentle tones. We agree with Boyd, that being prepared to listen and learn are excellent attitudes, but his vague attempts to establish these and similar virtues in hermeneutic theory are not plausible. Additionally, the current controversy about the use of human embryos in stem cell therapy research shows Boyd missing the opportunity to illustrate how conversation would improve understanding.

Pellegrino, Edmund D. Philosophy of medicine and medical ethics: a phenomenological perspective. *In:* Khushf, George, ed. Handbook of Bioethics: Taking Stock of the Field From a Philosophical Perspective. Dordrecht; Boston: Kluwer Academic, 2004: 183-202. NRCBL: 2.1; 1.1; 4.1.2.

Powers, Penny. Persuasion and coercion: a critical review of philosophical and empirical approaches. *HEC (Healthcare Ethics Committee) Forum* 2007 June; 19(2):125-143. NRCBL: 2.1; 1.1. SC: em; rv.

Rabe, Marianne. Ethik in der Pflegeausbildung [Ethics in nursing education]. *Ethik in der Medizin* 2006 December; 18(4): 379-384. NRCBL: 2.3; 4.1.3; 7.2.

Rudnick, Abraham. Processes and pitfalls of dialogical bioethics. *Health Care Analysis: An International Journal of Health Philosophy and Policy* 2007 June; 15(2): 123-135. NRCBL: 2.1; 1.1.
Abstract: Bioethics uses various theories, methods and institutions for its decision-making. Lately, a dialogical, i.e., dialogue-based, approach has been argued for in bioethics. The aim of this paper is to explore some of the decision-making processes that may be involved in this dialogical approach, as well as related pitfalls that may have to be addressed in order for this approach to be helpful, particularly in clinical ethics. Using informal logic, an analysis is presented of the notion of dialogue and of the stages of dialogical decision-making, and then processes and related pitfalls associated with these stages in the context of clinical ethics are examined. The results of this exploration are expected to facilitate the implementation and empirical testing of dialogical bioethics.

Sass, Hans-Martin. Fritz Jahr's 1927 concept of bioethics. *Kennedy Institute of Ethics Journal* 2007 December; 17(4): 279-295. NRCBL: 2.2; 1.1; 22.1; 16.1; 2.3.
Abstract: In 1927, Fritz Jahr, a Protestant pastor, philosopher, and educator in Halle an der Saale, published an article entitled "Bio-Ethics: A Review of the Ethical Relationships of Humans to Animals and Plants" and proposed a "Bioethical Imperative," extending Kant's moral imperative to all forms of life. Reviewing new physiological knowledge of his times and moral challenges associated with the development of secular and pluralistic societies, Jahr redefines moral obligations towards human and nonhuman forms of life, outlining the concept of bioethics as an academic discipline, principle, and virtue. Although he had no immediate long-lasting influence during politically and morally turbulent times, his argument that new science and technology requires new ethical and philosophical reflection and resolve may contribute toward clarification of terminology and of normative and practical visions of bioethics, including understanding of the geoethical dimensions of bioethics.

Sayson, Ciriaco M. Asian bioethics: theoretical background. *Asian Biotechnology and Development Review* 2006 November; 9(1): 7-11 [Online]. Accessed: http://www.openj-gate.org/articlelist.asp?LatestYear=2007&JCode=104346&year=2006&vol=9&issue=1&ICode=583649 [2008 February 29]. NRCBL: 2.1; 1.1.

Schmidt, Ulf. Turning the history of medical ethics from its head onto its feet: a critical commentary on Baker and McCullough. *Kennedy Institute of Ethics Journal* 2007 March; 17(1): 31-42. NRCBL: 2.1; 1.3.1; 2.2; 1.1; 7.1; 4.1.2; 20.5.1; 21.4. SC: an. Identifiers: Karl Brandt.
Abstract: The paper provides a critical commentary on the article by Baker and McCullough on Medical Ethics' Appropriation of Moral Philosophy. The author argues that Baker and McCullough offer a more "pragmatic" approach to the history of medical ethics that has the potential to enrich the bioethics field with a greater historical grounding and sound methodology. Their approach can help us to come to a more nuanced understanding about the way in which medical ethics has connected, disconnected, and reconnected with philosophical ideas throughout the centuries. The author points out that Baker and McCullough's model can run the danger of overemphasizing the role of medical ethicists whilst marginalizing the influence of philosophers and of other historical actors and forces. He critically reviews the two case studies on which Baker and McCullough focus and concludes that scholars need to bear in mind the levels of uncertainty and ambivalence that accompany the process of transformation and dissemination of moral values in medicine and medical practice.

Shahzad, Qaiser. Playing God and the ethics of divine names: an Islamic paradigm for biomedical ethics. *Bioethics* 2007 October; 21(8): 413-418. NRCBL: 2.1; 1.2; 1.1. Conference: Eighth World Congress of Bioethics: A Just and Healthy Society; Beijing, China; 2006 August 9.
Abstract: The notion of 'playing God' frequently comes to fore in discussions of bioethics, especially in religious contexts. The phrase has always been analyzed and discussed from Christian and secular standpoints. Two interpretations exist in the literature. The first one takes 'God' seriously and playing 'playfully'. It argues that this concept does state a principle but invokes a perspective on the world. The second takes both terms playfully. In the Islamic Intellectual tradition, the Sufi concept of 'adopting divine character traits' provides a legitimate paradigm for 'playing God'. This paradigm is interesting because here we take both terms 'God' and 'playing' seriously. It is significant for the development of biomedical

---

NRCBL: National Reference Center for Bioethics Literature Classification Scheme   See inside front cover for terms.

55

ethics in contemporary Islamic societies as it can open new vistas for viewing biotechnological developments.

Solomon, David. Domestic disarray and imperial ambition: contemporary applied ethics and the prospects for global bioethics. *In:* Engelhardt, H. Tristram, ed. Global Bioethics: The Collapse of Consensus. Salem, MA: M&M Scrivener Press, 2006: 335-361. NRCBL: 2.1; 1.1; 2.2; 21.1.

Spiess, Christian. Recognition and social justice: a Roman Catholic view of Christian bioethics of long-term care and community service. *Christian Bioethics* 2007 September-December;(13)3: 287-301. NRCBL: 2.1; 1.2; 1.1; 9.5.1; 9.5.2; 9.5.10.

Abstract: Contemporary Christian ethics encounters the challenge to communicate genuinely Christian normative orientations within the scientific debate in such a way as to render these orientations comprehensible, and to maintain or enhance their plausibility even for non-Christians. This essay, therefore, proceeds from a biblical motif, takes up certain themes from the Christian tradition (in particular the idea of social justice), and connects both with a compelling contemporary approach to ethics by secular moral philosophy, i.e. with Axel Honneth's reception of Hegel, as based on Hegel's theory of recognition. As a first step, elements of an ethics of recognition are developed on the basis of an anthropological recourse to the conditions of intersubjective encounters. These conditions are then brought to bear on the idea of social justice, as developed in the social-Catholic tradition, and as systematically explored in the Pastoral Letter of the United States Conference of Catholic Bishops, Economic Justice For All (1986). Proceeding from this basis, aspects of a Christian ethics of community service with regard to long-term care can be defined.

Steigleder, Klaus. Medizinethik und Philosophie [Medical ethics and philosophy]. *Ethik in der Medizin* 2006 December; 18(4): 310-314. NRCBL: 2.1; 1.1.

Sugarman, Jeremy. Roles of moral philosophy in appropriated bioethics: a response to Baker and McCullough. *Kennedy Institute of Ethics Journal* 2007 March; 17(1): 65-67. NRCBL: 2.1; 1.3.1; 2.2; 1.1; 4.1.2. SC: an.

Abstract: Strong arguments support the notion that much of modern bioethics is a result of appropriation rather than strict application of traditional moral philosophy. Nevertheless, it is important to recognize these sources and approaches associated with them, even when working with appropriated theories, since traditional ethical theory does and should influence modern bioethics.

Sulmasy, Daniel P. 'Reinventing' the rule of double effect. *In:* Steinbock, Bonnie, ed. The Oxford Handbook of Bioethics. Oxford; New York: Oxford University Press, 2007: 114-149. NRCBL: 2.1; 1.1.

Thiele, Felix. Bioethics: its foundation and application in political decision making. *In:* Machamer, Peter; Wolters, Gereon, eds. Science, Values, and Objectivity. Pittsburgh,

PA: University of Pittsburgh Press, 2004: 256-274. NRCBL: 2.1; 1.1; 5.3.

Thomasma, David C. Virtue theory in philosophy of medicine. *In:* Khushf, George, ed. Handbook of Bioethics: Taking Stock of the Field From a Philosophical Perspective. Dordrecht; Boston: Kluwer Academic, 2004: 89-120. NRCBL: 2.1; 1.1.

Thompson, Richard E. Look what's happened to medical ethics: broader horizons, updated ideas, fresh language. *Physician Executive* 2006 March-April; 32(2): 60-62. NRCBL: 2.1; 2.2; 1.1.

Tollefsen, Christopher. Religious reasons and public healthcare deliberations. *Christian Bioethics* 2007 May-August; (13)2: 139-157. NRCBL: 2.1; 1.2; 7.1; 1.1. SC: an.

Abstract: This paper critically explores the path of some of the controversies over public reason and religion through four distinct steps. The first part of this article considers the engagement of John Finnis and Robert P. George with John Rawls over the nature of public reason. The second part moves to the question of religion by looking at the engagement of Nicholas Wolterstorff with Rawls, Robert Audi, and others. Here the question turns specifically to religious reasons, and their permissible use by citizens in public debate and discourse. The third part engages Jürgen Habermas's argument that while citizens must be free to make religious arguments, still, there is an obligation of translation, and a motivational constraint on lawmakers. The final section argues that even though Habermas's proposal fails, nevertheless he recognizes a key difficulty for religious citizens in contemporary liberal polities. Restoration of a full role for religiously grounded justificatory reasons in public debate is one part of an adequate solution to this problem, but a second plank must be added to the solution: recognition that religious reasons can enter into public deliberation not just as first-order justifications of particular policies, but as second-order reasons, to be considered by any polity that respects its religious citizens and, more broadly, the good of religion.

Tong, Rosemarie. Feminist approaches to bioethics. *In:* Khushf, George, ed. Handbook of Bioethics: Taking Stock of the Field From a Philosophical Perspective. Dordrecht; Boston: Kluwer Academic, 2004: 143-161. NRCBL: 2.1; 1.1; 10.

Veatch, Robert M. Is bioethics applied ethics? *Kennedy Institute of Ethics Journal* 2007 March; 17(1): 1-2. NRCBL: 2.1; 1.3.1; 2.2; 1.1; 7.1; 4.1.2.

Widdershoven, Guy A.M. How to combine hermeneutics and Wide Reflective Equilibrium. *Medicine, Health Care and Philosophy* 2007 March; 10(1): 49-52. NRCBL: 2.1; 1.1. SC: an; em. Comments: Mette Ebbesen and Birthe D. Pedersen. Using empirical research to formulate normative ethical principles in biomedicine. Medicine, Health Care and Philosophy 2007 March; 10(1): 33-48.

Wildes, Kevin Wm. Global and particular bioethics. *In:* Engelhardt, H. Tristram, ed. Global Bioethics: The Collapse of Consensus. Salem, MA: M&M Scrivener Press, 2006: 362-379. NRCBL: 2.1; 1.1; 21.1.

Yalisove, Daniel. From the ivory tower to the trenches: teaching professional ethics to substance abuse counselors. *In:* Kleinig, John; Einstein, Stanley, eds. Ethical Challenges for Intervening in Drug Use: Policy, Research and Treatment Issues. Huntsville, TX: Office of International Criminal Justice; Criminal Justice Center, Sam Houston State University, 2006: 507-527. NRCBL: 2.3; 9.5.9; 17.1; 1.1; 1.3.1.

Zoloth, Laurie. I want you: notes toward a theory of hospitality. *In:* Eckenwiler, Lisa A.; Cohn, Felicia G., eds. The Ethics of Bioethics: Mapping the Moral Landscape. Baltimore, MD: Johns Hopkins University Press, 2007: 205-219. NRCBL: 2.1; 1.1.

BIOETHICS AND MEDICAL ETHICS/ RELIGIOUS ASPECTS

Aramesh, Kiarash. The influences of bioethics and Islamic jurisprudence on policy-making in Iran. *American Journal of Bioethics* 2007 October; 7(10): 42-44. NRCBL: 2.1; 1.2. Comments: Myra J. Christopher. "Show me" bioethics and politics. American Journal of Bioethics 2007 October; 7(10): 28-33.

Bedford-Strohm, Heinrich. Justice and long-term care: a theological ethical perspective. *Christian Bioethics* 2007 September-December; (13)3: 269-285. NRCBL: 2.1; 1.2; 1.1; 9.5.1; 9.5.2; 9.5.10.
Abstract: The relevance of justice for the current debate on long-term care is explored on the basis of demographic and economic data, especially in the U.S. and Germany. There is a justice question concerning the quality and availability of long-term care for different groups within society. Mapping the justice debate by discussing the two main opponents, John Rawls and Robert Nozick, the article identifies fundamental assumptions in both theories. An exploration of the biblical concept of the "option for the poor" and its influence on a new "ecumenical social teaching from below" leads to the conclusion that a Christian ethical account of long-term care will argue for a system that guarantees decent care to every citizen. The German model of Soziale Pflegeversicherung is presented as one possible option for putting this ethical guideline into political practice. In a final reflection, the role of religious affiliation for long-term care is discussed by looking at empirical data and by naming seven dimensions of faith-driven long-term care.

Boer, Theo A.; Schroten, Egbert. Life and death seen from a (Dutch) reformed position: a Calvinistic approach to bioethics. *In:* Østnor, Lars, ed. Bioetikk og teologi: Rapport fra Nordisk teologisk nettverk for bioetikks workshop I Stockholm 27.-29. September 1996. Oslo: Nord. teol. nettv. for bioetikk, 1996: 31-53. NRCBL: 2.1; 1.2; 20.1.

Boyle, Joseph. Casuistry. *In:* Khushf, George, ed. Handbook of Bioethics: Taking Stock of the Field From a Philosophical Perspective. Dordrecht; Boston: Kluwer Academic, 2004: 75-88. NRCBL: 2.1; 1.1; 1.2.

Burton, Olivette R. Why bioethics cannot figure out what to do with race. *American Journal of Bioethics* 2007 February; 7(2): 6-12. 20 refs. 9 fn. NRCBL: 2.1; 1.2; 13.1; 15.5.
Keywords: *bioethical issues; *bioethics; *blacks; *racial groups; capitalism; culture; eugenics; historical aspects; population control; religion; scientific misconduct; social discrimination; whites; Proposed Keywords: slavery; Keyword Identifiers: Seventeenth Century; Eighteenth Century; Nineteenth Century; Twentieth Century; Twenty-First Century; United States
Abstract: Race and religion are integral parts of bioethics. Harm and oppression, with the aim of social and political control, have been wrought in the name of religion against Blacks and people of color as embodied in the Ten Commandments, the Inquisition, and in the history of the Holy Crusades. Missionaries came armed with Judeo/Christian beliefs went to nations of people of color who had their own belief systems and forced change and caused untold harms because the indigenous belief systems were incompatible with their own. The indigenous people were denounced as ungodly, pagan, uncivilized, and savage. Hence, laws were enacted because of their perceived need to structure a sense of morality and to create and build a culture for these indigenous people of color. To date bioethics continues to be informed by a Western worldview that is Judeo/Christian in belief and orientation. However, missing from bioethical discourse in America is the historical influence of the Black Church as a cultural repository, which continues to influence the culture of Africans and Blacks. Cultural aspects of peoples of color are still largely ignored today. In attempting to deal with issues of race while steering clear of the religious and cultural impact of the Black Church, bioethics finds itself in the middle of a distressing situation: it simply cannot figure out what to do with race.

Capaldi, Nicholas. How philosophy and theology have undermined bioethics. *Christian Bioethics* 2007 January-April; (13)1: 53-66. NRCBL: 2.1; 1.2; 1.1; 7.1.
Abstract: This essay begins by distinguishing among the viewpoints of philosophy, theology, and religion; it then explores how each deals with "sin" in the bioethical context. The conclusions are that the philosophical and theological viewpoints are intellectually defective in that they cripple our ability to deal with normative issues, and are in the end unable to integrate Christian concepts like "sin" successfully into bioethics. Sin is predicated only of beings with free will, though only in Western Christianity must all sins be committed with knowledge and voluntarily. Without the notions of free will, sin, and a narrative of redemption, bioethics remains unable to provide itself with an adequate normative framework. Bioethics, and morality in general, remain a morass precisely because there has been a failure to translate Christian morality into fully secular and scientist terms.

NRCBL: National Reference Center for Bioethics Literature Classification Scheme   See inside front cover for terms.

57

Cherry, Mark J. Traditional Christian norms and the shaping of public moral life: how should Christians engage in bioethical debate within the public forum? *Christian Bioethics* 2007 May-August; (13)2: 129-138. NRCBL: 2.1; 1.2; 7.1; 4.4. SC: an.

Cohen, Eric. In whose image shall we die? *New Atlantis* 2007 Winter; 15: 21-39. NRCBL: 2.1; 1.2; 20.3.1.

Dan, Ovidiu. The philanthropy of the Orthodox Church: a Rumanian case study. *Christian Bioethics* 2007 September-December;(13)3: 303-307. NRCBL: 2.1; 1.2; 9.1. SC: cs.
   Abstract: On the basis of a definition of God as "love", human philanthropy is derived from Divine philanthropy, and therefore extends to all human beings. Because Divine philanthropy is most centrally expressed in Christ's incarnation and resurrection, Christ's identification with all who suffer presents the strongest motivation for human philanthropy. After a short review of the Romanian Orthodox Church's development after 1989, the author turns to his special case study, the Social-Medical Day-Care Christian Centre for older citizens. He describes the wan [sic; way] in which Church-based philanthropy can integrate social-medical with Christian pastoral care, and how this work draws the local communities into assuming a shared responsibility.

Delany, Mike. General medical practice: the problem of cooperation in evil. *In:* Watt, Helen, ed. Cooperation, Complicity and Conscience: Problems in Healthcare, Science, Law and Public Policy. London: Linacre Centre, 2005: 128-138. NRCBL: 2.1; 1.2; 12.1; 11.1; 14.1.

Delkeskamp-Hayes, Corinna. Resisting the therapeutic reduction: on the significance of sin. *Christian Bioethics* 2007 January-April; (13)1: 105-127. NRCBL: 2.1; 1.2; 4.2; 1.1; 7.1; 21.7.
   Abstract: Sin-talk, though politically incorrect, is indispensable. Placing human life under the "hermeneutic of sin" means acknowledging that one ought to aim flawlessly at God, and that one can fail in this endeavor. None of this can be appreciated within the contemporary post-Christian, mindset, which has attempted to reduce religion to morality and culture. In such a secular context, the guilt-feelings connected with the recognition of sin are considered to be harmful; the eternal benefit of a repentance is disregarded. Nevertheless, spirituality appears to have therapeutic benefits. Therefore attempts are made to re-locate within healthcare a religion shorn of its transcendent claims, so as then to harvest the benefits of a spirituality "saved from sin". This reduction of religiosity to its therapeutic function is nourished by a post-modern constructivist construal of religion. This article critically examines the dis-ingenuity marring such recasting, as well as the incoherence of related attempts to reduce transcendence to solidarity, and to re-shape the significance of religious rituals.

Dikova, Rossitza. Introduction [to issue on Philanthropy, Caritas, Diakonia — European approaches to Christians' service in the world]. *Christian Bioethics* 2007 September-December;(13)3: 245-250. NRCBL: 2.1; 1.2; 9.5.1; 9.5.2; 9.5.10; 1.1.

Engelhardt, H. Tristram. Why ecumenism fails: taking theological differences seriously. *Christian Bioethics* 2007 January-April; (13)1: 25-51. NRCBL: 2.1; 1.2; 1.1.
   Abstract: Contemporary Christians are separated by foundationally disparate understandings of Christianity itself. Christians do not share one theology, much less a common understanding of the significance of sin, suffering, disease, and death. These foundational disagreements not only stand as impediments to an intellectually defensible ecumenism, but they also form the underpinnings of major disputes in the culture wars, particularly as these are expressed in healthcare. There is not one Christian bioethics of sin, suffering, sickness, and death. In this article, the character of the moral-theological visions separating the various Christianities and thus their bioethics is examined. Particular emphasis is placed on the differences that set contemporary Western theology at odds with the theology of the Christianity of the first millennium. As is shown, the ground for this gulf lies in the divide between traditional and post-traditional views of the appropriate role of philosophy in theology, a difference rooted in disparate understandings of the meaning of church and of the meaning of the logos, the Son of God.

Fisher, Anthony. Cooperation in evil: understanding the issues. *In:* Watt, Helen, ed. Cooperation, Complicity and Conscience: Problems in Healthcare, Science, Law and Public Policy. London: Linacre Centre, 2005: 27-64. NRCBL: 2.1; 1.1; 1.2.

Frey, Christofer. Bioethics from the perspective of universalisation. *In:* Roetz, Heiner, ed. Cross-Cultural Issues in Bioethics: The Example of Human Cloning. New York: Rodopi, 2006: 341-361. NRCBL: 2.1; 1.2; 4.4; 14.5.

Gentry, Glenn. Rawls and religious community: ethical decision making in the public square. *Christian Bioethics* 2007 May-August; (13)2: 171-181. NRCBL: 2.1; 1.2; 7.1; 1.1. SC: an.
   Abstract: While most people may initially agree that justice is fairness, as an evangelical Protestant I argue that, for many religious comprehensive doctrines, the Rawlsean model does not possess the resources necessary to sustain tolerance in moral decision making. The weakness of Rawls's model centers on the reasonable priority of convictions that arise from private comprehensive doctrines. To attain a free and pluralistic society, people need resources sufficient to provide reasons to tolerate actions that are otherwise intolerable. In addition to arguing for the deficiency of the Rawlsean political model, I sketch out a preliminary model of ambassadorship that offers religious communities, and in particular Protestant evangelicals, the necessary resources to engage the broader society tolerantly while maintaining their religious convictions. As a citizen of the church and a member of another kingdom, Christians serve as ambassadors to those who are not of the heavenly kingdom. I take this model to be more ambitious than that of a sojourner who lives in the land but is isolated as much as possible from

society, while more modest than that of reconstructionists who seek to implement their own sacred law on all others.

Giampietro, Anthony E. Improving the Catholic approach to healthcare? [review of More Humane Medicine: A Liberal Catholic Bioethics, by James F. Drane]. *HEC (Healthcare Ethics Committee) Forum* 2007 September; 19(3): 261-270. NRCBL: 2.1; 1.2; 1.1.

Gormally, Luke. Why not dirty your hands? Or: on the supposed rightness of (sometimes) intentionally cooperating in wrongdoing. *In:* Watt, Helen, ed. Cooperation, Complicity and Conscience: Problems in Healthcare, Science, Law and Public Policy. London: Linacre Centre, 2005: 12-26. NRCBL: 2.1; 1.1; 1.2.

Gustafson, James M. Styles of religious reflection about medical ethics: further discussion. *In:* Østnor, Lars, ed. Bioetikk og teologi: Rapport fra Nordisk teologisk nettverk for bioetikks workshop I Stockholm 27.-29. September 1996. Oslo: Nord. teol. nettv. for bioetikk, 1996: 12-30. NRCBL: 2.1; 1.2; 3.1; 15.1. SC: an.

Haker, Hille. Medizinethik auf dem Weg ins 21. Jahrhundert - Bilanz und Zukunftsperspektiven aus Sicht der Katholischen Theologie [Medical ethics on the way to the 21st century - current and future prospects from the point of view of Catholic theology]. *Ethik in der Medizin* 2006 December; 18(4): 325-330. NRCBL: 2.1; 1.2.

Hamel, Ron. New directives for health care ethics? *Health Progress* 2007 January-February; 88(1): 4-5. NRCBL: 2.1; 2.2; 1.2.

Hanson, Stephen S. Moral acquaintances: Loewy, Wildes, and beyond. *HEC (Healthcare Ethics Committee) Forum* 2007 September; 19(3): 207-225. NRCBL: 2.1; 1.1; 1.2; 4.1.2; 21.7. SC: cs.

Hedayat, K.M. The possibility of a universal declaration of biomedical ethics. *Journal of Medical Ethics* 2007 January; 33(1): 17-20. NRCBL: 2.1; 1.2; 6.
Abstract: Statements on issues in biomedical ethics, purporting to represent international interests, have been put forth by numerous groups. Most of these groups are composed of thinkers in the tradition of European secularism, and do not take into account the values of other ethical systems. One fifth of the world's population is accounted for by Islam, which is a universal religion, with more than 1400 years of scholarship. Although many values are held in common by secular ethical systems and Islam, their inferences are different. The question, "Is it possible to derive a truly universal declaration of biomedical ethics?" is discussed here by examining the value and extent of personal autonomy in Western and Islamic biomedical ethical constructs. These constructs are then tested vis-à-vis the issue of abortion. It is concluded that having a universal declaration of biomedical ethics in practice is not possible, although there are many conceptual similarities and agreements between secular and Islamic value systems, unless a radical paradigm shift occurs in segments of the world's deliberative bodies. The appellation

"universal" should not be used on deliberative statements unless the ethical values of all major schools of thought are satisfied.

Knobel, Peter. An expanded approach to Jewish bioethics: a liberal/aggadic approach. *In:* Cutter, William, ed. Healing and the Jewish Imagination: Spiritual and Practical Perspectives on Judaism and Health. Woodstock, VT: Jewish Lights Pub., 2007: 171-183. NRCBL: 2.1; 1.2. SC: an.

Kreß, Hartmut. Ethik in der Medizin — Schlaglichter aus der Sicht protestantischer Ethik [Ethics in medicine — highlights from the point of view of Protestant ethics]. *Ethik in der Medizin* 2006 December; 18(4): 320-324. NRCBL: 2.1; 1.2.

Kuczewski, Mark G. Empirical metaethical absolutism in contemporary Catholic bioethics [review of Contemporary Catholic Health Care Ethics, by David F. Kelly]. *Medical Humanities Review* 2005 Spring-Fall; 19(1-2): 76-80. NRCBL: 2.1; 1.1; 1.2.

Levada, William. The magisterium's role in bioethics. *Origins* 2007 March 1; 36(37): 581-588. NRCBL: 2.1; 1.2; 4.4; 14.1.

Majumdar, Sisir K. History of evolution of the concept of medical ethics. *Bulletin of the Indian Institute of History of Medicine (Hyderabad)* 2003 January-June; 33(1): 17-31. NRCBL: 2.2; 2.1; 1.1; 1.3.1; 1.2.

McKay, Angela. Publicly accessible intuitions: "neutral reasons" and bioethics. *Christian Bioethics* 2007 May-August; (13)2: 183-197. NRCBL: 2.1; 1.2; 7.1; 1.1; 4.4; 20.7; 20.5.1; 8.1. SC: an.
Abstract: This article examines Leon Kass's contention that a choice for physician-assisted suicide is "undignified." Although Kass is Jewish rather than Christian, he argues for positions that most Christians share, and he argues for these positions without presupposing the truth of specific religious claims. I argue that although Kass has some important intuitions, he too readily assumes that these intuitions will be shared by his audience, and that this assumption diminishes the force of his argument. An examination of the limitations of Kass's argument is helpful insofar as it illustrates the real challenge faced by religious believers who wish to defend their beliefs in the "public forum." For it illustrates that what needs to be made "accessible" is the Judeo-Christian understanding of man and his place in the world. While I do not wish to claim that this task is impossible, I do think that it is far more difficult than most realize. Like all important tasks, however, unless we wrestle with the difficulties it raises, our arguments will strike many as unconvincing.

Müller, Denis. The original risk: overtheologizing ethics and undertheologizing sin. *Christian Bioethics* 2007 January-April; (13)1: 7-23. NRCBL: 2.1; 1.2; 1.1.
Abstract: The project of articulating a theological ethics on the basis of liturgical anthropology is bound to fail if the necessary consequence is that one has to quit the fo-

NRCBL: National Reference Center for Bioethics Literature Classification Scheme   See inside front cover for terms.

59

rum of critical modern rationality. The risk of Engel-hardt's approach is to limit rationality to a narrow vision of reason. Sin is not to be understood as the negation of human holiness, but as the negation of divine holiness. The only way to renew theological ethics is to understand sin as the anthropological and ethical expression of the biblical message of the justification by faith only. Sin is therefore a secondary category, which can only by inter-preted in light of the positive manifestation of liberation, justification, and grace. The central issue of Christian ethics is not ritual purity or morality, but experience, confession and recognition of our own injustice in our dealing with God and men.

Orr, Robert D. The role of Christian belief in public pol-icy. *Christian Bioethics* 2007 May-August; (13)2: 199-209. NRCBL: 2.1; 1.2; 7.1; 20.7; 20.5.1; 8.1; 12.3. SC: an.

Abstract: It seems intuitive to the believer that God in-tended through instruction in the Law to define morality, intended to lead humankind to "the right and the good." Further, God's love for humankind, exemplified by the incarnation, atonement and teachings of Jesus, and em-powered by the Holy Spirit, should lead to a better world. Indeed, the Christian worldview is a coherent and valid way to look at bioethical issues in public policy and at the bedside. Yet, as this paper explores, in a pluralistic soci-ety such as the United States, it is neither possible nor de-sirable for Christians to try to force their views on others. Still, it is obligatory for Christians to stand up and articu-late their views in the public square. We should try to per-suade others using either prudential or moral arguments. While we must be willing to live with "the will of the peo-ple," at the same time, we must not be intimidated into ac-cepting the position that our voice is not valid because it has a religious basis.

Padela, Aasim I. Islamic medical ethics: a primer. *Bioethics* 2007 March; 21(3): 169-178. NRCBL: 2.1; 4.1.2; 1.2. SC: rv.

Abstract: Modern medical practice is becoming increas-ingly pluralistic and diverse. Hence, cultural competency and awareness are given more focus in physician training seminars and within medical school curricula. A renewed interest in describing the varied ethical constructs of spe-cific populations has taken place within medical litera-ture. This paper aims to provide an overview of Islamic Medical Ethics. Beginning with a definition of Islamic Medical Ethics, the reader will be introduced to the scope of Islamic Medical Ethics literature, from that aimed at developing moral character to writings grounded in Is-lamic law. In the latter form, there is an attempt to derive an Islamic perspective on bioethical issues such as abor-tion, gender relations within the patient-doctor relation-ship, end-of-life care and euthanasia. It is hoped that the insights gained will aid both clinicians and ethicists to better understand the Islamic paradigm of medical ethics and thereby positively affect patient care.

Shahzad, Qaiser. Playing God and the ethics of divine names: an Islamic paradigm for biomedical ethics. *Bioethics* 2007 October; 21(8): 413-418. NRCBL: 2.1;

1.2; 1.1. Conference: Eighth World Congress of Bioethics: A Just and Healthy Society; Beijing, China; 2006 August 9.

Abstract: The notion of 'playing God' frequently comes to fore in discussions of bioethics, especially in religious contexts. The phrase has always been analyzed and dis-cussed from Christian and secular standpoints. Two inter-pretations exist in the literature. The first one takes 'God' seriously and playing 'playfully'. It argues that this con-cept does state a principle but invokes a perspective on the world. The second takes both terms playfully. In the Islamic Intellectual tradition, the Sufi concept of 'adopt-ing divine character traits' provides a legitimate para-digm for 'playing God'. This paradigm is interesting because here we take both terms 'God' and 'playing' seri-ously. It is significant for the development of biomedical ethics in contemporary Islamic societies as it can open new vistas for viewing biotechnological developments.

Sherlock, Richard. Bioethics in liberal regimes: a review of the President's Council. *Ethics and Medicine: An Inter-national Journal of Bioethics* 2007 Fall; 23(3): 169-188. 10 fn. NRCBL: 2.4; 2.2; 2.1; 1.2; 9.1; 14.1; 15.1; 15.2; 9.5.4; 10; 5.1. SC: rv.

Keywords: *advisory committees; *bioethical issues; *bio-ethics; biomedical enhancement; biotechnology; cloning; cultural pluralism; democracy; embryo research; embryonic stem cells; eugenics; euthanasia; genetic engineering; his-torical aspects; human dignity; human experimentation; morality; preimplantation diagnosis; public policy; religion; reproductive technologies; Proposed Keywords: publica-tions; Keyword Identifiers: *President's Council on Bio-ethics; *United States; National Bioethics Advisory Com-mission; National Commission for the Protection of Human Subjects and Biomedical and Behavioral Research; Presi-dent's Commission for the Study of Ethical Problems in Biomedical and Behavioral Research; Twentieth Century; Twenty-First Century

Smith, George P. Procreational autonomy or theological restraints. *In his:* The Christian Religion and Biotechnol-ogy: A Search for Principled Decision-making. Dordrecht: Springer, 2005: 61-84. NRCBL: 2.1; 1.2; 12.3; 14.4. SC: le.

Spiess, Christian. Recognition and social justice: a Ro-man Catholic view of Christian bioethics of long-term care and community service. *Christian Bioethics* 2007 Septem-ber-December;(13)3: 287-301. NRCBL: 2.1; 1.2; 1.1; 9.5.1; 9.5.2; 9.5.10.

Abstract: Contemporary Christian ethics encounters the challenge to communicate genuinely Christian normative orientations within the scientific debate in such a way as to render these orientations comprehensible, and to main-tain or enhance their plausibility even for non-Christians. This essay, therefore, proceeds from a biblical motif, takes up certain themes from the Christian tradition (in particular the idea of social justice), and connects both with a compelling contemporary approach to ethics by secular moral philosophy, i.e. with Axel Honneth's re-ception of Hegel, as based on Hegel's theory of recogni-tion. As a first step, elements of an ethics of recognition are developed on the basis of an anthropological recourse to the conditions of intersubjective encounters. These conditions are then brought to bear on the idea of social justice, as developed in the social-Catholic tradition, and as systematically explored in the Pastoral Letter of the United States Conference of Catholic Bishops, Economic Justice For All (1986). Proceeding from this basis, aspects of a Christian ethics of community service with regard to long-term care can be defined.

Tham, Joseph. Bioethics and anointing of the sick. *Linacre Quarterly* 2007 August; 74(3): 253-257. NRCBL: 2.1; 1.2; 4.4; 20.5.1.

Tollefsen, Christopher. Religious reasons and public healthcare deliberations. *Christian Bioethics* 2007 May-August; (13)2: 139-157. NRCBL: 2.1; 1.2; 7.1; 1.1. SC: an.

Abstract: This paper critically explores the path of some of the controversies over public reason and religion through four distinct steps. The first part of this article considers the engagement of John Finnis and Robert P. George with John Rawls over the nature of public reason. The second part moves to the question of religion by looking at the engagement of Nicholas Wolterstorff with Rawls, Robert Audi, and others. Here the question turns specifically to religious reasons, and their permissible use by citizens in public debate and discourse. The third part engages Jürgen Habermas's argument that while citizens must be free to make religious arguments, still, there is an obligation of translation, and a motivational constraint on lawmakers. The final section argues that even though Habermas's proposal fails, nevertheless he recognizes a key difficulty for religious citizens in contemporary liberal polities. Restoration of a full role for religiously grounded justificatory reasons in public debate is one part of an adequate solution to this problem, but a second plank must be added to the solution: recognition that religious reasons can enter into public deliberation not just as first-order justifications of particular policies, but as second-order reasons, to be considered by any polity that respects its religious citizens and, more broadly, the good of religion.

Vantsos, Miltiadis; Kiroudi, Marina. An Orthodox view of philanthropy and Church diaconia. *Christian Bioethics* 2007 September-December; (13)3: 251-268. NRCBL: 2.1; 1.2; 7.1; 9.5.10.

Abstract: According to Orthodox theology, philanthropy refers to the love of God toward man, which man is called to imitate by loving his neighbor as himself. This love consists not just in emotions but requires specific acts of philanthropy toward our fellow man in need. The church, in keeping the commandments of Christ, has developed throughout her history a rich philanthropic work. The diaconia of the church has taken many forms, thus responding to historical change and to the specific human needs at different times. Concentrating on diaconia for those who are in need of long-term care, this article presents the Orthodox view of the diaconia of the church, as realized through her own philanthropic organizations as well as through her very specific contribution to the

diaconia offered by state sponsored charitable institutions.

Vogel, Lawrence. Natural-law Judaism? The genesis of bioethics in Hans Jonas, Leo Strauss, and Leon Kass. *In:* Schweiker, William; Johnson, Michael A.; Jung, Kevin, eds. Humanity Before God: Contemporary Faces of Jewish, Christian, and Islamic Ethics. Minneapolis, MN: Fortress Press, 2006: 209-237. NRCBL: 2.1; 1.2; 14.1; 10. SC: an.

Winslade, William J.; Carson, Ronald A. Foreword: the role of religion in health law and policy. *Houston Journal of Health Law and Policy* 2006; 6(2): 245-248. NRCBL: 2.1; 1.2; 12.3; 20.5.1; 4.4.

BIOETHICS COMMISSIONS *See* BIOETHICS AND MEDICAL ETHICS/ COMMISSIONS

BIOLOGICAL WARFARE *See* WAR AND TERRORISM

BIOMEDICAL RESEARCH
See also BEHAVIORAL RESEARCH; HUMAN EXPERIMENTATION; JOURNALISM AND PUBLISHING; NANOTECHNOLOGY

Bainbridge, William Sims. Converging technologies and human destiny. *Journal of Medicine and Philosophy* 2007 May-June; 32(3): 197-216. NRCBL: 5.1; 1.1; 4.5.

Abstract: The rapid fertility decline in most advanced industrial nations, coupled with secularization and the disintegration of the family, is a sign that Western Civilization is beginning to collapse, even while radical religious movements pose challenges to Western dominance. Under such dire circumstances, it is pointless to be cautious about developing new Converging Technologies. Historical events are undermining the entire basis of ethical decision-making, so it is necessary to seek a new basis for ethics in the intellectual unification of science and the power to do good inherent in the related technological convergence. This article considers the uneasy relations between science and religion, in the context of fertility decline, and the prospects for developing a new and self-sustaining civilization based in a broad convergence of science and technology, coalescing around a core of nanotechnology, biotechnology, information technology, and cognitive technologies. It concludes with the suggestion that the new civilization should become interstellar.

Bauer, Keith A. Wired patients: implantable microchips and biosensors in patient care. *CQ: Cambridge Quarterly of Healthcare Ethics* 2007 Summer; 16(3): 281-290. NRCBL: 5.1; 9.7; 9.5.1; 4.4; 4.5; 8.1.

Bibbee, Jeffrey R.; Viens, A.M.; Moreno, Jonathan D.; Berger, Sam. The inseparability of religion and politics in the neoconservative critique of biotechnology. *American Journal of Bioethics* 2007 October; 7(10): 18-20; author reply W1-W3. NRCBL: 5.1; 1.2; 15.1; 21.1. Comments:

NRCBL: National Reference Center for Bioethics Literature Classification Scheme   See inside front cover for terms.

61

Jonathan D. Moreno and Sam Berger. Biotechnology and the new right: neoconservatism's red menace. *American Journal of Bioethics* 2007 October; 7(10): 7-13.

Burchell, Kevin. Boundary work, associative argumentation and switching in the advocacy of agricultural biotechnology. *Science as Culture* 2007 March; 16(1): 49-70. NRCBL: 5.1; 15.1; 1.3.11; 7.1. SC: em.

Campbell, Courtney S.; Clark, Lauren A.; Loy, David; Keenan, James F.; Matthews, Kathleen; Winograd, Terry; Zoloth, Laurie. The bodily incorporation of mechanical devices: ethical and religious issues (Part 1). *CQ: Cambridge Quarterly of Healthcare Ethics* 2007 Spring; 16(2): 229-239. NRCBL: 5.1; 4.4; 1.2; 5.4.

Campbell, Courtney S.; Clark, Lauren A.; Loy, David; Keenan, James F.; Matthews, Kathleen; Winograd, Terry; Zoloth, Laurie. The bodily incorporation of mechanical devices: ethical and religious issues (Part 2). *CQ: Cambridge Quarterly of Healthcare Ethics* 2007 Summer; 16(3): 268-280. NRCBL: 5.1; 9.7; 9.5.1; 4.4; 1.1; 1.2; 21.7. SC: rv.

Cline, Cheryl A.; Moreno, Jonathan D.; Berger, Sam. Biotechnology and the new right: a progressive red herring? *American Journal of Bioethics* 2007 October; 7(10): 15-17; author reply W1-W3. NRCBL: 5.1; 18.5.4; 15.1; 21.1. Comments: Jonathan D. Moreno and Sam Berger. Biotechnology and the new right: neoconservatism's red menace. American Journal of Bioethics 2007 October; 7(10): 7-13.

Conference Coordinators, Second National Bioethics Conference. Indian Journal of Medical Ethics. Moral and ethical imperatives of health care technologies: scientific, legal and socio-economic perspectives. *Indian Journal of Medical Ethics* 2007 January-March; 4(1): 35-37. NRCBL: 5.1; 2.1.

Dörner, Klaus. Bio-Medizin als Verarmung einer zukünftigen Medizin. *In:* Neuer-Miebach, Therese; Wunder, Michael, eds. Bio-Ethik und die Zukunft der Medizin. Bonn: Psychiatrie-Verlag, 1998: 16-19. NRCBL: 5.1.

Dupuy, Jean-Pierre. Some pitfalls in the philosophical foundations of nanoethics. *Journal of Medicine and Philosophy* 2007 May-June; 32(3): 237-261. NRCBL: 5.1; 1.1; 4.5.
  Abstract: If such a thing as nanoethics is possible, it can only develop by confronting the great questions of moral philosophy, thus avoiding the pitfalls so common to regional ethics. We identify and analyze some of these pitfalls: the restriction of ethics to prudence understood as rational risk management; the reduction of ethics to cost/benefit analysis; the confusion of technique with technology and of human nature with the human condition. Once these points have been clarified, it is possible to take up some weighty philosophical and metaphysical questions which are not new, but which need to be raised

anew with respect to nanotechnologies: the artificialization of nature; the question of limits; the role of religion; the finiteness of the human condition as something with a beginning and an end; the relationship between knowledge and know-how; the foundations of ethics.

Dusyk, Nichole. The political and moral economies of science: a case study of genomics in Canada and the United Kingdom. *Health Law Review* 2007; 15(3): 3-5. 8 fn. NRCBL: 5.1; 15.1. SC: cs.
  Keywords: *economics; *genomics; *government financing; *public policy; *research priorities; *research support; entrepreneurship; genetic patents; industry; international aspects; policy making; politics; science; trends; universities; Proposed Keywords: technology transfer; Keyword Identifiers: *Canada; *Great Britain

Gillick, Muriel R. The technological imperative and the battle for the hearts of America. *Perspectives in Biology and Medicine* 2007 Spring; 50(2): 276-294. NRCBL: 5.2; 9.3.1; 9.5.2; 9.5.1; 9.1. Identifiers: Left Ventricular Assist Device.

Graham, John; Hu, Jianhui. The risk-benefit balance in the United States: who decides? *Health Affairs* 2007 May-June; 26(3): 625-635. NRCBL: 5.2; 9.1.

Hansson, Sven Ove. The ethics of enabling technology. *CQ: Cambridge Quarterly of Healthcare Ethics* 2007 Summer; 16(3): 257-267. NRCBL: 5.1; 9.7; 9.5.1; 4.4; 1.1; 9.4. SC: an.

Harvey, Matthew. Citizens in defence of something called science. *Science as Culture* 2007 March; 16(1): 31-48. NRCBL: 5.1; 1.1; 15.1; 1.3.11; 1.3.9. SC: em.

Hinsch, Kathryn M.; Fiore, Robin N.; Moreno, Jonathan D.; Berger, Sam. Responding to neocon critiques of biotechnology: a progressive agenda. *American Journal of Bioethics* 2007 October; 7(10): 14-15; author reply W1-W3. NRCBL: 5.1; 21.1. Comments: Jonathan D. Moreno and Sam Berger. Biotechnology and the new right: neoconservatism's red menace. American Journal of Bioethics 2007 October; 7(10): 7-13.

Hivon, Myriam; Lehoux, Pascale; Denis, Jean-Louis; Tailliez, Stéphanie. Use of health technology assessment in decision making: coresponsibility of users and producers? *International Journal of Technology Assessment in Health Care* 2005 Spring; 21(2): 268-275. NRCBL: 5.2; 9.1; 5.3; 7.1. SC: em.

Holm, Søren. A rose by any other name... is the research/non-research distinction still important and relevant? *Theoretical Medicine and Bioethics* 2007; 28(3): 153-155. NRCBL: 5.1; 1.3.9.

Kaebnick, Gregory E. Small talk. *Hastings Center Report* 2007 January-February; 37(1): inside front cover. NRCBL: 5.1; 15.7.

Kass, Leon R. Science, religion, and the human future. *Commentary* 2007 April; 123(4): 36-48. NRCBL: 5.1; 1.2; 1.1; 15.1.

Keiper, Adam. Nanoethics as a discipline? *New Atlantis* Spring 2007; (16): 55-67. NRCBL: 5.1; 4.4.

Khushf, George. Open questions in the ethics of convergence. *Journal of Medicine and Philosophy* 2007 May-June; 32(3): 299-310. NRCBL: 5.1; 1.1; 4.5.
Abstract: After historically situating NBIC Convergence in the context of earlier bioethical debate on genetics, ten questions are raised in areas related to the ethics of Convergence, indicating where future research is needed.

Khushf, George. The ethics of NBIC convergence. *Journal of Medicine and Philosophy* 2007 May-June; 32(3): 185-196. NRCBL: 5.1; 4.5; 1.1; 5.4. Identifiers: Nanotechnology, Biomedicine, Information Technology, and Cognitive Science [NBIC].

Litton, Paul. "Nanoethics"? What's new? *Hastings Center Report* 2007 January-February; 37(1): 22-25. NRCBL: 5.1; 16.1; 5.3; 5.2.

McGee, Ellen M.; Maguire, Gerald Q., Jr. Becoming borg to become immortal: regulating brain implant technologies. *CQ: Cambridge Quarterly of Healthcare Ethics* 2007 Summer; 16(3): 291-302. 51 refs. 13 fn. NRCBL: 5.1; 9.7; 9.5.1; 4.4; 17.1; 4.5; 14.5; 5.3. SC: rv.
Keywords: *biomedical enhancement; *biomedical technologies; *brain; *government regulation; *guidelines; *medical devices; autonomy; cloning; freedom; informed consent; international aspects; investigational therapies; neurosciences; policy making; privacy; reproductive technologies; risks and benefits; science; self concept; technology assessment; Proposed Keywords: *implants; cybernetics; Keyword Identifiers: Food and Drug Administration; United States

Moreno, Jonathan D.; Berger, Sam. Biotechnology and the new right: neoconservatism's red menace. *American Journal of Bioethics* 2007 October; 7(10): 7-13. NRCBL: 5.1; 4.4; 15.1; 21.1. SC: an. Comments: American Journal of Bioethics 2007 October; 7(10): 14-27.

Moreno, Rui; Afonso, Susana. Ethical, legal and organizational issues in the ICU: prediction of outcome. *Current Opinion in Critical Care* 2006 December; 12(6): 619-623. NRCBL: 5.2; 9.1; 9.5.1.

Nordmann, Alfred. Knots and strands: an argument for productive disillusionment. *Journal of Medicine and Philosophy* 2007 May-June; 32(3): 217-236. NRCBL: 5.1; 1.1; 5.4.
Abstract: This article offers a contrast between European and US-American approaches to the convergence of enabling technologies and to associated issues. It identifies an apparently paradoxical situation in which regional differences produce conflicting claims to universality, each telling us what can and will happen to the benefit of humanity. Those who might mediate and negotiate these competing claims are themselves entangled in the various positions. A possible solution is offered, namely a universalizable strategy that aims to disentangle premature claims to unity and universality as in the case of the greater "efficiency" of nanomedicine. This is the strategy by which Science and Technologies Studies (STS) can analytically tease apart what it has helped produce and sustain in the first place. The virtues and limits of this strategy are briefly presented, deliberation and decision-making under conditions of productive disillusionment recommended.

Outomuro, Delia; Moreno, Jonathan D.; Berger, Sam. Critiques on biotechnology and the problem of pigeonholing philosophical thinking. *American Journal of Bioethics* 2007 October; 7(10): 25-27; author reply W1-W3. NRCBL: 5.1; 1.1; 21.1. Comments: Jonathan D. Moreno and Sam Berger. Biotechnology and the new right: neoconservatism's red menace. American Journal of Bioethics 2007 October; 7(10): 7-13.

Peterson, M. Should the precautionary principle guide our actions or our beliefs? *Journal of Medical Ethics* 2007 January; 33(1): 5-10. NRCBL: 5.2; 1.1.
Abstract: Two interpretations of the precautionary principle are considered. According to the normative (action-guiding) interpretation, the precautionary principle should be characterised in terms of what it urges doctors and other decision makers to do. According to the epistemic (belief-guiding) interpretation, the precautionary principle should be characterised in terms of what it urges us to believe. This paper recommends against the use of the precautionary principle as a decision rule in medical decision making, based on an impossibility theorem presented in Peterson (2005). However, the main point of the paper is an argument to the effect that decision theoretical problems associated with the precautionary principle can be overcome by paying greater attention to its epistemic dimension. Three epistemic principles inherent in a precautionary approach to medical risk analysis are characterised and defended.

Roongrerngsuke, Siriyupa; Phornprapha, Sarote. Perceptions of issues in biotechnology management in Thailand. *Eubios Journal of Asian and International Bioethics* 2007 November; 17(6): 185-190. NRCBL: 5.1; 1.3.2; 5.3; 15.1; 1.3.11. SC: em.

Sawyer, Robert J. Robot ethics [editorial]. *Science* 2007 November 16; 318(5853): 1037. NRCBL: 5.1.

Steinbrook, Robert. Guidance for guidelines. *New England Journal of Medicine* 2007 January 25; 356(4): 331-333. NRCBL: 5.2; 9.8; 9.4.

White, Gladys B.; Moreno, Jonathan D.; Berger, Sam. The sky is falling . . . or maybe not: the moral necessity of technology assessment. *American Journal of Bioethics* 2007 October; 7(10): 20-21; author reply W1-W3. NRCBL: 5.2; 5.4; 21.1. Comments: Jonathan D. Moreno and Sam Berger. Biotechnology and the new right: neoconservatism's red menace. American Journal of Bioethics 2007 October; 7(10): 7-13.

NRCBL: National Reference Center for Bioethics Literature Classification Scheme   See inside front cover for terms.

63

Wickins, Jeremy. The ethics of biometrics: the risk of social exclusion from the widespread use of electronic identification. *Science and Engineering Ethics* 2007 March; 13(1): 45-54. NRCBL: 5.1; 5.2; 1.3.12.

BIOMEDICAL RESEARCH/ RESEARCH ETHICS AND SCIENTIFIC MISCONDUCT
See also MALPRACTICE AND PROFESSIONAL MISCONDUCT

How to be good? Mentoring and training for ethical behaviour aren't all their cracked up to be [editorial]. *Nature* 2007 October 11; 449(7163): 638. NRCBL: 1.3.9; 7.2.

Japanese universities fire researchers for misconduct [news brief]. *Nature* 2007 January 4; 445(7123): 12. NRCBL: 1.3.9.

Leading by example [editorial]. *Nature* 2007 January 18; 445(7125): 229. NRCBL: 1.3.9. Identifiers: scientific misconduct.

Science at WHO and UNICEF: the corrosion of trust [editorial]. *Lancet* 2007 September 22-28; 370(9592): 1007. NRCBL: 1.3.9; 1.3.7; 5.3; 1.3.6.

When in doubt, disclose [editorial]. *Lancet* 2007 February 3-9; 369(9559): 344. NRCBL: 1.3.9; 18.6. Identifiers: Conflict of Interest Notification Study.

Acquavella, John F. Why focus only on financial interests? *Epidemiology* 2006 May; 17(3): 248-249. NRCBL: 1.3.9. SC: em.

Adrian, Manuella. Decisions involving research and ethics: misusing drug use(r) statistics. *In:* Kleinig, John; Einstein, Stanley, eds. Ethical Challenges for Intervening in Drug Use: Policy, Research and Treatment Issues. Huntsville, TX: Office of International Criminal Justice; Criminal Justice Center, Sam Houston State University, 2006: 217-258. NRCBL: 1.3.9; 9.5.9; 18.2; 7.1; 9.3.1.

Anderson, Melissa S. Collective openness and other recommendations for the promotion of research integrity. *Science and Engineering Ethics* 2007 December; 13(4): 387-394. NRCBL: 1.3.9. SC: em.

Anderson, Melissa S.; Martinson, Brian C.; De Vries, Raymond. Normative dissonance in science: results from a national survey of U.S. scientists. *Journal of Empirical Research on Human Research Ethics* 2007 December; 2(4): 3-14. NRCBL: 1.3.9; 7.3. SC: em.
Abstract: Norms of behavior in scientific research represent ideals to which most scientists subscribe. Our analysis of the extent of dissonance between these widely espoused ideals and scientists' perceptions of their own and others' behavior is based on survey responses from 3,247 mid- and early-career scientists who had research funding from the U.S. National Institutes of Health. We found substantial normative dissonance, particularly between espoused ideals and respondents' perceptions of other scientists' typical behavior. Also, respondents on average saw other scientists' behavior as more counternormative than normative. Scientists' views of their fields as cooperative or competitive were associated with their normative perspectives, with competitive fields showing more counternormative behavior. The high levels of normative dissonance documented here represent a persistent source of stress in science.

Anderson, Melissa S.; Ronning, Emily A.; de Vries, Raymond; Martinson, Brian C. The perverse effects of competition on scientists' work and relationships. *Science and Engineering Ethics* 2007 December; 13(4): 437-461. NRCBL: 1.3.9; 7.3. SC: em.

Antes, Alison L.; Brown, Ryan P.; Murphy, Stephen T.; Waples, Ethan P.; Mumford, Michael D.; Connelly, Shane; Devenport, Lynn D. Personality and ethical decision-making in research: the role of perceptions of self and others. *Journal of Empirical Research on Human Research Ethics* 2007 December; 2(4): 15-34. NRCBL: 1.3.9; 17.1; 9.4; 7.3. SC: em.
Abstract: This study examined basic personality characteristics, narcissism, and cynicism as predictors of ethical decision-making among graduate students training for careers in the sciences. Participants completed individual difference measures along with a scenario-based ethical decision-making measure that captures the complex, multifaceted nature of ethical decision-making in scientific research. The results revealed that narcissism and cynicism (individual differences influencing self-perceptions and perceptions of others) showed consistently negative relationships with aspects of ethical decision-making, whereas more basic personality characteristics (e.g., conscientiousness, agreeableness) were less consistent and weaker. Further analyses examined the relationship of personality to metacognitive reasoning strategies and social behavioral response patterns thought to underlie ethical decision-making. The findings indicated that personality was associated with many of these social-cognitive mechanisms which might, in part, explain the association between personality and ethical decisions.

Atkinson, Timothy N.; Gilleland, Diane S.; Pearson, Carolyn. The research environment norm inventory (RENI): a study of integrity in research administrative systems. *Accountability in Research* 2007 April-June; 14(2): 93-119. NRCBL: 1.3.9. SC: em.
Abstract: University research administrators have been generally ignored in basic studies of research integrity. Hensley (1986) noted that research administrators are "essential to the achievement of the specific missions of postsecondary institutions and to science and the academic infrastructure" (p. 47, 48). The following study sought to extend the scope of research on research integrity to research administrative structures with a new instrument called the Research Environment Norm Inventory or RENI. University research administrators and their professional association were targeted for data collection. Evidence suggested that research administration in the United States supports integrity in the research

environment through: (1) respect for community; (2) respect for institutional boundaries; (3) professionalism; (4) respect for authority structures; (5) sensitivity to system conflicts. The study suggested that integrity structures are dictated largely by the institutional settings and environments (Meyer and Rowan, 2006).

Blumsohn, Aubrey. The Gillberg Affair: profound ethical issues were smoothed over [letter]. *BMJ: British Medical Journal* 2007 September 29; 335(7621): 629. NRCBL: 1.3.9; 1.3.7. Comments: Jonathan Gornall. Hyperactivity in children: the Gillberg Affair. BMJ: British Medical Journal 2007 August 25; 335(7616): 370-373.

Böhme, Gernot. Rationalizing unethical medical research: taking seriously the case of Viktor von Weizsäcker. *In:* LaFleur, William R.; Böhme, Gernot; Shimazono, Susumu, eds. Dark Medicine: Rationalizing Medical Research. Bloomington: Indiana University Press, 2007: 15-29. NRCBL: 1.3.9; 18.1; 20.5.1; 1.3.5; 2.2; 21.4.

Bortolotti, Lisa; Heinrichs, Bert. Delimiting the concept of research: an ethical perspective. *Theoretical Medicine and Bioethics* 2007; 28(3): 157-179. NRCBL: 1.3.9; 1.1. SC: an.

Abstract: It is important to be able to offer an account of which activities count as scientific research, given our current interest in promoting research as a means to benefit humankind and in ethically regulating it. We attempt to offer such an account, arguing that we need to consider both the procedural and functional dimensions of an activity before we can establish whether it is a genuine instance of scientific research. By placing research in a broader schema of activities, the similarities and differences between research activities and other activities become visible. It is also easier to show why some activities that do not count as research can sometimes be confused with research and why some other activities can be regarded only partially as research. Although the concept of research is important to delimit a class of activities which we might be morally obliged to promote, we observe that the class of activities which are regarded as subject to ethical regulation is not exhausted by research activities. We argue that, whether they be research or not, all the activities that are likely to affect the rights and interests of the individuals involved and impact on the rights and interests of other individuals raise ethical issues and might be in need of ethical regulation.

Bosch, Xavier. Dealing with scientific misconduct [editorial]. *BMJ: British Medical Journal* 2007 September 15; 335(7619): 524-525. NRCBL: 1.3.9.

Boyd, Elizabeth; Bero, Lisa A. Defining financial conflicts and managing research relationships: an analysis of university conflict of interest committee decisions. *Science and Engineering Ethics* 2007 December; 13(4): 415-435. NRCBL: 1.3.9; 1.3.3; 5.3. SC: em. Identifiers: University of California.

Brown, James Robert. Self-censorship. *In:* Lemmens, Trudo; Waring, Duff R., eds. Law and Ethics in Biomedical

Research: Regulation, Conflict of Interest and Liability. Toronto; Buffalo: University of Toronto Press, 2006: 82-94. NRCBL: 1.3.9; 7.3.

Brumfiel, Geoff; Abbott, Alison; Cyranoski, David; Fuyuno, Ichiko; Giles, Jim; Odling-Smee, Lucy. Misconduct? It's all academic... [news]. *Nature* 2007 January 18; 445(7125): 240-214. NRCBL: 1.3.9; 1.3.3.

Brunekreef, Bert. He who pays the piper, calls the tune. . . . . *Epidemiology* 2006 May; 17(3): 246-247. NRCBL: 1.3.9. SC: em.

Bulger, Ruth Ellen. The responsible conduct of research, including responsible authorship and publication practices. *In:* Korthals, Michiel; Bogers, Robert J., eds. Ethics for Life Scientists. Dordrecht, The Netherlands: Springer, 2004: 55-62. NRCBL: 1.3.9; 1.3.7.

Butler, Declan. Long-held theory is in danger of losing its nerve. *Nature* 2007 September 13; 449(7159): 124-125. NRCBL: 1.3.9; 7.4. Identifiers: Henri Korn.

Camilleri, Michael; Dubnansky, Erin C.; Rustgi, Anil K. Conflicts of interest and disclosures in publications. *Clinical Gastroenterology and Hepatology* 2007 March; 5(3): 268-273. NRCBL: 1.3.9; 7.3; 1.3.7.

Catano, Victor M.; Turk, James. Fraud and misconduct in scientific research: a definition and procedures for investigation. *Medicine and Law: The World Association for Medical Law* 2007 September; 26(3): 465-476. NRCBL: 1.3.9; 7.3.

Abstract: Scientific fraud and misconduct appear to be on the rise throughout the scientific community. Whatever the reasons for fraud and whatever the number of cases, it is important that the academic research community consider this problem in a cool and rational manner, ensuring that allegations are dealt with through fair and impartial procedures. Increasingly, governments have either sought to regulate fraud and misconduct through legislation, or they have left it to universities and research institutions to deal with at the local level. The result has been less than uniform understanding of what constitutes scientific fraud and misconduct and a great deal of variance in procedures used to investigate such allegations. In this paper, we propose a standard definition of scientific fraud and misconduct and procedures for investigation based on natural justice and fairness. The issue of fraud and misconduct should not be left to government regulation by default. The standardized definition and procedures presented here should lead to more appropriate institutional responses in dealing with allegations of scientific fraud and misconduct.

Chadwick, Ruth. Science, context and professional ethics. *In:* Korthals, Michiel; Bogers, Robert J., eds. Ethics for Life Scientists. Dordrecht, The Netherlands: Springer, 2004: 175-182. NRCBL: 1.3.9; 1.3.1. SC: an.

Chang, Kenneth. Researcher cleared of misconduct, but case is still murky. *New York Times* 2007 February 13; p.

NRCBL: National Reference Center for Bioethics Literature Classification Scheme   See inside front cover for terms.

65

F4. NRCBL: 1.3.9. SC: po. Identifiers: Rusi P. Taleyarkhan; Purdue University.

Check, Erika. Transparency urged over research payments [news]. *Nature* 2007 August 16; 448(7155): 738. NRCBL: 1.3.9; 7.3; 9.7.

Chinn, John; Kulakowski, Elliott C. Conflict of interest in research. *In:* Kulakowski, Elliott C.; Chronister, Lynne U., eds. Research Administration and Management. Sudbury, MA: Jones and Bartlett, 2006: 511-521. NRCBL: 1.3.9; 18.2; 18.2.

Christie, Bryan. New helpline for those who blow whistle on research fraud. *BMJ: British Medical Journal* 2007 May 19; 334(7602): 1023. NRCBL: 1.3.9.

Cohen-Kohler, Jillian Clare; Esmail, Laura C. Scientific misconduct, the pharmaceutical industry, and the tragedy of institutions. *Medicine and Law: The World Association for Medical Law* 2007 September; 26(3): 431-446. NRCBL: 1.3.9; 9.7.
Abstract: This paper examines how current legislative and regulatory models do not adequately govern the pharmaceutical industry towards ethical scientific conduct. In the context of a highly profit-driven industry, governments need to ensure ethical and legal standards are not only in place for companies but that they are enforceable. We demonstrate with examples from both industrialized and developing countries how without sufficient controls, there is a risk that corporate behaviour will transgress ethical boundaries. We submit that there is a critical need for urgent drug regulatory reform. There must be robust regulatory structures in place which enforce corporate governance mechanisms to ensure that pharmaceutical companies maintain ethical standards in drug research and development and the marketing of pharmaceuticals. What is also needed is for the pharmaceutical industry to adopt authentic "corporate social responsibility" policies as current policies and practices are insufficient.

Corneliussen, Filippa. Adequate regulation, a stop-gap measure, or part of a package? Debates on codes of conduct for scientists could be diverting attention away from more serious questions. *EMBO Reports* 2006 July; 7 Special No: S50-S54. 19 refs. NRCBL: 1.3.9; 5.3; 6; 15.1.
Keywords: *biological sciences; *biological warfare; *bioterrorism; *codes of ethics; *professional ethics; *regulation; *researchers; *self regulation; biotechnology; disclosure; government regulation; guidelines; industry; international aspects; microbiology; policy making; professional organizations; public participation; public policy; recombinant DNA research; research ethics; review committees; Keyword Identifiers: National Institutes of Health; NIH Guidelines; United States

Cyranoski, David. Executed Chinese drug czar corrupted by system, observers say [news]. *Nature Medicine* 2007 August; 13(8): 889. NRCBL: 1.3.9; 18.2; 9.7. SC: le. Identifiers: China; State Food and Drug Administration [SFDA]; Zheng Xiaoyu.

Dahlquist, Gisela. Ethics in research: why and how? [editorial]. *Scandinavian Journal of Public Health* 2006; 34(5): 449-452. NRCBL: 1.3.9; 18.2; 18.6; 21.1.

Dalton, Rex. Passive-smoking study faces review [news]. *Nature* 2007 March 15; 446(7133): 242. NRCBL: 1.3.9; 9.5.9.

Daroff, Robert B. Scientific misconduct and breach of publication ethics: one editor's experience. *Medicine and Law: The World Association for Medical Law* 2007 September; 26(3): 527-533. NRCBL: 1.3.9; 1.3.7.
Abstract: I summarize my experience with scientific misconduct and breach of publication ethics during my 10 year term as Editor-in-Chief and my first 3 years as Scientific Integrity Advisor for Neurology, the official publication of the American Academy of Neurology. I describe in some detail the highly publicized, lengthy saga involving the accusation from a former colleague that James Abbs falsified data in an article published in Neurology. Nine years later, after numerous investigations and law suits, Abbs was found to have engaged in scientific misconduct which prompted the retraction of the article. Most of the problems I encountered were less complex and involved claims of plagiarism (regarded as "scientific misconduct") and self plagiarism (regarded as a "breach of publication ethics"). I conclude by providing helpful sources for editors in dealing with these infractions.

Davis, Mark S.; Riske-Morris, Michelle; Diaz, Sebastian R. Causal factors implicated in research misconduct: evidence from ORI case files. *Science and Engineering Ethics* 2007 December; 13(4): 395-414. NRCBL: 1.3.9. SC: em.

de Cock Buning, Tjard. Comments on Zwart: professional ethics and scholarly communication. *In:* Korthals, Michiel; Bogers, Robert J., eds. Ethics for Life Scientists. Dordrecht, The Netherlands: Springer, 2004: 81-84. NRCBL: 1.3.9.

Downie, Jocelyn. Grasping the nettle: confronting the issue of competing interests and obligations in health research policy. *In:* Flood, Colleen M., ed. Just Medicare: What's In, What's Out, How We Decide. Buffalo, NY: University of Toronto Press, 2006: 427-448. NRCBL: 1.3.9; 9.1; 5.3. SC: le. Identifiers: Canada.

Düwell, Marcus. Research as a challenge for ethical reflection. *In:* Korthals, Michiel; Bogers, Robert J., eds. Ethics for Life Scientists. Dordrecht, The Netherlands: Springer, 2004: 147-155. NRCBL: 1.3.9; 2.1; 1.1.

Dyer, Owen. GMC hearing against Andrew Wakefield opens [news]. *BMJ: British Medical Journal* 2007 July 14; 335(7610): 62-63. NRCBL: 1.3.9; 18.3; 18.5.3. Identifiers: Great Britain (United Kingdom); General Medical Council.

Dyer, Owen. Researcher accused of breaching research ethics faces GMC [news]. *BMJ: British Medical Journal*

2007 June 9; 334(7605): 1185. NRCBL: 1.3.9; 18.2. Identifiers: Great Britain (United Kingdom); General Medical Council; Tonmoy Sharma.

Epstein, Richard A. Conflicts of interest in health care: who guards the guardians? *Perspectives in Biology and Medicine* 2007 Winter; 50(1): 72-88. NRCBL: 1.3.9; 5.3; 7.3; 9.7; 1.3.2.

Faunce, Thomas Alured; Jefferys, Susannah. Whistleblowing and scientific misconduct: renewing legal and virtue ethics foundations. *Medicine and Law: The World Association for Medical Law* 2007 September; 26(3): 567-584. NRCBL: 1.3.9; 1.1; 2.1.

Abstract: Whistleblowing in relation to scientific research misconduct, despite the benefits of increased transparency and accountability it often has brought to society and the discipline of science itself, remains generally regarded as a pariah activity by many of the most influential relevant organizations. The motivations of whistleblowers and those supporting them continued to be questioned and their actions criticised by colleagues and management, despite statutory protections for reasonable disclosures appropriately made in good faith and for the public interest. One reason for this paradoxical position, explored here, is that whistle blowing concerning scientific misconduct lacks the policy support customarily derived from firm bioethical and jurisprudential foundations. Recommendations are made for altering this situation in the public interest.

Federation of American Societies for Experimental Biology [FASEB]; Brockway, Laura M.; Furcht, Leo T. Conflicts of interest in biomedical research — the FASEB guidelines. *FASEB Journal* 2006 December; 20(14): 2435-2438. NRCBL: 1.3.9; 7.3; 9.3.1; 6.

Finkel, Elizabeth. New misconduct rules aim to minister to an ailing system [news]. *Science* 2007 August 31; 317(5842): 1159. NRCBL: 1.3.9; 7.4. Identifiers: Australia.

Fisher, Richard. Fraudbusters. *New Scientist* 2007 November 10-16; 196(2629): 64-65. NRCBL: 1.3.9.

Fletcher, Robert H.; Black, Bert. "Spin" in scientific writing: scientific mischief and legal jeopardy. *Medicine and Law: The World Association for Medical Law* 2007 September; 26(3): 511-525. NRCBL: 1.3.9; 7.3; 1.3.7.

Abstract: In science, the data are supposed to speak for themselves. However, investigators have great latitude in how they report their results in the medical literature, even in an era of research protocols, pre-specified endpoints, reporting guidelines, and rigorous peer review. Authors' personal agendas, such as financial, personal, and intellectual conflicts of interest, can and sometimes do color how research results are described. Articles in peer-reviewed medical journals are the evidence base not only for the care of patients but also for legal decisions and the scientific record may be tailored for legal reasons as well. Journal editors preside over where and how the results of scientific research are published. We therefore suggest some actions that editors can take to foster a more trustworthy evidence base both for the care of patients and for legal decisions.

Friedberg, Errol C. Fraud in science — reflections on some whys and wherefores. *DNA Repair* 2006 March 7; 5(3): 291-293. NRCBL: 1.3.9; 1.3.7.

Funk, Carolyn L.; Barrett, Kirsten A.; Macrina, Francis L. Authorship and publication practices: evaluation of the effect of responsible conduct of research instruction to postdoctoral trainees. *Accountability in Research* 2007 October-December; 14(4): 269-305. NRCBL: 1.3.9; 1.3.7; 2.3.

Gert, Bernard. How common morality relates to business and the professions. *In:* Korthals, Michiel; Bogers, Robert J., eds. Ethics for Life Scientists. Dordrecht, The Netherlands: Springer, 2004: 129-139. NRCBL: 1.3.9; 1.1; 1.3.2.

Giles, Jim. Breeding cheats [news]. *Nature* 2007 January 18; 445(7125): 242-243. NRCBL: 1.3.9; 1.3.3.

Gilman, Paul. A conflict-of-interest policy for epidemiology. *Epidemiology* 2006 May; 17(3): 250-251. NRCBL: 1.3.9; 1.3.7. SC: em.

Glick, Michael. Scientific fraud — real consequences. *Journal of the American Dental Association* 2006 April; 137(4): 428, 430. NRCBL: 1.3.9; 1.3.7.

Goodyear, Michael. Free access to medical information: a moral right [letter]. *CMAJ/JAMC: Canadian Medical Association Journal* 2007 January 2; 176(1): 69. NRCBL: 1.3.9; 18.1.

Gornall, Jonathan. Hyperactivity in children: the Gillberg affair. *BMJ: British Medical Journal* 2007 August 25; 335(7616): 370-373. NRCBL: 1.3.9; 18.3; 17.4; 18.5.2. SC: le. Identifiers: Sweden; Christopher Gillberg.

Greif, Karen F.; Merz, Jon F. The darker side of science: scientific misconduct. *In their:* Current Controversies in the Biological Sciences: Case Studies of Policy Challenges from New Technologies. Cambridge, MA: MIT, 2007: 229-234. NRCBL: 1.3.9; 5.3.

Hanawalt, Philip C. Research collaborations: trial, trust, and truth. *Cell* 2006 September 8; 126(5): 823-825. NRCBL: 1.3.9; 1.3.7; 7.1; 7.3; 18.2.

Häyry, Matti. Bioscientists as ethical decision-makers. *In:* Korthals, Michiel; Bogers, Robert J., eds. Ethics for Life Scientists. Dordrecht, The Netherlands: Springer, 2004: 183-189. NRCBL: 1.3.9; 6; 1.1.

Heeger, Robert. Comments on Häyry: assessing bioscientific work from a moral point of view. *In:* Korthals, Michiel; Bogers, Robert J., eds. Ethics for Life Scientists. Dordrecht, The Netherlands: Springer, 2004: 191-193. NRCBL: 1.3.9; 6; 1.1.

NRCBL: National Reference Center for Bioethics Literature Classification Scheme   See inside front cover for terms.

67

Heinemann, Lothar A.J. Verlässlichkeit von Industrie- und öffentlich geförderten Studien. *In:* Shapiro, S.; Dinger, J.; Scriba, P., eds. Enabling Risk Assessment in Medicine: Farewell Symposium for Werner-Kari Raff. New Brunswick, NJ: Transaction Publishers, 2004: 51-66. NRCBL: 1.3.9; 1.3.2; 5.3; 9.7; 18.5.3.

Hofmann, Bjørn. That's not science! The role of moral philosophy in the science/non-science divide. *Theoretical Medicine and Bioethics* 2007; 28(3): 243-256. NRCBL: 1.3.9; 1.1; 5.1.
  Abstract: The science/non-science distinction has become increasingly blurred. This paper investigates whether recent cases of fraud in science can shed light on the distinction. First, it investigates whether there is an absolute distinction between science and non-science with respect to fraud, and in particular with regards to manipulation and fabrication of data. Finding that it is very hard to make such a distinction leads to the second step: scrutinizing whether there is a normative distinction between science and non-science. This is done by investigating one of the recent internationally famous frauds in science, the Sudbø case. This case demonstrates that moral norms are not only needed to regulate science because of its special characteristics, such as its potential for harm, but moral norms give science its special characteristics. Hence, moral norms are crucial in differentiating science from non-science. Although this does not mean that ethics can save the life of science, it can play a significant role in its resuscitation.

Holdstock, Douglas. A code of ethics for scientists [letter]. *Lancet* 2007 May 26 – June 1; 369(9575): 1789. NRCBL: 1.3.9; 2.4; 5.3; 21.3; 1.3.6. Comments: Ching Ling Pang. A code of ethics for scientists. Lancet 2007 March 31; 369(9567): 1068.

Hunter, Jennifer M. Plagiarism — does the punishment fit the crime? *Veterinary Anaesthesia and Analgesia* 2006 May; 33(3): 139-142. NRCBL: 1.3.9; 1.3.7.

Iannaccone, Philip M. An update on a misconduct investigation [letter]. *Science* 2007 August 17; 317(5840): 899. NRCBL: 1.3.9. Comments: J. Couzin. Truth and consequences. Science 2006 September 1: 1222.

Jesani, Amar. Response: questions of science, law and ethics. *Indian Journal of Medical Ethics* 2007 January-March; 4(1): 10-11. NRCBL: 1.3.9; 1.3.5; 7.4. SC: le.

Joly, Pierre-Benoit; Rip, Arie. A timely harvest. The public should be consulted on contentious research and development early enough for their opinions to influence the course of science and policy-making. *Nature* 2007 November 8; 450(7167): 174. NRCBL: 1.3.9; 5.3; 15.1; 1.3.11.

Jones, Nancy L. A code of ethics for the life sciences. *Science and Engineering Ethics* 2007 March; 13(1): 25-43. NRCBL: 1.3.9; 5.1.

Kaiser, Jocelyn. Privacy policies take a toll on research, survey finds [news]. *Science* 2007 November 16; 318(5853): 1049. NRCBL: 1.3.9; 8.4.

Kaiser, Jocelyn. Stung by controversy, biomedical groups urge consistent guidelines [news]. *Science* 2007 July 27; 317(5837): 441. NRCBL: 1.3.9.

Kanungo, R. Ethics in research [editorial]. *Indian Journal of Medical Microbiology* 2006 January; 24(1): 5-6. NRCBL: 1.3.9; 18.2; 18.5.9.

Kimmelman, Jonathan. Inventors as investigators: the ethics of patents in clinical trials. *Academic Medicine* 2007 January; 82(1): 24-31. NRCBL: 1.3.9; 5.3; 18.6; 8.1; 15.1.

Kondro, Wayne. Call for arm's-length national research integrity agency [news]. *CMAJ/JAMC: Canadian Medical Association Journal* 2007 March 13; 176(6): 749-750. NRCBL: 1.3.9; 5.3.

Kondro, Wayne; Hébert, Paul C. Research misconduct? What misconduct? = Inconduite scientifique? Quelle inconduite? [editorial]. *CMAJ/JAMC: Canadian Medical Association Journal* 2007 March 27; 176(7): 905, 907. NRCBL: 1.3.9.

Koper, Megan; Bubela, Tania; Caulfield, Timothy; Boon, Heather. Media portrayal of conflicts of interest in biomedical research. *Health Law Review* 2007; 15(3): 30-31. NRCBL: 1.3.9; 1.3.7. SC: em.

Kovac, Jeffrey. Moral rules, moral ideals, and use-inspired research. *Science and Engineering Ethics* 2007 June; 13(2): 159-169. NRCBL: 1.3.9; 1.1; 5.1.

Krimsky, Sheldon. Publication bias, data ownership, and the funding effect in science: threats to the integrity of biomedical research. *In:* Wagner, Wendy; Steinzor, Rena, eds. Rescuing Science from Politics: Regulation and the Distortion of Scientific Research. Cambridge: Cambridge University Press, 2006: 61-85. NRCBL: 1.3.9; 5.1; 5.3; 9.7.

Krimsky, Sheldon. The ethical and legal foundations of scientific 'conflict of interest'. *In:* Lemmens, Trudo; Waring, Duff R., eds. Law and Ethics in Biomedical Research: Regulation, Conflict of Interest and Liability. Toronto; Buffalo: University of Toronto Press, 2006: 63-81. NRCBL: 1.3.9; 7.3. SC: le.

Krimsky, Sheldon. When conflict-of-interest is a factor in scientific misconduct. *Medicine and Law: The World Association for Medical Law* 2007 September; 26(3): 447-463. NRCBL: 1.3.9; 7.3.
  Abstract: Under the guidelines adopted by the United States (U.S.) Office of Research Integrity (ORI), scientific misconduct is defined by one or more of three activities: fabrication of data, falsification of results, and plagiarism or the improper appropriation of other people's ideas or written work. This paper discusses whether three other breaches in scientific ethics, namely ghost

---

writing, fabricating credentials, and failure to disclose conflicts of interest, rise to the level of scientific misconduct. After discussing the funding effect in science, the paper argues that, like ghost writing and fabricated credentials, conflicts of interest can bias the outcome of research. Thus, lack of transparency to reviewers, journals and readers for conflicts of interest should be considered a form of scientific misconduct.

Krumholz, Harlan. What have we learnt from Vioxx? *BMJ: British Medical Journal* 2007 January 20; 334(7585): 120-123. NRCBL: 1.3.9; 5.3; 9.7; 18.1; 7.3; 1.3.7; 9.3.1.

Kubiak, Erik N.; Park, Samuel S.; Egol, Kenneth; Zuckerman, Joseph D.; Koval, Kenneth J. Increasingly conflicted: an analysis of conflicts of interest reported at the annual meetings of the Orthopaedic Trauma Association. *Bulletin (Hospital for Joint Diseases (New York, N.Y.))* 2006; 63(3-4): 83-87. NRCBL: 1.3.9; 1.3.7; 1.3.2; 9.3.1. SC: em.

Kulakowski, Elliott C. Dealing with allegations of research misconduct: the other side of responsible conduct of research. *In:* Kulakowski, Elliott C.; Chronister, Lynne U., eds. Research Administration and Management. Sudbury, MA: Jones and Bartlett, 2006: 617-624. NRCBL: 1.3.9; 18.2.

Kulynych, Jennifer. Intent to deceive: mental state and scienter in the new uniform federal definition of scientific misconduct. *Stanford Technology Law Review* 1998; 2: 35 p. [Online]. Accessed: http://stlr.stanford.edu/STLR/Articles/98_STLR_2/index.htm [2007 April 23]. NRCBL: 1.3.9; 18.3. SC: le.

Laine, Christine; Goodman, Steven N.; Griswold, Michael E.; Sox, Harold C. Reproducible research: moving toward research the public can really trust. *Annals of Internal Medicine* 2007 March 20; 146(6): 450-453. NRCBL: 1.3.9; 1.3.7.

Abstract: A community of scientists arrives at the truth by independently verifying new observations. In this time-honored process, journals serve 2 principal functions: evaluative and editorial. In their evaluative function, they winnow out research that is unlikely to stand up to independent verification; this task is accomplished by peer review. In their editorial function, they try to ensure transparent (by which we mean clear, complete, and unambiguous) and objective descriptions of the research. Both the evaluative and editorial functions go largely unnoticed by the public—the former only draws public attention when a journal publishes fraudulent research. However, both play a critical role in the progress of science. This paper is about both functions. We describe the evaluative processes we use and announce a new policy to help the scientific community evaluate, and build upon, the research findings that we publish.

Lemmens, Trudo. Commercialized medical research and the need for regulatory reform. *In:* Flood, Colleen M., ed.

Just Medicare: What's In, What's Out, How We Decide. Buffalo, NY: University of Toronto Press, 2006: 396-426. NRCBL: 1.3.9; 9.3.1; 5.3; 9.7. SC: le. Identifiers: Canada.

Lenzer, Jeanne. US Senate passes bill granting mandatory access to data [news]. *BMJ: British Medical Journal* 2007 November 3; 335(7626): 906. NRCBL: 1.3.9; 18.1.

Lexchin, Joel. The secret things belong unto the Lord our God: secrecy in the pharmaceutical arena. *Medicine and Law: The World Association for Medical Law* 2007 September; 26(3): 417-430. NRCBL: 1.3.9; 7.3; 9.7.

Abstract: Secrecy in the pharmaceutical arena has taken on more importance in the recent past as the pharmaceutical industry has assumed greater prominence in the funding of clinical research and has also become a funder of the agencies that are charged with regulating it. Governments have adopted a neo-liberal agenda that prioritizes private profit over public health and are therefore willing to let industry set the research agenda. As a result, secrecy, to protect intellectual property rights, is a major feature of clinical research. Secrecy also leads to biases in the published literature that conceal significant safety problems. Because regulators are now partially dependent on the pharmaceutical industry for their existence regulators are unwilling to challenge industry. By treating data on efficacy and safety as commercially confidential information they effectively collude with industry in denying health professionals and the public access to essential information to be able to use drugs appropriately.

Lind, Rebecca Ann; Lepper, Tammy Swenson. Sensitivity to research misconduct: a conceptual model. *Medicine and Law: The World Association for Medical Law* 2007 September; 26(3): 585-598. NRCBL: 1.3.9; 18.1.

Abstract: Ethical sensitivity research suggests techniques for assessing people's sensitivity to research misconduct (RM). Based on our prior work in assessing ethical sensitivity, we present a conceptual model for assessing RM sensitivity. We propose conceptual and operational definitions of RM sensitivity (RMsen), and consider how the construct could be measured. RMsen is conceptualized as a cognitive ability, a skill which can be learned and assessed. RMsen involves an awareness that the research situation presents the possibility for misconduct to occur, and that one may have to decide what is right or wrong in the situation. Indicators of RMsen can take many forms and represent multiple content domains and dimensions. Four main content domains of RMsen are situational characteristics, RM issues, consequences, and stakeholders. In addition, linkages are potential connections made among elements in the different content domains. Three dimensions applicable to assessing RMsen include time, breadth, and depth. Although our focus is on RMsen, we believe that our model and methods may be extended to assessing sensitivity to the responsible conduct of research.

Mabrouk, Patricia Ann. Introducing summer high school student-researchers to ethics in scientific research. *Journal of Chemical Education* 2007 June; 84(6): 952. NRCBL: 1.3.9. SC: cs.

NRCBL: National Reference Center for Bioethics Literature Classification Scheme   See inside front cover for terms.

69

Margetts, Barrie. Stopping the rot in nutrition science. *Public Health Nutrition* 2006 April; 9(2): 169-173. NRCBL: 1.3.9; 1.3.7.

Marusic, Ana; Katavic, Vedran; Marusic, Matko. Role of editors and journals in detecting and preventing scientific misconduct: strengths, weakness, opportunities, and threats. *Medicine and Law: The World Association for Medical Law* 2007 September; 26(3): 545-566. NRCBL: 1.3.9; 1.3.7.

Abstract: Scientific journals have a central place in protecting research integrity because published articles are the most visible documentation of research. We used SWOT analysis to audit (S)trengths and (W)eaknesses as internal and (O)pportunities and (T)hreats as external factors affecting journals' responsibility in addressing research integrity issues. Strengths include editorial independence, authority and expertise, power to formulate editorial policies, and responsibility for the integrity of published records. Weaknesses stem from having no mandate for legal action, reluctance to get involved, and lack of training. Opportunities for editors are new technologies for detecting misconduct, policies by editorial organization or national institutions, and greater transparency of published research. Editors face threats from the lack of legal regulation and culture of research integrity in academic communities, lack of support from stakeholders in scientific publishing, and different pressures. Journal editors cannot be the policing force of the scientific community but they should actively ensure the integrity of the scientific record.

Momen, Hooman; Gollogly, Laragh. Cross-cultural perspectives of scientific misconduct. *Medicine and Law: The World Association for Medical Law* 2007 September; 26(3): 409-416. NRCBL: 1.3.9; 21.7.

Abstract: The increasing globalization of scientific research lends urgency to the need for international agreement on the concepts of scientific misconduct. Universal spiritual and moral principles on which ethical standards are generally based indicate that it is possible to reach international agreement on the ethical principles underlying good scientific practice. Concordance on an operational definition of scientific misconduct that would allow independent observers to agree which behaviour constitutes misconduct is more problematic. Defining scientific misconduct to be universally recognized and universally sanctioned means addressing the broader question of ensuring that research is not only well-designed - and addresses a real need for better evidence - but that it is ethically conducted in different cultures. An instrument is needed to ensure that uneven ethical standards do not create unnecessary obstacles to research, particularly in developing countries.

Müller-Hill, Benno. The silence of the scholars. *In:* LaFleur, William R.; Böhme, Gernot; Shimazono, Susumu, eds. Dark Medicine: Rationalizing Medical Research. Bloomington: Indiana University Press, 2007: 57-62. NRCBL: 1.3.9; 1.3.7; 18.5.1; 21.4; 2.2.

Nagell, Hilde W. A penny for your thoughts — ethics in sponsored research. *In:* Gunning, Jennifer; Holm, Søren, eds. Ethics, law, and society. Volume 1. Aldershot, Hants, England; Burlington, VT: Ashgate, 2005: 45-60. NRCBL: 1.3.9; 7.3.

Neal, Joseph M.; Rathmell, James P. Scientific misconduct: no end in sight. *Regional Anesthesia and Pain Medicine* 2006 July-August; 31(4): 294-295. NRCBL: 1.3.9; 1.3.7.

Neill, Ushma S. Stop misbehaving! [editorial]. *Journal of Clinical Investigation* 2006 July; 116(7): 1740-1741. NRCBL: 1.3.9; 1.3.7. Identifiers: scientific misconduct.

Neutra, Raymond Richard; Cohen, Aaron; Fletcher, Tony; Michaels, David; Richter, Elihu D.; Soskolne, Colin L. Toward guidelines for the ethical reanalysis and reinterpretation of another's research. *Epidemiology* 2006 May; 17(3): 335-338. NRCBL: 1.3.9. SC: em.

Neutra, Raymond R. What to declare and why? *Epidemiology* 2006 May; 17(3): 244-245. NRCBL: 1.3.9; 1.3.7. SC: em.

Nieto, Antonio; Mazon, Angel; Pamies, Rafael; Linana, Juan J.; Lanuza, Amparo; Jiménez, Fernando Oliver; Medina-Hernandez, Alejandra; Nieto, F. Javier. Adverse effects of inhaled corticosteroids in funded and nonfunded studies. *Archives of Internal Medicine* 2007 October 22; 167(19): 2047-2053. NRCBL: 1.3.9; 5.3; 9.7; 18.1. SC: em.

Abstract: BACKGROUND: Evidence regarding the safety profile of drugs may vary depending on study sponsorship. We aimed to evaluate differences between studies funded by the pharmaceutical manufacturer of the drug (PF) and those with no pharmaceutical funding (NoPF) regarding the finding and interpretation of adverse effects of inhaled corticosteroids. METHODS: We assessed the safety reporting of inhaled corticosteroids in 275 PF and 229 NoPF studies identified by a MEDLINE search using prespecified criteria. RESULTS: Overall, the finding of statistically significant differences for adverse effects was significantly less frequent in PF (34.5%) than in NoPF (65.1%) studies (prevalence ratio, 0.53; 95% confidence interval, 0.44-0.64). This association became nonsignificant (prevalence ratio, 0.94; 95% confidence interval, 0.77-1.15) after controlling for design features (such as dose or use of parallel groups) that tended to be associated with less frequent finding of adverse effects and were more common in PF studies. Among studies finding a statistically significant increase in adverse effects associated with the study drug, the authors of PF articles concluded that the drug was "safe" more frequently than the authors of NoPF studies (prevalence ratio, 3.68; 95% confidence interval, 2.14-6.33). CONCLUSIONS: The type of funding may have determinant effects on the design of studies and on the interpretation of findings: funding by the industry is associated with design features less likely to lead to finding statistically significant adverse effects and with a more favorable clinical interpretation of such findings. Disclosure of conflicts of interest should be strengthened for a more balanced opinion on the safety of drugs.

Normile, Dennis. Japan Universities take action [news]. *Science* 2007 January 5; 315(5808): 26. NRCBL: 1.3.9.

Normile, Dennis. Osaka University researchers reject demand to retract Science paper. *Science* 2007 June 22; 316(5832): 1681. NRCBL: 1.3.9; 1.3.7. Identifiers: Japan.

Novack, Gary D. Research ethics. *The Ocular Surface* 2006 April; 4(2): 103-104. NRCBL: 1.3.9; 1.3.7. SC: rv.

Odling-Smee, Lucy; Giles, Jim; Fuyuno, Ichiko; Cyranoski, David; Marris, Emma. Where are they now? [news]. *Nature* 2007 January 18; 445(7125): 244-245. NRCBL: 1.3.9.

Ogbogu, Ubaka. Canada's approach to conflict-of-interest oversight [letter]. *CMAJ/JAMC: Canadian Medical Association Journal* 2007 August 14; 177(4): 375-376. NRCBL: 1.3.9; 5.3. Comments: Wayne Kondro. US proposes more stringent conflict-of-interest rules. CMAJ/JAMC: Canadian Medical Association Journal 2007 May 22; 176(11): 1571-1572.

Okike, Kanu; Kocher, Mininder S.; Mehlman, Charles T.; Bhandari, Mohit. Conflict of interest in orthopaedic research. An association between findings and funding in scientific presentations. *Journal of Bone and Joint Surgery. American volume* 2007 March; 89(3): 608-613. NRCBL: 1.3.9; 7.1; 1.3.2; 9.3.1; 9.8; 7.3.

Olsen, Jørn. Kafka's truth-seeking dogs. *Epidemiology* 2006 May; 17(3): 242-243. NRCBL: 1.3.9; 7.1. SC: em.

Pang, Ching Ling. A code of ethics for scientists. *Lancet* 2007 March 31-April 6; 369(9567): 1068. NRCBL: 1.3.9; 7.3.

Pascal, Chris B. Beyond the federal definition: other forms of misconduct. *In:* Kulakowski, Elliott C.; Chronister, Lynne U., eds. Research Administration and Management. Sudbury, MA: Jones and Bartlett, 2006: 523-530. NRCBL: 1.3.9; 5.3; 18.2.

Powell, Sean T.; Allison, Matthew A.; Kalichman, Michael W. Effectiveness of a responsible conduct of research course: a preliminary study. *Science and Engineering Ethics* 2007 June; 13(2): 249-264. NRCBL: 1.3.9; 7.2. SC: em.

Pryor, Erica R.; Habermann, Barbara; Broome, Marion E. Scientific misconduct from the perspective of research coordinators: a national survey. *Journal of Medical Ethics* 2007 June; 33(6): 365-369. NRCBL: 1.3.9. SC: em.

Abstract: OBJECTIVE: To report results from a national survey of coordinators and managers of clinical research studies in the US on their perceptions of and experiences with scientific misconduct. METHODS: Data were collected using the Scientific Misconduct Questionnaire-Revised. Eligible responses were received from 1645 of 5302 (31%) surveys sent to members of the Association of Clinical Research Professionals and to subscribers of Research Practitioner, published by the Center for Clinical Research Practice, between February 2004 and January 2005. Findings: Overall, the perceived frequency of misconduct was low. Differences were noted between workplaces with regard to perceived pressures on investigators and research coordinators, and on the effectiveness of the regulatory environment in reducing misconduct. First-hand experience with an incident of misconduct was reported by 18% of respondents. Those with first-hand knowledge of misconduct were more likely to report working in an academic medical setting, and to report that a typical research coordinator would probably do nothing if aware that a principal investigator or research staff member was involved in an incident of misconduct. CONCLUSION: These findings expand the knowledge on scientific misconduct by adding new information from the perspective of research coordinators. The findings provide some data supporting the influence of workplace climate on misconduct and also on the perceived effectiveness of institutional policies to reduce scientific misconduct.

Qiu, Jane. Chinese law aims to quell fear of failure [news]. *Nature* 2007 September 6; 449(7158): 12. NRCBL: 1.3.9; 7.4. Identifiers: China.

Reich, Eugenie Samuel. Misconduct report kept under wraps. *New Scientist* 2007 November 24-30; 196(2631): 16. NRCBL: 1.3.9; 7.4. Identifiers: Luk Van Parijs.

Resnik, David B. Some recent challenges to openness and freedom in scientific publication. *In:* Korthals, Michiel; Bogers, Robert J., eds. Ethics for Life Scientists. Dordrecht, The Netherlands: Springer, 2004: 85-99. NRCBL: 1.3.9; 1.3.7.

Revill, James; Dando, Malcolm R. A Hippocratic oath for life scientists. A Hippocratic-style oath in the life sciences could help to educate researchers about the dangers of dual-use research. *EMBO Reports* 2006 July; 7 Special No: S55-S60. NRCBL: 1.3.9; 6.

Richman, Vincent V.; Richman, Alex. Enhancing research integrity [letter]. *CMAJ/JAMC: Canadian Medical Association Journal* 2007 August 14; 177(4): 375. NRCBL: 1.3.9. Comments: Wayne Kondro and Paul Hébert. Research misconduct? What misconduct? CMAJ/JAMC: Canadian Medical Association Journal 2007 March 27; 176(7): 905.

Ropert-Coudert, Yan. Recognition could support a science code of conduct [letter]. *Nature* 2007 May 17; 447(7142): 259. NRCBL: 1.3.9.

Rust, Susanne; Spivak, Cary. Panel will investigate research firms' ethics: conflicts of interest exist, presidential candidate says. *Milwaukee Journal Sentinel: JS Online* 2007 May 5; 2 p. [Online]. Accessed: http://www.jsonline.

NRCBL: National Reference Center for Bioethics Literature Classification Scheme   See inside front cover for terms.

71

com/story/index.aspx?id=601378 [2007 May 7]. NRCBL: 1.3.9; 1.3.5. Identifiers: Dennis Kucinich; Constella Group.

Santosuosso, Amedeo; Sellaroli, Valentina; Fabio, Elisabetta. What constitutional protection for freedom of scientific research? *Journal of Medical Ethics* 2007 June; 33(6): 342-344. 11 refs. NRCBL: 1.3.9; 5.1; 15.1 15.5. SC: le; an.

> Keywords: *constitutional law; *freedom; *science; coercion; democracy; embryo research; embryonic stem cells; eugenics; genetic engineering; genetic research; human dignity; government financing; government regulation; legal rights; methods; patents; precautionary principle; public policy; reproductive technologies; Keyword Identifiers: Europe; North America

Sass, Hans-Martin. Ethical risk in medical research. *In:* Shapiro, S.; Dinger, J.; Scriba, P., eds. Enabling Risk Assessment in Medicine: Farewell Symposium for Werner-Kari Raff. New Brunswick, NJ: Transaction Publishers, 2004: 83-93. NRCBL: 1.3.9; 5.2; 18.2; 18.3.

Scheetz, Mary D. The Teaching Scholars Program: a proposed approach for promoting research integrity. *Medicine and Law: The World Association for Medical Law* 2007 September; 26(3): 599-614. NRCBL: 1.3.9; 18.6; 7.2.

> Abstract: All research environments are not created equal. They possess their own unique communication style, culture, and professional mores. Coupled with these distinct professional nuances is the fact that research collaborations today span not only a campus, but also the globe. While the opportunities for cross cultural collaborations are invaluable, they may present challenges that result in misunderstandings about how a research idea should be studied and the findings presented. Such misunderstandings are sometimes found at the center of research misconduct cases. And yet in light of highly visible cases of research misconduct, the attitude about ensuring research integrity remains rather opaque. This paper discusses the merits of the Teaching Scholars Program as a mechanism by which to promote research integrity. This paper will examine this education program against the backdrop of the US Office of Research Integrity (ORI), as an established office responsible for ensuring the integrity of federally funded biomedical and behavioral research.

Scholze, Simone. Setting standards for scientists. For almost ten years, COMEST has advised UNESCO on the formulation of ethical guidelines. *EMBO Reports* 2006 July; 7 Special No: S65-S67. NRCBL: 1.3.9; 6.

Seoul National University Investigation Committee. Summary of final report on Professor Hwang Woo-suk's research [press release]. Seoul, Korea: Seoul National University Investigation Committee, 2006 January 10; 4 p. [Online]. Accessed: http://www.useoul.edu/sk_board/boards/sk_news_read.jsp?board=11940&p_tid=86373&p_rel=null&id=63459 [2007 April 26]. NRCBL: 1.3.9; 14.5; 1.3.3.

Shapiro, Samuel. Integrity of independent and industry-sponsored studies. *In:* Shapiro, S.; Dinger, J.; Scriba, P., eds. Enabling Risk Assessment in Medicine: Farewell Symposium for Werner-Kari Raff. New Brunswick, NJ: Transaction Publishers, 2004: 67-82. NRCBL: 1.3.9; 1.3.2; 5.3; 18.2.

Shorvon, Simon. The prosecution of research — experience from Singapore. *Lancet* 2007 June 2-8; 369(9576): 1835-1837. NRCBL: 1.3.9; 21.7. SC: le. Identifiers: Great Britain (United Kingdom).

Siegel-Itzkovich, Judy. Israeli surgeons put their names on study they had not done. *BMJ: British Medical Journal* 2007 May 19; 334(7602): 1023. NRCBL: 1.3.9.

Spece, Roy G.; Bernstein, Carol. Scientific misconduct and liability for the acts of others. *Medicine and Law: The World Association for Medical Law* 2007 September; 26(3): 477-491. NRCBL: 1.3.9; 18.2. SC: cs.

> Abstract: We argue that two ambiguities in [U.S.] Public Health Service ("PHS") misconduct regulations make them so vague that they are unconstitutional and unfair: (1) they provide no guidance concerning when one can be held responsible for others' actions; and (2) they simultaneously are intended to allow misconduct findings only when there are "significant departure[s] from established practices of the relevant research community" but even if one complied with customary standards of practice in her research community, thus providing confusion rather than guidance. The effect of these ambiguities is not only to leave researchers without notice as to proscribed or prescribed conduct but also to give officials discretion to apply the regulations arbitrarily and discriminatorily. The regulations' effect is illustrated by applying them, hypothetically, to facts relating to the central charge in the misconduct case pressed by the University of Arizona in 1997 through 2003 against then Arizona Regents' Professor Marguerite Kay.

Spece, Roy G.; Bernstein, Carol. What is scientific misconduct, who has to (dis)prove it, and to what level of certainty? *Medicine and Law: The World Association for Medical Law* 2007 September; 26(3): 493-510. NRCBL: 1.3.9; 18.2. SC: le; cs.

> Abstract: This article traces the regulation of [U.S.] Public Health Service ("PHS")-funded research from changes begun with the proposal (1999) and then adoption (2000) of a basic, Uniform Federal ("research misconduct") Policy. It argues that the PHS misconduct regulations deny due process of law and are fundamentally unfair because they fail to specify the level of culpability for guilt, force accused researchers to prove that they are innocent, and, although admittedly quasi-criminal, adopt a standard of proof that tolerates nearly a 50 percent probability of false convictions. The regulations' infirmities will be demonstrated by applying them to facts relating to the central charge in the misconduct case pressed by the University of Arizona in 1997 through 2003 against then Arizona Regents' Professor Margue-

rite Kay, which facts are set forth in our companion piece in this theme issue.

Spier, Raymond E. Some thoughts on the 2007 World Conference on Research Integrity. *Science and Engineering Ethics* 2007 December; 13(4): 383-386. NRCBL: 1.3.9; 21.1.

Spitzer, Walter O. Minimizing bias and prejudice: special challenges for contractual research by academicians. *In:* Shapiro, S.; Dinger, J.; Scriba, P., eds. Enabling Risk Assessment in Medicine: Farewell Symposium for Werner-Kari Raff. New Brunswick, NJ: Transaction Publishers, 2004: 31-49. NRCBL: 1.3.9; 1.3.2; 5.3; 9.7; 18.5.3.

Stollorz, Volker. Öffentlich und Industrie-geförderte Studien in der Perzeption der Medien — ein Dilemma. *In:* Shapiro, S.; Dinger, J.; Scriba, P., eds. Enabling Risk Assessment in Medicine: Farewell Symposium for Werner-Kari Raff. New Brunswick, NJ: Transaction Publishers, 2004: 125-142. NRCBL: 1.3.9; 1.3.2; 1.3.7; 5.3; 9.7; 18.5.3.

Tabbara, Khalid; Al-Kawi, M. Zuheir. Ethics in medical research and publication. *Annals of Saudi Medicine* 2006 July-August; 26(4): 257-260. NRCBL: 1.3.9; 1.3.7; 18.2.

Tanne, Janice Hopkins. Investigators to review conflicts of interest at NIH [news]. *BMJ: British Medical Journal* 2007 April 14; 334(7597): 767. NRCBL: 1.3.9; 7.3.

Taylor, Patrick L. Rules of engagement. Is there an inherent conflict between public debate and free scientific inquiry? *Nature* 2007 November 8; 450(7167): 163-164. NRCBL: 1.3.9; 5.3; 15.1.

Thrush, Carol R.; Vander Putten, Jim; Rapp, Carla Gene; Pearson, L. Carolyn; Berry, Katherine Simms; O'Sullivan, Patricia S. Content validation of the Organizational Climate for Research Integrity (OCRI) Survey. *Journal of Empirical Research on Human Research Ethics* 2007 December; 2(4): 35-52. NRCBL: 1.3.9; 1.3.2; 18.2. SC: em.
Abstract: The purpose of this study was to develop and establish content validity of an instrument designed to measure the organizational climate for research integrity in academic health centers. Twenty-seven research integrity scholars and administrators evaluated 64 survey items for relevance and clarity, as well as overall comprehensiveness of the constructs that are measured (organizational inputs, structures, processes and outcomes). Both quantitative and qualitative methods were used, particularly content validity indices (CVI) and analyses of respondents' comments. The content validity index for the overall survey was initially high (CVI = .83) and improved (CVI = .90) when 17 marginal-to-poor items were removed. This study resulted in the Organizational Climate for Research Integrity (OCRI) survey, a 43-item fixed-response survey with established content validity.

Tierney, John. Are scientists playing God? It depends on your religion. *New York Times* 2007 November 20; p. F1, F2. NRCBL: 1.3.9; 14.5; 1.2; 15.1; 18.5.4.
Keywords: *cloning; *embryo research; *genetic engineering; *international aspects; *non-Western World; *religious ethics; *secularism; *Western World; biotechnology; Buddhist ethics; Christian ethics; embryonic stem cells; embryos; government regulation; Hindu ethics; legal aspects; public policy; reproductive technologies; Keyword Identifiers: Asia; Europe; North America; South America

Twombly, Renee. Goal of maintaining public's trust brings research groups together on conflict-of-interest guidelines [news]. *Journal of the National Cancer Institute* 2005 November 2; 97(21): 1560-1561. NRCBL: 1.3.9; 7.3; 5.3.

United States. Department of Energy [DOE]. Policy on research misconduct. *Federal Register* 2005 June 28; 70(123): 37010-37016 [Online]. Accessed: http://a257.g. akamaitech.net/7/257/2422/01jan20051800/edocket. access.gpo.gov/2005/pdf/05-12645.pdf [2007 April 24]. NRCBL: 1.3.9. SC: le.

United States. Department of Health and Human Services [DHHS]. Office of Research Integrity. Final rule on research misconduct: frequently asked questions and answers. Rockville, MD: Office of Research Integrity, 2005 June 13; 4 p. [Online]. Accessed: http://ori.dhhs. gov/policies/faq.shtml [2007 April 24]. NRCBL: 1.3.9; 5.3.

van den Belt, Henk. Comments on Bulger: the responsible conduct of research, including responsible authorship and publication practices. *In:* Korthals, Michiel; Bogers, Robert J., eds. Ethics for Life Scientists. Dordrecht, The Netherlands: Springer, 2004: 63-66. NRCBL: 1.3.9; 1.3.7.

Van Der Weyden, Martin B. Preventing and processing research misconduct: a new Australian code for responsible research: it all depends on compliance. *Medical Journal of Australia* 2006 May 1; 184(9): 430-431. NRCBL: 1.3.9; 1.3.7; 6.

van der Zijpp, Akke. Comments on Düwell: research as a challenge for ethical reflection. *In:* Korthals, Michiel; Bogers, Robert J., eds. Ethics for Life Scientists. Dordrecht, The Netherlands: Springer, 2004: 157-159. NRCBL: 1.3.9; 22.2; 1.1.

van Haselen, Robbert. Misconduct in CAM research: does it occur? *Complementary Therapies in Medicine* 2006 June; 14(2): 89-90. NRCBL: 1.3.9; 4.1.1; 7.1. Identifiers: complementary alternative medicine.

Veatch, Robert M. The roles of scientific and normative expertise in public policy formation: the anthrax vaccine case. *In:* Rasmussen, Lisa, ed. Ethics Expertise: History, Contemporary Perspectives, and Applications. Dordrecht: Springer, 2005: 211-225. NRCBL: 1.3.9; 5.3; 1.3.5; 2.1. SC: an.

NRCBL: National Reference Center for Bioethics Literature Classification Scheme   See inside front cover for terms.

73

von Elm, Erik. Research integrity: collaboration and research needed. *Lancet* 2007 October 20-26; 370(9596): 1403-1404. NRCBL: 1.3.9; 1.3.7; 21.1.

Wade, Nicholas. Panel says data is flawed in a major stem cell report. *New York Times* 2007 February 28; p. A15. NRCBL: 1.3.9; 1.3.7; 15.1. SC: po.

Wager, Elizabeth. What do journal editors do when they suspect research misconduct? *Medicine and Law: The World Association for Medical Law* 2007 September; 26(3): 535-544. NRCBL: 1.3.9; 1.3.7. SC: em.
Abstract: Several published guidelines urge journal editors to ensure that cases of suspected scientific misconduct are properly investigated. Using cases submitted to the Committee on Publication Ethics (COPE) I tried to discover what editors actually do when faced with such cases. Of the 79 cases referred to COPE between 1998 and 2003 relating to author misconduct, 33 related to redundant publication, 16 to unethical research, 13 to fabrication, 10 to clinical misconduct and 7 to plagiarism. Outcomes were reported in 49 cases. Authors were exonerated in 16 cases and reprimanded in another 17. An impasse (no or an unsatisfactory response) was reached in 16. Editors contacted the authors' institutions in 24 cases. Nearly half the cases (36) lasted over a year. This small survey highlights the difficulties faced by editors in pursuing cases of suspected misconduct and the need for better training and guidance for editors and more cooperation from institutions.

Watts, Geoff. Croatian academic is found guilty of plagiarism. *BMJ: British Medical Journal* 2007 May 26; 334(7603): 1077. NRCBL: 1.3.9; 1.3.7.

Wehling, Martin. Probleme Industrie-geförderter klinischer Studien im öffentlichen Bereich. *In:* Shapiro, S.; Dinger, J.; Scriba, P., eds. Enabling Risk Assessment in Medicine: Farewell Symposium for Werner-Kari Raff. New Brunswick, NJ: Transaction Publishers, 2004: 21-30. NRCBL: 1.3.9; 1.3.2; 5.3; 9.7; 18.2.

Weissmann, Gerald. Science fraud: from patchwork mouse to patchwork data. *FASEB Journal: Official Publication of the Federation of American Societies for Experimental Biology* 2006 April; 20(6): 587-590. NRCBL: 1.3.9; 22.2; 15.1.

Wellcome Trust. Statement on the handling of allegations of research misconduct. London: Wellcome Trust, 2005 November; 6 p. [Online]. Accessed: http://www.wellcome.ac.uk/doc_WTD002756.html [2007 April 18]. NRCBL: 1.3.9.

White, Caroline. Cancer expert attacks research paper [news]. *BMJ: British Medical Journal* 2007 September 8; 335(7618): 469. NRCBL: 1.3.9; 9.5.1; 9.7.

Wilcox, Allen J. On conflicts of interest. *Epidemiology* 2006 May; 17(3): 241. NRCBL: 1.3.9; 1.3.7. SC: em.

Willyard, Cassandra. Allegations of bias cloud conflicting reports on bisphenol A's effects [news]. *Nature Medicine* 2007 September; 13(9): 1002. NRCBL: 1.3.9; 1.3.5; 7.3.

Wilson, Kenneth; Schreier, Alan; Griffin, Angel; Resnik, David. Research records and the resolution of misconduct allegations at research universities. *Accountability in Research* 2007 January-March; 14(1): 57-71. NRCBL: 1.3.9. SC: em.

Young, Charles; Godlee, Fiona. Managing suspected research misconduct [editorial]. *BMJ: British Medical Journal* 2007 February 24; 334(7590): 378-379. NRCBL: 1.3.9; 1.3.7.

Zwart, Hub. Professional ethics and scholarly communication. *In:* Korthals, Michiel; Bogers, Robert J., eds. Ethics for Life Scientists. Dordrecht, The Netherlands: Springer, 2004: 67-80. NRCBL: 1.3.9.

BIOMEDICAL RESEARCH/ SOCIAL CONTROL OF SCIENCE AND TECHNOLOGY

California dreaming: universities should draw the line at certain types of support from the drug industry [editorial]. *Nature* 2007 July 26; 448(7152): 388. NRCBL: 5.3; 1.3.9; 7.3; 7.2; 9.7. Identifiers: University of California, San Francisco.

Federal funding for embryo research unlikely to rise after Bush. *BioEdge* 2007 August 22; 262: 4. NRCBL: 5.3; 18.5.4; 15.1;.

Abelson, Julia; Giacomini, Mita; Lehoux, Pascale; Gauvin, Francois-Pierre. Bringing 'the public' into health technology assessment and coverage policy decisions: from principles to practice. *Health Policy* 2007 June; 82(1): 37-50. NRCBL: 5.3; 5.2; 9.4; 9.1. Identifiers: Canada.

Brassington, Iain. On Heidegger, medicine, and the modernity of modern medical technology. *Medicine, Health Care and Philosophy* 2007 June; 10(2): 185-195. NRCBL: 5.3; 1.1. SC: an.
Abstract: This paper examines medicine's use of technology in a manner from a standpoint inspired by Heidegger's thinking on technology. In the first part of the paper, I shall suggest an interpretation of Heidegger's thinking on the topic, and attempt to show why he associates modern technology with danger. However, I shall also claim that there is little evidence that medicine's appropriation of modern technology is dangerous in Heidegger's sense, although there is no prima facie reason why it mightn't be. The explanation for this, I claim, is ethical. There is an initial attraction to the thought that Heidegger's thought echoes Kantian moral thinking, but I shall dismiss this. Instead, I shall suggest that the considerations that make modern technology dangerous for Heidegger are simply not in the character - the ethos - of

medicine properly understood. This is because there is a distinction to be drawn between chronological and historical modernity, and that even up-to-date medicine, empowered by technology, retains in its ethos crucial aspects of a historically pre-modern understanding of technology. A large part of the latter half of the paper will be concerned with explaining the difference.

Brody, Baruch. Intellectual property and biotechnology: the European debate. *Kennedy Institute of Ethics Journal* 2007 June; 17(2): 69-110. 48 refs. NRCBL: 5.3; 15.8; 4.4; 15.1; 22.2; 18.5.4; 21.1. SC: rv.

> Keywords: *biotechnology; *genetic patents; *moral policy; *patents; *policy making; advisory committees; body parts and fluids; commodification; cryopreservation; embryo research; embryonic stem cells; genetic research; genetic screening; genetically modified animals; historical aspects; industry; human body; human dignity; legal aspects; life; methods; property rights; public policy; sex preselection; sperm; trends; Proposed Keywords: mice; Keyword Identifiers: *European Patent Organization; *European Union; Europe; Council of Europe; European Parliament; Twentieth Century; Twenty-First Century

Abstract: The European patent system allows for the introduction of moral issues into decisions about the granting of patents. This feature has greatly impacted European debates about the patenting of biotechnology. This essay explores the European experience, in both the European Union and the European Patent Organization. It argues that there has been great confusion surrounding these issues primarily because the Europeans have not developed a general theory about when exclusion from patentability is the best social mechanism for dealing with morally offensive technologies.

Calabrese, Edward J. Elliott's ethics of expertise proposal and application: a dangerous precedent. *Science and Engineering Ethics* 2007 June; 13(2): 139-145. NRCBL: 5.3; 1.3.9; 1.3.7; 5.2; 8.3.1.

Campbell, Eric G.; Weissman, Joel S.; Ehringhaus, Susan; Rao, Sowmya R.; Moy, Beverly; Feibelmann, Sandra; Goold, Susan Dorr. Institutional academic-industry relationships. *JAMA: The Journal of the American Medical Association* 2007 October 17; 298(15): 1779-1786. NRCBL: 5.3; 7.3; 7.2. SC: em.

Abstract: CONTEXT: Institutional academic-industry relationships have the potential of creating institutional conflicts of interest. To date there are no empirical data to support the establishment and evaluation of institutional policies and practices related to managing these relationships. OBJECTIVE: To conduct a national survey of department chairs about the nature, extent, and consequences of institutional-academic industry relationships for medical schools and teaching hospitals. DESIGN, SETTING, AND PARTICIPANTS: National survey of department chairs in the 125 accredited allopathic medical schools and the 15 largest independent teaching hospitals in the United States, administered between February 2006 and October 2006. MAIN OUTCOME MEASURE: Types of relationships with industry. RESULTS: A total of 459 of 688 eligible department chairs completed the survey, yielding an overall response rate of 67%. Almost two-thirds (60%) of department chairs had some form of personal relationship with industry, including serving as a consultant (27%), a member of a scientific advisory board (27%), a paid speaker (14%), an officer (7%), a founder (9%), or a member of the board of directors (11%). Two-thirds (67%) of departments as administrative units had relationships with industry. Clinical departments were more likely than nonclinical departments to receive research equipment (17% vs 10%, P = .04), unrestricted funds (19% vs 3%, P .001), residency or fellowship training support (37% vs 2%, P .001), and continuing medial education support (65% vs 3%, P .001). However, nonclinical departments were more likely to receive funding from intellectual property licensing (27% vs 16%, P = .01). More than two-thirds of chairs perceived that having a relationship with industry had no effect on their professional activities, 72% viewed a chair's engaging in more than 1 industry-related activity (substantial role in a start-up company, consulting, or serving on a company's board) as having a negative impact on a department's ability to conduct independent unbiased research. CONCLUSION: Overall, institutional academic-industry relationships are highly prevalent and underscore the need for their active disclosure and management.

Corbellini, Gilberto. Scientists, bioethics and democracy: the Italian case and its meanings. *Journal of Medical Ethics* 2007 June; 33(6): 349-352. 4 refs. NRCBL: 5.3; 1.3.5; 1.3.9; 14.1; 15.1. SC: le.

> Keywords: *biotechnology; *democracy; *freedom; *government regulation; *politics; *science; animal cloning; bioethical issues; bioethics; embryo research; embryonic stem cells; evolution; genetic engineering; genetically modified organisms; information dissemination; legal aspects; political systems; preimplantation diagnosis; reproductive technologies; social impact; values; Western World; Proposed Keywords: technology; Keyword Identifiers: *Italy

Abstract: In June 2005, Italy held a referendum on repealing the law on medically assisted fertilization (Law 40/2004), which limits access to artificial reproduction to infertile couples, and prohibits the donation of gametes, the cryopreservation of embryos, preimplantation genetic diagnosis (PDG), and research on human embryos. The referendum was invalidated, and the law remained unchanged. The Italian political bioethical debate on assisted reproduction was manipulated by the Catholic Church, which distorted scientific data and issues at stake with the help of Catholic politicians and bioethicists. What happened in Italy shows that some perverse sociocultural political mechanisms are spreading the absurd and anti-historical view that scientific and technological advancements are threatening democracy and personal freedom. Scientists should not only contrast the political attempts at limiting freedom of scientific research, but also tell politicians, humanists and citizens that the invention of Western science with its view of scientific community as an "open society", contributed and still contributes, through scientific education, to the construction and maintaining of the moral and political values underlying Western democracies.

NRCBL: National Reference Center for Bioethics Literature Classification Scheme   See inside front cover for terms.

Dixon, Bernard. What do we need to say to each other? *New Science* 2007 January 6-12; 193(2585): 46-47. NRCBL: 5.3; 15.1.

Eaton, Lynne. Medical school accepts tobacco company funding for research [news]. *BMJ: British Medical Journal* 2007 March 10; 334(7592): 496. NRCBL: 5.3; 9.3.1; 1.3.9; 1.3.2; 9.5.9. Identifiers: University of Virginia.

Faunce, Thomas Alured. Global intellectual property protection of "innovative" pharmaceuticals: challenges for bioethics and health law. *In:* Bennett, Belinda; Tomossy, George F., eds. Globalization and Health: Challenges for Health Law and Bioethics. Dordrecht: Springer, 2006: 87-107. NRCBL: 5.3; 9.7; 9.3.1; 21.1; 9.1. SC: le. Identifiers: Trade related aspects of intellectual property rights (TRIPS).

Fielder, John. Ethics and the FDA. *IEEE Engineering in Medicine and Biology Magazine* 2006 July-August; 25(4): 13-17. NRCBL: 5.3; 9.1; 7.1.

Fisher, Morris A. Medicine and industry: a necessary but conflicted relationship. *Perspectives in Biology and Medicine* 2007 Winter; 50(1): 1-6. NRCBL: 5.3; 7.1; 9.7; 1.3.2; 1.3.9; 4.1.2.

Fossett, James W. Federalism by necessity: state and private support for human embryonic stem cell research. Rockefeller Institute Policy Brief 2007 August 9: 1-13 [Online]. Accessed: http://www.rockinst.org/WorkArea/showcontent.aspx?id=12064 [2007 October 25]. 24 fn. NRCBL: 5.3; 18.5.4; 15.1. SC: em; le.
  Keywords: *embryo research; *embryonic stem cells; *government financing; *private sector; *research support; *state government; federal government; government regulation; public policy; statistics; stem cells; Keyword Identifiers: *United States

Freckelton, Ian. Health practitioner regulation: emerging patterns and challenges for the age of globalization. *In:* Bennett, Belinda; Tomossy, George F., eds. Globalization and Health: Challenges for Health Law and Bioethics. Dordrecht: Springer, 2006: 187-206. NRCBL: 5.3; 7.3; 7.4; 18.1; 4.1.2; 4.1.1; 21.1. SC: le.

Gorovitz, Samuel. The past, present and future of human nature. *In:* Galston, Arthur W.; Peppard, Christiana Z., eds. Expanding Horizons in Bioethics. Dordrecht; Norwell, MA: Springer, 2005: 3-18. 25 refs. NRCBL: 5.3; 1.1; 2.1; 15.1.
  Keywords: *biotechnology; *human characteristics; advisory committees; autonomy; bioethical issues; biomedical enhancement; embryo research; evolution; forecasting; genetic determinism; genetic engineering; philosophy; public health; public policy; regulation; risks and benefits; science; stem cells; theology; uncertainty; values; Keyword Identifiers: New York State Task Force on Life and the Law; United States

Grimm, David. UC balks at campus-wide ban on tobacco money for research [news]. *Science* 2007 January 26;

315(5811): 447-448. NRCBL: 5.3; 1.3.3; 9.5.9. Identifiers: University of California.

Hoffmann, George R. Letter to the editor on ethics of expertise, informed consent, and hormesis. *Science and Engineering Ethics* 2007 June; 13(2): 135-137. NRCBL: 5.3; 8.3.1; 1.3.9.

Hsieh, Nien-hê. Property rights in crisis: managers and rescue. *In:* Santoro, Michael A.; Gorrie, Thomas M., eds. Ethics and the Pharmaceutical Industry. Cambridge; New York: Cambridge University Press, 2005: 379-385. NRCBL: 5.3; 9.5.6; 9.7; 21.1.

Irwin, Alan. The global context for risk governance: national regulatory policy in an international framework. *In:* Bennett, Belinda; Tomossy, George F., eds. Globalization and Health: Challenges for Health Law and Bioethics. Dordrecht: Springer, 2006: 71-85. NRCBL: 5.3; 5.2; 21.1; 15.1; 1.3.11. SC: cs; le. Identifiers: Great Britain (United Kingdom).

Jizba, Laurel. Ethics in grant funded academia: issues and questions. *Journal of Information Ethics* 2007 Spring; 16(1): 42-52. NRCBL: 5.3; 1.3.9; 1.3.3.

Kitcher, Phillip. Scientific research — who should govern? *NanoEthics* 2007 December; 1(3): 177-184. 8 fn. NRCBL: 5.3; 1.3.9; 15.1; 21.1. SC: an.
  Keywords: *democracy; *professional autonomy; *research priorities; *researchers; *science; *social control; biomedical research; decision making; economics; freedom; genetic research; government regulation; international aspects; policy making; self regulation; technical expertise; Keyword Identifiers: United States

Moran, Gordon. Rubber stamp-type decisions for funding of academic research: paradigms and conflicts of interest. *Journal of Information Ethics* 2007 Spring; 16(1): 53-58. NRCBL: 5.3; 1.3.9; 9.5.6.

Morgan, Derek. Regulating the bio-economy: a preliminary assessment of biotechnology and law. *In:* Bennett, Belinda; Tomossy, George F., eds. Globalization and Health: Challenges for Health Law and Bioethics. Dordrecht: Springer, 2006: 59-69. NRCBL: 5.3; 5.2; 21.1; 1.3.2. SC: le.

National Institutes of Health [NIH]. (United States). Department of Health and Human Services [DHHS]. Plan for implementation of Executive Order 13435: expanding approved stem cell lines in ethically responsible ways. Bethesda, MD: National Institutes of Health 2007 September 18; 25 p. [Online]. Accessed: http://stemcells.nih.gov/staticresources/policy/eo13435.pdf [2007 October 11]. NRCBL: 5.3; 18.5.4; 19.1; 15.1.

Persson, Anders; Hemlin, Sven; Welin, Stellan. Profitable exchanges for scientists: the case of Swedish human embryonic stem cell research. *Health Care Analysis: An International Journal of Health Philosophy and Policy*

2007 December; 15(4): 291-304. 30 refs. NRCBL: 5.3; 1.3.9; 18.5.4; 15.1; 9.3.1.

Keywords: *biomedical research; *embryo research; *embryonic stem cells; *industry; *research support; *researchers; authorship; biotechnology; cell lines; embryo disposition; entrepreneurship; genetic materials; genetic patents; motivation; politics; professional autonomy; property rights; research ethics committees; universities; Proposed Keywords: technology transfer; Keyword Identifiers: *Sweden; Cell Therapeutics; Ovacell

Abstract: In this article two inter-related issues concerning the ongoing commercialisation of biomedical research are analyzed. One aim is to explain how scientists and clinicians at Swedish public institutions can make profits, both commercially and scientifically, by controlling rare human biological material, like embryos and embryonic stem cell lines. This control in no way presupposes legal ownership or other property rights as an initial condition. We show how ethically sensitive material (embryos and stem cell lines) have been used in Sweden as a foundation for a commercial stem cell enterprise — despite all official Swedish strictures against commercialisation in this area. We also show how political decisions may amplify the value of controlling this kind of biological material. Another aim of the article is to analyze and discuss the meaning of this kind of academic commercial enterprise in a wider context of research funding strategies. A conclusion that is drawn is that the academic turn to commercial funding sources is dependent on the decline of public funding.

Stossel, Thomas P. Regulation of financial conflicts of interest in medical practice and medical research: a damaging solution in search of a problem. *Perspectives in Biology and Medicine* 2007 Winter; 50(1): 54-71. NRCBL: 5.3; 1.3.2; 1.3.9.

Tanne, Janice Hopkins. Group asks US institutes to reveal industry ties [news]. *BMJ: British Medical Journal* 2007 January 20; 334(7585): 115. NRCBL: 5.3; 7.3; 1.3.9; 9.7; 9.3.1.

Tucker, Jonathan B.; Hooper, Craig. Protein engineering: security implications; the increasing ability to manipulate protein toxins for hostile purposes has prompted calls for regulation. *EMBO Reports* 2006 July; 7 Special No: S14-S17. NRCBL: 5.3; 21.3; 1.3.9.

van Aken, Jan. When risk outweighs benefit. Dual-use research needs a scientifically sound risk-benefit analysis and legally binding biosecurity measures. *EMBO Reports* 2006 July; 7 Special No: S10-S13. 16 refs. NRCBL: 5.3; 21.3; 15.1.

Keywords: *biological sciences; *biological warfare; *bioterrorism; genomics; government regulation; influenza; international aspects; microbiology; publishing; recombinant DNA research; review committees; risk; risks and benefits; Proposed Keywords: crime; Keyword Identifiers: Biological and Toxin Weapons Convention

Waltz, Emily. Supreme Court boosts licensees in biotech patent battles [news]. *Nature Biotechnology* 2007 March;

25(3): 264-265. NRCBL: 5.3; 15.8. SC: le. Identifiers: MedImmune v. Genentech.

Keywords: *biotechnology; *legal aspects; *patents; *Supreme Court decisions; industry; universities; Proposed Keywords: licensure; Keyword Identifiers: *MedImmune v. Genentech; *United States

Weiss, Rick. House passes bill relaxing limits on stem cell research. *Washington Post* 2007 January 12; p. A4. NRCBL: 5.3; 15.1; 18.5.4. SC: po; le.

Keywords: *embryo research; *embryonic stem cells; *legislation; research support; government financing; politics; *public policy; embryo; federal government; Keyword Identifiers: *United States; *U.S. House of Representatives

BIOMEDICAL TECHNOLOGIES *See* ENHANCEMENT; GENETIC ENGINEERING AND BIOTECHNOLOGY; ORGAN AND TISSUE TRANSPLANTATION; REPRODUCTIVE TECHNOLOGIES

BLACKS AS RESEARCH SUBJECTS *See* HUMAN EXPERIMENTATION/ SPECIAL POPULATIONS

BLOOD BANKING, DONATION, AND TRANSFUSION
See also ORGAN AND TISSUE TRANSPLANTATION

Paid vs. unpaid donors. *Vox Sanguinis* 2006 January; 90(1): 63-70. NRCBL: 19.4; 9.3.1; 21.1; 19.5.

Chan, T.; Eckert, K.; Venesoen, P.; Leslie, K.; Chin-Yee, I. Consenting to blood: what do patients remember? *Transfusion Medicine* 2005 December; 15(6): 461-466. NRCBL: 19.4; 8.3.1; 8.1. SC: em.

Choudhury, L.P.; Tetali, S. Ethical challenges in voluntary blood donation in Kerala, India. *Journal of Medical Ethics* 2007 March; 33(3): 140-142. NRCBL: 19.4; 19.5.

Abstract: The National Blood Policy in India relies heavily on voluntary blood donors, as they are usually assumed to be associated with low levels of transfusion-transmitted infections (TTIs). In India, it is mandatory to test every unit of blood collected for hepatitis B, hepatitis C, HIV/AIDS, syphilis and malaria. Donors come to the blood bank with altruistic intentions. If donors test positive to any of the five infections, their blood is discarded. Although the blood policy advocates disclosure of TTI status, donors are not, in practice, informed about their results. The onus is on the donor to contact the blood bank. Out of approximately 16 000 donations in the past 2 years, 438 tested positive for TTI, including 107 for HIV. Only 20% of the donors contacted the blood bank; none of them were HIV positive. Disclosure by blood banks of TTI status by telephone or mail has resulted in serious consequences for some donors. Health providers face an ethical dilemma, in the absence of proper mechanisms in place for disclosure of test results, regarding notification to donors who may test positive but remain ignorant of

NRCBL: National Reference Center for Bioethics Literature Classification Scheme   See inside front cover for terms.

77

their TTI status. Given the high cost of neglecting to notify infected donors, the authors strongly recommend the use of rapid tests before collecting blood, instead of the current practice, which takes 3 h to obtain results, and disclosure of results directly to the donor by a counsellor, to avoid dropouts and to ensure confidentiality.

Gunning, Jennifer. Umbilical cord cell banking: a surprisingly controversial issue. *In:* Gunning, Jennifer; Holm, Søren, eds. Ethics, law, and society. Volume 2. Aldershot, Hants, England; Burlington, VT: Ashgate, 2006: 17-25. NRCBL: 19.4; 4.4; 18.5.2.

Gunning, Jennifer. Umbilical cord cell banking: an issue of self-interest versus altruism. *Medicine and Law: The World Association for Medical Law* 2007 December; 26(4): 769-780. NRCBL: 19.4; 19.5; 4.4.
Abstract: Stem cells from umbilical cord blood probably now form one of the most commonly banked types of human tissue. Originally stored for the treatment of haematological disorders these stem cells have now been found to be more versatile, even pluripotent, with potential for use in the treatment of a broader range of disorders and diseases and may be particularly valuable in cell therapy and regenerative medicine. This has led to the promotion of private storage of cord blood cells for autologous or family use and a rapidly growing private sector involvement. There is a growing tension between public and private banking and a number of ethical issues continue to be debated involving questions of regulation and quality assurance, ownership and commercialisation, and patenting. This paper aims to investigate some of these issues.

Kenny, Michael G. A question of blood, race, and politics. *Journal of the History of Medicine and Allied Sciences* 2006 October; 61(4): 456-491. NRCBL: 19.4; 9.5.4.

Kharaboyan, Linda; Knoppers, Bartha Maria; Avard, Denise; Nisker, Jeff. Understanding umbilical cord blood banking: what women need to know before deciding [editorial]. *Women's Health Issues* 2007 September-October; 17(5): 277-280. NRCBL: 19.4; 19.5; 9.5.5; 18.5.4; 8.4; 8.3.1.

Kuehn, Bridget M. Pediatrics group recommends public cord blood banking. *JAMA: The Journal of the American Medical Association* 2007 February 14; 297(6): 576. NRCBL: 19.4; 9.5.7. Identifiers: American Academy of Pediatrics.

McLachlan, Hugh V.; Swales, J. Kim. Altruism and blood donation. *In their:* From the Womb to the Tomb: Issues in Medical Ethics. Glasgow, Scotland: Humming Earth, 2007: 171-186. NRCBL: 19.4; 9.3.1; 19.5. SC: an.

Plant; Margo; Knoppers, Bartha Maria. Umbilical cord blood banking in Canada: socio-ethical and legal issues. *Health Law Journal* 2005; 13: 187-212. NRCBL: 19.4; 9.5.7; 8.3.2; 18.3; 9.5.4. SC: le; an.

Rapport, F.L.; Maggs, C.J. Titmuss and the gift relationship: altruism revisited. *Journal of Advanced Nursing* 2002 December; 40(5): 495-503. NRCBL: 19.4; 19.5.

Templeton, Allan; Braude, Peter. Umbilical cord blood banking and the RCOG. *Lancet* 2007 March 31-April 6; 369(9567): 1077. NRCBL: 19.4; 9.5.7; 9.8. Identifiers: Royal College of Obstetricians and Gynaecologists.

CAPITAL PUNISHMENT

Amnesty International. Execution by lethal injection — a quarter century of state poisoning. London: Amnesty International: 2007 October 4: 41 p. [Online]. Accessed: http://web.amnesty.org/library/pdf/ACT500072007ENGLISH/$File/ACT5000707.pdf [2007 October 9]. NRCBL: 20.6. SC: em. Identifiers: Great Britain (United Kingdom).

Black, Lee; Sade, Robert M. Lethal injection and physicians: state law vs medical ethics. *JAMA: The Journal of the American Medical Association* 2007 December 19; 298(23): 2779-2781. NRCBL: 20.6; 4.1.2; 1.3.1. SC: le.

Emery, Theo. U.S. judge blocks lethal injection in Tennessee. *New York Times* 2007 September 20; p. A14. NRCBL: 20.6. SC: po.

Kamerow, Douglas. Killing me softly. *BMJ: British Medical Journal* 2007 March 3; 334(7591): 454. NRCBL: 20.6; 4.1.2. Identifiers: lethal injection.

Koniaris, Leonidas G.; Sheldon, Jon P.; Zimmers, Teresa A. Can lethal injection for execution really be "fixed"? [commentary]. *Lancet* 2007 February 3-9; 369(9559): 352-353. NRCBL: 20.6; 9.7; 8.1.

Liang, Bryan A. Special doctor's docket. Lethal injection: policy considerations for medicine. *Journal of Clinical Anesthesia* 2006 September; 18(6): 466-470. NRCBL: 20.6; 7.1.

Liptak, Adam. Florida panel urges steps for painless executions. *New York Times* 2007 March 2; p. A12. NRCBL: 20.6. SC: po.

Matthews, Daryl; Wendler, Sheila. Ethical issues in the evaluation and treatment of death row inmates. *Current Opinion in Psychiatry* 2006 September; 19(5): 518-521. NRCBL: 20.6; 1.3.5; 9.5.1.

North Carolina Medical Board. Capital Punishment [position statement]. North Carolina Medical Board, adopted 2007 January; [Online]. Accessed: http://www.ncmedboard.org/Clients/NCBOM/Public/PublicMedia/capitalpunishment.htm [2007 February 9]. NRCBL: 20.6. SC: le.

Scott, Charles L. Psychiatry and the death penalty. *Psychiatric Clinics of North America* 2006 September; 29(3): 791-804. NRCBL: 20.6; 17.2; 7.1.

Varelius, Jukka. Execution by lethal injection, euthanasia, organ-donation and the proper goals of medicine. *Bioethics* 2007 March; 21(3): 140-149. NRCBL: 20.6; 9.7; 19.5; 20.5.1. SC: an.

Abstract: In a recent issue of this journal, David Silver and Gerald Dworkin discuss the physicians' role in execution by lethal injection. Dworkin concludes that discussion by stating that, at that point, he is unable to think of an acceptable set of moral principles to support the view that it is illegitimate for physicians to participate in execution by lethal injection that would not rule out certain other plausible moral judgements, namely that euthanasia is under certain conditions legitimate and that organ-donation surgery is sometimes permissible. This article draws attention to some problems in the views of Silver and Dworkin and suggests moral principles which support the three moral views just mentioned.

Weil, Elizabeth. The needle and the damage done: lethal injections are often botched and sometimes painful. Doctors don't want to administer them. Is it time to kill this form of execution? *New York Times Magazine* 2007 February 11; p. 46-51. NRCBL: 20.6. SC: po.

White, Caroline. Doctors who give lethal injections should be punished, says Amnesty [news]. *BMJ: British Medical Journal* 2007 October 6; 335(7622): 690. NRCBL: 20.6; 4.1.2. Identifiers: Amnesty International.

CARE FOR SPECIFIC GROUPS

See also HUMAN EXPERIMENTATION/ SPECIAL POPULATIONS

Access to health care for undocumented migrants in Europe [editorial]. *Lancet* 2007 December 22-2008 January 4; 370(9605): 2070. NRCBL: 9.5.1; 9.2; 4.1.2; 1.3.1.

HPV vaccine (Gardasil). *Health Care Ethics USA* 2007 Winter; 15(1): 15-16. NRCBL: 9.5.1; 9.7; 1.2; 9.5.5.

Abel, Gregory A.; Penson, Richard T.; Joffe, Steven; Schapira, Lidia; Chabner, Bruce A.; Lynch, Thomas J., Jr. Direct-to-consumer advertising in oncology. *Oncologist* 2006 February; 11(2): 217-226. NRCBL: 9.5.1; 9.7; 1.3.2; 8.1; 5.3.

Amundson, Ron; Tresky, Shari. On a bioethical challenge to disability rights. *Journal of Medicine and Philosophy* 2007 November-December; 32(6): 541-561. NRCBL: 9.5.1; 1.1; 2.1; 15.1.

Abstract: Tensions exist between the disability rights movement and the work of many bioethicists. These reveal themselves in a major recent book on bioethics and genetics, From Chance to Choice: Genetics and Justice. This book defends certain genetic policies against criticisms from disability rights advocates, in part by arguing that it is possible to accept both the genetic policies and the rights of people with impairments. However, a close reading of the book reveals a series of direct moral criticisms of the disability rights movement. The criticisms go beyond a defense of genetic policies from the criticisms of disability rights advocates. The disability rights movement is said not to have the same moral legitimacy as other civil rights movements, such as those for women or "racial" minorities. This paper documents, and in some cases shows the flaws within, these challenges to the disability rights movement.

Annas, George J. Cancer and the constitution — choice at life's end. *New England Journal of Medicine* 2007 July 26; 357(4):408-413. NRCBL: 9.5.1; 9.7. SC: an; le. Identifiers: Abigail Alliance; Food and Drug Administration.

Arber, Sara; McKinlay, John; Adams, Ann; Marceau, Lisa; Link, Carol; O'Donnell, Amy. Patient characteristics and inequalities in doctors' diagnostic and management strategies relating to CHD: a video-simulation experiment. *Social Science and Medicine* 2006 January; 62(1): 103-115. NRCBL: 9.5.1; 9.2; 10; 9.5.4; 9.5.2. SC: em. Identifiers: coronary heart disease.

Berlin, Jordan; Bruinooge, Suanna S.; Tannock, Ian F. Ethics in oncology: consulting for the investment industry. *Journal of Clinical Oncology* 2007 February 1; 25(4): 444-446. NRCBL: 9.5.1; 9.6; 7.1; 9.3.1; 18.2.

Bhandari, Mohit; Jönsson, Anders; Bühren, Volker. Conducting industry-partnered trials in orthopaedic surgery. *Injury* 2006 April; 37(4): 361-366. NRCBL: 9.5.1; 1.3.2; 9.7; 9.3.1.

Bowers, Libby. Ethical issues along the cancer continuum. *In:* Carroll-Johnson, Rose Mary; Gorman, Linda M.; Bush, Nancy Jo, eds. Psychosocial Nursing Care Along the Cancer Continuum. 2nd edition. Pittsburgh, PA: Oncology Nursing Society, 2006: 551-564. NRCBL: 9.5.1; 8.1; 20.5.1; 20.7; 9.4; 4.1.3.

Caplan, Arthur. Scared to death [op-ed]. *Free Inquiry* 2007 June-July; 27(4): 26. NRCBL: 9.5.1; 9.7; 9.3.1; 9.5.5.

Chiang, Hsien-Hsien; Chen, Mei-Bih; Sue, I-Ling. Self-state of nurses in caring for SARS survivors. *Nursing Ethics* 2007 January; 14(1): 18-26. NRCBL: 9.5.1; 7.1. SC: em.

Abstract: The aim of this study was to analyze nurses' experiences of role strain when taking care of patients with severe acute respiratory syndrome (SARS). We adopted an interpretive/ constructivist paradigm. Twenty-one nurses who had taken care of SARS patients were interviewed in focus groups. The data were analyzed using thematic analysis. The self-state of nurses during the SARS outbreak evolved into that of professional self as: (1) self-preservation; (2) self-mirroring; and (3) self-transcendence. The relationship between self-state and reflective practice is discussed.

Chou, Ann F.; Brown, Arleen F.; Jensen, Roxanne E.; Shih, Sarah; Pawlson, Greg; Scholle, Sarah Hudson. Gender and racial disparities in the management of diabetes mellitus among Medicare patients. *Women's Health Is-*

NRCBL: National Reference Center for Bioethics Literature Classification Scheme   See inside front cover for terms.

79

sues 2007 May-June; 17(3): 150-161. NRCBL: 9.5.1; 9.5.4; 9.5.5; 9.3.2. SC: em.

Cohen, J.S.; Erickson, J.M. Ethical dilemmas and moral distress in oncology nursing practice. *Clinical Journal of Oncology Nursing* 2006 December; 10(6): 775-780. NRCBL: 9.5.1; 4.1.3; 8.1.

Cole, Phillip. Human rights and the national interest: migrants, healthcare and social justice. *Journal of Medical Ethics* 2007 May; 33(5): 269-272. NRCBL: 9.5.1; 9.2; 21.1.

Abstract: The UK government has recently taken steps to exclude certain groups of migrants from free treatment under the National Health Service, most controversially from treatment for HIV. Whether this discrimination can have any coherent ethical basis is questioned in this paper. The exclusion of migrants of any status from any welfare system cannot be ethically justified because the distinction between citizens and migrants cannot be an ethical one.

Daneault, Serge; Lussier, Véronique; Mongeau, Suzanne; Hudon, Eveline; Paillé, Pierre; Dion, Dominique; Yelle, Louise. Primum non nocere: could the health care system contribute to suffering? In-depth study from the perspective of terminally ill cancer patients. *Canadian Family Physician* 2006 December; 52(12): 1575 e.1-5. NRCBL: 9.5.1; 20.5.1; 4.4. Note: Abstract in French.

Davidovitch, Nadav; Filc, Dani. Reconstructing data: evidence-based medicine and evidence-based public health in context. *Dynamis* 2006; 26: 287-306. NRCBL: 9.5.1; 9.1; 4.1.2; 7.1.

Day, Lisa. Family involvement in critical care: shortcomings of a utilitarian justification. *American Journal of Critical Care* 2006 March; 15(2): 223-225. NRCBL: 9.5.1; 8.1; 1.1.

de Vries, Jantina. The obesity epidemic: medical and ethical considerations. *Science and Engineering Ethics* 2007 March; 13(1): 55-67. NRCBL: 9.5.1; 9.5.7.

Dodds, Susan. Depending on care: recognition of vulnerability and the social contribution of care provision. *Bioethics* 2007 November; 21(9): 500-510. NRCBL: 9.5.1; 9.5.2; 10. SC: an.

Abstract: People who are paid to provide basic care for others are frequently undervalued, exploited and expected to reach often unrealistic standards of care. I argue that appropriate social recognition, support and fair pay for people who provide care for those who are disabled, frail and aged, or suffering ill health that impedes their capacity to negotiate daily activities without support, depends on a reconsideration of the paradigm of the citizen or and moral agent. I argue that by drawing on the ideas of human vulnerability and dependency as central to our personhood, a more realistic conception of selves, citizens and persons can be developed that better recognises the inevitability of human dependency and the social value of care work. I also indicate the significance of this

vulnerability-focussed view for ethical evaluation of the emotional aspects of care relationships.

Dubler, Nancy Neveloff; Kalkut, Gary E. Caring for VIPs in the hospital: the ethical thicket. *Israel Medical Association Journal* 2006 November; 8(11): 746-750. NRCBL: 9.5.1; 9.4; 8.4; 9.8; 20.5.1. SC: cs. Identifiers: Mickey Mantle.

Edwards, Steven D. Disablement and personal identity. *Medicine, Health Care and Philosophy* 2007 June; 10(2): 209-215. NRCBL: 9.5.1; 1.1; 4.4. SC: an.

Abstract: A number of commentators claim their disability to be a part of their identity. This claim can be labelled 'the identity claim'. It is the claim that disabling characteristics of persons can be identity-constituting. According to a central constraint on traditional discussions of personal identity over time, only essential properties can count as identity-constituting properties. By this constraint, contingent properties of persons (those they might not have instanced) cannot be identity-constituting. Viewed through the lens of traditional approaches to the problem of personal identity over time, disablement is most likely to be regarded as a contingent property of a person and not an essential one. Hence, on traditional approaches, the identity claim must be false. An alternative account of identity is sketched here. It is one which exploits the idea of narrative identity, and points to five basic features of personal existence. When accounts of identity are structured in relation to these five features, it is argued, disablement can be shown to be identity-constituting, and hence the identity claim can be accepted.

Ee, Pei-Lee; Kempen, Paul M. Elective surgery days after myocardial infarction: clinical and ethical considerations. *Journal of Clinical Anesthesia* 2006 August; 18(5): 363-366. NRCBL: 9.5.1; 8.1; 8.3.1. SC: cs.

Erde, Edmund L. Indecency/decency in cardiac surgery: a memoir of my education at a super-esteemed medical place. *Journal of Cardiac Surgery* 2007 January-February; 22(1): 43-48. NRCBL: 9.5.1; 7.1; 8.1; 8.3.1; 9.8.

Fiege, Angela B. Resident portfolio: a tale of two women. *Academic Emergency Medicine* 2006 September; 13(9): 989-990; discussion 990-992. NRCBL: 9.5.1; 9.5.5; 14.1; 12.5.1.

Fox, Marie; Thomson, Michael. Short changed? The law and ethics of male circumcision. *In:* Freeman, Michael, ed. Children's Health and Children's Rights. Boston: Martinus Nijhoff Publishers, 2006: 161-181. NRCBL: 9.5.1; 10; 21.1; 21.7. SC: le.

Gostin, Lawrence O.; DeAngelis, Catherine D. Mandatory HPV vaccination: public health vs. private wealth [editorial]. *JAMA: The Journal of the American Medical Association* 2007 May 2; 297(17): 1921-1923. NRCBL: 9.5.1; 9.5.5; 9.5.7; 9.7. SC: an.

Greif, Karen F.; Merz, Jon F. Emerging diseases: SARS and government responses. *In their:* Current Controversies in the Biological Sciences: Case Studies of Policy Challenges from New Technologies. Cambridge, MA: MIT, 2007: 255-266. NRCBL: 9.5.1; 9.1. Identifiers: severe acute respiratory syndrome.

Grewel, Hans. Behinderung und Philosophie: Ethik-Konzepte auf dem Prüfstand. *In:* Neuer-Miebach, Therese; Wunder, Michael, eds. Bio-Ethik und die Zukunft der Medizin. Bonn: Psychiatrie-Verlag, 1998: 87-105. NRCBL: 9.5.1; 1.1; 1.2.

Hawryluck, Laura. Ethics review: position papers and policies — are they really helpful to front-line ICU teams? *Critical Care* 2006; 10(6): 242. NRCBL: 9.5.1; 6. SC: rv. Identifiers: Canada.

Hinchley, Geoff; Patrick, Kirsten. Is infant male circumcision an abuse of the rights of the child? [debate]. *BMJ: British Medical Journal* 2007 December 8; 335(7631): 1180-1181. NRCBL: 9.5.1; 1.2; 10; 21.1; 1.2.

Hyatt, Adam. Medicinal marijuana and palliative care: carving a liberty interest out of the Glucksberg framework. *Fordham Urban Law Journal* 2006 November; 33(5): 1345-1367. NRCBL: 9.5.1; 4.4; 9.5.9; 9.7. SC: le. Identifiers: Compassionate Use Act of 1996 (California).

Infante, Peter F. The past suppression of industry knowledge of the toxicity of benzene to humans and potential bias in future benzene research. *International Journal of Occupational and Environmental Health* 2006 July-September; 12(3): 268-272. NRCBL: 9.5.1; 1.3.2; 8.3.1; 8.2; 16.1.

Jasen, Patricia. Breast cancer and the politics of abortion in the United States. *Medical History* 2005 October; 49(4): 423-444. NRCBL: 9.5.1; 9.5.5; 12.5.2. SC: rv.

Jones, James W.; McCullough, Laurence B.; Richman, Bruce W. Other people's money: ethics, finances, and bad outcomes. *Journal of Vascular Surgery* 2006 April; 43(4): 863-865. NRCBL: 9.5.1; 9.3.1; 1.3.2.

Kahn, Jeffrey. Baseball, alcohol and public health. *American Journal of Bioethics* 2007 July; 7(7): 3. NRCBL: 9.5.1; 9.5.9; 9.1.

Ladd, Paddy. Cochlear implantation, colonialism, and deaf rights. *In:* Komesaroff, Linda, ed. Surgical Consent: Bioethics and Cochlear Implantation. Washington, DC: Gallaudet University Press, 2007: 1-29. NRCBL: 9.5.1; 9.5.7; 15.5. Identifiers: Great Britain (United Kingdom).

Lane, Harlan. Ethnicity, ethics, and the deaf-world. *In:* Komesaroff, Linda, ed. Surgical Consent: Bioethics and Cochlear Implantation. Washington, DC: Gallaudet University Press, 2007: 42-69. NRCBL: 9.5.1; 21.7; 7.1; 15.5.

Liao, Lih Mei; Creighton, Sarah M. Requests for cosmetic genitoplasty: how should healthcare providers respond? *BMJ: British Medical Journal* 2007 May 26; 334(7603): 1090-1092. NRCBL: 9.5.1; 4.5; 10.

Lo, Bernard. Human papillomavirus vaccination programmes. *BMJ: British Medical Journal* 2007 August 25; 335(7616): 357-358. NRCBL: 9.5.1; 9.7; 9.5.5; 9.5.7.

Lyren, Anne; Leonard, Ethan. Vaccine refusal: issues for the primary care physician. *Clinical Pediatrics* 2006 June; 45(5): 399-404. NRCBL: 9.5.1; 9.7; 8.3.2; 8.3.4.

McCarthy, Patrick M.; Lamm, Richard D.; Sade, Robert M. Medical ethics collides with public policy: LVAD for a patient with leukemia. *Annals of Thoracic Surgery* 2005 September; 80(3): 793-798. NRCBL: 9.5.1; 9.2; 9.3.1; 9.4. Identifiers: Left Ventricular Assist Device.

McGee, Glenn; Johnson, Summer. Has the spread of HPV vaccine marketing conveyed immunity to common sense? [editorial]. *American Journal of Bioethics* 2007 July; 7(7): 1-2. NRCBL: 9.5.1; 9.3.1; 1.3.2; 9.1; 9.7.

McKneally, Martin F. Managing expectations and fear: invited commentary on "Indecency in cardiac surgery: a memoir of my education at a Super-Esteemed Medical Place (SEMP)," by Dr. Edmund Erde. *Journal of Cardiac Surgery* 2007 January-February; 22(1): 49-50. NRCBL: 9.5.1; 7.1; 8.1; 8.3.1; 9.8.

Memtsoudis, Stavros G.; Besculides, Melanie C.; Swamidoss, Cephas P. Do race, gender, and source of payment impact on anesthetic technique for inguinal hernia repair? *Journal of Clinical Anesthesia* 2006 August; 18(5): 328-333. NRCBL: 9.5.1; 9.3.1; 9.2; 9.8.

Mukherjee, Debjani; Levin, Rebecca L.; Heller, Wendy. The cognitive, emotional, and social sequelae of stroke: psychological and ethical concerns in post-stroke adaptation. *Topics in Stroke Rehabilitation* 2006 Fall; 13(4): 26-35. NRCBL: 9.5.1; 17.1; 4.4.

Nelson, W.; Pomerantz, A.; Howard, K.; Bushy, A. A proposed rural healthcare ethics agenda. *Journal of Medical Ethics* 2007 March; 33(3): 136-139. NRCBL: 9.5.1. SC: em. Identifiers: Coalition for Rural Health Care Ethics. Abstract: The unique context of the rural setting provides special challenges to furnishing ethical healthcare to its approximately 62 million inhabitants. Although rural communities are widely diverse, most have the following common features: limited economic resources, shared values, reduced health status, limited availability of and accessibility to healthcare services, overlapping professional-patient relationships and care giver stress. These rural features shape common healthcare ethical issues, including threats to confidentiality, boundary issues, professional-patient relationship and allocation of resources. To date, there exists a limited focus on rural healthcare ethics shown by the scarcity of rural healthcare ethics literature, rural ethics committees, rural focused ethics training and research on rural ethics issues. An interdisciplinary group of rural healthcare ethicists with backgrounds in medicine, nursing and philosophy was convened to explore the need for a rural healthcare ethics agenda. At the meeting, the Coalition for Rural Health

NRCBL: National Reference Center for Bioethics Literature Classification Scheme   See inside front cover for terms.

81

Care Ethics agreed to a definition of rural healthcare ethics and a broad-ranging rural ethics agenda with the ultimate goal of enhancing the quality of patient care in rural America. The proposed agenda calls for increasing awareness and understanding of rural healthcare ethics through the development of evidence — informed, rural-attuned research, scholarship and education in collaboration with rural healthcare professionals, healthcare institutions and the diverse rural population.

Padgett, Barry L.; Haas, Thomas. An ethical wrinkle on the face of therapy claims. *Plastic Surgical Nursing* 2004 July-September; 24(3): 123-126. NRCBL: 9.5.1; 9.7.

Redman, Barbara K. Responsibility for control; ethics of patient preparation for self-management of chronic disease. *Bioethics* 2007 June; 21(5): 243-250. NRCBL: 9.5.1; 1.1; 4.2; 8.1.
Abstract: Patient self-management (SM) of chronic disease is an evolving movement, with some forms documented as yielding important outcomes. Potential benefits from proper preparation and maintenance of patient SM skills include quality care tailored to the patient's preferences and life goals, and increase in skills in problem solving, confidence and success, generalizable to other parts of the patient's life. Four central ethical issues can be identified. 1) insufficient patient/family access to preparation that will optimize their competence to SM without harm to themselves, 2) lack of acknowledgement that an ethos of patient empowerment can mask transfer of responsibility beyond patient/family competency to handle that responsibility, 3) prevailing assumptions that preparation for SM cannot result in harm and that its main purpose is to deliver physician instructions, and 4) lack of standards for patient selection, which has the potential to exclude individuals who could benefit from learning to SM. Technology assessment offers one framework through which to examine available data about efficacy of patient SM and to answer the central question of what conditions must be put in place to optimize the benefits of SM while assuring that potential harms are controlled.

Schenkenberg, Thomas; Kochenour, Neil D.; Botkin, Jeffrey R. Ethical considerations in clinical care of the "VIP". *Journal of Clinical Ethics* 2007 Spring; 18(1): 56-63. NRCBL: 9.5.1; 9.8; 9.4.

Shewmon, D. Alan; De Giorgio, Christopher M. Early prognosis in anoxic coma: reliability and rationale. *Neurologic Clinics* 1989 November; 7(4): 823-843. NRCBL: 9.5.1; 20.5.1; 8.1.

Shiao, Judith Shu-Chu; Koh, David; Lo, Li-Hua; Lim, Meng-Kin; Guo, Yueliang Leon. Factors predicting nurses' consideration of leaving their job during the SARS outbreak. *Nursing Ethics* 2007 January; 14(1): 5-17. NRCBL: 9.5.1; 8.1; 7.1. SC: em. Identifiers: severe acute respiratory syndrome.
Abstract: Taiwan was affected by an outbreak of severe acute respiratory syndrome (SARS) in early 2003. A questionnaire survey was conducted to determine (1) the perceptions of risk of SARS infection in nurses; (2) the proportion of nurses considering leaving their job; and (3) work as well as non-work factors related to nurses' consideration of leaving their job because of the SARS outbreak. Nearly three quarters (71.9%) of the participants believed they were 'at great risk of exposure to SARS', 49.9% felt 'an increase in workload', and 32.4% thought that people avoided them because of their job; 7.6% of the nurses not only considered that they should not care for SARS patients but were looking for another job or considering resignation. The main predictors of nurses' consideration of leaving their job were shorter tenure, increased work stress, perceived risk of fatality from SARS, and affected social relationships. The findings are important in view of potential impending threats of pandemics such as avian influenza.

Smith, Valerie. A patient's perspective on moral issues and universal oral health care. *Journal of the American College of Dentists* 2007 Fall; 74(3): 27-31. NRCBL: 9.5.1; 4.1.1; 1.3.1; 8.1; 9.2.

Steward, Douglas O.; DeMarco, Joseph P. Rejoinder. *Journal of Bioethical Inquiry* 2006; 3(3): 137-138. NRCBL: 9.5.1; 8.3.4; 8.3.1. Comments: Response to: Malcolm Parker. Patients as rational traders: response to Stewart and DeMarco. Journal of Bioethical Inquiry 2006; 3(3): 133-136.

Sulmasy, Daniel P. Cancer care, money, and the value of life: whose justice? Which rationality? *Journal of Clinical Oncology* 2007 January 10; 25(2): 217-222. NRCBL: 9.5.1; 9.3.1; 9.2; 9.4; 7.1.

Tupasela, A. When legal worlds collide: from research to treatment in hereditary cancer prevention. *European Journal of Cancer Care* 2006 July; 15(3): 257-266. NRCBL: 9.5.1; 18.2; 18.5.9; 8.4; 18.3.

United States. Department of Defense [DOD]. Medical program support for detainee operations. Washington, DC: Department of Defense, [2310.08E], 2006 June 6; 10 p. [Online]. Accessed: http://www.dtic.mil/whs/directives/corres/pdf/ 231008p.pdf [2007 April 5]. NRCBL: 9.5.1; 1.3.5.

Van Vleet, Lee M. Between black and white. The gray area of ethics in EMS. *JEMS: A Journal of Emergency Medical Services* 2006 October; 31(10): 55-56, 58-63; quiz 64-65 [see correction in JEMS: A Journal of Emergency Medical Services 2006 December; 31(12): 16]. NRCBL: 9.5.1; 1.1; 2.1.

Vernick, William J. How long to postpone an operation after a myocardial infarction? When perioperative consultants contradict the literature, leaving the anesthesiologist in the middle. *Journal of Clinical Anesthesia* 2006 August; 18(5): 325-327. NRCBL: 9.5.1; 8.1; 8.3.1.

Vila, J.J.; Jimenez, F.J.; Inarrairaegui, M.; Prieto, C.; Nantes, O.; Borda, F. Informed consent document in gastrointestinal endoscopy: understanding and acceptance by patients. *Revista Española de Enfermedades Digestivas:*

Organo Oficial de la Sociedad Española de Patología Digestiva 2006 February; 98(2): 101-111. NRCBL: 9.5.1; 8.3.1.

Watnick, Suzanne. Obesity: a problem of Darwinian proportions? *Advances in Chronic Kidney Disease* 2006 October; 13(4): 428-432. NRCBL: 9.5.1; 8.1; 19.3; 9.4.

Weiss, Martin Meyer; Weiss, Peter D.; Weiss, Joseph B. Anthrax vaccine and public health policy. *American Journal of Public Health* 2007 November; 97(11): 1945-1951. NRCBL: 9.5.1; 9.7; 9.1.

Wynia, Matthew K. Mandating vaccination: what counts as a "mandate" in public health and when should they be used? *American Journal of Bioethics* 2007 December; 7(12): 2-6. NRCBL: 9.5.1; 9.7; 9.1.

CARE FOR SPECIFIC GROUPS/ AGED

Humane and compassionate elder care as a human right [editorial]. *Lancet* 2007 August 25-31; 370(9588): 629. NRCBL: 9.5.2; 21.1. SC: le.

Agich, George J. Reflections on the function of dignity in the context of caring for old people. *Journal of Medicine and Philosophy* 2007 September-October; 32(5): 483-494. NRCBL: 9.5.2.
    Abstract: This article accepts the proposition that old people want to be treated with dignity and that statements about dignity point to ethical duties that, if not independent of rights, at least enhance rights in ethically important ways. In contexts of policy and law, dignity can certainly have a substantive as well as rhetorical function. However, the article questions whether the concept of dignity can provide practical guidance for choosing among alternative approaches to the care of old people. The article explores the paradoxical relationship between the apparent lack of specific content in many conceptions of dignity and the broad utility that dignity appears to have as a concept expressive of shared social understandings about the status of old people.

Basta, Lofty L. Ethical issue: an oath for our time. *American Journal of Geriatric Cardiology* 2006 September-October; 15(5): 316-318. NRCBL: 9.5.2; 7.1; 8.1; 6.

Basta, Lofty L. Ethical issues in the management of geriatric cardiac patients: a 91-year old patient insists on an advance care plan that does not make sense. *American Journal of Geriatric Cardiology* 2004 September-October; 13(5): 276-277. NRCBL: 9.5.2; 9.6; 9.5.1.

Beaulieu, Marie; Leclerc, Nancy. Ethical and psycho-social issues raised by the practice in cases of mistreatment of older adults. *Journal of Gerontological Social Work* 2006 May; 46(3-4): 161-186. NRCBL: 9.5.2; 9.1; 4.1.2; 1.3.1; 8.1.

Boisaubin, Eugene V.; Chu, Adeline; Catalano, Janine M. Perceptions of long-term care, autonomy, and dignity, by residents, family and care-givers: the Houston experi-ence. *Journal of Medicine and Philosophy* 2007 September-October; 32(5): 447-464. NRCBL: 9.5.2; 9.5.1. SC: em.
    Abstract: Houston, Texas, is a major U.S. city with, like many, a growing aging population. The purpose of this study and ultimate book chapter is to explore the views and perceptions of long-term care (LTC) residents, family members and health care providers. Individuals primarily in independent living and group residential settings were interviewed and studied. Questions emphasized the concepts of personal autonomy, dignity, quality and location of care and decision making. Although a small sample of participants were involved, consistency was noted. Keeping the elderly in caring and loving home situations (theirs or family) was most preferred. Personal choice and independence were emphasized by residents, but family members needed to act as advocates. We also noted that the legal system emphasizes family control over individual decision making as competency declines with aging. Optimal personal decision making in the residents' best interest also became more difficult with loss of individual mental capacity

Chan, Ho Mun; Pang, Sam. Long-term care: dignity, autonomy, family integrity, and social sustainability: the Hong Kong experience. *Journal of Medicine and Philosophy* 2007 September-October; 32(5): 401-424. NRCBL: 9.5.2; 9.5.1.
    Abstract: This article reveals the outcome of a study on the perceptions of elders, family members, and healthcare professionals and administration providing care in a range of different long-term care facilities in Hong Kong with primary focus on the concepts of autonomy and dignity of elders, quality and location of care, decision making, and financing of long term care. It was found that aging in place and family care were considered the best approaches to long term care insofar as procuring and balancing the values of dignity, autonomy, family integrity and social sustainability were concerned. An elder having the final say was generally accepted. The results also initiated the importance of sharing of financial responsibility among elders, children and government albeit the emphasis was placed on individuals. Furthermore, dignity of elders was not considered purely a synonym of autonomy, but it had also to do with respect, family and social connections.

Cohen, Eric; Kass, Leon R. Cast me not off in old age. *Commentary* 2006 January; 121(1): 32-39. NRCBL: 9.5.2; 20.5.1; 20.5.4; 9.4; 20.3.3. Identifiers: euthanasia; assisted suicide.

Duhigg, Charles. At many homes, more profit and less nursing; insulated from lawsuits, private investors cut costs and staff. *New York Times* 2007 September 23; p. A1, A34, A35. NRCBL: 9.5.2; 9.3.1; 1.3.2. SC: po.

Elson, Paul. Do older adults presenting with memory complaints wish to be told if later diagnosed with Alzheimer's disease? *International Journal of Geriatric Psychiatry* 2006 May; 21(5): 419-425. NRCBL: 9.5.2; 8.3.1; 8.2; 17.1.

NRCBL: National Reference Center for Bioethics Literature Classification Scheme   See inside front cover for terms.

83

Eltis, Karen. Predicating dignity on autonomy? The need for further inquiry into the ethics of tagging and tracking dementia patients with GPS technology. *Elder Law Journal* 2005; 13(2): 387-415. NRCBL: 9.5.2; 17.1; 8.3.3; 4.3; 1.1. SC: le.

Engelhardt, H. Tristram. Long-term care: the family, post-modernity, and conflicting moral life-worlds. *Journal of Medicine and Philosophy* 2007 September-October; 32(5): 519-536. NRCBL: 9.5.2; 9.5.1.

Abstract: Long-term care is controversial because it involves foundational disputes. Some are moral-economic, bearing on whether the individual, the family, or the state is primarily responsible for long-term care, as well as on how one can establish a morally and financially sustainable long-term-care policy, given the moral hazard of people over-using entitlements once established, the political hazard of media democracies promising unfundable entitlements, the demographic hazard of relatively fewer workers to support those in need of long-term care, the moral hazard to responsibility of shifting accountability to third parties, and the bureaucratic hazard of moving from individual and family choice to bureaucratic oversight. These disputes are compounded by controversies regarding the nature of the family (Is it to be regarded primarily as a socio-biological category, a fundamental ontological category of social reality, or a construct resulting from the consent of the participants?), as well as its legal and moral autonomy and authority over its members. As the disputes show, there is no common understanding of respect and human dignity that will easily lead out of these disputes. The reflections on long-term care in this issue underscore the plurality of moralities defining bioethics.

Fahey, Charles J. The ethics of long-term care: recasting the policy discourse. *In:* Pruchno, Rachel A.; Smyer, Michael A., eds. Challenges of an Aging Society: Ethical Dilemmas, Political Issues. Baltimore: Johns Hopkins University Press, 2007: 52-73. NRCBL: 9.5.2; 9.3.1.

Fan, Ruiping. Which care? Whose responsibility? And why family? A Confucian account of long-term care for the elderly. *Journal of Medicine and Philosophy* 2007 September-October; 32(5): 495-517. NRCBL: 9.5.2; 1.2. Identifiers: China.

Abstract: Across the world, socio-economic forces are shifting the locus of long-term care from the family to institutional settings, producing significant moral, not just financial costs. This essay explores these costs and the distortions in the role of the family they involve. These reflections offer grounds for critically questioning the extent to which moral concerns regarding long-term care in Hong Kong and in mainland China are the same as those voiced in the United States, although family resemblances surely exist. Chinese moral values such as virtue and filial piety embedded in a Confucian moral and social context cannot be recast without distortion in terms of modern Western European notions. The essay concludes that the Confucian resources must be taken seriously in order to develop an authentic Chinese bioethics of long-term care and a defensible approach to long-term care policy for contemporary society in general and Chinese society in particular.

Gross, Jane. Aging and gay, and facing prejudice in twilight. *New York Times* 2007 October 9; p. A1, A25. NRCBL: 9.5.2; 10.

Häggström, Elisabeth; Kihlgren, Annica. Experiences of caregivers and relatives in public nursing homes. *Nursing Ethics* 2007 September; 14(5): 691-701. NRCBL: 9.5.2; 8.1. SC: em. Identifiers: Sweden.

Abstract: The aim of the present study was, by means of discussion highlighting ethical questions and moral reasonings, to increase understanding of the situations of caregivers and relatives of older persons living in a public nursing home in Sweden. The findings show that these circumstances can be better understood by considering two different perspectives: an individual perspective, which focuses on the direct contact that occurs among older people, caregivers and relatives; and a societal perspective, which focuses on the norms, values, rules and laws that govern a society. Relatives and caregivers thought that the politicians were sending out mixed messages: they were praising caregivers and relatives for their efforts, but at the same time the public health care sector was subjected to significant cutbacks in resources. Both caregivers and relatives were dissatisfied and frustrated with the present situation regarding the care of older persons in public nursing homes.

Illes, J.; Rosen, A.; Greicius, M.; Racine, E. Prospects for prediction: ethics analysis of neuroimaging in Alzheimer's disease. *Annals of the New York Academy of Sciences* 2007 February; 1097: 278-295. NRCBL: 9.5.2; 17.1; 9.5.1.

Lipman, Hannah I. Informed consent. *American Journal of Geriatric Cardiology* 2007 January-February; 16(1): 42-43. NRCBL: 9.5.2; 8.1; 8.3.1; 9.4.

Marik, Paul E. Should age limit admission to the intensive care unit? *American Journal of Hospice and Palliative Care* 2007 February-March; 24(1): 63-66. NRCBL: 9.5.2; 9.2; 9.4.

Mason, J.K.; Laurie, G.T. Treatment of the aged. *In their:* Mason and McCall Smith's Law and Medical Ethics. Seventh ed. Oxford; New York: Oxford University Press, 2005: 441-463. NRCBL: 9.5.2; 1.1; 8.3.4; 9.4. SC: le.

McCullough, Laurence B. Geroethics. *In:* Khushf, George, ed. Handbook of Bioethics: Taking Stock of the Field From a Philosophical Perspective. Dordrecht; Boston: Kluwer Academic, 2004: 507-523. NRCBL: 9.5.2; 1.1.

McKerlie, Dennis. Justice and the elderly. *In:* Steinbock, Bonnie, ed. The Oxford Handbook of Bioethics. Oxford; New York: Oxford University Press, 2007: 190-208. NRCBL: 9.5.2; 20.5.4; 1.1.

Moses, Sarah. A just society for the elderly: the importance of justice as participation. *Notre Dame Journal of Law, Ethics and Public Policy* 2007; 21(2): 335-362. NRCBL: 9.5.2; 1.1; 1.2; 9.4. Note: Symposium on Aging America.

Müller, Fernando Suárez. On futuristic gerontology: a philosophical evaluation of Aubrey de Grey's SENS project. *International Journal of Applied Philosophy* 2007 Fall; 21(2): 225-239. NRCBL: 9.5.2; 20.5.1; 4.5. SC: an. Identifiers: SENS (Strategies for Engineered Negligible Senescence).

O'Dowd, Adrian. Report highlights abuse of older people's human rights. *BMJ: British Medical Journal* 2007 August 25; 335(7616): 367. NRCBL: 9.5.2; 21.1. SC: le. Identifiers: Great Britain (United Kingdom).

Pérez-Cárceles, M.D.; Lorenzo, M.D.; Luna, A.; Osuna, E. Elderly patients also have rights. *Journal of Medical Ethics* 2007 December; 33(12): 712-716. NRCBL: 9.5.2; 8.3.1; 8.4; 8.2. SC: em. Identifiers: Spain.
  Abstract: BACKGROUND: Sharing information with relatives of elderly patients in primary care and in hospital has to fit into the complex set of obligations, justifications and pressures concerning the provision of information, and the results of some studies point to the need for further empirical studies exploring issues of patient autonomy, privacy and informed consent in the day-to-day care of older people. OBJECTIVES: To know the frequency with which "capable" patients over 65 years of age receive information when admitted to hospital, the information offered to the families concerned, the person who gives consent for medical intervention, and the degree of satisfaction with the information received and the healthcare provided. Method: A descriptive questionnaire given to 200 patients and 200 relatives during the patients' stay in hospital. RESULTS: Only 5% of patients confirmed that they had been asked whether information could be given to their relatives. A significantly higher proportion of relatives received information on the successive stages of the care offered than did patients themselves. As the age of the patients increased, so the number who were given information, understood the information and were asked for their consent for complementary tests decreased. The degree of satisfaction with the information offered was high for both patients and relatives (86.5% and 84%, respectively), despite the irregularities observed. CONCLUSIONS: The capacity of elderly patients to participate in the decision-making process is frequently doubted simply because they have reached a certain age and it is thought that relatives should act as their representatives. In Spain, the opinion of the family and doctors appears to play a larger role in making decisions than does the concept of patient autonomy.

Pishchita, A.N. Elderly patients as a vulnerable category of the population requiring special legal protection with respect to the provision of medical care. *European Journal of Health Law* 2007 December; 14(4): 349-354. NRCBL: 9.5.2; 8.3.3. SC: le. Identifiers: Russia.

Post, Stephen G. The aging society and the expansion of senility: biotechnological and treatment goals. *In:* Steinbock, Bonnie, ed. The Oxford Handbook of Bioethics. Oxford; New York: Oxford University Press, 2007: 304-323. NRCBL: 9.5.2; 20.3.1; 8.2; 20.5.1.

Proot, Ireen M.; ter Meulen, Ruud H.J.; Abu-Saad, Huda Huijer; Crebolder, Harry F.J.M. Supporting stroke patients' autonomy during rehabilitation. *Nursing Ethics* 2007 March; 14(2): 229-241. NRCBL: 9.5.2; 1.1. SC: em. Identifiers: Netherlands.
  Abstract: In a qualitative study, 22 stroke patients undergoing rehabilitation in three nursing homes were interviewed about constraints on and improvements in their autonomy and about approaches of health professionals regarding autonomy. The data were analysed using grounded theory, with a particular focus on the process of regaining autonomy. An approach by the health professionals that was responsive to changes in the patients' autonomy was found to be helpful for restoration of their autonomy. Two patterns in health professionals' approach appeared to be facilitatory: (1) from full support on admission through moderate support and supervision, to reduced supervision at discharge; and (2) from paternalism on admission through partial paternalism (regarding treatment) to shared decision making at discharge. The approach experienced by the patients did not always match their desires regarding their autonomy. Support and supervision were reduced over time, but paternalism was often continued too long. Additionally, the patients experienced a lack of information. Tailoring interventions to patients' progress in autonomy would stimulate their active participation in rehabilitation and in decision making, and would improve patients' preparation for autonomous living after discharge.

Robinson, L.; Hutchings, D.; Corner, L.; Beyer, F.; Dickinson, H.; Vanoli, A.; Finch, T.; Hughes, J.; Ballard, C.; May, C.; Bond, J. A systematic literature review of the effectiveness of non-pharmacological interventions to prevent wandering in dementia and evaluations of the ethical implications and acceptability of their use. *Health Technology Assessment* 2006 August; 10(26): iii-108. NRCBL: 9.5.2; 9.5.3; 17.3. SC: rv.

Sammet, Kai. Autonomy or protection from harm? Judgements of German courts on care for the elderly in nursing homes. *Journal of Medical Ethics* 2007 September; 33(9): 534-537. NRCBL: 9.5.2; 1.1; 17.3; 9.3.1. SC: le; cs.
  Abstract: The increase in life expectancy in developed countries has lead to an increase in the number of elderly people cared for in nursing homes. Given the physical frailty and deterioration of mental capacities in many of these residents, questions arise as to their autonomy and to their protection from harm. In 2005, one of the highest German courts, the Bundesgerichtshof (BGH) issued a seminal judgement that dealt with the obligations of nursing homes and with the preserving of autonomy and privacy in nursing home residents. An elderly woman had sustained a fracture of the neck of the femur during a fall. The health insurance company held that the nursing home

NRCBL: National Reference Center for Bioethics Literature Classification Scheme   See inside front cover for terms.

85

had breached its obligations to protect her from falling and sued the home for the hospital costs of her treatment. However, the BGH maintained that the case of the health insurance was not justified. It held that obligations of nursing homes have to be limited to normal arrangements within reasonable financial and personal effort, and that the autonomy of residents had to be protected from unnecessary interference. Permanent control or even restraining measures to reduce each risk would deprive the patient fully of her autonomy, and must therefore be hindered. Other judgments of other courts have emphasised the "protectionist" approach. The article deals with these different approaches and comments on both rationales. It will be shown that both approaches must be differentiated to establish fully autonomy and protection for nursing home residents.

Shim, Janet K.; Russ, Ann J.; Kaufman, Sharon R. Risk, life extension and the pursuit of medical possibility. *Sociology of Health and Illness* 2006 May; 28(4): 479-502. NRCBL: 9.5.2; 5.2; 4.4; 20.4.1.

Sloane, Peter; DeRenzo, Evan G. The case of Mr. A.B. *Journal of Clinical Ethics* 2007 Winter; 18(4): 399-401. NRCBL: 9.5.2; 8.3.1; 8.3.3; 20.3.3; 8.2. SC: cs.

Swidler, Robert N.; Seastrum, Terese; Shelton, Wayne. Difficult hospital inpatient discharge decisions: ethical, legal and clinical practice issues. *American Journal of Bioethics* 2007 March; 7(3): 23-28. NRCBL: 9.5.2; 9.2; 1.1; 9.3.1; 8.1. SC: le.

Vanlaere, Linus; Bouckaert, Filip; Gastmans, Chris. Care for suicidal older people: current clinical-ethical considerations. *Journal of Medical Ethics* 2007 July; 33(7): 376-381. NRCBL: 9.5.2; 20.7.
Abstract: This article opens by reviewing the state of the knowledge on the most current worldwide facts about suicide in older people. Next, a number of values that have a role in this problem are considered. Having a clear and current understanding of suicide and of the related self-held and social values forms the framework for a number of clinical–ethical recommendations for care practice. An important aspect of caring for older people with suicidal tendencies is to determine whether their primary care fosters self-esteem and affirms their dignity. In addition to providing a timely and appropriate diagnosis and treatment of suicidality, the caregiver is responsible for helping the patient to cope with stressful conditions, and for treating the patient with respect and consideration, thereby supporting the patient's dignity and giving the patient a reason to live. Paying attention to these central points will foster caring contact with suicidal older people.

Wah, Julia Tao Lai Po. Dignity in long-term care for older persons: a Confucian perspective. *Journal of Medicine and Philosophy* 2007 September-October; 32(5): 465-481. NRCBL: 9.5.2; 1.2. Identifiers: China.
Abstract: This article presents Mencius' concept of human dignity in the Chinese Confucian moral tradition, focused on the context of long-term care. The double nature of Mencius' notion of human dignity as an intrinsic quality of human beings qua being human is analyzed and contrasted with the dominant Western account of human dignity as grounded in personhood. Drawing on the heuristic force of an interview with an elder person in Hong Kong, the insights of the Mencian theory of human dignity are used to provide a moral foundation for long-term care for elder persons in a context of diminishing personhood and shrinking autonomy.

Wah, Julia Tao Lai Po; Chan, Ho Mun; Fan, Ruiping. Exploring the bioethics of long-term care. *Journal of Medicine and Philosophy* 2007 September-October; 32(5): 395-399. NRCBL: 9.5.2; 9.5.1.

Wells, Joseph K. Ethical dilemma and resolution: a case scenario. *Indian Journal of Medical Ethics* 2007 January-March; 4(1): 31-33. NRCBL: 9.5.2; 9.3.1; 1.1. SC: cs.

Werntoft, Elisbet; Hallberg, Ingalill R.; Edberg, Anna-Karin. Older people's reasoning about age-related prioritization in health care. *Nursing Ethics* 2007 May; 14(3): 399-412. NRCBL: 9.5.2; 9.4. SC: em. Identifiers: Sweden.
Abstract: The aim of this study was to describe the reasoning of people aged 60 years and over about prioritization in health care with regard to age and willingness to pay. Healthy people (n = 300) and people receiving continuous care and services (n = 146) who were between 60 and 101 years old were interviewed about their views on prioritization in health care. The transcribed interviews were analysed using manifest and latent qualitative content analysis. The participants' reasoning on prioritization embraced eight categories: feeling secure and confident in the health care system; being old means low priority; prioritization causes worries; using underhand means in order to be prioritized; prioritization as a necessity; being averse to anyone having precedence over others; having doubts about the distribution of resources; and buying treatment requires wealth.

Williams, Monique M. Invisible, unequal, and forgotten: health disparities in the elderly. *Notre Dame Journal of Law, Ethics and Public Policy* 2007; 21(2): 441-478. NRCBL: 9.5.2; 9.8; 9.2; 7.1.

Woolley, Douglas C.; Medvene, Louis J.; Kellerman, Rick D.; Base, Michelle; Mosack, Victoria. Do residents want automated external defibrillators in their retirement home? *Journal of the American Medical Directors Association* 2006 March; 7(3): 135-140. NRCBL: 9.5.2; 9.7. SC: em.

Yates, Ferdinand D., Jr. Holding the hospital hostage. *American Journal of Bioethics* 2007 March; 7(3): 36-37. NRCBL: 9.5.2; 1.1; 8.1; 9.1. Comments: Robert N. Swidler, Terese Seastrum, and Wayne Shelton. Difficult hospital inpatient discharge decisions: ethical, legal and clinical practice issues. American Journal of Bioethics 2007 March; 7(3): 23-28.

Zhai, Xiaomei; Qiu, Ren Zong. Perceptions of long-term care, autonomy, and dignity, by residents, family and caregivers: the Beijing experience. *Journal of Medicine and Philosophy* 2007 September-October; 32(5): 425-445. NRCBL: 9.5.2; 9.5.1. SC: em. Identifiers: China.

Abstract: This article documents the results of a study on the perceptions of long-term elder care in Beijing in the People's Republic of China by those most intimately involved. The study asked a sample of elderly, family members, and health care professionals, all of whom are involved in care at a variety of long-term care facilities in Beijing, about their perceptions of the care given at these facilities from their particular standpoints as regards issues such as the quality and ideal location of care, decision-making regarding the care receiving, who should be responsible for the financing of care, and the meaning of dignity for the elderly in these facilities. The results showed adherence to traditional family values at least on one level regarding the ideal location of care being with the family and in the home, but also the desire to pass on financing of long-term care facilities and the health care they provide the elderly on to the government. These results are not altogether surprising, but they also clearly demonstrate the larger conflict between traditional views about morality and economic considerations regarding health care financing in the China.

CARE FOR SPECIFIC GROUPS/ FETUSES
See also HUMAN EXPERIMENTATION/ SPECIAL POPULATIONS/ EMBRYOS AND FETUSES

Addelson, Kathryn Pyne. The emergence of the fetus. *In:* Mui, Constance L.; Murphy, Julien S., eds. Gender Struggles: Practical Approaches to Contemporary Feminism. Lanham, MD: Rowman & Littlefield Pub., 2002: 118-136. NRCBL: 9.5.8; 9.5.5; 10; 12.5.1.

American Pediatric Surgical Association Ethics and Advocacy Committee; Fallat, Mary E.; Caniano, Donna A.; Fecteau, Annie H. Ethics and the pediatric surgeon. *Journal of Pediatric Surgery* 2007 January; 42(1): 129-136; discussion 136. NRCBL: 9.5.8; 8.3.2; 9.5.1. SC: em.

Casper, Monica J. Fetal surgery then and now: there is too much emphasis on the fetus and not enough on the woman. *Conscience* 2007 Autumn; 28(3): 24-27. NRCBL: 9.5.8; 9.5.5.

Erikson, Susan L. Fetal views: histories and habits of looking at the fetus in Germany. *Journal of Medical Humanities* 2007 December; 28(4): 187-212. 28 refs. 34 fn. NRCBL: 9.5.8; 9.5.5; 7.1; 4.4; 14.1; 15.2.

Keywords: *fetuses; *historical aspects; *obstetrics; *prenatal diagnosis; attitudes; biomedical technologies; international aspects; mandatory programs; mass screening; metaphor; medicine; National Socialism; pregnant women; trends; Proposed Keywords: anatomy; photography; ultrasonography; Keyword Identifiers: *Germany; Fifteenth Century; Sixteenth Century; Seventeenth Century; Eighteenth Century; Nineteenth Century; Twentieth Century; Twenty-First Century

Abstract: This article examines historical and ideological trajectories that have made looking at the fetus via ultrasound a normal part of being pregnant for many women around the world. How did looking into so unlit a bodily space as the uterus become so natural? So everyday? So habit-forming? The answers lie in the convergence over time of technological hardware with knowledge practices that moved from medical to public domains. Germany serves as a site for an interrogation of how learned ways of thinking about anatomy, the development of technologies that "look," a privileging of the visual in medical domains, and seeing as metaphor for truth about health reinforced and normalized prenatal ultrasound use.

Ethics Group of the Newborn Drug Development Initiative; Baer, Gerri R.; Nelson, Robert M. Ethical challenges in neonatal research: summary report of the ethics group of the newborn drug development initiative. *Clinical Therapeutics* 2006 September; 28(9): 1399-1407. NRCBL: 9.5.8; 9.7; 18.2; 18.5.4.

Sayeed, Sadath A. Legal challenges at the limits of viability. *Medical Ethics Newsletter [Lahey Clinic]* 2007 Spring; 14(2): 6-8. NRCBL: 9.5.8; 4.4. SC: le.

Williams, Clare. Dilemmas in fetal medicine: premature application of technology or responding to women's choice? *Sociology of Health and Illness* 2006 January; 28(1): 1-20. NRCBL: 9.5.8; 9.5.5; 15.3; 5.1. SC: em. Identifiers: Great Britain (United Kingdom).

CARE FOR SPECIFIC GROUPS/ INDIGENTS

Feder, Judith; Pollitz, Karen. Reform's three essential elements: to be effective, insurance coverage must be adequate, affordable, and available. *Health Progress* 2007 May-June; 88(3): 30-31. NRCBL: 9.5.10; 9.3.1; 9.1.

Jones, James W.; McCullough, Laurence B.; Richman, Bruce W. My brother's keeper: uncompensated care for illegal immigrants. *Journal of Vascular Surgery* 2006 September; 44(3): 679-682. NRCBL: 9.5.10; 9.3.1; 8.1.

Kershaw, Sarah. U.S. rule limits emergency care for immigrants; shift in chemotherapy; New York used money under Medicaid for illegal residents. *New York Times* 2007 September 22; p. A1,A11. NRCBL: 9.5.10; 9.5.4; 9.3.1. SC: po; le.

Wen, Chuck K.; Hudak, Pamela L.; Hwang, Stephan W. Homeless people's perceptions of welcomeness and unwelcomeness in healthcare encounters. *JGIM: Journal of General Internal Medicine* 2007 July; 22(7): 1011-1017. NRCBL: 9.5.10. SC: em. Identifiers: Canada.

CARE FOR SPECIFIC GROUPS/ MENTALLY DISABLED
See also BEHAVIOR CONTROL; ELECTROCONVULSIVE THERAPY; INVOLUNTARY COMMITMENT; MENTAL HEALTH THERA-

NRCBL: National Reference Center for Bioethics Literature Classification Scheme   See inside front cover for terms.

87

PIES AND NEUROSCIENCES; PSYCHOPHAR-
MACOLOGY; PSYCHOTHERAPY

Ashley's treatment: unethical or compassionate? [edito-rial]. *Lancet* 2007 January 13-19; 369(9556): 80. NRCBL: 9.5.3; 8.3.2; 4.4.

Forever young: can stunting the growth of a disabled child ever be a good thing? [editorial]. *New Scientist* 2007 January 13-19; 193(2586): 3. NRCBL: 9.5.3; 9.5.7; 9.7.

Treatments to keep disabled girl small stir debate. *Washington Post* 2007 January 5; p. A2. NRCBL: 9.5.3; 8.3.2; 4.4. SC: po. Identifiers: Ashley; growth attenuation therapy.

Appelbaum, Kenneth L. Commentary: the use of restraint and seclusion in correctional mental health. *Journal of the American Academy of Psychiatry and the Law* 2007; 35(4): 431-435. NRCBL: 9.5.3; 1.3.5; 17.1; 17.8.

Baskin, Cyndy. Part I: Conceptualizing, framing and politicizing aboriginal ethics in mental health. *Journal of Ethics in Mental Health [electronic]* 2007 November; 2(2): 5 p. Accessed: http://www.jemh.ca [2008 January 24]. NRCBL: 9.5.3; 9.5.4. Identifiers: Canada.
  Abstract: This article looks at how mental health issues are conceptualized from the lens of Aboriginal world views. It refers to the legacy of colonization and the resulting historical trauma as the root of mental health "illnesses." But it also raises questions on how definitions of "mental illnesses" are arrived at by one's world view or lens. What may be seen as a mental health problem from one world view can be seen as a positive, healing spiritual experience from another.

Baskin, Cyndy. Part II: Working together in the circle: challenges and possibilities within mental health ethics. *Journal of Ethics in Mental Health [electronic]* 2007 November; 2(2): 4 p. Accessed: http://www.jemh.ca [2008 January 24]. NRCBL: 9.5.3; 8.1; 9.5.4. Identifiers: Canada.
  Abstract: This article explores how ethics are framed for both Aboriginal and non-Aboriginal helpers. It examines both the challenges and the possibilities of working in the area of mental health, particularly if one is looking at the world through only a Western lens. It finishes with a brief exploration of how the two groups of helpers – Aboriginal and Western - might be able to work together ethically.

Bersani, Hank, Jr.; Rotholz, David A.; Eidelman, Steven M.; Pierson, Joanna L.; Bradley, Valerie J.; Gomez, Sharon C.; Havercamp, Susan M.; Silverman, Wayne P.; Yeager, Mark H.; Morin, Diane; Wehmeyer, Michael L.; Carabello, Bernard J.; Croser, M. Doreen. Unjustifiable non-therapy: response to the issue of growth attenuation for young people on the basis of disability. *Intellectual and Developmental Disabilities* 2007 October; 45(5): 351-353. NRCBL: 9.5.3; 4.4; 8.3.2.

Buntinx, Wil H.E. Professional supports for persons with intellectual disability: products or relationships? *Ethics and Intellectual Disability* 2005 Winter; 8(2): 4-5. NRCBL: 9.5.3; 1.1.

Champion, Michael K. Commentary: seclusion and restraint in corrections — a time for change. *Journal of the American Academy of Psychiatry and the Law* 2007; 35(4): 426-430. NRCBL: 9.5.3; 1.3.5; 17.1; 9.8; 6.

Coombes, Rebecca. Ashley X: a difficult moral choice. Did the doctors and parents responsible for a severely disabled girl have the right to keep her small? *BMJ: British Medical Journal* 2007 January 13; 334(7584): 72-73. NRCBL: 9.5.3; 9.5.7.

Dawson, John; Szmukler, George. Fusion of mental health and incapacity legislation. *British Journal of Psychiatry* 2006 June; 188: 504-509. NRCBL: 9.5.3; 8.3.3; 17.7; 1.3.5.

Fontanarosa, Phil B.; Rennie, Drummond; DeAngelis, Catherine D. Access to care as a component of health system reform [editorial]. *JAMA: The Journal of the American Medical Association* 2007 March 14; 297(10): 1128-1130. NRCBL: 9.5.3.

Hale, Brenda. Justice and equality in mental health law: the European experience. *International Journal of Law and Psychiatry* 2007 January-February; 30(1): 18-28. NRCBL: 9.5.3; 9.2; 17.1; 17.8; 1.3.8; 21.1. SC: le.

Iversen, Knut Ivar; Høyer, Georg; Sexton, Harold C. Coercion and patient satisfaction on psychiatric acute wards. *International Journal of Law and Psychiatry* 2007 November-December; 30(6): 504-511. NRCBL: 9.5.3; 17.7; 17.8. SC: em.

Krahn, Gloria L.; Hammond, Laura; Turner, Anne. A cascade of disparities: health and health care access for people with intellectual disabilities. *Mental Retardation and Developmental Disabilities Research Reviews* 2006; 12(1): 70-82. NRCBL: 9.5.3; 9.2; 17.1; 21.1. SC: rv.

Ladd, Rosalind Ekman. Rights of the autistic child. *In:* Freeman, Michael, ed. Children's Health and Children's Rights. Boston: Martinus Nijhoff Publishers, 2006: 87-98. NRCBL: 9.5.3; 9.5.7; 8.3.2; 18.5.6; 21.1.

Leider, Robert. Quality of Life and Human Difference, edited by David Wasserman, Jerome Bickenbach, and Robert Wachbroit [book review]. *Ethics and Intellectual Disability* 2006 Spring; 9(2): 1, 4-5. NRCBL: 9.5.3; 4.4.

Marshall, Jessica. Operating in whose interest? *New Scientist* 2007 January 13-19; 193(2586): 6-7. NRCBL: 9.5.3; 9.5.7; 9.7.

McConville, Brad; Kelly, D. Clay. Cruel and unusual? Defining the conditions of confinement in the mentally ill. *Journal of the American Academy of Psychiatry and the*

Law 2007; 35(4): 533-534. NRCBL: 9.5.3; 1.3.5; 17.7. SC: le.

Meininger, Herman P. Embedded in professional practice: ethics at the 12th World Congress of IASSID. *Ethics and Intellectual Disability* 2005 Winter; 8(2): 1-3. NRCBL: 9.5.3; 1.1. Identifiers: International Association for the Scientific Study of Intellectual Disability.

Metzner, Jeffrey L.; Tardiff, Kenneth; Lion, John; Reid, William H.; Recupero, Patricia Ryan; Schetky, Diane H.; Edenfield, Bruce M.; Mattson, Marlin; Janofsky, Jeffrey S. Resource document on the use of restraint and seclusion in correctional mental health care. *Journal of the American Academy of Psychiatry and the Law* 2007; 35(4): 417-425. NRCBL: 9.5.3; 1.3.5; 17.1; 6.

Newsom, Robert W. Seattle syndrome: comments on the reaction to Ashley X. *Nursing Philosophy* 2007 October; 8(4): 291-294. NRCBL: 9.5.3; 9.5.7; 4.4.

O'Grady, John C. Commentary: a British perspective on the use of restraint and seclusion in correctional mental health care. *Journal of the American Academy of Psychiatry and the Law* 2007; 35(4): 439-443. NRCBL: 9.5.3; 1.3.5; 17.1; 17.8; 21.1.

Parkes, Georgina; Hall, Ian. Gender dysphoria and cross-dressing in people with intellectual disability: a literature review. *Mental Retardation* 2006 August; 44(4): 260-271. NRCBL: 9.5.3; 10; 8.3.1. SC: em.

Patfield, Martyn. The 'mentally disordered' provisions of the New South Wales Mental Health Act 1990: their ethical standing and effect on services. *Australasian Psychiatry* 2006 September; 14(3): 263-266. NRCBL: 9.5.3; 17.7; 17.8. SC: le.

Perske, Robert. The "big bang" theory and Down syndrome. *Ethics and Intellectual Disability* 2007 Winter; 10(1): 1, 4-6. NRCBL: 9.5.3.

Prior, Pauline M. Mentally disordered offenders and the European Court of Human Rights. *International Journal of Law and Psychiatry* 2007 November-December; 30(6): 546-557. NRCBL: 9.5.3; 1.3.5; 21.1; 17.7; 17.8. SC: le.

Robertson, Michael. Part I: psychiatrists and social justice — the concept of justice. *Journal of Ethics in Mental Health [electronic]* 2007 November; 2(2): 5 p. Accessed: http://www.jemh.ca [2008 January 24]. NRCBL: 9.5.3; 1.1; 1.3.1; 4.1.2; 9.4.
Abstract: These two papers consider the concept of social justice and the ethical obligations psychiatrists may have in its regard. In this first paper, the concept of social justice is defined in terms of the successful function of the social contract. Basic conceptions of justice are then considered.

Robertson, Michael. Part II: psychiatrists and social justice – when the social contract fails. *Journal of Ethics in Mental Health [electronic]* 2007 November; 2(2): 4 p. Accessed: http://www.jemh.ca [2008 January 24]. NRCBL: 9.5.3; 1.1; 1.3.1; 4.1.2; 9.4.
Abstract: This second paper explores psychiatrists' ethical obligations in the face of the failure of the social contract – inherent failures in distributive justice, the failure of the sovereign and the reconstitution of the social contract in post-conflict societies. Such situations present many sources of ethical tension between the professional ethical obligations of psychiatrists to their individual patients and to their society.

Sales, Becky; McKenzie, Nigel. Time to act on behalf of mentally disordered offenders. *BMJ: British Medical Journal* 2007 June 9; 334(7605): 1222. NRCBL: 9.5.3; 1.3.5.

Schmidt, Eric B. Making someone child-sized forever: ethical considerations in inhibiting the growth of a developmentally disabled child. *Clinical Ethics* 2007 March; 2(1): 46-49. NRCBL: 9.5.3; 4.4; 8.3.2; 9.5.7. SC: an.
Abstract: In a recent case, parents of a profoundly developmentally disabled child asked physicians to use high-dose oestrogen to inhibit the growth of their child in the interests of allowing better care of her as she ages. The physicians asked whether such an intervention would be ethically acceptable. Such an intervention would seem to violate the rights of the child to bodily integrity and to normal growth, making the intervention ethically objectionable. But in this paper, I argue that in some rare instances, a developmentally disabled child may have only a minimal right against interference with her growth. In those instances, parents may be acting ethically if they use medical interventions to inhibit the growth of their child for the purposes of facilitating better care. But they may so intervene only when the child's disabilities are so profound that the child has no personal interest in developing an adult size and when the intervention is the least intrusive means available for facilitating the care of the child.

Singer, Peter. A convenient truth [op-ed]. *New York Times* 2007 January 26; p. A21. NRCBL: 9.5.3; 4.4; 8.3.2. SC: po. Identifiers: Ashley; attenuated growth therapy.

Stansfield, Alison J.; Holland, A.J.; Clare, I.C.H. The sterilisation of people with intellectual disabilities in England and Wales during the period 1988 to 1999. *Journal of Intellectual Disability Research* 2007 August; 51(8): 569-570. NRCBL: 9.5.3; 11.3; 8.3.3; 2.2.

Vlach, David L.; Daniel, Anasseril. Commentary: evolving toward equivalency in correctional mental health care — a view from the maximum security trenches. *Journal of the American Academy of Psychiatry and the Law* 2007; 35(4): 436-438. NRCBL: 9.5.3; 1.3.5; 17.1; 17.8; 9.8.

Wikler, Daniel. Paternalism and the mildly retarded. *Philosophy and Public Affairs* 1979 Summer; 8(4): 377-392. NRCBL: 9.5.3; 1.1.

Wong, Sophia Isako. The moral personhood of individuals labeled "mentally retarded": a Rawlsian response to

NRCBL: National Reference Center for Bioethics Literature Classification Scheme   See inside front cover for terms.

89

Nussbaum. *Social Theory and Practice* 2007 October; 33(4): 579-594. NRCBL: 9.5.3; 4.4; 1.1.

CARE FOR SPECIFIC GROUPS/ MINORITIES

Armstron, Katrina; Ravenell, Karima L.; McMurphy, Suzanne; Putt, Mary. Racial/ethnic differences in physician distrust in the United States. *American Journal of Public Health* 2007 July; 97(7): 1282-1289. NRCBL: 9.5.4; 8.1. SC: em.

Abstract: OBJECTIVES: We examined the racial/ethnic and geographic variation in distrust of physicians in the United States. METHODS: We obtained data from the Community Tracking Study, analyzing 20 sites where at least 5% of the population was Hispanic and 5% was Black. RESULTS: In univariate analyses, Blacks and Hispanics reported higher levels of physician distrust than did Whites. Multivariate analyses, however, suggested a complex interaction among sociodemographic variables, city of residence, race/ethnicity, and distrust of physician. In general, lower socioeconomic status (defined as lower income, lower education, and no health insurance) was associated with higher levels of distrust, with men generally reporting more distrust than women. But the strength of these effects was modified by race/ethnicity. We present examples of individual cities in which Blacks reported consistently higher mean levels of distrust than did Whites, consistently lower mean levels of distrust than did Whites, or a mixed relationship dependent on socioeconomic status. In the same cities, Hispanics reported either consistently higher mean levels of distrust relative to Whites or a mixed relationship. CONCLUSIONS: Racial/ethnic differences in physician distrust are less uniform than previously hypothesized, with substantial geographic and individual variation present.

Bailey, James E.; Sprabery, Laura R. Inequitable funding may cause health care disparities [editorial]. *Archives of Internal Medicine* 2007 June 25; 167(12): 1226-1228. NRCBL: 9.5.4; 9.3.1.

Borry, Pascal; Schotsmans, Paul; Dierickx, Kris. Evidence-based medicine and its role in ethical decision-making. *Journal of Evaluation in Clinical Practice* 2006 June; 12(3): 306-311. NRCBL: 9.5.4; 9.1; 9.8.

Brown, Samuel L. Health policy and the politics of health care for African Americans. *In:* Livingston, Ivor Lensworth, ed. Praeger Handbook of Black American Health: Policies and Issues Behind Disparities in Health, Vol. II. 2nd edition. Westport, CT: Praeger Publishers, 2004: 685-700. NRCBL: 9.5.4; 9.2; 9.1. SC: le.

Casagrande, Sarah Stark; Gary, Tiffany L.; LaVeist, Thomas A.; Gaskin, Darrell J.; Cooper, Lisa A. Perceived discrimination and adherence to medical care in a racially integrated community. *JGIM: Journal of General Internal Medicine* 2007 March; 22(3): 389-395. NRCBL: 9.5.4. SC: em.

Dula, Annette. Whitewashing black health: lies, deceptions, assumptions, and assertions — and the disparities continue. *In:* Prograis, Lawrence J.; Pellegrino, Edmund D., eds. African American Bioethics: Culture, Race, and Identity. Washington, DC: Georgetown University Press, 2004: 47-65. NRCBL: 9.5.4; 9.4; 9.2. Conference: Symposium on African American Perspectives in Bioethics and Second Annual Conference on Health Disparities, held on September 23-24, 2004, at Georgetown University.

Duster, Troy. Medicalisation of race. *Lancet* 2007 February 24-March 2; 369(9562): 702-704. NRCBL: 9.5.4; 15.9; 9.7; 5.3; 7.1; 4.2.

Gray, Natalie; Bailie, Ross. Can human rights discourse improve the health of indigenous Australians? *Australian and New Zealand Journal of Public Health* 2006 October; 30(5): 448-452. NRCBL: 9.5.4; 9.5.10; 8.1; 21.1; 9.1.

Green, Alexander R.; Carney, Dana R.; Pallin, Daniel J.; Ngo, Long H.; Raymond, Kristal L.; Iezzoni, Lisa I.; Banaji, Mahzarin R. Implicit bias among physicians and its prediction of thrombolysis decisions for black and white patients. *JGIM: Journal of General Internal Medicine* 2007 September; 22(9): 1231-1238. NRCBL: 9.5.4; 9.7. SC: em.

Hasnain-Wynia, Romana; Baker, David W.; Nerenz, David; Feinglass, Joe; Beal, Anne C.; Landrum, Mary Beth; Behal, Raj; Weissman, Joel S. Disparities in health care are driven by where minority patients seek care. *Archives of Internal Medicine* 2007 June 25; 167(12): 1233-1239. NRCBL: 9.5.4; 9.8. SC: em.

Abstract: BACKGROUND: Racial/ethnic disparities in health care are well documented, but less is known about whether disparities occur within or between hospitals for specific inpatient processes of care. We assessed racial/ethnic disparities using the Hospital Quality Alliance Inpatient Quality of Care Indicators. METHODS: We performed an observational study using patient-level data for acute myocardial infarction (5 care measures), congestive heart failure (2 measures), community-acquired pneumonia (2 measures), and patient counseling (4 measures). Data were obtained from 123 hospitals reporting to the University HealthSystem Consortium from the third quarter of 2002 to the first quarter of 2005. A total of 320,970 patients 18 years or older were eligible for at least 1 of the 13 measures. RESULTS: There were consistent unadjusted differences between minority and nonminority patients in the quality of care across 8 of 13 quality measures (from 4.63 and 4.55 percentage points for angiotensin-converting enzyme inhibitors for acute myocardial infarction and congestive heart failure [P01] to 14.58 percentage points for smoking cessation counseling for pneumonia [P=.02]). Disparities were most pronounced for counseling measures. In multivariate models adjusted for individual patient characteristics and hospital effect, the magnitude of the disparities decreased substantially, yet remained significant for 3 of the 4 counseling measures; acute myocardial infarction (unadjusted, 9.00 [P001]; adjusted, 3.82 [P01]), congestive

---

heart failure (unadjusted, 8.45 [P=.02]; adjusted, 3.54 [P=.02]), and community-acquired pneumonia (unadjusted, 14.58 [P=.02]; adjusted, 4.96 [P=.01]). CONCLUSIONS: Disparities in clinical process of care measures are largely the result of differences in where minority and nonminority patients seek care. However, disparities in services requiring counseling exist within hospitals after controlling for site of care. Policies to reduce disparities should consider the underlying reasons for the disparities.

King, Patricia A. Race, equity, health policy, and the African American community. *In:* Prograis, Lawrence J.; Pellegrino, Edmund D., eds. African American Bioethics: Culture, Race, and Identity. Washington, DC: Georgetown University Press, 2004: 67-92. NRCBL: 9.5.4; 2.1; 9.4. Conference: Symposium on African American Perspectives in Bioethics and Second Annual Conference on Health Disparities, held on September 23-24, 2004, at Georgetown University.

Knudsen, Hannah K.; Ducharme, Lori J.; Roman, Paul M. Racial and ethnic disparities in SSRI availability in substance abuse treatment. *Psychiatric Services* 2007 January; 58(1): 55-62. NRCBL: 9.5.4; 9.5.9; 17.4. Identifiers: selective serotonin reuptake inhibitors.

Konotey-Ahulu, Felix I.D. Need for ethnic experts to tackle genetic public health [letter]. *Lancet* 2007 December 1-7; 370(9602): 1826-1827. NRCBL: 9.5.4; 15.3; 9.2; 15.2.

Livingston, Ivor Lensworth; Carter, J. Jacques. Eliminating racial and ethnic disparities in health: a framework for action. *In:* Livingston, Ivor Lensworth, ed. Praeger Handbook of Black American Health: Policies and Issues Behind Disparities in Health, Vol. II. 2nd edition. Westport, CT: Praeger Publishers, 2004: 835-861. NRCBL: 9.5.4; 9.1; 9.2.

Macintosh, Constance. Jurisdictional roulette: constitutional and structural barriers to aboriginal access to health. *In:* Flood, Colleen M., ed. Just Medicare: What's In, What's Out, How We Decide. Buffalo, NY: University of Toronto Press, 2006: 193-215. NRCBL: 9.5.4; 1.3.5; 9.1. SC: le. Identifiers: Canada.

Maze, Claire D. Martino. Registered nurses' willingness to serve populations on the periphery of society. *Journal of Nursing Scholarship* 2006; 38(3): 301-306. NRCBL: 9.5.4; 9.5.10; 4.1.3. SC: em.

Mehta, Pritti. Promoting equality and diversity in UK biomedical and clinical research. *Nature Reviews Genetics* 2006 September; 7(9): 668. NRCBL: 9.5.4; 9.2; 9.3.1.

Rogers, Naomi. Race and the politics of polio: Warm Springs, Tuskegee, and the March of Dimes. *American Journal of Public Health* 2007 May; 97(5): 784-795. NRCBL: 9.5.4; 7.1; 21.1.

Abstract: The Tuskegee Institute opened a polio center in 1941, funded by the March of Dimes. The center's founding was the result of a new visibility of Black polio survivors and the growing political embarrassment around the policy of the Georgia Warm Springs polio rehabilitation center, which Franklin Roosevelt had founded in the 1920s before he became president and which had maintained a Whites-only policy of admission. This policy, reflecting the ubiquitous norm of race-segregated health facilities of the era, was also sustained by a persuasive scientific argument about polio itself: that Blacks were not susceptible to the disease. After a decade of civil rights activism, this notion of polio as a White disease was challenged, and Black health professionals, emboldened by a new integrationist epidemiology, demanded that in polio, as in American medicine at large, health care should be provided regardless of race, color, or creed.

Saul, Stephanie. U.S. to review drug intended for one race. *New York Times* 2005 June 13: A1. NRCBL: 9.5.4; 9.7; 15.11.

Schenker, Yael; Wang, Frances; Selig, Sarah Jane; Ng, Rita; Fernandez, Alicia. The impact of language barriers on documentation of informed consent at a hospital with on-site interpreter services. *JGIM: Journal of General Internal Medicine* 2007 November; 22(Supplement 2): 294-299. NRCBL: 9.5.4; 8.3.1. SC: em.

Sheikh, Aziz; Esmail, Aneez. Should Muslims have faith based health services? *BMJ: British Medical Journal* 2007 January 13; 334(7584): 74-75. NRCBL: 9.5.4; 1.2.

Wasserman, J.; Flannery, M.A.; Clair, J.M. Raising the ivory tower: the production of knowledge and distrust of medicine among African Americans. *Journal of Medical Ethics* 2007 March; 33(3): 177-180. 50 refs. NRCBL: 9.5.4; 15.5; 18.5.1.
    Keywords: *blacks; *medicine; *trust; attitude of health personnel; culture; eugenics; genetic research; health care; health services accessibility; historical aspects; involuntary sterilization; mass media; medical education; non- therapeutic research; physicians; public policy; scientific misconduct; social discrimination; Keyword Identifiers: *United States; Nineteenth Century; Sims, J. Marion; Tuskegee Syphilis Study; Twentieth Century

Abstract: African American distrust of medicine has consequences for treatment seeking and healthcare behaviour. Much work has been done to examine acute events (eg, Tuskegee Syphilis Study) that have contributed to this phenomenon and a sophisticated bioethics discipline keeps watch on current practices by medicine. But physicians and clinicians are not the only actors in the medical arena, particularly when it comes to health beliefs and distrust of medicine. The purpose of this paper is to call attention not just to ethical shortcomings of the past, but to the structural contexts of those events and the contributions and responsibilities of popular media and academic disciplines in the production of (often mythic) knowledge. We argue that ignoring context and producing inaccurate work has real impacts on health and healthcare, particularly for African Americans, and thus engenders

NRCBL: National Reference Center for Bioethics Literature Classification Scheme   See inside front cover for terms.

91

ethical obligations incumbent on disciplines traditionally recognised as purely academic.

CARE FOR SPECIFIC GROUPS/ MINORS

Alderson, Priscilla; Hawthorne, Joanna; Killen, Margaret. The participation rights of premature babies. *In:* Freeman, Michael, ed. Children's Health and Children's Rights. Boston: Martinus Nijhoff Publishers, 2006: 31-50. NRCBL: 9.5.7; 9.5.8; 8.3.2; 21.1. Identifiers: United Nations Convention on the Rights of the Child.

Brabin, Loretta; Roberts, Stephen A.; Kitchener, Henry C. A semi-qualitative study of attitudes to vaccinating adolescents against human papillomavirus without parental consent. *BMC Public Health* 2007 February 9; 7: 20. NRCBL: 9.5.7; 9.1; 8.3.2.

Bridgeman, Jo. Caring for children with severe disabilities: boundaried and relational rights. *In:* Freeman, Michael, ed. Children's Health and Children's Rights. Boston: Martinus Nijhoff Publishers, 2006: 99-119. NRCBL: 9.5.7; 9.5.1; 20.4.2; 8.3.2; 21.1. SC: cs; le.

Carroll-Lind, Janis; Chapman, James W.; Gregory, Janet; Maxwell, Gabrielle. The key to the gatekeepers: passive consent and other ethical issues surrounding the rights of children to speak on issues that concern them. *Child Abuse and Neglect* 2006 September; 30(9): 979-989. NRCBL: 9.5.7; 8.3.2; 9.1.

Coleman, Gerald D. The irreversible disabling of a child: the "Ashley treatment". *National Catholic Bioethics Quarterly* 2007 Winter; 7(4): 711-728. 63 fn. NRCBL: 9.5.7; 9.5.1; 4.4; 2.4; 8.3.2; 1.2; 15.5. SC: cs.

Keywords: *children; *disabled persons; *human dignity; *patient care; brain pathology; choice behavior; clinical ethics committees; decision making; eugenics; parental consent; parents; physicians; quality of life; Roman Catholic ethics; sterilization; surgery; Proposed Keywords: *body height; *developmental disabilities; *growth attenuation; Keyword Identifiers: Children's Hospital (Seattle, WA)

Abstract: The controversial growth attenuation therapy for a severely disabled girl named Ashley is about intentional and deliberate medical acts of crippling. Does crippling children amount to a fundamental violation of their dignity and the oath of physicians to first do no harm? The way we formulate the dilemma, the picture we draw of its salient features, largely determines the conclusions we reach and the choices we make. It is intellectually important to try to view the events as the major participants view them.

Comeau, Pauline. Debate begins over public funding for HPV vaccine [news]. *CMAJ/JAMC: Canadian Medical Association Journal* 2007 March 27; 176(7): 913-914. NRCBL: 9.5.7; 9.3.1; 9.7.

Davies, Glanville; Poole, Richard F.; Akerman, Beverly R.; Gogol, Manfred. Reflections on the birth of conjoined twins [letters]. *CMAJ/JAMC: Canadian Medi-*

cal Association Journal 2007 November 6; 177(10): 1235-1236. NRCBL: 9.5.7; 9.3.1; 12.3.

Diller, Lawrence; Goldstein, Sam. Science, ethics, and the psychosocial treatment of ADHD [editorial]. *Journal of Attention Disorders* 2006 May; 9(4): 571-574. NRCBL: 9.5.7; 17.1.

English, Abigail. Health care for adolescents: ensuring access, protecting privacy. *Clearinghouse Review* 2005 July-August; 39: 217-218 [Online]. Accessed: http://www.povertylaw.org/clearinghouse-review/issues/2005/20050715/chr501090.pdf [2006 November 28]. NRCBL: 9.5.7; 9.3.1; 9.2; 8.3.2; 8.4. SC: le.

English, A.; Ford, C.A. More evidence supports the need to protect confidentiality in adolescent health care. *Journal of Adolescent Health* 2007 March; 40(3): 199-200. NRCBL: 9.5.7; 8.1; 8.4.

Ethics Working Group of Confederation of European Specialists in Paediatrics; Kurz, R.; Gill, D.; Mjones, S. Ethical issues in the daily medical care of children. *European Journal of Pediatrics* 2006 February; 165(2): 83-86. NRCBL: 9.5.7; 8.1; 8.3.2; 9.4.

Franklin, Anita; Sloper, Patricia. Listening and responding? Children's participation in health care within England. *In:* Freeman, Michael, ed. Children's Health and Children's Rights. Boston: Martinus Nijhoff Publishers, 2006: 11-29. NRCBL: 9.5.7; 8.3.2; 21.1.

Hentschel, Roland; Lindner, Katharina; Krueger, Markus; Reiter-Theil, Stella. Restriction of ongoing intensive care in neonates: a prospective study. *Pediatrics* 2006 August; 118(2): 563-569. NRCBL: 9.5.7; 9.2; 9.4; 8.1.

Iglehart, John K. Insuring all children: the new political imperative. *New England Journal of Medicine* 2007 July 5; 357(1): 70-76. NRCBL: 9.5.7; 9.3.1. SC: rv.

Jones, Melinda. Adolescent gender identity and the courts. *In:* Freeman, Michael, ed. Children's Health and Children's Rights. Boston: Martinus Nijhoff Publishers, 2006: 121-148. NRCBL: 9.5.7; 10; 8.3.2; 21.1. SC: cs; le. Identifiers: Australia.

Kennedy, Donald. Turning the tables with Mary Jane [editorial]. *Science* 2007 May 4; 316(5825): 661. NRCBL: 9.5.7; 9.7; 9.5.9. SC: le.

Kennedy, Evelyn P.; MacPhee, Cyndee. Access to confidential sexual health services. *Canadian Nurse* 2006 September; 102(7): 29-31. NRCBL: 9.5.7; 10; 8.4; 9.2; 4.1.3. SC: em.

Kmietowicz, Zosia. Doctors get advice on rights of children and young people [news]. *BMJ: British Medical Journal* 2007 September 29; 335(7621): 633. NRCBL: 9.5.7; 8.1.

Kumra, Sanjiv; Ashtari, Manzar; Anderson, Britt; Cervellione, Kelly L.; Kan, Li. Ethical and practical considerations in the management of incidental findings in pediatric MRI studies. *Journal of the American Academy of Child and Adolescent Psychiatry* 2006 August; 45(8): 1000-1006. NRCBL: 9.5.7; 18.2; 18.5.2; 9.8; 17.1.

Kuz, Kelly M. Young teenagers providing their own surgical consents: an ethical-legal dilemma for perioperative registered nurses. *Canadian Operating Room Nursing Journal* 2006 June; 24(2): 6-8, 10-11, 14-15. NRCBL: 9.5.7; 8.3.2.

Lantos, John D. At the Lok Nayak Hospital, Delhi. *Hastings Center Report* 2007 January-February; 37(1): 9. NRCBL: 9.5.7; 20.5.2.

Larcher, Vic. Ethical issues in child protection. *Clinical Ethics* 2007 December; 2(4): 208-212. NRCBL: 9.5.7; 9.1; 8.4; 1.3.5.
Abstract: The management of child protection concerns arouses strong emotions and controversies and creates ethical tensions for all concerned. This paper provides a rational analysis of some of the issues involved and suggests responses to them. The ethical and legal duties of health-care professionals are to act in the best interests of the child by safeguarding children and reporting concerns. But this may involve conflicts with parents and produce reluctance of professionals to become involved, especially in controversial types of abuse. Mandatory reporting of concerns might overcome such reluctance, but may be ineffective in the face of diagnostic uncertainties. Assembly of a stronger diagnostic evidence base would seem ethically justified, but organization of the necessary case controlled studies might be problematic. Even with a comprehensive evidence base, individual diagnoses of abuse will always involve value judgements that should be underpinned by effective training and assessment of core competencies of professionals. These manoeuvres are unlikely to prevent both justified and vexatious complaints, often in relation to breaches in professional duties or concerning professional misconduct. The tendency to blame experts may have contributed to a reluctance of other professionals to become involved, despite proposals for reforms in the expert witness and court systems. Current approaches to child protection may neither promote greater understanding nor be in the best interests of children. A revised social contract for the effective protection of children could include: a duty of care that adequately addresses the primacy of the child's welfare; the acquisition of a sound evidence base; professional transparency and accountability (but with protection from malicious and vexatious complaints); and a shift emphasis towards a more inquisitorial system that embraced the principles of truth and reconciliation.

Lee, Joyce M.; Howell, Joel D. Tall girls: the social shaping of a medical therapy. *Archives of Pediatrics and Adolescent Medicine* 2006 October; 160(10):1035-1039. NRCBL: 9.5.7; 4.2; 9.7.

Lehrer, Jocelyn A.; Pantell, Robert; Tebb, Kathleen; Shafer, Mary-Ann. Forgone health care among U.S. adolescents: associations between risk characteristics and confidentiality concern. *Journal of Adolescent Health* 2007 March; 40(3): 218-226. NRCBL: 9.5.7; 7.1; 8.1; 8.4; 8.3.4.

Louhiala, P. How tall is too tall? On the ethics of oestrogen treatment for tall girls. *Journal of Medical Ethics* 2007 January; 33(1): 48-50. NRCBL: 9.5.7; 4.2; 9.5.5; 9.7.
Abstract: Oestrogen treatment for girls, to prevent psychosocial problems due to extreme tallness, has been available for almost 50 years but uncertainty about its position prevails. The ethical problems of this treatment are focused on in this paper. After a brief overview on historical and medical aspects, ethical issues such as the general justification of oestrogen treatment, evaluation of its success and ethical concerns related to research in this subject are dealt with in detail.

Lundqvist, Anita; Nilstun, Tore. Human dignity in paediatrics: the effects of health care. *Nursing Ethics* 2007 March; 14(2): 215-228. NRCBL: 9.5.7; 4.4; 8.1. SC: em. Identifiers: Sweden.
Abstract: Human dignity is grounded in basic human attributes such as life and self-respect. When people cannot stand up for themselves they may lose their dignity towards themselves and others. The aim of this study was to elucidate if dignity remains intact for family members during care procedures in a children's hospital. A qualitative approach was adopted, using open non-participation observation. The findings indicate that dignity remains intact in family-centred care where all concerned parties encourage each other in a collaborative relationship. Dignity is shattered when practitioners care from their own perspective without seeing the individual in front of them. When there is a break in care, family members can restore their dignity because the interruption helps them to master their emotions. Family members' dignity is shattered and remains damaged when they are emotionally overwhelmed; they surrender themselves to practitioners' care, losing their self-esteem and self-respect.

Macer, Darryl; Toledano, Sarah Jane; de Castro, Leonardo D.; Siruno, Lalaine H. Republication: in that case. *Journal of Bioethical Inquiry* 2007; 4(3): 239, 241-244. NRCBL: 9.5.7; 9.1; 1.3.2. SC: cs. Identifiers: sponsorship of school health; obesity; poor nutrition. Note: Article originally published in the Journal of Bioethical Inquiry; 2007; 4(2): 157-158.

McKee, M. Diane; O'Sullivan, Lucia F.; Weber, Catherine M. Perspectives on confidential care for adolescent girls. *Annals of Family Medicine* 2006 November-December; 4(6): 519-526. NRCBL: 9.5.7; 8.4; 14.1.

Mills, Eithne. Parents, children, and medical treatment: legal rights and responsibilities. *In:* Komesaroff, Linda, ed. Surgical Consent: Bioethics and Cochlear Implantation. Washington, DC: Gallaudet University Press, 2007: 70-87. NRCBL: 9.5.7; 8.3.2; 11.3. SC: le. Identifiers: Australia.

NRCBL: National Reference Center for Bioethics Literature Classification Scheme							See inside front cover for terms.

93

Moss, Ralph W. Health checks, not shots: blanket vaccination against a sexually transmitted virus is the wrong way to protect women's health. *New Scientist* 2007 February 24-March 2; 193(2592): 20. NRCBL: 9.5.7; 9.5.1; 9.5.5; 10; 9.7.

Nobbs, Christopher. Probability potentiality. *CQ: Cambridge Quarterly of Healthcare Ethics* 2007 Spring; 16(2): 240-247. NRCBL: 9.5.7; 4.4; 20.5.2; 1.1.

Ogilvie, Gina S.; Remple, Valencia P.; Marra, Fawziah; McNeil, Shelly A.; Naus, Monika; Pielak, Karen L.; Ehlen, Thomas G.; Dobson, Simon R.; Money, Deborah M.; Patrick, David M. Parental intention to have daughters receive the human papillomavirus vaccine [letters and reply]. *CMAJ/JAMC: Canadian Medical Association Journal* 2007 December 4; 177(12): 1506-1512. NRCBL: 9.5.7; 9.5.1; 9.7; 9.5.5; 8.3.2. SC: em. Identifiers: Canada.

Pywell, Stephanie. Infant vaccination: a conflict of ethical imperatives? *In:* Garwood-Gowers, Austen; Tingle, John; Wheat, Kay, eds. Contemporary Issues in Healthcare Law and Ethics. Edinburgh; New York: Elsevier Butterworth-Heinemann, 2005: 213-232. NRCBL: 9.5.7; 9.7; 8.3.2. Identifiers: Great Britain (United Kingdom).

Rennie, Stuart; Muula, Adamson S.; Westreich, Daniel. Male circumcision and HIV prevention: ethical, medical and public health tradeoffs in low-income countries: ethical challenges surrounding the implementation of male circumcision as an HIV prevention strategy. *Journal of Medical Ethics* 2007 June; 33(6): 357-361. NRCBL: 9.5.7; 9.5.1; 9.5.6; 21.1.

Roy, Elizabeth; Samuels, Sumerlee. Ethical debating: therapy for adolescents with eating disorders. *Journal of Pediatric Nursing* 2006 April; 21(2): 161-166. NRCBL: 9.5.7; 17.2; 4.1.3.

Shenfield, Françoise; Sureau, Claude. The welfare of the child: whose responsibility? *In:* Shenfield, Françoise; Sureau, Claude, eds. Contemporary Ethical Dilemmas in Assisted Reproduction. Abingdon: Informa Healthcare, 2006: 73-83. NRCBL: 9.5.7; 7.1; 14.4.

Spinney, Laura. Therapy for autistic children causes outcry in France [comment]. *Lancet* 2007 August 25-31; 370(9588): 645-646. NRCBL: 9.5.7; 17.2; 17.3; 18.5.2; 18.4.

Udesky, Laurie. Push to mandate HPV vaccine triggers backlash in USA. *Lancet* 2007 March 24-30; 369(9566): 979-980. NRCBL: 9.5.7; 9.5.1; 10; 9.7; 9.3.1; 8.4; 9.1. SC: le.

Wright, C.M.; Waterston, A.J.R. Relationships between paediatricians and infant formula milk companies. *Archives of Disease in Childhood* 2006 May; 91(5): 383-385. NRCBL: 9.5.7; 7.3; 1.3.2.

CARE FOR SPECIFIC GROUPS/ SUBSTANCE ABUSERS

Baumrucker, Steven J. Ethics roundtable. Hospice and alcoholism. *American Journal of Hospice and Palliative Care* 2006 March-April; 23(2): 153-156. NRCBL: 9.5.9; 20.4.1; 8.1. SC: cs; le.

Benoit, Ellen; Magura, Stephen. Disability and substance user treatment/rehabilitation: ethical considerations. *In:* Kleinig, John; Einstein, Stanley, eds. Ethical Challenges for Intervening in Drug Use: Policy, Research and Treatment Issues. Huntsville, TX: Office of International Criminal Justice; Criminal Justice Center, Sam Houston State University, 2006: 153-170. NRCBL: 9.5.9; 9.4; 9.1; 4.2; 4.3; 9.3.1. SC: le.

Caplan, Arthur L. Ethical issues surrounding forced, mandated, or coerced treatment. *Journal of Substance Abuse Treatment* 2006 September; 31(2): 117-120. NRCBL: 9.5.9; 9.7; 8.3.4; 1.3.5; 1.1.

Charland, Louis C. Consent or coercion? Treatment referrals to Alcoholics Anonymous. *Journal of Ethics in Mental Health* 2007 April; 2(1): 3 p. Accessed: http://www.jemh.ca [2007 July 31]. NRCBL: 9.5.9; 8.3.4.

Clinton, Michael. On the colour of herring: response to commentary. *Journal of Ethics in Mental Health* 2007 April; 2(1): 3 p. [Online]. Accessed: http://www.jemh.ca [2007 July 31]. NRCBL: 9.5.9; 8.3.4.

Clinton, Michael. Should mental health professionals refer clients with substance use disorders to 12-step programs? *Journal of Ethics in Mental Health* 2007 April; 2(1): 4 p. [Online]. Accessed: http://www.jemh.ca [2007 July 31]. NRCBL: 9.5.9.

Abstract: Attendance at 12-step programs has become part of the orthodoxy of treating clients with substance abuse disorders. However, concerns have been raised about the assumptions on which 12-step programs are based. I argue that antirepresentationalism is the moral principle that underpins such concerns. After clarifying the principle of antirepresentationalism, I explore strategies for reconciling antirepresentationalism with 12-step programs. However, all the strategies I try fail. Consequently, I adopt an alternative way of thinking about antirepresentationalism that leaves mental health professionals free to refer clients to 12-step programs. However, such referrals can continue only at the cost of accepting objectionable assumptions about motivation, spirituality and human agency. Therefore, it might well be time to find an alternative to 12-step programs.

Cohen, Elliot D. Conceptualizing the professional relationship in drug user and alcohol misuser counseling. *In:* Kleinig, John; Einstein, Stanley, eds. Ethical Challenges for Intervening in Drug Use: Policy, Research and Treatment Issues. Huntsville, TX: Office of International Criminal Justice; Criminal Justice Center, Sam Houston State University, 2006: 367-382. NRCBL: 9.5.9; 8.1; 17.2; 8.4;

8.3.1. Identifiers: Alcoholics Anonymous; fiduciary model.

Coleman, Stephen. Ethical issues raised by non-punitive drug user policies. *In:* Kleinig, John; Einstein, Stanley, eds. Ethical Challenges for Intervening in Drug Use: Policy, Research and Treatment Issues. Huntsville, TX: Office of International Criminal Justice; Criminal Justice Center, Sam Houston State University, 2006: 99-119. NRCBL: 9.5.9; 9.1; 1.3.5.

Dagg, Paul K.B.; Hughes Julian C.; Sarkar, Sameer P. "Hey Bill, smoking is bad for you . . . ." [case study and commentaries]. *Journal of Ethics in Mental Health [electronic]* 2007 November; 2(2): 5 p. Accessed: http://www.jemh.ca [2008 January 24]. NRCBL: 9.5.9.

Einstein, Stanley. Drug users can't be treated, people can be! The creation and maintenance of ethical travesties, or at least dilemmas. *In:* Kleinig, John; Einstein, Stanley, eds. Ethical Challenges for Intervening in Drug Use: Policy, Research and Treatment Issues. Huntsville, TX: Office of International Criminal Justice; Criminal Justice Center, Sam Houston State University, 2006: 565-623. NRCBL: 9.5.9; 4.2; 7.1.

Estrin, Irene.; Sher, Leo. The constitutionality of random drug and alcohol testing of students in secondary schools. *International Journal of Adolescent Medicine and Health* 2006 January-March; 18(1): 21-25. NRCBL: 9.5.9; 9.5.7; 9.1. Identifiers: United States.

Gorman, Dennis M. Conflicts of interest in the evaluation and dissemination of drug use prevention programs. *In:* Kleinig, John; Einstein, Stanley, eds. Ethical Challenges for Intervening in Drug Use: Policy, Research and Treatment Issues. Huntsville, TX: Office of International Criminal Justice; Criminal Justice Center, Sam Houston State University, 2006: 171-187. NRCBL: 9.5.9; 9.1; 18.2; 9.7; 9.3.1.

Greif, Karen F.; Merz, Jon F. Concealing evidence: science, big business, and the tobacco industry. *In their:* Current Controversies in the Biological Sciences: Case Studies of Policy Challenges from New Technologies. Cambridge, MA: MIT, 2007: 205-228. NRCBL: 9.5.9; 1.3.2; 5.3; 7.3. SC: le.

Higgs, Peter; Moore, David; Aitken, Campbell. Engagement, reciprocity and advocacy: ethical harm reduction practice in research with injecting drug users. *Drug and Alcohol Review* 2006 September; 25(5): 419-423. NRCBL: 9.5.9; 7.1; 8.1.

Kasachkoff, Tziporah. Drug addiction and responsibility for health care of drug addicts. *In:* Kleinig, John; Einstein, Stanley, eds. Ethical Challenges for Intervening in Drug Use: Policy, Research and Treatment Issues. Huntsville, TX: Office of International Criminal Justice; Criminal Justice Center, Sam Houston State University, 2006: 189-204. NRCBL: 9.5.9; 9.4; 9.3.1. SC: an.

Kayser, Bengt; Mauron, Alexandre; Miah, Andy. Current anti-doping policy: a critical appraisal [debate]. *BMC Medical Ethics [electronic]* 2007; 8:2. 10 p. NRCBL: 9.5.9; 9.5.1; 15.1. Identifiers: World Anti Doping Agency [WADA].

Abstract: Background: Current anti-doping in competitive sports is advocated for reasons of fair-play and concern for the athlete's health. With the inception of the World Anti Doping Agency (WADA), anti-doping effort has been considerably intensified. Resources invested in anti-doping are rising steeply and increasingly involve public funding. Most of the effort concerns elite athletes with much less impact on amateur sports and the general public. Discussion: We review this recent development of increasingly severe anti-doping control measures and find them based on questionable ethical grounds. The ethical foundation of the war on doping consists of largely unsubstantiated assumptions about fairness in sports and the concept of a "level playing field". Moreover, it relies on dubious claims about the protection of an athlete's health and the value of the essentialist view that sports achievements reflect natural capacities. In addition, costly antidoping efforts in elite competitive sports concern only a small fraction of the population. From a public health perspective this is problematic since the high prevalence of uncontrolled, medically unsupervised doping practiced in amateur sports and doping-like behaviour in the general population (substance use for performance enhancement outside sport) exposes greater numbers of people to potential harm. In addition, anti-doping has pushed doping and doping-like behaviour underground, thus fostering dangerous practices such as sharing needles for injection. Finally, we argue that the involvement of the medical profession in doping and anti-doping challenges the principles of non-maleficience and of privacy protection. As such, current anti-doping measures potentially introduce problems of greater impact than are solved, and place physicians working with athletes or in anti-doping settings in an ethically difficult position. In response, we argue on behalf of enhancement practices in sports within a framework of medical supervision. Summary: Current anti-doping strategy is aimed at eradication of doping in elite sports by means of all-out repression, buttressed by a war-like ideology similar to the public discourse sustaining international efforts against illicit drugs. Rather than striving for eradication of doping in sports, which appears to be an unattainable goal, a more pragmatic approach aimed at controlled use and harm reduction may be a viable alternative to cope with doping and doping-like behaviour.

Kennett, Jeannette; Matthews, Stephen. The moral goal of treatment in cases of dual diagnosis. *In:* Kleinig, John; Einstein, Stanley, eds. Ethical Challenges for Intervening in Drug Use: Policy, Research and Treatment Issues. Huntsville, TX: Office of International Criminal Justice; Criminal Justice Center, Sam Houston State University, 2006: 409-436. NRCBL: 9.5.9; 17.1. SC: an.

Kleinig, John. Ethical issues on substance use intervention. *In:* Kleinig, John; Einstein, Stanley, eds. Ethical Chal-

NRCBL: National Reference Center for Bioethics Literature Classification Scheme   See inside front cover for terms.

95

lenges for Intervening in Drug Use: Policy, Research and Treatment Issues. Huntsville, TX: Office of International Criminal Justice; Criminal Justice Center, Sam Houston State University, 2006: 21-44. NRCBL: 9.5.9; 1.3.5; 8.3.1; 8.4.

Kleinig, John. Thinking ethically about needle and syringe programs. *In:* Kleinig, John; Einstein, Stanley, eds. Ethical Challenges for Intervening in Drug Use: Policy, Research and Treatment Issues. Huntsville, TX: Office of International Criminal Justice; Criminal Justice Center, Sam Houston State University, 2006: 121-132. NRCBL: 9.5.9; 9.1; 9.5.6. SC: an.

Kleinig, John. Thinking ethically about needle and syringe programs. *Substance Use and Misuse* 2006; 41(6-7): 815-825. NRCBL: 9.5.9; 9.1; 9.5.6.

Loue, Sana; Ioan, Beatrice. Legal and ethical issues in heroin diagnosis, treatment, and research. *Journal of Legal Medicine* 2007 April-June; 28(2): 193-221. NRCBL: 9.5.9; 18.3; 8.3.1. SC: le.

Morse, Stephen J. Medicine and morals, craving and compulsion. *In:* Kleinig, John; Einstein, Stanley, eds. Ethical Challenges for Intervening in Drug Use: Policy, Research and Treatment Issues. Huntsville, TX: Office of International Criminal Justice; Criminal Justice Center, Sam Houston State University, 2006: 323-339. NRCBL: 9.5.9; 4.3; 4.2; 1.3.5. SC: an.

Mukherjee, Raja; Eastman, Nigel; Turk, Jeremy; Hollins, Sheila. Fetal alcohol syndrome: law and ethics. *Lancet* 2007 April 7-13; 369(9568): 1149-1150. NRCBL: 9.5.9; 9.5.5; 9.5.8; 8.1. SC: le.

Neher, Jon O. You're fired. *Hastings Center Report* 2007 May-June; 37(3): 7-8. NRCBL: 9.5.9; 8.1. SC: cs. Identifiers: health professional's right to refuse to treat.

Pence, Gregory E. Kant on whether alcoholism is a disease. *In his:* The Elements of Bioethics. Boston: McGraw-Hill, 2007: 21-51. NRCBL: 9.5.9; 4.2; 15.6; 19.1; 4.4; 1.1.

Peters, Matthew J. Should smokers be refused surgery? *BMJ: British Medical Journal* 2007 January 6; 334(7583): 20-21. NRCBL: 9.5.9; 18.1; 9.4.

Phillips, Susan. Ethical decision-making when caring for the noncompliant patient. *Journal of Infusion Nursing* 2006 September-October; 29(5): 266-271. NRCBL: 9.5.9; 8.1; 4.1.3. SC: cs.

Presenza, Louis J. Naltrexone as a "mandate" or as a choice: comments on "Judicially mandated naltrexone use by criminal offenders: a legal analysis". *Journal of Substance Abuse Treatment* 2006 September; 31(2): 129-130. NRCBL: 9.5.9; 9.7; 8.3.4; 1.3.5. SC: le.

Reisman, Anna B. Saving Sylvia Cleary. *Hastings Center Report* 2007 July-August; 37(4): 9-10. NRCBL: 9.5.9; 8.1. SC: cs.

Rosenberg, Tina. Doctor or drug pusher? Pain is difficult to measure, and those who treat pain sufferers have to make highly subjective decisions about dosage levels of drugs that can be abused or even resold. When a doctor gets it wrong, is that bad medicine — or a drug felony? *New York Times Magazine* 2007 June 17; p. 48-55, 64, 68- 71. NRCBL: 9.5.9; 4.4; 8.1; 9.7; 1.3.5. SC: po.

Schomerus, G.; Matschinger, H.; Angermeyer, M.C. Alcoholism: illness beliefs and resource allocation preferences of the public. *Drug and Alcohol Dependence* 2006 May 20; 82(3): 204-210. NRCBL: 9.5.9; 9.2; 9.3.1; 9.4.

Walker, Robert; Logan, T.K.; Clark, James J.; Leukefeld, Carl. Informed consent to undergo treatment for substance abuse: a recommended approach. *Journal of Substance Abuse Treatment* 2005 December; 29(4): 241-251. NRCBL: 9.5.9; 8.3.1; 18.3.

Wallace, Barbara C. Ethical issues surrounding access to drug user counseling/treatment. *In:* Kleinig, John; Einstein, Stanley, eds. Ethical Challenges for Intervening in Drug Use: Policy, Research and Treatment Issues. Huntsville, TX: Office of International Criminal Justice; Criminal Justice Center, Sam Houston State University, 2006: 529-551. NRCBL: 9.5.9; 9.4; 1.3.5; 17.1.

Walsh, Adrian; Lynch, Tony. Drug user counseling remuneration and ethics. *In:* Kleinig, John; Einstein, Stanley, eds. Ethical Challenges for Intervening in Drug Use: Policy, Research and Treatment Issues. Huntsville, TX: Office of International Criminal Justice; Criminal Justice Center, Sam Houston State University, 2006: 453-466. NRCBL: 9.5.9; 17.1; 9.3.1. SC: an.

Wasserman, David. Addiction and disability: moral and policy issues. *In:* Kleinig, John; Einstein, Stanley, eds. Ethical Challenges for Intervening in Drug Use: Policy, Research and Treatment Issues. Huntsville, TX: Office of International Criminal Justice; Criminal Justice Center, Sam Houston State University, 2006: 133-152. NRCBL: 9.5.9; 4.2; 4.3; 9.1; 9.3.1; 9.4. SC: le.

Weisleder, Pedro. Inconsistency among American states on the age at which minors can consent to substance abuse treatment. *Journal for the American Academy of Psychiatry and the Law* 2007; 35(3): 317-322. NRCBL: 9.5.9; 8.3.2; 8.4. SC: le; rv.

Wiechelt, Shelly A. Ethical issues surrounding access to treatment for substance misuse. *In:* Kleinig, John; Einstein, Stanley, eds. Ethical Challenges for Intervening in Drug Use: Policy, Research and Treatment Issues. Huntsville, TX: Office of International Criminal Justice; Criminal Justice Center, Sam Houston State University, 2006: 553-563. NRCBL: 9.5.9; 9.4; 6; 9.3.1; 9.3.2.

Williams, Kylie; Cocking, Dean. Counselor-client relationships and professional role morality. *In:* Kleinig, John; Einstein, Stanley, eds. Ethical Challenges for Intervening in Drug Use: Policy, Research and Treatment Issues.

Huntsville, TX: Office of International Criminal Justice; Criminal Justice Center, Sam Houston State University, 2006: 341-365. NRCBL: 9.5.9; 8.1; 17.7; 1.3.5; 1.3.10. SC: an; le.

Zadunayski, Anna; Hicks, Matthew; Gibbard, Ben; Godlovitch, Glenys. Behind the screen: legal and ethical considerations in neonatal screening for prenatal exposure to alcohol. *Health Law Journal* 2006; 14: 105-127. NRCBL: 9.5.9; 9.5.8; 9.5.5; 8.3.4; 2.1. SC: le.

CARE FOR SPECIFIC GROUPS/ WOMEN

A necessary vaccine [editorial]. *New York Times* 2007 February 26; p. A20. NRCBL: 9.5.5; 9.7; 9.1; 9.5.7. SC: po. Identifiers: human papillomavirus vaccine (HPV).

Adams, Karen E. What's "normal": female genital mutilation, psychology, and body image. *Journal of the American Medical Women's Association* 2004 Summer; 59(3): 168-170. NRCBL: 9.5.5; 10; 21.7; 17.1. SC: cs.

Basu, Sanjay; Segraves, Britnye T.; Colgrove, James. Compulsory HPV vaccination [letter and reply]. *New England Journal of Medicine* 2007 March 8; 356(10): 1074-1075. NRCBL: 9.5.5; 8.3.4; 9.5.1; 9.5.7; 9.7.

Bergeron, Veronique. The ethics of cesarean section on maternal request: a feminist critique of the American College of Obstetricians and Gynecologists' position on patient-choice surgery. *Bioethics* 2007 November; 21(9): 478-487. NRCBL: 9.5.5; 1.1; 10.

Abstract: In recent years, the medical establishment has been speaking in favor of women's autonomy in childbirth by advocating cesarean delivery on maternal request (CDMR). This paper offers to look at the ethical dimension of CDMR through a feminist critique of the medicalization of childbirth and its influence on present-day medical ethics. I claim that the medicalization of childbirth reflects a sexist bias with regard to conceptions of the body and needs to be used with caution when applied to women's reproductive health. I then use this perspective to critically analyze the position of the American College of Obstetricians and Gynecologists (ACOG) on the ethics of decision-making in patient-choice surgery. I claim that informed consent cannot be meaningfully exercised unless women are made aware of the sexist underpinnings of the medical model of childbirth and its influence on the ethical reasoning of the American College of Obstetricians and Gynecologists. I also express concern about the effects of normalizing patient-choice cesarean sections on the choices available to pregnant women using as examples the institutional rules on mandatory cesarean sections for women with a previous cesarean delivery or breech presentation. I conclude with a call for more research into the real cost of convenience in CDMR, particularly as our increasingly strained publicly funded healthcare system would greatly benefit from the de-medicalization of normal body functions rather than an increased dependence on costly surgical technology.

Brown, Stephen D.; Truog, Robert D.; Johnson, Judith A.; Ecker, Jeffrey L. Do differences in the American Academy of Pediatrics and the American College of Obstetricians and Gynecologists positions on the ethics of maternal-fetal interventions reflect subtly divergent professional sensitivities to pregnant women and fetuses? *Pediatrics* 2006 April; 117(4): 1382-1387. NRCBL: 9.5.5; 9.5.8; 8.3.1; 8.3.4; 9.5.1. SC: le.

Chervenak, Frank A.; McCullough, Laurence B. Ethical issues and decision making in the management of diabetes in pregnancy. *In:* Langer, Oded, ed. The Diabetes in Pregnancy Dilemma: Leading Change with Proven Solutions. Lanham, MD: University Press of America, 2006: 23-33. NRCBL: 9.5.5; 1.1; 9.5.1; 9.5.8.

Chou, Ann F.; Scholle, Sarah Hudson; Weisman, Carol S.; Bierman, Arlene S.; Correa-de-Araujo, Rosaly; Mosca, Lori. Gender disparities in the quality of cardiovascular disease care in private managed care plans. *Women's Health Issues* 2007 May-June; 17(3): 120-130. NRCBL: 9.5.5; 9.5.1; 9.3.2. SC: em.

Chou, Ann F.; Wong, Lok; Weisman, Carol S.; Chan, Sophia; Bierman, Arlene S.; Correa-de-Araujo, Rosaly; Scholle, Sarah Hudson. Gender disparities in cardiovascular disease care among commercial and Medicare managed care plans. *Women's Health Issues* 2007 May-June; 17(3): 139-149. NRCBL: 9.5.5; 9.5.1; 9.3.2. SC: em.

Choudhury, Lincoln Priyadarshi; Kutty, V. Raman. Obstetric practices related to HIV in Kerala. *Indian Journal of Medical Ethics* 2007 January-March; 4(1): 12-15. NRCBL: 9.5.5; 9.5.6; 8.3.1; 9.8.

Christilaw, J.E. Cesarean section by choice: constructing a reproductive rights framework for the debate. *International Journal of Gynecology and Obstetrics* 2006 September; 94(3): 262-268. NRCBL: 9.5.5; 8.3.1; 9.2.

Dixon-Woods, Mary; Williams, Simon J.; Jackson, Clare J.; Akkad, Andrea; Kenyon, Sara; Habiba, Marwan. Why do women consent to surgery, even when they do not want to? An interactionist and Bourdieusian analysis. *Social Science and Medicine* 2006 June; 62(11): 2742-2753. NRCBL: 9.5.5; 8.3.1; 9.5.1; 1.1.

Dixon-Woods, Mary; Young, Bridget; Ross, Emma. Researching chronic childhood illness: the example of childhood cancer. *Chronic Illness* 2006 September; 2(3): 165-177. 68 refs. NRCBL: 9.5.5; 15.1; 18.5.2; 18.3.
Keywords: *cancer; *children; *chronically ill; *human experimentation; age factors; behavioral research; clinical trials; genetic research; informed consent; researcher subject relationship; risk; tissue donors

FIGO Committee for the Ethical Aspects of Human Reproduction and Women's Health. Ethical guidelines on obstetric fistula. *International Journal of Gynecology*

NRCBL: National Reference Center for Bioethics Literature Classification Scheme   See inside front cover for terms.

97

and Obstetrics 2006 August; 94(2): 174-175. NRCBL: 9.5.5; 14.1.

FIGO Committee for the Ethical Aspects of Human Reproduction and Women's Health. Safe motherhood. *International Journal of Gynecology and Obstetrics* 2006 August; 94(2): 167-168. NRCBL: 9.5.5; 14.1.

FIGO Committee for the Ethical Aspects of Human Reproduction and Women's Health; FIGO Committee on Women's Sexual and Reproductive Rights. Female genital cutting. *International Journal of Gynecology and Obstetrics* 2006 August; 94(2): 176-177. NRCBL: 9.5.5; 10; 9.5.7; 1.2.

Fremont, Allen M.; Coreea-de-Araujo, Rosaly; Hayes, Sharon. Gender disparities in managed care: it's time for action. *Women's Health Issues* 2007 May-June; 17(3): 116-119. NRCBL: 9.5.5; 9.3.2.

Gannon, Susanne; Müller-Rockstroh, Babette. In memory: women's experiences of (dangerous) breasts. *Philosophy in the Contemporary World* 2004 Spring-Summer; 11(1): 53-64. NRCBL: 9.5.5; 4.2; 4.4; 10.

Gass, C.W.J. It is the right of every anaesthetist to refuse to participate in a maternal-request caesarean section. *International Journal of Obstetric Anesthesia* 2006 January; 15(1): 33-35. NRCBL: 9.5.5; 9.5.8; 9.5.1; 8.1. SC: an.

Latham, Melanie. Cyberwoman and her surgeon in the twenty-first century. *In:* Garwood-Gowers, Austen; Tingle, John; Wheat, Kay, eds. Contemporary Issues in Healthcare Law and Ethics. Edinburgh; New York: Elsevier Butterworth-Heinemann, 2005: 233-249. NRCBL: 9.5.5; 1.1; 4.5; 9.5.1; 10. SC: le. Identifiers: Great Britain (United Kingdom).

Leavine, Barbara Ann. Court ordered cesareans: can a pregnant woman refuse? *Houston Law Review* 1992 Spring; 29(1): 185-213. NRCBL: 9.5.5; 1.1; 8.3.3; 8.3.4; 9.5.8. SC: le; rv.

Levy, Daniel R. The maternal-fetal conflict: the right of a woman to refuse a cesarean section versus the state's interest in saving the life of the fetus. *West Virginia Law Review* 2005 Fall; 108(1): 97-124. NRCBL: 9.5.5; 9.5.8; 8.3.4; 1.3.5.

Lindemann, Hilde. Breasts, wombs, and the body politic [review of Mass Hysteria: Medicine, Culture, and Mothers' Bodies, by Rebecca Kukla]. *Hastings Center Report* 2007 March-April; 37(2): 43-44. NRCBL: 9.5.5; 7.1; 10.

Lo, Bernard. Ethical issues in obstetrics and gynecology. *In his:* Resolving Ethical Dilemmas: A Guide for Clinicians. 3rd edition. Philadelphia, PA: Lippincott Williams and Wilkins, 2005: 249-255. NRCBL: 9.5.5; 9.5.8; 12.1; 14.1.

Macauley, Robert C. The role of substituted judgment in the aftermath of a suicide attempt. *Journal of Clinical Eth-*

ics 2007 Summer; 18(2): 111-118. NRCBL: 9.5.5; 20.7; 20.5.4; 8.1; 8.3.3. SC: cs.

Mansi, James A.; Franco, Eduardo L.; de Pokomandy, Alexandra; Spence, Andrea R.; Burchell, Ann N.; Trottier, Helen; Mayrand, Marie-Hélène; Lau, Susie; Ferenczy, Alex; Brophy, James M.; Cassels, Alan K.; Nisker, Jeff; Lippman, Abby; Boscoe, Madeline; Shimmin, Carolyn. Vaccination against human papillomavirus. *CMAJ/JAMC: Canadian Medical Association Journal* 2007 December 4; 177(12): 1524-1528. NRCBL: 9.5.5; 9.5.1; 9.5.7; 9.7.

Morris, Kelly. Issues on female genital mutilation/cutting — progress and parallels. *Lancet* 2006 December; 368(special issue): S64-S66. NRCBL: 9.5.5; 9.5.7; 10; 21.7; 21.1; 4.2; 9.5.1.

Moszynski, Peter. BMA backs police campaign against genital mutilation [news]. *BMJ: British Medical Journal* 2007 July 21; 335(7611): 116. NRCBL: 9.5.5; 9.5.7; 10; 21.7. Identifiers: British Medical Association.

Muula, A.S. Ethical and practical consideration of women choosing cesarean section deliveries without "medical indication" in developing countries. *Croatian Medical Journal* 2007 February; 48(1): 94-102. NRCBL: 9.5.5; 21.1; 9.2; 21.7.

Patni, Shalini; Wagstaff, John; Tofazzal, Nasima; Bonduelle, Myriam; Moselhi, Marsham; Kevelighan, Euan; Edwards, Steve. Metastatic unknown primary tumour presenting in pregnancy: a rarity posing an ethical dilemma. *Journal of Medical Ethics* 2007 August; 33(8): 442-443. NRCBL: 9.5.5; 9.5.1; 12.4.2. SC: cs.
Abstract: This brief report raises the ethical dilemma encountered by an obstetrician involved in the care of a pregnant woman with life-threatening disease. This is a particularly difficult issue if the maternal well-being is in conflict with the survival of the unborn child.

Powell, Tia. Commentary: support for case-based analysis in decision making after a suicide attempt. *Journal of Clinical Ethics* 2007 Summer; 18(2): 119-121. NRCBL: 9.5.5; 20.7; 20.5.4; 8.1; 8.3.3; 17.1. SC: cs. Comments: Robert C. Macauley. The role of substituted judgment in the aftermath of a suicide attempt. Journal of Clinical Ethics 2007 Summer; 18(2): 111-118.

Saunders, T.A.; Stein, D.J.; Dilger, J.P. Informed consent for labor epidurals: a survey of Society for Obstetric Anesthesia and Perinatology anesthesiologists from the United States. *International Journal of Obstetric Anesthesia* 2006 April; 15(2): 98-103. NRCBL: 9.5.5; 8.3.1; 7.1. SC: em.

Schultz, Jane E. Corpus interruptus: biotech drugs, insurance providers and the treatment of breast cancer. *Journal of Bioethical Inquiry* 2007; 4(2): 93-102. NRCBL: 9.5.5; 9.7; 9.5.1; 8.1; 9.3.1. SC: cs.

Abstract: In researching the biomedically-engineered drug Neulasta (filgrastim), a breast cancer patient becomes aware of the extent to which knowledge about the development and marketing of drugs influences her decisions with regard to treatment. Time spent on understanding the commercial interests of insurers and pharmaceutical companies initially thwarts but ultimately aids the healing process. This first-person narrative calls for physicians to recognize that the alignment of commercial interests transgresses the patient's humanity.

Schwartz, Linda; Woloshin, Steven. Participation in mammography screening: Women should be encouraged to decide what is right for them, rather than being told what to do. *BMJ: British Medical Journal* 2007 October 13; 335(7623): 731-732. NRCBL: 9.5.5; 8.1; 1.1.

Seymour, John. A question of autonomy? *In his:* Childbirth and the Law. Oxford; New York: Oxford University Press, 2000: 189-239. NRCBL: 9.5.5; 9.5.7; 1.1; 8.3.4. SC: an; le.

Szumacher, Ewa. The feminist approach in the decision-making process for treatment of women with breast cancer. *Annals of the Academy of Medicine, Singapore* 2006 September; 35(9): 655-661. NRCBL: 9.5.5; 9.5.4; 8.1; 10; 4.5; 9.5.1; 18.3. SC: rv.

Tong, Rosemarie. Gender-based disparities east/west: rethinking the burden of care in the United States and Taiwan. *Bioethics* 2007 November; 21(9): 488-499. NRCBL: 9.5.5; 10; 9.5.2; 21.1; 9.3.1. SC: rv.
Abstract: When feminist bioethicists express concerns about health-related gender disparities, they raise considerations about justice and gender that traditional bioethicists have either not raised or raised somewhat weakly. In this article, I first provide a feminist analysis of long-term healthcare by and for women in the United States and women in Taiwan. Next, I make the case that, on average, elderly US and Taiwanese women fare less well in long-term care contexts than do elderly US and Taiwanese men. Finally, I explore some suggested practical remedies to reduce gender disparities in long-term care contexts.

Turillazzi, E.; Fineschi, V. Female genital mutilation: the ethical impact of the new Italian law. *Journal of Medical Ethics* 2007 February; 33(2): 98-101. NRCBL: 9.5.5; 9.5.7; 10. SC: le.
Abstract: Despite global and local attempts to end female genital mutilation (FGM), the practice persists in some parts of the world and has spread to non-traditional countries through immigration. FGM is of varying degrees of invasiveness, but all forms raise health-related concerns that can be of considerable physical or psychological severity. FGM is becoming increasingly prohibited by law, both in countries where it is traditionally practised and in countries of immigration. Medical practice prohibits FGM. The Italian parliament passed a law prohibiting FGM, which has put in place a set of measures to prevent, to oppose and to suppress the practice of FGM as a violation of a person's fundamental rights to physical and mental integrity and to the health of women and girls. The Italian law not only treats new offences but also wants to deal with the problem in its entirety, providing important intervention in all the sectors. Different kinds of interventions are considered, starting with the development of informative campaigns, training of health workers, institution of a tollfree number, international cooperation programmes and the responsibility of the institution where the crime is committed. Particularly, the law recognises that doctors have a role in eliminating FGM by educating patients and communities.

Ward, C.M. The breast-implant controversy: a medico-moral critique. *British Journal of Plastic Surgery* 2001; 54(4): 352-357. NRCBL: 9.5.5; 4.5; 1.3.7; 7.3; 1.1. SC: le.

Weiniger, C.F.; Elchalal, U.; Sprung, C.L.; Weissman, C.; Matot, I. Holy consent — a dilemma for medical staff when maternal consent is withheld for emergency caesarean section. *International Journal of Obstetric Anesthesia* 2006 April; 15(2): 145-148. NRCBL: 9.5.5; 8.3.4; 1.2; 7.1.

CARING *See* NURSING ETHICS AND PHILOSOPHY; PHILOSOPHY OF MEDICINE; PATIENT RELATIONSHIPS

CHIMERAS AND HYBRIDS

An unwieldy hybrid [editorial]. *Nature* 2007 May 24; 447(7143): 353-354. NRCBL: 18.5.4; 15.1; 18.1; 22.1. SC: le.
Keywords: *chimeras; *embryo research; *legal aspects; genetically modified animals; government regulation; legislation; public policy; Keyword Identifiers: *Great Britain; Human Fertilisation and Embryology Authority (Great Britain)

Animal-human hybrid-embryo research [editorial]. *Lancet* 2007 September 15-21; 370(9591): 909. NRCBL: 15.1; 18.5.4; 22.2; 5.3; 18.6; 18.1; 22.1.

Avoiding a chimaera quagmire [editorial]. *Nature* 2007 January 4; 445(7123): 1. NRCBL: 15.1; 18.1; 22.1.
Keywords: *chimeras; *embryo research; *stem cells; human dignity; nuclear transfer techniques; primates; regulation; risk; Keyword Identifiers: Great Britain

Chimera research should be lightly regulated, not banned [editorial]. *Lancet* 2007 January 20-26; 369(9557): 164. NRCBL: 18.5.4; 15.1; 18.1; 22.1; 5.3; 1.3.7.
Keywords: *chimeras; *embryo research; *embryonic stem cells; *government regulation; legal aspects; mass media; public policy; Keyword Identifiers: *Great Britain; Human Fertilisation and Embryology Authority

Of animal eggs and human embryos [editorial]. *New York Times* 2007 September 24; p. A22. NRCBL: 15.1; 19.1; 14.6; 18.5.4; 18.1; 22.1. SC: po.
Keywords: *chimeras; *embryo research; *embryonic stem cells; alternatives; embryos; ovum donors; private sector; regulation; remuneration; scarcity; Keyword Identifiers: Great Britain; United States

Proyecto CHIMBRIDS: chimeras and hybrids in comparative European and international research-scientific, ethical, philosophical and legal aspects. *Revista de Derecho y Genoma Humano = Law and the Human Genome Review* 2007 July-December; (27): 227-243. 16 fn. NRCBL: 15.1; 18.1; 22.1; 21.1. SC: le; rv.

> Keywords: *chimeras; *international aspects; *moral policy; *public policy; animal experimentation; animal welfare; classification; comparative studies; embryo research; embryo transfer; embryos; ethical analysis; gene transfer techniques; genetic research; government regulation; legal aspects; moral status; nuclear transfer techniques; organ transplantation; research ethics; species specificity; stem cell transplantation; tissue transplantation; Keyword Identifiers: *European Union; Council of Europe; Europe

The use of human-animal hybrid embryos [news]. *ATLA: Alternatives to Laboratory Animals* 2007 March; 35(1): 8-9. 2 refs. NRCBL: 15.1; 18.1; 22.1; 18.5.4. SC: le.

> Keywords: *chimeras; *embryos; *research embryo creation; animal experimentation; biomedical research; embryonic stem cells; government regulation; Keyword Identifiers: Great Britain

Baylis, Françoise; Fenton, Andrew. Chimera research and stem cell therapies for human neurodegenerative disorders. *CQ: Cambridge Quarterly of Healthcare Ethics* 2007 Spring; 16(2): 195-208. 71 fn. NRCBL: 15.1; 18.1; 22.1; 17.1; 4.4; 18.5.4. SC: an.

> Keywords: *chimeras; *embryonic stem cells; *ethical analysis; *human dignity; *moral policy; *moral status; *primates; *stem cell transplantation; animal welfare; clinical trials; guidelines; human characteristics; speciesism; therapeutic research; risks and benefits; Proposed Keywords: *neurodegenerative diseases; Keyword Identifiers: National Academy of Sciences

Baylis, Françoise; Robert, Jason Scott. Part-human chimeras: worrying the facts, probing the ethics. *American Journal of Bioethics* 2007 May; 7(5): 41-45. 26 refs. NRCBL: 15.1; 18.1; 22.1. Comments: Comment on: Henry T. Greely, Mildred K. Cho, Linda F. Hogle, and Debra M. Satz. Thinking about the human neuron mouse. American Journal of Bioethics 2007 May; 7(5): 27-40.

> Keywords: *chimeras; animal experimentation; brain; embryonic stem cells; embryos; ethical analysis; government regulation; guidelines; human characteristics; legal aspects; moral policy; moral status; precautionary principle; public policy; research ethics; stem cell transplantation; terminology; Proposed Keywords: mice; species specificity; Keyword Identifiers: Canada; Canadian Institutes of Health Research; United States

Bobbert, Monika. Ethical questions concerning research on human embryos, embryonic stem cells and chimeras. *Biotechnology Journal* 2006 December; 1(12): 1352-1369. NRCBL: 18.5.4; 15.1; 22.1; 15.8; 4.4.

Cheshire, William P., Jr. The moral musings of a murine chimera. *American Journal of Bioethics* 2007 May; 7(5): 49-50. 6 refs. NRCBL: 15.1; 18.1; 22.1; 4.4. Comments: Comment on: Henry T. Greely, Mildred K. Cho, Linda F. Hogle, and Debra M. Satz. Thinking about the human neu-

ron mouse. American Journal of Bioethics 2007 May; 7(5): 27-40.

> Keywords: *chimeras; animal experimentation; brain; human characteristics; moral status; personhood; research ethics; self concept; stem cell transplantation; wedge argument; Proposed Keywords: *cognition; mice

Cohen, Cynthia B. Beyond the human neuron mouse to the NAS guidelines. *American Journal of Bioethics* 2007 May; 7(5): 46-49. 12 refs. NRCBL: 15.1; 18.1; 22.1; 18.5.4. Comments: Comment on: Henry T. Greely, Mildred K. Cho, Linda F. Hogle, and Debra M. Satz. Thinking about the human neuron mouse. American Journal of Bioethics 2007 May; 7(5): 27-40.

> Keywords: *chimeras; *guidelines; *stem cell transplantation; advisory committees; animal experimentation; brain; embryonic stem cells; embryos; fetuses; human characteristics; moral policy; policy analysis; policy making; precautionary principle; primates; public policy; regulation; research ethics; risks and benefits; Proposed Keywords: blastocysts; mice; species specificity; Keyword Identifiers: National Academies of Sciences; Stanford University; United States

Cole, Andrew. Scientists plead for right to create interspecies embryos [news]. *BMJ: British Medical Journal* 2007 June 23; 334(7607): 1294. NRCBL: 15.1; 18.1; 22.1; 18.5.4.

> Keywords: *chimeras; *research embryo creation; embryonic stem cells; government regulation; organizational policies; professional organizations; public policy; researchers; Keyword Identifiers: *Great Britain; Academy of Medical Sciences (Great Britain)

Davis, Bradley L. Compelled expression of the religiously forbidden: pharmacists, "duty to fill" statues, and the hybrid rights exception. *University of Hawai'i Law Review* 2006 Winter; 29(1): 97-121. NRCBL: 11.1; 4.1.1; 9.7. SC: le.

Day, Michael. UK may use hybrid embryos for research [news]. *BMJ: British Medical Journal* 2007 May 26; 334(7603): 1074. NRCBL: 18.5.4; 15.1; 18.1; 22.1.

> Keywords: *chimeras; *embryo research; *embryos; *legal aspects; *public policy; government; legislation; Keyword Identifiers: *Great Britain; Department of Health (Great Britain); *Human Tissue and Embryos Bill (Great Britain)

Eberl, Jason T. Creating non-human persons: might it be worth the risk? *American Journal of Bioethics* 2007 May; 7(5): 52-54. 11 refs. NRCBL: 15.1; 18.1; 22.1; 1.1; 4.4. Comments: Comment on: Henry T. Greely, Mildred K. Cho, Linda F. Hogle, and Debra M. Satz. Thinking about the human neuron mouse. American Journal of Bioethics 2007 May; 7(5): 27-40.

> Keywords: *chimeras; *moral status; *personhood; animal experimentation; brain; double effect; ethical analysis; human characteristics; philosophy; research ethics; risk; self concept; stem cell transplantation; Proposed Keywords: cognition; species specificity

Greely, Henry T.; Cho, Mildred K.; Hogle, Linda F.; Satz, Debra M. Response to open peer commentaries on "thinking about the human neuron mouse". *American*

Journal of Bioethics 2007 May; 7(5): W4-W6. 9 refs. NRCBL: 15.1; 18.1; 22.1; 4.4.

> Keywords: *chimeras; animal experimentation; animal rights; bioethics; brain; ethical analysis; human characteristics; moral status; personhood; research ethics; stem cell transplantation; wedge argument; Proposed Keywords: mice

Greely, Henry T.; Cho, Mildred K.; Hogle, Linda F.; Satz, Debra M. Thinking about the human neuron mouse. *American Journal of Bioethics* 2007 May; 7(5): 27-40. 46 refs. NRCBL: 15.1; 18.1; 22.1; 18.5.4; 4.4. SC: rv.

> Keywords: *chimeras; aborted fetuses; advisory committees; animal behavior; animal experimentation; animal welfare; biomedical research; body parts and fluids; brain; embryonic stem cells; embryos; fetal stem cells; government financing; guidelines; human characteristics; human dignity; legal aspects; moral policy; moral status; policy analysis; public policy; research ethics; research support; risks and benefits; stem cell transplantation; terminology; Proposed Keywords: cognition; mice; species specificity; Keyword Identifiers: Great Britain; National Academies of Science; National Institutes of Health; Stanford University; United States

Hoeyer, Klaus; Koch, Lene. The ethics of functional genomics: same, same, but different? *Trends in Biotechnology* 2006 September; 24(9): 387-389. 17 refs. NRCBL: 15.1; 18.1; 22.1;22.2; 4.4.

> Keywords: *chimeras; *genomics; *species specificity; animal experimentation; animal organs; animal rights; deontological ethics; DNA sequences; genetic engineering; genetically modified animals; human dignity; utilitarianism

Hopkin, Michael. UK set to reverse stance on research with chimeras [news]. *Nature Medicine* 2007 August; 13(8): 890-891. NRCBL: 18.5.4; 15.1; 18.1; 22.1; 18.6. SC: le. Identifiers: Human Fertilisation and Embryology Authority;.

> Keywords: *chimeras; *embryo research; legal aspects; embryonic stem cells; government regulation; Keyword Identifiers: *Great Britain

Hynes, Richard. Reply to 'UK set to reverse stance on research with chimeras' [letter]. *Nature Medicine* 2007 October; 13(10): 1133. NRCBL: 18.5.4; 15.1; 18.1; 22.1; 18.6. Comments: Comment on: Michael Hopkin. UK set to reverse stance on research with chimeras. Nature Medicine 2007 August; 13(8): 890-891.

> Keywords: *chimeras; *research embryo creation; *stem cells; *terminology; embryo research; editorial policies; public policy; Proposed Keywords: pluripotent stem cells; Keyword Identifiers: *Great Britain; Nature Medicine

Illes, Judy; Murphy, Emily R. Chimeras of nurture. *American Journal of Bioethics* 2007 May; 7(5):1-2. NRCBL: 17.1.

Kmietowicz, Zosia. Public support for hybrid embryos rises, poll shows [news]. *BMJ: British Medical Journal* 2007 September 8; 335(7618): 466-467. NRCBL: 15.1; 18.1; 18.5.4; 22.1. SC: em.

> Keywords: *chimeras; *public opinion; *research embryo creation; embryos; survey; Keyword Identifiers: Great Britain

Kmietowicz, Zosia. Regulator gives green light to using human-animal embryos [news]. *BMJ: British Medical Journal* 2007 September 15; 335(7619): 531. NRCBL: 15.1; 18.1; 22.1. SC: le.

Lavieri, Robert R. The ethical mouse: be not like Icarus. *American Journal of Bioethics* 2007 May; 7(5): 57-58. NRCBL: 15.1; 18.1; 22.1; 4.4. Comments: Comment on: Henry T. Greely, Mildred K. Cho, Linda F. Hogle, and Debra M. Satz. Thinking about the human neuron mouse. American Journal of Bioethics 2007 May; 7(5): 27-40.

> Keywords: *chimeras; animal experimentation; animal rights; brain; emotions; human characteristics; moral status; personhood; philosophy; research ethics; stem cell transplantation; Proposed Keywords: mice

Loike, John D.; Tendler, Moshe D. Ethical dilemmas in stem cell research: human-animal chimeras. *Tradition* 2007 Winter; 40(4): 28-49. 65 fn. NRCBL: 18.5.4; 15.1; 1.2; 18.1; 22.1.

> Keywords: *brain; *chimeras; *embryonic stem cells; *Jewish ethics; *stem cell transplantation; embryo research; embryos; human dignity; species specificity

Mamo, Laura. Negotiating conception: lesbians' hybrid-technological practices. *Science, Technology, and Human Values* 2007 May; 32(3): 369-393. NRCBL: 14.1; 10. SC: an.

Mayor, Susan. UK body wants consultation on human-animal hybrid research [news]. *BMJ: British Medical Journal* 2007 January 20; 334(7585): 112. NRCBL: 15.1; 18.1; 22.1.

> Keywords: *chimeras; *embryo research; *regulation; public policy; Keyword Identifiers: *Great Britain; *Human Fertilisation and Embryology Authority

Morris, Stephen G. Canada's Assisted Human Reproduction Act: a chimera of religion and politics. *American Journal of Bioethics* 2007 February; 7(2): 69-70. 3 refs. NRCBL: 14.5; 1.2. SC: le. Comments: Comment on: Timothy Caulfield and Tania Bubela. Why a criminal ban? Analyzing the arguments against somatic cell nuclear transfer in the Canadian parliamentary debate. American Journal of Bioethics 2007 February; 7(2): 51-61.

> Keywords: *cloning; *embryo research; *embryonic stem cells; *government regulation; *nuclear transfer techniques; *political activity; *public policy; *religion; embryos; criminal law; legislation; moral status; politics; reproductive technologies; Keyword Identifiers: *Canada; Right to Life Movement; United States

Mykitiuk, Roxanne; Nisker, Jeff; Bluhm, Robyn. The Canadian Assisted Human Reproduction Act: protecting women's health while potentially allowing human somatic cell nuclear transfer into non-human oocytes. *American Journal of Bioethics* 2007 February; 7(2): 71-73. 11 refs. NRCBL: 14.5; 18.5.4; 15.1. SC: le. Comments: Comment on: Timothy Caulfield and Tania Bubela. Why a criminal ban? Analyzing the arguments against somatic cell nuclear transfer in the Canadian parliamentary debate. American Journal of Bioethics 2007 February; 7(2): 51-61.

NRCBL: National Reference Center for Bioethics Literature Classification Scheme   See inside front cover for terms.

101

Keywords: *chimeras; *cloning; *embryo research; *government regulation; *legal aspects; *legislation; *nuclear transfer techniques; *ovum donors; *women's health; criminal law; embryonic stem cells; embryos; reproductive technologies; women; Keyword Identifiers: *Canada

O'Dowd, Adrian. MPs back creation of hybrid embryos [news]. *BMJ: British Medical Journal* 2007 April 14; 334(7597): 764. NRCBL: 15.1; 18.1; 22.2; 18.5.4.
Keywords: *chimeras; *research embryo creation; embryonic stem cells; government regulation; legal aspects; public policy; Keyword Identifiers: *Great Britain; Department of Health (Great Britain); House of Commons Select Committee on Science and Technology (Great Britain); Human Fertilisation and Embryology Authority

O'Dowd, Adrian. UK may allow creation of "cybrids" for stem cell research [news]. *BMJ: British Medical Journal* 2007 March 10; 334(7592): 495. NRCBL: 18.5.4; 15.1; 22.2; 14.5; 18.1; 22.1. SC: le.
Keywords: *chimeras; *embryo research; *embryonic stem cells; government regulation; public policy; Keyword Identifiers: *Great Britain

Pergament, Eugene. Controversies and challenges of array comparative genomic hybridization in prenatal genetic diagnosis. *Genetics in Medicine* 2007 September; 9(9): 596-599. NRCBL: 15.2; 15.3; 15.5.

Robert, Jason Scott. The science and ethics of making part-human animals in stem cell biology. *FASEB Journal: Official Publication of the Federation of American Societies for Experimental Biology* 2006 May; 20(7): 838-845. 50 refs. NRCBL: 15.1; 18.1; 22.1; 18.5.4; 5.3. SC: rv.
Keywords: *chimeras; *embryonic stem cells; brain; embryo research; ethical review; fetal stem cells; guidelines; historical aspects; human dignity; moral policy; moral status; primates; public policy; regulation; research design; research ethics; review committees; risks and benefits; species specificity; stem cell transplantation; terminology; Proposed Keywords: mice; Keyword Identifiers: National Academy of Sciences; Nineteenth Century; Twentieth Century; United States

Rollin, Bernard E. Of mice and men. *American Journal of Bioethics* 2007 May; 7(5): 55-57. 6 refs. NRCBL: 15.1; 18.1; 22.1; 4.4. Comments: Comment on: Henry T. Greely, Mildred K. Cho, Linda F. Hogle, and Debra M. Satz. Thinking about the human neuron mouse. American Journal of Bioethics 2007 May; 7(5): 27-40.
Keywords: *chimeras; animal cloning; animal experimentation; animal rights; brain; emotions; ethical analysis; genetically modified animals; human characteristics; moral status; personhood; public opinion; research ethics; stem cell transplantation; Proposed Keywords: mice; species specificity

Sagoff, Mark. Further thoughts about the human neuron mouse. *American Journal of Bioethics* 2007 May; 7(5): 51-52. 10 refs. NRCBL: 15.1; 18.1; 22.1; 4.4. Comments: Comment on: Henry T. Greely, Mildred K. Cho, Linda F. Hogle, and Debra M. Satz. Thinking about the human neuron mouse. American Journal of Bioethics 2007 May; 7(5): 27-40.
Keywords: *chimeras; animal behavior; animal experimentation; animal rights; brain; emotions; human characteris-

tics; literature; moral status; neurosciences; personhood; research ethics; stem cell transplantation; Proposed Keywords: mice

Sharp, Lesley A. Human, monkey, machine: the brave new world of human hybridity. *In her:* Bodies, Commodities, and Biotechnologies: Death, Mourning, and Scientific Desire in the Realm of Human Organ Transfer. New York: Columbia University Press, 2007: 77-115. NRCBL: 15.1; 18.1; 22.1.

Silver, Lee M. Human-animal combinations. *In his:* Challenging Nature: The Clash of Science and Spirituality at the New Frontiers of Life. New York: Ecco, 2006: 172-187, 385-386. 26 fn. NRCBL: 15.1; 18.1; 22.1; 19.1.
Keywords: *chimeras; animal organs; embryos; emotions; organ transplantation; political activity; stem cells; Proposed Keywords: species specificity; Keyword Identifiers: United States

Tabor, Holly K.; Cho, Mildred K. Ethical implications of array comparative genomic hybridization in complex phenotypes: points to consider in research. *Genetics in Medicine* 2007 September; 9(9): 626-631. NRCBL: 15.3; 15.1; 15.2; 18.2; 18.3.

CIVIL COMMITMENT *See* INVOLUNTARY COMMITMENT

CLINICAL ETHICISTS *See* ETHICISTS AND ETHICS COMMITTEES

CLINICAL ETHICS *See* BIOETHICS AND MEDICAL ETHICS; ETHICISTS AND ETHICS COMMITTEES; NURSING ETHICS AND PHILOSOPHY; PROFESSIONAL ETHICS

CLINICAL ETHICS COMMITTEES *See* ETHICISTS AND ETHICS COMMITTEES

CLINICAL TRIALS *See* BIOMEDICAL RESEARCH; HUMAN EXPERIMENTATION

CLONING
See also HUMAN EXPERIMENTATION/ SPECIAL POPULATIONS/ EMBRYOS AND FETUSES; REPRODUCTIVE TECHNOLOGIES

Annex: Relevant international and national documents. *In:* Vöneky, Silja; Wolfrum, Rüdiger, eds. Human Dignity and Human Cloning. Leiden: Nijhoff, 2004: 185-319. NRCBL: 14.5; 21.1.
Keywords: *cloning; *international aspects; embryo research; embryonic stem cells; human dignity; human experimentation; human rights; legal aspects; reproductive technologies; standards; Keyword Identifiers: Council of Europe; Declaration of Helsinki; European Convention on Human Rights and Biomedicine; European Union; Germany; International Convention Against the Reproductive Cloning of Human Beings; Universal Declaration on the Human Genome and Human Rights; United Nations; World Health Organization

Dolly's legacy: ten years on, mammalian cloning is moving forward with central societal issues remaining unresolved. Yet human reproductive cloning seems inevitable [editorial]. *Nature* 2007 February 22; 445(7130): 795. NRCBL: 14.5; 22.3.
> Keywords: *animal cloning; *cloning; human dignity; mass media; reproductive technologies; risk; risks and benefits; stem cells; trends; Proposed Keywords: research; sheep

Replicator review: Nature has implemented a peer-reviewed policy for strong claims [editorial]. *Nature* 2007 November 22; 450(7169): 457-458. NRCBL: 14.5; 1.3.7; 18.5.4; 15.1; 22.2.
> Keywords: *cloning; *nuclear transfer techniques; *peer review; editorial policies; embryos; primates; Keyword Identifiers: *Nature

Aramesh, Kiarash; Dabbagh, Soroush. An Islamic view to stem cell research and cloning: Iran's experience. *American Journal of Bioethics* 2007 February; 7(2): 62. 5 refs. NRCBL: 14.5; 18.5.4; 15.1; 1.2. Comments: Timothy Caulfield and Tania Bubela. Why a criminal ban? Analyzing the arguments against somatic cell nuclear transfer in the Canadian parliamentary debate. American Journal of Bioethics 2007 February; 7(2): 51-61.
> Keywords: *cloning; *embryo research; *embryonic stem cells; *Islamic ethics; reproductive technologies; Keyword Identifiers: *Iran

Arsanjani, Mahnoush. The negotiations on a treaty on cloning: some reflections. *In:* Vöneky, Silja; Wolfrum, Rüdiger, eds. Human Dignity and Human Cloning. Leiden: Nijhoff, 2004: 145-165. 27 fn. NRCBL: 14.5; 21.1.
> Keywords: *cloning; *international aspects; *policy making; consensus; cultural pluralism; dissent; embryo research; government regulation; human rights; legal aspects; reproductive technologies; Proposed Keywords: *negotiating; Keyword Identifiers: United Nations

Barilan, Y. Michael. The debate on human cloning: some contributions from the Jewish tradition. *In:* Roetz, Heiner, ed. Cross-Cultural Issues in Bioethics: The Example of Human Cloning. New York: Rodopi, 2006: 311-340. NRCBL: 14.5; 1.2; 14.5.

Becker, Gerhold K. Chinese ethics and human cloning: a view from Hong Kong. *In:* Roetz, Heiner, ed. Cross-Cultural Issues in Bioethics: The Example of Human Cloning. New York: Rodopi, 2006: 107-139. NRCBL: 14.5; 2.1; 21.7.

Bonnicksen, Andrea L. Therapeutic cloning: politics and policy. *In:* Steinbock, Bonnie, ed. The Oxford Handbook of Bioethics. Oxford; New York: Oxford University Press, 2007: 441-468. 94 refs. NRCBL: 14.5; 18.5.4; 15.1. SC: an; rv.
> Keywords: *cloning; *embryonic stem cells; *policy analysis; *policy making; advisory committees; cultural pluralism; embryo research; ethical analysis; ethics committees; federal government; government financing; government regulation; industry; international aspects; legal aspects; moral policy; nuclear transfer techniques; ovum donors; organizational policies; politics; professional organizations; research priorities; research support; resource allocation; state government; values; Keyword Identifiers: United States

Braune, Florian; Biller-Andorno, Nikola; Wiesemann, Claudia. Human reproductive cloning: a test case for individual rights? *In:* Roetz, Heiner, ed. Cross-Cultural Issues in Bioethics: The Example of Human Cloning. New York: Rodopi, 2006: 445-458. NRCBL: 14.5; 8.3.1; 8.1.

Brownsword, Roger. Cloning, zoning and the harm principle. *In:* McLean, Sheila A.M., ed. First Do No Harm: Law, Ethics, and Healthcare. Aldershot, England; Burlington, VT: Ashgate, 2006: 527-542. 43 fn. NRCBL: 14.5; 21.7. SC: an.
> Keywords: *cloning; *cultural pluralism; *ethical analysis; *regulation; beneficence; children; embryo research; embryonic stem cells; embryos; human dignity; human rights; international aspects; motivation; principle-based ethics; policy analysis; reproductive technologies; utilitarianism; values; Proposed Keywords: *harm

Carlson, Elof Axel. Cloning, stem cells, hyperbole, and cant. *In his:* Times of Triumph, Times of Doubt: Science and the Battle for Public Trust. Cold Spring Harbor, NY: Cold Spring Harbor Laboratory Press, 2006: 163-171. 9 fn. NRCBL: 14.5; 18.5.4; 15.1.
> Keywords: *cloning; *embryo research; *embryonic stem cells; motivation; reproductive technologies; stem cell transplantation; twinning; Proposed Keywords: blastocysts

Caulfield, Timothy. Human reproductive cloning: assessing the concerns. *In:* Eisen, Arri; Laderman, Gary, eds. Science, Religion, and Society: An Encyclopedia of History, Culture, and Controversy. Vol. 2. Armonk, NY: M.E. Sharpe, 2007: 795-802. 28 refs. NRCBL: 14.5.
> Keywords: *cloning; autonomy; embryo research; ethical analysis; government regulation; human dignity; international aspects; reproductive technologies

Caulfield, Timothy. Stem cells, clones, consensus, and the law. *In:* Knowles, Lori P.; Kaebnick, Gregory E., eds. Reprogenetics: Law, Policy, and Ethical Issues. Baltimore: Johns Hopkins University Press, 2007: 105-123. 73 refs. NRCBL: 14.5; 18.5.4; 15.1; 14.1. SC: an.
> Keywords: *cloning; *cultural pluralism; *embryo research; *genetic techniques; *policy making; *public opinion; *public policy; *regulation; *reproductive technologies; advisory committees; biotechnology; consensus; criminal law; dissent; embryonic stem cells; embryos; focus groups; government regulation; human dignity; international aspects; legal aspects; moral status; nuclear transfer techniques; policy analysis; religion; religious ethics; risks and benefits; survey; values; Keyword Identifiers: *United States; Canada; United Nations

Check, Erika. Dolly: a hard act to follow [news]. *Nature* 2007 February 22; 445(7130): 802. NRCBL: 14.5; 18.5.4; 15.1; 19.1; 22.3.
> Keywords: *animal cloning; adverse effects; methods; stem cells; trends; Proposed Keywords: sheep

NRCBL: National Reference Center for Bioethics Literature Classification Scheme   See inside front cover for terms.

103

Cibelli, Jose. A decade of cloning mystique. *Science* 2007 May 18; 316(5827): 990-992. 14 refs. NRCBL: 14.5; 22.2; 15.1.

> Keywords: *animal cloning; embryo transfer; methods; nuclear transfer techniques

Cibelli, Jose. Is therapeutic cloning dead? The ability to generate pluripotent stem cells directly from skin fibroblasts may render ethical debates over the use of human oocytes to create stem cells irrelevant. *Science* 2007 December 21; 318(5858): 1879-1880. 17 refs. NRCBL: 14.5; 18.5.4; 15.1.

> Keywords: *cloning; *stem cells; embryonic stem cells; genetic techniques; methods; nuclear transfer techniques; Proposed Keywords: *pluripotent stem cells

Cormick, Craig. Cloning goes to the movies = A clonagem vai ao cinema. *Historia, Ciencias, Saude — Manguinhos* 2006 October; 13(Supplement): 181-212. NRCBL: 14.5; 7.1; 15.1. Note: Abstract in Portuguese.

Cyranoski, David. Race to mimic human embryonic stem cells [news]. *Nature* 2007 November 22; 450(7169): 462-463. NRCBL: 14.5; 18.5.4; 15.1; 22.2.

> Keywords: *cloning; *embryonic stem cells; *methods; adult stem cells; embryo research; nuclear transfer techniques; primates; stem cell transplantation; Proposed Keywords: *pluripotent stem cells; regenerative medicine

de Melo-Martín, Immaculada. Cloning — or not — human beings. *In her:* Taking Biology Seriously: What Biology Can and Cannot Tell Us About Moral and Public Policy Issues. Lanham: Rowman and Littlefield, 2005: 45-61. 33 fn. NRCBL: 14.5. SC: an.

> Keywords: *cloning; *moral policy; biology; genetic determinism; genetic disorders; genetic identity; genotype; human dignity; human genome; motivation; phenotype; reproductive technologies

de Melo-Martín, Immaculada. Putting human cloning where it belongs. *In her:* Taking Biology Seriously: What Biology Can and Cannot Tell Us About Moral and Public Policy Issues. Lanham: Rowman and Littlefield, 2005: 63-76. 28 fn. NRCBL: 14.5. SC: an.

> Keywords: *cloning; *moral policy; biology; genetic disorders; genetic relatedness ties; homosexuality; infertility; motivation; reproductive technologies; resource allocation; risks and benefits; suffering

Döring, Ole. Culture and bioethics in the debate on the ethics of human cloning in China. *In:* Roetz, Heiner, ed. Cross-Cultural Issues in Bioethics: The Example of Human Cloning. New York: Rodopi, 2006: 77-105. NRCBL: 14.5; 2.1; 18.5.4; 21.7.

Eich, Thomas. The debate about human cloning among Muslim religious scholars since 1997. *In:* Roetz, Heiner, ed. Cross-Cultural Issues in Bioethics: The Example of Human Cloning. New York: Rodopi, 2006: 291-309. NRCBL: 14.5; 1.2; 2.2.

FIGO Committee for the Ethical Aspects of Human Reproduction and Women's Health; Serour, G.I. Human cloning. *International Journal of Gynecology and Obstetrics* 2006 June; 93(3): 282. NRCBL: 14.5; 18.5.4.

Gardner, Richard. Therapeutic and reproductive cloning — a scientific perspective. *In:* Gunning, Jennifer; Holm, Søren, eds. Ethics, law, and society. Volume 1. Aldershot, Hants, England; Burlington, VT: Ashgate, 2005: 9-16. 26 refs. NRCBL: 14.5.

> Keywords: *cloning; adult stem cells; embryo research; embryonic stem cells; reproductive technologies

Gratwohl, Alois. "Therapeutisches Klonen" aus der Sicht eines Klinikers. *In:* Schreiber, Hans-Peter, ed. Biomedizin und Ethik: Praxis, Recht, Moral. Basel; Boston: Birkhäuser, 2004: 23-28. NRCBL: 14.5; 18.5.4.

> Keywords: *cloning; *stem cell transplantation; adult stem cells; embryonic stem cells; reproductive technologies

Greif, Karen F.; Merz, Jon F. Brave new world revisited: human cloning and stem cells; the Asilomar Conference on Recombinant DNA: a model for self-regulation? *In their:* Current Controversies in the Biological Sciences: Case Studies of Policy Challenges from New Technologies. Cambridge, MA: MIT, 2007: 101-115. 54 refs. NRCBL: 14.5; 18.5.4; 15.1.

> Keywords: *cloning; *embryo research; *embryonic stem cells; *public policy; *recombinant DNA research; *regulation; advisory committees; government regulation; reproductive technologies; self regulation; Keyword Identifiers: *United States; Asilomar Conference; President's Council on Bioethics; Recombinant DNA Advisory Committee

Hall, Vanessa J.; Stojkovic, Petra; Stojkovic, Miodrag. Using therapeutic cloning to fight human disease: a conundrum or reality? *Stem Cells* 2006 July; 24(7): 1628-1637. 99 refs. NRCBL: 14.5; 18.5.4; 15.1; 18.2.

> Keywords: *cloning; *embryonic stem cells; *stem cell transplantation; adult stem cells; adverse effects; beginning of life; cell lines; commodification; in vitro fertilization; international aspects; legislation; ovum donors; reproductive technologies; research support; Proposed Keywords: parthenogenesis; Keyword Identifiers: Australia; Canada; Europe; Japan; Mexico

Haran, Joan. Managing the boundaries between maverick cloners and mainstream scientists: the life cycle of a news event in a contested field. *New Genetics and Society* 2007 August; 26(2): 203-219. 19 refs. NRCBL: 14.5; 1.3.7.

> Keywords: *cloning; *mass media; embryo transfer; journalism; peer review; public opinion; reproductive technologies; researchers; technical expertise; Proposed Keywords: hyperbole; public relations; Keyword Identifiers: *Zavos, Panos; Great Britain

Haraway, Donna J. Cloning mutts, saving tigers: ethical emergents in technocultural dog worlds. *In:* Franklin, Sarah; Lock, Margaret, eds. Remaking Life and Death: Toward an Anthropology of the Biosciences. Santa Fe: School of American Research Press; Oxford: James Currey, 2003: 293-327. 6 fn. NRCBL: 14.5; 15.1; 22.3.

> Keywords: *animal cloning; biotechnology; genetic diversity; industry; Proposed Keywords: *domestic animals; biodiversity; wild animals; zoo animals

Heng, Boon Chin; Tong, Guo Qing; Stojkovic, Miodrag. The egg-sharing model for human therapeutic cloning research: managing donor selection criteria, the proportion of shared oocytes allocated to research, and amount of financial subsidy given to the donor. *Medical Hypotheses* 2006; 66(5): 1022-1024. 13 refs. NRCBL: 14.5; 14.4; 18.1.

> Keywords: *cloning; *embryo research; *financial support; *incentives; *ovum; *ovum donors; *remuneration; *reproductive technologies; *resource allocation; age factors; patients; scarcity; standards

Hoffmann, Thomas Sören. Primordial ownership versus dispossession of the body. A contribution to the problem of cloning from the perspective of classical European philosophy of law. *In:* Roetz, Heiner, ed. Cross-Cultural Issues in Bioethics: The Example of Human Cloning. New York: Rodopi, 2006: 387-407. NRCBL: 14.5; 4.4; 14.1; 18.5.4; 1.1.

Holden, Constance. Former Hwang colleague faked monkey data, U.S. says [news]. *Science* 2007 January 19; 315(5810): 317. NRCBL: 14.5; 1.3.9; 15.1; 18.5.4; 22.2.

> Keywords: *animal cloning; *embryonic stem cells; *scientific misconduct; embryos; government regulation; nuclear transfer techniques; research institutes; researchers; universities; Keyword Identifiers: *Park, Jong Hyuk; Hwang, Woo Suk; Korea; Office of Research Integrity; Schatten, Gerald; United States; University of Pittsburgh

Horres, Robert; Ölschleger; Steineck, Christian. Cloning in Japan: public opinion, expert counselling, and bioethical reasoning. *In:* Roetz, Heiner, ed. Cross-Cultural Issues in Bioethics: The Example of Human Cloning. New York: Rodopi, 2006: 17-49. NRCBL: 14.5; 18.5.4; 15.1. SC: em.

Huxley, Andrew. The Pali Buddhist approach to human cloning. *In:* Vöneky, Silja; Wolfrum, Rüdiger, eds. Human Dignity and Human Cloning. Leiden: Nijhoff, 2004: 13-22. 16 fn. NRCBL: 14.5; 1.2.

> Keywords: *Buddhist ethics; *cloning; family relationship; reproductive technologies; Keyword Identifiers: Asia

Ilkilic, Ilhan. Human cloning as a challenge to traditional health care cultures. *In:* Roetz, Heiner, ed. Cross-Cultural Issues in Bioethics: The Example of Human Cloning. New York: Rodopi, 2006: 409-423. NRCBL: 14.5; 4.2; 21.7.

Jaenisch, Rudolf. Nuclear cloning, embryonic stem cells, and gene transfer. *In:* Rasko, John E.J.; O'Sullivan, Gabrielle M.; Ankeny, Rachel A., eds. The Ethics of Inheritable Genetic Modification: A Dividing Line? Cambridge: Cambridge University Press, 2006: 35-55. 43 fn. NRCBL: 14.5; 15.4; 18.5.4; 15.1.

> Keywords: *cloning; *embryonic stem cells; *gene therapy; *risks and benefits; adverse effects; animal cloning; chimeras; embryo disposition; embryo research; embryonic development; gene transfer techniques; human experimentation; nuclear transfer techniques; reproductive technologies

Jochemsen, Henk. Is cloning compatible with human rights and human dignity? *In:* Wagner, Teresa; Carbone,

Leslie, eds. Fifty Years after the Declaration: the United Nations' Record on Human Rights. Lanham, MD: University Press of America, 2001: 33-43. 15 fn. NRCBL: 14.5; 21.1; 4.4.

> Keywords: *cloning; *human dignity; embryonic stem cells; genetic engineering; germ cells; human rights; preimplantation diagnosis; reproductive technologies; research embryo creation

Joung, Phillan; Eggert, Marion. The cloning debate in South Korea. *In:* Roetz, Heiner, ed. Cross-Cultural Issues in Bioethics: The Example of Human Cloning. New York: Rodopi, 2006: 155-178. NRCBL: 14.5; 18.5.4; 15.1.

Kolata, Gina. Researcher who helped start stem cell war may now end it [news]. *New York Times* 2007 November 22; p. A1, A28. NRCBL: 14.5; 15.1; 18.5.4; 19.5. SC: po. Identifiers: James A. Thomson.

> Keywords: *adult stem cells; *embryonic stem cells; *methods; embryo research; researchers; Keyword Identifiers: *Thomson, James A.; University of Wisconsin; United States

Lysaught, M. Therese. Becoming one body: health care and cloning. *In:* Hauerwas, Stanley and Wells, Samuel, eds. The Blackwell Companion to Christian Ethics. Malden, MA : Blackwell, 2004: 263-275. NRCBL: 14.5; 1.2; 15.1.

Mameli, M. Reproductive cloning, genetic engineering and the autonomy of the child: the moral agent and the open future. *Journal of Medical Ethics* 2007 February; 33(2): 87-93. 18 refs. NRCBL: 14.5; 15.1; 1.1. SC: an.

> Keywords: *autonomy; *children; *cloning; *ethical analysis; *future generations; *genetic engineering; biotechnology; choice behavior; genetic determinism; genetic enhancement; moral obligations; parents; reproductive rights; reproductive technologies; self concept

Abstract: Some authors have argued that the human use of reproductive cloning and genetic engineering should be prohibited because these biotechnologies would undermine the autonomy of the resulting child. In this paper, two versions of this view are discussed. According to the first version, the autonomy of cloned and genetically engineered people would be undermined because knowledge of the method by which these people have been conceived would make them unable to assume full responsibility for their actions. According to the second version, these biotechnologies would undermine autonomy by violating these people's right to an open future. There is no evidence to show that people conceived through cloning and genetic engineering would inevitably or even in general be unable to assume responsibility for their actions; there is also no evidence for the claim that cloning and genetic engineering would inevitably or even in general rob the child of the possibility to choose from a sufficiently large array of life plans.

McLachlan, Hugh. Let's legalise cloning. *New Scientist* 2007 July 21-27; 195(2613): 20. NRCBL: 14.5.

> Keywords: *cloning; *public policy; freedom; legal aspects; reproductive technologies; risks and benefits

NRCBL: National Reference Center for Bioethics Literature Classification Scheme   See inside front cover for terms.

105

Miller, Henry I. Two views of the emperor's new clones [letter]. *Nature Biotechnology* 2007 March; 25(3): 281. NRCBL: 14.5; 22.3. Comments: The emperor's new clones. Nature Biotechnology 2007 January; 25(1): 1.
>   Keywords: *animal cloning; *food; *political activity; government regulation; mandatory programs; risks and benefits; Keyword Identifiers: Food and Drug Administration; United States

Morin, Karine; Green, Shane K. Professionalism in biomedical science. *American Journal of Bioethics* 2007 February; 7(2): 66-68. 6 refs. 2 fn. NRCBL: 14.5; 1.3.9; 18.5.4. Comments: Timothy Caulfield and Tania Bubela. Why a criminal ban? Analyzing the arguments against somatic cell nuclear transfer in the Canadian parliamentary debate. American Journal of Bioethics 2007 February; 7(2): 51-61.
>   Keywords: *cloning; *government regulation; *nuclear transfer techniques; *professional autonomy; *researchers; *self regulation; biomedical research; chimeras; criminal law; embryo research; guidelines; legal aspects; professional ethics; public opinion; reproductive technologies; research ethics; scientific misconduct; Keyword Identifiers: Canada

Morris, Jonathan. A brave new world of clones. *In his:* The Ethics of Biotechnology. Philadelphia: Chelsea Publishers, 2006: 30-55. NRCBL: 14.5; 22.1. SC: po.
>   Keywords: *animal cloning; *cloning; animal experimentation; domestic animals; historical aspects; methods; moral policy; political activity; public opinion; Keyword Identifiers: People for the Ethical Treatment of Animals (PETA)

Morris, Jonathan. Human cloning: should humans be cloned? *In his:* The Ethics of Biotechnology. Philadelphia: Chelsea Publishers, 2006: 56-82. NRCBL: 14.5; 14.1. SC: po.
>   Keywords: *cloning; eugenics; family relationship; genetic engineering; genetic enhancement; methods; moral policy; public policy; reproductive technologies; research embryo creation; risks and benefits

Neresini, Federico. Eve's sons. *New Genetics and Society* 2007 August; 26(2): 221-233. 33 refs. 7 fn. NRCBL: 14.5. Identifiers: Italy.
>   Keywords: *cloning; *mass media; morality; public opinion; reproductive technologies; self concept; Proposed Keywords: hyperbole; Keyword Identifiers: *Italy; *Raelians; Clonaid

Nor, Siti Nurani Mohd. The ethics of human cloning: with reference to the Malaysian bioethical discourse. *In:* Roetz, Heiner, ed. Cross-Cultural Issues in Bioethics: The Example of Human Cloning. New York: Rodopi, 2006: 215-246. NRCBL: 14.5; 1.2.

Oeming, Manfred. The Jewish perspective on cloning. *In:* Vöneky, Silja; Wolfrum, Rüdiger, eds. Human Dignity and Human Cloning. Leiden: Nijhoff, 2004: 35-45. 21 fn. NRCBL: 14.5; 1.2.
>   Keywords: *cloning; *Jewish ethics; embryo research; reproductive technologies

Pearson, Yvette. Never let me clone?: countering an ethical argument against the reproductive cloning of humans. *EMBO Reports* 2006 July; 7(7): 657-660. NRCBL: 14.5; 5.3; 15.4.

Pollard, Irina. Cloning technology. *In her:* Life, Love and Children: A Practical Introduction to Bioscience Ethics and Bioethics. Boston: Kluwer Academic Publishers, 2002: 145-155. 11 refs. (p. 237-238). NRCBL: 14.5.
>   Keywords: *cloning; embryo research; embryonic stem cells; moral policy; recombinant DNA research; reproductive technologies; stem cell transplantation; Proposed Keywords: regenerative medicine

Ratanakul, Pinit. Human cloning: Thai Buddhist perspectives. *In:* Roetz, Heiner, ed. Cross-Cultural Issues in Bioethics: The Example of Human Cloning. New York: Rodopi, 2006: 203-213. NRCBL: 14.5; 1.2; 18.5.4.

Renzong, Qiu. Cloning issues in China. *In:* Roetz, Heiner, ed. Cross-Cultural Issues in Bioethics: The Example of Human Cloning. New York: Rodopi, 2006: 51-75. NRCBL: 14.5; 18.5.4; 15.1.

Sachedina, Abdulaziz. The cultural and religious in Islamic biomedicine: the case of human cloning. *In:* Roetz, Heiner, ed. Cross-Cultural Issues in Bioethics: The Example of Human Cloning. New York: Rodopi, 2006: 263-290. NRCBL: 14.5; 1.2; 21.7.

Sadeghi, Mahmoud. Islamic perspectives on human cloning. *Human Reproduction and Genetic Ethics: An International Journal* 2007; 13(2): 32-40. NRCBL: 14.5; 1.2.

Schlieter, Jens. Some aspects of the Buddhist assessment of human cloning. *In:* Vöneky, Silja; Wolfrum, Rüdiger, eds. Human Dignity and Human Cloning. Leiden: Nijhoff, 2004: 23-33. 22 fn. NRCBL: 14.5; 1.2.
>   Keywords: *Buddhist ethics; *cloning; embryo research; human dignity; reproductive technologies; Keyword Identifiers: Asia

Schlieter, Jens. Some observations on Buddhist thoughts on human cloning. *In:* Roetz, Heiner, ed. Cross-Cultural Issues in Bioethics: The Example of Human Cloning. New York: Rodopi, 2006: 179-202. NRCBL: 14.5; 1.2; 18.5.4.

Schubert, David. Two views of the emperor's new clones [letter]. *Nature Biotechnology* 2007 March; 25(3): 282-283. 19 refs. NRCBL: 14.5; 22.3. Comments: The emperor's new clones [editorial]. Nature Biotechnology 2007 January; 25(1): 1.
>   Keywords: *animal cloning; *food; adverse effects; animal welfare; genetic diversity; genetically modified animals; genetically modified plants; government regulation; industry; mandatory programs; nuclear transfer techniques; Keyword Identifiers: Food and Drug Administration; United States

Shannon, Thomas A. Cloning, uniqueness, and individuality. *In:* Walter, James J.; Shannon, Thomas A., eds. Contemporary Issues in Bioethics: A Catholic Perspective. Lanham, MD: Rowman and Littlefield Publishers, 2005:

101-123. 47 fn. NRCBL: 14.5; 1.1; 1.2; 4.4; 15.1; 18.5.4. SC: an.

> Keywords: *cloning; *embryos; *genetic identity; *personhood; *philosophy; *theology; embryo research; ethical analysis; genetic engineering; human characteristics; human genome; Human Genome Project; nuclear transfer techniques; research priorities; social impact; twinning; Proposed Keywords: species specificity; Keyword Identifiers: Scotus, John Dunn

Shenfield, Françoise, Babinet, Charles; Teboul, Gérard. Human cloning: reproductive crime or therapeutic panacea — where are we now? *In:* Shenfield, Françoise; Sureau, Claude, eds. Contemporary Ethical Dilemmas in Assisted Reproduction. Abingdon: Informa Healthcare, 2006: 13-25. 47 refs. NRCBL: 14.5; 1.3.9.

> Keywords: *cloning; *reproductive technologies; embryonic stem cells; international aspects; legal aspects; ovum donors; public policy; research embryo creation; value of life; terminology; Keyword Identifiers: *Europe; European Group on Ethics in Science and New Technologies; European Society of Human Reproduction and Embryology; United Nations

Silver, Lee M. The politics of cloning. *In his:* Challenging Nature: The Clash of Science and Spirituality at the New Frontiers of Life. New York: Ecco, 2006: 125-146, 376-380. 62 fn. NRCBL: 14.5; 18.5.4; 15.1; 19.1.

> Keywords: *cloning; *embryo research; *embryos; *political activity; embryonic stem cells; international aspects; legislation; ovum donors; patents; public policy; terminology; Keyword Identifiers: *United States; Asia; Europe

Simitis, Spiros. A convention on cloning — annotations to an almost unsolvable dilemma. *In:* Vöneky, Silja; Wolfrum, Rüdiger, eds. Human Dignity and Human Cloning. Leiden: Nijhoff, 2004: 167-182. 51 fn. NRCBL: 14.5; 15.1; 21.1.

> Keywords: *cloning; *international aspects; *policy making; biotechnology; consensus; embryo research; government regulation; legal aspects; reproductive technologies; standards; Proposed Keywords: negotiating; Keyword Identifiers: European Union; United Nations

Spaemann, Robert. Christianity and western philosophy. *In:* Vöneky, Silja; Wolfrum, Rüdiger, eds. Human Dignity and Human Cloning. Leiden: Nijhoff, 2004: 47-51. 1 fn. NRCBL: 14.5; 1.2.

> Keywords: *Christian ethics; *cloning; embryo research; genetic intervention; human dignity; personhood; reproductive technologies

Starck, Christian. The human embryo is a person and not an object. *In:* Vöneky, Silja; Wolfrum, Rüdiger, eds. Human Dignity and Human Cloning. Leiden: Nijhoff, 2004: 63-67. 10 fn. NRCBL: 14.5; 4.4.

> Keywords: *cloning; embryo research; embryos; human dignity; legal aspects; personhood; reproductive technologies; Keyword Identifiers: Germany

Stoos, William Kevin. Who am I? Why am I? (The anguish of a clone). *Linacre Quarterly* 2007 May; 74(2): 171-173. NRCBL: 14.5; 1.2; 4.4.

Svenaeus, Fredrik. A Heideggerian defense of therapeutic cloning. *Theoretical Medicine and Bioethics* 2007; 28(1): 31-62. 42 refs. 48 fn. NRCBL: 14.5; 1.1; 4.4; 18.5.4; 15.1; 19.5. SC: an.

> Keywords: *cloning; *embryo research; *embryonic stem cells; *embryos; *ethical analysis; *moral status; *philosophy; beginning of life; in vitro fertilization; personhood; reproduction; Proposed Keywords: technology; Keyword Identifiers: *Heidigger, Martin; Habermas, Jürgen; Kant, Immanuel

Abstract: Debates about the legitimacy of embryonic stem-cell research have largely focused on the type of ethical value that should be accorded to the human embryo in vitro. In this paper, I try to show that, to broaden the scope of these debates, one needs to articulate an ontology that does not limit itself to biological accounts, but that instead focuses on the embryo's place in a totality of relevance surrounding and guiding a human practice. Instead of attempting to substantiate the ethical value of the embryo exclusively by pointing out that it has potentiality for personhood, one should examine the types of practices in which the embryo occurs and focus on the ends inherent to these practices. With this emphasis on context, it becomes apparent that the embryo's ethical significance can only be understood by elucidating the attitudes that are established towards it in the course of specific activities. The distinction between fertilized embryos and cloned embryos proves to be important in this contextual analysis, since, from the point of view of practice, the two types of embryos appear to belong to different human practices: (assisted) procreation and medical research, respectively. In my arguments, I highlight the concepts of practice, technology, and nature, as they have been analyzed in the phenomenological tradition, particularly by Martin Heidegger. I come to the conclusion that therapeutic cloning should be allowed, provided that it turns out to be a project that benefits medical science in its aim to battle diseases. Important precautions have to be taken, however, in order to safeguard the practice of procreation from becoming perverted by the aims and attitudes of medical science when the two practices intersect. The threat in question needs to be taken seriously, since it concerns the structure and goal of practices which are central to our very self understanding as human beings.

Swift, Jennifer. Eggs don't come cheap. *New Scientist* 2007 December 8-14; 196(2633): 22. NRCBL: 14.5; 14.2; 19.5; 18.5.3; 18.5.4; 15.1; 9.5.5.

> Keywords: *cloning; *ovum donors; *research embryo creation; *risk; *stem cells; adverse effects; conflict of interest; hormones; nontherapeutic research; remuneration; risks and benefits; women; Proposed Keywords: pluripotent stem cells

Tannert, Christof. The autonomy axiom and the cloning of humans. *Human Reproduction and Genetic Ethics: An International Journal* 2007; 13(1): 4-7. 7 refs. NRCBL: 14.5; 1.1; 14.1. SC: an.

> Keywords: *cloning; *moral policy; autonomy; beginning of life; cultural pluralism; embryo research; embryonic development; embryonic stem cells; ethical analysis; motivation; mother fetus relationship; reproductive technologies

---

NRCBL: National Reference Center for Bioethics Literature Classification Scheme   See inside front cover for terms.

107

van den Daele, Wolfgang. Moderne Tabus? — Zum Verbot des Klonens von Menschen. *In:* Schreiber, Hans-Peter, ed. Biomedizin und Ethik: Praxis, Recht, Moral. Basel; Boston: Birkhäuser, 2004: 77-83, 96. 5 refs. NRCBL: 14.5.

> Keywords: *cloning; human characteristics; human dignity; informal social control; regulation; reproductive technologies

Vitzthum, Wolfgang Graf. Back to Kant! An interjection in the debate on cloning and human dignity. *In:* Vöneky, Silja; Wolfrum, Rüdiger, eds. Human Dignity and Human Cloning. Leiden: Nijhoff, 2004: 87-106. 44 fn. NRCBL: 14.5; 4.4.

> Keywords: *cloning; *human dignity; Christian ethics; embryo research; government regulation; international aspects; legal aspects; philosophy; reproductive technologies; values; Keyword Identifiers: Germany; Kant, Immanuel

Wadman, Meredith. Dolly: a decade on. *Nature* 2007 February 22; 445(7130): 800-801. 4 refs. NRCBL: 14.5; 18.5.4; 15.1; 19.1.

> Keywords: *animal cloning; *cloning; *historical aspects; *research embryo creation; embryo research; embryonic stem cells; international aspects; mass media; public opinion; reproductive technologies; trends; Proposed Keywords: sheep; Keyword Identifiers: Twentieth Century; Twenty-First Century

Wessler, Heinz Werner. The charm of biotechnology: human cloning and Hindu bioethics in perspective. *In:* Roetz, Heiner, ed. Cross-Cultural Issues in Bioethics: The Example of Human Cloning. New York: Rodopi, 2006: 247-262. NRCBL: 14.5; 1.2; 18.5.4; 15.1.

Wexler, Barbara. Cloning. *In her:* Genetics and genetic engineering. 2005 ed. Detroit, MI: Thomson/Gale Group, 2006: 117-134. NRCBL: 14.5.

> Keywords: *cloning; adolescents; adults; animal cloning; embryos; genes; government regulation; international aspects; legislation; moral policy; morality; nuclear transfer techniques; public opinion; public policy; reproductive technologies; research embryo creation; state government; stem cells; Keyword Identifiers: United States

Wolfrum, Rüdiger; Vöneky, Silja. Who is protected by human rights conventions? Protection of the embryo vs. scientific freedom and public health. *In:* Vöneky, Silja; Wolfrum, Rüdiger, eds. Human Dignity and Human Cloning. Leiden: Nijhoff, 2004: 133-143. 43 fn. NRCBL: 14.5; 21.1.

> Keywords: *cloning; *human dignity; *human rights; *international aspects; embryos; freedom; government regulation; legal aspects; reproductive technologies; standards; Keyword Identifiers: Germany; United Nations; Universal Declaration of Human Rights

CLONING/ LEGAL ASPECTS

Annas, George J.; Andrews, Lori B.; Isasi, Rosario M. Protecting the endangered human: toward an international treaty prohibiting cloning and inheritable alterations. *In:* Gruskin, Sofia; Grodin, Michael A.; Annas, George J.; Marks, Stephen P., eds. Perspectives on Health and Human Rights. New York: Routledge, 2005: 135-162. 110 fn. NRCBL: 14.5; 15.1; 21.1; 4.4. SC: le.

> Keywords: *cloning; *genetic engineering; *germ cells; *international aspects; *regulation; human rights; legal aspects; nuclear transfer techniques; reproductive technologies; Keyword Identifiers: Convention on the Preservation of the Human Species; United States

Araujo, Robert John. The UN Declaration on Human Cloning: a survey and assessment of the debate. *National Catholic Bioethics Quarterly* 2007 Spring; 7(1): 129-149. NRCBL: 14.5; 18.2; 21.1; 4.4; 2.4. SC: le.

Blackford, Russell. Slippery slopes to slippery slopes: therapeutic cloning and the criminal law. *American Journal of Bioethics* 2007 February; 7(2): 63-64. 4 refs. NRCBL: 14.5; 15.1; 18.5.4. SC: le. Comments: Timothy Caulfield and Tania Bubela. Why a criminal ban? Analyzing the arguments against somatic cell nuclear transfer in the Canadian parliamentary debate. American Journal of Bioethics 2007 February; 7(2): 51-61.

> Keywords: *cloning; *embryo research; *legal aspects; *reproductive technologies; *wedge argument; criminal law; biomedical technologies; genetic engineering; government regulation; nuclear transfer techniques; risks and benefits; Proposed Keywords: logic; Keyword Identifiers: *Canada

Cameron, Nigel M. de S. The American debate on human cloning. *In:* Roetz, Heiner, ed. Cross-Cultural Issues in Bioethics: The Example of Human Cloning. New York: Rodopi, 2006: 363-386. NRCBL: 14.5; 18.5.4; 15.1. SC: le.

Casey, Michael. Cloning down under: an Australian reversal on embryo research. *New Atlantis* 2007 Winter; 15: 125-128. NRCBL: 14.5; 18.5.4; 1.3.5. SC: le.

> Keywords: *cloning; *embryo research; *embryonic stem cells; *embryos; *legal aspects; advisory committees; chimeras; government regulation; legislation; politics; public policy; research embryo creation; Keyword Identifiers: *Australia

Cash, Adrienne N. Attack of the clones: legislative approaches to human cloning in the United States. *Duke Law and Technology Review* 2005; 26: 14 p. [Online]. Accessed: http://www.law.duke.edu/journals/dltr/articles/pdf/2005dltr0026.pdf [2007 May 3]. 83 fn. NRCBL: 14.5. SC: le.

> Keywords: *cloning; *legal aspects; constitutional law; federal government; government regulation; legislation; reproductive technologies; state government; Keyword Identifiers: *United States

Caulfield, Timothy; Bubela, Tania. Why a criminal ban? Analyzing the arguments against somatic cell nuclear transfer in the Canadian parliamentary debate. *American Journal of Bioethics* 2007 February; 7(2): 51-61. 53 refs. 2 fn. NRCBL: 14.5; 18.5.4; 14.1; 4.4. SC: le.

> Keywords: *cloning; *embryo research; *embryonic stem cells; *government regulation; *legal aspects; *nuclear transfer techniques; *policy making; commodification; criminal law; cultural pluralism; embryos; international aspects; legislation; moral status; public opinion; public pol-

icy; reproductive technologies; Keyword Identifiers: *Canada

Abstract: Somatic cell nuclear transfer (SCNT) remains a controversial technique, one that has elicited a variety of regulatory responses throughout the world. On March 29, 2005, Canada's Assisted Human Reproduction Act came into force. This law prohibits a number of research activities, including SCNT. Given the pluralistic nature of Canadian society, the creation of this law stands as an interesting case study of the policy-making process and how and why a liberal democracy ends up making the relatively rare decision to use a statutory prohibition, backed by severe penalties, to stop a particular scientific activity. In this article, we provide a comprehensive and systematic legal analysis of the legislative process and parliamentary debates associated with the passage of this law.

Dreier, Horst. Does cloning violate the basic law's guarantee of human dignity? *In:* Vöneky, Silja; Wolfrum, Rüdiger, eds. Human Dignity and Human Cloning. Leiden: Nijhoff, 2004: 77-85. 2 fn. NRCBL: 14.5. SC: le.

Keywords: *cloning; *human dignity; *legal aspects; embryo research; moral status; reproductive technologies; Keyword Identifiers: *Germany

Ipsen, Jörn. Does the German basic law protect against human cloning? *In:* Vöneky, Silja; Wolfrum, Rüdiger, eds. Human Dignity and Human Cloning. Leiden: Nijhoff, 2004: 69-75. 19 fn. NRCBL: 14.5. SC: le.

Keywords: *cloning; *legal aspects; embryo research; embryos; genetic engineering; human dignity; reproductive technologies; Keyword Identifiers: *Germany

Kyo-hun, Chin. Current debates on 'human cloning' in Korea. *In:* Roetz, Heiner, ed. Cross-Cultural Issues in Bioethics: The Example of Human Cloning. New York: Rodopi, 2006: 141-153. NRCBL: 14.5; 18.5.4. SC: le.

Lilie, Hans. International legal limits to human cloning. *In:* Vöneky, Silja; Wolfrum, Rüdiger, eds. Human Dignity and Human Cloning. Leiden: Nijhoff, 2004: 125-132. 18 fn. NRCBL: 14.5; 21.1. SC: le.

Keywords: *cloning; *international aspects; *legal aspects; government regulation; human experimentation; informed consent; moral policy; reproductive technologies; risks and benefits; Keyword Identifiers: United Nations

Morioka, Masahiro. The ethics of human cloning and the sprout of human life. *In:* Roetz, Heiner, ed. Cross-Cultural Issues in Bioethics: The Example of Human Cloning. New York: Rodopi, 2006: 1-16. NRCBL: 14.5; 4.4; 18.5.4. SC: le.

Morris, Stephen G. Canada's Assisted Human Reproduction Act: a chimera of religion and politics. *American Journal of Bioethics* 2007 February; 7(2): 69-70. 3 refs. NRCBL: 14.5; 1.2. SC: le. Comments: Timothy Caulfield and Tania Bubela. Why a criminal ban? Analyzing the arguments against somatic cell nuclear transfer in the Canadian parliamentary debate. American Journal of Bioethics 2007 February; 7(2): 51-61.

Keywords: *cloning; *embryo research; *embryonic stem cells; *government regulation; *nuclear transfer techniques; *political activity; *public policy; *religion; embryos; criminal law; legislation; moral status; politics; reproductive technologies; Keyword Identifiers: *Canada; Right to Life Movement; United States

Mykitiuk, Roxanne; Nisker, Jeff; Bluhm, Robyn. The Canadian Assisted Human Reproduction Act: protecting women's health while potentially allowing human somatic cell nuclear transfer into non-human oocytes. *American Journal of Bioethics* 2007 February; 7(2): 71-73. 11 refs. NRCBL: 14.5; 18.5.4; 15.1. SC: le. Comments: Timothy Caulfield and Tania Bubela. Why a criminal ban? Analyzing the arguments against somatic cell nuclear transfer in the Canadian parliamentary debate. American Journal of Bioethics 2007 February; 7(2): 51-61.

Keywords: *chimeras; *cloning; *embryo research; *government regulation; *legal aspects; *legislation; *nuclear transfer techniques; *ovum donors; *women's health; criminal law; embryonic stem cells; embryos; reproductive technologies; women; Keyword Identifiers: *Canada

Stabile, Bonnie. Demographic profile of states with human cloning laws: morality policy meets political economy. *Politics and the Life Sciences* 2007 March; 26(1): 43-50. 31 refs. NRCBL: 14.5; 18.5.4; 15.1. SC: em; le.

Keywords: *cloning; *embryo research; *legal aspects; *legislation; *morality; *public policy; *socioeconomic factors; *state government; abortion; biotechnology; comparative studies; government regulation; policy analysis; policy making; politics; religion; reproductive technologies; survey; Proposed Keywords: demography; Keyword Identifiers: *United States

Stewart, Cameron. Recent developments. *Journal of Bioethical Inquiry* 2007; 4(3): 169-170. NRCBL: 14.5; 14.4; 20.7; 20.5.1; 12.4.4. SC: le.

Viens, A.M. Criminal law in the regulation of somatic cell nuclear transfer. *American Journal of Bioethics* 2007 February; 7(2): 73-75. 8 refs. NRCBL: 14.5. SC: le. Comments: Timothy Caulfield and Tania Bubela. Why a criminal ban? Analyzing the arguments against somatic cell nuclear transfer in the Canadian parliamentary debate. American Journal of Bioethics 2007 February; 7(2): 51-61.

Keywords: *cloning; *criminal law; *embryo research; *government regulation; *morality; *nuclear transfer techniques; *public policy; legal aspects; public opinion; Keyword Identifiers: Canada

CODES OF ETHICS

The Islamic Code of Medical Ethics. *World Medical Journal* 1982 September-October; 29(5): 78-80. NRCBL: 6; 2.1; 1.2. Note: Reprinted from SPEM Medical Times 1981 June; 16(6).

American Association of Electrodiagnostic Medicine. Guidelines for ethical behavior relating to clinical practice issues in electrodiagnostic medicine. *Muscle and Nerve*

NRCBL: National Reference Center for Bioethics Literature Classification Scheme   See inside front cover for terms.

Supplement 1999; 8: S43-S47. NRCBL: 6; 4.1.2; 1.3.1; 8.1.

American Dental Association [ADA]. Principles of Ethics and Code of Professional Conduct with official advisory opinions revised to January 2005. Chicago, IL: American Dental Association [ADA], 2005 January: 20 p. [Online]. Accessed: http://www.ada.org/prof/prac/law/code/ada_code.pdf [2007 June 19]. NRCBL: 6; 4.1.1; 1.3.1.

Australian College of Health Service Executives. Code of ethics. New South Wales, Australia: Australian College of Health Services Executives, n.d.; 1 p. [Online]. Accessed: http://www.achse.org.au/ethics.html [2007 April 30]. NRCBL: 6.

Baker, Robert. A history of codes of ethics for bioethicists. *In:* Eckenwiler, Lisa A.; Cohn, Felicia G., eds. The Ethics of Bioethics: Mapping the Moral Landscape. Baltimore, MD: Johns Hopkins University Press, 2007: 24-40. NRCBL: 6; 2.2.

Canadian Institutes of Health Research [CIHR]. Conflict of interest policy. Ottawa: Canadian Institutes of Health Research, 2000 June 26; 5 p. [Online]. Accessed: http://www.cihr-irsc.gc.ca/e/19039.html [2007 April 25]. NRCBL: 6; 1.3.9; 7.3.

Great Britain (United Kingdom). Department of Health. Research governance framework for health and social care: second edition. London: Department of Health, 2005 April; 49 p. [Online]. Accessed: http://www.dh.gov.uk/prod_consum_dh/idcplg?IdcService=GET_FILE&dID=26829&Rendition=Web [2007 April 25]. NRCBL: 6; 1.3.9.

Hasegawa, Thomas K.; Welie, Jos V.M. Role of codes of ethics in oral health care. *Journal of the American College of Dentists* 1998 Fall; 65(3): 12-14. NRCBL: 6; 4.1.1; 1.3.1.

Heymans, Regien; van der Arend, Arie; Gastmans, Chris. Dutch nurses' views on codes of ethics. *Nursing Ethics* 2007 March; 14(2): 156-170. NRCBL: 6; 4.1.3. SC: em. Identifiers: Netherlands.

Abstract: This study explored the experiences and views of Dutch nurses on the content, function, dissemination and implementation of their codes of ethics. A total of 39 participants, who differed in age, qualifications, length of work experience and health care setting, took part in focus groups. The findings revealed common unfamiliarity with and a rather implicit use of codes, and negative comments on the growing number of codes available in the Netherlands. Limited dissemination, implementation and functioning of codes of ethics were also identified. The findings were discussed using concepts from the literature, nursing practice and personal experience.

Islamic Organization for Medical Sciences [IOMS]. Islamic code of medical ethics: doctor's duty in war time.

Sulaibekhat, Kuwait: Islamic Organization for Medical Sciences, n.d.; 1 p. [Online]. Accessed: http://www.islamset.com/ethics/code/index.html [2007 April 19]. NRCBL: 6; 1.2; 21.2.

Kahnawake Schools Diabetes Prevention Project. Code of research ethics. Kahnawake Territory, Mohawk Nation via Quebec, Canada: Kahnawake Schools Diabetes Prevention Project, 2007; 31 p. [Online]. Accessed: http://www.ksdpp.org/i/ksdpp_code_of_research_ethics2007.pdf [2007 April 26]. NRCBL: 6; 1.3.9; 9.5.4.

Kannan, Ramya. Code of ethics for doctors not being enforced, says Anbumani [news]. *Hindu* 2007 January 1 [Online]. Accessed: http://www.thehindu.com/2007/01/31/stories/2007013108640100.htm [2007 January 31]. NRCBL: 6; 7.4. Identifiers: India.

Kenny, Nuala P. Codes and character: the pillars of professional ethics. *Journal of the American College of Dentists* 1998 Fall; 65(3): 5-8. NRCBL: 6; 4.1.1; 1.3.1.

Pentz, Rebecca D.; Joffe, Steven; Emanuel, Ezekiel J.; Schnipper, Lowell E.; Haskell, Charles M.; Tannock, Ian F. ASCO core values. *Journal of Clinical Oncology* 2006 December 20; 24(36): 5780-5782. NRCBL: 6; 9.5.1. Identifiers: American Society of Clinical Oncology.

Prescription Medicines Code of Practice Authority. Code of practice for the pharmaceutical industry 2006. London: Prescription Medicines Code of Practice Authority, 2006: 56 p. [Online]. Accessed: http://www.abpi.org.uk/links/assoc/PMCPA/Code06use.pdf [2007 April 17]. NRCBL: 6; 9.7.

Rx&D: Canada's Research-Based Pharmaceutical Companies. Rx&D code of conduct: 11 guiding principles. Ottawa: Rx&D: Canada's Research-Based Pharmaceutical Companies, 2006 Winter; 1 p. [Online]. Accessed: http://www.canadapharma.org/Industry_Publications/Code/GuidingPrinciples0512.pd f [2007 April 4]. NRCBL: 6; 9.7.

Stone, Julie. Evaluating the ethical and legal content of professional codes of ethics. *In:* Allsop, Judith; Saks, Mike, eds. Regulating the Health Professions. London; Thousand Oaks, CA: Sage Publications, 2002: 62-75. NRCBL: 6; 4.1.2; 1.3.1. SC: le.

World Medical Association [WMA]. World Medical Association Declaration of Helsinki: Ethical principles for medical research involving human subjects. *Bulletin of the World Health Organization* 2001; 79(4): 373. NRCBL: 6; 18.2; 21.1.

COMMISSIONS ON BIOETHICS *See* BIOETHICS AND MEDICAL ETHICS/ COMMISSIONS

CONFIDENTIALITY
See also AIDS/ CONFIDENTIALITY

Academy of Medical Sciences. Personal data for public good: using health information in medical research. London: Academy of Medical Sciences, 2006 January; 77 p. [Online]. Accessed: http://www.acmedsci.ac.uk/images/project/Personal.pdf [2007 April 3]. NRCBL: 8.4; 1.3.12; 18.1.

Acciani, Jennifer. Resident portfolio: breaking trust — a reflection on confidentiality and minors. *Academic Emergency Medicine* 2006 December; 13(12): 1339-1340. NRCBL: 8.4; 8.1; 7.2; 9.5.7.

American Civil Liberties Union [ACLU]. Answers to frequently asked questions about government access to personal medical information. New York: American Civil Liberties Union, 2003 May 30; 4 p. [Online]. Accessed: http://www.aclu.org/privacy/medical/15222res20030530.html [2007 April 26]. NRCBL: 8.4; 1.3.5; 1.3.12. SC: le. Identifiers: USA Patriot Act; HIPAA.

Aslam, S.A.; Colapinto, P.; Sheth, H.G.; Jain, R. Patient consultation survey in an ophthalmic outpatient department. *Journal of Medical Ethics* 2007 March; 33(3): 134-135. NRCBL: 8.4. SC: em.
Abstract: INTRODUCTION: Consultation methods differ between medical practitioners depending on the individual setting. However, the central tenet to the doctor-patient relationship is the issue of confidentiality. This prospective survey highlights patient attitudes towards consultation methods in the setting of an ophthalmic outpatient department. METHOD: Questionnaires were completed by 100 consecutive patients, who had been seen by an ophthalmologist in a single room, which had a joint doctor-patient consultation occurring simultaneously. RESULTS: Each question of all 100 questionnaires was completed. 58% of patients were not concerned about sharing a consultation room with another patient or doctor. However, this did not equate to the 49% of patients who were indifferent to discussing issues in the joint consultation room. The most common factor was the general issue of confidentiality. DISCUSSION: Ensuring total patient confidentiality may be deemed more necessary for certain medical specialties than for others, as seen in the practice of separate medical records in genitourinary medicine, for instance. However, with regard to patient consultations, the same level of confidentiality should be afforded across all specialties, and such factors should be borne in mind when planning outpatient clinics.

Auerbach, Rebecca S. New York's immediate need for a psychotherapist-patient privilege encompassing psychiatrists, psychologists, and social workers. *Albany Law Review* 2006; 69(3): 889-912. NRCBL: 8.4; 17.2; 8.1; 17.1. SC: le.

Bailey, Tracey M.; Penney, Steven. Healing, not squealing: recent amendments to Alberta's Health Information Act. *Health Law Review* 2007; 15(2): 3-14. NRCBL: 8.4; 1.3.5. SC: le.

Benziman, Uzi. Patient's condition — severe but stable. The press and the medical community: mutual expectations surrounding the health of national leaders. *Israel Medical Association Journal* 2006 November; 8(11): 763-767. NRCBL: 8.4; 21.1; 1.3.7. Identifiers: Israel; Ariel Sharon.

Beyleveld, Deryck. Conceptualising privacy in relation to medical research values. *In:* McLean, Sheila A.M., ed. First Do No Harm: Law, Ethics, and Healthcare. Aldershot, England; Burlington, VT: Ashgate, 2006: 151-163. NRCBL: 8.4; 18.1. SC: an; le.

Black, Lee; Anderson, Emily E. Physicians, patients and confidentiality: the role of physicians in electronic health records. *American Journal of Bioethics* 2007 March; 7(3): 50-51. NRCBL: 8.4; 1.3.12; 6. Comments: Mark A. Rothstein and Meghan K. Talbott. Compelled authorizations for disclosure of health records: magnitude and implications. American Journal of Bioethics 2007 March; 7(3): 38-45.

Brous, Edie A. HIPAA vs. law enforcement: A nurses' guide to managing conflicting responsibilities. *AJN: American Journal of Nursing* 2007 August; 107(8): 60-63. NRCBL: 8.4; 1.3.12; 1.3.5. SC: le. Identifiers: Health Insurance Portability and Accountability Act.

California Health Care Foundation; Bishop, Lynne "Sam"; Holmes, Bradford J.; Kelley, Christopher M.; Forrester Research, Inc. National consumer health privacy survey 2005: executive summary. Oakland, CA: California Health Care Foundation: 2005 November; 5 p. [Online]. Accessed: http://www.chcf.org/documents/ihealth/ConsumerPrivacy2005ExecSum.pdf [2007 October 1]. NRCBL: 8.4; 1.3.12. SC: em.

Canadian Institutes of Health Research [CIHR]. CIHR best practices for protecting privacy in health research. Ottawa: Canadian Institutes of Health Research, 2005 September; 161 p. [Online]. Accessed: http://www.cihr-irsc.gc.ca/e/documents/et_pbp_nov05_sept2005_e.pdf [2007 April 4]. NRCBL: 8.4; 18.1; 1.3.12.

Chaloner, C. Confidentiality. *In:* Holland, Stephen, ed. Introducing Nursing Ethics: Themes in Theory and Practice. Salisbury: APS, 2004: 65-90. NRCBL: 8.4; 4.1.3. SC: cs.

Coombes, Rebecca. Medical records: are patients' secrets up for grabs? *BMJ: British Medical Journal* 2007 January 6; 334(7583): 16-17. NRCBL: 8.4; 1.3.12; 8.2. Identifiers: Great Britain (United Kingdom).

Cross, Michael. House of cards. *BMJ: British Medical Journal* 2007 April 14; 334(7597): 772-773. NRCBL: 8.4; 1.3.12. Identifiers: electronic patient records.

Day, Michael. Hewitt says some Muslim GPs breach confidentiality [news]. *BMJ: British Medial Journal* 2007 April 7; 334(7596): 711. NRCBL: 8.4; 10; 1.2.

NRCBL: National Reference Center for Bioethics Literature Classification Scheme   See inside front cover for terms.

111

Deapen, Dennis. Cancer surveillance and information: balancing public health with privacy and confidentiality concerns (United States). *Cancer Causes and Control* 2006 June; 17(5): 633-637. NRCBL: 8.4; 9.1; 9.5.1.

Dickens, Bernard M. The doctor's duty of confidentiality: separating the rule from the expectations. *University of Toronto Medical Journal* 1999 December; 77(1): 40-43 [Online]. Accessed: http://www.utmj.org/issues/77.1/pdf/LawEthics77-1.pdf [2007 February 16]. NRCBL: 8.4; 8.1. SC: cs; le.

Draper, Heather; MacDiarmaid-Gordon, Adam; Strumidlo, Laura; Teuten, Bea; Update, Eleanor. Virtual Clinical Ethics Committee, case 5: can we give a son access to his mother's psychiatric notes? *Clinical Ethics* 2007 March; 2(1): 8-14. NRCBL: 8.4; 17.1; 9.6. SC: cs.

FIGO Committee for the Ethical Aspects of Human Reproduction and Women's Health. Confidentiality, privacy and security of patients' health care information. *International Journal of Gynecology and Obstetrics* 2006 May; 93(2): 184-186. NRCBL: 8.4; 9.5.5; 14.1; 7.1.

Fleetwood, Janet. STDs in patients with multiple partners: confidentiality. *American Family Physician* 2006 December 1; 74(11): 1963-1964. NRCBL: 8.4; 10; 9.5.1. SC: cs.

Fullbrook, Suzanne. Confidentiality. Part 3: Caldicott guardians and the control of data. *British Journal of Nursing* 2007 September 13-27; 16(16): 1008-1009. NRCBL: 8.4; 1.3.12. SC: le.

Fullbrook, Suzanne. Legal principles of confidentiality and other public interests:Part 1. *British Journal of Nursing* 2007 July 26 - August 8; 16(14): 874-875. NRCBL: 8.4. SC: le. Identifiers: Great Britain (United Kingdom).

Fullbrook, Suzanne. Regulatory codes of conduct and the common law. Part 2: confidentiality. *British Journal of Nursing* 2007 August 9 - September 12; 16(15): 946-947. NRCBL: 8.4; 6. SC: le.

Hartry, Nicola. Visually impaired drivers and public protection vs confidentiality. *British Journal of Nursing* 2007 February 22-March 7; 16(4): 226-230. NRCBL: 8.4; 9.5.1. SC: le.

Haynes, Charlotte L.; Cook, Gary A.; Jones, Michael A. Legal and ethical considerations in processing patient-identifiable data without parental consent: lessons learnt from developing a disease register. *Journal of Medical Ethics* 2007 May; 33(5): 302-307. NRCBL: 8.4; 18.2; 18.3. SC: le. Identifiers: Great Britain (United Kingdom).
Abstract: The legal requirements and justifications for collecting patient-identifiable data without patient consent were examined. The impetus for this arose from legal and ethical issues raised during the development of a population-based disease register. Numerous commentaries and case studies have been discussing the impact of the Data Protection Act 1998 (DPA1998) and Caldicott principles of good practice on the uses of personal data. But uncertainty still remains about the legal requirements for processing patient-identifiable data without patient consent for research purposes. This is largely owing to ignorance, or misunderstandings of the implications of the common law duty of confidentiality and section 60 of the Health and Social Care Act 2001. The common law duty of confidentiality states that patient-identifiable data should not be provided to third parties, regardless of compliance with the DPA1998. It is an obligation derived from case law, and is open to interpretation. Compliance with section 60 ensures that collection of patient-identifiable data without patient consent is lawful despite the duty of confidentiality. Fears regarding the duty of confidentiality have resulted in a common misconception that section 60 must be complied with. Although this is not the case, section 60 support does provide the most secure basis in law for collecting such data. Using our own experience in developing a disease register as a backdrop, this article will clarify the procedures, risks and potential costs of applying for section 60 support

Heikkinen, Anne M.; Wickström, Gustav J.; Leino-Kilpi, Helena; Katajisto, Jouko. Privacy and dual loyalties in occupational health practice. *Nursing Ethics* 2007 September; 14(5): 675-690. NRCBL: 8.4; 7.3; 16.3; 4.1.2; 4.1.3. SC: em. Identifiers: Finland.
Abstract: This survey set out to explore occupational health professionals' courses of action with respect to privacy in a situation of dual loyalty between employees and employers. A postal questionnaire was sent to randomly selected potential respondents. The overall response rate was 64%: 140 nurses and 94 physicians returned the questionnaire. Eight imaginary cases involving an ethical dilemma of privacy were presented to the respondents. Six different courses of action were constructed within the set alternatives proposed. The study indicated that privacy as an absolute value is not in the interest of either employees or employers. It also showed that, where dual loyalty is concerned, the most valid course of action in dealing with sensitive subjects such as drug and work community problems, sexual harassment and sick leave is to rely on tripartite co-operation. If they maintain their professional independence and impartiality, health professionals are well placed to succeed in this challenging task; if not, there are bound to be severe violations of privacy.

Herveg, Jean. The ban on processing medical data in European Law: consent and alternative solutions to legitimate processing of medical data in HealthGrid. *Studies in Health Technology and Informatics* 2006; 120: 107-116. NRCBL: 8.4; 1.3.12; 8.3.1. SC: le. Identifiers: Europe.

Hill, Kate. Consent, confidentiality and record keeping for the recording and usage of medical images. *Journal of Visual Communication in Medicine* 2006 June; 29(2): 76-79. NRCBL: 8.4; 1.3.12; 8.3.1; 7.4.

Hodge, James G., Jr. The flaw of informed consent. *American Journal of Bioethics* 2007 March; 7(3): 52-53.

NRCBL: 8.4; 8.3.1; 1.3.12. Comments: Mark A. Rothstein and Meghan K. Talbott. Compelled authorizations for disclosure of health records: magnitude and implications. American Journal of Bioethics 2007 March; 7(3): 38-45.

Hoffman, Sharona. Addressing privacy concerns through the Health Insurance Portability and Accountability Act privacy rule. *American Journal of Bioethics* 2007 March; 7(3): 48-49. NRCBL: 8.4; 1.3.12. SC: le. Comments: Mark A. Rothstein and Meghan K. Talbott. Compelled authorizations for disclosure of health records: magnitude and implications. American Journal of Bioethics 2007 March; 7(3): 38-45.

International Society of Nurses in Genetics [ISONG], Board of Directors. Privacy and confidentiality of genetic information: the role of the nurse [position statement]. Pittsburgh, PA: International Society of Nurses in Genetics, 2005 August 8: 2 p. [Online]. Accessed: http://www.isong.org/about/ps_privacy.cfm [2007 February 22]. 6 refs. NRCBL: 8.4; 15.1; 4.1.3; 6.
   Keywords: *genetic privacy; *nurse's role; genetic information; organizational policies; professional organizations

Jenkins, John G. GMC guidance on confidentiality [letter]. *BMJ: British Medical Journal* 2007 December 15; 335(7632): 1226. NRCBL: 8.4. Identifiers: General Medical Council; Great Britain (United Kingdom). Comments: Clare Dyer. Stringent restraints on use of patients' data are harming research. BMJ: British Medical Journal 2007 December 1; 335(7630): 1114-1115.

Jones, Christopher N. How much longer will patients trust us? [letter]. *BMJ: British Medical Journal* 2007 December 15; 335(7632): 1226. NRCBL: 8.4; 1.3.12. Comments: Annette Tuffs. One in 20 East German doctors spied on patients or colleagues. BMJ: British Medical Journal 2007 December 1; 335(7630): 1113.

Jørgensen, H.K.; Hartling, O.J. Anonymity in connection with sperm donation. *Medicine and Law: The World Association for Medical Law* 2007 March; 26(1): 137-143. NRCBL: 8.4; 14.6; 19.5. SC: an; le.

Kipnis, Kenneth. Medical confidentiality. *In:* Rhodes, Rosamond; Francis, Leslie P.; Silvers, Anita, eds. The Blackwell Guide to Medical Ethics. Malden, MA: Blackwell Pub., 2007: 104-127. NRCBL: 8.4.

Krishna, Rajeev; Kelleher, Kelly; Stahlberg, Eric. Patient confidentiality in the research use of clinical medical databases. *American Journal of Public Health* 2007 April; 97(4): 654-658. NRCBL: 8.4; 1.3.9; 1.3.12; 18.2.
   Abstract: Electronic medical record keeping has led to increased interest in analyzing historical patient data to improve care delivery. Such research use of patient data, however, raises concerns about confidentiality and institutional liability. Institutional review boards must balance patient data security with a researcher's ability to explore potentially important clinical relationships. We considered the issues involved when patient records from

health care institutions are used in medical research. We also explored current regulations on patient confidentiality, the need for identifying information in research, and the effectiveness of deidentification and data security. We will present an algorithm for researchers to use to think about the data security needs of their research, and we will introduce a vocabulary for documenting these techniques in proposals and publications.

Kruitenbrouwer, Frank. Private life: "frappez toujours". *In:* Gevers, J.K.M.; Hondius, E.H.; Hubben, J.H., eds. Health Law, Human Rights and the Biomedicine Convention: Essays in Honour of Henriette Roscam Abbing. Leiden; Boston: Martinus Nijhoff Publishers, 2005: 147-158. NRCBL: 8.4; 1.1; 1.3.12. SC: le.

Lothen-Kline, Christine; Howard, Donna E.; Hamburger, Ellen K.; Worrell, Kevin D.; Boekeloo, Bradley O. Truth and consequences: ethics, confidentiality, and disclosure in adolescent longitudinal prevention research. *Journal of Adolescent Health* 2003 November; 33(5): 385-394. NRCBL: 8.4; 9.5.7; 20.7; 18.2; 17.1; 7.1; 8.2. SC: em.

Mason, J.K.; Laurie, G.T. Medical confidentiality. *In their:* Mason and McCall Smith's Law and Medical Ethics. Seventh ed. Oxford; New York: Oxford University Press, 2005: 253-294. NRCBL: 8.4; 9.5.6; 9.5.7; 1.3.12; 18.1. SC: le.

Miller, Seumas. Privacy, confidentiality, and the treatment of drug addicts. *In:* Kleinig, John; Einstein, Stanley, eds. Ethical Challenges for Intervening in Drug Use: Policy, Research and Treatment Issues. Huntsville, TX: Office of International Criminal Justice; Criminal Justice Center, Sam Houston State University, 2006: 467-483. NRCBL: 8.4; 9.5.9; 4.2; 1.1; 9.5.6.

Mizani, Mehrdad A.; Baykal, N. A software platform to analyse the ethical issues of electronic patient privacy: the S3P example. *Journal of Medical Ethics* 2007 December; 33(12): 695-698. NRCBL: 8.4; 1.3.12.
   Abstract: Paper-based privacy policies fail to resolve the new changes posed by electronic healthcare. Protecting patient privacy through electronic systems has become a serious concern and is the subject of several recent studies. The shift towards an electronic privacy policy introduces new ethical challenges that cannot be solved merely by technical measures. Structured Patient Privacy Policy (S3P) is a software tool assuming an automated electronic privacy policy in an electronic healthcare setting. It is designed to simulate different access levels and rights of various professionals involved in healthcare in order to assess the emerging ethical problems. The authors discuss ethical issues concerning electronic patient privacy policies that have become apparent during the development and application of S3P.

Moore, Ilene N.; Snyder, Samuel Leason; Miller, Cynthia; An, Angel Qi; Blackford, Jennifer U.; Zhou, Chuan; Hickson, Gerald B. Confidentiality and privacy

NRCBL: National Reference Center for Bioethics Literature Classification Scheme   See inside front cover for terms.

113

in health care from the patient's perspective: does HIPAA help? *Health Matrix: The Journal of Law-Medicine* 2007 Spring; 17(2): 215-272. NRCBL: 8.4; 1.1. SC: em; le. Identifiers: Health Insurance Portability and Accountability Act.

National Health and Medical Research Council (Australia) [NHMRC]. Submission by the National Health and Medical Research Council to the review by the Federal Privacy Commissioner of the Private Sector Provisions of the Privacy Act 1988. Canberra: National Health and Medical Research Council, 2004 December 10; 39 p. [Online]. Accessed: http://www.nhmrc.gov.au/about/_files/psp.pdf [2007 April 18]. NRCBL: 8.4; 1.3.12.

National Health and Medical Research Council (Australia) [NHMRC]. The impact of privacy legislation on NHMRC stakeholders: comparative stakeholder analysis. Canberra: National Health and Medical Research Council, 2004 July; 51 p. [Online]. Accessed: http://www.nhmrc.gov.au/about/_files/st8.pdf [2007 April 18]. NRCBL: 8.4. SC: le; em.

Ness, Roberta B. Influence of the HIPAA Privacy Rule on health research. *JAMA: The Journal of the American Medical Association* 2007 November 14; 298(18): 2164-2170. NRCBL: 8.4; 18.2. SC: em; le. Identifiers: Health Insurance Portability and Accountability Act; Societies of Epidemiology Joint Policy Committee.
Abstract: CONTEXT: Anecdotal reports suggest that the Health Insurance Portability and Accountability Act Privacy Rule (HIPAA Privacy Rule) may be affecting health research in the United States. OBJECTIVE: To survey epidemiologists about their experiences with the HIPAA Privacy Rule. DESIGN, SETTING, AND PARTICIPANTS: Thirteen societies of epidemiology distributed a national Web-based survey; 2805 respondents accessed the survey Web site and 1527 eligible professionals anonymously answered questions. MAIN OUTCOME MEASURES: Responses related influences such as research delays and added cost after Privacy Rule implementation, frequency and type of Privacy Rule-related institutional review board modifications, level of difficulty obtaining deidentified data and waivers, experiences with multisite studies, and perceived participant privacy benefits under the rule. Respondents ranked their perceptions of Privacy Rule influence on 5-point Likert scales. RESULTS: A total of 875 (67.8%) respondents reported that the HIPAA Privacy Rule has made research more difficult at a level of 4 to 5 on a Likert scale, in which 5 indicates a great deal of added cost and time to study completion. A total of 684 (52.1%) of respondents identified a "most affected" protocol. Respondents indicated that the proportion of institutional review board applications in which the Privacy Rule had a negative influence on human subjects (participants) protection was significantly greater than the proportion in which it had a positive influence (P .001). CONCLUSION: In this national survey of clinical scientists, only a quarter perceived that the rule has enhanced participants' confidentiality and privacy, whereas the HIPAA Privacy Rule was perceived to have a substantial,

negative influence on the conduct of human subjects health research, often adding uncertainty, cost, and delay.

Ouellette, Alicia; Cohen, Beverly; Reider, Jacob. Practical, state, and federal limits on the scope of compelled disclosure of health records. *American Journal of Bioethics* 2007 March; 7(3): 46-48. NRCBL: 8.4; 1.3.12. SC: le. Comments: Mark A. Rothstein and Meghan K. Talbott. Compelled authorizations for disclosure of health records: magnitude and implications. American Journal of Bioethics 2007 March; 7(3): 38-45.

Page, Stacey A.; Mitchell, Ian. Patients' opinions on privacy, consent and the disclosure of health information for medical research. *Chronic Diseases in Canada* 2006; 27(2): 60-67. NRCBL: 8.4; 18.3; 18.5.1; 18.5.6; 18.2.

Pagliari, Claudia; Detmer, Don; Singleton, Peter. Potential of electronic personal health records. *BMJ: British Medical Journal* 2007 August 18; 335(7615): 330-333. NRCBL: 8.4; 1.3.12.

Palac, Diane M. Is it justified to breach confidentiality to protect a patient from abuse? *Medical Ethics Newsletter [Lahey Clinic]* 2007 Spring; 14(2): 3. NRCBL: 8.4; 8.3.3; 8.1; 17.1; 10. SC: cs.

Reches, Avinoam. Transparency with respect to the health of political leaders. *Israel Medical Association Journal* 2006 November; 8(11): 751-753. NRCBL: 8.4; 21.1. Identifiers: Israel; Ariel Sharon.

Reis, Shmuel; Biderman, Aya; Mitki, Revital; Borkan, Jeffrey M. Secrets in primary care: a qualitative exploration and conceptual model. *JGIM: Journal of General Internal Medicine* 2007 September; 22(9): 1246-1253. NRCBL: 8.4; 9.1. SC: em. Identifiers: Israel.

Riebschleger, Joanne; Nordquist, Gigi; Newman, Elana;. Mandated reporting. *Journal of Empirical Research on Human Research Ethics* 2007 March; 2(1): 61-64. NRCBL: 8.4; 18.3; 18.5.2.

Robinson, David J.; O'Neill, Desmond. Access to health care records after death: balancing confidentiality with appropriate disclosure. *JAMA: The Journal of the American Medical Association* 2007 February 14; 297(6): 634-636. 24 refs. NRCBL: 8.4; 15.1; 20.1.
Keywords: *access to information; *confidentiality; *death; *disclosure; *medical records; family members; genetic information; legal aspects; physician patient relationship; privacy; professional family relationship; Proposed Keywords: harm; Keyword Identifiers: United States

Rothstein, Mark A.; Talbott, Meghan K. Compelled authorizations for disclosure of health records: magnitude and implications. *American Journal of Bioethics* 2007 March; 7(3): 38-45. NRCBL: 8.4; 1.3.12.
Abstract: Each year individuals are required to execute millions of authorizations for the release of their health records as a condition of employment, applying for various types of insurance, and submitting claims for bene-

fits. Generally, there are no restrictions on the scope of information released pursuant to these compelled authorizations, and the development of a nationwide system of interoperable electronic health records will increase the amount of health information released. After quantifying the extent of these disclosures, this article discusses why it is important to limit disclosures of health information for nonmedical purposes as well as how it may be possible to do so.

Rothstein, Mark A.; Talbott, Meghan K. Response to open peer commentaries on "Compelled authorizations for disclosure of health records: magnitude and implications" [letter]. *American Journal of Bioethics* 2007 March; 7(3): W1-W3. NRCBL: 8.4; 6; 1.3.12.

Sade, Robert M.; Robinson, David J.; O'Neill, Desmond. Confidentiality of medical information after death [letter and reply]. *JAMA: The Journal of the American Medical Association* 2007 June 20; 297(23): 2585. NRCBL: 8.4; 20.1. Comments: D.J. Robinson and D. O'Neill. Access to health care records after death: balancing confidentiality with appropriate disclosure. JAMA: Journal of the American Medical Association 2007 February 14; 297(6): 634-636.

Shepherd, J.P.; Ho, M.; Shepherd, H.R.; Sivarajasingam, V. Confidential registration in health services: randomised controlled trial. *Emergency Medicine Journal* 2006 June; 23(6): 425-427. NRCBL: 8.4; 8.1; 18.2. SC: em. Identifiers: Great Britain (United Kingdom).

Shuchman, Miriam. FERPA, HIPAA, and the privacy of college students. *New England Journal of Medicine* 2007 July 12; 357(2): 109-110. NRCBL: 8.4; 9.5.7. SC: le. Identifiers: Family Educational Rights and Privacy Act; Health Insurance Portability and Accountability Act.

Sobel, Richard. The HIPAA paradox: the privacy rule that's not. *Hastings Center Report* 2007 July-August; 37(4): 40-50. NRCBL: 8.4; 8.3.1; 1.3.12; 5.3. SC: le. Identifiers: Health Insurance Portability and Accountability Act.

Stirewalt, Karolyn. To release or not to release? When is it all right for physicians who treat injured workers to release medical information without their consent? *Minnesota Medicine* 2007 September; 90(9): 52-53. NRCBL: 8.4; 16.3.

Stolberg, Sheryl Gay. President calls for genetic privacy bill. *New York Times* 2007 January 18; p. A14. NRCBL: 8.4; 15.1. SC: po; le. Identifiers: George W. Bush.
> Keywords: *genetic privacy; *legislation; politics; Keyword Identifiers: *United States

Tetali, Shailaja. The importance of patient privacy during a clinical examination. *Indian Journal of Medical Ethics* 2007 April-June; 4(2): 65. NRCBL: 8.4; 8.1.

Thomas, N.; Murray, E.; Rogstad, K.E. Confidentiality is essential if young people are to access sexual health ser-

vices. *International Journal of STD and AIDS* 2006 August; 17(8): 525-529. NRCBL: 8.4; 9.5.7; 10.

Weissbrodt, David; Pekin, Ferhat; Wilson, Amelia. Piercing the confidentiality veil: physician testimony in international criminal trials against perpetrators of torture. *Minnesota Journal of International Law* 2006 Winter; 15(1): 43-109. NRCBL: 8.4; 21.1; 21.4;. SC: an; cs; le.

Whiddett, Richard; Hunter, Inga; Engelbrecht, Judith; Handy, Jocelyn. Patients' attitudes towards sharing their health information. *International Journal of Medical Informatics* 2006 July; 75(7): 530-541. NRCBL: 8.4; 1.3.12; 1.3.5; 21.1.

Wolff, Katharina; Brun, Wibecke; Kvale, Gerd; Nordin, Karin. Confidentiality versus duty to inform — an empirical study on attitudes towards the handling of genetic information. *American Journal of Medical Genetics. Part A* 2007 January 15; 143(2): 142-148. NRCBL: 8.4; 15.3; 8.3.1; 8.2.

Wynia, Matthew K. Breaching confidentiality to protect the public: evolving standards of medical confidentiality for military detainees. *American Journal of Bioethics* 2007 August; 7(8): 1-5. NRCBL: 8.4; 9.5.1; 1.3.5.
> Abstract: Confidentiality is a core value in medicine and public health yet, like other core values, it is not absolute. Medical ethics has typically allowed for breaches of confidentiality when there is a credible threat of significant harm to an identifiable third party. Medical ethics has been less explicit in spelling out criteria for allowing breaches of confidentiality to protect populations, instead tending to defer these decisions to the law. But recently, issues in military detention settings have raised the profile of decisions to breach medical confidentiality in efforts to protect the broader population. National and international ethics documents say little about the confidentiality of detainee medical records. But initial decisions to use detainee medical records to help craft coercive interrogations led to widespread condemnation, and might have contributed to detainee health problems, such as a large number of suicide attempts several of which have been successful. More recent military guidance seems to reflect lessons learned from these problems and does more to protect detainee records. For the public health system, this experience is a reminder of the importance of confidentiality in creating trustworthy, and effective, means to protect the public's health.

CONTRACEPTION

Approval of nonprescription sale of Plan B muddies ethical waters. *ED Management* 2006 October; 18(10): 109-112. NRCBL: 11.1; 9.7; 9.5.1.

Catholic hospitals will comply with flawed law [news]. *America* 2007 October 15; 197(11): 7. NRCBL: 11.1; 9.7; 1.2; 9.1.

NRCBL: National Reference Center for Bioethics Literature Classification Scheme   See inside front cover for terms.

115

Connecticut Catholic hospitals will comply with Plan B law. *Origins* 2007 October 11; 37(18). NRCBL: 11.1; 1.2; 9.7; 9.5.5.

Emergency contraception: what's happening? *Health Care Ethics USA* 2007 Summer; 15(3): 20. NRCBL: 11.1; 9.7; 1.2.

American Life League. Emergency contraception: the morning-after pill. Stafford, VA: American Life League, 2006; 2 p. [Online]. Accessed: http://www.all.org/article.php?id=10130 [2007 April 9]. NRCBL: 11.1; 9.7; 1.2.

American Medical Women's Association [AMWA]. Emergency contraception [position statement]. Alexandria, VA: American Medical Women's Association, 1996 November: 4 p. [Online]. Accessed: http://www.amwa-doc.org/index.cfm?objectid=0EF88909-D567-0B25-531927EE4CC23EFB [2007 April 9]. NRCBL: 11.1; 9.7; 12.1; 11.4; 9.5.5.

American Public Health Association [APHA]. Support of public education about emergency contraception and reduction or elimination of barriers to access. Washington, DC: American Public Health Association, 2003 November 18, Policy No. 200315; 3 p. [Online]. Accessed: http://www.apha.org/advocacy/policy/policysearch/default.htm?id=1252 [2007 April 9]. NRCBL: 11.1; 9.7; 9.1; 9.5.5.

Austriaco, Nicanor Pier Giorgio. Is Plan B an abortifacient? A critical look at the scientific evidence. *National Catholic Bioethics Quarterly* 2007 Winter; 7(4): 703-707. NRCBL: 11.1; 9.7; 12.3.
  Abstract: On September 27, 2007, the Catholic bishops of Connecticut announced that they would allow the four Catholic hospitals in their state to comply with the state's emergency contraception law. The statement has generated much controversy and criticism from those who are convinced that Plan B is an abortifacient. This essay summarizes and critically reviews the scientific studies that have attempted to uncover the mechanism of action of levonorgestrel, the active drug in the contraceptive commonly known as Plan B. Mounting and recent evidence suggests that this emergency contraceptive has little or no effect on post-fertilization events.

Canadian Pharmacists Association. Emergency contraception now available from pharmacists. *Ottawa. Canadian Pharmacists Association.* 2005 April 19: 1 p. [Online]. Accessed: http://www.pharmacists.ca/content/about_cpha/whats_happening/news_releases/release_detail.cfm?release_id=122 [2007 August 28]. NRCBL: 11.1; 9.1; 12.1.

Center for Reproductive Rights. State trends in emergency contraception legislation. New York: Center for Reproductive Rights, 2006 January 10; 3 p [Online]. Accessed: http://www.reproductiverights.org/st_ec.html [2007 April 9]. NRCBL: 11.1; 9.7; 5.3; 9.1; 9.5.5. SC: le; em.

Chuang, Cynthia H.; Shank, Laura D. Availability of emergency contraception at rural and urban pharmacies in Pennsylvania. *Contraception* 2006 April; 73(4): 382-385. NRCBL: 11.1; 9.7; 9.2.

Cook, R.J.; Dickens, B.M.; Erdman, J.N. Emergency contraception, abortion and evidence-based law. *International Journal of Gynecology and Obstetrics* 2006 May; 93(2): 191-197. NRCBL: 11.1; 12.4.2; 12.5.1. SC: le.

Coons, Stephen Joel. Pharmacists' right of conscience: whose autonomy is it, anyway? [editorial]. *Clinical Therapeutics* 2005 June; 27(6): 924-925. NRCBL: 11.1; 9.7; 4.1.1.

Dailard, Cynthia; Richardson, Chinué Turner. Teenagers' access to confidential reproductive health services. *Guttmacher Report on Public Policy* 2005 November; 8(4): 6-11. NRCBL: 11.2; 12.4.1; 8.3.2; 8.4. SC: le.

Davis, Bradley L. Compelled expression of the religiously forbidden: pharmacists, "duty to fill" statutes, and the hybrid rights exception. *University of Hawai'i Law Review* 2006 Winter; 29(1): 97-121. NRCBL: 11.1; 4.1.1; 9.7. SC: le.

Davis, Thomas J., Jr. Plan B and the rout of religious liberty: reflection on the status of the law. *Ethics and Medics* 2007 December; 32(12): 1-4. NRCBL: 11.1; 9.1; 1.3.2; 1.3.5; 1.2. SC: le.

Douglas, Joshua A. When is a "minor" also an "adult"?: an adolescent's liberty interest in accessing contraceptives from public school distribution programs. *Willamette Law Review* 2007 Summer; 43(4): 545-576. NRCBL: 11.2; 8.3.1. SC: le.

Erdman, Joanna N.; Cook, Rebecca J. Protecting fairness in women's health: the case of emergency contraception. *In:* Flood, Colleen M., ed. Just Medicare: What's In, What's Out, How We Decide. Buffalo, NY: University of Toronto Press, 2006: 137-167. NRCBL: 11.1; 9.7; 4.1.1; 9.5.5. SC: le. Identifiers: Canada.

Fleming, John I.; Neville, Warwick; Pike, Gregory K. Another clash of orthodoxies: "the meaning of the universe" from "the other side of the pond." Abortion in the UK. Adelaide, Australia: Southern Cross Bioethics Institute, n.d.: 30 p. [Online]. Accessed: http://www.bioethics.org.au/docs/Other%20articles/MUNBY.PDF [2006 October 10]. NRCBL: 11.1; 9.7; 4.4; 12.4.2. SC: an; le.

Freeman, Michael. Rethinking Gillick. *In his:* Children's Health and Children's Rights. Boston: Martinus Nijhoff Publishers, 2006: 201-217. NRCBL: 11.2; 8.3.2. SC: cs; le. Identifiers: Gillick v. West Norfolk and Wisbech Area Health Authority; Great Britain (United Kingdom).

Gast, Kristen Marttila. Cold comfort pharmacy: pharmacist tort liability for conscientious refusals to dispense emergency contraception. *Texas Journal of Women and the*

Law 2007 Spring; 16(2): 149-184. NRCBL: 11.1; 9.7; 7.1; 11.4. SC: le.

Guilhem; Dirce; Azevedo, Anamaria Ferreira. Brazilian public policies for reproductive health: family planning, abortion and prenatal care. *Developing World Bioethics* 2007 August; 7(2): 68-77. NRCBL: 11.1; 9.5.5; 12.4.1; 12.5.1.

Abstract: This study is an ethical reflection on the formulation and application of public policies regarding reproductive health in Brazil. The Integral Assistance Program for Women's Health (PAISM) can be considered advanced for a country in development. Universal access for family planning is foreseen in the Brazilian legislation, but the services do not offer contraceptive methods for the population in a regular and consistent manner. Abortion is restricted by law to two cases: risk to the woman's life and rape. This reality favors the practice of unsafe abortion, which is the third largest cause of maternal death in Brazil. Legal abortion is regulated by the State and the procedure is performed in public health centers. However, there is resistance on the part of professionals to attend these women. Prenatal care is a priority strategy for promoting the quality of life of these women and of future generations. Nonetheless, it is still difficult for these women to access the prenatal care services and to have the required number of consultations. Moreover, managers and health professionals need to be made aware of the importance of implementing the actions indicated by the public policies in the area of sexual and reproductive health, favoring respect for autonomy in a context of personal freedom.

Hale, B. Culpability and blame after pregnancy loss. *Journal of Medical Ethics* 2007 January; 33(1): 24-27. NRCBL: 11.4; 1.1.

Abstract: The problem of feeling guilty about a pregnancy loss is suggested to be primarily a moral matter and not a medical or psychological one. Two standard approaches to women who blame themselves for a loss are first introduced, characterised as either psychologistic or deterministic. Both these approaches are shown to underdetermine the autonomy of the mother by depending on the notion that the mother is not culpable for the loss if she "could not have acted otherwise". The inability to act otherwise is explained as not being as strong a determinant of culpability as it may seem at first. Instead, people's culpability for a bad turn of events implies strongly that they have acted for the wrong reasons, which is probably not true in the case of women who have experienced a loss of pregnancy. The practical conclusion of this paper is that women who feel a sense of guilt in the wake of their loss have a good reason to reject both the psychologistic and the deterministic approaches to their guilt-that they are justified in feeling upset about what has gone wrong, even responsible for the life of the child, but are not culpable for the unfortunate turn of events.

Harris, John. The facts of life: controversial medical issues in the curriculum. *In:* Wellington, J.J., ed. Controversial Issues in the Curriculum. Oxford: Basil Blackwell, 1986: 99-108. NRCBL: 11.2; 12.3; 1.3.3. SC: le.

Hondius, Ewoud. The Kelly case — compensation for undue damage for wrongful treatment. *In:* Gevers, J.K.M.; Hondius, E.H.; Hubben, J.H., eds. Health Law, Human Rights and the Biomedicine Convention: Essays in Honour of Henriette Roscam Abbing. Leiden; Boston: Martinus Nijhoff Publishers, 2005: 105-116. NRCBL: 11.4; 8.5; 4.4. SC: an; le.

International Consortium for Emergency Contraception. EC status and availability by country. New York: International Consortium for Emergency Contraception, n.d.; 10 p. [Online]. Accessed: http://www.cecinfo.org/database/pill/viewAllCountry.php [2007 April 10]. NRCBL: 11.1; 9.7; 21.1; 5.3. SC: em. Identifiers: emergency contraception.

Kouri, Robert P. Achieving reproductive rights: access to emergency oral contraception and abortion in Quebec. *In:* Flood, Colleen M., ed. Just Medicare: What's In, What's Out, How We Decide. Buffalo, NY: University of Toronto Press, 2006: 168-190. NRCBL: 11.1; 9.7; 12.4.2; 9.2. SC: le.

Langlois, Natalie. Life-sustaining treatment law: a model for balancing a woman's reproductive rights with a pharmacist's conscientious objection. *Boston College Law Review* 2006 July; 47(4): 815-852. NRCBL: 11.1; 9.7; 4.1.1; 14.1; 8.1. SC: le.

Lee, Tricia K. Fujikawa. Emergency contraception in religious hospitals: the struggle between religious freedom and personal autonomy. *University of Hawai'i Law Review* 2004 Winter; 27(1): 65-109. NRCBL: 11.1; 9.7; 1.2; 4.1.1; 9.5.5. SC: le.

Lewin, Tamar. Court says health coverage may bar birth-control pills. *New York Times* 2007 March 17; p. A11. NRCBL: 11.1; 9.7; 9.5.5. SC: po; le. Identifiers: Union Pacific Railroad Company.

Mason, J.K.; Laurie, G.T. The control of fertility. *In their:* Mason and McCall Smith's Law and Medical Ethics. Seventh ed. Oxford; New York: Oxford University Press, 2005: 120-167. NRCBL: 11.1; 4.4; 9.7; 11.2; 11.3; 12.4.2; 7.1; 21.1. SC: le.

Murphy, Timothy F. When 'emergency contraception' is neither. *American Journal of Bioethics* 2007 August; 7(8): W7. NRCBL: 11.1; 9.7.

Oleson, Christopher; Rhonheimer, Martin; Cole, Basil. More on the contraceptive choice [letters and reply]. *National Catholic Bioethics Quarterly* 2007 Winter; 7(4): 649-654. NRCBL: 11.1; 1.2; 4.4; 9.5.6. Comments: Martin Rhonheimer. Contraceptive choice, condom use, and moral arguments based on nature: a reply to Christopher Oleson. National Catholic Bioethics Quarterly 2007 Summer; 7(2): 273-291.

Ranney, Megan L.; Gee, Erin M.; Merchant, Roland C. Nonprescription availability of emergency contraception

NRCBL: National Reference Center for Bioethics Literature Classification Scheme   See inside front cover for terms.

117

in the United States: current status, controversies, and impact on emergency medicine practice. *Annals of Emergency Medicine* 2006 May; 47(5): 461-471. NRCBL: 11.1; 9.7. SC: rv.

Schneider, A. Patrick. Emergency contraception (EC) for victims of rape: ten myths. *Linacre Quarterly* 2007 August; 74(3): 181-203. NRCBL: 11.1; 9.7; 1.2; 12.1.

Smearman, Claire A. Drawing the line: the legal, ethical and public policy implications of refusal clauses for pharmacists. *Arizona Law Review* 2006 Fall; 48(3): 469-540. NRCBL: 11.1; 9.7; 4.1.1; 10; 1.2; 8.1. SC: le.

Sparrow, Margaret June. Abortion politics and the impact on reproductive health care [editorial]. *Australian and New Zealand Journal of Obstetrics and Gynaecology* 2005 December; 45(6): 471-473. NRCBL: 11.1; 9.7; 12.5.1; 12.1.

Temkin, Elizabeth. Contraceptive equity: the birth control center of the International Workers Order. *American Journal of Public Health* 2007 October; 97(10): 1737-1745. NRCBL: 11.1; 9.3.1; 9.7. SC: le.

Wilan, Ken. Susan Wood. *Nature Biotechnology* 2007 May 25(5): 495. NRCBL: 11.1; 5.3; 1.3.5.

Williams, Anne. The morning-after pill. *Human Reproduction and Genetic Ethics: An International Journal* 2007; 13(1): 8-36. NRCBL: 11.1; 9.7; 12.1; 9.5.1; 10; 8.3.2; 1.1.

Zezima, Katie. Not all are pleased at plan to offer birth control at Maine middle school. *New York Times* 2007 October 21; p. A22. NRCBL: 11.2. SC: po.

COST OF HEALTH CARE *See* HEALTH CARE ECONOMICS

CRYOBANKING OF SPERM, OVA, AND EMBRYOS

Chan, Sarah; Quigley, Muireann. Frozen embryos, genetic information and reproductive rights. *Bioethics* 2007 October; 21(8): 439-448. 36 fn. NRCBL: 14.6; 4.4; 14.1; 14.4; 15.1. SC: cs; an. Conference: Eighth World Congress of Bioethics: A Just and Healthy Society; Beijing, China; 2006 August 9.

    Keywords: *cryopreservation; *dissent; *embryo disposition; *embryos; *ethical analysis; *genetic materials; *genetic relatedness ties; *in vitro fertilization; *parents; *property rights; donors; embryo transfer; genetic information; germ cells; informed consent; legal aspects; philosophy; reproductive rights; Proposed Keywords: exceptionalism; Keyword Identifiers: Evans v. Americus Healthcare Ltd.; Great Britain

    Abstract: Recent ethical and legal challenges have arisen concerning the rights of individuals over their IVF embryos, leading to questions about how, when the wishes of parents regarding their embryos conflict, such situations ought to be resolved. A notion commonly invoked

in relation to frozen embryo disputes is that of reproductive rights: a right to have (or not to have) children. This has sometimes been interpreted to mean a right to have, or not to have, one's own genetic children. But can such rights legitimately be asserted to give rise to claims over embryos? We examine the question of property in genetic material as applied to gametes and embryos, and whether rights over genetic information extend to grant control over IVF embryos. In particular we consider the purported right not to have one's own genetically related children from a property-based perspective. We argue that even if we concede that such (property) rights do exist, those rights become limited in scope and application upon engaging in reproduction. We want to show that once an IVF embryo is created for the purpose of reproduction, any right not to have genetically-related children that may be based in property rights over genetic information is ceded. There is thus no right to prevent one's IVF embryos from being brought to birth on the basis of a right to avoid having one's own genetic children. Although there may be reproductive rights over gametes and embryos, these are not grounded in genetic information.

Douglas, Gillian. Who has the right to determine the fate of their embryos? *In:* Gunning, Jennifer; Holm, Søren, eds. Ethics, law, and society. Volume 1. Aldershot, Hants, England; Burlington, VT: Ashgate, 2005: 265-268. NRCBL: 14.6; 4.4. SC: le.

Dyer, Clare. Woman loses final round of battle to use her frozen embryos at European court [news]. *BMJ: British Medical Journal* 2007 April 21; 334(7598): 818. NRCBL: 14.6; 14.4; 21.1. SC: le. Identifiers: Natallie Evans.

European Parliament. European Parliament resolution on the trade in human egg cells. Strasbourg: European Parliament 2005 March 10, P6_TA(2005)0074; 3 p. [Online]. Accessed: http://www.europarl.europa.eu/sides/getDoc.do?pubRef=-//EP//NONSGML+TA+P6-TA-2005-0074+0+DOC+PDF+V0//EN [2007 April 11]. NRCBL: 14.6; 19.5; 9.3.1. SC: le.

Heng, Boon Chin. Disparity in medical fees for donor and self freeze-thaw embryo transfer cycle — a covert form of embryo commercialization? [letter]. *Developing World Bioethics* 2007 April; 7(1): 49-50. NRCBL: 14.6; 14.4; 9.3.1.

Heng, Boon Chin. Donation of surplus frozen embryos for stem cell research or fertility treatment — should medical professionals and healthcare institutions be allowed to exercise undue influence on the informed decision of their former patients? *Journal of Assisted Reproduction and Genetics* 2006 September-October; 23(9-10): 381-382. NRCBL: 14.6; 15.1; 18.5.4; 7.1; 8.3.1.

Heng, B.C. Ethical issues in transnational "mail order" oocyte donation. *International Journal of Gynecology and Obstetrics* 2006 December; 95(3): 302-304. NRCBL: 14.6; 21.1; 9.3.1.

Lett, Dan. Health Canada dithers while "fertility preservations" proceed [news]. *CMAJ/JAMC: Canadian Medical Association Journal* 2007 July 17; 177(2): 135-136. NRCBL: 14.6; 9.5.5.

Lockwood, Gillian M. Whose embryos are they anyway? *In:* Shenfield, Françoise; Sureau, Claude, eds. Contemporary Ethical Dilemmas in Assisted Reproduction. Abingdon: Informa Healthcare, 2006:3-11. NRCBL: 14.6; 14.4. SC: le. Identifiers: Human Fertilisation and Embryology Act 1990 (HFEA); United Kingdom (Great Britain).

Motluk, Alison. Crisis of trust over sperm bank errors: a new register of DNA from donors and their offspring is exposing major gaps in US sperm bank regulation. *New Scientist* 2007 August 11-17; 195(2616): 6-7. NRCBL: 14.6; 14.2; 15.3.
> Keywords: *biological specimen banks; *genetic screening; *semen donors; *sperm; *standards; artificial insemination; industry; informed consent; medical errors; registries; regulation; Keyword Identifiers: *United States; Great Britain

Nowak, Rachel. Egg-freezing: a reproductive revolution [news]. *New Scientist* 2007 March 24-30; 193(2596): 8-9. NRCBL: 14.6; 14.4.

Serebrovska, Zoya; Serebrovskaya, Tatiana; Di Pietro, Maria Luisa; Pyle, Rebecca. Fertility restoration by the cryopreservation of oocytes and ovarian tissue from the position of biomedical ethics: a review. *Fiziolohichnyi Zhurnal* 2006; 52(6): 101-108. NRCBL: 14.6; 14.1; 9.5.1; 9.5.5. Note: Abstract in Russian.

CULTURAL PLURALISM

Capaldi, Nicholas. Manifesto: moral diversity in health care ethics. *In:* Engelhardt, H. Tristram, ed. Global Bioethics: The Collapse of Consensus. Salem, MA: M&M Scrivener Press, 2006: 131-153. NRCBL: 21.7; 1.1; 1.2; 9.1.

Fritz, K. Cultural diversity. *In:* Holland, Stephen, ed. Introducing Nursing Ethics: Themes in Theory and Practice. Salisbury: APS, 2004: 151-170. NRCBL: 21.7; 9.5.4. SC: cs.

Nie, Jing-Bao; Campbell, Alastair V. Multiculturalism and Asian bioethics: cultural war or creative dialogue? *Journal of Bioethical Inquiry* 2007; 4(3): 163-167. NRCBL: 21.7; 2.1. Identifiers: Korea.

Pellegrino, Edmund D. Culture and bioethics: where ethics and mores meet. *In:* Prograis, Lawrence J.; Pellegrino, Edmund D., eds. African American Bioethics: Culture, Race, and Identity. Washington, DC: Georgetown University Press, 2004: ix-xxi. NRCBL: 21.7; 2.1. Conference: Symposium on African American Perspectives in Bioethics and Second Annual Conference on Health Disparities, held on September 23-24, 2004, at Georgetown University.

Powell, Tia. Cultural context in medical ethics: lessons from Japan. *Philosophy, Ethics, and Humanities in Medicine [electronic]* 2006; 1(4): 7 p. Accessed: http://www.peh-med.com/content/1/1/4 [2006 May 15]. NRCBL: 21.7; 7.1; 2.2; 1.1; 8.3.1; 18.5.5; 18.5.9.

Schmidt, Kurt W. Lost in translation — bridging gaps through procedural norms: comments on the papers of Capaldi and Tao. *In:* Engelhardt, H. Tristram, ed. Global Bioethics: The Collapse of Consensus. Salem, MA: M&M Scrivener Press, 2006: 180-206. NRCBL: 21.7; 1.1; 8.1; 8.2. Identifiers: Nicholas Capaldi; Julia Tao Lai-Po-wah.

Stephens, Carolyn; Porter, John; Nettleton, Clive; Willis, Ruth. UN Declaration on the Rights of Indigenous Peoples [letter]. *Lancet* 2007 November 24-30; 370(9601): 1756. NRCBL: 21.7; 21.1; 9.2.

Valaitis, J.A. Cultural diversity in health care: interpersonal and ethical considerations. *WMJ: Official Publication of the State Medical Society of Wisconsin* 2006 June; 105(4): 12-15. NRCBL: 21.7; 8.1; 9.4.

Widdows, Heather. Is global ethics moral neo-colonialism? An investigation of the issue in the context of bioethics. *Bioethics* 2007 July; 21(6): 305-315. NRCBL: 21.7; 1.1.
> Abstract: This paper considers the possibility and desirability of global ethics in light of the claim that 'global ethics' in any form is not global, but simply the imposition of one form of local ethics – Western ethics – and, as such, a form of moral neo-colonialism. The claim that any form of global ethics is moral neo-colonialism is outlined using the work of a group of 'developing world bioethicists' who are sceptical of the possibility of global ethics. The work of virtue ethicists is then introduced and compared to the position of the developing world bioethicists in order to show that the divide between 'Western' and 'non-Western' ethics is exaggerated. The final section of the paper turns to the practical arena and considers the question of global ethics in light of practical issues in bioethics. The paper concludes that practical necessity is driving the creation of global ethics and thus the pertinent question is no longer 'Whether global ethics?', but 'Why global ethics?'.

DEATH AND DYING

Bleich, J. David. Cadavers on display. *Tradition* 2007 Spring; 40(1): 87-97. NRCBL: 20.1; 7.2; 1.2.

Garment, Ann; Lederer, Susan; Rogers, Naomi; Boult, Lisa. Let the dead teach the living: the rise of body bequeathal in 20th century America. *Academic Medicine* 2007 October; 82(10): 1000-1005. NRCBL: 20.1; 8.3.3; 19.5; 2.2; 7.1.

Jones, D. Gareth. Anatomical investigations and their ethical dilemmas. *Clinical Anatomy* 2007 April; 20(3): 338-343. NRCBL: 20.1; 7.2; 18.2; 18.5.4.

NRCBL: National Reference Center for Bioethics Literature Classification Scheme   See inside front cover for terms.

119

Khong, T.Y.; Tanner, Alison R. Foetal and neonatal autopsy rates and use of tissue for research: the influence of 'organ retention' controversy and new consent process. *Journal of Paediatrics and Child Health* 2006 June; 42(6): 366-369. NRCBL: 20.1; 8.3.3; 19.5; 18.1; 18.2.

Masterton, Malin; Helgesson, Gert; Höglund, Anna T.; Hansson, Mats G. Queen Christina's moral claim on the living: justification of a tenacious moral intuition. *Medicine, Health Care and Philosophy* 2007 September; 10(3): 321-327. NRCBL: 20.1; 1.1; 18.5.1.

Abstract: In the long-running debate on the interest of the dead, Joan C. Callahan argues against such interests and although Søren Holm for practical reasons is prepared to consider posthumous interests, he does not see any moral basis to support such interests. He argues that the whole question is irresolvable, yet finds privacy interests where Tutankhamen is concerned. Callahan argues that there can be reasons to hold on to the fiction that there are posthumous interests, namely if it is comforting for the living and instrumental for society. Thus, despite arguing against the position that the dead have any interests or for any moral basis for such interests, these "interests" are still taken into consideration in the end. This shows the unsatisfactory basis of their positions and indicates the tenacity of the moral intuition that the dead can have moral claims on the living. One example of a posthumous interest is the interest in one's good name. Here we argue that it is an interest of moral significance. This implies that if individuals restrict use of their sample when they are still alive, those restrictions apply after their death. Further, it implies that one should be concerned with the reputation of historic persons. Research that defeats these interests calls for justification. We have suggested two lines of thinking along which such a discussion could go: investigating the truth-value of the good name and the relevance of bringing it into possible disrepute.

Rodgers, M.E. Human bodies, inhuman uses: public reactions and legislative responses to the scandals of bodysnatching. *In:* Garwood-Gowers, Austen; Tingle, John; Wheat, Kay, eds. Contemporary Issues in Healthcare Law and Ethics. Edinburgh; New York: Elsevier Butterworth-Heinemann, 2005: 151-172. NRCBL: 20.1; 7.1; 18.1. SC: le. Identifiers: Great Britain (United Kingdom).

van Leeuwen, Evert; Kimsma, Gerrit. Public policy and ending lives. *In:* Rhodes, Rosamond; Francis, Leslie P.; Silvers, Anita, eds. The Blackwell Guide to Medical Ethics. Malden, MA: Blackwell Pub., 2007: 220-237. NRCBL: 20.1; 1.1; 20.2.1; 12.1; 20.5.2; 20.5.1. SC: le.

DEATH AND DYING/ ATTITUDES TO DEATH

See also ADVANCE DIRECTIVES; ASSISTED SUICIDE; EUTHANASIA AND ALLOWING TO DIE

Ackerman, Felicia Nimue. Patient and family decisions about life-extension and death. *In:* Rhodes, Rosamond;

Francis, Leslie P.; Silvers, Anita, eds. The Blackwell Guide to Medical Ethics. Malden, MA: Blackwell Pub., 2007: 52-68. NRCBL: 20.3.3; 20.5.1; 4.4; 20.5.4.

Bito, Seiji; Matsumura, Shinji; Singer, Marjorie Kagawa; Meredith, Lisa S.; Fukuhara, Shunichi; Wenger, Neil S. Acculturation and end-of-life decision making: comparison of Japanese and Japanese-American focus groups. *Bioethics* 2007 June; 21(5): 251-262. NRCBL: 20.3.3; 21.7; 20.4.1; 20.5.1. SC: em.

Abstract: Variation in decision-making about end-of-life care among ethnic groups creates clinical conflicts. In order to understand changes in preferences for end-of-life care among Japanese who immigrate to the United States, we conducted 18 focus groups with 122 participants: 65 English-speaking Japanese Americans, 29 Japanese-speaking Japanese Americans and 28 Japanese living in Japan. Negative feelings toward living in adverse health states and receiving life-sustaining treatment in such states permeated all three groups. Fear of being meiwaku, a physical, psychological or financial caregiving burden on loved ones, was a prominent concern. They preferred to die pokkuri (popping off) before they become end stage or physically frail. All groups preferred group-oriented decision-making with family. Although advance directives were generally accepted, Japanese participants saw written directives as intrusive whereas Japanese Americans viewed them mainly as tools to reduce conflict created by dying person's wishes and a family's kazoku no jo — responsibility to sustain the dying patient. These findings suggest that in the United States Japanese cultural values concerning end-of-life care and decision-making process are largely preserved.

Borasio, G.D.; Weltermann, B.; Voltz, R.; Reichmann, H.; Zierz, S. Einstellungen zur patientenbetreuung in der letzten lebensphase. Eine umfrage bei neurologischen chefärzten = Attitudes towards patient care at the end of life. A survey of directors of neurological departments. *Nervenarzt* 2004 December; 75(12): 1187-1193. NRCBL: 20.3.2; 20.7; 20.5.1; 20.4.1. SC: em.

Davies, Douglas. Cheating death: the invisibles. *New Scientist* 2007 October 13-19; 195(2625): 48-49. NRCBL: 20.3.1; 9.5.2.

Hamric, Ann B.; Blackhall, Leslie, J. Nurse-physician perspectives on the care of dying patients in intensive care units: collaboration, moral distress, and ethical climate. *Critical Care Medicine* 2007 February; 35(2): 422-429. NRCBL: 20.3.2; 4.1.3; 20.4.1; 7.3.

Hershenov, David B. Death, dignity, and degradation. *Public Affairs Quarterly* 2007 January; 21(1): 21-36. NRCBL: 20.3.1; 4.4; 1.1.

Hsin, Dena Hsin-Chen; Macer, Darryl. Comparisons of life images and end-of-life attitudes between the elderly in Taiwan and New Zealand. *Journal of Nursing Research* 2006 September; 14(3): 198-208. NRCBL: 20.3.1; 1.2; 20.4.1; 21.1.

ISDOC Study Group; Costantini, M.; Morasso, G.; Montella, M.; Borgia, P.; Cecioni, R.; Beccaro, M.; Sguazzotti, E.; Bruzzi, P. Diagnosis and prognosis disclosure among cancer patients. Results from an Italian mortality follow-back survey. *Annals of Oncology* 2006 May; 17(5): 853-859. NRCBL: 20.3.2; 8.2; 9.5.1. Identifiers: Italy

Karlsson, Marit; Milberg, Anna; Strang, Peter. Dying with dignity according to Swedish medical students. *Supportive Care in Cancer* 2006 April; 14(4): 334-339. NRCBL: 20.3.2; 20.4.1. SC: em. Identifiers: Sweden.

Kessler, David A.; Levy, Douglas A. Direct-to-consumer advertising: is it too late to manage the risks? *Annals of Family Medicine* 2007 January/February; 5(1): 4-5. NRCBL: 20.3.2; 9.7; 1.3.7; 9.8.

Leget, Carlo. Retrieving the ars moriendi tradition. *Medicine, Health Care and Philosophy* 2007 September; 10(3): 313-319. NRCBL: 20.3.1; 20.4.1.
  Abstract: North Atlantic culture lacks a commonly shared view on dying well that helps the dying, their social environment and caregivers to determine their place and role, interpret death and deal with the process of ethical deliberation. What is lacking nowadays, however, has been part of Western culture in medieval times and was known as the ars moriendi (art of dying well) tradition. In this paper an updated version of this tradition is presented that meets the demands of present day secularized and multiform society. Five themes are central to the new art of dying: autonomy and the self, pain control and medical intervention, attachment and relations, life balance and guilt, death and afterlife. The importance of retrieving the ancient ars moriendi outreaches the boundaries of palliative medicine, since it deals with issues that play a central role in every context of medical intervention and treatment.

Song, John; Ratner, Edward R.; Bartels, Dianne M.; Alderton, Lucy; Hudson, Brenda; Ahluwalia, Jasjit S. Experiences with and attitudes toward death and dying among homeless persons. *JGIM: Journal of General Internal Medicine* 2007 April; 22(4): 427-434. NRCBL: 20.3.1; 9.5.10. SC: em.

DEATH AND DYING/ DETERMINATION OF DEATH

Brain death revisited. *Health Care Ethics USA* 2007 Winter; 15(1): 15. NRCBL: 20.2.1; 1.2.

Anderson, Janice A.; Vernaglia, Lawrence W.; Morrigan, Shirley P. Refusal of brain death diagnosis: the health lawyers' perspective. *JONA's Healthcare Law, Ethics, and Regulation* 2007 July-September; 9(3): 90-92. NRCBL: 20.2.1; 1.2. SC: le.

Austriaco, Nicanor Pier Giorgio. Is the brain-dead patient really dead? *Studia Moralia* 2003 December; 41(2): 277-308. NRCBL: 20.2.1; 19.5; 1.2.

Bagheri, A. Individual choice in the definition of death. *Journal of Medical Ethics* 2007 March; 33(3): 146-149. NRCBL: 20.2.1; 19.5; 1.1. SC: le. Identifiers: Japan.
  Abstract: While there are numerous doubts, controversies and lack of consensus on alternative definitions of human death, it is argued that it is more ethical to allow people to choose either cessation of cardio-respiratory function or loss of entire brain function as the definition of death based on their own views. This paper presents the law of organ transplantation in Japan, which allows people to decide whether brain death can be used to determine their death in agreement with their family. Arguably, Japan could become a unique example of individual choice in the definition of death if the law is revised to allow individuals choose definition of death independently of their family. It suggests that such an approach is one of the reasonable policy options a country can adopt for legislation on issues related to the definition of death.

Bard, Terry R. Refusal of brain death diagnosis: a rabbi's response. *JONA's Healthcare Law, Ethics, and Regulation* 2007 July-September; 9(3): 92-94. NRCBL: 20.2.1; 1.2.

Baron, Leonard; Shemie, Sam D.; Teitelbaum, Jeannie; Doig, Christopher James. Brief review: history, concept and controversies in the neurological determination of death. *Canadian Journal of Anaesthesia* 2006 June; 53(6): 602-608. NRCBL: 20.2.1.

Beard, Edward L. Jr.; Johnson, Larry W. Conversations in ethics. *JONA's Healthcare Law, Ethics, and Regulation* 2007 July-September; 9(3): 95-96. NRCBL: 20.2.1; 1.2; 20.5.1; 9.3.1; 8.1. SC: cs.

Bernat, James L. Ethical issues in brain death and multiorgan transplantation. *Neurologic Clinics* 1989 November; 7(4): 715-728. NRCBL: 20.2.1; 19.5; 9.5.5; 1.2; 18.5.7.

Bosek, Marcia Sue DeWolf. Refusal of brain death diagnosis: the case. *JONA's Healthcare Law, Ethics, and Regulation* 2007 July-September; 9(3): 87. NRCBL: 20.2.1; 1.2. SC: cs.

Bosek, Marcia Sue DeWolf. Refusal of brain death diagnosis: the ethicist's response. *JONA's Healthcare Law, Ethics, and Regulation* 2007 July-September; 9(3): 87-90. NRCBL: 20.2.1; 1.2.

Brown, Grattan T. Reading the signs of death: a theological analysis. *National Catholic Bioethics Quarterly* 2007 Autumn; 7(3): 467-476. NRCBL: 20.2.1; 1.2.

Diamond, Eugene F. John Paul II and brain death. *National Catholic Bioethics Quarterly* 2007 Autumn; 7(3): 491-497. NRCBL: 20.2.1; 20.5.1; 1.2; 4.4.

Doig, Christopher James; Young, Kimberly; Teitelbaum, Jeannie; Shemie, Sam D. Brief survey: determining brain death in Canadian intensive care units = Enquête ponctuelle: la détermination de la mort encéphalique dans les units de soins intensifs au Canada.

NRCBL: National Reference Center for Bioethics Literature Classification Scheme   See inside front cover for terms.

121

Canadian Journal of Anaesthesia 2006 June; 53(6): 609-612. NRCBL: 20.2.1; 9.5.1. SC: em. Note: Abstract in French.

DuBois, James. Avoiding common pitfalls in the determination of death. *National Catholic Bioethics Quarterly* 2007 Autumn; 7(3): 545-559. NRCBL: 20.2.1; 19.5; 1.2; 4.4.

Eberl, Jason T. Dualist and animalist perspectives on death: a comparison with Aquinas. *National Catholic Bioethics Quarterly* 2007 Autumn; 7(3): 477-489. NRCBL: 20.2.1; 1.1; 4.4; 1.2.

Eberl, Jason T. The end of a human person's life. *In his:* Thomistic Principles and Bioethics. London; New York: Routledge, 2006: 43-61. NRCBL: 20.2.1; 1.1. SC: an.

Glannon, Walter. Brain death. *In his:* Bioethics and the Brain. New York: Oxford University Press, 2007: 148-177. NRCBL: 20.2.1; 4.4; 20.5.1. SC: an. Identifiers: Nancy Cruzan; Terri Schiavo.

Griniezakis, Archimandrite Makarios. Legal and ethical issues associated with brain death. *Ethics and Medicine: An International Journal of Bioethics* 2007 Summer; 23(2): 113-117. NRCBL: 20.2.1; 20.2.2; 20.5.4; 8.3.3; 19.5; 1.1; 4.4. SC: le.

Hinkley, Charles C. Defining death. *In his:* Moral Conflicts of Organ Retrieval: A Case for Constructive Pluralism. Amsterdam; New York: Rodopi, 2005: 91-103. NRCBL: 20.2.1.

Hornby, Karen; Shemie, Sam D.; Teitelbaum, Jeanni; Doig, Christopher. Variability in hospital-based brain death guidelines in Canada. *Canadian Journal of Anaesthesia* 2006 June; 53(6): 613-619. NRCBL: 20.2.1; 19.5. SC: em.

Hostetter, Larry. Higher-brain death: a critique. *National Catholic Bioethics Quarterly* 2007 Autumn; 7(3): 499-504. NRCBL: 20.2.1; 1.2; 4.4. SC: an.

Hughes, James. Cheating death: vital signs. *New Scientist* 2007 October 13-19; 195(2625): 44-45. NRCBL: 20.2.1.

Lock, Margaret. Inventing a new death and making it believable. *In:* van Dongen, Els; Fainzang, Sylvie, eds. Lying and Illness: Power and Performance. Amsterdam: Het Spinhuis; Piscataway, NJ: Transaction Publishers, 2005: 12-35. NRCBL: 20.2.1; 19.5.

Lock, Margaret. On making up the good-as-dead in a utilitarian world. *In:* Franklin, Sarah; Lock, Margaret, eds. Remaking Life and Death: Toward an Anthropology of the Biosciences. Santa Fe: School of American Research Press; Oxford: James Currey, 2003: 165-192. NRCBL: 20.2.1; 19.5; 2.2. SC: an.

Machado, Calixto; Kerein, Julius; Ferrer, Yazmina; Portela, Liana; de la C. García, Maria; Manero, José

M. The concept of brain death did not evolve to benefit organ transplants. *Journal of Medical Ethics* 2007 April; 33(4): 197-200. NRCBL: 20.2.1; 19.5; 2.2.

Abstract: Although it is commonly believed that the concept of brain death (BD) was developed to benefit organ transplants, it evolved independently. Transplantation owed its development to advances in surgery and immunosuppressive treatment; BD owed its origin to the development of intensive care. The first autotransplant was achieved in the early 1900s, when studies of increased intracranial pressure causing respiratory arrest with preserved heartbeat were reported. Between 1902 and 1950, the BD concept was supported by the discovery of EEG, Crile's definition of death, the use of EEG to demonstrate abolition of brain potentials after ischaemia, and Crafoord's statement that death was due to cessation of blood flow. Transplantation saw the first xenotransplant in humans and the first unsuccessful kidney transplant from a cadaver. In the 1950s, circulatory arrest in coma was identified by angiography, and the death of the nervous system and coma dépassé were described. Murray performed the first successful kidney transplant. In the 1960s, the BD concept and organ transplants were instantly linked when the first kidney transplant using a brain-dead donor was performed; Schwab proposed to use EEG in BD; the Harvard Committee report and the Sydney Declaration appeared; the first successful kidney, lung and pancreas transplants using cadaveric (not brain-dead) donors were achieved; Barnard performed the first human heart transplant. This historical review demonstrates that the BD concept and organ transplantation arose separately and advanced in parallel, and only began to progress together in the late 1960s. Therefore, the BD concept did not evolve to benefit transplantation.

Machado, Calixto; Korein, J.; Ferrer, Y.; Portela, L.; de la C. García, M.; Chinchilla, M.; Machado, Y.; Machado, Y.; Manero, J.M. The Declaration of Sydney on human death. *Journal of Medical Ethics* 2007 December; 33(12): 699-703. NRCBL: 20.2.1. SC: rv.

Abstract: On 5 August 1968, publication of the Harvard Committee's report on the subject of "irreversible coma" established a standard for diagnosing death on neurological grounds. On the same day, the 22nd World Medical Assembly met in Sydney, Australia, and announced the Declaration of Sydney, a pronouncement on death, which is less often quoted because it was overshadowed by the impact of the Harvard Report. To put those events into present-day perspective, the authors reviewed all papers published on this subject and the World Medical Association web page and documents, and corresponded with Dr A G Romualdez, the son of Dr A Z Romualdez. There was vast neurological expertise among some of the Harvard Committee members, leading to a comprehensible and practical clinical description of the brain death syndrome and the way to diagnose it. This landmark account had a global medical and social impact on the issue of human death, which simultaneously lessened reception of the Declaration of Sydney. Nonetheless, the Declaration of Sydney faced the main conceptual and philosophical issues on human death in a bold and forthright manner.

This statement differentiated the meaning of death at the cellular and tissue levels from the death of the person. This was a pioneering view on the discussion of human death, published as early as in 1968, that should be recognised by current and future generations.

Mason, J.K.; Laurie, G.T. The diagnosis of death. *In their:* Mason and McCall Smith's Law and Medical Ethics. Seventh ed. Oxford; New York: Oxford University Press, 2005: 464-476. NRCBL: 20.2.1; 20.2.2. SC: le.

Mistry, Parul R. Donation after cardiac death: an overview. *Mortality* 2006 May; 11(2): 182-195. NRCBL: 20.2.1; 19.5; 2.2. SC: rv. Identifiers: DCD Protocol; Dead Donor Rule.

Pennsylvania Catholic Health Association. Draft principles and guidelines for non-heart-beating organ donation. *National Catholic Bioethics Quarterly* 2007 Autumn; 7(3): 563-566. NRCBL: 20.2.1; 19.5; 1.2; 6.

Sachedina, Abdulaziz. Brain death in Islamic jurisprudence. Charlottesville, VA: University of Virginia, n.d.: 7 p. [Online]. Accessed: http://people.virginia.edu/~aas/article/article6.htm [2007 April 2]. NRCBL: 20.2.1; 1.2. SC: an.

Travaline, John M. Understanding brain death diagnosis — II. *Ethics and Medics* 2007 April; 32(4): 3-4. NRCBL: 20.2.1; 4.4; 1.2.

Whetstine, Leslie; Streat, Stephen; Darwin, Mike; Crippen, David. Pro/con ethics debate: when is dead really dead? *Critical Care (London, England)* 2005; 9(6): 538-542. NRCBL: 20.2.1; 19.5.

Wijdicks, Eelco F.M. The clinical criteria of brain death throughout the world: why has it come to this? = Les critères cliniques de mort encéphalique à travers le monde: pour quoi en arriver là [editorial]. *Canadian Journal of Anaesthesia* 2006 June; 53(6): 540-543. NRCBL: 20.2.1; 21.1.

Yamaori, Tetsuo. Strategies for survival versus accepting impermanence: rationalizing brain death and organ transplantation today. *In:* LaFleur, William R.; Böhme, Gernot; Shimazono, Susumu, eds. Dark Medicine: Rationalizing Medical Research. Bloomington: Indiana University Press, 2007: 165-179. NRCBL: 20.2.1; 19.2; 20.3.1.

Youngner, Stuart J. The definition of death. *In:* Steinbock, Bonnie, ed. The Oxford Handbook of Bioethics. Oxford; New York: Oxford University Press, 2007: 285-303. NRCBL: 20.2.1; 21.7.

DEATH AND DYING/ TERMINAL CARE

Abelson, Reed. A chance to pick hospice, and still hope to live. *New York Times* 2007 February 10; p. A1, C4. NRCBL: 20.4.1; 9.3.1. SC: po.

Aminoff, Bechor Zvi. The new Israeli law "The Dying Patient" and Relief of Suffering Units. *American Journal of Hospice and Palliative Care* 2007 February-March; 24(1): 54-58. NRCBL: 20.4.1; 4.4; 17.1.

Bach, John R. Palliative care becomes 'uninformed euthanasia' when patients are not offered noninvasive life preserving options. *Journal of Palliative Care* 2007 Autumn; 23(3): 181-184. NRCBL: 20.4.1; 20.5.1; 9.8.

Barilan, Y. Michael. The new Israeli law on the care of the terminally ill: conceptual innovations waiting for implementation. *Perspectives in Biology and Medicine* 2007 Autumn; 50(4): 557-571. NRCBL: 20.4.1; 1.2; 20.5.3; 20.5.2; 20.5.4; 1.2. SC: le.

Beck, Natalia Vonnegut. Building bridges: the Protestant perspective. *In:* Puchalski, Christina M., ed. A Time for Listening and Caring: Spirituality and the Care of the Chronically Ill and Dying. Oxford; New York: Oxford University Press, 2006: 155-170. NRCBL: 20.4.1; 1.2; 20.5.4. SC: cs.

Bentley, Philip J. The shattered vessel: the dying person in Jewish law and ethics. *Loyola University Chicago Law Journal* 2006 Winter; 37(2): 433-454. NRCBL: 20.4.1; 1.2; 4.4; 20.51.

Bouchal, Shelley Raffin. Moral meaning of caring for the dying. *In:* Johnston, Nancy E.; Scholler-Jaquish, Alwilda, eds. Meaning in Suffering: Caring Practices in the Health Professions. Madison: University of Wisconsin Press, 2007: 232-275. NRCBL: 20.4.1; 4.1.3; 7.1; 4.4. SC: em.

Byock, Ira R. To life! Reflections on spirituality, palliative practice, and politics. *American Journal of Hospice and Palliative Care* 2006 December-2007 January; 23(6): 436-438. NRCBL: 20.4.1; 20.3.1; 1.1; 1.2.

Casarett, David J.; Quill, Timothy E. "I'm not ready for hospice": strategies for timely and effective hospice decisions. *Annals of Internal Medicine* 2007 March 20; 146(6): 443-449. NRCBL: 20.4.1.
Abstract: Hospice programs offer unique benefits for patients who are near the end of life and their families, and growing evidence indicates that hospice can provide high-quality care. Despite these benefits, many patients do not enroll in hospice, and those who enroll generally do so very late in the course of their illness. Some barriers to hospice referral arise from the requirements of hospice eligibility, which will be difficult to eliminate without major changes to hospice organization and financing. However, the challenges of discussing hospice create other barriers that are more easily remedied. The biggest communication barrier is that physicians are often unsure of how to talk with patients clearly and directly about their poor prognosis and limited treatment options (both requirements of hospice referral) without depriving them of hope. This article describes a structured strategy for discussing hospice, based on techniques of effective communication that physicians use in other "bad news"

NRCBL: National Reference Center for Bioethics Literature Classification Scheme   See inside front cover for terms.

123

situations. This strategy can make hospice discussions both more compassionate and more effective.

Chochinov, Harvey Max. Dying, dignity, and new horizons in palliative end-of-life care. *CA: A Cancer Journal for Clinicians* 2006 March-April; 56(2): 84-103. NRCBL: 20.4.1; 4.4; 1.2; 9.1.

Cohen, Lewis M.; Moss, Alvin H.; Weisbord, Steven D.; Germain, Michael J. Renal palliative care. *Journal of Palliative Medicine* 2006 August; 9(4): 977-992. NRCBL: 20.4.1; 4.4; 19.3; 8.3.4; 8.1.

Collins, Niamh; Phelan, Dermot; Marsh, Brian; Sprung, Charles L. End-of-life care in the intensive care unit: the Irish Ethicus data. *Critical Care and Resuscitation* 2006 December; 8(4): 315-320. NRCBL: 20.4.1; 20.5.1; 9.5.1. SC: em. Identifiers: European Ethicus Study.

Dy, Sydney; Lynn, Joanne. Getting services right for those sick enough to die. *BMJ: British Medical Journal* 2007 March 10; 334(7592): 511-513. NRCBL: 20.4.1; 21.1. SC: rv.

Engström, Joakim; Bruno, Erik; Holm, Birgitta; Hellzén, Ove. Palliative sedation at end of life — a systematic literature review. *European Journal of Oncology Nursing* 2007 February; 11(1): 26-35. NRCBL: 20.4.1; 4.4; 4.1.3.

Flory, James; Emanuel, Ezekiel. Recent history of end-of-life care and implications for the future. *In:* Galston, Arthur W.; Peppard, Christiana Z., eds. Expanding Horizons in Bioethics. Dordrecht; Norwell, MA: Springer, 2005: 161-182. NRCBL: 20.4.1; 20.5.1; 9.3.1; 9.5.2. SC: em.

Fromme, Erik K.; Tilden, Virginia P.; Drach, Linda L.; Tolle, Susan W. Increased family reports of pain or distress in dying Oregonians: 1996 to 2002. *Journal of Palliative Medicine* 2004 June; 7(3): 431-442. NRCBL: 20.4.1; 4.4. Identifiers: Oregon.

Gazelle, Gail. Understanding hospice — an underutilized option for life's final chapter. *New England Journal of Medicine* 2007 July 26; 357(4): 321-324. NRCBL: 20.4.1. SC: cs; rv.

Georges, Jean-Jacques; Onwuteaka-Philipsen, Bregje D.; van der Heide, Agnes; van der Wal, G.; van der Maas, P.J. Physicians' opinions on palliative care and euthanasia in the Netherlands. *Journal of Palliative Medicine* 2006 October; 9(5): 1137-1144. NRCBL: 20.4.1; 20.5.1; 4.1.2. SC: em.

Giordano, James. Hospice, palliative care, and pain medicine: meeting the obligations of non-abandonment and preserving the personal dignity of terminally ill patients. *Delaware Medical Journal* 2006 November; 78(11): 419-422. NRCBL: 20.4.1; 4.4.

Griffith, Richard. Controlled drugs and the principle of double effect. *British Journal of Community Nursing* 2006 August; 11(8): 352, 354-357. NRCBL: 20.4.1; 9.7; 4.4; 1.1; 4.1.3.

Han, Sung-Suk. Ethical issues in nursing care at the end of life. *Dolentium Hominum* 2007; 22(2): 28-32. NRCBL: 20.4.1; 4.1.3; 4.4; 1.1.

Hasan, Yusuf; Salaam, Yusef. Faith and Islamic issues at the end of life. *In:* Puchalski, Christina M., ed. A Time for Listening and Caring: Spirituality and the Care of the Chronically Ill and Dying. Oxford; New York: Oxford University Press, 2006: 183-192. NRCBL: 20.4.1; 1.2.

Jauhar, Sandeep. Between comfort and care, a blurry line. *New York Times* 2007 September 18; p. F5. NRCBL: 20.4.1; 9.3.1; 20.3.2. SC: po.

Kmietowicz, Zosia. Dying patients are often not told of the closeness of death [news]. *BMJ: British Medical Journal* 2007 December 8; 335(7631): 1176. NRCBL: 20.4.1; 8.2.

Kwak, Jung; Salmon, Jennifer R. Attitudes and preferences of Korean-American older adults and caregivers on end-of-life care. *Journal of the American Geriatrics Society* 2007 November; 55(11): 1867-1872. NRCBL: 20.4.1; 20.5.4; 20.3.3; 21.7. SC: em.

Legemaate; Johan; Verkerk, Marian; van Wijlick, Eric; de Graeff, Alexander. Palliative sedation in The Netherlands: starting-points and contents of a national guideline. *European Journal of Health Law* 2007 April; 14(1): 61-73. NRCBL: 20.4.1; 20.5.1; 9.8. SC: le.

Lewis, William R.; Luebke, Donna L.; Johnson, Nancy J.; Harrington, Michael D.; Costantini, Ottorino; Aulisio, Mark P. Withdrawing implantable defibrillator shock therapy in terminally ill patients. *American Journal of Medicine* 2006 October; 119(10): 892-896. NRCBL: 20.4.1; 20.5.1; 4.4; 9.4.

Linder, John F.; Meyers, Frederick J. Palliative care for prison inmates: "don't let me die in prison". *JAMA: The Journal of the American Medical Association* 2007 August 22-29; 298(8): 894-901. NRCBL: 20.4.1; 9.5.1.
Abstract: The number of older inmates in US correctional facilities is increasing and with it the need for quality palliative health care services. Morbidity and mortality are high in this population. Palliative care in the correctional setting includes most of the challenges faced in the free-living community and several unique barriers to inmate care. Successful models of hospice care in prisons have been established and should be disseminated and evaluated. This article highlights why the changing demographics of prison populations necessitates hospice in this setting and highlights many of the barriers that correctional and consulting physicians face while providing palliative care. Issues specific to palliative care and hospice in prison include palliative care standards, inmate-physician and inmate-family relationships, confidential-

ity, interdisciplinary care, do-not-resuscitate orders and advance medical directives, medical parole, and the use of inmate volunteers in prison hospice programs. We also include practical recommendations to community-based physicians working with incarcerated or recently released prisoners and describe solutions that can be implemented on an individual and systems basis.

Löfmark, Rurik; Nilstun, T.; Bolmsjö, I Ågren. From cure to palliation: concept, decision and acceptance. *Journal of Medical Ethics* 2007 December; 33(12): 685-688. NRCBL: 20.4.1; 4.1.1. SC: em. Identifiers: Sweden.

Abstract: The aim of this paper is to present and discuss nurses' and physicians' comments in a questionnaire relating to patients' transition from curative treatment to palliative care. The four-page questionnaire relating to experiences of and attitudes towards communication, decision-making, documentation and responsibility of nurses and physicians and towards the competence of patients was developed and sent to a random sample of 1672 nurses and physicians of 10 specialties. The response rate was 52% (n = 844), and over one-third made comments. The respondents differed in their comments about three areas: the concept of palliative care, experiences of unclear decision-making and difficulties in acceptance of the patient's situation. The responses are analysed in terms of four ethical theories: virtue ethics, deontology, consequentialism and casuistry. Many virtues considered to be appropriate for healthcare personnel to possess were invoked. Compassion, honesty, justice and prudence are especially important. However, principles of medical ethics, such as the deontological principle of respect for self-determination and the consequence of avoidance of harm, are also implied. Casuistry may be particularly helpful in analysing certain areas of difficulty-namely, what is meant by "palliative care", decision-making and accepting the patient's situation. Keeping a patient in a state of uncertainty often causes more suffering than necessary. Communication among the staff and with patients must be explicit. Many of the staff have not had adequate training in communicating with patients who are at the end of their life. Time for joint reflection has to be regained, and training in decision-making is essential. In our opinion, palliative care in Sweden is in need of improvement.

Malakoff, Marion. Palliative care/physician-assisted dying: alternative or continuing care? *Care Management Journals* 2006 Spring; 7(1): 41-44. NRCBL: 20.4.1; 20.7; 20.5.1; 8.1. SC: le.

Marco, Catherine A.; Schears, Raquel M. Death, dying, and last wishes. *Emergency Medicine Clinics of North America* 2006 November; 24(4): 969-987. NRCBL: 20.4.1; 20.5.1; 8.1; 9.4.

McNeil, Donald G., Jr. Drugs banned, world's poor suffer in pain. *New York Times* 2007 September 10; p. A1. A14, A15. NRCBL: 20.4.1; 9.7; 4.4. SC: po.

McNeil, Donald G., Jr. In India, a quest to ease the pain of the dying. *New York Times* 2007 September 11; p. F1, F5. NRCBL: 20.4.1; 9.7; 4.4. SC: po.

McNeil, Donald G., Jr. Japanese slowly shedding their misgivings about the use of painkilling drugs. *New York Times* 2007 September 10; p. A15. NRCBL: 20.4.1; 4.4; 9.7. SC: po.

Muller, David. GOMER. *Health Affairs* 2007 May-June; 26(3): 831-835. NRCBL: 20.4.1. Identifiers: "Get Out of My Emergency Room".

Mysorekar, Uma. Spirituality in palliative care — a Hindu perspective. *In:* Puchalski, Christina M., ed. A Time for Listening and Caring: Spirituality and the Care of the Chronically Ill and Dying. Oxford; New York: Oxford University Press, 2006: 171-182. NRCBL: 20.4.1; 1.2.

O'Connell, Laurence J. Spirituality in palliative care: an ethical imperative. *In:* Puchalski, Christina M., ed. A Time for Listening and Caring: Spirituality and the Care of the Chronically Ill and Dying. Oxford; New York: Oxford University Press, 2006: 27-38. NRCBL: 20.4.1; 1.2. SC: cs.

O'Gorman, Mary Lou. Spirituality in end-of-life care from a Catholic perspective: reflections of a hospital chaplain. *In:* Puchalski, Christina M., ed. A Time for Listening and Caring: Spirituality and the Care of the Chronically Ill and Dying. Oxford; New York: Oxford University Press, 2006: 139-154. NRCBL: 20.4.1; 1.2; 4.4; 20.5.4. SC: cs.

Olthuis, Gert; Leget, Carlo; Dekkers, Wim. Why hospice nurses need high self-esteem. *Nursing Ethics* 2007 January; 14(1): 62-71. NRCBL: 20.4.1; 4.1.3.

Abstract: This article discusses the relationship between personal and professional qualities in hospice nurses. We examine the notion of self-esteem in personal and professional identity. The focus is on two questions: (1) what is self-esteem, and how is it related to personal identity and its moral dimension? and (2) how do self-esteem and personal identity relate to the professional identity of nurses? We demonstrate it is important that the moral and personal goals in nurses' life coincide. If nurses' personal view of the good life is compatible with their experiences and feelings as professionals, this improves their performance as nurses. We also discuss how good nursing depends on the responses that nurses receive from patients, colleagues and family; they make nurses feel valued as persons and enable them to see the value of the work they do.

Palmer, Robert Chi-Noodin; Palmer, Marianne Leslie. Ojibwe beliefs and rituals in end-of-life care. *In:* Puchalski, Christina M., ed. A Time for Listening and Caring: Spirituality and the Care of the Chronically Ill and Dying. Oxford; New York: Oxford University Press, 2006: 215-225. NRCBL: 20.4.1; 1.2; 20.3.1.

Rapgay, Lobsang. A Buddhist approach to end-of-life care. *In:* Puchalski, Christina M., ed. A Time for Listening

NRCBL: National Reference Center for Bioethics Literature Classification Scheme   See inside front cover for terms.

125

and Caring: Spirituality and the Care of the Chronically Ill and Dying. Oxford; New York: Oxford University Press, 2006: 131-137. NRCBL: 20.4.1; 1.2; 20.3.1.

Sharp, Helen M. Ethical issues in the management of dysphagia after stroke. *Topics in Stroke Rehabilitation* 2006 Fall; 13(4): 18-25. NRCBL: 20.4.1; 20.5.1. SC: an.

Song, John; Bartels, Dianne M.; Ratner, Edward R.; Alderton, Lucy; Hudson, Brenda; Ahluwalia, Jasjit S. Dying on the streets: homeless persons' concerns and desires about end of life care. *JGIM: Journal of General Internal Medicine* 2007 April; 22(4): 435-441. NRCBL: 20.4.1; 9.5.10. SC: em.

Terry, W.; Olson, L.G.; Ravenscroft, P.; Wilss, L.; Boulton-Lewis, G. Hospice patients' views on research in palliative care. *Internal Medicine Journal* 2006 July; 36(7): 406-413. NRCBL: 20.4.1; 8.3.1; 18.5.7. SC: em.

Tuffrey-Wijne, Irene; Bernal, Jane; Butler, Gary; Hollins, Sheila; Curfs, Leopold. Using nominal group technique to investigate the views of people with intellectual disabilities on end-of-life care provision. *Journal of Advanced Nursing* 2007 April; 58(1): 80-89. NRCBL: 20.4.1; 9.5.3. SC: em.

United States. Veterans Health Administration. National Ethics Committee. The ethics of palliative sedation as a therapy of last resort. *American Journal of Hospice and Palliative Care* 2006 December-2007 January; 23(6): 483-491. NRCBL: 20.4.1; 4.4; 20.5.1; 9.4; 7.1.

Walter, James J. Terminal sedation: a Catholic perspective. *In:* Walter, James J.; Shannon, Thomas A., eds. Contemporary Issues in Bioethics: A Catholic Perspective. Lanham, MD: Rowman and Littlefield Publishers, 2005: 225-229. NRCBL: 20.4.1; 1.2; 9.7; 20.5.1.

Wright, Alexis A.; Katz, Ingrid T. Letting go of the rope — aggressive treatment, hospice care, and open access. *New England Journal of Medicine* 2007 July 26; 357(4): 324-327. NRCBL: 20.4.1; 9.3.1; 9.5.2; 9.7. SC: cs.

Zucker, David J.; Taylor, Bonita E. Spirituality, suffering, and prayerful presence within Jewish tradition. *In:* Puchalski, Christina M., ed. A Time for Listening and Caring: Spirituality and the Care of the Chronically Ill and Dying. Oxford; New York: Oxford University Press, 2006: 193-214. NRCBL: 20.4.1; 1.2; 4.4.

DEATH AND DYING/ TERMINAL CARE/ MINORS

Banerjee, Neela. A place to turn when a newborn is fated to die. *New York Times* 2007 March 13; p. A1, A14. NRCBL: 20.4.2; 20.5.2. SC: po.

McConnell, Yarrow; Frager, Gerri; Levetown, Marcia. Decision making in pediatric palliative care. *In:* Carter, Brian S.; Levetown, Marcia, eds. Palliative Care for In-

fants, Children, and Adolescents: A Practical Handbook. Baltimore: Johns Hopkins University Press, 2004: 69-111. NRCBL: 20.4.2; 8.3.2; 8.1; 20.5.2. SC: cs.

Ramnarayan, Padmanabhan; Craig, Finella; Petros, Andy; Pierce, Christine. Characteristics of deaths occurring in hospitalised children: changing trends. *Journal of Medical Ethics* 2007 May; 33(5): 255-260. NRCBL: 20.4.2; 20.3.2. SC: em. Identifiers: Hong Kong.
Abstract: BACKGROUND: Despite a gradual shift in the focus of medical care among terminally ill patients to a palliative model, studies suggest that many children with life-limiting chronic illnesses continue to die in hospital after prolonged periods of inpatient admission and mechanical ventilation. OBJECTIVES: To (1) examine the characteristics and location of death among hospitalised children, (2) investigate yearwise trends in these characteristics and (3) test the hypothesis that professional ethical guidance from the UK Royal College of Paediatrics and Child Health (1997) would lead to significant changes in the characteristics of death among hospitalised children. METHODS: Routine administrative data from one large tertiary-level UK children's hospital was examined over a 7-year period (1997-2004) for children aged 0-18 years. Demographic details, location of deaths, source of admission (within hospital vs external), length of stay and final diagnoses (International Classification of Diseases-10 codes) were studied. Statistical significance was tested by the Kruskal-Wallis analysis of ranks and median test (non-parametric variables), chi(2) test (proportions) and Cochran-Armitage test (linear trends). RESULTS: Of the 1127 deaths occurring in hospital over the 7-year period, the majority (57.7%) were among infants. The main diagnoses at death included congenital malformations (22.2%), perinatal diseases (18.1%), cardiovascular disorders (14.9%) and neoplasms (12.4%). Most deaths occurred in an intensive care unit (ICU) environment (85.7%), with a significant increase over the years (80.1% in 1997 to 90.6% in 2004). There was a clear increase in the proportion of admissions from in-hospital among the ICU cohort (14.8% in 1998 to 24.8% in 2004). Infants with congenital malformations and perinatal conditions were more likely to die in an ICU (OR 2.42, 95% CI 1.65 to 3.55), and older children with malignancy outside the ICU (OR 6.5, 95% CI 4.4 to 9.6). Children stayed for a median of 13 days interquartile range 4.0-23.25 days) on a hospital ward before being admitted to an ICU where they died. CONCLUSIONS: A greater proportion of hospitalised children are dying in an ICU environment. Our experience indicates that professional ethical guidance by itself may be inadequate in reversing the trends observed in this study.

Strong, Carson; Feudtner, Chris; Carter, Brian S.; Rushton, Cynda H. Goals, values, and conflict resolution. *In:* Carter, Brian S.; Levetown, Marcia, eds. Palliative Care for Infants, Children, and Adolescents: A Practical Handbook. Baltimore: Johns Hopkins University Press, 2004: 23-43. NRCBL: 20.4.2; 8.1; 4.4. SC: an, cs.

DELIVERY OF HEALTH CARE *See* CARE FOR SPECIFIC GROUPS

DETERMINATION OF DEATH *See* DEATH AND DYING/ DETERMINATION OF DEATH

DISCLOSURE *See* CONFIDENTIALITY; INFORMED CONSENT; HUMAN EXPERIMENTATION/ INFORMED CONSENT; TRUTH DISCLOSURE

DISTRIBUTIVE JUSTICE *See* RESOURCE ALLOCATION

DIVERSITY *See* CULTURAL PLURALISM

DNA FINGERPRINTING *See* GENETIC PRIVACY

DONATION *See* BLOOD BANKING, DONATION, AND TRANSFUSION; ORGAN AND TISSUE TRANSPLANTATION/ DONATION AND PROCUREMENT

DRUG INDUSTRY

Addicted to secrecy: sealed drug documents should be opened up [editorial]. *Nature* 2007 April 19; 446(7138): 832. NRCBL: 9.7; 1.3.2; 18.6. SC: le.

Ethics Committee to review physician-drug industry ties [news]. *Minnesota Medicine* 2007 June; 90(6): 23. NRCBL: 9.7; 7.3.

Flogging Gardasil. In its rush to market its human papillomavirus vaccine, Merck forgot to make a strong and compelling case for compulsory immunization [editorial]. *Nature Biotechnology* 2007 March; 25(3): 261. NRCBL: 9.7; 9.5.1; 9.5.5; 9.5.7.

Probity gone nuts [editorial]. *Nature Biotechnology* 2007 May 25(5): 483. NRCBL: 9.7.

Afif, Maryam T. Prescription ethics: can states protect pharmacists who refuse to dispense contraceptive prescriptions? *Pace Law Review* 2005 Fall; 26(1): 243-272. NRCBL: 9.7; 8.1; 11.1.

Agnello, Vincent. Commentary: Vioxx and public policy? Taking the right steps. *Organizational Ethics: Healthcare, Business, and Policy* 2006 Fall-Winter; 3(2): 125-130. NRCBL: 9.7; 1.3.9; 1.3.2; 1.3.5. Identifiers: Food and Drug Administration; Merck and Co., Inc. Comments: Patricia M. Tereskerz. Financial conflict and Vioxx: a public policy case study. Organizational Ethics: Healthcare, Business, and Policy 2006 Fall-Winter; 3(2): 112-119.

Aldhous, Peter. Prescribed opinions. *New Science* 2007 January 6-12; 193(2585): 17. NRCBL: 9.7; 1.3.7.

Almirall, Nat. The ethics of engagement with the pharmaceutical industry. *Michigan Medicine* 2006 January-February; 105(1): 10-12. NRCBL: 9.7; 7.3.

Ausman, James I. I told you it was going to happen . . . Part II [editorial]. *Surgical Neurology* 2006 May; 65(5): 520-521. NRCBL: 9.7; 1.3.2; 9.3.1. Identifiers: Medtronics.

Avorn, Jerry. Paying for drug approvals — who's using whom. *New England Journal of Medicine* 2007 April 26; 356(17): 1697-1700. NRCBL: 9.7; 1.3.2; 7.3. SC: em. Identifiers: Food and Drug Administration.

Bakalar, Nicholas. Review finds drug makers issue more positive studies. *New York Times* 2007 February 27; p. F7. NRCBL: 9.7; 1.3.9. SC: po; em.

Bardes, Charles L. Ethics and prescribing: the clinician's perspective. *In:* Santoro, Michael A.; Gorrie, Thomas M., eds. Ethics and the Pharmaceutical Industry. Cambridge; New York: Cambridge University Press, 2005: 136-152. NRCBL: 9.7; 4.1.1; 9.3.1; 1.3.2.

Bibbins-Domingo, Kirsten; Fernandez, Alicia; Kahn, Jonathan D.; Temple, Robert; Stockbridge, Norman L. BiDil for heart failure in black patients [letters and response]. *Annals of Internal Medicine* 2007 August 7; 147(3): 214-216. 15 refs. NRCBL: 9.7; 9.5.4; 15.11.
   Keywords: *blacks; *drugs; *heart diseases; biomedical research; drug industry; genetic ancestry; government regulation; patents; pharmacogenetics; racial groups; selection for treatment; Proposed Keywords: *drug approval; drug therapy; Keyword Identifiers: *BiDil; *Food and Drug Administration; United States

Brennan, Troyen A.; Mello, Michelle M. Sunshine laws and the pharmaceutical industry [editorial]. *JAMA: The Journal of the American Medical Association* 2007 March 21; 297(11): 1255-1257. NRCBL: 9.7; 7.3; 8.2. Comments: Ross, Joseph S., et al. Pharmaceutical company payments to physicians: early experiences with disclosure laws in Vermont and Minnesota. JAMA 2007 March 21; 297(11): 1216-1223.

Brown, Hannah. Sweetening the pill. *BMJ: British Medical Journal* 2007 March 31; 334(7595): 664-666. NRCBL: 9.7; 1.3.2. Identifiers: pharmaceutical industry; direct to consumer advertising.

Burton, Bob. Diabetes expert accuses drug company of "intimidation" [news]. *BMJ: British Medical Journal* 2007 December 1; 335(7630): 1113. NRCBL: 9.7; 7.3.

Burton, Bob. Industry loses bid to block disclosure of doctor's gifts [news]. *BMJ: British Medical Journal* 2007 July 7; 335(7609): 12. NRCBL: 9.7; 7.3; 1.3.2. Identifiers: Australia.

Burton, Bob. Roche fined over "extravagant" meals for doctors [news]. *BMJ: British Medical Journal* 2007 Feb-

NRCBL: National Reference Center for Bioethics Literature Classification Scheme   See inside front cover for terms.

127

ruary 24; 334(7590): 384. NRCBL: 9.7; 1.3.2; 9.3.1. SC: le. Identifiers: Australia.

Cahana, Alex; Mauron, Alexandre. The story of Vioxx — no pain and a lot of gain: ethical concerns regarding conduct of the pharmaceutical industry. *Journal of Anesthesia* 2006; 20(4): 348-351. NRCBL: 9.7; 4.4; 1.3.5; 1.3.9.

Campbell, Eric G. Doctors and drug companies — scrutinizing influential relationships. *New England Journal of Medicine* 2007 November 1; 357(18): 1796-1797. NRCBL: 9.7; 1.3.2; 7.3; 9.3.1.

Campbell, Eric G.; Gruen, Russell L.; Mountford, James; Miller, Lawrence G.; Cleary, Paul D.; Blumenthal, David. A national survey of physician-industry relationships. *New England Journal of Medicine* 2007 April 26; 356(17): 1742-1750. NRCBL: 9.7; 7.3. SC: em.

Abstract: BACKGROUND: Relationships between physicians and pharmaceutical, medical device, and other medically related industries have received considerable attention in recent years. We surveyed physicians to collect information about their financial associations with industry and the factors that predict those associations. METHODS: We conducted a national survey of 3167 physicians in six specialties (anesthesiology, cardiology, family practice, general surgery, internal medicine, and pediatrics) in late 2003 and early 2004. The raw response rate for this probability sample was 52%, and the weighted response rate was 58%. RESULTS: Most physicians (94%) reported some type of relationship with the pharmaceutical industry, and most of these relationships involved receiving food in the workplace (83%) or receiving drug samples (78%). More than one third of the respondents (35%) received reimbursement for costs associated with professional meetings or continuing medical education, and more than one quarter (28%) received payments for consulting, giving lectures, or enrolling patients in trials. Cardiologists were more than twice as likely as family practitioners to receive payments. Family practitioners met more frequently with industry representatives than did physicians in other specialties, and physicians in solo, two-person, or group practices met more frequently with industry representatives than did physicians practicing in hospitals and clinics. CONCLUSIONS: The results of this national survey indicate that relationships between physicians and industry are common and underscore the variation among such relationships according to specialty, practice type, and professional activities.

Cardarelli, Robert; Licciardone, John C.; Taylor, Lockwood G. A cross-sectional evidence-based review of pharmaceutical promotional marketing brochures and their underlying studies: is what they tell us important and true? *BMC Family Practice* 2006 March 3; 7: 13-18. NRCBL: 9.7; 1.3.2; 7.3; 7.1.

Carlat, Daniel. Dr. drug rep: during a year of being paid to give talks to doctors about an antidepressant, a psychiatrist

comes to terms with the fact that taking pharmaceutical money can cloud your judgment. *New York Times Magazine* 2007 November 25; p. 64-69. NRCBL: 9.7; 1.3.2; 9.3.1. SC: po.

Carreyrou, John. Inside Abbott's tactics to protect AIDS drug: older pill's price hike helps sales of flagship; a probe in Illinois. *Wall Street Journal* 2007 January 3; p. A1, A10. NRCBL: 9.7; 9.3.1. SC: po. Identifiers: Abbott Laboratories; Kaletra; Norvir.

Chandrasekhar, Charu A. Rx for drugstore discrimination: challenging pharmacy refusals to dispense prescription contraceptives under state public accommodations laws. *Albany Law Review* 2006; 70(1): 55-115. NRCBL: 9.7; 11.1; 9.2. SC: le.

Charatan, Fred. Drug company payments to doctors still hard to access [news]. *BMJ: British Medical Journal* 2007 March 31; 334(7595): 655. NRCBL: 9.7; 7.3.

Charatan, Fred. Drug makers end free lunches [news]. *BMJ: British Medical Journal* 2007 January 13; 334(7584): 64-65. NRCBL: 9.7; 1.3.2; 6; 9.3.1.

Charo, R. Alta. Politics, parents, and prophylaxis: mandating HPV vaccination in the United States. *New England Journal of Medicine* 2007 May 10; 356(19): 1905-1908. NRCBL: 9.7; 9.5.1; 9.5.7. SC: an; le.

Chimonas, Susan; Brennan, Troyen A.; Rothman, David J. Physicians and drug representatives: exploring the dynamics of the relationship. *JGIM: Journal of General Internal Medicine* 2007 February; 22(2): 184-190. NRCBL: 9.7; 1.3.2; 7.3. SC: em.

Collier, Joe. Inside big pharma's box of tricks [review of BBC program Panorama: The Secrets of the Drug Trials]. *BMJ: British Medical Journal* 2007 January 27; 334(7586): 209. NRCBL: 9.7; 18.5.2; 18.5.6; 1.3.9; 1.3.2; 1.3.7. Identifiers: British Broadcasting Corporation.

Coombes, Rebecca. Cancer drugs: swallowing big pharma's line? *BMJ: British Medical Journal* 2007 May 19; 334(7602): 1034-1035. NRCBL: 9.7; 1.3.7; 9.5.1.

Coombes, Rebecca. Life saving treatment or giant experiment. *BMJ: British Medial Journal* 2007 April 7; 334(7596): 721-723. NRCBL: 9.7; 9.5.1; 9.5.7.

Cooper, R.J.; Bissell, P.; Wingfield, J. A new prescription for empirical ethics research in pharmacy: a critical review of the literature. *Journal of Medical Ethics* 2007 February; 33(2): 82-86. NRCBL: 9.7; 4.1.1. SC: em; rv.

Abstract: Empirical ethics research is increasingly valued in bioethics and healthcare more generally, but there remain as yet under-researched areas such as pharmacy, despite the increasingly visible attempts by the profession to embrace additional roles beyond the supply of medicines. A descriptive and critical review of the extant empirical pharmacy ethics literature is provided here. A chronological change from quantitative to qualitative approaches is highlighted in this review, as well as differing theoretical approaches such as cognitive moral development and the four principles of biomedical ethics. Research with pharmacy student cohorts is common, as is representation from American pharmacists. Many examples of ethical problems are identified, as well as commercial and legal influences on ethical understanding and decision making. In this paper, it is argued that as pharmacy seeks to develop additional roles with concomitant ethical responsibilities, a new prescription is needed for empirical ethics research in pharmacy-one that embraces an agenda of systematic research using a plurality of methodological and theoretical approaches to better explore this under-researched discipline.

Daniels, Norman; Sabin, James E.; Teagarden, J. Russell. Who should get access to which drugs? An ethical template for pharmacy benefits. *In:* Santoro, Michael A.; Gorrie, Thomas M., eds. Ethics and the Pharmaceutical Industry. Cambridge; New York: Cambridge University Press, 2005: 206-224. NRCBL: 9.7; 9.3.2; 9.4; 1.3.2.

Davis, Joel J. Consumers' preferences for the communication of risk information in drug advertising. *Health Affairs* 2007 May-June; 26(3): 863-870. NRCBL: 9.7; 1.3.2; 8.3.1. SC: em.

Day, Lisa. Industry gifts to healthcare providers: are the concerns serious? *American Journal of Critical Care* 2006 September; 15(5): 510-513. NRCBL: 9.7; 1.3.2; 9.3.1.

Day, Michael. Who's funding WHO? WHO guidelines state that it will not accept money from drug companies, but how rigorous is it enforcing this? *BMJ: British Medical Journal* 2007 February 17; 334(7589): 338-340. NRCBL: 9.7; 9.3.1; 21.1. Identifiers: World Health Organization.

DeMaria, Anthony N. Your soul for a pen? *Journal of the American College of Cardiology* 2007 March 20; 49(11): 1220-1222. NRCBL: 9.7; 1.3.2; 9.3.1.

Diaz-Navarlaz, T.; Segui-Gomez, M. Commentary on Armitage G (2005) Drug errors, qualitative research and some reflections on ethics. Journal of Clinical Nursing 14, 869-875. *Journal of Clinical Nursing* 2006 September; 15(9): 1208-1209; discussion 1209. NRCBL: 9.7; 7.1; 9.8.

Donohue, Julie M.; Cevasco, Marisa; Rosenthal, Meredith B. A decade of direct-to-consumer advertising of prescription drugs. *New England Journal of Medicine* 2007 August 16; 357(7): 673-681. NRCBL: 9.7; 8.1; 1.3.2. SC: em.

Dresser, Rebecca. The curious case of off-label use. *Hastings Center Report* 2007 May-June; 37(3): 9-11. NRCBL: 9.7.

Drews, Jürgen. Drug research: between ethical demands and economic constraints. *In:* Santoro, Michael A.; Gorrie, Thomas M., eds. Ethics and the Pharmaceutical Industry. Cambridge; New York: Cambridge University Press, 2005: 21-36. NRCBL: 9.7; 1.3.9; 4.1.1; 1.3.2; 9.3.1; 9.5.6.

Dumit, Joseph; Greenslit, Nathan. Informed health and ethical identity management. *Culture, Medicine and Psychiatry* 2006 June; 30(2): 127-134. NRCBL: 9.7; 1.3.2; 9.3.1.

Dyer, Clare. US parents take government to court over MMR vaccine claims [news]. *BMJ: British Medical Journal* 2007 June 16; 334(7606): 1241. NRCBL: 9.7; 1.3.5.

Egan, Erin A. Who should regulate the practice of medicine? *Journal of Opioid Management* 2005 March-April; 1(1): 11-12. NRCBL: 9.7; 20.7; 4.4; 7.1; 1.3.5.

Epstein, Richard A. Influence of pharmaceutical funding on the conclusions of meta-analyses [editorial]. *BMJ: British Medical Journal* 2007 December 8; 335(7631): 1167. NRCBL: 9.7; 18.2.

Evans, Emily W. Conscientious objection: a pharmacist's right or professional negligence? *American Journal of Health-System Pharmacy* 2007 January 15; 64(2): 139-141. NRCBL: 9.7; 8.1; 4.1.1.

Falcón, M.; Martinez-Cánovas, F.J.; Pérez-Carceles, M.D.; Osuna, E.; Luna, A. Ethical problems related to information and pharmaceutical care in Spain. *Medicine and Law: The World Association for Medical Law* 2007 March; 26(1): 85-93. NRCBL: 9.7. SC: le.
  Abstract: This paper represents a reflection on the limits and objectives of the information pharmacists should offer in pharmacies. The obligation of a pharmacist to follow the patient's therapeutic progress makes it necessary to integrate this figure into an ethical-legal framework and to define the objective of the health-related information offered, taking into account the patient's welfare and constitutional rights.

Ferris, Lorraine E.; Naylor, C. David. Promoting integrity in industry-sponsored clinical drug trials: conflict of interest for Canadian academic health sciences centres. *In:* Lemmens, Trudo; Waring, Duff R., eds. Law and Ethics in Biomedical Research: Regulation, Conflict of Interest and Liability. Toronto; Buffalo: University of Toronto Press, 2006: 95-131. NRCBL: 9.7; 18.2; 1.3.9; 7.3.

Finucane, Thomas E.; Peterson, Eric D.; Boyce, Kurt; Overstreet, Karen; Sapers, Benjamin L.; Steinman, Michael A.; Chren, Mary-Margaret; Landefeld, C. Seth; Bero, Lisa A. The promotion of Gabapentin [letters and reply]. *Annals of Internal Medicine* 2007 February 20; 146(4): 312-314. NRCBL: 9.7; 1.3.2; 1.3.7; 1.3.9.

Freeman, Robert A. Industry perspectives on equity, access, and corporate social responsibility: a view from the inside. *In:* Cohen, Jillian Clare; Illingworth, Patricia; Schüklenk, Udo, eds. The Power of Pills: Social, Ethical and Legal Issues in Drug Development, Marketing, and Pricing. London; Ann Arbor, MI: Pluto, 2006: 65-73. NRCBL: 9.7; 5.3; 9.3.1.

NRCBL: National Reference Center for Bioethics Literature Classification Scheme   See inside front cover for terms.

129

Frosch, Dominick L.; Krueger, Patrick M.; Hornik, Robert C.; Cronholm, Peter F.; Barg, Frances K. Creating demand for prescription drugs: a content analysis of television direct-to-consumer advertising. *Annals of Family Medicine* 2007 January/February; 5(1): 6-12. NRCBL: 9.7; 1.3.7; 9.1; 1.3.2.

Fudin, Jeffrey. Blowing the whistle: a pharmacist's vexing experience unraveled. *American Journal of Health-System Pharmacy* 2006 November 15; 63(22): 2262-2265. NRCBL: 9.7; 7.3; 9.8.

Fugh-Berman, Adriane; Shahram, Ahari. Following the script: how drug reps make friends and influence doctors. *PLoS Medicine* 2007 April; 4(4): e150 [Online]. Accessed:http://medicine.plosjournals.org/perlserv/?request =get-document&doi=10.1371%2Fjournal.pmed. 0040150 [2007 August 27]. NRCBL: 9.7; 7.3; 9.3.1; 7.1.

Gathii, James Thuo. Third world perspectives on global pharmaceutical access. *In:* Santoro, Michael A.; Gorrie, Thomas M., eds. Ethics and the Pharmaceutical Industry. Cambridge; New York: Cambridge University Press, 2005: 336-351. NRCBL: 9.7; 21.1; 9.2; 5.3; 9.3.1; 9.5.6; 18.5.1.

Giles, Jim. Court case to reclaim confidential data. *Nature* 2007 April 19; 446(7138): 838-839. NRCBL: 9.7; 1.3.2; 18.6. SC: le.

Giles, Jim. Say no to lunch [commentary]. *New Scientist* 2007 April 28-May 4; 194(2601): 18. NRCBL: 9.7; 1.3.2; 7.3.

Giles, Jim. US vaccines on trial over link to autism. *New Scientist* 2007 June 23-29; 194(2609): 6-7. NRCBL: 9.7; 1.3.9; 9.5.7.

Grande, David. Prescriber profiling: time to call it quits [editorial]. *Annals of Internal Medicine* 2007 May 15; 146(10): 751-752. NRCBL: 9.7; 1.3.2; 9.3.1.

Great Britain (United Kingdom). Secretary of State for Health. Government response to the health committee's report on the influence of the pharmaceutical industry. London: Secretary of State for Health, 2005 September; 24 p. [Online]. Accessed: http://www.dh.gov.uk/prod_consum_dh/groups/dh_digitalassets/@dh/@en/documents/digitalasset/dh_4118608.pdf [2007 April 18]. NRCBL: 9.7; 5.3.

Greene, Jeremy A. Pharmaceutical marketing research and the prescribing physician. *Annals of Internal Medicine* 2007 May 15; 146(10): 742-748. NRCBL: 9.7; 1.3.2; 9.3.1.

Abstract: Surveillance of physicians' prescribing patterns and the accumulation and sale of these data for pharmaceutical marketing are currently the subjects of legislation in several states and action by state and national medical associations. Contrary to common perception, the growth of the health care information organization industry has not been limited to the past decade but has been building slowly over the past 50 years, beginning in the 1940s when growth in the prescription drug market fueled industry interest in understanding and influencing prescribing patterns. The development of this surveillance system was not simply imposed on the medical profession by the pharmaceutical industry but was developed through the interactions of pharmaceutical salesmen, pharmaceutical marketers, academic researchers, individual physicians, and physician organizations. Examination of the role of physicians and physician organizations in the development of prescriber profiling is directly relevant to the contemporary policy debate surrounding this issue.

Greene, Jeremy A. Pharmaceuticals and the economy of medical knowledge. *Chronicle of Higher Education* 2007 November 30; 54(14): B12-B13. NRCBL: 9.7; 5.3; 1.3.2; 1.3.5; 18.1.

Griffin, Leslie C. Conscience and emergency contraception. *Houston Journal of Health Law and Policy* 2006; 6(2): 299-318. NRCBL: 9.7; 13.1; 9.2; 1.2. SC: le.

Harris, Gardiner. Lawmaker calls for registry of drug firms paying doctors. *New York Times* 2007 August 4; p. A9. NRCBL: 9.7; 1.3.2; 9.3.1. SC: po.

Hazaray, Neil F. Do the benefits outweigh the risks? The legal, business, and ethical ramifications of pulling a blockbuster drug off the market. *Indiana Health Law Review* 2007; 4(1): 115-150. NRCBL: 9.7; 9.8; 9.2. SC: le.

Howell, Jonathan V. Direct to consumer advertising: the world of the market place [letter]. *BMJ: British Medical Journal* 2007 October 6; 335(7622): 683-684. NRCBL: 9.7; 1.3.2. Comments: Nicola Magrini, and Maria Font. Direct to consumer advertising of drugs in Europe. BMJ: British Medical Journal 2007 September 15; 335(7619): 526.

Hughes, Virginia. Mercury rising: parents of autistic children are mounting a vicious campaign against scientists who refute the link between vaccines and autism. *Nature Medicine* 2007 August; 13(8): 896-897. NRCBL: 9.7; 9.5.1; 21.1. SC: le.

Idänpään-Heikkilä, Juhana E.; Fluss, Sev. Emerging international norms for clinical testing: good clinical trial practice. *In:* Santoro, Michael A.; Gorrie, Thomas M., eds. Ethics and the Pharmaceutical Industry. Cambridge; New York: Cambridge University Press, 2005: 37-47. NRCBL: 9.7; 18.5.9; 18.2; 21.1.

Jacobson, Peter D.; Parmet, Wendy E. A new era of unapproved drugs: the case of Abigail Alliance v Von Eschenbach. *JAMA: The Journal of the American Medical Association* 2007 January 10; 297(2): 205-208. NRCBL: 9.7; 18.5.7. SC: le.

Jagadeesh, N. Narco analysis leads to more questions than answers. *Indian Journal of Medical Ethics* 2007 January-March; 4(1): 9. NRCBL: 9.7; 1.3.5; 21.4; 7.4. SC: le.

Kahn, Jeffrey. What vaccination programs mean for research [editorial]. *American Journal of Bioethics* 2007 March; 7(3): 5-10. NRCBL: 9.7; 9.1; 18.5.2; 18.6.

Kassirer, Jerome P. Pharmaceutical ethics? [review of Ethics and the Pharmaceutical Industry, edited by Michael A. Santoro and Thomas M. Gorrie]. *Open Medicine [electronic]* 2007; 1(1): 58-59. Accessed: http://www.openmedicine.ca [2007 April 19]. NRCBL: 9.7.

Kent, Alastair; Mintzes, Barbara. Should patient groups accept money from drug companies? [debate]. *BMJ: British Medial Journal* 2007 May 5; 334(7600): 934-935. NRCBL: 9.7; 9.3.1; 9.5.1. SC: an.

Kieve, Millie. Falling on deaf ears [comment]. *New Scientist* 2007 September 15-21; 195(2621): 24. NRCBL: 9.7; 1.3.9; 7.3; 9.8.

Kmietowicz, Zosia. Doctors threaten to withdraw subscription to GMC [news]. *BMJ: British Medical Journal* 2007 July 7; 335(7609): 14. NRCBL: 9.7; 1.3.2. Identifiers: Great Britain (United Kingdom); General Medical Council.

Kmietowicz, Zosia. Repeal law that puts "FDA on the payroll of the industry" [news]. *BMJ: British Medical Journal* 2007 March 3; 334(7591): 447. NRCBL: 9.7; 5.3; 9.3.1. SC: le.

Koski, Edward Greg. Renegotiating the grand bargain: balancing prices, profits, people, and principles. *In:* Santoro, Michael A.; Gorrie, Thomas M., eds. Ethics and the Pharmaceutical Industry. Cambridge; New York: Cambridge University Press, 2005: 393-403. NRCBL: 9.7; 1.3.2; 5.3; 18.6; 4.1.1.

Lenzer, Jeanne. Drug company tries to suppress internal memos [news]. *BMJ: British Medical Journal* 2007 January 13; 334(7584): 59. NRCBL: 9.7; 1.3.2. SC: le.

Lesser, Eugene A.; Starr, Jennifer; Kong, Xuan; Megerian, J. Thomas; Gozani, Shai N. Point-of-service nerve conduction studies: an example of industry-driven disruptive innovation in health care. *Perspectives in Biology and Medicine* 2007 Winter; 50(1): 40-53. NRCBL: 9.7; 7.4; 7.3.

Light, Terry R. Orthopaedic gifts: opportunities and obligations. *Journal of Bone and Joint Surgery. American volume* 2006 November; 88(11): 2521-2526. NRCBL: 9.7; 7.1; 1.3.2; 9.3.1; 4.1.2; 21.1.

Liss, Howard. Publication bias in the pulmonary/allergy literature: effect of pharmaceutical company sponsorship. *The Israel Medical Association Journal* 2006 July; 8(7): 451-454. NRCBL: 9.7; 18.2; 1.3.7.

Luna, Florencia. Assumptions in the "standard of care" debate. *In:* Cohen, Jillian Clare; Illingworth, Patricia; Schüklenk, Udo, eds. The Power of Pills: Social, Ethical and Legal Issues in Drug Development, Marketing, and Pricing. London; Ann Arbor, MI: Pluto, 2006: 215-223. NRCBL: 9.7; 18.5.9; 18.2; 9.5.6; 21.1.

Lybecker, Kristina M. Social, ethical, and legal issues in drug development, marketing, and pricing policies: setting priorities: pharmaceuticals as private organizations and the duty to make money/maximize profits. *In:* Cohen, Jillian Clare; Illingworth, Patricia; Schüklenk, Udo, eds. The Power of Pills: Social, Ethical and Legal Issues in Drug Development, Marketing, and Pricing. London; Ann Arbor, MI: Pluto, 2006: 25-31. NRCBL: 9.7; 5.3; 9.3.1; 1.3.2.

Magrini, Nicola; Font, Maria. Direct to consumer advertising of drugs in Europe [editorial]. *BMJ: British Medical Journal* 2007 September 15; 335(7619): 526. NRCBL: 9.7; 1.3.2; 1.3.7; 21.1.

Marco, Catherine A.; Moskop, John C.; Solomon, Robert C.; Geiderman, Joel M.; Larkin, Gregory L. Gifts to physicians from the pharmaceutical industry: an ethical analysis. *Annals of Emergency Medicine* 2006 November; 48(5): 513-521. NRCBL: 9.7; 7.1; 1.3.2; 9.3.1.

Marshall, Ian E. Physicians and the pharmaceutical industry: a symbiotic relationship? *In:* Cohen, Jillian Clare; Illingworth, Patricia; Schüklenk, Udo, eds. The Power of Pills: Social, Ethical and Legal Issues in Drug Development, Marketing, and Pricing. London; Ann Arbor, MI: Pluto, 2006: 57-64. NRCBL: 9.7; 1.3.2; 9.3.1; 6. SC: le.

Martin, Emily. Pharmaceutical virtue. *Culture, Medicine and Psychiatry* 2006 June; 30(2): 157-174. NRCBL: 9.7; 17.4; 1.3.2.

McCarthy, Michael. US campaign tackles drug company influence over doctors. *Lancet* 2007 March 3-9; 369(9563): 730. NRCBL: 9.7; 1.3.2; 9.3.1; 9.8; 1.3.9.

McNeill, Paul M.; Kerridge, Ian H.; Arciuli, Catherine; Henry, David A.; Macdonald, Graham J.; Day, Richard O.; Hill, Suzanne R. Gifts, drug samples and other items given to medical specialists by pharmaceutical companies. *Journal of Bioethical Inquiry* 2006; 3(3): 139-148. NRCBL: 9.7; 7.3; 6. SC: em.

McNeill, P.M.; Kerridge, I.H.; Henry, D.A.; Stokes, B.; Hill, S.R.; Newby, D.; Macdonald, G.J; Day, R.O.; Maguire, J.; Henderson, K.M. Giving and receiving of gifts between pharmaceutical companies and medical specialists in Australia. *Internal Medicine Journal* 2006 September; 36(9): 571-578. NRCBL: 9.7; 1.3.2; 9.3.1.

Metzl, Jonathan M. If direct-to-consumer advertisements come to Europe: lessons from the USA. *Lancet* 2007 February 24-March 2; 369(9562): 704-706. NRCBL: 9.7; 1.3.2; 4.2; 7.1; 8.1; 21.1.

NRCBL: National Reference Center for Bioethics Literature Classification Scheme   See inside front cover for terms.

131

Mills, Ann; Werhane, Patricia; Gorman, Michael. The pharmaceutical industry and its obligations in the developing world. *In:* Cohen, Jillian Clare; Illingworth, Patricia; Schüklenk, Udo, eds. The Power of Pills: Social, Ethical and Legal Issues in Drug Development, Marketing, and Pricing. London; Ann Arbor, MI: Pluto, 2006: 32-40. NRCBL: 9.7; 21.1. SC: an.

Mitton, Craig R.; McMahon, Meghan; Morgan, Steve; Gibson, Jennifer. Centralized drug review processes: are they fair? *Social Science and Medicine* 2006 July; 63(1): 200-211. NRCBL: 9.7; 5.3; 21.1.

Mohan, Bannur Muthai. Misconceptions about narco analysis. *Indian Journal of Medical Ethics* 2007 January-March; 4(1): 7-8. NRCBL: 9.7; 1.3.5; 21.4; 7.4. SC: le.

Moran, Nuala. UK parses merits of value-based drug pricing. *Nature Biotechnology* 2007 April; 25(4): 369-370. NRCBL: 9.7; 9.3.1.

Morgan, Steven G. Direct-to-consumer advertising and expenditure on prescription drugs: a comparison of experiences in the United States and Canada. *Open Medicine [electronic]* 2007; 1(1): 37-45. Accessed: http://www.openmedicine.ca [2007 April 19]. NRCBL: 9.7; 1.3.2; 21.1.

Morris, Albert W., Jr.; Gadson, Sandra L.; Burroughs, Valentine. "For the good of the patient," survey of the physicians of the National Medical Association regarding perceptions of DTC advertising, Part II, 2006. *Journal of the National Medical Association* 2007 March; 99(3): 287-293. NRCBL: 9.7; 7.1; 1.3.2; 1.3.7. SC: em.

Moynihan, Ray. Attempt to undermine European ban on advertising drugs fails in France [news]. *BMJ: British Medical Journal* 2007 February 10; 334(7588): 279. NRCBL: 9.7; 1.3.2.

Moynihan, Ray. Direct to consumer advertising should not come to Europe. *BMJ: British Medical Journal* 2007 May 19; 334(7602): 1025. NRCBL: 9.7; 1.3.2.

Moynihan, Ray. EC report on drug advertising found to be "biased". *BMJ: British Medical Journal* 2007 June 23; 334(7607): 1290. NRCBL: 9.7; 1.3.2. Identifiers: European Commission.

Moynihan, Ray. Healthcare giant advertises to children in Australia's classrooms [news]. *BMJ: British Medical Journal* 2007 September 29; 335(7621): 637. NRCBL: 9.7; 1.3.2; 9.5.7.

Mueller, Paul S.; Hook, C.Christopher; Litin, Scott C. Physician preferences and attitudes regarding industry support of CME programs. *American Journal of Medicine* 2007 March; 120(3): 281-285. NRCBL: 9.7; 7.1; 7.2; 9.3.1; 1.3.2. SC: em.

Murphy, M. Dianne; Goldkind, Sara F. The regulatory and ethical challenges of pediatric research. *In:* Santoro, Michael A.; Gorrie, Thomas M., eds. Ethics and the Pharmaceutical Industry. Cambridge; New York: Cambridge University Press, 2005: 48-67. NRCBL: 9.7; 18.5.2; 18.2; 18.6.

Oldani, Michael. Can doctors take back the script? Understanding the total system of prescription generation. *Atrium* 2007 Summer; 4: 15-17, 28. NRCBL: 9.7; 1.3.2; 9.3.1.

Parmet, Wendy E. Pharmaceuticals, public health, and the law: a public health perspective. *In:* Cohen, Jillian Clare; Illingworth, Patricia; Schüklenk, Udo, eds. The Power of Pills: Social, Ethical and Legal Issues in Drug Development, Marketing, and Pricing. London; Ann Arbor, MI: Pluto, 2006: 77-87. NRCBL: 9.7; 9.5.1; 9.1; 21.1.

Paul, Norbert W.; Fangerau, Heiner. Why should we bother? Ethical and social issues in individualized medicine. *Current Drug Targets* 2006 December; 7(12): 1721-1727. NRCBL: 9.7; 15.1; 15.3; 1.1.

Pearson, Helen. Cancer patients opt for unapproved drug [news]. *Nature* 2007 March 29; 446(7135): 474-475. NRCBL: 9.7; 9.5.1; 18.5.7.

Peppin, Patricia. Directing consumption: direct-to-consumer advertising and global public health. *In:* Bennett, Belinda; Tomossy, George F., eds. Globalization and Health: Challenges for Health Law and Bioethics. Dordrecht: Springer, 2006: 109-128. NRCBL: 9.7; 1.3.2; 5.3; 21.1; 9.1; 8.1. SC: le; rv. Identifiers: Canada; New Zealand; Australia.

Peppin, Patricia. The power of illusion and the illusion of power: direct-to-consumer advertising and Canadian health care. *In:* Flood, Colleen M., ed. Just Medicare: What's In, What's Out, How We Decide. Buffalo, NY: University of Toronto Press, 2006: 355-378. NRCBL: 9.7; 1.3.2. SC: le.

Pharmaceutical Research and Manufacturers of America [PhRMA]. PhRMA guiding principles: direct to consumer advertisements about prescription medicines. Washington, DC: Pharmaceutical Research and Manufacturers of America, 2005 November; 10 p. [Online]. Accessed: http://www.phrma.org/files/DTCGuidingprinciples.pdf [2007 April 30]. NRCBL: 9.7; 1.3.2; 17.3; 17.4.

Pinto, Sharrel L.; Lipowski, Earlene; Segal, Richard; Kimberlin, Carole; Algina, James. Physicians' intent to comply with the American Medical Association's guidelines on gifts from the pharmaceutical industry. *Journal of Medical Ethics* 2007 June; 33(6): 313-319. NRCBL: 9.7; 1.3.2; 7.3; 9.3.1. SC: em.
  Abstract: OBJECTIVE: To identify factors that predict physicians' intent to comply with the American Medical Association's (AMA's) ethical guidelines on gifts from the pharmaceutical industry. METHODS: A survey was designed and mailed in June 2004 to a random sample of

850 physicians in Florida, USA, excluding physicians with inactive licences, incomplete addresses, addresses in other states and pretest participants. Factor analysis extracted six factors: attitude towards following the guidelines, subjective norms (eg, peers, patients, etc), facilitating conditions (eg, knowledge of the guidelines, etc), profession-specific precedents (eg, institution's policies, etc), individual-specific precedents (physicians' own discretion, policies, etc) and intent. Multivariate regression modelling was conducted. RESULTS: Surveys were received from 213 physicians representing all specialties, with a net response rate of 25.5%. 62% (n = 133) of respondents were aware of the guidelines; 50% (n = 107) had read them. 48% (n = 102) thought that following the guidelines would increase physicians' credibility and professional image; 68% (n = 145) agreed that it was important to do so. Intent to comply was positively associated with attitude, subjective norms, facilitators and sponsorship of continuing medical education (CME) events, while individual-specific precedents had a negative relationship with intent to comply. Predictors of intent (R(2) = 0.52, p) were attitude, subjective norms, the interaction term (attitude and subjective norms), sponsorship of CME events and individual-specific precedents. CONCLUSIONS: Physicians are more likely to follow the AMA guidelines if they have positive attitudes towards the guidelines, greater subjective norms, fewer expectations of CME sponsorship and fewer individual-specific precedents. Physicians believing that important individuals or organisations expect them to comply with the guidelines are more likely to express intent, despite having fewer beliefs that positive outcomes would result through compliance.

Pitts, Peter J. Settling for second best? [letter]. *Nature Biotechnology* 2007 July; 25(7): 715-716. NRCBL: 9.7; 1.3.9; 5.3. Identifiers: U.S. Food and Drug Administration [FDA]. Comments: Probity gone nuts. Nature Biotechnology 2007 May; 25(5): 483.

Pray, W. Steven. Ethical, scientific, and educational concerns with unproven medications. *American Journal of Pharmaceutical Education* 2006 December 15; 70(6): 141. NRCBL: 9.7; 9.8; 9.5.1; 4.1.1.

Resnik, David B. Access to medications and global justice. *In:* Cohen, Jillian Clare; Illingworth, Patricia; Schüklenk, Udo, eds. The Power of Pills: Social, Ethical and Legal Issues in Drug Development, Marketing, and Pricing. London; Ann Arbor, MI: Pluto, 2006: 88-97. NRCBL: 9.7; 21.1; 5.3.

Roehr, Bob. More than 90% of US doctors receive drug company favours. *BMJ: British Medical Journal* 2007 April 28; 334(7599): 869. NRCBL: 9.7; 7.3.

Roper, James E. Commentary on "Financial conflict and Vioxx: a public policy case study" by Tereskerz. *Organizational Ethics: Healthcare, Business, and Policy* 2006 Fall-Winter; 3(2): 120-124. NRCBL: 9.7; 1.3.9; 1.3.2; 1.3.5. Identifiers: Food and Drug Administration [FDA]; Merck and Co., Inc. Comments: Patricia M. Tereskerz. Financial

conflict and Vioxx: a public policy case study. Organizational Ethics: Healthcare, Business, and Policy 2006 Fall-Winter; 3(2): 112-119.

Rosenthal, Meredith B.; Donohue, Julie M. Direct-to-consumer advertising of prescription drugs: a policy dilemma. *In:* Santoro, Michael A.; Gorrie, Thomas M., eds. Ethics and the Pharmaceutical Industry. Cambridge; New York: Cambridge University Press, 2005: 169-183. NRCBL: 9.7; 1.3.2; 9.1.

Ross, David B. The FDA and the case of ketek. *New England Journal of Medicine* 2007 April 19; 356(16): 1601-1604. NRCBL: 9.7; 1.3.5; 1.3.9. Identifiers: Food and Drug Administration; telithromycin.

Ross, Joseph S.; Lackner, Josh E.; Lurie, Peter; Gross, Cary P.; Wolfe, Sidney; Krumholz, Harlan M. Pharmaceutical company payments to physicians: early experiences with disclosure laws in Vermont and Minnesota. *JAMA: The Journal of the American Medical Association* 2007 March 21; 297(11): 1216-1223. NRCBL: 9.7; 7.3; 8.2. SC: em.

Abstract: CONTEXT: Recent legislation in 5 states and the District of Columbia mandated state disclosure of payments made to physicians by pharmaceutical companies. In 2 of these states, Vermont and Minnesota, payment disclosures are publicly available. OBJECTIVES: To determine the accessibility and quality of the data available in Vermont and Minnesota and to describe the prevalence and magnitude of disclosed payments. DESIGN AND SETTING: Cross-sectional analysis of publicly available data from July 1, 2002, through June 30, 2004, in Vermont and from January 1, 2002, through December 31, 2004, in Minnesota. MAIN OUTCOME MEASURES: Accessibility and quality of disclosure data and the number, value, and type of payments of $100 or more to physicians. RESULTS: Access to payment data required extensive negotiation with the Office of the Vermont Attorney General and manual photocopying of individual disclosure forms at Minnesota's State Board of Pharmacy. In Vermont, 61% of payments were not released to the public because pharmaceutical companies designated them as trade secrets and 75% of publicly disclosed payments were missing information necessary to identify the recipient. In Minnesota, 25% of companies reported in each of the 3 years. In Vermont, among 12,227 payments totaling $2.18 million publicly disclosed, there were 2416 payments of $100 or more to physicians; total, $1.01 million; median payment, $177 (range, $100-$20,000). In Minnesota, among 6946 payments totaling $30.96 million publicly disclosed, there were 6238 payments of $100 or more to physicians; total, $22.39 million; median payment, $1000 (range, $100-$922,239). Physician-specific analyses were possible only in Minnesota, identifying 2388 distinct physicians who received payment of $100 or more; median number of payments received, 1 (range, 1-88) and the median amount received, $1000 (range, $100-$1,178,203). CONCLUSIONS: The Vermont and Minnesota laws requiring disclosure of payments do not provide easy

NRCBL: National Reference Center for Bioethics Literature Classification Scheme   See inside front cover for terms.

133

access to payment information for the public and are of limited quality once accessed. However, substantial numbers of payments of $100 or more were made to physicians by pharmaceutical companies.

Roy, Nobhojit; Madhiwalla, Neha; Pai, Sanjay A. Drug promotional practices in Mumbai: a qualitative study. *Indian Journal of Medical Ethics* 2007 April-June; 4(2): 57-61. NRCBL: 9.7; 1.3.2; 9.3.1. SC: em.

Sade, Robert M.; Grande, David; Gorske, Arnold L.; Campbell, Eric G. A national survey of physician-industry relationships [letters and reply]. *New England Journal of Medicine* 2007 August 2; 357(5): 507-508. NRCBL: 9.7; 1.3.2; 9.3.1; 7.3.

Sandberg, David E. Growth attenuation in developmental disabilities. *Growth, Genetics and Hormones* 2007 March; 23(1): 12-13. NRCBL: 9.7; 9.5.7. Comments: D.F. Gunther, and D.S. Diekema. Attenuating growth in children with profound developmental disability: a new approach to an old dilemma. Archives of Pediatric and Adolescent Medicine 2006 October 160(10): 1013-1017.

Saul, Stephanie. Doctors and drug makers: a move to end cozy ties. *New York Times* 2007 February 12; p. C10. NRCBL: 9.7; 1.3.2; 9.3.1. SC: po.

Selgelid, Michael J. Ethics and drug resistance. *Bioethics* 2007 May; 21(4): 218-229. NRCBL: 9.7; 1.3.2; 7.1; 9.5.1. SC: an.
Abstract: This paper reviews the dynamics behind, and ethical issues associated with, the phenomenon of drug resistance. Drug resistance is an important ethical issue partly because of the severe consequences likely to result from the increase in drug resistant pathogens if more is not done to control them. Drug resistance is also an ethical issue because, rather than being a mere quirk of nature, the problem is largely a product of drug distribution. Drug resistance results from the over-consumption of antibiotics by the wealthy; and it, ironically, results from the under-consumption of antibiotics, usually by the poor or otherwise marginalized. In both kinds of cases the phenomenon of drug resistance illustrates why health (care) — at least in the context of infectious disease — should be treated as a (global) public good. The point is that drug resistance involves 'externalities' affecting third parties. When one patient develops a resistant strain of disease because of her over- or under-consumption of medication, this more dangerous malady poses increased risk to others. The propriety of free-market distribution of goods subject to externalities is famously dubious — given that the 'efficiency' rationale behind markets assumes an absence of externalities. Market failure in the context of drug resistance is partly revealed by the fact that no new classes of antibiotics have been developed since 1970. I conclude by arguing that the case of drug resistance reveals additional reasons — to those traditionally appealed to by bioethicists — for treating health care as something special when making policy decisions about its distribution.

Shuchman, Miriam. Drug risks and free speech — can Congress ban consumer drug ads? *New England Journal of Medicine* 2007 May 31; 356(22): 2236-2239. NRCBL: 9.7; 1.3.2; 9.3.1. SC: le.

Smith, Richard. Curbing the influence of the drug industry: a British view. *PLoS Medicine* 2005 September; 2(9): e241 (0821-0823). NRCBL: 9.7; 1.3.2; 5.3.

Soreth, Janice; Cox, Edward; Kweder, Sandra; Jenkins, John; Galson, Steven. Ketek — the FDA perspective. *New England Journal of Medicine* 2007 April 19; 356(16): 1675-1676. NRCBL: 9.7; 1.3.5; 1.3.9. Identifiers: Food and Drug Administration.

Spurgeon, David. New York Times reveals payments to doctors by drug firms [news]. *BMJ: British Medical Journal* 2007 March 31; 334(7595): 655. NRCBL: 9.7; 7.3.

Stange, Kurt C. In this issue: doctor-patient and drug company-patient communication. *Annals of Family Medicine* 2007 January/February; 5(1): 2-3. NRCBL: 9.7; 8.1; 1.3.7.

Steinberg, Brian. New medical-device ads; old concerns: Can a knee implant be sold this way, and should it be? *Wall Street Journal* 2007 April 10; p. B6. NRCBL: 9.7; 8.1; 1.3.2. SC: po.

Stewart, Alexandra M. Mandating HPV vaccination: private rights, public good [letter]. *New England Journal of Medicine* 2007 May 10; 356(19): 1998-1999. NRCBL: 9.7; 9.5.1; 9.5.5. SC: le.

Strumolo, Adaline R. Prescription privacy. *American Journal of Law and Medicine* 2007; 33(4): 705-708. NRCBL: 9.7; 8.4. SC: le.

Sugarman, Stephen D. Cases in vaccine court — legal battles over vaccines and autism. *New England Journal of Medicine* 2007 September 27; 357(13): 1275-1277. NRCBL: 9.7; 9.5.7. SC: le.

Tanne, Janice Hopkins. Drug advertisements in US paint a "black and white scenario" [news]. *BMJ: British Medical Journal* 2007 February 10; 334(7588): 279. NRCBL: 9.7; 1.3.2; 1.3.7.

Tanne, Janice Hopkins. FDA places "black box" warnings on anaemia drugs amid reports of incentives to doctors. *BMJ: British Medical Journal* 2007 May 19; 334(7602): 1022. NRCBL: 9.7; 7.3.

Tanne, Janice Hopkins. US campaign aims to end industry gifts and speaking fees [news]. *BMJ: British Medical Journal* 2007 February 24; 334(7590): 385. NRCBL: 9.7; 1.3.2; 9.3.1.

Tanne, Janice Hopkins. US guidelines often influenced by industry [news]. *BMJ: British Medical Journal* 2007 January 27; 334(7586): 171. NRCBL: 9.7; 18.2; 9.8; 1.3.9; 7.3.

Tereskerz, Patricia M. Financial conflict and Vioxx: a public policy case study. *Organizational Ethics: Healthcare, Business, and Policy* 2006 Fall-Winter; 3(2): 112-119. NRCBL: 9.7; 1.3.9; 1.3.2; 1.3.5. SC: cs. Comments: Organizational Ethics: Healthcare, Business, and Policy 2006 Fall-Winter; 3(2): 120-130.

Theofrastous, Theodore C. Session 8: Canada and U.S. approaches to health care: how the Canadian and U.S. political, regulatory, and legal systems impact health care. *Canadian - United States Law Journal* 2005; 31: 269-280. NRCBL: 9.7; 1.3.9; 5.1; 5.3; 8.3.1; 21.1. SC: le.

Toop, Les; Mangin, Dee. Industry funded patient information and the slippery slope to New Zealand. *BMJ: British Medical Journal* 2007 October 6; 335(7622): 694-695. NRCBL: 9.7; 1.3.2.

Triggle, David J. Treating desires not diseases: a pill for every ill and an ill for every pill? *Drug Discovery Today* 2007 February; 12(3-4): 161-166. NRCBL: 9.7; 1.3.2; 4.2; 4.3; 1.3.7; 9.3.1.

United States. Food and Drug Administration [FDA]. Expanded access to investigational drugs for treatment use. *Federal Register* 2006 December 14; 71(240): 75147-75168 [Online]. Accessed: http://www.fda.gov/ OHRMS/DOCKET/98fr/06-9684.pdf [2007 October 10]. NRCBL: 9.7; 5.2; 5.3; 1.3.5.

United States. Food and Drug Administration [FDA]. Requirements on content and format of labeling for human prescription drug and biological products and draft guidances and two guidances for industry on the content and format of labeling for human prescription drug and biological products; final rule and notices [21 CFR Parts 201, 314, and 601]. *Federal Register* 2006 January 24; 71(15): 3922-3997. NRCBL: 9.7; 5.3.

Vastag, Brian. US aims to tighten rules on direct-to-consumer drug ads. *Nature Biotechnology* 2007 March; 25(3): 267. NRCBL: 9.7; 1.3.2. SC: le.

Werhane, Patricia H.; Gorman, Michael E. Intellectual property rights, access to life-enhancing drugs, and corporate moral responsibilities. *In:* Santoro, Michael A.; Gorrie, Thomas M., eds. Ethics and the Pharmaceutical Industry. Cambridge; New York: Cambridge University Press, 2005: 260-281. NRCBL: 9.7; 1.1; 5.3; 1.3.2; 21.1; 9.5.6. SC: le.

Wingfield, Joy. Researching the chemists: towards an integrated research agenda: conference report. *Clinical Ethics* 2007 March; 2(1): 42-44. NRCBL: 9.7; 2.3. Conference: Researching the "Chemists": A one day conference for researchers in healthcare law, healthcare ethics and pharmacy practice; held at Manchester, UK; 29 June 2006.

Wingfield, Joy. You pays your money and you takes your choice? *In:* Gunning, Jennifer; Holm, Søren, eds. Ethics,

law, and society. Volume 2. Aldershot, Hants, England; Burlington, VT: Ashgate, 2006: 287-289. NRCBL: 9.7; 1.3.2.

Wong, D.; Kyle, G. Some ethical considerations for the "off-label" use of drugs such as Avastin. *British Journal of Ophthalmology* 2006 October; 90(10): 1218-1219. NRCBL: 9.7; 9.8; 9.5.1.

Yeates, Neil. Health Canada's new standards on conflict of interest. *CMAJ/JAMC: Canadian Medical Association Journal* 2007 October 9; 177(8): 900. NRCBL: 9.7; 1.3.9; 7.3.

DURABLE POWER OF ATTORNEY *See* ADVANCE DIRECTIVES

ECONOMICS *See* GENETIC SCREENING/ ECONOMIC ASPECTS; HEALTH CARE ECONOMICS; ORGAN AND TISSUE TRANSPLANTATION/ DONATION AND PROCUREMENT/ ECONOMIC ASPECTS

EDUCATION *See* BIOETHICS AND MEDICAL ETHICS/ EDUCATION; MEDICAL EDUCATION

ELECTROCONVULSIVE THERAPY
See also BEHAVIOR CONTROL; CARE FOR SPECIFIC GROUPS/ MENTALLY DISABLED; MENTAL HEALTH THERAPIES AND NEUROSCIENCES

Illes, Judy; Gallo, Marisa; Kirschen, Matthew P. An ethics perspective on transcranial magnetic stimulation (TMS) and human neuromodulation. *Behavioural Neurology* 2006; 17(3-4): 149-157. NRCBL: 17.5; 8.3.1; 8.3.3.

Jones, Leslie Sargent. The ethics of transcranial magnetic stimulation [letter and reply]. *Science* 2007 March 23; 315(5819): 1663-1664. NRCBL: 17.5; 18.1.

Newell, Elizabeth R. Competency, consent, and electroconvulsive therapy: a mentally ill prisoner's right to refuse invasive medical treatment in Oregon's criminal justice system. *Lewis and Clark Law Review* 2005 Winter; 9(4): 1019-1045. NRCBL: 17.5; 17.8; 1.3.5; 8.3.3; 8.3.4; 9.5.1. SC: le.

Racine, Eric; Waldman, Sarah; Palmour, Nicole; Risse, David; Illes, Judy. "Currents of hope": neurostimulation techniques in U.S. and U.K. print media. *CQ: Cambridge Quarterly of Healthcare Ethics* 2007 Summer; 16(3): 312-316. NRCBL: 17.5; 21.1; 1.3.7. SC: em.

EMBRYOS *See* CARE FOR SPECIFIC GROUPS/ FETUSES; CRYOBANKING OF SPERM, OVA AND EMBRYOS; HUMAN EXPERIMENTATION/ SPECIAL POPULATIONS/ EMBRYOS AND FETUSES

NRCBL: National Reference Center for Bioethics Literature Classification Scheme   See inside front cover for terms.

135

ENHANCEMENT

Enhancing, not cheating. A broad debate about the use of drugs that improve cognition for both the healthy and the ill is needed [editorial]. *Nature* 2007 November 15; 450(7168): 320. NRCBL: 4.5; 9.7; 17.4.

Agar, Nicholas. Whereto transhumanism? The literature reaches a critical mass. *Hastings Center Report* 2007 May-June; 37(3): 12-17. NRCBL: 4.5; 4.4; 14.1; 1.1.

Bolt, L.L.E. True to oneself? Broad and narrow ideas on authenticity in the enhancement debate. *Theoretical Medicine and Bioethics* 2007; 28(4): 285-300. NRCBL: 4.5; 1.1; 4.4; 17.4.
Abstract: Our knowledge of the human brain and the influence of pharmacological substances on human mental functioning is expanding. This creates new possibilities to enhance personality and character traits. Psycho- pharmacological enhancers, as well as other enhancement technologies, raise moral questions concerning the boundary between clinical therapy and enhancement, risks and safety, coercion and justice. Other moral questions include the meaning and value of identity and authenticity, the role of happiness for a good life, or the perceived threats to humanity. Identity and authenticity are central in the debate on psychopharmacological enhancers. In this paper, I first describe the concerns at issue here as extensively propounded by Carl Elliott. Next, I address David DeGrazia's theory, which holds that there are no fundamental identity-related and authenticity-related arguments against enhancement technologies. I argue, however, that DeGrazia's line of reasoning does not succeed in settling these concerns. His conception of identity does not seem able to account for the importance we attach to personal identity in cases where personal identity is changed through enhancement technology. Moreover, his conception of authenticity does not explain the reason why we find inauthentic values objectionable. A broader approach to authenticity can make sense of concerns about changes in personal identity by means of enhancement technologies.

Burke, Michael. What would happen if a 'woman' outpaced the winner of the gold medal in the 'men's' one hundred meters? female sport, drugs and the transgressive cyborg body. *Philosophy in the Contemporary World* 2004 Spring-Summer; 11(1): 33-41. NRCBL: 4.5; 5.1; 9.5.1; 10; 4.1.1. SC: an.

Carson, P.A.; Holt, J. Ethics of studies involving human volunteers. II. Relevance and practical implementation for cosmetic scientists. *Journal of Cosmetic Science* 2006 May-June; 57(3): 223-231. NRCBL: 4.5; 18.1. SC: em.

Chatterjee, Anjan. "Cosmetic neurology" and the problem of pain. *Cerebrum: The DANA Forum on Brain Science* 2007 July: 6 p. [Online]. Accessed: http://www.dana.org/news/cerebrum/detail.aspx?id=8794 [20070920]. NRCBL: 4.5; 4.4; 17.1.

Clark, Andy. Re-inventing ourselves: the plasticity of embodiment, sensing, and mind. *Journal of Medicine and Philosophy* 2007 May-June; 32(3): 263-282. NRCBL: 4.5; 1.1; 5.1.
Abstract: Recent advances in cognitive science and cognitive neuroscience open up new vistas for human enhancement. Central to much of this work is the idea of new human-machine interfaces (in general) and new brain-machine interfaces (in particular). But despite the increasing prominence of such ideas, the very idea of such an interface remains surprisingly under-explored. In particular, the notion of human enhancement suggests an image of the embodied and reasoning agent as literally extended or augmented, rather than the more conservative image of a standard (non-enhanced) agent using a tool via some new interface. In this essay, I explore this difference, and attempt to lay out some of the conditions under which the more radical reading (positing brand new integrated agents or systemic wholes) becomes justified. I adduce some empirical evidence suggesting that the radical result is well within our scientific reach. The main reason why this is so has less to do with the advancement of our science (though that certainly helps) than with our native biological plasticity. We humans, I shall try to show, are biologically disposed towards literal (and repeated) episodes of sensory re-calibration, of bodily re-configuration and of mental extension. Such potential for literal and repeated re-configuration is the mark of what I shall call "profoundly embodied agency," contrasting it with a variety of weaker (less philosophically and scientifically interesting) understandings of the nature and importance of embodiment for minds and persons. The article ends by relating the image of profound embodiment to some questions (and fears) concerning converging technologies for improving human performance.

Deane-Drummond, Celia. Future perfect? God, the transhuman future and the quest for immortality. *In:* Deane-Drummond, Celia; Scott, Peter Manley, eds. Future Perfect?: God, Medicine and Human Identity. New York: T and T Clark International, 2006: 168-182. NRCBL: 4.5; 5.2; 1.2; 1.1.

Doyle, Jacqueline. Surgical solution becoming acceptable, as for birth [letter]. *BMJ: British Medical Journal* 2007 June 9; 334(7605): 1179-1180. NRCBL: 4.5; 9.5.1; 9.5.5. Comments: Lih Mei Liao and Sarah M. Creighton. Requests for cosmetic genitoplasty: how should healthcare providers respond? BMJ: British Medical Journal 2007 May 26; 334(7603): 1090-1092.

Görman, Ulf. Never too late to live a little longer? The quest for extended life and immortality — some ethical considerations. *In:* Deane-Drummond, Celia; Scott, Peter Manley, eds. Future Perfect?: God, Medicine and Human Identity. New York: T and T Clark International, 2006: 143-154. NRCBL: 4.5; 4.2; 4.4; 20.5.1.

Hughes, James; Bostrum, Nick; Agar, Nicholas. Human vs. posthuman [letters and reply]. *Hastings Center Report*

2007 September-October; 37(5): 4-6. NRCBL: 4.5; 4.4; 1.1. Comments: Nicholas Agar. Whereto transhumanism? The literature reaches a critical mass. Hastings Center Report 2007 May-June; 37(3): 12-17.

Murray, Thomas H. Enhancement. *In:* Steinbock, Bonnie, ed. The Oxford Handbook of Bioethics. Oxford; New York: Oxford University Press, 2007: 491-515. 20 refs. NRCBL: 4.5; 15.1; 17.1. SC: an; rv.
   Keywords: *enhancement technologies; *ethical analysis; *moral policy; adults; biotechnology; children; doping in sports; drugs; freedom; hormones; human characteristics; justice; moral complicity; motivation; parents; public policy; regulation; Proposed Keywords: body height; therapeutics

Plotz, David. The ethics of enhancement: we can make ourselves stronger, fast, smarter. Should we? *Slate Magazine* 2003 March 12: 3 p. [Online]. Accessed: http://www.slate.com/id/2079310 [2007 April 12]. NRCBL: 4.5; 15.1.

Roco, Mihail C. Progress in governance of converging technologies integrated from the nanoscale. *Annals of the New York Academy of Sciences* 2006 December; 1093: 1-23. NRCBL: 4.5; 5.3; 21.1.

Sahakian, Barbara; Morein-Zamir, Sharon. Professor's little helper. The use of cognitive-enhancing drugs by both ill and healthy individuals raises ethical questions that should not be ignored. *Nature* 2007 December 20-27; 450(7173): 1157-1159. NRCBL: 4.5; 17.2.

Sheldon, Tony. Cosmetic surgery gets under Dutch skin. *BMJ: British Medical Journal* 2007 September 15; 335(7619): 541. NRCBL: 4.5; 10; 9.5.5.

Tomasini, Floris. Imagining human enhancement: whose future, which rationality? *Theoretical Medicine and Bioethics* 2007; 28(6): 497-507. NRCBL: 4.5; 15.5. SC: an.

Tuffs, Annette. German doctors may have to report patients who have piercings and beauty treatments [news]. *BMJ: British Medical Journal* 2007 November 3; 335(7626): 905. NRCBL: 4.5; 8.4; 9.3.1.

Twine, Richard. Thinking across species — a critical bioethics approach to enhancement. *Theoretical Medicine and Bioethics* 2007; 28(6): 509-523. NRCBL: 4.5; 22.1; 1.1; 2.1.

Waters, Brent. Saving us from ourselves: Christology, anthropology and the seduction of posthuman medicine. *In:* Deane-Drummond, Celia; Scott, Peter Manley, eds. Future Perfect?: God, Medicine and Human Identity. New York: T and T Clark International, 2006: 183-195. NRCBL: 4.5; 1.2; 1.1; 1.3.1.

Wilson, James. Transhumanism and moral equality. *Bioethics* 2007 October; 21(8): 419-425. NRCBL: 4.5;

1.1. Conference: Eighth World Congress of Bioethics: A Just and Healthy Society; Beijing, China; 2006 August 9.
   Abstract: Conservative thinkers such as Francis Fukuyama have produced a battery of objections to the transhumanist project of fundamentally enhancing human capacities. This article examines one of these objections, namely that by allowing some to greatly extend their capacities, we will undermine the fundamental moral equality of human beings. I argue that this objection is groundless: once we understand the basis for human equality, it is clear that anyone who now has sufficient capacities to count as a person from the moral point of view will continue to count as one even if others are fundamentally enhanced; and it is mistaken to think that a creature which had even far greater capacities than an unenhanced human being should count as more than an equal from the moral point of view.

ETHICISTS AND ETHICS COMMITTEES
See also BIOETHICS AND MEDICAL ETHICS/ COMMISSIONS

American Psychological Association. Ethics Committee. Report of the Ethics Committee, 2005. *American Psychologist* 2006 July-August; 61(5): 522-529. NRCBL: 9.6; 2.3; 17.1.

Anderson-Shaw, Lisa; Ahrens, William; Fetzer, Marny. Ethics consultation in the emergency department. *JONA's Healthcare Law, Ethics, and Regulation* 2007 January-March; 9(1): 32-35. NRCBL: 9.6; 9.5.1. SC: em.
   Abstract: Clinical ethics teams exist in various forms and have assisted care providers for several decades. Our clinical ethics service at an urban, tertiary, teaching hospital provides ethics consultation to care providers, patients, and their family members. Scenarios prompting an ethics consultation may be complex, often involving social, cultural, and fiscal components. Because patients who receive an ethics consultation often require a lengthy hospital stay, our group searched for unique identifiers in a patient's presentation to facilitate earlier and, potentially, more effective interventions. Of particular interest to our group was the presentation of these patients to our institution from the emergency department (ED). Our group's subjective experience indicated that factors requiring ethics consultation were often present very early during hospitalization. A retrospective medical record review of a convenience sample of 50 records of patients who had received a formal clinical ethics consult within a 14-month timeframe was done. Those patients who were admitted to the hospital via the ED and subsequently received an ethics consultation were identified. The critical issues prompting the ethics consult were then evaluated. Eighteen (35%) of the study patients were originally admitted through the ED. Results showed that the ethical issue(s) that prompted the clinical ethics consult was regularly identifiable in the ED. Our study results indicate that issues prompting ethics consults may potentially be identified as patients present to the ED. Rapid and effective interventions proscribed through institutional policy guidelines could greatly assist nurses and other ED

NRCBL: National Reference Center for Bioethics Literature Classification Scheme   See inside front cover for terms.

137

providers in identifying these at-risk patients upon entry of the ED. Such a policy would ultimately benefit both patient and provider.

Arnold, Robert; Aulisio, Mark; Begler, Ann; Seltzer, Deborah. A commentary on Caplan and Bergman: ethics mediation — questions for the future. *Journal of Clinical Ethics* 2007 Winter; 18(4): 350-354. NRCBL: 9.6; 8.1; 20.5.1. Comments: Arthur L. Caplan and Edward J. Bergman. Beyond Schiavo. Journal of Clinical Ethics 2007 Winter; 18(4): 340-345.

Baylis, Françoise. Of courage, honor, and integrity. *In:* Eckenwiler, Lisa A.; Cohn, Felicia G., eds. The Ethics of Bioethics: Mapping the Moral Landscape. Baltimore, MD: Johns Hopkins University Press, 2007: 193-204. NRCBL: 9.6; 1.1.

Bockenheimer-Lucius, Gisela. Ethikberatung und Ethik-Komitee im Altenpflegeheim (EKA) - Herausforderung und Chance für eine ethische Entscheidungskultur = Ethics committee in a long-term care facility - a challenge and a chance for an ethical decision-making culture. *Ethik in der Medizin* 2007 December; 19(4): 320-330. NRCBL: 9.6; 9.5.2; 17.4.

Bockenheimer-Lucius, Gisela; May, Arnd T. Ethikberatung - Ethik-Komitee in Einrichtungen der stationären Altenhilfe (EKA). Eckpunkte für ein Curriculum [Ethics consultation - ethics committees in geriatric care institutions. Key elements of a curriculum]. *Ethik in der Medizin* 2007 December; 19(4): 331-339. NRCBL: 9.6; 9.5.2; 7.2.

Bosk, Charles L. Disinterested commitment as moral heroism. *Atrium* 2007 Summer; 4: 1-4. NRCBL: 9.6; 1.3.1.

Breier-Mackie, Sarah. Who is the clinical ethicist? *Gastroenterology Nursing: the Official Journal of the Society of Gastroenterology Nurses and Associates* 2006 January-February; 29(1): 70-72. NRCBL: 9.6; 2.1; 4.1.3.

Cheng-tek Tai, Michael. Clinical ethics consultation — a checklist approach from Asian perspective. *Formosan Journal of Medical Humanities* 2007 July; 8(1-2): 21-25. NRCBL: 9.6; 1.1; 21.7. SC: cs.

Chwang, Eric; Landy, David C.; Sharp, Richard R. Views regarding the training of ethics consultants: a survey of physicians caring for patients in ICU. *Journal of Medical Ethics* 2007 June; 33(6): 320-324. NRCBL: 9.6; 7.3; 9.1. SC: em.
  Abstract: BACKGROUND: Despite the expansion of ethics consultation services, questions remain about the aims of clinical ethics consultation, its methods and the expertise of those who provide such services. OBJECTIVE: To describe physicians' expectations regarding the training and skills necessary for ethics consultants to contribute effectively to the care of patients in intensive care unit (ICU). DESIGN: Mailed survey. PARTICIPANTS: Physicians responsible for the care of at least 10 patients in ICU over a 6-month period at a 921-bed private teaching hospital with an established ethics consul-

tation service. 69 of 92 (75%) eligible physicians responded. Measurements: Importance of specialised knowledge and skills for ethics consultants contributing to the care of patients in ICU; need for advanced disciplinary training; expectations regarding formal-training programmes for ethics consultants. RESULTS: Expertise in ethics was described most often as important for ethics consultants taking part in the care of patients in ICU, compared with expertise in law (p.03), religious traditions (p.001), medicine (p.001) and conflict-mediation techniques (p.001). When asked about the formal training consultants should possess, however, physicians involved in the care of patients in ICU most often identified advanced medical training as important. CONCLUSIONS: Although many physicians caring for patients in ICU believe ethics consultants must possess non-medical expertise in ethics and law if they are to contribute effectively to patient care, these physicians place a very high value on medical training as well, suggesting a "medicine plus one" view of the training of an ideal ethics consultant. As ethics consultation services expand, clear expectations regarding the training of ethics consultants should be established.

Cohn, Felicia; Goodman-Crews, Paula; Rudman, William; Schneiderman, Lawrence J.; Waldman, Ellen. Proactive ethics consultation in the ICU: a comparison of value perceived by healthcare professionals and recipients. *Journal of Clinical Ethics* 2007 Summer; 18(2): 140-147. NRCBL: 9.6; 8.1. SC: em.

Community-State Partnerships to Improve End-of-life Care. How regional long-term care ethics committees improve end-of-life care. Kansas City, MO: State Initiatives Policy Brief, Community State Partnerships to Improve End-of-Life Care 2000; (6): 104 [Online]. Accessed: http://www.rwjf.org [2006 September 28]. NRCBL: 9.6; 20.4.1; 20.5.1.

Cranford, Ronald E. The neurologist as ethics consultant and as a member of the institutional ethics committee: the neuroethicist. *Neurologic Clinics* 1989 November; 7(4): 697-713. NRCBL: 9.6; 17.1; 20.2.1; 20.5.1.

de Freitas, Genival Fernandes; Oguisso, Taka; Merighi, Miriam Aparecida Barbosa. Ethical events in nursing: daily activities of nurse managers and nursing ethics committee members. *Revista Latino-Americana Enfermagem* 2006 July-August; 14(4): 497-502. NRCBL: 9.6; 4.1.3; 7.2; 8.1. Identifiers: Brazil.

Diekema, Douglas S. The armchair ethicist: it's all about location. *Journal of Clinical Ethics* 2007 Fall; 18(3): 227-232. NRCBL: 9.6; 4.1.2; 20.5.1; 20.5.4. Comments: Jaklin Eliott and Ian Olver. Autonomy and the family as (in)appropriate surrogates for DNR decisions: a qualitative analysis of dying cancer patients' talk. Journal of Clinical Ethics 2007 Fall; 18(3): 206-218.

Dörries, Andrea; Hespe-Jungesblut, Katharina. Die Implementierung klinischer Ethikberatung in Deutschland

— Ergebnisse einer bundesweiten Umfrage bei Krankenhäusern [The implementation of clinical ethics consultation in Germany — results of a nationwide survey in hospitals]. *Ethik in der Medizin* 2007 June; 19(2): 148-156. NRCBL: 9.6. SC: em.

Dubler, Nancy Neveloff; Blustein, Jeffrey. Credentialing ethics consultants: an invitation to collaboration. *American Journal of Bioethics* 2007 February; 7(2): 35-37. NRCBL: 9.6; 7.2. Comments: Ellen Fox, Sarah Myers, and Robert Pearlman. Ethics consultation in United States hospitals: a national survey. American Journal of Bioethics 2007 February; 7(2): 13-25.

Dudzinski, Denise M. Education to dispel the myth. *American Journal of Bioethics* 2007 February; 7(2): 39-40. NRCBL: 9.6; 7.2. Comments: Ellen Fox, Sarah Myers, and Robert Pearlman. Ethics consultation in United States hospitals: a national survey. American Journal of Bioethics 2007 February; 7(2): 13-25.

Fiester, Autumn. Mediation and moral aporia. *Journal of Clinical Ethics* 2007 Winter; 18(4): 355-356. NRCBL: 9.6; 1.1. Identifiers: Terri Schiavo.

Fiester, Autumn. The failure of the consult model: why "mediation" should replace "consultation". *American Journal of Bioethics* 2007 February; 7(2): 31-32. NRCBL: 9.6. Comments: Ellen Fox, Sarah Myers, and Robert Pearlman. Ethics consultation in United States hospitals: a national survey. American Journal of Bioethics 2007 February; 7(2): 13-25.

Fleming, David A. Responding to ethical dilemmas in nursing homes: do we always need an "ethicist"? *HEC (Healthcare Ethics Committee) Forum* 2007 September; 19(3): 245-259. NRCBL: 9.6; 9.5.2; 20.5.1; 1.1. SC: an.

Ford, Paul J. Professional clinical ethicist: knowing why and limits. *Journal of Clinical Ethics* 2007 Fall; 18(3): 243-246. NRCBL: 9.6; 8.1; 9.8.

Ford, Paul J.; Boissy, Adrienne R. Different questions, different goals. *American Journal of Bioethics* 2007 February; 7(2): 46-47. NRCBL: 9.6. Comments: Ellen Fox, Sarah Myers, and Robert Pearlman. Ethics consultation in United States hospitals: a national survey. American Journal of Bioethics 2007 February; 7(2): 13-25.

Fox, Ellen; Myers, Sarah; Pearlman, Robert A. Ethics consultation in United States hospitals: a national survey. *American Journal of Bioethics* 2007 February; 7(2): 13-25. NRCBL: 9.6. SC: em.
Abstract: Context: Although ethics consultation is commonplace in United States (U.S.) hospitals, descriptive data about this health service are lacking. Objective: To describe the prevalence, practitioners, and processes of ethics consultation in U.S. hospitals. Design: A 56-item phone or questionnaire survey of the "best informant" within each hospital. Participants: Random sample of 600 U.S. general hospitals, stratified by bed size. Results: The response rate was 87.4%. Ethics consultation services (ECSs) were found in 81% of all general hospitals in the U.S., and in 100% of hospitals with more than 400 beds. The median number of consults performed by ECSs in the year prior to survey was 3. Most individuals performing ethics consultation were physicians (34%), nurses (31%), social workers (11%), or chaplains (10%). Only 41% had formal supervised training in ethics consultation. Consultation practices varied widely both within and between ECSs. For example, 65% of ECSs always made recommendations, whereas 6% never did. These findings highlight a need to clarify standards for ethics consultation practices.

Fox, Ellen; Myers, Sarah; Pearlman, Robert A. Response to open peer commentaries on "Ethics Consultation in U.S. Hospitals: A National Survey" [letter]. *American Journal of Bioethics [Online].* 2007 February; 7(2): W1-W3. NRCBL: 9.6.

Frewer, Andreas; Fahr, Uwe. Clinical ethics and confidentiality: opinions of experts and ethics committees. *HEC (Healthcare Ethics Committee) Forum* 2007 December; 19(4): 277-291. NRCBL: 9.6; 8.4. SC: em. Identifiers: Germany.

Gordijn, Bert. Genetic diagnosis, confidentiality and counseling: an ethics committee's potential deliberations about the do's and don'ts. *HEC (Healthcare Ethics Committee) Forum* 2007 December; 19(4): 303-312. NRCBL: 9.6; 15.3; 8.4; 15.2; 1.1.

Gordon, Elisa J. A better way to evaluate clinical ethics consultations? An ecological approach. *American Journal of Bioethics* 2007 February; 7(2): 26-29. NRCBL: 9.6. Comments: Ellen Fox, Sarah Myers, and Robert Pearlman. Ethics consultation in United States hospitals: a national survey. American Journal of Bioethics 2007 February; 7(2): 13-25.

Heeley, Gerry. A system's transition to next generation model of ethics. *Health Care Ethics USA* 2007 Fall; 15(4): 2-4. NRCBL: 9.6.

Hurst, Samia A.; Reiter-Theil, Stella; Perrier, Arnaud; Forde, Reidun; Slowther, Anne-Marie; Pegoraro, Renzo; Danis, Marion. Physicians' access to ethics support services in four European countries. *Health Care Analysis: An International Journal of Health Philosophy and Policy* 2007 December; 15(4): 321-335. NRCBL: 9.6; 21.1. SC: em. Identifiers: Italy; Norway; Switzerland; Great Britain (United Kingdom).
Abstract: Clinical ethics support services are developing in Europe. They will be most useful if they are designed to match the ethical concerns of clinicians. We conducted a cross-sectional mailed survey on random samples of general physicians in Norway, Switzerland, Italy, and the UK, to assess their access to different types of ethics support services, and to describe what makes them more likely to have used available ethics support. Respondents reported access to formal ethics support services such as

NRCBL: National Reference Center for Bioethics Literature Classification Scheme   See inside front cover for terms.

139

clinical ethics committees (23%), consultation in individual cases (17.6%), and individual ethicists (8.8%), but also to other kinds of less formal ethics support (23.6%). Access to formal ethics support services was associated with work in urban hospitals. Informal ethics resources were more evenly distributed. Although most respondents (81%) reported that they would find help useful in facing ethical difficulties, they reported having used the available services infrequently (14%). Physicians with greater confidence in their knowledge of ethics (P=0.001), or who had had ethics courses in medical school (P=0.006), were more likely to have used available services. Access to help in facing ethical difficulties among general physicians in the surveyed countries is provided by a mix of official ethics support services and other resources. Developing ethics support services may benefit from integration of informal services. Development of ethics education in medical school curricula could lead to improved physicians sensitivity to ethical difficulties and greater use of ethics support services. Such support services may also need to be more proactive in making their help available.

Hurst, S.A.; Perrier, A.; Pegoraro, R.; Reiter-Theil, S.; Forde, R.; Slowther, A.-M.; Garrett-Mayer, E.; Danis, M. Ethical difficulties in clinical practice: experiences of European doctors. *Journal of Medical Ethics* 2007 January; 33(1): 51-57. NRCBL: 9.6. SC: em.

Abstract: BACKGROUND: Ethics support services are growing in Europe to help doctors in dealing with ethical difficulties. Currently, insufficient attention has been focused on the experiences of doctors who have faced ethical difficulties in these countries to provide an evidence base for the development of these services. METHODS: A survey instrument was adapted to explore the types of ethical dilemma faced by European doctors, how they ranked the difficulty of these dilemmas, their satisfaction with the resolution of a recent ethically difficult case and the types of help they would consider useful. The questionnaire was translated and given to general internists in Norway, Switzerland, Italy and the UK. RESULTS: Survey respondents (n=656, response rate 43%) ranged in age from 28 to 82 years, and averaged 25 years in practice. Only a minority (17.6%) reported having access to ethics consultation in individual cases. The ethical difficulties most often reported as being encountered were uncertain or impaired decision-making capacity (94.8%), disagreement among caregivers (81.2%) and limitation of treatment at the end of life (79.3%). The frequency of most ethical difficulties varied among countries, as did the type of issue considered most difficult. The types of help most often identified as potentially useful were professional reassurance about the decision being correct (47.5%), someone capable of providing specific advice (41.1%), help in weighing outcomes (36%) and clarification of the issues (35.9%). Few of the types of help expected to be useful varied among countries. CONCLUSION: Cultural differences may indeed influence how doctors perceive ethical difficulties. The type of help needed, however, did not vary markedly. The general structure of ethics support services would not have to be radically altered to suit cultural variations among the surveyed countries.

Illhardt, Franz-Josef. Conflict between a patient's family and the medical team. *HEC (Healthcare Ethics Committee) Forum* 2007 December; 19(4): 381-388. NRCBL: 9.6; 8.3.3; 20.5.1. SC: cs.

Iltis, Ana Smith. Bioethical expertise in health care organizations. *In:* Rasmussen, Lisa, ed. Ethics Expertise: History, Contemporary Perspectives, and Applications. Dordrecht: Springer, 2005: 259-267. NRCBL: 9.6; 2.1. SC: an.

Johnson, Larry W. Practice pointers for the nurse leader: lessons in conducting an ethics consult. *JONA's Healthcare Law, Ethics, and Regulation* 2007 July-September; 9(3): 97-99. NRCBL: 9.6.

Jotkowitz, Alan B. Ethics consultation: whose ethics? *American Journal of Bioethics* 2007 February; 7(2): 41-42. NRCBL: 9.6. Comments: Ellen Fox, Sarah Myers, and Robert Pearlman. Ethics consultation in United States hospitals: a national survey. American Journal of Bioethics 2007 February; 7(2): 13-25.

Kaebnick, Gregory E. The problem with trust and sympathy [editorial]. *Hastings Center Report* 2007 March-April; 37(2): 2. NRCBL: 9.6; 7.3; 5.3.

Kirby, Jeff; Simpson, Christy. An innovative, inclusive process for meso-level health policy development. *HEC (Healthcare Ethics Committee) Forum* 2007 June; 19(2):161-176. NRCBL: 9.6; 9.1; 1.3.2. SC: an.

Klitzman, Robert. Additional implications of a national survey on ethics consultation in United States hospitals. *American Journal of Bioethics* 2007 February; 7(2): 47-48. NRCBL: 9.6; 18.2. Comments: Ellen Fox, Sarah Myers, and Robert Pearlman. Ethics consultation in United States hospitals: a national survey. American Journal of Bioethics 2007 February; 7(2): 13-25.

Kuczewski, Mark G. Ethics committees and case consultation: theory and practice. *In:* Khushf, George, ed. Handbook of Bioethics: Taking Stock of the Field From a Philosophical Perspective. Dordrecht; Boston: Kluwer Academic, 2004: 315-334. NRCBL: 9.6; 7.2.

Lebeer, Guy. Clinical ethics committees in Europe — assistance in medical decisions, fora for democratic debates, or bodies to monitor basic rights? *In:* Gunning, Jennifer; Holm, Søren, eds. Ethics, law, and society. Volume 1. Aldershot, Hants, England; Burlington, VT: Ashgate, 2005: 65-72. NRCBL: 9.6.

McDaniel, Charlotte. Melding or meddling: compliance and ethics programs. *HEC (Healthcare Ethics Committee) Forum* 2007 June; 19(2): 97-107. NRCBL: 9.6; 9.8; 9.1; 1.3.2. SC: an.

McLean, Sheila A.M. What and who are clinical ethics committees for? [editorial]. *Journal of Medical Ethics*

2007 September; 33(9): 497-500. NRCBL: 9.6. Identifiers: Great Britain (United Kingdom).

Meyers, Christopher. Clinical ethics consulting and conflict of interest structurally intertwined. *Hastings Center Report* 2007 March-April; 37(2): 32-40. NRCBL: 9.6; 7.3; 5.3.

Abstract: Clinical ethical consultants are subject to an unavoidable conflict of interest. Their work requires that they be independent, but incentives attached to their role chip relentlessly at independence. This is a problem without any solution, but it can at least be ameliorated through careful management.

Mielke, Jens. Clinical ethics in the developing world: a case in point: in Zimbabwe. *Formosan Journal of Medical Humanities* 2007 July; 8(1-2): 15-20. NRCBL: 9.6; 9.1; 21.1. SC: cs.

Morreim, E. Haavi. Ties without tethers: bioethics corporate relations in the AbioCor artificial heart trial. *In:* Eckenwiler, Lisa A.; Cohn, Felicia G., eds. The Ethics of Bioethics: Mapping the Moral Landscape. Baltimore, MD: Johns Hopkins University Press, 2007: 181-190. NRCBL: 9.6; 1.3.2; 9.7; 9.3.1; 18.5.1; 19.2. SC: cs.

National Center for Ethics in Health Care (United States). IntegratedEthics [Integrated Ethics]: IntegratredEthics Toolkit — A Manual for the IntegratedEthics Program Officer; Ethics Consultation Toolkit — A Manual for the Ethics Consultation Coordinator; Ethical Leadership Toolkit — A Manual for the Ethical Leadership Coordinator; Preventive Ethics Toolkit — A Manual for the Preventive Ethics Coordinator. Washington, DC: National Center for Ethics in Health Care, Veterans Health Administration, 2007: multiple pages in 4 volumes. NRCBL: 9.6; 2.1.

Neitzke, Gerald. Confidentiality, secrecy, and privacy in ethics consultation. *HEC (Healthcare Ethics Committee) Forum* 2007 December; 19(4): 293-302. NRCBL: 9.6; 8.4; 1.1. Identifiers: Germany.

Otto, Sheila. Memento . . . life imitates art: the request for an ethics consultation. *Journal of Clinical Ethics* 2007 Fall; 18(3): 247-251. NRCBL: 9.6; 8.3.3; 8.1; 9.5.3. SC: cs.

Parker, Lisa S. Ethical expertise, maternal thinking, and the work of clinical ethicists. *In:* Rasmussen, Lisa, ed. Ethics Expertise: History, Contemporary Perspectives, and Applications. Dordrecht: Springer, 2005: 165-207. NRCBL: 9.6; 2.1; 9.5.5; 9.5.7. SC: an.

Parsi, Kayhan; Kuczewski, Mark G. Failure to thrive: can education save the life of ethics consultation? *American Journal of Bioethics* 2007 February; 7(2): 37-39. NRCBL: 9.6; 7.2. Comments: Ellen Fox, Sarah Myers, and Robert Pearlman. Ethics consultation in United States hospitals: a national survey. American Journal of Bioethics 2007 February; 7(2): 13-25.

Quist, Norman. Hope, uncertainty, and lacking mechanisms. *Journal of Clinical Ethics* 2007 Winter; 18(4): 357-361. NRCBL: 9.6; 8.1; 20.3.3; 20.5.1. Identifiers: Terri Schiavo. Comments: Arthur L. Caplan and Edmund J. Bergman. Beyond Schiavo. Journal of Clinical Ethics 2007 Winter; 18(4): 340-345.

Racine, Eric. HEC member perspectives on the case analysis process: a qualitative multi-site study. *HEC (Healthcare Ethics Committee) Forum* 2007 September; 19(3): 185-206. NRCBL: 9.6; 1.1; 7.3. SC: em.

Richter, Gerd. Greater patient, family and surrogate involvement in clinical ethics consultation: the model of clinical ethics liaison service as a measure for preventive ethics. *HEC (Healthcare Ethics Committee) Forum* 2007 December; 19(4): 327-340. NRCBL: 9.6; 8.4; 8.3.3. SC: rv. Identifiers: Germany.

Russell, Barbara J.; Pape, Deborah A. Ethics consultation: continuing its analysis. *Journal of Clinical Ethics* 2007 Fall; 18(3): 235-242. NRCBL: 9.6; 7.2; 8.1; 8.3.4; 8.2.

Säfken, Christian; Frewer, Andreas. The duty to warn and clinical ethics: legal and ethical aspects of confidentiality and HIV/AIDS. *HEC (Healthcare Ethics Committee) Forum* 2007 December; 19(4): 313-326. NRCBL: 9.6; 8.4; 9.5.6. SC: le.

Schmidt, Kurt W.; Frewer, Andreas. Current problems of clinical ethics: confidentiality and end-of-life decisions — is silence always golden? Introduction. *HEC (Healthcare Ethics Committee) Forum* 2007 December; 19(4): 273-276. NRCBL: 9.6; 8.4.

Scofield, Giles R. The war on error. *American Journal of Bioethics* 2007 February; 7(2): 44-45. NRCBL: 9.6. Comments: Ellen Fox, Sarah Myers, and Robert Pearlman. Ethics consultation in United States hospitals: a national survey. American Journal of Bioethics 2007 February; 7(2): 13-25.

Silberman, Jordan; Morrison, Wynne; Feudtner, Chris. Pride and prejudice: how might ethics consultation services minimize bias? *American Journal of Bioethics* 2007 February; 7(2): 32-34. NRCBL: 9.6. Comments: Ellen Fox, Sarah Myers, and Robert Pearlman. Ethics consultation in United States hospitals: a national survey. American Journal of Bioethics 2007 February; 7(2): 13-25.

Smith, Martin L.; Weise, Kathryn L. The goals of ethics consultation: rejecting the role of "ethics police". *American Journal of Bioethics* 2007 February; 7(2): 42-44. NRCBL: 9.6; 9.8. Comments: Ellen Fox, Sarah Myers, and Robert Pearlman. Ethics consultation in United States hospitals: a national survey. American Journal of Bioethics 2007 February; 7(2): 13-25.

NRCBL: National Reference Center for Bioethics Literature Classification Scheme   See inside front cover for terms.

141

Sokol, Daniel K. No patient is an island. *BMJ: British Medical Journal* 2007 September 15; 335(7619): 568. NRCBL: 9.6; 1.3.7; 17.1; 8.3.3; 8.3.4. Identifiers: Inside the Ethics Committee, BBC Radio 4 series.

Spike, Jeffrey P. Memory identity and capacity. *Journal of Clinical Ethics* 2007 Fall; 18(3): 252-255. NRCBL: 9.6; 8.1; 8.3.3. SC: cs. Comments: Sheila Otto. Memento . . . life imitates art: the request for an ethics consultation. Journal of Clinical Ethics 2007 Fall; 18(3): 247-251.

Spike, Jeffrey P. Who's guarding the henhouse? Ramifications of the Fox study. *American Journal of Bioethics* 2007 February; 7(2): 48-50. NRCBL: 9.6; 7.2. Identifiers: American Society for Bioethics and Humanities [ASBH]. Comments: Ellen Fox, Sarah Myers, and Robert Pearlman. Ethics consultation in United States hospitals: a national survey. American Journal of Bioethics 2007 February; 7(2): 13-25.

Wear, Stephen. Ethical expertise in the clinical setting. *In:* Rasmussen, Lisa, ed. Ethics Expertise: History, Contemporary Perspectives, and Applications. Dordrecht: Springer, 2005: 243-258. NRCBL: 9.6; 2.1.

Williamson, Laura. Empirical assessments of clinical ethics services: implications for clinical ethics committees. *Clinical Ethics* 2007 December; 2(4): 187-192. NRCBL: 9.6; 7.1. SC: em.

Abstract: The need to evaluate the performance of clinical ethics services is widely acknowledged although work in this area is more developed in the United States. In the USA many studies that assess clinical ethics services have utilized empirical methods and assessment criteria. The value of these approaches is thought to rest on their ability to measure the value of services in a demonstrable fashion. However, empirical measures tend to lack ethical content, making their contribution to developments in ethical governance unclear. The steady increase of clinical ethics committees in the UK must be accompanied by efforts to evaluate their performance. As part of this evaluative work it is important to examine how the practice of clinical ethics committees can be informed by empirical measures.

Yarborough, Mark; Sharp, Richard R. Bioethics consultation and patient advocacy organizations: expanding the dialogue about professional conflicts of interest. *CQ: Cambridge Quarterly of Healthcare Ethics* 2007 Winter; 16(1): 74-81. NRCBL: 9.6; 1.3.1; 9.3.1; 7.1; 7.3.

Zaner, Richard M. A comment on community consultation. *American Journal of Bioethics* 2007 February; 7(2): 29-30. NRCBL: 9.6. Comments: Ellen Fox, Sarah Myers, and Robert Pearlman. Ethics consultation in United States hospitals: a national survey. American Journal of Bioethics 2007 February; 7(2): 13-25.

ETHICS COMMITTEES *See* ETHICISTS AND ETHICS COMMITTEES; HUMAN EXPERIMEN-

TATION/ ETHICS COMMITTEES AND POLICY GUIDELINES

EUGENICS

Sterile thinking. *New Atlantis* 2007 Winter; 15: 143. NRCBL: 15.5; 11.3.

> Keywords: *eugenics; *historical aspects; *involuntary sterilization; mentally disabled persons; National Socialism; prisoners; Keyword Identifiers: *United States; Germany; Indiana; Twentieth Century

Bayertz, Kurt. Eugenik. *In:* Schreiber, Hans-Peter, ed. Biomedizin und Ethik: Praxis, Recht, Moral. Basel; Boston: Birkhäuser, 2004: 72-76. NRCBL: 15.5.

> Keywords: *eugenics; genetic intervention; germ cells; historical aspects; Keyword Identifiers: Europe; United States; Nineteenth Century; Twentieth Century

Boas, Franz. Eugenics. *In his:* Anthropology and Modern Life. New Brunswick, NJ: Transaction Publishers, 2004: 106-121. 2 fn. (p. 329-330). NRCBL: 15.5; 1.3.1.

> Keywords: *anthropology; *eugenics; behavioral genetics; culture; genetic determinism

Brüne, Martin. On human self-domestication, psychiatry, and eugenics. *Philosophy, Ethics, and Humanities in Medicine [electronic]* 2007; (2)21: 9 p. Accessed: http://www.peh-med.com/content/pdf/1747-5341-2-21.pdf [2007 December 18]. 62 refs. NRCBL: 15.5; 3.1; 15.6; 17.1; 2.2. SC: an.

> Keywords: *eugenics; *psychiatry; attitude of health personnel; evolution; historical aspects; human characteristics; mental disorders; National Socialism; philosophy; physicians; racial groups; Proposed Keywords: Darwinism; Keyword Identifiers: Germany; Twentieth Century

Abstract: The hypothesis that anatomically modern homo sapiens could have undergone changes akin to those observed in domesticated animals has been contemplated in the biological sciences for at least 150 years. The idea had already plagued philosophers such as Rousseau, who considered the civilisation of man as going against human nature, and eventually "sparked over" to the medical sciences in the late 19th and early 20th century. At that time, human "self-domestication" appealed to psychiatry, because it served as a causal explanation for the alleged degeneration of the "erbgut" (genetic material) of entire populations and the presumed increase of mental disorders. Consequently, Social Darwinists emphasised preventing procreation by people of "lower genetic value" and positively selecting favourable traits in others. Both tendencies culminated in euthanasia and breeding programs ("Lebensborn") during the Nazi regime in Germany. Whether or not domestication actually plays a role in some anatomical changes since the late Pleistocene period is, from a biological standpoint, contentious, and the currently resurrected debate depends, in part, on the definitional criteria applied. However, the example of human self-domestication may illustrate that scientific ideas, especially when dealing with human biology, are prone to misuse, particularly if "is" is confused with "ought", i.e., if moral principles are deduced from biological facts. Although such naturalistic fallacies appear to be banned, modern genetics may, at least in theory, pose similar ethical problems to medicine, including psychiatry. In times during which studies into the genetics of psychiatric disorders are scientifically more valued than studies into environmental causation of disorders (which is currently the case), the prospects of genetic therapy may be tempting to alter the human genome in patients, probably at costs that no-one can foresee. In the case of "self- domestication", it is proposed that human characteristics resembling domesticated traits in animals should be labelled "domestication-like", or better, objectively described as genuine adaptations to sedentism.

Carlson, Elof Axel. Heroes with feet of clay: Francis Galton and Harry Clay Sharp. *In his:* Times of Triumph, Times of Doubt: Science and the Battle for Public Trust. Cold Spring Harbor, NY: Cold Spring Harbor Laboratory Press, 2006: 55-63. 16 fn. NRCBL: 15.5; 2.2; 7.4.
> Keywords: *eugenics; *historical aspects; famous persons; involuntary sterilization; physicians; prisoners; professional misconduct; Keyword Identifiers: *Galton, Francis; *Sharp, Harry Clay; Twentieth Century; United States

Carlson, Elof Axel. The banality of evil: the careers of Charles Davenport and Harry Laughlin. *In his:* Times of Triumph, Times of Doubt: Science and the Battle for Public Trust. Cold Spring Harbor, NY: Cold Spring Harbor Laboratory Press, 2006: 39-54. 32 fn. NRCBL: 15.5; 2.2.
> Keywords: *eugenics; *historical aspects; famous persons; genetics; involuntary sterilization; researchers; Keyword Identifiers: *Davenport, Charles; *Laughlin, Harry; Eugenics Record Office; Twentieth Century; United States

Cleminson, Richard. "A century of civilization under the influence of eugenics": Dr. Enrique Diego Madrazo, socialism and scientific progress. *Dynamis* 2006; 26: 221-251. NRCBL: 15.5; 1.3.5; 21.1.

Daub, Ute. Soziologische Überlegungen zur Funktion der Eugenik in der neuen politischen Ökonomie. *In:* Neuer-Miebach, Therese; Wunder, Michael, eds. Bio-Ethik und die Zukunft der Medizin. Bonn: Psychiatrie-Verlag, 1998: 139-154. 45 fn. NRCBL: 15.5; 21.1.
> Keywords: *economics; *eugenics; political systems; Proposed Keywords: *sociology

Deane-Drummond, Celia. Living in the shadow of eugenics. *In her:* Genetics and Christian Ethics. Cambridge: Cambridge University Press, 2006: 55-75. 48 fn. NRCBL: 15.5; 2.2.
> Keywords: *eugenics; *historical aspects; clinical genetics; international aspects; involuntary sterilization; National Socialism; politics; population genetics; public policy; researchers; terminology; Keyword Identifiers: Germany; Great Britain; Latin America; Scandinavia; United States

Dolan, Deborah V. Psychiatry, psychology, and human sterilization then and now: "therapeutic" or in the social interest? *Ethical Human Psychology and Psychiatry* 2007; 9(2): 99-108. NRCBL: 15.5; 11.3; 17.1; 2.2.

Falk, Raphael. Nervous diseases and eugenics of the Jews: a view from 1918. *Korot* 2003-2004; 17: 23-46, ix-x. NRCBL: 15.5; 15.11.

Gerson, Michael. The eugenics temptation [op-ed]. *Washington Post* 2007 October 24; p. A19. NRCBL: 15.5. SC: po.
> Keywords: *eugenics; attitudes; behavioral genetics; congenital disorders; disabled persons; Down syndrome; homosexuality; intelligence; population genetics; prenatal diagnosis; racial groups; researchers; selective abortion; trends; Keyword Identifiers: United States; Watson, James

Indiana. Senate. Senate Resolution 0091. A concurrent resolution to mark the centennial of Indiana's 1907 eugenical sterilization law and to express the regret of the Senate and House of Representatives of the 115th Indiana General Assembly for Indiana's experience with eugenics. Indiana: General Assembly 2007: 3 p. [Online]. Accessed: http://www.in.gov/legislative/bills/2007/SRESP/SC0091.html [2008 February 15]. NRCBL: 15.5; 1.3.5; 4.4.

Jackson, Chuck. Waste and whiteness: Zora Neale Hurston and the politics of eugenics. *African American Review* 2000 Winter; 34(4): 639-660. 44 refs. 27 fn. NRCBL: 15.5; 7.1; 4.4.
> Keywords: *eugenics; *literature; blacks; metaphor; mothers; sex factors; social discrimination; whites; Keyword Identifiers: *Seraph on the Suwanee (Hurston. Z.N.); Hurston, Zora Neale

Kirby, David A. The devil in our DNA: a brief history of eugenics in science fiction films. *Literature and Medicine* 2007 Spring; 26(1): 83-108. NRCBL: 15.5; 7.1; 15.11; 5.3; 3.2; 15.7.

Kline, Wendy. A new deal for the child: Ann Cooper Hewitt and sterilization in the 1930s. *In:* Currell, Susan; Cogdell, Christina, eds. Popular Eugenics: National Efficiency and American Mass Culture in the 1930s. Athens, OH: Ohio University Press, 2006: 17-43. 81 fn. NRCBL: 15.5; 11.3; 8.3.4; 2.2.
> Keywords: *eugenics; *historical aspects; *involuntary sterilization; child abuse; dissent; family; famous persons; females; international aspects; legal liability; mass media; mental competence; mentally disabled persons; minors; mothers; parent-child relationship; parental consent; physicians; public opinion; public policy; professional organizations; sexuality; social control; state government; values; Keyword Identifiers: *Cooper Hewitt, Ann; *United States; American Eugenics Society; California; Hoover, Herbert; Roosevelt, Franklin; San Francisco; Twentieth Century

Kröner, Hans-Peter. Eugenik: Zur Geschichte bio-medizinischer Utopien. *In:* Neuer-Miebach, Therese; Wunder, Michael, eds. Bio-Ethik und die Zukunft der Medizin. Bonn: Psychiatrie-Verlag, 1998: 20-30. 24 fn. NRCBL: 15.5; 2.2.
> Keywords: *eugenics; *historical aspects; international aspects; National Socialism; Keyword Identifiers: Germany; Great Britain; Nineteenth Century; Twentieth Century; United States

---

NRCBL: National Reference Center for Bioethics Literature Classification Scheme   See inside front cover for terms.

143

Kusmer, Ken. Indiana apologizes for role in eugenics. *ABC News* 2007 April 13: 2 p. [Online]. Accessed: http://abcnews.go.com/US/wireStory?id=3036919 [2008 February 21]. NRCBL: 15.5; 1.3.5; 4.4.

Laing, Jacqueline. The prohibition on eugenics and reproductive autonomy. *University of New South Wales Law Journal* 2006; 29(2): 261-265. 11 fn. NRCBL: 15.5; 21.1. SC: le.
  Keywords: *eugenics; *reproductive rights; *reproductive technologies; regulation; reproduction; disabled persons; Keyword Identifiers: Universal Declaration of Human Rights

Lombardo, Paul A.; Dorr, Gregory M. Eugenics, medical education, and the public health service: another perspective on the Tuskegee syphilis experiment. *Bulletin of the History of Medicine* 2006 Summer; 80(2): 291-316. NRCBL: 15.5; 7.2; 18.5.1; 18.3; 10; 9.1.

O'Reilly, Keviin B. [sic; Kevin]. Confronting eugenics: does the now discredited practice have relevance to today's technology? *AMnews* 2007 July 9: 6 p. [Online]. Accessed: http://www.ama-assn.org/amednews/2007/07/09/prsa0709.htm [2008 February 12]. NRCBL: 15.5; 11.3; 1.3.5. SC: le.

Ogino, Miho. Eugenics, reproductive technologies, and the feminist dilemma in Japan. *In:* LaFleur, William R.; Böhme, Gernot; Shimazono, Susumu, eds. Dark Medicine: Rationalizing Medical Research. Bloomington: Indiana University Press, 2007: 223-232. 12 refs.; 1 fn. NRCBL: 15.5; 14.1; 10; 11.3; 12.5.1; 15.2.
  Keywords: *abortion; *eugenics; *prenatal diagnosis; *reproductive technologies; *women's rights; attitudes; disabled persons; family planning; feminist ethics; historical aspects; informal social control; international aspects; legal aspects; legislation; physicians; political activity; preimplantation diagnosis; reproductive rights; statistics; surrogate mothers; trends; Keyword Identifiers: *Eugenic Protection Act (Japan); *Japan; Twentieth Century; Twenty-First Century; World War II

Parker, Michael. The best possible child. *Journal of Medical Ethics* 2007 May; 33(5): 279-283. 21 refs. NRCBL: 15.5; 14.1; 1.1; 15.2; 14.4; 15.3; 9.5.1. SC: an.
  Keywords: *beneficence; *embryo transfer; *ethical analysis; *preimplantation diagnosis; *prenatal diagnosis; *quality of life; *reproduction; autonomy; choice behavior; disabled persons; ethical relativism; genetic disorders; genetic enhancement; in vitro fertilization; moral obligations; parents; refusal to treat; uncertainty
  Abstract: Julian Savulescu argues for two principles of reproductive ethics: reproductive autonomy and procreative beneficence, where the principle of procreative beneficence is conceptualised in terms of a duty to have the child, of the possible children that could be had, who will have the best opportunity of the best life. Were it to be accepted, this principle would have significant implications for the ethics of reproductive choice and, in particular, for the use of prenatal testing and other reproductive technologies for the avoidance of disability, and for enhancement. In this paper, it is argued that this principle should

be rejected, and it is concluded that while potential parents do have important obligations in relation to the foreseeable lives of their future children, these obligations are not best captured in terms of a duty to have the child with the best opportunity of the best life.

Paul, Diane B. On drawing lessons from the history of eugenics. *In:* Knowles, Lori P.; Kaebnick, Gregory E., eds. Reprogenetics: Law, Policy, and Ethical Issues. Baltimore: Johns Hopkins University Press, 2007: 3-19. 64 refs. NRCBL: 15.5; 15.1; 2.2.
  Keywords: *eugenics; *genetic engineering; *freedom; *historical aspects; *informal social control; *reproductive technologies; *values; autonomy; attitudes; coercion; contraception; directive counseling; disabled persons; feminism; genetic services; public policy; reproduction; reproductive rights; researchers; sterilization; trends; voluntary programs; women; Proposed Keywords: ectogenesis; transhumanism; Keyword Identifiers: Twentieth Century; Twenty-First Century

Pence, Gregory E. Are genetic abortions eugenic? *In his:* The Elements of Bioethics. Boston: McGraw-Hill, 2007: 172-202. NRCBL: 15.5; 12.1; 9.5.3; 4.4; 15.3; 12.4.2; 15.2. SC: an; cs.

Pollack, Robert E. Von der religiösen Pflicht, die Natur und den Staat zu hinterfragen: D. Bonhoeffer über den Schutz und die Würde des menschlichen Lebens im Zusammenhang mit der modernen genetischen Medizin. *In:* Gestrich, Christof; Neugebauer, Johannes, eds. Der Wert menschlichen Lebens: medizinische Ethik bei Karl Bonhoeffer und Dietrich Bonhoeffer. Berlin: Wichern-Verlag, 2006: 66-97. NRCBL: 15.5; 1.2; 2.2; 21.4.

Rafter, Nicole H. Claims-making and socio-cultural context in the first U.S. eugenics campaign. *Social Problems* 1992 February; 39(1): 17-34. NRCBL: 15.5; 9.5.5; 9.5.3.

Reggiani, Andés H. "Drilling eugenics into people's minds": expertise, public opinion, and biopolitics in Alexis Carrel's Man, the Unknown. *In:* Currell, Susan; Cogdell, Christina, eds. Popular Eugenics: National Efficiency and American Mass Culture in the 1930s. Athens, OH: Ohio University Press, 2006: 70-90. 51 fn. NRCBL: 15.5.
  Keywords: *eugenics; *historical aspects; animal experimentation; attitudes; euthanasia; famous persons; human experimentation; involuntary sterilization; mass media; mentally disabled persons; National Socialism; physicians; researchers; social discrimination; socioeconomic factors; Keyword Identifiers: *Man, the Unknown (Carrel, Alexis); Lindbergh, Charles; Rockefeller Institute; Twentieth Century; United States

Revie, Linda. "More than just boots! The eugenic and commercial concerns behind A. R. Kaufman's birth controlling activities". *Canadian Bulletin of Medical History* 2006; 23(1): 119-143. NRCBL: 15.5; 11.1; 1.3.2; 11.3.

Rydell, Robert; Cogdell, Christina; Largent, Mark. The Nazi eugenics exhibit in the United States, 1934-43. *In:* Currell, Susan; Cogdell, Christina, eds. Popular Eugenics: National Efficiency and American Mass Culture in the

1930s. Athens, OH: Ohio University Press, 2006: 359-384. 57 fn. NRCBL: 15.5; 11.3; 8.3.4; 21.4; 1.3.5; 2.2.
Keywords: *eugenics; *historical aspects; *National Socialism; involuntary sterilization; legal aspects; mentally disabled persons; prisoners; professional organizations; state government; Proposed Keywords: *exhibits; museums; Keyword Identifiers: *United States; Germany; Twentieth Century

Savulescu, Julian. In defence of procreative beneficence. *Journal of Medical Ethics* 2007 May; 33(5): 284-288. 17 refs. NRCBL: 15.5; 1.1; 14.1; 14.4; 15.2; 15.3; 9.5.1. SC: an.
Keywords: *beneficence; *ethical analysis; *preimplantation diagnosis; *quality of life; *reproduction; autonomy; behavioral genetics; choice behavior; disabled persons; ethical relativism; genetic disorders; genetic enhancement; genetic predisposition; genetic screening; in vitro fertilization; justice; moral obligations; parents; risks and benefits; uncertainty

Selgelid, Michael J. Neugenics? *Monash Bioethics Review* 2000 October; 19(4): 9-33. NRCBL: 15.5; 15.2; 15.3; 5.3; 2.2; 15.10. SC: an.

Weiss, Sheila Faith. Human genetics and politics as mutually beneficial resources: the case of the Kaiser Wilhelm Institute for Anthropology, Human Heredity and Eugenics during the Third Reich. *Journal of the History of Biology* 2006 Spring; 39(1): 41-88. NRCBL: 15.5; 1.3.5; 15.11; 2.2.

Wray, Matt. Three generations of imbeciles are enough: American eugenics and poor white trash. *In his:* Not Quite White: White Trash and the Boundaries of Whiteness. Durham: Duke University Press, 2006: 65-95. References embedded at back of book. NRCBL: 15.5; 2.2; 7.1; 11.3; 8.3.4; 9.5.10.
Keywords: *eugenics; *historical aspects; *involuntary sterilization; *poverty; *whites; attitudes; genetics; legal aspects; mentally retarded persons; pedigree; prisoners; professional organizations; public policy; racial groups; social discrimination; social problems; social sciences; stereotyping; socioeconomic factors; Supreme Court decisions; terminology; women; Proposed Keywords: *social class; rural population; Keyword Identifiers: *United States; Buck v. Bell; Davenport, Charles; Eugenic Records Office; Europe; Galton, Francis; Nineteenth Century; Twentieth Century

EUTHANASIA AND ALLOWING TO DIE
See also ADVANCE DIRECTIVES; ASSISTED SUICIDE; DEATH AND DYING

Dying in America, post Schiavo. *Health Care Ethics USA* 2007 Summer; 15(3): 20-21. NRCBL: 20.5.1; 20.7.

Euthanasia in Belgium up by 10% [news]. *BMJ: British Medial Journal* 2007 April 7; 334(7596): 714. NRCBL: 20.5.1.

Italy: doctor cleared in right-to-die case. *New York Times* 2007 March 7; p. A9. NRCBL: 20.5.1; 8.3.4. SC: po. Identifiers: Mario Riccio; Piergiorgio Welby.

Special report: Switzerland: an appointment with death. *Update (International Task Force on Euthanasia and Assisted Suicide)* 2007; 21(1): 4 p. [Online]. Accessed: http://www.internationaltaskforce.org/ina40.htm#2 [2008 February 4]. NRCBL: 20.5.1; 20.7; 4.4.

Ackerman, Felicia Nimue. Death is a punch in the jaw: life-extension and its discontents. *In:* Steinbock, Bonnie, ed. The Oxford Handbook of Bioethics. Oxford; New York: Oxford University Press, 2007: 324-348. NRCBL: 20.5.1; 9.5.2; 4.5.

Anwar, Rahij; Ahmed, Azeem. Who is responsible for "do not resuscitate" status in patients with broken hips? *BMJ: British Medical Journal* 2007 January 20; 334(7585): 155. NRCBL: 20.5.1; 9.8; 9.5.2. Identifiers: Great Britain (United Kingdom).

Babylon, Debra M.; Monk-Turner, Elizabeth. Should incurable patients be allowed to die? *Omega* 2006; 53(4): 311-319. NRCBL: 20.5.1; 7.1. SC: em; rv.

Baggs, Judith Gedney. Prognostic information provided during family meetings in the intensive care unit. *Critical Care Medicine* 2007 February; 35(2): 646-647. NRCBL: 20.5.1; 8.1; 9.2; 9.4.

Bendiane, Marc-Karim; Bouhnik, Anne-Deborah; Favre, Roger; Galinier,Anne; Obadia, Yolande; Moatti, Jean-Paul; Peretti-Watel, Patrick. Morphine prescription in end-of-life care and euthanasia: French home nurses' opinions. *Journal of Opioid Management* 2007 January-February; 3(1): 21-26. NRCBL: 20.5.1; 20.4.1; 9.7; 4.4.

Berner, Yitshal N. Non-benefit of active nutritional support in advanced dementia. *Israel Medical Association Journal* 2006 July; 8(7): 505-506. NRCBL: 20.5.1.

Bito, Seiji; Asai, Atsushi. Attitudes and behaviors of Japanese physicians concerning withholding and withdrawal of life-sustaining treatment for end-of-life patients: results from an Internet survey. *BMC Medical Ethics* 2007; 8:7, 9 p. [Online]. Accessed: http://www.biomedcentral.com/1472-6939/8/7 [2007 July 20]. NRCBL: 20.5.1; 9.5.2. SC: em.
Abstract: BACKGROUND: Evidence concerning how Japanese physicians think and behave in specific clinical situations that involve withholding or withdrawal of medical interventions for end-of-life or frail elderly patients is yet insufficient. METHODS: To analyze decisions and actions concerning the withholding/withdrawal of life-support care by Japanese physicians, we conducted cross-sectional web-based internet survey presenting three scenarios involving an elderly comatose patient following a severe stroke. Volunteer physicians were recruited for the survey through mailing lists and medical journals. The respondents answered questions concerning attitudes and behaviors regarding decision-making for the withholding/withdrawal of life-support care, namely, the initiation/withdrawal of tube feeding

NRCBL: National Reference Center for Bioethics Literature Classification Scheme   See inside front cover for terms.

145

and respirator attachment. RESULTS: Of the 304 responses analyzed, a majority felt that tube feeding should be initiated in these scenarios. Only 18% felt that a respirator should be attached when the patient had severe pneumonia and respiratory failure. Over half the respondents felt that tube feeding should not be withdrawn when the coma extended beyond 6 months. Only 11% responded that they actually withdrew tube feeding. Half the respondents perceived tube feeding in such a patient as a "life-sustaining treatment," whereas the other half disagreed. Physicians seeking clinical ethics consultation supported the withdrawal of tube feeding (OR, 6.4; 95% CI, 2.5-16.3; P 0.001). CONCLUSION: Physicians tend to harbor greater negative attitudes toward the withdrawal of life-support care than its withholding. On the other hand, they favor withholding invasive life-sustaining treatments such as the attachment of a respirator over less invasive and long-term treatments such as tube feeding. Discrepancies were demonstrated between attitudes and actual behaviors. Physicians may need systematic support for appropriate decision-making for end-of-life care.

Boles, Jean-Michel. End of life in the intensive care unit: from practice to law. What do the lawmakers tell the caregivers? A new series in Intensive Care Medicine. *Intensive Care Medicine* 2006 July; 32(7): 955-957. NRCBL: 20.5.1; 1.3.5; 21.1; 9.4.

Bookman, Kelly; Abbott, Jean. Ethics seminars: withdrawal of treatment in the emergency department — when and how? *Academic Emergency Medicine* 2006 December; 13(12): 1328-1332. NRCBL: 20.5.1; 9.5.1; 9.4; 20.5.4.

Boz, Bora; Acar, Kemalettin; Ergin, Ahmet; Kurtulus, Ayse; Ergin, Nesrin; Oguzhanoglu, Nalan. Effect of locus of control on acceptability of euthanasia among medical students and residents in Denizli, Turkey. *Journal of Palliative Care* 2007 Winter; 23(4): 286-290. NRCBL: 20.5.1; 7.2. SC: em.

Brannigan, Michael C. On medical futility: considerations and guidelines. *Missouri Medicine* 2006 March-April; 103(2): 113-117. NRCBL: 20.5.1; 9.4.

Breitbart, William. What can we learn from the death of Terri Schiavo? *Palliative and Supportive Care* 2005 March; 3(1): 1-3. NRCBL: 20.5.1; 20.3.2; 4.4.

British Medical Association [BMA]. Ethics Department. Treatment of patients in persistent vegetative state. Guidance from the BMA's Medical Ethics Department. London: British Medical Association, October 2007: 6 p. [Online]. Accessed: http://www.bma.org.uk/ap.nsf/AttachmentsByTitle/PDFpvs2007/$FILE/PVS07.pdf [2008 January 14]. NRCBL: 20.5.1.

Brotherton, Alisa M.; Abbott, Janice; Hurley, Margaret A.; Aggett, Peter J. Home percutaneous endoscopic gastrostomy feeding: perceptions of patients, carers, nurses and dietitians. *Journal of Advanced Nursing* 2007

August; 59(4): 388-397. NRCBL: 20.5.1; 8.1. SC: em. Identifiers: Great Britain (United Kingdom).

Button, James. My name is Dr. John Elliott and I'm about to die, with my head held high. *Sydney Morning Herald* 2007 January 26; 4 p. [Online]. Accessed: http://www.smh.com.au/news/world/my-name-is-dr-john-elliott-and-im-about-to-die-with-my-head-heldhigh/2007/01/26/1169788692086.html [2008 February 8]. NRCBL: 20.5.1; 20.7; 4.4.

Campbell, Tom. Euthanasia as a human right. *In:* McLean, Sheila A.M., ed. First Do No Harm: Law, Ethics, and Healthcare. Aldershot, England; Burlington, VT: Ashgate, 2006: 447-459. NRCBL: 20.5.1; 4.4; 21.1. SC: an.

Canadian Researchers at the End of Life Network; Heyland, Daren K.; Frank, Chris; Groll, Dianne; Pichora, Deb; Dodek, Peter; Rocker, Graeme; Gafni, Amiram. Understanding cardiopulmonary resuscitation decision making: perspectives of seriously ill hospitalized patients and family members. *Chest* 2006 August; 130(2): 419-428. NRCBL: 20.5.1; 9.4; 8.3.1; 9.5.2; .1.

Cantor, Norman L. On hastening death without violating legal and moral prohibitions. *Loyola University Chicago Law Journal* 2006 Winter; 37(2): 407-431. NRCBL: 20.5.1; 20.4.1; 9.7.

Caplan, Arthur L.; Bergman, Edward J. Beyond Schiavo. *Journal of Clinical Ethics* 2007 Winter; 18(4): 340-345. NRCBL: 20.5.1; 20.5.3; 9.6; 8.1; 20.3.3. SC: an. Identifiers: Terri Schiavo.

Cheng, Guang-Shing. Compromise [case study]. *Hastings Center Report* 2007 September-October; 37(5): 8-9. NRCBL: 20.5.1; 8.1; 9.4. SC: cs.

Clarkson, Frederick. Tragedy on the national stage: conservative intervention into the Terri Schiavo case was a disservice to everybody. *Conscience* 2007 Autumn; 28(3): 35-38. NRCBL: 20.5.1; 7.1.

Cohen, J.; Marcoux, I.; Bilsen, J.; Deboosere, P.; van der Wal, G.; Deliens, L. Trends in acceptance of euthanasia among the general public in 12 European countries (1981-1999). *European Journal of Public Health* 2006 December; 16(6): 663-669. NRCBL: 20.5.1; 4.4; 7.1; 2.2; 21.1. SC: em.

Collins, N.; Phelan, D.; Carton, E. End of life in ICU — care of the dying or 'pulling the plug'? *Irish Medical Journal* 2006 April; 99(4): 112-114. NRCBL: 20.5.1; 20.4.1. SC: em.

Cosgriff, JoAnne Alissi; Pisani, Margaret; Bradley, Elizabeth H.; O'Leary, John R.; Fried, Terri R. The association between treatment preferences and trajectories of care at the end-of-life. *JGIM: Journal of General Internal*

Medicine 2007 November; 22(11): 1566-1571. NRCBL: 20.5.1; 20.4.1. SC: em.

Crippen, David. Medical treatment for the terminally ill: the 'risk of unacceptable badness'. *Critical Care (London, England)* 2005 August; 9(4): 317-318. NRCBL: 20.5.1; 4.4; 9.4.

Curtis, J. Randall. Interventions to improve care during withdrawal of life-sustaining treatments. *Journal of Palliative Medicine* 2005; 8(Supplement 1): S116-S131. NRCBL: 20.5.1; 18.5.7; 4.4; 9.8. SC: rv.

Day, Lisa. Questions concerning the goodness of hastening death. *American Journal of Critical Care* 2006 May; 15(3): 312-314. NRCBL: 20.5.1; 20.4.1; 7.1; 20.3.2.

De Gendt, Cindy; Bilsen, Johan; Stichele, Robert Vander; Van Den Noortgate, Nele; Lambert, Margareta; Deliens, Luc. Nurses' involvement in 'do not resuscitate' decisions on acute elder care wards. *Journal of Advanced Nursing* 2007 February; 57(4): 404-409. NRCBL: 20.5.1; 7.1; 9.5.2. SC: em. Identifiers: Belgium.

Doig, Christopher; Murray, Holt; Bellomo, Rinaldo; Kuiper, Michael; Costa, Rubens; Azoulay, Elie; Crippen, David. Ethics roundtable debate: patients and surrogates want 'everything done' — what does 'everything' mean? *Critical Care* 2006; 10(5): 231. NRCBL: 20.5.1; 9.4; 8.1; 9.5.1; 21.1. SC: cs. Identifiers: Canada; Australia; Netherlands; France; Brazil.

Doyal, Len. The futility of opposing the legalisation of non-voluntary and voluntary euthanasia. *In:* McLean, Sheila A.M., ed. First Do No Harm: Law, Ethics, and Healthcare. Aldershot, England; Burlington, VT: Ashgate, 2006: 461-477. NRCBL: 20.5.1; 8.1; 20.7. SC: an.

Dubler, Nancy Neveloff. Commentary on "Beyond Schiavo": beyond theory. *Journal of Clinical Ethics* 2007 Winter; 18(4): 346-349. NRCBL: 20.5.1; 9.6; 8.1; 8.3.4; 20.3.3. SC: cs. Comments: Arthur L. Caplan and Edward J. Bergman. Beyond Schiavo. Journal of Clinical Ethics 2007 Winter; 18(4): 340-345.

Dunlop, John. Permissibility to stop man's ventilator on his request. *Ethics and Medicine: An International Journal of Bioethics* 2007 Spring; 23(1): 15-17. NRCBL: 20.5.1; 8.3.4. SC: cs.

Eberl, Jason T. Issues and the end of human life: PVS patients, euthanasia, and organ donation. *In his:* Thomistic Principles and Bioethics. London; New York: Routledge, 2006: 95-127. NRCBL: 20.5.1; 4.4; 20.7; 19.5; 20.2.1. SC: an. Identifiers: permanent vegetative state.

Egan, Danielle. Cheating death: we're going to live forever. *New Scientist* 2007 October 13-19; 195(2625): 46. NRCBL: 20.5.1; 4.5.

Elder, John. Illegal book heads through the Internet gateway. *Age (Australia)* 2007 April 1; 2 p. [Online]. Accessed: http://www.theage.com.au/news/national/ illegal-book-heads-through-internet-gateway/ 2007/03/31/1174761817937.html [2008 February 8]. NRCBL: 20.5.1; 9.7; 20.7. Identifiers: The Peaceful Pill; euthanasia.

Eliot, Jaklin; Olver, Ian. Autonomy and the family as (in)appropriate surrogates for DNR decisions: a qualitative analysis of dying cancer patients' talk. *Journal of Clinical Ethics* 2007 Fall; 18(3): 206-218. NRCBL: 20.5.1; 8.3.3; 20.5.4; 20.3.3. SC: cs. Identifiers: Australia.

Eliott, Jaklin; Olver, Ian. Response from Eliott and Olver. *Journal of Clinical Ethics* 2007 Fall; 18(3): 233-234. NRCBL: 20.5.1; 8.1; 20.3.3. Comments: James L. Nelson and Hilde Lindemann. What families say about surrogacy: a response to "Autonomy and the family as (in)appropriate surrogates for DNR decisions." Journal of Clinical Ethics 2007 Fall; 18(3): 219-226.

England, Ruth; England, Tim; Coggon, John. The ethical and legal implications of deactivating an implantable cardioverter-defibrillator in a patient with terminal cancer. *Journal of Medical Ethics* 2007 September; 33(9): 538-540. NRCBL: 20.5.1; 9.7. SC: an.
Abstract: In this paper, the ethical and legal issues raised by the deactivation of implantable cardioverter-defibrillators (ICDs) in patients with terminal cancer is considered. It is argued that the ICD cannot be well described either as a treatment or as a non-treatment option, and thus raises complex questions regarding how rules governing deactivation should be framed. A new category called "integral devices" is proposed. Integral devices require their own special rules, reflecting their position as a "halfway house" between a form of treatment and a part of the body. The practical problems faced by doctors working in palliative medicine with regard to the deactivation of ICDs are also considered.

EURELD (European End-of-life Decision) Consortium; van Delden, Johannes J.M.; Löfmark, Rurik; Deliens, Luc; Bosshard, Georg; Norup, Michael; Cecioni, Riccardo; van der Heide, Agnes. Do-not-resuscitate decisions in six European countries. *Critical Care Medicine* 2006 June; 34(6): 1686-1690. NRCBL: 20.5.1; 21.1. SC: em. Identifiers: Belgium; Denmark; Italy; Netherlands; Sweden; Switzerland.

Ewanchuk, Mark; Brindley, Peter G. Perioperative do-not-resuscitate orders — doing 'nothing' when 'something' can be done. *Critical Care (London, England)* 2006; 10(4): 219. NRCBL: 20.5.1; 7.1; 9.4.

Finlay, Ilora. Crossing the 'bright line' — difficult decisions at the end of life. *Clinical Medicine (London, England)* 2006 July-August; 6(4): 398-402. NRCBL: 20.5.1; 8.1; 9.4; 7.1.

Fins, Joseph J. The minimally conscious state: ethics and diagnostic nosology. *Medical Ethics Newsletter [Lahey Clinic]* 2007 Fall; 14(3): 1-2, 5. NRCBL: 20.5.1; 20.2.1.

NRCBL: National Reference Center for Bioethics Literature Classification Scheme   See inside front cover for terms.

147

Fox, Ellen; Daskal, Frona C.; Stocking, Carol. Ethics consultants' recommendations for life-prolonging treatment of patients in a persistent vegetative state: a follow-up study. *Journal of Clinical Ethics* 2007 Spring; 18(1): 64-71. NRCBL: 20.5.1; 9.6; 8.1.

Fried, Terri R.; O'Leary, John; Van Ness, Peter; Fraenkel, Liana. Inconsistency over time in the preferences of older persons with advanced illness for life-sustaining treatment. *Journal of the American Geriatrics Society* 2007 July; 55(7): 1007-1014. NRCBL: 20.5.1; 9.5.2. SC: em.

Fried, Terri R.; Van Ness, Peter H.; Byers, Amy L.; Towle, Virginia R.; O'Leary, John R.; Dubin, Joel A. Changes in preferences for life-sustaining treatment among older persons with advanced illness. *JGIM: Journal of General Internal Medicine* 2007 April; 22(4): 495-501. NRCBL: 20.5.1; 9.5.2. SC: em.

Friedman, Sandra; Gilmore, Dana. Factors that impact resuscitation preferences for young people with severe developmental disabilities. *Intellectual and Developmental Disabilities* 2007 April; 45(2): 90-97. NRCBL: 20.5.1; 8.3.3. SC: em. Note: Abstract in French.

Galanakis, E.; Dimoliatis, I.D.K. Early European attitudes towards "good death": Eugenios Voulgaris, Treatise on euthanasia, St Petersburg, 1804. *Medical Humanities* 2007 June; 33(1): 1-4. NRCBL: 20.5.1; 2.2.
  Abstract: Eugenios Voulgaris (Corfu, Greece, 1716; St Petersburg, Russia, 1806) was an eminent theologian and scholar, and bishop of Kherson, Ukraine. He copiously wrote treatises in theology, philosophy and sciences, greatly influenced the development of modern Greek thought, and contributed to the perception of Western thought throughout the Eastern Christian world. In his Treatise on euthanasia (1804), Voulgaris tried to moderate the fear of death by exalting the power of faith and trust in the divine providence, and by presenting death as a universal necessity, a curative physician and a safe harbour. Voulgaris presented his views in the form of a consoling sermon, abundantly enriched with references to classical texts, the Bible and the Church Fathers, as well as to secular sources, including vital statistics from his contemporary England and France. Besides euthanasia, he introduced terms such as dysthanasia, etoimothanasia and prothanasia. The Treatise on euthanasia is one of the first books, if not the very first, devoted to euthanasia in modern European thought and a remarkable text for the study of the very early European attitudes towards "good death". In the Treatise, euthanasia is clearly meant as a spiritual preparation and reconciliation with dying rather than a physician-related mercy killing, as the term progressed to mean during the 19th and the 20th centuries. This early text is worthy of study not only for the historian of medical ethics or of religious ethics, but for everybody who is trying to courageously confront death, either in private or in professional settings.

Georges, Jean-Jacques.; Onwuteaka-Philipsen, Bregje D.; Muller, MartienT.; Van Der Wal, Gerrit.; Van Der Heide, Agnes; Van Der Maas, Paul J. Relatives' perspective on the terminally ill patients who died after euthanasia or physician-assisted suicide: a retrospective cross-sectional interview study in the Netherlands. *Death Studies* 2007 January-February; 31(1): 1-15. NRCBL: 20.5.1; 20.3.3; 20.7; 4.4.

Gillett, Grant; Chisholm, Nick. Locked in syndrome, PVS and ethics at the end of life. *Journal of Ethics in Mental Health [electronic]* 2007 November; 2(2): 6 p. Accessed: http://www.jemh.ca [2008 January 24]. NRCBL: 20.5.1.

Goldblatt, David; Greenlaw, Jane. Starting and stopping the ventilator for patients with amyotrophic lateral sclerosis. *Neurologic Clinics* 1989 November; 7(4): 789-806. NRCBL: 20.5.1; 9.4. SC: cs; em. Identifiers: Lou Gehrig's disease.

Goodman, Kenneth W.; Allen, Bill; Cerminara, Kathy L.; Fiore, Robin N.; Moseley, Ray; Mulvey, Ben; Spike, Jeffrey; Walker, Robert M. Florida bioethics leaders' analysis on HB701. Miami: University of Miami, 2005 March 7; 7 p. [Online]. Accessed: http://www6.miami.edu/ethics/schiavo/pdf_files/030805-HB701-EthicsAnalysis.pdf [2007 April 5]. NRCBL: 20.5.1; 2.4. SC: an. Identifiers: Theresa Marie Schiavo.

Grayling, A.C. Cheating death: will we have to choose who lives? *New Scientist* 2007 October 13-19; 195(2625): 47. NRCBL: 20.5.1; 20.7.

Gunning, Karel. Euthanasia and the United Nations' Universal Declaration of Human Rights. *In:* Wagner, Teresa; Carbone, Leslie, eds. Fifty Years after the Declaration: the United Nations' Record on Human Rights. Lanham, MD: University Press of America, 2001: 17-23. NRCBL: 20.5.1; 2.2; 8.1; 20.7; 21.1.

Hildén, Hanna-Mari; Honkasalo, Marja-Liisa. Unethical bunglers or humane professionals? Discussions in the media of end-of-life treatment decisions. *Communication and Medicine* 2006; 3(2): 125-134. NRCBL: 20.5.1; 8.1; 7.1; 9.4.

Hofmann, Paul B.; Schneiderman, Laurence J. Futility, in short [letter and reply]. *Hastings Center Report* 2007 July-August; 37(4): 8. NRCBL: 20.5.1; 9.4; 8.1. Comments: Paul B. Hofmann and Laurence J. Schneiderman. Physicians should not always pursue a good "clinical" outcome. Hastings Center Report 2007 May-June 37(3): inside back cover.

Hofmann, Paul B.; Schneiderman, Lawrence J. Physicians should not always pursue a good "clinical" outcome. *Hastings Center Report* 2007 May-June; 37(3): inside back cover. NRCBL: 20.5.1; 9.4; 8.1. SC: an.

Jecker, Nancy S. Medical futility: a paradigm analysis. *HEC (Healthcare Ethics Committee) Forum* 2007 March; 19(1): 13-32. NRCBL: 20.5.1; 2.2; 4.4. SC: an; cs.

Jennings, Beth. The politics of end-of-life decision-making: computerised decision-support tools, physicians' jurisdiction and morality. *Sociology of Health and Illness* 2006 April; 28(3): 350-375. NRCBL: 20.5.1; 9.4; 8.1; 1.3.12; 7.1; 5.3. SC: cs; em.

Kapp, Marshall B. Pain control for dying patients: hastening death or ensuring comfort? *Journal of Opioid Management* 2006 May-June; 2(3): 128-129. NRCBL: 20.5.1; 9.7; 4.4; 20.4.1.

Karenberg, Axel. Neurosciences and the Third Reich — Introduction. *Journal of the History of the Neurosciences* 2006 September; 15(3): 168-172. NRCBL: 20.5.1; 9.5.3; 21.3; 1.3.5.

Kelley, Amy S.; Gold, Heather T.; Roach, Keith W.; Fins, Joseph J. Differential medical and surgical house staff involvement in end-of-life decisions: a retrospective chart review. *Journal of Pain and Symptom Management* 2006 August; 32(2): 110-117. NRCBL: 20.5.1; 7.1; 8.1; 7.2.

Kimsma, Geritt K.; van Leeuwen; Evert. The role of family in euthanasia decision making. *HEC (Healthcare Ethics Committee) Forum* 2007 December; 19(4): 365-373. NRCBL: 20.5.1; 8.3.3.

Komatsu, Yoshihiko. The age of a "revolutionized human body" and the right to die. *In:* LaFleur, William R.; Böhme, Gernot; Shimazono, Susumu, eds. Dark Medicine: Rationalizing Medical Research. Bloomington: Indiana University Press, 2007: 180-200. NRCBL: 20.5.1; 4.4.

Kwok, Timothy; Twinn, Sheila; Yan, Elsie. The attitudes of Chinese family caregivers of older people with dementia towards life sustaining treatments. *Journal of Advanced Nursing* 2007 May; 58(3): 256-262. NRCBL: 20.5.1; 8.1; 17.1; 9.5.2; 20.3.3. SC: em. Identifiers: Hong Kong.

Lacey, Debra. End-of-Life decision making for nursing home residents with dementia: a survey of nursing home social services staff. *Health and Social Work* 2006 August; 31(3): 189-199. NRCBL: 20.5.1; 9.5.2; 9.5.3; 20.5.4; 20.3.2.

Lemiengre, Joke; de Casterlé, Bernadette Dierckx; Van Craen, Katleen; Schotsmans, Paul; Gastmans, Chris. Institutional ethics policies on medical end-of-life decisions: a literature review. *Health Policy* 2007 October; 83(2-3): 131-143. NRCBL: 20.5.1; 7.1; 9.1. SC: em.

Lemiengre, Joke; Dierckx de Casterlé, Bernadette; Verbeke, Geert; Guisson, Catherine; Schotsmans, Paul; Gastmans, Chris. Ethics policies on euthanasia in hospitals — a survey in Flanders (Belgium). *Health Policy* 2007 December; 84(2-3): 170-180. NRCBL: 20.5.1; 9.1; 7.1. SC: em.

Lett, Dan. Manitoba physicians consider DNR guidelines [news]. *CMAJ/JAMC: Canadian Medical Association Journal* 2007 January 30; 176(3): 310-311. NRCBL: 20.5.1.

Level of Care Study Investigators; Canadian Critical Care Trials Group; Cook, Deborah; Rocker, Graeme; Marshall, John; Griffith, Lauren; McDonald, Ellen; Guyatt, Gordon. Levels of care in the intensive care unit: a research program. *American Journal of Critical Care* 2006 May; 15(3): 269-379. NRCBL: 20.5.1; 20.5.4; 4.4; 7.1; 8.1.

Lewis, Milton James. Medicine and euthanasia. *In his:* Medicine and Care of the Dying: A Modern History. Oxford; New York: Oxford University Press, 2007: 198-228. NRCBL: 20.5.1; 2.2.

Logan, Lara. Switzerland's suicide tourists. *CBSNews.com* 2003 July 23: 2 p. [Online]. Accessed: http://www.cbsnews.com/stories/2003/02/12/60II/printable540332.shtml [2008 February 7]. NRCBL: 20.5.1; 20.7; 4.4. SC: po.

McCarron, Mary; McCallion, Philip. End-of-life care challenges for persons with intellectual disability and dementia: making decisions about tube feeding. *Intellectual and Developmental Disabilities* 2007 April; 45(2): 128-131. NRCBL: 20.5.1; 8.3.3; 9.5.3.

McGrath, Pam D.; Forrester, Kim. Ethico-legal issues in relation to end-of-life care and institutional mental health. *Australian Health Review* 2006 August; 30(3): 286-297. NRCBL: 20.5.1; 20.4.1; 9.5.3; 17.8.

Meyer, Erin K.G.; AuBuchon, James P. Conflicting duties: an ethical dilemma in transfusion medicine. *Medical Ethics Newsletter [Lahey Clinic]* 2007 Fall; 14(3): 3,7. NRCBL: 20.5.1; 19.4; 19.5; 8.1. SC: cs.

Michalsen, Andrej; Reinhart, Konrad. "Euthanasia": a confusing term, abused under the Nazi regime and misused in present end-of-life debate. *Intensive Care Medicine* 2006 September; 32(9): 1304-1310. NRCBL: 20.5.1; 15.5; 1.3.5; 21.4; 21.1.

Mitchell, Susan L.; Teno, Joan M.; Intrator, Orna; Feng, Zhanlian; Mor, Vincent. Decisions to forgo hospitalization in advance dementia: a nationwide study. *Journal of the American Geriatrics Society* 2007 March; 55(3): 432-438. NRCBL: 20.5.1; 17.1; 9.5.2. SC: em.

Mohindra, Raj. Obligations to treat, personal autonomy, and artificial nutrition and hydration. *Clinical Medicine* 2006 May-June; 6(3): 271-273. NRCBL: 20.5.1; 7.1; 8.1; 8.3.4.

NRCBL: National Reference Center for Bioethics Literature Classification Scheme   See inside front cover for terms.

149

Mohindra, R.K. Medical futility: a conceptual model. *Journal of Medical Ethics* 2007 February; 33(2): 71-75. NRCBL: 20.5.1; 9.4; 8.1; 9.2. SC: an.

Abstract: This paper introduces the medical factual matrix as a new and potentially valuable tool in medical ethical analysis. Using this tool it demonstrates the idea that a defined medical intervention can only be meaningfully declared futile in relation to a defined goal(s) of treatment. It argues that a declaration of futility made solely in relation to a defined medical intervention is inchoate. It recasts the definition of goal futility as an intervention that cannot alter the probability of the existence of the important outcome states that might flow from a defined intervention. The idea of value futility and the extent of physician obligations in futile situations are also addressed. It also examines the source of substantive conflicts which commonly arise within the doctor-patient relationship and the ensuing power relations that operate between doctor and patient when questions of futility arise.

Morgenstern, L.; Laquer, M.; Treyzon, L. Ethical challenges of percutaneous endoscopic gastrostomy. *Surgical Endoscopy* 2005 March; 19(3): 398-400. NRCBL: 20.5.1; 9.5.1; 9.6. SC: cs; em.

Morris, Marilyn C.; Fischbach, Ruth L.; Nelson, Robert M.; Schleien, Charles L. A paradigm for inpatient resuscitation research with an exception from informed consent. *Critical Care Medicine* 2006 October; 34(10): 2567-2575. NRCBL: 20.5.1; 8.3.2.

Moselli, N.M.; Debernardi, F.; Piovano, F. Forgoing life sustaining treatments: differences and similarities between North America and Europe. *Acta Anaesthesiologica Scandinavica* 2006 November; 50(10): 1177-1186. NRCBL: 20.5.1; 9.4; 20.5.4; 21.1.

Mostert, Mark P. Cultures of death, old and new. *Human Life Review* 2007 Fall; 33(4): 54-65. 29 fn. NRCBL: 20.5.1; 9.5.1; 2.2; 15.5; 21.4.

Keywords: *disabled persons; *eugenics; *involuntary euthanasia; *Killing; *National Socialism; historical aspects; involuntary sterilization; legal aspects; moral complicity; physicians; quality of life; value of life; Keyword Identifiers: *Germany; Twentieth Century

Murphy, Donald J.; Santilli, Sara. Elderly patients' preferences for long-term life support. *Archives of Family Medicine* 1998 September; 7(5): 484-488. NRCBL: 20.5.1; 9.5.2. SC: em.

Neil, David A.; Coady, C.A.J.; Thompson, J.; Kuhse, H. End-of-life decisions in medical practice: a survey of doctors in Victoria (Australia). *Journal of Medical Ethics* 2007 December; 33(12): 721-725. NRCBL: 20.5.1; 20.3.2. SC: em.

Abstract: OBJECTIVES: To discover the current state of opinion and practice among doctors in Victoria, Australia, regarding end-of-life decisions and the legalisation of voluntary euthanasia. Longitudinal comparison with similar 1987 and 1993 studies. Design and PARTICI-PANTS: Cross-sectional postal survey of doctors in Victoria. RESULTS: 53% of doctors in Victoria support the legalisation of voluntary euthanasia. Of doctors who have experienced requests from patients to hasten death, 35% have administered drugs with the intention of hastening death. There is substantial disagreement among doctors concerning the definition of euthanasia. CONCLUSIONS: Disagreement among doctors concerning the meaning of the term euthanasia may contribute to misunderstanding in the debate over voluntary euthanasia. Among doctors in Victoria, support for the legalisation of voluntary euthanasia appears to have weakened slightly over the past 17 years. Opinion on this issue is sharply polarised.

Nelson, James L.; Lindemann, Hilde. What families say about surrogacy: a response to "Autonomy and the family as (in)appropriate surrogates for DNR decisions". *Journal of Clinical Ethics* 2007 Fall; 18(3): 219-226. NRCBL: 20.5.1; 20.3.3; 8.3.3; 20.5.4. Comments: Jaklin Eliott and Ian Olver. Autonomy and the family as (in)appropriate surrogates for DNR decisions: a qualitative analysis of dying cancer patients' talk. Journal of Clinical Ethics 2007 Fall; 18(3): 206-218.

Netherlands. Ministry of Health, Welfare and Sport. Regional Euthanasia Review Committees. Annual report 2005. Arnhem: Regional Euthanasia Review Committees, 2006 April; 30 p. [Online]. Accessed: http://www.toetsingscommissieseuthanasie.nl/Images/Annuall%20Report%202005%20English_tcm12-2439.pdf [2007 April 17]. NRCBL: 20.5.1.

Neumärker, Klaus-Jürgen. Karl Bonhoeffers Entscheidungen zur Zwangssterilisation und Euthanasie. Versuch einer ethischen Beurteilung unter Berücksichtigung D. Bonhoeffers. *In:* Gestrich, Christof; Neugebauer, Johannes, eds. Der Wert menschlichen Lebens: medizinische Ethik bei Karl Bonhoeffer und Dietrich Bonhoeffer. Berlin: Wichern-Verlag, 2006: 33-65. NRCBL: 20.5.1; 11.3; 17.1; 2.2; 15.5; 21.4; 2.2.

Nicolasora, Nelson; Pannala, Rahul; Mountantonakis, Stavros; Shanmugam, Baia; DeGirolamo, Angela; Amoateng-Adjepong, Yaw.; Manthous, Constantine A. If asked, hospitalized patients will choose whether to receive life-sustaining therapies. *Journal of Hospital Medicine* 2006 May; 1(3): 161-167. NRCBL: 20.5.1; 8.3.1; 9.4; 20.5.4. SC: em.

O'Rourke, Kevin. Ethical reflection continues post-Schiavo. *Health Care Ethics USA* 2007 Winter; 15(1): 13-14. NRCBL: 20.5.1.

Oh, Do-Youn; Kim, Jee-Hyun; Kim, Dong-Wan; Im, Seock-Ah; Kim, Tae-You; Heo, Dae Seog; Bang, Yung-Jue; Kim, Noe Kyeong. CPR or DNR? End-of-life decision in Korean cancer patients: a single center's experience. *Supportive Care in Cancer* 2006 February; 14(2): 103-108. NRCBL: 20.5.1; 9.5.1. SC: em.

Oliver, D. A perspective on euthanasia. *British Journal of Cancer* 2006 October 23; 95(8): 953-954. NRCBL: 20.5.1; 20.7; 7.1. SC: em.

Peiffer, Jürgen. Phases in the postwar German reception of the "Euthanasia Program" (1939-1945) involving the killing of the mentally disabled and its exploitation by neuroscientists. *Journal of the History of the Neurosciences* 2006 September; 15(3): 210-244. NRCBL: 20.5.1; 21.4; 1.3.5; 2.2; 9.5.3.

Pence, Gregory E. Is there a duty to die? *In his:* The Elements of Bioethics. Boston: McGraw-Hill, 2007: 233-262. NRCBL: 20.5.1; 9.4; 9.3.1; 9.5.2; 17.1; 20.5.4. SC: cs.

Phillips, Helen. Impossible awakenings. *New Scientist* 2007 July 7-13; 195(2611): 40-43. NRCBL: 20.5.1.

Pollard, Irina. The state of wellbeing: on the end of life care and euthanasia. *In her:* Life, Love and Children: A Practical Introduction to Bioscience Ethics and Bioethics. Boston: Kluwer Academic Publishers, 2002: 97-103. NRCBL: 20.5.1.

Reiling, Jennifer. Euthanasia as a romantic motive. *JAMA: The Journal of the American Medical Association* 2007 November 7; 298(17): 2076. NRCBL: 20.5.1. Identifiers: The Fruit of the Tree by Edith Wharton.

Reinders, Hans S. Euthanasia and disability: comments on the Terry Schiavo case. *Ethics and Intellectual Disability* 2005 Summer; 9(1): 6-7. NRCBL: 20.5.1; 9.5.3.

Rietjens, Judith A.C.; Bilsen, Johan; Fischer, Susanne; Van Der Heide, Agnes; Van Der Maas, Paul J.; Miccinesi, Guido; Norup, Michael; Onwuteaka-Philipsen, Bregje D.; Vrakking, Astrid M.; Van Der Wal, Gerrit. Using drugs to end life without an explicit request of the patient. *Death Studies* 2007 March; 31(3): 205-221. NRCBL: 20.5.1; 8.1; 9.7. SC: em. Identifiers: Belgium; Denmark; Sweden; Netherlands.

Robinson, Ellen M.; Phipps, Marion; Purtilo, Ruth B.; Tsoumas, Angelica; Hamel-Nardozzi, Marguerite. Complexities in decision making for persons with disabilities nearing end of life. *Topics in Stroke Rehabilitation* 2006 Fall; 13(4): 54-67. NRCBL: 20.5.1; 2.1; 8.3.4. SC: cs.

Rocker, Graeme. Life-support limitation in the pre-hospital setting. *Intensive Care Medicine* 2006 October; 32(10): 1464-1466. NRCBL: 20.5.1; 7.1; 9.4.

Rousseau, Paul. Allegations of euthanasia. *American Journal of Hospice and Palliative Care* 2006 October-November; 23(5): 422-423. NRCBL: 20.5.1; 7.1; 20.3.2.

Rubin, Susan B. If we think it's futile, can't we just say no? *HEC (Healthcare Ethics Committee) Forum* 2007 March; 19(1): 45-65. NRCBL: 20.5.1; 8.1; 7.3; 20.3.2; 20.3.3.

Salomon, F. Leben erhalten und Sterben ermöglichen. Entscheidungskonflikte in der Intensivmedizin = Saving life and permitting death. Decision conflicts in intensive medicine. *Der Anaesthesist* 2006 January; 55(1): 64-69. NRCBL: 20.5.1. SC: cs.

Schaffer, Marjorie A. Ethical problems in end-of-life decisions for elderly Norwegians. *Nursing Ethics* 2007 March; 14(2): 242-257. NRCBL: 20.5.1; 9.5.2. SC: em.
Abstract: Norwegian health professionals, elderly people and family members experience ethical problems involving end-of-life decision making for elders in the context of the values of Norwegian society. This study used ethical inquiry and qualitative methodology to conduct and analyze interviews carried out with 25 health professionals, six elderly people and five family members about the ethical problems they encountered in end-of-life decision making in Norway. All three participant groups experienced ethical problems involving the adequacy of health care for elderly Norwegians. Older people were concerned about being a burden to their families at the end of their life. However, health professionals wished to protect families from the burden of difficult decisions regarding health care for elderly parents at the end of life. Strategies are suggested for dialogue about end-of-life decisions and the integration of palliative care approaches into health care services for frail elderly people.

Schleger, Heidi Albisser. „Alter" und „Kosten" - Faktoren bei Therapieentscheidungen am Lebensende? Eine Analyse informeller Wissensstrukturen bei Ärzten und Pflegenden = "Age" and "Costs" - factors in treatment decisions at the end-of-life? An analysis of informal knowledge structures of doctors and nurses. *Ethik in der Medizin* 2007 June; 19(2): 103-119. NRCBL: 20.5.1; 9.5.2; 9.3.1. SC: em.

Schneiderman, Lawrence J. Effect of ethics consultations in the intensive care unit. *Critical Care Medicine* 2006 November; 34(11, Supplement): S359-S363. NRCBL: 20.5.1; 9.6; 8.1; 9.4.

Schonfeld, Toby L.; Romberger, Debra J.; Hester, D. Micah; Shannon, Sarah E. Resuscitating a bad patient. *Hastings Center Report* 2007 January-February; 37(1): 14-16. NRCBL: 20.5.1; 8.1; 17.1; 8.3.3. SC: cs.

Shakespeare, Jocasta. A date with death [interview]. *Sunday Times Magazine (London)* 2006 April 16: 5 p. [Online]. Accessed: http://www.timesonline.co.uk/tol/life_and_style/article702621.ece?token=null&offset=24 [2008 February 7]. NRCBL: 20.5.1; 20.7; 4.4. Identifiers: Ludwig A. Minelli, General Secretary of Dignitas.

Sham, C.O.; Cheng, Y.W.; Ho, K.W.; Lai, P.H.; Lo, L.W.; Wan, H.L.; Wong, C.Y.; Yeung, Y.N.; Yuen, S.H.; Wong, A.Y. Do-not-resuscitate decision: the attitudes of medical and non-medical students. *Journal of Medical Ethics* 2007 May; 33(5): 261-265. NRCBL: 20.5.1; 20.3.2. SC: em. Identifiers: Hong Kong.

NRCBL: National Reference Center for Bioethics Literature Classification Scheme   See inside front cover for terms.

151

Abstract: OBJECTIVES: To study the attitudes of both medical and non-medical students towards the do-not-resuscitate (DNR) decision in a university in Hong Kong, and the factors affecting their attitudes. METHODS: A questionnaire-based survey conducted in the campus of a university in Hong Kong. Preferences and priorities of participants on cardiopulmonary resuscitation in various situations and case scenarios, experience of death and dying, prior knowledge of DNR and basic demographic data were evaluated. RESULTS: A total of 766 students participated in the study. There were statistically significant differences in their DNR decisions in various situations between medical and non-medical students, clinical and preclinical students, and between students who had previously experienced death and dying and those who had not. A prior knowledge of DNR significantly affected DNR decision, although 66.4% of non-medical students and 18.7% of medical students had never heard of DNR. 74% of participants from both medical and non-medical fields considered the patient's own wish as the most important factor that the healthcare team should consider when making DNR decisions. Family wishes might not be decisive on the choice of DNR. CONCLUSIONS: Students in medical and non-medical fields held different views on DNR. A majority of participants considered the patient's own wish as most important in DNR decisions. Family wishes were considered less important than the patient's own wishes.

Shapiro, Dvorah S.; Friedmann, Reuven. To feed or not to feed the terminal demented patient — is there any question? *Israel Medical Association Journal* 2006 July; 8(7): 507-508. NRCBL: 20.5.1; 17.1; 9.5.2.

Sheldon, Tony. Incidence of euthanasia in the Netherlands falls [news]. *BMJ: British Medical Journal* 2007 May 26; 334(7603): 1075. NRCBL: 20.5.1. SC: em.

Sibbald, Robert; Downar, James; Hawryluck, Laura. Perceptions of "futile care" among caregivers in intensive care units. *CMAJ/JAMC: Canadian Medical Association Journal* 2007 November 6; 177(10): 1201-1208. NRCBL: 20.5.1; 20.3.2. SC: em.

Sidhu, Navdeep S.; Dunkley, Margaret E.; Egan, Melinda J. "Not-for-resuscitation" orders in Australian public hospitals: policies, standardised order forms and patient information leaflets. *Medical Journal of Australia* 2007 January 15; 186(2): 72-75. NRCBL: 20.5.1; 8.1.

Sizemore, Rebecca. Separating medical and ethical: helping families determine the best interests of loved ones. *Dimensions of Critical Care Nursing* 2006 September-October; 25(5): 216-220. NRCBL: 20.5.1; 20.3.2; 8.1; 9.4; 4.4.

Smith, Craig S. France: doctors petition for euthanasia. *New York Times* 2007 March 10; p. A6. NRCBL: 20.5.1. SC: po.

Smith, Wesley J. "We never say no." The right-to-die movement abandons pretense. *Weekly Standard* 2006 April 27: 4 p. [Online]. Accessed: http://www.weeklystandard.com/Content/Public/Articles/000/000/012/124abkbr.asp [2008 February 7]. NRCBL: 20.5.1; 20.7; 4.4.

Sweet, Victoria. Code pearl. *Health Affairs* 2008 January-February; 27(1): 216-220. NRCBL: 20.5.1. SC: cs.

Thomas, George. Response: such neat resolutions are not possible in India. *Indian Journal of Medical Ethics* 2007 January-March; 4(1): 34. NRCBL: 20.5.1; 9.4; 9.3.1. SC: cs.

Tonelli, Mark R. What medical futility means to clinicians. *HEC (Healthcare Ethics Committee) Forum* 2007 March; 19(1): 83-93. NRCBL: 20.5.1; 8.1.

Trotter, Griffin. Editorial introduction: futility in the 21st century. *HEC (Healthcare Ethics Committee) Forum* 2007 March; 19(1): 1-12. NRCBL: 20.5.1; 2.2. SC: an.

van Delden, Johannes J.M. Terminal sedation: source of a restless ethical debate. *Journal of Medical Ethics* 2007 April; 33(4): 187-188. NRCBL: 20.5.1; 9.7; 20.4.1.

Van Dijk, Yehuda; Sonnenblick, Moshe. Enteral feeding in terminal dementia — a dilemma without a consensual solution. *Israel Medical Association Journal* 2006 July; 8(7): 503-504. NRCBL: 20.5.1; 17.1; 9.5.2.

Vincent, Jean-Louis; Brun-Buisson, Christian; Niederman, Michael; Haenni, Christian; Harbarth, Stephan; Sprumont, Dominique; Valencia, Mauricio; Torres, Antoni. Ethics roundtable debate: a patient dies from an ICU-acquired infection related to methicillin-resistant Staphylococcus aureus — how do you defend your case and your team? *Critical Care* 2005 February; 9(1): 5-9. NRCBL: 20.5.1; 9.5.1; 21.1. SC: cs.

Walters, LeRoy. Der Widerstand Paul Braunes und des Bonhoefferkreises gegen die "Euthanasie" - Programm der Nationalsozialisten. *In:* Gestrich, Christof; Neugebauer, Johannes, eds. Der Wert menschlichen Lebens: medizinische Ethik bei Karl Bonhoeffer und Dietrich Bonhoeffer. Berlin: Wichern-Verlag, 2006: 98-146. NRCBL: 20.5.1; 2.2; 15.5; 21.4.

Werner, Wolfgang Franz. "Euthanasie" und Widerstand in der Rheinprovinz. *In:* Faust, Anselm, ed. Verfolgung und Widerstand im Rheinland und in Westfalen 1933-1945. Köln: W. Kohlhammer, 1992: 224-233. 7 refs. NRCBL: 20.5.1; 20.5.2; 1.3.5; 2.2; 15.5; 21.4.
    Keywords: *active euthanasia; *historical aspects; *killing; *mentally disabled persons; *National Socialism, adults; attitude of health personnel; children; eugenics; public opinion; religion; Keyword Identifiers: *Germany; *Twentieth Century

White, Douglas B.; Braddock, Clarence H., III; Bereknyei, Sylvia; Curtis, J. Randall. Toward shared decision making at the end of life in intensive care units: opportunities for improvement. *Archives of Internal*

Medicine 2007 March 12; 167(5): 461-467. NRCBL: 20.5.1; 8.1; 8.3.3. SC: em.

White, Douglas B.; Engelberg, Ruth A.; Wenrich, Marjorie D.; Lo, Bernard; Curtis, J. Randall. Prognostication during physician-family discussions about limiting life support in intensive care units. *Critical Care Medicine* 2007 February; 35(2): 442-448. NRCBL: 20.5.1; 8.1; 7.1; 9.4.

Whyte, J. Treatments to enhance recovery from the vegetative and minimally conscious states: ethical issues surrounding efficacy studies. *American Journal of Physical Medicine and Rehabilitation* 2007 February; 86(2): 86-92. NRCBL: 20.5.1; 18.2; 18.3; 18.5.7.

Wu, Eugene B. The ethics of implantable devices. *Journal of Medical Ethics* 2007 September; 33(9): 532-533. NRCBL: 20.5.1; 8.3.4; 9.7.

EUTHANASIA AND ALLOWING TO DIE/ LEGAL ASPECTS

France: doctor convicted of euthanasia but avoids prison. *New York Times* 2007 March 16; p. A6. NRCBL: 20.5.1. SC: po; le. Identifiers: Laurence Tamois.

France: doctor convicted of euthanasia but avoids prison. *New York Times* 2007 March 16; p. A6. NRCBL: 20.5.1. SC: po; le. Identifiers: Laurence Tamois.

Italy: No disciplinary action for doctor in right-to-die case. *New York Times* 2007 February 2; p. A8. NRCBL: 20.5.1; 7.3. SC: po; le. Identifiers: Piergiorgio Welby; Mario Riccio.

Anscombe, G.E.M. Murder and the morality of euthanasia. *In her:* Human Life, Action and Ethics: Essays. Exeter: Imprint Academic, 2005: 261-277. NRCBL: 20.5.1; 1.1; 1.2. SC: le.

Beresford, H. Richard. Legal aspects of termination of treatment decisions. *Neurologic Clinics* 1989 November; 7(4): 775-787. NRCBL: 20.5.1. SC: le.

Blakely, Gillian; Millward, Jennifer. Moral dilemmas associated with the withdrawal of artificial hydration. *British Journal of Nursing* 2007 August 9 - September 12; 16(15): 916-919. NRCBL: 20.5.1. SC: le.

Brock, Dan W.; Truog, Robert D.; Brett, Allan S.; Frader, Joel; Downie, Jocelyn. Withholding and withdrawing life-sustaining treatment. *In:* Baylis, Françoise; Downie, Jocelyn; Freedman, Benjamin; Hoffmaster, Barry; Sherwin, Susan, eds. Health Care Ethics in Canada. Toronto: Harcourt Brace Canada, 1995: 487-525. NRCBL: 20.5.1; 9.4; 8.1; 20.5.4. SC: le. Note: Section includes: Introduction — Forgoing life-sustaining food and water: is it killing? / Dan W. Brock — The problem with futility / Robert D. Truog, Allan S. Brett and Joel Frader —

"Where there is a will, there may be a better way": legislating advance directives / Jocelyn Downie.

Burt, Robert A. Law's effect on the quality of end-of-life care: lessons from the Schiavo case. *Critical Care Medicine* 2006 November; 34(11, Supplement): S348-S354. NRCBL: 20.5.1; 8.3.3; 7.1; 8.1. SC: le.

Calabresi, Steven G. The Terri Schiavo case: in defense of the special law enacted by Congress and President Bush. *Northwestern University Law Review* 2006; 100(1): 151-170. NRCBL: 20.5.1. SC: le.

Cantor, Norman L. On hastening death without violating legal and moral prohibitions. *Specialty Law Digest: Health Care Law* 2007 June; (338): 9-31. NRCBL: 20.5.1. SC: le.

Council of Europe. Parliamentary Assembly. Recommendations 1418(1999) on protection of the human rights and dignity of the terminally ill and the dying. Strasbourg, France: Council of Europe 1999 June 25; 5 p. [Online]. Accessed: http://assembly.coe.int//main.asp?link=http://assembly.coe.int/documents/adoptedtext/ta99/erec1418.htm [2007 April 11]. NRCBL: 20.5.1; 21.1. SC: le.

Council of Europe. Parliamentary Assembly. Committee on Legal Affairs and Human Rights; McNamara, Kevin. Euthanasia: opinion. Strasbourg, France: Council of Europe, 2003 September 23; 10 p. [Online]. Accessed: http://assembly.coe.int/main.asp?Link=/documents/workingdocs/doc04/edoc9923.htm [2007 April 12]. NRCBL: 20.5.1. SC: le.

Cranford, Ronald. Terri Schiavo was not disabled. *Ethics and Intellectual Disability* 2005 Summer; 9(1): 1, 4-5. NRCBL: 20.5.1. SC: le.

Dubler, Nancy Neveloff. The legal aspects of end-of-life decision making. *In:* Pruchno, Rachel A.; Smyer, Michael A., eds. Challenges of an Aging Society: Ethical Dilemmas, Political Issues. Baltimore: Johns Hopkins University Press, 2007: 19-33. NRCBL: 20.5.1. SC: le.

Dyer, Clare. Dying woman seeks backing to hasten death [news]. *BMJ: British Medical Journal* 2007 February 17; 334(7589): 329. NRCBL: 20.5.1; 20.4.1. SC: le. Identifiers: Kelly Taylor; Identifiers: Great Britain (United Kingdom).

Eidelman, Steve; Drake, Steve. Not yet dead. *Ethics and Intellectual Disability* 2005 Summer; 9(1): 1-3. NRCBL: 20.5.1. SC: le. Identifiers: Terri Schiavo.

Fine, Robert L. Tackling medical futility in Texas [letter]. *New England Journal of Medicine* 2007 October 11; 357(15): 1558-1559. NRCBL: 20.5.1. SC: le. Comments: R.D. Truog. Tackling medical futility in Texas. New England Journal of Medicine 2007; 357: 1-3.

Fullbrook, Suzanne. End-of-life issues: common law and the Mental Capacity Act 2005. *British Journal of Nursing*

NRCBL: National Reference Center for Bioethics Literature Classification Scheme   See inside front cover for terms.

153

2007 July 12-25; 16(13): 816-818. NRCBL: 20.5.1; 8.3.3. SC: le. Identifiers: Great Britain (United Kingdom).

Hall, Mark A.; Bobinski, Mary Anne; Orentlicher, David. The right and "duty" to die. *In their:* Bioethics and Public Health Law. New York: Aspen Publishers, 2005: 221-338. NRCBL: 20.5.1; 20.7; 20.5.2; 8.3.3. SC: le; cs.

Halliday, Samantha. Regulating active voluntary euthanasia: what can England and Wales learn from Belgium and the Netherlands? *In:* Garwood-Gowers, Austen; Tingle, John; Wheat, Kay, eds. Contemporary Issues in Healthcare Law and Ethics. Edinburgh; New York: Elsevier Butterworth-Heinemann, 2005: 269-301. NRCBL: 20.5.1; 20.7. SC: le.

Henry, Maureen. Update on end-of-life issues in Utah. *Utah Bar Journal* 2006 January-February; 19: 6-10. NRCBL: 20.5.1; 4.4; 7.1; 9.4. SC: le.

Kakuk, Peter. The slippery slope of the middle ground: reconsidering euthanasia in Britain. *HEC (Healthcare Ethics Committee) Forum* 2007 June; 19(2):145-159. NRCBL: 20.5.1; 1.1. SC: cs; le.

Keown, John. Defending the Council of Europe's opposition to euthanasia. *In:* McLean, Sheila A.M., ed. First Do No Harm: Law, Ethics, and Healthcare. Aldershot, England; Burlington, VT: Ashgate, 2006: 479-494. NRCBL: 20.5.1; 21.1. SC: an; le. Identifiers: Marty Report.

King, Patricia A.; Areen, Judith; Gostin, Lawrence O. Death and dying. *In their:* Law, Medicine and Ethics. New York: Foundation Press, 2006: 370-508. NRCBL: 20.5.1; 20.2.1; 4.4; 20.7; 8.3.3. SC: le.

Kirk, Kenneth C. The Alaska Health Care Decisions Act, analyzed. *Alaska Law Review* 2005 December; 22(2): 213-253. NRCBL: 20.5.1; 20.5.4. SC: cs; le.

Kollas, Chad D.; Boyer-Kollas, Beth. Closing the Schiavo case: an analysis of legal reasoning. *Journal of Palliative Medicine* 2006 October; 9(5): 1145-1163. NRCBL: 20.5.1; 20.5.4. SC: le; rv.

Lo, Bernard. Decisions about life-sustaining interventions. *In his:* Resolving Ethical Dilemmas: A Guide for Clinicians. 3rd edition. Philadelphia, PA: Lippincott Williams and Wilkins, 2005: 103-152. NRCBL: 20.5.1; 20.3.2; 9.6. SC: le.

Long, Robert Emmet. Nancy Cruzan and the "right to die". *In his:* Suicide. New York: H.W. Wilson, 1995: 54-75. NRCBL: 20.5.1; 8.3.3; 8.3.4; 2.2. SC: le.

Manaouil, C.; Gignon, M.; Decourcelle, M.; Jardé, O. Law, ethics and medicine: a new legal frame for end of life in France. *Journal of Medical Ethics* 2007 May; 33(5): 278. NRCBL: 20.5.1. SC: le.

Mason, J.K.; Laurie, G.T. Euthanasia. *In their:* Mason and McCall Smith's Law and Medical Ethics. Seventh ed.

Oxford; New York: Oxford University Press, 2005: 598-647. NRCBL: 20.5.1; 8.1; 8.3.4; 20.7. SC: le. Identifiers: Great Britain (United Kingdom); Netherlands.

Mason, J.K.; Laurie, G.T. Medical futility. *In their:* Mason and McCall Smith's Law and Medical Ethics. Seventh ed. Oxford; New York: Oxford University Press, 2005: 539-597. NRCBL: 20.5.1; 4.4; 9.4; 20.3.2; 20.5.2. SC: le.

McLean, Sheila A.M. From Bland to Burke: the law and politics of assisted nutrition and hydration. *In:* McLean, Sheila A.M., ed. First Do No Harm: Law, Ethics, and Healthcare. Aldershot, England; Burlington, VT: Ashgate, 2006: 431-446. NRCBL: 20.5.1; 4.4. SC: le. Identifiers: Airedale NHS Trust v. Bland; R (on the application of Burke) v. General Medical Council.

Miller, Geoffrey. Ten days in Texas. *Hastings Center Report* 2007 July-August; 37(4): inside back cover. NRCBL: 20.5.1; 9.4; 20.5.4. SC: le. Identifiers: futility.

Naudts, Kris; Ducatelle, Caroline; Kovacs, Jozsef; Laurens, Kristin; van den Eynde, Frederique; van Heeringen, Cornelis. Euthanasia: the role of the psychiatrist. *British Journal of Psychiatry* 2006 May; 188: 405-409. NRCBL: 20.5.1; 4.3; 17.1; 21.1. SC: le.

Noah, Barbara A. The role of religion in the Schiavo controversy. *Houston Journal of Health Law and Policy* 2006; 6(2): 319-346. NRCBL: 20.5.1; 1.2. SC: le; cs.

Oregon. Department of Human Services. Public Health Division. Death with Dignity Act history. Salem: Department of Human Services, n.d.; 1 p. [Online]. Accessed: http://egov.oregon.gov/DHS/ph/pas/docs/History.pdf [2007 April 12]. NRCBL: 20.5.1. SC: le.

Paris, John. A life too burdensome. *Tablet* 2007 January 6; 261(8672): 10-11. NRCBL: 20.5.1; 8.1; 1.2. SC: le. Identifiers: Italy.

Paris, J.J.; Billinngs, J.A.; Cummings, B.; Moreland, M.P. Howe v. MGH and Hudson v. Texas Children's Hospital: two approaches to resolving family-physician disputes in end-of-life care. *Journal of Perinatology* 2006 December; 26(12): 726-729. NRCBL: 20.5.1; 9.4; 8.1; 20.5.2;. SC: le; cs.

Perry, Joshua E. Biopolitics at the bedside: proxy wars and feeding tubes. *Journal of Legal Medicine* 2007 April-June; 28(2): 171-192. NRCBL: 20.5.1; 8.3.2; 2.1; 2.4. SC: le.

Petsko, Gregory A. A matter of life and death. *Genome Biology* 2005; 6(5): 109.1-109.3. NRCBL: 20.5.1; 21.1. SC: le.

Seale, Clive. National survey of end-of-life decisions made by UK medical practitioners. *Palliative Medicine* 2006 January; 20(1): 3-10. NRCBL: 20.5.1; 20.5.2; 21.1. SC: le.

Sheperd, Lois. Terri Shiavo: unsettling the settled. *Loyola University Chicago Law Journal* 2006 Winter; 37(2): 297-341. NRCBL: 20.5.1; 8.3.3; 8.3.4; 20.5.4. SC: le.

Skene, Loane. Life-prolonging treatment and patients' legal rights. *In:* McLean, Sheila A.M., ed. First Do No Harm: Law, Ethics, and Healthcare. Aldershot, England; Burlington, VT: Ashgate, 2006: 421-429. NRCBL: 20.5.1; 8.3.4; 9.2. SC: le.

Smith, George P. A compassionate death. *In his:* The Christian Religion and Biotechnology: A Search for Principled Decision-making. Dordrecht: Springer, 2005: 189-230. NRCBL: 20.5.1; 1.1; 9.4; 20.6; 20.7. SC: le.

Spurgeon, Brad. Doctors sign petition calling for euthanasia to be decriminalised [news]. *BMJ: British Medical Journal* 2007 March 17; 334(7593): 555. NRCBL: 20.5.1; 4.1.2. SC: le. Identifiers: France.

Starks, Helene; Back, Anthony L.; Pearlman, Roberta A.; Koenig, Barbara A.; Hsu, Clarissa; Gordon,Judith R.; Bharucha, Ashok J. Family member involvement in hastened death. *Death Studies* 2007 February; 31(2): 105-130. NRCBL: 20.5.1; 9.4; 20.3.3; 1.1. SC: cs; le.

Tucker, Kathryn L. Privacy and dignity at the end of life: protecting the right of Montanans to choose aid in dying. *Montana Law Review* 2007 Summer; 68(2): 317-333. NRCBL: 20.5.1; 4.4; 8.4; 20.7. SC: le.

Weyers, Heleen. Legal recognition of the right to die. *In:* Garwood-Gowers, Austen; Tingle, John; Wheat, Kay, eds. Contemporary Issues in Healthcare Law and Ethics. Edinburgh; New York: Elsevier Butterworth-Heinemann, 2005: 252-267. NRCBL: 20.5.1; 21.1. SC: le. Identifiers: Netherlands; Belgium; Denmark.

Wilcockson, Michael. Euthanasia and doctors' ethics. *In his:* Issues of Life and Death. London: Hodder and Stoughton Educational, 1999: 57-71. NRCBL: 20.5.1; 20.5.2; 20.7; 8.1; 1.1; 1.2. SC: cs; le.

Yacoub, Ahmed Abdel Aziz. Euthanasia. *In his:* The Fiqh of Medicine: Responses in Islamic Jurisprudence to Developments in Medical Science. London, UK: Ta-Ha Publishers Ltd., 2001: 159-201. NRCBL: 20.5.1; 1.2; 20.2.1. SC: le.

EUTHANASIA AND ALLOWING TO DIE/ MINORS

April, Carolyn; Parker, Michael. End of life decision-making in neonatal care. *Journal of Medical Ethics* 2007 March; 33(3): 126-127. NRCBL: 20.5.2.

Beals, Daniel A. Permissibility to stop off-label use of expensive drug treatment for child? *Ethics and Medicine: An International Journal of Bioethics* 2007 Fall; 23(3): 141-144. NRCBL: 20.5.2; 9.6; 1.2. SC: cs.

Brazier, Margaret; Archard, David. Letting babies die [editorial]. *Journal of Medical Ethics* 2007 March; 33(3): 125-126. NRCBL: 20.5.2.

Browning, David M.; Meyer, Elaine C.; Brodsky, Dara; Truog, Robert D. Reflections on love, fear, and specializing in the impossible. *Journal of Clinical Ethics* 2007 Winter; 18(4): 373-376. NRCBL: 20.5.2; 20.4.2; 20.3.3; 4.1.2. Comments: Annie Janvier. How much emotion is enough? Journal of Clinical Ethics 2007 Winter; 18(4): 362-365.

Brunnquell, Donald. Case report: parental request for life-prolonging interventions. *HEC (Healthcare Ethics Committee) Forum* 2007 December; 19(4): 375-376. NRCBL: 20.5.2; 8.3.2. SC: cs.

Burns, Jeffrey. Ask the ethicist. Does anyone actually invoke their hospital futility policy? *Advances in Neonatal Care: Official Journal of the National Association of Neonatal Nurses* 2006 April; 6(2): 66-67. NRCBL: 20.5.2; 8.3.2; 9.4. SC: cs.

Canadian Paediatric Society. Use of anencephalic newborns as organ donors: Position Statement B 2005-01. *Paediatrics and Child Health* 2005 July-August; 10(6): 335-337 (English); 339-341 (French) [Online]. Accessed: http://www.cps.ca/english/statements/B/B05-01.pdf [2007 January 15]. NRCBL: 20.5.2; 19.5; 9.5.7. Note: Available in French at http://www.cps.ca/Francais/enonces/B/B05-01.pdf.

Carnevale, Franco A. The birth of tragedy in pediatrics: a phronetic conception of bioethics. *Nursing Ethics* 2007 September; 14(5): 571-582. NRCBL: 20.5.2; 1.1; 2.1; 8.3.2; 9.6.
Abstract: Accepted standards of parental decisional autonomy and child best interests do not address adequately the complex moral problems involved in the care of critically ill children. A growing body of moral discourse is calling for the recognition of ;'tragedy' in selected human problems. A tragic dilemma is an irresolvable dilemma with forced terrible alternatives, where even the virtuous agent inescapably emerges with ;'dirty hands'. The shift in moral framework described here recognizes that the form of conduct called for by tragic dilemmas is the practice of phronesis. The phronetic agent has acquired a capacity to discern good agency in tragic circumstances. This discernment is practiced through the artful creation of moral narratives: stories that convey that which is morally meaningful in a particular situation; that is, stories that are ;'meaning making'. The phronetic agent addresses tragic dilemmas involving children as a narrator of contextualized temporal embodied human (counter) stories.

Costeloe, Kate. Euthanasia in neonates. *BMJ: British Medial Journal* 2007 May 5; 334(7600): 912-913. NRCBL: 20.5.2.

Cremer, R.; Binoche, A.; Noizet, O.; Fourier, C.; Leteurtre, S.; Moutel, G.; Leclerc, F. Are the GFRUP's

NRCBL: National Reference Center for Bioethics Literature Classification Scheme   See inside front cover for terms.

155

recommendations for withholding or withdrawing treatments in critically ill children applicable? Results of a two-year survey. *Journal of Medical Ethics* 2007 March; 33(3): 128-133. NRCBL: 20.5.2. SC: em. Identifiers: France; Groupe Francophone de Réanimation ed Urgence Pédiatriques.

Abstract: OBJECTIVE: To evaluate feasibility of the guidelines of the Groupe Francophone de Réanimation et Urgence Pédiatriques (French-speaking group of paediatric intensive and emergency care; GFRUP) for limitation of treatments in the paediatric intensive care unit (PICU). DESIGN: A 2-year prospective survey. SETTING: A 12-bed PICU at the Hôpital Jeanne de Flandre, Lille, France. PATIENTS: Were included when limitation of treatments was expected. RESULTS: Of 967 children admitted, 55 were included with a 2-day median delay. They were younger than others (24 v 60 months), had a higher paediatric risk of mortality (PRISM) score (14 v 4), and a higher paediatric overall performance category (POPC) score at admission (2 v 1); all p.002. 34 (50% of total deaths) children died. A limitation decision was made without meeting for 7 children who died: 6 received do-not-resuscitate orders (DNROs) and 1 received withholding decision. Decision-making meetings were organised for 31 children, and the following decisions were made: 12 DNROs (6 deaths and 6 survivals), 4 withholding (1 death and 3 survivals), with 14 withdrawing (14 deaths) and 1 continuing treatment (survival). After limitation, 21 (31% of total deaths) children died and 10 survived (POPC score 4). 13 procedures were interrupted because of death and 11 because of clinical improvement (POPC score 4). Parents' opinions were obtained after 4 family conferences (for a total of 110 min), 3 days after inclusion. The first meeting was planned for 6 days after inclusion and held on the 7th day after inclusion; 80% of parents were immediately informed of the decision, which was implemented after half a day. CONCLUSIONS: GFRUPs procedure was applicable in most cases. The main difficulties were anticipating the correct date for the meeting and involving nurses in the procedure. Children for whom the procedure was interrupted because of clinical improvement and who survived in poor condition without a formal decision pointed out the need for medical criteria for questioning, which should lead to a formal decision-making process.

Euronic Study Group; Cuttini, Marina; Casotto, Veronica; Orzalesi, Marcello. Ethical issues in neonatal intensive care and physicians' practices: a European perspective. *Acta Paediatrica Supplement* 2006 July; 95(452): 42-46. NRCBL: 20.5.2; 9.4; 21.1.

Faulkner, Janet. Conjoined twins: the ethics of separation. *Midwives: the Official Journal of the Royal College of Midwives* 2006 March; 9(3): 86-87. NRCBL: 20.5.2; 1.1.

FIGO Committee for the Ethical Aspects of Human Reproduction and Women's Health. Ethical guidelines on resuscitation of newborns. *International Journal of Gynecology and Obstetrics* 2006 August; 94(2): 169-171. NRCBL: 20.5.2; 14.1.

Fortune, Peter-Marc. Euthanasia in neonates: are we asking the right questions? [letter]. *BMJ: British Medical Journal* 2007 May 26; 334(7603): 1072. NRCBL: 20.5.2. Comments: Kate Costeloe. Euthanasia in neonates. BMJ: British Medical Journal 2007 May 5; 334(7600): 912-913.

Frader, Joel E. Discontinuing artificial fluids and nutrition: discussions with children's families. *Hastings Center Report* 2007 January-February; 37(1): inside back cover. NRCBL: 20.5.2; 20.5.1.

French LATASAMU Group; Ferrand, Edouard; Marty, Jean. Prehospital withholding and withdrawal of life-sustaining treatments. The French LATASAMU survey. *Intensive Care Medicine* 2006 October; 32(10): 1498-1505. NRCBL: 20.5.2; 7.1; 20.4.1; 8.3.1. SC: em.

Friedman, Sandra L. Parent resuscitation preferences for young people with severe developmental disabilities. *Journal of the American Medical Directors Association* 2006 February; 7(2): 67-72. NRCBL: 20.5.2; 20.5.4; 8.3.2; 9.5.3. SC: em.

Howe, Edmund G. When family members disagree. *Journal of Clinical Ethics* 2007 Winter; 18(4): 331-339. NRCBL: 20.5.2; 20.3.3; 20.4.2; 20.5.4; 9.6. SC: cs.

Isaacs, David; Kilham, Henry; Gordon, Adrienne; Jeffery, Heather; Tarnow-Mordi, William; Woolnough, Janet; Hamblin, Julie; Tobin, Bernadette. Withdrawal of neonatal mechanical ventilation against the parents' wishes. *Journal of Paediatrics and Child Health* 2006 May; 42(5): 311-315. NRCBL: 20.5.2; 8.3.2; 9.4; 8.1; 1.1.

Jaing, Tang-Her; Tsay, Pei-Kwei; Fang, En-Chen; Yang, Shu-Ho; Chen, Shih-Hsiang; Yang, Chao-Ping; Hung, lou-Jih. "Do-not-resuscitate" orders in patients with cancer at a children's hospital in Taiwan. *Journal of Medical Ethics* 2007 April; 33(4): 194-196. NRCBL: 20.5.2; 9.5.7. SC: em.

Abstract: OBJECTIVES: To quantify the use of do-not-resuscitate (DNR) orders in a tertiary-care children's hospital and to characterise the circumstances in which such orders are written. DESIGN: Retrospective study conducted in a 500-bed children's hospital in Taiwan. PATIENTS: The course of 101 patients who died between January 2002 and December 2005 was reviewed. The following data were collected: age at death, gender, disease and its status, place of death and survival. There were 59 males and 42 females with a median age of 103 months (range 1-263 months). 50 children had leukaemias, and 51 had malignancies other than leukaemia. The t test and the chi(2) test were applied as appropriate. RESULTS: The study found that 44% of patient deaths occurred in the paediatric oncology ward; 29% of patient deaths occurred in the intensive care unit; and 28% of patients died in their home or at another hospital. Other findings included the following: 46 of 101 (46%) patients died after attempted cardiopulmonary resuscitation and 55 (54%) died with a DNR order in effect. The mean age at death was 9.8 years in both groups with or without DNR orders. CONCLUSIONS: From the study of patient deaths in this tertiary-care children's hospital, it was concluded that an explicit DNR order is now the rule rather than the exception, with more DNR orders being written for patients who have been ill longer, who have solid tumours, who are not in remission and who are in the ward.

Janvier, Annie. How much emotion is enough? *Journal of Clinical Ethics* 2007 Winter; 18(4): 362-365. NRCBL: 20.5.2; 8.3.2; 20.4.2. SC: cs.

Jonas, Monique. The Baby MB case: medical decision making in the context of uncertain infant suffering. *Journal of Medical Ethics* 2007 September; 33(9): 541-544. NRCBL: 20.5.2; 8.3.2; 4.4. SC: an. Identifiers: Great Britain (United Kingdom).

Kaposy, Chris. Can infants have interests in continued life? *Theoretical Medicine and Bioethics* 2007; 28(4): 301-330. NRCBL: 20.5.2; 1.1; 4.4; 12.3; 22.1. SC: an.
Abstract: The philosophers Peter Singer and Jeff McMahan hold variations of the view that infant interests in continued life are suspect because infants lack the cognitive complexity to anticipate the future. Since infants cannot see themselves as having a future, Singer argues that the future cannot have value for them, and McMahan argues that the future can only have minimal value for an infant. This paper critically analyzes these arguments and defends the view that infants can have interests in continuing to live. Even though infants themselves lack a strong psychological connection to the future, others who are involved in an infant's life can anticipate, on an infant's behalf, the kind of future that awaits the infant, and on the basis of this insight judge that continuing to live would be in the infant's interests. After defending this position, I argue that this position on the interests of infants in continued life does not commit one to opposing abortion, and it does not commit one to the view that our ethical obligations to protect the lives of sentient animals are the same as our ethical obligations to protect infant lives.

Kipnis, Kenneth. Harm and uncertainty in newborn intensive care. *Theoretical Medicine and Bioethics* 2007; 28(5): 393-412. NRCBL: 20.5.2; 8.3.2; 8.3.4.
Abstract: There is a broadly held view that neonatologists are ethically obligated to act to override parental nontreatment decisions for imperiled premature newborns when there is a reasonable chance of a good outcome. It is argued here that three types of uncertainty undercut any such general obligation: (1) the vagueness of the boundary at which an infant's deficits become so intolerable that death could be reasonably preferred; (2) the uncertainty about whether aggressive treatment will result in the survival of a reasonably healthy child or, alternatively, the survival of a child with intolerable deficits; and (3) the inability to determine an acceptable ratio between the likelihoods of those two outcomes. It is argued that the broadly held view accords insufficient weight to the fact that newborn intensive care increases the likelihood of harm to the child by effecting survival with intolerable deficits. Though treatment may offer a reasonable chance of a good outcome, it is argued that there are situations in which neonatologists should nonetheless defer to parental nontreatment decisions.

Kon, Alexander A. Neonatal euthanasia is unsupportable: the Groningen Protocol should be abandoned. *Theoretical Medicine and Bioethics* 2007; 28(5): 453-463. NRCBL: 20.5.2; 4.4. SC: an. Identifiers: Netherlands.
Abstract: The growing support for voluntary active euthanasia (VAE) is evident in the recently approved Dutch Law on Termination of Life on Request. Indeed, the debate over legalized VAE has increased in European countries, the United States, and many other nations over the last several years. The proponents of VAE argue that when a patient judges that the burdens of living outweigh the benefits, euthanasia can be justified. If some adults suffer to such an extent that VAE is justified, then one may conclude that some children suffer to this extent as well. In an attempt to alleviate the suffering of extremely ill neonates, the University Medical Center Groningen developed a protocol for neonatal euthanasia. In this article, I first present the ethical justifications for VAE and discuss how these arguments relate to euthanizing ill neonates. I then argue that, even if one accepts the justification for VAE in adults, neonatal euthanasia cannot be supported, primarily because physicians and parents can never accurately assess the suffering of children. I argue that without the testament of the patient herself as to the nature and magnitude of her suffering, physicians can never accurately weigh the benefits and burdens of a child's life, and therefore any such system would condemn to death some children whose suffering is not unbearable. I conclude that because the primary duty of physicians is to never harm their patients, neonatal euthanasia cannot be supported.

Krill, Edward J. What parents face with their child's life-threatening illness: comment on "How much emotion is enough?" and "Real life informs consent". *Journal of Clinical Ethics* 2007 Winter; 18(4): 369-372. NRCBL: 20.5.2; 20.4.2; 20.3.3. Comments: Annie Janvier. How much emotion is enough? Journal of Clinical Ethics 2007 Winter; 18(4): 362-365; Felicia Cohn. Real life informs consent. Journal of Clinical Ethics 2007 Winter; 18(4): 366-368.

Krug, E.F., III. Law and ethics at the border of viability. *Journal of Perinatology* 2006 June; 26(6): 321-324. NRCBL: 20.5.2; 9.4; 8.3.2; 9.5.7.

Larcher, Vic; Hird, Michael F. Withholding and withdrawing neonatal intensive care. *Current Paediatrics* 2002 December; 12(6): 470-475. NRCBL: 20.5.2; 8.1. SC: le. Identifiers: Great Britain (United Kingdom).

Mannaerts, Debbie; Mortier, Freddy. Minors and euthanasia. *In:* Freeman, Michael, ed. Children's Health and Children's Rights. Boston: Martinus Nijhoff Publishers, 2006: 255-277. NRCBL: 20.5.2; 20.5.1; 8.3.2. SC: le. Identifiers: Belgium; Netherlands.

NRCBL: National Reference Center for Bioethics Literature Classification Scheme   See inside front cover for terms.

157

McGraw, Melanie P.; Perlman, Jeffrey; Chervenak, Frank A.; McCullough, Laurence B. Clinical concepts of futility and ethically justified limits on neonatal care: a case presentation of an infant with Apgar scores of 0 at 1, 5, and 10 minutes. *American Journal of Perinatology* 2006 April; 23(3): 159-162. NRCBL: 20.5.2; 9.4; 7.1. SC: cs.

Mercurio, M.R. Parental authority, patient's best interest and refusal of resuscitation at borderline gestational age. *Journal of Perinatology* 2006 August; 26(8): 452-457. NRCBL: 20.5.2; 9.5.8; 8.3.2; 8.3.4.

New York. Supreme Court. Queens County. Matter of Long Island Jewish Medical Center, Petitioner. Baby Doe, a Minor Patient, Respondent. [Date of Decision: 28 February 1996]. West's New York Supplement, 2d Series, 1996; 641: 989-992. NRCBL: 20.5.2; 20.2. SC: le.
  Abstract: Court Decision: 641 New York Supplement, 2d Series 989; 28 Feb 1996 (date of decision). The Supreme Court of New York authorized a hospital to withdraw artificial respiratory support from a brain dead infant. After several days of care and testing, two hospital physicians certified the five-month-old baby on artificial respiratory support was brain dead. A medical expert retained by the infant's parents, an anesthesiologist father and an attorney mother, concurred with that medical assessment. The court held that the conclusion of all three medical experts was consistent with the definition of death set forth under state law. The court also held that the hospital adequately notified the parents of their baby's status, and that the hospital fulfilled its obligations as set forth in the hospital's administrative manual. However, because the hospital did not have a written policy regarding "reasonable accommodation" of the individual's religious or moral objection, the court reviewed the facts to determine this issue. Finding that the hospital did, in fact, reasonably accommodate parents' religious and moral beliefs, the court granted an order authorizing the hospital's withdrawal of the baby's artificial respiratory support.

Paris, J.J.; Graham, N.; Schreiber, M.D.; Goodwin, M. Approaches to end-of-life decision-making in the NICU: insights from Dostoevsky's The Grand Inquisitor. *Journal of Perinatology* 2006 July; 26(7): 389-391. NRCBL: 20.5.2; 9.4; 7.1.

Poland, Susan. Court clarifies role of guardians in foregoing life-support. *Ethics and Intellectual Disability* 2005 Winter; 8(2): 3, 7. NRCBL: 20.5.2; 9.5.3. SC: le.

Porta, Nicolas; Frader, Joel. Withholding hydration and nutrition in newborns. *Theoretical Medicine and Bioethics* 2007; 28(5): 443-451. NRCBL: 20.5.2.
  Abstract: In the twenty-first century, decisions to withhold or withdraw life-supporting measures commonly precede death in the neonatal intensive care unit without major ethical controversy. However, caregivers often feel much greater turmoil with regard to stopping medical hydration and nutrition than they do when considering discontinuation of mechanical ventilation or circulatory support. Nevertheless, forgoing medical fluids and food

represents a morally acceptable option as part of a carefully developed palliative care plan considering the infant's prognosis and the burdens of continued treatment. Decisions to stop any form of life support should focus on the clinical circumstances, not the means used to sustain life.

Reitman, James. Shall we prolong life in order to give a patient time to decide about faith? *Today's Christian Doctor* 2007 Spring; 38(1): 28-29. NRCBL: 20.5.2; 1.2; 8.1; 9.6.

Richards, N. Life or death decisions in the NICU. *Journal of Perinatology* 2006 April; 26(4): 248-251. NRCBL: 20.5.2; 8.3.2; 4.4; 9.4. Identifiers: neonatal intensive care unit.

Schexnayder, Stephen M.; Hester, D. Micah. A new perspective on community consultation in pediatric resuscitation research. *Critical Care Medicine* 2006 October; 34(10): 2684-2685. NRCBL: 20.5.2; 8.3.2.

Sharma, Renuka M. The ethics of birth and death: gender infanticide in India. *Journal of Bioethical Inquiry* 2007; 4(3): 181-192. NRCBL: 20.5.2; 10; 14.3; 1.1.
  Abstract: This paper discusses the persistent devaluation of the girl child in India and the link between the entrenched perception of female valuelessness and the actual practice of infanticide of girl babies or foetuses. It seeks to place female infanticide, or 'gendercide,' within the context of Western-derived conceptions of ethics, justice and rights. To date, current ethical theories and internationally purveyed moral frameworks, as well as legal and political declarations, have fallen short of an adequate moral appraisal of infanticide. This paper seeks to rethink the issue.

Singer, Peter. Treating (or not) the tiniest babies. *Free Inquiry* 2007 June-July; 27(4): 20-21. NRCBL: 20.5.2; 9.5.7; 8.3.2; 9.4.

Steger, Florian. Neuropathological research at the "Deutsche Forschungsanstalt fuer Psychiatrie" (German Institute for Psychiatric Research) in Munich (Kaiser-Wilhelm-Institute). Scientific utilization of children's organs from the "Kinderfachabteilungen" (Children's Special Departments) at Bavarian State Hospitals. *Journal of the History of the Neurosciences* 2006 September; 15(3): 173-185. NRCBL: 20.5.2; 9.5.3; 9.5.7; 21.4; 1.3.5; 2.2.

Strong, Carson. Case commentary: parental request for life-prolonging interventions. *HEC (Healthcare Ethics Committee) Forum* 2007 December; 19(4): 377-380. NRCBL: 20.5.2; 8.3.2. SC: cs.

Tai, Michael Cheng-tek; Hill, Donald. A Confucian perspective on bioethical principles in ethics consultation. *Clinical Ethics* 2007 December; 2(4): 201-207. NRCBL: 20.5.2; 9.6; 1.1; 21.7. SC: cs.
  Abstract: With the rapid development of biotechnology, the physician is now more able to keep a patient's life going indefinitely on a life support system. The question of

whether we should switch off the machine often arises when, according to the medical prognosis, there is no hope of recovery, or in a no-win situation where you are 'damned if you do and damned if you don't'. In a case which seems without hope, the dilemma of whether to prolong a life or let it go disturbs many people, including health professionals as well as the family of the patient. In this painful situation, an ethics consultant who has received intensive training can help the concerned parties to arrive at what may be the best decision. How do Asians, especially those living in countries influenced by Confucian teachings, reach their answers? Three aspects are usually considered: (1) motivation and situation; (2) reasonableness and propriety; and (3) lawfulness and legality. More specifically, three questions are deliberated, as follows. (1) Where an action has already been taken, what motivated it and in what situation? Or, where a decision has still to be made, what should motivate it, and what are the relevant features of the situation? (2) Was the attempted resolution of the dilemma, or is its prospective resolution, reasonable and in accordance with traditional principles of ethical behaviour? (3) Was the action taken lawful, or would the intended action be lawful? This approach to finding an answer has been practised for centuries in Confucian society. But what is legal may not always be reasonable, what is reasonable may not always be compassionate, and what is compassionate may not always be either legal or reasonable. Principles to guide decision-making are therefore called for. This article, written by Michael Cheng-tek Tai in collaboration with Donald Hill, discusses the Confucian method of solving a problem and examines its principal features and how they are applied in ethics consultations. The article is followed by a series of questions and answers and a commentary by Donald Hill.

Templeton, Sarah-Kate. Doctors: let us kill disabled babies. *TimesOnline (London)* 2006 November 5: 2 p. [Online]. Accessed: http://www.timesonline.co.uk/tol/news/uk/article625477.ece [2007 December 12]. NRCBL: 20.5.2; 7.1; 9.4; 8.3.2.

Templeton, Sarah-Kate; Swinford, Steven. Haunted mother who backs mercy killing. *TimesOnline (London)* 2006 November 5: 2 p. [Online]. Accessed: http://www.timesonline.co.uk/tol/news/uk/article625550.ece [2007 December 12]. NRCBL: 20.5.2; 8.1; 9.4; 8.3.2.

Thomas, Florian P.; Beres, Alana; Shevell, Michael I. "A cold wind coming": Heinrich Gross and child euthanasia in Vienna. *Journal of Child Neurology* 2006 April; 21(4): 342-348. NRCBL: 20.5.2; 18.5.2; 1.3.9; 15.5; 2.2.

Truog, Robert D. Tackling medical futility in Texas. *New England Journal of Medicine* 2007 July 5; 357(1): 1-3. NRCBL: 20.5.2; 9.4; 8.1. SC: cs. Identifiers: Emilio Gonzales.

Veatch, Robert M. Court authorizes withdrawing of ventilator and nutrition. *Ethics and Intellectual Disability* 2006 Spring; 9(2): 1, 3-4. NRCBL: 20.5.2; 9.5.3. SC: le.

Vrakking, Astrid M.; van der Heide, Agnes; Onwuteaka-Philipsen, Bregje D.; van der Maas, Paul J.; van der Wal, Gerrit. Regulating physician-assisted dying for minors in the Netherlands: views of paediatricians and other physicians. *Acta Paediatrica* 2007 January; 96(1): 117-121. NRCBL: 20.5.2; 7.1; 20.3.2. SC: em.

Zamperetti, N.; Proietti, R. End of life in the ICU: laws, rules and practices: the situation in Italy. *Intensive Care Medicine* 2006 October; 32(10): 1620-1622. NRCBL: 20.5.2; 1.3.5.

EUTHANASIA AND ALLOWING TO DIE/ PHILOSOPHICAL ASPECTS

Anscombe, G.E.M. Murder and the morality of euthanasia. *In her:* Human Life, Action and Ethics: Essays. Exeter: Imprint Academic, 2005: 261-277. NRCBL: 20.5.1; 1.1; 1.2. SC: le.

Asscher, Joachim. Killing and letting die: the similarity criterion. *Journal of Applied Philosophy* 2007 August; 24(3): 271-282. NRCBL: 20.5.1; 1.1. SC: an.
Abstract: Applied ethics engages with concrete moral issues. This engagement involves the application of philosophical tools. When the philosophical tools used in applied ethics are problematic, conclusions about applied problems can become skewed. In this paper, I focus on problems with the idea that comparison cases must be exactly alike, except for the moral issue at hand. I argue that this idea has skewed the debate regarding the moral distinction between killing and letting die. I begin with problems that can arise from attempts to produce comparison cases that are exactly alike, except for the moral issue at hand. I then argue that attempts to produce such examples are doomed to failure. Finally, I argue that abandoning concerns about similarity advances the debate regarding the moral distinction between killing and letting die.

Ball, Susan C. Nurse-patient advocacy and the right to die. *Journal of Psychosocial Nursing and Mental Health Services* 2006 December; 44(12): 36-42. NRCBL: 20.5.1; 4.1.3; 8.1.

Bendiane, Marc-Karim; Galinier, A.; Favre, R.; Ribiere, C.; Lapiana, J.-M.; Obadia, Y.; Peretti-Watel, Patrick. French district nurses' opinions towards euthanasia, involvement in end-of-life care and nurse-patient relationship: a national phone survey. *Journal of Medical Ethics* 2007 December; 33(12): 708-711. NRCBL: 20.5.1; 4.1.3; 8.1. SC: em.
Abstract: OBJECTIVES: To assess French district nurses' opinions towards euthanasia and to study factors associated with these opinions, with emphasis on attitudes towards terminal patients. DESIGN AND SETTING: An anonymous telephone survey carried out in 2005 among a national random sample of French district nurses. PARTICIPANTS: District nurses currently delivering home care who have at least 1 year of professional experience. Of 803 district nurses contacted, 602 agreed

NRCBL: National Reference Center for Bioethics Literature Classification Scheme   See inside front cover for terms.

159

to participate (response rate 75%). MAIN OUTCOME MEASURES: Opinion towards the legalisation of euthanasia (on a five-point Likert scale from "strongly agree" to "strongly disagree"), attitudes towards terminal patients (discussing end-of-life issues with them, considering they should be told their prognosis, valuing the role of advance directives and surrogates). RESULTS: Overall, 65% of the 602 nurses favoured legalising euthanasia. Regarding associated factors, this proportion was higher among those who discuss end-of-life issues with terminal patients (70%), who consider competent patients should always be told their prognosis (81%) and who value the role of advance directives and surrogates in end-of-life decision-making for incompetent patients (68% and 77% respectively). Women and older nurses were less likely to favour legalising euthanasia, as were those who believed in a god who masters their destiny. CONCLUSIONS: French nurses are more in favour of legalising euthanasia than French physicians; these two populations contrast greatly in the factors associated with this support. Further research is needed to investigate how and to what extent such attitudes may affect nursing practice and emotional well-being in the specific context of end-of-life home care.

Benedict, Susan; Caplan, Arthur; Page, Traute Lafrenz. Duty and 'euthanasia': the nurses of Meseritz-Obrawalde. *Nursing Ethics* 2007 November; 14(6): 781-794. NRCBL: 20.5.1; 4.1.3; 21.4; 2.2. Identifiers: Germany.

Bosek, Marcia Sue DeWolf; Stammer, Karen. Ethical commitments during desperate times. *JONA's Healthcare Law, Ethics, and Regulation* 2006 October-December; 8(4): 123-128. NRCBL: 20.5.1; 4.1.3. Identifiers: Hurricane Katrina; Louisiana.

Carmel, Sara; Werner, Perla; Ziedenberg, Hanna. Physicians' and nurses' preferences in using life-sustaining treatments. *Nursing Ethics* 2007 September; 14(5): 665-674. NRCBL: 20.5.1; 4.1.2; 4.1.3. SC: em.
   Abstract: This study examined why intensive care unit (ICU) nurses experience difficulties in respecting the wishes of patients in end-of-life care in Japan. A questionnaire survey was conducted with ICU nurses working in Japanese university hospitals. The content of their narratives was analyzed concerning the reasons why the nurses believed that patients' wishes were not respected. The most commonly stated reason was that patients' wishes were impossible to realize, followed by the fact that decision making was performed by others, regardless of whether the patients' wishes were known, if the death was sudden, and time constraints. Many nurses wanted to respect the wishes of dying patients, but they questioned how patients die in ICUs and were therefore faced with ethical dilemmas. However, at the same time, many of the nurses realized that respecting patients' wishes about end-of-life care in an ICU would be difficult and that being unable to respect these wishes would often be unavoidable. The results thus suggest that there has been insufficient discussion about respecting the wishes of patients undergoing intensive care.

Clarke, C. Ethics and the end of life. *In:* Holland, Stephen, ed. Introducing Nursing Ethics: Themes in Theory and Practice. Salisbury: APS, 2004: 49-64. NRCBL: 20.5.1; 4.1.3. SC: cs.

Cook, Deborah; Rocker, Graeme; Giacomini, Mita; Sinuff, Tasnim; Heyland, Daren. Understanding and changing attitudes toward withdrawal and withholding of life support in the intensive care unit. *Critical Care Medicine* 2006 November; 34(11, Supplement): S317-S323. NRCBL: 20.5.1; 20.3.2; 1.1; 8.1; 9.4.

Dennis, William J. What is death with dignity? *Ethics and Medics* 2007 August; 32(8): 1-2. NRCBL: 20.5.1; 4.4; 1.2; 1.1. Identifiers: Jack Kevorkian.

Ferrell, Betty R. Understanding the moral distress of nurses witnessing medically futile care. *Oncology Nursing Forum* 2006 September 1; 33(5): 922-930. NRCBL: 20.5.1; 4.1.3; 7.1; 1.1.

Garcia, J.L.A. Health versus harm: euthanasia and physicians' duties. *Journal of Medicine and Philosophy* 2007 January-February; 32(1): 7-24. NRCBL: 20.5.1; 1.1; 1.3.1. SC: an.
   Abstract: This essay rebuts Gary Seay's efforts to show that committing euthanasia need not conflict with a physician's professional duties. First, I try to show how his misunderstanding of the correlativity of rights and duties and his discussion of the foundation of moral rights undermine his case. Second, I show aspects of physicians' professional duties that clash with euthanasia, and that attempts to avoid this clash lead to absurdities. For professional duties are best understood as deriving from professional virtues and the commitments and purposes with which the professional as such ought to act, and there is no plausible way in which her death can be seen as advancing the patient's medical welfare. Third, I argue against Prof. Seay's assumption that apparent conflicts among professional duties must be resolved through "balancing" and argue that, while the physician's duty to extend life is continuous with her duty to protect health, any duty to relieve pain is subordinate to these. Finally, I show that what is morally determinative here, as throughout the moral life, is the agent's intention and that Prof. Seay's implicitly preferred consequentialism threatens not only to distort moral thinking but would altogether undermine the medical (and any other) profession and its internal ethics.

Gastmans, Chris. The care perspective in healthcare ethics. *In:* Davis, Anne J.; Tschudin, Verena; de Raeve, Louise, eds. Essentials of Teaching and Learning in Nursing Ethics: Perspectives and Methods. New York: Churchill Livingstone Elsevier, 2006: 135-148. NRCBL: 20.5.1; 4.1.3; 4.1.1. SC: cs.

Gavrin, Jonathan R. Ethical considerations at the end of life in the intensive care unit. *Critical Care Medicine* 2007 February; 35(2, Supplement): S85-S94. NRCBL: 20.5.1; 1.1; 20.4.1; 7.1; 9.4.

SC (Subject Captions):   an=analytical   cs=case studies   em=empirical   le=legal   po=popular   rv=review

Gedge, E.; Giacomini, M.; Cook, D. Withholding and withdrawing life support in critical care settings: ethical issues concerning consent. *Journal of Medical Ethics* 2007 April; 33(4): 215-218. NRCBL: 20.5.1; 8.3.1; 1.1. SC: an.

Abstract: The right to refuse medical intervention is well established, but it remains unclear how best to respect and exercise this right in life support. Contemporary ethical guidelines for critical care give ambiguous advice, largely because they focus on the moral equivalence of withdrawing and withholding care without confronting the very real differences regarding who is aware and informed of intervention options and how patient values are communicated and enacted. In withholding care, doctors typically withhold information about interventions judged too futile to offer. They thus retain greater decision-making burden (and power) and face weaker obligations to secure consent from patients or proxies. In withdrawing care, there is a clearer imperative for the doctor to include patients (or proxies) in decisions, share information and secure consent, even when continued life support is deemed futile. How decisions to withhold and withdraw life support differ ethically in their implications for positive versus negative interpretations of patient autonomy, imperatives for consent, definitions of futility and the subjective evaluation of (and submission to) benefits and burdens of life support in critical care settings are explored. Professional reflection is required to respond to trends favouring a more positive interpretation of patient autonomy in the context of life support decisions in critical care. Both the bioethics and critical care communities should investigate the possibilities and limits of growing pressure for doctors to disclose their reasoning or seek patient consent when decisions to withhold life support are made.

Hov, Reidun; Hedelin, Birgitta; Athlin, Elsy. Being an intensive care nurse related to questions of withholding or withdrawing curative treatment. *Journal of Clinical Nursing* 2007 January; 16(1): 203-211. NRCBL: 20.5.1; 7.1; 4.1.3; 8.1.

Howsepian, A.A. Cerebral neurophysiology, 'Libetian' action, and euthanasia. *Ethics and Medicine: An International Journal of Bioethics* 2007 Summer; 23(2): 103-111. NRCBL: 20.5.1; 17.1; 1.1; 4.4. SC: an. Identifiers: Benjamin Libet.

Huxtable, Richard; Möller, Maaike. 'Setting a principled boundary'? Euthanasia as a response to 'life fatigue'. *Bioethics* 2007 March; 21(3): 117-126. NRCBL: 20.5.1; 4.4; 1.1. SC: an, cs.

Abstract: The Dutch case of Brongersma presents novel challenges to the definition and evaluation of voluntary euthanasia since it involved a doctor assisting the suicide of an individual who was (merely?) 'tired of life'. Legal officials had called on the courts to 'set a principled boundary', excluding such cases from the scope of permissible voluntary euthanasia, but they arguably failed. This failure is explicable, however, since the case seems justifiable by reference to the two major principles in favour of that practice, respect for autonomy and benefi-

cence. Ultimately, it will be argued that those proponents of voluntary euthanasia who are wary of its use in such circumstances may need to draw upon 'practical' objections, in order to erect an otherwise arbitrary perimeter. Furthermore, it will be suggested that the issues raised by the case are not peculiarly Dutch in nature and that, therefore, there are lessons here for other jurisdictions too.

Hyde, Michael J.; McSpiritt, Sarah. Coming to terms with perfection: the case of Terri Schiavo. *Quarterly Journal of Speech* 2007 May; 93 (2): 150-178. NRCBL: 20.5.1; 4.4; 1.1.

Kaczor, Christopher. Philosophy and theology: the authority of Pope John Paul II allocution; is ANH required for PVS patients?; papal allocution and Catholic tradition; human life as intrinsic good;. *National Catholic Bioethics Quarterly* 2007 Autumn; 7(3): 595-605. NRCBL: 20.5.1; 1.1; 1.2; 4.4. Identifiers: artificially administered nutrition and hydration; persistent vegetative state.

Kakuk, Peter. The slippery slope of the middle ground: reconsidering euthanasia in Britain. *HEC (Healthcare Ethics Committee) Forum* 2007 June; 19(2):145-159. NRCBL: 20.5.1; 1.1. SC: cs; le.

Kinoshita, Satomi. Respecting the wishes of patients in intensive care units. *Nursing Ethics* 2007 September; 14(5): 651-664. NRCBL: 20.5.1; 4.1.3. SC: em. Identifiers: Japan.

Abstract: This study examined why intensive care unit (ICU) nurses experience difficulties in respecting the wishes of patients in end-of-life care in Japan. A questionnaire survey was conducted with ICU nurses working in Japanese university hospitals. The content of their narratives was analyzed concerning the reasons why the nurses believed that patients' wishes were not respected. The most commonly stated reason was that patients' wishes were impossible to realize, followed by the fact that decision making was performed by others, regardless of whether the patients' wishes were known, if the death was sudden, and time constraints. Many nurses wanted to respect the wishes of dying patients, but they questioned how patients die in ICUs and were therefore faced with ethical dilemmas. However, at the same time, many of the nurses realized that respecting patients' wishes about end-of-life care in an ICU would be difficult and that being unable to respect these wishes would often be unavoidable. The results thus suggest that there has been insufficient discussion about respecting the wishes of patients undergoing intensive care.

Körner, U.; Bondolfi, A.; Bühler, E.; Macfie, J.; Meguid, M.M.; Messing, B.; Oehmichen, F.; Valentini, L.; Allison, S.P. Ethical and legal aspects of enteral nutrition. *Clinical Nutrition* 2006 April; 25(2): 196-202. NRCBL: 20.5.1; 1.1; 2.1; 7.1.

Kumas, Gülsah; Öztunç, Gürsel; Alparslan, Z. Nazan. Intensive care unit nurses' opinions about euthanasia.

NRCBL: National Reference Center for Bioethics Literature Classification Scheme   See inside front cover for terms.

161

Nursing Ethics 2007 September; 14(5): 637-650. NRCBL: 20.5.1; 4.1.3; 7.1. SC: em. Identifiers: Turkey.
Abstract: This study was conducted to gain opinions about euthanasia from nurses who work in intensive care units. The research was planned as a descriptive study and conducted with 186 nurses who worked in intensive care units in a university hospital, a public hospital, and a private not-for-profit hospital in Adana, Turkey, and who agreed to complete a questionnaire. Euthanasia is not legal in Turkey. One third (33.9%) of the nurses supported the legalization of euthanasia, whereas 39.8% did not. In some specific circumstances, 44.1% of the nurses thought that euthanasia was being practiced in our country. The most significant finding was that these Turkish intensive care unit nurses did not overwhelmingly support the legalization of euthanasia. Those who did support it were inclined to agree with passive rather than active euthanasia (P = 0.011).

Lippert-Rasmussen, Kasper. Why killing some people is more seriously wrong than killing others. *Ethics* 2007 July; 117(4): 716-738. NRCBL: 20.5.1; 1.1; 4.4. SC: an.

Lustig, B. Andrew. Death, dying, euthanasia, and palliative care: perspectives from philosophy of medicine and ethics. *In:* Khushf, George, ed. Handbook of Bioethics: Taking Stock of the Field From a Philosophical Perspective. Dordrecht; Boston: Kluwer Academic, 2004: 441-471. NRCBL: 20.5.1; 1.1; 4.1.2; 4.4; 20.7. SC: rv.

McCabe, Helen. Nursing involvement in euthanasia: how sound is the philosophical support? *Nursing Philosophy* 2007 July; 8(3): 167-175. NRCBL: 20.5.1; 1.1; 4.1.3.

McCabe, Helen. Nursing involvement in euthanasia:a 'nursing-as-healing-praxis' approach. *Nursing Philosophy* 2007 July; 8(3): 176-186. NRCBL: 20.5.1; 1.1; 4.1.3.

Noakes, J.; Pridham, G. The 'euthanasie' programme 1939-1945. *In their:* Nazism, 1919-1945, Volume 3: Foreign Policy, War and Racial Extermination: A Documentary Reader. Exeter: University of Exeter Press, 1998-2001: 389-440. NRCBL: 20.5.1; 2.2; 4.1.2; 21.4.

Papadimitriou, John D.; Skiadas, Panayiotis; Mavrantonis, Constantino S.; Polimeropoulos, Vassilis; Papadimitriou, Dimitris J.; Papacostas, Kyriaki J. Euthanasia and suicide in antiquity: viewpoint of the dramatists and philosophers. *Journal of the Royal Society of Medicine* 2007 January; 100(1): 25-28. NRCBL: 20.5.1; 20.7; 7.1; 2.2; 1.1.

Pijnenburg, Martien A.M.; Leget, Carlo. Who wants to live forever? Three arguments against extending the human lifespan. *Journal of Medical Ethics* 2007 October; 33(10): 585-587. NRCBL: 20.5.1; 1.1; 4.4; 4.5. SC: an.
Abstract: The wish to extend the human lifespan has a long tradition in many cultures. Optimistic views of the possibility of achieving this goal through the latest developments in medicine feature increasingly in serious scientific and philosophical discussion. The authors of this paper argue that research with the explicit aim of extend-

ing the human lifespan is both undesirable and morally unacceptable. They present three serious objections, relating to justice, the community and the meaning of life.

Ray, Ratna; Raju, Mohan. Attitude towards euthanasia in relation to death anxiety among a sample of 343 nurses in India. *Psychological Reports* 2006 August; 99(1): 20-26. NRCBL: 20.5.1; 20.3.2; 4.1.3. SC: em.

Reiter-Theil, Stella; Mertz, Marcel; Meyer-Zehnder, Barbara. The complex roles of relatives in end-of-life decision-making: an ethical analysis. *HEC (Healthcare Ethics Committee) Forum* 2007 December; 19(4): 341-364. NRCBL: 20.5.1; 8.3.3; 1.1; 20.3.3. SC: em; cs.

Rich, Ben A. Causation and intent: persistent conundrums in end-of-life care. *CQ: Cambridge Quarterly of Healthcare Ethics* 2007 Winter; 16(1): 63-73. NRCBL: 20.5.1; 1.1.

Salladay, Susan A. Life and death disagreements [interview]. *Journal of Christian Nursing* 2007 January-March; 24(1): 38-40. NRCBL: 20.5.1; 1.2; 1.1; 4.4. Identifiers: Terri Schiavo; David C. Gibbs; Robert Orr.

Scherer, Yvonne.; Jezewski, Mary Ann; Graves, Brian; Wu, Yow-Wu; Bu, Xiaoyan. Advance directives and end-of-life decision making: survey of critical care nurses' knowledge, attitude, and experience. *Critical Care Nurse* 2006 August; 26(4): 30-40. NRCBL: 20.5.1; 20.5.4; 7.1; 4.1.3; 20.3.2. SC: em.

Shaw, David. The body as unwarranted life support: a new perspective on euthanasia. *Journal of Medical Ethics* 2007 September; 33(9): 519-521. NRCBL: 20.5.1; 1.1; 20.7. SC: an.

Shiffrin, Seana Valentine. Autonomy, beneficence, and the permanently demented. *In:* Burley, Justine, ed. Dworkin and His Critics: With Replies by Dworkin. Malden, MA: Blackwell Pub., 2004: 195-217. NRCBL: 20.5.1; 1.1; 20.5.4; 17.1; 8.3.3. SC: an.

Slosar, John Paul. Medical futility in the post-modern context. *HEC (Healthcare Ethics Committee) Forum* 2007 March; 19(1): 67-82. NRCBL: 20.5.1; 1.1; 7.1.

Smith, George P. A compassionate death. *In his:* The Christian Religion and Biotechnology: A Search for Principled Decision-making. Dordrecht: Springer, 2005: 189-230. NRCBL: 20.5.1; 1.1; 9.4; 20.6; 20.7. SC: le.

Snodgrass, Mary Ellen. Nazi nurses. *In her:* Historical Encyclopedia of Nursing. Santa Barbara, CA: ABC-CLIO, 1999: 191-194. NRCBL: 20.5.1; 2.2; 4.1.3; 21.4.

Solomon, Lewis D. Life extension: public policy aspects. *In his:* The Quest for Human Longevity: Science, Business, and Public Policy. New Brunswick, NJ: Transaction Publishers, 2006: 165-187. NRCBL: 20.5.1; 1.1; 4.5; 5.3.

Spielthenner, Georg. Ordinary and extraordinary means of treatment. *Ethics and Medicine: An International Journal of Bioethics* 2007 Fall; 23(3): 145-158. NRCBL: 20.5.1; 4.1.2; 1.1; 1.2. SC: rv.

Spurgeon, Brad. Doctors sign petition calling for euthanasia to be decriminalised [news]. *BMJ: British Medical Journal* 2007 March 17; 334(7593): 555. NRCBL: 20.5.1; 4.1.2; 20.3.2. SC: le. Identifiers: France.

Starks, Helene; Back, Anthony L.; Pearlman, Roberta A.; Koenig, Barbara A.; Hsu, Clarissa; Gordon,Judith R.; Bharucha, Ashok J. Family member involvement in hastened death. *Death Studies* 2007 February; 31(2): 105-130. NRCBL: 20.5.1; 9.4; 20.3.3; 1.1. SC: cs; le.

Strous, Rael D. Hitler's psychiatrists: healers and researchers turned executioners and its relevance today. *Harvard Review of Psychiatry* 2006 January-February; 14(1): 30-37. NRCBL: 20.5.1; 4.1.2; 17.1; 2.2; 21.4; 7.4.

Tomlinson, Thomas. Futility beyond CPR: the case of dialysis. *HEC (Healthcare Ethics Committee) Forum* 2007 March; 19(1): 33-43. NRCBL: 20.5.1; 1.1.

Tsaloglidou, Areti; Rammos, Kyriakos; Kiriklidis, Konstantinos; Zourladani, Athanasia; Matziari, Chrysoula. Nurses' ethical decision-making role in artificial nutritional support. *British Journal of Nursing* 2007 September 13-27; 16(16): 996-998. NRCBL: 20.5.1; 4.1.3. SC: em.

van Bruchem-van de Scheur, G.G.; van der Arend, Arie J.G.; Spreeuwenberg, Cor; Abu-Saad, Huda Huijer; ter Meulen, Ruud H.J. Euthanasia and physician-assisted suicide in the Dutch homecare sector: the role of the district nurse. *Journal of Advanced Nursing* 2007 April; 58(1): 44-52. NRCBL: 20.5.1; 20.7; 4.1.3; 9.5.1. SC: em.

Wilcockson, Michael. Euthanasia and doctors' ethics. *In his:* Issues of Life and Death. London: Hodder and Stoughton Educational, 1999: 57-71. NRCBL: 20.5.1; 20.5.2; 20.7; 8.1; 1.1; 1.2. SC: cs; le.

EUTHANASIA AND ALLOWING TO DIE/ RELIGIOUS ASPECTS

Extraordinary measures. Perpetuating a vegetative, unresponsive life may not in every case protect human dignity. *Christian Century* 2007 October 16; 124(21): 5. NRCBL: 20.5.1; 1.2.

Italy: Cardinal says patient should have the right to die. *New York Times* 2007 January 23; p. A9. NRCBL: 20.5.1; 1.2. SC: po. Identifiers: Piergiorgio Welby; Carlo Maria Martini.

Vatican clarifies position on artificial nutrition [news]. *America* 2007 October 1; 197(9): 6. NRCBL: 20.5.1; 1.2.

Anglican Diocese of Sydney. Social Issues Executive; Cameron, Andrew; Nodder, Tracy; Watts, Lisa. Euthanasia and the abandonment of life. Social Issues Executive #057 2007 February 9: 4 p. [Online]. Accessed: http://your.sydneyanglicans.net/socialissues/057_euthanasia_and_the_abandonment_of_life [2007 February 12]. NRCBL: 20.5.1; 20.7; 1.2. Identifiers: Dr. John Elliott; Dignitas. Identifiers: Australia.

Anscombe, G.E.M. Murder and the morality of euthanasia. *In her:* Human Life, Action and Ethics: Essays. Exeter: Imprint Academic, 2005: 261-277. NRCBL: 20.5.1; 1.1; 1.2. SC: le.

Bayley, Carol; Cardone, Joseph; Harvey, John Collins; O'Brien, Daniel; Panicola, Michael; Repenshek, Mark; Sheehan, Myles; Worsley, Stephen. Sampling of responses to the CDF statement on nutrition and hydration. *Health Care Ethics USA* 2007 Fall; 15(4): 8-14. NRCBL: 20.5.1; 1.2. Identifiers: Congregation for the Doctrine of the Faith.

Berger, Jeffrey T. When surrogates' responsibilities and religious concerns intersect. *Journal of Clinical Ethics* 2007 Winter; 18(4): 391-393. NRCBL: 20.5.1; 8.3.3; 1.2; 8.3.4. SC: cs. Comments: Craig D. Blinderman. Jewish law and end-of-life decision making: a case report. Journal of Clinical Ethics 2007 Winter; 18(4): 384-390.

Blinderman, Craig D. Jewish law and end-of-life decision making: a case report. *Journal of Clinical Ethics* 2007 Winter; 18(4): 384-390. NRCBL: 20.5.1; 1.2; 20.5.4; 20.4.1. SC: cs.

Brusco, Angelo. Treating and caring. *Dolentium Hominum* 2007; 22(2): 58-60. NRCBL: 20.5.1; 4.4; 1.2.

Catholic Church. Congregatio Pro Doctrina Fidei = Congregation for the Doctrine of the Faith. Commentary [on Responses to Certain Questions of the United States Conference of Catholic Bishops Concerning Artificial Nutrition and Hydration]. Rome: Congregation for the Doctrine of the Faith 2007 August 1: 4 p. [Online]. Accessed: http://www.vatican.va/roman_curia/ congregations/cfaith/documents/rc_con/cfaith_doc_20070801_nota-commento_en.html [2007 December 13]. NRCBL: 20.5.1; 1.2.

Catholic Church. Congregatio Pro Doctrina Fidei = Congregation for the Doctrine of the Faith. Commentary on responses to questions on nutrition and hydration. *Origins* 2007 September 27; 37(16): 242-245. NRCBL: 20.5.1; 1.2.

Catholic Church. Congregatio Pro Doctrina Fidei = Congregation for the Doctrine of the Faith. Responses to certain questions of the United States Conference of Catholic Bishops concerning artificial nutrition and hydration. Rome: Congregation for the Doctrine of the Faith, 2007 August 1: 1 p. [Online]. Accessed: http://www.vatican.va/roman_curia/congregations/

NRCBL: National Reference Center for Bioethics Literature Classification Scheme   See inside front cover for terms.

163

cfaith/documents/rc_con_cfaith_doc_20070801_riposte-usa_en.htm [2007 December 13]. NRCBL: 20.5.1; 1.2.

Catholic Church. Congregatio Pro Doctrina Fidei = Congregation for the Doctrine of the Faith. Responses to certain questions of the USCCB concerning artificial nutrition and hydration. *Ethics and Medics* 2007 November; 32(11): 1-3. NRCBL: 20.5.1; 1.2. Identifiers: United States Conference of Catholic Bishops.

Catholic Church. Congregatio Pro Doctrina Fidei = Congregation for the Doctrine of the Faith; Levada, William; Amato, Angelo. Responses to certain questions concerning artificial nutrition and hydration. *Origins* 2007 September 27; 37(16): 241-242. NRCBL: 20.5.1; 1.2.

Catholic Church. Pontifical Academy for Life = Pontificia Academia pro vita; World Federation of Catholic Medical Associations = Fédération Internationale des Associations Médicales Catholiques. Joint Statement on the Vegetative State: scientific and ethical problems related to the vegetative state. Vatican City: Pontifical Academy for Life 2004 March 20: 3 p. [Online]. Accessed: http://www.vatican.va/roman_curia/pontifical_academis/acdlife/documents/rc_pont-acd_life_doc20040320_joint-statement-veget-state_en.html [2007 December 13]. NRCBL: 20.5.1; 1.2. Conference: International Congress on "Life-Sustaining Treatments and Vegetative State: Scientific Advances and Ethical Dilemmas" Rome, Italy; 2004 March 10-17.

Daleiden, Joseph L. Euthanasia. *In his:* The Science of Morality: The Individual, Community, and Future Generations. Amherst, NY: Prometheus Books, 1998: 373-409. NRCBL: 20.5.1; 1.2; 20.7; 20.5.2. SC: an.

Daly, Daniel J. Prudence and the debate on death and dying; in the Catholic theological tradition, temporal life is not the highest good. *Health Progress* 2007 September-October; 88(5): 49-54. NRCBL: 20.5.1; 1.2.

Davis, Dena S. A tale of two daughters: Jewish law and end-of-life decision making. *Journal of Clinical Ethics* 2007 Winter; 18(4): 394-395. NRCBL: 20.5.1; 8.3.3; 1.2. SC: cs. Comments: Craig D. Blinderman. Jewish law and end-of-life decision making: a case report. Journal of Clinical Ethics 2007 Winter; 18(4): 384-390.

Dennis, William J. What is death with dignity? *Ethics and Medics* 2007 August; 32(8): 1-2. NRCBL: 20.5.1; 4.4; 1.2; 1.1. Identifiers: Jack Kevorkian.

Diamond, Eugene F. Catholic health care decision making [editorial]. *Linacre Quarterly* 2007 May; 74(2): 92-93. NRCBL: 20.5.1; 1.2.

Dorff, Elliot N. Judaism and ethical issues in end of life care. *In:* Eisen, Arri; Laderman, Gary, eds. Science, Religion, and Society: An Encyclopedia of History, Culture, and Controversy. Vol. 2. Armonk, NY: M.E. Sharpe, 2007: 712-719. NRCBL: 20.5.1; 1.2; 20.4.1; 19.1.

Dunlop, John. A good death [commentary]. *Ethics and Medicine: An International Journal of Bioethics* 2007 Summer; 23(2): 69-75. NRCBL: 20.5.1; 20.1; 1.2. SC: cs.

Fisher, Ian. Pope's death is drawn into euthanasia debate. *New York Times* 2007 September 28; p. A6. NRCBL: 20.5.1; 1.2. SC: po. Identifiers: Pope John Paul II.

Gesundheit, Benjamin; Steinberg, Avraham; Glick, Shimon; Or, Reuven; Jotkovitz, Alan. Euthanasia: an overview and the Jewish perspective. *Cancer Investigation* 2006 October; 24(6): 621-629. NRCBL: 20.5.1; 1.2; 7.2. SC: rv.

Hamel, Ron. The CDF statement on artificial nutrition and hydration: what should we make of it? *Health Care Ethics USA* 2007 Fall; 15(4): 5-7. NRCBL: 20.5.1; 1.2. Identifiers: Congregation for the Doctrine of the Faith.

Hardt, John J.; O'Rourke, Kevin D. Nutrition and hydration: the CDF response, in perspective. *Health Progress* 2007 November-December; 88(6): 44-47. NRCBL: 20.5.1; 1.2. Identifiers: Congregation for the Doctrine of the Faith.

Heneghan, Tom. Does Italy have its own "Terry Schiavo case?" Reuters. FaithWorld Blog 2007 October 24: 2 p. [Online]. Accessed: http://blogs.reuters.com/faithworld/2007/10/24/does-italy-have-itsown-terry-schiavo-case/ [2007 October 26]. NRCBL: 20.5.1; 1.2; 1.3.5.

Hong, Suk Young. Patients in a vegetative state and the quality of life. *Dolentium Hominum* 2007; 22(2): 22-27. NRCBL: 20.5.1; 1.2; 2.2; 4.4.

John Paul II, Pope. Address To the Participants to the International Congress "Life-Sustaining Treatments and Vegetative State: Scientific Advances and Ethical Dilemmas" [official English version]. Vatican City: Magisterium 2004 March 20; 5 p. [Online]. Accessed: http://www.academiavita.org/template.jsp?sez=DocumentiMagistero&pag=papi/gp_sv/gp_sv&lang=english [2004 March 29]. NRCBL: 20.5.1; 1.2; 4.4; 20.4.1. Note: Published in L'Osservatore Romano (English edition), 2004 March 31: 5. Conference: International Congress, Life-Sustaining Treatments and Vegetative State: Scientific Advances and Ethical Dilemmas; Rome, Italy; 2004 March 17-20; Pontifical Academy for Life and the International Federation of Catholic Medical Associations.

Kaczor, Christopher. Philosophy and theology: the authority of Pope John Paul II allocution; is ANH required for PVS patients?; papal allocution and Catholic tradition; human life as intrinsic good;. *National Catholic Bioethics Quarterly* 2007 Autumn; 7(3): 595-605. NRCBL: 20.5.1; 1.1; 1.2; 4.4. Identifiers: artificially administered nutrition and hydration; persistent vegetative state.

Kaplan, Kalman J. Zeno, Job and Terry Schiavo: the right to die versus the right to life. *Ethics and Medicine: An In-*

ternational *Journal of Bioethics* 2007 Summer; 23(2): 95-102. NRCBL: 20.5.1; 1.2; 4.4; 20.7. SC: an.

Kinlaw, Kathy. Prolonging living and dying. *In:* Eisen, Arri; Laderman, Gary, eds. Science, Religion, and Society: An Encyclopedia of History, Culture, and Controversy. Vol. 2. Armonk, NY: M.E. Sharpe, 2007: 731-738. NRCBL: 20.5.1; 20.5.4; 1.2; 20.4.1.

Langan, John. Catholic perspectives on nutrition. *Ethics and Intellectual Disability* 2005 Summer; 9(1): 3-4. NRCBL: 20.5.1; 1.2.

Linacre Institute. Catholic medical decision-making on the concept of futility. *Linacre Quarterly* 2007 August; 74(3): 258-262. NRCBL: 20.5.1; 1.2; 9.4.

National Catholic Bioethics Center. Brief comments on the CDF responses [from Statement of the NCBC On the CDF's "Responses to Certain Questions Concerning Artificial Nutrition and Hydration"]. *Ethics and Medics* 2007 November; 32(11): 3-4. NRCBL: 20.5.1; 1.2. Identifiers: Congregation for the Doctrine of the Faith.

Neugebauer, Matthias. Der theologische Lebensbegriff Dietrich Bonhoeffers im Lichte aktueller Fragen um Euthanasie, Sterbehilfe und Zwangssterilisation. *In:* Gestrich, Christof; Neugebauer, Johannes, eds. Der Wert menschlichen Lebens: medizinische Ethik bei Karl Bonhoeffer und Dietrich Bonhoeffer. Berlin: Wichern-Verlag, 2006: 147-165. NRCBL: 20.5.1; 1.2; 11.3; 15.5; 21.4; 2.2.

Noah, Barbara A. The role of religion in the Schiavo controversy. *Houston Journal of Health Law and Policy* 2006; 6(2): 319-346. NRCBL: 20.5.1; 1.2. SC: le; cs.

O'Rourke, Kevin D. Artificial nutrition and hydration and the Catholic tradition: the Terri Schiavo case had even members of Congress debating the issue. *Health Progress* 2007 May-June; 88(3): 50-54. NRCBL: 20.5.1; 1.2.

Panicola, Michael R. Making decisions about medically administered nutrition and hydration. *In:* Hamel, Ronald, ed. Making Health Care Decisions: A Catholic Guide. Liguori, MO: Liguori Publications, 2006: 109-126. NRCBL: 20.5.1; 1.2. Identifiers: Pope John Paul II.

Paris, John. A life too burdensome. *Tablet* 2007 January 6; 261(8672): 10-11. NRCBL: 20.5.1; 8.1; 1.2. SC: le. Identifiers: Italy.

Pence, Gregory E. Terri Schiavo: when does personhood end? *In his:* The Elements of Bioethics. Boston: McGraw-Hill, 2007: 137-171. NRCBL: 20.5.1; 4.4; 1.2; 20.2.1. SC: cs.

Providence Center for Health Care Ethics. A primer for understanding the CDF's Responses regarding ANH for the PVS patient. *Health Care Ethics USA* 2007 Fall; 15(4): 15-9. NRCBL: 20.5.1; 1.2. Identifiers: Congregation for

the Doctrine of the Faith; artificial nutrition and hydration; persistent vegetative state.

Rosner, Fred. Commentary on "Jewish law and end-of-life decision making". *Journal of Clinical Ethics* 2007 Winter; 18(4): 396-398. NRCBL: 20.5.1; 1.2. SC: cs. Comments: Craig D. Blinderman. Jewish law and end-of-life decision making: a case report. Journal of Clinical Ethics 2007 Winter; 18(4): 384-390.

Salladay, Susan A. Life and death disagreements [interview]. *Journal of Christian Nursing* 2007 January-March; 24(1): 38-40. NRCBL: 20.5.1; 1.2; 1.1; 4.4. Identifiers: Terri Schiavo; David C. Gibbs; Robert Orr.

Shannon, Thomas A.; Walter, James J. Artificial nutrition and hydration: assessing the papal statement. *In their:* Contemporary Issues in Bioethics: A Catholic Perspective. Lanham, MD: Rowman and Littlefield Publishers, 2005: 257-261. NRCBL: 20.5.1; 1.2. Identifiers: Pope John Paul II.

Shannon, Thomas A.; Walter, James J. Assisted nutrition and hydration and the Catholic tradition: the case of Terri Schiavo. *In their:* Contemporary Issues in Bioethics: A Catholic Perspective. Lanham, MD: Rowman and Littlefield Publishers, 2005: 269-280. NRCBL: 20.5.1; 1.2.

Shannon, Thomas A.; Walter, James J. Implications of the papal allocution on feeding tubes. *In their:* Contemporary Issues in Bioethics: A Catholic Perspective. Lanham, MD: Rowman and Littlefield Publishers, 2005: 263-268. NRCBL: 20.5.1; 1.2. Identifiers: Pope John Paul II.

Shannon, Thomas A.; Walter, James J. The PVS patient and the forgoing/withdrawing of medical nutrition and hydration. *In their:* Contemporary Issues in Bioethics: A Catholic Perspective. Lanham, MD: Rowman and Littlefield Publishers, 2005: 231-256. NRCBL: 20.5.1; 1.2; 4.4. SC: em.

Sparks, Richard C. Making decisions about end-of-life care. *In:* Hamel, Ronald, ed. Making Health Care Decisions: A Catholic Guide. Liguori, MO: Liguori Publications, 2006: 73-90. NRCBL: 20.5.1; 1.2; 4.4; 20.5.4; 20.7.

Spielthenner, Georg. Ordinary and extraordinary means of treatment. *Ethics and Medicine: An International Journal of Bioethics* 2007 Fall; 23(3): 145-158. NRCBL: 20.5.1; 4.1.2; 1.1; 1.2. SC: rv.

Stonington, Scott; Ratanakul, Pinit. Is there a global bioethics? End-of-life in Thailand and the case for local difference. California: Pacific Rim Research Program (University of California, Multi-Campus Research Unit), 2006: 7 p. [Online]. Accessed: http://repositories.cdlib.org/cgi/viewcontent.cgi?article=1021&context=pacrim [2007 October 1]. NRCBL: 20.5.1; 1.2; 2.1; 21.7. SC: cs.

NRCBL: National Reference Center for Bioethics Literature Classification Scheme   See inside front cover for terms.

165

Sujdak Mackiewicz, Birgitta N. Artificial nutrition and hydration: advancing the conversation. *Health Care Ethics USA* 2007 Summer; 15(3): 9-11. NRCBL: 20.5.1; 1.2.

Sullivan, Scott M. The development and nature of the ordinary/extraordinary means distinction in the Roman Catholic tradition. *Bioethics* 2007 September; 21(7): 386-397. NRCBL: 20.5.1; 1.2.

Abstract: In the Roman Catholic tradition the nature of the ordinary/extraordinary means distinction is best understood in light of its historical development. The moralist tradition that reared and nurtured this distinction implicitly developed a set of general criteria to distinguish the extraordinary from the ordinary. These criteria, conjoined with the context within which they were understood, can play an important role in refereeing the contemporary debate over the aggressiveness of medical treatment and the extent of one's moral obligation.

Wilcockson, Michael. Euthanasia and doctors' ethics. *In his:* Issues of Life and Death. London: Hodder and Stoughton Educational, 1999: 57-71. NRCBL: 20.5.1; 20.5.2; 20.7; 8.1; 1.1; 1.2. SC: cs; le.

Yacoub, Ahmed Abdel Aziz. Euthanasia. *In his:* The Fiqh of Medicine: Responses in Islamic Jurisprudence to Developments in Medical Science. London, UK: Ta-Ha Publishers Ltd., 2001: 159-201. NRCBL: 20.5.1; 1.2; 20.2.1. SC: le.

FETUSES *See* CARE FOR SPECIFIC GROUPS/ FETUSES; HUMAN EXPERIMENTATION/ SPECIAL POPULATIONS/ EMBRYOS AND FETUSES

FOREIGN NATIONALS *See* HUMAN EXPERIMENTATION/ SPECIAL POPULATIONS/ FOREIGN NATIONALS

GENE THERAPY

Gene therapy trials for cystic fibrosis to begin in U.K. [news]. *GeneWatch* 2007 January-February; 20(1): 19. NRCBL: 15.4; 18.1; 9.5.1.

Beutler, Ernest. Lysosomal storage diseases: natural history and ethical and economic aspects. *Molecular Genetics and Metabolism* 2006 July; 88(3): 208-215. NRCBL: 15.4; 9.3.1; 9.4.

Deane-Drummond, Celia. Gene therapies. *In her:* Genetics and Christian Ethics. Cambridge: Cambridge University Press, 2006: 124-159. 71 fn. NRCBL: 15.4; 1.2; 14.1; 18.2.
  Keywords: *gene therapy; advisory committees; Christian ethics; clinical trials; cloning; embryonic stem cells; ethical analysis; fetal therapy; future generations; gene transfer techniques; genetic disorders; genetic engineering; genetic enhancement; germ cells; pharmacogenetics; risks and benefits; virtues; Proposed Keywords: transhumanism; Keyword Identifiers: Great Britain; Human Fertilisation and Embryology Authority

European Commission. Directorate-General Enterprise and Industry. Human tissue engineering and beyond: proposal for a community regulatory framework on advanced therapies. Brussels: European Commission, 2005 May 4; 15 p. [Online]. Accessed: http://ec.europa.eu/enterprise/pharmaceuticals/advtherapies/docs/consultatiopaper-advancedtherapies-2005-may-04.pdf [2007 April 5]. NRCBL: 15.4; 19.1; 9.7.

Fox, Jeffrey L. FDA clarifies stance on long-term follow-up for gene therapy clinical trials [news brief]. *Nature Biotechnology* 2007 February; 25(2): 153. NRCBL: 15.4; 18.2. SC: le.
  Keywords: *clinical trials; *gene therapy; *guidelines; *legal aspects; federal government; government regulation; Keyword Identifiers: *Food and Drug Administration; *United States

Fuchs, Michael. Gene therapy. An ethical profile of a new medical territory. *Journal of Gene Medicine* 2006 November; 8(11): 1358-1362. NRCBL: 15.4; 18.5.1; 18.2.

Gelsinger, Paul L. Uninformed consent: the case of Jesse Gelsinger. *In:* Lemmens, Trudo; Waring, Duff R., eds. Law and Ethics in Biomedical Research: Regulation, Conflict of Interest and Liability. Toronto; Buffalo: University of Toronto Press, 2006: 12-32. 14 refs. NRCBL: 15.4; 18.3; 1.3.9; 7.3; 8.5; 18.6.
  Keywords: *clinical trials; *gene therapy; *informed consent; adverse effects; altruism; conflict of interest; legal liability; nontherapeutic research; research related injuries; research subjects; Keyword Identifiers: *Gelsinger, Jesse; *University of Pennsylvania; Food and Drug Administration; United States

Hughes, Virginia. Therapy on trial. *Nature Medicine* 2007 September; 13(9): 1008-1009. NRCBL: 15.4; 18.2; 18.3; 18.6.
  Keywords: *adverse effects; *clinical trials; *gene therapy; *risks and benefits; death; gene transfer techniques; government regulation; industry; informed consent; international aspects; research subjects; research support; therapeutic misconception; Proposed Keywords: rheumatoid arthritis; Keyword Identifiers: *Mohr, Jolee; Food and Drug Administration; Gelsinger, Jesse; Targeted Genetics Corp.; United States

Kaiser, Jocelyn. Death prompts a review of gene therapy vector [news]. *Science* 2007 August 3; 317(5838): 580. NRCBL: 15.4; 18.5.1; 20.1.
  Keywords: *adverse effects; *clinical trials; *gene therapy; death; gene transfer techniques; industry; research subjects; selection of subjects; Proposed Keywords: arthritis; Keyword Identifiers: Mohr, Jolee; Targeted Genetics Corp.; United States

Kaiser, Jocelyn. Gene transfer an unlikely contributor to patient's death [news]. *Science* 2007 December 7; 318(5856): 1535. NRCBL: 15.4; 18.5.1; 20.1.
  Keywords: *adverse effects; *clinical trials; *gene therapy; *research subjects; advisory committees; death; industry; Keyword Identifiers: *Mohr, Jolee; Recombinant DNA Advisory Committee; Targeted Genetics Corp.

Kaiser, Jocelyn. Questions remain on cause of death in arthritis trial [news]. *Science* 2007 September 21; 317(5845): 1665. NRCBL: 15.4; 18.5.1; 18.3.

Keywords: *adverse effects; *clinical trials; *gene therapy; death; drugs; industry; informed consent; research subjects; Proposed Keywords: arthritis; Keyword Identifiers: *Mohr, Jolee; Targeted Genetics Corp.

Karpin, Isabel; Mykitiuk, Roxanne. Regulating inheritable genetic modification, or policing the fertile scientific imagination? A feminist legal response. *In:* Rasko, John E.J.; O'Sullivan, Gabrielle M.; Ankeny, Rachel A., eds. The Ethics of Inheritable Genetic Modification: A Dividing Line? Cambridge: Cambridge University Press, 2006: 193-222. 79 fn. NRCBL: 15.4; 9.5.8; 14.1; 15.1; 10; 5.3. SC: an; le.

Keywords: *feminist ethics; *fetal therapy; *gene therapy; *genetic intervention; *government regulation; *legal aspects; chimeras; cloning; embryos; family; freedom; gene pool; genetic determinism; germ cells; human experimentation; preimplantation diagnosis; prenatal diagnosis; public policy; reproduction; reproductive technologies; social impact; women; Keyword Identifiers: *Australia; *Canada

Keim, Brandon. Gene therapy trials on trial: the unfortunate tale of Jolee Mohr. *GeneWatch* 2007 September-October; 20(5): 10-11. NRCBL: 15.4; 18.5.1; 20.1.

Kim, Scott Y.H. Assessing and communicating the risks and benefits of gene transfer clinical trials. *Current Opinion in Molecular Therapeutics* 2006 October; 8(5): 384-389. NRCBL: 15.4; 18.2; 18.3.

Ledford, Heidi. Death in gene therapy trial raises questions about private IRBs [news]. *Nature Biotechnology* 2007 October; 25(10): 1067. NRCBL: 15.4; 18.2; 18.5.1; 20.1.

Keywords: *adverse effects; *clinical trials; *conflict of interest; *ethical review; *gene therapy; *research ethics committees; consent forms; death; evaluation; industry; research design; research subjects; research support; Keyword Identifiers: Mohr, Jolee; Targeted Genetics Corp.; United States

Nycum, Gillian; Reid, Lynette. The harm-benefit trade-off in "bad deal" trials. *Kennedy Institute of Ethics Journal* 2007 December; 17(4): 321-350. 80 refs. 11 fn. NRCBL: 15.4; 18.2; 5.2; 1.1. SC: an.

Keywords: *clinical trials; *evaluation; *gene therapy; *gene transfer techniques; *nontherapeutic research; *risks and benefits; adverse effects; altruism; biomedical research; cancer; emotions; ethical review; human experimentation; motivation; prognosis; research design; research ethics committees; research ethics; research subjects; selection of subjects; terminally ill; therapeutic misconception; uncertainty; vulnerable populations; Proposed Keywords: *phase I clinical trials; glioblastoma

Abstract: This paper examines the nature of the harm-benefit tradeoff in early clinical research for interventions that involve remote possibility of direct benefit and likelihood of direct harms to research participants with fatal prognoses, by drawing on the example of gene transfer trials for glioblastoma multiforme. We argue that the appeal made by the component approach to clinical equi-poise fails to account fully for the nature of the harm-benefit tradeoff—individual harm for social benefit—that would be required to justify such research. An analysis of what we label "collateral affective benefits," such as the experience of hope or exercise of altruism, shows that the existence of these motivations reinforces rather than mitigates the necessity of justification by reference to social benefit. Evaluations of social benefit must be taken seriously in the research ethics review process to avoid the exploitation of research participants' motivations of hope or altruism and to avoid the possibility of inadvertent exploitation of high-risk research participants and the harms that would associate with such exploitation.

Parker, H. Stewart. Reply to "Poor trial design leaves gene therapy death a mystery" [letter]. *Nature Medicine* 2007 November; 13(11): 1276. NRCBL: 15.4; 18.2. Comments: Meredith Wadman. Poor trial design leaves gene therapy death a mystery. Nature Medicine 2007 October: 13(10): 1124.

Keywords: *adverse effects; *clinical trials; *gene therapy; *research design; advisory committees; death; industry; peer review; research subjects; Proposed Keywords: rheumatoid arthritis; Keyword Identifiers: *Mohr, Jolee; National Institutes of Health; Recombinant DNA Advisory Committee; Targeted Genetics Corp.; United States

Pollack, Andrew. Death in gene therapy treatment is still unexplained. *New York Times* 2007 September 18; p. A22. NRCBL: 15.4; 18.5.1; 18.2. SC: po.

Keywords: *adverse effects; *clinical trials; *gene therapy; *research subjects; death; gene transfer techniques; industry; nontherapeutic research; review committees; selection of subjects; uncertainty; Proposed Keywords: arthritis; Keyword Identifiers: *Mohr, Jolee; Targeted Genetics Corp.; University of Chicago Medical Center

Pollack, Andrew. Gene therapy study to resume after woman's death [news]. *New York Times* 2007 November 26; p. A16. NRCBL: 15.4; 18.5.1; 18.3. SC: po.

Keywords: *gene therapy; *phase I clinical trials; *research subjects; adverse effects; consent forms; death; federal government; government regulation; industry; investigational drugs; Keyword Identifiers: *Mohr, Jolee; Targeted Genetics Corp.; United States

Rasko, John E.J.; O'Sullivan, Gabrielle M.; Ankeny, Rachel A. Is inheritable genetic modification the new dividing line? *In:* Rasko, John E.J.; O'Sullivan, Gabrielle M.; Ankeny, Rachel A., eds. The Ethics of Inheritable Genetic Modification: A Dividing Line? Cambridge: Cambridge University Press, 2006: 1-15. 50 fn. NRCBL: 15.4; 15.1.

Keywords: *gene therapy; *genetic engineering; *germ cells; adverse effects; clinical trials; ethical analysis; future generations; gene transfer techniques; historical aspects; human experimentation; reproductive technologies; scientific misconduct; terminology; Keyword Identifiers: Twentieth Century

Scully, Jackie Leach. Inheritable genetic modification and disability: normality and identity. *In:* Rasko, John E.J.; O'Sullivan, Gabrielle M.; Ankeny, Rachel A., eds. The Ethics of Inheritable Genetic Modification: A Dividing

NRCBL: National Reference Center for Bioethics Literature Classification Scheme   See inside front cover for terms.

167

Line? Cambridge: Cambridge University Press, 2006: 175-192. 37 fn. NRCBL: 15.4; 15.1; 9.5.1. SC: an.

Keywords: *disabled persons; *gene therapy; *genetic engineering; *germ cells; *normality; *self concept; choice behavior; chronically ill; culture; ethical analysis; genetic diversity; hearing disorders; moral policy parents; patient participation; reproduction; suffering; wedge argument

Tanne, Janice Hopkins. US gene therapy trial is to restart, despite patient's death. *BMJ: British Medical Journal* 2007 December 8; 335(7631): 1172-1173. NRCBL: 15.4; 18.2.

Thomson, Mary M. Bringing research into therapy: liability anyone? *In:* Lemmens, Trudo; Waring, Duff R., eds. Law and Ethics in Biomedical Research: Regulation, Conflict of Interest and Liability. Toronto; Buffalo: University of Toronto Press, 2006: 183-205. 54 fn. NRCBL: 15.4; 18.3; 1.3.9; 7.3; 8.5; 18.6. SC: le.

Keywords: *clinical trials; *gene therapy; *legal liability; adverse effects; conflict of interest; death; disclosure; gene transfer techniques; informed consent; negligence; research ethics committees; research institutes; research subjects; research support; researchers; risk; Keyword Identifiers: *Canada; *United States; Dent, James; Gelsinger, Jesse

Tong, Rosemarie. Traditional and feminist bioethical perspectives on gene transfer: is inheritable genetic modification really the problem? *In:* Rasko, John E.J.; O'Sullivan, Gabrielle M.; Ankeny, Rachel A., eds. The Ethics of Inheritable Genetic Modification: A Dividing Line? Cambridge: Cambridge University Press, 2006: 159-173. 55 fn. NRCBL: 15.4; 15.1; 1.1; 10. SC: an.

Keywords: *gene therapy; *feminist ethics; *genetic engineering; *moral policy; bioethics; choice behavior; disadvantaged persons; embryos; ethical analysis; fetal therapy; gene transfer techniques; genetic services; health services accessibility; pregnant women; preimplantation diagnosis; risks and benefits; treatment refusal; women

Wadman, Meredith. Poor trial design leaves gene therapy death a mystery [news]. *Nature Medicine* 2007 October; 13(10): 1124. NRCBL: 15.4; 18.2.

Keywords: *adverse effects; *clinical trials; *gene therapy; *research design; death; drugs; research subjects; uncertainty; Proposed Keywords: rheumatoid arthritis; Keyword Identifiers: *Mohr, Jolee; Targeted Genetics Corp.; United States

Walter, James J. Theological perspectives on cancer genetics and gene therapy: the Roman Catholic tradition. *In:* Walter, James J.; Shannon, Thomas A., eds. Contemporary Issues in Bioethics: A Catholic Perspective. Lanham, MD: Rowman and Littlefield Publishers, 2005: 199-207. 35 fn. NRCBL: 15.4; 1.2; 15.1.

Keywords: *gene therapy; *genetic intervention; *Roman Catholic ethics; *theology; cancer; genetic enhancement; germ cells; human genome; natural law; risks and benefits

Weiss, Rick. Death points to risks in research; one woman's experience in gene therapy trial highlights weaknesses in the patient safety net. *Washington Post* 2007 August 6; p. A1, A7. NRCBL: 15.4; 18.3; 18.5.1. SC: po.

Keywords: *adverse effects; *phase I clinical trials; *gene therapy; *research subjects; consent forms; contracts; death; industry; investigational drugs; physicians; remuneration; research ethics committees; selection of subjects; Keyword Identifiers: *Mohr, Jolee; Targeted Genetics Corp.

Weiss, Rick. Suspended gene therapy test had drawn early questions. *Washington Post* 2007 July 28; p. A9. NRCBL: 15.4; 18.1. SC: po.

Keywords: *adverse effects; *clinical trials; *gene therapy; advisory committees; consent forms; death; disclosure; ethical review; industry; methods; nontherapeutic research; research subjects; risk; Proposed Keywords: arthritis; Keyword Identifiers: *Mohr, Jolee; Recombinant DNA Advisory Committee; Targeted Genetics Corp.; United States

GENETIC COUNSELING

See also GENETIC SCREENING; SEX DETERMINATION

Mohler would favor altering 'gay' fetus [news]. *Christian Century* 2007 April 3; 124(7): 15. NRCBL: 15.2; 10; 9.5.8. Identifiers: R. Albert Mohler, Jr.; Southern Baptist Theological Seminary.

Alliman, S.; McCarthy Veach, P.; Bartels, D.; Bower, M.; James, C.; LeRoy, B. Ethical and professional challenges in clinical practice: a comparative analysis of Australian and U.S. genetic counselors. *Journal of Genetic Counseling* 2007 December; 16(6): 687. NRCBL: 15.2; 1.3.1; 21.1. SC: em.

Alsulaiman, Ayman; Hewison, J. Attitudes to prenatal testing and termination of pregnancy in Saudi Arabia. *Community Genetics* 2007; 10(3): 169-173. 23 refs. NRCBL: 15.2; 12.5.2. SC: em.

Keywords: *attitudes; *genetic disorders; *parents; *prenatal diagnosis; *selective abortion; fathers; mothers; questionnaires; survey; Keyword Identifiers: *Saudi Arabia

American Society for Reproductive Medicine [ASRM]. Practice Committee; Society for Assisted Reproductive Technology [SART]. Practice Committee. Preimplantation genetic diagnosis. *Fertility and Sterility* 2006 November; 86(5, Supplement): S257-S258. NRCBL: 15.2; 14.4.

Arbeitsgruppe Pränataldiagnostik; Merkel, Reinhard. Das „Strudlhof"-Symposion. Konsensus-Statement: Bedingungen spezieller pränataler genetischer Diagnostik [The "Strudlhof" symposium. Consensus statement: conditions for special prenatal genetic diagnosis]. *Ethik in der Medizin* 2007 September; 19(3): 221-225. NRCBL: 15.2; 9.5.5. SC: le.

Keywords: *prenatal diagnosis; autonomy; consensus; decision making; legal aspects; pregnant women; risks and benefits; selective abortion; Keyword Identifiers: Austria

Asch, Adrienne; Wasserman, Davis. A response to Nelson and Mahowald. *CQ: Cambridge Quarterly of Healthcare Ethics* 2007 Fall; 16(4): 468-475. NRCBL: 15.2; 4.4; 1.1; 7.1. SC: an. Identifiers: prenatal testing for selection against disabilities.

Austin, Jehannine C.; Honer, William G. The genomic era and serious mental illness: a potential application for psychiatric genetic counseling. *Psychiatric Services* 2007 February; 58(2): 254-261. 94 refs. NRCBL: 15.2; 17.1.
Keywords: *genetic counseling; *mentally ill persons; family members; genetic predisposition; patient care team; patient education; referral and consultation; stigmatization

Autti-Rämö, Ilona; Mäkelä, Marjukka. Screening for fetal abnormalities: from a health technology assessment report to a national statute. *International Journal of Technology Assessment in Health Care* 2007 Fall; 23(4): 436-442. 17 refs. NRCBL: 15.2; 15.3; 5.2; 9.1; 7.1. SC: em.
Keywords: *chromosome abnormalities; *congenital disorders; *mass screening; *policy making; *prenatal diagnosis; *public policy; *technology assessment; attitudes; genetic services; government regulation; health services; survey accessibility;; Keyword Identifiers: *Finland

Baggot, Paddy Jim; Baggot, M.G. Obstetric genetic counseling for lethal anomalies. *Linacre Quarterly* 2007 February; 74(1): 60-67. NRCBL: 15.2; 1.2.

Baldwin, Thomas. Choosing who: what is wrong with making better children? *In:* Spencer, J.R.; du Bois-Pedain, Antje, eds. Freedom and Responsibility in Reproductive Choice. Portland: Hart Pub., 2006: 15-30. 13 fn. NRCBL: 15.2; 1.1; 14.1; 15.5. SC: an.
Keywords: *beneficence; *ethical analysis; *eugenics; *genetic intervention; *moral policy; *reproduction; autonomy; choice behavior; ecology; deontological ethics; genetic disorders; genetic screening; growth disorders; homosexuality; moral obligations; parents; preimplantation diagnosis; prenatal diagnosis; quality of life; reproductive rights; teleological ethics; value of life

Barfield, Raymond C.; Kodish, Eric. Pediatric ethics in the age of molecular medicine. *Pediatric Clinics of North America* 2006 August; 53(4): 639-648. NRCBL: 15.2; 15.3; 9.5.7; 18.5.4.

Bauer, Patricia E. What's lost in prenatal testing [op-ed]. *Washington Post* 2007 January 14; p. B7. NRCBL: 15.2; 9.5.3. SC: po.
Keywords: *prenatal diagnosis; *Down syndrome; organizational policies; professional organizations; obstetrics and gynecology; selective abortion; practice guidelines; Proposed Keywords: mental retardation; Keyword Identifiers: American College of Obstetricians and Gynecologists

Beeson, Diane; Lippman, Abby. Egg harvesting for stem cell research: medical risks and ethical problems. *Reproductive Biomedicine Online* 2006 October; 13(4): 573-579. 50 refs. NRCBL: 15.2; 18.5.4; 9.5.5; 14.4; 14.5.
Keywords: *embryonic stem cells; *ovum donors; *research embryo creation; *risk; cloning; conflict of interest; remuneration; reproduction; researchers; women's health; Proposed Keywords: tissue harvesting

Borkenhagen, A.; Brähler, E.; Wisch, S.; Stöbel-Richter, Y.; Strauss, B.; Kentenich, H. Attitudes of German infertile couples towards preimplantation genetic diagnosis for different uses: a comparison to international studies.

Human Reproduction 2007 July; 22(7): 2051-2057. NRCBL: 15.2; 14.4. SC: em.

Braude, Peter; Flinter, Frances. Use and misuse of preimplantation genetic testing. *BMJ: British Medical Journal* 2007 October 13; 335(7623): 752-754. NRCBL: 15.2; 14.4.

Burke, Wylie; Press, Nancy. Ethical obligations and counseling challenges in cancer genetics. *Journal of the National Comprehensive Cancer Network: JNCCN* 2006 February; 4(2): 185-191. 38 refs. NRCBL: 15.2; 15.3; 9.5.1; 8.4. SC: cs.
Keywords: *cancer; *genetic counseling; *genetic predisposition; adults; autonomy; beneficence; clinical genetics; duty to warn; family members; genetic screening; justice; minors; moral obligations; physician patient relationship; primary health care; professional family relationship; truth disclosure; uncertainty

Callus, Thérèse. Pre-implantation genetic diagnosis — towards a principled construction of law? *In:* Garwood-Gowers, Austen; Tingle, John; Wheat, Kay, eds. Contemporary Issues in Healthcare Law and Ethics. Edinburgh; New York: Elsevier Butterworth-Heinemann, 2005: 133-147. 22 refs. NRCBL: 15.2; 14.4; 4.4. SC: le.
Keywords: *preimplantation diagnosis; *legal aspects; children; embryos; family members; genetic disorders; legislation; regulation; siblings; value of life; Proposed Keywords: tissue typing; Keyword Identifiers: *Great Britain; Human Fertilisation and Embryology Act 1990 (Great Britain); Human Fertilisation and Embryology Authority

Carlson, Elof Axel. Prenatal diagnosis and an alleged eugenics through the back door. *In his:* Times of Triumph, Times of Doubt: Science and the Battle for Public Trust. Cold Spring Harbor, NY: Cold Spring Harbor Laboratory Press, 2006: 155-162. 10 fn. NRCBL: 15.2; 12.3; 15.5.
Keywords: *eugenics; *prenatal diagnosis; congenital disorders; genetic counseling; genetic disorders; preimplantation diagnosis; risk; selective abortion; social impact

Case, Amy P.; Ramadhani, Tunu A.; Canfield, Mark A.; Wicklund, Catherine A. Awareness and attitudes regarding prenatal testing among Texas women of childbearing age. *Journal of Genetic Counseling* 2007 October; 16(5): 655-661. NRCBL: 15.2. SC: em.

Chachkin, Carolyn Jacobs. What potent blood: non-invasive prenatal genetic diagnosis and the transformation of modern prenatal care. *American Journal of Law and Medicine* 2007; 33(1): 9-53. 306 fn. NRCBL: 15.2; 9.8; 9.3.1.
Keywords: *genetic disorders; *methods; *prenatal diagnosis; costs and benefits; genetic screening; health insurance; insurance coverage; pregnant women; risks and benefits; selective abortion; social impact; standards; Proposed Keywords: predictive value of tests; prenatal care; Keyword Identifiers: Medicaid; United States

Civin, Curt I.; Rao, Mahendra S. How many human embryonic stem cell lines are sufficient? A. U.S. perspective. *Stem Cells* 2006 April; 24(4): 800-803 [Online]. http://www.StemCells.com/cgi/content/full/24/4/800

NRCBL: National Reference Center for Bioethics Literature Classification Scheme   See inside front cover for terms.

169

[2007 December 4]. 30 refs. NRCBL: 15.2; 18.5.4; 18.6; 5.3.

Keywords: *cell lines; *embryo research; *embryonic stem cells; beginning of life; cloning; donors; embryo disposition; genetic diversity; government regulation; informed consent; research support; stem cell transplantation; Keyword Identifiers: *United States

Davey, Angela; Newson, Ainsley; O'Leary, Peter. Communication of genetic information within families: the case for familial comity. *Journal of Bioethical Inquiry* 2006; 3(3): 161-166. NRCBL: 15.2; 15.3; 8.4. SC: cs.

de Melo-Martín, Immaculada; Rosenwaks, Zev; Fins, Joseph J. New methods for deriving embryonic stem cell lines: are the ethical problems solved? *Fertility and Sterility* 2006 November; 86(5): 1330-1332. NRCBL: 15.2; 18.5.4; 15.3.

de Montalembert, Mariane de; Bonnet, Doris; Lena-Russo, Danielle; Briard, Marie Louise. Ethical aspects of neonatal screening for sickle cell disease in Western European countries. *Acta Paediatrica* 2005 May; 94(5): 528-530. 16 refs. NRCBL: 15.2; 15.3; 9.1; 14.1; 21.1; 21.7.

Keywords: *genetic screening; *immigrants; *mass screening; *newborns; *sickle cell anemia; costs and benefits; disclosure; genetic counseling; parents; risks and benefits; stigmatization; Keyword Identifiers: *Europe

de Wert, Guido; Geraedts, Joep P.M. Preimplantation genetic diagnosis for hereditary disorders that do not show a simple Mendelian pattern: an ethical exploration. *In:* Shenfield, Françoise; Sureau, Claude, eds. Contemporary Ethical Dilemmas in Assisted Reproduction. Abingdon: Informa Healthcare, 2006: 85-98. 39 refs. NRCBL: 15.2; 14.4; 15.1.

Keywords: *genetic disorders; *preimplantation diagnosis; adults; children; genetic counseling; late-onset disorders; prenatal diagnosis

de Wert, Guido; Liebaers, Inge; Van de Velde, Hilde. The future (r)evolution of preimplantation genetic diagnosis/human leukocyte antigen testing: ethical reflections. *Stem Cells* 2007 September; 25: 2167-2172 [Online]. Accessed: http://www.StemCells.com/cgi/content/full/25/9/2167 [2007 December 4]. 47 refs. NRCBL: 15.2; 14.4; 18.5.4.

Keywords: *embryonic stem cells; *preimplantation diagnosis; *moral policy; *stem cells; *tissue typing; adverse effects; classification; family members; ethical analysis; forecasting; motivation; reproduction; research embryo creation; risks and benefits; stem cell transplantation; trends; Proposed Keywords: *hematopoietic stem cells

Deane-Drummond, Celia. Genetic counselling. *In her:* Genetics and Christian Ethics. Cambridge: Cambridge University Press, 2006: 101-123. 42 fn. NRCBL: 15.2; 1.2; 14.4.

Keywords: *genetic counseling; Christian ethics; disabled persons; precautionary principle; preimplantation diagnosis; prenatal diagnosis; reproductive technologies; risk; selective abortion; trust; Keyword Identifiers: Great Britain; Human Fertilisation and Embryology Authority

Dery, Anat Mishori; Carmi, Rivka; Vardi, Ilana Shoham. Different perceptions and attitudes regarding prenatal testing among service providers and consumers in Israel. *Community Genetics* 2007; 10(4): 242-251. 37 refs. NRCBL: 15.2. SC: em.

Keywords: *attitude of health personnel; *attitudes; *comparative studies; *pregnant women; *prenatal diagnosis; congenital disorders; Jews; patient satisfaction; questionnaires; socioeconomic factors; Keyword Identifiers: *Israel

Doerflinger, Richard M. Washington insider: House passes amended Genetic Nondiscrimination Bill; continued impasse on stem cell legislation, new executive order; defeat of deceptive human cloning bill; Supreme Court decision on partial-birth abortion. *National Catholic Bioethics Quarterly* 2007 Autumn; 7(3): 455-463. 21 fn. NRCBL: 15.2; 8.4; 18.5.4; 14.5. SC: le.

Keywords: *cloning; *embryo research; *embryonic stem cells; *genetic discrimination; *legislation; abortion; adult stem cells; cell lines; federal government; government financing; government regulation; politics; reproductive technologies; research support; Keyword Identifiers: *U. S. Congress; *United States; Genetic Information Nondiscrimination Act; Human Cloning Prohibition Act; Partial Birth Abortion Ban Act; Stem Cell Research Enhancement Act

Ehrich, Kathryn; Farsides, Bobbie; Williams, Clare; Scott, Rosamund. Testing the embryo, testing the fetus. *Clinical Ethics* 2007 December; 2(4): 181-186. 40 refs. NRCBL: 15.2; 7.1; 12.1; 14.4. SC: em.

Keywords: *attitude of health personnel; *beginning of life; *embryos; *fetuses; *moral status; *preimplantation diagnosis; *prenatal diagnosis; attitudes; comparative studies; embryo disposition; embryo transfer; embryonic development; genetic counseling; genetic disorders; in vitro fertilization; interviews; researchers; selective abortion; survey; value of life; Keyword Identifiers: Great Britain

Abstract: This paper stems from an ethnographic, multidisciplinary study that explored the views and experiences of practitioners and scientists on social, ethical and clinical dilemmas encountered when working in the area of pre-implantation genetic diagnosis for serious genetic disorders. We focus here on staff perceptions and experiences of working with embryos and helping women/couples to make choices that will result in selecting embryos for transfer and disposal of 'affected' embryos, compared to the termination of affected pregnancies following prenatal diagnosis. Analysis and discussion of our data led us to consider the possible advantages of pre-implantation genetic diagnosis and whether a gradualist account of the embryo's and fetus's moral status can account for all of these, particularly since a gradualist account concentrates on the significance of time (developmental stage) and makes no comment as to the significance of place (in vitro, in utero).

Freeman, Michael. Saviour siblings. *In:* McLean, Sheila A.M., ed. First Do No Harm: Law, Ethics, and Healthcare. Aldershot, England; Burlington, VT: Ashgate, 2006: 389-406. 131 fn. NRCBL: 15.2; 14.4. SC: an; rv.

Keywords: *ethical analysis; *preimplantation diagnosis; *reproduction; *siblings; *tissue typing; autonomy; commodification; directed donation; embryo transfer; legal aspects; moral obligations; parents; reproductive rights; risks and benefits; tissue donors; wedge argument; Keyword Identifiers: Australia; Great Britain; R (Quintavalle) v. Human Fertilisation and Embryology Authority

Gavaghan, Colin. Right problem, wrong solution: a pro-choice response to "expressivist" concerns about preimplantation genetic diagnosis. *CQ: Cambridge Quarterly of Healthcare Ethics* 2007 Winter; 16(1): 20-34. 50 fn. NRCBL: 15.2; 9.5.1; 4.2; 12.5.1; 14.4. SC: an.
Keywords: *disabled persons; *preimplantation diagnosis; autonomy; ethical analysis; eugenics; genetic disorders; genetic predisposition to disease; genetic screening; health care delivery; obligations of society; public policy; regulation; prenatal diagnosis; reproductive rights; selective abortion; social impact; Keyword Identifiers: Great Britain; Human Fertilisation and Embryology Authority

Gilbar, Roy. Communicating genetic information in the family: the familial relationship as the forgotten factor. *Journal of Medical Ethics* 2007 July; 33(7): 390-393. 29 refs. NRCBL: 15.2; 8.4; 1.1; 8.1.
Keywords: *communication; *disclosure; *duty to warn; *family members; *family relationship; *genetic counseling; *genetic privacy; confidentiality; informed consent; physician patient relationship; professional family relationship; right not to know; utilitarianism

Abstract: Communicating genetic information to family members has been the subject of an extensive debate recently in bioethics and law. In this context, the extent of the relatives' right to know and not to know is examined. The mainstream in the bioethical literature adopts a liberal perception of patient autonomy and offers a utilitarian mechanism for solving familial tensions over genetic information. This reflects a patient-centred approach in which disclosure without consent is justified only to prevent serious harm or death to others. Based on a legal and bioethical analysis on the one hand, and an examination of empirical studies on the other, this paper advocates the adoption of a relational perception of autonomy, which, in the context of genetics, takes into account the effect that any decision—whether to disclose or not to disclose—will have on the familial relationship and the dynamics of the particular family. Adding this factor to the criteria usually advocated by lawyers and ethicists will facilitate reaching a sensitive decision, which recognises the various interests of family members beyond the risk to physical health. Taking this factor into account will require a process of deliberation both between doctors and patients, and in the family. It will also require a relaxation of medical confidentiality, as the family rather than the patient is gradually perceived as the unit of care. Moreover, adopting such a relational approach will accord with current views of doctors and patients who base their decision primarily on the nature of the familial relationship.

Goldman, Bruce. The first cut. *Nature* 2007 February 1; 445(7127): 479-480. 7 refs. NRCBL: 15.2; 14.4. SC: em.

Keywords: *preimplantation diagnosis; *risks and benefits; chromosome disorders; embryo transfer; empirical research; in vitro fertilization

Gosden, Roger. Genetic test may lead to waste of healthy embryos [letter]. *Nature* 2007 March 22; 446(7134): 372. NRCBL: 15.2; 14.4.
Keywords: *preimplantation diagnosis; chromosome abnormalities; embryo transfer; genetic screening; risks and benefits; Proposed Keywords: predictive value of tests

Grace, Jan; Drakeley, Andrew. Preimplantation genetic diagnosis. *British Journal of Hospital Medicine* 2006 April; 67(4): 197-199. NRCBL: 15.2; 14.4.

Gruen, Lori; Grabel, Laura. Concise review: scientific and ethical roadblocks to human embryonic stem cell therapy. *Stem Cells* 2006 October; 24(10): 2162-2169 [Online]. Accessed: http://stemcell.alphamedpress.org/cgi/content/full/24/10/2162 [2007 December 4]. 72 refs. NRCBL: 15.2; 18.5.4; 5.3; 18.6. SC: rv.
Keywords: *embryo research; *embryonic stem cells; *stem cell transplantation; advisory committees; cell lines; federal government; research support; state government; Proposed Keywords: altered nuclear transfer; blastocysts

Guilam, Maria Cristina R.; Corrêa, Marilena C.D.V. Risk, medicine and women: a case study on prenatal genetic counselling in Brazil. *Developing World Bioethics* 2007 August; 7(2): 78-85. 19 fn. NRCBL: 15.2; 12.4.1.
Keywords: *congenital disorders; *genetic counseling; *prenatal diagnosis; *selective abortion; anencephaly; autonomy; communication; decision making; developing countries; Down syndrome; genetic services; illegal abortion; legal aspects; physician's role; pregnant women; public hospitals; reproductive rights; risk; women's rights; Keyword Identifiers: *Brazil; Rio de Janeiro

Abstract: Genetic counselling is an important aspect of prenatal care in many developed countries. This tendency has also begun to emerge in Brazil, although few medical centres offer this service. Genetic counselling provides prenatal risk control through a process of individual decision-making based on medical information, in a context where diagnostic and therapeutic possibilities overlap. Detection of severe foetal anomalies can lead to a decision involving possible termination of pregnancy. This paper focuses on medical and legal consequences of the detection of severe foetal anomalies, mainly anencephaly and Down syndrome, and in light of the fact that abortion is illegal in Brazil. The discussion is based on the literature and empirical research at a high-complexity public hospital in Rio de Janeiro.

Harrel, T. Recontacting former patients regarding BRCA1/2 rearrangement testing: opinions and practices of genetics professionals. *Journal of Genetic Counseling* 2007 December; 16(6): 670. NRCBL: 15.2; 15.3; 9.5.5. SC: em.

Harvey, Erin K.; Fogel, Chana E.; Peyrot, Mark; Christensen, Kurt D.; Terry, Sharon F.; McInerney, Joseph D. Providers' knowledge of genetics: a survey of 5915 individuals and families with genetic conditions. *Ge-*

NRCBL: National Reference Center for Bioethics Literature Classification Scheme   See inside front cover for terms.

171

netics in Medicine 2007 May; 9(5): 259-267. 16 refs. NRCBL: 15.2; 15.1; 7.2. SC: em.

> Keywords: *genetic counseling; *genetics; *health personnel; *professional competence; evaluation studies; family members; genetic disorders; medical education; patient education; patients; primary health care; survey

Hashiloni-Dolev, Yael. "Wrongful life", in the eyes of the law, the counselors and the disabled. *In her:* A Life (Un)Worthy of Living: Reproductive Genetics in Israel and Germany. Dordrecht: Springer, 2007: 119-130. 1 fn. NRCBL: 15.2; 12.5.1; 21.1. SC: le.

> Keywords: *attitude of health personnel; *genetic counseling; *legal aspects; *wrongful life; cross cultural comparison; disabled persons; genetic screening; patient advocacy; public opinion; quality of life; value of life; Keyword Identifiers: *Germany; *Israel

Hashiloni-Dolev, Yael. Genetic counselors' moral practices. *In her:* A Life (Un)Worthy of Living: Reproductive Genetics in Israel and Germany. Dordrecht: Springer, 2007: 63-81. 4 fn. NRCBL: 15.2; 4.1.1; 1.3.1; 12.5.1; 21.1. SC: em.

> Keywords: *attitude of health personnel; *genetic counseling; *professional ethics; *selective abortion; age factors; congenital disorders; cross cultural comparison; directive counseling; disabled persons; eugenics; interviews; medical specialties; prenatal diagnosis; religion; sex factors; statistics; Keyword Identifiers: *Germany; *Israel

Hashiloni-Dolev, Yael. The conflicts between individuals, families and society, as well as between different family members, embodied in reproductive genetics. *In her:* A Life (Un)Worthy of Living: Reproductive Genetics in Israel and Germany. Dordrecht: Springer, 2007: 131-146. 1 fn. NRCBL: 15.2; 12.5.1; 14.4; 8.1; 21.1.

> Keywords: *attitude of health personnel; *genetic counseling; *preimplantation diagnosis; *prenatal diagnosis; choice behavior; cross cultural comparison; culture; disabled persons; eugenics; genetic disorders; public opinion; selective abortion; Keyword Identifiers: *Germany; *Israel

Heyman, Bob; Hundt, Gillian; Sandall, Jane; Spencer, Kevin; Williams, Clare; Grellier, Rachel; Pitson, Laura. On being at higher risk: a qualitative study of prenatal screening for chromosomal anomalies. *Social Science and Medicine* 2006 May; 62(10): 2360-2372. NRCBL: 15.2; 8.3.1; 9.5.5.

Hiraki, Susan; Ormond, Kelly E.; Kim, Katherine; Ross, Lainie F. Attitudes of genetic counselors towards expanding newborn screening and offering predictive genetic testing to children. *American Journal of Medical Genetics. Part A* 2006 November 1; 140(21): 2312-2319. NRCBL: 15.2; 15.3; 9.5.7.

Hoffman, Jan. Where risk and choice and hope converge, a guiding voice. *New York Times* 2007 September 18; p. F5, F10. NRCBL: 15.2. SC: po.

> Keywords: *genetic counseling; *prenatal diagnosis; chromosome disorders; communication; Hispanic Americans; socioeconomic factors; Keyword Identifiers: New York City

Holm, Søren. Wrongful life, the welfare principle and the non-identity problem: some further complications. *In:* McLean, Sheila A.M., ed. First Do No Harm: Law, Ethics, and Healthcare. Aldershot, England; Burlington, VT: Ashgate, 2006: 407-414. NRCBL: 15.2; 4.4; 11.4. SC: le.

Howe, Edmund G. "I'm still glad you were born" — careproviders and genetic counseling. *Journal of Clinical Ethics* 2007 Summer; 18(2): 99-110. 43 fn. NRCBL: 15.2; 8.1; 8.2; 8.4.

> Keywords: *directive counseling; *genetic counseling; *professional patient relationship; caring; choice behavior; communication; emotions; empirical research; family members; genetic screening; health personnel; parents; preimplantation diagnosis; prenatal diagnosis; psychological stress; risks and benefits

Hudson, Kathy; Baruch, Susannah; Javitt, Gail. Genetic testing of human embryos: ethical challenges and policy choices. *In:* Galston, Arthur W.; Peppard, Christiana Z., eds. Expanding Horizons in Bioethics. Dordrecht; Norwell, MA: Springer, 2005: 103-122. 47 refs. NRCBL: 15.2; 14.4; 5.3. SC: rv.

> Keywords: *preimplantation diagnosis; embryo transfer; embryos; evaluation studies; federal government; genetic screening; government regulation; insurance; legal aspects; moral policy; motivation; policy making; public participation; public policy; risks and benefits; self regulation; standards; social impact; state government; value of life; Proposed Keywords: insurance coverage; Keyword Identifiers: Food and Drug Administration; United States

International Society of Nurses in Genetics [ISONG]. Provision of quality genetic services and care: building a multidisciplinary, collaborative approach among genetic nurses and genetic counselors [position statement]. Pittsburgh, PA: International Society of Nurses in Genetics, 2006 November 1: 2 p. [Online]. Accessed: http://www.isong.org/about/ps_multidisciplinarygeneticcare.cfm [2007 February 22]. NRCBL: 15.2; 15.1; 4.1.3; 6.

International Society of Nurses of Genetics [ISONG]. Genetic Counseling for Vulnerable Populations: The Role of Nursing: Position Statement. Pittsburgh, PA: International Society of Nurses of Genetics, 2002 October 10: 4 p. [Online]. Accessed: http://www.isong.org/about/ps_vulnerable.cfm [2007 February 22]. 12 refs. NRCBL: 15.2; 8.3.1; 8.3.3; 9.5.3; 9.5.7; 4.1.3; 6.

> Keywords: *genetic counseling; *nurse's role; *vulnerable populations; codes of ethics; nursing ethics; organizational policies; Proposed Keywords: societies

Iredale, Rachel; Longley, Marcus; Thomas, Christian; Shaw, Anita. What choices should we be able to make about designer babies? A Citizens' Jury of young people in South Wales. *Health Expectations* 2006 September; 9(3): 207-217. NRCBL: 15.2; 15.3; 19.5; 14.3.

Kaimal, Girija; Steinberg, Annie G.; Ennis, Sara; Harasink, Sue Moyer; Ewing, Rachel; Li, Yuelin. Parental narratives about genetic testing for hearing loss: a one year follow up study. *Journal of Genetic Counseling*

2007 December; 16(6): 775-787. NRCBL: 15.2; 9.5.7. SC: em.

Klitzman, Robert; Thorne, Deborah; Williamson, Jennifer; Chung, Wendy; Marder, Karen. Decision-making about reproductive choices among individuals at-risk for Huntington's disease. *Journal of Genetic Counseling* 2007 June; 16(3): 347-362. NRCBL: 15.2; 14.1; 15.3; 15.1; 14.4. SC: em.

Korenromp, Marijke J.; Page-Christiaens, Godelieve C.M.L.; van den Bout, Jan; Mulder, Eduard J.H.; Visser, Gerard H.A. Maternal decision to terminate pregnancy in case of Down syndrome. *American Journal of Obstetrics and Gynecology* 2007 February; 196(2): 149. NRCBL: 15.2; 12.1; 9.5.5; 9.4.

Krones, Tanja; Schlüter, Elmar; Neuwohner, Elke; El Ansari, Susan; Wissner, Thomas; Richter, Gerd. What is the preimplantation embryo? *Social Science and Medicine* 2006 July; 63(1): 1-20. NRCBL: 15.2; 14.4.

Latimer, Joanna. Becoming in-formed: genetic counselling, ambiguity and choice. *Health Care Analysis: An International Journal of Health Philosophy and Policy* 2007 March; 15(1): 13-23. 37 refs. 3 fn. NRCBL: 15.2; 14.1.
    Keywords: *genetic counseling; *parents; *patient participation; *professional family relationship; children; choice behavior; congenital disorders; diagnosis; parental consent; pedigree; professional role; reproduction; uncertainty
Abstract: The paper presents findings from an ethnography of dysmorphology, a specialism in genetic medicine, to explore genetic counselling as a process through which parents 'become informed.' Current professional and policy debate over the use of genetic technology in medicine emphasises the need for informed choice making, and for genetic services that provide parents with what is referred to as 'non-directive genetic counselling.' In the paper the process of becoming informed is shown to be very specific and to have its own effects. Specifically, genetics is performed in dysmorphology as a space of ambiguity and uncertainty. In addition, parents are engaged by the clinic as participants in the very processes through which their child, and perhaps their family, are clinically classified. The paper examines the effects of parents' immersion in this clinical space of deferral to suggest how the need for reproductive choice, and calculation, is predicated upon clinical processes that shift parents between the experience of definition and uncertainty. The paper thus troubles simple stories about autonomous and informed choice, particularly reproductive choice, as icons of contemporary versions of what it is to be fully human.

Lenhard, Wolfgang; Breitenach, Erwin; Ebert, Harald; Schindelhauer-Deutscher, H. Joachim; Zang, Klaus D.; Henn, Wolfram. Attitudes of mothers towards their child with Down syndrome before and after the introduction of prenatal diagnosis. *Intellectual and Developmental Disabilities* 2007 April; 45(2): 98-102. 12 refs. NRCBL: 15.2; 8.1; 9.5.3. SC: em. Note: Abstract in French.

    Keywords: *attitudes; *Down syndrome; *mothers; *parent child relationship; *prenatal diagnosis; children; emotions; questionnaires; time factors

Leroi, Armand Marie. The future of neo-eugenics: now that many people approve the elimination of certain genetically defective fetuses, is society closer to screening all fetuses for all known mutations? *EMBO Reports* 2006 December; 7(12): 1184-1187. NRCBL: 15.2; 15.1; 15.5; 12.1; 9.5.8.

Lisker, Rubén; Carnevale, Alessandra. Changing opinions of Mexican geneticists on ethical issues. *Archives of Medical Research* 2006 August; 37(6): 794-803. NRCBL: 15.2; 8.4; 8.3.1; 9.2; 15.3. SC: em.

MacKenzie, Catriona. Feminist bioethics and genetic termination. *Bioethics* 2007 November; 21(9): 515-516. NRCBL: 15.2; 1.1; 10; 12.5.1; 15.3. Comments: Angela Thachuk. The space in between: narratives of silence and genetic terminations. Bioethics 2007 November; 21(9): 511-514.
    Abstract: A brief discussion of how relational autonomy, phenomenological theories of embodiment and narrative approaches to clinical ethics can open up the space for more subtle feminist ethical reflection about genetic termination.

Mahowald, Mary Briody. Preconception and prenatal decisions. *In her:* Bioethics and Women: Across the Life Span. Oxford; New York: Oxford University Press, 2006: 73-91, 250-251. 26 fn. NRCBL: 15.2; 14.1; 14.4.
    Keywords: *counseling; *preimplantation diagnosis; *prenatal diagnosis; *reproduction; *women; autonomy; beneficence; cystic fibrosis; disabled persons; disclosure; embryos; fetuses; genetic disorders; genetic predisposition; genetic screening; moral status; paternity; phenylketonuria; professional patient relationship; racial groups; reproductive technologies; sex determination; spousal consent; Proposed Keywords: preconception care

Mahowald, Mary B. Prenatal testing for selection against disabilities. *CQ: Cambridge Quarterly of Healthcare Ethics* 2007 Fall; 16(4): 457-462. NRCBL: 15.2; 4.4; 1.1. SC: an.

Nelson, James Lindemann. Synecdoche and stigma. *CQ: Cambridge Quarterly of Healthcare Ethics* 2007 Fall; 16(4): 475-478. NRCBL: 15.2; 4.4; 1.1; 7.1; 12.1. SC: an. Identifiers: prenatal testing for selection against disabilities.

Nelson, James Lindemann. Testing, terminating, and discriminating. *CQ: Cambridge Quarterly of Healthcare Ethics* 2007 Fall; 16(4): 462-468. NRCBL: 15.2; 4.4; 1.1; 12.1. SC: an. Identifiers: prenatal testing for selection against disabilities.

Ozakinci, Gozde; Humphris, Gerry; Steel, Michael. Provision of breast cancer risk information to women at the lower end of the familial risk spectrum. *Community Genetics* 2007; 10(1): 41-44. 16 refs. NRCBL: 15.2; 8.1. SC: em.

NRCBL: National Reference Center for Bioethics Literature Classification Scheme   See inside front cover for terms.

173

Keywords: *breast cancer; *communication; *genetic information; *information dissemination; genetic counseling; risk; survey; women; Keyword Identifiers: Great Britain

Paonessa, Louis. Straightening out your heir: on the constitutionality of regulating the use of preimplantation technologies to select preembryos or modify the genetic profile thereof based on expected sexual orientation. *Rutgers Computer and Technology Law Journal* 2007; 33(2): 331-366. NRCBL: 15.2; 14.3; 14.4; 5.3.

Pattinson, Shaun D. Designing donors. *In:* Gunning, Jennifer; Holm, Søren, eds. Ethics, law, and society. Volume 1. Aldershot, Hants, England; Burlington, VT: Ashgate, 2005: 251-256. 6 refs. NRCBL: 15.2; 14.4; 19.5. SC: le.
   Keywords: *directed donation; *legal aspects; *preimplantation diagnosis; *reproduction; *siblings; embryo transfer; genetic disorders; genetic screening; motivation; thalassemia; Proposed Keywords: *tissue donors; *tissue typing; anemia; Keyword Identifiers: *Great Britain; Human Fertilization and Embryology Authority

Penasa, Simone. The issue of constitutional law legitimacy on "human assisted reproduction" between reasonableness of the choices and effectiveness of the protection of all involved subjects. *Revista de Derecho y Genoma Humano = Law and the Human Genome Review* 2006 July-December; (25): 117-137. 53 fn. NRCBL: 15.2; 14.4; 14.1. SC: le.
   Keywords: *constitutional law; *legal aspects; *preimplantation diagnosis; *reproductive technologies; Keyword Identifiers: *Italy

Pergament, Eugene. Controversies and challenges of array comparative genomic hybridization in prenatal genetic diagnosis. *Genetics in Medicine* 2007 September; 9(9): 596-599. NRCBL: 15.2; 15.3; 15.5.

Perry, Sandy; Woodall, Angela L.; Pressman, Eva K. Association of ultrasound findings with decision to continue Down syndrome pregnancies. *Community Genetics* 2007; 10(4): 227-230. 5 refs. NRCBL: 15.2; 12.5.2. SC: em.
   Keywords: *childbirth; *choice behavior; *Down syndrome; *pregnant women; *prenatal diagnosis; *selective abortion; decision making; evaluation studies; Proposed Keywords: retrospective studies; ultrasonography

Pirzadeh, Sara M.; McCarthy Veach, Patricia; Bartels, Dianne M.; Kao, Juihsien; LeRoy, Bonnie S. A national survey of genetic counselors' personal values. *Journal of Genetic Counseling* 2007 December; 16(6): 763-773. NRCBL: 15.2; 1.1. SC: em.

Polansky, Samara. Overcoming the obstacles: a collaborative approach to informed consent in prenatal genetic screening. *Health Law Journal* 2006; 14: 21-43. 106 fn. NRCBL: 15.2; 8.3.1; 12.5.3; 8.1. SC: le.
   Keywords: *females; *informed consent; *prenatal diagnosis; *sex determination; *sex preselection; autonomy; coercion; comprehension; congenital disorders; disclosure; genetic counseling; genetic screening; legal aspects; physician patient relationship; pregnant women; preimplantation

diagnosis; selective abortion; standards; wrongful life; Keyword Identifiers: *Canada

Reis, Linda M.; Baumiller, Robert; Scrivener, William; Yager, Geoffrey; Warren, Nancy Steinberg. Spiritual assessment in genetic counseling. *Journal of Genetic Counseling* 2007 February; 16(1): 41-52. 45 refs. NRCBL: 15.2; 1.2; 4.1.1. SC: em.
   Keywords: *genetic counseling; *knowledge, attitudes, practice; *spirituality; health personnel; professional patient relationship

Resnik, David B.; Vorhaus, Daniel B. Genetic modification and genetic determinism. *Philosophy, Ethics and Humanities in Medicine [electronic]* 2006; 1(9): 11 p. Accessed: http://www.peh-med.com/content/1/1/9 n.d. NRCBL: 15.2; 14.5; 15.4; 2.2. SC: an.

Resta, Robert G. Defining and redefining the scope and goals of genetic counseling. *American Journal of Medical Genetics. Part C, Seminars in Medical Genetics* 2006 November 15; 142(4): 269-275. NRCBL: 15.2; 8.1.

Sawyer, Susan M.; Cerritelli, Belinda; Carter, Lucy S.; Cooke, Mary; Glazner, Judith A.; Massie, John. Changing their minds with time: a comparison of hypothetical and actual reproductive behaviors in parents of children with cystic fibrosis. *Pediatrics* 2006 September; 118(3): e649-e656. NRCBL: 15.2; 12.5.2.

Sermon, K.D.; Michiels, A.; Harton, G.; Moutou, C.; Repping, S.; Scriven, P.N.; SenGupta, S.; Traeger-Synodinos, J.; Vesela, K.; Viville, S.; Wilton, L.; Harper, J.C. ESHRE PGD Consortium data collection VI: cycles from January to December 2003 with pregnancy follow-up to October 2004. *Human Reproduction* 2007 February; 22(2): 323-326. NRCBL: 15.2; 14.4. SC: em.

Shuster, Evelyne. Microarray genetic screening: a prenatal roadblock for life? *Lancet* 2007 February 10-16; 369(9560): 526-529. 31 refs. NRCBL: 15.2; 7.1.
   Keywords: *genetic disorders; *genetic techniques; *methods; *prenatal diagnosis; *technology assessment; autonomy; costs and benefits; decision making; disclosure; eugenics; fetuses; genetic counseling; genetic screening; informed consent; justice; normality; philosophy; parents; physicians; predictive value of tests; preimplantation diagnosis; public policy; risks and benefits; selective abortion; standards; trends; values; wrongful life

Silver, R.; Bernhardt, B.; Wilfond, B.; Geller, G. Genetic counselors' experiences of moral value conflicts with clients. *Journal of Genetic Counseling* 2007 December; 16(6): 690. NRCBL: 15.2; 8.1. SC: em.

Silversides, Ann. The wide gap between genetic research and clinical needs [news]. *CMAJ/JAMC: Canadian Medical Association Journal* 2007 January 30; 176(3): 315-316. 2 refs. NRCBL: 15.2; 15.3.
   Keywords: *clinical genetics; *genetic research; *genetic services; *government financing; *research support; *scarcity; education; health personnel; internship and residency;

referral and consultation; Proposed Keywords: waiting lists; Keyword Identifiers: *Canada

Simpson, Bob. Negotiating the therapeutic gap: prenatal diagnostics and termination of pregnancy in Sri Lanka. *Journal of Bioethical Inquiry* 2007; 4(3): 207-215. NRCBL: 15.2; 15.3; 12.5.2; 12.3.

Abstract: In Sri Lanka, termination of pregnancy, other than in extreme circumstances, is strictly illegal. Among the public and large sections of the medical community there is widespread support for some degree of liberalization of the law, particularly where this relates to serious genetic conditions which can be identified prenatally. Tension emerges out of a publicly maintained conservatism on issues of abortion on the one hand and a growing disconnection from unregulated practices of termination in the private sector on the other. Social science approaches have much to contribute when understanding the 'therapeutic gap' that opens up and, in particular, the way that local ideas of fate, destiny and how suffering might be ameliorated become blended with the predictive power of genetic testing.

Skene, Loane. Should the law limit genetic tests on embryos and foetuses? *University of New South Wales Law Journal* 2006; 29(2): 250-253. 4 fn. NRCBL: 15.2; 15.5; 5.3; 1.3.5. SC: le.
Keywords: *legal aspects; *preimplantation diagnosis; *prenatal diagnosis; eugenics; genetic screening; international aspects; legislation; Keyword Identifiers: Australia; Germany; Italy

Snelling, Jeanne. Implications for providers and patients: a comment on the regulatory framework for preimplantation genetic diagnosis in New Zealand. *Medical Law International* 2007; 8(1): 23-49. NRCBL: 15.2; 14.4; 5.3. SC: le; an.

Soini, S. Preimplantation genetic diagnosis (PGD) in Europe: diversity of legislation a challenge to the community and its citizens. *Medicine and Law: The World Association for Medical Law* 2007 June; 26(2): 309-323. 40 fn. NRCBL: 15.2; 14.4; 21.1. SC: le.
Keywords: *international aspects; *legal aspects; *preimplantation diagnosis; cross cultural comparison; decision making; embryos; eugenics; genetic counseling; genetic disorders; guidelines; motivation; regulation; risk; sex preselection; Proposed Keywords: tissue typing; Keyword Identifiers: *Europe; European Union

Abstract: Preimplantation genetic diagnosis (PGD) aims to safeguard the reproductive confidence of couples who have an increased risk of having a child with a serious hereditary disease. Non-directive genetic counselling is an essential part of PGD. Lately, performance of PGD for some new and non-medical indications, such as selecting for a tissue-matching embryo for a saviour sibling, or sex-selection for family-balancing, has raised ethical concerns. Who decides when to perform PGD, and for which conditions? The European member states have very diverse regulation on PGD. Some countries totally ban PGD, while the others keep close track of the new applications. The people in need of PGD seek it in the other member states. These cross-border treatments cause psy-

chological stress and pose many so far unresolved legal questions. The individuals need more information about all the aspects of PGD. This article analyses contemporary indications for PGD in Europe and relevant ethical discussion, and second, shows the diversity in regulation and reflects the consequences thereof.

Stretton, Dean. Harriton v Stephens; Waller v James: wrongful life and the logic of non-existence. *Melbourne University Law Review* 2006 December; 30(3): 972-1001. 208 fn. NRCBL: 15.2; 1.1; 8.4; 12.4.2. SC: le.
Keywords: *congenital disorders; *legal aspects; *legal liability; *negligence; *wrongful life; duty to warn; genetic disorders; physicians; prenatal diagnosis; rubella; Keyword Identifiers: *Australia; *Harriton v. Stephens; *Waller v. James

Thachuk, Angela. The space in between: narratives of silence and genetic terminations. *Bioethics* 2007 November; 21(9): 511-514. NRCBL: 15.2; 15.3; 1.1; 12.5.1. SC: an.

Abstract: In North America, prenatal testing and genetic terminations are becoming clinically normalized. Yet despite this implied social acceptance, open discussions surrounding genetic terminations remain taboo and silenced. Women are socially isolated, their experiences kept secret, and their grief disenfranchised. The lack of social consensus regarding genetic terminations, the valorization of scientific knowledge, and the bioethical framing of the issue as a matter of personal choice and autonomy collectively serve to reify this silence. In many respects genetic screening offers a form of technological surveillance procuring security from the unwanted kind of child. Yet the manner in which 'the unwanted kind of child' is understood varies from context to context. While we carry with us the consequences of decisions made elsewhere, the institutionalized discourses upon which these decisions are made are not always so readily transportable. One must somehow reconcile 'the unwanted kind of child' of the biomedical model with 'the unwanted kind of child' who was to be a member of one's family. In this paper, my intention is not to engage in the broader debate surrounding prenatal testing and genetic terminations. Rather, I employ my clinical encounters with these practices to illustrate the absence of an ethical language that might do justice to the experiences such practices construct. The limitations of a bioethical discourse that remains abstracted from lived experience are discussed.

Tonti-Filippini, Nicholas. Reproductive discrimination. *University of New South Wales Law Journal* 2006; 29(2): 254-260. 12 fn. NRCBL: 15.2; 15.5; 21.1. SC: le.
Keywords: *disabled persons; *genetic intervention; *reproductive rights; *social discrimination; eugenics; genetic disorders; guidelines; international aspects; preimplantation diagnosis; prenatal diagnosis; reproductive technologies; Keyword Identifiers: Australia; Europe; European Convention on Human Rights and Biomedicine; National Health and Medical Research Council (Australia); United Nations; Universal Declaration on the Human Genome and Human Rights

NRCBL: National Reference Center for Bioethics Literature Classification Scheme   See inside front cover for terms.

175

Tuohey, John F. Screening for aneuploidy: a complex ethical issue. *Health Care Ethics USA* 2007 Spring; 15(2): 4-8. 25 refs. NRCBL: 15.2; 1.2; 9.5.5; 9.5.8.
   Keywords: *chromosome abnormalities; *prenatal diagnosis; Down syndrome; fetal development; moral obligations; pregnant women; Roman Catholic ethics; risk

Turone, Fabio. Court upholds demand for preimplantation genetic diagnosis [news]. *BMJ: British Medical Journal* 2007 October 6; 335(7622): 687. NRCBL: 15.2; 14.4. SC: le. Identifiers: Italy.

Wade, Christopher H.; Wilfond, Benjamin S. Ethical and clinical practice considerations for genetic counselors related to direct-to-consumer marketing of genetic tests. *American Journal of Medical Genetics. Part C, Seminars in Medical Genetics* 2006 November 15; 142(4): 284-292; discussion 293. NRCBL: 15.2; 15.3; 8.1.

Wasserman, David; Asch, Adrienne. Reply to Nelson. *CQ: Cambridge Quarterly of Healthcare Ethics* 2007 Fall; 16(4): 478-482. NRCBL: 15.2; 4.4; 1.1; 7.1; 12.1. SC: an. Identifiers: prenatal testing for selection against disabilities.

Woltanski, A.; Cragun, R.; Myers, M.; Cragun, D. Views on religion and abortion: a comparison of genetic counselors and the general population. *Journal of Genetic Counseling* 2007 December; 16(6): 686. NRCBL: 15.2; 1.2; 12.3. SC: em.

Wüstner, Kerstin; Heinze, Ulrich. Attitudes towards preimplantation genetic diagnosis — a German and Japanese comparison. *New Genetics and Society* 2007 April; 26(1): 1-27. NRCBL: 15.2; 14.4; 21.1. SC: em.

Yuen, R.K.N.; Lam, S.T.S.; Allison, D. Bioethics and prenatal diagnosis of foetal diseases. *Hong Kong Medical Journal* 2006 December; 12(6): 488. NRCBL: 15.2; 2.1; 7.2.

Zeiler, Kristin. Shared decision-making, gender and new technologies. *Medicine, Health Care and Philosophy* 2007 September; 10(3): 279-287. 21 refs. 5 fn. NRCBL: 15.2; 14.4; 1.1; 8.1; 10. SC: an.
   Keywords: *autonomy; *clinical genetics; *decision making; *embryo transfer; *ethical analysis; *in vitro fertilization; *interpersonal relations; *men; *preimplantation diagnosis; *reproductive medicine; *women; choice behavior; dissent; empirical research; genetic counseling; genetic disorders; genetic screening; health personnel; intention; married persons; philosophy; physician patient relationship; prenatal diagnosis; professional role; reproductive technologies; selective abortion; Proposed Keywords: *sex factors; cooperative behavior
   Abstract: Much discussion of decision-making processes in medicine has been patient-centred. It has been assumed that there is, most often, one patient. Less attention has been given to shared decision-making processes where two or more patients are involved. This article aims to contribute to this special area. What conditions need to be met if decision-making can be said to be shared? What is a shared decision-making process and

what is a shared autonomous decision-making process? Why make the distinction? Examples are drawn from the area of new reproductive medicine and clinical genetics. Possible gender-differences in shared decision-making are discussed.

GENETIC DATABASES AND BIOBANKS

A new tool to promote research: the Law on Biomedical Research [editorial]. *Revista de Derecho y Genoma Humano = Law and the Human Genome Review* 2007 January-June; (26): 17-20. NRCBL: 15.1. SC: le.
   Keywords: *biological specimen banks; *biomedical research; *genetic databases; *genetic research; *legal aspects; *legislation; advisory committees; cloning;; economics; embryo research; embryonic stem cells; genetic engineering; human dignity; nuclear transfer techniques; Keyword Identifiers: *Spain; Council of Europe

Genome abuse [editorial]. *Nature* 2007 September 27; 449(7161): 377. NRCBL: 15.1; 1.3.5; 1.3.12. Identifiers: France; Great Britain (United Kingdom).

Aray-Blais, Christiane; Patenaude, Johane. Biobanking primer: down to basics [letter]. *Science* 2007 May 11; 316(5826): 830. NRCBL: 15.1; 1.3.12; 18.2.
   Keywords: *biological specimen banks; *genetic databases; ethical review; genetic research; informed consent; population genetics; research ethics committees; research subjects

Árnason, Gardar. Second thoughts on biobanks: the Icelandic experience. *In:* Einsiedel, Edna; Timmermans, Frank, eds. Crossing Over: Genomics in the Public Arena. Calgary, Alberta, Canada: University of Calgary Press, 2005: 193-203. 21 refs.; 6 fn. NRCBL: 15.1; 1.3.12. Conference: Essays from the conference held Apr. 25-27, 2003, Kananaskis, Alta.
   Keywords: *databases; *genetic databases; advisory committees; biological specimen banks; committee membership; confidentiality; conflict of interest; donors; genetic research; government; health services research; guidelines; informed consent; interinstitutional relations; legislation; medical records; pedigree; population genetics; presumed consent; privacy; public health; public policy; regulation; research ethics; research subjects; standards; Keyword Identifiers: *Iceland; Declaration of Helsinki; deCode Genetics; Health Sector Database (Iceland); National Bioethics Committee (Iceland); Nuremberg Code

Bauer, Martin W. The public career of the 'gene' — trends in public sentiments from 1946 to 2002. *New Genetics and Society* 2007 April; 26(1): 29-45. NRCBL: 15.1; 1.3.11; 2.2; 15.5; 1.3.12. SC: em. Identifiers: Great Britain (United Kingdom).

Campbell, Alastair V. The ethical challenges of biobanks: safeguarding altruism and trust. *In:* McLean, Sheila A.M., ed. First Do No Harm: Law, Ethics, and Healthcare. Aldershot, England; Burlington, VT: Ashgate, 2006: 203-214. 30 fn. NRCBL: 15.1; 1.3.12. SC: rv.
   Keywords: *biological specimen banks; *genetic databases; access to information; altruism; autonomy; blood specimen collection; confidentiality; databases; decision making; disclosure; ethical review; industry; informed consent; interna-

tional aspects; medical records; population genetics; public health; regulation; research findings; research subjects; risk; tissue donors; trust

Cotton, Richard G.H.; Sallée, Clémentine; Knoppers, Bartha M. Locus-specific databases: from ethical principles to practice. *Human Mutation* 2005 November; 26(5): 489-493. NRCBL: 15.1; 1.3.12; 18.3; 8.4. SC: em.

Couzin, Jennifer. Kaiser to set up gene bank [news]. *Science* 2007 February 23; 315(5815): 1067. NRCBL: 15.1; 1.3.12.

Cullen, Rowena; Marshall, Stephen. Genetic research and genetic information: a health information professional's perspective on the benefits and risks. *Health Information and Libraries Journal* 2006 December; 23(4): 275-282. NRCBL: 15.1; 1.3.12; 15.10; 15.3; 15.4; 9.7.

Ducournau, Pascal. The viewpoint of DNA donors on the consent procedure. *New Genetics and Society* 2007 April; 26(1): 105-116. NRCBL: 15.1; 1.3.12; 18.3; 19.5. SC: em.

Einsiedel, Edna; Sheremeta, Lorraine. Biobanks and the challenges of commercialization. *In:* Sensen, Chrisopher W., ed. Handbook of Genome Research: Genomics, Proteomics, Metabolomics, Bioinformatics, Ethical and Legal Issues. Volume 2. Weinheim: Wiley-VCH, 2005: 537-559. 112 refs. NRCBL: 15.1; 1.3.12; 15.8; 9.3.1. SC: rv.
  Keywords: *biological specimen banks; *genetic databases; *population genetics; benefit sharing; biotechnology; commerce; genetic materials; genetic patents; genomics; human genome; industry; international aspects; private sector; property rights; public opinion; public participation; public sector; regulation; Proposed Keywords: genetic resources; Keyword Identifiers: Myriad Genetics Inc.

Eischen, Kyle. Commercializing Iceland: biotechnology, culture, and the information society. *In:* Mehta, Michael D., ed. Biotechnology Unglued: Science, Society and Social Cohesion. Vancouver: UBC Press, 2005: 95-116. 48 refs. 2 fn. NRCBL: 15.1; 1.3.12; 9.3.1; 15.11; 13.1; 5.3.
  Keywords: *genetic databases; *genetic research; *social impact; biotechnology; culture; genetic ancestry; genetic privacy; geographic factors; industry; legal aspects; population genetics; socioeconomic factors; Proposed Keywords: commerce; Keyword Identifiers: *Iceland; *deCode Genetics

Engels, Eve-Marie. Biobanken für die medizinische Forschung: Probleme und Potenzial. *In:* Schreiber, Hans-Peter, ed. Biomedizin und Ethik: Praxis, Recht, Moral. Basel; Boston: Birkhäuser, 2004: 29-40, 95-96. 9 refs. NRCBL: 15.1; 1.3.12; 18.1.
  Keywords: *biological specimen banks; *genetic databases; benefit sharing; biomedical research; confidentiality; genetic information; informed consent; international aspects; organization and administration; property rights; regulation; tissue donors

Gamero, Joaquin J.; Romero, Jose-Luis; Peralta, Juan-Luis; Carvalho, Mónica; Corte-Real, Francisco. Spanish public awareness regarding DNA profile data-

bases in forensic genetics: what type of DNA profiles should be included? *Journal of Medical Ethics* 2007 October; 33(10): 598-604. 52 refs. NRCBL: 15.1; 1.3.5; 1.3.12; 1.3.5; 8.4; 18.3. SC: em.
  Keywords: *biological specimen banks; *DNA fingerprinting; *forensic genetics; *genetic databases; *public opinion; *public policy; genetic privacy; government regulation; informed consent; law enforcement; legislation; policy making; predictive value of tests; prisoners; public participation; survey; Keyword Identifiers: *Spain; European Union

Abstract: The importance of non-codifying DNA polymorphism for the administration of justice is now well known. In Spain, however, this type of test has given rise to questions in recent years: (a) Should consent be obtained before biological samples are taken from an individual for DNA analysis? (b) Does society perceive these techniques and methods of analysis as being reliable? (c) There appears to be lack of knowledge concerning the basic norms that regulate databases containing private or personal information and the protection that information of this type must be given. This opinion survey and the subsequent analysis of the results in ethical terms may serve to reveal the criteria and the degree of information that society has with regard to DNA databases. In the study, 73.20% (SE 1.12%) of the population surveyed was in favour of specific legislation for computer files in which DNA analysis results for forensic purposes are stored.

Gerlach, Neil. Biotechnology and social control: the Canadian DNA data bank. *In:* Mehta, Michael D., ed. Biotechnology Unglued: Science, Society and Social Cohesion. Vancouver: UBC Press, 2005: 117-132. 40 refs. 2 fn. NRCBL: 15.1; 1.3.12; 1.3.5.
  Keywords: *DNA fingerprinting; *genetic databases; legislation; public policy; Keyword Identifiers: *Canada

Gesche, Astrid H. Genetic testing and human genetic databases. *In:* Betta, Michela, ed. The Moral, Social, and Commercial Imperatives of Genetic Testing and Screening: the Australian Case. Dordrecht: Springer, 2006: 71-94. 84 fn. NRCBL: 15.1; 1.3.5; 1.3.12; 15.3; 18.3. SC: rv.
  Keywords: *genetic databases; DNA fingerprinting; genetic information; genetic privacy; genetic research; genetic screening; law enforcement; private sector; property rights; public sector; refusal to participate; registries; regulation; research subjects; Proposed Keywords: classification; tissue donors; Keyword Identifiers: Australia

Gevers, Sjef. Human tissue research, with particular reference to DNA banking. *In:* Gevers, J.K.M.; Hondius, E.H.; Hubben, J.H., eds. Health Law, Human Rights and the Biomedicine Convention: Essays in Honour of Henriette Roscam Abbing. Leiden; Boston: Martinus Nijhoff Publishers, 2005: 231-243. 25 fn. NRCBL: 15.1; 1.3.12.
  Keywords: *biological specimen banks; *genetic databases; family members; government regulation; guidelines; human experimentation; informed consent; legal aspects; research subjects; Proposed Keywords: tissue donors; Keyword Identifiers: *European Convention on Human Rights and Biomedicine; Europe

NRCBL: National Reference Center for Bioethics Literature Classification Scheme   See inside front cover for terms.

177

Godard, Béatrice. Involving communities: a matter of trust and communication. *In:* Einsiedel, Edna; Timmermans, Frank, eds. Crossing Over: Genomics in the Public Arena. Calgary, Alberta, Canada: University of Calgary Press, 2005: 87-98. 33 refs., 2 fn. NRCBL: 15.1; 1.3.12; 15.11; 5.1. Conference: Essays from the conference held Apr. 25-27, 2003, Kananaskis, Alta.

> Keywords: *genetic databases; *genetic research; *population genetics; *public participation; communication; confidentiality; genetic diversity; genomics; information dissemination; public opinion; research design; research subjects; selection of subjects; trust; Proposed Keywords: research findings; Keyword Identifiers: *Cartagene Project; *Quebec

Godard, Béatrice; Marshall, Jennifer; Laberge, Claude. Community engagement in genetic research: results of the first public consultation for the Quebec CARTaGENE project. *Community Genetics* 2007; 10(3): 147-158. 28 refs. NRCBL: 15.1; 1.3.12; 18.1. SC: em.

> Keywords: *genetic databases; *genetic research; *public opinion; *public participation; biological specimen banks; choice behavior; confidentiality; donors; duty to recontact; ethnic groups; focus groups; genetic privacy; knowledge, attitudes, practice; public sector; private sector; questionnaires; research subjects; researcher subject relationship; risks and benefits; Proposed Keywords: research findings; Keyword Identifiers: *Quebec; *Quebec CARTaGENE

Gulcher, Jeff; Stefansson, Kari. The Icelandic healthcare database: a tool to create knowledge, a social debate, and a bioethical and privacy challenge [editorial]. *Medscape Molecular Medicine* 1999: 5 p. [Online]. Accessed: http://www.medscape.com/viewarticle/414505 [2008 March 4]. NRCBL: 15.1; 8.4; 1.3.12.

Heinrichs, Bert. A comparative analysis of selected European guidelines and recommendations for biobanks with special regard to the research / non-research distinction. *Revista de Derecho y Genoma Humano = Law and the Human Genome Review* 2007 July-December; (27): 205-224. 24 fn. NRCBL: 15.1; 1.3.12; 18.2; 19.5; 21.1.

> Keywords: *biological specimen banks; *genetic databases; *genetic research; *guidelines; advisory committees; comparative studies; genetic information; informed consent; international aspects; organizational policies; public policy; research subjects; tissue donors; Keyword Identifiers: *Europe; Council of Europe; European Group on Ethics in Science and New Technologies; German National Ethics Council; Human Genetics Commission (Great Britain)

Helft, Paul R.; Champion, Victoria L.; Eckles, Rachael; Johnson, Cynthia S.; Meslin, Eric M. Cancer patients' attitudes toward future research uses of stored human biological materials. *Journal of Empirical Research on Human Research Ethics* 2007 September; 2(3): 15-22. 15 refs. NRCBL: 19.5; 18.3; 9.5.1; 15.1. SC: em.

> Keywords: *attitudes; *biological specimen banks; *biomedical research; *donors; *genetic materials; *informed consent; *patients; cancer; privacy; racial groups; research subjects; survey; time factors; Proposed Keywords: *tissue donors; Keyword Identifiers: Indiana University Cancer Center

Abstract: The policy debate concerning informed consent for future, unspecified research of stored human biological materials (HBM) would benefit from an understanding of the attitudes of individuals who contribute tissue specimens to HBM repositories. Cancer patients who contributed leftover tissue to the Indiana University Cancer Center Tissue Bank under such conditions were recruited for a mail survey study of their attitudes. Our findings suggest that a clear majority of subjects would permit unlimited future research on stored HBMs without re-contact and reconsent, and a significant minority appear to desire ongoing control over future research uses of their tissue. These differences merit further investigation and suggest that a policy of blanket consent for all future, unspecified research would be premature.

Helgesson, Gert; Dillner, Joakim; Carlson, Joyce; Bartram, Claus R.; Hansson, Mats G. Ethical framework for previously collected biobank samples [letter]. *Nature Biotechnology* 2007 September; 25(9): 973-976. 16 refs. NRCBL: 15.1; 1.3.12.

> Keywords: *biological specimen banks; *biomedical research; *genetic research; *informed consent; *presumed consent; *standards; *tissue donors; access to information; confidentiality; duty to recontact; ethical review; genetic databases; guidelines; legal aspects; refusal to participate; regulation; research findings; research subjects; time factors

Irving, Louise; Harris, John. Biobanking. *In:* Steinbock, Bonnie, ed. The Oxford Handbook of Bioethics. Oxford; New York: Oxford University Press, 2007: 240-257. 20 refs. NRCBL: 15.1; 1.3.12. SC: an; rv.

> Keywords: *biological specimen banks; *genetic databases; *genetic information; *moral policy; biomedical research; common good; confidentiality; genetic discrimination; genetic privacy; genetic research; historical aspects; informed consent; international aspects; moral obligations; policy analysis; property rights; public health; public policy; research subjects; risks and benefits; tissue donors; trends; Proposed Keywords: exceptionalism

Jayaraman, K.S. Database targets Parsi genes [news]. *Nature* 2007 March 29; 446(7135): 475. NRCBL: 15.1; 1.3.12; 15.11. Identifiers: India.

> Keywords: *ethnic groups; *genetic databases; *population genetics; access to information; commerce; industry; informed consent; international aspects; medical records; pedigree; research subjects; Keyword Identifiers: *India; *Parsis

Kaiser, Matthias. Practical ethics in search of a toolbox: ethics of science and technology at the crossroads. *In:* Gunning, Jennifer; Holm, Søren, eds. Ethics, law, and society. Volume 2. Aldershot, Hants, England; Burlington, VT: Ashgate, 2006: 35-44. 38 refs. NRCBL: 15.1; 1.3.12; 22.3. SC: an.

> Keywords: *biotechnology; *democracy; *decision making; *ethical analysis; *genetically modified organisms; *policy analysis; *policy making; *public policy; advisory committees; consensus; cultural pluralism; ethics committees; genetically modified food; guidelines; international aspects; public participation; risks and benefits; standards; technology assessment; theoretical models; Proposed Keywords: stakeholders

Kaye, D.H. Bioethics, bench, and bar: selected arguments in Landry v. Attorney General. *Jurimetrics* 2000 Winter; 40(2): 193-216. 119 fn. NRCBL: 15.1; 1.3.12; 1.3.5; 8.4. SC: le.

> Keywords: *DNA fingerprinting; *genetic databases; *legal aspects; biological specimen banks; blood specimen collection; family members; genetic privacy; genetic research; government regulation; law enforcement; prisoners; research subjects; Keyword Identifiers: *Landry v. Attorney General; *Massachusetts; Federal Policy (Common Rule) for the Protection of Human Subjects 1991; Nuremberg Code; United States

Kaye, D.H.; Smith, Michael E. DNA identification databases: legality, legitimacy, and the case for population-wide coverage. *Wisconsin Law Review* 2003; 2003(3): 413-459. 153 fn. NRCBL: 15.1; 1.3.5. 1.3.12; 8.4. SC: le.

> Keywords: *DNA fingerprinting; *genetic databases; *law enforcement; *legal aspects; biological specimen banks; constitutional law; costs and benefits; criminal law; genetic privacy; justice; population genetics; prisoners; public opinion; racial groups; Supreme Court decisions; Keyword Identifiers: *United States

Knoppers, Bartha Maria; Sallée, Clémentine. Ethical aspects of genome research and banking. *In:* Sensen, Chrisopher W., ed. Handbook of Genome Research: Genomics, Proteomics, Metabolomics, Bioinformatics, Ethical and Legal Issues. Volume 2. Weinheim: Wiley-VCH, 2005: 509-536. 65 refs. NRCBL: 15.1; 1.3.12. SC: rv.

> Keywords: *genetic databases; *genetic research; access to information; advisory committees; benefit sharing; biological specimen banks; cloning; disclosure; duty to warn; economics; ethical review; family members; gene therapy; genetic patents; genetic privacy; germ cells; guidelines; human rights; informed consent; international aspects; legal aspects; property rights; regulation; research ethics; research subjects; right not to know; trends; Proposed Keywords: tissue donors

Krimsky, Sheldon; Simoncelli, Tania. Genetic privacy: new frontiers. *GeneWatch* 2007 September-October; 20(5): 3-10. NRCBL: 15.1; 1.3.5; 8.4; 1.3.12. SC: le. Identifiers: DNA Fingerprints Act of 2005.

Mc Fleming, Jennifer. The governance of human genetic research databases in mental health research. *International Journal of Law and Psychiatry* 2007 May-June; 30(3): 182-190. 45 refs. NRCBL: 15.1; 1.3.12; 15.6; 8.4; 17.1; 18.3. SC: le.

> Keywords: *genetic databases; *genetic research; *mental health; biological specimen banks; confidentiality; disclosure; empirical research; genetic information; genetic privacy; guidelines; informed consent; legal aspects; mentally ill persons; property rights; regulation; research subjects; schizophrenia; stigmatization; vulnerable populations; Proposed Keywords: clinical utility; exceptionalism; research findings; Keyword Identifiers: Australia

Merlo, D.F.; Knudsen, L.E.; Matusiewicz, K.; Niebrój, L.; Vähäkangas, K.H. Ethics in studies on children and environmental health. *Journal of Medical Ethics* 2007 July; 33(7): 408-413. 58 refs. NRCBL: 18.5.2; 16.1; 18.2; 18.3; 18.5.4; 15.1; 1.3.12.

> Keywords: *adolescents; *biological specimen banks; *children; *nontherapeutic research; age factors; autonomy; biomedical research; consent forms; genetic databases; genetic research; guidelines; health hazards; informed consent; minors; parental consent; public health; refusal to participate; research ethics committees; research subjects; standards; tissue donors; vulnerable populations; Proposed Keywords: *environmental health; Keyword Identifiers: Europe

Abstract: Children, because of age-related reasons, are a vulnerable population, and protecting their health is a social, scientific and emotional priority. The increased susceptibility of children and fetuses to environmental (including genotoxic) agents has been widely discussed by the scientific community. Children may experience different levels of chemical exposure than adults, and their sensitivity to chemical toxicities may be increased or decreased in comparison with adults. Such considerations also apply to unborn (fetal exposure) and newborn (neonatal exposure) children. Therefore, research on children is necessary in both clinical and environmental fields, to provide age-specific relevant data regarding the efficacy and safety of medical treatments, and regarding the assessment of risk from unintended environmental exposure. In this context, the stakeholders are many, including children and their parents, physicians and public health researchers, and the society as a whole, with its ethical, regulatory, administrative and political components. The important ethical issues are information of participants and consent to participate. Follow-up and protection of data (samples and information derived from samples) should be discussed in the context of biobanks, where children obtain individual rights when they become adults. It is important to realise that there are highly variable practices within European countries, which may have, in the past, led to differences in practical aspects of research in children. A number of recommendations are provided for research with children and environmental health. Environmental research with children should be scientifically justified, with sound research questions and valid study protocols of sufficient statistical power, ensuring the autonomy of the child and his/her family at the time of the study and later in life, if data and samples are used for follow-up studies. When children are enrolled, we recommend a consent dyad, including (1) parental (or legal guardian) informed consent and (2) the child's assent and/or informed consent from older minors. For evaluation of the studies including children, a paediatrician should always be involved in the research ethics committee.

Milanovic, Fabien; Pontille, David; Cambon-Thomsen, Anne. Biobanking and data sharing: a plurality of exchange regimes. *Genomics, Society and Policy* 2007 April; 3(1): 17-30. 24 fn. NRCBL: 15.1; 1.3.12; 5.3. SC: em.

> Keywords: *access to information; *biological specimen banks; *genetic databases; *genetic materials; *interinstitutional relations; *institutional policies; benefit sharing; biomedical research; clinical genetics; contracts; genetic research; hospitals; industry; interviews; interprofessional relations; laboratories; organization and administration;

NRCBL: National Reference Center for Bioethics Literature Classification Scheme   See inside front cover for terms.

179

physicians; property rights; research institutes; researchers; science; theoretical models; Proposed Keywords: *cooperative behavior; genetic resources; stakeholders; Keyword Identifiers: France

Abstract: Key activities in biomedicine and related research rely on collections of biological samples and related files. Access to such resources in industry and in academic contexts has become strategic and represents a central issue in the general framework of rising patenting practices and in debates about the knowledge economy. It raises important issues concerning the organisation of scientific and medical work, the outline of data-sharing guidelines, and science policy's contribution to the elaboration of an adapted framework. This paper presents an ethnographic study of three French human biobanks. Building on field work (participant observation and in-depth interviews), the study focuses on data access in the concrete practices in biobanks. The paper develops a perspective based on an analysis of different exchange regimes. We argue that access practices are submitted to the different regimes that can coexist and be articulated within the daily activities of each biobank. We also discuss how this perspective can further our understanding of biomedical research, and how it might inform data access policy.

Motluk, Alison. Crisis of trust over sperm bank errors: a new register of DNA from donors and their offspring is exposing major gaps in US sperm bank regulation. *New Scientist* 2007 August 11-17; 195(2616): 6-7. NRCBL: 14.6; 14.2; 15.3.

Keywords: *biological specimen banks; *genetic screening; *semen donors; *sperm; *standards; artificial insemination; industry; informed consent; medical errors; registries; regulation; Keyword Identifiers: *United States; Great Britain

Myskja, Bjørn K. Lay expertise: why involve the public in biobank governance? *Genomics, Society and Policy* 2007 April; 3(1): 1-16. 52 fn. NRCBL: 15.1; 1.3.12; 5.1.

Keywords: *genetic databases; *policy making; *public participation; *social control; *technical expertise; democracy; deontological ethics; focus groups; genetic research; international aspects; methods; politics; science; technology; trust; Proposed Keywords: stakeholders

Abstract: Key to concerns about public involvement in technology governance is the concept of lay expertise, the idea that lay people possess some kind of special knowledge that neither trained experts in technology, ethics and social sciences nor professional politicians possess. There are at least four different meanings of "lay expert": (1) Lay people who are educated into quasi-experts on a particular issue or technology; (2) Lay people who turn themselves into experts in order to challenge scientific experts; (3) Lay people with particular knowledge based on tradition and experience; (4) Lay people who represent an alternative perspective to expert views because they are non-experts. The challenge is that lay people are often ignorant in the relevant matters and wish to leave governance to experts. Still, there are normative reasons for lay engagement, either as stakeholders or as citizens in a deliberative democracy. According to the second approach, political decisions should be based on an inclusive open debate aimed at the better argument, providing lay people a crucial role in governance. In order to include lay people without making them hostage to experts, politicians or interest groups, we can engage them in focus group interviews which are analysed by social scientists and included in the interdisciplinary debate in journals and political forums.

National Institutes of Health [NIH] (United States). Stem Cell Information [government document]. Bethesda, MD: National Institutes of Health [NIH] (United States), 2007 May 4 and 2007 September 11: [34 p.] [Online]. Accessed: http://stemcells.nih.gov/staticresources/research/registry/PDFs/Registry.pdf [2007 October 2]. NRCBL: 18.5.4; 15.1; 5.3. SC: em.

Keywords: *biological specimen banks; *cell lines; *embryo research; *embryonic stem cells; *federal government; *government financing; *public policy; *research support; guidelines; international aspects; registries; Keyword Identifiers: *National Institutes of Health; *United States

Palmer, Lyle J. UK Biobank: bank on it. *Lancet* 2007 June 16-22; 369(9578): 1980-1981. NRCBL: 15.1; 1.3.12; 21.1; 9.8.

Pentz, Rebecca D.; Billot, Laurent; Wendler, David. Research on stored biological samples: views of African American and White American cancer patients. *American Journal of Medical Genetics. Part A* 2006 April 1; 140(7): 733-739. 14 refs. NRCBL: 15.1; 1.3.12; 19.5; 18.3; 9.5.4. SC: em.

Keywords: *attitudes; *biological specimen banks; *biomedical research; *blacks; *cancer; *minority groups; *patients; *whites; body parts and fluids; comparative studies; confidentiality; consent forms; donors; duty to recontact; genetic databases; informed consent; refusal to participate; socioeconomic factors; survey; time factors; Proposed Keywords: *tissue donors; Keyword Identifiers: Georgia

Petersen, Alan. 'Biobanks' "engagements": engendering trust or engineering consent? *Genomics, Society and Policy* 2007 April; 3(1): 31-43. 56 fn. NRCBL: 15.1; 1.3.12; 5.1.

Keywords: *genetic databases; *international aspects; *policy making; *public participation; *trust; administrators; biotechnology; genetic research; genetically modified organisms; informed consent; population genetics; nanotechnology; politics; presumed consent; program descriptions; researchers; risk; science; social control; Keyword Identifiers: *UK Biobank; Great Britain

Abstract: The rapid development of biobanks internationally reflects the considerable expectations attached to the exploitation of genetics knowledge. However, establishing consent and legitimacy for the new generation of biobanks is not without its challenges because they tend to be prospective in nature, involving the collection of DNA, personal medical and lifestyle data generally held over a very long period of time for unspecified research purposes. Thus far, biobanks have tended to be established ahead of wide-ranging debate about their broad implications. Making specific reference to the 'engagement' processes employed by UK Biobank during its establishment phase, this article focuses on the politics of 'public engagement'. It examines the context of arguments for 'public engagement', drawing attention to how particular approaches to 'engagement' within biobank projects may serve to limit debate on substantive questions arising from their development. Unless biobanks' 'engagements' substantially involve publics in deliberations about their development, it is argued, publics are likely to become distrustful of projects and perhaps resist other similar population-based health initiatives in the future.

Prainsack, Barbara. Research populations: biobanks in Israel. *New Genetics and Society* 2007 April; 26(1): 85-103. NRCBL: 15.1; 19.5; 15.3; 1.3.12; 18.2. SC: em.

Roche, Patricia A.; Annas, George J. New genetic privacy concerns: DNA donors may give up more than they realize. *GeneWatch* 2007 January-February; 20(1): 14-17. NRCBL: 15.1; 1.3.12.

Roscam Abbing, Henriette. Human tissue research, individual rights and bio-banks. *In:* Gunning, Jennifer; Holm, Søren, eds. Ethics, law, and society. Volume 2. Aldershot, Hants, England; Burlington, VT: Ashgate, 2006: 7-15. 10 fn. NRCBL: 15.1; 1.3.12.
> Keywords: *biological specimen banks; *genetic databases; *genetic research; biomedical research; commodification; confidentiality; family members; genetic information; human rights; informed consent; nontherapeutic research; population genetics; presumed consent; research subjects; Proposed Keywords: third-party research subjects; tissue donors

Royal College of Obstetricians and Gynaecologists (Great Britain). RCOG response to MRC consultation on the Code of Practice for the use of human stem cell lines. London: Royal College of Obstetricians and Gynaecologists, 2004 May 27; 3 p. [Online]. Accessed: http://www.rcog.org.uk/index.asp?PageID=1349 [2007 April 5]. NRCBL: 18.5.4; 15.1; 6. Identifiers: Great Britain (United Kingdom).
> Keywords: *biological specimen banks; *cell lines; *embryonic stem cells; *guidelines; *standards; embryo research; ethical review; fetal research; informed consent; organizational policies; physicians; professional organizations; stem cells; Keyword Identifiers: *Royal College of Obstetricians and Gynaecologists (Great Britain); Great Britain

Rushlow, Jenny. Rapid DNA database expansion and disparate minority impact. *GeneWatch* 2007 July-August; 20(4): 3-11. 48 fn. NRCBL: 15.1; 1.3.5; 1.3.12; 9.5.4; 15.11. SC: le; em.
> Keywords: *DNA fingerprinting; *genetic databases; *law enforcement; *legislation; *minority groups; *social impact; *state government; adults; biological specimen banks; evaluation studies; federal government; legal aspects; legal rights; minors; racial groups; social discrimination; statistics; trends; Proposed Keywords: crime; Keyword Identifiers: *United States

Stein, Rob. 'Embryo bank' stirs ethics fears; firm lets clients pick among fertilized eggs. *Washington Post* 2007 January 6; p. A1, A8. NRCBL: 14.4; 14.6; 15.5. SC: po.
> Keywords: *embryo transfer; *remuneration; commodification; ovum donors; semen donors; cryopreservation; *embryo; eugenics; *biological specimen banks

Swede, Helen; Stone, Carol L.; Norwood, Alyssa R. National population-based biobanks for genetic research. *Genetics in Medicine* 2007 March; 9(3): 141-149. 73 refs. NRCBL: 15.1; 1.3.12; 19.5; 21.1; 18.3. SC: rv.
> Keywords: *biological specimen banks; *genetic databases; *population genetics; *international aspects; *public policy; *standards; confidentiality; donors; federal government; genetic privacy; genetic research; guidelines; informed consent; policy making; public health; public participation; research subjects; research support; state government; trends; Keyword Identifiers: *United States

Tai, Michael Cheng-tek. The debate on establishing a biobank in Taiwan. *Asian Biotechnology and Development Review* 2006 November; 9(1): 31-36 [Online]. Accessed. http://www.openj-gate.org/articlelist.asp?LatestYear=2007&JCode=104346&year=2006&vol=9&issue=1&ICode=583649 [2008 February 29]. NRCBL: 15.1; 1.3.12; 4.4; 15.11.

UK Biobank. UK Biobank ethics and governance framework [EGF]: Version 2.0. London: UK Biobank, 2006 July; 20 p. [Online]. Accessed: http://www.ukbiobank.ac.uk/docs/EGF_Version2_July%2006%20most%20uptodate.pdf [2007 April 3]. NRCBL: 15.1; 1.3.12; 18.3.
> Keywords: *biological specimen banks; *genetic databases; access to information; accountability; benefit sharing; biomedical research; confidentiality; donors; ethical review; financial support; information dissemination; informed consent; medical records; organization and administration; refusal to participate; remuneration; research subjects; selection of subjects; standards; Keyword Identifiers: *Great Britain; *UK Biobank

Wallace, Helen. The UK National DNA Database. Balancing crime detection, human rights and privacy. *EMBO Reports* 2006 July; 7 Special No: S26-S30. 24 refs. NRCBL: 15.1; 1.3.5; 1.3.12; 8.4.
> Keywords: *biological specimen banks; *DNA fingerprinting; *forensic genetics; *genetic databases; *law enforcement; accountability; behavioral genetics; donors; genetic privacy; genetic research; human rights; public policy; Proposed Keywords: crime; Keyword Identifiers: *Great Britain

Watts, Geoff. Genes on ice. *BMJ: British Medical Journal* 2007 March 31; 334(7595): 662-663. NRCBL: 15.1; 1.3.12; 21.1.
> Keywords: *genetic databases; *international aspects; access to information; biological specimen banks; genetic privacy; genetic research; industry; informed consent; population genetics; private sector; public sector; research subjects; Keyword Identifiers: Asia; Australia; deCode Genetics; Europe; Great Britain; Iceland; North America; UK Biobank

Watts, Geoff. UK Biobank gets 10% response rate as it starts recruiting volunteers in Manchester [news]. *BMJ: British Medical Journal* 2007 March 31; 334(7595): 659. NRCBL: 15.1; 1.3.12.

NRCBL: National Reference Center for Bioethics Literature Classification Scheme   See inside front cover for terms.

181

Keywords: *biological specimen banks; *genetic databases; donors; genetic research; informed consent; research design; research subjects; selection of subjects; statistics; Proposed Keywords: *pilot projects; Keyword Identifiers: *Great Britain; *UK Biobank; Manchester

Wylie, Jean E.; Mineau, Geraldine P. Biomedical databases: protecting privacy and promoting research. *Trends in Biotechnology* 2003 March; 21(3): 113-116. 23 refs. NRCBL: 15.1; 1.3.12; 1.3.9; 15.10; 8.4.
Keywords: *biomedical research; *databases; *genetic research; biological specimen banks; computers; confidentiality; epidemiology; genetic databases; industry; medical records; privacy; regulation; standards; Keyword Identifiers: deCode Genetics; Iceland; United States; Utah

GENETIC ENGINEERING AND BIOTECHNOLOGY

Biotechnology. *In:* Marks, Stephen P., ed. Health and Human Rights: Basic International Documents. Cambridge, MA: Harvard University, Francois-Xavier Bagnoud Center for Health and Human Rights, 2004: 281-304. NRCBL: 15.1; 21.1.
Keywords: *biotechnology; *cloning; *genetic information; *guidelines; *human genome; *human rights; *international aspects; *social control; genetic databases; genetic research; genetic screening; reproductive technologies; Keyword Identifiers: *Draft International Convention on the Prohibition of All Forms of Human Cloning; *European Convention on Human Rights and Biomedicine; *International Declaration on Human Genetic Data; *Universal Declaration on the Human Genome and Human Rights; UNESCO

Acharya, Tara; Daar, Abdallah S.; Dowdeswell, Elizabeth; Singer, Peter A. Better global governance to promote genomics for development. Toronto, Canada: University of Toronto, Joint Centre for Bioethics, 2004; 10 p. [Online]. Accessed: http://www.utoronto.ca/jcb/genomics/documents/CGI-paper.pdf [2007 April 16]. 19 refs. NRCBL: 15.1; 5.3; 21.1. SC: le.
Keywords: *biotechnology; *developing countries; *genomics; *international aspects; access to information; government; health priorities; industry; justice; interinstitutional relations; policy making; public health; regulation; research priorities; risks and benefits; universities; Proposed Keywords: biodiversity; stakeholders; technology transfer; Keyword Identifiers: *Global Genomics Initiative

Altieri, Miguel Angel. The myths of biotechnology: some ethical questions. *In:* Serageldin, I.; Martin-Brown, J., eds. Proceedings of an associated event of the 5th Annual World Bank Conference on Environmentally and Socially Sustainable Development.Washington, D.C.: World Bank, 1998: 53- 58 [Online]. Accessed: http://nature.berkeley.edu/~agroeco3/the_myths.html [2007 April 24]. 25 refs. NRCBL: 1.3.11; 15.1; 16.1; 5.3.
Keywords: *agriculture; *biotechnology; *genetically modified plants; developing countries; ecology; genetic engineering; industry; international aspects; property rights; risks and benefits; Proposed Keywords: biodiversity

Annas, George J.; Andrews, Lori B.; Isasi, Rosario M. Protecting the endangered human: toward an international treaty prohibiting cloning and inheritable alterations. *In:* Gruskin, Sofia; Grodin, Michael A.; Annas, George J.; Marks, Stephen P., eds. Perspectives on Health and Human Rights. New York: Routledge, 2005: 135-162. 110 fn. NRCBL: 14.5; 15.1; 21.1; 4.4. SC: le.
Keywords: *cloning; *genetic engineering; *germ cells; *international aspects; *regulation; human rights; legal aspects; nuclear transfer techniques; reproductive technologies; Keyword Identifiers: Convention on the Preservation of the Human Species; United States

Armstrong, Susan J.; Botzler, Richard G. Animals and biotechnology. *In their:* The Animal Ethics Reader. London; New York: Routledge, 2003: 311-377. 20 refs. 12 fn. NRCBL: 15.1; 22.1; 1.3.11; 22.2.
Keywords: *animal welfare; *biotechnology; *genetically modified animals; agriculture; animal experimentation; animal rights; ethical analysis; methods; regulation; researchers; risks and benefits; suffering; utilitarianism; Proposed Keywords: domestic animals; laboratory animals

Bayertz, Kurt; Schmidt, Kurt W. Testing genes and constructing humans — ethics and genetics. *In:* Khushf, George, ed. Handbook of Bioethics: Taking Stock of the Field From a Philosophical Perspective. Dordrecht; Boston: Kluwer Academic, 2004: 415-438. 59 refs. 4 fn. NRCBL: 15.1; 15.3; 1.1. SC: an; rv.
Keywords: *biotechnology; *ethical analysis; *gene therapy; *genetic engineering; *genetic enhancement; *genetic screening; *moral policy; autonomy; confidentiality; diagnosis; eugenics; future generations; genetic information; genetic predisposition; germ cells; human characteristics; medicine; philosophy; physician patient relationship; population genetics; professional family relationship; prenatal diagnosis; principle-based ethics; right not to know; risks and benefits; social impact

Baylis, Françoise; Robert, Jason Scott. Radical rupture: exploring biological sequelae of volitional inheritable genetic modification. *In:* Rasko, John E.J.; O'Sullivan, Gabrielle M.; Ankeny, Rachel A., eds. The Ethics of Inheritable Genetic Modification: A Dividing Line? Cambridge: Cambridge University Press, 2006: 131-148. NRCBL: 15.1. SC: an.
Keywords: *genetic engineering; anthropology; chimeras; evolution; future generations; gene therapy; genetic enhancement; genetic intervention; germ cells; human experimentation; human genome; pedigree; risks and benefits; social impact

Berry, Roberta M. Can bioethics speak to politics about the prospect of inheritable genetic modification? If so, what might it say? *In:* Rasko, John E.J.; O'Sullivan, Gabrielle M.; Ankeny, Rachel A., eds. The Ethics of Inheritable Genetic Modification: A Dividing Line? Cambridge: Cambridge University Press, 2006: 243-277. NRCBL: 15.1; 2.1. SC: an.
Keywords: *bioethics; *clinical genetics; *ethical analysis; *genetic engineering; *policy making; behavioral genetics; biotechnology; choice behavior; coercion; common good; decision making; deontological ethics; eugenics; genetic disorders; germ cells; hearing disorders; historical aspects;

Huntington disease; parents; politics; policy analysis; public policy; reproductive technologies; utilitarianism; Keyword Identifiers: Twentieth Century; United States

Betta, Michela. Diagnostic knowledge in the genetic economy and commerce. *In:* Betta, Michela, ed. The Moral, Social, and Commercial Imperatives of Genetic Testing and Screening: the Australian Case. Dordrecht: Springer, 2006: 25-52. 88 fn. NRCBL: 15.1; 15.3; 19.4; 1.3.2. SC: an.

Keywords: *biotechnology; *economics; *genetic research; *genetic screening; *policy making; blood banks; commerce; cord blood; diagnosis; disclosure; embryo research; embryonic stem cells; entrepreneurship; genetic databases; genetic discrimination; genetic disorders; genetic information; genetic predisposition; genomics; government financing; government regulation; industry; insurance; legal aspects; patents; pharmacogenetics; policy analysis; public policy; regulation; research support; researchers; science; trends; Keyword Identifiers: Australia

Bharadwaj, Minakshi. Looking back, looking beyond: revisiting the ethics of genome generation. *Journal of Biosciences* 2006 March; 31(1): 167-176. 42 refs. NRCBL: 15.1; 1.3.11; 2.1; 15.3; 15.10; 15.5. SC: rv.

Keywords: *biotechnology; *genetic engineering; *genetic research; *genomics; benefit sharing; bioethics; cloning; developing countries; economics; embryo research; ethical analysis; genes; genetic determinism; genetic information; genetic patents; genetic screening; genetically modified food; genetically modified plants; Human Genome Project; international aspects; Proposed Keywords: exceptionalism; genetic resources; stakeholders

Boyle, Joseph. Genetics, medicine, and the human person: the papal theology. *In:* Monsour, H. Daniel, ed. Ethics and the New Genetics: An Integrated Approach. Toronto: University of Toronto Press, 2007: 134-142. 21 fn. NRCBL: 15.1; 1.2.

Keywords: *clinical genetics; *genetic engineering; *Roman Catholic ethics; gene therapy; human dignity; risks and benefits; theology; Keyword Identifiers: *Pope John Paul II; *Pope Pius XII

Brody, Baruch. Intellectual property and biotechnology: the European debate. *Kennedy Institute of Ethics Journal* 2007 June; 17(2): 69-110. 48 refs. NRCBL: 5.3; 15.8; 4.4; 15.1; 22.2; 18.5.4; 21.1. SC: rv.

Keywords: *biotechnology; *genetic patents; *moral policy; *patents; *policy making; advisory committees; body parts and fluids; commodification; cryopreservation; embryo research; embryonic stem cells; genetic research; genetic screening; genetically modified animals; historical aspects; industry; human body; human dignity; legal aspects; life; methods; property rights; public policy; sex preselection; sperm; trends; Proposed Keywords: mice; Keyword Identifiers: *European Patent Organization; *European Union; Europe; Council of Europe; European Parliament; Twentieth Century; Twenty-First Century

Abstract: The European patent system allows for the introduction of moral issues into decisions about the granting of patents. This feature has greatly impacted European debates about the patenting of biotechnology. This essay explores the European experience, in both the European Union and the European Patent Organization.

It argues that there has been great confusion surrounding these issues primarily because the Europeans have not developed a general theory about when exclusion from patentability is the best social mechanism for dealing with morally offensive technologies.

Burgess, Michael M. Ethical analysis of representation in the governance of biotechnology. *In:* Einsiedel, Edna; Timmermans, Frank, eds. Crossing Over: Genomics in the Public Arena. Calgary, Alberta, Canada: University of Calgary Press, 2005: 157-172. 30 refs.; 7 fn. NRCBL: 15.1; 5.3. Conference: Essays from the conference held Apr. 25-27, 2003, Kananaskis, Alta.

Keywords: *biotechnology; *genetic research; *policy making; *social control; ethical analysis; focus groups; genomics; government; industry; international aspects; political activity; public participation; regulation; research priorities; researchers; risks and benefits; Proposed Keywords: consumer advocacy; stakeholders

Caplan, Arthur L.; Baruch, Susannah; Schmidt, Harald; Jennings, Bruce; Bonnicksen, Andrea; Greenfield, Debra; Baylis, Françoise; Robertson, John A.; Fleck, Leonard M.; Furger, Franco; Fukuyama, Francis. Needed: a modest proposal [letters and replies]. *Hastings Center Report* 2007 November-December; 37(6): 4-11. NRCBL: 14.4; 14.1; 18.5.4; 15.1; 14.2; 19.5; 9.3.1. Comments: Comments on: Furger, Franco; Fukuyama, Francis. A proposal for modernizing the regulation of human biotechnologies. Hastings Center Report 2007 July-August 37(4): 16-20; Fossett, James W. Managing reproductive pluralism: the case for decentralized governance. Hastings Center Report 2007 July-August 37(4): 24-35; Fleck, Leonard M. Can we trust "democratic deliberation"? Hastings Center Report 2007 July-August 37(4): 22-25; Robertson, John A. The virtues of muddling through. Hastings Center Report 2007 July-August 37(4): 26-28; Johnston, Josephine. Tied up in nots over genetic parentage. Hastings Center Report 2007 July-August 37(4): 28-31.

Keywords: *biotechnology; *government regulation; *public policy; *reproductive technologies; advisory committees; bioethical issues; chimeras; common good; embryo research; embryonic stem cells; federal government; freedom; genetic engineering; legal aspects; ovum donors; policy making; public participation; remuneration

Caulfield, Timothy. Popular media, biotechnology, and the "cycle of hype". *Houston Journal of Health Law and Policy* 2005; 5(2): 213-233. 104 fn. NRCBL: 15.1; 1.3.7.

Keywords: *biotechnology; *mass media; genetic research; risks and benefits; Proposed Keywords: *hyperbole

Comber, Julie; Griffin, Gilly. Genetic engineering and other factors that might affect human-animal interactions in the research setting. *Journal of Applied Animal Welfare Science* 2007 June; 10(3): 267-277. 35 refs. NRCBL: 15.1; 22.1.

Keywords: *animal welfare; *genetically modified animals; agriculture; animal behavior; animal cloning; animal experimentation; attitudes; guidelines; researchers; Proposed

NRCBL: National Reference Center for Bioethics Literature Classification Scheme   See inside front cover for terms.

183

Keywords: domestic animals; laboratory animals; Keyword Identifiers: Canada; Canadian Council on Animal Care

Corbellini, Gilberto. Scientists, bioethics and democracy: the Italian case and its meanings. *Journal of Medical Ethics* 2007 June; 33(6): 349-352. 4 refs. NRCBL: 5.3; 1.3.5; 1.3.9; 14.1; 15.1. SC: le.
> Keywords: *biotechnology; *democracy; *freedom; *government regulation; *politics; *science; animal cloning; bioethical issues; bioethics; embryo research; embryonic stem cells; evolution; genetic engineering; genetically modified organisms; information dissemination; legal aspects; political systems; preimplantation diagnosis; reproductive technologies; social impact; values; Western World; Proposed Keywords: technology; Keyword Identifiers: *Italy

Abstract: In June 2005, Italy held a referendum on repealing the law on medically assisted fertilization (Law 40/2004), which limits access to artificial reproduction to infertile couples, and prohibits the donation of gametes, the cryopreservation of embryos, preimplantation genetic diagnosis (PDG), and research on human embryos. The referendum was invalidated, and the law remained unchanged. The Italian political bioethical debate on assisted reproduction was manipulated by the Catholic Church, which distorted scientific data and issues at stake with the help of Catholic politicians and bioethicists. What happened in Italy shows that some perverse sociocultural political mechanisms are spreading the absurd and anti-historical view that scientific and technological advancements are threatening democracy and personal freedom. Scientists should not only contrast the political attempts at limiting freedom of scientific research, but also tell politicians, humanists and citizens that the invention of Western science with its view of scientific community as an "open society", contributed and still contributes, through scientific education, to the construction and maintaining of the moral and political values underlying Western democracies.

Deane-Drummond, Celia. Biotechnology and theology. *In:* Eisen, Arri; Laderman, Gary, eds. Science, Religion, and Society: An Encyclopedia of History, Culture, and Controversy. Vol. 2. Armonk, NY: M.E. Sharpe, 2007: 780-786. 10 refs. NRCBL: 15.1; 1.2.
> Keywords: *biotechnology; *theology; animal cloning; genetic engineering; genetic research; genetically modified organisms; human genome

Deane-Drummond, Celia. Genetics and environmental concern. *In her:* Genetics and Christian Ethics. Cambridge: Cambridge University Press, 2006: 220-244. 70 fn. NRCBL: 15.1; 1.2; 16.1; 22.2.
> Keywords: *animal rights; *ecology; *genetic engineering; *genetically modified animals; *theology; agriculture; Christian ethics; covenant; genetically modified plants; philosophy; speciesism; virtues

Egorova, Yulia. 'Up in the sky': human and social sciences' responses to genetics. *In:* Gunning, Jennifer; Holm, Søren, eds. Ethics, law, and society. Volume 2. Aldershot, Hants, England; Burlington, VT: Ashgate, 2006: 45-53. 33 fn. NRCBL: 15.1; 1.1. SC: an.
> Keywords: *biotechnology; *genetic research; *social impact; cloning; culture; genetic identity; genetic intervention; genetic patents; human genome; metaphor; philosophy; postmodernism; reproductive technologies; social sciences; Proposed Keywords: transhumanism

Ehrenfeld, David. Unethical contexts for ethical questions. *In:* Galston, Arthur W.; Peppard, Christiana Z., eds. Expanding Horizons in Bioethics. Dordrecht; Norwell, MA: Springer, 2005: 19-34. 41 refs. 42 fn. NRCBL: 15.1; 5.3; 15.7; 22.1.
> Keywords: *biotechnology; agriculture; animal welfare; cloning; ecology; genetic engineering; genetic patents; industry; genetically modified animals; genetically modified food; genetically modified plants; reproductive technologies; science; technology; values

Einsiedel, Edna. Telling technological tales: the media and the evolution of biotechnology. *In:* Einsiedel, Edna; Timmermans, Frank, eds. Crossing Over: Genomics in the Public Arena. Calgary, Alberta, Canada: University of Calgary Press, 2005: 143-154. 40 refs. NRCBL: 15.1; 1.3.7. Conference: Essays from the conference held Apr. 25-27, 2003, Kananaskis, Alta.
> Keywords: *biotechnology; *mass media; cloning; genetic research; genetically modified food; industry; information dissemination; international aspects; Internet; marketing; public opinion; risks and benefits; social impact; Keyword Identifiers: Great Britain; United States

Eischen, Kyle. Commercializing Iceland: biotechnology, culture, and the information society. *In:* Mehta, Michael D., ed. Biotechnology Unglued: Science, Society and Social Cohesion. Vancouver: UBC Press, 2005: 95-116. 48 refs. 2 fn. NRCBL: 15.1; 1.3.12; 9.3.1; 15.11; 13.1; 5.3.
> Keywords: *genetic databases; *genetic research; *social impact; biotechnology; culture; genetic ancestry; genetic privacy; geographic factors; industry; legal aspects; population genetics; socioeconomic factors; Proposed Keywords: commerce; Keyword Identifiers: *Iceland; *deCode Genetics

Fiester, Autumn. Casuistry and the moral continuum: evaluating animal biotechnology. *Politics and the Life Sciences* 2006 March-September; 25(1-2): 15-22. 22 refs. NRCBL: 15.1; 22.2; 14.5; 19.5. SC: an; cs.
> Keywords: *animal cloning; *animal organs; *casuistry; *ethical analysis; *genetically modified animals; animal rights; animal welfare; biotechnology; drugs; organ transplantation; risks and benefits; teleological ethics; Proposed Keywords: biodiversity; goats; species specificity; swine; zoonoses

Fleck, Leonard M. Can we trust "democratic deliberation"? *Hastings Center Report* 2007 July-August; 37(4): 22-25. 4 refs. NRCBL: 21.1; 14.1; 15.1; 5.3. SC: an; le. Comments: Comment on: Franco Furger and Francis Fukuyama. A proposal for modernizing the regulation of human biotechnologies. Hastings Center Report 2007 July-August; 37(4): 16-20.
> Keywords: *biotechnology; *democracy; *policy making; *public participation; *reproductive technologies; accountability; bioethical issues; cloning; cultural pluralism; embryo research; embryonic stem cells; freedom; genetic engineering; government financing; government regulation;

policy analysis; preimplantation diagnosis; public policy; regulation; Keyword Identifiers: *United States

Fossett, James W. Managing reproductive pluralism: the case for decentralized governance. *Hastings Center Report* 2007 July-August; 37(4): 20-22. NRCBL: 14.1; 15.1; 5.3. SC: an; le. Comments: Comment on: Franco Furger and Francis Fukuyama. A proposal for modernizing the regulation of human biotechnologies. Hastings Center Report 2007 July-August; 37(4): 16-20.
> Keywords: *biotechnology; *cultural pluralism; *government regulation; *public policy; *reproductive technologies; bioethical issues; cloning; decision making; democracy; dissent; embryo research; embryonic stem cells; embryos; federal government; moral status; policy analysis; preimplantation diagnosis; reproductive medicine; self regulation; state government; Keyword Identifiers: *United States; Great Britain

Fox, Dov. Silver spoons and golden genes: genetic engineering and the egalitarian ethos. *American Journal of Law and Medicine* 2007; 33(4): 567-623. NRCBL: 15.1; 15.5; 1.1; 15.2; 14.6.

Franklin, Sarah. Ethical biocapital: new strategies of cell culture. *In:* Franklin, Sarah; Lock, Margaret, eds. Remaking Life and Death: Toward an Anthropology of the Biosciences. Santa Fe: School of American Research Press; Oxford: James Currey, 2003: 97-127. 20 fn. NRCBL: 15.1; 1.3.11; 14.5; 22.3.
> Keywords: *animal cloning; *biotechnology; *genetic engineering; *industry; *stem cells; capitalism; cell lines; embryo research; embryonic stem cells; gene therapy; genetically modified food; genetically modified organisms; nuclear transfer techniques; patents; stem cell transplantation; trust; Proposed Keywords: domestic animals; regenerative medicine; sheep; Keyword Identifiers: Geron Corp.; Great Britain; United States

Furger, Franco; Fukuyama, Francis. A proposal for modernizing the regulation of human biotechnologies. *Hastings Center Report* 2007 July-August; 37(4): 16-20. NRCBL: 14.1; 15.1; 5.3; 15.2; 14.4. SC: an; le.
> Keywords: *biotechnology; *government regulation; *reproductive technologies; accountability; biomedical research; cloning; commerce; embryo research; embryonic stem cells; embryos; freedom; genetic engineering; germ cells; guidelines; industry; judicial role; legal aspects; political activity; policy analysis; politics; preimplantation diagnosis; public participation; public policy; quality of health care; standards; Keyword Identifiers: *United States

Fürst, Gebhard. The (im)perfect human — his own creator? Bioethics and genetics at the beginning of life. *In:* Sensen, Chrisopher W., ed. Handbook of Genome Research: Genomics, Proteomics, Metabolomics, Bioinformatics, Ethical and Legal Issues. Volume 2. Weinheim: Wiley-VCH, 2005: 561-569. 3 refs. NRCBL: 15.1; 1.2; 14.5; 18.5.4.
> Keywords: *cloning; *genetic engineering; *human dignity; *Roman Catholic ethics; biotechnology; embryo research; embryos; eugenics; genetic research; international aspects; personhood; quality of life; regulation; reproductive technologies; value of life; Proposed Keywords: biological sciences; Keyword Identifiers: Germany

Gerlach, Neil. Biotechnology and social control: the Canadian DNA data bank. *In:* Mehta, Michael D., ed. Biotechnology Unglued: Science, Society and Social Cohesion. Vancouver: UBC Press, 2005: 117-132. 40 refs. 2 fn. NRCBL: 15.1; 1.3.12; 1.3.5.
> Keywords: *DNA fingerprinting; *genetic databases; legislation; public policy; Keyword Identifiers: *Canada

Goldman, Michael A. Calamity gene: when biotechnology spins out of control [review of Next, by Michael Crichton]. *Nature* 2007 February 22; 445(7130): 819-820. NRCBL: 15.1; 2.1.
> Keywords: *biotechnology; genes; genetically modified animals; genetic patents; hyperbole; industry; literature; universities; Keyword Identifiers: *Next (Crichton, Michael)

Gorovitz, Samuel. The past, present and future of human nature. *In:* Galston, Arthur W.; Peppard, Christiana Z., eds. Expanding Horizons in Bioethics. Dordrecht; Norwell, MA: Springer, 2005: 3-18. 25 refs. NRCBL: 5.3; 1.1; 2.1; 15.1.
> Keywords: *biotechnology; *human characteristics; advisory committees; autonomy; bioethical issues; biomedical enhancement; embryo research; evolution; forecasting; genetic determinism; genetic engineering; philosophy; public health; public policy; regulation; risks and benefits; science; stem cells; theology; uncertainty; values; Keyword Identifiers: New York State Task Force on Life and the Law; United States

Graham, Elaine. In whose image? Representations of technology and the 'ends' of humanity. *In:* Deane-Drummond, Celia; Scott, Peter Manley, eds. Future Perfect?: God, Medicine and Human Identity. New York: T and T Clark International, 2006: 56-69. 33 fn. NRCBL: 15.1; 4.5; 1.2.
> Keywords: *biomedical enhancement; *Christian ethics; *genetic engineering; *theology; biotechnology; human dignity; personhood; philosophy; values; Proposed Keywords: *technology; *transhumanism

Graham, Gordon. Human nature and the human condition. *In:* Deane-Drummond, Celia; Scott, Peter Manley, eds. Future Perfect?: God, Medicine and Human Identity. New York: T and T Clark International, 2006: 33-44. 11 fn. NRCBL: 15.1.
> Keywords: *genetic engineering; *human characteristics; biotechnology; genetic enhancement; human genome; life extension; philosophy; precautionary principle; religious ethics; risks and benefits; sociobiology; theology; Proposed Keywords: transhumanism

Gunderson, Martin. Seeking perfection: a Kantian look at human genetic engineering. *Theoretical Medicine and Bioethics* 2007; 28(2): 87-102. 24 refs. 50 fn. NRCBL: 15.1; 1.1. SC: an.
> Keywords: *ethical analysis; *ethical theory; *gene therapy; *genetic engineering; *genetic enhancement; *philosophy; autonomy; freedom; future generations; germ cells; government regulation; human dignity; moral obligations; moral policy; paternalism; public policy; risks and benefits; socioeconomic factors; Proposed Keywords: happiness; Keyword Identifiers: *Kant, Immanuel

NRCBL: National Reference Center for Bioethics Literature Classification Scheme   See inside front cover for terms.

185

Abstract: It is tempting to argue that Kantian moral philosophy justifies prohibiting both human germ-line genetic engineering and non-therapeutic genetic engineering because they fail to respect human dignity. There are, however, good reasons for resisting this temptation. In fact, Kant's moral philosophy provides reasons that support genetic engineering-even germ-line and non-therapeutic. This is true of Kant's imperfect duties to seek one's own perfection and the happiness of others. It is also true of the categorical imperative. Kant's moral philosophy does, however, provide limits to justifiable genetic engineering.

Henaghan, Mark. The 'do no harm' principle and the genetic revolution in New Zealand. *In:* McLean, Sheila A.M., ed. First Do No Harm: Law, Ethics, and Healthcare. Aldershot, England; Burlington, VT: Ashgate, 2006: 511-526. 68 fn. NRCBL: 15.1; 14.1; 15.2; 14.4; 15.3; 21.1.

Keywords: *genetic engineering; *genetic services; *preimplantation diagnosis; *public policy; *regulation; advisory committees; beneficence; biotechnology; clinical trials; cloning; common good; embryos; ethical analysis; ethics committees; genetic screening; genetically modified organisms; germ cells; government regulation; guidelines; legal aspects; precautionary principle; reproductive technologies; risks and benefits; sex preselection; tissue typing; Keyword Identifiers: *New Zealand; National Ethics Committee on Assisted Human Reproduction (New Zealand)

Hohlfeld, Rainer. Politische Ökonomie und Bio-Medizin. *In:* Neuer-Miebach, Therese; Wunder, Michael, eds. Bio-Ethik und die Zukunft der Medizin. Bonn: Psychiatrie-Verlag, 1998: 44-59. 38 fn. NRCBL: 15.1; 1.3.2; 4.5.

Keywords: *biotechnology; *genomics; drug industry; economics; eugenics; genetic determinism; genetic enhancement; genetic predisposition; Human Genome Project; synthetic biology; Keyword Identifiers: Novartis

Holland, Suzanne. Market transactions in reprogenetics: a case for regulation. *In:* Knowles, Lori P.; Kaebnick, Gregory E., eds. Reprogenetics: Law, Policy, and Ethical Issues. Baltimore: Johns Hopkins University Press, 2007: 89-104. 20 refs. 6 fn. NRCBL: 14.1; 15.2; 14.4; 5.3. SC: an.

Keywords: *commerce; *freedom; *genetic engineering; *genetic services; *germ cells; *ovum donors; *public policy; *regulation; *remuneration; *reproductive technologies; commodification; common good; democracy; embryo research; embryos; federal government; genetic relatedness ties; government financing; health facilities; in vitro fertilization; industry; justice; moral policy; policy analysis; politics; preimplantation diagnosis; private sector; public opinion; reproductive rights; socioeconomic factors; surrogate mothers; values; women; Keyword Identifiers: *United States; Canada; Great Britain

Indigenous Peoples Council on Biocolonialism. Indigenous people, genes and genetics: what indigenous people should know about biocolonialism: a primer and resource guide. Nixon, NV: Indigenous Peoples Council on Biocolonialism, 2000 June; 25 p. [Online]. Accessed: http://www.ipcb.org/publications/primers/htmls/ipgg.html [2007 April 23]. 2 refs. NRCBL: 15.11; 18.5.9.

Keywords: *American Indians; *genetic engineering; *genetic research; *indigenous populations; agriculture; biological specimen banks; biotechnology; commerce; eugenics; genes; genetic ancestry; genetic databases; genetic discrimination; genetic materials; genetic patents; genetically modified organisms; genetics; health hazards; Human Genome Diversity Project; informed consent; international aspects; political activity; population genetics; public policy; research ethics; research priorities; research subjects; Proposed Keywords: *genetic resources; biodiversity; colonialism

Johnston, Josephine. Tied up in nots over genetic parentage. *Hastings Center Report* 2007 July-August; 37(4): 28-31. 3 refs. NRCBL: 14.1; 15.1; 18.5.2; 18.5.4; 5.3. SC: an; le. Comments: Comment on: Franco Furger and Francis Fukuyama. A proposal for modernizing the regulation of human biotechnologies. Hastings Center Report 2007 July-August; 37(4): 16-20.

Keywords: *biotechnology; *genetic relatedness ties; *government regulation; *reproductive technologies; cloning; family relationship; genetic techniques; moral policy; ovum donors; policy analysis; public policy; Keyword Identifiers: United States

Johnston, Josephine; Wasunna, Angela A. Patents, biomedical research, and treatments: examining concerns, canvassing solutions. *Hastings Center Report* 2007 January-February; 37(1): S2-S35. 183 fn. NRCBL: 15.8; 21.1; 9.3.1; 18.5.4; 9.5.6; 18.6. SC: an; rv.

Keywords: *biomedical research; *biomedical technologies; *biotechnology; *genetic materials; *international aspects; *patents; *policy analysis; *property rights; benefit sharing; developing countries; drug industry; drugs; economics; embryonic stem cells; genetic patents; genetic screening; government financing; guidelines; health care delivery; health services accessibility; historical aspects; HIV infections; incentives; legal aspects; moral policy; poliomyelitis; private sector; public policy; public sector; research support; trends; Proposed Keywords: licensure; vaccines; Keyword Identifiers: National Institutes of Health; Organization for Economic Cooperation and Development; South Africa; Trade-Related Aspects of Intellectual Property Rights (TRIPS); United States

Juengst, Eric T. "Alter-ing" the human species? Misplaced essentialism in science policy. *In:* Rasko, John E.J.; O'Sullivan, Gabrielle M.; Ankeny, Rachel A., eds. The Ethics of Inheritable Genetic Modification: A Dividing Line? Cambridge: Cambridge University Press, 2006: 149-158. 30 fn. NRCBL: 15.1. SC: an.

Keywords: *genetic engineering; *moral policy; future generations; gene pool; gene therapy; genetic enhancement; genetic research; germ cells; human characteristics; human experimentation; human genome; human rights; international aspects; policy analysis; risks and benefits; social control; Proposed Keywords: *species specificity

Kaebnick, Gregory. Putting concerns about nature in context: the case of agricultural biotechnology. *Perspectives in Biology and Medicine* 2007 Autumn; 50(4): 572-584. NRCBL: 15.1; 1.3.11; 1.1.

Kaiser, Matthias. Practical ethics in search of a toolbox: ethics of science and technology at the crossroads. *In:*

Gunning, Jennifer; Holm, Søren, eds. Ethics, law, and society. Volume 2. Aldershot, Hants, England; Burlington, VT: Ashgate, 2006: 35-44. 38 refs. NRCBL: 15.1; 1.3.12; 22.3. SC: an.

Keywords: *biotechnology; *democracy; *decision making; *ethical analysis; *genetically modified organisms; *policy analysis; *policy making; *public policy; advisory committees; consensus; cultural pluralism; ethics committees; genetically modified food; guidelines; international aspects; public participation; risks and benefits; standards; technology assessment; theoretical models; Proposed Keywords: stakeholders

Kenny, Denis. Inheritable genetic modification as moral responsibility in a creative universe. *In:* Rasko, John E.J.; O'Sullivan, Gabrielle M.; Ankeny, Rachel A., eds. The Ethics of Inheritable Genetic Modification: A Dividing Line? Cambridge: Cambridge University Press, 2006: 77-102. 46 fn. NRCBL: 15.1; 1.1; 5.3.

Keywords: *genetic engineering; *morality; biotechnology; cloning; evolution; future generations; germ cells; human characteristics; postmodernism; risks and benefits

Kleinman, Daniel Lee; Kinchy, Abby J. Against the neoliberal steamroller? The Biosafety Protocol and the social regulation of agricultural biotechnologies. *Agriculture and Human Values* 2007 Summer; 24(2): 195-206. 46 refs. 5 fn. NRCBL: 15.7; 1.3.11; 15.1.

Keywords: *agriculture; *biotechnology; *genetically modified food; *genetically modified plants; *international aspects; *policy making; biodiversity; ecology; genetically modified organisms; precautionary principle; regulation; risk; socioeconomic factors; Keyword Identifiers: *Cartagena Protocol on Biosafety

Krimsky, Sheldon. The profit of scientific discovery and its normative implications. *Chicago-Kent Law Review* 1999; 75(1): 15-39. 164 fn. NRCBL: 15.8; 1.3.7; 1.3.9; 2.2; 5.3. SC: le; rv.

Keywords: *biomedical research; *biotechnology; *conflict of interest; *genetic patents; *legal aspects; authorship; cell lines; disclosure; editorial policies; entrepreneurship; federal government; genes; genetic engineering; genetic research; genetic screening; genetically modified organisms; government financing; government regulation; industry; interinstitutional relations; journalism; public policy; publishing; recombinant DNA research; research support; science; Supreme Court decisions; universities; Keyword Identifiers: *United States;

Lassen, Jesper; Gjerris, Mickey; Sandoe, Peter. After Dolly — ethical limits to the use of biotechnology on farm animals. *Theriogenology* 2006 March 15; 65(5): 992-1004. 14 refs. NRCBL: 22.3; 14.5; 15.1. SC: em.

Keywords: *animal cloning; *animal welfare; *biotechnology; *genetically modified animals; agriculture; animal experimentation; animal organs; attitudes; focus groups; food; genetically modified organisms; interviews; public opinion; risk; wedge argument; Proposed Keywords: *domestic animals; laboratory animals; Keyword Identifiers: *Denmark; European Union

Lee, Robert. GM resistant: Europe and the WTO panel dispute on biotech products. *In:* Gunning, Jennifer; Holm, Søren, eds. Ethics, law, and society. Volume 1. Aldershot,

Hants, England; Burlington, VT: Ashgate, 2005: 131-140. 36 fn. NRCBL: 15.1; 1.3.11; 15.7; 21.1.

Keywords: *genetically modified food; *genetically modified organisms; *regulation; biotechnology; commerce; risk; Keyword Identifiers: *European Union; *World Trade Organization; Europe

Létourneau, Lyne. The regulation of animal biotechnology: at the crossroads of law and ethics. *In:* Einsiedel, Edna; Timmermans, Frank, eds. Crossing Over: Genomics in the Public Arena. Calgary, Alberta, Canada: University of Calgary Press, 2005: 173-192. 49 refs.; 2 fn. NRCBL: 15.1; 22.2; 22.3; 1.3.11; 5.1; 5.3. Conference: Essays from the conference held Apr. 25-27, 2003, Kananaskis, Alta.

Keywords: *animal rights; *animal welfare; *genetically modified animals; *moral policy; *regulation; advisory committees; agriculture; animal experimentation; animal organs; biotechnology; ethical analysis; guidelines; legislation; organ transplantation; philosophy; public policy; standards; Proposed Keywords: domestic animals; laboratory animals; Keyword Identifiers: Canada; Canadian Council on Animal Care; Committee to Consider the Ethical Implications of Emerging Technologies in the Breeding of Farm Animals (Great Britain); Great Britain

Mackie, Jocelyn E.; Taylor, Andrew D.; Finegold, David L.; Daar, Abdallah S.; Singer, Peter A. Lessons on ethical decision making from the bioscience industry. *PLoS Medicine* 2006 May; 3(5): e129. 4 refs. NRCBL: 15.1; 1.3.2; 2.1. SC: em.

Keywords: *biotechnology; *business ethics; *ethics consultation; *industry; *institutional ethics; *organizational policies; administrators; advisory committees; agriculture; case studies; conflict of interest; decision making; disclosure; drug industry; employment; entrepreneurship; ethicists; evaluation; interinstitutional relations; interviews; organization and administration; Proposed Keywords: stakeholders

Mameli, M. Reproductive cloning, genetic engineering and the autonomy of the child: the moral agent and the open future. *Journal of Medical Ethics* 2007 February; 33(2): 87-93. 18 refs. NRCBL: 14.5; 15.1. SC: an.

Keywords: *autonomy; *children; *cloning; *ethical analysis; *future generations; *genetic engineering; biotechnology; choice behavior; genetic determinism; genetic enhancement; moral obligations; parents; reproductive rights; reproductive technologies; self concept

Abstract: Some authors have argued that the human use of reproductive cloning and genetic engineering should be prohibited because these biotechnologies would undermine the autonomy of the resulting child. In this paper, two versions of this view are discussed. According to the first version, the autonomy of cloned and genetically engineered people would be undermined because knowledge of the method by which these people have been conceived would make them unable to assume full responsibility for their actions. According to the second version, these biotechnologies would undermine autonomy by violating these people's right to an open future. There is no evidence to show that people conceived through cloning and genetic engineering would inevitably or even in general be unable to assume responsibility for their actions; there is also no evidence for the claim

NRCBL: National Reference Center for Bioethics Literature Classification Scheme   See inside front cover for terms.

187

that cloning and genetic engineering would inevitably or even in general rob the child of the possibility to choose from a sufficiently large array of life plans.

Marks, Stephen P. Tying Prometheus down: human rights issues of human genetic manipulation. *In:* Gruskin, Sofia; Grodin, Michael A.; Annas, George J.; Marks, Stephen P., eds. Perspectives on Health and Human Rights. New York: Routledge, 2005: 163-178. 57 fn. NRCBL: 15.1; 4.4; 21.1.

> Keywords: *genetic engineering; *human rights; *international aspects; autonomy; cloning; human dignity; legal aspects; regulation

Meilaender, Gilbert. Genes as resources. *Hedgehog Review* 2002 Fall [Online]. Accessed: http://www.virginia.edu/iasc/hedgehog.html [2006 September 25]. 20 refs. NRCBL: 15.1; 7.1; 4.4; 1.2.

> Keywords: *genetic engineering; *human genome; biotechnology; commodification; dehumanization; embryo research; future generations; genes; genetic resources; justice; love; moral obligations; personhood; theology

Memis, Tekin. Debate on patentability of biotechnological studies in Turkey. *Revista de Derecho y Genoma Humano = Law and the Human Genome Review* 2007 January-June; (26): 121-135. 26 fn. NRCBL: 15.8. SC: le.

> Keywords: *biotechnology; *genetic materials; *genetic patents; *legal aspects; developing countries; genetic research; genetic resources; industry; international aspects; Keyword Identifiers: *Turkey; European Union

Mepham, Ben. Food ethics. *In:* Gunning, Jennifer; Holm, Søren, eds. Ethics, law, and society. Volume 1. Aldershot, Hants, England; Burlington, VT: Ashgate, 2005: 141-151. 34 fn. NRCBL: 15.1; 1.3.11; 15.7.

> Keywords: *biotechnology; *food; *ethical analysis; *genetically modified organisms; agriculture; animal welfare; autonomy; beneficence; ecology; justice; theoretical models; Proposed Keywords: biodiversity

Novas, Carlos. What is the bioscience industry doing to address the ethical issues it faces? *PLoS Medicine* 2006 May; 3(5): e142. 5 refs. NRCBL: 15.1; 1.3.2. SC: em.

> Keywords: *biotechnology; *business ethics; *drug industry; *industry; *institutional ethics; *organizational policies; administrators; ethicists; ethics consultation; interinstitutional relations

O'Sullivan, Gabrielle M. Ethics and welfare issues in animal genetic modification. *In:* Rasko, John E.J.; O'Sullivan, Gabrielle M.; Ankeny, Rachel A., eds. The Ethics of Inheritable Genetic Modification: A Dividing Line? Cambridge: Cambridge University Press, 2006: 103-129. 97 fn. NRCBL: 15.1; 14.5; 22.2; 22.3. SC: an.

> Keywords: *animal welfare; *genetic engineering; *genetically modified animals; *risks and benefits; animal cloning; animal experimentation; ethical analysis; human experimentation; moral policy; trends; Proposed Keywords: species specificity

Obasogie, Osagie. Racial alchemy. It may not be long before genetic skin-lightening treatments are on sale, so it's time to stop pretending colour prejudice isn't a problem.

New Scientist 2007 August 18-24; 195(2617): 17. NRCBL: 15.11; 9.7.

> Keywords: *biotechnology; *genetic ancestry; *racial groups; population genetics; social discrimination; Proposed Keywords: *cosmetic techniques; *skin pigmentation

Palmer, Julie Gage. Governmental regulation of genetic technology, and the lessons learned. *In:* Knowles, Lori P.; Kaebnick, Gregory E., eds. Reprogenetics: Law, Policy, and Ethical Issues. Baltimore: Johns Hopkins University Press, 2007: 20-63. 46 refs. NRCBL: 15.1; 15.4; 18.5.4; 18.6; 5.3. SC: le; rv.

> Keywords: *federal government; *freedom; *gene therapy; *genetic engineering; *government regulation; *historical aspects; *public policy; adverse effects; advisory committees; biotechnology; clinical trials; cloning; decision making; embryo research; ethical review; eugenics; germ cells; guidelines; human experimentation; justice; parents; recombinant DNA research; research priorities; research subjects; researchers; reproductive technologies; resource allocation; rights; self regulation; social worth; Keyword Identifiers: *United States; Asilomar Conference; Department of Health and Human Services; Federal Policy (Common Rule) for the Protection of Human Subjects; Food and Drug Administration; National Commission for the Protection of Human Subjects; National Institutes of Health; Points to Consider in Human Somatic Cell Therapy and Gene Therapy; President's Commission for the Study of Ethical Problems; Recombinant DNA Advisory Committee; Twentieth Century; Twenty-First Century; Working Group on Human Gene Therapy

Parens, Erik; Knowles, Lori P. Reprogenetics and public policy: reflections and recommendations. *In:* Knowles, Lori P.; Kaebnick, Gregory E., eds. Reprogenetics: Law, Policy, and Ethical Issues. Baltimore: Johns Hopkins University Press, 2007: 253-294. 96 refs., 7 fn. NRCBL: 14.1; 18.5.4; 15.1. SC: rv.

> Keywords: *embryo research; *genetic engineering; *policy making; *public policy; *regulation; *reproductive technologies; adverse effects; advisory committees; cloning; embryonic stem cells; embryos; federal government; freedom; government regulation; historical aspects; human experimentation; international aspects; ovum donors; preimplantation diagnosis; professional organizations; recombinant DNA research; risk; social impact; state government; Proposed Keywords: tissue typing; Keyword Identifiers: Canada; Ethics Advisory Board; Food and Drug Administration; Great Britain; Human Embryo Research Panel; National Bioethics Advisory Commission; Recombinant DNA Advisory Committee; Reprogenetics Technology Board; Twentieth Century; Twenty-First Century; United States; Warnock Committee

Paul, Diane B. On drawing lessons from the history of eugenics. *In:* Knowles, Lori P.; Kaebnick, Gregory E., eds. Reprogenetics: Law, Policy, and Ethical Issues. Baltimore: Johns Hopkins University Press, 2007: 3-19. 64 refs. NRCBL: 15.5; 15.1; 2.2.

> Keywords: *eugenics; *genetic engineering; *freedom; *historical aspects; *informal social control; *reproductive technologies; *values; autonomy; attitudes; coercion; contraception; directive counseling; disabled persons; feminism; genetic services; public policy; reproduction; reproductive rights; researchers; sterilization; trends; voluntary programs; women; Proposed Keywords: ectogenesis;

transhumanism; Keyword Identifiers: Twentieth Century; Twenty-First Century

Persson, Anders. Research ethics and the development of medical biotechnology. *Xenotransplantation* 2006 November; 13(6): 511-513. NRCBL: 15.1; 15.4; 5.3; 18.2; 19.1; 22.2.

Pilarski, Linda M.; Mehta, Michael D.; Caulfield, Timothy; Kaler, Karan V.I.S.; Backhouse, Christopher J. Microsystems and nanoscience for biomedical applications: a view to the future. *In:* Hunt, Geoffrey; Mehta, Michael D., eds. Nanotechnology: Risk, Ethics and Law. London; Sterling, VA: Earthscan, 2006: 35-42. 15 refs. NRCBL: 5.4; 5.3; 15.1.
> Keywords: *biomedical technologies; *biotechnology; *nanotechnology; genetic screening; genetic services; genomics; marketing; pharmacogenetics; policy making; public participation; risks and benefits; social impact; values

Pivetti, Monica. Natural and unnatural: activists' representations of animal biotechnology. *New Genetics and Society* 2007 August; 26(2): 137-157. 75 refs. 5 fn. NRCBL: 15.1; 22.2; 22.1. SC: em.
> Keywords: *animal rights; *animal welfare; *attitudes; *biotechnology; *genetically modified animals; animal experimentation; animal organs; emotions; focus groups; moral policy; organ transplantation; political activity; public opinion; utilitarianism; Keyword Identifiers: *Italy; Anti-Vivisection League (Italy); Centre for Animal Aid (Italy); National Foundation for the Protection of Animals (Italy)

Pollack, Andrew. Round 2 [two] for biotech beets; after delay over safety fears, engineered crop will be planted [news]. *New York Times* 2007 November 27; p. C1, C2. NRCBL: 15.1; 1.3.11. SC: po.
> Keywords: *genetically modified plants; agriculture; genetically modified food; industry; trends; Keyword Identifiers: United States

Prusak, Bernard G.; Malmqvist, Erik; Fenton, Elizabeth. Back to the future: Habermas's The Future of Human Nature [letters and reply]. *Hastings Center Report* 2007 March-April; 37(2): 4-6. NRCBL: 15.1; 15.5; 4.5; 1.1. Comments: Comment on: Elizabeth Fenton. Liberal eugenics and human nature: against Habermas. Hastings Center Report 2006 November-December; 36(6): 35-42.
> Keywords: *autonomy; *eugenics; *genetic engineering; *genetic enhancement; *human characteristics; freedom; parent child relationship; parents; Keyword Identifiers: *The Future of Human Nature (Habermas, Jürgen)

Rasko, John E.J.; O'Sullivan, Gabrielle M.; Ankeny, Rachel A. Is inheritable genetic modification the new dividing line? *In:* Rasko, John E.J.; O'Sullivan, Gabrielle M.; Ankeny, Rachel A., eds. The Ethics of Inheritable Genetic Modification: A Dividing Line? Cambridge: Cambridge University Press, 2006: 1-15. 50 fn. NRCBL: 15.4; 15.1.
> Keywords: *gene therapy; *genetic engineering; *germ cells; adverse effects; clinical trials; ethical analysis; future generations; gene transfer techniques; historical aspects; hu-

man experimentation; reproductive technologies; scientific misconduct; terminology; Keyword Identifiers: Twentieth Century

Rehmann-Sutter, Christoph. Controlling bodies and creating monsters: popular perceptions of genetic modifications. *In:* Rasko, John E.J.; O'Sullivan, Gabrielle M.; Ankeny, Rachel A., eds. The Ethics of Inheritable Genetic Modification: A Dividing Line? Cambridge: Cambridge University Press, 2006: 57-76. 41 fn. NRCBL: 15.1; 1.3.7; 14.5; 7.1.
> Keywords: *genetic engineering; *mass media; adverse effects; chimeras; cloning; eugenics; gene therapy; genetic identity; germ cells; human genome; Human Genome Project; literature; philosophy; public opinion; Keyword Identifiers: *The Island of Dr. Moreau (motion picture); Frankenstein (Shelley, M.); Gattaca (motion picture); The 6th Day (motion picture)

Robertson, John A. The virtues of muddling through. *Hastings Center Report* 2007 July-August; 37(4): 26-28. NRCBL: 14.1; 15.1; 5.3. SC: an; le. Comments: Comment on: Franco Furger and Francis Fukuyama. A proposal for modernizing the regulation of human biotechnologies Hastings Center Report 2007 July-August; 37(4): 16-20.
> Keywords: *biotechnology; *policy making; *reproductive technologies; advisory committees; cultural pluralism; democracy; dissent; federal government; freedom; genetic engineering; government regulation; policy analysis; preimplantation diagnosis; public participation; public policy; Keyword Identifiers: *United States

Rollin, Bernard. Ethics, biotechnology, and animals. *In:* Waldau, Paul; Patton, Kimberly, eds. A Communion of Subjects: Animals in Religion, Science, and Ethics. New York: Columbia University Press, 2006: 519-532. 19 refs. NRCBL: 15.1; 1.3.11; 14.5; 22.1; 22.2; 22.3. SC: an; rv.
> Keywords: *animal cloning; *animal welfare; *genetically modified animals; *moral policy; agriculture; animal experimentation; biotechnology; cloning; commodification; ecology; ethical analysis; gene pool; genetic diversity; genetic engineering; genetic research; industry; literature; religion; risks and benefits; suffering; Proposed Keywords: animal production; domestic animals; laboratory animals

Schmidt, Charlie. Negotiating the RNAi patent thicket. *Nature Biotechnology* 2007 March; 25(3): 273-275. 1 ref. NRCBL: 15.8.
> Keywords: *biotechnology; *genetic patents; drugs; economics; genetic techniques; industry; legal aspects; Proposed Keywords: licensure; Keyword Identifiers: United States

Scully, Jackie Leach. Inheritable genetic modification and disability: normality and identity. *In:* Rasko, John E.J.; O'Sullivan, Gabrielle M.; Ankeny, Rachel A., eds. The Ethics of Inheritable Genetic Modification: A Dividing Line? Cambridge: Cambridge University Press, 2006: 175-192. 37 fn. NRCBL: 15.4; 15.1; 9.5.1. SC: an.
> Keywords: *disabled persons; *gene therapy; *genetic engineering; *germ cells; *normality; *self concept; choice behavior; chronically ill; culture; ethical analysis; genetic diversity; hearing disorders; moral policy parents; patient participation; reproduction; suffering; wedge argument

NRCBL: National Reference Center for Bioethics Literature Classification Scheme   See inside front cover for terms.

189

Silver, Lee M. The battle for mother nature's genes. *In his:* Challenging Nature: The Clash of Science and Spirituality at the New Frontiers of Life. New York: Ecco, 2006: 278-293, 397-399. 44 fn. NRCBL: 15.1; 1.3.11.

> Keywords: *genetic engineering; *genetically modified food; *genetically modified plants; genetically modified organisms; genetic patents; microbiology; public opinion; recombinant DNA research; risk; Keyword Identifiers: Europe; United States

Simon, Jürgen. Biotechnology and law: biotechnology patents. Special considerations on the inventions with human material. *Revista de Derecho y Genoma Humano = Law and the Human Genome Review* 2006 July-December; (25): 139-159. 83 fn. NRCBL: 15.8; 15.1; 21.1. SC: le.

> Keywords: *biotechnology; *cloning; *genetic materials; *genetic patents; *legal aspects; *stem cells; DNA sequences; embryo research; genes; genetic engineering; germ cells; international aspects; reproductive technologies; Keyword Identifiers: Europe; United States

Smith, George P. Freedom of scientific investigation. *In his:* The Christian Religion and Biotechnology: A Search for Principled Decision-making. Dordrecht: Springer, 2005: 85-148. 406 fn. NRCBL: 15.1; 1.3.9; 14.1.

> Keywords: *biotechnology; *freedom; *science; abortion; advisory committees; cloning; embryo disposition; embryo research; embryonic stem cells; eugenics; federal government; fetal research; gene therapy; genetic discrimination; genetic engineering; genetic patents; genetic privacy; genetic research; genetic screening; genetically modified organisms; government regulation; legal aspects; politics; recombinant DNA research; reproductive technologies; Proposed Keywords: parthenogenesis; Keyword Identifiers: *United States; Diamond v. Chakrabarty; Great Britain; Human Fertilisation and Embryology Act 1990 (Great Britain); Patent and Trademark Office; President's Council on Bioethics

Tierney, John. Are scientists playing God? It depends on your religion. *New York Times* 2007 November 20; p. F1, F2. NRCBL: 1.3.9; 14.5; 1.2; 15.1; 18.5.4.

> Keywords: *cloning; *embryo research; *genetic engineering; *international aspects; *non-Western World; *religious ethics; *secularism; *Western World; biotechnology; Buddhist ethics; Christian ethics; embryonic stem cells; embryos; government regulation; Hindu ethics; legal aspects; public policy; reproductive technologies; Keyword Identifiers: Asia; Europe; North America; South America

Tong, Rosemarie. Traditional and feminist bioethical perspectives on gene transfer: is inheritable genetic modification really the problem? *In:* Rasko, John E.J.; O'Sullivan, Gabrielle M.; Ankeny, Rachel A., eds. The Ethics of Inheritable Genetic Modification: A Dividing Line? Cambridge: Cambridge University Press, 2006: 159-173. 55 fn. NRCBL: 15.4; 15.1; 1.1; 10. SC: an.

> Keywords: *gene therapy; *feminist ethics; *genetic engineering; *moral policy; bioethics; choice behavior; disadvantaged persons; embryos; ethical analysis; fetal therapy; gene transfer techniques; genetic services; health services accessibility; preimplantation diagnosis; pregnant women; risks and benefits; treatment refusal; women

Turney, Jon. Inhuman, superhuman, or posthuman? Images of genetic futures. *In:* Einsiedel, Edna; Timmermans, Frank, eds. Crossing Over: Genomics in the Public Arena. Calgary, Alberta, Canada: University of Calgary Press, 2005: 225-235. 26 refs.; 1 fn. NRCBL: 15.1; 3.1; 15.5. Conference: Essays from the conference held Apr. 25-27, 2003, Kananaskis, Alta.

> Keywords: *evolution; *forecasting; *genetic engineering; biotechnology; eugenics; genetics; historical aspects; literature; philosophy; public opinion; risks and benefits; social impact; Proposed Keywords: *transhumanism; Keyword Identifiers: Nineteenth Century; Twentieth Century

Twine, Richard. Animal genomics and ambivalence: a sociology of animal bodies in agricultural biotechnology. *Genomics, Society and Policy* 2007 August; 3(2): 99-117. 62 fn. NRCBL: 15.1; 22.3; 1.3.11. SC: an.

> Keywords: *animal welfare; *genomics; agriculture; animal cloning; biodiversity; biotechnology; chimeras; economics; food; genetic databases; international aspects; regulation; social sciences; trends; Proposed Keywords: *animal production; *domestic animals

Abstract: How may emergent biotechnologies impact upon our relations with other animals? To what extent are any changes indicative of new relations between society and nature? This paper critically explores which sociological tools can contribute to an understanding of the technologisation of animal bodies. By drawing upon interview data with animal scientists I argue that such technologies are being partly shaped by broader changes in agriculture. The complexity of genomics trajectories in animal science is partly fashioned through the deligitimisation of the productivist paradigm but continue to sit in tension around particular conceptions of sustainability in farm animal production. In spite of this deligitimisation process genomics is now being framed in the context of a new productivism (termed the livestock revolution) bound up in projected global changes in animal consumption during the first half of the 21st century. This potentially jars against both social trends that seek to re-enchant animal life and sustainability discourses which include social and environmental contexts. Nevertheless the possibility of a new productivism is supported by various interconnected trends including the emergence of a discourse of the 'bioeconomy' and a liberal regulatory apparatus for farm animal breeding technologies. Ultimately an understanding of the possibility of emerging new bio-capitalisations on animal life should be set in a broader context of competing agricultural paradigms as well as ongoing tensions over 'naturalness' in human/animal relations.

Tyshenko, Michael G. Management of natural and bioterrorism induced pandemics. *Bioethics* 2007 September; 21(7): 364-369. 31 fn. NRCBL: 21.3; 9.1; 15.1; 5.3.

> Keywords: *biological warfare; *biotechnology; *bioterrorism; *genetic engineering; historical aspects; public health; resource allocation; risk; social control; world health

Abstract: A recent approach for bioterrorism risk management calls for stricter regulations over biotechnology as a way to control subversion of technology that may be used to create a man-made pandemic. This approach is

---

largely unworkable given the increasing pervasiveness of molecular techniques and tools throughout society. Emerging technology has provided the tools to design much deadlier pathogens but concomitantly the ability to respond to emerging pandemics to reduce mortality has also improved significantly in recent decades. In its historical context determining just how 'risky' biological weapons is an important consideration for decision making and resource allocation. Management should attempt to increase capacity, share resources, provide accurate infectious disease reporting, deliver information transparency and improve communications to help mitigate the magnitude of future pandemics.

Walter, James J. Perspectives on medical ethics: biotechnology and genetic medicine. *In:* Walter, James J.; Shannon, Thomas A., eds. Contemporary Issues in Bioethics: A Catholic Perspective. Lanham, MD: Rowman and Littlefield Publishers, 2005: 181-197. 49 fn. NRCBL: 15.1; 1.2.

Keywords: *biotechnology; *clinical genetics; *Roman Catholic ethics; *theology; bioethics; cloning; eugenics; genetic engineering; genotype; health priorities; human dignity; Human Genome Project; justice; natural law; phenotype; recombinant DNA research; Keyword Identifiers: *Curran, Charles

Waltz, Emily. Supreme Court boosts licensees in biotech patent battles [news]. *Nature Biotechnology* 2007 March; 25(3): 264-265. NRCBL: 5.3; 15.8. SC: le. Identifiers: MedImmune v. Genentech.

Keywords: *biotechnology; *legal aspects; *patents; *Supreme Court decisions; industry; universities; Proposed Keywords: licensure; Keyword Identifiers: *MedImmune v. Genentech; *United States

Wexler, Barbara. Ethical issues and public opinion. *In her:* Genetics and genetic engineering. 2005 ed. Detroit, MI: Thomson/Gale Group, 2006: 163-173. NRCBL: 15.1; 15.3. SC: em.

Keywords: *genetic engineering; *genetic research; *genetic screening; *public opinion; access to information; biotechnology; cloning; disclosure; employment; eugenics; genetic discrimination; genetic information; Human Genome Project; insurance selection bias; preimplantation diagnosis; prenatal diagnosis; reproductive technologies; research priorities; risks and benefits; social impact; survey; Keyword Identifiers: United States

Wexler, Barbara. Genetic engineering and biotechnology. *In her:* Genetics and genetic engineering. 2005 ed. Detroit, MI: Thomson/Gale Group, 2006: 135-161. NRCBL: 15.1; 1.3.11.

Keywords: *biotechnology; *genetic engineering; *genetically modified food; *genetically modified plants; agriculture; DNA fingerprinting; doping in sports; ecology; gene therapy; genetic enhancement; health hazards; international aspects; nanotechnology; public opinion; recombinant DNA research; regulation; risks and benefits; stem cells; Proposed Keywords: forensic genetics; Keyword Identifiers: United States

Wolpert, Lewis. Is cell science dangerous? *Journal of Medical Ethics* 2007 June; 33(6): 345-348. 7 refs.

NRCBL: 15.1; 5.3; 1.3.9; 14.4; 14.5; 15.2; 15.4; 15.5. SC: an.

Keywords: *genetic engineering; *genetic research; *science; behavioral genetics; bioethical issues; decision making; democracy; embryo research; embryonic stem cells; eugenics; freedom; genetically modified organisms; mass media; moral obligations; policy making; prenatal diagnosis; public participation; reproductive technologies; researchers; risk; Proposed Keywords: technology

Zwart, Hub. Statements, declarations and the problems of ethical expertise [editorial]. *Genomics, Society and Policy* 2007 April; 3(1): ii-iv. 2 refs. NRCBL: 15.1; 2.1; 5.1; 21.1.

Keywords: *bioethical issues; *bioethics; *biotechnology; *genetics; *guidelines; *international aspects; *policy making; *public participation; *social control; *technical expertise; cloning; ethicists; ethics committees; genetic databases; genetic research; human genome; human rights;; Keyword Identifiers: Human Genome Organization (HUGO); UNESCO; Universal Declaration on Bioethics and Human Rights; Universal Declaration on the Human Genome and Human Rights

Zycinski, Joseph. Ethics in medical technologies: the Roman Catholic viewpoint. *Journal of Clinical Neuroscience* 2006 June; 13(5): 518-523. 20 refs. NRCBL: 18.5.4; 15.1; 1.2; 4.4.

Keywords: *biotechnology; *embryo research; *Roman Catholic ethics; ecology; embryonic stem cells; eugenics; human dignity; quality of life; value of life; values; Proposed Keywords: altered nuclear transfer

Zycinski, J. Bioethics, technology and human dignity: the Roman Catholic viewpoint. *Acta Neurochirurgica Supplement* 2006; 98: 1-7. 25 refs. NRCBL: 18.5.4; 15.1; 1.2; 4.4.

Keywords: *biotechnology; *embryos; *human dignity; *research embryo creation; *Roman Catholic ethics; bioethics; biomedical technologies; embryonic stem cells; evolution; moral status; nuclear transfer techniques; quality of life; value of life

GENETIC ENHANCEMENT

Bayertz, Kurt; Schmidt, Kurt W. Testing genes and constructing humans — ethics and genetics. *In:* Khushf, George, ed. Handbook of Bioethics: Taking Stock of the Field From a Philosophical Perspective. Dordrecht; Boston: Kluwer Academic, 2004: 415-438. 59 refs. 4 fn. NRCBL: 15.1; 15.3; 1.1. SC: an; rv.

Keywords: *biotechnology; *ethical analysis; *gene therapy; *genetic engineering; *genetic enhancement; *genetic screening; *moral policy; autonomy; confidentiality; diagnosis; eugenics; future generations; genetic information; genetic predisposition; germ cells; human characteristics; medicine; philosophy; physician patient relationship; population genetics; professional family relationship; prenatal diagnosis; principle-based ethics; right not to know; risks and benefits; social impact

Betta, Michela. From destiny to freedom? On human nature and liberal eugenics in the age of genetic manipulation. *In:* Betta, Michela, ed. The Moral, Social, and Commercial Imperatives of Genetic Testing and Screening: the Australian Case. Dordrecht: Springer, 2006: 3-24. 53 fn. NRCBL: 15.1; 1.1; 15.5; 4.5. SC: an.

NRCBL: National Reference Center for Bioethics Literature Classification Scheme   See inside front cover for terms.

191

Keywords: *eugenics; *freedom; *genetic enhancement; *genetic intervention; *philosophy; children; ethical theory; future generations; genetic determinism; genetic identity; human characteristics; moral policy; parent child relationship; parents; preimplantation diagnosis; regulation; Keyword Identifiers: Australia; Habermas, Jurgen

Borenstein, Jason. Shaping our future: the implications of genetic enhancement. *Human Reproduction and Genetic Ethics: An International Journal* 2007; 13(2): 4-15. NRCBL: 15.1; 4.5; 14.1; 1.1.

Bortolotti, Lisa; Harris, John. Disability, enhancement and the harm-benefit continuum. *In:* Spencer, J.R.; du Bois-Pedain, Antje, eds. Freedom and Responsibility in Reproductive Choice. Portland: Hart Pub., 2006: 31-49. 29 fn. NRCBL: 15.1; 1.1; 14.1; 4.5. SC: an.
  Keywords: *beneficence; *disabled persons; *ethical analysis; *genetic intervention; *moral policy; *reproduction; *risks and benefits; choice behavior; congenital disorders; moral obligations; normality; parents; precautionary principle; preimplantation diagnosis; selective abortion; socioeconomic factors; Proposed Keywords: transhumanism

Elliott, Carl. The mixed promise of genetic medicine. *New England Journal of Medicine* 2007 May 17; 356(20): 2024-2025. 4 refs. NRCBL: 15.1; 4.5; 5.3; 15.5.
  Keywords: *biomedical enhancement; *genetic enhancement; attitudes; cosmetic surgery; eugenics; philosophy; Keyword Identifiers: United States

Graham, Elaine. In whose image? Representations of technology and the 'ends' of humanity. *In:* Deane-Drummond, Celia; Scott, Peter Manley, eds. Future Perfect?: God, Medicine and Human Identity. New York: T and T Clark International, 2006: 56-69. 33 fn. NRCBL: 15.1; 4.5; 1.2.
  Keywords: *biomedical enhancement; *Christian ethics; *genetic engineering; *theology; biotechnology; human dignity; personhood; philosophy; values; Proposed Keywords: *technology; *transhumanism

Gunderson, Martin. Seeking perfection: a Kantian look at human genetic engineering. *Theoretical Medicine and Bioethics* 2007; 28(2): 87-102. 24 refs. 50 fn. NRCBL: 15.1; 1.1. SC: an.
  Keywords: *ethical analysis; *ethical theory; *gene therapy; *genetic engineering; *genetic enhancement; *philosophy; autonomy; freedom; future generations; germ cells; government regulation; human dignity; moral obligations; moral policy; paternalism; public policy; risks and benefits; socioeconomic factors; Proposed Keywords: happiness; Keyword Identifiers: *Kant, Immanuel

Abstract: It is tempting to argue that Kantian moral philosophy justifies prohibiting both human germ-line genetic engineering and non-therapeutic genetic engineering because they fail to respect human dignity. There are, however, good reasons for resisting this temptation. In fact, Kant's moral philosophy provides reasons that support genetic engineering-even germ-line and non-therapeutic. This is true of Kant's imperfect duties to seek one's own perfection and the happiness of others. It is also true of the categorical imperative. Kant's moral

philosophy does, however, provide limits to justifiable genetic engineering.

Hohlfeld, Rainer. Politische Ökonomie und Bio-Medizin. *In:* Neuer-Miebach, Therese; Wunder, Michael, eds. Bio-Ethik und die Zukunft der Medizin. Bonn: Psychiatrie-Verlag, 1998: 44-59. 38 fn. NRCBL: 15.1; 1.3.2; 4.5.
  Keywords: *biotechnology; *genomics; drug industry; economics; eugenics; genetic determinism; genetic enhancement; genetic predisposition; Human Genome Project; synthetic biology; Keyword Identifiers: Novartis

Holm, Søren. The nature of human welfare. *In:* Deane-Drummond, Celia; Scott, Peter Manley, eds. Future Perfect?: God, Medicine and Human Identity. New York: T and T Clark International, 2006: 45-55. 30 fn. NRCBL: 15.1; 1.1; 4.5. SC: an.
  Keywords: *ethical analysis; *genetic enhancement; *human characteristics; bioethics; biotechnology; disabled persons; gene therapy; germ cells; intelligence; philosophy; teleological ethics; Proposed Keywords: *transhumanism

Junker-Kenny, Maureen. Genetic perfection, or fulfilment of creation in Christ? *In:* Deane-Drummond, Celia; Scott, Peter Manley, eds. Future Perfect?: God, Medicine and Human Identity. New York: T and T Clark International, 2006: 155-167. 28 fn. NRCBL: 15.1; 1.2; 4.5.
  Keywords: *Christian ethics; *genetic enhancement; *theology; human characteristics; philosophy

Kline, A. David. Giftedness, humility and genetic enhancement. *Human Reproduction and Genetic Ethics: An International Journal* 2007; 13(2): 16-21. NRCBL: 15.1; 4.5.

Lock, Margaret. Utopias of health eugenics, and germline engineering. *In:* Nichter, Mark and Lock, Margaret, eds. New Horizons in Medical Anthropology: Essays in Honour of Charles Leslie. London: Routledge, 2002: 240-266. NRCBL: 15.1; 1.1; 4.5; 15.3; 15.4; 15.5; 15.10.

Peters, Ted. Perfect humans or trans-humans? *In:* Deane-Drummond, Celia; Scott, Peter Manley, eds. Future Perfect?: God, Medicine and Human Identity. New York: T and T Clark International, 2006: 15-32. 61 fn. NRCBL: 15.1; 4.5; 1.2; 15.4.
  Keywords: *Christian ethics; *genetic enhancement; *theology; attitudes to death; biomedical enhancement; gene therapy; health; justice; neurosciences; personhood; philosophy; Proposed Keywords: *transhumanism

Prusak, Bernard G.; Malmqvist, Erik; Fenton, Elizabeth. Back to the future: Habermas's The Future of Human Nature [letters and reply]. *Hastings Center Report* 2007 March-April; 37(2): 4-6. NRCBL: 15.1; 15.5; 4.5; 1.1. Comments: Comment on: Elizabeth Fenton. Liberal eugenics and human nature: against Habermas. Hastings Center Report 2006 November-December; 36(6): 35-42.
  Keywords: *autonomy; *eugenics; *genetic engineering; *genetic enhancement; *human characteristics; freedom; parent child relationship; parents; Keyword Identifiers: *The Future of Human Nature (Habermas, Jürgen)

Savulescu, Julian. Genetic interventions and the ethics of enhancement of human beings. *In:* Steinbock, Bonnie, ed. The Oxford Handbook of Bioethics. Oxford; New York: Oxford University Press, 2007: 516-535. 41 refs. NRCBL: 15.1; 4.5. SC: an; rv.

> Keywords: *ethical analysis; *genetic enhancement; *moral policy; adults; autonomy; behavioral genetics; beneficence; decision making; enhancement technologies; eugenics; genetic discrimination; genetic engineering; guidelines; human characteristics; justice; minors; public policy; reproduction; risk and benefits

Smith, George P. Genetic enhancement. *In his:* The Christian Religion and Biotechnology: A Search for Principled Decision-making. Dordrecht: Springer, 2005: 149-188. 285 fn. NRCBL: 15.1; 4.5; 14.1; 15.5; 15.9.

> Keywords: *eugenics; *genetic enhancement; autonomy; biotechnology; cloning; genetic counseling; genetic engineering; genetic screening; government regulation; historical aspects; involuntary sterilization; legal aspects; reproductive rights; reproductive technologies; sociobiology; trends; Keyword Identifiers: Europe; United States

GENETIC PATENTS *See* PATENTS

GENETIC PRIVACY

Alpert, Sheri. Brain privacy: how can we protect it? *American Journal of Bioethics* 2007 September; 7(9): 70-73. 11 refs. NRCBL: 17.1; 8.4; 15.1. SC: le. Comments: Comment on: Stacey A. Tovino. Functional neuroimaging and the law: trends and directions for future scholarship. American Journal of Bioethics 2007 September; 7(9): 44-56.

> Keywords: *access to information; *brain; *exceptionalism; *government regulation; *neurosciences; *privacy; disclosure; federal government; genetic information; legislation; presumed consent; state government; Proposed Keywords: *brain imaging; magnetic resonance imaging; Keyword Identifiers: *United States; Health Insurance Portability and Accountability Act (HIPAA)

Barash, Carol Isaacson. Threats to privacy protection [letter]. *Science* 2007 November 9; 318(5852): 913-914. NRCBL: 15.1; 8.4. Comments: Comment on: William W. Lowrance and Francis S. Collins. Identifiability in genomic research. Science 2007 August 3; 317(5838): 600-602.

Barnoy, Sivia; Tabak, Nili. Israeli nurses and genetic information disclosure. *Nursing Ethics* 2007 May; 14(3): 280-294. 37 refs. NRCBL: 15.1; 4.1.3; 8.4; 8.1; 15.3. SC: em.

> Keywords: *attitude of health personnel; *disclosure; *family members; *genetic information; *genetic privacy; *nurse's role' *nurses; family relationship; genetic screening; interprofessional relations; international aspects; knowledge, attitudes, practice; legal aspects; nurse patient relationship; patients; professional family relationship; questionnaires; Proposed Keywords: self disclosure; Keyword Identifiers: *Israel

Abstract: The debate continues about whether people have a duty to pass on the positive results of their genetic tests to relatives who are at risk from the same disease, and, should they refuse, whether physicians and genetic counselors then have the duty to do so. To date, the role and views of nurses in this debate have not been investigated. In our study, a sample of Israeli nurses, untrained in genetics, were asked for their theoretical opinions and what practical steps they would take in the case of patients' refusal to disclose. The nurses were very sure that patients should inform their families but were equally sure that nurses must respect their decision to disclose or not. Few said they would take practical steps to disclose information if the patient objected. The authors believe that the most useful and appropriate role for nurses in this field is in working to bring about co-operation between patients and family members.

Cotton, Richard G.H.; Sallée, Clémentine; Knoppers, Bartha M. Locus-specific databases: from ethical principles to practice. *Human Mutation* 2005 November; 26(5): 489-493. NRCBL: 15.1; 1.3.12; 18.3; 8.4. SC: em.

Duster, Troy. Differential trust in DNA forensics: grounded assessment or inexplicable paranoia? *GeneWatch* 2007 January-February; 20(1): 3-10. NRCBL: 15.1; 1.3.5; 15.11.

Gamero, Joaquin J.; Romero, Jose-Luis; Peralta, Juan-Luis; Carvalho, Mónica; Corte-Real, Francisco. Spanish public awareness regarding DNA profile databases in forensic genetics: what type of DNA profiles should be included? *Journal of Medical Ethics* 2007 October; 33(10): 598-604. 52 refs. NRCBL: 15.1; 1.3.5; 1.3.12; 1.3.5; 8.4; 18.3. SC: em.

> Keywords: *biological specimen banks; *DNA fingerprinting; *forensic genetics; *genetic databases; *public opinion; *public policy; genetic privacy; government regulation; informed consent; law enforcement; legislation; policy making; predictive value of tests; prisoners; public participation; survey; Keyword Identifiers: *Spain; European Union

Abstract: The importance of non-codifying DNA polymorphism for the administration of justice is now well known. In Spain, however, this type of test has given rise to questions in recent years: (a) Should consent be obtained before biological samples are taken from an individual for DNA analysis? (b) Does society perceive these techniques and methods of analysis as being reliable? (c) There appears to be lack of knowledge concerning the basic norms that regulate databases containing private or personal information and the protection that information of this type must be given. This opinion survey and the subsequent analysis of the results in ethical terms may serve to reveal the criteria and the degree of information that society has with regard to DNA databases. In the study, 73.20% (SE 1.12%) of the population surveyed was in favour of specific legislation for computer files in which DNA analysis results for forensic purposes are stored.

Gerlach, Neil. Biotechnology and social control: the Canadian DNA data bank. *In:* Mehta, Michael D., ed. Biotechnology Unglued: Science, Society and Social

---

NRCBL: National Reference Center for Bioethics Literature Classification Scheme   See inside front cover for terms.

193

Cohesion. Vancouver: UBC Press, 2005: 117-132. 40 refs. 2 fn. NRCBL: 15.1; 1.3.12; 1.3.5.

Keywords: *DNA fingerprinting; *genetic databases; legislation; public policy; Keyword Identifiers: *Canada

Gilbar, Roy. Communicating genetic information in the family: the familial relationship as the forgotten factor. *Journal of Medical Ethics* 2007 July; 33(7): 390-393. 29 refs. NRCBL: 15.2; 8.4; 1.1; 8.1.

Keywords: *communication; *disclosure; *duty to warn; *family members; *family relationship; *genetic counseling; *genetic privacy; confidentiality; informed consent; physician patient relationship; professional family relationship; right not to know; utilitarianism

Abstract: Communicating genetic information to family members has been the subject of an extensive debate recently in bioethics and law. In this context, the extent of the relatives' right to know and not to know is examined. The mainstream in the bioethical literature adopts a liberal perception of patient autonomy and offers a utilitarian mechanism for solving familial tensions over genetic information. This reflects a patient-centred approach in which disclosure without consent is justified only to prevent serious harm or death to others. Based on a legal and bioethical analysis on the one hand, and an examination of empirical studies on the other, this paper advocates the adoption of a relational perception of autonomy, which, in the context of genetics, takes into account the effect that any decision—whether to disclose or not to disclose—will have on the familial relationship and the dynamics of the particular family. Adding this factor to the criteria usually advocated by lawyers and ethicists will facilitate reaching a sensitive decision, which recognises the various interests of family members beyond the risk to physical health. Taking this factor into account will require a process of deliberation both between doctors and patients, and in the family. It will also require a relaxation of medical confidentiality, as the family rather than the patient is gradually perceived as the unit of care. Moreover, adopting such a relational approach will accord with current views of doctors and patients who base their decision primarily on the nature of the familial relationship.

Gilbar, Roy. Patient autonomy and relatives' right to know genetic information. *Medicine and Law: The World Association for Medical Law* 2007 December; 26(4): 677-697. NRCBL: 15.1; 8.4; 1.1. SC: le; an.

Abstract: One of the most difficult issues doctors face is a conflict between their professional duties. Such a conflict may arise when doctors know that information has implications not only for patients but also for family members but their duty of confidentiality prevents them from disclosing it. A comparative analysis of English and Israeli medical law reveals that the doctors' duty is based on two principles: a liberal perception of patient autonomy and an overriding utilitarian principle of prevention of harm. However, socio-medical research indicates that these principles do not entirely reflect the views of patients and doctors and are too narrow to deal with the complex situations in practice. Thus, it is argued that the doctor's legal duty of confidentiality should be reconsidered and quali-

fied when it concerns the family. It is suggested that if medical law seeks to recognize the various interests family members have in genetic information then we should consider a different approach, founded on a relational interpretation of autonomy and communitarian notions of solidarity and moral responsibility. This approach perceives confidentiality and privacy as embracing the family unit, based on the view that close relatives are not entirely outside the private sphere of the individual but rather are integral to his or her identity. Thus, to the utilitarian mechanism available in medical law this approach adds a social criterion: The effect any decision (to disclose or not to disclose) will have on the familial relationship and on the dynamics of the particular family. This will provide a more flexible and workable alternative for doctors to resolve familial tensions over access to genetic information.

Gulcher, Jeff; Stefansson, Kari. The Icelandic healthcare database: a tool to create knowledge, a social debate, and a bioethical and privacy challenge [editorial]. *Medscape Molecular Medicine* 1999: 5 p. [Online]. Accessed: http://www.medscape.com/viewarticle/414505 [2008 March 4]. NRCBL: 15.1; 8.4; 1.3.12.

Holm, Søren; Ashcroft, Richard. Should genetic information be disclosed to insurers? [debate]. *BMJ: British Medical Journal* 2007 June 9; 334(7605): 1196-1197. NRCBL: 15.1; 9.3.1; 15.3; 8.4. SC: an.

Keywords: *disclosure; *genetic information; *insurance; genetic discrimination; genetic privacy; genetic screening; government regulation; Keyword Identifiers: Great Britain

International Society of Nurses in Genetics [ISONG], Board of Directors. Privacy and confidentiality of genetic information: the role of the nurse [position statement]. Pittsburgh, PA: International Society of Nurses in Genetics, 2005 August 8: 2 p. [Online]. Accessed: http://www.isong.org/about/ps_privacy.cfm [2007 February 22]. 6 refs. NRCBL: 8.4; 15.1; 4.1.3; 6.

Keywords: *genetic privacy; *nurse's role; genetic information; organizational policies; professional organizations

Kaye, D.H. Bioethics, bench, and bar: selected arguments in Landry v. Attorney General. *Jurimetrics* 2000 Winter; 40(2): 193-216. 119 fn. NRCBL: 15.1; 1.3.12; 1.3.5; 8.4. SC: le.

Keywords: *DNA fingerprinting; *genetic databases; *legal aspects; biological specimen banks; blood specimen collection; family members; genetic privacy; genetic research; government regulation; law enforcement; prisoners; research subjects; Keyword Identifiers: *Landry v. Attorney General; *Massachusetts; Federal Policy (Common Rule) for the Protection of Human Subjects 1991; Nuremberg Code; United States

Kaye, D.H.; Smith, Michael E. DNA identification databases: legality, legitimacy, and the case for population-wide coverage. *Wisconsin Law Review* 2003; 2003(3): 413-459. 153 fn. NRCBL: 15.1; 1.3.5. 1.3.12; 8.4. SC: le.

Keywords: *DNA fingerprinting; *genetic databases; *law enforcement; *legal aspects; biological specimen banks; constitutional law; costs and benefits; criminal law; genetic privacy; justice; population genetics; prisoners; public opinion; racial groups; Supreme Court decisions; Keyword Identifiers: *United States

Kohut, Kelly; Manno, Michael; Gallinger, Steven; Esplen, Mary Jane. Should healthcare providers have a duty to warn family members of individuals with an HNPCC-causing mutation? A survey of patients from the Ontario Familial Colon Cancer Registry. *Journal of Medical Genetics* 2007 June; 44(6): 404-407. 30 refs. NRCBL: 15.1; 8.4. SC: em. Identifiers: hereditary non-polyposis colorectal cancer.

> Keywords: *duty to warn; *family members; *genetic information; *genetic predisposition; *genetic screening; *health personnel; attitudes; genetic counseling; genetic privacy; informed consent; legal aspects; patients; registries; risk; self disclosure; survey; Proposed Keywords: colon cancer; Keyword Identifiers: Ontario

Krimsky, Sheldon; Simoncelli, Tania. Genetic privacy: new frontiers. *GeneWatch* 2007 September-October; 20(5): 3-10. NRCBL: 15.1; 1.3.5; 8.4; 1.3.12. SC: le. Identifiers: DNA Fingerprints Act of 2005.

Lowrence, William W.; Collins, Francis S. Identifiability in genomic research. *Science* 2007 August 3; 317(5838): 600-602. 25 refs. NRCBL: 15.1; 8.4.

> Keywords: *access to information; *genetic databases; *genetic privacy; *genetic research; *genomics; biological specimen banks; confidentiality; ethical review; government regulation; human experimentation; human genome; information dissemination; informed consent; international aspects; legal aspects; research subjects; Keyword Identifiers: Health Insurance Portability and Accessibility Act (HIPAA); United States

Lucassen, Anneke; Clarke, Angus. Should families own genetic information [debate]. *BMJ: British Medical Journal* 2007 July 7; 335(7609): 22-23. NRCBL: 15.8; 1.3.12; 8.4.

> Keywords: *disclosure; *family members; *genetic information; *genetic privacy; *property rights; altruism; autonomy; confidentiality; genetic counseling; genetic screening; pedigree; professional family relationship; risk

Marshall, Eliot. Sequencers of a famous genome confront privacy issues [news]. *Science* 2007 March 30; 315(5820): 1780. NRCBL: 15.10; 8.4.

> Keywords: *disclosure; *DNA sequences; *famous persons; *genetic privacy; *genome mapping; *human genome; access to information; genetic databases; genetic predisposition; industry; informed consent; researchers; Proposed Keywords: self disclosure; Keyword Identifiers: *Watson, James; Church, George; Venter, J. Craig

Mc Fleming, Jennifer. The governance of human genetic research databases in mental health research. *International Journal of Law and Psychiatry* 2007 May-June; 30(3): 182-190. 45 refs. NRCBL: 15.1; 1.3.12; 15.6; 8.4; 17.1; 18.3. SC: le.

> Keywords: *genetic databases; *genetic research; *mental health; biological specimen banks; confidentiality; disclosure; empirical research; genetic information; genetic privacy; guidelines; informed consent; legal aspects; mentally ill persons; property rights; regulation; research subjects;

schizophrenia; stigmatization; vulnerable populations; Proposed Keywords: clinical utility; exceptionalism; research findings; Keyword Identifiers: Australia

Moore, Molly; Gavard, Corinne. French plan to screen DNA of visa-seekers draws anger [news]. *Washington Post* 2007 September 21; p. A14. NRCBL: 15.1; 1.3.5; 15.3. SC: po.

> Keywords: *DNA fingerprinting; *genetic relatedness ties; *immigrants; *public policy; ethnic groups; family members; genetic ancestry; international aspects; legal aspects; pedigree; politics; racial groups; social discrimination; Proposed Keywords: travel; Keyword Identifiers: *France

Moore, Solomon. DNA exoneration leads to change in legal system; states pass new laws; new police procedures — prisoners gain evidence access [news]. *New York Times* 2007 October 1; p. A1, A22. NRCBL: 15.1; 1.3.5. SC: le; po.

> Keywords: *DNA fingerprinting; *law enforcement; *legal aspects; *state government; access to information; laboratories; legislation; prisoners; standards; Proposed Keywords: *forensic genetics; Keyword Identifiers: *United States

Nyrhinen, Tarja; Hietala, Marja; Puukka, Pauli; Leino-Kilpi, Helena. Privacy and equality in diagnostic genetic testing. *Nursing Ethics* 2007 May; 14(3): 295-308. 37 refs. NRCBL: 15.3; 8.4; 9.5.7. SC: em.

> Keywords: *attitudes; *genetic privacy; *genetic screening; adults; attitude of health personnel; children; comparative studies; diagnosis; evaluation studies; education; genetic counseling; genetic discrimination; genetic information; genetic services; justice; knowledge, attitudes, practice; nurses; parents; patients; physicians; questionnaires; Keyword Identifiers: *Finland

Abstract: This study aimed to determine the extent to which the principles of privacy and equality were observed during diagnostic genetic testing according to views held by patients or child patients' parents (n = 106) and by staff (n = 162) from three Finnish university hospitals. The data were collected through a structured questionnaire and analysed using the SAS 8.1 statistical software. In general, the two principles were observed relatively satisfactorily in clinical practice. According to patients/parents, equality in the post-analytic phase and, according to staff, privacy in the pre-analytic phase, involved the greatest ethical problems. The two groups differed in their views concerning pre-analytic privacy. Although there were no major problems regarding the two principles, the differences between the testing phases require further clarification. To enhance privacy protection and equality, professionals need to be given more genetics/ethics training, and patients individual counselling by genetics units staff, giving more consideration to patients' world-view, the purpose of the test and the test result.

Pirakitikulr, Darlyn; Bursztajn, Harold J. Pride and prejudice: avoiding genetic gossip in the age of genetic testing. *Journal of Clinical Ethics* 2007 Summer; 18(2): 156-161. 21 fn. NRCBL: 15.3; 8.1; 8.4.

> Keywords: *genetic privacy; *genetic screening; disclosure; employment; genetic discrimination; genetic information; informed consent; insurance; legal aspects; legislation;

NRCBL: National Reference Center for Bioethics Literature Classification Scheme   See inside front cover for terms.

195

medical records; Keyword Identifiers: Health Insurance Portability and Accountability Act 1996; United States

Preston, Julia. U.S. set to begin a vast expansion of DNA sampling; big effect on immigrants; law to cover most people detained or arrested by federal agents. *New York Times* 2007 February 5; p. A1, A15. NRCBL: 15.1; 1.3.5. SC: po; le.

Keywords: *DNA fingerprinting; *legal aspects; *law enforcement; *legislation; immigrants; genetic databases; federal government; Keyword Identifiers: *United States; Department of Justice; Federal Bureau of Investigation; Violence against Women Act

Pullman, D.; Hodgkinson, K. Genetic knowledge and moral responsibility: ambiguity at the interface of genetic research and clinical practice. *Clinical Genetics* 2006 March; 69(3): 199-203. 32 refs. NRCBL: 15.1; 15.3; 8.4; 9.8. SC: cs; em.

Keywords: *clinical genetics; *duty to recontact; *duty to warn; *family members; *genetic research; *research findings; *research subjects; confidentiality; disclosure; genetic counseling; genetic privacy; research ethics; Keyword Identifiers: Newfoundland

Robinson, David J.; O'Neill, Desmond. Access to health care records after death: balancing confidentiality with appropriate disclosure. *JAMA: The Journal of the American Medical Association* 2007 February 14; 297(6): 634-636. 24 refs. NRCBL: 8.4; 15.1; 20.1.

Keywords: *access to information; *confidentiality; *death; *disclosure; *medical records; family members; genetic information; legal aspects; physician patient relationship; privacy; professional family relationship; Proposed Keywords: harm; Keyword Identifiers: United States

Romeo-Malanda, Sergio; Nicol, Dianne. Protection of genetic data in medical genetics: a legal analysis in the European context. *Revista de Derecho y Genoma Humano = Law and the Human Genome Review* 2007 July-December; (27): 97-134. 119 fn. NRCBL: 15.3; 8.4; 15.2; 21.1. SC: le.

Keywords: *genetic information; *genetic privacy; *genetic screening; *legal aspects; access to information; anonymous testing; autonomy; confidentiality; duty to warn; exceptionalism; family members; genetic counseling; genetic research; government regulation; informed consent; legal rights; medical records; public opinion; research subjects; right not to know; risks and benefits; Keyword Identifiers: *Europe; Council of Europe

Roscam Abbing, Henriette D.C. Pharmacogenetics: a new challenge for health law. *Medicine and Law: The World Association for Medical Law* 2007 December; 26(4): 781-789. NRCBL: 15.1; 9.7; 8.3.1; 8.4. SC: le.

Abstract: Developments in pharmacogenetics make it possible to determine the genetic factors that influence variations in response to medicine. Differences in response to medication may be related to the genetic characteristics of the individual, to the genetic make-up of the diseased tissue or to both. Advantages include optimal therapeutic effect, safe medication, minimised side-effects, and development of medication for small groups of patients. Strict adherence to patients' rights and to the

medical professional standard must prevent negative effects of pharmacogenetics on individual rights, notably the right (not) to know, to privacy and informed consent. Use of pharmacogenetics by third parties for non-health related purposes may bring about a disproportionate intrusion of the privacy of an individual; it may result in barriers for accessing primary social goods, and it may be a disincentive for the individual to have a pharmacogenetic analysis performed for individual health care purposes or to participate in a drug trial. Medical examinations before employment must be justified by the health requirements unavoidably inherent to the job (their objective being the protection of health and not the financial interests of the employer). In a system that relies on private insurance for having access to primary social goods (health, disability — and life insurance), the use and the outcome of a pharmacogenetic analysis for the purpose of differentiation between insurance candidates on the basis of their "risk-profile" must be restricted; where appropriate measures should take into account justified interests of the insurance company to prevent adverse selection. Current measures in several European countries are not effective enough to meet the concerns specifically inherent to pahrmacogenetics [sic; pharmacogenetics]. Human rights principles must be at the basis of national and European policies for providing adequate protection against disproportionate intrusion into private life, for guaranteeing equity in access to health care and accessibility of other primary social goods.

Rushlow, Jenny. Rapid DNA database expansion and disparate minority impact. *GeneWatch* 2007 July-August; 20(4): 3-11. 48 fn. NRCBL: 15.1; 1.3.5; 1.3.12; 9.5.4; 15.11. SC: le; em.

Keywords: *DNA fingerprinting; *genetic databases; *law enforcement; *legislation; *minority groups; *social impact; *state government; adults; biological specimen banks; evaluation studies; federal government; legal aspects; legal rights; minors; racial groups; social discrimination; statistics; trends; Proposed Keywords: crime; Keyword Identifiers: *United States

Sellaroli, Valentina; Cucca, Francesco; Santosuosso, Amedeo. Shared genetic data and the rights of involved people. *Revista de Derecho y Genoma Humano = Law and the Human Genome Review* 2007 January-June; (26): 193-231. 44 fn. NRCBL: 15.1; 8.4; 1.3.5. SC: le.

Keywords: *DNA fingerprinting; *genetic databases; *genetic information; *genetic privacy; *genetic screening; *legal aspects; access to information; exceptionalism; family members; genetic ancestry; genetic predisposition; international aspects; law enforcement; medical records; pregnant women; rights; Keyword Identifiers: *Europe; United States

Stolberg, Sheryl Gay. President calls for genetic privacy bill. *New York Times* 2007 January 18; p. A14. NRCBL: 8.4; 15.1. SC: po; le. Identifiers: George W. Bush.

Keywords: *genetic privacy; *legislation; politics; Keyword Identifiers: *United States

Taylor, Mark J. Data protection, shared (genetic) data and genetic discrimination. *Medical Law International* 2007; 8(1): 51-77. NRCBL: 15.1; 8.4; 13.1. SC: le.

United States. Congress. Senate. A bill to prohibit discrimination on the basis of genetic information with respect to health insurance and employment. Washington, DC: U.S. G.P.O., 2007. 80 p. [Online]. Accessed: http://frwebgate.access.gpo.gov/cgi-bin/useftp.cgi?IPaddress=162.140.64.21&filename=s358is.pdf&directory=/diskb/wais/data/110_cong_bills [2007 February 20]. NRCBL: 15.3; 9.3.1; 8.4. SC: le. Identifiers: Genetic Information Nondiscrimination Act of 2007. Note: S. 358., 110th Congress, 1st session. Introduced by Sen. Snowe, January 22, 2007. Referred to the Committee on Health, Education, Labor, and Pensions.

> Keywords: *genetic discrimination; *genetic information; *genetic privacy; *genetic screening; *legal aspects; employment; family members; health insurance; government regulation; Keyword Identifiers: *United States

Varma, Sumeeta; Ratterman, Allison Griffin. Sharing data, DNA and tissue samples. *Journal of Empirical Research on Human Research Ethics* 2007 March; 2(1): 97-100. 4 refs. NRCBL: 19.5; 15.1; 1.3.12; 8.4.

> Keywords: *access to information; *children; *confidentiality; *DNA sequences; *genetic databases; *genetic information; *genetic privacy; *genetic research; *publishing; adolescents; biological specimen banks; cancer; disclosure; donors; empirical research; federal government; genetic predisposition; government regulation; human genome; Internet; legal aspects; parental consent; research ethics committees; research subjects; risk; risks and benefits; state government; Proposed Keywords: genome-wide association studies; tissue donors; Keyword Identifiers: Health Insurance Portability and Accountability Act (HIPAA); United States

Wallace, Helen. The UK National DNA Database. Balancing crime detection, human rights and privacy. *EMBO Reports* 2006 July; 7 Special No: S26-S30. 24 refs. NRCBL: 15.1; 1.3.5; 1.3.12; 8.4.

> Keywords: *biological specimen banks; *DNA fingerprinting; *forensic genetics; *genetic databases; *law enforcement; accountability; behavioral genetics; donors; genetic privacy; genetic research; human rights; public policy; Proposed Keywords: crime; Keyword Identifiers: *Great Britain

Weisbrot, David. The imperative of the "new genetics": challenges for ethics, law, and social policy. *In:* Betta, Michela, ed. The Moral, Social, and Commercial Imperatives of Genetic Testing and Screening: the Australian Case. Dordrecht: Springer, 2006: 95-124. 101 fn. NRCBL: 15.1; 8.4. SC: le.

> Keywords: *genetic information; *genetic privacy; *legal aspects; *public policy; access to information; advisory committees; disclosure; DNA fingerprinting; employment; family members; genetic discrimination; genetic markers; informed consent; insurance; international aspects; legislation; regulation; Keyword Identifiers: *Australia; Australian Law Reform Commission; Genetic Privacy Act (Australia)

Williams, Robin; Johnson, Paul. Forensic DNA databasing: a European perspective. Interim report. Durham, Great Britain: University of Durham, 2005 June; 143 p. [Online]. Accessed: http://www.dur.ac.uk/resources/sass/Williams%20and%20Johnson%20Interim%20Report%202005-1.pdf [2007 April 25]. Refs., p. 133-143. NRCBL: 15.1; 1.3.5; 21.1.

> Keywords: *DNA fingerprinting; *genetic databases; *law enforcement; biological specimen banks; international aspects; legal aspects; organization and administration; regulation; trends; Proposed Keywords: forensic genetics; Keyword Identifiers: *European Union; Europe

Wylie, Jean E.; Mineau, Geraldine P. Biomedical databases: protecting privacy and promoting research. *Trends in Biotechnology* 2003 March; 21(3): 113-116. 23 refs. NRCBL: 15.1; 1.3.12; 1.3.9; 15.10; 8.4.

> Keywords: *biomedical research; *databases; *genetic research; biological specimen banks; computers; confidentiality; epidemiology; genetic databases; industry; medical records; privacy; regulation; standards; Keyword Identifiers: deCode Genetics; Iceland; United States; Utah

Zali, Mohammad Reza; Shahraz, Saeed; Borzabadi, Shokoufeh. Bioethics in Iran: legislation as the main problem. *Archives of Iranian Medicine* 2002 July; 5(3): 7 p. [Online]. Accessed: http://www.ac.ir./AIM/0253/0253136.htm [2008 March 31]. NRCBL: 15.1; 15.2; 8.3.1; 8.4; 1.2; 21.1.

GENETIC RESEARCH

See also GENETICALLY MODIFIED ORGANISMS AND FOOD; GENOME MAPPING AND SEQUENCING

A new tool to promote research: the Law on Biomedical Research [editorial]. *Revista de Derecho y Genoma Humano = Law and the Human Genome Review* 2007 January-June; (26): 17-20. NRCBL: 15.1. SC: le.

> Keywords: *biological specimen banks; *biomedical research; *genetic databases; *genetic research; *legal aspects; *legislation; advisory committees; cloning;; economics; embryo research; embryonic stem cells; genetic engineering; human dignity; nuclear transfer techniques; Keyword Identifiers: *Spain; Council of Europe

Animal-human hybrid-embryo research [editorial]. *Lancet* 2007 September 15-21; 370(9591): 909. NRCBL: 15.1; 18.5.4; 22.2; 5.3; 18.6; 18.1; 22.1.

Do-it-yourself science: how much involvement can patient advocates have in genetics? [editorial]. *Nature* 2007 October 18; 449(7164): 755-756. NRCBL: 15.1; 15.10; 15.11.

> Keywords: *genetic research; *patient advocacy; personal genomics; clinical genetics

Meanings of 'life' [editorial]. *Nature* 2007 June 28; 447(7148): 1031-1032. NRCBL: 15.1.

> Keywords: *life; *synthetic biology; genetic patents; genome

Proyecto CHIMBRIDS: chimeras and hybrids in comparative European and international research-scientific, ethical, philosophical and legal aspects. *Revista de Derecho y Genoma Humano = Law and the Human Genome Review* 2007 July-December; (27): 227-243. 16 fn. NRCBL: 15.1; 18.1; 22.1; 21.1. SC: le; rv.

NRCBL: National Reference Center for Bioethics Literature Classification Scheme   See inside front cover for terms.

197

Keywords: *chimeras; *international aspects; *moral policy; *public policy; animal experimentation; animal welfare; classification; comparative studies; embryo research; embryo transfer; embryos; ethical analysis; gene transfer techniques; genetic research; government regulation; legal aspects; moral status; nuclear transfer techniques; organ transplantation; research ethics; species specificity; stem cell transplantation; tissue transplantation; Keyword Identifiers: *European Union; Council of Europe; Europe

Aldhous, Peter. Angry reception greets patent for synthetic life [news]. *New Scientist* 2007 June 16-22; 194(2608): 13. NRCBL: 15.8.
Keywords: *genetic patents; *genome; *synthetic biology; industry; microbiology; Keyword Identifiers: Celera Genomics; Venter, Craig

Árnason, Arnar; Simpson, Bob. Refractions through culture: the new genomics in Iceland. *Ethnos* 2003 December; 68(4): 533-553. 52 refs. 8 fn. NRCBL: 15.11; 15.1; 1.3.12; 13.1.
Keywords: *culture; *genetic databases; *genetic research; *industry; *information dissemination; attitudes; biological specimen banks; biotechnology; genetic identity; genomics; mass media; medical records; metaphor; pedigree; politics; population genetics; public opinion; public participation; presumed consent; public policy; regulation; research subjects; researchers; risks and benefits; Proposed Keywords: persuasive communication; Keyword Identifiers: *deCode Genetics; *Iceland

Baylis, Françoise; Fenton, Andrew. Chimera research and stem cell therapies for human neurodegenerative disorders. *CQ: Cambridge Quarterly of Healthcare Ethics* 2007 Spring; 16(2): 195-208. 71 fn. NRCBL: 15.1; 18.1; 22.1; 17.1; 4.4; 18.5.4. SC: an.
Keywords: *chimeras; *embryonic stem cells; *ethical analysis; *human dignity; *moral policy; *moral status; *primates; *stem cell transplantation; animal welfare; clinical trials; guidelines; human characteristics; speciesism; therapeutic research; risks and benefits; Proposed Keywords: *neurodegenerative diseases; Keyword Identifiers: National Academy of Sciences

Bennett, Gaymon. Genetics, society, and spirituality. *In:* Eisen, Arri; Laderman, Gary, eds. Science, Religion, and Society: An Encyclopedia of History, Culture, and Controversy. Vol. 2. Armonk, NY: M.E. Sharpe, 2007: 763-779. 18 refs. NRCBL: 15.1; 1.2; 15.5.
Keywords: *genetic research; *genetics; eugenics; genetic determinism; genetic identity; genetic intervention; genetic variation; genotype; historical aspects; human dignity; human genome; Human Genome Project; phenotype; population genetics; risks and benefits; spirituality; trends

Betta, Michela. Diagnostic knowledge in the genetic economy and commerce. *In:* Betta, Michela, ed. The Moral, Social, and Commercial Imperatives of Genetic Testing and Screening: the Australian Case. Dordrecht: Springer, 2006: 25-52. 88 fn. NRCBL: 15.1; 15.3; 19.4; 1.3.2. SC: an.
Keywords: *biotechnology; *economics; *genetic research; *genetic screening; *policy making; blood banks; commerce; cord blood; diagnosis; disclosure; embryo research; embryonic stem cells; entrepreneurship; genetic databases; genetic discrimination; genetic disorders; genetic

information; genetic predisposition; genomics; government financing; government regulation; industry; insurance; legal aspects; patents; pharmacogenetics; policy analysis; public policy; regulation; research support; researchers; science; trends; Keyword Identifiers: Australia

Bharadwaj, Minakshi. Looking back, looking beyond: revisiting the ethics of genome generation. *Journal of Biosciences* 2006 March; 31(1): 167-176. 42 refs. NRCBL: 15.1; 1.3.11; 2.1; 15.3; 15.10; 15.5. SC: rv.
Keywords: *biotechnology; *genetic engineering; *genetic research; *genomics; benefit sharing; bioethics; cloning; developing countries; economics; embryo research; ethical analysis; genes; genetic determinism; genetic information; genetic patents; genetic screening; genetically modified food; genetically modified plants; Human Genome Project; international aspects; Proposed Keywords: exceptionalism; genetic resources; stakeholders

Brown, Barry F.; Sawa, Russell J. Key issues in genetic research, testing, and patenting. *In:* Monsour, H. Daniel, ed. Ethics and the New Genetics: An Integrated Approach. Toronto: University of Toronto Press, 2007: 143-164. 33 fn. NRCBL: 15.1; 1.2; 15.3; 15.8. SC: cs; rv.
Keywords: *clinical genetics; *genetic patents; *genetic research; *genetic screening; autonomy; case studies; chromosome abnormalities; commerce; common good; embryo research; gene therapy; genetic disorders; genetic engineering; genetic predisposition; genetically modified organisms; health care reform; human dignity; human genome; Human Genome Project; legal aspects; prenatal diagnosis; risks and benefits; selective abortion; Keyword Identifiers: Canada

Bubela, Tania M.; Caulfield, Timothy. Media representations of genetic research. *In:* Einsiedel, Edna; Timmermans, Frank, eds. Crossing Over: Genomics in the Public Arena. Calgary, Alberta, Canada: University of Calgary Press, 2005: 117-130. 51 refs., 1 fn. NRCBL: 15.1; 1.3.7. Conference: Essays from the conference held Apr. 25-27, 2003, Kananaskis, Alta.
Keywords: *genetic research; *mass media; biotechnology; conflict of interest; editorial policies; genetic determinism; genetic engineering; journalism; public opinion; researchers; risks and benefits; social impact; trust; Proposed Keywords: publication bias

Burgess, Michael M. Ethical analysis of representation in the governance of biotechnology. *In:* Einsiedel, Edna; Timmermans, Frank, eds. Crossing Over: Genomics in the Public Arena. Calgary, Alberta, Canada: University of Calgary Press, 2005: 157-172. 30 refs.; 7 fn. NRCBL: 15.1; 5.3. Conference: Essays from the conference held Apr. 25-27, 2003, Kananaskis, Alta.
Keywords: *biotechnology; *genetic research; *policy making; *social control; ethical analysis; focus groups; genomics; government; industry; international aspects; political activity; public participation; regulation; research priorities; researchers; risks and benefits; Proposed Keywords: consumer advocacy; stakeholders

Caulfield, Timothy. The media, marketing, and genetic services. *In:* Flood, Colleen M., ed. Just Medicare: What's In, What's Out, How We Decide. Buffalo, NY: University of Toronto Press, 2006: 379-395. 55 fn. NRCBL: 1.3.7; 1.3.9; 15.1.

Keywords: *genetic research; *genetic services; *marketing; *mass media; biomedical research; biotechnology; government financing; industry; journalism; policy making; public participation; research priorities; research support; researchers; science; technology assessment; Keyword Identifiers: Canada

Comber, Julie; Griffin, Gilly. Genetic engineering and other factors that might affect human-animal interactions in the research setting. *Journal of Applied Animal Welfare Science* 2007 June; 10(3): 267-277. 35 refs. NRCBL: 15.1; 22.1.

Keywords: *animal welfare; *genetically modified animals; agriculture; animal behavior; animal cloning; animal experimentation; attitudes; guidelines; researchers; Proposed Keywords: domestic animals; laboratory animals; Keyword Identifiers: Canada; Canadian Council on Animal Care

Condit, Celeste M. Lay people actively process messages about genetic research. *In:* Einsiedel, Edna; Timmermans, Frank, eds. Crossing Over: Genomics in the Public Arena. Calgary, Alberta, Canada: University of Calgary Press, 2005: 131-141. 22 refs.; 1 fn. NRCBL: 15.1; 1.3.7. Conference: Essays from the conference held Apr. 25-27, 2003, Kananaskis, Alta.

Keywords: *genetic determinism; *genetic research; *mass media; *public opinion; alcohol abuse; behavioral genetics; comprehension; empirical research; genetic ancestry; genetic predisposition; genetics; racial groups; social impact

Condit, C.M.; Parrott, R.L.; Bates, B.R.; Bevan, J.L.; Achter, P.J. Exploration of the impact of messages about genes and race on lay attitudes. *Clinical Genetics* 2004 November; 66(5): 402-408. 45 refs. NRCBL: 15.11; 9.5.4. SC: em.

Keywords: *attitudes; *genetic ancestry; *genetic discrimination; *genetic research; *information dissemination; *public opinion; *racial groups; *social impact; blacks; mass media; minority groups; social discrimination; survey; whites; Keyword Identifiers: United States

Couzin, Jennifer; Kaiser, Jocelyn. Closing the net on common disease genes. *Science* 2007 May 11; 316(5826): 820-822. NRCBL: 15.1.

Keywords: *genetic markers; *genetic predisposition; *genetic research; *human genome; *methods; access to information; clinical genetics; federal government; genetic databases; genetic screening; genetic techniques; informed consent; population genetics; public policy; research design; research subjects; trends; Proposed Keywords: *genome-wide association studies; Keyword Identifiers: National Institutes of Health; United States

Cullen, Rowena; Marshall, Stephen. Genetic research and genetic information: a health information professional's perspective on the benefits and risks. *Health Information and Libraries Journal* 2006 December; 23(4): 275-282. NRCBL: 15.1; 1.3.12; 15.10; 15.3; 15.4; 9.7.

Egorova, Yulia. 'Up in the sky': human and social sciences' responses to genetics. *In:* Gunning, Jennifer; Holm, Søren, eds. Ethics, law, and society. Volume 2. Aldershot, Hants, England; Burlington, VT: Ashgate, 2006: 45-53. 33 fn. NRCBL: 15.1; 1.1. SC: an.

Keywords: *biotechnology; *genetic research; *social impact; cloning; culture; genetic identity; genetic intervention; genetic patents; human genome; metaphor; philosophy; postmodernism; reproductive technologies; social sciences; Proposed Keywords: transhumanism

Egorova, Yulia. The meanings of science: conversations with geneticists. *Health Care Analysis: An International Journal of Health Philosophy and Policy* 2007 March; 15(1): 51-58. 16 refs. NRCBL: 15.1; 5.1; 21.1. SC: em.

Keywords: *attitudes; *genetic research; *genetics; *researchers; clinical genetics; culture; evolution; genetic determinism; genetic engineering; Human Genome Project; interdisciplinary communication; international aspects; interviews; mass media; personhood; philosophy; professional role; religion; science; secularism; self concept; technical expertise; Keyword Identifiers: Great Britain; Russia; United States

Abstract: It is often suggested in the mass media and popular academic literature that scientists promote a secular and reductionist understanding of the implications of the life sciences for the concept of being human. Is adhering to this view considered to be one of the components of the notion of being a good scientist? This paper explores responses of geneticists interviewed in the UK, the USA and Russia about the cultural meanings of their work. When discussing this question the interviewees distinguished between their 'personal' and 'professional' views. When talking as 'lay people' they demonstrated a wide range of opinions none of which was perceived as incompatible with scientific practice. When talking as 'scientists' the respondents stressed that the cultural implications of their research were not a matter of their professional concern. It is suggested that these two trends in their answers could be explained by scientists tending to relegate the implications of their work to the realm of the social which they construe as divorced from scientific practice.

Eischen, Kyle. Commercializing Iceland: biotechnology, culture, and the information society. *In:* Mehta, Michael D., ed. Biotechnology Unglued: Science, Society and Social Cohesion. Vancouver: UBC Press, 2005: 95-116. 48 refs. 2 fn. NRCBL: 15.1; 1.3.12; 9.3.1; 15.11; 13.1; 5.3.

Keywords: *genetic databases; *genetic research; *social impact; biotechnology; culture; genetic ancestry; genetic privacy; geographic factors; industry; legal aspects; population genetics; socioeconomic factors; Proposed Keywords: commerce; Keyword Identifiers: *Iceland; *deCode Genetics

Farrell, David Blake; De Neeve, Eileen. Commercialization of human genetic research. *In:* Monsour, H. Daniel, ed. Ethics and the New Genetics: An Integrated Approach. Toronto: University of Toronto Press, 2007: 58-75. 46 fn. NRCBL: 15.1; 15.8; 9.3.1. SC: rv.

Keywords: *commerce; *genetic patents; *genetic research; *government; *human genome; *industry; commodification; community consent; conflict of interest; economics; genetic databases; indigenous populations; interinstitutional relations; population genetics; private sector; property rights; public policy; public sector; research support; risks and benefits; social impact; trends; universities; utilitarianism; values

NRCBL: National Reference Center for Bioethics Literature Classification Scheme   See inside front cover for terms.

199

Gevers, Sjef. Human tissue research, with particular reference to DNA banking. *In:* Gevers, J.K.M.; Hondius, E.H.; Hubben, J.H., eds. Health Law, Human Rights and the Biomedicine Convention: Essays in Honour of Henriette Roscam Abbing. Leiden; Boston: Martinus Nijhoff Publishers, 2005: 231-243. 25 fn. NRCBL: 15.1; 1.3.12.

Keywords: *biological specimen banks; *genetic databases; family members; government regulation; guidelines; human experimentation; informed consent; legal aspects; research subjects; Proposed Keywords: tissue donors; Keyword Identifiers: *European Convention on Human Rights and Biomedicine; Europe

Godard, Béatrice. Involving communities: a matter of trust and communication. *In:* Einsiedel, Edna; Timmermans, Frank, eds. Crossing Over: Genomics in the Public Arena. Calgary, Alberta, Canada: University of Calgary Press, 2005: 87-98. 33 refs., 2 fn. NRCBL: 15.1; 1.3.12; 15.11; 5.1. Conference: Essays from the conference held Apr. 25-27, 2003, Kananaskis, Alta.

Keywords: *genetic databases; *genetic research; *population genetics; *public participation; communication; confidentiality; genetic diversity; genomics; information dissemination; public opinion; research design; research subjects; selection of subjects; trust; Proposed Keywords: research findings; Keyword Identifiers: *Cartagene Project; *Quebec

Godard, Béatrice; Marshall, Jennifer; Laberge, Claude. Community engagement in genetic research: results of the first public consultation for the Quebec CARTaGENE project. *Community Genetics* 2007; 10(3): 147-158. 28 refs. NRCBL: 15.1; 1.3.12; 18.1. SC: em.

Keywords: *genetic databases; *genetic research; *public opinion; *public participation; biological specimen banks; choice behavior; confidentiality; donors; duty to recontact; ethnic groups; focus groups; genetic privacy; knowledge, attitudes, practice; public sector; private sector; questionnaires; research subjects; researcher subject relationship; risks and benefits; Proposed Keywords: research findings; Keyword Identifiers: *Quebec; *Quebec CARTaGENE

Green, Ronald M. From genome to brainome: charting the lessons learned. *In:* Illes, Judy, ed. Neuroethics: Defining the Issues in Theory, Practice, and Policy. New York: Oxford University Press, 2006: 105-121. 82 refs. NRCBL: 15.1; 17.1. SC: an; rv.

Keywords: *comparative study; *genetic research; *neurosciences; biomedical enhancement; biomedical technologies; brain; coercion; diagnosis; DNA fingerprinting; ethical analysis; eugenics; family members; genetic determinism; genetic diversity; genetic identity; genetic information; genetic screening; Human Genome Project; informed consent; justice; normality; personhood; philosophy; prenatal diagnosis; privacy; research subjects; risks and benefits; terminology; uncertainty; Proposed Keywords: brain imaging

Hausman, Daniel M. Group risks, risks to groups, and group engagement in genetics research. *Kennedy Institute of Ethics Journal* 2007 December; 17(4): 351-369. 12 refs. 1 fn. NRCBL: 15.1; 18.2; 18.3; 5.2; 15.11. SC: an.

Keywords: *genetic research; *moral policy; *research subjects; *risk; *vulnerable populations; anthropology; classification; community consent; ethical analysis; ethnic groups; genetic ancestry; genetic epidemiology; indigenous populations; moral obligations; policy making; public participation; racial groups; research findings; social discrimination; stigmatization; Proposed Keywords: *community participation; *harm; *population groups; *third-party research subjects

Abstract: This essay distinguishes between two kinds of group harms: harms to individuals in virtue of their membership in groups and harms to "structured" groups that have a continuing existence, an organization, and interests of their own. Genetic research creates risks of causing both kinds of group harms, and engagement with the groups at risk can help to mitigate those harms. The two kinds of group harms call for different kinds of group engagement.

Heinrichs, Bert. A comparative analysis of selected European guidelines and recommendations for biobanks with special regard to the research / non-research distinction. *Revista de Derecho y Genoma Humano = Law and the Human Genome Review* 2007 July-December; (27): 205-224. 24 fn. NRCBL: 15.1; 1.3.12; 18.2; 19.5; 21.1.

Keywords: *biological specimen banks; *genetic databases; *genetic research; *guidelines; advisory committees; comparative studies; genetic information; informed consent; international aspects; organizational policies; public policy; research subjects; tissue donors; Keyword Identifiers: *Europe; Council of Europe; European Group on Ethics in Science and New Technologies; German National Ethics Council; Human Genetics Commission (Great Britain)

Helgesson, Gert; Dillner, Joakim; Carlson, Joyce; Bartram, Claus R.; Hansson, Mats G. Ethical framework for previously collected biobank samples [letter]. *Nature Biotechnology* 2007 September; 25(9): 973-976. 16 refs. NRCBL: 15.1; 1.3.12.

Keywords: *biological specimen banks; *biomedical research; *genetic research; *informed consent; *presumed consent; *standards; *tissue donors; access to information; confidentiality; duty to recontact; ethical review; genetic databases; guidelines; legal aspects; refusal to participate; regulation; research findings; research subjects; time factors

Hoedemaekers, Rogeer; Gordijn, Bert; Pijnenburg, Martien. Solidarity and justice as guiding principles in genomic research. *Bioethics* 2007 July; 21(6): 342-350. 25 fn. NRCBL: 15.10; 15.1; 1.3.12; 1.1; 9.1; 18.3; 18.5.1. SC: an.

Keywords: *ethical analysis; *future generations; *genetic research; *justice; *moral obligations; *moral policy; *nontherapeutic research; *obligations to society; *presumed consent; altruism; autonomy; biological specimen banks; communitarianism; ethical review; financial support; genetic databases; genomics; health priorities; industry; informed consent; obligations of society; public policy; research design; research priorities; research subjects; research support; resource allocation; risks and benefits; theoretical models

Abstract: In genomic research the ideal standard of free, informed, prior and explicit consent is sometimes difficult to apply. This has raised concern that important genomic research will be restricted. Different consent procedures have therefore been proposed. This paper explicitly examines the question how, in genomic research, the principles of solidarity and justice can be used to justify forms of diminished individual control over personal data and bio-samples. After a discussion of the notions of solidarity and justice and how they can be related to health care and genomic research, we examine how and in which situations these notions can form a strong moral basis for demanding certain financial sacrifices. Then we examine when these principles can justify consent procedures which diverge from the ideal standard. Because much genomic research is not expected to lead to immediate (clinical) benefits we also discuss the question of whether we can be obliged to make any sacrifices for future (not yet existing) patients. We conclude with the formulation of a number of conditions that have to be met before autonomy sacrifices can be reasonably demanded in genomic research.

Huntington Study Group. Event Monitoring Committee; Erwin, Cheryl; Hersch, Steven. Monitoring reportable events and unanticipated problems: the PHAROS and PREDICT studies of Huntington disease. *IRB: Ethics and Human Research* 2007 May-June; 29(3): 11-16. 18 fn. NRCBL: 18.2; 15.1. Identifiers: Prospective Huntington At Risk Observational Study; Neurobiological Predictors of Huntington Disease [PREDICT-HD].

Keywords: *adverse effects; *clinical trials; *data monitoring committees; *genetic research; *Huntington disease; *research subjects; *risk; government regulation; information dissemination; multicenter studies; nontherapeutic research; research design; research ethics committees; research findings; risks and benefits; Proposed Keywords: *observation; prospective studies; Keyword Identifiers: *Neurobiological Predictors of Huntington Disease [PREDICT-HD]; *Prospective Huntington At Risk Observational Study; United States

Indigenous Peoples Council on Biocolonialism. Indigenous people, genes and genetics: what indigenous people should know about biocolonialism: a primer and resource guide. Nixon, NV: Indigenous Peoples Council on Biocolonialism, 2000 June; 25 p. [Online]. Accessed: http://www.ipcb.org/publications/primers/htmls/ipgg.html [2007 April 23]. 2 refs. NRCBL: 15.11; 18.5.9.

Keywords: *American Indians; *genetic engineering; *genetic research; *indigenous populations; agriculture; biological specimen banks; biotechnology; commerce; eugenics; genes; genetic ancestry; genetic databases; genetic discrimination; genetic materials; genetic patents; genetically modified organisms; genetics; health hazards; Human Genome Diversity Project; informed consent; international aspects; political activity; population genetics; public policy; research ethics; research priorities; research subjects; Proposed Keywords: *genetic resources; biodiversity; colonialism

Institute of Medicine (United States) [IOM]. Board on Health Sciences Policy. Committee on Assessing Interactions Among Social, Behavioral, and Genetic Factors in Health; Hernandez, Lyla M.; Blazer, Dan G. Ethical, legal, and social implications. *In their:* Genes, Behavior, and the Social Environment: Moving Beyond the Nature/Nurture Debate. Washington, DC: National Academies Press, 2006: 202-218. 43 refs., 3 fn. NRCBL: 15.1; 9.1; 7.1; 15.6; 16.1.

Keywords: *genetic predisposition; *genetic research; *health promotion; access to information; behavioral genetics; confidentiality; decision making; disclosure; ecology; genetic disorders; genetic patents; genomics; informed consent; interdisciplinary communication; occupational exposure; policy making; population genetics; public participation; public policy; research ethics committees; research priorities; research subjects; risk; socioeconomic factors; vulnerable populations; Proposed Keywords: genetic epidemiology; research findings; Keyword Identifiers: National Institutes of Health; United States

Juengst, Eric T. Population genetic research and screening: conceptual and ethical issues. *In:* Steinbock, Bonnie, ed. The Oxford Handbook of Bioethics. Oxford; New York: Oxford University Press, 2007: 471-490. 66 refs. NRCBL: 15.11. SC: an; rv.

Keywords: *genetic research; *genetic screening; *population genetics; autonomy; benefit sharing; clinical genetics; community consent; cultural pluralism; decision making; ethnic groups; eugenics; gene pool; genetic ancestry; genetic discrimination; genetic diversity; genetic information; genetic resources; genotype; genomics; goals; international aspects; justice; mass screening; moral policy; phenotype; policy analysis; preventive medicine; property rights; public health; public policy; racial groups; research subjects; values

Kaiser, Jocelyn. Attempt to patent artificial organism draws a protest [news]. *Science* 2007 June 15; 316(5831): 1557. NRCBL: 15.8.

Keywords: *genetic patents; *synthetic biology; biotechnology; microbiology; Keyword Identifiers: Venter, J. Craig

Kettis-Lindblad, Åsa; Ring, Lena; Viberth, Eva; Hansson, Mats G. Genetic research and donation of tissue samples to biobanks. What do potential sample donors in the Swedish general public think? *European Journal of Public Health* 2006 August; 16(4): 433-440. NRCBL: 15.1; 19.1; 18.5.1; 18.3.

Kettner, Matthias. Die Herstellung einer öffentlichen Hegemonie. Humangenomforschung in der deutschen und der US-amerikanischen Presse [The making of a public hegemony. Human genome research in the German and US press, by Jürgen Gerhards and Mike Steffen Schäfer] [book review]. *Ethik in der Medizin* 2007 June; 19(2): 167-168. NRCBL: 15.1; 1.3.7; 21.1.

Keywords: *genetic research; *human genome; *mass media; *public opinion; international aspects; Keyword Identifiers: *Germany; *United States; Austria; France; Great Britain

Khoury, Muin J.; Gwinn, Marta; Bowen, Scott J.; Westfall, John M.; Mold, James; Fagnan, Lyle; Jones, Loretta; Wells, Kenneth B. Genomics and public health research [letter and replies]. *JAMA: The Journal of the American Medical Association* 2007 June 6; 297(21): 2347-2348. NRCBL: 15.1; 9.1.

Knoppers, Bartha Maria; Joly, Yoly; Simard, Jacques; Durocher, Francine. The emergence of an ethical duty to

NRCBL: National Reference Center for Bioethics Literature Classification Scheme   See inside front cover for terms.

201

disclose genetic research results: international perspectives. *European Journal of Human Genetics* 2006 November; 14(11): 1170-1178 [see correction in: European Journal of Human Genetics 2006 December; 14(12): 1322]. NRCBL: 15.1; 15.11; 18.3; 18.5.1; 21.1; 8.2. SC: le.

Knoppers, Bartha Maria; Sallée, Clémentine. Ethical aspects of genome research and banking. *In:* Sensen, Chrisopher W., ed. Handbook of Genome Research: Genomics, Proteomics, Metabolomics, Bioinformatics, Ethical and Legal Issues. Volume 2. Weinheim: Wiley-VCH, 2005: 509-536. 65 refs. NRCBL: 15.1; 1.3.12. SC: rv.

Keywords: *genetic databases; *genetic research; access to information; advisory committees; benefit sharing; biological specimen banks; cloning; disclosure; duty to warn; economics; ethical review; family members; gene therapy; genetic patents; genetic privacy; germ cells; guidelines; human rights; informed consent; international aspects; legal aspects; property rights; regulation; research ethics; research subjects; right not to know; trends; Proposed Keywords: tissue donors

Lipworth, W.; Ankeny, R.; Kerridge, I. Consent in crisis: the need to reconceptualize consent to tissue banking research. *Internal Medicine Journal* 2006 February; 36(2): 124-128. NRCBL: 15.1; 19.5; 18.3.

Lowrence, William W.; Collins, Francis S. Identifiability in genomic research. *Science* 2007 August 3; 317(5838): 600-602. 25 refs. NRCBL: 15.1; 8.4.

Keywords: *access to information; *genetic databases; *genetic privacy; *genetic research; *genomics; biological specimen banks; confidentiality; ethical review; government regulation; human experimentation; human genome; information dissemination; informed consent; international aspects; legal aspects; research subjects; Keyword Identifiers: Health Insurance Portability and Accessibility Act (HIPAA); United States

MacIntosh, Constance. Indigenous self-determination and research on human genetic material: a consideration of the relevance of debates on patents and informed consent, and the political demands on researchers. *Health Law Journal* 2005; 13: 213-251. 195 fn. NRCBL: 15.11; 15.8; 18.3; 13.1. SC: le.

Keywords: *genetic databases; *genetic diversity; *genetic materials; *genetic patents; *genetic research; *HapMap Project; *Human Genome Diversity Project; *indigenous populations; *informed consent; *international aspects; *population genetics; American Indians; anthropology; autonomy; benefit sharing; biological specimen banks; culture; ethnic groups; genetic identity; human rights; industry; legal aspects; moral policy; public policy; research subjects; Keyword Identifiers: *Genographic Project; Canada; European Union; Hagahai; National Geographic Society; National Institutes of Health; Patent Act; United States

MacKellar, Calum. Ethics and genetics of human behaviour [commentary]. *Ethics and Medicine: An International Journal of Bioethics* 2007 Spring; 23(1): 7-9. 6 fn. NRCBL: 15.6; 4.3.

Keywords: *behavioral genetics; *genetic research; eugenics; genetic determinism; mental disorders; risks and benefits; social discrimination; stigmatization

Malek, Janet. Understanding risks and benefits in research on reproductive genetic technologies. *Journal of Medicine and Philosophy* 2007 July-August; 32(4): 339-358. 22 refs. 9 fn. NRCBL: 15.1; 14.3; 15.2; 14.4; 18.6. SC: an.

Keywords: *genetic research; *genetic techniques; *reproductive technologies; *research subjects; *risks and benefits; adverse effects; beneficence; children; disabled persons; donors; economics; ethnic groups; ethical analysis; evaluation; family members; future generations; gene therapy; genetic engineering; genetic information; germ cells; guidelines; human experimentation; innovative therapies; men; preimplantation diagnosis; regulation; research ethics committees; stigmatization; women; wrongful life; Proposed Keywords: *third-party research subjects

Abstract: Research protocols must have a reasonable balance of risks and anticipated benefits to be ethically and legally acceptable. This article explores three characteristics of research on reproductive genetic technologies that complicate the assessment of the risk-benefit ratio for such research. First, a number of different people may be affected by a research protocol, raising the question of who should be considered to be the subject of reproductive genetic research. Second, such research could involve a wide range of possible harms and benefits, making the evaluation and comparison of those harms and benefits a challenging task. Finally, the risk-benefit ratio for this type of research is difficult to estimate because such research can have unpredictable, long-term implications. The article aims to facilitate the assessment of risk-benefit ratios in research on reproductive genetic technologies by proposing and defending some guidelines for dealing with each of these complicating factors.

Malek, Janet; Kopelman, Loretta M. The well-being of subjects and other parties in genetic research and testing. *Journal of Medicine and Philosophy* 2007 July-August; 32(4): 311-319. 8 refs. NRCBL: 15.1; 15.3; 15.11; 18.6. SC: an.

Keywords: *genetic research; *genetic screening; *risks and benefits; decision making; ethical analysis; ethnic groups; evaluation; exceptionalism; family members; gene therapy; genetic ancestry; genetic information; genetic materials; research subjects; right not to know; stigmatization; Proposed Keywords: harm; third-party research subjects

Matsui, Kenji; Lie, Reidar K.; Kita, Yoshikuni. Two methods of obtaining informed consent in a genetic epidemiological study: effects on understanding. *Journal of Empirical Research on Human Research Ethics* 2007 September; 2(3): 39-48. 23 refs. NRCBL: 15.1; 18.2; 18.3; 7.1. SC: em.

Keywords: *comprehension; *genetic research; *informed consent; *methods; *research subjects; comparative studies; consent forms; epidemiology; motivation; Proposed Keywords: *genetic epidemiology; Keyword Identifiers: Japan

Abstract: This study evaluated the effect on participant understanding and participation rates of two different approaches to obtaining informed consent, using 2,192 actual research subjects in a genetic cohort study. One group received the routine approach consisting of written materials and an oral explanation. The other group received a more intense approach consisting of educational lectures and group meetings in addition to the routine approach. Subjects in the intense approach group were relatively more likely to read some or all of the explanatory material. Those in the intense group who did not read the material were more likely than those in the routine group to express uncertainty about their understanding of the research. Those in the intense group who read the material perceived that they had a higher level of understanding of the research and this was associated with a higher frequency of volunteering to participate. In contrast, subjects in the routine group were less likely to read the written material, but ironically more likely to assume that they understood what the research was about. These rather paradoxical findings raised questions about what motivates potential research subjects to become sufficiently engaged to seek actual understanding of the research before volunteering.

Matthews, Robert. Are you looking at me? Medical researchers keen to scour patients' data for insights into disease should get consent first or risk coming seriously unstuck. *New Scientist* 2007 August 4-10; 195(2615): 18. NRCBL: 18.3; 8.4; 15.1; 1.3.12.

> Keywords: *biomedical research; *databases; *genetic databases; *genetic research; *informed consent; *presumed consent; confidentiality; epidemiology; international aspects; medical records; registries; Keyword Identifiers: Great Britain; UK Biobank

Mayor, Susan. UK body wants consultation on human-animal hybrid research [news]. *BMJ: British Medical Journal* 2007 January 20; 334(7585): 112. NRCBL: 15.1; 18.1; 22.1.

> Keywords: *chimeras; *embryo research; *regulation; public policy; Keyword Identifiers: *Great Britain; *Human Fertilisation and Embryology Authority

Mc Fleming, Jennifer. The governance of human genetic research databases in mental health research. *International Journal of Law and Psychiatry* 2007 May-June; 30(3): 182-190. 45 refs. NRCBL: 15.1; 1.3.12; 15.6; 8.4; 17.1; 18.3. SC: le.

> Keywords: *genetic databases; *genetic research; *mental health; biological specimen banks; confidentiality; disclosure; empirical research; genetic information; genetic privacy; guidelines; informed consent; legal aspects; mentally ill persons; property rights; regulation; research subjects; schizophrenia; stigmatization; vulnerable populations; Proposed Keywords: clinical utility; exceptionalism; research findings; Keyword Identifiers: Australia

McCarty, Catherine A.; Nair, Anuradha; Austin, Diane M.; Giampietro, Philip F. Informed consent and subject motivation to participate in a large, population-based genomics study: the Marshfield Clinic Personalized Medicine Research Project. *Community Genetics* 2007; 10(1): 2-9. 23 refs. NRCBL: 18.3; 15.1; 15.11; 1.3.12. SC: em.

> Keywords: *genetic databases; *genetic research; *informed consent; *motivation; *population genetics; *research subjects; biological specimen banks; coercion; donors; genomics; questionnaires; researchers; Keyword Identifiers: Wisconsin

Miles, Steven H. Human genomic research ethics: changing the rules. *In:* Cohen, Jillian Clare; Illingworth, Patricia; Schüklenk, Udo, eds. The Power of Pills: Social, Ethical and Legal Issues in Drug Development, Marketing, and Pricing. London; Ann Arbor, MI: Pluto, 2006: 203-214. 66 refs. NRCBL: 15.1; 21.1.

> Keywords: *benefit sharing; *developing countries; *genetic research; *international aspects; advisory committees; genetic databases; genetic diversity; genomics; human experimentation; human genome; justice; population genetics; property rights; research ethics;; research support; Proposed Keywords: biodiversity; Keyword Identifiers: Convention on Biological Diversity; Human Genome Organization (HUGO); United States

Ormondroyd, E.; Moynihan, C.; Watson, M.; Foster, C.; Davolls, S.; Ardern-Jones, A.; Eeles, R. Disclosure of genetic research results after the death of the patient participant: a qualitative study of the impact on relatives. *Journal of Genetic Counseling* 2007 August; 16(4): 527-538. 38 refs. NRCBL: 15.1; 8.1; 8.2; 18.5.1; 20.1. SC: em.

> Keywords: *BRCA2 genes; *cancer; *disclosure; *family members; *genetic counseling; *genetic predisposition; *genetic research; *research findings; *truth disclosure; breast cancer; death; genetic screening; informed consent; interviews; ovarian cancer; prostate cancer; qualitative research; research subjects

Paradies, Yin C.; Montoya, Michael J.; Fullerton, Stephanie M. Racialized genetics and the study of complex diseases: the thrifty genotype revisited. *Perspectives in Biology and Medicine* 2007 Spring; 50(2): 203-227. 143 refs. NRCBL: 15.1; 9.5.4; 15.11. SC: an; rv.

> Keywords: *diabetes; *ethnic groups; *genetic ancestry; *genetic predisposition; *genetic research; *population genetics; *racial groups; genetic diversity; genotype; health status; Proposed Keywords: *genetic epidemiology

Pentz, Rebecca D.; Billot, Laurent; Wendler, David. Research on stored biological samples: views of African American and White American cancer patients. *American Journal of Medical Genetics. Part A* 2006 April 1; 140(7): 733-739. 14 refs. NRCBL: 15.1; 1.3.12; 19.5; 18.3; 9.5.4. SC: em.

> Keywords: *attitudes; *biological specimen banks; *biomedical research; *blacks; *cancer; *minority groups; *patients; *whites; body parts and fluids; comparative studies; confidentiality; consent forms; donors; duty to recontact; genetic databases; informed consent; refusal to participate; socioeconomic factors; survey; time factors; Proposed Keywords: *tissue donors; Keyword Identifiers: Georgia

Persson, Anders. Research ethics and the development of medical biotechnology. *Xenotransplantation* 2006 November; 13(6): 511-513. NRCBL: 15.1; 15.4; 5.3; 18.2; 19.1; 22.2.

NRCBL: National Reference Center for Bioethics Literature Classification Scheme   See inside front cover for terms.

203

Phillips, Peter W.B.; Einsiedel, Edna. The future of genomics. *In:* Einsiedel, Edna; Timmermans, Frank, eds. Crossing Over: Genomics in the Public Arena. Calgary, Alberta, Canada: University of Calgary Press, 2005: 239-247. NRCBL: 15.1.

> Keywords: *genetic research; *genomics; biotechnology; commerce; decision making; interdisciplinary communication; policy making; private sector; public sector; research priorities; research support; risks and benefits; social sciences

Prainsack, Barbara. Research populations: biobanks in Israel. *New Genetics and Society* 2007 April; 26(1): 85-103. NRCBL: 15.1; 19.5; 15.3; 1.3.12; 18.2. SC: em.

Pullman, D.; Hodgkinson, K. Genetic knowledge and moral responsibility: ambiguity at the interface of genetic research and clinical practice. *Clinical Genetics* 2006 March; 69(3): 199-203. 32 refs. NRCBL: 15.1; 15.3; 8.4; 9.8. SC: cs; em.

> Keywords: *clinical genetics; *duty to recontact; *duty to warn; *family members; *genetic research; *research findings; *research subjects; confidentiality; disclosure; genetic counseling; genetic privacy; research ethics; Keyword Identifiers: Newfoundland

Rohter, Larry. In the Amazon, giving blood but getting nothing. *New York Times* 2007 June 20; p. A1, A4. NRCBL: 18.5.9; 19.4; 9.3.1; 1.3.9; 15.1; 15.11. SC: po. Identifiers: Karitiana Indians; Brazil.

> Keywords: *blood specimen collection; *commerce; *genetic materials; *genetic research; *indigenous populations; American Indians; culture; deception; informed consent; international aspects; population genetics; property rights; scientific misconduct; Proposed Keywords: tissue donors; Keyword Identifiers: *Brazil; *Karitiana Indians

Roscam Abbing, Henriette. Human tissue research, individual rights and bio-banks. *In:* Gunning, Jennifer; Holm, Søren, eds. Ethics, law, and society. Volume 2. Aldershot, Hants, England; Burlington, VT: Ashgate, 2006: 7-15. 10 fn. NRCBL: 15.1; 1.3.12.

> Keywords: *biological specimen banks; *genetic databases; *genetic research; biomedical research; commodification; confidentiality; family members; genetic information; human rights; informed consent; nontherapeutic research; population genetics; presumed consent; research subjects; Proposed Keywords: third-party research subjects; tissue donors

Rothstein, M.A.; Epps, P.G. Pharmacogenomics and the (ir)relevance of race. *Pharmacogenomics Journal* 2001; 1(2): 104-108. 29 refs. NRCBL: 15.11; 9.7; 15.1.

> Keywords: *genetic ancestry; *genetic research; *pharmacogenetics; *racial groups; biomedical research; clinical genetics; ethnic groups; genetic diversity; historical aspects; research subjects; selection of subjects; trends; Proposed Keywords: classification

Rusnak, A.J.; Chudley, A.E. Stem cell research: cloning, therapy and scientific fraud. *Clinical Genetics* 2006 October; 70(4): 302-305. NRCBL: 15.1; 14.5; 1.3.9; 18.5.1.

Schneider, Ingrid. Oocyte donation for reproduction and research cloning — the perils of commodification and the need for European and international regulation. *Revista de Derecho y Genoma Humano = Law and the Human Genome Review* 2006 July-December; (25): 205-241. 79 refs. NRCBL: 14.1; 14.5; 18.3.1; 21.1; 5.3. SC: le.

> Keywords: *cloning; *commodification; *embryo research; *genetic research; *ovum donors; *regulation; *remuneration; *reproductive technologies; advertising; embryo disposition; germ cells; in vitro fertilization; informed consent; international aspects; nuclear transfer techniques; therapeutic misconception; Keyword Identifiers: Europe

Schroeder, D. Benefit sharing: it's time for a definition. *Journal of Medical Ethics* 2007 April; 33(4): 205-209. 29 refs. NRCBL: 15.1; 1.1; 21.1. SC: an.

> Keywords: *benefit sharing; *genetic materials; *genetic research; *justice; *terminology; biological specimen banks; developing countries; donors; ethical analysis; ethics; genetic databases; guidelines; human genome; informed consent; international aspects; law; property rights; research subjects; Proposed Keywords: *genetic resources; biodiversity

Abstract: Benefit sharing has been a recurrent theme in international debates for the past two decades. However, despite its prominence in law, medical ethics and political philosophy, the concept has never been satisfactorily defined. In this conceptual paper, a definition that combines current legal guidelines with input from ethics debates is developed. Philosophers like boxes; protective casings into which they can put concisely-defined concepts. Autonomy is the human capacity for self-determination; beneficence denotes the virtue of good deeds, coercion is the intentional threat of harm and so on. What about benefit sharing? Does the concept have a box and are the contents clearly defined? The answer to this question has to be no. The concept of benefit sharing is almost unique in that various disciplines use it regularly without precise definitions. In this article, a definition for benefit sharing is provided, to eliminate unnecessary ambiguity.

Sharp, Richard R.; Foster, Morris W. Grappling with groups: protecting collective interests in biomedical research. *Journal of Medicine and Philosophy* 2007 July-August; 32(4): 321-337. 56 refs. 7 fn. NRCBL: 15.11; 15.1; 18.6. SC: an.

> Keywords: *genetic diversity; *genetic research; *Human Genome Diversity Project; *research; biomedical research; community consent; cultural pluralism; decision making; dissent; donors; ethnic groups; genetic ancestry; genetic materials; guidelines; indigenous populations; international aspects; justice; population genetics; public participation; research design; research subjects; researcher subject relationship; risks and benefits; vulnerable populations; Proposed Keywords: *population groups; harm; social identification; third-party research subjects

Abstract: Strategies for protecting historically disadvantaged groups have been extensively debated in the context of genetic variation research, making this a useful starting point in examining the protection of social groups from harm resulting from biomedical research. We analyze research practices developed in response to concerns about the involvement of indigenous communities in studies of genetic variation and consider their potential application in other contexts. We highlight several conceptual ambiguities and practical challenges associated with the protection of group interests and argue that protectionist strategies developed in the context of genetic research will not be easily adapted to other types of research in which social groups are placed at risk. We suggest that it is this set of conceptual and practical issues that philosophers, ethicists, and others should focus on in their efforts to protect identifiable social groups from harm resulting from biomedical research.

Sheldon, Jane P.; Jayaratne, Toby Epstein; Feldbaum, Merle B.; DiNardo, Courtney D.; Petty, Elizabeth M. Applications and implications of advances in human genetics: perspectives from a group of Black Americans. *Community Genetics* 2007; 10(2): 82-92. 39 refs. NRCBL: 15.11. SC: em.

> Keywords: *attitudes; *blacks; *genetic ancestry; *genetic predisposition; *genetic research; *racial groups; behavioral genetics; ethnic groups; genetic disorders; intelligence; interviews; risks and benefits; questionnaires; social discrimination; social impact; stigmatization; trust; violence; whites; Keyword Identifiers: United States

Shelton, B.L. Consent and consultation in genetic research on American Indians and Alaskan Natives. Nixon, NV: Indigenous Peoples Council on Biocolonialism, 2002; 3 p. [Online]. Accessed: http://www.ipcb.org/publications/briefing_papers/files/consent.html [2007 April 24]. NRCBL: 15.11; 18.3; 18.5.9.

> Keywords: *American Indians; *genetic research; community consent; government; indigenous populations; informed consent; policy making; public participation; refusal to participate; research subjects; Keyword Identifiers: United States

Sherwood, Mylaina L.; Buchinsky, Farrel J.; Quigley, Matthew R.; Donfack, Joseph; Choi, Sukgi S.; Conley, Stephen F.; Derkay, Craig S.; Myer, Charles M., III; Ehrlich, Garth D.; Post, J. Christopher. Unique challenges of obtaining regulatory approval for a multicenter protocol to study the genetics of RRP and suggested remedies. *Otolaryngology and Head and Neck Surgery* 2006 August; 135(2): 189-196. NRCBL: 15.1; 18.2; 18.3; 18.5.1.

Silversides, Ann. The wide gap between genetic research and clinical needs [news]. *CMAJ/JAMC: Canadian Medical Association Journal* 2007 January 30; 176(3): 315-316. 2 refs. NRCBL: 15.2; 15.3.

> Keywords: *clinical genetics; *genetic research; *genetic services; *government financing; *research support; *scarcity; education; health personnel; internship and residency; referral and consultation; Proposed Keywords: waiting lists; Keyword Identifiers: *Canada

Storz, Philipp; Kolpatzik, Kai; Perleth, Matthias; Klein, Silvia; Häussler, Bertram. Future relevance of genetic testing: a systematic horizon scanning analysis. *International Journal of Technology Assessment in Health Care* 2007 Fall; 23(4): 495-504. 17 refs. NRCBL: 15.3; 5.2; 7.1. SC: em.

> Keywords: *genetic epidemiology; *genetic research; *genetic screening; *research priorities; databases; genetic dis-

orders; genetic predisposition; literature; survey; technology assessment; Keyword Identifiers: Germany

Swede, Helen; Stone, Carol L.; Norwood, Alyssa R. National population-based biobanks for genetic research. *Genetics in Medicine* 2007 March; 9(3): 141-149. 73 refs. NRCBL: 15.1; 1.3.12; 19.5; 21.1; 18.3. SC: rv.

> Keywords: *biological specimen banks; *genetic databases; *population genetics; *international aspects; *public policy; *standards; confidentiality; donors; federal government; genetic privacy; genetic research; guidelines; informed consent; policy making; public health; public participation; research subjects; research support; state government; trends; Keyword Identifiers: *United States

Treloar, Susan A.; Morley, Katherine I.; Taylor, Sandra D.; Hall, Wayne D. Why do they do it? A pilot study towards understanding participant motivation and experience in a large genetic epidemiological study of endometriosis. *Community Genetics* 2007; 10(2): 61-71. 49 refs. NRCBL: 15.1; 18.3; 9.5.5. SC: em.

> Keywords: *attitudes; *genetic predisposition to disease; *genetic research; *motivation; *research subjects; donors; blood specimen collection; epidemiology; family members; genetic privacy; interviews; nontherapeutic research; women; Proposed Keywords: *endometriosis; *genetic epidemiology; pilot projects; Keyword Identifiers: Australia

United States. Congress. Senate. A bill to secure the promise of personalized medicine for all Americans by expanding and accelerating genomics research and initiatives to improve the accuracy of disease diagnosis, increase the safety of drugs, and identify novel treatments. Washington, DC: U.S. G.P.O., 2007. 32 p. [Online]. Accessed: http://frwebgate.access.gpo.gov/cgi-bin/useftp.cgi?IPaddress=162.140.64.21&filename=s976is.pdf&directoRY=/diskb/wais/data/110_cong_bills [2007 April 4]. NRCBL: 15.1; 9.7; 19.5. SC: le. Identifiers: Genomics and Personalized Medicine Act of 2007. Note: S. 976, 110th Congress, 1st session. Introduced by Sen. Obama on March 23, 2007. Referred to the Committee on Health, Education, Labor, and Pensions.

> Keywords: *genetic research; *genomics; *legal aspects; *pharmacogenetics; *public policy; advertising; biological specimen banks; drugs; genetic screening; guidelines; research support; standards; Keyword Identifiers: *United States

Varma, Sumeeta; Ratterman, Allison Griffin. Sharing data, DNA and tissue samples. *Journal of Empirical Research on Human Research Ethics* 2007 March; 2(1): 97-100. 4 refs. NRCBL: 19.5; 15.1; 1.3.12; 8.4.

> Keywords: *access to information; *children; *confidentiality; *DNA sequences; *genetic databases; *genetic information; *genetic privacy; *genetic research; *publishing; adolescents; biological specimen banks; cancer; disclosure; donors; empirical research; federal government; genetic predisposition; government regulation; human genome; Internet; legal aspects; parental consent; research ethics committees; research subjects; risk; risks and benefits; state government; Proposed Keywords: genome-wide association studies; tissue donors; Keyword Identifiers: Health Insurance Portability and Accountability Act (HIPAA); United States

NRCBL: National Reference Center for Bioethics Literature Classification Scheme   See inside front cover for terms.

205

Wexler, Barbara. Ethical issues and public opinion. *In her:* Genetics and genetic engineering. 2005 ed. Detroit, MI: Thomson/Gale Group, 2006: 163-173. NRCBL: 15.1; 15.3. SC: em.

Keywords: *genetic engineering; *genetic research; *genetic screening; *public opinion; access to information; biotechnology; cloning; disclosure; employment; eugenics; genetic discrimination; genetic information; Human Genome Project; insurance selection bias; preimplantation diagnosis; prenatal diagnosis; reproductive technologies; research priorities; risks and benefits; social impact; survey; Keyword Identifiers: United States

Wolpert, Lewis. Is cell science dangerous? *Journal of Medical Ethics* 2007 June; 33(6): 345-348. 7 refs. NRCBL: 15.1; 5.3; 1.3.9; 14.4; 14.5; 15.2; 15.4; 15.5. SC: an.

Keywords: *genetic engineering; *genetic research; *science; behavioral genetics; bioethical issues; decision making; democracy; embryo research; embryonic stem cells; eugenics; freedom; genetically modified organisms; mass media; moral obligations; policy making; prenatal diagnosis; public participation; reproductive technologies; researchers; risk; Proposed Keywords: technology

Wylie, Jean E.; Mineau, Geraldine P. Biomedical databases: protecting privacy and promoting research. *Trends in Biotechnology* 2003 March; 21(3): 113-116. 23 refs. NRCBL: 15.1; 1.3.12; 1.3.9; 15.10; 8.4.

Keywords: *biomedical research; *databases; *genetic research; biological specimen banks; computers; confidentiality; epidemiology; genetic databases; industry; medical records; privacy; regulation; standards; Keyword Identifiers: deCode Genetics; Iceland; United States; Utah

Zwart, Nijmegen Hub. Genomics and self-knowledge: implications for societal research and debate. *New Genetics and Society* 2007 August; 26(2): 181-202. 45 refs. 11 fn. NRCBL: 15.1; 5.1; 15.10; 1.1.

Keywords: *genetic determinism; *genomics; *human genome; *Human Genome Project; *social impact; *trends; biotechnology; genetic engineering; genetic information; metaphor; philosophy; Proposed Keywords: transhumanism; Keyword Identifiers: Jurassic Park (Crichton, M.)

GENETIC SCREENING
See also DNA FINGERPRINTING; GENETIC COUNSELING; GENOME MAPPING AND SEQUENCING

Myriad Genetic launches direct-to-consumer advertising of breast cancer gene test in Northeastern cities [news]. *Kaiser Daily Women's Health Policy Report* 2007 September 11: 2 p. [Online]. Accessed: http://kaisernetwork.org/daily_reports/print_report.cfm?DR_ID=47409&dr_cat=2 [2007 September 11]. NRCBL: 15.3; 5.3. Identifiers: BRCA1; BRCA2; Myriad Genetic Laboratories; Boston; Hartford, CN; New York City; Providence, RI; BRCAnalysis.

Aitken, Maryanne; Metcalfe, Sylvia. The social imperative for community genetic screening: an Australian perspective. *In:* Betta, Michela, ed. The Moral, Social, and Commercial Imperatives of Genetic Testing and Screening: the Australian Case. Dordrecht: Springer, 2006: 165-184. 109 fn. NRCBL: 15.3.

Keywords: *genetic screening; attitudes; genetic carriers; genetic disorders; genetic markers; genetic predisposition; guidelines; mass screening; newborns; pregnant women; prenatal diagnosis; program descriptions; public opinion; risks and benefits; voluntary programs; Keyword Identifiers: *Australia

Anido, Aimee; Carlson, Lisa M.; Sherman, Stephanie L. Attitudes toward fragile X mutation carrier testing from women identified in a general population survey. *Journal of Genetic Counseling* 2007 February; 16(1): 97-104. 17 refs. NRCBL: 15.3. SC: em.

Keywords: *attitudes; *chromosome abnormalities; *genetic carriers; *genetic screening; *women; focus groups; interviews; motivation; prenatal diagnosis; selective abortion; survey; Proposed Keywords: *fragile X syndrome

Ashley, Benedict M.; deBlois, Jean K.; O'Rourke, Kevin D. Genetic screening and counseling. *In their:* Health Care Ethics: A Catholic Theological Analysis. 5th edition. Washington, DC: Georgetown University Press, 2006: 98-103. NRCBL: 15.3; 15.2; 1.2.

Keywords: *genetic counseling; *genetic screening; *Roman Catholic ethics; eugenics; mandatory testing; mass screening; moral obligations; newborns; parents; prenatal diagnosis; selective abortion

Austin, Jehannine C.; Smith, Geoffrey N.; Honer, William G. The genomic era and perceptions of psychotic disorders: genetic risk estimation, associations with reproductive decisions and views about predictive testing. *American Journal of Medical Genetics. Part B, Neuropsychiatric Genetics* 2006 December 5; 141(8): 926-928. NRCBL: 15.3; 15.6; 14.1.

Bird, Stephanie J. Genetic testing for neurologic diseases: a rose with thorns. *Neurologic Clinics* 1989 November; 7(4): 859-870. NRCBL: 15.3; 17.1; 8.4; 15.2.

Blase, Terri; Martinez, Ariadna; Grody, Wayne W.; Schimmenti, Lisa; Palmer, Christina G.S. Sharing GJB2/GJB6 genetic test information with family members. *Journal of Genetic Counseling* 2007 June; 16(3): 313-324. NRCBL: 15.3; 8.1; 8.2; 15.2. SC: em.

Borry, P.; Stultiens, L.; Nys, H.; Cassiman, J.-J.; Dierickx, K. Presymptomatic and predictive genetic testing in minors: a systematic review of guidelines and position papers. *Clinical Genetics* 2006 November; 70(5): 374-381. NRCBL: 15.3; 9.5.7.

British Council Switzerland; University of Basel, Institute for Applied Ethics and Medical Ethics. Conflicts of interest: ethics and predictive medicine. Bern and Basel: British Council Switzerland; Institute of Applied Ethics and Medical Ethics, 2003 February; 72 p. [Online]. Accessed: http://www.britishcouncil.ch/governance/Genetic%20and%20Ethics.pdf [2007 April 30]. NRCBL: 15.3;

7.3; 9.5.1; 9.4; 9.5.7. Conference: Seminar: Conflicts of Interest: Ethics and Predictive Medicine; Basel, Switzerland; 20-21 February 2003; British Council Switzerland and University of Basel, Institute of Applied Ethics and Medical Ethics.

British In Vitro Diagnostics Association [BIVDA]. Genetic testing: the difference diagnostics can make [position paper 6]. London: British In Vitro Diagnostics Association, 2004 November; 10 p. [Online]. Accessed: http://www.bivda.co.uk/Portals/0/Positionpaper6-genetics3.pdf [2007 April 17]. NRCBL: 15.3; 6.

Brown, David. For the first time, FDA recommends gene testing. *Washington Post* 2007 August 17; p. A10. NRCBL: 15.3; 5.3; 9.1; 9.7. SC: po.
> Keywords: *drugs; *genetic screening; *government regulation; pharmacogenetics; Proposed Keywords: *drug prescriptions; drug labeling; Keyword Identifiers: *Food and Drug Administration; *Warfarin; United States

Burke, Wylie; Psaty, Bruce M. Personalized medicine in the era of genomics. *JAMA: The Journal of the American Medical Association* 2007 October 10; 298(14): 1682-1684. 25 refs. NRCBL: 15.3; 15.1.
> Keywords: *clinical genetics; *genetic predisposition; *genomics; genetic screening; preventive medicine; risk; trends; Proposed Keywords: *personalized medicine

Burke, Wylie; Zimmern, Ronald L.; Kroese, Mark. Defining purpose: a key step in genetic test evaluation. *Genetics in Medicine* 2007 October; 9(10): 675-681. 45 refs. NRCBL: 15.3. SC: rv.
> Keywords: *genetic screening; *evaluation studies; *goals; classification; clinical genetics; decision making; genetic services; health care delivery; predictive value of tests; standards

Calsbeek, Hiske; Morren, Mattijn; Bensing, Jozien; Rijken, Mieke. Knowledge and attitudes towards genetic testing: a two year follow-up study in patients with asthma, diabetes mellitus and cardiovascular disease. *Journal of Genetic Counseling* 2007 August; 16(4): 493-504. 19 refs. NRCBL: 15.3; 9.5.1. SC: em.
> Keywords: *chronically ill; *genetic screening; *knowledge, attitudes, practice; genetic counseling; genetics; patients; survey; Proposed Keywords: asthma; cardiovascular diseases; diabetes mellitus; follow-up studies; Keyword Identifiers: Netherlands

Caruso, Denise. Genetic tests offer promise, but raise questions too. *New York Times* 2007 February 18; p. BU5. NRCBL: 15.3. SC: po.

Chen, Lei-Shih; Goodson, Patricia. Factors affecting decisions to accept or decline cystic fibrosis carrier testing/screening: a theory-guided systematic review. *Genetics in Medicine* 2007 July; 9(7): 442-450. NRCBL: 15.3. SC: em.

Coghlan, Andy. Genetic testing: an informed choice? *New Scientist* 2007 October 6-12; 195(2624): 8-9. NRCBL: 15.3; 1.3.2; 9.7.

Collins, John A. Preimplantation genetic screening in older mothers [editorial]. *New England Journal of Medicine* 2007 July 5; 357(1); 61-63. 12 refs. NRCBL: 15.3; 14.4.
> Keywords: *age factors; *chromosome abnormalities; *genetic screening; *preimplantation diagnosis; genetic disorders; in vitro fertilization; pregnant women; risks and benefits; Proposed Keywords: pregnancy outcome

Comfort, Nathaniel. "Polyhybrid heterogeneous bastards": promoting medical genetics in America in the 1930s and 1940s. *Journal of the History of Medicine and Allied Sciences* 2006 October; 61(4): 415-455. NRCBL: 15.3; 15.1; 15.11; 15.5; 2.2.

Copelovitch, Lawrence; Kaplan, Bernard S. Is genetic testing of healthy pre-symptomatic children with possible Alport syndrome ethical? *Pediatric Nephrology* 2006 April; 21(4): 455-456. NRCBL: 15.3; 9.5.7.

Council of Europe. Working Party on Human Genetics [CDBI-CO-GT4]. Working document on the applications of genetics for health purposes. Strasbourg: Council of Europe. Working Party on Human Genetics (CDBI-CO-GT4), 2003 February 7: 8 p. [Online]. Accessed: http://www.coe.int/t/e/legal_affairs/legal_co-operation/bioethics/activities/human_genetics/INF(2003)3E_Wkgdoc_genetics.pdf [2007 March 5]. NRCBL: 15.3; 9.5.1; 8.3.1; 8.3.3; 18.2.
> Keywords: *genetic screening; biological specimen banks; family members; gene therapy; genetic privacy; genetic research; genetic services; informed consent; public policy; Keyword Identifiers: *Council of Europe

Couzin, Jennifer. Amid debate, gene-based cancer test approved [news]. *Science* 2007 February 16; 315(5814): 924. NRCBL: 15.3; 21.1; 5.3.

de Melo-Martín, Immaculada. Genetic testing: the appropriate means for a desired goal? *Journal of Bioethical Inquiry* 2006; 3(3): 167-177. NRCBL: 15.3; 9.1; 5.3; 15.2; 9.5.1. SC: an.

Deane-Drummond, Celia. Genetic testing and screening. *In her:* Genetics and Christian Ethics. Cambridge: Cambridge University Press, 2006: 76-100. 39 fn. NRCBL: 15.3; 1.3.12; 14.4; 15.1; 15.2.
> Keywords: *genetic screening; advisory committees; children; genetic databases; genetic disorders; population genetics; preimplantation diagnosis; prenatal diagnosis; public policy; Keyword Identifiers: Great Britain; Human Genetics Commission (Great Britain); Human Fertilisation and Embryology Authority; UK Biobank

DiMichele, D.; Chuansumrit, A.; London, A.J.; Thompson, A.R.; Cooper, C.G.; Killian, R.M.; Ross, L.F.; Lillicrap, D.; Kimmelman, J. Ethical issues in haemophilia. *Haemophilia* 2006 July; 12(Supplement 3): 30-35. NRCBL: 15.3; 15.4; 9.2; 21.1; 9.5.1.

Duncan, R.E.; Delatycki, M.B. Predictive genetic testing in young people for adult-onset conditions: where is the empirical evidence? *Clinical Genetics* 2006 January;

NRCBL: National Reference Center for Bioethics Literature Classification Scheme   See inside front cover for terms.

207

69(1): 8-16; discussion 17-20. 46 refs. NRCBL: 15.3; 9.5.7.

> Keywords: *genetic screening; *late-onset disorders; *minors; adolescents; children; genetic counseling; genetic predisposition; empirical research; risks and benefits

Erde, Edmund L.; McCormack, Michael K.; Steer, Robert A.; Ciervo, Carman A., Jr.; McAbee, Gary N. Patient confidentiality vs disclosure of inheritable risk: a survey-based study. *Journal of the American Osteopathic Association* 2006 October; 106(10): 615-620. NRCBL: 15.3; 8.4; 8.2; 7.1; 5.3. SC: em.

Fox, Jeffrey L. Feds eye genetic testing [news]. *Nature Biotechnology* 2007 December; 25(12): 1340. NRCBL: 15.3; 5.3.

> Keywords: *genetic screening; *public policy; advertising; advisory committees; federal government; genetic services; government regulation; industry; Keyword Identifiers: *Secretary's Advisory Committee on Genetics, Health, and Society (SACGHS); Food and Drug Administration; Department of Health and Human Services; United States

Freeman, Bradley D.; Kennedy, Carie R.; Coopersmith, Craig M.; Zehnbauer, Barbara A.; Buchman, Timothy G. Genetic research and testing in critical care: surrogates' perspective. *Critical Care Medicine* 2006 April; 34(4): 986-994. NRCBL: 15.3; 8.3.3; 18.2.

Gagen, Wendy Jane; Bishop, Jeffrey P. Ethics, justification and the prevention of spina bifida. *Journal of Medical Ethics* 2007 September; 33(9): 501-507. 42 refs. NRCBL: 15.3; 15.2; 9.1. SC: an; em.

> Keywords: *attitudes; *prenatal diagnosis; *spina bifida; attitude of health personnel; authorship; congenital disorders; costs and benefits; disabled persons; editorial policies; ethical analysis; eugenics; historical aspects; mass screening; methods; neural tube defects; paternalism; preventive medicine; public health; public policy; risks and benefits; selective abortion; survey; utilitarianism; Keyword Identifiers: Great Britain; Lancet; National Health Service; Twentieth Century

Abstract: During the 1970s, prenatal screening technologies were in their infancy, but were being swiftly harnessed to uncover and prevent spina bifida. The historical rise of this screening process and prevention programme is analysed in this paper, and the role of ethical debates in key studies, editorials and letters reported in the Lancet, and other related texts and governmental documents between 1972 and 1983, is considered. The silence that surrounded rigorous ethical debate served to highlight where discussion lay—namely, within the justifications offered for the prevention of spina bifida, and the efficacy and benefits of screening. In other words, the ethical justification for screening and prevention of spina bifida, when the authors are not explicitly interested in ethics, is considered. These justifications held certain notions of disability as costly to society, with an imperative to screen and prevent spina bifida for the good of society.

Genetics and Public Policy Center. FDA Regulation of Genetic Tests. Washington, D.C.: Genetics and Public Pol-

icy Center 2007 September 27: 2 p. [Online]. Accessed: http://www.dnapolicy.org/images/issuebriefspdfs/FDA_Regulation_of_Genetic_Test_Issue_Brief.pdf [2008 January 7]. NRCBL: 15.3; 18.6; 5.3; 18.2.

> Keywords: *genetic screening; *government regulation; federal government; genetic services; laboratories; medical devices; Keyword Identifiers: *Food and Drug Administration; *United States

German Society of Human Genetics [DGfH]. Committee for Public Relations and Ethical Issues. Statement on population screening for heterozygotes [position statement]. *Medizinische Genetik* 1991; 3(2): 11-12. [Online]. Accessed: http://www.medgenetik.de/sonderdruck/en/Heterozygote_screening.pdf [2006 July 31]. NRCBL: 15.3; 15.2; 14.1.

Gilani, Ahmed I.; Jadoon, Atif S.; Qaiser, Rabia; Nasim, Sana; Meraj, Riffat; Nasir, Nosheen; Naqvi, Fizza F.; Latif, Zafar; Memon, Muhammad A.; Menezes, Esme V.; Malik, Imran; Memon, Muhammad Z.; Kazim, Syed F.; Ahmad, Usman. Attitudes towards genetic diagnosis in Pakistan: a survey of medical and legal communities and parents of thalassemic children. *Community Genetics* 2007; 10(3): 140-146. 30 refs. NRCBL: 15.3. SC: em.

> Keywords: *attitudes; *genetic screening; *lawyers; *medical students; *parents; *physicians; *prenatal diagnosis; *selective abortion; *thalassemia; children; choice behavior; comparative studies; costs and benefits; genetic carriers; genetic disorders; mandatory programs; mass screening; prognosis; quality of life; survey; voluntary programs; Keyword Identifiers: *Pakistan

Godard, Béatrice; Pratte, Annabelle; Dumont, Martine; Simard-Lebrun, Adèle; Simard, Jacques. Factors associated with an individual's decision to withdraw from genetic testing for breast and ovarian cancer susceptibility: implications for counseling. *Genetic Testing* 2007 Spring; 11(1): 45-54. 53 refs. NRCBL: 15.3; 15.2. SC: em.

> Keywords: *breast cancer; *choice behavior; *genetic counseling; *genetic screening; *refusal to participate; ethnic groups; genetic predisposition; motivation; pedigree; psychology; Proposed Keywords: *ovarian cancer; BRCA1 genes; BRCA2 genes

Gupta, Jyotsna Agnihotri. Private and public eugenics: genetic testing and screening in India. *Journal of Bioethical Inquiry* 2007; 4(3): 217-228. NRCBL: 15.3; 15.5; 15.2; 9.1; 8.4; 8.3.1. SC: em.

Abstract: Epidemiologists and geneticists claim that genetics has an increasing role to play in public health policies and programs in the future. Within this perspective, genetic testing and screening are instrumental in avoiding the birth of children with serious, costly or untreatable disorders. This paper discusses genetic testing and screening within the framework of eugenics in the health care context of India. Observations are based on literature review and empirical research using qualitative methods. I distinguish 'private' from 'public' eugenics. I refer to the practice of prenatal diagnosis as an aspect of private eugenics, when the initiative to test comes from the pregnant woman herself. Public eugenics involves testing initiated by the state or medical profession through (more or less) obligatory testing programmes. To illustrate these concepts I discuss the management of thalassaemia which I see as an example of private eugenics that is moving into the sphere of public eugenics. I then discuss the recently launched newborn screening programme as an example of public eugenics. I use Foucault's concepts of power and governmentality to explore the thin line separating individual choice and overt or covert coercion, and between private and public eugenics. We can expect that the use of genetic testing technology will have serious and far-reaching implications for cultural perceptions regarding health and disease and women's experience of pregnancy, besides creating new ethical dilemmas and new professional and parental responsibilities. Therefore, culturally sensitive health literacy programmes to empower the public and sensitise professionals need attention.

Gustafson, Shanna L.; Gettig, Elizabeth A.; Watt-Morse, Margaret; Krishnamurti, Lakshmanan. Health beliefs among African American women regarding genetic testing and counseling for sickle cell disease. *Genetics in Medicine* 2007 May; 9(5): 303-310. 51 refs. NRCBL: 15.3; 9.5.4; 15.2; 9.5.5. SC: em.
> Keywords: *attitudes; *blacks; *genetic counseling; *genetic screening; *sickle cell anemia; *women; knowledge, attitudes, practice; population genetics; questionnaires; risks and benefits; survey; Keyword Identifiers: United States

Harmon, Amy. Cancer free, but weighing a mastectomy. *New York Times* 2007 September 16; p. A1, A20, A21. NRCBL: 15.3; 8.3.1; 9.5.5. SC: po.
> Keywords: *breast cancer; *choice behavior; *genetic predisposition; *surgery; age factors; family members; genetic counseling; genetic screening; pedigree; preventive medicine; risk; women; Proposed Keywords: *mastectomy; BRCA1 genes; BRCA2 genes

Harris, L.; Yashar, B.; Burmeister, M. Genetic testing for bipolar disorder: exploring patients' attitudes and receptivity. *Journal of Genetic Counseling* 2007 December; 16(6): 678. NRCBL: 15.3; 17.1. SC: em.

Hoffman, K.; Thomas, S.B.; Gettig, E.; Grubs, R.E.; Krishnamurti, L.; Butler, J. Assessing the attitudes and beliefs of African-Americans toward newborn screening and sickle cell disease. *Journal of Genetic Counseling* 2007 December; 16(6): 674. NRCBL: 15.3; 9.5.4; 15.2; 9.5.7. SC: em.

Hogben, Susan; Boddington, Paula. The rhetorical construction of ethical positions: policy recommendations for nontherapeutic genetic testing in childhood. *Communication and Medicine* 2006; 3(2): 135-146. NRCBL: 15.3; 9.5.7; 1.3.5; 21.1.

Huang, Mei-Chih; Lee, Chia-Kuei; Lin, Shio-Jean; Lu, I-Chen. A survey of parental consent process for newborn screening in Taiwan. *Acta Paediatrica Taiwanica = Tai-*

wan er ke yi xue hui za zhi 2005 November-December; 46(6): 361-369. NRCBL: 15.3; 8.3.2;. SC: em.

International Society of Nurses in Genetics [ISONG]. Informed Decision-Making and Consent: The Role of Nursing: Position Statement [Revised]. Pittsburgh, PA: International Society of Nurses in Genetics, 2005 April 4: 2 p. [Online]. Accessed: http://www.isong.org/about/ps_consent.cfm [2007 February 22]. 2 refs. NRCBL: 15.3; 8.3.1; 4.1.3; 6.
> Proposed Keywords: *genetic screening; *informed consent; *nurse's role; genetic services; organizational policies; professional organizations

Joint Committee on Medical Genetics; Farndon, Peter A. Recording, using and sharing genetic information and test results: consent is the key in all medical specialties. *Clinical Medicine* 2006 May-June; 6(3): 236-238. NRCBL: 15.3; 8.4; 8.3.1.

Klitzman, Robert; Thorne, Deborah; Williamson, Jennifer; Marder, Karen. The roles of family members, health care workers, and others in decision-making processes about genetic testing among individuals at risk for Huntington disease. *Genetics in Medicine* 2007 June; 9(6): 358-371. 79 refs. NRCBL: 15.3; 8.1; 8.3.1. SC: em.
> Keywords: *decision making; *family members; *genetic screening; *health personnel; *Huntington disorder; choice behavior; coercion; family relationship; genetic carriers; genetic counseling; interviews; patient care team; patients; professional role; psychology; referral and consultation; risk

Konda, Vani; Huo, Dezheng; Hermes, Gretchen; Liu, Michael; Patel, Roshan; Rubin, David T. Do patients with inflammatory bowel disease want genetic testing? *Inflammatory Bowel Disease* 2006 June; 12(6): 497-502. 15 refs. NRCBL: 15.3; 8.3.1; 9.5.1. SC: em.
> Keywords: *attitudes; *genetic screening; *patients; family members; genetic counseling; genetic predisposition; questionnaires; survey; uncertainty; Proposed Keywords: *inflammatory bowel disease; Keyword Identifiers: University of Chicago Inflammatory Bowel Disease Center

Koopmans, Joy; Hiraki, Susan; Ross, Laine Friedman. Attitudes and beliefs of pediatricians and genetic counselors regarding testing and screening for CF and G6PD: implications for policy. *American Journal of Medical Genetics. Part A* 2006 November 1; 140(21): 2305-2311. NRCBL: 15.3; 15.2; 9.1; 9.5.7.

Kosunen, Tiina. Ethical implications of genetic testing and screening [abstract]. *In:* Østnor, Lars, ed. Bioetikk og teologi: Rapport fra Nordisk teologisk nettverk for bioetikks workshop i Stockholm 27.-29. September 1996. Oslo: Nord. teol. nettv. for bioetikk, 1996: 107-115. Refs., p. 110-115. NRCBL: 15.3.
> Keywords: *genetic screening; ethical analysis

Langston, Anne L.; Johnston, Marie; Robertson, Clare; Campbell, Marion K.; Entwistle, Vikky A.; Marteau, Theresa M., McCallum, Marilyn; Ralston, Stuart H. Protocol for stage 1 of the GaP study (Genetic

NRCBL: National Reference Center for Bioethics Literature Classification Scheme										See inside front cover for terms.

209

testing acceptability for Paget's disease of bone): an interview study about genetic testing and preventive treatment: would relatives of people with Paget's disease want testing and treatment if they were available? *BMC Health Services Research* 2006 June 8; 6: 71: 9 p. NRCBL: 15.3; 15.2.

Lavery, J.V.; Slutsky, A.S. Substitute decisions about genetic testing in critical care research: a glimpse behind the curtain. *Critical Care Medicine* 2006 April; 34(4): 1257-1259. NRCBL: 15.3; 8.3.3; 18.2.

Lenzer, Jeanne. Advert for breast cancer gene test triggers inquiry [news]. *BMJ: British Medical Journal* 2007 September 22; 335(7620): 579. NRCBL: 15.3; 1.3.2; 1.3.7.
   Keywords: *advertising; *breast cancer; *genetic screening; genetic predisposition; genetic services; industry; mass media; ovarian cancer; Keyword Identifiers: Myriad Genetics; United States

Levy, Douglas E.; Youatt, Emily J; Shields, Alexandra E. Primary care physicians' concerns about offering a genetic test to tailor smoking cessation treatment. *Genetics in Medicine* 2007 December; 9(12): 842-849. NRCBL: 15.3; 9.5.9; 9.1. SC: em.

Lippman, Abby; Wertz, Dorothy C.; Fletcher, John C.; Nolan, Kathleen. Genetics. *In:* Baylis, Françoise; Downie, Jocelyn; Freedman, Benjamin; Hoffmaster, Barry; Sherwin, Susan, eds. Health Care Ethics in Canada. Toronto: Harcourt Brace Canada, 1995: 367-410. Includes references. NRCBL: 15.3; 15.2. Note: Section includes: Introduction — Prenatal genetic testing and screening: constructing needs and reinforcing inequities / Abby Lippman — A critique of some feminist challenges to prenatal diagnosis / Dorothy C. Wertz and John C. Fletcher — First fruits: genetic screening / Kathleen Nolan.
   Keywords: *genetic screening; *prenatal diagnosis; disabled persons; eugenics; feminist ethics; genetic counseling; genetic disorders; preimplantation diagnosis; selective abortion; social control; stigmatization; women; Keyword Identifiers: Canada

Lo, Bernard. Testing for genetic conditions. *In his:* Resolving Ethical Dilemmas: A Guide for Clinicians. 3rd edition. Philadelphia, PA: Lippincott Williams and Wilkins, 2005: 272-279. 42 refs. NRCBL: 15.3; 15.2.
   Keywords: *genetic screening; confidentiality; DNA fingerprinting; genetic counseling; genetic determinism; genetic disorders; informed consent; risks and benefits

Malpas, Phillipa. Predictive genetic testing in children and respect for autonomy. *In:* Freeman, Michael, ed. Children's Health and Children's Rights. Boston: Martinus Nijhoff Publishers, 2006: 297-309. 32 refs. 2 fn. NRCBL: 15.3; 9.5.7; 1.1.
   Keywords: *autonomy; *children; *genetic screening; *late-onset disorders; *risks and benefits; adolescents; adoption; age factors; disclosure; genetic disorders; genetic predisposition; guidelines; international aspects; parental consent

Mettner, Jeanne. Code YOU: will genetic testing make personalized medicine a reality? *Minnesota Medicine* 2007 May; 90(5): 26-29. NRCBL: 15.3; 15.1; 9.1.

Nyrhinen, Tarja; Hietala, Marja; Puukka, Pauli; Leino-Kilpi, Helena. Consequences as ethical issues in diagnostic genetic testing — a comparison of the perceptions of patients/parents and personnel. *New Genetics and Society* 2007 April; 26(1): 47-63. NRCBL: 15.3; 15.2. SC: em. Identifiers: Finland.

Offit, Kenneth; Kohut, Kelly; Clagett, Bartholt; Wadsworth, Eve A.; Lafaro, Kelly J.; Cummings, Shelly; White, Melody; Sagi, Michal; Bernstein, Donna; Davis, Jessica G. Cancer genetic testing and assisted reproduction. *Journal of Clinical Oncology* 2006 October 10; 24(29): 4775-4782. NRCBL: 15.3; 9.5.1; 14.1; 15.2.

Ormond, K.E.; Iris, M.; Banuvar, S.; Minogue, J.; Annas, G.J.; Elias, S. What do patients prefer: informed consent models for genetic carrier testing. *Journal of Genetic Counseling* 2007 August; 16(4): 539-550. 22 refs. NRCBL: 15.3; 8.3.1; 15.2. SC: em.
   Keywords: *choice behavior; *genetic carriers; *genetic screening; *informed consent; attitudes; communication; decision making; focus groups; genetic counseling; genetic disorders; knowledge, attitudes, practice; mass screening; questionnaires; parents; patients; pregnant women; prenatal diagnosis; selective abortion

Palombo, Enzo A.; Bhave, Mrinal. Genetically transformed healthcare: healthy children and parents. *In:* Betta, Michela, ed. The Moral, Social, and Commercial Imperatives of Genetic Testing and Screening: the Australian Case. Dordrecht: Springer, 2006: 185-199. 57 fn. NRCBL: 15.3.
   Keywords: *genetic screening; children; ethnic groups; genetic counseling; genetic disorders; genetic intervention; genetic predisposition; genetic privacy; insurance; mass screening; methods; preimplantation diagnosis; prenatal diagnosis; Keyword Identifiers: Australia

Phelps, Ceri; Wood, F.; Bennett, P.; Brain, K.; Gray, J. Knowledge and expectations of women undergoing cancer genetic risk assessment: a qualitative analysis of free-text questionnaire comments. *Journal of Genetic Counseling* 2007 August; 16(4): 505-514. 50 refs. NRCBL: 15.3; 15.2; 9.5.1; 9.5.5. SC: em.
   Keywords: *breast cancer; *genetic counseling; *genetic predisposition; *genetic screening; *knowledge, attitudes, practice; *ovarian cancer; *women; mass screening; questionnaires; risk; Keyword Identifiers: Wales

Plass, A.M.C. Informed consent for newborn screening? *Community Genetics* 2007; 10(4): 262-263. NRCBL: 15.3; 9.5.6; 8.3.2. SC: em. Comments: Comment on: E.P. Parsons, J.T. King, J.A. Israel, and D.M. Bradley. Mothers' accounts of screening newborn babies in Wales (UK). Midwifery 2007; 23: 59-65.
   Keywords: *attitudes; *genetic screening; *mothers; *mass screening; *newborns; *parental consent; *patient satisfaction; survey; Keyword Identifiers: Wales

---

Power, Tara E.; Adams, Paul C.; Barton, James C.; Acton, Ronald T.; Howe, Edmund, III; Palla, Shana; Walker, Ann P.; Anderson, Roger; Harrison, Barbara. Psychosocial impact of genetic testing for hemochromatosis in the HEIRS study: a comparison of participants recruited in Canada and in the United States. *Genetic Testing* 2007 Spring; 11(1): 55-64. 31 refs. NRCBL: 15.3; 21.1. SC: em. Identifiers: Hemochromatosis and Iron Overload Screening.
> Keywords: *genetic screening; *international aspects; *psychological stress; comparative studies; mass screening; questionnaires; psychology; Keyword Identifiers: *Canada; *United States

Remennick, Larissa. The quest for the perfect baby: why do Israeli women seek prenatal genetic testing? *Sociology of Health and Illness* 2006 January; 28(1): 21-53. 49 refs. 2 fn. NRCBL: 15.3; 15.2; 9.5.5. SC: em.
> Keywords: *attitudes; *choice behavior; *pregnant women; *prenatal diagnosis; genetic carriers; genetic disorders; genetic screening; interviews; Jews; socioeconomic factors; Keyword Identifiers: *Israel

Richards, F.H. Maturity of judgement in decision making for predictive testing for nontreatable adult-onset neurogenetic conditions: a case against predictive testing of minors. *Clinical Genetics* 2006 November; 70(5): 396-401. NRCBL: 15.3; 7.1; 9.4; 9.5.7; 17.1.

Rowley, Emma. On doing 'being ordinary': women's accounts of BRCA testing and maternal responsibility. *New Genetics and Society* 2007 December; 26(3): 241-250. NRCBL: 15.3; 9.5.5. SC: em. Identifiers: Great Britain (United Kingdom).

Satia, Jessie A.; McRitchie, Susan; Kupper, Lawrence L.; Halbert, Chanita Hughes. Genetic testing for colon cancer among African-Americans in North Carolina. *Preventive Medicine* 2006 January; 42(1): 51-59. NRCBL: 15.3; 9.5.4; 9.5.1.

Schaller, Jean; Moser, Hugo; Begleiter, Michael L.; Edwards, Janice. Attitudes of families affected by adrenoleukodystrophy toward prenatal diagnosis, presymptomatic and carrier testing, and newborn screening. *Genetic Testing* 2007 Fall; 11(3): 296-302. 40 refs. NRCBL: 15.3; 15.2; 9.5.7. SC: em.
> Keywords: *attitudes; *genetic carriers; *genetic disorders; *genetic screening; *prenatal diagnosis; family members; newborns; preimplantation diagnosis; survey; sex factors

Schiltz, Elizabeth R. The disabled Jesus: a parent looks at the logic behind prenatal testing and stem cell research. *America* 2007 March 12; 196(9): 16-18. NRCBL: 15.3; 18.5.4; 1.2.
> Keywords: *embryo research; *embryonic stem cells; *prenatal diagnosis; *Roman Catholic ethics; *value of life; adult stem cells; attitudes; choice behavior; congenital disorders; Down syndrome; embryos; eugenics; mentally disabled persons; preimplantation diagnosis; quality of life; risks and benefits; selective abortion; Proposed Keywords: autistic disorder

Schmidt, Eric B. The parental obligation to expand a child's range of open futures when making genetic trait selections for their child. *Bioethics* 2007, May; 21(4): 191-197. 15 fn. NRCBL: 15.3; 1.1; 14.1. SC: an.
> Keywords: *children; *ethical analysis; *genetic intervention; *moral obligations; *parents; autonomy; beneficence; decision making; congenital disorders; disabled persons; genetic disorders; freedom; genetic enhancement; hearing disorders; preimplantation diagnosis; reproduction; selective abortion; standards

Abstract: As parents become increasingly able to make genetic trait selections on behalf of their children, they will need ethical guidance in deciding what genetic traits to select. Dena Davis has argued that parents act unethically if they make selections that constrain their child's range of futures. But some selections may expand the child's range of futures. And other selections may shift the child's range of futures, without either constraining or expanding that range. I contend that not only would parents act unethically if they make selections that constrain the range of their child's futures, they would act unethically if they make selections that shift the range of their child's futures, because selections that shift the range of the child's futures would allow parents to over-determine their child's futures. Thus, I contend that parents would act ethically only if they make selections that expand their child's range of futures.

Scully, Jackie Leach; Porz, Rouven; Rehmann-Sutter, Christoph. 'You don't make genetic test decisions from one day to the next' — using time to preserve moral space. *Bioethics* 2007, May; 21(4):208-217. 16 fn. NRCBL: 15.3; 8.1; 15.2. SC: an; em.
> Keywords: *decision making; *empirical research; *ethical analysis; *genetic screening; *patients; *prenatal diagnosis; *time factors; age factors; autonomy; cancer; choice behavior; emotions; genetic counseling; Huntington disease; informed consent; interviews; mental competence; narrative ethics; qualitative research; theoretical models; uncertainty; Keyword Identifiers: Switzerland

Abstract: The part played by time in ethics is often taken for granted, yet time is essential to moral decision making. This paper looks at time in ethical decisions about having a genetic test. We use a patient-centred approach, combining empirical research methods with normative ethical analysis to investigate the patients' experience of time in (i) prenatal testing of a foetus for a genetic condition, (ii) predictive or diagnostic testing for breast and colon cancer, or (iii) testing for Huntington's disease (HD). We found that participants often manipulated their experience of time, either using a stepwise process of microdecisions to extend it or, under the time pressure of pregnancy, changing their temporal 'depth of field'. We discuss the implications of these strategies for normative concepts of moral agency, and for clinical ethics.

Shostak, Sara; Ottman, Ruth. Ethical, legal, and social dimensions of epilepsy genetics. *Epilepsia* 2006 October; 47(10): 1595-1602. NRCBL: 15.3; 15.11; 8.2; 8.4.

Shute, Nancy. Unraveling your DNA's secrets. Do-it-yourself genetic tests promise to reveal your risk of coming down with a disease. But do they really deliver?

NRCBL: National Reference Center for Bioethics Literature Classification Scheme   See inside front cover for terms.

211

U.S. News and World Report 2007 January 8; 142(1): 50-54, 57-58. NRCBL: 15.3; 15.2; 9.7; 5.3; 7.1; 8.4.

Skirton, Heather; Frazier, Lorraine Q.; Calvin, Amy O.; Cohen, Marlene Z. A legacy for the children — attitudes of older adults in the United Kingdom to genetic testing. *Journal of Clinical Nursing* 2006 May; 15(5): 565-573. NRCBL: 15.3. SC: em.

Sleeboom-Faulkner, Margaret. Predictive genetic testing in Asia: social science perspectives on the bioethics of choice. *Journal of Bioethical Inquiry* 2007; 4(3): 193-195. NRCBL: 15.3; 5.3; 7.1; 21.7.

Sleeboom-Faulkner, Margaret. Social-science perspectives on bioethics: predictive genetic testing (PGT) in Asia. *Journal of Bioethical Inquiry* 2007; 4(3): 197-206. NRCBL: 15.3; 15.2; 21.7; 2.1; 5.3; 7.1. SC: an.

Storz, Philipp; Kolpatzik, Kai; Perleth, Matthias; Klein, Silvia; Häussler, Bertram. Future relevance of genetic testing: a systematic horizon scanning analysis. *International Journal of Technology Assessment in Health Care* 2007 Fall; 23(4): 495-504. 17 refs. NRCBL: 15.3; 5.2; 7.1. SC: em.
Keywords: *genetic epidemiology; *genetic research; *genetic screening; *research priorities; databases; genetic disorders; genetic predisposition; literature; survey; technology assessment; Keyword Identifiers: Germany

Tabor, Holly K.; Cho, Mildred K. Ethical implications of array comparative genomic hybridization in complex phenotypes: points to consider in research. *Genetics in Medicine* 2007 September; 9(9): 626-631. NRCBL: 15.3; 15.1; 15.2; 18.2; 18.3.

Tan, Eng-King; Lee, Jennie; Hunter, Christine; Shinawi, Lina; Fook-Chong, S.; Jankovic, Joseph. Comparing knowledge and attitudes towards genetic testing in Parkinson's disease in an American and Asian population. *Journal of the Neurological Sciences* 2007 January 31; 252(2): 113-120. NRCBL: 15.3; 9.5.4; 15.11.

Tauer, Carol A. Making decisions about genetic testing. *In:* Hamel, Ronald, ed. Making Health Care Decisions: A Catholic Guide. Liguori, MO: Liguori Publications, 2006: 37-53. 6 refs. NRCBL: 15.3; 1.2.
Keywords: *genetic screening; *Roman Catholic ethics; adults; genetic carriers; minors; pharmacogenetics; prenatal diagnosis

Tercyak, Kenneth P.; Peshkin, Beth N.; Wine, Lauren A.; Walker, Leslie R. Interest of adolescents in genetic testing for nicotine addiction susceptibility. *Preventive Medicine* 2006 January; 42(1): 60-65. NRCBL: 15.3; 9.5.9; 9.5.7.

Turney, Lyn. Essentially whose? Genetic testing and the ownership of genetic information. *In:* Betta, Michela, ed. The Moral, Social, and Commercial Imperatives of Genetic Testing and Screening: the Australian Case.

Dordrecht: Springer, 2006: 237-245. 16 fn. NRCBL: 15.3; 4.4.
Keywords: *genetic information; *genetic screening; *paternity; *property rights; access to information; children; DNA fingerprinting; genetic databases; genetic privacy; genetic relatedness ties; genetic research; mandatory testing; men; population genetics; Keyword Identifiers: Australia

Wakefield, Claire E.; Kasparian, Nadine A.; Meiser, Bettina; Homewood, Judi; Kirk, Judy; Tucker, Kathy. Attitudes toward genetic testing for cancer risk after genetic counseling and decision support: a qualitative comparison between hereditary cancer types. *Genetic Testing* 2007 Winter; 11(4): 401-411. 37 refs. NRCBL: 15.3; 15.2; 9.5.5; 15.1; 9.5.1. SC: em.
Keywords: *attitudes; *breast cancer; *cancer; *genetic predisposition; *genetic screening; *motivation; *ovarian cancer; *risks and benefits; decision making; genetic counseling; patients; questionnaires; risk; women; Proposed Keywords: *colon cancer; Keyword Identifiers: Australia

White, Mary Terrell. Uncertainty and moral judgment: the limits of reason in genetic decision making. *Journal of Clinical Ethics* 2007 Summer; 18(2): 148-155. 23 fn. NRCBL: 15.3; 15.2; 8.1.
Keywords: *genetic counseling; *genetic screening; *uncertainty; decision making; directive counseling; genetic information; prenatal diagnosis; psychological stress; risk

Zimmerman, Richard K.; Tabbarah, Melissa; Nowalk, Mary P.; Raymund, Mahlon; Jewell, Ilene K.; Wilson, Stephen A; Ricci, Edmund M. Racial differences in beliefs about genetic screening among patients at inner-city neighborhood health centers. *Journal of the National Medical Association* 2006 March; 98(3): 370-377. NRCBL: 15.3; 9.5.4. SC: em.

Zuckerman, Shachar; Lahad, Amnon; Shmueli, Amir; Zimran, Ari; Peleg, Leah; Orr-Urtreger, Avi; Levy-Lahad, Ephrat; Sagi, Michal. Carrier screening for Gaucher disease: lessons for low-penetrance, treatable diseases. *JAMA: The Journal of the American Medical Association* 2007 September 19; 298(11): 1281-1290. 50 refs. NRCBL: 15.3; 15.2; 9.1. SC: em.
Keywords: *genetic carriers; *genetic screening; *Jews; *statistics; genetic counseling; genetic disorders; genetic services; interviews; mass screening; prenatal diagnosis; risks and benefits; selective abortion; Proposed Keywords: *Gaucher disease; Keyword Identifiers: *Israel

Abstract: CONTEXT: The aim of carrier screening is to prevent severe, untreatable genetic disease by identifying couples at risk before the birth of an affected child, and providing such couples with options for reproductive outcomes for affected pregnancies. Gaucher disease (GD) is an autosomal recessive storage disorder, relatively frequent in Ashkenazi Jews. Carrier screening for GD is controversial because common type 1 GD is often asymptomatic and effective treatment exists. However, screening is offered to Ashkenazi Jews worldwide and has been offered in Israel since 1995. OBJECTIVE: To examine the scope and outcomes of nationwide GD screening. DESIGN, SETTING, AND PARTICIPANTS: All Israeli genetic centers provided data on the number of individuals screened for GD, the number of carriers identified, the number of carrier couples identified, and the mutations identified in these couples between January 1, 1995, and March 31, 2003. Carrier couples were interviewed via telephone between January 21, 2003, and August 31, 2004, using a structured questionnaire for relevant outcome measures. MAIN OUTCOME MEASURES: Screening scope (number of testing centers, tested individuals, and carrier couples), screening process (type of pretest and posttest consultations), and screening outcomes (utilization of prenatal diagnosis and pregnancy terminations). RESULTS: Between January 1, 1995, and March 31, 2003, 10 of 12 Israeli genetic centers (83.3%) offered carrier screening. Carrier frequency was 5.7%, and 83 carrier couples were identified among an estimated 28,893 individuals screened. There were 82 couples at risk for offspring with type 1 GD. Seventy of 82 couples (85%) were at risk for asymptomatic or mildly affected offspring and 12 of 82 couples (15%) were at risk for moderately affected offspring. At postscreening, 65 interviewed couples had 90 pregnancies, and prenatal diagnosis was performed in 68 pregnancies (76%), detecting 16 fetuses with GD (24%). Pregnancies were terminated in 2 of 13 fetuses (15%) predicted to be asymptomatic or mildly affected and 2 of 3 fetuses (67%) with predicted moderate disease. There were significantly fewer pregnancy terminations in couples who in addition to genetic counseling had medical counseling with a GD expert (1 of 13 [8%] vs 3 of 3 with no medical counseling [100%], P = .007). CONCLUSIONS: In this study of GD screening among Ashkenazi Jewish couples in Israel, most couples did not terminate affected pregnancies, although screening was associated with a few pregnancy terminations. The main possible benefit was providing couples with knowledge and control. The divergence of these outcomes from stated goals of screening programs is likely to confront carrier screening programs for low-penetrance diseases.

GENETIC SCREENING/ LEGAL ASPECTS

Discriminating on genes: the United States is belatedly establishing necessary protections in law. Others, take note [editorial]. *Nature* 2007 July 5; 448(7149): 2. NRCBL: 15.3. SC: le.
> Keywords: *genetic discrimination; *genetic screening; *government regulation; employment; genetic information; insurance; international aspects; legislation; Keyword Identifiers: *Genetic Information Nondiscrimination Act; *United States; European Union

Messing with home brews. Political moves to expand FDA oversight to home brews are a bad idea. [editorial]. *Nature Biotechnology* 2007 March; 25(3): 262. NRCBL: 15.3; 5.3. SC: le.
> Keywords: *genetic screening; *government regulation; *nutrition; genetic services; marketing; nutrigenomics; standards; Proposed Keywords: predictive value of tests; quackery; Keyword Identifiers: *Food and Drug Administration; United States

American Society of Human Genetics [ASHG]. Statement from the American Society of Human Genetics

(ASHG): the Board of Directors of the American Society of Human Genetics has endorsed Senate Bill 318, the Genetic Nondiscrimination and Health Insurance and Employment Act. Bethesda, MD: American Society of Human Genetics, 2001 December 18: 1 p. [Online]. Accessed: http://genetics.faseb.org/genetics/ashg/pubs/policy/pol-47.htm [2006 July 26]. NRCBL: 15.3; 9.3.1. SC: le. Note: Statement from J.A. Boughman, Ph.D., Executive Vice President of ASHG.
> Keywords: *genetic discrimination; *genetic screening; *government regulation; *legislation; employment; federal government; health insurance; organizational policies; professional organizations; state government; Keyword Identifiers: *American Society of Human Genetics; *United States

Ballard, Rebecca. You get a line, I'll get a pole, we'll go fish'n in the plaintiff's gene pool. *Defense Counsel Journal* 2007 January; 74(1): 22-34. NRCBL: 15.3; 8.4. SC: le.

Berg, Jonathan S.; French, Shannon L.; McCullough, Laurence B.; Kleppe, Soledad; Sutton, Vernon R.; Gunn, Sheila K.; Karaviti, Lefkothea P. Ethical and legal implications of genetic testing in androgen insensitivity syndrome. *Journal of Pediatrics* 2007 April; 150(4): 434-438. NRCBL: 15.3; 8.4; 7.1; 10. SC: le; cs.

Epstein, Richard A. The social response to genetic conditions: beware of the antidiscrimination law. *Health Affairs* 2007 September-October; 26(5): 1249-1252. NRCBL: 15.3; 16.3; 9.3.1. SC: le. Identifiers: Genetic Information Nondiscrimination Act of 2007.

Genetics and Public Policy Center. Who regulates genetic tests? Washington, DC: Genetics and Public Policy 2007 September 27: 2p [Online]. Accessed: http://www.dnapolicy.org/images/issuesbriefspdfs/Who_Regulates_Genetic_Tests_Issue_Brief.pdf [2008 January 7]. NRCBL: 15.3; 5.3. SC: le.
> Keywords: *genetic screening; *government regulation; *standards; federal government; genetic services; laboratories; legal aspects; medical devices; state government; Keyword Identifiers: *Department of Health and Human Services; United States

Hogarth, Stuart; Melzer, David; Zimmern, Ron. The regulation of commercial genetic testing services in the UK: a briefing for the Human Genetics Commission. Cambridge: Department of Public Health and Primary Care, Cambridge University, 2005: 23 p. [Online]. Accessed: http://www.phpc.cam.ac.uk/epg/dtc.pdf [2007 April 17]. NRCBL: 15.3; 5.3. SC: le.

Holden, Constance. Long-awaited genetic nondiscrimination bill headed for easy passage [news]. *Science* 2007 May 4; 316(5825): 676. NRCBL: 15.3; 8.4; 9.3.1. SC: le.
> Keywords: *genetic discrimination; *government regulation; *legal aspects; *legislation; employment; genetic information; genetic privacy; insurance; Keyword Identifiers: *Genetic Information Nondiscrimination Act; *United States; U.S. House of Representatives

NRCBL: National Reference Center for Bioethics Literature Classification Scheme   See inside front cover for terms.

213

Hudson, Kathy L. Prohibiting genetic discrimination. *New England Journal of Medicine* 2007 May 17; 356(20): 2021-2023. 5 refs. NRCBL: 15.3; 9.3.1; 8.4; 16.3. SC: le.

Keywords: *genetic discrimination; *genetic screening; *government regulation; *legislation; access to information; attitude of health personnel; employment; genetic privacy; federal government; genetic research; health insurance; legal aspects; public opinion; state government; trust; Keyword Identifiers: *Genetic Information Nondiscrimination Act; *United States; Americans with Disabilities Act 1990; Health Insurance Portability and Accountability Act 1996

Ireni-Saban, Liza. Embracing personal and community empowerment: genetic information policy making in Israel. *Eubios Journal of Asian and International Bioethics* 2007 November; 17(6): 181-184. NRCBL: 15.3; 8.4; 18.5.1; 18.3. SC: le.

Javitt, Gail H. In search of a coherent framework: options for FDA oversight of genetics tests. *Food and Drug Law Journal* 2007; 62(4): 617-652. 193 fn. NRCBL: 15.3; 18.6; 5.3; 18.2. SC: le; rv.

Keywords: *genetic screening; *government regulation; *standards; exceptionalism; federal government; genetic services; industry; laboratories; legal aspects; legislation; medical devices; Keyword Identifiers: *Food and Drug Administration; United States

Kaye, Jane. Testing times: what is the legal situation when an adolescent wants a genetic test? *Clinical Ethics* 2007 December; 2(4): 176-180. 40 refs. NRCBL: 15.3; 9.5.7; 8.3.1; 4.3. SC: le. Identifiers: Gillick principles (assessment of competence).

Keywords: *adolescents; *competence; *genetic screening; *informed consent; *legal aspects; age factors; decision making; diagnosis; genetic counseling; parental consent; professional family relationship; Keyword Identifiers: *England; Gillick v. West Norfolk and Wisbech AHA

Abstract: Clinicians, as well as other health-care professionals in genetics clinics, may find themselves in the position where they must consider whether it would be appropriate to offer a diagnostic genetic test to an adolescent. While a clinician's decision to offer a diagnostic genetic test may be straightforward in clinical terms, the dynamics of family interaction and circumstances may make the decision-making process more complicated. Disagreement between parent and child place clinicians in a difficult position and they must be clear about the scope of their professional responsibility and obligations, to both parents and the adolescent. The purpose of this paper is to discuss the Gillick principles and statutory requirements regarding the genetic testing of adolescents. While I will discuss the clinician's obligations, these legal requirements also have applicability to other health-care professionals, such as genetic counsellors, working in genetics clinics.

Makdisi, June Mary Zekan. The protection of embryonic life in the European Council's Convention on Biomedicine. *National Catholic Bioethics Quarterly* 2007 Spring; 7(1): 31-39. NRCBL: 15.3; 15.2; 18.5.4; 4.4; 21.1; 14.4; 19.1. SC: le.

Romeo-Malanda, Sergio; Nicol, Dianne. Protection of genetic data in medical genetics: a legal analysis in the European context. *Revista de Derecho y Genoma Humano = Law and the Human Genome Review* 2007 July-December; (27): 97-134. 119 fn. NRCBL: 15.3; 8.4; 15.2; 21.1. SC: le.

Keywords: *genetic information; *genetic privacy; *genetic screening; *legal aspects; access to information; anonymous testing; autonomy; confidentiality; duty to warn; exceptionalism; family members; genetic counseling; genetic research; government regulation; informed consent; legal rights; medical records; public opinion; research subjects; right not to know; risks and benefits; Keyword Identifiers: *Europe; Council of Europe

Slaughter, Louise M. Your genes and privacy [editorial]. *Science* 2007 May 11; 316(5826): 797. NRCBL: 15.3; 9.3.1; 8.4; 16.3. SC: le.

Keywords: *employment; *genetic discrimination; *genetic information; *government regulation; *health insurance; *legislation; genetic predisposition; genetic privacy; genetic research; genetic screening; insurance selection bias; research subjects; Keyword Identifiers: *Genetic Information Nondiscrimination Act; *United States; U.S. House of Representatives

United States. Congress. House. A bill to prohibit discrimination on the basis of genetic information with respect to health insurance and employment. Washington, DC: U.S. G.P.O., 2006. 81 p. [Online]. Accessed: http://frwebgate.access.gpo.gov/cgi-bin/useftp.cgi?IPaddress=162.140.64.21&filename=h493ih.pdf&directory=/diskb/wais/data/110_cong_bills [2007 January 20]. NRCBL: 15.3; 8.4. SC: le. Identifiers: Genetic Information Nondiscrimination Act of 2007. Note: H.R. 493, 110th Congress, 1st session. Introduced by Rep. Slaughter on January 16, 2007. Referred to the Committee on Education and Labor, Committee on Energy and Commerce, and the Committee on Ways and Means.

Keywords: *employment; *genetic discrimination; *genetic information; *genetic screening; *health insurance; *legal aspects; confidentiality; family members; genetic privacy; genetic services; government regulation; Keyword Identifiers: *United States

United States. Congress. Senate. A bill to prohibit discrimination on the basis of genetic information with respect to health insurance and employment. Washington, DC: U.S. G.P.O., 2007. 80 p. [Online]. Accessed: http://frwebgate.access.gpo.gov/cgi-bin/useftp.cgi?IPaddress=162.140.64.21&filename=s358is.pdf&directory=/diskb/wais/data/110_cong_bills [2007 February 20]. NRCBL: 15.3; 9.3.1; 8.4. SC: le. Identifiers: Genetic Information Nondiscrimination Act of 2007. Note: S. 358., 110th Congress, 1st session. Introduced by Sen. Snowe, January 22, 2007. Referred to the Committee on Health, Education, Labor, and Pensions.

Keywords: *genetic discrimination; *genetic information; *genetic privacy; *genetic screening; *legal aspects; employment; family members; health insurance; government regulation; Keyword Identifiers: *United States

Van Hoyweghen, Ine; Horstman, Klasien; Schepers, Rita. Genetic 'risk carriers' and lifestyle 'risk takers'. Which risks deserve our legal protection in insurance? *Health Care Analysis: An International Journal of Health Philosophy and Policy* 2007 September; 15(3): 179-193. 49 refs.; 7 fn. NRCBL: 15.3; 8.4; 9.3.1. SC: an; em; le.

Keywords: *genetic discrimination; *genetic information; *genetic predisposition; *genetic screening; *insurance; *insurance selection bias; *legislation; *risk;*self induced illness; genetic privacy; government regulation; policy analysis; smoking; Proposed Keywords: *exceptionalism; *life style; Keyword Identifiers: *Belgium

Abstract: Over the past years, one of the most contentious topics in policy debates on genetics has been the use of genetic testing in insurance. In the rush to confront concerns about potential abuses of genetic information, most countries throughout Europe and the US have enacted genetics-specific legislation for insurance. Drawing on current debates on the pros and cons of a genetics-specific legislative approach, this article offers empirical insight into how such legislation works out in insurance practice. To this end, ethnographic fieldwork was done in the underwriting departments of Belgian insurance companies. Belgium was one of the first European countries introducing genetics-specific legislation in insurance. Although this approach does not allow us to speak in terms of ' the causal effects of the law', it enables us to point to some developments in insurance practice that are quite different than the law's original intentions. It will not only become clear that the Belgian genetics-specific legislation does not offer adequate solutions to the underlying issues it was intended for. We will also show that, while the legislation's focus has been on the inadmissibility of genetic discrimination, at the same time differences are made in the insurance appraisal within the group of the asymptomatic ill. In other words, by giving exclusive legal protection to the group of genetic risks, other non-genetic risk groups are unintendedly being under-protected. From a policy point of view, studying genetics-specific legislation is especially valuable because it forces us to return to first principles: Which risks deserve our legal protection in insurance? Who do we declare our solidarity with?

Wadman, Meredith. US genetics bill blocked again [news]. *Nature* 2007 August 9; 448(7154): 631. NRCBL: 15.3. SC: le.

Keywords: *genetic discrimination; *genetic information; *legislation; employment; federal government; genetic screening; government regulation; insurance; Keyword Identifiers: *Genetic Information Nondiscrimination Act; *U.S. Senate; *United States

GENETIC SCREENING/ SOCIOECONOMIC ASPECTS

Discriminating on genes: the United States is belatedly establishing necessary protections in law. Others, take note [editorial]. *Nature* 2007 July 5; 448(7149): 2. NRCBL: 15.3. SC: le.

Keywords: *genetic discrimination; *genetic screening; *government regulation; employment; genetic information;

insurance; international aspects; legislation; Keyword Identifiers: *Genetic Information Nondiscrimination Act; *United States; European Union

Pulling rank: why should US military personnel be singled out for genetic discrimination? [editorial]. *Nature* 2007 August 30; 448(7157): 969. NRCBL: 15.3; 1.3.5.

American Society of Human Genetics [ASHG]. Statement from the American Society of Human Genetics (ASHG): the Board of Directors of the American Society of Human Genetics has endorsed Senate Bill 318, the Genetic Nondiscrimination and Health Insurance and Employment Act. Bethesda, MD: American Society of Human Genetics, 2001 December 18: 1 p. [Online]. Accessed: http://genetics.faseb.org/genetics/ashg/pubs/policy/pol-47.htm [2006 July 26]. NRCBL: 15.3; 9.3.1. SC: le. Note: Statement from J.A. Boughman, Ph.D., Executive Vice President of ASHG.

Keywords: *genetic discrimination; *genetic screening; *government regulation; *legislation; employment; federal government; health insurance; organizational policies; professional organizations; state government; Keyword Identifiers: *American Society of Human Genetics; *United States

Avard, Denise; Kharaboyan, Linda; Knoppers, Bartha. Newborn screening for sickle cell disease: socioethical implications. *In:* McLean, Sheila A.M., ed. First Do No Harm: Law, Ethics, and Healthcare. Aldershot, England; Burlington, VT: Ashgate, 2006: 495-509. 84 fn. NRCBL: 15.3; 9.5.7; 9.1. SC: rv.

Keywords: *genetic screening; *mass screening; *newborns; *sickle cell anemia; diagnosis; disclosure; ethnic groups; genetic carriers; genetic counseling; genetic privacy; incidental findings; parental consent; policy making; prevalence; program descriptions; racial groups; risks and benefits; social discrimination

Baron, Roberta H. Genetic susceptibility testing: issues and psychosocial implications. *In:* Carroll-Johnson, Rose Mary; Gorman, Linda M.; Bush, Nancy Jo, eds. Psychosocial Nursing Care Along the Cancer Continuum. 2nd edition. Pittsburgh, PA: Oncology Nursing Society, 2006: 499-509. 34 refs. NRCBL: 15.3; 15.2.

Keywords: *genetic screening; attitudes; cancer; confidentiality; genetic counseling; genetic discrimination; genetic predisposition; nurse's role; prenatal diagnosis; psychological stress

Borry, Pascal; Fryns, Jean-Pierre; Schotsmans, Paul; Dierickx, Kris. Carrier testing in minors: a systematic review of guidelines and position papers. *European Journal of Human Genetics : EJHG* 2006 February; 14(2): 133-138. NRCBL: 15.3; 9.3.1; 8.3.1.

de Melo-Martín, Immaculada. Genetic information and moral obligations. *In her:* Taking Biology Seriously: What Biology Can and Cannot Tell Us About Moral and Public Policy Issues. Lanham: Rowman and Littlefield, 2005: 83-103. 48 fn. NRCBL: 15.3; 15.2; 1.1.

Keywords: *genetic information; *genetic predisposition; *genetic screening; *moral policy; autonomy; beneficence;

NRCBL: National Reference Center for Bioethics Literature Classification Scheme   See inside front cover for terms.

215

biology; choice behavior; disclosure; family members; genetic counseling; genetic discrimination; genetic disorders; genetic engineering; genetic research; genotype; phenotype; prenatal diagnosis; reproduction; reproductive technologies; right not to know; uncertainty; Proposed Keywords: *predictive value of tests

de Melo-Martín, Immaculada. Moral obligations, genetic information, and social context. *In her:* Taking Biology Seriously: What Biology Can and Cannot Tell Us About Moral and Public Policy Issues. Lanham: Rowman and Littlefield, 2005: 105-127. 66 fn. NRCBL: 15.3; 15.2; 1.1. SC: an.

> Keywords: *genetic information;*genetic screening; *moral obligations; *moral policy; *right not to know; autonomy; biology; choice behavior; comprehension; disabled persons; disclosure; family members; genetic carriers; genetic counseling; genetic services; justice; health services accessibility; minority groups; prenatal diagnosis; reproduction; social discrimination; social impact; women

English, Veronica; Gardner, Jessica; Romano-Critchley, Gillian; Sommerville, Ann. Genetics and insurance. *Journal of Medical Ethics* 2001 June; 27(3): 204. 8 refs. NRCBL: 15.3; 9.3.1; 8.4.

> Keywords: *genetic information; *genetic screening; *insurance; advisory committees; international aspects; legal aspects; self regulation; Keyword Identifiers: *Great Britain; Association of British Insurers; Genetics and Insurance Committee (Great Britain); Human Genetics Commission (Great Britain)

Epstein, Richard A. The social response to genetic conditions: beware of the antidiscrimination law. *Health Affairs* 2007 September-October; 26(5): 1249-1252. NRCBL: 15.3; 16.3; 9.3.1. SC: le. Identifiers: Genetic Information Nondiscrimination Act of 2007.

European Alliance of Patient and Parent Organizations for Genetics Services and Innovation in Medicine [EAGS]; Dutch Genetic Alliance [VSOP]; European Organisation for Rare Diseases [EURODIS]; European Federation of Pharmaceutical Industries and Associations [EFPIA]. Report: Workshop 'Genetic Testing: Challenges for Society'. Brussels, Belgium: European Federation of Pharmaceutical Industries and Associations (EFPIA), 2001 September 24: 13 p. [Online]. Accessed: http://www.egaweb.org/documents/24092001GENTEST [2007 May 1]. NRCBL: 15.3; 8.4; 9.3.1; 13.1. Conference: Genetic Testing: Challenges for Society; Brussels, Belgium; 2001 September 24; European patients' organisations: European Alliance of Patient and Parent Organisations for Genetic Services and Innovation in Medicine (EAGS), Dutch Genetic Alliance (VSOP), and European Organisation for Rare Diseases (EURORDIS), and the European Federation of Pharmaceutical Industries and Associations (EFPIA),.

Evans, James P. Health care in the age of genetic medicine. *JAMA: The Journal of the American Medical Association* 2007 December 12; 298(22): 2670-2672. 6 refs. NRCBL: 15.3; 15.1; 9.1.

> Keywords: *clinical genetics; *health care delivery; economics; genetic discrimination; genetic predisposition; genetic screening; genomics; health insurance; insurance coverage; justice; pharmacogenetics; preventive medicine; public health; resource allocation; trends; Proposed Keywords: personalized medicine; universal coverage; Keyword Identifiers: United States

Gesche, Astrid H. Protecting the vulnerable: genetic testing and screening for parentage, immigration, and aboriginality. *In:* Betta, Michela, ed. The Moral, Social, and Commercial Imperatives of Genetic Testing and Screening: the Australian Case. Dordrecht: Springer, 2006: 221-236. 81 fn. NRCBL: 15.3; 15.11.

> Keywords: *genetic ancestry; *genetic screening; *vulnerable populations; children; DNA fingerprinting; genetic discrimination; genetic relatedness ties; immigrants; indigenous populations; informed consent; parental consent; paternity; population genetics; public policy; racial groups; Keyword Identifiers: Australia

Gewin, Virginia. Crunch time for multiple-gene tests. *Nature* 2007 January 25; 445(7126): 354-355. NRCBL: 15.3; 9.3.1.

Great Britain (United Kingdom). Department of Health. Concordat and moratorium on genetics and insurance. London: Department of Health, 2005 March; 6 p. [Online]. Accessed: http://www.dh.gov.uk/prod_consum_dh/idcplg?IdcService=GET_FILE&dID=384&Rendition=Web [2007 April 30]. NRCBL: 15.3; 15.1; 9.3.1; 5.3.

> Keywords: *genetic information; *genetic screening; *insurance; *regulation; *standards; access to information; advisory committees; disclosure; government; industry; interinstitutional relations; public policy voluntary programs; Keyword Identifiers: *Association of British Insurers; *Great Britain; Genetics and Insurance Committee (Great Britain); Human Genetics Commission (Great Britain)

Holden, Constance. Long-awaited genetic nondiscrimination bill headed for easy passage [news]. *Science* 2007 May 4; 316(5825): 676. NRCBL: 15.3; 8.4; 9.3.1. SC: le.

> Keywords: *genetic discrimination; *government regulation; *legal aspects; *legislation; employment; genetic information; genetic privacy; insurance; Keyword Identifiers: *Genetic Information Nondiscrimination Act; *United States; U.S. House of Representatives

Hudson, Kathy L. Prohibiting genetic discrimination. *New England Journal of Medicine* 2007 May 17; 356(20): 2021-2023. 5 refs. NRCBL: 15.3; 9.3.1; 8.4; 16.3. SC: le.

> Keywords: *genetic discrimination; *genetic screening; *government regulation; *legislation; access to information; attitude of health personnel; employment; genetic privacy; federal government; genetic research; health insurance; legal aspects; public opinion; state government; trust; Keyword Identifiers: *Genetic Information Nondiscrimination Act; *United States; Americans with Disabilities Act 1990; Health Insurance Portability and Accountability Act 1996

Jamieson, Suzanne. Genetic information and the Australian labour movement. *In:* Betta, Michela, ed. The Moral,

---

Social, and Commercial Imperatives of Genetic Testing and Screening: the Australian Case. Dordrecht: Springer, 2006: 211-220. 49 fn. NRCBL: 15.3; 16.3.

Keywords: *employment; *genetic discrimination; *genetic information; *regulation; advisory committees; genetic screening; industry; Proposed Keywords: *labor unions;; Keyword Identifiers: *Australia; Australian Law Reform Commission

Kieran, Shannon; Loescher, Lois J.; Lim, Kyung Hee. The role of financial factors in acceptance of clinical BRCA genetic testing. *Genetic Testing* 2007 Spring; 11(1): 101-110. 40 refs. NRCBL: 15.3; 9.3.1; 9.5.5. SC: em.

Keywords: *choice behavior; *genetic screening; *health insurance; breast cancer; decision making; economics; genetic predisposition; genetic privacy; risk; survey; women; Proposed Keywords: *BRCA1 genes; *BRCA2 genes; *insurance coverage; ovarian cancer

Kopelman, Loretta M. Using the best interests standard to decide whether to test children for untreatable, late-onset genetic diseases. *Journal of Medicine and Philosophy* 2007 July-August; 32(4): 375-394. 43 refs. 13 fn. NRCBL: 15.3; 8.3.2. SC: an.

Keywords: *children; *decision making; *guideline adherence; *genetic screening; *late-onset disorders; *parental consent; *standards; autonomy; dementia; dissent; ethical analysis; genetic discrimination; genetic disorders; genetic predisposition; guidelines; Huntington disease; minors; moral policy; parent child relationship; physicians; policy analysis; professional organizations; risks and benefits; truth disclosure; Proposed Keywords: Alzheimer disease

Abstract: A new analysis of the Best Interests Standard is given and applied to the controversy about testing children for untreatable, severe late-onset genetic diseases, such as Huntington's disease or Alzheimer's disease. A professional consensus recommends against such predictive testing, because it is not in children's best interest. Critics disagree. The Best Interests Standard can be a powerful way to resolve such disputes. This paper begins by analyzing its meaning into three necessary and jointly sufficient conditions showing it: 1. is an "umbrella" standard, used differently in different contexts, 2. has objective and subjective features, 3. is more than people's intuitions about how to rank potential benefits and risks in deciding for others but also includes evidence, established rights, duties and thresholds of acceptable care, and 4. can have different professional, medical, moral and legal uses, as in this dispute. Using this standard, support is given for the professional consensus based on concerns about discrimination, analogies to adult choices, consistency with clinical judgments for adults, and desires to preserve of an open future for children. Support is also given for parents' legal authority to decide what genetic tests to do.

Lewis, M. Jane; Peterson, Susan K. Perceptions of genetic testing for cancer predisposition among Ashkenazi Jewish women. *Community Genetics* 2007; 10(2): 72-81. 42 refs. NRCBL: 15.3; 9.5.5. SC: em.

Keywords: *attitudes; *cancer; *genetic predisposition; *genetic screening; *Jews; focus groups; genetic discrimination; genetic privacy; mass screening; risks and benefits;

women; Proposed Keywords: BRCA1 genes; BRCA2 genes; Keyword Identifiers: New Jersey

McLean, Sheila A.M.; Mason, J. Kenyon. Our inheritance, our future: their rights? *In:* Freeman, Michael, ed. Children's Health and Children's Rights. Boston: Martinus Nijhoff Publishers, 2006: 279-296. 22 refs. 27 fn. NRCBL: 15.3; 9.5.7; 15.2.

Keywords: *children; *genetic screenings; *genetic services; *mass screening; *prenatal diagnosis; *public policy; cystic fibrosis; Down syndrome; eugenics; genetic carriers; genetic discrimination; genetic privacy; hearing disorders; human rights; late-onset disorders; newborns; risks and benefits; sickle cell anemia; thalassemia; Keyword Identifiers: *Great Britain; National Health Service

Nyrhinen, Tarja; Hietala, Marja; Puukka, Pauli; Leino-Kilpi, Helena. Privacy and equality in diagnostic genetic testing. *Nursing Ethics* 2007 May; 14(3): 295-308. 37 refs. NRCBL: 15.3; 8.4; 9.5.7. SC: em.

Keywords: *attitudes; *genetic privacy; *genetic screening; adults; attitude of health personnel; children; comparative studies; diagnosis; evaluation studies; education; genetic counseling; genetic discrimination; genetic information; genetic services; justice; knowledge, attitudes, practice; nurses; parents; patients; physicians; questionnaires; Keyword Identifiers: *Finland

Abstract: This study aimed to determine the extent to which the principles of privacy and equality were observed during diagnostic genetic testing according to views held by patients or child patients' parents (n = 106) and by staff (n = 162) from three Finnish university hospitals. The data were collected through a structured questionnaire and analysed using the SAS 8.1 statistical software. In general, the two principles were observed relatively satisfactorily in clinical practice. According to patients/parents, equality in the post-analytic phase and, according to staff, privacy in the pre-analytic phase, involved the greatest ethical problems. The two groups differed in their views concerning pre-analytic privacy. Although there were no major problems regarding the two principles, the differences between the testing phases require further clarification. To enhance privacy protection and equality, professionals need to be given more genetics/ethics training, and patients individual counselling by genetics units staff, giving more consideration to patients' world-view, the purpose of the test and the test result.

Pennicuik, Susan. The Australian law reform inquiry into genetic commission testing — a worker's perspective. *In:* Betta, Michela, ed. The Moral, Social, and Commercial Imperatives of Genetic Testing and Screening: the Australian Case. Dordrecht: Springer, 2006: 201-210. 2 fn. NRCBL: 15.3; 16.3.

Keywords: *employment; *genetic screening; *occupational exposure; advisory committees; coercion; genetic discrimination; genetic predisposition; genetic privacy; regulation; Proposed Keywords: workers' compensation; Keyword Identifiers: *Australia; Australian Law Reform Commission

Pennock, Robert T. Pre-existing conditions: genetic testing, causation, and the justice of medical insurance. *In:*

NRCBL: National Reference Center for Bioethics Literature Classification Scheme   See inside front cover for terms.

217

Rhodes, Rosamond; Francis, Leslie P.; Silvers, Anita, eds. The Blackwell Guide to Medical Ethics. Malden, MA: Blackwell Pub., 2007: 407-424. NRCBL: 15.3; 9.3.1. SC: an.

Pirakitikulr, Darlyn; Bursztajn, Harold J. Pride and prejudice: avoiding genetic gossip in the age of genetic testing. *Journal of Clinical Ethics* 2007 Summer; 18(2): 156-161. 21 fn. NRCBL: 15.3; 8.1; 8.4.
> Keywords: *genetic privacy; *genetic screening; disclosure; employment; genetic discrimination; genetic information; informed consent; insurance; legal aspects; legislation; medical records; Keyword Identifiers: Health Insurance Portability and Accountability Act 1996; United States

Pollack, Andrew. A genetic test that very few need, marketed to the masses. *New York Times* 2007 September 11; p. C3. NRCBL: 15.3; 9.7; 9.3.1. SC: po.
> Keywords: *advertising; *genetic screening; *industry; *mass media; breast cancer; genetic services; women; Keyword Identifiers: *Myriad Genetics Inc.

Quinlivan, Julie A.; Suriadi, Christine. Attitudes of new mothers towards genetics and newborn screening. *Journal of Psychosomatic Obstetrics and Gynaecology* 2006 March; 27(1): 67-72. 22 refs. NRCBL: 15.3; 9.5.5; 9.5.7. SC: em.
> Keywords: *attitudes; *genetic screening; *mass screening; *mothers; *newborns; genetic discrimination; interviews; parental consent; questionnaires; Keyword Identifiers: Australia

Rogowski, Wolf. Current impact of gene technology on healthcare: a map of economic assessments. *Health Policy* 2007 February; 80(2): 340-357. NRCBL: 15.3; 9.3.1; 15.1; 9.7; 7.1. SC: em.

Rose, Nikolas S. At genetic risk. *In his:* Politics of Life Itself: Biomedicine, Power, and Subjectivity in the Twenty-first Century. Princeton: Princeton University Press, 2007: 106-130, 280-283. 25 fn. NRCBL: 15.3; 1.1; 4.4; 15.2. SC: an.
> Keywords: *genetic identity; *genetic predisposition; *genetic screening; *philosophy; *risk; adults; autonomy; behavioral genetics; children; choice behavior; confidentiality; disclosure; education; employment; eugenics; family members; family relationship; genetic counseling; genetic information; genetic determinism; genetic discrimination; genetic disorders; genetic relatedness ties; government regulation; health; historical aspects; Huntington disease; international aspects; insurance; moral obligations; patient participation; patients; personhood; professional patient relationship; reproduction; rights; self concept; Keyword Identifiers: Great Britain; Internet; United States

Silver, Ken; Sharp, Richard R. Ethical considerations in testing workers for the -Glu69 marker of genetic susceptibility to chronic beryllium disease. *Journal of Occupational and Environmental Medicine* 2006 April; 48(4): 434-443. NRCBL: 15.3; 16.3.

Slaughter, Louise M. Your genes and privacy [editorial]. *Science* 2007 May 11; 316(5826): 797. NRCBL: 15.3; 9.3.1; 8.4; 16.3. SC: le.
> Keywords: *employment; *genetic discrimination; *genetic information; *government regulation; *health insurance; *legislation; genetic predisposition; genetic privacy; genetic research; genetic screening; insurance selection bias; research subjects; Keyword Identifiers: *Genetic Information Nondiscrimination Act; *United States; U.S. House of Representatives

Sui, Suli; Sleeboom-Faulkner, Margaret. Commercial genetic testing in mainland China: social, financial and ethical issues. *Journal of Bioethical Inquiry* 2007; 4(3): 229-237. NRCBL: 15.3; 1.3.2; 5.3; 15.2; 8.4; 9.3.1.
> Abstract: This paper provides an empirical account of commercial genetic predisposition testing in mainland China, based on interviews with company mangers, regulators and clients, and literature research during fieldwork in mainland China from July to September 2006. This research demonstrates that the commercialization of genetic testing and the lack of adequate regulation have created an environment in which dubious advertising practices and misleading and unprofessional medical advice are commonplace. The consequences of these ethically problematic activities for the users of predictive tests are, as yet, unknown. The paper concludes with a bioethical and social science perspective on the social and ethical issues raised by the dissemination and utilization of genetic testing in mainland China.

UK Cystic Fibrosis Database Steering Committee; Sims, Erika J.; Mugford, Miranda; Clark, Allan; Aitken, David; McCormick, Jonathan; Mehta, Gita; Mehta, Anil. Economic implications of newborn screening for cystic fibrosis: a cost of illness retrospective cohort study. *Lancet* 2007 April 7-13; 369(9568): 1187-1195. NRCBL: 15.3; 9.5.7; 9.5.1; 9.3.1. SC: em.
> Abstract: BACKGROUND: Newborn screening for cystic fibrosis might not be introduced if implementation and running costs are perceived as prohibitive. Compared with clinical diagnosis, newborn screening is associated with clinical benefit and reduced treatment needs. We estimate the potential savings in treatment costs attributable to newborn screening. METHODS: Using the UK Cystic Fibrosis Database, we used a prevalence strategy to undertake a cost of illness retrospective snapshot cohort study. We estimated yearly costs of long-term therapies and intravenous antibiotics for 184 patients who were diagnosed as a result of screening as newborn babies, and 950 patients who were clinically diagnosed aged 1-9 years in 2002. Costs of adding cystic fibrosis screening to an established newborn screening service in Scotland were adjusted to 2002 prices and applied to the UK as a whole. Costs were recalculated in US$. FINDINGS: Cost of therapy for patients diagnosed by newborn screening was significantly lower than equivalent therapies for clinically diagnosed patients: mean ($7228 vs $12 008, 95% CI of difference -6736 to -2028, p.0001) and median ($352 vs $2442, -1916 to -180, p.0001). When we limited the clinically diagnosed group to only those diagnosable with a 31 cystic fibrosis transmembrane regulator muta-

tion assay and assumed similar disease progression in the clinically diagnosed group as in the newborn screening group, we showed that mean ($3,397,344) or median ($947,032) drug cost savings could have offset the estimated cost of adding cystic fibrosis to a UK national newborn screening service ($2,971,551). INTERPRETATION: Including indirect costs savings, newborn screening for cystic fibrosis might have even greater financial benefits to society than our estimate shows. Clinical, social, and now economic evidence suggests that universal newborn screening programmes for cystic fibrosis should be adopted internationally.

United States. Congress. House. A bill to prohibit discrimination on the basis of genetic information with respect to health insurance and employment. Washington, DC: U.S. G.P.O., 2006. 81 p. [Online]. Accessed: http://frwebgate.access.gpo.gov/cgi-bin/useftp.cgi?IPaddress=162.140.64.21&filename=h493ih.pdf&directory=/diskb/wais/data/110_cong_bills [2007 January 20]. NRCBL: 15.3; 8.4. SC: le. Identifiers: Genetic Information Nondiscrimination Act of 2007. Note: H.R. 493, 110th Congress, 1st session. Introduced by Rep. Slaughter on January 16, 2007. Referred to the Committee on Education and Labor, Committee on Energy and Commerce, and the Committee on Ways and Means.

> Keywords: *employment; *genetic discrimination; *genetic information; *genetic screening; *health insurance; *legal aspects; confidentiality; family members; genetic privacy; genetic services; government regulation; Keyword Identifiers: *United States

United States. Congress. Senate. A bill to prohibit discrimination on the basis of genetic information with respect to health insurance and employment. Washington, DC: U.S. G.P.O., 2007. 80 p. [Online]. Accessed: http://frwebgate.access.gpo.gov/cgi-bin/useftp.cgi?IPaddress=162.140.64.21&filename=s358is.pdf&directory=/diskb/wais/data/110_cong_bills [2007 February 20]. NRCBL: 15.3; 9.3.1; 8.4. SC: le. Identifiers: Genetic Information Nondiscrimination Act of 2007. Note: S. 358., 110th Congress, 1st session. Introduced by Sen. Snowe, January 22, 2007. Referred to the Committee on Health, Education, Labor, and Pensions.

> Keywords: *genetic discrimination; *genetic information; *genetic privacy; *genetic screening; *legal aspects; employment; family members; health insurance; government regulation; Keyword Identifiers: *United States

Van Hoyweghen, Ine; Horstman, Klasien; Schepers, Rita. Genetic 'risk carriers' and lifestyle 'risk takers'. Which risks deserve our legal protection in insurance? *Health Care Analysis: An International Journal of Health Philosophy and Policy* 2007 September; 15(3): 179-193. 49 refs.; 7 fn. NRCBL: 15.3; 8.4; 9.3.1. SC: an; em; le.

> Keywords: *genetic discrimination; *genetic information; *genetic predisposition; *genetic screening; *insurance; *insurance selection bias; *legislation; *risk;*self induced illness; genetic privacy; government regulation; policy analysis; smoking; Proposed Keywords: *exceptionalism; *life style; Keyword Identifiers: *Belgium

Abstract: Over the past years, one of the most contentious topics in policy debates on genetics has been the use of genetic testing in insurance. In the rush to confront concerns about potential abuses of genetic information, most countries throughout Europe and the US have enacted genetics-specific legislation for insurance. Drawing on current debates on the pros and cons of a genetics-specific legislative approach, this article offers empirical insight into how such legislation works out in insurance practice. To this end, ethnographic fieldwork was done in the underwriting departments of Belgian insurance companies. Belgium was one of the first European countries introducing genetics-specific legislation in insurance. Although this approach does not allow us to speak in terms of ' the causal effects of the law', it enables us to point to some developments in insurance practice that are quite different than the law's original intentions. It will not only become clear that the Belgian genetics-specific legislation does not offer adequate solutions to the underlying issues it was intended for. We will also show that, while the legislation's focus has been on the inadmissibility of genetic discrimination, at the same time differences are made in the insurance appraisal within the group of the asymptomatic ill. In other words, by giving exclusive legal protection to the group of genetic risks, other non-genetic risk groups are unintendedly being under-protected. From a policy point of view, studying genetics-specific legislation is especially valuable because it forces us to return to first principles: Which risks deserve our legal protection in insurance? Who do we declare our solidarity with?

Wadman, Meredith. US genetics bill blocked again [news]. *Nature* 2007 August 9; 448(7154): 631. NRCBL: 15.3. SC: le.

> Keywords: *genetic discrimination; *genetic information; *legislation; employment; federal government; genetic screening; government regulation; insurance; Keyword Identifiers: *Genetic Information Nondiscrimination Act; *U.S. Senate; *United States

Wei, S.; Quigg, M.H.; Monaghan, Kristin G. Is cystic fibrosis carrier screening cost effective? *Community Genetics* 2007; 10(2): 103-109. 14 refs. NRCBL: 15.3; 9.5.1; 9.3.1. SC: em.

> Keywords: *costs and benefits; *cystic fibrosis; *genetic carriers; *genetic screening; ethnic groups; mass screening; prenatal diagnosis; racial groups; statistics; women; Keyword Identifiers: Michigan

Weisbrot, David; Opeskin, Brian. Insurance and genetics: regulating a private market in the public interest. *In:* Betta, Michela, ed. The Moral, Social, and Commercial Imperatives of Genetic Testing and Screening: the Australian Case. Dordrecht: Springer, 2006: 125-164. 135 fn. NRCBL: 15.3; 9.3.1.

> Keywords: *genetic discrimination; *genetic information; *insurance; *insurance selection bias; *regulation; advisory committees; disabled persons; genetic privacy; genetic screening; legal aspects; legislation; private sector; risk; Keyword Identifiers: *Australia; Australian Law Reform Commission

NRCBL: National Reference Center for Bioethics Literature Classification Scheme   See inside front cover for terms.

219

Wexler, Barbara. Genetic testing. *In her:* Genetics and genetic engineering. 2005 ed. Detroit, MI: Thomson/Gale Group, 2006: 83-98. NRCBL: 15.3.
Keywords: *genetic screening; adults; advertising; children; genetic carriers; genetic counseling; genetic discrimination; genetic disorders; genetic predisposition; genetic services; government regulation; incidental findings; late-onset disorders; legislation; mass screening; newborns; population genetics; preimplantation diagnosis; prenatal diagnosis; psychological stress; Proposed Keywords: predictive value of tests

GENETIC TESTING *See* GENETIC SCREENING

GENETICALLY MODIFIED ORGANISMS AND FOOD

Directive action required: Europe's handling of applications to grow genetically modified crops amounts to bad governance [editorial]. *Nature* 2007 December 13; 450(7172): 921. NRCBL: 15.1; 1.3.11; 5.3; 21.1.
Keywords: *genetically modified plants; *politics; agriculture; genetically modified food; industry; regulation; risks and benefits; Keyword Identifiers: *European Union; Europe

Abbott, Alison; Schiermeier, Quirin. Showdown for Europe. The European Union is set to make a landmark decision on genetically modified crops. *Nature* 2007 December 13; 450(7172): 928-929. NRCBL: 15.1; 1.3.11; 5.3; 21.1.
Keywords: *genetically modified plants; agriculture; genetically modified food; regulation; Keyword Identifiers: *European Union; Europe

Altieri, Miguel Angel. The myths of biotechnology: some ethical questions. *In:* Serageldin, I.; Martin-Brown, J., eds. Proceedings of an associated event of the 5th Annual World Bank Conference on Environmentally and Socially Sustainable Development.Washington, D.C.: World Bank, 1998: 53-58 [Online]. Accessed: http://nature.berkeley.edu/~agroeco3/the_myths.html [2007 April 24]. 25 refs. NRCBL: 1.3.11; 15.1; 16.1; 5.3.
Keywords: *agriculture; *biotechnology; *genetically modified plants; developing countries; ecology; genetic engineering; industry; international aspects; property rights; risks and benefits; Proposed Keywords: biodiversity

Andrée, Peter. The biopolitics of genetically modified organisms in Canada. *Journal of Canadian Studies* 2002 Fall; 37(3): 162-191. 76 refs. 15 fn. NRCBL: 15.1; 1.3.11; 5.3. SC: rv.
Keywords: *genetically modified food; *genetically modified plants; *government regulation; *politics; biotechnology; genetically modified organisms; policy making; precautionary principle; public policy; risk; science; Keyword Identifiers: *Canada

Armstrong, Susan J.; Botzler, Richard G. Animals and biotechnology. *In their:* The Animal Ethics Reader. London; New York: Routledge, 2003: 311-377. 20 refs. 12 fn. NRCBL: 15.1; 22.1; 1.3.11; 22.2.
Keywords: *animal welfare; *biotechnology; *genetically modified animals; agriculture; animal experimentation; animal rights; ethical analysis; methods; regulation; researchers; risks and benefits; suffering; utilitarianism; Proposed Keywords: domestic animals; laboratory animals

Bauer, Martin W. The public career of the 'gene' — trends in public sentiments from 1946 to 2002. *New Genetics and Society* 2007 April; 26(1): 29-45. NRCBL: 15.1; 1.3.11; 2.2; 15.5; 1.3.12. SC: em. Identifiers: Great Britain (United Kingdom).

Bharadwaj, Minakshi. Looking back, looking beyond: revisiting the ethics of genome generation. *Journal of Biosciences* 2006 March; 31(1): 167-176. 42 refs. NRCBL: 15.1; 1.3.11; 2.1; 15.3; 15.10; 15.5. SC: rv.
Keywords: *biotechnology; *genetic engineering; *genetic research; *genomics; benefit sharing; bioethics; cloning; developing countries; economics; embryo research; ethical analysis; genes; genetic determinism; genetic information; genetic patents; genetic screening; genetically modified food; genetically modified plants; Human Genome Project; international aspects; Proposed Keywords: exceptionalism; genetic resources; stakeholders

Bratspies, Rebecca M. Glowing in the dark: how America's first transgenic animal escaped regulation. *Minnesota Journal of Law, Science and Technology* 2004-2005; 6(2): 457-504. 200 fn. NRCBL: 15.1; 22.3; 5.3. SC: le.
Keywords: *genetically engineered animals; *government regulation; *legal aspects; agriculture; commerce; ecology; federal government; risk; state government; Proposed Keywords: *fish; Keyword Identifiers: *Food and Drug Administration; *United States; Environmental Protection Agency; Fish and Wildlife Service; Food, Drug, and Cosmetic Act

Brown, Marilyn J.; Murray, Kathleen A. Phenotyping of genetically engineered mice: humane, ethical, environmental, and husbandry issues. *ILAR Journal: Institute of Laboratory Animal Resources* 2006; 47(2): 118-123. NRCBL: 15.1; 22.3; 22.1. SC: em.

Carlson, Elof Axel. Genetically modified foods — as usual. *In his:* Times of Triumph, Times of Doubt: Science and the Battle for Public Trust. Cold Spring Harbor, NY: Cold Spring Harbor Laboratory Press, 2006: 129-137. 11 fn. NRCBL: 15.1; 1.3.11; 15.7.
Keywords: *genetically modified food; agriculture; industry; regulation; risk

Carter, Lucy. A case for a duty to feed the hungry: GM plants and the third world. *Science and Engineering Ethics* 2007 March; 13(1): 69-82. 14 refs. NRCBL: 15.1; 1.3.11. SC: an.
Keywords: *developing countries; *genetically modified food; *genetically modified plants; *moral obligations; *precautionary principle; *risks and benefits; ecology; nutrition; Proposed Keywords: malnutrition

Charles, Dan. Transgenic hay mowed [news]. *Science* 2007 May 11; 316(5826): 815. NRCBL: 15.1; 1.3.11; 15.7. SC: le.
Keywords: *genetically modified plants; *legal aspects; agriculture; ecology; federal government; government regula-

---

tion; risk; Keyword Identifiers: *United States; Department of Agriculture

Charles, Dan. U.S. courts say transgenic crops need tighter scrutiny [news]. *Science* 2007 February 23; 315(5815): 1069. NRCBL: 15.7; 15.1; 1.3.11.

Comber, Julie; Griffin, Gilly. Genetic engineering and other factors that might affect human-animal interactions in the research setting. *Journal of Applied Animal Welfare Science* 2007 June; 10(3): 267-277. 35 refs. NRCBL: 15.1; 22.1.

> Keywords: *animal welfare; *genetically modified animals; agriculture; animal behavior; animal cloning; animal experimentation; attitudes; guidelines; researchers; Proposed Keywords: domestic animals; laboratory animals; Keyword Identifiers: Canada; Canadian Council on Animal Care

Deane-Drummond, Celia. Genetics and environmental concern. *In her:* Genetics and Christian Ethics. Cambridge: Cambridge University Press, 2006: 220-244. 70 fn. NRCBL: 15.1; 1.2; 16.1; 22.2.

> Keywords: *animal rights; *ecology; *genetic engineering; *genetically modified animals; *theology; agriculture; Christian ethics; covenant; genetically modified plants; philosophy; speciesism; virtues

Ferretti, Maria Paola. Why public participation in risk regulation? The case of authorizing GMO products in European Union. *Science as Culture* 2007 December; 16(4): 377-395. NRCBL: 15.1; 1.3.11; 5.2; 5.3; 7.1. Identifiers: genetically modified organisms.

Fiester, Autumn. Casuistry and the moral continuum: evaluating animal biotechnology. *Politics and the Life Sciences* 2006 March-September; 25(1-2): 15-22. 22 refs. NRCBL: 15.1; 22.2; 14.5; 19.5. SC: an; cs.

> Keywords: *animal cloning; *animal organs; *casuistry; *ethical analysis; *genetically modified animals; animal rights; animal welfare; biotechnology; drugs; organ transplantation; risks and benefits; teleological ethics; Proposed Keywords: biodiversity; goats; species specificity; swine; zoonoses

Fox, Jeffrey L. US courts thwart GM alfalfa and turf grass [news]. *Nature Biotechnology* 2007 April; 25(4): 367-368. NRCBL: 15.1; 1.3.11; 15.7. SC: le.

> Keywords: *genetically modified plants; *legal aspects; agriculture; government regulation; risk; Keyword Identifiers: *United States; Department of Agriculture

Fox, Marie. Exposing harm: the erasure of animal bodies in healthcare law. *In:* McLean, Sheila A.M., ed. First Do No Harm: Law, Ethics, and Healthcare. Aldershot, England; Burlington, VT: Ashgate, 2006: 543-559. 86 fn. NRCBL: 22.1; 15.1; 19.1; 22.2. SC: an; le.

> Keywords: *animal welfare; *genetically modified animals; *legal aspects; animal experimentation; animal organs; biotechnology; ethical theory; organ transplantation; speciesism; suffering; Proposed Keywords: *harm; laboratory animals; Keyword Identifiers: Great Britain; Nuffield Council on Bioethics

Franklin, Sarah. Ethical biocapital: new strategies of cell culture. *In:* Franklin, Sarah; Lock, Margaret, eds. Remak-

ing Life and Death: Toward an Anthropology of the Biosciences. Santa Fe: School of American Research Press; Oxford: James Currey, 2003: 97-127. 20 fn. NRCBL: 15.1; 1.3.11; 14.5; 22.3.

> Keywords: *animal cloning; *biotechnology; *genetic engineering; *industry; *stem cells; capitalism; cell lines; embryo research; embryonic stem cells; gene therapy; genetically modified food; genetically modified organisms; nuclear transfer techniques; patents; stem cell transplantation; trust; Proposed Keywords: domestic animals; regenerative medicine; sheep; Keyword Identifiers: Geron Corp.; Great Britain; United States

Greif, Karen F.; Merz, Jon F. Science misunderstood: genetically modified organisms and international trade. *In their:* Current Controversies in the Biological Sciences: Case Studies of Policy Challenges from New Technologies. Cambridge, MA: MIT, 2007: 267-287. 71 refs. NRCBL: 15.1; 1.3.11; 1.3.2; 21.1.

> Keywords: *genetically modified plants; *international aspects; agriculture; commerce; ecology; genetically modified food; genetically modified organisms; legal aspects; regulation; risks and benefits; Keyword Identifiers: Europe; United States

Hauskeller, Michael. The reification of life. *Genomics, Society and Policy* 2007 August; 3(2): 70-81. 27 fn. NRCBL: 22.1; 1.1; 15.1. SC: an.

> Keywords: *animal rights; *animal welfare; *commodification; *genetically modified animals; *moral policy; animal experimentation; biotechnology; genetic patents; Proposed Keywords: laboratory animals

Abstract: What's wrong – fundamentally wrong – with the way animals are treated (…) isn't the pain, the suffering, isn't the deprivation. (…) The fundamental wrong is the system that allows us to view animals as our resources, here for us – to be eaten, or surgically manipulated, or exploited for sport or money.' Tom Regan made this claim 20 years ago. What he maintains is basically that the fundamental wrong is not the suffering we inflict on animals but the way we look at them. What we do to them, what we believe we are allowed to do to them, is dependent on how we perceive or conceptualize them. We not only treat them as resources but prior to this we already think of them as resources, and when we look at them, all we tend to see is resources. In our perception of them they exist not for themselves but 'for us'. But obviously it can only be fundamentally wrong in a moral sense to view them that way if it is wrong in a factual sense, that is, if animals are in fact not 'for us'. But is it wrong?

Holloway, Lewis; Morris, Carol. Exploring biopower in the regulation of farm animal bodies: genetic policy interventions in UK livestock. *Genomics, Society and Policy* 2007 August; 3(2): 82-98. 60 fn. NRCBL: 15.1; 22.3; 1.3.11. SC: an; cs.

> Keywords: *animal welfare; *food; *genetically modified animals; agriculture; biotechnology; case studies; commerce; genetic resources; genomics; policy analysis; public policy; Proposed Keywords: *animal production; *domestic animals; Keyword Identifiers: *Great Britain

Abstract: This paper explores the analytical relevance of Foucault's notion of biopower in the context of regulating and managing non-human lives and populations, spe-

NRCBL: National Reference Center for Bioethics Literature Classification Scheme   See inside front cover for terms.

221

cifically those animals that are the focus of livestock breeding based on genetic techniques. The concept of biopower is seen as offering theoretical possibilities precisely because it is concerned with the regulation of life and of populations. The paper approaches the task of testing the 'analytic mettle' of biopower through an analysis of four policy documents concerned with farm animal genetics: the UK's National Scrapie Plan (2003); the UK National Action Plan on Farm Animal Genetic Resources (2006); the Agriculture and Environment Biotechnology Committee's report on Animals and Biotechnology (2002); and the Farm Animal Welfare Council's report on the Welfare Implications of Animal Breeding and Breeding Technologies in Commercial Agriculture (2004). Of interest is whether and how the four policy case studies articulate a form of biopower in relation to human-livestock animal relations in the context of genetic approaches to livestock breeding, and how biopower is variably expressed in relation to the different policy issues addressed. In concluding, the paper considers the overall applicability and relevance of biopower in the context of regulating animal lives within livestock breeding, highlighting both possibilities and limitations, and offers suggestions for taking forward research on livestock populations from a neo-Foucaultian perspective.

Jefferson, Valeria. The ethical dilemma of genetically modified food. *Journal of Environmental Health* 2006 July-August; 69(1): 33-34. NRCBL: 15.1; 1.3.11; 15.7; 21.1.

Kaebnick, Gregory. Putting concerns about nature in context: the case of agricultural biotechnology. *Perspectives in Biology and Medicine* 2007 Autumn; 50(4): 572-584. NRCBL: 15.1; 1.3.11; 1.1.

Kaiser, Matthias. Practical ethics in search of a toolbox: ethics of science and technology at the crossroads. *In:* Gunning, Jennifer; Holm, Søren, eds. Ethics, law, and society. Volume 2. Aldershot, Hants, England; Burlington, VT: Ashgate, 2006: 35-44. 38 refs. NRCBL: 15.1; 1.3.12; 22.3. SC: an.
> Keywords: *biotechnology; *democracy; *decision making; *ethical analysis; *genetically modified organisms; *policy analysis; *policy making; *public policy; advisory committees; consensus; cultural pluralism; ethics committees; genetically modified food; guidelines; international aspects; public participation; risks and benefits; standards; technology assessment; theoretical models; Proposed Keywords: stakeholders

Kanter, James. Proposed ban on genetically modified corn in Europe [news]. *New York Times* 2007 November 23; p. C3. NRCBL: 15.1; 1.3.11; 5.3; 21.1. SC: po; le.
> Keywords: *genetically modified plants; *government regulation; agriculture; ecology; genetically modified food; industry; policy making; precautionary principle; Keyword Identifiers: *European Union; Europe

Kleinman, Daniel Lee; Kinchy, Abby J. Against the neoliberal steamroller? The Biosafety Protocol and the social regulation of agricultural biotechnologies. *Agriculture*

and Human Values 2007 Summer; 24(2): 195-206. 46 refs. 5 fn. NRCBL: 15.7; 1.3.11; 15.1.
> Keywords: *agriculture; *biotechnology; *genetically modified food; *genetically modified plants; *international aspects; *policy making; biodiversity; ecology; genetically modified organisms; precautionary principle; regulation; risk; socioeconomic factors; Keyword Identifiers: *Cartagena Protocol on Biosafety

Lassen, Jesper; Gjerris, Mickey; Sandoe, Peter. After Dolly — ethical limits to the use of biotechnology on farm animals. *Theriogenology* 2006 March 15; 65(5): 992-1004. 14 refs. NRCBL: 22.3; 14.5; 15.1. SC: em.
> Keywords: *animal cloning; *animal welfare; *biotechnology; *genetically modified animals; agriculture; animal experimentation; animal organs; attitudes; focus groups; food; genetically modified organisms; interviews; public opinion; risk; wedge argument; Proposed Keywords: *domestic animals; laboratory animals; Keyword Identifiers: *Denmark; European Union

Ledford, Heidi. Out of bounds. *Nature* 2007 January 11; 445(7124): 132-133. NRCBL: 15.1; 1.3.11; 15.7; 21.1.

Lee, Robert. GM resistant: Europe and the WTO panel dispute on biotech products. *In:* Gunning, Jennifer; Holm, Søren, eds. Ethics, law, and society. Volume 1. Aldershot, Hants, England; Burlington, VT: Ashgate, 2005: 131-140. 36 fn. NRCBL: 15.1; 1.3.11; 15.7; 21.1.
> Keywords: *genetically modified food; *genetically modified organisms; *regulation; biotechnology; commerce; risk; Keyword Identifiers: *European Union; *World Trade Organization; Europe

Létourneau, Lyne. The regulation of animal biotechnology: at the crossroads of law and ethics. *In:* Einsiedel, Edna; Timmermans, Frank, eds. Crossing Over: Genomics in the Public Arena. Calgary, Alberta, Canada: University of Calgary Press, 2005: 173-192. 49 refs.; 2 fn. NRCBL: 15.1; 22.2; 22.3; 1.3.11; 5.1; 5.3. Conference: Essays from the conference held Apr. 25-27, 2003, Kananaskis, Alta.
> Keywords: *animal rights; *animal welfare; *genetically modified animals; *moral policy; *regulation; advisory committees; agriculture; animal experimentation; animal organs; biotechnology; ethical analysis; guidelines; legislation; organ transplantation; philosophy; public policy; standards; Proposed Keywords: domestic animals; laboratory animals; Keyword Identifiers: Canada; Canadian Council on Animal Care; Committee to Consider the Ethical Implications of Emerging Technologies in the Breeding of Farm Animals (Great Britain); Great Britain

Mepham, Ben. Food ethics. *In:* Gunning, Jennifer; Holm, Søren, eds. Ethics, law, and society. Volume 1. Aldershot, Hants, England; Burlington, VT: Ashgate, 2005: 141-151. 34 fn. NRCBL: 15.1; 1.3.11; 15.7.
> Keywords: *biotechnology; *food; *ethical analysis; *genetically modified organisms; agriculture; animal welfare; autonomy; beneficence; ecology; justice; theoretical models; Proposed Keywords: biodiversity

Morris, Jonathan. GM foods: what are they doing to our dinner? *In his:* The Ethics of Biotechnology. Philadelphia: Chelsea Publishers, 2006: 104-130. NRCBL: 15.1; 1.3.11. SC: po.

Keywords: *genetically modified food; disclosure; ecology; genetic engineering; genetically modified organisms; moral policy; public policy; risks and benefits

O'Sullivan, Gabrielle M. Ethics and welfare issues in animal genetic modification. *In:* Rasko, John E.J.; O'Sullivan, Gabrielle M.; Ankeny, Rachel A., eds. The Ethics of Inheritable Genetic Modification: A Dividing Line? Cambridge: Cambridge University Press, 2006: 103-129. 97 fn. NRCBL: 15.1; 14.5; 22.2; 22.3. SC: an.

Keywords: *animal welfare; *genetic engineering; *genetically modified animals; *risks and benefits; animal cloning; animal experimentation; ethical analysis; human experimentation; moral policy; trends; Proposed Keywords: species specificity

Pearce, David. Economics and genetic diversity. *Futures* 1987; 19(6): 710-712. 3 refs. NRCBL: 15.1; 16.1; 9.3.1; 1.3.11.

Keywords: *economics; *genetic diversity; *international aspects; commerce; developing countries; ecology; gene pool; property rights; Proposed Keywords: *biodiversity; *genetic resources; Keyword Identifiers: United States

Pivetti, Monica. Natural and unnatural: activists' representations of animal biotechnology. *New Genetics and Society* 2007 August; 26(2): 137-157. 75 refs. 5 fn. NRCBL: 15.1; 22.2; 22.1. SC: em.

Keywords: *animal rights; *animal welfare; *attitudes; *biotechnology; *genetically modified animals; animal experimentation; animal organs; emotions; focus groups; moral policy; organ transplantation; political activity; public opinion; utilitarianism; Keyword Identifiers: *Italy; Anti-Vivisection League (Italy); Centre for Animal Aid (Italy); National Foundation for the Protection of Animals (Italy)

Pollack, Andrew. Round 2 [two] for biotech beets; after delay over safety fears, engineered crop will be planted [news]. *New York Times* 2007 November 27; p. C1, C2. NRCBL: 15.1; 1.3.11. SC: po.

Keywords: *genetically modified plants; agriculture; genetically modified food; industry; trends; Keyword Identifiers: United States

Rollin, Bernard. Ethics, biotechnology, and animals. *In:* Waldau, Paul; Patton, Kimberly, eds. A Communion of Subjects: Animals in Religion, Science, and Ethics. New York: Columbia University Press, 2006: 519-532. 19 refs. NRCBL: 15.1; 1.3.11; 14.5; 22.1; 22.2; 22.3. SC: an; rv.

Keywords: *animal cloning; *animal welfare; *genetically modified animals; *moral policy; agriculture; animal experimentation; biotechnology; cloning; commodification; ecology; ethical analysis; gene pool; genetic diversity; genetic engineering; genetic research; industry; literature; religion; risks and benefits; suffering; Proposed Keywords: animal production; domestic animals; laboratory animals

Silver, Lee M. The battle for mother nature's genes. *In his:* Challenging Nature: The Clash of Science and Spirituality at the New Frontiers of Life. New York: Ecco, 2006: 278-293, 397-399. 44 fn. NRCBL: 15.1; 1.3.11.

Keywords: *genetic engineering; *genetically modified food; *genetically modified plants; genetically modified organisms; genetic patents; microbiology; public opinion; re-

combinant DNA research; risk; Keyword Identifiers: Europe; United States

Twine, Richard. Animal genomics and ambivalence: a sociology of animal bodies in agricultural biotechnology. *Genomics, Society and Policy* 2007 August; 3(2): 99-117. 62 fn. NRCBL: 15.1; 22.3; 1.3.11. SC: an.

Keywords: *animal welfare; *genomics; agriculture; animal cloning; biodiversity; biotechnology; chimeras; economics; food; genetic databases; international aspects; regulation; social sciences; trends; Proposed Keywords: *animal production; *domestic animals

Abstract: How may emergent biotechnologies impact upon our relations with other animals? To what extent are any changes indicative of new relations between society and nature? This paper critically explores which sociological tools can contribute to an understanding of the technologisation of animal bodies. By drawing upon interview data with animal scientists I argue that such technologies are being partly shaped by broader changes in agriculture. The complexity of genomics trajectories in animal science is partly fashioned through the deligitimisation of the productivist paradigm but continue to sit in tension around particular conceptions of sustainability in farm animal production. In spite of this deligitimisation process genomics is now being framed in the context of a new productivism (termed the livestock revolution) bound up in projected global changes in animal consumption during the first half of the 21st century. This potentially jars against both social trends that seek to re-enchant animal life and sustainability discourses which include social and environmental contexts. Nevertheless the possibility of a new productivism is supported by various interconnected trends including the emergence of a discourse of the 'bioeconomy' and a liberal regulatory apparatus for farm animal breeding technologies. Ultimately an understanding of the possibility of emerging new bio-capitalisations on animal life should be set in a broader context of competing agricultural paradigms as well as ongoing tensions over 'naturalness' in human/animal relations.

Van Dooren, Thom. Terminated seed: death, proprietary kinship and the production of (bio)wealth. *Science as Culture* 2007 March; 16(1): 71-93. NRCBL: 15.1; 5.3; 1.3.11; 21.1. Identifiers: Genetic Use Restriction Technologies.

Veeman, Michele; Adamowicz, Wiktor; Hu, Wuyang; Hünnemeyer, Anne. Canadian attitudes to genetically modified food. *In:* Einsiedel, Edna; Timmermans, Frank, eds. Crossing Over: Genomics in the Public Arena. Calgary, Alberta, Canada: University of Calgary Press, 2005: 99-113. 18 refs.; 4 fn. NRCBL: 15.1; 1.3.11. Conference: Essays from the conference held Apr. 25-27, 2003, Kananaskis, Alta.

Keywords: *genetically modified food; *public opinion; agriculture; animal welfare; choice behavior; ecology; focus groups; genetically modified organisms; health hazards; industry; information dissemination; knowledge, attitudes, practice; political activity; risks and benefits; survey; trust; Keyword Identifiers: *Canada

NRCBL: National Reference Center for Bioethics Literature Classification Scheme   See inside front cover for terms.

223

Wells, D.J.; Playle, L.C.; Enser, W.E.; Flecknell, P.A.; Gardiner, M.A.; Holland, J.; Howard, B.R.; Hubrecht, R.; Humphreys, K.R.; Jackson, I.J.; Lane, N.; Maconochie, M.; Mason, G.; Morton, D.B.; Raymond, R.; Robinson, V.; Smith, J.A.; Watt, N. Assessing the welfare of genetically altered mice. *Lab Animal* 2006 April; 40(2): 111-114. 8 refs. NRCBL: 15.1; 22.1.

> Keywords: *animal welfare; *genetically modified animals; advisory committees; ethical review; guidelines; Proposed Keywords: *mice; Keyword Identifiers: Great Britain

Wexler, Barbara. Genetic engineering and biotechnology. *In her:* Genetics and genetic engineering. 2005 ed. Detroit, MI: Thomson/Gale Group, 2006: 135-161. NRCBL: 15.1; 1.3.11.

> Keywords: *biotechnology; *genetic engineering; *genetically modified food; *genetically modified plants; agriculture; DNA fingerprinting; doping in sports; ecology; gene therapy; genetic enhancement; health hazards; international aspects; nanotechnology; public opinion; recombinant DNA research; regulation; risks and benefits; stem cells; Proposed Keywords: forensic genetics; Keyword Identifiers: United States

Wickson, Fern. From risk to uncertainty in the regulation of GMOs: social theory and Australian practice. *New Genetics and Society* 2007 December; 26(3): 325-339. NRCBL: 15.7; 5.3; 15.1.

GENETICS AND GENOMICS

. . .but not as we know it. Synthetic life is on the way, and we need to think about the consequences [editorial]. *New Scientist* 2007 October 20-26; 196(2626): 5. NRCBL: 15.1.

Andorno, Roberto. Human dignity and the UNESCO Declaration on the Human Genome. *In:* Gunning, Jennifer; Holm, Søren, eds. Ethics, law, and society. Volume 1. Aldershot, Hants, England; Burlington, VT: Ashgate, 2005: 73-81. 27 fn. NRCBL: 15.1; 21.1.

> Keywords: *human genome; *human dignity; *international aspects; genetic privacy; human rights; legal aspects; Keyword Identifiers: *Declaration on the Human Genome (UNESCO)

Ashley, Benedict M.; de Blois, Jean K.; O'Rourke, Kevin D. Genetic intervention. *In their:* Health Care Ethics: A Catholic Theological Analysis. 5th edition. Washington, DC: Georgetown University Press, 2006: 94-98. NRCBL: 15.1; 1.2.

> Keywords: *genetic intervention; *Roman Catholic ethics; chimeras; eugenics; recombinant DNA research; sex preselection

Beck, Matthias. Illness, disease and sin: the connection between genetics and spirituality. *Christian Bioethics* 2007 January-April; (13)1: 67-89. NRCBL: 15.1; 1.2; 4.2; 7.1.

> Abstract: The New Testament, while rejecting any superficial connection between illness and sin, does not reject a possible connection between illness and a person's relationship with God. An example can be seen in the story of the young blind man who was healed (St. John 9:3). His blindness does not result from any fault he or his parents had committed but apparently from God's wish to reveal his own healing power. The inner blindness of the Pharisees is a different type of blindness far more difficult to heal. The blind young man was actually healed, not only in body but also in soul. Such miraculous healings are rare nowadays. However, if one takes a closer look at modern genetics and psycho-neuro-immunological findings, one may come to a better understanding of how miracle healings are linked to man's inner life and therefore also to his religiousness. Many diseases have genetic backgrounds. Defective genes, however, do not necessarily lead to subsequent illness. Genes have to be switched on or off. Only activated genes trigger pathological change. The human brain and all of man's thinking and feeling are intimately connected with such activations. We may thus conclude that both inner life and religious outlook on life are relevant to the origin and development of diseases.

Beene-Harris, Rosalyn Y.; Wang, Catharine; Bach, Janice V. Barriers to access: results from focus groups to identify genetic service needs in the community. *Community Genetics* 2007; 10(1): 10-17. NRCBL: 15.1; 9.3.1; 9.5.1; 9.5.4; 15.2; 15.3. SC: em.

> Keywords: *genetic services; *health services accessibility; focus groups; genetic counseling; genetic disorders; health facilities; health insurance; health personnel; knowledge, attitudes, practice; minority groups; parents; patient satisfaction; public opinion; public policy; state government; Proposed Keywords: *needs assessment; Keyword Identifiers: Michigan

Betta, Michela. Self-knowledge and self-care in the age of genetic manipulation. *In:* Betta, Michela, ed. The Moral, Social, and Commercial Imperatives of Genetic Testing and Screening: the Australian Case. Dordrecht: Springer, 2006: 249-256. 10 fn. NRCBL: 15.1; 1.1.

> Keywords: *genetic intervention; *philosophy; *self concept; autonomy; cloning; genetic enhancement; genetic screening; human body; sexuality; Proposed Keywords: aging; Keyword Identifiers: Foucault, Michel

Burley, Justine. Morality and the "new genetics". *In her:* Dworkin and His Critics: With Replies by Dworkin. Malden, MA: Blackwell Pub., 2004: 170-192. NRCBL: 15.1; 14.5; 14.1; 1.1. SC: an; cs. Identifiers: Ronald Dworkin; moral free-fall hypothesis.

Caplan, Arthur L.; Curry, David R. Leveraging genetic resources or moral blackmail? Indonesia and avian flu virus sample sharing [editorial]. *American Journal of Bioethics* 2007 November; 7(11): 1-2. 4 refs. NRCBL: 15.1; 9.1; 9.7; 21.1.

> Keywords: *genetic resources; *influenza; *international aspects; *property rights; *public policy; *world health; benefit sharing; developing countries; industry; obligations to society; public health; vaccines; Proposed Keywords: *viruses; communicable disease control; Keyword Identifiers: Indonesia; World Health Organization

Castle, David; Cline, Cheryl; Daar, Abdallah S.; Tsamis, Charoula; Singer, Peter A. The ethics of nutrigenomic tests and information. *In their:* Science, Society, and the Supermarket: The Opportunities and Challenges of Nutrigenomics. Hoboken, NJ: Wiley-Interscience, 2007: 49-75. 46 refs. NRCBL: 15.1; 15.3.

Keywords: *genetic information; *genetic screening; *nutrigenomics; adolescents; adults; autonomy; children; clinical genetics; confidentiality; duty to warn; employment; ethical analysis; family members; genetic discrimination; genetic privacy; genetic research; government regulation; informed consent; insurance; international aspects; pedigree; predisposition to disease

Castle, David; Cline; Cheryl; Daar, Abdallah S.; Tsamis, Charoula; Singer, Peter A. Nutrigenomics: justice, equity, and access. *In their:* Science, Society, and the Supermarket: The Opportunities and Challenges of Nutrigenomics. Hoboken, NJ: Wiley-Interscience, 2007: 133-151. 54 refs. NRCBL: 15.1; 1.1; 21.1.

Keywords: *international aspects; *justice; *nutrigenomics; access to information; developing countries; economics; genetic databases; genetic patents; genetic research; genetic services; health disparities; health services accessibility; industry; property rights; public health; risks and benefits; Proposed Keywords: developed countries; personalized medicine; world health

Chapman, Elizabeth. The social and ethical implications of changing medical technologies: the views of people living with genetic conditions. *Journal of Health Psychology* 2002 March; 7(2): 195-206. 34 refs. NRCBL: 15.1; 4.4; 15.2. SC: em.

Keywords: *attitudes; *genetic disorders; *prenatal diagnosis; cystic fibrosis; disabled persons; genetic screening; Huntington disease; quality of life; selective abortion; Keyword Identifiers: Great Britain

Collins, Francis S.; Manolio, Teri A. Necessary but not sufficient. *Nature* 2007 January 18; 445(7125): 259. NRCBL: 15.1; 7.1; 15.11; 16.1; 19.1.

Dunston, Georgia M.; Royal, Charmaine D.M. The human genome: implications for the health of African Americans. *In:* Livingston, Ivor Lensworth, ed. Praeger Handbook of Black American Health: Policies and Issues Behind Disparities in Health, Vol. II. 2nd edition. Westport, CT: Praeger Publishers, 2004: 757-775. 83 refs; 5 Web site citations. NRCBL: 15.11; 9.5.4.

Keywords: *blacks; *genetic diversity; *genetic epidemiology; *genomics; *health disparities; *human genome; genetic ancestry; genetic research; genetic services; genome mapping; health services accessibility; population genetics; racial groups; Keyword Identifiers: *United States; Africa; Genomic Research in the African Diaspora (GRAD) Project; National Human Genome Center; Howard University

Dusyk, Nichole. The political and moral economies of science: a case study of genomics in Canada and the United Kingdom. *Health Law Review* 2007; 15(3): 3-5. 8 fn. NRCBL: 5.1; 15.1. SC: cs.

Keywords: *economics; *genomics; *government financing; *public policy; *research priorities; *research support; entrepreneurship; genetic patents; industry; international

aspects; policy making; politics; science; trends; universities; Proposed Keywords: technology transfer; Keyword Identifiers: *Canada; *Great Britain

Egorova, Yulia. The meanings of genetics: science and the concepts of personhood. *Health Care Analysis: An International Journal of Health Philosophy and Policy* 2007 March; 15(1): 1-3. 4 refs. NRCBL: 15.1; 4.4.

Keywords: *genetics; biotechnology; personhood; self concept

FitzGerald, Kevin; Royal, Charmaine. Race, genetics, and ethics. *In:* Prograis, Lawrence J.; Pellegrino, Edmund D., eds. African American Bioethics: Culture, Race, and Identity. Washington, DC: Georgetown University Press, 2004: 137-151. 38 fn. NRCBL: 15.11; 9.5.4. Conference: Symposium on African American Perspectives in Bioethics and Second Annual Conference on Health Disparities, held on September 23-24, 2004, at Georgetown University.

Keywords: *genetic ancestry; *genomics; *racial groups; biomedical research; blacks; drugs; ethnic groups; health care delivery; health disparities; justice; pharmacogenetics; social discrimination; Proposed Keywords: classification; Keyword Identifiers: BiDil; United States

Grüber, Katrin. Abschied vom Gendogma — für eine verantwortbare medizinische Forschung. *In:* Neuer-Miebach, Therese; Wunder, Michael, eds. Bio-Ethik und die Zukunft der Medizin. Bonn: Psychiatrie-Verlag, 1998: 120-129. 10 fn. NRCBL: 15.1; 4.2; 5.1.

Keywords: *biomedical research; *genetic determinism; biotechnology; disease; science

Harmon, Amy. Facing life with a lethal gene: a young woman's DNA test reveals an inevitably grim fate. *New York Times* 2007 March 18; p. A1, A26, A27. NRCBL: 15.1; 15.3; 8.1. SC: po. Identifiers: Katharine Moser; Huntington's disease.

Harvey, Erin K.; Fogel, Chana E.; Peyrot, Mark; Christensen, Kurt D.; Terry, Sharon F.; McInerney, Joseph D. Providers' knowledge of genetics: a survey of 5915 individuals and families with genetic conditions. *Genetics in Medicine* 2007 May; 9(5): 259-267. 16 refs. NRCBL: 15.2; 15.1; 7.2. SC: em.

Keywords: *genetic counseling; *genetics; *health personnel; *professional competence; evaluation studies; family members; genetic disorders; medical education; patient education; patients; primary health care; survey

Harvey, Matthew. Animal genomics in science, social science and culture. *Genomics, Society and Policy* 2007 August; 3(2): 1-28. 156 fn. NRCBL: 15.1; 22.1; 5.1. SC: rv.

Keywords: *animal welfare; *genomics; *species specificity; animal experimentation; animal organs; animal rights; biodiversity; biotechnology; chimeras; food; genetically modified animals; humanities; personhood; regulation; social sciences; Proposed Keywords: animal production; classification; domestic animals; laboratory animals; wild animals

Abstract: Animals are commonplace in genomic research, yet to date there has been little direct interrogation

NRCBL: National Reference Center for Bioethics Literature Classification Scheme   See inside front cover for terms.

225

of the position, role and construction of animals in the otherwise flourishing social science of genomics. Following a brief discussion of this omission, I go on to suggest that there is much of interest for the social sciences and the humanities in this field of science. I show that animal genomics not only updates and extends established debates about the use of animals in science and society, but also raises novel issues and promotes new ways of thinking about what animals are, and the social and biological relationships between animals and humans. Organising the science of interest into six themes (sameness, difference and classification; crossing boundaries; the maintenance of borders; farmyard supermodels; laboratory supermodels; knowing, relating and looking at animals), for each I review some of the science that is being done, some of the conceptual issues that are raised, and some of the social science that is or could be done. I conclude by briefly considering the development of socially responsive policies for animal genomics.

Hendriks, Aart. Protection against genetic discrimination and the Biomedicine Convention. *In:* Gevers, J.K.M.; Hondius, E.H.; Hubben, J.H., eds. Health Law, Human Rights and the Biomedicine Convention: Essays in Honour of Henriette Roscam Abbing. Leiden; Boston: Martinus Nijhoff Publishers, 2005: 207-218. 49 fn. NRCBL: 15.1; 15.3.
> Keywords: *genetic discrimination; genetic information; government regulation; human dignity; human rights; justice; legal aspects; terminology; Keyword Identifiers: *European Convention on Human Rights and Biomedicine; Council of Europe; Europe

Henn, Wolfram. Auf dem Weg zur „ökonomischen Indikation" zum Schwangerschaftsabbruch bei therapierbaren Erbleiden? = Towards terminations of pregnancy due to the therapy costs of treatable hereditary diseases? *Ethik in der Medizin* 2007 June; 19(2): 120-127. 20 refs. NRCBL: 15.1; 9.3.1; 9.4; 12.5.1; 15.2.
> Keywords: *economics; *genetic carriers; *genetic disorders; *prenatal diagnosis; *reproduction; *resource allocation; *selective abortion; children; coercion; health insurance reimbursement; legal aspects; parents; patient care; reproductive rights; Proposed Keywords: Gaucher disease; Keyword Identifiers: Germany

International Society of Nurses in Genetics [ISONG]. Access to Genomic Healthcare: The Role of the Nurse: Position Statement. Pittsburgh, PA: International Society of Nurses in Genetics, 2003 September 9: 4 p. [Online]. Accessed: http://www.isong.org/about/ps_genomic.cfm [2007 February 22]. 10 refs. NRCBL: 15.1; 4.1.3; 6.
> Keywords: *genetic services; *nurse's role; codes of ethics; health services accessibility; nursing ethics; organizational policies; professional organizations; Keyword Identifiers: *International Society of Nurses in Genetics [ISONG]

Keller, Johannes. In genes we trust: the biological component of psychological essentialism and its relationship to mechanisms of motivated social cognition. *Journal of Personality and Social Psychology* 2005 April; 88(4): 686-702. NRCBL: 15.1; 15.11; 17.1; 1.1.

King, Nancy M.P. Genes and Tourette syndrome: scientific, ethical, and social implications. *Advances in Neurology* 2006; 99: 144-147. NRCBL: 15.1; 15.3; 15.10.

King, Patricia A.; Areen Judith; Gostin, Lawrence O. The human genome: pathways to health. *In their:* Law, Medicine and Ethics. New York: Foundation Press, 2006: 1-111. Includes references. NRCBL: 15.1; 15.3.
> Keywords: *genetics; *genomics; *human genome; advisory committees; autonomy; beneficence; commerce; disclosure; duty to warn; ethical theory; eugenics; family members; gene therapy; genetic ancestry; genetic databases; genetic discrimination; genetic enhancement; genetic information; genetic patents; genetic privacy; genetic research; genetic screening; Human Genome Project; human rights; justice; legal aspects; moral policy; pharmacogenetics; professional organizations; public health; public policy; regulation; social impact; Proposed Keywords: exceptionalism; Keyword Identifiers: United States

Koeman, Jan H. Comments on Korthals: new public responsibilities for life scientists. *In:* Korthals, Michiel; Bogers, Robert J., eds. Ethics for Life Scientists. Dordrecht, The Netherlands: Springer, 2004: 171-174. NRCBL: 15.1; 1.3.9.

Korthals, Michiel. New public responsibilities for life scientists. *In:* Korthals, Michiel; Bogers, Robert J., eds. Ethics for Life Scientists. Dordrecht, The Netherlands: Springer, 2004: 163-170. NRCBL: 15.1; 1.3.9; 5.3; 1.1.

Lebacqz, Karen. Choosing our children: the uneasy alliance of law and ethics in John Robertson's thought. *In:* Galston, Arthur W.; Peppard, Christiana Z., eds. Expanding Horizons in Bioethics. Dordrecht; Norwell, MA: Springer, 2005: 123-139. 29 refs. 64 fn. NRCBL: 15.1; 1.1; 14.1; 15.2. SC: an.
> Keywords: *freedom; *ethical analysis; *justice; *genetic intervention; *prenatal diagnosis; *reproductive rights; *rights; *values; choice behavior; cultural diversity; disabled persons; feminist ethics; genetic disorders; genetic enhancement; genetic information; legal aspects; morality; parents; reproductive technologies; selective abortion; sex determination; social discrimination; social control; women's rights; Proposed Keywords: deafness; Keyword Identifiers: *Robertson, John; United States

Loike, John D.; Tendler, Moshe D. Molecular genetics, evolution, and Torah principles. *Torah u-Madda Journal* 2006-2007; 14: 173-192. NRCBL: 15.1; 1.2; 3.1.

Maher, Brendan. His daughter's DNA. *Nature* 2007 October 18; 449(7164): 772-776. 4 refs. NRCBL: 15.1; 15.10. Identifiers: Hugh Rienhoff.
> Keywords: *personal genomics; *family members; *genetic disorders; children; diagnosis; DNA sequences; fathers

Mason, J.K.; Laurie, G.T. Genetic information and the law. *In their:* Mason and McCall Smith's Law and Medical Ethics. Seventh ed. Oxford; New York: Oxford University Press, 2005: 206-252. 226 fn. NRCBL: 15.1; 15.2; 15.3; 14.5; 15.4. SC: le.

Keywords: *genetic information; *legal aspects; cloning; employment; family members; gene therapy; genetic counseling; genetic discrimination; genetic disorders; genetic privacy; genetic research; genetic screening; insurance; prenatal diagnosis; right not to know; Keyword Identifiers: Great Britain

McGuire, Amy L.; Cho, Mildred K.; McGuire, Sean E.; Caulfield, Timothy. The future of personal genomics. *Science* 2007 September 21; 317(5845): 1687. 15 refs. NRCBL: 15.1; 15.3.
> Keywords: *clinical genetics; *genome mapping; *genomics; costs and benefits; forecasting; genetic counseling; genetic predisposition; genetic screening; human genome; industry; justice; practice guidelines; risks and benefits; trends; Proposed Keywords: *personalized medicine

Moscarillo, T.J.; Holt, H.; Perman, M.; Goldberg, S.; Cortellini, L.; Stoler, J.M.; DeJong, W.; Miles, B.J.; Albert, M.S.; Go, R.C.P.; Blacker, Deborah. Knowledge of and attitudes about Alzheimer disease genetics: report of a pilot survey and two focus groups. *Community Genetics* 2007; 10(2): 97-102. 48 refs. NRCBL: 15.1; 9.5.2; 15.3. SC: em.
> Keywords: *Alzheimer disease; *family members; *genetic predisposition; *genetic screening; *knowledge, attitudes, practice; focus groups; genetic research; survey; Proposed Keywords: pilot projects; Keyword Identifiers: Alabama; Massachusetts

National Reference Center for Bioethics Literature. News from the National Reference Center for Bioethics Literature (NRCBL) and the National Information Resource on Ethics and Human Genetics (NIREHG). *Kennedy Institute of Ethics Journal* 2007 December; 17(4): 399-403. NRCBL: 2.1; 1.3.12; 15.1.
> Keywords: *bioethical issues; *bioethics; *databases; *genetics; *information dissemination; *literature; *terminology; eugenics; genetic patents; human experimentation; professional organizations; publishing; Proposed Keywords: *abstracting and indexing; *information services; publications; Keyword Identifiers: *Bioethics Thesaurus for Genetics; *ETHXWeb; *GenETHX; *National Information Resource on Ethics and Genetics; *National Reference Center for Bioethics Literature; Bioethics Thesaurus; Georgetown University; Kennedy Institute of Ethics

Petersen, Alan. The genetic conception of health: is it as radical as claimed? *Health (London)* 2006 October; 10(4): 481-500. NRCBL: 15.1; 4.2; 9.1.

Pollard, Irina. The recombinant DNA technologies. *In her:* Life, Love and Children: A Practical Introduction to Bioscience Ethics and Bioethics. Boston: Kluwer Academic Publishers, 2002: 127-143. 17 refs. (p. 236-237). NRCBL: 15.1.
> Proposed Keywords: *genetic engineering; *genetic research; eugenics; gene therapy; genetic enhancement; genetic patents; genetic screening; genetically modified organisms; germ cells; mass screening; moral policy; Human Genome Diversity Project; Human Genome Project; prenatal diagnosis; recombinant DNA research

Rapp, Rayna. Cell life and death, child life and death: genomic horizons, genetic diseases, family stories. *In:* Franklin, Sarah; Lock, Margaret, eds. Remaking Life and Death: Toward an Anthropology of the Biosciences. Santa Fe: School of American Research Press; Oxford: James Currey, 2003: 129-164. 17 fn. NRCBL: 15.1; 8.1.
> Keywords: *genomics; attitudes; biotechnology; clinical genetics; economics; gene therapy; genetic determinism; genetic disorders; genetic intervention; genetic materials; genetic patents; genetic research; genetic screening; life; mass media; metaphor; patient advocacy; philosophy; quality of life; research priorities; researchers; stem cells; Proposed Keywords: foundations; hyperbole

Richards, Martin. Genes, genealogies, and paternity: making babies in the twenty-first century. *In:* Spencer, J.R.; du Bois-Pedain, Antje, eds. Freedom and Responsibility in Reproductive Choice. Portland: Hart Pub., 2006: 53-72. 56 fn. NRCBL: 15.1; 14.2; 14.4.
> Keywords: *DNA fingerprinting; *fathers; *parent child relationship; *paternity; artificial insemination; confidentiality; disclosure; genetic identity; genetic relatedness ties; married persons; methods; reproductive technologies; semen donors; single persons

Robert, Jason Scott. Gene maps, brain scans, and psychiatric nosology. *CQ: Cambridge Quarterly of Healthcare Ethics* 2007 Spring; 16(2): 209-218. 44 fn. NRCBL: 17.1; 4.4; 15.1; 1.1; 4.1.2.
> Keywords: *classification; *genomics; *neurosciences; *psychiatric diagnosis; *philosophy; behavioral genetics; brain; genotype; mental disorders; phenotype; psychiatry; schizophrenia

Salter, Frank K. On the ethics of defending genetic interests. *In his:* On Genetic Interests: Family, Ethnicity, and Humanity in an Age of Mass Migration. New Brunswick, NJ: Transaction Publishers, 2007: 283-323. 72 fn. NRCBL: 15.1; 1.1.
> Keywords: *ethical theory; *morality; *sociobiology; *utilitarianism; behavioral genetics; ethnic groups; evolution; family members; freedom; justice; moral obligations; philosophy; religion; rights; Proposed Keywords: species specificity

Secretariat of the Convention on Biological Diversity. Bonn guidelines on access to genetic resources and fair and equitable sharing of the benefits arising out of their utilization. Montreal: Secretariat of the Convention on Biological Diversity, 2002: 20 p. [Online]. Accessed: http://www.biodiv.org/doc/publications/cbd-bonn-gdls-en.pdf [2007 April 16]. NRCBL: 15.1; 15.3; 15.8; 18.5.1; 18.2.
> Keywords: *benefit sharing; *genetic materials; *guidelines; *international aspects; developing countries; ecology; indigenous populations; informed consent; regulation; voluntary programs; Proposed Keywords: *biodiversity; *genetic resources; stakeholders; Keyword Identifiers: *Convention on Biological Diversity

Sharp, Michael. The effect of genetic determinism and exceptionalism on law and policy. *Health Law Review* 2007; 15(3): 16-18. 11 fn. NRCBL: 15.1. SC: le.
> Keywords: *genetic determinism; *genetic information; *policy making; *public policy; attitudes; biotechnology;

NRCBL: National Reference Center for Bioethics Literature Classification Scheme   See inside front cover for terms.

227

genetic materials; genetic screening; genomics; government regulation; human genome; international aspects; legal aspects; risks and benefits; social impact; Proposed Keywords: exceptionalism

Shaw, Alison. The contingency of the 'genetic link' in the construction of kinship and inheritance — an anthropological perspective. *In:* Spencer, J.R.; du Bois-Pedain, Antje, eds. Freedom and Responsibility in Reproductive Choice. Portland: Hart Pub., 2006: 73-90. 77 fn. NRCBL: 15.1; 14.1.

Keywords: *culture; *family relationship; *genetic relatedness ties; anthropology; ethnic groups; genetic services; reproductive technologies; Keyword Identifiers: Great Britain; Pakistan

Simpson, Bob. On parrots and thorns: Sri Lankan perspective on genetics, science and personhood. *Health Care Analysis: An International Journal of Health Philosophy and Policy* 2007 March; 15(1): 41-49. 32 refs. 7 fn. NRCBL: 15.1; 1.2.

Keywords: *culture; *genetics; *non-Western World; anthropology; biotechnology; Buddhist ethics; developing countries; genetic determinism; genomics; knowledge, attitudes, practice; personhood; science; suffering; Keyword Identifiers: * Sri Lanka

Abstract: This paper addresses the issue of how the scientific discourse of genetics is expressed in local idioms. The examples used are taken from fieldwork conducted in Sri Lanka and relate principally to Sinhala Buddhist attempts to socialise 'big science.' The paper explores idioms of both nature and nurture in local imagery and narratives and draws attention to the rhetorical dimensions of genetic discourses when used in context. The article concludes with a preliminary attempt to identify the ways in which explanations of genetic causality are aligned with notions of karma in the explanation of illness and misfortune.

Skene, Loane. Theft of DNA: do we need a new criminal offence? *In:* Gunning, Jennifer; Holm, Søren, eds. Ethics, law, and society. Volume 1. Aldershot, Hants, England; Burlington, VT: Ashgate, 2005: 85-94. 36 fn. NRCBL: 15.1; 7.4; 8.3.1. SC: le.

Keywords: *criminal law; *genetic information; *genetic materials; *legal liability; *property rights; advisory committees; commodification; DNA fingerprinting; genetic privacy; genetic screening; informed consent; publishing; Proposed Keywords: *theft; Keyword Identifiers: Australia; Australian Law Reform Commission; Great Britain; Human Genetics Commission (Great Britain)

Smith, Rachel A. Picking a frame for communicating about genetics: stigmas or challenges. *Journal of Genetic Counseling* 2007 June; 16(3): 289-298. NRCBL: 15.1; 15.3; 8.1.

Wang, Grace; Watts, Carolyn. The role of genetics in the provision of essential public health services. *American Journal of Public Health* 2007 April; 97(4): 620-625. 45 refs. NRCBL: 15.1; 1.3.5; 9.1.

Keywords: *genetic services; *public health; congenital disorders; evaluation studies; genetic privacy; government financing; health hazards; health promotion; health services

accessibility; information dissemination; institutional policies; mass screening; program descriptions; public policy; quality of health care; standards; state government; survey; Keyword Identifiers: United States

Abstract: States include genetics services among their public health programs, but budget shortfalls raise the question, is genetics an essential part of public health? We used the Essential Services of Public Health consensus statement and data from state genetics plans to analyze states' public health genetics programs. Public health genetics programs fulfill public health obligations: birth defects surveillance and prevention programs protect against environmental hazards, newborn screening programs prevent injuries, and clinical genetics programs ensure the quality and accessibility of health services. These programs fulfill obligations by providing 4 essential public health services, and they could direct future efforts toward privacy policies, research on communications, and rigorous evaluations.

Warnock, Mary. The limits of rights-based discourse. *In:* Spencer, J.R.; du Bois-Pedain, Antje, eds. Freedom and Responsibility in Reproductive Choice. Portland: Hart Pub., 2006: 3-14. 17 fn. NRCBL: 15.1; 14.1; 15.2; 14.4.

Keywords: *genetic intervention; *reproductive rights; *reproductive technologies; *rights; choice behavior; eugenics; health care; homosexuality; human rights; international aspects; legal rights; obligations of society; parents; philosophy; preimplantation diagnosis; refusal to treat; social discrimination; Proposed Keywords: tissue typing; Keyword Identifiers: Great Britain

Welsh, Ian; Plows, Alexandra; Evans, Robert. Human rights and genomics: science, genomics and social movements at the 2004 London Social Forum. *New Genetics and Society* 2007 August; 26(2): 123-135. 43 refs. NRCBL: 15.1; 21.1.

Keywords: *genomics; *human rights; *international aspects; *political activity; organizational policies; private sector; public participation; Proposed Keywords: nongovernmental organizations (NGOs)

Widdows, Heather. Conceptualising the self in the genetic era. *Health Care Analysis: An International Journal of Health Philosophy and Policy* 2007 March; 15(1): 5-12. 27 refs. 5 fn. NRCBL: 15.1; 1.1. SC: an.

Keywords: *autonomy; *communitarianism; *genetics; *self concept; *philosophy; bioethics; community consent; confidentiality; ethical theory; family members; freedom; genetic information; genetic patents; genetic privacy; genetic research; informed consent; freedom; literature; morality; population genetics; theoretical models; Keyword Identifiers: Murdoch, Iris

Abstract: This paper addresses the impact of genetic advances and understandings on our concept of the self and the individual. In particular it focuses on conceptions of the 'autonomous individual' in the post-Enlightenment tradition and in bioethics. It considers the ascendancy of the autonomous individual as the model of the self and describes the erosion of substantial concepts of the self and the reduction of the self to "the will"—with the accompanying values of freedom, choice and autonomy. This conception of the self as an isolated, autonomous individual, characterised by acts of 'will' is then critiqued drawing on both theoretical sources, particularly the work of Iris Murdoch, and practical sources, namely the difficulties raised by genetics.

Willett, Walter C.; Blot, William J.; Colditz, Graham A.; Folsom, Aaron R.; Henderson, Brian E.; Stampfer, Meir J. Merging and emerging cohorts: necessary but not sufficient. *Nature* 2007 January 18; 445(7125): 257-258. NRCBL: 15.1; 7.1; 15.11; 16.1; 19.1.

Zeiler, Kristin. Who am I? When do "I" become another? An analytic exploration of identities, sameness and difference, genes and genomes. *Health Care Analysis: An International Journal of Health Philosophy and Policy* 2007 March; 15(1): 25-32. 14 refs. 3 fn. NRCBL: 15.1; 1.1; 4.4; 15.2; 14.4; 15.4. SC: an.

   Keywords: *gene therapy; *genetic identity; *germ cells; *personhood; *philosophy; *preimplantation diagnosis; *time factors; DNA fingerprinting; embryo transfer; embryos; genes; human genome; paternity; self concept; twinning

Abstract: What is the impact of genetics and genomics on issues of identity and what do we mean when we speak of identity? This paper explores how certain concepts of identity used in philosophy can be brought together in a multi-layered concept of identity. It discusses the concepts of numerical, qualitative, personal and genetic identity-over-time as well as rival concepts of genomic identity-over-time. These are all understood as layers in the multi-layered concept of identity. Furthermore, the paper makes it clear that our understanding of genomic identity and the importance attached to genomic sameness-over-time matters for the ethical questions raised by certain new gene technologies.

GENETICS AND HUMAN ANCESTRY

Árnason, Arnar; Simpson, Bob. Refractions through culture: the new genomics in Iceland. *Ethnos* 2003 December; 68(4): 533-553. 52 refs. 8 fn. NRCBL: 15.11; 15.1; 1.3.12; 13.1.

   Keywords: *culture; *genetic databases; *genetic research; *industry; *information dissemination; attitudes; biological specimen banks; biotechnology; genetic identity; genomics; mass media; medical records; metaphor; pedigree; politics; population genetics; public opinion; public participation; presumed consent; public policy; regulation; research subjects; researchers; risks and benefits; Proposed Keywords: persuasive communication; Keyword Identifiers: *deCode Genetics; *Iceland

Bibbins-Domingo, Kirsten; Fernandez, Alicia. BiDil for heart failure in black patients: implications of the U.S. Food and Drug Administration approval. *Annals of Internal Medicine* 2007 January 2; 146(1): 52-56. 49 refs. NRCBL: 15.11; 5.3; 9.5.4; 9.7.

   Keywords: *blacks; *pharmacogenetics; clinical trials; drugs; federal government; health disparities; heart diseases; public policy; racial groups; Proposed Keywords: *drug approval; Keyword Identifiers: *BiDil; *Food and Drug Administration; United States

Bolnick, Deborah A.; Fullwiley, Duana; Duster, Troy; Cooper, Richard S.; Fujimura, Joan H.; Kahn, Jona-than; **Kaufman, Jay S.; Marks, Jonathan; Morning, Ann; Nelson, Alondra; Ossorio, Pilar; Reardon, Jenny; Reverby, Susan M.; TallBear, Kimberly.** The science and business of genetic ancestry testing. *Science* 2007 October 19; 318(5849): 399-400. 21 refs. NRCBL: 15.11; 15.3.

   Keywords: *commerce; *genetic ancestry; *genetic screening; ethnic groups; genetic diversity; industry; racial groups; risks and benefits; Proposed Keywords: haplotypes

Coghlan, Andy. A subtle key to human diversity. *New Scientist* 2007 January 13-19; 193(2586): 8. NRCBL: 15.11; 1.3.9. Identifiers: gene expression.

Condit, C.M.; Parrott, R.L.; Bates, B.R.; Bevan, J.L.; Achter, P.J. Exploration of the impact of messages about genes and race on lay attitudes. *Clinical Genetics* 2004 November; 66(5): 402-408. 45 refs. NRCBL: 15.11; 9.5.4. SC: em.

   Keywords: *attitudes; *genetic ancestry; *genetic discrimination; *genetic research; *information dissemination; *public opinion; *racial groups; *social impact; blacks; mass media; minority groups; social discrimination; survey; whites; Keyword Identifiers: United States

Davis, Dena S. The changing face of "misidentified paternity". *Journal of Medicine and Philosophy* 2007 July-August; 32(4): 359-373. 36 refs. NRCBL: 15.11; 14.1; 15.1; 15.2. SC: rv.

   Keywords: *genetic ancestry; *genetic screening; *paternity; *pedigree; attitudes; confidentiality; disclosure; ethnic groups; family relationship; genetic counseling; genetic identity; genetic services; incidental findings; informed consent; Internet; population genetics; racial groups; risks and benefits; trends

Abstract: Advances in genetic research and technology can have a profound impact on identity and family dynamics when genetic findings disrupt deeply held assumptions about the nuclear family. Ancestry tracing and paternity testing present parallel risks and opportunities. As these latter uses are now available over the internet directly to the consumer, bypassing the genetic counselor, consumers need adequate warning when making use of these new modalities.

Dunston, Georgia M.; Royal, Charmaine D.M. The human genome: implications for the health of African Americans. *In:* Livingston, Ivor Lensworth, ed. Praeger Handbook of Black American Health: Policies and Issues Behind Disparities in Health, Vol. II. 2nd edition. Westport, CT: Praeger Publishers, 2004: 757-775. 83 refs; 5 Web site citations. NRCBL: 15.11; 9.5.4.

   Keywords: *blacks; *genetic diversity; *genetic epidemiology; *genomics; *health disparities; *human genome; genetic ancestry; genetic research; genetic services; genome mapping; health services accessibility; population genetics; racial groups; Keyword Identifiers: *United States; Africa; Genomic Research in the African Diaspora (GRAD) Project; National Human Genome Center; Howard University

Elliott, Carl. Ethnicity, citizenship, family: identity after the HGP. Minneapolis: University of Minnesota, Center for Bioethics, n.d.; 14 p. [Online]. Accessed: http://www.

NRCBL: National Reference Center for Bioethics Literature Classification Scheme   See inside front cover for terms.

229

bioethics.umn.edu/genetics_and_identity/docs/gen-grant.pdf [2007 April 24]. 43 refs. NRCBL: 15.11; 15.10; 18.5.1.

Keywords: *ethnic groups; *genetic ancestry; *genetic diversity; *genetic identity; *genetic relatedness ties; *Human Genome Project; *population genetics; family; genetic information; racial groups; self concept; social discrimination; Proposed Keywords: classification; Keyword Identifiers: United States

Fausto-Sterling, Anne. Refashioning race: DNA and the politics of health care. *Differences: A Journal of Feminist Cultural Studies* 2004 Fall; 15(3): 1-37. NRCBL: 15.11; 3.1; 15.1.

FitzGerald, Kevin; Royal, Charmaine. Race, genetics, and ethics. *In:* Prograis, Lawrence J.; Pellegrino, Edmund D., eds. African American Bioethics: Culture, Race, and Identity. Washington, DC: Georgetown University Press, 2004: 137-151. 38 fn. NRCBL: 15.11; 9.5.4. Conference: Symposium on African American Perspectives in Bioethics and Second Annual Conference on Health Disparities, held on September 23-24, 2004, at Georgetown University.

Keywords: *genetic ancestry; *genomics; *racial groups; biomedical research; blacks; drugs; ethnic groups; health care delivery; health disparities; justice; pharmacogenetics; social discrimination; Proposed Keywords: classification; Keyword Identifiers: BiDil; United States

Fullwiley, Duana. The molecularization of race: institutionalizing human difference in pharmacogenetics practice. *Science as Culture* 2007 March; 16(1): 1-30. NRCBL: 15.11; 9.7; 9.5.4.

Harmon, Amy. In DNA era, new worries about prejudice. *New York Times* 2007 November 11; p. A1, A26. NRCBL: 15.11. SC: po.

Keywords: *genetic ancestry; *genetic discrimination; *genetic diversity; *racial groups; behavioral genetics; genetic determination; genetic research; intelligence; population genetics; social impact

Harmon, Amy. My genome, myself: seeking clues in DNA. *New York Times* 2007 November 17; p. A1, A16. NRCBL: 15.11; 15.3. SC: po.

Keywords: *DNA fingerprinting; *genetic ancestry; *genetic predisposition; *genetic services; industry; pedigree; self concept

Indigenous Peoples Council on Biocolonialism. Indigenous people, genes and genetics: what indigenous people should know about biocolonialism: a primer and resource guide. Nixon, NV: Indigenous Peoples Council on Biocolonialism, 2000 June; 25 p. [Online]. Accessed: http://www.ipcb.org/publications/primers/htmls/ipgg.html [2007 April 23]. 2 refs. NRCBL: 15.11; 18.5.9.

Keywords: *American Indians; *genetic engineering; *genetic research; *indigenous populations; agriculture; biological specimen banks; biotechnology; commerce; eugenics; genes; genetic ancestry; genetic databases; genetic discrimination; genetic materials; genetic patents; genetically modified organisms; genetics; health hazards; Human Genome Diversity Project; informed consent; inter-

national aspects; political activity; population genetics; public policy; research ethics; research priorities; research subjects; Proposed Keywords: *genetic resources; biodiversity; colonialism

International HapMap Consortium; Rotimi, Charles; Leppert, Mark; Matsuda, Ichiro; Zeng, Changqing; Zhang, Houcan; Adebamowo, Clement; Ajayi, Ike; Aniagwu, Toyin; Dixon, Missy; Fukushima, Yoshimitsu; Macer, Darryl; Marshall, Patricia; Nkwodimmah, Chibuzor; Peiffer, Andy; Royal, Charmaine; Suda, Eiko; Zhao, Hui; Wang, Vivian Ota; McEwen, Jean. Community engagement and informed consent in the international HapMap project. *Community Genetics* 2007; 10(3): 186-198. 20 refs. NRCBL: 15.11; 15.10; 18.3. SC: em.

Keywords: *genetic diversity; *HapMap Project; *informed consent; *international aspects; *population genetics; *public participation; blood specimen collection; donors; genetic databases; genetic research; incentives; program descriptions; public opinion; racial groups; research subjects; risks and benefits; Keyword Identifiers: China; Japan; National Institutes of Health; Nigeria; United States

Juengst, Eric T. Population genetic research and screening: conceptual and ethical issues. *In:* Steinbock, Bonnie, ed. The Oxford Handbook of Bioethics. Oxford; New York: Oxford University Press, 2007: 471-490. 66 refs. NRCBL: 15.11. SC: an; rv.

Keywords: *genetic research; *genetic screening; *population genetics; autonomy; benefit sharing; clinical genetics; community consent; cultural pluralism; decision making; ethnic groups; eugenics; gene pool; genetic ancestry; genetic discrimination; genetic diversity; genetic information; genetic resources; genotype; genomics; goals; international aspects; justice; mass screening; moral policy; phenotype; policy analysis; preventive medicine; property rights; public health; public policy; racial groups; research subjects; values

Lee, S.S.-J. The ethical implications of stratifying by race in pharmacogenomics. *Clinical Pharmacology and Therapeutics* 2007 January; 81(1): 122-125. NRCBL: 15.11; 9.7.

Lillquist, Erik; Sullivan, Charles A. The law and genetics of racial profiling in medicine. *Harvard Civil Rights-Civil Liberties Law Review* 2004 Summer; 39(2): 391-483. 491 fn. NRCBL: 15.11; 9.5.4. SC: le; rv.

Keywords: *biomedical research; *genetic ancestry; *health disparities; *legal aspects; *medicine; *racial groups; alternatives; clinical trials; constitutional law; federal government; genetic disorders; genetic predisposition; genetic research; genetic screening; government regulation; health promotion; legal rights; minority groups; patient care; population genetics; public policy; social discrimination; socioeconomic factors; Proposed Keywords: classification; environmental health; Keyword Identifiers: *United States; First Amendment; Fourteenth Amendment

MacIntosh, Constance. Indigenous self-determination and research on human genetic material: a consideration of the relevance of debates on patents and informed consent, and the political demands on researchers. *Health Law*

Journal 2005; 13: 213-251. 195 fn. NRCBL: 15.11; 15.8; 18.3; 13.1. SC: le.
  Keywords: *genetic databases; *genetic diversity; *genetic materials; *genetic patents; *genetic research; *HapMap Project; *Human Genome Diversity Project; *indigenous populations; *informed consent; *international aspects; *population genetics; American Indians; anthropology; autonomy; benefit sharing; biological specimen banks; culture; ethnic groups; genetic identity; human rights; industry; legal aspects; moral policy; public policy; research subjects; Keyword Identifiers: *Genographic Project; Canada; European Union; Hagahai; National Geographic Society; National Institutes of Health; Patent Act; United States

Mangon, R. The medical (ir)relevance of race and ethnicity in a multiethnic society. *Community Genetics* 2007; 10(3): 199. 4 refs. NRCBL: 15.11; 9.7; 9.5.4. Comments: Troy Duster. Medicalisation of race. Lancet 2007 February 24 - March 7; 369(9562): 702-704.
  Keywords: *ethnic groups; *genetic ancestry; *medicine; *racial groups; biomedical research; clinical trials; drug industry; drugs; genetic diversity; genetic predisposition; mortality; morbidity; pharmacogenetics; research design; selection of subjects; socioeconomic factors; Proposed Keywords: drug approval; Keyword Identifiers: Bidil; Food and Drug Administration; United States

Mastroianni, George R. Kurt Gottschaldt's ambiguous relationship with National Socialism. *History of Psychology* 2006 February; 9(1): 38-54. NRCBL: 15.11; 15.6; 1.3.5; 2.2.

McMurray, David L., Jr. Genomics and ethnicity: using a tool in the U.S. Environmental Protection Agency's environmental justice toolkit. *Journal of Health Care Law and Policy* 2007; 10(1): 187-214. 233 fn. NRCBL: 15.11; 16.3; 9.7; 15.1; 9.5.4.
  Keywords: *ethnic groups; *genetic ancestry; *genetic predisposition; *health hazards; *justice; *public policy; ecology; genomics; government regulation; health disparities; indigents; minority groups; pharmacogenetics; population genetics; public health; social discrimination; toxicity; Proposed Keywords: *environmental health; *toxicogenetics; genetic epidemiology; Keyword Identifiers: *Environmental Protection Agency; United States

Nixon, Ron. DNA tests find branches but few roots. *New York Times* 2007 November 25; p. A1, A7. NRCBL: 15.11; 15.3; 1.3.2. SC: po.
  Keywords: *blacks; *DNA fingerprinting; *genetic ancestry; *genetic services; *pedigree; economics; industry; methods; population genetics; Keyword Identifiers: *United States; Africa; Gates, Henry Louis

Obasogie, Osagie. Racial alchemy. It may not be long before genetic skin-lightening treatments are on sale, so it's time to stop pretending colour prejudice isn't a problem. *New Scientist* 2007 August 18-24; 195(2617): 17. NRCBL: 15.11; 9.7.
  Keywords: *biotechnology; *genetic ancestry; *racial groups; population genetics; social discrimination; Proposed Keywords: *cosmetic techniques; *skin pigmentation

Pigliucci, Massimo; Kaplan, Jonathan. On the concept of biological race and its applicability to humans. *Philoso-*

phy of Science 2003 December; 70(5): 1161-1172. 33 refs. NRCBL: 15.11; 3.1.
  Keywords: *racial groups; biological sciences; ecology; evolution; genetic ancestry; genetic diversity; genotype; geographic factors; phenotype; population genetics

Po, Alain Li Wan. Personalised medicine: who is an Asian? *Lancet* 2007 May 26 - June 1; 369(9575): 1770-1771. NRCBL: 15.11; 9.7; 9.5.4.

Reardon, Jenny. Decoding race and human difference in a genomic age. *Differences: A Journal of Feminist Cultural Studies* 2004 Fall; 15(3): 38-65. NRCBL: 15.11; 3.1.

Rose, Nikolas S. Race in the age of genomic medicine. *In his:* Politics of Life Itself: Biomedicine, Power, and Subjectivity in the Twenty-first Century. Princeton: Princeton University Press, 2007: 155-186, 287-291. 48 fn. NRCBL: 15.11. SC: an.
  Keywords: *ethnic groups; *genetic ancestry; *population genetics; *racial groups; biomedical research; blacks; classification; eugenics; genetic identity; genetic screening; genomics; HapMap Project; health disparities; health services research; historical aspects; international aspects; Jews; pharmacogenetics; public policy; social discrimination; social dominance; Proposed Keywords: colonialism; Keyword Identifiers: China; Europe; Nineteenth Century; Twentieth Century; United States

Rothstein, M.A.; Epps, P.G. Pharmacogenomics and the (ir)relevance of race. *Pharmacogenomics Journal* 2001; 1(2): 104-108. 29 refs. NRCBL: 15.11; 9.7; 15.1.
  Keywords: *genetic ancestry; *genetic research; *pharmacogenetics; *racial groups; biomedical research; clinical genetics; ethnic groups; genetic diversity; historical aspects; research subjects; selection of subjects; trends; Proposed Keywords: classification

Schwartz, John. DNA pioneer's genome blurs race lines. *New York Times* 2007 December 12; p. A22. NRCBL: 15.11. SC: po.
  Keywords: *genetic ancestry; *racial groups; blacks; famous persons; genetic research; genetic screening; intelligence; researchers; social discrimination; whites; Proposed Keywords: predictive value of tests; Keyword Identifiers: *Watson, James D.

Sharp, Richard R.; Foster, Morris W. Grappling with groups: protecting collective interests in biomedical research. *Journal of Medicine and Philosophy* 2007 July-August; 32(4): 321-337. 56 refs. 7 fn. NRCBL: 15.11; 15.1; 18.6. SC: an.
  Keywords: *genetic diversity; *genetic research; *Human Genome Diversity Project; *research; biomedical research; community consent; cultural pluralism; decision making; dissent; donors; ethnic groups; genetic ancestry; genetic materials; guidelines; indigenous populations; international aspects; justice; population genetics; public participation; research design; research subjects; researcher subject relationship; risks and benefits; vulnerable populations; Proposed Keywords: *population groups; harm; social identification; third-party research subjects
Abstract: Strategies for protecting historically disadvantaged groups have been extensively debated in the context of genetic variation research, making this a useful

NRCBL: National Reference Center for Bioethics Literature Classification Scheme   See inside front cover for terms.

231

starting point in examining the protection of social groups from harm resulting from biomedical research. We analyze research practices developed in response to concerns about the involvement of indigenous communities in studies of genetic variation and consider their potential application in other contexts. We highlight several conceptual ambiguities and practical challenges associated with the protection of group interests and argue that protectionist strategies developed in the context of genetic research will not be easily adapted to other types of research in which social groups are placed at risk. We suggest that it is this set of conceptual and practical issues that philosophers, ethicists, and others should focus on in their efforts to protect identifiable social groups from harm resulting from biomedical research.

Sheldon, Jane P.; Jayaratne, Toby Epstein; Feldbaum, Merle B.; DiNardo, Courtney D.; Petty, Elizabeth M. Applications and implications of advances in human genetics: perspectives from a group of Black Americans. *Community Genetics* 2007; 10(2): 82-92. 39 refs. NRCBL: 15.11. SC: em.
> Keywords: *attitudes; *blacks; *genetic ancestry; *genetic predisposition; *genetic research; *racial groups; behavioral genetics; ethnic groups; genetic disorders; intelligence; interviews; risks and benefits; questionnaires; social discrimination; social impact; stigmatization; trust; violence; whites; Keyword Identifiers: United States

Shelton, B.L. Consent and consultation in genetic research on American Indians and Alaskan Natives. Nixon, NV: Indigenous Peoples Council on Biocolonialism, 2002; 3 p. [Online]. Accessed: http://www.ipcb.org/ publications/briefing_papers/files/consent.html [2007 April 24]. NRCBL: 15.11; 18.3; 18.5.9.
> Keywords: *American Indians; *genetic research; community consent; government; indigenous populations; informed consent; policy making; public participation; refusal to participate; research subjects; Keyword Identifiers: United States

Temple, Robert; Stockbridge, Norman L. BiDil for heart failure in black patients: the U.S. Food and Drug Administration perspective. *Annals of Internal Medicine* 2007 January 2; 146(1): 52-62. 28 refs. NRCBL: 15.11; 5.3; 9.5.4; 9.7.
> Keywords: *blacks; *pharmacogenetics; clinical trials; drugs; federal government; heart diseases; public policy; racial groups; Proposed Keywords: *drug approval; Keyword Identifiers: *BiDil; *Food and Drug Administration; United States

GENOCIDE *See* TORTURE, GENOCIDE, AND WAR CRIMES

GENOME MAPPING AND SEQUENCING
See also GENETIC RESEARCH; GENETIC SCREENING; GENETICALLY MODIFIED ORGANISMS AND FOOD

Aldhous, Peter. Your own book of life. *New Scientist* 2007 September 8-14; 195(2620): 8-11. NRCBL: 15.10; 8.4; 9.7.
> Keywords: *genetic services; *genome mapping; *trends; DNA sequences; genetic predisposition; genomics; human genome; industry; Proposed Keywords: *personalized medicine; Keyword Identifiers: Venter, Craig

Brenner, Steven E. Common sense for our genomes. *Nature* 2007 October 18; 449(7164): 783-784. 5 refs. NRCBL: 15.10; 1.3.12; 15.11.
> Keywords: *genetic databases; *genetic diversity; *personal genomics; famous persons; research support; researchers; Keyword Identifiers: *Genome Commons; Venter, J. Craig; Watson, James

Butcher, James. Kari Stefánsson: a general of genetics. *Lancet* 2007 January 27 - February 2; 369(9558): 267. NRCBL: 15.10; 1.3.12; 13.1.

Check, Erika. Celebrity genomes alarm researchers [news]. *Nature* 2007 May 24; 447(7143): 358-359. NRCBL: 15.10; 1.3.12.
> Keywords: *famous persons; *genome mapping; *human genome; *researchers; *selection of subjects; DNA sequences; family members; genetic databases; genetic privacy; genomics; government financing; Human Genome Project; medical records; private sector; public sector; research design; research subjects; research support; Proposed Keywords: personalized medicine; Keyword Identifiers: Church, George; National Human Genome Research Institute; Venter, J. Craig; Watson, James

Greif, Karen F.; Merz, Jon F. Big science: the Human Genome Project and the public funding of science. *In their:* Current Controversies in the Biological Sciences: Case Studies of Policy Challenges from New Technologies. Cambridge, MA: MIT, 2007: 17-34. 37 refs. NRCBL: 15.10.
> Keywords: *government financing; *Human Genome Project; *research support; federal government; genetic patents; historical aspects; industry; international aspects; private sector; program descriptions; public sector; Keyword Identifiers: *United States; National Institutes of Health; Twentieth Century

Hayden, Erika Check. Personalized genomes go mainstream [news]. *Nature* 2007 November 1; 450(7166): 11. NRCBL: 15.10; 9.7; 8.4.

Henderson, Lesley; Kitzinger, Jenny. Orchestrating a science 'event': the case of the Human Genome Project. *New Genetics and Society* 2007 April; 26(1): 65-83. NRCBL: 15.10; 1.3.7. SC: em. Identifiers: Great Britain (United Kingdom).

Hoedemaekers, Rogeer; Gordijn, Bert; Pijnenburg, Martien. Solidarity and justice as guiding principles in genomic research. *Bioethics* 2007 July; 21(6): 342-350. 25 fn. NRCBL: 15.10; 15.1; 1.3.12; 1.1; 9.1; 18.3; 18.5.1. SC: an.
> Keywords: *ethical analysis; *future generations; *genetic research; *justice; *moral obligations; *moral policy; *nontherapeutic research; *obligations to society; *presumed consent; altruism; autonomy; biological specimen

banks; communitarianism; ethical review; financial support; genetic databases; genomics; health priorities; industry; informed consent; obligations of society; public policy; research design; research priorities; research subjects; research support; resource allocation; risks and benefits; theoretical models

Abstract: In genomic research the ideal standard of free, informed, prior and explicit consent is sometimes difficult to apply. This has raised concern that important genomic research will be restricted. Different consent procedures have therefore been proposed. This paper explicitly examines the question how, in genomic research, the principles of solidarity and justice can be used to justify forms of diminished individual control over personal data and bio-samples. After a discussion of the notions of solidarity and justice and how they can be related to health care and genomic research, we examine how and in which situations these notions can form a strong moral basis for demanding certain financial sacrifices. Then we examine when these principles can justify consent procedures which diverge from the ideal standard. Because much genomic research is not expected to lead to immediate (clinical) benefits we also discuss the question of whether we can be obliged to make any sacrifices for future (not yet existing) patients. We conclude with the formulation of a number of conditions that have to be met before autonomy sacrifices can be reasonably demanded in genomic research.

Marshall, Eliot. Sequencers of a famous genome confront privacy issues [news]. *Science* 2007 March 30; 315(5820): 1780. NRCBL: 15.10; 8.4.

Keywords: *disclosure; *DNA sequences; *famous persons; *genetic privacy; *genome mapping; *human genome; access to information; genetic databases; genetic predisposition; industry; informed consent; researchers; Proposed Keywords: self disclosure; Keyword Identifiers: *Watson, James; Church, George; Venter, J. Craig

Mayor, Susan. Genome sequence of one person is published for first time [news]. *BMJ: British Medical Journal* 2007 September 15; 335(7619): 530-531. NRCBL: 15.10.

McLean, Margaret R. Religion, ethics, and the Human Genome Project. *In:* Eisen, Arri; Laderman, Gary, eds. Science, Religion, and Society: An Encyclopedia of History, Culture, and Controversy. Vol. 2. Armonk, NY: M.E. Sharpe, 2007: 787-794. 15 refs. NRCBL: 15.10; 1.2.

Keywords: *Human Genome Project; genetic research; genetic screening; human dignity; pharmacogenetics;; Keyword Identifiers: NIH-DOE Working Group on Ethical, Legal, and Social Implications (ELSI)

Wexler, Barbara. The Human Genome Project. *In her:* Genetics and genetic engineering. 2005 ed. Detroit, MI: Thomson/Gale Group, 2006: 99-116. NRCBL: 15.10.

Keywords: *Human Genome Project; genetic diversity; genetic patents; genome; genomics; government financing; HapMap Project; historical aspects; international aspects; mass media; private sector; public sector; research support; researchers; Keyword Identifiers: Department of Energy; National Human Genome Research Institute; National Institutes of Health; Twentieth Century; United States

GENOME SEQUENCING *See* GENOME MAPPING AND SEQUENCING

GENOMICS *See* GENETICS AND GENOMICS

HEALTH CARE
See also CARE FOR SPECIFIC GROUPS; HEALTH CARE ECONOMICS; HEALTH CARE QUALITY; RESOURCE ALLOCATION; RIGHT TO HEALTH CARE

AMA unveils health care ethics program, toolkit. *Healthcare Benchmarks and Quality Improvement* 2006 January; 13(1): 9-11. NRCBL: 9.1; 1.3.2; 9.8.

Due process — right to medical access — Supreme Court of Canada holds that ban on private health insurance violates Quebec charter of human rights and freedoms — Chaoulli v. Quebec (Attorney General), 2005 S.C.C. 35, 29272, [2005] S.C.J. No. 33 QUICKLAW (June 9, 2005). *Harvard Law Review* 2005 December; 119(2): 677-684. NRCBL: 9.1; 9.3.1. SC: le.

Adams, Samantha; de Bont, Antoinette. Information Rx: prescribing good consumerism and responsible citizenship. *Health Care Analysis: An International Journal of Health Philosophy and Policy* 2007 December; 15(4): 273-290. NRCBL: 9.1; 1.3.12; 8.1.

Abstract: Recent medical informatics and sociological literature has painted the image of a new type of patient — one that is reflexive and informed, with highly specified information needs and perceptions, as well as highly developed skills and tactics for acquiring information. Patients have been re-named "reflexive consumers." At the same time, literature about the questionable reliability of web-based information has suggested the need to create both user tools that have pre-selected information and special guidelines for individuals to use to check the individual characteristics of the information they encounter. In this article, we examine suggestions that individuals must be assisted in developing skills for "reflexive consumerism" and what these particular skills should be. Using two types of data (discursive data from websites and promotional items, and supplementary data from interviews and ethnographic observations carried out with those working to sustain these initiatives), we examine how users are directly addressed and discussed. We argue that these initiatives prescribe skills and practices that extend beyond finding and assessing information on the internet and demonstrate that they include ideals of consumerism and citizenship.

American Medical Association. Institute for Ethics. Ethical Force Program; Levine, Mark A.; Wynia, Matthew K.; Schyve, Paul M.; Teagarden, J. Russell; Fleming, David A.; Donohue, Sharon King; Anderson, Ron J.; Sabin, James; Emanuel, Ezekiel J. Improving access to health care: a consensus ethical framework to guide proposals for reform. *Hastings Center Report* 2007 September-October; 37(5): 14-19. NRCBL: 9.1; 9.2; 9.3.1; 9.4.

NRCBL: National Reference Center for Bioethics Literature Classification Scheme   See inside front cover for terms.

233

Banja, John; Eig, Jennifer; Williams, Mark V. Discharge dilemmas as system failures. *American Journal of Bioethics* 2007 March; 7(3): 29-31. NRCBL: 9.1; 8.1; 9.4. Comments: Robert N. Swidler, Terese Seastrum, and Wayne Shelton. Difficult hospital inpatient discharge decisions: ethical, legal and clinical practice issues. American Journal of Bioethics 2007 March; 7(3): 23-28.

Barnes, Charles. Why compliance programs fail: economics, ethics, and the role of leadership. *HEC (Healthcare Ethics Committee) Forum* 2007 June; 19(2): 109-123. NRCBL: 9.1; 1.3.2; 9.8; 1.3.1; 1.1. SC: an. Identifiers: Enron.

Battin, Margaret P.; Francis, Leslie P.; Jacobson, Jay A.; Smith, Charles B. The patient as victim and vector: the challenge of infectious disease for bioethics. *In:* Rhodes, Rosamond; Francis, Leslie P.; Silvers, Anita, eds. The Blackwell Guide to Medical Ethics. Malden, MA: Blackwell Pub., 2007: 269-286. NRCBL: 9.1; 9.5.6; 1.1.

Belde, David. Toward a "total organizational ethic" in health care ethics. *Health Care Ethics USA* 2007 Spring; 15(2): 9-11. NRCBL: 9.1; 1.3.2; 1.2; 9.6.

Brody, Howard. Ethics, justice, and health reform. *In:* Engström, Timothy H.; Robison, Wade L., eds. Health Care Reform: Ethics and Politics. Rochester, NY: University of Rochester Press, 2006: 40-66. NRCBL: 9.1; 7.1; 1.1.

Cahill, Lisa Sowle. Global health and Catholic social commitment. *Health Progress* 2007 May-June; 88(3): 55-57. NRCBL: 9.1; 1.2; 21.1.

Catholic Health Association of the United States. Report on a theological dialogue on the Principle of Cooperation: executive summary. *National Catholic Bioethics Quarterly* 2007 Winter; 7(4): 773-776. NRCBL: 9.1; 4.1.1; 1.2.

Churchill, Larry R. Preparing for the next health care reform: notes for an interim ethics. *In:* Engström, Timothy H.; Robison, Wade L., eds. Health Care Reform: Ethics and Politics. Rochester, NY: University of Rochester Press, 2006: 195-208. NRCBL: 9.1; 9.3.1; 7.1.

Cooper, Robert W.; Frank, Garry L.; Gouty, Carol Ann; Hansen, Mary C. Key ethical issues encountered in healthcare organizations: perceptions of nurse executives. *JONA: The Journal of Nursing Administration* 2002 June; 32(6): 331-337. NRCBL: 9.1; 9.8; 1.3.2; 4.1.3.

Daniels, Norman. Fairness and national health care reform. *In:* Engström, Timothy H.; Robison, Wade L., eds. Health Care Reform: Ethics and Politics. Rochester, NY: University of Rochester Press, 2006: 240-263. NRCBL: 9.1; 9.3.1; 9.4; 1.1.

Darr, Kurt. Virtue ethics: worth another look. *Hospital Topics* 2006 Fall; 84(4): 29-31. NRCBL: 9.1; 1.3.2; 1.1.

de Melo-Martín, Immaculada. The promise of the human papillomavirus vaccine does not confer immunity against ethical reflection. *Oncologist* 2006 April; 11(4): 393-396. NRCBL: 9.1; 9.5.1; 10; 9.5.7; 8.3.2.

Degnin, Francis Dominic; Wood, Donna J. Levinas and society's most vulnerable: a philosopher's view of the business of healthcare. *Organizational Ethics: Healthcare, Business, and Policy* 2007 Spring-Summer; 4(1): 65-80. NRCBL: 9.1; 1.1; 1.3.2; 4.1.1.

Dekkers, Wim; Gordijn, Bert. Practical wisdom in medicine and health care [editorial]. *Medicine, Health Care and Philosophy* 2007 September; 10(3): 231-232. NRCBL: 9.1; 1.1; 4.1.1; 9.8; 20.1.

Edmondson, Ricca; Pearce, Jane. The practice of health care: wisdom as a model. *Medicine, Health Care and Philosophy* 2007 September; 10(3): 233-244. NRCBL: 9.1; 1.1; 9.8; 17.1.
Abstract: Reasoning and judgement in health care entail complex responses to problems whose demands typically derive from several areas of specialism at once. We argue that current evidence- or value-based models of health care reasoning, despite their virtues, are insufficient to account for responses to such problems exhaustively. At the same time, we offer reasons for contending that health professionals in fact engage in forms of reasoning of a kind described for millennia under the concept of wisdom. Wisdom traditions refer to forms of deliberation which combine knowledge, reflection and life experience with social, emotional and ethical capacities. Wisdom is key in dealing with problems which are vital to human affairs but lack prescribed solutions. Uncertainty and fluidity must be tolerated in seeking to resolve them. We illustrate the application of wisdom using cases in psychiatry, where non-technical aspects of problems are often prominent and require more systematic analysis than conventional approaches offer, but we argue that our thesis applies throughout the health care field. We argue for the relevance of a threefold model of reasoning to modern health care situations in which multifaceted teamwork and complex settings demand wise judgement. A model based on practical wisdom highlights a triadic process with features activating capacities of the self (professional), other (patient and/or carers and/or colleagues) and aspects of the problem itself. Such a framework could be used to develop current approaches to health care based on case review and experiential learning.

Emanuel, Ezekiel J. What cannot be said on television about health care [commentary]. *JAMA: The Journal of the American Medical Association* 2007 May 16; 297(19): 2131-2133. NRCBL: 9.1; 5.2; 9.3.1; 9.4; 9.8. SC: rv.

Engström, Timothy H.; Richter, Gerd. Citizens and customers: establishing the ethical foundations of the German and U.S. health care systems. *In:* Engström, Timothy H.; Robison, Wade L., eds. Health Care Reform: Ethics and

Politics. Rochester, NY: University of Rochester Press, 2006: 166-186. NRCBL: 9.1; 9.3.1; 7.1; 21.1.

Frenk, Julio. Ethical considerations in health systems. *In:* Marinker, Marshall, ed. Constructive Conversations about Health: Policy and Values. Oxford; Seattle: Radcliffe Pub., 2006: 165-175. NRCBL: 9.1; 9.3.1; 9.4.

George, Tom; Van Oeveren, Edward L.; Gostin, Lawrence O. Using law to facilitate healthier lifestyles [letters and reply]. *JAMA: The Journal of the American Medical Association* 2007 May 9; 297(18): 1981-1983. NRCBL: 9.1. SC: le. Comments: Lawrence O. Gostin. Law as a tool to facilitate healthier lifestyles and prevent obesity. JAMA: The Journal of the American Medical Association 2007 January 3; 297(1): 87-90.

Glaser, John W. Catholic health ministry: fruit on the diseased tree of U.S. health care. *Health Care Ethics USA* 2007 Winter; 15(1): 2-4. NRCBL: 9.1; 1.2; 1.3.2.

Gordon, Elisa J.; Wolf, Michael S.; Volandes, Angelo E.; Paasche-Orlow, Michael K. Beyond the basics: designing a comprehensive response to low health literacy. *American Journal of Bioethics* 2007 November; 7(11): 11-13; author reply W1-W2. NRCBL: 9.1; 8.1. Comments: Angelo E. Volandes and Michael K. Paasche-Orlow. Health literacy, health inequality and a just healthcare system. American Journal of Bioethics 2007 November; 7(11): 5-10.

Gostin, Lawrence O. Law as a tool to facilitate healthier lifestyles and prevent obesity. *JAMA: The Journal of the American Medical Association* 2007 January 3; 297(1): 87-90. NRCBL: 9.1; 9.5.1. SC: le.

Gostin, Lawrence O. Meeting the survival needs of the world's least healthy people: a proposed model for global health governance. *JAMA: The Journal of the American Medical Association* 2007 July 11; 298(2): 225-228. NRCBL: 9.1; 21.1.

Gostin, Lawrence O. Why rich countries should care about the world's least healthy people [commentary]. *JAMA: The Journal of the American Medical Association* 2007 July 4; 298(1): 89-92. NRCBL: 9.1; 21.1.

Gostin, Lawrence O. Why should we care about social justice? *Hastings Center Report* 2007 July-August; 37(4): 3. NRCBL: 9.1; 21.1; 1.1.

Hammerly, Milt. Disruptive germination in health care. *Health Care Ethics USA* 2007 Spring; 15(2): 1-3. NRCBL: 9.1; 1.2; 9.3.1.

Harrington, John. Globalization and English medical law: strains and contradictions. *In:* Bennett, Belinda; Tomossy, George F., eds. Globalization and Health: Challenges for Health Law and Bioethics. Dordrecht: Springer, 2006: 169-185. NRCBL: 9.1; 1.3.5; 21.1; 19.5. SC: le; rev.

Hathaway, Andrew D. Ushering in another harm reduction era? Discursive authenticity, drug policy and research. *Drug and Alcohol Review* 2005 November; 24(6): 549-550. NRCBL: 9.1; 9.5.9.

Holm, Søren. Can politics be taken out of the (English) NHS? [editorial]. *Journal of Medical Ethics* 2007 October; 33(10): 559. NRCBL: 9.1. Identifiers: Great Britain (United Kingdom); National Health Service.

Holm, Søren. Policy-making in pluralistic societies. *In:* Steinbock, Bonnie, ed. The Oxford Handbook of Bioethics. Oxford; New York: Oxford University Press, 2007: 153-174. NRCBL: 9.1; 1.1; 2.1.

John, S.D. How to take deontological concerns seriously in risk-cost-benefit analysis: a re-interpretation of the precautionary principle. *Journal of Medical Ethics* 2007 April; 33(4): 221-224. NRCBL: 9.1; 1.1. SC: an.
   Abstract: In this paper the coherence of the precautionary principle as a guide to public health policy is considered. Two conditions that any account of the principle must meet are outlined, a condition of practicality and a condition of publicity. The principle is interpreted in terms of a tripartite division of the outcomes of action (good outcomes, normal bad outcomes and special bad outcomes). Such a division of outcomes can be justified on either "consequentialist" or "deontological" grounds. In the second half of the paper, it is argued that the precautionary principle is not necessarily opposed to risk-cost-benefit analysis, but, rather, should be interpreted as suggesting a lowering of our epistemic standards for assessing evidence that there is a link between some policy and "special bad" outcomes. This suggestion is defended against the claim that it mistakes the nature of statistical testing and against the charge that it is unscientific or antiscientific, and therefore irrational.

Jotkowitz, Alan; Porath, Avi; Volandes, Angelo E.; Paasche-Orlow, Michael K. Health literacy, access to care and outcomes of care. *American Journal of Bioethics* 2007 November; 7(11): 25-27; author reply W1-W2. NRCBL: 9.1; 8.1; 9.2; 9.8. Comments: Angelo E. Volandes and Michael K. Paasche-Orlow. Health literacy, health inequality and a just healthcare system. American Journal of Bioethics 2007 November; 7(11): 5-10.

Kaleebu, Pontiano. HIV vaccine trials in Uganda: personal experience as an investigator. *In:* AIDS Vaccine Handbook, 2nd edition: Global Perspectives. New York: AIDS Vaccine Advocacy Coalition, 2005: 145-151 [Online]. Accessed: http://www.avac.org/pdf/primer2/AVH_CH21.pdf [2006 March 8]. NRCBL: 9.1; 9.7; 9.5.6; 18.5.9.

Kapp, Marshall B. Patient autonomy in the age of consumer-driven health care: informed consent and informed choice. *Journal of Legal Medicine* 2007 January-March; 28(1): 91-117. NRCBL: 9.1; 8.3.1; 1.1.

NRCBL: National Reference Center for Bioethics Literature Classification Scheme   See inside front cover for terms.

235

Krohmal, Benjamin J.; Emanuel, Ezekiel J. Tiers without tears: the ethics of a two-tier health care system. *In:* Steinbock, Bonnie, ed. The Oxford Handbook of Bioethics. Oxford; New York: Oxford University Press, 2007: 175-189. NRCBL: 9.1; 9.3.1; 9.4.

Lanoix, Monique. When cure entails care. *American Journal of Bioethics* 2007 March; 7(3): 34-36. NRCBL: 9.1; 8.1. Comments: Robert N. Swidler, Terese Seastrum, and Wayne Shelton. Difficult hospital inpatient discharge decisions: ethical, legal and clinical practice issues. American Journal of Bioethics 2007 March; 7(3): 23-28.

Levitt, Cheryl. Spirituality and family medicine. *In:* Meier, Augustine; O'Connor, Thomas St. James; VanKatwyk, Peter, eds. Spirituality and Health: Multidisciplinary Explorations. Waterloo, Ont.: Wilfrid Laurier University Press, 2005: 61-72. NRCBL: 9.1; 1.2; 8.1; 9.5.1.

Levy, Barry S. Health and peace. *Croatian Medical Journal* 2002 April; 43(2):114-116. NRCBL: 9.1; 21.1.

Lewis, Steven; Southern, Danielle A.; Maxwell, Colleen J.; Dunn, James R.; Noseworthy, Tom W.; Ghali, William A. What prosperous, highly educated Americans living in Canada think of the Canadian and US health care systems. *Open Medicine [electronic]* 2007; 1(2): E68-E74. NRCBL: 9.1; 9.2; 9.3.1; 21.1. SC: em.

London, Leslie. 'Issues of equity are also issues of rights': lessons from experiences in Southern Africa. *BMC Public Health* 2007 January 26; 7:14. NRCBL: 9.1; 21.1; 9.2.

Loughlin, Michael. A platitude too far: 'evidence-based ethics'. Commentary on Borry (2006), Evidence-based medicine and its role in ethical decision-making. Journal of Evaluation in Clinical Practice 12, 306-311. *Journal of Evaluation in Clinical Practice* 2006 June; 12(3): 312-318 [see correction in: J Eval Clin Pract. 2006 August; 12(4): 471]. NRCBL: 9.1; 9.8; 9.4.

Ludwick, Ruth; Silva, Mary Cipriano. What would you do? Ethics and infection control. *Online Journal of Issues in Nursing [electronic]* 2007; 12(1): 8 p. Accessed: http://www.nursingworld.org/ojin/tocv12n1.htm [2007 March 15]. NRCBL: 9.1; 4.1.3; 9.5.1.

May, Thomas; Silverman, Ross D. Free-riding, fairness and the rights of minority groups in exemption from mandatory childhood vaccination. *Human Vaccines* 2005 January-February; 1(1): 12-15. NRCBL: 9.1; 9.7; 8.3.2; 8.3.4; 9.5.4; 1.2; 9.5.7. SC: le.

McIntyre, Di; Whitehead, Margaret; Gilson, Lucy; Dahlgren, Göran; Tang, Shenglan. Equity impacts of neoliberal reforms: what should the policy responses be? *International Journal of Health Services* 2007; 37(4): 693-709. NRCBL: 9.1; 9.5.3.

Miller, Vail M.; Volandes, Angelo E.; Paasche-Orlow, Michael K. Poor eHealth literacy and consumer-directed health plans: a recipe for market failure. *American Journal of Bioethics* 2007 November; 7(11): 20-22; author reply W1-W2. NRCBL: 9.1; 8.1; 1.3.12; 9.3.1. Comments: Angelo E. Volandes and Michael K. Paasche-Orlow. Health literacy, health inequality and a just healthcare system. American Journal of Bioethics 2007 November; 7(11): 5-10.

Milwaukee Guild of the Catholic Medical Association. Checklist for Catholic hospitals. *Linacre Quarterly* 2007 May; 74(2): 159-163. NRCBL: 9.1; 1.2.

Novak, David. A Jewish argument for socialized medicine. *In his:* The Sanctity of Human Life. Washington, DC: Georgetown University Press, 2007: 91-110. NRCBL: 9.1; 1.2; 4.1.2.

Nuzzo, Jennifer B. HHS proposes changes to federal quarantine regulations. *Biosecurity and bioterrorism: Biodefense Strategy, Practice, and Science* 2006; 4(1): 11-12. NRCBL: 9.1; 1.3.5.

Nuzzo, Jennifer B.; Henderson, Donald A.; O'Toole, Tara; Inglesby, Thomas V. Comments from the Center for Biosecurity of UPMC on proposed revisions to federal quarantine rules. *Biosecurity and Bioterrorism* 2006; 4(2): 204-206. NRCBL: 9.1; 1.3.5.

Orlikoff, James E.; Totten, Mary K. Conflict of interest and governance. New approaches for a new healthcare environment. *Healthcare Executive* 2006 September-October; 21(5): 52, 54. NRCBL: 9.1; 1.3.2; 7.1.

Parmet, Wendy E. Legal power and legal rights — isolation and quarantine in the case of drug-resistant tuberculosis. *New England Journal of Medicine* 2007 August 2; 357(5): 433-435. NRCBL: 9.1; 9.5.1. SC: le.

Raspe, Heiner. Individuelle Gesundheitsleistungen in der vertragsärztlichen Versorgung - Eine medizinethische Diskussion = Individual health services within Germany's statutory health insurance system: ethical considerations. *Ethik in der Medizin* 2007 March; 19(1): 24-38. NRCBL: 9.1; 9.3.1; 8.1; 9.8.

Rathert, Cheryl; Fleming, David A. Ethical climates of HCOs and end-of-life moral conflict in care terms. *Organizational Ethics: Healthcare, Business, and Policy* 2006 Fall-Winter; 3(2): 101-111. NRCBL: 9.1; 1.3.2; 20.4.1; 20.5.1.

Reckless, Ian. Patients and doctors: rights and responsibilities in the NHS. *Clinical Medicine* 2005 September-October; 5(5): 499-500. NRCBL: 9.1; 8.1; 1.3.2. Identifiers: National Health Service; Great Britain (United Kingdom).

Resnik, D.B. Responsibility for health: personal, social, and environmental. *Journal of Medical Ethics* 2007 August; 33(8): 444-445. NRCBL: 9.1.

Abstract: Most of the discussion in bioethics and health policy concerning social responsibility for health has focused on society's obligation to provide access to healthcare. While ensuring access to healthcare is an important social responsibility, societies can promote health in many other ways, such as through sanitation, pollution control, food and drug safety, health education, disease surveillance, urban planning and occupational health. Greater attention should be paid to strategies for health promotion other than access to healthcare, such as environmental and public health and health research.

Robison, Wade L. The moral crisis in health care. *In:* Engström, Timothy H.; Robison, Wade L., eds. Health Care Reform: Ethics and Politics. Rochester, NY: University of Rochester Press, 2006: 13-39. NRCBL: 9.1; 9.4; 9.3.1; 7.3.

Schillinger, Dean; Volandes, Angelo E.; Paasche-Orlow, Michael K. Literacy and health communication: reversing the 'inverse care law'. *American Journal of Bioethics* 2007 November; 7(11): 15-18; author reply W1-W2. NRCBL: 9.1; 8.1. Comments: Angelo E. Volandes and Michael K. Paasche-Orlow. Health literacy, health inequality and a just healthcare system. American Journal of Bioethics 2007 November; 7(11): 5-10.

Schneider, Carl E. The cash nexus. *Hastings Center Report* 2007 July-August; 37(4): 11-12. NRCBL: 9.1; 1.3.2; 4.1.2; 9.3.1. SC: le. Identifiers: consumer-directed health care.

Schüklenk, Udo; Bello, Braimoh. Globalization and health: a developing world perspective on ethical and policy issues. *In:* Bennett, Belinda; Tomossy, George F., eds. Globalization and Health: Challenges for Health Law and Bioethics. Dordrecht: Springer, 2006: 13-25. NRCBL: 9.1; 21.1. SC: rv.

Schüklenk, Udo; Gartland, K.M.A. Confronting an influenza pandemic: ethical and scientific issues. *Biochemical Society Transactions* 2006 December; 34(Pt 6): 1151-1154. NRCBL: 9.1; 7.1; 21.1.

Seedhouse, David. Ethics and health promotion. *In his:* Health Promotion: Philosophy, Prejudice, and Practice. 2nd edition. New York: J. Wiley, 2004: 197-213. NRCBL: 9.1.

Semin, Semih; Aras, Sahbal. Bioethics and Turkey: crossroads and challenges. *Politics and the Life Sciences* 2007 March; 26(1): 2-9. NRCBL: 9.1; 2.1.

Singer, Lawrence E. Does mission matter? *Houston Journal of Health Law and Policy* 2006; 6(2): 347-377. NRCBL: 9.1; 1.2; 7.1. SC: le.

Smith, Melissa. Patients and doctors: rights and responsibilities in the NHS (2). *Clinical Medicine* 2005 September-October; 5(5): 501-502. NRCBL: 9.1; 8.1; 1.3.2. Identifiers: National Health Service; Great Britain (United Kingdom).

Smyth, Tim. Bioethics, politics and policy development. *Monash Bioethics Review* 2000 October; 19(4): 38-45. NRCBL: 9.1; 21.1; 5.3; 18.6; 2.1; 2.3. SC: an. Conference: Symposium on the Role of Bioethics and Bioethicists; Sydney, Australia; 2000 July 7; Australian Bioethics Association and the Australian Institute of Health, Law and Ethics at the University of Sydney.

Steckel, Cynthia M. Mandatory influenza immunization for health care workers — an ethical discussion. *AAOHN Journal* 2007 January; 55(1): 34-39. NRCBL: 9.1; 7.1.

Storch, Janet. Building moral communities in health care [editorial]. *Nursing Ethics* 2007 September; 14(5): 569-570. NRCBL: 9.1; 4.1.3.

Tengland, Per-Anders. Empowerment: a goal or a means for health promotion? *Medicine, Health Care and Philosophy* 2007 June; 10(2): 197-207. NRCBL: 9.1; 1.1; 4.4.

Abstract: Empowerment is a concept that has been much used and discussed for a number of years. However, it is not always explicitly clarified what its central meaning is. The present paper intends to clarify what empowerment means, and relate it to the goals of health promotion. The paper starts with the claim that health-related quality of life is the ultimate general goal for health promotion, and continues by briefly presenting definitions of some central concepts: "welfare", "health" and "quality of life". Several suggestions as to what empowerment is are then discussed: autonomy, freedom, knowledge, self-esteem, self-confidence, and control over health or life. One conclusion of this discussion is that empowerment can be seen as a complex goal which includes aspects of the three central concepts welfare, health and quality of life. To the extent that the empowerment goals aimed at are health-related, it is concluded that empowerment is a legitimate goal for health promotion. But empowerment is not only a goal, it can also be described as a process or as an approach. This process, or approach, in a fundamental way involves the participants in problem formulation, decision making and action, which means that the experts have to relinquish some of their control and power.

Todres, Les; Galvin, Kathleen; Dahlberg, Karin. Lifeworld-led healthcare: revisiting a humanizing philosophy that integrates emerging trends. *Medicine, Health Care and Philosophy* 2007 March; 10(1): 53-63. NRCBL: 9.1; 1.1.

Abstract: In this paper, we describe the value and philosophy of lifeworld-led care. Our purpose is to give a philosophically coherent foundation for lifeworld-led care and its core value as a humanising force that moderates technological progress. We begin by indicating the timeliness of these concerns within the current context of citizen-oriented, participative approaches to healthcare. We believe that this context is in need of a deepening philosophy if it is not to succumb to the discourses of mere consumerism. We thus revisit the potential of Husserl's

NRCBL: National Reference Center for Bioethics Literature Classification Scheme   See inside front cover for terms.

237

notion of the lifeworld and how lifeworld-led care could provide important ideas and values that are central to the humanisation of healthcare practice. This framework provides a synthesis of the main arguments of the paper and is finally expressed in a model of lifeworld-led care that includes its core value, core perspectives, relevant indicative methodologies and main benefits. The model is offered as a potentially broad-based approach for integrating many existing practices and trends. In the spirit of Husserl's interest in both commonality and variation, we highlight the central, less contestable foundations of lifeworld-led care, without constraining the possible varieties of confluent practices.

Trachtman, Howard; Volandes, Angelo E.; Paasche-Orlow, Michael K. Illiteracy ain't what it used to be. *American Journal of Bioethics* 2007 November; 7(11): 27-28; author reply W1-W2. NRCBL: 9.1; 8.1. Comments: Angelo E. Volandes and Michael K. Paasche-Orlow. Health literacy, health inequality and a just healthcare system. American Journal of Bioethics 2007 November; 7(11): 5-10.

Travaline, John M. Medicine: Notes and abstracts. *National Catholic Bioethics Quarterly* 2007 Winter; 7(4): 793-808. NRCBL: 9.1; 14.2; 14.4; 9.7; 12.; 20.5.1; 20.5.2; 18.2.

Tuohey, John F. A matrix for ethical decision making in a pandemic: the Oregon Tool for emergency preparedness. *Health Progress* 2007 November-December; 88(6): 20-25. NRCBL: 9.1; 9.4.

Uscher-Pines, Lori; Duggan, Patrick S.; Garoon, Joshua P.; Karron, Ruth A.; Faden, Ruth R. Planning for an influenza pandemic: social justice and disadvantaged groups. *Hastings Center Report* 2007 July-August; 37(4): 32-39. NRCBL: 9.1; 21.1; 9.5.4.

Volandes, Angelo E.; Paasche-Orlow, Michael K. Health literacy, health inequality and a just healthcare system. *American Journal of Bioethics* 2007 November; 7(11): 5-10. NRCBL: 9.1; 8.1; 1.1; 1.3.12; 9.3.1. Comments: American Journal of Bioethics 2007 November; 7(11): 11-30.

Wall, Sarah. Organizational ethics, change, and stakeholder involvement: a survey of physicians. *HEC (Healthcare Ethics Committee) Forum* 2007 September; 19(3): 227-243. NRCBL: 9.1; 1.3.2; 7.3. SC: em.

Warwick, P. Health care policy. *In:* Holland, Stephen, ed. Introducing Nursing Ethics: Themes in Theory and Practice. Salisbury: APS, 2004: 189-208. NRCBL: 9.1; 1.3.2; 6; 4.1.3. SC: cs.

HEALTH CARE/ HEALTH CARE ECONOMICS
See also RESOURCE ALLOCATION

Andereck, William S. Commodified care. *CQ: Cambridge Quarterly of Healthcare Ethics* 2007 Fall; 16(4): 398-406. NRCBL: 9.3.1; 8.1; 9.8; 1.1. SC: an.

Andereck, William S.; Jonsen, Albert R. Conclusion [commercialism in medicine]. *CQ: Cambridge Quarterly of Healthcare Ethics* 2007 Fall; 16(4): 439-442. NRCBL: 9.3.1; 7.3; 1.3.2; 4.1.2; 1.3.1; 1.1; 8.1.

Beaudin, David J. Ethical funds for physicians [letter]. *CMAJ/JAMC: Canadian Medical Association Journal* 2007 August 14; 177(4): 375. NRCBL: 9.3.1; 7.3; 1.3.2.

Bloche, M. Gregg. Health care for all? *New England Journal of Medicine* 2007 September 20; 357(12): 1173-1175. NRCBL: 9.3.1; 9.2.

Brett, Allan S. Two-tiered health care: a problematic double standard [editorial]. *Archives of Internal Medicine* 2007 March 12; 167(5): 430-432. NRCBL: 9.3.1; 9.2; 9.4.

Browne, Andrew. In China, preventive medicine pits doctor against system; hospitals see threat to profit, bonuses; Dr. Hu's house call. *Wall Street Journal* 2007 January 16; p. A1, A18. NRCBL: 9.3.1; 7.3; 7.1. SC: po. Identifiers: Loudi, China; Hu Weimin.

Burkett, Levi. Medical tourism; concerns, benefits, and the American legal perspective. *Journal of Legal Medicine* 2007 April-June; 28(2): 223-245. NRCBL: 9.3.1; 21.1; 9.2; 9.8. SC: le.

Chen, Xiao-Yang. Defensive medicine or economically motivated corruption? A Confucian reflection on physician care in China today. *Journal of Medicine and Philosophy* 2007 November-December; 32(6): 635-648. NRCBL: 9.3.1; 1.1; 8.5.

Abstract: In contemporary China, physicians tend to require more diagnostic work-ups and prescribe more expensive medications than are clearly medically indicated. These practices have been interpreted as defensive medicine in response to a rising threat of potential medical malpractice lawsuits. After outlining recent changes in Chinese malpractice law, this essay contends that the overuse of expensive diagnostic and therapeutic interventions cannot be attributed to malpractice concerns alone. These practice patterns are due as well, if not primarily, to the corruption of medical decision-making by physicians being motivated to earn supplementary income, given the constraints of an ill-structured governmental policy by the over-use of expensive diagnostic and therapeutic interventions. To respond to these difficulties of Chinese health care policy, China will need not only to reform the particular policies that encourage these behaviors, but also to nurture a moral understanding that can place the pursuit of profit within the pursuit of virtue. This can be done by drawing on Confucian moral resources that integrate the pursuit of profit within an appreciation of benevolence. It is this Confucian moral account that can formulate a medical care policy suitable to China's contemporary market economy.

Choudhry, Sujit; Choudhry, Niteesh K.; Brown, Adalsteinn D. The legal regulation of referral incentives: physician kickbacks and physician self-referral. *In:* Flood, Colleen M., ed. Just Medicare: What's In, What's Out, How We Decide. Buffalo, NY: University of Toronto Press, 2006: 261-280. NRCBL: 9.3.1; 7.3; 1.3.2. SC: le. Identifiers: Canada.

Churchill, Larry R. The hegemony of money: commercialism and professionalism in American medicine. *CQ: Cambridge Quarterly of Healthcare Ethics* 2007 Fall; 16(4): 407-414. NRCBL: 9.3.1; 7.3; 1.3.2; 4.1.2; 1.3.1; 1.1. SC: an.

Craig, Amber; Bollinger, Dan. Of waste and want: a nationwide survey of Medicaid funding for medically unnecessary, non-therapeutic circumcision. *In:* Denniston, George C.; Gallo, Pia Grassivaro; Hodges, Frederick M.; Milos, Marilyn Fayre; Viviani, Franco, eds. Bodily Integrity and the Politics of Circumcision: Culture, Controversy, and Change. New York: Springer, 2006: 233-246. NRCBL: 9.3.1; 9.5.10; 10; 9.5.7; 7.1. SC: em.

Daniels, Norman. Rescuing universal health care. *Hastings Center Report* 2007 March-April; 37(2): 3. NRCBL: 9.3.1; 7.1.

Davis, Karen. Uninsured in America: problems and possible solutions. *BMJ: British Medical Journal* 2007 February 17; 334(7589): 346-348. NRCBL: 9.3.1.

Denny, Colleen C.; Emanuel, Ezekiel J.; Pearson, Steven D. Why well-insured patients should demand value-based insurance benefits [commentary]. *JAMA: The Journal of the American Medical Association* 2007 June 13; 297(22): 2515-2518. NRCBL: 9.3.1; 5.2; 9.4.

Emanuel, Ezekiel J.; Fuchs, Victor R. Beyond healthcare band-aids [op-ed]. *Washington Post* 2007 February 7; p. A17. NRCBL: 9.3.1; 9.1. SC: po.

Fins, Joseph J. Commercialism in the clinic: finding balance in medical professionalism. *CQ: Cambridge Quarterly of Healthcare Ethics* 2007 Fall; 16(4): 425-432. NRCBL: 9.3.1; 7.3; 1.3.2; 4.1.2; 1.3.1; 8.1. SC: an.

Graham, Jane H. Community care or therapeutic stalking: two sides of the same coin? *Journal of Psychosocial Nursing and Mental Health Services* 2006 August; 44(8): 41-47. NRCBL: 9.3.1; 7.1; 17.8.

Hamilton, Geert Jim. Equal access and financing of health services in Europe. *In:* Gevers, J.K.M.; Hondius, E.H.; Hubben, J.H., eds. Health Law, Human Rights and the Biomedicine Convention: Essays in Honour of Henriette Roscam Abbing. Leiden; Boston: Martinus Nijhoff Publishers, 2005: 61-76. NRCBL: 9.3.1; 9.2; 21.1. SC: le.

Jonsen, Albert R. Guest editorial: a note on the notion of commercialism. *CQ: Cambridge Quarterly of Healthcare*

Ethics 2007 Fall; 16(4): 368-373. NRCBL: 9.3.1; 1.3.2; 1.1; 7.3; 4.1.2; 1.3.1.

Kassirer, Jerome P. Commercialism and medicine: an overview. *CQ: Cambridge Quarterly of Healthcare Ethics* 2007 Fall; 16(4): 377-386. NRCBL: 9.3.1; 1.3.2; 7.3; 4.1.2; 1.3.1. SC: rv.

Kemble, Sarah; King, Susanne L.; Fleck, Leonard M. The price of compromise: the Massachusetts health care reform [letters]. *Hastings Center Report* 2007 January-February; 37(1): 4-7. NRCBL: 9.3.1.

Kenny, Nuala; Chafe, Roger. Pushing right against the evidence: turbulent times for Canadian health care. *Hastings Center Report* 2007 September-October; 37(5): 24-26. NRCBL: 9.3.1. SC: le.

Krohmal, Benjamin J.; Emanuel, Ezekiel J. Access and ability to pay: the ethics of a tiered health care system. *Archives of Internal Medicine* 2007 March 12; 167(5): 433-437. NRCBL: 9.3.1; 9.2; 9.4.

Latham, Stephen R. Justice and the financing of health care. *In:* Rhodes, Rosamond; Francis, Leslie P.; Silvers, Anita, eds. The Blackwell Guide to Medical Ethics. Malden, MA: Blackwell Pub., 2007: 341-353. NRCBL: 9.3.1; 1.1.

Lehmann, Lisa Soleymani; Swartz, Katherine; Chin, Michael; Angell, Marcia; Daniels, Norman; Brock, Dan; Relman, Bud; Fein, Rashi. Harvard Medical School public forum: insuring the uninsured: does Massachusetts have the right model? 17 May 2007. *Journal of Clinical Ethics* 2007 Fall; 18(3): 270-293. NRCBL: 9.3.1; 9.3.2; 9.2; 9.4. SC: em.

Nation, George A., III. Obscene contracts: the doctrine of unconscionability and hospital billing of the uninsured. *Kentucky Law Journal* 2005-2006; 94(1): 101-137. NRCBL: 9.3.1; 1.1; 1.3.2; 9.1. SC: cs; le.

Needleman, Jacob. A philosopher's reflection on commercialism in medicine. *CQ: Cambridge Quarterly of Healthcare Ethics* 2007 Fall; 16(4): 433-438. NRCBL: 9.3.1; 7.3; 1.3.2; 4.1.2; 1.3.1; 1.1; 8.1. SC: an.

Nichols, Len M. The moral case for covering children (and everyone else): policy analysis and evaluation cannot rest until there is real health care justice throughout the entire land. *Health Affairs* 2007 March-April; 26(2): 405-407. NRCBL: 9.3.1; 9.5.7; 1.2.

Rajczi, Alex. A critique of the innovation argument against a national health program. *Bioethics* 2007 July; 21(6): 316-323. NRCBL: 9.3.1; 9.2. SC: an.
Abstract: President Bush and his Council of Economic Advisors have claimed that the US shouldn't adopt a national health program because doing so would slow innovation in health care. Some have attacked this argument by challenging its moral claim that innovativeness is a good ground for choosing between health care systems.

NRCBL: National Reference Center for Bioethics Literature Classification Scheme   See inside front cover for terms.

239

This reply is misguided. If we want to refute the argument from innovation, we have to undercut the premise that seems least controversial – the premise that our current system produces more innovation than a national health program would. I argue that this premise is false. The argument requires clarifying the concept 'national health program' and examining various theories of human well-being.

Relman, Arnold S. Medical professionalism in a commercialized health care market. *JAMA: The Journal of the American Medical Association* 2007 December 12; 298(22): 2668-2670. NRCBL: 9.3.1; 4.1.2; 1.3.1.

Relman, Arnold S. The problem of commercialism in medicine. *CQ: Cambridge Quarterly of Healthcare Ethics* 2007 Fall; 16(4): 375-376. NRCBL: 9.3.1; 1.3.2; 7.1; 7.3; 4.1.2; 1.3.1.

Rodwin, Marc A. Medical commerce, physician entrepreneurialism, and conflicts of interest. *CQ: Cambridge Quarterly of Healthcare Ethics* 2007 Fall; 16(4): 387-397. NRCBL: 9.3.1; 7.3; 1.3.2; 7.1. SC: rv; le.

Ruger, Jennifer Prah. Health, health care, and incompletely theorized agreements: a normative theory of health policy decision making. *Journal of Health Politics, Policy and Law* 2007 February; 32(1): 51-87. NRCBL: 9.3.1; 9.1. SC: an.
Abstract: The years 2003-2004 marked the tenth anniversary of the rapid rise and demise of the Clinton administration's health reform efforts. Health reform may again be a political issue in the 2008 congressional and presidential elections. However, analysts still disagree over why large-scale health reform efforts continue to fail in the American political landscape. This article presents a normative theory for analyzing federal health policy decision making in the United States. This theory states that values and norms, particularly their level of generality, and the social agreement or lack thereof around them have a central role in understanding health policy reform. This theory does not attempt to arrive at a single unified framework for explaining health policy reform, and it recognizes the complementary roles of political science and economic explanations. Nonetheless, it argues that unarticulated values and norms have a critical role to play in health-policy making and reform; this role has been inadequately studied and has lacked a theoretical framework. Within this perspective, this article argues that policy goals, which require individuals to make financial commitments (e.g., tax contributions) in the form of redistributing resources for implementation (e.g., universal health insurance), should be analyzed within a normative framework that evaluates individuals' ethical commitments to making such sacrifices that are beyond their self-interest. The distribution of public moral norms, their degree of internalization, and the social consensus, or lack thereof, that applies to them must be objects of study in the effort to better understand health policy reform. By emphasizing these factors, this approach offers findings distinct from those provided by existing
analyses, and the article concludes with prescriptions for future health reform efforts.

Ruger, J.P. The moral foundations of health insurance. *Quarterly Journal of Medicine* 2007; 100(1): 5 p. [Online]. Accessed: http://papers.ssrn.com/sol3/papers.cfm?abstract_id=957971 [2007 November 12]. NRCBL: 9.3.1; 1.1; 9.2.

Saniotis, Arthur. Changing ethics in medical practice: a Thai perspective. *Indian Journal of Medical Ethics* 2007 January-March; 4(1): 24-25. NRCBL: 9.3.1; 4.5; 21.6; 4.1.2.

Saver, Richard S. The costs of avoiding physician conflicts of interest: a cautionary tale of gainsharing regulation. *In:* Flood, Colleen M., ed. Just Medicare: What's In, What's Out, How We Decide. Buffalo, NY: University of Toronto Press, 2006: 281-306. NRCBL: 9.3.1; 7.3; 1.3.2. SC: le. Identifiers: Canada.

Schneiderman, Lawrence J. The media and the medical market. *CQ: Cambridge Quarterly of Healthcare Ethics* 2007 Fall; 16(4): 420-424. NRCBL: 9.3.1; 1.3.2; 1.3.7; 9.7.

Snyder, Lois; Neubauer, Richard L. Pay-for-performance principles that promote patient-centered care: an ethics manifesto. *Annals of Internal Medicine* 2007 December 4; 147(11): 792-794. NRCBL: 9.3.1; 9.8; 8.1; 7.1.
Abstract: Pay-for-performance programs are growing, but little evidence exists on their effectiveness or on their potential unintended consequences and effects on the patient-physician relationship. Pay-for-performance has the potential to help improve the quality of care, if it can be aligned with the goals of medical professionalism. Initiatives that provide incentives for a few specific elements of a single disease or condition, however, may neglect the complexity of care for the whole patient, especially the elderly patient with multiple chronic conditions. Such programs could also result in the deselection of patients, "playing to the measures" rather than focusing on the patient as a whole, and misalignment of perceptions between physicians and patients. The primary focus of the quality movement in health care should not be on "pay for" or "performance" based on limited measures, but rather on the patient. The American College of Physicians hopes to move the pay-for-performance debate forward with a patient-centered focus — one that puts the needs and interests of the patient first — as these programs evolve.

Taylor, Robert M. Ethical aspects of medical economics. *Neurologic Clinics* 1989 November; 7(4): 883-900. NRCBL: 9.3.1; 19.4. SC: rv.

United States. Department of Health and Human Services [DHHS]; United States. Centers for Medicare and Medicaid Services. Medicare and Medicaid programs; hospital conditions of participation: patients' rights. Final rule. *Federal Register* 2006 December 8; 71(236): 71377-71428. NRCBL: 9.3.1; 9.5.2; 9.6; 8.1; 9.2.

United States. Department of Health and Human Services [DHHS]; United States. Centers for Medicare and Medicaid Services. Medicare program; physicians referrals to health care entities with which they have financial relationships; exceptions for certain electronic prescribing and electronic health records arrangements; final rule. *Federal Register* 2006 August 8; 71(152): 45139-45171. NRCBL: 9.3.1; 9.5.2; 1.3.12; 5.3. SC: le.

United States. Department of Health and Human Services [DHHS]. Office of Inspector General. Medicare and state health care programs: fraud and abuse; safe harbors for certain electronic prescribing and electronic health records arrangements under the anti-kickback statute; final rule. *Federal Register* 2006 August 8; 71(152): 45109-45137. NRCBL: 9.3.1; 9.5.2; 1.3.12; 5.3. SC: le.

Van Rosendaal, Guido M.A. Queue jumping: social justice and the doctor-patient relationship [editorial]. *Canadian Family Physician* 2006 December; 52(12): 1525-1526. NRCBL: 9.3.1; 8.1; 7.1.

Weiner, Rory B. A cooperative beneficence approach to health care reform. *In:* Engström, Timothy H.; Robison, Wade L., eds. Health Care Reform: Ethics and Politics. Rochester, NY: University of Rochester Press, 2006: 209-239. NRCBL: 9.3.1; 1.1; 9.1.

Wells, David A.; Ross, Joseph S.; Detsky, Allan S. What is different about the market for health care? *JAMA: The Journal of the American Medical Association* 2007 December 19; 298(23): 2785-2787. NRCBL: 9.3.1.

White, Lawrence W. Corporatization of health care. *In:* Engström, Timothy H.; Robison, Wade L., eds. Health Care Reform: Ethics and Politics. Rochester, NY: University of Rochester Press, 2006: 99-115. NRCBL: 9.3.1; 8.1; 1.1.

HEALTH CARE/.../ MANAGED CARE PROGRAMS

Berwick, Donald M.; Kaplan, Madge. 'What's the ethics of that?' A conversation with Thomas O. Pyle. *Health Affairs* 2008 January-February; 27(1): 143-150. NRCBL: 9.3.2.

Cohen, Julie; Marecek, Jeanne; Gillham, Jane. Is three a crowd? Clients, clinicians, and managed care. *American Journal of Orthopsychiatry* 2006 April; 76(2): 251-259. NRCBL: 9.3.2; 17.2; 8.1.

Patel, Mitesh S.; Chernew, Michael E. The impact of the adoption of gag laws on trust in the patient-physician relationship. *Journal of Health Politics, Policy and Law* 2007 October; 32(5): 819-842. NRCBL: 9.3.2; 8.1; 8.2. SC: em; le.

Abstract: Physician organizations, policy makers, and patient advocates have expressed concern that health plans have contractually limited the freedom of physicians to communicate with their patients. In response, many states have adopted gag laws that limit the ability of managed care contracts to restrict patient-physician communication. We examine the impact of these laws on patient trust in the physician. We analyzed patients' ratings of trust in their physicians in states before and after adoption of gag laws. Individuals in states that had such laws throughout the study period were used as the comparison group. The analysis is based on a nationally representative sample of adults obtained from the 1996-1997 and 1998-1999 Community Tracking Study Household Surveys. After adjustment for patient characteristics, it was estimated that the adoption of gag laws had no statistically significant impact on trust in the physician for the average patient. However, the adoption of gag laws is estimated to have increased trust in the physician by a modest amount (25 percent of a standard deviation) for health maintenance organization (HMO) enrollees who did not have a usual source of care. Gag laws may assure HMO enrollees without a usual source of care that their physicians are free to speak candidly about treatment options. This does not necessarily imply that physicians are prohibited from speaking freely in the absence of such laws, but gag laws indicate concerns (justified or not) that patients have about unrestricted communication with their health care providers.

Reinhardt, Uwe E. A social contract for twenty-first century health care: three-tier health care with bounty hunting. *In:* Engström, Timothy H.; Robison, Wade L., eds. Health Care Reform: Ethics and Politics. Rochester, NY: University of Rochester Press, 2006: 67-98. NRCBL: 9.3.2; 9.1.

Sabin, James E.; Cochran, David. Confronting trade-offs in health care: Harvard Pilgrim Health Care's organizational ethics program. *Health Affairs* 2007 July-August; 26(4): 1129-1134. NRCBL: 9.3.2; 1.3.2; 9.4; 9.7.

Teagarden, J. Russell; Wynia, Matthew K. Ensuring fairness in coverage decisions: applying the American Medical Association Ethical Force Program's consensus report to managed care pharmacy. *American Journal of Health-System Pharmacy* 2006 September 15; 63(18): 1749-1754. NRCBL: 9.3.2; 9.7; 9.2; 9.4.

HEALTH CARE/ HEALTH CARE QUALITY

Baumrucker, Steven J. A medical error leads to tragedy: how do we inform the patient? *American Journal of Hospice and Palliative Care* 2006 October-November; 23(5): 417-421. NRCBL: 9.8; 7.1; 8.1; 20.4.1. SC: cs.

Bird, Chloe E.; Fremont, Allen M.; Bierman, Arlene S.; Wickstrom, Steve; Shah, Mona; Rector, Thomas; Horstman, Thomas; Escarce, José J. Does quality of care for cardiovascular disease and diabetes differ by gender for enrollees in managed care plans? *Women's Health Issues* 2007 May-June; 17(3): 131-138. NRCBL: 9.8; 10; 9.5.5; 9.5.1; 9.3.2. SC: em.

Blustein, Jeffrey. Doctoring and self-forgiveness. *In:* Walker, Rebecca L.; Ivanhoe, Philip J., eds. Working Vir-

NRCBL: National Reference Center for Bioethics Literature Classification Scheme   See inside front cover for terms.

241

tue: Virtue Ethics and Contemporary Moral Problems. Oxford: Clarendon, 2007: 87-111. NRCBL: 9.8; 8.1; 7.1. SC: an.

Boyle, Dennis; O'Connell, Daniel; Platt, Frederic W.; Albert, Richard K. Disclosing errors and adverse events in the intensive care unit. *Critical Care Medicine* 2006 May; 34(5): 1532-1537. NRCBL: 9.8; 8.2; 9.5.1. SC: cs; rv.

Bryant, Rosemary. Contradictions in the concept of professional culpability. *Health Care Analysis: An International Journal of Health Philosophy and Policy* 2007 June; 15(2): 137-152. NRCBL: 9.8; 1.3.1; 7.1; 8.5.

Abstract: Increasing recognition of adverse events in health care is wide spread. Implementing improved system arrangements, which prevent adverse events taking place rather than focussing on individual culpability is increasingly being recognised as a more effective preventative strategy. But does such a perspective mean individual health practitioners remain accountable for their practice? This article explores the philosophical, psychological and professional contradictions inherent in attempting to understand where the responsibility for our actions lies and concludes by arguing that while the case for the system approach to adverse event reduction is strong, the notion of individual professional culpability needs to be maintained.

Callens, Stefaan; Volbragt, Ilse; Nys, Herman. Legal thoughts on the implications of cost-reducing guidelines for the quality of health care. *Health Policy* 2007 March; 80(3): 422-431. NRCBL: 9.8; 9.2; 9.3.1; 21.1; 8.5. Identifiers: Europe; Belgium; The Netherlands; France.

Candib, Lucy M. How turning a QI project into "research" almost sank a great program. *Hastings Center Report* 2007 January-February; 37(1): 26-30. NRCBL: 9.8; 18.1; 18.2.

Dawson, Liza; Hyder, Adnan A. Understanding the 'de jure' standard of care for research: a reply to Faust [letter]. *Developing World Bioethics* 2007 April; 7(1): 46-47. NRCBL: 9.8; 9.4; 21.7. Comments: Hally S. Faust. Is a national standard of care always the right one? Developing World Bioethics 2007 April; 7(1): 45-46.

Delbanco, Tom; Bell, Sigall K. Guilty, afraid and alone — struggling with medical error. *New England Journal of Medicine* 2007 October 25; 357(17): 1682-1683. NRCBL: 9.8; 7.1; 8.5.

Dyer, Owen. US medical authorities are accused of failing to act over doctors in Guantanamo [news]. *BMJ: British Medical Journal* 2007 September 15; 335(7619): 530. NRCBL: 9.8; 4.1.2.

Faust, Halley S. Is a national standard of care always the right one? [letter]. *Developing World Bioethics* 2007 April; 7(1): 45-46. NRCBL: 9.8; 9.4; 21.7. Comments: Adnan A. Hyder and Liza Dawson. Defining standard of care in the developing world: the intersection of interna-

tional research ethics and health systems analysis. developing World Bioethics 2005; 5: 142-152.

Fein, Stephanie P.; Hilborne, Lee H.; Spiritus, Eugene M.; Seymann, Gregory B.; Keenan, Craig R.; Shojania, Kaveh G.; Kagawa-Singer, Marjorie; Wenger, Neil S. The many faces of error disclosure: a common set of elements and a definition. *JGIM: Journal of General Internal Medicine* 2007 June; 22(6): 755-761. NRCBL: 9.8; 8.2. SC: em.

Flemons, W. Ward; Davies, Jan M.; MacLeod, Bruce. Disclosing medical errors [letter]. *CMAJ/JAMC: Canadian Medical Association Journal* 2007 November 6; 177(10): 1236. NRCBL: 9.8; 8.2.

Fox, Renée C. Toward an ethics of iatrogenesis. *In:* LaFleur, William R.; Böhme, Gernot; Shimazono, Susumu, eds. Dark Medicine: Rationalizing Medical Research. Bloomington: Indiana University Press, 2007: 149-164. NRCBL: 9.8; 9.1; 5.2.

Gallagher, Thomas H.; Studdert, David; Levinson, Wendy. Disclosing harmful medical errors to patients. *New England Journal of Medicine* 2007 June 28; 356(26): 2713-2719. NRCBL: 9.8; 8.2. SC: rv. Identifiers: National Quality Forum.

Garau, J. Impact of antibiotic restrictions: the ethical perspective. *Clinical Microbiology and Infection* 2006 August; 12(Supplement 5): 16-24. NRCBL: 9.8; 9.7.

Garbutt, Jane; Brownstein, DenaR.; Klein, Eileen J.; Waterman, Amy; Krauss, MelissaJ.; Marcuse,Edgar K.; Hazel, Erik; Dunagan, Wm. Claiborne; Fraser, Victoria; Gallagher, Thomas H. Reporting and disclosing medical errors: pediatricians' attitudes and behaviors. *Archives of Pediatrics and Adolescent Medicine* 2007 February; 161(2): 179-185. NRCBL: 9.8; 8.2; 7.1. SC: em.

Genuis, Stephen J. Diagnosis: contemporary medical hubris; Rx: a tincture of humility. *Journal of Evaluation in Clinical Practice* 2006 February; 12(1): 24-30. NRCBL: 9.8; 9.1; 7.1; 8.5.

Gerber, Andreas; Hentzelt, Frieder; Lauterbach, Karl W. Can evidence-based medicine implicitly rely on current concepts of disease or does it have to develop its own definition? *Journal of Medical Ethics* 2007 July; 33(7): 394-399. NRCBL: 9.8; 1.1; 4.2.

Abstract: Decisions in healthcare are made against the background of cultural and philosophical definitions of disease, sickness and illness. These concepts or definitions affect both health policy (macro level) and research (meso level), as well as individual encounters between patients and physicians (micro level). It is therefore necessary for evidence-based medicine to consider whether any of the definitions underlying research prior to the hierarchisation of knowledge are indeed compatible with its own epistemological principles.

Graber, Mark A. "Can I have that drug I saw on TV?" Justice, cost-effectiveness, and the ethics of prescribing. *JAAPA: Official Journal of the American Academy of Physician Assistants* 2006 July; 19(7): 48-49. NRCBL: 9.8; 8.1; 9.7.

Grady, Christine. Quality improvement and ethical oversight [editorial]. *Annals of Internal Medicine* 2007 May 1; 146(9): 680-681. NRCBL: 9.8; 18.6.

Hauswald, Mark; Wells, Robert J.; Candib, Lucy M. Overseeing quality improvement [letters and reply]. *Hastings Center Report* 2007 July-August; 37(4): 6-8. NRCBL: 9.8; 18.1; 18.2. Comments: Lucy M. Candib. How turning a QI project into "research" almost sank a great program. Hastings Center Report 2007 January-February; 37(1): 26-30.

Huddle, Thomas S. The limits of objective assessment of medical practice. *Theoretical Medicine and Bioethics* 2007; 28(6): 487-496. NRCBL: 9.8. SC: an.
   Abstract: Medical work is increasingly being subjected to objective assessment as those who pay for it seek to grasp the quality of that work and how best to improve it. While objective measures have a role in the assessment of health care, I argue that this role is currently overestimated and that no human practice such as medicine can be fully comprehended by objective assessment. I suggest that the character of practices, in which formalizations are combined with judgment, requires that valid assessment involve the perspective of the skilled practitioner. Relying exclusively on objective measures in assessing health care will not only distort our assessments of it but lead to damage as the incentives of health care workers are directed away from the important aspects of their work that are not captured by objective measures.

Itoh, Kenji; Andersen, Henning Boje; Madsen, Marlene Dyrlov; Østergaard, Doris; Ikeno, Masaaki. Patient views of adverse events: comparisons of self-reported healthcare staff attitudes with disclosure of accident information. *Applied Ergonomics* 2006 July; 37(4): 513-523. NRCBL: 9.8; 8.2; 21.1.

Jansen-van der Weide, Marijke Catharina; Onwuteaka-Philipsen, Bregje Dorien; van der Wal, Gerrit. Quality of consultation and the project 'Support and Consultation on Euthanasia in the Netherlands' (SCEN). *Health Policy* 2007 January; 80(1): 97-106. NRCBL: 9.8; 7.3; 20.7; 8.1; 20.5.1; 7.1. SC: em.

Johnson, Summer. Making up is hard to do [review of After Harm: Medical Error and the Ethics of Forgiveness, by Nancy Berlinger]. *Hastings Center Report* 2007 March-April; 37(2): 45-46. NRCBL: 9.8; 7.1; 8.1.

Jones, James W.; McCullough, Laurence B. Ethics of over-scheduling: when enough becomes too much. *Journal of Vascular Surgery* 2007 March; 45(3): 635-636. NRCBL: 9.8; 4.1.2; 8.1. SC: cs.

Kaldjian, Lauris C.; Jones, Elizabeth W.; Rosenthal, Gary E. Facilitating and impeding factors for physicians' error disclosure: a structured literature review. *Joint Commission Journal on Quality and Patient Safety / Joint Commission Resources* 2006 April; 32(4): 188-198. NRCBL: 9.8; 8.2. SC: em.

Kaldjian, Lauris C.; Jones, Elizabeth W.; Wu, Barry J.; Forman-Hoffman, Valerie L.; Levi, Benjamin H.; Rosenthal, Gary E. Disclosing medical errors to patients: attitudes and practices of physicians and trainees. *JGIM: Journal of General Internal Medicine* 2007 July; 22(7): 988-996 [see correction in JGIM: Journal of General Internal Medicine 2007 September; 22(9): 1384]. NRCBL: 9.8; 8.2. SC: em.

Kamerow, Douglas. Great health care, guaranteed. *BMJ: British Medical Journal* 2007 May 26; 334(7603): 1086. NRCBL: 9.8; 9.3.1; 1.3.2.

Luce, John M. Acknowledging our mistakes. *Critical Care Medicine* 2006 May; 34(5): 1575-1576. NRCBL: 9.8; 8.2.

Lynn, Joanne; Baily, Mary Ann; Bottrell, Melissa; Jennings, Bruce; Levine, Robert J.; Davidoff, Frank; Casarett, David; Corrigan, Janet; Fox, Ellen; Wynia, Matthew K.; Agich, George J.; O'Kane, Margaret; Speroff, Theodore; Schyve, Paul; Batalden, Paul; Tunis, Sean; Berlinger, Nancy; Cronenwett, Linda; Fitzmaurice, J. Michael; Neveloff Dubler, Nancy; James, Brent. The ethics of using quality improvement methods in health care. *Annals of Internal Medicine* 2007 May 1; 146(9): 666-673. NRCBL: 9.8; 18.6.
   Abstract: Quality improvement (QI) activities can improve health care but must be conducted ethically. The Hastings Center convened leaders and scholars to address ethical requirements for QI and their relationship to regulations protecting human subjects of research. The group defined QI as systematic, data-guided activities designed to bring about immediate improvements in health care delivery in particular settings and concluded that QI is an intrinsic part of normal health care operations. Both clinicians and patients have an ethical responsibility to participate in QI, provided that it complies with specified ethical requirements. Most QI activities are not human subjects research and should not undergo review by an institutional review board; rather, appropriately calibrated supervision of QI activities should be part of professional supervision of clinical practice. The group formulated a framework that would use key characteristics of a project and its context to categorize it as QI, human subjects research, or both, with the potential of a customized institutional review board process for the overlap category. The group recommended a period of innovation and evaluation to refine the framework for ethical conduct of QI and to integrate that framework into clinical practice.

Magill, Gerard. Ethical and policy issues related to medical error and patient safety. *In:* McLean, Sheila A.M., ed. First Do No Harm: Law, Ethics, and Healthcare. Alder-

---

NRCBL: National Reference Center for Bioethics Literature Classification Scheme   See inside front cover for terms.

243

shot, England; Burlington, VT: Ashgate, 2006: 101-116. NRCBL: 9.8; 7.1; 16.3.

Mohammadi, S. Mehrdad; Mohammadi, S. Farzad; Hedges, Jerris R. Conceptualizing a quality plan for healthcare: a philosophical reflection on the relevance of the health profession to society. *Health Care Analysis: An International Journal of Health Philosophy and Policy* 2007 December; 15(4): 337-361. NRCBL: 9.8; 1.1; 4.2; 4.4.

Abstract: Today, health systems around the world are under pressure to create greater value for patients and society; increasing access, improving client orientation and responsiveness, reducing medical errors and safety, restraining utilization via managed care, and implementing priority allocation of resources for high-burden health problems are examples of strategies towards this end. The quality paradigm by virtue of its strategic consumer focus and its methods for achieving operational excellence has proved an effective approach for creating higher value in many sectors. If applied in a deliberate and holistic manner, the quality paradigm can bring about a more cost-effective organization of the health systems. In this article, we apply quality concepts to healthcare in a conceptual format; we characterize the health system's customers and outputs with their quality dimensions. The product of this effort is a blueprint for a customer-driven health system which identifies six types of customers, nine types of outputs and the associated operations. As a preliminary step, a new analysis and definition of health and disease is provided. Rethinking the structure of health system in this manner and the related conceptual model can guide medical research, health sciences education, and health services policy, and help the practitioner to integrate all modern trends in healthcare delivery.

Moskop, John C.; Geiderman, Joel M.; Hobgood, Cherri D.; Larkin, Gregory L. Emergency physicians and disclosure of medical errors. *Annals of Emergency Medicine* 2006 November; 48(5): 523-531. NRCBL: 9.8; 7.1.

Pear, Robert. Medicare says it won't cover "preventable" hospital errors. *New York Times* 2007 August 19; p A1, A20. NRCBL: 9.8; 7.1; 8.5; 9.3.1; 9.5.2. SC: po.

Scheirton, Linda S.; Mu, K.; Lohman, H.; Cochran, T.M. Error and patient safety: ethical analysis of cases in occupational and physical therapy practice. *Medicine, Health Care and Philosophy* 2007 September; 10(3): 301-311. NRCBL: 9.8; 1.1; 4.1.1.

Abstract: Compared to other health care professions such as medicine, nursing and pharmacy, few studies have been conducted to examine the nature of practice errors in occupational and physical therapy. In an ongoing study to determine root causes, typographies and impact of occupational and physical therapy error on patients, focus group interviews have been conducted across the United States. A substantial number of harmful practice errors and/or other patient safety events (deviations or acci-

dents) have been identified. Often these events have had moral dimensions that troubled the therapist involved. In this article, six of these transcribed cases are analyzed, using predominant bioethical theories, ethical principles and professional codes of ethics. The cases and their analyses are intended to be exemplary, improving the readers' ability to discern and critically address similar such events. Several patient safety strategies are suggested that might have prevented the events described in these cases.

Silversides, Ann. Slouching toward disclosure. *CMAJ/JAMC: Canadian Medical Association Journal* 2007 November 20; 177(11): 1342-1343. NRCBL: 9.8; 8.2.

van der Wal, Gerrit. Quality of care, patient safety, and the role of the patient. *In:* Gevers, J.K.M.; Hondius, E.H.; Hubben, J.H., eds. Health Law, Human Rights and the Biomedicine Convention: Essays in Honour of Henriette Roscam Abbing. Leiden; Boston: Martinus Nijhoff Publishers, 2005: 77-92. NRCBL: 9.8; 7.1; 8.5.

Voss Horrell, Sarah C.; MacLean, William E., Jr.; Conley, Virginia M. Patient and parent/guardian perspectives on the health care of adults with mental retardation. *Mental Retardation* 2006 August; 44(4): 239-248. NRCBL: 9.8; 8.1; 9.5.3. SC: em.

Wade, Derick. Ethics of collecting and using healthcare data. *BMJ: British Medical Journal* 2007 June 30; 334(7608): 1330-1331. NRCBL: 9.8; 1.3.9; 1.3.12; 8.4.

HEALTH CARE RATIONING See RESOURCE ALLOCATION

HEALTH CARE RIGHTS See RIGHT TO HEALTH CARE

HEALTH, CONCEPT OF
See also MENTAL HEALTH, CONCEPT OF

Cannon, Geoffrey. Out of the box. *Public Health Nutrition* 2006 April; 9(2): 174-177. NRCBL: 4.2; 5.1; 9.1; 21.1; 2.1. Identifiers: ethics in nutrition science.

Davis, Kathy. Rethinking "normal" [reviews of No Child Left Different, edited by Sharon Olfman; Cutting to the Core:Exploring the Ethics of Contested Surgeries, edited by David Benatar; Surgically Shaping Children: Technology, Ethics, and the Pursuit of Normality, edited by Erik Parens]. *Hastings Center Report* 2007 May-June; 37(3): 44-47. NRCBL: 4.2; 9.5.7; 9.7; 9.5.9.

Meilaender, Gilbert. Fitness fixation: why health is not a civic virtue. *Christian Century* 2007 October 16; 124(21): 8-9. NRCBL: 4.2; 9.3.1; 9.5.1.

Metzl, Jonathan M.; Herzig, Rebecca M. Medicalisation in the 21st century: introduction. *Lancet* 2007 February 24-March 2; 369(9562):697-698. NRCBL: 4.2; 7.1; 9.7. Identifiers: Ivan Illich.

Nordenfelt, Lennart. The logic of health concepts. *In:* Khushf, George, ed. Handbook of Bioethics: Taking Stock of the Field From a Philosophical Perspective. Dordrecht; Boston: Kluwer Academic, 2004: 205-222. NRCBL: 4.2. SC: rv.

Rose, Nikolas. Beyond medicalisation. *Lancet* 2007 February 24-March 2; 369(9562): 700-702. NRCBL: 4.2; 7.1; 4.4.

Savulescu, Julian. Autonomy, the good life, and controversial choices. *In:* Rhodes, Rosamond; Francis, Leslie P.; Silvers, Anita, eds. The Blackwell Guide to Medical Ethics. Malden, MA: Blackwell Pub., 2007: 17-37. NRCBL: 4.2; 1.1.

Tarzian, Anita J. Disability and slippery slopes [Perspective]. *Hastings Center Report* 2007 September-October; 37(5): inside back cover. NRCBL: 4.2; 8.3.2; 9.5.1.

Tengland, Per-Anders. A two-dimensional theory of health. *Theoretical Medicine and Bioethics* 2007; 28(4): 257-284. NRCBL: 4.2; 1.1.

Abstract: The starting point for the contemporary debate about theories of health should be the holistic theory of Lennart Nordenfelt, claims George Khushf, not the refuted theory of Christopher Boorse. The present paper is an attempt to challenge Nordenfelt and to present an alternative theory to his and other theories, including Boorse's. The main problems with Nordenfelt's theory are that it is relativistic, that it leads to counter-intuitive results as to what goals can count as healthy, that it focuses on the wrong kind of abilities, that it makes measuring health extra difficult, and that it does not give us a sufficient account of health, at most a necessary one. The alternative theory proposed is two-dimensional. First, health is to have developed the abilities and dispositions that members of one's culture typically develop, and be able to use them, in acceptable circumstances; and second, health is to experience positive moods and sensations, the kinds that have internal causes. The theory solves the problems attached to Nordenfelt's theory by not being individual relativistic, by eliminating the goals in the definition, by giving an alternative interpretation of "ability," by making health easier to measure, and by adding the dimension of well-being that, together with health as ability, not only gives us a necessary, but also a sufficient, account of health.

Tomes, Nancy. Patient empowerment and the dilemmas of late-modern medicalisation. *Lancet* 2007 February 24-March 2; 369(9562): 698-700. NRCBL: 4.2; 7.1; 8.1.

HEMODIALYSIS *See* ORGAN AND TISSUE TRANSPLANTATION

HISTORY OF BIOETHICS *See* BIOETHICS AND MEDICAL ETHICS/ HISTORY

HOSPICES *See* DEATH AND DYING/ TERMINAL CARE

HOSPITAL ETHICS COMMITTEES *See* ETHICISTS AND ETHICS COMMITTEES

HUMAN EXPERIMENTATION
See also AIDS/ HUMAN EXPERIMENTATION; BEHAVIORAL RESEARCH; BIOMEDICAL RESEARCH

Blustein, J. The history and moral foundations of human-subject research. *American Journal of Physical Medicine and Rehabilitation* 2007 February; 86(2): 82-85. NRCBL: 18.1; 18.2; 1.1; 7.1; 1.3.5; 21.4.

Bracken, Wendy; Simon, Gayle; Cox, Susan; McDonald, Michael; Fitzgerald, Maureen. Protecting researchers. *Journal of Empirical Research on Human Research Ethics* 2007 March; 2(1): 93-96. NRCBL: 18.1.

Brassington, Iain. John Harris' argument for a duty to research. *Bioethics* 2007 March; 21(3): 160-168. NRCBL: 18.1; 1.1. SC: an.

Abstract: John Harris suggests that participation in or support for research, particularly medical research, is a moral duty. One kind of defence of this position rests on an appeal to the past, and produces two arguments. The first of these arguments is that it is unfair to accept the benefits of research without contributing something back in the form of support for, or participation in, research. A second argument is that we have a social duty to maintain those practices and institutions that sustain us, such as those which contribute to medical knowledge. This argument is related to the first, but it does not rely so heavily on fairness. Another kind of defence of the duty to research rests on an appeal to the future benefits of research: research is an effective way to discharge a duty to rescue others from serious illness or death, therefore we have a duty to research. I suggest that all three of Harris' lines fail to provide a compelling duty to research and spell out why. Moreover, not only do the lines of argument fail in their own terms: in combination, they turn out to be antagonistic to the very position that Harris wants to defend. While it is not my intention here to deny that there might be a duty to research, I claim that Harris' argument for the existence of such a duty is not the best way to establish it.

Brody, Baruch. The ethics of controlled clinical trials. *In:* Khushf, George, ed. Handbook of Bioethics: Taking Stock of the Field From a Philosophical Perspective. Dordrecht; Boston: Kluwer Academic, 2004: 337-352. NRCBL: 18.1; 18.2.

Crooks, Glenna M. The rights of patients to participate in clinical research. *In:* Santoro, Michael A.; Gorrie, Thomas M., eds. Ethics and the Pharmaceutical Industry. Cambridge; New York: Cambridge University Press, 2005: 97-108. NRCBL: 18.1; 9.7; 18.3; 18.6; 18.2.

Decullier, Evelyne; Chapuis, François. Impact of funding on biomedical research: a retrospective cohort study.

NRCBL: National Reference Center for Bioethics Literature Classification Scheme   See inside front cover for terms.

245

BMC Public Health 2006 June 22; 6: 165. NRCBL: 18.1; 5.3; 9.3.1; 18.2.

Deming, Nicole; Fryer-Edwards, Kelly; Dudzinski, Denise; Starks, Helene; Culver, Julie; Hopley, Elizabeth; Robins, Lynne; Burke, Wylie. Incorporating principles and practical wisdom in research ethics education: a preliminary study. *Academic Medicine* 2007 January; 82(1): 18-23. NRCBL: 18.1; 2.3. SC: em.

Emanuel, Ezekiel J.; Miller, Franklin G. Money and distorted ethical judgments about research: ethical assessment of the TeGenero TGN1412 trial. *American Journal of Bioethics* 2007 February; 7(2): 76-81. NRCBL: 18.1; 9.7; 18.3; 9.3.1; 1.3.9; 7.3.

> Abstract: The recent TeGenero phase I trial of a novel monoclonal antibody in healthy volunteers produced a drastic inflammatory reaction in participants receiving the experimental agent. Commentators on the ethics of the research have focused considerable attention on the role of financial considerations: the for-profit status of the biotechnology company and Contract Research Organization responsible respectively for sponsoring and conducting the trial and the amount of monetary compensation to participants. We argue that these financial considerations are largely irrelevant and distort ethical appraisal of this tragic research. Except for administering the antibody to all 6 participants nearly simultaneously, the trial appears to fulfill all of the critical ethical requirements for clinical research—social value, scientific validity, fair subject selection, favorable risk-benefit ratio, independent review, informed consent, and respect for enrolled participants.

Ferguson, Pamela R. Human 'guinea pigs': why patients participate in clinical trials. *In:* McLean, Sheila A.M., ed. First Do No Harm: Law, Ethics, and Healthcare. Aldershot, England; Burlington, VT: Ashgate, 2006: 165-185. NRCBL: 18.1; 7.1. SC: em.

Frewer, Andreas. Medical research, morality, and history: the German journal Ethik and the limits of human experimentation. *In:* LaFleur, William R.; Böhme, Gernot; Shimazono, Susumu, eds. Dark Medicine: Rationalizing Medical Research. Bloomington: Indiana University Press, 2007: 30-45. NRCBL: 18.1; 2.2; 10; 18.2; 1.3.5; 15.5; 21.4.

Grady, Christine. Ethical principles in clinical research. *In:* Gallin, John I.; Ognibene, Frederick P., eds. Principles and Practice of Clinical Research. 2nd edition. Oxford: Academic, 2007: 15-26. NRCBL: 18.1; 2.1; 18.2.

Hale, Benjamin. Risk, judgment and fairness in research incentives. *American Journal of Bioethics* 2007 February; 7(2): 82-83. NRCBL: 18.1; 9.7; 1.3.9; 7.3. Comments: Ezekiel J. Emanuel and Franklin G. Miller. Money and distorted ethical judgments about research: ethical assessment of the TeGenero TGN1412 trial. American Journal of Bioethics 2007 February; 7(2): 76-81.

Hardell, Lennart; Walker, Martin J.; Walhjalt, Bo; Friedman, Lee S.; Richter, Elihu D. Secret ties to industry and conflicting interests in cancer research. *American Journal of Industrial Medicine* 2007 March; 50(3): 227-233 [See correction in: American Journal of Industrial Medicine 2007 March; 50(3): 234]. NRCBL: 18.1; 1.3.2; 9.5.1; 7.1; 9.3.1.

Hyder, Adnan A.; Harrison, Rachel A.; Kass, Nancy; Maman, Suzanne. A case study of research ethics capacity development in Africa. *Academic Medicine* 2007 July; 82(7): 675-683. NRCBL: 18.1; 21.1; 2.3. SC: em; cs.

Khalil, Susan S.; Silverman, Henry J.; Raafat, May; El-Kamary, Samer; El-Setouhy, Maged. Attitudes, understanding, and concerns regarding medical research amongst Egyptians: a qualitative pilot study. *BMC Medical Ethics [electronic]* 2007; 8(9): 12 p. Accessed: http://www.biomedcentral.com/content/pdf/1472-6939-8-9.pdf [2007 December 18]. NRCBL: 18.1; 18.3. SC: em.

Kimmelman, Jonathan. The therapeutic misconception at 25: treatment, research, and confusion. *Hastings Center Report* 2007 November-December; 37(6): 36-42. NRCBL: 18.1; 18.3; 8.2; 5.2.

> Abstract: "Therapeutic misconception" has been misconstrued, and some of the newer, mistaken interpretations are troublesome. They exaggerate the distinction between research and treatment revealing problems in the foundations of research ethics and possibly weakening informed consent.

King, Patricia A.; Areen, Judith; Gostin, Lawrence O. The human body. *In their:* Law, Medicine and Ethics. New York: Foundation Press, 2006: 208-369. NRCBL: 18.1; 4.4; 2.2; 18.2. SC: le.

Klitzman, Robert; Albala, Ilene; Siragusa, Joseph; Nelson, Kristen N.; Appelbaum, Paul S. The reporting of monetary compensation in research articles. *Journal of Empirical Research on Human Research Ethics* 2007 December; 2(4): 61-67. NRCBL: 18.1; 7.3; 9.3.1; 8.2; 1.3.7; 18.2. SC: em.

> Abstract: Study participant compensation is of increasing concern, yet few investigations have explored it; none have examined whether published journal articles report it. Medline searches for articles in six areas—HIV, substance abuse (heroin and cocaine), depression, essential hypertension, and cardiac surgery—reveal very low mention of payment (0–32.1%). Of 207 articles, only 13.5% mentioned financial compensation in any way, and only 11.1% listed amounts. Of the 207 studies, 92 involved more than minimal risk interventions, but were not more likely to mention compensation. Studies that included substance users were significantly more likely than others to mention payment (p .001). These overall low rates are concerning as they can hamper evaluation of ethical issues, and impact study replicability. Publication requirements should consider discussion of compensation.

Kvochak, Patricia A. Legal issues. *In:* Gallin, John I.; Ognibene, Frederick P., eds. Principles and Practice of Clinical Research. 2nd edition. Oxford: Academic, 2007: 109-120. NRCBL: 18.1; 1.3.9; 7.3; 8.4; 18.3. SC: le.

LaFleur, William R. Refusing utopia's bait: research, rationalizations, and Hans Jonas. *In:* LaFleur, William R.; Böhme, Gernot; Shimazono, Susumu, eds. Dark Medicine: Rationalizing Medical Research. Bloomington: Indiana University Press, 2007: 233-245. NRCBL: 18.1; 1.3.9; 21.4.

Lenzer, Jeanne. Nigeria files criminal charges against Pfizer. *BMJ: British Medical Journal* 2007 June 9; 334(7605): 1181. NRCBL: 18.1; 1.3.8; 9.7. SC: le.

Liao, S. Matthew; Goldschmidt-Clermont, Pascal J.; Sugarman, Jeremy. Ethical and policy issues relating to progenitor-cell-based strategies for prevention of atherosclerosis. *Journal of Medical Ethics* 2007 November; 33(11): 643-646. NRCBL: 18.1; 9.1; 9.4; 19.5. Note: A report of the Working Group on Ethics of Progenitor Cellbased Strategies for Disease Prevention.

Abstract: OBJECTIVE: To examine important ethical and societal issues relating to the use of progenitor-cell-based strategies for disease prevention, particularly atherosclerosis. BACKGROUND: Several nascent lines of evidence suggest the feasibility of using progenitor cells to reverse the health consequence of atherosclerosis. Such potential uses of progenitor cells are scientifically exciting, yet they raise important ethical and societal issues. METHOD: The Working Group on Ethics of Progenitor Cell-based Strategies for Disease Prevention met to discuss the relevant issues. Several drafts of a report were then circulated to the entire Working Group for comments until a consensus was reached. RESULTS: Scientific evidence suggests the appropriateness of using progenitor-cell-based strategies for some rare conditions involving atherosclerosis, but additional preclinical data are needed for other, more prevalent conditions before human trials begin. All such trials raise a set of ethical issues, especially since trials aimed at prevention rather than treatment may involve persons who do not yet have disease but will be exposed to the risks of interventions. In addition, enrolment in prevention trials may be hazardous and harmful if participants erroneously believe experimental interventions will necessarily prevent disease. Finally, given the high prevalence of atherosclerosis, there are some important public policy implications of taking such an approach to prevention, including the sources of progenitor cells for such interventions as well as the allocation of health resources. CONCLUSION: Potential uses of progenitor-cell-based strategies for preventing atherosclerosis must be considered in the context of a range of social and ethical issues.

London, Alex John. Two dogmas of research ethics and the integrative approach to human-subjects research. *Journal of Medicine and Philosophy* 2007 March-April; 32(2): 99-116. NRCBL: 18.1; 1.1; 18.2. SC: an.

Abstract: This article argues that lingering uncertainty about the normative foundations of research ethics is perpetuated by two unfounded dogmas of research ethics. The first dogma is that clinical research, as a social activity, is an inherently utilitarian endeavor. The second dogma is that an acceptable framework for research ethics must impose constraints on this endeavor whose moral force is grounded in role-related obligations of either physicians or researchers. This article argues that these dogmas are common to traditional articulations of the equipoise requirement and to recently articulated alternatives, such as the non-exploitation approach. Moreover, important shortcomings of these approaches can be traced to their acceptance of these dogmas. After highlighting these shortcomings, this article illustrates the benefits of rejecting these dogmas by sketching the broad outlines of an alternative called the "integrative approach" to clinical research.

Madsen, S.M.; Holm, S.; Riis, P. Attitudes towards clinical research among cancer trial participants and non-participants: an interview study using a grounded theory approach. *Journal of Medical Ethics* 2007 April; 33(4): 234-240. NRCBL: 18.1; 18.3; 18.5.3. SC: em.

Abstract: The attitudes of women patients with cancer were explored when they were invited to participate in one of three randomised trials that included chemotherapy at two university centres and a satellite centre. Fourteen patients participating in and 15 patients declining trials were interviewed. Analysis was based on the constant comparative method. Most patients voiced positive attitudes towards clinical research, believing that trials are necessary for further medical development, and most spontaneously argued that participation is a moral obligation. Most trial decliners, however, described a radical change in focus as they faced the actual personal choice. Almost no one got an impression of clinical equipoise between treatments in the trials, and most patients expressed discomfort with randomisation. A patient's choice to participate was mainly determined by whether the primary focus was on treatment effect or on adverse effects. Both knowledge about and feelings towards trials originated mostly from the media, although paradoxically the media were largely seen as untrustworthy. Mistrust was shown towards the pharmaceutical industry, and although most patients originally trusted that doctors primarily pursued the interest of patients, they did not trust the adequacy of doctors or industry in maintaining self-regulation. Thus, public control measures were judged to be essential.

Maschke, Karen J. The pressure to tolerate risk in human subjects research [review of Lesser Harms: The Morality of Risk in Medical Research by Sydney A. Halpern]. *Medical Humanities Review* 2005 Spring-Fall; 19(1-2): 39-44. NRCBL: 18.1; 1.3.9; 18.6; 18.2; 9.5.7.

Mason, J.K.; Laurie, G.T. Biomedical human research and experimentation. *In their:* Mason and McCall Smith's Law and Medical Ethics. Seventh ed. Oxford; New York: Oxford University Press, 2005: 648-684. NRCBL: 18.1; 1.3.9; 9.3.1; 18.2; 18.3. SC: le.

NRCBL: National Reference Center for Bioethics Literature Classification Scheme   See inside front cover for terms.

247

Nussenblatt, Robert B.; Gottesman, Michael M. Rules to prevent conflict of interest for clinical investigators conducting human subjects research. *In:* Gallin, John I.; Ognibene, Frederick P., eds. Principles and Practice of Clinical Research. 2nd edition. Oxford: Academic, 2007: 121-127. NRCBL: 18.1; 1.3.9; 7.3; 18.2.

Phillips, Trisha B. Money, advertising and seduction in human subjects research. *American Journal of Bioethics* 2007 February; 7(2): 88-90. NRCBL: 18.1; 18.2; 9.3.1; 1.3.9; 7.3. Comments: Ezekiel J. Emanuel and Franklin G. Miller. Money and distorted ethical judgments about research: ethical assessment of the TeGenero TGN1412 trial. American Journal of Bioethics 2007 February; 7(2): 76-81.

Piccart, Martine; Goldhirsch, Aron. Keeping faith with trial volunteers: how best to serve patients' interests in large clinical trials? [commentary]. *Nature* 2007 March 8; 446(7132): 137-138. NRCBL: 18.1; 1.3.9; 9.7.

Plemmons, Dena K.; Kalichman, Michael W. Reported goals for knowledge to be learned in responsible conduct of research courses. *Journal of Empirical Research on Human Research Ethics* 2007 June; 2(2): 57-66. NRCBL: 18.1; 1.3.9; 7.2. SC: em.

Abstract: Education in responsible conduct of research (RCR) has been a required part of training for students on U.S. National Institutes of Health (NIH) training grants for over 15 years. However, there is little evidence of commonly accepted goals for RCR instruction, making it difficult to assess effectiveness. As part of a larger study examining RCR instructors' goals for RCR education, this report focuses on those reported goals categorized as knowledge. To identify RCR instructors, e-mail requests were sent to the 116 recipients of NIH training grants awarded in 2000. Of 67 verified RCR instructors, 50 (75% response rate) from 37 different institutions were successfully interviewed. Despite a shared sense of the basics to be taught in RCR courses, these instructors were diverse in their views and understanding of goals for RCR education. This diversity suggests a challenge to be overcome not only for improving the effectiveness of RCR education, but also for attempts to assess that effectiveness.

Qiu, Jane. To walk again. *New Scientist* 2007 November 10-16; 196(2629): 57-59. NRCBL: 18.1; 18.5.1. Identifiers: China.

Resnik, David B. Intentional exposure studies of environmental agents on human subjects: assessing benefits and risks. *Accountability in Research* 2007 January-March; 14(1): 35-55. NRCBL: 18.1; 16.1; 18.2; 22.2.

Richardson, A.; Sitton-Kent, L. Research ethics. *In:* Holland, Stephen, ed. Introducing Nursing Ethics: Themes in Theory and Practice. Salisbury: APS, 2004: 131-150. NRCBL: 18.1; 4.1.3. SC: cs.

Shamoo, Adil E.; Schwartz, Jack. Universal and uniform protections of human subjects in research. *American Journal of Bioethics* 2007 December; 7(12): 7-9. NRCBL: 18.1; 18.6.

Shamoo, Adil; Woeckner, Elizabeth. Ethical flaws in the TeGenero trial. *American Journal of Bioethics* 2007 February; 7(2): 90-92. NRCBL: 18.1; 18.6; 9.7; 9.3.1; 1.3.9; 7.3. Comments: Ezekiel J. Emanuel and Franklin G. Miller. Money and distorted ethical judgments about research: ethical assessment of the TeGenero TGN1412 trial. American Journal of Bioethics 2007 February; 7(2): 76-81.

Shapshay, Sandra; Pimple, Kenneth D. Participation in biomedical research is an imperfect moral duty: a response to John Harris. *Journal of Medical Ethics* 2007 July; 33(7): 414-417. NRCBL: 18.1; 1.1; 1.3.9. SC: an. Comments: John Harris. Scientific research is a moral duty. Journal of Medical Ethics 2005; 31: 242-248.

Abstract: In his paper "Scientific research is a moral duty", John Harris argues that individuals have a moral duty to participate in biomedical research by volunteering as research subjects. He supports his claim with reference to what he calls the principle of beneficence as embodied in the "rule of rescue" (the moral obligation to prevent serious harm), and the principle of fairness embodied in the prohibition on "free riding" (we are obliged to share the sacrifices that make possible social practices from which we benefit). His view that biomedical research is an important social good is agreed upon, but it is argued that Harris succeeds only in showing that such participation and support is a moral good, among many other moral goods, while failing to show that there is a moral duty to participate in biomedical research in particular. The flaws in Harris's arguments are detailed here, and it is shown that the principles of beneficence and fairness yield only a weaker discretionary or imperfect obligation to help others in need and to reciprocate for sacrifices that others have made for the public good. This obligation is discretionary in the sense that the individuals are free to choose when, where, and how to help others in need and reciprocate for earlier sacrifices. That Harris has not succeeded in claiming a special status for biomedical research among all other social goods is shown here.

Singapore. Bioethics Advisory Committee. Personal Information in Biomedical Research: a report by the Bioethics Advisory Committee. Bioethics Advisory Committee 2007 May: 48 p.; Appendices 141 p. NRCBL: 18.1; 8.4; 18.3; 1.3.12.

Spielman, Bethany. Faulty premise, premature conclusion: that money was extraneous to the research ethics of the TGN1412 study. *American Journal of Bioethics* 2007 February; 7(2): 93-94. NRCBL: 18.1; 18.6; 9.3.1. Comments: Ezekiel J. Emanuel and Franklin G. Miller. Money and distorted ethical judgments about research: ethical assessment of the TeGenero TGN1412 trial. American Journal of Bioethics 2007 February; 7(2): 76-81.

SC (Subject Captions):   an=analytical   cs=case studies   em=empirical   le=legal   po=popular   rv=review

Svensson, Sara; Hansson, Sven Ove. Protecting people in research: a comparison between biomedical and traffic research. *Science and Engineering Ethics* 2007 March; 13(1): 99-115. NRCBL: 18.1; 1.3.9; 18.3.

United States. Environmental Protection Agency [EPA]. Science Advisory Board. Comments on the use of Data from the Testing of Human Subjects: A Report by the Science Advisory Board and the FIFRA Scientific Advisory Panel. Washington, DC: Environmental Protection Agency, 2000 September 11: 40 p. [Online]. Accessed: http://www.epa.gov/sab/pdf/ec0017.pdf [2007 October 24]. NRCBL: 18.1; 18.5.1. Identifiers: Federal Insecticide, Fungicide, and Rodenticide Act.

HUMAN EXPERIMENTATION/ ETHICS COMMITTEES AND POLICY GUIDELINES

Safeguarding clinical trials [editorial]. *Nature Medicine* 2007 February; 13(2): 107. NRCBL: 18.2; 18.6; 21.1. Identifiers: European Union; Clinical Trial Directive.

Aagaard-Hansen, Jens; Johansen, Maria Vang.; Riis, Pols. Research ethical challenges in cross-disciplinary and cross-cultural health research: the diversity of codes. *Danish Medical Bulletin* 2004 February; 51(1): 117-120. NRCBL: 18.2; 18.6; 21.7; 6.

Akabayashi, Akira; Slingsby, Brian T.; Nagao, Noriko; Kai, Ichiro; Sato, Hajime. An eight-year follow-up national study of medical school and general hospital ethics committees in Japan. *BMC Medical Ethics* 2007; 8(8), 8 p. [Online]. Accessed: http://www.biomedcentral.com/1472-6939-8-8 [2007 July 20]. NRCBL: 18.2; 9.6; 8.3.4; 1.2. SC: em.

Abstract: BACKGROUND: Ethics committees and their system of research protocol peer-review are currently used worldwide. To ensure an international standard for research ethics and safety, however, data is needed on the quality and function of each nation's ethics committees. The purpose of this study was to describe the characteristics and developments of ethics committees established at medical schools and general hospitals in Japan. METHODS: This study consisted of four national surveys sent twice over a period of eight years to two separate samples. The first target was the ethics committees of all 80 medical schools and the second target was all general hospitals with over 300 beds in Japan (n = 1457 in 1996 and n = 1491 in 2002). Instruments contained four sections: (1) committee structure, (2) frequency of annual meetings, (3) committee function, and (4) existence of a set of guidelines for the refusal of blood transfusion by Jehovah's Witnesses. RESULTS: Committee structure was overall interdisciplinary. Frequency of annual meetings increased significantly for both medical school and hospital ethics committees over the eight years. The primary activities for medical school and hospital ethics committees were research protocol reviews and policy making. Results also showed a significant increase in the use of ethical guidelines, particularly those related to the refusal of blood transfusion by Jehovah's Witnesses,

among both medical school and hospital ethics committees. CONCLUSION: Overall findings indicated a greater recognized degree of responsibilities and an increase in workload for Japanese ethics committees.

American Association of University Professors. Protecting human beings: institutional review boards and social science research. *Academe* 2001 May/June; 87(3): 55-67. NRCBL: 18.2; 1.3.9; 1.3.3; 1.3.5.

Anderson, Melissa S.; Horn, Aaron S.; Risbey, Kelly R.; Ronning, Emily A.; De Vries, Raymond; Martinson, Brian C. What do mentoring and training in the responsible conduct of research have to do with scientists' misbehavior? Findings from a national survey of NIH-funded scientists. *Academic Medicine* 2007 September; 82(9): 853-860. NRCBL: 18.2; 2.3; 1.3.9. SC: em.

Anscombe, G.E.M. Sins of omission? The non-treatment of controls in clinical trials. *In her:* Human Life, Action and Ethics: Essays. Exeter: Imprint Academic, 2005: 286-291. NRCBL: 18.2; 1.1.

Barata, Paula C.; Gucciardi, Enza; Ahmad, Farah; Stewart, Donna E. Cross-cultural perspectives on research participation and informed consent. *Social Science and Medicine* 2006 January; 62(2): 479-490. NRCBL: 18.2; 18.3; 21.7. SC: cs; em.

Becker, Gary J. Financial relationships with industry and device research involving non-Food and Drug Administration-approved use: a perspective. *Radiology* 2006 June; 239(3): 626-628. NRCBL: 18.2; 9.3.1; 1.3.2; 7.3; 18.6.

Bhutta, Zulfiqar Ahmed. Ethics in international health research: a perspective from the developing world. *Bulletin of the World Health Organization* 2002; 80(2): 114-120. NRCBL: 18.2; 18.5.9; 18.6; 18.3; 2.1.

Boudoulas, Harisios. Ethics in biomedical research. *Hellenic Journal of Cardiology* 2006 May-June; 47(3): 193. NRCBL: 18.2; 1.3.9.

Bramstedt, Katrina A.; Ford, Paul J. Protecting human subjects in neurosurgical trials: the challenge of psychogenic dystonia. *Contemporary Clinical Trials* 2006 April; 27(2): 161-164. NRCBL: 18.2; 18.3; 18.5.6; 7.1; 17.1.

Brandt, Michelle L. IRB burden studied in cost analysis. *Stanford Report* 2003 August 6: 3 p. [Online]. Accessed: http://news.service.standford.edu/news/2003/august6/humphreys.html [2007 December 13]. NRCBL: 18.2; 9.3.1; 9.5.1; 18.5.1.

Branson, Richard D.; Davis, Kenneth, Jr.; Butler, Karyn L. African Americans' participation in clinical research: importance, barriers, and solutions. *American Journal of Surgery* 2007 January; 193(1): 32-39; discussion 40. NRCBL: 18.2; 18.5.1.

Braunschweiger, Paul; Goodman, Kenneth W. The CITI program: an international online resource for educa-

NRCBL: National Reference Center for Bioethics Literature Classification Scheme   See inside front cover for terms.

249

tion in human subjects protection and the responsible conduct of research. *Academic Medicine* 2007 September; 82(9): 861-864. NRCBL: 18.2; 2.3.

Brown, Stephen D.; Daly, Jennifer C.; Kalish, Leslie A.; McDaniel, Samuel A. Financial disclosures of scientific papers presented at the 2003 RSNA Annual Meeting: association with reporting of non-Food and Drug Administration-approved uses of industry products. *Radiology* 2006 June; 239(3): 849-855. NRCBL: 18.2; 9.3.1; 1.3.2; 7.3; 18.6; 1.3.7.

Brown, Susan. Use of research-ethics boards is growing in Africa, study finds [news]. *Chronicle of Higher Education* 2007 February 2; 53(22): A13. NRCBL: 18.2; 1.3.9; 21.1.

Bulger, Ruth Ellen; Heitman, Elizabeth. Expanding responsible conduct of research instruction across the university. *Academic Medicine* 2007 September; 82(9): 876-878. NRCBL: 18.2; 2.3; 1.3.9.

Carlson, Robert V.; van Ginneken, Nadja H.; Pettigrew, Luisa M.; Davies, Alan; Boyd, Kenneth M.; Webb, David J. The three official language versions of the Declaration of Helsinki: what's lost in translation? *Journal of Medical Ethics* 2007 September; 33(9): 545-548. NRCBL: 18.2; 21.1.

Abstract: Background: The Declaration of Helsinki, the World Medical Association's (WMA's) statement of ethical guidelines regarding medical research, is published in the three official languages of the WMA: English, French and Spanish. Methods: A detailed comparison of the three official language versions was carried out to determine ways in which they differed and ways in which the wording of the three versions might illuminate the interpretation of the document. Results: There were many minor linguistic differences between the three versions. However, in paragraphs 1, 6, 29, 30 and in the note of clarification to paragraph 29, there were differences that could be considered potentially significant in their ethical relevance. Interpretation: Given the global status of the Declaration of Helsinki and the fact that it is translated from its official versions into many other languages for application to the ethical conduct of research, the differences identified are of concern. It would be best if such differences could be eliminated but, at the very least, a commentary to explain any differences that are unavoidable on the basis of language or culture should accompany the Declaration of Helsinki. This evidence further strengthens the case for international surveillance of medical research ethics as has been proposed by the WMA.

Carson, P.A.; Holt, J. Ethics of studies involving human volunteers. I. Historical background. *Journal of Cosmetic Science* 2006 May-June; 57(3): 215-221. NRCBL: 18.2; 18.3; 18.5.1; 18.6.

Castellano, Marlene Brant. Ethics of Aboriginal research. *Journal of Aboriginal Health* 2004 January; 1(1):

98-114 [Online]. Accessed: http://www.naho.ca/english/pdf/journal_p98-114.pdf [2007 March 29]. NRCBL: 18.2; 18.5.9; 21.7; 18.6.

Catholic Medical Association; National Catholic Bioethics Center. Catholic principles and guidelines for clinical research. *National Catholic Bioethics Quarterly* 2007 Spring; 7(1): 153-165. NRCBL: 18.2; 18.5.1; 1.2; 2.4.

Center for Advanced Study. Center for Advanced Study Project Steering Committee; Gunsalus, C.K.; Bruner, Edward M.; Burbules, Nicholas C.; Dash, Leon; Finkin, Matthew; Goldberg, Joseph P.; Greenough, William; Miller, Gregory A.; Pratt, Michael G.; Iriye, Masumi; Aronson, Deb. Illinois White Paper. Improving the system for protecting human subjects: counteracting IRB "mission creep". University of Illinois at Urbana-Champaign. Center for Advanced Study 2005 November 17: 32 p. [Online]. Accessed: http://www.law.uiuc.edu/conferences/whitepaper/whitepaper.pdf [2007 December 13]. NRCBL: 18.2. SC: rv. Conference: Human Subject Protection Regulations and Research Outside the Biomedical Sphere; Champaign, Illinois; 2003 April 11-12; University of Illinois Center for Advanced Study, Colleges of Law and Liberal Arts and Sciences, the Vice Chancellor for Research.

Clarke, Amanda. Qualitative interviewing: encountering ethical issues and challenges. *Nurse Researcher* 2006; 13(4): 19-29. NRCBL: 18.2; 18.5.7; 18.3; 8.1.

Clinton, William J. Memorandum of March 27, 1997 — strengthened protections for human subjects of classified research. *Federal Register* 1997 May 13; 62(92): 26367-26372 [Online]. Accessed: http://frwebgate.access.gpo.gov/cgi-bin/multidb.cgi [2005 December 28]. NRCBL: 18.2; 18.3; 18.6.

Cohen, Patricia. As ethics panels expand, no research field is exempt. *New York Times* 2007 February 28; p. A15. NRCBL: 18.2. SC: po.

Colt, Henri G.; Mulnard, Ruth A. Writing an application for a human subjects institutional review board. *Chest* 2006 November; 130(5): 1605-1607. NRCBL: 18.2.

Consortium to Examine Clinical Research Ethics; Speckman, Jeanne L.; Byrne, Margaret M.; Gerson, Jason; Getz, Kenneth; Wangsmo, Gary; Muse, Carianne T.; Sugarman, Jeremy. Determining the costs of institutional review boards. *IRB: Ethics and Human Research* 2007 March-April; 29(2): 7-13. NRCBL: 18.2; 9.3.1. SC: em.

Corbie-Smith, Giselle M.; Durant, Raegan W.; St George, Diane Marie M. Investigators' assessment of NIH mandated inclusion of women and minorities in research. *Contemporary Clinical Trials* 2006 December; 27(6): 571-579. NRCBL: 18.2; 18.5.3; 1.3.5; 18.6.

Davies, Hugh. Ethical reflections on Edward Jenner's experimental treatment. *Journal of Medical Ethics* 2007 March; 33(3): 174-176. NRCBL: 18.2; 2.2; 9.5.1; 9.7.

Abstract: In 1798 Dr Edward Jenner published his famous account of "vaccination". Some claim that a Research Ethics Committee, had it existed in the 1790s, might have rejected his work. I provide the historical context of his work and argue that it addressed a major risk to the health of the community, and, given the devastating nature of smallpox and the significant risk of variolation, the only alternative preventative measure, Jenner's study had purpose, justification and a base in the practice of the day.

Dawson, Angus J.; Yentis, Steve M. Contesting the science/ethics distinction in the review of clinical research. *Journal of Medical Ethics* 2007 March; 33(3): 165-167. NRCBL: 18.2. SC: an.

Abstract: Recent policy in relation to clinical research proposals in the UK has distinguished between two types of review: scientific and ethical. This distinction has been formally enshrined in the recent changes to research ethics committee (REC) structure and operating procedures, introduced as the UK response to the EU Directive on clinical trials. Recent reviews and recommendations have confirmed the place of the distinction and the separate review processes. However, serious reservations can be mounted about the science/ethics distinction and the policy of separate review that has been built upon it. We argue here that, first, the science/ethics distinction is incoherent, and, second, that RECs should not only be permitted to consider a study's science, but that they have an obligation do so.

de Melo-Martín, Immaculada; Palmer, Larry I.; Fins, Joseph J. Developing a research ethics consultation service to foster responsive and responsible clinical research. *Academic Medicine* 2007 September; 82(9): 900-904. NRCBL: 18.2; 9.6.

De Ville, Kenneth; Hassler, Gregory; Lewis, Michael J. Rejuvenating a foundering institutional review board: one institution's story. *Academic Medicine* 2007 January; 82(1): 11-17. NRCBL: 18.2; 18.6.

DeIorio, Nicole M.; McClure, Katie B.; Nelson, Maria; McConnell, K. John; Schmidt, Terri A. Ethics committee experience with emergency exception from informed consent protocols. *Journal of Empirical Research on Human Research Ethics* 2007 September; 2(3): 23-30. NRCBL: 18.2; 18.3; 18.6. SC: em.

Abstract: Since 1996, U.S. federal regulations allow research without informed consent to study emergency conditions, if there is currently no satisfactory treatment for the condition, no time to obtain advance consent from the patient or representative, and if there is community involvement through a public disclosure and community consultation process. REB experiences since then are unknown. We surveyed REB chairpersons at the 126 United States medical schools to quantify reviewed protocols and identify attitudes about the rule, to better understand the rule's impact on REBs. Sixty-nine surveys were returned (55%). Fifty-two respondents reviewing human research had heard of the Rule. Forty-eight percent (25/52) had reviewed such a study; 40% of those had rejected at least one. Seventy-eight percent believe the rule protects human subjects, and 88% feel prepared to implement them. REB views differed from public opinion on how best to enact notification and consultation.

DeMets, David L.; Fost, Norman; Powers, Madison. An Institutional Review Board dilemma: responsible for safety monitoring but not in control. *Clinical Trials* 2006; 3(2): 142-148. NRCBL: 18.2; 18.6.

Ding, Eric L.; Powe, Neil R.; Manson, JoAnn E.; Sherber, Noëlle S.; Braunstein, Joel B. Sex differences in perceived risks, distrust, and willingness to participate in clinical trials. *Archives of Internal Medicine* 2007 May 14; 167(9): 905-912. NRCBL: 18.2; 18.5.1; 18.5.3; 10. SC: em.

Abstract: Background: Multiple sex differences exist in cardiovascular disease burden and treatment efficacies; adequate participation of both sexes is crucial to clinical research. Methods: A multicenter, double-blind, randomized study evaluated sex and trial scenarios on willingness to participate (WTP) in cardiovascular prevention trials and examined sex differences in perceived risks and distrust. Hypothetical trial scenarios randomized multifactorial vignettes of adverse effects, trial durations, sponsors, financial incentives, and conflicts of interest. Results: With 783 participants across 13 clinical centers, women showed lower distrust of medical researchers, perceived greater risk of myocardial infarction, and perceived greater risk of harm from trial participation than men. Men had 15% greater WTP than women (33.1% vs 28.7%; relative risk [RR], 1.15; 95% confidence interval [CI], 1.02-1.31); adjusting for explanatory mediators, we found that sex differences in perceived risks and benefits explained the sex gap in WTP. Although greater perceived probability of harm (RR, 0.41; 95% CI, 0.23-0.72), health benefit (RR, 2.99; 95% CI, 1.63-5.46), and quality of care (RR, 1.71; 95% CI, 1.12-2.61) strongly predicted WTP (for perceived probabilities 80% vs %) similarly in both sexes, and perceptions of distrust and myocardial infarction risk predicted WTP differently between sexes (P.01 for interactions), age, history of coronary artery disease, hypertension, and diabetes mellitus increased WTP in men but not in women (P.05 for sex interactions). Compared with no financial conflict, disclosure of investigator patent ownership increased WTP in women, while it decreased WTP in men (P = .02 for sex interaction). Monetary incentives were overall more effective on WTP in women (P = .03 for sex interaction). Conclusions: In this multicenter study, women perceived greater risk of harm and myocardial infarction and showed lower WTP in cardiovascular prevention trials. Evidence underscores the importance of sex in influencing clinical trial enrollment.

Dingwall, Robert. An exercise in fatuity: research governance and the emasculation of HSR [editorial]. *Journal of Health Services Research and Policy* 2006 October; 11(4):

NRCBL: National Reference Center for Bioethics Literature Classification Scheme   See inside front cover for terms.

251

193-194. NRCBL: 18.2; 5.3; 9.3.1; 9.2; 9.4. Identifiers: health services research.

Dinnett, Eleanor M.; Mungall, Moira M.B.; Kent, Jane A.; Ronald, Elizabeth S.; McIntyre, Karen E.; Anderson, Elizabeth; Gaw, Allan. Unblinding of trial participants to their treatment allocation: lessons from the Prospective Study of Pravastatin in the Elderly at Risk (PROSPER). *Clinical Trials* 2005 June; 2(3): 254-259. NRCBL: 18.2; 18.5.7. SC: em.

Djulbegovic, Benjamin. Articulating and responding to uncertainties in clinical research. *Journal of Medicine and Philosophy* 2007 March-April; 32(2): 79-98. NRCBL: 18.2; 1.1. SC: an.

Abstract: This paper introduces taxonomy of clinical uncertainties and argues that the choice of scientific method should match the underlying level of uncertainty. Clinical trial is one of these methods aiming to resolve clinical uncertainties. Whenever possible these uncertainties should be quantified. The paper further shows that the still ongoing debate about the usage of "equipoise" vs. "uncertainty principle" vs. "indifference" as an entry criterion to clinical trials actually refers to the question "whose uncertainty counts". This question is intimately linked to the control of research agenda, which is not quantifiable and hence is not solvable to equal acceptability to all interested parties. The author finally shows that there is a predictable relation between [acknowledgement of] uncertainty (the moral principle) on which trials are based and the ultimate outcomes of clinical trials. That is, [acknowledgement of] uncertainty determines a pattern of success in medicine and drives clinical discoveries.

Duval, Gordon. The benefits and threats of research partnerships with industry. *Critical Care (London, England)* 2005 August; 9(4): 309-310. NRCBL: 18.2; 9.7; 9.3.1.

Dyrbye, Liselotte N.; Thomas, Matthew R.; Mechaber, Alex J.; Eacker, Anne; Harper, William; Massie, F. Stanford; Power, David V.; Shanafelt, Tait D. Medical education research and IRB review: an analysis and comparison of the IRB review process at six institutions. *Academic Medicine* 2007 July; 82(7): 654-660. NRCBL: 18.2; 7.2. SC: em.

Effa, Pierre; Massougbodji, Achille; Ntoumi, Francine; Hirsch, François; Debois, Henri; Vicari, Marissa; Derme, Assetou; Ndemanga-Kamoune, Jacques; Nguembo, Joseph; Impouma, Benido; Akué, Jean-Paul; Ehouman, Armand; Dieye, Alioune; Kilama, Wen. Ethics committees in western and central Africa: concrete foundations. *Developing World Bioethics* 2007 December; 7(3): 136-142. NRCBL: 18.2; 18.5.9; 21.1.

Abstract: The involvement of developing countries in international clinical trials is necessary for the development of appropriate medicines for local populations. However, the absence of appropriate structures for ethical review represents a barrier for certain countries. Currently there is very little information available on existing structures dedicated to ethics in western and central Africa. This article briefly describes historical milestones in the development of networks dedicated to capacity building in ethical review in these regions and outlines the major conclusions of two workshops on this issue, which were held in September and October 2002 in Libreville, Gabon, and Paris, France. The workshops were the culmination of collaboration between the African Malaria Network Trust (AMANET) and the Pan African Bioethics Initiative (PABIN). They produced an update on ethics organizations with regard to mission, function, activities, members, and contact people, in eight countries within the regions discussed. As a result of the commitment of mandated delegates, a further prominent outcome followed these workshops: the creation of national structures, where none existed before, dedicated to the ethical review of clinical trials.

Elsayed, Dya Eldin M.; Kass, Nancy E. Assessment of the ethical review process in Sudan. *Developing World Bioethics* 2007 December; 7(3): 143-148. NRCBL: 18.2; 21.1. SC: em.

Abstract: The ethical review process is an important component of contemporary health research worldwide. Sudan started an ethical review process rather late in comparison with other countries. In this study, we evaluate the structure and functions of existing ethics review committees. We also explore the knowledge and attitudes of Sudanese researchers toward the ethical review process and their experience with existing ethics review committees. There are four ethics review committees in the country; these committees have no institutional regulations to govern their functions. Furthermore, Sudan also lacks national guidelines. Ethical reviews are carried out primarily for studies seeking international funding and are almost always governed by the funding agencies' requirements. Nearly half of respondents (46.3%) knew about the existence of research ethics committees in Sudan. Researchers reported a variety of experiences with the ethical review process; most of them were unable to define 'ethics committee'.

Emanuel, Ezekiel J.; Lemmens, Trudo; Elliot, Carl. Should society allow research ethics boards to be run as for-profit enterprises? *PLoS Medicine* 2006 July; 3(7): e309. NRCBL: 18.2; 9.3.1; 1.3.2.

Epstein, M.; Wingate, D.L. Is the NHS research ethics committees system to be outsourced to a low-cost offshore call centre? Reflections on human research ethics after the Warner Report. *Journal of Medical Ethics* 2007 January; 33(1): 45-47. NRCBL: 18.2. Identifiers: Great Britain (United Kingdom); National Health Service.

Abstract: The recently published Report of the AHAG on the Operation of NHS Research Ethics Committees (the Warner Report) advocates major reforms of the NHS research ethics committees system. The main implications of the proposed changes and their probable effects on the major stakeholders are described.

Espirit Group; Pace, Christine; Grady, Christine; Wendler, David; Bebchuk, Judith D.; Tavel, Jorge A.; McNay, Laura A.; Forster, Heidi P.; Killen, Jack; Emanuel, Ezekiel J. Post-trial access to tested interventions: the views of IRB/REC chair, investigators, and research participants in a multinational HIV/AIDS study. *AIDS Research and Human Retroviruses* 2006 September; 22(9): 837-841. NRCBL: 18.2; 1.3.2; 9.7; 9.5.6.

Expert Scientific Group on Phase One Clinical Trials. Expert scientific group on phase one clinical trials: final report. London: Expert Scientific Group on Phase One Clinical Trials, 2006 November 30; 106 p. [Online]. Accessed: http://www.dh.gov.uk/prod_consum_dh/idcplg? IdcService=GET_FILE&dID=136063&Rendition=Web [2007 April 23]. NRCBL: 18.2.

Falusi, Adeyinka G.; Olopade, Olufunmilayo I.; Olopade, Christopher O. Establishment of a standing ethics/institutional review board in a Nigerian university: a blueprint for developing countries. *Journal of Empirical Research on Human Research Ethics* 2007 March; 2(1): 21-30. NRCBL: 18.2; 18.5.9; 21.1. Identifiers: Nigeria.
Abstract: An ethics/institutional review board(IRB) was established according to International standards at the University of Ibadan in Nigeria. To achieve this, a private-public partnership was developed to support a review of prevailing practice and the development of necessary infrastructure for an effective IRB. An internationally registered and well-constituted IRB with a federal-wide assurance (FWA) from the National Institute of Health in the United States was established within a year. Over a 3-year period, the number of proposals reviewed increased by 150% while time to approval decreased by 62%. International collaboration and external research funding has increased substantially. These findings support our initial supposition that the development of a properly functioning IRB can be a catalyst for increased research productivity at academic centers in developing countries while ensuring the protection of vulnerable human research subjects. The University of Ibadan is now assisting other academic Institutions in Nigeria and sub-Saharan Africa with the establishment of their own IRBs.

Feder, Barnaby J. F.D.A. weighs flexibility in trials of heart treatment. *New York Times* 2007 September 21; p. A15. NRCBL: 18.2; 5.3. SC: po. Identifiers: Food and Drug Administration.

Fernandez, Conrad V. Our moral obligations in caring for patients with orphan cancers [editorial]. *CMAJ/JAMC: Canadian Medical Association Journal* 2007 January 30; 176(3): 297, 299. NRCBL: 18.2; 18.5.2; 9.3.1; 9.5.7.

Fost, Norman; Levine, Robert J. The dysregulation of human subjects research [editorial]. *JAMA: The Journal of the American Medical Association* 2007 November 14; 298(18): 2196-2198. NRCBL: 18.2.

Fox, R.M. Debate: should Australia move towards a centralized ethics committees system? The case for. *Internal Medicine Journal* 2005 April; 35(4): 247-248. NRCBL: 18.2.

Freemantle, Nick; Calvert, Mel. Composite and surrogate outcomes in randomized controlled trials. *BMJ: British Medical Journal* 2007 April 14; 334(7597): 756-757. NRCBL: 18.2; 9.7; 21.1.

Fuchs, Thomas. Ethical issues in neuroscience. *Current Opinion in Psychiatry* 2006 November; 19(6): 600-607. NRCBL: 18.2; 18.5.1; 17.1; 17.6.

Gabriele, Edward F. Belmont as parable: research leadership and the spirit of integrity. *In:* Kulakowski, Elliott C.; Chronister, Lynne U., eds. Research Administration and Management. Sudbury, MA: Jones and Bartlett, 2006: 473-480. NRCBL: 18.2; 1.1; 18.6. Identifiers: Belmont Report.

Garattini, Silvio; Bertelé, Vittorio. Non-inferiority trials are unethical because they disregard patients' interests. *Lancet* 2007 December 1-7; 370(9602): 1875-1877. NRCBL: 18.2; 9.7. SC: an.

Geissler, P.W.; Pool, R. Popular concerns about medical research projects in sub-Saharan Africa — a critical voice in debates about medical research ethics [editorial]. *Tropical Medicine and International Health* 2006 July; 11(7): 975-982. NRCBL: 18.2; 7.1; 21.1.

Gifford, Fred. Pulling the plug on clinical equipoise: a critique of Miller and Weijer. *Kennedy Institute of Ethics Journal* 2007 September; 17(3): 203-226. NRCBL: 18.2; 4.1.2; 1.1. SC: an.
Abstract: As clinicians, researchers, bioethicists, and members of society, we face a number of moral dilemmas concerning randomized clinical trials. How we manage the starting and stopping of such trials — how we conceptualize what evidence is sufficient for these decisions — has implications for both our obligations to trial participants and for the nature and security of the resultant medical knowledge. One view of how this is to be done, "clinical equipoise," recently has been given an extended defense by Paul Miller and Charles Weijer in their article "Rehabilitating Equipoise." The present paper critiques this position and Miller and Weijer's defense of it. I argue that their attempted rehabilitation fails. Their analysis suffers from a number of confusions, as well as a failure to make crucial distinctions, adequately to clarify key concepts, or to think through exactly what needs to be established to justify their claim. We are left with little reason to uphold the clinical equipoise criterion.

Gifford, Fred. So-called "clinical equipoise" and the argument from design. *Journal of Medicine and Philosophy* 2007 March-April; 32(2): 135-150. NRCBL: 18.2; 1.1. SC: an.
Abstract: In this article, I review and expand upon arguments showing that Freedman's so-called "clinical equipoise" criterion cannot serve as an appropriate guide and justification for the moral legitimacy of carrying out randomized clinical trials. At the same time, I try to explain

NRCBL: National Reference Center for Bioethics Literature Classification Scheme   See inside front cover for terms.

253

why this approach has been given so much credence despite compelling arguments against it, including the fact that Freedman's original discussion framed the issues in a misleading way, making certain things invisible: Clinical equipoise is conflated with community equipoise, and several versions of each are also conflated. But a misleading impression is given that, rather than distinct criteria being arbitrarily conflated, a puzzle is solved and a number of features unified. Various issues are pushed under the rug, hiding flaws of the "clinical equipoise" approach and thus deceiving us into thinking that we have a solution when we do not. Particularly significant is the ignoring of the crucial distinction between the individual patient decision and the policy decision.

Gifford, Fred. Taking equipoise seriously: the failure of clinical or community equipoise to resolve the ethical dilemmas in randomized clinical trials. *In:* Kincaid, Harold; McKitrick, Jennifer, eds. Establishing Medical Reality: Essays in the Metaphysics and Epistemology of Biomedical Science. Dordrecht, The Netherlands: Springer, 2007: 215-233. NRCBL: 18.2; 1.1. SC: an.

Gonzalez, Luis S., 3rd; Miller, Stephanie; Barnhart, Donna; Leifheit, Michael. Institutional review board approval of projects presented as posters at an ASHP midyear clinical meeting. *American Journal of Health-System Pharmacy* 2005 September 15; 62(18): 1890-1893. NRCBL: 18.2; 9.7; 9.8.

Goodman, Steven N. Ethics and evidence in clinical trials [editorial]. *Clinical Trials* 2005 June; 2(3): 195-196. NRCBL: 18.2.

Goodman, Steven N. Stopping at nothing? Some dilemmas of data monitoring in clinical trials [commentary]. *Annals of Internal Medicine* 2007 June 19; 146(12): 882-887. NRCBL: 18.2; 7.1.

Abstract: This commentary reviews the argument that clinical trials with data monitoring committees that use statistical stopping guidelines should generally not be stopped early for large observed efficacy differences because efficacy estimates may be exaggerated and there is minimal information on treatment harms. Overall, the average of estimates from trials that use these boundaries differs minimally from the true value. Estimates from a given trial that seem implausibly high can be moderated by using Bayesian methods. Data monitoring committees are not ethically required to precisely estimate a large efficacy difference if that difference differs convincingly from zero, and the requirement to detect harms and balance efficacy against harm depends on whether the nature of the harm is known or unknown before the trial.

Goodyear, Michael D.E.; Krleza-Jeric, Karmela; Lemmens, Trudo. The Declaration of Helsinki: mosaic tablet, dynamic document, or dinosaur [editorial]. *BMJ: British Medical Journal* 2007 September 29; 335(7621): 624-625. NRCBL: 18.2; 6; 18.5.9.

Graham, David Y.; Yamaoka, Yoshio. Ethical considerations of comparing sequential and traditional anti-helicobacter pylori therapy [letter]. *Annals of Internal Medicine* 2007 September 18; 147(6): 434-436. NRCBL: 18.2; 9.7.

Great Britain (United Kingdom). National Health Service. Health Service Guidelines: ethics committee review of multi-centre research: establishment of multi-centre research ethics committees: HSG (97)23. London: National Health Service (NHS): 1997 April 14; 3 p. [Online]. Accessed: http://www.dh.gov.uk/en/Publicationsandstatistics/Lettersandcirculars/Healthserviceguidelines/DH_4018331 [2007 October 9]. NRCBL: 18.2.

Green, David; Cushman, Mary; Dermond, Norma; Johnson, Eric A.; Castro, Cecilia; Arnett, Donna; Hill, Joel; Manolio, Teri A. Obtaining informed consent for genetic studies: the multiethnic study of atherosclerosis. *American Journal of Epidemiology* 2006 November 1; 164(9): 845-851. NRCBL: 18.2; 18.5.1; 15.3; 18.3; 8.3.4.

Greene, Sarah M.; Geiger, Ann M. A review finds that multicenter studies face substantial challenges but strategies exist to achieve Institutional Review Board approval. *Journal of Clinical Epidemiology* 2006 August; 59(8): 784-790. NRCBL: 18.2; 18.1; 18.6. SC: em.

Greene, Sarah M.; Geiger, Ann M.; Harris, Emily L.; Altschuler, Andrea; Nekhlyudov, Larissa; Barton, Mary B.; Rolnick, Sharon J.; Elmore, Joann G.; Fletcher, Suzanne. Impact of IRB requirements on a multicenter survey of prophylactic mastectomy outcomes. *Annals of Epidemiology* 2006 April; 16(4): 275-278. NRCBL: 18.2; 18.5.3; 9.5.1; 7.1.

Griffin, Joan M.; Struve, James K.; Collins, Dorothea; Liu, An; Nelson, David B.; Bloomfield, Hanna E. Long term clinical trials: how much information do participants retain from the informed consent process? *Contemporary Clinical Trials* 2006 October; 27(5): 441-448. NRCBL: 18.2; 18.3.

Grove, Matthew L. Trials and electronic records: a frightening industry proposal [letter]. *BMJ: British Medical Journal* 2007 June 16; 334(7606): 1236. NRCBL: 18.2; 1.3.12. Comments: J. Butcher. UK will lose clinical trials if electronic records system is delayed, ABPI warns. BMJ: British Medical Journal 2007 June 2; 334(7604): 1132.

Gunsalus, C. Kristina. Human subject protections: some thoughts on costs and benefits in the humanistic disciplines. *In:* Galston, Arthur W.; Peppard, Christiana Z., eds. Expanding Horizons in Bioethics. Dordrecht; Norwell, MA: Springer, 2005: 35-58. NRCBL: 18.2; 1.3.3; 18.6; 7.1.

Hadskis, Michael R. Giving voice to research participants: should IRBs hear from research participant representatives? *Accountability in Research* 2007 July-September; 14(3): 155-177. NRCBL: 18.2; 18.6.

SC (Subject Captions): an=analytical cs=case studies em=empirical le=legal po=popular rv=review

Abstract: The current decision-making model for the review of human research contains inadequate mechanisms to ensure that the interests and perspectives of research participants are considered by Institutional Review Boards, whose decisions may profoundly affect the safety and well-being of participants. As a result, this model is far from being optimized to realize Institutional Review Boards' principal mandate and undermines the credibility of the research review process. This article proposes a procedural mechanism that would ameliorate these systemic deficiencies by allowing "research participant representatives" to give voice to participants during the research review process.

Hardy, Pollyahanna.; Clemens, Felicity. Stopping a randomized trial early: from protocol to publication. Commentary to Thome at al.: outcome of extremely preterm infants randomized at birth to different PaCO2 targets during the first seven days of life (Biology of the Neonate 2006; 90: 218-225). *Biology of the Neonate* 2006; 90(4): 226-228. NRCBL: 18.2; 18.6.

Harkness, Jon; Lederer, Susan E.; Wikler, Daniel. Laying ethical foundations for clinical research: Public Health Classics. *Bulletin of the World Health Organization* 2001; 79(4): 365-372. NRCBL: 18.2; 18.6; 18.3; 2.2. Identifiers: Henry K. Beecher.

Hausman, Daniel M. Third-party risks in research: should IRBs address them? *IRB: Ethics and Human Research* 2007 May-June; 29(3): 1-5. NRCBL: 18.2; 18.3.

Heaven, Ben; Murtagh, Madeleine; Rapley, Tim; May, Carl; Graham, Ruth; Kaner, Eileen; Thomson, Richard. Patients or research subjects? A qualitative study of participation in a randomised controlled trial of a complex intervention. *Patient Education and Counseling* 2006 August; 62(2): 260-270. NRCBL: 18.2; 18.5.1; 18.3; 8.1.

Heilig, Charles M.; Weijer, Charles. A critical history of individual and collective ethics in the lineage of Lellouch and Schwartz. *Clinical Trials* 2005 June; 2(3): 244-253. NRCBL: 18.2; 2.2; 8.1; 6. SC: rv.

Heitman, Elizabeth; Olsen, Cara H.; Anestidou, Lida; Bulger, Ruth Ellen. New graduate students' baseline knowledge of the responsible conduct of research. *Academic Medicine* 2007 September; 82(9): 838-845. NRCBL: 18.2; 2.3. SC: em.

Hemelaar, Joris. Minimising risk in first-in-man trials. *Lancet* 2007 May 5-11; 369(9572): 1496-1497. NRCBL: 18.2; 5.2; 18.6; 1.3.12.

Holtedahl, Knut A., Meland, Eivind. Drug trials in general practice: time for a quality check before recruiting patients [letter]. *BMJ: British Medical Journal* 2007 July 7; 335(7609): 7. NRCBL: 18.2; 9.7. Identifiers: Norway.

Huang, David T.; Hadian, Mehrnaz. Bench-to-bedside review: human subjects research — are more standards

needed? *Critical Care* 2006; 10(6): 244. NRCBL: 18.2; 18.6; 18.3.

Hunter, David. Efficiency and the proposed reforms to the NHS research ethics system. *Journal of Medical Ethics* 2007 November; 33(11): 651-654. NRCBL: 18.2. Identifiers: Great Britain (United Kingdom); National Health Service.

Abstract: Significant changes are proposed for the research ethics system governing the review of the conduct of medical research in the UK. This paper examines these changes and whether they will meet the aimed-for goal of improving the efficiency of the research ethics system. The author concludes that, unfortunately, they will not and thus should be rejected.

Hunter, D. Proportional ethical review and the identification of ethical issues. *Journal of Medical Ethics* 2007 April; 33(4): 241-245. NRCBL: 18.2. SC: an. Identifiers: Great Britain (United Kingdom).

Abstract: Presently, there is a movement in the UK research governance framework towards what is referred to as proportional ethical review. Proportional ethical review is the notion that the level of ethical review and scrutiny given to a research project ought to reflect the level of ethical risk represented by that project. Relatively innocuous research should receive relatively minimal review and relatively risky research should receive intense scrutiny. Although conceptually attractive, the notion of proportional review depends on the possibility of effectively identifying the risks and ethical issues posed by an application with some process other than a full review by a properly constituted research ethics committee. In this paper, it is argued that this cannot be achieved and that the only appropriate means of identifying risks and ethical issues is consideration by a full committee. This implies that the suggested changes to the National Health Service research ethics system presently being consulted on should be strenuously resisted.

Huntington Study Group. Event Monitoring Committee; Erwin, Cheryl; Hersch, Steven. Monitoring reportable events and unanticipated problems: the PHAROS and PREDICT studies of Huntington disease. *IRB: Ethics and Human Research* 2007 May-June; 29(3): 11-16. 18 fn. NRCBL: 18.2; 15.1. Identifiers: Prospective Huntington At Risk Observational Study; Neurobiological Predictors of Huntington Disease [PREDICT-HD].

Keywords: *adverse effects; *clinical trials; *data monitoring committees; *genetic research; *Huntington disease; *research subjects; *risk; government regulation; information dissemination; multicenter studies; nontherapeutic research; research design; research ethics committees; research findings; risks and benefits; Proposed Keywords: *observation; prospective studies; Keyword Identifiers: *Neurobiological Predictors of Huntington Disease [PREDICT-HD]; *Prospective Huntington At Risk Observational Study; United States

INSECT Study Group; Seiler, C.M.; Kellmeyer, P.; Kienle, P.; Büchler, M.W.; Knaebel, H.-P. Assessment of the ethical review process for non-pharmacological

NRCBL: National Reference Center for Bioethics Literature Classification Scheme   See inside front cover for terms.

255

multicentre studies in Germany on the basis of a random-ised surgical trial. *Journal of Medical Ethics* 2007 February; 33(2): 113-118. NRCBL: 18.2. SC: em. Identifiers: German Surgical Society; INterrupted or continuous Slowly absorbable sutures-Evaluation of abdominal Closure Techniques.

Abstract: OBJECTIVE: To examine the current ethical review process (ERP) of ethics committees in a non-phar-macological trial from the perspective of a clinical inves-tigator. DESIGN: Prospective collection of data at the Study Centre of the German Surgical Society on the dura-tion, costs and administrative effort of the ERP of a ran-domised controlled multicentre surgical INSECT Trial (INterrupted or continuous Slowly absorbable su-tures-Evaluation of abdominal Closure Techniques Trial, ISRCTN 24023541) between November 2003 and May 2005. SETTING: Germany. PARTICIPANTS: 18 ethics committees, including the ethics committee handling the primary approval, responsible overall for 32 clinical sites throughout Germany. 8 ethics committees were located at university medical schools (MSU) and 10 at medical chambers. Duration was measured as days between sub-mission and receipt of final approval, costs in euros and administrative effort by calculation of the product of the total number of different types of documents and the mean number of copies required (primary approval act-ing as the reference standard). RESULTS: The duration of the ERP ranged from 1 to 176 (median 31) days. The median duration was 26 days at MSUs compared with 34 days at medical chambers. The total cost was euro2947. 1 of 8 ethics committees at universities (euro250) and 8 of 10 at medical chambers charged a median fee of euro162 (mean euro269.70). The administrative effort for primary approval was 30. Four ethics committees required a higher administrative effort for secondary approval (37, 39, 42 and 104). CONCLUSION: The ERP for non-phar-macological multicentre trials in Germany needs im-provement. The administrative process has to be standardised: the application forms and the number and content of the documents required should be identical or at least similar. The fees charged vary considerably and are obviously too high for committees located at medical chambers. However, the duration of the ERP was, with some exceptions, excellent. A centralised ethics committee in Germany for multicentre trials such as the INSECT Trial can simplify the ERP for clinical investigators in and outside the country.

Jester, Penelope M.; Tilden, Samuel J.; Li, Yufeng; Whitley, Richard J.; Sullender, Wayne M. Regulatory challenges: lessons from recent West Nile virus trials in the United States. *Contemporary Clinical Trials* 2006 June; 27(3): 254-259. NRCBL: 18.2; 18.5.1; 18.6.

Kalichman, Michael W. Responding to challenges in ed-ucating for the responsible conduct of research. *Academic Medicine* 2007 September; 82(9): 870-875. NRCBL: 18.2; 2.3.

Kalichman, Michael W.; Plemmons, Dena K. Reported goals for responsible conduct of research courses. *Aca-demic Medicine* 2007 September; 82(9): 846-852. NRCBL: 18.2; 2.3. SC: em.

Karbwang, Juntra; Crawley, Francis P. Need to strengthen ethics committees. SciDevNet: Science and De-velopment Network 2007 November 12: 2 p. [Online]. Ac-cessed: http://www.scidev.net/dossiers/index.cfm? fuseaction=dossierreaditem&dossier=5&t ype=3& itemid=687&language=1 [2008 February 21]. NRCBL: 18.2; 18.3; 18.5.9; 8.1.

Karunaratne, A.S.; Myles P.S.; Ago M.J.; Komesaroff, P.A. Communication deficiencies in research and monitor-ing by ethics committees. *Internal Medicine Journal* 2006 February; 36(2): 86-91. NRCBL: 18.2; 18.3; 8.1. SC: em. Identifiers: Australia.

Kass, Nancy E.; Hyder, Adnan Ali; Ajuwon, Ademola; Appiah-Poku, John; Barsdorf, Nicola; Elsayed, Dya Eldin; Mokhachane, Mantoa; Mupenda, Bavon; Ndebele, Paul; Ndossi, Godwin; Sikateyo, Bornwell; Tangwa, Godfrey; Tindana, Paulline. The structure and function of research ethics committees in Africa: a case study. *PLoS Medicine* 2007 January; 4(1): e3: 0026-0031 [Online]. Accessed: http://www.plos.org/press/plme-04-01-Kass.pdf [2007 January 25]. NRCBL: 18.2; 18.5.9; 21.1. SC: cs.

Kass, N.E.; Myers, R.; Fuchs, E.J.; Carson, K.A.; Flexner, C. Balancing justice and autonomy in clinical re-search with healthy volunteers. *Clinical Pharmacology and Therapeutics* 2007 August; 82(2): 219-227. NRCBL: 18.2; 18.5.1; 1.1; 18.3.

Keim, Brandon. Tied up in red tape, European trials shut down [news]. *Nature Medicine* 2007 February; 13(2): 110. NRCBL: 18.2; 18.6; 21.1.

Khoo, Chong-Yew. Ethical issues in ophthalmology and vision research. *Annals of the Academy of Medicine, Sin-gapore* 2006 July; 35(7): 512-516. NRCBL: 18.2; 18.5.1; 8.4; 21.1; 18.6.

Kohane, Isaac C.; Mandl, Kenneth D.; Taylor, Patrick L.; Holm, Ingrid A.; Nigrin, Daniel J.; Kunkel, Louis M. Reestablishing the researcher-patient compact. *Science* 2007 May 11; 316(5826): 836-837. NRCBL: 18.2; 8.1; 18.3.

Kojima, Somei; Waikagul, Jitra; Rojekittikhun, Wichit; Keicho, Naoto. The current situation regarding the establishment of national ethical guidelines for bio-medical research in Thailand and its neighboring coun-tries. *Southeast Asian Journal of Tropical Medicine and Public Health* 2005 May; 36(3): 728-732. 11 refs. NRCBL: 18.2; 18.3; 15.1; 6. SC: rv.

Keywords: *guidelines; *human experimentation; ethical review; ethics committees; genetic research; informed con-sent; international aspects; Keyword Identifiers: *Cambo-dia; *Myanmar; *Thailand

Kozanczyn, Christa; Collins, Katie; Fernandez, Conrad V. Offering results to research subjects: U.S. institutional review board policy. *Accountability in Research* 2007 October-December; 14(4): 255-267. NRCBL: 18.2. SC: em.

Krimsky, Sheldon; Simoncelli, Tania. Testing pesticides in humans: of mice and men divided by ten. *JAMA: The Journal of the American Medical Association* 2007 June 6; 297(21): 2405-2407. NRCBL: 18.2; 16.1. Identifiers: Environmental Protection Agency.

Krishna, Anurag. The ethics of research in children [editorial]. *Indian Pediatrics* 2005 May; 42(5): 419-423. NRCBL: 18.2; 18.5.2; 18.5.9.

Krousel-Wood, Marie; Muntner, Paul; Jannu, Ann; Hyre, Amanda; Breault, Joseph. Does waiver of written informed consent from the institutional review board affect response rate in a low-risk research study? *Journal of Investigative Medicine* 2006 May; 54(4): 174-179. NRCBL: 18.2; 18.3; 18.5.7. SC: em.

Kukla, Rebecca. Resituating the principle of equipoise: justice and access to care in non-ideal conditions. *Kennedy Institute of Ethics Journal* 2007 September; 17(3): 171-202. NRCBL: 18.2; 4.1.2; 1.1; 9.8; 9.5.6; 9.5.10; 18.5.9. SC: an.
Abstract: The principle of equipoise traditionally is grounded in the special obligations of physician- investigators to provide research participants with optimal care. This grounding makes the principle hard to apply in contexts with limited health resources, to research that is not directed by physicians, or to non-therapeutic research. I propose a different version of the principle of equipoise that does not depend upon an appeal to the Hippocratic duties of physicians and that is designed to be applicable within a wider range of research contexts and types—including health services research and research on social interventions. I consider three examples of ethically contentious research trials conducted in three different social settings. I argue that in each case my version of the principle of equipoise provides more plausible and helpful guidance than does the traditional version of the principle.

Kulynych, Jennifer J. The regulation of MR neuro- imaging research: disentangling the Gordian knot. *American Journal of Law and Medicine* 2007; 33(2-3): 295-317. NRCBL: 18.2; 18.5.6. Identifiers: magnetic resonance imaging.

Lawrence Livermore National Laboratory. Institutional Review Board. New investigator instructions: required reading for all new investigators. Livermore, CA: Lawrence Livermore National Laboratory, 2006 May 9; 4 p. [Online]. Accessed: http://www.llnl.gov/HumanSubjects/pi_instructions.html [2007 April 24]. NRCBL: 18.2; 1.3.9.

Ledford, Heidi. Trial and error: the ethics committees that oversee research done in humans have been attacked from all sides. *Nature* 2007 August 2; 448(7153): 530-532. NRCBL: 18.2; 1.3.2; 1.3.9; 18.6.

Lemaire, François. Do all types of human research need ethics committee approval? *American Journal of Respiratory and Critical Care Medicine* 2006 August 15; 174(4): 363-364. NRCBL: 18.2; 18.3; 21.1.

Lockwood, Michael; Anscombe, G.E.M. Sins of omission? The non-treatment of controls in clinical trials. *Proceedings of the Aristotelian Society, Supplementary Volumes* 1983; 57: 207-227. NRCBL: 18.2; 1.1.

Lubowitz, James H. Randomize, then consent: a strategy for improving patient acceptance of participation in randomized controlled trials. *Arthroscopy* 2006 September; 22(9): 1007-1008. NRCBL: 18.2; 18.3; 18.5.1; 9.5.1.

Luna, Florencia. Research in developing countries. *In:* Steinbock, Bonnie, ed. The Oxford Handbook of Bioethics. Oxford; New York: Oxford University Press, 2007: 621-647. NRCBL: 18.2; 2.2; 18.5.9.

Luna, Florencia. Social science research and respect for persons. *In:* Luna, Florencia; Herissone-Kelly, Peter, ed. Bioethics and Vulnerability: a Latin American View. Amsterdam; New York: Rodopi, 2006: 73-85. NRCBL: 18.2; 18.3; 7.1; 18.5.1.

Macduff, Colin; McKie, Andrew; Martindale, Sheelagh; Rennie, Anne Marie; West, Bernice; Wilcock, Sylvia. A novel framework for reflecting on the functioning of research ethics review panels. *Nursing Ethics* 2007 January; 14(1): 99-116. NRCBL: 18.2. SC: em.
Abstract: In the past decade structures and processes for the ethical review of UK health care research have undergone rapid change. Although this has focused users' attention on the functioning of review committees, it remains rare to read a substantive view from the inside. This article presents details of processes and findings resulting from a novel structured reflective exercise undertaken by a newly formed research ethics review panel in a university school of nursing and midwifery. By adopting and adapting some of the knowledge to be found in the art and science of malt whisky tasting, a framework for critical reflection is presented and applied. This enables analysis of the main contemporary issues for a review panel that is primarily concerned with research into nursing education and practice. In addition to structuring the panel's own literary narrative, the framework also generates useful visual representation for further reflection. Both the analysis of issues and the framework itself are presented as of potential value to all nurses, health care professionals and educationalists with an interest in ethical review.

MacNeil, S. Danielle; Fernandez, Conrad V. Attitudes of research ethics board chairs toward disclosure of research results to participants: results of a national survey.

NRCBL: National Reference Center for Bioethics Literature Classification Scheme   See inside front cover for terms.

Journal of Medical Ethics 2007 September; 33(9): 549-553. NRCBL: 18.2; 8.2. SC: em.

Abstract: Background: The offer of aggregate study results to research participants following study completion is increasingly accepted as a means of demonstrating greater respect for participants. The attitudes of research ethics board (REB) chairs towards this practice, although integral to policy development, are unknown. Objectives: To determine the attitudes of REB chairs and the practices of REBs with respect to disclosure of results to research participants. Design: A postal questionnaire was distributed to the chairs of English-language university-based REBs in Canada. In total, 88 REB chairs were eligible. The questionnaire examined respondents' attitudes towards offering participants completed study results, methods for delivering this information, and barriers to disclosing results. Findings: The response rate was 89.8%. Chairs were highly supportive (94.8%) of offering results to research participants. Only 19.5% of chairs responded that a policy or guideline that governed the return of research results to participants existed at their institution. Most chairs (72.0%) supported the idea of their REB instituting a set of guidelines recommending that researchers offer results to participants in a lay format. Chairs identified the major impediments to the implementation of programmes offering to return results to participants as being financial cost (57.5%) and retaining contact with research participants (78.1%). Conclusions: University-based REB chairs overwhelmingly support the offer of research results to participants. This is incongruent with the frequent lack of existing REB guidelines recommending this practice. REBs should support guidelines that diminish identified barriers and promote consistency in offering to return results.

Macrina, Francis L. Scientific societies and promotion of the responsible conduct of research: codes, policies, and education. *Academic Medicine* 2007 September; 82(9): 865-869. NRCBL: 18.2; 2.3; 7.3.

Maggon, Krishan. Regulatory reforms and GCP clinical trials with new drugs in India. *Clinical Trials (London, England)* 2004; 1(5): 461-467. NRCBL: 18.2; 9.7; 18.3; 18.5.9.

Maloney, Dennis M. Case study: institutional review board (IRB) must review hundreds of protocols — again. *Human Research Report* 2007 November; 22(11): 6-7. NRCBL: 18.2; 18.5.5.

Maloney, Dennis M. Case study: university forms two new institutional review boards (IRBs) to rereview suspended studies. *Human Research Report* 2007 June; 22(6): 6-7. NRCBL: 18.2; 1.3.3; 18.6.

Maloney, Dennis M. Case study: university says it is doing what it can to earn right to resume research. *Human Research Report* 2007 July; 22(7): 6-7. NRCBL: 18.2; 1.3.3; 18.6.

Maloney, Dennis M. Changes for expedited reviews by institutional review boards (IRBs). *Human Research Report* 2007 December; 22(12): 1-2. NRCBL: 18.2.

Maloney, Dennis M. Final guidance on adverse events for institutional review boards (IRBs). *Human Research Report* 2007 March; 22(3): 1-2. NRCBL: 18.2; 18.6.

Maloney, Dennis M. Institutional review board (IRB) accused of relying too much on subcommittee [case study]. *Human Research Report* 2007 February; 22(2): 6-7. NRCBL: 18.2; 18.3.

Maloney, Dennis M. Institutional review boards (IRBs), privacy rule, and subject recruitment. *Human Research Report* 2007 July; 22(7): 1-2. NRCBL: 18.2; 8.4; 18.3.

Maloney, Dennis M. IRB members must receive continuing education on protection of human subjects. *Human Research Report* 2007 May; 22(5): 6-7. NRCBL: 18.2; 18.6.

Maloney, Dennis M. University explains how it will strengthen its support for institutional review boards (IRBs) [case study]. *Human Research Report* 2007 October; 22(10): 6-7. NRCBL: 18.2; 1.3.3. SC: cs.

Maloney, Dennis M. University says its institutional review board (IRB) policies and procedures were just misunderstood. *Human Research Report* 2007 January; 22(1): 6-7. NRCBL: 18.2; 18.5.5.

Mann, Howard. Deception in the single-blind run-in phase of clinical trials. *IRB: Ethics and Human Research* 2007 March-April; 29(2): 14-17. NRCBL: 18.2; 18.6. Identifiers: Office for Human Research Protections; Food and Drug Administration.

Mann, Howard. Evaluation of research design by research ethics committees: misleading reassurance and the need for substantive reforms. *American Journal of Bioethics* 2007 February; 7(2): 84-86. NRCBL: 18.2; 18.1; 9.7; 1.3.9; 7.3. Comments: Ezekiel J. Emanuel and Franklin G. Miller. Money and distorted ethical judgments about research: ethical assessment of the TeGenero TGN1412 trial. American Journal of Bioethics 2007 February; 7(2): 76-81.

Mano, Max S.; Rosa, Daniela D.; Dal Lago, Lissandra. Multinational clinical trials in oncology and post-trial benefits for host countries: where do we stand? *European Journal of Cancer* 2006 November; 42(16): 2675-2677. NRCBL: 18.2; 18.5.9; 18.6; 21.1.

Mayer, Musa. Listen to all the voices: an advocate's perspective on early access to investigational therapies. *Clinical Trials* 2006; 3(2): 149-153. NRCBL: 18.2; 18.5.1; 18.6; 9.7.

Mayers, Douglas L.; Chung, Jain; Kohlbrenner, Veronika M.; Hall, David B.; DeMasi, Ralph A.; Neubacher, Dietmar; Buss, Neil E.; Salgo, Miklos P. Seeking ethical designs for HIV clinical trials in treatment-experienced patients: an industry perspective. *AIDS Research and Human Retroviruses* 2006 November; 22(11): 1110-1112. NRCBL: 18.2; 18.5.1; 1.3.2; 18.6.

McClure, Katie B.; Delorio, Nicole M.; Schmidt, Terri A.; Chiodo, Gary; Gorman, Paul. A qualitative study of institutional review board members' experience reviewing research proposals using emergency exception from informed consent. *Journal of Medical Ethics* 2007 May; 33(5): 289-293. NRCBL: 18.2; 18.3. SC: em.

Abstract: BACKGROUND: Emergency exception to informed consent regulation was introduced to provide a venue to perform research on subjects in emergency situations before obtaining informed consent. For a study to proceed, institutional review boards (IRBs) need to determine if the regulations have been met. AIM: To determine IRB members' experience reviewing research protocols using emergency exception to informed consent. METHODS: This qualitative research used semistructured telephone interviews of 10 selected IRB members from around the US in the fall of 2003. IRB members were chosen as little is known about their views of exception to consent, and part of their mandate is the protection of human subjects in research. Interview questions focused on the length of review process, ethical and legal considerations, training provided to IRB members on the regulations, and experience using community consultation and notification. Content analysis was performed on the transcripts of interviews. To ensure validity, data analysis was performed by individuals with varying backgrounds: three emergency physicians, an IRB member and a layperson. RESULTS: Respondents noted that: (1) emergency exception to informed consent studies require lengthy review; (2) community consultation and notification regulations are vague and hard to implement; (3) current regulations, if applied correctly, protect human subjects; (4) legal counsel is an important aspect of reviewing exception to informed-consent protocols; and (5) IRB members have had little or no formal training in these regulations, but are able to access materials needed to review such protocols. CONCLUSIONS: This preliminary study suggests that IRB members find emergency exception to informed consent studies take longer to review than other protocols, and that community consultation and community notification are the most difficult aspect of the regulations with which to comply but that they adequately protect human subjects.

McDonald, Katherine; Hernandez, Brigida; Plemmons, Dena; Simmerling, Mary. Privacy in organizational research. *Journal of Empirical Research on Human Research Ethics* 2007 March; 2(1): 69-73. NRCBL: 18.2; 8.4; 1.3.2; 18.5.1.

McMichael, Anthony J.; Bambrick, Hilary J. Greenhouse-gas costs of clinical trials. *Lancet* 2007 May 12–18; 369(9573): 1584-1585. NRCBL: 18.2; 16.1. SC: em.

McRae, Andrew D.; Weijer, Charles. Lessons from everyday lives: a moral justification for acute care research. *Critical Care Medicine* 2002 May; 30(5): 1146-1151. NRCBL: 18.2; 1.1; 18.3. Identifiers: Canada.

Menikoff, Jerry. Toward a general theory of research ethics. *Hastings Center Report* 2007 May-June; 37(3): 3. NRCBL: 18.2; 18.5.2; 18.5.1.

Merritt, Maria; Grady, Christine. Reciprocity and post-trial access for participants in antiretroviral therapy trials. *AIDS* 2006 September 11; 20(14): 1791-1794. NRCBL: 18.2; 9.2; 9.7; 9.5.6.

Messer, Neil. Medicine, science and virtue. *In:* Deane-Drummond, Celia; Scott, Peter Manley, eds. Future Perfect?: God, Medicine and Human Identity. New York: T and T Clark International, 2006: 113-125. NRCBL: 18.2; 9.5.1; 1.1; 1.2.

Mielke, J.; Ndebele, P. Making research ethics review work in Zimbabwe — the case for investment in local capacity. *Central African Journal of Medicine* 2004 November-December; 50(11-12): 115-119. NRCBL: 18.2; 18.5.9; 9.6.

Miller, Franklin G.; Brody, Howard. Clinical equipoise and the incoherence of research ethics. *Journal of Medicine and Philosophy* 2007 March-April; 32(2): 151-165. NRCBL: 18.2; 1.1. SC: an.

Abstract: The doctrine of clinical equipoise is appealing because it appears to permit physicians to maintain their therapeutic obligation to offer optimal medical care to patients while conducting randomized controlled trials (RCTs). The appearance, however, is deceptive. In this article we argue that clinical equipoise is defective and incoherent in multiple ways. First, it conflates the sound methodological principle that RCTs should begin with an honest null hypothesis with the questionable ethical norm that participants in these trials should never be randomized to an intervention known to be inferior to standard treatment. Second, the claim that RCTs preserve the therapeutic obligation of physicians misrepresents the patient-centered orientation of medical care. Third, the appeal to clinical equipoise as a basic principle of risk-benefit assessment for RCTs is incoherent. Finally, the difficulties with clinical equipoise cannot be resolved by viewing it as a presumptive principle subject to exceptions. In the final sections of the article, we elaborate on the non-exploitation framework for the ethics clinical research and indicate issues that warrant further inquiry.

Miller, Franklin G.; Campbell, Eric G.; Vogeli, Christine; Weissman, Joel S. Financial relationships of institutional review board members [letter and reply]. *New England Journal of Medicine* 2007 March 1; 356(9): 965. NRCBL: 18.2; 5.3; 1.3.9; 9.3.1; 9.7.

Miller, Franklin G.; Fins, Joseph J. Protecting human subjects in brain research: a pragmatic approach. *In:* Illes, Judy, ed. Neuroethics: Defining the Issues in Theory, Practice, and Policy. New York: Oxford University Press, 2006: 123-140. NRCBL: 18.2; 1.1; 17.1; 4.4; 18.5.6.

Miller, Franklin G.; Wertheimer, Alan. Facing up to paternalism in research ethics. *Hastings Center Report* 2007 May-June; 37(3): 24-34. NRCBL: 18.2; 18.3; 9.1; 1.1.

NRCBL: National Reference Center for Bioethics Literature Classification Scheme   See inside front cover for terms.

259

Miller, Paul B.; Weijer, Charles. Equipoise and the duty of care in clinical research: a philosophical response to our critics. *Journal of Medicine and Philosophy* 2007 March-April; 32(2): 117-133. NRCBL: 18.2; 1.1. SC: an.

Abstract: Franklin G. Miller and colleagues have stimulated renewed interest in research ethics through their work criticizing clinical equipoise. Over three years and some twenty articles, they have also worked to articulate a positive alternative view on norms governing the conduct of clinical research. Shared presuppositions underlie the positive and critical dimensions of Miller and colleagues' work. However, recognizing that constructive contributions to the field ought to enjoy priority, we presently scrutinize the constructive dimension of their work. We argue that it is wanting in several respects.

Miller, Paul B.; Weijer, Charles. Revisiting equipoise; a response to Gifford. *Kennedy Institute of Ethics Journal* 2007 September; 17(3): 227-246. NRCBL: 18.2; 4.1.2; 1.1. SC: an.

Abstract: The authors respond to objections Fred Gifford has raised against their paper "Rehabilitating Equipoise." They situate this exchange in the wider context of recent debate over equipoise, highlighting substantial points of agreement between themselves and Gifford. The authors offer a brief restatement of "Rehabilitating Equipoise" in which they amplify some of its core arguments. They then assess Gifford's objections. Finding each to be unfounded, they argue that there is no justification for "pulling the plug" on clinical equipoise.

Miossec, Marie; Miossec, Pierre. New regulatory rules for clinical trials in the United States and the European Union: key points and comparisons. *Arthritis and Rheumatism* 2006 December; 54(12): 3735-3740. NRCBL: 18.2; 21.1; 1.3.5.

Mitchell, Peter. Critics pan timid European response to TeGenero disaster. *Nature Biotechnology* 2007 May 25(5): 485-486. NRCBL: 18.2; 18.6. Identifiers: TGN1412; European Medicines Evaluation Agency.

Moerman, C.J.; Haafkens, J.A.; Söderström, M.; Rásky, É.; Maguire, P.; Maschewsky-Schneider, U.; Norstedt, M.; Hahn, D.; Reinerth, H.; McKevitt, M. Gender equality in the work of local research ethics committees in Europe: a study of practice of five countries. *Journal of Medical Ethics* 2007 February; 33(2): 107-112. NRCBL: 18.2; 10; 18.5.5; 21.1. SC: em; rv. Identifiers: Austria; Germany; Ireland; Netherlands; Sweden.

Abstract: BACKGROUND: Funding organisations and research ethics committees (RECs) should play a part in strengthening attention to gender equality in clinical research. In the research policy of European Union (EU), funding measures have been taken to realise this, but such measures are lacking in the EU policy regarding RECs. OBJECTIVE: To explore how RECs in Austria, Germany, Ireland, The Netherlands and Sweden deal with gender equality issues by asking two questions: (1) Do existing procedures promote representation of women and gender expertise in the committee? (2) How are sex and gender issues dealt with in protocol evaluation? METHODS: Two RECs were selected from each country. Data were obtained through interviews with key informants and content analysis of relevant documents (regulations, guidelines and review tools in use in 2003). RESULTS: All countries have rules (mostly informal) to ensure the presence of women on RECs; gender expertise is not required. Drug study protocols are carefully evaluated, sometimes on a formal basis, as regards the inclusion of women of childbearing age. The reason for excluding either one of the sexes or including specific groups of women or making a gender-specific risk-benefit analysis are investigated by some RECs. Such measures are, however, neither defined in the regulations nor integrated in review tools. CONCLUSIONS: The RECs investigated in five European member states are found to pay limited attention to gender equality in their working methods and, in particular in protocol evaluation. Policy and regulations of EU are needed to strengthen attention to gender equality in the work of RECs.

Moodley, Kaymanthri; Myer, Landon. Health research ethics committees in South Africa 12 years into democracy. *BMC Medical Ethics [electronic]* 2007; 8(1): 8 p. Accessed: http://www.biomedcentral.com/1472-6939/8/1 [2007 February 21]. NRCBL: 18.2. SC: em.

Moran, Maureen B. Ethical issues in research with human subjects. *Journal of the American Dietetic Association* 2006 September; 106(9): 1346, 1348. NRCBL: 18.2; 18.3; 8.1; 8.4.

Moreno, Jonathan D. Stumbling toward bioethics: human experiments policy and the early Cold War. *In:* LaFleur, William R.; Böhme, Gernot; Shimazono, Susumu, eds. Dark Medicine: Rationalizing Medical Research. Bloomington: Indiana University Press, 2007: 138-146. NRCBL: 18.2; 18.3; 6; 18.5.1; 1.3.5; 16.2; 21.3; 2.2.

Mueller, Paul S.; Montori, Victor M.; Bassler, Dirk; Koenig, Barbara A.; Guyatt, Gordon H. Ethical issues in stopping randomized trials early because of apparent benefit. *Annals of Internal Medicine* 2007 June 19; 146(12): 878-881. NRCBL: 18.2.

Abstract: Stopping randomized trials early because of an apparent benefit is becoming more common. To protect and promote the interests of trial participants, investigators may feel obligated to stop a trial early because of the apparent benefit of a study treatment (compared with placebo or other treatment). There are, however, serious ethical problems with doing so. Truncated trials systematically overestimate treatment effects; in cases where the number of accrued outcome events is small, the overestimation may be very large. Generating seriously inflated estimates of treatment effect violates the ethical research requirement of scientific validity. Subsequent use of inflated estimates to inform clinical decision making and practice guidelines violates the ethical requirements of social value and a favorable risk–benefit ratio. Researchers should ensure that a large number of outcome events accrues before stopping a trial and then continue recruitment to assess whether positive trends continue. This can balance the need to protect research participants with the ethical requirements of scientific validity, social value, and a favorable risk–benefit ratio.

Musschenga, A.W.; van Luijn, H.E.M.; Keus, R.B.; Aaronson, N.K. Are risks and benefits of oncological research protocols both incommensurable and incompensable? *Accountability in Research* 2007 July-September; 14(3): 179-196. NRCBL: 18.2; 5.2; 18.5.1. SC: em; an. Identifiers: Netherlands.

Abstract: Institutional review boards (IRBs) are legally required to determine whether the balance between the risks and benefits (the risk-benefit ratio or RBR) of a proposed study is "reasonable" or "proportional". This obligation flows from their duty to protect the interests of research subjects. It has been argued that it is difficult, perhaps even impossible for IRBs to determine the RBR of studies, because the risks and benefits are not only heterogeneous, but also incommensurable. After arguing that the relevant meaning of incommensurability is incomparability, we discuss whether the risks of participating in a trial and the benefits are comparable. We conclude that at least the risks and the benefits to participants are comparable. In the last section we show that the main problem of RBR analyses is that of interpersonal incompensability. IRBs have to assume that risks to research subjects be compensated by benefits to others. The question is: To what extent? When does it become unreasonable to ask that patients accept the risks of participating in a trial for the benefit of science and/or future patients?

National Institutes of Health [NIH] (United States); United States. Department of Health and Human Services [DHHS]. Office for Human Research Protections [OHRP]; Association of American Medical Colleges [AAMC]; American Society of Clinical Oncology [ASCO]. Alternative Models of IRB Review Workshop Summary Report, November 17-18, 2005. [Rockville, MD]: Office of Human Research Protections, [2007 April]: 7 p. [Online]. Accessed: http://www.hhs.gov/ohrp/sachrp/documents/AltModIRB.pdf [2007 April 30]. NRCBL: 18.2. Conference: National Conference on Alternative IRB Models: Optimizing Human Subjects Protections; 2006 November 19-21, 2006; Washington, DC.

Newcombe, J.P.; Kerridge, I.H. Assessment by human research ethics committees of potential conflicts of interest arising from pharmaceutical sponsorship of clinical research. *Internal Medicine Journal* 2007 January; 37(1): 12-17. NRCBL: 18.2; 9.7; 1.3.2; 18.5.1.

Noble, John H., Jr. Declaration of Helsinki: dead [letter]. *BMJ: British Medical Journal* 2007 October 13; 335(7623): 736. NRCBL: 18.2; 6. Comments: Michael D.E. Goodyear, Karmela Krieza-Jeric, and Trudo Lemmens. The Declaration of Helsinki. BMJ: British Medical Journal 2007 September 29; 624-625.

O'Beirne; Maeve; Stingl, Michael; Hayward, Sarah. Who reviews the projects of unaffiliated researchers for ethics? A case study from Alberta. *CQ: Cambridge Quarterly of Healthcare Ethics* 2007 Summer; 16(3): 346-355. NRCBL: 18.2; 18.6.

Panda, Mukta; Heath, Gregory W.; Desbiens, Norman A.; Moffitt, Benjamin. Research status of case reports for medical school institutional review boards [letter]. *JAMA: The Journal of the American Medical Association* 2007 September 19; 298(11): 1277-1278. NRCBL: 18.2; 7.2. SC: em.

Pandya, Dipak P.; Dave, Jay. Protection of human subjects in clinical research: the pitfalls in clinical research. *Comprehensive Therapy* 2005 Spring; 31(1): 72-77. NRCBL: 18.2; 18.3.

Parvizi, Javad; Tarity, T. David; Conner, Kyle; Smith, J. Bruce. Institutional review board approval: why it matters. *Journal of Bone and Joint Surgery. American volume* 2007 February; 89(2): 418-426. NRCBL: 18.2; 18.3; 7.1.

Pentz, Rebecca D.; Flamm, Anne L.; Sugarman, Jeremy; Cohen, Marlene Z.; Xu, Zhiheng; Herbst, Roy S.; Abbruzzese, James L. Who should go first in trials with scarce agents? The views of potential participants. *IRB: Ethics and Human Research* 2007 July-August; 29(4): 1-6. NRCBL: 18.2; 9.7; 18.5.1. SC: em.

Perkins, Alexis C.; Choi, Joanna Mimi; Kimball, Alexa B. Reporting of ethical review of clinical research submitted to the Journal of the American Academy of Dermatology. *Journal of the American Academy of Dermatology* 2007 February; 56(2): 279-284. NRCBL: 18.2; 18.3; 7.1; 1.3.7.

Perlis, Clifford S.; Harwood, Michael; Perlis, Roy H. Extent and impact of industry sponsorship conflicts of interest in dermatology research. *Journal of the American Academy of Dermatology* 2005 June; 52(6): 967-971. NRCBL: 18.2; 1.3.7; 7.3; 9.7; 1.3.2. SC: em; rv.

Pimple, Kenneth D. Ethical issues in drug user treatment research. *In:* Kleinig, John; Einstein, Stanley, eds. Ethical Challenges for Intervening in Drug Use: Policy, Research and Treatment Issues. Huntsville, TX: Office of International Criminal Justice; Criminal Justice Center, Sam Houston State University, 2006: 205-216. NRCBL: 18.2; 1.3.9; 18.3; 9.5.9. Identifiers: Belmont Report.

Poff, Deborah. Community-based REBS: the experience of the British Columbia medical services foundation. *NCEHR Communique CNERH* 2006 Spring; 14(1): 24-25. NRCBL: 18.2; 18.6.

Quest, Dale. Case vignette 1: a randomized double-blind double-dummy cross-over study of oral hexylinsulin monoconjugate 2 [PEGInsulin} versus insulin lispro for postprandial glycæmic control in adult patients with Type

NRCBL: National Reference Center for Bioethics Literature Classification Scheme   See inside front cover for terms.

261

2 diabetes mllitus. *NCEHR Communique CNERH* 2005 Spring; 13(1): 7-8. NRCBL: 18.2; 7.2. SC: cs.

Quest, Dale. Case vignette 2: a phase 2, randomized, double-blind, placebo-controlled study of DPE6591A in rheumatoid arthritis patients. *NCEHR Communique CNERH* 2005 Spring; 13(1): 9-10. NRCBL: 18.2; 7.2. SC: cs.

Quest, Dale. Case vignette 3: a multi-centered trial to compare TCT vs Clozapine for treatment-resistant schizophrenia. *NCEHR Communique CNERH* 2005 Spring; 13(1): 11-12. NRCBL: 18.2; 7.2. SC: cs.

Rabin, Cheryl; Tabak, Nili. Healthy participants in phase I clinical trials: the quality of their decision to take part. *Journal of Clinical Nursing* 2006 August; 15(8): 971-979. NRCBL: 18.2; 18.3; 18.5.1.

Racine, Eric; Illes, Judy. Neuroethical responsibilities. *Canadian Journal of Neurological Sciences* 2006 August; 33(3): 269-277, 260-268. NRCBL: 18.2; 18.5.1; 18.6.

Robinson, Louise; Murdoch-Eaton, Deborah; Carter, Yvonne. NHS research ethics committees — still need more common sense and less bureaucracy [editorial]. *BMJ: British Medical Journal* 2007 July 7; 335(7609): 6. NRCBL: 18.2. Identifiers: Great Britain (United Kingdom); National Health Service.

Sade, Robert M. Reports of clinical trials: ethical aspects. *Journal of Thoracic and Cardiovascular Surgery* 2006 August; 132(2): 245-246. NRCBL: 18.2; 18.5.1; 1.3.7.

Saunders, John. More guidelines on research ethics? With its new research ethics guidelines, the UK Royal College of Physicians continues a useful tradition of providing guidance to medical researchers. [editorial]. *Journal of Medical Ethics* 2007 December; 33(12): 683-684. NRCBL: 18.2. Identifiers: Great Britain (United Kingdom).

Sayers, G.M. Should research ethics committees be told how to think? *Journal of Medical Ethics* 2007 January; 33(1): 39-42. NRCBL: 18.2. SC: an.
  Abstract: Research ethics committees (RECs) are charged with providing an opinion on whether research proposals are ethical. These committees are overseen by a central office that acts for the Department of Health and hence the State. An advisory group has recently reported back to the Department of Health, recommending that it should deal with (excessive) inconsistency in the decisions made by different RECs. This article questions the desirability and feasibility of questing for consistent ethical decisions.

Schmelzer, Marilee. Institutional review boards: friend, not foe. *Gastroenterology Nursing: the Official Journal of the Society of Gastroenterology Nurses and Associates* 2006 January-February; 29(1): 80-81. NRCBL: 18.2; 18.3.

Schonfeld, Toby; Gordon, Bruce; Amoura, Jean; Brown, Joseph Spencer. Money matters. *American Jour-*

nal of Bioethics 2007 February; 7(2): 86-88. NRCBL: 18.2; 9.3.1; 1.3.9; 7.3. Comments: Ezekiel J. Emanuel and Franklin G. Miller. Money and distorted ethical judgments about research: ethical assessment of the TeGenero TGN1412 trial. American Journal of Bioethics 2007 February; 7(2): 76-81.

Schuppli, C.A.; Fraser, D. Factors influencing the effectiveness of research ethics committees. *Journal of Medical Ethics* 2007 May; 33(5): 294-301. NRCBL: 18.2; 9.8. SC: em. Identifiers: Canada.
  Abstract: Research ethics committees - animal ethics committees (AECs) for animal-based research and institutional research boards (IRBs) for human subjects - have a key role in research governance, but there has been little study of the factors influencing their effectiveness. The objectives of this study were to examine how the effectiveness of a research ethics committee is influenced by committee composition and dynamics, recruitment of members, workload, participation level and member turnover. As a model, 28 members of AECs at four universities in western Canada were interviewed. Committees were selected to represent variation in the number and type of protocols reviewed, and participants were selected to include different types of committee members. We found that a bias towards institutional or scientific interests may result from (1) a preponderance of institutional and scientist members, (2) an intimidating atmosphere for community members and other minority members, (3) recruitment of community members who are affiliated with the institution and (4) members joining for reasons other than to fulfil the committee mandate. Thoroughness of protocol review may be influenced by heavy workloads, type of review process and lack of full committee participation. These results, together with results from the literature on research ethics committees, suggested potential ways to improve the effectiveness of research ethics committees.

Schwartz, Joan P. Integrity in research: individual and institutional responsibility. *In:* Gallin, John I.; Ognibene, Frederick P., eds. Principles and Practice of Clinical Research. 2nd edition. Oxford: Academic, 2007: 39-46. NRCBL: 18.2; 1.3.9; 18.5.1.

Scotland. Scottish Executive. Health Department. Chief Scientist Office. Scottish Ethics Advisory Group [SEAG]. Consultation Report: Review of NHS Research Ethics Committees. Edinburgh: Scottish Executive, 2006 January 12: 21 p. [Online]. Accessed: http://www. scotland.gov.uk/Publications/2006/01/12093352/0 [2007 May 7]. NRCBL: 18.2. Identifiers: National Health Service [NHS]; Warner Report.

Scotland. Scottish Executive. Health Department. Chief Scientist Office.; Nolan, Moira; Hinds, Alison. Review of the NHS Research Ethics Committee System [letter with Consultation attachments]. Edinburgh: Scottish Executive, 2005 June 30: 10 p. [Online]. Accessed: http://www.scotland.gov.uk/Publications/2005/07/

0194908/49116 [2007 May 7]. NRCBL: 18.2. Identifiers: National Health Service; Warner Report.

Shamoo, Adil E. Deregulating low-risk research. *Chronicle of Higher Education* 2007 August 3; 53(48): B16. NRCBL: 18.2; 18.6.

Shragge, Jeremy E. A graduate student perspective on the accreditation of programs ensuring ethical research with humans. *NCEHR Communique CNERH* 2006 Spring; 14(1): 21-22. NRCBL: 18.2; 9.8. Identifiers: Canada.

Shuchman, Miriam. Commercializing clinical trials — risks and benefits of the CRO boom. *New England Journal of Medicine* 2007 October 4; 357(14): 1365-1368. NRCBL: 18.2; 1.3.9; 1.3.2; 9.7. Identifiers: Contract Research Organizations.

Shweder, Richard A. Protecting human subjects and preserving academic freedom: prospects at the University of Chicago. *American Ethnologists* 2006 November; 33(4): 507-518. NRCBL: 18.2; 1.3.3; 1.3.9.

Sieber, Joan E. Institutional introspection [editorial]. *Journal of Empirical Research on Human Research Ethics* 2007 December; 2(4): 1-2. NRCBL: 18.2; 1.3.9; 7.3.

Sinclair, Andrew H.; Schofield, Peter R. Human embryonic stem cell research: an Australian perspective. *Cell* 2007 January 26; 128(2): 221-223. NRCBL: 18.2; 18.5.4; 15.1; 1.3.5.

Singer, Eleanor; Bossarte, Robert M. Incentives for survey participation when are they "coercive"? *American Journal of Preventive Medicine* 2006 November; 31(5): 411-418. NRCBL: 18.2; 18.3; 9.3.1; 18.6; 9.1.

Slaven, Marcia Jacobson. First impressions: the experiences of a community member on a research ethics committee. *IRB: Ethics and Human Research* 2007 May-June; 29(3): 17-19. NRCBL: 18.2.

Slutsman, Julia; Buchanan, David; Grady, Christine. Ethical issues in cancer chemoprevention trials: considerations for IRBs and investigators. *IRB: Ethics and Human Research* 2007 March-April; 29(2): 1-6. NRCBL: 18.2; 18.3.

Snowdon, Claire; Garcia, Jo; Elbourne, Diana. Making sense of randomization: responses of parents of critically ill babies to random allocation of treatment in a clinical trial. *Social Science and Medicine* 1997 November; 45(9): 1337-1355. NRCBL: 18.2; 18.5.2; 18.3; 9.4.

Sobolski, Gregory K.; Flores, Leonardo; Emanuel, Ezekiel J. Institutional review board review of multicenter studies [letter]. *Annals of Internal Medicine* 2007 May 15; 146(10): 759. NRCBL: 18.2.

Society for Adolescent Medicine. Guidelines for adolescent health research. 1995. *Journal of Adolescent Health*

2003 November; 33(5): 410-415. NRCBL: 18.2; 18.5.2; 18.3; 8.3.2; 5.2.

Society for Clinical Trials; Dickersin, Kay; Davis, Barry R.; Dixon, Dennis O.; George, Stephen L.; Hawkins, Barbara S.; Lachin, John; Peduzzi, Peter; Pocock, Stuart. The Society for Clinical Trials supports United States legislation mandating trials registration. Position paper. *Clinical Trials (London, England)* 2004; 1(5): 417-420. NRCBL: 18.2; 1.3.5; 18.6.

Steneck, Nicholas H.; Bulger, Ruth Ellen. The history, purpose, and future of instruction in the responsible conduct of research. *Academic Medicine* 2007 September; 82(9): 829-834. NRCBL: 18.2; 18.6; 2.3; 2.2.

Stone, Judy. Ethical issues in human subjects research. *In her:* Conducting Clinical Research: A Practical Guide for Physicians, Nurses, Study Coordinators, and Investigators. Cumberland, MD: Mountainside MD Press, 2006: 145-172. NRCBL: 18.2; 5.2; 18.3; 2.2; 9.3.1. SC: cs.

Swazo, Norman K. Research integrity and rights of indigenous peoples: appropriating Foucault's critique of knowledge/power. *Studies in History and Philosophy of Biological and Biomedical Sciences* 2005 September; 36(3): 568-584. NRCBL: 18.2; 18.5.9; 21.1.

ter Meulen, Ruud. Ethical issues of evidence-based medicine. *In:* Gunning, Jennifer; Holm, Søren, eds. Ethics, law, and society. Volume 1. Aldershot, Hants, England; Burlington, VT: Ashgate, 2005: 51-58. NRCBL: 18.2; 9.1; 8.1; 7.1.

Tod, A.M.; Nicolson, P.; Allmark, P. Ethical review of health service research in the UK: implications for nursing. *Journal of Advanced Nursing* 2002 November; 40(4): 379-386. NRCBL: 18.2; 9.1; 8.4; 18.3. Identifiers: Great Britain (United Kingdom).

Turale, Sue. Reflections on the ethics involved in international research. *Nursing and Health Sciences* 2006 September; 8(3): 131-132. NRCBL: 18.2; 18.5.9; 21.7; 21.1.

United States. Department of Education [DOE]. Protection of human subjects; proposed rule. *Federal Register* 1997 May 22; 62(99): 28155-28159 [Online]. Accessed: http://frwebgate.access.gpo.gov/cgi-bin/multidb.cgi [2005 December 27]. NRCBL: 18.2; 18.5.2; 18.6.

United States. Department of Health and Human Services [DHHS]. Office for Human Research Protections [OHRP]. Guidance on reporting and reviewing adverse events and unanticipated problems involving risks to subjects or others: Draft - October 11, 2005. Rockville, MD: Office for Human Research Protections, 2005 October 11: 25 p. [Online]. Accessed: http://www.hhs.gov/ohrp/requests/aerg.pdf [2007 March 10]. NRCBL: 18.2.

United States. Department of Health and Human Services [DHHS]. Office of the Secretary. Protection of hu-

---

NRCBL: National Reference Center for Bioethics Literature Classification Scheme   See inside front cover for terms.

263

man research subjects: delay of effective date. *Federal Register* 2001 March 19; 66(53): 15352 [Online]. Accessed: http://frwebgate.access.gpo.gov/cgi-bin/multidb.cgi [2005 December 27]. NRCBL: 18.2.

United States. Food and Drug Administration [FDA]. Strengthening the regulation of clinical trials and bioresearch monitoring. *FDA Consumer* 2006 November- December; 40(6): 35. NRCBL: 18.2; 1.3.5; 5.2.

United States. Food and Drug Administration [FDA]; Center for Drug Evaluation and Research [CDER], Office of the Commissioner; Center for Biologics Evaluation and Research [CBER]; Center for Devices and Radiological Health [CDRH];Good Clinical Practice Program [GCPP] (United States). Adverse Event Reporting — Improving Human Subject Protection. Guidance for Clinical Investigators, Sponsors and IRBs [draft guidance]. Rockville, MD: Food and Drug Administration 2007 April: 8 p. [Online]. Accessed: http://www.clinicalresearchresources.com/images/fdaguidanceadvreport.pdf [2007 September 12]. NRCBL: 18.2; 9.7; 18.6.

United States. Food and Drug Administration [FDA]; Center for Drug Evaluation and Research [CDER]; Center for Biologics Evaluation and Research [CBER]; Center for Devices and Radiological Health [CDRH] (United States). Protecting the Rights, Safety, and Welfare of Study Subjects — Supervisory Responsibilities of Investigators. Guidance for Industry [draft guidance]. Rockville, MD: Food and Drug Administration 2007 May: 16 p. [Online]. Accessed: http://www.clinicalresearchresources.com/images/fdaguidancestudysub.pdf [2007 September 12]. NRCBL: 18.2; 18.6.

Van Denend, Toni; Finlayson, Marcia. Ethical decision making in clinical research: application of CELIBATE. *American Journal of Occupational Therapy* 2007 January-February; 61(1): 92-95. NRCBL: 18.2; 16.3; 9.4; 4.1.1.

van Luijn, H.E.M.; Musschenga, A.W.; Keus, R.B.; Aaronson, N.K. Evaluating the risks and benefits of phase II and III clinical cancer trials: a look at institutional review board members in the Netherlands. *IRB: Ethics and Human Research* 2007 January-February; 29(1): 13-17. NRCBL: 18.2; 1.3.9. SC: em.

van Veen, E.-B.; Riegman, P.H.J.; Dinjens, W.N.M.; Lam, K.H.; Oomen, M.H.A.; Spatz, A.; Mager, R.; Ratcliffe, C.; Knox, K.; Kerr, D.; van Damme, B.; van de Vijver, M.; van Boven, H.; Morente, M.M.; Alonso, S.; Kerjaschki, D.; Pammer, J.; Lopez-Guerrero, J.A.; Llombart Bosch, A.; Carbone, A.; Gloghini, A.; Teodorovic, I.; Isabelle, M.; Passioukov, A.; Lejeune, S.; Therasse, P.; Oosterhuis, J.W. TuBaFrost 3: regulatory and ethical issues on the exchange of residual tissue for research across Europe. *European Journal of Cancer* 2006 November; 42(17): 2914-2923. NRCBL: 18.2; 18.5.9; 21.1; 18.6.

Vanderwel, Marianne. Accreditation: the application of quality principles to the protection of human research subjects. *NCEHR Communique CNERH* 2006 Spring; 14(1): 9-11. NRCBL: 18.2; 9.8. Identifiers: Canada.

Vasgird, Daniel R. Prevention over cure: the administrative rationale for education in the responsible conduct of research. *Academic Medicine* 2007 September; 82(9): 835-837. NRCBL: 18.2; 18.6; 2.3.

Veatch, Robert M. The irrelevance of equipoise. *Journal of Medicine and Philosophy* 2007 March-April; 32(2): 167-183. NRCBL: 18.2; 1.1; 18.3. SC: an.
Abstract: It is commonly believed in research ethics that some form of equipoise is a necessary condition for justifying randomized clinical trials, that without it clinicians are violating the moral duty to do what is best for the patient. Recent criticisms have shown how complex the concept of equipoise is, but often retain the commitment to some form of equipoise for randomization to be justified. This article rejects that claim. It first asks for what one should be equally poised (scientific or clinical equipoise), then asks who should be equally poised (scientist, clinician, or subject), and finally asks why any of these players need be equally poised between treatment options. The article argues that only the subject's evaluation of the options is morally relevant and that even the subject need not be equally poised or indifferent between the options in order to volunteer for randomization. All that is needed is adequately informed, free, and unexploited consent. It concludes equipoise is irrelevant.

Weijer, Charles; Miller, P.B. Refuting the net risks test: a response to Wendler and Miller's "Assessing research risks systematically". *Journal of Medical Ethics* 2007 August; 33(8): 487-490. NRCBL: 18.2. SC: an. Comments: D. Wendler and F.G. Miller. Assessing research risks systematically: the net risks test. Journal of Medical Ethics 2007 August; 33(8): 481-486.
Abstract: Earlier in the pages of this journal (p 481), Wendler and Miller offered the "net risks test" as an alternative approach to the ethical analysis of benefits and harms in research. They have been vocal critics of the dominant view of benefit-harm analysis in research ethics, which encompasses core concepts of duty of care, clinical equipoise and component analysis. They had been challenged to come up with a viable alternative to component analysis which meets five criteria. The alternative must (1) protect research subjects; (2) allow clinical research to proceed; (3) explain how physicians may offer trial enrolment to their patients; (4) address the challenges posed by research containing a mixture of interventions and (5) define ethical standards according to which the risks and potential benefits of research may be consistently evaluated. This response argues that the net risks test meets none of these criteria and concludes that it is not a viable alternative to component analysis.

Weinfurt, Kevin P.; Allsbrook, Jennifer S.; Friedman, Joëlle Y.; Dinan, Michaela A.; Hall, Mark A.; Schulman, Kevin A.; Sugarman, Jeremy. Developing model language for disclosing financial interests to potential clinical research participants. *IRB: Ethics and Human Research* 2007 January-February; 29(1): 1-5. NRCBL: 18.2; 1.3.9.

Wendler, D.; Miller, F.G. Assessing research risks systematically: the net risks test. *Journal of Medical Ethics* 2007 August; 33(8): 481-486. NRCBL: 18.2. SC: an.

Abstract: Dual-track assessment directs research ethics committees (RECs) to assess the risks of research interventions based on the unclear distinction between therapeutic and non-therapeutic interventions. The net risks test, in contrast, relies on the clinically familiar method of assessing the risks and benefits of interventions in comparison to the available alternatives and also focuses attention of the RECs on the central challenge of protecting research participants.

Werner, Michael J.; Price, Elizabeth. Managing conflicts of interest: a survival guide for biotechs. *Nature Biotechnology* 2007 February; 25(2): 161-163. NRCBL: 18.2; 1.3.9.

Wichman, Alison. Institutional review boards. *In:* Gallin, John I.; Ognibene, Frederick P., eds. Principles and Practice of Clinical Research. 2nd edition. Oxford: Academic, 2007: 47-58. NRCBL: 18.2; 2.2.

Williamson, Graham R.; Prosser, Sue. Action research: politics, ethics and participation. *Journal of Advanced Nursing* 2002 December; 40(5): 587-593. NRCBL: 18.2.

Wisely, J.; Lilleyman, J. Implementing the district hospital recommendations for the National Health Service Research Ethics Service in England. *Journal of Medical Ethics* 2007 March; 33(3): 168. NRCBL: 18.2.

Wolf, Leslie E.; Zandecki, Jolanta. Conflicts of interest in research: how IRBs address their own conflicts. *IRB: Ethics and Human Research* 2007 January-February; 29(1): 6-12. NRCBL: 18.2; 1.3.9; 7.3.

Wolinsky, Howard. The battle of Helsinki: two troublesome paragraphs in the Declaration of Helsinki are causing a furore over medical research ethics. *EMBO Reports* 2006 July; 7(7): 670-672. NRCBL: 18.2; 21.1; 1.3.5.

Yank, Veronia; Rennie, Drummond; Bero, Lisa A. Financial ties and concordance between results and conclusions in meta-analyses: retrospective cohort study. *BMJ: British Medical Journal* 2007 December 8; 335(7631): 1202-1205. NRCBL: 18.2; 9.7; 1.3.2. SC: em.

HUMAN EXPERIMENTATION/ . . . / LEGAL ASPECTS

Appleton, J.; Caan, W.; Cowley, S.; Kendall, S. Busting the bureaucracy: lessons from research governance in primary care. *Community Practitioner* 2007 February; 80(2): 29-32. NRCBL: 18.2; 18.6; 9.5.1. SC: cs; le. Identifiers: Great Britain (United Kingdom).

Birmingham, Karen; Frumston, Michael. Avon longitudinal study of parents and children (ALSPAC): ethical process. *In:* Gunning, Jennifer; Holm, Søren, eds. Ethics, law, and society. Volume 2. Aldershot, Hants, England; Burlington, VT: Ashgate, 2006: 65-74. NRCBL: 18.2; 18.5.2; 18.3. SC: le.

Carvalho, Fatima Lampreia. Regulation of clinical research and bioethics in Portugal. *Bioethics* 2007 June; 21(5): 290-302. NRCBL: 18.2; 9.6. SC: le.

Abstract: This article presents an overview of the Portuguese transposition of the European Directive on Good Clinical Practice (2001/20/E) concerning scientific and academic debates on bioethics and clinical investigation. Since the Directive was transposed into Portuguese law by its National Assembly, the bureaucracy of clinical trials has been ever more complex. Despite demands for swift application processes by the Pharmaceutical industry, supported by the European Parliament, the Directive's transcription to the national law has not always delivered the expected outcome. However, this has led to an increased number of applications for clinical trials in Portuguese hospitals. In this article I revise bioethical publications and decree-laws enabling an informed appraisal of the anxieties and prospects for the implementation of the clinical trials Directive in Portugal. This article also places the European Directive in the field of sociology of bioethics, arguing that Portuguese bioethical institutions differ from those of the US, and also from Northern European counterparts. The main divergence is that those people in Portugal who claim expertise in 'legal' bioethics do not dominate either the bureaucratic structure of research or ethics committees for health. Even experts in the applied ethics field now claim that 'professional bioethicists do not exist'. The recent creation of a national Ethics Committee for Clinical Investigation (CEIC) in line with the European Directive on Good Clinical Practice (GCP) will not change the present imbalance between different professional jurisdictions in the national bioethical debate in Portugal.

Chalmers, Don. International medical research regulation: from ethics to law. *In:* McLean, Sheila A.M., ed. First Do No Harm: Law, Ethics, and Healthcare. Aldershot, England; Burlington, VT: Ashgate, 2006: 81-100. NRCBL: 18.2; 2.1; 2.2; 15.1; 21.1. SC: le. Identifiers: Australia.

Douglas, Thomas M. Ethics committees and the legality of research. *Journal of Medical Ethics* 2007 December; 33(12): 732-736. NRCBL: 18.2. SC: an; le. Identifiers: New Zealand.

Abstract: One role of research ethics committees (RECs) is to assess the ethics of proposed health research. In some countries, RECs are also instructed to assess its legality. However, in other countries they are explicitly instructed not to do so. In this paper, I defend the claim that

---

NRCBL: National Reference Center for Bioethics Literature Classification Scheme   See inside front cover for terms.

public policy should instruct RECs not to assess the legality of proposed research ("the Claim"). I initially defend a presumption in favour of the Claim, citing reasons for making research institutions solely responsible for assessing the legality of their own research. I then consider three arguments against the Claim which may over-ride this presumption-namely, that policy should instruct RECs to assess the legality of research because (1) doing so would minimise the costs of assessing the legality of research, (2) whether research is legal may partly determine whether it is ethical and (3) whether research is legal may constitute evidence for whether it is ethical. I reject the first two arguments and note that whether the third succeeds depends on the answer to a more fundamental question about the appropriate nature of REC ethical deliberation. I end with a brief discussion of this question, tentatively concluding that the third argument also fails.

Hambruger, Philip. The new censorship: institutional review boards. Supreme Court Review 2004 October: 271-354 [Online]. Accessed: http://papers.ssrn.com/sol3/papers.cfm?abstract_id=721363 [2007 December 13]. NRCBL: 18.2; 1.3.8; 18.1. SC: le. Note: Public Law and Legal Theory Working Paper Series: no.95. The Law School, University of Chicago. May 2005.

Hyman, David A. Institutional review boards: is this the least we can do? *Northwestern University Law Review* 2007; 101(2): 749-773. NRCBL: 18.2; 18.6. SC: le.

Kesselheim, Aaron S.; Mello, Michelle M. Confidentiality laws and secrecy in medical research: improving public access to data on drug safety. Concealing clinical trial data from public scrutiny has implications for Americans' health. *Health Affairs* 2007 March-April; 26(2): 483-491. NRCBL: 18.2; 8.4; 9.7. SC: le.

Lötjönen, Salla. Research on human subjects. *In:* Gevers, J.K.M.; Hondius, E.H.; Hubben, J.H., eds. Health Law, Human Rights and the Biomedicine Convention: Essays in Honour of Henriette Roscam Abbing. Leiden; Boston: Martinus Nijhoff Publishers, 2005: 175-190. NRCBL: 18.2; 18.6; 1.3.9; 21.1. SC: le.

Maloney, Dennis M. Court says actions of institutional review board (IRB) members were ethically wrong. *Human Research Report* 2007 April; 22(4): 8. NRCBL: 18.2; 16.1; 18.5.2. SC: le. Identifiers: Grimes v. Kennedy Krieger Institute (KKI), (associated with Johns Hopkins University), (Part 6).

Maloney, Dennis M. Court says institutional review boards (IRBs) are not objective enough to protect human subjects. *Human Research Report* 2007 March; 22(3): 8. NRCBL: 18.2; 18.5.2; 18.3. SC: le. Identifiers: Grimes v. Kennedy Krieger Institute (KKI), (associated with Johns Hopkins University), (Part 5).

Maloney, Dennis M. In court: legal principles for protecting human subjects. *Human Research Report* 2007 October; 22(10): 8. NRCBL: 18.2; 18.3. SC: le. Identifiers:

Grimes v. Kennedy Krieger Institute (KKI), (associated with Johns Hopkins University), (Part 11).

Moore, Mary E.; Berk, Stephen N.; Freedman, Benjamin; Salisbury, David A.; Schechter, Martin T. Research involving human subjects. *In:* Baylis, Françoise; Downie, Jocelyn; Freedman, Benjamin; Hoffmaster, Barry; Sherwin, Susan, eds. Health Care Ethics in Canada. Toronto: Harcourt Brace Canada, 1995: 319-364. NRCBL: 18.2; 18.3; 18.5.1; 9.5.6; 4.1.1. SC: le. Note: Section includes: Introduction — Halushka v. University of Saskatchewan et al. / Saskatchewan Court of Appeal — Ethical considerations encountered in a study of acupuncture — a reappraisal / Mary E. Moore and Stephen N. Berk — Equipoise and the ethics of clinical research / Benjamin Freedman — AIDS trials, civil liberties and the social control of therapy: should we embrace new drugs with open arms? / David A. Salisbury and Martin T. Schechter.

Osborn, Andrew. Three Russian doctors face trial for vaccine tests [news]. *BMJ: British Medial Journal* 2007 April 21; 334(7598): 817. NRCBL: 18.2; 9.7; 18.5.2. SC: le.

Pullen, Sophie. Research on people: ethical considerations. *In:* Pullen, Sophie; Gray, Carol, eds. Ethics, Law and the Veterinary Nurse. New York: Elsevier Butterworth Heinemann, 2006: 117-128. NRCBL: 18.2. SC: le.

Romeo-Casabona, Carlos; Nicolas, Pilar. Research ethics committees in Spain. *In:* Beyleveld, D.; Townend, D.; Wright, J., eds. Research Ethics Committees, Data Protection and Medical Research in European Countries. Hants, England; Burlington, VT: Ashgate; 2005: 233-244. NRCBL: 18.2. SC: le.

Slocum, J. Michael. Legal issues in clinical trials. *In:* Kulakowski, Elliott C.; Chronister, Lynne U., eds. Research Administration and Management. Sudbury, MA: Jones and Bartlett, 2006: 189-206. NRCBL: 18.2; 9.7; 18.6; 18.3. SC: le.

Society for Adolescent Medicine; Santelli, John S.; Smith Rogers, Audrey; Rosenfeld, Walter D.; DuRant, Robert H.; Dubler, Nancy; Morreale, Madlyn; English, Abigail; Lyss, Sheryl; Wimberly, Yolanda; Schissel, Anna. Guidelines for adolescent health research. A position paper of the Society for Adolescent Medicine. *Journal of Adolescent Health* 2003 November; 33(5): 396-409. NRCBL: 18.2; 18.5.2; 18.3; 8.3.2; 8.4; 18.4. SC: le.

Tonti-Filippini, Nicholas. The need for ethics committees, and their role and function. *National Catholic Bioethics Quarterly* 2007 Winter; 7(4): 749-769. NRCBL: 18.2; 1.2; 4.1.1; 18.6; 18.3; 18.3. SC: le.
  Abstract: The search for truth is not the sole end of science. Science serves humanity, not humanity science. Science must never forget that the human being is not a mere means to scientific ends, but the reason for and goal of research. The central function of bioethics committees is to guide the development of medical science so that it genuinely seeks knowledge within the context of recog-

nizing that each human being is created in God's own image and likeness and that no member of the human family may be used or treated merely as an object of use.

United States. Department of Health and Human Services [DHHS]. Protecting Personal Health Information in Research: Understanding the HIPAA Privacy Rule. Washington, DC: Department of Health and Human Services (HHS), 2003: 33 p. NRCBL: 18.2; 8.4. SC: le. Identifiers: Health Insurance Portability and Accountability Act.

United States. Department of Health and Human Services [DHHS]. Protection of human research subjects; notice of proposed rule making. *Federal Register* 2001 July 6; 66(130): 35576-35580 [Online]. Accessed: http://frwebgate.access.gpo.gov/cgi-bin/multidb.cgi [2005 December 27]. NRCBL: 18.2; 18.6; 18.5.2; 18.5.3; 18.5.4; 18.5.6. SC: le.

United States. Department of Health and Human Services [DHHS]. Office for Human Research Protections [OHRP]. Part 46 - Protection of human subjects [Revised 2005 June 23; Effective 2005 June 23]. Code of Federal Regulations: Title 45: Public Welfare 2005 October 1: 117-134 [Online]. Accessed: http://www.hhs.gov/ohrp/humansubjects/guidance/45cfr46.htm [2007 April 25]. NRCBL: 18.2. SC: le.

United States. Environmental Protection Agency [EPA]. Protections for subjects in human research; proposed rule. *Federal Register* 2005 September 12; 70(175): 53837-53866 [Online]. Accessed: http://www.frwebgate.access.gpo.gov/cgi-bin/multidb.cgi [2005 December 27]. NRCBL: 18.2; 16.1; 18.5.2; 18.5.3; 18.5.4; 18.5.5; 18.6. SC: le.

Wilson, Roxanne M. Litigating on the new frontier: inroads on the duties of sponsors and investigators in clinical trials. *Federation of Defense and Corporate Counsel Quarterly* 2005 Fall; 56(1): 49-75. NRCBL: 18.2; 18.3. SC: le.

HUMAN EXPERIMENTATION/ INFORMED CONSENT
See also INFORMED CONSENT

Uninformed consent? The US should revamp rules on informed consent to ensure that people have all the information and support they need before deciding to enroll in clinical trials [editorial]. *Nature Medicine* 2007 September; 13(9): 999. NRCBL: 18.3; 18.6; 15.4.
> Keywords: *clinical trials; *gene therapy; *government regulation; *informed consent; adverse effects; conflict of interest; consent forms; death; disclosure; industry; physicians; research ethics committees; research subjects; research support; researchers; risks and benefits; therapeutic misconception; Proposed Keywords: rheumatoid arthritis; Keyword Identifiers: *United States; Mohr, Jolee; Targeted Genetics Corp.

Adams, Mary; Fischler, Ira; Simmerling, Mary;. Deception. *Journal of Empirical Research on Human Research Ethics* 2007 March; 2(1): 87-91. NRCBL: 18.3; 8.2.

Alzheimer's Association. National Board of Directors. Ethical issues in dementia research (with special emphasis on "informed consent"). Chicago, IL: Alzheimer's Association, 1997 May; 2 p. [Online]. Accessed: http://www.alz.org/national/documents/statements_ethicalissues.pdf [2007 September 17]. NRCBL: 18.3; 18.5.6.

American College of Epidemiology Policy Committee; Ness, Roberta B. Biospecimen "ownership": point. *Cancer Epidemiology, Biomarkers and Prevention* 2007 February; 16(2): 188-189. NRCBL: 18.3; 15.11; 18.6; 4.4.

Anderson, Andrea. Ethicists balk at new emergency trials that skip informed consent [news]. *Nature Medicine* 2007 July; 13(7): 765. NRCBL: 18.3; 18.5.1; 18.6.

Baylis, Françoise; Ram, Natalie. Eligibility of cryopreserved human embryos for stem cell research in Canada. *JOGC: Journal of Obstetrics and Gynaecology Canada = JOGC:Journal d'Obstétrique et Gynécologie du Canada* 2005 October; 27(10): 949-955. NRCBL: 18.3; 18.5.4; 15.1; 14.6. SC: em. Note: Abstract in French.

Biros, Michelle H. Research without consent: exception from and waiver of informed consent in resuscitation research. *Science and Engineering Ethics* 2007 September; 13(3): 361-369. NRCBL: 18.3; 18.5.1; 18.6.

Biros, Michelle H. The ethics of research in emergency medicine. *Science and Engineering Ethics* 2007 September; 13(3): 279-280. NRCBL: 18.3; 18.5.1.

Black, Betty S.; Kass, Nancy E.; Fogarty, Linda A.; Rabins, Peter V. Informed consent for dementia research: the study enrollment encounter. *IRB: Ethics and Human Research* 2007 July-August; 29(4): 7-14. NRCBL: 18.3; 18.5.6; 18.5.7. SC: em.

Blackwood, Bronagh. Informed consent for research in critical care: implications for nursing. *Nursing in Critical Care* 2006 July-August; 11(4): 151-153. NRCBL: 18.3; 18.2; 9.5.1; 4.1.3.

Booth, Malcolm G. Informed consent in emergency research: a contradiction in terms. *Science and Engineering Ethics* 2007 September; 13(3): 351-359. NRCBL: 18.3; 18.5.1.

Bowling, Ann; Rowe, Gene. "You decide doctor". What do patient preference arms in clinical trials really mean? [editorial]. *Journal of Epidemiology and Community Health* 2005 November; 59(11): 914-915. NRCBL: 18.3; 8.1.

Bradburn, Norman; Simon, Gayle; Bankowski, Susan Burner; Beattie, Elizabeth; Buckwalter, Kathleen; Clark, Laura; Diehl, Dawn. Informed consent. *Journal*

NRCBL: National Reference Center for Bioethics Literature Classification Scheme   See inside front cover for terms.

267

of Empirical Research on Human Research Ethics 2007 March; 2(1): 75-82. NRCBL: 18.3; 18.5.2; 18.5.6.

Breese, Peter E.; Burman, William J.; Goldberg, Stefan; Weis, Stephen E. Education level, primary language, and comprehension of the informed consent process. *Journal of Empirical Research on Human Research Ethics* 2007 December; 2(4): 69-79. NRCBL: 18.3; 21.7. SC: em.

Abstract: To information on how persons from diverse backgrounds experience the informed consent process, we surveyed adults with a wide variety of educational levels and different primary languages (English, Spanish, or Vietnamese) who had recently enrolled in a study requiring written informed consent. Of the 100 participants, 62 were non-White, 43 had less than a high school education, and 60 had a primary language other than English. The median score for comprehension was 62% (IQR 50–76%); the median satisfaction score was 86% (IQR 71–100%). In multivariate analysis, only educational level was significantly associated with comprehension and satisfaction with the informed consent process (p 0.001). Comprehension and satisfaction with the informed consent process were markedly lower among persons with lower educational levels.

Breese, Peter; Rietmeijer, Cornelis; Burman, William. Content among locally approved HIPAA authorization forms for research. *Journal of Empirical Research on Human Research Ethics* 2007 March; 2(1): 43-46. NRCBL: 18.3; 18.6; 8.4. SC: le. Identifiers: Health Insurance Portability and Accountability Act.

Abstract: This study was designed to access differences in the content of HIPAA authorization forms now required for clinical research. Authorization forms were collected from 111 institutions, including academic medical centers and commercial Institutional Review Boards. The requirement for an element covering the use of information acquired was fulfilled in 95% of the forms, and 100% had a statement fulfilling the core requirement of a description of the data to be collected. However, only 19% distinguished between entities that could see personal identifiers versus aggregate data. Significant differences existed in how long the disclosure agreement would remain in effect, and complex legalistic language was common. Thus, while research authorization forms technically met the requirements, the complex language and confusion over personal identifiers may raise concerns in prospective research participants.

Buchanan, David; Miller, Franklin G. Justice in research on human subjects. *In:* Rhodes, Rosamond; Francis, Leslie P.; Silvers, Anita, eds. The Blackwell Guide to Medical Ethics. Malden, MA: Blackwell Pub., 2007: 373-392. NRCBL: 18.3; 18.2; 1.1.

Cahana, Alex; Romagnioli, Simone. Not all placebos are the same: a debate on the ethics of placebo use in clinical trials versus clinical practice. *Journal of Anesthesia* 2007; 21(1): 102-105. NRCBL: 18.3; 18.2.

Catania, Joseph A.; Wolf, Leslie E.; Wertleib, Stacey; Lo, Bernard; Henne, Jeff. Research participants' perceptions of the Certificate of Confidentiality's assurances and limitations. *Journal of Empirical Research on Human Research Ethics* 2007 December; 2(4): 53-59. NRCBL: 18.3; 8.4; 18.2. SC: em.

Abstract: The certificate of confidentiality (COC) provides additional protections to personal and sensitive research data. COC guarantees are not absolute and investigators are obligated to inform potential participants of COC limitations. The present study utilized qualitative and partnership methodology to examine participants' (N = 24) perceptions of COC assurances and limitations in the context of a hypothetical study on depression. Although some participants were comforted by COC assurances, a majority of participants had confidentiality/privacy concerns specifically with COC passages concerning federal audits and legal reporting requirements. As one respondent noted, "Why is it that you guys don't have to turn the records over to the court unless I say so ... but you have to give them over to the government? . . . I don't know about what is goin' on." Our findings underscore the need for larger quantitative investigations to examine the negative and positive impact of COCs on research participation and response bias.

Chenaud, Catherine; Merlani, Paolo; Luyasu, Samuel; Ricou, Bara. Informed consent for research obtained during the intensive care unit stay. *Critical Care* 2006; 10(6): R170. NRCBL: 18.3; 9.5.1. SC: em.

Choi, Joanna M.; Salter, Sharon A.; Kimball, Alexa B. Innovative care, medical research, and the ethics of informed consent. *Journal of the American Academy of Dermatology* 2007 February; 56(2): 330-332. NRCBL: 18.3; 18.5.1; 7.1; 8.1; 8.3.1.

Christopher, Paul P.; Foti, Mary Ellen; Roy-Bujnowski, Kristen; Appelbaum, Paul S. Consent form readability and educational levels of potential participants in mental health research. *Psychiatric Services* 2007 February; 58(2): 227-232. NRCBL: 18.3; 18.5.6. SC: em.

Coats, T.J. Consent for emergency care research: the Mental Capacity Act 2005. *Emergency Medicine Journal* 2006 December; 23(12): 893-894. NRCBL: 18.3; 18.5.6; 9.5.1. SC: le. Identifiers: United Kingdom (Great Britain).

Cohen, A.T.; Maillardet, L.M.A. Are placebo-controlled trials ethical in areas where current guidelines recommend therapy? Yes. *Journal of Thrombosis and Haemostasis* 2006 October; 4(10): 2130-2132. NRCBL: 18.3; 18.2. SC: an.

Cooper, Matthew. Sharing data and results in ethnographic research: why this should not be an ethical imperative. *Journal of Empirical Research on Human Research Ethics* 2007 March; 2(1): 3-19. NRCBL: 18.3; 8.2; 1.3.1; 18.6; 18.5.9. SC: cs; rv.

Abstract: Researchers recently have argued that offering to share research results with study participants should be an "ethical imperative." This article considers that suggestion in light of the practice of ethnographic, particularly anthropological, research. Sharing results is discussed in relation to several issues, e.g., whether it occurs during or after completion of a project, whether the research is long-term, the complexities involved in depositing field materials in archives, the changing politics of ethnographic research, research not concerned with communities, situations in which participants and the anthropologist may be in danger, and changing styles of ethnographic research. I argue that, ideally, sharing should be a regular component of ethnographic research but should not be an ethical requirement. Given the complexity, variety and changing political contexts of ethnographic research, implementing such a requirement would often be practically impossible and sometimes would be inadvisable. I recommend instead that research ethics boards educate themselves about the nature of ethnographic research. Further, they should approach decision making on the issue of data or results sharing on a case-by-case basis. For researchers, I recommend that discussion of data and result sharing should become part of the education of all ethnographers and that discussion of the issue should be fostered.

Corrigan, Oonagh. Empty ethics: the problem with informed consent. *Sociology of Health and Illness* 2003 November; 25(7): 768-792. NRCBL: 18.3; 18.2. SC: em.

Cox, A.C.; Fallowfield, L.J.; Jenkins, V.A. Communication and informed consent in phase 1 trials: a review of the literature. *Supportive Care in Cancer* 2006 April; 14(4): 303-309. NRCBL: 18.3; 18.2; 9.5.1. SC: rv.

Dew, Rachel E. Informed consent for research in borderline personality disorder [debate]. *BMC Medical Ethics [electronic]* 2007; 8:4. 4 p. NRCBL: 18.3; 17.1; 18.2.

Abstract: Background: Previous research on informed consent for research in psychiatric patients has centered on disorders that affect comprehension and appreciation of risks. Little has been written about consent to research in those subjects with Borderline Personality Disorder, a prevalent and disabling condition. Discussion: Despite apparently intact cognition and comprehension of risks, a borderline subject may deliberately choose self-harm in order to fulfill abnormal psychological needs, or due to suicidality. Alternatively, such a subject may refuse enrollment due to transference or the desire to harm him or herself. Such phenomena could be precipitated or prevented by the interpersonal dynamics of the informed consent encounter. Summary: Caution should be exercised in obtaining informed consent for research from subjects with Borderline Personality Disorder. A literature review and recommendations for future research are discussed.

Dickert, Neal W.; Sugarman, Jeremy. Getting the ethics right regarding research in the emergency setting: lessons from the PolyHeme Study. *Kennedy Institute of Ethics Journal* 2007 June; 12(2): 153-169. NRCBL: 18.3; 18.2; 18.6; 19.4. SC: rv.

Abstract: Research in emergency settings (RES) has become a major public issue with urgent policy implications. Significant attention has focused recently on RES in response to the trial of PolyHeme, a synthetic blood substitute, in trauma victims in hemorrhagic shock. Unfortunately, the discussion of the PolyHeme trial in the popular and scholarly press leaves important questions unanswered. This paper articulates three important lessons from the PolyHeme trial that have significant policy implications. First, the RES regulations should be re-visited, particularly the requirement that existing treatments be unproven or unsatisfactory in order for research to be acceptable without consent. Second, further conceptual and empirical scholarship is needed to accomplish the goal of effectively involving communities. Third, a more subtle analysis is needed regarding how to balance the needs of maintaining public trust and protecting confidential trade information in the context of RES.

Dressler, Lynn G. Biospecimen "ownership": counterpoint. *Cancer Epidemiology, Biomarkers and Prevention* 2007 February; 16(2): 190-191. NRCBL: 18.3; 15.11; 18.6.

Feuchtbaum, Lisa; Cunningham, George; Sciortino, Stan. Questioning the need for informed consent: a case study of California's experience with a pilot newborn screening research project. *Journal of Empirical Research on Human Research Ethics* 2007 September; 2(3): 3-14. 31 refs. NRCBL: 18.3; 18.5.2; 15.3; 18.6; 8.3.2; 15.1. SC: cs; an; em.

Keywords: *genetic research; *genetic screening; *mandatory testing; *mass screening; *newborns; *public health; attitudes; autonomy; blood specimen collection; communitarianism; genetic disorders; interviews; moral policy; policy analysis; pregnant women; public policy; refusal to participate; risks and benefits; utilitarianism; Proposed Keywords: pilot projects; Keyword Identifiers: California

Abstract: California provides mandatory newborn screening for disorders that cause irreversible, severe disabilities if not identified and treated early in life. Parental consent is not required. In 2001, the Genetic Disease Branch was mandated to pilot test a new technology that could identify many additional disorders using the same blood specimen already collected. Study participation required informed consent, which was obtained for 47% of births during the study timeframe. The inability of hospitals to carry out the consent procedure for all newborns resulted in denial of testing and missed cases. If informed consent were waived, all newborns could have been tested. Several empirical questions are posed and each is examined from the perspective of society, the parents and the newborn. It is concluded that the legitimate needs of society and the interests of newborns should not be sacrificed to respond to the autonomy interests of the few parents who did not wish their infant to participate in the study, and that in the future, parental consent should be waived for projects evaluating new screening technologies.

Gawande, Atul. A lifesaving checklist [op-ed]. *New York Times* 2007 December 30; p. WK8. NRCBL: 18.3; 8.4;

NRCBL: National Reference Center for Bioethics Literature Classification Scheme   See inside front cover for terms.

269

9.8; 5.1. SC: po; em. Identifiers: Johns Hopkins University; Office for Human Research Protections [OHRP].

Goldim, José Roberto; Clotet, Joaquim; Ribeiro, Jorge Pinto. Adequacy of informed consent in research carried out in Brazil. *Eubios Journal of Asian and International Bioethics* 2007 November; 17(6): 177-180. NRCBL: 18.3; 9.6. SC: em.

Hamilton, S.; Hepper, J.; Hanby, A.; Hewison, J. Consent gained from patients after breast surgery for the use of surplus tissue in research: an exploration. *Journal of Medical Ethics* 2007 April; 33(4): 229-233. NRCBL: 18.3; 19.5; 9.5.5. SC: em.

Abstract: OBJECTIVES: (1) To investigate the quality of consent gained for the use in research of tissue that is surplus after surgery. (2) To compare the use of two consent forms: a simple locally introduced form and a more complex centrally instigated form. (3) To discuss the attitudes of patients towards the use of their surplus tissue in research. DESIGN: Data were collected through interviews and analysed with a combination of quantitative and qualitative analytical techniques. Participants and SETTING: Patients of the breast care unit at a teaching hospital were interviewed at home or in a quiet room at the hospital. RESULTS: 57 people were interviewed out of 81 approached, between October 2003 and March 2004. Most participants had a poor level of knowledge about the consent they had given, but reported being happy about having given it. The patients who had signed the locally introduced form had considerably more knowledge than those who had signed the centrally instigated form (z = -2.56; p.05). Participants considered being well informed to be less important than believing that their opinions were valued and respected. CONCLUSIONS: The findings suggest that traditional models of informed consent are not universally applicable and, in this case, seem to overstate what people wish to know. The simple consent form achieved a better quality of informed consent and provided a better model of practice than the complex form, and it seemed that a focused approach to consent seeking is more effective and acceptable than more complex approaches.

Hem, Marit Helene; Heggen, Kristin; Ruyter, Knut W. Questionable requirement for consent in observational research in psychiatry. *Nursing Ethics* 2007 January; 14(1): 41-53. NRCBL: 18.3; 18.5.6; 8.3.3. SC: em. Identifiers: Norway.

Abstract: Informed consent represents a cornerstone of the endeavours to make health care research ethically acceptable. Based on experience of qualitative research on power dynamics in nursing care in acute psychiatry, we show that the requirement for informed consent may be practised in formalistic ways that legitimize the researcher's activities without taking the patient's changing perception of the situation sufficiently into account. The presentation of three patient case studies illustrates a diversity of issues that the researcher must consider in each situation. We argue for the necessity of researchers to base their judgement on a complex set of competencies. Consciousness of research ethics must be combined with knowledge of the challenges involved in research methodology in qualitative research and familiarity with the therapeutic arena in which the research is being conducted. The article shows that the alternative solution is not simple but must emphasize the researcher's ability to doubt and be based on an awareness of the researcher's fallibility.

Huntington, Ian; Robinson, Walter. The many ways of saying yes and no: reflections on the research coordinator's role in recruiting research participants and obtaining informed consent. *IRB: Ethics and Human Research* 2007 May-June; 29(3): 6-10. NRCBL: 18.3; 18.2.

Ilfeld, Brian M. Informed consent for medical research: an ethical imperative. *Regional Anesthesia and Pain Medicine* 2006 July-August; 31(4): 353-357. NRCBL: 18.3; 18.2.

Illes, J.; Chin, V. Trust and reciprocity: foundational principles for human subjects imaging research. *Canadian Journal of Neurological Sciences* 2007 February; 34(1): 3-4. NRCBL: 18.3; 18.2; 8.1.

Iorio, A.; Agnelli, G. Are placebo-controlled trials ethical in areas where current guidelines recommend therapy? No. *Journal of Thrombosis and Haemostasis* 2006 October; 4(10): 2133-2136. NRCBL: 18.3; 18.2. SC: an.

Iwanoswski, Piotr S. Informed consent procedure for clinical trials in emergency settings: the Polish perspective. *Science and Engineering Ethics* 2007 September; 13(3): 333-336. NRCBL: 18.3; 18.5.1.

Karlawish, Jason. Research on cognitively impaired adults. *In:* Steinbock, Bonnie, ed. The Oxford Handbook of Bioethics. Oxford; New York: Oxford University Press, 2007: 597-620. NRCBL: 18.3; 8.3.3.

Kim, Scott Y.; Kieburtz, Karl. Appointing a proxy for research consent after one develops dementia: the need for further study [editorial]. *Neurology* 2006 May 9; 66(9): 1298-1299 [Online]. Accessed: http://www.neurology.org/ [2007 April 16]. NRCBL: 18.3; 18.5.6; 18.2.

Kimberly, Michael B.; Hoehn, K. Sarah; Feudtner, Chris; Nelson, Robert M.; Schreiner, Mark. Variation in standards of research compensation and child assent practices: a comparison of 69 institutional review board-approved informed permission and assent forms for 3 multicenter pediatric clinical trials. *Pediatrics* 2006 May; 117(5): 1706-1711. NRCBL: 18.3; 9.3.1; 18.2; 18.5.2. SC: em; rv.

Kompanje, Erwin J.O. 'No time to be lost!' Ethical considerations on consent for inclusion in emergency pharmacological research in severe traumatic brain injury in the European Union. *Science and Engineering Ethics* 2007 September; 13(3): 371-381. NRCBL: 18.3; 18.5.1; 9.7; 21.1. SC: em.

SC (Subject Captions):   an=analytical   cs=case studies   em=empirical   le=legal   po=popular   rv=review

Krosin, Michael T.; Klitzman, Robert; Levin, Bruce; Cheng, Jianfeng; Ranney, Megan L. Problems in comprehension of informed consent in rural and peri-urban Mali, West Africa. *Clinical Trials (London, England)* 2006; 3(3): 306-313. NRCBL: 18.3; 18.5.9. SC: em.

Länsimies-Antikainen, Helena; Pietilä, Anna-Maija; Laitinen, Tomi; Schwab, Ursula; Rauramaa, Rainer; Länsimies, Esko. Evaluation of informed consent: a pilot study. *Journal of Advanced Nursing* 2007 July; 59(2): 146-154. NRCBL: 18.3. SC: em. Identifiers: Finland.

Luce, John M. California's new law allowing surrogate consent for clinical research involving subjects with impaired decision-making capacity. *Intensive Care Medicine* 2003 June; 29(6): 1024-1025. NRCBL: 18.3; 18.2; 18.6. SC: le.

Maloney, Dennis M. Case study: university says it will modify many human subject protection procedures. *Human Research Report* 2007 August; 22(8): 6-7. NRCBL: 18.3; 18.2; 18.5.5; 18.6. SC: cs.

Maloney, Dennis M. In court: court says nontherapeutic research requires special informed consent measures. *Human Research Report* 2007 November; 22(11): 8. NRCBL: 18.3; 18.5.2; 15.4. SC: le.
> Keywords: *informed consent; *legal aspects; *nontherapeutic research; disclosure; gene therapy; children; research subjects; researcher subject relationship; Keyword Identifiers: *Grimes v. Kennedy Krieger Institute; Gelsinger, Jesse

Maloney, Dennis M. In court: researchers didn't tell subject's mother of high lead levels in house until after blood tests. *Human Research Report* 2007 July; 22(7): 8. NRCBL: 18.3; 16.1; 18.5.2. SC: le. Identifiers: Grimes v Kennedy Krieger Institute (KKI), (associated with Johns Hopkins University), (Part 9).

Maloney, Dennis M. IRBs have some leeway on methods of informed consent. *Human Research Report* 2007 August; 22(8): 3. NRCBL: 18.3; 18.5.2; 18.2.

Marshall, Jennifer; Martin, Toby; Downie, Jocelyn; Malisza, Krisztina. A comprehensive analysis of MRI research risks: in support of full disclosure. *Canadian Journal of Neurological Sciences* 2007 February; 34(1): 11-17. NRCBL: 18.3; 18.2; 18.5.1.

Matthews, Robert. Are you looking at me? Medical researchers keen to scour patients' data for insights into disease should get consent first or risk coming seriously unstuck. *New Scientist* 2007 August 4-10; 195(2615): 18. NRCBL: 18.3; 8.4; 15.1; 1.3.12.
> Keywords: *biomedical research; *databases; *genetic databases; *genetic research; *informed consent; *presumed consent; confidentiality; epidemiology; international aspects; medical records; registries; Keyword Identifiers: Great Britain; UK Biobank

May, Thomas; Craig, J.M.; Spellecy, Ryan. IRBs, hospital ethics committees, and the need for "translational in-

formed consent". *Academic Medicine* 2007 July; 82(7): 670-674. NRCBL: 18.3; 18.2; 8.3.1; 9.6.

McCarty, Catherine A.; Nair, Anuradha; Austin, Diane M.; Giampietro, Philip F. Informed consent and subject motivation to participate in a large, population-based genomics study: the Marshfield Clinic Personalized Medicine Research Project. *Community Genetics* 2007; 10(1): 2-9. 23 refs. NRCBL: 18.3; 15.1; 15.11; 1.3.12. SC: em.
> Keywords: *genetic databases; *genetic research; *informed consent; *motivation; *population genetics; *research subjects; biological specimen banks; coercion; donors; genomics; questionnaires; researchers; Keyword Identifiers: Wisconsin

Mello, Michelle M.; Joffe, Steven. Compact versus contract: industry sponsors' obligations to their research subjects. *New England Journal of Medicine* 2007 June 28; 356(26): 2737-2743. NRCBL: 18.3; 5.3; 9.7; 18.2. SC: le.

Miller, F.G.; Kaptchuk, T.J. Acupuncture trials and informed consent. *Journal of Medical Ethics* 2007 January; 33(1): 43-44. NRCBL: 18.3; 4.1.1.
> Abstract: Participants are often not informed by investigators who conduct randomised, placebo-controlled acupuncture trials that they may receive a sham acupuncture intervention. Instead, they are told that one or more forms of acupuncture are being compared in the study. This deceptive disclosure practice lacks a compelling methodological rationale and violates the ethical requirement to obtain informed consent. Participants in placebo-controlled acupuncture trials should be provided an accurate disclosure regarding the use of sham acupuncture, consistent with the practice of placebo-controlled drug trials.

Miller, Robin L.; Forte, Draco; Wilson, Bianca Della; Greene, George J. Protecting sexual minority youth from research risks: conflicting perspectives. *American Journal of Community Psychology* 2006 June; 37(3-4): 341-348. NRCBL: 18.3; 8.3.2; 18.2; 18.5.2; 18.6; 10.

Miller, Tina; Bell, Linda. Consenting to what? Issues of access, gate-keeping and 'informed' consent. *In:* Mauthner, Melanie; Birch, Maxine; Jessop, Julie; Miller, Tina, eds. Ethics in Qualitative Research. London; Thousand Oaks, CA: Sage Publications Ltd., 2002: 53-69. NRCBL: 18.3; 18.5.2; 10; 21.7.

Newton, Sam K.; Appiah-Poku, John. The perspectives of researchers on obtaining informed consent in developing countries. *Developing World Bioethics* 2007 April; 7(1): 19-24. NRCBL: 18.3; 18.5.9; 21.1. SC: em.
> Abstract: Background: The doctrine of informed consent (IC) exists to protect individuals from exploitation or harm. This study into IC was carried out to investigate how different researchers perceived the process whereby researchers obtained consent. It also examined researchers' perspectives on what constituted IC, and how different settings influenced the process. Methods: The study recorded in-depth interviews with 12 lecturers and five doctoral students, who had carried out research in devel-

NRCBL: National Reference Center for Bioethics Literature Classification Scheme   See inside front cover for terms.

271

oping countries, at a leading school of public health in the United Kingdom. A purposive, snowballing approach was used to identify interviewees. Results: Although the concept and application of the doctrine of IC should have been the same, irrespective of where the research was carried out, the process of obtaining it had to be different. The setting had to be taken into consideration and the autonomy of the subject had to be respected at all times. In areas of high illiteracy, and where understanding of the subject was likely to be a problem, there was an added responsibility placed on the researcher to devise innovative ways of carrying out the study, taking into consideration the peculiarities of the environment. Conclusion: The ethical issues for IC were the same, irrespective of where the research was conducted. However, because the backgrounds, setting, and knowledge of populations differed, there was the need to be similarly sensitive in obtaining consent. The problems of obtaining genuine IC were not limited to developing countries.

Revheim, Nadine; Javitt, Daniel C. Reading disability goes beyond consent forms [letter]. *Psychiatric Services* 2007 April; 58(4): 566. NRCBL: 18.3; 18.5.6.

Rivera, Roberto; Borasky, David; Rice, Robert; Carayon, Florence; Wong, Emelita. Informed consent: an international researchers' perspective. *American Journal of Public Health* 2007 January; 97(1): 25-30. NRCBL: 18.3; 21.1. SC: em.
Abstract: We reported 164 researchers' recommendations for information that should be included in the informed consent process. These recommendations were obtained during training workshops conducted in Africa, Europe, and the United States. The 8 elements of informed consent of the US Code of Federal Regulations were used to identify 95 items of information ("points"), most related to benefits and research description. Limited consensus was found among the 3 workshops: of the 95 points, only 27 (28%) were identified as useful by all groups. These points serve as a springboard for identifying information applicable in different geographic areas and indicate the need for involving a variety of individuals and stakeholders, with different research and cultural perspectives, in the development of informed consent, particularly for research undertaken in international settings.

Rózynska, Joanna; Czarkowski, Marek. Emergency research without consent under Polish Law. *Science and Engineering Ethics* 2007 September; 13(3): 337-350. NRCBL: 18.3; 18.5.1. SC: le.

Saks, Elyn R.; Dunn, Laura B.; Marshall, Barbara J.; Nayak, Gauri V.; Golshan, Shahrokh; Jeste, Dilip V. The California Scale of Appreciation: a new instrument to measure the appreciation component of capacity to consent to research. *American Journal of Geriatric Psychiatry* 2002 March-April; 10(2): 166-174. NRCBL: 18.3; 18.5.6. SC: em.

Sammons, Helen M.; Atkinson, Maria; Choonara, Imti; Stephenson, Terence. What motivates British parents to consent for research? A questionnaire study. *BMC Pediatrics* 2007 March 9; 7: 12. NRCBL: 18.3; 18.5.2; 18.2.

Schellings, Ron; Kessels, Alfons G.; ter Riet, Gerben; Knottnerus, J. André; Sturmans, Ferd. Randomized consent designs in randomized controlled trials: systematic literature search. *Contemporary Clinical Trials* 2006 August; 27(4): 320-332. NRCBL: 18.3; 18.2. SC: rv.

Shaibu, Sheila. Ethical and cultural considerations in informed consent in Botswana. *Nursing Ethics* 2007 July; 14(4): 503-509. NRCBL: 18.3; 8.4; 18.5.9; 21.7. SC: em.
Abstract: Reflections on my experience of conducting research in Botswana are used to highlight tensions and conflicts that arise from adhering to the western conceptualization of bioethics and the need to be culturally sensitive when carrying out research in one's own culture. Cultural practices required the need to exercise discretionary judgement guided by respect for the culture and decision-making protocols of the research participants. Ethical challenges that arose are discussed. The brokerage role of nurse educators and leaders in contextualizing western bioethics is emphasized.

Sheremeta, Lorraine. Public meets private: challenges for informed consent and umbilical cord blood banking in Canada. *Health Law Review* 2007; 15(2): 23-29. NRCBL: 18.3; 19.4. SC: le.

Sieber, Joan E. Respect for persons and informed consent: a moving target [editorial]. *Journal of Empirical Research on Human Research Ethics* 2007 September; 2(3): 1-2. NRCBL: 18.3.

Slaughter, Susan; Cole, Dixie; Jennings, Eileen; Reimer, Marlene A. Consent and assent to participate in research from people with dementia. *Nursing Ethics* 2007 January; 14(1): 27-40. NRCBL: 18.3; 18.5.6; 18.5.7; 8.3.3.
Abstract: Conducting research with vulnerable populations involves careful attention to the interests of individuals. Although it is generally understood that informed consent is a necessary prerequisite to research participation, it is less clear how to proceed when potential research participants lack the capacity to provide this informed consent. The rationale for assessing the assent or dissent of vulnerable individuals and obtaining informed consent by authorized representatives is discussed. Practical guidelines for recruitment of and data collection from people in the middle or late stage of dementia are proposed. These guidelines were used by research assistants in a minimal risk study.

Snowden, Claire; Elbourne, Diana; Garcia, Jo. Declining enrolment in a clinical trial and injurious misconceptions: is there a flipside to the therapeutic misconception? *Clinical Ethics* 2007 December; 2(4): 193-200. NRCBL: 18.3; 18.5.2; 8.3.2; 7.1; 18.2. SC: em. Identifiers: CANDA Trial.
Abstract: The term 'therapeutic misconception' (TM) was introduced in 1982 to conceptualize how some psy-

chiatry trial participants perceived and interpreted their involvement in research. TM has since been identified in many settings and is a major component in research ethics discussions. A qualitative study included a subgroup of interviews with five parents (two couples, one mother) who declined to enrol their baby in a neonatal trial. Analysis suggested the possibility of a counterpart to TM which, given the original terminology, we term the 'injurious misconception' (IM). While TM is closely linked to the elision of care and research, and involves an overstated sense of benefit and protection, IM may be a product of a particularly keen and discomforting sense of distinctions between care and research and a correspondingly over-stated sense of risk and threat.

Sugarman, Jeremy. Examining the provisions for research without consent in the emergency setting. *Hastings Center Report* 2007 January-February; 37(1): 12-13. NRCBL: 18.3; 18.2; 18.6.

Sugarman, Jeremy; Roter, Debra; Cain, Carole; Wallace, Roberta; Schmechel, Don; Welsh-Bohmer, Kathleen A. Proxies and consent discussions for dementia research. *Journal of the American Geriatrics Society* 2007 April; 55(4): 556-561. NRCBL: 18.3; 18.5.6; 18.5.7. SC: em.

Vaslef, Steven N.; Cairns, Charles B.; Falletta, John M. Ethical and regulatory challenges associated with the exception from informed consent requirements for emergency research: from experimental design to institutional review board approval. *Archives of Surgery* 2006 October; 141(10): 1019-1023; discussion 1024. NRCBL: 18.3; 18.6; 18.2; 9.5.1. SC: em.

Waring, Duff R.; Glass, Kathleen Cranley. Legal liability for harm to research participants: the case of placebo-controlled trials. *In:* Lemmens, Trudo; Waring, Duff R., eds. Law and Ethics in Biomedical Research: Regulation, Conflict of Interest and Liability. Toronto; Buffalo: University of Toronto Press, 2006: 206-227. NRCBL: 18.3; 8.5. SC: le.

Winau, Rolf. Experimentation on humans and informed consent: how we arrived where we are. *In:* LaFleur, William R.; Böhme, Gernot; Shimazono, Susumu, eds. Dark Medicine: Rationalizing Medical Research. Bloomington: Indiana University Press, 2007: 46-56. NRCBL: 18.3; 2.2.

HUMAN EXPERIMENTATION/ REGULATION

Board games: the way research on human subjects is overseen in the United States requires reform [editorial]. *Nature* 2007 August 2; 448(7153): 511-512. NRCBL: 18.6; 18.2.

Risk, consent and IRB models. *Protecting Human Subjects* 2007 November (15): 1, 4-5. NRCBL: 18.6; 18.2; 18.3. Identifiers: U.S. Department of Health and Human Ser-

vices Secretary's Advisory Committee on Human Research Protections [SACHRP].

Blumenthal, Daniel S. A community coalition board creates a set of values for community-based research. *Preventing Chronic Disease* 2006 January; 3(1): 7 p. NRCBL: 18.6.

Caron-Flinterman, J. Francisca; Broerse, Jacqueline E.W.; Bunders, Joske F.G. Patient partnership in decision-making on biomedical research: changing the network. *Science, Technology, and Human Values* 2007 May; 32(3): 339-368. NRCBL: 18.6; 8.1. SC: em.

Cash, Richard A. Research ethics involves continuous learning. *Indian Journal of Medical Ethics* 2007 April-June; 4(2): 82-83. NRCBL: 18.6; 18.2; 7.1.

Chen, Donna T.; Jones, Loretta; Gelberg, Lillian. Ethics of clinical research within a community-academic partnered participatory framework. *Ethnicity and Disease* 2006 Winter; 16(1 Supplement 1): S118-S135. NRCBL: 18.6; 9.5.4.

Drazen, Jeffrey M.; Morrissey, Stephen; Curfman, Gregory D. Open clinical trials. *New England Journal of Medicine* 2007 October 25; 357(17): 1756-1757. NRCBL: 18.6; 9.7.

Emanuel, Ezekiel J.; Grady, Christine. Four paradigms of clinical research and research oversight. *CQ: Cambridge Quarterly of Healthcare Ethics* 2007 Winter; 16(1): 82-96. NRCBL: 18.6; 2.2; 8.1.

Epstein, Miran. Clinical trials in the developing world [letter]. *Lancet* 2007 June 2-8; 369(9576): 1859. NRCBL: 18.6; 18.5.9; 21.1; 1.3.2. Comments: Strengthening clinical research in India. Lancet 2007 April 14; 369(9569): 1233.

Glass, Kathleen Cranley. Question and challenges in the governance of research involving humans: a Canadian perspective. *In:* Lemmens, Trudo; Waring, Duff R., eds. Law and Ethics in Biomedical Research: Regulation, Conflict of Interest and Liability. Toronto; Buffalo: University of Toronto Press, 2006: 35-46. NRCBL: 18.6; 1.3.9; 7.3.

Glass, Kathleen Cranley; Kaufert, Joseph. Research ethics review and aboriginal community values: can the two be reconciled? *Journal of Empirical Research on Human Research Ethics* 2007 June; 2(2): 25-40. NRCBL: 18.6; 18.5.1; 18.2.

Abstract: Contemporary Research Ethics Review Committees (RECs) are heavily influenced by the established academic or health care institutional frameworks in which they operate, sharing a cultural, methodological and ethical perspective on the conduct of research involving humans. The principle of autonomous choice carries great weight in what is a highly individualistic decision-making process in medical practice and research. This assumes that the best protection lies in the ability of patients or research participants to make competent, vol-

NRCBL: National Reference Center for Bioethics Literature Classification Scheme   See inside front cover for terms.

273

untary, informed choices, evaluating the risks and bene-fits from a personal perspective. Over the past two decades, North American and international indigenous researchers, policy makers and communities have identi-fied key issues of relevance to them, but ignored by most institutional or university-based RECs. They critique the current research review structure, and propose changes on a variety of levels in an attempt to develop more com-munity sensitive research ethics review processes. In do-ing so, they have emphasized recognition of collective rights including community consent. Critics see alterna-tive policy guidelines and community-based review bod-ies as challenging the current system of ethics review. Some view them as reflecting a fundamental difference in values. In this paper, we explore these developments in the context of the political, legal and ethical frameworks that have informed REC review. We examine the process and content of these frameworks and ask how this con-trasts with emerging Aboriginal proposals for commu-nity-based research ethics review. We follow this with recommendations on how current REC review models might accommodate the requirements of both communi-ties and RECs.

Halwani, Sana. Her Majesty's research subjects: liability of the crown in research involving humans. *In:* Lemmens, Trudo; Waring, Duff R., eds. Law and Ethics in Biomedical Research: Regulation, Conflict of Interest and Liability. Toronto; Buffalo: University of Toronto Press, 2006: 206-227. NRCBL: 18.6; 8.5; 18.2. SC: le. Identifiers: Canada.

International Committee of Medical Journal Editors. Clinical trial registration: looking back and moving ahead. *Lancet* 2007 June 9-15; 369(9577): 1909-1911. NRCBL: 18.6; 1.3.7; 1.3.9; 1.3.12; 21.1. SC: em.

Jalali, Rakesh; Howie, Stephen. Conduct of clinical trials in developing countries [letters]. *Lancet* 2007 August 18-24; 370(9587): 562. NRCBL: 18.6; 18.5.9; 18.2; 9.7.

Kimmelman, Jonathan. Missing the forest: further thoughts on the ethics of bystander risk in medical re-search. *CQ: Cambridge Quarterly of Healthcare Ethics* 2007 Fall; 16(4): 483-490. NRCBL: 18.6; 18.2; 18.5.1; 15.1; 7.1. SC: an.

King, Jean. Accepting tobacco industry money for re-search: has anything changed now that harm reduction is on the agenda? *Addiction* 2006 August; 101(8): 1067-1069. NRCBL: 18.6; 1.3.9; 1.3.2; 18.2; 9.3.1.

Laine, Christine; Horton, Richard; DeAngelis, Catherine D.; Drazen, Jeffrey M.; Frizelle, Frank A.; Godlee, Fiona; Haug, Charlotte; Hébert, Paul C.; Kotzin, Sheldon; Marusic, Ana; Sahni, Peush; Schroeder, Torben V.; Sox, Harold C.; Van Der Weyden, Martin B.; Verheugt, Freek W.A. Clinical trial registration: looking back and moving ahead [editorial]. *Annals of Internal Medicine* 2007 August 21; 147(4): 275-277. NRCBL: 18.6; 1.3.7; 1.3.9; 1.3.12; 21.1.

Laine, Christine; Horton, Richard; DeAngelis, Catherine D.; Drazen, Jeffrey M.; Frizelle, Frank A.; Godlee, Fiona; Haug, Charlotte; Hébert, Paul C.; Kotzin, Sheldon; Marusic, Ana; Sahni, Peush; Schroeder, Torben V.; Sox, Harold C.; Van Der Weyden, Martin B.; Verheugt, Freek W.A. Clinical trial registration: looking back and moving ahead [editorial]. *BMJ: British Medical Journal* 2007 June 9; 334(7605): 1177-1178. NRCBL: 18.6; 1.3.7; 1.3.9; 1.3.12; 21.1.

Laine, Christine; Horton, Richard; DeAngelis, Catherine D.; Drazen, Jeffrey M.; Frizelle, Frank A.; Godlee, Fiona; Haug, Charlotte; Hébert, Paul C.; Kotzin, Sheldon; Marusic, Ana; Sahni, Peush; Schroeder, Torben V.; Sox, Harold C.; Van Der Weyden, Martin B.; Verheugt, Freek W.A. Clinical trial registration: looking back and moving ahead [editorial]. *JAMA: The Journal of the American Medical Association* 2007 July 4; 298(1): 93-94. NRCBL: 18.6; 1.3.7; 1.3.9; 1.3.12; 21.1.

Laine, Christine; Horton, Richard; DeAngelis, Catherine D.; Drazen, Jeffrey M.; Frizelle, Frank A.; Godlee, Fiona; Haug, Charlotte; Hébert, Paul C.; Kotzin, Sheldon; Marusic, Ana; Sahni, Peush; Schroeder, Torben V.; Sox, Harold C.; Van Der Weyden, Martin B.; Verheught, Freek W.A. Clinical trial registration: looking back and moving ahead [edito-rial]. *CMAJ/JAMC: Canadian Medical Association Jour-nal* 2007 July 3; 177(1): 57-58. NRCBL: 18.6; 1.3.7; 1.3.9; 1.3.12; 21.1.

Laine, Christine; Horton, Richard; DeAngelis, Catherine D.; Drazen, Jeffrey M.; Frizelle, Frank A.; Godlee, Fiona; Haug, Charlotte; Hébert, Paul C.; Kotzin, Sheldon; Marusic, Ana; Sahni, Peush; Schroeder, Torben V.; Sox, Harold C.; Van Der Weyden, Martin B.; Verheugt, Freek W.A. Clinical tri-als registration: looking back and moving ahead [edito-rial]. *New England Journal of Medicine* 2007 June 28; 356(26): 2734-2736. NRCBL: 18.6; 1.3.7; 1.3.9; 1.3.12; 21.1.

Lemmens, Trudo; Miller, Paul B. The human subjects trade: ethical, legal, and regulatory remedies to deal with recruitment incentives and to protect scientific integrity. *In:* Lemmens, Trudo; Waring, Duff R., eds. Law and Ethics in Biomedical Research: Regulation, Conflict of Interest and Liability. Toronto; Buffalo: University of Toronto Press, 2006: 132-179. NRCBL: 18.6; 18.2; 1.3.9; 7.3; 18.5.6. SC: le.

Levin, Leonard A.; Palmer, Julie Gage. Institutional re-view boards should require clinical trial registration. *Ar-chives of Internal Medicine* 2007 August 13-27; 167(15): 1576-1580. NRCBL: 18.6; 1.3.12; 18.2; 18.3. SC: le.

Lopus, Jane S.; Grimes, Paul W.; Becker, William E.;Pearson, Rodney A. Effects of human subjects requirements on classroom research: multidisciplinary evidence. *Journal of Empirical Research on Human Research Ethics* 2007 September; 2(3): 69-77. NRCBL: 18.6; 18.2; 7.2. SC: em.

Abstract: Professors who include their students as subjects in classroom-based research projects typically must submit to a review by their university's research ethics committee (REC) even in cases which present only minimal risks, and when the investigation is intended for evaluation of teaching approaches only, and not for publication. Results of a web-based survey with 378 respondents indicate that the perceived costs of the review process may outweigh the perceived benefits to subjects. A logistic regression analysis identifies the time it takes to complete the review application, the time it takes to receive a response, and the necessity of revising a project as significant factors in respondents viewing the REC process as a barrier to research. Instituting policies of expedited review for minimal-risk classroom research and exempting evaluations that are not to be published, both of which are permitted under the current regulations, would decrease burdens on both researchers and REC members, and foster improvement of teaching.

Maloney, Dennis M. Another institution ordered to halt human subjects research projects. *Human Research Report* 2007 November; 22(11): 1-2. NRCBL: 18.6; 18.2. Identifiers: Saint John's Health System of Anderson, Indiana.

Maloney, Dennis M. Case study: university is allowed to resume its human subjects research projects. *Human Research Report* 2007 December; 22(12): 6-7. NRCBL: 18.6; 18.5.5; 18.3; 18.2.

Maloney, Dennis M. Federal office orders college to halt all federally-supported human research. *Human Research Report* 2007 September; 22(9): 1-2. NRCBL: 18.6; 18.2; 18.3. Identifiers: Bluefield State College.

Mastroianni, Anna; Kahn, Jeffrey. Swinging on the pendulum: shifting views of justice in human subjects research. *In:* Lemmens, Trudo; Waring, Duff R., eds. Law and Ethics in Biomedical Research: Regulation, Conflict of Interest and Liability. Toronto; Buffalo: University of Toronto Press, 2006: 47-60. NRCBL: 18.6; 18.3; 1.1.

Meier, Barry. Participants left uninformed in some halted medical trials. *New York Times* 2007 October 30; p. C1, C2. NRCBL: 18.6; 18.3; 9.7. SC: po.

Mills, Peter. Recent issues in assisted reproduction: evolutions in science, law and ethics. *In:* Gunning, Jennifer; Holm, Søren, eds. Ethics, law, and society. Volume 1. Aldershot, Hants, England; Burlington, VT: Ashgate, 2005: 23-31. NRCBL: 18.6; 18.5.4. SC: le.

Ploem, Corrette. Freedom of research and its relation to the right to privacy. *In:* Gevers, J.K.M.; Hondius, E.H.; Hubben, J.H., eds. Health Law, Human Rights and the Biomedicine Convention: Essays in Honour of Henriette

Roscam Abbing. Leiden; Boston: Martinus Nijhoff Publishers, 2005: 159-173. NRCBL: 18.6; 8.4; 1.1; 1.3.9; 21.1. SC: le.

Racher, Frances .E. The evolution of ethics for community practice. *Journal of Community Health Nursing* 2007 Spring; 24(1): 65-76. NRCBL: 18.6; 4.1.3; 9.1.

Resnik, David B. The new EPA regulations for protecting human subjects: haste makes waste. *Hastings Center Report* 2007 January-February; 37(1): 17-21. NRCBL: 18.6; 16.1; 18.2; 15.1. SC: le. Identifiers: Environmental Protection Agency.

Scott, Christopher Thomas; Baker, Monya. Overhauling clinical trials. *Nature Biotechnology* 2007 March; 25(3): 287-292. NRCBL: 18.6.

Shore, Nancy. Community-based participatory research and the ethics review process. *Journal of Empirical Research on Human Research Ethics* 2007 March; 2(1): 31-41. NRCBL: 18.6; 18.2; 7.1; 7.3.

Abstract: This exploratory study examines the experiences of community-based participatory researchers' (CBPR) with the IRB. CBPR is oftentimes applied to non-clinical questions where academic researchers collaborate with community partners to address local concerns. Constant Comparative Method guided the analysis of ten CBPR interviews. The interview questions included: How does your conceptualization of research coincide with the regulations' definition? How are community partners involved in the IRB process? What are the benefits/challenges of the IRB process? And, what recommendations do you have to strengthen the IRB process? The article concludes with suggestions for IRB reviewers and CBPR partners on how to facilitate the review of CBPR projects.

Shrader-Frechette, Kristin. EPA's 2006 human-subjects rule for pesticide experiments. *Accountability in Research* 2007 October-December; 14(4): 211-254. NRCBL: 18.6; 16.1; 18.5.2. SC: le. Identifiers: Environmental Protection Agency.

Stone, Judy. Society and politics. *In her:* Conducting Clinical Research: A Practical Guide for Physicians, Nurses, Study Coordinators, and Investigators. Cumberland, MD: Mountainside MD Press, 2006: 173-194. NRCBL: 18.6; 18.5.3; 1.2; 18.5.1; 21.1.

Taylor, Holly A. Moving beyond compliance: measuring ethical quality to enhance the oversight of human subjects research. *IRB: Ethics and Human Research* 2007 September-October; 29(5): 9-14. NRCBL: 18.6; 18.2.

United States. Department of Health and Human Services [DHHS]. Office for Human Research Protections [OHRP]. Guidance on reporting incidents to OHRP. Washington, DC: Office for Human Research Protections, 2005 May 27; 6 p. [Online]. Accessed: http://www. hhs.gov/ohrp/policy/incidreport_ohrp.html [2007 April 24]. NRCBL: 18.6; 18.2.

NRCBL: National Reference Center for Bioethics Literature Classification Scheme   See inside front cover for terms.

275

Watson, Rory. Developing countries need stronger research guidelines [news]. *BMJ: British Medical Journal* 2007 May 26; 334(7603): 1076. NRCBL: 18.6; 21.1.

Wibulpolprasert, Suwit; Moosa, Sheena; Satyanarayana, K.; Samarage, Sarath; Tangcharoensathien, Viroj. WHO's web-based public hearings: hijacked by pharma? [letter]. *Lancet* 2007 November 24-30; 370(9601): 1754. NRCBL: 18.6; 21.1; 9.7; 7.1; 9.3.1; 1.3.12. Identifiers: World Health Organization.

Zettler, Patricia; Wolf, Leslie E.; Lo, Bernard. Establishing procedures for institutional oversight of stem cell research. *Academic Medicine* 2007 January; 82(1): 6-10. NRCBL: 18.6; 18.5.4; 18.5.1; 15.1.

HUMAN EXPERIMENTATION/ SPECIAL POPULATIONS

Baader, Gerhard. Menschenversuche in der Medizin. *In:* Neuer-Miebach, Therese; Wunder, Michael, eds. Bio-Ethik und die Zukunft der Medizin. Bonn: Psychiatrie-Verlag, 1998: 31-43. 33 fn. NRCBL: 18.5.1; 2.2; 15.5; 21.4.
> Keywords: *eugenics; *National Socialism; *nontherapeutic human experimentation; active euthanasia; historical aspects; involuntary sterilization; killing; misconduct; physicians; Keyword Identifiers: *Germany; Twentieth Century

Beger, H.G.; Arbogast, R. The art of surgery in the 21st century: based on natural sciences and new ethical dimensions. *Langenbeck's Archives of Surgery* 2006 April; 391(2): 143-148. NRCBL: 18.5.1; 9.5.1.

Bleyer, W. Archie; Tejeda, Heriberto A.; Murphy, Sharon B.; Brawley, Otis W.; Smith, Malcolm A.; Ungersleider, Richard S. Equal participation of minority patients in U.S. national pediatric cancer clinical trials. *Journal of Pediatric Hematology and Oncology* 1997 September-October; 19(5): 423-427. NRCBL: 18.5.1; 9.5.1; 18.2; 18.5.2. SC: em.

Brawley, Otis W.; Tejeda, Heriberto. Minority inclusion in clinical trials: issues and potential strategies. *Journal of National Cancer Institute Monographs* 1995; 17: 55-57. NRCBL: 18.5.1; 18.2.

Burroughs, Valentine J. Racial and ethnic inclusiveness in clinical trials. *In:* Santoro, Michael A.; Gorrie, Thomas M., eds. Ethics and the Pharmaceutical Industry. Cambridge; New York: Cambridge University Press, 2005: 80-96, 421-425. NRCBL: 18.5.1; 15.1; 9.7.
> Keywords: *clinical trials; *ethnic groups; *pharmacogenetics; *racial groups; culture; drug industry; drugs; genetic ancestry; health hazards; minority groups; public policy; Keyword Identifiers: BiDil; United States

Carlson, Elof Axel. Medical deception and syphilis. *In his:* Times of Triumph, Times of Doubt: Science and the Battle for Public Trust. Cold Spring Harbor, NY: Cold Spring Harbor Laboratory Press, 2006: 141-153. NRCBL: 18.5.1; 18.3; 10. Identifiers: Tuskegee Syphilis Study of Untreated Syphilis in the Negro Male.

Croft, Jason R.; Festinger, David S.; Dugosh, Karen L.; Marlowe, Douglas B.; Rosenwasser, Beth J. Does size matter?:salience of follow-up payments in drug abuse research. *IRB: Ethics and Human Research* 2007 July-August; 29(4): 15-19. NRCBL: 18.5.1; 9.3.1. SC: em.

Denny, Colleen C.; Grady, Christine. Clinical research with economically disadvantaged populations. *Journal of Medical Ethics* 2007 July; 33(7): 382-385 [see correction in Journal of Medical Ethics 2007 August; 33(8): 496]. NRCBL: 18.5.1. SC: an.
> Abstract: Concerns about exploiting the poor or economically disadvantaged in clinical research are widespread in the bioethics community. For some, any research that involves economically disadvantaged individuals is de facto ethically problematic. The economically disadvantaged are thought of as "venerable" [sic; "vulnerable"]to exploitation, impaired decision making, or both, thus requiring either special protections or complete exclusion from research. A closer examination of the worries about vulnerabilities among the economically disadvantaged reveals that some of these worries are empirically or logically untenable, while others can be better resolved by improved study designs than by blanket exclusion of poorer individuals from research participation. The scientific objective to generate generalisable results and the ethical objective to fairly distribute both the risks and benefits of research oblige researchers not to unnecessarily bar economically disadvantaged subjects from clinical research participation.

Dickinson, Frederick R. Biohazard: Unit 731 in postwar Japanese politics of national "forgetfulness". *In:* LaFleur, William R.; Böhme, Gernot; Shimazono, Susumu, eds. Dark Medicine: Rationalizing Medical Research. Bloomington: Indiana University Press, 2007: 85-104. NRCBL: 18.5.1; 18.5.8; 18.5.9; 21.3; 2.2; 1.3.5. SC: rv.

Farmer, Deborah F.; Jackson, Sharon A.; Camacho, Fabian; Hall, Mark A. Attitudes of African American and low socioeconomic status white women toward medical research. *Journal of Health Care for the Poor and Underserved* 2007 February; 18(1): 85-99. NRCBL: 18.5.1; 18.5.3; 8.1. SC: em.

Fendrich, Michael; Lippert, Adam M.; Johnson, Timothy P. Respondent reactions to sensitive questions. *Journal of Empirical Research on Human Research Ethics* 2007 September; 2(3): 31-37. NRCBL: 18.5.1; 18.3; 10. SC: em.
> Abstract: We administered debriefing probes to gauge respondent discomfort in reaction to sensitive questions. These probes assessed respondents' own reactions to being asked to report on substance use (subjective discomfort), as well as their beliefs about the reaction of others (projective discomfort). We investigated whether a sample of men from the general population were more uncomfortable with questions about drug use than a sample

of men who have sex with men (MSM) surveyed from the same city (Chicago). We also investigated whether those who disclosed drug use on the survey experienced higher levels of discomfort. Contrary to opinions often expressed as research ethics committee (REC) recommendations, questions about drug use do not generate much subjective discomfort. MSM did not differ from the general population with respect to subjective discomfort. General population males did, however, report higher levels of "drug specific" projective discomfort. Respondents disclosing recent drug use reported higher levels of subjective discomfort. Implications for the REC practice, researcher and REC education, and directions for future research are discussed.

Fischler, Ira; Simmerling, Mary; Fitzgerald, Maureen; Weitlauf, Julie; Frayne, Susan M.; Lee, Tina; Ruzek, Josef; Finney, John; Thrailkill, Ann; Newman, Elana. Trauma research. *Journal of Empirical Research on Human Research Ethics* 2007 March; 2(1): 51-59. NRCBL: 18.5.1; 17.1; 18.3; 18.5.3; 18.2; 18.5.8; 17.7. SC: cs.

Gefenas, Eugenijus. Balancing ethical principles in emergency medicine research. *Science and Engineering Ethics* 2007 September; 13(3): 281-288. NRCBL: 18.5.1; 18.2; 18.3; 21.1.

Godlaski, Theodore M.; Johnson, Jeannette; Haring, Rodney. Reflections on ethical issues in research with aboriginal peoples. *In:* Kleinig, John; Einstein, Stanley, eds. Ethical Challenges for Intervening in Drug Use: Policy, Research and Treatment Issues. Huntsville, TX: Office of International Criminal Justice; Criminal Justice Center, Sam Houston State University, 2006: 281-305. NRCBL: 18.5.1; 18.2; 21.7; 1.2; 21.4.

Halila, Ritva. Assessing the ethics of medical research in emergency settings: how do international regulations work in practice? *Science and Engineering Ethics* 2007 September; 13(3): 305-313. NRCBL: 18.5.1; 18.2; 18.6; 21.1.

Hoag, Hannah. Rules tightened for aboriginal studies [news]. *Nature* 2007 May 17; 447(7142): 241. NRCBL: 18.5.1; 18.4. Identifiers: Canada.

Hussain-Gambles, Mah; Atkin, Karl; Leese, Brenda. South Asian participation in clinical trials: the views of lay people and health professionals. *Health Policy* 2006 July; 77(2): 149-165. NRCBL: 18.5.1; 21.7; 1.2.; 18.2; 7.1. SC: em. Identifiers: Great Britain (United Kingdom).

Iserson, Kenneth V. Has emergency medicine research benefitted patients? An ethical question. *Science and Engineering Ethics* 2007 September; 13(3): 289-295. NRCBL: 18.5.1.

Kafarowski, Joanna. The woman/gender questions: best practices of conducting research with indigenous peoples in Canada. *NCEHR Communique CNERH* 2006 Spring; 14(1): 18-20. NRCBL: 18.5.1; 18.5.3.

Kelty, Miriam; Bates, Angela; Pinn, Vivian W. National Institutes of Health policy on the inclusion of women and minorities as subjects in clinical research. *In:* Gallin, John I.; Ognibene, Frederick P., eds. Principles and Practice of Clinical Research. 2nd edition. Oxford: Academic, 2007: 129-142. NRCBL: 18.5.1; 18.2; 18.5.3.

Landeweer, Elleke; Berghmans, Ron; Abma, Tineke; Widdershoven, Guy. Coercive treatment in mental hospitals: legal regulations and experiences in the Netherlands. *EACME Newsletter* 2007 July; (17): 3 p. [Online]. Accessed: http://www.eacmeweb.com/newsletter/n17.htm [2007 August 15]. NRCBL: 18.5.1; 17.7; 18.6. SC: le; em.

Lindee, Susan. Experimental injury: wound ballistics and aviation medicine in mid-century America. *In:* LaFleur, William R.; Böhme, Gernot; Shimazono, Susumu, eds. Dark Medicine: Rationalizing Medical Research. Bloomington: Indiana University Press, 2007: 121-137. NRCBL: 18.5.1; 1.3.5; 22.2; 2.2.

Mason, J.K.; Laurie, G.T. Research on children, fetuses and embryos. *In their:* Mason and McCall Smith's Law and Medical Ethics. Seventh ed. Oxford; New York: Oxford University Press, 2005: 685-709. 113 refs. NRCBL: 18.5.1; 18.5.2; 18.5.4; 15.1; 18.3; 21.1. SC: le.
    Keywords: *children; *embryo research; *fetal research; *human experimentation; *public policy; aborted fetuses; advisory committees; embryonic stem cells; embryos; fetuses; legal aspects; moral status; nontherapeutic research; parental consent; stem cells; therapeutic research; Keyword Identifiers: Great Britain

McCallum, Jan M.; Arekere, Dhananjaya M.; Green, B. Lee; Katz, Ralph V.; Rivers, Brian M. Awareness and knowledge of the U.S. Public Health Service syphilis study at Tuskegee: implications for biomedical research. *Journal of Health Care the Poor and Underserved* 2006 November; 17(4): 716-733. NRCBL: 18.5.1; 18.3; 10; 18.6; 2.2. SC: rv.

McManus, John; McClinton, Annette; Gerhardt, Robert; Morris, Michael. Performance of ethical military research is possible: on and off the battlefield. *Science and Engineering Ethics* 2007 September; 13(3): 297-303. NRCBL: 18.5.1; 18.5.8; 18.3; 18.6.

Moreno, Jonathan. Secret state experiments and medical ethics. *In:* Galston, Arthur W.; Peppard, Christiana Z., eds. Expanding Horizons in Bioethics. Dordrecht; Norwell, MA: Springer, 2005: 59-69. NRCBL: 18.5.1; 18.6; 1.3.5.

Ozdogan, Mustafa; Samur, Mustafa; Artac, Mehmet; Yildiz, Mustafa; Savas, Burhan; Bozcuk, Hakan Sat. Factors related to truth-telling practice of physicians treating patients with cancer in Turkey. *Journal of Palliative Medicine* 2006 October; 9(5): 1114-1119. NRCBL: 18.5.1; 8.2; 8.1; 9.5.1. SC: em.

Panikkar, Bindu; Brugge, Doug. The ethical issues in uranium mining research in the Navajo nation. *Account-*

NRCBL: National Reference Center for Bioethics Literature Classification Scheme   See inside front cover for terms.

277

ability in Research 2007 April-June; 14(2): 121-153. NRCBL: 18.5.1; 1.3.9; 16.3; 18.2. SC: em.

Abstract: We explore the experience of Navajo communities living under the shadow of nuclear age fallout who were subjects of five decades of research. In this historical analysis of public health (epidemiological) research conducted in the Navajo lands since the inception of uranium mining from the 1950s until the end of the 20th century, we analyze the successes and failures in the research initiatives conducted on Navajo lands, the ethical breaches, and the harms and benefits that this research has brought about to the community. We discuss how scientific and moral uncertainty, lack of full stakeholder participation and community wide outreach and education can impact ethical decisions made in research.

Racine, Eric; Illes, Judy. Emerging ethical challenges in advanced neuroimaging research: review, recommendations and research agenda. *Journal of Empirical Research on Human Research Ethics* 2007 June; 2(2): 1-10. NRCBL: 18.5.1; 17.1; 8.4; 18.2; 18.3.

Abstract: The dynamic and ever-evolving nature of neuroimaging research creates important ethical challenges. New domains of neuroscience research and improving technological capabilities in neuroimaging have expanded the scope of studies that probe the biology of the social and ethical brain, the range of eligible volunteers for research, and the extent of academic-industry relationships. Accordingly, challenges in informed consent and subject protection are surfacing. In this context, we provide an overview of the current landscape for neuroimaging and discuss specific research ethics topics arising from it. We suggest preliminary approaches to tackle current issues, and identify areas for further collaboration between neuroimagers and institutional review boards (research ethics committee).

Ritter, Alison J.; Fry, Craig L.; Swan, Amy. The ethics of reimbursing drug users for public health research interviews: what price are we prepared to pay? [editorial]. *International Journal of Drug Policy* 2003 February; 14(1): 1-3. NRCBL: 18.5.1; 9.3.1; 18.2; 18.3.

Samuels, Allison. Brutal case studies. A new book documents a true ethics horror story. *Newsweek* 2007 February 12; 149(7): 49. NRCBL: 18.5.1; 18.3; 2.2. SC: po. Identifiers: Medical Apartheid: The Dark History of Medical Experimentation on Black Americans from Colonial Times to the Present, by Harriet Washington.

Schüklenk, Udo; Ashcroft, Richard. HIV vaccine trials: reconsidering the therapeutic misconception and the question of what constitutes trial related injuries [editorial]. *Developing World Bioethics* 2007 December; 7(3): ii-iv. NRCBL: 18.5.1; 9.5.6; 18.2; 21.1.

Streeter, Oscar E.; Cuyjet, Aloysius B.; Norris, Keith; Hylton, Kevin. Issues surrounding the involvement of African Americans in clinical trials and other research. *In:* Livingston, Ivor Lensworth, ed. Praeger Handbook of Black American Health: Policies and Issues Behind Dis-

parities in Health, Vol. II. 2nd edition. Westport, CT: Praeger Publishers, 2004: 808-834. NRCBL: 18.5.1; 9.5.1; 9.5.6; 10; 18.3.

Tejeda, Heriberto A.; Green, Sylvan B.; Trimble, Edward L.; Ford, Leslie; High, Joseph L.; Ungersleider, Richard S.; Friedman, Michael A.; Brawley, Otis W. Representation of African Americans, Hispanics, and Whites in National Cancer Institute treatment trials. *Journal of National Cancer Institute* 1996 June 19; 88(12): 812-816. NRCBL: 18.5.1; 9.5.1; 18.2.

Thomas, Charles R., Jr.; Pinto, Harlan A.; Roach, Mack, III; Vaughn, Clarence B. Participation in clinical trials: is it state-of the art treatment for African Americans and other people of color? *Journal of National Medical Association* 1994; 86(3): 177-182. NRCBL: 18.5.1; 9.5.1; 18.2.

Tsuneishi, Kei-chi. Unit 731 and the human skulls discovered in 1989: physicians carrying out organized crimes. *In:* LaFleur, William R.; Böhme, Gernot; Shimazono, Susumu, eds. Dark Medicine: Rationalizing Medical Research. Bloomington: Indiana University Press, 2007: 73-84. NRCBL: 18.5.1; 18.5.8; 18.5.9; 21.3; 2.2.

United States. Food and Drug Administration [FDA]. Human drugs and biologics; determination that informed consent is not feasible or is contrary to the best interests of recipients; revocation of 1990 interim final rule; establishment of new interim final rule. *Federal Register* 1999 October 5; 64(192): 54180-54189 [Online]. Accessed: http://www.fda.gov/oc/gcp/preambles/64fr54180.html [2005 December 27]. NRCBL: 18.5.1; 18.5.8; 18.2; 18.6; 9.7. SC: le.

Welch, H. Gilbert; Woloshin, Steven; Schwartz, Lisa M. How two studies on cancer screening led to two results. *New York Times* 2007 March 13; p. F5, F8. NRCBL: 18.5.1; 5.2; 4.4. SC: po; em.

Wendler, David; Kington, Raynard; Madans, Jennifer; Van Wye, Gretchen; Christ-Schmidt, Heidi; Pratt, Laura A.; Brawley, Otis W.; Gross, Cary P.; Emanuel, Ezekiel. Are racial and ethnic minorities less willing to participate in health research? *PLoS Medicine* 2006 February; 3(2): e19: 0201-0210. NRCBL: 18.5.1; 18.3. SC: em; rv.

Wnukiewicz-Kozlowka, Agata. The admissibility of research in emergency medicine. *Science and Engineering Ethics* 2007 September; 13(3): 315-332. NRCBL: 18.5.1; 18.6; 21.1. SC: le. Identifiers: Poland.

Wong-Kim, Evaon; Song, Young; Vasgird, Daniel. Cultural competence of researchers. *Journal of Empirical Research on Human Research Ethics* 2007 March; 2(1): 83-85. NRCBL: 18.5.1; 18.6; 21.7.

HUMAN EXPERIMENTATION/ . . . / AGED AND TERMINALLY ILL

Adamson, Peter C.; Paradis, Carmen; Smith, Martin L. All for one, or one for all? [case study and commentary]. *Hastings Center Report* 2007 July-August; 37(4): 13-15. NRCBL: 18.5.7; 18.5.2; 5.2; 18.2; 1.3.2; 9.5.1. SC: cs. Identifiers: Penelope London; experimental drug; Neotropix, Inc.

Bender, Shira; Flicker, Lauren; Rhodes, Rosamond. Access for the terminally ill to experimental medical innovations: a three-pronged threat. *American Journal of Bioethics* 2007 October; 7(10): 3-6. NRCBL: 18.5.7; 1.1; 9.7.

Bristol, Nellie. Should terminally ill patients have access to phase I drugs? *Lancet* 2007 March 10-16; 369(9564): 815-816. NRCBL: 18.5.7; 9.7; 9.2; 5.3. SC: le.

Byock, Ira. Palliative care and the ethics of research: medicare, hospice, and phase I trials. *Journal of Supportive Oncology* 2003 July-August; 1(2): 139-141. NRCBL: 18.5.7; 20.4.1; 9.3.1; 9.5.2.

Casarett, David. Ethical considerations in end-of-life care and research. *Journal of Palliative Medicine* 2005; 8(Supplement 1): S148-S160. NRCBL: 18.5.7; 20.4.1; 4.4; 18.3; 18.5.6; 8.3.3.

Dyer, Clare. Husband says judge's ruling on wife's treatment was "inhumane" [news]. *BMJ: British Medical Journal* 2007 January 27; 334(7586): 176. NRCBL: 18.5.7; 20.5.1. SC: le. Identifiers: Great Britain (United Kingdom).

Freireich, Emil; Gesme, Dean. Should terminally ill patients have the right to take drugs that pass phase 1 testing? [debate]. *BMJ: British Medical Journal* 2007 September 8; 335(7618): 478-479. NRCBL: 18.5.7; 18.2; 9.7; 1.3.2.

Johnston, Therese E. Issues surrounding protection and assent in pediatric research. *Pediatric Physical Therapy* 2006 Summer; 18(2): 133-140. NRCBL: 18.5.7; 18.2; 18.3; 18.5.2.

Kendall, Marilyn; Harris, Fiona; Boyd, Kirsty; Sheikh, Aziz; Murray, Scott A.; Brown, Duncan; Mallinson, Ian; Kearney, Nora; Worth, Allison. Key challenges and ways forward in researching the "good death": qualitative in-depth interview and focus group study. *BMJ: British Medical Journal* 2007 March 10; 334(7592): 521-524. NRCBL: 18.5.7; 20.4.1. SC: em. Identifiers: Scotland.

Natale, JoAnne E.; Joseph, Jill G.; Pretzlaff, Robert K.; Silber, Tomas J.; Guerguerian, Anne-Marie. Clinical trials in pediatric traumatic brain injury: unique challenges and potential responses. *Developmental Neuroscience* 2006; 28(4-5): 276-290. NRCBL: 18.5.7; 18.2; 18.5.6.

Siu, Lillian L. Clinical trials in the elderly — a concept comes of age. *New England Journal of Medicine* 2007 April 12; 356(15): 1575-1576. NRCBL: 18.5.7; 9.5.1.

Williams, Charlotte J.; Shuster, John L.; Clay, Olivio J.; Burgio, Kathryn L. Interest in research participation among hospice patients, caregivers, and ambulatory senior citizens: practical barriers or ethical constraints? *Journal of Palliative Medicine* 2006 August; 9(4): 968-974. NRCBL: 18.5.7; 20.4.1; 18.5.1. SC: em.

Workman, Stephen. Researching a good death [editorial]. *BMJ: British Medical Journal* 2007 March 10; 334(7592): 485-486. NRCBL: 18.5.7; 20.4.1; 8.2. SC: em.

HUMAN EXPERIMENTATION/ . . . / EMBRYOS AND FETUSES
See also CLONING

Amniotic fluid supplies 'repair kit' for later life. *New Scientist* 2007 January 13-19; 193(2586): 9. NRCBL: 18.5.4; 15.1; 19.1.

An inconvenient truth: research on human embryonic stem cells must go on [editorial]. *Nature* 2007 November 29; 450(7170): 585-586. NRCBL: 18.5.4; 15.1; 18.5.1.
> Keywords: *adult stem cells; *embryo research; *embryonic stem cells; international aspects; regulation

Britain to let women donate eggs for research. *New York Times* 2007 February 22; p. A13. NRCBL: 18.5.4; 19.5; 14.6. SC: po.

Criteria creep: the politically motivated extension of US stem-cell registry makes no scientific sense [editorial]. *Nature* 2007 October 18; 449(7164): 756. NRCBL: 18.5.4; 15.1.
> Keywords: *embryonic stem cells; *pluripotent stem cells; *public policy; *registries; adult stem cells; cell lines; embryo research; federal government; research support; Keyword Identifiers: United States; National Institutes of Health

The long and winding road [editorial]. *Nature* 2007 September 27; 449(7161): 377. NRCBL: 18.5.4; 18.5.1; 15.1; 19.5. Identifiers: Germany; human embryonic stem-cell research.

American Society of Human Genetics [ASHG]. Statement on stem cell research. Bethesda, MD: American Society of Human Genetics, 2001 August 27: 1 p. [Online]. Accessed: http://genetics.faseb.org/genetics/ashg/pubs/policy/pol-44.htm [2006 July 26]. NRCBL: 18.5.4; 15.1; 6.
> Keywords: *embryo research; *embryonic stem cells; *stem cells; advisory committees; government financing; organizational policies; professional organizations; public policy; Keyword Identifiers: United States

Baker, Monya. Stem cells by any other name [news]. *Nature* 2007 September 27; 449(7161): 389. NRCBL: 18.5.4; 15.1; 1.3.12; 19.5. Identifiers: Human Pluripotent Stem Cell Registry.

NRCBL: National Reference Center for Bioethics Literature Classification Scheme   See inside front cover for terms.

279

Bass, Sarah Bauerle. Why can't a fetus be more like a sperm? The women's role in fetal issue research and how women are left out of the discussion. *Gender Issues* 2001 Winter; 19(1): 19-32. NRCBL: 18.5.4; 12.1; 10; 15.1.

Bobrow, James C. The ethics and politics of stem cell research. *Transactions of the American Ophthalmological Society* 2005; 103: 138-141; discussion 141-142. NRCBL: 18.5.4; 15.1; 18.5.1; 18.6.

Brainard, Jeffrey. California stem-cell researchers ponder next steps after court victory. *Chronicle of Higher Education* 2007 June 1; 53(39): A20. NRCBL: 18.5.4; 15.1; 1.3.9.

    Keywords: *government financing; *research support; *state government; *stem cells; *universities; biomedical research; conflict of interest; embryo research; laboratories; legal aspects; patient advocacy; politics; public policy; research institutes; Keyword Identifiers: *California; Stanford University; University of California, San Francisco

Brainard, Jeffrey. NIH director calls for easing administration's stem-cell restrictions [news]. *Chronicle of Higher Education* 2007 March 30; 53(30): A26. NRCBL: 18.5.4; 15.1; 5.3.

    Keywords: *embryo research; *embryonic stem cells; *government financing; *public policy; *research support; cell lines; government regulation; Keyword Identifiers: *United States; *Zerhouni, Elias; National Institutes of Health

Brown, Susan. China challenges the west in stem-cell research: unconstrained by public debate, cities like Shanghai and Beijing lure scientists with new laboratories and grants. *Chronicle of Higher Education* 2007 April 13; 53(32): A14-16, A18. NRCBL: 18.5.4; 15.1; 21.1; 1.3.9.

Brown, Susan. International group proposes guidelines for embryonic-stem-cell research. *Chronicle of Higher Education* 2007 February 16; 53(24): A21. NRCBL: 18.5.4; 15.1; 18.2.

Burke, William; Pullicino, Patrick; Richard, Edward J. The biological basis of the oocyte assisted reprogramming (OAR) hypothesis: is it an ethical procedure for making embryonic stem cells? *Linacre Quarterly* 2007 August; 74(3): 204-212. NRCBL: 18.5.4; 15.1.

Canadian Institutes of Health Research [CIHR]. Human pluripotent stem cell research: recommendations for CIHR-funded research: report of the ad hoc Working Group on Stem Cell Research. Ottawa: Canadian Institutes of Health Research, 2002 January; 21 p. [Online]. Accessed: http://www.cihr-irsc.gc.ca/e/1489.html [2007 April 19]. NRCBL: 18.5.4; 15.1; 5.3.

    Keywords: *embryo research; *embryonic stem cells; *government financing; *guidelines; *public policy; *research support; cell lines; commerce; confidentiality; disclosure; ethical review; germ cells; informed consent; policy making; public participation; review committees; standards; stem cell transplantation; stem cells; Keyword Identifiers: *Canada; *Canadian Institutes of Health Research

Chekar, Choon Key; Kitzinger, Jenny. Science, patriotism and discourses of nation and culture: reflections on the South Korean stem cell breakthroughs and scandals. *New Genetics and Society* 2007 December; 26(3): 289-307. NRCBL: 18.5.4; 15.1; 14.5; 1.3.9. Identifiers: Woo Suk Hwang.

Condic, Maureen L. What we know about embryonic stem cells. *First Things* 2007 January; (169): 25-29. NRCBL: 18.5.4; 15.1; 14.5; 18.5.1; 22.2; 19.1.

Cyranoski, David. Stem-cell fraudster 'is working in Thailand' [news]. *Nature* 2007 September 27; 449(7161): 387. NRCBL: 18.5.4; 15.1; 14.5; 18.1; 22.1. Identifiers: Woo Suk Hwang.

Daley, George Q.; Ahrlund-Richter, Lars; Auerbach, Jonathan M.; Benvenisty, Nissim; Charo, R. Alta; Chen, Grace; Deng, Hong-kui; Goldstein, Lawrence S.; Hudson, Kathy L.; Hyun, Insoo; Junn, Sung Chull; Love, Jane; Lee, Eng Hin; McLaren, Anne; Mummery, Christine L.; Nakatsuji, Norio; Racowsky, Catherine; Rooke, Heather; Rossant, Janet; Schöler, Hans R.; Solbakk, Jan Helge; Taylor, Patrick; Trounson, Alan O.; Weissman, Irving L.; Wilmut, Ian; Yu, John; Zoloth, Laurie. The ISSCR guidelines for human embryonic stem cell research. *Science* 2007 February 2; 315(5812): 603-604. 12 refs. NRCBL: 18.5.4; 15.1.

    Keywords: *embryo research; *embryonic stem cells; *guidelines; *professional organizations; access to information; chimeras; cloning; donors; editorial policies; ethical review; germ cells; guideline adherence; informed consent; international aspects; regulation; remuneration; researchers; Proposed Keywords: pluripotent stem cells; Keyword Identifiers: *International Society for Stem Cell Research; National Academy of Sciences; United States

Day, Michael. UK may use hybrid embryos for research [news]. *BMJ: British Medical Journal* 2007 May 26; 334(7603): 1074. NRCBL: 18.5.4; 15.1; 18.1; 22.1.

    Keywords: *chimeras; *embryo research; *embryos; *legal aspects; *public policy; government; legislation; Keyword Identifiers: *Great Britain; Department of Health (Great Britain); *Human Tissue and Embryos Bill (Great Britain)

Downey, Robin; Geransar, Rose; Einsiedel, Edna. Angles of vision: stakeholders and human embryonic stem cell policy development. *In:* Einsiedel, Edna; Timmermans, Frank, eds. Crossing Over: Genomics in the Public Arena. Calgary, Alberta, Canada: University of Calgary Press, 2005: 61-84. 69 refs. NRCBL: 18.5.4; 15.1; 19.5; 5.3. Conference: Essays from the conference held Apr. 25-27, 2003, Kananaskis, Alta.

    Keywords: *embryo research; *embryonic stem cells; *policy making; *public policy; advisory committees; biomedical research; cloning; democracy; government financing; government regulation; guidelines; historical aspects; international aspects; legal aspects; legislation; patient advocacy; political activity; research support; researchers; Proposed Keywords: *stakeholders; lobbying; Keyword Identifiers: *Canada; Canadian Institutes of Health Research; Royal Commission on New Reproductive Technologies; Right to Life Movement; Twentieth Century

Ecker, Jeffrey L.; O'Rourke, Patricia Pearl; Lott, Jason P.; Savulescu, Julian. An immodest proposal: banking embryonic stem cells for solid organ transplantation is problematic and premature. *American Journal of Bioethics* 2007 August; 7(8): 48-50; author reply W4-W6. 7 refs. NRCBL: 18.5.4; 15.1; 19.5. Comments: Jason P. Lott and Julian Savulescu. Towards a global human embryonic stem cell bank. American Journal of Bioethics 2007 August; 7(8): 37-44.
>   Keywords: *embryonic stem cells; *tissue banks; donors; incentives; mandatory programs; moral policy; organ transplantation; remuneration; resource allocation; risks and benefits; scarcity; standards; stem cell transplantation; utilitarianism; Proposed Keywords: embryo donation

European Society of Human Genetics [ESHG]. EU Temporary Committee on Genetics Recap Overview by the European Society of Human Genetics. Vienna, Austria: European Society of Human Genetics, 2001: 2 p. [Online]. Accessed: http://www.eshg.org/Fiorireporthistory.pdf [2007 February 8]. NRCBL: 18.5.4; 15.1; 14.5. Identifiers: European Union.

European Society of Human Genetics [ESHG]. Letter to members of the European Parliament from the European Society of Human Genetics. Re: Fiori Report on the ethical, legal, economic and social implications of human genetics. Vienna, Austria: European Society of Human Genetics, [2001]: 3 p. [Online]. Accessed: http://www.eshg.org/ESHGlettertoMEPs.pdf [2007 May 1]. NRCBL: 18.5.4; 15.1; 14.5; 15.2.

European Society of Human Genetics [ESHG]. Report on the ethical, legal, economic and social implications of human genetics by the temporary committee on human genetics and other new technologies in modern medicine [of the European Parliament]: changes recommended by the European Society of Human Genetics. Vienna, Austria: European Society of Human Genetics, 2001 November 8: 6 p. [Online]. Accessed: http://www.eshg.org/ESHGrecchangesFiori.pdf [2007 February 8]. NRCBL: 18.5.4; 15.1; 14.5.

Feresin, Emiliano. Italian bioethics committee in uproar [news]. *Nature* 2007 October 25; 449(7165): 955. NRCBL: 18.5.4; 15.1; 2.4.

FIGO Committee for the Ethical Aspects of Human Reproduction and Women's Health; Serour, G.I. Embryo research. *International Journal of Gynecology and Obstetrics* 2006 May; 93(2): 182-183. NRCBL: 18.5.4; 15.1; 14.1.
>   Keywords: *embryo research; embryo disposition; embryonic stem cells; in vitro fertilization; informed consent; ovum donors; professional organizations; research embryo creation; Keyword Identifiers: International Federation of Gynecology and Obstetrics

Franklin, Sarah. Embryonic economies: the double reproductive value of stem cells. *Biosocieties* 2006 March;

1(1): 71-90. 97 refs. NRCBL: 18.5.4; 15.1; 19.5; 14.1; 18.3.
>   Keywords: *embryo disposition; *embryo research; *embryonic stem cells; *in vitro fertilization; biological specimen banks; cell lines; donors; economics; motivation; public policy; regulation; trends; Proposed Keywords: *embryo donation; Keyword Identifiers: Great Britain

Genetics Committee of the Society of Obstetricians and Gynaecologists of Canada; Wilson, R. Douglas; Desilets, Valerie; Gagnon, Alain; Summers, Anne; Wyatt, Philip; Allen, Victoria; Langlois, Sylvie. Present role of stem cells for fetal genetic therapy = Rôle actuel des cellules souches en matière de thérapie génique fetale. *JOGC: Journal of Obstetrics and Gynaecology Canada = JOGC: Journal d'Obstétrique et Gynécologie du Canada* 2005 November; 27(11): 1038-1047. NRCBL: 18.5.4; 15.1; 14.4; 19.5.

Giordano, Simona; Cappato, Marco. Scientific freedom [editorial]. *Journal of Medical Ethics* 2007 June; 33(6): 311-312. NRCBL: 18.5.4; 15.1; 1.3.9.
>   Keywords: *embryo research; *embryonic stem cells; *freedom; *public policy; *research support; biomedical research; cloning; democracy; government financing; government regulation; international aspects; legal aspects; religion; science; Keyword Identifiers: *European Union

Great Britain (United Kingdom). Department of Health. UK Stem Cell Initiative. UK stem cell initiative: report and recommendations. London: UK Stem Cell Initiative, 2005 November; 118 p. [Online]. Accessed: http://www.advisorybodies.doh.gov.uk/uksci/ukscireportnov05.pdf [2007 April 26]. NRCBL: 18.5.4; 5.3; 15.1; 19.1; 19.5.

Hayflick, L. The limited in vitro lifetime of human diploid cell strains. *Experimental Cell Research* 1965; 37: 614-636. NRCBL: 18.5.4; 15.1; 14.6.

Hinkley, Charles C. Stem cell research. *In his:* Moral Conflicts of Organ Retrieval: A Case for Constructive Pluralism. Amsterdam; New York: Rodopi, 2005: 127-133. NRCBL: 18.5.4; 15.1;.

Holden, Constance. U.K. takes eggstra time [news]. *Science* 2007 January 19; 315(5810): 317. NRCBL: 18.5.4; 5.3; 15.1. Identifiers: Human Fertilisation and Embryology Authority [HFEA]; Great Britain (United Kingdom).

Johnson, Martin H. Regulating the science and therapeutic application of human embryo research: managing the tension between biomedical creativity and public concern. *In:* Spencer, J.R.; du Bois-Pedain, Antje, eds. Freedom and Responsibility in Reproductive Choice. Portland: Hart Pub., 2006: 91-106. 35 fn. NRCBL: 18.5.4; 14.1; 15.1.
>   Keywords: *embryo research; *regulation; *reproductive technologies; cloning; government regulation; physicians; preimplantation diagnosis; professional autonomy; reproduction; researchers; risk; self regulation; sexuality; Keyword Identifiers: Great Britain; Human Fertilisation and Embryology Act 1990 (Great Britain); Human Fertilisation and Embryology Authority

NRCBL: National Reference Center for Bioethics Literature Classification Scheme   See inside front cover for terms.

281

Kaczor, Christopher Robert. An ethical assessment of Bush's guidelines for stem cell research. *In his:* The Edge of Life: Human Dignity and Contemporary Bioethics. Dordrecht: Springer, 2005: 83-96. Includes references. NRCBL: 18.5.4; 15.1. SC: an.
　　Keywords: *double effect; *embryo research; *embryonic stem cells; *ethical analysis; *government financing; *moral complicity; *moral policy; *public policy; *value of life; adult stem cells; allowing to die; beginning of life; cell lines; common good; embryos; federal government; guidelines; human dignity; killing; moral status; parental consent; personhood; Roman Catholic ethics; Keyword Identifiers: Bush, George; United States

Kahn, Jeffrey P. Organs and stem cells: policy lessons and cautionary tales. *Hastings Center Report* 2007 March-April; 37(2): 11-12. 4 fn. NRCBL: 18.5.4; 15.1; 19.5; 19.6.
　　Keywords: *embryonic stem cells; *organ transplantation; *property rights; *public policy; *social control; biomedical technologies; body parts and fluids; cadavers; consensus; determination of death; embryo research; embryos; government regulation; kidneys; living donors; moral status; organ donors; policy making; remuneration; resource allocation; Proposed Keywords: stakeholders; Keyword Identifiers: United Network for Organ Sharing; United States

Kim, Tae-gyu. Korea mulls allowing research using cloned embryos [news]. *Korea Times* 2007 January 29: 2 p. [Online]. Accessed: http://times.hankooki.com/service/print/Print.php?po=times.hankooki.com/1page/2 00701/kt2007011917571310230.htm [2007 February 6]. NRCBL: 18.5.4; 15.1; 14.5.

Kimmelman, Jonathan; Lott, Jason P.; Savulescu, Julian. Towards a global human embryonic stem cell bank: differential termination. *American Journal of Bioethics* 2007 August; 7(8): 52-53; author reply W4-W6. 6 refs. NRCBL: 18.5.4; 15.1; 19.5. Comments: Jason P. Lott and Julian Savulescu. Towards a global human embryonic stem cell bank. American Journal of Bioethics 2007 August; 7(8): 37-44.
　　Keywords: *embryonic stem cells; *tissue banks; donors; incentives; international aspects; justice; minority groups; moral policy; remuneration; stem cell transplantation; Proposed Keywords: embryo donation

Kitzinger, Jenny; Williams, Clare. Forecasting the future: legitimizing hope and calming fears in the embryo stem-cell debate. *In:* Deane-Drummond, Celia; Scott, Peter Manley, eds. Future Perfect?: God, Medicine and Human Identity. New York: T and T Clark International, 2006: 129-142. 14 fn. NRCBL: 18.5.4; 15.1; 1.3.7. SC: em.
　　Keywords: *embryo research; *embryonic stem cells; *mass media; biotechnology; editorial policies; genetic research; journalism; survey; Keyword Identifiers: Great Britain

Langer, Gary; Lyerly, Anne Drapkin; Faden, Ruth. Counting on embryos [letter and reply]. *Science* 2007 October 26; 318(5850): 566, 568. NRCBL: 18.5.4; 15.1; 19.5; 14.6. Comments: A.D. Lyerly and R.R. Faden. Willingness to donate frozen embryos for stem cell research. Science 2007 July 6; 317(5834): 46-47.

Lanza, Robert. Stem cell breakthrough: don't forget the ethics [letter]. *Science* 2007 December 21; 318(5858): 1865. 4 refs. NRCBL: 18.5.4; 15.1.
　　Keywords: *stem cells; chimeras; embryo research; embryonic stem cells; genetic techniques; methods; risk; Proposed Keywords: *pluripotent stem cells

Levin, Yuval. A middle ground for stem cells [op-ed]. *New York Times* 2007 January 19; p. A23. NRCBL: 18.5.4; 15.1. SC: po.
　　Keywords: *embryonic stem cells; *embryo research; *public policy; embryo; value of life; research support; government financing; Keyword Identifiers: *United States

Longaker, Michael T.; Baker, Laurence C.; Greely, Henry T. Proposition 71 and CIRM — assessing the return on investment. *Nature Biotechnology* 2007 May; 25(5): 513-521. NRCBL: 18.5.4; 15.1; 5.3. Identifiers: California Institute for Regenerative Medicine.

Lott, Jason P.; Savulescu, Julian. Towards a global human embryonic stem cell bank. *American Journal of Bioethics* 2007 August; 7(8): 37-44. 27 refs. NRCBL: 18.5.4; 15.1; 19.5; 14.5; 9.3.1. SC: an. Comments: American Journal of Bioethics 2007 August; 7(8): 45-53.
　　Keywords: *embryonic stem cells; *tissue banks; cell lines; cloning; donors; embryo disposition; ethnic groups; economics; financial support; in vitro fertilization; incentives; informed consent; international aspects; justice; mandatory programs; minority groups; moral policy; nuclear transfer techniques; organ transplantation; policy analysis; racial groups; remuneration; resource allocation; scarcity; social discrimination; standards; stem cell transplantation; transplant recipients; Proposed Keywords: embryo donation; haplotypes; tissue typing

Abstract: An increasingly unbridgeable gap exists between the supply and demand of transplantable organs. Human embryonic stem cell technology could solve the organ shortage problem by restoring diseased or damaged tissue across a range of common conditions. However, such technology faces several largely ignored immunological challenges in delivering cell lines to large populations. We address some of these challenges and argue in favor of encouraging contribution or intentional creation of embryos from which widely immunocompatible stem cell lines could be derived. Further, we argue that current immunological constraints in tissue transplantation demand the creation of a global stem cell bank, which may hold particular promise for minority populations and other sub-groups currently marginalized from organ procurement and allocation systems. Finally, we conclude by offering a number of practical and ethically oriented recommendations for constructing a human embryonic stem cell bank that we hope will help solve the ongoing organ shortage problem.

Lyerly, Anne Drapkin; Faden, Ruth R. Willingness to donate frozen embryos for stem cell research. *Science* 2007 July 6; 317(5834): 46-47. 17 refs. NRCBL: 18.5.4; 15.1; 19.5; 14.6. SC: em.

Keywords: *attitudes; *cryopreservation; *donors; *embryo disposition; *embryo research; *embryonic stem cells; *embryos; *patients; biomedical research; embryo transfer; in vitro fertilization; informed consent; moral status; public opinion; survey; Keyword Identifiers: *United States; Australia; Great Britain

Master, Zubin; Williams-Jones, Bryn; Lott, Jason P.; Savulescu, Julian. The global HLA banking of embryonic stem cells requires further scientific justification. *American Journal of Bioethics* 2007 August; 7(8): 45-46; author reply W4-W6. 7 refs. NRCBL: 18.5.4; 15.1; 19.5. Comments: Jason P. Lott and Julian Savulescu. Towards a global human embryonic stem cell bank. American Journal of Bioethics 2007 August; 7(8): 37-44.

Keywords: *embryonic stem cells; *tissue banks; cloning; donors; incentives; international aspects; justice; mandatory programs; moral policy; nuclear transfer techniques; organ transplantation; presumed consent; risks and benefits; scarcity; stem cell transplantation; voluntary programs; Proposed Keywords: embryo donation

McCartney, James J. Embryonic stem cell research and respect for human life: philosophical and legal reflections. *Albany Law Review* 2002; 65(3): 597-624. NRCBL: 18.5.4; 15.1; 14.5; 5.3.

McKechnie, L.; Gill, A.B. Consent for neonatal research. *Archives of Disease in Childhood. Fetal and Neonatal Edition* 2006 September; 91(5): F374-F376. NRCBL: 18.5.4; 18.3; 18.2.

McKneally, Martin. Controversies in cardiothoracic surgery: should therapeutic cloning be supported to provide stem cells for cardiothoracic surgery research and treatment? [debate]. *Journal of Thoracic and Cardiovascular Surgery* 2006 May; 131(5): 937-940. NRCBL: 18.5.4; 15.1; 14.5; 9.5.1.

McLeod, Carolyn; Baylis, Françoise. Donating fresh versus frozen embryos to stem cell research: in whose interests. *Bioethics* 2007 November; 21(9): 465-477. NRCBL: 18.5.4; 10; 14.4; 14.6; 19.5.

Abstract: Some stem cell researchers believe that it is easier to derive human embryonic stem cells from fresh rather than frozen embryos and they have had in vitro fertilization (IVF) clinicians invite their infertility patients to donate their fresh embryos for research use. These embryos include those that are deemed 'suitable for transfer' (i.e. to the woman's uterus) and those deemed unsuitable in this regard. This paper focuses on fresh embryos deemed suitable for transfer – hereafter 'fresh embryos'– which IVF patients have good reason not to donate. We explain why donating them to research is not in the self-interests specifically of female IVF patients. Next, we consider the other-regarding interests of these patients and conclude that while fresh embryo donation may serve those interests, it does so at unnecessary cost to patients' self-interests. Lastly, we review some of the potential barriers to the autonomous donation of fresh embryos to research and highlight the risk that female IVF patients invited to donate these embryos will misunderstand key aspects of the donation decision, be coerced to donate, or be exploited in the consent process. On the basis of our analysis, we conclude that patients should not be asked to donate their fresh embryos to stem cell research.

Mitchell, Peter. EU cell therapy legislation. *Nature Biotechnology* 2007 June; 25(6): 614. NRCBL: 18.5.4; 15.1; 5.3; 21.1. SC: le. Identifiers: European Union.

Morris, Jonathan. Stem cell research. *In his:* The Ethics of Biotechnology. Philadelphia: Chelsea Publishers, 2006: 83-104. NRCBL: 18.5.4; 15.1. SC: po.

Keywords: *embryo research; *embryonic stem cells; abortion; federal government; government financing; legal aspects; moral policy; religious ethics; risks and benefits; stem cell transplantation; Keyword Identifiers: United States

Murray, Fiona. The stem cell market — patents and the pursuit of scientific progress. *New England Journal of Medicine* 2007 June 7; 356(23): 2341-2343. 4 refs. NRCBL: 18.5.4; 15.1; 1.3.9; 15.8; 19.1.

Keywords: *biomedical research; *embryonic stem cells; *patents; *research support; *universities; access to information; cell lines; commerce; contracts; embryo research; industry; legal aspects; private sector; publishing; science; Proposed Keywords: technology transfer; Keyword Identifiers: *Wisconsin Alumni Research Foundation; United States; University of Wisconsin

National Institutes of Health [NIH] (United States). Stem Cell Information [government document]. Bethesda, MD: National Institutes of Health [NIH] (United States), 2007 May 4 and 2007 September 11: [34 p.] [Online]. Accessed: http://stemcells.nih.gov/staticresources/research/registry/PDFs/Registry.pdf [2007 October 2]. NRCBL: 18.5.4; 15.1; 5.3. SC: em.

Keywords: *biological specimen banks; *cell lines; *embryo research; *embryonic stem cells; *federal government; *government financing; *public policy; *research support; guidelines; international aspects; registries; Keyword Identifiers: *National Institutes of Health; *United States

Onder, Robert. "People need a fairy tale": the embryonic stem cell and cloning debate in Missouri. *Missouri Medicine* 2006 March-April; 103(2): 106-111. NRCBL: 18.5.4; 14.5; 15.1. SC: em.

Outomuro, Delia; Lott, James P.; Savulescu, Julian. Moral dilemmas around a global human embryonic stem cell bank. *American Journal of Bioethics* 2007 August; 7(8): 47-48; author reply W4-W6. 10 refs. NRCBL: 18.5.4; 15.1; 19.5. Comments: Jason P. Lott and Julian Savulescu. Towards a global human embryonic stem cell bank. American Journal of Bioethics 2007 August; 7(8): 37-44.

Keywords: *embryonic stem cells; *tissue banks; donors; embryos; health services accessibility; incentives; international aspects; justice; mandatory programs; minority groups; moral policy; moral status; organ transplantation; presumed consent; remuneration; required request; risks and benefits; scarcity; stem cell transplantation; voluntary programs; Proposed Keywords: embryo donation

---

NRCBL: National Reference Center for Bioethics Literature Classification Scheme   See inside front cover for terms.

283

Rosen, Michael R. Are stem cells drugs? The regulation of stem cell research and development. *Circulation* 2006 October 31; 114(18): 1992-2000. NRCBL: 18.5.4; 15.1; 18.6.

Royal College of Obstetricians and Gynaecologists (Great Britain). RCOG response to MRC consultation on the Code of Practice for the use of human stem cell lines. London: Royal College of Obstetricians and Gynaecologists, 2004 May 27; 3 p. [Online]. Accessed: http://www. rcog.org.uk/index.asp?PageID=1349 [2007 April 5]. NRCBL: 18.5.4; 15.1; 6.

Keywords: *biological specimen banks; *cell lines; *embryonic stem cells; *guidelines; *standards; embryo research; ethical review; fetal research; informed consent; organizational policies; physicians; professional organizations; stem cells; Keyword Identifiers: *Royal College of Obstetricians and Gynaecologists (Great Britain); Great Britain

Salter, Brian. Bioethics, politics and the moral economy of human embryonic stem cell science: the case of the European Union's Sixth Framework Programme. *New Genetics and Society* 2007 December; 26(3): 269-288. NRCBL: 18.5.4; 15.1; 21.1.

Sass, Hans-Martin. Let probands and patients decide about moral risk: stem cell research and medical treatment. *In:* Roetz, Heiner, ed. Cross-Cultural Issues in Bioethics: The Example of Human Cloning. New York: Rodopi, 2006: 425-444. NRCBL: 18.5.4; 15.1; 14.5.

Schwartz, Peter. Stem cells: biopsy on frozen embryos [letter]. *Hastings Center Report* 2007 January-February; 37(1): 7-8. NRCBL: 18.5.4; 15.1; 14.6.

Keywords: *cryopreservation; *embryo research; *embryonic stem cells; *embryos; in vitro fertilization; Proposed Keywords: blastomeres

Somerville, Margaret A. The importance of empirical research in bioethics: the case of human embryo stem cell research = Importance de la recherche empirique en bioéthique: cas de la recherche sur les cellules souches embryonnaires humaines [editorial]. *JOGC: Journal of Obstetrics and Gynaecology Canada = JOGC: Journal d'Obstétrique et Gynécologie du Canada* 2005 October; 27(10): 929-932. NRCBL: 18.5.4; 15.1; 18.3; 2.1.

Taylor, Patrick L. Research sharing, ethics and public benefit. *Nature Biotechnology* 2007 April; 25(4): 398-401. 36 refs. NRCBL: 18.5.4; 15.1; 1.3.9.

Keywords: *biomedical research; *embryonic stem cells; *interprofessional relations; beneficence; biotechnology; embryo research; genetic research; guidelines; industry; informed consent; international aspects; justice; obligations to society; patents; research ethics; research subjects; researchers; universities; Proposed Keywords: *cooperative behavior; Keyword Identifiers: International Society for Stem Cell Research

te Braake, Trees. Research on human embryos. *In:* Gevers, J.K.M.; Hondius, E.H.; Hubben, J.H., eds. Health Law, Human Rights and the Biomedicine Convention: Es-

says in Honour of Henriette Roscam Abbing. Leiden; Boston: Martinus Nijhoff Publishers, 2005: 191-203. 20 fn. NRCBL: 18.5.4; 15.1.

Keywords: *embryo research; cloning; fetal research; genetic engineering; guidelines; human dignity; human experimentation; in vitro fertilization; value of life; Keyword Identifiers: *Council of Europe; Ad Hoc Committee of Experts on Progress in the Medical Sciences (CAHBI); Europe; European Convention on Human Rights and Biomedicine

Trivedi, Bijal. Researchers detour around stem-cell rules. *Chronicle of Higher Education* 2007 October 3; 54(6): A12-A15. NRCBL: 18.5.4; 15.1; 18.6; 19.5.

Keywords: *embryonic stem cells; *ovum donors; *remuneration; embryo disposition; embryo research; government regulation; in vitro fertilization; indigents; ovum; researchers; scarcity; women; Keyword Identifiers: Great Britain; United States

United States. Office of the President. Fact Sheet: Embryonic Stem Cell Research. *Washington, DC: The White House, Office of the Press Secretary,* 2001 August 9; 3 p. [Online]. Accessed: http://www.whitehouse.gov/news/releases/2001/08/text/20010809-1.html [2001 August 10]. NRCBL: 18.5.4; 15.1; 18.6. SC: po.

Keywords: *embryo research; *embryonic stem cells; *federal government; *government financing; *public policy; *research support; adult stem cells; advisory committees; cell lines; guidelines; Keyword Identifiers: *United States; President's Council on Bioethics

United States. Office of the President. Domestic Policy Council. Advancing stem cell science without destroying human life. Washington, DC: The White House, 2007 January 9: 64 p. [Online]. Accessed: http://www. whitehouse.gov/infocus/healthcare/stemcell_010907.pdf [2007 January 24]. 130 refs. NRCBL: 18.5.4; 15.1; 18.5.1; 14.5.

Keywords: *embryo research; *legal aspects; *public policy; *stem cells; adult stem cells; alternatives; cloning; embryo disposition; embryonic stem cells; federal government; government financing; methods; research support; value of life; Keyword Identifiers: *United States; President's Council on Bioethics

Vogel, Gretchen. Still waiting for cybrids. *Science* 2007 September 14; 317(5844): 1483. NRCBL: 18.5.4; 15.1; 22.2; 18.6. Identifiers: Great Britain (United Kingdom).

Wainwright, Steven; Williams, Clare; Michael, Mike; Farsides, Bobbie; Cribb, Alan. Remaking the body? Scientists' genetic discourses and practices as examples of changing expectations on embryonic stem cell therapy for diabetes. *New Genetics and Society* 2007 December; 26(3): 251-268. NRCBL: 18.5.4; 15.1.

Weiss, Rick. Stem cells created with no harm to human embryos; but concerns are raised about the technique. *Washington Post* 2006 August 24; p. A3. NRCBL: 18.5.4; 19.1; 15.1. SC: po.

Keywords: *embryonic stem cells; research embryo creation; *methods; public policy; research support; government financing; Keyword Identifiers: United States

World Federation of Neurology; Rosenberg, Roger N. World Federation of Neurology position paper on human stem cell research. *Journal of the Neurological Sciences* 2006 April 15; 243(1-2): 1-2. NRCBL: 18.5.4; 15.1; 6.

Yoshimura, Yasunori. Bioethical aspects of regenerative and reproductive medicine. *Human Cell* 2006 May; 19(2): 83-86. NRCBL: 18.5.4; 14.5; 15.1; 18.2; 14.1.

Zycinski, J. Bioethics, technology and human dignity: the Roman Catholic viewpoint. *Acta Neurochirurgica Supplement* 2006; 98: 1-7. 25 refs. NRCBL: 18.5.4; 15.1; 1.2; 4.4.
    Keywords: *biotechnology; *embryos; *human dignity; *research embryo creation; *Roman Catholic ethics; bioethics; biomedical technologies; embryonic stem cells; evolution; moral status; nuclear transfer techniques; quality of life; value of life

Zycinski, Joseph. Ethics in medical technologies: the Roman Catholic viewpoint. *Journal of Clinical Neuroscience* 2006 June; 13(5): 518-523. 20 refs. NRCBL: 18.5.4; 15.1; 1.2; 4.4.
    Keywords: *biotechnology; *embryo research; *Roman Catholic ethics; ecology; embryonic stem cells; eugenics; human dignity; quality of life; value of life; values; Proposed Keywords: altered nuclear transfer

HUMAN EXPERIMENTATION/ . . . /
EMBRYOS AND FETUSES/ LEGAL ASPECTS
See also CLONING

An unwieldy hybrid [editorial]. *Nature* 2007 May 24; 447(7143): 353-354. NRCBL: 18.5.4; 15.1; 18.1; 22.1. SC: le.
    Keywords: *chimeras; *embryo research; *legal aspects; genetically modified animals; government regulation; legislation; public policy; Keyword Identifiers: *Great Britain; Human Fertilisation and Embryology Authority (Great Britain)

German stem-cell law should change, says ethics council [news]. *Nature* 2007 July 26; 448(7152): 399. NRCBL: 18.5.4; 15.1. SC: le.
    Keywords: *embryo research; *embryonic stem cells; *government regulation; *legal aspects; advisory committees; cell lines; public policy; Keyword Identifiers: *Germany; National Ethics Council (Germany)

Burns, Lawrence. Overstating the ban, ignoring the compromise. *American Journal of Bioethics* 2007 February; 7(2): 65-66. 4 refs. NRCBL: 18.5.4; 15.1; 14.5. SC: le. Comments: Timothy Caulfield and Tania Bubela. Why a criminal ban? Analyzing the arguments against somatic cell nuclear transfer in the Canadian parliamentary debate. American Journal of Bioethics 2007 February; 7(2): 51-61.
    Keywords: *cloning; *embryo research; *legal aspects; *nuclear transfer techniques; *public policy; abortion; consensus; embryonic stem cells; embryos; government regulation; moral status; political activity; reproductive technologies; wedge argument; Keyword Identifiers: Canada

Doerflinger, Richard M. Washington insider: 2006 in Congress; Senate Hearing on Misrepresentations in stem cell research; Opening battle of 2007; Genetic Nondiscrimination Bill may see action. *National Catholic Bioethics Quarterly* 2007 Spring; 7(1): 15-21. NRCBL: 18.5.4; 15.1; 18.5.1; 1.2; 15.3. SC: le. Identifiers: H.R. 810 — Stem Cell Research Enhancement Act.

Green, Shane K.; Lott, Jason P.; Savulescu, Julian. Is Canada's stem cell legislation unwittingly discriminatory? *American Journal of Bioethics* 2007 August; 7(8): 50-52; author reply W4-W6. 4 refs. NRCBL: 18.5.4; 15.1; 19.5. SC: le. Comments: Jason P. Lott and Julian Savulescu. Towards a global human embryonic stem cell bank. American Journal of Bioethics 2007 August; 7(8): 37-44.
    Keywords: *embryonic stem cells; *legislation; *tissue banks; cell lines; cloning; donors; government regulation; guidelines; in vitro fertilization; incentives; international aspects; justice; minority groups; moral policy; nuclear transfer techniques; public policy; remuneration; scarcity; social discrimination; Proposed Keywords: embryo donation; haplotypes; Keyword Identifiers: *Canada

Grubb, Andrew. Regulating reprogenetics in the United Kingdom. *In:* Knowles, Lori P.; Kaebnick, Gregory E., eds. Reprogenetics: Law, Policy, and Ethical Issues. Baltimore: Johns Hopkins University Press, 2007: 144-177. 30 refs., 31 fn. NRCBL: 18.5.4; 15.1; 14.1; 14.5; 15.2; 14.4. SC: le.
    Keywords: *cloning; *embryo research; *legislation; *policy making; *preimplantation diagnosis; *public policy; *regulation; *reproductive technologies; advisory committees; embryonic stem cells; embryos; nuclear transfer techniques; organization and administration; standards; Proposed Keywords: licensure; tissue typing; Keyword Identifiers: *Great Britain; *Human Fertilisation and Embryology Authority; Human Fertilisation and Embryology Act 1990 (Great Britain); Warnock Committee

Guenin, Louis M. A proposed stem cell research policy. *Stem Cells* 2005 September; 23(8): 1023-1027. NRCBL: 18.5.4; 15.1; 18.6. SC: le.

Holden, Constance. Scientists protest 'misrepresentation' as Senate vote looms [news]. *Science* 2007 January 19; 315(5810): 315-316. NRCBL: 18.5.4; 15.1; 5.3. SC: le.

Holt, Rush. How should government regulate stem-cell research? Views from a scientist- legislator. *In:* Santoro, Michael A.; Gorrie, Thomas M., eds. Ethics and the Pharmaceutical Industry. Cambridge; New York: Cambridge University Press, 2005: 109-122, 427-431. 53 refs. NRCBL: 18.5.4; 15.1. SC: le.
    Keywords: *embryo research; *legal aspects; *stem cells; cloning; embryonic stem cells; federal government; government financing; in vitro fertilization; international aspects; legislation; public policy; reproductive technologies; state government; stem cell transplantation; Keyword Identifiers: *United States; Europe; Asia

Martínez, Jaime Vidal. Biomedical research with human embryos: changes in the legislation on assisted reproduction in Spain. *Revista de Derecho y Genoma Humano =*

NRCBL: National Reference Center for Bioethics Literature Classification Scheme   See inside front cover for terms.

285

Law and the Human Genome Review 2006 July-December; (25): 161-182. 81 fn. NRCBL: 18.5.4; 14.1; 15.1. SC: le.

    Keywords: *embryo research; *legal aspects; *reproductive technologies; biomedical research; embryonic stem cells; embryos; fetal research; legislation; genetic engineering; legal rights; property rights; Keyword Identifiers: *Spain

Mitka, Mike. Stem cell legislation [news]. *JAMA: The Journal of the American Medical Association* 2007 February 14; 297(6): 581. NRCBL: 18.5.4; 15.1. SC: le.

    Keywords: *embryo research; *embryonic stem cells; *government financing; *legislation; federal government; politics; public policy; research support; Keyword Identifiers: *Stem Cell Research Enhancement Act; *U.S. Congress; *United States

O'Dowd, Adrian. UK may allow creation of "cybrids" for stem cell research [news]. *BMJ: British Medical Journal* 2007 March 10; 334(7592): 495. NRCBL: 18.5.4; 15.1; 22.2; 14.5; 18.1; 22.1. SC: le.

    Keywords: *chimeras; *embryo research; *embryonic stem cells; government regulation; public policy; Keyword Identifiers: *Great Britain

O'Neil, Graeme. Australia ends stem-cell cloning ban [news brief]. *Nature Biotechnology* 2007 February; 25(2): 153. NRCBL: 18.5.4; 15.1; 14.5. SC: le.

    Keywords: *cloning; *embryo research; *embryonic stem cells; *embryos; *legal aspects; government regulation; Keyword Identifiers: *Australia

Ostrer, H.; Wilson, D.I.; Hanley, N.A. Human embryo and early fetus research. *Clinical Genetics* 2006 August; 70(2): 98-107. NRCBL: 18.5.4; 15.1; 18.6.

Ries, Nola M. Regulation of human stem cell research in Japan and Canada: a comparative analysis. *University of New Brunswick Law Journal = Revue de Droit de L'université du Nouveau Brunswick* 2005; 54: 62-74. 68 fn. NRCBL: 18.5.4; 15.1; 18.6. SC: le.

    Keywords: *embryo research; *embryonic stem cells; *government regulation; *international aspects; cloning; cross-cultural comparison; embryos; ethical review; government financing; guidelines; legal aspects; public policy; reproductive technologies; research support; stem cells; Keyword Identifiers: *Canada; *Japan

Schreiber, Hans-Peter. Embryonen- und Stammzellforschung. *In:* Schreiber, Hans-Peter, ed. Biomedizin und Ethik: Praxis, Recht, Moral. Basel; Boston: Birkhäuser, 2004: 84-87, 96. 3 refs. NRCBL: 18.5.4; 15.1. SC: le.

    Keywords: *embryo research; *embryonic stem cells; adult stem cells; cloning; international aspects; legal aspects; Keyword Identifiers: France; Great Britain; Sweden; Switzerland

Spar, Debora; Harrington, Anna. Selling stem cell science: how markets drive law along the technological frontier. *American Journal of Law and Medicine* 2007; 33(4): 541-565. NRCBL: 18.5.4; 15.1; 18.6; 5.3; 11.1; 14.4. SC: le.

Takala, Tuija; Häyry, Matti. Benefiting from past wrongdoing, human embryonic stem cell lines, and the fra-

gility of the German legal position. *Bioethics* 2007 March; 21(3): 150-159. 50 fn. NRCBL: 18.5.4; 15.1; 19.5; 1.1. SC: an; le.

    Keywords: *cell lines; *embryo research; *embryonic stem cells; *ethical analysis; *legislation; *moral complicity; *moral policy; aborted fetuses; commerce; commodification; dehumanization; embryo disposition; embryos; government regulation; historical aspects; in vitro fertilization; international aspects; morality; National Socialism; ovum donors; policy analysis; public policy; scientific misconduct; Keyword Identifiers: *Germany; *Stem Cell Act 2002 (Germany)

Abstract: This paper examines the logic and morality of the German Stem Cell Act of 2002. After a brief description of the law's scope and intent, its ethical dimensions are analysed in terms of symbolic threats, indirect consequences, and the encouragement of immorality. The conclusions are twofold. For those who want to accept the law, the arguments for its rationality and morality can be sound. For others, the emphasis on the uniqueness of the German experience, the combination of absolute and qualified value judgments, and the lingering questions of indirect encouragement of immoral activities will probably be too much.

United States. Congress. House. A bill to derive human pluripotent stem cell lines using techniques that do not harm human embryos. Washington, DC: U.S. G.P.O., 2007. 4 p. [Online]. Accessed: http://frwebgate.access. gpo.gov/cgi-bin/useftp.cgi?IPaddress=162.140.64.21& filename=h322ih.pdf&directory=/diskb/wais/data/ 110_cong_bills [17 January 2007]. NRCBL: 18.5.4; 15.1; 4.4. SC: le. Identifiers: Alternative Pluripotent Stem Cell Therapies Enhancement Act of 2007. Note: H.R. 322, 110th Congress, 1st session. Introduced by Rep. Bartlett on January 9, 2007. Referred to the Committee on Energy and Commerce.

    Keywords: *embryos; *legal aspects; *stem cells; alternatives; cell lines; embryo research; methods; Keyword Identifiers: *United States

United States. Congress. House. An act to amend the Public Health Service Act to provide for human embryonic stem cell research. Washington, DC: U.S. G.P.O., 2007. 4 p. [Online]. Accessed: http://frwebgate.access.gpo.gov/ cgi-bin/useftp.cgi?IPaddress=162.140.64.21&filename= h3eh.pdf&directory=/diskb/wais/data/110_cong_bills [2007 January 17]. NRCBL: 18.5.4; 15.1. SC: le. Identifiers: Stem Cell Research Enhancement Act of 2007. Note: H.R. 3, 110th Congress, 1st session. Introduced by Ms. DeGette and others, January 5, 2007. Passed the House of Representatives January 11, 2007.

    Keywords: *embryo research; *embryonic stem cells; *legal aspects; Keyword Identifiers: *United States

United States. Congress. Senate. A bill to amend the Public Health Service Act to provide for human embryonic stem cell research. Washington, DC: U.S. G.P.O., 2007. 3 p. [Online]. Accessed: http://frwebgate.access.gpo.gov/ cgi-bin/useftp.cgi?IPaddress=162.140.64.21&filename= s5pcs.pdf&directory=/diskb/wais/data/110_cong_bills [2007 April 10]. NRCBL: 18.5.4; 15.1. SC: le. Identifiers:

Stem Cell Research Enhancement Act of 2007. Note: S. 5, 110th Congress, 1st session. Introduced by Sen. Reid, January 4, 2007. Read the second time and placed on calendar, January 8, 2007.

Keywords: *embryo research; *embryonic stem cells; *legal aspects; embryo disposition; Keyword Identifiers: *United States

United States. Congress. Senate. A bill to derive human pluripotent stem cell lines using techniques that do not knowingly harm embryos. Washington, DC: U.S. G.P.O., 2007. 5 p. [Online]. Accessed: http://frwebgate.access. gpo.gov/cgi-bin/useftp.cgi?IPaddress=162.140.64.21& filename=s51is.pdf&directory=/diskb/wais/data/ 110_cong_bills [2007 January 17]. NRCBL: 18.5.4; 15.1; 4.4. SC: le. Identifiers: Pluripotent Stem Cell Therapy Enhancement Act of 2007. Note: S. 51, 110th Congress, 1st session. Introduced by Sen. Isakson on January 4, 2007. Referred to the Committee on Health, Education, Labor, and Pensions.

Keywords: *cell lines; *legal aspects; *research embryo creation; *stem cells; alternatives; embryonic stem cells; public policy; Keyword Identifiers: *United States

United States. Congress. Senate. A bill to intensify research to derive human pluripotent stem cell lines. Washington, DC: U.S. G.P.O., 2007. 6 p. [Online]. Accessed: http://frwebgate.access.gpo.gov/cgi-bin/useftp.cgi? IPaddress=162.140.64.21&filename=s30hds.pdf& directory=/diskb/wais/data/110_cong_bills [2007 April 10]. NRCBL: 18.5.4; 15.1. SC: le. Identifiers: Hope Offered through Principled and Ethical Stem Cell Research Act; HOPE Act. Note: S. 30, 110th Congress, 1st session. Introduced by Sen. Coleman, March 29, 2007.

Keywords: *legal aspects; *stem cells; alternatives; biomedical research; embryo research; embryos; guidelines; public policy; Keyword Identifiers: *United States

United States. Congress. Senate. A bill to provide increased Federal funding for stem cell research, to expand the number of embryonic stem cell lines available for Federally funded research, to provide ethical guidelines for stem cell research, to derive human pluripotent stem cell lines using techniques that do not create an embryo or embryos for research or knowingly harm embryos, and for other purposes. Washington, DC: U.S. G.P.O., 2007. 15 p. [Online]. Accessed: http://frwebgate.access.gpo.gov/cgi-bin/useftp.cgi?IPaddress=162.140.64.21&filename= s363is.pdf&directory=/diskb/wais/data/110_cong_bills [2007 April 4]. NRCBL: 18.5.4; 15.1. SC: le. Identifiers: Hope Offered through Principled, Ethically-Sound Stem Cell Research Act; HOPE Act. Note: S. 363, 110th Congress, 1st session. Introduced by Sen. Coleman on January 23, 2007. Referred to the Committee on Health, Education, Labor, and Pensions.

Keywords: *embryonic stem cells; *embryo research; *government financing; *legal aspects; *research support; *stem cells; adult stem cells; alternatives; cell lines; embryos; informed consent; research ethics committees; Keyword Identifiers: *United States

Wadman, Meredith. Stem-cell issue moves up the US agenda [news]. *Nature* 2007 April 19; 446(7138): 842. NRCBL: 18.5.4; 15.1. SC: le.

Keywords: *embryo research; *embryonic stem cells; *government financing; *legal aspects; *research support; cell lines; federal government; legislation; politics; Keyword Identifiers: *United States; Stem Cell Research Enhancement Act 2007; U.S. House of Representative; U.S. Senate

HUMAN EXPERIMENTATION/ . . . / EMBR. & FETUSES/ PHIL. & RELIG. ASPECTS
See also CLONING

Agar, Nicholas. Embryonic potential and stem cells. *Bioethics* 2007, May; 21(4): 198-207. 23 fn. NRCBL: 18.5.4; 1.1; 4.4; 15.1; 19.5. SC: an.

Keywords: *embryo research; *embryonic stem cells; *embryos; *ethical analysis; *moral status; beginning of life; cloning; embryo disposition; embryonic development; in vitro fertilization; intention; nuclear transfer techniques; philosophy; Proposed Keywords: *blastocysts

Abstract: This paper examines three arguments that use the concept of potential to identify embryos that are morally suitable for embryonic stem cell research (ESCR). According to the first argument, due to Ronald Green, the fact that they are scheduled for disposal makes embryos left over from IVF treatments morally appropriate for research. Paul McHugh argues that embryos created by somatic cell nuclear transfer differ from those that result directly from the meeting of sperm and egg in having potential especially conducive to the therapeutic use of their stem cells. I reject both of these arguments. According to the way of making distinctions in embryonic potential that I defend, it is the absence of a functional relationship with a womb that marks embryos morally suitable for ESCR.

Balint, John A. Ethical issues in stem cell research. *Albany Law Review* 2002; 65(3): 729-742. NRCBL: 18.5.4; 15.1; 14.5; 1.2.

Barnes, Richard. Stem cell research funding: testimony. *Origins* 2007 March 15; 36(39): 616-620. 10 refs. NRCBL: 18.5.4; 15.1; 1.2; 9.3.1.

Keywords: *embryo research; *embryonic stem cells; *government financing; *research support; *Roman Catholic ethics; *state government; adult stem cells; biotechnology; cloning; conflict of interest; economics; embryos; industry; legislation; public opinion; public policy; value of life; Keyword Identifiers: *New York; United States

Brown, Mark T. The potential of the human embryo. *Journal of Medicine and Philosophy* 2007 November-December; 32(6): 585-618. NRCBL: 18.5.4; 4.4; 1.1; 15.1. SC: an.

Abstract: A higher order potential analysis of moral status clarifies the issues that divide Human Being Theorists who oppose embryo research from Person Theorists who favor embryo research. Higher order potential personhood is transitive if it is active, identity preserving and morally relevant. If the transition from the Second Order Potential of the embryo to the First Order Potential

NRCBL: National Reference Center for Bioethics Literature Classification Scheme   See inside front cover for terms.

287

of an infant is transitive, opponents of embryo research make a powerful case for the moral status of the embryo. If it is intransitive, then the Person Theorist can draw lines between levels of moral status that permit embryo research to proceed.

Byrnes, W. Malcolm. Partial trajectory: the story of the altered nuclear transfer-oocyte assisted reprogramming (ANT-OAR) proposal. *Linacre Quarterly* 2007 February; 74(1): 50-59. NRCBL: 18.5.4; 15.1; 1.2.

Catholic Church. Catholic Bishops' Conference of England and Wales Linacre Centre for Healthcare Ethics. Catholic Bishops' Conference of England and Wales and Linacre Centre for Healthcare Ethics joint response to the Human Tissue and Embryos (Draft) Bill. London: Linacre Centre for Healthcare Ethics. 2007 June 20: 5 p. [Online]. Accessed: http://www.linacre.org/Linacre%20 Joint%20submission%20on%20Human%20Tissue%20 and %20Embryos%20_draft_%20Bill.pdf [2007 July 12]. NRCBL: 18.5.4; 19.5; 1.2.

Condic, Maureen L.; Furton, Edward J. Harvesting embryonic stem cells from deceased human embryos. *National Catholic Bioethics Quarterly* 2007 Autumn; 7(3): 507-525. 35 fn. NRCBL: 18.5.4; 15.1; 19.5; 20.2.1; 14.4; 18.3; 1.2.
  Keywords: *death; *determination of death; *embryo disposition; *embryonic stem cells; *embryos; *moral policy; cryopreservation; embryo research; human dignity; in vitro fertilization; methods; model legislation; moral complicity; parental consent; public policy; Roman Catholic ethics; Proposed Keywords: *embryo death

Deckers, J. Why two arguments from probability fail and one argument from Thomson's analogy of the violinist succeeds in justifying embryo destruction in some situations. *Journal of Medical Ethics* 2007 March; 33(3): 160-164. 10 refs. NRCBL: 18.5.4; 14.5; 1.1; 15.1; 4.4. SC: an.
  Keywords: *embryo research; *embryos; *ethical analysis; *killing; *moral status; abortion; advisory committees; beginning of life; embryo disposition; legal aspects; philosophy; public policy; value of life; Proposed Keywords: probability; spontaneous abortion; Keyword Identifiers: Chief Medical Officer's Expert Group (Great Britain); Great Britain; House of Lords Select Committee on Medical Ethics; Thomson, Judith
Abstract: The scope of embryo research in the UK has been expanded by the Human Fertilisation and Embryology (Research Purposes) Regulations 2001. Two advisory bodies — the Chief Medical Officer's Expert Group and the House of Lords' Select Committee — presented various arguments in favour of embryo research. One of these is the view that, just as lottery tickets have relatively little value before the draw because of the low probability of their being the winning ticket, early embryos have relatively little value because of the presumed low probability that they will mature into more developed embryos. This (first) argument from probability is questioned in this paper, as well as the contention that allowing embryo destruction is incompatible with the view that embryos have full moral status. Although I challenge Savulescu's view that early embryos should be entered into a lottery in which they are subjected to the probability of being destroyed (the second argument from probability), a revised version of Thomson's analogy of the famous violinist defies the view that the position that the embryo has full moral status is incompatible with qualified support for embryo destruction.

Doerflinger, Richard M. Washington insider: 2006 in Congress; Senate Hearing on Misrepresentations in stem cell research; Opening battle of 2007; Genetic Nondiscrimination Bill may see action. *National Catholic Bioethics Quarterly* 2007 Spring; 7(1): 15-21. NRCBL: 18.5.4; 15.1; 18.5.1; 1.2; 15.3. SC: le. Identifiers: H.R. 810 — Stem Cell Research Enhancement Act.

Green, Ronald M. Can we develop ethically universal embryonic stem-cell lines? *Nature Reviews Genetics* 2007 June; 8(6): 480-485. NRCBL: 18.5.4; 15.1; 19.5; 4.4; 15.4. SC: em.

Hagen, John D., Jr. Bentham's mummy and stem cells. *America* 2007 May 14; 196(17): 12-14. NRCBL: 18.5.4; 15.1; 1.1; 4.4.

Hurlbut, William B. Stem cells, embryos, and ethics: is there a way forward? *Update (Loma Linda Center)* 2007 January 21(3): 1-10. 5 refs. NRCBL: 18.5.4; 15.1; 4.4; 1.2. Note: Adapted from the Health and Faith Forum: Bioethics and Wholeness Grand Rounds presentation, 2007 January 10.
  Keywords: *beginning of life; *embryo research; *embryonic stem cells; *embryos; *moral policy; alternatives; cloning; government financing; in vitro fertilization; moral status; nuclear transfer techniques; public policy; twinning; value of life; Proposed Keywords: regenerative medicine; Keyword Identifiers: United States

Jersild, Paul. Theological and moral reflections on stem cell research. *Journal of Lutheran Ethics [electronic]* 2007 March; 7(3): 4 p. Accessed: http://www.elca.org/jle/article.asp?k=705 [2007 February 28]. 7 fn. NRCBL: 18.5.4; 15.1; 1.2.
  Keywords: *embryonic stem cells; *embryos; *Protestant ethics; *value of life; beginning of life; embryo research; ethical analysis; moral status; personhood; Roman Catholic ethics; stem cell transplantation; theology

Johnson, Luke. Embryonic stem cell research: a legitimate application of just-war theory? *Ethics and Medicine: An International Journal of Bioethics* 2007 Spring; 23(1): 19-30. 14 refs. 7 fn. NRCBL: 18.5.4; 15.1; 21.2; 1.2; 1.1. SC: an.
  Keywords: *embryo research; *embryonic stem cells; *ethical theory; *moral policy; *war; embryos; intention; killing; moral status; rights

Lysaught, M. Therese. Making decisions about embryonic stem cell research. *In:* Hamel, Ronald, ed. Making Health Care Decisions: A Catholic Guide. Liguori, MO: Liguori Publications, 2006: 19-36. 7 refs. NRCBL: 18.5.4; 1.2; 15.1.

Keywords: *embryo research; *embryonic stem cells; *Roman Catholic ethics; beginning of life; cloning; embryos; justice; moral status; personhood; value of life

Mauceri, Joseph M. Evolution and the embryo: the evidence for special creation. *Linacre Quarterly* 2007 February; 74(1): 30-49. NRCBL: 18.5.4; 3.2; 15.1; 14.5; 1.2.

Napier, Stephen. Human embryos as human subjects. *Ethics and Medics* 2007 September; 32(9): 3-4. NRCBL: 18.5.4; 15.1; 18.6; 1.2; 1.1.

Northcott, Michael S. In the waters of Babylon: the moral geography of the embryo. *In:* Deane-Drummond, Celia; Scott, Peter Manley, eds. Future Perfect?: God, Medicine and Human Identity. New York: T and T Clark International, 2006: 73-86. NRCBL: 18.5.4; 15.1; 14.4; 14.5; 4.4; 1.2.

Novak, David. On the use of embryonic stem cells. *In his:* The Sanctity of Human Life. Washington, DC: Georgetown University Press, 2007: 1-89. 202 fn. NRCBL: 18.5.4; 15.1; 1.1; 1.2; 4.4; 12.3. SC: an.
  Keywords: *embryo research; *embryonic stem cells; *embryos; *Jewish ethics; *moral policy; *moral status; *philosophy; *public policy; *theology; abortion; beginning of life; cultural pluralism; fetuses; killing; morality; natural law; personhood; policy analysis; politics; secularism; value of life; Proposed Keywords: embryo death; embryonic development

Parker, C. Ethics for embryos. *Journal of Medical Ethics* 2007 October; 33(10): 614-616. 3 refs. NRCBL: 18.5.4; 4.4; 15.1; 19.5. SC: an.
  Keywords: *embryo research; *embryonic stem cells; *embryos; *moral status; beginning of life; cloning; embryonic development; ethical analysis; personhood; rights; value of life; Proposed Keywords: blastocysts
Abstract: This paper responds to DW Brock's technically strong case for the use of human embryonic stem cells in medical research. His main issue in this context is the question of whether it is moral to destroy viable human embryos. He offers a number of reasons to support his view that it is moral to destroy them, but his use of conceptual arguments is not adequate to secure his position. The purpose and scope of this paper is wholly concerned with his arguments rather than with the conclusion that it is justifiable to destroy human embryos. The author proceeds through his variety of arguments and offers reasons for rejecting them. The author concludes that Brock has not shown that it is moral to destroy viable human embryos.

Pruss, Alexander R. Cooperation with past evil and use of cell-lines derived from aborted fetuses. *In:* Watt, Helen, ed. Cooperation, Complicity and Conscience: Problems in Healthcare, Science, Law and Public Policy. London: Linacre Centre, 2005: 89-104. NRCBL: 18.5.4; 15.1; 19.5; 1.2.

Scolding, Neil. Cooperation problems in science: use of embryonic/fetal material. *In:* Watt, Helen, ed. Cooperation, Complicity and Conscience: Problems in Healthcare,

Science, Law and Public Policy. London: Linacre Centre, 2005: 105-117. NRCBL: 18.5.4; 15.1; 19.5; 1.2.

Shimazono, Susumu. Why we must be prudent in research using human embryos: differing views of human dignity. *In:* LaFleur, William R.; Böhme, Gernot; Shimazono, Susumu, eds. Dark Medicine: Rationalizing Medical Research. Bloomington: Indiana University Press, 2007: 201-222. NRCBL: 18.5.4; 4.4; 1.2.

Takala, Tuija; Häyry, Matti. Benefiting from past wrongdoing, human embryonic stem cell lines, and the fragility of the German legal position. *Bioethics* 2007 March; 21(3): 150-159. 50 fn. NRCBL: 18.5.4; 15.1; 19.5; 1.1. SC: an; le.
  Keywords: *cell lines; *embryo research; *embryonic stem cells; *ethical analysis; *legislation; *moral complicity; *moral policy; aborted fetuses; commerce; commodification; dehumanization; embryo disposition; embryos; government regulation; historical aspects; in vitro fertilization; international aspects; morality; National Socialism; ovum donors; policy analysis; public policy; scientific misconduct; Keyword Identifiers: *Germany; *Stem Cell Act 2002 (Germany)
Abstract: This paper examines the logic and morality of the German Stem Cell Act of 2002. After a brief description of the law's scope and intent, its ethical dimensions are analysed in terms of symbolic threats, indirect consequences, and the encouragement of immorality. The conclusions are twofold. For those who want to accept the law, the arguments for its rationality and morality can be sound. For others, the emphasis on the uniqueness of the German experience, the combination of absolute and qualified value judgments, and the lingering questions of indirect encouragement of immoral activities will probably be too much.

Tangwa, Godfrey B. Moral status of embryonic stem cells: perspective of an African villager. *Bioethics* 2007 October; 21(8): 449-457. 6 fn. NRCBL: 18.5.4; 15.1; 1.1; 1.3.1; 4.4; 21.7. SC: an. Conference: Eighth World Congress of Bioethics: A Just and Healthy Society; Beijing, China; 2006 August 9.
  Keywords: *embryonic stem cells; *embryos; *moral status; *value of life; beginning of life; cloning; ethical analysis; embryo research; morality; non-Western World; personhood; teleological ethics; wedge argument; Proposed Keywords: blastocysts; Keyword Identifiers: Africa
Abstract: One of the most important as well as most awesome achievements of modern biotechnology is the possibility of cloning human embryonic stem cells, if not human beings themselves. The possible revolutionary role of such stem cells in curative, preventive and enhancement medicine has been voiced and chorused around the globe. However, the question of the moral status of embryonic stem cells has not been clearly and unequivocally answered. Taking inspiration from the African adage that 'the hand that reaches beneath the incubating hen is not guiltless', I attempt answering this question, from the background of traditional African moral sensibility and sensitivity. I reach the following conclusions. Stem cells in themselves do not have human status and therefore lack moral worth/value. Embryos do

NRCBL: National Reference Center for Bioethics Literature Classification Scheme   See inside front cover for terms.

289

have human status and a morally significant line cannot be drawn between human embryos and other human beings. What is morally at stake in stem cell research is therefore the question of the source of derivation or generation of the cells, not of the cells as such.

Thompson, Clive. How to farm stem cells without losing your soul. *Wired Magazine* 2005 June; 13.06: 4 p. [Online]. Accessed: http://www.wired.com/wired/archive/13.06/stemcells.html [2006 October 12]. NRCBL: 18.5.4; 1.2; 15.1.

> Keywords: *embryonic stem cells; *methods; *research embryo creation; advisory committees; alternatives; beginning of life; Christian ethics; dissent; embryos; Keyword Identifiers: *Hurlbut, William; President's Council on Bioethics

Tong, Rosemary. Stem-cell research and the affirmation of life. *Conscience* 2007 Autumn; 28(3): 19-23. NRCBL: 18.5.4; 15.1; 4.4; 1.2; 7.1.

United States. Congress. House. A bill to derive human pluripotent stem cell lines using techniques that do not harm human embryos. Washington, DC: U.S. G.P.O., 2007. 4 p. [Online]. Accessed: http://frwebgate.access.gpo.gov/cgi-bin/useftp.cgi?IPaddress=162.140.64.21&filename=h322ih.pdf&directory=/diskb/wais/data/110_cong_bills [17 January 2007]. NRCBL: 18.5.4; 15.1; 4.4. SC: le. Identifiers: Alternative Pluripotent Stem Cell Therapies Enhancement Act of 2007. Note: H.R. 322, 110th Congress, 1st session. Introduced by Rep. Bartlett on January 9, 2007. Referred to the Committee on Energy and Commerce.

> Keywords: *embryos; *legal aspects; *stem cells; alternatives; cell lines; embryo research; methods; Keyword Identifiers: *United States

United States. Congress. Senate. A bill to derive human pluripotent stem cell lines using techniques that do not knowingly harm embryos. Washington, DC: U.S. G.P.O., 2007. 5 p. [Online]. Accessed: http://frwebgate.access.gpo.gov/cgi-bin/useftp.cgi?IPaddress=162.140.64.21&filename=s51is.pdf&directory=/diskb/wais/data/110_cong_bills [2007 January 17]. NRCBL: 18.5.4; 15.1; 4.4. SC: le. Identifiers: Pluripotent Stem Cell Therapy Enhancement Act of 2007. Note: S. 51, 110th Congress, 1st session. Introduced by Sen. Isakson on January 4, 2007. Referred to the Committee on Health, Education, Labor, and Pensions.

> Keywords: *cell lines; *legal aspects; *research embryo creation; *stem cells; alternatives; embryonic stem cells; public policy; Keyword Identifiers: *United States

Walter, James J. A Catholic reflection on embryonic stem cell research. *In:* Walter, James J.; Shannon, Thomas A., eds. Contemporary Issues in Bioethics: A Catholic Perspective. Lanham, MD: Rowman and Littlefield Publishers, 2005: 91-99. 21 fn. NRCBL: 18.5.4; 1.2; 15.1.

> Keywords: *embryo research; *embryonic stem cells; *Roman Catholic ethics; embryos; government financing; moral status; public policy; research support; theology

Wang, Yanguang. The moral status of the human embryo in Chinese stem cell research. *Asian Biotechnology and Development Review* 2006 November; 9(1): 45-63 [Online]. Accessed: http://www.openj-gate.org/articlelist.asp?LatestYear=2007&JCode=104346&year=2006&vol=9&issue=1&ICode=583649 [2008 February 29]. NRCBL: 18.5.4; 19.1; 1.1; 18.6.

HUMAN EXPERIMENTATION/ . . . / FOREIGN NATIONALS

Angell, Marcia. Cross-cultural considerations in medical ethics: the case of human subjects research. *In:* Galston, Arthur W.; Peppard, Christiana Z., eds. Expanding Horizons in Bioethics. Dordrecht; Norwell, MA: Springer, 2005: 71-84. NRCBL: 18.5.9; 18.6; 5.3; 21.7.

Caniza, Miguela A.; Clara, Wilfrido; Maron, Gabriela; Navarro-Marin, Jose Ernesto; Rivera, Roberto; Howard, Scott C.; Camp, Jonathan; Barfield, Raymond C. Establishment of ethical oversight of human research in El Salvador: lessons learned. *Lancet Oncology* 2006 December; 7(12): 1027-1033. NRCBL: 18.5.9; 18.2; 2.4.

Dickert, N.; DeRiemer, K.; Duffy, P.E.; Garcia-Garcia, L.; Mutabingwa, T.K.; Sina, B.J.; Tindana, P.; Lie, R. Ancillary-care responsibilities in observational research: two cases, two issues. *Lancet* 2007 March 10-16; 369(9564): 874-877. NRCBL: 18.5.9; 18.2; 18.6; 9.8.

Dunbar, Terry; Scrimgeour, Margaret. Ethics in indigenous research — connecting with community. *Journal of Bioethical Inquiry* 2006; 3(3): 179-185. NRCBL: 18.5.9; 18.6; 18.2; 18.3.

Ford, Jolyon; Tomossy, George. Clinical trials in developing countries: the plaintiff's challenge. *Law, Social Justice and Global Development* 2004 June 4; (1): 17 p. [Online]. Accessed: http://www2.warwick.ac.uk/fac/soc/law/elj/lgd/2004_1/ford/ [2007 April 4]. NRCBL: 18.5.9. SC: le.

Ghayur, Muhammad N.; Ghayur, Ayesha; Janssen, Luke J. State of clinical research ethics in Pakistan [letter]. *Nature Medicine* 2007 September; 13(9): 1011. NRCBL: 18.5.9; 18.2; 18.3. Comments: Cassandra Willyard. Pfizer lawsuit spotlights ethics of developing world clinical trials. Nature Medicine 2007 July; 13(7): 763.

Henderson, Gail E.; Corneli, Amy L.; Mahoney, David B.; Nelson, Daniel K.; Mwansambo, Charles. Applying research ethics guidelines: the view from a sub-Saharan research ethics committee. *Journal of Empirical Research on Human Research Ethics* 2007 June; 2(2): 41-48. NRCBL: 18.5.9; 18.2; 18.3; 18.6; 21.7. SC: em. Identifiers: Malawi.

> Abstract: Considerable variation has been demonstrated in applying regulations across research ethics committees (RECs) in the U.S., U.K., and European nations. With the

rise of international research collaborations, RECs in developing countries apply a variety of international regulations. We conducted a qualitative descriptive pilot study with members of the national REC in Malawi to determine criteria they use to review research, and their views on international collaborations. Qualitative content analysis demonstrated that international guidelines are interpreted in light of local African conditions such that emphasis is placed on examining benefit to the community and ensuring the informed consent process translates concepts in locally-meaningful ways. Members suggest that RECs often must comply with regulations that do not fit local conditions. Recommendations are provided for improving such international collaborations.

Horn, Lyn. Research vulnerability: an illustrative case study from the South African mining industry. *Developing World Bioethics* 2007 December; 7(3): 119-127. NRCBL: 18.5.9. SC: cs.

Abstract: The concept of 'vulnerability' is well established within the realm of research ethics and most ethical guidelines include a section on 'vulnerable populations'. However, the term 'vulnerability', used within a human research context, has received a lot of negative publicity recently and has been described as being simultaneously 'too broad' and 'too narrow'. The aim of the paper is to explore the concept of research vulnerability by using a detailed case study - that of mineworkers in post-apartheid South Africa. In particular, the usefulness of Kipnis's taxonomy of research vulnerability will be examined. In recent years the volume of clinical research on human subjects in South Africa has increased significantly. The HIV and TB pandemics have contributed to this increase. These epidemics have impacted negatively on the mining industry; and mining companies have become increasingly interested in research initiatives that address these problems. This case study explores the potential research vulnerability of mineworkers in the context of the South African mining industry and examines measures that can reduce this vulnerability.

Jesani, Amar; Coutinho, Lester. AIDS vaccine trials in India: ethical benchmarks and unanswered questions [editorial]. *Indian Journal of Medical Ethics* 2007 January-March; 4(1): 2-3. NRCBL: 18.5.9; 9.5.6; 9.7; 15.1; 18.2. Identifiers: Targeted Genetics.

Kilmarx, Peter H.; Ramjee, Gita; Kitayaporn, Dwip; Kunasol, Prayura. Protection of human subjects' rights in HIV-preventive clinical trials in Africa and Asia: experiences and recommendations. *AIDS* 2001; 15 (suppl.5): S73-S79. NRCBL: 18.5.9; 9.5.6; 18.3; 18.6.

Kimmelman, Jonathan. Clinical trials and SCID row: the ethics of phase 1 trials in the developing world. *Developing World Bioethics* 2007 December; 7(3): 128-135. 43 fn. NRCBL: 18.5.9; 15.4. SC: an.

Keywords: *clinical trials; *developing countries; *disadvantaged; *gene therapy; *gene transfer techniques; drugs; economics; genetic disorders; guideline adherence; guidelines; health care delivery; hemophilia; HIV infections; international aspects; justice; nontherapeutic research; research design; research ethics; socioeconomic factors; standards; therapeutic misconception; Proposed Keywords: *phase I clinical trials; adenosine deaminase; severe combined immunodeficiency; Keyword Identifiers: Africa; Council for International Organizations of Medical Sciences; Declaration of Helsinki; Italy

Abstract: Relatively little has been written about the ethics of conducting early phase clinical trials involving subjects from the developing world. Below, I analyze ethical issues surrounding one of gene transfer's most widely praised studies conducted to date: in this study, Italian investigators recruited two subjects from the developing world who were ineligible for standard of care because of economic considerations. Though the study seems to have rendered a cure in these two subjects, it does not appear to have complied with various international guidelines that require that clinical trials conducted in the developing world be responsive to their populations' health needs. Nevertheless, policies devised to address large scale, late stage trials, such as the AZT short-course placebo trials, map somewhat awkwardly to early phase studies. I argue that interest in conducting translational research in the developing world, particularly in the context of hemophilia trials, should motivate more rigorous ethical thinking around clinical trials involving economically disadvantaged populations.

Lertsithichai, Panuwat. Health research, fair benefits and access to medicines. *Journal of the Medical Association of Thailand = Chotmaihet Thangphaet* 2006 April; 89(4): 558-564. NRCBL: 18.5.9; 9.2; 2.4; 18.2.

Manafa, Ogenna; Lindegger, Graham; Ijsselmuiden, Carel. Informed consent in an antiretroviral trial in Nigeria. *Indian Journal of Medical Ethics* 2007 January-March; 4(1): 26-30. NRCBL: 18.5.9; 18.3; 9.5.6; 9.7. SC: em.

Newton, Sam K.; Appiah-Poku, John. Opinions of researchers based in the UK on recruiting subjects from developing countries into randomized controlled trials. *Developing World Bioethics* 2007 December; 7(3): 149-156. NRCBL: 18.5.9; 18.3. SC: em.

Abstract: BACKGROUND: Explaining technical terms in consent forms prior to seeking informed consent to recruit into trials can be challenging in developing countries, and more so when the studies are randomized controlled trials. This study was carried out to examine the opinions of researchers on ways of dealing with these challenges in developing countries. METHODS: Recorded in-depth interviews with 12 lecturers and five doctoral students, who had carried out research in developing countries, at a leading school of public health in the United Kingdom. A purposive, snowballing approach was used to identify interviewees. RESULTS: Researchers were divided on the feasibility of explaining technical trials in illiterate populations; the majority of them held the view that local analogies could be used to explain these technical terms. Others were of the opinion that this could not be done since it was too difficult to explain technical trials, such as randomized controlled trials, even to people in developed countries. CONCLUSION: Researchers acknowledged the difficulty in explaining randomized controlled trials but it was also their perception that this was an important part of the ethics of the work of scientific research involving human subjects. These difficulties notwithstanding, efforts should be made to ensure that subjects have sufficient understanding to consent, taking into account the fact that peculiar situations in developing countries might compound this difficulty.

Rohter, Larry. In the Amazon, giving blood but getting nothing. *New York Times* 2007 June 20; p. A1, A4. NRCBL: 18.5.9; 19.4; 9.3.1; 1.3.9; 15.1; 15.11. SC: po.
　　Keywords: *blood specimen collection; *commerce; *genetic materials; *genetic research; *indigenous populations; American Indians; culture; deception; informed consent; international aspects; population genetics; property rights; scientific misconduct; Proposed Keywords: tissue donors; Keyword Identifiers: *Brazil; *Karitiana Indians

Salvi, Vinita; Damania, K. HIV, research, ethics and women. *Journal of Postgraduate Medicine* 2006 July-September; 52(3): 161-162. NRCBL: 18.5.9; 18.5.3; 18.3; 9.5.6.

Terrell White, Mary. A right to benefit from international research: a new approach to capacity building in less-developed countries. *Accountability in Research* 2007 April-June; 14(2): 73-92. NRCBL: 18.5.9; 18.2.
　　Abstract: This article proposes a means by which benefits provided in international research collaborations might be employed to strengthen health care, research, and other capacities in less-developed countries. The Declaration of Helsinki and CIOMS Guidelines define certain expectations of benefits, but these requirements are ambiguous, logistically problematic, and studies suggest they are inconsistently upheld. Drawing on the principle of respect for persons, a right to benefit from hosting externally-sponsored research is proposed. This right guarantees host communities benefits of a certain value, the nature and use of which is controlled by indigenous personnel. Suggestions are made as to how implementation of this right, using structured incentives, may systematically promote capacity building in host communities.

Tomossy, George F.; Ford, Jolyon. Globalization and clinical trials: compensating subjects in developing countries. *In:* Bennett, Belinda; Tomossy, George F., eds. Globalization and Health: Challenges for Health Law and Bioethics. Dordrecht: Springer, 2006: 27-45. NRCBL: 18.5.9; 21.1; 9.3.1. SC: le; rv.

Willyard, Cassandra. Pfizer lawsuit spotlights ethics of developing world clinical trials [news]. *Nature Medicine* 2007 July; 13(7): 763. NRCBL: 18.5.9; 1.3.2; 9.7. SC: le. Identifiers: Nigeria.

HUMAN EXPERIMENTATION/ . . . / MENTALLY DISABLED

Anderson, Kelly K.; Mukherjee, Som D. The need for additional safeguards in the informed consent process in schizophrenia research. *Journal of Medical Ethics* 2007 November; 33(11): 647-650. NRCBL: 18.5.6; 18.3.
　　Abstract: The process of obtaining informed consent to participate in a clinical study presents many challenges for research conducted in a population of patients with schizophrenia. Morally valid, informed consent must include information sharing, decisional capacity, and capacity for voluntarism. This paper examines the unique features of schizophrenia that may threaten each of these elements of informed consent, and it proposes additional safeguards in the process of gaining informed consent from individuals with schizophrenia in order to maximise the decision-making potential of this patient population.

Brashler, Rebecca. Ethics, family caregivers, and stroke. *Topics in Stroke Rehabilitation* 2006 Fall; 13(4): 11-17. NRCBL: 18.5.6; 9.5.1; 8.1.

Dewing, Jan. From ritual to relationship: a person-centered approach to consent in qualitative research with older people who have dementia. *Dementia: The International Journal of Social Research and Practice* 2002 June; 1(2): 157-171. NRCBL: 18.5.6; 4.4; 18.3; 18.5.7.

Hellström, Ingrid; Nolan, Mike; Nordenfelt, Lennart; Lundh, Ulla. Ethical and methodological issues in interviewing persons with dementia. *Nursing Ethics* 2007 September; 14(5): 608-619. NRCBL: 18.5.6; 18.5.7; 18.3.
　　Abstract: People with dementia have previously not been active participants in research, with ethical difficulties often being cited as the reason for this. A wider inclusion of people with dementia in research raises several ethical and methodological challenges. This article adds to the emerging debate by reflecting on the ethical and methodological issues raised during an interview study involving people with dementia and their spouses. The study sought to explore the impact of living with dementia. We argue that there is support for the inclusion of people with dementia in research and that the benefits of participation usually far outweigh the risks, particularly when a ;safe context' has been created. The role of gatekeepers as potentially responsible for excluding people with dementia needs further consideration, with particular reference to the appropriateness of viewing consent as a primarily cognitive, universalistic and exclusionary event as opposed to a more particularistic, inclusive and context relevant process.

Iacono, Teresa. Ethical challenges and complexities of including people with intellectual disability as participants in research. *Journal of Intellectual and Developmental Disability* 2006 September; 31(3): 173-179; discussion 180-191. NRCBL: 18.5.6; 18.3. Identifiers: Australia.

Lanter, Jennifer. Clinical research with cognitively impaired subjects: issues for nurses. *Dimensions of Critical*

Care Nursing 2006 March-April; 25(2): 89-92. NRCBL: 18.5.6; 4.1.3; 18.3; 18.6.

Leidinger, Friedrich. Müssen Demenzkranke ein "Sonderopfer für die Forschung" bringen? — Für eine neue Wissenschaft von der Demenz! *In:* Neuer-Miebach, Therese; Wunder, Michael, eds. Bio-Ethik und die Zukunft der Medizin. Bonn: Psychiatrie-Verlag, 1998: 106-119. NRCBL: 18.5.6.

Maloney, Dennis M. Research with adult subjects who have impaired decision-making capacity. *Human Research Report* 2007 October; 22(10): 1-2. NRCBL: 18.5.6; 18.2.

McHale, Jean. Law reform, clinical research and adults without mental capacity — much needed clarification or a recipe for further uncertainty? *In:* McLean, Sheila A.M., ed. First Do No Harm: Law, Ethics, and Healthcare. Aldershot, England; Burlington, VT: Ashgate, 2006: 215-233. NRCBL: 18.5.6; 18.3. SC: le. Identifiers: Great Britain (United Kingdom).

Pence, Gregory E. Can research be just on people with schizophrenia? *In his:* The Elements of Bioethics. Boston: McGraw-Hill, 2007: 203-232. NRCBL: 18.5.6; 18.3; 18.2; 2.2; 18.5.1.

Reid, Clare L.; Menon, David K. Researching incapacity: time to get our acts together [letter]. *BMJ: British Medical Journal* 2007 September 1; 335(7617): 415. NRCBL: 18.5.6; 18.3. SC: le. Identifiers: Great Britain (United Kingdom).

Savage, Teresa A. Ethical issues in research with patients who have experienced stroke. *Topics in Stroke Rehabilitation* 2006 Fall; 13(4): 1-10. NRCBL: 18.5.6; 18.3; 18.2.

Siegel, Paul E.; Ellis, Norman R. Note on the recruitment of subjects for mental retardation research. *American Journal of Mental Deficiency* 1985 January; 89(4): 431-433. NRCBL: 18.5.6; 18.2; 18.6.

HUMAN EXPERIMENTATION/ . . . / MINORS

Alderson, Priscilla. Ethics. *In:* Fraser, Sandy; Lewis, Vicky; Ding, Sharon; Kellett, Mary; Robinson, Chris, eds. Doing Research with Children and Young People. London; Thousand Oaks, CA: Sage Publications, 2004: 97-112. NRCBL: 18.5.2; 18.2; 18.3; 1.3.9.

Anand, K.J.S.; Aranda, Jacob V.; Berde, Charles B.; Buckman, ShaAvhrée; Capparelli, Edmund V.; Carlo, Waldemar A.; Hummel, Patricia; Lantos, John; Johnston, C. Celeste; Lehr, Victoria Tutag; Lynn, Anne M.; Maxwell, Lynne G.; Oberlander, Tim F.; Raju, Tonse N.K.; Soriano, Sulpicio G.; Taddio, Anna; Walco, Gary A. Analgesia and anesthesia for neonates: study design and ethical issues. *Clinical Therapeutics* 2005 June; 27(6): 814-843. NRCBL: 18.5.2; 18.2; 18.6; 4.4. SC: rv.

Austin, Joan K. Ethical issues related to the increased emphasis on children participating in research. *Chronic Illness* 2006 September; 2(3): 181-182. NRCBL: 18.5.2; 18.2.

Banister, Elizabeth; Leadbeater, Bonnie; Benoit, Cecilia; Jansson, Michael; Marshall, Anne; Riecken, Ted. Ethical issues in community-based research with children and youth. *NCEHR Communique CNERH* 2006 Spring; 14(1): 23-24. NRCBL: 18.5.2; 18.3; 18.6. Identifiers: Canada.

Brody, Janet L.; Scherer, David G.; Annett, Robert D.; Turner, Charles; Dalen, Jeanne. Family and physician influence on asthma research participation decisions for adolescents: the effects of adolescent gender and research risk. *Pediatrics* 2006 August; 118(2): e356-e362. NRCBL: 18.5.2; 18.3; 9.4.

Coch, Donna. Neuroimaging research with children: ethical issues and case scenarios. *Journal of Moral Education* 2007 March; 36(1): 1-18. NRCBL: 18.5.2; 18.4; 17.1; 1.3.1; 18.6. SC: cs.

Downie, Jocelyn; Schmidt, Matthais; Kenny, Nuala; D'Arcy, Ryan; Hadskis, Michael; Marshall, Jennifer. Paediatric MRI research ethics: the priority issues. *Journal of Bioethical Inquiry* 2007; 4(2): 85-91. NRCBL: 18.5.2; 8.3.2; 8.4; 5.3.

Abstract: Abstract In this paper, we first briefly describe neuroimaging technology, our reasons for studying magnetic resonance imaging (MRI) technology, and then provide a discussion of what we have identified as priority issues for paediatric MRI research. We examine the issues of respectful involvement of children in the consent process as well as privacy and confidentiality for this group of MRI research participants. In addition, we explore the implications of unexpected findings for paediatric MRI research participants. Finally, we explore the ethical issues concerning advances in functional MRI. This paper aims to provide a clear description of priority paediatric MRI research ethics issues to make some preliminary recommendations regarding next steps.

Dyer, Owen. Andrew Wakefield is accused of paying children for blood [news]. *BMJ: British Medical Journal* 2007 July 21; 335(7611): 118-119. NRCBL: 18.5.2; 1.3.9; 18.2; 19.5; 9.3.1. SC: le.

Evans, Jane; Simon, Gayle. Family Educational Rights and Privacy Act (FERPA). *Journal of Empirical Research on Human Research Ethics* 2007 March; 2(1): 101-104. NRCBL: 18.5.2; 1.3.3; 8.4; 1.3.12.

Fisher, Celia B.; Kornetsky, Susan Z.; Prentice, Ernest D. Determining risk in pediatric research with no prospect of direct benefit: time for a national consensus on the interpretation of federal regulations. *American Journal of Bioethics* 2007 March; 7(3): 5-10. NRCBL: 18.5.2; 18.6; 18.2. SC: le. Identifiers: United States.

NRCBL: National Reference Center for Bioethics Literature Classification Scheme   See inside front cover for terms.

Abstract: United States federal regulations for pediatric research with no prospect of direct benefit restrict institutional review board (IRB) approval to procedures presenting: 1) no more than "minimal risk" (§ 45CFR46.404); or 2) no more than a "minor increase over minimal risk" if the research is commensurate with the subjects' previous or expected experiences and intended to gain vitally important information about the child's disorder or condition (§ 45CFR46.406) (DHHS 2001). During the 25 years since their adoption, these regulations have helped IRBs balance subject protections with the pursuit of scientific knowledge to advance children's welfare. At the same time, inconsistency in IRB application of these regulations to pediatric protocols has been widespread, in part because of the ambiguity of the regulatory language. During the past decade, three federally-charged committees have addressed these ambiguities: 1) the National Human Research Protections Advisory Committee (NHRPAC) (Washington, DC), 2) the Institute of Medicine (IOM) Committee on the Ethical Conduct of Clinical Research Involving Children (Washington, DC); and 3) the United States Department of Health and Human Services Secretary's Advisory Committee for Human Research Protections (SACHRP) (Washington, DC). The committees have reached similar conclusions on interpretation of language within regulations § § 45CFR46.404 and 406; these conclusions are remarkably consistent with recent international recommendations and those of the original National Commission for the Protection of Human Subjects of Biomedical and Behavioral Research (1977) report from which current regulations are based. Drawing on the committees' public reports, this article identifies the ethical issues posed by ambiguities in regulatory language, summarizes the committees' deliberations, and calls for a national consensus on recommended criteria.

Flotte, Terence R.; Frentzen, Barbara; Humphries, Margaret R.; Rosenbloom, Arlan L. Recent developments in the protection of pediatric research subjects. *Journal of Pediatrics* 2006 September; 149(3): 285-286. NRCBL: 18.5.2; 18.2.

Frith, Lucy. Researching chronic childhood illness: autonomy or beneficence? *Chronic Illness* 2006 September; 2(3): 178-180. NRCBL: 18.5.2; 18.3; 1.1. Identifiers: Great Britain (United Kingdom).

Glaser, Nicole; Kuppermann, Nathan; Marcin, James; Schalick, Walton O. A comment on "The risky business of assessing research risk". *American Journal of Bioethics* 2007 November; 7(11): W5-W6. NRCBL: 18.5.2; 18.2. Comments: A.A. Kon. The risky business of assessing research risk. American Journal of Bioethics 2007; 7(3): 21-22.

Hagger, Lynn; Woods, Simon. Children and research: a risk of double jeopardy? *In:* Freeman, Michael, ed. Children's Health and Children's Rights. Boston: Martinus Nijhoff Publishers, 2006: 51-72. NRCBL: 18.5.2; 8.3.2. SC: cs; le.

Hartman, Rhonda Gay. Word from the academies: a primer for legal policy analysis regarding adolescent research participation. *Rutgers Journal of Law and Public Policy* 2006 Fall; 4(1): 152-199. NRCBL: 18.5.2; 18.2; 18.3; 8.3.2. SC: le.

Hazen, Rebecca; Greenley, Rachel Neff; Drotar, Dennis; Kodish, Eric. Recommending randomized trials for pediatric Leukemia: observer and physician report of recommendations. *Journal of Empirical Research on Human Research Ethics* 2007 June; 2(2): 49-55. NRCBL: 18.5.2; 4.1.2; 8.1; 18.3; 9.5.1. SC: em.
Abstract: Physicians' presentation of treatment options in a non-coercive manner is critical for informed consent for participation in randomized clinical trials (RCTs). This study examined discrepancies between observer and physician report of treatment recommendations for pediatric leukemia RCTs. This study also assessed relationships between recommendations and decisions to participate in RCTs. Participants were 104 parents of children with leukemia and the treating physicians. Measures included observations of informed consent conferences (ICCs), physician report of treatment recommendations, and parent report of trial participation. Observation revealed that physicians recommended RCTs in 38% of ICCs, while physicians reported recommending RCTs in 73% of ICCs. Treatment recommendations were unrelated to decisions to participate in RCTs. Results highlight the importance of enhancing parent-physician communication regarding RCT participation.

Hunter, David; Pierscionek, Barbara K. Children, Gillick competency and consent for involvement in research. *Journal of Medical Ethics* 2007 November; 33(11): 659-662. NRCBL: 18.5.2; 18.3. SC: an.
Abstract: This paper looks at the issue of consent from children and whether the test of Gillick competency, applied in medical and healthcare practice, ought to extend to participation in research. It is argued that the relatively broad usage of the test of Gillick competency in the medical context should not be considered applicable for use in research. The question of who would and could determine Gillick competency in research raises further concerns relating to the training of the researcher to make such a decision as well as to the obvious issue of the researcher's personal interest in the project and possibility of benefiting from the outcome. These could affect the judgment of Gillick competency if the researcher is charged with making this decision. The above notwithstanding, there are two exceptional research situations in which Gillick competency might be legitimately applied: (1) when the research is likely to generate significant advantages for the participants while exposing them to relatively minor risks, and (2) when it is likely to generate great societal benefit, pose minimal risks for the participants and yet raise parental objection. In both cases, to ensure that autonomy is genuinely respected and to protect against personal interest, Gillick competency should be assessed by an individual who has no interest or involvement in the research.

Iltis, Ana. Pediatric research posing a minor increase over minimal risk and no prospect of direct benefit: challenging 45 CFR 46.406. *Accountability in Research* 2007 January-March; 14(1): 19-34. NRCBL: 18.5.2. SC: le.

Iltis, Ana S.; DeVader, Shannon; Matsuo, Hisako. Payments to children and adolescents enrolled in research: a pilot study. *Pediatrics* 2006 October; 118(4): 1546-1552. NRCBL: 18.5.2; 18.3; 9.3.1.

Joffe, Steven; Fernandez, Conrad V.; Pentz, Rebecca D.; Ungar, David R.; Mathew, N. Ajoy; Turner, Curtis W.; Alessandri, Angela J.; Woodman, Catherine L.; Singer, Dale A.; Kodish, Eric. Involving children with cancer in decision-making about research participation. *Journal of Pediatrics* 2006 December; 149(6): 862-868. NRCBL: 18.5.2; 9.5.1; 18.3; 8.1.

John, Jill E. The child's right to participate in research: myth or misconception? *British Journal of Nursing* 2007 February 8-21; 16(3): 157-160. NRCBL: 18.5.2; 18.3; 8.3.2.

Johnson, Jeannette L.; Vandermark, Nancy R. Ethics in prevention research with children. *In:* Kleinig, John; Einstein, Stanley, eds. Ethical Challenges for Intervening in Drug Use: Policy, Research and Treatment Issues. Huntsville, TX: Office of International Criminal Justice; Criminal Justice Center, Sam Houston State University, 2006: 259-279. NRCBL: 18.5.2; 9.5.9; 18.2; 18.3; 21.7.

Kassam-Adams, Nancy; Newman, Elana. The reactions to research participation questionnaires for children and for parents (RRPQ-C and RRPQ-P). *General Hospital Psychiatry* 2002 September-October; 24(5): 336-342. NRCBL: 18.5.2; 18.3; 18.4. SC: em.

Kon, Alexander A. The risky business of assessing research risk. *American Journal of Bioethics* 2007 March; 7(3): 21-22. NRCBL: 18.5.2; 18.2; 18.3. Comments: Celia B. Fisher, Susan Z. Kornetsky, and Ernest D. Prentice. Determining risk in pediatric research with no prospect of direct benefit: time for a national consensus on the interpretation of federal regulations. American Journal of Bioethics 2007 March; 7(3): 5-10.

Kopelman, Loretta M. When can children with conditions be in no-benefit, higher-hazard pediatric studies? *American Journal of Bioethics* 2007 March; 7(3): 15-17. NRCBL: 18.5.2; 18.1; 18.2; 18.6. Comments: Celia B. Fisher, Susan Z. Kornetsky, and Ernest D. Prentice. Determining risk in pediatric research with no prospect of direct benefit: time for a national consensus on the interpretation of federal regulations. American Journal of Bioethics 2007 March; 7(3): 5-10.

Kutty, V. Raman. The study served no purpose. *Indian Journal of Medical Ethics* 2007 April-June; 4(2): 78. NRCBL: 18.5.2; 18.2. SC: cs. Comments: Response to: Mala Ramanthan and Amar Jesani. Ethics in nutrition in-tervention research. Indian Journal of Medical Ethics 2007 April-June; 4(2): 76-77.

Lantos, John D. Research in wonderland: does "minimal risk" mean whatever an institutional review board says it means? *American Journal of Bioethics* 2007 March; 7(3): 11-12. NRCBL: 18.5.2; 18.2; 18.3; 19.5; 18.5.4; 18.1. Comments: Celia B. Fisher, Susan Z. Kornetsky, and Ernest D. Prentice. Determining risk in pediatric research with no prospect of direct benefit: time for a national consensus on the interpretation of federal regulations. American Journal of Bioethics 2007 March; 7(3): 5-10.

Lynch, Margaret A.; Glaser, Danya; Prior, Vivien; Inwood, Vivien. Following up children who have been abused: ethical considerations for research design. *Child Psychology and Psychiatry Review* 1999 May; 4(2): 68-75. NRCBL: 18.5.2; 10; 18.3; 9.5.7; 9.1; 18.2. SC: rv; em. Identifiers: Great Britain (United Kingdom).

Magnus, David. Playing it safe [editorial]. *American Journal of Bioethics* 2007 March; 7(3): 1-2. NRCBL: 18.5.2; 18.6.

Maloney, Dennis M. Court says institutional review board (IRB) abdicated duty to protect children as subjects. *Human Research Report* 2007 January; 22(1): 8. NRCBL: 18.5.2; 18.2; 18.3. SC: le. Identifiers: Grimes v. Kennedy Krieger Institute (KKI), Part (3).

Maloney, Dennis M. Final guidance issued on protection of children as research subjects. *Human Research Report* 2007 February; 22(2): 1-2. NRCBL: 18.5.2; 18.2.

Maloney, Dennis M. In court: former research subjects charge researchers and their institutions with negligence. *Human Research Report* 2007 August; 22(8): 8. NRCBL: 18.5.2; 16.1; 18.3. SC: le. Identifiers: Grimes v. Kennedy Krieger Institute (KKI), (associated with Johns Hopkins University), (Part 10).

Maloney, Dennis M. In court: informed consent form lacked crucial information, says court. *Human Research Report* 2007 June; 22(6): 8. NRCBL: 18.5.2; 16.1; 18.3. SC: le. Identifiers: Grimes v. Kennedy Krieger Institute (KKI),(associated with Johns Hopkins University), (Part 8).

Maloney, Dennis M. In court: researchers and a history of unethical behavior. *Human Research Report* 2007 December; 22(12): 8. NRCBL: 18.5.2; 18.3; 18.4; 1.3.9. SC: le. Identifiers: Grimes v. Kennedy Krieger Institute (KKI), (associated with Johns Hopkins University), (Part 14); Wendell Johnson.

Maloney, Dennis M. Informed consent issue dwarfed by ethics of research study itself. *Human Research Report* 2007 February; 22(2): 8. NRCBL: 18.5.2; 18.3. SC: le. Identifiers: Grimes v. Kennedy Krieger Institute (KKI), (associated with Johns Hopkins University), (Part 4).

NRCBL: National Reference Center for Bioethics Literature Classification Scheme   See inside front cover for terms.

295

Maloney, Dennis M. Study's children lived in housing with various levels of possible lead exposure. *Human Research Report* 2007 May; 22(5): 8. NRCBL: 18.5.2; 16.1; 18.3. SC: le. Identifiers: Grimes v. Kennedy Krieger Institute (KKI), (associated with Johns Hopkins University), (Part 7).

Martin, Rebecca A.; Robert, Jason Scott. Is risky pediatric research without prospect of direct benefit ever justified? *American Journal of Bioethics* 2007 March; 7(3): 12-15. 12 refs. NRCBL: 18.5.2; 18.2; 15.1; 18.3. Comments: Celia B. Fisher, Susan Z. Kornetsky, and Ernest D. Prentice. Determining risk in pediatric research with no prospect of direct benefit: time for a national consensus on the interpretation of federal regulations. American Journal of Bioethics 2007 March; 7(3): 5-10.
>   Keywords: *children; *clinical trials; *genetic disorders; *nontherapeutic research; *risk; *terminally ill; brain; government regulation; informed consent; investigational therapies; parental consent; pediatrics; research ethics committees; research subjects; stem cell transplantation; therapeutic misconception; vulnerable populations; Proposed Keywords: Batten disease; Keyword Identifiers: United States

Merlo, D.F.; Knudsen, L.E.; Matusiewicz, K.; Niebrój, L.; Vähäkangas, K.H. Ethics in studies on children and environmental health. *Journal of Medical Ethics* 2007 July; 33(7): 408-413. 58 refs. NRCBL: 18.5.2; 16.1; 18.2; 18.3; 18.5.4; 15.1; 1.3.12.
>   Keywords: *adolescents; *biological specimen banks; *children; *nontherapeutic research; age factors; autonomy; biomedical research; consent forms; genetic databases; genetic research; guidelines; health hazards; informed consent; minors; parental consent; public health; refusal to participate; research ethics committees; research subjects; standards; tissue donors; vulnerable populations; Proposed Keywords: *environmental health; Keyword Identifiers: Europe

Abstract: Children, because of age-related reasons, are a vulnerable population, and protecting their health is a social, scientific and emotional priority. The increased susceptibility of children and fetuses to environmental (including genotoxic) agents has been widely discussed by the scientific community. Children may experience different levels of chemical exposure than adults, and their sensitivity to chemical toxicities may be increased or decreased in comparison with adults. Such considerations also apply to unborn (fetal exposure) and newborn (neonatal exposure) children. Therefore, research on children is necessary in both clinical and environmental fields, to provide age-specific relevant data regarding the efficacy and safety of medical treatments, and regarding the assessment of risk from unintended environmental exposure. In this context, the stakeholders are many, including children and their parents, physicians and public health researchers, and the society as a whole, with its ethical, regulatory, administrative and political components. The important ethical issues are information of participants and consent to participate. Follow-up and protection of data (samples and information derived from samples) should be discussed in the context of biobanks, where children obtain individual rights when they become adults. It is important to realise that there are highly variable practices within European countries, which may have, in the past, led to differences in practical aspects of research in children. A number of recommendations are provided for research with children and environmental health. Environmental research with children should be scientifically justified, with sound research questions and valid study protocols of sufficient statistical power, ensuring the autonomy of the child and his/her family at the time of the study and later in life, if data and samples are used for follow-up studies. When children are enrolled, we recommend a consent dyad, including (1) parental (or legal guardian) informed consent and (2) the child's assent and/or informed consent from older minors. For evaluation of the studies including children, a paediatrician should always be involved in the research ethics committee.

Millum, Joseph; Emanuel, Ezekiel J. The ethics of international research with abandoned children. Research with abandoned children does not necessarily involve exploitation. *Science* 2007 December 21; 318(5858): 1874-1875. NRCBL: 18.5.2; 18.5.9; 21.1.

Nelson, Robert M. Including children in research: participation or exploitation? *In:* Santoro, Michael A.; Gorrie, Thomas M., eds. Ethics and the Pharmaceutical Industry. Cambridge; New York: Cambridge University Press, 2005: 68-79. NRCBL: 18.5.2; 9.7; 18.2; 18.6; 18.3.

Pasternak, Ryan H.; Geller, Gail; Parrish, Catherine; Cheng, Tina L. Adolescent and parent perceptions on youth participation in risk behavior research. *Archives of Pediatrics and Adolescent Medicine* 2006 November; 160(11): 1159-1166. NRCBL: 18.5.2; 17.3; 18.3.

Ramanthan, Mala; Jesani, Amar. Ethics in nutrition intervention research. *Indian Journal of Medical Ethics* 2007 April-June; 4(2): 76-77. NRCBL: 18.5.2; 18.2. SC: cs.

Ravindran, G.D. The study was unjustified and fallacious. *Indian Journal of Medical Ethics* 2007 April-June; 4(2): 81. NRCBL: 18.5.2; 18.2. SC: cs. Comments: Response to: Mala Ramanthan and Amar Jesani. Ethics in nutrition intervention research. Indian Journal of Medical Ethics 2007 April-June; 4(2): 76-77.

Resnik, David B. Are the new EPA regulations concerning intentional exposure studies involving children overprotective? *IRB: Ethics and Human Research* 2007 September-October; 29(5): 15-19. NRCBL: 18.5.2; 18.6; 16.1. Identifiers: Environmental Protection Agency.

Resnik, David B.; Wing, Steven. Lessons learned from the children's environmental exposure research study. *American Journal of Public Health* 2007 March; 97(3): 414-418. NRCBL: 18.5.2; 16.1; 9.5.4; 18.3; 1.3.9.

Rosner, Fred. Medical research in children: ethical issues. *Cancer Investigation* 2006 March; 24(2): 218-220. NRCBL: 18.5.2; 18.6; 1.2. SC: rv.

Ross, Lainie Friedman; Philipson, Louis H.; Voltarelli, Julio C.; Couri, Carlos E.B.; Stracieri, Ana B.P.L.; Oliveira, Maria C.; Moraes, Daniela A.; Fabiano, Pieroni; Coutinho, Marina; Malmegrim, Kelen C.R.; Foss-Freitas, Maria C.; Simões, Belinda P.; Foss, Milton C.; Squiers, Elizabeth; Burt, Richard K. Ethics of hematopoietic stem cell transplantation in type 1 diabetes mellitus [letter and reply]. *JAMA: The Journal of the American Medical Association* 2007 July 18; 298(3): 285-286. NRCBL: 18.5.2; 18.5.1; 19.1. Identifiers: Brazil.

Sax, Joanna K. Reforming FDA policy for pediatric testing: challenges and changes in the wake of studies using antidepressant drugs. *Indiana Health Law Review* 2007; 4(1): 61-84. NRCBL: 18.5.2; 9.7; 17.4. SC: le.

Shatrugna, Veena. An extremely cynical study. *Indian Journal of Medical Ethics* 2007 April-June; 4(2): 79-80. NRCBL: 18.5.2; 18.2. SC: cs. Comments: Response to: Mala Ramanthan and Amar Jesani. Ethics in nutrition intervention research. Indian Journal of Medical Ethics 2007 April-June; 4(2): 76-77.

Singh, Jerome Amir; Abdool Karim, Salim S.; Abdool Karim, Quarrisha; Mlisana, Koleka; Williamson, Carolyn; Gray, Clive; Govender, Michelle; Gray, Andrew. Enrolling adolescents in research on HIV and other sensitive issues: lessons from South Africa. *PLoS Medicine* 2006 July; 3(7): e180 (0984-0988). NRCBL: 18.5.2; 18.5.1; 9.5.6.

Slack, Catherine; Strode, Ann; Fleischer, Theodore; Gray, Glenda; Ranchod, Chitra. Enrolling adolescents in HIV vaccine trials: reflections on legal complexities from South Africa. *BMC Medical Ethics* 2007; 8:5; 8 p. [Online]. Accessed: http://www.biomedcentral.com/1472-6939/8/5 [2007 June 18]. NRCBL: 18.5.2; 9.5.6; 9.5.7; 18.3. SC: le.

Abstract: Background: South Africa is likely to be the first country in the world to host an adolescent HIV vaccine trial. Adolescents may be enrolled in late 2007. In the development and review of adolescent HIV vaccine trial protocols there are many complexities to consider, and much work to be done if these important trials are to become a reality. Discussion: This article sets out essential requirements for the lawful conduct of adolescent research in South Africa including compliance with consent requirements, child protection laws, and processes for the ethical and regulatory approval of research. Summary: This article outlines likely complexities for researchers and research ethics committees, including determining that trial interventions meet current risk standards for child research. Explicit recommendations are made for role-players in other jurisdictions who may also be planning such trials. This article concludes with concrete steps for implementing these important trials in South Africa and other jurisdictions, including planning for consent processes; delineating privacy rights; compiling information necessary for ethics committees to assess risks to child participants; training trial site staff to recognize when disclosures trig mandatory reporting response; networking among relevant ethics committees; and lobbying the National Regulatory Authority for guidance.

Spriggs, Merle. When "risk" and "benefit" are open to interpretation - as is generally the case. *American Journal of Bioethics* 2007 March; 7(3): 17-19. NRCBL: 18.5.2; 18.6; 18.3. Identifiers: Kennedy Krieger Institute [KKI]. Comments: Celia B. Fisher, Susan Z. Kornetsky, and Ernest D. Prentice. Determining risk in pediatric research with no prospect of direct benefit: time for a national consensus on the interpretation of federal regulations. American Journal of Bioethics 2007 March; 7(3): 5-10.

Taylor, Holly A. Instead of revising half the story, why not rewrite the whole thing? *American Journal of Bioethics* 2007 March; 7(3): 19-21. NRCBL: 18.5.2; 18.2. Comments: Celia B. Fisher, Susan Z. Kornetsky, and Ernest D. Prentice. Determining risk in pediatric research with no prospect of direct benefit: time for a national consensus on the interpretation of federal regulations. American Journal of Bioethics 2007 March; 7(3): 5-10.

United States. Department of Health and Human Services [DHHS]. Office for Human Research Protections [OHRP]. Special protections for children as research subjects. Children involved as subjects in research: guidance on the HHS 45CFR 46.407 ("407") review process. Washington, DC: Office for Human Research Protections, 2005 May 26; 9 p. [Online]. Accessed: http://www.hhs.gov/ohrp/children/guidance_407process.html [2007 April 24]. NRCBL: 18.5.2; 18.2.

Wendler, David; Varma, Sumeeta. Minimal risk in pediatric research. *Journal of Pediatrics* 2006 December; 149(6): 855-861. NRCBL: 18.5.2; 18.6; 5.2.

HUMAN EXPERIMENTATION/ . . . / PRISONERS

Caplan, Arthur. The ethics of evil: the challenge of the lessons of the Nazi medical experiments. *In:* LaFleur, William R.; Böhme, Gernot; Shimazono, Susumu, eds. Dark Medicine: Rationalizing Medical Research. Bloomington: Indiana University Press, 2007: 63-72. NRCBL: 18.5.5; 21.4; 1.3.9; 2.1; 2.2.

Gostin, Lawrence O. Biomedical research involving prisoners: ethical values and legal regulation. *JAMA: The Journal of the American Medical Association* 2007 February 21; 297(7): 737-740. NRCBL: 18.5.5. SC: le; rv.

Grady, Denise. White doctors, black subjects: abuse disguised as research [review of Medical Apartheid: The Dark History of Medical Experimentation on Black Americans from Colonial Times to the Present by Harriet A. Washington]. *New York Times* 2007 January 23; p. F5, F8. NRCBL: 18.5.5; 18.6; 18.5.1. SC: po.

NRCBL: National Reference Center for Bioethics Literature Classification Scheme   See inside front cover for terms.

Lerner, Barron H. Subjects or objects? Prisoners and human experimentation. *New England Journal of Medicine* 2007 May 3; 356(18): 1806-1807. NRCBL: 18.5.5; 18.3.

Maloney, Dennis M. Impermissible research with prisoners and generally improper protocol reviews [case study]. *Human Research Report* 2007 April; 22(4): 6-7. NRCBL: 18.5.5; 18.2.

HUMAN EXPERIMENTATION/ . . . / WOMEN

Agarwal, Sanjay K.; Estrada, Sylvia; Foster, Warren G.; Wall, L. Lewis; Brown, Doug; Revis, Elaine S.; Rodriguez, Suzanne. What motivates women to take part in clinical and basic science endometriosis research? *Bioethics* 2007 June; 21(5): 263-269. NRCBL: 18.5.3; 18.1. SC: em.
Abstract: BACKGROUND: The objective of this study was to identify factors motivating women to take part in endometriosis research and to determine if these factors differ for women participating in clinical versus basic science studies. METHODS: A consecutive series of 24 women volunteering for participation in endometriosis-related research were asked to indicate, in their own words, why they chose to volunteer. In addition, the women were asked to rate, on a scale of 0 to 10, sixteen potentially motivating factors. The information was gathered in the form of an anonymous self-administered questionnaire. RESULTS: Strong motivating factors (mean score 8) included potential benefit to other women's health, improvement to one's own condition, and participation in scientific advancement. Weak motivating factors (mean score 3) included financial compensation, making one's doctor happy, and use of 'natural' products. No difference was detected between clinical and basic science study participants. CONCLUSION: This study is the first study to specifically investigate the factors that motivate women to take part in endometriosis research. Understanding why women choose to take part in such research is important to the integrity of the informed consent process. The factors most strongly motivating women to participate in endometriosis research related to improving personal or public health; the weakest, to financial compensation and pleasing the doctor.

Hess, Rosanna F. Postabortion research: methodological and ethical issues. *Qualitative Health Research* 2006 April; 16(4): 580-587. NRCBL: 18.5.3; 12.5.1; 18.2.

Kenyon, S.; Dixon-Woods, M.; Jackson, C.J.; Windridge, K.; Pitchforth, E. Participating in a trial in a critical situation: a qualitative study in pregnancy. *Quality and Safety in Health Care* 2006 April; 15(2): 98-101. NRCBL: 18.5.3; 18.3; 18.5.4. SC: em.

Kopelman, Loretta M. Clinical trials for breast cancer and informed consent: how women helped make research a cooperative venture. *In:* Rawlinson, Mary C.; Lundeen, Shannon, eds. The Voice of Breast Cancer in Medicine and Bioethics. Dordrecht, Netherlands: Springer, 2006: 133-161. NRCBL: 18.5.3; 18.3; 2.2; 18.2.

Kuppermann, Miriam; Learman, Lee A.; Gates, Elena; Gregorich, Steven E.; Nease, Robert F., Jr.; Lewis, James; Washington, A. Eugene. Beyond race or ethnicity and socioeconomic status: predictors of prenatal testing for Down syndrome. *Obstetrics and Gynecology* 2006 May; 107(5): 1087-1097. [see correction in: Obstetrics and Gynecology 2006 August; 108(2): 453]. NRCBL: 18.5.3; 18.5.1; 15.2. SC: em.

Mahowald, Mary Briody. Research issues. *In her:* Bioethics and Women: Across the Life Span. Oxford; New York: Oxford University Press, 2006: 214-229, 263-265. 24 fn. NRCBL: 18.5.3; 14.5.
   Keywords: *cloning; *human experimentation; *pregnant women; *women; cord blood; cryopreservation; embryo research; ethical review; fetal therapy; informed consent; ovum; paternalism; randomized clinical trials; reproductive technologies; research subjects; selection of subjects; stem cells

Manderson, Lenore; Kelaher, Margaret; Williams, Gail; Shannon, Cindy. The politics of community: negotiation and consultation in research on women's health. *Human Organization* 1998 Summer; 57(2): 222-229. NRCBL: 18.5.3; 18.6; 18.5.1. Identifiers: Australia.

McCullough, Laurence B.; Coverdale, John H.; Chervenak, Frank A. Preventive ethics for including women of childbearing potential in clinical trials. *American Journal of Obstetrics and Gynecology* 2006 May; 194(5): 1221-1227. NRCBL: 18.5.3; 18.3; 18.2; 18.5.2.

National Institutes of Health [NIH] (United States). NIH Tracking/Inclusion Committee; Pinn, Vivian W.; Roth, Carl; Bates, Angela C.; Caban, Carlos; Jarema, Kim. Monitoring Adherence to the NIH Policy on the Inclusion of Women and Minorities as Subjects in Clinical Research. Comprehensive Report: Tracking of Human Subjects Research as Reported in Fiscal Year 2004 and Fiscal Year 2005. Bethesda, MD: National Institutes of Health [NIH], 2006: 147 p. NRCBL: 18.5.3; 18.5.1; 18.2. SC: em.

Shakur, Haleema; Roberts, Ian; Barnetson, Lin; Coats, Tim. Clinical trials in emergency situations [editorial]. *BMJ: British Medical Journal* 2007 January 27; 334(7586): 165-166. NRCBL: 18.5.3; 9.5.1.

Uhl, K.; Parekh, A.; Kweder, S. Females in clinical studies: where are we going? *Clinical Pharmacology and Therapeutics* 2007 April; 81(4): 600-602. NRCBL: 18.5.3; 18.2; 9.3.1; 10.

Wild, Verina. Plädoyer für einen Einschluss schwangerer Frauen in Arzneimittelstudien = Towards the inclusion of pregnant women in drug trials. *Ethik in der Medizin* 2007 March; 19(1): 7-23. NRCBL: 18.5.3; 9.7; 18.2. SC: em.

HUMAN RIGHTS *See* INTERNATIONAL HEALTH AND HUMAN RIGHTS

HYBRIDS *See* CHIMERAS AND HYBRIDS

IMMUNIZATION *See* CARE FOR SPECIFIC GROUPS; DRUG INDUSTRY; PUBLIC HEALTH

IN VITRO FERTILIZATION
See also REPRODUCTIVE TECHNOLOGIES

California Supreme Court to hear case involving physicians who refused to perform IVF because of religious beliefs. *Kaiser Daily Women's Health Policy Reports* 2007 August 3: 2 p. [Online]. Accessed: http://kaisernetwork. org/daily_reports/rep_index.cfm?DR_ID=46645 [2007 August 3]. NRCBL: 14.4; 1.2; 7.1; 10; 9.5.5. SC: le. Identifiers: Guadalupe Benitez v. North Coast Women's Care Medical Group; Christine Brody, M.D.; Douglas Fenton, M.D.; Guadalupe Benitez; same-sex partners; California anti-discrimination laws.

Ethical ruling and matured eggs offer hope for fertility [news brief]. *Nature* 2007 July 5; 448(7149): 13. NRCBL: 14.4; 14.6. Identifiers: Canada; Israel.

Adam, G.M. Assisted human reproduction — legal rights of the unborn in respect of avoidable damage. *Medicine and Law: The World Association for Medical Law* 2007 June; 26(2): 325-337. NRCBL: 14.4; 9.5.8. SC: le.
Abstract: The author describes various risks to the foetus arising from assisted reproduction technology (ART). These risks are examined from the legal viewpoint, especially considering the rights of the foetus as interpreted in a number of jurisdictions. He distinguishes between the avoidable and inherent risks to the foetus resulting from ART and the potential hazards of ART relevant to criminal law. The basic internationally accepted conventions on foetal rights are compared relative to decisions in a number of cases heard and decided.

Allahbadia, Gautam N.; Kaur, Kulvinder. Accreditation, supervision, and regulation of ART clinics in India — a distant dream? *Journal of Assisted Reproduction and Genetics* 2003 July; 20(7): 276-280. NRCBL: 14.4; 9.8; 21.7. SC: le.

American Society for Reproductive Medicine [ASRM]. Practice Committee; Society for Assisted Reproductive Technology [SART]. Practice Committee. Elements to be considered in obtaining informed consent for ART. *Fertility and Sterility* 2006 November; 86(5, Supplement): S272-S273. NRCBL: 14.4; 8.3.1; 14.6.

American Society for Reproductive Medicine [ASRM]. Practice Committee; Society for Assisted Reproductive Technology [SART]. Practice Committee. 2006 guidelines for gamete and embryo donation. *Fertility and Sterility* 2006 November; 86(5, Supplement): S38-S50. NRCBL: 14.4; 7.1; 15.3; 17.1.

Beem, Penelope; Morgan, Derek. What's love got to do with it? Regulating reproductive technologies and second hand emotions. *In:* McLean, Sheila A.M., ed. First Do No Harm: Law, Ethics, and Healthcare. Aldershot, England; Burlington, VT: Ashgate, 2006: 369-388. NRCBL: 14.4.

SC: le. Identifiers: Australia; Canada; Ireland; New Zealand; Great Britain (United Kingdom).

Birrittieri, Cara. How IVF changed my life. *Lancet* 2006 December; 368(special issue): S58. NRCBL: 14.4; 14.1. SC: cs; po.

Blyth, Eric; Farrand, Abigail. Reproductive tourism — a price worth paying for reproductive autonomy? *Critical Social Policy* 2005; 25(1): 91-114 [Online]. Accessed: http://csp.sagepub.com/cgi/reprint/25/1/91 [13 March 2007]. NRCBL: 14.4; 14.6; 21.1. SC: rv.

Boada, M.; Veiga, A.; Barri, P.N. Spanish regulations on assisted reproduction techniques. *Journal of Assisted Reproduction and Genetics* 2003 July; 20(7): 271-275. NRCBL: 14.4; 14.6; 14.2; 14.5. SC: le.

British Broadcasting Corporation [BBC]. Half-price IVF offered for eggs [news]. *London: BBC News.* 2007 September 13; 1 p. [Online]. Accessed: http://newsvote. bbc.co.uk/mpapps/pagetools/print/news.bbc.co.uk/2/hi/ uk_news/england/6992642.stm [2007 September 14]. NRCBL: 14.4; 9.3.1; 9.5.10; 14.6; 18.5.4. SC: po.

Caplan, Arthur L.; Baruch, Susannah; Schmidt, Harald; Jennings, Bruce; Bonnicksen, Andrea; Greenfield, Debra; Baylis, Françoise; Robertson, John A.; Fleck, Leonard M.; Furger, Franco; Fukuyama, Francis. Needed: a modest proposal [letters and replies]. *Hastings Center Report* 2007 November-December; 37(6): 4-11. NRCBL: 14.4; 14.1; 18.5.4; 15.1; 14.2; 19.5; 9.3.1. Comments: Furger, Franco; Fukuyama, Francis. A proposal for modernizing the regulation of human biotechnologies. Hastings Center Report 2007 July-August 37(4): 16-20; Fossett, James W. Managing reproductive pluralism: the case for decentralized governance. Hastings Center Report 2007 July-August 37(4): 24-35; Fleck, Leonard M. Can we trust "democratic deliberation"? Hastings Center Report 2007 July-August 37(4):22-25; Robertson, John A. The virtues of muddling through. Hastings Center Report 2007 July-August 37(4):26-28; Johnston, Josephine. Tied up in nots over genetic parentage. Hastings Center Report 2007 July-August 37(4):28-31.
Keywords: *biotechnology; *government regulation; *public policy; *reproductive technologies; advisory committees; bioethical issues; chimeras; common good; embryo research; embryonic stem cells; federal government; freedom; genetic engineering; legal aspects; ovum donors; policy making; public participation; remuneration

Carlson, Elof Axel. Assisted reproduction and the argument of playing God. *In his:* Times of Triumph, Times of Doubt: Science and the Battle for Public Trust. Cold Spring Harbor, NY: Cold Spring Harbor Laboratory Press, 2006: 173-183. NRCBL: 14.4; 1.2.

Chung, Lisa Hird. Free trade in human reproductive cells: a solution to procreative tourism and the unregulated Internet. *Minnesota Journal of International Law* 2006

NRCBL: National Reference Center for Bioethics Literature Classification Scheme   See inside front cover for terms.

299

Winter; 15(1): 263-296. NRCBL: 14.4; 14.6; 1.3.12; 9.3.1; 21.1. SC: le.

Cioffi, Alfred. The Church and assisted procreation: cautions for the "infertile" couple. *Ethics and Medics* 2007 October; 32(10): 1-4. NRCBL: 14.4; 1.2.

Colliton, William F., Jr. In vitro fertilization and the wisdom of the Roman Catholic church. *Linacre Quarterly* 2007 February; 74(1): 10-29. NRCBL: 14.4; 1.2; 18.5.4; 12.3.

Covington, Sharon N.; Gibbons, William E. What is happening to the price of eggs? *Fertility and Sterility* 2007 May; 87(5): 1001-1004. NRCBL: 14.4; 14.6; 9.3.1; 19.5; 9.5.5.

Cutas, Daniela. Postmenopausal motherhood: immoral, illegal? A case study. *Bioethics* 2007 October; 21(8): 458-463. NRCBL: 14.4; 9.5.2; 9.5.5. SC: an; cs; sc. Identifiers: Romania; Adrianna Iliescu. Conference: Eighth World Congress of Bioethics: A Just and Healthy Society; Beijing, China; 2006 August 9.
   Abstract: The paper explores the ethics of post-menopausal motherhood by looking at the case of Adriana Iliescu, the oldest woman ever to have given birth (so far). To this end, I will approach the three most common objections brought against the mother and/or against the team of healthcare professionals who made it happen: the age of the mother, the fact that she is single, the appropriateness of her motivation and of that of the medical team.

Davis, Dena S. The puzzle of IVF. *Houston Journal of Health Law and Policy* 2006; 6(2): 275-297. NRCBL: 14.4; 4.4; 14.1; 12.3; 11.1; 12.5.1. SC: le.

Eaton, Lynn. Controversial embryo bill receives second reading in Lords. *BMJ: British Medical Journal* 2007 November 24; 335(7629): 1069. NRCBL: 14.4; 10. SC: le. Identifiers: Great Britain (United Kingdom); House of Lords.

Eaton, Lynn. Fertilisation authority raids controversial fertility clinics [news]. *BMJ: British Medical Journal* 2007 January 20; 334(7585): 115. NRCBL: 14.4. Identifiers: Great Britain (United Kingdom); Human Fertilisation and Embryology Authority.

European Society of Human Reproduction and Embryology [ESHRE]. Task Force on Ethics and Law; Pennings, Guido; de Wert, G.; Shenfield, F.; Cohen, J.; Tarlatzis, B.; Devroey, P. ESHRE Task Force on Ethics and Law 12: Oocyte donation for non-reproductive purposes. *Human Reproduction* 2007 May; 22(5): 1210-1213. NRCBL: 14.4; 14.6; 18.3; 4.4; 18.5.3.

Ferber, Deborah Sarah. As sure as eggs? Responses to an ethical question posed by Abramov, Elchalal, and Schenker. *Journal of Clinical Ethics* 2007 Spring; 18(1): 35-48. NRCBL: 14.4; 9.7; 14.2; 18.5.3.

Ferber, Deborah Sarah. Some reflections on IVF, emotions, and patient autonomy. *Journal of Clinical Ethics* 2007 Spring; 18(1): 53-55. NRCBL: 14.4; 8.1.

Ferrell, Robyn. Reproducing technology. *In her:* Copula: Sexual Technologies, Reproductive Powers. Albany: State University of New York Press, 2006: 37-47. NRCBL: 14.4; 10.

Frassoni, F. The laws covering in vitro fertilization and embryo research in Italy. *Bone Marrow Transplantation* 2006 July; 38(1): 5-6. NRCBL: 14.4; 1.3.5; 15.1; 18.5.4.

Gleicher, N.; Weghofer, A.; Barad, D. On the benefit of assisted reproduction techniques, a comparison of the USA and Europe [letter]. *Human Reproduction* 2007 February; 22(2): 624-626. NRCBL: 14.4; 21.1. SC: em.

Greif, Karen F.; Merz, Jon F. Manufacturing children: assisted reproductive technologies and self-regulation by scientists and clinicians. *In their:* Current Controversies in the Biological Sciences: Case Studies of Policy Challenges from New Technologies. Cambridge, MA: MIT, 2007: 77-99. NRCBL: 14.4; 5.3; 9.3.1; 14.1; 15.2. SC: le.

Heng, B.C. The advent of international 'mail-order' egg donation. *BJOG: An International Journal of Obstetrics and Gynaecology* 2006 November; 113(11): 1225-1227. NRCBL: 14.4; 9.3.1; 21.1.

Holbrook, Daniel. All embryos are equal?: Issues in pre-implantation genetic diagnosis, IVF implantation, embryonic stem cell research, and therapeutic cloning. *International Journal of Applied Philosophy* 2007 Spring; 21(1): 43-53. 12 fn. NRCBL: 14.4; 15.2; 18.5.4; 15.1; 14.6; 14.5; 4.4. SC: an.
   Keywords: *beginning of life; *embryos; *moral status; cloning; cryopreservation; embryo disposition; embryo research; embryonic stem cells; in vitro fertilization; moral obligations; moral policy; personhood; preimplantation diagnosis; reproductive technologies; value of life

Katayama, Alyce C. U.S. ART practitioners soon to begin their forced march into a regulated future. *Journal of Assisted Reproduction and Genetics* 2003 July; 20(7): 265-270. NRCBL: 14.4; 14.6. SC: le.

Klagsbrun, Francine. Multiple choice. Human reproductive technology enables many people to have children, but it poses dangers. *Moment* 1999 August; 24(4): 24-25. NRCBL: 14.4; 1.2.

Kok, Jeroen D. Is subfertility a medical condition? *Journal of Clinical Ethics* 2007 Spring; 18(1): 49-52. NRCBL: 14.4; 18.5.3; 4.2.

Luna, Florencia. Assisted reproduction and local experiences: women and context in Latin America. *In:* Luna, Florencia; Herissone-Kelly, Peter, ed. Bioethics and Vulnerability: a Latin American View. Amsterdam; New York: Rodopi, 2006: 63-71. NRCBL: 14.4; 9.5.5; 10; 9.2.

Mackenzie, Robin. Regulating reprogenetics: strategic sacralisation and semantic massage. *Health Care Analysis: An International Journal of Health Philosophy and Policy* 2007 December; 15(4): 305-319. 89 refs. NRCBL: 14.4; 1.2; 10; 15.1; 18.5.4.

Keywords: *embryo research; *feminist ethics; *regulation; *reproductive technologies; *values; anthropology; advisory committees; biotechnology; commodification; decision making; donors; embryonic stem cells; genes; in vitro fertilization; legal aspects; surrogate mothers; value of life; women; Proposed Keywords: embryo donation; Keyword Identifiers: Great Britain; Warnock Committee

Abstract: This paper forms part of the feminist critique of the regulatory consequences of biomedicine's systematic exclusion of the role of women's bodies in the development of reprogenetic technologies. I suggest that strategic use of notions of the sacred to decontextualise and delimit disagreement fosters this marginalisation. Here conceptions of the sacred and sacralisation afford a means by which pragmatic consensus over regulation may be achieved, through the deployment of a bricolage of dense images associated with cultural loyalties to solidify support or exclude contradictory elements. Hence an explicit renegotiation of the symbolic order structuring salient debates is necessary to disrupt and enrich the entrenched and exclusionary dominant discourse over reprogenetic regulation. I draw upon previous analyses of strategic rhetoric associated with the regulation of infertility treatment and embryo research in the United Kingdom, the cultural anthropology of biomedicine and feminist ethnographies of reprogenetics to illustrate these claims.

Madden, Deirdre. Assisted reproduction in the Republic of Ireland — a legal quagmire. *In:* Gunning, Jennifer; Holm, Søren, eds. Ethics, law, and society. Volume 2. Aldershot, Hants, England; Burlington, VT: Ashgate, 2006: 27-34. NRCBL: 14.4; 4.4. SC: le.

Mahowald, Mary Briody; Sherwin, Susan; Overall, Christine. Assisted reproductive technologies. *In:* Baylis, Françoise; Downie, Jocelyn; Freedman, Benjamin; Hoffmaster, Barry; Sherwin, Susan, eds. Health Care Ethics in Canada. Toronto: Harcourt Brace Canada, 1995: 449-485. NRCBL: 14.4; 14.2; 9.5.5; 10; 2.4. SC: le. Note: Section includes: Introduction — Fertility enhancement and the right to have a baby / Mary Briody Mahowald — New reproductive technologies / Susan Sherwin — Surrogate motherhood / Christine Overall — Proceed with care: final report of the Royal Commission on New Reproductive Technologies.

Mayor, Susan. UK study will reimburse part of cost of IVF to women who donate eggs for research [news]. *BMJ: British Medical Journal* 2007 September 22; 335(7620): 581. NRCBL: 14.4; 9.3.1; 14.5; 18.5.4; 15.1; 19.1; 19.5.

Keywords: *embryonic stem cells; *in vitro fertilization; *incentives; *ovum donors; *remuneration; cloning; embryo research; public policy; research support; Keyword Identifiers: *Medical Research Council (Great Britain); Great Britain; National Health Service; Newcastle Fertility Centre

Montagut, Jacques; Menezo, Yves. How to legislate in human reproduction: the French experience. *Journal of Assisted Reproduction and Genetics* 2003 July; 20(7): 287-289. NRCBL: 14.4; 9.8. SC: le.

Moses, Lyria Bennett. Understanding legal responses to technological change: the example of in vitro fertilization. *Minnesota Journal of Law, Science and Technology* 2004-2005; 6(2): 505-618. NRCBL: 14.4; 2.2; 21.1. SC: le.

Nelson, Erin L. Legal and ethical issues in ART "outcomes" research. *Health Law Journal* 2005; 13: 165-186. NRCBL: 14.4; 18.6; 18.5.3; 18.5.4; 18.5.2; 18.3. SC: le.

Ng, Ernest Hung Yu; Liu, Athena; Chan, Celia H.Y.; Chan, Cecilia Lai Wan; Yeung, William Shu Biu; Ho, Pak Chung. Regulating reproductive technology in Hong Kong. *Journal of Assisted Reproduction and Genetics* 2003 July; 20(7): 281-286. NRCBL: 14.4; 9.8; 2.4. SC: le.

Pence, Gregory E. Emotivism and banning some conceptions. *In his:* The Elements of Bioethics. Boston: McGraw-Hill, 2007: 109-136. NRCBL: 14.4; 7.1; 11.3; 1.1; 14.2; 15.1; 14.5. SC: cs.

Petersen, Kerry. The rights of donor-conceived children to know the identity of their donor: the problem of the known unknowns and the unknown unknowns. *In:* Bennett, Belinda; Tomossy, George F., eds. Globalization and Health: Challenges for Health Law and Bioethics. Dordrecht: Springer, 2006: 151-167. NRCBL: 14.4; 8.4; 14.2; 9.5.7; 21.1; 19.1. SC: le; rv.

Powledge, Tabitha M. Looking at ART: is it time to scrutinize assisted reproduction? *Scientific American* 2002 April; 286(4): 20, 23. NRCBL: 14.4; 14.5; 14.6; 18.6.

Rabin, Roni Caryn. As demand for donor eggs soars, high prices stir ethical concerns. *New York Times* 2007 May 15; p. F6. NRCBL: 14.4; 14.6; 9.3.1; 19.5; 9.5.5. SC: po.

Reid, Lynette; Ram, Natalie; Brown, Blake. Compensation for gamete donation: the analogy with jury duty. *CQ: Cambridge Quarterly of Healthcare Ethics* 2007 Winter; 16(1): 35-43. NRCBL: 14.4; 19.5; 9.3.1; 7.1.

Schenker, Joseph G. Legal aspects of ART practice in Israel. *Journal of Assisted Reproduction and Genetics* 2003 July; 20(7): 250-259. NRCBL: 14.4; 14.2; 14.5. SC: le.

Serour, Gamal I. Religious perspectives on ethical issues in assisted reproductive technologies. *In:* Shenfield, Françoise; Sureau, Claude, eds. Contemporary Ethical Dilemmas in Assisted Reproduction. Abingdon: Informa Healthcare, 2006: 99-114. NRCBL: 14.4; 1.2; 14.3.

Sforza, Teri. Thousands of human eggs may be mssing. Bankrupt egg-donor registry says fertility doctors may have transferred eggs without permission. *Orange County Register* 2007 July 23: 10 p. [Online]. Accessed: http://

NRCBL: National Reference Center for Bioethics Literature Classification Scheme   See inside front cover for terms.

301

www.ocregister.com/eggs-options-doctors-1782305-embryos-egg [2007 July 24]. NRCBL: 14.4; 14.1; 7.1; 8.3.1.

Shanley, Mary L. The Baby Business: How Money, Science and Politics Drive the Commerce of Conception, by Debora L. Spar [book review]. *DePaul Journal of Health Care Law* 2007; 10(4): 557-566. NRCBL: 14.4; 15.2.

Short, Robert. HFEA wants greater use of single embryo transfers in IVF [news]. *BMJ: British Medical Journal* 2007 April 14; 334(7597): 766. NRCBL: 14.4. Identifiers: Great Britain (United Kingdom).

Smajdor, Anna. State-funded IVF will make us rich . . . or will it? *Journal of Medical Ethics* 2007 August; 33(8): 468-469. NRCBL: 14.4; 9.3.1. Identifiers: Great Britain (United Kingdom).

Society for Assisted Reproductive Technology [SART]. Practice Committee; American Society for Reproductive Medicine [ASRM]. Practice Committee. Guidelines on number of embryos transferred. *Fertility and Sterility* 2006 November; 86(5, Supplement): S51-S52. NRCBL: 14.4; 9.5.5; 8.1; 9.8.

Society of Obstetricians and Gynecologists of Canada; Canadian Fertility and Andrology Society; Min, Jason K.; Claman, Paul; Hughes, Ed. Guidelines for the number of embryos to transfer following in vitro fertilization. *Journal of Obstetrics and Gynaecology Canada* 2006 September; 28(9): 799-813. NRCBL: 14.4.

Spar, Debora. The egg trade — making sense of the market for human oocytes. *New England Journal of Medicine* 2007 March 29; 356(13): 1289-1291. NRCBL: 14.4; 14.6; 9.3.1; 18.5.4; 15.1; 19.5.

Stein, Rob. 'Embryo bank' stirs ethics fears; firm lets clients pick among fertilized eggs. *Washington Post* 2007 January 6; p. A1, A8. NRCBL: 14.4; 14.6; 15.5. SC: po.
   Keywords: *embryo transfer; *remuneration; commodification; ovum donors; semen donors; cryopreservation; *embryo; eugenics; *biological specimen banks

Steinbock, Bonnie. Defining parenthood. *In:* Freeman, Michael, ed. Children's Health and Children's Rights. Boston: Martinus Nijhoff Publishers, 2006: 311-334. NRCBL: 14.4; 14.2; 10; 8.3.2. SC: cs; le.

Stillman, Robert J. A 47-year-old woman with fertility problems who desires a multiple pregnancy. *JAMA: The Journal of the American Medical Association* 2007 February 28; 297(8): 858-867. NRCBL: 14.4; 9.5.2. SC: rv; cs.

Sutcliffe, Alastair G.; Ludwig, Michael. Outcome of assisted reproduction. *Lancet* 2007 July 28-August 3; 370(9584): 351-359. NRCBL: 14.4; 9.5.7. SC: em.
   Abstract: In-vitro fertilisation has been done for nearly 30 years; in developed countries at least 1% of births are from assisted reproductive therapies (ART). These children now represent a substantial proportion of the population but little is known about their health. Some of the morbidity associated with ART does not result from the techniques but from the underlying health risks of being subfertile. Much of the amplified risk associated with ART is related to high birth order. However, risk of intrauterine and subsequent perinatal complications is enhanced after ART, and urogenital malformations can be present in boys, even in singleton infants. No increase in discord or other difficulties within families has been recorded. Long-term follow-up of children born after ART to reproductive age and beyond is necessary.

Takeshita, Naoki; Hanaoka, Kanako; Shibui, Yukihiro; Jinnai, Hikoyoshi; Abe, Yuji; Kubo, Harumi. Regulating assisted reproductive technologies in Japan. *Journal of Assisted Reproduction and Genetics* 2003 July; 20(7): 260-264. NRCBL: 14.4; 15.2. SC: le.

Ten, C.L. A child's right to a father. *Monash Bioethics Review* 2000 October; 19(4): 33-37. NRCBL: 14.4; 1.1; 10; 14.1. SC: le. Identifiers: Australia; Sex Discrimination Act.

Turone, Fabio. New reproduction law reduces success rate [news]. *BMJ: British Medical Journal* 2007 July 14; 335(7610): 62. NRCBL: 14.4. SC: le.

Wolf, Don P. An opinion on regulating the assisted reproductive technologies. *Journal of Assisted Reproduction and Genetics* 2003 July; 20(7): 290-292. NRCBL: 14.4; 9.8. SC: le.

INCOMPETENTS *See* INFORMED CONSENT/ INCOMPETENTS

INDIGENTS *See* CARE FOR SPECIFIC GROUPS/ INDIGENTS

INFANTICIDE *See* EUTHANASIA AND ALLOWING TO DIE/ MINORS

INFANTS *See* CARE FOR SPECIFIC GROUPS/ MINORS; EUTHANASIA AND ALLOWING TO DIE/ MINORS; HUMAN EXPERIMENTATION/ SPECIAL POPULATIONS/ MINORS

INFORMATICS *See* TELEMEDICINE AND INFORMATICS

INFORMED CONSENT
See also HUMAN EXPERIMENTATION/ INFORMED CONSENT; TREATMENT REFUSAL

Ågård, Anders; Löfmark, Rurik; Edvardsson, Nils; Ekman, Inger. Views of patients with heart failure about their role in the decision to start implantable cardioverter-defibrillator treatment: prescription rather than participation. *Journal of Medical Ethics* 2007 September; 33(9): 514-518. NRCBL: 8.3.1; 9.5.1. SC: em.
   Abstract: Background: There is a shortage of reports on what potential recipients of implantable cardioverter-

defibrillators (ICDs) need to be informed about and what role they can and want to play in the decision-making process when it comes to whether or not to implant an ICD. Aims: To explore how patients with heart failure and previous episodes of malignant arrhythmia experience and view their role in the decision to initiate ICD treatment. Patients and methods: A qualitative content analysis of semistructured interviews was used. The study population consisted of 31 outpatients with moderate heart failure at the time of their first ICD implantation. Setting: The study was performed at Sahlgrenska University Hospital, Göteborg, Sweden. Results: None of the respondents had discussed the alternative option of receiving treatment with anti-arrhythmic drugs, the estimated risk of a fatal arrhythmia, or the expected time of survival from heart failure in itself. Even so, very little criticism was directed at the lack of information or the lack of participation in the decision-making process. The respondents felt that they had to rely on the doctors' recommendation when it comes to such a complex and important decision. None of them regretted implantation of the ICD. Conclusions: The respondents were confronted by a matter of fact. They needed an ICD and were given an offer they could not refuse, simply because life was precious to them. Being able to give well-informed consent seemed to be a matter of less importance for them.

Albera, R.; Argentero, P.; Bonziglia, S.; De Andreis, M.; Preti, G.; Palonta, F.; Canale, A. Informed consent in ENT. Patient's judgement about a specific consensus form. *Acta Otorhinolaryngologica Italica: Organo Ufficiale della Società Italiana di Otorinolaringologia e Chirurgia Cervico-facciale* 2005 October; 25(5): 304-311 [Online]. Accessed: http://www.actaitalica.it/ [2007 April 13]. NRCBL: 8.3.1;8.1; 9.5.1. SC: em.

Appelbaum, Paul S. Comment on the case of Mr. A.B. *Journal of Clinical Ethics* 2007 Winter; 18(4): 402-403. NRCBL: 8.3.1; 8.2; 20.3.3; 20.3.2. Comments: Peter Sloane and Evan G. DeRenzo. The case of Mr. A.B. Journal of Clinical Ethics 2007 Winter; 18(4): 399-401.

Archard, David. Informed consent and the grounds of autonomy. *In:* Nys, Thomas; Denier, Yvonne; Vandevelde, Toon, eds. Autonomy and Paternalism: Reflections on the Theory and Practice of Health Care. Leuven; Dudley, MA: Peeters, 2007: 113-128. NRCBL: 8.3.1; 1.1; 4.4.

Brands, W.G. The standard for the duty to inform patients about risks: from the responsible dentist to the reasonable patient. *British Dental Journal* 2006 August 26; 201(4): 207-210. NRCBL: 8.3.1; 8.1; 4.1.1.

Brennan, Patricia A.W. The medical and ethical aspects of photography in the sexual assault examination: why does it offend? *Journal of Clinical Forensic Medicine* 2006 May; 13(4): 194-202. NRCBL: 8.3.1; 10; 8.4; 7.1; 9.2.

British Medical Association [BMA]. Medical Ethics Department. Human Tissue Legislation. Guidance from the BMA's Medical Ethics Department. London: British Medical Association, 2006 September: 8 p. [Online]. Accessed: http://www.bma.org.uk/ap.nsf/AttachmentsByTitle/PDFHumantissue/$FILE/HumanTissueLegislation.pdf [2007 August 13]. NRCBL: 8.3.1; 19.5; 18.5; 20.1. SC: le. Identifiers: Human Tissue Act 2004; Human Tissue (Scotland) Act; Human Tissue Act 1961; Anatomy Act 1984; Human Organ Transplants Act 1989; Human Tissue Act (Northern Ireland) 1962; Human Organ Transplants (Northern Ireland) Order 1989; Anatomy (Northern Ireland) Order 1992; Human Tissue Authority.

Cahill, H. Consent. *In:* Holland, Stephen, ed. Introducing Nursing Ethics: Themes in Theory and Practice. Salisbury: APS, 2004: 91-110. NRCBL: 8.3.1. SC: cs.

Canada. Supreme Court; Ontario. Court of Appeal; Jecker, Nancy S.; Loewy, Erich H. Consent. *In:* Baylis, Françoise; Downie, Jocelyn; Freedman, Benjamin; Hoffmaster, Barry; Sherwin, Susan, eds. Health Care Ethics in Canada. Toronto: Harcourt Brace Canada, 1995: 201-230. NRCBL: 8.3.1; 1.1. SC: le. Note: Section includes: Introduction — Reibl v. Hughes / Supreme Court of Canada — Zampara v. Brisson / Ontario Court of Appeal — Being a burden on others / Nancy S. Jecker — Changing one's mind: when is Odysseus to be believed? / Erich H. Loewy.

Clarke, Steve; Levy, Neil. On the competence of substance users to consent to treatment programs. *In:* Kleinig, John; Einstein, Stanley, eds. Ethical Challenges for Intervening in Drug Use: Policy, Research and Treatment Issues. Huntsville, TX: Office of International Criminal Justice; Criminal Justice Center, Sam Houston State University, 2006: 309-322. NRCBL: 8.3.1; 4.3; 9.5.9. SC: an.

Coggon, John. Varied and principled understandings of autonomy in English law: justifiable inconsistency or blinkered moralism? *Health Care Analysis: An International Journal of Health Philosophy and Policy* 2007 September; 15(3): 235-255. NRCBL: 8.3.1; 1.1; 8.3.4. SC: an; le.

Abstract: Autonomy is a concept that holds much appeal to social and legal philosophers. Within a medical context, it is often argued that it should be afforded supremacy over other concepts and interests. When respect for autonomy merely requires non-intervention, an adult's right to refuse treatment is held at law to be absolute. This apparently simple statement of principle does not hold true in practice. This is in part because an individual must be found to be competent to make a valid refusal of consent to medical treatment, and capacity to decide is not an absolute concept. But further to this, I argue that there are three relevant understandings of autonomy within our society, and each can demand in differing cases that different courses of action be followed. Judges, perhaps inadvertently, have been able to take advantage of the equivocal nature of the concept to come tacitly to decisions that reflect their own moral judgments of patients or

NRCBL: National Reference Center for Bioethics Literature Classification Scheme   See inside front cover for terms.

303

decisions made in particular cases. The result is the inconsistent application of principle. I ask whether this is an unforeseen outcome or if it reflects a wilful disregard for equal treatment in favour of silent moral judgments in legal cases. Whatever the cause, I suggest that once this practice is seen to occur, acceptable justification of it in some cases is difficult to find.

Corfield, Lorraine F. To inform or not to inform: how should the surgeon proceed when the patient refuses to discuss surgical risk? *Journal of Vascular Surgery* 2006 July; 44(1): 219-221. NRCBL: 8.3.1; 9.5.1; 8.1. SC: cs.

Derse, Arthur R.; Easton, Raul B.; Graber, Mark A.; Monnahan, Jay; Hughes, Jason. Is patients' time too valuable for informed consent? *American Journal of Bioethics* 2007 December; 7(12): 45-46; author reply W3-W4. NRCBL: 8.3.1. Comments: Raul B. Easton, Mark A. Graber, Jay Monnahan, and Jason Hughes. Defining the scope of implied consent in the emergency department. American Journal of Bioethics 2007 December; 7(12): 35-38.

Easton, Raul B.; Graber, Mark A.; Monnahan, Jay; Hughes, Jason. Defining the scope of implied consent in the emergency department. *American Journal of Bioethics* 2007 December; 7(12): 35-38. NRCBL: 8.3.1; 9.5.1. SC: em. Comments: American Journal of Bioethics 2007 December; 7(12): 39-54.

Egonsson, Dan. Hypothetical approval in prudence and medicine. *Medicine, Health Care and Philosophy* 2007 September; 10(3): 245-252. NRCBL: 8.3.1; 1.1; 8.3.4; 17.1; 20.5.1.
Abstract: We often assume that hypothetical approval - either in the form of preferences or consent - under ideal conditions adds to the legitimacy of an arrangement or act. I want to show that this assumption, reasonable as it may seem, will also give rise to ethical problems. I focus on three problem areas: prudence, euthanasia and coercive psychiatric treatment . If we are to count as prudentially or morally relevant those preferences you would have if you were informed and rational, we will run into difficulties in all these areas if your actual and rational preferences are at variance with each other. In the prudential sphere we may question the personal value of satisfying preferences that a person does not actually have. In this case our problem concerns the point of satisfying a rational preference in conflict with an actual one. In the cases of euthanasia and coercive care it concerns instead whether it would be morally right to do such a thing. I doubt there is a simple solution to our problem. In this paper at most I prepare the way for a solution or for wiser decisions in the hard cases, by pointing out what they will have to deal with.

Eonas, Anthony; McCoy, John D.; Eaton, Silviya H.M. Medical informed consent: clarity or confusion? *Journal of Hospital Marketing and Public Relations* 2006; 16(1-2): 69-88. NRCBL: 8.3.1; 8.1. SC: le.

Fisher-Jeffes, Lisa; Barton, Charlotte; Finlay, Fiona. Clinicans' knowledge of informed consent. *Journal of Medical Ethics* 2007 March; 33(3): 181-184. NRCBL: 8.3.1; 7.1. SC: em. Identifiers: Great Britain (United Kingdom).
Abstract: OBJECTIVE: To audit doctors' knowledge of informed consent. DESIGN: 10 consent scenarios with "true", "false", or "don't know" answers were completed by doctors who care for children at a large district general hospital. These questions tested clinicians' knowledge of who could give consent in different clinical situations. SETTING: Royal United Hospital, Bath, UK. RESULTS: 51 doctors participated (25 paediatricians and 26 other clinicians). Paediatricians scored higher than other clinicians (average correct response 69% v 49%). Only 36% (9/25) of paediatricians and 8% (2/26) of other clinicians realised that the biological father of a child born before 1 December 2003 needed a court order or a parental responsibility agreement to acquire parental responsibility, and thus be able to consent on behalf of his child, if he was not married to the child's mother. Non-paediatric clinicians were unsure or incorrect when tested on situations where people with parental responsibility do not agree, or where young people (years), who are Fraser competent do not want to consult their parents. Most clinicians did not know that the parents of a 20-year-old man with severe learning difficulties are unable to consent to surgery on his behalf, and many non-paediatricians were unclear on who could give consent when a child lived with foster parents. CONCLUSION: Clinicians who obtain consent for the treatment of children need to increase their knowledge on who is able to give informed consent to ensure best (legal and safe) practice.

Foote, Robert L.; Brown, Paul D.; Garces, Yolanda I.; Okuno, Scott H.; Miller, Robert C.; Strome, Scott E. Informed consent in advanced laryngeal cancer. *Head and Neck* 2007 March; 29(3): 230-235. NRCBL: 8.3.1; 9.5.1; 8.1.

Fullbrook, Suzanne. Autonomy and care: acting in a person's best interets [sic: interest]. *British Journal of Nursing* 2007 February 22-March 7; 16(4): 236-237. NRCBL: 8.3.1; 1.1; 4.1.3.

Fullbrook, Suzanne. Consent: the issue of rights and responsibilities for the health worker. *British Journal of Nursing* 2007 March 8-21; 16(5): 318-319. NRCBL: 8.3.1; 8.3.2; 8.3.3. SC: le. Identifiers: Great Britain (United Kingdom).

Geppert, Cynthia M.A.; Abbott, Christopher. Voluntarism in consultation psychiatry: the forgotten capacity. *American Journal of Psychiatry* 2007 March; 164(3): 409-413. NRCBL: 8.3.1; 17.1; 8.3.3.

Giordano, James; Ernst, E. Informed consent: a potential dilemma for complementary medicine [letter and reply]. *Journal of Manipulative and Physiological*

Therapeutics 2004 November-December; 27(9): 596-597. NRCBL: 8.3.1; 4.1.1.

González San Segundo, Carmen; Santos Miranda, Juan A. Informed consent in radiation oncology: is consenting easier than informing? *Clinical and Translational Oncology* 2006 November; 8(11): 802-804. NRCBL: 8.3.1; 8.2; 8.1; 9.5.1.

Great Britain (United Kingdom). Department of Health. Reference guide to consent for examination or treatment. London: Department of Health. 2001 April 6; 30 p. [Online]. Accessed: http://www.dh.gov.uk/prod_consum_dh/idcplg?IdcService=GET_FILE&dID=29069 &Rendition=Web [2007 April 12]. NRCBL: 8.3.1; 8.3.4; 15.1; 19.5. SC: po.

Hart, Dieter. Patient information on drug therapy. A problem of medical malpractice law: between product safety and user safety. *European Journal of Health Law* 2007 April; 14(1): 47-59. NRCBL: 8.3.1; 9.7; 8.5. SC: le.

Hayes, Margaret Oot. Prisoners and autonomy: implications for the informed consent process with vulnerable populations. *Journal of Forensic Nursing* 2006 Summer; 2(2): 84-89. NRCBL: 8.3.1; 9.5.1; 1.3.5.

Hunt, Linda M.; de Voogd, Katherine B. Are good intentions good enough?: Informed consent without trained interpreters. *JGIM: Journal of General Internal Medicine* 2007 May; 22(5): 598-605. NRCBL: 8.3.1; 9.5.4; 15.2. SC: em.

Iserson, Kenneth V.; Easton, Raul B.; Graber, Mark A.; Monnahan, Jay; Hughes, Jason. The three faces of "yes": consent for emergency department procedures. *American Journal of Bioethics* 2007 December; 7(12): 42-45; author reply W3-W4. NRCBL: 8.3.1; 9.5.1; 8.3.3; 20.5.4. Comments: Raul B. Easton, Mark A. Graber, Jay Monnahan, and Jason Hughes. Defining the scope of implied consent in the emergency department. American Journal of Bioethics 2007 December; 7(12): 35-38.

Issa, M.M.; Setzer, E.; Charaf, C.; Webb, A.L.; Derico, R.; Kimberl, I.J.; Fink, A.S. Informed versus uninformed consent for prostate surgery: the value of electronic consents. *Journal of Urology* 2006 August; 176(2): 694-699; discussion 699. NRCBL: 8.3.1; 9.5.1; 1.3.12. SC: em.

Kristinsson, Sigurdur. Autonomy and informed consent: a mistaken association? *Medicine, Health Care and Philosophy* 2007 September; 10(3): 253-264. NRCBL: 8.3.1; 1.1; 18.3. SC: an. Identifiers: Belmont Report.
Abstract: For decades, the greater part of efforts to improve regulatory frameworks for research ethics has focused on informed consent procedures; their design, codification and regulation. Why is informed consent thought to be so important? Since the publication of the Belmont Report in 1979, the standard response has been that obtaining informed consent is a way of treating individuals as autonomous agents. Despite its political suc-

cess, the philosophical validity of this Belmont view cannot be taken for granted. If the Belmont view is to be based on a conception of autonomy that generates moral justification, it will either have to be reinterpreted along Kantian lines or coupled with a something like Mill's conception of individuality. The Kantian interpretation would be a radical reinterpretation of the Belmont view, while the Millian justification is incompatible with the liberal requirement that justification for public policy should be neutral between controversial conceptions of the good. This consequence might be avoided by replacing Mill's conception of individuality with a procedural conception of autonomy, but I argue that the resulting view would in fact fail to support a non-Kantian, autonomy-based justification of informed consent. These difficulties suggest that insofar as informed consent is justified by respect for persons and considerations of autonomy, as the Belmont report maintained, the justification should be along the lines of Kantian autonomy and not individual autonomy.

Krizova, Eva; Simek, Jiri. Theory and practice of informed consent in the Czech Republic. *Journal of Medical Ethics* 2007 May; 33(5): 273-277. NRCBL: 8.3.1. SC: em.
Abstract: The large-scale change of Czech society since 1989 has involved the democratic transformation of the health system. To empower the patient was one important goal of the healthcare reform launched immediately after the Velvet Revolution. The process has been enhanced by the accession of the Czech Republic to the European Union and the adoption of important European conventions regulating the area. The concept of informed consent and a culture of negotiation are being inserted into a traditionally paternalistic culture. Our article describes the current situation on the issue of the communication of information on state of health and treatment, and on the question of the participation of the patient in decisions on treatment. We present empirical results of a public opinion survey on this issue. The results show a still prevailing submissive attitude towards the physicians, despite the fact that the concept of informed consent has become more and more publicly familiar (42% of respondents gave the completely correct answer regarding informed consent). The impact of age, education and sex on answers to the questionnaire was analysed. Men, younger and more educated respondents were more likely to show the autonomous attitude, whereas women, older and less educated people tended to show the traditional submissive attitude. Further, our article raises the question of the cultural and historical background within which the current ethically and legally binding norms (products of western democracies, in fact) are interpreted. The question is how far cultural modifications are tolerable in the practical implementation of universal ethical constructs (informed consent).

Kukla, Rebecca. How do patients know? *Hastings Center Report* 2007 September-October; 37(5): 27-35. NRCBL: 8.3.1; 8.1; 1.1.
Abstract: The way patients make health care decisions is much more complicated than is often recognized. Patient autonomy allows both that patients will sometimes defer

NRCBL: National Reference Center for Bioethics Literature Classification Scheme   See inside front cover for terms.

305

to clinicians and that they should sometimes be active inquirers, ready to question their clinicians and do some independent research. At the same time, patients' active inquiry requires clinicians' support.

Ladas, Spiros D. Informed consent: still far from ideal? *Digestion* 2006; 73(2-3): 187-188. NRCBL: 8.3.1; 8.1.

Langworthy, Jennifer M.; le Fleming, Christine. Consent or submission? The practice of consent within UK chiropractic. *Journal of Manipulative and Physiological Therapeutics* 2005 January; 28(1): 15-24. NRCBL: 8.3.1; 4.1.1; 8.5. SC: em; le.

Lee, K. Jane.; Havens, Peter L.; Sato, Thomas T.; Hoffman, George M.; Leuthner, Steven R. Assent for treatment: clinician knowledge, attitudes, and practice. *Pediatrics* 2006 August; 118(2): 723-730. NRCBL: 8.3.1; 7.1; 8.1.

Lo, Bernard. Ethical issues in surgery. *In his:* Resolving Ethical Dilemmas: A Guide for Clinicians. 3rd edition. Philadelphia, PA: Lippincott Williams and Wilkins, 2005: 243-248. NRCBL: 8.3.1; 8.3.4; 8.1 9.5.1.

Lyden, Martin. Capacity issues related to the health care proxy. *Mental Retardation* 2006 August; 44(4): 272-282. NRCBL: 8.3.1; 8.3.3. Identifiers: New York.

Mantese, Theresamarie; Pfeiffer, Christine; McClinton, Jacquelyn. Cosmetic surgery and informed consent: legal and ethical considerations. *Michigan Bar Journal* 2006 January; 85(1): 26-29 [Online]. Accessed: http://www.michbar.org/journal/pdf/pdf4article957.pdf [2007 May 11]. NRCBL: 8.3.1; 9.5.1.

Mason, J.K.; Laurie, G.T. Consent to treatment. *In their:* Mason and McCall Smith's Law and Medical Ethics. Seventh ed. Oxford; New York: Oxford University Press, 2005: 348-411. NRCBL: 8.3.1; 8.3.2; 8.3.3; 8.3.4; 9.5.5; 9.5.6. SC: le.

Masood, Junaid; Hafeez, Azhar; Wiseman, Oliver; Hill, James T. Informed consent: are we deluding ourselves? A randomized controlled study. *BJU International* 2007 January; 99(1): 4-5. NRCBL: 8.3.1; 9.5.1; 8.1.

McCullough, Laurence B.; McGuire, Amy L.; Whitney, Simon N.; Easton, Raul B.; Graber, Mark A.; Monnahan, Jay; Hughes, Jason. Consent: informed, simple, implied and presumed. *American Journal of Bioethics* 2007 December; 7(12): 49-50; author reply W3-W4. NRCBL: 8.3.1. Comments: Raul B. Easton, Mark A. Graber, Jay Monahan, and Jason Hughes. Defining the scope of implied consent in the emergency department. American Journal of Bioethics 2007 December; 7(12): 35-38.

Mishra, Pankaj Kumar; Ozalp, Faruk; Gardner,Roy S.; Arangannal, Arul; Murday, Andrew. Informed consent in cardiac surgery: is it truly informed? *Journal of Cardiovascular Medicine* 2006 September; 7(9): 675-681. NRCBL: 8.3.1; 9.5.1; 7.1; 8.1.

Moskop, John C.; Easton, Raul B.; Graber, Mark A.; Monnahan, Jay; Hughes, Jason. Information disclosure and consent: patient preferences and provider responsibilities. *American Journal of Bioethics* 2007 December; 7(12): 47-49; author reply W3-W4. NRCBL: 8.3.1; 9.5.1. Comments: Raul B. Easton, Mark A. Graber, Jay Monnahan, and Jason Hughes. Defining the scope of implied consent in the emergency department. American Journal of Bioethics 2007 December; 7(12): 35-38.

Murphy, Dominic; Dandeker, Christopher; Horn, Oded; Hotopf, Matthew; Hull, Lisa; Jones, Margaret; Marteau, Theresa; Rona, Roberto; Wessely, Simon. UK armed forces responses to an informed consent policy for anthrax vaccination: a paradoxical effect? *Vaccine* 2006 April 12; 24(16): 3109-3114. NRCBL: 8.3.1; 9.1; 9.5.1.

Najjar, Dany; Srinivasan, M.; Hammersmith, Kristin M.; Cohen, Elisabeth J.; Rapuano, Christopher J.; Laibson, Peter R. Informed consent for Creutzfeldt-Jakob disease after corneal transplantation [letters and reply]. *Cornea* 2005 January; 24(1): 121-122. NRCBL: 8.3.1; 9.5.1.

O'Brien, C.M.; Thorburn, T.G.; Sibbel-Linz, A.; McGregor, A.D. Consent for plastic surgical procedures. *Journal of Plastic, Reconstructive and Aesthetic Surgery* 2006; 59(9): 983-989. NRCBL: 8.3.1; 8.2; 9.5.1; 5.2. SC: em; le. Identifiers: Great Britain (United Kingdom).

O'Connor, Annette M.; Wennberg, John E.; Legare, France; Llewellyn-Thomas, Hilary A.; Moulton, Benjamin W.; Sepucha, Karen R.; Sodano, Andrea G.; King, Jaime S. Toward the 'tipping point': decision aids and informed patient choice. *Health Affairs* 2007 May-June; 26(3): 716-725. NRCBL: 8.3.1; 5.2.

Palmboom, G.G.; Willems, D.L.; Janssen, N.B.A.T.; de Haes, J.C.J.M. Doctor's views on disclosing or withholding information on low risks of complication. *Journal of Medical Ethics* 2007 February; 33(2): 67-70. NRCBL: 8.3.1; 8.2. SC: em.

Abstract: BACKGROUND: More and more quantitative information is becoming available about the risks of complications arising from medical treatment. In everyday practice, this raises the question whether each and every risk, however low, should be disclosed to patients. What could be good reasons for doing or not doing so? This will increasingly become a dilemma for practitioners. OBJECTIVE: To report doctors' views on whether to disclose or withhold information on low risks of complications. METHODS: In a qualitative study design, 37 respondents (gastroenterologists and gynaecologists or obstetricians) were included. Focus group interviews were held with 22 respondents and individual in-depth interviews with 15. RESULTS: Doctors have doubts about disclosing or withholding information on complication risk, especially in a risk range of 1 in 200 to 1 in 10,000. Their considerations on whether to disclose or to withhold information depend on a complicated mix of patient and doctor-associated reasons; on medical and personal considerations; and on the kind and purpose of intervention. DISCUSSION: Even though the degree of a risk is important in a doctor's considerations, the severity of the possible complications and patients' wishes and competencies have an important role as well. Respondents said that low risks should always be communicated when there are alternatives for the intervention or when the patient may prevent or mitigate the risk. When the appropriateness of disclosing risks is doubtful, doctors should always tell their patients that no intervention is without risk, give them the opportunity to gather all the information they need or want, and enable them to detect a complication at an early stage.

Parsons, Brian; Kennedy, Miriam. A review of recorded information given to patients starting to take clozapine and the development of guidelines on disclosure, a key component of informed consent. *Journal of Medical Ethics* 2007 October; 33(10): 564-567. NRCBL: 8.3.1; 8.3.3; 17.4. SC: em.

Abstract: Clozapine is a very effective drug with both significant benefits and significant risks in treatment-resistant schizophrenia. Informed consent is generally accepted as both desirable and necessary in order to ensure that the patient's human rights and dignity are respected. Disclosure is a key element of informed consent. It is unclear if the adequate documentation of disclosure is standard practice before initiation of clozapine. The aim of this study was to assess the adequacy of the documentation of disclosure in consent to clozapine treatment in an adult mental health service and to develop guidelines on disclosure. The method was a retrospective analysis of charts of patients given clozapine who received the drug through the pharmacy of a single North Dublin psychiatric hospital. Results show that current practice has evident gaps. The professional, ethical and legal issues are discussed.

Pozgar, George D. Patient consent. *In his:* Legal Aspects of Health Care Administration. 9th edition. Sudbury, MA: Jones and Bartlett Publishers, 2004: 313-334. NRCBL: 8.3.1; 8.3.4; 8.3.2. SC: le; cs.

Raja, Kavitha. Patients' perspectives on medical information: results of a formal survey. *Indian Journal of Medical Ethics* 2007 January-March; 4(1): 16-17. NRCBL: 8.3.1; 1.1. SC: em.

Recupero, Patricia R.; Rainey, Samara E. Informed consent to e-therapy. *American Journal of Psychotherapy* 2005; 59(4): 319-331. NRCBL: 8.3.1; 17.1; 1.3.12; 8.1.

Reider, Alan E.; Dahlinghaus, Andrew B. The impact of new technology on informed consent. *Comprehensive Ophthalmology Update* 2006 November-December; 7(6): 299-302. NRCBL: 8.3.1; 8.2; 8.5; 9.5.1; 4.5. SC: le.

Reiheld, Alison; Easton, Raul B.; Graber, Mark A.; Monnahan, Jay; Hughes, Jason. Consent by survey: losing autonomy one percentage point at a time. *American Journal of Bioethics* 2007 December; 7(12): 53-54; author reply W3-W4,. NRCBL: 8.3.1. Comments: Raul B. Easton, Mark A. Graber, Jay Monnahan, and Jason Hughes. Defining the scope of implied consent in the emergency department. American Journal of Bioethics 2007 December; 7(12): 35-38.

Richards, R. Jason. How we got where we are: a look at informed consent in Colorado — past, present, and future. *Northern Illinois University Law Review* 2005 Fall; 26(1): 69-99. NRCBL: 8.3.1; 1.1; 2.2. SC: le.

Sakaguchi, Misa; Maeda, Shoichi. Informed consent for anesthesia: survey of current practices in Japan. *Journal of Anesthesia* 2005; 19(4): 315-319. NRCBL: 8.3.1; 9.5.1.

Sanz-Ortiz, Jaime. Informed consent and sedation. *Clinical and Translational Oncology* 2006 February; 8(2): 94-97. NRCBL: 8.3.1; 8.1; 9.7; 20.4.1. SC: le. Identifiers: Spain.

Schachter, Debbie; Kleinman, Irwin. Psychiatrists' documentation of informed consent: a representative survey. *Canadian Journal of Psychiatry* 2006 June; 51(7): 438-444. NRCBL: 8.3.1; 17.1; 8.1.

Schneider, Carl E. Void for vagueness. *Hastings Center Report* 2007 January-February; 37(1): 10-11. NRCBL: 8.3.1; 8.1; 5.3. SC: le.

Siddiky, Abul. Should junior doctors be obtaining consent? *British Journal of Hospital Medicine* 2006 November; 67(11): M214. NRCBL: 8.3.1; 7.1.

Slowther, Anne-Marie. The concept of autonomy and its interpretation in health care. *Clinical Ethics* 2007 December; 2(4): 173-175. NRCBL: 8.3.1; 1.1; 8.1.

Triner, Wayne; Jacoby, Liva; Shelton, Wayne; Burk, Mathew; Imarenakhue, Samual; Watt, James; Larkin, Gregory; McGee, Glenn. Exception from informed consent enrollment in emergency medical research: attitudes and awareness. *Academic Emergency Medicine* 2007 February; 14(2): 187-191. NRCBL: 8.3.1; 9.5.1.

van den Brink-Muinen, Atie; van Dulmen, Sandra M.; de Haes, Hanneke C.J.M.; Visser, Adriaan Ph.; Schellevis, F.G.; Bensing, J.M. Has patients' involvement in the decision-making process changed over time? *Health Expectations* 2006 December; 9(4): 333-342. NRCBL: 8.3.1; 8.1; 9.5.1; 9.5.2.

Veatch, Robert M.; Easton, Raul B.; Graber, Mark A.; Monnahan, Jay; Hughes, Jason. Implied, presumed, and waived-consent: the relative moral wrongs of under- and over-informing. *American Journal of Bioethics* 2007 December; 7(12): 39-41; author reply W3-W4. NRCBL: 8.3.1. Comments: Raul B. Easton, Mark A. Graber, Jay

NRCBL: National Reference Center for Bioethics Literature Classification Scheme   See inside front cover for terms.

307

Monnahan, and Jason Hughes. Defining the scope of implied consent in the emergency department. American Journal of Bioethics 2007 December; 7(12): 35-38.

Verheijde, Joseph L.; Rady, Mohamed Y.; McGregor, Joan L.; Easton, Raul B.; Graber, Mark A.; Monnahan, Jay; Hughes, Jason. Defining the scope of implied consent in the emergency department: shortchanging patient's right to self determination. *American Journal of Bioethics* 2007 December; 7(12): 51-52; author reply W3-W4. NRCBL: 8.3.1; 9.5.1. Comments: Raul B. Easton, Mark A. Graber, Jay Monnahan, and Jason Hughes. Defining the scope of implied consent in the emergency department. American Journal of Bioethics 2007 December; 7(12): 35-38.

Wear, Stephen. Informed consent. *In:* Khushf, George, ed. Handbook of Bioethics: Taking Stock of the Field From a Philosophical Perspective. Dordrecht; Boston: Kluwer Academic, 2004: 251-290. NRCBL: 8.3.1; 18.3; 1.1. SC: em; rv.

Weinstein, James N.; Clay, Kate; Morgan, Tamara S. Informed patient choice: patient-centered valuing of surgical risks and benefits. *Health Affairs* 2007 May-June; 26(3): 726-730. NRCBL: 8.3.1; 5.2.

Wilson, James. Is respect for autonomy defensible? *Journal of Medical Ethics* 2007 June; 33(6): 353-356. NRCBL: 8.3.1; 1.1. SC: an.

Abstract: Three main claims are made in this paper. First, it is argued that Onora O'Neill has uncovered a serious problem in the way medical ethicists have thought about both respect for autonomy and informed consent. Medical ethicists have tended to think that autonomous choices are intrinsically worthy of respect, and that informed consent procedures are the best way to respect the autonomous choices of individuals. However, O'Neill convincingly argues that we should abandon both these thoughts. Second, it is argued that O'Neill's proposed solution to this problem is inadequate. O'Neill's approach requires that a more modest view of the purpose of informed consent procedures be adopted. In her view, the purpose of informed consent procedures is simply to avoid deception and coercion, and the ethical justification for informed consent derives from a different ethical principle, which she calls principled autonomy. It is argued that contrary to what O'Neill claims, the wrongness of coercion cannot be derived from principled autonomy, and so its credentials as a justification for informed consent procedures is weak. Third, it is argued that we do better to rethink autonomy and informed consent in terms of respecting persons as ends in themselves, and a characteristically liberal commitment to allowing individuals to make certain categories of decisions for themselves.

Wirtz, Veronika; Cribb, Alan; Barber, Nick. The use of informed consent for medication treatment in hospital: a qualitative study of the views of doctors and nurses. *Clini-* *cal Ethics* 2007 March; 2(1): 36-41. NRCBL: 8.3.1; 9.7; 7.1. SC: em; le.

Abstract: The use of informed consent for surgery or research has been widely studied; however, its use in other areas of clinical practice has received less attention. This study investigates how doctors and nurses understand informed consent in relation to the prescription and administration of medicines in secondary care. It uses a qualitative analysis of semi-structured in-depth interviews with 19 doctors and 6 nurses recruited from various specialties in a teaching hospital. The results indicate a striking gap between official and actual standards of practice. Providing information, assuring adherence and communication about potential treatment harms were raised as key issues instead of the principal goals of informed consent. Rather than simply treating these findings as support for a 'deficit' account of professionalism, the paper concludes that we need a richer and more grounded account of exactly when hospital medication decisions need to be subjected to the highest standards of informed consent.

Woodrow, Susannah R.; Jenkins, Anthony P. How thorough is the process of informed consent prior to outpatient gastroscopy? A study of practice in a United Kingdom District Hospital. *Digestion* 2006; 73(2-3): 189-197. NRCBL: 8.3.1; 8.1; 9.5.1.

Yang, Julia A.; Kombarakaran, Francis A. A practitioner's response to the new health privacy regulations. *Health and Social Work* 2006 May; 31(2): 129-136. NRCBL: 8.3.1; 8.2; 8.4; 1.3.5.

INFORMED CONSENT/ INCOMPETENTS

Alonzi, Andrew; Pringle, Mike. Mental Capacity Act 2005 should guide doctors to help protect vulnerable people [editorial]. *BMJ: British Medical Journal* 2007 November 3; 335(7626): 898. NRCBL: 8.3.3; 8.1. SC: le. Identifiers: Great Britain (United Kingdom).

Amar, Jonathan. Medical decision on behalf of incompetent patients: federal court upholds law allowing medical decisions for incompetent patients — Doe ex re. Tarlow v. District of Columbia. *American Journal of Law and Medicine* 2007; 33(4): 703-705. NRCBL: 8.3.3. SC: le.

Appelbaum, Paul S. Assessment of patients' competence to consent to treatment. *New England Journal of Medicine* 2007 November 1; 357(18): 1834-1840. NRCBL: 8.3.3; 8.3.4; 8.1. SC: cs; rv.

Aveyard, Helen; Woolliams, Mary. In whose best interests? Nurses' experiences of the administration of sedation in general medical wards in England: an application of the critical incident technique. *International Journal of Nursing Studies* 2006 November; 43(8): 929-939. NRCBL: 8.3.3; 4.4; 8.1; 9.7; 9.5.1. SC: em.

Brady Wagner, Lynne C.; Stein, Joel. Failure to achieve assent in a communicative patient: what are the caregiver's obligations? *Topics in Stroke Rehabilitation* 2006 Fall; 13(4): 36-41. NRCBL: 8.3.3; 8.3.4. SC: cs.

Brock, Dan W. Patient competence and surrogate decision-making. *In:* Rhodes, Rosamond; Francis, Leslie P.; Silvers, Anita, eds. The Blackwell Guide to Medical Ethics. Malden, MA: Blackwell Pub., 2007: 128-141. NRCBL: 8.3.3.

Broström, Linus; Johansson, Mats; Nielsen, Morten Klemme. "What the patient would have decided": a fundamental problem with the substituted judgment standard. *Medicine, Health Care and Philosophy* 2007 September; 10(3): 265-278. NRCBL: 8.3.3; 1.1.

Abstract: Decision making for incompetent patients is a much-discussed topic in bioethics. According to one influential decision making standard, the substituted judgment standard, the decision that ought to be made for the incompetent patient is the decision the patient would have made, had he or she been competent. Although the merits of this standard have been extensively debated, some important issues have not been sufficiently explored. One fundamental problem is that the substituted judgment standard, as commonly formulated, is indeterminate in content and thus offers the surrogate little or no guidance. What the standard does not specify is just how competent one should imagine the patient to be, and what else one ought to envision about the patient's hypothetical outlook and the circumstances surrounding his or her decision making. The article discusses this problem of underdetermined decision conditions.

Cherney, Leora Reiff. Ethical issues involving the right hemisphere stroke patient: to treat or not to treat? *Topics in Stroke Rehabilitation* 2006 Fall; 13(4): 47-53. NRCBL: 8.3.3; 8.3.4; 4.4.

Childress, James F. Must we always respect religious belief? *Hastings Center Report* 2007 January-February; 37(1): 3. NRCBL: 8.3.3; 8.1; 1.2.

Devereux, John. Continuing conundrums in competency. *In:* McLean, Sheila A.M., ed. First Do No Harm: Law, Ethics, and Healthcare. Aldershot, England; Burlington, VT: Ashgate, 2006: 235-253. NRCBL: 8.3.3. SC: le.

Dimond, Bridgit. Mental capacity and decision making: defining capacity. *British Journal of Nursing* 2007 October 11-24; 16(18): 1138-1139. NRCBL: 8.3.3. SC: le. Identifiers: Great Britain (United Kingdom).

Dimond, Bridgit. The Mental Capacity Act 2005 and decision-making: advocacy. *British Journal of Nursing* 2007 December 13-2008 January 9; 16(22): 1414-1416. NRCBL: 8.3.3. SC: le. Identifiers: Great Britain (United Kingdom).

Dimond, Bridgit. The Mental Capacity Act 2005 and decision-making: best interests. *British Journal of Nursing* 2007 October 25-November 7; 16(19): 1208-1210. NRCBL: 8.3.3. SC: le. Identifiers: Great Britain (United Kingdom).

Dimond, Bridgit. The Mental Capacity Act 2005: lasting power of attorney. *British Journal of Nursing* 2007 November 8-21; 16(20): 1284-1285. NRCBL: 8.3.3; 20.5.4. SC: le. Identifiers: Great Britain (United Kingdom).

Dimond, Bridgit. The Mental Capacity Act 2005: the new Court of Protection. *British Journal of Nursing* 2007 November 22-December 12; 16(21): 1328-1330. NRCBL: 8.3.3; 20.5.4. SC: le. Identifiers: Great Britain (United Kingdom).

Fraleigh, Anna Schork. An alternative to guardianship: should Michigan statutorily allow acute-care hospitals to make medical treatment decisions for incompetent patients who have neither identifiable surrogates nor advance directives? *University of Detroit Mercy Law Review* 1999 Summer; 76(4): 1079-1134. NRCBL: 8.3.3; 8.4; 9.5.2; 17.1; 20.5.4. SC: le; an.

Fulford, K.W.M. (Bill); Thornton, Tim; Graham, George. It's the law! Rationality and consent as a case study in values and mental health law. *In their:* Oxford Textbook of Philosophy and Psychiatry. Oxford; New York: Oxford University Press, 2006: 539-563. NRCBL: 8.3.3; 17.1; 20.5.4; 4.3. SC: le; cs.

Fullbrook, Suzanne. Best interests. An holistic approach: part 2(b). *British Journal of Nursing* 2007 June 28-July 11; 16(12): 746-747. NRCBL: 8.3.3. SC: le. Identifiers: Great Britain (United Kingdom).

Fullbrook, Suzanne. Best interest. A review of the legal principles involved: Part 2(a). *British Journal of Nursing* 2007 June 14-27; 16(11): 682-683. NRCBL: 8.3.3. SC: le. Identifiers: Great Britain (United Kingdom).

Fullbrook, Suzanne. Consent and capacity: principles of the Mental Capacity Act 2005. *British Journal of Nursing* 2007 April 12-25; 16(7): 412-413. NRCBL: 8.3.3. SC: le. Identifiers: Great Britain (United Kingdom).

Fullbrook, Suzanne; Sanders, Karen. Consent and capacity 2: the Mental Capacity Act 2005 and 'living wills'. *British Journal of Nursing* 2007 April 26-May 9; 16(8): 474-475. NRCBL: 8.3.3; 20.5.4. SC: le. Identifiers: Great Britain (United Kingdom).

Fullbrook, Suzanne; Sanders, Karen. Consent and capacity: other aspects of the Mental Capacity Act. *British Journal of Nursing* 2007 May 10-23; 16(9): 538-539. NRCBL: 8.3.3; 20.5.4. SC: le. Identifiers: Great Britain (United Kingdom).

Garwood-Gowers, Austen. The proper limits for medical intervention that harms the therapeutic interests of incompetents. *In:* Garwood-Gowers, Austen; Tingle, John; Wheat, Kay, eds. Contemporary Issues in Healthcare Law and Ethics. Edinburgh; New York: Elsevier Butterworth-Heinemann, 2005: 191-211. NRCBL: 8.3.3; 4.3; 18.5.6; 19.5. SC: an; le. Identifiers: Great Britain (United Kingdom).

NRCBL: National Reference Center for Bioethics Literature Classification Scheme   See inside front cover for terms.

309

Great Britain (United Kingdom). Department of Health. Bournewood briefing sheet. London: Department of Health, 2006 June; 8 p. [Online]. Accessed: http://www.dh.gov.uk/prod_consum_dh/idcplg?IdcService=GET_FILE&dID=13928&Rendition=Web [2007 April 17]. NRCBL: 8.3.3; 9.5.1; 17.8.

Haque, Omar Sultan; Bursztajn, Harold. Decision-making capacity, memory and informed consent, and judgment at the boundaries of the self. *Journal of Clinical Ethics* 2007 Fall; 18(3): 256-261. NRCBL: 8.3.3; 8.1; 8.3.1. SC: cs. Comments: Sheila Otto. Memento . . . life imitates art: the request for an ethics consultation. Journal of Clinical Ethics 2007 Fall; 18(3): 247-251.

Hattab, Jocelyn Y.; Kohn, Yoav. Informed consent in child psychiatry -- a theoretical review. *Journal of Ethics in Mental Health [electronic]* 2007 November; 2(2): 6 p. Accessed: http://www.jemh.ca [2008 January 24]. NRCBL: 8.3.3; 9.5.3; 9.5.7.

Abstract: In this theoretical review we examine the issue of informed consent in child psychiatry. We describe the development of the concept of informed consent in the history of medicine and review the limited research on its application in child psychiatry. We analyze special features of informed consent unique to our field, such as the capacity of the child to give consent, the status of the "mature minor", the special situation of the child within the family, the place of informed consent in psychotherapy, and the ability of child psychiatrists to give full information prior to consent. We conclude that children, even under the legal age, should be part of the process of giving consent to treatment. On the other hand the complex process of obtaining consent should be aimed at achieving real involvement of patients and families and not merely adhering to formal requirements.

Hubben, Joep H. Decisions on competency and professional standards. *In:* Gevers, J.K.M.; Hondius, E.H.; Hubben, J.H., eds. Health Law, Human Rights and the Biomedicine Convention: Essays in Honour of Henriette Roscam Abbing. Leiden; Boston: Martinus Nijhoff Publishers, 2005: 93-103. NRCBL: 8.3.3. SC: le.

Johnston, Carolyn; Liddle, Jane. The Mental Capacity Act 2005: a new framework for healthcare decision making. *Journal of Medical Ethics* 2007 February; 33(2): 94-97. NRCBL: 8.3.3; 8.4; 17.1. SC: le. Identifiers: Mental Capacity Act; Great Britain (United Kingdom).

Abstract: The Mental Capacity Act received Royal Assent on 7 April 2005, and it will be implemented in 2007. The Act defines when someone lacks capacity and it supports people with limited decision-making ability to make as many decisions as possible for themselves. The Act lays down rules for substitute decision making. Someone taking decisions on behalf of the person lacking capacity must act in the best interests of the person concerned and choose the options least restrictive of his or her rights and freedoms. Decision making will be allowed without any formal procedure unless specific provisions apply, such as a written advance decision, lasting

powers of attorney or a decision by the court of protection.

Li, L.L.-M.; Cheong, K.Y.P.; Yaw, L.K.; Liu, E.H.C. The accuracy of surrogate decisions in intensive care scenarios. *Anaesthesia and Intensive Care* 2007 February; 35(1): 46-51. NRCBL: 8.3.3; 9.5.1; 9.4; 7.1.

Lo, Bernard. Shared decision making. *In his:* Resolving Ethical Dilemmas: A Guide for Clinicians. 3rd edition. Philadelphia, PA: Lippincott Williams and Wilkins, 2005: 57-102. NRCBL: 8.3.3; 8.3.4; 20.5.4.

Macklin, Ruth; Weir, Robert F.; Paris, John J.; Crone, Robert K.; Reardon, Frank; Jecker, Nancy S. Children and the elderly: who should decide? *In:* Baylis, Françoise; Downie, Jocelyn; Freedman, Benjamin; Hoffmaster, Barry; Sherwin, Susan, eds. Health Care Ethics in Canada. Toronto: Harcourt Brace Canada, 1995: 277-318. NRCBL: 8.3.3; 8.3.2; 8.1; 20.5.2; 1.1. SC: le. Note: Section includes: Introduction — Deciding for others / Ruth Macklin — Procedure: criteria, options, and recommendations / Robert F. Weir — Physicians' refusal of requested treatment: the case of Baby L / John J. Paris, Robert K. Crone and Frank Reardon — The role of intimate others in medical decision making / Nancy S. Jecker.

Martin, Adrienne M. Tales publicly allowed: competence, capacity, and religious belief. *Hastings Center Report* 2007 January-February; 37(1): 33-40. NRCBL: 8.3.3; 8.1; 1.2; 1.1.

Mendelson, Danuta. Roman concept of mental capacity to make end-of-life decisions. *International Journal of Law and Psychiatry* 2007 May-June; 30(3): 201-212. NRCBL: 8.3.3; 8.3.4; 20.5.1; 1.1; 2.2.

Oyebode, Femi. The Mental Capacity Act 2005. *Clinical Medicine* 2006 March-April; 6(2): 130-131. NRCBL: 8.3.3; 20.5.4; 18.5.6. Identifiers: Great Britain (United Kingdom).

Shickle, Darren. The Mental Capacity Act 2005. *Clinical medicine* 2006 March-April; 6(2): 169-173. NRCBL: 8.3.3; 20.5.4; 18.5.6. Identifiers: Great Britain (United Kingdom).

Slowther, Anne-Marie. Determining best interests in patients who lack capacity to decide for themselves. *Clinical Ethics* 2007 March; 2(1): 19-21. NRCBL: 8.3.3; 20.5.1.

Smith, David H.; Davis, Dena S.; Cohen, Cynthia B.; Martin, Adrienne M. Taking religion seriously [letters and reply]. *Hastings Center Report* 2007 July-August; 37(4): 4-6. NRCBL: 8.3.3; 8.1; 1.2; 1.1. Comments: Adrienne M. Martin. Tales publicly allowed: competence, capacity, and religious belief. Hastings Center Report 2007 January-February; 37(1): 33-40.

Stein, Joel; Brady Wagner, Lynne C. Is informed consent a "yes or no" response? Enhancing the shared decision-making process for persons with aphasia. *Topics in Stroke Rehabilitation* 2006 Fall; 13(4): 42-46. NRCBL: 8.3.3; 8.3.4.

Stewart, Cameron. Recent developments. *Journal of Bioethical Inquiry* 2007; 4(2): 81-84. NRCBL: 8.3.3; 20.5.3; 14.5; 18.5.4. SC: le.

Straus, Sharon; Stelfox, Tom. Whose life is it anyway? Capacity and consent in Canada. *CMAJ/JAMC: Canadian Medical Association Journal* 2007 November 20; 177(11): 1329. NRCBL: 8.3.3; 20.5.1. SC: cs.

Torke, Alexia M.; Alexander, G. Caleb; Lantos, John; Siegler, Mark. The physician-surrogate relationship. *Archives of Internal Medicine* 2007 June 11; 167(11): 1117-1121. NRCBL: 8.3.3; 8.1.

Abstract: The physician-patient relationship is a cornerstone of the medical encounter and has been analyzed extensively. But in many cases, this relationship is altered because patients are unable to make decisions for themselves. In such cases, physicians rely on surrogates, who are often asked to "speak for the patient." This view overlooks the fundamental fact that the surrogate decision maker cannot be just a passive spokesperson for the patient but is also an active agent who develops a complex relationship with the physician. Although there has been much analysis of the ethical guidelines by which surrogates should make decisions, there has been little previous analysis of the special features of the physician-surrogate relationship. Such an analysis seems crucial as the population ages and life-sustaining technologies improve, which is likely to make surrogate decision making even more common. We outline key issues affecting the physician-surrogate relationship and provide guidance for physicians who are making decisions with surrogates.

Varma, Sumeeta; Wendler, David. Medical decision making for patients without surrogates. *Archives of Internal Medicine* 2007 September 10; 167(16): 1711-1715. NRCBL: 8.3.3.

Abstract: Patients who lose decision-making capacity and lack advance directives and next of kin present a quandary for physicians. Current mechanisms for making treatment decisions for these patients rely on decision makers, such as courts, public guardians, committees, and physicians, who typically do not have sufficient knowledge to predict the patients' preferences. Thus, these mechanisms likely yield decisions that are inconsistent with patients' treatment preferences in many cases. A population-based treatment indicator is a computer-based tool that predicts which treatment a given patient would prefer based on the treatment preferences of similar patients in similar situations. A recent analysis suggests that a population-based treatment indicator could predict patient preferences as accurately as patient-appointed surrogates and next of kin. This analysis suggests that a population-based treatment indicator may provide a mechanism to respect the treatment preferences of patients without surrogates and ensure that their treatment preferences are respected as much as the preferences of patients who have surrogates. Collection of data on patients' treatment preferences, especially those without surrogates, incorporation of these data into a treatment indicator, and exploration of ways to implement this approach for patients without surrogates are called for.

Vig, Elizabeth K.; Starks, Helene; Taylor, Janelle S.; Hopley, Elizabeth K.; Fryer-Edwards, Kelly. Surviving surrogate decision-making: what helps and hampers the experience of making medical decisions for others. *JGIM: Journal of General Internal Medicine* 2007 September; 22(9): 1274-1279. NRCBL: 8.3.3. SC: em.

White, Douglas B.; Curtis, J. Randall; Wolf, Leslie E.; Prendergast, Thomas J.; Taichman, Darren B.; Kuniyoshi, Gary; Acerra, Frank; Lo, Bernard; Luce, John M. Life support for patients without a surrogate decision maker: who decides? *Annals of Internal Medicine* 2007 July 3; 147(1): 34-40. NRCBL: 8.3.3; 20.5.1. SC: em.

Abstract: Background: Physicians in intensive care units have withdrawn life support in incapacitated patients who lack surrogate decision makers and advance directives, yet little is known about how often this occurs or under what circumstances. Objective: To determine the proportion of deaths in intensive care units that occur in patients who lack decision-making capacity and a surrogate and the process that physicians use to make these decisions. Design: Multicenter, prospective cohort study. Setting: Intensive care units of 7 medical centers in 2004 to 2005. Patients: 3011 consecutive critically ill adults. Measurements: Attending physicians completed a questionnaire about the decision-making process for each incapacitated patient without a surrogate or advance directive for whom they considered limiting life support. Results: Overall, 5.5% (25 of 451 patients) of deaths in intensive care units occurred in incapacitated patients who lacked a surrogate decision maker and an advance directive. This percentage ranged from 0% to 27% across the 7 centers. Physicians considered limiting life support in 37 such patients or would have considered it if a surrogate had been available. In 6 patients, there was prospective hospital review of the decision, and in 1 patient, there was court review. In the remaining 30 patients, the decision was made by the intensive care unit team alone or by the intensive care unit team plus another attending physician. The authors found wide variability in hospital policies, professional society guidelines, and state laws regarding who should make life-support decisions for this patient population. Thirty-six of 37 life-support decisions were made in a manner inconsistent with American College of Physicians guidelines for judicial review. Limitations: The results are based on physicians' self-reported practices and may not match actual practices. The number of incapacitated patients without surrogates in the study is small. Conclusions: Incapacitated patients without surrogates accounted for approximately 1 in 20 deaths in intensive care units. Most life-support decisions were made by physicians without institutional or judicial review.

NRCBL: National Reference Center for Bioethics Literature Classification Scheme   See inside front cover for terms.

311

Wilson, Naomi. Professionals' experiences of addressing ethical issues in services for people with intellectual disabilities: a brief report. *Ethics and Intellectual Disability* 2005 Winter; 8(2): 5-7. NRCBL: 8.3.3; 9.5.3; 1.1.

Wong, J.G.; Clare, I.C.H.; Gunn, M.J.; Holland, A.J. Capacity to make health care decisions: its importance in clinical practice. *Psychological Medicine* 1999 March; 29(2): 437-446. NRCBL: 8.3.3; 20.5.4. SC: le; rv.

Wrigley, A. Proxy consent: moral authority misconceived. *Journal of Medical Ethics* 2007 September; 33(9): 527-531. NRCBL: 8.3.3. SC: an; le.
Abstract: The Mental Capacity Act 2005 has provided unified scope in the British medical system for proxy consent with regard to medical decisions, in the form of a lasting power of attorney. While the intentions are to increase the autonomous decision making powers of those unable to consent, the author of this paper argues that the whole notion of proxy consent collapses into a paternalistic judgement regarding the other person's best interests and that the new legislation introduces only an advisor, not a proxy with the moral authority to make treatment decisions on behalf of another. The criticism is threefold. First, there is good empirical evidence that people are poor proxy decision makers as regards accurately representing other people's desires and wishes, and this is therefore a pragmatically inadequate method of gaining consent. Second, philosophical theory explaining how we represent other people's thought processes indicates that we are unlikely ever to achieve accurate simulations of others' wishes in making a proxy decision. Third, even if we could accurately simulate other people's beliefs and wishes, the current construction of proxy consent in the Mental Capacity Act means that it has no significant ethical authority to match that of autonomous decision making. Instead, it is governed by a professional, paternalistic, best-interests judgement that undermines the intended role of a proxy decision maker. The author argues in favour of clearly adopting the paternalistic best-interests option and viewing the proxy as solely an advisor to the professional medical team in helping make best-interests judgements.

INFORMED CONSENT/ MINORS

Alderson, Priscilla. Consent to surgery for deaf children: making informed decisions. *In:* Komesaroff, Linda, ed. Surgical Consent: Bioethics and Cochlear Implantation. Washington, DC: Gallaudet University Press, 2007: 30-41. NRCBL: 8.3.2; 9.5.7; 8.3.1; 2.1.

Cashmore, Judy. Ethical issues concerning consent in obtaining children's reports on their experience of violence. *Child Abuse and Neglect* 2006 September; 30(9): 969-977. NRCBL: 8.3.2; 9.5.7; 9.1; 18.2; 18.3; 18.5.2.

Cohn, Felicia. Real life informs consent. *Journal of Clinical Ethics* 2007 Winter; 18(4): 366-368. NRCBL: 8.3.2; 8.3.1; 9.6; 20.5.2.

Furton, Edward J. Morality is not a medical problem. *Ethics and Medics* 2007 July; 32(7): 3-4. NRCBL: 8.3.2; 9.5.1; 9.7; 9.5.5; 9.5.7; 8.3.4; 1.2.

Great Britain (United Kingdom). Department of Health. Consent — What You Have a Right to Expect: A Guide for Parents. London: Department of Health, 2001 July; 11 p. NRCBL: 8.3.2.

Halpern, Jodi. Let's value, but not idealize, emotions. *Journal of Clinical Ethics* 2007 Winter; 18(4): 380-383. NRCBL: 8.3.2; 20.5.2; 20.3.3. Comments: Annie Janvier. How much suffering is enough? Journal of Clinical Ethics 2007 Winter; 18(4): 362-365.

Hester, D. Micah. Interests and neonates: there is more to the story than we explicitly acknowledge. *Theoretical Medicine and Bioethics* 2007; 28(5): 357-372. NRCBL: 8.3.2; 9.5.7.
Abstract: Although there are many different moral arguments concerning the use of Best Interests in neonatal decision-making, there seems in practice a firm commitment to application of the concept. And yet, there is still little reflection given by practitioners about what employing a Best Interest determination means in infant care. The following lays out a comprehensive taxonomy of interest-sources in order to provide for more robust considerations of what constitutes best interests of/for neonates.

Hickey, Kathryn. Minors rights in medical decision making. *JONA's Healthcare Law, Ethics, and Regulation* 2007 July-September; 9(3): 100-106. NRCBL: 8.3.2; 1.2; 8.3.4; 8.4; 18.5.2; 18.2.
Abstract: In the past, minors were not considered legally capable of making medical decisions and were viewed as incompetent because of their age. The authority to consent or refuse treatment for a minor remained with a parent or guardian. This parental authority was derived from the constitutional right to privacy regarding family matters, common law rule, and a general presumption that parents or guardians will act in the best interest of their incompetent child. However, over the years, the courts have gradually recognized that children younger than 18 years who show maturity and competence deserve a voice in determining their course of medical treatment. This article will explore the rights and interests of minors, parents, and the state in medical decision making and will address implications for nursing administrators and leaders.

Higginson, Jason D. Emotion, suffering, and hope: commentary on "How much suffering is enough?". *Journal of Clinical Ethics* 2007 Winter; 18(4): 377-379. NRCBL: 8.3.2; 20.5.2; 20.4.2; 20.3.3; 4.2. Comments: Annie Janvier. How much suffering is enough? Journal of Clinical Ethics 2007 Winter; 18(4): 362-365.

Kopelman, Loretta M.; Kopelman, Arthur E. Using a new analysis of the best interests standard to address cultural disputes: whose data, which values? *Theoretical*

Medicine and Bioethics 2007; 28(5): 373-391. NRCBL: 8.3.2; 8.3.4; 20.5.2; 21.7. SC: cs.

Abstract: Clinicians sometimes disagree about how much to honor surrogates' deeply held cultural values or traditions when they differ from those of the host country. Such a controversy arose when parents requested a cultural accommodation to let their infant die by withdrawing life saving care. While both the parents and clinicians claimed to be using the Best Interests Standard to decide what to do, they were at an impasse. This standard is analyzed into three necessary and jointly sufficient conditions and used to resolve the question of how much to accommodate cultural preferences and how to treat this infant. The extreme versions of absolutism and relativism are rejected. Properly understood, the Best Interests Standard can serve as a powerful tool in settling disputes about how to make good decisions for those who cannot decide for themselves.

Liao, S. Matthew; Savulescu, Julian; Sheehan, Mark. The Ashley treatment: best interests, convenience, and parental decision-making. *Hastings Center Report* 2007 March-April; 37(2): 16-20. NRCBL: 8.3.2; 9.5.7; 10.

Lo, Bernard. Ethical issues in pediatrics. *In his:* Resolving Ethical Dilemmas: A Guide for Clinicians. 3rd edition. Philadelphia, PA: Lippincott Williams and Wilkins, 2005: 235-242. NRCBL: 8.3.2; 8.3.4; 9.5.7.

Slonina, Mary Irene. State v. physicians et al.: legal standards guiding the mature minor doctrine and the bioethical judgment of pediatricians in life-sustaining medical treatment. *Health Matrix: The Journal of Law-Medicine* 2007 Winter; 17(1): 181-214. NRCBL: 8.3.2; 20.4.2; 20.5.2. SC: le.

Sobsey, Dick. Growth attenuation and indirect-benefit rationale. *Ethics and Intellectual Disability* 2007 Winter; 10(1): 1,2, 7-8. NRCBL: 8.3.2; 9.5.3; 1.1.

Stultiëns, Loes; Goffin, Tom; Borry, Pascal; Dierickx, Kris; Nys, Herman. Minors and informed consent: a comparative approach. *European Journal of Health Law* 2007 April; 14(1): 21-46. NRCBL: 8.3.2; 21.1. SC: le; rv.

Wilfond, Benjamin S. The Ashley case: the public response and policy implications. *Hastings Center Report* 2007 September-October; 37(5): 12-13. NRCBL: 8.3.2; 9.5.7. SC: le.

Wright, Wendy; Staible, Nancy. HPV mandates: parents trump politics. *Ethics and Medics* 2007 July; 32(7): 1-3. NRCBL: 8.3.2; 9.5.1; 9.7; 9.5.5; 9.5.7; 8.3.4. Identifiers: human papillomavirus.

INSTITUTIONAL REVIEW BOARDS *See* HUMAN EXPERIMENTATION/ ETHICS COMMITTEES AND POLICY GUIDELINES

INTERNATIONAL HEALTH AND HUMAN RIGHTS

See also TORTURE, GENOCIDE AND WAR CRIMES; WAR AND TERRORISM

Bennett, Belinda. Globalising the body: globalisation and reproductive rights. *University of New South Wales Law Journal* 2006; 29(2): 266-271. NRCBL: 21.1; 4.4; 14.1; 2.1.

Bibeau, Gilles; Pedersen, Duncan. A return to scientific racism in medical social sciences: the case of sexuality and the AIDS epidemic in Africa. *In:* Nichter, Mark and Lock, Margaret, eds. New Horizons in Medical Anthropology: Essays in Honour of Charles Leslie. London: Routledge, 2002: 141-171. NRCBL: 21.1; 9.5.4; 9.5.6; 9.1; 21.7; 15.1.

Borst-Eisler, Els. Th role of public debate and politics in the implementation of the Convention. *In:* Gevers, J.K.M.; Hondius, E.H.; Hubben, J.H., eds. Health Law, Human Rights and the Biomedicine Convention: Essays in Honour of Henriette Roscam Abbing. Leiden; Boston: Martinus Nijhoff Publishers, 2005: 247-254. NRCBL: 21.1; 18.6; 5.3. SC: le.

Charo, R. Alta. The endarkenment. *In:* Eckenwiler, Lisa A.; Cohn, Felicia G., eds. The Ethics of Bioethics: Mapping the Moral Landscape. Baltimore, MD: Johns Hopkins University Press, 2007: 95-107. NRCBL: 21.1; 2.4.

DeCamp, Matthew. Scrutinizing global short-term medical outreach. *Hastings Center Report* 2007 November-December; 37(6): 21-23. NRCBL: 21.1; 7.2; 18.6; 18.5.9; 2.1; 9.1.

Denniston, George C. Human rights advances in the United States. *In:* Denniston, George C.; Gallo, Pia Grassivaro; Hodges, Frederick M.; Milos, Marilyn Fayre; Viviani, Franco, eds. Bodily Integrity and the Politics of Circumcision: Culture, Controversy, and Change. New York: Springer, 2006: 189-201. NRCBL: 21.1; 10; 9.5.7; 6. SC: le.

Fleck, Leonard M. Can we trust "democratic deliberation"? *Hastings Center Report* 2007 July-August; 37(4): 22-25. 4 refs. NRCBL: 21.1; 14.1; 15.1; 5.3. SC: an; le. Comments: Franco Furger and Francis Fukuyama. A proposal for modernizing the regulation of human biotechnologies. Hastings Center Report 2007 July-August; 37(4): 16-20.

Keywords: *biotechnology; *democracy; *policy making; *public participation; *reproductive technologies; accountability; bioethical issues; cloning; cultural pluralism; embryo research; embryonic stem cells; freedom; genetic engineering; government financing; government regulation; policy analysis; preimplantation diagnosis; public policy; regulation; Keyword Identifiers: *United States

Gadd, Elaine. The global significance of the Convention on Human Rights and Biomedicine. *In:* Gevers, J.K.M.; Hondius, E.H.; Hubben, J.H., eds. Health Law, Human

NRCBL: National Reference Center for Bioethics Literature Classification Scheme   See inside front cover for terms.

313

Rights and the Biomedicine Convention: Essays in Honour of Henriette Roscam Abbing. Leiden; Boston: Martinus Nijhoff Publishers, 2005: 35-46. NRCBL: 21.1; 2.1. SC: le.

Gostin, Lawrence O. The international health regulations: a new paradigm for global health governance? *In:* McLean, Sheila A.M., ed. First Do No Harm: Law, Ethics, and Healthcare. Aldershot, England; Burlington, VT: Ashgate, 2006: 59-79. NRCBL: 21.1; 9.1. SC: le. Identifiers: World Health Organization.

Gruskin, Sofia; Tarantola, Daniel. Health and human rights. *In:* Detels, Roger; McEwen, James; Beaglehole, Robert; Tanaka, Heizo, eds. Oxford Textbook of Public Health. Fourth edition. Oxford; New York: Oxford University Press, 2004: 311-335. NRCBL: 21.1; 9.2; 9.1.

Jayasinghe, Saroj. Faith-based NGOs and healthcare in poor countries: a preliminary exploration of ethical issues. *Journal of Medical Ethics* 2007 November; 33(11): 623-626. NRCBL: 21.1; 1.2; 9.1.

Abstract: An increasing number of non-governmental organisations (NGOs) provide humanitarian assistance, including healthcare. Some faith-based NGOs combine proselytising work with humanitarian aid. This can result in ethical dilemmas that are rarely discussed in the literature. The article explores several ethical issues, using four generic activities of faith-based NGOs: (1) It is discriminatory to deny aid to a needy community because it provides less opportunity for proselytising work. Allocating aid to a community with fewer health needs but potential for proselytising work is unjust, since it neither maximises welfare (utilitarianism) nor assists the most needy (egalitarianism). (2) Faith-based-NGOs may state that proselytising work combined with humanitarian assistance improves spiritual wellbeing and overall benefit. However, proselytising work creates religious doubts, which could transiently decrease wellbeing. (3) Proselytising work is unlikely to be a perceived need of the population and, if carried out without consent, breaches the principle of autonomy. Such work also exploits the vulnerability of disaster victims. (4) Governments that decline the assistance of a faith-based NGO involved in proselytising work may deprive the needy of aid. Three strategies are proposed: (a) Increase knowledge to empower communities, individuals and governments; information on NGOs could be provided through an accessible register that discloses objectives, funding sources and intended spiritual activities. (b) Clearly demarcate between humanitarian aid from proselytising work, by setting explicit guidelines for humanitarian assistance. (c) Strengthen self-regulation by modifying the Code of Conduct of the Red Cross to state criteria for selecting communities for assistance and procedures for proselytising work.

Macklin, Ruth. Global health. *In:* Steinbock, Bonnie, ed. The Oxford Handbook of Bioethics. Oxford; New York: Oxford University Press, 2007: 696-720. NRCBL: 21.1; 1.1; 9.1.

Novotny, Thomas E.; Mordini, Emilio; Chadwick, Ruth; Pedersen, J. Martin; Fabbri, Fabrizio; Lie, Reidar; Thanachaiboot, Natapong; Mossialos, Elias; Permanand, Govin. Bioethical implications of globalization: an international consortium project of the European Commission. *PLoS Medicine* 2006 February; 3(2); e43: 0173-0176. 16 refs. NRCBL: 21.1; 9.1; 1.3.12; 15.1; 9.3.1; 21.3.

Keywords: *communicable diseases; *international aspects; *policy making; *public health; *trends; access to information; biomedical research; bioterrorism; commerce; developing countries; genetic predisposition; genetic screening; genetically modified plants; genomics; health care delivery; immigrants; Internet; population genetics; property rights; resource allocation; social impact; social problems; Proposed Keywords: *technology; *world health; developed countries; genetic resources; travel; Keyword Identifiers: *Bioethical Implications of Globalization Project; *European Commission; Europe; European Union

Orbinski, James; Beyrer, Chris; Singh, Sonal. Violations of human rights: health practitioners as witnesses. *Lancet* 2007 August 25-31; 370(9588): 698-704. NRCBL: 21.1; 21.2; 21.4; 9.5.6; 9.2; 7.1.

Orentlicher, David. Bioethics and society: from the ivory tower to the state house. *In:* Eckenwiler, Lisa A.; Cohn, Felicia G., eds. The Ethics of Bioethics: Mapping the Moral Landscape. Baltimore, MD: Johns Hopkins University Press, 2007: 74-82. NRCBL: 21.1; 5.3. SC: le.

Parker, Lisa S. Bioethics as activism. *In:* Eckenwiler, Lisa A.; Cohn, Felicia G., eds. The Ethics of Bioethics: Mapping the Moral Landscape. Baltimore, MD: Johns Hopkins University Press, 2007: 144-157. NRCBL: 21.1; 2.1; 1.1.

Pogge, Thomas. Montréal statement on the human right to essential medicines. *CQ: Cambridge Quarterly of Healthcare Ethics* 2007 Winter; 16(1): 97-108. NRCBL: 21.1; 9.7; 5.3.

Robertson, Geoffrey. Health and human rights series [comment]. *Lancet* 2007 August 4-10; 370(9585): 368-369. NRCBL: 21.1; 21.2; 21.4; 9.2; 8.4; 1.3.6. Identifiers: International Committee of the Red Cross.

Roucounas, Emmanuel. The Biomedicine Convention in Relation to other international instruments. *In:* Gevers, J.K.M.; Hondius, E.H.; Hubben, J.H., eds. Health Law, Human Rights and the Biomedicine Convention: Essays in Honour of Henriette Roscam Abbing. Leiden; Boston: Martinus Nijhoff Publishers, 2005: 23-34. NRCBL: 21.1; 2.1. SC: le.

Rowson, Richard. Nurses' difficulties with rights. *Nursing Ethics* 2007 November; 14(6): 838-840. NRCBL: 21.1; 4.1.3.

Sharma, Shridhar. Human rights in psychiatric care: an Asian perspective. *Acta Psychiatrica Scandinavica* 2000 January; 101(399 Supplement): 97-101. NRCBL: 21.1; 17.1.

Sibley, Robert. When healers become killers: the doctor as terrorist. *CMAJ/JAMC: Canadian Medical Association* 2007 September 11; 177(6): 688. NRCBL: 21.1; 7.4.

Singh, Jerome Amir; Govender, Michelle; Mills, Edward J. Do human rights matter to health? *Lancet* 2007 August 11-17; 370(9586): 521-527. NRCBL: 21.1; 9.2. SC: le. Identifiers: India; South Africa.

Abstract: Legal instruments and litigation as a way to enforce the rights to life and to health is a relatively new strategy that is increasingly common. We show how legal measures have been used to attain health and human rights with case examples from India and South Africa that resulted in large public-health benefits.

Sokalska, Maria E. Implementation of the Convention in central Europe: the case of Poland. *In:* Gevers, J.K.M.; Hondius, E.H.; Hubben, J.H., eds. Health Law, Human Rights and the Biomedicine Convention: Essays in Honour of Henriette Roscam Abbing. Leiden; Boston: Martinus Nijhoff Publishers, 2005: 255-268. NRCBL: 21.1; 9.2. SC: le; cs.

Taylor, Allyn L.; Bettcher, Douglas W.; Fluss, Sev S.; DeLand, Katherine; Yach, Derek. International health instruments: an overview. *In:* Detels, Roger; McEwen, James; Beaglehole, Robert; Tanaka, Heizo, eds. Oxford Textbook of Public Health. Fourth edition. Oxford; New York: Oxford University Press, 2004: 359-386. NRCBL: 21.1; 9.1; 7.1. SC: le.

Turner, Leigh. Global health inequalities and bioethics. *In:* Eckenwiler, Lisa A.; Cohn, Felicia G., eds. The Ethics of Bioethics: Mapping the Moral Landscape. Baltimore, MD: Johns Hopkins University Press, 2007: 229-240. NRCBL: 21.1; 9.4; 9.2; 9.5.10; 2.1.

Ventura, Carla A. Arena; Mendes, Isabel Amelia Costa; Trevizan, Maria Auxiliadora. Psychiatric nursing care in Brazil: legal and ethical aspects. *Medicine and Law: The World Association for Medical Law* 2007 December; 26(4): 829-840. NRCBL: 21.1; 17.7; 4.1.3. SC: le.

Abstract: Human rights, considered as rights inherent to all human beings, must be respected unconditionally, especially during health care delivery. These rights became actually protected by International Law when the UN was created in 1945 and, later, when the Universal Declaration of Human Rights was issued in 1948, giving rise to various subsequent treaties. Based on the historical evolution of Human Rights in the international sphere, associated with the principles of constitutional, penal and civil law and psychiatric patient rights in Brazil, we aim to understand some dilemmas of psychiatric nursing care: individuals' rights as psychiatric patients, hospitalization and nursing professionals' practice. In their practice, nurses attempt to conciliate patients' rights with their legal role and concerns with high-quality psychiatric care. In coping with these dilemmas, these professionals are active in three spheres: as health care providers, as employees of a health organization and as citizens.

White, Mary Terrell. Bioethics without a map. *Medical Humanities Review* 2005 Spring-Fall; 19(1-2): 9-12. NRCBL: 21.1; 21.7; 9.2; 7.2. Identifiers: Uganda.

Zuckerberg, Joaquin. International human rights for mentally ill persons: the Ontario experience. *International Journal of Law and Psychiatry* 2007 November-December; 30(6): 512-529. NRCBL: 21.1; 9.5.3; 9.2; 17.1. SC: le.

INTERNATIONAL MIGRATION OF HEALTH CARE PROFESSIONALS

Benatar, Solomon. An examination of ethical aspects of migration and recruitment of health care professionals from developing countries. *Clinical Ethics* 2007 March; 2(1): 2-7. NRCBL: 21.6; 7.1.

Cutcliffe, John R.; Yarbrough, Susan. Globalization, commodification and mass transplant of nurses: Part 2. *British Journal of Nursing* 2007 August 9 - September 12; 16(15): 926-930. NRCBL: 21.6.

Dwyer, James. What's wrong with the global migration of health care professionals? Individual rights and international justice. *Hastings Center Report* 2007 September-October; 37(5): 36-43. NRCBL: 21.6; 21.1.

Abstract: When health care workers migrate from poor countries to rich countries, they are exercising an important human right and helping rich countries fulfill obligations of social justice. They are also, however, creating problems of social justice in the countries they leave. Solving these problems requires balancing social needs against individual rights and studying the relationship of social justice to international justice.

INVOLUNTARY COMMITMENT

Arikan, Rasim; Appelbaum, Paul S.; Sercan, Mustafa; Turkcan, Solmaz; Satmis, Nevzat; Polat, Aslihan. Civil commitment in Turkey: reflections on a bill drafted by psychiatrists. *International Journal of Law and Psychiatry* 2007 January-February; 30(1): 29-35. NRCBL: 17.7; 17.8. SC: le.

Bauer, Arie; Rosca, Paula; Grinshpoon, Alexander; Khawaled, Razek; Mester, Roberto; Yoffe, Rinat; Ponizovsky, Alexander M. Trends in involuntary psychiatric hospitalization in Israel 1991-2000. *International Journal of Law and Psychiatry* 2007 January-February; 30(1): 60-70. NRCBL: 17.7. SC: em.

Brooks, Robert A. Psychiatrists' opinions about involuntary civil commitment: results of a national survey. *Journal of the American Academy of Psychiatry and the Law* 2007; 35(2): 219-228. NRCBL: 17.7; 4.1.1. SC: em.

Brooks, Robert A. U.S. psychiatrists' beliefs and wants about involuntary civil commitment grounds. *International Journal of Law and Psychiatry* 2006 January-February; 29(1): 13-21. NRCBL: 17.7. SC: em.

NRCBL: National Reference Center for Bioethics Literature Classification Scheme   See inside front cover for terms.

315

Butcher, James. Controversial mental health bill reaches the finishing line [commentary]. *Lancet* 2007 July 14-20; 370(9582): 117-118. NRCBL: 17.7; 4.3. Identifiers: Great Britain (United Kingdom).

Byatt, Nancy; Pinals, Debra; Arikan, Rasim. Involuntary hospitalization of medical patients who lack decisional capacity: an unresolved issue. *Psychosomatics* 2006 September-October; 47(5): 443-448. NRCBL: 17.7.

Dreßing, Harald. Compulsory admission and compulsory treatment in psychiatry. *In:* Schramme, Thomas; Thome, Johannes, eds. Philosophy and Psychiatry. Berlin; New York: De Gruyter, 2004: 353-356. NRCBL: 17.7; 17.8; 1.1.

Dyer, Clare. Community treatment orders stay in mental health bill. *BMJ: British Medical Journal* 2007 June 23; 334(7607): 1293. NRCBL: 17.7. SC: le. Identifiers: Great Britain (United Kingdom).

Dyer, Clare. Government suffers defeat in the House of Lords over Mental Health Bill for England and Wales. *BMJ: British Medical Journal* 2007 February 24; 334(7590): 384-385. NRCBL: 17.7. SC: le.

Dyer, Clare. Government suffers its first defeat over Mental Health Bill [news]. *BMJ: British Medical Journal* 2007 January 20; 334(7585): 113. NRCBL: 17.7; 17.8; 8.3.4. SC: le. Identifiers: Great Britain (United Kingdom).

Griffith, Richard. Authorizing the deprivation of liberty of incapable adults in institutions. *British Journal of Community Nursing* 2006 December; 11(12): 538-541. NRCBL: 17.7; 8.3.3; 17.8.

Kaltiala-Heino, Riittakerttu; Fröjd, Sari. Severe mental disorder as a basic commitment criterion for minors. *International Journal of Law and Psychiatry* 2007 January-February; 30(1): 81-94. NRCBL: 17.7; 9.5.3; 9.5.7.

Kuosmanen, Lauri; Hätönen, Heli; Malkavaara, Heikki; Kylmä, Jari; Välimäki, Maritta. Deprivation of liberty in psychiatry hospital care: the patient's perspective. *Nursing Ethics* 2007 September; 14(5): 597-607. NRCBL: 17.7. SC: em. Identifiers: Finland.
Abstract: Deprivation of liberty in psychiatric hospitals is common world-wide. The aim of this study was to find out whether patients had experienced deprivation of their liberty during psychiatric hospitalization and to explore their views about it. Patients (n = 51) in two acute psychiatric inpatient wards were interviewed in 2001. They were asked to describe in their own words their experiences of being deprived of their liberty. The data were analysed by inductive content analysis. The types of deprivation of liberty in psychiatric hospital care reported by these patients were: restrictions on leaving the ward and on communication, confiscation of property, and various coercive measures. The patients' experiences of being deprived of their liberty were negative, although some saw the rationale for using these interventions, considering them as part of hospital care.

Lee, James E.; Kelly, D. Clay. Constitutional challenge to grave disability. *Journal of the American Academy of Psychiatry and the Law* 2007; 35(4): 534-535. NRCBL: 17.7; 4.3; 9.5.3. SC: le.

Linhorst, Donald. Individual rights, coercion, and empowerment. *In his:* Empowering People with Severe Mental Illness: A Practical Guide. New York: Oxford University Press, 2006: 40-64. NRCBL: 17.7; 8.3.4; 17.8. SC: le.

Niveau, G.; Materi, J. Psychiatric commitment: over 50 years of case law from the European Court of Human Rights. *European Psychiatry* 2007 January; 22(1): 59-67. NRCBL: 17.7; 6. SC: le. Note: Table 1.is updated in this article. An earlier verison was published in European Psychiatry 2006 October; 21(7): 429.

Niveau, G.; Materi, J. Psychiatric commitment: over 50 years of case law from the European Court of Human Rights. *European Psychiatry* 2006 October; 21(7): 427-435. NRCBL: 17.7; 6. SC: le. Note: See European Psychiatry 2007 January; 22(1): 61 for updated version of Table 1.

Pescosolido, Bernice A; Fettes, Danielle L.; Martin, Jack K.; Monahan, John; McLeod, Jane D. Perceived dangerousness of children with mental health problems and support for coerced treatment. *Psychiatric Services* 2007 May; 58(5): 619-625. NRCBL: 17.7; 4.3; 8.3.2. SC: em.

Schramme, Thomas. Coercive threats and offers in psychiatry. *In:* Schramme, Thomas; Thome, Johannes, eds. Philosophy and Psychiatry. Berlin; New York: De Gruyter, 2004: 357-369. NRCBL: 17.7; 17.8.

Segal, Steven P.; Tauber, Alfred I.; Zanni, Guido R.; Stavis, Paul F. Revisiting Hume's law [comment and reply]. *American Journal of Bioethics* 2007 November; 7(11): 43-45; author reply W3-W4. NRCBL: 17.7. Comments: Guido R. Zanni and Paul F. Stavis. The effectiveness and ethical justification of psychiatric outpatient commitment. American Journal of Bioethics 2007 November; 7(11): 31-41.

Spellecy, Ryan; Zanni, Guido R.; Stavis, Paul F. Psychiatric outpatient commitment: one tool along a continuum [comment and reply]. *American Journal of Bioethics* 2007 November; 7(11): 45-47; author reply W3-W4. NRCBL: 17.7. Comments: Guido R. Zanni and Paul F. Stavis. The effectiveness and ethical justification of psychiatric outpatient commitment. American Journal of Bioethics 2007 November; 7(11): 31-41.

Zanni, Guido R.; Stavis, Paul F. The effectiveness and ethical justification of psychiatric outpatient commitment. *American Journal of Bioethics* 2007 November; 7(11): 31-41. NRCBL: 17.7; 9.3.1. SC: em; le. Comments: American Journal of Bioethics 2007 November; 7(11): 42-47.

Zion, Deborah; Jureidini, Jon; Newman, Louise; Kyambi, Sarah; Zion, Deborah;. Republication: in that case [case study and commentaries]. *Journal of Bioethical Inquiry* 2006; 3(3): 193-202. NRCBL: 17.7; 1.3.5; 9.5.4; 4.3; 4.1.1; 8.3.3; 10; 17.1; 17.8. SC: cs; le.

JOURNALISM AND PUBLISHING
See also BIOMEDICAL RESEARCH

The plagiarism policy of the American Journal of Nursing. *AJN: American Journal of Nursing* 2007 July; 107(7): 78-79. NRCBL: 1.3.7. SC: em.

Who is accountable? [editorial]. *Nature* 2007 November 1; 450(7166): 1. NRCBL: 1.3.7; 1.3.9.

Anaissie, E.J.; Segal, B.H.; Graybill, J.R.; Arndt, C.; Perfect, J.R.; Kleinberg, M.; Pappas, P.; Benjamin, D.; Rubin, R.; Aberg, J.A.; Adderson, E.E.; Adler-Shohet, F.C.; Akan, H.; Akova, M.; Almyroudis, N.G.; Alexander, B.D.; Andes, D.; Arrieta, A.; Baddley, J.W.; Barron, M.A.; et al. Clinical research in the lay press: irresponsible journalism raises a huge dose of doubt. *Clinical Infectious Diseases* 2006 October 15; 43(8): 1031-1039 [see correction in Clinical Infectious Diseases 2007 March 15; 44(6): 894]. NRCBL: 1.3.7; 9.7; 18.2; 1.3.5; 7.1. SC: cs.

Ancker, Jessica S.; Flanagin, Annette. A comparison of conflict of interest policies at peer-reviewed journals in different scientific disciplines. *Science and Engineering Ethics* 2007 June; 13(2): 147-157. NRCBL: 1.3.7; 1.3.9. SC: em.

Batt, Sharon; Braun, Joshua A. Limits on autonomy: political meta-narratives and health stories in the media. *American Journal of Bioethics* 2007 August; 7(8): 23-25; author reply W1-W2. NRCBL: 1.3.7; 1.1; 9.1; 21.1. Comments: Joshua A. Braun. The imperatives of narrative: health interest groups and morality in network news. American Journal of Bioethics 2007 August; 7(8): 6-14.

Benham, Bryan; Clark, Dale; Francis, Leslie P. Authorship: credit, responsibility, and accountability. *In:* Kulakowski, Elliott C.; Chronister, Lynne U., eds. Research Administration and Management. Sudbury, MA: Jones and Bartlett, 2006: 501-510. NRCBL: 1.3.7; 18.1.

Benítez-Bribiesca, Luis; Modiano-Esquenazi, Marcos. Ethics of scientific publication after the human stem cell scandal [editorial]. *Archives of Medical Research* 2006 May; 37(4): 423-424. NRCBL: 1.3.7; 1.3.9.

Bevan, Joan C.; Miller, Donald R. Medical journals and cross-cultural research ethics. *Canadian Journal of Anaesthesia = Journal Canadien d'Anesthésie* 2005 December; 52(10): 1009-1016. NRCBL: 1.3.7; 21.7; 18.2.

Bogod, D.G. The editor as umpire: clinical trial registration and dispute resolution. *Anaesthesia* 2006 December; 61(12): 1133-1135. NRCBL: 1.3.7; 18.2; 18.6.

Brand, Richard A.; Jacobs, Joshua J.; Heckman, James D. Professionalism in publishing. *Journal of Bone and Joint Surgery. American volume* 2006 November; 88(11): 2323-2325. NRCBL: 1.3.7; 7.1.

Braun, Joshua A. The imperatives of narrative: health interest groups and morality in network news. *American Journal of Bioethics* 2007 August; 7(8): 6-14. NRCBL: 1.3.7; 9.1; 1.1; 2.1. Comments: American Journal of Bioethics 2007 August: 15-25.

Abstract: This article examines some of the story conventions of network television news to explain the ways in which healthcare interest groups develop and maintain their presence in this medium - a process that has significant implications for public understanding of healthcare issues, and therefore to bioethics. The article is divided into three sections. The first section focuses on three major normative conventions of television news: adherence to a simple narrative structure, the balance ethic, and avoidance of the "think-piece" and outlines the basic strategies available to interest groups for exploiting these normative conventions. Section two introduces three case studies of organizations and individuals who have run high-profile media campaigns. Section three explores the implications for bioethics of the observations made in this article.

Caulfield, Timothy. The media, marketing, and genetic services. *In:* Flood, Colleen M., ed. Just Medicare: What's In, What's Out, How We Decide. Buffalo, NY: University of Toronto Press, 2006: 379-395. 55 fn. NRCBL: 1.3.7; 1.3.9; 15.1.

Keywords: *genetic research; *genetic services; *marketing; *mass media; biomedical research; biotechnology; government financing; industry; journalism; policy making; public participation; research priorities; research support; researchers; science; technology assessment; Keyword Identifiers: Canada

Chambers, Tod; Braun, Joshua A. It's narrative all the way down. *American Journal of Bioethics* 2007 August; 7(8): 15-16; author reply W1-W2. NRCBL: 1.3.7; 2.1. Comments: Joshua A. Braun. The imperatives of narrative: health interest groups and morality in network news. American Journal of Bioethics 2007 August; 7(8): 6-14.

de Melo-Martín, Immaculada; Intemann, Kristen. Authors' financial interests should be made known to manuscript reviewers. *Nature* 2007 July 12; 448(7150): 129. NRCBL: 1.3.7; 1.3.9; 7.3.

Derbyshire, Stuart W.G. Medical journals: past their sell by date? *BMJ: British Medical Journal* 2007 January 6; 334(7583): 45. NRCBL: 1.3.7; 1.3.9; 9.7.

Friedman, Lee S.; Richter, Elihu D. Excessive and disproportionate advertising in peer-reviewed journals. *International Journal of Occupational and Environmental Health : Official Journal of the International Commission on Occupational Health* 2006 January-March; 12(1): 59-64. NRCBL: 1.3.7; 1.3.2; 1.3.9.

NRCBL: National Reference Center for Bioethics Literature Classification Scheme   See inside front cover for terms.

317

Godlee, Fiona. Ethical assets at the BMJ [editorial]. *BMJ: British Medical Journal* 2007 February 24; 334(7590): 374. NRCBL: 1.3.7; 1.3.9.

Goldacre, Ben. Why don't journalists mention the data? *BMJ: British Medical Journal* 2007 June 16; 334(7606): 1249. NRCBL: 1.3.7; 1.3.9; 18.1.

Gornall, Jonathan. Duplicate publication: a bitter dispute. *BMJ: British Medial Journal* 2007 April 7; 334(7596): 717-720. NRCBL: 1.3.7; 1.3.9; 14.1. Identifiers: Korea.

Grieger, Maria Christina Anna. Authorship: an ethical dilemma of science. *Sao Paulo Medical Journal = Revista Paulista de Medicina* 2005 September 1; 123(5): 242-246. NRCBL: 1.3.7; 1.3.9. SC: rv.

Holaday, Margot; Yost, Tracey E. Authorship credit and ethical guidelines. *Counseling and Values* 1995 October; 40(1): 24-31. NRCBL: 1.3.7; 1.3.9; 7.3. SC: rv.

Hong, Mi-Kung; Bero, Lisa A. Tobacco industry sponsorship of a book and conflict of interest. *Addiction* 2006 August; 101(8): 1202-1211. NRCBL: 1.3.7; 1.3.2; 9.3.1; 1.3.7.

Hren, Darko; Sambunjak, Dario; Ivanis, Ana; Marusic, Matko; Marusic, Ana. Perceptions of authorship criteria: effects of student instruction and scientific experience. *Journal of Medical Ethics* 2007 July; 33(7): 428-432. NRCBL: 1.3.7; 1.3.9. SC: em.
Abstract: Objective: To analyse medical students', graduate students' and doctors' and medical teachers' perceptions of research contributions as criteria for authorship in relation to the authorship criteria defined by the International Committee of Medical Journal Editors (ICMJE). Design: Medical students with (n = 152) or without (n = 85) prior instruction on ICMJE criteria, graduate students/doctors (n = 125) and medical teachers (n = 112) rated the importance of 11 contributions as authorship qualifications. They also reported single contributions eligible for authorship, as well as acceptable combinations of two or three qualifying contributions. Results: Conception and design, Analysis and interpretation and Drafting of article formed the most important cluster in all four groups. Students without prior instruction rated Critical revision and Final approval lower than the other three groups. "Final approval" was a part of the least important cluster in all groups except among students with instruction. Conclusions: Conception and design, Analysis and interpretation and Drafting of article were recognised as the most important of the ICMJE criteria by all participants. They can be considered independent of previous instruction or experience. Final approval and Critical revision should be actively taught as important authorship criteria to future scientists.

Jayaraman, K.S. Indian scientists battle journal retraction [news]. *Nature* 2007 June 14; 447(7146): 764. NRCBL: 1.3.7; 1.3.9.

Johnson, Claire. Repetitive, duplicate, and redundant publications: a review for authors and readers. *Journal of Manipulative and Physiological Therapies* 2006 September; 29(7): 505-509. NRCBL: 1.3.7.

Johnson, Jonas T.; Niparko, John K.; Levine, Paul A.; Kennedy, David W.; Rudy, Susan F.; Weber, Pete; Weber, Randal S.; Benninger, Michael S.; Rosenfeld, Richard M.; Ruben, Robert J.; Smith, Richard J.H.; Sataloff, Robert Thayer; Weir, Neil. Standards for ethical publication. *American Journal of Otolaryngology* 2007 January-February; 28(1): 1-2. NRCBL: 1.3.7; 1.3.9.

Johnson, Jonas T.; Niparko, John K.; Levine, Paul A.; Kennedy, David W.; Rudy, Susan F.; Weber, Peter C.; Weber, Randal S.; Benniger, Michael S.; Rosenfeld, Richard J.; Ruben, Robert J.; Smith, Richard J.H.; Sataloff, Robert Thayer; Weir, Neil. Standards for ethical publication. *Ear, Nose, and Throat Journal* 2006 December; 85(12): 792, 795. NRCBL: 1.3.7; 1.3.9.

Kaebnick, Gregory E. What should HCR publish? [editorial]. *Hastings Center Report* 2007 November-December; 37(6): 2. NRCBL: 1.3.7; 2.1.

Kassirer, Jerome P. Assault on editorial independence: improprieties of the Canadian Medical Association [editorial]. *Journal of Medical Ethics* 2007 February; 33(2): 63-66. NRCBL: 1.3.7.

Katavic, Vedran. Five-year report of Croatian Medical Journal's research integrity editor — policy, policing, or policing policy. *Croatian Medical Journal* 2006 April; 47(2): 220-227. NRCBL: 1.3.7; 1.3.9.

Kenny, Nuala P.; McMahon, Meghan; Flood, Colleen M.; Braun, Joshua A. Canadian media and health policy research: the limits of stories. *American Journal of Bioethics* 2007 August; 7(8): 19-21; author reply W1-W2. NRCBL: 1.3.7; 9.1. Comments: Joshua A. Braun. The imperatives of narrative: health interest groups and morality in network news. American Journal of Bioethics 2007 August; 7(8): 6-14.

Klugman, Craig M.; Braun, Joshua A. Buying the fourth estate. *American Journal of Bioethics* 2007 August; 7(8): 16-18; author reply W1-W2. NRCBL: 1.3.7; 2.1. Comments: Joshua A. Braun. The imperatives of narrative: health interest groups and morality in network news. American Journal of Bioethics 2007 August; 7(8): 6-14.

Koplewicz, Harold S. Conflict of interest in the eyes of the beholder. *Journal of Child and Adolescent Psychopharmacology* 2006 October; 16(5): 511-512. NRCBL: 1.3.7; 9.5.5; 17.4; 9.3.1; 9.7.

Light, Donald W. Review of Richard Smith, The Trouble with Medical Journals [book review]. *American Journal of Bioethics* 2007 May; 7(5): 61-63. NRCBL: 1.3.7; 1.3.12. SC: rv.

Luft, Harold S.; Flood, Ann Barry; Escarce, José J. New policy on disclosures at Health Services Research. *Health Services Research* 2006 October; 41(5): 1721-1732. NRCBL: 1.3.7; 7.3; 9.7; 1.3.2; 9.3.1.

Lumley, Judith; Daly, Jeanne. Authors' misconduct in the firing line. *Australian and New Zealand Journal of Public Health* 2006 October; 30(5): 403-404. NRCBL: 1.3.7; 7.3; 1.3.9.

Marusic, Ana; Bates, Tamara; Anic, Ante; Marusic, Matko. How the structure of contribution disclosure statements affects validity of authorship: a randomized study in a general medical journal. *Current Medical Research and Opinion* 2006 June; 22(6): 1035-1044. NRCBL: 1.3.7; 1.3.9; 8.2. SC: em.

McKneally, Martin. Put my name on that paper: reflections on the ethics of authorship [editorial]. *Journal of Thoracic and Cardiovascular Surgery* 2006 March; 131(3): 517-519. NRCBL: 1.3.7; 1.3.9; 8.2.

Medical Research Council [MRC] (Great Britain). MRC guidance on open access to published research. London: Medical Research Council, 2006 October 1; 2 p. [Online]. Accessed: http://www.mrc.ac.uk/consumption/groups/public/documents/content/mrc002548.pdf [2007 April 4]. NRCBL: 1.3.7; 1.3.9; 1.3.12. Identifiers: Great Britain (United Kingdom).

Moffatt, Barton; Elliott, Carl. Ghost marketing: pharmaceutical companies and ghostwritten journal articles. *Perspectives in Biology and Medicine* 2007 Winter; 50(1): 18-31. NRCBL: 1.3.7; 9.7.

Möller, Hans-Jürgen. Ethical aspects of publishing [editorial]. *World Journal of Biological Psychiatry* 2006; 7(2): 66-69. NRCBL: 1.3.7; 1.3.9; 7.4.

Monaghan, Peter. Panel warns psychological journals about corporate influence. *Chronicle of Higher Education* 2007 December 21; 54(17): A12. NRCBL: 1.3.7; 9.7.

Naparstek, Yaakov. Ariel Sharon's illness: should we dedicate a medical journal issue to a single case study? *Israel Medical Association Journal* 2006 November; 8(11): 739-740. NRCBL: 1.3.7; 9.5.1; 9.4; 21.1. SC: cs.

O'Malley, Patricia. Pharmaceutical advertising and clinical nurse specialist practice. *Clinical Nurse Specialist CNS.* 2006 January-February; 20(1): 13-15. NRCBL: 1.3.7; 9.7; 8.1; 7.1.

Parsi, Kayhan; Braun, Joshua A. Media and health: are bioethicists just another interest group? *American Journal of Bioethics* 2007 August; 7(8): 18-19; author reply W1-W2. NRCBL: 1.3.7; 2.1. Comments: Joshua A. Braun. The imperatives of narrative: health interest groups and morality in network news. American Journal of Bioethics 2007 August; 7(8): 6-14.

Price, Connie C.; Braun, Joshua A. Cinematic thinking: narratives and bioethics unbound. *American Journal of Bioethics* 2007 August; 7(8): 21-23; author reply W1-W2. NRCBL: 1.3.7; 2.1. Comments: Joshua A. Braun. The imperatives of narrative: health interest groups and morality in network news. American Journal of Bioethics 2007 August; 7(8): 6-14.

Prideaux, David; Rogers, Wendy. Audit or research: the ethics of publication. *Medical Education* 2006 June; 40(6): 497-499. NRCBL: 1.3.7; 18.2; 18.6.

Probst, Janice C. Prisoners' dilemma: the importance of negative results. *Family Medicine* 2006 November-December; 38(10): 742-743. NRCBL: 1.3.7; 7.1.

Saver, Cynthia. Legal and ethical aspects of publishing. *AORN Journal* 2006 October; 84(4): 571-5 [see correction in: AORN Journal 2006 December; 84(6): 950]. NRCBL: 1.3.7.

Schüklenk, Udo. More on publication ethics [editorial]. *Bioethics* 2007 March; 21(3): ii. NRCBL: 1.3.7.

Schultz, Heather Yarnall; Blalock, Elisabeth. Transparency is the key to the relationship between biomedical journals and medical writers. *Journal of Investigative Dermatology* 2007 April; 127(4): 735-737. NRCBL: 1.3.7; 1.3.9; 8.2.

Smith, Richard. Lapses at the New England Journal of Medicine. *Journal of the Royal Society of Medicine* 2006 August; 99(8): 380-382. NRCBL: 1.3.7; 9.7; 18.2.

Smith, Richard. The highly profitable but unethical business of publishing medical research. *Journal of Royal Society of Medicine* 2006 September; 99(9): 452-456. NRCBL: 1.3.7; 1.3.9; 9.3.1; 1.3.2.

Smith, Richard; Williams, Gareth. Should medical journals carry drug advertising? [debate]. *BMJ: British Medical Journal* 2007 July 14; 335(7610): 74-75. NRCBL: 1.3.7; 9.7; 1.3.2.

Stobbart, L.; et al. "We saw human guinea pigs explode". *BMJ: British Medical Journal* 2007 March 17; 334(7593): 566-567. NRCBL: 1.3.7; 1.3.9; 18.1. Identifiers: Great Britain (United Kingdom); TGN1412 trial.

Turpin, David L. Policies for biomedical journals address ethics, confidentiality, and corrections. *American Journal of Orthodontics and Dentofacial Orthopedics* 2006 December; 130(6): 693-695. NRCBL: 1.3.7; 8.4; 7.1.

Upshur, Ross; Buetow, Stephen; Loughlin,Michael; Miles, Andrew. Can academic and clinical journals be in financial conflict of interest situations? The case of evidence-based incorporated. *Journal of Evaluation in Clinical Practice* 2006 August; 12(4): 405-409. NRCBL: 1.3.7; 1.3.9; 1.3.2; 5.3; 7.3.

NRCBL: National Reference Center for Bioethics Literature Classification Scheme   See inside front cover for terms.

319

Vaithianathan, Rhema. Better the devil you know than the doctor you don't: is advertising drugs to doctors more harmful than advertising to patients? *Journal of Health Services Research and Policy* 2006 October; 11(4): 235-239. NRCBL: 1.3.7; 7.1; 1.3.2; 9.7.

Woolley, Karen L. Goodbye Ghostwriters!: How to work ethically and efficiently with professional medical writers. *Chest* 2006 September; 130(3): 921-923. NRCBL: 1.3.7; 7.1; 7.3.

Yoshikawa, Thomas T.; Ouslander, Joseph G. Integrity in publishing: update on policies and statements on authorship, duplicate publications and conflict of interest [editorial]. *Journal of the American Geriatrics Society* 2007 February; 55(2): 155-157. NRCBL: 1.3.7; 1.3.9.

JUSTICE *See* RESOURCE ALLOCATION; RIGHT TO HEALTH CARE

LEGAL ASPECTS *See* ABORTION/ LEGAL ASPECTS; AIDS/ LEGAL ASPECTS; BIOETHICS AND MEDICAL ETHICS/ LEGAL ASPECTS; CLONING/ LEGAL ASPECTS; EUTHANASIA AND ALLOWING TO DIE/ LEGAL ASPECTS; GENETIC SCREENING/ LEGAL ASPECTS; HUMAN EXPERIMENTATION/ ETHICS COMMITTEES AND POLICY GUIDELINES/ LEGAL ASPECTS; HUMAN EXPERIMENTATION/ SPECIAL POPULATIONS/ EMBRYOS AND FETUSES/ LEGAL ASPECTS; ORGAN AND TISSUE TRANSPLANTATION/ DONATION AND PROCUREMENT/ LEGAL ASPECTS

LIVING WILLS *See* ADVANCE DIRECTIVES

MALPRACTICE AND PROFESSIONAL MISCONDUCT
See also BIOMEDICAL RESEARCH/ RESEARCH ETHICS AND SCIENTIFIC MISCONDUCT

Barber, Christopher F. Abuse by care professionals. Part 1: an introduction. *British Journal of Nursing* 2007 August 9 - September 12; 16(15): 938-940. NRCBL: 7.4; 21.1.

Barber, Christopher F. Abuse by care professionals. Part 2: a behavioural assessment. *British Journal of Nursing* 2007 September 13-27; 16(16): 1023-1025. NRCBL: 7.4.

Cuperus-Bosma, Jacquelyne M.; Hout, Fredericus A.G.; Hubben, Joep H.; van der Wal, Gerrit. Views of physicians, disciplinary board members and practicing lawyers on the new statutory disciplinary system for health care in the Netherlands. *Health Policy* 2006 July; 77(2): 202-211. NRCBL: 7.4; 7.3; 7.1. SC: em; le.

Draper, Heather. Paternity fraud and compensation for misattributed paternity. *Journal of Medical Ethics* 2007 August; 33(8): 475-480. 23 refs. NRCBL: 8.5; 14.1; 15.3; 15.1; 1.3.5. SC: an.
Keywords: *compensation; *deception; *DNA fingerprinting; *fathers; *fraud; *paternity; autonomy; children; financial support; genetic relatedness ties; legal aspects; marital relationship; mothers
Abstract: Claims for reimbursement of child support, the reversal of property settlements and compensation can arise when misattributed paternity is discovered. The ethical justifications for such claims seem to be related to the financial cost of bringing up children, the absence of choice about taking on these expenses, the hard work involved in child rearing, the emotional attachments that are formed with children, the obligation of women to make truthful claims about paternity, and the deception involved in infidelity. In this paper it is argued that there should not be compensation for infidelity and that reimbursement is appropriate where the claimant has made child support payments but has not taken on the social role of father. Where the claimant's behaviour suggests a social view of fatherhood, on the other hand, claims for compensation are less coherent. Where the genetic model of fatherhood dominates, the "other" man (the woman's lover and progenitor of the children) might also have a claim for the loss of the benefits of fatherhood. It is concluded that claims for reimbursement and compensation in cases of misattributed paternity produce the same distorted and thin view of what it means to be a father that paternity testing assumes, and underscores a trend that is not in the interests of children.

Dyer, Clare. Doctors lose power to run their profession [news]. *BMJ: British Medical Journal* 2007 March 3; 334(7591): 441. NRCBL: 7.4; 9.8; 4.1.3.

Dyer, Clare. GMC to introduce "plea bargaining" for less serious misconduct cases. *BMJ: British Medical Journal* 2007 April 14; 334(7597): 763. NRCBL: 7.4. Identifiers: Great Britain (United Kingdom); General Medical Council.

Dyer, Owen. GMC clears GP accused of giving "junk science" evidence [news]. *BMJ: British Medical Journal* 2007 September 1; 335(7617): 416-417. NRCBL: 7.4. SC: le. Identifiers: Great Britain (United Kingdom); General Medical Council.

Fan, Ruiping. Corrupt practices in Chinese medical care: the root in public policies and a call for Confucian-market approach. *Kennedy Institute of Ethics Journal* 2007 June; 17(2): 111-131. NRCBL: 7.4; 1.1; 1.3.2; 2.1. SC: an.
Abstract: This paper argues that three salient corrupt practices that mark contemporary Chinese health care, namely the over-prescription of indicated drugs, the prescription of more expensive forms of medication and more expensive diagnostic work-ups than needed, and illegal cash payments to physicians-i.e., red packages-result not from the introduction of the market to China, but from two clusters of circumstances. First, there has been a loss of the Confucian appreciation of the proper role of financial reward for good health care. Second, misguided governmental policies have distorted the behavior of

physicians and hospitals. The distorting policies include (1) setting very low salaries for physicians, (2) providing bonuses to physicians and profits to hospitals from the excessive prescription of drugs and the use of more expensive drugs and unnecessary expensive diagnostic procedures, and (3) prohibiting payments by patients to physicians for higher quality care. The latter problem is complicated by policies that do not allow the use of governmental insurance and funds from medical savings accounts in private hospitals as well as other policies that fail to create a level playing field for both private and government hospitals. The corrupt practices currently characterizing Chinese health care will require not only abolishing the distorting governmental policies but also drawing on Confucian moral resources to establish a rightly directed appreciation of the proper place of financial reward in the practice of medicine.

Johnson, Edward L., Jr.; Johnson, Larry W. Conversations in ethics. *JONA's Healthcare Law, Ethics, and Regulation* 2007 October-December; 9(4): 117-118. NRCBL: 7.4; 9.5.9; 8.4; 7.3. SC: cs.

Jones, James W.; McCullough, Laurence B. Ethics of unprofessional behavior that disrupts: crossing the line. *Journal of Vascular Surgery* 2007 February; 45(2): 433-435. NRCBL: 7.4; 7.3. SC: cs.

Kassirer, Jerome P. Professional societies and industry supports: what is the quid pro quo? *Perspectives in Biology and Medicine* 2007 Winter; 50(1): 7-17. NRCBL: 7.4; 9.7; 9.3.1.

Magnavita, Nicola. The unhealthy physician. *Journal of Medical Ethics* 2007 April; 33(4): 210-214. NRCBL: 7.4; 16.3; 9.1; 8.1.
  Abstract: BACKGROUND: Physicians, if affected by transmissible or impairing diseases, could be hazardous for third persons. AIM: To solve the apparent chasm between patient's and sick worker's rights, a consensus-building process leading to hospital-wide policies is the better alternative to individual decision making. CONCLUSIONS: Policies have to balance the rights of the sick worker, the right of the other workers, patients and customers, and society's expectations.

Montrose, J.L. Is negligence an ethical or a sociological concept? *Modern Law Review* 1958 May; 21(3): 259-264. NRCBL: 8.5; 7.1. SC: le. Identifiers: Great Britain (United Kingdom).

Siegel-Itzkovich, Judy. Doctor's licence suspended after he admitted removing hundreds of ova without consent [news]. *BMJ: British Medical Journal* 2007 March 17; 334(7593): 557. NRCBL: 7.4; 14.4; 19.3; 8.3.1; 14.6. SC: le. Identifiers: Israel.

Webster, Paul. Canadian soldiers and doctors face torture allegations. *Lancet* 2007 April 28 - May 4; 369(9571): 1419-1420. NRCBL: 7.4; 21.4.

Yacoub, Ahmed Abdel Aziz. The rights and responsibilities of patients and those who treat them. *In his:* The Fiqh of Medicine: Responses in Islamic Jurisprudence to Developments in Medical Science. London, UK: Ta-Ha Publishers Ltd., 2001: 100-126. NRCBL: 8.5; 1.2. SC: le.

MANAGED CARE PROGRAMS *See* HEALTH CARE/ HEALTH CARE ECONOMICS/ MANAGED CARE PROGRAMS

MASS SCREENING *See* PUBLIC HEALTH

MEDICAL EDUCATION
See also BIOETHICS AND MEDICAL ETHICS/ EDUCATION

Educational advantage. *Journal of Empirical Research on Human Research Ethics* 2007 March; 2(1): 47-48. NRCBL: 7.2; 18.2; 18.6; 1.3.1; 8.4. SC: le.
  Abstract: The research articles in the March 2007 issue of JERHRE explore two major topics: • Research methods wherein the investigator does not have unilateral control over the setting, as in ethnography and community-based participatory research, raise special problems of ethical oversight and problem solving. • Introducing new ethical oversight, whether developing an effective ethics committee in a developing country or implementing HIPAA requirements in human research, call for mindfulness of ethical objectives rather than simple rule following. JERHRE has an advantage in the ethics education arena. Lecturing about what should be done is an ineffective way to change people's hearts and minds, much less their behavior. In contrast, JERHRE provides concepts and methods that learners can use to discover for themselves what should be done. In the process, learners discover that what they should do is synonymous with what is in their best interests. Such is the persuasive power of evidence-based ethical problem solving.

Austin, Wendy. The Terminal: a tale of virtue. *Nursing Ethics* 2007 January; 14(1): 54-61. NRCBL: 7.2; 1.1.
  Abstract: The movie, The Terminal, is used to illustrate MacIntyre's description of virtue ethics. The terminal is a mythical tale about a traveler, Viktor Navorski, who is stranded by circumstances in a New York airport. Viktor is a person who, without a strict reliance on duty or rules, has developed the disposition to act well despite variation in his circumstances. His character is revealed in contrast to that of three other characters: a cleaner, a flight attendant and the airport manager. Stories like this one may be a good way to open dialogue among clinicians about being virtuous as a practitioner. Such dialogue may make striving to be virtuous an acceptable goal for practitioners and less like an idealistic, pseudo-goal for those aiming for sainthood.

Austin, Zubin; Collins, David; Remillard, Alfred; Kelcher, Sheila; Chui, Stephanie. Influence of attitudes toward curriculum on dishonest academic behavior. *American Journal of Pharmaceutical Education* 2006 June 15; 70(3): 50. NRCBL: 7.2; 7.4; 9.7.

NRCBL: National Reference Center for Bioethics Literature Classification Scheme   See inside front cover for terms.

321

Baxter, Pamela E.; Boblin, Sheryl L. The moral development of baccalaureate nursing students: understanding unethical behavior in classroom and clinical settings. *Journal of Nursing Education* 2007 January; 46(1): 20-27. NRCBL: 7.2; 1.3.1; 4.1.3.

Beckett, Alesha; Gilbertson, Sarah; Greenwood, Sallie. Doing the right thing: nursing students, relational practice, and moral agency. *Journal of Nursing Education* 2007 January; 46(1): 28-32. NRCBL: 7.2; 4.1.3.

Begley, Ann M. Creative approaches to ethics: poetry, prose and dialogue. *In:* Tschudin, Verena, ed. Approaches to Ethics: Nursing Beyond Boundaries. New York: Butterworth-Heinemann, 2003: 125-135. NRCBL: 7.2; 2.1; 7.1.

Bercovitch, Lionel; Long, Thomas P. Dermatoethics: a curriculum in bioethics and professionalism for dermatology residents at Brown Medical School. *Journal of the American Academy of Dermatology* 2007 April; 56(4): 679-682. NRCBL: 7.2; 2.3.

Boyd, J. Wesley; Himmelstein, David U.; Lasser, Karen; McCormick, Danny; Bor, David H.; Cutrona, Sarah L.; Woolhandler, Steffie. U.S. medical students' knowledge about the military draft, the Geneva Conventions, and military medical ethics. *International Journal of Health Services* 2007; 37(4): 643-650. NRCBL: 7.2; 1.3.5; 21.1; 21.4. SC: em.

Buchanan, David; Witlen, Renee. Balancing service and education: ethical management of student-run clinics. *Journal of Health Care for the Poor and Underserved* 2006 August; 17(3): 477-485. NRCBL: 7.2; 2.3; 9.5.10; 9.8.

Buchner, Benedikt. Industry-sponsored medical education — in the quest for professional integrity and legal certainty. *European Journal of Health Law* 2007 December; 14(4): 313-319. NRCBL: 7.2; 7.3. SC: le.

Carline, Jan D.; O'Sullivan, Patricia S.; Gruppen, Larry D.; Richardson-Nassif, Karen. Crafting successful relationships with the IRB. *Academic Medicine* 2007 October; 82(10, Supplement): S57-S60. NRCBL: 7.2; 18.2. SC: em.

Carpenter, Robert O.; Spooner, John; Arbogast, Patrick G.; Tarpley, John L.; Griffin, Marie R.; Lomis, Kimberly D. Work hours restrictions as an ethical dilemma for residents: a descriptive survey of violation types and frequency. *Current Surgery* 2006 November-December; 63(6): 448-455. NRCBL: 7.2; 7.1; 16.3. SC: em.

Casada, Jane P.; Willis, David O.; Butters, Janice M. An investigation of dental student values. *Journal of the American College of Dentists* 1998 Fall; 65(3): 36-40. NRCBL: 7.2; 4.1.1; 1.3.1.

Chen, Daniel; Lew, Robert; Hershman, Warren; Orlander, Jay. A cross-sectional measurement of medical student empathy. *JGIM: Journal of General Internal Medicine* 2007 October; 22(10): 1434-1438. NRCBL: 7.2; 8.1. SC: em.

Chiaramonte, Gabrielle R.; Friend, Ronald. Medical students' and residents' gender bias in the diagnosis, treatment, and interpretation of coronary heart disease symptoms. *Health Psychology* 2006 May; 25(3): 255-266. NRCBL: 7.2; 9.5.5; 10; 8.1.

Chiong, Winston. Justifying patient risks associated with medical education. *JAMA: The Journal of the American Medical Association* 2007 September 5; 298(9): 1046-1048. NRCBL: 7.2; 8.1; 9.1; 9.8.

Cooper, Richard A.; Tauber, Alfred I. Values and ethics: a collection of curricular reforms for a new generation of physicians. *Academic Medicine* 2007 April; 82(4): 321-323. NRCBL: 7.2; 2.1; 2.3; 4.1.2; 1.3.1.

DeRosa, G. Paul. Professionalism and virtues. *Clinical Orthopaedics and Related Research* 2006 August; 449: 28-33. NRCBL: 7.2; 2.3; 1.1; 4.1.2.

Doukas, David J. The medical-social education compact and the medical learner. *In:* Kenny, Nuala; Shelton, Wayne, eds. Lost Virtue: Professional Character Development in Medical Education. Amsterdam; Oxford: Elsevier, 2006: 185-209. NRCBL: 7.2; 4.1.2; 1.3.1; 6.

Elcin, Melih; Odabasi, Orhan; Gokler, Bahar; Sayek, Isender; Akova, Murat; Kiper, Nural. Developing and evaluating professionalism. *Medical Teacher* 2006 February; 28(1): 36-39. NRCBL: 7.2; 4.1.2; 1.3.1.

Fiester, Autumn. Why the clinical ethics we teach fails patients. *Academic Medicine* 2007 July; 82(7): 684-689. NRCBL: 7.2; 2.1; 1.1.

FIGO Committee for the Ethical Aspects of Human Reproduction and Women's Health. Ethical issues in medical education: gifts and obligations. *International Journal of Gynecology and Obstetrics* 2006 May; 93(2): 189-190. NRCBL: 7.2; 6.

Finkel, Alan G. Conflict of interest or productive collaboration? The pharma: academic relationship and its implications for headache medicine. *Headache* 2006 July-August; 46(7): 1181-1185. NRCBL: 7.2; 9.7; 1.3.2; 9.3.1.

Fitz, Matthew M.; Homan, David; Reddy, Shalini; Griffith, Charles H.; Baker, Elizabeth; Simpson, Kevin P. The hidden curriculum: medical students' changing opinions toward the pharmaceutical industry. *Academic Medicine* 2007 October; 82(10, Supplement): S1-S3. NRCBL: 7.2; 9.7; 1.3.2; 9.3.1. SC: em.

Frohna, Alice. Medical students' professionalism. *Medical Teacher* 2006 February; 28(1): 1-2. NRCBL: 7.2; 4.1.2; 1.3.1.

Gallagher, Ann. The teaching of nursing ethics: content and method. Promoting ethical competence. *In:* Davis, Anne J.; Tschudin, Verena; de Raeve, Louise, eds. Essentials of Teaching and Learning in Nursing Ethics: Perspectives and Methods. New York: Churchill Livingstone Elsevier, 2006: 223-239. NRCBL: 7.2; 1.1; 4.1.3. SC: an.

Garetto, Lawrence P.; Senour, Wendy. Using an ethics across the curriculum strategy in dental education. *Journal of the American College of Dentists* 2006 Winter; 73(4): 33-35. NRCBL: 7.2; 4.1.1; 1.3.1. Identifiers: Indiana University School of Dentistry.

Goodman, Robert L. Medical education and the pharmaceutical industry. *Perspectives in Biology and Medicine* 2007 Winter; 50(1): 32-39. NRCBL: 7.2; 9.7; 1.3.3.

Görgülü, Refia Selma; Dinç, Leyla. Ethics in Turkish nursing education programs. *Nursing Ethics* 2007 November; 14(6): 741-752. NRCBL: 7.2; 2.3; 4.1.3. SC: em.

Graham, Bruce S. Educating dental students about oral health care access disparities. *Journal of Dental Education* 2006 November; 70(11): 1208-1211. NRCBL: 7.2; 9.2; 4.1.1.

Green, Stephen A. The ethical commitments of academic faculty in psychiatric education. *Academic Psychiatry* 2006 January-February; 30(1): 48-54 [Online]. Accessed: http://ap.psychiatryonline.org/ [2007 April 13]. NRCBL: 7.2; 17.2; 4.1.2.

Hamilton, Patricia. Ethical dilemmas in training tomorrow's doctors. *Paediatric Respiratory Reviews* 2006 June; 7(2): 129-134. NRCBL: 7.2; 8.1; 9.5.7; 9.8.

Iramaneerat, Cherdsak. Moral education in medical schools. *Journal of the Medical Association of Thailand* 2006 November; 89(11): 1987-1993. NRCBL: 7.2; 4.1.2; 1.1.

Ives, J. Kant, curves and medical learning practice: a reply to Le Morvan and Stock. *Journal of Medical Ethics* 2007 February; 33(2): 119-122. NRCBL: 7.2; 8.1; 1.1. SC: an. Comments: P. Le Morvan and B. Stock. Medical learning curves and the Kantian ideal. Journal of Medical Ethics 2005 September; 31(9): 513-518.
Abstract: In a recent paper published in the Journal of Medical Ethics, Le Morvan and Stock claim that the kantian ideal of treating people always as ends in themselves and never merely as a means is in direct and insurmountable conflict with the current medical practice of allowing practitioners at the bottom of their "learning curve" to "practise their skills" on patients. In this response, I take up the challenge they issue is and try to reconcile this conflict. The kantian ideal offered in the paper is an incomplete characterisation of Kant's moral philosophy, and the formula of humanity is considered in isolation without taking into account other salient kantian principles. I also suggest that their argument based on "necessary for the patient" assumes too narrow a reading

of "necessary". This reply is intended as an extension to, rather than a criticism of, their work.

Jakusovaite, Irayda; Bankauskaite, Vaida. Teaching ethics in a masters program in public health in Lithuania. *Journal of Medical Ethics* 2007 July; 33(7): 423-432. NRCBL: 7.2; 2.3; 9.1.
Abstract: This article aims to present 10 years of experience of teaching ethics in a Masters Program in Public Health in Lithuania, and to discuss the content, skills, teaching approach and tools of this programme. In addition, the article analyses the links between ethics and law, identifies the challenges of the teaching process and suggests future teaching strategies. The important role of teaching ethics in countries that are in transition owing to a radically changing value system is emphasised.

Johnston, Carolyn; Haughton, Peter. Medical students' perceptions of their ethics teaching. *Journal of Medical Ethics* 2007 July; 33(7): 418-422. NRCBL: 7.2; 2.3. SC: em.
Abstract: The teaching of ethics in UK medical schools has recently been reviewed, from the perspective of the teachers themselves. A questionnaire survey of medical undergraduates at King's College London School of Medicine provides useful insight into the students' perception of ethics education, what they consider to be the value of learning ethics and law, and how engaged they feel with the subject.

Kanter, Steven L.; Wimmers, Paul F.; Levine, Arthur S. In-depth learning: one school's initiatives to foster integration of ethics, values, and the human dimensions of medicine. *Academic Medicine* 2007 April; 82(4): 405-409. NRCBL: 7.2; 2.1; 2.3; 4.1.2; 1.3.1.

Kim, Jung-Ran; Fisher, Murray J.; Elliott, Doug. Undergraduate nursing students' knowledge and attitudes towards organ donation in Korea: implications for education. *Nurse Education Today* 2006 August; 26(6): 465-474. NRCBL: 7.2; 19.5; 20.2.1; 7.1. SC: em.

Klingberg-Allvin, Marie; Van Tam, Vu; Nga, Nguyen Tha; Ransjo-Arvidson, Anna-Berit; Johansson, Annika. Ethics of justice and ethics of care. Values and attitudes among midwifery students on adolescent sexuality and abortion in Vietnam and their implications for midwifery education: a survey by questionnaire and interview. *International Journal of Nursing Studies* 2007 January; 44(1): 37-46. NRCBL: 7.2; 9.5.7; 10; 12.5.2.

Kon, Alexander A. Resident-generated versus instructor-generated cases in ethics and professionalism training. *Philosophy, Ethics and Humanities in Medicine [electronic]* 2006; 1(10): 6 p. Accessed: http://www.peh-med.com/content/1/1/10 [2006 August 15]. NRCBL: 7.2; 2.3. SC: em.

Langone, Melissa. Promoting integrity among nursing students. *Journal of Nursing Education* 2007 January; 46(1): 45-47. NRCBL: 7.2; 4.1.3.

Lantz, Cheryl M. Teaching spiritual care in a public institution: legal implications, standards of practice, and ethical obligations. *Journal of Nursing Education* 2007 January; 46(1): 33-38. NRCBL: 7.2; 1.2; 4.1.3. SC: le.

Leino-Kilpi, Helena. [Education in nursing ethics research] [editorial]. *Nursing Ethics* 2007 July; 14(4): 443-444. NRCBL: 7.2; 4.1.3.

Lim, E.-C.; Seet, R.C.S. Attitudes of medical students to placebo therapy. *Internal Medicine Journal* 2007 March; 37(3): 156-160. NRCBL: 7.2; 18.3; 8.2. SC: em.

Lipscomb, Martin; Snelling, Paul C. Moral content and assignment marking: an exploratory study. *Nurse Education Today* 2006 August; 26(6): 457-464. NRCBL: 7.2; 1.1; 4.1.3; 7.1; 7.3.

Lown, Beth A.; Chou, Calvin L.; Clark, William D.; Haidet, Paul; White, Maysel Kemp; Krupat, Edward; Pelletier, Stephen; Weissmann, Peter; Anderson, M. Brownell. Caring attitudes in medical education: perceptions of deans and curriculum leaders. *JGIM: Journal of General Internal Medicine* 2007 November; 22(11): 1514-1522. NRCBL: 7.2; 4.1.1; 4.1.2; 8.1. SC: em.

Mangan, Katherine. Medical schools stop using dogs and pigs in teaching: training of future doctors now largely depends on new technologies rather than lab animals. *Chronicle of Higher Education* 2007 October 12; 54(7): A12. NRCBL: 7.2; 22.2.

Nash, David A. "The profession of dentistry:" the University of Kentucky's curriculum in professional ethics. *Journal of the American College of Dentists* 1996 Spring; 63(1): 25-29. NRCBL: 7.2; 4.1.1; 1.3.1.

O'Donnell, Charlie. Medical training: cooperation problems and solutions. *In:* Watt, Helen, ed. Cooperation, Complicity and Conscience: Problems in Healthcare, Science, Law and Public Policy. London: Linacre Centre, 2005: 118-127. NRCBL: 7.2; 1.2.

Polifroni, E. Carol. Ethical knowing and nursing education [editorial]. *Journal of Nursing Education* 2007 January; 46(1): 3. NRCBL: 7.2; 4.1.3.

Rabi, Suzanne M.; Patton, Lynn R.; Fjortoft, Nancy; Zgarrick, David P. Characteristics, prevalence, attitudes, and perceptions of academic dishonesty among pharmacy students. *American Journal of Pharmaceutical Education* 2006 August 15; 70(4): 73. NRCBL: 7.2; 7.4; 9.7.

Resnick, Andrew S.; Mullen, James L.; Kaiser, Larry R.; Morris, Jon B. Patterns and predictions of resident misbehavior — a 10-year retrospective look. *Current Surgery* 2006 November-December; 63(6): 418-425. NRCBL: 7.2; 7.1; 7.3; 9.8. SC: em.

Schildmann, Jan; Steger, Florian; Vollmann, Jochen. „Aufklärung im ärztlichen Alltag" - ein Lehrmodul zur integrierten Bearbeitung medizinethischer und -historischer Aspekte im neuen Querschnittsbereich GTE = "Informed consent" - an integrated teaching module on ethical and historical aspects for the new subject "history, theory, ethics of medicine". *Ethik in der Medizin* 2007 September; 19(3): 187-199. NRCBL: 7.2; 8.3.1. SC: em.

Shapiro, Johanna; Rucker, Lloyd; Robitshek, Daniel. Teaching the art of doctoring: an innovative medical student elective. *Medical Teacher* 2006 February; 28(1): 30-35. NRCBL: 7.2; 4.1.2; 1.3.1.

Slováčková, Birgita; Slováček, Ladislav. Moral judgement competence and moral attitudes of medical students. *Nursing Ethics* 2007 May; 14(3): 320-328. NRCBL: 7.2; 1.3.1. SC: em. Identifiers: Czech Republic.
Abstract: A cross-sectional study explored the moral judgement competence and moral attitudes of 310 Czech and Slovak and 70 foreign national students at the Medical Faculty of Charles University in Hradec Králové, Czech Republic. Lind's Moral Judgement Test was used to evaluate moral judgement competence and moral attitudes depending on factors such as age, number of semesters of study, sex, nationality and religion. Moral judgement competence decreased significantly in the Czech and Slovak medical students as they grew older; in medical students from other countries it did not significantly increase. The influence of other factors (sex, nationality and religion) on moral judgement competence was not proven in either the Czech and Slovak or the foreign national medical students. Moral attitudes do not change; the Czech and Slovak as well as the foreign students preferred the post-conventional levels of moral judgement (Kohlberg's 5th and 6th stages). The fact that the Czech and Slovak students' moral judgement competence decreased with age and number of semesters of study completed is not an optimistic sign: medical students who had undergone a lower number of semesters of study were morally more competent.

Sporrong, Sofia Kälvemark; Arnetz, Bengt; Hansson, Mats G.; Westerholm, Peter; Höglund, Anna T. Developing ethical competence in health care organizations. *Nursing Ethics* 2007 November; 14(6): 825-837. NRCBL: 7.2; 4.1.1. SC: em. Identifiers: Sweden.

Suchman, Anthony L. Advancing humanism in medical education. *JGIM: Journal of General Internal Medicine* 2007 November; 22(11): 1630-1631. NRCBL: 7.2; 1.1; 7.1.

Talbott, John A.; Mallott, David B. Professionalism, medical humanism, and clinical bioethics: the new wave — does psychiatry have a role? *Journal of Psychiatric Practice* 2006 November; 12(6): 384-390. NRCBL: 7.2; 4.1.2; 1.3.1; 17.1; 7.1.

Toliušiene, Jolanta; Peicius, Eimantas. Changes in nursing ethics education in Lithuania. *Nursing Ethics* 2007 November; 14(6): 753-757. NRCBL: 7.2; 2.3; 4.1.3.

van Hooft, Stan. Socratic dialogue: an example. *In:* Tschudin, Verena, ed. Approaches to Ethics: Nursing Be-

---

yond Boundaries. New York: Butterworth-Heinemann, 2003: 115-123. NRCBL: 7.2; 1.1; 10.

Vanlaere, Linus; Gastmans, Chris. Ethics in nursing education: learning to reflect on care practices. *Nursing Ethics* 2007 November; 14(6): 758-766. NRCBL: 7.2; 2.3; 1.1; 4.1.3.

Veatch, Robert M. Character formation in professional education: a word of caution. *In:* Kenny, Nuala; Shelton, Wayne, eds. Lost Virtue: Professional Character Development in Medical Education. Amsterdam; Oxford: Elsevier, 2006: 29-45. NRCBL: 7.2; 1.1; 4.1.2. SC: an.

Vezeau, Toni M. Teaching professional values in a BSN program. *International Journal of Nursing Education Scholarship* 2006; 3: Article 25. NRCBL: 7.2; 4.1.3; 8.1.

Whiting, Demian. Inappropriate attitudes, fitness to practise and the challenges facing medical educators. *Journal of Medical Ethics* 2007 November; 33(11): 667-670. NRCBL: 7.2; 9.8; 2.3.
  Abstract: The author outlines a number of reasons why morally inappropriate attitudes may give rise to concerns about fitness to practise. He argues that inappropriate attitudes may raise such concerns because they can lead to harmful behaviours (such as a failure to give proper care or treatment), and because they are often themselves harmful (both because of the offence that they can cause and because of the unhealthy pall that they may cast over relations between healthcare practitioners and patients). He also outlines some of the challenges that the cultivation and assessment of attitudes in students raise for medical educators and some of the ways in which those challenges may be approached and possibly overcome.

Wiesing, Urban. Ethical aspects of limiting residents' work hours. *Bioethics* 2007 September; 21(7): 398-405. NRCBL: 7.2; 1.1; 7.1; 16.3. SC: an.
  Abstract: Definition of the problem: The regulation of residents' work hours involves several ethical conflicts which need to be systematically analysed and evaluated. Arguments and conclusion: The most important ethical principle when regulating work hours is to avoid the harm resulting from the over-work of physicians and from an excessive division of labour. Additionally, other ethical principles have to be taken into account, in particular the principles of nonmaleficence and beneficence for future patients and for physicians. The article presents arguments for balancing the relevant ethical principles and analyses the structural difficulties that occur unavoidably in any regulation of the complex activities of physicians.

Wolfberg, Adam J. The patient as ally — learning the pelvic examination. *New England Journal of Medicine* 2007 March 1; 356(9): 889-890. NRCBL: 7.2; 9.5.5; 8.3.1; 8.1.

Yarbrough, Susan; Klotz, Linda. Incorporating cultural issues in education for ethical practice. *Nursing Ethics* 2007 July; 14(4): 492-502. NRCBL: 7.2; 4.1.3; 21.7. SC: cs.

Abstract: The population of most non-dominant ethnic groups in the USA is growing dramatically. Faculty members are challenged to develop curricula that adequately prepare our future nurses. An increased focus on clinical ethics has resulted from the use of sophisticated technology, changes in health care financing, an increasing elderly population and the shift of care from inpatient to outpatient settings. Nurses frequently face situations demanding resolution of ethical dilemmas involving cultural differences. Nursing curricula must include content on both ethics and cultural sensitivity. Active student participation is an important element providing a foundation for ethical practice. A proposed educational format was introduced with graduating baccalaureate students. In a pilot study, curricular content on cultural sensitivity and ethical practice was taught in separate modules. Students were then asked to identify and problem solve an ethical dilemma involving patients and professional caregivers from vastly different cultures. Course faculty members provided discussion questions to guide the students' thinking.

MEDICAL ERRORS *See* HEALTH CARE/ HEALTH CARE QUALITY

MEDICAL ETHICS *See* BIOETHICS AND MEDICAL ETHICS

MENTAL HEALTH, CONCEPT OF
See also MENTAL HEALTH THERAPIES AND NEUROSCIENCES

Barry, Colleen L. The political evolution of mental health parity. *Harvard Review of Psychiatry* 2006 July-August; 14(4): 185-194. NRCBL: 4.3; 9.3.1; 1.3.2; 1.3.5.

Kennett, Jeanette. Mental disorder, moral agency and the self. *In:* Steinbock, Bonnie, ed. The Oxford Handbook of Bioethics. Oxford; New York: Oxford University Press, 2007: 90-113. NRCBL: 4.3; 1.1.

Ritchie, Karen; Freedman, Benjamin; Culver, Charles M.; Ferrell, Richard B.; Green, Ronald M. Competence and mental illness. *In:* Baylis, Françoise; Downie, Jocelyn; Freedman, Benjamin; Hoffmaster, Barry; Sherwin, Susan, eds. Health Care Ethics in Canada. Toronto: Harcourt Brace Canada, 1995: 231-273. NRCBL: 4.3; 8.3.3; 8.3.4; 17.5; 10. SC: le. Note: Section includes: Introduction — The little woman meets son of DSM-III / Karen Ritchie — Competence, marginal and otherwise: concepts and ethics / Benjamin Freedman — In the matter of The Mental Health Act, R.S.O. 1980, c. 262 as amended, and in the Matter of KV, a Patient at a Hospital in Ontario — ECT and special problems of informed consent / Charles M. Culver, Richard B. Ferrell and Ronald M. Green.

Rosen, Jeffrey. The brain on the stand: how neuroscience is transforming the legal system. *New York Times Magazine* 2007 March 11; p. 48-53, 70, 77, 82, 83. NRCBL: 4.3; 17.1; 1.3.8; 1.3.5. SC: po; le.

NRCBL: National Reference Center for Bioethics Literature Classification Scheme   See inside front cover for terms.

325

Wiggins, Osborne P.; Schwartz, Michael Alan. Philosophical issues in psychiatry. *In:* Khushf, George, ed. Handbook of Bioethics: Taking Stock of the Field From a Philosophical Perspective. Dordrecht; Boston: Kluwer Academic, 2004: 473-488. NRCBL: 4.3; 1.1; 17.1.

MENTAL HEALTH THERAPIES AND NEUROSCIENCES

See also BEHAVIOR CONTROL; CARE FOR SPECIFIC GROUPS/ MENTALLY DISABLED; ELECTROCONVULSIVE THERAPY; HUMAN EXPERIMENTATION/ SPECIAL POPULATIONS/ MENTALLY DISABLED; INVOLUNTARY COMMITMENT; MENTAL HEALTH, CONCEPT OF; PSYCHOPHARMACOLOGY; PSYCHOTHERAPY

Alpert, Sheri. Brain privacy: how can we protect it? *American Journal of Bioethics* 2007 September; 7(9): 70-73. 11 refs. NRCBL: 17.1; 8.4; 15.1. SC: le. Comments: Stacey A. Tovino. Functional neuroimaging and the law: trends and directions for future scholarship. American Journal of Bioethics 2007 September; 7(9): 44-56.
> Keywords: *access to information; *brain; *exceptionalism; *government regulation; *neurosciences; *privacy; disclosure; federal government; genetic information; legislation; presumed consent; state government; Proposed Keywords: *brain imaging; magnetic resonance imaging; Keyword Identifiers: *United States; Health Insurance Portability and Accountability Act (HIPAA)

Alpert, Sheri. Total information awareness: forgotten but not gone: lessons for neuroethics. *American Journal of Bioethics* 2007 May; 7(5): 24-26. NRCBL: 17.1; 1.3.12; 1.3.5. Identifiers: Defense Advance Projects Research Agency (DARPA). Comments: Turhan Canli, Susan Brandon, William Casebeer, Philip J. Crowley, Don DuRousseau, Henry T. Greely, and Alvaro Pascual-Leone. Neuroethics and national security. American Journal of Bioethics 2007 May; 7(5): 3-13.

Ani, Cornelius; Ani, Obeagaeli. Institutional racism: editorial is unduly provocative [letter]. *BMJ: British Medical Journal* 2007 April 14; 334(7597): 761. NRCBL: 17.1; 9.5.4. Comments: Kwame McKenzie and Kamaldeep Bhui. Institutional racism in mental health care. British Medical Journal 2007 March 31; 334(7595): 649-650.

Ashcroft, Richard; Campbell, Alastair V.; Capps, Ben. Ethical aspects of developments in neuroscience and drug addiction. London: Foresight Brain Science, Addiction and Drugs project, v. 1.0; n.d.: 65 p. [Online]. Accessed: http://www.foresight.gov.uk/Previous_Projects/Brain_Science_Addiction_and_Drugs/Reports_and_Publications/ScienceReviews/Ethics.pdf [2007 April 25]. Refs., p. 53-65. NRCBL: 17.1;15.1; 9.5.9.
> Keywords: *drug abuse; *neurosciences; autonomy; behavior control; behavioral genetics; biomedical enhancement; coercion; confidentiality; genetic engineering; genetic research; genetic screening; immunization; informed consent; preimplantation diagnosis; psychoactive drugs

Banja, John. Personhood: elusive but not illusory. *American Journal of Bioethics* 2007 January; 7(1): 60-62. NRCBL: 17.1; 4.4; 1.1. Comments: Martha J. Farah and Andrea S. Heberlein. Personhood and neuroscience: naturalizing or nihilating? American Journal of Bioethics 2007 January; 7(1): 37-48.

Bard, Jennifer S. Learning from law's past: a call for caution in incorporating new innovations in neuroscience. *American Journal of Bioethics* 2007 September; 7(9): 73-75. NRCBL: 17.1. SC: le. Comments: Stacey A. Tovino. Functional neuroimaging and the law: trends and directions for future scholarship. American Journal of Bioethics 2007 September; 7(9): 44-56.

Begley, Sharon. New ethical minefield: drugs to boost memory and sharpen attention. *Wall Street Journal* 2004 October 1: B1. NRCBL: 17.1; 4.5.

Biever, Celeste. Uproar flares over Alzheimer's tags [news]. *New Scientist* 2007 May 19-25; 194(2604): 14. NRCBL: 17.1; 9.5.2; 8.4.

Blackford, Russell. Differing vulnerabilities: the moral significance of Lockean personhood. *American Journal of Bioethics* 2007 January; 7(1): 70-71. NRCBL: 17.1; 4.4; 1.1. Comments: Martha J. Farah and Andrea S. Heberlein. Personhood and neuroscience: naturalizing or nihilating? American Journal of Bioethics 2007 January; 7(1): 37-48.

Blank, Robert H. Policy implications of the new neuroscience. *CQ: Cambridge Quarterly of Healthcare Ethics* 2007 Spring; 16(2): 169-180. NRCBL: 17.1; 4.4; 5.3; 17.3.

Bolton, Derek. What's the problem? A response to "secular humanism and scientific psychiatry". *Philosophy, Ethics, and Humanities in Medicine [electronic]* 2006; 1(6): 2 p. Accessed: http://www.peh-med.com/content/1/1/6 [2006 May 15]. NRCBL: 17.1; 7.1. Comments: Thomas Szasz. Secular humanism and "scientific psychiatry." Philosophy, Ethics, and Humanities [electronic] 2006; 1(5): 2 p.

Brüggenmann, Bernd. Ethische Aspekte der Frühintervention und Akutbehandlung schizophrener Störungen = Ethics of early intervention and acute treatment of schizophrenic disorders. *Ethik in der Medizin* 2007 June; 19(2): 91-102. NRCBL: 17.1; 1.1.

Buford, Chris; Allhoff, Fritz. Neuroscience and metaphysics (redux). *American Journal of Bioethics* 2007 January; 7(1): 58-60. NRCBL: 17.1; 4.4. Comments: Martha J. Farah and Andrea S. Heberlein. Personhood and neuroscience: naturalizing or nihilating? American Journal of Bioethics 2007 January; 7(1): 37-48.

Buller, Tom. Brains, lies, and psychological explanations. *In:* Illes, Judy, ed. Neuroethics: Defining the Issues in Theory, Practice, and Policy. New York: Oxford University Press, 2006: 51-60. NRCBL: 17.1; 4.4.

Burke, Greg F. Medicine: mental health and war; dementia and vasectomy; oocyte donation and sale; pedophilia; benefits of rest; diabetes mellitus; abortion and breast cancer; spirituality and health; euthanasia in the Netherlands; the Supreme Court decision on partial-birth abortion. *National Catholic Bioethics Quarterly* 2007 Autumn; 7(3): 579-594. NRCBL: 17.1; 21.2; 14.4; 10; 12.1; 20.7.

Canli, Turhan; Brandon, Susan; Casebeer, William; Crowley, Philip J.; DuRousseau, Don; Greely, Henry T.; Pascual-Leone, Alvaro. Neuroethics and national security. *American Journal of Bioethics* 2007 May; 7(5): 3-13. NRCBL: 17.1; 5.1; 1.3.8; 1.3.5; 21.3.

Canli, Turhan; Brandon, Susan; Casebeer, William; Crowley, Philip J.; DuRousseau, Don; Greely, Henry T.; Pascual-Leone, Alvaro. Response to open peer commentaries on "neuroethics and national security". *American Journal of Bioethics* 2007 May; 7(5): W1-W3. NRCBL: 17.1; 21.1.

Caplan, Arthur L. Bioethics and the brain [book review]. *New England Journal of Medicine* 2007 June 28; 356(26): 2758-2759. NRCBL: 17.1.

Carter, Adrian; Hall, Wayne. The social implications of neurobiological explanations of resistible compulsions. *American Journal of Bioethics* 2007 January; 7(1): 15-17. NRCBL: 17.1; 9.5.9; 17.3; 7.1. Comments: Steven E. Hyman. The neurobiology of addiction: implications for voluntary control of behavior. American Journal of Bioethics 2007 January; 7(1): 8-11.

Charland, Louis C. Affective neuroscience and addiction. *American Journal of Bioethics* 2007 January; 7(1): 20-21. NRCBL: 17.1; 9.5.9. Comments: Steven E. Hyman. The neurobiology of addiction: implications for voluntary control of behavior. American Journal of Bioethics 2007 January; 7(1): 8-11.

Chatterjee, Anjan. Cosmetic neurology and cosmetic surgery: parallels, predictions, and challenges. *CQ: Cambridge Quarterly of Healthcare Ethics* 2007 Spring; 16(2): 129-137. NRCBL: 17.1; 4.4; 4.5; 5.2; 7.1; 9.5.1.

Cheshire, William P. Can grey voxels resolve neuroethical dilemmas? *Ethics and Medicine: An International Journal of Bioethics* 2007 Fall; 23(3): 135-140. NRCBL: 17.1; 4.4. SC: an.

Cheshire, William P. Glimpsing the grey marble. *Ethics and Medicine: An International Journal of Bioethics* 2007 Summer; 23(2): 119-121. NRCBL: 17.1; 4.4; 7.1.

Cheshire, William P., Jr. The matter of the brightened grey. *Ethics and Medicine: An International Journal of Bioethics* 2007 Spring; 23(1): 35-38. NRCBL: 17.1; 4.4; 4.5; 17.4.

Churchland, Patricia Smith. Moral decision-making and the brain. *In:* Illes, Judy, ed. Neuroethics: Defining the Issues in Theory, Practice, and Policy. New York: Oxford University Press, 2006: 3-16. NRCBL: 17.1; 4.4.

Churchland, Patricia Smith. The necessary-and-sufficient boondoggle. *American Journal of Bioethics* 2007 January; 7(1): 54-55. NRCBL: 17.1; 4.4. Comments: Martha J. Farah and Andrea S. Heberlein. Personhood and neuroscience: naturalizing or nihilating? American Journal of Bioethics 2007 January; 7(1): 37-48.

Clark, Thomas W. Review of Walter Glannon, Bioethics and the Brain [book review]. *American Journal of Bioethics* 2007 May; 7(5): 59-60. NRCBL: 17.1; 2.1. SC: rv.

Cochrane, Thomas I. Brain disease or moral condition? Wrong question. *American Journal of Bioethics* 2007 January; 7(1): 24-25. NRCBL: 17.1; 9.5.9. Comments: Steven E. Hyman. The neurobiology of addiction: implications for voluntary control of behavior. American Journal of Bioethics 2007 January; 7(1): 8-11.

Coghlan, Andy. Bipolar disorder: young and moody or mentally ill? [news]. *New Scientist* 2007 May 19-25; 194(2604): 6-7. NRCBL: 17.1; 9.5.7; 17.4. SC: em.

Cohen, Peter J. Addiction, molecules and morality: disease does not obviate responsibility. *American Journal of Bioethics* 2007 January; 7(1): 21-23. NRCBL: 17.1; 9.5.1; 9.5.9. Comments: Steven E. Hyman. The neurobiology of addiction: implications for voluntary control of behavior. American Journal of Bioethics 2007 January; 7(1): 8-11.

Damasio, Antonio. Neuroscience and ethics: intersections [commentary]. *American Journal of Bioethics* 2007 January; 7(1): 3-7. NRCBL: 17.1; 1.1.

Dees, Richard H. Better brains, better selves? The ethics of neuroenhancements. *Kennedy Institute of Ethics Journal* 2007 December; 17(4): 371-395. NRCBL: 17.1; 4.5; 4.4; 5.2. SC: an.
Abstract: The idea of enhancing our mental functions through medical means makes many people uncomfortable. People have a vague feeling that altering our brains tinkers with the core of our personalities and the core of ourselves. It changes who we are, and doing so seems wrong, even if the exact reasons for the unease are difficult to define. Many of the standard arguments against neuroenhancements—that they are unsafe, that they violate the distinction between therapy and enhancements, that they undermine equality, and that they will be used coercively—fail to show why the use of any such technologies is wrong in principle. Two other objections—the arguments that such changes undermine our integrity and that they prevent us from living authentic lives—will condemn only a few of the uses that are proposed. The result is that very few uses of these drugs are morally suspect and that most uses are morally permissible.

NRCBL: National Reference Center for Bioethics Literature Classification Scheme   See inside front cover for terms.

327

Doucet, Hubert. Anthropological challenges raised by neuroscience: some ethical reflections. *CQ: Cambridge Quarterly of Healthcare Ethics* 2007 Spring; 16(2): 219-226. NRCBL: 17.1; 4.4; 7.1.

Downie, Jocelyn; Marshall, Jennifer. Pediatric neuroimaging ethics. *CQ: Cambridge Quarterly of Healthcare Ethics* 2007 Spring; 16(2): 147-160. NRCBL: 17.1; 4.4; 9.5.7; 5.2; 8.4; 9.4.

Downie, Jocelyn; Murphy, Ronalda. Inadmissible, eh? *American Journal of Bioethics* 2007 September; 7(9): 67-69. NRCBL: 17.1; 1.3.5. SC: le. Identifiers: Canada; R v. Béland. Comments: Stacey A. Tovino. Functional neuroimaging and the law: trends and directions for future scholarship.American Journal of Bioethics 2007 September; 7(9): 44-56.

Dyer, Clare. Mental health act becomes law after concessions are made [news]. *BMJ: British Medical Journal* 2007 July 14; 335(7610): 65. NRCBL: 17.1; 1.3.5; 17.7. SC: le.

Eaton, Margaret L.; Illes, Judy. Commercializing cognitive neurotechnology — the ethical terrain. *Nature Biotechnology* 2007 April; 25(4): 393-397. NRCBL: 17.1; 4.4; 8.4; 5.1; 9.3.1.

Egan, Erin A. Neuroimaging as evidence. *American Journal of Bioethics* 2007 September; 7(9): 62-63. NRCBL: 17.1; 1.3.5. SC: le. Comments: Stacey A. Tovino. Functional neuroimaging and the law: trends and directions for future scholarship. American Journal of Bioethics 2007 September; 7(9): 44-56.

Ellilä, Heikki; Välimäki, Maritta; Warne, Tony; Sourander, Andre. Ideology of nursing care in child psychiatric inpatient treatment. *Nursing Ethics* 2007 September; 14(5): 583-596. NRCBL: 17.1; 9.5.7; 4.1.3; 8.1. SC: em. Identifiers: Finland.
Abstract: Research on nursing ideology and the ethics of child and adolescent psychiatric nursing care is limited. The aim of this study was to describe and explore the ideological approaches guiding psychiatric nursing in child and adolescent psychiatric inpatient wards in Finland, and discuss the ethical, theoretical and practical concerns related to nursing ideologies. Data were collected by means of a national questionnaire survey, which included one open-ended question seeking managers' opinions on the nursing ideology used in their area of practice. Questionnaires were sent to all child and adolescent psychiatric inpatient wards (n = 69) in Finland; 61 ward managers responded. Data were analysed by qualitative and quantitative content analysis. Six categories — family centred care, individual care, milieu centred care, integrated care, educational care and psychodynamic care — were formed to specify ideological approaches used in inpatient nursing. The majority of the wards were guided by two or more approaches. Nursing models, theories and codes of ethics were almost totally ignored in the ward managers' ideological descriptions.

Esiri, Margaret. Why do research on human brains? *In:* Gunning, Jennifer; Holm, Søren, eds. Ethics, law, and society. Volume 1. Aldershot, Hants, England; Burlington, VT: Ashgate, 2005: 33-39. NRCBL: 17.1; 18.1; 19.5; 18.3.

Evers, Kathinka. Perspectives on memory manipulation: using beta-blockers to cure post-traumatic stress disorder. *CQ: Cambridge Quarterly of Healthcare Ethics* 2007 Spring; 16(2): 138-146. NRCBL: 17.1; 4.4; 9.7.

Farah, Martha J.; Heberlein, Andrea S. Personhood and neuroscience: naturalizing or nihilating? *American Journal of Bioethics* 2007 January; 7(1): 37-48. NRCBL: 17.1; 4.4; 1.1.
Abstract: Personhood is a foundational concept in ethics, yet defining criteria have been elusive. In this article we summarize attempts to define personhood in psychological and neurological terms and conclude that none manage to be both specific and non-arbitrary. We propose that this is because the concept does not correspond to any real category of objects in the world. Rather, it is the product of an evolved brain system that develops innately and projects itself automatically and irrepressibly onto the world whenever triggered by stimulus features such as a human-like face, body, or contingent patterns of behavior. We review the evidence for the existence of an autonomous person network in the brain and discuss its implications for the field of ethics and for the implicit morality of everyday behavior.

Farah, Martha J.; Heberlein, Andrea S. Response to open peer commentaries on "Personhood and neuroscience: naturalizing or nihilating?": getting personal [letter]. *American Journal of Bioethics* 2007 January; 7(1): W1-W4. NRCBL: 17.1; 4.4; 1.1.

Farah, Martha J.; Noble, Kimberly G.; Hurt, Hallam. Poverty, privilege, and brain development: empirical findings and ethical implications. *In:* Illes, Judy, ed. Neuroethics: Defining the Issues in Theory, Practice, and Policy. New York: Oxford University Press, 2006: 277-287. NRCBL: 17.1; 4.4; 9.5.10. SC: em.

Farah, Martha J.; Wolpe, Paul Root; Caplan, Arthur. Brain research and neuroethics. *In:* Gunning, Jennifer; Holm, Søren, eds. Ethics, law, and society. Volume 1. Aldershot, Hants, England; Burlington, VT: Ashgate, 2005: 261-264. NRCBL: 17.1; 18.4.

Fergusson, Andrew. Neuroethics: the new frontier. *Ethics and Medicine: An International Journal of Bioethics* 2007 Spring; 23(1): 31-33. NRCBL: 17.1; 4.4; 4.5. Conference: Report on the 13th Annual Conference on Bioethics; Deerfield, Illinois; Trinity International University; 2006 July 13-15; sponsored by the Center for Bioethics and Human Dignity.

Fins, Joseph J. Border zones of consciousness: another immigration debate? *American Journal of Bioethics* 2007 January; 7(1): 51-54. NRCBL: 17.1; 4.4; 20.5.1. Comments: Martha J. Farah and Andrea S. Heberlein.

Personhood and neuroscience: naturalizing or nihilating? American Journal of Bioethics 2007 January; 7(1): 37-48.

Fins, Joseph J.; Rezai, Ali R.; Greenberg, Benjamin D. Psychosurgery: avoiding an ethical redux while advancing a therapeutic future. *Neurosurgery* 2006 October; 59(4): 713-716. NRCBL: 17.6; 7.1.

Foddy, Bennett; Savulescu, Julian. Addiction is not an affliction: addictive desires are merely pleasure-oriented desires. *American Journal of Bioethics* 2007 January; 7(1): 29-32. NRCBL: 17.1; 9.5.9. Comments: Steven E. Hyman. The neurobiology of addiction: implications for voluntary control of behavior. American Journal of Bioethics 2007 January; 7(1): 8-11.

Ford, Paul J. Neurosurgical implants: clinical protocol considerations. *CQ: Cambridge Quarterly of Healthcare Ethics* 2007 Summer; 16(3): 308-311. NRCBL: 17.1; 4.5; 18.5.6; 18.3; 18.2.

Ford, Paul J.; DeMarco, Joseph P. Brains, ethics, and elective surgeries: emerging ethics consultation. *Ethics and Medicine: An International Journal of Bioethics* 2007 Spring; 23(1): 39-45. NRCBL: 17.1; 4.4; 4.5; 9.6. SC: cs.

Ford, Paul J.; Henderson, Jaimie M. Functional neurosurgical intervention: neuroethics in the operating room. *In:* Illes, Judy, ed. Neuroethics: Defining the Issues in Theory, Practice, and Policy. New York: Oxford University Press, 2006: 213-228. NRCBL: 17.1; 4.4; 8.3.1; 18.3.

Ford, Paul J.; Kubu, Cynthia S. Ameliorating and exacerbating: surgical "prosthesis" in addiction. *American Journal of Bioethics* 2007 January; 7(1): 32-34. NRCBL: 17.1; 9.5.9; 17.5; 17.3. Comments: Steven E. Hyman. The neurobiology of addiction: implications for voluntary control of behavior. American Journal of Bioethics 2007 January; 7(1): 8-11.

Foster, Kenneth R. Engineering the brain. *In:* Illes, Judy, ed. Neuroethics: Defining the Issues in Theory, Practice, and Policy. New York: Oxford University Press, 2006: 185-199. NRCBL: 17.1; 4.4; 18.4.

Fulford, K.W.M. (Bill); Thornton, Tim; Graham, George. From bioethics to values-based practice in psychiatric diagnosis. *In their:* Oxford Textbook of Philosophy and Psychiatry. Oxford; New York: Oxford University Press, 2006: 585-608. NRCBL: 17.1; 4.3; 2.1.

Fulford, K.W.M. (Bill); Thornton, Tim; Graham, George. Tools of the trade: an introduction to psychiatric ethics. *In their:* Oxford Textbook of Philosophy and Psychiatry. Oxford; New York: Oxford University Press, 2006: 469-497. NRCBL: 17.1; 7.2; 8.3.3; 8.4.

Fulford, K.W.M. (Bill); Thornton, Tim; Graham, George. Values in psychiatric psychiatry. *In their:* Oxford Textbook of Philosophy and Psychiatry. Oxford; New

York: Oxford University Press, 2006: 564-584. NRCBL: 17.1; 4.3. SC: cs.

Gazzaniga, Michael S. Facts, fictions and the future of neuroethics. *In:* Illes, Judy, ed. Neuroethics: Defining the Issues in Theory, Practice, and Policy. New York: Oxford University Press, 2006: 141-148. NRCBL: 17.1; 4.4.

Glannon, Walter. Neurosurgery, psychosurgery, and neurostimulation. *In his:* Bioethics and the Brain. New York: Oxford University Press, 2007: 116-147. NRCBL: 17.6; 5.2; 19.1; 17.3; 1.3.12; 8.3.3. SC: an.

Glannon, Walter. Persons, metaphysics and ethics. *American Journal of Bioethics* 2007 January; 7(1): 68-69. NRCBL: 17.1; 4.4; 1.1. Comments: Martha J. Farah and Andrea S. Heberlein. Personhood and neuroscience: naturalizing or nihilating? American Journal of Bioethics 2007 January; 7(1): 37-48.

Glover-Thomas, Nicola. A new 'new' Mental Health Act? Reflections on the proposed amendments to the Mental Health Act 1983. *Clinical Ethics* 2007 March; 2(1): 28-31. NRCBL: 17.1; 4.3; 8.1; 17.1. SC: le. Identifiers: Great Britain (United Kingdom); 'Bournewood' proposals.
  Abstract: Since 1998, several attempts have been made to reform the existing mental health legislation - the Mental Health Act 1983. However, all efforts thus far have been resoundingly rejected by mental health charities, psychiatrists and related professions. Following the Government's decision to abandon the draft Mental Health Bill in March 2006, plans to introduce new legislation designed to amend the existing 1983 Act have been published. This shorter bill was introduced before Parliament in November 2006. The amendments focused on six key policy areas including supervised community treatment, the nearest relative, the definition of mental disorder and detention criteria. It is also intended that the Mental Capacity Act 2005 will be amended to bridge the present 'Bournewood' gap.

Grainger-Monsen, Maren: Karetsky, Kim. The mind in the movies: a neuroethical analysis of the portrayal of the mind in popular media. *In:* Illes, Judy, ed. Neuroethics: Defining the Issues in Theory, Practice, and Policy. New York: Oxford University Press, 2006: 297-311. NRCBL: 17.1; 4.4; 7.1.

Greely, Henry. On neuroethics [editorial]. *Science* 2007 October 26; 318(5850): 533. NRCBL: 17.1; 4.4.

Greely, Henry T. The social effects of advances in neuroscience: legal problems, legal perspectives. *In:* Illes, Judy, ed. Neuroethics: Defining the Issues in Theory, Practice, and Policy. New York: Oxford University Press, 2006: 245-263. NRCBL: 17.1; 4.4; 14.5; 1.3.5.

Grey, Betsy J. Neuroscience, emotional harm, and emotional distress tort claims. *American Journal of Bioethics* 2007 September; 7(9): 65-67. NRCBL: 17.1. SC: le. Comments: Stacey A. Tovino. Functional neuroimaging and the

NRCBL: National Reference Center for Bioethics Literature Classification Scheme   See inside front cover for terms.

329

law: trends and directions for future scholarship. American Journal of Bioethics 2007 September; 7(9): 44-56.

Grey, William; Hall, Wayne; Carter, Adrian. Persons and personification. *American Journal of Bioethics* 2007 January; 7(1): 57-58. NRCBL: 17.1; 4.4. Comments: Martha J. Farah and Andrea S. Heberlein. Personhood and neuroscience: naturalizing or nihilating? American Journal of Bioethics 2007 January; 7(1): 37-48.

Hassert, Derrick L. Neuroethics and the person: should neurological and cognitive criteria be used to define human value? *Ethics and Medicine: An International Journal of Bioethics* 2007 Spring; 23(1): 47-55. NRCBL: 17.1; 4.4; 4.5; 1.1. SC: an.

Hewitt, J.L.; Edwards, S.D. Moral perspectives on the prevention of suicide in mental health settings. *Journal of Psychiatric and Mental Health Nursing* 2006 December; 13(6): 665-672. NRCBL: 17.1; 20.7; 1.1; 4.1.3; 1.1. SC: cs.

Hyman, Steven E. The neurobiology of addiction: implications for voluntary control of behavior. *American Journal of Bioethics* 2007 January; 7(1): 8-11. NRCBL: 17.1; 9.5.9; 17.4.
    Abstract: There continues to be a debate on whether addiction is best understood as a brain disease or a moral condition. This debate, which may influence both the stigma attached to addiction and access to treatment, is often motivated by the question of whether and to what extent we can justly hold addicted individuals responsible for their actions. In fact, there is substantial evidence for a disease model, but the disease model per se does not resolve the question of voluntary control. Recent research at the intersection of neuroscience and psychology suggests that addicted individuals have substantial impairments in cognitive control of behavior, but this "loss of control" is not complete or simple. Possible mechanisms and implications are briefly reviewed.

Illes, Judy. Ipsa scientia potestas est (Knowledge is power) [editorial]. *American Journal of Bioethics* 2007 January; 7(1): 1-2. NRCBL: 17.1.

Illes, Judy. Not forgetting forgetting [editorial]. *American Journal of Bioethics* 2007 September; 7(9): 1-2. NRCBL: 17.1; 4.4.

Illes, Judy; Bird, Stephanie J. Neuroethics: a modern context for ethics in neuroscience. *Trends in Neurosciences* 2006 September; 29(9): 511-517. NRCBL: 17.1; 4.4; 22. SC: rv.

Illes, Judy; Murphy, Emily R. Chimeras of nurture. *American Journal of Bioethics* 2007 May; 7(5):1-2. NRCBL: 17.1.

Illes, Judy; Racine, Eric; Krischen, Matthew P. A picture is worth 1000 words, but which 1000? *In:* Illes, Judy, ed. Neuroethics: Defining the Issues in Theory, Practice,

and Policy. New York: Oxford University Press, 2006: 149-168. NRCBL: 17.1; 4.4. SC: em.

Jaworska, Agnieszka. Ethical dilemmas in neurodegenerative disease: respecting patients at the twilight of agency. *In:* Illes, Judy, ed. Neuroethics: Defining the Issues in Theory, Practice, and Policy. New York: Oxford University Press, 2006: 87-101. NRCBL: 17.1; 4.4; 1.1.

Johnson, Kevin A.; Kozel, F. Andrew; Laken, Steven J.; George, Mark S. The neuroscience of functional magnetic resonance imaging fMRI for deception detection. *American Journal of Bioethics* 2007 September; 7(9): 58-60. NRCBL: 17.1. Comments: Stacey A. Tovino. Functional neuroimaging and the law: trends and directions for future scholarship. American Journal of Bioethics 2007 September; 7(9): 44-56.

Jones, Dan. The depths of disgust. Is there wisdom to be found in repugnance? Or is disgust 'the nastiest of all emotions', offering nothing but support to prejudice? *Nature* 2007 June 14; 447(7146): 768-771. NRCBL: 17.1; 2.1; 3.1.

Jones, D. Gareth. Neuroscience and the modification of human beings. *In:* Deane-Drummond, Celia; Scott, Peter Manley, eds. Future Perfect?: God, Medicine and Human Identity. New York: T and T Clark International, 2006: 87-99. NRCBL: 17.1; 4.5; 4.4; 1.2.

Jones, Roland; Kingdon, David. Council of Europe recommendation on human rights and psychiatry: a major opportunity for mental health services [editorial]. *European Psychiatry* 2005 November; 20(7): 461-464. NRCBL: 17.1; 17.7; 1.3.5; 6; 21.1.

Justo, Luis; Erazun, Fabiana. Neuroethics and human rights. *American Journal of Bioethics* 2007 May; 7(5): 16-18. NRCBL: 17.1; 21.1; 1.3.5. Comments: Turhan Canli, Susan Brandon, William Casebeer, Philip J. Crowley, Don DuRousseau, Henry T.Greely, and Alvaro Pascual-Leone. Neuroethics and national security. American Journal of Bioethics 2007 May; 7(5): 3-13.

Klitzman, Robert. Clinicians, patients, and the brain. *In:* Illes, Judy, ed. Neuroethics: Defining the Issues in Theory, Practice, and Policy. New York: Oxford University Press, 2006: 229-241. NRCBL: 17.1; 4.4; 15.3; 8.4; 8.1.

Kulynych, Jennifer J. Some thoughts about the evaluation of non-clinical functional magnetic resonance imaging. *American Journal of Bioethics* 2007 September; 7(9): 57-58. NRCBL: 17.1; 5.3. Comments: Stacey A. Tovino. Functional neuroimaging and the law: trends and directions for future scholarship. American Journal of Bioethics 2007 September; 7(9): 44-56.

Levy, Neil. Rethinking neuroethics in the light of the extended mind thesis. *American Journal of Bioethics* 2007 September; 7(9): 3-11. NRCBL: 17.1; 4.4; 1.1; 4.5; 17.4.

Abstract: The extended mind thesis is the claim that mental states extend beyond the skulls of the agents whose states they are. This seemingly obscure and bizarre claim has far-reaching implications for neuroethics, I argue. In the first half of this article, I sketch the extended mind thesis and defend it against criticisms. In the second half, I turn to its neuroethical implications. I argue that the extended mind thesis entails the falsity of the claim that interventions into the brain are especially problematic just because they are internal interventions, but that many objections to such interventions rely, at least in part, on this claim. Further, I argue that the thesis alters the focus of neuroethics, away from the question of whether we ought to allow interventions into the mind, and toward the question of which interventions we ought to allow and under what conditions. The extended mind thesis dramatically expands the scope of neuroethics: because interventions into the environment of agents can count as interventions into their minds, decisions concerning such interventions become questions for neuroethics.

Levy, Neil. The social: a missing term in the debate over addiction and voluntary control. *American Journal of Bioethics* 2007 January; 7(1): 35-36. NRCBL: 17.1; 9.5.9. Comments: Steven E. Hyman. The neurobiology of addiction: implications for voluntary control of behavior. American Journal of Bioethics 2007 January; 7(1): 8-11.

Loeben, Gregory; Stoehr, James D. Normative judgments, responsibility and executive function. *American Journal of Bioethics* 2007 January; 7(1): 27-29. NRCBL: 17.1; 9.5.9. Comments: Steven E. Hyman. The neurobiology of addiction: implications for voluntary control of behavior. American Journal of Bioethics 2007 January; 7(1): 8-11.

Lunstroth, John; Goldman, Jan. Ethical intelligence from neuroscience: is it possible? *American Journal of Bioethics* 2007 May; 7(5): 18-20. NRCBL: 17.1; 1.3.5; 21.1. Comments: Turhan Canli, Susan Brandon, William Casebeer, Philip J. Crowley, Don DuRousseau, Henry T. Greely, and Alvaro Pascual-Leone. Neuroethics and national security. American Journal of Bioethics 2007 May; 7(5): 3-13.

Madueme, Hans. Addiction as an amoral condition? The case remains unproven. *American Journal of Bioethics* 2007 January; 7(1): 25-27. NRCBL: 17.1; 9.5.9. Comments: Steven E. Hyman. The neurobiology of addiction: implications for voluntary control of behavior. American Journal of Bioethics 2007 January; 7(1): 8-11.

Manning, Christopher L. Institutional racism: article too strong? I think not. *BMJ: British Medical Journal* 2007 April 14; 334(7597): 761. NRCBL: 17.1; 9.5.4. Comments: Kwame McKenzie and Kamaldeep Bhui. Institutional racism in mental health care. British Medical Journal 2007 March 31; 334(7595): 649-650.

Mason, J.K.; Laurie, G.T. Mental health and human rights. *In their:* Mason and McCall Smith's Law and Medi-

cal Ethics. Seventh ed. Oxford; New York: Oxford University Press, 2005: 710-739. NRCBL: 17.1; 8.3.1; 4.3; 17.8; 21.5. SC: le.

Matthews, Eric. Is autonomy relevant to psychiatric ethics? *In:* Nys, Thomas; Denier, Yvonne; Vandevelde, Toon, eds. Autonomy and Paternalism: Reflections on the Theory and Practice of Health Care. Leuven; Dudley, MA: Peeters, 2007: 129-146. NRCBL: 17.1; 1.1; 8.1; 17.8.

McKenzie, Kwame; Bhui, Kamaldeep. Institutional racism in mental health care. *BMJ: British Medical Journal* 2007 March 31; 334(7595): 649-650. NRCBL: 17.1; 9.5.4. Identifiers: Great Britain (United Kingdom).

McMonagle, Ethan. Functional neuroimaging and the law: a Canadian perspective. *American Journal of Bioethics* 2007 September; 7(9): 69-70. NRCBL: 17.1; 1.3.5. SC: le. Identifiers: Canada; Charter of Rights and Freedoms. Comments: Stacey A. Tovino. Functional neuroimaging and the law: trends and directions for future scholarship. American Journal of Bioethics 2007 September; 7(9): 44-56.

Meghani, Zahra. Is personhood an illusion? *American Journal of Bioethics* 2007 January; 7(1): 62-63. NRCBL: 17.1; 4.4; 1.1. Comments: Martha J. Farah and Andrea S. Heberlein. Personhood and neuroscience: naturalizing or nihilating? American Journal of Bioethics 2007 January; 7(1): 37-48.

Meyers, Christopher. Personhood: empirical thing or rational concept? *American Journal of Bioethics* 2007 January; 7(1): 63-65. NRCBL: 17.1; 4.4; 1.1. Comments: Martha J. Farah and Andrea S. Heberlein. Personhood and neuroscience: naturalizing or nihilating? American Journal of Bioethics 2007 January; 7(1): 37-48.

Michaels, Mark H. Ethical considerations in writing psychological assessment reports. *Journal of Clinical Psychology* 2006 January; 62(1): 47-58. NRCBL: 17.1; 1.3.7; 1.3.12; 8.4.

Morris, Stephen G. Neuroscience and the free will conundrum. *American Journal of Bioethics* 2007 May; 7(5): 20-22. NRCBL: 17.1; 1.1. Comments: Turhan Canli, Susan Brandon, William Casebeer, Philip J. Crowley, Don DuRousseau, Henry T. Greely, and Alvaro Pascual-Leone. Neuroethics and national security. American Journal of Bioethics 2007 May; 7(5): 3-13.

Morse, Stephen J. Moral and legal responsibility and the new neuroscience. *In:* Illes, Judy, ed. Neuroethics: Defining the Issues in Theory, Practice, and Policy. New York: Oxford University Press, 2006: 33-50. NRCBL: 17.1; 4.4.

Morse, Stephen J. Voluntary control of behavior and responsibility. *American Journal of Bioethics* 2007 January; 7(1): 12-13. NRCBL: 17.1; 9.5.9. Comments: Steven E. Hyman. The neurobiology of addiction: implications for

NRCBL: National Reference Center for Bioethics Literature Classification Scheme   See inside front cover for terms.

331

voluntary control of behavior. American Journal of Bioethics 2007 January; 7(1): 8-11.

Nelson, James Lindemann. Illusions about persons. *American Journal of Bioethics* 2007 January; 7(1): 65-66. NRCBL: 17.1; 4.4; 1.1. Comments: Martha J. Farah and Andrea S. Heberlein. Personhood and neuroscience: naturalizing or nihilating? American Journal of Bioethics 2007 January; 7(1): 37-48.

Northoff, Georg. The influence of brain implants on personal identity and personality — a combined theoretical and empirical investigation in 'neuroethics'. *In:* Schramme, Thomas; Thome, Johannes, eds. Philosophy and Psychiatry. Berlin; New York: De Gruyter, 2004: 326-344. NRCBL: 17.1; 4.4; 17.6.

Ozgen, C. Ethics in the use of new medical technologies for neurosurgery: "Islamic viewpoint". *Acta Neurochirurgica. Supplement* 2006; 98: 13-17. NRCBL: 17.1; 9.5.1; 4.5; 1.2. Identifiers: Islam.

Parens, Erik. Creativity, gratitude, and the enhancement debate. *In:* Illes, Judy, ed. Neuroethics: Defining the Issues in Theory, Practice, and Policy. New York: Oxford University Press, 2006: 75-86. NRCBL: 17.1; 4.4; 4.5.

Perring, Christian. Against scientism, for personhood. *American Journal of Bioethics* 2007 January; 7(1): 67-68. NRCBL: 17.1; 4.4; 1.1. Comments: Martha J. Farah and Andrea S. Heberlein. Personhood and neuroscience: naturalizing or nihilating? American Journal of Bioethics 2007 January; 7(1): 37-48.

Phelps, Elizabeth A. The neuroscience of a person network. *American Journal of Bioethics* 2007 January; 7(1): 49-54. NRCBL: 17.1. Comments: Martha J. Farah and Andrea S. Heberlein. Personhood and neuroscience: naturalizing or nihilating? American Journal of Bioethics 2007 January; 7(1): 37-48.

Potter, Nancy Nyguist; Zanni, Guido R.; Stavis, Paul F. Querying the "community" in community mental health. *American Journal of Bioethics* 2007 November; 7(11): 42-43; author reply W3-W4. NRCBL: 17.1; 17.7. Comments: Guido R. Zanni and Paul F. Stavis. The effectiveness and ethical justification of psychiatric outpatient commitment. American Journal of Bioethics 2007 November; 7(11): 31-41.

Powell, Tia. Wrestling satan and conquering dopamine: addiction and free will. *American Journal of Bioethics* 2007 January; 7(1): 14-15. NRCBL: 17.1; 9.5.9. Comments: Steven E. Hyman. The neurobiology of addiction: implications for voluntary control of behavior. American Journal of Bioethics 2007 January; 7(1): 8-11.

Racine, Eric. Identifying challenges and conditions for the use of neuroscience in bioethics. *American Journal of Bioethics* 2007 January; 7(1): 74-75. NRCBL: 17.1; 4.4; 1.1. Comments: Martha J. Farah and Andrea S. Heberlein.

Personhood and neuroscience: naturalizing or nihilating? American Journal of Bioethics 2007 January; 7(1): 37-48.

Racine, Eric; van der Loos, Hz Adriaan; Illes, Judy. Internet marketing of neuroproducts: new practices and healthcare policy changes. *CQ: Cambridge Quarterly of Healthcare Ethics* 2007 Spring; 16(2): 181-194. NRCBL: 17.1; 4.4; 1.3.12; 1.3.2; 5.3; 8.1. SC: em.

Radden, Jennifer. Virtue ethics as professional ethics: the case of psychiatry. *In:* Walker, Rebecca L.; Ivanhoe, Philip J., eds. Working Virtue: Virtue Ethics and Contemporary Moral Problems. Oxford: Clarendon, 2007: 113-134. NRCBL: 17.1; 4.1.2; 1.3.1. SC: an.

Rappaport, Z.H. Robotics and artificial intelligence: Jewish ethical perspectives. *Acta Neurochirurgica. Supplement* 2006; 98: 9-12. NRCBL: 17.1; 17.6; 17.5; 1.2; 5.3. Identifiers: Judaism.

Resnik, David B. Neuroethics, national security and secrecy. *American Journal of Bioethics* 2007 May; 7(5): 14-15. NRCBL: 17.1; 21.3; 1.3.9. Comments: Turhan Canli, Susan Brandon, William Casebeer, Philip J. Crowley, Don DuRousseau, Henry T.Greely, and Alvaro Pascual-Leone. Neuroethics and national security. American Journal of Bioethics 2007 May; 7(5): 3-13.

Richardson, Genevra. Balancing autonomy and risk: a failure of nerve in England and Wales? *International Journal of Law and Psychiatry* 2007 January-February; 30(1): 71-80. NRCBL: 17.1; 17.8; 21.1; 1.1. SC: le.

Robert, Jason Scott. Gene maps, brain scans, and psychiatric nosology. *CQ: Cambridge Quarterly of Healthcare Ethics* 2007 Spring; 16(2): 209-218. 44 fn. NRCBL: 17.1; 4.4; 15.1; 1.1; 4.1.2.
   Keywords: *classification; *genomics; *neurosciences; *psychiatric diagnosis; *philosophy; behavioral genetics; brain; genotype; mental disorders; phenotype; psychiatry; schizophrenia

Roberts, Laura Weiss; Coverdale, John; Louie, Alan K. Philanthropy, ethics, and leadership in academic psychiatry. *Academic Psychiatry* 2006 July-August; 30(4): 269-272. NRCBL: 17.1; 1.3.3; 9.3.1.

Robertson, Michael; Walter, Garry. A critical reflection on utilitarianism as the basis for psychiatric ethics. Part I: Utilitarianism as ethical theory. *Journal of Ethics in Mental Health [electronic]* 2007 April; 2(1): 4 p. Accessed: http://www.jemh.ca [2007 July 31]. NRCBL: 17.1; 2.1; 4.1.2.
   Abstract: Utilitarianism is one of the "grand Enlightenment" moral philosophies. It provides a means of evaluating the ethical implications of common and unusual situations faced by psychiatrists, and offers a logical and ostensibly scientific method of moral justification and action. In this first of our two papers, we trace the evolution of utilitarianism into a contemporary moral theory and review the main theoretical critiques. In the second paper we contextualize utilitarianism in psychiatry and con-

sider its function within the realm of the professional ethics of psychiatrist as physician, before applying it to two dilemmas faced by psychiatrists as individuals and as members of a profession. We conclude that psychiatry must search beyond utilitarianism in grappling with everyday clinical scenarios.

Robertson, Michael; Walter, Gary. A critical reflection on utilitarianism as the basis for psychiatric ethics. Part II: Utilitarianism and psychiatry. *Journal of Ethics in Mental Health* 2007 April; 2(1): 4 p. Accessed: http://www.jemh. ca [2007 July 31]. NRCBL: 17.1; 2.1; 4.1.2.

Abstract: In this second paper we contextualize utilitarianism to the craft of psychiatry and consider its function within the realm of the professional ethics of psychiatrists as physicians. We then apply it to two dilemmas faced by psychiatrists as individuals and as members of a profession. We conclude that psychiatry must search beyond utilitarianism in grappling with everyday clinical scenarios.

Roelcke, Volker. Psychiatrie und Nervenheilkunde im Nationalsozialismus: Ärztliches Verhalten zwischen Bewährung und Versagen. *In:* Gestrich, Christof; Neugebauer, Johannes, eds. Der Wert menschlichen Lebens: medizinische Ethik bei Karl Bonhoeffer und Dietrich Bonhoeffer. Berlin: Wichern-Verlag, 2006: 13-32. NRCBL: 17.1; 15.5; 2.2; 11.3; 20.5.1; 21.4; 2.2.

Rosenberg, Leah; Gehrie, Eric. Against the use of medical technologies for military or national security interests. *American Journal of Bioethics* 2007 May; 7(5): 22-24. NRCBL: 17.1; 2.1; 8.1; 1.3.5; 21.2. Comments: Turhan Canli, Susan Brandon, William Casebeer, Philip J. Crowley, Don DuRousseau, Henry T.Greely, and Alvaro Pascual-Leone. Neuroethics and national security. American Journal of Bioethics 2007 May; 7(5): 3-13.

Roskies, Adina. A case study of neuroethics: the nature of moral judgment. *In:* Illes, Judy, ed. Neuroethics: Defining the Issues in Theory, Practice, and Policy. New York: Oxford University Press, 2006: 17-32. NRCBL: 17.1; 4.4.

Roskies, Adina L. The illusion of personhood. *American Journal of Bioethics* 2007 January; 7(1): 55-57. NRCBL: 17.1; 4.4; 1.1. Comments: Martha J. Farah and Andrea S. Heberlein. Personhood and neuroscience: naturalizing or nihilating? American Journal of Bioethics 2007 January; 7(1): 37-48.

Sachs, Greg A.; Cassel, Christine K. Ethical aspects of dementia. *Neurologic Clinics* 1989 November; 7(4): 845-858. NRCBL: 17.1; 9.5.2; 8.3.3. SC: cs.

Sadler, John Z. The rhetorician's craft, distinctions in science, and political morality. *Philosophy, Ethics, and Humanities in Medicine [electronic]* 2006; 1(7): 2 p. Accessed: http://www.peh-med.com/content/1/1/7 [2006 August 15]. NRCBL: 17.1; 5.3. Comments: Thomas Szasz. Secular humanism and "scientific psychiatry." Phi-

losophy, Ethics, and Humanities [electronic] 2006; 1(5): 2 p.

Sagoff, Mark. A transcendental argument for the concept of personhood in neuroscience. *American Journal of Bioethics* 2007 January; 7(1): 72-73. NRCBL: 17.1; 4.4; 1.1. Comments: Martha J. Farah and Andrea S. Heberlein. Personhood and neuroscience: naturalizing or nihilating? American Journal of Bioethics 2007 January; 7(1): 37-48.

Salman, Rustam Al-Shahi, Stone, Jon; Warlow, Charles. What do patients think about appearing in neurology "grand rounds"? *Journal of Neurology, Neurosurgery, and Psychiatry* 2007 May; 78(5): 454-456. NRCBL: 17.1; 7.2; 8.1. SC: em.

Schmidt, Charles. Putting the brakes on psychosis. *Science* 2007 May 18; 316(5827): 976-977. NRCBL: 17.1; 17.3; 17.4.

Sen, Piyal; Gordon, Harvey; Adshead, Gwen; Irons, Ashley. Ethical dilemmas in forensic psychiatry: two illustrative cases. *Journal of Medical Ethics* 2007 June; 33(6): 337-341. NRCBL: 17.1; 1.3.1; 1.3.5; 2.1; 4.1.2; 6; 8.1; 8.3.3; 8.3.4; 17.3. SC: cs.

Abstract: One approach to the analysis of ethical dilemmas in medical practice uses the "four principles plus scope" approach. These principles are: respect for autonomy, beneficence, non-maleficence and justice, along with concern for their scope of application. However, conflicts between the different principles are commonplace in psychiatric practice, especially in forensic psychiatry, where duties to patients often conflict with duties to third parties such as the public. This article seeks to highlight some of the specific ethical dilemmas encountered in forensic psychiatry: the excessive use of segregation for the protection of others, the ethics of using mechanical restraint when clinically beneficial and the use of physical treatment without consent. We argue that justice, as a principle, should be paramount in forensic psychiatry, and that there is a need for a more specific code of ethics to cover specialised areas of medicine like forensic psychiatry. This code should specify that in cases of conflict between different principles, justice should gain precedence over the other principles.

Sheridan, Kimberly; Zinchenko, Elena; Gardner, Howard. Neuroethics in education. *In:* Illes, Judy, ed. Neuroethics: Defining the Issues in Theory, Practice, and Policy. New York: Oxford University Press, 2006: 265-275. NRCBL: 17.1; 4.4; 1.3.3.

Sheth, Hiten G.; Sheth, A.G. Frequent attenders to ophthalmic accident and emergency departments [letter]. *Journal of Medical Ethics* 2007 August; 33(8): 496. NRCBL: 17.1; 9.4.

Snead, Carter. Neuroimaging, entrapment, and the predisposition to crime. *American Journal of Bioethics* 2007 September; 7(9): 60-61. NRCBL: 17.1; 1.3.5. SC: le. Comments: Stacey A. Tovino. Functional neuroimaging and the

NRCBL: National Reference Center for Bioethics Literature Classification Scheme   See inside front cover for terms.

333

law: trends and directions for future scholarship. American Journal of Bioethics 2007 September; 7(9): 44-56.

Srebnik, Debra S.; Russo, Joan. Consistency of psychiatric crisis care with advance directive instructions. *Psychiatric Services* 2007 September; 58(9): 1157-1163. NRCBL: 17.1; 8.3.3; 8.3.4. SC: em.

Stephenson, James A.; Staal, Mark A. An ethical decision-making model for operational psychology. *Ethics and Behavior* 2007; 17(1): 61-82. NRCBL: 17.1; 1.3.5. SC: cs.
  Abstract: Operational psychology is an emerging subdiscipline that has enhanced the U.S. military's combat capabilities during the Global War on Terrorism. What makes this subdiscipline unique is its use of psychological principles and skills to improve a commander's decision making as it pertains to conducting combat (or related operations). Due to psychology's expanding role in combat support, psychologists are being confronted with challenges that require the application of their professional ethics in areas in which little if any guidance has been provided. Operational psychologists are at the forefront of this expansion. Accordingly, they need a decision model to assist them in this complex dynamic environment. To this end, this article reviews various decision models and ethical frameworks, selects the most appropriate, and then applies it to the challenges faced by operational psychologists. A naturalistic decision model that integrates rational and intuitive elements is recommended.

Steven, Megan S.; Pascual-Leone, Alvaro. Transcranial magnetic stimulation and the human brain: an ethical evaluation. *In:* Illes, Judy, ed. Neuroethics: Defining the Issues in Theory, Practice, and Policy. New York: Oxford University Press, 2006: 201-211. NRCBL: 17.1; 4.4; 1.1.

Stocking, Carol B.; Houghham, Gavin W.; Danner, Deborah D.; Patterson, Marian B.; Whitehouse, Peter J.; Sachs, Greg A. Empirical assessment of a research advance directive for persons with dementia and their proxies. *Journal of the American Geriatrics Society* 2007 October; 55(10): 1609-1612. NRCBL: 17.1; 8.3.3; 8.3.4; 18.5.6; 18.3. SC: em.

Swartz, Marvin S.; Swanson, Jeffrey W. Psychiatric advance directives and recovery-oriented care. *Psychiatric Services* 2007 September; 58(9): 1164. NRCBL: 17.1; 8.3.3; 8.3.4.

Szasz, Thomas. Secular humanism and "scientific psychiatry". *Philosophy, Ethics, and Humanities in Medicine [electronic]* 2006; 1(5): 5 p. Accessed: http://www.peh-med.com/content/1/1/5 [2006 July 19]. NRCBL: 17.7; 17.4; 4.3; 1.3.5; 1.1; 7.1. SC: an.

Van Citters, Aricca D.; Naidoo, Umadevi; Foti, Mary Ellen. Using a hypothetical scenario to inform psychiatric advance directives. *Psychiatric Services* 2007 November; 58(11): 1467-1471. NRCBL: 17.1; 8.3.3; 8.3.4. SC: em.

van der Loos, H.F. Machiel. Design and engineering ethics considerations for neurotechnologies. *CQ: Cambridge Quarterly of Healthcare Ethics* 2007 Summer; 16(3): 303-307. NRCBL: 17.1; 4.5; 7.1; 1.1; 5.1.

Viens, A.M. Addiction, responsibility and moral psychology. *American Journal of Bioethics* 2007 January; 7(1): 17-20. NRCBL: 17.1; 9.5.9. Comments: Steven E. Hyman. The neurobiology of addiction: implications for voluntary control of behavior. American Journal of Bioethics 2007 January; 7(1): 8-11.

Viens, A.M. The use of functional neuroimaging technology in the assessment of loss and damages in tort law. *American Journal of Bioethics* 2007 September; 7(9): 63-65. NRCBL: 17.1. SC: le. Comments: Stacey A. Tovino. Functional neuroimaging and the law: trends and directions for future scholarship. American Journal of Bioethics 2007 September; 7(9): 44-56.

Widdershoven, Guy; Berghmans, Ron. Coercion and pressure in psychiatry: lessons from Ulysses. *Journal of Medical Ethics* 2007 October; 33(10): 560-563. NRCBL: 17.1; 8.3.4.
  Abstract: Coercion and pressure in mental healthcare raise moral questions. This article focuses on moral questions raised by the everyday practice of pressure and coercion in the care for the mentally ill. In view of an example from literature—the story of Ulysses and the Sirens—several ethical issues surrounding this practice of care are discussed. Care giver and patient should be able to express feelings such as frustration, fear and powerlessness, and attention must be paid to those feelings. In order to be able to evaluate the intervention, one has to be aware of the variety of goals the intervention can aim at. One also has to be aware of the variety of methods of intervention, each with its own benefits and drawbacks. Finally, an intervention requires a context of care and responsibility, along with good communication and fair treatment before, during and after the use of coercion and pressure.

Wildeman, Sheila. Access to treatment of serious mental illness: enabling choice or enabling treatment? *In:* Flood, Colleen M., ed. Just Medicare: What's In, What's Out, How We Decide. Buffalo, NY: University of Toronto Press, 2006: 231-257. NRCBL: 17.1; 17.8; 9.8; 17.7. SC: le. Identifiers: Canada.

Wilder, Christine M.; Elbogen, Eric B.; Swartz, Marvin S.; Swanson, Jeffrey W.; Van Dorn, Richard A. Effect of patients' reasons for refusing treatment on implementing psychiatric advance directives. *Psychiatric Services* 2007 October; 58(10): 1348-1350. NRCBL: 17.1; 8.3.3; 8.3.4. SC: em.

Winslade, William J. Severe brain injury: recognizing the limits of treatment and exploring the frontiers of research. *CQ: Cambridge Quarterly of Healthcare Ethics* 2007 Spring; 16(2): 161-168. NRCBL: 17.1; 4.4; 20.5.1.

Wolpe, Paul Root. Religious responses to neuroscientific questions. *In:* Illes, Judy, ed. Neuroethics: Defining the Issues in Theory, Practice, and Policy. New York: Oxford University Press, 2006: 289-296. NRCBL: 17.1; 4.4; 1.2.

Zoloth, Laurie. Being in the world: neuroscience and the ethical agent. *In:* Illes, Judy, ed. Neuroethics: Defining the Issues in Theory, Practice, and Policy. New York: Oxford University Press, 2006: 61-73. NRCBL: 17.1; 4.4; 1.1.

MENTALLY DISABLED *See* CARE FOR SPECIFIC GROUPS/ MENTALLY DISABLED; HUMAN EXPERIMENTATION/ SPECIAL POPULATIONS/ MENTALLY DISABLED; INFORMED CONSENT/ INCOMPETENTS

MENTALLY HANDICAPPED *See* CARE FOR SPECIFIC GROUPS/ MENTALLY DISABLED; HUMAN EXPERIMENTATION/ SPECIAL POPULATIONS/ MENTALLY DISABLED; INFORMED CONSENT/ INCOMPETENTS

MENTALLY ILL *See* CARE FOR SPECIFIC GROUPS/ MENTALLY DISABLED; HUMAN EXPERIMENTATION/ SPECIAL POPULATIONS/ MENTALLY DISABLED; INFORMED CONSENT/ INCOMPETENTS; TREATMENT REFUSAL/ MENTALLY ILL

MERCY KILLING *See* EUTHANASIA AND ALLOWING TO DIE

MIGRATION OF HEALTH CARE PROFESSIONALS *See* INTERNATIONAL MIGRATION OF HEALTH CARE PROFESSIONALS

MINORITIES *See* CARE FOR SPECIFIC GROUPS/ MINORITIES

MINORS *See* CARE FOR SPECIFIC GROUPS/ MINORS; DEATH AND DYING/ TERMINAL CARE FOR MINORS; EUTHANASIA AND ALLOWING TO DIE/ MINORS; HUMAN EXPERIMENTATION/ SPECIAL POPULATIONS/ MINORS; INFORMED CONSENT/ MINORS

MISCONDUCT *See* BIOMEDICAL RESEARCH/ RESEARCH ETHICS AND SCIENTIFIC MISCONDUCT; MALPRACTICE AND PROFESSIONAL MISCONDUCT

MORAL AND RELIGIOUS ASPECTS *See* ABORTION/ MORAL AND RELIGIOUS ASPECTS; HUMAN EXPERIMENTATION/ SPECIAL POPULATIONS/ EMBRYOS AND FETUSES/ PHILOSOPHICAL AND RELIGIOUS ASPECTS

NANOTECHNOLOGY

Enough talk already: governments should act on researchers' attempts to engage the public over nanotechnology [editorial]. *Nature* 2007 July 5; 448(7149): 1-2. NRCBL: 5.4; 5.3.

Allhoff, Fritz. On the autonomy and justification of nanoethics. *NanoEthics* 2007 December; 1(3): 185-210. NRCBL: 5.4; 1.1; 1.3.2; 2.1; 16.1; 17.1. SC: an; rv.

Bowman, Diana M.; Hodge, Graeme A. Editorial - governing nanotechnology: more than a small matter? *NanoEthics* 2007 December; 1(3): 239-241. NRCBL: 5.4; 5.3.

Bruce, Donald. Ethical and social issues in nanobiotechnologies: Nano2Life provides a European ethical 'think tank' for research in biology at the nanoscale. *EMBO Reports* 2006 August; 7(8): 754-758. NRCBL: 5.4; 4.4; 21.1; 15.1.

Bruce, Donald. Faster, higher, stronger. *Nano Now!* 2007 February; 1(1): 18-19 [Online]. Accessed: http://www.nanonow.co.uk/nanonow_issue1.pdf [2007 March 2]. NRCBL: 5.4; 4.5. SC: an. Identifiers: human performance enhancement.

Burri, Regula Valérie. Deliberating risks under uncertainty: experience, trust, and attitudes in a Swiss nanotechnology stakeholder discussion group. *NanoEthics* 2007 August; 1(2): 143-154. NRCBL: 5.4; 5.2. SC: em.

Dorbeck-Jung, Bärbel R. What can prudent public regulators learn from the United Kingdom government's nanotechnological regulatory activities? *NanoEthics* 2007 December; 1(3): 257-270. NRCBL: 5.4; 5.3.

Doubleday, Robert. The laboratory revisited: academic science and the responsible development of nanotechnology. *NanoEthics* 2007 August; 1(2): 167-176. NRCBL: 5.4; 5.2; 5.3. SC: em.

Edwards, Steven A. Fear of nano: dangers and ethical challenges. *In his:* The Nanotech Pioneers: Where Are They Taking Us? Weinheim: Wiley-VCH, 2006: 197-229. NRCBL: 5.4; 5.3; 5.1; 16.1; 20.5.1; 4.4; 21.3.

European Commission. European Group on Ethics in Science and New Technologies. Opinion on the ethical aspects of nanomedicine. Brussels. Belgium: The European Commission, European Group on Ethics in Science and New technologies to the European Commission: 2007 January 17; 164 p. [Online]. Accessed: http://ec.europa.eu/european_group_ethics/activities/docs/opinion_21_nano_en.pd f [2007 January 30]. NRCBL: 5.4; 5.3; 21.1. Note: Opinion No. 21.

Fisher, Erik. Ethnographic invention: probing the capacity of laboratory decisions. *NanoEthics* 2007 August; 1(2): 155-165. NRCBL: 5.4; 5.2; 5.3. SC: em.

NRCBL: National Reference Center for Bioethics Literature Classification Scheme   See inside front cover for terms.

335

Grunwald, Armin; Julliard, Yannick. Nanotechnology — steps toward understanding human beings as technology? *NanoEthics* 2007 August; 1(2): 77-87. NRCBL: 5.4; 1.1; 4.4; 4.5. SC: an.

Hermerén, Göran. Challenges in the evaluation of nanoscale research: ethical aspects. *NanoEthics* 2007 December; 1(3): 223-237. NRCBL: 5.4. SC: an; rv.

Kearnes, Matthew; Wynne, Brian. On nanotechnology and ambivalence: the politics of enthusiasm. *NanoEthics* 2007 August; 1(2): 131-142. NRCBL: 5.4; 5.2. SC: an.

Kjølberg, Kamilla; Wickson, Fern. Social and ethical interactions with nano: mapping the early literature. *NanoEthics* 2007 August; 1(2): 89-104. NRCBL: 5.4; 1.1; 5.2. SC: em; rv.

Lenk, Christian; Biller-Andorno, Nikola. Nanomedicine — emerging or re-emerging ethical issues? A discussion of four ethical themes. *Medicine, Health Care and Philosophy* 2007 June; 10(2): 173-184. NRCBL: 5.4; 4.4; 4.5; 5.2; 9.4. SC: rv.

Abstract: Nanomedicine plays a prominent role among emerging technologies. The spectrum of potential applications is as broad as it is promising. It includes the use of nanoparticles and nanodevices for diagnostics, targeted drug delivery in the human body, the production of new therapeutic materials as well as nanorobots or nanoprotheses. Funding agencies are investing large sums in the development of this area, among them the European Commission, which has launched a large network for life-sciences related nanotechnology. At the same time government agencies as well as the private sector are putting forward reports of working groups that have looked into the promises and risks of these developments. This paper will begin with an introduction to the central ethical themes as identified by selected reports from Europe and beyond. In a next step, it will analyse the most frequently invoked ethical concerns-risk assessment and management, the issues of human identity and enhancement, possible implications for civil liberties (e.g. nanodevices that might be used for covert surveillance), and concerns about equity and fair access. Although it seems that the main ethical issues are not unique to nano- technologies, the conclusion will argue against shrugging them off as non-specific items that have been considered before in the context of other biomedical technologies, such as gene therapy or xenotransplantation. Rather, the paper will call on ethicists to help foster a rational, fair and participatory discourse on the different potential applications of nanotechnologies in medicine, which can form the basis for informed and responsible societal and political decisions.

Lin, Patrick. Nanotechnology bound: evaluating the case for more regulation. *NanoEthics* 2007 August; 1(2): 105-122. NRCBL: 5.4; 5.2; 5.3; 16.1. SC: an; le.

Lupton, M. Nanotechnology — salvation or damnation for humans? *Medicine and Law: The World Association for Medical Law* 2007 June; 26(2): 349-362. 38 fn. NRCBL: 5.4; 4.4; 15.1; 20.5.1. SC: le.

Keywords: *nanotechnology; cloning; embryo research; gene therapy; genetic engineering; international aspects; life extension; medicine; personhood; privacy; regulation; risks and benefits

Abstract: Nanotechnology is a term derived from the Greek word nanos, meaning dwarf. It is used to describe activities at the level of atoms and molecules. The application of this technology is aimed at controlling and manipulating the physical properties of materials with single molecule precision. Scientists use the technology to build working devices, systems and materials, molecule by molecule. This enables them to exploit the unique and powerful electrical, physical and chemical properties found at that scale. Nanotech holds the potential to revolutionise medicine, electronics and chemistry. Nanomedicine would facilitate the repair and improvement of the human body from the inside out, with a precision and delicacy far greater than the finest surgical instruments permit. Problem areas stemming from the technology include the following:- Who will benefit - just the rich or the poor as well? This paper will explore the role of law, ethics and suitable control mechanisms to limit the dangers and maximise the benefits of nanotechnology for society, especially in the field of medicine.

Mody, Cyrus; McCray, Patrick; Roberts, Jody; Berne, Rosalyn; Lin, Patrick; Keiper, Adam. Debating nanoethics [letters and reply]. *New Atlantis* 2007 Summer; (17): 5-14. NRCBL: 5.4; 4.4. Comments: Adam Keiper. Nanoethics as a discipline? New Atlantis 2007 Spring; (16): 55-67.

Petersen, Alan; Anderson, Alison. A question of balance or blind faith?: scientists' and science policymakers' representations of the benefits and risks of nanotechnologies. *NanoEthics* 2007 December; 1(3): 243-256. NRCBL: 5.4; 5.2. SC: em.

Pilarski, Linda M.; Mehta, Michael D.; Caulfield, Timothy; Kaler, Karan V.I.S.; Backhouse, Christopher J. Microsystems and nanoscience for biomedical applications: a view to the future. *In:* Hunt, Geoffrey; Mehta, Michael D., eds. Nanotechnology: Risk, Ethics and Law. London; Sterling, VA: Earthscan, 2006: 35-42. 15 refs. NRCBL: 5.4; 5.3; 15.1.

Keywords: *biomedical technologies; *biotechnology; *nanotechnology; genetic screening; genetic services; genomics; marketing; pharmacogenetics; policy making; public participation; risks and benefits; social impact; values

Rogers-Hayden, Tee; Mohr, Alison; Pidgeon, Nick. Introduction: engaging with nanotechnology — engaging differently? *NanoEthics* 2007 August; 1(2): 123-130. NRCBL: 5.4; 5.2. SC: an.

Sheremeta, Lorraine. Nanotechnologies and the ethical conduct of research involving human subjects. *In:* Hunt, Geoffrey; Mehta, Michael D., eds. Nanotechnology: Risk,

---

Ethics and Law. London; Sterling, VA: Earthscan, 2006: 247-258. NRCBL: 5.4; 18.2; 15.3; 9.7. Identifiers: Canada; Tri-Council Policy Statement on the Ethical Conduct for Research Involving Humans.

Shrader-Frechette, Kristin. Nanotoxicology and ethical conditions for informed consent. *NanoEthics* 2007 March; 1(1): 47-56. NRCBL: 5.4; 8.3.1; 16.1. SC: an.

Thurs, Daniel Patrick. No longer academic: models of commercialization and the construction of a nanotech industry. *Science as Culture* 2007 June; 16(2): 169-186. NRCBL: 5.4; 5.3; 1.3.2.

Toumey, Chris. Privacy in the shadow of nanotechnology. *NanoEthics* 2007 December; 1(3): 211-222. NRCBL: 5.4; 8.4. SC: an.

van den Hoven, Jeroen; Vermaas, Pieter E. Nanotechnology and privacy: on continuous surveillance outside the panopticon. *Journal of Medicine and Philosophy* 2007 May-June; 32(3): 283-297. NRCBL: 5.4; 1.1; 8.4.
Abstract: We argue that nano-technology in the form of invisible tags, sensors, and Radio Frequency Identity Chips (RFIDs) will give rise to privacy issues that are in two ways different from the traditional privacy issues of the last decades. One, they will not exclusively revolve around the idea of centralization of surveillance and concentration of power, as the metaphor of the Panopticon suggests, but will be about constant observation at decentralized levels. Two, privacy concerns may not exclusively be about constraining information flows but also about designing of materials and nano-artifacts such as chips and tags. We begin by presenting a framework for structuring the current debates on privacy, and then present our arguments.

NATIVE AMERICANS AS RESEARCH SUBJECTS *See* HUMAN EXPERIMENTATION/ SPECIAL POPULATIONS

NEUROSCIENCES *See* MENTAL HEALTH THERAPIES AND NEUROSCIENCES

NONTHERAPEUTIC HUMAN EXPERIMENTATION *See* HUMAN EXPERIMENTATION

NURSE PATIENT RELATIONSHIP *See* NURSING ETHICS AND PHILOSOPHY; PATIENT RELATIONSHIPS

NURSING CARE *See* CARE FOR SPECIFIC GROUPS; DEATH AND DYING/ TERMINAL CARE; NURSING ETHICS AND PHILOSOPHY

NURSING ETHICS AND PHILOSOPHY
See also BIOETHICS AND MEDICAL ETHICS; CODES OF ETHICS; PROFESSIONAL ETHICS

Arman, Maria; Rehnsfeldt, Arne. The presence of love in ethical caring. *Nursing Forum* 2006 January-March; 41(1): 4-12. NRCBL: 4.1.3; 1.1; 8.1.

Austin, Wendy; Bergum, Vangie; Dossetor, John. Relational ethics: an action ethic as a foundation for health care. *In:* Tschudin, Verena, ed. Approaches to Ethics: Nursing Beyond Boundaries. New York: Butterworth-Heinemann, 2003: 45-52. NRCBL: 4.1.3; 1.1; 4.4.

Badger, James M.; O'Connor, Bonnie. Moral discord, cognitive coping strategies, and medical intensive care unit nurses: insights from a focus group study. *Critical Care Nursing Quarterly* 2006 April-June; 29(2): 147-151. NRCBL: 4.1.3; 1.1; 7.3; 8.1.

Barazzetti, Gaia; Radaelli, Stefania; Sala, Roberta. Autonomy, responsibility and the Italian Code of Deontology for Nurses. *Nursing Ethics* 2007 January; 14(1): 83-98. NRCBL: 4.1.3; 1.1; 6.
Abstract: This article is a first assessment of the Italian Code of deontology for nurses (revised in 1999) on the basis of data collected from focus groups with nurses taking part in the Ethical Codes in Nursing (ECN) project. We illustrate the professional context in which the Code was introduced and explain why the 1999 revision was necessary in the light of changes affecting the Italian nursing profession. The most remarkable findings concern professional autonomy and responsibility, and how the Code is thought of as a set of guidelines for nursing practice. We discuss these issues, underlining that the 1999 Code represents a valuable instrument for ethical reflection and examination, a stimulus for putting the moral sense of the nursing profession into action, and that it represents a new era for professional nursing practice in Italy. The results of the analysis also deserve further qualitative study and future consideration.

Berggren, Ingela; Severinsson, Elisabeth. The significance of nurse supervisors' different ethical decision-making styles. *Journal of Nursing Management* 2006 November; 14(8): 637-643. NRCBL: 4.1.3; 7.3.

Breier-Mackie, Sarah. Medical ethics and nursing ethics: is there really any difference? *Gastroenterology Nursing: the Official Journal of the Society of Gastroenterology Nurses and Associates* 2006 March-April; 29(2): 182-183. NRCBL: 4.1.3; 4.1.2.

Bunch, Eli Haugen. Norway: some ethical challenges faced by health providers who work with first-generation immigrant men from Pakistan diagnosed with type 2 diabetes. *In:* Davis, Anne J.; Tschudin, Verena; de Raeve, Louise, eds. Essentials of Teaching and Learning in Nursing Ethics: Perspectives and Methods. New York: Churchill Livingstone Elsevier, 2006: 281-289. NRCBL: 4.1.3; 9.5.1; 21.7; 9.2; 1.1.

Dahlqvist, Vera; Eriksson, Sture; Glasberg, Ann-Louise; Lindahl, Elisabeth; Lützén, Kim; Strandberg, Gunilla; Söderberg, Anna; Sørlie, Venke; Norberg,

---

NRCBL: National Reference Center for Bioethics Literature Classification Scheme   See inside front cover for terms.

337

Astrid. Development of the perceptions of conscience questionnaire. *Nursing Ethics* 2007 March; 14(2): 181-193. NRCBL: 4.1.3. SC: em.

Abstract: Health care often involves ethically difficult situations that may disquiet the conscience. The purpose of this study was to develop a questionnaire for identifying various perceptions of conscience within a framework based on the literature and on explorative interviews about perceptions of conscience (Perceptions of Conscience Questionnaire). The questionnaire was tested on a sample of 444 registered nurses, enrolled nurses, nurses' assistants and physicians. The data were analysed using principal component analysis to explore possible dimensions of perceptions of conscience. The results showed six dimensions, found also in theory and empirical health care studies. Conscience was perceived as authority, a warning signal, demanding sensitivity, an asset, a burden and depending on culture. We conclude that the Perceptions of Conscience Questionnaire is valid for assessing some perceptions of conscience relevant to health care providers.

Davis, Anne J. An ethical voice for nurses — is anybody listening? [letter]. *Nursing Ethics* 2007 March; 14(2): 264. NRCBL: 4.1.3; 1.1.

Davis, Anne J. International nursing ethics: context and concerns. *In:* Tschudin, Verena, ed. Approaches to Ethics: Nursing Beyond Boundaries. New York: Butterworth-Heinemann, 2003: 95-104. NRCBL: 4.1.3; 21.1; 2.2.

Davis, Anne J.; Fowler, Marsha. Caring and caring ethics depicted in selected literature: what we know and what we need to ask. *In:* Davis, Anne J.; Tschudin, Verena; de Raeve, Louise, eds. Essentials of Teaching and Learning in Nursing Ethics: Perspectives and Methods. New York: Churchill Livingstone Elsevier, 2006: 165-179. NRCBL: 4.1.3; 1.1; 7.2. SC: an.

Davis, Anne J.; Tschudin, Verena; de Raeve, Louise. The future: teaching nursing ethics. *In their:* Essentials of Teaching and Learning in Nursing Ethics: Perspectives and Methods. New York: Churchill Livingstone Elsevier, 2006: 339-352. NRCBL: 4.1.3; 7.2.

de Raeve, Louise. A critique of virtue ethics. *In:* Davis, Anne J.; Tschudin, Verena; de Raeve, Louise, eds. Essentials of Teaching and Learning in Nursing Ethics: Perspectives and Methods. New York: Churchill Livingstone Elsevier, 2006: 109-122. NRCBL: 4.1.3; 1.1.

de Raeve, Louise. Teaching virtue ethics. *In:* Davis, Anne J.; Tschudin, Verena; de Raeve, Louise, eds. Essentials of Teaching and Learning in Nursing Ethics: Perspectives and Methods. New York: Churchill Livingstone Elsevier, 2006: 123-134. NRCBL: 4.1.3; 1.1; 7.2.

de Raeve, Louise. Virtue ethics. *In:* Davis, Anne J.; Tschudin, Verena; de Raeve, Louise, eds. Essentials of Teaching and Learning in Nursing Ethics: Perspectives

and Methods. New York: Churchill Livingstone Elsevier, 2006: 97-108. NRCBL: 4.1.3; 1.1.

Dinç, Leyla. Turkey: teaching ethics in Turkish nursing education programmes. *In:* Davis, Anne J.; Tschudin, Verena; de Raeve, Louise, eds. Essentials of Teaching and Learning in Nursing Ethics: Perspectives and Methods. New York: Churchill Livingstone Elsevier, 2006: 271-280. NRCBL: 4.1.3; 7.2; 10.

Dobrowolska, Beata; Wronska, Irena; Fidecki, Wieslaw; Wysokinski, Mariusz. Moral obligations of nurses based on the ICN, UK, Irish and Polish codes of ethics for nurses. *Nursing Ethics* 2007 March; 14(2): 171-180. NRCBL: 4.1.3; 6; 21.1. Identifiers: International Council of Nurses [ICN]; Great Britain (United Kingdom); Iceland; Poland.

Abstract: A code of professional conduct is a collection of norms appropriate for the nursing profession and should be the point of reference for all decisions made during the care process. Codes of ethics for nurses are formulated by members of national nurses' organizations. These codes can be considered to specify general norms that function in the relevant society, adjusting them to the character of the profession and enriching them with rules signifying the essence of nursing professionalism. The aim of this article is to present a comparative analysis of codes of ethics for nurses: the ICN's Code of ethics for nurses, the UK's Code of professional conduct, the Irish Code of professional conduct for each nurse and midwife, and the Polish Code of professional ethics for nurses and midwives. This analysis allows the identification of common elements in the professional ethics of nurses in these countries.

Drought, Theresa. The application of principle-based ethics to nursing practice and management: implications for the education of nurses. *In:* Davis, Anne J.; Tschudin, Verena; de Raeve, Louise, eds. Essentials of Teaching and Learning in Nursing Ethics: Perspectives and Methods. New York: Churchill Livingstone Elsevier, 2006: 81-96. NRCBL: 4.1.3; 1.1; 7.2; 9.1; 1.3.2. SC: an; cs.

Edwards, Steven. A principle-based approach to nursing ethics. *In:* Davis, Anne J.; Tschudin, Verena; de Raeve, Louise, eds. Essentials of Teaching and Learning in Nursing Ethics: Perspectives and Methods. New York: Churchill Livingstone Elsevier, 2006: 55-66. NRCBL: 4.1.3; 1.1; 2.1. SC: an.

Flaming, Don. The ethics of Foucault and Ricoeur: an underrepresented discussion in nursing. *Nursing Inquiry* 2006 September; 13(3): 220-227. NRCBL: 4.1.3; 1.1.

Fowler, Marsha. Religious and clinical ethics. *In:* Davis, Anne J.; Tschudin, Verena; de Raeve, Louise, eds. Essentials of Teaching and Learning in Nursing Ethics: Perspectives and Methods. New York: Churchill Livingstone Elsevier, 2006: 37-48. NRCBL: 4.1.3; 1.2; 1.1.

Fowler, Marsha. Social ethics, the profession and society. *In:* Davis, Anne J.; Tschudin, Verena; de Raeve, Louise, eds. Essentials of Teaching and Learning in Nursing Ethics: Perspectives and Methods. New York: Churchill Livingstone Elsevier, 2006: 27-36. NRCBL: 4.1.3; 6; 7.1.

Fry, Sara T. Nursing ethics. *In:* Khushf, George, ed. Handbook of Bioethics: Taking Stock of the Field From a Philosophical Perspective. Dordrecht; Boston: Kluwer Academic, 2004: 489-505. NRCBL: 4.1.3; 2.2.

Fullbrook, Suzanne. Best interests: a review of issues that affect nurses' decision making. *British Journal of Nursing* 2007 May 24 - June 13; 16(10): 600-601. NRCBL: 4.1.3; 8.1; 8.3.3. SC: le.

Fullbrook, Suzanne. Common law and a duty of care: the application of principles. *British Journal of Nursing* 2007 September 27 - October 10; 16(17):1074-1075. NRCBL: 4.1.3; 8.1. SC: le.

Gallagher, Ann. The respectful nurse. *Nursing Ethics* 2007 May; 14(3): 360-371. NRCBL: 4.1.3; 8.1.
  Abstract: Respect is much referred to in professional codes, in health policy documents and in everyday conversation. What respect means and what it requires in everyday contemporary nursing practice is less than clear. Prescriptions in professional codes are insufficient, given the complexity and ambiguity of everyday nursing practice. This article explores the meaning and requirements of respect in relation to nursing practice. Fundamentally, respect is concerned with value: where ethical value or worth is present, respect is indicated. Raz has argued that the two ways of encountering value are to respect and to engage with it. The former requires acknowledgement and preservation. Respect in nursing practice necessarily requires also engagement. Respect is an active value and can be conceptualized within the context of virtue ethics as a hybrid virtue having both intellectual and ethical components. Examples from the literature are provided to illustrate situations where the respectful nurse requires these components or capabilities.

Garzón, Nelly. Colombia: social justice in nursing ethics. *In:* Davis, Anne J.; Tschudin, Verena; de Raeve, Louise, eds. Essentials of Teaching and Learning in Nursing Ethics: Perspectives and Methods. New York: Churchill Livingstone Elsevier, 2006: 241-250. NRCBL: 4.1.3; 1.1; 7.2; 9.5.1; 9.2.

Gasull, Maria. Spain: professionalism and issues within nursing between nursing and other health professions. *In:* Davis, Anne J.; Tschudin, Verena; de Raeve, Louise, eds. Essentials of Teaching and Learning in Nursing Ethics: Perspectives and Methods. New York: Churchill Livingstone Elsevier, 2006: 313-321. NRCBL: 4.1.3; 7.2; 7.3.

Haegert, Sandy. Whose culture? An attempt at raising a culturally sensitive ethical awareness. *In:* Tschudin, Verena, ed. Approaches to Ethics: Nursing Beyond Boundaries. New York: Butterworth-Heinemann, 2003: 83-93. NRCBL: 4.1.3; 21.7. Identifiers: Africa.

Hirschfield, Miriam. An international perspective. *In:* Davis, Anne J.; Tschudin, Verena; de Raeve, Louise, eds. Essentials of Teaching and Learning in Nursing Ethics: Perspectives and Methods. New York: Churchill Livingstone Elsevier, 2006: 325-337. NRCBL: 4.1.3; 21.1.

Horton, Khim; Tschudin, Verena; Forget, Armorel. The value of nursing: a literature review. *Nursing Ethics* 2007 November; 14(6): 716-740. NRCBL: 4.1.3; 1.3.1; 21.1.

Johnson, Martin; Haigh, Carol; Yates-Bolton, Natalie. Valuing of altruism and honesty in nursing students: a two-decade replication study. *Journal of Advanced Nursing* 2007 February; 57(4): 366-374. NRCBL: 4.1.3; 7.2. SC: em. Identifiers: Great Britain (United Kingdom).

Juthberg, Christina; Eriksson, Sture; Norberg, Astrid; Sundin, Karin. Perceptions of conscience in relation to stress of conscience. *Nursing Ethics* 2007 May; 14(3): 329-343. NRCBL: 4.1.3; 9.5.2. SC: em. Identifiers: Sweden.
  Abstract: Every day situations arising in health care contain ethical issues influencing care providers' conscience. How and to what extent conscience is influenced may differ according to how conscience is perceived. This study aimed to explore the relationship between perceptions of conscience and stress of conscience among care providers working in municipal housing for elderly people. A total of 166 care providers were approached, of which 146 (50 registered nurses and 96 nurses' aides/enrolled nurses) completed a questionnaire containing the Perceptions of Conscience Questionnaire and the Stress of Conscience Questionnaire. A multivariate canonical correlation analysis was conducted. The first two functions emerging from the analysis themselves explained a noteworthy amount of the shared variance (25.6% and 17.8%). These two dimensions of the relationship were interpreted either as having to deaden one's conscience relating to external demands in order to be able to collaborate with coworkers, or as having to deaden one's conscience relating to internal demands in order to uphold one's identity as a 'good' health care professional.

Kidd, J.; Finlayson, M. Navigating uncharted water: research ethics and emotional engagement in human inquiry. *Journal of Psychiatric and Mental Health Nursing* 2006 August; 13(4): 423-428. NRCBL: 4.1.3; 7.1; 17.1; 4.3; 18.5.6.

Kim, Yong-Soon; Park, Jin-Hee; Han, Sung-Suk. Differences in moral judgment between nursing students and qualified nurses. *Nursing Ethics* 2007 May; 14(3): 309-319. NRCBL: 4.1.3; 1.3.1; 7.2. SC: em. Identifiers: Korea.
  Abstract: This longitudinal study examined how nursing students' moral judgment changes after they become qualified nurses working in a hospital environment. The

---

NRCBL: National Reference Center for Bioethics Literature Classification Scheme   See inside front cover for terms.

339

sample used was a group of 80 nursing students attending a university in Suwon, Korea, between 2001 and 2003. By using a Korean version of the Judgment About Nursing Decisions questionnaire, an instrument used in nursing care research, moral judgment scores based on Ketefian's six nursing dilemmas were determined. The results were as follows: (1) the qualified nurses had significantly higher idealistic moral judgment scores than the nursing students; (2) the qualified nurses showed significantly higher realistic moral judgment scores than the nursing students; and (3) when comparing idealistic and realistic moral judgment scores, both the qualified nurses and the nursing students had higher scores for idealistic moral judgment. Further study is recommended to examine changes in moral judgment.

Kirkham, Sheryl R.; Browne, Annette J. Toward a critical theoretical interpretation of social justice discourses in nursing. *ANS: Advances in Nursing Science* 2006 October-December; 29(4): 324-339. NRCBL: 4.1.3; 1.1.

Kolmer, D.M. Beneken Genaamd; Tellings, A.; Garretsen, H.F.L.; Bongers, I.M.B. Communalization of health care: how to do it properly. *Medicine and Law: The World Association for Medical Law* 2007 March; 26(1): 53-68. NRCBL: 4.1.3; 1.1.; 9.1.
Abstract: Communalization of health care refers to the increasing responsibility of citizens to look after their ill or handicapped fellow members of society and to provide care to them. Governments in Western Europe more and more develop health care policies directed at communalization of health care. The article discusses the care responsibilities of individuals based on the views of the philosophers Buber, Levinas, and Ricoeur and on the views of the family therapist Nagy. The care responsibilities of states are discussed in terms of the views of the political philosophers Rawls and Daniels and these are linked to right liberal, left liberal, and Christian-democrat views on care responsibilities of states. Thereupon, four criteria for a proper communalization of health care are proposed and different forms of health care policies with respect to communalization of care are assessed. In the last section, we look closely at several measures in the just reformed Dutch health care system and discuss how far these measures meet our criteria for a proper communalization. We focus in this section on the effects of these measures on family care because more and more family care plays an important role in good functioning of the health care system.

Konishi, Emiko; Davis, Anne J. Japan: the teaching of nursing ethics in Japan. *In:* Davis, Anne J.; Tschudin, Verena; de Raeve, Louise, eds. Essentials of Teaching and Learning in Nursing Ethics: Perspectives and Methods. New York: Churchill Livingstone Elsevier, 2006: 251-260. NRCBL: 4.1.3; 7.2; 1.1.

Kostas-Polston, Elizabeth A.; Hayden, Susan J. Living ethics: contributing to knowledge building through qualitative inquiry. *Nursing Science Quarterly* 2006 October; 19(4): 304-310. NRCBL: 4.1.3; 1.1; 8.1.

Laabs, Carolyn Ann. Primary care nurse practitioners' integrity when faced with moral conflict. *Nursing Ethics* 2007 November; 14(6): 795-809. NRCBL: 4.1.3; 9.1. SC: em.

Leners, Debra Woodward; Roehrs, Carol; Piccone, Adam Vincent. Tracking the development of professional values in undergraduate nursing students. *Journal of Nursing Education* 2006 December; 45(12): 504-511. NRCBL: 4.1.3; 7.2. SC: em.

Liaschenko, Joan; Peter, Elizabeth. Feminist ethics. *In:* Tschudin, Verena, ed. Approaches to Ethics: Nursing Beyond Boundaries. New York: Butterworth-Heinemann, 2003: 33-43. NRCBL: 4.1.3; 1.1; 10.

Liaschenko, Joan; Peter, Elizabeth. Feminist ethics: a way of doing ethics. *In:* Davis, Anne J.; Tschudin, Verena; de Raeve, Louise, eds. Essentials of Teaching and Learning in Nursing Ethics: Perspectives and Methods. New York: Churchill Livingstone Elsevier, 2006: 181-190. NRCBL: 4.1.3; 10; 1.1. SC: an.

Lindh, Inga-Britt; Severinsson, Elisabeth; Berg, Agneta. Moral responsibility: a relational way of being. *Nursing Ethics* 2007 March; 14(2): 129-140. NRCBL: 4.1.3; 7.2. SC: em. Identifiers: Sweden.
Abstract: This article reports a study exploring the meaning of the complex phenomenon of moral responsibility in nursing practice. Each of three focus groups with a total of 14 student nurses were conducted twice to gather their views on moral responsibility in nursing practice. The data were analysed by qualitative thematic content analysis. Moral responsibility was interpreted as a relational way of being, which involved guidance by one's inner compass composed of ideals, values and knowledge that translate into a striving to do good. It was concluded that, if student nurses are to continue striving to do good in a way that respects themselves and other people, it is important that they do not feel forced to compromise their values. Instead they should be given space and encouragement in their endeavours to do good in a relational way that advances nursing as a moral practice.

Lindsay, Robin; Graham, Helen. Relational narratives: solving an ethical dilemma concerning an individual's insurance policy. *In:* Tschudin, Verena, ed. Approaches to Ethics: Nursing Beyond Boundaries. New York: Butterworth-Heinemann, 2003: 53-60. NRCBL: 4.1.3; 1.1; 9.3.1. SC: cs.

Lundmark, Mikael. Vocation in theology-based nursing theories. *Nursing Ethics* 2007 November; 14(6): 767-780. NRCBL: 4.1.3; 1.2.

McNamee, Michael John. Nursing Schadenfreude: the culpability of emotional construction. *Medicine, Health Care and Philosophy* 2007 September; 10(3): 289-299. NRCBL: 4.1.3; 1.1; 4.1.2. SC: an.
Abstract: The purpose of this paper is to examine the concept of Schadenfreude - the pleasure felt at another's mis-

fortune - and to argue that feeling it in the course of health care work, as elsewhere, is evidence of a deficient character. In order to show that Schadenfreude is an objectionable emotion in health care work, I first offer some conceptual remarks about emotions generally and their differential treatment in Kantian and Aristotelian thought. Second, I argue that an appreciation of the rationality of the emotions is crucial to our self-understanding as persons in general and nurses in particular. Third, I present a critique of Portmann's (2000, When Bad Things Happen to Other People . London: Routledge) defence of Schadenfreude with examples from both nursing and medical scenarios. Specifically, I show how his exculpation of the emotion in terms of low self-esteem and a commitment to justice are not compelling. I argue that we are active in the construction of our emotional experiences of Schadenfreude , how we may indeed ,nurse' the emotion, and thus become culpable for them in ethical terms.

Memarian, Robabeh; Salsali, Mahvash; Vanaki, Zohreh; Ahmadi, Fazlolah; Hajizadeh, Ebrahim. Professional ethics as an important factor in clinical competency in nursing. *Nursing Ethics* 2007 March; 14(2): 203-214. NRCBL: 4.1.3; 1.3.1; 9.8. SC: em. Identifiers: Iran.

Abstract: It is imperative to understand the factors that influence clinical competency. Consequently, it is essential to study those that have an impact on the process of attaining clinical competency. A grounded theory approach was adopted for this study. Professional competency empowers nurses and enables them to fulfill their duties effectively. Internal and external factors were identified as affecting clinical competency. A total of 36 clinical nurses, nurse educators, hospital managers and members of the Nursing Council in Tehran participated in this research. Data were obtained by semistructured interviews. Personal factors and useful work experience were considered to be significant, based on knowledge and skills, ethical conduct, professional commitment, self-respect and respect for others, as well as from effective relationships, interest, responsibility and accountability. Effective management, education systems and technology were named as influential environmental factors. Personal and environmental factors affect clinical competency. Ethical persons are responsible and committed to their work, acquiring relevant work experience. A suitable work environment that is structured and ordered also encourages an ethical approach by nurses.

Olsen, Douglas. Editorial comment: nursing and other health care disciplines have a longstanding tradition of conscientious objection. *Nursing Ethics* 2007 May; 14(3): 277-279. NRCBL: 4.1.3; 8.1.

Paley, John. Past caring: the limitations of one-to-one ethics. *In:* Davis, Anne J.; Tschudin, Verena; de Raeve, Louise, eds. Essentials of Teaching and Learning in Nursing Ethics: Perspectives and Methods. New York: Churchill Livingstone Elsevier, 2006: 149-164. NRCBL: 4.1.3; 1.1. SC: an.

Pang, Samantha Mei-che. The principle-based approach to nursing ethics: a critical analysis. *In:* Davis, Anne J.; Tschudin, Verena; de Raeve, Louise, eds. Essentials of Teaching and Learning in Nursing Ethics: Perspectives and Methods. New York: Churchill Livingstone Elsevier, 2006: 67-79. NRCBL: 4.1.3; 1.1. SC: an; cs.

Peter, Elizabeth. Feminist ethics: a critique. *In:* Davis, Anne J.; Tschudin, Verena; de Raeve, Louise, eds. Essentials of Teaching and Learning in Nursing Ethics: Perspectives and Methods. New York: Churchill Livingstone Elsevier, 2006: 191-201. NRCBL: 4.1.3; 1.1; 10. SC: an.

Råholm, Maj-Britt. Caritative caring ethics: a description reflected through the Aristotelian terms phronesis, techne and episteme. *In:* Tschudin, Verena, ed. Approaches to Ethics: Nursing Beyond Boundaries. New York: Butterworth-Heinemann, 2003: 13-23. NRCBL: 4.1.3; 1.1. SC: an.

Rozsos, Elizabeth. Hungary: nursing and nursing ethics after the Communist era. *In:* Davis, Anne J.; Tschudin, Verena; de Raeve, Louise, eds. Essentials of Teaching and Learning in Nursing Ethics: Perspectives and Methods. New York: Churchill Livingstone Elsevier, 2006: 301-312. NRCBL: 4.1.3; 1.1; 21.1; 17.1; 21.4.

Scott, P. Anne. Virtue, nursing and the moral domain of practice. *In:* Tschudin, Verena, ed. Approaches to Ethics: Nursing Beyond Boundaries. New York: Butterworth-Heinemann, 2003: 25-32. NRCBL: 4.1.3; 1.1.

Shirley, Jamie L. Limits of autonomy in nursing's moral discourse. *ANS: Advances in Nursing Science* 2007 January-March; 30(1): 14-25. NRCBL: 4.1.3; 1.1; 8.1.

Solomon, Margot R.; DeNatale, Mary Lou. Academic dishonesty and professional practice: a convocation. *Nurse Educator* 2000 November-December; 25(6): 270-271. NRCBL: 4.1.3; 1.3.1; 2.3; 7.2; 8.2; 1.3.3.

Tabak, Nili. Israel: teaching nursing ethics in Israel: ancient values meet modern healthcare problems. *In:* Davis, Anne J.; Tschudin, Verena; de Raeve, Louise, eds. Essentials of Teaching and Learning in Nursing Ethics: Perspectives and Methods. New York: Churchill Livingstone Elsevier, 2006: 261-270. NRCBL: 4.1.3; 1.2; 7.2; 20.4.1.

Tang, Ping Fen; Johansson, Camilla; Wadensten, Barbro; Wenneberg, Stig; Ahlström, Gerd. Chinese nurses' ethical concerns in a neurological ward. *Nursing Ethics* 2007 November; 14(6): 810-824. NRCBL: 4.1.3; 9.5.1; 9.8. SC: em.

Toiviainen, Leila. 'The Globalisation of Nursing: Ethical, Legal and Political Issues' University of Surrey 10-11 July 2006: a summary of the deliberations of the concurrent working groups. *Nursing Ethics* 2007 March; 14(2): 258-263. NRCBL: 4.1.3; 21.1.

NRCBL: National Reference Center for Bioethics Literature Classification Scheme   See inside front cover for terms.

341

Torjuul, Kirsti; Elstad, Ingunn; Sørlie, Venke. Compassion and responsibility in surgical care. *Nursing Ethics* 2007 July; 14(4): 522-534. NRCBL: 4.1.3; 4.1.1; 8.1. SC: em. Identifiers: Norway.

Abstract: Ten nurses at a university hospital in Norway were interviewed as part of a comprehensive investigation into the narratives of nurses and physicians about being in ethically difficult situations in surgical units. The transcribed interview texts were subjected to a phenomenological-hermeneutic interpretation. The main theme in the narratives was being close to and moved by the suffering of patients and relatives. The nurses' responsibility for patients and relatives was expressed as a commitment to act, and they needed to ask themselves whether their responsibility had been fulfilled, that nothing had been left undone, overlooked or neglected, before they could leave the unit. When there was confirmation by the patients, relatives, colleagues and themselves that the needs of patients and relatives had been attended to in a morally and professionally satisfying manner, this increased the nurses' confidence and satisfaction in their work, and their strength to live with the burden of being in ethically difficult situations.

Tschudin, Verena. Narrative ethics. *In her:* Approaches to Ethics: Nursing Beyond Boundaries. New York: Butterworth-Heinemann, 2003: 61-72. NRCBL: 4.1.3; 1.1.

Uhrenfeldt, Lisbeth; Hall, Elisabeth O.C. Clinical wisdom among proficient nurses. *Nursing Ethics* 2007 May; 14(3): 387-398. NRCBL: 4.1.3; 8.1. SC: em. Identifiers: Denmark.

Abstract: This article examines clinical wisdom, which has emerged from a broader study about nurse managers' influence on proficient registered nurse turnover and retention. The purpose of the study was to increase understanding of proficient nurses' experience and clinical practice by giving voice to the nurses themselves, and to look for differences in their practice. This was a qualitative study based on semistructured interviews followed by analysis founded on Gadamerian hermeneutics. The article describes how proficient nurses experience their practice. Proficient practice constitutes clinical wisdom based on responsibility, thinking and ethical discernment, and a drive for action. The study showed that poor working conditions cause proficient nurses to regress to non-proficient performance. Further studies are recommended to allow deeper searching into the area of working conditions and their relationship to lack of nurse proficiency.

van Hooft, Stan. Caring and ethics in nursing. *In:* Tschudin, Verena, ed. Approaches to Ethics: Nursing Beyond Boundaries. New York: Butterworth-Heinemann, 2003: 1-12. NRCBL: 4.1.3; 1.1. SC: an.

Vivaldelli, Joan. Therapeutic reciprocity: "A union through pain.". *AJN: American Journal of Nursing* 2007 July; 107(7): 74, 76. NRCBL: 4.1.3; 8.1.

Zuzelo, Patti Rager. Exploring the moral distress of registered nurses. *Nursing Ethics* 2007 May; 14(3): 344-359. NRCBL: 4.1.3; 7.3; 8.1. SC: em.

Abstract: Registered nurses (RNs) employed in an urban medical center in the USA identified moral distress as a practice concern. This study describes RNs' moral distress and the frequency of morally distressing events. Data were collected using the Moral Distress Scale and an open-ended questionnaire. The instruments were distributed to direct-care-providing RNs; 100 responses were returned. Morally distressing events included: working with staffing levels perceived as 'unsafe', following families' wishes for patient care even though the nurse disagreed with the plan, and continuing life support for patients owing to family wishes despite patients' poor prognoses. One high frequency distressing event was carrying out orders for unnecessary tests and treatments. Qualitative data analysis revealed that the nurses sought support and information from nurse managers, chaplaincy services and colleagues. The RNs requested further information on biomedical ethics, suggested ethics rounds, and requested a non-punitive environment surrounding the initiation of ethics committee consultations.

ORGAN AND TISSUE TRANSPLANTATION
See also BLOOD BANKING, DONATION, AND TRANSFUSION

Badzek, Laurie A.; Cline, Heather S.; Moss, Alvin H.; Hines, Stephen C. Inappropriate use of dialysis for some elderly patients: nephrology nurses' perceptions and concerns. *Nephrology Nursing Journal* 2000 October; 27(5): 462-470; discussion 471-472. NRCBL: 19.3; 9.5.2; 4.4; 4.1.3.

Botkin, Jeffrey R.; Munger, Mark A.; Shea, Patrick A.; Coffin, Cheryl; Mineau, Geraldine P. Management of human tissue resources for research in academic medical centers: points to consider. *In:* Kulakowski, Elliott C.; Chronister, Lynne U., eds. Research Administration and Management. Sudbury, MA: Jones and Bartlett, 2006: 567-575. NRCBL: 19.1; 19.5; 18.3.

Caplan, Arthur L.; Perry, Constance; Plante, Lauren A.; Saloma, Joseph; Batzer, Frances R. Moving the womb. *Hastings Center Report* 2007 May-June; 37(3): 18-20. NRCBL: 19.1; 9.5.5; 14.1. Identifiers: uterus transplant.

Charatan, Fred. Organ recipients may die when insurance for drugs runs out [news]. *BMJ: British Medical Journal* 2007 March 17; 334(7593): 556. NRCBL: 19.3; 9.5.7; 9.3.1; 9.7; 20.1.

Cheema, Puneet; Mehta, Paulette. Pediatric stem cell transplantation ethical concerns. *In:* Mehta, P., ed. Pediatric Stem Cell Transplantation. Sudbury: Jones and Bartlet Publishers, 2004: p. 91-98. NRCBL: 19.1; 8.3.2; 19.4; 19.5; 18.5.4; 9.5.7; 1.1.

Cohen, Joshua T.; Neumann, Peter J. Dialysis facility ownership and epoetin dosing in hemodialysis patients: a medical economic perspective. *American Journal of Kidney Diseases* 2007 September; 50(3): 362-365. NRCBL: 19.3; 5.3; 9.7; 9.3.1. SC: em.

Crippin, Jeffrey S. Treatment of hepatocellular carcinoma after transplantation and human rights. *Hepatology* 2007 February; 45(2): 263-265. NRCBL: 19.1; 9.5.1; 9.2.

Daar, Abdallah S. The case for a regulated system of living kidney sales. *Nature Clinical Practice. Nephrology* 2006 November; 2(11): 600-601. NRCBL: 19.3; 19.5; 9.3.1.

Danovitch, Gabriel M.; Bunnapradist, Suphamai. Allocating deceased donor kidneys: maximizing years of life. *American Journal of Kidney Diseases* 2007 February; 49(2): 180-182. NRCBL: 19.3; 19.6.

Danovitch, G.M. A kidney for all ages. *American Journal of Transplantation* 2006 June; 6(6): 1267-1268. NRCBL: 19.3; 9.5.2; 19.6.

Einollahi, Behzad; Nourbala, Mohammad-Hossein; Bahaeloo-Horeh, Saeid; Assari, Shervin; Lessan-Pezeshki, Mahboob; Simforoosh, Naser. Deceased-donor kidney transplantation in Iran: trends, barriers and opportunities. *Indian Journal of Medical Ethics* 2007 April-June; 4(2): 70-72. NRCBL: 19.3; 19.5; 20.2.1. SC: le.

Fortin, Marie-Chantal; Roigt, Delphine; Doucet, Hubert. What should we do with patients who buy a kidney overseas? *Journal of Clinical Ethics* 2007 Spring; 18(1): 23-34. NRCBL: 19.3; 19.5; 8.1; 21.1; 9.3.1. SC: le.

Freeman, Michael; Jaoudé, Pauline Abou. Justifying surgery's last taboo: the ethics of face transplants. *Journal of Medical Ethics* 2007 February; 33(2): 76-81. NRCBL: 19.1; 9.5.1; 4.4; 8.1. SC: an.
  Abstract: Should face transplants be undertaken? This article examines the ethical problems involved from the perspective of the recipient, looking particularly at the question of identity, the donor and the donor's family, and the disfigured community and society more generally. Concern is expressed that full face transplants are going ahead.

Gaston, R.S.; Danovitch, G.M.; Epstein, R.A.; Kahn, J.P.; Matas, A.J.; Schnitzler, M.A. Limiting financial disincentives in live organ donation: a rational solution to the kidney shortage. *American Journal of Transplantation* 2006 November; 6(11): 2548-2555. NRCBL: 19.3; 19.5; 9.3.1.

Gillett, Grant. The use of human tissue. *Journal of Bioethical Inquiry* 2007; 4(2): 119-127. NRCBL: 19.1; 19.5; 8.3.1; 18.3; 18.2.
  Abstract: Abstract The use of human tissue raises ethical issues of great concern to health care professionals, biomedical researchers, ethics committees, tissue banks and policy makers because of the heightened importance given to informed consent and patient autonomy. The debate has been intensified by high profile scandals such as the "baby hearts" debacle and revelations about the retention of human brains in neuropathology laboratories worldwide. Respect for patient's rights seems, however, to impede research and development of clinical knowledge in contemporary health care. The Common clinical endeavour argument and a Presumption for beneficial use argument suggest that the use of tissues for research and teaching in contemporary health care can respect patients and their values in multicultural communities where there are provisions for oversight and for opting not to contribute, both of which should respect the diverse views of different ethnic or cultural groups.

Haddad, Haissam. Cardiac retransplantation: an ethical dilemma. *Current Opinion in Cardiology* 2006 March; 21(2): 118-119. NRCBL: 19.2; 19.5; 19.6.

Hippen, B.E.; Gaston, R.S. The conspicuous costs of more of the same. *American Journal of Transplantation* 2006 July; 6(7): 1503-1504. NRCBL: 19.3; 9.4; 19.5.

Howe, Edmund G. Taking patients' values seriously. *Journal of Clinical Ethics* 2007 Spring; 18(1): 4-11. NRCBL: 19.3; 19.5; 8.1; 14.4; 8.2.

Hyde, Merv; Power, Des. Some ethical dimensions of cochlear implantation for deaf children and their families. *Journal of Deaf Studies and Deaf Education* 2006 Winter; 11(1): 102-111. NRCBL: 19.1; 9.5.7; 19.6; 21.7.

Kasiske, Bertram L. Dialysis facility ownership and epoetin dosing in hemodialysis patients: a US physician perspective. *American Journal of Kidney Diseases* 2007 September; 50(3): 354-357. NRCBL: 19.3; 5.3; 9.7; 9.3.1. SC: em.

Kim, S. Joseph; Gordon, Elisa J.; Powe, Neil R. The economics and ethics of kidney transplantation: perspectives in 2006. *Current Opinion in Nephrology and Hypertension* 2006 November; 15(6): 593-598. NRCBL: 19.3; 9.3.1; 9.2.

Langlands, Nicola. Life after death: my life after a heart and lung transplant. *In:* Gunning, Jennifer; Holm, Søren, eds. Ethics, law, and society. Volume 2. Aldershot, Hants, England; Burlington, VT: Ashgate, 2006: 291-294. NRCBL: 19.2. SC: cs.

Lazarus, J. Michael; Hakim, Raymond M. Dialysis facility ownership and epoetin dosing in hemodialysis patients: a dialysis provider's perspective. *American Journal of Kidney Diseases* 2007 September; 50(3): 366-370. NRCBL: 19.3; 5.3; 9.7; 9.3.1. SC: em.

Lerner, Barron H. Hero or victim? Barney Clark and the technological imperative. *In his:* When Illness Goes Public: Celebrity Patients and How We Look at Medicine. Baltimore: Johns Hopkins University Press, 2006: 180-200. NRCBL: 19.2; 18.3; 4.4; 5.1. SC: cs.

NRCBL: National Reference Center for Bioethics Literature Classification Scheme   See inside front cover for terms.

343

MacDougall, Iain C. Dialysis facility ownership and epoetin dosing in hemodialysis patients: a view from Europe. *American Journal of Kidney Diseases* 2007 September; 50(3): 358-361. NRCBL: 19.3; 5.3; 9.7; 9.3.1. SC: em.

Martínez-Alarcón, L.; Ríos, A.; Conesa, C.; Alcaraz, J.; González, M.J.; Ramírez, P.; Parrilla, P. Attitude of kidney patients on the transplant waiting list toward related-living donation. A reason for the scarce development of living donation in Spain. *Clinical Transplantation* 2006 November-December; 20(6): 719-724. NRCBL: 19.3; 19.5.

Mazaris, Evangelos; Papalois, Vassilios E. Ethical issues in living donor kidney transplantation. *Experimental and Clinical Transplantation* 2006 December; 4(2): 485-497. NRCBL: 19.3; 19.5; 9.3.1.

Meckler, Laura. More kidneys for transplants may go to young; policy to stress benefit to patient over length of time on wait list. *Wall Street Journal* 2007 March 10; p. A1, A7. NRCBL: 19.3; 19.6; 9.5.7. SC: po.

Moosa, M.R.; Kidd, M. The dangers of rationing dialysis treatment: the dilemma facing a developing country. *Kidney International* 2006 September; 70(6): 1107-1114. NRCBL: 19.3; 9.2; 9.4.

Morgan, Myfanwy; Hooper, Richard; Mayblin, Maya; Jones, Roger. Attitudes to kidney donation and registering as a donor among ethnic groups in the UK. *Journal of Public Health* 2006 September; 28(3): 226-234. NRCBL: 19.3; 19.5; 9.5.4. SC: em.

Murphy, Fiona; Byrne, Gobnait. Ethical issues regarding live kidney transplantation. *British Journal of Nursing* 2007 October 25-November 7; 16(19): 1224-1229. NRCBL: 19.3; 19.5.

Northern Ireland Targeting Social Need Renal Group; Kee, Frank; Reaney, Elizabeth; Savage, Gerard; O'Reilly, Dermot; Patterson, Chris; Maxwell, Peter; Fogarty, Damian. Are gatekeepers to renal services referring patients equitably? *Journal of Health Services Research and Policy* 2007 January; 12(1): 36-41. NRCBL: 19.3; 9.5.1; 9.2; 9.4.

Pearson, Helen. Infertility researchers target uterus transplant [news]. *Nature* 2007 February 1; 445(7127): 466-467. NRCBL: 19.1; 14.1; 18.5.3.

Pollack, Andrew. The dialysis business: fair treatment? *New York Times* 2007 September 16; p. BU1, BU10. NRCBL: 19.3; 1.3.2; 9.5.1; 9.3.1. SC: po. Identifiers: DaVita; Kent J. Thiry.

Preminger, Beth A.; Fins, Joseph J. Face transplantation: an extraordinary case with lessons for ordinary practice. *Plastic and Reconstrive Surgery* 2006 September 15; 118(4): 1073-1074. NRCBL: 19.1; 4.4; 9.5.1.

Reach, G. Innovative therapies: some ethical considerations. *Diabetes and Metabolism* 2006 December; 32(5, Part 2): 527-531. NRCBL: 19.1; 1.1; 5.2.

Ringel, Steven P. Autonomy and ars moriendi. *Neurology* 2006 September 26; 67(6): 1101-1102. NRCBL: 19.2; 8.3.1; 8.3.2; 8.3.4; 20.5.1.

Rohrich, Rod J.; Longaker, Michael T.; Cunningham, Bruce. On the ethics of composite tissue allotransplantation (facial transplantation) [editorial]. *Plastic and Reconstructive Surgery* 2006 May; 117(6): 2071-2073. NRCBL: 19.1; 9.7; 8.3.1.

Royal College of Surgeons of England. Working party. Facial transplantation: working party report. London: Royal College of Surgeons of England, 2006 November; 48 p. [Online]. Accessed: http://www.rcseng.ac.uk/publications/docs/facial_transplant_report_2006.html/pdffile/downloadFile [2007 April 12]. NRCBL: 19.1; 19.6.

Sheldon, Tony. Holland bans private stem cell therapy [news]. *BMJ: British Medical Journal* 2007 January 6; 334(7583): 12. NRCBL: 19.1; 15.1; 18.5; 9.5.1. SC: le. Identifiers: Netherlands.

Stein, Rob. First U.S. uterus transplant planned; some experts say risk isn't justified. *Washington Post* 2007 January 15; p. A1, A9. NRCBL: 19.1; 9.5.5; 14.1. SC: po.

Steinberg, David. How much risk can medicine allow a willing altruist? *Journal of Clinical Ethics* 2007 Spring; 18(1): 12-17. NRCBL: 19.3; 19.5; 8.1.

Stewart, Cameron. Introduction: the human body — The Land That Time Forgot. *Journal of Bioethical Inquiry* 2007; 4(2): 117-118. NRCBL: 19.1; 2.1. SC: le.

Swindell, J.S. Facial allograft transplantation, personal identity and subjectivity. *Journal of Medical Ethics* 2007 August; 33(8): 449-453. NRCBL: 19.1; 1.1; 4.4. SC: an.
  Abstract: An analysis of the identity issues involved in facial allograft transplantation is provided in this paper. The identity issues involved in organ transplantation in general, under both theoretical accounts of personal identity and subjective accounts provided by organ recipients, are examined. It is argued that the identity issues involved in facial allograft transplantation are similar to those involved in organ transplantation in general, but much stronger because the face is so closely linked with personal identity. Recipients of facial allograft transplantation have the potential to feel that their identity is a mix between their own and the donor's, and the donor's family is potentially likely to feel that their loved one "lives on". It is also argued that facial allograft transplantation allows the recipients to regain an identity, because they can now be seen in the social world. Moreover, they may regain expressivity, allowing for them to be seen even more by others, and to regain an identity to an even greater extent. Informing both recipients and donors about the role that identity plays in facial allograft

transplantation could enhance the consent process for facial allograft transplantation and donation.

Talone, Patricia A. Making decisions about organ transplantation. *In:* Hamel, Ronald, ed. Making Health Care Decisions: A Catholic Guide. Liguori, MO: Liguori Publications, 2006: 55-72. NRCBL: 19.1; 1.2; 8.3.1; 9.3.1; 19.5.

Taylor, Robert S. Self-ownership and transplantable human organs. *Public Affairs Quarterly* 2007 January; 21(1): 89-107. NRCBL: 19.1; 19.5; 19.6; 1.1; 4.4; 9.3.1.

Thamer, Mae; Zhang, Yi. Dialysis facility ownership and epoetin dosing in patients receiving hemodialysis: the authors respond. *American Journal of Kidney Diseases* 2007 October; 50(4): 538-541. NRCBL: 19.3; 5.3; 9.7; 9.3.1. SC: em.

Waterman, A.D.; Schenk, E.A.; Barrett, A.C.; Waterman, B.M.; Rodrigue, J.R.; Woodle, E.S.; Shenoy, S.; Jendrisak, M.; Schnitzler, M. Incompatible kidney donor candidates' willingness to participate in donor-exchange and non-directed donation. *American Journal of Transplantation* 2006 July; 6(7): 1631-1638. NRCBL: 19.3; 19.5; 19.6.

Weiner, Daniel E.; Levey, Andrew S. Dialysis facility ownership and epoetin dosing in hemodialysis patients: an overview. *American Journal of Kidney Diseases* 2007 September; 50(3): 349-353. NRCBL: 19.3; 5.3; 9.7; 9.3.1. SC: em.

ORGAN AND TISSUE TRANSPLANTATION/ ALLOCATION

Brudney, Daniel. Are alcoholics less deserving of liver transplants? *Hastings Center Report* 2007 January-February; 37(1): 41-47. NRCBL: 19.6; 9.5.9.

Chandler, Jennifer A. Priority systems in the allocation of organs for transplant: should we reward those who have previously agreed to donate? *Health Law Journal* 2005; 13: 99-138. NRCBL: 19.6; 19.5; 9.4. SC: le; an.

de Beaufort, Inez; Meulenberg, Frans. The dangers of triage by television. *BMJ: British Medical Journal* 2007 June 9; 334(7605): 1194-1195. NRCBL: 19.5; 1.3.7; 9.4.

Draper, Heather; MacDiarmaid-Gordon, Adam; Strumidlo, Laura; Teuten, Bea; Updale, Eleanor. Virtual clinical ethics committee, case 8/case 4 vol 2: should non-medical circumstances determine whether a child is placed on the transplant register when there is a risk of wasting a scarce organ? *Clinical Ethics* 2007 December; 2(4): 166-172. NRCBL: 19.6; 20.5.2; 19.2; 9.6. SC: cs.

Kerr, Cathel. Hoax raises awareness about organ shortages [news]. *CMAJ/JAMC: Canadian Medical Association Journal* 2007 July 17; 177(2): 135. NRCBL: 19.6; 19.5.

Magnus, David; Tabor, Holly; Karkazis, Katrina. Transplants for developmentally delayed children. *Ethics and Intellectual Disability* 2007 Winter; 10(1): 3-4. NRCBL: 19.6; 9.5.3; 7.2.

Neuberger, James; Gimson, Alexander. Selfless adults and split donor livers [comment]. *Lancet* 2007 July 28-August 3; 370(9584): 299-300. NRCBL: 19.6; 19.5; 1.1; 9.5.7. SC: em.

Pennings, Guido. Directed organ donation: discrimination or autonomy? *Journal of Applied Philosophy* 2007; 24(1): 41-49. NRCBL: 19.6; 1.1; 19.5.
Abstract: Numerous measures have been proposed to change the collection procedure in order to increase the supply of organ donations. One such proposal is to give the candidate donors the right to direct their organs to groups of recipients characterised by specific features like sex, age, disease and geographic location. Four possible justifications for directed donation of organs are considered: the utilitarian benefit, the egalitarian principle of justice, the maximin principle of justice and the autonomy principle. It is concluded that none of these principles justifies the acceptance of designated donations. When potentially life-saving resources are distributed, only a pure egalitarian distribution is in agreement with the principle of justice.

Tuffs, Annette. Media claim allocations of organs to Saudi patients was unfair [news]. *BMJ: British Medical Journal* 2007 September 29; 335(7621): 634. NRCBL: 19.6; 19.1; 19.3; 21.1.

United Network for Organ Sharing [UNOS]. Ethics Committee. Preferred status for organ donors: a report of the United Network for Organ Sharing Ethics Committee. Richmond, VA: United Network for Organ Sharing [UNOS], 1993 June 30: 4 p. [Online]. Accessed: http://www.unos.org/Resources/bioethics.asp?index=5 [2007 May 8]. NRCBL: 19.6.

United Network for Organ Sharing [UNOS]. Ethics Committee, Organ Procurement and Transplantation Network [OPTN]. Allocation of Organs from Non-Directed Living Donors. Richmond, VA: United Network for Organ Sharing, 2002 June: 1 p. [Online]. Accessed: http://www.unos.org/Resources/bioethics.asp?index=9 [2007 May 8]. NRCBL: 19.6.

Veatch, Robert M.; Balint, John A.; Glannon, Walter; Cohen, Peter J.; Brudney, Daniel. Just deserts? [letters and reply]. *Hastings Center Report* 2007 May-June; 37(3): 4-6. NRCBL: 19.6; 9.5.9. Comments: Daniel Brudney. Are alcoholics less deserving of liver transplants? Hastings Center Report 2007 January-February; 37(1): 41-47.

Weimer, David L. Public and private regulation of organ transplantation: liver allocation and the final rule. *Journal of Health Politics, Policy and Law* 2007 February; 32(1): 9-49. NRCBL: 19.6; 5.3. SC: le.

NRCBL: National Reference Center for Bioethics Literature Classification Scheme   See inside front cover for terms.

345

Abstract: The allocation of cadaveric organs for transplantation in the United States is governed by a process of private regulation. Through the Organ Procurement and Transplantation Network (OPTN), stakeholders and public representatives determine the substantive content of allocation rules. Between 1994 and 2000 the U.S. Department of Health and Human Services conducted a rule making to define more clearly the public and private roles in the determination of organ allocation policy. Several prominent liver transplant centers that were losing market share as a result of the proliferation of transplant centers used the rule making as a vehicle for challenging the local priority for organ allocation inherent in the OPTN rules. The process leading to the final rule provides a window on the politics of organ allocation. It also facilitates an assessment of the strengths and weaknesses of private rule making. Overall, private rule making appears to be relatively effective in tapping the technical expertise and tacit knowledge of stakeholders to allow for the adaptation of rules in the face of changing technology and information. However, the particular system of representation employed may give less influence to some stakeholders than they would have in public regulatory arenas, giving them an incentive to seek public rule making as a remedy for their persistent losses within the framework of private rule making.

ORGAN AND TISSUE TRANSPLANTATION/ DONATION AND PROCUREMENT

China: doctors agree not to take organs from prisoners. *New York Times* 2007 October 6; p. A9. NRCBL: 19.5; 18.5.5. SC: po.

Presumed consent. *ATLA:Alternatives to Laboratory Animals* 2007 August; 35(4): 379. NRCBL: 19.5; 8.3.1. Identifiers: Great Britain (United Kingdom).

Two sample policies: donation after cardiac death. *Health Care Ethics USA* 2007 Summer; 15(3): 12-18. NRCBL: 19.5; 1.2; 20.2.1.

Aulisio, Mark P.; Devita, Michael; Luebke, Donna. Taking values seriously: ethical challenges in organ donation and transplantation for critical care professionals. *Critical Care Medicine* 2007 February; 35(2, Supplement): S95-S101. NRCBL: 19.5; 1.1; 19.1; 20.1.

Banasik, Miroslaw. Living donor transplantation — the real gift of life. Procurement and the ethical assessment. *Annals of Transplantation* 2006; 11(1): 4-6. NRCBL: 19.5; 19.3.

Bartley, Richard. Is prognosis key in donation? [letter]. *BMJ: British Medical Journal* 2007 June 9; 334(7605): 1179. NRCBL: 19.5. Comments: Veronica English. Is presumed consent the answer to organ shortages? Yes. BMJ: British Medical Journal 2007 May 26; 334(7603): 1088.

Bayley, Carol. Back to basics: examining the assumptions of donation after cardiac death. *Health Care Ethics USA* 2007 Summer; 15(3): 2-4. NRCBL: 19.5; 20.2.1.

Bellomo, Rinaldo; Zamperetti, Nereo. Defining the vital condition for organ donation [commentary]. *Philosophy, Ethics, and Humanities in Medicine [electronic]* 2007 November 19; 2(27): 3 p. Accessed: http://www.peh-med.com/ [2008 January 24]. NRCBL: 19.5; 20.2.1. Comments: J.L. Verheijde, M.Y. Rady, and J. McGregor. Recovery of transplantable organs after cardiac or circulatory death: transforming the paradigm for the ethics of organ donation. Philosophy, Ethics, and Humanities in Medicine 2007 May 22; 2:8.

Björkman, Barbro. Different types — different rights: distinguishing between different perspectives on ownership of biological material. *Science and Engineering Ethics* 2007 June; 13(2): 221-233. 34 refs. 18 fn. NRCBL: 19.5; 4.4; 15.1. SC: an.
    Keywords: *body parts and fluids; *ethical analysis; *genetic materials; *property rights; biotechnology; commodification; donors; hearts; kidneys; organ donation; rights; standards; remuneration; stem cells; tissue donation; transplant recipients

Brazier, Margaret; Quigley, Muireann. Deceased organ donation: in praise of pragmatism [editorial]. *Clinical Ethics* 2007 December; 2(4): 164-165. NRCBL: 19.5; 20.1; 7.1.

Brennan, Patricia. Public solicitation of organs on the Internet: ethical and policy issues. *Journal of Emergency Nursing* 2006 April; 32(2): 191-193. NRCBL: 19.5; 1.3.12; 19.3.

Brody, Jane E. The solvable problem of organ shortages. *New York Times* 2007 August 28; p F7. NRCBL: 19.5; 20.2.1. SC: po.

Canova, Daniele; De Bona, Manuela; Ruminati, Rino; Ermani, Mario; Naccarato, Remo; Burra, Patrizia. Understanding of and attitudes to organ donation and transplantation: a survey among Italian university students. *Clinical Transplantation* 2006 May-June; 20(3): 307-312. NRCBL: 19.5; 19.1. SC: em.

Chianchiano, D. The Uniform Anatomical Gift Act and organ donation in the United States. *Advances in Chronic Kidney Disease* 2006 April; 13(2): 189-191. NRCBL: 19.5. SC: le.

Cronin, Antonia J. Transplants save lives, defending the double veto does not: a reply to Wilkinson. *Journal of Medical Ethics* 2007 April; 33(4): 219-220. NRCBL: 19.5; 8.3.1. SC: an.

de Beaufort, Inez; Meulenberg, Frans. The dangers of triage by television. *BMJ: British Medical Journal* 2007 June 9; 334(7605): 1194-1195. NRCBL: 19.5; 1.3.7; 9.4.

DeJohn, Carla; Zwischenberger, Joseph B. Ethical implications of extracorporeal interval support for organ retrieval (EISOR) [editorial]. *ASAIO Journal: American Society for Artificial Internal Organs* 2006 March-April; 52(2): 119-122. NRCBL: 19.5; 20.4.1; 1.2.

Donatelli, Luke A.; Geocadin, Romergryko G.; Williams, Michael A. Ethical issues in critical care and cardiac arrest: clinical research, brain death, and organ donation. *Seminars in Neurology* 2006 September; 26(4): 452-459. NRCBL: 19.5; 8.3.1; 20.2.1; 20.4.1; 20.5.1.

DuBois, James M. Donation after cardiac death: a reply to Bayley and Gallagher. *Health Care Ethics USA* 2007 Summer; 15(3): 7-8. NRCBL: 19.5; 20.2.1.

DuBois, James M.; Anderson, Emily E. Attitudes toward death criteria and organ donation among healthcare personnel and the general public. *Progress in Transplantation* 2006 March; 16(1): 65-73. NRCBL: 19.5; 8.3.1; 20.2.1; 20.3.1; 20.3.2. SC: rv.

DuBois, James M.; DeVita, Michael. Donation after cardiac death in the United States: how to move forward. *Critical Care Medicine* 2006 December; 34(12): 3045-3047. NRCBL: 19.5; 20.1.

Dyer, Clare. UK considers moving to new system to increase organ donation [news]. *BMJ: British Medical Journal* 2007 September 29; 335(7621): 634-635. NRCBL: 19.5.

Dyer, Owen. Inquiry will study removal of Sellafield workers' body parts. *BMJ: British Medical Journal* 2007 April 28; 334(7599): 868. NRCBL: 19.5; 8.3.3; 16.2.

English, Veronica; Wright, Linda. Is presumed consent the answer to organ shortages? [debate]. *BMJ: British Medical Journal* 2007 May 26; 334(7603): 1088-1089. NRCBL: 19.5; 8.3.1.

Farrugia, Albert. When do tissues and cells become products? — Regulatory oversight of emerging biological therapies. *Cell and Tissue Banking* 2006; 7(4): 325-335. NRCBL: 19.5; 18.6. Identifiers: Australia; Canada.

Fujita, Misao; Akabayashi, Akira; Slingsby, Brian Taylor; Kosugi, Shinji; Fujimoto, Yasuhiro; Tanaka, Koichi. A model of donors' decision-making in adult-to-adult living donor liver transplantation in Japan: having no choice. *Liver Transplantation* 2006 May; 12(5): 768-774. NRCBL: 19.5; 9.4; 8.3.1.

Gallagher, John A. Donation after cardiac death: an ethical reflection on the development of a protocol. *Health Care Ethics USA* 2007 Summer; 15(3): 5-6. NRCBL: 19.5; 20.2.1; 8.3.1; 1.2; 9.7.

Glanville, A.R. Ethical and equity issues in lung transplantation and lung volume reduction surgery. *Chronic Respiratory Disease.* 2006; 3(1): 53-58. NRCBL: 19.5; 19.6; 19.1; 9.2. Identifiers: Australia.

Goz, Fugen; Goz, Mustafa; Erkan, Medine. Knowledge and attitudes of medical, nursing, dentistry and health technician students towards organ donation: a pilot study. *Journal of Clinical Nursing* 2006 November; 15(11): 1371-1375. NRCBL: 19.5; 7.2; 19.1. SC: em.

Grasser, Phyllis L. Donation after cardiac death: major ethical issues. *National Catholic Bioethics Quarterly* 2007 Autumn; 7(3): 527-543. NRCBL: 19.5; 9.6; 1.2; 20.2.1; 20.3.3.

Griffith, Richard. Legal requirements for donating and retaining organs: the Human Tissue Act. *British Journal of Community Nursing* 2006 October; 11(10): 446-449. NRCBL: 19.5; 1.3.5; 19.1.

Hanto, Douglas W. Ethical challenges posed by the solicitation of deceased and living organ donors. *New England Journal of Medicine* 2007 March 8; 356(10): 1062-1066. NRCBL: 19.5; 19.6.

Harris, John. Mark Anthony or Macbeth: some problems concerning the dead and the incompetent when it comes to consent. *In:* McLean, Sheila A.M., ed. First Do No Harm: Law, Ethics, and Healthcare. Aldershot, England; Burlington, VT: Ashgate, 2006: 287-301. NRCBL: 19.5; 8.3.1; 20.1.

Healy, G.W. Moral and legal aspects of transplantation: prisoners or death convicts as donors. *Transplantation Proceedings* 1998 November; 30(7): 3653-3654. NRCBL: 19.5; 1.3.5; 9.5.1.

Helft, Paul R.; Champion, Victoria L.; Eckles, Rachael; Johnson, Cynthia S.; Meslin, Eric M. Cancer patients' attitudes toward future research uses of stored human biological materials. *Journal of Empirical Research on Human Research Ethics* 2007 September; 2(3): 15-22. 15 refs. NRCBL: 19.5; 18.3; 9.5.1; 15.1. SC: em.

Keywords: *attitudes; *biological specimen banks; *biomedical research; *donors; *genetic materials; *informed consent; *patients; cancer; privacy; racial groups; research subjects; survey; time factors; Proposed Keywords: *tissue donors; Keyword Identifiers: Indiana University Cancer Center

Abstract: The policy debate concerning informed consent for future, unspecified research of stored human biological materials (HBM) would benefit from an understanding of the attitudes of individuals who contribute tissue specimens to HBM repositories. Cancer patients who contributed leftover tissue to the Indiana University Cancer Center Tissue Bank under such conditions were recruited for a mail survey study of their attitudes. Our findings suggest that a clear majority of subjects would permit unlimited future research on stored HBMs without re-contact and reconsent, and a significant minority appear to desire ongoing control over future research uses of their tissue. These differences merit further investigation and suggest that a policy of blanket consent for all future, unspecified research would be premature.

Helft, Paul R.; Daugherty, Christopher K. Are we taking without giving in return? The ethics of research-related biopsies and the benefits of clinical trial participation. *Journal of Clinical Oncology* 2006 October 20; 24(30): 4793-4795. NRCBL: 19.5; 18.1; 18.2; 18.3; 9.5.1.

NRCBL: National Reference Center for Bioethics Literature Classification Scheme   See inside front cover for terms.

347

Hilhorst, Medard T.; Kranenburg, Leonieke W.; Busschbach, Jan J.V. Should health care professionals encourage living kidney donation? *Medicine, Health Care and Philosophy* 2007 March; 10(1): 81-90. NRCBL: 19.5; 19.3; 8.1. SC: an.

Abstract: Living kidney donation provides a promising opportunity in situations where the scarcity of cadaveric kidneys is widely acknowledged. While many patients and their relatives are willing to accept its benefits, others are concerned about living kidney programs; they appear to feel pressured into accepting living kidney transplantations as the only proper option for them. As we studied the attitudes and views of patients and their relatives, we considered just how actively health care professionals should encourage living donation. We argue that active interference in peoples' personal lives is justified - if not obligatory. First, we address the ambiguous ideals of non-directivity and value neutrality in counselling. We describe the main pitfalls implied in these concepts, and conclude that these concepts cannot account for the complex reality of living donation and transplantation. We depict what is required instead as truthful information and context-relative counselling. We then consider professional interference into personal belief systems. We argue that individual convictions are not necessarily strong, stable, or deep. They may be flawed in many ways. In order to justify interference in peoples' personal lives, it is crucial to understand the structure of these convictions. Evidence suggests that both patients and their relatives have attitudes towards living kidney donation that are often open to change and, accordingly, can be influenced. We show how ethical theories can account for this reality and can help us to discern between justified and unjustified interference. We refer to Stephen Toulmin's model of the structure of logical argument, the Rawlsian model of reflective equilibrium, and Thomas Nagel's representation of the particularistic position.

Hirshberg, Boaz. Can we justify living donor islet transplantation? *Current Diabetes Reports* 2006 August; 6(4): 307-309. NRCBL: 19.5; 19.1.

Holmes, Anne. Add carrying card to QOF [letter]. *BMJ: British Medical Journal* 2007 June 9; 334(7605): 1179. NRCBL: 19.5. Identifiers: Quality and Outcomes Framework.

Howard, R.J. We have an obligation to provide organs for transplantation after we die. *American Journal of Transplantation* 2006 August; 6(8): 1786-1789. NRCBL: 19.5; 20.1; 1.1.

Joffe, Ari R. The ethics of donation and transplantation: are definitions of death being distorted for organ transplantation? *Philosophy, Ethics, and Humanities in Medicine [electronic]* 2007 November 25; 2(28): 7 p. Accessed: http://www.peh-med.com/ [2008 January 24]. NRCBL: 19.5; 20.2.1. Comments: S.D. Shemie. Clarifying the paradigm for the ethics of donation and transplantation: was 'dead' really so clear before organ donation? Philosophy, Ethics, and Humanities in Medicine 2007; 2(18).

Lo, Bernard. Ethical issues in organ transplantation. *In his:* Resolving Ethical Dilemmas: A Guide for Clinicians. 3rd edition. Philadelphia, PA: Lippincott Williams and Wilkins, 2005: 264-271. NRCBL: 19.5; 20.2.1; 1.1.

Mandell, M. Susan;; Zamudio, Stacy; Seem, Debbie; McGaw, Lin J.; Wood, Geri; Liehr, Patricia.; Ethier, Angela; D'Alessandro, Anthony M. National evaluation of healthcare provider attitudes toward organ donation after cardiac death. *Critical Care Medicine* 2006 December; 34(12): 2952-2958. NRCBL: 19.5; 7.1; 20.3.2.

Mayer, Susan. HFEA allows women to donate their eggs for research [news]. *BMJ: British Medical Journal* 2007 March 3; 334(7591): 445. NRCBL: 19.5; 14.6; 14.5; 1.3.9; 18.1. Identifiers: Great Britain (United Kingdom); Human Fertilsation and Embryology Authority.

Meckler, Laura. What living organ donors need to know; even as transplants surge, data on long-term impact on givers remain scant. *Wall Street Journal* 2007 January 30; p. B1, B13. NRCBL: 19.5; 9.5.1. SC: po; em.

Munson, Ronald. Organ transplantation. *In:* Steinbock, Bonnie, ed. The Oxford Handbook of Bioethics. Oxford; New York: Oxford University Press, 2007: 211-239. NRCBL: 19.5; 19.3; 4.4.

Muula, A.S.; Mfutso-Bengo, J.M. Responsibilities and obligations of using human research specimens transported across national boundaries. *Journal of Medical Ethics* 2007 January; 33(1): 35-38. NRCBL: 19.5; 1.3.9; 9.5.6; 18.3; 21.1. Identifiers: Malawi.

Abstract: Research collaboration beyond national jurisdiction is one aspect of the globalisation of health research. It has potential to complement researchers in terms of research skills, equipment and lack of adequate numbers of potential research subjects. Collaboration at an equal level of partnership though desirable, may not be practicable. Sometimes, human research specimens must be transported from one country to other. Where this occurs, there should be clear understanding between the collaborating research institutions regarding issues of access and control of the specimens as well as the duration of storage of specimens. The researchers have the duty to inform the research participants about specimen storage and transport across national boundaries. While obtaining informed consent from study subjects if specimens are to be stored beyond the life of the present study could be the ideal, there still remains significant challenges in a multi-cultural world.

O'Neil, Peter. China's doctors signal retreat on organ harvest. *CMAJ/JAMC: Canadian Medical Association Journal* 2007 November 20; 177(11): 1341. NRCBL: 19.5; 1.3.5; 20.6.

Olsen, Douglas P. Arranging live organ donation over the Internet: is it ethical? Each nurse must decide. *AJN: American Journal of Nursing* 2007 March; 107(3): 69-72. NRCBL: 19.5; 19.3; 1.3.12.

Øverland, Gerhard. Survival lotteries reconsidered. *Bioethics* 2007 September; 21(7): 355-363. NRCBL: 19.5; 2.1. SC: an.

Abstract: In 1975 John Harris envisaged a survival lottery to redistribute organs from one to a greater number in order to reduce number of deaths as a consequence of organ failure. In this paper I reach a conclusion about when running a survival lottery is permissible by looking at the reason prospective participants have for allowing the procedure from a contractual perspective. I identify three versions of the survival lottery. In a National Lottery, everyone within a jurisdiction is a candidate for being a donor for everyone else, disregarding all differences between individuals' eventual possibility of needing an organ. In a Group Specific Lottery, it is a question of running a lottery among members of a specific group who share the same probability of getting organ failure. In a Local Lottery one randomises among individuals who are already in need of a new organ but who happen to be compatible and in need of different organs. While the first is vulnerable to considerations of fairness, it is difficult to perceive a feasible way to implement the second option that does not come with a host of unwelcome consequences. I argue, however, that it is permissible to run Local Lotteries.

Parry, Jane. A matter of life and death. China is moving towards changing its transplantation practices in a bid to gain wider international acceptance [news]. *BMJ: British Medical Journal* 2007 November 10; 335(7627): 961. NRCBL: 19.5; 21.1.

Pitchers, M.; Stokes, A.; Lonsdale, R.; Premachandra, D.J.; Edwards, D.R. Research tissue banking in otolaryngology: organization, methods and uses, with reference to practical, ethical and legal issues. *Journal of Laryngology and Otology* 2006 June; 120(6): 433-438. NRCBL: 19.5; 18.5.1; 9.5.1; 18.3.

Quigley, Muireann. Property and the body: applying Honoré. *Journal of Medical Ethics* 2007 November; 33(11): 631-634. NRCBL: 19.5; 1.1; 4.4. SC: an.

Abstract: This paper argues that the new commercial and quasi-commercial activities of medicine, scientists, pharmaceutical companies and industry with regard to human tissue has given rise to a whole new way of valuing our bodies. It is argued that a property framework may be an effective and constructive method of exploring issues arising from this. The paper refers to A M Honoré's theory of ownership and aims to show that we have full liberal ownership of our own bodies and as such can be considered to be self-owners.

Rady, Mohamed Y.; Verheijde, Joseph L.; Spital, Aaron. Ethically increasing the supply of transplantable organs [letters]. *Annals of Internal Medicine* 2007 April 3; 146(7): 537-538. NRCBL: 19.5; 20.2.1.

Ríos, A.; Ramírez, P.; Martínez, L.; Montoya, M.J.; Lucas, D.; Alcaraz, J.; Rodríguez, M.M.; Rodríguez, J.M.; Parrilla, P. Are personnel in transplant hospitals in favor of cadaveric organ donation? Multivariate attitudinal study in a hospital with a solid organ transplant program. *Clinical Transplantation* 2006 November-December; 20(6): 743-754. NRCBL: 19.5; 20.3.2; 20.1; 7.1. SC: em.

Roff, Sue Rabbitt. Self-interest, self-abnegation and self-esteem: towards a new moral economy of non-directed kidney donation. *Journal of Medical Ethics* 2007 August; 33(8): 437-441. NRCBL: 19.5; 19.3. Identifiers: Great Britain (United Kingdom).

Abstract: As of September 2006, non-directed donation of kidneys and other tissues and organs is permitted in the UK under the new Human Tissue Acts. At the same time as making provision for psychiatric and clinical assessment of so-called "altruistic" donations to complete strangers, the Acts intensify assessments required for familial, genetically related donations, which will now require the same level as genetically unrelated but "emotionally" connected donations by locally based independent assessors reporting to the newly constituted Human Tissue Authority. But there will also need to be considerable reflection on the criteria for "stranger donation", which may lead us to a new understanding of the moral economy of altruistic organ donation, no matter how mixed the motives of the donor may be. This paper looks at some of the issues that will have to be accommodated in such a framework.

Ross, Lainie Friedman; Siegler, Mark; Thistlethwaite, J. Richard, Jr. We need a registry of living kidney donors. *Hastings Center Report* 2007 November-December; 37(6): Inside back cover. NRCBL: 19.5; 19.3.

Rushton, Cynda Hylton. Donation after cardiac death: ethical implications and implementation strategies. *AACN Advanced Critical Care* 2006 July-September; 17(3): 345-349. NRCBL: 19.5; 20.1.

Sanner, Margareta A. Giving and taking — to whom and from whom? People's attitudes toward transplantation of organs and tissues from different sources. *Clinical Transplantation* 1998 December; 12(6): 530-537. NRCBL: 19.5; 19.1; 22.2. SC: em. Identifiers: Sweden.

Sanner, Margareta A. People's attitudes and reactions to organ donation. *Mortality* 2006 May; 11(2): 133-150. NRCBL: 19.5; 7.1; 20.3.1. SC: em.

Satel, Sally. Desperately seeking a kidney. What you learn about people — and yourself — when you need them to donate an organ. *New York Times Magazine* 2007 December 16; p. 62-67. NRCBL: 19.5; 19.3. SC: po.

Schauenburg, H.; Hildebrandt, A. Public knowledge and attitudes on organ donation do not differ in Germany and Spain. *Transplantation Proceedings* 2006 June; 38(5): 1218-1220. NRCBL: 19.5; 20.1; 21.1. SC: em.

Scott, Larry D. Living donor liver transplant — is the horse already out of the barn? *The American Journal of Gastroenterology* 2006 April; 101(4): 686-688. NRCBL: 19.5; 19.1; 8.3.1. SC: cs.

NRCBL: National Reference Center for Bioethics Literature Classification Scheme   See inside front cover for terms.

349

Spital, Aaron; Snyder, Deborah J.; Miller, Franklin J.; Rosenstein, Donald L.; Domingo, Angela F.; Salvana, Edsel Maurice T.; Rady, Mohamed Y.; Verheijde, Joseph L.; McGregor, Joan; Hanto, Douglas W. Solicitation of deceased and living organ donors [letters and reply]. *New England Journal of Medicine* 2007 June 7; 356(23): 2427-2429. NRCBL: 19.5.

Spital, Aaron; Steinberg, David. Living donor list exchanges disadvantage blood-group-O recipients [letter and reply]. *American Journal of Kidney Diseases* 2005 May; 45(5): 962. NRCBL: 19.5; 19.4.

Sque, Magi; Payne, Sheila; Clark, Jill Macleod. Gift of life or sacrifice?: key discourses to understanding organ donor families' decision-making. *Mortality* 2006 May; 11(2): 117-132. NRCBL: 19.5; 8.1; 7.1. SC: em.

Stein, Rob. New trend in organ donation raises questions; as alternative approach becomes more frequent, doctors worry that it puts donors at risk. *Washington Post* 2007 March 18;. NRCBL: 19.5; 5.2. SC: po; em.

Stein, Rob. New zeal in organ procurement raises fears; donation groups say they walk a fine line, but critics see potential for abuses. *Washington Post* 2007 September 13; p. A1, A9. NRCBL: 19.5. SC: po.

Steinberg, David. Reply to Valapour, "Living donor transplantation: the perfect balance of public oversight and medical responsibility". *Journal of Clinical Ethics* 2007 Spring; 18(1): 21-22. NRCBL: 19.5; 8.1.

Steinbrook, Robert. Organ donation after death. *New England Journal of Medicine* 2007 July 19; 357(3): 209-213. NRCBL: 19.5; 20.2.1. SC: rv.

Sulmasy, Dan. DCD policies: the devil is always in the details. *Health Care Ethics USA* 2007 Fall; 15(4): 22. NRCBL: 19.5; 20.2.1; 1.2. Identifiers: donation after cardiac death.

United Network for Organ Sharing [UNOS]. Board of Directors. Directed Donation. Richmond, VA: United Network for Organ Sharing, 1996 June: 2 p. [Online]. Accessed: http://www.unos.org/Resources/bioethics.asp?index=10 [2007 May 8]. NRCBL: 19.5; 19.6.

United Network for Organ Sharing [UNOS]. Ethics Committee. Payment Subcommittee. Financial incentives for organ donation. A report of the payment subcommittee. Richmond, VA: United Network for Organ Sharing, 1993 June 30: 4 p. [Online]. Accessed: http://www.unos.org/Resources/bioethics.asp?index=3 [2007 May 8]. NRCBL: 19.5.

United Network for Organ Sharing [UNOS]. Research to Practice Steering Committee; Metzger, Robert A.; Taylor, Gloria J.; McGaw, Lin J.; Weber, Phyllis G.; Delmonico, Francis L.; Prottas, Jeffrey M. Research to practice: a national consensus conference. *Progress in Transplantation* 2005 December; 15(4): 379-384. NRCBL: 19.5; 8.3.1; 8.1; 20.1.

United States. Department of Health and Human Services [DHHS]; United States. Centers for Medicare and Medicaid Services. Medicare and Medicaid programs; conditions for coverage for organ procurement organizations (OPOs). Final rule. *Federal Register* 2006 May 31; 71(104): 30981-31054. NRCBL: 19.5; 1.3.5; 19.1.

Upshur, Ross E.G.; Lavery, James V.; Tindana, Paulina O. Taking tissue seriously means taking communities seriously. *BMC Medical Ethics* 2007; 8(11): 1-6 [Online]. Accessed: http://www.biomedcentral.com/content/pdf/1472-6939-8-11.pdf [2008 January 24]. NRCBL: 19.5; 18.5.9; 18.6; 18.3; 21.1. Identifiers: Tonga; India; Uganda.
Abstract: BACKGROUND: Health research is increasingly being conducted on a global scale, particularly in the developing world to address leading causes of morbidity and mortality. While research interest has increased, building scientific capacity in the developing world has not kept pace. This often leads to the export of human tissue (defined broadly) from the developing to the developed world for analysis. These practices raise a number of important ethical issues that require attention. DISCUSSION: In the developed world, there is great heterogeneity of regulatory practices regarding human tissues. In this paper, we outline the salient ethical issues raised by tissue exportation, review the current ethical guidelines and norms, review the literature on what is known empirically about perceptions and practices with respect to tissue exportation from the developing to the developed world, set out what needs to be known in terms of a research agenda, and outline what needs to be done immediately in terms of setting best practices. We argue that the current status of tissue exportation is ambiguous and requires clarification lest problems that have plagued the developed world occur in the context of global heath research with attendant worsening of inequities. Central to solutions to current ethical concerns entail moving beyond concern with individual level consent and embracing a robust interaction with communities engaged in research. CONCLUSION: Greater attention to community engagement is required to understand the diverse issues associated with tissue exportation.

Valapour, Maryam. Living donor transplantation: the perfect balance of public oversight and medical responsibility. *Journal of Clinical Ethics* 2007 Spring; 18(1): 18-20. NRCBL: 19.5; 19.3.

Varma, Sumeeta; Ratterman, Allison Griffin. Sharing data, DNA and tissue samples. *Journal of Empirical Research on Human Research Ethics* 2007 March; 2(1): 97-100. 4 refs. NRCBL: 19.5; 15.1; 1.3.12; 8.4.
Keywords: *access to information; *children; *confidentiality; *DNA sequences; *genetic databases; *genetic information; *genetic privacy; *genetic research; *publishing; adolescents; biological specimen banks; cancer; disclosure; donors; empirical research; federal government; genetic predisposition; government regulation; human genome; Internet; legal aspects; parental consent; research ethics

committees; research subjects; risk; risks and benefits; state government; Proposed Keywords: genome-wide association studies; tissue donors; Keyword Identifiers: Health Insurance Portability and Accountability Act (HIPAA); United States

Wilkinson, T.M. Individual and family decisions about organ donation. *Journal of Applied Philosophy* 2007; 24(1): 26-40. NRCBL: 19.5; 8.1; 1.1; 4.4. SC: an.

Abstract: This paper examines, from a philosophical point of view, the ethics of the role of the family and the deceased in decisions about organ retrieval. The paper asks: Who, out of the individual and the family, should have the ultimate power to donate or withhold organs? On the side of respecting the wishes of the deceased individual, the paper considers and rejects arguments by analogy with bequest and from posthumous bodily integrity. It develops an argument for posthumous autonomy based on the liberal idea of self-development and argues that this establishes a right of veto over donation. It claims, however, that whether the family's power to veto would conflict with posthumous autonomy rights depends on how it comes about. On the side of respecting the family's wishes, the paper first considers an argument from family distress. This supports a contingent, non-rights-based reason for the family's power that is trumped by the deceased's rights. It then outlines and criticises an argument based on family autonomy. The conclusion is that the individual has the right to veto the family's wish to donate and that, while the family has no right to veto the individual's wishes to donate, it can legitimately acquire this power and has done so in practice.

Wilson, Penelope; Sexton, Wendy; Singh, Andrea; Smith, Melissa; Durham, Stephanie; Cowie, Anne; Fritschi, Lin. Family experiences of tissue donation in Australia. *Progress in Transplantation* 2006 March; 16(1): 52-56. NRCBL: 19.5; 8.3.3; 20.1; 8.1. SC: em.

Yacoub, Ahmed Abdel Aziz. Transplantation. *In his:* The Fiqh of Medicine: Responses in Islamic Jurisprudence to Developments in Medical Science. London, UK: Ta-Ha Publishers Ltd., 2001: 254-280. NRCBL: 19.5; 1.2; 20.2.1; 18.5.4; 15.1.

ORGAN AND TISSUE TRANSPLANTATION/ . . . / ECONOMIC ASPECTS

Legal and illegal organ donation [editorial]. *Lancet* 2007 June 9-15; 369(9577): 1901. NRCBL: 19.5; 9.3.1; 21.1. SC: le.

Arnold, Robert; Bartlett, Steven; Bernat, James; Colonna, John; Dafoe, Donald; Dubler, Nancy; Gruber, Scott; Kahn, Jeffrey; Luskin, Richard; Nathan, Howard; Orloff, Susan; Prottas, Jeffrey; Shapiro, Robyn; Ricordi, Camillo; Youngner, Stuart; Delmonico, Francis L. Financial incentives for cadaver organ donation: an ethical reappraisal. *Transplantation* 2002 April 27; 73(8): 1361-1367. NRCBL: 19.5; 4.4; 9.3.1; 19.1; 20.1.

Batson, Andrew; Oster, Shai. China reconsiders fairness of "transplant tourism"; foreigners pay more for scarce organs; Israelis debate reform. *Wall Street Journal* 2007 April 6; p. A1, A9. NRCBL: 19.5; 19.6; 21.1; 9.3.1. SC: po. Identifiers: China; Israel.

Baylis, Françoise; McLeod, C. The stem cell debate continues: the buying and selling of eggs for research. *Journal of Medical Ethics* 2007 December; 33(12): 726-731. 41 refs. NRCBL: 19.5; 14.5; 1.3.9; 18.5.4; 15.1; 18.5.3; 9.3.1. SC: an.

Keywords: *embryo research; *embryonic stem cells; *guidelines; *ovum donors; *remuneration; cloning; embryo disposition; ethical analysis; disadvantaged persons; in vitro fertilization; incentives; international aspects; organizational policies; ovum; patients; professional organizations; research subjects; risk; socioeconomic factors; women; Keyword Identifiers: *ISSCR Guidelines; *International Society for Stem Cell Research

Boulware, L.E.; Troll, M.U.; Wang, N.Y.; Powe, N.R. Public attitudes toward incentives for organ donation: a national study of different racial/ethnic and income groups. *American Journal of Transplantation* 2006 November; 6(11): 2774-2785. NRCBL: 19.5; 3.1; 9.3.1.

Brown, Susan; Glenn, David. The true price of a human organ: economists and surgeons debate on whether legalizing the sale of body parts will help or harm. *Chronicle of Higher Education* 2007 March 23; 53(29): A12-A15. NRCBL: 19.5; 19.6; 21.1; 1.3.2; 9.3.1. SC: le.

Buciuniene, Ilona; Stonienė, Laimutė; Blazeviciene, Aurelija; Kazlauskaite, Ruta; Skudiene, Vida. Blood donors' motivation and attitude to non-remunerated blood donation in Lithuania. *BMC Public Health* 2006 June 22; 6: 166-173. NRCBL: 19.5; 19.4; 9.3.1. SC: em.

Clay, Megan; Block, Walter. A free market for human organs. *Journal of Social, Political and Economic Studies* 2002 Summer; 27(2): 227-236. NRCBL: 19.5; 9.3.1. SC: an; le.

Coleman, Gerald D. Organ donation: charity or commerce? *America* 2007 March 5; 196(8): 22-24. NRCBL: 19.5; 1.2; 9.3.1.

Council of Europe. Parliamentary Assembly. Trafficking in organs in Europe. Recommendation 1611 (2003). Strasbourg, France: Council of Europe (21st Sitting), 2003 June 25: 4 p. [Online]. Accessed: http://assembly.coe.int/Documents/AdoptedText/ta03/EREC1611.htm [2007 May 1]. NRCBL: 19.5; 9.3.1; 21.1.

Dickenson, Donna L. Tissue economies: biomedicine and commercialization [review of Tissue Economics: Blood, Organs, and Cell Lines in Late Capitalism by Catherine Waldby and Robert Mitchell]. *Perspectives in Biology and Medicine* 2007 Spring; 50(2): 308-311. NRCBL: 19.5; 4.4; 9.3.1.

NRCBL: National Reference Center for Bioethics Literature Classification Scheme   See inside front cover for terms.

351

Dobson, Roger. WHO reports on the growing commercial trade in transplant organs [news]. *BMJ: British Medical Journal* 2007 November 17; 335(7628): 1013. NRCBL: 19.5; 21.1; 9.3.1. Identifiers: World Health Organization.

Dykstra, Alyssa. Should incentives be used to increase organ donation? *Plastic Surgical Nursing* 2004 April-June; 24(2): 70-74. NRCBL: 19.5; 1.1; 9.3.1.

Epstein, Miran. The ethics of poverty and the poverty of ethics: the case of Palestinian prisoners in Israel seeking to sell their kidneys in order to feed their children. *Journal of Medical Ethics* 2007 August; 33(8): 473-474. NRCBL: 19.5; 9.5.1; 19.3; 9.3.1.
  Abstract: Bioethical arguments conceal the coercion underlying the choice between poverty and selling ones organs.

Fox, M.D. The price is wrong: the moral cost of living donor inducements. *American Journal of Transplantation* 2006 November; 6(11): 2529-2530. NRCBL: 19.5; 9.3.1; 1.1.

Friedman, E.A.; Friedman, A.L. Payment for donor kidneys: pros and cons. *Kidney International* 2006 February 15; 69(5): 960-962 [Online]. Accessed: http://www.nature.com/ki/journal/v69/n6/pdf/5000262a.pdf [2006 October 3]. NRCBL: 19.5; 19.3; 9.3.1. SC: an.

Garden, Rebecca; Murphree, Hyon Joo Yoo. Class and ethnicity in the global market for organs: the case of Korean cinema. *Journal of Medical Humanities* 2007 December; 28(4): 213-229. NRCBL: 19.5; 4.4; 7.1; 21.1; 9.3.1.
  Abstract: While organ transplantation has been established in the medical imagination since the 1960s, this technology is currently undergoing a popular re-imagination in the era of global capitalism. As transplantation procedures have become routine in medical centers in non-Western and developing nations and as organ sales and transplant tourism become increasingly common, organs that function as a material resource increasingly derive from subaltern bodies. This essay explores this development as represented in Korean filmmaker Park Chan-wook's 2002 Sympathy for Mr. Vengeance, focusing on the ethnic and class characteristics of the global market in organs and possible modes of counter-logic to transplant technologies and related ethical discourses.

Griffin, Anne. Kidneys on demand. *BMJ: British Medical Journal* 2007 March 10; 334(7592): 502-505. NRCBL: 19.5; 9.3.1; 19.6; 9.1; 1.3.5; 19.3. SC: le; rv. Identifiers: Iran.

Hinkley, Charles C. The selling of organs. *In his:* Moral Conflicts of Organ Retrieval: A Case for Constructive Pluralism. Amsterdam; New York: Rodopi, 2005: 105-115. NRCBL: 19.5; 9.3.1.

Jafarey, Aamir; Thomas, George; Ahmad, Aasim; Srinivasan, Sandhya. Asia's organ farms [editorial]. *Indian Journal of Medical Ethics* 2007 April-June; 4(2): 52-53. NRCBL: 19.5; 9.3.1.

Khalili, Mohammed I. Organ trading in Jordan: bad news, good news. *Politics and the Life Sciences* 2007 March; 26(1): 12-14. NRCBL: 19.5; 9.3.1; 21.1.

Meckler, Laura. Kidney shortage inspires a radical idea: organ sales. *Wall Street Journal* 2007 November 13; p. A1 [Online]. Accessed: http://online.wsj.com/public/article_print/SB119490273908090431.html [2007 November 13]. NRCBL: 19.5; 9.3.1; 19.3; 19.6. SC: po.

Pence, Gregory E. Kant's critique of adult organ donation. *In his:* The Elements of Bioethics. Boston: McGraw-Hill, 2007: 52-80. NRCBL: 19.5; 1.1; 9.3.1.

Porter, Susan. Organ transplants. Part two: Questions and controversy. *Ohio State Medical Journal* 1984 January; 80(1): 33, 37, 39. NRCBL: 19.5; 9.3.1; 1.3.2; 19.6; 21.1.

Rai, Mohammad A.; Afzal, Omer. Organs in the bazaar: the end of the beginning? *Politics and the Life Sciences* 2007 March; 26(1): 10-11. NRCBL: 19.5; 9.3.1; 21.1. Identifiers: Pakistan.

Rothman, S.M.; Rothman, D.J. The hidden cost of organ sale. *American Journal of Transplantation* 2006 July; 6(7): 1524-1528. NRCBL: 19.5; 1.1; 9.3.1.

Scheper-Hughes, Nancy. Rotten trade: millennial capitalism, human values and global justice in organs trafficking. *Journal of Human Rights* 2003 June; 2(2): 197-226. NRCBL: 19.5; 19.3; 2.1; 9.3.1; 21.1. SC: cs.

Scheper-Hughes, Nancy. The ends of the body: commodity fetishism and the global traffic in organs. *SAIS Review* 2002 Winter-Spring; 22(1): 61-80 [Online]. Accessed: http://muse.jhu.edu/journals/sais_review/v022/22.1scheper.pdf [2007 February 21]. NRCBL: 19.5; 21.1; 19.3; 9.3.1. SC: cs.

Sheldon, Tony. Undertakers offer cash incentive for organ donation in an attempt to encourage donors. *BMJ: British Medical Journal* 2007 June 2; 334(7604): 1131. NRCBL: 19.5; 9.3.1. Identifiers: Netherlands.

Shroff, Sunil. Working towards ethical organ transplants. *Indian Journal of Medical Ethics* 2007 April-June; 4(2): 68-69. NRCBL: 19.5; 9.3.1; 7.1. SC: le.

Steiner, Hillel. The right to trade in human body parts. *In:* Seglow, Jonathan ed. The Ethics of Altruism. London: Frank Cass, 2004: 187-193. NRCBL: 19.5; 9.3.1; 1.1; 4.4. SC: an.

Sun, Chiao-Yin.; Lee, Chin-Chan; Chang, Chiz-Tzung; Hung, Cheng-Chih; Wu, Mai-Szu. Commercial cadaveric renal transplant: an ethical rather than medical issue. *Clinical Transplantation* 2006 May-June; 20(3): 340-345. NRCBL: 19.5; 20.1; 1.3.2; 9.3.1.

Taylor, J.S. A "queen of hearts" trial of organ markets: why Scheper-Hughes's objections to markets in human organs fail. *Journal of Medical Ethics* 2007 April; 33(4): 201-204. NRCBL: 19.5; 9.3.1; 19.3; 1.1. SC: an.

Abstract: Nancy Scheper-Hughes is one of the most prominent critics of markets in human organs. Unfortunately, Scheper-Hughes rejects the view that markets should be used to solve the current (and chronic) shortage of transplant organs without engaging with the arguments in favour of them. Scheper-Hughes's rejection of such markets is of especial concern, given her influence over their future, for she holds, among other positions, the status of an adviser to the World Health Organization (Geneva) on issues related to global transplantation. Given her influence, it is important that Scheper-Hughes's moral condemnation of markets in human organs be subject to critical assessment. Such critical assessment, however, has not generally been forthcoming. A careful examination of Scheper-Hughes's anti-market stance shows that it is based on serious mischaracterisations of both the pro-market position and the medical and economic realities that underlie it. In this paper, the author will expose and correct these mischaracterisations and, in so doing, show that her objections to markets in human organs are unfounded.

Vathsala, A. Commercial renal transplantation — body parts for sale [editorial]. *Annals of the Academy of Medicine, Singapore* 2006 April; 35(4): 227-228. NRCBL: 19.5; 19.3; 9.3.1. SC: le.

Waltz, Emily. The body snatchers: rising demand has created a thriving market for human body parts — and not all of it above ground [news]. *Nature Medicine* 2006 May; 12(5): 487-488. NRCBL: 19.5; 4.4; 9.3.1; 18.1; 20.1. SC: le.

Warburg, Ronnie. Renal transplantation: living donors and markets for body parts — Halakha in concert with halakhic policy or public policy? *Tradition* 2007 Summer; 40(2): 14-48. NRCBL: 19.5; 1.2; 1.3.2; 9.3.1.

Watts, Jonathan. China introduces new rules to deter human organ trade. *Lancet* 2007 June 9-15; 369(9577): 1917-1918. NRCBL: 19.5; 9.3.1; 21.1. SC: le.

Zarocostas, John. UN calls for tougher rules to prevent sale of children's organs [news]. *BMJ: British Medical Journal* 2007 March 31; 334(7595): 656. NRCBL: 19.5; 19.2; 9.5.7; 9.3.1; 21.1.

ORGAN AND TISSUE TRANSPLANTATION/ . . . / LEGAL ASPECTS

Infected patient's lawyer says risk wasn't disclosed. *New York Times* 2007 November 18; p. A35. NRCBL: 19.5; 18.3; 9.5.6. SC: po; le. Identifiers: HIV; hepatitis.

Legal and illegal organ donation [editorial]. *Lancet* 2007 June 9-15; 369(9577): 1901. NRCBL: 19.5; 9.3.1; 21.1. SC: le.

Abadie, Alberto; Gay, Sebastien. The impact of presumed consent legislation on cadaveric organ donation: a cross-country study. *Journal of Health Economics* 2006 July; 25(4): 599-620. NRCBL: 19.5; 8.3.1; 20.1; 21.1. SC: rv; le.

Archibold, Randal C. 2 accused of trading in cadaver parts. *New York Times* 2007 March 8; p. A20. NRCBL: 19.5. SC: po; le.

Arnold, Brent. Legal solutions to Ontario's organ shortage: redrawing the boundaries of consent. *Health Law Journal* 2005; 13: 139-163. NRCBL: 19.5; 8.3.1; 8.3.3; 20.3.1. SC: le.

Barshes, Neal R.; Hacker, Carl S.; Freeman, Richard B., Jr.; Vierling, John M.; Goss, John A. Justice, administrative law, and the transplant clinician: the ethical and legislative basis of a national policy on donor liver allocation. *Journal of Contemporary Health Law and Policy* 2007 Spring; 23(2): 200-230. NRCBL: 19.5; 2.1; 1.1; 19.6. SC: le.

Bell, M.D. Emergency medicine, organ donation and the Human Tissue Act. *Emergency Medicine Journal* 2006 November; 23(11): 824-827. NRCBL: 19.5; 20.1; 8.3.1. SC: le. Identifiers: Great Britain (United Kingdom).

Bucklin, Leonard H. Woe unto those who request consent: ethical and legal considerations in rejecting a deceased's anatomical gift because there is no consent by the survivors. *North Dakota Law Review* 2002 October; 78: 323-354. NRCBL: 19.5; 20.1; 8.3.1; 8.3.3. SC: le.

Clay, Megan; Block, Walter. A free market for human organs. *Journal of Social, Political and Economic Studies* 2002 Summer; 27(2): 227-236. NRCBL: 19.5; 9.3.1. SC: an; le.

Cook, Kristin. Familial consent for registered organ donors: a legally rejected concept. *Health Matrix: The Journal of Law-Medicine* 2007 Winter; 17(1): 117-145. NRCBL: 19.5; 20.3.3. SC: le.

DeVita, Michael A.; Caplan, Arthur L. Caring for organs or for patients? Ethical concerns about the Uniform Anatomical Gift Act (2006). *Annals of Internal Medicine* 2007 December 18; 147(12): 876-879. NRCBL: 19.5; 20.4.1. SC: le.

Abstract: In 2006, the National Conference of Commissioners on Uniform State Laws rewrote the Uniform Anatomical Gift Act. To overcome the problem of family members prohibiting organ donation from their deceased loved ones even when a donor card existed, the commissioners modified the act to prevent end-of-life care from precluding organ donation. An unintended consequence of the new wording creates the potential for end-of-life care that prioritizes care of the potential donor organs over care and comfort of the dying person. The commissioners have now revised the act, but the original version has already been legislated in many states, with others poised to follow. To protect dying patients' wishes about their end-of-life care, states that have legislated or are considering the original act must replace it with the re-

NRCBL: National Reference Center for Bioethics Literature Classification Scheme   See inside front cover for terms.

353

vised version. A long-term and important ethical precept must stand: Care of dying patients takes precedence over organs. Another laudable goal must be promoted as well: Organ donation is an important part of end-of-life care.

Feifer, Jason. Paying big to be a donor; gifting an organ can be costly. Would a tax break cross a moral line? *Washington Post* 2007 March 20; p. F1, F4. NRCBL: 19.5; 9.3.1. SC: po; le.

Glazier, Alexandra K.; Sasjack, Scott. Should it be illicit to solicit? A legal analysis of policy options to regulate solicitation of organs for transplant. *Health Matrix: The Journal of Law-Medicine* 2007 Winter; 17(1): 63-99. NRCBL: 19.5; 19.6; 4.4. SC: le.

Greif, Karen F.; Merz, Jon F. Who lives and who dies? Organ transplantation. *In their:* Current Controversies in the Biological Sciences: Case Studies of Policy Challenges from New Technologies. Cambridge, MA: MIT, 2007: 329-365. NRCBL: 19.5; 19.1; 20.2.1. SC: le.

Griffin, Anne. Kidneys on demand. *BMJ: British Medical Journal* 2007 March 10; 334(7592): 502-505. NRCBL: 19.5; 9.3.1; 19.6; 9.1; 1.3.5; 19.3. SC: rv. Identifiers: Iran.

Hall, Mark A.; Bobinski, Mary Anne; Orentlicher, David. Organ transplantation: the control, use, and allocation of body parts. *In their:* Bioethics and Public Health Law. New York: Aspen Publishers, 2005: 339-395. NRCBL: 19.5; 4.4; 9.6. SC: le; cs.

Jacob, Marie-Andree. Frail connections: legal and psychiatric knowledge practices in U.S. adjudication over organ donations by children and incompetent adults. *In:* Freeman, Michael, ed. Children's Health and Children's Rights. Boston: Martinus Nijhoff Publishers, 2006: 219-252. NRCBL: 19.5; 8.3.2; 8.3.3; 17.1. SC: cs; le.

Kirby, Neil. Treatment or crime? the status of stem cell therapies and research in South African law. *Medicine and Law: The World Association for Medical Law* 2007 March; 26(1): 95-115. 54 fn. NRCBL: 19.5; 18.3; 8.3.2; 4.4; 18.5.4; 15.1. SC: le.
  Keywords: *cord blood; *fetal stem cells; *legal aspects; *parental consent; biomedical research; DNA; legislation; newborns; property rights; stem cell transplantation; Proposed Keywords: *South Africa
Abstract: The author develops a thorough analysis of current and proposed South African law in relation to the harvesting and use of stem cells. He begins with the question of ownership of the umbilical cord at birth and afterwards. The problems of informed consent in these situations are discussed. Changes in the law in South Africa, now in progress, should ameliorate some of the difficulties.

Mason, J.K.; Laurie, G.T. The donation of organs and transplantation. *In their:* Mason and McCall Smith's Law and Medical Ethics. Seventh ed. Oxford; New York: Oxford University Press, 2005: 477-510. NRCBL: 19.5; 9.5.8; 22.2; 20.1. SC: le.

McLean, Sheila A.M.; Campbell, Alastair; Gutridge, Kerry; Harper, Helen. Human tissue legislation and medical practice: a benefit or a burden? *Medical Law International* 2007; 8(1): 1-21. NRCBL: 19.5; 4.4; 5.3. SC: le; em.

Mertes, Heidi.; Pennings, G. Oocyte donation for stem cell research. *Human Reproduction* 2007 March; 22(3): 629-634. NRCBL: 19.5; 14.6; 9.5.5; 18.5.4; 15.1; 4.4; 18.3. SC: le.

Nowenstein, Graciela. Nemo censetur ignorare legem? Presumed consent to organ donation in France, from Parliament to hospitals. *In:* Garwood-Gowers, Austen; Tingle, John; Wheat, Kay, eds. Contemporary Issues in Healthcare Law and Ethics. Edinburgh; New York: Elsevier Butterworth-Heinemann, 2005: 173-188. NRCBL: 19.5; 8.3.1. SC: le.

Nys, Herman. Organ transplantation. *In:* Gevers, J.K.M.; Hondius, E.H.; Hubben, J.H., eds. Health Law, Human Rights and the Biomedicine Convention: Essays in Honour of Henriette Roscam Abbing. Leiden; Boston: Martinus Nijhoff Publishers, 2005: 219-230. NRCBL: 19.5; 21.1. SC: le.

Otlowski, Margaret. Donor perspectives on issues associated with donation of genetic samples and information: an Australian viewpoint. *Journal of Bioethical Inquiry* 2007; 4(2): 135-150. NRCBL: 19.5; 15.1; 15.7; 18.6; 18.2; 18.3. SC: le.
  Abstract: This paper provides a legal overview of key issues associated with donation of genetic samples and information from a donor perspective. In particular, it addresses the property status of samples as well as issues in respect of consent, privacy, commercialisation and benefit sharing. The paper highlights the need for appropriate protection and safeguards for individuals, but also, importantly, for understanding what donors actually think and want in terms of genetic research and the use of their samples and information. The paper seeks to emphasise the importance of transparency and accountability in the conduct of research in order to maximise donor participation and confidence and public trust in general.

Peterson, Kathryn E. My father's eyes and my mother's heart: the due process rights of the next of kin in organ donation. *Valparaiso University Law Review* 2005 Fall; 40(1): 169. NRCBL: 19.5; 8.3.1; 7.1; 20.1. SC: le; cs. Identifiers: National Organ Transplantation Act; Uniform Anatomical Gift Act.

Pishchita, A. Presumed consent in the law of the Russian Federation on transplanting organs. *Medicine and Law: The World Association for Medical Law* 2007 March; 26(1): 179-188. NRCBL: 19.5; 8.3.1; 8.3.4. SC: le.
  Abstract: In this article, the author defines and discusses various concepts of consent in relation to organ transplantation. Beginning with the law of the Russian Federation, he highlights the benefits and shortcomings of basic provisions and parameters of consent in Russian law. The

situation is [sic; in] some other countries is reviewed from a comparative aspect. With the object of making transplantation more widespread, liberal interpretations of the rules are to be encouraged.

Pugliese, Elizabeth. Organ trafficking and the TVPA: why one word makes a difference in international enforcement efforts. *Journal of Contemporary Health Law and Policy* 2007 Fall; 24(1): 181-208. NRCBL: 19.5; 21.1; 4.4; 1.3.5. SC: le; an. Identifiers: Federal Trafficking Victims Protection Act.

Rady, Mohamed Y.; Verheijde, Joseph L.; McGregor, Joan. Organ donation after circulatory death: the forgotten donor? *Critical Care* 2006; 10(5): 166. NRCBL: 19.5; 20.1. SC: le.

Resnicoff, Steven H. Supplying human body parts: a Jewish law perspective. *DePaul Law Review* 2005-2006; 55: 851-874. NRCBL: 19.5; 20.1; 1.2. SC: le.

Sanders, Karen; Fullbrook, Suzanne. Autonomy and care: respecting the wishes of the deceased patient. *British Journal of Nursing* 2007 March 22-April 11; 16(6): 360-361. NRCBL: 19.5; 1.1; 8.3.1; 8.3.3; 20.1. SC: le. Identifiers: Human Tissue Act 2004.

Sangster, Catriona. 'Cooling corpses': Section 43 of the Human Tissue Act 2004 and organ donation. *Clinical Ethics* 2007 March; 2(1): 23-27. NRCBL: 19.5; 20.2.1; 8.3.1. SC: le.

Abstract: In an attempt to increase the number of organs available for transplantation, section 43 of the Human Tissue Act 2004 provides, for the first time, a statutory basis for the non-consensual preservation of organs. However, several issues arise out of the terminology of the section relating to where the preservation steps can be carried out and, indeed, what preservation steps can be performed which may affect the success of this attempt to increase the organ donor pool.

Sarig, Merav. Israeli surgeon is arrested for suspected organ trafficking [news]. *BMJ: British Medical Journal* 2007 May 12; 334(7601): 973. NRCBL: 19.5. SC: le.

Keywords: Israel

Shroff, Sunil. Working towards ethical organ transplants. *Indian Journal of Medical Ethics* 2007 April-June; 4(2): 68-69. NRCBL: 19.5; 9.3.1; 7.1. SC: le.

Skene, Loane. Legal rights of human bodies, body parts and tissue. *Journal of Bioethical Inquiry* 2007; 4(2): 129-133. NRCBL: 19.5; 18.3; 18.2; 18.6. SC: le.

Tuffs, Annette. German council demands opt-out system for transplants [news]. *BMJ: British Medical Journal* 2007 May 12; 334(7601): 973. NRCBL: 19.5. SC: le.

Keywords: Germany

Underwood, J.C.E. The impact on histopathology practice of new human tissue legislation in the UK. *Histopathology* 2006 September; 49(3): 221-228. NRCBL: 19.5; 19.1; 20.1; 18.3. SC: le.

Vathsala, A. Commercial renal transplantation — body parts for sale [editorial]. *Annals of the Academy of Medicine, Singapore* 2006 April; 35(4): 227-228. NRCBL: 19.5; 19.3; 9.3.1. SC: le.

Waltz, Emily. The body snatchers: rising demand has created a thriving market for human body parts — and not all of it above ground [news]. *Nature Medicine* 2006 May; 12(5): 487-488. NRCBL: 19.5; 4.4; 9.3.1; 18.1; 20.1. SC: le.

Waltz, Emily. Tracking down tissues. *Nature Biotechnology* 2007 November; 25(11): 1204-1206. NRCBL: 19.5; 9.3.1. SC: le.

Watts, Jonathan. China introduces new rules to deter human organ trade. *Lancet* 2007 June 9-15; 369(9577): 1917-1918. NRCBL: 19.5; 9.3.1; 21.1. SC: le.

ORGAN AND TISSUE TRANSPLANTATION/ XENOTRANSPLANTATION

Allan, Jonathan S.; Aluwihare, A.P.R.; Bach, Fritz H.; Caplan, Arthur; Chapman, Louisa; Dickens, Bernard M.; Fishman, Jay A.; Groth, C.G.; Breimer, M.E.; Menache, Andr?; Morris, Peter J.; van Rongen, Eric. Round Table Discussion: animal-to-human organ transplants. *Bulletin of the World Health Organization* 1999; 77(1): 62-81. NRCBL: 19.1; 15.1; 18.5.1; 22.2; 22.1; 18.2; 18.3; 18.6; 1.1; 5.2; 21.1; 19.5; 19.3. SC: le.

Brown, Nik. Xenotransplantation: normalizing disgust. *Science as Culture* 1999 September; 8(3): 327-355. NRCBL: 19.1; 1.3.7; 4.2; 22.2.

Daar, A.S. Animal-to-human organ transplants — a solution or a new problem? *Bulletin of the World Health Organization* 1999; 77(1): 54-61. NRCBL: 19.1; 7.1; 18.2; 18.5.1; 18.6; 18.5.9; 22.2.

Fovargue, Sara. Consenting to bio-risk: xenotransplantation and the law. *Legal Studies* 2005 September; 25(3): 404-430. NRCBL: 19.1; 18.3; 18.6; 22.2. SC: le. Identifiers: Great Britain (United Kingdom).

George, James F. Xenotransplantation: an ethical dilemma. *Current Opinion in Cardiology* 2006 March; 21(2): 138-141. NRCBL: 19.1; 22.2.

Hinkley, Charles C. Xenografts. *In his:* Moral Conflicts of Organ Retrieval: A Case for Constructive Pluralism. Amsterdam; New York: Rodopi, 2005: 117-126. NRCBL: 19.1; 22.2.

Hughes, Jonathan. Justice and third party risk: the ethics of xenotransplantation. *Journal of Applied Philosophy* 2007 May; 24(2): 151-168. NRCBL: 19.1; 19.5; 22.1; 1.1; 22.2. SC: an. Identifiers: Robert Veatch.

Abstract: The question of when it is permissible to inflict risks on others without their consent is one that we all face in our everyday lives, but which is often brought to

NRCBL: National Reference Center for Bioethics Literature Classification Scheme   See inside front cover for terms.

355

our attention in contexts of technological innovation and scientific uncertainty. Xenotransplantation, the transplantation of organs or tissues from animals to humans, has the potential to save or improve the lives of many patients but gives rise to the possibility of infectious agents being transferred from donor animals into the human population. As well as being an important ethical issue in its own right it therefore provides a useful vehicle for exploring the more general question of how to balance the benefits of a practice against the risks to third parties. This paper focuses on the Rawlsian, justice-based analysis of the risks of xenotransplantation proposed by Robert Veatch. It argues that Veatch is right to take considerations of distributive justice into account, but that his particular approach is flawed. It is hoped that consideration of Veatch's arguments, and of the underlying assumptions will suggest better ways of executing a justice-based approach.

McLean, Sheila; Williamson, Laura. The demise of UKXIRA and the regulation of solid-organ xenotransplantation in the UK [editorial]. *Journal of Medical Ethics* 2007 July; 33(7): 373-375. NRCBL: 19.1; 22.2.

O'Neill, Robert D. Xenotransplantation: the solution to the shortage of human organs for transplantation? *Mortality* 2006 May; 11(2): 211-231. NRCBL: 19.1; 22.2; 19.6; 2.2; 1.1.

ORGAN DONATION *See* ORGAN AND TISSUE TRANSPLANTATION/ DONATION AND PROCUREMENT

OVA *See* CRYOBANKING OF SPERM, OVA AND EMBRYOS

OVUM DONORS *See* REPRODUCTIVE TECHNOLOGIES

PAIN AND PAIN CARE *See* QUALITY AND VALUE OF LIFE

PALLIATIVE CARE *See* DEATH AND DYING/ TERMINAL CARE

PARENTAL CONSENT *See* EUTHANASIA AND ALLOWING TO DIE/ MINORS; CARE FOR SPECIFIC GROUPS/ MINORS; HUMAN EXPERIMENTATION/ SPECIAL POPULATIONS/ MINORS; INFORMED CONSENT/ MINORS

PATENTS

Aldhous, Peter. Angry reception greets patent for synthetic life [news]. *New Scientist* 2007 June 16-22; 194(2608): 13. NRCBL: 15.8.
    Keywords: *genetic patents; *genome; *synthetic biology; industry; microbiology; Keyword Identifiers: Celera Genomics; Venter, Craig

Baumgartner, Christoph. Exclusion by inclusion? On difficulties with regard to an effective ethical assessment of patenting in the field of agricultural bio-technology. *Journal of Agricultural and Environmental Ethics* 2006; 19(6): 521-539. NRCBL: 15.8; 1.3.11. SC: an; le.

Bergman, Karl; Graff, Gregory D. The global stem cell patent landscape: implications for efficient technology transfer and commercial development. *Nature Biotechnology* 2007 April; 25(4): 419-425. 28 refs. NRCBL: 15.8; 15.1. SC: em.
    Keywords: *international aspects; *patents; *stem cells; access to information; databases; embryonic stem cells; industry; interinstitutional relations; private sector; public sector; research institutes; statistics; universities; Proposed Keywords: *technology transfer; licensure; Keyword Identifiers: United States

Calvert, Jane. Patenting genomic objects: genes, genomes, function and information. *Science as Culture* 2007 June; 16(2): 207-223. NRCBL: 15.8; 5.1; 3.1.

Caruso, Denise. Someone (other than you) may own your genes [op-ed]. *New York Times* 2007 January 28; p. BU3. NRCBL: 15.8; 5.3. SC: po.
    Keywords: *genes; *patents; *property rights; legal aspects; industry; biotechnology

Caulfield, Timothy; Bubela, Tania; Murdoch, C.J. Myriad and the mass media: the covering of a gene patent controversy. *Genetics in Medicine* 2007 December; 9(12): 850-855. NRCBL: 15.8; 1.3.7; 1.3.2; 9.5.5; 21.1. SC: em.

Caulfield, Timothy; von Tigerstrom, Barbara. Globalization and biotechnology policy: the challenges created by gene patents and cloning technologies. *In:* Bennett, Belinda; Tomossy, George F., eds. Globalization and Health: Challenges for Health Law and Bioethics. Dordrecht: Springer, 2006: 129-149. NRCBL: 15.8; 14.5; 21.1; 9.3.1; 5.3. SC: le; rv. Identifiers: Trade related aspects of intellectual property rights (TRIPS).

Check, Erika. Patenting the obvious? *Nature* 2007 May 3; 447(7140): 16-17. NRCBL: 15.8; 18.5.4; 15.1. SC: le.
    Keywords: *embryonic stem cells; *legal aspects; biomedical research; biotechnology; economics; methods; public policy; research support; researchers; social impact; state government; universities; Proposed Keywords: licensure; technology transfer; Keyword Identifiers: *Patent and Trademark Office; *United States; *Wisconsin; *Wisconsin Alumni Research Foundation; California

Chin, Andrew. Research in the shadow of DNA patents. *Journal of the Patent and Trademark Office Society* 2005 November; 87(11): 846-906. 317 fn. NRCBL: 15.8; 15.1. SC: em; le.
    Keywords: *DNA; *genetic patents; *legal aspects; *risks and benefits; biotechnology; DNA sequences; guidelines; human genome; industry; legislation; policy analysis; public policy; recombinant DNA research; Keyword Identifiers: *Patent and Trademark Office; *United States; Patent Act

Commission on Intellectual Property Rights [CIPR]; Thambisetty, Sivaramjani. Human genome patents and developing countries. Study paper 10. London: Commission on Intellectual Property Rights, 2001; 70 p. [Online]. Accessed: http://www.iprcommission.org/papers/pdfs/study_papers/10_human_genome_patents.pdf [2007 April 16]. 141 fn. NRCBL: 15.8; 15.10. SC: le.

> Keywords: *developing countries; *genetic materials; *genetic patents; benefit sharing; community consent; donors; genetic research; human genome; industry; informed consent; international aspects; legal aspects; research subjects; Western World; Proposed Keywords: genetic resources

Deane-Drummond, Celia. Gene patenting. *In her:* Genetics and Christian Ethics. Cambridge: Cambridge University Press, 2006: 160-190. 84 fn. NRCBL: 15.8; 1.2. SC: an.

> Keywords: *genetic patents; benefit sharing; Christian ethics; common good; genetic materials; embryonic stem cells; international aspects; justice; legal aspects; philosophy; theology; virtues; Keyword Identifiers: Europe; European Biotechnology Patent Directive; Human Genome Organization (HUGO); United States

European Society of Human Genetics [ESHG]. EPO upholds limited patent on BRCA2 gene: singling out an ethnic group is a 'dangerous precedent' says European Society of Human Genetics. *Vienna, Austria: European Society of Human Genetics,* 2005 July 1: 2 p. [Online]. Accessed: http://www.eshg.org/ESHGPressRelease01July2005.pdf [2007 February 8]. NRCBL: 15.8; 15.3; 15.11; 21.1. Identifiers: European Patent Office.

European Society of Human Genetics [ESHG]. Geneticists oppose singling out Jewish women in European breast cancer patent [press release]. *Vienna, Austria: European Society of Human Genetics,* 2005 June 15: 3 p. [Online]. Accessed: http://www.eshg.org/PressReleaseESHG15-06-2005.pdf [2007 February 8]. NRCBL: 15.8; 15.3; 15.11; 21.1.

Garforth, Kathryn. Health care and access to patented technologies. *Health Law Journal* 2005; 13: 77-97. NRCBL: 15.8; 9.2. SC: le; cs.

Gold, E. Richard; Bubela, Tania; Miller, Fiona A.; Nicol, Dianne; Piper, Tina. Gene patents — more evidence needed, but policymakers must act [letter]. *Nature Biotechnology* 2007 April; 25(4): 388-389. NRCBL: 15.8. Comments: Timothy Caulfield, Robert M. Cook-Deegan, F. Scott Kieff, and John P. Walsh. Evidence and anecdotes: an analysis of human gene patenting controversies Nature Biotechnology 2006 September; 24(9): 1091-1094.

> Keywords: *access to information; *biomedical research; *genes; *genetic patents; *interinstitutional relations; empirical research; guidelines; industry; international aspects; justice; legal aspects; organizational policies; policy making; Proposed Keywords: licensure; technology transfer

Greif, Karen F.; Merz, Jon F. Who owns the genome? The patenting of human genes; Who owns life? Mr. Moore's spleen; The Canavan disease case. *In their:* Current Controversies in the Biological Sciences: Case Stud-

ies of Policy Challenges from New Technologies. Cambridge, MA: MIT, 2007: 49-76. 42 refs. NRCBL: 15.8. SC: le.

> Keywords: *genetic patents; *legal aspects; biomedical research; cell lines; clinical genetics; genes; genetic research; genetic screening; genetic services; international aspects; patients; patients' rights; physicians; property rights; recombinant DNA research; research subjects; researchers; Supreme Court decisions; Proposed Keywords: Canavan disease; licensure; tissue donors; Keyword Identifiers: *United States; Diamond v. Chakrabarty; Greenberg v. Miiami Children's Hospital Research Institute; Moore v. Regents of the University of California; Patent and Trademark Office

Gulbrandsen, Carl. WARF's licensing policy for ES cell lines [letter]. *Nature Biotechnology* 2007 April; 25(4): 387-388. 1 ref. NRCBL: 15.8; 18.5.4; 15.1. Identifiers: Wisconsin Alumni Research Foundation. Comments: Burning bridges. Nature Biotechnology 2007 January; 25(1): 2.

> Keywords: *cell lines; *embryonic stem cells; biomedical research; embryo research; industry; interinstitutional relations; organizational policies; property rights; research support; universities; Proposed Keywords: *licensure; foundations; technology transfer; Keyword Identifiers: *Wisconsin Alumni Research Foundation; United States; University of Wisconsin

Hansson, Mats G.; Helgesson, Gert; Wessman, Richard; Jaenisch, Rudolf. Commentary: isolated stem cells — patentable as cultural artifacts? *Stem Cells* 2007 June; 25(6): 1507-1510 [Online]. Accessed: http://www.StemCells.com/cgi/content/full/25/6/1507 [2007 December 3]. 31 refs. NRCBL: 15.8; 15.1; 18.5.4; 1.1.

> Keywords: *embryonic stem cells; *patents; adult stem cells; commodification; donors; embryo research; embryos; informed consent; international aspects; legal aspects; moral status; public policy; Keyword Identifiers: Europe

Holden, Constance. Prominent researchers join the attack on stem cell patents [news]. *Science* 2007 July 13; 317(5835): 187. NRCBL: 15.8; 18.5.4; 15.1.

> Keywords: *cell lines; *embryonic stem cells; *legal aspects; *patents; attitudes; researchers; universities; Proposed Keywords: consumer advocacy; Keyword Identifiers: *Wisconsin Alumni Research Foundation; Patent and Trademark Office; Thomson, James; University of Wisconsin; United States

Holden, Constance. U.S. Patent Office casts doubt on Wisconsin stem cell patents [news]. *Science* 2007 April 13; 316(5822): 182. NRCBL: 15.8; 15.1; 18.5.4. Identifiers: Wisconsin Alumni Research Foundation.

> Keywords: *cell lines; *embryonic stem cells; *patents; industry; legal aspects; universities; Proposed Keywords: foundations; licensure; technology transfer; Keyword Identifiers: *Patent and Trademark Office; *Wisconsin Alumni Research Foundation; United States

Hopkins, Michael M.; Mahdi, Surya; Patel, Pari; Thomas, Sandy M. DNA patenting: the end of an era? *Nature Biotechnology* 2007 February; 25(2): 185-188. 18 refs. NRCBL: 15.8; 21.1. SC: em.

NRCBL: National Reference Center for Bioethics Literature Classification Scheme   See inside front cover for terms.

357

Keywords: *DNA sequences; *genetic patents; *international aspects; *statistics; *trends; federal government; industry; legal aspects; social impact; standards; universities; Keyword Identifiers: *Europe; *Japan; *United States; European Patent Office; Japan Patent Office; Patent and Trademark Office

Johnston, Josephine; Wasunna, Angela A. Patents, biomedical research, and treatments: examining concerns, canvassing solutions. *Hastings Center Report* 2007 January-February; 37(1): S2-S35. 183 fn. NRCBL: 15.8; 21.1; 9.3.1; 18.5.4; 9.5.6; 18.6. SC: an; rv.

Keywords: *biomedical research; *biomedical technologies; *biotechnology; *genetic materials; *international aspects; *patents; *policy analysis; *property rights; benefit sharing; developing countries; drug industry; drugs; economics; embryonic stem cells; genetic patents; genetic screening; government financing; guidelines; health care delivery; health services accessibility; historical aspects; HIV infections; incentives; legal aspects; moral policy; poliomyelitis; private sector; public policy; public sector; research support; trends; Proposed Keywords: licensure; vaccines; Keyword Identifiers: National Institutes of Health; Organization for Economic Cooperation and Development; South Africa; Trade-Related Aspects of Intellectual Property Rights (TRIPS); United States

Kaiser, Jocelyn. Attempt to patent artificial organism draws a protest [news]. *Science* 2007 June 15; 316(5831): 1557. NRCBL: 15.8.

Keywords: *genetic patents; *synthetic biology; biotechnology; microbiology; Keyword Identifiers: Venter, J. Craig

Kintisch, Eli. New cell rules [news]. *Science* 2007 January 26; 315(5811): 449. NRCBL: 15.8; 18.5.4. Identifiers: Wisconsin Alumni Research Foundation.

Klein, Roger D. Gene patents and genetic testing in the United States. As genetic testing moves into mainstream medicine, its restriction by gene patent holders will have far-reaching, detrimental effects on the healthcare system. *Nature Biotechnology* 2007 September; 25(9): 989-991. 23 refs. NRCBL: 15.8; 15.3.

Keywords: *genetic patents; *genetic screening; DNA sequences; economics; genes; legal aspects; recombinant DNA research; social impact; Supreme Court decisions; Keyword Identifiers: Diamond v. Charkrabarty; Diamond v. Diehr; Patent and Trademark Office; United States

Krimsky, Sheldon. The profit of scientific discovery and its normative implications. *Chicago-Kent Law Review* 1999; 75(1): 15-39. 164 fn. NRCBL: 15.8; 1.3.7; 1.3.9; 2.2; 5.3. SC: le; rv.

Keywords: *biomedical research; *biotechnology; *conflict of interest; *genetic patents; *legal aspects; authorship; cell lines; disclosure; editorial policies; entrepreneurship; federal government; genes; genetic engineering; genetic research; genetic screening; genetically modified organisms; government financing; government regulation; industry; interinstitutional relations; journalism; public policy; publishing; recombinant DNA research; research support; science; Supreme Court decisions; universities; Keyword Identifiers: *United States;

Lucassen, Anneke; Clarke, Angus. Should families own genetic information [debate]. *BMJ: British Medical Journal* 2007 July 7; 335(7609): 22-23. NRCBL: 15.8; 1.3.12; 8.4.

Keywords: *disclosure; *family members; *genetic information; *genetic privacy; *property rights; altruism; autonomy; confidentiality; genetic counseling; genetic screening; pedigree; professional family relationship; risk

Macer, D.R.J. Patent or perish? An ethical approach to patenting human genes and proteins. *Pharmacogenomics Journal* 2002; 2(6): 361-366. 46 refs. NRCBL: 15.8. SC: an.

Keywords: *genes; *genetic materials; *genetic patents; *moral policy; *standards; beneficence; benefit sharing; biomedical research; biotechnology; DNA sequences; economics; ethical analysis; ethical review; genomics; government; government regulation; human experimentation; human genome; industry; international aspects; justice; publishing; scientific misconduct; universities; Proposed Keywords: biodiversity; genetic resources

Memis, Tekin. Debate on patentability of biotechnological studies in Turkey. *Revista de Derecho y Genoma Humano = Law and the Human Genome Review* 2007 January-June; (26): 121-135. 26 fn. NRCBL: 15.8. SC: le.

Keywords: *biotechnology; *genetic materials; *genetic patents; *legal aspects; developing countries; genetic research; genetic resources; industry; international aspects; Keyword Identifiers: *Turkey; European Union

Paradise, Jordan; Janson, Christopher. Decoding the research exemption. *Nature Reviews Genetics* 2006 February; 7(2): 148-154. 71 refs. NRCBL: 15.8; 5.3. SC: le.

Keywords: *DNA sequences; *genes; *genetic patents; *legal aspects; genetic privacy; genetic research; international aspects; Proposed Keywords: BRCA1 genes; BRCA2 genes; Keyword Identifiers: Unied States

Resnik, David B. Embryonic stem cell patents and human dignity. *Health Care Analysis: An International Journal of Health Philosophy and Policy* 2007 September; 15(3): 211-222. 41 refs. NRCBL: 15.8; 18.5.4; 15.1; 4.4. SC: an.

Keywords: *embryonic stem cells; *human dignity; *moral policy; *moral status; *patents; beginning of life; biotechnology; embryos; ethical analysis; genetic patents; genetically modified organisms; guidelines; historical aspects; human characteristics; legal aspects; natural law; stem cells; terminology; value of life; Proposed Keywords: *multipotent stem cells; *pluripotent stem cells; *totipotent stem cells; blastocysts; classification; Keyword Identifiers: Europe; Twentieth Century; United States

Abstract: This article examines the assertion that human embryonic stem cells patents are immoral because they violate human dignity. After analyzing the concept of human dignity and its role in bioethics debates, this article argues that patents on human embryos or totipotent embryonic stem cells violate human dignity, but that patents on pluripotent or multipotent stem cells do not. Since patents on pluripotent or multipotent stem cells may still threaten human dignity by encouraging people to treat embryos as property, patent agencies should carefully monitor and control these patents to ensure that patents are not inadvertently awarded on embryos or totipotent stem cells.

Resnik, David B. The human genome: common resource but not common heritage. *In:* Korthals, Michiel; Bogers, Robert J., eds. Ethics for Life Scientists. Dordrecht, The Netherlands: Springer, 2004: 197-210. NRCBL: 15.8; 15.10; 1.3.9. SC: an; le. Identifiers: Moore vs. Regents of the University of California (1990); Greenberg v. Miami Children's Hospital Research Institute (2003).

Rutz, Berthold; Yeats, Siobhan. Patents: patenting of stem cell related inventions in Europe. *Biotechnology Journal* 2006 April; 1(4): 384-387. 14 refs. NRCBL: 15.8; 21.1. SC: le.
　　　Keywords: *embryonic stem cells; *legal aspects; *patents; *regulation; international aspects; Keyword Identifiers: *Europe; European Group on Ethics in Science and New Technologies; European Patent Convention; European Patent Office; European Union

Schmidt, Charlie. Negotiating the RNAi patent thicket. *Nature Biotechnology* 2007 March; 25(3): 273-275. 1 ref. NRCBL: 15.8.
　　　Keywords: *biotechnology; *genetic patents; drugs; economics; genetic techniques; industry; legal aspects; Proposed Keywords: licensure; Keyword Identifiers: United States

Simon, Jürgen. Biotechnology and law: biotechnology patents. Special considerations on the inventions with human material. *Revista de Derecho y Genoma Humano = Law and the Human Genome Review* 2006 July-December; (25): 139-159. 83 fn. NRCBL: 15.8; 15.1; 21.1. SC: le.
　　　Keywords: *biotechnology; *cloning; *genetic materials; *genetic patents; *legal aspects; *stem cells; DNA sequences; embryo research; genes; genetic engineering; germ cells; international aspects; reproductive technologies; Keyword Identifiers: Europe; United States

Singeo, Lindsey. The patentability of the native Hawaiian genome. *American Journal of Law and Medicine* 2007; 33(1): 119-139. NRCBL: 15.8; 15.10; 1.3.12; 13.1; 21.7; 18.3; 8.4. SC: le.

Straus, Joseph. Patentierung von Leben? *In:* Schreiber, Hans-Peter, ed. Biomedizin und Ethik: Praxis, Recht, Moral. Basel; Boston: Birkhäuser, 2004: 50-55. NRCBL: 15.8. SC: le.
　　　Keywords: *genetic patents; *legal aspects; biotechnology; DNA sequences; genetically modified organisms; microbiology; Keyword Identifiers: Europe

Sulmasy, Daniel P. Who owns the human genome? *In:* Monsour, H. Daniel, ed. Ethics and the New Genetics: An Integrated Approach. Toronto: University of Toronto Press, 2007: 123-133. 18 fn. NRCBL: 15.8; 1.1; 4.4. SC: an; rv.
　　　Keywords: *genes; *genetic patents; *human genome; *property rights; body parts and fluids; casuistry; genetic information; genetic privacy; genetically modified organisms; human dignity; justice; legal aspects; philosophy

United States. Congress. House. A bill to amend title 35, United States Code, to prohibit the patenting of human genetic material. Washington, DC: U.S. G.P.O., 2007. 15 p. [Online]. Accessed: http://frwebgate.access.gpo.gov/cgi-bin/useftp.cgi?IPaddress=162.140.64.21&filename=h977ih.pdf&directory=/diskb/wais/data/110_cong_bills [2007 April 4]. NRCBL: 15.8. SC: le. Identifiers: Genomic Research and Accessibility Act. Note: H.R.977, 110th Congress, 1st session. Introduced by Rep. Becerra on February 9, 2007. Referred to the Committee on the Judiciary.
　　　Keywords: *genetic patents; *legal aspects; DNA sequences; Keyword Identifiers: *United States

Van Overwalle, Geertrui; van Zimmeren, Esther; Verbeure, Birgit; Matthijs, Gert. Models for facilitating access to patents on genetic inventions. *Nature Reviews Genetics* 2006 February; 7(2): 143-148. 47 refs. NRCBL: 15.8; 5.3.
　　　Keywords: *access to information; *genetic patents; biotechnology; clinical genetics; genetic research; international aspects; legal aspects; Proposed Keywords: *organizational models; licensure; Keyword Identifiers: United States

Watson, Rory. Scientists welcome ruling on patent on breast cancer gene [news]. *BMJ: British Medical Journal* 2007 October 13; 335(7623): 740-741. NRCBL: 15.8; 9.5.5.

Watts, Geoff. The locked code. *BMJ: British Medical Journal* 2007 May 19; 334(7602): 1032-1033. NRCBL: 15.8. SC: le. Identifiers: DNA; gene patents.

World Intellectual Property Organization [WIPO]. Study takes critical look at benefit sharing of genetic resources and traditional knowledge [press release]. Kuala Lumpur/Geneva/Nairobi: World Intellectual Property Organization, 2004 February 10; 4 p. [Online]. Accessed: http://www.wipo.int/edocs/prdocs/en/2004/wipo_pr_2004_373.html [2007 April 16]. NRCBL: 15.8; 22.1; 15.1; 9.7; 15.10. SC: le.
　　　Keywords: *benefit sharing; *genetic patents; *international aspects; drugs; ecology; food; genetic diversity; genetic materials; property rights; voluntary programs; Proposed Keywords: *biodiversity; *genetic resources; Keyword Identifiers: Africa; India; United Nations; United Nations Environment Program; World Intellectual Property Organization

PATERNALISM *See* PATIENT RELATIONSHIPS

PATIENT ACCESS TO RECORDS *See* CONFIDENTIALITY; TRUTH DISCLOSURE

PATIENT CARE *See* CARE FOR SPECIFIC GROUPS; DEATH AND DYING/ TERMINAL CARE; PATIENT RELATIONSHIPS

PATIENT RELATIONSHIPS
See also CARE FOR SPECIFIC GROUPS; PROFESSIONAL ETHICS

An erosion of conscientious objection? *Health Care Ethics USA* 2007 Spring; 15(2): 15. NRCBL: 8.1; 1.2; 4.1.2. SC: em.

NRCBL: National Reference Center for Bioethics Literature Classification Scheme　　　　See inside front cover for terms.

359

Adams, Jared R.; Drake, Robert E.; Wolford, George L. Shared decision-making preferences of people with severe mental illness. *Psychiatric Services* 2007 September; 58(9): 1219-1221. NRCBL: 8.1; 17.1. SC: em.

Andereck, William. From patient to consumer in the medical marketplace. *CQ: Cambridge Quarterly of Healthcare Ethics* 2007 Winter; 16(1): 109-113. NRCBL: 8.1; 9.3.1; 1.1; 4.4.

Arroll, Bruce; Falloon, Karen. Should doctors go to patients' funerals? *BMJ: British Medical Journal* 2007 June 23; 334(7607): 1322. NRCBL: 8.1; 20.1.

Austin, Larry J. Religious bias colors physicians' views [letter]. *Archives of Internal Medicine* 2007 October 8; 167(18): 2007. NRCBL: 8.1; 1.2. Comments: F.A. Curlin, S.A. Sellergren, J.D. Lantos, and M.H. Chin. Physicians' observations and interpretations of the influence of religion and spirituality on health. Archives of Internal Medicine 2007; 167(7): 649-654.

Banja, John D.; Volandes, Angelo E.; Paasche-Orlow, Michael K. My what? *American Journal of Bioethics* 2007 November; 7(11): 13-15; author reply W1-W2. NRCBL: 8.1; 9.1. Comments: Angelo E. Volandes and Michael K. Paasche-Orlow. Health literacy, health inequality and a just healthcare system. American Journal of Bioethics 2007 November; 7(11): 5-10.

Beach, Mary Catherine; Duggan, Patrick S.; Cassel, Christine K.; Geller, Gail. What does 'respect' mean? Exploring the moral obligation of health professionals to respect patients. *JGIM: Journal of General Internal Medicine* 2007 May; 22(5): 692-695. NRCBL: 8.1; 1.1; 4.1.2.

Bensing, Jozien. Bridging the gap. The separate worlds of evidence-based medicine and patient-centered medicine. *Patient Education and Counseling* 2000 January; 39(1): 17-25. NRCBL: 8.1; 9.4; 18.2.

Berlinger, Nancy. Martin Luther at the bedside: conscientious objection and community. *Hastings Center Report* 2007 March-April; 37(2): inside back cover. NRCBL: 8.1; 4.1.2; 11.1; 12.3; 20.4.1; 20.7.

Berry, Philip A. The absence of sadness: darker reflections on the doctor-patient relationship. *Journal of Medical Ethics* 2007 May; 33(5): 266-268. NRCBL: 8.1; 20.4.1.
  Abstract: Recognising a diminution in his emotional response to patients' deaths, the author analyses in detail his internal reactions in an attempt to understand what he believes is a common phenomenon among doctors. He identifies factors that may erode the connection between patient and physician: an instinct to separate oneself from another's suffering, professional unease in the case of therapeutic failure, the atrophying effect of perceived hopelessness, insincerities in the establishment of the initial relationship, and an inability to imbue the sedated or unconscious patient with human qualities. He concludes

that recognition of these negative influences, without necessarily changing behaviours that are natural, may be a first step towards protecting doctors against what might be an otherwise insidious process of dehumanisation.

Bosek, Marcia Sue DeWolf. When respecting patient autonomy may not be in the patient's best interest. *JONA's Healthcare Law, Ethics, and Regulation* 2007 April-June; 9(2): 46-49. NRCBL: 8.1; 1.1. SC: cs.

Bowers, Len. On conflict, containment and the relationship between them. *Nursing Inquiry* 2006 September; 13(3): 172-180. NRCBL: 8.1; 8.3.4; 17.2; 4.3.

Brewster, Luke P.; Bennett, Barry K.; Gamelli, Richard L. Application of rehabilitation ethics to a selected burn patient population's perspective. *Journal of the American College of Surgeons* 2006 November; 203(5): 766-771. NRCBL: 8.1; 1.1; 8.3.1; 9.5.1.

Brody, Howard; Whitney, Simon N.; McCullough, Laurence B. Transparency and self-censorship in shared decision-making. *American Journal of Bioethics* 2007 July; 7(7): 44-46; author reply W1-W3. NRCBL: 8.1; 8.3.1. Comments: Simon N. Whitney and Laurence B. McCullough. Physicians' silent decisions: because patient autonomy does not always come first. American Journal of Bioethics 2007 July; 7(7): 33-38.

Browne, Alister; Browne, Katharine. Morality, prudential rationality, and cheating. *CQ: Cambridge Quarterly of Healthcare Ethics* 2007 Winter; 16(1): 53-62. NRCBL: 8.1; 1.1; 9.3.1; 9.4.

Burns, Tom; Shaw, Joanne. Is it acceptable for people to be paid to adhere to medication?[forum]. *BMJ: British Medical Journal* 2007 August 4; 335(7613): 232-233. NRCBL: 8.1; 9.3.1; 17.8; 17.4.

Camann, William. It is the right of every anaesthetist to refuse to participate in a maternal-request caesarean section. *International Journal of Obstetric Anesthesia* 2006 January; 15(1): 35-37. NRCBL: 8.1; 9.5.5; 9.5.8; 9.5.1. SC: an.

Centor, Robert M. Seek first to understand. *Philosophy, Ethics, and Humanities in Medicine [electronic]* 2007 November 28; 2(29): 2 p. Accessed: http://www.peh-med.com/ [2008 January 24]. NRCBL: 8.1; 9.8. Comments: Susan H. McDaniel, Howard B. Beckman, Diane S. Morse, Jordan Silberman, David B. Seaburn, and Ronald M. Epstein. Physician self-disclosure in primary care visits: enough about you, what about me? Archives of Internal Medicine 2007 June 25; 167(12): 1321-1326.

Chochinov, Harvey Max. Dignity and the essence of medicine: the A, B, C, and D of dignity conserving care. *BMJ: British Medical Journal* 2007 July 28; 335(7612): 184-187. NRCBL: 8.1; 4.1.2; 7.1.

Cochrane, Thomas I. Religious delusions and the limits of spirituality in decision-making. *American Journal of*

Bioethics 2007 July; 7(7): 14-15. NRCBL: 8.1; 1.2; 8.3.3. Comments: Mark G. Kuczewski. Talking about spirituality in the clinical setting: can being professional require being personal? American Journal of Bioethics 2007 July; 7(7): 4-11.

Cohen, Cynthia B. Ways of being personal and not being personal about religious beliefs in the clinical setting. *American Journal of Bioethics* 2007 July; 7(7): 16-18. NRCBL: 8.1; 1.2; 8.3.1; 9.6. Comments: Mark G. Kuczewski. Talking about spirituality in the clinical setting: can being professional require being personal? American Journal of Bioethics 2007 July; 7(7): 4-11.

Collins, Kenneth. Maimonides and the ethics of patient autonomy. *Israel Medical Association Journal* 2007 January; 9(1): 55-58. NRCBL: 8.1; 1.1; 1.2.

Coulter, Angela; Ellins, Jo. Effectiveness of strategies for informing, educating, and involving patients. *BMJ: British Medical Journal* 2007 July 7; 335(7609): 24-27. NRCBL: 8.1; 9.5.1; 9.8.

Curlin, Farr A.; Chin, Marshall H.; Sellergren, Sarah A.; Roach, Chad J.; Lantos, John D. The association of physicians' religious characteristics with their attitudes and self-reported behaviors regarding religion and spirituality in the clinical encounter. *Medical Care* 2006 May; 44(5): 446-453. NRCBL: 8.1; 1.2. SC: em.

Curlin, Farr A.; Lawrence, Ryan E.; Chin, Marshall H.; Lantos, John D. Religion, conscience, and controversial clinical practices. *New England Journal of Medicine* 2007 February 8; 356(6): 593-600. NRCBL: 8.1; 1.2; 8.2; 9.6; 4.1.2.

Curlin, Farr A.; Roach, Chad J. By intuitions differently formed: how physicians assess and respond to spiritual issues in the clinical encounter. *American Journal of Bioethics* 2007 July; 7(7): 19-20. NRCBL: 8.1; 1.2. Comments: Mark G. Kuczewski. Talking about spirituality in the clinical setting: can being professional require being personal. American Journal of Bioethics 2007 July; 7(7): 4-11.

Dees, Richard H.; Volandes, Angelo E.; Paasche-Orlow, Michael K. Health literacy and autonomy. *American Journal of Bioethics* 2007 November; 7(11): 22-23; author reply W1-W2. NRCBL: 8.1; 9.1; 1.1. Comments: Angelo E. Volandes and Michael K. Paasche-Orlow. Health literacy, health inequality and a just healthcare system. American Journal of Bioethics 2007 November; 7(11): 5-10.

Detmer, Don E.; Singleton, Peter; Ratzan, Scott C. The need for better health information: advancing the informed patient in Europe. *In:* Santoro, Michael A.; Gorrie, Thomas M., eds. Ethics and the Pharmaceutical Industry. Cambridge; New York: Cambridge University Press, 2005: 196-205. NRCBL: 8.1; 8.2; 9.7; 1.3.2.

DiSilvestro, Russell. What's wrong with deliberately proselytizing patients? *American Journal of Bioethics* 2007 July; 7(7): 22-24. NRCBL: 8.1; 1.2. Comments: Mark G. Kuczewski. Talking about spirituality in the clinical setting: can being professional require being personal. American Journal of Bioethics 2007 July; 7(7): 4-11.

Duggan, Patrick S.; Geller, Gail; Cooper, Lisa A.; Beach, Mary Catherine. The moral nature of patient-centeredness: is it "just the right thing to do"? *Patient Counselling and Health Education* 2006 August; 62(2): 271-276. NRCBL: 8.1; 7.1; 1.1.

Ehrenstein, Boris P.; Hanses, Frank; Salzberger, Bernd. Influenza pandemic and professional duty: family or patients first? A survey of hospital employees. *BMC Public Health* 2006 December 28; 6: 311. NRCBL: 8.1; .1; 9.2; 16.3.

Eisen, Arri. The challenge of spirituality in the clinic: symptom of a larger syndrome. *American Journal of Bioethics* 2007 July; 7(7): 12-13. NRCBL: 8.1; 1.2; 5.1. Comments: Mark G. Kuczewski. Talking about spirituality in the clinical setting: can being professional require being personal? American Journal of Bioethics 2007 July; 7(7): 4-11.

Epstein, Miran. Legal and institutional fictions in medical ethics: a common, and yet largely overlooked, phenomenon: a theoretical platform for a much-needed change in the provision of healthcare based on restoring the autonomy of doctor-patient relationships. *Journal of Medical Ethics* 2007 June; 33(6): 362-364. NRCBL: 8.1; 2.1; 5.3; 8.3.1; 18.1; 19.5; 20.5.1.

Evans, H.M. Do patients have duties? *Journal of Medical Ethics* 2007 December; 33(12): 689-694. NRCBL: 8.1; 1.1; 9.1; 18.1. SC: an.
   Abstract: The notion of patients' duties has received periodic scholarly attention but remains overwhelmed by attention to the duties of healthcare professionals. In a previous paper the author argued that patients in publicly funded healthcare systems have a duty to participate in clinical research, arising from their debt to previous patients. Here the author proposes a greatly extended range of patients' duties grounding their moral force distinctively in the interests of contemporary and future patients, since medical treatment offered to one patient is always liable to be an opportunity cost (however justifiable) in terms of medical treatment needed by other patients. This generates both negative and positive duties. Ten duties-enjoining obligations ranging from participation in healthcare schemes to promoting one's own earliest recovery from illness-are proposed. The characteristics of these duties, including their basis, moral force, extent and enforceability, are considered. They are tested against a range of objections-principled, societal, epistemological and practical-and found to survive. Finally, the paper suggests that these duties could be thought to reinforce a regrettably adversarial characteristic, shared with rights-based approaches, and that a pref-

NRCBL: National Reference Center for Bioethics Literature Classification Scheme   See inside front cover for terms.

361

erable alternative might be sought through the (here unexplored) notion of a "virtuous patient" contributing to a problem-solving partnership with the clinician. However, in defining and giving content to that partnership, there is a clear role for most, if not all, of the proposed duties; their value thus extends beyond the adversarial context in which they might first be thought to arise.

Fioriglio, Gianluigi; Szolovits, Peter. Copy fees and patients' rights to obtain a copy of their medical records: from law to reality. *AMIA Annual Symposium Proceedings* 2005: 251-255. NRCBL: 8.1; 9.2; 9.3.1; 1.3.12.

Fontana, Nicholas. A question of professionalism: are we treating patients as people or procedures? *Journal of the Michigan Dental Association* 2006 October; 88(10): 28-30. NRCBL: 8.1; 7.1; 4.1.1.

Fraenkel, Liana; McGraw, Sarah. What are the essential elements to enable patient participation in medical decision making? *JGIM: Journal of General Internal Medicine* 2007 May; 22(5): 614-619. NRCBL: 8.1; 8.3.1. SC: em.

Friele, Roland D.; Sluijs, Emmy M. Patient expectations of fair complaint handling in hospitals: empirical data. *BMC Health Services Research* 2006 August 18; 6: 106: 9 p. NRCBL: 8.1; 7.1; 9.8. SC: em.

Gallant, Mae H.; Beaulieu, Marcia C.; Carnevale, Franco A. Partnership: an analysis of the concept within the nurse-client relationship. *Journal of Advanced Nursing* 2002 October; 40(2): 149-157. NRCBL: 8.1; 4.1.3.

Geier, G. Richard. Professionalism, ethics, and trust. *Minnesota Medicine* 2007 June; 90(6): 20. NRCBL: 8.1; 7.1; 9.7.

Goldberg, Daniel S.; Brody, Howard. Spirituality: respect but don't reveal. *American Journal of Bioethics* 2007 July; 7(7): 21-22. NRCBL: 8.1; 1.2. Comments: Mark G. Kuczewski. Talking about spirituality in the clinical setting: can being professional require being personal. American Journal of Bioethics 2007 July; 7(7): 4-11.

Goldberg, Daniel S.; Volandes, Angelo E.; Paasche-Orlow, Michael K. Justice, health literacy and social epidemiology. *American Journal of Bioethics* 2007 November; 7(11): 18-20; author reply W1-W2. NRCBL: 8.1; 1.1; 7.1. Comments: Angelo E. Volandes and Michael K. Paasche-Orlow. Health literacy, health inequality and a just healthcare system. American Journal of Bioethics 2007 November; 7(11): 5-10.

Grindrod, Eirlys; Gardiner, Esther. Withdrawal of consent during surgery. *Journal of Perioperative Practice* 2006 September; 16(9): 418, 420. NRCBL: 8.1; 9.5.1; 4.1.3. SC: cs.

Groopman, Leonard C.; Miller, Franklin G.; Fins, Joseph J. The patient's work. *CQ: Cambridge Quarterly of Healthcare Ethics* 2007 Winter; 16(1): 44-52. NRCBL: 8.1; 8.3.1; 4.1.2; 4.2; 1.1.

Hall, Judith A.; Horgan, Terrence G.; Stein, Terry S.; Roter, Debra L. Liking in the physician-patient relationship. *Patient Education and Counseling* 2002 September; 48(1): 69-77. NRCBL: 8.1; 7.1.

Hall, Mark A.; Bobinski, Mary Anne; Orentlicher, David. The treatment relationship. *In their:* Bioethics and Public Health Law. New York: Aspen Publishers, 2005: 91-219. NRCBL: 8.1; 8.3.1; 18.3; 9.2; 9.4. SC: le; cs.

Heath, Iona; Nessa, John. Objectification of physicians and loss of therapeutic power. *Lancet* 2007 March 17-23; 369(9565): 886-887. NRCBL: 8.1; 4.1.2.

Henderson, Amanda; Van Eps, Mary Ann; Pearson, Kate; James, Catherine; Henderson, Peter; Osborne, Yvonne. 'Caring for' behaviours that indicate to patients that nurses 'care about' them. *Journal of Advanced Nursing* 2007 October; 60(2): 146-153. NRCBL: 8.1; 4.1.3. SC: em.

Higginson, Irene J.; Hall, S. Rediscovering dignity at the bedside. *BMJ: British Medical Journal* 2007 July 28; 335(7612): 167. NRCBL: 8.1; 7.2.

Hilliard, Marie T. The duty to care: when health care workers face personal risk. *National Catholic Bioethics Quarterly* 2007 Winter; 7(4): 673-682. NRCBL: 8.1; 9.1; 4.1.1; 1.1.
Abstract: A pandemic due to the avian flu virus (H5N1) is possible, and if it occurs, the event will not be unfamiliar to health care workers. History provides us with numerous examples. In the twentieth century alone, there were three pandemics, the largest being the 1918 "Spanish" influenza pandemic, in which forty to fifty million people died worldwide within one year. Five hundred thousand persons died in the United States alone. Such crises have generated heroic responses by health care workers. The question that arises today is whether such heroism will prevail in the face of varying perceptions concerning the duty of health care workers to care.

Howell, Joel D. Trust and the Tuskegee experiments. *In:* Duffin, Jacalyn, ed. Clio in the Clinic: History in Medical Practice. New York: Oxford University Press, 2005: 213-225. NRCBL: 8.1; 8.2; 18.5.1; 18.3; 10; 7.2.

Hurst, Samia A.; Volandes, Angelo E.; Paasche-Orlow, Michael K. De-clustering national and international inequality. *American Journal of Bioethics* 2007 November; 7(11): 24-25; author reply W1-W2. NRCBL: 8.1; 9.1; 21.1. Comments: Angelo E. Volandes and Michael K. Paasche-Orlow. Health literacy, health inequality and a just healthcare system. American Journal of Bioethics 2007 November; 7(11): 5-10.

Jagsi, Reshma. Conflicts of interest and the physician-patient relationship in the era of direct-to-patient advertising.

Journal of Clinical Oncology 2007 March 1; 25(7): 902-905. NRCBL: 8.1; 1.3.7; 9.7; 7.1.

Jensen, Norman. Empathy and patient-physician conflicts: exploring respect. *JGIM: Journal of General Internal Medicine* 2007 October; 22(10): 1485. NRCBL: 8.1.

Kaba, R.; Sooriakumaran, P. The evolution of the doctor-patient relationship. *International Journal of Surgery* 2007 February; 5(1): 57-65. NRCBL: 8.1; 7.1.

Kara, Mahmut Alpertunga. Applicability of the principle of respect for autonomy: the perspective of Turkey. *Journal of Medical Ethics* 2007 November; 33(11): 627-630. NRCBL: 8.1; 1.1; 21.7.
  Abstract: Turkey has a complex character, which has differences from the Western world or Eastern Asia as well as common points. Even after more than a century of efforts to modernise and integrate with the West, Turkish society has values that are different from those of the West, as well as having Western values. It is worth questioning whether ordinary Turkish people show an individualistic character. The principle of respect for individual autonomy arises from a perception of oneself as an individual, and the person's situation may affect the applicability of the principle. Patients who perceive themselves to be members of a community rather than free persons and who prefer to participate in the common decisions of the community and to consider the common interest and the common value system of the community concerning problems of their life (except healthcare or biomedical research) rather than to decide as independent, rational individuals may not be competent to make an autonomous choice. Expectations that such patients will behave as autonomous individuals may be unjustified. The family, rather than the patient, may take a primary role in decisions. A flexible system considering cultural differences in the concept of autonomy may be more feasible than a system following strict universal norms.

Karsjens, Kari L.; Whitney, Simon N.; McCullough, Laurence B. Exploring the nature of physician intent in "silent decisions". *American Journal of Bioethics* 2007 July; 7(7): 42-44; author reply W1-W3. NRCBL: 8.1; 8.3.1. Comments: Simon N. Whitney and Laurence B. McCullough. Physicians' silent decisions: because patient autonomy does not always come first. American Journal of Bioethics 2007 July; 7(7): 33-38.

Kirklin, D. Framing, truth telling and the problem with non-directive counselling. *Journal of Medical Ethics* 2007 January; 33(1): 58-62. NRCBL: 8.1; 1.1; 8.2; 9.1; 17.1.
  Abstract: In this paper several reasons as to why framing issues should be of greater interest to both medical ethicists and healthcare professionals are suggested: firstly, framing can help in explaining health behaviours that can, from the medical perspective, appear perverse; secondly, framing provides a way of describing the internal structure of ethical arguments; and thirdly, an understanding of framing issues can help in identifying clinical practices, such as non-directive counselling, which may, inadvertently, be failing to meet their own stated ethical aims. The effect of framing on how individuals interpret information and how healthcare choices are influenced by framing are described. Next, the role of framing in ethical discourse is discussed with specific reference to Judith Jarvis Thomson's philosophical mind experiment about abortion and the violinist. Finally, the implications of this analysis are examined for the practice of non-directive counselling, which aims at communicating information in a neutral, value-free way and thereby protecting patient autonomy.

Kissoon, Niranjan. Bench-to-bedside review: humanism in pediatric critical care medicine - a leadership challenge. *Critical Care (London, England)* 2005 August; 9(4): 371-375. NRCBL: 8.1; 7.1; 9.5.7.

Klitzman, Robert. Pleasing doctors: when it gets in the way. *BMJ: British Medical Journal* 2007 September 8; 335(7618): 514. NRCBL: 8.1.

Kon, Alexander A.; Whitney, Simon N.; McCullough, Laurence B. Silent decisions or veiled paternalism? Physicians are not experts in judging character. *American Journal of Bioethics* 2007 July; 7(7): 40-42; author reply W1-W3. NRCBL: 8.1; 8.3.1. Comments: Simon N. Whitney and Laurence B. McCullough. Physicians' silent decisions: because patient autonomy does not always come first. American Journal of Bioethics 2007 July; 7(7): 33-38.

Kothari, Sunil; Kirschner, Kristi L. Abandoning the golden rule: the problem with "putting ourselves in the patient's place". *Topics in Stroke Rehabilitation* 2006 Fall; 13(4): 68-73. NRCBL: 8.1; 4.4; 9.5.1. SC: em.

Kuczewski, Mark G. Talking about spirituality in the clinical setting: can being professional require being personal? *American Journal of Bioethics* 2007 July; 7(7): 4-11. NRCBL: 8.1; 1.2; 8.3.1; 9.6. SC: cs. Comments: American Journal of Bioethics 2007 July; 7(7): 12-32.
  Abstract: Spirituality or religion often presents as a foreign element to the clinical environment, and its language and reasoning can be a source of conflict there. As a result, the use of spirituality or religion by patients and families seems to be a solicitation that is destined to be unanswered and seems to open a distance between those who speak this language and those who do not. I argue that there are two promising approaches for engaging such language and helping patients and their families to productively engage in the decision-making process. First, patient-centered interviewing techniques can be employed to explore the patient's religious or spiritual beliefs and successfully translate them into choices. Second, and more radically, I suggest that in some more recalcitrant conflicts regarding treatment plans, resolution may require that clinicians become more involved, personally engaging in discussion and disclosure of religious and spiritual worldviews. I believe that both these approaches are supported by rich models of informed consent such as the transparency model and identify considerations and circumstances that can justify such personal disclosures. I conclude by offering some

NRCBL: National Reference Center for Bioethics Literature Classification Scheme   See inside front cover for terms.

363

considerations for curbing potential unprofessional excesses or abuses in discussing spirituality and religion with patients.

Kullnat, Megan Wills. Boundaries. *JAMA: The Journal of the American Medical Association* 2007 January 24-31; 297(4): 343-344. NRCBL: 8.1.

Lo, Bernard. The doctor-patient relationship. *In his:* Resolving Ethical Dilemmas: A Guide for Clinicians. 3rd edition. Philadelphia, PA: Lippincott Williams and Wilkins, 2005: 153-182. NRCBL: 8.1; 8.4; 18.3; 7.3.

Lyckholm, Laurie; Quillin, John. Equanimity abandoned? *American Journal of Bioethics* 2007 July; 7(7): 31-32. NRCBL: 8.1; 1.2. Comments: Mark G. Kuczewski. Talking about spirituality in the clinical setting: can being professional require being personal. American Journal of Bioethics 2007 July; 7(7): 4-11.

MacDonald, Hannah. Relational ethics and advocacy in nursing: literature review. *Journal of Advanced Nursing* 2007 January; 57(2): 119-126. NRCBL: 8.1; 4.1.3; 9.1.

Macdonald, Marilyn. Origins of difficulty in the nurse-patient encounter. *Nursing Ethics* 2007 July; 14(4): 510-521. NRCBL: 8.1; 4.1.3. SC: em. Identifiers: Canada.
  Abstract: The purpose of this study was to look beyond the patient as the source of difficulty and to examine the context of care encounters for factors that contributed to the construction of difficulty in the nurse-patient encounter. The study explains the origins of difficulty in the nurse-patient encounter. This explanation broadens the thinking limits previously imposed by locating difficulty within the individual. Key elements of this explanation are: knowing the patient minimizes the likelihood of difficulty in the encounter; and families, availability of supplies and equipment, who is working, and care space changes are contextual factors that contribute to the construction of difficulty in the nurse-patient encounter. Awareness of these findings has implications for the strategies nurses employ in difficult encounters.

Messikomer, Carla M. "Our options have changed. . . we will not call you back": communicating with my primary care physician. *Perspectives in Biology and Medicine* 2007 Summer; 50(3): 435-443. NRCBL: 8.1.

Moser, Albine; Houtepen, Rob; Widdershoven, Guy. Patient autonomy in nurse-led shared care: a review of theoretical and empirical literature. *Journal of Advanced Nursing* 2007 February; 57(4): 357-365. NRCBL: 8.1; 1.1; 4.1.3.

Mujovic-Zornic, Hajrija. Legislation and patients' rights: some necessary remarks. *Medicine and Law: The World Association for Medical Law* 2007 December; 26(4): 709-719. NRCBL: 8.1; 21.1. SC: le.
  Abstract: The essence of a patient's rights and legislation framework requires an answer to the question on how legislation can work towards better defining, respecting, protecting and effectiveness of these rights. First, it is necessary to give a short introduction to patients' rights, their definition and different classifications. In the long list of human rights, patients' rights obviously take one of the very important places. Human life and health are the values, which, in comparison with all other human values, are considered as values of the highest rank. Patients' rights represent a legal expression of something, which every person basically and naturally expects from a doctor, medical staff, and from a health care system in general. The subject of the second part of this paper presents the intention, scope and conception of necessary legislation. How should it be considered - in a wider sense or as a special law? Some theoretical and practical questions regarding interaction between medical ethics regulation, confidentiality, and legislation are discussed as well. In the European context there are numerous examples of laws with the specific purpose of protecting patients' rights. Special attention and critical review will be paid to the situation of patients' rights in Serbia. The paper concludes with the point that the role of legislation is evidently important, but the traditional view should be replaced with a new one, due to the reason that modem [sic; modern] health law puts the protection of patients' rights on a higher level. De lege lata, the whole system of health law in its diversity (civil, penal and administrative) is characterized by better understanding of rights, duties and legal relations, either through regulation or the protection of patients' rights.

Myers, Richard S. US law and conscientious objection in healthcare. *In:* Watt, Helen, ed. Cooperation, Complicity and Conscience: Problems in Healthcare, Science, Law and Public Policy. London: Linacre Centre, 2005: 296-315. NRCBL: 8.1; 9.1; 1.2. SC: le; an.

Myser, Catherine. The challenges of amnesia in assessing capacity, assigning a proxy, and deciding to forego life-prolonging medical treatment. *Journal of Clinical Ethics* 2007 Fall; 18(3): 262-269. NRCBL: 8.1; 8.3.3; 20.5.1. SC: cs. Comments: Sheila Otto. Memento . . . life imitates art: the request for an ethics consultation. Journal of Clinical Ethics 2007 Fall; 18(3): 247-251.

Neal, Karama C.; Volandes, Angelo E.; Paasche-Orlow, Michael K. Health literacy: more than a one-way street. *American Journal of Bioethics* 2007 November; 7(11): 29-30; author reply W1-W2. NRCBL: 8.1; 9.1. Comments: Angelo E. Volandes and Michael K. Paasche-Orlow. Health literacy, health inequality and a just healthcare system. American Journal of Bioethics 2007 November; 7(11): 5-10.

Nordby, Halvor. Meaning and normativity in nurse-patient interaction. *Nursing Philosophy* 2007 January; 8(1): 16-27. NRCBL: 8.1.

Omonzejele, Peter F. Obligation of non-maleficence: moral dilemma in physician-patient relationship. *Journal of Medicine and Biomedical Research* 2005 June; 4(1): 7 p. [Online]. Accessed: http://www.bioline.org.br/titles?id=

jm&year=2005&vol=4&num=01&keys=V4N1 [2008 February 14]. NRCBL: 8.1; 1.1; 20.5.1; 20.2.1. SC: cs.

Paris, John J.; Moreland, Michael P.; Whitney, Simon N.; McCullough, Laurence B. Silence is not always golden in medical decision-making. *American Journal of Bioethics* 2007 July; 7(7): 39-40; author reply W1-W3. NRCBL: 8.1; 8.3.1. Comments: Simon N. Whitney and Laurence B. McCullough. Physicians' silent decisions: because patient autonomy does not always come first. American Journal of Bioethics 2007 July; 7(7): 33-38.

Parker, Malcolm. Patients as rational traders: response to Stewart and DeMarco. *Journal of Bioethical Inquiry* 2006; 3(3): 133-136. NRCBL: 8.1; 8.3.4; 9.5.1. SC: an. Comments: D.O. Stewart and J.P. DeMarco. An economic theory of patient decision-making. Journal of Bioethical Inquiry 2005; 2(3): 153-164.

Abstract: Stewart and DeMarco's economic theory of patient decision-making applied to the case of diabetes is flawed by clinical inaccuracies and an unrealistic depiction of patients as rational traders. The theory incorrectly represents patients' struggles to optimize their management as calculated trade-offs against the costs of care, and gives an unrealistic, inflexible account of such costs. It imputes to physicians the view that their patients' lack of compliance is unreasonable, but physicians are accustomed to the variety of human factors which contribute to suboptimal compliance, and work with patients to minimize their influence. By depicting patients as rational traders rather than human beings with a range of motivations and burdens, the economic theory distorts the proper function of informed consent.

Peniston, Reginald L. Does an African American perspective alter clinical ethical decision making at the bedside? *In:* Prograis, Lawrence J.; Pellegrino, Edmund D., eds. African American Bioethics: Culture, Race, and Identity. Washington, DC: Georgetown University Press, 2004: 127-136. NRCBL: 8.1; 2.1; 21.7. Conference: Symposium on African American Perspectives in Bioethics and Second Annual Conference on Health Disparities, held on September 23-24, 2004, at Georgetown University.

Pozgar, George D. Patient rights and responsibilities. *In his:* Legal Aspects of Health Care Administration. 9th edition. Sudbury, MA: Jones and Bartlett Publishers, 2004: 365-374. NRCBL: 8.1; 9.2; 8.3.4. SC: le; cs.

Reeves, Roy R.; Douglas, Sharon P.; Garner, Rosa T.; Reynolds, Marti D.; Silvers, Anita. The individual rights of the difficult patient [case study and commentaries]. *Hastings Center Report* 2007 March-April; 37(2): 13-15. NRCBL: 8.1; 9.5.1; 1.1. SC: cs.

Rentmeester, Christy A. Should a good healthcare professional be (at least a little) callous? *Journal of Medicine and Philosophy* 2007 January-February; 32(1): 43-64. NRCBL: 8.1; 4.1.2. SC: an.

Abstract: The term "callous" has not, to this point, been studied empirically or considered philosophically in the context of healthcare professionalism. It should be, however, because its uses seem peculiar. Sometimes "callous" is used to suggest that becoming callous confers a benefit of some protection against emotional distress, which might be considered expedient in the healthcare work environment. But, "callous" also refers to a person's unappealing demeanor of hardened insensitivity. The tension between these different moral connotations of "callous" prompts several empirical, psychological, and moral questions; I introduce and entertain a few here. I also suggest a distinction between callousness and inurement and argue for why this distinction is important to appreciate and uphold in health professions education.

Robinson, Mary R.; Thiel, Mary Martha; Meyer, Elaine C. On being a spiritual care generalist. *American Journal of Bioethics* 2007 July; 7(7): 24-26. NRCBL: 8.1; 1.2; 8.3.3. Comments: Mark G. Kuczewski. Talking about spirituality in the clinical setting: can being professional require being personal? American Journal of Bioethics 2007 July; 7(7): 4-11.

Rose, Donald N. Respect for patient autonomy in forensic psychiatric nursing. *Journal of Forensic Nursing* 2005 Spring; 1(1): 23-27. NRCBL: 8.1; 1.1; 4.1.3; 17.1.

Rosenthal, M. Sara. Patient misconceptions and ethical challenges in radioactive iodine scanning and therapy. *Journal of Nuclear Medicine Technology* 2006 September; 34(3): 143-150; quiz 151-152. NRCBL: 8.1; 1.3.12; 9.7; 8.3.1; 8.4; 2.1; 16.2.

Schmidt, Harald. Patients' charters and health responsibilities. *BMJ: British Medical Journal* 2007 December 8; 335(7631): 1187-1189. NRCBL: 8.1; 9.1; 9.3.1.

Schwartz, Peter H.; Whitney, Simon N.; McCullough, Laurence B. Silence about screening. *American Journal of Bioethics* 2007 July; 7(7): 46-48; author reply W1-W3. NRCBL: 8.1; 8.3.1; 9.1. Comments: Simon N. Whitney and Laurence B. McCullough. Physicians' silent decisions: because patient autonomy does not always come first. American Journal of Bioethics 2007 July; 7(7): 33-38.

Segal, Judy Z. "Compliance" to "concordance": a critical view. *Journal of Medical Humanities* 2007 June; 28(2): 81-96. NRCBL: 8.1; 9.7. SC: an.

Abstract: Advocates of "concordance" describe it as a new model of shared decision-making between physicians and patients based on a partnership of equals. "Concordance" is meant to make obsolete the notion of "compliance," in which patients are seen as, ideally, following doctors' orders. This essay offers a critical view of concordance, arguing that the literature itself on concordance, including materials at the web site of Medicines Partnership, the implementation arm in Great Britain of the concordance model, is full of contradiction; concordance, in fact, harbors an ideology of compliance. The essay suggests that an improvement in patient medication use will more likely come from a frank consideration of the relation of compliance issues and commercial ones,

---

NRCBL: National Reference Center for Bioethics Literature Classification Scheme   See inside front cover for terms.

365

and that a key question across domains is, "how are patients/health agents/consumers persuaded to acquire certain drugs and take them as directed?"

Sellman, Derek. Trusting patients, trusting nurses. *Nursing Philosophy* 2007 January; 8(1): 28-36. NRCBL: 8.1.

Slieper, Chad F.; Hyle, Laurel R.; Rodriguez, Maria Alma. Difficult discharge: lessons from the oncology setting. *American Journal of Bioethics* 2007 March; 7(3): 31-32. NRCBL: 8.1; 9.4; 9.5.1. Comments: Robert N. Swidler, Terese Seastrum, and Wayne Shelton. Difficult hospital inpatient discharge decisions: ethical, legal and clinical practice issues. American Journal of Bioethics 2007 March; 7(3): 23-28.

Slieper, Chad F.; Wasson, Katherine; Ramondetta, Lois M. From technician to professional: integrating spirituality into medical practice. *American Journal of Bioethics* 2007 July; 7(7): 26-27. NRCBL: 8.1; 1.2; 7.1. Comments: Mark G. Kuczewski. Talking about spirituality in the clinical setting: can being professional require being personal. American Journal of Bioethics 2007 July; 7(7): 4-11.

Sloan, Richard P. Ethical problems. *In his:* Blind Faith: The Unholy Alliance of Religion and Medicine. New York: St. Martin's Press, 2006: 181-206. NRCBL: 8.1; 1.2; 8.4.

Smith, Alexander K.; Davis, Roger B.; Krakauer, Eric L. Differences in the quality of the patient-physician relationship among terminally ill African-American and white patients: impact on advance care planning and treatment preferences. *JGIM: Journal of General Internal Medicine* 2007 November; 22(11): 1579-1582. NRCBL: 8.1; 9.5.4; 20.4.1; 20.5.1. SC: em.

Sokol, Daniel K. What would you do, doctor? *BMJ: British Medial Journal* 2007 April 21; 334(7598): 853. NRCBL: 8.1.

Sprague, Stuart. What part of spirituality don't you understand? *American Journal of Bioethics* 2007 July; 7(7): 28-29. NRCBL: 8.1; 1.2. Comments: Mark G. Kuczewski. Talking about spirituality in the clinical setting: can being professional require being personal. American Journal of Bioethics 2007 July; 7(7): 4-11.

Stotland, Nada L.; Ross, Lainie F.; Clayton, Ellen W.; Mishtal, Joanna Z.; Chavkin, Wendy; Zarate, Victor; O'Connell, Patrick; Mistrot, Jacques; Parson, Kenneth C.; Curlin, Farr A.; Lawrence, Ryan E.; Lantos, John D. Religion, conscience, and controversial clinical practices [letters and reply]. *New England Journal of Medicine* 2007 May 3; 356(18): 1889-1892. NRCBL: 8.1; 1.2; 4.1.2; 8.2; 9.1. Identifiers: Chile.

Strong, P.M.; Davis, A.G. Roles, role formats and medical encounters: a cross-cultural analysis of staff-client relationships in children's clinics. *Sociological Review* 1977

November; 25(4): 775-800. NRCBL: 8.1; 7.1; 9.5.7; 9.5.10. SC: em.

Swota, Alissa Hurwitz. Changing policy to reflect a concern for patients who sign out against medical advice. *American Journal of Bioethics* 2007 March; 7(3): 32-34. NRCBL: 8.1; 9.3.1. Comments: Robert N. Swidler, Terese Seastrum, and Wayne Shelton. Difficult hospital inpatient discharge decisions: ethical, legal and clinical practice issues. American Journal of Bioethics 2007 March; 7(3): 23-28.

Tao Lai-Po-wah, Julia. A Confucian approach to a "shared family decision model" in health care: reflections on moral pluralism. *In:* Engelhardt, H. Tristram, ed. Global Bioethics: The Collapse of Consensus. Salem, MA: M&M Scrivener Press, 2006: 154-179. NRCBL: 8.1; 1.1; 2.1; 21.7.

Tuffs, Annette. German doctors: public enemy number one? *BMJ: British Medical Journal* 2007 May 26; 334(7603): 1087. NRCBL: 8.1; 1.3.7; 7.1.

Watt, Helen. Cooperation problems in care of suicidal patients. *In her:* Cooperation, Complicity and Conscience: Problems in Healthcare, Science, Law and Public Policy. London: Linacre Centre, 2005: 139-147. NRCBL: 8.1; 20.7; 1.2.

Whitney, Simon N.; McCullough, Laurence B. Physicians' silent decisions: because patient autonomy does not always come first. *American Journal of Bioethics* 2007 July; 7(7): 33-38. NRCBL: 8.1; 8.3.1; 1.1. Comments: American Journal of Bioethics 2007 July; 7(7): 39-48.

Wicclair, Mark R. Professionalism, religion and shared decision-making. *American Journal of Bioethics* 2007 July; 7(7): 29-31. NRCBL: 8.1; 1.2; 8.3.1. Comments: Mark G. Kuczewski. Talking about spirituality in the clinical setting: can being professional require being personal? American Journal of Bioethics 2007 July; 7(7): 4-11.

Wirtz, Veronika; Cribb, Alan; Barber, Nick. Patient-doctor decision-making about treatment within the consultation — a critical analysis of models. *Social Science and Medicine* 2006 January; 62(1): 116-124. NRCBL: 8.1; 7.3.

Zaner, Richard. Physicians and patients in relation: clinical interpretation and dialogues of trust. *In:* Khushf, George, ed. Handbook of Bioethics: Taking Stock of the Field From a Philosophical Perspective. Dordrecht; Boston: Kluwer Academic, 2004: 223-250. NRCBL: 8.1; 1.1.

PATIENTS' RIGHTS See CARE FOR SPECIFIC GROUPS; CONFIDENTIALITY; INFORMED CONSENT; RIGHT TO HEALTH CARE; TREATMENT REFUSAL; TRUTH DISCLOSURE

PERSONHOOD *See* QUALITY AND VALUE OF
LIFE

PHARMACOGENETICS

Bibbins-Domingo, Kirsten; Fernandez, Alicia. BiDil
for heart failure in black patients: implications of the U.S.
Food and Drug Administration approval. *Annals of Internal Medicine* 2007 January 2; 146(1): 52-56. 49 refs.
NRCBL: 15.11; 5.3; 9.5.4; 9.7.
    Keywords: *blacks; *pharmacogenetics; clinical trials;
    drugs; federal government; health disparities; heart diseases; public policy; racial groups; Proposed Keywords:
    *drug approval; Keyword Identifiers: *BiDil; *Food and
    Drug Administration; United States

Burroughs, Valentine J. Racial and ethnic inclusiveness
in clinical trials. *In:* Santoro, Michael A.; Gorrie, Thomas
M., eds. Ethics and the Pharmaceutical Industry. Cambridge; New York: Cambridge University Press, 2005:
80-96, 421-425. NRCBL: 18.5.1; 15.1; 9.7.
    Keywords: *clinical trials; *ethnic groups;
    *pharmacogenetics; *racial groups; culture; drug industry;
    drugs; genetic ancestry; health hazards; minority groups;
    public policy; Keyword Identifiers: BiDil; United States

DeCamp, Matthew; Buchanan, Allen. Pharmacogenomics, ethical and regulatory issues. *In:* Steinbock,
Bonnie, ed. The Oxford Handbook of Bioethics. Oxford;
New York: Oxford University Press, 2007: 536-568. 96
refs. 10 fn. NRCBL: 15.1; 9.7; 5.3. SC: an; rv.
    Keywords: *moral policy; *pharmacogenetics; clinical genetics; clinical trials; community consent; confidentiality;
    drugs; economics; ethical analysis; ethnic groups; genetic
    ancestry; genetic databases; genetic diversity; genetic information; genetic research; genetic screening; genetic services; health services accessibility; informed consent;
    justice; marketing; property rights; population genetics; privacy; racial groups; regulation; research subjects; risks and
    benefits; standards; Proposed Keywords: community
    participation; exceptionalism

Fox, Jeffrey L. Despite glacial progress, US government
signals support for personalized medicine. *Nature Biotechnology* 2007 May 25(5): 489-490. NRCBL: 15.1; 9.7.

Holm, Søren. Pharmacogenetics and global (in)justice.
In: Cohen, Jillian Clare; Illingworth, Patricia; Schüklenk,
Udo, eds. The Power of Pills: Social, Ethical and Legal Issues in Drug Development, Marketing, and Pricing. London; Ann Arbor, MI: Pluto, 2006: 98-105. 21 refs.
NRCBL: 15.1; 9.7; 21.1. SC: an.
    Keywords: *developing countries; *drug industry; *ethical
    analysis; *international aspects; *justice;
    *pharmacogenetics; *regulation; clinical genetics; clinical
    trials; economics; health care delivery; informed consent;
    marketing; risks and benefits; social impact; trends;
    Proposed Keywords: world health

Human Genome Organisation. Ethics Committee.
HUGO statement on pharmacogenomics (PGx): solidarity,
equity and governance [position statement]. *Genomics,
Society and Policy* 2007 April; 3(1): 44-47. 1 fn. NRCBL:
15.1; 9.7.

Keywords: *pharmacogenetics; *organizational policies;
access to information; clinical genetics; drugs; ethics committees; family members; genetic research; genetic services;
goals; international aspects; justice; population genetics;
public participation; research priorities; risks and benefits;
social impact; standards; Proposed Keywords: stakeholders;
Keyword Identifiers: *Human Genome Organization
(HUGO)

Jones, David S.; Perlis, Roy H. Pharmacogenetics, race,
and psychiatry: prospects and challenges. *Harvard Review
of Psychiatry* 2006 March-April; 14(2): 92-108. NRCBL:
15.1; 9.7; 17.4.

Mayor, Susan. Fitting the drug to the patient. *BMJ: British
Medical Journal* 2007 March 3; 334(7591): 452-453. 2
refs. NRCBL: 15.1; 9.7.
    Keywords: *pharmacogenetics; drug industry; forecasting;
    marketing; social impact; trends

Nielsen, Louise Fuks; Møldrup, Claus. Lay perspective
on pharmacogenetics and its application to future drug
treatment: a Danish quantitative survey. *New Genetics and
Society* 2007 December; 26(3): 309-324. NRCBL: 15.1;
9.7. SC: em.

Patowary, S. Pharmacogenomics — therapeutic and ethical issues. *Kathmandu University Medical Journal
(KUMJ)* 2005 October-December; 3(4): 428-430.
NRCBL: 15.1; 9.7; 15.11.

Roscam Abbing, Henriette D.C. Pharmacogenetics: a
new challenge for health law. *Medicine and Law: The
World Association for Medical Law* 2007 December;
26(4): 781-789. NRCBL: 15.1; 9.7; 8.3.1; 8.4. SC: le.
    Abstract: Developments in pharmacogenetics make it
    possible to determine the genetic factors that influence
    variations in response to medicine. Differences in response to medication may be related to the genetic characteristics of the individual, to the genetic make-up of the
    diseased tissue or to both. Advantages include optimal
    therapeutic effect, safe medication, minimised side-effects, and development of medication for small groups of
    patients. Strict adherence to patients' rights and to the
    medical professional standard must prevent negative effects of pharmacogenetics on individual rights, notably
    the right (not) to know, to privacy and informed consent.
    Use of pharmacogenetics by third parties for non-health
    related purposes may bring about a disproportionate intrusion of the privacy of an individual; it may result in
    barriers for accessing primary social goods, and it may be
    a disincentive for the individual to have a pharmacogenetic analysis performed for individual health care purposes or to participate in a drug trial. Medical examinations before employment must be justified by the health
    requirements unavoidably inherent to the job (their objective being the protection of health and not the financial
    interests of the employer). In a system that relies on private insurance for having access to primary social goods
    (health, disability — and life insurance), the use and the
    outcome of a pharmacogenetic analysis for the purpose of
    differentiation between insurance candidates on the basis
    of their "risk-profile" must be restricted; where appropri-

NRCBL: National Reference Center for Bioethics Literature Classification Scheme   See inside front cover for terms.

367

ate measures should take into account justified interests of the insurance company to prevent adverse selection. Current measures in several European countries are not effective enough to meet the concerns specifically inherent to pahrmacogenetics [sic; pharmacogenetics]. Human rights principles must be at the basis of national and European policies for providing adequate protection against disproportionate intrusion into private life, for guaranteeing equity in access to health care and accessibility of other primary social goods.

Rothstein, M.A.; Epps, P.G. Pharmacogenomics and the (ir)relevance of race. *Pharmacogenomics Journal* 2001; 1(2): 104-108. 29 refs. NRCBL: 15.11; 9.7; 15.1.
> Keywords: *genetic ancestry; *genetic research; *pharmacogenetics; *racial groups; biomedical research; clinical genetics; ethnic groups; genetic diversity; historical aspects; research subjects; selection of subjects; trends; Proposed Keywords: classification

Temple, Robert; Stockbridge, Norman L. BiDil for heart failure in black patients: the U.S. Food and Drug Administration perspective. *Annals of Internal Medicine* 2007 January 2; 146(1): 52-62. 28 refs. NRCBL: 15.11; 5.3; 9.5.4; 9.7.
> Keywords: *blacks; *pharmacogenetics; clinical trials; drugs; federal government; heart diseases; public policy; racial groups; Proposed Keywords: *drug approval; Keyword Identifiers: *BiDil; *Food and Drug Administration; United States

United States. Congress. Senate. A bill to secure the promise of personalized medicine for all Americans by expanding and accelerating genomics research and initiatives to improve the accuracy of disease diagnosis, increase the safety of drugs, and identify novel treatments. Washington, DC: U.S. G.P.O., 2007. 32 p. [Online]. Accessed: http://frwebgate.access.gpo.gov/cgi-bin/useftp.cgi?IPaddress=162.140.64.21&filename=s976is.pdf&directory=/diskb/wais/data/110_cong_bills [2007 April 4]. NRCBL: 15.1; 9.7; 19.5. SC: le. Identifiers: Genomics and Personalized Medicine Act of 2007. Note: S. 976, 110th Congress, 1st session. Introduced by Sen. Obama on March 23, 2007. Referred to the Committee on Health, Education, Labor, and Pensions.
> Keywords: *genetic research; *genomics; *legal aspects; *pharmacogenetics; *public policy; advertising; biological specimen banks; drugs; genetic screening; guidelines; research support; standards; Keyword Identifiers: *United States

PHARMACOGENOMICS *See* PHARMACOGENETICS

PHILOSOPHICAL ASPECTS *See* BIOETHICS AND MEDICAL ETHICS/ PHILOSOPHICAL ASPECTS; EUTHANASIA AND ALLOWING TO DIE/ PHILOSOPHICAL ASPECTS

PHILOSOPHY *See* BIOETHICS AND MEDICAL ETHICS/ PHILOSOPHICAL ASPECTS; EUTHANASIA AND ALLOWING TO DIE/

PHILOSOPHICAL ASPECTS; NURSING ETHICS AND PHILOSOPHY; PHILOSOPHY OF MEDICINE

PHILOSOPHY OF MEDICINE

Adams, Marcus P.; Lawrence, Ryan E.; Curlin, Farr A. Conscience and conflict. *American Journal of Bioethics* 2007 December; 7(12): 28-29; author reply W1-W2. NRCBL: 4.1.2; 1.2. Comments: Ryan E. Lawrence and Farr A. Curlin. Clash of definitions: controversies about conscience in medicine. American Journal of Bioethics 2007 December; 7(12): 10-14.

Baier, Annette C. Trust, suffering, and the Aesculapian virtues. *In:* Walker, Rebecca L.; Ivanhoe, Philip J., eds. Working Virtue: Virtue Ethics and Contemporary Moral Problems. Oxford: Clarendon, 2007: 135-153. NRCBL: 4.1.2; 1.3.1; 20.5.1; 4.4; 1.1.

Barfield, Raymond; Lawrence, Ryan E.; Curlin, Farr A. Conscience is the means by which we engage the moral dimension of medicine. *American Journal of Bioethics* 2007 December; 7(12): 26-27; author reply W1-W2. NRCBL: 4.1.2. Comments: Ryan E. Lawrence and Farr A. Curlin. Clash of definitions: controversies about conscience in medicine. American Journal of Bioethics 2007 December; 7(12): 10-14.

Bebeau, Muriel J. Evidence-based character development. *In:* Kenny, Nuala; Shelton, Wayne, eds. Lost Virtue: Professional Character Development in Medical Education. Amsterdam; Oxford: Elsevier, 2006: 47-86. NRCBL: 4.1.2; 7.2. SC: em.

Birnbacher, Dieter. Die Grenzen der Philosophie und die Grenzen des Lebens [The boundaries of philosophy and the limits of life]. *Ethik in der Medizin* 2006 December; 18(4): 315-319. NRCBL: 4.1.2; 1.1; 2.1; 4.4; 20.2.1.

Campbell, Eric G.; Regan, Susan; Gruen, Russell L.; Ferris, Timothy G.; Rao, Sowmya R.; Cleary, Paul D.; Blumenthal, David. Professionalism in medicine: results of a national survey of physicians. *Annals of Internal Medicine* 2007 December 4; 147(11): 795-802. NRCBL: 4.1.2; 1.3.1. SC: em.
> Abstract: BACKGROUND: The prospect of improving care through increasing professionalism has been gaining momentum among physician organizations. Although there have been efforts to define and promote professionalism, few data are available on physician attitudes toward and conformance with professional norms. OBJECTIVE: To ascertain the extent to which practicing physicians agree with and act consistently with norms of professionalism. DESIGN: National survey using a stratified random sample. SETTING: Medical care in the United States. PARTICIPANTS: 3504 practicing physicians in internal medicine, family practice, pediatrics, surgery, anesthesiology, and cardiology. MEASUREMENTS: Attitudes and behaviors were assessed by using

indicators for each domain of professionalism developed by the American College of Physicians and the American Board of Internal Medicine. Of the eligible sampled physicians, 1662 responded, yielding a 58% weighted response rate (adjusting for noneligible physicians). RESULTS: Ninety percent or more of the respondents agreed with specific statements about principles of fair distribution of finite resources, improving access to and quality of care, managing conflicts of interest, and professional self-regulation. Twenty-four percent disagreed that periodic recertification was desirable. Physician behavior did not always reflect the standards they endorsed. For example, although 96% of respondents agreed that physicians should report impaired or incompetent colleagues to relevant authorities, 45% of respondents who encountered such colleagues had not reported them. LIMITATIONS: Our measures of behavior did not capture all activities that may reflect on the norms in question. Furthermore, behaviors were self-reported, and the results may not be generalizable to physicians in specialties not included in the study. CONCLUSION: Physicians agreed with standards of professional behavior promulgated by professional societies. Reported behavior, however, did not always conform to those norms.

Colman, Richard D.; Caine, Jane A. Role of the doctor: to care for patients' wellbeing [letters]. *BMJ: British Medical Journal* 2007 December 8; 335(7631): 1169. NRCBL: 4.1.2; 1.3.1; 1.3.2. Comments: Fiona Godlee. The role of the doctor. BMJ: British Medical Journal 2007 November 17; 335(7628): Editor's choice.

Cook, E. David; Lawrence, Ryan E.; Curlin, Farr A. Always let your conscience be your guide. *American Journal of Bioethics* 2007 December; 7(12): 17-19; author reply W1-W2. NRCBL: 4.1.2; 7.2. Comments: Ryan E. Lawrence and Farr A. Curlin. Clash of definitions: controversies about conscience in medicine. American Journal of Bioethics 2007 December; 7(12): 10-14.

Cruess, Sylvia R. Professionalism and medicine's social contract with society. *Clinical Orthopaedics and Related Research* 2006 August; 449: 170-176. NRCBL: 4.1.2; 1.1; 7.1; 9.3.1.

DasGupta, Sayantani. The doctor's wife. *Hastings Center Report* 2007 March-April; 37(2): 7-8. NRCBL: 4.1.2; 8.1.

Denier, Yvonne. Autonomy in dependence: a defence of careful solidarity. *In:* Nys, Thomas; Denier, Yvonne; Vandevelde, Toon, eds. Autonomy and Paternalism: Reflections on the Theory and Practice of Health Care. Leuven; Dudley, MA: Peeters, 2007: 93-111. NRCBL: 4.1.2; 1.1; 8.1; 8.3.1.

Elliott, Carl. Disillusioned doctors. *In:* Kenny, Nuala; Shelton, Wayne, eds. Lost Virtue: Professional Character Development in Medical Education. Amsterdam; Oxford: Elsevier, 2006: 87-97. NRCBL: 4.1.2.

Emerson, Claudia I.; Daar, Abdallah S.; Lawrence, Ryan E.; Curlin, Farr A. Defining conscience and acting conscientiously. *American Journal of Bioethics* 2007 December; 7(12): 19-21; author reply W1-W2. NRCBL: 4.1.2. Comments: Ryan E. Lawrence and Farr A. Curlin. Clash of definitions: controversies about conscience in medicine. American Journal of Bioethics 2007 December; 7(12): 10-14.

Epstein, Ronald M. Mindful practice and the tacit ethics of the moment. *In:* Kenny, Nuala; Shelton, Wayne, eds. Lost Virtue: Professional Character Development in Medical Education. Amsterdam; Oxford: Elsevier, 2006: 115-144. NRCBL: 4.1.2; 7.2; 8.1.

Fulford, K.W.M. (Bill). Ten principles of values-based medicine (VBM). *In:* Schramme, Thomas; Thome, Johannes, eds. Philosophy and Psychiatry. Berlin; New York: De Gruyter, 2004: 50-80. NRCBL: 4.1.2; 2.1; 9.8; 17.1.

Genuis, S.J. Dismembering the ethical physician. *Postgraduate Medical Journal* 2006 April; 82(966): 233-238. NRCBL: 4.1.2; 18.6; 2.1.

Glenn, Linda MacDonald; Boyce, Jeanann; Lawrence, Ryan E.; Curlin, Farr A. The Tao of conscience: conflict and resolution. *American Journal of Bioethics* 2007 December; 7(12): 33-34; author reply W1-W2. NRCBL: 4.1.2; 1.2. Comments: Ryan E. Lawrence and Farr A. Curlin. Clash of definitions: controversies about conscience in medicine. American Journal of Bioethics 2007 December; 7(12): 10-14.

Jones, James W.; McCullough, Laurence B.; Richman, Bruce W. Ethics and professionalism: do we need yet another surgeons' charter? *Journal of Vascular Surgery* 2006 October; 44(4): 903-906. NRCBL: 4.1.2; 1.3.1; 6.

Kenny, Nuala. Searching for doctor good: virtues for the twenty-first century. *In:* Kenny, Nuala; Shelton, Wayne, eds. Lost Virtue: Professional Character Development in Medical Education. Amsterdam; Oxford: Elsevier, 2006: 211-233. NRCBL: 4.1.2; 1.1; 1.3.1; 7.2.

Kinghorn, Warren A.; McEvoy, Matthew D.; Michel, Andrew; Balboni, Michael. Professionalism in modern medicine: does the emperor have any clothes? *Academic Medicine* 2007 January; 82(1): 40-45. NRCBL: 4.1.2; 1.3.1; 1.1; 7.2.

Kittay, Eva F. Beyond autonomy and paternalism: the caring transparent self. *In:* Nys, Thomas; Denier, Yvonne; Vandevelde, Toon, eds. Autonomy and Paternalism: Reflections on the Theory and Practice of Health Care. Leuven; Dudley, MA: Peeters, 2007: 23-70. NRCBL: 4.1.2; 1.1; 8.3.1; 8.3.3; 9.1. SC: an.

Klahr, Saulo. One physician's exploration of the ethics in the practice of medicine. *Kidney International* 2006 August; 70(4): 613-614. NRCBL: 4.1.2; 1.3.1.

NRCBL: National Reference Center for Bioethics Literature Classification Scheme   See inside front cover for terms.

369

Ladd, Rosalind Ekman; Lawrence, Ryan E.; Curlin, Farr A. Some reflections on conscience. *American Journal of Bioethics* 2007 December; 7(12): 32-33; author reply W1-W2. NRCBL: 4.1.2. Comments: Ryan E. Lawrence and Farr A. Curlin. Clash of definitions: controversies about conscience in medicine. American Journal of Bioethics 2007 December; 7(12): 10-14.

LaFollette, Hugh; Lawrence, Ryan E.; Curlin, Farr A. The physician's conscience. *American Journal of Bioethics* 2007 December; 7(12): 15-17; author reply W1-W2. NRCBL: 4.1.2; 1.2. Comments: Ryan E. Lawrence and Farr A. Curlin. Clash of definitions: controversies about conscience in medicine. American Journal of Bioethics 2007 December; 7(12): 10-14.

Lawrence, Ryan E.; Curlin, Farr A. Clash of definitions: controversies about conscience in medicine. *American Journal of Bioethics* 2007 December; 7(12): 10-14. NRCBL: 4.1.2; 1.2. Comments: American Journal of Bioethics 2007 December; 7(12): 15-34.

Le Coz, Pierre; Tassy, Sebastien. The philosophical moment of the medical decision: revisiting emotions felt, to improve ethics of future decisions. *Journal of Medical Ethics* 2007 August; 33(8): 470-472. NRCBL: 4.1.2; 1.1; 8.1.

Abstract: The present investigation looks for a solution to the problem of the influence of feelings and emotions on our ethical decisions. This problem can be formulated in the following way. On the one hand, emotions (fear, pity and so on) can alter our sense of discrimination and lead us to make our wrong decisions. On the other hand, it is known that lack of sensitivity can alter our judgment and lead us to sacrifice basic ethical principles such as autonomy, beneficence, non-maleficence and justice. Only emotions can turn a decision into an ethical one, but they can also turn it into an unreasonable one. To avoid this contradiction, suggest integrating emotions with the decisional factors of the process of "retrospective thinking". During this thinking, doctors usually try to identify the nature and impact of feelings on the decision they have just made. In this retrospective moment of analysis of the decision, doctors also question themselves on the feelings they did not experience. They do this to estimate the consequences of this lack of feeling on the way they behaved with the patient.

Mann, Karen V. Learning and teaching in professional character development. *In:* Kenny, Nuala; Shelton, Wayne, eds. Lost Virtue: Professional Character Development in Medical Education. Amsterdam; Oxford: Elsevier, 2006: 145-183. NRCBL: 4.1.2; 7.2.

McCullough, Laurence B. The ethical concept of medicine as a profession: its origins in modern medical ethics and implications for physicians. *In:* Kenny, Nuala; Shelton, Wayne, eds. Lost Virtue: Professional Character Development in Medical Education. Amsterdam; Oxford: Elsevier, 2006: 17-27. NRCBL: 4.1.2; 2.2; 7.2. SC: an.

Night, Susan S.; Lawrence, Ryan E.; Curlin, Farr A. Negotiating the tension between two integrities: a richer perspective on conscience. *American Journal of Bioethics* 2007 December; 7(12): 24-26; author reply W1-W2. NRCBL: 4.1.2. Comments: Ryan E. Lawrence and Farr A. Curlin. Clash of definitions: controversies about conscience in medicine. American Journal of Bioethics 2007 December; 7(12): 10-14.

Orr, Robert D.; Lawrence, Ryan E.; Curlin, Farr A. The role of moral complicity in issues of conscience. *American Journal of Bioethics* 2007 December; 7(12): 23-24; author reply W1-W2. NRCBL: 4.1.2. Comments: Ryan E. Lawrence and Farr A. Curlin. Clash of definitions: controversies about conscience in medicine. American Journal of Bioethics 2007 December; 7(12): 10-14.

Parke, David W.; Durfee, David A.; Zacks, Charles M.; Orloff, Paul N., eds. Ethics in ophthalmology. *In their:* The Profession of Ophthalmology: Practice Management, Ethics, and Advocacy. San Francisco: American Academy of Ophthalmology, 2005: 167- 247. NRCBL: 4.1.2; 2.1; 6; 2.2; 9.5.1. SC: cs.

Pellegrino, Edmund D. Character formation and the making of good physicians [M575]. *In:* Kenny, Nuala and Shelton, Wayne, eds. Lost Virtue: Professional Character Development in Medical Education. Amsterdam; Oxford: Elsevier, 2006: 1-15. NRCBL: 4.1.2; 1.1; 1.3.1; 7.2; 8.1.

Pellegrino, Edmund D. Professing medicine, virtue based ethics, and the retrieval of professionalism. *In:* Walker, Rebecca L.; Ivanhoe, Philip J., eds. Working Virtue: Virtue Ethics and Contemporary Moral Problems. Oxford: Clarendon, 2007: 61-85. NRCBL: 4.1.2; 1.1; 1.3.1. SC: an.

Petrova, Mila; Dale, Jeremy; Fulford, Bill (KWM). Values-based practice in primary care: easing the tensions between individual values, ethical principles and best evidence. *British Journal of General Practice* 2006 September; 56(530): 703-709. NRCBL: 4.1.2; 8.1; 7.1.

Rees, Charlotte E.; Knight, Lynn V. The trouble with assessing students' professionalism: theoretical insights from sociocognitive psychology. *Academic Medicine* 2007 January; 82(1): 46-50. NRCBL: 4.1.2; 1.3.1; 7.2; 2.1. SC: em.

Rhodes, Rosamond. The professional responsibilities of medicine. *In:* Rhodes, Rosamond; Francis, Leslie P.; Silvers, Anita, eds. The Blackwell Guide to Medical Ethics. Malden, MA: Blackwell Pub., 2007: 71-87. NRCBL: 4.1.2; 7.1.

Rhodes, Rosamond; Smith, Lawrence G. Molding professional character. *In:* Kenny, Nuala; Shelton, Wayne, eds. Lost Virtue: Professional Character Development in Medical Education. Amsterdam; Oxford: Elsevier, 2006: 99-114. NRCBL: 4.1.2; 7.2. SC: an.

Salomon, Fred; Ziegler, Andrea. Moral und Abhängigkeit - Ethische Entscheidungskonflikte im hierarchischen System Krankenhaus = Morals and dependency - ethical conflicts in the hierarchical system of a hospital. *Ethik in der Medizin* 2007 September; 19(3): 174-186. NRCBL: 4.1.2; 7.3; 9.1; 1.3.2.

Savulescu, Julian; Lawrence, Ryan E.; Curlin, Farr A. The proper place of values in the delivery of medicine. *American Journal of Bioethics* 2007 December; 7(12): 21-22; author reply W1-W2. NRCBL: 4.1.2; 1.3.1. Comments: Ryan E. Lawrence and Farr A. Curlin. Clash of definitions: controversies about conscience in medicine. American Journal of Bioethics 2007 December; 7(12): 10-14.

Sox, Harold C. Medical professionalism and the parable of the craft guilds [editorial]. *Annals of Internal Medicine* 2007 December; 147(11): 809-810. NRCBL: 4.1.2; 1.3.1.

Stark, Patsy; Roberts, Chris; Newble, David; Bax, Nigel. Discovering professionalism through guided reflection. *Medical Teacher* 2006 February; 28(1): e25-e31. NRCBL: 4.1.2; 1.3.1; 7.2.

Stevens, David. Medical martyrdom? *Today's Christian Doctor* 2007 Fall; 38(3): 18-21. NRCBL: 4.1.2; 1.2. SC: le.

Wicclair, Mark R.; Lawrence, Ryan E.; Curlin, Farr A. The moral significance of claims of conscience in healthcare. *American Journal of Bioethics* 2007 December; 7(12): 30-31; author reply W1-W2. NRCBL: 4.1.2. Comments: Ryan E. Lawrence and Farr A. Curlin. Clash of definitions: controversies about conscience in medicine. American Journal of Bioethics 2007 December; 7(12): 10-14.

PHILOSOPHY OF NURSING *See* NURSING ETHICS AND PHILOSOPHY

PHYSICIAN PATIENT RELATIONSHIP *See* BIOETHICS AND MEDICAL ETHICS; PATIENT RELATIONSHIPS

PREIMPLANTATION DIAGNOSIS *See* GENETIC COUNSELING; GENETIC SCREENING

PRENATAL DIAGNOSIS *See* GENETIC COUNSELING; GENETIC SCREENING; SEX DETERMINATION

PRIORITIES IN HEALTH CARE *See* RESOURCE ALLOCATION

PRISONERS *See* HUMAN EXPERIMENTATION/ SPECIAL POPULATIONS/ PRISONERS; TREATMENT REFUSAL

PRIVACY *See* CONFIDENTIALITY; GENETIC PRIVACY

PRIVILEGED COMMUNICATION *See* CONFIDENTIALITY

PROCUREMENT *See* ORGAN AND TISSUE TRANSPLANTATION/ DONATION AND PROCUREMENT

PROFESSIONAL ETHICS
See also BIOETHICS AND MEDICAL ETHICS; CODES OF ETHICS; NURSING ETHICS AND PHILOSOPHY

Arman, Maria; Rehnsfeldt, Arne; Oberle, Kathleen. The 'little extra' that alleviates suffering. *Nursing Ethics* 2007 May; 14(3): 372-384; discussion: 384-386. NRCBL: 4.1.1; 4.1.3; 8.1; 4.4. SC: em.
   Abstract: Nursing, or caring science, is mainly concerned with developing knowledge of what constitutes ideal, good health care for patients as whole persons, and how to achieve this. The aim of this study was to find clinical empirical indications of good ethical care and to investigate the substance of ideal nursing care in praxis. A hermeneutic method was employed in this clinical study, assuming the theoretical perspective of caritative caring and ethics of the understanding of life. The data consisted of two Socratic dialogues: one with nurses and one with nursing students, and interviews with two former patients. The empirical data are first described from a phenomenological approach. Observations of caregivers offering 'the little extra' were taken to confirm that patients were 'being seen', not from the perspective of an ideal nursing model, but from that of interaction as a fellow human being. The study provides clinical evidence that, as an ontological response to suffering, 'symbolic acts' such as giving the 'little extra' may work to bridge gaps in human interaction. The fact that 'little things' have the power to preserve dignity and make patients feel they are valued offers hope. Witnessing benevolent acts also paves the way for both patients and caregivers to increase their understanding of life.

Beemsterboer, Phyllis L. Developing an ethic of access to care in dentistry. *Journal of Dental Education* 2006 November; 70(11): 1212-1216. NRCBL: 4.1.1; 7.2.

Benner, Patricia; Wrubel, Judith. Response to: 'Edwards, Benner and Wrubel on caring' by S. Horrocks (2002) Journal of Advanced Nursing 40, 36-41. *Journal of Advanced Nursing* 2002 October; 40(1): 45-47. NRCBL: 4.1.1; 1.1; 4.1.3.

Botto, Ronald W. Addressing the marketplace mentality and improving professionalism in dental education: response to Richard Masella's "Renewing professionalism in dental education". *Journal of Dental Education* 2007 February; 71(2): 217-221. NRCBL: 4.1.1; 7.2.

Campbell, N. Ethics in South African dentistry 2006. *SADJ: Journal of the South African Dental Association* 2006 July; 61(6): 240; discussion 242. NRCBL: 4.1.1; 7.4; 8.1; 1.3.2; 6.

NRCBL: National Reference Center for Bioethics Literature Classification Scheme   See inside front cover for terms.

371

Card, Robert F. Response to commentators on "Conscientious objection and emergency contraception": sex, drugs and the rocky role of Levonorgestrel [letter]. *American Journal of Bioethics* 2007 October; 7(10): W4-W6. NRCBL: 4.1.1; 9.7; 11.1.

Catalanotto, Frank A.; Patthoff, Donald E.; Gray, Carolyn F. Ethics of access to oral health care: an introduction to the special issue. *Journal of Dental Education* 2006 November; 70(11): 1117-1119. NRCBL: 4.1.1; 9.2.

Chambers, David W. Ethics summit I: assembling the ethical community. *Journal of the American College of Dentists* 1998 Fall; 65(3): 9-11. NRCBL: 4.1.1; 1.3.1.

Chambers, David W. Moral communities. *Journal of Dental Education* 2006 November; 70(11): 1226-1234. NRCBL: 4.1.1.

Cohen, Michael H. Legal and ethical issues relating to use of complementary therapies in pediatric hematology/oncology. *Journal of Pediatric Hematology/Oncology* 2006 March; 28(3): 190-193. NRCBL: 4.1.1; 8.1; 8.3.2; 9.5.1. SC: le; rv.

Davis, Michael. Eighteen rules for writing a code of professional ethics. *Science and Engineering Ethics* 2007 June; 13(2): 171-189. NRCBL: 1.3.1; 6.

Dharamsi, Shafik. Building moral communities? First, do no harm. *Journal of Dental Education* 2006 November; 70(11): 1235-1240. NRCBL: 4.1.1; 9.2.

Fan, Ruiping; Holliday, Ian. Which medicine? Whose standard? Critical reflections on medical integration in China. *Journal of Medical Ethics* 2007 August; 33(8): 454-461. NRCBL: 4.1.1; 21.1.
Abstract: There is a prevailing conviction that if traditional medicine (TRM) or complementary and alternative medicine (CAM) are integrated into healthcare systems, modern scientific medicine (MSM) should retain its principal status. This paper contends that this position is misguided in medical contexts where TRM is established and remains vibrant. By reflecting on the Chinese policy on three entrenched forms of TRM (Tibetan, Mongolian and Uighur medicines) in western regions of China, the paper challenges the ideology of science that lies behind the demand that all traditional forms of medicine be evaluated and reformed according to MSM standards. Tibetan medicine is used as a case study to indicate the falsity of a major premise of the scientific ideology. The conclusion is that the proper integrative system for TRM and MSM is a dual standard based system in which both TRM and MSM are free to operate according to their own medical standards.

Henley, Lesley D.; Frank, Denise M. Reporting ethical protections in physical therapy research. *Physical Therapy* 2006 April; 86(4): 499-509. NRCBL: 4.1.1; 18.2; 18.3; 1.3.7. SC: em.

Hogan, Carol. Conscience clauses and the challenge of cooperation in a pluralistic society. Sacramento: California Catholic Conference, 2003 February; 12 p. [Online]. Accessed: http://www.cacatholic.org/rfconscience.html [2007 April 4]. NRCBL: 4.1.1; 1.2; 12.4.3; 11.3.

Horrocks, Stephen. Edwards, Benner, and Wrubel on caring. *Journal of Advanced Nursing* 2002 October; 40(1): 36-41. NRCBL: 4.1.1; 1.1; 4.1.3.

Kenny, Belinda; Lincoln, Michelle; Balandin, Susan. A dynamic model of ethical reasoning in speech pathology. *Journal of Medical Ethics* 2007 September; 33(9): 508-513. NRCBL: 4.1.1. SC: em.
Abstract: Ten new graduate speech pathologists recounted their experiences in managing workplace ethical dilemmas in semi-structured interviews. Their stories were analysed for elements that described the nature and management of the ethical dilemmas. Ethical reasoning themes were generated to reflect the participants' approaches to managing these dilemmas. Finally, a conceptual model, the Dynamic Model of Ethical Reasoning, was developed. This model incorporates the elements of awareness, independent problem solving, supported problem solving, and decision and outcome evaluation. Features of the model demonstrate the complexity of ethical reasoning and the challenges that new graduates encounter when managing ethical dilemmas. The results have implications for preparing new graduates to manage ethical dilemmas in the workplace.

Kirkpatrick, William J.; Reamer, Frederic G.; Sykulski, Marilyn. Social work ethics audits in health care settings: a case study. *Health and Social Work* 2006 August; 31(3): 225-228. NRCBL: 1.3.1; 2.3. SC: cs.

Kotsirilos, Vicki; Hassed, Craig S.; Arnold, Peter C.; Kerridge, Ian H.; McPhee, John R. Ethical and legal issues at the interface of complementary and conventional medicine [letters and reply]. *Medical Journal of Australia* 2004 November 15; 181(10): 581-582. NRCBL: 4.1.1.

Largent, Beverly A. When is it proper to refer a patient receiving public aid to another dentist? *Journal of the American Dental Association* 2006 March; 137(3): 395-396. NRCBL: 4.1.1; 9.3.1; 9.5.10; 7.3.

Long, J. Michael. Student views of professional ethics. *Journal of the American College of Dentists* 1996 Spring; 63(1): 37-42. NRCBL: 4.1.1; 1.3.1; 7.2; 9.3.1.

Masella, Richard S. Renewing professionalism in dental education: overcoming the market environment. *Journal of Dental Education* 2007 February; 71(2): 205-216. NRCBL: 4.1.1; 7.2.

McGrath, Pam; Henderson, David; Holewa, Hamish. Patient-centred care: qualitative findings on health professionals' understanding of ethics in acute medicine. *Journal of Bioethical Inquiry* 2006; 3(3): 149-160. NRCBL: 4.1.1; 9.6; 7.1; 8.1. SC: cs.

Abstract: In recent years the literature on bioethics has begun to pose the sociological challenge of how to explore organisational processes that facilitate a systemic response to ethical concerns. The present discussion seeks to make a contribution to this important new direction in ethical research by presenting findings from an Australian pilot study. The research was initiated by the Clinical Ethics Committee of Redland Hospital at Bayside Health Service District in Queensland, Australia, and explores health professionals' understanding of the nature of ethics and their experience with ethical decision-making within an acute medical ward. This study focuses on the actual experience, understanding and attitudes of clinical professionals in a general medical ward. In particular, the discussion explores the specific findings from the study concerned with how a multi-disciplinary team of health professionals define and operationalise the notion of ethics in an acute ward hospital setting. The key issue reported is that health professionals are not only able to clearly articulate notions of ethics, but that the notions expressed by a multi-disciplinary diversity of participants share a common definitional concept of ethics as patient-centred care. The central finding is that all professional groups indicated that there is a guiding principle to address their ethical sense of the 'good' or the 'ought' and that is to act in a way that furthered the interests of patients and their families. The findings affirm the importance of a sociological perspective as a productive new direction in bioethical research.

Mertz, Marcel. Complementary and alternative medicine: the challenges of ethical justification. *Medicine, Health Care and Philosophy* 2007 September; 10(3): 329-345. NRCBL: 4.1.1; 1.1.

Abstract: With the prevalence of complementary and alternative medicine (CAM) increasing in western societies, questions of the ethical justification of these alternative health care approaches and practices have to be addressed. In order to evaluate philosophical reasoning on this subject, it is of paramount importance to identify and analyse possible arguments for the ethical justification of CAM considering contemporary biomedical ethics as well as more fundamental philosophical aspects. Moreover, it is vital to provide adequate analytical instruments for this task, such as separating, CAM as belief system' and ,CAM as practice'. Findings show that beneficence and non-maleficence are central issues for an ethical justification of CAM as practice, while freedom of thought and religion are central to CAM as belief system. Many justification strategies have limitations and qualifications that have to be taken into account. Singularly descriptive premises in an argument often prove to be more problematic than universal ethical principles. Thus, non-ethical issues related to a general philosophical underpinning - e.g. epistemology, semantics, and ontology - are highly relevant for determining a justification strategy, especially when strong metaphysical assumptions are involved. Even if some values are shared with traditional biomedicine, axiological differences have to be considered as well. Further research should be done about specific CAM positions. These could be combined with applied qualitative social research methods.

Newburger, Amy E.; Caplan, Arthur L. Taking ethics seriously in cosmetic dermatology. *Archives of Dermatology* 2006 December; 142(12): 1641-1642. NRCBL: 4.1.1; 7.1; 1.3.2; 7.2; 9.3.1.

Nyika, Aceme. Ethical and regulatory issues surrounding African traditional medicine in the context of HIV/AIDS. *Developing World Bioethics* 2007 April; 7(1): 25-34. NRCBL: 4.1.1; 9.5.6; 21.7.

Abstract: It has been estimated that more than 80% of people in Africa use traditional medicine (TM). With the HIV/AIDS epidemic claiming many lives in Africa, the majority of people affected rely on TM mainly because it is relatively affordable and available to the poor populations who cannot afford orthodox medicine. Whereas orthodox medicine is practiced under stringent regulations and ethical guidelines emanating from The Nuremburg Code,1 African TM seems to be exempt from such scrutiny. Although recently there have been calls for TM to be incorporated into the health care system, less emphasis has been placed on ethical and regulatory issues. In this paper, an overview of the use of African TM in general, and for HIV/AIDS in particular, is given, followed by a look at: (i) the relative laxity in the application of ethical standards and regulatory requirements with regards to TM; (ii) the importance of research on TM in order to improve and demystify its therapeutic qualities; (iii) the need to tailor-make intellectual property laws to protect traditional knowledge and biodiversity. A framework of partnerships involving traditional healers' associations, scientists, policy makers, patients, community leaders, members of the communities, and funding organizations is suggested as a possible method to tackle these issues. It is hoped that this paper will stimulate objective and constructive debate that could enhance the protection of patients' welfare.

O'Toole, Brian. Four ways we approach ethics. *Journal of Dental Education* 2006 November; 70(11): 1152-1158. NRCBL: 4.1.1; 2.1.

Oguamanam, Chidi. Biomedical orthodoxy and complementary and alternative medicine: ethical challenges of integrating medical cultures. *Journal of Alternative and Complementary Medicine* 2006 July-August; 12(6): 577-581. NRCBL: 4.1.1; 21.7; 8.1; 7.1.

Ozar, David T. Conflicting values in oral health care. *Journal of the American College of Dentists* 1998 Fall; 65(3): 15-18. NRCBL: 4.1.1; 1.3.1.

Paley, John. Caring as a slave morality: Nietzschean themes in nursing ethics. *Journal of Advanced Nursing* 2002 October; 40(1): 25-35. NRCBL: 4.1.1; 1.1; 4.1.3.

Patthoff, Donald. Defining the ethical organization in oral health care. *Journal of the American College of Dentists* 1998 Fall; 65(3): 24-26. NRCBL: 4.1.1; 1.3.1.

NRCBL: National Reference Center for Bioethics Literature Classification Scheme   See inside front cover for terms.

373

Patthoff, Donald E. How did we get here? Where are we going? Hopes and gaps in access to oral health care. *Journal of Dental Education* 2006 November; 70(11): 1125-1132. NRCBL: 4.1.1; 9.2; 9.3.1.

Patthoff, Donald E. The need for dental ethicists and the promise of universal patient acceptance: response to Richard Masella's "Renewing professionalism in dental education". *Journal of Dental Education* 2007 February; 71(2): 222-226. NRCBL: 4.1.1; 7.2.

Peltier, Bruce. Response to unethical behavior in oral health care. *Journal of the American College of Dentists* 1998 Fall; 65(3): 19-23. NRCBL: 4.1.1; 1.3.1; 7.4.

Poulis, Ioannis. Bioethics and physiotherapy [editorial]. *Journal of Medical Ethics* 2007 August; 33(8): 435-436. NRCBL: 4.1.1; 9.5.1; 8.1.

Rapport, Frances L. Response to: 'Caring as a slave morality: Nietzschean themes in nursing ethics' by J. Paley (2002) Journal of Advanced Nursing 40, 25-35. *Journal of Advanced Nursing* 2002 October; 40(1): 42-44. NRCBL: 4.1.1; 1.1; 4.1.3.

Schwab, Abraham. Getting rid of heroes. *Atrium* 2007 Summer; 4: 25-28. NRCBL: 1.3.1; 4.1.2; 18.4; 18.5.2.

Smith, Jayne L.; Cervero, Ronald M.; Valentine, Thomas. Impact of commercial support on continuing pharmacy education. *Journal of Continuing Education in the Health Professions* 2006 Fall; 26(4): 302-312. NRCBL: 4.1.1; 7.2; 1.3.2; 7.3. SC: em.

Tangwa, Godfrey B. How not to compare western scientific medicine with African traditional medicine. *Developing World Bioethics* 2007 April; 7(1): 41-44. NRCBL: 4.1.1; 9.5.6; 21.7; 21.1. Comments: Aceme Nyika. Ethical and regulatory issues surrounding African traditional medicine in the context of HIV/AIDS. Developing World Bioethics 2007 April; 7(1): 25-34.
Abstract: In his commentary on Aceme Nyika's paper 'Ethical and Regulatory Issues Surrounding African Traditional Medicine in the Context of HIV/AIDS',1 Godfrey B. Tangwa charges the author with inappropriately using expressions, terminology and criteria of evaluation appropriate in Western scientific medicine to judge African traditional medicine (TM). He seriously frowns on Nyika's suggestion that African TM needs to be incorporated into, and subjected to the canons of Western scientific medicine. Such a suggestion, he believes, is a prescription for invasion, colonization and exploitation so characteristic of the relationship between Africa and the Western world. However, he thinks that African TM is quite compatible with Western scientific medicine.

Van Bogaert, Donna Knapp. Ethical considerations in African traditional medicine: a response to Nyika. *Developing World Bioethics* 2007 April; 7(1): 35-40. NRCBL: 4.1.1; 9.5.6; 21.7. Comments: Aceme Nyika. Ethical and regulatory issues surrounding African traditional medicine

in the context of HIV/AIDS. Developing World Bioethics 2007 April; 7(1): 25-34.
Abstract: Like other so-called 'parallel' practices in medicine, traditional medicine (TM) does not avoid criticism or even rejection. Nyika's article 'Ethical and Regulatory Issues Surrounding African Traditional Medicine in the Context of HIV/AIDS' looks at some of the issues from a traditional Western ethical perspective and suggests that it should be rejected. I respond to this article agreeing with Nyika's three major criticisms: lack of informed consent, confidentiality and paternalism. However, as traditional healers are consulted by over 70% of South Africans before any other type of healthcare professional, a blanket negation of TM is not possible, nor is it politically feasible. A pragmatic approach would be to work within the current structures for positive change. I point out that, as all cultural practices do, TM will change over time. Yet, until some regulations and change occur, the problem of harm to patients remains a major concern.

Weaver, Kathryn. Ethical sensitivity: state of knowledge and needs for further research. *Nursing Ethics* 2007 March; 14(2): 141-155. NRCBL: 1.3.1; 4.1.3. SC: em; rv.
Abstract: Ethical sensitivity was introduced to caring science to describe the first component of decision making in professional practice; that is, recognizing and interpreting the ethical dimension of a care situation. It has since been conceptualized in various ways by scholars of professional disciplines. While all have agreed that ethical sensitivity is vital to practice, there has been no consensus regarding its definition, its characteristics, the conditions needed for it to occur, or the outcomes to professionals and society. The purpose of this article is to explore the meaning of the concept of ethical sensitivity based on a review of the professional literature of selected disciplines. Qualitative content analysis of the many descriptors found within the literature was conducted to enhance understanding of the concept and identify its essential characteristics. Ethical sensitivity is considered to be an emerging concept with potential utility in research and practice.

Weaver, Kathryn; Morse, Janice M. Pragmatic utility: using analytical questions to explore the concept of ethical sensitivity. *Research and Theory for Nursing Practice* 2006 Fall; 20(3): 191-214. NRCBL: 4.1.1; 1.1; 4.4.

Yamalik, Nermin. The responsibilities and rights of dental professionals 1. Introduction. *International Dental Journal* 2006 April; 56(2): 109-111. NRCBL: 4.1.1; 7.1; 9.2. Identifiers: Turkey.

Zarkowski, Pamela. Professional promises: summary and next steps. *Journal of Dental Education* 2006 November; 70(11): 1241-1245. NRCBL: 4.1.1; 9.2.

PROFESSIONAL MISCONDUCT *See* BIOMEDICAL RESEARCH/ RESEARCH ETHICS AND SCIENTIFIC MISCONDUCT; MALPRACTICE AND PROFESSIONAL MISCONDUCT

PROFESSIONAL PATIENT RELATIONSHIP

See CARE FOR SPECIFIC GROUPS; NURSING ETHICS AND PHILOSOPHY; PATIENT RELATIONSHIPS; PROFESSIONAL ETHICS

PROFESSIONAL PROFESSIONAL RELATIONSHIP

Caldicott, Catherine V. "Sweeping up after the parade": professional, ethical, and patient care implications of "turfing". *Perspectives in Biology and Medicine* 2007 Winter; 50(1): 136-149. NRCBL: 7.3; 8.1; 7.2.

Chervenak, Frank A.; McCullough, Laurence B.; Baril, Thomas E., Sr. Ethics, a neglected dimension of power relationships of physician leaders. *American Journal of Obstetrics and Gynecology* 2006 September; 195(3): 651-656. NRCBL: 7.3; 7.1; 1.3.2.

Cosgrove, Lisa; Krimsky, Sheldon; Vijayaraghavan, Manisha; Schneider, Lisa. Financial ties between DSM-IV panel members and the pharmaceutical industry. *Psychotherapy and Psychosomatics* 2006; 75(3): 154-160. NRCBL: 7.3; 17.1; 9.7; 9.3.1.

Cram, Peter; Rosenthal; Gary E. Physician-owned specialty hospitals and coronary revascularization utilization: too much of a good thing? *JAMA: The Journal of the American Medical Association* 2007 March 7; 297(9): 998-999. NRCBL: 7.3; 9.3.1; 9.5.1.

Davis, Anne J.; Konishi, Emiko. Whistleblowing in Japan. *Nursing Ethics* 2007 March; 14(2): 194-202. NRCBL: 7.3; 4.1.3; 9.8. SC: em.

Abstract: This article, written from research data, focuses on the possible meaning of the data rather than on detailed statistical reporting. It defines whistleblowing as an act of the international nursing ethical ideal of advocacy, and places it in the larger context of professional responsibility. The experiences, actions, and ethical positions of 24 Japanese nurses regarding whistleblowing or reporting a colleague for wrongdoing provide the data. Of these respondents, similar in age, educational level and clinical experience, 10 had previously reported another nurse and 12 had reported a physician for a wrongful act. These data raise questions about overt actions to expose a colleague in a culture that values group loyalty and saving face. Additional research is needed for an in-depth understanding of whistleblowing, patient advocacy and professional responsibility across cultures, especially those that value group loyalty, saving face and similar concepts to the Japanese Ishin Denshin, where the value is on implicit understanding requiring indirect communication. Usually, being direct and openly discussing sensitive topics is not valued in Japan because such behavior disrupts the most fundamental value, harmony (wa).

Debruin, Debra A. Ethics on the inside? *In:* Eckenwiler, Lisa A.; Cohn, Felicia G., eds. The Ethics of Bioethics: Mapping the Moral Landscape. Baltimore, MD: Johns Hopkins University Press, 2007: 161-169. NRCBL: 7.3; 21.1; 1.1.

Duvall, David G. Conflict of interest or ideological divide: the need for ongoing collaboration between physicians and industry. *Current Medical Research and Opinion* 2006 September; 22(9): 1807-1812. NRCBL: 7.3; 1.3.2; 9.7.

Giles, Jim. Drug firms accused of biasing doctors' training [news]. *Nature* 2007 November 22; 450(7169): 464-465. NRCBL: 7.3; 7.2; 9.7.

Kerridge, I.; Maguire, J.; Newby, D.; McNeill, P.M.; Henry, D.; Hill, S.; Day, R.; Macdonald, G.; Stokes, B.; Henderson, K. Cooperative partnerships or conflict-of-interest? A national survey of interaction between the pharmaceutical industry and medical organizations. *Internal Medicine Journal* 2005 April; 35(4): 206-210. NRCBL: 7.3; 9.7; 1.3.2; 9.3.1. SC: em. Identifiers: Australia.

Komesaroff, P. Ethical issues in the relationships involving medicine and industry: evolving problems require evolving. *Internal Medicine Journal* 2005 April; 35(4): 203-205. NRCBL: 7.3; 1.3.2; 9.3.1; 9.7. Identifiers: Australia.

Kwiecinski, Maureen. Limiting conflicts of interest arising from physician investment in specialty hospitals. *Specialty Law Digest: Health Care Law* 2006 January; (321). NRCBL: 7.3; 9.3.1. SC: le.

Lo, Bernard. Conflicts of interest. *In his:* Resolving Ethical Dilemmas: A Guide for Clinicians. 3rd edition. Philadelphia, PA: Lippincott Williams and Wilkins, 2005: 183-232. NRCBL: 7.3; 9.4; 8.2; 7.4; 9.5.9.

McKneally, Martin F. Beyond disclosure: managing conflicts of interest to strengthen trust in our profession. *Journal of Thoracic and Cardiovascular Surgery* 2007 February; 133(2): 300-302. NRCBL: 7.3; 8.2; 9.3.1; 9.5.1; 1.3.7.

Parker, Lisa S.; Satkoske, Valerie B. Conflicts of interest: are informed consent an appropriate model and disclosure an appropriate remedy? *Journal of the American College of Dentists* 2007 Summer; 74(2): 19-26. NRCBL: 7.3; 4.1.1; 1.3.1; 8.2.

Rentmeester, Christy A. "Why aren't you doing what we want?" Cultivating collegiality and communication between specialist and generalist physicians and residents. *Journal of Medical Ethics* 2007 May; 33(5): 308-310. NRCBL: 7.3.

Abstract: Developing residents' communication skills has been a goal of residency training programmes since the Accreditation Council for Graduate Medical Education codified it as a core competency. In this article, a case that features problematic communication between a generalist and specialist physician is drawn upon, and it is suggested how their communication might become open and effective through a practice of reason exchange. This is a practice of giving reasons, listening to reasons given

NRCBL: National Reference Center for Bioethics Literature Classification Scheme   See inside front cover for terms.

375

by others, evaluating reasons and deciding which particulars of situations constitute reasons to act and reasons how to act. Drawing on recent literature in teaching communication to radiology residents, it is proposed that practices of reason exchange are part of the skill set generally referred to as "negotiation skills" that should be cultivated in all residents. Particularly, in cases in which generalist and specialist physicians disagree about the reasons to do something, not do something or do something this way or that way, how well physicians are trained to practice reason exchange depends on whether they can communicate effectively and negotiate disagreement collegially.

Rich, Karen L. Using Buddhist Sangha as a model of communitarianism in nursing. *Nursing Ethics* 2007 July; 14(4): 466-477. NRCBL: 7.3; 1.2; 1.3.1; 4.1.3.
Abstract: In spite of a continuing long and rich history of caring for patients, many nurses have not been satisfied with their work. One cause among others for this dissatisfaction is that nurses often do not care for one another. The philosophy of a Buddhist Sangha, or community, is similar to the philosophy of western communitarian ethics. Both philosophies emphasize the importance of people working together harmoniously towards a common good. In this article, unsatisfactory nurse-nurse relationships have been considered and a model for communitarian nursing practice has been suggested based on a Buddhist Sangha.

Rorty, Mary V.; Mills, Ann E.; Werhane, Patricia H. Institutional practices, ethics, and the physician. *In:* Rhodes, Rosamond; Francis, Leslie P.; Silvers, Anita, eds. The Blackwell Guide to Medical Ethics. Malden, MA: Blackwell Pub., 2007: 180-197. NRCBL: 7.3.

Schwartz, Barry. The evolving relationship between specialists and general dentists: practical and ethical challenges. *Journal of the American College of Dentists* 2007 Spring; 74(1): 22-26. NRCBL: 7.3; 4.1.1.

Standridge, John B. Of doctor conventions and drug companies. *Family Medicine* 2006 July-August; 38(7): 518-520. NRCBL: 7.3; 9.7; 9.3.1; 1.3.2.

Storch, Janet L.; Kenny, Nuala. Shared moral work of nurses and physicians. *Nursing Ethics* 2007 July; 14(4): 478-491. NRCBL: 7.3; 4.1.2; 4.1.3.
Abstract: Physicians and nurses need to sustain their unique strengths and work in true collaboration, recognizing their interdependence and the complementarity of their knowledge, skills and perspectives, as well as their common moral commitments. In this article, challenges often faced by both nurses and physicians in working collaboratively are explored with a focus on the ways in which each profession's preparation for practice has differed over time, including shifts in knowledge development and codes of ethics guiding their practice. A call for envisioning their practice as shared moral work as well as practical strategies to begin that work are offered as a basis for reflection towards enhanced nurse-physician relationships.

Tanne, Janice Hopkins. US companies are fined for payments to surgeons [news]. *BMJ: British Medical Journal* 2007 November 24; 335(7629): 1065. NRCBL: 7.3; 9.7. SC: le.

Tattersall, Martin H.N.; Kerridge, Ian H. Doctors behaving badly? [editorial]. *Medical Journal of Australia* 2006 September 18; 185(6): 299-300 [see correction in: Medical Journal of Australia 2006 November 20; 185(10): 576]. NRCBL: 7.3; 9.7; 9.3.1; 1.3.2.

Turton, Frederick E.; Snyder, Lois. Physician-industry relations [letter]. *Annals of Internal Medicine* 2007 March 20; 146(6): 469. NRCBL: 7.3; 1.3.2; 9.3.1.

PROLONGATION OF LIFE *See* EUTHANASIA AND ALLOWING TO DIE

PROXY DECISION MAKING *See* ADVANCE DIRECTIVES; EUTHANASIA AND ALLOWING TO DIE; INFORMED CONSENT/ INCOMPETENTS; INFORMED CONSENT/ MINORS

PSYCHOPHARMACOLOGY
See also BEHAVIOR CONTROL; CARE FOR SPECIFIC GROUPS/ MENTALLY DISABLED; MENTAL HEALTH THERAPIES AND NEUROSCIENCES

Bell, Jennifer A.; Henry, Michael; Fishman, Jennifer R.; Youngner, Stuart J. Preventing post-traumatic stress disorder or pathologizing bad memories? *American Journal of Bioethics* 2007 September; 7(9): 29-30; author reply W1-W3. NRCBL: 17.4; 2.4. Comments: Michael Henry, Jennifer R. Fishman, and Stuart J. Youngner. Propranolol and the prevention of post-traumatic stress disorder: is it wrong to erase the "sting" of bad memories? American Journal of Bioethics 2007 September 7(9): 12-20.

Claassen, Dirk. Financial incentives for antipsychotic depot medication: ethical issues. *Journal of Medical Ethics* 2007 April; 33(4): 189-193. NRCBL: 17.4; 9.3.1. SC: em.
Abstract: BACKGROUND: Giving money as a direct incentive for patients in exchange for depot medication has proved beneficial in some clinical cases in assertive outreach (AO). However, ethical concerns around this practice have been raised, and will be analysed in more detail here. Method: Ethical concern voiced in a survey of all AO teams in England were analysed regarding their content. These were grouped into categories. RESULTS: 53 of 70 team managers mentioned concerns, many of them serious and expressing a negative attitude towards giving money for depot adherence. Four broad categories of ethical concern following Christensen's concept were distinguished: valid consent and refusal (n = 5), psychiatric paternalism (n = 31), resource allocation (n = 4), organisational relationships (n = 2), with a residual category others and unspecified (n = 11). DISCUSSION: The main concerns identified are discussed on the background of existing ethical theories in healthcare and the

specific problems of community mental health and AO. Points for practice are derived from this discussion. A way forward is outlined that includes informed consent and an operational policy in the use of incentives, further randomised controlled trials and qualitative studies, and continuing discussions with all stakeholders, especially service users.

Cosgrove, Lisa; Bursztajn, Harold J. Undoing undue industry influence: lessons from psychiatry as psychopharmacology. *Organizational Ethics: Healthcare, Business, and Policy* 2006 Fall-Winter; 3(2): 131-133. NRCBL: 17.4; 1.3.2; 7.3; 17.1.

Craigie, Jillian; Henry, Michael; Fishman, Jennifer R.; Youngner, Stuart J. Propranolol, cognitive biases, and practical decision-making. *American Journal of Bioethics* 2007 September; 7(9): 31-32; author reply W1-W3. NRCBL: 17.4; 2.4. Comments: Michael Henry, Jennifer R. Fishman, and Stuart J. Youngner. Propranolol and the prevention of post-traumatic stress disorder: is it wrong to erase the "sting" of bad memories? American Journal of Bioethics 2007 September 7(9): 12-20.

Dekkers, Wim; Rikkert, Marcel Olde. Memory enhancing drugs and Alzheimer's disease: enhancing the self or preventing the loss of it? *Medicine, Health Care and Philosophy* 2007 June; 10(2): 141-151. NRCBL: 17.4; 1.1; 4.5; 4.4; 9.5.2. SC: an.

Abstract: In this paper we analyse some ethical and philosophical questions related to the development of memory enhancing drugs (MEDs) and anti-dementia drugs. The world of memory enhancement is coloured by utopian thinking and by the desire for quicker, sharper, and more reliable memories. Dementia is characterized by decline, fragility, vulnerability, a loss of the most important cognitive functions and even a loss of self. While MEDs are being developed for self-improvement, in Alzheimer's Disease (AD) the self is being lost. Despite this it is precisely those patients with AD and other forms of dementia that provide the subjects for scientific research on memory improvement. Biomedical research in the field of MEDs and anti- dementia drugs appears to provide a strong impetus for rethinking what we mean by 'memory', 'enhancement', 'therapy', and 'self'. We conclude (1) that the enhancement of memory is still in its infancy, (2) that current MEDs and anti-dementia drugs are at best partially and minimally effective under specific conditions, (3) that 'memory' and 'enhancement' are ambiguous terms, (4) that there is no clear-cut distinction between enhancement and therapy, and (5) that the research into MEDs and anti-dementia drugs encourages a reductionistic view of the human mind and of the self.

Diehm, Alexander; Ebsen, Ingwer. Ansätze zur „heimärztlichen Versorgung" und die geplante Pflegereform - Rechtliche Aspekte dargestellt am Beispiel der Psychopharmakaversorgung = The medical treatment in nursing homes and plans for a legislative reform - legal aspects with particular reference to supply of psycho-

tropic drugs. *Ethik in der Medizin* 2007 December; 19(4): 301-312. NRCBL: 17.4; 9.5.2. SC: le.

Elliott, Carl. Against happiness [review of Against Depression, by Peter D. Kramer]. *Medicine, Health Care and Philosophy* 2007 June; 10(2): 167-171. NRCBL: 17.4; 4.5.

Ghaemi, S. Nassir; Goodwin, Frederick K. The ethics of clinical innovation in psychopharmacology: challenging traditional bioethics. *Philosophy, Ethics, and Humanities in Medicine [electronic]* 2007 November 8; 2(26): 8 p. Accessed: http://www.peh-med.com/ [2008 January 24]. NRCBL: 17.4; 18.5.1; 9.5.1; 18.3; 2.1. SC: cs; rv.

Glannon, Walter. Pharmacological and psychological interventions. *In his:* Bioethics and the Brain. New York: Oxford University Press, 2007: 76-115. NRCBL: 17.4; 8.2; 17.3. SC: an.

Hall, Wayne; Carter, Adrian; Henry, Michael; Fishman, Jennifer R.; Youngner, Stuart J. Debunking alarmist objections to the pharmacological prevention of PTSD. *American Journal of Bioethics* 2007 September; 7(9): 23-25; author reply W1-W3. NRCBL: 17.4; 2.4. Comments: Michael Henry, Jennifer R. Fishman, and Stuart J. Youngner. Propranolol and the prevention of post-traumatic stress disorder: is it wrong to erase the "sting" of bad memories? American Journal of Bioethics 2007 September 7(9): 12-20.

Hawthorne, Susan. ADHD drugs: values that drive the debates and decisions. *Medicine, Health Care and Philosophy* 2007 June; 10(2): 129-140. NRCBL: 17.4; 17.3. SC: an.

Abstract: Use of medication for treatment of ADHD (or its historical precursors) has been debated for more than forty years. Reasons for the ongoing differences of opinion are analyzed by exploring some of the arguments for and against considering ADHD a mental disorder. Relative to two important DSM criteria - that a mental disorder causes some sort of harm to the individual and that a mental disorder is the manifestation of a dysfunction in the individual - ADHD's classification as a mental disorder is found to be contentiously value-laden. The disagreements spill over to reasoning regarding appropriate management, because justification for a drug prescription is in part predicated on the idea that the drugs manage mental disorders. These debates do not appear to be nearing resolution, so individuals offering advice, or trying to decide whether ADHD drugs are appropriate for themselves or their children, may find it helpful to compare the values underlying various perspectives with their own.

Henry, Michael; Fishman, Jennifer R.; Youngner, Stuart J. Propranolol and the prevention of post-traumatic stress disorder: is it wrong to erase the "sting" of bad memories? *American Journal of Bioethics* 2007 September; 7(9): 12-20. NRCBL: 17.4; 4.4; 2.4; 8.3.3; 18.3. SC: le. Comments: American Journal of Bioethics 2007 September; 7(9): 21-42.

Abstract: The National Institute of Mental Health (Bethesda, MD) reports that approximately 5.2 million Americans experience post-traumatic stress disorder (PTSD) each year. PTSD can be severely debilitating and diminish quality of life for patients and those who care for them. Studies have indicated that propranolol, a beta-blocker, reduces consolidation of emotional memory. When administered immediately after a psychic trauma, it is efficacious as a prophylactic for PTSD. Use of such memory-altering drugs raises important ethical concerns, including some futuristic dystopias put forth by the President's Council on Bioethics. We think that adequate informed consent should facilitate ethical research using propranolol and, if it proves efficacious, routine treatment. Clinical evidence from studies should certainly continue to evaluate realistic concerns about possible ill effects of diminishing memory. If memory-attenuating drugs prove effective, we believe that the most immediate social concern is the over-medicalization of bad memories, and its subsequent exploitation by the pharmaceutical industry.

Hinton, Jeremy; Forrest, Robert. Involuntary non-emergent psychotropic medication. *Journal for the American Academy of Psychiatry and the Law* 2007; 35(3): 396-398. NRCBL: 17.4; 8.3.4; 17.8. SC: le.

Hurley, Elisa A.; Henry, Michael; Fishman, Jennifer R.; Youngner, Stuart J. The moral costs of prophylactic propranolol. *American Journal of Bioethics* 2007 September; 7(9): 35-36; author reply W1-W3. NRCBL: 17.4. Comments: Michael Henry, Jennifer R. Fishman, and Stuart J. Youngner. Propranolol and the prevention of post-traumatic stress disorder: is it wrong to erase the "sting" of bad memories? American Journal of Bioethics 2007 September 7(9): 12-20.

Kabasenche, William P.; Henry, Michael; Fishman, Jennifer R.; Youngner, Stuart J. Emotions, memory suppression, and identity. *American Journal of Bioethics* 2007 September; 7(9): 33-34; author reply W1-W3. NRCBL: 17.4. Comments: Michael Henry, Jennifer R. Fishman, and Stuart J. Youngner. Propranolol and the prevention of post-traumatic stress disorder: is it wrong to erase the "sting" of bad memories? American Journal of Bioethics 2007 September 7(9): 12-20.

Kolber, Adam; Henry, Michael; Fishman, Jennifer R.; Youngner, Stuart J. Clarifying the debate over therapeutic forgetting. *American Journal of Bioethics* 2007 September; 7(9): 25-27; author reply W1-W3. NRCBL: 17.4; 2.4. Comments: Michael Henry, Jennifer R. Fishman, and Stuart J. Youngner. Propranolol and the prevention of post-traumatic stress disorder: is it wrong to erase the "sting" of bad memories? American Journal of Bioethics 2007 September 7(9): 12-20.

Liao, S. Matthew; Wasserman, David T.; Henry, Michael; Fishman, Jennifer R.; Youngner, Stuart J. Neuroethical concerns about moderating traumatic memories. *American Journal of Bioethics* 2007 September; 7(9): 38-40; author reply W1-W3. NRCBL: 17.4; 4.4; 2.4. Comments: Michael Henry, Jennifer R. Fishman, and Stuart J. Youngner. Propranolol and the prevention of post-traumatic stress disorder: is it wrong to erase the "sting" of bad memories? American Journal of Bioethics 2007 September 7(9): 12-20.

McCullough, Laurence B.; Coverdale, John H.; Chervenak, Frank A. Constructing a systematic review for argument-based clinical ethics literature: the example of concealed medications. *Journal of Medicine and Philosophy* 2007 January-February; 32(1): 65-76. NRCBL: 17.4; 8.2. SC: em; rv.

Abstract: The clinical ethics literature is striking for the absence of an important genre of scholarship that is common to the literature of clinical medicine: systematic reviews. As a consequence, the field of clinical ethics lacks the internal, corrective effect of review articles that are designed to reduce potential bias. This article inaugurates a new section of the annual "Clinical Ethics" issue of the Journal of Medicine and Philosophy on systematic reviews. Using recently articulated standards for argument-based normative ethics, we provide a systematic review of the literature on concealed medication for the management of psychiatric disorders. Four steps are completed: identify a focused question; conduct a literature search using key terms relevant to the focused question; assess the adequacy of the argument-based methods of the papers identified; and identify conclusions drawn in each paper and whether they apply to the focused question. We identified seven papers and provide an assessment of them. While none of the papers fully meet the standards of argument-based ethics, they did provide rationales for the use of concealed medications, with the important requirement such a practice be accountable in explicit organizational policy to prevent abuse of patients with mental illness or dementia.

Pantel, Johannes; Haberstroh, Julia. Psychopharmakaverordnung im Altenpflegeheim - Zwischen indikationsgeleiteter Therapie und „Chemical Restraint" = Psychotropic drug use in nursing homes - between adequate care and "chemical restraint". *Ethik in der Medizin* 2007 December; 19(4): 258-269. NRCBL: 17.4; 9.5.2.

Rosenberg, Leah B.; Henry, Michael; Fishman, Jennifer R.; Youngner, Stuart J. Necessary forgetting: on the use of propranolol in post-traumatic stress disorder management. *American Journal of Bioethics* 2007 September; 7(9): 27-28; author reply W1-W3. NRCBL: 17.4; 2.4. Comments: Michael Henry, Jennifer R. Fishman, and Stuart J. Youngner. Propranolol and the prevention of post-traumatic stress disorder: is it wrong to erase the "sting" of bad memories? American Journal of Bioethics 2007 September 7(9): 12-20.

Sade, Robert M.; Henry, Michael; Fishman, Jennifer R.; Youngner, Stuart J. On moralizing and hidden agendas: the pot and the kettle in political bioethics. *American Journal of Bioethics* 2007 September; 7(9): 42-43; author reply W1-W3. NRCBL: 17.4; 2.4; 1.3.2. Comments: Michael Henry, Jennifer R. Fishman, and Stuart J. Youngner. Propranolol and the prevention of post-traumatic stress disorder: is it wrong to erase the "sting" of bad memories? *American Journal of Bioethics* 2007 September 7(9): 12-20.

Schermer, M.H.N. Brave New World versus Island — utopian and dystopian views on psychopharmacology. *Medicine, Health Care and Philosophy* 2007 June; 10(2): 119-128. NRCBL: 17.4; 4.5; 7.1. SC: an.

Abstract: Aldous Huxley's Brave New World is a famous dystopia, frequently called upon in public discussions about new biotechnology. It is less well known that 30 years later Huxley also wrote a utopian novel, called Island. This paper will discuss both novels focussing especially on the role of psychopharmacological substances. If we see fiction as a way of imagining what the world could look like, then what can we learn from Huxley's novels about psychopharmacology and how does that relate to the discussion in the ethical and philosophical literature on this subject? The paper argues that in the current ethical discussion the dystopian vision on psychopharmacology is dominant, but that a comparison between Brave New World and Island shows that a more utopian view is possible as well. This is illustrated by a discussion of the issue of psychopharmacology and authenticity. The second part of the paper draws some further conclusions for the ethical debate on psychopharmacology and human enhancement, by comparing the novels not only with each other, but also with our present reality. It is claimed that the debate should not get stuck in an opposition of dystopian and utopian views, but should address important issues that demand attention in our real world: those of evaluation and governance of enhancing psychopharmacological substances in democratic, pluralistic societies.

Schulte, Peter F.J.; Stienen, Juan J.; Bogers, Jan; Cohen, Dan; van Dijk, Daniel; Lionarons, Wendell H.; Sanders, Sophia S.; Heck, Adolph H. Compulsory treatment with clozapine: a retrospective long-term cohort study. *International Journal of Law and Psychiatry* 2007 November-December; 30(6): 539-545. NRCBL: 17.4; 8.3.4; 9.5.3; 18.5.6. SC: em.

Sernyak, Michael; Rosenheck, Robert. Experience of VA psychiatrists with pharmaceutical detailing of antipsychotic medications. *Psychiatric Services* 2007 October; 58(10): 1292-1296. NRCBL: 17.4; 1.3.2; 9.7. SC: em.

Svenaeus, Fredrik. Do antidepressants affect the self? A phenomenological approach. *Medicine, Health Care and Philosophy* 2007 June; 10(2): 153-166. NRCBL: 17.4; 1.1; 4.4. SC: an.

Abstract: In this paper, I explore the questions of how and to what extent new antidepressants (selective serotonin-reuptake inhibitors, or SSRIs) could possibly affect the self. I do this by way of a phenomenological approach, using the works of Martin Heidegger and Thomas Fuchs to analyze the roles of attunement and embodiment in normal and abnormal ways of being-in-the-world. The nature of depression and anxiety disorders - the diagnoses for which treatment with antidepressants is most commonly indicated - is also explored by way of this phenomenological approach, as are the basic structures of self-being. Special attention is paid in the analysis to the moods of boredom, anxiety and grief, since they play fundamental roles in depression and anxiety disorders and since their intensity and frequency appear to be modulated by antidepressants. My conclusion is that the effect of these drugs on the self can be thought of in terms of changes in self-feeling, or, more precisely, self-vibration of embodiment. I present the idea of a spectrum of bodily resonance, which extends from the normal resonance of the lived body, in which the body is able to pick up a wide range of different moods; continuing over various kinds of sensitivities, preferences and idiosyncrasies, in which certain moods are favored over others; to cases that we unreservedly label pathologies, in which the body is severely out of tune, or even devoid of tune and thus useless as a tool of resonance. Different cultures and societies favor slightly differently attuned self-styles as paradigmatic of the normal and good life, and the popularity of the SSRIs can therefore be explained, not only by defects of embodiment, but also by the presence of certain cultural norms in our contemporary society.

Svenaeus, Fredrik. Psychopharmacology and the self: an introduction to the theme [editorial]. *Medicine, Health Care and Philosophy* 2007 June; 10(2): 115-117. NRCBL: 17.4; 4.5.

Synofzik, Matthis; Maetzler, Walter. Wie sollen wir Patienten mit Demenz behandeln? Die ethisch problematische Funktion der Antidementiva = How should we treat dementia patients? The ethically problematic function of antidementia drugs. *Ethik in der Medizin* 2007 December; 19(4): 270-280. NRCBL: 17.4; 17.1; 9.5.2.

Tenenbaum, Evelyn M.; Reese, Brian; Henry, Michael; Fishman, Jennifer R.; Youngner, Stuart J. Memory-altering drugs: shifting the paradigm of informed consent. *American Journal of Bioethics* 2007 September; 7(9): 40-42; author reply W1-W3. NRCBL: 17.4; 18.3; 18.5.6. Comments: Michael Henry, Jennifer R. Fishman, and Stuart J. Youngner. Propranolol and the prevention of post-traumatic stress disorder: is it wrong to erase the "sting" of bad memories? American Journal of Bioethics 2007 September 7(9): 12-20.

Tovino, Stacey A. Functional neuroimaging and the law: trends and directions for future scholarship. *American Journal of Bioethics* 2007 September; 7(9): 44-56. NRCBL: 17.4; 4.4; 1.3.2; 1.3.5; 5.3; 8.4. SC: le. Comments: American Journal of Bioethics 2007 September 7(9): 57-75.

Abstract: Under the umbrella of the burgeoning neurotransdisciplines, scholars are using the principles and research methodologies of their primary and secondary fields to examine developments in neuroimaging, neuromodulation and psychopharmacology. The path for

NRCBL: National Reference Center for Bioethics Literature Classification Scheme   See inside front cover for terms.

379

advanced scholarship at the intersection of law and neuroscience may clear if work across the disciplines is collected and reviewed and outstanding and debated issues are identified and clarified. In this article, I organize, examine and refine a narrow class of the burgeoning neurotransdiscipline scholarship; that is, scholarship at the interface of law and functional magnetic resonance imaging (fMRI).

Trachtman, Howard; Henry, Michael; Fishman, Jennifer R.; Youngner, Stuart J. Spinoza's passions. *American Journal of Bioethics* 2007 September; 7(9): 21-23; author reply W1-W3. NRCBL: 17.4; 1.1. Comments: Michael Henry, Jennifer R. Fishman, and Stuart J. Youngner. Propranolol and the prevention of post-traumatic stress disorder: is it wrong to erase the "sting" of bad memories? American Journal of Bioethics 2007 September 7(9): 12-20.

Trost, Bernd. Ethische Probleme der Pflegenden im Altenpflegeheim - Nachdenkliches zur Psychopharmakaverordnung aus Sicht eines Heimleitenden = Ethical questions in a long term care nursing home — Reflections of an administrator considering psychotropic prescriptions. *Ethik in der Medizin* 2007 December; 19(4): 281-288. NRCBL: 17.4; 9.5.2.

Warnick, Jason E.; Henry, Michael; Fishman, Jennifer R.; Youngner, Stuart J. Propranolol and its potential inhibition of positive post-traumatic growth. *American Journal of Bioethics* 2007 September; 7(9): 37-38; author reply W1-W3. NRCBL: 17.4; 2.4. Comments: Michael Henry, Jennifer R. Fishman, and Stuart J. Youngner. Propranolol and the prevention of post-traumatic stress disorder: is it wrong to erase the "sting" of bad memories? American Journal of Bioethics 2007 September 7(9): 12-20.

PSYCHOTHERAPY
See also CARE FOR SPECIFIC GROUPS/ MENTALLY DISABLED; INVOLUNTARY COMMITMENT; MENTAL HEALTH THERAPIES AND NEUROSCIENCES

Arboleda-Flórez, Julio E. The ethics of forensic psychiatry. *Current Opinion in Psychiatry* 2006 September; 19(5): 544-546. NRCBL: 17.2; 4.3; 1.3.5. SC: le.

Bloom, Alexandra. The ostrich raises its head: "knowing" and moral accountability in the practice of psychotherapy. *Women and Therapy* 1999; 22(2): 7-20. NRCBL: 17.2; 1.1; 10.

Brabender, V. The ethical group psychotherapist: a coda. *International Journal of Group Psychotherapy* 2007 January; 57(1): 41-47; discussion 49-59. NRCBL: 17.2; 4.1.2; 9.4; 8.1.

Bruns, Cindy M.; Lesko, Teresa M. In the belly of the beast: morals, ethics, and feminist psychotherapy with women in prison. *Women and Therapy* 1999; 22(2): 69-85. NRCBL: 17.2; 1.1; 10. 17.8.

Clemens, Norman A. When colleagues go astray . . . *Journal of Psychiatric Practice* 2007 January; 13(1): 40-43. NRCBL: 17.2; 7.4; 7.3; 8.1.

Cust, Kenneth. Philosophers return to the agora. *In:* Rasmussen, Lisa, ed. Ethics Expertise: History, Contemporary Perspectives, and Applications. Dordrecht: Springer, 2005: 227-241. NRCBL: 17.2; 1.3.1; 2.1. SC: cs.

Danzinger, Paula R.; Welfel, Elizabeth Reynolds. The impact of managed care on mental health counselors: a survey of perceptions, practices, and compliance with ethical standards. *Journal of Mental Health Counseling* 2001 April; 23(2): 137-150. NRCBL: 17.2; 1.3.2; 8.3.1; 6; 8.4. SC: em.

Debiak, Dennis. Attending to diversity in group psychotherapy: an ethical imperative. *International Journal of Group Psychotherapy* 2007 January; 57(1): 1-12; discussion 49-59, 61-66. NRCBL: 17.2; 8.1; 7.1; 7.2; 10. SC: cs.

Fennig, Silvana; Secker, Aya; Treves, Ilan; Ben Yakar, Motti; Farina, Jorje; Roe, David; Levkovitz, Yechiel; Fennig, Shmuel. Ethical dilemmas in psychotherapy: comparison between patients, therapists and laypersons. *Israel Journal of Psychiatry and Related Sciences* 2005; 42(4): 251-257. NRCBL: 17.2; 8.4; 8.1. SC: em; cs.

Gottdiener, William H. Is harm reduction psychotherapy ethical? *In:* Kleinig, John; Einstein, Stanley, eds. Ethical Challenges for Intervening in Drug Use: Policy, Research and Treatment Issues. Huntsville, TX: Office of International Criminal Justice; Criminal Justice Center, Sam Houston State University, 2006: 91-98. NRCBL: 17.2; 9.5.9.

Greenberg, Stuart A.; Shuman, Daniel W. Irreconcilable conflict between therapeutic and forensic roles. *Professional Psychology: Research and Practice* 1997 February; 28(1): 50-57. NRCBL: 17.2; 1.3.5. SC: le.

Hajdin, Mane. The prohibition of sexual relationships between drug users and their counselors: is it justified? *In:* Kleinig, John; Einstein, Stanley, eds. Ethical Challenges for Intervening in Drug Use: Policy, Research and Treatment Issues. Huntsville, TX: Office of International Criminal Justice; Criminal Justice Center, Sam Houston State University, 2006: 437-452. NRCBL: 17.2; 10; 7.4; 9.5.9; 1.3.5. SC: an.

Johnson, W. Brad; Bacho, Roderick; Heim, Mark; Ralph, John. Multiple-role dilemmas for military mental health care providers. *Military Medicine* 2006 April; 171(4): 311-315. NRCBL: 17.2; 7.3; 1.3.5; 8.4.

Lowe, Jennifer; Pomerantz, Andrew M.; Pettibone, Jon C. The influence of payment method on psychologists' diagnostic decisions: expanding the range of presenting problems. *Ethics and Behavior* 2007; 17(1): 83-93. NRCBL: 17.2; 9.3.2; 8.3.1; 8.2; 7.1. SC: em.

Abstract: Previous research (Kielbasa, Pomerantz, Krohn, and Sullivan, 2004; Pomerantz and Segrist, 2006) indicates that when psychologists consider a client with symptoms of depression or anxiety, payment method significantly influences diagnostic decisions. This study extends the scope of the previous research to consider clients with symptoms of social phobia and attention deficit hyperactivity disorder (ADHD). Psychologists in independent practice responded to vignettes of clients whose descriptions deliberately included subclinical impairment. Half of the participants were told that the clients would pay via managed care; the other half were told that the clients would pay out-of-pocket. Confirming previous studies, payment method had a highly significant impact on diagnosis such that compared to out-of-pocket clients, managed care clients were much more likely to be assigned Diagnostic and Statistical Manual of Mental Disorders (4th ed. [DSM-IV]; American Psychiatric Association, 1994) diagnoses. Ethical implications relate to informed consent, accuracy and truthfulness in diagnosis, and psychologists' integrity.

Mangione, Lorraine; Forti, Rosalind; Iacuzzi, Catherine M. Ethics and endings in group psychotherapy: saying good-bye and saying it well. *International Journal of Group Psychotherapy* 2007 January; 57(1): 25-40; discussion 49-59, 61-66. NRCBL: 17.2; 8.1; 7.1.

Merlino, Joseph P. Psychoanalysis and ethics — relevant then, essential now. *Journal of the American Academy of Psychoanalysis and Dynamic Psychiatry* 2006 Summer; 34(2): 231-247. NRCBL: 17.2; 2.2.

Pawelzik, Markus; Prinz, Aloys. The moral economics of psychotherapy. *In:* Schramme, Thomas; Thome, Johannes, eds. Philosophy and Psychiatry. Berlin; New York: De Gruyter, 2004: 370-386. NRCBL: 17.2; 1.1; 8.1; 9.3.1.

Riggs, Billy J. Ethical considerations of integrating spiritual direction into psychotherapy. *Journal of Pastoral Care and Counseling* 2006 Winter; 60(4): 353-362. NRCBL: 17.2; 1.2.

Roberts, Laura Weiss; Johnson, Mark E.; Brems, Christiane; Warner, Teddy D. Preferences of Alaska and New Mexico psychiatrists regarding professionalism and ethics training. *Academic Psychiatry* 2006 May-June; 30(3): 200-204. NRCBL: 17.2; 7.2; 2.3.

Schmidt-Felzmann, Heike. Authority and influence in the psychotherapeutic relationship. *In:* Nys, Thomas; Denier, Yvonne; Vandevelde, Toon, eds. Autonomy and Paternalism: Reflections on the Theory and Practice of Health Care. Leuven; Dudley, MA: Peeters, 2007: 167-180. NRCBL: 17.2; 1.1; 8.1.

Weiss Roberts, Laura; Coverdale, John; Louie, Alan. Professionalism and the ethics-related roles of academic psychiatrists [editorial]. *Academic Psychiatry* 2005 November-December; 29(5): 413-415 [Online]. Accessed: http://ap.psychiatryonline.org/ [2007 April 13]. NRCBL: 17.2; 7.2; 4.1.2.

PUBLIC HEALTH
See also AIDS; HEALTH CARE

The ethics of public health [editorial]. *Lancet* 2007 December 1-7; 370(9602): 1801. NRCBL: 9.1; 4.1.1; 7.1; 1.3.2.

Bennett, Belinda. Travel in a small world: SARS, globalization and public health laws. *In:* Bennett, Belinda; Tomossy, George F., eds. Globalization and Health: Challenges for Health Law and Bioethics. Dordrecht: Springer, 2006: 1-12. NRCBL: 9.1; 21.1; 5.3. SC: le. Identifiers: severe acute respiratory syndrome.

Berkman, Alan; Susser, Ezra. Paternalism and the public's health. *Chronic Illness* 2006 March; 2(1): 17-18. NRCBL: 9.1; 9.5.1; 8.1; 1.1.

Calman, K.C.; Downie, R.S. Ethical principles and ethical issues in public health. *In:* Detels, Roger; McEwen, James; Beaglehole, Robert; Tanaka, Heizo, eds. Oxford Textbook of Public Health. Fourth edition. Oxford; New York: Oxford University Press, 2004: 387-399. NRCBL: 9.1; 4.2; 2.1.

Clancy, Anne; Svensson, Tommy. 'Faced' with responsibility: Levinasian ethics and the challenges of responsibility in Norwegian public health nursing. *Nursing Philosophy* 2007 July; 8(3): 158-166. NRCBL: 9.1; 1.1; 4.1.3.

Fairchild, Amy L.; Alkon, Ava. Back to the future? Diabetes, HIV, and the boundaries of public health. *Journal of Health Politics, Policy and Law* 2007 August; 32(4): 561-593. NRCBL: 9.1; 9.5.1; 9.5.6.

Gostin, Lawrence O. "Police" powers and public health paternalism: HIV and diabetes surveillance. *Hastings Center Report* 2007 March-April; 37(2): 9-10. NRCBL: 9.1; 9.5.6; 9.5.1; 8.4.

Gostin, Lawrence O. A theory and definition of public health law. *Journal of Health Care Law and Policy* 2007; 10(1): 1-12. NRCBL: 9.1; 1.1. SC: le.

Hall, Mark A.; Bobinski, Mary Anne; Orentlicher, David. Public health law. *In their:* Bioethics and Public Health Law. New York: Aspen Publishers, 2005: 519-588. NRCBL: 9.1; 8.3.4; 8.4; 17.8; 9.5.6; 9.7. SC: le; cs.

Hasman, Andreas. Restrictive or engaging: redefining public health promotion. *In:* Gunning, Jennifer; Holm, Søren, eds. Ethics, law, and society. Volume 2. Aldershot, Hants, England; Burlington, VT: Ashgate, 2006: 77-83. NRCBL: 9.1. SC: le.

Hodge, James G. Jr.; Gostin, Lawrence O.; Vernick, Jon S. The Pandemic and All-Hazards Preparedness Act:

NRCBL: National Reference Center for Bioethics Literature Classification Scheme   See inside front cover for terms.

381

improving public health emergency response. *JAMA: The Journal of the American Medical Association* 2007 April 18; 297(15): 1708-1711. NRCBL: 9.1; 1.3.12; 8.4; 9.4; 9.7. SC: le.

Hunter, Nan D. "Public-private" health law: multiple directions in public health. *Journal of Health Care Law and Policy* 2007; 10(1): 89-119. NRCBL: 9.1; 1.3.5. SC: le.

Jones, Marian Moser; Bayer, Ronald. Paternalism and its discontents: motorcycle helmet laws, libertarian values, and public health. *American Journal of Public Health* 2007 February; 97(2): 208-217. NRCBL: 9.1; 7.1.
  Abstract: The history of motorcycle helmet legislation in the United States reflects the extent to which concerns about individual liberties have shaped the public health debate. Despite overwhelming epidemiological evidence that motorcycle helmet laws reduce fatalities and serious injuries, only 20 states currently require all riders to wear helmets. During the past 3 decades, federal government efforts to push states toward enactment of universal helmet laws have faltered, and motorcyclists' advocacy groups have been successful at repealing state helmet laws. This history raises questions about the possibilities for articulating an ethics of public health that would call upon government to protect citizens from their own choices that result in needless morbidity and suffering.

Kahn, Jeffrey P. Why public health and politics don't mix [editorial]. *American Journal of Bioethics* 2007 November; 7(11): 3-4. NRCBL: 9.1; 21.1.

Kahn, Jeffrey; Mastroianni, Anna. The implications of public health for bioethics. *In:* Steinbock, Bonnie, ed. The Oxford Handbook of Bioethics. Oxford; New York: Oxford University Press, 2007: 671-695. NRCBL: 9.1; 2.1; 9.5.6; 9.4; 15.1; 8.4.

Kenny, Nuala P.; Melnychuk, Ryan M.; Asada, Yukiko. The promise of public health: ethical reflections. *Canadian Journal of Public Health* 2006 September-October; 97(5): 402-404. NRCBL: 9.1; 1.3.5.

Laaser, Ulrich; Donev, Donco; Bjegovic, Vesna; Sarolli, Ylli. Public health and peace [editorial]. *Croatian Medical Journal* 2002 April; 43(2):107-113. NRCBL: 9.1; 21.1.

Lang, Slobodan; Kovacic, Luka; Šogoric, Selma; Brborovic, Ognjen. Challenges of goodness III: public health facing war. *Croatian Medical Journal* 2002 April; 43(2): 156-165. NRCBL: 9.1; 21.1; 1.1. SC: em.

Lo, Bernard. Ethical issues in public health emergencies. *In his:* Resolving Ethical Dilemmas: A Guide for Clinicians. 3rd edition. Philadelphia, PA: Lippincott Williams and Wilkins, 2005: 280-284. NRCBL: 9.1; 8.3.4.

Mariner, Wendy K. Medicine and public health: crossing legal boundaries. *Journal of Health Care Law and Policy* 2007; 10(1): 121-151. NRCBL: 9.1; 9.5.1; 8.4; 8.1. SC: cs; le.

Markel, Howard; Gostin, Lawrence O.; Fidler, David P. Extensively drug-resistant tuberculosis: an isolation order, public health powers, and a global crisis [commentary]. *JAMA: The Journal of the American Medical Association* 2007 July 4; 298(1): 83-86. NRCBL: 9.1.

Mason, J.K.; Laurie, G.T. Public health and the state/patient relationships. *In their:* Mason and McCall Smith's Law and Medical Ethics. Seventh ed. Oxford; New York: Oxford University Press, 2005: 29-47. NRCBL: 9.1; 9.5.6; 21.1; 8.1. SC: le. Identifiers: Great Britain (United Kingdom); Human Rights Act of 1998.

Meier, Benjamin Mason; Mori, Larisa M. The highest attainable standard: advancing a collective human right to public health. *Columbia Human Rights Law Review* 2005 Fall; 37(1): 101-147. NRCBL: 9.1; 9.2; 21.1. SC: le.

Nash, Robert. Health-care workers in influenza pandemics [comment]. *Lancet* 2007 July 28-August 3; 370(9584): 300-301. NRCBL: 9.1; 8.1; 9.5.1; 21.1. SC: em.

Ortendahl, M. Risk in public health and clinical work [letter]. *Journal of Medical Ethics* 2007 April; 33(4): 246. NRCBL: 9.1; 5.2.

Parmet, Wendy E. Public health and constitutional law: recognizing the relationship. *Journal of Health Care Law and Policy* 2007; 10(1): 13-25. NRCBL: 9.1; 1.3.8. SC: le.

Rest, Kathleen M.; Halpern, Michael H. Politics and the erosion of federal scientific capacity: restoring scientific integrity to public health science. *American Journal of Public Health* 2007 November; 97(11): 1939-1944. NRCBL: 9.1; 5.3; 21.1.

Richards, Edward P. Public health law as administrative law: example lessons. *Journal of Health Care Law and Policy* 2007; 10(1): 61-88. NRCBL: 9.1; 1.3.8. SC: le.

Roemer, Ruth; Roemer, Milton I. Comparative national public health legislation. *In:* Detels, Roger; McEwen, James; Beaglehole, Robert; Tanaka, Heizo, eds. Oxford Textbook of Public Health. Fourth edition. Oxford; New York: Oxford University Press, 2004: 337-357. NRCBL: 9.1; 16.1; 9.7; 9.8; 21.1. SC: le.

Weed, Douglas L. Ethics and philosophy of public health. *In:* Khushf, George, ed. Handbook of Bioethics: Taking Stock of the Field From a Philosophical Perspective. Dordrecht; Boston: Kluwer Academic, 2004: 525-547. NRCBL: 9.1; 1.1; 4.2.

West, Robin L. Social justice, public health, and constitutional authority [review of Social Justice: The Moral Foundations of Public Health and Health Policy, by Madison Powers and Ruth Faden]. *DePaul Journal of Health Care Law* 2007; 10(4): 567-585. NRCBL: 9.1; 1.1. SC: le.

Wynia, Matthew K. Ethics and public health emergencies: restrictions on liberty [editorial]. *American Journal of Bioethics* 2007 February; 7(2): 1-5. NRCBL: 9.1; 1.1.

Abstract: Responses to public health emergencies can entail difficult decisions about restricting individual liberties to prevent the spread of disease. The quintessential example is quarantine. While isolating sick patients tends not to provoke much concern, quarantine of healthy people who only might be infected often is controversial. In fact, as the experience with severe acute respiratory syndrome (SARS) shows, the vast majority of those placed under quarantine typically don't become ill. Efforts to enforce involuntary quarantine through military or police powers also can backfire, stoking both panic and disease spread. Yet quarantine is part of a limited arsenal of options when effective treatment or prophylaxis is not available, and some evidence suggests it can be effective, especially when it is voluntary, home-based and accompanied by extensive outreach, communication and education efforts. Even assuming that quarantine is medically effective, however, it still must be ethically justified because it creates harms for many of those affected. Moreover, ethical principles of reciprocity, transparency, nondiscrimination and accountability should guide any implementation of quarantine.

PUBLISHING *See* JOURNALISM AND PUBLISHING

QUALITY AND VALUE OF LIFE

Abrams, Frederick R. Colorado revised statutes in support of palliative care limiting criminal liability. *Journal of Palliative Medicine* 2006 December; 9(6): 1254-1256. NRCBL: 4.4; 20.4.1; 20.7; 20.5.1. SC: le.

Alton, David. The paramount human right: the right to life. *In:* Wagner, Teresa; Carbone, Leslie, eds. Fifty Years after the Declaration: the United Nations' Record on Human Rights. Lanham, MD: University Press of America, 2001: 11-16. NRCBL: 4.4; 12.3; 21.1.

Baruch, Jay. Doctor versus patient: pain management in the ED. *Atrium* 2007 Summer; 4: 10-13. NRCBL: 4.4; 9.5.1.

Bayertz, Kurt. Zur Idee der Menschenwürde. *In:* Schreiber, Hans-Peter, ed. Biomedizin und Ethik: Praxis, Recht, Moral. Basel; Boston: Birkhäuser, 2004: 63-66. NRCBL: 4.4.

Bindig, Todd S. Confusion about speciesism and moral status. *Linacre Quarterly* 2007 May; 74(2): 145-155. NRCBL: 4.4; 1.2; 22.1.

Bortolotti, Lisa. Disputes over moral status: philosophy and science in the future of bioethics. *Health Care Analysis: An International Journal of Health Philosophy and Policy* 2007 June; 15(2): 153-158. NRCBL: 4.4; 1.1.
Abstract: Various debates in bioethics have been focused on whether non-persons, such as marginal humans or non-human animals, deserve respectful treatment. It has been argued that, where we cannot agree on whether these individuals have moral status, we might agree that they have symbolic value and ascribe to them moral value

in virtue of their symbolic significance. In the paper I resist the suggestion that symbolic value is relevant to ethical disputes in which the respect for individuals with no intrinsic moral value is in conflict with the interests of individuals with intrinsic moral value. I then turn to moral status and discuss the suitability of personhood as a criterion. There some desiderata for a criterion for moral status: it should be applicable on the basis of our current scientific knowledge; it should have a solid ethical justification; and it should be in line with some of our moral intuitions and social practices. Although it highlights an important connection between the possession of some psychological properties and eligibility for moral status, the criterion of personhood does not meet the desiderata above. I suggest that all intentional systems should be credited with moral status in virtue of having preferences and interests that are relevant to their well-being.

Bozzato, Gianni. "Lay" reduction of the human-embryo individual. *Linacre Quarterly* 2007 May; 74(2): 122-134. NRCBL: 4.4; 1.2.

Bras, Marijana; Loncar, Zoran; Fingler, Mira. The relief of pain as a human right. *Psychiatria Danubina* 2006 June; 18(1-2): 108-110. NRCBL: 4.4; 21.1; 6. Identifiers: Osijek Declaration on the Rights of Patients with Chronic Pain.

Brazier, Margot. Human(s) (as) medicine(s). *In:* McLean, Sheila A.M., ed. First Do No Harm: Law, Ethics, and Healthcare. Aldershot, England; Burlington, VT: Ashgate, 2006: 187-202. NRCBL: 4.4; 19.1; 9.3.1; 20.1; 14.4.

Briggs, Sheila. The ethics of life: medical advances have exposed inconsistencies in the Roman Catholic hierarchy's position on life. *Conscience* 2007 Autumn; 28(3): 14-18. NRCBL: 4.4; 1.2; 20.5.1; 20.1. Identifiers: Terri Schiavo; natural death.

Cohen, Alfred. The valance of pain in Jewish thought and practice. *Journal of Halacha and Contemporary Society* 2007 Spring; (53): 25-52. NRCBL: 4.4; 1.2.

Community-State Partnerships to Improve End-of-Life Care. Advances in state pain policy and medical practice. Kansas City, MO: State Initiatives Policy Brief, Community State Partnerships to Improve End-of-Life Care 1999 April; (4): 1-8 [Online]. Accessed: http://www.practicalbioethics.org/FileUploads/SI_4.pdf [2006 September 28]. NRCBL: 4.4; 20.4.1; 5.3. SC: le.

Cooper, Rachel. Can it be a good thing to be deaf? *Journal of Medicine and Philosophy* 2007 November-December; 32(6): 563-583. NRCBL: 4.4; 1.1; 9.5.1; 18.4. SC: an.
Abstract: Increasingly, Deaf activists claim that it can be good to be Deaf. Still, much of the hearing world remains unconvinced, and continues to think of deafness in negative terms. I examine this debate and argue that to determine whether it can be good to be deaf it is necessary to examine each claimed advantage or disadvantage of being deaf, and then to make an overall judgment regarding

NRCBL: National Reference Center for Bioethics Literature Classification Scheme   See inside front cover for terms.

383

the net cost or benefit. On the basis of such a survey I conclude that being deaf may plausibly be a good thing for some deaf people but not for others.

de Graeff, Alexander; Dean, Mervyn. Palliative sedation therapy in the last weeks of life: a literature review and recommendations for standards. *Journal of Palliative Medicine* 2007 February; 10(1): 67-85. NRCBL: 4.4; 20.4.1; 9.

De Grazia, David. Must we have full moral status throughout our existence? A reply to Alfonso Gomez-Lobo. *Kennedy Institute of Ethics Journal* 2007 December; 17(4): 297-310. NRCBL: 4.4; 1.1; 12.3; 22.1. SC: an.

Abstract: Those who are morally opposed to abortion generally make several pivotal assumptions. This paper focuses on the assumption that we have full moral status throughout our existence. Coupled with the assumption that we come into existence at conception, the assumption about moral status entails that all human fetuses have full moral status, including a right to life. Is the assumption about moral status correct? In addressing this question, I respond to several arguments advanced, in this journal and other venues, by Alfonso Gómez-Lobo. Gómez-Lobo's reasoning resolves into two basic arguments: (1) an appeal to the practical necessity of early moral protection and (2) an appeal to our kind membership and potentiality. I respond to these in turn before offering further reflections.

Deckers, Jan. Are those who subscribe to the view that early embryos are persons irrational and inconsistent? A reply to Brock. *Journal of Medical Ethics* 2007 February; 33(2): 102-106. NRCBL: 4.4; 1.1. SC: an. Comments: D.W. Brock. Is a consensus possible on stem cell research? Journal of Medical Ethics 2006 January; 32(1): 36- 42.

Abstract: Dan Brock has asserted that those who claim that the early embryo has full moral status are not consistent, and that the rationality of such a position is dubious when it is adopted from a religious perspective. I argue that both claims are flawed. Starting with the second claim, which is grounded in Brock's moral abstolutist position, I argue that Brock has provided no argument on why the religious position should be less rational than the secular position. With regard to the first claim, I argue that those who hold the view that the early embryo has full moral status can be consistent even if they do not oppose sexual reproduction, even if they do not grieve as much over the loss of embryos as over the loss of other humans, even if they prefer to save one child instead of 100 embryos in the event of fire, and even if they do not accept racism and sexism.

Deckers, Jan. Why Eberl is wrong. Reflections on the beginning of personhood. *Bioethics* 2007 June; 21(5): 270-282. NRCBL: 4.4; 1.1. SC: an. Comments: Jason T. Eberl. A Thomistic perspective on the beginning of personhood: redux. Bioethics 2007 June; 21(5): 283-289.

Abstract: In a paper published in Bioethics, Jason Eberl has argued that early embryos are not persons and should not be granted the status possessed by them. Eberl bases this position upon the following claims: (1) The early em-bryo has a passive potentiality for development into a person. (2) The early embryo has not established both 'unique genetic identity' and 'ongoing ontological identity', which are necessary conditions for ensoulment. (3) The early embryo has a low probability of developing into a more developed embryo. This paper examines these claims. I argue against (1) that a plausible view is that the early embryo has an active potentiality to grow into a more developed embryo. Against (2), I argue that neither 'unique genetic identity' nor 'ongoing ontological identity' are necessary conditions for ensoulment, and that 'ongoing ontological identity' is established between early embryos and more developed embryos. Against (3), I argue that the fact that the early embryo has a low probability of developing into a more developed embryo, if true, does not warrant the conclusion that the early embryo is not a person. If Eberl is right that the human soul is that which organises the activities of a human being and that ensouled humans are persons, embryos are persons from conception.

Diamond, Sheila M. Response to Todd Bindig's "Confusion about speciesism and moral status". *Linacre Quarterly* 2007 May; 74(2): 156-158. NRCBL: 4.4; 1.2; 22.1. Comments: Todd S. Bindig. Confusion about speciesism and moral status. Linacre Quarterly 2007 May; 74(2): 156-158.

Dolgin, Janet L. New terms for an old debate: embryos, dying, and the "culture wars". *Houston Journal of Health Law and Policy* 2006; 6(2): 249-273. NRCBL: 4.4; 12.5.1; 12.3; 20.5.1; 20.3.1. SC: le; an.

Dombrowski, Dan. Personhood and life issues: a Catholic view. *Conscience* 2007 Autumn; 28(3): 30-33. NRCBL: 4.4; 1.2; 20.5.1; 22.1.

Durante, Chris. Persons, identities, and medical ethics [review of Human Identity and Bioethics, by David DeGrazia]. *Hastings Center Report* 2007 March-April; 37(2): 47. NRCBL: 4.4; 2.1.

Eberl, Jason T. A Thomistic perspective on the beginning of personhood: redux. *Bioethics* 2007 June; 21(5): 283-289. NRCBL: 4.4; 1.1. SC: an. Comments: Jan Deckers. Why Eberl is wrong. Reflections on the beginning of personhood. Bioethics 2007 June; 21(5): 270-282.

Eberl, Jason T. The beginning of a human person's life. *In his:* Thomistic Principles and Bioethics. London; New York: Routledge, 2006: 23-42. NRCBL: 4.4; 15.1; 1.1. SC: an.

Gibson, Susanne. Uses of respect and uses of the human embryo. *Bioethics* 2007 September; 21(7): 370-378. NRCBL: 4.4; 1.1; 18.5.4. SC: an; le. Identifiers: Great Britain (United Kingdom); Human Fertilisation and Embryology Act (HFEA) 1990; Report of the Committee of Inquiry into Human Fertilisation and Embryology (Warnock Report).

Abstract: In most parts of the world, research on the human embryo is subject to tight controls. In the United

Kingdom it is restricted by means of both a fourteen-day time limit and the permitted purposes of the research. One of the ways in which the argument for these restrictions has been put is in terms of respect. That is, the human embryo is said to be the kind of thing that is worthy of a measure of respect such that there are limits to what can be done to it. This paper considers some of the ways in which this principle of respect has been understood as well as some objections to the very idea that research resulting in the destruction of the human embryo can claim to show that embryo respect. It will be argued that an account of 'respectful destruction' can be articulated on the grounds of our shared finitude as human moral agents, and in particular on the grounds of our shared lack of certainty regarding the moral status of the embryo.

Giordano, James. Cassandra's curse: interventional pain management, policy and preserving meaning against a market mentality. *Pain Physician* 2006 July; 9(3): 167-169. NRCBL: 4.4; 7.1; 9.3.1; 1.3.2.

Gómez-Lobo, Alfonso. A note on metaphysics and embryology [letter]. *Theoretical Medicine and Bioethics* 2007; 28(4): 331-335. NRCBL: 4.4; 1.1. Comments: Carson Strong. Preembryo personhood: an assessment of the President's Council arguments. Theoretical Medicine and Bioethics 2006; 27(5): 433-453.

Gómez-Lobo, Alfonso. Inviolability at any age. *Kennedy Institute of Ethics Journal* 2007 December; 17(4): 311-320. NRCBL: 4.4; 1.1; 12.3; 22.1. SC: an.
Abstract: This paper starts from three assumptions: that we are essentially human organisms, that we start to exist at conception, and that we retain our identity throughout our lives. The identity claim provides the background to argue that it is irrational for a person to claim that it would be impermissible to kill her now but permissible to have killed her at an earlier age. The notion of "full moral status" as an ascertainable property is questioned and shown to be dependent on previously accepted moral norms. It is concluded that the exclusion of the very young from the scope of the norm o f common morality that prohibits the killing of the innocent amounts to discrimination on the basis of age.

Haas, John M. Person and human being in the UNESCO Declaration on Bioethics and Human Rights. *National Catholic Bioethics Quarterly* 2007 Spring; 7(1): 41-50. NRCBL: 4.4; 21.1; 2.1; 1.2.

Hamel, Ron. Human dignity: an "energizing vision" of health care. *Health Progress* 2007 May-June; 88(3): 12-13. NRCBL: 4.4; 1.2; 9.1.

Hartman, Rhonda Gay. The face of dignity: principled oversight of biomedical innovation. *Santa Clara Law Review* 2007; 47(1): 55-91. NRCBL: 4.4; 19.1; 5.3; 18.1; 1.1. Identifiers: face transplants; facial transplants.

Hewson, Barbara. Dancing on the head of a pin? Foetal life and the European convention. *Feminist Legal Studies* 2005; 13(3): 363-375. NRCBL: 4.4; 9.5.8; 12.4.2.

Hirskyj, Peter. QALY: an ethical issue that dare not speak its name. *Nursing Ethics* 2007 January; 14(1): 72-82. NRCBL: 4.4; 9.4; 4.1.3. Identifiers: Great Britain (United Kingdom).
Abstract: The current British Government's policy towards resource allocation for health care has been informed by the commissioned Wanless Report. This makes a case for the use of quality adjusted life years (QALYs) to form a rationale for resourcing health care and has implications for the staff and patients who work in and use the health service. This article offers a definition of the term 'QALY' and considers some of the strengths and weaknesses of this approach to resource distribution. An account is also given of an alternative formula, the DALY (disability adjusted life years), which can address some of the problems that are associated with QALYs. The values of the public, patients and nurses are identified and linked to the potential effect of a QALY formula. The implications of QALY use are applied to the health care of patients and a discussion is offered with regard to whether this method of resource allocation can be considered as just.

Honnefelder, Ludger. Science, law and ethics: the Biomedicine Convention as an ethico-legal response to current scientific challenges. *In:* Gevers, J.K.M.; Hondius, E.H.; Hubben, J.H., eds. Health Law, Human Rights and the Biomedicine Convention: Essays in Honour of Henriette Roscam Abbing. Leiden; Boston: Martinus Nijhoff Publishers, 2005: 13-22. NRCBL: 4.4; 21.1. SC: le.

Horne, Benjamin D. Human at conception: the 14th amendment and the acquisition of personhood. *Human Life Review* 2007 Summer; 33(3): 73-81. NRCBL: 4.4; 12.4.2; 9.5.5. SC: le.

Janvier, Annie; Bauer, Karen Lynn; Lantos, John D. Are newborns morally different from older children? *Theoretical Medicine and Bioethics* 2007; 28(5): 413-425. NRCBL: 4.4; 8.3.2; 20.5.2.
Abstract: Policies and position statements regarding decision-making for extremely premature babies exist in many countries and are often directive, focusing on parental choice and expected outcomes. These recommendations often state survival and handicap as reasons for optional intervention. The fact that such outcome statistics would not justify such approaches in other populations suggests that some other powerful factors are at work. The value of neonatal intensive care has been scrutinized far more than intensive care for older patients and suggests that neonatal care is held to a higher standard of justification. The relative value placed on the life of newborns, in particular the preterm, is less than expected by any objective medical data or any prevailing moral frameworks about the value of individual lives. Why do we feel less obligated to treat the premature baby? Do we put newborns in a special and lesser moral category? We explore this question from a legal and ethical perspective and offer several hypotheses pertaining to personhood, reproductive choices, "precious children," and probable evolutionary and anthropological factors.

NRCBL: National Reference Center for Bioethics Literature Classification Scheme   See inside front cover for terms.

Johnson, Kirk. Proposed Colorado measure on rights for human eggs. *New York Times* 2007 November 18; p. A23. NRCBL: 4.4. SC: po; le.

Kaczor, Christopher Robert. All human beings are persons. *In his:* The Edge of Life: Human Dignity and Contemporary Bioethics. Dordrecht: Springer, 2005: 41-66. NRCBL: 4.4.

Kaczor, Christopher Robert. How is the dignity of the person as agent recognized?: distinguishing intention from foresight. *In his:* The Edge of Life: Human Dignity and Contemporary Bioethics. Dordrecht: Springer, 2005: 67-81. NRCBL: 4.4.

Kaiser, Jocelyn. Issues with tissues [news brief]. *Science* 2007 June 29; 316(5833): 1829. NRCBL: 4.4; 19.5. SC: le.

Kass, Leon R. The right to life and human dignity. *New Atlantis* 2007 Spring; (16): 23-40. NRCBL: 4.4; 20.5.1; 20.7; 1.1. Identifiers: Thomas Hobbes.

Kolber, Adam J. Pain detection and the privacy of subjective experience. *American Journal of Law and Medicine* 2007; 33(2-3): 433-456. NRCBL: 4.4.

Lee, Dong-Ik. The bioethics of 'quality of life' and 'sanctity of life'. *Dolentium Hominum* 2007; 22(2): 33-39. NRCBL: 4.4; 1.2; 2.1; 6.

Lizza, John P. Potentiality and human embryos. *Bioethics* 2007 September; 21(7): 379-385. NRCBL: 4.4; 1.1; 1.3.5; 18.5.4. SC: an.
Abstract: Consideration of the potentiality of human embryos to develop characteristics of personhood, such as intellect and will, has figured prominently in arguments against abortion and the use of human embryos for research. In particular, such consideration was the basis for the call of the US President's Council on Bioethics for a moratorium on stem cell research on human embryos. In this paper, I critique the concept of potentiality invoked by the Council and offer an alternative account. In contrast to the Council's view that an embryo's potentiality is determined by definition and is not affected by external conditions that may prevent certain possibilities from ever being realized, I propose an empirically grounded account of potentiality that involves an assessment of the physical and decisional conditions that may restrict an embryo's possibilities. In my view, some human embryos lack the potentiality to become a person that other human embryos have. Assuming for the sake of argument that the potential to become a person gives a being special moral status, it follows that some human embryos lack this status. This argument is then used to support Gene Outka's suggestion that it is morally permissible to experiment on 'spare' frozen embryos that are destined to be destroyed.

Mason, J.K.; Laurie, G.T. The body as property. *In their:* Mason and McCall Smith's Law and Medical Ethics. Seventh ed. Oxford; New York: Oxford University Press, 2005: 511-538. NRCBL: 4.4; 5.3; 1.3.9; 15.8; 19.1; 20.1.

SC: le. Identifiers: Great Britain (United Kingdom); The Nuffield Report.

McLachlan, Hugh V.; Swales, J. Kim. Embryology and human cloning. *In their:* From the Womb to the Tomb: Issues in Medical Ethics. Glasgow, Scotland: Humming Earth, 2007: 21-83. 50 refs. 35 fn. NRCBL: 4.4; 1.2; 14.5; 18.5.4; 15.1.
Keywords: *cloning; *embryo research; *embryos; advisory committees; beginning of life; embryonic stem cells; genetic diversity; human dignity; legal aspects; moral status; personhood; Protestant ethics; public policy; reproductive technologies; twinning; value of life; Keyword Identifiers: Church of Scotland; European Union; Great Britain; Human Fertilisation and Embryology Act 1990 (Great Britain); Quintavalle v. Human Fertilisation and Embryology Authority; Warnock Committee

Molyneux, David. "And how is life going for you?" — an account of subjective welfare in medicine. *Journal of Medical Ethics* 2007 October; 33(10): 568-572. NRCBL: 4.4; 1.1; 4.1.2.
Abstract: The dominant account of welfare in medicine is an objective one; welfare consists of certain favoured health states, or in having needs satisfied, or in certain capabilities and functionings. By contrast, I present a subjective account of welfare, suggested initially by LW Sumner and called "authentic happiness". The adoption of such an account of welfare within medicine offers several advantages over other subjective and objective accounts, and systematises several intuitions about patient-centredness and autonomy. Subjective accounts of welfare are unpopular because of their implications for justice and the autonomy of the healthcare professional. This account of welfare, however, seems to have the resources to resist these criticisms.

Murdoch, C.J. Kant's antipode: Nietzche's transvaluation of human dignity and its "implications" for biotechnology policy. *Health Law Review* 2007; 15(3): 24-26. NRCBL: 4.4; 1.1; 2.1.

Nota, L.; Ferrari, L.; Soresi, S.; Wehmeyer, M. Self-determination, social abilities and the quality of life of people with intellectual disability. *Journal of Intellectual Disability Research* 2007 November; 51(11): 850-865. NRCBL: 4.4; 9.5.3. SC: em.

Ohlin, Jens David. Is the concept of the person necessary for human rights? *Columbia Law Review* 2005 January; 105(1): 209-249. NRCBL: 4.4; 4.3.

Peters, Philip G., Jr. The ambiguous meaning of human conception. *U.C. Davis Law Review* 2006 November; 40(1): 199-228. NRCBL: 4.4; 18.5.4; 15.1; 18.6. SC: le.

Ralston, D. Christopher; Ho, Justin. Disability, humanity, and personhood: a survey of moral concepts. *Journal of Medicine and Philosophy* 2007 November-December; 32(6): 619-633. NRCBL: 4.4; 1.1; 4.2; 9.5.1. SC: an.
Abstract: Three of the articles included in this issue of the Journal of Medicine and Philosophy - Ron Amundson and Shari Tresky's "On a Bioethical Challenge to Dis-

ability Rights"; Rachel Cooper's "Can It Be a Good Thing to Be Deaf?"; and Mark T. Brown's "The Potential of the Human Embryo" - interact (in various ways) with the concepts of disability, humanity, and personhood and their normative dimensions. As one peruses these articles, it becomes apparent that terms like "disability," "human being," and "person" carry with them great normative significance. There is, however, much disagreement concerning both the definition and the extension of such terms. This is significant because different terms and definitions are associated with different sets of normative requirements. In what follows we reconstruct the argument of each of the articles, and then offer some brief critical analysis intended to stimulate further thought about and discussion of the issues that each raises.

Raposo, Vera Lúcia; Osuna, Eduardo. Embryo dignity: the status and juridical protection of the in vitro embryo. *Medicine and Law: The World Association for Medical Law* 2007 December; 26(4): 737-746. NRCBL: 4.4; 14.6; 18.5.4. SC: le.

Abstract: In the context of research and reproduction, the status of the human in vitro embryo ranges from being regarded as a person to being regarded as mere property. As regards the first view, one extreme of the spectrum for offering possible legal protection considers that the embryo constitutes a legal person from the moment of conception. For opponents of this view life is a continuum that runs from conception until death. In this process one of the most important stages is birth, the reason being that birth represents the transition between a potential person and a person. The term "embryo" is used to express the being that exists after fusion of the egg and a spermatozoon during the process of embryogenesis until it reaches eight weeks, after which time it is termed a foetus. The embryo's life is recognized as a constitutional value which deserves juridical protection, but not as a person. It only becomes a person with birth.

Reiman, Jeffrey. Being fair to future people: the non-identity problem in the original position. *Philosophy and Public Affairs* 2007 Winter; 35(1): 69-92. NRCBL: 4.4; 1.1. SC: an.

Richards, Janet Radcliffe. Selling organs, gametes, and surrogacy services. *In:* Rhodes, Rosamond; Francis, Leslie P.; Silvers, Anita, eds. The Blackwell Guide to Medical Ethics. Malden, MA: Blackwell Pub., 2007: 254-268. NRCBL: 4.4; 14.6; 14.2; 1.1; 9.3.1; 19.1.

Ross, Lainie Friedman. The moral status of the newborn and its implications for medical decision making. *Theoretical Medicine and Bioethics* 2007 October; 28(5): 349-355. NRCBL: 4.4; 8.3.2; 20.5.2.

Savell, Kristin. The legal significance of birth. *University of New South Wales Law Journal* 2006; 29(2): 200-206. NRCBL: 4.4; 1.3.5. SC: le.

Schefer, Markus. Geltung der Grundrechte vor der Geburt. *In:* Schreiber, Hans-Peter, ed. Biomedizin und

Ethik: Praxis, Recht, Moral. Basel; Boston: Birkhäuser, 2004: 43-49, 96. 3 refs. NRCBL: 4.4; 12.4.1; 15.1. SC: le.

Keywords: *embryos; *fetuses; *legal rights; abortion; biotechnology; embryo research; genetic intervention; Keyword Identifiers: Europe; United States

Schicktanz, Silke. Why the way we consider the body matters — reflections on four bioethical perspectives on the human body. *Philosophy, Ethics, and Humanities in Medicine [electronic]* 2007 December 4; 2(30): 12 p. Accessed: http://www.peh-med.com/ [2008 January 24]. NRCBL: 4.4; 1.1.

Shannon, Thomas A.; Wolter, Allan B. Reflections on the moral status of the pre-embryo. *In:* Walter, James J.; Shannon, Thomas A., eds. Contemporary Issues in Bioethics: A Catholic Perspective. Lanham, MD: Rowman and Littlefield Publishers, 2005: 67-90. NRCBL: 4.4; 1.2.

Silver, Lee M. The embryonic soul. *In his:* Challenging Nature: The Clash of Science and Spirituality at the New Frontiers of Life. New York: Ecco, 2006: 98-124. NRCBL: 4.4; 1.2. SC: an.

Snead, O. Carter. Assessing the Universal Declaration on Bioethics and Human Rights: implications for human dignity and the respect for human life. *National Catholic Bioethics Quarterly* 2007 Spring; 7(1): 53-71. NRCBL: 4.4; 21.1; 2.4.

Steinberg, Douglas. Consciousness is missing — and so is research. After the Schiavo controversy in the USA, obstacles still hinder the study of people with little or no awareness. *EMBO Reports* 2005 November; 6(11): 1009-1011. NRCBL: 4.4; 20.5.1; 7.1; 9.4; 18.5.6.

Steinbock, Bonnie. Moral status, moral value, and human embryos: implications for stem cell research. *In:* Steinbock, Bonnie, ed. The Oxford Handbook of Bioethics. Oxford; New York: Oxford University Press, 2007: 416-440. 43 refs. 12 fn. NRCBL: 4.4; 18.5.4; 15.1. SC: an; rv.

Keywords: *embryo research; *embryonic stem cells; *embryos; *moral policy; *moral status; abortion; beginning of life; cadavers; ethical analysis; human characteristics; killing; nuclear transfer techniques; personhood; public policy; research embryo creation; species specificity; stem cells; twinning; values; Proposed Keywords: altered nuclear transfer; embryo death; pluripotent stem cells; sentience

Strong, Carson. Embryology, metaphysics, and common sense: a response to Gómez-Lobo [letter]. *Theoretical Medicine and Bioethics* 2007; 28(4): 337-340. NRCBL: 4.4; 1.1. Comments: Alfonso Gómez-Lobo. A note on metaphysics and embryology. Theoretical Medicine and Bioethics 2007; 28(4): 331-335.

Takala, Tuija. Concepts of "person" and "liberty," and their implications to our fading notions of autonomy. *Journal of Medical Ethics* 2007 April; 33(4): 225-228. NRCBL: 4.4; 1.1; 2.1; 8.1. SC: an.

NRCBL: National Reference Center for Bioethics Literature Classification Scheme   See inside front cover for terms.

387

Abstract: It is commonly held that respect for autonomy is one of the most important principles in medical ethics. However, there are a number of interpretations as to what that respect actually entails in practice and a number of constraints have been suggested even on our self-regarding choices. These limits are often justified in the name of autonomy. In this paper, it is argued that these different interpretations can be explained and understood by looking at the discussion from the viewpoints of positive and negative liberty and the various notions of a "person" that lay beneath. It will be shown how all the appeals to positive liberty presuppose a particular value system and are therefore problematic in multicultural societies.

Talseth, Anne-Grethe; Gilje, Fredricka. Unburdening suffering: responses of psychiatrists to patients' suicide deaths. *Nursing Ethics* 2007 September; 14(5): 620-636. NRCBL: 4.4; 20.7; 17.1; 8.1. SC: em. Identifiers: Norway.
Abstract: The research questions was: 'How do psychiatrists describe their responses to patients' suicidal deaths in the light of a published model of consolation?' The textual data (n = 5) was a subset of a larger (n = 19) study. Thematic analysis showed a main theme, 'unburdening grief', and six themes. Embedded in the results is a story about suffering that reveals that, through ethical reflectiveness, a meaning of suffering can be recreated that unburdens grief and opens up new understandings with and among disciplines. This can help to prepare health professionals to respond to people who suffer because of suicidal death.

Taylor, Joseph G. NICE, Alzheimer's and QALY. *Clinical Ethics* 2007 March; 2(1): 50-54. NRCBL: 4.4; 9.4; 17.1; 9.5.2; 1.1. Identifiers: National Institute for Health and Clinical Excellence; quality-adjusted life year.
Abstract: The introduction of National Institute for Health and Clinical Excellence (NICE) guidance on Alzheimer's medication in November 2006 will have a significant effect on the treatment of patients, and is opposed by the Royal College of Psychiatrists and many charities dealing with the elderly. The use of the Quality-Adjusted Life Year (QALY) in the guidance formulation is much debated due to questions of ageism. This article seeks to examine the basis of these accusations and whether NICE can be justified using a utilitarian calculation in a principally egalitarian system such as the NHS.

Verhovek, Sam Howe. Parents halt growth of severely disabled girl. *Seattle Times* 2007 January 4; 3 p. [Online]. Accessed: http://seattletimes.nwsource.com [2007 January 8]. NRCBL: 4.4; 9.5.3; 8.3.2. SC: po.

Walter, James J. The meaning and validity of quality of life judgments in contemporary Roman Catholic medical ethics. *In:* Walter, James J.; Shannon, Thomas A., eds. Contemporary Issues in Bioethics: A Catholic Perspective. Lanham, MD: Rowman and Littlefield Publishers, 2005: 209-221. NRCBL: 4.4; 1.2.

Watt, Helen. Embryos and pseudoembryos: parthenotes, reprogrammed oocytes and headless clones. *Journal of Medical Ethics* 2007 September; 33(9): 554-556. 9 refs. NRCBL: 4.4; 14.5; 18.1; 18.5.4; 15.1; 19.1; 22.1. SC: an.
Keywords: *cloning; *embryos; *moral status; *ovum; *research embryo creation; anencephaly; chimeras; embryonic stem cells; genetic engineering; human dignity; methods; nuclear transfer techniques; Proposed Keywords: *parthenogenesis; altered nuclear transfer

Wilkinson, Stephen. Separating conjoined twins: the case of Ladan and Lelah Bijani. *In:* Gunning, Jennifer; Holm, Søren, eds. Ethics, law, and society. Volume 1. Aldershot, Hants, England; Burlington, VT: Ashgate, 2005: 257-260. NRCBL: 4.4; 1.1; 18.3. SC: cs.

Yu, Erika; Fan, Ruiping. A Confucian view of personhood and bioethics. *Journal of Bioethical Inquiry* 2007; 4(3): 171-179. NRCBL: 4.4; 1.1; 1.2; 2.1; 8.3.1; 21.7.
Abstract: This paper focuses on Confucian formulations of personhood and the implications they may have for bioethics and medical practice. We discuss how an appreciation of the Confucian concept of personhood can provide insights into the practice of informed consent and, in particular, the role of family members and physicians in medical decision-making in societies influenced by Confucian culture. We suggest that Western notions of informed consent appear ethically misguided when viewed from a Confucian perspective.

Zucchi, Pierluigi; Honings, Bonifacio. Cancer pain, death and dying. *Dolentium Hominum* 2007; 22(2): 61-72. NRCBL: 4.4; 20.4.1; 1.2.

RATIONING OF HEALTH CARE *See* RESOURCE ALLOCATION

RECOMBINANT DNA RESEARCH *See* GENETICALLY MODIFIED ORGANISMS AND FOOD

REGULATION *See* ABORTION/ LEGAL ASPECTS; BIOETHICS AND MEDICAL ETHICS/ LEGAL ASPECTS; CLONING/ LEGAL ASPECTS; EUTHANASIA AND ALLOWING TO DIE/ LEGAL ASPECTS; GENETIC SCREENING/ LEGAL ASPECTS; HUMAN EXPERIMENTATION/ ETHICS COMMITTEES AND POLICY GUIDELINES/ LEGAL ASPECTS; HUMAN EXPERIMENTATION/ REGULATION; ORGAN AND TISSUE TRANSPLANTATION/ DONATION AND PROCUREMENT/ LEGAL ASPECTS

RELIGIOUS ASPECTS *See* ABORTION/ MORAL AND RELIGIOUS ASPECTS; BIOETHICS AND MEDICAL ETHICS/ RELIGIOUS ASPECTS; EUTHANASIA AND ALLOWING TO DIE/ RELIGIOUS ASPECTS; HUMAN EXPERIMENTATION/ SPECIAL POP-

ULATIONS/ EMBRYOS AND FETUSES/
PHILOSOPHICAL AND RELIGIOUS ASPECTS

RENAL DIALYSIS *See* ORGAN AND TISSUE
TRANSPLANTATION

REPRODUCTION *See* REPRODUCTIVE
TECHNOLOGIES

REPRODUCTIVE TECHNOLOGIES
See also ARTIFICIAL INSEMINATION AND
SURROGATE MOTHERS; CLONING; CRYO-
BANKING OF SPERM, OVA, AND EMBRYOS;
IN VITRO FERTILIZATION; SEX DETERMIN-
ATION

**American Society for Reproductive Medicine [ASRM].
Ethics Committee.** Access to fertility treatment by gays,
lesbians, and unmarried persons. *Fertility and Sterility*
2006 November; 86(5): 1333-1335. NRCBL: 14.1; 10.

**American Society for Reproductive Medicine [ASRM].
Practice Committee.** Definition of "experimental". *Fer-
tility and Sterility* 2006 November; 86(5, Supplement):
S123. NRCBL: 14.1; 1.3.7; 1.3.9.

Ankeny, Rachel A. Individual responsibility and reproduction. *In:* Rhodes, Rosamond; Francis, Leslie P.; Silvers,
Anita, eds. The Blackwell Guide to Medical Ethics.
Malden, MA: Blackwell Pub., 2007: 38-51. NRCBL: 14.1;
1.1.

Baines, Jennifer A. Gamete donors and mistaken identities: the importance of genetic awareness and proposals favoring donor identity disclosure for children born from
gamete donations in the United States. *Family Court Review* 2007 January; 45(1): 116-132. NRCBL: 14.1; 8.4;
14.4; 21.1. SC: le.

Baird, Patricia A. The evolution of public policy on
reprogenetics in Canada. *In:* Knowles, Lori P.; Kaebnick,
Gregory E., eds. Reprogenetics: Law, Policy, and Ethical
Issues. Baltimore: Johns Hopkins University Press, 2007:
178-193. 23 refs., 1 fn. NRCBL: 14.1; 15.1.
   Keywords: *public policy; *regulation; *reproductive technologies; advisory committees; cloning; embryo research;
   federal government; genetic engineering; historical aspects;
   justice; legislation; policy making; public participation; religion; surrogate mothers; values; Proposed Keywords:
   licensure; Keyword Identifiers: *Canada; Royal Commission on New Reproductive Technologies (Canada);
   Twentieth Century; United States

Bard, Jennifer S. Immaculate gestation? How will ectogenesis change current paradigms of social relationships
and values? *In:* Gelfand, Scott; Shook, John R., eds. Ectogenesis: Artificial Womb Technology and the Future of
Human Reproduction. Amsterdam; New York: Editions
Rodopi, B.V., 2006: 149-157. NRCBL: 14.1; 19.1; 14.6;
5.1.

Benatar, David. Grim news from the original position: a
reply to Professor Doyal. *Journal of Medical Ethics* 2007
October; 33(10): 577. NRCBL: 14.1; 1.1; 4.4. SC: an.
   Abstract: In his review of my book, Better never to have
   been, Len Doyal suggests, contrary to my view, that rational beings in the original position might prefer coming
   into existence to the alternative of never existing, if their
   lives were to include enough good and not too much bad.
   I argue, in response, that Professor Doyal fails to make
   his case.

Bennett, Rebecca; Harris, John. Reproductive choice.
In: Rhodes, Rosamond; Francis, Leslie P.; Silvers, Anita,
eds. The Blackwell Guide to Medical Ethics. Malden, MA:
Blackwell Pub., 2007: 201-219. NRCBL: 14.1; 1.1; 15.2;
14.5.

Berger, Abi. Seeds of discontent. *BMJ: British Medical
Journal* 2007 June 23; 334(7607): 1323. NRCBL: 14.1.

**Blyth, Eric; Crawshaw, Marilyn; Haase, Jean; Speirs,
Jennifer.** The implications of adoption for donor offspring
following donor-assisted conception. *Child and Family
Social Work* 2001; 6(4): 295-304. NRCBL: 14.1; 14.4;
14.2; 15.1; 8.4. Identifiers: Great Britain (United
Kingdom).

Bomhoff, Jacco; Zucca, Lorenzo. European Court of Human Rights. The tragedy of Ms Evans: conflicts and
incommensurability of rights, Evans v. the United Kingdom, Fourth Section Judgment of 7 March 2006, application no. 6339/05. *European Constitutional Law Review*
2006 October; 2(3): 424-442. NRCBL: 14.1; 14.4; 14.6.
SC: le.

Bonnicksen, Andrea L. Oversight of assisted reproductive technologies: the last twenty years. *In:* Knowles, Lori
P.; Kaebnick, Gregory E., eds. Reprogenetics: Law, Policy,
and Ethical Issues. Baltimore: Johns Hopkins University
Press, 2007: 64-86. 47 refs. NRCBL: 14.1; 14.5; 18.5.4;
15.1; 2.2. SC: le.
   Keywords: *embryo research; *historical aspects; *regulation; *reproductive technologies; advisory committees;
   cloning; embryonic stem cells; ethical review; federal government; gene therapy; government regulation; guidelines;
   human experimentation; legal aspects; nuclear transfer techniques; policy making; preimplantation diagnosis; professional organizations; recombinant DNA research; self
   regulation; state government; standards; Keyword Identifiers: *United States; Department of Health and Human Services; Department of Health, Education, and Welfare; Food
   and Drug Administration; Human Embryo Research Panel;
   Twentieth Century; U.S. Congress

Bourg, Claudine. Ethical dilemmas in medically assisted
procreation: a psychological perspective. *Human Reproduction and Genetic Ethics: An International Journal*
2007; 13(2): 22-31. NRCBL: 14.1; 14.2; 14.4.

Cannold, Leslie. Women, ectogenesis, and ethical theory.
In: Gelfand, Scott; Shook, John R., eds. Ectogenesis: Artificial Womb Technology and the Future of Human Repro-

duction. Amsterdam; New York: Editions Rodopi, B.V., 2006: 47-58. NRCBL: 14.1; 19.1; 1.1; 12.3.

Casal, Paula; Williams, Andrew. Equality of resources and procreative justice. *In:* Burley, Justine, ed. Dworkin and His Critics: With Replies by Dworkin. Malden, MA: Blackwell Pub., 2004: 150-169. NRCBL: 14.1; 1.3.5. SC: an; cs.

Chadwick, Ruth. Reproductive autonomy — a special issue [editorial]. *Bioethics* 2007 July; 21(6): 304. NRCBL: 14.1; 1.1.

Cook, Rebecca; Pretorius, Rika. Duties to implement reproductive rights: the case of adolescents. *In:* Dennerstein, Lorraine, ed. Women's Rights and Bioethics. Paris: UNESCO, 2000: 175-189. NRCBL: 14.1; 9.5.7. SC: le.

Deane-Drummond, Celia. Women and genetic technologies. *In her:* Genetics and Christian Ethics. Cambridge: Cambridge University Press, 2006: 191-219. 60 fn. NRCBL: 14.1; 1.2; 15.1.
   Keywords: *feminist ethics; *reproductive technologies; *theology; *women; autonomy; caring; Christian ethics; ethical theory; family relationship; genetic counseling; genetic disorders; genetic services; genetic intervention; justice; natural law; selective abortion; sociobiology; virtues

Dickens, Bernard M. Conscientious objection: a shield or a sword? *In:* McLean, Sheila A.M., ed. First Do No Harm: Law, Ethics, and Healthcare. Aldershot, England; Burlington, VT: Ashgate, 2006: 337-351. NRCBL: 14.1; 12.4.3; 12.3; 8.1; 11.1; 11.3; 14.6. SC: le.

Diniz, Debora; Perea, Juan-Guillermo Figueroa; Luna, Florencia. Reproductive health ethics: Latin American perspectives [editorial]. *Developing World Bioethics* 2007 August; 7(2): ii-iii. NRCBL: 14.1; 2.1; 14.2; 12.3; 21.1.

Doyal, Len. Is human existence worth its consequent harm? [review of Better Never to Have Been: The Harm of Coming Into Existence, by David Benatar]. *Journal of Medical Ethics* 2007 October; 33(10): 573-576. NRCBL: 14.1; 1.1; 4.4.
   Abstract: Benatar argues that it is better never to have been born because of the harms always associated with human existence. Non-existence entails no harm, along with no experience of the absence of any benefits that existence might offer. Therefore, he maintains that procreation is morally irresponsible, along with the use of reproductive technology to have children. Women should seek termination if they become pregnant and it would be better for potential future generations if humans become extinct as soon as humanely possible. These views are challenged by the argument that while decisions not to procreate may be rational on the grounds of the harm that might occur, it may equally rational to gamble under certain circumstances that future children would be better-off experiencing the harms and benefits of life rather than never having the opportunity of experiencing anything. To the degree that Benatar's arguments

preclude the potential rationality of any such gamble, their moral relevance to concrete issues concerning human reproduction is weakened. However, he is right to emphasise the importance of foreseen harm when decisions are made to attempt to have children.

Dyer, Clare. Fertility group calls for investigation into UK regulator after search warrants declared illegal. *BMJ: British Medical Journal* 2007 July 7; 335(7609): 10. NRCBL: 14.1; 2.4; 1.3.7. SC: le. Identifiers: British Fertililty Society; Mohammed Taraniss; Human Fertilisation and Embryology Authority(HFEA).

Ferrell, Robyn. Brave new world. *In her:* Copula: Sexual Technologies, Reproductive Powers. Albany: State University of New York Press, 2006: 21-36. NRCBL: 14.1; 10; 4.4.

FIGO Committee for the Ethical Aspects of Human Reproduction and Women's Health. Ethical guidelines on iatrogenic and self-induced infertility. *International Journal of Gynecology and Obstetrics* 2006 August; 94(2): 172-173. NRCBL: 14.1; 9.5.5.

FIGO Committee for the Ethical Aspects of Human Reproduction and Women's Health. HIV and fertility treatment. *International Journal of Gynecology and Obstetrics* 2006 May; 93(2): 187-188. NRCBL: 14.1; 9.5.6; 9.5.5.

Fossett, James W. Managing reproductive pluralism: the case for decentralized governance. *Hastings Center Report* 2007 July-August; 37(4): 20-22. NRCBL: 14.1; 15.1; 5.3. SC: an; le. Comments: Franco Furger and Francis Fukuyama. A proposal for modernizing the regulation of human biotechnologies. Hastings Center Report 2007 July-August; 37(4): 16-20.
   Keywords: *biotechnology; *cultural pluralism; *government regulation; *public policy; *reproductive technologies; bioethical issues; cloning; decision making; democracy; dissent; embryo research; embryonic stem cells; embryos; federal government; moral status; policy analysis; preimplantation diagnosis; reproductive medicine; self regulation; state government; Keyword Identifiers: *United States; Great Britain

Furger, Franco; Fukuyama, Francis. A proposal for modernizing the regulation of human biotechnologies. *Hastings Center Report* 2007 July-August; 37(4): 16-20. NRCBL: 14.1; 15.1; 5.3; 15.2; 14.4. SC: an; le.
   Keywords: *biotechnology; *government regulation; *reproductive technologies; accountability; biomedical research; cloning; commerce; embryo research; embryonic stem cells; embryos; freedom; genetic engineering; germ cells; guidelines; industry; judicial role; legal aspects; political activity; policy analysis; politics; preimplantation diagnosis; public participation; public policy; quality of health care; standards; Keyword Identifiers: *United States

Gelfand, Scott. Ectogenesis and the ethics of care. *In:* Gelfand, Scott; Shook, John R., eds. Ectogenesis: Artificial Womb Technology and the Future of Human Reproduction. Amsterdam; New York: Editions Rodopi, B.V., 2006: 89-107. NRCBL: 14.1; 1.1; 5.3; 12.3; 19.1.

Great Britain (United Kingdom). Human Fertilisation and Embryology Authority [HFEA]. The HFEA guide to infertility. London: Human Fertilisation and Embryology Authority, 2007-2008; 51 p. [Online]. Accessed: http://www.hfea.gov.uk/docs/Guide2.pdf [2007 April 5]. NRCBL: 14.1; 14.2; 14.4; 14.6; 9.5.5.

Hall, Mark A.; Bobinski, Mary Anne; Orentlicher, David. Reproductive rights and genetic technologies. *In their:* Bioethics and Public Health Law. New York: Aspen Publishers, 2005: 397-518. 9 fn. NRCBL: 14.1; 15.1. SC: le.

> Keywords: *government regulation; *legal aspects; *legal rights; *reproductive rights; *reproductive technologies; abortion; cesarean section; cloning; constitutional law; contraception; drug abuse; embryo disposition; eugenics; federal government; fetuses; gene therapy; genetic counseling; genetic disorders; genetic engineering; genetic predisposition; genetic screening; in vitro fertilization; informed consent; involuntary sterilization; living wills; mental competence; mentally retarded persons; minors; ovum donors; parent-child relationship; personhood; pregnant women; prisoners; privacy; spousal consent; state government; Supreme Court decisions; surrogate mothers; treatment refusal; Keyword Identifiers: *United States; Buck v. Bell; Griswold v. Connecticut; Ferguson v. City of Charleston; In re A.C.; Planned Parenthood of Southeastern Pennsylvania v. Casey; Roe v. Wade; Skinner v. Oklahoma; Stenberg v. Carhart; Whitner v. South Carolina

Hallamaa, Jaana. Human reproductive technology - is there a Lutheran perspective? *In:* Østnor, Lars, ed. Bioetikk og teologi: Rapport fra Nordisk teologisk nettverk for bioetikks workshop i Stockholm 27.-29. September 1996. Oslo: Nord. teol. nettv. for bioetikk, 1996: 76-89. NRCBL: 14.1; 1.2; 2.1.

Heng, B.C. Should fertility specialists refer local patients abroad for shared or commercialized oocyte donation? *Fertility and Sterility* 2007 January; 87(1): 6-7. NRCBL: 14.1; 21.1; 8.1; 7.1; 9.3.1.

Himmel, Wolfgang; Michelmann, Hans Wilhelm. Access to genetic material: reproductive technologies and bioethical issues. *Reproductive Biomedicine Online* 2007 September; 15 (Suppl. 1): 18-24. 33 refs. NRCBL: 14.1; 18.5.4; 15.1; 14.4; 15.4.

> Keywords: *embryo disposition; *embryo research; *embryos; *germ cells; *reproductive technologies; age factors; beginning of life; cryopreservation; embryo transfer; embryonic stem cells; freedom; genetic materials; in vitro fertilization; informed consent; legal aspects; moral status; oocyte donation; preimplantation diagnosis; public opinion; regulation; selection for treatment; sex preselection; Proposed Keywords: blastocysts; Keyword Identifiers: Germany; Embryo Protection Act (Germany)

Hitchen, Lisa. Keeping the scientists in step with society [news]. *BMJ: British Medical Journal* 2007 May 26; 334(7603): 1079. NRCBL: 14.1; 2.4; 14.4; 15.1; 15.2; 22.2. Identifiers: Great Britain (United Kingdom).

Ho, Dien. Leaving people alone: liberalism, ectogenesis, and the limits of medicine. *In:* Gelfand, Scott; Shook, John R., eds. Ectogenesis: Artificial Womb Technology and the Future of Human Reproduction. Amsterdam; New York: Editions Rodopi, B.V., 2006: 139-147. NRCBL: 14.1; 19.1; 1.1.

Holland, Suzanne. Market transactions in reprogenetics: a case for regulation. *In:* Knowles, Lori P.; Kaebnick, Gregory E., eds. Reprogenetics: Law, Policy, and Ethical Issues. Baltimore: Johns Hopkins University Press, 2007: 89-104. 20 refs. 6 fn. NRCBL: 14.1; 15.2; 14.4; 5.3; 9.3.1. SC: an.

> Keywords: *commerce; *freedom; *genetic engineering; *genetic services; *germ cells; *ovum donors; *public policy; *regulation; *remuneration; *reproductive technologies; commodification; common good; democracy; embryo research; embryos; federal government; genetic relatedness ties; government financing; health facilities; in vitro fertilization; industry; justice; moral policy; policy analysis; politics; preimplantation diagnosis; private sector; public opinion; reproductive rights; socioeconomic factors; surrogate mothers; values; women; Keyword Identifiers: *United States; Canada; Great Britain

Johnson, Susan C. New reproductive technologies — discourse and dilemma. *In:* Garwood-Gowers, Austen; Tingle, John; Wheat, Kay, eds. Contemporary Issues in Healthcare Law and Ethics. Edinburgh; New York: Elsevier Butterworth-Heinemann, 2005: 115-131. NRCBL: 14.1; 1.1; 5.3; 15.1. SC: le. Identifiers: Great Britain (United Kingdom).

Johnston, Josephine. Tied up in nots over genetic parentage. *Hastings Center Report* 2007 July-August; 37(4): 28-31. 3 refs. NRCBL: 14.1; 15.1; 18.5.2; 18.5.4; 5.3. SC: an; le. Comments: Franco Furger and Francis Fukuyama. A proposal for modernizing the regulation of human biotechnologies. Hastings Center Report 2007 July-August; 37(4): 16-20.

> Keywords: *biotechnology; *genetic relatedness ties; *government regulation; *reproductive technologies; cloning; family relationship; genetic techniques; moral policy; ovum donors; policy analysis; public policy; Keyword Identifiers: United States

Kabir, M.; az-Zubair, Banu. Who is a parent? Parenthood in Islamic ethics. *Journal of Medical Ethics* 2007 October; 33(10): 605-609. 33 refs. NRCBL: 14.1; 1.2; 15.1.

> Keywords: *Islamic ethics; *parent child relationship; *reproductive technologies; adoption; beginning of life; bioethical issues; cross-cultural comparison; embryo transfer; fathers; fetuses; genetic relatedness ties; in vitro fertilization; minors; mothers; ovum donors; paternity; rights; Western World; Proposed Keywords: marriage

Kaczor, Christopher Robert. Could artificial wombs end the abortion debate? *In his:* The Edge of Life: Human Dignity and Contemporary Bioethics. Dordrecht: Springer, 2005: 105-131. NRCBL: 14.1; 19.1.

Kaczor, Christopher Robert. Moral absolutism and ectopic pregnancy. *In his:* The Edge of Life: Human Dig-

---

NRCBL: National Reference Center for Bioethics Literature Classification Scheme   See inside front cover for terms.

391

nity and Contemporary Bioethics. Dordrecht: Springer, 2005: 97-103. NRCBL: 14.1; 1.1; 9.5.5.

Kennedy, Holly P.; Renfrew, Mary J.; Madi, Banyana C.; Opoku, Dora; Thompson, Joyce B. The conduct of ethical research collaboration across international and culturally diverse communities. *Midwifery* 2006 June; 22(2): 100-107. NRCBL: 14.1; 9.5.5; 18.2; 18.5.3; 21.7; 21.1.

King, Patricia A.; Areen, Judith; Gostin, Lawerence O. Reproduction and the new genetics. *In their:* Law, Medicine and Ethics. New York: Foundation Press, 2006: 509-654. NRCBL: 14.1; 4.4; 12.4.2; 14.4; 14.2; 14.5; 15.2; 15.3. SC: le.

Knowles, Lori P. The governance of reprogenetic technology: international models. *In:* Knowles, Lori P.; Kaebnick, Gregory E., eds. Reprogenetics: Law, Policy, and Ethical Issues. Baltimore: Johns Hopkins University Press, 2007: 127-143. 32 refs. 6 fn. NRCBL: 14.1; 18.5.4; 15.1; 14.5; 21.1; 5.3. SC: rv.
  Keywords: *embryo research; *international aspects; *public policy; *regulation; *reproductive technologies; advisory committees; biotechnology; chimeras; cloning; consensus; cross cultural comparison; dissent; donors; embryonic stem cells; embryos; genetic engineering; government regulation; human dignity; in vitro fertilization; informed consent; legal aspects; legislation; moral status; policy making; public opinion; trends; Proposed Keywords: organizational models; Keyword Identifiers: Asia; Australia; Canada; Europe; Human Fertilisation and Embryology Authority; United States

Landau, Ruth. Assisted human reproduction: lessons of the Canadian experience. *Philosophy and Public Policy Quarterly* 2007 Winter-Spring; 27(1-2): 18-23. NRCBL: 14.1. SC: le. Identifiers: Assisted Human Reproduction Act.

Macklin, Ruth. Reproductive rights and health in the developing world. *In:* Galston, Arthur W.; Peppard, Christiana Z., eds. Expanding Horizons in Bioethics. Dordrecht; Norwell, MA: Springer, 2005: 87-101. NRCBL: 14.1; 21.1; 13.3; 1.2; 9.2.

Mamo, Laura. Negotiating conception: lesbians' hybrid-technological practices. *Science, Technology, and Human Values* 2007 May; 32(3): 369-393. NRCBL: 14.1; 10. SC: an.

Mason, J.K.; Laurie, G.T. The management of infertility and childlessness. *In their:* Mason and McCall Smith's Law and Medical Ethics. Seventh ed. Oxford; New York: Oxford University Press, 2005: 71-119. NRCBL: 14.1; 12.1; 14.2; 14.4. SC: le. Identifiers: Great Britain (United Kingdom); HFEA (Human Fertilisation and Embryology Authority).

Masson, Judith. Parenting by being; parenting by doing — in search of principles for founding families. *In:* Spencer, J.R.; du Bois-Pedain, Antje, eds. Freedom and

Responsibility in Reproductive Choice. Portland: Hart Pub., 2006: 131-155. NRCBL: 14.1; 14.4. SC: le.

McDougall, Rosalind. Parental virtue: a new way of thinking about the morality of reproductive actions. *Bioethics* 2007, May; 21(4): 181-190. NRCBL: 14.1; 1.1. SC: an; cs.
  Abstract: In this paper I explore the potential of virtue ethical ideas to generate a new way of thinking about the ethical questions surrounding the creation of children. Applying ideas from neo-Aristotelian virtue ethics to the parental sphere specifically, I develop a framework for the moral assessment of reproductive actions that centres on the concept of parental virtue. I suggest that the character traits of the good parent can be used as a basis for determining the moral permissibility of a particular reproductive action. I posit three parental virtues and argue that we can see the moral status of a reproductive action as determined by the relationship between such an action and (at least) these virtues. Using a case involving selection for deafness, I argue that thinking in terms of the question 'would a virtuous parent do this?' when morally assessing reproductive action is a viable and useful way of thinking about issues in reproductive ethics.

Mittra, James. Marginalising 'eugenic anxiety' through a rhetoric of 'liberal choice': a critique of the House of Commons Select Committee Report on reproductive technologies. *New Genetics and Society* 2007 August; 26(2): 159-179. 53 refs. NRCBL: 14.1; 15.5; 15.2. SC: le.
  Keywords: *eugenics; *freedom; *government regulation; *preimplantation diagnosis; *regulation; *reproductive technologies; embryos; genetic intervention; genetic services; legislation; prenatal diagnosis; public policy; reproductive rights; Keyword Identifiers: *Great Britain; *House of Commons Select Committee on Science and Technology (Great Britain); Human Fertilisation and Embryology Act 1990 (Great Britain; Human Fertilisation and Embryology Authority

Murphy, Julien S. Is pregnancy necessary? Feminist concerns about ectogenesis. *In:* Gelfand, Scott; Shook, John R., eds. Ectogenesis: Artificial Womb Technology and the Future of Human Reproduction. Amsterdam; New York: Editions Rodopi, B.V., 2006: 27-46. NRCBL: 14.1; 19.1; 10.

Murtagh, Ged M. Ethical reflection on the harm in reproductive decision-making. *Journal of Medical Ethics* 2007 December; 33(12): 717-720. 13 refs. NRCBL: 14.1; 1.1; 4.4; 11.1; 15.2. SC: an.
  Keywords: *autonomy; *congenital disorders; *ethical analysis; *moral obligations; *reproduction; choice behavior; decision making; disabled persons; ethical relativism; future generations; genetic disorders; genetic screening; obligations to society; parents; prenatal diagnosis; quality of life; reproductive rights; reproductive technologies; selective abortion; suffering; wrongful life; Proposed Keywords: harm
  Abstract: Advances in reproductive technologies continue to present ethical problems concerning their implementation and use. These advances have preoccupied bioethicists in their bid to gauge our moral responsibili-

ties and obligations when making reproductive decisions. The aim of this discussion is to highlight the importance of a sensibility to differences in moral perspective as part of our ethical inquiry in these matters. Its focal point is the work of John Harris, who has consistently addressed the ethical issues raised by advancing reproductive technologies. The discussion is aimed at a central tenet of Harris's position on reproductive decision-making-namely, that in some instances, giving birth to a worthwhile life may cause harm and will therefore be morally wrong. It attempts to spell out some of the implications of Harris's position that the author takes to involve a misplaced generality. To support this claim, some examples are explored that demonstrate the variety of ways in which concepts (such as harm) may manifest themselves as moral considerations within the context of reproductive decision-making. The purpose is to demonstrate that Harris's general conception of the moral limits of reproductive autonomy obscures the issues raised by particular cases, which in themselves may reveal important directions for our ethical inquiry.

New Zealand. Parliament. Human Assisted Reproductive Technology Bill. Wellington, New Zealand: Government Printer, 2004. No.195-3 [Online]. Accessed: http://www.parliament.nz/NR/rdonlyres/65E86459-2539-4FC3-B6F5-03440792D310/4617 2/DBHOH_BILL_24_49999999999997.pdf [2007 March 14]. NRCBL: 14.1; 9.6; 14.2; 14.4; 14.6. SC: le.

Panforti, Maria Donata; Serio, Mario. Genetics and artificial procreation in Italy. *In:* Meulders-Klein, Marie-Thérèse; Deech, Ruth; Vlaardingerbroek, Paul, eds. Biomedicine, the Family and Human Rights. New York: Kluwer Law International, 2002: 253-257. NRCBL: 14.1; 15.1. SC: le.
  Keywords: *legal aspects; *reproductive technologies; cloning; embryo disposition; genetic engineering; genetic screening; informed consent; Keyword Identifiers: *Italy

Parens, Erik; Knowles, Lori P. Reprogenetics and public policy: reflections and recommendations. *In:* Knowles, Lori P.; Kaebnick, Gregory E., eds. Reprogenetics: Law, Policy, and Ethical Issues. Baltimore: Johns Hopkins University Press, 2007: 253-294. 96 refs., 7 fn. NRCBL: 14.1; 18.5.4; 15.1. SC: rv.
  Keywords: *embryo research; *genetic engineering; *policy making; *public policy; *regulation; *reproductive technologies; adverse effects; advisory committees; cloning; embryonic stem cells; embryos; federal government; freedom; government regulation; historical aspects; human experimentation; international aspects; ovum donors; preimplantation diagnosis; professional organizations; recombinant DNA research; risk; social impact; state government; Proposed Keywords: tissue typing; Keyword Identifiers: Canada; Ethics Advisory Board; Food and Drug Administration; Great Britain; Human Embryo Research Panel; National Bioethics Advisory Commission; Recombinant DNA Advisory Committee; Reprogenetics Technology Board; Twentieth Century; Twenty-First Century; United States; Warnock Committee

Paxson, Heather. Reproduction as spiritual kin work: orthodoxy, IVF, and the moral economy of motherhood in Greece. *Culture, Medicine and Psychiatry* 2006 December; 30(4): 481-505. NRCBL: 14.1; 14.4; 1.2; 9.5.5.

Pearson, Yvette E. Storks, cabbage patches, and the right to procreate. *Journal of Bioethical Inquiry* 2007; 4(2): 105-115. NRCBL: 14.1; 1.1. SC: an.
  Abstract: In this paper I examine the prevailing assumption that there is a right to procreate and question whether there exists a coherent notion of such a right. I argue that we should question any and all procreative activities, not just alternative procreative means and contexts. I suggest that clinging to the assumption of a right to procreate prevents serious scrutiny of reproductive behavior and that, instead of continuing to embrace this assumption, attempts should be made to provide a proper foundation for it. I argue that the focus of procreative activities and discourse on reproductive ethics should be on obligations instead of rights, as rights talk tends to obfuscate recognition of obligations toward others, particularly those who bear the most significant burdens of the procreative process. I examine some possible foundations of a right to procreate as well as John Robertson's thoughtful account of "procreative liberty" but conclude that at the present time there exists no compelling account of a right to procreate. Finally, I conclude that in the absence of a satisfactory account of a right to procreate, we should refrain from grounding practices or polices on the assumption that there is such a right.

Pence, Gregory. What's so good about natural motherhood? (In praise of unnatural gestation). *In:* Gelfand, Scott; Shook, John R., eds. Ectogenesis: Artificial Womb Technology and the Future of Human Reproduction. Amsterdam; New York: Editions Rodopi, B.V., 2006: 77-88. NRCBL: 14.1; 19.1.

Pennings, Guido. International parenthood via procreative tourism. *In:* Shenfield, Françoise; Sureau, Claude, eds. Contemporary Ethical Dilemmas in Assisted Reproduction. Abingdon: Informa Healthcare, 2006: 43-56. NRCBL: 14.1; 21.1. SC: le.

Pollard, Irina. Procreative technology: achievements and desired goals. *In her:* Life, Love and Children: A Practical Introduction to Bioscience Ethics and Bioethics. Boston: Kluwer Academic Publishers, 2002: 105-125. NRCBL: 14.1; 7.1; 14.2; 14.4.

Raskin, Joyce M.; Mazor, Nadav A. The artificial womb and human subject research. *In:* Gelfand, Scott; Shook, John R., eds. Ectogenesis: Artificial Womb Technology and the Future of Human Reproduction. Amsterdam; New York: Editions Rodopi, B.V., 2006: 159-182. NRCBL: 14.1; 19.1; 18.5.4; 1.1; 4.4; 12.1.

Robertson, John A. The virtues of muddling through. *Hastings Center Report* 2007 July-August; 37(4): 26-28. NRCBL: 14.1; 15.1; 5.3. SC: an; le. Comments: Franco Furger and Francis Fukuyama. A proposal for modernizing

NRCBL: National Reference Center for Bioethics Literature Classification Scheme   See inside front cover for terms.

393

the regulation of human biotechnologies Hastings Center Report 2007 July-August; 37(4): 16-20.

Keywords: *biotechnology; *policy making; *reproductive technologies; advisory committees; cultural pluralism; democracy; dissent; federal government; freedom; genetic engineering; government regulation; policy analysis; preimplantation diagnosis; public participation; public policy; Keyword Identifiers: *United States

Sander-Staudt, Maureen. Of machine born? A feminist assessment of ectogenesis and artificial wombs. *In:* Gelfand, Scott; Shook, John R., eds. Ectogenesis: Artificial Womb Technology and the Future of Human Reproduction. Amsterdam; New York: Editions Rodopi, B.V., 2006: 109-128. NRCBL: 14.1; 19.1; 10; 1.1.

Schneider, Ingrid. Oocyte donation for reproduction and research cloning — the perils of commodification and the need for European and international regulation. *Revista de Derecho y Genoma Humano = Law and the Human Genome Review* 2006 July-December; (25): 205-241. 79 refs. NRCBL: 14.1; 14.5; 18.3.1; 21.1; 5.3; 9.3.1. SC: le.

Keywords: *cloning; *commodification; *embryo research; *genetic research; *ovum donors; *regulation; *remuneration; *reproductive technologies; advertising; embryo disposition; germ cells; in vitro fertilization; informed consent; international aspects; nuclear transfer techniques; therapeutic misconception; Keyword Identifiers: Europe

Shannon, Thomas A. Reproductive technologies: ethical and religious issues. *In:* Walter, James J.; Shannon, Thomas A., eds. Contemporary Issues in Bioethics: A Catholic Perspective. Lanham, MD: Rowman and Littlefield Publishers, 2005: 125-143. NRCBL: 14.1; 1.2; 2.2.

Sheldon, Sally. Reproductive technologies and the legal determination of fatherhood. *Feminist Legal Studies* 2005; 13(3): 349-362. NRCBL: 14.1. SC: le. Identifiers: Great Britain (United Kingdom); Human Fertilisation and Embryology Act 1990.

Shields, Wayne C.; Jordan, Beth. Adding value to reproductive health research: communicating about the moral dimensions of science. *Contraception* 2006 September; 74(3): 199-200. NRCBL: 14.1; 1.3.9; 1.1; 5.3.

Shildrick, Margrit. Reconfiguring the bioethics of reproduction. *Philosophy in the Contemporary World* 2004 Spring-Summer; 11(1): 75-83. NRCBL: 14.1; 10; 2.1; 1.1.

Singer, Peter; Wells, Deane. Ectogenesis. *In:* Gelfand, Scott; Shook, John R., eds. Ectogenesis: Artificial Womb Technology and the Future of Human Reproduction. Amsterdam; New York: Editions Rodopi, B.V., 2006: 9-25. NRCBL: 14.1; 19.1.

Smajdor, Anna. The moral imperative for ectogenesis. *CQ: Cambridge Quarterly of Healthcare Ethics* 2007 Summer; 16(3): 336-345. NRCBL: 14.1; 1.1; 9.2; 9.4; 9.5.5; 5.2; 9.5.7; 19.1. SC: an.

Society for Assisted Reproductive Technology [SART]. Practice Committee; American Society for Reproduc- tive Medicine **[ASRM]. Practice Committee.** Revised minimum standards for practices offering assisted reproductive technologies. *Fertility and Sterility* 2006 November; 86(5, Supplement): S53-S56. NRCBL: 14.1; 14.4; 7.1; 9.8.

Soini, Sirpa; Ibarreta, Dolores; Anastasiadou, Violetta; Aymé, Ségolène; Braga, Suzanne; Cornel, Martina; Coviello, Domenico A.; Evers-Kiebooms, Gerry; Geraedts, Joep; Gianaroli, Luca; Harper, Joyce; Kosztolanyi, György; Lundin, Kersti; Rodrigues-Cerezo, Emilio; Sermon, Karen; Sequeiros, Jorge; Tranebjaerg, Lisbeth; Kääriäinen, Helena. The interface between assisted reproductive technologies and genetics: technical, social, ethical, and legal issues. *European Journal of Human Genetics* 2006 May; 14(5): 588-645. NRCBL: 14.1; 15.3; 15.2; 21.1.

Steinbock, Bonnie. Defining personhood. *In:* Spencer, J.R.; du Bois-Pedain, Antje, eds. Freedom and Responsibility in Reproductive Choice. Portland: Hart Pub., 2006: 107-128. NRCBL: 14.1; 14.2; 14.4; 4.4. SC: le.

Storrow, Richard F. The bioethics of prospective parenthood: in pursuit of the proper standard for gatekeeping in infertility clinics. *Cardozo Law Review* 2007 April; 28(5):101-138. NRCBL: 14.1; 8.1; 9.2; 9.4; 21.1. SC: le.

Sutherland, Elaine E. Is there a right not to procreate? *In:* McLean, Sheila A.M., ed. First Do No Harm: Law, Ethics, and Healthcare. Aldershot, England; Burlington, VT: Ashgate, 2006: 319-336. NRCBL: 14.1; 12.4.1; 11.4; 14.6. SC: le.

Tollefsen, Christopher. Sic et non: some disputed questions in reproductive ethics. *In:* Khushf, George, ed. Handbook of Bioethics: Taking Stock of the Field From a Philosophical Perspective. Dordrecht; Boston: Kluwer Academic, 2004: 381-413. 97 refs. 12 fn. NRCBL: 14.1; 4.4; 12.3; 14.5; 15.1; 1.1. SC: an; rv.

Keywords: *autonomy; *beginning of life; *cloning; *embryos; *ethical analysis; *ethical theory; *fetuses; *moral policy; *personhood; *philosophy; *reproduction; *reproductive technologies; abortion; childbirth; chimeras; common good; contraception; deontological ethics; embryo disposition; embryo research; embryonic stem cells; feminist ethics; fetal research; freedom; genetic engineering; intention; justice; killing; marriage; moral status; mother fetus relationship; pregnant women; principle-based ethics; preimplantation diagnosis; reproductive rights; sexuality; utilitarianism; wedge argument; value of life; values

Tong, Rosemarie. Out of body gestation: in whose best interests? *In:* Gelfand, Scott; Shook, John R., eds. Ectogenesis: Artificial Womb Technology and the Future of Human Reproduction. Amsterdam; New York: Editions Rodopi, B.V., 2006: 59-76. NRCBL: 14.1; 19.1; 1.1; 10.

Tong, Rosemarie. Out-of-body gestation: in whose best interests? *Philosophy in the Contemporary World* 2004 Spring-Summer; 11(1): 65-74. NRCBL: 14.1; 19.1; 1.1; 10.

SC (Subject Captions): an=analytical   cs=case studies   em=empirical   le=legal   po=popular   rv=review

White, Gladys B. The development of reprogenetic policy and practice in the United States: looking to the United Kingdom. *In:* Knowles, Lori P.; Kaebnick, Gregory E., eds. Reprogenetics: Law, Policy, and Ethical Issues. Baltimore: Johns Hopkins University Press, 2007: 240-252. 15 refs. NRCBL: 14.1; 15.1; 21.1.
> Keywords: *policy making; *public policy; *regulation; *reproductive technologies; advisory committees; cross cultural comparison; genetic intervention; guidelines; public participation; Proposed Keywords: licensure; Keyword Identifiers: *United States; Great Britain; Human Fertilisation and Embryology Authority; President's Council on Bioethics

Woolfrey, Joan. Ectogenesis: liberation, technological tyranny, or just more of the same? *In:* Gelfand, Scott; Shook, John R., eds. Ectogenesis: Artificial Womb Technology and the Future of Human Reproduction. Amsterdam; New York: Editions Rodopi, B.V., 2006: 129-139. NRCBL: 14.1; 19.1; 1.1.

Yacoub, Ahmed Abdel Aziz. Reproduction and cloning. *In his:* The Fiqh of Medicine: Responses in Islamic Jurisprudence to Developments in Medical Science. London, UK: Ta-Ha Publishers Ltd., 2001: 233-253. 66 fn. NRCBL: 14.1; 1.2; 14.2; 14.4; 14.5.
> Keywords: *cloning; *Islamic ethics; *reproduction; *reproductive technologies; beginning of life; embryo research; embryo transfer; in vitro fertilization; infertility; posthumous conception; surrogate mothers

Young, Alison Harvison. Possible policy strategies for the United States: comparative lessons. *In:* Knowles, Lori P.; Kaebnick, Gregory E., eds. Reprogenetics: Law, Policy, and Ethical Issues. Baltimore: Johns Hopkins University Press, 2007: 226-239. 15 refs. NRCBL: 14.1; 2.4; 14.2; 15.1; 21.1. SC: le.
> Keywords: *legal aspects; *policy making; *public policy; *regulation; *reproductive technologies; advisory committees; cloning; contracts; cross cultural comparison; genetic intervention; government regulation; guidelines; mass media; surrogate mothers; torts; Keyword Identifiers: *United States; Canada

Zypries, Brigitte. From procreation to generation? Constitutional and legal-political issues in bioethics. *In:* Vöneky, Silja; Wolfrum, Rüdiger, eds. Human Dignity and Human Cloning. Leiden: Nijhoff, 2004: 107-121. 3 fn. NRCBL: 14.1; 15.2; 14.4; 14.5; 18.5.4. SC: le.
> Keywords: *embryo research; *legal aspects; *preimplantation diagnosis; *reproductive technologies; bioethical issues; cloning; embryonic stem cells; embryos; human dignity; legal rights; value of life; values; Keyword Identifiers: *Germany

RESEARCH *See* BEHAVIORAL RESEARCH; BIOMEDICAL RESEARCH; GENETIC RESEARCH; HUMAN EXPERIMENTATION

RESEARCH ETHICS *See* ANIMAL EXPERIMENTATION; BIOMEDICAL RESEARCH/ RESEARCH ETHICS AND SCIENTIFIC MISCONDUCT; HUMAN EXPERIMENTATION

RESEARCH ETHICS COMMITTEES *See* HUMAN EXPERIMENTATION/ ETHICS COMMITTEES AND POLICY GUIDELINES

RESOURCE ALLOCATION
See also HEALTH CARE ECONOMICS

Celebrities in the ED: managers often face both ethical and operational challenges. *ED Management* 2006 December; 18(12): 133-135. NRCBL: 9.4; 9.5.1.

Allan, Robert. Rationing of medical care by age [editorial]. *Clinical Medicine (London, England)* 2006 July-August; 6(4): 329-330. NRCBL: 9.4; 9.5.2; 9.2.

Alvarez, Allen Andrew A. Threshold considerations in fair allocation of health resources: justice beyond scarcity. *Bioethics* 2007 October; 21(8): 426-438. NRCBL: 9.4; 1.1; 21.1; 9.3.1. SC: an. Conference: Eighth World Congress of Bioethics: A Just and Healthy Society; Beijing, China; 2006 August 9.
> Abstract: Application of egalitarian and prioritarian accounts of health resource allocation in low-income countries have both been criticized for implying distribution outcomes that allow decreasing/undermining health gains and for tolerating unacceptable standards of health care and health status that result from such allocation schemes. Insufficient health care and severe deprivation of health resources are difficult to accept even when justified by aggregative efficiency or legitimized by fair deliberative process in pursuing equality and priority oriented outcomes. I affirm the sufficientarian argument that, given extreme scarcity of public health resources in low-income countries, neither health status equality between populations nor priority for the worse off is normatively adequate. Nevertheless, the threshold norm alone need not be the sole consideration when a country's total health budget is extremely scarce. Threshold considerations are necessary in developing a theory of fair distribution of health resources that is sensitive to the lexically prior norm of sufficiency. Based on the intuition that shares must not be taken away from those who barely achieve a minimal level of health, I argue that assessments based on standards of minimal physical/mental health must be developed to evaluate the sufficiency of the total resources of health systems in low-income countries prior to pursuing equality, priority, and efficiency based resource allocation. I also begin to examine how threshold sensitive health resource assessment could be used in the Philippines.

Bakalar, Nicholas. All breast cancer patients are not treated the same. *New York Times* 2007 January 23; p. F7. NRCBL: 9.4; 9.5.5; 9.5.4; 9.5.10. SC: po; em.

Banja, John D. The role of feelings in making moral decisions. *Case Manager* 2006 September-October; 17(5): 17-19, 62. NRCBL: 9.4; 1.1; 7.1.

Baxter, Nancy N. Equal for whom? Addressing disparities in the Canadian medical system must become a national

NRCBL: National Reference Center for Bioethics Literature Classification Scheme   See inside front cover for terms.

395

priority. *CMAJ/JAMC: Canadian Medical Association Journal* 2007 December 4; 177(12): 1522-1523. NRCBL: 9.4; 9.2; 9.5.5; 9.5.2; 9.5.4; 10.

Bierman, Arlene S. Sex matters: gender disparities in quality and outcomes of care. *CMAJ/JAMC: Canadian Medical Association Journal* 2007 December 4; 177(12): 1520-1521. NRCBL: 9.4; 9.2; 9.5.5; 10.

Blachar, Yoram; Borow, Malke. The health of leaders: information, interpretation and the media. *Israel Medical Association Journal* 2006 November; 8(11): 741-743. NRCBL: 9.4; 21.1; 1.3.7. Identifiers: Israel; Ariel Sharon.

Bond, Claire. Lieberman Award; Section 15 of the Charter and the allocation of resources in health care: a comment on Auton v. British Columbia. *Health Law Journal* 2005; 13: 253-271. NRCBL: 9.4; 9.5.3; 17.3. SC: le; cs.

Brindle, David. Seeing red [interview]. *BMJ: British Medical Journal* 2007 May 12; 334(7601): 976-977. NRCBL: 9.4; 9.3.2; 18.5.1. Identifiers: Great Britain (United Kingdom); National Health Service; Julian Tudor Hart.

Brock, Dan. Ethical issues in the use of cost effectiveness analysis for the prioritization of health resources. *In:* Khushf, George, ed. Handbook of Bioethics: Taking Stock of the Field From a Philosophical Perspective. Dordrecht; Boston: Kluwer Academic, 2004: 353-380. NRCBL: 9.4; 1.1.

Burgess, Diana J.; van Ryn, Michelle; Crowley-Matoka, Megan; Malat, Jennifer. Understanding the provider contribution to race/ethnicity disparities in pain treatment: insights from dual process models of stereotyping. *Pain Medicine* 2006 March-April; 7(2): 119-134. NRCBL: 9.4; 4.4; 9.5.4; 9.7.

Cappelen, Alexander W.; Norheim, Ole Frithjof. Responsibility, fairness and rationing in health care. *Health Policy* 2006 May; 76(3): 312-319. NRCBL: 9.4; 9.3.1; 5.2. SC: an.

Cassel, Christine K.; Brennan, Troyen E. Managing medical resources: return to the commons? [commentary]. *JAMA: The Journal of the American Medical Association* 2007 June 13; 297(22): 2518-2521. NRCBL: 9.4; 9.3.1. Identifiers: Tragedy of the Commons; Garrett Hardin.

Clarke, C. Resource allocation. *In:* Holland, Stephen, ed. Introducing Nursing Ethics: Themes in Theory and Practice. Salisbury: APS, 2004: 171-188. NRCBL: 9.4; 4.1.3; 8.1. SC: cs.

Claxton, Karl; Culyer, Anthony J. Rights, responsibilities and NICE: a rejoinder to Harris. *Journal of Medical Ethics* 2007 August; 33(8): 462-464. NRCBL: 9.4. Identifiers: Great Britain (United Kingdom); National Institute for Clinical Excellence. Comments: J. Harris. NICE is not cost-effective Journal of Medical Ethics 2006; 32: 378-380.

Abstract: Harris' reply to our defence of the National Institute for Clinical Excellence's (NICE) current cost-effectiveness procedures contains two further errors. First, he wrongly draws a conclusion from the fact that NICE does not and cannot evaluate all possible uses of healthcare resources at any one time and generally cannot know which National Health Service (NHS) activities would be displaced or which groups of patients would have to forgo health benefits: the inference is that no estimate is or can be made by NICE of the benefits to be forgone. This is a non-sequitur. Second, he asserts that it is a flaw at the heart of the use of quality-adjusted life years (QALYs) as an outcome measure that comparisons between people need to be made. Such comparisons do indeed have to be made, but this is not a consequence of the choice of any particular outcome measure, be it the QALY or anything else.

Douw, Karla; Vondeling, Hindrik; Oortwijn, Wija. Priority setting for horizon scanning of new health technologies in Denmark: views of health care stakeholders and health economists. *Health Policy* 2006 May; 76(3): 334-345. NRCBL: 9.4; 5.2; 7.1. SC: em.

Dyer, Clare. Charity challenges decision to refuse drug to 84 year old [news]. *BMJ: British Medical Journal* 2007 July 14; 335(7610): 64-65. NRCBL: 9.4; 9.7; 9.5.2.

Engelman, Michal; Johnson, Summer. Population aging and international development: addressing competing claims of distributive justice. *Developing World Bioethics* 2007 April; 7(1): 8-18. NRCBL: 9.4; 9.5.2; 21.1. SC: an.
Abstract: To date, bioethics and health policy scholarship has given little consideration to questions of aging and intergenerational justice in the developing world. Demographic changes are precipitating rapid population aging in developing nations, however, and ethical issues regarding older people's claim to scarce healthcare resources must be addressed. This paper posits that the traditional arguments about generational justice and age-based rationing of healthcare resources, which were developed primarily in more industrialized nations, fail to adequately address the unique challenges facing older persons in developing nations. Existing philosophical approaches to age-based resource allocation underemphasize the importance of older persons for developing countries and fail to adequately consider the rights and interests of older persons in these settings. Ultimately, the paper concludes that the most appropriate framework for thinking about generational justice in developing nations is a rights-based approach that allows for the interests of all age groups, including the oldest, to be considered in the determination of health resource allocation.

Fan, Eddy; Needham, Dale M. Deciding who to admit to a critical care unit [editorial]. *BMJ: British Medical Journal* 2007 December 1; 335(7630): 1103-1104. NRCBL: 9.4; 20.5.1.

Fleck, Leonard M. Just caring: the challenges of priority-setting in public health. *In:* Rhodes, Rosamond; Francis, Leslie P.; Silvers, Anita, eds. The Blackwell Guide to Medical Ethics. Malden, MA: Blackwell Pub., 2007: 323-340. NRCBL: 9.4; 9.1.

Flum, David R.; Khan, Tipu V.; Dellinger, E. Patchen. Toward the rational and equitable use of bariatric surgery. *JAMA: The Journal of the American Medical Association* 2007 September 26; 298(12): 1442-1444. NRCBL: 9.4; 9.5.1.

Foster, Charles. Simple rationality? The law of healthcare resource allocation in England. *Journal of Medical Ethics* 2007 July; 33(7): 404-407. NRCBL: 9.4. SC: le.
Abstract: This paper examines the law relating to healthcare resource allocation in England. The National Health Service (NHS) Act 1977 does not impose an absolute duty to provide specified healthcare services. The courts will only interfere with a resource allocation decision made by an NHS body if that decision is frankly irrational (or where the decision infringes the principle of proportionality when a right under the European Convention on Human Rights (ECHR) is engaged). Such irrationality is very difficult to establish. The ECHR has made no significant contribution to domestic English law in the arena of healthcare provision. The decision of the European Court in the Yvonne Watts case establishes that, in relation to the question of entitlement to seek treatment abroad at the expense of the NHS, a clinical judgment about the urgency of treatment trumps an administrative decision about waiting list targets. That decision goes against the grain of domestic law about healthcare allocation, but is not likely to have wide ramifications in domestic law.

Fowler, Robert A.; Sabur, Natasha; Li, Ping; Juurlink, David N.; Pinto, Ruxandra; Hladunewich, Michelle A.; Adhikari, Neill K.J.; Sibbald, William J.; Martin, Claudio M. Sex- and age-based differences in the delivery and outcomes of critical care. *CMAJ/JAMC: Canadian Medical Association Journal* 2007 December 4; 177(12): 1513-1519. NRCBL: 9.4; 9.2; 9.5.2; 9.5.5; 10. SC: em.

Francis, Leslie P. Discrimination in medical practice: justice and the obligations of health care providers to disadvantaged patients. *In:* Rhodes, Rosamond; Francis, Leslie P.; Silvers, Anita, eds. The Blackwell Guide to Medical Ethics. Malden, MA: Blackwell Pub., 2007: 162-179. NRCBL: 9.4; 1.1; 9.5.2; 9.5.3; 9.5.4; 9.5.5; 9.5.7; 9.5.10.

Guo-ping, Wang. Ethical evaluation of decision-making for distribution of health resources in China. *Medicine and Law: The World Association for Medical Law* 2007 June; 26(2): 283-289. NRCBL: 9.4.
Abstract: Since distribution of health resources involves various aspects of ethics, the evaluation of ethical problems should be emphasised in health decisions using criteria of fairness and fundamental principles of ethics correctly understood and chosen in order to solve the real conflicts evident in the distribution of health resources and to enable fair and reasonable distribution of health resources.

Harris, John. NICE rejoinder [comment]. *Journal of Medical Ethics* 2007 August; 33(8): 467. NRCBL: 9.4. Identifiers: Great Britain (United Kingdom); National Institute for Clinical Excellence. Comments: K. Claxton and A.J. Culyer. Wickedness or folly? The ethics of NICE's decisions. Journal of Medical Ethics 2006; 32: 375.

Howe, Edmund G. How should careproviders respond when the medical system leaves a patient short? *Journal of Clinical Ethics* 2007 Fall; 18(3): 195-205. NRCBL: 9.4; 8.1; 8.2; 8.4.

Hunter, David. Am I my brother's gatekeeper? Professional ethics and the prioritisation of healthcare. *Journal of Medical Ethics* 2007 September; 33(9): 522-526. NRCBL: 9.4; 4.1.2. SC: an.
Abstract: At the 5th International Conference on Priorities in Health Care in Wellington, New Zealand, 2004, one resonating theme was that for priority setting to be effective, it has to include clinicians in both decision making and the enforcement of those decisions. There was, however, a disturbing undertone to this theme, namely that doctors, in particular, were unjustifiably thwarting good systems of prioritising scarce healthcare resources. This undertone seems unfair precisely because doctors may, and in some cases do, feel obligated by their professional ethics to remain uninvolved either in deciding priorities and in some cases in enforcing them. I will argue that the professional role of a doctor ought not be considered inconsistent with the role of a priority setter or enforcer, as long as one crucial element is in place, a rationally coherent and broadly justifiable regime for prioritising healthcare. Given this I conclude both that prioritisation and doctoring are not incompatible under certain conditions, and that the education of healthcare professionals ought to include material on distributive justice in healthcare.

Hurst, Samia A.; Danis, Marion. A framework for rationing by clinical judgment. *Kennedy Institute of Ethics Journal* 2007 September; 17(3): 247-266. NRCBL: 9.4; 8.1; 4.1.2; 1.1. SC: an.
Abstract: Although rationing by clinical judgment is controversial, its acceptability partly depends on how it is practiced. In this paper, rationing by clinical judgment is defined in three different circumstances that represent increasingly wider circles of resource pools in which the rationing decision takes place: triage during acute shortage, comparison to other potential patients in a context of limited but not immediately strained resources, and determination of whether expected benefit of an intervention is deemed sufficient to warrant its cost by reference to published population based thresholds. Notions of procedural justice are applied along with an analytical framework of six minimal requisites in order to facilitate fair bedside rationing: (1) a closed system that offers reciprocity, (2) attention to general concerns of justice, (3) respect for individual variations, (4) application of a consistent process, (5) explicitness, and (6) review of

---

NRCBL: National Reference Center for Bioethics Literature Classification Scheme   See inside front cover for terms.

397

decisions. The process could be monitored for its applicability and appropriateness.

Jacobs, Lesley A. Justice in health care: can Dworkin justify universal access? *In:* Burley, Justine, ed. Dworkin and His Critics: With Replies by Dworkin. Malden, MA: Blackwell Pub., 2004: 134-149. NRCBL: 9.4; 1.1; 9.3.1. SC: an.

Kapiriri, Lydia; Norheim, Ole Frithjof; Martin, Douglas K. Priority setting at the micro-, meso- and macro-levels in Canada, Norway and Uganda. *Health Policy* 2007 June; 82(1): 78-94. NRCBL: 9.4; 21.1. SC: em.

Katz, Leo. Choice, consent, and cycling: the hidden limitations of consent. *Michigan Law Review* 2006 February; 104(4): 627-670. NRCBL: 9.4; 1.1; 8.3.1; 16.1. SC: le.

Lauridsen, S.M.R.; Norup, M.S.; Rossel, P.J.H. The secret art of managing healthcare expenses: investigating implicit rationing and autonomy in public healthcare systems. *Journal of Medical Ethics* 2007 December; 33(12): 704-707. NRCBL: 9.4; 9.3.1. SC: an.
Abstract: Rationing healthcare is a difficult task, which includes preventing patients from accessing potentially beneficial treatments. Proponents of implicit rationing argue that politicians cannot resist pressure from strong patient groups for treatments and conclude that physicians should ration without informing patients or the public. The authors subdivide this specific programme of implicit rationing, or "hidden rationing", into local hidden rationing, unsophisticated global hidden rationing and sophisticated global hidden rationing. They evaluate the appropriateness of these methods of rationing from the perspectives of individual and political autonomy and conclude that local hidden rationing and unsophisticated global hidden rationing clearly violate patients' individual autonomy, that is, their right to participate in medical decision-making. While sophisticated global hidden rationing avoids this charge, the authors point out that it nonetheless violates the political autonomy of patients, that is, their right to engage in public affairs as citizens. A defence of any of the forms of hidden rationing is therefore considered to be incompatible with a defence of autonomy.

Leget, Carlo; Hoedemaekers, R. Teaching medical students about fair distribution of healthcare resources. *Journal of Medical Ethics* 2007 December; 33(12): 737-741. NRCBL: 9.4; 7.2. SC: an. Identifiers: Dunning Committee; Netherlands.
Abstract: Healthcare package decisions are complex. Different judgements about effectiveness, cost-effectiveness and disease burden influence the decision-making process. Moreover, different concepts of justice generate different ideas about fair distribution of healthcare resources. This paper presents a decision model that is used in medical school in order to familiarise medical students with the different concepts of justice and the ethical dimension of making concrete choices. The model is based on the four-stage decision model developed in the Netherlands by the Dunning Committee and the discussion

that followed its presentation in 1991. Having to deal with 10 medical services, students working with the model learn to discern and integrate four different ideas of distributive justice that are integrated in a flow chart: libertarian, communitarian, egalitarian and utilitarian.

Levin, Philip D.; Sprung, Charles L. Intensive care triage — the hardest rationing decision of them all. *Critical Care Medicine* 2006 April; 34(4): 1250-1251. NRCBL: 9.4; 9.5.1; 21.1.

Marcus, Robert; Firth, John. Should you tell patients about beneficial treatments that they cannot have? [debate]. *BMJ: British Medial Journal* 2007 April 21; 334(7598): 826-827. NRCBL: 9.4; 8.2; 9.3.2; 9.7; 7.3.

Mason, J.K.; Laurie, G.T. Health resources and dilemmas in treatment. *In their:* Mason and McCall Smith's Law and Medical Ethics. Seventh ed. Oxford; New York: Oxford University Press, 2005: 412-440. NRCBL: 9.4; 4.4; 9.3.1; 9.5.1. SC: le. Identifiers: Great Britain (United Kingdom); NICE (National Institute for Health and Clinical Excellence).

McLachlan, Hugh V.; Swales, J. Kim. Health and health care — justice, rights and equality. *In their:* From the Womb to the Tomb: Issues in Medical Ethics. Glasgow, Scotland: Humming Earth, 2007: 187-249. NRCBL: 9.4; 9.1; 9.3.1; 14.1; 9.5.9. SC: an. Identifiers: Julian LeGrand.

Menon, Devidas; Stafinski, Tania; Martin, Douglas. Priority-setting for healthcare: who, how, and is it fair? *Health Policy* 2007 December; 84(2-3): 220-233. NRCBL: 9.4; 9.3.1; 7.1. SC: em. Identifiers: Alberta, Canada.

Menzel, Paul. Allocation of scarce resources. *In:* Rhodes, Rosamond; Francis, Leslie P.; Silvers, Anita, eds. The Blackwell Guide to Medical Ethics. Malden, MA: Blackwell Pub., 2007: 305-322. NRCBL: 9.4; 1.1.

Moskop, John C.; Iserson, Kenneth V. Triage in medicine, Part II: underlying values and principles. *Annals of Emergency Medicine* 2007 March; 49(3): 282-287. NRCBL: 9.4; 1.1.

Ozar, David T. Basic oral health needs: a public priority. *Journal of Dental Education* 2006 November; 70(11): 1159-1165. NRCBL: 9.4; 9.2; 9.1; 4.1.1.

Quigley, Muireann. A NICE fallacy. *Journal of Medical Ethics* 2007 August; 33(8): 465-466. NRCBL: 9.4. Identifiers: Great Britain (United Kingdom); National Institute for Health and Clinical Excellence. Comments: K. Claxton and A.J. Culyer. Wickedness or folly? The ethics of NICE's decisions. Journal of Medical Ethics 2006; 32: 375.
Abstract: A response is given to the claim by Claxton and Culyer, who stated that the policies of the National Institute for Health and Clinical Excellence (NICE) do not evaluate patients rather than treatments. The argument is made that the use of values such as quality of life and

life-years is ethically dubious when used to choose which patients ought to receive treatments in the National Health Service (NHS).

Rameix, Suzanne; Durand-Zaleski, Isabelle. Justice and the allocation of healthcare. *In:* Marinker, Marshall, ed. Constructive Conversations about Health: Policy and Values. Oxford; Seattle: Radcliffe Pub., 2006: 177-183. NRCBL: 9.4; 9.3.1; 21.1.

Rosén, Per. Public dialogue on healthcare prioritisation. *Health Policy* 2006 November; 79(1): 107-116. NRCBL: 9.4; 9.2; 7.1; 21.1. SC: em. Identifiers: Sweden.

Salek, Sam. Health economics and access to treatment. *In:* Gunning, Jennifer; Holm, Søren, eds. Ethics, law, and society. Volume 1. Aldershot, Hants, England; Burlington, VT: Ashgate, 2005: 59-64. NRCBL: 9.4; 9.3.1; 9.2.

Schramme, Thomas. The significance of the concept of disease for justice in health care. *Theoretical Medicine and Bioethics* 2007; 28(2): 121-135. NRCBL: 9.4; 1.1; 4.2. SC: an.

Abstract: In this paper, I want to scrutinise the value of utilising the concept of disease for a theory of distributive justice in health care. Although many people believe that the presence of a disease-related condition is a prerequisite of a justified claim on health care resources, the impact of the philosophical debate on the concept of disease is still relatively minor. This is surprising, because how we conceive of disease determines the amount of justified claims on health care resources. Therefore, the severity of scarcity depends on our interpretation of the concept of disease. I want to defend a specific combination of a theory of disease with a theory of distributive justice. A naturalist account of disease, together with sufficientarianism, is able to perform a gate-keeping function regarding entitlements to medical treatment. Although this combination cannot solve all problems of justice in health care, it may inform rationing decisions as well.

Silvers, Anita. Judgment and justice: evaluating health care for chronically ill and disabled patients. *In:* Rhodes, Rosamond; Francis, Leslie P.; Silvers, Anita, eds. The Blackwell Guide to Medical Ethics. Malden, MA: Blackwell Pub., 2007: 354-372. NRCBL: 9.4; 4.4; 9.5.2; 9.5.3.

Task Force on Values, Ethics, and Rationing in Critical Care (VERICC); Truog, Robert D.; Brock, Dan W.; Cook, Deborah J.; Danis, Marion; Luce, John M.; Rubenfeld, Gordon D.; Levy, Mitchell M. Rationing in the intensive care unit. *Critical Care Medicine* 2006 April; 34(4): 958-963; quiz 971. NRCBL: 9.4; 9.5.1; 7.2.

Ubel, Peter A. Confessions of a bedside rationer: commentary on Hurst and Danis. *Kennedy Institute of Ethics Journal* 2007 September; 17(3): 267-269. NRCBL: 9.4; 8.1; 9.3.1; 1.1. SC: cs.

Abstract: Samia Hurst and Marion Danis provide a thoughtful framework for how to judge the morality of bedside rationing decisions. In this commentary, I applaud Hurst and Danis for advancing the level of debate about bedside rationing. But when I attempt to apply the framework to my own clinical practice, I conclude that the framework comes up short.

Vannelli, Alberto; Battaglia, Luigi; Poiasina, Elia; Belli, Filiberto; Bonfanti, Giuliano; Gallino, Gianfrancesco; Vitellaro, Marco; De Dosso, Sara; Leo, Ermanno. The art of decision-making in surgery. To what extent does economics influence choice? *Chirurgia Italiana* 2006 November-December; 58(6): 717-722. NRCBL: 9.4; 9.3.1; 9.5.1.

Weiss, Gail Garfinkel. What would you do? New issues in medical ethics. *Medical Economics* 2006 August 18; 83(16): 56-61, 63-64. NRCBL: 9.4; 9.3.1; 7.1; 14.1.

RESUSCITATION ORDERS *See* EUTHANASIA AND ALLOWING TO DIE

RIGHT TO DIE *See* ASSISTED SUICIDE; EUTHANASIA AND ALLOWING TO DIE

RIGHT TO HEALTH CARE

Access to investigational drugs in the USA [editorial]. *Lancet* 2007 August 18-24; 370(9587): 540. NRCBL: 9.2; 9.7; 18.5.7; 20.5.1. SC: le. Identifiers: Abigail Burroughs; The Abigail Alliance for Better Access to Developmental Drugs.

Annas, George J. The right to health and the nevirapine case in South Africa. *In:* Gruskin Sofia; Grodin, Michael A.; Annas, George J.; Marks, Stephen P., eds. Perspectives on health and Human Rights. New York: Routledge, 2005: 497-505. NRCBL: 9.2; 9.5.6; 9.7; 9.5.7. SC: le.

Anya, Ike; Virgilio, Antonia; Defilippi, Loris; Moschochoritis, Konstantinos; Ravinetto, Raffaella; den Otter, Joost J.; Tavenier, Daniel. Right to health care for vulnerable migrants [letters]. *Lancet* 2007 September 8-14; 370(9590): 827-828. NRCBL: 9.2; 9.5.10; 21.1. Comments: Vulnerable migrants have a right to health. Lancet 2007 July 7; 370: 2.

Austin, Wendy. Using the human rights paradigm in health ethics: the problems and the possibilities. *In:* Tschudin, Verena, ed. Approaches to Ethics: Nursing Beyond Boundaries. New York: Butterworth-Heinemann, 2003: 105-114. NRCBL: 9.2; 21.1. SC: cs.

Baker, Brook K. Placing access to essential medicines on the human rights agenda. *In:* Cohen, Jillian Clare; Illingworth, Patricia; Schüklenk, Udo, eds. The Power of Pills: Social, Ethical and Legal Issues in Drug Development, Marketing, and Pricing. London; Ann Arbor, MI: Pluto, 2006: 239-248. NRCBL: 9.2; 9.7; 21.1; 9.5.6.

Beyrer, Chris; Villar, Juan Carlos; Suwanvanichkij, Voravit; Singh, Sonal; Baral, Stefan D.; Mills, Edward J. Neglected diseases, civil conflicts, and the right to

NRCBL: National Reference Center for Bioethics Literature Classification Scheme   See inside front cover for terms.

399

health. *Lancet* 2007 August 18-24; 370(9587): 619-627. NRCBL: 9.2; 21.1; 7.1.

Abstract: Neglected diseases remain one of the largest causes of disease and mortality. In addition to the difficulties in provision of appropriate drugs for specific diseases, many other factors contribute to the prevalence of such diseases and the difficulties in reducing their burden. We address the role that poor governance and politically motivated oppression have on the epidemiology of neglected diseases. We give case examples including filariasis in eastern Burma and vector-borne diseases (Chagas' disease, leishmaniasis, and yellow fever) in Colombia, we show the links between systematic human rights violations and the effects of infectious disease on health. We also discuss the role of researchers in advocating for and researching within oppressed populations.

Bilmore, Isabel. The "right to health" according to WHO. *In:* Wagner, Teresa; Carbone, Leslie, eds. Fifty Years after the Declaration: the United Nations' Record on Human Rights. Lanham, MD: University Press of America, 2001: 25-31. NRCBL: 9.2; 21.1; 14.1; 10; 21.1.

Birmontiene, Toma. The influence of the rulings of the Constitutional Court on the development of health law in Lithuania. *European Journal of Health Law* 2007 December; 14(4): 321-333. NRCBL: 9.2; 9.7; 1.3.2. SC: le.

Braveman, Paula. Health disparities and health equity: concepts and measurement. *Annual Review of Public Health* 2006; 27: 167-194. NRCBL: 9.2; 9.1; 9.5.1. SC: rv.

Catalanotto, Frank A. A welcome to the workshop on "Professional Promises: Hopes and Gaps in Access to Oral Health Care". *Journal of Dental Education* 2006 November; 70(11): 1120-1124. NRCBL: 9.2; 4.1.1.

Chambers, David W. Access denied; invalid password. *Journal of Dental Education* 2006 November; 70(11): 1146-1151. NRCBL: 9.2; 4.1.1.

Chandler, Michelle. The rights of the medically uninsured: an analysis of social justice and disparate health outcomes. *Journal of Health and Social Policy* 2006; 21(3): 17-36. NRCBL: 9.2; 9.3.1; 1.1; 9.2; 9.8.

Corsino, Bruce V.; Patthoff, Donald E., Jr. The Ethical and practical aspects of acceptance and Universal Patient Acceptance. *Journal of Dental Education* 2006 November; 70(11): 1198-1201. NRCBL: 9.2; 2.1; 4.1.1.

Crall, James J. Access to oral health care: professional and societal considerations. *Journal of Dental Education* 2006 November; 70(11): 1133-1138. NRCBL: 9.2; 9.5.7; 9.5.2; 4.1.1.

de Groot, Rolf. Right to health care and scarcity of resources. *In:* Gevers, J.K.M.; Hondius, E.H.; Hubben, J.H., eds. Health Law, Human Rights and the Biomedicine Convention: Essays in Honour of Henriette Roscam Abbing.

Leiden; Boston: Martinus Nijhoff Publishers, 2005: 49-59. NRCBL: 9.2; 9.4; 21.1. SC: le.

Dostal, O. Patient rights protection in the Czech Republic: challenges of a transition from communism to a modern legal system. *Medicine and Law: The World Association for Medical Law* 2007 March; 26(1): 75-84. NRCBL: 9.2; 8.3.1; 8.4; 8.5; 21.1. SC: le.

Abstract: The post-Communist countries in Central Europe, including the Czech Republic, underwent a rapid transformation of their legal systems, within which the concept of patient rights passed through revolutionary changes. This process however often left significant gaps in patient rights protection. There are practical difficulties for patients in defending their rights before the courts, such as problems with obtaining evidence and independent expert opinions, long delays and high costs of court proceedings, strict burden of proof rules and low compensation levels. Modern patient rights often collide with the systems of health care provision that are still unprepared for patient autonomy and responsibility. The experience gained in the transition process might be applicable also to other countries that undergo changes from traditional to modern system of patient rights protection.

Dute, Jos. The leading principles of the Convention on Human Rights and Biomedicine. *In:* Gevers, J.K.M.; Hondius, E.H.; Hubben, J.H., eds. Health Law, Human Rights and the Biomedicine Convention: Essays in Honour of Henriette Roscam Abbing. Leiden; Boston: Martinus Nijhoff Publishers, 2005: 3-12. NRCBL: 9.2; 1.1; 21.1. SC: le.

Evans, Caswell A. Eliminating oral health disparities: ethics workshop reactor comments. *Journal of Dental Education* 2006 November; 70(11): 1180-1183. NRCBL: 9.2; 4.1.1.

Forman, Lisa. Trade rules, intellectual property, and the right to health. *Ethics and International Affairs* 2007 Fall; 21(3): 337-357. NRCBL: 9.2; 5.3; 21.1.

Garetto, Lawrence P.; Yoder, Karen M. Basic oral health needs: a professional priority. *Journal of Dental Education* 2006 November; 70(11): 1166-1169. NRCBL: 9.2; 4.1.1.

Gruskin, Sofia, Mills, Edward J.; Tarantola, Daniel. History, principles, and practice of health and human rights. *Lancet* 2007 August 4-10; 370(9585): 449-455. NRCBL: 9.2; 21.1; 2.2. SC: le.

Abstract: Individuals and populations suffer violations of their rights that affect health and wellbeing. Health professionals have a part to play in reduction and prevention of these violations and ensuring that health-related policies and practices promote rights. This needs efforts in terms of advocacy, application of legal standards, and public-health programming. We discuss the changing views of human rights in the context of the HIV/AIDS epidemic and propose further development of the right to health by increased practice, evidence, and action.

Haiun, Delphine. The Israeli Patients' Rights Law: a discourse analysis of some main values. *Korot* 2003-2004; 17: 97-124, xi-xii. NRCBL: 9.2; 8.4; 4.4; 1.3.5. SC: le.

Hall, Peter Lawrence. Health care for refused asylum seekers in the UK [comment]. *Lancet* 2007 August 11-17; 370(9586): 466-467. NRCBL: 9.2; 21.1.

Hall, Peter L.; Sheather, Julian C. Asylum seekers' health rights: BMA is in denial [letter and reply]. *BMJ: British Medical Journal* 2007 September 29; 335(7621): 629-630. NRCBL: 9.2; 9.5.1. Identifiers: British Medical Association.

Heath, Iona. Let's get tough on the causes of health inequality. *BMJ: British Medical Journal* 2007 June 23; 334(7607): 1301. NRCBL: 9.2.

Hunt, Paul. Right to the highest attainable standard of health. *Lancet* 2007 August 4-10; 370(9585): 369-371. NRCBL: 9.2; 9.1; 9.8; 21.1. SC: le. Identifiers: International Covenant on Economic, Social and Cultural Rights; General Comment 14.

Kisely, Stephen; Smith, Mark; Lawrence, David; Cox, Martha; Campbell, Leslie Anne; Maaten, Sarah. Inequitable access for mentally ill patients to some medically necessary procedures. *CMAJ/JAMC: Canadian Medical Association Journal* 2007 March 13; 176(6): 779-784. NRCBL: 9.2; 9.3.1; 17.1. SC: em.

Largent, Beverly A. Reaction to Universal Patient Acceptance: the perspective of a private practice dentist. *Journal of Dental Education* 2006 November; 70(11): 1202-1207. NRCBL: 9.2; 4.1.1; 9.3.1; 9.5.10.

Loewy, Erich H.; Loewy, Roberta Springer. Framing issues in health care: do American ideals demand basic health care and other social necessities for all? *Health Care Analysis: An International Journal of Health Philosophy and Policy* 2007 December; 15(4): 261-271. NRCBL: 9.2; 1.1; 1.3.1; 1.3.5. SC: an.

Abstract: This paper argues for the necessity of universal health care (as well as universal free education) using a different argument than most that have been made heretofore. It is not meant to conflict with but to strengthen the arguments previously made by others. Using the second paragraph of the Declaration of Independence and the Preamble to the Constitution we argue that universal health care in this day and age has become a necessary condition if the ideals of life, liberty and the pursuit of happiness are to be more than an empty promise and if the discussion of "promoting of general welfare" in the preamble is to have any meaning.

Mangalore, Roshni; Knapp, Martin. Equity in mental health. *Epidemiologia e Psichiatria Sociale* 2006 October-December; 15(4): 260-266. NRCBL: 9.2; 17.1; 9.3.1; 1.1.

Mason, J.K.; Laurie, G.T. Health rights and obligations in the European Union. *In their:* Mason and McCall

Smith's Law and Medical Ethics. Seventh ed. Oxford; New York: Oxford University Press, 2005: 48-70. NRCBL: 9.2; 5.1; 21.1. SC: le.

Mouradian, Wendy E. Band-Aid solutions to the dental access crisis: conceptually flawed — a response to Dr. David H. Smith. *Journal of Dental Education* 2006 November; 70(11): 1174-1179. NRCBL: 9.2; 9.5.7; 4.1.1. Comments: David H. Smith. Band-Aid solutions to problems of access: their origins and limits. Journal of Dental Education 2006 November; 70(11): 1170-1173.

Newdick, Christopher. The positive side of healthcare rights. *In:* McLean, Sheila A.M., ed. First Do No Harm: Law, Ethics, and Healthcare. Aldershot, England; Burlington, VT: Ashgate, 2006: 573-586. NRCBL: 9.2; 9.3.1; 9.4. SC: le.

Niveau, Gérard. Relevance and limits of the principle of "equivalence of care" in prison medicine. *Journal of Medical Ethics* 2007 October; 33(10): 610-613. NRCBL: 9.2; 1.3.5; 9.5.1. SC: rv.

Abstract: The principle of "equivalence of care" in prison medicine is a principle by which prison health services are obliged to provide prisoners with care of a quality equivalent to that provided for the general public in the same country. It is cited in numerous national and international directives and recommendations. The principle of equivalence is extremely relevant from the point of view of normative ethics but requires adaptation from the point of view of applied ethics. From a clinical point of view, the principle of equivalence is often insufficient to take account of the adaptations necessary for the organization of care in a correctional setting. The principle of equivalence is cost-effective in general, but has to be overstepped to ensure the humane management of certain special cases.

Nys, Herman; Stultiëns, Loes; Borry, Pascal; Goffin, Tom; Dierickx, Kris. Patient rights in EU Member States after the ratification of the Convention on Human Rights and Biomedicine. *Health Policy* 2007 October; 83(2-3): 223-235. NRCBL: 9.2; 21.1. SC: le.

O'Toole, Brian. Promoting access to oral health care: more than professional ethics is needed. *Journal of Dental Education* 2006 November; 70(11): 1217-1220. NRCBL: 9.2; 4.1.1.

Ozar, David T. Applying systems thinking to oral health care: commentary on Dr. Patricia H. Werhane's article. *Journal of Dental Education* 2006 November; 70(11): 1196-1197. NRCBL: 9.2; 4.1.1; 9.1; 1.3.2. Comments: Patricia H. Werhane. Access, responsibility, and funding: a systems thinking approach to universal access to oral health. Journal of Dental Education 2006 November; 70(11): 1184-1195.

Ozar, David T. Ethics, access, and care. *Journal of Dental Education* 2006 November; 70(11): 1139-1145. NRCBL: 9.2; 4.1.1; 2.1.

NRCBL: National Reference Center for Bioethics Literature Classification Scheme   See inside front cover for terms.

401

Peltier, Bruce. Codes and colleagues: is there support for Universal Patient Acceptance. *Journal of Dental Education* 2006 November; 70(11): 1221-1225. NRCBL: 9.2; 6; 4.1.1.

Ruger, J.P. Ethics and governance of global health inequalities. *Journal of Epidemiology and Community Health* 2006 November; 60(11): 998-1003 [Online]. Accessed: http://jech.bmj.com/ cgi/reprint/60/11/998 [2007 November 13]. NRCBL: 9.2; 1.1; 9.4; 21.1. SC: rv.

Ruger, J.P. Rethinking equal access: agency, quality, and norms. *Global Public Health* 2007 January; 2(1): 78-96. NRCBL: 9.2; 1.1; 9.3.1; 9.4; 9.8. SC: an.

Selby, Patricia L. On whose conscience? Patient rights disappear under broad protective measures for conscientious objectors in health care. *University of Detroit Mercy Law Review* 2006 Summer; 83(4): 507-543. NRCBL: 9.2; 9.1; 8.1; 1.3.5; 1.2. SC: le.

Smith, David H. Band-Aid solutions to problems of access: their origins and limits. *Journal of Dental Education* 2006 November; 70(11): 1170-1173. NRCBL: 9.2; 1.3.5; 4.1.1.

Sreenivasan, Gopal. Health care and equality of opportunity. *Hastings Center Report* 2007 March-April; 37(2): 21-31. NRCBL: 9.2; 9.4; 9.3.1; 1.1; 7.1. SC: an.
  Abstract: One widely accepted way of justifying universal access to health care is to argue that access to health care is necessary to ensure health, which is necessary to provide equality of opportunity. But the evidence on the social determinants of health undermines this argument.

Torres, Mary Ann. The human right to health, national courts, and access to HIV/AIDS treatment: a case study from Venezuela. *In:* Gruskin Sofia; Grodin, Michael A.; Annas, George J.; Marks, Stephen P., eds. Perspectives on health and Human Rights. New York: Routledge, 2005: 507-516. NRCBL: 9.2; 9.5.6; 9.7. SC: le; cs. Identifiers: Cruz Bermúdez v. Ministerio de Sanidad y Asistencia Social.

Tuohey, John F. Making access a priority: ethics has a vital role in fostering collaboration in health care. *Health Progress* 2007 March-April; 88(2): 67-72. NRCBL: 9.2; 9.3.1; 1.2.

United Nations. Committee on Economic, Social, and Cultural Rights. General comment no. 14 (2000): the right to the highest attainable standard of health (Article 12 of the International Covenant on Economic, Social, and Cultural Rights). *In:* Gruskin Sofia; Grodin, Michael A.; Annas, George J.; Marks, Stephen P., eds. Perspectives on health and Human Rights. New York: Routledge, 2005: 473-495. NRCBL: 9.2; 21.1. SC: le.

Vallgårda, Signild. When are health inequalities a political problem? *European Journal of Public Health* 2006 December; 16(6): 615-616. NRCBL: 9.2; 21.1; 9.1.

Werhane, Patricia H. Access, responsibility, and funding: a systems thinking approach to universal access to oral health. *Journal of Dental Education* 2006 November; 70(11): 1184-1195. NRCBL: 9.2; 4.1.1; 9.1; 1.3.2.

Wilmot, Stephen. A fair range of choice: justifying maximum patient choice in the British National Health Service. *Health Care Analysis: An International Journal of Health Philosophy and Policy* 2007 June; 15(2): 59-72. NRCBL: 9.2; 1.1; 9.1; 9.3.1. SC: an.
  Abstract: In this paper I put forward an ethical argument for the provision of extensive patient choice by the British National Health Service. I base this argument on traditional liberal rights to freedom of choice, on a welfare right to health care, and on a view of health as values-based. I argue that choice, to be ethically sustainable on this basis, must be values-based and rational. I also consider whether the British taxpayer may be persuadable with regard to the moral acceptability of patient choice, making use of Rawls' theory of political liberalism in this context. I identify issues that present problems in terms of public acceptance of choice, and also identify a boundary issue with regard to public health choices as against individual choices.

World Health Organization [WHO]. Commission on Social Determinants of Health; Marmot, Michael. Achieving health equity: from root causes to fair outcomes. *Lancet* 2007 September 29-October 5; 370(9593): 1153-1163. NRCBL: 9.2; 1.1; 7.1; 21.1.

Yates, Tom; Crane, Rosie; Burnett, Angela. Rights and the reality of healthcare charging in the United Kingdom. *Medicine, Conflict and Survival* 2007 October-December; 23(4): 297-304. NRCBL: 9.2; 21.1.

RIGHTS *See* INTERNATIONAL HEALTH AND HUMAN RIGHTS; RIGHT TO HEALTH CARE

SCIENCE AND TECHNOLOGY *See* BIOMEDICAL RESEARCH/ SOCIAL CONTROL OF SCIENCE AND TECHNOLOGY; NANOTECHNOLOGY

SCIENTIFIC MISCONDUCT *See* BIOMEDICAL RESEARCH/ RESEARCH ETHICS AND SCIENTIFIC MISCONDUCT

SEX DETERMINATION
See also GENETIC COUNSELING; GENETIC SCREENING

Bandyopadhyay, Sutapa; Singh, Amarjeet. History of son preference and sex selection in India and in the west. *Bulletin of the Indian Institute of History of Medicine (Hyderabad)* 2003 July-December; 33(2): 149-167. NRCBL: 14.3; 10; 21.1; 15.2; 15.9. SC: le.

Dahl, E.; Beutel, M.; Brosig, B.; Grüssner, S.; Stöbel-Richter, Y.; Tinneberg, H.-R.; Brähler, E. Social sex selection and the balance of the sexes: empirical evidence from Germany, the UK, and the US. *Journal of Assisted Reproduction and Genetics* 2006 July-August; 23(7-8): 311-318. NRCBL: 14.3; 15.2; 21.1.

Grady, Denise. Girl or boy? As fertility technology advances, so does an ethical debate. *New York Times* 2007 February 6; p. F5, F10. NRCBL: 14.3; 14.4; 15.2. SC: po.
Keywords: *sex preselection; methods; organizational policies; professional organizations; international aspects; attitude of health personnel; Keyword Identifiers: American College of Obstetricians and Gynecologists; American Society for Reproductive Medicine

Grazi, Richard V.; Wolowelsky, Joel B. Addressing the idiosyncratic needs of Orthodox Jewish couples requesting sex selection by preimplantation genetic diagnosis (PGD). *Journal of Assisted Reproduction and Genetics* 2006 November-December; 23(11-12): 421-425. NRCBL: 14.3; 15.2; 1.2.

Heng, Boon Chin. Regulated family balancing by equalizing the sex-ratio of gender-selected births. *Journal of Assisted Reproduction and Genetics* 2006 July-August; 23(7-8): 319-320. NRCBL: 14.3; 15.2.

Herissone-Kelly, Peter. Parental love and the ethics of sex selection. *CQ: Cambridge Quarterly of Healthcare Ethics* 2007 Summer; 16(3): 326-335. NRCBL: 14.3; 1.1.

Herissone-Kelly, Peter. The "parental love" objection to nonmedical sex selection: deepening the argument. *CQ: Cambridge Quarterly of Healthcare Ethics* 2007 Fall; 16(4): 446-455. NRCBL: 14.3; 4.4; 1.1. SC: an.

Kluge, Eike-Henner W. Sex selection: some ethical and policy considerations. *Health Care Analysis: An International Journal of Health Philosophy and Policy* 2007 June; 15(2): 73-89. NRCBL: 14.3; 1.3.1. SC: an.
Abstract: Sex selection, which refers to the attempt to choose or control the sex of a child prior to its birth, has become the subject of increasing ethical scrutiny and many jurisdictions have criminalized it except for serious sex-linked diseases or conditions that cannot easily be ameliorated or remedied. This paper argues that such a blanket prohibition is ethically unwarranted because it is based on a flawed understanding of the difference between sexist values and mere sex-oriented preferences. It distinguishes between ethics and public policy, and suggests a way of allowing preference-based sex selection as a matter of public policy without permitting value-based sex selection. It further argues that medically-based sex selection should be publicly funded but that preference-base sex selection should not be paid for by society, and that the prohibition against value-based sex selection should be enforced through legislation that controls the licensing of health care facilities and through disciplinary procedures against health care professionals.

Kusum, K. Sex selection in India. *In:* Dennerstein, Lorraine, ed. Women's Rights and Bioethics. Paris: UNESCO, 2000: 50-58. NRCBL: 14.3; 10. SC: an; cs; le.

Long, Angela M. Why criminalizing sex selection techniques is unjust: an argument challenging conventional wisdom. *Health Law Journal* 2006; 14: 69-104. 149 fn. NRCBL: 14.3; 1.3.5; 15.2; 14.4. SC: le.
Keywords: *criminal law; *females; *justice; *legal aspects; *sex determination; *sex preselection; attitudes; advisory committees; freedom; infanticide; international aspects; justice; legal rights; males; moral policy; preimplantation diagnosis; prenatal diagnosis; selective abortion; social discrimination; Keyword Identifiers: *Canada; Assisted Human Reproduction Act (Canada); Canadian Charter of Rights and Freedoms; Royal Commission on New Reproductive Technologies (Canada)

Rogers, Wendy; Ballantyne, Angela; Draper, Heather. Is sex-selective abortion morally justified and should it be prohibited? *Bioethics* 2007 November; 21(9): 520-524. NRCBL: 14.3; 12.5.1. SC: an.
Abstract: In this paper we argue that sex-selective abortion (SSA) cannot be morally justified and that it should be prohibited. We present two main arguments against SSA. First, we present reasons why the decision for a woman to seek SSA in cultures with strong son-preference cannot be regarded as autonomous on either a narrow or a broad account of autonomy. Second, we identify serious harms associated with SSA including perpetuation of discrimination against women, disruption to social and familial networks, and increased violence against women. For these reasons, SSA should be prohibited by law, and such laws should be enforced. Finally, we describe additional strategies for decreasing son-preference. Some of these strategies rely upon highlighting the disadvantages of women becoming scarce, such as lack of brides and daughters-in-law to care for elderly parents. We should, however, be cautious not to perpetuate the view that the purpose of women is to be the consorts for, and carers of, men, and the providers of children. Arguments against SSA should be located within a concerted effort to ensure greater, deeper social and cultural equality between the sexes.

Schroeder, Doris. Editorial: rights and procreative liberty. *CQ: Cambridge Quarterly of Healthcare Ethics* 2007 Summer; 16(3): 325. NRCBL: 14.3; 1.1; 5.3.

Scully, Jackie Leach; Banks, Sarah; Shakespeare, Tom W. Chance, choice and control: lay debate on prenatal social sex selection. *Social Science and Medicine* 2006 July; 63(1): 21-31. NRCBL: 14.3.

Seavilleklein, Victoria; Sherwin, Susan. The myth of the gendered chromosome: sex selection and the social interest. *CQ: Cambridge Quarterly of Healthcare Ethics* 2007 Winter; 16(1): 7-19. 32 fn. NRCBL: 14.3; 15.1; 7.1; 10; 9.8. SC: an.
Keywords: *sex determination; *sex preselection; *moral obligations; advertising; ethical analysis; females; industry; males; methods; moral policy; parents; preimplantation diagnosis; prenatal diagnosis; public policy; reproductive technologies; selective abortion; sexuality; Western World

NRCBL: National Reference Center for Bioethics Literature Classification Scheme   See inside front cover for terms.

403

Tuffs, Annette. Doctors protest about fetal sex tests in early pregnancy. *BMJ: British Medial Journal* 2007 April 7; 334(7596): 712. NRCBL: 14.3. Identifiers: Germany.

Wolowelsky, Joel B.; Grazi, Richard V.; Brander, Kenneth; Freundel, Barry; Friedman, Michelle; Goldberg, Judah; Greenberger, Ben; Kaplan, Feige; Reichman, Edward; Zimmerman, Deena R. Sex selection and Halakhic ethics: a contemporary discussion. *Tradition* 2007 Spring; 40(1): 45-78. 60 fn. NRCBL: 14.3; 1.2; 15.2; 14.4.
> Keywords: *Jewish ethics; *preimplantation diagnosis; *sex determination; *sex preselection; advisory committees; embryo transfer; eugenics; family planning; marital relationship; motivation; parent child relationship; parents; prenatal diagnosis; public policy; reproduction; reproductive technologies; semen donors; Keyword Identifiers: Israel; United States

Zilberberg, Julie. Sex selection and restricting abortion and sex determination. *Bioethics* 2007 November; 21(9): 517-519. NRCBL: 14.3; 1.1; 10; 15.2; 12.4.2; 12.5.1.
> Abstract: Sex selection in India and China is fostered by a limiting social structure that disallows women from performing the roles that men perform, and relegates women to a lower status level. Individual parents and individual families benefit concretely from having a son born into the family, while society, and girls and women as a group, are harmed by the widespread practice of sex selection. Sex selection reinforces oppression of women and girls. Sex selection is best addressed by ameliorating the situations of women and girls, increasing their autonomy, and elevating their status in society. One might argue that restricting or prohibiting abortion, prohibiting sex selection, and prohibiting sex determination would eliminate sex selective abortion. But this decreases women's autonomy rather than increases it. Such practices will turn underground. Sex selective infanticide, and slower death by long term neglect, could increase. If abortion is restricted, the burden is placed on women seeking abortions to show that they have a legally acceptable or legitimate reason for a desired abortion, and this seriously limits women's autonomy. Instead of restricting abortion, banning sex selection, and sex determination, it is better to address the practice of sex selection by elevating the status of women and empowering women so that giving birth to a girl is a real and positive option, instead of a detriment to the parents and family as it is currently. But, if a ban on sex selective abortion or a ban on sex determination is indeed instituted, then wider social change promoting women's status in society should be instituted simultaneously.

SEX PRESELECTION *See* SEX DETERMINATION

SEXUALITY
See also MALPRACTICE AND PROFESSIONAL MISCONDUCT

Ben-Asher, Noa. The necessity of sex change: a struggle for intersex and transsex liberties. *Harvard Journal of Law and Gender* 2006 Winter; 29(1): 51-98. NRCBL: 10; 8.3.1; 8.3.2; 9.5.1. SC: le.

George, Alison. Body swap. *New Scientist* 2007 April 21-27; 194(2600): 40-43. NRCBL: 10; 8.3.2; 9.5.7; 9.7. Identifiers: gender identity disorder.

Gurney, Karen. Sex and the surgeon's knife: the family court's dilemma... informed consent and the specter of iatrogenic harm to children with intersex characteristics. *American Journal of Law and Medicine* 2007; 33(4): 625-661. NRCBL: 10; 8.3.2; 9.5.7. SC: le.

Hale, C. Jacob. Ethical problems with the mental health evaluation standards of care for adult gender variant prospective patients. *Perspectives in Biology and Medicine* 2007 Autumn; 50(4): 491-505. NRCBL: 10; 1.1; 9.8; 14.1; 17.1. SC: an.

Maharaj, N.R.; Dhai, A.; Wiersma, R.; Moodley, J. Intersex conditions in children and adolescents: surgical, ethical, and legal considerations. *Journal of Pediatric and Adolescent Gynecology* 2005 December; 18(6): 399-402. NRCBL: 10; 9.5.7; 9.5.1. SC: le; cs.

Manson, H. The role of the 'lifestyle' label and negative bias in the allocation of health resources for erectile dysfunction drugs: an ethics-based appraisal. *International Journal of Impotence Research* 2006 January-February; 18(1): 98-103.[see corrections in: International Journal of Impotence Research 2006 March-April; 18(2): 221]. NRCBL: 10; 9.7; 4.5; 9.5.2.

Parker, Richard G. Sexuality, health and human rights [editorial]. *American Journal of Public Health* 2007 June; 97(6): 972-973. NRCBL: 10; 9.5.6; 21.1.

Reis, Elizabeth. Divergence or disorder: the politics of naming intersex. *Perspectives in Biology and Medicine* 2007 Autumn; 50(4): 535-543. NRCBL: 10; 4.2. SC: an.

Steinmetzer, Jan; Groß, Dominik; Duncker, Tobias Heinrich. Ethische Fragen im Umgang mit transidenten Personen - Limitierende Faktoren des gegenwärtigen Konzepts von "Transsexualität" = Ethical problems concerning transgender persons: limiting factors of present concepts of "transsexualism". *Ethik in der Medizin* 2007 March; 19(1): 39-54. NRCBL: 10; 9.5.1.

Wiesen, Jonathan; Kulak, David. "Male and female He created them:" revisiting gender assignment and treatment in intersex children. *Journal of Halacha and Contemporary Society* 2007 Fall; (54): 5-29. NRCBL: 10; 9.5.7.

SOCIAL ASPECTS *See* ABORTION/ SOCIAL ASPECTS

SOCIAL CONTROL OF SCIENCE AND TECHNOLOGY *See* BIOMEDICAL RESEARCH/ SOCIAL CONTROL OF SCIENCE AND TECHNOLOGY

SOCIAL JUSTICE *See* RESOURCE ALLOCATION; RIGHT TO HEALTH CARE

SOCIOECONOMIC ASPECTS *See* GENETIC SCREENING/ SOCIOECONOMIC ASPECTS

SOCIOLOGY OF MEDICINE

American College of Obstetricians and Gynecologists. Committee on Ethics. The limits of conscientious refusal in reproductive medicine. *ACOG Committee Opinion* 2007 November; (385): 1-6 [Online]. Accessed: http://www.acog.org/from_home/publications/ethics/co385.pdf [2008 March 25]. NRCBL: 7.1; 8.1; 8.3.1;.

American Medical Association. Council on Ethical and Judicial Affairs; Taub, Sara; Morin, Karine; Goldrich, Michael S.; Ray, Priscilla; Benjamin, Regina. Physician health and wellness. *Occupational Medicine (Oxford, England)* 2006 March; 56(2): 77-82. NRCBL: 7.1; 7.4; 9.5.9.

American Society of Clinical Oncology. Interactions with the investment industry: practical and ethical implications. *Journal of Clinical Oncology* 2007 January 20; 25(3): 338-340. NRCBL: 7.1; 9.3.1; 1.3.2.

Barilan, Y. Michael. Contemporary art and ethics of anatomy. *Perspectives in Biology and Medicine* 2007 Winter; 50(1): 104-123. NRCBL: 7.1.

Barker, P. Mental health nursing: the craft of the impossible? *Journal of Psychiatric and Mental Health Nursing* 2006 August; 13(4): 385-387. NRCBL: 7.1; 4.1.3; 17.1; 18.2.

Bouknight, Heyward H., III. Between the scalpel and the lie: comparing theories of physician accountability for misrepresentations of experience and competence. *Washington and Lee Law Review* 2003 Fall; 60(4): 1515-1560. NRCBL: 7.1; 8.3.1; 8.1; 9.8; 8.5. SC: le.

Charon, Rita. The bioethics of narrative medicine. *In her:* Narrative Medicine: Honoring the Stories of Illness. New York: Oxford University Press, 2006: 203-218. NRCBL: 7.1; 2.1; 1.1; 8.1; 18.3; 20.4.1.

Ebbesen, Mette; Pedersen, Birthe D. Empirical investigation of the ethical reasoning of physicians and molecular biologists — the importance of the four principles of biomedical ethics. *Philosophy, Ethics, and Humanities in Medicine [electronic]* 2007 October 25; 2(23): 16 p. Accessed: http://www.peh-med.com/ [2008 January 24]. NRCBL: 7.1; 1.1; 1.3.9; 8.1. SC: em. Identifiers: Tom. L Beauchamp; James F. Childress; Denmark.

Endacott, Ruth; Wood, Anita; Judd, Fiona; Hulbert, Carol; Thomas, Ben; Grigg, Margaret. Impact and management of dual relationships in metropolitan, regional and rural mental health practice. *Australian and New Zea-*land Journal of Psychiatry* 2006 November-December; 40(11-12): 987-994. NRCBL: 7.1; 7.3; 8.1; 17.1.

Freeman, Jeanine. The ethics of epidemics — when healing puts the doctor's life in danger. *Iowa Medicine* 2006 January-February; 96(1): 10-11. NRCBL: 7.1; 6.

Garlin, Amy B.; Goldschmidt, Ronald. Pregnant physicians and infectious disease risk. *American Family Physician* 2007 January 1; 75(1): 112, 114. NRCBL: 7.1; 14.1; 8.1; 16.3.

International Dual Loyalty Working Group; Physicians for Human Rights; University of Cape Town. Health Services Faculty. Dual loyalty and human rights in health professional practice: proposed guidelines and institutional mechanisms. Boston: Physicians for Human Rights, 2002; 136 p. [Online]. Accessed: http://physiciansforhumanrights.org/library/documents/reports/report-2002-duelloyalty.pdf [2007 April 25]. NRCBL: 7.1; 21.1.

King, Patricia A.; Areen, Judith; Gostin, Lawrence O. Private control of science and medicine. *In their:* Law, Medicine and Ethics. New York: Foundation Press, 2006: 112-207. NRCBL: 7.1; 6; 8.2; 8.3.1; 18.2; 16.3; 2.1. SC: le.

Kory, Deborah. How psychologists aid torture. AlterNet (Independent Media Institute) 2007 June 27: 5 p. Accessed: http://www.alternet.org/story/ 55308 [2007 June 27]. NRCBL: 7.1; 21.4; 17.1; 4.1.1.

McLean, Michelle; Naidoo, Soornarain S. Medical students' views on the white coat: a South African perspective on ethical issues. *Ethics and Behavior* 2007 December; 17(4): 387-402. NRCBL: 7.1; 7.2. SC: em.
Abstract: There is a debate regarding the use of the white coat, a traditional symbol of the medical profession, by students. In a study evaluating final-year South African medical students' perceptions, the white coat was associated with traditional symbolic values (e.g., trust) and had practical uses (e.g., identification). The coat was generally perceived to evoke positive emotions in patients, but some recognized that it may cause anxiety or mistrust. Donning a white coat generally implied a responsibility to the profession. For a few, without the coat, patients would not cooperate, resulting in some perceiving no need to be distinguished from qualified practitioners. There was thus some evidence of entitled (vs. earned) respect. In the light of the underresourced health care setting in which these students learn clinical medicine, we recommend that students be able to recognize the potential for unprofessional or unethical behavior. Students should also be able to identify role models.

Qureshi, K.; Gershon, R.R.M.; Sherman, M.F.; Straub, T.; Gebbie, E.; McCollum, M.; Erwin, M.J.; Morse, S.S. Health care workers' ability and willingness to report to duty during catastrophic disasters. *Journal of Urban Health: Bulletin of the New York Academy of Medicine* 2005 September; 82(3): 378-388. NRCBL: 7.1; 4.1.1.

NRCBL: National Reference Center for Bioethics Literature Classification Scheme   See inside front cover for terms.

405

Rees, C.E.; Knight, L.V. "The stroke is eighty nine": understanding unprofessional behaviour through physician-authored prose. *Medical Humanities* 2007 June; 33(1): 38-43. NRCBL: 7.1; 7.4; 9.8.

Abstract: The unprofessional behaviour of medics is explored through their depiction in two physician-authored books—the novel Bodies and the autobiography Bedside stories: confessions of a junior doctor. Using the Integrative Model of Behavioural Prediction, not only the range and nature of professionalism lapses outlined in these books but also the reasons behind such unprofessional behaviours are examined. The books contained examples of lapses in professionalism outlined in research investigating the unprofessional behaviour of medical students, such as communication violations, objectification of patients and causing harm to patients. More interestingly, various reasons behind lapses in professionalism were found. Most examples of unprofessional behaviour were unintentional acts and therefore due to environmental constraints and skill deficits. Seemingly intentional acts were largely influenced by normative beliefs—that is, people feeling pressurised to act unprofessionally. Further research is needed to examine the depiction of lapses in professionalism in a wider range of physician-authored prose.

Rie, Michael A.; Kofke, W. Andrew. Nontherapeutic quality improvement: the conflict of organizational ethics and societal rule of law. *Critical Care Medicine* 2007 February; 35(2, Supplement): S66-S84. NRCBL: 7.1; 4.1.2; 9.3.1; 9.2; 9.4; 18.2.

Rubin, M.H. Is there a doctor in the house? *Journal of Medical Ethics* 2007 March; 33(3): 158-159. NRCBL: 7.1; 8.1; 4.1.2.

SAEM Ethics Committee; Schears, R.M.; Watters, A.; Schmidt, T.A.; Marco, C.A.; Larkin, G.L.; Marshall, J.P.; Mason, J.D.; McBeth, B.D.; Mello, M.J. The Society for Academic Emergency Medicine position on ethical relationships with the biomedical industry. *Academic Emergency Medicine* 2007 February; 14(2): 179-181. NRCBL: 7.1; 9.5.1; 1.3.2.

Schwenzer, Karen J.; Wang, Lijuan. Assessing moral distress in respiratory care practitioners. *Critical Care Medicine* 2006 December; 34(12): 2967-2973. NRCBL: 7.1; 4.1.3.

Sokol, Daniel K. Virulent epidemics and scope of healthcare workers' duty of care. *Emerging Infectious Diseases* 2006 August; 12(8): 1238-1241. NRCBL: 7.1; 9.1; 8.1.

Solskone, Colin L.; Light, Andrew. Towards ethics guidelines for environmental epidemiologists. *Science of the Total Environment* 1996; 184(1-2): 137-147. NRCBL: 7.1; 16.1; 6; 4.1.1.

Steers, Edward, Jr. Dr. Mudd and the "colored" witnesses. *Civil War History* 2000; 46(4): 324-336. NRCBL: 7.1; 9.2.

Thornton, Tim. Tacit knowledge as the unifying factor in evidence based medicine and clinical judgement. *Philosophy, Ethics, and Humanities in Medicine [electronic]* 2006; 1(2): 10 p. Accessed: http://www.peh-med.com/content/1/1/2 [2006 June 15]. NRCBL: 7.1; 7.2; 1.1; 18.1. SC: an.

Van Groenou, Aneema A.; Bakes, Katherine Mary. Art, Chaos, Ethics, and Science (ACES): a doctoring curriculum for emergency medicine. *Annals of Emergency Medicine* 2006 November; 48(5): 532-537. NRCBL: 7.1; 7.2.

Williams, Hywel C.; Naldi, Luigi; Paul, Carle; Vahlquist, Anders; Schroter, Sara; Jobling, Ray. Conflicts of interest in dermatology. *Acta Dermato-Venereologica* 2006; 86(6): 485-497. NRCBL: 7.1; 1.3.2; 9.3.1; 1.3.7.

Wolf, J.H.; Foley, P. Hans Gerhard Creutzfeldt (1885-1964): a life in neuropathology. *Journal of Neural Transmission* 2005 August; 112(8): I-XCVII. NRCBL: 7.1; 9.5.1; 17.1.

Xenakis, B.G. Stephen. Military medical ethics under attack. *Journal of Ambulatory Care Management* 2006 October-December; 29(4): 342-344. NRCBL: 7.1; 1.3.5; 21.1; 21.4.

SPECIAL POPULATIONS *See* CARE FOR SPECIFIC GROUPS; HUMAN EXPERIMENTATION/ SPECIAL POPULATIONS

SPERM *See* CRYOBANKING OF SPERM, OVA AND EMBRYOS

STEM CELL RESEARCH

An inconvenient truth: research on human embryonic stem cells must go on [editorial]. *Nature* 2007 November 29; 450(7170): 585-586. NRCBL: 18.5.4; 15.1; 18.5.1.
Keywords: *adult stem cells; *embryo research; *embryonic stem cells; international aspects; regulation

Avoiding a chimaera quagmire [editorial]. *Nature* 2007 January 4; 445(7123): 1. NRCBL: 15.1; 18.1; 22.1.
Keywords: *chimeras; *embryo research; *stem cells; human dignity; nuclear transfer techniques; primates; regulation; risk; Keyword Identifiers: Great Britain

Criteria creep: the politically motivated extension of US stem-cell registry makes no scientific sense [editorial]. *Nature* 2007 October 18; 449(7164): 756. NRCBL: 18.5.4; 15.1.
Keywords: *embryonic stem cells; *pluripotent stem cells; *public policy; *registries; adult stem cells; cell lines; embryo research; federal government; research support; Keyword Identifiers: United States; National Institutes of Health

German stem-cell law should change, says ethics council [news]. *Nature* 2007 July 26; 448(7152): 399. NRCBL: 18.5.4; 15.1. SC: le.
Keywords: *embryo research; *embryonic stem cells; *government regulation; *legal aspects; advisory commit-

tees; cell lines; public policy; Keyword Identifiers: *Germany; National Ethics Council (Germany)

New Jersey Catholic hospitals support stem cell research, promote cord blood donation. *Health Care Ethics USA* 2007 Winter; 15(1): 15. NRCBL: 18.5.1; 15.1; 19.4; 9.5.7; 1.2.

> Keywords: *adult stem cells; *cord blood; *institutional policies; *religious hospitals; biomedical research; blood banks; blood donation; Roman Catholic ethics; Keyword Identifiers: *New Jersey

Timeout, not the final buzzer, in the stem cell debate [editorial]. *Lancet* 2007 December 1-7; 370(9602): 1802. NRCBL: 18.5.1; 15.1; 18.5.4.

Agar, Nicholas. Embryonic potential and stem cells. *Bioethics* 2007, May; 21(4): 198-207. 23 fn. NRCBL: 18.5.4; 1.1; 4.4; 15.1; 19.5. SC: an.

> Keywords: *embryo research; *embryonic stem cells; *embryos; *ethical analysis; *moral status; beginning of life; cloning; embryo disposition; embryonic development; in vitro fertilization; intention; nuclear transfer techniques; philosophy; Proposed Keywords: *blastocysts

Abstract: This paper examines three arguments that use the concept of potential to identify embryos that are morally suitable for embryonic stem cell research (ESCR). According to the first argument, due to Ronald Green, the fact that they are scheduled for disposal makes embryos left over from IVF treatments morally appropriate for research. Paul McHugh argues that embryos created by somatic cell nuclear transfer differ from those that result directly from the meeting of sperm and egg in having potential especially conducive to the therapeutic use of their stem cells. I reject both of these arguments. According to the way of making distinctions in embryonic potential that I defend, it is the absence of a functional relationship with a womb that marks embryos morally suitable for ESCR.

American Society of Human Genetics [ASHG]. Statement on stem cell research. Bethesda, MD: American Society of Human Genetics, 2001 August 27: 1 p. [Online]. Accessed: http://genetics.faseb.org/genetics/ashg/pubs/policy/pol-44.htm [2006 July 26]. NRCBL: 18.5.4; 15.1; 6.

> Keywords: *embryo research; *embryonic stem cells; *stem cells; advisory committees; government financing; organizational policies; professional organizations; public policy; Keyword Identifiers: United States

Aramesh, Kiarash; Dabbagh, Soroush. An Islamic view to stem cell research and cloning: Iran's experience. *American Journal of Bioethics* 2007 February; 7(2): 62. 5 refs. NRCBL: 14.5; 18.5.4; 15.1; 1.2. Comments: Comment on: Timothy Caulfield and Tania Bubela. Why a criminal ban? Analyzing the arguments against somatic cell nuclear transfer in the Canadian parliamentary debate. American Journal of Bioethics 2007 February; 7(2): 51-61.

> Keywords: *cloning; *embryo research; *embryonic stem cells; *Islamic ethics; reproductive technologies; Keyword Identifiers: *Iran

Atala, Anthony. Stem cells — brave new world? *Journal of Urology* 2005 December; 174(6): 2085. NRCBL: 15.1; 18.5.4; 18.5.1.

> Keywords: *stem cells; embryonic stem cells; embryo research; federal government; pluripotent stem cells; research support; Keyword Identifiers: United States

Baker, Monya. Stem cells by any other name [news]. *Nature* 2007 September 27; 449(7161): 389. NRCBL: 18.5.4; 15.1; 1.3.12; 19.5. Identifiers: Human Pluripotent Stem Cell Registry.

Balint, John A. Ethical issues in stem cell research. *Albany Law Review* 2002; 65(3): 729-742. NRCBL: 18.5.4; 15.1; 14.5; 1.2.

Barnes, Richard. Stem cell research funding: testimony. *Origins* 2007 March 15; 36(39): 616-620. 10 refs. NRCBL: 18.5.4; 15.1; 1.2; 9.3.1.

> Keywords: *embryo research; *embryonic stem cells; *government financing; *research support; *Roman Catholic ethics; *state government; adult stem cells; biotechnology; cloning; conflict of interest; economics; embryos; industry; legislation; public opinion; public policy; value of life; Keyword Identifiers: *New York; United States

Baylis, Françoise; Fenton, Andrew. Chimera research and stem cell therapies for human neurodegenerative disorders. *CQ: Cambridge Quarterly of Healthcare Ethics* 2007 Spring; 16(2): 195-208. 71 fn. NRCBL: 15.1; 18.1; 22.1; 17.1; 4.4; 18.5.4. SC: an.

> Keywords: *chimeras; *embryonic stem cells; *ethical analysis; *human dignity; *moral policy; *moral status; *primates; *stem cell transplantation; animal welfare; clinical trials; guidelines; human characteristics; speciesism; therapeutic research; risks and benefits; Proposed Keywords: *neurodegenerative diseases; Keyword Identifiers: National Academy of Sciences

Baylis, Françoise; McLeod, C. The stem cell debate continues: the buying and selling of eggs for research. *Journal of Medical Ethics* 2007 December; 33(12): 726-731. 41 refs. NRCBL: 19.5; 14.5; 1.3.9; 18.5.4; 15.1; 18.5.3; 9.3.1. SC: an.

> Keywords: *embryo research; *embryonic stem cells; *guidelines; *ovum donors; *remuneration; cloning; embryo disposition; ethical analysis; disadvantaged persons; in vitro fertilization; incentives; international aspects; organizational policies; ovum; patients; professional organizations; research subjects; risk; socioeconomic factors; women; Keyword Identifiers: *ISSCR Guidelines; *International Society for Stem Cell Research

Baylis, Françoise; Ram, Natalie. Eligibility of cryopreserved human embryos for stem cell research in Canada. *JOGC: Journal of Obstetrics and Gynaecology Canada = JOGC:Journal d'Obstétrique et Gynécologie du Canada* 2005 October; 27(10): 949-955. NRCBL: 18.3; 18.5.4; 15.1; 14.6. SC: em. Note: Abstract in French.

Beeson, Diane; Lippman, Abby. Egg harvesting for stem cell research: medical risks and ethical problems. *Reproductive Biomedicine Online* 2006 October; 13(4): 573-579. 50 refs. NRCBL: 15.2; 18.5.4; 9.5.5; 14.4; 14.5.

NRCBL: National Reference Center for Bioethics Literature Classification Scheme   See inside front cover for terms.

407

Keywords: *embryonic stem cells; *ovum donors; *research embryo creation; *risk; cloning; conflict of interest; remuneration; reproduction; researchers; women's health; Proposed Keywords: tissue harvesting

Bell, Leanne; Devaney, Sarah. Gaps and overlaps: improving the current regulation of stem cells in the UK. *Journal of Medical Ethics* 2007 November; 33(11): 621-622. NRCBL: 18.5.1; 15.1; 19.1. SC: le.

Benítez-Bribiesca, Luis; Modiano-Esquenazi, Marcos. Ethics of scientific publication after the human stem cell scandal [editorial]. *Archives of Medical Research* 2006 May; 37(4): 423-424. NRCBL: 1.3.7; 1.3.9.

Bergman, Karl; Graff, Gregory D. The global stem cell patent landscape: implications for efficient technology transfer and commercial development. *Nature Biotechnology* 2007 April; 25(4): 419-425. 28 refs. NRCBL: 15.8; 15.1. SC: em.
Keywords: *international aspects; *patents; *stem cells; access to information; databases; embryonic stem cells; industry; interinstitutional relations; private sector; public sector; research institutes; statistics; universities; Proposed Keywords: *technology transfer; licensure; Keyword Identifiers: United States

Bobbert, Monika. Ethical questions concerning research on human embryos, embryonic stem cells and chimeras. *Biotechnology Journal* 2006 December; 1(12): 1352-1369. NRCBL: 18.5.4; 15.1; 22.1; 15.8; 4.4.

Bobrow, James C. The ethics and politics of stem cell research. *Transactions of the American Ophthalmological Society* 2005; 103: 138-141; discussion 141-142. NRCBL: 18.5.4; 15.1; 18.5.1; 18.6.

Brainard, Jeffrey. California stem-cell researchers ponder next steps after court victory. *Chronicle of Higher Education* 2007 June 1; 53(39): A20. NRCBL: 18.5.4; 15.1; 1.3.9.
Keywords: *government financing; *research support; *state government; *stem cells; *universities; biomedical research; conflict of interest; embryo research; laboratories; legal aspects; patient advocacy; politics; public policy; research institutes; Keyword Identifiers: *California; Stanford University; University of California, San Francisco

Brainard, Jeffrey. NIH director calls for easing administration's stem-cell restrictions [news]. *Chronicle of Higher Education* 2007 March 30; 53(30): A26. NRCBL: 18.5.4; 15.1.
Keywords: *embryo research; *embryonic stem cells; *government financing; *public policy; *research support; cell lines; government regulation; Keyword Identifiers: *United States; *Zerhouni, Elias; National Institutes of Health

Brown, Susan. China challenges the west in stem-cell research: unconstrained by public debate, cities like Shanghai and Beijing lure scientists with new laboratories and grants. *Chronicle of Higher Education* 2007 April 13; 53(32): A14-16, A18. NRCBL: 18.5.4; 15.1; 21.1; 1.3.9.

Brown, Susan. International group proposes guidelines for embryonic-stem-cell research. *Chronicle of Higher Education* 2007 February 16; 53(24): A21. NRCBL: 18.5.4; 15.1; 18.2.

Burke, William; Pullicino, Patrick; Richard, Edward J. The biological basis of the oocyte assisted reprogramming (OAR) hypothesis: is it an ethical procedure for making embryonic stem cells? *Linacre Quarterly* 2007 August; 74(3): 204-212. NRCBL: 18.5.4; 15.1.

Canadian Institutes of Health Research [CIHR]. Human pluripotent stem cell research: recommendations for CIHR-funded research: report of the ad hoc Working Group on Stem Cell Research. Ottawa: Canadian Institutes of Health Research, 2002 January; 21 p. [Online]. Accessed: http://www.cihr-irsc.gc.ca/e/1489.html [2007 April 19]. NRCBL: 18.5.4; 15.1; 5.3.
Keywords: *embryo research; *embryonic stem cells; *government financing; *guidelines; *public policy; *research support; cell lines; commerce; confidentiality; disclosure; ethical review; germ cells; informed consent; policy making; public participation; review committees; standards; stem cell transplantation; stem cells; Keyword Identifiers: *Canada; *Canadian Institutes of Health Research

Canadian Institutes of Health Research [CIHR]. Updated guidelines for human pluripotent stem cell research. Ottawa: Canadian Institutes of Health Research, 2006 June 28; 8 p. [Online]. Accessed: http://www.cihr-irsc.gc.ca/e/31488.html [2007 April 23]. NRCBL: 18.5.4; 15.1; 5.3.
Keywords: *biomedical research; *cell lines; *embryo research; *embryonic stem cells; *guidelines; *public policy; *stem cells; confidentiality; conflict of interest; directed donation; donors; ethical review; fetal research; industry; informed consent; international aspects; registries; remuneration; refusal to participate; research embryo creation; research support; researchers; Keyword Identifiers: *Canada; *Canadian Institutes of Health Research; Tri-Council Policy Statement on Ethical Conduct for Research Involving Humans

Carlson, Elof Axel. Cloning, stem cells, hyperbole, and cant. *In his:* Times of Triumph, Times of Doubt: Science and the Battle for Public Trust. Cold Spring Harbor, NY: Cold Spring Harbor Laboratory Press, 2006: 163-171. 9 fn. NRCBL: 14.5; 18.5.4; 15.1.
Keywords: *cloning; *embryo research; *embryonic stem cells; motivation; reproductive technologies; stem cell transplantation; twinning; Proposed Keywords: blastocysts

Caulfield, Timothy. Stem cells, clones, consensus, and the law. *In:* Knowles, Lori P.; Kaebnick, Gregory E., eds. Reprogenetics: Law, Policy, and Ethical Issues. Baltimore: Johns Hopkins University Press, 2007: 105-123. 73 refs. NRCBL: 14.5; 18.5.4; 15.1; 14.1. SC: an.
Keywords: *cloning; *cultural pluralism; *embryo research; *genetic techniques; *policy making; *public opinion; *public policy; *regulation; *reproductive technologies; advisory committees; biotechnology; consensus; criminal law; dissent; embryonic stem cells; embryos; focus groups; government regulation; human dignity; international aspects; legal aspects; moral status; nuclear transfer

techniques; policy analysis; religion; religious ethics; risks and benefits; survey; values; Keyword Identifiers: *United States; Canada; United Nations

Cheema, Puneet; Mehta, Paulette. Pediatric stem cell transplantation ethical concerns. *In:* Mehta, P., ed. Pediatric Stem Cell Transplantation. Sudbury: Jones and Bartlet Publishers, 2004: p. 91-98. NRCBL: 19.1; 8.3.2; 19.4; 19.5; 18.5.4; 9.5.7; 1.1.

Chekar, Choon Key; Kitzinger, Jenny. Science, patriotism and discourses of nation and culture: reflections on the South Korean stem cell breakthroughs and scandals. *New Genetics and Society* 2007 December; 26(3): 289-307. NRCBL: 18.5.4; 15.1; 14.5; 1.3.9. Identifiers: Woo Suk Hwang.

Cibelli, Jose. Is therapeutic cloning dead? The ability to generate pluripotent stem cells directly from skin fibroblasts may render ethical debates over the use of human oocytes to create stem cells irrelevant. *Science* 2007 December 21; 318(5858): 1879-1880. 17 refs. NRCBL: 14.5; 18.5.4; 15.1.
  Keywords: *cloning; *stem cells; embryonic stem cells; genetic techniques; methods; nuclear transfer techniques; Proposed Keywords: *pluripotent stem cells

Civin, Curt I.; Rao, Mahendra S. How many human embryonic stem cell lines are sufficient? A. U.S. perspective. *Stem Cells* 2006 April; 24(4): 800-803 [Online]. http://www.StemCells.com/cgi/content/full/24/4/800 [2007 December 4]. 30 refs. NRCBL: 15.2; 18.5.4; 18.6; 5.3.
  Keywords: *cell lines; *embryo research; *embryonic stem cells; beginning of life; cloning; donors; embryo disposition; genetic diversity; government regulation; informed consent; research support; stem cell transplantation; Keyword Identifiers: *United States

Cohen, Cynthia B. Beyond the human neuron mouse to the NAS guidelines. *American Journal of Bioethics* 2007 May; 7(5): 46-49. 12 refs. NRCBL: 15.1; 18.1; 22.1; 18.5.4. Comments: Comment on: Henry T. Greely, Mildred K. Cho, Linda F. Hogle, and Debra M. Satz. Thinking about the human neuron mouse. American Journal of Bioethics 2007 May; 7(5): 27-40.
  Keywords: *chimeras; *guidelines; *stem cell transplantation; advisory committees; animal experimentation; brain; embryonic stem cells; embryos; fetuses; human characteristics; moral policy; policy analysis; policy making; precautionary principle; primates; public policy; regulation; research ethics; risks and benefits; Proposed Keywords: blastocysts; mice; species specificity; Keyword Identifiers: National Academies of Sciences; Stanford University; United States

Colman, Alan; Burley, Justine. Stem cells: recycling the abnormal [news]. *Nature* 2007 June 7; 447(7145): 649-650. NRCBL: 18.5.1; 18.5.4; 15.1.
  Keywords: *embryonic stem cells; *methods; *nuclear transfer techniques; chromosome abnormalities; cloning; embryonic development; ovum donors; primates; Proposed Keywords: blastocysts; mice

Condic, Maureen L. What we know about embryonic stem cells. *First Things* 2007 January; (169): 25-29. NRCBL: 18.5.4; 15.1; 14.5; 18.5.1; 22.2; 19.1.

Condic, Maureen L.; Furton, Edward J. Harvesting embryonic stem cells from deceased human embryos. *National Catholic Bioethics Quarterly* 2007 Autumn; 7(3): 507-525. 35 fn. NRCBL: 18.5.4; 15.1; 19.5; 20.2.1; 14.4; 18.3; 1.2.
  Keywords: *death; *determination of death; *embryo disposition; *embryonic stem cells; *embryos; *moral policy; cryopreservation; embryo research; human dignity; in vitro fertilization; methods; model legislation; moral complicity; parental consent; public policy; Roman Catholic ethics; Proposed Keywords: *embryo death

Cyranoski, David. Race to mimic human embryonic stem cells [news]. *Nature* 2007 November 22; 450(7169): 462-463. NRCBL: 14.5; 18.5.4; 15.1; 22.2.
  Keywords: *cloning; *embryonic stem cells; *methods; adult stem cells; embryo research; nuclear transfer techniques; primates; stem cell transplantation; Proposed Keywords: *pluripotent stem cells; regenerative medicine

Cyranoski, David. Stem-cell fraudster 'is working in Thailand' [news]. *Nature* 2007 September 27; 449(7161): 387. NRCBL: 18.5.4; 15.1; 14.5; 18.1; 22.1. Identifiers: Woo Suk Hwang.

Daley, George Q.; Ahrlund-Richter, Lars; Auerbach, Jonathan M.; Benvenisty, Nissim; Charo, R. Alta; Chen, Grace; Deng, Hong-kui; Goldstein, Lawrence S.; Hudson, Kathy L.; Hyun, Insoo; Junn, Sung Chull; Love, Jane; Lee, Eng Hin; McLaren, Anne; Mummery, Christine L.; Nakatsuji, Norio; Racowsky, Catherine; Rooke, Heather; Rossant, Janet; Schöler, Hans R.; Solbakk, Jan Helge; Taylor, Patrick; Trounson, Alan O.; Weissman, Irving L.; Wilmut, Ian; Yu, John; Zoloth, Laurie. The ISSCR guidelines for human embryonic stem cell research. *Science* 2007 February 2; 315(5812): 603-604. 12 refs. NRCBL: 18.5.4; 15.1. Identifiers: International Society for Stem Cell Research.
  Keywords: *embryo research; *embryonic stem cells; *guidelines; *professional organizations; access to information; chimeras; cloning; donors; editorial policies; ethical review; germ cells; guideline adherence; informed consent; international aspects; regulation; remuneration; researchers; Proposed Keywords: pluripotent stem cells; Keyword Identifiers: *International Society for Stem Cell Research; National Academy of Sciences; United States

de Melo-Martín, Immaculada; Rosenwaks, Zev; Fins, Joseph J. New methods for deriving embryonic stem cell lines: are the ethical problems solved? *Fertility and Sterility* 2006 November; 86(5): 1330-1332. NRCBL: 15.2; 18.5.4; 15.3.

de Wert, Guido; Liebaers, Inge; Van de Velde, Hilde. The future (r)evolution of preimplantation genetic diagnosis/human leukocyte antigen testing: ethical reflections. *Stem Cells* 2007 September; 25: 2167-2172 [Online]. Accessed: http://www.StemCells.com/cgi/content/full/25/9/

NRCBL: National Reference Center for Bioethics Literature Classification Scheme   See inside front cover for terms.

409

2167 [2007 December 4]. 47 refs. NRCBL: 15.2; 14.4; 18.5.4.

> Keywords: *embryonic stem cells; *preimplantation diagnosis; *moral policy; *stem cells; *tissue typing; adverse effects; classification; family members; ethical analysis; forecasting; motivation; reproduction; research embryo creation; risks and benefits; stem cell transplantation; trends; Proposed Keywords: *hematopoietic stem cells

Dickens, B.M.; Cook, R.J. Acquiring human embryos for stem-cell research. *International Journal of Gynecology and Obstetrics* 2007 January; 96(1): 67-71. NRCBL: 18.5.4; 15.1; 18.3; 9.5.5.

Doerflinger, Richard M. Washington insider: House passes amended Genetic Nondiscrimination Bill; continued impasse on stem cell legislation, new executive order; defeat of deceptive human cloning bill; Supreme Court decision on partial-birth abortion. *National Catholic Bioethics Quarterly* 2007 Autumn; 7(3): 455-463. 21 fn. NRCBL: 15.2; 8.4; 18.5.4; 14.5. SC: le.

> Keywords: *cloning; *embryo research; *embryonic stem cells; *genetic discrimination; *legislation; abortion; adult stem cells; cell lines; federal government; government financing; government regulation; politics; reproductive technologies; research support; Keyword Identifiers: *U. S. Congress; *United States; Genetic Information Nondiscrimination Act; Human Cloning Prohibition Act; Partial Birth Abortion Ban Act; Stem Cell Research Enhancement Act

Doerflinger, Richard M. Washington insider: 2006 in Congress; Senate Hearing on Misrepresentations in stem cell research; Opening battle of 2007; Genetic Nondiscrimination Bill may see action. *National Catholic Bioethics Quarterly* 2007 Spring; 7(1): 15-21. NRCBL: 18.5.4; 15.1; 18.5.1; 1.2; 15.3. SC: le. Identifiers: H.R. 810 — Stem Cell Research Enhancement Act.

Downey, Robin; Geransar, Rose; Einsiedel, Edna. Angles of vision: stakeholders and human embryonic stem cell policy development. *In:* Einsiedel, Edna; Timmermans, Frank, eds. Crossing Over: Genomics in the Public Arena. Calgary, Alberta, Canada: University of Calgary Press, 2005: 61-84. 69 refs. NRCBL: 18.5.4; 15.1; 19.5; 5.3. Conference: Essays from the conference held Apr. 25-27, 2003, Kananaskis, Alta.

> Keywords: *embryo research; *embryonic stem cells; *policy making; *public policy; advisory committees; biomedical research; cloning; democracy; government financing; government regulation; guidelines; historical aspects; international aspects; legal aspects; legislation; patient advocacy; political activity; research support; researchers; Proposed Keywords: *stakeholders; lobbying; Keyword Identifiers: *Canada; Canadian Institutes of Health Research; Royal Commission on New Reproductive Technologies; Right to Life Movement; Twentieth Century

Eberl, Jason T. Issues at the beginning of human life: abortion, embryonic stem cell research, and cloning. *In his:* Thomistic Principles and Bioethics. London; New York: Routledge, 2006: 62-94. 19 fn. NRCBL: 12.3; 1.1; 4.4; 14.5; 18.5.4; 15.1. SC: an.

> Keywords: *abortion; *beginning of life; *cloning; *embryo research; *embryos; *moral status; abortifacients; adult stem cells; embryo disposition; fetal stem cells; in vitro fertilization; moral complicity; natural law; personhood; philosophy; reproductive technologies; Proposed Keywords: blastocysts; Keyword Identifiers: *Thomas Aquinas

Ecker, Jeffrey L.; O'Rourke, Patricia Pearl; Lott, Jason P.; Savulescu, Julian. An immodest proposal: banking embryonic stem cells for solid organ transplantation is problematic and premature. *American Journal of Bioethics* 2007 August; 7(8): 48-50; author reply W4-W6. 7 refs. NRCBL: 18.5.4; 15.1; 19.5. Comments: Comment on: Jason P. Lott and Julian Savulescu. Towards a global human embryonic stem cell bank. American Journal of Bioethics 2007 August; 7(8): 37-44.

> Keywords: *embryonic stem cells; *tissue banks; donors; incentives; mandatory programs; moral policy; organ transplantation; remuneration; resource allocation; risks and benefits; scarcity; standards; stem cell transplantation; utilitarianism; Proposed Keywords: embryo donation

Feldmann, Robert E., Jr.; Mattern, Rainer. The human brain and its neural stem cells postmortem: from dead brains to live therapy. *International Journal of Legal Medicine* 2006 July; 120(4): 201-211. NRCBL: 18.5.1; 15.1; 17.1; 20.1.

Fielder, John. Sex and stem cell research. *IEEE Engineering in Medicine and Biology Magazine* 2006 November-December; 25(6): 96-98. NRCBL: 18.5.4; 15.1; 14.1; 10.

Fossett, James W. Federalism by necessity: state and private support for human embryonic stem cell research. Rockefeller Institute Policy Brief 2007 August 9: 1-13 [Online]. Accessed: http://www.rockinst.org/WorkArea/showcontent.aspx?id=12064 [2007 October 25]. 24 fn. NRCBL: 5.3; 18.5.4; 15.1. SC: em; le.

> Keywords: *embryo research; *embryonic stem cells; *government financing; *private sector; *research support; *state government; federal government; government regulation; public policy; statistics; stem cells; Keyword Identifiers: *United States

Franklin, Sarah. Embryonic economies: the double reproductive value of stem cells. *Biosocieties* 2006 March; 1(1): 71-90. 97 refs. NRCBL: 18.5.4; 15.1; 19.5; 14.1; 18.3.

> Keywords: *embryo disposition; *embryo research; *embryonic stem cells; *in vitro fertilization; biological specimen banks; cell lines; donors; economics; motivation; public policy; regulation; trends; Proposed Keywords: *embryo donation; Keyword Identifiers: Great Britain

Franklin, Sarah. Ethical biocapital: new strategies of cell culture. *In:* Franklin, Sarah; Lock, Margaret, eds. Remaking Life and Death: Toward an Anthropology of the Biosciences. Santa Fe: School of American Research Press; Oxford: James Currey, 2003: 97-127. 20 fn. NRCBL: 15.1; 1.3.11; 14.5; 22.3.

> Keywords: *animal cloning; *biotechnology; *genetic engineering; *industry; *stem cells; capitalism; cell lines; embryo research; embryonic stem cells; gene therapy; genetically modified food; genetically modified organisms;

nuclear transfer techniques; patents; stem cell transplantation; trust; Proposed Keywords: domestic animals; regenerative medicine; sheep; Keyword Identifiers: Geron Corp.; Great Britain; United States

Genetics Committee of the Society of Obstetricians and Gynaecologists of Canada; Wilson, R. Douglas; Desilets, Valerie; Gagnon, Alain; Summers, Anne; Wyatt, Philip; Allen, Victoria; Langlois, Sylvie. Present role of stem cells for fetal genetic therapy = Rôle actuel des cellules souches en matière de thérapie génique fœtale. *JOGC: Journal of Obstetrics and Gynaecology Canada = JOGC: Journal d'Obstétrique et Gynécologie du Canada* 2005 November; 27(11): 1038-1047. NRCBL: 18.5.4; 15.1; 15.4; 19.5.

Giacomini, Mita; Baylis, Françoise; Robert, Jason. Banking on it: public policy and the ethics of stem cell research and development. *Social Science and Medicine* 2007 October; 65(7): 1490-1500. NRCBL: 18.5.4; 15.1; 18.5.1; 5.2; 5.3.

Glenn, David. Questions plague landmark paper on versatility of adult stem cells. *Chronicle of Higher Education* 2007 March 9; 53(27): A22. NRCBL: 18.5.1; 15.1; 1.3.9; 1.3.7.

Gratwohl, Alois. "Therapeutisches Klonen" aus der Sicht eines Klinikers. *In:* Schreiber, Hans-Peter, ed. Biomedizin und Ethik: Praxis, Recht, Moral. Basel; Boston: Birkhäuser, 2004: 23-28. NRCBL: 14.5; 18.5.4.
   Keywords: *cloning; *stem cell transplantation; adult stem cells; embryonic stem cells; reproductive technologies

Great Britain (United Kingdom). Department of Health. UK Stem Cell Initiative. UK stem cell initiative: report and recommendations. London: UK Stem Cell Initiative, 2005 November; 118 p. [Online]. Accessed: http://www.advisorybodies.doh.gov.uk/uksci/uksci-reportnov05.pdf [2007 April 26]. NRCBL: 18.5.4; 5.3; 15.1; 19.1; 19.5.

Green, Ronald M. Can we develop ethically universal embryonic stem-cell lines? *Nature Reviews Genetics* 2007 June; 8(6): 480-485. NRCBL: 18.5.4; 15.1; 19.5; 4.4; 15.4. SC: em.

Green, Shane K.; Lott, Jason P.; Savulescu, Julian. Is Canada's stem cell legislation unwittingly discriminatory? *American Journal of Bioethics* 2007 August; 7(8): 50-52; author reply W4-W6. 4 refs. NRCBL: 18.5.4; 15.1; 19.5. SC: le. Comments: Comment on: Jason P. Lott and Julian Savulescu. Towards a global human embryonic stem cell bank. American Journal of Bioethics 2007 August; 7(8): 37-44.
   Keywords: *embryonic stem cells; *legislation; *tissue banks; cell lines; cloning; donors; government regulation; guidelines; in vitro fertilization; incentives; international aspects; justice; minority groups; moral policy; nuclear transfer techniques; public policy; remuneration; scarcity; social discrimination; Proposed Keywords: embryo donation; haplotypes; Keyword Identifiers: *Canada

Greif, Karen F.; Merz, Jon F. Brave new world revisited: human cloning and stem cells; the Asilomar Conference on Recombinant DNA: a model for self-regulation? *In their:* Current Controversies in the Biological Sciences: Case Studies of Policy Challenges from New Technologies. Cambridge, MA: MIT, 2007: 101-115. 54 refs. NRCBL: 14.5; 18.5.4; 15.1.
   Keywords: *cloning; *embryo research; *embryonic stem cells; *public policy; *recombinant DNA research; *regulation; advisory committees; government regulation; reproductive technologies; self regulation; Keyword Identifiers: *United States; Asilomar Conference; President's Council on Bioethics; Recombinant DNA Advisory Committee

Gruen, Lori; Grabel, Laura. Concise review: scientific and ethical roadblocks to human embryonic stem cell therapy. *Stem Cells* 2006 October; 24(10): 2162-2169 [Online]. Accessed: http://stemcell.alphamedpress.org/cgi/content/full/24/10/2162 [2007 December 4]. 72 refs. NRCBL: 15.2; 18.5.4; 5.3; 18.6. SC: rv.
   Keywords: *embryo research; *embryonic stem cells; *stem cell transplantation; advisory committees; cell lines; federal government; research support; state government; Proposed Keywords: altered nuclear transfer; blastocysts

Guenin, Louis M. A proposed stem cell research policy. *Stem Cells* 2005 September; 23(8): 1023-1027. NRCBL: 18.5.4; 15.1; 18.6. SC: le.

Hagen, John D., Jr. Bentham's mummy and stem cells. *America* 2007 May 14; 196(17): 12-14. NRCBL: 18.5.4; 15.1; 1.1; 4.4.

Hall, Vanessa J.; Stojkovic, Petra; Stojkovic, Miodrag. Using therapeutic cloning to fight human disease: a conundrum or reality? *Stem Cells* 2006 July; 24(7): 1628-1637. 99 refs. NRCBL: 14.5; 18.5.4; 15.1; 18.2.
   Keywords: *cloning; *embryonic stem cells; *stem cell transplantation; adult stem cells; adverse effects; beginning of life; cell lines; commodification; in vitro fertilization; international aspects; legislation; ovum donors; reproductive technologies; research support; Proposed Keywords: parthenogenesis; Keyword Identifiers: Australia; Canada; Europe; Japan; Mexico

Hansson, Mats G.; Helgesson, Gert; Wessman, Richard; Jaenisch, Rudolf. Commentary: isolated stem cells — patentable as cultural artifacts? *Stem Cells* 2007 June; 25(6): 1507-1510 [Online]. Accessed: http://www.StemCells.com/cgi/content/full/25/6/1507 [2007 December 3]. 31 refs. NRCBL: 15.8; 15.1; 18.5.4; 1.1.
   Keywords: *embryonic stem cells; *patents; adult stem cells; commodification; donors; embryo research; embryos; informed consent; international aspects; legal aspects; moral status; public policy; Keyword Identifiers: Europe

Heng, Boon Chin. Donation of surplus frozen embryos for stem cell research or fertility treatment — should medical professionals and healthcare institutions be allowed to exercise undue influence on the informed decision of their former patients? *Journal of Assisted Reproduction and Genetics* 2006 September-October; 23(9-10): 381-382. NRCBL: 14.6; 15.1; 18.5.4; 7.1; 8.3.1.

NRCBL: National Reference Center for Bioethics Literature Classification Scheme   See inside front cover for terms.

411

Hinkley, Charles C. Stem cell research. *In his:* Moral Conflicts of Organ Retrieval: A Case for Constructive Pluralism. Amsterdam; New York: Rodopi, 2005: 127-133. NRCBL: 18.5.4; 15.1;.

Holbrook, Daniel. All embryos are equal?: Issues in pre-implantation genetic diagnosis, IVF implantation, embryonic stem cell research, and therapeutic cloning. *International Journal of Applied Philosophy* 2007 Spring; 21(1): 43-53. 12 fn. NRCBL: 14.4; 15.2; 18.5.4; 15.1; 14.6; 14.5; 4.4. SC: an.

> Keywords: *beginning of life; *embryos; *moral status; cloning; cryopreservation; embryo disposition; embryo research; embryonic stem cells; in vitro fertilization; moral obligations; moral policy; personhood; preimplantation diagnosis; reproductive technologies; value of life

Holden, Constance. Prominent researchers join the attack on stem cell patents [news]. *Science* 2007 July 13; 317(5835): 187. NRCBL: 15.8; 18.5.4; 15.1.

> Keywords: *cell lines; *embryonic stem cells; *legal aspects; *patents; attitudes; researchers; universities; Proposed Keywords: consumer advocacy; Keyword Identifiers: *Wisconsin Alumni Research Foundation; Patent and Trademark Office; Thomson, James; University of Wisconsin; United States

Holden, Constance. U.S. Patent Office casts doubt on Wisconsin stem cell patents [news]. *Science* 2007 April 13; 316(5822): 182. NRCBL: 15.8; 15.1; 18.5.4. Identifiers: Wisconsin Alumni Research Foundation.

> Keywords: *cell lines; *embryonic stem cells; *patents; industry; legal aspects; universities; Proposed Keywords: foundations; licensure; technology transfer; Keyword Identifiers: *Patent and Trademark Office; *Wisconsin Alumni Research Foundation; United States

Holt, Rush. How should government regulate stem-cell research? Views from a scientist- legislator. *In:* Santoro, Michael A.; Gorrie, Thomas M., eds. Ethics and the Pharmaceutical Industry. Cambridge; New York: Cambridge University Press, 2005: 109-122, 427-431. 53 refs. NRCBL: 18.5.4; 15.1. SC: le.

> Keywords: *embryo research; *legal aspects; *stem cells; cloning; embryonic stem cells; federal government; government financing; in vitro fertilization; international aspects; legislation; public policy; reproductive technologies; state government; stem cell transplantation; Keyword Identifiers: *United States; Europe; Asia

Hurlbut, William B. Stem cells, embryos, and ethics: is there a way forward? *Update (Loma Linda Center)* 2007 January 21(3): 1-10. 5 refs. NRCBL: 18.5.4; 15.1; 4.4; 1.2. Note: Adapted from the Health and Faith Forum: Bioethics and Wholeness Grand Rounds presentation, 2007 January 10.

> Keywords: *beginning of life; *embryo research; *embryonic stem cells; *embryos; *moral policy; alternatives; cloning; government financing; in vitro fertilization; moral status; nuclear transfer techniques; public policy; twinning; value of life; Proposed Keywords: regenerative medicine; Keyword Identifiers: United States

Hynes, Richard. Reply to 'UK set to reverse stance on research with chimeras' [letter]. *Nature Medicine* 2007 October; 13(10): 1133. NRCBL: 18.5.4; 15.1; 18.1; 22.1; 18.6. Comments: Comment on: Michael Hopkin. UK set to reverse stance on research with chimeras. Nature Medicine 2007 August; 13(8): 890-891.

> Keywords: *chimeras; *research embryo creation; *stem cells; *terminology; embryo research; editorial policies; public policy; Proposed Keywords: pluripotent stem cells; Keyword Identifiers: *Great Britain; Nature Medicine

Jaenisch, Rudolf. Nuclear cloning, embryonic stem cells, and gene transfer. *In:* Rasko, John E.J.; O'Sullivan, Gabrielle M.; Ankeny, Rachel A., eds. The Ethics of Inheritable Genetic Modification: A Dividing Line? Cambridge: Cambridge University Press, 2006: 35-55. 43 fn. NRCBL: 14.5; 15.4; 18.5.4; 15.1.

> Keywords: *cloning; *embryonic stem cells; *gene therapy; *risks and benefits; adverse effects; animal cloning; chimeras; embryo disposition; embryo research; embryonic development; gene transfer techniques; human experimentation; nuclear transfer techniques; reproductive technologies

Jersild, Paul. Theological and moral reflections on stem cell research. *Journal of Lutheran Ethics [electronic]* 2007 March; 7(3): 4 p. Accessed: http://www.elca.org/jle/article.asp?k=705 [2007 February 28]. 7 fn. NRCBL: 18.5.4; 15.1; 1.2.

> Keywords: *embryonic stem cells; *embryos; *Protestant ethics; *value of life; beginning of life; embryo research; ethical analysis; moral status; personhood; Roman Catholic ethics; stem cell transplantation; theology

Johnson, Luke. Embryonic stem cell research: a legitimate application of just-war theory? *Ethics and Medicine: An International Journal of Bioethics* 2007 Spring; 23(1): 19-30. 14 refs. 7 fn. NRCBL: 18.5.4; 15.1; 21.2; 1.2; 1.1. SC: an.

> Keywords: *embryo research; *embryonic stem cells; *ethical theory; *moral policy; *war; embryos; intention; killing; moral status; rights

Kaczor, Christopher Robert. An ethical assessment of Bush's guidelines for stem cell research. *In his:* The Edge of Life: Human Dignity and Contemporary Bioethics. Dordrecht: Springer, 2005: 83-96. Includes references. NRCBL: 18.5.4; 15.1. SC: an.

> Keywords: *double effect; *embryo research; *embryonic stem cells; *ethical analysis; *government financing; *moral complicity; *moral policy; *public policy; *value of life; adult stem cells; allowing to die; beginning of life; cell lines; common good; embryos; federal government; guidelines; human dignity; killing; moral status; parental consent; personhood; Roman Catholic ethics; Keyword Identifiers: Bush, George; United States

Kahn, Jeffrey P. Organs and stem cells: policy lessons and cautionary tales. *Hastings Center Report* 2007 March-April; 37(2): 11-12. 4 fn. NRCBL: 18.5.4; 15.1; 19.5; 19.6.

> Keywords: *embryonic stem cells; *organ transplantation; *property rights; *public policy; *social control; biomedical technologies; body parts and fluids; cadavers; consensus; determination of death; embryo research; embryos; government regulation; kidneys; living donors; moral status; organ

donors; policy making; remuneration; resource allocation; Proposed Keywords: stakeholders; Keyword Identifiers: United Network for Organ Sharing; United States

Kimmelman, Jonathan; Lott, Jason P.; Savulescu, Julian. Towards a global human embryonic stem cell bank: differential termination. *American Journal of Bioethics* 2007 August; 7(8): 52-53; author reply W4-W6. 6 refs. NRCBL: 18.5.4; 15.1; 19.5. Comments: Comment on: Jason P. Lott and Julian Savulescu. Towards a global human embryonic stem cell bank. American Journal of Bioethics 2007 August; 7(8): 37-44.

> Keywords: *embryonic stem cells; *tissue banks; donors; incentives; international aspects; justice; minority groups; moral policy; remuneration; stem cell transplantation; Proposed Keywords: embryo donation

Kirby, Neil. Treatment or crime? the status of stem cell therapies and research in South African law. *Medicine and Law: The World Association for Medical Law* 2007 March; 26(1): 95-115. 54 fn. NRCBL: 19.5; 18.3; 8.3.2; 4.4; 18.5.4; 15.1. SC: le.

> Keywords: *cord blood; *fetal stem cells; *legal aspects; *parental consent; biomedical research; DNA; legislation; newborns; property rights; stem cell transplantation; Proposed Keywords: *South Africa

Abstract: The author develops a thorough analysis of current and proposed South African law in relation to the harvesting and use of stem cells. He begins with the question of ownership of the umbilical cord at birth and afterwards. The problems of informed consent in these situations are discussed. Changes in the law in South Africa, now in progress, should ameliorate some of the difficulties.

Kitzinger, Jenny; Williams, Clare. Forecasting the future: legitimizing hope and calming fears in the embryo stem-cell debate. *In:* Deane-Drummond, Celia; Scott, Peter Manley, eds. Future Perfect?: God, Medicine and Human Identity. New York: T and T Clark International, 2006: 129-142. 14 fn. NRCBL: 18.5.4; 15.1; 1.3.7. SC: em.

> Keywords: *embryo research; *embryonic stem cells; *mass media; biotechnology; editorial policies; genetic research; journalism; survey; Keyword Identifiers: Great Britain

Kolata, Gina. Researcher who helped start stem cell war may now end it [news]. *New York Times* 2007 November 22; p. A1, A28. NRCBL: 14.5; 15.1; 18.5.4; 19.5. SC: po. Identifiers: James A. Thomson.

> Keywords: *adult stem cells; *embryonic stem cells; *methods; embryo research; researchers; Keyword Identifiers: *Thomson, James A.; University of Wisconsin; United States

Kolata, Gina. Scientists bypass need for embryo to get stem cells; method using human skin is seen as defusing the debate over ethics [news]. *New York Times* 2007 November 21; p. A1, A23. NRCBL: 18.5.1; 15.1; 19.1; 18.5.4. SC: po. Identifiers: Japan; Wisconsin; Shinya Yamanaka; James A. Thomson.

Keywords: *adult stem cells; *methods; cloning; embryonic stem cells; politics; researchers; Keyword Identifiers: Japan; Thomson, James A.; United States; Yamanaka, Shinya

Lanza, Robert. Stem cell breakthrough: don't forget the ethics [letter]. *Science* 2007 December 21; 318(5858): 1865. 4 refs. NRCBL: 18.5.4; 15.1.

> Keywords: *stem cells; chimeras; embryo research; embryonic stem cells; genetic techniques; methods; risk; Proposed Keywords: *pluripotent stem cells

Levin, Yuval. A middle ground for stem cells [op-ed]. *New York Times* 2007 January 19; p. A23. NRCBL: 18.5.4; 15.1. SC: po.

> Keywords: *embryonic stem cells; *embryo research; *public policy; embryo; value of life; research support; government financing; Keyword Identifiers: *United States

Little, Melissa; Hall, Wayne; Orlandi, Amy. Delivering on the promise of human stem-cell research. What are the real barriers? *EMBO Reports* 2006 December; 7(12): 1188-1192. NRCBL: 18.5.4; 15.1; 18.6.

Loike, John D.; Tendler, Moshe D. Ethical dilemmas in stem cell research: human-animal chimeras. *Tradition* 2007 Winter; 40(4): 28-49. 65 fn. NRCBL: 18.5.4; 15.1; 1.2; 18.1; 22.1.

> Keywords: *brain; *chimeras; *embryonic stem cells; *Jewish ethics; *stem cell transplantation; embryo research; embryos; human dignity; species specificity

Lott, Jason P.; Savulescu, Julian. Towards a global human embryonic stem cell bank. *American Journal of Bioethics* 2007 August; 7(8): 37-44. 27 refs. NRCBL: 18.5.4; 15.1; 19.5; 14.5; 9.3.1. SC: an. Comments: See comments in American Journal of Bioethics 2007 August; 7(8): 45-53.

> Keywords: *embryonic stem cells; *tissue banks; cell lines; cloning; donors; embryo disposition; ethnic groups; economics; financial support; in vitro fertilization; incentives; informed consent; international aspects; justice; mandatory programs; minority groups; moral policy; nuclear transfer techniques; organ transplantation; policy analysis; racial groups; remuneration; resource allocation; scarcity; social discrimination; standards; stem cell transplantation; transplant recipients; Proposed Keywords: embryo donation; haplotypes; tissue typing

Abstract: An increasingly unbridgeable gap exists between the supply and demand of transplantable organs. Human embryonic stem cell technology could solve the organ shortage problem by restoring diseased or damaged tissue across a range of common conditions. However, such technology faces several largely ignored immunological challenges in delivering cell lines to large populations. We address some of these challenges and argue in favor of encouraging contribution or intentional creation of embryos from which widely immunocompatible stem cell lines could be derived. Further, we argue that current immunological constraints in tissue transplantation demand the creation of a global stem cell bank, which may hold particular promise for minority populations and other sub-groups currently marginalized from organ procurement and allocation systems. Finally, we conclude by offering a number of practical and ethically

NRCBL: National Reference Center for Bioethics Literature Classification Scheme   See inside front cover for terms.

413

oriented recommendations for constructing a human embryonic stem cell bank that we hope will help solve the ongoing organ shortage problem.

Lyerly, Anne Drapkin; Faden, Ruth R. Willingness to donate frozen embryos for stem cell research. *Science* 2007 July 6; 317(5834): 46-47. 17 refs. NRCBL: 18.5.4; 15.1; 19.5; 14.6. SC: em.

> Keywords: *attitudes; *cryopreservation; *donors; *embryo disposition; *embryo research; *embryonic stem cells; *embryos; *patients; biomedical research; embryo transfer; in vitro fertilization; informed consent; moral status; public opinion; survey; Keyword Identifiers: *United States; Australia; Great Britain

Lysaught, M. Therese. Making decisions about embryonic stem cell research. *In:* Hamel, Ronald, ed. Making Health Care Decisions: A Catholic Guide. Liguori, MO: Liguori Publications, 2006: 19-36. 7 refs. NRCBL: 18.5.4; 1.2; 15.1.

> Keywords: *embryo research; *embryonic stem cells; *Roman Catholic ethics; beginning of life; cloning; embryos; justice; moral status; personhood; value of life

Master, Zubin; McLeod, Marcus; Mendez, Ivar. Benefits, risks and ethical considerations in translation of stem cell research to clinical applications in Parkinson's disease. *Journal of Medical Ethics* 2007 March; 33(3): 169-173. 30 refs. NRCBL: 18.5.1; 18.5.4; 15.1; 19.1. SC: an.

> Keywords: *clinical trials; *research subjects; *risks and benefits; *stem cell transplantation; adverse effects; embryonic stem cells; ethical review; therapeutic research; placebos; research design; surgery; Proposed Keywords: sham surgery; Keyword Identifiers: *Parkinson disease

Abstract: Stem cells are likely to be used as an alternate source of biological material for neural transplantation to treat Parkinson's disease in the not too distant future. Among the several ethical criteria that must be fulfilled before proceeding with clinical research, a favourable benefit to risk ratio must be obtained. The potential benefits to the participant and to society are evaluated relative to the risks in an attempt to offer the participants a reasonable choice. Through examination of preclinical studies transplanting stem cells in animals and the transplantation of fetal tissue in patients with Parkinson's disease, a current set of potential benefits and risks for neural transplantation of stem cells in clinical research of Parkinson's disease are derived. The potential benefits to research participants undergoing stem cell transplantation are relief of parkinsonian symptoms and decreasing doses of parkinsonian drugs. Transplantation of stem cells as a treatment for Parkinson's disease may benefit society by providing knowledge that can be used to help determine better treatments in the future. The risks to research participants undergoing stem cell transplantation include tumour formation, inappropriate stem cell migration, immune rejection of transplanted stem cells, haemorrhage during neurosurgery and postoperative infection. Although some of these risks are general to neurosurgical transplantation and may not be reduced for participants, the potential risk of tumour formation and inappropriate stem cell migration must be minimised before obtaining a favourable potential benefit to risk calculus and to provide participants with a reasonable choice before they enroll in clinical studies.

Master, Zubin; Williams-Jones, Bryn; Lott, Jason P.; Savulescu, Julian. The global HLA banking of embryonic stem cells requires further scientific justification. *American Journal of Bioethics* 2007 August; 7(8): 45-46; author reply W4-W6. 7 refs. NRCBL: 18.5.4; 15.1; 19.5. Comments: Comment on: Jason P. Lott and Julian Savulescu. Towards a global human embryonic stem cell bank. American Journal of Bioethics 2007 August; 7(8): 37-44.

> Keywords: *embryonic stem cells; *tissue banks; cloning; donors; incentives; international aspects; justice; mandatory programs; moral policy; nuclear transfer techniques; organ transplantation; presumed consent; risks and benefits; scarcity; stem cell transplantation; voluntary programs; Proposed Keywords: embryo donation

McCartney, James J. Embryonic stem cell research and respect for human life: philosophical and legal reflections. *Albany Law Review* 2002; 65(3): 597-624. NRCBL: 18.5.4; 15.1; 14.5; 5.3.

McKneally, Martin. Controversies in cardiothoracic surgery: should therapeutic cloning be supported to provide stem cells for cardiothoracic surgery research and treatment? [debate]. *Journal of Thoracic and Cardiovascular Surgery* 2006 May; 131(5): 937-940. NRCBL: 18.5.4; 15.1; 14.5; 9.5.1.

McLeod, Carolyn; Baylis, Françoise. Donating fresh versus frozen embryos to stem cell research: in whose interests. *Bioethics* 2007 November; 21(9): 465-477. NRCBL: 18.5.4; 10; 14.4; 14.6; 19.5.

> Abstract: Some stem cell researchers believe that it is easier to derive human embryonic stem cells from fresh rather than frozen embryos and they have had in vitro fertilization (IVF) clinicians invite their infertility patients to donate their fresh embryos for research use. These embryos include those that are deemed 'suitable for transfer' (i.e. to the woman's uterus) and those deemed unsuitable in this regard. This paper focuses on fresh embryos deemed suitable for transfer – hereafter 'fresh embryos'– which IVF patients have good reason not to donate. We explain why donating them to research is not in the self-interests specifically of female IVF patients. Next, we consider the other-regarding interests of these patients and conclude that while fresh embryo donation may serve those interests, it does so at unnecessary cost to patients' self-interests. Lastly, we review some of the potential barriers to the autonomous donation of fresh embryos to research and highlight the risk that female IVF patients invited to donate these embryos will misunderstand key aspects of the donation decision, be coerced to donate, or be exploited in the consent process. On the basis of our analysis, we conclude that patients should not be asked to donate their fresh embryos to stem cell research.

Mertes, Heidi.; Pennings, G. Oocyte donation for stem cell research. *Human Reproduction* 2007 March; 22(3):

SECTION I   STEM CELL RESEARCH

629-634. NRCBL: 19.5; 14.6; 9.5.5; 18.5.4; 15.1; 4.4; 18.3. SC: le.

Mitka, Mike. Stem cell legislation [news]. *JAMA: The Journal of the American Medical Association* 2007 February 14; 297(6): 581. NRCBL: 18.5.4; 15.1. SC: le.
Keywords: *embryo research; *embryonic stem cells; *government financing; *legislation; federal government; politics; public policy; research support; Keyword Identifiers: *Stem Cell Research Enhancement Act; *U.S. Congress; *United States

Morris, Jonathan. Stem cell research. *In his:* The Ethics of Biotechnology. Philadelphia: Chelsea Publishers, 2006: 83-104. NRCBL: 18.5.4; 15.1. SC: po.
Keywords: *embryo research; *embryonic stem cells; abortion; federal government; government financing; legal aspects; moral policy; religious ethics; risks and benefits; stem cell transplantation; Keyword Identifiers: United States

Murray, Fiona. The stem cell market — patents and the pursuit of scientific progress. *New England Journal of Medicine* 2007 June 7; 356(23): 2341-2343. 4 refs. NRCBL: 18.5.4; 15.1; 1.3.9; 15.8; 19.1;.
Keywords: *biomedical research; *embryonic stem cells; *patents; *research support; *universities; access to information; cell lines; commerce; contracts; embryo research; industry; legal aspects; private sector; publishing; science; Proposed Keywords: technology transfer; Keyword Identifiers: *Wisconsin Alumni Research Foundation; United States; University of Wisconsin

National Institutes of Health [NIH] (United States). Stem Cell Information [government document]. Bethesda, MD: National Institutes of Health [NIH] (United States), 2007 May 4 and 2007 September 11: [34 p.] [Online]. Accessed: http://stemcells.nih.gov/staticresources/research/registry/PDFs/Registry.pdf [2007 October 2]. NRCBL: 18.5.4; 15.1; 5.3. SC: em.
Keywords: *biological specimen banks; *cell lines; *embryo research; *embryonic stem cells; *federal government; *government financing; *public policy; *research support; guidelines; international aspects; registries; Keyword Identifiers: *National Institutes of Health; *United States

National Institutes of Health [NIH]. (United States). Department of Health and Human Services [DHHS]. Plan for implementation of Executive Order 13435: expanding approved stem cell lines in ethically responsible ways. Bethesda, MD: National Institutes of Health 2007 September 18; 25 p. [Online]. Accessed: http://stemcells.nih.gov/staticresources/policy/eo13435.pdf [2007 October 11]. NRCBL: 5.3; 18.5.4; 19.1.

Novak, David. On the use of embryonic stem cells. *In his:* The Sanctity of Human Life. Washington, DC: Georgetown University Press, 2007: 1-89. 202 fn. NRCBL: 18.5.4; 15.1; 1.1; 1.2; 4.4; 12.3. SC: an.
Keywords: *embryo research; *embryonic stem cells; *embryos; *Jewish ethics; *moral policy; *moral status; *philosophy; *public policy; *theology; abortion; beginning of life; cultural pluralism; fetuses; killing; morality; natural law; personhood; policy analysis; politics; secularism; value

of life; Proposed Keywords: embryo death; embryonic development

O'Dowd, Adrian. UK may allow creation of "cybrids" for stem cell research [news]. *BMJ: British Medical Journal* 2007 March 10; 334(7592): 495. NRCBL: 18.5.4; 15.1; 22.2; 14.5; 18.1; 22.1. SC: le.
Keywords: *chimeras; *embryo research; *embryonic stem cells; government regulation; public policy; Keyword Identifiers: *Great Britain

O'Neil, Graeme. Australia ends stem-cell cloning ban [news brief]. *Nature Biotechnology* 2007 February; 25(2): 153. NRCBL: 18.5.4; 15.1; 14.5. SC: le.
Keywords: *cloning; *embryo research; *embryonic stem cells; *embryos; *legal aspects; government regulation; Keyword Identifiers: *Australia

Okarma, Tom. Adult stem cells won't do. *New Scientist* 2007 March 10-16; 193(2594): 20. NRCBL: 18.5.1; 15.1; 19.1; 18.5.4.
Keywords: *embryo research; *embryonic stem cells; *government financing; *research support; adult stem cells; federal government; government regulation; industry; politics; public policy; stem cell transplantation; Keyword Identifiers: *United States; National Institutes of Health

Onder, Robert. "People need a fairy tale": the embryonic stem cell and cloning debate in Missouri. *Missouri Medicine* 2006 March-April; 103(2): 106-111. NRCBL: 18.5.4; 14.5; 15.1. SC: em.

Outomuro, Delia; Lott, James P.; Savulescu, Julian. Moral dilemmas around a global human embryonic stem cell bank. *American Journal of Bioethics* 2007 August; 7(8): 47-48; author reply W4-W6. 10 refs. NRCBL: 18.5.4; 15.1; 19.5. Comments: Comment on: Jason P. Lott and Julian Savulescu. Towards a global human embryonic stem cell bank. American Journal of Bioethics 2007 August; 7(8): 37-44.
Keywords: *embryonic stem cells; *tissue banks; donors; embryos; health services accessibility; incentives; international aspects; justice; mandatory programs; minority groups; moral policy; moral status; organ transplantation; presumed consent; remuneration; required request; risks and benefits; scarcity; stem cell transplantation; voluntary programs; Proposed Keywords: embryo donation

Parry, Sarah. (Re)constructing embryos in stem cell research: exploring the meaning of embryos for people involved in fertility treatments. *Social Science and Medicine* 2006 May; 62(10): 2349-2359. NRCBL: 18.5.4; 15.1; 14.1; 9.5.5; 18.3.

Persson, Anders; Hemlin, Sven; Welin, Stellan. Profitable exchanges for scientists: the case of Swedish human embryonic stem cell research. *Health Care Analysis: An International Journal of Health Philosophy and Policy* 2007 December; 15(4): 291-304. 30 refs. NRCBL: 5.3; 1.3.9; 18.5.4; 15.1.
Keywords: *biomedical research; *embryo research; *embryonic stem cells; *industry; *research support; *researchers; authorship; biotechnology; cell lines; embryo disposition; entrepreneurship; genetic materials; genetic

NRCBL: National Reference Center for Bioethics Literature Classification Scheme   See inside front cover for terms.

415

patents; motivation; politics; professional autonomy; property rights; research ethics committees; universities; Proposed Keywords: technology transfer; Keyword Identifiers: *Sweden; Cell Therapeutics; Ovacell

Abstract: In this article two inter-related issues concerning the ongoing commercialisation of biomedical research are analyzed. One aim is to explain how scientists and clinicians at Swedish public institutions can make profits, both commercially and scientifically, by controlling rare human biological material, like embryos and embryonic stem cell lines. This control in no way presupposes legal ownership or other property rights as an initial condition. We show how ethically sensitive material (embryos and stem cell lines) have been used in Sweden as a foundation for a commercial stem cell enterprise — despite all official Swedish strictures against commercialisation in this area. We also show how political decisions may amplify the value of controlling this kind of biological material. Another aim of the article is to analyze and discuss the meaning of this kind of academic commercial enterprise in a wider context of research funding strategies. A conclusion that is drawn is that the academic turn to commercial funding sources is dependent on the decline of public funding.

Petri, J. Thomas. Altered nuclear transfer, gift, and mystery: an Aristotelian-Thomistic response to David L. Schindler. *National Catholic Bioethics Quarterly* 2007 Winter; 7(4): 729-747. 84 fn. NRCBL: 15.1; 18.5.4; 1.1; 1.2; 4.4.
   Keywords: *beginning of life; *nuclear transfer techniques; *philosophy; *research embryo creation; *pluripotent stem cells; *Roman Catholic ethics; embryos; methods; personhood; theology; Proposed Keywords: *altered nuclear transfer; Keyword Identifiers: *Aristotle; *Thomas Aquinas

Abstract: The leaders of the resistance against not only ANT-OAR and ANT-Cdx2 but any ANT procedure are two editors of Communio, David Schindler and Adrian Walker. Both scholars offer what they hold to be an Aristotelian-Thomistic objection to any ANT procedure. While many of their intuitions resonate with Aristotelians and Thomists, I do not believe they have represented either the Philosopher or the Common Doctor accurately. This article focuses almost exclusively on the Schindler's reading of Aristotle and St. Thomas Aquinas to show why his reading of them cannot be used to mount an effective objection against ANT on strictly Aristotelian-Thomistic grounds.

Pittman, Larry J. Embryonic stem cell research and religion: the ban on federal funding as a violation of the establishment clause. *University of Pittsburgh Law Review* 2006 Fall; 68(1): 131-190. NRCBL: 18.5.4; 15.1; 1.2; 5.3.

Prentice, David A. The whole truth about stem cells and relevant therapies. *Today's Christian Doctor* 2007 Summer; 38(2): 21-24. NRCBL: 18.5.1; 18.5.4; 15.1; 14.5.
   Keywords: *adult stem cells; *embryo research; *embryonic stem cells; *stem cell transplantation; cloning

Prentice, David A.; Tarne, Gene. Treating diseases with adult stem cells [letter]. *Science* 2007 January 19; 315(5810): 328. 8 refs. NRCBL: 18.5.1; 15.1; 5.3.
   Keywords: *adult stem cells; cancer; clinical trials; embryonic stem cells; government regulation; stem cell transplantation; treatment outcome; Proposed Keywords: *therapeutics; Keyword Identifiers: Food and Drug Administration; United States

Rao, Mahendra S. Are there morally acceptable alternatives to blastocyst derived ESC? *Journal of Cellular Biochemistry* 2006 August 1; 98(5): 1054-1061. NRCBL: 15.1; 18.5.4; 9.3.1; 19.1.

Rao, Mahendra S. Mired in the quagmire of uncertainty: The "catch-22" of embryonic stem cell research. *Stem Cells and Development* 2006 August; 15(4): 492-496. NRCBL: 18.5.4; 15.1; 15.8; 5.3; 18.2; 9.3.1.

Resnik, David B. Embryonic stem cell patents and human dignity. *Health Care Analysis: An International Journal of Health Philosophy and Policy* 2007 September; 15(3): 211-222. 41 refs. NRCBL: 15.8; 18.5.4; 15.1; 4.4. SC: an.
   Keywords: *embryonic stem cells; *human dignity; *moral policy; *moral status; *patents; beginning of life; biotechnology; embryos; ethical analysis; genetic patents; genetically modified organisms; guidelines; historical aspects; human characteristics; legal aspects; natural law; stem cells; terminology; value of life; Proposed Keywords: *multipotent stem cells; *pluripotent stem cells; *totipotent stem cells; blastocysts; classification; Keyword Identifiers: Europe; Twentieth Century; United States

Abstract: This article examines the assertion that human embryonic stem cells patents are immoral because they violate human dignity. After analyzing the concept of human dignity and its role in bioethics debates, this article argues that patents on human embryos or totipotent embryonic stem cells violate human dignity, but that patents on pluripotent or multipotent stem cells do not. Since patents on pluripotent or multipotent stem cells may still threaten human dignity by encouraging people to treat embryos as property, patent agencies should carefully monitor and control these patents to ensure that patents are not inadvertently awarded on embryos or totipotent stem cells.

Ries, Nola M. Regulation of human stem cell research in Japan and Canada: a comparative analysis. *University of New Brunswick Law Journal = Revue de Droit de L'universite du Nouveau Brunswick* 2005; 54: 62-74. 68 fn. NRCBL: 18.5.4; 15.1; 18.6. SC: le.
   Keywords: *embryo research; *embryonic stem cells; *government regulation; *international aspects; cloning; cross-cultural comparison; embryos; ethical review; government financing; guidelines; legal aspects; public policy; reproductive technologies; research support; stem cells; Keyword Identifiers: *Canada; *Japan

Robert, Jason Scott. The science and ethics of making part-human animals in stem cell biology. *FASEB Journal: Official Publication of the Federation of American Societies for Experimental Biology* 2006 May; 20(7): 838-845. 50 refs. NRCBL: 15.1; 18.1; 22.1; 18.5.4; 5.3. SC: rv.

Keywords: *chimeras; *embryonic stem cells; brain; embryo research; ethical review; fetal stem cells; guidelines; historical aspects; human dignity; moral policy; moral status; primates; public policy; regulation; research design; research ethics; review committees; risks and benefits; species specificity; stem cell transplantation; terminology; Proposed Keywords: mice; Keyword Identifiers: National Academy of Sciences; Nineteenth Century; Twentieth Century; United States

Romano, Gaetano. Perspectives and controversies in the field of stem cell research. *Drug News and Perspectives* 2006 September; 19(7): 433-439. NRCBL: 18.5.4; 15.1; 18.5.1; 18.2.

Rosen, Michael R. Are stem cells drugs? The regulation of stem cell research and development. *Circulation* 2006 October 31; 114(18): 1992-2000. NRCBL: 18.5.4; 15.1; 18.6.

Ross, Lainie Friedman; Philipson, Louis H.; Voltarelli, Julio C.; Couri, Carlos E.B.; Stracieri, Ana B.P.L.; Oliveira, Maria C.; Moraes, Daniela A.; Fabiano, Pieroni; Coutinho, Marina; Malmegrim, Kelen C.R.; Foss-Freitas, Maria C.; Simões, Belinda P.; Foss, Milton C.; Squiers, Elizabeth; Burt, Richard K. Ethics of hematopoietic stem cell transplantation in type 1 diabetes mellitus [letter and reply]. *JAMA: The Journal of the American Medical Association* 2007 July 18; 298(3): 285-286. NRCBL: 18.5.2; 18.5.1; 19.1. Identifiers: Brazil.

Royal College of Obstetricians and Gynaecologists (Great Britain). RCOG response to MRC consultation on the Code of Practice for the use of human stem cell lines. London: Royal College of Obstetricians and Gynaecologists, 2004 May 27; 3 p. [Online]. Accessed: http://www.rcog.org.uk/index.asp?PageID=1349 [2007 April 5]. NRCBL: 18.5.4; 15.1; 6. Identifiers: Great Britain (United Kingdom).
Keywords: *biological specimen banks; *cell lines; *embryonic stem cells; *guidelines; *standards; embryo research; ethical review; fetal research; informed consent; organizational policies; physicians; professional organizations; stem cells; Keyword Identifiers: *Royal College of Obstetricians and Gynaecologists (Great Britain); Great Britain

Rusnak, A.J.; Chudley, A.E. Stem cell research: cloning, therapy and scientific fraud. *Clinical Genetics* 2006 October; 70(4): 302-305. NRCBL: 15.1; 14.5; 1.3.9; 18.5.1.

Rutz, Berthold; Yeats, Siobhan. Patents: patenting of stem cell related inventions in Europe. *Biotechnology Journal* 2006 April; 1(4): 384-387. 14 refs. NRCBL: 15.8; 21.1. SC: le.
Keywords: *embryonic stem cells; *legal aspects; *patents; *regulation; international aspects; Keyword Identifiers: *Europe; European Group on Ethics in Science and New Technologies; European Patent Convention; European Patent Office; European Union

Salter, Brian. Bioethics, politics and the moral economy of human embryonic stem cell science: the case of the Eu-

ropean Union's Sixth Framework Programme. *New Genetics and Society* 2007 December; 26(3): 269-288. NRCBL: 18.5.4; 15.1; 21.1.

Salter, Brian; Salter, Charlotte. Bioethics and the global moral economy: the cultural politics of human embryonic stem cell science. *Science, Technology and Human Values* 2007 September; 32(5): 554-581. NRCBL: 2.1; 18.5.4; 15.1; 21.1. SC: an; em.

Sass, Hans-Martin. Let probands and patients decide about moral risk: stem cell research and medical treatment. *In:* Roetz, Heiner, ed. Cross-Cultural Issues in Bioethics: The Example of Human Cloning. New York: Rodopi, 2006: 425-444. NRCBL: 18.5.4; 15.1; 14.5.

Schiltz, Elizabeth R. The disabled Jesus: a parent looks at the logic behind prenatal testing and stem cell research. *America* 2007 March 12; 196(9): 16-18. NRCBL: 15.3; 18.5.4; 1.2.
Keywords: *embryo research; *embryonic stem cells; *prenatal diagnosis; *Roman Catholic ethics; *value of life; adult stem cells; attitudes; choice behavior; congenital disorders; Down syndrome; embryos; eugenics; mentally disabled persons; preimplantation diagnosis; quality of life; risks and benefits; selective abortion; Proposed Keywords: autistic disorder

Schwartz, Peter. Stem cells: biopsy on frozen embryos [letter]. *Hastings Center Report* 2007 January-February; 37(1): 7-8. NRCBL: 18.5.1; 15.1; 14.6.
Keywords: *cryopreservation; *embryo research; *embryonic stem cells; *embryos; in vitro fertilization; Proposed Keywords: blastomeres

Sheldon, Tony. Holland bans private stem cell therapy [news]. *BMJ: British Medical Journal* 2007 January 6; 334(7583): 12. NRCBL: 19.1; 15.1; 18.5; 9.5.1. SC: le. Identifiers: Netherlands.

Simon, Jürgen. Biotechnology and law: biotechnology patents. Special considerations on the inventions with human material. *Revista de Derecho y Genoma Humano = Law and the Human Genome Review* 2006 July-December; (25): 139-159. 83 fn. NRCBL: 15.8; 15.1; 21.1. SC: le.
Keywords: *biotechnology; *cloning; *genetic materials; *genetic patents; *legal aspects; *stem cells; DNA sequences; embryo research; genes; genetic engineering; germ cells; international aspects; reproductive technologies; Keyword Identifiers: Europe; United States

Sinclair, Andrew H.; Schofield, Peter R. Human embryonic stem cell research: an Australian perspective. *Cell* 2007 January 26; 128(2): 221-223. NRCBL: 18.2; 18.5.4; 15.1; 1.3.5.

Smith, Shane; Neaves, William; Teitelbaum, Steven; Prentice, David A.; Tarne, Gene. Adult versus embryonic stem cells: treatments [letter and reply]. *Science* 2007 June 8; 316(5830): 1422-1423. 22 refs. NRCBL: 18.5.1; 15.1; 18.5.4; 18.2.

NRCBL: National Reference Center for Bioethics Literature Classification Scheme   See inside front cover for terms.

417

Keywords: *adult stem cells; *embryonic stem cells; *stem cell transplantation; *treatment outcome; clinical trials; embryo research; Proposed Keywords: *therapeutics

Somerville, Margaret A. The importance of empirical research in bioethics: the case of human embryo stem cell research = Importance de la recherche empirique en bioéthique: cas de la recherche sur les cellules souches embryonnaires humaines [editorial]. *JOGC: Journal of Obstetrics and Gynaecology Canada = JOGC: Journal d'Obstétrique et Gynécologie du Canada* 2005 October; 27(10): 929-932. NRCBL: 18.5.4; 15.1; 18.3; 2.1.

Song, Sang-yong. The rise and fall of embryonic stem cell research in Korea. *Asian Biotechnology and Development Review* 2006 November; 9(1): 65-73 [Online]. Accessed: http://www.openj-gate.org/articlelist.asp?LatestYear= 2007&JCode=104346&year=2006&vol=9&issue=1 &ICode=583649 [2008 February 29]. NRCBL: 18.5.4; 15.1; 18.6; 18.2.

Spar, Debora; Harrington, Anna. Selling stem cell science: how markets drive law along the technological frontier. *American Journal of Law and Medicine* 2007; 33(4): 541-565. NRCBL: 18.5.4; 15.1; 18.6; 5.3; 11.1; 14.4. SC: le.

Steinbock, Bonnie. Moral status, moral value, and human embryos: implications for stem cell research. *In:* Steinbock, Bonnie, ed. The Oxford Handbook of Bioethics. Oxford; New York: Oxford University Press, 2007: 416-440. 43 refs. 12 fn. NRCBL: 4.4; 18.5.4; 15.1. SC: an; rv.
  Keywords: *embryo research; *embryonic stem cells; *embryos; *moral policy; *moral status; abortion; beginning of life; cadavers; ethical analysis; human characteristics; killing; nuclear transfer techniques; personhood; public policy; research embryo creation; species specificity; stem cells; twinning; values; Proposed Keywords: altered nuclear transfer; embryo death; pluripotent stem cells; sentience

Swift, Jennifer. Eggs don't come cheap. *New Scientist* 2007 December 8-14; 196(2633): 22. NRCBL: 14.5; 14.2; 19.5; 18.5.3; 18.5.4; 15.1; 9.5.5.
  Keywords: *cloning; *ovum donors; *research embryo creation; *risk; *stem cells; adverse effects; conflict of interest; hormones; nontherapeutic research; remuneration; risks and benefits; women; Proposed Keywords: pluripotent stem cells

Takala, Tuija; Häyry, Matti. Benefiting from past wrongdoing, human embryonic stem cell lines, and the fragility of the German legal position. *Bioethics* 2007 March; 21(3): 150-159. 50 fn. NRCBL: 18.5.4; 15.1; 19.5; 1.1. SC: an; le.
  Keywords: *cell lines; *embryo research; *embryonic stem cells; *ethical analysis; *legislation; *moral complicity; *moral policy; aborted fetuses; commerce; commodification; dehumanization; embryo disposition; embryos; government regulation; historical aspects; in vitro fertilization; international aspects; morality; National Socialism; ovum donors; policy analysis; public policy; scientific misconduct; Keyword Identifiers: *Germany; *Stem Cell Act 2002 (Germany)

Abstract: This paper examines the logic and morality of the German Stem Cell Act of 2002. After a brief description of the law's scope and intent, its ethical dimensions are analysed in terms of symbolic threats, indirect consequences, and the encouragement of immorality. The conclusions are twofold. For those who want to accept the law, the arguments for its rationality and morality can be sound. For others, the emphasis on the uniqueness of the German experience, the combination of absolute and qualified value judgments, and the lingering questions of indirect encouragement of immoral activities will probably be too much.

Tangwa, Godfrey B. Moral status of embryonic stem cells: perspective of an African villager. *Bioethics* 2007 October; 21(8): 449-457. 6 fn. NRCBL: 18.5.4; 15.1; 1.1; 1.3.1; 4.4; 21.7. SC: an. Conference: Eighth World Congress of Bioethics: A Just and Healthy Society; Beijing, China; 2006 August 9.
  Keywords: *embryonic stem cells; *embryos; *moral status; *value of life; beginning of life; cloning; ethical analysis; embryo research; morality; non-Western World; personhood; teleological ethics; wedge argument; Proposed Keywords: blastocysts; Keyword Identifiers: Africa
Abstract: One of the most important as well as most awesome achievements of modern biotechnology is the possibility of cloning human embryonic stem cells, if not human beings themselves. The possible revolutionary role of such stem cells in curative, preventive and enhancement medicine has been voiced and chorused around the globe. However, the question of the moral status of embryonic stem cells has not been clearly and unequivocally answered. Taking inspiration from the African adage that 'the hand that reaches beneath the incubating hen is not guiltless', I attempt answering this question, from the background of traditional African moral sensibility and sensitivity. I reach the following conclusions. Stem cells in themselves do not have human status and therefore lack moral worth/value. Embryos do have human status and a morally significant line cannot be drawn between human embryos and other human beings. What is morally at stake in stem cell research is therefore the question of the source of derivation or generation of the cells, not of the cells as such.

Thompson, Clive. How to farm stem cells without losing your soul. *Wired Magazine* 2005 June; 13.06: 4 p. [Online]. Accessed: http://www.wired.com/wired/archive/ 13.06/stemcells.html [2006 October 12]. NRCBL: 18.5.4; 1.2; 15.1.
  Keywords: *embryonic stem cells; *methods; *research embryo creation; advisory committees; alternatives; beginning of life; Christian ethics; dissent; embryos; Keyword Identifiers: *Hurlbut, William; President's Council on Bioethics

Tong, Rosemary. Stem-cell research and the affirmation of life. *Conscience* 2007 Autumn; 28(3): 19-23. NRCBL: 18.5.4; 15.1; 4.4; 1.2; 7.1.

Trivedi, Bijal. Researchers detour around stem-cell rules. *Chronicle of Higher Education* 2007 October 3; 54(6): A12-A15. NRCBL: 18.5.4; 15.1; 18.6; 19.5.

Keywords: *embryonic stem cells; *ovum donors; *remuneration; embryo disposition; embryo research; government regulation; in vitro fertilization; indigents; ovum; researchers; scarcity; women; Keyword Identifiers: Great Britain; United States

United States. Congress. House. A bill to derive human pluripotent stem cell lines using techniques that do not harm human embryos. Washington, DC: U.S. G.P.O., 2007. 4 p. [Online]. Accessed: http://frwebgate.access. gpo.gov/cgi-bin/useftp.cgi?IPaddress=162.140.64.21& filename=h322ih.pdf&directory=/diskb/wais/data/ 110_cong_bills [17 January 2007]. NRCBL: 18.5.4; 15.1; 4.4. SC: le. Identifiers: Alternative Pluripotent Stem Cell Therapies Enhancement Act of 2007. Note: H.R. 322, 110th Congress, 1st session. Introduced by Rep. Bartlett on January 9, 2007. Referred to the Committee on Energy and Commerce.

Keywords: *embryos; *legal aspects; *stem cells; alternatives; cell lines; embryo research; methods; Keyword Identifiers: *United States

United States. Congress. House. An act to amend the Public Health Service Act to provide for human embryonic stem cell research. Washington, DC: U.S. G.P.O., 2007. 4 p. [Online]. Accessed: http://frwebgate.access.gpo.gov/ cgi-bin/useftp.cgi?IPaddress=162.140.64.21&filename= h3eh.pdf&directory=/diskb/wais/data/110_cong_bills [2007 January 17]. NRCBL: 18.5.4; 15.1. SC: le. Identifiers: Stem Cell Research Enhancement Act of 2007. Note: H.R. 3, 110th Congress, 1st session. Introduced by Ms. DeGette and others, January 5, 2007. Passed the House of Representatives January 11, 2007.

Keywords: *embryo research; *embryonic stem cells; *legal aspects; Keyword Identifiers: *United States

United States. Congress. Senate. A bill to amend the Public Health Service Act to provide for human embryonic stem cell research. Washington, DC: U.S. G.P.O., 2007. 3 p. [Online]. Accessed: http://frwebgate.access.gpo.gov/ cgi-bin/useftp.cgi?IPaddress=162.140.64.21&filename= s5pcs.pdf&directory=/diskb/wais/data/110_cong_bills [2007 April 10]. NRCBL: 18.5.4; 15.1. SC: le. Identifiers: Stem Cell Research Enhancement Act of 2007. Note: S. 5, 110th Congress, 1st session. Introduced by Sen. Reid, January 4, 2007. Read the second time and placed on calendar, January 8, 2007.

Keywords: *embryo research; *embryonic stem cells; *legal aspects; embryo disposition; Keyword Identifiers: *United States

United States. Congress. Senate. A bill to derive human pluripotent stem cell lines using techniques that do not knowingly harm embryos. Washington, DC: U.S. G.P.O., 2007. 5 p. [Online]. Accessed: http://frwebgate.access. gpo.gov/cgi-bin/useftp.cgi?IPaddress=162.140.64.21& filename=s51is.pdf&directory=/diskb/wais/data/ 110_cong_bills [2007 January 17]. NRCBL: 18.5.4; 15.1; 4.4. SC: le. Identifiers: Pluripotent Stem Cell Therapy Enhancement Act of 2007. Note: S. 51, 110th Congress, 1st session. Introduced by Sen. Isakson on January 4, 2007.

Referred to the Committee on Health, Education, Labor, and Pensions.

Keywords: *cell lines; *legal aspects; *research embryo creation; *stem cells; alternatives; embryonic stem cells; public policy; Keyword Identifiers: *United States

United States. Congress. Senate. A bill to intensify research to derive human pluripotent stem cell lines. Washington, DC: U.S. G.P.O., 2007. 6 p. [Online]. Accessed: http://frwebgate.access.gpo.gov/cgi-bin/useftp.cgi? IPaddress=162.140.64.21&filename=s30hds.pdf& directory=/diskb/wais/data/110_cong_bills [2007 April 10]. NRCBL: 18.5.4; 15.1. SC: le. Identifiers: Hope Offered through Principled and Ethical Stem Cell Research Act; HOPE Act. Note: S. 30, 110th Congress, 1st session. Introduced by Sen. Coleman, March 29, 2007.

Keywords: *legal aspects; *stem cells; alternatives; biomedical research; embryo research; embryos; guidelines; public policy; Keyword Identifiers: *United States

United States. Congress. Senate. A bill to provide increased Federal funding for stem cell research, to expand the number of embryonic stem cell lines available for Federally funded research, to provide ethical guidelines for stem cell research, to derive human pluripotent stem cell lines using techniques that do not create an embryo or embryos for research or knowingly harm embryos, and for other purposes. Washington, DC: U.S. G.P.O., 2007. 15 p. [Online]. Accessed: http://frwebgate.access.gpo.gov/ cgi-bin/useftp.cgi?IPaddress=162.140.64.21&filename= s363is.pdf&directory=/diskb/wais/data/110_cong_bills [2007 April 4]. NRCBL: 18.5.4; 15.1. SC: le. Identifiers: Hope Offered through Principled, Ethically-Sound Stem Cell Research Act; HOPE Act. Note: S. 363, 110th Congress, 1st session. Introduced by Sen. Coleman on January 23, 2007. Referred to the Committee on Health, Education, Labor, and Pensions.

Keywords: *embryonic stem cells; *embryo research; *government financing; *legal aspects; *research support; *stem cells; adult stem cells; alternatives; cell lines; embryos; informed consent; research ethics committees; Keyword Identifiers: *United States

United States. Office of the President. Fact Sheet: Embryonic Stem Cell Research. *Washington, DC: The White House, Office of the Press Secretary,* 2001 August 9; 3 p. [Online]. Accessed: http://www.whitehouse.gov/news/releases/2001/08/text/20010809-1.html [2001 August 10]. NRCBL: 18.5.4; 15.1; 18.6. SC: po.

Keywords: *embryo research; *embryonic stem cells; *federal government; *government financing; *public policy; *research support; adult stem cells; advisory committees; cell lines; guidelines; Keyword Identifiers: *United States; President's Council on Bioethics

United States. Office of the President. Domestic Policy Council. Advancing stem cell science without destroying human life. Washington, DC: The White House, 2007 January 9: 64 p. [Online]. Accessed: http://www.whitehouse. gov/infocus/healthcare/stemcell_010907.pdf [2007 January 24]. 130 refs. NRCBL: 18.5.4; 15.1; 18.5.1; 14.5.

NRCBL: National Reference Center for Bioethics Literature Classification Scheme   See inside front cover for terms.

419

Keywords: *embryo research; *legal aspects; *public policy; *stem cells; adult stem cells; alternatives; cloning; embryo disposition; embryonic stem cells; federal government; government financing; methods; research support; value of life; Keyword Identifiers: *United States; President's Council on Bioethics

Wade, Nicholas. Panel says data is flawed in a major stem cell report. *New York Times* 2007 February 28; p. A15. NRCBL: 1.3.9; 1.3.7; 15.1. SC: po.

Wadman, Meredith. Stem-cell issue moves up the US agenda [news]. *Nature* 2007 April 19; 446(7138): 842. NRCBL: 18.5.4; 15.1. SC: le.
Keywords: *embryo research; *embryonic stem cells; *government financing; *legal aspects; *research support; cell lines; federal government; legislation; politics; Keyword Identifiers: *United States; Stem Cell Research Enhancement Act 2007; U.S. House of Representative; U.S. Senate

Wainwright, Steven; Williams, Clare; Michael, Mike; Farsides, Bobbie; Cribb, Alan. Remaking the body? Scientists' genetic discourses and practices as examples of changing expectations on embryonic stem cell therapy for diabetes. *New Genetics and Society* 2007 December; 26(3): 251-268. NRCBL: 18.5.4; 15.1.

Walter, James J. A Catholic reflection on embryonic stem cell research. *In:* Walter, James J.; Shannon, Thomas A., eds. Contemporary Issues in Bioethics: A Catholic Perspective. Lanham, MD: Rowman and Littlefield Publishers, 2005: 91-99. 21 fn. NRCBL: 18.5.4; 1.2; 15.1.
Keywords: *embryo research; *embryonic stem cells; *Roman Catholic ethics; embryos; government financing; moral status; public policy; research support; theology

Wang, Yanguang. The moral status of the human embryo in Chinese stem cell research. *Asian Biotechnology and Development Review* 2006 November; 9(1): 45-63 [Online]. Accessed: http://www.openj-gate.org/articlelist.asp?LatestYear=2007&JCode=104346&year=2006&vol=9&issue=1&ICode=583649 [2008 February 29]. NRCBL: 18.5.4; 19.1; 1.1; 18.6.

Weiss, Rick. House passes bill relaxing limits on stem cell research. *Washington Post* 2007 January 12; p. A4. NRCBL: 5.3; 15.1; 18.5.4. SC: po; le.
Keywords: *embryo research; *embryonic stem cells; *legislation; research support; government financing; politics; *public policy; embryo; federal government; Keyword Identifiers: *United States; *U.S. House of Representatives

Weiss, Rick. Stem cells created with no harm to human embryos; but concerns are raised about the technique. *Washington Post* 2006 August 24; p. A3. NRCBL: 18.5.4; 19.1; 15.1. SC: po.
Keywords: *embryonic stem cells; research embryo creation; *methods; public policy; research support; government financing; Keyword Identifiers: United States

Whittaker, Peter. Stem cells, patents and ethics. *In:* Gunning, Jennifer; Holm, Søren, eds. Ethics, law, and society.

Volume 1. Aldershot, Hants, England; Burlington, VT: Ashgate, 2005: 17-21. NRCBL: 18.5.1; 18.5.4; 5.3; 15.8.
Keywords: *patents; *stem cells; biotechnology; cloning; embryo research; embryonic stem cells; legal aspects; Keyword Identifiers: European Biotechnology Patent Directive; European Union

Winickoff, David E. Governing stem cell research in California and the USA: towards a social infrastructure. *Trends in Biotechnology* 2006 September; 24(9): 390-394. NRCBL: 18.5.4; 15.1; 18.6; 5.3.

World Federation of Neurology; Rosenberg, Roger N. World Federation of Neurology position paper on human stem cell research. *Journal of the Neurological Sciences* 2006 April 15; 243(1-2): 1-2. NRCBL: 18.5.4; 15.1; 6.

Zettler, Patricia; Wolf, Leslie E.; Lo, Bernard. Establishing procedures for institutional oversight of stem cell research. *Academic Medicine* 2007 January; 82(1): 6-10. NRCBL: 18.6; 18.5.4; 18.5.1; 15.1.

SUBSTANCE ABUSERS *See* CARE FOR SPECIFIC GROUPS/ SUBSTANCE ABUSERS

SUICIDE *See* ASSISTED SUICIDE

SURROGATE DECISION MAKING *See* EUTHANASIA AND ALLOWING TO DIE; INFORMED CONSENT

SURROGATE MOTHERS *See* ARTIFICIAL INSEMINATION AND SURROGATE MOTHERS

TECHNOLOGIES, BIOMEDICAL *See* ORGAN AND TISSUE TRANSPLANTATION; REPRODUCTIVE TECHNOLOGIES

TELEMEDICINE AND INFORMATICS

Ball, Douglas E.; Tisocki, Klara; Herxheimer, Andrew. Advertising and disclosure of funding on patient organisation websites: a cross-sectional survey. *BMC Public Health* 2006 August 3; 6: 201. NRCBL: 1.3.12; 1.3.2; 9.3.1; 9.7.

Brownstein, John S.; Cassa, Christopher A.; Kohane, Isaac S.; Mandl, Kenneth D. Reverse geocoding: concerns about patient confidentiality in the display of geospatial health data. *AMIA Annual Symposium Proceedings* 2005: 905. NRCBL: 1.3.12; 1.3.9; 8.4.

Dickens, B.M.; Cook, R.J. Legal and ethical issues in telemedicine and robotics. *International Journal of Gynecology and Obstetrics* 2006 July; 94(1): 73-78. NRCBL: 1.3.12; 1.3.5; 21.1; 9.1.

Gavel, Ylva; Andersson, Per-Olov; Knutsson, Gun-Brit. Euroethics — a database network on biomedical ethics. *Health Information and Libraries Journal* 2006 September; 23(3): 169-178. NRCBL: 1.3.12; 2.3; 21.1; 2.1.

Hongladarom, Soraj. Ethics of bioinformatics: a convergence between bioethics and computer ethics. *Asian Biotechnology and Development Review* 2006 November; 9(1): 37-44 [Online]. Accessed: http://www.openj-gate.org/articlelist.asp?LatestYear=2007&JCode=104346&year=2006&vol=9&issue=1&ICode=583649 [2008 February 29]. NRCBL: 1.3.12; 2.1; 1.3.1.

Huang, Nicole; Shih, Shu-Fang; Chang, Hsing-Yi; Chou, Yiing-Jeng. Record linkage research and informed consent: who consents? *BMC Health Services Research* 2007 February 12; 7: 18. NRCBL: 1.3.12; 8.3.1.

Kondro, Wayne. American Medical Association boards implantable chip wagon [news]. *CMAJ/JAMC: Canadian Medical Association Journal* 2007 August 14; 177(4): 331-332. NRCBL: 1.3.12; 5.2; 8.4.

McCubbin, Caroline N. Legal and ethico-legal issues in e-healthcare research projects in the UK. *Social Science and Medicine* 2006 June; 62(11): 2768-2773. NRCBL: 1.3.12; 18.2; 18.6. Identifiers: Great Britain (United Kingdom)

Owen, Paul. Government gives ground on NHS database. *Guardian Unlimited* 2006 December 18: 2 p. [Online]. Accessed: http://politics.guardian.co.uk./print/0,,329665933-110251,00.html [2006 December 18]. NRCBL: 1.3.12; 8.4; 8.3.1. Identifiers: National Health Service; Great Britain (United Kingdom).

Stanberry, Benedict. Legal and ethical aspects of telemedicine. *Journal of Telemedicine and Telecare* 2006; 12(4): 166-175. NRCBL: 1.3.12; 9.1. SC: le.

Torrance, Rebecca J.; Lasome, Caterina E.M.; Agazio, Janice B. Ethics and computer-mediated communication: implications for practice and policy. *JONA: The Journal of Nursing Administration* 2002 June; 32(6): 346-353. NRCBL: 1.3.12; 9.1; 1.1.

Winfield, Marlene. For patients' sake, don't boycott e-health records. *BMJ: British Medical Journal* 2007 July 21; 335(7611): 158. NRCBL: 1.3.12; 9.1.

TERMINAL CARE *See* DEATH AND DYING/ TERMINAL CARE

TERMINALLY ILL *See* DEATH AND DYING/ TERMINAL CARE; HUMAN EXPERIMENTATION/ SPECIAL POPULATIONS/ AGED AND TERMINALLY ILL

TERRORISM *See* WAR AND TERRORISM

TEST TUBE FERTILIZATION *See* IN VITRO FERTILIZATION

THERAPEUTIC RESEARCH *See* HUMAN EXPERIMENTATION

THIRD PARTY CONSENT *See* HUMAN EXPERIMENTATION/ INFORMED CONSENT; INFORMED CONSENT

TISSUE DONATION *See* ORGAN AND TISSUE TRANSPLANTATION/ DONATION AND PROCUREMENT

TORTURE, GENOCIDE, AND WAR CRIMES
See also WAR AND TERRORISM

Adler, Robert. Unwitting accomplices. *New Scientist* 2007 September 29-October 5; 195(2623): 18. NRCBL: 21.4; 4.1.1. Identifiers: American Psychological Association.

Brumfiel, Geoff. Interrogation comes under fire [news]. *Nature* 2007 January 25; 445(7126): 349. NRCBL: 21.4.

Carlson, Elof Axel. Evil at its worst: Nazi medicine and biology. *In his:* Times of Triumph, Times of Doubt: Science and the Battle for Public Trust. Cold Spring Harbor, NY: Cold Spring Harbor Laboratory Press, 2006: 21-37. 38 fn. NRCBL: 21.4; 2.2; 7.4; 15.5.
   Keywords: *eugenics; *historical aspects; *Holocaust; *National Socialism; *physicians; *researchers; killing; involuntary euthanasia; involuntary sterilization; nontherapeutic human experimentation; moral complicity; professional misconduct; Keyword Identifiers: Germany; Twentieth Century; Nuremberg Trials

Dossey, Larry. Where were the doctors? Torture and the betrayal of medicine. *Explore (NY)* 2006 November-December; 2(6): 473-481. NRCBL: 21.4; 21.1; 6. SC: rv.

Glenn, David. A policy on torture roils psychologists' annual meeting. *Chronicle of Higher Education* 2007 September 7; 54(2): A16-A17. NRCBL: 21.4; 6; 1.3.1; 17.1.

Langston, Edward L.; Burns-Cox, C.J.; Halpin, David; Frost, C. Stephen; Hall, Peter. Ethical treatment of military detainees [letters]. *Lancet* 2007 December 15-21; 370(9604): 1999-2000. NRCBL: 21.4; 7.3; 21.1.

Luban, David. Torture and the professions [commentary]. *Criminal Justice Ethics* 2007 Summer-Fall; 26(2): 2, 58-66. NRCBL: 21.4; 1.3.1; 1.3.8; 17.1.

Mayer, Jane. The experiment. *The New Yorker* 2005 July 11 and 18; 61-71. NRCBL: 21.4; 21.2. SC: po. Identifiers: Guantanamo Bay; psychological torture and interrogation techniques.

Mor-Yosef, Shlomo; Weiss, Yuval; Birnbaum, Yair C. Problems and questions regarding the treatment of political leaders. *Israel Medical Association Journal* 2006 November; 8(11): 754-756. NRCBL: 21.4; 9.4; 8.4. Identifiers: Israel; Ariel Sharon.

Nathanson, Vivienne. Indian doctors are promised help to fight abuse of prisoners [news]. *BMJ: British Medical*

NRCBL: National Reference Center for Bioethics Literature Classification Scheme   See inside front cover for terms.

421

Journal 2007 May 12; 334(7601): 972. NRCBL: 21.4. Identifiers: India

Nicholl, David J.; Jenkins, Trefor; Miles, Steven H.; Hopkins, William; Siddiqui, Adnan; Boulton, Frank, et. al. Biko to Guantanamo: 30 years of medical involvement in torture [letter]. *Lancet* 2007 September 8-14; 370(9590): 823. NRCBL: 21.4; 7.4.

Vedantam, Shankar. APA rules on interrogation abuse; psychologists' group bars member participation in certain techniques. *Washington Post* 2007 August 20; p. A3. NRCBL: 21.4; 17.1. SC: po. Identifiers: American Psychological Association.

Wessely, Simon. When doctors become terrorists. *New England Journal of Medicine* 2007 August 16; 357(7): 635-637. NRCBL: 21.4; 4.1.2; 7.4.

Winik, Lyric Wallwork. The 'good' Nazi doctor: according to the 'ways of Auschwitz'. *Moment* 1999 October; 24(5): 60-77. NRCBL: 21.4; 1.3.5; 20.5.1; 7.1. Identifiers: Hans Münch.

TRANSPLANTATION *See* ORGAN AND TISSUE TRANSPLANTATION

TREATMENT REFUSAL
See also ADVANCE DIRECTIVES; EUTHANASIA AND ALLOWING TO DIE; INFORMED CONSENT

Right to refuse treatment: prisoner's claim that conditioning eligibility for parole on taking potentially medically inappropriate medication violated his due process rights is not frivolous. *Journal of the American Academy of Psychiatry and the Law* 2007; 35(2): 260-262. NRCBL: 8.3.4; 17.4; 1.3.5. SC: le.

Cranston, Robert. The doctor who wanted to die. *Today's Christian Doctor* 2007 Fall; 38(3): 29-30. NRCBL: 8.3.4; 20.7; 1.2.

Crosby, Sondra S.; Apovian, Caroline M.; Grodin, Michael A. Hunger strikes, force-feeding, and physicians' responsibilities. *JAMA: The Journal of the American Medical Association* 2007 August 1; 298(5): 563-566. NRCBL: 21.5; 4.1.2; 8.1.

Dute, Joseph. Case of Ciorap v. Moldova, 19 June 2007, no. 12066/02 (Fourth Section) ECHR 2007/11. *European Journal of Health Law* 2007 December; 14(4): 381-384. NRCBL: 21.5. SC: le.

Fullbrook, Suzanne. Death by denomination: a Jehovah Witness's right to die. *British Journal of Nursing* 2007 November 22-December 12; 16(21): 1306-1307. NRCBL: 8.3.4; 19.4; 1.2. SC: le.

Hord, Jeffrey D.; Rehman, Waqas; Hannon, Patricia; Anderson-Shaw, Lisa; Schmidt, Mary Lou. Do parents have the right to refuse standard treatment for their child with favorable-prognosis cancer? Ethical and legal concerns. *Journal of Clinical Oncology* 2006 December 1; 24(34): 5454-5456. NRCBL: 8.3.4; 8.3.2; 9.5.1; 9.5.7. SC: le.

Jones, James W.; McCullough, Laurence B.; Richman, Bruce W. Painted into a corner: unexpected complications in treating a Jehovah's Witness. *Journal of Vascular Surgery* 2006 August; 44(2): 425-428. NRCBL: 8.3.4; 19.4; 1.2.

Levins, Susan. Teen leaves "his only hope" behind in U.S. After 20 months, 14-year-old with leukemia returns home, saying no more chemotherapy or bone marrow transplants. *Washington Post* 2007 January 11; p. B1, B5. NRCBL: 8.3.4; 9.5.7. SC: po. Identifiers: Kabir Sekhri; India.

Levy, Y. Practical aspects of the issue of patients refusing medical care. *Medicine and Law: The World Association for Medical Law* 2007 March; 26(1): 23-31. NRCBL: 8.3.4; 8.3.2; 1.2; 19.4; 20.5.1. SC: le.

Mertl, Steve. B.C. seized three ailing sextuplets; blood transfusions carried out before parents could challenge move in court. *Toronto Star* 2007 February 1; 2 p. [Online]. Accessed: http://www.thestar.com/printArticle/177069 [2007 February 5]. NRCBL: 8.3.4; 8.3.2; 1.2; 19.4. SC: po. Identifiers: British Columbia; Canada.

Niemann, Ulrich; Tag, Brigitte. Amputation bei einer Patientin mit einer Psychose in der Vorgeschichte? [Amputation in a patient with a history of psychosis]. *Ethik in der Medizin* 2007 June; 19(2): 128-138. NRCBL: 8.3.4; 17.1.

Olsen, Douglas P. Unwanted treatment: what are the ethical implications? *AJN: American Journal of Nursing* 2007 September; 107(9): 51-53. NRCBL: 8.3.4; 8.3.2. Identifiers: Starchild Abraham Cherrix; Virginia.

Orr; Robert D.; Craig, Debra. Old enough [case study and commentaries]. *Hastings Center Report* 2007 November-December; 37(6): 15-16. NRCBL: 8.3.4; 8.3.2; 19.4; 8.1; 1.2. SC: cs; le.

Paris, John J.; Schreiber, Michael D.; Moreland, Michael P. Parental refusal of medical treatment for a newborn. *Theoretical Medicine and Bioethics* 2007; 28(5): 427-441. NRCBL: 8.3.4; 8.3.2; 20.5.2.
  Abstract: When there is a conflict between parents and the physician over appropriate care due to an infant whose decision prevails? What standard, if any, should guide such decisions? This article traces the varying standards articulated over the past three decades from the proposal in Duff and Campbell's 1973 essay that these decisions are best left to the parents to the Baby Doe Regs of the 1980s which required every life that could be salvaged be continued. We conclude with support for the policy articulated in the 2007 guidelines of the American

Academy of Pediatrics on non-intervention or withdrawal of intensive care for high-risk newborns.

Pence, Gregory E. Treating Jehovah's Witnesses professionally. *In his:* The Elements of Bioethics. Boston: McGraw-Hill, 2007: 263-279. NRCBL: 8.3.4; 1.2; 8.3.2; 9.5.7.

Rothman, Marc D.; Van Ness, Peter H.; O'Leary, John R.; Fried, Terri R. Refusal of medical and surgical interventions by older persons with advanced chronic disease. *JGIM: Journal of General Internal Medicine* 2007 July; 22(7): 982-987. NRCBL: 8.3.4; 9.5.2. SC: em.

Russell, Barbara. The crucible of anorexia nervosa. *Journal of Ethics in Mental Health [electronic]* 2007 November; 2(2): 6 p. Accessed: http://www.jemh.ca [2008 January 24]. NRCBL: 8.3.4; 9.5.1; 9.5.5.

Abstract: Anorexia nervosa (AN) is a very serious condition because of the suffering and loss of life that it causes. However, the wishes of the people directly involved can be strongly opposed. The person with severe AN may not want treatment, yet her family beseeches professionals to unilaterally intervene and clinical teams are divided over the defensibility of involuntary hospitalization and treatment. The metaphor of a crucible is used in this paper to help identify how much is at stake and how much is in conflict when someone has AN. Frank (2004) cautions against ethical analyses that rely mostly on substantive principles or rules and institutional conflict resolution procedures. This paper applies his heuristic concepts of "ethics-as-substance" and "ethics-as-process" to a prototypical AN case to illustrate how process activities can expand understanding of, and responsiveness to, those who are living with this dire condition or those who are obligated to help.

Rutecki, Greg. Permissibility to accept refusal of potentially life-saving treatment. *Ethics and Medicine: An International Journal of Bioethics* 2007 Summer; 23(2): 77-80. NRCBL: 8.3.4; 21.7. SC: cs.

Sayers, Gwen M.; Gabe, Simon M. Restraint in order to feed: justifying a lawful policy for the UK. *European Journal of Health Law* 2007 April; 14(1): 3-20. NRCBL: 21.5; 8.3.4; 20.5.1; 21.4; 4.4. SC: le.

Vialettes, B.; Samuelian-Massat, C.; Valéro, R.; Béliard, S. The refusal of treatment in anorexia nervosa, an ethical conflict with three characters: "the girl, the family and the medical profession". Discussion in a French legislative context. *Diabetes and Metabolism* 2006 September; 32(4): 306-311. NRCBL: 8.3.4; 8.3.2; 20.5.2; 8.1; 4.1.2; 1.1. SC: le. Note: Abstract in French.

White, Ben; Willmott, Lindy. Will you do as I ask? Compliance with instructions about health care in Queensland. *Queensland University of Technology Law and Justice Journal* 2004; 4(1): 77-87. NRCBL: 8.3.4; 8.3.3; 20.5.4. SC: le.

TREATMENT REFUSAL/ MENTALLY ILL

Bober, Daniel I.; Pinals, Debra A. Prisoner's rights and deliberate indifference. *Journal for the American Academy of Psychiatry and the Law* 2007; 35(3): 388-391. NRCBL: 17.8; 1.3.5; 20.7. SC: le.

Fennell, Philip. Reducing rights in the name of convention compliance: mental health law reform and the new human rights agenda. *In:* Gunning, Jennifer; Holm, Søren, eds. Ethics, law, and society. Volume 2. Aldershot, Hants, England; Burlington, VT: Ashgate, 2006: 125-133. NRCBL: 17.8; 8.3.4. SC: le.

Gottstein, James B. Psychiatrists' failure to inform: is there substantial financial exposure? *Ethical Human Psychology and Psychiatry* 2007; 9(2): 117-125. NRCBL: 17.8; 8.3.4; 17.4; 1.3.5. SC: le.

Harcourt Bernard E. The mentally ill, behind bars [op-ed]. *New York Times* 2007 January 15; p. A15. NRCBL: 17.8. SC: po; em.

Herbel, Bryon L.; Stelmach, Hans. Involuntary medication treatment for competency restoration of 22 defendants with delusional disorder. *Journal of the American Academy of Psychiatry and the Law* 2007; 35(1): 47-59. NRCBL: 17.8; 1.3.5; 8.3.3; 8.3.4; 9.5.3; 17.1. SC: em.

Legemaate, Johan. Psychiatry and human rights. *In:* Gevers, J.K.M.; Hondius, E.H.; Hubben, J.H., eds. Health Law, Human Rights and the Biomedicine Convention: Essays in Honour of Henriette Roscam Abbing. Leiden; Boston: Martinus Nijhoff Publishers, 2005: 119-130. NRCBL: 17.8; 21.1. SC: le.

Lo, Bernard. Ethical issues in psychiatry. *In his:* Resolving Ethical Dilemmas: A Guide for Clinicians. 3rd edition. Philadelphia, PA: Lippincott Williams and Wilkins, 2005: 256-263. NRCBL: 17.8; 8.3.4; 8.4; 17.2; 4.3.

Mayman, Daniel M.; Guyer, Melvin. Defendants' constitutional rights against forced medication in Sell hearings. *Journal of the American Academy of Psychiatry and the Law* 2007; 35(1): 118-120. NRCBL: 17.8; 1.3.5; 8.3.3; 8.3.4; 17.4. SC: le. Identifiers: U.S. v. Rivera-Guerrero.

Young, John; Saleh, Fabien M. The prisoner's right to treatment. *Journal for the American Academy of Psychiatry and the Law* 2007; 35(3): 384-386. NRCBL: 17.8; 1.3.5. SC: le.

TRUTH DISCLOSURE

Doctors who fail their patients [editorial]. *New York Times* 2007 February 13; p. A22. NRCBL: 8.2; 11.2; 20.5.1.

Ambrose, Ella Grace. Placebos: the nurse and the iron pills. *Journal of Medical Ethics* 2007 June; 33(6): 325-328. NRCBL: 8.2; 4.1.3; 21.1. Identifiers: Africa.
Abstract: In sub-Saharan Africa, a nurse gives iron pills as placebos to terminally ill patients. She tells them, act-

ing in what she believes is in their best interests, "these will make you feel better". The patients believe it will help their AIDS and their well-being improves. Do the motive and the patient's positive outcome in well-being make the deceit justifiable when other issues such as consent, autonomy and potential consequences regarding the patient and the wider community are considered? Is there a difference between lying and non-lying deception when the end result is the same? The patients feel better, but at what cost if the deceit was found out? It will be argued that although the actions of the nurse are understandable and to some extent defensible, they are unethical. It is not ethically acceptable to take away the patient's autonomy and risk the health of the community even though the risk of deceit being discovered is a small one.

Ausman, James I. Trust, malpractice, and honesty in medicine: should doctors say they are sorry? *Surgical neurology* 2006 July; 66(1): 105-106. NRCBL: 8.2; 8.1; 8.5.

Ayers, Tressie A. Dutchyn. A partnership in like-minded thinking generating hopefulness in persons with cancer. *Medicine, Health Care and Philosophy* 2007 March; 10(1): 65-80. NRCBL: 8.2; 9.5.1; 1.1; 8.1; 20.4.1.
Abstract: A conceptual model of a partnership in 'like-minded thinking' consists of the following components: a relationship, a shared goal with mutual agreement to work toward that goal, and reciprocal encouragement between two people. A like-minded alliance is a relationship that offers support while at the same time encourages hope and establishes a reciprocating emotional attitude of hopefulness. The discussion focuses on the principles of such a model that is designed primarily as a lay intervention for anyone who has a close friend with cancer and who wants to assist the friend in maintaining a hopeful attitude in the face of illness. While this model is not directed at healthcare professionals it may be transferable into psychosocial interventions to assist persons toward sustaining hopefulness in the context of the cancer trajectory. Much has been written in the literature about how hopelessness spawns despair for individuals who have cancer and in those near the end of life; it may even create a desire for hastened death (Breitbart W., Heller K.S.: 2003, 'Reframing Hope: Meaning-Centered Care for Patients Near the End of Life'. Journal of Palliative Medicine 6, 979-988; Jones J.M., Huggins M.A., Rydall A.C., Rodin G.M.: 2003, 'Symptomatic distress, hopelessness, and the desire for hastened death in hospitalized cancer patients', Journal of Psychosomatic Research 55, 411-418). Therefore, the aim of this paper is to explore how like-minded thinking for a person with cancer and his or her support person provides a framework for a personal shared worldview that is hope-based, meaningful and coherent.

Back, Anthony L.; Arnold, R.M. Discussing prognosis: "how much do you want to know?" talking to patients who are prepared for explicit information. *Journal of Clinical Oncology.* 2006 September 1; 24(25): 4209-4213. NRCBL: 8.2; 8.1; 9.5.1; 8.3.1.

Back, Anthony L.; Arnold, R.M. Discussing prognosis: "how much do you want to know?" talking to patients who do not want information or who are ambivalent. *Journal of Clinical Oncology* 2006 September 1; 24(25): 4214-4217. NRCBL: 8.2; 8.1; 9.5.1.

Blazekovic-Milakovic, Sanja; Matijasevic, Ivana; Stojanovic-Spehar, Stanislava; Supe, Svjetlana. Family physicians' views on disclosure of a diagnosis of cancer and care of terminally ill patients in Croatia. *Psychiatria Danubina* 2006 June; 18(1-2): 19-29. NRCBL: 8.2; 8.1; 20.4.1; 9.5.1. SC: em.

Brennan, Troyen Anthony. Ethics of disclosure following a medical injury: time for reform? *In:* Rhodes, Rosamond; Francis, Leslie P.; Silvers, Anita, eds. The Blackwell Guide to Medical Ethics. Malden, MA: Blackwell Pub., 2007: 393-406. NRCBL: 8.2; 8.5. SC: le.

Capozzi, James D.; Rhodes, Rosamond. A family's request for deception. *Journal of Bone and Joint Surgery. American Volume* 2006 April; 88(4): 906-908. NRCBL: 8.2; 8.1; 21.7. SC: cs.

Derksen, E.; Vernooij-Dassen, M.; Gillissen, F.; Olde Rikkert, M.; Scheltens, P. Impact of diagnostic disclosure in dementia on patients and carers: qualitative case series analysis. *Aging and Mental Health* 2006 September; 10(5): 525-531. NRCBL: 8.2; 8.3.3; 9.5.2; 8.1; 17.1.

Evans, H.M.; Hungin, A.P.S. Uncomfortable implications: placebo equivalence in drug management of a functional illness. *Journal of Medical Ethics* 2007 November; 33(11): 635-638. NRCBL: 8.2; 9.7. SC: an.
Abstract: Using a fictional but representative general practice consultation, involving the diagnosis of irritable bowel syndrome in a patient who is anxious for some relief from the discomfort his condition entails, this paper argues that when both (a) a drug fails to out-perform placebo and (b) the condition in question is a functional illness with no demonstrable underlying pathology, then the action of the drug is not only no better than placebo, and it is also no different from it either. The paper also argues that, in the circumstances of the consultation described, it is striking that current governance deems it ethical for a practitioner to prescribe either a drug or a placebo, both of which appear to rely for their effectiveness on a measure of concealment on the part of the doctor, yet deems it unethical for a practitioner openly to prescribe a harmless and enjoyable substance which (in equivalent conditions of transparency and information) is likely to be no less effective than either drug or placebo and is also likely to be better-tolerated and cheaper than the drug.

Fainzang, Sylvie. When doctors and patients lie to each other: lying and power within the doctor-patient relationship. *In:* van Dongen, Els; Fainzang, Sylvie, eds. Lying and Illness: Power and Performance. Amsterdam: Het Spinhuis; Piscataway, NJ: Transaction Publishers, 2005: 36-55. NRCBL: 8.2; 8.1; 7.1.

Helgesson, Gert; Eriksson, Stefan; Swartling, Ulrica. Limited relevance of the right not to know — reflections on a screening study. *Accountability in Research* 2007 July-September; 14(3): 197-209. NRCBL: 8.2; 9.5.7; 18.5.2; 9.1. SC: an; em. Identifiers: Sweden.

Abstract: The right not to know personal health-related information has been included in prominent human rights documents and subsequently in national legislation since the middle of the 1990s. Apart from situations where another life is at stake, the right not to know has in these documents been formulated as if it should have precedence over other interests. This article argues against giving the right not to know such a prominent position. It does so by questioning the ethical relevance of the concept for both theoretical and empirical reasons. The main focus of the article is on empirical data from a prospective population screening for Type I diabetes. Data indicate that research participants are not as autonomous as is generally assumed by the defenders of the right not to know.

Higgs, Roger. Truth telling. *In:* Rhodes, Rosamond; Francis, Leslie P.; Silvers, Anita, eds. The Blackwell Guide to Medical Ethics. Malden, MA: Blackwell Pub., 2007: 88-103. NRCBL: 8.2.

Hinks, Timothy S. Deceiving patients: ends never justify means [letter]. *BMJ: British Medical Journal* 2007 May 26; 334(7603): 1072. NRCBL: 8.2. Comments: Daniel K. Sokol. Can deceiving patients be morally acceptable? BMJ: British Medical Journal 2007 May 12; 334(7601): 984-986.

Jotkowitz, A.; Glick, Shimon; Gezundheit, B. Truthtelling in a culturally diverse world. *Cancer Investigation* 2006 December; 24(8): 786-789. NRCBL: 8.2; 8.1; 7.1; 21.7.

Kendall, Sharon. Being asked not to tell: nurses' experiences of caring for cancer patients not told their diagnosis. *Journal of Clinical Nursing* 2006 September; 15(9): 1149-1157. NRCBL: 8.2; 9.5.1; 7.1; 8.1; 21.7.

Kirklin, D. Truth telling, autonomy and the role of metaphor. *Journal of Medical Ethics* 2007 January; 33(1): 11-14. NRCBL: 8.2. SC: an.

Abstract: This paper examines the potential role of metaphors in helping healthcare professionals to communicate honestly with patients and in helping patients gain a richer and more nuanced understanding of what is being explained. One of the ways in which doctors and nurses may intentionally, or unintentionally, avoid telling the truth to patients is either by using metaphors that obscure the truth or by failing to deploy appropriately powerful and revealing metaphors in their discussions. This failure to tell the truth may partly account for the observation by clinicians that patients sometimes make decisions that, from the perspective of their clinician, and given all that the clinician knows, seem unwise. For example, patients with advanced cancer may choose to undergo further, aggressive, treatment despite the fact that they are likely to accrue little or no benefit as a result. While acknowledging that the immediate task of telling patients the truth can be difficult for all those concerned, I argue that the long-term consequences of denying patients autonomy at the end of life can be harmful to patients and can leave doctors and nurses distressed and confused.

Kolber, Adam J. A limited defense of clinical placebo deception. *Yale Law and Policy Review* 2007; 26: 75-134 [Online]. Accessed: http://papers.ssrn.com/sol3/papers.cfm?abstract_id=967563 [2008 April 10]. NRCBL: 8.2; 7.1; 8.1; 8.3.1. SC: le.

Lee, A.; Wu, H.Y. Diagnosis disclosure in cancer patients — when the family says "no!". *Singapore Medical Journal* 2002 October; 43(10): 533-538. NRCBL: 8.2; 8.1; 20.3.1; 20.3.3; 21.1. SC: rv.

Mack, Jennifer W.; Wolfe, Joanne; Grier, Holocombe E.; Cleary, Paul D.; Weeks, Jane C. Communication about prognosis between parents and physicians of children with cancer: parent preferences and the impact of prognostic information. *Journal of Clinical Oncology* 2006 November 20; 24(33): 5265-5270. NRCBL: 8.2; 8.3.2; 9.5.1; 9.5.7. SC: em.

Miller, Franklin G.; Wendler, David; Swartzman, Leora C. Deception in research on the placebo effect. *PLoS Medicine* 2005 September; 2(9): e262 (0853-0859). NRCBL: 8.2; 18.3; 8.1.

Moynihan, Timothy J.; Schapira, Lidia. Preparing ourselves, our trainees, and our patients: a commentary on truthtelling. *Journal of Clinical Oncology* 2007 February 1; 25(4): 456-457. NRCBL: 8.2; 8.1; 7.2.

Nelson, William A.; Campfield, Justin. Ethical implications of transparency. Valid justification is required when withholding information. *Healthcare Executive* 2006 November-December; 21(6): 33-34. NRCBL: 8.2; 7.1; 8.3.1; 9.3.1.

Öksüzoglu, Berna; Abali, Hüseyin; Bakar, Meral; Yildirim, Nuriye; Zengin, Nurullah. Disclosure of cancer diagnosis to patients and their relatives in Turkey: views of accompanying persons and influential factors in reaching those views. *Tumori* 2006 January-February; 92(1): 62-66. NRCBL: 8.2; 9.5.1; 8.1. SC: em.

Oliffe, John; Thorne, Sally; Hislop, T.Gregory; Armstrong, Elizabeth-Anne. "Truth telling" and cultural assumptions in an era of informed consent. *Family and Community Health* 2007 January-March; 30(1): 5-15. NRCBL: 8.2; 8.3.1; 8.1; 9.5.1; 21.7.

Page, Cameron. Hope. *Hastings Center Report* 2007 November-December; 37(6): 12. NRCBL: 8.2; 8.1; 9.5.6. SC: cs.

Pence, Gregory E. Lying to patients and ethical relativism. *In his:* The Elements of Bioethics. Boston: McGraw-Hill, 2007: 1-20. NRCBL: 8.2; 1.1; 4.1.2.

NRCBL: National Reference Center for Bioethics Literature Classification Scheme   See inside front cover for terms.

Richards, Tessa. My illness, my record. *BMJ: British Medical Journal* 2007 March 10; 334(7592): 510. NRCBL: 8.2. Identifiers: Great Britain.

Rockwell, Lindsay E. Truthtelling. *Journal of Clinical Oncology* 2007 February 1; 25(4): 454-455. NRCBL: 8.2; 9.5.1; 20.3.2.

Sheu, Shuh-Jen; Huang, Shu-He; Tang, Fu-In; Huang, Song-Lih. Ethical decision making on truth telling in terminal cancer: medical students' choices between patient autonomy and family paternalism. *Medical Education* 2006 June; 40(6): 590-598. NRCBL: 8.2; 7.2; 8.1; 9.5.1.

Sokol, Daniel K. Can deceiving patients be morally acceptable? *BMJ: British Medical Journal* 2007 May 12; 334(7601): 984-986. NRCBL: 8.2. SC: cs.

Sokol, Daniel K. How the doctor's nose has shortened over time; a historical overview of the truth-telling debate in the doctor-patient relationship. *Journal of the Royal Society of Medicine* 2006 December; 99(12): 632-636. NRCBL: 8.2; 8.1; 2.2.

Srivastava, Ranjana. The art of letting go. *New England Journal of Medicine* 2007 July 5; 357(1): 3-5. NRCBL: 8.2; 20.5.1; 9.4. SC: cs.

Waite, Michael. To tell the truth: the ethical and legal implications of disclosure of medical error. *Health Law Journal* 2005; 13: 1-33. NRCBL: 8.2; 7.2; 8.5; 9.8. SC: le.

Wright, Linda; MacRae, Susan; Gordon, Debra; Elliot, Esther; Dixon, David; Abbey, Susan; Richarson, Robert. Disclosure of misattributed paternity: issues involved in the discovery of unsought information. *Seminars in Dialysis* 2002 June; 15(3): 202-206. NRCBL: 8.2; 19.5. SC: cs.

UMBILICAL CORD BLOOD *See* BLOOD BANKING, DONATION, AND TRANSFUSION

VALUE OF LIFE *See* QUALITY AND VALUE OF LIFE

WAR AND TERRORISM
See also TORTURE, GENOCIDE, AND WAR CRIMES

Beshai, James A.; Tushup, Richard J. Sanctity of human life in war: ethics and post traumatic stress disorder. *Psychological Reports* 2006 February; 98(1): 217-225. NRCBL: 21.2; 4.4; 7.1; 17.2.

Daccord, Yves. ICRC and confidentiality [letter]. *Lancet* 2007 September 8-14; 370(9590): 823-824. NRCBL: 21.2; 21.4; 8.4. Identifiers: International Committee of the Red Cross. Comments: Geoffrey Robertson. Health and human rights series. Lancet 2007 August 4; 370: 368-369.

Federation of American Scientists [FAS]. Working Group on BW [Biological Warfare]; Stockholm International Peace Research Institute [SIPRI]; Verification Research, Training and Information Centre [VERTIC]; International Network of Engineers and Scientists for Global Responsibility[INES]' Acronym Institute for Disarmament Diplomacy; Sunshine Project; Pax Christi International; Physicians for Social Responsibility; 20/20 Vision. Draft recommendations for a code of conduct for biodefense programs. Federation of American Scientists 2002 November; 2 p. [Online]. Accessed: http://www.fas.org/bwc/papers/code.pdf [2007 May 11]. NRCBL: 21.3; 1.3.5; 6; 21.1.

Green, Colin; Khan, Asad; Karmi, Ghada; Burns-Cox, Chris; Birnstingl, Martin; Halpin, David; Summerfield, Derek. Medical ethical violations in Gaza. *Lancet* 2007 December 22-2008 January 4; 370(9605): 2102. NRCBL: 21.2; 21.4; 7.4; 7.3.

Greif, Karen F.; Merz, Jon F. Science in the national interest: bioterrorism and civil liberties. *In their:* Current Controversies in the Biological Sciences: Case Studies of Policy Challenges from New Technologies. Cambridge, MA: MIT, 2007: 235-254. NRCBL: 21.3; 21.1.

Guillemin, Jeanne. Scientists and the history of biological weapons. A brief historical overview of the development of biological weapons in the twentieth century. *EMBO Reports* 2006 July; 7 Special No: S45-S49. NRCBL: 21.3; 1.3.9.

Holmes, D.; Perron, A. Violating ethics: unlawful combatants, national security and health professionals. *Journal of Medical Ethics* 2007 March; 33(3): 143-145. NRCBL: 21.2; 21.4; 4.1.2; 4.1.3.

Hurst, G. Cameron. Biological weapons: the United States and the Korean War. *In:* LaFleur, William R.; Böhme, Gernot; Shimazono, Susumu, eds. Dark Medicine: Rationalizing Medical Research. Bloomington: Indiana University Press, 2007: 105-120. NRCBL: 21.3; 21.2; 1.3.5.

Kaebnick, Gregory E. Secrets and open societies [editorial]. *Hastings Center Report* 2007 May-June; 37(3): 2. NRCBL: 21.3; 1.3.9; 1.3.7; 5.3. Identifiers: national security.

Kmietowicz, Zosia. Doctors condemn use of drugs as weapon [news]. *BMJ: British Medical Journal* 2007 May 26; 334(7603): 1073. NRCBL: 21.3; 1.3.5; 9.7.

Marks, Jonathan H. The bioethics of war [review of Oath Betrayed: Torture, Medical Complicity and the War on Terror, by Steven H. Miles; Bioethics and Armed Conflicts: Moral Dilemmas of Medicine and War, by Michael L. Gross]. *Hastings Center Report* 2007 March-April; 37(2): 41-42. NRCBL: 21.2; 21.4.

Miller, Seumas; Selgelid, Michael. Ethical and philosophical consideration of the dual-use dilemma in the biological sciences. *Science and Engineering Ethics* 2007 December; 13(4): 523-580. 164 refs. NRCBL: 21.3; 1.3.9; 15.1; 15.7. SC: an.
Keywords: *biological sciences; *biomedical research; *bioterrorism; *ethical analysis; *moral policy; biological warfare; decision making; freedom; genetically modified organisms; government regulation; guidelines; information dissemination; international aspects; microbiology; philosophy; policy analysis; property rights; public health; public policy; publishing; researchers; risk; risks and benefits; science; self regulation; synthetic biology; Proposed Keywords: viruses; Keyword Identifiers: Biological Weapons Convention; National Biodefense Analysis and Countermeasures Center; United States

Moreno, Jonathan D. Bioethics and bioterrorism. *In:* Steinbock, Bonnie, ed. The Oxford Handbook of Bioethics. Oxford; New York: Oxford University Press, 2007: 721-733. NRCBL: 21.3; 2.1; 15.1.

Moreno, Jonathan D.; Selgelid, Michael J. The dual-use dilemma [letter and reply]. *Hastings Center Report* 2007 September-October; 37(5): 6-7. NRCBL: 21.3; 1.3.7; 1.3.9; 1.3.5. Comments: Michael J. Selgelid. A tale of two studies: ethics, bioterrorism, and the censorship of science. Hastings Center Report 2007 May-June; 37(3): 35-43.

Sanders, Cheryl J. Religion and ethical decision making in the African American community: bioterrorism and the black postal workers. *In:* Prograis, Lawrence J.; Pellegrino, Edmund D., eds. African American Bioethics: Culture, Race, and Identity. Washington, DC: Georgetown University Press, 2004: 93-104. NRCBL: 21.3; 1.2; 1.1; 9.5.4. Conference: Symposium on African American Perspectives in Bioethics and Second Annual Conference on Health Disparities, held on September 23-24, 2004, at Georgetown University.

Schmaltz, Florian. Neurosciences and research on chemical weapons of mass destruction in Nazi Germany. *Journal of the History of the Neurosciences* 2006 September; 15(3): 186-209. NRCBL: 21.2; 21.3; 1.3.5.

Selgelid, Michael J. A tale of two studies: ethics, bioterrorism, and the censorship of science. *Hastings Center Report* 2007 May-June; 37(3): 35-43. NRCBL: 21.3; 1.3.7; 1.3.5.

Tyshenko, Michael G. Management of natural and bioterrorism induced pandemics. *Bioethics* 2007 September; 21(7): 364-369. 31 fn. NRCBL: 21.3; 9.1; 15.1; 5.3.
Keywords: *biological warfare; *biotechnology; *bioterrorism; *genetic engineering; historical aspects; public health; resource allocation; risk; social control; world health
Abstract: A recent approach for bioterrorism risk management calls for stricter regulations over biotechnology as a way to control subversion of technology that may be used to create a man-made pandemic. This approach is largely unworkable given the increasing pervasiveness of molecular techniques and tools throughout society. Emerging technology has provided the tools to design much deadlier pathogens but concomitantly the ability to respond to emerging pandemics to reduce mortality has also improved significantly in recent decades. In its historical context determining just how 'risky' biological weapons is an important consideration for decision making and resource allocation. Management should attempt to increase capacity, share resources, provide accurate infectious disease reporting, deliver information transparency and improve communications to help mitigate the magnitude of future pandemics.

Wright, Samuel L. Protecting the right to medical treatment from the war on terror: the status of Guantanamo Bay detainees. *Journal of Legal Medicine* 2007 January-March; 28(1): 135-149. NRCBL: 21.2; 9.2.

WITHHOLDING TREATMENT *See* EUTHANASIA AND ALLOWING TO DIE

WOMEN *See* CARE FOR SPECIFIC GROUPS/ WOMEN; HUMAN EXPERIMENTATION/ SPECIAL POPULATIONS/ WOMEN

WRONGFUL BIRTH *See* CONTRACEPTION

XENOTRANSPLANTATION *See* ORGAN AND TISSUE TRANSPLANTATION/ XENOTRANSPLANTATION

SECTION II:
PERIODICAL LITERATURE
AND ESSAYS

AUTHOR INDEX

Section II: Periodical Literature and Essays
Author Index

A

Aagaard-Hansen, Jens; Johansen, Maria Vang; Riis, Pols. Research ethical challenges in cross-disciplinary and cross-cultural health research: the diversity of codes. *Danish Medical Bulletin* 2004 February; 51(1): 117-120. Subject: 18.2

Abadie, Alberto; Gay, Sebastien. The impact of presumed consent legislation on cadaveric organ donation: a cross-country study. *Journal of Health Economics* 2006 July; 25(4): 599-620. Subject: 19.5

Abbott, Alison; Schiermeier, Quirin. Showdown for Europe. The European Union is set to make a landmark decision on genetically modified crops. *Nature* 2007 December 13; 450(7172): 928-929. Subject: 15.1

Abel, Gregory A.; Penson, Richard T.; Joffe, Steven; Schapira, Lidia; Chabner, Bruce A.; Lynch, Thomas J., Jr. Direct-to-consumer advertising in oncology. *Oncologist* 2006 February; 11(2): 217-226. Subject: 9.5.1

Abelson, Julia; Giacomini, Mita; Lehoux, Pascale; Gauvin, Francois-Pierre. Bringing 'the public' into health technology assessment and coverage policy decisions: from principles to practice. *Health Policy* 2007 June; 82(1): 37-50. Subject: 5.3

Abelson, Reed. A chance to pick hospice, and still hope to live. *New York Times* 2007 February 10; p. A1, C4. Subject: 20.4.1

Abrams, Frederick R. Colorado revised statutes in support of palliative care limiting criminal liability. *Journal of Palliative Medicine* 2006 December; 9(6): 1254-1256. Subject: 4.4

Academy of Medical Sciences. Personal data for public good: using health information in medical research. London: Academy of Medical Sciences, 2006 January; 77 p. [Online]. Accessed: http://www.acmedsci.ac.uk/images/project/Personal.pdf [2007 April 3]. Subject: 8.4

Acciani, Jennifer. Resident portfolio: breaking trust — a reflection on confidentiality and minors. *Academic Emergency Medicine* 2006 December; 13(12): 1339-1340. Subject: 8.4

Acharya, Tara; Daar, Abdallah S.; Dowdeswell, Elizabeth; Singer, Peter A. Better global governance to promote genomics for development. Toronto, Canada: University of Toronto, Joint Centre for Bioethics, 2004; 10 p. [Online]. Accessed: http://www.utoronto.ca/jcb/genomics/documents/CGI-paper.pdf [2007 April 16]. Subject: 15.1

Ackerman, Felicia Nimue. Death is a punch in the jaw: life-extension and its discontents. *In:* Steinbock, Bonnie, ed. The Oxford Handbook of Bioethics. Oxford; New York: Oxford University Press, 2007: 324-348. Subject: 20.5.1

Ackerman, Felicia Nimue. Patient and family decisions about life-extension and death. *In:* Rhodes, Rosamond; Francis, Leslie P.; Silvers, Anita, eds. The Blackwell Guide to Medical Ethics. Malden, MA: Blackwell Pub., 2007: 52-68. Subject: 20.3.3

Acquavella, John F. Why focus only on financial interests? *Epidemiology* 2006 May; 17(3): 248-249. Subject: 1.3.9

Adam, G.M. Assisted human reproduction — legal rights of the unborn in respect of avoidable damage. *Medicine and Law: The World Association for Medical Law* 2007 June; 26(2): 325-337. Subject: 14.4

Adams, Jared R.; Drake, Robert E.; Wolford, George L. Shared decision-making preferences of people with severe mental illness. *Psychiatric Services* 2007 September; 58(9): 1219-1221. Subject: 8.1

Adams, John. Prescribing: the ethical dimension. *Nurse Prescriber [electronic]* 2004 July; 1(7): 3 p. Accessed: http://journals.cambridge.org/action/displayJournal?jid=NPR [2007 May 11]. Subject: 2.1

Adams, Karen E. What's "normal": female genital mutilation, psychology, and body image. *Journal of the American Medical Women's Association* 2004 Summer; 59(3): 168-170. Subject: 9.5.5

Adams, Marcus P.; Lawrence, Ryan E.; Curlin, Farr A. Conscience and conflict. *American Journal of Bioethics* 2007 December; 7(12): 28-29; author reply W1-W2. Subject: 4.1.2

See SUBJECT HEADING KEY FOR SECTION II on inside back cover.

431

Adams, Mary; Fischler, Ira; Simmerling, Mary;. Deception. *Journal of Empirical Research on Human Research Ethics* 2007 March; 2(1): 87-91. Subject: 18.3

Adams, Samantha; de Bont, Antoinette. Information Rx: prescribing good consumerism and responsible citizenship. *Health Care Analysis: An International Journal of Health Philosophy and Policy* 2007 December; 15(4): 273-290. Subject: 9.1

Adamson, Peter C.; Paradis, Carmen; Smith, Martin L. All for one, or one for all? [case study and commentary]. *Hastings Center Report* 2007 July-August; 37(4): 13-15. Subject: 18.5.7

Addelson, Kathryn Pyne. The emergence of the fetus. *In:* Mui, Constance L.; Murphy, Julien S., eds. Gender Struggles: Practical Approaches to Contemporary Feminism. Lanham, MD: Rowman & Littlefield Pub., 2002: 118-136. Subject: 9.5.8

Adkins, Jason A. Meet me at the (West Coast) hotel: the Lochner Era and the demise of Roe v. Wade. *Specialty Law Digest: Health Care Law* 2007 July; (339): 9-43. Subject: 12.4.1

Adler, Robert. Unwitting accomplices. *New Scientist* 2007 September 29-October 5; 195(2623): 18. Subject: 21.4

Adrian, Manuella. Decisions involving research and ethics: misusing drug use(r) statistics. *In:* Kleinig, John; Einstein, Stanley, eds. Ethical Challenges for Intervening in Drug Use: Policy, Research and Treatment Issues. Huntsville, TX: Office of International Criminal Justice; Criminal Justice Center, Sam Houston State University, 2006: 217-258. Subject: 1.3.9

Afif, Maryam T. Prescription ethics: can states protect pharmacists who refuse to dispense contraceptive prescriptions? *Pace Law Review* 2005 Fall; 26(1): 243-272. Subject: 9.7

Agar, Nicholas. Embryonic potential and stem cells. *Bioethics* 2007, May; 21(4): 198-207. Subject: 18.5.4

Agar, Nicholas. Whereto transhumanism? The literature reaches a critical mass. *Hastings Center Report* 2007 May-June; 37(3): 12-17. Subject: 4.5

Ågård, Anders; Löfmark, Rurik; Edvardsson, Nils; Ekman, Inger. Views of patients with heart failure about their role in the decision to start implantable cardioverter-defibrillator treatment: prescription rather than participation. *Journal of Medical Ethics* 2007 September; 33(9): 514-518. Subject: 8.3.1

Agarwal, Sanjay K.; Estrada, Sylvia; Foster, Warren G.; Wall, L. Lewis; Brown, Doug; Revis, Elaine S.; Rodriguez, Suzanne. What motivates women to take part in clinical and basic science endometriosis research? *Bioethics* 2007 June; 21(5): 263-269. Subject: 18.5.3

Agich, George J. Autonomy as a problem for clinical ethics. *In:* Nys, Thomas; Denier, Yvonne; Vandevelde, Toon,

eds. Autonomy and Paternalism: Reflections on the Theory and Practice of Health Care. Leuven; Dudley, MA: Peeters, 2007: 71-91. Subject: 2.1

Agich, George J. Reflections on the function of dignity in the context of caring for old people. *Journal of Medicine and Philosophy* 2007 September-October; 32(5): 483-494. Subject: 9.5.2

Agnello, Vincent. Commentary: Vioxx and public policy? Taking the right steps. *Organizational Ethics: Healthcare, Business, and Policy* 2006 Fall-Winter; 3(2): 125-130. Subject: 9.7

Aitken, Maryanne; Metcalfe, Sylvia. The social imperative for community genetic screening: an Australian perspective. *In:* Betta, Michela, ed. The Moral, Social, and Commercial Imperatives of Genetic Testing and Screening: the Australian Case. Dordrecht: Springer, 2006: 165-184. Subject: 15.3

Akabayshi, Akira; Slingsby, Brian T.; Nagao, Noriko; Kai, Ichiro; Sato, Hajime. An eight-year follow-up national study of medical school and general hospital ethics committees in Japan. *BMC Medical Ethics* 2007; 8(8), 8 p. [Online]. Accessed: http://www.biomedcentral.com/1472-6939-8-8 [2007 July 20]. Subject: 18.2

Albera, R.; Argentero, P.; Bonziglia, S.; De Andreis, M.; Preti, G.; Palonta, F.; Canale, A. Informed consent in ENT. Patient's judgement about a specific consensus form. *Acta Otorhinolaryngologica Italica: Organo Ufficiale della Società Italiana di Otorinolaringologia e Chirurgia Cervico-facciale* 2005 October; 25(5): 304-311 [Online]. Accessed: http://www.actaitalica.it/ [2007 April 13]. Subject: 8.3.1

Alderson, Priscilla. Consent to surgery for deaf children: making informed decisions. *In:* Komesaroff, Linda, ed. Surgical Consent: Bioethics and Cochlear Implantation. Washington, DC: Gallaudet University Press, 2007: 30-41. Subject: 8.3.2

Alderson, Priscilla. Ethics. *In:* Fraser, Sandy; Lewis, Vicky; Ding, Sharon; Kellett, Mary; Robinson, Chris, eds. Doing Research with Children and Young People. London; Thousand Oaks, CA: Sage Publications, 2004: 97-112. Subject: 18.5.2

Alderson, Priscilla; Hawthorne, Joanna; Killen, Margaret. The participation rights of premature babies. *In:* Freeman, Michael, ed. Children's Health and Children's Rights. Boston: Martinus Nijhoff Publishers, 2006: 31-50. Subject: 9.5.7

Aldhous, Peter. Angry reception greets patent for synthetic life [news]. *New Scientist* 2007 June 16-22; 194(2608): 13. Subject: 15.8

Aldhous, Peter. Prescribed opinions. *New Science* 2007 January 6-12; 193(2585): 17. Subject: 9.7

Aldhous, Peter. Your own book of life. *New Scientist* 2007 September 8-14; 195(2620): 8-11. Subject: 15.10

Subject = NRCBL Primary Classification Number; see inside front cover.

Allahbadia, Gautam N.; Kaur, Kulvinder. Accreditation, supervision, and regulation of ART clinics in India — a distant dream? *Journal of Assisted Reproduction and Genetics* 2003 July; 20(7): 276-280. Subject: 14.4

Allan, Jonathan S.; Aluwihare, A.P.R.; Bach, Fritz H.; Caplan, Arthur; Chapman, Louisa; Dickens, Bernard M.; Fishman, Jay A.; Groth, C.G.; Breimer, M.E.; Menache, Andr?; Morris, Peter J.; van Rongen, Eric. Round Table Discussion: animal-to-human organ transplants. *Bulletin of the World Health Organization* 1999; 77(1): 62-81. Subject: 19.1

Allan, Robert. Rationing of medical care by age [editorial]. *Clinical Medicine (London, England)* 2006 July-August; 6(4): 329-330. Subject: 9.4

Allen, Charlotte F. Back off! I'm not dead yet. I don't want a living will. Why should I? [op-ed]. *Washington Post* 2007 October 14; p. B1, B4. Subject: 20.5.4

Allen, Colin; Bekoff, Marc. Animal minds, cognitive ethology, and ethics. *Journal of Ethics* 2007; 11(3): 299-317. Subject: 22.1

Allhoff, Fritz. On the autonomy and justification of nanoethics. *NanoEthics* 2007 December; 1(3): 185-210. Subject: 5.4

Alliman, S.; McCarthy Veach, P.; Bartels, D.; Bower, M.; James, C.; LeRoy, B. Ethical and professional challenges in clinical practice: a comparative analysis of Australian and U.S. genetic counselors. *Journal of Genetic Counseling* 2007 December; 16(6): 687. Subject: 15.2

Almirall, Nat. The ethics of engagement with the pharmaceutical industry. *Michigan Medicine* 2006 January-February; 105(1): 10-12. Subject: 9.7

Alonzi, Andrew; Pringle, Mike. Mental Capacity Act 2005 should guide doctors to help protect vulnerable people [editorial]. *BMJ: British Medical Journal* 2007 November 3; 335(7626): 898. Subject: 8.3.3

Alpert, Joseph S. Ethical precepts for cardiologists. *Current Cardiology Reports* 2005 January; 7(1): 1-2. Subject: 2.1

Alpert, Sheri. Brain privacy: how can we protect it? *American Journal of Bioethics* 2007 September; 7(9): 70-73. Subject: 17.1

Alpert, Sheri. Total information awareness: forgotten but not gone: lessons for neuroethics. *American Journal of Bioethics* 2007 May; 7(5): 24-26. Subject: 17.1

Alsulaiman, Ayman; Hewison, J. Attitudes to prenatal testing and termination of pregnancy in Saudi Arabia. *Community Genetics* 2007; 10(3): 169-173. Subject: 15.2

Altieri, Miguel Angel. The myths of biotechnology: some ethical questions. *In:* Serageldin, I.; Martin-Brown, J., eds. Proceedings of an associated event of the 5th Annual World Bank Conference on Environmentally and Socially Sustainable Development. Washington, D.C.: World Bank, 1998: 53- 58 [Online]. Accessed: http://nature.berkeley. edu/~agroeco3/the_myths.html [2007 April 24]. Subject: 1.3.11

Altman, Dennis. Taboos and denial in government responses. *International Affairs* 2006 March; 82(2): 257-268. Subject: 9.5.6

Alton, David. The paramount human right: the right to life. *In:* Wagner, Teresa; Carbone, Leslie, eds. Fifty Years after the Declaration: the United Nations' Record on Human Rights. Lanham, MD: University Press of America, 2001: 11-16. Subject: 4.4

Alvarez, Allen Andrew A. Threshold considerations in fair allocation of health resources: justice beyond scarcity. *Bioethics* 2007 October; 21(8): 426-438. Subject: 9.4

Alzheimer's Association. National Board of Directors. Ethical issues in dementia research (with special emphasis on "informed consent"). Chicago, IL: Alzheimer's Association, 1997 May; 2 p. [Online]. Accessed: http://www.alz. org/national/documents/statements_ethicalissues.pdf [2007 September 17]. Subject: 18.3

Amar, Jonathan. Medical decision on behalf of incompetent patients: federal court upholds law allowing medical decisions for incompetent patients — Doe ex re. Tarlow v. District of Columbia. *American Journal of Law and Medicine* 2007; 33(4): 703-705. Subject: 8.3.3

Ambrose, Ella Grace. Placebos: the nurse and the iron pills. *Journal of Medical Ethics* 2007 June; 33(6): 325-328. Subject: 8.2

American Association for Laboratory Animal Science. AALAS position statement on the humane care and use of laboratory animals. *Comparative Medicine* 2006 December; 56(6): 534. Subject: 22.2

American Association of Electrodiagnostic Medicine. Guidelines for ethical behavior relating to clinical practice issues in electrodiagnostic medicine. *Muscle and Nerve Supplement* 1999; 8: S43-S47. Subject: 6

American Association of University Professors. Protecting human beings: institutional review boards and social science research. *Academe* 2001 May/June; 87(3): 55-67. Subject: 18.2

American Civil Liberties Union [ACLU]. Answers to frequently asked questions about government access to personal medical information. New York: American Civil Liberties Union, 2003 May 30; 4 p. [Online]. Accessed: http://www.aclu.org/privacy/medical/ 15222res20030530.html [2007 April 26]. Subject: 8.4

American College of Epidemiology Policy Committee; Ness, Roberta B. Biospecimen "ownership": point. *Cancer Epidemiology, Biomarkers and Prevention* 2007 February; 16(2): 188-189. Subject: 18.3

American College of Obstetricians and Gynecologists. Committee on Ethics. The limits of conscientious refusal in reproductive medicine. *ACOG Committee Opinion* 2007 November; (385): 1-6 [Online]. Accessed:

See SUBJECT HEADING KEY FOR SECTION II on inside back cover.

433

http://www.acog.org/from_home/publications/ethics/co385.pdf [2008 March 25]. Subject: 7.1

American Dental Association [ADA]. Principles of Ethics and Code of Professional Conduct with official advisory opinions revised to January 2005. Chicago, IL: American Dental Association [ADA], 2005 January: 20 p. [Online]. Accessed: http://www.ada.org/prof/prac/law/code/ada_code.pdf [2007 June 19]. Subject: 6

American Life League. Emergency contraception: the morning-after pill. Stafford, VA: American Life League, 2006; 2 p. [Online]. Accessed: http://www.all.org/article.php?id=10130 [2007 April 9]. Subject: 11.1

American Medical Association. Council on Ethical and Judicial Affairs; Taub, Sara; Morin, Karine; Goldrich, Michael S.; Ray, Priscilla; Benjamin, Regina. Physician health and wellness. *Occupational Medicine (Oxford, England)* 2006 March; 56(2): 77-82. Sub- ject: 7.1

American Medical Association. Institute for Ethics. Ethical Force Program; Levine, Mark A.; Wynia, Matthew K.; Schyve, Paul M.; Teagarden, J. Russell; Fleming, David A.; Donohue, Sharon King; Anderson, Ron J.; Sabin, James; Emanuel, Ezekiel J. Improving access to health care: a consensus ethical framework to guide proposals for reform. *Hastings Center Report* 2007 September-October; 37(5): 14-19. Subject: 9.1

American Medical Women's Association [AMWA]. Emergency contraception [position statement]. Alexandria, VA: American Medical Women's Association, 1996 November: 4 p. [Online]. Accessed: http://www.amwa-doc.org/index.cfm?objectid=0EF88909-D567-0B25-531927EE4CC23EFB [2007 April 9]. Subject: 11.1

American Nurses Association [ANA]. Ethics and human rights position statements: assisted suicide. Silver Spring, MD: American Nurses Association, 1994 December 8: 5 p. [Online]. Accessed: http://www.nursingworld.org/readroom/position/ethics/etsuic.htm [2007 April 19]. Subject: 20.7

American Pediatric Surgical Association Ethics and Advocacy Committee; Fallat, Mary E.; Caniano, Donna A.; Fecteau, Annie H. Ethics and the pediatric surgeon. *Journal of Pediatric Surgery* 2007 January; 42(1): 129-136; discussion 136. Subject: 9.5.8

American Psychological Association. Ethics Committee. Report of the Ethics Committee, 2005. *American Psychologist* 2006 July-August; 61(5): 522-529. Subject: 9.6

American Public Health Association [APHA]. Support of public education about emergency contraception and reduction or elimination of barriers to access. Washington, DC: American Public Health Association, 2003 November 18, Policy No. 200315; 3 p. [Online]. Accessed: http://www.apha.org/advocacy/policy/policysearch/default.htm?id=1252 [2007 April 9]. Subject: 11.1

American Society for Reproductive Medicine [ASRM]. Ethics Committee. Access to fertility treatment by gays, lesbians, and unmarried persons. *Fertility and Sterility* 2006 November; 86(5): 1333-1335. Subject: 14.1

American Society for Reproductive Medicine [ASRM]. Practice Committee. Definition of "experimental". *Fertility and Sterility* 2006 November; 86(5, Supplement): S123. Subject: 14.1

American Society for Reproductive Medicine [ASRM]. Practice Committee; Society for Assisted Reproductive Technology [SART]. Practice Committee. Elements to be considered in obtaining informed consent for ART. *Fertility and Sterility* 2006 November; 86(5, Supplement): S272-S273. Subject: 14.4

American Society for Reproductive Medicine [ASRM]. Practice Committee; Society for Assisted Reproductive Technology [SART]. Practice Committee. Preimplantation genetic diagnosis. *Fertility and Sterility* 2006 November; 86(5, Supplement): S257-S258. Subject: 15.2

American Society for Reproductive Medicine [ASRM]. Practice Committee; Society for Assisted Reproductive Technology [SART]. Practice Committee. 2006 guidelines for gamete and embryo donation. *Fertility and Sterility* 2006 November; 86(5, Supplement): S38-S50. Subject: 14.4

American Society of Clinical Oncology. Interactions with the investment industry: practical and ethical implications. *Journal of Clinical Oncology* 2007 January 20; 25(3): 338-340. Subject: 7.1

American Society of Human Genetics [ASHG]. Statement from the American Society of Human Genetics (ASHG): the Board of Directors of the American Society of Human Genetics has endorsed Senate Bill 318, the Genetic Nondiscrimination and Health Insurance and Employment Act. Bethesda, MD: American Society of Human Genetics, 2001 December 18: 1 p. [Online]. Accessed: http://genetics.faseb.org/genetics/ashg/pubs/policy/pol-47.htm [2006 July 26]. Subject: 15.3

American Society of Human Genetics [ASHG]. Statement on stem cell research. Bethesda, MD: American Society of Human Genetics, 2001 August 27: 1 p. [Online]. Accessed: http://genetics.faseb.org/genetics/ashg/pubs/policy/pol-44.htm [2006 July 26]. Subject: 18.5.4

Aminoff, Bechor Zvi. The new Israeli law "The Dying Patient" and Relief of Suffering Units. *American Journal of Hospice and Palliative Care* 2007 February-March; 24(1): 54-58. Subject: 20.4.1

Amnesty International. Execution by lethal injection — a quarter century of state poisoning. London: Amnesty International: 2007 October 4: 41 p. [Online]. Accessed: http://web.amnesty.org/library/pdf/ACT500072007ENGLISH/$File/ACT5000707.pdf [2007 October 9]. Subject: 20.6

Amundson, Ron; Tresky, Shari. On a bioethical challenge to disability rights. *Journal of Medicine and Philosophy* 2007 November-December; 32(6): 541-561. Subject: 9.5.1

Anaissie, E.J.; Segal, B.H.; Graybill, J.R.; Arndt, C.; Perfect, J.R.; Kleinberg, M.; Pappas, P.; Benjamin, D.; Rubin, R.; Aberg, J.A.; Adderson, E.E.; Adler-Shohet, F.C.; Akan, H.; Akova, M.; Almyroudis, N.G.; Alexander, B.D.; Andes, D.; Arrieta, A.; Baddley, J.W.; Barron, M.A.; et al. Clinical research in the lay press: irresponsible journalism raises a huge dose of doubt. *Clinical Infectious Diseases* 2006 October 15; 43(8): 1031-1039 [see correction in Clinical Infectious Diseases 2007 March 15; 44(6): 894]. Subject: 1.3.7

Anand, K.J.S.; Aranda, Jacob V.; Berde, Charles B.; Buckman, ShaAvhrée; Capparelli, Edmund V.; Carlo, Waldemar A.; Hummel, Patricia; Lantos, John; Johnston, C. Celeste; Lehr, Victoria Tutag; Lynn, Anne M.; Maxwell, Lynne G.; Oberlander, Tim F.; Raju, Tonse N.K.; Soriano, Sulpicio G.; Taddio, Anna; Walco, Gary A. Analgesia and anesthesia for neonates: study design and ethical issues. *Clinical Therapeutics* 2005 June; 27(6): 814-843. Subject: 18.5.2

Ancker, Jessica S.; Flanagin, Annette. A comparison of conflict of interest policies at peer-reviewed journals in different scientific disciplines. *Science and Engineering Ethics* 2007 June; 13(2): 147-157. Subject: 1.3.7

Andereck, William. From patient to consumer in the medical marketplace. *CQ: Cambridge Quarterly of Healthcare Ethics* 2007 Winter; 16(1): 109-113. Subject: 8.1

Andereck, William S. Commodified care. *CQ: Cambridge Quarterly of Healthcare Ethics* 2007 Fall; 16(4): 398-406. Subject: 9.3.1

Andereck, William S.; Jonsen, Albert R. Conclusion [commercialism in medicine]. *CQ: Cambridge Quarterly of Healthcare Ethics* 2007 Fall; 16(4): 439-442. Subject: 9.3.1

Anderson, Andrea. Ethicists balk at new emergency trials that skip informed consent [news]. *Nature Medicine* 2007 July; 13(7): 765. Subject: 18.3

Anderson, Janice A.; Vernaglia, Lawrence W.; Morrigan, Shirley P. Refusal of brain death diagnosis: the health lawyers' perspective. *JONA's Healthcare Law, Ethics, and Regulation* 2007 July-September; 9(3): 90-92. Subject: 20.2.1

Anderson, Kelly K.; Mukherjee, Som D. The need for additional safeguards in the informed consent process in schizophrenia research. *Journal of Medical Ethics* 2007 November; 33(11): 647-650. Subject: 18.5.6

Anderson, Melissa S. Collective openness and other recommendations for the promotion of research integrity. *Science and Engineering Ethics* 2007 December; 13(4): 387-394. Subject: 1.3.9

Anderson, Melissa S.; Horn, Aaron S.; Risbey, Kelly R.; Ronning, Emily A.; De Vries, Raymond; Martinson, Brian C. What do mentoring and training in the responsible conduct of research have to do with scientists' misbehavior? Findings from a national survey of NIH-funded scientists. *Academic Medicine* 2007 September; 82(9): 853-860. Subject: 18.2

Anderson, Melissa S.; Martinson, Brian C.; De Vries, Raymond. Normative dissonance in science: results from a national survey of U.S. scientists. *Journal of Empirical Research on Human Research Ethics* 2007 December; 2(4): 3-14. Subject: 1.3.9

Anderson, Melissa S.; Ronning, Emily A.; de Vries, Raymond; Martinson, Brian C. The perverse effects of competition on scientists' work and relationships. *Science and Engineering Ethics* 2007 December; 13(4): 437-461. Subject: 1.3.9

Anderson-Shaw, Lisa; Ahrens, William; Fetzer, Marny. Ethics consultation in the emergency department. *JONA's Healthcare Law, Ethics, and Regulation* 2007 January-March; 9(1): 32-35. Subject: 9.6

Andorno, Roberto. Human dignity and the UNESCO Declaration on the Human Genome. *In:* Gunning, Jennifer; Holm, Søren, eds. Ethics, law, and society. Volume 1. Aldershot, Hants, England; Burlington, VT: Ashgate, 2005: 73-81. Subject: 15.1

Andorno, R. Global bioethics at UNESCO: in defence of the Universal Declaration on Bioethics and Human Rights. *Journal of Medical Ethics* 2007 March; 33(3): 150-154. Subject: 2.1

Andre, Judith. Learning to listen: second-order moral perception and the work of bioethics. *In:* Eckenwiler, Lisa A.; Cohn, Felicia G., eds. The Ethics of Bioethics: Mapping the Moral Landscape. Baltimore, MD: Johns Hopkins University Press, 2007: 220-228. Subject: 2.1

Andrée, Peter. The biopolitics of genetically modified organisms in Canada. *Journal of Canadian Studies* 2002 Fall; 37(3): 162-191. Subject: 15.1

Angell, Marcia. Cross-cultural considerations in medical ethics: the case of human subjects research. *In:* Galston, Arthur W.; Peppard, Christiana Z., eds. Expanding Horizons in Bioethics. Dordrecht; Norwell, MA: Springer, 2005: 71-84. Subject: 18.5.9

Anglican Diocese of Sydney. Social Issues Executive; Cameron, Andrew; Nodder, Tracy; Watts, Lisa. Euthanasia and the abandonment of life. Social Issues Executive #057 2007 February 9: 4 p. [Online]. Accessed: http://your.sydneyanglicans.net/socialissues/057_euthanasia_and_the_abandonment_of_life [2007 February 12]. Subject: 20.5.1

Ani, Cornelius; Ani, Obeagaeli. Institutional racism: editorial is unduly provocative [letter]. *BMJ: British Medical Journal* 2007 April 14; 334(7597): 761. Subject: 17.1

See SUBJECT HEADING KEY FOR SECTION II on inside back cover.

435

Anido, Aimee; Carlson, Lisa M.; Sherman, Stephanie L. Attitudes toward fragile X mutation carrier testing from women identified in a general population survey. *Journal of Genetic Counseling* 2007 February; 16(1): 97-104. Subject: 15.3

Ankeny, Rachel A. Individual responsibility and reproduction. *In:* Rhodes, Rosamond; Francis, Leslie P.; Silvers, Anita, eds. The Blackwell Guide to Medical Ethics. Malden, MA: Blackwell Pub., 2007: 38-51. Subject: 14.1

Annas, George J. Cancer and the constitution — choice at life's end. *New England Journal of Medicine* 2007 July 26; 357(4):408-413. Subject: 9.5.1

Annas, George J. The right to health and the nevirapine case in South Africa. *In:* Gruskin Sofia; Grodin, Michael A.; Annas, George J.; Marks, Stephen P., eds. Perspectives on health and Human Rights. New York: Routledge, 2005: 497-505. Subject: 9.2

Annas, George J. The Supreme Court and abortion rights. *New England Journal of Medicine* 2007 May 24; 356(21): 2201-2207. Subject: 12.4.1

Annas, George J.; Andrews, Lori B.; Isasi, Rosario M. Protecting the endangered human: toward an international treaty prohibiting cloning and inheritable alterations. *In:* Gruskin, Sofia; Grodin, Michael A.; Annas, George J.; Marks, Stephen P., eds. Perspectives on Health and Human Rights. New York: Routledge, 2005: 135-162. Subject: 14.5

Anscombe, G.E.M. Murder and the morality of euthanasia. *In her:* Human Life, Action and Ethics: Essays. Exeter: Imprint Academic, 2005: 261-277. Subject: 20.5.1

Anscombe, G.E.M. Sins of omission? The non-treatment of controls in clinical trials. *In her:* Human Life, Action and Ethics: Essays. Exeter: Imprint Academic, 2005: 286-291. Subject: 18.2

Ansell, Nicola; Van Blerk, Lorraine. Joining the conspiracy? Negotiating ethics and emotions in researching (around) AIDS in Southern Africa. *Ethics Place and Environment* 2005 March; 8(1): 61-82. Subject: 9.5.6

Antes, Alison L.; Brown, Ryan P.; Murphy, Stephen T.; Waples, Ethan P.; Mumford, Michael D.; Connelly, Shane; Devenport, Lynn D. Personality and ethical decision-making in research: the role of perceptions of self and others. *Journal of Empirical Research on Human Research Ethics* 2007 December; 2(4): 15-34. Subject: 1.3.9

Anwar, Rahij; Ahmed, Azeem. Who is responsible for "do not resuscitate" status in patients with broken hips? *BMJ: British Medical Journal* 2007 January 20; 334(7585): 155. Subject: 20.5.1

Anya, Ike; Virgilio, Antonia; Defilippi, Loris; Moschochoritis, Konstantinos; Ravinetto, Raffaella; den Otter, Joost J.; Tavenier, Daniel. Right to health care for vulnerable migrants [letters]. *Lancet* 2007 September 8-14; 370(9590): 827-828. Subject: 9.2

Appel, Jacob M. A suicide right for the mentally ill? A Swiss case opens a new debate. *Hastings Center Report* 2007 May-June; 37(3): 21-23. Subject: 20.7

Appelbaum, Kenneth L. Commentary: the use of restraint and seclusion in correctional mental health. *Journal of the American Academy of Psychiatry and the Law* 2007; 35(4): 431-435. Subject: 9.5.3

Appelbaum, Paul S. Assessment of patients' competence to consent to treatment. *New England Journal of Medicine* 2007 November 1; 357(18): 1834-1840. Subject: 8.3.3

Appelbaum, Paul S. Comment on the case of Mr. A.B. *Journal of Clinical Ethics* 2007 Winter; 18(4): 402-403. Subject: 8.3.1

Appleton, J.; Caan, W.; Cowley, S.; Kendall, S. Busting the bureaucracy: lessons from research governance in primary care. *Community Practitioner* 2007 February; 80(2): 29-32. Subject: 18.2

April, Carolyn; Parker, Michael. End of life decision-making in neonatal care. *Journal of Medical Ethics* 2007 March; 33(3): 126-127. Subject: 20.5.2

Aramesh, Kiarash. The influences of bioethics and Islamic jurisprudence on policy-making in Iran. *American Journal of Bioethics* 2007 October; 7(10): 42-44. Subject: 2.1

Aramesh, Kiarash; Dabbagh, Soroush. An Islamic view to stem cell research and cloning: Iran's experience. *American Journal of Bioethics* 2007 February; 7(2): 62. Subject: 14.5

Araujo, Robert John. The UN Declaration on Human Cloning: a survey and assessment of the debate. *National Catholic Bioethics Quarterly* 2007 Spring; 7(1): 129-149. Subject: 14.5

Aray-Blais, Christiane; Patenaude, Johane. Biobanking primer: down to basics [letter]. *Science* 2007 May 11; 316(5826): 830. Subject: 15.1

Arbeitsgruppe Pränataldiagnostik; Merkel, Reinhard. Das „Strudlhof"-Symposion. Konsensus-Statement: Bedingungen spezieller pränataler genetischer Diagnostik [The "Strudlhof" symposium. Consensus statement: conditions for special prenatal genetic diagnosis]. *Ethik in der Medizin* 2007 September; 19(3): 221-225. Subject: 15.2

Arber, Sara; McKinlay, John; Adams, Ann; Marceau, Lisa; Link, Carol; O'Donnell, Amy. Patient characteristics and inequalities in doctors' diagnostic and management strategies relating to CHD: a video-simulation experiment. *Social Science and Medicine* 2006 January; 62(1): 103-115. Subject: 9.5.1

Arboleda-Flórez, Julio E. The ethics of forensic psychiatry. *Current Opinion in Psychiatry* 2006 September; 19(5): 544-546. Subject: 17.2

Archard, David. Informed consent and the grounds of autonomy. *In:* Nys, Thomas; Denier, Yvonne; Vandevelde, Toon, eds. Autonomy and Paternalism: Reflections on the

Theory and Practice of Health Care. Leuven; Dudley, MA: Peeters, 2007: 113-128. Subject: 8.3.1

Archibold, Randal C. 2 accused of trading in cadaver parts. *New York Times* 2007 March 8; p. A20. Subject: 19.5

Arikan, Rasim; Appelbaum, Paul S.; Sercan, Mustafa; Turkcan, Solmaz; Satmis, Nevzat; Polat, Aslihan. Civil commitment in Turkey: reflections on a bill drafted by psychiatrists. *International Journal of Law and Psychiatry* 2007 January-February; 30(1): 29-35. Subject: 17.7

Arman, Maria; Rehnsfeldt, Arne. The presence of love in ethical caring. *Nursing Forum* 2006 January-March; 41(1): 4-12. Subject: 4.1.3

Arman, Maria; Rehnsfeldt, Arne; Oberle, Kathleen. The 'little extra' that alleviates suffering. *Nursing Ethics* 2007 May; 14(3): 372-384; discussion: 384-386. Subject: 4.1.1

Armstron, Katrina; Ravenell, Karima L.; McMurphy, Suzanne; Putt, Mary. Racial/ethnic differences in physician distrust in the United States. *American Journal of Public Health* 2007 July; 97(7): 1282-1289. Subject: 9.5.4

Armstrong, Susan J.; Botzler, Richard G. Animal experimentation. *In their:* The Animal Ethics Reader. London; New York: Routledge, 2003: 245-307. Subject: 22.2

Armstrong, Susan J.; Botzler, Richard G. Animals and biotechnology. *In their:* The Animal Ethics Reader. London; New York: Routledge, 2003: 311-377. Subject: 15.1

Árnason, Arnar; Simpson, Bob. Refractions through culture: the new genomics in Iceland. *Ethnos* 2003 December; 68(4): 533-553. Subject: 15.11

Árnason, Gardar. Second thoughts on biobanks: the Icelandic experience. *In:* Einsiedel, Edna; Timmermans, Frank, eds. Crossing Over: Genomics in the Public Arena. Calgary, Alberta, Canada: University of Calgary Press, 2005: 193-203. Subject: 15.1

Árnason, Vilhjálmur. The global and the local: fruitful tensions in medical ethics. *Ethik in der Medizin* 2006 December; 18(4): 385-389. Subject: 2.1

Arnold, Brent. Legal solutions to Ontario's organ shortage: redrawing the boundaries of consent. *Health Law Journal* 2005; 13: 139-163. Subject: 19.5

Arnold, Robert; Aulisio, Mark; Begler, Ann; Seltzer, Deborah. A commentary on Caplan and Bergman: ethics mediation — questions for the future. *Journal of Clinical Ethics* 2007 Winter; 18(4): 350-354. Subject: 9.6

Arnold, Robert; Bartlett, Steven; Bernat, James; Colonna, John; Dafoe, Donald; Dubler, Nancy; Gruber, Scott; Kahn, Jeffrey; Luskin, Richard; Nathan, Howard; Orloff, Susan; Prottas, Jeffrey; Shapiro, Robyn; Ricordi, Camillo; Youngner, Stuart; Delmonico, Francis L. Financial incentives for cadaver organ donation: an ethical reappraisal. *Transplantation* 2002 April 27; 73(8): 1361-1367. Subject: 19.5

Arras, John D. The way we reason now: reflective equilibrium in bioethics. *In:* Steinbock, Bonnie, ed. The Oxford Handbook of Bioethics. Oxford; New York: Oxford University Press, 2007: 46-71. Subject: 2.1

Arroll, Bruce; Falloon, Karen. Should doctors go to patients' funerals? *BMJ: British Medical Journal* 2007 June 23; 334(7607): 1322. Subject: 8.1

Arsanjani, Mahnoush. The negotiations on a treaty on cloning: some reflections. *In:* Vöneky, Silja; Wolfrum, Rüdiger, eds. Human Dignity and Human Cloning. Leiden: Nijhoff, 2004: 145-165. Subject: 14.5

Asch, Adrienne; Wasserman, Davis. A response to Nelson and Mahowald. *CQ: Cambridge Quarterly of Healthcare Ethics* 2007 Fall; 16(4): 468-475. Subject: 15.2

Ashcraft, Mark H.; Krause, Jeremy A. Social and behavioral researchers' experiences with their IRBs. *Ethics and Behavior* 2007; 17(1): 1-17. Subject: 18.4

Ashcroft, Richard; Campbell, Alastair V.; Capps, Ben. Ethical aspects of developments in neuroscience and drug addiction. London: Foresight Brain Science, Addiction and Drugs project, v. 1.0; n.d.: 65 p. [Online]. Accessed: http://www.foresight.gov.uk/Previous_Projects/Brain_Science_Addiction_and_Drugs/Reports_and_Publications/ScienceReviews/Ethics.pdf [2007 April 25]. Subject: 17.1

Ashley, Benedict M.; de Blois, Jean K.; O'Rourke, Kevin D. Genetic intervention. *In their:* Health Care Ethics: A Catholic Theological Analysis. 5th edition. Washington, DC: Georgetown University Press, 2006: 94-98. Subject: 15.1

Ashley, Benedict M.; deBlois, Jean K.; O'Rourke, Kevin D. Genetic screening and counseling. *In their:* Health Care Ethics: A Catholic Theological Analysis. 5th edition. Washington, DC: Georgetown University Press, 2006: 98-103. Subject: 15.3

Aslam, S.A.; Colapinto, P.; Sheth, H.G.; Jain, R. Patient consultation survey in an ophthalmic outpatient department. *Journal of Medical Ethics* 2007 March; 33(3): 134-135. Subject: 8.4

Asscher, Joachim. Killing and letting die: the similarity criterion. *Journal of Applied Philosophy* 2007 August; 24(3): 271-282. Subject: 20.5.1

Atala, Anthony. Stem cells — brave new world? *Journal of Urology* 2005 December; 174(6): 2085. Subject: 15.1

Atkinson, Timothy N.; Gilleland, Diane S.; Pearson, Carolyn. The research environment norm inventory (RENI): a study of integrity in research administrative systems. *Accountability in Research* 2007 April-June; 14(2): 93-119. Subject: 1.3.9

Audi, Robert. Preventing abortion as a test case for the justifiability of violence. *Journal of Ethics* 1997; 1(2): 141-163. Subject: 12.3

See SUBJECT HEADING KEY FOR SECTION II on inside back cover.

Auerbach, Rebecca S. New York's immediate need for a psychotherapist-patient privilege encompassing psychiatrists, psychologists, and social workers. *Albany Law Review* 2006; 69(3): 889-912. Subject: 8.4

Aulisio, Mark P.; Devita, Michael; Luebke, Donna. Taking values seriously: ethical challenges in organ donation and transplantation for critical care professionals. *Critical Care Medicine* 2007 February; 35(2, Supplement): S95-S101. Subject: 19.5

Ausman, James I. I told you it was going to happen . . . Part II [editorial]. *Surgical Neurology* 2006 May; 65(5): 520-521. Subject: 9.7

Ausman, James I. Trust, malpractice, and honesty in medicine: should doctors say they are sorry? *Surgical neurology* 2006 July; 66(1): 105-106. Subject: 8.2

Austin, Jehannine C.; Honer, William G. The genomic era and serious mental illness: a potential application for psychiatric genetic counseling. *Psychiatric Services* 2007 February; 58(2): 254-261. Subject: 15.2

Austin, Jehannine C.; Smith, Geoffrey N.; Honer, William G. The genomic era and perceptions of psychotic disorders: genetic risk estimation, associations with reproductive decisions and views about predictive testing. *American Journal of Medical Genetics. Part B, Neuropsychiatric Genetics* 2006 December 5; 141(8): 926-928. Subject: 15.3

Austin, Joan K. Ethical issues related to the increased emphasis on children participating in research. *Chronic Illness* 2006 September; 2(3): 181-182. Subject: 18.5.2

Austin, Larry J. Religious bias colors physicians' views [letter]. *Archives of Internal Medicine* 2007 October 8; 167(18): 2007. Subject: 8.1

Austin, Wendy. The ethics of everyday practice: healthcare environments as moral communities. *ANS: Advances in Nursing Science* 2007 January-March; 30(1): 81-88. Subject: 2.1

Austin, Wendy. The Terminal: a tale of virtue. *Nursing Ethics* 2007 January; 14(1): 54-61. Subject: 7.2

Austin, Wendy. Using the human rights paradigm in health ethics: the problems and the possibilities. *In:* Tschudin, Verena, ed. Approaches to Ethics: Nursing Beyond Boundaries. New York: Butterworth-Heinemann, 2003: 105-114. Subject: 9.2

Austin, Wendy; Bergum, Vangie; Dossetor, John. Relational ethics: an action ethic as a foundation for health care. *In:* Tschudin, Verena, ed. Approaches to Ethics: Nursing Beyond Boundaries. New York: Butterworth-Heinemann, 2003: 45-52. Subject: 4.1.3

Austin, Zubin; Collins, David; Remillard, Alfred; Kelcher, Sheila; Chui, Stephanie. Influence of attitudes toward curriculum on dishonest academic behavior. *American Journal of Pharmaceutical Education* 2006 June 15; 70(3): 50. Subject: 7.2

Australian College of Health Service Executives. Code of ethics. New South Wales, Australia: Australian College of Health Services Executives, n.d.; 1 p. [Online]. Accessed: http://www.achse.org.au/ethics.html [2007 April 30]. Subject: 6

Austriaco, Nicanor Pier Giorgio. Is Plan B an abortifacient? A critical look at the scientific evidence. *National Catholic Bioethics Quarterly* 2007 Winter; 7(4): 703-707. Subject: 11.1

Austriaco, Nicanor Pier Giorgio. Is the brain-dead patient really dead? *Studia Moralia* 2003 December; 41(2): 277-308. Subject: 20.2.1

Austria. Bioethikkommission beim Bundeskanzleramt Austria. Bioethics Commission at the Federal Chancellery. Praimplantationsdiagnostik (PID): Bericht der Bioethikkommission beim Bundeskanzleramt [Preimplantation genetic diagnosis [PGD]: report of the Bioethics Commission of the Federal Chancellery]. Vienna, Austria: Bioethikkommission beim Bundeskanzleramt, 2004 July: 71 p. Refs.: 68-71. Subject: 14.2

Austria. Bioethikkommission beim Bundeskanzleramt Austria. Bioethics Commission at the Federal Chancellery. Preimplantation genetic diagnosis (PGD): report of the Bioethics Commission at the Federal Chancellery. Vienna, Austria: Bioethikkommission beim Bundeskanzleramt, 2004 July: 67 p. Refs.: 64-67. Subject: 14.2

Autti-Rämö, Ilona; Mäkelä, Marjukka. Screening for fetal abnormalities: from a health technology assessment report to a national statute. *International Journal of Technology Assessment in Health Care* 2007 Fall; 23(4): 436-442. Subject: 15.2

Avard, Denise; Kharaboyan, Linda; Knoppers, Bartha. Newborn screening for sickle cell disease: socio-ethical implications. *In:* McLean, Sheila A.M., ed. First Do No Harm: Law, Ethics, and Healthcare. Aldershot, England; Burlington, VT: Ashgate, 2006: 495-509. Subject: 15.3

Aveyard, Helen; Woolliams, Mary. In whose best interests? Nurses' experiences of the administration of sedation in general medical wards in England: an application of the critical incident technique. *International Journal of Nursing Studies* 2006 November; 43(8): 929-939. Subject: 8.3.3

Avorn, Jerry. Paying for drug approvals — who's using whom. *New England Journal of Medicine* 2007 April 26; 356(17): 1697-1700. Subject: 9.7

Ayers, Tressie A. Dutchyn. A partnership in like-minded thinking generating hopefulness in persons with cancer. *Medicine, Health Care and Philosophy* 2007 March; 10(1): 65-80. Subject: 8.2

Azira bt Tengku Zainuden, Tengku Noor. Partial-birth abortion from the perspective of Malaysian criminal law.

Formosan Journal of Medical Humanities 2007 July; 8(1-2): 1-13. NRCBL: 12.4.3.. Partial-birth abortion from the perspective of Malaysian criminal law. *Formosan Journal of Medical Humanities* 2007 July; 8(1-2): 1-13. Subject: 12.4.3

B

Baader, Gerhard. Menschenversuche in der Medizin. *In:* Neuer-Miebach, Therese; Wunder, Michael, eds. Bio-Ethik und die Zukunft der Medizin. Bonn: Psychiatrie-Verlag, 1998: 31-43. Subject: 18.5.1

Babylon, Debra M.; Monk-Turner, Elizabeth. Should incurable patients be allowed to die? *Omega* 2006; 53(4): 311-319. Subject: 20.5.1

Bach, John R. Palliative care becomes 'uninformed euthanasia' when patients are not offered noninvasive life preserving options. *Journal of Palliative Care* 2007 Autumn; 23(3): 181-184. Subject: 20.4.1

Back, Anthony L.; Arnold, R.M. Discussing prognosis: "how much do you want to know?" talking to patients who are prepared for explicit information. *Journal of Clinical Oncology.* 2006 September 1; 24(25): 4209-4213. Subject: 8.2

Back, Anthony L.; Arnold, R.M. Discussing prognosis: "how much do you want to know?" talking to patients who do not want information or who are ambivalent. *Journal of Clinical Oncology* 2006 September 1; 24(25): 4214-4217. Subject: 8.2

Backlar, Patricia. Genes and behavior: will genetic information change the way we see ourselves? *Community Mental Health Journal* 1996 June; 32: 205-209. Subject: 15.6

Badger, James M.; O'Connor, Bonnie. Moral discord, cognitive coping strategies, and medical intensive care unit nurses: insights from a focus group study. *Critical Care Nursing Quarterly* 2006 April-June; 29(2): 147-151. Subject: 4.1.3

Badzek, Laurie A.; Cline, Heather S.; Moss, Alvin H.; Hines, Stephen C. Inappropriate use of dialysis for some elderly patients: nephrology nurses' perceptions and concerns. *Nephrology Nursing Journal* 2000 October; 27(5): 462-470; discussion 471-472. Subject: 19.3

Baggot, Paddy Jim. Hard cases do not justify partial birth abortion. *Linacre Quarterly* 2007 August; 74(3): 248-252. Subject: 12.1

Baggot, Paddy Jim; Baggot, M.G. Obstetric genetic counseling for lethal anomalies. *Linacre Quarterly* 2007 February; 74(1): 60-67. Subject: 15.2

Baggs, Judith Gedney. Prognostic information provided during family meetings in the intensive care unit. *Critical Care Medicine* 2007 February; 35(2): 646-647. Subject: 20.5.1

Bagheri, A. Individual choice in the definition of death. *Journal of Medical Ethics* 2007 March; 33(3): 146-149. Subject: 20.2.1

Bahadur, Guido. Till death do us part: to be or not to be . . . a parent after one's death? *In:* Shenfield, Françoise; Sureau, Claude, eds. Contemporary Ethical Dilemmas in Assisted Reproduction. Abingdon: Informa Healthcare, 2006: 29-42. Subject: 14.2

Baier, Annette C. Trust, suffering, and the Aesculapian virtues. *In:* Walker, Rebecca L.; Ivanhoe, Philip J., eds. Working Virtue: Virtue Ethics and Contemporary Moral Problems. Oxford: Clarendon, 2007: 135-153. Subject: 4.1.2

Bailey, James E.; Sprabery, Laura R. Inequitable funding may cause health care disparities [editorial]. *Archives of Internal Medicine* 2007 June 25; 167(12): 1226-1228. Subject: 9.5.4

Bailey, Tracey M.; Penney, Steven. Healing, not squealing: recent amendments to Alberta's Health Information Act. *Health Law Review* 2007; 15(2): 3-14. Subject: 8.4

Bainbridge, William Sims. Converging technologies and human destiny. *Journal of Medicine and Philosophy* 2007 May-June; 32(3): 197-216. Subject: 5.1

Baines, Jennifer A. Gamete donors and mistaken identities: the importance of genetic awareness and proposals favoring donor identity disclosure for children born from gamete donations in the United States. *Family Court Review* 2007 January; 45(1): 116-132. Subject: 14.1

Baird, Patricia A. The evolution of public policy on reprogenetics in Canada. *In:* Knowles, Lori P.; Kaebnick, Gregory E., eds. Reprogenetics: Law, Policy, and Ethical Issues. Baltimore: Johns Hopkins University Press, 2007: 178-193. Subject: 14.1

Bakalar, Nicholas. All breast cancer patients are not treated the same. *New York Times* 2007 January 23; p. F7. Subject: 9.4

Bakalar, Nicholas. Review finds drug makers issue more positive studies. *New York Times* 2007 February 27; p. F7. Subject: 9.7

Baker, Brook K. Placing access to essential medicines on the human rights agenda. *In:* Cohen, Jillian Clare; Illingworth, Patricia; Schüklenk, Udo, eds. The Power of Pills: Social, Ethical and Legal Issues in Drug Development, Marketing, and Pricing. London; Ann Arbor, MI: Pluto, 2006: 239-248. Subject: 9.2

Baker, Monya. Stem cells by any other name [news]. *Nature* 2007 September 27; 449(7161): 389. Subject: 18.5.4

Baker, Robert. A history of codes of ethics for bioethicists. *In:* Eckenwiler, Lisa A.; Cohn, Felicia G., eds. The Ethics of Bioethics: Mapping the Moral Landscape. Baltimore, MD: Johns Hopkins University Press, 2007: 24-40. Subject: 6

See SUBJECT HEADING KEY FOR SECTION II on inside back cover.

439

Baker, Robert B.; McCullough, Laurence B. The relationship between moral philosophy and medical ethics reconsidered. *Kennedy Institute of Ethics Journal* 2007 September; 17(3): 271-276. Subject: 2.1

Baker, Robert; McCullough, Laurence. Medical ethics' appropriation of moral philosophy: the case of the sympathetic and the unsympathetic physician. *Kennedy Institute of Ethics Journal* 2007 March; 17(1): 3-22. Subject: 2.1

Baldwin, Ann; Bekoff, Marc. Too stressed to work: scientists must provide lab animals with decent living conditions or accept that their results could be useless [commentary]. *New Scientist* 2007 June 2-8; 194(2606): 24. Subject: 22.2

Baldwin, Thomas. Choosing who: what is wrong with making better children? *In:* Spencer, J.R.; du Bois-Pedain, Antje, eds. Freedom and Responsibility in Reproductive Choice. Portland: Hart Pub., 2006: 15-30. Subject: 15.2

Balint, John A. Ethical issues in stem cell research. *Albany Law Review* 2002; 65(3): 729-742. Subject: 18.5.4

Ball, Douglas E.; Tisocki, Klara; Herxheimer, Andrew. Advertising and disclosure of funding on patient organisation websites: a cross-sectional survey. *BMC Public Health* 2006 August 3; 6: 201. Subject: 1.3.12

Ball, Susan C. Nurse-patient advocacy and the right to die. *Journal of Psychosocial Nursing and Mental Health Services* 2006 December; 44(12): 36-42. Subject: 20.5.1

Ballard, Rebecca. You get a line, I'll get a pole, we'll go fish'n in the plaintiff's gene pool. *Defense Counsel Journal* 2007 January; 74(1): 22-34. Subject: 15.3

Balls, Michael; Combes, Robert. Putting replacement first [editorial]. *ATLA: Alternatives to Laboratory Animals* 2007 June; 35(3): 297-298. Subject: 22.2

Banasik, Miroslaw. Living donor transplantation — the real gift of life. Procurement and the ethical assessment. *Annals of Transplantation* 2006; 11(1): 4-6. Subject: 19.5

Bandyopadhyay, Sutapa; Singh, Amarjeet. History of son preference and sex selection in India and in the west. *Bulletin of the Indian Institute of History of Medicine (Hyderabad)* 2003 July-December; 33(2): 149-167. Subject: 14.3

Banerjee, Albert; Birenbaum-Carmeli, Daphna. Ordering suicide: media reporting of family assisted suicide in Britain. *Journal of Medical Ethics* 2007 November; 33(11): 639-642. Subject: 20.7

Banerjee, Neela. A place to turn when a newborn is fated to die. *New York Times* 2007 March 13; p. A1, A14. Subject: 20.4.2

Banister, Elizabeth; Leadbeater, Bonnie; Benoit, Cecilia; Jansson, Michael; Marshall, Anne; Riecken, Ted. Ethical issues in community-based research with children and youth. *NCEHR Communique CNERH* 2006 Spring; 14(1): 23-24. Subject: 18.5.2

Banja, John. Personhood: elusive but not illusory. *American Journal of Bioethics* 2007 January; 7(1): 60-62. Subject: 17.1

Banja, John D. The role of feelings in making moral decisions. *Case Manager* 2006 September-October; 17(5): 17-19, 62. Subject: 9.4

Banja, John D.; Volandes, Angelo E.; Paasche-Orlow, Michael K. My what? *American Journal of Bioethics* 2007 November; 7(11): 13-15; author reply W1-W2. Subject: 8.1

Banja, John; Eig, Jennifer; Williams, Mark V. Discharge dilemmas as system failures. *American Journal of Bioethics* 2007 March; 7(3): 29-31. Subject: 9.1

Barash, Carol Isaacson. Threats to privacy protection [letter]. *Science* 2007 November 9; 318(5852): 913-914. Subject: 15.1

Barata, Paula C.; Gucciardi, Enza; Ahmad, Farah; Stewart, Donna E. Cross-cultural perspectives on research participation and informed consent. *Social Science and Medicine* 2006 January; 62(2): 479-490. Subject: 18.2

Barazzetti, Gaia; Radaelli, Stefania; Sala, Roberta. Autonomy, responsibility and the Italian Code of Deontology for Nurses. *Nursing Ethics* 2007 January; 14(1): 83-98. Subject: 4.1.3

Barber, Christopher F. Abuse by care professionals. Part 1: an introduction. *British Journal of Nursing* 2007 August 9 - September 12; 16(15): 938-940. Subject: 7.4

Barber, Christopher F. Abuse by care professionals. Part 2: a behavioural assessment. *British Journal of Nursing* 2007 September 13-27; 16(16): 1023-1025. Subject: 7.4

Bard, Jennifer S. Immaculate gestation? How will ectogenesis change current paradigms of social relationships and values? *In:* Gelfand, Scott; Shook, John R., eds. Ectogenesis: Artificial Womb Technology and the Future of Human Reproduction. Amsterdam; New York: Editions Rodopi, B.V., 2006: 149-157. Subject: 14.1

Bard, Jennifer S. Learning from law's past: a call for caution in incorporating new innovations in neuroscience. *American Journal of Bioethics* 2007 September; 7(9): 73-75. Subject: 17.1

Bard, Terry R. Refusal of brain death diagnosis: a rabbi's response. *JONA's Healthcare Law, Ethics, and Regulation* 2007 July-September; 9(3): 92-94. Subject: 20.2.1

Bardes, Charles L. Ethics and prescribing: the clinician's perspective. *In:* Santoro, Michael A.; Gorrie, Thomas M., eds. Ethics and the Pharmaceutical Industry. Cambridge; New York: Cambridge University Press, 2005: 136-152. Subject: 9.7

Barfield, Raymond C.; Kodish, Eric. Pediatric ethics in the age of molecular medicine. *Pediatric Clinics of North America* 2006 August; 53(4): 639-648. Subject: 15.2

Subject = NRCBL Primary Classification Number; see inside front cover.

Barfield, Raymond; Lawrence, Ryan E.; Curlin, Farr A. Conscience is the means by which we engage the moral dimension of medicine. *American Journal of Bioethics* 2007 December; 7(12): 26-27; author reply W1-W2. Subject: 4.1.2

Barilan, Y. Michael. Contemporary art and ethics of anatomy. *Perspectives in Biology and Medicine* 2007 Winter; 50(1): 104-123. Subject: 7.1

Barilan, Y. Michael. The debate on human cloning: some contributions from the Jewish tradition. *In:* Roetz, Heiner, ed. Cross-Cultural Issues in Bioethics: The Example of Human Cloning. New York: Rodopi, 2006: 311-340. Subject: 14.5

Barilan, Y. Michael. The new Israeli law on the care of the terminally ill: conceptual innovations waiting for implementation. *Perspectives in Biology and Medicine* 2007 Autumn; 50(4): 557-571. Subject: 20.4.1

Barker, P. Mental health nursing: the craft of the impossible? *Journal of Psychiatric and Mental Health Nursing* 2006 August; 13(4): 385-387. Subject: 7.1

Barnes, Charles. Why compliance programs fail: economics, ethics, and the role of leadership. *HEC (Healthcare Ethics Committee) Forum* 2007 June; 19(2): 109-123. Subject: 9.1

Barnes, Richard. Stem cell research funding: testimony. *Origins* 2007 March 15; 36(39): 616-620. Subject: 18.5.4

Barnoy, Sivia; Tabak, Nili. Israeli nurses and genetic information disclosure. *Nursing Ethics* 2007 May; 14(3): 280-294. Subject: 15.1

Baron, Leonard; Shemie, Sam D.; Teitelbaum, Jeannie; Doig, Christopher James. Brief review: history, concept and controversies in the neurological determination of death. *Canadian Journal of Anaesthesia* 2006 June; 53(6): 602-608. Subject: 20.2.1

Baron, Roberta H. Genetic susceptibility testing: issues and psychosocial implications. *In:* Carroll-Johnson, Rose Mary; Gorman, Linda M.; Bush, Nancy Jo, eds. Psychosocial Nursing Care Along the Cancer Continuum. 2nd edition. Pittsburgh, PA: Oncology Nursing Society, 2006: 499-509. Subject: 15.3

Barry, Colleen L. The political evolution of mental health parity. *Harvard Review of Psychiatry* 2006 July-August; 14(4): 185-194. Subject: 4.3

Barshes, Neal R.; Hacker, Carl S.; Freeman, Richard B., Jr.; Vierling, John M.; Goss, John A. Justice, administrative law, and the transplant clinician: the ethical and legislative basis of a national policy on donor liver allocation. *Journal of Contemporary Health Law and Policy* 2007 Spring; 23(2): 200-230. Subject: 19.5

Barsky, Lauren E.; Donner, William R. Lessons from the field: human needs often complicate ethical duties in disaster research. Hurricane Katrina investigation. *Pro-tecting Human Subjects* 2007 November (15): 18-19. Subject: 18.4

Bartley, Richard. Is prognosis key in donation? [letter]. *BMJ: British Medical Journal* 2007 June 9; 334(7605): 1179. Subject: 19.5

Baruch, Jay. Doctor versus patient: pain management in the ED. *Atrium* 2007 Summer; 4: 10-13. Subject: 4.4

Baskin, Cyndy. Part I: Conceptualizing, framing and politicizing aboriginal ethics in mental health. *Journal of Ethics in Mental Health [electronic]* 2007 November; 2(2): 5 p. Accessed: http://www.jemh.ca [2008 January 24]. Subject: 9.5.3

Baskin, Cyndy. Part II: Working together in the circle: challenges and possibilities within mental health ethics. *Journal of Ethics in Mental Health [electronic]* 2007 November; 2(2): 4 p. Accessed: http://www.jemh.ca [2008 January 24]. Subject: 9.5.3

Bass, Sarah Bauerle. Why can't a fetus be more like a sperm? The women's role in fetal issue research and how women are left out of the discussion. *Gender Issues* 2001 Winter; 19(1): 19-32. Subject: 18.5.4

Basta, Lofty L. Dying on one's own terms. *American Journal of Geriatric Cardiology* 2006 July-August; 15(4): 250-252. Subject: 20.7

Basta, Lofty L. Ethical issue: an oath for our time. *American Journal of Geriatric Cardiology* 2006 September-October; 15(5): 316-318. Subject: 9.5.2

Basta, Lofty L. Ethical issues in the management of geriatric cardiac patients: a 91-year old patient insists on an advance care plan that does not make sense. *American Journal of Geriatric Cardiology* 2004 September-October; 13(5): 276-277. Subject: 9.5.2

Basu, Sanjay; Segraves, Britnye T.; Colgrove, James. Compulsory HPV vaccination [letter and reply]. *New England Journal of Medicine* 2007 March 8; 356(10): 1074-1075. Subject: 9.5.5

Batson, Andrew; Oster, Shai. China reconsiders fairness of "transplant tourism"; foreigners pay more for scarce organs; Israelis debate reform. *Wall Street Journal* 2007 April 6; p. A1, A9. Subject: 19.5

Batt, Sharon; Braun, Joshua A. Limits on autonomy: political meta-narratives and health stories in the media. *American Journal of Bioethics* 2007 August; 7(8): 23-25; author reply W1-W2. Subject: 1.3.7

Battin, Margaret P.; Francis, Leslie P.; Jacobson, Jay A.; Smith, Charles B. The patient as victim and vector: the challenge of infectious disease for bioethics. *In:* Rhodes, Rosamond; Francis, Leslie P.; Silvers, Anita, eds. The Blackwell Guide to Medical Ethics. Malden, MA: Blackwell Pub., 2007: 269-286. Subject: 9.1

Battin, Margaret P.; van der Heide, Agnes; Ganzini, Linda; van der Wal, Gerrit; Onwuteaka-Philipsen, Bregje D. Legal physician-assisted dying in Oregon and

See SUBJECT HEADING KEY FOR SECTION II on inside back cover.

441

the Netherlands: evidence concerning the impact on patients in "vulnerable" groups. *Journal of Medical Ethics* 2007 October; 33(10): 591-597. Subject: 20.7

Bauer, Arie; Rosca, Paula; Grinshpoon, Alexander; Khawaled, Razek; Mester, Roberto; Yoffe, Rinat; Ponizovsky, Alexander M. Trends in involuntary psychiatric hospitalization in Israel 1991-2000. *International Journal of Law and Psychiatry* 2007 January-February; 30(1): 60-70. Subject: 17.7

Bauer, Keith A. Wired patients: implantable microchips and biosensors in patient care. *CQ: Cambridge Quarterly of Healthcare Ethics* 2007 Summer; 16(3): 281-290. Subject: 5.1

Bauer, Martin W. The public career of the 'gene' — trends in public sentiments from 1946 to 2002. *New Genetics and Society* 2007 April; 26(1): 29-45. Subject: 15.1

Bauer, Patricia E. What's lost in prenatal testing [op-ed]. *Washington Post* 2007 January 14; p. B7. Subject: 15.2

Baumgartner, Christoph. Exclusion by inclusion? On difficulties with regard to an effective ethical assessment of patenting in the field of agricultural bio-technology. *Journal of Agricultural and Environmental Ethics* 2006; 19(6): 521-539. Subject: 15.8

Baumrucker, Steven J. A medical error leads to tragedy: how do we inform the patient? *American Journal of Hospice and Palliative Care* 2006 October-November; 23(5): 417-421. Subject: 9.8

Baumrucker, Steven J. Durable power of attorney versus the advance directive: who wins, who suffers? *American Journal of Hospice and Palliative Care* 2007 February-March; 24(1): 68-73. Subject: 20.5.4

Baumrucker, Steven J. Ethics roundtable. Hospice and alcoholism. *American Journal of Hospice and Palliative Care* 2006 March-April; 23(2): 153-156. Subject: 9.5.9

Bavastro, Paolo. Europäische Initiative gegen Bio-Ethik und deren Folgen. *In:* Neuer-Miebach, Therese; Wunder, Michael, eds. Bio-Ethik und die Zukunft der Medizin. Bonn: Psychiatrie-Verlag, 1998: 155-158. Subject: 2.1

Baxter, Nancy N. Equal for whom? Addressing disparities in the Canadian medical system must become a national priority. *CMAJ/JAMC: Canadian Medical Association Journal* 2007 December 4; 177(12): 1522-1523. Subject: 9.4

Baxter, Pamela E.; Boblin, Sheryl L. The moral development of baccalaureate nursing students: understanding unethical behavior in classroom and clinical settings. *Journal of Nursing Education* 2007 January; 46(1): 20-27. Subject: 7.2

Bayer, Ronald. Ethics and public policy: engaging the moral challenges posed by AIDS. *AIDS Patient Care and STDs* 2006 July; 20(7): 456-460. Subject: 9.5.6

Bayertz, Kurt. Eugenik. *In:* Schreiber, Hans-Peter, ed. Biomedizin und Ethik: Praxis, Recht, Moral. Basel; Boston: Birkhäuser, 2004: 72-76. Subject: 15.5

Bayertz, Kurt. Struggling for consensus and living without it: the construction of a common European bioethics. *In:* Engelhardt, H. Tristram, ed. Global Bioethics: The Collapse of Consensus. Salem, MA: M&M Scrivener Press, 2006: 207-237. Subject: 2.1

Bayertz, Kurt. Zur Idee der Menschenwürde. *In:* Schreiber, Hans-Peter, ed. Biomedizin und Ethik: Praxis, Recht, Moral. Basel; Boston: Birkhäuser, 2004: 63-66. Subject: 4.4

Bayertz, Kurt; Schmidt, Kurt W. Testing genes and constructing humans — ethics and genetics. *In:* Khushf, George, ed. Handbook of Bioethics: Taking Stock of the Field From a Philosophical Perspective. Dordrecht; Boston: Kluwer Academic, 2004: 415-438. Subject: 15.1

Bayley, Carol. Back to basics: examining the assumptions of donation after cardiac death. *Health Care Ethics USA* 2007 Summer; 15(3): 2-4. Subject: 19.5

Bayley, Carol; Cardone, Joseph; Harvey, John Collins; O'Brien, Daniel; Panicola, Michael; Repenshek, Mark; Sheehan, Myles; Worsley, Stephen. Sampling of responses to the CDF statement on nutrition and hydration. *Health Care Ethics USA* 2007 Fall; 15(4): 8-14. Subject: 20.5.1

Baylis, Françoise. Of courage, honor, and integrity. *In:* Eckenwiler, Lisa A.; Cohn, Felicia G., eds. The Ethics of Bioethics: Mapping the Moral Landscape. Baltimore, MD: Johns Hopkins University Press, 2007: 193-204. Subject: 9.6

Baylis, Françoise; Fenton, Andrew. Chimera research and stem cell therapies for human neurodegenerative disorders. *CQ: Cambridge Quarterly of Healthcare Ethics* 2007 Spring; 16(2): 195-208. Subject: 15.1

Baylis, Françoise; McLeod, C. The stem cell debate continues: the buying and selling of eggs for research. *Journal of Medical Ethics* 2007 December; 33(12): 726-731. Subject: 19.5

Baylis, Françoise; Ram, Natalie. Eligibility of cryopreserved human embryos for stem cell research in Canada. *JOGC: Journal of Obstetrics and Gynaecology Canada = JOGC:Journal d'Obstétrique et Gynécologie du Canada* 2005 October; 27(10): 949-955. Subject: 18.3

Baylis, Françoise; Robert, Jason Scott. Part-human chimeras: worrying the facts, probing the ethics. *American Journal of Bioethics* 2007 May; 7(5): 41-45. Subject: 15.1

Baylis, Françoise; Robert, Jason Scott. Radical rupture: exploring biological sequelae of volitional inheritable genetic modification. *In:* Rasko, John E.J.; O'Sullivan, Gabrielle M.; Ankeny, Rachel A., eds. The Ethics of Inheritable Genetic Modification: A Dividing Line? Cam-

---

bridge: Cambridge University Press, 2006: 131-148. Subject: 15.1

Bayne, Kathryn A.; Harkness, John E. Welfare of research animals. *In:* Kulakowski, Elliott C.; Chronister, Lynne U., eds. Research Administration and Management. Sudbury, MA: Jones and Bartlett, 2006: 577-581. Subject: 22.2

Beach, Mary Catherine; Duggan, Patrick S.; Cassel, Christine K.; Geller, Gail. What does 'respect' mean? Exploring the moral obligation of health professionals to respect patients. *JGIM: Journal of General Internal Medicine* 2007 May; 22(5): 692-695. Subject: 8.1

Beals, Daniel A. Permissibility to stop off-label use of expensive drug treatment for child? *Ethics and Medicine: An International Journal of Bioethics* 2007 Fall; 23(3): 141-144. Subject: 20.5.2

Beard, Edward L. Jr.; Johnson, Larry W. Conversations in ethics. *JONA's Healthcare Law, Ethics, and Regulation* 2007 July-September; 9(3): 95-96. Subject: 20.2.1

Beauchamp, Tom L. History and theory in "applied ethics". *Kennedy Institute of Ethics Journal* 2007 March; 17(1): 55-64. Subject: 2.1

Beauchamp, Tom L.; DeGrazia, David. Principles and principlism. *In:* Khushf, George, ed. Handbook of Bioethics: Taking Stock of the Field From a Philosophical Perspective. Dordrecht; Boston: Kluwer Academic, 2004: 55-74. Subject: 2.1

Beaudin, David J. Ethical funds for physicians [letter]. *CMAJ/JAMC: Canadian Medical Association Journal* 2007 August 14; 177(4): 375. Subject: 9.3.1

Beaulieu, Marie; Leclerc, Nancy. Ethical and psychosocial issues raised by the practice in cases of mistreatment of older adults. *Journal of Gerontological Social Work* 2006 May; 46(3-4): 161-186. Subject: 9.5.2

Bebeau, Muriel J. Evidence-based character development. *In:* Kenny, Nuala; Shelton, Wayne, eds. Lost Virtue: Professional Character Development in Medical Education. Amsterdam; Oxford: Elsevier, 2006: 47-86. Subject: 4.1.2

Beck, Matthias. Illness, disease and sin: the connection between genetics and spirituality. *Christian Bioethics* 2007 January-April; (13)1: 67-89. Subject: 15.1

Beck, Natalia Vonnegut. Building bridges: the Protestant perspective. *In:* Puchalski, Christina M., ed. A Time for Listening and Caring: Spirituality and the Care of the Chronically Ill and Dying. Oxford; New York: Oxford University Press, 2006: 155-170. Subject: 20.4.1

Becker, Gary J. Financial relationships with industry and device research involving non-Food and Drug Administration-approved use: a perspective. *Radiology* 2006 June; 239(3): 626-628. Subject: 18.2

Becker, Gerhold K. Chinese ethics and human cloning: a view from Hong Kong. *In:* Roetz, Heiner, ed. Cross-Cul-

tural Issues in Bioethics: The Example of Human Cloning. New York: Rodopi, 2006: 107-139. Subject: 14.5

Beckett, Alesha; Gilbertson, Sarah; Greenwood, Sallie. Doing the right thing: nursing students, relational practice, and moral agency. *Journal of Nursing Education* 2007 January; 46(1): 28-32. Subject: 7.2

Beckmann, Jan P. Ethik in der Medizin in Aus- und Weiterbildung aus der Sicht der Philosophie [Ethics in medicine in education and continuing education from the philosophical point of view]. *Ethik in der Medizin* 2006 December; 18(4): 369-373. Subject: 2.3

Bedford-Strohm, Heinrich. Justice and long-term care: a theological ethical perspective. *Christian Bioethics* 2007 September-December; (13)3: 269-285. Subject: 2.1

Beem, Penelope; Morgan, Derek. What's love got to do with it? Regulating reproductive technologies and second hand emotions. *In:* McLean, Sheila A.M., ed. First Do No Harm: Law, Ethics, and Healthcare. Aldershot, England; Burlington, VT: Ashgate, 2006: 369-388. Subject: 14.4

Beemsterboer, Phyllis L. Developing an ethic of access to care in dentistry. *Journal of Dental Education* 2006 November; 70(11): 1212-1216. Subject: 4.1.1

Beene-Harris, Rosalyn Y.; Wang, Catharine; Bach, Janice V. Barriers to access: results from focus groups to identify genetic service needs in the community. *Community Genetics* 2007; 10(1): 10-17. Subject: 15.1

Beeson, Diane; Lippman, Abby. Egg harvesting for stem cell research: medical risks and ethical problems. *Reproductive Biomedicine Online* 2006 October; 13(4): 573-579. Subject: 15.2

Beger, H.G.; Arbogast, R. The art of surgery in the 21st century: based on natural sciences and new ethical dimensions. *Langenbeck's Archives of Surgery* 2006 April; 391(2): 143-148. Subject: 18.5.1

Begley, Ann M. Creative approaches to ethics: poetry, prose and dialogue. *In:* Tschudin, Verena, ed. Approaches to Ethics: Nursing Beyond Boundaries. New York: Butterworth-Heinemann, 2003: 125-135. Subject: 7.2

Begley, Sharon. New ethical minefield: drugs to boost memory and sharpen attention. *Wall Street Journal* 2004 October 1: B1. Subject: 17.1

Belde, David. Toward a "total organizational ethic" in health care ethics. *Health Care Ethics USA* 2007 Spring; 15(2): 9-11. Subject: 9.1

Belgium. Ministry of Justice. The Belgian Act on Euthanasia of May 28th, 2002. *Ethical Perspectives* 2002; 9(2-3): 182-188 [Online]. Accessed: http://www.kuleuven.ac.be/cbmer/viewpic.php?LAN=E&TABLE=DOCS&ID=23 [2007 April 17]. Subject: 20.7

Bell, Jennifer A.; Henry, Michael; Fishman, Jennifer R.; Youngner, Stuart J. Preventing post-traumatic stress disorder or pathologizing bad memories? *American Jour-*

See SUBJECT HEADING KEY FOR SECTION II on inside back cover.

443

nal of Bioethics 2007 September; 7(9): 29-30; author reply W1-W3. Subject: 17.4

Bell, Leanne; Devaney, Sarah. Gaps and overlaps: improving the current regulation of stem cells in the UK. *Journal of Medical Ethics* 2007 November; 33(11): 621-622. Subject: 18.5.1

Bell, M.D. Emergency medicine, organ donation and the Human Tissue Act. *Emergency Medicine Journal* 2006 November; 23(11): 824-827. Subject: 19.5

Bellomo, Rinaldo; Zamperetti, Nereo. Defining the vital condition for organ donation [commentary]. *Philosophy, Ethics, and Humanities in Medicine [electronic]* 2007 November 19; 2(27): 3 p. Accessed: http://www.peh-med. com/ [2008 January 24]. Subject: 19.5

Benatar, David. Grim news from the original position: a reply to Professor Doyal. *Journal of Medical Ethics* 2007 October; 33(10): 577. Subject: 14.1

Benatar, David. Unscientific ethics: science and selective ethics. *Hastings Center Report* 2007 January-February; 37(1): 30-32. Subject: 22.2

Benatar, D. Moral theories may have some role in teaching applied ethics. *Journal of Medical Ethics* 2007 November; 33(11): 671-672. Subject: 2.3

Benatar, Solomon. An examination of ethical aspects of migration and recruitment of health care professionals from developing countries. *Clinical Ethics* 2007 March; 2(1): 2-7. Subject: 21.6

Benatar, Solomon R. Achieving gold standards in ethics and human rights in medical practice. *PLoS Medicine* 2005 August; 2(8): e260. Subject: 9.5.6

Bender, Shira; Flicker, Lauren; Rhodes, Rosamond. Access for the terminally ill to experimental medical innovations: a three-pronged threat. *American Journal of Bioethics* 2007 October; 7(10): 3-6. Subject: 18.5.7

Bendiane, Marc-Karim; Bouhnik, Anne-Deborah; Favre, Roger; Galinier,Anne; Obadia, Yolande; Moatti, Jean-Paul; Peretti-Watel, Patrick. Morphine prescription in end-of-life care and euthanasia: French home nurses' opinions. *Journal of Opioid Management* 2007 January-February; 3(1): 21-26. Subject: 20.5.1

Bendiane, Marc-Karim; Galinier, A.; Favre, R.; Ribiere, C.; Lapiana, J.-M.; Obadia, Y.; Peretti-Watel, Patrick. French district nurses' opinions towards euthanasia, involvement in end-of-life care and nurse-patient relationship: a national phone survey. *Journal of Medical Ethics* 2007 December; 33(12): 708-711. Subject: 20.5.1

Benedict, Susan; Caplan, Arthur; Page, Traute Lafrenz. Duty and 'euthanasia': the nurses of Meseritz-Obrawalde. *Nursing Ethics* 2007 November; 14(6): 781-794. Subject: 20.5.1

Benham, Bryan; Clark, Dale; Francis, Leslie P. Authorship: credit, responsibility, and accountability. *In:* Kulakowski, Elliott C.; Chronister, Lynne U., eds. Research

Administration and Management. Sudbury, MA: Jones and Bartlett, 2006: 501-510. Subject: 1.3.7

Benítez-Bribiesca, Luis; Modiano-Esquenazi, Marcos. Ethics of scientific publication after the human stem cell scandal [editorial]. *Archives of Medical Research* 2006 May; 37(4): 423-424. Subject: 1.3.7

Benner, Patricia; Wrubel, Judith. Response to: 'Edwards, Benner and Wrubel on caring' by S. Horrocks (2002) Journal of Advanced Nursing 40, 36-41. *Journal of Advanced Nursing* 2002 October; 40(1): 45-47. Subject: 4.1.1

Bennett, Belinda. Globalising the body: globalisation and reproductive rights. *University of New South Wales Law Journal* 2006; 29(2): 266-271. Subject: 21.1

Bennett, Belinda. Travel in a small world: SARS, globalization and public health laws. *In:* Bennett, Belinda; Tomossy, George F., eds. Globalization and Health: Challenges for Health Law and Bioethics. Dordrecht: Springer, 2006: 1-12. Subject: 9.1

Bennett, Gaymon. Genetics, society, and spirituality. *In:* Eisen, Arri; Laderman, Gary, eds. Science, Religion, and Society: An Encyclopedia of History, Culture, and Controversy. Vol. 2. Armonk, NY: M.E. Sharpe, 2007: 763-779. Subject: 15.1

Bennett, Rebecca. Routine antenatal HIV testing and informed consent: an unworkable marriage? *Journal of Medical Ethics* 2007 August; 33(8): 446-448. Subject: 9.5.6

Bennett, Rebecca; Harris, John. Reproductive choice. *In:* Rhodes, Rosamond; Francis, Leslie P.; Silvers, Anita, eds. The Blackwell Guide to Medical Ethics. Malden, MA: Blackwell Pub., 2007: 201-219. Subject: 14.1

Benoit, Ellen; Magura, Stephen. Disability and substance user treatment/rehabilitation: ethical considerations. *In:* Kleinig, John; Einstein, Stanley, eds. Ethical Challenges for Intervening in Drug Use: Policy, Research and Treatment Issues. Huntsville, TX: Office of International Criminal Justice; Criminal Justice Center, Sam Houston State University, 2006: 153-170. Subject: 9.5.9

Bensing, Jozien. Bridging the gap. The separate worlds of evidence-based medicine and patient-centered medicine. *Patient Education and Counseling* 2000 January; 39(1): 17-25. Subject: 8.1

Bentley, Philip J. The shattered vessel: the dying person in Jewish law and ethics. *Loyola University Chicago Law Journal* 2006 Winter; 37(2): 433-454. Subject: 20.4.1

Benziman, Uzi. Patient's condition — severe but stable. The press and the medical community: mutual expectations surrounding the health of national leaders. *Israel Medical Association Journal* 2006 November; 8(11): 763-767. Subject: 8.4

Subject = NRCBL Primary Classification Number; see inside front cover.

Ben-Asher, Noa. The necessity of sex change: a struggle for intersex and transsex liberties. *Harvard Journal of Law and Gender* 2006 Winter; 29(1): 51-98. Subject: 10

Bercovitch, Lionel; Long, Thomas P. Dermatoethics: a curriculum in bioethics and professionalism for dermatology residents at Brown Medical School. *Journal of the American Academy of Dermatology* 2007 April; 56(4): 679-682. Subject: 7.2

Beresford, H. Richard. Legal aspects of termination of treatment decisions. *Neurologic Clinics* 1989 November; 7(4): 775-787. Subject: 20.5.1

Berg, Jonathan S.; French, Shannon L.; McCullough, Laurence B.; Kleppe, Soledad; Sutton, Vernon R.; Gunn, Sheila K.; Karaviti, Lefkothea P. Ethical and legal implications of genetic testing in androgen insensitivity syndrome. *Journal of Pediatrics* 2007 April; 150(4): 434-438. Subject: 15.3

Berger, Abi. Seeds of discontent. *BMJ: British Medical Journal* 2007 June 23; 334(7607): 1323. Subject: 14.1

Berger, Jeffrey T. When surrogates' responsibilities and religious concerns intersect. *Journal of Clinical Ethics* 2007 Winter; 18(4): 391-393. Subject: 20.5.1

Bergeron, Veronique. The ethics of cesarean section on maternal request: a feminist critique of the American College of Obstetricians and Gynecologists' position on patient-choice surgery. *Bioethics* 2007 November; 21(9): 478-487. Subject: 9.5.5

Berggren, Ingela; Severinsson, Elisabeth. The significance of nurse supervisors' different ethical decision-making styles. *Journal of Nursing Management* 2006 November; 14(8): 637-643. Subject: 4.1.3

Berghmans, Ron L.P.; Widdershoven, Guy A.M. Physician-assisted dying in the Netherlands. *EACME Newsletter* 2007 July; (17): 3 p. [Online]. Accessed: http://www.eacmeweb.com/newsletter/n17.htm [2007 August 15]. Subject: 20.7

Berghmans, R.; de Jong, Johan; Tibben, A.; de Wert, G. Genetics of alcoholism: ethical and societal implications. *EACME Newsletter* 2007 July; (17): 2 p. [Online]. Accessed: http://www.eacmeweb.com/newsletter/n17.htm [2007 August 15]. Subject: 15.6

Bergman, Karl; Graff, Gregory D. The global stem cell patent landscape: implications for efficient technology transfer and commercial development. *Nature Biotechnology* 2007 April; 25(4): 419-425. Subject: 15.8

Bergmeier, L. Animal welfare is not just another bureaucratic hoop [letter]. *Nature* 2007 July 19; 448(7151): 251. Subject: 22.2

Bergner, Daniel. Death in the family: Booth Gardner, a former governor of Washington State who has Parkinson's, is urgently lobbying for a doctor-assisted-suicide law. His son is among those fighting him every step of the way. *New York Times Magazine* 2007 December 2; p. 38-45, 60, 76, 78, 80. Subject: 20.7

Berkman, Alan; Susser, Ezra. Paternalism and the public's health. *Chronic Illness* 2006 March; 2(1): 17-18. Subject: 9.1

Berlin, Jordan; Bruinooge, Suanna S.; Tannock, Ian F. Ethics in oncology: consulting for the investment industry. *Journal of Clinical Oncology* 2007 February 1; 25(4): 444-446. Subject: 9.5.1

Berlinger, Nancy. Martin Luther at the bedside: conscientious objection and community. *Hastings Center Report* 2007 March-April; 37(2): inside back cover. Subject: 8.1

Bermúdez, José Luis. Thinking without words: an overview for animal ethics. *Journal of Ethics* 2007; 11(3): 319-335. Subject: 22.1

Bernat, James L. Ethical issues in brain death and multiorgan transplantation. *Neurologic Clinics* 1989 November; 7(4): 715-728. Subject: 20.2.1

Berner, Yitshal N. Non-benefit of active nutritional support in advanced dementia. *Israel Medical Association Journal* 2006 July; 8(7): 505-506. Subject: 20.5.1

Berry, Philip A. The absence of sadness: darker reflections on the doctor-patient relationship. *Journal of Medical Ethics* 2007 May; 33(5): 266-268. Subject: 8.1

Berry, Roberta M. Can bioethics speak to politics about the prospect of inheritable genetic modification? If so, what might it say? *In:* Rasko, John E.J.; O'Sullivan, Gabrielle M.; Ankeny, Rachel A., eds. The Ethics of Inheritable Genetic Modification: A Dividing Line? Cambridge: Cambridge University Press, 2006: 243-277. Subject: 15.1

Bersani, Hank, Jr.; Rotholz, David A.; Eidelman, Steven M.; Pierson, Joanna L.; Bradley, Valerie J.; Gomez, Sharon C.; Havercamp, Susan M.; Silverman, Wayne P.; Yeager, Mark H.; Morin, Diane; Wehmeyer, Michael L.; Carabello, Bernard J.; Croser, M. Doreen. Unjustifiable non-therapy: response to the issue of growth attenuation for young people on the basis of disability. *Intellectual and Developmental Disabilities* 2007 October; 45(5): 351-353. Subject: 9.5.3

Berwick, Donald M.; Kaplan, Madge. 'What's the ethics of that?' A conversation with Thomas O. Pyle. *Health Affairs* 2008 January-February; 27(1): 143-150. Subject: 9.3.2

Beshai, James A.; Tushup, Richard J. Sanctity of human life in war: ethics and post traumatic stress disorder. *Psychological Reports* 2006 February; 98(1): 217-225. Subject: 21.2

Betta, Michela. Diagnostic knowledge in the genetic economy and commerce. *In:* Betta, Michela, ed. The Moral, Social, and Commercial Imperatives of Genetic Testing and Screening: the Australian Case. Dordrecht: Springer, 2006: 25-52. Subject: 15.1

See SUBJECT HEADING KEY FOR SECTION II on inside back cover.

Betta, Michela. From destiny to freedom? On human nature and liberal eugenics in the age of genetic manipulation. *In:* Betta, Michela, ed. The Moral, Social, and Commercial Imperatives of Genetic Testing and Screening: the Australian Case. Dordrecht: Springer, 2006: 3-24. Subject: 15.1

Betta, Michela. Self-knowledge and self-care in the age of genetic manipulation. *In:* Betta, Michela, ed. The Moral, Social, and Commercial Imperatives of Genetic Testing and Screening: the Australian Case. Dordrecht: Springer, 2006: 249-256. Subject: 15.1

Beutler, Ernest. Lysosomal storage diseases: natural history and ethical and economic aspects. *Molecular Genetics and Metabolism* 2006 July; 88(3): 208-215. Subject: 15.4

Bevan, Joan C.; Miller, Donald R. Medical journals and cross-cultural research ethics. *Canadian Journal of Anaesthesia = Journal Canadien d'Anesthésie* 2005 December; 52(10): 1009-1016. Subject: 1.3.7

Beyleveld, Deryck. Conceptualising privacy in relation to medical research values. *In:* McLean, Sheila A.M., ed. First Do No Harm: Law, Ethics, and Healthcare. Aldershot, England; Burlington, VT: Ashgate, 2006: 151-163. Subject: 8.4

Beyrer, Chris; Villar, Juan Carlos; Suwanvanichkij, Voravit; Singh, Sonal; Baral, Stefan D.; Mills, Edward J. Neglected diseases, civil conflicts, and the right to health. *Lancet* 2007 August 18-24; 370(9587): 619-627. Subject: 9.2

Bhandari, Mohit; Jönsson, Anders; Bühren, Volker. Conducting industry-partnered trials in orthopaedic surgery. *Injury* 2006 April; 37(4): 361-366. Subject: 9.5.1

Bharadwaj, Minakshi. Looking back, looking beyond: revisiting the ethics of genome generation. *Journal of Biosciences* 2006 March; 31(1): 167-176. Subject: 15.1

Bhutta, Zulfiqar Ahmed. Ethics in international health research: a perspective from the developing world. *Bulletin of the World Health Organization* 2002; 80(2): 114-120. Subject: 18.2

Bianchi, Susan B. Living will for a handicapped child [letter]. *Health Affairs* 2007 September-October; 26(5): 1507. Subject: 20.5.4

Bibbee, Jeffrey R.; Viens, A.M.; Moreno, Jonathan D.; Berger, Sam. The inseparability of religion and politics in the neoconservative critique of biotechnology. *American Journal of Bioethics* 2007 October; 7(10): 18-20; author reply W1-W3. Subject: 5.1

Bibbins-Domingo, Kirsten; Fernandez, Alicia. BiDil for heart failure in black patients: implications of the U.S. Food and Drug Administration approval. *Annals of Internal Medicine* 2007 January 2; 146(1): 52-56. Subject: 15.11

Bibbins-Domingo, Kirsten; Fernandez, Alicia; Kahn, Jonathan D.; Temple, Robert; Stockbridge, Norman L.

BiDil for heart failure in black patients [letters and response]. *Annals of Internal Medicine* 2007 August 7; 147(3): 214-216. Subject: 9.7

Bibeau, Gilles; Pedersen, Duncan. A return to scientific racism in medical social sciences: the case of sexuality and the AIDS epidemic in Africa. *In:* Nichter, Mark and Lock, Margaret, eds. New Horizons in Medical Anthropology: Essays in Honour of Charles Leslie. London: Routledge, 2002: 141-171. Subject: 21.1

Bierman, Arlene S. Sex matters: gender disparities in quality and outcomes of care. *CMAJ/JAMC: Canadian Medical Association Journal* 2007 December 4; 177(12): 1520-1521. Subject: 9.4

Biever, Celeste. Uproar flares over Alzheimer's tags [news]. *New Scientist* 2007 May 19-25; 194(2604): 14. Subject: 17.1

Biller-Andorno, Nikola. Epilogue: cross-cultural discourse in bioethics: it's a small world after all. *In:* Roetz, Heiner, ed. Cross-Cultural Issues in Bioethics: The Example of Human Cloning. New York: Rodopi, 2006: 459-463. Subject: 2.1

Biller-Andorno, Nikola. The global, the local, and the parochial - a commentary on Vilhjálmur Árnason. *Ethik in der Medizin* 2006 December; 18(4): 390-392. Subject: 2.1

Bilmore, Isabel. The "right to health" according to WHO. *In:* Wagner, Teresa; Carbone, Leslie, eds. Fifty Years after the Declaration: the United Nations' Record on Human Rights. Lanham, MD: University Press of America, 2001: 25-31. Subject: 9.2

Bindig, Todd S. Confusion about speciesism and moral status. *Linacre Quarterly* 2007 May; 74(2): 145-155. Subject: 4.4

Bird, Chloe E.; Fremont, Allen M.; Bierman, Arlene S.; Wickstrom, Steve; Shah, Mona; Rector, Thomas; Horstman, Thomas; Escarce, José J. Does quality of care for cardiovascular disease and diabetes differ by gender for enrollees in managed care plans? *Women's Health Issues* 2007 May-June; 17(3): 131-138. Subject: 9.8

Bird, Stephanie J. Genetic testing for neurologic diseases: a rose with thorns. *Neurologic Clinics* 1989 November; 7(4): 859-870. Subject: 15.3

Birmingham, Karen; Frumston, Michael. Avon longitudinal study of parents and children (ALSPAC): ethical process. *In:* Gunning, Jennifer; Holm, Søren, eds. Ethics, law, and society. Volume 2. Aldershot, Hants, England; Burlington, VT: Ashgate, 2006: 65-74. Subject: 18.2

Birmontiene, Toma. The influence of the rulings of the Constitutional Court on the development of health law in Lithuania. *European Journal of Health Law* 2007 December; 14(4): 321-333. Subject: 9.2

Birnbacher, Dieter. Die Grenzen der Philosophie und die Grenzen des Lebens [The boundaries of philosophy and the limits of life]. *Ethik in der Medizin* 2006 December; 18(4): 315-319. Subject: 4.1.2

Birnbacher, Dieter; Dabrock, Peter; Taupitz, Jochen; Vollmann, Jochen. Wie sollen Ärzte mit Patientenverfügungen umgehen? Ein Vorschlag aus interdisziplinärer Sicht [How should physicians deal with advance directives? A proposal from an interdisciplinary point of view]. *Ethik in der Medizin* 2007 June; 19(2): 139-147. Subject: 20.5.4

Biros, Michelle H. Research without consent: exception from and waiver of informed consent in resuscitation research. *Science and Engineering Ethics* 2007 September; 13(3): 361-369. Subject: 18.3

Biros, Michelle H. The ethics of research in emergency medicine. *Science and Engineering Ethics* 2007 September; 13(3): 279-280. Subject: 18.3

Birrittieri, Cara. How IVF changed my life. *Lancet* 2006 December; 368(special issue): S58. Subject: 14.4

Bito, Seiji; Asai, Atsushi. Attitudes and behaviors of Japanese physicians concerning withholding and withdrawal of life-sustaining treatment for end-of-life patients: results from an Internet survey. *BMC Medical Ethics* 2007; 8:7, 9 p. [Online]. Accessed: http://www.biomedcentral.com/1472-6939/8/7 [2007 July 20]. Subject: 20.5.1

Bito, Seiji; Matsumura, Shinji; Singer, Marjorie Kagawa; Meredith, Lisa S.; Fukuhara, Shunichi; Wenger, Neil S. Acculturation and end-of-life decision making: comparison of Japanese and Japanese-American focus groups. *Bioethics* 2007 June; 21(5): 251-262. Subject: 20.3.3

Björkman, Barbro. Different types — different rights: distinguishing between different perspectives on ownership of biological material. *Science and Engineering Ethics* 2007 June; 13(2): 221-233. Subject: 19.5

Blachar, Yoram; Borow, Malke. The health of leaders: information, interpretation and the media. *Israel Medical Association Journal* 2006 November; 8(11): 741-743. Subject: 9.4

Black, Betty S.; Kass, Nancy E.; Fogarty, Linda A.; Rabins, Peter V. Informed consent for dementia research: the study enrollment encounter. *IRB: Ethics and Human Research* 2007 July-August; 29(4): 7-14. Subject: 18.3

Black, Lee; Anderson, Emily E. Physicians, patients and confidentiality: the role of physicians in electronic health records. *American Journal of Bioethics* 2007 March; 7(3): 50-51. Subject: 8.4

Black, Lee; Sade, Robert M. Lethal injection and physicians: state law vs medical ethics. *JAMA: The Journal of the American Medical Association* 2007 December 19; 298(23): 2779-2781. Subject: 20.6

Blackford, Russell. Differing vulnerabilities: the moral significance of Lockean personhood. *American Journal of Bioethics* 2007 January; 7(1): 70-71. Subject: 17.1

Blackford, Russell. Slippery slopes to slippery slopes: therapeutic cloning and the criminal law. *American Journal of Bioethics* 2007 February; 7(2): 63-64. Subject: 14.5

Blackmer, Jeff. Clarification of the CMA's position concerning induced abortion [letter]. *CMAJ/JAMC: Canadian Medical Association Journal* 2007 April 24; 176(9): 1310. Subject: 12.1

Blacksher, Erika. Bioethics and politics: a values analysis of the mission of the Center for Practical Bioethics. *American Journal of Bioethics* 2007 October; 7(10): 34-36. Subject: 2.1

Blackwood, Bronagh. Informed consent for research in critical care: implications for nursing. *Nursing in Critical Care* 2006 July-August; 11(4): 151-153. Subject: 18.3

Blakely, Gillian; Millward, Jennifer. Moral dilemmas associated with the withdrawal of artificial hydration. *British Journal of Nursing* 2007 August 9 - September 12; 16(15): 916-919. Subject: 20.5.1

Blank, Robert H. Policy implications of the new neuroscience. *CQ: Cambridge Quarterly of Healthcare Ethics* 2007 Spring; 16(2): 169-180. Subject: 17.1

Blase, Terri; Martinez, Ariadna; Grody, Wayne W.; Schimmenti, Lisa; Palmer, Christina G.S. Sharing GJB2/GJB6 genetic test information with family members. *Journal of Genetic Counseling* 2007 June; 16(3): 313-324. Subject: 15.3

Blazekovic-Milakovic, Sanja; Matijasevic, Ivana; Stojanovic-Spehar, Stanislava; Supe, Svjetlana. Family physicians' views on disclosure of a diagnosis of cancer and care of terminally ill patients in Croatia. *Psychiatria Danubina* 2006 June; 18(1-2): 19-29. Subject: 8.2

Bleich, J. David. Cadavers on display. *Tradition* 2007 Spring; 40(1): 87-97. Subject: 20.1

Bleyer, W. Archie; Tejeda, Heriberto A.; Murphy, Sharon B.; Brawley, Otis W.; Smith, Malcolm A.; Ungersleider, Richard S. Equal participation of minority patients in U.S. national pediatric cancer clinical trials. *Journal of Pediatric Hematology and Oncology* 1997 September-October; 19(5): 423-427. Subject: 18.5.1

Blinderman, Craig D. Jewish law and end-of-life decision making: a case report. *Journal of Clinical Ethics* 2007 Winter; 18(4): 384-390. Subject: 20.5.1

Bloche, M. Gregg. Health care for all? *New England Journal of Medicine* 2007 September 20; 357(12): 1173-1175. Subject: 9.3.1

Bloom, Alexandra. The ostrich raises its head: "knowing" and moral accountability in the practice of psychotherapy. *Women and Therapy* 1999; 22(2): 7-20. Subject: 17.2

Blumenthal, Daniel S. A community coalition board creates a set of values for community-based research. *Preventing Chronic Disease* 2006 January; 3(1): 7 p. Subject: 18.6

See SUBJECT HEADING KEY FOR SECTION II on inside back cover.

447

Blumsohn, Aubrey. The Gillberg Affair: profound ethical issues were smoothed over [letter]. *BMJ: British Medical Journal* 2007 September 29; 335(7621): 629. Subject: 1.3.9

Blustein, Jeffrey. Doctoring and self-forgiveness. *In:* Walker, Rebecca L.; Ivanhoe, Philip J., eds. Working Virtue: Virtue Ethics and Contemporary Moral Problems. Oxford: Clarendon, 2007: 87-111. Subject: 9.8

Blustein, J. The history and moral foundations of human-subject research. *American Journal of Physical Medicine and Rehabilitation* 2007 February; 86(2): 82-85. Subject: 18.1

Blyth, Eric; Crawshaw, Marilyn; Haase, Jean; Speirs, Jennifer. The implications of adoption for donor offspring following donor-assisted conception. *Child and Family Social Work* 2001; 6(4): 295-304. Subject: 14.1

Blyth, Eric; Farrand, Abigail. Reproductive tourism—a price worth paying for reproductive autonomy? *Critical Social Policy* 2005; 25(1): 91-114 [Online]. Accessed: http://csp.sagepub.com/cgi/reprint/25/1/91 [13 March 2007]. Subject: 14.4

Boada, M.; Veiga, A.; Barri, P.N. Spanish regulations on assisted reproduction techniques. *Journal of Assisted Reproduction and Genetics* 2003 July; 20(7): 271-275. Subject: 14.4

Boas, Franz. Eugenics. *In his:* Anthropology and Modern Life. New Brunswick, NJ: Transaction Publishers, 2004: 106-121. Subject: 15.5

Bobbert, Monika. Ethical questions concerning research on human embryos, embryonic stem cells and chimeras. *Biotechnology Journal* 2006 December; 1(12): 1352-1369. Subject: 18.5.4

Bober, Daniel I.; Pinals, Debra A. Prisoner's rights and deliberate indifference. *Journal for the American Academy of Psychiatry and the Law* 2007; 35(3): 388-391. Subject: 17.8

Bobrow, James C. The ethics and politics of stem cell research. *Transactions of the American Ophthalmological Society* 2005; 103: 138-141; discussion 141-142. Subject: 18.5.4

Bockenheimer-Lucius, Gisela. Behandlungsbegrenzung durch eine Patientenverfügung - im individuellen Fall auch mit Blick auf neue therapeutische Möglichkeiten! = Limitation of treatment by an advanced directive - in the individual case also with regard to the therapeutic possibilities! *Ethik in der Medizin* 2007 March; 19(1): 5-6. Subject: 20.5.4

Bockenheimer-Lucius, Gisela. Ethikberatung und Ethik-Komitee im Altenpflegeheim (EKA) - Herausforderung und Chance für eine ethische Entscheidungskultur = Ethics committee in a long-term care facility - a challenge and a chance for an ethical deci-

sion-making culture. *Ethik in der Medizin* 2007 December; 19(4): 320-330. Subject: 9.6

Bockenheimer-Lucius, Gisela; May, Arnd T. Ethikberatung - Ethik-Komitee in Einrichtungen der stationären Altenhilfe (EKA). Eckpunkte für ein Curriculum [Ethics consultation - ethics committees in geriatric care institutions. Key elements of a curriculum]. *Ethik in der Medizin* 2007 December; 19(4): 331-339. Subject: 9.6

Bodger, Jessica Ansley. Taking the sting out of reporting requirements: reproductive health clinics and the constitutional right to informational privacy. *Duke Law Journal* 2006 November; 56(2): 583-609. Subject: 12.4.2

Boer, Theo A.; Schroten, Egbert. Life and death seen from a (Dutch) reformed position: a Calvinistic approach to bioethics. *In:* Østnor, Lars, ed. Bioetikk og teologi: Rapport fra Nordisk teologisk nettverk for bioetikks workshop i Stockholm 27.-29. September 1996. Oslo: Nord. teol. nettv. for bioetikk, 1996: 31-53. Subject: 2.1

Bogod, D.G. The editor as umpire: clinical trial registration and dispute resolution. *Anaesthesia* 2006 December; 61(12): 1133-1135. Subject: 1.3.7

Böhme, Gernot. Rationalizing unethical medical research: taking seriously the case of Viktor von Weizsäcker. *In:* LaFleur, William R.; Böhme, Gernot; Shimazono, Susumu, eds. Dark Medicine: Rationalizing Medical Research. Bloomington: Indiana University Press, 2007: 15-29. Subject: 1.3.9

Boisaubin, Eugene V.; Chu, Adeline; Catalano, Janine M. Perceptions of long-term care, autonomy, and dignity, by residents, family and care-givers: the Houston experience. *Journal of Medicine and Philosophy* 2007 September-October; 32(5): 447-464. Subject: 9.5.2

Boles, Jean-Michel. End of life in the intensive care unit: from practice to law. What do the lawmakers tell the caregivers? A new series in Intensive Care Medicine. *Intensive Care Medicine* 2006 July; 32(7): 955-957. Subject: 20.5.1

Bolks, Sean M.; Evans, Diana; Polinard J.L.; Wrinkle, Robert D. Core beliefs and abortion attitudes: a look at Latinos. *Social Science Quarterly* 2000; 81(1): 253-260. Subject: 12.5.2

Bolnick, Deborah A.; Fullwiley, Duana; Duster, Troy; Cooper, Richard S.; Fujimura, Joan H.; Kahn, Jonathan; Kaufman, Jay S.; Marks, Jonathan; Morning, Ann; Nelson, Alondra; Ossorio, Pilar; Reardon, Jenny; Reverby, Susan M.; TallBear, Kimberly. The science and business of genetic ancestry testing. *Science* 2007 October 19; 318(5849): 399-400. Subject: 15.11

Bolt, L.L.E. True to oneself? Broad and narrow ideas on authenticity in the enhancement debate. *Theoretical Medicine and Bioethics* 2007; 28(4): 285-300. Subject: 4.5

Bolton, Derek. What's the problem? A response to "secular humanism and scientific psychiatry". *Philosophy, Ethics, and Humanities in Medicine [electronic]* 2006; 1(6): 2

p. Accessed: http://www.peh-med.com/content/1/1/6 [2006 May 15]. Subject: 17.1

Bomhoff, Jacco; Zucca, Lorenzo. European Court of Human Rights. The tragedy of Ms Evans: conflicts and incommensurability of rights, Evans v. the United Kingdom, Fourth Section Judgment of 7 March 2006, application no. 6339/05. *European Constitutional Law Review* 2006 October; 2(3): 424-442. Subject: 14.1

Bond, Claire. Lieberman Award; Section 15 of the Charter and the allocation of resources in health care: a comment on Auton v. British Columbia. *Health Law Journal* 2005; 13: 253-271. Subject: 9.4

Bonnicksen, Andrea L. Oversight of assisted reproductive technologies: the last twenty years. *In:* Knowles, Lori P.; Kaebnick, Gregory E., eds. Reprogenetics: Law, Policy, and Ethical Issues. Baltimore: Johns Hopkins University Press, 2007: 64-86. Subject: 14.1

Bonnicksen, Andrea L. Therapeutic cloning: politics and policy. *In:* Steinbock, Bonnie, ed. The Oxford Handbook of Bioethics. Oxford; New York: Oxford University Press, 2007: 441-468. Subject: 14.5

Bookman, Kelly; Abbott, Jean. Ethics seminars: withdrawal of treatment in the emergency department — when and how? *Academic Emergency Medicine* 2006 December; 13(12): 1328-1332. Subject: 20.5.1

Boon, Mieke. Comments on Thompson: research ethics for animal biotechnology. *In:* Korthals, Michiel; Bogers, Robert J., eds. Ethics for Life Scientists. Dordrecht, The Netherlands: Springer, 2004: 121-125. Subject: 22.2

Boonstra, Heather D.; Gold, Rachel Benson; Richards, Cory L.; Finer, Lawrence B. Abortion in women's lives. New York: Guttmacher Institute, 2006; 44 p. [Online]. Accessed: http://www.guttmacher.org/pubs/2006/05/04/AiWL.pdf [2007 April 3]. Subject: 12.5.1

Booth, Malcolm G. Informed consent in emergency research: a contradiction in terms. *Science and Engineering Ethics* 2007 September; 13(3): 351-359. Subject: 18.3

Borasio, G.D.; Weltermann, B.; Voltz, R.; Reichmann, H.; Zierz, S. Einstellungen zur patientenbetreuung in der letzten lebensphase. Eine umfrage bei neurologischen chefärzten = Attitudes towards patient care at the end of life. A survey of directors of neurological departments. *Nervenarzt* 2004 December; 75(12): 1187-1193. Subject: 20.3.2

Borenstein, Jason. Shaping our future: the implications of genetic enhancement. *Human Reproduction and Genetic Ethics: An International Journal* 2007; 13(2): 4-15. Subject: 15.1

Borkenhagen, A.; Brähler, E.; Wisch, S.; Stöbel-Richter, Y.; Strauss, B.; Kentenich, H. Attitudes of German infertile couples towards preimplantation genetic diagnosis for different uses: a comparison to international studies.

Human Reproduction 2007 July; 22(7): 2051-2057. Subject: 15.2

Borry, Pascal; Fryns, Jean-Pierre; Schotsmans, Paul; Dierickx, Kris. Carrier testing in minors: a systematic review of guidelines and position papers. *European Journal of Human Genetics : EJHG* 2006 February; 14(2): 133-138. Subject: 15.3

Borry, Pascal; Schotsmans, Paul; Dierickx, Kris. Evidence-based medicine and its role in ethical decision-making. *Journal of Evaluation in Clinical Practice* 2006 June; 12(3): 306-311. Subject: 9.5.4

Borry, P.; Stultiens, L.; Nys, H.; Cassiman, J.-J.; Dierickx, K. Presymptomatic and predictive genetic testing in minors: a systematic review of guidelines and position papers. *Clinical Genetics* 2006 November; 70(5): 374-381. Subject: 15.3

Borst-Eisler, Els. Th role of public debate and politics in the implementation of the Convention. *In:* Gevers, J.K.M.; Hondius, E.H.; Hubben, J.H., eds. Health Law, Human Rights and the Biomedicine Convention: Essays in Honour of Henriette Roscam Abbing. Leiden; Boston: Martinus Nijhoff Publishers, 2005: 247-254. Subject: 21.1

Bortolotti, Lisa. Disputes over moral status: philosophy and science in the future of bioethics. *Health Care Analysis: An International Journal of Health Philosophy and Policy* 2007 June; 15(2): 153-158. Subject: 4.4

Bortolotti, Lisa; Harris, John. Disability, enhancement and the harm-benefit continuum. *In:* Spencer, J.R.; du Bois-Pedain, Antje, eds. Freedom and Responsibility in Reproductive Choice. Portland: Hart Pub., 2006: 31-49. Subject: 15.1

Bortolotti, Lisa; Heinrichs, Bert. Delimiting the concept of research: an ethical perspective. *Theoretical Medicine and Bioethics* 2007; 28(3): 157-179. Subject: 1.3.9

Bosch, Xavier. Dealing with scientific misconduct [editorial]. *BMJ: British Medical Journal* 2007 September 15; 335(7619): 524-525. Subject: 1.3.9

Bosek, Marcia Sue DeWolf. Refusal of brain death diagnosis: the case. *JONA's Healthcare Law, Ethics, and Regulation* 2007 July-September; 9(3): 87. Subject: 20.2.1

Bosek, Marcia Sue DeWolf. Refusal of brain death diagnosis: the ethicist's response. *JONA's Healthcare Law, Ethics, and Regulation* 2007 July-September; 9(3): 87-90. Subject: 20.2.1

Bosek, Marcia Sue DeWolf. When respecting patient autonomy may not be in the patient's best interest. *JONA's Healthcare Law, Ethics, and Regulation* 2007 April-June; 9(2): 46-49. Subject: 8.1

Bosek, Marcia Sue DeWolf; Stammer, Karen. Ethical commitments during desperate times. *JONA's Healthcare Law, Ethics, and Regulation* 2006 October-December; 8(4): 123-128. Subject: 20.5.1

See SUBJECT HEADING KEY FOR SECTION II on inside back cover.

449

Bosk, Charles L. Disinterested commitment as moral heroism. *Atrium* 2007 Summer; 4: 1-4. Subject: 9.6

Botkin, Jeffrey R.; Munger, Mark A.; Shea, Patrick A.; Coffin, Cheryl; Mineau, Geraldine P. Management of human tissue resources for research in academic medical centers: points to consider. *In:* Kulakowski, Elliott C.; Chronister, Lynne U., eds. Research Administration and Management. Sudbury, MA: Jones and Bartlett, 2006: 567-575. Subject: 19.1

Botto, Ronald W. Addressing the marketplace mentality and improving professionalism in dental education: response to Richard Masella's "Renewing professionalism in dental education". *Journal of Dental Education* 2007 February; 71(2): 217-221. Subject: 4.1.1

Bouchal, Shelley Raffin. Moral meaning of caring for the dying. *In:* Johnston, Nancy E.; Scholler-Jaquish, Alwilda, eds. Meaning in Suffering: Caring Practices in the Health Professions. Madison: University of Wisconsin Press, 2007: 232-275. Subject: 20.4.1

Boudoulas, Harisios. Ethics in biomedical research. *Hellenic Journal of Cardiology* 2006 May-June; 47(3): 193. Subject: 18.2

Bouknight, Heyward H., III. Between the scalpel and the lie: comparing theories of physician accountability for misrepresentations of experience and competence. *Washington and Lee Law Review* 2003 Fall; 60(4): 1515-1560. Subject: 7.1

Boulware, L.E.; Troll, M.U.; Wang, N.Y.; Powe, N.R. Public attitudes toward incentives for organ donation: a national study of different racial/ethnic and income groups. *American Journal of Transplantation* 2006 November; 6(11): 2774-2785. Subject: 19.5

Bourg, Claudine. Ethical dilemmas in medically assisted procreation: a psychological perspective. *Human Reproduction and Genetic Ethics: An International Journal* 2007; 13(2): 22-31. Subject: 14.1

Bowers, Len. On conflict, containment and the relationship between them. *Nursing Inquiry* 2006 September; 13(3): 172-180. Subject: 8.1

Bowers, Libby. Ethical issues along the cancer continuum. *In:* Carroll-Johnson, Rose Mary; Gorman, Linda M.; Bush, Nancy Jo, eds. Psychosocial Nursing Care Along the Cancer Continuum. 2nd edition. Pittsburgh, PA: Oncology Nursing Society, 2006: 551-564. Subject: 9.5.1

Bowling, Ann; Rowe, Gene. "You decide doctor". What do patient preference arms in clinical trials really mean? [editorial]. *Journal of Epidemiology and Community Health* 2005 November; 59(11): 914-915. Subject: 18.3

Bowman, Diana M.; Hodge, Graeme A. Editorial - governing nanotechnology: more than a small matter? *NanoEthics* 2007 December; 1(3): 239-241. Subject: 5.4

Boyd, Elizabeth; Bero, Lisa A. Defining financial conflicts and managing research relationships: an analysis of university conflict of interest committee decisions. *Science and Engineering Ethics* 2007 December; 13(4): 415-435. Subject: 1.3.9

Boyd, J. Wesley; Himmelstein, David U.; Lasser, Karen; McCormick, Danny; Bor, David H.; Cutrona, Sarah L.; Woolhandler, Steffie. U.S. medical students' knowledge about the military draft, the Geneva Conventions, and military medical ethics. *International Journal of Health Services* 2007; 37(4): 643-650. Subject: 7.2

Boyd, Kenneth. Medical ethics: Hippocratic and democratic ideals. *In:* McLean, Sheila A.M., ed. First Do No Harm: Law, Ethics, and Healthcare. Aldershot, England; Burlington, VT: Ashgate, 2006: 29-38. Subject: 2.1

Boyle, Dennis; O'Connell, Daniel; Platt, Frederic W.; Albert, Richard K. Disclosing errors and adverse events in the intensive care unit. *Critical Care Medicine* 2006 May; 34(5): 1532-1537. Subject: 9.8

Boyle, Joseph. Casuistry. *In:* Khushf, George, ed. Handbook of Bioethics: Taking Stock of the Field From a Philosophical Perspective. Dordrecht; Boston: Kluwer Academic, 2004: 75-88. Subject: 2.1

Boyle, Joseph. Genetics, medicine, and the human person: the papal theology. *In:* Monsour, H. Daniel, ed. Ethics and the New Genetics: An Integrated Approach. Toronto: University of Toronto Press, 2007: 134-142. Subject: 15.1

Boyle, Joseph. The bioethics of global biomedicine: a natural law reflection. *In:* Engelhardt, H. Tristram, ed. Global Bioethics: The Collapse of Consensus. Salem, MA: M&M Scrivener Press, 2006: 300-334. Subject: 2.1

Boz, Bora; Acar, Kemalettin; Ergin, Ahmet; Kurtulus, Ayse; Ergin, Nesrin; Oguzhanoglu, Nalan. Effect of locus of control on acceptability of euthanasia among medical students and residents in Denizli, Turkey. *Journal of Palliative Care* 2007 Winter; 23(4): 286-290. Subject: 20.5.1

Bozzato, Gianni. "Lay" reduction of the human-embryo individual. *Linacre Quarterly* 2007 May; 74(2): 122-134. Subject: 4.4

Brabender, V. The ethical group psychotherapist: a coda. *International Journal of Group Psychotherapy* 2007 January; 57(1): 41-47; discussion 49-59. Subject: 17.2

Brabin, Loretta; Roberts, Stephen A.; Kitchener, Henry C. A semi-qualitative study of attitudes to vaccinating adolescents against human papillomavirus without parental consent. *BMC Public Health* 2007 February 9; 7: 20. Subject: 9.5.7

Bracken, Wendy; Simon, Gayle; Cox, Susan; McDonald, Michael; Fitzgerald, Maureen. Protecting researchers. *Journal of Empirical Research on Human Research Ethics* 2007 March; 2(1): 93-96. Subject: 18.1

Bradburn, Norman; Simon, Gayle; Bankowski, Susan Burner; Beattie, Elizabeth; Buckwalter, Kathleen; Clark, Laura; Diehl, Dawn. Informed consent. *Journal* *of Empirical Research on Human Research Ethics* 2007 March; 2(1): 75-82. Subject: 18.3

Brady Wagner, Lynne C.; Stein, Joel. Failure to achieve assent in a communicative patient: what are the caregiver's obligations? *Topics in Stroke Rehabilitation* 2006 Fall; 13(4): 36-41. Subject: 8.3.3

Brainard, Jeffrey. California stem-cell researchers ponder next steps after court victory. *Chronicle of Higher Education* 2007 June 1; 53(39): A20. Subject: 18.5.4

Brainard, Jeffrey. NIH director calls for easing administration's stem-cell restrictions [news]. *Chronicle of Higher Education* 2007 March 30; 53(30): A26. Subject: 18.5.4

Bramstedt, Katrina A.; Ford, Paul J. Protecting human subjects in neurosurgical trials: the challenge of psychogenic dystonia. *Contemporary Clinical Trials* 2006 April; 27(2): 161-164. Subject: 18.2

Brand, Richard A.; Jacobs, Joshua J.; Heckman, James D. Professionalism in publishing. *Journal of Bone and Joint Surgery. American volume* 2006 November; 88(11): 2323-2325. Subject: 1.3.7

Brands, W.G. The standard for the duty to inform patients about risks: from the responsible dentist to the reasonable patient. *British Dental Journal* 2006 August 26; 201(4): 207-210. Subject: 8.3.1

Brandt, Michelle L. IRB burden studied in cost analysis. *Stanford Report* 2003 August 6: 3 p. [Online]. Accessed: http://news.service.standford.edu/news/2003/august6/humphreys.html [2007 December 13]. Subject: 18.2

Brannigan, Michael C. On medical futility: considerations and guidelines. *Missouri Medicine* 2006 March-April; 103(2): 113-117. Subject: 20.5.1

Branson, Richard D.; Davis, Kenneth, Jr.; Butler, Karyn L. African Americans' participation in clinical research: importance, barriers, and solutions. *American Journal of Surgery* 2007 January; 193(1): 32-39; discussion 40. Subject: 18.2

Bras, Marijana; Loncar, Zoran; Fingler, Mira. The relief of pain as a human right. *Psychiatria Danubina* 2006 June; 18(1-2): 108-110. Subject: 4.4

Brashler, Rebecca. Ethics, family caregivers, and stroke. *Topics in Stroke Rehabilitation* 2006 Fall; 13(4): 11-17. Subject: 18.5.6

Brassington, Iain. John Harris' argument for a duty to research. *Bioethics* 2007 March; 21(3): 160-168. Subject: 18.1

Brassington, Iain. On Heidegger, medicine, and the modernity of modern medical technology. *Medicine, Health Care and Philosophy* 2007 June; 10(2): 185-195. Subject: 5.3

Bratspies, Rebecca M. Glowing in the dark: how America's first transgenic animal escaped regulation. *Minnesota*

Journal of Law, Science and Technology 2004-2005; 6(2): 457-504. Subject: 15.1

Braude, Peter; Flinter, Frances. Use and misuse of pre-implantation genetic testing. *BMJ: British Medical Journal* 2007 October 13; 335(7623): 752-754. Subject: 15.2

Braun, Joshua A. The imperatives of narrative: health interest groups and morality in network news. *American Journal of Bioethics* 2007 August; 7(8): 6-14. Subject: 1.3.7

Braune, Florian; Biller-Andorno, Nikola; Wiesemann, Claudia. Human reproductive cloning: a test case for individual rights? *In:* Roetz, Heiner, ed. Cross-Cultural Issues in Bioethics: The Example of Human Cloning. New York: Rodopi, 2006: 445-458. Subject: 14.5

Braunschweiger, Paul; Goodman, Kenneth W. The CITI program: an international online resource for education in human subjects protection and the responsible conduct of research. *Academic Medicine* 2007 September; 82(9): 861-864. Subject: 18.2

Braveman, Paula. Health disparities and health equity: concepts and measurement. *Annual Review of Public Health* 2006; 27: 167-194. Subject: 9.2

Brawley, Otis W.; Tejeda, Heriberto. Minority inclusion in clinical trials: issues and potential strategies. *Journal of National Cancer Institute Monographs* 1995; 17: 55-57. Subject: 18.5.1

Brazier, Margaret; Archard, David. Letting babies die [editorial]. *Journal of Medical Ethics* 2007 March; 33(3): 125-126. Subject: 20.5.2

Brazier, Margaret; Quigley, Muireann. Deceased organ donation: in praise of pragmatism [editorial]. *Clinical Ethics* 2007 December; 2(4): 164-165. Subject: 19.5

Brazier, Margot. Human(s) (as) medicine(s). *In:* McLean, Sheila A.M., ed. First Do No Harm: Law, Ethics, and Healthcare. Aldershot, England; Burlington, VT: Ashgate, 2006: 187-202. Subject: 4.4

Breen, John M.; Scaperlanda, Michael A. Never get out'a the boat: Stenberg v. Carhart and the future of American law. *Connecticut Law Review* 2006 November; 39(1): 297-323. Subject: 12.4.2

Breese, Peter E.; Burman, William J.; Goldberg, Stefan; Weis, Stephen E. Education level, primary language, and comprehension of the informed consent process. *Journal of Empirical Research on Human Research Ethics* 2007 December; 2(4): 69-79. Subject: 18.3

Breese, Peter; Rietmeijer, Cornelis; Burman, William. Content among locally approved HIPAA authorization forms for research. *Journal of Empirical Research on Human Research Ethics* 2007 March; 2(1): 43-46. Subject: 18.3

Breger, Marshall. Freedom to choose. For Jewish women, a pro-choice stance on abortion often reflects their political values and culture as much as their views on reproductive

See SUBJECT HEADING KEY FOR SECTION II on inside back cover.

451

freedom. *Moment* 1999 August; 24(4): 28-29. Subject: 12.3

Breier-Mackie, Sarah. Medical ethics and nursing ethics: is there really any difference? *Gastroenterology Nursing: the Official Journal of the Society of Gastroenterology Nurses and Associates* 2006 March-April; 29(2): 182-183. Subject: 4.1.3

Breier-Mackie, Sarah. Who is the clinical ethicist? *Gastroenterology Nursing: the Official Journal of the Society of Gastroenterology Nurses and Associates* 2006 January-February; 29(1): 70-72. Subject: 9.6

Breitbart, William. What can we learn from the death of Terri Schiavo? *Palliative and Supportive Care* 2005 March; 3(1): 1-3. Subject: 20.5.1

Brennan, Patricia. Public solicitation of organs on the Internet: ethical and policy issues. *Journal of Emergency Nursing* 2006 April; 32(2): 191-193. Subject: 19.5

Brennan, Patricia A.W. The medical and ethical aspects of photography in the sexual assault examination: why does it offend? *Journal of Clinical Forensic Medicine* 2006 May; 13(4): 194-202. Subject: 8.3.1

Brennan, Troyen Anthony. Ethics of disclosure following a medical injury: time for reform? *In:* Rhodes, Rosamond; Francis, Leslie P.; Silvers, Anita, eds. The Blackwell Guide to Medical Ethics. Malden, MA: Blackwell Pub., 2007: 393-406. Subject: 8.2

Brennan, Troyen A.; Mello, Michelle M. Sunshine laws and the pharmaceutical industry [editorial]. *JAMA: The Journal of the American Medical Association* 2007 March 21; 297(11): 1255-1257. Subject: 9.7

Brenner, Steven E. Common sense for our genomes. *Nature* 2007 October 18; 449(7164): 783-784. Subject: 15.10

Brett, Allan S. Two-tiered health care: a problematic double standard [editorial]. *Archives of Internal Medicine* 2007 March 12; 167(5): 430-432. Subject: 9.3.1

Brewster, Luke P.; Bennett, Barry K.; Gamelli, Richard L. Application of rehabilitation ethics to a selected burn patient population's perspective. *Journal of the American College of Surgeons* 2006 November; 203(5): 766-771. Subject: 8.1

Bridgeman, Jo. Caring for children with severe disabilities: boundaried and relational rights. *In:* Freeman, Michael, ed. Children's Health and Children's Rights. Boston: Martinus Nijhoff Publishers, 2006: 99-119. Subject: 9.5.7

Briggs, Sheila. The ethics of life: medical advances have exposed inconsistencies in the Roman Catholic hierarchy's position on life. *Conscience* 2007 Autumn; 28(3): 14-18. Subject: 4.4

Brindle, David. Seeing red [interview]. *BMJ: British Medical Journal* 2007 May 12; 334(7601): 976-977. Subject: 9.4

Bristol, Nellie. Should terminally ill patients have access to phase I drugs? *Lancet* 2007 March 10-16; 369(9564): 815-816. Subject: 18.5.7

British Broadcasting Corporation [BBC]. Half-price IVF offered for eggs [news]. *London: BBC News.* 2007 September 13; 1 p. [Online]. Accessed: http://newsvote. bbc.co.uk/mpapps/pagetools/print/news.bbc.co.uk/2/hi/ uk_news/en gland/6992642.stm [2007 September 14]. Subject: 14.4

British Council Switzerland; University of Basel, Institute for Applied Ethics and Medical Ethics. Conflicts of interest: ethics and predictive medicine. Bern and Basel: British Council Switzerland; Institute of Applied Ethics and Medical Ethics, 2003 February; 72 p. [Online]. Accessed: http://www.britishcouncil.ch/governance/Genetic %20and%20Ethics.pdf [2007 April 30]. Subject: 15.3

British In Vitro Diagnostics Association [BIVDA]. Genetic testing: the difference diagnostics can make [position paper 6]. London: British In Vitro Diagnostics Association, 2004 November; 10 p. [Online]. Accessed: http:// www.bivda.co.uk/Portals/0/Positionpaper6-genetics3.pdf [2007 April 17]. Subject: 15.3

British Medical Association [BMA]. Ethics Department. Treatment of patients in persistent vegetative state. Guidance from the BMA's Medical Ethics Department. London: British Medical Association, October 2007: 6 p. [Online]. Accessed: http://www.bma.org.uk/ap.nsf/ AttachmentsByTitle/PDFpvs2007/$FILE/PVS07.pdf [2008 January 14]. Subject: 20.5.1

British Medical Association [BMA]. Medical Ethics Department. Human Tissue Legislation. Guidance from the BMA's Medical Ethics Department. London: British Medical Association, 2006 September: 8 p. [Online]. Accessed: http://www.bma.org.uk/ap.nsf/ AttachmentsByTitle/PDFHumantissue/ $FILE/ HumanTissueLegislation.pdf [2007 August 13]. Subject: 8.3.1

Brock, Dan. Ethical issues in the use of cost effectiveness analysis for the prioritization of health resources. *In:* Khushf, George, ed. Handbook of Bioethics: Taking Stock of the Field From a Philosophical Perspective. Dordrecht; Boston: Kluwer Academic, 2004: 353-380. Subject: 9.4

Brock, Dan W. Patient competence and surrogate decision-making. *In:* Rhodes, Rosamond; Francis, Leslie P.; Silvers, Anita, eds. The Blackwell Guide to Medical Ethics. Malden, MA: Blackwell Pub., 2007: 128-141. Subject: 8.3.3

Brock, Dan W.; Truog, Robert D.; Brett, Allan S.; Frader, Joel; Downie, Jocelyn. Withholding and withdrawing life-sustaining treatment. *In:* Baylis, Françoise; Downie, Jocelyn; Freedman, Benjamin; Hoffmaster, Barry; Sherwin, Susan, eds. Health Care Ethics in Canada. Toronto: Harcourt Brace Canada, 1995: 487-525. Subject: 20.5.1

Brody, Baruch. Intellectual property and biotechnology: the European debate. *Kennedy Institute of Ethics Journal* 2007 June; 17(2): 69-110. Subject: 5.3

Brody, Baruch. The ethics of controlled clinical trials. *In:* Khushf, George, ed. Handbook of Bioethics: Taking Stock of the Field From a Philosophical Perspective. Dordrecht; Boston: Kluwer Academic, 2004: 337-352. Subject: 18.1

Brody, Howard. Ethics, justice, and health reform. *In:* Engström, Timothy H.; Robison, Wade L., eds. Health Care Reform: Ethics and Politics. Rochester, NY: University of Rochester Press, 2006: 40-66. Subject: 9.1

Brody, Howard; Whitney, Simon N.; McCullough, Laurence B. Transparency and self-censorship in shared decision-making. *American Journal of Bioethics* 2007 July; 7(7): 44-46; author reply W1-W3. Subject: 8.1

Brody, Jane E. The solvable problem of organ shortages. *New York Times* 2007 August 28; p F7. Subject: 19.5

Brody, Janet L.; Scherer, David G.; Annett, Robert D.; Turner, Charles; Dalen, Jeanne. Family and physician influence on asthma research participation decisions for adolescents: the effects of adolescent gender and research risk. *Pediatrics* 2006 August; 118(2): e356-e362. Subject: 18.5.2

Brody, Julia Green; Morello-Frosch, Rachel; Brown, Phil; Rudel, Ruthann A.; Altman, Rebecca Gasior; Frye, Margaret; Osimo, Cheryl A.; Pérez, Carla; Seryak, Liesel M. "Is it safe?": new ethics for reporting personal exposures to environmental chemicals. *American Journal of Public Health* 2007 September; 97(9): 1547-1554. Subject: 16.1

Brooks, Robert A. Psychiatrists' opinions about involuntary civil commitment: results of a national survey. *Journal of the American Academy of Psychiatry and the Law* 2007; 35(2): 219-228. Subject: 17.7

Brooks, Robert A. U.S. psychiatrists' beliefs and wants about involuntary civil commitment grounds. *International Journal of Law and Psychiatry* 2006 January-February; 29(1): 13-21. Subject: 17.7

Broom, Alex. Ethical issues in social research. *Complementary Therapies in Medicine* 2006 June; 14(2): 151-156. Subject: 18.4

Broström, Linus; Johansson, Mats; Nielsen, Morten Klemme. "What the patient would have decided": a fundamental problem with the substituted judgment standard. *Medicine, Health Care and Philosophy* 2007 September; 10(3): 265-278. Subject: 8.3.3

Brotherton, Alisa M.; Abbott, Janice; Hurley, Margaret A.; Aggett, Peter J. Home percutaneous endoscopic gastrostomy feeding: perceptions of patients, carers, nurses and dietitians. *Journal of Advanced Nursing* 2007 August; 59(4): 388-397. Subject: 20.5.1

Brous, Edie A. HIPAA vs. law enforcement: A nurses' guide to managing conflicting responsibilities. *AJN:* *American Journal of Nursing* 2007 August; 107(8): 60-63. Subject: 8.4

Brown, Barry F.; Sawa, Russell J. Key issues in genetic research, testing, and patenting. *In:* Monsour, H. Daniel, ed. Ethics and the New Genetics: An Integrated Approach. Toronto: University of Toronto Press, 2007: 143-164. Subject: 15.1

Brown, David. For the first time, FDA recommends gene testing. *Washington Post* 2007 August 17; p. A10. Subject: 15.3

Brown, Grattan T. Reading the signs of death: a theological analysis. *National Catholic Bioethics Quarterly* 2007 Autumn; 7(3): 467-476. Subject: 20.2.1

Brown, Hannah. Abortion round the world. *BMJ: British Medical Journal* 2007 November 17; 335(7628): 1018-1019. Subject: 12.5.2

Brown, Hannah. Sweetening the pill. *BMJ: British Medical Journal* 2007 March 31; 334(7595): 664-666. Subject: 9.7

Brown, James Robert. Self-censorship. *In:* Lemmens, Trudo; Waring, Duff R., eds. Law and Ethics in Biomedical Research: Regulation, Conflict of Interest and Liability. Toronto; Buffalo: University of Toronto Press, 2006: 82-94. Subject: 1.3.9

Brown, Marilyn J.; Murray, Kathleen A. Phenotyping of genetically engineered mice: humane, ethical, environmental, and husbandry issues. *ILAR Journal: Institute of Laboratory Animal Resources* 2006; 47(2): 118-123. Subject: 15.1

Brown, Mark T. The potential of the human embryo. *Journal of Medicine and Philosophy* 2007 November-December; 32(6): 585-618. Subject: 18.5.4

Brown, Nik. Xenotransplantation: normalizing disgust. *Science as Culture* 1999 September; 8(3): 327-355. Subject: 19.1

Brown, R.W.; Jewell, R.T.; Rous, J.J. Abortion decisions among Hispanic women along the Texas-Mexico border. *Social Science Quarterly* 2000; 81(1): 237-252. Subject: 12.5.2

Brown, Samuel L. Health policy and the politics of health care for African Americans. *In:* Livingston, Ivor Lensworth, ed. Praeger Handbook of Black American Health: Policies and Issues Behind Disparities in Health, Vol. II. 2nd edition. Westport, CT: Praeger Publishers, 2004: 685-700. Subject: 9.5.4

Brown, Stephen D.; Daly, Jennifer C.; Kalish, Leslie A.; McDaniel, Samuel A. Financial disclosures of scientific papers presented at the 2003 RSNA Annual Meeting: association with reporting of non-Food and Drug Administration-approved uses of industry products. *Radiology* 2006 June; 239(3): 849-855. Subject: 18.2

Brown, Stephen D.; Truog, Robert D.; Johnson, Judith A.; Ecker, Jeffrey L. Do differences in the American

See SUBJECT HEADING KEY FOR SECTION II on inside back cover.

453

Academy of Pediatrics and the American College of Obstetricians and Gynecologists positions on the ethics of maternal-fetal interventions reflect subtly divergent professional sensitivities to pregnant women and fetuses? *Pediatrics* 2006 April; 117(4): 1382-1387. Subject: 9.5.5

Brown, Susan. China challenges the west in stem-cell research: unconstrained by public debate, cities like Shanghai and Beijing lure scientists with new laboratories and grants. *Chronicle of Higher Education* 2007 April 13; 53(32): A14-16, A18. Subject: 18.5.4

Brown, Susan. International group proposes guidelines for embryonic-stem-cell research. *Chronicle of Higher Education* 2007 February 16; 53(24): A21. Subject: 18.5.4

Brown, Susan. Use of research-ethics boards is growing in Africa, study finds [news]. *Chronicle of Higher Education* 2007 February 2; 53(22): A13. Subject: 18.2

Brown, Susan; Glenn, David. The true price of a human organ: economists and surgeons debate on whether legalizing the sale of body parts will help or harm. *Chronicle of Higher Education* 2007 March 23; 53(29): A12-A15. Subject: 19.5

Browne, Alister; Browne, Katharine. Morality, prudential rationality, and cheating. *CQ: Cambridge Quarterly of Healthcare Ethics* 2007 Winter; 16(1): 53-62. Subject: 8.1

Browne, Andrew. In China, preventive medicine pits doctor against system; hospitals see threat to profit, bonuses; Dr. Hu's house call. *Wall Street Journal* 2007 January 16; p. A1, A18. Subject: 9.3.1

Browning, David M.; Meyer, Elaine C.; Brodsky, Dara; Truog, Robert D. Reflections on love, fear, and specializing in the impossible. *Journal of Clinical Ethics* 2007 Winter; 18(4): 373-376. Subject: 20.5.2

Brownstein, John S.; Cassa, Christopher A.; Kohane, Isaac S.; Mandl, Kenneth D. Reverse geocoding: concerns about patient confidentiality in the display of geospatial health data. *AMIA Annual Symposium Proceedings* 2005: 905. Subject: 1.3.12

Brownsword, Roger. Cloning, zoning and the harm principle. *In:* McLean, Sheila A.M., ed. First Do No Harm: Law, Ethics, and Healthcare. Aldershot, England; Burlington, VT: Ashgate, 2006: 527-542. Subject: 14.5

Bruce, Donald. Ethical and social issues in nanobiotechnologies: Nano2Life provides a European ethical 'think tank' for research in biology at the nanoscale. *EMBO Reports* 2006 August; 7(8): 754-758. Subject: 5.4

Bruce, Donald. Faster, higher, stronger. *Nano Now!* 2007 February; 1(1): 18-19 [Online]. Accessed: http://www.nanonow.co.uk/nanonow_issue1.pdf [2007 March 2]. Subject: 5.4

Bruckner, Donald W. Considerations on the morality of meat consumption: hunted-game versus farm-raised animals. *Journal of Social Philosophy* 2007 Summer; 38(2): 311-330. Subject: 22.3

Brudney, Daniel. Are alcoholics less deserving of liver transplants? *Hastings Center Report* 2007 January-February; 37(1): 41-47. Subject: 19.6

Brüggenmann, Bernd. Ethische Aspekte der Frühintervention und Akutbehandlung schizophrener Störungen = Ethics of early intervention and acute treatment of schizophrenic disorders. *Ethik in der Medizin* 2007 June; 19(2): 91-102. Subject: 17.1

Brumfiel, Geoff. Interrogation comes under fire [news]. *Nature* 2007 January 25; 445(7126): 349. Subject: 21.4

Brumfiel, Geoff; Abbott, Alison; Cyranoski, David; Fuyuno, Ichiko; Giles, Jim; Odling-Smee, Lucy. Misconduct? It's all academic . . . [news]. *Nature* 2007 January 18; 445(7125): 240-214. Subject: 1.3.9

Brüne, Martin. On human self-domestication, psychiatry, and eugenics. *Philosophy, Ethics, and Humanities in Medicine [electronic]* 2007; (2)21: 9 p. Accessed: http://www.peh-med.com/content/pdf/1747-5341-2-21.pdf [2007 December 18]. Subject: 15.5

Brunekreef, Bert. He who pays the piper, calls the tune. . . . *Epidemiology* 2006 May; 17(3): 246-247. Subject: 1.3.9

Brunnquell, Donald. Case report: parental request for life-prolonging interventions. *HEC (Healthcare Ethics Committee) Forum* 2007 December; 19(4): 375-376. Subject: 20.5.2

Bruns, Cindy M.; Lesko, Teresa M. In the belly of the beast: morals, ethics, and feminist psychotherapy with women in prison. *Women and Therapy* 1999; 22(2): 69-85. Subject: 17.2

Brusco, Angelo. Treating and caring. *Dolentium Hominum* 2007; 22(2): 58-60. Subject: 20.5.1

Bryant, Rosemary. Contradictions in the concept of professional culpability. *Health Care Analysis: An International Journal of Health Philosophy and Policy* 2007 June; 15(2): 137-152. Subject: 9.8

Bubela, Tania M.; Caulfield, Timothy. Media representations of genetic research. *In:* Einsiedel, Edna; Timmermans, Frank, eds. Crossing Over: Genomics in the Public Arena. Calgary, Alberta, Canada: University of Calgary Press, 2005: 117-130. Subject: 15.1

Buchanan, Allen. Social moral epistemology and the role of bioethicists. *In:* Eckenwiler, Lisa A.; Cohn, Felicia G., eds. The Ethics of Bioethics: Mapping the Moral Landscape. Baltimore, MD: Johns Hopkins University Press, 2007: 288-296. Subject: 2.1

Buchanan, David; Miller, Franklin G. Justice in research on human subjects. *In:* Rhodes, Rosamond; Francis, Leslie P.; Silvers, Anita, eds. The Blackwell Guide to Medical Ethics. Malden, MA: Blackwell Pub., 2007: 373-392. Subject: 18.3

Buchanan, David; Witlen, Renee. Balancing service and education: ethical management of student-run clinics.

Journal of Health Care for the Poor and Underserved 2006 August; 17(3): 477-485. Subject: 7.2

Buchner, Benedikt. Industry-sponsored medical education — in the quest for professional integrity and legal certainty. *European Journal of Health Law* 2007 December; 14(4): 313-319. Subject: 7.2

Buciuniene, Ilona; Stonienë, Laimutë; Blazeviciene, Aurelija; Kazlauskaite, Ruta; Skudiene, Vida. Blood donors' motivation and attitude to non-remunerated blood donation in Lithuania. *BMC Public Health* 2006 June 22; 6: 166-173. Subject: 19.5

Buck, Victoria. Who will start the 3Rs ball rolling for animal welfare? [letter]. *Nature* 2007 April 19; 446(7138): 856. Subject: 22.2

Bucklin, Leonard H. Woe unto those who request consent: ethical and legal considerations in rejecting a deceased's anatomical gift because there is no consent by the survivors. *North Dakota Law Review* 2002 October; 78: 323-354. Subject: 19.5

Buford, Chris; Allhoff, Fritz. Neuroscience and metaphysics (redux). *American Journal of Bioethics* 2007 January; 7(1): 58-60. Subject: 17.1

Bulger, Ruth Ellen. The responsible conduct of research, including responsible authorship and publication practices. *In:* Korthals, Michiel; Bogers, Robert J., eds. Ethics for Life Scientists. Dordrecht, The Netherlands: Springer, 2004: 55-62. Subject: 1.3.9

Bulger, Ruth Ellen; Heitman, Elizabeth. Expanding responsible conduct of research instruction across the university. *Academic Medicine* 2007 September; 82(9): 876-878. Subject: 18.2

Buller, Tom. Brains, lies, and psychological explanations. *In:* Illes, Judy, ed. Neuroethics: Defining the Issues in Theory, Practice, and Policy. New York: Oxford University Press, 2006: 51-60. Subject: 17.1

Bunch, Eli Haugen. Norway: some ethical challenges faced by health providers who work with first-generation immigrant men from Pakistan diagnosed with type 2 diabetes. *In:* Davis, Anne J.; Tschudin, Verena; de Raeve, Louise, eds. Essentials of Teaching and Learning in Nursing Ethics: Perspectives and Methods. New York: Churchill Livingstone Elsevier, 2006: 281-289. Subject: 4.1.3

Buntinx, Wil H.E. Professional supports for persons with intellectual disability: products or relationships? *Ethics and Intellectual Disability* 2005 Winter; 8(2): 4-5. Subject: 9.5.3

Burchell, Kevin. Boundary work, associative argumentation and switching in the advocacy of agricultural biotechnology. *Science as Culture* 2007 March; 16(1): 49-70. Subject: 5.1

Burgess, Diana J.; van Ryn, Michelle; Crowley-Matoka, Megan; Malat, Jennifer. Understanding the provider contribution to race/ethnicity disparities in pain treatment: insights from dual process models of stereotyping. *Pain Medicine* 2006 March-April; 7(2): 119-134. Subject: 9.4

Burgess, Michael M. Ethical analysis of representation in the governance of biotechnology. *In:* Einsiedel, Edna; Timmermans, Frank, eds. Crossing Over: Genomics in the Public Arena. Calgary, Alberta, Canada: University of Calgary Press, 2005: 157-172. Subject: 15.1

Burgess-Jackson, Keith. Doing right by our animal companions. *Journal of Ethics* 1998; 2(2): 159-185. Subject: 22.1

Burke, Greg F. Medicine: mental health and war; dementia and vasectomy; oocyte donation and sale; pedophilia; benefits of rest; diabetes mellitus; abortion and breast cancer; spirituality and health; euthanasia in the Netherlands; the Supreme Court decision on partial-birth abortion. *National Catholic Bioethics Quarterly* 2007 Autumn; 7(3): 579-594. Subject: 17.1

Burke, John; Diehl, Dawn; Durosinmi, Brenda; McGinnis, Troy A. The privacy of stigmatized persons. *Journal of Empirical Research on Human Research Ethics* 2007 March; 2(1): 65-67. Subject: 18.4

Burke, Michael. What would happen if a 'woman' outpaced the winner of the gold medal in the 'men's' one hundred meters? female sport, drugs and the transgressive cyborg body. *Philosophy in the Contemporary World* 2004 Spring-Summer; 11(1): 33-41. Subject: 4.5

Burke, William; Pullicino, Patrick; Richard, Edward J. The biological basis of the oocyte assisted reprogramming (OAR) hypothesis: is it an ethical procedure for making embryonic stem cells? *Linacre Quarterly* 2007 August; 74(3): 204-212. Subject: 18.5.4

Burke, Wylie; Press, Nancy. Ethical obligations and counseling challenges in cancer genetics. *Journal of the National Comprehensive Cancer Network: JNCCN* 2006 February; 4(2): 185-191. Subject: 15.2

Burke, Wylie; Psaty, Bruce M. Personalized medicine in the era of genomics. *JAMA: The Journal of the American Medical Association* 2007 October 10; 298(14): 1682-1684. Subject: 15.3

Burke, Wylie; Zimmern, Ronald L.; Kroese, Mark. Defining purpose: a key step in genetic test evaluation. *Genetics in Medicine* 2007 October; 9(10): 675-681. Subject: 15.3

Burkett, Levi. Medical tourism; concerns, benefits, and the American legal perspective. *Journal of Legal Medicine* 2007 April-June; 28(2): 223-245. Subject: 9.3.1

Burley, Justine. Morality and the "new genetics". *In her:* Dworkin and His Critics: With Replies by Dworkin. Malden, MA: Blackwell Pub., 2004: 170-192. Subject: 15.1

Burns, Jeffrey. Ask the ethicist. Does anyone actually invoke their hospital futility policy? *Advances in Neonatal*

See SUBJECT HEADING KEY FOR SECTION II on inside back cover.

455

Care: Official Journal of the National Association of Neonatal Nurses 2006 April; 6(2): 66-67. Subject: 20.5.2

Burns, Lawrence. Overstating the ban, ignoring the compromise. *American Journal of Bioethics* 2007 February; 7(2): 65-66. Subject: 18.5.4

Burns, Tom; Shaw, Joanne. Is it acceptable for people to be paid to adhere to medication?[forum]. *BMJ: British Medical Journal* 2007 August 4; 335(7613): 232-233. Subject: 8.1

Burri, Regula Valérie. Deliberating risks under uncertainty: experience, trust, and attitudes in a Swiss nanotechnology stakeholder discussion group. *NanoEthics* 2007 August; 1(2): 143-154. Subject: 5.4

Burroughs, Valentine J. Racial and ethnic inclusiveness in clinical trials. *In:* Santoro, Michael A.; Gorrie, Thomas M., eds. Ethics and the Pharmaceutical Industry. Cambridge; New York: Cambridge University Press, 2005: 80-96, 421-425. Subject: 18.5.1

Burt, Robert A. Law's effect on the quality of end-of-life care: lessons from the Schiavo case. *Critical Care Medicine* 2006 November; 34(11, Supplement): S348-S354. Subject: 20.5.1

Burton, Bob. Diabetes expert accuses drug company of "intimidation" [news]. *BMJ: British Medical Journal* 2007 December 1; 335(7630): 1113. Subject: 9.7

Burton, Bob. Industry loses bid to block disclosure of doctor's gifts [news]. *BMJ: British Medical Journal* 2007 July 7; 335(7609): 12. Subject: 9.7

Burton, Bob. Roche fined over "extravagant" meals for doctors [news]. *BMJ: British Medical Journal* 2007 February 24; 334(7590): 384. Subject: 9.7

Burton, Olivette R. Why bioethics cannot figure out what to do with race. *American Journal of Bioethics* 2007 February; 7(2): 6-12. Subject: 2.1

Butcher, James. Controversial mental health bill reaches the finishing line [commentary]. *Lancet* 2007 July 14-20; 370(9582): 117-118. Subject: 17.7

Butcher, James. Kari Stefánsson: a general of genetics. *Lancet* 2007 January 27 - February 2; 369(9558): 267. Subject: 15.10

Butler, Declan. Long-held theory is in danger of losing its nerve. *Nature* 2007 September 13; 449(7159): 124-125. Subject: 1.3.9

Button, James. Dealing in the desire for death. *Sydney Morning Herald* 2007 February 3; 4 p. [Online]. Accessed: http://www.smh.com.au/news/world/dealing-in-the-desire-for-death/2007/02/02/116 9919531030.html [2008 February 8]. Subject: 20.7

Button, James. My name is Dr. John Elliott and I'm about to die, with my head held high. *Sydney Morning Herald* 2007 January 26; 4 p. [Online]. Accessed: http://www.smh.com.au/news/world/my-name-is-dr-john-elliott-and-im-about-to-die-with-my-head-heldhigh/2007/01/26/1169788692086.html [2008 February 8]. Subject: 20.5.1

Byatt, Nancy; Pinals, Debra; Arikan, Rasim. Involuntary hospitalization of medical patients who lack decisional capacity: an unresolved issue. *Psychosomatics* 2006 September-October; 47(5): 443-448. Subject: 17.7

Byock, Ira. Palliative care and the ethics of research: medicare, hospice, and phase I trials. *Journal of Supportive Oncology* 2003 July-August; 1(2): 139-141. Subject: 18.5.7

Byock, Ira R. To life! Reflections on spirituality, palliative practice, and politics. *American Journal of Hospice and Palliative Care* 2006 December-2007 January; 23(6): 436-438. Subject: 20.4.1

Byrne, Margaret. Against Bioethics, by Jonathan Baron [book review]. *DePaul Journal of Health Care Law* 2007; 10(4): 535-542. Subject: 2.1

Byrnes, W. Malcolm. Partial trajectory: the story of the altered nuclear transfer-oocyte assisted reprogramming (ANT-OAR) proposal. *Linacre Quarterly* 2007 February; 74(1): 50-59. Subject: 18.5.4

C

Cahana, Alex; Mauron, Alexandre. The story of Vioxx — no pain and a lot of gain: ethical concerns regarding conduct of the pharmaceutical industry. *Journal of Anesthesia* 2006; 20(4): 348-351. Subject: 9.7

Cahana, Alex; Romagnioli, Simone. Not all placebos are the same: a debate on the ethics of placebo use in clinical trials versus clinical practice. *Journal of Anesthesia* 2007; 21(1): 102-105. Subject: 18.3

Cahill, H. Consent. *In:* Holland, Stephen, ed. Introducing Nursing Ethics: Themes in Theory and Practice. Salisbury: APS, 2004: 91-110. Subject: 8.3.1

Cahill, H.; Holland, S. Abortion and other 'beginning of life' issues. *In:* Holland, Stephen, ed. Introducing Nursing Ethics: Themes in Theory and Practice. Salisbury: APS, 2004: 29-48. Subject: 12.1

Cahill, Lisa Sowle. Global health and Catholic social commitment. *Health Progress* 2007 May-June; 88(3): 55-57. Subject: 9.1

Calabrese, Edward J. Elliott's ethics of expertise proposal and application: a dangerous precedent. *Science and Engineering Ethics* 2007 June; 13(2): 139-145. Subject: 5.3

Calabresi, Steven G. The Terri Schiavo case: in defense of the special law enacted by Congress and President Bush. *Northwestern University Law Review* 2006; 100(1): 151-170. Subject: 20.5.1

Caldicott, Catherine V. "Sweeping up after the parade": professional, ethical, and patient care implications of "turfing". *Perspectives in Biology and Medicine* 2007 Winter; 50(1): 136-149. Subject: 7.3

California Health Care Foundation; Bishop, Lynne "Sam"; Holmes, Bradford J.; Kelley, Christopher M.; Forrester Research, Inc. National consumer health privacy survey 2005: executive summary. Oakland, CA: California Health Care Foundation: 2005 November; 5 p. [Online]. Accessed: http://www.chcf.org/documents/ ihealth/ConsumerPrivacy2005ExecSum.pdf [2007 October 1]. Subject: 8.4

Callens, Stefaan; Volbragt, Ilse; Nys, Herman. Legal thoughts on the implications of cost-reducing guidelines for the quality of health care. *Health Policy* 2007 March; 80(3): 422-431. Subject: 9.8

Callus, Thérèse. Pre-implantation genetic diagnosis — towards a principled construction of law? *In:* Garwood-Gowers, Austen; Tingle, John; Wheat, Kay, eds. Contemporary Issues in Healthcare Law and Ethics. Edinburgh; New York: Elsevier Butterworth-Heinemann, 2005: 133-147. Subject: 15.2

Calman, K.C.; Downie, R.S. Ethical principles and ethical issues in public health. *In:* Detels, Roger; McEwen, James; Beaglehole, Robert; Tanaka, Heizo, eds. Oxford Textbook of Public Health. Fourth edition. Oxford; New York: Oxford University Press, 2004: 387-399. Subject: 9.1

Calsbeek, Hiske; Morren, Mattijn; Bensing, Jozien; Rijken, Mieke. Knowledge and attitudes towards genetic testing: a two year follow-up study in patients with asthma, diabetes mellitus and cardiovascular disease. *Journal of Genetic Counseling* 2007 August; 16(4): 493-504. Subject: 15.3

Calvert, Jane. Patenting genomic objects: genes, genomes, function and information. *Science as Culture* 2007 June; 16(2): 207-223. Subject: 15.8

Camann, William. It is the right of every anaesthetist to refuse to participate in a maternal-request caesarean section. *International Journal of Obstetric Anesthesia* 2006 January; 15(1): 35-37. Subject: 8.1

Cameron, Nigel M. de S. The American debate on human cloning. *In:* Roetz, Heiner, ed. Cross-Cultural Issues in Bioethics: The Example of Human Cloning. New York: Rodopi, 2006: 363-386. Subject: 14.5

Camilleri, Michael; Dubnansky, Erin C.; Rustgi, Anil K. Conflicts of interest and disclosures in publications. *Clinical Gastroenterology and Hepatology* 2007 March; 5(3): 268-273. Subject: 1.3.9

Campbell, Alastair V. The ethical challenges of biobanks: safeguarding altruism and trust. *In:* McLean, Sheila A.M., ed. First Do No Harm: Law, Ethics, and Healthcare. Aldershot, England; Burlington, VT: Ashgate, 2006: 203-214. Subject: 15.1

Campbell, Courtney S.; Clark, Lauren A.; Loy, David; Keenan, James F.; Matthews, Kathleen; Winograd, Terry; Zoloth, Laurie. The bodily incorporation of mechanical devices: ethical and religious issues (Part 1). *CQ:*

Cambridge Quarterly of Healthcare Ethics 2007 Spring; 16(2): 229-239. Subject: 5.1

Campbell, Courtney S.; Clark, Lauren A.; Loy, David; Keenan, James F.; Matthews, Kathleen; Winograd, Terry; Zoloth, Laurie. The bodily incorporation of mechanical devices: ethical and religious issues (Part 2). *CQ: Cambridge Quarterly of Healthcare Ethics* 2007 Summer; 16(3): 268-280. Subject: 5.1

Campbell, Eric G. Doctors and drug companies — scrutinizing influential relationships. *New England Journal of Medicine* 2007 November 1; 357(18): 1796-1797. Subject: 9.7

Campbell, Eric G.; Gruen, Russell L.; Mountford, James; Miller, Lawrence G.; Cleary, Paul D.; Blumenthal, David. A national survey of physician-industry relationships. *New England Journal of Medicine* 2007 April 26; 356(17): 1742-1750. Subject: 9.7

Campbell, Eric G.; Regan, Susan; Gruen, Russell L.; Ferris, Timothy G.; Rao, Sowmya R.; Cleary, Paul D.; Blumenthal, David. Professionalism in medicine: results of a national survey of physicians. *Annals of Internal Medicine* 2007 December 4; 147(11): 795-802. Subject: 4.1.2

Campbell, Eric G.; Weissman, Joel S.; Ehringhaus, Susan; Rao, Sowmya R.; Moy, Beverly; Feibelmann, Sandra; Goold, Susan Dorr. Institutional academic-industry relationships. *JAMA: The Journal of the American Medical Association* 2007 October 17; 298(15): 1779-1786. Subject: 5.3

Campbell, N. Ethics in South African dentistry 2006. *SADJ: Journal of the South African Dental Association* 2006 July; 61(6): 240; discussion 242. Subject: 4.1.1

Campbell, Tom. Euthanasia as a human right. *In:* McLean, Sheila A.M., ed. First Do No Harm: Law, Ethics, and Healthcare. Aldershot, England; Burlington, VT: Ashgate, 2006: 447-459. Subject: 20.5.1

Canada. Supreme Court; Brock, Dan W.; Callahan, Daniel. Euthanasia and assisted suicide. *In:* Baylis, Françoise; Downie, Jocelyn; Freedman, Benjamin; Hoffmaster, Barry; Sherwin, Susan, eds. Health Care Ethics in Canada. Toronto: Harcourt Brace Canada, 1995: 527-571. Subject: 20.7

Canada. Supreme Court; Ontario. Court of Appeal; Jecker, Nancy S.; Loewy, Erich H. Consent. *In:* Baylis, Françoise; Downie, Jocelyn; Freedman, Benjamin; Hoffmaster, Barry; Sherwin, Susan, eds. Health Care Ethics in Canada. Toronto: Harcourt Brace Canada, 1995: 201-230. Subject: 8.3.1

Canadian Council on Animal Care [CCAC]. Facts and figures: CCAC animal use survey 2002. Ottawa, Canada: Canadian Council on Animal Care, 2002: 13 p. [Online]. Accessed: http://www.ccac.ca/en/Publications/Facts_Figures/ pdfs/aus2002en-all.pdf [2007 April 26]. Subject: 22.2

See SUBJECT HEADING KEY FOR SECTION II on inside back cover.

Canadian Institutes of Health Research [CIHR]. CIHR best practices for protecting privacy in health research. Ottawa: Canadian Institutes of Health Research, 2005 September; 161 p. [Online]. Accessed: http://www.cihr-irsc.gc.ca/e/documents/et_pbp_nov05_sept2005_e.pdf [2007 April 4]. Subject: 8.4

Canadian Institutes of Health Research [CIHR]. Conflict of interest policy. Ottawa: Canadian Institutes of Health Research, 2000 June 26; 5 p. [Online]. Accessed: http://www.cihr-irsc.gc.ca/e/19039.html [2007 April 25]. Subject: 6

Canadian Institutes of Health Research [CIHR]. Human pluripotent stem cell research: recommendations for CIHR-funded research: report of the ad hoc Working Group on Stem Cell Research. Ottawa: Canadian Institutes of Health Research, 2002 January; 21 p. [Online]. Accessed: http://www.cihr-irsc.gc.ca/e/1489.html [2007 April 19]. Subject: 18.5.4

Canadian Institutes of Health Research [CIHR]. Updated guidelines for human pluripotent stem cell research. Ottawa: Canadian Institutes of Health Research, 2006 June 28; 8 p. [Online]. Accessed: http://www.cihr-irsc.gc.ca/e/31488.html [2007 April 23]. Subject: 18.5.4

Canadian Paediatric Society. Use of anencephalic newborns as organ donors: Position Statement B 2005-01. *Paediatrics and Child Health* 2005 July-August; 10(6): 335-337 (English); 339-341 (French) [Online]. Accessed: http://www.cps.ca/english/statements/B/B05-01.pdf [2007 January 15]. Subject: 20.5.2

Canadian Pharmacists Association. Emergency contraception now available from pharmacists. *Ottawa. Canadian Pharmacists Association.* 2005 April 19: 1 p. [Online]. Accessed: http://www.pharmacists.ca/content/about_cpha/whats_happening/news_releases/release_detail.cfm?release_id=122 [2007 August 28]. Subject: 11.1

Canadian Researchers at the End of Life Network; Heyland, Daren K.; Frank, Chris; Groll, Dianne; Pichora, Deb; Dodek, Peter; Rocker, Graeme; Gafni, Amiram. Understanding cardiopulmonary resuscitation decision making: perspectives of seriously ill hospitalized patients and family members. *Chest* 2006 August; 130(2): 419-428. Subject: 20.5.1

Candib, Lucy M. How turning a QI project into "research" almost sank a great program. *Hastings Center Report* 2007 January-February; 37(1): 26-30. Subject: 9.8

Caniza, Miguela A.; Clara, Wilfrido; Maron, Gabriela; Navarro-Marin, Jose Ernesto; Rivera, Roberto; Howard, Scott C.; Camp, Jonathan; Barfield, Raymond C. Establishment of ethical oversight of human research in El Salvador: lessons learned. *Lancet Oncology* 2006 December; 7(12): 1027-1033. Subject: 18.5.9

Canli, Turhan. When genes and brains unite: ethical implications of genomic neuroimaging. *In:* Illes, Judy, ed.

Neuroethics: Defining the Issues in Theory, Practice, and Policy. New York: Oxford University Press, 2006: 169-183. Subject: 15.6

Canli, Turhan; Brandon, Susan; Casebeer, William; Crowley, Philip J.; DuRousseau, Don; Greely, Henry T.; Pascual-Leone, Alvaro. Neuroethics and national security. *American Journal of Bioethics* 2007 May; 7(5): 3-13. Subject: 17.1

Canli, Turhan; Brandon, Susan; Casebeer, William; Crowley, Philip J.; DuRousseau, Don; Greely, Henry T.; Pascual-Leone, Alvaro. Response to open peer commentaries on "neuroethics and national security". *American Journal of Bioethics* 2007 May; 7(5): W1-W3. Subject: 17.1

Cannold, Leslie. Women, ectogenesis, and ethical theory. *In:* Gelfand, Scott; Shook, John R., eds. Ectogenesis: Artificial Womb Technology and the Future of Human Reproduction. Amsterdam; New York: Editions Rodopi, B.V., 2006: 47-58. Subject: 14.1

Cannon, Geoffrey. Out of the box. *Public Health Nutrition* 2006 April; 9(2): 174-177. Subject: 4.2

Canova, Daniele; De Bona, Manuela; Ruminati, Rino; Ermani, Mario; Naccarato, Remo; Burra, Patrizia. Understanding of and attitudes to organ donation and transplantation: a survey among Italian university students. *Clinical Transplantation* 2006 May-June; 20(3): 307-312. Subject: 19.5

Cantor, Norman L. On hastening death without violating legal and moral prohibitions. *Loyola University Chicago Law Journal* 2006 Winter; 37(2): 407-431. Subject: 20.5.1

Cantor, Norman L. On hastening death without violating legal and moral prohibitions. *Specialty Law Digest: Health Care Law* 2007 June; (338): 9-31. Subject: 20.5.1

Capaldi, Nicholas. How philosophy and theology have undermined bioethics. *Christian Bioethics* 2007 January-April; (13)1: 53-66. Subject: 2.1

Capaldi, Nicholas. Manifesto: moral diversity in health care ethics. *In:* Engelhardt, H. Tristram, ed. Global Bioethics: The Collapse of Consensus. Salem, MA: M&M Scrivener Press, 2006: 131-153. Subject: 21.7

Caplan, Arthur. Scared to death [op-ed]. *Free Inquiry* 2007 June-July; 27(4): 26. Subject: 9.5.1

Caplan, Arthur. The ethics of evil: the challenge of the lessons of the Nazi medical experiments. *In:* LaFleur, William R.; Böhme, Gernot; Shimazono, Susumu, eds. Dark Medicine: Rationalizing Medical Research. Bloomington: Indiana University Press, 2007: 63-72. Subject: 18.5.5

Caplan, Arthur L. Bioethics and the brain [book review]. *New England Journal of Medicine* 2007 June 28; 356(26): 2758-2759. Subject: 17.1

Caplan, Arthur L. Ethical issues surrounding forced, mandated, or coerced treatment. *Journal of Substance*

Abuse Treatment 2006 September; 31(2): 117-120. Subject: 9.5.9

Caplan, Arthur L.; Baruch, Susannah; Schmidt, Harald; Jennings, Bruce; Bonnicksen, Andrea; Greenfield, Debra; Baylis, Françoise; Robertson, John A.; Fleck, Leonard M.; Furger, Franco; Fukuyama, Francis. Needed: a modest proposal [letters and replies]. *Hastings Center Report* 2007 November-December; 37(6): 4-11. Subject: 14.4

Caplan, Arthur L.; Bergman, Edward J. Beyond Schiavo. *Journal of Clinical Ethics* 2007 Winter; 18(4): 340-345. Subject: 20.5.1

Caplan, Arthur L.; Curry, David R. Leveraging genetic resources or moral blackmail? Indonesia and avian flu virus sample sharing [editorial]. *American Journal of Bioethics* 2007 November; 7(11): 1-2. Subject: 15.1

Caplan, Arthur L.; Perry, Constance; Plante, Lauren A.; Saloma, Joseph; Batzer, Frances R. Moving the womb. *Hastings Center Report* 2007 May-June; 37(3): 18-20. Subject: 19.1

Caplan, Arthur; Marino, Thomas A. The role of scientists in the beginning-of-life debate: a 25-year retrospective. *Perspectives in Biology and Medicine* 2007 Autumn; 50(4): 603-613. Subject: 12.4.2

Capozzi, James D.; Rhodes, Rosamond. A family's request for deception. *Journal of Bone and Joint Surgery. American Volume* 2006 April; 88(4): 906-908. Subject: 8.2

Cappelen, Alexander W.; Norheim, Ole Frithjof. Responsibility, fairness and rationing in health care. *Health Policy* 2006 May; 76(3): 312-319. Subject: 9.4

Capron, Alexander Morgan. Imagining a new world: using internationalism to overcome the 10/90 gap in bio- ethics. *Bioethics* 2007 October; 21(8): 409-412. Subject: 2.1

Card, Robert F. Response to commentators on "Conscientious objection and emergency contraception": sex, drugs and the rocky role of Levonorgestrel [letter]. *American Journal of Bioethics* 2007 October; 7(10): W4-W6. Subject: 4.1.1

Cardarelli, Robert; Licciardone, John C.; Taylor, Lockwood G. A cross-sectional evidence-based review of pharmaceutical promotional marketing brochures and their underlying studies: is what they tell us important and true? *BMC Family Practice* 2006 March 3; 7: 13-18. Subject: 9.7

Carlat, Daniel. Dr. drug rep: during a year of being paid to give talks to doctors about an antidepressant, a psychiatrist comes to terms with the fact that taking pharmaceutical money can cloud your judgment. *New York Times Magazine* 2007 November 25; p. 64-69. Subject: 9.7

Carline, Jan D.; O'Sullivan, Patricia S.; Gruppen, Larry D.; Richardson-Nassif, Karen. Crafting successful relationships with the IRB. *Academic Medicine* 2007 October; 82(10, Supplement): S57-S60. Subject: 7.2

Carlson, Elof Axel. Assisted reproduction and the argument of playing God. *In his:* Times of Triumph, Times of Doubt: Science and the Battle for Public Trust. Cold Spring Harbor, NY: Cold Spring Harbor Laboratory Press, 2006: 173-183. Subject: 14.4

Carlson, Elof Axel. Cloning, stem cells, hyperbole, and cant. *In his:* Times of Triumph, Times of Doubt: Science and the Battle for Public Trust. Cold Spring Harbor, NY: Cold Spring Harbor Laboratory Press, 2006: 163-171. Subject: 14.5

Carlson, Elof Axel. Evil at its worst: Nazi medicine and biology. *In his:* Times of Triumph, Times of Doubt: Science and the Battle for Public Trust. Cold Spring Harbor, NY: Cold Spring Harbor Laboratory Press, 2006: 21-37. Subject: 21.4

Carlson, Elof Axel. Genetically modified foods — as usual. *In his:* Times of Triumph, Times of Doubt: Science and the Battle for Public Trust. Cold Spring Harbor, NY: Cold Spring Harbor Laboratory Press, 2006: 129-137. Subject: 15.1

Carlson, Elof Axel. Heroes with feet of clay: Francis Galton and Harry Clay Sharp. *In his:* Times of Triumph, Times of Doubt: Science and the Battle for Public Trust. Cold Spring Harbor, NY: Cold Spring Harbor Laboratory Press, 2006: 55-63. Subject: 15.5

Carlson, Elof Axel. Medical deception and syphilis. *In his:* Times of Triumph, Times of Doubt: Science and the Battle for Public Trust. Cold Spring Harbor, NY: Cold Spring Harbor Laboratory Press, 2006: 141-153. Subject: 18.5.1

Carlson, Elof Axel. Prenatal diagnosis and an alleged eugenics through the back door. *In his:* Times of Triumph, Times of Doubt: Science and the Battle for Public Trust. Cold Spring Harbor, NY: Cold Spring Harbor Laboratory Press, 2006: 155-162. Subject: 15.2

Carlson, Elof Axel. The banality of evil: the careers of Charles Davenport and Harry Laughlin. *In his:* Times of Triumph, Times of Doubt: Science and the Battle for Public Trust. Cold Spring Harbor, NY: Cold Spring Harbor Laboratory Press, 2006: 39-54. Subject: 15.5

Carlson, Robert V.; van Ginneken, Nadja H.; Pettigrew, Luisa M.; Davies, Alan; Boyd, Kenneth M.; Webb, David J. The three official language versions of the Declaration of Helsinki: what's lost in translation? *Journal of Medical Ethics* 2007 September; 33(9): 545-548. Subject: 18.2

Carmel, Sara; Werner, Perla; Ziedenberg, Hanna. Physicians' and nurses' preferences in using life-sustaining treatments. *Nursing Ethics* 2007 September; 14(5): 665-674. Subject: 20.5.1

Carnevale, Franco A. The birth of tragedy in pediatrics: a phronetic conception of bioethics. *Nursing Ethics* 2007 September; 14(5): 571-582. Subject: 20.5.2

See SUBJECT HEADING KEY FOR SECTION II on inside back cover.

459

Caron-Flinterman, J. Francisca; Broerse, Jacqueline E.W.; Bunders, Joske F.G. Patient partnership in decision-making on biomedical research: changing the network. *Science, Technology, and Human Values* 2007 May; 32(3): 339-368. Subject: 18.6

Carpenter, Robert O.; Spooner, John; Arbogast, Patrick G.; Tarpley,John L.; Griffin, Marie R.; Lomis, Kimberly D. Work hours restrictions as an ethical dilemma for residents: a descriptive survey of violation types and frequency. *Current Surgery* 2006 November-December; 63(6): 448-455. Subject: 7.2

Carranza, María. The therapeutic exception: abortion, sterilization and medical necessity in Costa Rica. *Developing World Bioethics* 2007 August; 7(2): 55-63. Subject: 12.4.1

Carreyrou, John. Inside Abbott's tactics to protect AIDS drug: older pill's price hike helps sales of flagship; a probe in Illinois. *Wall Street Journal* 2007 January 3; p. A1, A10. Subject: 9.7

Carroll-Lind, Janis; Chapman, James W.; Gregory, Janet; Maxwell, Gabrielle. The key to the gatekeepers: passive consent and other ethical issues surrounding the rights of children to speak on issues that concern them. *Child Abuse and Neglect* 2006 September; 30(9): 979-989. Subject: 9.5.7

Carruthers, Peter. Invertebrate minds: a challenge for ethical theory. *Journal of Ethics* 2007; 11(3): 275-297. Subject: 22.1

Carson, P.A.; Holt, J. Ethics of studies involving human volunteers. II. Relevance and practical implementation for cosmetic scientists. *Journal of Cosmetic Science* 2006 May-June; 57(3): 223-231. Subject: 4.5

Carson, P.A.; Holt, J. Ethics of studies involving human volunteers. I. Historical background. *Journal of Cosmetic Science* 2006 May-June; 57(3): 215-221. Subject: 18.2

Carson, Ronald A. Engaged humanities: moral work in the precincts of medicine. *Perspectives in Biology and Medicine* 2007 Summer; 50(3): 321-333. Subject: 2.2

Carter, Adrian; Hall, Wayne. The social implications of neurobiological explanations of resistible compulsions. *American Journal of Bioethics* 2007 January; 7(1): 15-17. Subject: 17.1

Carter, Lucy. A case for a duty to feed the hungry: GM plants and the third world. *Science and Engineering Ethics* 2007 March; 13(1): 69-82. Subject: 15.1

Caruso, Denise. Genetic tests offer promise, but raise questions too. *New York Times* 2007 February 18; p. BU5. Subject: 15.3

Caruso, Denise. Someone (other than you) may own your genes [op-ed]. *New York Times* 2007 January 28; p. BU3. Subject: 15.8

Carvalho, Fatima Lampreia. Regulation of clinical research and bioethics in Portugal. *Bioethics* 2007 June; 21(5): 290-302. Subject: 18.2

Casada, Jane P.; Willis, David O.; Butters, Janice M. An investigation of dental student values. *Journal of the American College of Dentists* 1998 Fall; 65(3): 36-40. Subject: 7.2

Casagrande, Sarah Stark; Gary, Tiffany L.; LaVeist, Thomas A.; Gaskin, Darrell J.; Cooper, Lisa A. Perceived discrimination and adherence to medical care in a racially integrated community. *JGIM: Journal of General Internal Medicine* 2007 March; 22(3): 389-395. Subject: 9.5.4

Casal, Paula; Williams, Andrew. Equality of resources and procreative justice. *In:* Burley, Justine, ed. Dworkin and His Critics: With Replies by Dworkin. Malden, MA: Blackwell Pub., 2004: 150-169. Subject: 14.1

Casarett, David. Ethical considerations in end-of-life care and research. *Journal of Palliative Medicine* 2005; 8(Supplement 1): S148-S160. Subject: 18.5.7

Casarett, David J.; Quill, Timothy E. "I'm not ready for hospice": strategies for timely and effective hospice decisions. *Annals of Internal Medicine* 2007 March 20; 146(6): 443-449. Subject: 20.4.1

Case, Amy P.; Ramadhani, Tunu A.; Canfield, Mark A.; Wicklund, Catherine A. Awareness and attitudes regarding prenatal testing among Texas women of childbearing age. *Journal of Genetic Counseling* 2007 October; 16(5): 655-661. Subject: 15.2

Case, Gretchen. The few and the proud. *Atrium* 2007 Summer; 4: 9. Subject: 14.2

Casey, Michael. Cloning down under: an Australian reversal on embryo research. *New Atlantis* 2007 Winter; 15: 125-128. Subject: 14.5

Cash, Adrienne N. Attack of the clones: legislative approaches to human cloning in the United States. *Duke Law and Technology Review* 2005; 26: 14 p. [Online]. Accessed: http://www.law.duke.edu/journals/dltr/articles/pdf/2005dltr0026.pdf [2007 May 3]. Subject: 14.5

Cash, Richard A. Research ethics involves continuous learning. *Indian Journal of Medical Ethics* 2007 April-June; 4(2): 82-83. Subject: 18.6

Cashmore, Judy. Ethical issues concerning consent in obtaining children's reports on their experience of violence. *Child Abuse and Neglect* 2006 September; 30(9): 969-977. Subject: 8.3.2

Casper, Monica J. Fetal surgery then and now: there is too much emphasis on the fetus and not enough on the woman. *Conscience* 2007 Autumn; 28(3): 24-27. Subject: 9.5.8

Cassel, Christine K.; Brennan, Troyen E. Managing medical resources: return to the commons? [commentary]. *JAMA: The Journal of the American Medical Association* 2007 June 13; 297(22): 2518-2521. Subject: 9.4

Cassell, Eric J. Unanswered questions: bioethics and human relationships. *Hastings Center Report* 2007 September-October; 37(5): 20-23. Subject: 2.1

Castellano, Marlene Brant. Ethics of Aboriginal research. *Journal of Aboriginal Health* 2004 January; 1(1): 98-114 [Online]. Accessed: http://www.naho.ca/english/pdf/journal_p98-114.pdf [2007 March 29]. Subject: 18.2

Castle, David; Cline, Cheryl; Daar, Abdallah S.; Tsamis, Charoula; Singer, Peter A. The ethics of nutrigenomic tests and information. *In their:* Science, Society, and the Supermarket: The Opportunities and Challenges of Nutrigenomics. Hoboken, NJ: Wiley-Interscience, 2007: 49-75. Subject: 15.1

Castle, David; Cline; Cheryl; Daar, Abdallah S.; Tsamis, Charoula; Singer, Peter A. Nutrigenomics: justice, equity, and access. *In their:* Science, Society, and the Supermarket: The Opportunities and Challenges of Nutrigenomics. Hoboken, NJ: Wiley-Interscience, 2007: 133-151. Subject: 15.1

Catalanotto, Frank A. A welcome to the workshop on "Professional Promises: Hopes and Gaps in Access to Oral Health Care". *Journal of Dental Education* 2006 November; 70(11): 1120-1124. Subject: 9.2

Catalanotto, Frank A.; Patthoff, Donald E.; Gray, Carolyn F. Ethics of access to oral health care: an introduction to the special issue. *Journal of Dental Education* 2006 November; 70(11): 1117-1119. Subject: 4.1.1

Catania, Joseph A.; Wolf, Leslie E.; Wertleib, Stacey; Lo, Bernard; Henne, Jeff. Research participants' perceptions of the Certificate of Confidentiality's assurances and limitations. *Journal of Empirical Research on Human Research Ethics* 2007 December; 2(4): 53-59. Subject: 18.3

Catano, Victor M.; Turk, James. Fraud and misconduct in scientific research: a definition and procedures for investigation. *Medicine and Law: The World Association for Medical Law* 2007 September; 26(3): 465-476. Subject: 1.3.9

Catholic Church. Catholic Bishops' Conference of England and Wales Linacre Centre for Healthcare Ethics. Catholic Bishops' Conference of England and Wales and Linacre Centre for Healthcare Ethics joint response to the Human Tissue and Embryos (Draft) Bill. London: Linacre Centre for Healthcare Ethics. 2007 June 20: 5 p. [Online]. Accessed: http://www.linacre.org/Linacre%20Joint%20submission%20on%20Human%20Tissue%20and%20Embryos%20_draft_%20Bill.pdf [2007 July 12]. Subject: 18.5.4

Catholic Church. Congregatio Pro Doctrina Fidei = Congregation for the Doctrine of the Faith. Commentary [on Responses to Certain Questions of the United States Conference of Catholic Bishops Concerning Artificial Nutrition and Hydration]. Rome: Congregation for the Doctrine of the Faith 2007 August 1: 4 p. [Online]. Accessed: http://www.vatican.va/roman_curia/ congrega-tions/cfaith/documents/rc_con/cfaith_doc_20070801_nota-commento_en.html [2007 December 13]. Subject: 20.5.1

Catholic Church. Congregatio Pro Doctrina Fidei = Congregation for the Doctrine of the Faith. Commentary on responses to questions on nutrition and hydration. *Origins* 2007 September 27; 37(16): 242-245. Subject: 20.5.1

Catholic Church. Congregatio Pro Doctrina Fidei = Congregation for the Doctrine of the Faith. Responses to certain questions of the United States Conference of Catholic Bishops concerning artificial nutrition and hydration. Rome: Congregation for the Doctrine of the Faith, 2007 August 1: 1 p. [Online]. Accessed: http://www.vatican.va/roman_curia/congregations/cfaith/ documents/rc_con_cfaith_doc_20070801_riposte-usa_en.htm [2007 December 13]. Subject: 20.5.1

Catholic Church. Congregatio Pro Doctrina Fidei = Congregation for the Doctrine of the Faith. Responses to certain questions of the USCCB concerning artificial nutrition and hydration. *Ethics and Medics* 2007 November; 32(11): 1-3. Subject: 20.5.1

Catholic Church. Congregatio Pro Doctrina Fidei = Congregation for the Doctrine of the Faith; Levada, William; Amato, Angelo. Responses to certain questions concerning artificial nutrition and hydration. *Origins* 2007 September 27; 37(16): 241-242. Subject: 20.5.1

Catholic Church. Pontifical Academy for Life = Pontificia Academia pro vita; World Federation of Catholic Medical Associations = Fédération Internationale des Associations Médicales Catholiques. Joint Statement on the Vegetative State: scientific and ethical problems related to the vegetative state. Vatican City: Pontifical Academy for Life 2004 March 20: 3 p. [Online]. Accessed: http://www.vatican.va/roman_curia/pontifical_academis/acdlife/documents/rc_pont-acd_life_doc20040320_joint-statement-veget-state_en.html [2007 December 13]. Subject: 20.5.1

Catholic Health Association of the United States. Report on a theological dialogue on the Principle of Cooperation: executive summary. *National Catholic Bioethics Quarterly* 2007 Winter; 7(4): 773-776. Subject: 9.1

Catholic Medical Association; National Catholic Bioethics Center. Catholic principles and guidelines for clinical research. *National Catholic Bioethics Quarterly* 2007 Spring; 7(1): 153-165. Subject: 18.2

Caulfield, Timothy. Human reproductive cloning: assessing the concerns. *In:* Eisen, Arri; Laderman, Gary, eds. Science, Religion, and Society: An Encyclopedia of History, Culture, and Controversy. Vol. 2. Armonk, NY: M.E. Sharpe, 2007: 795-802. Subject: 14.5

Caulfield, Timothy. Popular media, biotechnology, and the "cycle of hype". *Houston Journal of Health Law and Policy* 2005; 5(2): 213-233. Subject: 15.1

See SUBJECT HEADING KEY FOR SECTION II on inside back cover.

461

Caulfield, Timothy. Stem cells, clones, consensus, and the law. *In:* Knowles, Lori P.; Kaebnick, Gregory E., eds. Reprogenetics: Law, Policy, and Ethical Issues. Baltimore: Johns Hopkins University Press, 2007: 105-123. Subject: 14.5

Caulfield, Timothy. The media, marketing, and genetic services. *In:* Flood, Colleen M., ed. Just Medicare: What's In, What's Out, How We Decide. Buffalo, NY: University of Toronto Press, 2006: 379-395. Subject: 1.3.7

Caulfield, Timothy; Bubela, Tania. Why a criminal ban? Analyzing the arguments against somatic cell nuclear transfer in the Canadian parliamentary debate. *American Journal of Bioethics* 2007 February; 7(2): 51-61. Subject: 14.5

Caulfield, Timothy; Bubela, Tania; Murdoch, C.J. Myriad and the mass media: the covering of a gene patent controversy. *Genetics in Medicine* 2007 December; 9(12): 850-855. Subject: 15.8

Caulfield, Timothy; von Tigerstrom, Barbara. Globalization and biotechnology policy: the challenges created by gene patents and cloning technologies. *In:* Bennett, Belinda; Tomossy, George F., eds. Globalization and Health: Challenges for Health Law and Bioethics. Dordrecht: Springer, 2006: 129-149. Subject: 15.8

Center for Advanced Study. Center for Advanced Study Project Steering Committee; Gunsalus, C.K.; Bruner, Edward M.; Burbules, Nicholas C.; Dash, Leon; Finkin, Matthew; Goldberg, Joseph P.; Greenough, William; Miller, Gregory A.; Pratt, Michael G.; Iriye, Masumi; Aronson, Deb. Illinois White Paper. Improving the system for protecting human subjects: counteracting IRB "mission creep". University of Illinois at Urbana-Champaign. Center for Advanced Study 2005 November 17: 32 p. [Online]. Accessed: http://www.law.uiuc.edu/conferences/whitepaper/whitepaper.pdf [2007 December 13]. Subject: 18.2

Center for Reproductive Rights. State trends in emergency contraception legislation. New York: Center for Reproductive Rights, 2006 January 10; 3 p [Online]. Accessed: http://www.reproductiverights.org/st_ec.html [2007 April 9]. Subject: 11.1

Center for Reproductive Rights. The world's abortion laws. New York, NY: Center for Reproductive Rights, 2005 April: 4 p. [Online]. Accessed: http://www.reproductiverights.org/pub_fac_abortion_laws.html [2007 April 2]. Subject: 12.4.1

Centor, Robert M. Seek first to understand. *Philosophy, Ethics, and Humanities in Medicine [electronic]* 2007 November 28; 2(29): 2 p. Accessed: http://www.peh-med.com/ [2008 January 24]. Subject: 8.1

Chachkin, Carolyn Jacobs. What potent blood: non-invasive prenatal genetic diagnosis and the transformation of modern prenatal care. *American Journal of Law and Medicine* 2007; 33(1): 9-53. Subject: 15.2

Chadwick, Ruth. Reproductive autonomy — a special issue [editorial]. *Bioethics* 2007 July; 21(6): 304. Subject: 14.1

Chadwick, Ruth. Science, context and professional ethics. *In:* Korthals, Michiel; Bogers, Robert J., eds. Ethics for Life Scientists. Dordrecht, The Netherlands: Springer, 2004: 175-182. Subject: 1.3.9

Chalmers, Don. International medical research regulation: from ethics to law. *In:* McLean, Sheila A.M., ed. First Do No Harm: Law, Ethics, and Healthcare. Aldershot, England; Burlington, VT: Ashgate, 2006: 81-100. Subject: 18.2

Chaloner, C. Confidentiality. *In:* Holland, Stephen, ed. Introducing Nursing Ethics: Themes in Theory and Practice. Salisbury: APS, 2004: 65-90. Subject: 8.4

Chambers, David W. Access denied; invalid password. *Journal of Dental Education* 2006 November; 70(11): 1146-1151. Subject: 9.2

Chambers, David W. Ethics summit I: assembling the ethical community. *Journal of the American College of Dentists* 1998 Fall; 65(3): 9-11. Subject: 4.1.1

Chambers, David W. Moral communities. *Journal of Dental Education* 2006 November; 70(11): 1226-1234. Subject: 4.1.1

Chambers, Tod. The virtue of attacking the bioethicist. *In:* Eckenwiler, Lisa A.; Cohn, Felicia G., eds. The Ethics of Bioethics: Mapping the Moral Landscape. Baltimore, MD: Johns Hopkins University Press, 2007: 281-287. Subject: 2.1

Chambers, Tod; Braun, Joshua A. It's narrative all the way down. *American Journal of Bioethics* 2007 August; 7(8): 15-16; author reply W1-W2. Subject: 1.3.7

Champion, Michael K. Commentary: seclusion and restraint in corrections — a time for change. *Journal of the American Academy of Psychiatry and the Law* 2007; 35(4): 426-430. Subject: 9.5.3

Chan, Ho Mun; Pang, Sam. Long-term care: dignity, autonomy, family integrity, and social sustainability: the Hong Kong experience. *Journal of Medicine and Philosophy* 2007 September-October; 32(5): 401-424. Subject: 9.5.2

Chan, K.Y.; Reidpath, D.D. Future research on structural and institutional forms of HIV discrimination. *AIDS Care* 2005 July; 17(Supplement 2): S215-S218. Subject: 9.5.6

Chan, K.Y.; Reidpath, D.D. Methodological considerations in the measurement of institutional and structural forms of HIV discrimination. *AIDS Care* 2005 July; 17(Supplement 2): S205-S213. Subject: 9.5.6

Chan, Sarah; Quigley, Muireann. Frozen embryos, genetic information and reproductive rights. *Bioethics* 2007 October; 21(8): 439-448. Subject: 14.6

Subject = NRCBL Primary Classification Number; see inside front cover.

Chan, T.; Eckert, K.; Venesoen, P.; Leslie, K.; Chin-Yee, I. Consenting to blood: what do patients remember? *Transfusion Medicine* 2005 December; 15(6): 461-466. Subject: 19.4

Chandler, Jennifer A. Priority systems in the allocation of organs for transplant: should we reward those who have previously agreed to donate? *Health Law Journal* 2005; 13: 99-138. Subject: 19.6

Chandler, Michelle. The rights of the medically uninsured: an analysis of social justice and disparate health outcomes. *Journal of Health and Social Policy* 2006; 21(3): 17-36. Subject: 9.2

Chandna, Alka; Stephens, Martin L.; Runkle, Deborah; Pippin, John J.; Greek, Ray; Perry, Seth. What have we learned from the use of animals in scientific research? [letters and reply]. *Chronicle of Higher Education* 2007 March 23; 53(29): B17-B18. Subject: 22.2

Chandrasekhar, Charu A. Rx for drugstore discrimination: challenging pharmacy refusals to dispense prescription contraceptives under state public accommodations laws. *Albany Law Review* 2006; 70(1): 55-115. Subject: 9.7

Chang, Kenneth. Researcher cleared of misconduct, but case is still murky. *New York Times* 2007 February 13; p. F4. Subject: 1.3.9

Chapman, Elizabeth. The social and ethical implications of changing medical technologies: the views of people living with genetic conditions. *Journal of Health Psychology* 2002 March; 7(2): 195-206. Subject: 15.1

Charatan, Fred. Drug company payments to doctors still hard to access [news]. *BMJ: British Medical Journal* 2007 March 31; 334(7595): 655. Subject: 9.7

Charatan, Fred. Drug makers end free lunches [news]. *BMJ: British Medical Journal* 2007 January 13; 334(7584): 64-65. Subject: 9.7

Charatan, Fred. Organ recipients may die when insurance for drugs runs out [news]. *BMJ: British Medical Journal* 2007 March 17; 334(7593): 556. Subject: 19.3

Charland, Louis C. Affective neuroscience and addiction. *American Journal of Bioethics* 2007 January; 7(1): 20-21. Subject: 17.1

Charland, Louis C. Consent or coercion? Treatment referrals to Alcoholics Anonymous. *Journal of Ethics in Mental Health* 2007 April; 2(1): 3 p. Accessed: http://www.jemh.ca [2007 July 31]. Subject: 9.5.9

Charles, Dan. Transgenic hay mowed [news]. *Science* 2007 May 11; 316(5826): 815. Subject: 15.1

Charles, Dan. U.S. courts say transgenic crops need tighter scrutiny [news]. *Science* 2007 February 23; 315(5815): 1069. Subject: 15.7

Charo, R. Alta. Politics, parents, and prophylaxis: mandating HPV vaccination in the United States. *New England Journal of Medicine* 2007 May 10; 356(19): 1905-1908. Subject: 9.7

Charo, R. Alta. The endarkenment. *In:* Eckenwiler, Lisa A.; Cohn, Felicia G., eds. The Ethics of Bioethics: Mapping the Moral Landscape. Baltimore, MD: Johns Hopkins University Press, 2007: 95-107. Subject: 21.1

Charo, R. Alta. The partial death of abortion rights. *New England Journal of Medicine* 2007 May 24; 356(21): 2125-2128. Subject: 12.4.2

Charon, Rita. The bioethics of narrative medicine. *In her:* Narrative Medicine: Honoring the Stories of Illness. New York: Oxford University Press, 2006: 203-218. Subject: 7.1

Chatterjee, Anjan. Cosmetic neurology and cosmetic surgery: parallels, predictions, and challenges. *CQ: Cambridge Quarterly of Healthcare Ethics* 2007 Spring; 16(2): 129-137. Subject: 17.1

Chatterjee, Anjan. "Cosmetic neurology" and the problem of pain. *Cerebrum: The DANA Forum on Brain Science* 2007 July: 6 p. [Online]. Accessed: http://www.dana.org/news/cerebrum/detail.aspx?id=8794 [20070920]. Subject: 4.5

Check, Erika. Celebrity genomes alarm researchers [news]. *Nature* 2007 May 24; 447(7143): 358-359. Subject: 15.10

Check, Erika. Dolly: a hard act to follow [news]. *Nature* 2007 February 22; 445(7130): 802. Subject: 14.5

Check, Erika. Patenting the obvious? *Nature* 2007 May 3; 447(7140): 16-17. Subject: 15.8

Check, Erika. Transparency urged over research payments [news]. *Nature* 2007 August 16; 448(7155): 738. Subject: 1.3.9

Cheema, Puneet; Mehta, Paulette. Pediatric stem cell transplantation ethical concerns. *In:* Mehta, P., ed. Pediatric Stem Cell Transplantation. Sudbury: Jones and Bartlet Publishers, 2004: p. 91-98. Subject: 19.1

Chekar, Choon Key; Kitzinger, Jenny. Science, patriotism and discourses of nation and culture: reflections on the South Korean stem cell breakthroughs and scandals. *New Genetics and Society* 2007 December; 26(3): 289-307. Subject: 18.5.4

Chen, Daniel; Lew, Robert; Hershman, Warren; Orlander, Jay. A cross-sectional measurement of medical student empathy. *JGIM: Journal of General Internal Medicine* 2007 October; 22(10): 1434-1438. Subject: 7.2

Chen, Donna T.; Jones, Loretta; Gelberg, Lillian. Ethics of clinical research within a community-academic partnered participatory framework. *Ethnicity and Disease* 2006 Winter; 16(1 Supplement 1): S118-S135. Subject: 18.6

Chen, Lei-Shih; Goodson, Patricia. Factors affecting decisions to accept or decline cystic fibrosis carrier test-

See SUBJECT HEADING KEY FOR SECTION II on inside back cover.

463

ing/screening: a theory-guided systematic review. *Genetics in Medicine* 2007 July; 9(7): 442-450. Subject: 15.3

Chen, Xiao-Yang. Defensive medicine or economically motivated corruption? A Confucian reflection on physician care in China today. *Journal of Medicine and Philosophy* 2007 November-December; 32(6): 635-648. Subject: 9.3.1

Chenaud, Catherine; Merlani, Paolo; Luyasu, Samuel; Ricou, Bara. Informed consent for research obtained during the intensive care unit stay. *Critical Care* 2006; 10(6): R170. Subject: 18.3

Cheng, Guang-Shing. Compromise [case study]. *Hastings Center Report* 2007 September-October; 37(5): 8-9. Subject: 20.5.1

Cheng-tek Tai, Michael. Clinical ethics consultation — a checklist approach from Asian perspective. *Formosan Journal of Medical Humanities* 2007 July; 8(1-2): 21-25. Subject: 9.6

Cherney, Leora Reiff. Ethical issues involving the right hemisphere stroke patient: to treat or not to treat? *Topics in Stroke Rehabilitation* 2006 Fall; 13(4): 47-53. Subject: 8.3.3

Cherry, Mark J. Preserving the possibility for liberty in health care. *In:* Engelhardt, H. Tristram, ed. Global Bioethics: The Collapse of Consensus. Salem, MA: M&M Scrivener Press, 2006: 95-130. Subject: 2.1

Cherry, Mark J. Traditional Christian norms and the shaping of public moral life: how should Christians engage in bioethical debate within the public forum? *Christian Bioethics* 2007 May-August; (13)2: 129-138. Subject: 2.1

Chervenak, Frank A.; McCullough, Laurence B. Ethical issues and decision making in the management of diabetes in pregnancy. *In:* Langer, Oded, ed. The Diabetes in Pregnancy Dilemma: Leading Change with Proven Solutions. Lanham, MD: University Press of America, 2006: 23-33. Subject: 9.5.5

Chervenak, Frank A.; McCullough, Laurence B.; Baril, Thomas E., Sr. Ethics, a neglected dimension of power relationships of physician leaders. *American Journal of Obstetrics and Gynecology* 2006 September; 195(3): 651-656. Subject: 7.3

Cheshire, William P. Can grey voxels resolve neuro-ethical dilemmas? *Ethics and Medicine: An International Journal of Bioethics* 2007 Fall; 23(3): 135-140. Subject: 17.1

Cheshire, William P. Glimpsing the grey marble. *Ethics and Medicine: An International Journal of Bioethics* 2007 Summer; 23(2): 119-121. Subject: 17.1

Cheshire, William P., Jr. The matter of the brightened grey. *Ethics and Medicine: An International Journal of Bioethics* 2007 Spring; 23(1): 35-38. Subject: 17.1

Cheshire, William P., Jr. The moral musings of a murine chimera. *American Journal of Bioethics* 2007 May; 7(5): 49-50. Subject: 15.1

Chianchiano, D. The Uniform Anatomical Gift Act and organ donation in the United States. *Advances in Chronic Kidney Disease* 2006 April; 13(2): 189-191. Subject: 19.5

Chiang, Hsien-Hsien; Chen, Mei-Bih; Sue, I-Ling. Self-state of nurses in caring for SARS survivors. *Nursing Ethics* 2007 January; 14(1): 18-26. Subject: 9.5.1

Chiaramonte, Gabrielle R.; Friend, Ronald. Medical students' and residents' gender bias in the diagnosis, treatment, and interpretation of coronary heart disease symptoms. *Health Psychology* 2006 May; 25(3): 255-266. Subject: 7.2

Childress, James F. Mentoring in bioethics: possibilities and problems. *In:* Eckenwiler, Lisa A.; Cohn, Felicia G., eds. The Ethics of Bioethics: Mapping the Moral Landscape. Baltimore, MD: Johns Hopkins University Press, 2007: 260-269. Subject: 2.3

Childress, James F. Methods in bioethics. *In:* Steinbock, Bonnie, ed. The Oxford Handbook of Bioethics. Oxford; New York: Oxford University Press, 2007: 15-45. Subject: 2.1

Childress, James F. Must we always respect religious belief? *Hastings Center Report* 2007 January-February; 37(1): 3. Subject: 8.3.3

Chimonas, Susan; Brennan, Troyen A.; Rothman, David J. Physicians and drug representatives: exploring the dynamics of the relationship. *JGIM: Journal of General Internal Medicine* 2007 February; 22(2): 184-190. Subject: 9.7

Chin, Andrew. Research in the shadow of DNA patents. *Journal of the Patent and Trademark Office Society* 2005 November; 87(11): 846-906. Subject: 15.8

Chinn, John; Kulakowski, Elliott C. Conflict of interest in research. *In:* Kulakowski, Elliott C.; Chronister, Lynne U., eds. Research Administration and Management. Sudbury, MA: Jones and Bartlett, 2006: 511-521. Subject: 1.3.9

Chiong, Winston. Justifying patient risks associated with medical education. *JAMA: The Journal of the American Medical Association* 2007 September 5; 298(9): 1046-1048. Subject: 7.2

Chochinov, Harvey Max. Dignity and the essence of medicine: the A, B, C, and D of dignity conserving care. *BMJ: British Medical Journal* 2007 July 28; 335(7612): 184-187. Subject: 8.1

Chochinov, Harvey Max. Dying, dignity, and new horizons in palliative end-of-life care. *CA: A Cancer Journal for Clinicians* 2006 March-April; 56(2): 84-103. Subject: 20.4.1

Choi, Joanna M.; Salter, Sharon A.; Kimball, Alexa B. Innovative care, medical research, and the ethics of informed consent. *Journal of the American Academy of Dermatology* 2007 February; 56(2): 330-332. Subject: 18.3

Chou, Ann F.; Brown, Arleen F.; Jensen, Roxanne E.; Shih, Sarah; Pawlson, Greg; Scholle, Sarah Hudson. Gender and racial disparities in the management of diabetes mellitus among Medicare patients. *Women's Health Issues* 2007 May-June; 17(3): 150-161. Subject: 9.5.1

Chou, Ann F.; Scholle, Sarah Hudson; Weisman, Carol S.; Bierman, Arlene S.; Correa-de-Araujo, Rosaly; Mosca, Lori. Gender disparities in the quality of cardiovascular disease care in private managed care plans. *Women's Health Issues* 2007 May-June; 17(3): 120-130. Subject: 9.5.5

Chou, Ann F.; Wong, Lok; Weisman, Carol S.; Chan, Sophia; Bierman, Arlene S.; Correa-de-Araujo, Rosaly; Scholle, Sarah Hudson. Gender disparities in cardiovascular disease care among commercial and Medicare managed care plans. *Women's Health Issues* 2007 May-June; 17(3): 139-149. Subject: 9.5.5

Choudhry, Sujit; Choudhry, Niteesh K.; Brown, Adalsteinn D. The legal regulation of referral incentives: physician kickbacks and physician self-referral. *In:* Flood, Colleen M., ed. Just Medicare: What's In, What's Out, How We Decide. Buffalo, NY: University of Toronto Press, 2006: 261-280. Subject: 9.3.1

Choudhury, Lincoln Priyadarshi; Kutty, V. Raman. Obstetric practices related to HIV in Kerala. *Indian Journal of Medical Ethics* 2007 January-March; 4(1): 12-15. Subject: 9.5.5

Choudhury, L.P.; Tetali, S. Ethical challenges in voluntary blood donation in Kerala, India. *Journal of Medical Ethics* 2007 March; 33(3): 140-142. Subject: 19.4

Christie, Bryan. New helpline for those who blow whistle on research fraud. *BMJ: British Medical Journal* 2007 May 19; 334(7602): 1023. Subject: 1.3.9

Christie, Timothy; Asrat, Getnet A.; Jiwani, Bashir; Maddix, Thomas; Montaner, Julio S.G. Exploring disparities between global HIV/AIDS funding and recent tsunami relief efforts: an ethical analysis. *Developing World Bioethics* 2007 April; 7(1): 1-7. Subject: 9.5.6

Christilaw, J.E. Cesarean section by choice: constructing a reproductive rights framework for the debate. *International Journal of Gynecology and Obstetrics* 2006 September; 94(3): 262-268. Subject: 9.5.5

Christopher, Myra J. "Show me" bioethics and politics. *American Journal of Bioethics* 2007 October; 7(10): 28-33. Subject: 2.1

Christopher, Paul P.; Foti, Mary Ellen; Roy-Bujnowski, Kristen; Appelbaum, Paul S. Consent form readability and educational levels of potential participants in mental health research. *Psychiatric Services* 2007 February; 58(2): 227-232. Subject: 18.3

Chuang, Cynthia H.; Shank, Laura D. Availability of emergency contraception at rural and urban pharmacies in Pennsylvania. *Contraception* 2006 April; 73(4): 382-385. Subject: 11.1

Chung, Lisa Hird. Free trade in human reproductive cells: a solution to procreative tourism and the unregulated Internet. *Minnesota Journal of International Law* 2006 Winter; 15(1): 263-296. Subject: 14.4

Churchill, Larry R. Preparing for the next health care reform: notes for an interim ethics. *In:* Engström, Timothy H.; Robison, Wade L., eds. Health Care Reform: Ethics and Politics. Rochester, NY: University of Rochester Press, 2006: 195-208. Subject: 9.1

Churchill, Larry R. The hegemony of money: commercialism and professionalism in American medicine. *CQ: Cambridge Quarterly of Healthcare Ethics* 2007 Fall; 16(4): 407-414. Subject: 9.3.1

Churchland, Patricia Smith. Moral decision-making and the brain. *In:* Illes, Judy, ed. Neuroethics: Defining the Issues in Theory, Practice, and Policy. New York: Oxford University Press, 2006: 3-16. Subject: 17.1

Churchland, Patricia Smith. The necessary-and-sufficient boondoggle. *American Journal of Bioethics* 2007 January; 7(1): 54-55. Subject: 17.1

Chwang, Eric; Landy, David C.; Sharp, Richard R. Views regarding the training of ethics consultants: a survey of physicians caring for patients in ICU. *Journal of Medical Ethics* 2007 June; 33(6): 320-324. Subject: 9.6

Cibelli, Jose. A decade of cloning mystique. *Science* 2007 May 18; 316(5827): 990-992. Subject: 14.5

Cibelli, Jose. Is therapeutic cloning dead? The ability to generate pluripotent stem cells directly from skin fibroblasts may render ethical debates over the use of human oocytes to create stem cells irrelevant. *Science* 2007 December 21; 318(5858): 1879-1880. Subject: 14.5

Cioffi, Alfred. The Church and assisted procreation: cautions for the "infertile" couple. *Ethics and Medics* 2007 October; 32(10): 1-4. Subject: 14.4

Civin, Curt I.; Rao, Mahendra S. How many human embryonic stem cell lines are sufficient? A. U.S. perspective. *Stem Cells* 2006 April; 24(4): 800-803 [Online]. http://www.StemCells.com/cgi/content/full/24/4/800 [2007 December 4]. Subject: 15.2

Claassen, Dirk. Financial incentives for antipsychotic depot medication: ethical issues. *Journal of Medical Ethics* 2007 April; 33(4): 189-193. Subject: 17.4

Clancy, Anne; Svensson, Tommy. 'Faced' with responsibility: Levinasian ethics and the challenges of responsibility in Norwegian public health nursing. *Nursing Philosophy* 2007 July; 8(3): 158-166. Subject: 9.1

Clark, Andy. Re-inventing ourselves: the plasticity of embodiment, sensing, and mind. *Journal of Medicine and Philosophy* 2007 May-June; 32(3): 263-282. Subject: 4.5

See SUBJECT HEADING KEY FOR SECTION II on inside back cover.

465

Clark, Thomas W. Review of Walter Glannon, Bioethics and the Brain [book review]. *American Journal of Bioethics* 2007 May; 7(5): 59-60. Subject: 17.1

Clark, Tom C. Religion, morality and abortion: a constitutional appraisal. *Loyola University Los Angeles Law Review* 1969; 2: 1-11. Subject: 12.3

Clarke, Amanda. Qualitative interviewing: encountering ethical issues and challenges. *Nurse Researcher* 2006; 13(4): 19-29. Subject: 18.2

Clarke, C. Ethics and the end of life. *In:* Holland, Stephen, ed. Introducing Nursing Ethics: Themes in Theory and Practice. Salisbury: APS, 2004: 49-64. Subject: 20.5.1

Clarke, C. Resource allocation. *In:* Holland, Stephen, ed. Introducing Nursing Ethics: Themes in Theory and Practice. Salisbury: APS, 2004: 171-188. Subject: 9.4

Clarke, Steve; Levy, Neil. On the competence of substance users to consent to treatment programs. *In:* Kleinig, John; Einstein, Stanley, eds. Ethical Challenges for Intervening in Drug Use: Policy, Research and Treatment Issues. Huntsville, TX: Office of International Criminal Justice; Criminal Justice Center, Sam Houston State University, 2006: 309-322. Subject: 8.3.1

Clarkson, Frederick. Tragedy on the national stage: conservative intervention into the Terri Schiavo case was a disservice to everybody. *Conscience* 2007 Autumn; 28(3): 35-38. Subject: 20.5.1

Claudot, Frédérique; Alla, François; Ducrocq, Xavier; Coudane, Henry. Teaching ethics in Europe. *Journal of Medical Ethics* 2007 August; 33(8): 491-495. Subject: 2.3

Claxton, Karl; Culyer, Anthony J. Rights, responsibilities and NICE: a rejoinder to Harris. *Journal of Medical Ethics* 2007 August; 33(8): 462-464. Subject: 9.4

Clay, Megan; Block, Walter. A free market for human organs. *Journal of Social, Political and Economic Studies* 2002 Summer; 27(2): 227-236. Subject: 19.5

Cleaton-Jones, Peter. The first randomised trial of male circumcision for preventing HIV: what were the ethical issues? *PLoS Medicine* 2005 November; 2(11): e287: 1073-1075. Subject: 9.5.6

Clemens, Norman A. When colleagues go astray . . . . *Journal of Psychiatric Practice* 2007 January; 13(1): 40-43. Subject: 17.2

Cleminson, Richard. "A century of civilization under the influence of eugenics": Dr. Enrique Diego Madrazo, socialism and scientific progress. *Dynamis* 2006; 26: 221-251. Subject: 15.5

Cline, Cheryl A.; Moreno, Jonathan D.; Berger, Sam. Biotechnology and the new right: a progressive red herring? *American Journal of Bioethics* 2007 October; 7(10): 15-17; author reply W1-W3. Subject: 5.1

Clinton, Michael. On the colour of herring: response to commentary. *Journal of Ethics in Mental Health* 2007 April; 2(1): 3 p. [Online]. Accessed: http://www.jemh.ca [2007 July 31]. Subject: 9.5.9

Clinton, Michael. Should mental health professionals refer clients with substance use disorders to 12-step programs? *Journal of Ethics in Mental Health* 2007 April; 2(1): 4 p. [Online]. Accessed: http://www.jemh.ca [2007 July 31]. Subject: 9.5.9

Clinton, William J. Memorandum of March 27, 1997 — strengthened protections for human subjects of classified research. *Federal Register* 1997 May 13; 62(92): 26367-26372 [Online]. Accessed: http://frwebgate. access.gpo.gov/cgi-bin/multidb.cgi [2005 December 28]. Subject: 18.2

Clouser, K. Danner; Gert, Bernard. Common morality. *In:* Khushf, George, ed. Handbook of Bioethics: Taking Stock of the Field From a Philosophical Perspective. Dordrecht; Boston: Kluwer Academic, 2004: 121-141. Subject: 2.1

Coats, T.J. Consent for emergency care research: the Mental Capacity Act 2005. *Emergency Medicine Journal* 2006 December; 23(12): 893-894. Subject: 18.3

Coch, Donna. Neuroimaging research with children: ethical issues and case scenarios. *Journal of Moral Education* 2007 March; 36(1): 1-18. Subject: 18.5.2

Cochrane, Thomas I. Brain disease or moral condition? Wrong question. *American Journal of Bioethics* 2007 January; 7(1): 24-25. Subject: 17.1

Cochrane, Thomas I. Religious delusions and the limits of spirituality in decision-making. *American Journal of Bioethics* 2007 July; 7(7): 14-15. Subject: 8.1

Coggon, John. Varied and principled understandings of autonomy in English law: justifiable inconsistency or blinkered moralism? *Health Care Analysis: An International Journal of Health Philosophy and Policy* 2007 September; 15(3): 235-255. Subject: 8.3.1

Coghlan, Andy. A subtle key to human diversity. *New Scientist* 2007 January 13-19; 193(2586): 8. Subject: 15.11

Coghlan, Andy. Bipolar disorder: young and moody or mentally ill? [news]. *New Scientist* 2007 May 19-25; 194(2604): 6-7. Subject: 17.1

Coghlan, Andy. Genetic testing: an informed choice? *New Scientist* 2007 October 6-12; 195(2624): 8-9. Subject: 15.3

Coghlan, Andy. Pro-choice? Pro-life? No choice. *New Scientist* 2007 October 20-26; 196(2626): 8-9. Subject: 12.4.1

Cohen, Alfred. The valance of pain in Jewish thought and practice. *Journal of Halacha and Contemporary Society* 2007 Spring; (53): 25-52. Subject: 4.4

Cohen, Andrew I. Contractarianism, other-regarding attitudes, and the moral standing of nonhuman animals. *Journal of Applied Philosophy* 2007 May; 24(2): 188-201. Subject: 22.1

Subject = NRCBL Primary Classification Number; see inside front cover.

Cohen, A.T.; Maillardet, L.M.A. Are placebo-controlled trials ethical in areas where current guidelines recommend therapy? Yes. *Journal of Thrombosis and Haemostasis* 2006 October; 4(10): 2130-2132. Subject: 18.3

Cohen, Cynthia B. Beyond the human neuron mouse to the NAS guidelines. *American Journal of Bioethics* 2007 May; 7(5): 46-49. Subject: 15.1

Cohen, Cynthia B. Philosophical challenges to the use of advance directives. *In:* Khushf, George, ed. Handbook of Bioethics: Taking Stock of the Field From a Philosophical Perspective. Dordrecht; Boston: Kluwer Academic, 2004: 291-314. Subject: 20.5.4

Cohen, Cynthia B. Ways of being personal and not being personal about religious beliefs in the clinical setting. *American Journal of Bioethics* 2007 July; 7(7): 16-18. Subject: 8.1

Cohen, Elliot D. Conceptualizing the professional relationship in drug user and alcohol misuser counseling. *In:* Kleinig, John; Einstein, Stanley, eds. Ethical Challenges for Intervening in Drug Use: Policy, Research and Treatment Issues. Huntsville, TX: Office of International Criminal Justice; Criminal Justice Center, Sam Houston State University, 2006: 367-382. Subject: 9.5.9

Cohen, Eric. In whose image shall we die? *New Atlantis* 2007 Winter; 15: 21-39. Subject: 2.1

Cohen, Eric; Kass, Leon R. Cast me not off in old age. *Commentary* 2006 January; 121(1): 32-39. Subject: 9.5.2

Cohen, Jon. Brazil, Thailand override big pharma patents. *Science* 2007 May 11; 316(5826): 816. Subject: 9.5.6

Cohen, Jon. Feud over AIDS vaccine trials leads prominent Italian researchers to court. *Science* 2007 August 10; 317(5839): 738-739. Subject: 9.5.6

Cohen, Jon. NIH to end chimp breeding for research [news]. *Science* 2007 June 1; 316(5829): 1265. Subject: 22.2

Cohen, Jon. The endangered lab chimp [news]. *Science* 2007 January 26; 315(5811): 450-452. Subject: 22.2

Cohen, Joshua T.; Neumann, Peter J. Dialysis facility ownership and epoetin dosing in hemodialysis patients: a medical economic perspective. *American Journal of Kidney Diseases* 2007 September; 50(3): 362-365. Subject: 19.3

Cohen, Julie; Marecek, Jeanne; Gillham, Jane. Is three a crowd? Clients, clinicians, and managed care. *American Journal of Orthopsychiatry* 2006 April; 76(2): 251-259. Subject: 9.3.2

Cohen, J.; Marcoux, I.; Bilsen, J.; Deboosere, P.; van der Wal, G.; Deliens, L. Trends in acceptance of euthanasia among the general public in 12 European countries (1981-1999). *European Journal of Public Health* 2006 December; 16(6): 663-669. Subject: 20.5.1

Cohen, J.S.; Erickson, J.M. Ethical dilemmas and moral distress in oncology nursing practice. *Clinical Journal of Oncology Nursing* 2006 December; 10(6): 775-780. Subject: 9.5.1

Cohen, Lewis M.; Moss, Alvin H.; Weisbord, Steven D.; Germain, Michael J. Renal palliative care. *Journal of Palliative Medicine* 2006 August; 9(4): 977-992. Subject: 20.4.1

Cohen, Michael H. Legal and ethical issues relating to use of complementary therapies in pediatric hematology/oncology. *Journal of Pediatric Hematology/Oncology* 2006 March; 28(3): 190-193. Subject: 4.1.1

Cohen, Patricia. As ethics panels expand, no research field is exempt. *New York Times* 2007 February 28; p. A15. Subject: 18.2

Cohen, Peter J. Addiction, molecules and morality: disease does not obviate responsibility. *American Journal of Bioethics* 2007 January; 7(1): 21-23. Subject: 17.1

Cohen-Kohler, Jillian Clare; Esmail, Laura C. Scientific misconduct, the pharmaceutical industry, and the tragedy of institutions. *Medicine and Law: The World Association for Medical Law* 2007 September; 26(3): 431-446. Subject: 1.3.9

Cohn, Felicia. Real life informs consent. *Journal of Clinical Ethics* 2007 Winter; 18(4): 366-368. Subject: 8.3.2

Cohn, Felicia; Goodman-Crews, Paula; Rudman, William; Schneiderman, Lawrence J.; Waldman, Ellen. Proactive ethics consultation in the ICU: a comparison of value perceived by healthcare professionals and recipients. *Journal of Clinical Ethics* 2007 Summer; 18(2): 140-147. Subject: 9.6

Cole, Andrew. Botched abortions kill more than 66,000 women each year [news]. *BMJ: British Medical Journal* 2007 October 27; 335(7625): 845. Subject: 12.1

Cole, Andrew. Scientists plead for right to create interspecies embryos [news]. *BMJ: British Medical Journal* 2007 June 23; 334(7607): 1294. Subject: 15.1

Cole, Phillip. Human rights and the national interest: migrants, healthcare and social justice. *Journal of Medical Ethics* 2007 May; 33(5): 269-272. Subject: 9.5.1

Coleman, Gerald D. Organ donation: charity or commerce? *America* 2007 March 5; 196(8): 22-24. Subject: 19.5

Coleman, Gerald D. The irreversible disabling of a child: the "Ashley treatment". *National Catholic Bioethics Quarterly* 2007 Winter; 7(4): 711-728. Subject: 9.5.7

Coleman, Stephen. Ethical issues raised by non-punitive drug user policies. *In:* Kleinig, John; Einstein, Stanley, eds. Ethical Challenges for Intervening in Drug Use: Policy, Research and Treatment Issues. Huntsville, TX: Office of International Criminal Justice; Criminal Justice Center, Sam Houston State University, 2006: 99-119. Subject: 9.5.9

See SUBJECT HEADING KEY FOR SECTION II on inside back cover.

467

Collier, Joe. Inside big pharma's box of tricks [review of BBC program Panorama: The Secrets of the Drug Trials]. *BMJ: British Medical Journal* 2007 January 27; 334(7586): 209. Subject: 9.7

Collins, Francis S.; Manolio, Teri A. Necessary but not sufficient. *Nature* 2007 January 18; 445(7125): 259. Subject: 15.1

Collins, John A. Preimplantation genetic screening in older mothers [editorial]. *New England Journal of Medicine* 2007 July 5; 357(1); 61-63. Subject: 15.3

Collins, Kenneth. Maimonides and the ethics of patient autonomy. *Israel Medical Association Journal* 2007 January; 9(1): 55-58. Subject: 8.1

Collins, Niamh; Phelan, Dermot; Marsh, Brian; Sprung, Charles L. End-of-life care in the intensive care unit: the Irish Ethicus data. *Critical Care and Resuscitation* 2006 December; 8(4): 315-320. Subject: 20.4.1

Collins, N.; Phelan, D.; Carton, E. End of life in ICU — care of the dying or 'pulling the plug'? *Irish Medical Journal* 2006 April; 99(4): 112-114. Subject: 20.5.1

Colliton, William F., Jr. In vitro fertilization and the wisdom of the Roman Catholic church. *Linacre Quarterly* 2007 February; 74(1): 10-29. Subject: 14.4

Colman, Alan; Burley, Justine. Stem cells: recycling the abnormal [news]. *Nature* 2007 June 7; 447(7145): 649-650. Subject: 18.5.1

Colman, Richard D.; Caine, Jane A. Role of the doctor: to care for patients' wellbeing [letters]. *BMJ: British Medical Journal* 2007 December 8; 335(7631): 1169. Subject: 4.1.2

Colt, Henri G.; Mulnard, Ruth A. Writing an application for a human subjects institutional review board. *Chest* 2006 November; 130(5): 1605-1607. Subject: 18.2

Comber, Julie; Griffin, Gilly. Genetic engineering and other factors that might affect human-animal interactions in the research setting. *Journal of Applied Animal Welfare Science* 2007 June; 10(3): 267-277. Subject: 15.1

Comeau, Pauline. Debate begins over public funding for HPV vaccine [news]. *CMAJ/JAMC: Canadian Medical Association Journal* 2007 March 27; 176(7): 913-914. Subject: 9.5.7

Comfort, Nathaniel. "Polyhybrid heterogeneous bastards": promoting medical genetics in America in the 1930s and 1940s. *Journal of the History of Medicine and Allied Sciences* 2006 October; 61(4): 415-455. Subject: 15.3

Commission on Intellectual Property Rights [CIPR]; Thambisetty, Sivaramjani. Human genome patents and developing countries. Study paper 10. London: Commission on Intellectual Property Rights, 2001; 70 p. [Online]. Accessed: http://www.iprcommission.org/papers/pdfs/study_papers/10_human_genome_patents.pdf [2007 April 16]. Subject: 15.8

Community-State Partnerships to Improve End-of-life Care. How regional long-term care ethics committees improve end-of-life care. Kansas City, MO: State Initiatives Policy Brief, Community State Partnerships to Improve End-of-Life Care 2000; (6): 104 [Online]. Accessed: http://www.rwjf.org [2006 September 28]. Subject: 9.6

Community-State Partnerships to Improve End-of-Life Care. Advances in state pain policy and medical practice. Kansas City, MO: State Initiatives Policy Brief, Community State Partnerships to Improve End-of-Life Care 1999 April; (4): 1-8 [Online]. Accessed: http://www.practicalbioethics.org/FileUploads/SI_4.pdf [2006 September 28]. Subject: 4.4

Condic, Maureen L. What we know about embryonic stem cells. *First Things* 2007 January; (169): 25-29. Subject: 18.5.4

Condic, Maureen L.; Furton, Edward J. Harvesting embryonic stem cells from deceased human embryos. *National Catholic Bioethics Quarterly* 2007 Autumn; 7(3): 507-525. Subject: 18.5.4

Condit, Celeste. How culture and science make race "genetic": motives and strategies for discrete categorization of the continuous and heterogeneous. *Literature and Medicine* 2007 Spring; 26(1): 240-268. Subject: 15.11

Condit, Celeste M. Lay people actively process messages about genetic research. *In:* Einsiedel, Edna; Timmermans, Frank, eds. Crossing Over: Genomics in the Public Arena. Calgary, Alberta, Canada: University of Calgary Press, 2005: 131-141. Subject: 15.1

Condit, C.M.; Parrott, R.L.; Bates, B.R.; Bevan, J.L.; Achter, P.J. Exploration of the impact of messages about genes and race on lay attitudes. *Clinical Genetics* 2004 November; 66(5): 402-408. Subject: 15.11

Conference Coordinators, Second National Bioethics Conference. Indian Journal of Medical Ethics. Moral and ethical imperatives of health care technologies: scientific, legal and socio-economic perspectives. *Indian Journal of Medical Ethics* 2007 January-March; 4(1): 35-37. Subject: 5.1

Consortium to Examine Clinical Research Ethics; Speckman, Jeanne L.; Byrne, Margaret M.; Gerson, Jason; Getz, Kenneth; Wangsmo, Gary; Muse, Carianne T.; Sugarman, Jeremy. Determining the costs of institutional review boards. *IRB: Ethics and Human Research* 2007 March-April; 29(2): 7-13. Subject: 18.2

Cook, Deborah; Rocker, Graeme; Giacomini, Mita; Sinuff, Tasnim; Heyland, Daren. Understanding and changing attitudes toward withdrawal and withholding of life support in the intensive care unit. *Critical Care Medicine* 2006 November; 34(11, Supplement): S317-S323. Subject: 20.5.1

Cook, E. David; Lawrence, Ryan E.; Curlin, Farr A. Always let your conscience be your guide. *American Jour-*

nal of Bioethics 2007 December; 7(12): 17-19; author reply W1-W2. Subject: 4.1.2

Cook, Kristin. Familial consent for registered organ donors: a legally rejected concept. *Health Matrix: The Journal of Law-Medicine* 2007 Winter; 17(1): 117-145. Subject: 19.5

Cook, Rebecca; Pretorius, Rika. Duties to implement reproductive rights: the case of adolescents. *In:* Dennerstein, Lorraine, ed. Women's Rights and Bioethics. Paris: UNESCO, 2000: 175-189. Subject: 14.1

Cook, R.J.; Dickens, B.M.; Erdman, J.N. Emergency contraception, abortion and evidence-based law. *International Journal of Gynecology and Obstetrics* 2006 May; 93(2): 191-197. Subject: 11.1

Cook, R.J.; Ortega-Ortiz, A.; Romans, S.; Ross, L.E. Legal abortion for mental health indications. *International Journal of Gynecology and Obstetrics* 2006 November; 95(2): 185-190. Subject: 12.4.2

Coombes, Rebecca. Ashley X: a difficult moral choice. Did the doctors and parents responsible for a severely disabled girl have the right to keep her small? *BMJ: British Medical Journal* 2007 January 13; 334(7584): 72-73. Subject: 9.5.3

Coombes, Rebecca. Bad blood. *BMJ: British Medical Journal* 2007 April 28; 334(7599): 879-880. Subject: 9.5.6

Coombes, Rebecca. Cancer drugs: swallowing big pharma's line? *BMJ: British Medical Journal* 2007 May 19; 334(7602): 1034-1035. Subject: 9.7

Coombes, Rebecca. Life saving treatment or giant experiment. *BMJ: British Medial Journal* 2007 April 7; 334(7596): 721-723. Subject: 9.7

Coombes, Rebecca. Medical records: are patients' secrets up for grabs? *BMJ: British Medical Journal* 2007 January 6; 334(7583): 16-17. Subject: 8.4

Coons, Stephen Joel. Pharmacists' right of conscience: whose autonomy is it, anyway? [editorial]. *Clinical Therapeutics* 2005 June; 27(6): 924-925. Subject: 11.1

Cooper, Matthew. Sharing data and results in ethnographic research: why this should not be an ethical imperative. *Journal of Empirical Research on Human Research Ethics* 2007 March; 2(1): 3-19. Subject: 18.3

Cooper, Rachel. Can it be a good thing to be deaf? *Journal of Medicine and Philosophy* 2007 November-December; 32(6): 563-583. Subject: 4.4

Cooper, Richard A.; Tauber, Alfred I. Values and ethics: a collection of curricular reforms for a new generation of physicians. *Academic Medicine* 2007 April; 82(4): 321-323. Subject: 7.2

Cooper, Robert W.; Frank, Garry L.; Gouty, Carol Ann; Hansen, Mary C. Key ethical issues encountered in healthcare organizations: perceptions of nurse executives.

JONA: The Journal of Nursing Administration 2002 June; 32(6): 331-337. Subject: 9.1

Cooper, R.J.; Bissell, P.; Wingfield, J. A new prescription for empirical ethics research in pharmacy: a critical review of the literature. *Journal of Medical Ethics* 2007 February; 33(2): 82-86. Subject: 9.7

Copelovitch, Lawrence; Kaplan, Bernard S. Is genetic testing of healthy pre-symptomatic children with possible Alport syndrome ethical? *Pediatric Nephrology* 2006 April; 21(4): 455-456. Subject: 15.3

Corbellini, Gilberto. Scientists, bioethics and democracy: the Italian case and its meanings. *Journal of Medical Ethics* 2007 June; 33(6): 349-352. Subject: 5.3

Corbie-Smith, Giselle M.; Durant, Raegan W.; St George, Diane Marie M. Investigators' assessment of NIH mandated inclusion of women and minorities in research. *Contemporary Clinical Trials* 2006 December; 27(6): 571-579. Subject: 18.2

Corfield, Lorraine F. To inform or not to inform: how should the surgeon proceed when the patient refuses to discuss surgical risk? *Journal of Vascular Surgery* 2006 July; 44(1): 219-221. Subject: 8.3.1

Cormick, Craig. Cloning goes to the movies = A clonagem vai ao cinema. *Historia, Ciencias, Saude — Manguinhos* 2006 October; 13(Supplement): 181-212. Subject: 14.5

Corneliussen, Filippa. Adequate regulation, a stop-gap measure, or part of a package? Debates on codes of conduct for scientists could be diverting attention away from more serious questions. *EMBO Reports* 2006 July; 7 Special No: S50-S54. Subject: 1.3.9

Corrigan, Oonagh. Empty ethics: the problem with informed consent. *Sociology of Health and Illness* 2003 November; 25(7): 768-792. Subject: 18.3

Corsino, Bruce V.; Patthoff, Donald E., Jr. The Ethical and practical aspects of acceptance and Universal Patient Acceptance. *Journal of Dental Education* 2006 November; 70(11): 1198-1201. Subject: 9.2

Cosgriff, JoAnne Alissi; Pisani, Margaret; Bradley, Elizabeth H.; O'Leary, John R.; Fried, Terri R. The association between treatment preferences and trajectories of care at the end-of-life. *JGIM: Journal of General Internal Medicine* 2007 November; 22(11): 1566-1571. Subject: 20.5.1

Cosgrove, Lisa; Bursztajn, Harold J. Undoing undue industry influence: lessons from psychiatry as psycho- pharmacology. *Organizational Ethics: Healthcare, Business, and Policy* 2006 Fall-Winter; 3(2): 131-133. Subject: 17.4

Cosgrove, Lisa; Krimsky, Sheldon; Vijayaraghavan, Manisha; Schneider, Lisa. Financial ties between DSM-IV panel members and the pharmaceutical industry. *Psychotherapy and Psychosomatics* 2006; 75(3): 154-160. Subject: 7.3

See SUBJECT HEADING KEY FOR SECTION II on inside back cover.

469

Costa, Rosely Gomes. Racial classification regarding semen donor selection in Brazil. *Developing World Bioethics* 2007 August; 7(2): 104-111. Subject: 14.2

Costeloe, Kate. Euthanasia in neonates. *BMJ: British Medial Journal* 2007 May 5; 334(7600): 912-913. Subject: 20.5.2

Cotton, Richard G.H.; Sallée, Clémentine; Knoppers, Bartha M. Locus-specific databases: from ethical principles to practice. *Human Mutation* 2005 November; 26(5): 489-493. Subject: 15.1

Coulter, Angela; Ellins, Jo. Effectiveness of strategies for informing, educating, and involving patients. *BMJ: British Medical Journal* 2007 July 7; 335(7609): 24-27. Subject: 8.1

Council of Europe. Parliamentary Assembly. Recommendations 1418(1999) on protection of the human rights and dignity of the terminally ill and the dying. Strasbourg, France: Council of Europe 1999 June 25; 5 p. [Online]. Accessed: http://assembly.coe.int//main.asp?link= http://assembly.coe.int/documents/adopte dtext/ta99/erec1418.htm [2007 April 11]. Subject: 20.5.1

Council of Europe. Parliamentary Assembly. Trafficking in organs in Europe. Recommendation 1611 (2003). Stras- bourg, France: Council of Europe (21st Sitting), 2003 June 25: 4 p. [Online]. Accessed: http://assembly. coe.int/Documents/AdoptedText/ta03/EREC1611.htm [2007 May 1]. Subject: 19.5

Council of Europe. Parliamentary Assembly. Committee on Legal Affairs and Human Rights; McNamara, Kevin. Euthanasia: opinion. Strasbourg, France: Council of Europe, 2003 September 23; 10 p. [Online]. Accessed: http://assembly.coe.int/main.asp?Link=/documents/workingdocs/doc04/edoc9923.htm [2007 April 12]. Subject: 20.5.1

Council of Europe. Working Party on Human Genetics [CDBI-CO-GT4]. Working document on the applications of genetics for health purposes. Strasbourg: Council of Europe. Working Party on Human Genetics (CDBI-CO-GT4), 2003 February 7: 8 p. [Online]. Accessed: http://www.coe.int/t/e/legal_affairs/legal_co-operation/bioethics/activities/human_genetics/INF(2003)3E_Wkgdoc_genetics.pdf [2007 March 5]. Subject: 15.3

Couzin, Jennifer. Amid debate, gene-based cancer test approved [news]. *Science* 2007 February 16; 315(5814): 924. Subject: 15.3

Couzin, Jennifer. Kaiser to set up gene bank [news]. *Science* 2007 February 23; 315(5815): 1067. Subject: 15.1

Couzin, Jennifer; Kaiser, Jocelyn. Closing the net on common disease genes. *Science* 2007 May 11; 316(5826): 820-822. Subject: 15.1

Covington, Sharon N.; Gibbons, William E. What is happening to the price of eggs? *Fertility and Sterility* 2007 May; 87(5): 1001-1004. Subject: 14.4

Cowley, Christopher. Why medical ethics should not be taught by philosophers. *Discourse* 2005 Autumn; 5(1): 50-63 [Online]. Accessed: http://prs.heacademy.ac.uk/publications/autumn2005.pdf [2006 December 6]. Subject: 2.3

Cox, A.C.; Fallowfield, L.J.; Jenkins, V.A. Communication and informed consent in phase 1 trials: a review of the literature. *Supportive Care in Cancer* 2006 April; 14(4): 303-309. Subject: 18.3

Craig, Alexa; Cronin, Beth; Eward, William; Metz, James; Murray, Logan; Rose, Gail; Suess, Eric; Vergara, Maria E. Attitudes toward physician-assisted suicide among physicians in Vermont. *Journal of Medical Ethics* 2007 July; 33(7): 400-403. Subject: 20.7

Craig, Amber; Bollinger, Dan. Of waste and want: a nationwide survey of Medicaid funding for medically unnecessary, non-therapeutic circumcision. *In:* Denniston, George C.; Gallo, Pia Grassivaro; Hodges, Frederick M.; Milos, Marilyn Fayre; Viviani, Franco, eds. Bodily Integrity and the Politics of Circumcision: Culture, Controversy, and Change. New York: Springer, 2006: 233-246. Subject: 9.3.1

Craigie, Jillian; Henry, Michael; Fishman, Jennifer R.; Youngner, Stuart J. Propranolol, cognitive biases, and practical decision-making. *American Journal of Bioethics* 2007 September; 7(9): 31-32; author reply W1-W3. Subject: 17.4

Crall, James J. Access to oral health care: professional and societal considerations. *Journal of Dental Education* 2006 November; 70(11): 1133-1138. Subject: 9.2

Cram, Peter; Rosenthal; Gary E. Physician-owned specialty hospitals and coronary revascularization utilization: too much of a good thing? *JAMA: The Journal of the American Medical Association* 2007 March 7; 297(9): 998-999. Subject: 7.3

Cranford, Ronald. Terri Schiavo was not disabled. *Ethics and Intellectual Disability* 2005 Summer; 9(1): 1, 4-5. Subject: 20.5.1

Cranford, Ronald E. The neurologist as ethics consultant and as a member of the institutional ethics committee: the neuroethicist. *Neurologic Clinics* 1989 November; 7(4): 697-713. Subject: 9.6

Cranston, Robert. The doctor who wanted to die. *Today's Christian Doctor* 2007 Fall; 38(3): 29-30. Subject: 8.3.4

Cremer, R.; Binoche, A.; Noizet, O.; Fourier, C.; Leteurtre, S.; Moutel, G.; Leclerc, F. Are the GFRUP's recommendations for withholding or withdrawing treatments in critically ill children applicable? Results of a two-year survey. *Journal of Medical Ethics* 2007 March; 33(3): 128-133. Subject: 20.5.2

Crippen, David. Medical treatment for the terminally ill: the 'risk of unacceptable badness'. *Critical Care (London, England)* 2005 August; 9(4): 317-318. Subject: 20.5.1

Subject = NRCBL Primary Classification Number; see inside front cover.

Crippin, Jeffrey S. Treatment of hepatocellular carcinoma after transplantation and human rights. *Hepatology* 2007 February; 45(2): 263-265. Subject: 19.1

Croft, Jason R.; Festinger, David S.; Dugosh, Karen L.; Marlowe, Douglas B.; Rosenwasser, Beth J. Does size matter?:salience of follow-up payments in drug abuse research. *IRB: Ethics and Human Research* 2007 July-August; 29(4): 15-19. Subject: 18.5.1

Cronin, Antonia J. Transplants save lives, defending the double veto does not: a reply to Wilkinson. *Journal of Medical Ethics* 2007 April; 33(4): 219-220. Subject: 19.5

Crook, Jamie. Balancing intellectual property protection with the human right to health. *Berkeley Journal of International Law* 2005; 23(3): 524-550. Subject: 9.5.6

Crooks, Glenna M. The rights of patients to participate in clinical research. *In:* Santoro, Michael A.; Gorrie, Thomas M., eds. Ethics and the Pharmaceutical Industry. Cambridge; New York: Cambridge University Press, 2005: 97-108. Subject: 18.1

Crosby, Sondra S.; Apovian, Caroline M.; Grodin, Michael A. Hunger strikes, force-feeding, and physicians' responsibilities. *JAMA: The Journal of the American Medical Association* 2007 August 1; 298(5): 563-566. Subject: 21.5

Cross, Michael. House of cards. *BMJ: British Medical Journal* 2007 April 14; 334(7597): 772-773. Subject: 8.4

Cruess, Sylvia R. Professionalism and medicine's social contract with society. *Clinical Orthopaedics and Related Research* 2006 August; 449: 170-176. Subject: 4.1.2

Cullen, Rowena; Marshall, Stephen. Genetic research and genetic information: a health information professional's perspective on the benefits and risks. *Health Information and Libraries Journal* 2006 December; 23(4): 275-282. Subject: 15.1

Cuperus-Bosma, Jacquelyne M.; Hout, Fredericus A.G.; Hubben, Joep H.; van der Wal, Gerrit. Views of physicians, disciplinary board members and practicing lawyers on the new statutory disciplinary system for health care in the Netherlands. *Health Policy* 2006 July; 77(2): 202-211. Subject: 7.4

Curlin, Farr A.; Chin, Marshall H.; Sellergren, Sarah A.; Roach, Chad J.; Lantos, John D. The association of physicians' religious characteristics with their attitudes and self-reported behaviors regarding religion and spirituality in the clinical encounter. *Medical Care* 2006 May; 44(5): 446-453. Subject: 8.1

Curlin, Farr A.; Lawrence, Ryan E.; Chin, Marshall H.; Lantos, John D. Religion, conscience, and controversial clinical practices. *New England Journal of Medicine* 2007 February 8; 356(6): 593-600. Subject: 8.1

Curlin, Farr A.; Roach, Chad J. By intuitions differently formed: how physicians assess and respond to spiritual issues in the clinical encounter. *American Journal of Bioethics* 2007 July; 7(7): 19-20. Subject: 8.1

Curtis, J. Randall. Interventions to improve care during withdrawal of life-sustaining treatments. *Journal of Palliative Medicine* 2005; 8(Supplement 1): S116-S131. Subject: 20.5.1

Cust, Kenneth. Philosophers return to the agora. *In:* Rasmussen, Lisa, ed. Ethics Expertise: History, Contemporary Perspectives, and Applications. Dordrecht: Springer, 2005: 227-241. Subject: 17.2

Cutas, Daniela. Postmenopausal motherhood: immoral, illegal? A case study. *Bioethics* 2007 October; 21(8): 458-463. Subject: 14.4

Cutcliffe, John R.; Yarbrough, Susan. Globalization, commodification and mass transplant of nurses: Part 2. *British Journal of Nursing* 2007 August 9 - September 12; 16(15): 926-930. Subject: 21.6

Cutter, Mary Ann G. Expert moral choice in medicine: a study of uncertainty and locality. *In:* Rasmussen, Lisa, ed. Ethics Expertise: History, Contemporary Perspectives, and Applications. Dordrecht: Springer, 2005: 125-137. Subject: 2.1

Cyranoski, David. Executed Chinese drug czar corrupted by system, observers say [news]. *Nature Medicine* 2007 August; 13(8): 889. Subject: 1.3.9

Cyranoski, David. Race to mimic human embryonic stem cells [news]. *Nature* 2007 November 22; 450(7169): 462-463. Subject: 14.5

Cyranoski, David. Stem-cell fraudster 'is working in Thailand' [news]. *Nature* 2007 September 27; 449(7161): 387. Subject: 18.5.4

D

Daar, Abdallah S. The case for a regulated system of living kidney sales. *Nature Clinical Practice. Nephrology* 2006 November; 2(11): 600-601. Subject: 19.3

Daar, A.S. Animal-to-human organ transplants — a solution or a new problem? *Bulletin of the World Health Organization* 1999; 77(1): 54-61. Subject: 19.1

Daccord, Yves. ICRC and confidentiality [letter]. *Lancet* 2007 September 8-14; 370(9590): 823-824. Subject: 21.2

Dagg, Paul K.B.; Hughes Julian C.; Sarkar, Sameer P. "Hey Bill, smoking is bad for you . . . ." [case study and commentaries]. *Journal of Ethics in Mental Health [electronic]* 2007 November; 2(2): 5 p. Accessed: http://www.jemh.ca [2008 January 24]. Subject: 9.5.9

Daher, Michel. Current trends in medical ethics education. *Journal Medical Libanais* 2006 July-September; 54(3): 121-123. Subject: 2.3

Dahl, E.; Beutel, M.; Brosig, B.; Grüssner, S.; Stöbel-Richter, Y.; Tinneberg, H.-R.; Brähler, E. Social sex selection and the balance of the sexes: empirical evi-

See SUBJECT HEADING KEY FOR SECTION II on inside back cover.

dence from Germany, the UK, and the US. *Journal of Assisted Reproduction and Genetics* 2006 July-August; 23(7-8): 311-318. Subject: 14.3

Dahlquist, Gisela. Ethics in research: why and how? [editorial]. *Scandinavian Journal of Public Health* 2006; 34(5): 449-452. Subject: 1.3.9

Dahlqvist, Vera; Eriksson, Sture; Glasberg, Ann-Louise; Lindahl, Elisabeth; Lützén, Kim; Strandberg, Gunilla; Söderberg, Anna; Sørlie, Venke; Norberg, Astrid. Development of the perceptions of conscience questionnaire. *Nursing Ethics* 2007 March; 14(2): 181-193. Subject: 4.1.3

Dailard, Cynthia; Richardson, Chinué Turner. Teenagers' access to confidential reproductive health services. *Guttmacher Report on Public Policy* 2005 November; 8(4): 6-11. Subject: 11.2

Daleiden, Joseph L. Abortion. *In his:* The Science of Morality: The Individual, Community, and Future Generations. Amherst, NY: Prometheus Books, 1998: 348-372. Subject: 12.1

Daleiden, Joseph L. Euthanasia. *In his:* The Science of Morality: The Individual, Community, and Future Generations. Amherst, NY: Prometheus Books, 1998: 373-409. Subject: 20.5.1

Daley, George Q.; Ahrlund-Richter, Lars; Auerbach, Jonathan M.; Benvenisty, Nissim; Charo, R. Alta; Chen, Grace; Deng, Hong-kui; Goldstein, Lawrence S.; Hudson, Kathy L.; Hyun, Insoo; Junn, Sung Chull; Love, Jane; Lee, Eng Hin; McLaren, Anne; Mummery, Christine L.; Nakatsuji, Norio; Racowsky, Catherine; Rooke, Heather; Rossant, Janet; Schöler, Hans R.; Solbakk, Jan Helge; Taylor, Patrick; Trounson, Alan O.; Weissman, Irving L.; Wilmut, Ian; Yu, John; Zoloth, Laurie. The ISSCR guidelines for human embryonic stem cell research. *Science* 2007 February 2; 315(5812): 603-604. Subject: 18.5.4

Dalton, Rex. Passive-smoking study faces review [news]. *Nature* 2007 March 15; 446(7133): 242. Subject: 1.3.9

Daly, Daniel J. Prudence and the debate on death and dying; in the Catholic theological tradition, temporal life is not the highest good. *Health Progress* 2007 September-October; 88(5): 49-54. Subject: 20.5.1

Damasio, Antonio. Neuroscience and ethics: intersections [commentary]. *American Journal of Bioethics* 2007 January; 7(1): 3-7. Subject: 17.1

Dan, Ovidiu. The philanthropy of the Orthodox Church: a Rumanian case study. *Christian Bioethics* 2007 September-December;(13)3: 303-307. Subject: 2.1

Dandie, Geoff. Report of the International Consensus meeting on carbon dioxide euthanasia of laboratory animals. *ANZCCART News (Australian and New Zealand Council for the Care of Animals in Research and Teaching)* 2006; 19(2): 1-7 [Online]. Accessed: http://www.adelaide.

edu.au/ANZCCART/news/AN19_2.pdf [2008 February 15]. Subject: 22.1

Daneault, Serge; Lussier, Véronique; Mongeau, Suzanne; Hudon, Eveline; Paillé, Pierre; Dion, Dominique; Yelle, Louise. Primum non nocere: could the health care system contribute to suffering? In-depth study from the perspective of terminally ill cancer patients. *Canadian Family Physician* 2006 December; 52(12): 1575 e.1-5. Subject: 9.5.1

Daniels, Ken; Meadows, Letitia. Sharing information with adults conceived as a result of donor insemination. *Human Fertility* 2006 June; 9(2): 93-99. Subject: 14.2

Daniels, Norman. Fairness and national health care reform. *In:* Engström, Timothy H.; Robison, Wade L., eds. Health Care Reform: Ethics and Politics. Rochester, NY: University of Rochester Press, 2006: 240-263. Subject: 9.1

Daniels, Norman. Rescuing universal health care. *Hastings Center Report* 2007 March-April; 37(2): 3. Subject: 9.3.1

Daniels, Norman; Sabin, James E.; Teagarden, J. Russell. Who should get access to which drugs? An ethical template for pharmacy benefits. *In:* Santoro, Michael A.; Gorrie, Thomas M., eds. Ethics and the Pharmaceutical Industry. Cambridge; New York: Cambridge University Press, 2005: 206-224. Subject: 9.7

Danovitch, Gabriel M.; Bunnapradist, Suphamai. Allocating deceased donor kidneys: maximizing years of life. *American Journal of Kidney Diseases* 2007 February; 49(2): 180-182. Subject: 19.3

Danovitch, G.M. A kidney for all ages. *American Journal of Transplantation* 2006 June; 6(6): 1267-1268. Subject: 19.3

Danzinger, Paula R.; Welfel, Elizabeth Reynolds. The impact of managed care on mental health counselors: a survey of perceptions, practices, and compliance with ethical standards. *Journal of Mental Health Counseling* 2001 April; 23(2): 137-150. Subject: 17.2

Daroff, Robert B. Scientific misconduct and breach of publication ethics: one editor's experience. *Medicine and Law: The World Association for Medical Law* 2007 September; 26(3): 527-533. Subject: 1.3.9

Darr, Kurt. Virtue ethics: worth another look. *Hospital Topics* 2006 Fall; 84(4): 29-31. Subject: 9.1

DasGupta, Sayantani. The doctor's wife. *Hastings Center Report* 2007 March-April; 37(2): 7-8. Subject: 4.1.2

Daub, Ute. Soziologische Überlegungen zur Funktion der Eugenik in der neuen politischen Ökonomie. *In:* Neuer-Miebach, Therese; Wunder, Michael, eds. Bio-Ethik und die Zukunft der Medizin. Bonn: Psychiatrie-Verlag, 1998: 139-154. Subject: 15.5

Davey, Angela; Newson, Ainsley; O'Leary, Peter. Communication of genetic information within families: the case for familial comity. *Journal of Bioethical Inquiry* 2006; 3(3): 161-166. Subject: 15.2

Davey, Gareth; Wu, Zhihui. Attitudes in China toward the use of animals in laboratory research. *ATLA: Alternatives to Laboratory Animals* 2007 June; 35(3): 313-316. Subject: 22.2

Davidovitch, Nadav; Filc, Dani. Reconstructing data: evidence-based medicine and evidence-based public health in context. *Dynamis* 2006; 26: 287-306. Subject: 9.5.1

Davies, Douglas. Cheating death: the invisibles. *New Scientist* 2007 October 13-19; 195(2625): 48-49. Subject: 20.3.1

Davies, Glanville; Poole, Richard F.; Akerman, Beverly R.; Gogol, Manfred. Reflections on the birth of conjoined twins [letters]. *CMAJ/JAMC: Canadian Medical Association Journal* 2007 November 6; 177(10): 1235-1236. Subject: 9.5.7

Davies, Hugh. Ethical reflections on Edward Jenner's experimental treatment. *Journal of Medical Ethics* 2007 March; 33(3): 174-176. Subject: 18.2

Davis, Anne J. An ethical voice for nurses — is anybody listening? [letter]. *Nursing Ethics* 2007 March; 14(2): 264. Subject: 4.1.3

Davis, Anne J. International nursing ethics: context and concerns. *In:* Tschudin, Verena, ed. Approaches to Ethics: Nursing Beyond Boundaries. New York: Butterworth-Heinemann, 2003: 95-104. Subject: 4.1.3

Davis, Anne J.; Fowler, Marsha. Caring and caring ethics depicted in selected literature: what we know and what we need to ask. *In:* Davis, Anne J.; Tschudin, Verena; de Raeve, Louise, eds. Essentials of Teaching and Learning in Nursing Ethics: Perspectives and Methods. New York: Churchill Livingstone Elsevier, 2006: 165-179. Subject: 4.1.3

Davis, Anne J.; Konishi, Emiko. Whistleblowing in Japan. *Nursing Ethics* 2007 March; 14(2): 194-202. Subject: 7.3

Davis, Anne J.; Tschudin, Verena; de Raeve, Louise. The future: teaching nursing ethics. *In their:* Essentials of Teaching and Learning in Nursing Ethics: Perspectives and Methods. New York: Churchill Livingstone Elsevier, 2006: 339-352. Subject: 4.1.3

Davis, A. Surrogates and outcast mothers: racism and reproductive politics in the nineties. *In:* James, J., ed. The Angela Y. Davis Reader. Malden, MA: Blackwell, 1998: 210-221. Subject: 14.2

Davis, Bradley L. Compelled expression of the religiously forbidden: pharmacists, "duty to fill" statutes, and the hybrid rights exception. *University of Hawai'i Law Review* 2006 Winter; 29(1): 97-121. Subject: 11.1

Davis, Dena S. A tale of two daughters: Jewish law and end-of-life decision making. *Journal of Clinical Ethics* 2007 Winter; 18(4): 394-395. Subject: 20.5.1

Davis, Dena S. The changing face of "misidentified paternity". *Journal of Medicine and Philosophy* 2007 July-August; 32(4): 359-373. Subject: 15.11

Davis, Dena S. The puzzle of IVF. *Houston Journal of Health Law and Policy* 2006; 6(2): 275-297. Subject: 14.4

Davis, Gayle; Davidson, Roger. "A fifth freedom" or "hideous atheistic expediency"? The medical community and abortion law reform in Scotland, c.1960-1975. *Medical History* 2006 January; 50(1): 29-48. Subject: 12.4.1

Davis, Joel J. Consumers' preferences for the communication of risk information in drug advertising. *Health Affairs* 2007 May-June; 26(3): 863-870. Subject: 9.7

Davis, John K. Intuition and the junctures of judgment in decision procedures for clinical ethics. *Theoretical Medicine and Bioethics* 2007; 28(1): 1-30. Subject: 2.1

Davis, John K. Precedent, autonomy, advance directives, and end-of-life care. *In:* Steinbock, Bonnie, ed. The Oxford Handbook of Bioethics. Oxford; New York: Oxford University Press, 2007: 349-374. Subject: 20.5.4

Davis, Karen. Uninsured in America: problems and possible solutions. *BMJ: British Medical Journal* 2007 February 17; 334(7589): 346-348. Subject: 9.3.1

Davis, Kathy. Rethinking "normal" [reviews of No Child Left Different, edited by Sharon Olfman; Cutting to the Core:Exploring the Ethics of Contested Surgeries, edited by David Benatar; Surgically Shaping Children: Technology, Ethics, and the Pursuit of Normality, edited by Erik Parens]. *Hastings Center Report* 2007 May-June; 37(3): 44-47. Subject: 4.2

Davis, Mark S.; Riske-Morris, Michelle; Diaz, Sebastian R. Causal factors implicated in research misconduct: evidence from ORI case files. *Science and Engineering Ethics* 2007 December; 13(4): 395-414. Subject: 1.3.9

Davis, Michael. Eighteen rules for writing a code of professional ethics. *Science and Engineering Ethics* 2007 June; 13(2): 171-189. Subject: 1.3.1

Davis, Thomas J., Jr. Plan B and the rout of religious liberty: reflection on the status of the law. *Ethics and Medics* 2007 December; 32(12): 1-4. Subject: 11.1

Dawson, Angus J.; Yentis, Steve M. Contesting the science/ethics distinction in the review of clinical research. *Journal of Medical Ethics* 2007 March; 33(3): 165-167. Subject: 18.2

Dawson, John; Szmukler, George. Fusion of mental health and incapacity legislation. *British Journal of Psychiatry* 2006 June; 188: 504-509. Subject: 9.5.3

Dawson, Liza; Hyder, Adnan A. Understanding the 'de jure' standard of care for research: a reply to Faust [letter]. *Developing World Bioethics* 2007 April; 7(1): 46-47. Subject: 9.8

See SUBJECT HEADING KEY FOR SECTION II on inside back cover.

Day, Lisa. Family involvement in critical care: shortcomings of a utilitarian justification. *American Journal of Critical Care* 2006 March; 15(2): 223-225. Subject: 9.5.1

Day, Lisa. Industry gifts to healthcare providers: are the concerns serious? *American Journal of Critical Care* 2006 September; 15(5): 510-513. Subject: 9.7

Day, Lisa. Questions concerning the goodness of hastening death. *American Journal of Critical Care* 2006 May; 15(3): 312-314. Subject: 20.5.1

Day, Michael. Hewitt says some Muslim GPs breach confidentiality [news]. *BMJ: British Medial Journal* 2007 April 7; 334(7596): 711. Subject: 8.4

Day, Michael. Number of sperm donors rises in UK despite removal of anonymity [news]. *BMJ: British Medical Journal* 2007 May 12; 334(7601): 971. Subject: 14.2

Day, Michael. UK may use hybrid embryos for research [news]. *BMJ: British Medical Journal* 2007 May 26; 334(7603): 1074. Subject: 18.5.4

Day, Michael. Who's funding WHO? WHO guidelines state that it will not accept money from drug companies, but how rigorous is it enforcing this? *BMJ: British Medical Journal* 2007 February 17; 334(7589): 338-340. Subject: 9.7

de Beaufort, Inez; Meulenberg, Frans. The dangers of triage by television. *BMJ: British Medical Journal* 2007 June 9; 334(7605): 1194-1195. Subject: 19.5

de Cock Buning, Tjard. Comments on Zwart: professional ethics and scholarly communication. *In:* Korthals, Michiel; Bogers, Robert J., eds. Ethics for Life Scientists. Dordrecht, The Netherlands: Springer, 2004: 81-84. Subject: 1.3.9

de Costa, Caroline; de Costa, Naomi. Medical abortion and the law. *University of New South Wales Law Journal* 2006; 29(2): 218-223. Subject: 12.1

de Freitas, Genival Fernandes; Oguisso, Taka; Merighi, Miriam Aparecida Barbosa. Ethical events in nursing: daily activities of nurse managers and nursing ethics committee members. *Revista Latino-Americana Enfermagem* 2006 July-August; 14(4): 497-502. Subject: 9.6

De Gendt, Cindy; Bilsen, Johan; Stichele, Robert Vander; Van Den Noortgate, Nele; Lambert, Margareta; Deliens, Luc. Nurses' involvement in 'do not resuscitate' decisions on acute elder care wards. *Journal of Advanced Nursing* 2007 February; 57(4): 404-409. Subject: 20.5.1

de Graeff, Alexander; Dean, Mervyn. Palliative sedation therapy in the last weeks of life: a literature review and recommendations for standards. *Journal of Palliative Medicine* 2007 February; 10(1): 67-85. Subject: 4.4

De Grazia, David. Must we have full moral status throughout our existence? A reply to Alfonso Gomez-Lobo. *Kennedy Institute of Ethics Journal* 2007 December; 17(4): 297-310. Subject: 4.4

de Groot, Rolf. Right to health care and scarcity of resources. *In:* Gevers, J.K.M.; Hondius, E.H.; Hubben, J.H., eds. Health Law, Human Rights and the Biomedicine Convention: Essays in Honour of Henriette Roscam Abbing. Leiden; Boston: Martinus Nijhoff Publishers, 2005: 49-59. Subject: 9.2

De Jonge, Christopher; Barratt, Christopher L. Gamete donation: a question of anonymity. *Fertility and Sterility* 2006 February; 85(2): 500-501. Subject: 14.2

de la Gorgendière, Louise. Rights and wrongs: HIV/AIDS research in Africa. *Human Organization* 2005 Summer; 64(2): 166-178. Subject: 9.5.6

de Melo-Martín, Immaculada. Cloning — or not — human beings. *In her:* Taking Biology Seriously: What Biology Can and Cannot Tell Us About Moral and Public Policy Issues. Lanham: Rowman and Littlefield, 2005: 45-61. Subject: 14.5

de Melo-Martín, Immaculada. Genetic information and moral obligations. *In her:* Taking Biology Seriously: What Biology Can and Cannot Tell Us About Moral and Public Policy Issues. Lanham: Rowman and Littlefield, 2005: 83-103. Subject: 15.3

de Melo-Martín, Immaculada. Genetic testing: the appropriate means for a desired goal? *Journal of Bioethical Inquiry* 2006; 3(3): 167-177. Subject: 15.3

de Melo-Martín, Immaculada. Moral obligations, genetic information, and social context. *In her:* Taking Biology Seriously: What Biology Can and Cannot Tell Us About Moral and Public Policy Issues. Lanham: Rowman and Littlefield, 2005: 105-127. Subject: 15.3

de Melo-Martín, Immaculada. Putting human cloning where it belongs. *In her:* Taking Biology Seriously: What Biology Can and Cannot Tell Us About Moral and Public Policy Issues. Lanham: Rowman and Littlefield, 2005: 63-76. Subject: 14.5

de Melo-Martín, Immaculada. The promise of the human papillomavirus vaccine does not confer immunity against ethical reflection. *Oncologist* 2006 April; 11(4): 393-396. Subject: 9.1

de Melo-Martín, Immaculada. When is biology destiny? Biological determinism and social responsibility. *Philosophy of Science* 2003 December; 70(5): 1184-1194. Subject: 15.6

de Melo-Martín, Immaculada; Intemann, Kristen. Authors' financial interests should be made known to manuscript reviewers. *Nature* 2007 July 12; 448(7150): 129. Subject: 1.3.7

de Melo-Martín, Immaculada; Palmer, Larry I.; Fins, Joseph J. Developing a research ethics consultation service to foster responsive and responsible clinical research. *Academic Medicine* 2007 September; 82(9): 900-904. Subject: 18.2

Subject = NRCBL Primary Classification Number; see inside front cover.

de Melo-Martín, Immaculada; Rosenwaks, Zev; Fins, Joseph J. New methods for deriving embryonic stem cell lines: are the ethical problems solved? *Fertility and Sterility* 2006 November; 86(5): 1330-1332. Subject: 15.2

de Montalembert, Mariane de; Bonnet, Doris; Lena-Russo, Danielle; Briard, Marie Louise. Ethical aspects of neonatal screening for sickle cell disease in Western European countries. *Acta Paediatrica* 2005 May; 94(5): 528-530. Subject: 15.2

de Raeve, Louise. A critique of virtue ethics. *In:* Davis, Anne J.; Tschudin, Verena; de Raeve, Louise, eds. Essentials of Teaching and Learning in Nursing Ethics: Perspectives and Methods. New York: Churchill Livingstone Elsevier, 2006: 109-122. Subject: 4.1.3

de Raeve, Louise. Teaching virtue ethics. *In:* Davis, Anne J.; Tschudin, Verena; de Raeve, Louise, eds. Essentials of Teaching and Learning in Nursing Ethics: Perspectives and Methods. New York: Churchill Livingstone Elsevier, 2006: 123-134. Subject: 4.1.3

de Raeve, Louise. Virtue ethics. *In:* Davis, Anne J.; Tschudin, Verena; de Raeve, Louise, eds. Essentials of Teaching and Learning in Nursing Ethics: Perspectives and Methods. New York: Churchill Livingstone Elsevier, 2006: 97-108. Subject: 4.1.3

de Roubaix, J.A.M; van Niekerk, A.A. Separation-survivability — the elusive moral cut-off point? *South African Medical Journal = Suid-afrikaanse Tydskrif vir Geneeskunde* 2006 July; 96(7): 623-626. Subject: 12.3

de Roubaix, M. Ten years hence — has the South African Choice on Termination of Pregnancy Act, Act 92 of 1996, realised its aims? A moral-critical evaluation. *Medicine and Law: The World Association for Medical Law* 2007 March; 26(1): 145-177. Subject: 12.4.2

De Ville, Kenneth; Hassler, Gregory; Lewis, Michael J. Rejuvenating a foundering institutional review board: one institution's story. *Academic Medicine* 2007 January; 82(1): 11-17. Subject: 18.2

de Vries, Jantina. The obesity epidemic: medical and ethical considerations. *Science and Engineering Ethics* 2007 March; 13(1): 55-67. Subject: 9.5.1

De Vries, Raymond G.; Turner, Leigh; Orfali, Kristina; Bosk, Charles L. Social science and bioethics: morality from the ground up. *Clinical Ethics* 2007 March; 2(1): 33-35. Subject: 2.1

de Wert, Guido; Geraedts, Joep P.M. Preimplantation genetic diagnosis for hereditary disorders that do not show a simple Mendelian pattern: an ethical exploration. *In:* Shenfield, Françoise; Sureau, Claude, eds. Contemporary Ethical Dilemmas in Assisted Reproduction. Abingdon: Informa Healthcare, 2006: 85-98. Subject: 15.2

de Wert, Guido; Liebaers, Inge; Van de Velde, Hilde. The future (r)evolution of preimplantation genetic diagnosis/human leukocyte antigen testing: ethical reflections.

Stem Cells 2007 September; 25: 2167-2172 [Online]. Accessed: http://www.StemCells.com/cgi/content/full/25/9/2167 [2007 December 4]. Subject: 15.2

de Zulueta, Paquita; Boulton, Mary. Routine antenatal HIV testing: the responses and perceptions of pregnant women and the viability of informed consent. A qualitative study. *Journal of Medical Ethics* 2007 June; 33(6): 329-336. Subject: 9.5.6

Deane-Drummond, Celia. Biotechnology and theology. *In:* Eisen, Arri; Laderman, Gary, eds. Science, Religion, and Society: An Encyclopedia of History, Culture, and Controversy. Vol. 2. Armonk, NY: M.E. Sharpe, 2007: 780-786. Subject: 15.1

Deane-Drummond, Celia. Future perfect? God, the transhuman future and the quest for immortality. *In:* Deane-Drummond, Celia; Scott, Peter Manley, eds. Future Perfect?: God, Medicine and Human Identity. New York: T and T Clark International, 2006: 168-182. Subject: 4.5

Deane-Drummond, Celia. Gene patenting. *In her:* Genetics and Christian Ethics. Cambridge: Cambridge University Press, 2006: 160-190. Subject: 15.8

Deane-Drummond, Celia. Gene therapies. *In her:* Genetics and Christian Ethics. Cambridge: Cambridge University Press, 2006: 124-159. Subject: 15.4

Deane-Drummond, Celia. Genetic counselling. *In her:* Genetics and Christian Ethics. Cambridge: Cambridge University Press, 2006: 101-123. Subject: 15.2

Deane-Drummond, Celia. Genetic testing and screening. *In her:* Genetics and Christian Ethics. Cambridge: Cambridge University Press, 2006: 76-100. Subject: 15.3

Deane-Drummond, Celia. Genetics and environmental concern. *In her:* Genetics and Christian Ethics. Cambridge: Cambridge University Press, 2006: 220-244. Subject: 15.1

Deane-Drummond, Celia. Living in the shadow of eugenics. *In her:* Genetics and Christian Ethics. Cambridge: Cambridge University Press, 2006: 55-75. Subject: 15.5

Deane-Drummond, Celia. Women and genetic technologies. *In her:* Genetics and Christian Ethics. Cambridge: Cambridge University Press, 2006: 191-219. Subject: 14.1

Deapen, Dennis. Cancer surveillance and information: balancing public health with privacy and confidentiality concerns (United States). *Cancer Causes and Control* 2006 June; 17(5): 633-637. Subject: 8.4

Debiak, Dennis. Attending to diversity in group psychotherapy: an ethical imperative. *International Journal of Group Psychotherapy* 2007 January; 57(1): 1-12; discussion 49-59, 61-66. Subject: 17.2

Debruin, Debra A. Ethics on the inside? *In:* Eckenwiler, Lisa A.; Cohn, Felicia G., eds. The Ethics of Bioethics: Mapping the Moral Landscape. Baltimore, MD: Johns Hopkins University Press, 2007: 161-169. Subject: 7.3

See SUBJECT HEADING KEY FOR SECTION II on inside back cover.

475

DeCamp, Matthew. Scrutinizing global short-term medical outreach. *Hastings Center Report* 2007 November-December; 37(6): 21-23. Subject: 21.1

DeCamp, Matthew; Buchanan, Allen. Pharmacogenomics, ethical and regulatory issues. *In:* Steinbock, Bonnie, ed. The Oxford Handbook of Bioethics. Oxford; New York: Oxford University Press, 2007: 536-568. Subject: 15.1

Deckers, Jan. Are those who subscribe to the view that early embryos are persons irrational and inconsistent? A reply to Brock. *Journal of Medical Ethics* 2007 February; 33(2): 102-106. Subject: 4.4

Deckers, Jan. Why Eberl is wrong. Reflections on the beginning of personhood. *Bioethics* 2007 June; 21(5): 270-282. Subject: 4.4

Deckers, J. Why two arguments from probability fail and one argument from Thomson's analogy of the violinist succeeds in justifying embryo destruction in some situations. *Journal of Medical Ethics* 2007 March; 33(3): 160-164. Subject: 18.5.4

Decullier, Evelyne; Chapuis, François. Impact of funding on biomedical research: a retrospective cohort study. *BMC Public Health* 2006 June 22; 6: 165. Subject: 18.1

Dees, Richard H. Better brains, better selves? The ethics of neuroenhancements. *Kennedy Institute of Ethics Journal* 2007 December; 17(4): 371-395. Subject: 17.1

Dees, Richard H.; Volandes, Angelo E.; Paasche-Orlow, Michael K. Health literacy and autonomy. *American Journal of Bioethics* 2007 November; 7(11): 22-23; author reply W1-W2. Subject: 8.1

Degnin, Francis Dominic; Wood, Donna J. Levinas and society's most vulnerable: a philosopher's view of the business of healthcare. *Organizational Ethics: Healthcare, Business, and Policy* 2007 Spring-Summer; 4(1): 65-80. Subject: 9.1

DeIorio, Nicole M.; McClure, Katie B.; Nelson, Maria; McConnell, K. John; Schmidt, Terri A. Ethics committee experience with emergency exception from informed consent protocols. *Journal of Empirical Research on Human Research Ethics* 2007 September; 2(3): 23-30. Subject: 18.2

DeJohn, Carla; Zwischenberger, Joseph B. Ethical implications of extracorporeal interval support for organ retrieval (EISOR) [editorial]. *ASAIO Journal: American Society for Artificial Internal Organs* 2006 March-April; 52(2): 119-122. Subject: 19.5

Dekkers, Wim; Gordijn, Bert. Practical wisdom in medicine and health care [editorial]. *Medicine, Health Care and Philosophy* 2007 September; 10(3): 231-232. Subject: 9.1

Dekkers, Wim; Rikkert, Marcel Olde. Memory enhancing drugs and Alzheimer's disease: enhancing the self or preventing the loss of it? *Medicine, Health Care and Philosophy* 2007 June; 10(2): 141-151. Subject: 17.4

Delaney, Martin. AIDS activism and the pharmaceutical industry. *In:* Santoro, Michael A.; Gorrie, Thomas M., eds. Ethics and the Pharmaceutical Industry. Cambridge; New York: Cambridge University Press, 2005: 300-325. Subject: 9.5.6

Delany, Mike. General medical practice: the problem of cooperation in evil. *In:* Watt, Helen, ed. Cooperation, Complicity and Conscience: Problems in Healthcare, Science, Law and Public Policy. London: Linacre Centre, 2005: 128-138. Subject: 2.1

Delbanco, Tom; Bell, Sigall K. Guilty, afraid and alone — struggling with medical error. *New England Journal of Medicine* 2007 October 25; 357(17): 1682-1683. Subject: 9.8

Delkeskamp-Hayes, Corinna. Implementing health care rights versus health care cultures: the limits of tolerance, Kant's rationality, and the moral pitfalls of international bioethics standardization. *In:* Engelhardt, H. Tristram, ed. Global Bioethics: The Collapse of Consensus. Salem, MA: M&M Scrivener Press, 2006: 50-94. Subject: 2.1

Delkeskamp-Hayes, Corinna. Resisting the therapeutic reduction: on the significance of sin. *Christian Bioethics* 2007 January-April; (13)1: 105-127. Subject: 2.1

Delkeskamp-Hayes, Corinna. Societal consensus and the problem of consent: refocusing the problem of ethics expertise in liberal democracies. *In:* Rasmussen, Lisa, ed. Ethics Expertise: History, Contemporary Perspectives, and Applications. Dordrecht: Springer, 2005: 139-163. Subject: 2.1

Delpierre, Cyrille; Cuzin, Lise; Lert, France. Routine testing to reduce late HIV diagnosis in France. *BMJ: British Medical Journal* 2007 June 30; 334(7608): 1354-1356. Subject: 9.5.6

DeMarco, Donald. Fetal pain: real or relative? *Human Life Review* 2007 Winter; 33(1): 68-71. Subject: 12.4.2

DeMaria, Anthony N. Your soul for a pen? *Journal of the American College of Cardiology* 2007 March 20; 49(11): 1220-1222. Subject: 9.7

DeMets, David L.; Fost, Norman; Powers, Madison. An Institutional Review Board dilemma: responsible for safety monitoring but not in control. *Clinical Trials* 2006; 3(2): 142-148. Subject: 18.2

Deming, Nicole; Fryer-Edwards, Kelly; Dudzinski, Denise; Starks, Helene; Culver, Julie; Hopley, Elizabeth; Robins, Lynne; Burke, Wylie. Incorporating principles and practical wisdom in research ethics education: a preliminary study. *Academic Medicine* 2007 January; 82(1): 18-23. Subject: 18.1

Denier, Yvonne. Autonomy in dependence: a defence of careful solidarity. *In:* Nys, Thomas; Denier, Yvonne; Vandevelde, Toon, eds. Autonomy and Paternalism: Reflections on the Theory and Practice of Health Care.

Subject = NRCBL Primary Classification Number; see inside front cover.

Leuven; Dudley, MA: Peeters, 2007: 93-111. Subject: 4.1.2

Dennis, William J. What is death with dignity? *Ethics and Medics* 2007 August; 32(8): 1-2. Subject: 20.5.1

Dennison, Tania; Leach, Matthew. Animal research, ethics and law. *In:* Pullen, Sophie; Gray, Carol, eds. Ethics, Law and the Veterinary Nurse. New York: Elsevier Butterworth Heinemann, 2006: 103-116. Subject: 22.2

Denniston, George C. Human rights advances in the United States. *In:* Denniston, George C.; Gallo, Pia Grassivaro; Hodges, Frederick M.; Milos, Marilyn Fayre; Viviani, Franco, eds. Bodily Integrity and the Politics of Circumcision: Culture, Controversy, and Change. New York: Springer, 2006: 189-201. Subject: 21.1

Denny, Colleen C.; Emanuel, Ezekiel J.; Pearson, Steven D. Why well-insured patients should demand value-based insurance benefits [commentary]. *JAMA: The Journal of the American Medical Association* 2007 June 13; 297(22): 2515-2518. Subject: 9.3.1

Denny, Colleen C.; Grady, Christine. Clinical research with economically disadvantaged populations. *Journal of Medical Ethics* 2007 July; 33(7): 382-385 [see correction in Journal of Medical Ethics 2007 August; 33(8): 496]. Subject: 18.5.1

Derbyshire, Stuart W.G. Medical journals: past their sell by date? *BMJ: British Medical Journal* 2007 January 6; 334(7583): 45. Subject: 1.3.7

Derksen, E.; Vernooij-Dassen, M.; Gillissen, F.; Olde Rikkert, M.; Scheltens, P. Impact of diagnostic disclosure in dementia on patients and carers: qualitative case series analysis. *Aging and Mental Health* 2006 September; 10(5): 525-531. Subject: 8.2

DeRosa, G. Paul. Professionalism and virtues. *Clinical Orthopaedics and Related Research* 2006 August; 449: 28-33. Subject: 7.2

Derse, Arthur R. The evolution of medical ethics education at the Medical College of Wisconsin. *WMJ: Official publication of the State Medical Society of Wisconsin* 2006 June; 105(4): 18-20. Subject: 2.3

Derse, Arthur R.; Easton, Raul B.; Graber, Mark A.; Monnahan, Jay; Hughes, Jason. Is patients' time too valuable for informed consent? *American Journal of Bioethics* 2007 December; 7(12): 45-46; author reply W3-W4. Subject: 8.3.1

Dery, Anat Mishori; Carmi, Rivka; Vardi, Ilana Shoham. Different perceptions and attitudes regarding prenatal testing among service providers and consumers in Israel. *Community Genetics* 2007; 10(4): 242-251. Subject: 15.2

Detmer, Don E.; Singleton, Peter; Ratzan, Scott C. The need for better health information: advancing the informed patient in Europe. *In:* Santoro, Michael A.; Gorrie, Thomas M., eds. Ethics and the Pharmaceutical Industry.

Cambridge; New York: Cambridge University Press, 2005: 196-205. Subject: 8.1

Deutsche Gesellschaft für Humangenetik = German Society of Human Genetics [DGfH]. Committee for Public Relations and Ethical Issues. Statement on the revision of section 218 of the German Penal Code with elimination of the so-called embryopathic indication for terminating pregnancy [position statement]. *Medizinische Genetik* 1995; 7: 360-361 (3 p.) [Online]. Accessed: http://www.medgenetik.de/sonderdruck/en/218_ Revision.pdf [2007 February 12]. Subject: 12.4.2

Devereux, John. Continuing conundrums in competency. *In:* McLean, Sheila A.M., ed. First Do No Harm: Law, Ethics, and Healthcare. Aldershot, England; Burlington, VT: Ashgate, 2006: 235-253. Subject: 8.3.3

DeVita, Michael A.; Caplan, Arthur L. Caring for organs or for patients? Ethical concerns about the Uniform Anatomical Gift Act (2006). *Annals of Internal Medicine* 2007 December 18; 147(12): 876-879. Subject: 19.5

Dew, Rachel E. Informed consent for research in borderline personality disorder [debate]. *BMC Medical Ethics [electronic]* 2007; 8:4. 4 p. Subject: 18.3

Dewing, Jan. From ritual to relationship: a person-centered approach to consent in qualitative research with older people who have dementia. *Dementia: The International Journal of Social Research and Practice* 2002 June; 1(2): 157-171. Subject: 18.5.6

Dharamsi, Shafik. Building moral communities? First, do no harm. *Journal of Dental Education* 2006 November; 70(11): 1235-1240. Subject: 4.1.1

Diamond, Eugene F. Catholic health care decision making [editorial]. *Linacre Quarterly* 2007 May; 74(2): 92-93. Subject: 20.5.1

Diamond, Eugene F. John Paul II and brain death. *National Catholic Bioethics Quarterly* 2007 Autumn; 7(3): 491-497. Subject: 20.2.1

Diamond, Sheila M. Response to Todd Bindig's "Confusion about speciesism and moral status". *Linacre Quarterly* 2007 May; 74(2): 156-158. Subject: 4.4

Diaz-Navarlaz, T.; Segui-Gomez, M. Commentary on Armitage G (2005) Drug errors, qualitative research and some reflections on ethics. Journal of Clinical Nursing 14, 869-875. *Journal of Clinical Nursing* 2006 September; 15(9): 1208-1209; discussion 1209. Subject: 9.7

Dickens, Bernard M. Conscientious objection: a shield or a sword? *In:* McLean, Sheila A.M., ed. First Do No Harm: Law, Ethics, and Healthcare. Aldershot, England; Burlington, VT: Ashgate, 2006: 337-351. Subject: 14.1

Dickens, Bernard M. The doctor's duty of confidentiality: separating the rule from the expectations. *University of Toronto Medical Journal* 1999 December; 77(1): 40-43 [Online]. Accessed: http://www.utmj.org/issues/77.1/pdf/LawEthics77-1.pdf [2007 February 16]. Subject: 8.4

See SUBJECT HEADING KEY FOR SECTION II on inside back cover.

477

Dickens, B.M.; Cook, R.J. Acquiring human embryos for stem-cell research. *International Journal of Gynecology and Obstetrics* 2007 January; 96(1): 67-71. Subject: 18.5.4

Dickens, B.M.; Cook, R.J. Legal and ethical issues in telemedicine and robotics. *International Journal of Gynecology and Obstetrics* 2006 July; 94(1): 73-78. Subject: 1.3.12

Dickenson, Donna L. Tissue economies: biomedicine and commercialization [review of Tissue Economics: Blood, Organs, and Cell Lines in Late Capitalism by Catherine Waldby and Robert Mitchell]. *Perspectives in Biology and Medicine* 2007 Spring; 50(2): 308-311. Subject: 19.5

Dickert, Neal W.; Sugarman, Jeremy. Getting the ethics right regarding research in the emergency setting: lessons from the PolyHeme Study. *Kennedy Institute of Ethics Journal* 2007 June; 12(2): 153-169. Subject: 18.3

Dickert, N.; DeRiemer, K.; Duffy, P.E.; Garcia-Garcia, L.; Mutabingwa, T.K.; Sina, B.J.; Tindana, P.; Lie, R. Ancillary-care responsibilities in observational research: two cases, two issues. *Lancet* 2007 March 10-16; 369(9564): 874-877. Subject: 18.5.9

Dickinson, Frederick R. Biohazard: Unit 731 in postwar Japanese politics of national "forgetfulness". *In:* LaFleur, William R.; Böhme, Gernot; Shimazono, Susumu, eds. Dark Medicine: Rationalizing Medical Research. Bloomington: Indiana University Press, 2007: 85-104. Subject: 18.5.1

Diehm, Alexander; Ebsen, Ingwer. Ansätze zur „heimärztlichen Versorgung" und die geplante Pflegereform - Rechtliche Aspekte dargestellt am Beispiel der Psychopharmakaversorgung = The medical treatment in nursing homes and plans for a legislative reform - legal aspects with particular reference to supply of psychotropic drugs. *Ethik in der Medizin* 2007 December; 19(4): 301-312. Subject: 17.4

Diekema, Douglas S. The armchair ethicist: it's all about location. *Journal of Clinical Ethics* 2007 Fall; 18(3): 227-232. Subject: 9.6

Dieterle, J.M. Physician assisted suicide: a new look at the arguments. *Bioethics* 2007 March; 21(3): 127-139. Subject: 20.7

Dikova, Rossitza. Introduction [to issue on Philanthropy, Caritas, Diakonia — European approaches to Christians' service in the world]. *Christian Bioethics* 2007 September-December;(13)3: 245-250. Subject: 2.1

Diller, Lawrence; Goldstein, Sam. Science, ethics, and the psychosocial treatment of ADHD [editorial]. *Journal of Attention Disorders* 2006 May; 9(4): 571-574. Subject: 9.5.7

DiMichele, D.; Chuansumrit, A.; London, A.J.; Thompson, A.R.; Cooper, C.G.; Killian, R.M.; Ross, L.F.; Lillicrap, D.; Kimmelman, J. Ethical issues in haemophilia. *Haemophilia* 2006 July; 12(Supplement 3): 30-35. Subject: 15.3

Dimitrov, Borislav D.; Glutnikova, Zlatka; St. Dimitrova, Bogdana. Education and practice of medical ethics in Bulgaria after political and socio-economic changes in the 90's [commentary]. *Ethics and Medicine: An International Journal of Bioethics* 2007 Spring; 23(1): 11-14. Subject: 2.1

Dimond, Bridgit. Mental capacity and decision making: defining capacity. *British Journal of Nursing* 2007 October 11-24; 16(18): 1138-1139. Subject: 8.3.3

Dimond, Bridgit. The Mental Capacity Act 2005 and decision-making: advocacy. *British Journal of Nursing* 2007 December 13-2008 January 9; 16(22): 1414-1416. Subject: 8.3.3

Dimond, Bridgit. The Mental Capacity Act 2005 and decision-making: best interests. *British Journal of Nursing* 2007 October 25-November 7; 16(19): 1208-1210. Subject: 8.3.3

Dimond, Bridgit. The Mental Capacity Act 2005: lasting power of attorney. *British Journal of Nursing* 2007 November 8-21; 16(20): 1284-1285. Subject: 8.3.3

Dimond, Bridgit. The Mental Capacity Act 2005: the new Court of Protection. *British Journal of Nursing* 2007 November 22-December 12; 16(21): 1328-1330. Subject: 8.3.3

Dinç, Leyla. Turkey: teaching ethics in Turkish nursing education programmes. *In:* Davis, Anne J.; Tschudin, Verena; de Raeve, Louise, eds. Essentials of Teaching and Learning in Nursing Ethics: Perspectives and Methods. New York: Churchill Livingstone Elsevier, 2006: 271-280. Subject: 4.1.3

Ding, Eric L.; Powe, Neil R.; Manson, JoAnn E.; Sherber, Noëlle S.; Braunstein, Joel B. Sex differences in perceived risks, distrust, and willingness to participate in clinical trials. *Archives of Internal Medicine* 2007 May 14; 167(9): 905-912. Subject: 18.2

Dingwall, Robert. An exercise in fatuity: research governance and the emasculation of HSR [editorial]. *Journal of Health Services Research and Policy* 2006 October; 11(4): 193-194. Subject: 18.2

Diniz, Debora. Selective abortion in Brazil: the anencephaly case. *Developing World Bioethics* 2007 August; 7(2): 64-67. Subject: 12.4.1

Diniz, Debora; Perea, Juan-Guillermo Figueroa; Luna, Florencia. Reproductive health ethics: Latin American perspectives [editorial]. *Developing World Bioethics* 2007 August; 7(2): ii-iii. Subject: 14.1

Dinnett, Eleanor M.; Mungall, Moira M.B.; Kent, Jane A.; Ronald, Elizabeth S.; McIntyre, Karen E.; Anderson, Elizabeth; Gaw, Allan. Unblinding of trial participants to their treatment allocation: lessons from the Prospective Study of Pravastatin in the Elderly at Risk (PROS-

PER). *Clinical Trials* 2005 June; 2(3): 254-259. Subject: 18.2

DiSilvestro, Russell. What's wrong with deliberately proselytizing patients? *American Journal of Bioethics* 2007 July; 7(7): 22-24. Subject: 8.1

Dissanayake, V.H.W; Lanerolle, R.D.; Mendis, N. Research ethics and ethical review committees in Sri Lanka: a 25 year journey. *Ceylon Medical Journal* 2006 September; 51(3): 110-113. Subject: 2.4

Dixon, Bernard. What do we need to say to each other? *New Science* 2007 January 6-12; 193(2585): 46-47. Subject: 5.3

Dixon-Mueller, Ruth; Germain, Adrienne. HIV testing: the mutual rights and responsibilities of partners. *Lancet* 2007 December 1-7; 370(9602): 1808-1809. Subject: 9.5.6

Dixon-Woods, Mary; Williams, Simon J.; Jackson, Clare J.; Akkad, Andrea; Kenyon, Sara; Habiba, Marwan. Why do women consent to surgery, even when they do not want to? An interactionist and Bourdieusian analysis. *Social Science and Medicine* 2006 June; 62(11): 2742-2753. Subject: 9.5.5

Dixon-Woods, Mary; Young, Bridget; Ross, Emma. Researching chronic childhood illness: the example of childhood cancer. *Chronic Illness* 2006 September; 2(3): 165-177. Subject: 9.5.5

Djulbegovic, Benjamin. Articulating and responding to uncertainties in clinical research. *Journal of Medicine and Philosophy* 2007 March-April; 32(2): 79-98. Subject: 18.2

Dobrowolska, Beata; Wronska, Irena; Fidecki, Wieslaw; Wysokinski, Mariusz. Moral obligations of nurses based on the ICN, UK, Irish and Polish codes of ethics for nurses. *Nursing Ethics* 2007 March; 14(2): 171-180. Subject: 4.1.3

Dobson, Roger. WHO reports on the growing commercial trade in transplant organs [news]. *BMJ: British Medical Journal* 2007 November 17; 335(7628): 1013. Subject: 19.5

Dodds, Susan. Depending on care: recognition of vulnerability and the social contribution of care provision. *Bioethics* 2007 November; 21(9): 500-510. Subject: 9.5.1

Doerflinger, Richard M. Washington insider: 2006 in Congress; Senate Hearing on Misrepresentations in stem cell research; Opening battle of 2007; Genetic Nondiscrimination Bill may see action. *National Catholic Bioethics Quarterly* 2007 Spring; 7(1): 15-21. Subject: 18.5.4

Doerflinger, Richard M. Washington insider: House passes amended Genetic Nondiscrimination Bill; continued impasse on stem cell legislation, new executive order; defeat of deceptive human cloning bill; Supreme Court decision on partial-birth abortion. *National Catholic Bio-*

ethics Quarterly 2007 Autumn; 7(3): 455-463. Subject: 15.2

Doig, Christopher James; Young, Kimberly; Teitelbaum, Jeannie; Shemie, Sam D. Brief survey: determining brain death in Canadian intensive care units = Enquête ponctuelle: la détermination de la mort encéphalique dans les units de soins intensifs au Canada. *Canadian Journal of Anaesthesia* 2006 June; 53(6): 609-612. Subject: 20.2.1

Doig, Christopher; Murray, Holt; Bellomo, Rinaldo; Kuiper, Michael; Costa, Rubens; Azoulay, Elie; Crippen, David. Ethics roundtable debate: patients and surrogates want 'everything done' — what does 'everything' mean? *Critical Care* 2006; 10(5): 231. Subject: 20.5.1

Dolan, Deborah V. Psychiatry, psychology, and human sterilization then and now: "therapeutic" or in the social interest? *Ethical Human Psychology and Psychiatry* 2007; 9(2): 99-108. Subject: 15.5

Dolgin, Janet L. New terms for an old debate: embryos, dying, and the "culture wars". *Houston Journal of Health Law and Policy* 2006; 6(2): 249-273. Subject: 4.4

Dombrowski, Dan. Personhood and life issues: a Catholic view. *Conscience* 2007 Autumn; 28(3): 30-33. Subject: 4.4

Donatelli, Luke A.; Geocadin, Romergryko G.; Williams, Michael A. Ethical issues in critical care and cardiac arrest: clinical research, brain death, and organ donation. *Seminars in Neurology* 2006 September; 26(4): 452-459. Subject: 19.5

Donohue, Julie M.; Cevasco, Marisa; Rosenthal, Meredith B. A decade of direct-to-consumer advertising of prescription drugs. *New England Journal of Medicine* 2007 August 16; 357(7): 673-681. Subject: 9.7

Dorbeck-Jung, Bärbel R. What can prudent public regulators learn from the United Kingdom government's nanotechnological regulatory activities? *NanoEthics* 2007 December; 1(3): 257-270. Subject: 5.4

Dorff, Elliot N. Judaism and ethical issues in end of life care. *In:* Eisen, Arri; Laderman, Gary, eds. Science, Religion, and Society: An Encyclopedia of History, Culture, and Controversy. Vol. 2. Armonk, NY: M.E. Sharpe, 2007: 712-719. Subject: 20.5.1

Döring, Ole. Culture and bioethics in the debate on the ethics of human cloning in China. *In:* Roetz, Heiner, ed. Cross-Cultural Issues in Bioethics: The Example of Human Cloning. New York: Rodopi, 2006: 77-105. Subject: 14.5

Dörner, Klaus. Bio-Medizin als Verarmung einer zukünftigen Medizin. *In:* Neuer-Miebach, Therese; Wunder, Michael, eds. Bio-Ethik und die Zukunft der Medizin. Bonn: Psychiatrie-Verlag, 1998: 16-19. Subject: 5.1

See SUBJECT HEADING KEY FOR SECTION II on inside back cover.

Dörries, Andrea; Hespe-Jungesblut, Katharina. Die Implementierung klinischer Ethikberatung in Deutschland — Ergebnisse einer bundesweiten Umfrage bei Krankenhäusern [The implementation of clinical ethics consultation in Germany — results of a nationwide survey in hospitals]. *Ethik in der Medizin* 2007 June; 19(2): 148-156. Subject: 9.6

Dossey, Larry. Where were the doctors? Torture and the betrayal of medicine. *Explore (NY)* 2006 November-December; 2(6): 473-481. Subject: 21.4

Dostal, O. Patient rights protection in the Czech Republic: challenges of a transition from communism to a modern legal system. *Medicine and Law: The World Association for Medical Law* 2007 March; 26(1): 75-84. Subject: 9.2

Doubleday, Robert. The laboratory revisited: academic science and the responsible development of nano- technology. *NanoEthics* 2007 August; 1(2): 167-176. Subject: 5.4

Doucet, Hubert. Anthropological challenges raised by neuroscience: some ethical reflections. *CQ: Cambridge Quarterly of Healthcare Ethics* 2007 Spring; 16(2): 219-226. Subject: 17.1

Douglas, Gillian. Who has the right to determine the fate of their embryos? *In:* Gunning, Jennifer; Holm, Søren, eds. Ethics, law, and society. Volume 1. Aldershot, Hants, England; Burlington, VT: Ashgate, 2005: 265-268. Subject: 14.6

Douglas, Joshua A. When is a "minor" also an "adult"?: an adolescent's liberty interest in accessing contraceptives from public school distribution programs. *Willamette Law Review* 2007 Summer; 43(4): 545-576. Subject: 11.2

Douglas, Kate. Just like us. Humans have rights, other animals don't — no matter how human-like they are. *New Scientist* 2007 June 2-8; 194(2606): 46-49. Subject: 22.1

Douglas, Thomas M. Ethics committees and the legality of research. *Journal of Medical Ethics* 2007 December; 33(12): 732-736. Subject: 18.2

Doukas, David J. The medical-social education compact and the medical learner. *In:* Kenny, Nuala; Shelton, Wayne, eds. Lost Virtue: Professional Character Development in Medical Education. Amsterdam; Oxford: Elsevier, 2006: 185-209. Subject: 7.2

Douw, Karla; Vondeling, Hindrik; Oortwijn, Wija. Priority setting for horizon scanning of new health technologies in Denmark: views of health care stakeholders and health economists. *Health Policy* 2006 May; 76(3): 334-345. Subject: 9.4

Downey, Robin; Geransar, Rose; Einsiedel, Edna. Angles of vision: stakeholders and human embryonic stem cell policy development. *In:* Einsiedel, Edna; Timmermans, Frank, eds. Crossing Over: Genomics in the Public Arena. Calgary, Alberta, Canada: University of Calgary Press, 2005: 61-84. Subject: 18.5.4

Downie, Jocelyn. Grasping the nettle: confronting the issue of competing interests and obligations in health research policy. *In:* Flood, Colleen M., ed. Just Medicare: What's In, What's Out, How We Decide. Buffalo, NY: University of Toronto Press, 2006: 427-448. Subject: 1.3.9

Downie, Jocelyn; Marshall, Jennifer. Pediatric neuroimaging ethics. *CQ: Cambridge Quarterly of Healthcare Ethics* 2007 Spring; 16(2): 147-160. Subject: 17.1

Downie, Jocelyn; Murphy, Ronalda. Inadmissible, eh? *American Journal of Bioethics* 2007 September; 7(9): 67-69. Subject: 17.1

Downie, Jocelyn; Schmidt, Matthais; Kenny, Nuala; D'Arcy, Ryan; Hadskis, Michael; Marshall, Jennifer. Paediatric MRI research ethics: the priority issues. *Journal of Bioethical Inquiry* 2007; 4(2): 85-91. Subject: 18.5.2

Doyal, Len. Is human existence worth its consequent harm? [review of Better Never to Have Been: The Harm of Coming Into Existence, by David Benatar]. *Journal of Medical Ethics* 2007 October; 33(10): 573-576. Subject: 14.1

Doyal, Len. The futility of opposing the legalisation of non-voluntary and voluntary euthanasia. *In:* McLean, Sheila A.M., ed. First Do No Harm: Law, Ethics, and Healthcare. Aldershot, England; Burlington, VT: Ashgate, 2006: 461-477. Subject: 20.5.1

Doyle, Jacqueline. Surgical solution becoming acceptable, as for birth [letter]. *BMJ: British Medical Journal* 2007 June 9; 334(7605): 1179-1180. Subject: 4.5

Draper, Heather. Paternity fraud and compensation for misattributed paternity. *Journal of Medical Ethics* 2007 August; 33(8): 475-480. Subject: 8.5

Draper, Heather; MacDiarmaid-Gordon, Adam; Strumidlo, Laura; Teuten, Bea; Updale, Eleanor. Virtual clinical ethics committee, case 8/case 4 vol 2: should non-medical circumstances determine whether a child is placed on the transplant register when there is a risk of wasting a scarce organ? *Clinical Ethics* 2007 December; 2(4): 166-172. Subject: 19.6

Draper, Heather; MacDiarmaid-Gordon, Adam; Strumidlo, Laura; Teuten, Bea; Update, Eleanor. Virtual Clinical Ethics Committee, case 5: can we give a son access to his mother's psychiatric notes? *Clinical Ethics* 2007 March; 2(1): 8-14. Subject: 8.4

Drazen, Jeffrey M. Government in medicine [editorial]. *New England Journal of Medicine* 2007 May 24; 356(21): 2195. Subject: 12.4.1

Drazen, Jeffrey M.; Morrissey, Stephen; Curfman, Gregory D. Open clinical trials. *New England Journal of Medicine* 2007 October 25; 357(17): 1756-1757. Subject: 18.6

Dreier, Horst. Does cloning violate the basic law's guarantee of human dignity? *In:* Vöneky, Silja; Wolfrum,

Rüdiger, eds. Human Dignity and Human Cloning. Leiden: Nijhoff, 2004: 77-85. Subject: 14.5

Dreßing, Harald. Compulsory admission and compulsory treatment in psychiatry. *In:* Schramme, Thomas; Thome, Johannes, eds. Philosophy and Psychiatry. Berlin; New York: De Gruyter, 2004: 353-356. Subject: 17.7

Dresser, Rebecca. Protecting women from their abortion choices. *Hastings Center Report* 2007 November-December; 37(6): 13-14. Subject: 12.4.2

Dresser, Rebecca. The curious case of off-label use. *Hastings Center Report* 2007 May-June; 37(3): 9-11. Subject: 9.7

Dressler, Lynn G. Biospecimen "ownership": counterpoint. *Cancer Epidemiology, Biomarkers and Prevention* 2007 February; 16(2): 190-191. Subject: 18.3

Drews, Jürgen. Drug research: between ethical demands and economic constraints. *In:* Santoro, Michael A.; Gorrie, Thomas M., eds. Ethics and the Pharmaceutical Industry. Cambridge; New York: Cambridge University Press, 2005: 21-36. Subject: 9.7

Drought, Theresa. The application of principle-based ethics to nursing practice and management: implications for the education of nurses. *In:* Davis, Anne J.; Tschudin, Verena; de Raeve, Louise, eds. Essentials of Teaching and Learning in Nursing Ethics: Perspectives and Methods. New York: Churchill Livingstone Elsevier, 2006: 81-96. Subject: 4.1.3

Dubler, Nancy Neveloff. Commentary on "Beyond Schiavo": beyond theory. *Journal of Clinical Ethics* 2007 Winter; 18(4): 346-349. Subject: 20.5.1

Dubler, Nancy Neveloff. The legal aspects of end-of-life decision making. *In:* Pruchno, Rachel A.; Smyer, Michael A., eds. Challenges of an Aging Society: Ethical Dilemmas, Political Issues. Baltimore: Johns Hopkins University Press, 2007: 19-33. Subject: 20.5.1

Dubler, Nancy Neveloff; Blustein, Jeffrey. Credentialing ethics consultants: an invitation to collaboration. *American Journal of Bioethics* 2007 February; 7(2): 35-37. Subject: 9.6

Dubler, Nancy Neveloff; Kalkut, Gary E. Caring for VIPs in the hospital: the ethical thicket. *Israel Medical Association Journal* 2006 November; 8(11): 746-750. Subject: 9.5.1

DuBois, James. Avoiding common pitfalls in the determination of death. *National Catholic Bioethics Quarterly* 2007 Autumn; 7(3): 545-559. Subject: 20.2.1

DuBois, James M. Donation after cardiac death: a reply to Bayley and Gallagher. *Health Care Ethics USA* 2007 Summer; 15(3): 7-8. Subject: 19.5

DuBois, James M.; Anderson, Emily E. Attitudes toward death criteria and organ donation among healthcare personnel and the general public. *Progress in Transplantation* 2006 March; 16(1): 65-73. Subject: 19.5

DuBois, James M.; DeVita, Michael. Donation after cardiac death in the United States: how to move forward. *Critical Care Medicine* 2006 December; 34(12): 3045-3047. Subject: 19.5

Ducournau, Pascal. The viewpoint of DNA donors on the consent procedure. *New Genetics and Society* 2007 April; 26(1): 105-116. Subject: 15.1

Dudzinski, Denise M. Education to dispel the myth. *American Journal of Bioethics* 2007 February; 7(2): 39-40. Subject: 9.6

Duggan, Patrick S.; Geller, Gail; Cooper, Lisa A.; Beach, Mary Catherine. The moral nature of patient-centeredness: is it "just the right thing to do"? *Patient Counselling and Health Education* 2006 August; 62(2): 271-276. Subject: 8.1

Duhigg, Charles. At many homes, more profit and less nursing; insulated from lawsuits, private investors cut costs and staff. *New York Times* 2007 September 23; p. A1, A34, A35. Subject: 9.5.2

Dula, Annette. Whitewashing black health: lies, deceptions, assumptions, and assertions — and the disparities continue. *In:* Prograis, Lawrence J.; Pellegrino, Edmund D., eds. African American Bioethics: Culture, Race, and Identity. Washington, DC: Georgetown University Press, 2004: 47-65. Subject: 9.5.4

Dumit, Joseph; Greenslit, Nathan. Informed health and ethical identity management. *Culture, Medicine and Psychiatry* 2006 June; 30(2): 127-134. Subject: 9.7

Dunbar, Terry; Scrimgeour, Margaret. Ethics in indigenous research — connecting with community. *Journal of Bioethical Inquiry* 2006; 3(3): 179-185. Subject: 18.5.9

Duncan, R.E.; Delatycki, M.B. Predictive genetic testing in young people for adult-onset conditions: where is the empirical evidence? *Clinical Genetics* 2006 January; 69(1): 8-16; discussion 17-20. Subject: 15.3

Dunlop, John. A good death [commentary]. *Ethics and Medicine: An International Journal of Bioethics* 2007 Summer; 23(2): 69-75. Subject: 20.5.1

Dunlop, John. Permissibility to stop man's ventilator on his request. *Ethics and Medicine: An International Journal of Bioethics* 2007 Spring; 23(1): 15-17. Subject: 20.5.1

Dunston, Georgia M.; Royal, Charmaine D.M. The human genome: implications for the health of African Americans. *In:* Livingston, Ivor Lensworth, ed. Praeger Handbook of Black American Health: Policies and Issues Behind Disparities in Health, Vol. II. 2nd edition. Westport, CT: Praeger Publishers, 2004: 757-775. Subject: 15.11

Dupuy, Jean-Pierre. Some pitfalls in the philosophical foundations of nanoethics. *Journal of Medicine and Philosophy* 2007 May-June; 32(3): 237-261. Subject: 5.1

Durante, Chris. Persons, identities, and medical ethics [review of Human Identity and Bioethics, by David

See SUBJECT HEADING KEY FOR SECTION II on inside back cover.

481

DeGrazia]. *Hastings Center Report* 2007 March-April; 37(2): 47. Subject: 4.4

Duster, Troy. Differential trust in DNA forensics: grounded assessment or inexplicable paranoia? *GeneWatch* 2007 January-February; 20(1): 3-10. Subject: 15.1

Duster, Troy. Medicalisation of race. *Lancet* 2007 February 24-March 2; 369(9562): 702-704. Subject: 9.5.4

Dusyk, Nichole. The political and moral economies of science: a case study of genomics in Canada and the United Kingdom. *Health Law Review* 2007; 15(3): 3-5. Subject: 5.1

Dute, Jos. The leading principles of the Convention on Human Rights and Biomedicine. *In:* Gevers, J.K.M.; Hondius, E.H.; Hubben, J.H., eds. Health Law, Human Rights and the Biomedicine Convention: Essays in Honour of Henriette Roscam Abbing. Leiden; Boston: Martinus Nijhoff Publishers, 2005: 3-12. Subject: 9.2

Dute, Joseph. Case of Ciorap v. Moldova, 19 June 2007, no. 12066/02 (Fourth Section) ECHR 2007/11. *European Journal of Health Law* 2007 December; 14(4): 381-384. Subject: 21.5

Duttge, Gunnar. Zukunftsperspektiven der Medizinethik — aus Sicht des Rechts [Future prospects of medical ethics — from the legal point of view]. *Ethik in der Medizin* 2006 December; 18(4): 331-336. Subject: 2.1

Duval, Gordon. The benefits and threats of research partnerships with industry. *Critical Care (London, England)* 2005 August; 9(4): 309-310. Subject: 18.2

Duvall, David G. Conflict of interest or ideological divide: the need for ongoing collaboration between physicians and industry. *Current Medical Research and Opinion* 2006 September; 22(9): 1807-1812. Subject: 7.3

Düwell, Marcus. Research as a challenge for ethical reflection. *In:* Korthals, Michiel; Bogers, Robert J., eds. Ethics for Life Scientists. Dordrecht, The Netherlands: Springer, 2004: 147-155. Subject: 1.3.9

Dworkin, Gerald. Physician-assisted death: the state of the debate. *In:* Steinbock, Bonnie, ed. The Oxford Handbook of Bioethics. Oxford; New York: Oxford University Press, 2007: 375-392. Subject: 20.7

Dwyer, James. What's wrong with the global migration of health care professionals? Individual rights and international justice. *Hastings Center Report* 2007 September-October; 37(5): 36-43. Subject: 21.6

Dy, Sydney; Lynn, Joanne. Getting services right for those sick enough to die. *BMJ: British Medical Journal* 2007 March 10; 334(7592): 511-513. Subject: 20.4.1

Dyer, Clare. BMA gives advice on withdrawing treatment [news]. *BMJ: British Medical Journal* 2007 April 7; 334(7596): 711. Subject: 20.5.4

Dyer, Clare. Charity challenges decision to refuse drug to 84 year old [news]. *BMJ: British Medical Journal* 2007 July 14; 335(7610): 64-65. Subject: 9.4

Dyer, Clare. Community treatment orders stay in mental health bill. *BMJ: British Medical Journal* 2007 June 23; 334(7607): 1293. Subject: 17.7

Dyer, Clare. Dignitas is forced to offer its services from a former factory [news]. *BMJ: British Medical Journal* 2007 December 8; 335(7631): 1176. Subject: 20.7

Dyer, Clare. Doctors lose power to run their profession [news]. *BMJ: British Medical Journal* 2007 March 3; 334(7591): 441. Subject: 7.4

Dyer, Clare. Dying woman seeks backing to hasten death [news]. *BMJ: British Medical Journal* 2007 February 17; 334(7589): 329. Subject: 20.5.1

Dyer, Clare. Experts clash over reducing abortion limit [news]. *BMJ: British Medical Journal* 2007 October 20; 335(7624): 789. Subject: 12.1

Dyer, Clare. Fertility group calls for investigation into UK regulator after search warrants declared illegal. *BMJ: British Medical Journal* 2007 July 7; 335(7609): 10. Subject: 14.1

Dyer, Clare. Girl carrying anencephalic fetus is granted right to travel. *BMJ: British Medical Journal* 2007 May 19; 334(7602): 1026. Subject: 12.4.2

Dyer, Clare. GMC guidance on conscience goes too far, says BMA [news]. *BMJ: British Medical Journal* 2007 October 6; 335(7622): 688. Subject: 2.1

Dyer, Clare. GMC to introduce "plea bargaining" for less serious misconduct cases. *BMJ: British Medical Journal* 2007 April 14; 334(7597): 763. Subject: 7.4

Dyer, Clare. Government suffers defeat in the House of Lords over Mental Health Bill for England and Wales. *BMJ: British Medical Journal* 2007 February 24; 334(7590): 384-385. Subject: 17.7

Dyer, Clare. Government suffers its first defeat over Mental Health Bill [news]. *BMJ: British Medical Journal* 2007 January 20; 334(7585): 113. Subject: 17.7

Dyer, Clare. Husband says judge's ruling on wife's treatment was "inhumane" [news]. *BMJ: British Medical Journal* 2007 January 27; 334(7586): 176. Subject: 18.5.7

Dyer, Clare. Mental health act becomes law after concessions are made [news]. *BMJ: British Medical Journal* 2007 July 14; 335(7610): 65. Subject: 17.1

Dyer, Clare. Patients win right to have their advance decisions honoured by medical staff [news]. *BMJ: British Medical Journal* 2007 October 6; 335(7622): 688-689. Subject: 20.5.4

Dyer, Clare. UK considers moving to new system to increase organ donation [news]. *BMJ: British Medical Journal* 2007 September 29; 335(7621): 634-635. Subject: 19.5

Dyer, Clare. US parents take government to court over MMR vaccine claims [news]. *BMJ: British Medical Journal* 2007 June 16; 334(7606): 1241. Subject: 9.7

Dyer, Clare. Woman loses final round of battle to use her frozen embryos at European court [news]. *BMJ: British Medical Journal* 2007 April 21; 334(7598): 818. Subject: 14.6

Dyer, Owen. Andrew Wakefield is accused of paying children for blood [news]. *BMJ: British Medical Journal* 2007 July 21; 335(7611): 118-119. Subject: 18.5.2

Dyer, Owen. GMC clears GP accused of giving "junk science" evidence [news]. *BMJ: British Medical Journal* 2007 September 1; 335(7617): 416-417. Subject: 7.4

Dyer, Owen. GMC hearing against Andrew Wakefield opens [news]. *BMJ: British Medical Journal* 2007 July 14; 335(7610): 62-63. Subject: 1.3.9

Dyer, Owen. Inquiry will study removal of Sellafield workers' body parts. *BMJ: British Medical Journal* 2007 April 28; 334(7599): 868. Subject: 19.5

Dyer, Owen. Researcher accused of breaching research ethics faces GMC [news]. *BMJ: British Medical Journal* 2007 June 9; 334(7605): 1185. Subject: 1.3.9

Dyer, Owen. US medical authorities are accused of failing to act over doctors in Guantanamo [news]. *BMJ: British Medical Journal* 2007 September 15; 335(7619): 530. Subject: 9.8

Dykstra, Alyssa. Should incentives be used to increase organ donation? *Plastic Surgical Nursing* 2004 April-June; 24(2): 70-74. Subject: 19.5

Dyrbye, Liselotte N.; Thomas, Matthew R.; Mechaber, Alex J.; Eacker, Anne; Harper, William; Massie, F. Stanford; Power, David V.; Shanafelt, Tait D. Medical education research and IRB review: an analysis and comparison of the IRB review process at six institutions. *Academic Medicine* 2007 July; 82(7): 654-660. Subject: 18.2

Dzur, Albert W.; Levin, Daniel. The primacy of the public: in support of bioethics commissions as deliberative forums. *Kennedy Institute of Ethics Journal* 2007 June; 17(2): 133-142. Subject: 2.4

E

Easton, Raul B.; Graber, Mark A.; Monnahan, Jay; Hughes, Jason. Defining the scope of implied consent in the emergency department. *American Journal of Bioethics* 2007 December; 7(12): 35-38. Subject: 8.3.1

Eaton, Lynn. Controversial embryo bill receives second reading in Lords. *BMJ: British Medical Journal* 2007 November 24; 335(7629): 1069. Subject: 14.4

Eaton, Lynn. Fertilisation authority raids controversial fertility clinics [news]. *BMJ: British Medical Journal* 2007 January 20; 334(7585): 115. Subject: 14.4

Eaton, Lynne. Medical school accepts tobacco company funding for research [news]. *BMJ: British Medical Journal* 2007 March 10; 334(7592): 496. Subject: 5.3

Eaton, Margaret L.; Illes, Judy. Commercializing cognitive neurotechnology — the ethical terrain. *Nature Biotechnology* 2007 April; 25(4): 393-397. Subject: 17.1

Ebbesen, Mette; Pedersen, Birthe D. Empirical investigation of the ethical reasoning of physicians and molecular biologists — the importance of the four principles of biomedical ethics. *Philosophy, Ethics, and Humanities in Medicine [electronic]* 2007 October 25; 2(23): 16 p. Accessed: http://www.peh-med.com/ [2008 January 24]. Subject: 7.1

Ebbesen, Mette; Pedersen, Birthe D. Using empirical research to formulate normative ethical principles in biomedicine. *Medicine, Health Care and Philosophy* 2007 March; 10(1): 33-48. Subject: 2.1

Eberl, Jason T. A Thomistic perspective on the beginning of personhood: redux. *Bioethics* 2007 June; 21(5): 283-289. Subject: 4.4

Eberl, Jason T. Creating non-human persons: might it be worth the risk? *American Journal of Bioethics* 2007 May; 7(5): 52-54. Subject: 15.1

Eberl, Jason T. Dualist and animalist perspectives on death: a comparison with Aquinas. *National Catholic Bioethics Quarterly* 2007 Autumn; 7(3): 477-489. Subject: 20.2.1

Eberl, Jason T. Issues and the end of human life: PVS patients, euthanasia, and organ donation. *In his:* Thomistic Principles and Bioethics. London; New York: Routledge, 2006: 95-127. Subject: 20.5.1

Eberl, Jason T. Issues at the beginning of human life: abortion, embryonic stem cell research, and cloning. *In his:* Thomistic Principles and Bioethics. London; New York: Routledge, 2006: 62-94. Subject: 12.3

Eberl, Jason T. The beginning of a human person's life. *In his:* Thomistic Principles and Bioethics. London; New York: Routledge, 2006: 23-42. Subject: 4.4

Eberl, Jason T. The end of a human person's life. *In his:* Thomistic Principles and Bioethics. London; New York: Routledge, 2006: 43-61. Subject: 20.2.1

Echevarria, Laura. Personally opposed, but. *Human Life Review* 2007 Spring; 33(2): 30-38. Subject: 12.5.1

Ecker, Jeffrey L.; O'Rourke, Patricia Pearl; Lott, Jason P.; Savulescu, Julian. An immodest proposal: banking embryonic stem cells for solid organ transplantation is problematic and premature. *American Journal of Bioethics* 2007 August; 7(8): 48-50; author reply W4-W6. Subject: 18.5.4

Edmondson, Ricca; Pearce, Jane. The practice of health care: wisdom as a model. *Medicine, Health Care and Philosophy* 2007 September; 10(3): 233-244. Subject: 9.1

See SUBJECT HEADING KEY FOR SECTION II on inside back cover.

Edwards, Steven. A principle-based approach to nursing ethics. *In:* Davis, Anne J.; Tschudin, Verena; de Raeve, Louise, eds. Essentials of Teaching and Learning in Nursing Ethics: Perspectives and Methods. New York: Churchill Livingstone Elsevier, 2006: 55-66. Subject: 4.1.3

Edwards, Steven A. Fear of nano: dangers and ethical challenges. *In his:* The Nanotech Pioneers: Where Are They Taking Us? Weinheim: Wiley-VCH, 2006: 197-229. Subject: 5.4

Edwards, Steven D. Disablement and personal identity. *Medicine, Health Care and Philosophy* 2007 June; 10(2): 209-215. Subject: 9.5.1

Ee, Pei-Lee; Kempen, Paul M. Elective surgery days after myocardial infarction: clinical and ethical considerations. *Journal of Clinical Anesthesia* 2006 August; 18(5): 363-366. Subject: 9.5.1

Effa, Pierre; Massougbodji, Achille; Ntoumi, Francine; Hirsch, François; Debois, Henri; Vicari, Marissa; Derme, Assetou; Ndemanga-Kamoune, Jacques; Nguembo, Joseph; Impouma, Benido; Akué, Jean-Paul; Ehouman, Armand; Dieye, Alioune; Kilama, Wen. Ethics committees in western and central Africa: concrete foundations. *Developing World Bioethics* 2007 December; 7(3): 136-142. Subject: 18.2

Egan, Danielle. Cheating death: we're going to live forever. *New Scientist* 2007 October 13-19; 195(2625): 46. Subject: 20.5.1

Egan, Erin A. Neuroimaging as evidence. *American Journal of Bioethics* 2007 September; 7(9): 62-63. Subject: 17.1

Egan, Erin A. Who should regulate the practice of medicine? *Journal of Opioid Management* 2005 March-April; 1(1): 11-12. Subject: 9.7

Egonsson, Dan. Hypothetical approval in prudence and medicine. *Medicine, Health Care and Philosophy* 2007 September; 10(3): 245-252. Subject: 8.3.1

Egorova, Yulia. 'Up in the sky': human and social sciences' responses to genetics. *In:* Gunning, Jennifer; Holm, Søren, eds. Ethics, law, and society. Volume 2. Aldershot, Hants, England; Burlington, VT: Ashgate, 2006: 45-53. Subject: 15.1

Egorova, Yulia. The meanings of genetics: science and the concepts of personhood. *Health Care Analysis: An International Journal of Health Philosophy and Policy* 2007 March; 15(1): 1-3. Subject: 15.1

Egorova, Yulia. The meanings of science: conversations with geneticists. *Health Care Analysis: An International Journal of Health Philosophy and Policy* 2007 March; 15(1): 51-58. Subject: 15.1

Ehlrich, J. Shoshanna. Shifting boundaries: abortion, criminal culpability and the indeterminate legal status of adolescents. *Wisconsin Women's Law Journal* 2003 Spring; 18(1): 77-116. Subject: 12.4.2

Ehrenfeld, David. Unethical contexts for ethical questions. *In:* Galston, Arthur W.; Peppard, Christiana Z., eds. Expanding Horizons in Bioethics. Dordrecht; Norwell, MA: Springer, 2005: 19-34. Subject: 15.1

Ehrenstein, Boris P.; Hanses, Frank; Salzberger, Bernd. Influenza pandemic and professional duty: family or patients first? A survey of hospital employees. *BMC Public Health* 2006 December 28; 6: 311. Subject: 8.1

Ehrich, Kathryn; Farsides, Bobbie; Williams, Clare; Scott, Rosamund. Testing the embryo, testing the fetus. *Clinical Ethics* 2007 December; 2(4): 181-186. Subject: 15.2

Ehrlich, Joseph B. Schiavo: cold justice. Did the courts pursue, in view of the high court's decision in Troxell v. Granville, the correct conclusion to the Schiavo case? *The Journal of Law in Society Wayne State University Law School* 2005 Fall; 7(1): 1-15. Subject: 20.5.4

Eich, Thomas. The debate about human cloning among Muslim religious scholars since 1997. *In:* Roetz, Heiner, ed. Cross-Cultural Issues in Bioethics: The Example of Human Cloning. New York: Rodopi, 2006: 291-309. Subject: 14.5

Eidelman, Steve; Drake, Steve. Not yet dead. *Ethics and Intellectual Disability* 2005 Summer; 9(1): 1-3. Subject: 20.5.1

Einollahi, Behzad; Nourbala, Mohammad-Hossein; Bahaeloo-Horeh, Saeid; Assari, Shervin; Lessan-Pezeshki, Mahboob; Simforoosh, Naser. Deceased-donor kidney transplantation in Iran: trends, barriers and opportunities. *Indian Journal of Medical Ethics* 2007 April- June; 4(2): 70-72. Subject: 19.3

Einsiedel, Edna. Telling technological tales: the media and the evolution of biotechnology. *In:* Einsiedel, Edna; Timmermans, Frank, eds. Crossing Over: Genomics in the Public Arena. Calgary, Alberta, Canada: University of Calgary Press, 2005: 143-154. Subject: 15.1

Einsiedel, Edna; Sheremeta, Lorraine. Biobanks and the challenges of commercialization. *In:* Sensen, Chrisopher W., ed. Handbook of Genome Research: Genomics, Proteomics, Metabolomics, Bioinformatics, Ethical and Legal Issues. Volume 2. Weinheim: Wiley-VCH, 2005: 537-559. Subject: 15.1

Einstein, Stanley. Drug users can't be treated, people can be! The creation and maintenance of ethical travesties, or at least dilemmas. *In:* Kleinig, John; Einstein, Stanley, eds. Ethical Challenges for Intervening in Drug Use: Policy, Research and Treatment Issues. Huntsville, TX: Office of International Criminal Justice; Criminal Justice Center, Sam Houston State University, 2006: 565-623. Subject: 9.5.9

Eischen, Kyle. Commercializing Iceland: biotechnology, culture, and the information society. *In:* Mehta, Michael D., ed. Biotechnology Unglued: Science, Society and So-

---

Subject = NRCBL Primary Classification Number; see inside front cover.

cial Cohesion. Vancouver: UBC Press, 2005: 95-116. Subject: 15.1

Eisen, Arri. The challenge of spirituality in the clinic: symptom of a larger syndrome. *American Journal of Bioethics* 2007 July; 7(7): 12-13. Subject: 8.1

Elamon, J. A situational analysis of HIV/AIDS-related discrimination in Kerala, India. *AIDS Care* 2005 July; 17(Supplement 2): S141-S151. Subject: 9.5.6

Elcin, Melih; Odabasi, Orhan; Gokler, Bahar; Sayek, Isender; Akova, Murat; Kiper, Nural. Developing and evaluating professionalism. *Medical Teacher* 2006 February; 28(1): 36-39. Subject: 7.2

Elder, John. Illegal book heads through the Internet gateway. *Age (Australia)* 2007 April 1; 2 p. [Online]. Accessed: http://www.theage.com.au/news/national/illegal-book-heads-through-internet-gate way/2007/03/31/1174761817937.html [2008 February 8]. Subject: 20.5.1

Eliot, Jaklin; Olver, Ian. Autonomy and the family as (in)appropriate surrogates for DNR decisions: a qualitative analysis of dying cancer patients' talk. *Journal of Clinical Ethics* 2007 Fall; 18(3): 206-218. Subject: 20.5.1

Eliott, Jaklin; Olver, Ian. Response from Eliott and Olver. *Journal of Clinical Ethics* 2007 Fall; 18(3): 233-234. Subject: 20.5.1

Ellilä, Heikki; Välimäki, Maritta; Warne, Tony; Sourander, Andre. Ideology of nursing care in child psychiatric inpatient treatment. *Nursing Ethics* 2007 September; 14(5): 583-596. Subject: 17.1

Elliott, Carl. Against happiness [review of Against Depression, by Peter D. Kramer]. *Medicine, Health Care and Philosophy* 2007 June; 10(2): 167-171. Subject: 17.4

Elliott, Carl. Disillusioned doctors. *In:* Kenny, Nuala; Shelton, Wayne, eds. Lost Virtue: Professional Character Development in Medical Education. Amsterdam; Oxford: Elsevier, 2006: 87-97. Subject: 4.1.2

Elliott, Carl. Ethnicity, citizenship, family: identity after the HGP. Minneapolis: University of Minnesota, Center for Bioethics, n.d.; 14 p. [Online]. Accessed: http://www.bioethics.umn.edu/genetics_and_identity/docs/gen-grant.pdf [2007 April 24]. Subject: 15.11

Elliott, Carl. The mixed promise of genetic medicine. *New England Journal of Medicine* 2007 May 17; 356(20): 2024-2025. Subject: 15.1

Elliott, Carl. The tyranny of expertise. *In:* Eckenwiler, Lisa A.; Cohn, Felicia G., eds. The Ethics of Bioethics: Mapping the Moral Landscape. Baltimore, MD: Johns Hopkins University Press, 2007: 43-46. Subject: 2.1

Elsayed, Dya Eldin M.; Kass, Nancy E. Assessment of the ethical review process in Sudan. *Developing World Bioethics* 2007 December; 7(3): 143-148. Subject: 18.2

Elson, Paul. Do older adults presenting with memory complaints wish to be told if later diagnosed with Alzheimer's disease? *International Journal of Geriatric Psychiatry* 2006 May; 21(5): 419-425. Subject: 9.5.2

Eltis, Karen. Predicating dignity on autonomy? The need for further inquiry into the ethics of tagging and tracking dementia patients with GPS technology. *Elder Law Journal* 2005; 13(2): 387-415. Subject: 9.5.2

Emanuel, Ezekiel J. Researching a bioethical question. *In:* Gallin, John I.; Ognibene, Frederick P., eds. Principles and Practice of Clinical Research. 2nd edition. Oxford: Academic, 2007: 27-38. Subject: 2.1

Emanuel, Ezekiel J. What cannot be said on television about health care [commentary]. *JAMA: The Journal of the American Medical Association* 2007 May 16; 297(19): 2131-2133. Subject: 9.1

Emanuel, Ezekiel J.; Fuchs, Victor R. Beyond healthcare band-aids [op-ed]. *Washington Post* 2007 February 7; p. A17. Subject: 9.3.1

Emanuel, Ezekiel J.; Grady, Christine. Four paradigms of clinical research and research oversight. *CQ: Cambridge Quarterly of Healthcare Ethics* 2007 Winter; 16(1): 82-96. Subject: 18.6

Emanuel, Ezekiel J.; Lemmens, Trudo; Elliot, Carl. Should society allow research ethics boards to be run as for-profit enterprises? *PLoS Medicine* 2006 July; 3(7): e309. Subject: 18.2

Emanuel, Ezekiel J.; Miller, Franklin G. Money and distorted ethical judgments about research: ethical assessment of the TeGenero TGN1412 trial. *American Journal of Bioethics* 2007 February; 7(2): 76-81. Subject: 18.1

Emerson, Claudia I.; Daar, Abdallah S.; Lawrence, Ryan E.; Curlin, Farr A. Defining conscience and acting conscientiously. *American Journal of Bioethics* 2007 December; 7(12): 19-21; author reply W1-W2. Subject: 4.1.2

Emery, Theo. U.S. judge blocks lethal injection in Tennessee. *New York Times* 2007 September 20; p. A14. Subject: 20.6

Endacott, Ruth; Wood, Anita; Judd, Fiona; Hulbert, Carol; Thomas, Ben; Grigg, Margaret. Impact and management of dual relationships in metropolitan, regional and rural mental health practice. *Australian and New Zealand Journal of Psychiatry* 2006 November-December; 40(11-12): 987-994. Subject: 7.1

Engelhardt, H. Tristram. Bioethics as politics: a critical reassessment. *In:* Eckenwiler, Lisa A.; Cohn, Felicia G., eds. The Ethics of Bioethics: Mapping the Moral Landscape. Baltimore, MD: Johns Hopkins University Press, 2007: 118-133. Subject: 2.1

Engelhardt, H. Tristram. Long-term care: the family, post-modernity, and conflicting moral life-worlds. *Journal of Medicine and Philosophy* 2007 September-October; 32(5): 519-536. Subject: 9.5.2

See SUBJECT HEADING KEY FOR SECTION II on inside back cover.

485

Engelhardt, H. Tristram. The search for global morality: bioethics, the culture wars, and moral diversity. *In:* Engelhardt, H. Tristram, ed. Global Bioethics: The Collapse of Consensus. Salem, MA: M&M Scrivener Press, 2006: 18-49. Subject: 2.1

Engelhardt, H. Tristram. Why ecumenism fails: taking theological differences seriously. *Christian Bioethics* 2007 January-April; (13)1: 25-51. Subject: 2.1

Engelman, Michal; Johnson, Summer. Population aging and international development: addressing competing claims of distributive justice. *Developing World Bioethics* 2007 April; 7(1): 8-18. Subject: 9.4

Engels, Eve-Marie. Biobanken für die medizinische Forschung: Probleme und Potenzial. *In:* Schreiber, Hans-Peter, ed. Biomedizin und Ethik: Praxis, Recht, Moral. Basel; Boston: Birkhäuser, 2004: 29-40, 95-96. Subject: 15.1

England, Ruth; England, Tim; Coggon, John. The ethical and legal implications of deactivating an implantable cardioverter-defibrillator in a patient with terminal cancer. *Journal of Medical Ethics* 2007 September; 33(9): 538-540. Subject: 20.5.1

English, Abigail. Health care for adolescents: ensuring access, protecting privacy. *Clearinghouse Review* 2005 July-August; 39: 217-218 [Online]. Accessed: http://www.povertylaw.org/clearinghouse-review/issues/2005/20050715/chr501090.pdf [2006 November 28]. Subject: 9.5.7

English, A.; Ford, C.A. More evidence supports the need to protect confidentiality in adolescent health care. *Journal of Adolescent Health* 2007 March; 40(3): 199-200. Subject: 9.5.7

English, Veronica; Gardner, Jessica; Romano-Critchley, Gillian; Sommerville, Ann. Genetics and insurance. *Journal of Medical Ethics* 2001 June; 27(3): 204. Subject: 15.3

English, Veronica; Hamm, Danielle; Harrison, Caroline; Mussell, Rebecca; Sheather, Julian; Sommerville, Ann. Ethics briefings. *Journal of Medical Ethics* 2007 July; 33(7): 433-434 [see correction Journal of Medical Ethics 2007 October; 33(10): 620]. Subject: 2.1

English, Veronica; Hamm, Danielle; Harrison, Caroline; Sheather, Julian; Sommerville, Ann. Ethics briefings. *Journal of Medical Ethics* 2007 February; 33(2): 123-124. Subject: 2.1

English, Veronica; Mussell, Rebecca; Sheather, Julian; Sommerville, Ann. Autonomy and its limits: what place for the public good? *In:* McLean, Sheila A.M., ed. First Do No Harm: Law, Ethics, and Healthcare. Aldershot, England; Burlington, VT: Ashgate, 2006: 117-130. Subject: 2.1

English, Veronica; Wright, Linda. Is presumed consent the answer to organ shortages? [debate]. *BMJ: British Medical Journal* 2007 May 26; 334(7603): 1088-1089. Subject: 19.5

Engström, Joakim; Bruno, Erik; Holm, Birgitta; Hellzén, Ove. Palliative sedation at end of life — a systematic literature review. *European Journal of Oncology Nursing* 2007 February; 11(1): 26-35. Subject: 20.4.1

Engström, Timothy H.; Richter, Gerd. Citizens and customers: establishing the ethical foundations of the German and U.S. health care systems. *In:* Engström, Timothy H.; Robison, Wade L., eds. Health Care Reform: Ethics and Politics. Rochester, NY: University of Rochester Press, 2006: 166-186. Subject: 9.1

Eonas, Anthony; McCoy, John D.; Eaton, Silviya H.M. Medical informed consent: clarity or confusion? *Journal of Hospital Marketing and Public Relations* 2006; 16(1-2): 69-88. Subject: 8.3.1

Epstein, Miran. Clinical trials in the developing world [letter]. *Lancet* 2007 June 2-8; 369(9576): 1859. Subject: 18.6

Epstein, Miran. Legal and institutional fictions in medical ethics: a common, and yet largely overlooked, phenomenon: a theoretical platform for a much-needed change in the provision of healthcare based on restoring the autonomy of doctor-patient relationships. *Journal of Medical Ethics* 2007 June; 33(6): 362-364. Subject: 8.1

Epstein, Miran. The ethics of poverty and the poverty of ethics: the case of Palestinian prisoners in Israel seeking to sell their kidneys in order to feed their children. *Journal of Medical Ethics* 2007 August; 33(8): 473-474. Subject: 19.5

Epstein, M.; Wingate, D.L. Is the NHS research ethics committees system to be outsourced to a low-cost offshore call centre? Reflections on human research ethics after the Warner Report. *Journal of Medical Ethics* 2007 January; 33(1): 45-47. Subject: 18.2

Epstein, Richard A. Conflicts of interest in health care: who guards the guardians? *Perspectives in Biology and Medicine* 2007 Winter; 50(1): 72-88. Subject: 1.3.9

Epstein, Richard A. Influence of pharmaceutical funding on the conclusions of meta-analyses [editorial]. *BMJ: British Medical Journal* 2007 December 8; 335(7631): 1167. Subject: 9.7

Epstein, Richard A. The social response to genetic conditions: beware of the antidiscrimination law. *Health Affairs* 2007 September-October; 26(5): 1249-1252. Subject: 15.3

Epstein, Ronald M. Mindful practice and the tacit ethics of the moment. *In:* Kenny, Nuala; Shelton, Wayne, eds. Lost Virtue: Professional Character Development in Medical Education. Amsterdam; Oxford: Elsevier, 2006: 115-144. Subject: 4.1.2

Erde, Edmund L. Indecency/decency in cardiac surgery: a memoir of my education at a super-esteemed medical place. *Journal of Cardiac Surgery* 2007 January-February; 22(1): 43-48. Subject: 9.5.1

Erde, Edmund L.; McCormack, Michael K.; Steer, Robert A.; Ciervo, Carman A., Jr.; McAbee, Gary N. Patient confidentiality vs disclosure of inheritable risk: a survey-based study. *Journal of the American Osteopathic Association* 2006 October; 106(10): 615-620. Subject: 15.3

Erdman, Joanna N.; Cook, Rebecca J. Protecting fairness in women's health: the case of emergency contraception. *In:* Flood, Colleen M., ed. Just Medicare: What's In, What's Out, How We Decide. Buffalo, NY: University of Toronto Press, 2006: 137-167. Subject: 11.1

Erikson, Susan L. Fetal views: histories and habits of looking at the fetus in Germany. *Journal of Medical Humanities* 2007 December; 28(4): 187-212. Subject: 9.5.8

Ernst, Erik; Ingerslev, Hans Jakob.; Schou, Ole; Stoltenberg, Meredin. Attitudes among sperm donors in 1992 and 2002: a Danish questionnaire survey. *Acta obstetricia et gynecologica Scandinavica* 2007 March; 86(3): 327-333. Subject: 14.2

Esiri, Margaret. Why do research on human brains? *In:* Gunning, Jennifer; Holm, Søren, eds. Ethics, law, and society. Volume 1. Aldershot, Hants, England; Burlington, VT: Ashgate, 2005: 33-39. Subject: 17.1

Espirit Group; Pace, Christine; Grady, Christine; Wendler, David; Bebchuk, Judith D.; Tavel, Jorge A.; McNay, Laura A.; Forster, Heidi P.; Killen, Jack; Emanuel, Ezekiel J. Post-trial access to tested interventions: the views of IRB/REC chair, investigators, and research participants in a multinational HIV/AIDS study. *AIDS Research and Human Retroviruses* 2006 September; 22(9): 837-841. Subject: 18.2

Estrin, Irene.; Sher, Leo. The constitutionality of random drug and alcohol testing of students in secondary schools. *International Journal of Adolescent Medicine and Health* 2006 January-March; 18(1): 21-25. Subject: 9.5.9

Ethics Group of the Newborn Drug Development Initiative; Baer, Gerri R.; Nelson, Robert M. Ethical challenges in neonatal research: summary report of the ethics group of the newborn drug development initiative. *Clinical Therapeutics* 2006 September; 28(9): 1399-1407. Subject: 9.5.8

Ethics Working Group of Confederation of European Specialists in Paediatrics; Kurz, R.; Gill, D.; Mjones, S. Ethical issues in the daily medical care of children. *European Journal of Pediatrics* 2006 February; 165(2): 83-86. Subject: 9.5.7

EURELD (European End-of-life Decision) Consortium; van Delden, Johannes J.M.; Löfmark, Rurik; Deliens, Luc; Bosshard, Georg; Norup, Michael; Cecioni, Riccardo; van der Heide, Agnes. Do-not-resuscitate decisions in six European countries. *Critical Care Medicine* 2006 June; 34(6): 1686-1690. Subject: 20.5.1

Euronic Study Group; Cuttini, Marina; Casotto, Veronica; Orzalesi, Marcello. Ethical issues in neonatal intensive care and physicians' practices: a European perspective. *Acta Paediatrica Supplement* 2006 July; 95(452): 42-46. Subject: 20.5.2

European Alliance of Patient and Parent Organizations for Genetics Services and Innovation in Medicine [EAGS]; Dutch Genetic Alliance [VSOP]; European Organisation for Rare Diseases [EURODIS]; European Federation of Pharmaceutical Industries and Associations [EFPIA]. Report: Workshop 'Genetic Testing: Challenges for Society'. Brussels, Belgium: European Federation of Pharmaceutical Industries and Associations (EFPIA), 2001 September 24: 13 p. [Online]. Accessed: http://www.egaweb.org/documents/24092001GENTEST [2007 May 1]. Subject: 15.3

European Commission of Human Rights. Brüggemann v. Germany [Date of Decision: 12 July 1977]. European Human Rights Reports 1977; 3: 244-258. Subject: 12.4.4

European Commission. Directorate-General Enterprise and Industry. Human tissue engineering and beyond: proposal for a community regulatory framework on advanced therapies. Brussels: European Commission, 2005 May 4; 15 p. [Online]. Accessed: http://ec.europa.eu/enterprise/pharmaceuticals/advtherapies/docs/consultatiopaper-advancedtherapies-2005-may-04.pdf [2007 April 5]. Subject: 15.4

European Commission. European Group on Ethics in Science and New Technologies. Opinion on the ethical aspects of nanomedicine. Brussels. Belgium: The European Commission, European Group on Ethics in Science and New technologies to the European Commission: 2007 January 17; 164 p. [Online]. Accessed: http://ec.europa.eu/european_group_ethics/activities/docs/opinion_21_nano_en.pd f [2007 January 30]. Subject: 5.4

European Parliament. European Parliament resolution on the trade in human egg cells. Strasbourg: European Parliament 2005 March 10, P6_TA(2005)0074; 3 p. [Online]. Accessed: http://www.europarl.europa.eu/sides/getDoc.do?pubRef=-//EP//NONSGML+TA+P6-TA-20 05-0074+0+DOC+PDF+V0//EN [2007 April 11]. Subject: 14.6

European Society of Human Genetics [ESHG]. EPO upholds limited patent on BRCA2 gene: singling out an ethnic group is a 'dangerous precedent' says European Society of Human Genetics. *Vienna, Austria: European Society of Human Genetics,* 2005 July 1: 2 p. [Online]. Accessed: http://www.eshg.org/ESHGPressRelease01July2005.pdf [2007 February 8]. Subject: 15.8

European Society of Human Genetics [ESHG]. EU Temporary Committee on Genetics Recap Overview by the European Society of Human Genetics. Vienna, Austria: European Society of Human Genetics, 2001: 2 p. [Online]. Accessed: http://www.eshg.org/Fiorireporthistory.pdf [2007 February 8]. Subject: 18.5.4

See SUBJECT HEADING KEY FOR SECTION II on inside back cover.

European Society of Human Genetics [ESHG]. Geneticists oppose singling out Jewish women in European breast cancer patent [press release]. *Vienna, Austria: European Society of Human Genetics,* 2005 June 15: 3 p. [Online]. Accessed: http://www.eshg.org/PressReleaseESHG15-06-2005.pdf [2007 February 8]. Subject: 15.8

European Society of Human Genetics [ESHG]. Letter to members of the European Parliament from the European Society of Human Genetics. Re: Fiori Report on the ethical, legal, economic and social implications of human genetics. Vienna, Austria: European Society of Human Genetics, [2001]: 3 p. [Online]. Accessed: http://www.eshg.org/ESHGlettertoMEPs.pdf [2007 May 1]. Subject: 18.5.4

European Society of Human Genetics [ESHG]. Report on the ethical, legal, economic and social implications of human genetics by the temporary committee on human genetics and other new technologies in modern medicine [of the European Parliament]: changes recommended by the European Society of Human Genetics. Vienna, Austria: European Society of Human Genetics, 2001 November 8: 6 p. [Online]. Accessed: http://www.eshg.org/ESHGrecchangesFiori.pdf [2007 February 8]. Subject: 18.5.4

European Society of Human Reproduction and Embryology [ESHRE]. Task Force on Ethics and Law; Pennings, Guido; de Wert, G.; Shenfield, F.; Cohen, J.; Tarlatzis, B.; Devroey, P. ESHRE Task Force on Ethics and Law 12: Oocyte donation for non-reproductive purposes. *Human Reproduction* 2007 May; 22(5): 1210-1213. Subject: 14.4

Evans, Caswell A. Eliminating oral health disparities: ethics workshop reactor comments. *Journal of Dental Education* 2006 November; 70(11): 1180-1183. Subject: 9.2

Evans, Emily W. Conscientious objection: a pharmacist's right or professional negligence? *American Journal of Health-System Pharmacy* 2007 January 15; 64(2): 139-141. Subject: 9.7

Evans, H.M. Do patients have duties? *Journal of Medical Ethics* 2007 December; 33(12): 689-694. Subject: 8.1

Evans, H.M.; Hungin, A.P.S. Uncomfortable implications: placebo equivalence in drug management of a functional illness. *Journal of Medical Ethics* 2007 November; 33(11): 635-638. Subject: 8.2

Evans, James P. Health care in the age of genetic medicine. *JAMA: The Journal of the American Medical Association* 2007 December 12; 298(22): 2670-2672. Subject: 15.3

Evans, Jane; Simon, Gayle. Family Educational Rights and Privacy Act (FERPA). *Journal of Empirical Research on Human Research Ethics* 2007 March; 2(1): 101-104. Subject: 18.5.2

Evers, Kathinka. Perspectives on memory manipulation: using beta-blockers to cure post-traumatic stress disorder.

CQ: Cambridge Quarterly of Healthcare Ethics 2007 Spring; 16(2): 138-146. Subject: 17.1

Ewanchuk, Mark; Brindley, Peter G. Perioperative do-not-resuscitate orders — doing 'nothing' when 'something' can be done. *Critical Care (London, England)* 2006; 10(4): 219. Subject: 20.5.1

Expert Scientific Group on Phase One Clinical Trials. Expert scientific group on phase one clinical trials: final report. London: Expert Scientific Group on Phase One Clinical Trials, 2006 November 30; 106 p. [Online]. Accessed: http://www.dh.gov.uk/prod_consum_dh/idcplg?IdcService=GET_FILE&dID=136063&Rendition=Web [2007 April 23]. Subject: 18.2

F

Faber, Berit A. Bioethics in Europe. *In:* Gunning, Jennifer; Holm, Søren, eds. Ethics, law, and society. Volume 1. Aldershot, Hants, England; Burlington, VT: Ashgate, 2005: 41-44. Subject: 2.1

Faden, Ruth R.; Duggan, Patrick S.; Karron, Ruth. Who pays to stop a pandemic? [op-ed]. *New York Times* 2007 February 9; p. A19. Subject: 22.3

Fahey, Charles J. The ethics of long-term care: recasting the policy discourse. *In:* Pruchno, Rachel A.; Smyer, Michael A., eds. Challenges of an Aging Society: Ethical Dilemmas, Political Issues. Baltimore: Johns Hopkins University Press, 2007: 52-73. Subject: 9.5.2

Fainzang, Sylvie. When doctors and patients lie to each other: lying and power within the doctor-patient relationship. *In:* van Dongen, Els; Fainzang, Sylvie, eds. Lying and Illness: Power and Performance. Amsterdam: Het Spinhuis; Piscataway, NJ: Transaction Publishers, 2005: 36-55. Subject: 8.2

Fairchild, Amy L.; Alkon, Ava. Back to the future? Diabetes, HIV, and the boundaries of public health. *Journal of Health Politics, Policy and Law* 2007 August; 32(4): 561-593. Subject: 9.1

Falcón, M.; Martinez-Cánovas, F.J.; Pérez-Carceles, M.D.; Osuna, E.; Luna, A. Ethical problems related to information and pharmaceutical care in Spain. *Medicine and Law: The World Association for Medical Law* 2007 March; 26(1): 85-93. Subject: 9.7

Falk, Raphael. Nervous diseases and eugenics of the Jews: a view from 1918. *Korot* 2003-2004; 17: 23-46, ix-x. Subject: 15.5

Falusi, Adeyinka G.; Olopade, Olufunmilayo I.; Olopade, Christopher O. Establishment of a standing ethics/institutional review board in a Nigerian university: a blueprint for developing countries. *Journal of Empirical Research on Human Research Ethics* 2007 March; 2(1): 21-30. Subject: 18.2

Subject = NRCBL Primary Classification Number; see inside front cover.

Fan, Eddy; Needham, Dale M. Deciding who to admit to a critical care unit [editorial]. *BMJ: British Medical Journal* 2007 December 1; 335(7630): 1103-1104. Subject: 9.4

Fan, Ruiping. Bioethics: globalization, communitization, or localization? *In:* Engelhardt, H. Tristram, ed. Global Bioethics: The Collapse of Consensus. Salem, MA: M&M Scrivener Press, 2006: 271-299. Subject: 2.1

Fan, Ruiping. Corrupt practices in Chinese medical care: the root in public policies and a call for Confucian-market approach. *Kennedy Institute of Ethics Journal* 2007 June; 17(2): 111-131. Subject: 7.4

Fan, Ruiping. Which care? Whose responsibility? And why family? A Confucian account of long-term care for the elderly. *Journal of Medicine and Philosophy* 2007 September-October; 32(5): 495-517. Subject: 9.5.2

Fan, Ruiping; Holliday, Ian. Which medicine? Whose standard? Critical reflections on medical integration in China. *Journal of Medical Ethics* 2007 August; 33(8): 454-461. Subject: 4.1.1

Farah, Martha J.; Heberlein, Andrea S. Personhood and neuroscience: naturalizing or nihilating? *American Journal of Bioethics* 2007 January; 7(1): 37-48. Subject: 17.1

Farah, Martha J.; Heberlein, Andrea S. Response to open peer commentaries on "Personhood and neuroscience: naturalizing or nihilating?": getting personal [letter]. *American Journal of Bioethics* 2007 January; 7(1): W1-W4. Subject: 17.1

Farah, Martha J.; Noble, Kimberly G.; Hurt, Hallam. Poverty, privilege, and brain development: empirical findings and ethical implications. *In:* Illes, Judy, ed. Neuroethics: Defining the Issues in Theory, Practice, and Policy. New York: Oxford University Press, 2006: 277-287. Subject: 17.1

Farah, Martha J.; Wolpe, Paul Root; Caplan, Arthur. Brain research and neuroethics. *In:* Gunning, Jennifer; Holm, Søren, eds. Ethics, law, and society. Volume 1. Aldershot, Hants, England; Burlington, VT: Ashgate, 2005: 261-264. Subject: 17.1

Farmer, Deborah F.; Jackson, Sharon A.; Camacho, Fabian; Hall, Mark A. Attitudes of African American and low socioeconomic status white women toward medical research. *Journal of Health Care for the Poor and Underserved* 2007 February; 18(1): 85-99. Subject: 18.5.1

Farquhar, Carey; John-Stewart, Grace C.; John, Francis N.; Kabura, Marjory N.; Kiarie, James N. Pediatric HIV type 1 vaccine trial acceptability among mothers in Kenya. *AIDS Research and Human Retroviruses* 2006 June; 22(6): 491-495. Subject: 9.5.6

Farrell, David Blake; De Neeve, Eileen. Commercialization of human genetic research. *In:* Monsour, H. Daniel, ed. Ethics and the New Genetics: An Integrated Approach.

Toronto: University of Toronto Press, 2007: 58-75. Subject: 15.1

Farrugia, Albert. When do tissues and cells become products? — Regulatory oversight of emerging biological therapies. *Cell and Tissue Banking* 2006; 7(4): 325-335. Subject: 19.5

Faulkner, Janet. Conjoined twins: the ethics of separation. *Midwives: the Official Journal of the Royal College of Midwives* 2006 March; 9(3): 86-87. Subject: 20.5.2

Faunce, Thomas Alured. Global intellectual property protection of "innovative" pharmaceuticals: challenges for bioethics and health law. *In:* Bennett, Belinda; Tomossy, George F., eds. Globalization and Health: Challenges for Health Law and Bioethics. Dordrecht: Springer, 2006: 87-107. Subject: 5.3

Faunce, Thomas Alured; Jefferys, Susannah. Whistle-blowing and scientific misconduct: renewing legal and virtue ethics foundations. *Medicine and Law: The World Association for Medical Law* 2007 September; 26(3): 567-584. Subject: 1.3.9

Faust, Halley S. Is a national standard of care always the right one? [letter]. *Developing World Bioethics* 2007 April; 7(1): 45-46. Subject: 9.8

Fausto-Sterling, Anne. Refashioning race: DNA and the politics of health care. *Differences: A Journal of Feminist Cultural Studies* 2004 Fall; 15(3): 1-37. Subject: 15.11

Feder, Barnaby J. F.D.A. weighs flexibility in trials of heart treatment. *New York Times* 2007 September 21; p. A15. Subject: 18.2

Feder, Judith; Pollitz, Karen. Reform's three essential elements: to be effective, insurance coverage must be adequate, affordable, and available. *Health Progress* 2007 May-June; 88(3): 30-31. Subject: 9.5.10

Federation of American Scientists [FAS]. Working Group on BW [Biological Warfare]; Stockholm International Peace Research Institute [SIPRI]; Verification Research, Training and Information Centre [VERTIC]; International Network of Engineers and Scientists for Global Responsibility[INES]' Acronym Institute for Disarmament Diplomacy; Sunshine Project; Pax Christi International; Physicians for Social Responsibility; 20/20 Vision. Draft recommendations for a code of conduct for biodefense programs. Federation of American Scientists 2002 November; 2 p. [Online]. Accessed: http://www.fas.org/bwc/papers/code.pdf [2007 May 11]. Subject: 21.3

Federation of American Societies for Experimental Biology [FASEB]; Brockway, Laura M.; Furcht, Leo T. Conflicts of interest in biomedical research — the FASEB guidelines. *FASEB Journal* 2006 December; 20(14): 2435-2438. Subject: 1.3.9

See SUBJECT HEADING KEY FOR SECTION II on inside back cover.

489

Feifer, Jason. Paying big to be a donor; gifting an organ can be costly. Would a tax break cross a moral line? *Washington Post* 2007 March 20; p. F1, F4. Subject: 19.5

Fein, Stephanie P.; Hilborne, Lee H.; Spiritus, Eugene M.; Seymann, Gregory B.; Keenan, Craig R.; Shojania, Kaveh G.; Kagawa-Singer, Marjorie; Wenger, Neil S. The many faces of error disclosure: a common set of elements and a definition. *JGIM: Journal of General Internal Medicine* 2007 June; 22(6): 755-761. Subject: 9.8

Feldmann, Robert E., Jr.; Mattern, Rainer. The human brain and its neural stem cells postmortem: from dead brains to live therapy. *International Journal of Legal Medicine* 2006 July; 120(4): 201-211. Subject: 18.5.1

Fendrich, Michael; Lippert, Adam M.; Johnson, Timothy P. Respondent reactions to sensitive questions. *Journal of Empirical Research on Human Research Ethics* 2007 September; 2(3): 31-37. Subject: 18.5.1

Fennell, Philip. Reducing rights in the name of convention compliance: mental health law reform and the new human rights agenda. *In:* Gunning, Jennifer; Holm, Søren, eds. Ethics, law, and society. Volume 2. Aldershot, Hants, England; Burlington, VT: Ashgate, 2006: 125-133. Subject: 17.8

Fenner, Dagmar. Ist die Institutionalisierung und Legalisierung der Suizidbeihilfe gefährlich? Eine kritische Analyze der Gegenargumente = Is the institutionalization and legalization of assistance to suicide dangerous? A critical analysis of counterarguments. *Ethik in der Medizin* 2007 September; 19(3): 200-214. Subject: 20.7

Fennig, Silvana; Secker, Aya; Treves, Ilan; Ben Yakar, Motti; Farina, Jorje; Roe, David; Levkovitz, Yechiel; Fennig, Shmuel. Ethical dilemmas in psychotherapy: comparison between patients, therapists and laypersons. *Israel Journal of Psychiatry and Related Sciences* 2005; 42(4): 251-257. Subject: 17.2

Ferber, Deborah Sarah. As sure as eggs? Responses to an ethical question posed by Abramov, Elchalal, and Schenker. *Journal of Clinical Ethics* 2007 Spring; 18(1): 35-48. Subject: 14.4

Ferber, Deborah Sarah. Some reflections on IVF, emotions, and patient autonomy. *Journal of Clinical Ethics* 2007 Spring; 18(1): 53-55. Subject: 14.4

Feresin, Emiliano. Italian bioethics committee in uproar [news]. *Nature* 2007 October 25; 449(7165): 955. Subject: 18.5.4

Ferguson, Pamela R. Human 'guinea pigs': why patients participate in clinical trials. *In:* McLean, Sheila A.M., ed. First Do No Harm: Law, Ethics, and Healthcare. Aldershot, England; Burlington, VT: Ashgate, 2006: 165-185. Subject: 18.1

Fergusson, Andrew. Neuroethics: the new frontier. *Ethics and Medicine: An International Journal of Bioethics* 2007 Spring; 23(1): 31-33. Subject: 17.1

Fernandez, Conrad V. Our moral obligations in caring for patients with orphan cancers [editorial]. *CMAJ/JAMC: Canadian Medical Association Journal* 2007 January 30; 176(3): 297, 299. Subject: 18.2

Ferreira, N. Latest legal and social developments in the euthanasia debate: bad moral consciences and political unrest. *Medicine and Law: The World Association for Medical Law* 2007 June; 26(2): 387-407. Subject: 20.7

Ferrell, Betty R. Understanding the moral distress of nurses witnessing medically futile care. *Oncology Nursing Forum* 2006 September 1; 33(5): 922-930. Subject: 20.5.1

Ferrell, Robyn. Brave new world. *In her:* Copula: Sexual Technologies, Reproductive Powers. Albany: State University of New York Press, 2006: 21-36. Subject: 14.1

Ferrell, Robyn. Reproducing technology. *In her:* Copula: Sexual Technologies, Reproductive Powers. Albany: State University of New York Press, 2006: 37-47. Subject: 14.4

Ferretti, Maria Paola. Why public participation in risk regulation? The case of authorizing GMO products in European Union. *Science as Culture* 2007 December; 16(4): 377-395. Subject: 15.1

Ferris, Lorraine E.; Naylor, C. David. Promoting integrity in industry-sponsored clinical drug trials: conflict of interest for Canadian academic health sciences centres. *In:* Lemmens, Trudo; Waring, Duff R., eds. Law and Ethics in Biomedical Research: Regulation, Conflict of Interest and Liability. Toronto; Buffalo: University of Toronto Press, 2006: 95-131. Subject: 9.7

Feuchtbaum, Lisa; Cunningham, George; Sciortino, Stan. Questioning the need for informed consent: a case study of California's experience with a pilot newborn screening research project. *Journal of Empirical Research on Human Research Ethics* 2007 September; 2(3): 3-14. Subject: 18.3

Fiege, Angela B. Resident portfolio: a tale of two women. *Academic Emergency Medicine* 2006 September; 13(9): 989-990; discussion 990-992. Subject: 9.5.1

Fielder, John. Ethics and the FDA. *IEEE Engineering in Medicine and Biology Magazine* 2006 July-August; 25(4): 13-17. Subject: 5.3

Fielder, John. Sex and stem cell research. *IEEE Engineering in Medicine and Biology Magazine* 2006 November-December; 25(6): 96-98. Subject: 18.5.4

Fielder, Odicie; Altice, Frederick L. Attitudes toward and beliefs about prenatal HIV testing policies and mandatory HIV testing of newborns among drug users. *AIDS and Public Policy Journal* 2005 Fall-Winter; 20(3-4): 74-91. Subject: 9.5.6

Fiester, Autumn. Casuistry and the moral continuum: evaluating animal biotechnology. *Politics and the Life Sci-*

ences 2006 March-September; 25(1-2): 15-22. Subject: 15.1

Fiester, Autumn. Mediation and moral aporia. *Journal of Clinical Ethics* 2007 Winter; 18(4): 355-356. Subject: 9.6

Fiester, Autumn. The failure of the consult model: why "mediation" should replace "consultation". *American Journal of Bioethics* 2007 February; 7(2): 31-32. Subject: 9.6

Fiester, Autumn. Why the clinical ethics we teach fails patients. *Academic Medicine* 2007 July; 82(7): 684-689. Subject: 7.2

FIGO Committee for the Ethical Aspects of Human Reproduction and Women's Health. Confidentiality, privacy and security of patients' health care information. *International Journal of Gynecology and Obstetrics* 2006 May; 93(2): 184-186. Subject: 8.4

FIGO Committee for the Ethical Aspects of Human Reproduction and Women's Health. Ethical guidelines on iatrogenic and self-induced infertility. *International Journal of Gynecology and Obstetrics* 2006 August; 94(2): 172-173. Subject: 14.1

FIGO Committee for the Ethical Aspects of Human Reproduction and Women's Health. Ethical guidelines on obstetric fistula. *International Journal of Gynecology and Obstetrics* 2006 August; 94(2): 174-175. Subject: 9.5.5

FIGO Committee for the Ethical Aspects of Human Reproduction and Women's Health. Ethical guidelines on resuscitation of newborns. *International Journal of Gynecology and Obstetrics* 2006 August; 94(2): 169-171. Subject: 20.5.2

FIGO Committee for the Ethical Aspects of Human Reproduction and Women's Health. Ethical issues in medical education: gifts and obligations. *International Journal of Gynecology and Obstetrics* 2006 May; 93(2): 189-190. Subject: 7.2

FIGO Committee for the Ethical Aspects of Human Reproduction and Women's Health. HIV and fertility treatment. *International Journal of Gynecology and Obstetrics* 2006 May; 93(2): 187-188. Subject: 14.1

FIGO Committee for the Ethical Aspects of Human Reproduction and Women's Health. Safe motherhood. *International Journal of Gynecology and Obstetrics* 2006 August; 94(2): 167-168. Subject: 9.5.5

FIGO Committee for the Ethical Aspects of Human Reproduction and Women's Health; FIGO Committee on Women's Sexual and Reproductive Rights. Female genital cutting. *International Journal of Gynecology and Obstetrics* 2006 August; 94(2): 176-177. Subject: 9.5.5

FIGO Committee for the Ethical Aspects of Human Reproduction and Women's Health; Serour, G.I. Embryo research. *International Journal of Gynecology and Obstetrics* 2006 May; 93(2): 182-183. Subject: 18.5.4

FIGO Committee for the Ethical Aspects of Human Reproduction and Women's Health; Serour, G.I. Human cloning. *International Journal of Gynecology and Obstetrics* 2006 June; 93(3): 282. Subject: 14.5

Fine, Robert L. Tackling medical futility in Texas [letter]. *New England Journal of Medicine* 2007 October 11; 357(15): 1558-1559. Subject: 20.5.1

Finkel, Alan G. Conflict of interest or productive collaboration? The pharma: academic relationship and its implications for headache medicine. *Headache* 2006 July-August; 46(7): 1181-1185. Subject: 7.2

Finkel, Elizabeth. New misconduct rules aim to minister to an ailing system [news]. *Science* 2007 August 31; 317(5842): 1159. Subject: 1.3.9

Finlay, Ilora. Crossing the 'bright line' — difficult decisions at the end of life. *Clinical Medicine (London, England)* 2006 July-August; 6(4): 398-402. Subject: 20.5.1

Finnis, John. "A vote decisive for . . . a more restrictive law". *In:* Watt, Helen, ed. Cooperation, Complicity and Conscience: Problems in Healthcare, Science, Law and Public Policy. London: Linacre Centre, 2005: 269-295. Subject: 12.4.1

Finnis, John. Restricting legalised abortion is not intrinsically unjust. *In:* Watt, Helen, ed. Cooperation, Complicity and Conscience: Problems in Healthcare, Science, Law and Public Policy. London: Linacre Centre, 2005: 209-245. Subject: 12.4.1

Fins, Joseph J. Border zones of consciousness: another immigration debate? *American Journal of Bioethics* 2007 January; 7(1): 51-54. Subject: 17.1

Fins, Joseph J. Commercialism in the clinic: finding balance in medical professionalism. *CQ: Cambridge Quarterly of Healthcare Ethics* 2007 Fall; 16(4): 425-432. Subject: 9.3.1

Fins, Joseph J. The minimally conscious state: ethics and diagnostic nosology. *Medical Ethics Newsletter [Lahey Clinic]* 2007 Fall; 14(3): 1-2, 5. Subject: 20.5.1

Fins, Joseph J.; Rezai, Ali R.; Greenberg, Benjamin D. Psychosurgery: avoiding an ethical redux while advancing a therapeutic future. *Neurosurgery* 2006 October; 59(4): 713-716. Subject: 17.6

Finucane, Thomas E.; Peterson, Eric D.; Boyce, Kurt; Overstreet, Karen; Sapers, Benjamin L.; Steinman, Michael A.; Chren, Mary-Margaret; Landefeld, C. Seth; Bero, Lisa A. The promotion of Gabapentin [letters and reply]. *Annals of Internal Medicine* 2007 February 20; 146(4): 312-314. Subject: 9.7

Fioriglio, Gianluigi; Szolovits, Peter. Copy fees and patients' rights to obtain a copy of their medical records: from law to reality. *AMIA Annual Symposium Proceedings* 2005: 251-255. Subject: 8.1

See SUBJECT HEADING KEY FOR SECTION II on inside back cover.

491

Fischer, Henry W. Protecting human subjects from themselves . . . after the disaster. *Protecting Human Subjects* 2007 November (15): 20-21. Subject: 18.4

Fischler, Ira; Simmerling, Mary; Fitzgerald, Maureen; Weitlauf, Julie; Frayne, Susan M.; Lee, Tina; Ruzek, Josef; Finney, John; Thrailkill, Ann; Newman, Elana. Trauma research. *Journal of Empirical Research on Human Research Ethics* 2007 March; 2(1): 51-59. Subject: 18.5.1

Fish, Mark. The health professional and the dying patient. *In:* Gunning, Jennifer; Holm, Søren, eds. Ethics, law, and society. Volume 1. Aldershot, Hants, England; Burlington, VT: Ashgate, 2005: 239-241. Subject: 20.7

Fisher, Anthony. Cooperation in evil: understanding the issues. *In:* Watt, Helen, ed. Cooperation, Complicity and Conscience: Problems in Healthcare, Science, Law and Public Policy. London: Linacre Centre, 2005: 27-64. Subject: 2.1

Fisher, Celia B.; Kornetsky, Susan Z.; Prentice, Ernest D. Determining risk in pediatric research with no prospect of direct benefit: time for a national consensus on the interpretation of federal regulations. *American Journal of Bioethics* 2007 March; 7(3): 5-10. Subject: 18.5.2

Fisher, Erik. Ethnographic invention: probing the capacity of laboratory decisions. *NanoEthics* 2007 August; 1(2): 155-165. Subject: 5.4

Fisher, Ian. Pope's death is drawn into euthanasia debate. *New York Times* 2007 September 28; p. A6. Subject: 20.5.1

Fisher, Morris A. Medicine and industry: a necessary but conflicted relationship. *Perspectives in Biology and Medicine* 2007 Winter; 50(1): 1-6. Subject: 5.3

Fisher, Richard. Fraudbusters. *New Scientist* 2007 November 10-16; 196(2629): 64-65. Subject: 1.3.9

Fisher-Jeffes, Lisa; Barton, Charlotte; Finlay, Fiona. Clincians' knowledge of informed consent. *Journal of Medical Ethics* 2007 March; 33(3): 181-184. Subject: 8.3.1

Fitz, Matthew M.; Homan, David; Reddy, Shalini; Griffith, Charles H.; Baker, Elizabeth; Simpson, Kevin P. The hidden curriculum: medical students' changing opinions toward the pharmaceutical industry. *Academic Medicine* 2007 October; 82(10, Supplement): S1-S3. Subject: 7.2

FitzGerald, Kevin; Royal, Charmaine. Race, genetics, and ethics. *In:* Prograis, Lawrence J.; Pellegrino, Edmund D., eds. African American Bioethics: Culture, Race, and Identity. Washington, DC: Georgetown University Press, 2004: 137-151. Subject: 15.11

Flagel, David C.; Best, Lisa A.; Hunter, Aren C. Perceptions of stress among students participating in psychology research: a Canadian survey. *Journal of Empirical Research on Human Research Ethics* 2007 September; 2(3): 61-67. Subject: 18.4

Flaming, Don. The ethics of Foucault and Ricoeur: an underrepresented discussion in nursing. *Nursing Inquiry* 2006 September; 13(3): 220-227. Subject: 4.1.3

Fleck, Leonard M. Can we trust "democratic deliberation"? *Hastings Center Report* 2007 July-August; 37(4): 22-25. Subject: 21.1

Fleck, Leonard M. Just caring: the challenges of priority-setting in public health. *In:* Rhodes, Rosamond; Francis, Leslie P.; Silvers, Anita, eds. The Blackwell Guide to Medical Ethics. Malden, MA: Blackwell Pub., 2007: 323-340. Subject: 9.4

Fleetwood, Janet. STDs in patients with multiple partners: confidentiality. *American Family Physician* 2006 December 1; 74(11): 1963-1964. Subject: 8.4

Fleming, David A. Responding to ethical dilemmas in nursing homes: do we always need an "ethicist"? *HEC (Healthcare Ethics Committee) Forum* 2007 September; 19(3): 245-259. Subject: 9.6

Fleming, John I.; Neville, Warwick; Pike, Gregory K. Another clash of orthodoxies: "the meaning of the universe" from "the other side of the pond." Abortion in the UK. Adelaide, Australia: Southern Cross Bioethics Institute, n.d.: 30 p. [Online]. Accessed: http://www.bioethics.org.au/docs/Other%20articles/MUNBY.PDF [2006 October 10]. Subject: 11.1

Flemons, W. Ward; Davies, Jan M.; MacLeod, Bruce. Disclosing medical errors [letter]. *CMAJ/JAMC: Canadian Medical Association Journal* 2007 November 6; 177(10): 1236. Subject: 9.8

Fletcher, Robert H.; Black, Bert. "Spin" in scientific writing: scientific mischief and legal jeopardy. *Medicine and Law: The World Association for Medical Law* 2007 September; 26(3): 511-525. Subject: 1.3.9

Flood, Patrick J. Is international law on the side of the unborn child? *National Catholic Bioethics Quarterly* 2007 Spring; 7(1): 73-95. Subject: 12.4.2

Flory, James; Emanuel, Ezekiel. Recent history of end-of-life care and implications for the future. *In:* Galston, Arthur W.; Peppard, Christiana Z., eds. Expanding Horizons in Bioethics. Dordrecht; Norwell, MA: Springer, 2005: 161-182. Subject: 20.4.1

Flotte, Terence R.; Frentzen, Barbara; Humphries, Margaret R.; Rosenbloom, Arlan L. Recent developments in the protection of pediatric research subjects. *Journal of Pediatrics* 2006 September; 149(3): 285-286. Subject: 18.5.2

Flum, David R.; Khan, Tipu V.; Dellinger, E. Patchen. Toward the rational and equitable use of bariatric surgery. *JAMA: The Journal of the American Medical Association* 2007 September 26; 298(12): 1442-1444. Subject: 9.4

Foddy, Bennett; Savulescu, Julian. Addiction is not an affliction: addictive desires are merely pleasure-oriented

---

desires. *American Journal of Bioethics* 2007 January; 7(1): 29-32. Subject: 17.1

Fontana, Nicholas. A question of professionalism: are we treating patients as people or procedures? *Journal of the Michigan Dental Association* 2006 October; 88(10): 28-30. Subject: 8.1

Fontanarosa, Phil B.; Rennie, Drummond; DeAngelis, Catherine D. Access to care as a component of health system reform [editorial]. *JAMA: The Journal of the American Medical Association* 2007 March 14; 297(10): 1128-1130. Subject: 9.5.3

Foote, Robert L.; Brown, Paul D.; Garces, Yolanda I.; Okuno, Scott H.; Miller, Robert C.; Strome, Scott E. Informed consent in advanced laryngeal cancer. *Head and Neck* 2007 March; 29(3): 230-235. Subject: 8.3.1

Ford, Jolyon; Tomossy, George. Clinical trials in developing countries: the plaintiff's challenge. *Law, Social Justice and Global Development* 2004 June 4; (1): 17 p. [Online]. Accessed: http://www2.warwick.ac.uk/fac/soc/law/elj/lgd/2004_1/ford/ [2007 April 4]. Subject: 18.5.9

Ford, Mary. The consent model of pregnancy: deadlock undiminished. *McGill Law Journal* 2005; 50: 619-666. Subject: 12.4.2

Ford, Paul J. Neurosurgical implants: clinical protocol considerations. *CQ: Cambridge Quarterly of Healthcare Ethics* 2007 Summer; 16(3): 308-311. Subject: 17.1

Ford, Paul J. Professional clinical ethicist: knowing why and limits. *Journal of Clinical Ethics* 2007 Fall; 18(3): 243-246. Subject: 9.6

Ford, Paul J.; Boissy, Adrienne R. Different questions, different goals. *American Journal of Bioethics* 2007 February; 7(2): 46-47. Subject: 9.6

Ford, Paul J.; DeMarco, Joseph P. Brains, ethics, and elective surgeries: emerging ethics consultation. *Ethics and Medicine: An International Journal of Bioethics* 2007 Spring; 23(1): 39-45. Subject: 17.1

Ford, Paul J.; Henderson, Jaimie M. Functional neurosurgical intervention: neuroethics in the operating room. *In:* Illes, Judy, ed. Neuroethics: Defining the Issues in Theory, Practice, and Policy. New York: Oxford University Press, 2006: 213-228. Subject: 17.1

Ford, Paul J.; Kubu, Cynthia S. Ameliorating and exacerbating: surgical "prosthesis" in addiction. *American Journal of Bioethics* 2007 January; 7(1): 32-34. Subject: 17.1

Forman, Lisa. Claiming equity and justice in health: the role of the South African right to health in ensuring access to HIV/AIDS treatment. *In:* Flood, Colleen M., ed. Just Medicare: What's In, What's Out, How We Decide. Buffalo, NY: University of Toronto Press, 2006: 80-104. Subject: 9.5.6

Forman, Lisa. Trade rules, intellectual property, and the right to health. *Ethics and International Affairs* 2007 Fall; 21(3): 337-357. Subject: 9.2

Forsythe, Clarke D. A lack of prudence. *Human Life Review* 2007 Fall; 33(4): 15-21. Subject: 12.4.2

Fortin, Marie-Chantal; Roigt, Delphine; Doucet, Hubert. What should we do with patients who buy a kidney overseas? *Journal of Clinical Ethics* 2007 Spring; 18(1): 23-34. Subject: 19.3

Fortune, Peter-Marc. Euthanasia in neonates: are we asking the right questions? [letter]. *BMJ: British Medical Journal* 2007 May 26; 334(7603): 1072. Subject: 20.5.2

Fossett, James W. Federalism by necessity: state and private support for human embryonic stem cell research. Rockefeller Institute Policy Brief 2007 August 9: 1-13 [Online]. Accessed: http://www.rockinst.org/WorkArea/showcontent.aspx?id=12064 [2007 October 25]. Subject: 5.3

Fossett, James W. Managing reproductive pluralism: the case for decentralized governance. *Hastings Center Report* 2007 July-August; 37(4): 20-22. Subject: 14.1

Fossett, James W.; Ouellette, Alicia R.; Philpott, Sean; Magnus, David; McGee, Glenn. Federalism and bioethics: states and moral pluralism. *Hastings Center Report* 2007 November-December; 37(6): 24-35. Subject: 2.1

Fost, Norman; Levine, Robert J. The dysregulation of human subjects research [editorial]. *JAMA: The Journal of the American Medical Association* 2007 November 14; 298(18): 2196-2198. Subject: 18.2

Foster, Charles. Simple rationality? The law of healthcare resource allocation in England. *Journal of Medical Ethics* 2007 July; 33(7): 404-407. Subject: 9.4

Foster, Kenneth R. Engineering the brain. *In:* Illes, Judy, ed. Neuroethics: Defining the Issues in Theory, Practice, and Policy. New York: Oxford University Press, 2006: 185-199. Subject: 17.1

Fovargue, Sara. Consenting to bio-risk: xenotransplantation and the law. *Legal Studies* 2005 September; 25(3): 404-430. Subject: 19.1

Fowler, Marsha. Religious and clinical ethics. *In:* Davis, Anne J.; Tschudin, Verena; de Raeve, Louise, eds. Essentials of Teaching and Learning in Nursing Ethics: Perspectives and Methods. New York: Churchill Livingstone Elsevier, 2006: 37-48. Subject: 4.1.3

Fowler, Marsha. Social ethics, the profession and society. *In:* Davis, Anne J.; Tschudin, Verena; de Raeve, Louise, eds. Essentials of Teaching and Learning in Nursing Ethics: Perspectives and Methods. New York: Churchill Livingstone Elsevier, 2006: 27-36. Subject: 4.1.3

Fowler, Marsha; Tschudin, Verena. Ethics in nursing: an historical perspective. *In:* Davis, Anne J.; Tschudin, Verena; de Raeve, Louise, eds. Essentials of Teaching and

See SUBJECT HEADING KEY FOR SECTION II on inside back cover.

493

Learning in Nursing Ethics: Perspectives and Methods. New York: Churchill Livingstone Elsevier, 2006: 13-25. Subject: 2.2

Fowler, Robert A.; Sabur, Natasha; Li, Ping; Juurlink, David N.; Pinto, Ruxandra; Hladunewich, Michelle A.; Adhikari, Neill K.J.; Sibbald, William J.; Martin, Claudio M. Sex- and age-based differences in the delivery and outcomes of critical care. *CMAJ/JAMC: Canadian Medical Association Journal* 2007 December 4; 177(12): 1513-1519. Subject: 9.4

Fox, Daniel M. Selective appropriation, medical ethics, and health politics: the complementarity of Baker, McCullough, and me. *Kennedy Institute of Ethics Journal* 2007 March; 17(1): 23-30. Subject: 2.1

Fox, Dov. Silver spoons and golden genes: genetic engineering and the egalitarian ethos. *American Journal of Law and Medicine* 2007; 33(4): 567-623. Subject: 15.1

Fox, Ellen; Daskal, Frona C.; Stocking, Carol. Ethics consultants' recommendations for life-prolonging treatment of patients in a persistent vegetative state: a follow-up study. *Journal of Clinical Ethics* 2007 Spring; 18(1): 64-71. Subject: 20.5.1

Fox, Ellen; Myers, Sarah; Pearlman, Robert A. Ethics consultation in United States hospitals: a national survey. *American Journal of Bioethics* 2007 February; 7(2): 13-25. Subject: 9.6

Fox, Ellen; Myers, Sarah; Pearlman, Robert A. Response to open peer commentaries on "Ethics Consultation in U.S. Hospitals: A National Survey" [letter]. *American Journal of Bioethics [Online].* 2007 February; 7(2): W1-W3. Subject: 9.6

Fox, Jeffrey L. Despite glacial progress, US government signals support for personalized medicine. *Nature Biotechnology* 2007 May 25(5): 489-490. Subject: 15.1

Fox, Jeffrey L. Feds eye genetic testing [news]. *Nature Biotechnology* 2007 December; 25(12): 1340. Subject: 15.3

Fox, Jeffrey L. FDA clarifies stance on long-term follow-up for gene therapy clinical trials [news brief]. *Nature Biotechnology* 2007 February; 25(2): 153. Subject: 15.4

Fox, Jeffrey L. US courts thwart GM alfalfa and turf grass [news]. *Nature Biotechnology* 2007 April; 25(4): 367-368. Subject: 15.1

Fox, Marie. Exposing harm: the erasure of animal bodies in healthcare law. *In:* McLean, Sheila A.M., ed. First Do No Harm: Law, Ethics, and Healthcare. Aldershot, England; Burlington, VT: Ashgate, 2006: 543-559. Subject: 22.1

Fox, Marie; Thomson, Michael. Short changed? The law and ethics of male circumcision. *In:* Freeman, Michael, ed. Children's Health and Children's Rights. Boston: Martinus Nijhoff Publishers, 2006: 161-181. Subject: 9.5.1

Fox, M.D. The price is wrong: the moral cost of living donor inducements. *American Journal of Transplantation* 2006 November; 6(11): 2529-2530. Subject: 19.5

Fox, Renée C. Toward an ethics of iatrogenesis. *In:* LaFleur, William R.; Böhme, Gernot; Shimazono, Susumu, eds. Dark Medicine: Rationalizing Medical Research. Bloomington: Indiana University Press, 2007: 149-164. Subject: 9.8

Fox, R.M. Debate: should Australia move towards a centralized ethics committees system? The case for. *Internal Medicine Journal* 2005 April; 35(4): 247-248. Subject: 18.2

Frader, Joel E. Discontinuing artificial fluids and nutrition: discussions with children's families. *Hastings Center Report* 2007 January-February; 37(1): inside back cover. Subject: 20.5.2

Fraenkel, Liana; McGraw, Sarah. What are the essential elements to enable patient participation in medical decision making? *JGIM: Journal of General Internal Medicine* 2007 May; 22(5): 614-619. Subject: 8.1

Fraleigh, Anna Schork. An alternative to guardianship: should Michigan statutorily allow acute-care hospitals to make medical treatment decisions for incompetent patients who have neither identifiable surrogates nor advance directives? *University of Detroit Mercy Law Review* 1999 Summer; 76(4): 1079-1134. Subject: 8.3.3

Francescotti, Robert. Animal mind and animal ethics: an introduction. *Journal of Ethics* 2007; 11(3): 239-252. Subject: 22.1

Francis, Leslie P. Discrimination in medical practice: justice and the obligations of health care providers to disadvantaged patients. *In:* Rhodes, Rosamond; Francis, Leslie P.; Silvers, Anita, eds. The Blackwell Guide to Medical Ethics. Malden, MA: Blackwell Pub., 2007: 162-179. Subject: 9.4

Frank, Arthur W. Social bioethics and the critique of autonomy. *Health* 2000 July; 4(3): 378-394 [Online]. Accessed: http://hea.sagepub.com/cgi/reprint/4/3/378 [2007 May 10]. Subject: 2.1

Franklin, Anita; Sloper, Patricia. Listening and responding? Children's participation in health care within England. *In:* Freeman, Michael, ed. Children's Health and Children's Rights. Boston: Martinus Nijhoff Publishers, 2006: 11-29. Subject: 9.5.7

Franklin, Sarah. Embryonic economies: the double reproductive value of stem cells. *Biosocieties* 2006 March; 1(1): 71-90. Subject: 18.5.4

Franklin, Sarah. Ethical biocapital: new strategies of cell culture. *In:* Franklin, Sarah; Lock, Margaret, eds. Remaking Life and Death: Toward an Anthropology of the Biosciences. Santa Fe: School of American Research Press; Oxford: James Currey, 2003: 97-127. Subject: 15.1

Frassoni, F. The laws covering in vitro fertilization and embryo research in Italy. *Bone Marrow Transplantation* 2006 July; 38(1): 5-6. Subject: 14.4

Freckelton, Ian. Health practitioner regulation: emerging patterns and challenges for the age of globalization. *In:* Bennett, Belinda; Tomossy, George F., eds. Globalization and Health: Challenges for Health Law and Bioethics. Dordrecht: Springer, 2006: 187-206. Subject: 5.3

Freeman, Bradley D.; Kennedy, Carie R.; Coopersmith, Craig M.; Zehnbauer, Barbara A.; Buchman, Timothy G. Genetic research and testing in critical care: surrogates' perspective. *Critical Care Medicine* 2006 April; 34(4): 986-994. Subject: 15.3

Freeman, Jeanine. The ethics of epidemics — when healing puts the doctor's life in danger. *Iowa Medicine* 2006 January-February; 96(1): 10-11. Subject: 7.1

Freeman, Michael. Rethinking Gillick. *In his:* Children's Health and Children's Rights. Boston: Martinus Nijhoff Publishers, 2006: 201-217. Subject: 11.2

Freeman, Michael. Saviour siblings. *In:* McLean, Sheila A.M., ed. First Do No Harm: Law, Ethics, and Healthcare. Aldershot, England; Burlington, VT: Ashgate, 2006: 389-406. Subject: 15.2

Freeman, Michael; Jaoudé, Pauline Abou. Justifying surgery's last taboo: the ethics of face transplants. *Journal of Medical Ethics* 2007 February; 33(2): 76-81. Subject: 19.1

Freeman, Robert A. Industry perspectives on equity, access, and corporate social responsibility: a view from the inside. *In:* Cohen, Jillian Clare; Illingworth, Patricia; Schüklenk, Udo, eds. The Power of Pills: Social, Ethical and Legal Issues in Drug Development, Marketing, and Pricing. London; Ann Arbor, MI: Pluto, 2006: 65-73. Subject: 9.7

Freemantle, Nick; Calvert, Mel. Composite and surrogate outcomes in randomized controlled trials. *BMJ: British Medical Journal* 2007 April 14; 334(7597): 756-757. Subject: 18.2

Freireich, Emil; Gesme, Dean. Should terminally ill patients have the right to take drugs that pass phase 1 testing? [debate]. *BMJ: British Medical Journal* 2007 September 8; 335(7618): 478-479. Subject: 18.5.7

Fremont, Allen M.; Coreea-de-Araujo, Rosaly; Hayes, Sharon. Gender disparities in managed care: it's time for action. *Women's Health Issues* 2007 May-June; 17(3): 116-119. Subject: 9.5.5

French LATASAMU Group; Ferrand, Edouard; Marty, Jean. Prehospital withholding and withdrawal of life-sustaining treatments. The French LATASAMU survey. *Intensive Care Medicine* 2006 October; 32(10): 1498-1505. Subject: 20.5.2

Frenk, Julio. Ethical considerations in health systems. *In:* Marinker, Marshall, ed. Constructive Conversations about Health: Policy and Values. Oxford; Seattle: Radcliffe Pub., 2006: 165-175. Subject: 9.1

Frewer, Andreas. Medical research, morality, and history: the German journal Ethik and the limits of human experimentation. *In:* LaFleur, William R.; Böhme, Gernot; Shimazono, Susumu, eds. Dark Medicine: Rationalizing Medical Research. Bloomington: Indiana University Press, 2007: 30-45. Subject: 18.1

Frewer, Andreas; Fahr, Uwe. Clinical ethics and confidentiality: opinions of experts and ethics committees. *HEC (Healthcare Ethics Committee) Forum* 2007 December; 19(4): 277-291. Subject: 9.6

Frey, Christofer. Bioethics from the perspective of universalisation. *In:* Roetz, Heiner, ed. Cross-Cultural Issues in Bioethics: The Example of Human Cloning. New York: Rodopi, 2006: 341-361. Subject: 2.1

Fried, Terri R.; O'Leary, John; Van Ness, Peter; Fraenkel, Liana. Inconsistency over time in the preferences of older persons with advanced illness for life-sustaining treatment. *Journal of the American Geriatrics Society* 2007 July; 55(7): 1007-1014. Subject: 20.5.1

Fried, Terri R.; Van Ness, Peter H.; Byers, Amy L.; Towle, Virginia R.; O'Leary, John R.; Dubin, Joel A. Changes in preferences for life-sustaining treatment among older persons with advanced illness. *JGIM: Journal of General Internal Medicine* 2007 April; 22(4): 495-501. Subject: 20.5.1

Friedberg, Errol C. Fraud in science — reflections on some whys and wherefores. *DNA Repair* 2006 March 7; 5(3): 291-293. Subject: 1.3.9

Friedman, E.A.; Friedman, A.L. Payment for donor kidneys: pros and cons. *Kidney International* 2006 February 15; 69(5): 960-962 [Online]. Accessed: http://www.nature.com/ki/journal/v69/n6/pdf/5000262a.pdf [2006 October 3]. Subject: 19.5

Friedman, Lee S.; Richter, Elihu D. Excessive and disproportionate advertising in peer-reviewed journals. *International Journal of Occupational and Environmental Health : Official Journal of the International Commission on Occupational Health* 2006 January-March; 12(1): 59-64. Subject: 1.3.7

Friedman, Sandra L. Parent resuscitation preferences for young people with severe developmental disabilities. *Journal of the American Medical Directors Association* 2006 February; 7(2): 67-72. Subject: 20.5.2

Friedman, Sandra; Gilmore, Dana. Factors that impact resuscitation preferences for young people with severe developmental disabilities. *Intellectual and Developmental Disabilities* 2007 April; 45(2): 90-97. Subject: 20.5.1

Friele, Roland D.; Sluijs, Emmy M. Patient expectations of fair complaint handling in hospitals: empirical data. *BMC Health Services Research* 2006 August 18; 6: 106: 9 p. Subject: 8.1

See SUBJECT HEADING KEY FOR SECTION II on inside back cover.

495

Frith, Lucy. Researching chronic childhood illness: autonomy or beneficence? *Chronic Illness* 2006 September; 2(3): 178-180. Subject: 18.5.2

Fritz, K. Cultural diversity. *In:* Holland, Stephen, ed. Introducing Nursing Ethics: Themes in Theory and Practice. Salisbury: APS, 2004: 151-170. Subject: 21.7

Frohna, Alice. Medical students' professionalism. *Medical Teacher* 2006 February; 28(1): 1-2. Subject: 7.2

Fromme, Erik K.; Tilden, Virginia P.; Drach, Linda L.; Tolle, Susan W. Increased family reports of pain or distress in dying Oregonians: 1996 to 2002. *Journal of Palliative Medicine* 2004 June; 7(3): 431-442. Subject: 20.4.1

Frosch, Dominick L.; Krueger, Patrick M.; Hornik, Robert C.; Cronholm, Peter F.; Barg, Frances K. Creating demand for prescription drugs: a content analysis of television direct-to-consumer advertising. *Annals of Family Medicine* 2007 January/February; 5(1): 6-12. Subject: 9.7

Fry, Sara T. Nursing ethics. *In:* Khushf, George, ed. Handbook of Bioethics: Taking Stock of the Field From a Philosophical Perspective. Dordrecht; Boston: Kluwer Academic, 2004: 489-505. Subject: 4.1.3

Fry-Revere, Sigrid. Legal trends in bioethics. *Journal of Clinical Ethics* 2007 Spring; 18(1): 72-90. Subject: 2.1

Fry-Revere, Sigrid. Legal trends in bioethics. *Journal of Clinical Ethics* 2007 Summer; 18(2): 162-188. Subject: 2.1

Fry-Revere, Sigrid; Koshy, Sheeba. Legal trends in bioethics. *Journal of Clinical Ethics* 2007 Fall; 18(3): 294-328. Subject: 2.1

Fry-Revere, Sigrid; Koshy, Sheeba; Leppard IV, John. Legal trends in bioethics. *Journal of Clinical Ethics* 2007 Winter; 18(4): 404-424. Subject: 2.1

Fuchs, Michael. Gene therapy. An ethical profile of a new medical territory. *Journal of Gene Medicine* 2006 November; 8(11): 1358-1362. Subject: 15.4

Fuchs, Thomas. Ethical issues in neuroscience. *Current Opinion in Psychiatry* 2006 November; 19(6): 600-607. Subject: 18.2

Fuchs, Ursel. Bürger gegen Bio-Ethik: Internationale Initiative gegen die geplante Bio-Ethik-Konvention. *In:* Neuer-Miebach, Therese; Wunder, Michael, eds. Bio-Ethik und die Zukunft der Medizin. Bonn: Psychiatrie-Verlag, 1998: 165-166. Subject: 2.1

Fuchs, Ursel. Experten entscheiden. Unter sich. Und über uns. Die Bio-Ethik-Konvention geht alle an. *In:* Neuer-Miebach, Therese; Wunder, Michael, eds. Bio-Ethik und die Zukunft der Medizin. Bonn: Psychiatrie-Verlag, 1998: 130-138. Subject: 2.1

Fudin, Jeffrey. Blowing the whistle: a pharmacist's vexing experience unraveled. *American Journal of Health-System Pharmacy* 2006 November 15; 63(22): 2262-2265. Subject: 9.7

Fugh-Berman, Adriane; Shahram, Ahari. Following the script: how drug reps make friends and influence doctors. *PLoS Medicine* 2007 April; 4(4): e150 [Online]. Accessed:http://medicine.plosjournals.org/perlserv/?request =get-document&doi=10.1371%2Fjournal.pmed. 0040150 [2007 August 27]. Subject: 9.7

Fujita, Misao; Akabayashi, Akira; Slingsby, Brian Taylor; Kosugi, Shinji; Fujimoto, Yasuhiro; Tanaka, Koichi. A model of donors' decision-making in adult-to-adult living donor liver transplantation in Japan: having no choice. *Liver Transplantation* 2006 May; 12(5): 768-774. Subject: 19.5

Fulford, K.W.M. (Bill). Ten principles of values-based medicine (VBM). *In:* Schramme, Thomas; Thome, Johannes, eds. Philosophy and Psychiatry. Berlin; New York: De Gruyter, 2004: 50-80. Subject: 4.1.2

Fulford, K.W.M. (Bill); Thornton, Tim; Graham, George. From bioethics to values-based practice. *In their:* Oxford Textbook of Philosophy and Psychiatry. Oxford; New York: Oxford University Press, 2006: 498-538. Subject: 2.1

Fulford, K.W.M. (Bill); Thornton, Tim; Graham, George. From bioethics to values-based practice in psychiatric diagnosis. *In their:* Oxford Textbook of Philosophy and Psychiatry. Oxford; New York: Oxford University Press, 2006: 585-608. Subject: 17.1

Fulford, K.W.M. (Bill); Thornton, Tim; Graham, George. It's the law! Rationality and consent as a case study in values and mental health law. *In their:* Oxford Textbook of Philosophy and Psychiatry. Oxford; New York: Oxford University Press, 2006: 539-563. Subject: 8.3.3

Fulford, K.W.M. (Bill); Thornton, Tim; Graham, George. Tools of the trade: an introduction to psychiatric ethics. *In their:* Oxford Textbook of Philosophy and Psychiatry. Oxford; New York: Oxford University Press, 2006: 469-497. Subject: 17.1

Fulford, K.W.M. (Bill); Thornton, Tim; Graham, George. Values in psychiatric psychiatry. *In their:* Oxford Textbook of Philosophy and Psychiatry. Oxford; New York: Oxford University Press, 2006: 564-584. Subject: 17.1

Fullbrook, Suzanne. Autonomy and care: acting in a person's best interets [sic: interest]. *British Journal of Nursing* 2007 February 22-March 7; 16(4): 236-237. Subject: 8.3.1

Fullbrook, Suzanne. Best interests: a review of issues that affect nurses' decision making. *British Journal of Nursing* 2007 May 24 - June 13; 16(10): 600-601. Subject: 4.1.3

Fullbrook, Suzanne. Best interests. An holistic approach: part 2(b). *British Journal of Nursing* 2007 June 28-July 11; 16(12): 746-747. Subject: 8.3.3

Subject = NRCBL Primary Classification Number; see inside front cover.

Fullbrook, Suzanne. Best interest. A review of the legal principles involved: Part 2(a). *British Journal of Nursing* 2007 June 14-27; 16(11): 682-683. Subject: 8.3.3

Fullbrook, Suzanne. Common law and a duty of care: the application of principles. *British Journal of Nursing* 2007 September 27 - October 10; 16(17):1074-1075. Subject: 4.1.3

Fullbrook, Suzanne. Confidentiality. Part 3: Caldicott guardians and the control of data. *British Journal of Nursing* 2007 September 13-27; 16(16): 1008-1009. Subject: 8.4

Fullbrook, Suzanne. Consent and capacity: principles of the Mental Capacity Act 2005. *British Journal of Nursing* 2007 April 12-25; 16(7): 412-413. Subject: 8.3.3

Fullbrook, Suzanne. Consent: the issue of rights and responsibilities for the health worker. *British Journal of Nursing* 2007 March 8-21; 16(5): 318-319. Subject: 8.3.1

Fullbrook, Suzanne. Death by denomination: a Jehovah Witness's right to die. *British Journal of Nursing* 2007 November 22-December 12; 16(21): 1306-1307. Subject: 8.3.4

Fullbrook, Suzanne. End-of-life issues: common law and the Mental Capacity Act 2005. *British Journal of Nursing* 2007 July 12-25; 16(13): 816-818. Subject: 20.5.1

Fullbrook, Suzanne. Legal principles of confidentiality and other public interests:Part 1. *British Journal of Nursing* 2007 July 26 - August 8; 16(14): 874-875. Subject: 8.4

Fullbrook, Suzanne. Regulatory codes of conduct and the common law. Part 2: confidentiality. *British Journal of Nursing* 2007 August 9 - September 12; 16(15): 946-947. Subject: 8.4

Fullbrook, Suzanne; Sanders, Karen. Consent and capacity 2: the Mental Capacity Act 2005 and 'living wills'. *British Journal of Nursing* 2007 April 26-May 9; 16(8): 474-475. Subject: 8.3.3

Fullbrook, Suzanne; Sanders, Karen. Consent and capacity: other aspects of the Mental Capacity Act. *British Journal of Nursing* 2007 May 10-23; 16(9): 538-539. Subject: 8.3.3

Fullwiley, Duana. The molecularization of race: institutionalizing human difference in pharmacogenetics practice. *Science as Culture* 2007 March; 16(1): 1-30. Subject: 15.11

Funk, Carolyn L.; Barrett, Kirsten A.; Macrina, Francis L. Authorship and publication practices: evaluation of the effect of responsible conduct of research instruction to postdoctoral trainees. *Accountability in Research* 2007 October-December; 14(4): 269-305. Subject: 1.3.9

Furger, Franco; Fukuyama, Francis. A proposal for modernizing the regulation of human biotechnologies. *Hastings Center Report* 2007 July-August; 37(4): 16-20. Subject: 14.1

Fürst, Gebhard. The (im)perfect human — his own creator? Bioethics and genetics at the beginning of life. *In:* Sensen, Chrisopher W., ed. Handbook of Genome Research: Genomics, Proteomics, Metabolomics, Bioinformatics, Ethical and Legal Issues. Volume 2. Weinheim: Wiley-VCH, 2005: 561-569. Subject: 15.1

Furton, Edward J. Morality is not a medical problem. *Ethics and Medics* 2007 July; 32(7): 3-4. Subject: 8.3.2

G

Gaber, Tarek A-Z.K. Medico-legal and ethical aspects in the management of wandering patients following brain injury: questionnaire survey. *Disability and Rehabilitation* 2006 November 30; 28(22): 1413-1416. Subject: 17.3

Gabriele, Edward F. Belmont as parable: research leadership and the spirit of integrity. *In:* Kulakowski, Elliott C.; Chronister, Lynne U., eds. Research Administration and Management. Sudbury, MA: Jones and Bartlett, 2006: 473-480. Subject: 18.2

Gadd, Elaine. The global significance of the Convention on Human Rights and Biomedicine. *In:* Gevers, J.K.M.; Hondius, E.H.; Hubben, J.H., eds. Health Law, Human Rights and the Biomedicine Convention: Essays in Honour of Henriette Roscam Abbing. Leiden; Boston: Martinus Nijhoff Publishers, 2005: 35-46. Subject: 21.1

Gagen, Wendy Jane; Bishop, Jeffrey P. Ethics, justification and the prevention of spina bifida. *Journal of Medical Ethics* 2007 September; 33(9): 501-507. Subject: 15.3

Galanakis, E.; Dimoliatis, I.D.K. Early European attitudes towards "good death": Eugenios Voulgaris, Treatise on euthanasia, St Petersburg, 1804. *Medical Humanities* 2007 June; 33(1): 1-4. Subject: 20.5.1

Gallagher, Ann. The respectful nurse. *Nursing Ethics* 2007 May; 14(3): 360-371. Subject: 4.1.3

Gallagher, Ann. The teaching of nursing ethics: content and method. Promoting ethical competence. *In:* Davis, Anne J.; Tschudin, Verena; de Raeve, Louise, eds. Essentials of Teaching and Learning in Nursing Ethics: Perspectives and Methods. New York: Churchill Livingstone Elsevier, 2006: 223-239. Subject: 7.2

Gallagher, John A. Donation after cardiac death: an ethical reflection on the development of a protocol. *Health Care Ethics USA* 2007 Summer; 15(3): 5-6. Subject: 19.5

Gallagher, Romayne. Does knowledge of ethics and end-of-life issues inform choices in advance care planning scenarios? [letter]. *Journal of the American Geriatrics Society* 2007 October; 55(10): 1695-1696. Subject: 20.5.4

Gallagher, Thomas H.; Studdert, David; Levinson, Wendy. Disclosing harmful medical errors to patients. *New England Journal of Medicine* 2007 June 28; 356(26): 2713-2719. Subject: 9.8

Gallant, Mae H.; Beaulieu, Marcia C.; Carnevale, Franco A. Partnership: an analysis of the concept within

See SUBJECT HEADING KEY FOR SECTION II on inside back cover.

497

the nurse-client relationship. *Journal of Advanced Nursing* 2002 October; 40(2): 149-157. Subject: 8.1

Galvão, Pedro. Boonin on the future-like-ours argument against abortion. *Bioethics* 2007 July; 21(6): 324-328. Subject: 12.3

Gamero, Joaquin J.; Romero, Jose-Luis; Peralta, Juan-Luis; Carvalho, Mónica; Corte-Real, Francisco. Spanish public awareness regarding DNA profile databases in forensic genetics: what type of DNA profiles should be included? *Journal of Medical Ethics* 2007 October; 33(10): 598-604. Subject: 15.1

Gannon, Susanne; Müller-Rockstroh, Babette. In memory: women's experiences of (dangerous) breasts. *Philosophy in the Contemporary World* 2004 Spring-Summer; 11(1): 53-64. Subject: 9.5.5

Ganz, Freda DeKeyser; Musgrave, Catherine F. Israeli critical care nurses' attitudes toward physician-assisted dying. *Heart and Lung* 2006 November-December; 35(6): 412-422. Subject: 20.7

Ganzini, Linda; Beer, Tomasz M.; Brouns, Matthew C. Views on physician-assisted suicide among family members of Oregon cancer patients. *Journal of Pain and Symptom Management* 2006 September; 32(3): 230-236. Subject: 20.7

Garattini, Silvio; Bertelé, Vittorio. Non-inferiority trials are unethical because they disregard patients' interests. *Lancet* 2007 December 1-7; 370(9602): 1875-1877. Subject: 18.2

Garau, J. Impact of antibiotic restrictions: the ethical perspective. *Clinical Microbiology and Infection* 2006 August; 12(Supplement 5): 16-24. Subject: 9.8

Garber, Mandy; Hanusa, Barbara H.; Switzer, Galen E.; Mellors, John; Arnold, Robert M. HIV-infected African Americans are willing to participate in HIV treatment trials. *JGIM: Journal of General Internal Medicine* 2007 January; 22(1): 17-42. Subject: 9.5.6

Garbutt, Jane; Brownstein, DenaR.; Klein, Eileen J.; Waterman, Amy; Krauss, MelissaJ.; Marcuse,Edgar K.; Hazel, Erik; Dunagan, Wm. Claiborne; Fraser, Victoria; Gallagher, Thomas H. Reporting and disclosing medical errors: pediatricians' attitudes and behaviors. *Archives of Pediatrics and Adolescent Medicine* 2007 February; 161(2): 179-185. Subject: 9.8

Garcia, Jorge L.A. Revisiting African American perspectives on biomedical ethics: distinctiveness and other questions. *In:* Prograis, Lawrence J.; Pellegrino, Edmund D., eds. African American Bioethics: Culture, Race, and Identity. Washington, DC: Georgetown University Press, 2004: 1-23. Subject: 2.1

Garcia, J.L.A. Health versus harm: euthanasia and physicians' duties. *Journal of Medicine and Philosophy* 2007 January-February; 32(1): 7-24. Subject: 20.5.1

Garden, Rebecca; Murphree, Hyon Joo Yoo. Class and ethnicity in the global market for organs: the case of Korean cinema. *Journal of Medical Humanities* 2007 December; 28(4): 213-229. Subject: 19.5

Gardner, Richard. Therapeutic and reproductive cloning — a scientific perspective. *In:* Gunning, Jennifer; Holm, Søren, eds. Ethics, law, and society. Volume 1. Aldershot, Hants, England; Burlington, VT: Ashgate, 2005: 9-16. Subject: 14.5; 14.1

Garetto, Lawrence P.; Senour, Wendy. Using an ethics across the curriculum strategy in dental education. *Journal of the American College of Dentists* 2006 Winter; 73(4): 33-35. Subject: 7.2

Garetto, Lawrence P.; Yoder, Karen M. Basic oral health needs: a professional priority. *Journal of Dental Education* 2006 November; 70(11): 1166-1169. Subject: 9.2

Garforth, Kathryn. Health care and access to patented technologies. *Health Law Journal* 2005; 13: 77-97. Subject: 15.8

Garlin, Amy B.; Goldschmidt, Ronald. Pregnant physicians and infectious disease risk. *American Family Physician* 2007 January 1; 75(1): 112, 114. Subject: 7.1

Garment, Ann; Lederer, Susan; Rogers, Naomi; Boult, Lisa. Let the dead teach the living: the rise of body bequeathal in 20th century America. *Academic Medicine* 2007 October; 82(10): 1000-1005. Subject: 20.1

Garner, Samual A. Dear bioethics, the country needs you. *American Journal of Bioethics* 2007 October; 7(10): 38-39. Subject: 2.1

Garwood-Gowers, Austen. The proper limits for medical intervention that harms the therapeutic interests of incompetents. *In:* Garwood-Gowers, Austen; Tingle, John; Wheat, Kay, eds. Contemporary Issues in Healthcare Law and Ethics. Edinburgh; New York: Elsevier Butterworth-Heinemann, 2005: 191-211. Subject: 8.3.3

Garzón, Nelly. Colombia: social justice in nursing ethics. *In:* Davis, Anne J.; Tschudin, Verena; de Raeve, Louise, eds. Essentials of Teaching and Learning in Nursing Ethics: Perspectives and Methods. New York: Churchill Livingstone Elsevier, 2006: 241-250. Subject: 4.1.3

Gass, C.W.J. It is the right of every anaesthetist to refuse to participate in a maternal-request caesarean section. *International Journal of Obstetric Anesthesia* 2006 January; 15(1): 33-35. Subject: 9.5.5

Gast, Kristen Marttila. Cold comfort pharmacy: pharmacist tort liability for conscientious refusals to dispense emergency contraception. *Texas Journal of Women and the Law* 2007 Spring; 16(2): 149-184. Subject: 11.1

Gastmans, Chris. The care perspective in healthcare ethics. *In:* Davis, Anne J.; Tschudin, Verena; de Raeve, Louise, eds. Essentials of Teaching and Learning in Nursing Ethics: Perspectives and Methods. New York: Churchill Livingstone Elsevier, 2006: 135-148. Subject: 20.5.1

Gaston, R.S.; Danovitch, G.M.; Epstein, R.A.; Kahn, J.P.; Matas, A.J.; Schnitzler, M.A. Limiting financial disincentives in live organ donation: a rational solution to the kidney shortage. *American Journal of Transplantation* 2006 November; 6(11): 2548-2555. Subject: 19.3

Gasull, Maria. Spain: professionalism and issues within nursing between nursing and other health professions. *In:* Davis, Anne J.; Tschudin, Verena; de Raeve, Louise, eds. Essentials of Teaching and Learning in Nursing Ethics: Perspectives and Methods. New York: Churchill Livingstone Elsevier, 2006: 313-321. Subject: 4.1.3

Gathii, James Thuo. Third world perspectives on global pharmaceutical access. *In:* Santoro, Michael A.; Gorrie, Thomas M., eds. Ethics and the Pharmaceutical Industry. Cambridge; New York: Cambridge University Press, 2005: 336-351. Subject: 9.7

Gavaghan, Colin. Right problem, wrong solution: a pro-choice response to "expressivist" concerns about preimplantation genetic diagnosis. *CQ: Cambridge Quarterly of Healthcare Ethics* 2007 Winter; 16(1): 20-34. Subject: 15.2

Gavel, Ylva; Andersson, Per-Olov; Knutssøn, Gun-Brit. Euroethics — a database network on biomedical ethics. *Health Information and Libraries Journal* 2006 September; 23(3): 169-178. Subject: 1.3.12

Gavrin, Jonathan R. Ethical considerations at the end of life in the intensive care unit. *Critical Care Medicine* 2007 February; 35(2, Supplement): S85-S94. Subject: 20.5.1

Gawande, Atul. A lifesaving checklist [op-ed]. *New York Times* 2007 December 30; p. WK8. Subject: 18.3

Gazelle, Gail. Understanding hospice — an underutilized option for life's final chapter. *New England Journal of Medicine* 2007 July 26; 357(4): 321-324. Subject: 20.4.1

Gazzaniga, Michael S. Facts, fictions and the future of neuroethics. *In:* Illes, Judy, ed. Neuroethics: Defining the Issues in Theory, Practice, and Policy. New York: Oxford University Press, 2006: 141-148. Subject: 17.1

Gbadegesin, Segun. The moral weight of culture in ethics. *In:* Prograis, Lawrence J.; Pellegrino, Edmund D., eds. African American Bioethics: Culture, Race, and Identity. Washington, DC: Georgetown University Press, 2004: 25-45. Subject: 2.1

Gedge, E.; Giacomini, M.; Cook, D. Withholding and withdrawing life support in critical care settings: ethical issues concerning consent. *Journal of Medical Ethics* 2007 April; 33(4): 215-218. Subject: 20.5.1

Gefenas, Eugenijus. Balancing ethical principles in emergency medicine research. *Science and Engineering Ethics* 2007 September; 13(3): 281-288. Subject: 18.5.1

Geier, G. Richard. Professionalism, ethics, and trust. *Minnesota Medicine* 2007 June; 90(6): 20. Subject: 8.1

Geissler, P.W.; Pool, R. Popular concerns about medical research projects in sub-Saharan Africa — a critical voice in debates about medical research ethics [editorial]. *Tropical Medicine and International Health* 2006 July; 11(7): 975-982. Subject: 18.2

Gelfand, Scott. Ectogenesis and the ethics of care. *In:* Gelfand, Scott; Shook, John R., eds. Ectogenesis: Artificial Womb Technology and the Future of Human Reproduction. Amsterdam; New York: Editions Rodopi, B.V., 2006: 89-107. Subject: 14.1

Gelsinger, Paul L. Uninformed consent: the case of Jesse Gelsinger. *In:* Lemmens, Trudo; Waring, Duff R., eds. Law and Ethics in Biomedical Research: Regulation, Conflict of Interest and Liability. Toronto; Buffalo: University of Toronto Press, 2006: 12-32. Subject: 15.4

Genetics and Public Policy Center. FDA Regulation of Genetic Tests. Washington, D.C.: Genetics and Public Policy Center 2007 September 27: 2 p. [Online]. Accessed: http://www.dnapolicy.org/images/issuebriefspdfs/ FDA_Regulation_of_Genetic_Test_Issue_Brief.pdf [2008 January 7]. Subject: 15.3

Genetics and Public Policy Center. Who regulates genetic tests? Washington, DC: Genetics and Public Policy 2007 September 27: 2p [Online]. Accessed: http://www. dnapolicy.org/images/issuesbriefspdfs/Who_Regulates_ Genetic_Tests_Issue_Brief.pdf [2008 January 7]. Subject: 15.3

Genetics Committee of the Society of Obstetricians and Gynaecologists of Canada; Wilson, R. Douglas; Desilets, Valerie; Gagnon, Alain; Summers, Anne; Wyatt, Philip; Allen, Victoria; Langlois, Sylvie. Present role of stem cells for fetal genetic therapy = Rôle actuel des cellules souches en matière de thérapie génique fœtale. *JOGC: Journal of Obstetrics and Gynaecology Canada = JOGC: Journal d'Obstétrique et Gynécologie du Canada* 2005 November; 27(11): 1038-1047. Subject: 18.5.4

Gentry, Glenn. Rawls and religious community: ethical decision making in the public square. *Christian Bioethics* 2007 May-August; (13)2: 171-181. Subject: 2.1

Genuis, Stephen J. Diagnosis: contemporary medical hubris; Rx: a tincture of humility. *Journal of Evaluation in Clinical Practice* 2006 February; 12(1): 24-30. Subject: 9.8

Genuis, S.J. Dismembering the ethical physician. *Postgraduate Medical Journal* 2006 April; 82(966): 233-238. Subject: 4.1.2

George, Alison. Body swap. *New Scientist* 2007 April 21-27; 194(2600): 40-43. Subject: 10

George, James F. Xenotransplantation: an ethical dilemma. *Current Opinion in Cardiology* 2006 March; 21(2): 138-141. Subject: 19.1

George, Tom; Van Oeveren, Edward L.; Gostin, Lawrence O. Using law to facilitate healthier lifestyles [letters and reply]. *JAMA: The Journal of the American Medical Association* 2007 May 9; 297(18): 1981-1983. Subject: 9.1

See SUBJECT HEADING KEY FOR SECTION II on inside back cover.

499

Georges, Jean-Jacques; Onwuteaka-Philipsen, Bregje D.; van der Heide, Agnes; van der Wal, G.; van der Maas, P.J. Physicians' opinions on palliative care and euthanasia in the Netherlands. *Journal of Palliative Medicine* 2006 October; 9(5): 1137-1144. Subject: 20.4.1

Georges, Jean-Jacques.; Onwuteaka-Philipsen, Bregje D.; Muller, Martien T.; Van Der Wal, Gerrit.; Van Der Heide, Agnes; Van Der Maas, Paul J. Relatives' perspective on the terminally ill patients who died after euthanasia or physician-assisted suicide: a retrospective cross-sectional interview study in the Netherlands. *Death Studies* 2007 January-February; 31(1): 1-15. Subject: 20.5.1

Geppert, Cynthia M.A.; Abbott, Christopher. Voluntarism in consultation psychiatry: the forgotten capacity. *American Journal of Psychiatry* 2007 March; 164(3): 409-413. Subject: 8.3.1

Gerber, Andreas; Hentzelt, Frieder; Lauterbach, Karl W. Can evidence-based medicine implicitly rely on current concepts of disease or does it have to develop its own definition? *Journal of Medical Ethics* 2007 July; 33(7): 394-399. Subject: 9.8

Gerlach, Neil. Biotechnology and social control: the Canadian DNA data bank. *In:* Mehta, Michael D., ed. Biotechnology Unglued: Science, Society and Social Cohesion. Vancouver: UBC Press, 2005: 117-132. Subject: 15.1

German Society of Human Genetics [DGfH]. Committee for Public Relations and Ethical Issues. Statement on population screening for heterozygotes [position statement]. *Medizinische Genetik* 1991; 3(2): 11-12. [Online]. Accessed: http://www.medgenetik.de/sonderdruck/en/Heterozygote_screening.pdf [2006 July 31]. Subject: 15.3

Gerson, Michael. The eugenics temptation [op-ed]. *Washington Post* 2007 October 24; p. A19. Subject: 15.5

Gert, Bernard. How common morality relates to business and the professions. *In:* Korthals, Michiel; Bogers, Robert J., eds. Ethics for Life Scientists. Dordrecht, The Netherlands: Springer, 2004: 129-139. Subject: 1.3.9

Gesche, Astrid H. Genetic testing and human genetic databases. *In:* Betta, Michela, ed. The Moral, Social, and Commercial Imperatives of Genetic Testing and Screening: the Australian Case. Dordrecht: Springer, 2006: 71-94. Subject: 15.1

Gesche, Astrid H. Protecting the vulnerable: genetic testing and screening for parentage, immigration, and aboriginality. *In:* Betta, Michela, ed. The Moral, Social, and Commercial Imperatives of Genetic Testing and Screening: the Australian Case. Dordrecht: Springer, 2006: 221-236. Subject: 15.3

Gesundheit, Benjamin; Steinberg, Avraham; Glick, Shimon; Or, Reuven; Jotkovitz, Alan. Euthanasia: an overview and the Jewish perspective. *Cancer Investigation* 2006 October; 24(6): 621-629. Subject: 20.5.1

Gevers, Sjef. Evaluation of the Dutch legislation on euthanasia and assisted suicide. *European Journal of Health Law* 2007 December; 14(4): 369-379. Subject: 20.7

Gevers, Sjef. Human tissue research, with particular reference to DNA banking. *In:* Gevers, J.K.M.; Hondius, E.H.; Hubben, J.H., eds. Health Law, Human Rights and the Biomedicine Convention: Essays in Honour of Henriette Roscam Abbing. Leiden; Boston: Martinus Nijhoff Publishers, 2005: 231-243. Subject: 15.1

Gewin, Virginia. Crunch time for multiple-gene tests. *Nature* 2007 January 25; 445(7126): 354-355. Subject: 15.3

Ghaemi, S. Nassir; Goodwin, Frederick K. The ethics of clinical innovation in psychopharmacology: challenging traditional bioethics. *Philosophy, Ethics, and Humanities in Medicine [electronic]* 2007 November 8; 2(26): 8 p. Accessed: http://www.peh-med.com/ [2008 January 24]. Subject: 17.4

Ghayur, Muhammad N.; Ghayur, Ayesha; Janssen, Luke J. State of clinical research ethics in Pakistan [letter]. *Nature Medicine* 2007 September; 13(9): 1011. Subject: 18.5.9

Giacomini, Mita; Baylis, Françoise; Robert, Jason. Banking on it: public policy and the ethics of stem cell research and development. *Social Science and Medicine* 2007 October; 65(7): 1490-1500. Subject: 18.5.4

Giampietro, Anthony E. Improving the Catholic approach to healthcare? [review of More Humane Medicine: A Liberal Catholic Bioethics, by James F. Drane]. *HEC (Healthcare Ethics Committee) Forum* 2007 September; 19(3): 261-270. Subject: 2.1

Gianelli, Diane M.; Davis, F. Daniel. News from the President's Council on Bioethics. *Kennedy Institute of Ethics Journal* 2007 December; 17(4): 397-398. Subject: 2.4

Gibson, Susanne. Uses of respect and uses of the human embryo. *Bioethics* 2007 September; 21(7): 370-378. Subject: 4.4

Gifford, Fred. Pulling the plug on clinical equipoise: a critique of Miller and Weijer. *Kennedy Institute of Ethics Journal* 2007 September; 17(3): 203-226. Subject: 18.2

Gifford, Fred. So-called "clinical equipoise" and the argument from design. *Journal of Medicine and Philosophy* 2007 March-April; 32(2): 135-150. Subject: 18.2

Gifford, Fred. Taking equipoise seriously: the failure of clinical or community equipoise to resolve the ethical dilemmas in randomized clinical trials. *In:* Kincaid, Harold; McKitrick, Jennifer, eds. Establishing Medical Reality: Essays in the Metaphysics and Epistemology of Biomedical Science. Dordrecht, The Netherlands: Springer, 2007: 215-233. Subject: 18.2

Gilam, Lynn. What is bioethics all about? *Monash Bioethics Review* 2000 October; 19(4): 51-54. Subject: 2.1

Gilani, Ahmed I.; Jadoon, Atif S.; Qaiser, Rabia; Nasim, Sana; Meraj, Riffat; Nasir, Nosheen; Naqvi,

Fizza F.; Latif, Zafar; Memon, Muhammad A.; Menezes, Esme V.; Malik, Imran; Memon, Muhammad Z.; Kazim, Syed F.; Ahmad, Usman. Attitudes towards genetic diagnosis in Pakistan: a survey of medical and legal communities and parents of thalassemic children. *Community Genetics* 2007; 10(3): 140-146. Subject: 15.3

Gilbar, Roy. Communicating genetic information in the family: the familial relationship as the forgotten factor. *Journal of Medical Ethics* 2007 July; 33(7): 390-393. Subject: 15.2

Gilbar, Roy. Patient autonomy and relatives' right to know genetic information. *Medicine and Law: The World Association for Medical Law* 2007 December; 26(4): 677-697. Subject: 15.1

Giles, Jim. Breeding cheats [news]. *Nature* 2007 January 18; 445(7125): 242-243. Subject: 1.3.9

Giles, Jim. Court case to reclaim confidential data. *Nature* 2007 April 19; 446(7138): 838-839. Subject: 9.7

Giles, Jim. Drug firms accused of biasing doctors' training [news]. *Nature* 2007 November 22; 450(7169): 464-465. Subject: 7.3

Giles, Jim. Say no to lunch [commentary]. *New Scientist* 2007 April 28-May 4; 194(2601): 18. Subject: 9.7

Giles, Jim. US vaccines on trial over link to autism. *New Scientist* 2007 June 23-29; 194(2609): 6-7. Subject: 9.7

Gillett, Grant. The use of human tissue. *Journal of Bioethical Inquiry* 2007; 4(2): 119-127. Subject: 19.1

Gillett, Grant; Chisholm, Nick. Locked in syndrome, PVS and ethics at the end of life. *Journal of Ethics in Mental Health [electronic]* 2007 November; 2(2): 6 p. Accessed: http://www.jemh.ca [2008 January 24]. Subject: 20.5.1

Gillick, Muriel R. The technological imperative and the battle for the hearts of America. *Perspectives in Biology and Medicine* 2007 Spring; 50(2): 276-294. Subject: 5.2

Gilling-Smith, Carole. Risking parenthood? Serious viral illness, parenting and the welfare of the child. *In:* Shenfield, Françoise; Sureau, Claude, eds. Contemporary Ethical Dilemmas in Assisted Reproduction. Abingdon: Informa Healthcare, 2006: 57-69. Subject: 9.5.6

Gilman, Paul. A conflict-of-interest policy for epidemiology. *Epidemiology* 2006 May; 17(3): 250-251. Subject: 1.3.9

Giordano, James. Cassandra's curse: interventional pain management, policy and preserving meaning against a market mentality. *Pain Physician* 2006 July; 9(3): 167-169. Subject: 4.4

Giordano, James. Hospice, palliative care, and pain medicine: meeting the obligations of non-abandonment and preserving the personal dignity of terminally ill patients. *Delaware Medical Journal* 2006 November; 78(11): 419-422. Subject: 20.4.1

Giordano, James; Ernst, E. Informed consent: a potential dilemma for complementary medicine [letter and reply]. *Journal of Manipulative and Physiological Therapeutics* 2004 November-December; 27(9): 596-597. Subject: 8.3.1

Giordano, Simona; Cappato, Marco. Scientific freedom [editorial]. *Journal of Medical Ethics* 2007 June; 33(6): 311-312. Subject: 18.5.4

Glannon, Walter. Brain death. *In his:* Bioethics and the Brain. New York: Oxford University Press, 2007: 148-177. Subject: 20.2.1

Glannon, Walter. Neurosurgery, psychosurgery, and neurostimulation. *In his:* Bioethics and the Brain. New York: Oxford University Press, 2007: 116-147. Subject: 17.6

Glannon, Walter. Persons, metaphysics and ethics. *American Journal of Bioethics* 2007 January; 7(1): 68-69. Subject: 17.1

Glannon, Walter. Pharmacological and psychological interventions. *In his:* Bioethics and the Brain. New York: Oxford University Press, 2007: 76-115. Subject: 17.4

Glanville, A.R. Ethical and equity issues in lung transplantation and lung volume reduction surgery. *Chronic Respiratory Disease.* 2006; 3(1): 53-58. Subject: 19.5

Glaser, John W. Catholic health ministry: fruit on the diseased tree of U.S. health care. *Health Care Ethics USA* 2007 Winter; 15(1): 2-4. Subject: 9.1

Glaser, Nicole; Kuppermann, Nathan; Marcin, James; Schalick, Walton O. A comment on "The risky business of assessing research risk". *American Journal of Bioethics* 2007 November; 7(11): W5-W6. Subject: 18.5.2

Glass, Kathleen Cranley. Question and challenges in the governance of research involving humans: a Canadian perspective. *In:* Lemmens, Trudo; Waring, Duff R., eds. Law and Ethics in Biomedical Research: Regulation, Conflict of Interest and Liability. Toronto; Buffalo: University of Toronto Press, 2006: 35-46. Subject: 18.6

Glass, Kathleen Cranley; Kaufert, Joseph. Research ethics review and aboriginal community values: can the two be reconciled? *Journal of Empirical Research on Human Research Ethics* 2007 June; 2(2): 25-40. Subject: 18.6

Glazier, Alexandra K.; Sasjack, Scott. Should it be illicit to solicit? A legal analysis of policy options to regulate solicitation of organs for transplant. *Health Matrix: The Journal of Law-Medicine* 2007 Winter; 17(1): 63-99. Subject: 19.5

Gleicher, N.; Weghofer, A.; Barad, D. On the benefit of assisted reproduction techniques, a comparison of the USA and Europe [letter]. *Human Reproduction* 2007 February; 22(2): 624-626. Subject: 14.4

Glenn, David. A policy on torture roils psychologists' annual meeting. *Chronicle of Higher Education* 2007 September 7; 54(2): A16-A17. Subject: 21.4

See SUBJECT HEADING KEY FOR SECTION II on inside back cover.

501

Glenn, David. Questions plague landmark paper on versatility of adult stem cells. *Chronicle of Higher Education* 2007 March 9; 53(27): A22. Subject: 18.5.1

Glenn, Linda MacDonald; Boyce, Jeanann; Lawrence, Ryan E.; Curlin, Farr A. The Tao of conscience: conflict and resolution. *American Journal of Bioethics* 2007 December; 7(12): 33-34; author reply W1-W2. Subject: 4.1.2

Glick, Michael. Scientific fraud — real consequences. *Journal of the American Dental Association* 2006 April; 137(4): 428, 430. Subject: 1.3.9

Glick, Shimon; Jotkowitz, Alan. Compromise and dialogue in bioethical disputes. *American Journal of Bioethics* 2007 October; 7(10): 36-38. Subject: 2.1

Glover-Thomas, Nicola. A new 'new' Mental Health Act? Reflections on the proposed amendments to the Mental Health Act 1983. *Clinical Ethics* 2007 March; 2(1): 28-31. Subject: 17.1

Godard, Béatrice. Involving communities: a matter of trust and communication. *In:* Einsiedel, Edna; Timmermans, Frank, eds. Crossing Over: Genomics in the Public Arena. Calgary, Alberta, Canada: University of Calgary Press, 2005: 87-98. Subject: 15.1

Godard, Béatrice; Marshall, Jennifer; Laberge, Claude. Community engagement in genetic research: results of the first public consultation for the Quebec CARTaGENE project. *Community Genetics* 2007; 10(3): 147-158. Subject: 15.1

Godard, Béatrice; Pratte, Annabelle; Dumont, Martine; Simard-Lebrun, Adèle; Simard, Jacques. Factors associated with an individual's decision to withdraw from genetic testing for breast and ovarian cancer susceptibility: implications for counseling. *Genetic Testing* 2007 Spring; 11(1): 45-54. Subject: 15.3

Godlaski, Theodore M.; Johnson, Jeannette; Haring, Rodney. Reflections on ethical issues in research with aboriginal peoples. *In:* Kleinig, John; Einstein, Stanley, eds. Ethical Challenges for Intervening in Drug Use: Policy, Research and Treatment Issues. Huntsville, TX: Office of International Criminal Justice; Criminal Justice Center, Sam Houston State University, 2006: 281-305. Subject: 18.5.1

Godlee, Fiona. Ethical assets at the BMJ [editorial]. *BMJ: British Medical Journal* 2007 February 24; 334(7590): 374. Subject: 1.3.7

Gold, E. Richard; Bubela, Tania; Miller, Fiona A.; Nicol, Dianne; Piper, Tina. Gene patents — more evidence needed, but policymakers must act [letter]. *Nature Biotechnology* 2007 April; 25(4): 388-389. Subject: 15.8

Goldacre, Ben. Why don't journalists mention the data? *BMJ: British Medical Journal* 2007 June 16; 334(7606): 1249. Subject: 1.3.7

Goldberg, Daniel S.; Brody, Howard. Spirituality: respect but don't reveal. *American Journal of Bioethics* 2007 July; 7(7): 21-22. Subject: 8.1

Goldberg, Daniel S.; Volandes, Angelo E.; Paasche-Orlow, Michael K. Justice, health literacy and social epidemiology. *American Journal of Bioethics* 2007 November; 7(11): 18-20; author reply W1-W2. Subject: 8.1

Goldberg, Jordan. The Commerce Clause and federal abortion law: why progressives might be tempted to embrace federalism. *Fordham Law Review* 2006 October; 75(1): 301-354. Subject: 12.4.1

Goldblatt, David; Greenlaw, Jane. Starting and stopping the ventilator for patients with amyotrophic lateral sclerosis. *Neurologic Clinics* 1989 November; 7(4): 789-806. Subject: 20.5.1

Goldim, José Roberto; Clotet, Joaquim; Ribeiro, Jorge Pinto. Adequacy of informed consent in research carried out in Brazil. *Eubios Journal of Asian and International Bioethics* 2007 November; 17(6): 177-180. Subject: 18.3

Goldman, Bruce. The first cut. *Nature* 2007 February 1; 445(7127): 479-480. Subject: 15.2

Goldman, Michael A. Calamity gene: when biotechnology spins out of control [review of Next, by Michael Crichton]. *Nature* 2007 February 22; 445(7130): 819-820. Subject: 15.1

Gómez-Lobo, Alfonso. A note on metaphysics and embryology [letter]. *Theoretical Medicine and Bioethics* 2007; 28(4): 331-335. Subject: 4.4

Gómez-Lobo, Alfonso. Inviolability at any age. *Kennedy Institute of Ethics Journal* 2007 December; 17(4): 311-320. Subject: 4.4

González San Segundo, Carmen; Santos Miranda, Juan A. Informed consent in radiation oncology: is consenting easier than informing? *Clinical and Translational Oncology* 2006 November; 8(11): 802-804. Subject: 8.3.1

Gonzalez, Luis S., 3rd; Miller, Stephanie; Barnhart, Donna; Leifheit, Michael. Institutional review board approval of projects presented as posters at an ASHP midyear clinical meeting. *American Journal of Health-System Pharmacy* 2005 September 15; 62(18): 1890-1893. Subject: 18.2

Goodman, Kenneth W.; Allen, Bill; Cerminara, Kathy L.; Fiore, Robin N.; Moseley, Ray; Mulvey, Ben; Spike, Jeffrey; Walker, Robert M. Florida bioethics leaders' analysis on HB701. Miami: University of Miami, 2005 March 7; 7 p. [Online]. Accessed: http://www6.miami.edu/ethics/schiavo/pdf_files/030805-HB701-EthicsAnalysis.pdf [2007 April 5]. Subject: 20.5.1

Goodman, Robert L. Medical education and the pharmaceutical industry. *Perspectives in Biology and Medicine* 2007 Winter; 50(1): 32-39. Subject: 7.2

Goodman, Steven N. Ethics and evidence in clinical trials [editorial]. *Clinical Trials* 2005 June; 2(3): 195-196. Subject: 18.2

Goodman, Steven N. Stopping at nothing? Some dilemmas of data monitoring in clinical trials [commentary]. *Annals of Internal Medicine* 2007 June 19; 146(12): 882-887. Subject: 18.2

Goodyear, Michael. Free access to medical information: a moral right [letter]. *CMAJ/JAMC: Canadian Medical Association Journal* 2007 January 2; 176(1): 69. Subject: 1.3.9

Goodyear, Michael D.E.; Krleza-Jeric, Karmela; Lemmens, Trudo. The Declaration of Helsinki: mosaic tablet, dynamic document, or dinosaur [editorial]. *BMJ: British Medical Journal* 2007 September 29; 335(7621): 624-625. Subject: 18.2

Gordijn, Bert. Genetic diagnosis, confidentiality and counseling: an ethics committee's potential deliberations about the do's and don'ts. *HEC (Healthcare Ethics Committee) Forum* 2007 December; 19(4): 303-312. Subject: 9.6

Gordon, Elisa J. A better way to evaluate clinical ethics consultations? An ecological approach. *American Journal of Bioethics* 2007 February; 7(2): 26-29. Subject: 9.6

Gordon, Elisa J.; Wolf, Michael S.; Volandes, Angelo E.; Paasche-Orlow, Michael K. Beyond the basics: designing a comprehensive response to low health literacy. *American Journal of Bioethics* 2007 November; 7(11): 11-13; author reply W1-W2. Subject: 9.1

Görgülü, Refia Selma; Dinç, Leyla. Ethics in Turkish nursing education programs. *Nursing Ethics* 2007 November; 14(6): 741-752. Subject: 7.2

Gormally, Luke. Why not dirty your hands? Or: on the supposed rightness of (sometimes) intentionally cooperating in wrongdoing. *In:* Watt, Helen, ed. Cooperation, Complicity and Conscience: Problems in Healthcare, Science, Law and Public Policy. London: Linacre Centre, 2005: 12-26. Subject: 2.1

Gorman, Dennis M. Conflicts of interest in the evaluation and dissemination of drug use prevention programs. *In:* Kleinig, John; Einstein, Stanley, eds. Ethical Challenges for Intervening in Drug Use: Policy, Research and Treatment Issues. Huntsville, TX: Office of International Criminal Justice; Criminal Justice Center, Sam Houston State University, 2006: 171-187. Subject: 9.5.9

Görman, Ulf. Never too late to live a little longer? The quest for extended life and immortality — some ethical considerations. *In:* Deane-Drummond, Celia; Scott, Peter Manley, eds. Future Perfect?: God, Medicine and Human Identity. New York: T and T Clark International, 2006: 143-154. Subject: 4.5

Gornall, Jonathan. Duplicate publication: a bitter dispute. *BMJ: British Medial Journal* 2007 April 7; 334(7596): 717-720. Subject: 1.3.7

Gornall, Jonathan. Hyperactivity in children: the Gillberg affair. *BMJ: British Medical Journal* 2007 August 25; 335(7616): 370-373. Subject: 1.3.9

Gornall, Jonathan. Where do we draw the line? Numerous attempts have been made to change the rules on abortion since it was legalised 40 years ago. *BMJ: British Medical Journal* 2007 February 10; 334(7588): 285-289. Subject: 12.4.1

Gorovitz, Samuel. The centrality of marginalization. *Monash Bioethics Review* 2000 October; 19(4): 49-51. Subject: 2.1

Gorovitz, Samuel. The past, present and future of human nature. *In:* Galston, Arthur W.; Peppard, Christiana Z., eds. Expanding Horizons in Bioethics. Dordrecht; Norwell, MA: Springer, 2005: 3-18. Subject: 5.3

Gosden, Roger. Genetic test may lead to waste of healthy embryos [letter]. *Nature* 2007 March 22; 446(7134): 372. Subject: 15.2

Gostin, Lawrence O. "Police" powers and public health paternalism: HIV and diabetes surveillance. *Hastings Center Report* 2007 March-April; 37(2): 9-10. Subject: 9.1

Gostin, Lawrence O. A theory and definition of public health law. *Journal of Health Care Law and Policy* 2007; 10(1): 1-12. Subject: 9.1

Gostin, Lawrence O. Abortion politics: clinical freedom, trust in the judiciary, and the autonomy of women. *JAMA: The Journal of the American Medical Association* 2007 October 3; 298(13): 1562-1564. Subject: 12.4.2

Gostin, Lawrence O. Biomedical research involving prisoners: ethical values and legal regulation. *JAMA: The Journal of the American Medical Association* 2007 February 21; 297(7): 737-740. Subject: 18.5.5

Gostin, Lawrence O. Global climate change: the Roberts Court and environmental justice. *Hastings Center Report* 2007 September-October; 37(5): 10-11. Subject: 16.1

Gostin, Lawrence O. Law as a tool to facilitate healthier lifestyles and prevent obesity. *JAMA: The Journal of the American Medical Association* 2007 January 3; 297(1): 87-90. Subject: 9.1

Gostin, Lawrence O. Meeting the survival needs of the world's least healthy people: a proposed model for global health governance. *JAMA: The Journal of the American Medical Association* 2007 July 11; 298(2): 225-228. Subject: 9.1

Gostin, Lawrence O. The international health regulations: a new paradigm for global health governance? *In:* McLean, Sheila A.M., ed. First Do No Harm: Law, Ethics, and Healthcare. Aldershot, England; Burlington, VT: Ashgate, 2006: 59-79. Subject: 21.1

See SUBJECT HEADING KEY FOR SECTION II on inside back cover.

Gostin, Lawrence O. Why rich countries should care about the world's least healthy people [commentary]. *JAMA: The Journal of the American Medical Association* 2007 July 4; 298(1): 89-92. Subject: 9.1

Gostin, Lawrence O. Why should we care about social justice? *Hastings Center Report* 2007 July-August; 37(4): 3. Subject: 9.1

Gostin, Lawrence O.; DeAngelis, Catherine D. Mandatory HPV vaccination: public health vs. private wealth [editorial]. *JAMA: The Journal of the American Medical Association* 2007 May 2; 297(17): 1921-1923. Subject: 9.5.1

Gottdiener, William H. Is harm reduction psychotherapy ethical? *In:* Kleinig, John; Einstein, Stanley, eds. Ethical Challenges for Intervening in Drug Use: Policy, Research and Treatment Issues. Huntsville, TX: Office of International Criminal Justice; Criminal Justice Center, Sam Houston State University, 2006: 91-98. Subject: 17.2

Gottstein, James B. Psychiatrists' failure to inform: is there substantial financial exposure? *Ethical Human Psychology and Psychiatry* 2007; 9(2): 117-125. Subject: 17.8

Goz, Fugen; Goz, Mustafa; Erkan, Medine. Knowledge and attitudes of medical, nursing, dentistry and health technician students towards organ donation: a pilot study. *Journal of Clinical Nursing* 2006 November; 15(11): 1371-1375. Subject: 19.5

Graber, Mark A. "Can I have that drug I saw on TV?" Justice, cost-effectiveness, and the ethics of prescribing. *JAAPA: Official Journal of the American Academy of Physician Assistants* 2006 July; 19(7): 48-49. Subject: 9.8

Grace, Jan; Drakeley, Andrew. Preimplantation genetic diagnosis. *British Journal of Hospital Medicine* 2006 April; 67(4): 197-199. Subject: 15.2

Grady, Christine. Ethical principles in clinical research. *In:* Gallin, John I.; Ognibene, Frederick P., eds. Principles and Practice of Clinical Research. 2nd edition. Oxford: Academic, 2007: 15-26. Subject: 18.1

Grady, Christine. Quality improvement and ethical oversight [editorial]. *Annals of Internal Medicine* 2007 May 1; 146(9): 680-681. Subject: 9.8

Grady, Denise. Girl or boy? As fertility technology advances, so does an ethical debate. *New York Times* 2007 February 6; p. F5, F10. Subject: 14.3

Grady, Denise. White doctors, black subjects: abuse disguised as research [review of Medical Apartheid: The Dark History of Medical Experimentation on Black Americans from Colonial Times to the Present by Harriet A. Washington]. *New York Times* 2007 January 23; p. F5, F8. Subject: 18.5.5

Graham, Bruce S. Educating dental students about oral health care access disparities. *Journal of Dental Education* 2006 November; 70(11): 1208-1211. Subject: 7.2

Graham, David Y.; Yamaoka, Yoshio. Ethical considerations of comparing sequential and traditional anti-helicobacter pylori therapy [letter]. *Annals of Internal Medicine* 2007 September 18; 147(6): 434-436. Subject: 18.2

Graham, Elaine. In whose image? Representations of technology and the 'ends' of humanity. *In:* Deane-Drummond, Celia; Scott, Peter Manley, eds. Future Perfect?: God, Medicine and Human Identity. New York: T and T Clark International, 2006: 56-69. Subject: 15.1

Graham, Gordon. Human nature and the human condition. *In:* Deane-Drummond, Celia; Scott, Peter Manley, eds. Future Perfect?: God, Medicine and Human Identity. New York: T and T Clark International, 2006: 33-44. Subject: 15.1

Graham, Jane H. Community care or therapeutic stalking: two sides of the same coin? *Journal of Psychosocial Nursing and Mental Health Services* 2006 August; 44(8): 41-47. Subject: 9.3.1

Graham, John; Hu, Jianhui. The risk-benefit balance in the United States: who decides? *Health Affairs* 2007 May-June; 26(3): 625-635. Subject: 5.2

Grainger-Monsen, Maren: Karetsky, Kim. The mind in the movies: a neuroethical analysis of the portrayal of the mind in popular media. *In:* Illes, Judy, ed. Neuroethics: Defining the Issues in Theory, Practice, and Policy. New York: Oxford University Press, 2006: 297-311. Subject: 17.1

Grande, David. Prescriber profiling: time to call it quits [editorial]. *Annals of Internal Medicine* 2007 May 15; 146(10): 751-752. Subject: 9.7

Grasser, Phyllis L. Donation after cardiac death: major ethical issues. *National Catholic Bioethics Quarterly* 2007 Autumn; 7(3): 527-543. Subject: 19.5

Gratwohl, Alois. "Therapeutisches Klonen" aus der Sicht eines Klinikers. *In:* Schreiber, Hans-Peter, ed. Biomedizin und Ethik: Praxis, Recht, Moral. Basel; Boston: Birkhäuser, 2004: 23-28. Subject: 14.5

Graumann, Sigrid. Ethik in der Medizin und ihre Aufgaben in der Politik [Ethics in medicine and its role in politics]. *Ethik in der Medizin* 2006 December; 18(4): 359-363. Subject: 2.1

Gray, Glenda. When bodies remember [review of Experiences and Politics of AIDS in South Africa, by Didier Fassin]. *New England Journal of Medicine* 2007 October 25; 357(17): 1783-1784. Subject: 9.5.6

Gray, Natalie; Bailie, Ross. Can human rights discourse improve the health of indigenous Australians? *Australian and New Zealand Journal of Public Health* 2006 October; 30(5): 448-452. Subject: 9.5.4

Grayling, A.C. Cheating death: will we have to choose who lives? *New Scientist* 2007 October 13-19; 195(2625): 47. Subject: 20.5.1

Subject = NRCBL Primary Classification Number; see inside front cover.

Grazi, Richard V.; Wolowelsky, Joel B. Addressing the idiosyncratic needs of Orthodox Jewish couples requesting sex selection by preimplantation genetic diagnosis (PGD). *Journal of Assisted Reproduction and Genetics* 2006 November-December; 23(11-12): 421-425. Subject: 14.3

Great Britain (United Kingdom). Department of Health. Bournewood briefing sheet. London: Department of Health, 2006 June; 8 p. [Online]. Accessed: http://www.dh.gov.uk/prod_consum_dh/idcplg?IdcService=GET_FILE&dID=13928&Rendition=Web [2007 April 17]. Subject: 8.3.3

Great Britain (United Kingdom). Department of Health. Concordat and moratorium on genetics and insurance. London: Department of Health, 2005 March; 6 p. [Online]. Accessed: http://www.dh.gov.uk/prod_consum_dh/idcplg?IdcService=GET_FILE&dID=384&Rendition=Web [2007 April 30]. Subject: 15.3

Great Britain (United Kingdom). Department of Health. Consent — What You Have a Right to Expect: A Guide for Parents. London: Department of Health, 2001 July; 11 p. Subject: 8.3.2

Great Britain (United Kingdom). Department of Health. Reference guide to consent for examination or treatment. London: Department of Health. 2001 April 6; 30 p. [Online]. Accessed: http://www.dh.gov.uk/prod_consum_dh/idcplg?IdcService=GET_FILE&dID=29069&Rendition=Web [2007 April 12]. Subject: 8.3.1

Great Britain (United Kingdom). Department of Health. Research governance framework for health and social care: second edition. London: Department of Health, 2005 April; 49 p. [Online]. Accessed: http://www.dh.gov.uk/prod_consum_dh/idcplg?IdcService=GET_FILE&dID=26829&Rendition=Web [2007 April 25]. Subject: 6

Great Britain (United Kingdom). Department of Health. UK Stem Cell Initiative. UK stem cell initiative: report and recommendations. London: UK Stem Cell Initiative, 2005 November; 118 p. [Online]. Accessed: http://www.advisorybodies.doh.gov.uk/uksci/uksci-reportnov05.pdf [2007 April 26]. Subject: 18.5.4

Great Britain (United Kingdom). Human Fertilisation and Embryology Authority [HFEA]. The HFEA guide to infertility. London: Human Fertilisation and Embryology Authority, 2007-2008; 51 p. [Online]. Accessed: http://www.hfea.gov.uk/docs/Guide2.pdf [2007 April 5]. Subject: 14.1

Great Britain (United Kingdom). National Health Service. Health Service Guidelines: ethics committee review of multi-centre research: establishment of multi-centre research ethics committees: HSG (97)23. London: National Health Service (NHS): 1997 April 14; 3 p. [Online]. Accessed: http://www.dh.gov.uk/en/Publicationsandstatistics/Lettersandcirculars/Healthserviceguidelines/DH_4018331 [2007 October 9]. Subject: 18.2

Great Britain (United Kingdom). Secretary of State for Health. Government response to the health committee's report on the influence of the pharmaceutical industry. London: Secretary of State for Health, 2005 September; 24 p. [Online]. Accessed: http://www.dh.gov.uk/prod_consum_dh/groups/dh_digitalassets/@dh/@en/documents/digitalasset/dh_4118608.pdf [2007 April 18]. Subject: 9.7

Greely, Henry. On neuroethics [editorial]. *Science* 2007 October 26; 318(5850): 533. Subject: 17.1

Greely, Henry T. The social effects of advances in neuroscience: legal problems, legal perspectives. *In:* Illes, Judy, ed. Neuroethics: Defining the Issues in Theory, Practice, and Policy. New York: Oxford University Press, 2006: 245-263. Subject: 17.1

Greely, Henry T.; Cho, Mildred K.; Hogle, Linda F.; Satz, Debra M. Response to open peer commentaries on "thinking about the human neuron mouse". *American Journal of Bioethics* 2007 May; 7(5): W4-W6. Subject: 15.1

Greely, Henry T.; Cho, Mildred K.; Hogle, Linda F.; Satz, Debra M. Thinking about the human neuron mouse. *American Journal of Bioethics* 2007 May; 7(5): 27-40. Subject: 15.1

Green, Alexander R.; Carney, Dana R.; Pallin, Daniel J.; Ngo, Long H.; Raymond, Kristal L.; Iezzoni, Lisa I.; Banaji, Mahzarin R. Implicit bias among physicians and its prediction of thrombolysis decisions for black and white patients. *JGIM: Journal of General Internal Medicine* 2007 September; 22(9): 1231-1238. Subject: 9.5.4

Green, Colin; Khan, Asad; Karmi, Ghada; Burns-Cox, Chris; Birnstingl, Martin; Halpin, David; Summerfield, Derek. Medical ethical violations in Gaza. *Lancet* 2007 December 22-2008 January 4; 370(9605): 2102. Subject: 21.2

Green, David; Cushman, Mary; Dermond, Norma; Johnson, Eric A.; Castro, Cecilia; Arnett, Donna; Hill, Joel; Manolio, Teri A. Obtaining informed consent for genetic studies: the multiethnic study of atherosclerosis. *American Journal of Epidemiology* 2006 November 1; 164(9): 845-851. Subject: 18.2

Green, Ronald M. Can we develop ethically universal embryonic stem-cell lines? *Nature Reviews Genetics* 2007 June; 8(6): 480-485. Subject: 18.5.4

Green, Ronald M. From genome to brainome: charting the lessons learned. *In:* Illes, Judy, ed. Neuroethics: Defining the Issues in Theory, Practice, and Policy. New York: Oxford University Press, 2006: 105-121. Subject: 15.1

Green, Shane K.; Lott, Jason P.; Savulescu, Julian. Is Canada's stem cell legislation unwittingly discriminatory?

See SUBJECT HEADING KEY FOR SECTION II on inside back cover.

505

American Journal of Bioethics 2007 August; 7(8): 50-52; author reply W4-W6. Subject: 18.5.4

Green, Stephen A. The ethical commitments of academic faculty in psychiatric education. *Academic Psychiatry* 2006 January-February; 30(1): 48-54 [Online]. Accessed: http://ap.psychiatryonline.org/ [2007 April 13]. Subject: 7.2

Greenberg, Stuart A.; Shuman, Daniel W. Irreconcilable conflict between therapeutic and forensic roles. *Professional Psychology: Research and Practice* 1997 February; 28(1): 50-57. Subject: 17.2

Greene, Jeremy A. Pharmaceutical marketing research and the prescribing physician. *Annals of Internal Medicine* 2007 May 15; 146(10): 742-748. Subject: 9.7

Greene, Jeremy A. Pharmaceuticals and the economy of medical knowledge. *Chronicle of Higher Education* 2007 November 30; 54(14): B12-B13. Subject: 9.7

Greene, Michael F. The intimidation of American physicians: banning partial-birth abortion. *New England Journal of Medicine* 2007 May 24; 356(21): 2128-2129. Subject: 12.4.1

Greene, Sarah M.; Geiger, Ann M. A review finds that multicenter studies face substantial challenges but strategies exist to achieve Institutional Review Board approval. *Journal of Clinical Epidemiology* 2006 August; 59(8): 784-790. Subject: 18.2

Greene, Sarah M.; Geiger, Ann M.; Harris, Emily L.; Altschuler, Andrea; Nekhlyudov, Larissa; Barton, Mary B.; Rolnick, Sharon J.; Elmore, Joann G.; Fletcher, Suzanne. Impact of IRB requirements on a multicenter survey of prophylactic mastectomy outcomes. *Annals of Epidemiology* 2006 April; 16(4): 275-278. Subject: 18.2

Greif, Karen F.; Merz, Jon F. Big science: the Human Genome Project and the public funding of science. *In their:* Current Controversies in the Biological Sciences: Case Studies of Policy Challenges from New Technologies. Cambridge, MA: MIT, 2007: 17-34. Subject: 15.10

Greif, Karen F.; Merz, Jon F. Brave new world revisited: human cloning and stem cells; the Asilomar Conference on Recombinant DNA: a model for self-regulation? *In their:* Current Controversies in the Biological Sciences: Case Studies of Policy Challenges from New Technologies. Cambridge, MA: MIT, 2007: 101-115. Subject: 14.5

Greif, Karen F.; Merz, Jon F. Concealing evidence: science, big business, and the tobacco industry. *In their:* Current Controversies in the Biological Sciences: Case Studies of Policy Challenges from New Technologies. Cambridge, MA: MIT, 2007: 205-228. Subject: 9.5.9

Greif, Karen F.; Merz, Jon F. Emerging diseases: SARS and government responses. *In their:* Current Controversies in the Biological Sciences: Case Studies of Policy Chal-

lenges from New Technologies. Cambridge, MA: MIT, 2007: 255-266. Subject: 9.5.1

Greif, Karen F.; Merz, Jon F. Manufacturing children: assisted reproductive technologies and self-regulation by scientists and clinicians. *In their:* Current Controversies in the Biological Sciences: Case Studies of Policy Challenges from New Technologies. Cambridge, MA: MIT, 2007: 77-99. Subject: 14.4

Greif, Karen F.; Merz, Jon F. Protecting the public: the FDA and new AIDS drugs. *In their:* Current Controversies in the Biological Sciences: Case Studies of Policy Challenges from New Technologies. Cambridge, MA: MIT, 2007: 117-147. Subject: 9.5.6

Greif, Karen F.; Merz, Jon F. Science in the national interest: bioterrorism and civil liberties. *In their:* Current Controversies in the Biological Sciences: Case Studies of Policy Challenges from New Technologies. Cambridge, MA: MIT, 2007: 235-254. Subject: 21.3

Greif, Karen F.; Merz, Jon F. Science misunderstood: genetically modified organisms and international trade. *In their:* Current Controversies in the Biological Sciences: Case Studies of Policy Challenges from New Technologies. Cambridge, MA: MIT, 2007: 267-287. Subject: 15.1

Greif, Karen F.; Merz, Jon F. The darker side of science: scientific misconduct. *In their:* Current Controversies in the Biological Sciences: Case Studies of Policy Challenges from New Technologies. Cambridge, MA: MIT, 2007: 229-234. Subject: 1.3.9

Greif, Karen F.; Merz, Jon F. Who lives and who dies? Organ transplantation. *In their:* Current Controversies in the Biological Sciences: Case Studies of Policy Challenges from New Technologies. Cambridge, MA: MIT, 2007: 329-365. Subject: 19.5

Greif, Karen F.; Merz, Jon F. Who owns the genome? The patenting of human genes; Who owns life? Mr. Moore's spleen; The Canavan disease case. *In their:* Current Controversies in the Biological Sciences: Case Studies of Policy Challenges from New Technologies. Cambridge, MA: MIT, 2007: 49-76. Subject: 15.8; 14.4

Grewel, Hans. Behinderung und Philosophie: Ethik-Konzepte auf dem Prüfstand. *In:* Neuer-Miebach, Therese; Wunder, Michael, eds. Bio-Ethik und die Zukunft der Medizin. Bonn: Psychiatrie-Verlag, 1998: 87-105. Subject: 9.5.1

Grey, Betsy J. Neuroscience, emotional harm, and emotional distress tort claims. *American Journal of Bioethics* 2007 September; 7(9): 65-67. Subject: 17.1

Grey, William; Hall, Wayne; Carter, Adrian. Persons and personification. *American Journal of Bioethics* 2007 January; 7(1): 57-58. Subject: 17.1

Grieger, Maria Christina Anna. Authorship: an ethical dilemma of science. *Sao Paulo Medical Journal = Revista Paulista de Medicina* 2005 September 1; 123(5): 242-246. Subject: 1.3.7

Griffin, Anne. Kidneys on demand. *BMJ: British Medical Journal* 2007 March 10; 334(7592): 502-505. Subject: 19.5

Griffin, Joan M.; Struve, James K.; Collins, Dorothea; Liu, An; Nelson, David B.; Bloomfield, Hanna E. Long term clinical trials: how much information do participants retain from the informed consent process? *Contemporary Clinical Trials* 2006 October; 27(5): 441-448. Subject: 18.2

Griffin, Leslie C. Conscience and emergency contraception. *Houston Journal of Health Law and Policy* 2006; 6(2): 299-318. Subject: 9.7

Griffith, Ezra E.H. Personal narrative and an African American perspective on medical ethics. *In:* Prograis, Lawrence J.; Pellegrino, Edmund D., eds. African American Bioethics: Culture, Race, and Identity. Washington, DC: Georgetown University Press, 2004: 105-125. Subject: 2.1

Griffith, Richard. Authorizing the deprivation of liberty of incapable adults in institutions. *British Journal of Community Nursing* 2006 December; 11(12): 538-541. Subject: 17.7

Griffith, Richard. Controlled drugs and the principle of double effect. *British Journal of Community Nursing* 2006 August; 11(8): 352, 354-357. Subject: 20.4.1

Griffith, Richard. Legal requirements for donating and retaining organs: the Human Tissue Act. *British Journal of Community Nursing* 2006 October; 11(10): 446-449. Subject: 19.5

Grimm, David. UC balks at campus-wide ban on tobacco money for research [news]. *Science* 2007 January 26; 315(5811): 447-448. Subject: 5.3

Grindrod, Eirlys; Gardiner, Esther. Withdrawal of consent during surgery. *Journal of Perioperative Practice* 2006 September; 16(9): 418, 420. Subject: 8.1

Griniezakis, Archimandrite Makarios. Legal and ethical issues associated with brain death. *Ethics and Medicine: An International Journal of Bioethics* 2007 Summer; 23(2): 113-117. Subject: 20.2.1

Groopman, Leonard C.; Miller, Franklin G.; Fins, Joseph J. The patient's work. *CQ: Cambridge Quarterly of Healthcare Ethics* 2007 Winter; 16(1): 44-52. Subject: 8.1

Gross, Jane. Aging and gay, and facing prejudice in twilight. *New York Times* 2007 October 9; p. A1, A25. Subject: 9.5.2

Gross, Jed Adam; Moreno, Jonathan D.; Berger, Sam. Gray, not red: the hue of neoconservative bioethics. *American Journal of Bioethics* 2007 October; 7(10): 22-25; author reply W1-W3. Subject: 2.1

Grove, Matthew L. Trials and electronic records: a frightening industry proposal [letter]. *BMJ: British Medical Journal* 2007 June 16; 334(7606): 1236. Subject: 18.2

Grubb, Andrew. Regulating reprogenetics in the United Kingdom. *In:* Knowles, Lori P.; Kaebnick, Gregory E., eds. Reprogenetics: Law, Policy, and Ethical Issues. Baltimore: Johns Hopkins University Press, 2007: 144-177. Subject: 18.5.4

Grüber, Katrin. Abschied vom Gendogma — für eine verantwortbare medizinische Forschung. *In:* Neuer-Miebach, Therese; Wunder, Michael, eds. Bio-Ethik und die Zukunft der Medizin. Bonn: Psychiatrie-Verlag, 1998: 120-129. Subject: 15.1

Gruen, Lori; Grabel, Laura. Concise review: scientific and ethical roadblocks to human embryonic stem cell therapy. *Stem Cells* 2006 October; 24(10): 2162-2169 [Online]. Accessed: http://stemcell.alphamedpress.org/cgi/content/full/24/10/2162 [2007 December 4]. Subject: 15.2

Grunwald, Armin; Julliard, Yannick. Nanotechnology — steps toward understanding human beings as technology? *NanoEthics* 2007 August; 1(2): 77-87. Subject: 5.4

Gruskin, Sofia, Mills, Edward J.; Tarantola, Daniel. History, principles, and practice of health and human rights. *Lancet* 2007 August 4-10; 370(9585): 449-455. Subject: 9.2

Gruskin, Sofia; Tarantola, Daniel. Health and human rights. *In:* Detels, Roger; McEwen, James; Beaglehole, Robert; Tanaka, Heizo, eds. Oxford Textbook of Public Health. Fourth edition. Oxford; New York: Oxford University Press, 2004: 311-335. Subject: 21.1

Guenin, Louis M. A proposed stem cell research policy. *Stem Cells* 2005 September; 23(8): 1023-1027. Subject: 18.5.4

Guevin, Benedict M. Extraordinary treatment or suicide? *Ethics and Medics* 2007 May; 32(5): 1-2. Subject: 20.7

Guilam, Maria Cristina R.; Corrêa, Marilena C.D.V. Risk, medicine and women: a case study on prenatal genetic counselling in Brazil. *Developing World Bioethics* 2007 August; 7(2): 78-85. Subject: 15.2

Guilhem; Dirce; Azevedo, Anamaria Ferreira. Brazilian public policies for reproductive health: family planning, abortion and prenatal care. *Developing World Bioethics* 2007 August; 7(2): 68-77. Subject: 11.1

Guillemin, Jeanne. Scientists and the history of biological weapons. A brief historical overview of the development of biological weapons in the twentieth century. *EMBO Reports* 2006 July; 7 Special No: S45-S49. Subject: 21.3

Gulbrandsen, Carl. WARF's licensing policy for ES cell lines [letter]. *Nature Biotechnology* 2007 April; 25(4): 387-388. Subject: 15.8

Gulcher, Jeff; Stefansson, Kari. The Icelandic healthcare database: a tool to create knowledge, a social debate, and a bioethical and privacy challenge [editorial]. *Medscape Molecular Medicine* 1999: 5 p. [Online]. Accessed: http://www.medscape.com/viewarticle/414505 [2008 March 4]. Subject: 15.1

See SUBJECT HEADING KEY FOR SECTION II on inside back cover.

507

Gunderson, Martin. Seeking perfection: a Kantian look at human genetic engineering. *Theoretical Medicine and Bioethics* 2007; 28(2): 87-102. Subject: 15.1

Gunning, Jennifer. Umbilical cord cell banking: a surprisingly controversial issue. *In:* Gunning, Jennifer; Holm, Søren, eds. Ethics, law, and society. Volume 2. Aldershot, Hants, England; Burlington, VT: Ashgate, 2006: 17-25. Subject: 19.4

Gunning, Jennifer. Umbilical cord cell banking: an issue of self-interest versus altruism. *Medicine and Law: The World Association for Medical Law* 2007 December; 26(4): 769-780. Subject: 19.4

Gunning, Karel. Euthanasia and the United Nations' Universal Declaration of Human Rights. *In:* Wagner, Teresa; Carbone, Leslie, eds. Fifty Years after the Declaration: the United Nations' Record on Human Rights. Lanham, MD: University Press of America, 2001: 17-23. Subject: 20.5.1

Gunsalus, C. Kristina. Human subject protections: some thoughts on costs and benefits in the humanistic disciplines. *In:* Galston, Arthur W.; Peppard, Christiana Z., eds. Expanding Horizons in Bioethics. Dordrecht; Norwell, MA: Springer, 2005: 35-58. Subject: 18.2

Guo-ping, Wang. Ethical evaluation of decision-making for distribution of health resources in China. *Medicine and Law: The World Association for Medical Law* 2007 June; 26(2): 283-289. Subject: 9.4

Gupta, Jyotsna Agnihotri. Private and public eugenics: genetic testing and screening in India. *Journal of Bioethical Inquiry* 2007; 4(3): 217-228. Subject: 15.3

Gurney, Karen. Sex and the surgeon's knife: the family court's dilemma . . . informed consent and the specter of iatrogenic harm to children with intersex characteristics. *American Journal of Law and Medicine* 2007; 33(4): 625-661. Subject: 10

Gustafson, James M. Styles of religious reflection about medical ethics: further discussion. *In:* Østnor, Lars, ed. Bioetikk og teologi: Rapport fra Nordisk teologisk nettverk for bioetikks workshop i Stockholm 27.-29. September 1996. Oslo: Nord. teol. nettv. for bioetikk, 1996: 12-30. Subject: 2.1

Gustafson, Shanna L.; Gettig, Elizabeth A.; Watt-Morse, Margaret; Krishnamurti, Lakshmanan. Health beliefs among African American women regarding genetic testing and counseling for sickle cell disease. *Genetics in Medicine* 2007 May; 9(5): 303-310. Subject: 15.3

H

Haas, John M. Person and human being in the UNESCO Declaration on Bioethics and Human Rights. *National Catholic Bioethics Quarterly* 2007 Spring; 7(1): 41-50. Subject: 4.4

Hackam, Daniel G. Translating animal research into clinical benefit [editorial]. *BMJ: British Medical Journal* 2007 January 27; 334(7586): 163-164. Subject: 22.2

Haddad, Haissam. Cardiac retransplantation: an ethical dilemma. *Current Opinion in Cardiology* 2006 March; 21(2): 118-119. Subject: 19.2

Hadskis, Michael R. Giving voice to research participants: should IRBs hear from research participant representatives? *Accountability in Research* 2007 July-September; 14(3): 155-177. Subject: 18.2

Haegert, Sandy. Whose culture? An attempt at raising a culturally sensitive ethical awareness. *In:* Tschudin, Verena, ed. Approaches to Ethics: Nursing Beyond Boundaries. New York: Butterworth-Heinemann, 2003: 83-93. Subject: 4.1.3

Hagen, John D., Jr. Bentham's mummy and stem cells. *America* 2007 May 14; 196(17): 12-14. Subject: 18.5.4

Hagger, Lynn; Woods, Simon. Children and research: a risk of double jeopardy? *In:* Freeman, Michael, ed. Children's Health and Children's Rights. Boston: Martinus Nijhoff Publishers, 2006: 51-72. Subject: 18.5.2

Häggström, Elisabeth; Kihlgren, Annica. Experiences of caregivers and relatives in public nursing homes. *Nursing Ethics* 2007 September; 14(5): 691-701. Subject: 9.5.2

Haiun, Delphine. The Israeli Patients' Rights Law: a discourse analysis of some main values. *Korot* 2003-2004; 17: 97-124, xi-xii. Subject: 9.2

Hajdin, Mane. The prohibition of sexual relationships between drug users and their counselors: is it justified? *In:* Kleinig, John; Einstein, Stanley, eds. Ethical Challenges for Intervening in Drug Use: Policy, Research and Treatment Issues. Huntsville, TX: Office of International Criminal Justice; Criminal Justice Center, Sam Houston State University, 2006: 437-452. Subject: 17.2

Haker, Hille. Medizinethik auf dem Weg ins 21. Jahrhundert - Bilanz und Zukunftsperspektiven aus Sicht der Katholischen Theologie [Medical ethics on the way to the 21st century - current and future prospects from the point of view of Catholic theology]. *Ethik in der Medizin* 2006 December; 18(4): 325-330. Subject: 2.1

Hale, Benjamin. Risk, judgment and fairness in research incentives. *American Journal of Bioethics* 2007 February; 7(2): 82-83. Subject: 18.1

Hale, Brenda. Justice and equality in mental health law: the European experience. *International Journal of Law and Psychiatry* 2007 January-February; 30(1): 18-28. Subject: 9.5.3

Hale, B. Culpability and blame after pregnancy loss. *Journal of Medical Ethics* 2007 January; 33(1): 24-27. Subject: 11.4

Hale, C. Jacob. Ethical problems with the mental health evaluation standards of care for adult gender variant prospective patients. *Perspectives in Biology and Medicine* 2007 Autumn; 50(4): 491-505. Subject: 10

Halila, Ritva. Assessing the ethics of medical research in emergency settings: how do international regulations work in practice? *Science and Engineering Ethics* 2007 September; 13(3): 305-313. Subject: 18.5.1

Hall, Judith A.; Horgan, Terrence G.; Stein, Terry S.; Roter, Debra L. Liking in the physician-patient relationship. *Patient Education and Counseling* 2002 September; 48(1): 69-77. Subject: 8.1

Hall, Mark A.; Bobinski, Mary Anne; Orentlicher, David. Organ transplantation: the control, use, and allocation of body parts. *In their:* Bioethics and Public Health Law. New York: Aspen Publishers, 2005: 339-395. Subject: 19.5

Hall, Mark A.; Bobinski, Mary Anne; Orentlicher, David. Public health law. *In their:* Bioethics and Public Health Law. New York: Aspen Publishers, 2005: 519-588. Subject: 9.1

Hall, Mark A.; Bobinski, Mary Anne; Orentlicher, David. Reproductive rights and genetic technologies. *In their:* Bioethics and Public Health Law. New York: Aspen Publishers, 2005: 397-518. Subject: 14.1

Hall, Mark A.; Bobinski, Mary Anne; Orentlicher, David. The right and "duty" to die. *In their:* Bioethics and Public Health Law. New York: Aspen Publishers, 2005: 221-338. Subject: 20.5.1

Hall, Mark A.; Bobinski, Mary Anne; Orentlicher, David. The treatment relationship. *In their:* Bioethics and Public Health Law. New York: Aspen Publishers, 2005: 91-219. Subject: 8.1

Hall, Peter Lawrence. Health care for refused asylum seekers in the UK [comment]. *Lancet* 2007 August 11-17; 370(9586): 466-467. Subject: 9.2

Hall, Peter L.; Sheather, Julian C. Asylum seekers' health rights: BMA is in denial [letter and reply]. *BMJ: British Medical Journal* 2007 September 29; 335(7621): 629-630. Subject: 9.2

Hall, Vanessa J.; Stojkovic, Petra; Stojkovic, Miodrag. Using therapeutic cloning to fight human disease: a conundrum or reality? *Stem Cells* 2006 July; 24(7): 1628-1637. Subject: 14.5

Hall, Wayne; Carter, Adrian; Henry, Michael; Fishman, Jennifer R.; Youngner, Stuart J. Debunking alarmist objections to the pharmacological prevention of PTSD. *American Journal of Bioethics* 2007 September; 7(9): 23-25; author reply W1-W3. Subject: 17.4

Hallamaa, Jaana. Human reproductive technology - is there a Lutheran perspective? *In:* Østnor, Lars, ed. Bioetikk og teologi: Rapport fra Nordisk teologisk nettverk for bioetikks workshop i Stockholm 27.-29. September 1996. Oslo: Nord. teol. nettv. for bioetikk, 1996: 76-89. Subject: 14.1

Halliday, Samantha. Regulating active voluntary euthanasia: what can England and Wales learn from Belgium and the Netherlands? *In:* Garwood-Gowers, Austen; Tingle, John; Wheat, Kay, eds. Contemporary Issues in Healthcare Law and Ethics. Edinburgh; New York: Elsevier Butterworth-Heinemann, 2005: 269-301. Subject: 20.5.1

Halpern, Jodi. Let's value, but not idealize, emotions. *Journal of Clinical Ethics* 2007 Winter; 18(4): 380-383. Subject: 8.3.2

Halpern, Scott D.; Metkus, Thomas S.; Fuchs, Barry D.; Ward, Nicholas S.; Siegel, Mark D.; Luce, John M.; Curtis, J. Randall. Nonconsented human immunodeficiency virus testing among critically ill patients: intensivists' practices and the influence of state laws. *Archives of Internal Medicine* 2007 November 26; 167(21): 2323-2328. Subject: 9.5.6

Halwani, Sana. Her Majesty's research subjects: liability of the crown in research involving humans. *In:* Lemmens, Trudo; Waring, Duff R., eds. Law and Ethics in Biomedical Research: Regulation, Conflict of Interest and Liability. Toronto; Buffalo: University of Toronto Press, 2006: 206-227. Subject: 18.6

Hambruger, Philip. The new censorship: institutional review boards. Supreme Court Review 2004 October: 271-354 [Online]. Accessed: http://papers.ssrn.com/sol3/papers.cfm?abstract_id=721363 [2007 December 13]. Subject: 18.2

Hamel, Ron. Human dignity: an "energizing vision" of health care. *Health Progress* 2007 May-June; 88(3): 12-13. Subject: 4.4

Hamel, Ron. New directives for health care ethics? *Health Progress* 2007 January-February; 88(1): 4-5. Subject: 2.1

Hamel, Ron. The CDF statement on artificial nutrition and hydration: what should we make of it? *Health Care Ethics USA* 2007 Fall; 15(4): 5-7. Subject: 20.5.1

Hamer, Ian. Pastor, the gene made me do it! *Concordia Journal* 1997 January; 23: 18-26. Subject: 15.6

Hamilton, Geert Jim. Equal access and financing of health services in Europe. *In:* Gevers, J.K.M.; Hondius, E.H.; Hubben, J.H., eds. Health Law, Human Rights and the Biomedicine Convention: Essays in Honour of Henriette Roscam Abbing. Leiden; Boston: Martinus Nijhoff Publishers, 2005: 61-76. Subject: 9.3.1

Hamilton, Patricia. Ethical dilemmas in training tomorrow's doctors. *Paediatric Respiratory Reviews* 2006 June; 7(2): 129-134. Subject: 7.2

Hamilton, S.; Hepper, J.; Hanby, A.; Hewison, J. Consent gained from patients after breast surgery for the use of surplus tissue in research: an exploration. *Journal of Medical Ethics* 2007 April; 33(4): 229-233. Subject: 18.3

See SUBJECT HEADING KEY FOR SECTION II on inside back cover.

509

Hammerly, Milt. Disruptive germination in health care. *Health Care Ethics USA* 2007 Spring; 15(2): 1-3. Subject: 9.1

Hammill, M.; Burgoine, K., Farrell, F.; Hemelaar, J.; Patel, G.; Welchew, D.E.; Jaffe, H.W. Time to move towards opt-out testing for HIV in the UK. *BMJ: British Medical Journal* 2007 June 30; 334(7608): 1352-1354. Subject: 9.5.6

Hamric, Ann B.; Blackhall, Leslie, J. Nurse-physician perspectives on the care of dying patients in intensive care units: collaboration, moral distress, and ethical climate. *Critical Care Medicine* 2007 February; 35(2): 422-429. Subject: 20.3.2

Han, Sung-Suk. Ethical issues in nursing care at the end of life. *Dolentium Hominum* 2007; 22(2): 28-32. Subject: 20.4.1

Hanawalt, Philip C. Research collaborations: trial, trust, and truth. *Cell* 2006 September 8; 126(5): 823-825. Subject: 1.3.9

Hanna, Kathi E. A brief history of public debate about reproductive technologies: politics and commissions. *In:* Knowles, Lori P.; Kaebnick, Gregory E., eds. Reprogenetics: Law, Policy, and Ethical Issues. Baltimore: Johns Hopkins University Press, 2007: 197-225. Subject: 2.4

Hanson, Stephen S. Moral acquaintances: Loewy, Wildes, and beyond. *HEC (Healthcare Ethics Committee) Forum* 2007 September; 19(3): 207-225. Subject: 2.1

Hansson, Mats G.; Helgesson, Gert; Wessman, Richard; Jaenisch, Rudolf. Commentary: isolated stem cells — patentable as cultural artifacts? *Stem Cells* 2007 June; 25(6): 1507-1510 [Online]. Accessed: http://www.StemCells.com/cgi/content/full/25/6/1507 [2007 December 3]. Subject: 15.8

Hansson, Mats G.; Kihlbom, Ulrik; Tuvemo, Torsten; Olsen, Leif A.; Rodriguez, Alina. Ethics takes time, but not that long. *BMC Medical Ethics [electronic]* 2007; 8:6. 7 p. Subject: 2.1

Hansson, Sven Ove. The ethics of enabling technology. *CQ: Cambridge Quarterly of Healthcare Ethics* 2007 Summer; 16(3): 257-267. Subject: 5.1

Hanto, Douglas W. Ethical challenges posed by the solicitation of deceased and living organ donors. *New England Journal of Medicine* 2007 March 8; 356(10): 1062-1066. Subject: 19.5

Haque, Omar Sultan; Bursztajn, Harold. Decision-making capacity, memory and informed consent, and judgment at the boundaries of the self. *Journal of Clinical Ethics* 2007 Fall; 18(3): 256-261. Subject: 8.3.3

Haran, Joan. Managing the boundaries between maverick cloners and mainstream scientists: the life cycle of a news event in a contested field. *New Genetics and Society* 2007 August; 26(2): 203-219. Subject: 14.5

Haraway, Donna J. Cloning mutts, saving tigers: ethical emergents in technocultural dog worlds. *In:* Franklin, Sarah; Lock, Margaret, eds. Remaking Life and Death: Toward an Anthropology of the Biosciences. Santa Fe: School of American Research Press; Oxford: James Currey, 2003: 293-327. Subject: 14.5

Harcourt Bernard E. The mentally ill, behind bars [op-ed]. *New York Times* 2007 January 15; p. A15. Subject: 17.8

Hardell, Lennart; Walker, Martin J.; Walhjalt, Bo; Friedman, Lee S.; Richter, Elihu D. Secret ties to industry and conflicting interests in cancer research. *American Journal of Industrial Medicine* 2007 March; 50(3): 227-233 [See correction in: American Journal of Industrial Medicine 2007 March; 50(3): 234]. Subject: 18.1

Hardt, John J.; O'Rourke, Kevin D. Nutrition and hydration: the CDF response, in perspective. *Health Progress* 2007 November-December; 88(6): 44-47. Subject: 20.5.1

Hardy, Pollyahanna.; Clemens, Felicity. Stopping a randomized trial early: from protocol to publication. Commentary to Thome at al.: outcome of extremely preterm infants randomized at birth to different PaCO2 targets during the first seven days of life (Biology of the Neonate 2006; 90: 218-225). *Biology of the Neonate* 2006; 90(4): 226-228. Subject: 18.2

Harkness, Jon; Lederer, Susan E.; Wikler, Daniel. Laying ethical foundations for clinical research: Public Health Classics. *Bulletin of the World Health Organization* 2001; 79(4): 365-372. Subject: 18.2

Harmon, Amy. Cancer free, but weighing a mastectomy. *New York Times* 2007 September 16; p. A1, A20, A21. Subject: 15.3

Harmon, Amy. Facing life with a lethal gene: a young woman's DNA test reveals an inevitably grim fate. *New York Times* 2007 March 18; p. A1, A26, A27. Subject: 15.1

Harmon, Amy. In DNA era, new worries about prejudice. *New York Times* 2007 November 11; p. A1, A26. Subject: 15.11

Harmon, Amy. My genome, myself: seeking clues in DNA. *New York Times* 2007 November 17; p. A1, A16. Subject: 15.11

Harrel, T. Recontacting former patients regarding BRCA1/2 rearrangement testing: opinions and practices of genetics professionals. *Journal of Genetic Counseling* 2007 December; 16(6): 670. Subject: 15.2

Harrington, John. Globalization and English medical law: strains and contradictions. *In:* Bennett, Belinda; Tomossy, George F., eds. Globalization and Health: Challenges for Health Law and Bioethics. Dordrecht: Springer, 2006: 169-185. Subject: 9.1

Harris, Gardiner. Lawmaker calls for registry of drug firms paying doctors. *New York Times* 2007 August 4; p. A9. Subject: 9.7

Harris, John. Mark Anthony or Macbeth: some problems concerning the dead and the incompetent when it comes to consent. *In:* McLean, Sheila A.M., ed. First Do No Harm: Law, Ethics, and Healthcare. Aldershot, England; Burlington, VT: Ashgate, 2006: 287-301. Subject: 19.5

Harris, John. NICE rejoinder [comment]. *Journal of Medical Ethics* 2007 August; 33(8): 467. Subject: 9.4

Harris, John. The facts of life: controversial medical issues in the curriculum. *In:* Wellington, J.J., ed. Controversial Issues in the Curriculum. Oxford: Basil Blackwell, 1986: 99-108. Subject: 11.2

Harris, L.; Yashar, B.; Burmeister, M. Genetic testing for bipolar disorder: exploring patients' attitudes and receptivity. *Journal of Genetic Counseling* 2007 December; 16(6): 678. Subject: 15.3

Hart, Dieter. Patient information on drug therapy. A problem of medical malpractice law: between product safety and user safety. *European Journal of Health Law* 2007 April; 14(1): 47-59. Subject: 8.3.1

Harte, Colin. The opening up of a discussion: a response to John Finnis. *In:* Watt, Helen, ed. Cooperation, Complicity and Conscience: Problems in Healthcare, Science, Law and Public Policy. London: Linacre Centre, 2005: 246-268. Subject: 12.4.1

Hartman, Rhonda Gay. The face of dignity: principled oversight of biomedical innovation. *Santa Clara Law Review* 2007; 47(1): 55-91. Subject: 4.4

Hartman, Rhonda Gay. Word from the academies: a primer for legal policy analysis regarding adolescent research participation. *Rutgers Journal of Law and Public Policy* 2006 Fall; 4(1): 152-199. Subject: 18.5.2

Hartry, Nicola. Visually impaired drivers and public protection vs confidentiality. *British Journal of Nursing* 2007 February 22-March 7; 16(4): 226-230. Subject: 8.4

Harvey, Erin K.; Fogel, Chana E.; Peyrot, Mark; Christensen, Kurt D.; Terry, Sharon F.; McInerney, Joseph D. Providers' knowledge of genetics: a survey of 5915 individuals and families with genetic conditions. *Genetics in Medicine* 2007 May; 9(5): 259-267. Subject: 15.2

Harvey, Matthew. Animal genomics in science, social science and culture. *Genomics, Society and Policy* 2007 August; 3(2): 1-28. Subject: 15.1

Harvey, Matthew. Citizens in defence of something called science. *Science as Culture* 2007 March; 16(1): 31-48. Subject: 5.1

Hasan, Yusuf; Salaam, Yusef. Faith and Islamic issues at the end of life. *In:* Puchalski, Christina M., ed. A Time for Listening and Caring: Spirituality and the Care of the Chronically Ill and Dying. Oxford; New York: Oxford University Press, 2006: 183-192. Subject: 20.4.1

Hasegawa, Thomas K.; Welie, Jos V.M. Role of codes of ethics in oral health care. *Journal of the American College of Dentists* 1998 Fall; 65(3): 12-14. Subject: 6

Hashiloni-Dolev, Yael. "Wrongful life", in the eyes of the law, the counselors and the disabled. *In her:* A Life (Un)Worthy of Living: Reproductive Genetics in Israel and Germany. Dordrecht: Springer, 2007: 119-130. Subject: 15.2

Hashiloni-Dolev, Yael. Abortions on embryopathic grounds: policy and practice in Israel and Germany. *In her:* A Life (Un)Worthy of Living: Reproductive Genetics in Israel and Germany. Dordrecht: Springer, 2007: 83-104. Subject: 12.5.1

Hashiloni-Dolev, Yael. Genetic counselors' moral practices. *In her:* A Life (Un)Worthy of Living: Reproductive Genetics in Israel and Germany. Dordrecht: Springer, 2007: 63-81. Subject: 15.2

Hashiloni-Dolev, Yael. Sex chromosome anomalies (SCAs) in Israel and Germany: assessing "birth defects" and medical risks according to the importance of fertility. *In her:* A Life (Un)Worthy of Living: Reproductive Genetics in Israel and Germany. Dordrecht: Springer, 2007: 105-117. Subject: 12.5.1

Hashiloni-Dolev, Yael. The conflicts between individuals, families and society, as well as between different family members, embodied in reproductive genetics. *In her:* A Life (Un)Worthy of Living: Reproductive Genetics in Israel and Germany. Dordrecht: Springer, 2007: 131-146. Subject: 15.2

Hasman, Andreas. Restrictive or engaging: redefining public health promotion. *In:* Gunning, Jennifer; Holm, Søren, eds. Ethics, law, and society. Volume 2. Aldershot, Hants, England; Burlington, VT: Ashgate, 2006: 77-83. Subject: 9.1

Hasnain-Wynia, Romana; Baker, David W.; Nerenz, David; Feinglass, Joe; Beal, Anne C.; Landrum, Mary Beth; Behal, Raj; Weissman, Joel S. Disparities in health care are driven by where minority patients seek care. *Archives of Internal Medicine* 2007 June 25; 167(12): 1233-1239. Subject: 9.5.4

Hassert, Derrick L. Neuroethics and the person: should neurological and cognitive criteria be used to define human value? *Ethics and Medicine: An International Journal of Bioethics* 2007 Spring; 23(1): 47-55. Subject: 17.1

Hatfield, Amy J.; Kelley, Shana D. Case study: lessons learned through digitizing the National Commission for the Protection of Human Subjects of Biomedical and Behavioral Research collection. *Journal of the Medical Library Association* 2007 July; 95(3): 267-270. Subject: 2.1

Hathaway, Andrew D. Ushering in another harm reduction era? Discursive authenticity, drug policy and research. *Drug and Alcohol Review* 2005 November; 24(6): 549-550. Subject: 9.1

See SUBJECT HEADING KEY FOR SECTION II on inside back cover.

511

Hattab, Jocelyn Y.; Kohn, Yoav. Informed consent in child psychiatry — a theoretical review. *Journal of Ethics in Mental Health [electronic]* 2007 November; 2(2): 6 p. Accessed: http://www.jemh.ca [2008 January 24]. Subject: 8.3.3

Hatziavramidis, Katie. Parental involvement law for abortion in the United States and the United Nations conventions on the rights of the child: can international law secure the right to choose for minors? *Texas Journal of Women and the Law* 2007 Spring; 16(2): 185-204. Subject: 12.4.2

Hauskeller, Michael. The reification of life. *Genomics, Society and Policy* 2007 August; 3(2): 70-81. Subject: 22.1

Hausman, Daniel M. Group risks, risks to groups, and group engagement in genetics research. *Kennedy Institute of Ethics Journal* 2007 December; 17(4): 351-369. Subject: 15.1

Hausman, Daniel M. Third-party risks in research: should IRBs address them? *IRB: Ethics and Human Research* 2007 May-June; 29(3): 1-5. Subject: 18.2

Hauswald, Mark; Wells, Robert J.; Candib, Lucy M. Overseeing quality improvement [letters and reply]. *Hastings Center Report* 2007 July-August; 37(4): 6-8. Subject: 9.8

Hawryluck, Laura. Ethics review: position papers and policies — are they really helpful to front-line ICU teams? *Critical Care* 2006; 10(6): 242. Subject: 9.5.1

Hawthorne, Susan. ADHD drugs: values that drive the debates and decisions. *Medicine, Health Care and Philosophy* 2007 June; 10(2): 129-140. Subject: 17.4

Hayden, Erika Check. Personalized genomes go mainstream [news]. *Nature* 2007 November 1; 450(7166): 11. Subject: 15.10

Hayes, Margaret Oot. Prisoners and autonomy: implications for the informed consent process with vulnerable populations. *Journal of Forensic Nursing* 2006 Summer; 2(2): 84-89. Subject: 8.3.1

Hayflick, L. The limited in vitro lifetime of human diploid cell strains. *Experimental Cell Research* 1965; 37: 614-636. Subject: 18.5.4

Haynes, Charlotte L.; Cook, Gary A.; Jones, Michael A. Legal and ethical considerations in processing patient-identifiable data without parental consent: lessons learnt from developing a disease register. *Journal of Medical Ethics* 2007 May; 33(5): 302-307. Subject: 8.4

Häyry, Matti. Bioscientists as ethical decision-makers. *In:* Korthals, Michiel; Bogers, Robert J., eds. Ethics for Life Scientists. Dordrecht, The Netherlands: Springer, 2004: 183-189. Subject: 1.3.9

Hazaray, Neil F. Do the benefits outweigh the risks? The legal, business, and ethical ramifications of pulling a blockbuster drug off the market. *Indiana Health Law Review* 2007; 4(1): 115-150. Subject: 9.7

Hazen, Rebecca; Greenley, Rachel Neff; Drotar, Dennis; Kodish, Eric. Recommending randomized trials for pediatric Leukemia: observer and physician report of recommendations. *Journal of Empirical Research on Human Research Ethics* 2007 June; 2(2): 49-55. Subject: 18.5.2

Healy, G.W. Moral and legal aspects of transplantation: prisoners or death convicts as donors. *Transplantation Proceedings* 1998 November; 30(7): 3653-3654. Subject: 19.5

Heath, Iona. Let's get tough on the causes of health inequality. *BMJ: British Medical Journal* 2007 June 23; 334(7607): 1301. Subject: 9.2

Heath, Iona; Nessa, John. Objectification of physicians and loss of therapeutic power. *Lancet* 2007 March 17-23; 369(9565): 886-887. Subject: 8.1

Heaven, Ben; Murtagh, Madeleine; Rapley, Tim; May, Carl; Graham, Ruth; Kaner, Eileen; Thomson, Richard. Patients or research subjects? A qualitative study of participation in a randomised controlled trial of a complex intervention. *Patient Education and Counseling* 2006 August; 62(2): 260-270. Subject: 18.2

Hedayat, K.M. The possibility of a universal declaration of biomedical ethics. *Journal of Medical Ethics* 2007 January; 33(1): 17-20. Subject: 2.1

Heeger, Robert. Comments on Häyry: assessing bioscientific work from a moral point of view. *In:* Korthals, Michiel; Bogers, Robert J., eds. Ethics for Life Scientists. Dordrecht, The Netherlands: Springer, 2004: 191-193. Subject: 1.3.9

Heeley, Gerry. A system's transition to next generation model of ethics. *Health Care Ethics USA* 2007 Fall; 15(4): 2-4. Subject: 9.6

Heikkinen, Anne M.; Wickström, Gustav J.; Leino-Kilpi, Helena; Katajisto, Jouko. Privacy and dual loyalties in occupational health practice. *Nursing Ethics* 2007 September; 14(5): 675-690. Subject: 8.4

Heilig, Charles M.; Weijer, Charles. A critical history of individual and collective ethics in the lineage of Lellouch and Schwartz. *Clinical Trials* 2005 June; 2(3): 244-253. Subject: 18.2

Heinemann, Lothar A.J. Verlässlichkeit von Industrie- und öffentlich geförderten Studien. *In:* Shapiro, S.; Dinger, J.; Scriba, P., eds. Enabling Risk Assessment in Medicine: Farewell Symposium for Werner-Kari Raff. New Brunswick, NJ: Transaction Publishers, 2004: 51-66. Subject: 1.3.9

Heinrichs, Bert. A comparative analysis of selected European guidelines and recommendations for biobanks with special regard to the research / non-research distinction. *Revista de Derecho y Genoma Humano = Law and the Human Genome Review* 2007 July-December; (27): 205-224. Subject: 15.1

Heitman, Elizabeth; Olsen, Cara H.; Anestidou, Lida; Bulger, Ruth Ellen. New graduate students' baseline knowledge of the responsible conduct of research. *Academic Medicine* 2007 September; 82(9): 838-845. Subject: 18.2

Helft, Paul R.; Champion, Victoria L.; Eckles, Rachael; Johnson, Cynthia S.; Meslin, Eric M. Cancer patients' attitudes toward future research uses of stored human biological materials. *Journal of Empirical Research on Human Research Ethics* 2007 September; 2(3): 15-22. Subject: 19.5

Helft, Paul R.; Daugherty, Christopher K. Are we taking without giving in return? The ethics of research-related biopsies and the benefits of clinical trial participation. *Journal of Clinical Oncology* 2006 October 20; 24(30): 4793-4795. Subject: 19.5

Helgesson, Gert; Dillner, Joakim; Carlson, Joyce; Bartram, Claus R.; Hansson, Mats G. Ethical framework for previously collected biobank samples [letter]. *Nature Biotechnology* 2007 September; 25(9): 973-976. Subject: 15.1

Helgesson, Gert; Eriksson, Stefan; Swartling, Ulrica. Limited relevance of the right not to know — reflections on a screening study. *Accountability in Research* 2007 July-September; 14(3): 197-209. Subject: 8.2

Hellström, Ingrid; Nolan, Mike; Nordenfelt, Lennart; Lundh, Ulla. Ethical and methodological issues in interviewing persons with dementia. *Nursing Ethics* 2007 September; 14(5): 608-619. Subject: 18.5.6

Helmchen, Hanfried. Ethics as a focus of controversy in postmodern antagonisms. *In:* Schramme, Thomas; Thome, Johannes, eds. Philosophy and Psychiatry. Berlin; New York: De Gruyter, 2004: 347-351. Subject: 2.1

Hem, Marit Helene; Heggen, Kristin; Ruyter, Knut W. Questionable requirement for consent in observational research in psychiatry. *Nursing Ethics* 2007 January; 14(1): 41-53. Subject: 18.3

Hemelaar, Joris. Minimising risk in first-in-man trials. *Lancet* 2007 May 5-11; 369(9572): 1496-1497. Subject: 18.2

Henaghan, Mark. The 'do no harm' principle and the genetic revolution in New Zealand. *In:* McLean, Sheila A.M., ed. First Do No Harm: Law, Ethics, and Healthcare. Aldershot, England; Burlington, VT: Ashgate, 2006: 511-526. Subject: 15.1

Henderson, Amanda; Van Eps, Mary Ann; Pearson, Kate; James, Catherine; Henderson, Peter; Osborne, Yvonne. 'Caring for' behaviours that indicate to patients that nurses 'care about' them. *Journal of Advanced Nursing* 2007 October; 60(2): 146-153. Subject: 8.1

Henderson, Gail E.; Corneli, Amy L.; Mahoney, David B.; Nelson, Daniel K.; Mwansambo, Charles. Applying research ethics guidelines: the view from a sub-Saharan research ethics committee. *Journal of Empirical Research on Human Research Ethics* 2007 June; 2(2): 41-48. Subject: 18.5.9

Henderson, Lesley; Kitzinger, Jenny. Orchestrating a science 'event': the case of the Human Genome Project. *New Genetics and Society* 2007 April; 26(1): 65-83. Subject: 15.10

Hendriks, Aart. Protection against genetic discrimination and the Biomedicine Convention. *In:* Gevers, J.K.M.; Hondius, E.H.; Hubben, J.H., eds. Health Law, Human Rights and the Biomedicine Convention: Essays in Honour of Henriette Roscam Abbing. Leiden; Boston: Martinus Nijhoff Publishers, 2005: 207-218. Subject: 15.1

Heneghan, Tom. Does Italy have its own "Terry Schiavo case?" Reuters. FaithWorld Blog 2007 October 24: 2 p. [Online]. Accessed: http://blogs.reuters.com/faithworld/2007/10/24/does-italy-have-its own-terry-schiavo-case/ [2007 October 26]. Subject: 20.5.1

Heng, Boon Chin. Disparity in medical fees for donor and self freeze-thaw embryo transfer cycle — a covert form of embryo commercialization? [letter]. *Developing World Bioethics* 2007 April; 7(1): 49-50. Subject: 14.6

Heng, Boon Chin. Donation of surplus frozen embryos for stem cell research or fertility treatment — should medical professionals and healthcare institutions be allowed to exercise undue influence on the informed decision of their former patients? *Journal of Assisted Reproduction and Genetics* 2006 September-October; 23(9-10): 381-382. Subject: 14.6

Heng, Boon Chin. Regulated family balancing by equalizing the sex-ratio of gender-selected births. *Journal of Assisted Reproduction and Genetics* 2006 July-August; 23(7-8): 319-320. Subject: 14.3

Heng, Boon Chin; Tong, Guo Qing; Stojkovic, Miodrag. The egg-sharing model for human therapeutic cloning research: managing donor selection criteria, the proportion of shared oocytes allocated to research, and amount of financial subsidy given to the donor. *Medical Hypotheses* 2006; 66(5): 1022-1024. Subject: 14.5

Heng, B.C. Ethical issues in transnational "mail order" oocyte donation. *International Journal of Gynecology and Obstetrics* 2006 December; 95(3): 302-304. Subject: 14.6

Heng, B.C. Should fertility specialists refer local patients abroad for shared or commercialized oocyte donation? *Fertility and Sterility* 2007 January; 87(1): 6-7. Subject: 14.1

Heng, B.C. The advent of international 'mail-order' egg donation. *BJOG: An International Journal of Obstetrics and Gynaecology* 2006 November; 113(11): 1225-1227. Subject: 14.4

Henley, Lesley D.; Frank, Denise M. Reporting ethical protections in physical therapy research. *Physical Therapy* 2006 April; 86(4): 499-509. Subject: 4.1.1

See SUBJECT HEADING KEY FOR SECTION II on inside back cover.

Henn, Wolfram. Auf dem Weg zur „ökonomischen Indikation" zum Schwangerschaftsabbruch bei therapierbaren Erbleiden? = Towards terminations of pregnancy due to the therapy costs of treatable hereditary diseases? *Ethik in der Medizin* 2007 June; 19(2): 120-127. Subject: 15.1

Hennig, Wolfgang. Bioethics in China: although national guidelines are in place, their implementation remains difficult. *EMBO Reports* 2006 September; 7(9): 850-854. Subject: 2.1

Henry, Maureen. Update on end-of-life issues in Utah. *Utah Bar Journal* 2006 January-February; 19: 6-10. Subject: 20.5.1

Henry, Michael; Fishman, Jennifer R.; Youngner, Stuart J. Propranolol and the prevention of post-traumatic stress disorder: is it wrong to erase the "sting" of bad memories? *American Journal of Bioethics* 2007 September; 7(9): 12-20. Subject: 17.4

Hentschel, Roland; Lindner, Katharina; Krueger, Markus; Reiter-Theil, Stella. Restriction of ongoing intensive care in neonates: a prospective study. *Pediatrics* 2006 August; 118(2): 563-569. Subject: 9.5.7

Herbel, Bryon L.; Stelmach, Hans. Involuntary medication treatment for competency restoration of 22 defendants with delusional disorder. *Journal of the American Academy of Psychiatry and the Law* 2007; 35(1): 47-59. Subject: 17.8

Herissone-Kelly, Peter. Parental love and the ethics of sex selection. *CQ: Cambridge Quarterly of Healthcare Ethics* 2007 Summer; 16(3): 326-335. Subject: 14.3

Herissone-Kelly, Peter. The "parental love" objection to nonmedical sex selection: deepening the argument. *CQ: Cambridge Quarterly of Healthcare Ethics* 2007 Fall; 16(4): 446-455. Subject: 14.3

Hermerén, Göran. Challenges in the evaluation of nanoscale research: ethical aspects. *NanoEthics* 2007 December; 1(3): 223-237. Subject: 5.4

Hershberger, Patricia; Klock, Susan C.; Barnes, Randall B. Disclosure decisions among pregnant women who received donor oocytes: a phenomenological study. *Fertility and Sterility* 2007 February; 87(2): 288-296. Subject: 14.2

Hershenov, David B. Death, dignity, and degradation. *Public Affairs Quarterly* 2007 January; 21(1): 21-36. Subject: 20.3.1

Herveg, Jean. The ban on processing medical data in European Law: consent and alternative solutions to legitimate processing of medical data in HealthGrid. *Studies in Health Technology and Informatics* 2006; 120: 107-116. Subject: 8.4

Hess, Rosanna F. Postabortion research: methodological and ethical issues. *Qualitative Health Research* 2006 April; 16(4): 580-587. Subject: 18.5.3

Hester, D. Micah. Interests and neonates: there is more to the story than we explicitly acknowledge. *Theoretical Medicine and Bioethics* 2007; 28(5): 357-372. Subject: 8.3.2

Hewitt, J.L.; Edwards, S.D. Moral perspectives on the prevention of suicide in mental health settings. *Journal of Psychiatric and Mental Health Nursing* 2006 December; 13(6): 665-672. Subject: 17.1

Hewson, Barbara. Abortion in Poland: a new human rights ruling. *Conscience* 2007 Summer; 28(2): 34-35. Subject: 12.4.2

Hewson, Barbara. Dancing on the head of a pin? Foetal life and the European convention. *Feminist Legal Studies* 2005; 13(3): 363-375. Subject: 4.4

Hewson, Caroline J. Veterinarians who swear: animal welfare and the veterinary oath. *Canadian Veterinary Journal. La Revue Vétérinaire Canadienne* 2006 August; 47(8): 807-811. Subject: 22.1

Heyman, Bob; Hundt, Gillian; Sandall, Jane; Spencer, Kevin; Williams, Clare; Grellier, Rachel; Pitson, Laura. On being at higher risk: a qualitative study of prenatal screening for chromosomal anomalies. *Social Science and Medicine* 2006 May; 62(10): 2360-2372. Subject: 15.2

Heymans, Regien; van der Arend, Arie; Gastmans, Chris. Dutch nurses' views on codes of ethics. *Nursing Ethics* 2007 March; 14(2): 156-170. Subject: 6

Hickey, Kathryn. Minors rights in medical decision making. *JONA's Healthcare Law, Ethics, and Regulation* 2007 July-September; 9(3): 100-106. Subject: 8.3.2

Hicks, Madelyn Hsiao-Rei. Physician-assisted suicide: a review of the literature concerning practical and clinical implications for UK doctors. *BMC Family Practice* 2006 June 22; 7: 39-55. Subject: 20.7

Higginson, Irene J.; Hall, S. Rediscovering dignity at the bedside. *BMJ: British Medical Journal* 2007 July 28; 335(7612): 167. Subject: 8.1

Higginson, Jason D. Emotion, suffering, and hope: commentary on "How much suffering is enough?". *Journal of Clinical Ethics* 2007 Winter; 18(4): 377-379. Subject: 8.3.2

Higgs, Peter; Moore, David; Aitken, Campbell. Engagement, reciprocity and advocacy: ethical harm reduction practice in research with injecting drug users. *Drug and Alcohol Review* 2006 September; 25(5): 419-423. Subject: 9.5.9

Higgs, Roger. Truth telling. *In:* Rhodes, Rosamond; Francis, Leslie P.; Silvers, Anita, eds. The Blackwell Guide to Medical Ethics. Malden, MA: Blackwell Pub., 2007: 88-103. Subject: 8.2

Hil, Richard; Hindmarsch, Richard. Body talk: genetic screening as a device of crime regulation. *In:* Betta, Michela, ed. The Moral, Social, and Commercial Imperatives of Genetic Testing and Screening: the Australian Case. Dordrecht: Springer, 2006: 55-70. Subject: 15.6

Hildén, Hanna-Mari; Honkasalo, Marja-Liisa. Unethical bunglers or humane professionals? Discussions in the media of end-of-life treatment decisions. *Communication and Medicine* 2006; 3(2): 125-134. Subject: 20.5.1

Hilhorst, Medard T.; Kranenburg, Leonieke W.; Busschbach, Jan J.V. Should health care professionals encourage living kidney donation? *Medicine, Health Care and Philosophy* 2007 March; 10(1): 81-90. Subject: 19.5

Hill, Kate. Consent, confidentiality and record keeping for the recording and usage of medical images. *Journal of Visual Communication in Medicine* 2006 June; 29(2): 76-79. Subject: 8.4

Hilliard, Marie T. The duty to care: when health care workers face personal risk. *National Catholic Bioethics Quarterly* 2007 Winter; 7(4): 673-682. Subject: 8.1

Himmel, Wolfgang; Michelmann, Hans Wilhelm. Access to genetic material: reproductive technologies and bioethical issues. *Reproductive Biomedicine Online* 2007 September; 15 (Suppl. 1): 18-24. Subject: 14.1

Hinchley, Geoff; Patrick, Kirsten. Is infant male circumcision an abuse of the rights of the child? [debate]. *BMJ: British Medical Journal* 2007 December 8; 335(7631): 1180-1181. Subject: 9.5.1

Hinkley, Charles C. Defining death. *In his:* Moral Conflicts of Organ Retrieval: A Case for Constructive Pluralism. Amsterdam; New York: Rodopi, 2005: 91-103. Subject: 20.2.1

Hinkley, Charles C. Stem cell research. *In his:* Moral Conflicts of Organ Retrieval: A Case for Constructive Pluralism. Amsterdam; New York: Rodopi, 2005: 127-133. Subject: 18.5.4

Hinkley, Charles C. The selling of organs. *In his:* Moral Conflicts of Organ Retrieval: A Case for Constructive Pluralism. Amsterdam; New York: Rodopi, 2005: 105-115. Subject: 19.5

Hinkley, Charles C. Xenografts. *In his:* Moral Conflicts of Organ Retrieval: A Case for Constructive Pluralism. Amsterdam; New York: Rodopi, 2005: 117-126. Subject: 19.1

Hinks, Timothy S. Deceiving patients: ends never justify means [letter]. *BMJ: British Medical Journal* 2007 May 26; 334(7603): 1072. Subject: 8.2

Hinsch, Kathryn M.; Fiore, Robin N.; Moreno, Jonathan D.; Berger, Sam. Responding to neocon critiques of biotechnology: a progressive agenda. *American Journal of Bioethics* 2007 October; 7(10): 14-15; author reply W1-W3. Subject: 5.1

Hinton, Jeremy; Forrest, Robert. Involuntary non-emergent psychotropic medication. *Journal for the American Academy of Psychiatry and the Law* 2007; 35(3): 396-398. Subject: 17.4

Hippen, B.E.; Gaston, R.S. The conspicuous costs of more of the same. *American Journal of Transplantation* 2006 July; 6(7): 1503-1504. Subject: 19.3

Hiraki, Susan; Ormond, Kelly E.; Kim, Katherine; Ross, Lainie F. Attitudes of genetic counselors towards expanding newborn screening and offering predictive genetic testing to children. *American Journal of Medical Genetics. Part A* 2006 November 1; 140(21): 2312-2319. Subject: 15.2

Hirschfield, Miriam. An international perspective. *In:* Davis, Anne J.; Tschudin, Verena; de Raeve, Louise, eds. Essentials of Teaching and Learning in Nursing Ethics: Perspectives and Methods. New York: Churchill Livingstone Elsevier, 2006: 325-337. Subject: 4.1.3

Hirshberg, Boaz. Can we justify living donor islet transplantation? *Current Diabetes Reports* 2006 August; 6(4): 307-309. Subject: 19.5

Hirskyj, Peter. QALY: an ethical issue that dare not speak its name. *Nursing Ethics* 2007 January; 14(1): 72-82. Subject: 4.4

Hitchen, Lisa. Keeping the scientists in step with society [news]. *BMJ: British Medical Journal* 2007 May 26; 334(7603): 1079. Subject: 14.1

Hivon, Myriam; Lehoux, Pascale; Denis, Jean-Louis; Tailliez, Stéphanie. Use of health technology assessment in decision making: coresponsibility of users and producers? *International Journal of Technology Assessment in Health Care* 2005 Spring; 21(2): 268-275. Subject: 5.2

Ho, Dien. Leaving people alone: liberalism, ectogenesis, and the limits of medicine. *In:* Gelfand, Scott; Shook, John R., eds. Ectogenesis: Artificial Womb Technology and the Future of Human Reproduction. Amsterdam; New York: Editions Rodopi, B.V., 2006: 139-147. Subject: 14.1

Hoag, Hannah. Rules tightened for aboriginal studies [news]. *Nature* 2007 May 17; 447(7142): 241. Subject: 18.5.1

Hodge, James G. Jr.; Gostin, Lawrence O.; Vernick, Jon S. The Pandemic and All-Hazards Preparedness Act: improving public health emergency response. *JAMA: The Journal of the American Medical Association* 2007 April 18; 297(15): 1708-1711. Subject: 9.1

Hodge, James G., Jr. The flaw of informed consent. *American Journal of Bioethics* 2007 March; 7(3): 52-53. Subject: 8.4

Hoedemaekers, Rogeer; Gordijn, Bert; Pijnenburg, Martien. Solidarity and justice as guiding principles in genomic research. *Bioethics* 2007 July; 21(6): 342-350. Subject: 15.10

Hoeyer, Klaus; Koch, Lene. The ethics of functional genomics: same, same, but different? *Trends in Biotechnology* 2006 September; 24(9): 387-389. Subject: 15.1

See SUBJECT HEADING KEY FOR SECTION II on inside back cover.

515

Hoffenberg, Raymond. Advance healthcare directives. *Clinical Medicine* 2006 May-June; 6(3): 231-233. Subject: 20.5.4

Hoffman, Jan. Where risk and choice and hope converge, a guiding voice. *New York Times* 2007 September 18; p. F5, F10. Subject: 15.2

Hoffman, K.; Thomas, S.B.; Gettig, E.; Grubs, R.E.; Krishnamurti, L.; Butler, J. Assessing the attitudes and beliefs of African-Americans toward newborn screening and sickle cell disease. *Journal of Genetic Counseling* 2007 December; 16(6): 674. Subject: 15.3

Hoffman, Sharona. Addressing privacy concerns through the Health Insurance Portability and Accountability Act privacy rule. *American Journal of Bioethics* 2007 March; 7(3): 48-49. Subject: 8.4

Hoffmann, George R. Letter to the editor on ethics of expertise, informed consent, and hormesis. *Science and Engineering Ethics* 2007 June; 13(2): 135-137. Subject: 5.3

Hoffmann, Thomas Sören. Primordial ownership versus dispossession of the body. A contribution to the problem of cloning from the perspective of classical European philosophy of law. *In:* Roetz, Heiner, ed. Cross-Cultural Issues in Bioethics: The Example of Human Cloning. New York: Rodopi, 2006: 387-407. Subject: 14.5

Höfling, Wolfram. Das "Menschenrechts-übereinkommen zur Bio-Medizin" und die Grund- und Menschenrechte. *In:* Neuer-Miebach, Therese; Wunder, Michael, eds. Bio-Ethik und die Zukunft der Medizin. Bonn: Psychiatrie-Verlag, 1998: 72-86. Subject: 2.1

Hofmann, Bjørn. That's not science! The role of moral philosophy in the science/non-science divide. *Theoretical Medicine and Bioethics* 2007; 28(3): 243-256. Subject: 1.3.9

Hofmann, Paul B.; Schneiderman, Laurence J. Futility, in short [letter and reply]. *Hastings Center Report* 2007 July-August; 37(4): 8. Subject: 20.5.1

Hofmann, Paul B.; Schneiderman, Lawrence J. Physicians should not always pursue a good "clinical" outcome. *Hastings Center Report* 2007 May-June; 37(3): inside back cover. Subject: 20.5.1

Hogan, Carol. Conscience clauses and the challenge of cooperation in a pluralistic society. Sacramento: California Catholic Conference, 2003 February; 12 p. [Online]. Accessed: http://www.cacatholic.org/rfconscience.html [2007 April 4]. Subject: 4.1.1

Hogarth, Stuart; Melzer, David; Zimmern, Ron. The regulation of commercial genetic testing services in the UK: a briefing for the Human Genetics Commission. Cambridge: Department of Public Health and Primary Care, Cambridge University, 2005: 23 p. [Online]. Accessed: http://www.phpc.cam.ac.uk/epg/dtc.pdf [2007 April 17]. Subject: 15.3

Hogben, Susan; Boddington, Paula. The rhetorical construction of ethical positions: policy recommendations for nontherapeutic genetic testing in childhood. *Communication and Medicine* 2006; 3(2): 135-146. Subject: 15.3

Hohlfeld, Rainer. Politische Ökonomie und Bio-Medizin. *In:* Neuer-Miebach, Therese; Wunder, Michael, eds. Bio-Ethik und die Zukunft der Medizin. Bonn: Psychiatrie-Verlag, 1998: 44-59. Subject: 15.1

Holaday, Margot; Yost, Tracey E. Authorship credit and ethical guidelines. *Counseling and Values* 1995 October; 40(1): 24-31. Subject: 1.3.7

Holbrook, Daniel. All embryos are equal?: Issues in pre-implantation genetic diagnosis, IVF implantation, embryonic stem cell research, and therapeutic cloning. *International Journal of Applied Philosophy* 2007 Spring; 21(1): 43-53. Subject: 14.4

Holden, Constance. Former Hwang colleague faked monkey data, U.S. says [news]. *Science* 2007 January 19; 315(5810): 317. Subject: 14.5

Holden, Constance. Long-awaited genetic nondiscrimination bill headed for easy passage [news]. *Science* 2007 May 4; 316(5825): 676. Subject: 15.3

Holden, Constance. Prominent researchers join the attack on stem cell patents [news]. *Science* 2007 July 13; 317(5835): 187. Subject: 15.8

Holden, Constance. Scientists protest 'misrepresentation' as Senate vote looms [news]. *Science* 2007 January 19; 315(5810): 315-316. Subject: 18.5.4

Holden, Constance. U.K. takes eggstra time [news]. *Science* 2007 January 19; 315(5810): 317. Subject: 18.5.4

Holden, Constance. U.S. Patent Office casts doubt on Wisconsin stem cell patents [news]. *Science* 2007 April 13; 316(5822): 182. Subject: 15.8

Holdstock, Douglas. A code of ethics for scientists [letter]. *Lancet* 2007 May 26 – June 1; 369(9575): 1789. Subject: 1.3.9

Holland, Suzanne. Market transactions in reprogenetics: a case for regulation. *In:* Knowles, Lori P.; Kaebnick, Gregory E., eds. Reprogenetics: Law, Policy, and Ethical Issues. Baltimore: Johns Hopkins University Press, 2007: 89-104. Subject: 14.1

Holland, S. Theories, principles, and types of arguments. *In his:* Introducing Nursing Ethics: Themes in Theory and Practice. Salisbury: APS, 2004: 1-28. Subject: 2.1

Holloway, Lewis; Morris, Carol. Exploring biopower in the regulation of farm animal bodies: genetic policy interventions in UK livestock. *Genomics, Society and Policy* 2007 August; 3(2): 82-98. Subject: 15.1

Holm, Søren. A rose by any other name... is the research/non-research distinction still important and relevant? *Theoretical Medicine and Bioethics* 2007; 28(3): 153-155. Subject: 5.1

Subject = NRCBL Primary Classification Number; see inside front cover.

Holm, Søren. Can politics be taken out of the (English) NHS? [editorial]. *Journal of Medical Ethics* 2007 October; 33(10): 559. Subject: 9.1

Holm, Søren. Pharmacogenetics and global (in)justice. *In:* Cohen, Jillian Clare; Illingworth, Patricia; Schüklenk, Udo, eds. The Power of Pills: Social, Ethical and Legal Issues in Drug Development, Marketing, and Pricing. London; Ann Arbor, MI: Pluto, 2006: 98-105. Subject: 15.1

Holm, Søren. Policy-making in pluralistic societies. *In:* Steinbock, Bonnie, ed. The Oxford Handbook of Bioethics. Oxford; New York: Oxford University Press, 2007: 153-174. Subject: 9.1

Holm, Søren. The nature of human welfare. *In:* Deane-Drummond, Celia; Scott, Peter Manley, eds. Future Perfect?: God, Medicine and Human Identity. New York: T and T Clark International, 2006: 45-55. Subject: 15.1

Holm, Søren. Wrongful life, the welfare principle and the non-identity problem: some further complications. *In:* McLean, Sheila A.M., ed. First Do No Harm: Law, Ethics, and Healthcare. Aldershot, England; Burlington, VT: Ashgate, 2006: 407-414. Subject: 15.2

Holm, Søren; Ashcroft, Richard. Should genetic information be disclosed to insurers? [debate]. *BMJ: British Medical Journal* 2007 June 9; 334(7605): 1196-1197. Subject: 15.1

Holm, Søren; Bortolotti, Lisa. Large scale surveys for policy formation and research—a study in inconsistency. *Theoretical Medicine and Bioethics* 2007; 28(3): 205-220. Subject: 18.4

Holm, S.; Takala, T. High hopes and automatic escalators: a critique of some new arguments in bioethics. *Journal of Medical Ethics* 2007 January; 33(1): 1-4. Subject: 2.1

Holmes, Anne. Add carrying card to QOF [letter]. *BMJ: British Medical Journal* 2007 June 9; 334(7605): 1179. Subject: 19.5

Holmes, D.; Perron, A. Violating ethics: unlawful combatants, national security and health professionals. *Journal of Medical Ethics* 2007 March; 33(3): 143-145. Subject: 21.2

Holt, Rush. How should government regulate stem-cell research? Views from a scientist-legislator. *In:* Santoro, Michael A.; Gorrie, Thomas M., eds. Ethics and the Pharmaceutical Industry. Cambridge; New York: Cambridge University Press, 2005: 109-122, 427-431. Subject: 18.5.4

Holtedahl, Knut A., Meland, Eivind. Drug trials in general practice: time for a quality check before recruiting patients [letter]. *BMJ: British Medical Journal* 2007 July 7; 335(7609): 7. Subject: 18.2

Hondius, Ewoud. The Kelly case — compensation for undue damage for wrongful treatment. *In:* Gevers, J.K.M.; Hondius, E.H.; Hubben, J.H., eds. Health Law, Human Rights and the Biomedicine Convention: Essays in Honour

of Henriette Roscam Abbing. Leiden; Boston: Martinus Nijhoff Publishers, 2005: 105-116. Subject: 11.4

Hong, Mi-Kung; Bero, Lisa A. Tobacco industry sponsorship of a book and conflict of interest. *Addiction* 2006 August; 101(8): 1202-1211. Subject: 1.3.7

Hong, Suk Young. Patients in a vegetative state and the quality of life. *Dolentium Hominum* 2007; 22(2): 22-27. Subject: 20.5.1

Hongladarom, Soraj. Ethics of bioinformatics: a convergence between bioethics and computer ethics. *Asian Biotechnology and Development Review* 2006 November; 9(1): 37-44 [Online]. Accessed: http://www.openj-gate.org/articlelist.asp?LatestYear=2007&JCode=104346&year=2006&vol=9&issue=1&ICode=583649 [2008 February 29]. Subject: 1.3.12

Honnefelder, Ludger. Science, law and ethics: the Biomedicine Convention as an ethico-legal response to current scientific challenges. *In:* Gevers, J.K.M.; Hondius, E.H.; Hubben, J.H., eds. Health Law, Human Rights and the Biomedicine Convention: Essays in Honour of Henriette Roscam Abbing. Leiden; Boston: Martinus Nijhoff Publishers, 2005: 13-22. Subject: 4.4

Hooft, Stan van. Bioethics and caring. *In his:* Caring about Health. Aldershot, England; Burlington, VT: Ashgate, 2006: 27-39. Subject: 2.1

Hopkin, Michael. UK set to reverse stance on research with chimeras [news]. *Nature Medicine* 2007 August; 13(8): 890-891. Subject: 18.5.4

Hopkins, Michael M.; Mahdi, Surya; Patel, Pari; Thomas, Sandy M. DNA patenting: the end of an era? *Nature Biotechnology* 2007 February; 25(2): 185-188. Subject: 15.8

Hord, Jeffrey D.; Rehman, Waqas; Hannon, Patricia; Anderson-Shaw, Lisa; Schmidt, Mary Lou. Do parents have the right to refuse standard treatment for their child with favorable-prognosis cancer? Ethical and legal concerns. *Journal of Clinical Oncology* 2006 December 1; 24(34): 5454-5456. Subject: 8.3.4

Horn, Lyn. Research vulnerability: an illustrative case study from the South African mining industry. *Developing World Bioethics* 2007 December; 7(3): 119-127. Subject: 18.5.9

Hornby, Karen; Shemie, Sam D.; Teitelbaum, Jeanni; Doig, Christopher. Variability in hospital-based brain death guidelines in Canada. *Canadian Journal of Anaesthesia* 2006 June; 53(6): 613-619. Subject: 20.2.1

Horne, Benjamin D. Human at conception: the 14th amendment and the acquisition of personhood. *Human Life Review* 2007 Summer; 33(3): 73-81. Subject: 4.4

Horres, Robert; Ölschleger; Steineck, Christian. Cloning in Japan: public opinion, expert counselling, and bioethical reasoning. *In:* Roetz, Heiner, ed. Cross-Cultural

See SUBJECT HEADING KEY FOR SECTION II on inside back cover.

517

Issues in Bioethics: The Example of Human Cloning. New York: Rodopi, 2006: 17-49. Subject: 14.5

Horrocks, Stephen. Edwards, Benner, and Wrubel on caring. *Journal of Advanced Nursing* 2002 October; 40(1): 36-41. Subject: 4.1.1

Horst, Jason M. The meaning of "life": the morning-after pill, the question of when life begins, and judicial review. *Texas Journal of Women and the Law* 2007 Spring; 16(2): 205. Subject: 12.1

Horton, Khim; Tschudin, Verena; Forget, Armorel. The value of nursing: a literature review. *Nursing Ethics* 2007 November; 14(6): 716-740. Subject: 4.1.3

Hostetter, Larry. Higher-brain death: a critique. *National Catholic Bioethics Quarterly* 2007 Autumn; 7(3): 499-504. Subject: 20.2.1

Hov, Reidun; Hedelin, Birgitta; Athlin, Elsy. Being an intensive care nurse related to questions of withholding or withdrawing curative treatment. *Journal of Clinical Nursing* 2007 January; 16(1): 203-211. Subject: 20.5.1

Howard, R.J. We have an obligation to provide organs for transplantation after we die. *American Journal of Transplantation* 2006 August; 6(8): 1786-1789. Subject: 19.5

Howe, Edmund G. How should careproviders respond when the medical system leaves a patient short? *Journal of Clinical Ethics* 2007 Fall; 18(3): 195-205. Subject: 9.4

Howe, Edmund G. Taking patients' values seriously. *Journal of Clinical Ethics* 2007 Spring; 18(1): 4-11. Subject: 19.3

Howe, Edmund G. "I'm still glad you were born"—careproviders and genetic counseling. *Journal of Clinical Ethics* 2007 Summer; 18(2): 99-110. Subject: 15.2

Howe, Edmund G. When family members disagree. *Journal of Clinical Ethics* 2007 Winter; 18(4): 331-339. Subject: 20.5.2

Howell, Joel D. Trust and the Tuskegee experiments. *In:* Duffin, Jacalyn, ed. Clio in the Clinic: History in Medical Practice. New York: Oxford University Press, 2005: 213-225. Subject: 8.1

Howell, Jonathan V. Direct to consumer advertising: the world of the market place [letter]. *BMJ: British Medical Journal* 2007 October 6; 335(7622): 683-684. Subject: 9.7

Howsepian, A.A. Cerebral neurophysiology, 'Libetian' action, and euthanasia. *Ethics and Medicine: An International Journal of Bioethics* 2007 Summer; 23(2): 103-111. Subject: 20.5.1

Hren, Darko; Sambunjak, Dario; Ivanis, Ana; Marusic, Matko; Marusic, Ana. Perceptions of authorship criteria: effects of student instruction and scientific experience. *Journal of Medical Ethics* 2007 July; 33(7): 428-432. Subject: 1.3.7

Hsieh, Nien-hê. Property rights in crisis: managers and rescue. *In:* Santoro, Michael A.; Gorrie, Thomas M., eds.

Ethics and the Pharmaceutical Industry. Cambridge; New York: Cambridge University Press, 2005: 379-385. Subject: 5.3

Hsin, Dena Hsin-Chen; Macer, Darryl. Comparisons of life images and end-of-life attitudes between the elderly in Taiwan and New Zealand. *Journal of Nursing Research* 2006 September; 14(3): 198-208. Subject: 20.3.1

Huang, David T.; Hadian, Mehrnaz. Bench-to-bedside review: human subjects research — are more standards needed? *Critical Care* 2006; 10(6): 244. Subject: 18.2

Huang, Mei-Chih; Lee, Chia-Kuei; Lin, Shio-Jean; Lu, I-Chen. A survey of parental consent process for newborn screening in Taiwan. *Acta Paediatrica Taiwanica = Taiwan er ke yi xue hui za zhi* 2005 November-December; 46(6): 361-369. Subject: 15.3

Huang, Nicole; Shih, Shu-Fang; Chang, Hsing-Yi; Chou, Yiing-Jeng. Record linkage research and informed consent: who consents? *BMC Health Services Research* 2007 February 12; 7: 18. Subject: 1.3.12

Hubben, Joep H. Decisions on competency and professional standards. *In:* Gevers, J.K.M.; Hondius, E.H.; Hubben, J.H., eds. Health Law, Human Rights and the Biomedicine Convention: Essays in Honour of Henriette Roscam Abbing. Leiden; Boston: Martinus Nijhoff Publishers, 2005: 93-103. Subject: 8.3.3

Huddle, Thomas S. The limits of objective assessment of medical practice. *Theoretical Medicine and Bioethics* 2007; 28(6): 487-496. Subject: 9.8

Hudson, Kathy L. Prohibiting genetic discrimination. *New England Journal of Medicine* 2007 May 17; 356(20): 2021-2023. Subject: 15.3

Hudson, Kathy; Baruch, Susannah; Javitt, Gail. Genetic testing of human embryos: ethical challenges and policy choices. *In:* Galston, Arthur W.; Peppard, Christiana Z., eds. Expanding Horizons in Bioethics. Dordrecht; Norwell, MA: Springer, 2005: 103-122. Subject: 15.2

Huff, Sarah A. The abortion crisis in Peru: finding a woman's right to obtain safe and legal abortions in the convention on the elimination of all forms of discrimination against women. *Boston College International and Comparative Law Review* 2007 Winter; 30(1): 237-248. Subject: 12.4.2

Hughes, James. Cheating death: vital signs. *New Scientist* 2007 October 13-19; 195(2625): 44-45. Subject: 20.2.1

Hughes, James; Bostrum, Nick; Agar, Nicholas. Human vs. posthuman [letters and reply]. *Hastings Center Report* 2007 September-October; 37(5): 4-6. Subject: 4.5

Hughes, Jonathan. Justice and third party risk: the ethics of xenotransplantation. *Journal of Applied Philosophy* 2007 May; 24(2): 151-168. Subject: 19.1

Hughes, Virginia. Mercury rising: parents of autistic children are mounting a vicious campaign against scientists who refute the link between vaccines and autism. *Nature Medicine* 2007 August; 13(8): 896-897. Subject: 9.7

Hughes, Virginia. Therapy on trial. *Nature Medicine* 2007 September; 13(9): 1008-1009. Subject: 15.4

Human Genome Organisation. Ethics Committee. HUGO statement on pharmacogenomics (PGx): solidarity, equity and governance [position statement]. *Genomics, Society and Policy* 2007 April; 3(1): 44-47. Subject: 15.1

Hunt, Linda M.; de Voogd, Katherine B. Are good intentions good enough?: Informed consent without trained interpreters. *JGIM: Journal of General Internal Medicine* 2007 May; 22(5): 598-605. Subject: 8.3.1

Hunt, Paul. Right to the highest attainable standard of health. *Lancet* 2007 August 4-10; 370(9585): 369-371. Subject: 9.2

Hunter, David. Am I my brother's gatekeeper? Professional ethics and the prioritisation of healthcare. *Journal of Medical Ethics* 2007 September; 33(9): 522-526. Subject: 9.4

Hunter, David. Efficiency and the proposed reforms to the NHS research ethics system. *Journal of Medical Ethics* 2007 November; 33(11): 651-654. Subject: 18.2

Hunter, David; Pierscionek, Barbara K. Children, Gillick competency and consent for involvement in research. *Journal of Medical Ethics* 2007 November; 33(11): 659-662. Subject: 18.5.2

Hunter, D. Proportional ethical review and the identification of ethical issues. *Journal of Medical Ethics* 2007 April; 33(4): 241-245. Subject: 18.2

Hunter, Jennifer M. Plagiarism — does the punishment fit the crime? *Veterinary Anaesthesia and Analgesia* 2006 May; 33(3): 139-142. Subject: 1.3.9

Hunter, Nan D. "Public-private" health law: multiple directions in public health. *Journal of Health Care Law and Policy* 2007; 10(1): 89-119. Subject: 9.1

Hunter, Nan D. Justice Blackmun, abortion, and the myth of medical independence. *Brooklyn Law Review* 2006 Fall; 72(1): 147-197. Subject: 12.4.2

Huntington Study Group. Event Monitoring Committee; Erwin, Cheryl; Hersch, Steven. Monitoring reportable events and unanticipated problems: the PHAROS and PREDICT studies of Huntington disease. *IRB: Ethics and Human Research* 2007 May-June; 29(3): 11-16. Subject: 18.2

Huntington, Ian; Robinson, Walter. The many ways of saying yes and no: reflections on the research coordinator's role in recruiting research participants and obtaining informed consent. *IRB: Ethics and Human Research* 2007 May-June; 29(3): 6-10. Subject: 18.3

Hurlbut, William B. Stem cells, embryos, and ethics: is there a way forward? *Update (Loma Linda Center)* 2007 January 21(3): 1-10. Subject: 18.5.4

Hurley, Elisa A.; Henry, Michael; Fishman, Jennifer R.; Youngner, Stuart J. The moral costs of prophylactic propranolol. *American Journal of Bioethics* 2007 September; 7(9): 35-36; author reply W1-W3. Subject: 17.4

Hurst, G. Cameron. Biological weapons: the United States and the Korean War. *In:* LaFleur, William R.; Böhme, Gernot; Shimazono, Susumu, eds. Dark Medicine: Rationalizing Medical Research. Bloomington: Indiana University Press, 2007: 105-120. Subject: 21.3

Hurst, Samia A.; Danis, Marion. A framework for rationing by clinical judgment. *Kennedy Institute of Ethics Journal* 2007 September; 17(3): 247-266. Subject: 9.4

Hurst, Samia A.; Reiter-Theil, Stella; Perrier, Arnaud; Forde, Reidun; Slowther, Anne-Marie; Pegoraro, Renzo; Danis, Marion. Physicians' access to ethics support services in four European countries. *Health Care Analysis: An International Journal of Health Philosophy and Policy* 2007 December; 15(4): 321-335. Subject: 9.6

Hurst, Samia A.; Volandes, Angelo E.; Paasche-Orlow, Michael K. De-clustering national and international inequality. *American Journal of Bioethics* 2007 November; 7(11): 24-25; author reply W1-W2. Subject: 8.1

Hurst, S.A.; Perrier, A.; Pegoraro, R.; Reiter-Theil, S.; Forde, R.; Slowther, A.-M.; Garrett-Mayer, E.; Danis, M. Ethical difficulties in clinical practice: experiences of European doctors. *Journal of Medical Ethics* 2007 January; 33(1): 51-57. Subject: 9.6

Hussain-Gambles, Mah; Atkin, Karl; Leese, Brenda. South Asian participation in clinical trials: the views of lay people and health professionals. *Health Policy* 2006 July; 77(2): 149-165. Subject: 18.5.1

Huxley, Andrew. The Pali Buddhist approach to human cloning. *In:* Vöneky, Silja; Wolfrum, Rüdiger, eds. Human Dignity and Human Cloning. Leiden: Nijhoff, 2004: 13-22. Subject: 14.5

Huxtable, Richard; Möller, Maaike. 'Setting a principled boundary'? Euthanasia as a response to 'life fatigue'. *Bioethics* 2007 March; 21(3): 117-126. Subject: 20.5.1

Hyatt, Adam. Medicinal marijuana and palliative care: carving a liberty interest out of the Glucksberg framework. *Fordham Urban Law Journal* 2006 November; 33(5): 1345-1367. Subject: 9.5.1

Hyde, Merv; Power, Des. Some ethical dimensions of cochlear implantation for deaf children and their families. *Journal of Deaf Studies and Deaf Education* 2006 Winter; 11(1): 102-111. Subject: 19.1

Hyde, Michael J.; McSpiritt, Sarah. Coming to terms with perfection: the case of Terri Schiavo. *Quarterly Journal of Speech* 2007 May; 93 (2): 150-178. Subject: 20.5.1

Hyder, Adnan A.; Harrison, Rachel A.; Kass, Nancy; Maman, Suzanne. A case study of research ethics capacity development in Africa. *Academic Medicine* 2007 July; 82(7): 675-683. Subject: 18.1

See SUBJECT HEADING KEY FOR SECTION II on inside back cover.

519

Hyman, David A. Institutional review boards: is this the least we can do? *Northwestern University Law Review* 2007; 101(2): 749-773. Subject: 18.2

Hyman, Steven E. The neurobiology of addiction: implications for voluntary control of behavior. *American Journal of Bioethics* 2007 January; 7(1): 8-11. Subject: 17.1

Hynes, Richard. Reply to 'UK set to reverse stance on research with chimeras' [letter]. *Nature Medicine* 2007 October; 13(10): 1133. Subject: 18.5.4

I

Iacono, Teresa. Ethical challenges and complexities of including people with intellectual disability as participants in research. *Journal of Intellectual and Developmental Disability* 2006 September; 31(3): 173-179; discussion 180-191. Subject: 18.5.6

Iannaccone, Philip M. An update on a misconduct investigation [letter]. *Science* 2007 August 17; 317(5840): 899. Subject: 1.3.9

Idänpään-Heikkilä, Juhana E.; Fluss, Sev. Emerging international norms for clinical testing: good clinical trial practice. *In:* Santoro, Michael A.; Gorrie, Thomas M., eds. Ethics and the Pharmaceutical Industry. Cambridge; New York: Cambridge University Press, 2005: 37-47. Subject: 9.7

Iglehart, John K. Insuring all children: the new political imperative. *New England Journal of Medicine* 2007 July 5; 357(1): 70-76. Subject: 9.5.7

Ilfeld, Brian M. Informed consent for medical research: an ethical imperative. *Regional Anesthesia and Pain Medicine* 2006 July-August; 31(4): 353-357. Subject: 18.3

Ilkilic, Ilhan. Human cloning as a challenge to traditional health care cultures. *In:* Roetz, Heiner, ed. Cross-Cultural Issues in Bioethics: The Example of Human Cloning. New York: Rodopi, 2006: 409-423. Subject: 14.5

Illes, Judy. Ipsa scientia potestas est (Knowledge is power) [editorial]. *American Journal of Bioethics* 2007 January; 7(1): 1-2. Subject: 17.1

Illes, Judy. Not forgetting forgetting [editorial]. *American Journal of Bioethics* 2007 September; 7(9): 1-2. Subject: 17.1

Illes, Judy; Bird, Stephanie J. Neuroethics: a modern context for ethics in neuroscience. *Trends in Neurosciences* 2006 September; 29(9): 511-517. Subject: 17.1

Illes, Judy; Gallo, Marisa; Kirschen, Matthew P. An ethics perspective on transcranial magnetic stimulation (TMS) and human neuromodulation. *Behavioural Neurology* 2006; 17(3-4): 149-157. Subject: 17.5

Illes, Judy; Murphy, Emily R. Chimeras of nurture. *American Journal of Bioethics* 2007 May; 7(5):1-2. Subject: 17.1

Illes, Judy; Racine, Eric; Krischen, Matthew P. A picture is worth 1000 words, but which 1000? *In:* Illes, Judy, ed. Neuroethics: Defining the Issues in Theory, Practice, and Policy. New York: Oxford University Press, 2006: 149-168. Subject: 17.1

Illes, J.; Chin, V. Trust and reciprocity: foundational principles for human subjects imaging research. *Canadian Journal of Neurological Sciences* 2007 February; 34(1): 3-4. Subject: 18.3

Illes, J.; Rosen, A.; Greicius, M.; Racine, E. Prospects for prediction: ethics analysis of neuroimaging in Alzheimer's disease. *Annals of the New York Academy of Sciences* 2007 February; 1097: 278-295. Subject: 9.5.2

Illhardt, Franz-Josef. Conflict between a patient's family and the medical team. *HEC (Healthcare Ethics Committee) Forum* 2007 December; 19(4): 381-388. Subject: 9.6

Iltis, Ana. Pediatric research posing a minor increase over minimal risk and no prospect of direct benefit: challenging 45 CFR 46.406. *Accountability in Research* 2007 January-March; 14(1): 19-34. Subject: 18.5.2

Iltis, Ana Smith. Bioethical expertise in health care organizations. *In:* Rasmussen, Lisa, ed. Ethics Expertise: History, Contemporary Perspectives, and Applications. Dordrecht: Springer, 2005: 259-267. Subject: 9.6

Iltis, Ana S.; DeVader, Shannon; Matsuo, Hisako. Payments to children and adolescents enrolled in research: a pilot study. *Pediatrics* 2006 October; 118(4): 1546-1552. Subject: 18.5.2

Indiana. Senate. Senate Resolution 0091. A concurrent resolution to mark the centennial of Indiana's 1907 eugenical sterilization law and to express the regret of the Senate and House of Representatives of the 115th Indiana General Assembly for Indiana's experience with eugenics. Indiana: General Assembly 2007: 3 p. [Online]. Accessed: http://www.in.gov/legislative/bills/2007/SRESP/SC0091.html [2008 February 15]. Subject: 15.5

Indigenous Peoples Council on Biocolonialism. Indigenous people, genes and genetics: what indigenous people should know about biocolonialism: a primer and resource guide. Nixon, NV: Indigenous Peoples Council on Biocolonialism, 2000 June; 25 p. [Online]. Accessed: http://www.ipcb.org/publications/primers/htmls/ipgg.html [2007 April 23]. Subject: 15.11

Infante, Peter F. The past suppression of industry knowledge of the toxicity of benzene to humans and potential bias in future benzene research. *International Journal of Occupational and Environmental Health* 2006 July-September; 12(3): 268-272. Subject: 9.5.1

INSECT Study Group; Seiler, C.M.; Kellmeyer, P.; Kienle, P.; Büchler, M.W.; Knaebel, H.-P. Assessment of the ethical review process for non-pharmacological multi-centre studies in Germany on the basis of a randomised surgical trial. *Journal of Medical Ethics* 2007 February; 33(2): 113-118. Subject: 18.2

Institute of Medicine (United States) [IOM]. Board on Health Sciences Policy. Committee on Assessing Interactions Among Social, Behavioral, and Genetic Factors in Health; Hernandez, Lyla M.; Blazer, Dan G. Ethical, legal, and social implications. *In their:* Genes, Behavior, and the Social Environment: Moving Beyond the Nature/Nurture Debate. Washington, DC: National Academies Press, 2006: 202-218. Subject: 15.1

International Committee of Medical Journal Editors. Clinical trial registration: looking back and moving ahead. *Lancet* 2007 June 9-15; 369(9577): 1909-1911. Subject: 18.6

International Consortium for Emergency Contraception. EC status and availability by country. New York: International Consortium for Emergency Contraception, n.d.; 10 p. [Online]. Accessed: http://www.cecinfo.org/database/pill/viewAllCountry.php [2007 April 10]. Subject: 11.1

International Dual Loyalty Working Group; Physicians for Human Rights; University of Cape Town. Health Services Faculty. Dual loyalty and human rights in health professional practice: proposed guidelines and institutional mechanisms. Boston: Physicians for Human Rights, 2002; 136 p. [Online]. Accessed: http://physiciansforhumanrights.org/library/documents/reports/report-2002-duell oyalty.pdf [2007 April 25]. Subject: 7.1

International HapMap Consortium; Rotimi, Charles; Leppert, Mark; Matsuda, Ichiro; Zeng, Changqing; Zhang, Houcan; Adebamowo, Clement; Ajayi, Ike; Aniagwu, Toyin; Dixon, Missy; Fukushima, Yoshimitsu; Macer, Darryl; Marshall, Patricia; Nkwodimmah, Chibuzor; Peiffer, Andy; Royal, Charmaine; Suda, Eiko; Zhao, Hui; Wang, Vivian Ota; McEwen, Jean. Community engagement and informed consent in the international HapMap project. *Community Genetics* 2007; 10(3): 186-198. Subject: 15.11

International Planned Parenthood Federation [IPPF]. Death and denial: unsafe abortion and poverty. London: International Planned Parenthood Federation, 2006 January: 18 p. [Online]. Accessed: http://content.ippf.org/output/ORG/files/13108.pdf [2007 April 30]. Subject: 12.5.1

International Society of Nurses in Genetics [ISONG]. Access to Genomic Healthcare: The Role of the Nurse: Position Statement. Pittsburgh, PA: International Society of Nurses in Genetics, 2003 September 9: 4 p. [Online]. Accessed: http://www.isong.org/about/ps_genomic.cfm [2007 February 22]. Subject: 15.1

International Society of Nurses in Genetics [ISONG]. Informed Decision-Making and Consent: The Role of Nursing: Position Statement [Revised]. Pittsburgh, PA: International Society of Nurses in Genetics, 2005 April 4: 2 p. [Online]. Accessed: http://www.isong.org/about/ps_consent.cfm [2007 February 22]. Subject: 15.3

International Society of Nurses in Genetics [ISONG]. Provision of quality genetic services and care: building a multidisciplinary, collaborative approach among genetic nurses and genetic counselors [position statement]. Pittsburgh, PA: International Society of Nurses in Genetics, 2006 November 1: 2 p. [Online]. Accessed: http://www.isong.org/about/ps_multidisciplinarygeneticcare.cfm [2007 February 22]. Subject: 15.2

International Society of Nurses in Genetics [ISONG], Board of Directors. Privacy and confidentiality of genetic information: the role of the nurse [position statement]. Pittsburgh, PA: International Society of Nurses in Genetics, 2005 August 8: 2 p. [Online]. Accessed: http://www.isong.org/about/ps_privacy.cfm [2007 February 22]. Subject: 8.4

International Society of Nurses of Genetics [ISONG]. Genetic Counseling for Vulnerable Populations: The Role of Nursing: Position Statement. Pittsburgh, PA: International Society of Nurses of Genetics, 2002 October 10: 4 p. [Online]. Accessed: http://www.isong.org/about/ps_vulnerable.cfm [2007 February 22]. Subject: 15.2

Iorio, A.; Agnelli, G. Are placebo-controlled trials ethical in areas where current guidelines recommend therapy? No. *Journal of Thrombosis and Haemostasis* 2006 October; 4(10): 2133-2136. Subject: 18.3

Ipsen, Jörn. Does the German basic law protect against human cloning? *In:* Vöneky, Silja; Wolfrum, Rüdiger, eds. Human Dignity and Human Cloning. Leiden: Nijhoff, 2004: 69-75. Subject: 14.5

Iramaneerat, Cherdsak. Moral education in medical schools. *Journal of the Medical Association of Thailand* 2006 November; 89(11): 1987-1993. Subject: 7.2

Iredale, Rachel; Longley, Marcus; Thomas, Christian; Shaw, Anita. What choices should we be able to make about designer babies? A Citizens' Jury of young people in South Wales. *Health Expectations* 2006 September; 9(3): 207-217. Subject: 15.2

Ireni-Saban, Liza. Embracing personal and community empowerment: genetic information policy making in Israel. *Eubios Journal of Asian and International Bioethics* 2007 November; 17(6): 181-184. Subject: 15.3

Irving, Kate. Governing the conduct of conduct: are restraints inevitable? *Journal of Advanced Nursing* 2002 November; 40(4): 405-412. Subject: 17.3

Irving, Louise; Harris, John. Biobanking. *In:* Steinbock, Bonnie, ed. The Oxford Handbook of Bioethics. Oxford; New York: Oxford University Press, 2007: 240-257. Subject: 15.1

Irwin, Alan. The global context for risk governance: national regulatory policy in an international framework. *In:* Bennett, Belinda; Tomossy, George F., eds. Globalization and Health: Challenges for Health Law and Bioethics. Dordrecht: Springer, 2006: 71-85. Subject: 5.3

See SUBJECT HEADING KEY FOR SECTION II on inside back cover.

521

Isaacs, David; Kilham, Henry; Gordon, Adrienne; Jeffery, Heather; Tarnow-Mordi, William; Woolnough, Janet; Hamblin, Julie; Tobin, Bernadette. Withdrawal of neonatal mechanical ventilation against the parents' wishes. *Journal of Paediatrics and Child Health* 2006 May; 42(5): 311-315. Subject: 20.5.2

ISDOC Study Group; Costantini, M.; Morasso, G.; Montella, M.; Borgia, P.; Cecioni, R.; Beccaro, M.; Sguazzotti, E.; Bruzzi, P. Diagnosis and prognosis disclosure among cancer patients. Results from an Italian mortality follow-back survey. *Annals of Oncology* 2006 May; 17(5): 853-859. Subject: 20.3.2

Iserson, Kenneth V. Has emergency medicine research benefitted patients? An ethical question. *Science and Engineering Ethics* 2007 September; 13(3): 289-295. Subject: 18.5.1

Iserson, Kenneth V.; Easton, Raul B.; Graber, Mark A.; Monnahan, Jay; Hughes, Jason. The three faces of "yes": consent for emergency department procedures. *American Journal of Bioethics* 2007 December; 7(12): 42-45; author reply W3-W4. Subject: 8.3.1

Islamic Organization for Medical Sciences [IOMS]. Islamic code of medical ethics: doctor's duty in war time. Sulaibekhat, Kuwait: Islamic Organization for Medical Sciences, n.d.; 1 p. [Online]. Accessed: http://www.islamset.com/ethics/code/index.html [2007 April 19]. Subject: 6

Issa, M.M.; Setzer, E.; Charaf, C.; Webb, A.L.; Derico, R.; Kimberl, I.J.; Fink, A.S. Informed versus uninformed consent for prostate surgery: the value of electronic consents. *Journal of Urology* 2006 August; 176(2): 694-699; discussion 699. Subject: 8.3.1

Ito, Shigehiko. Beyond standing: a search for a new solution in animal welfare. *Santa Clara Law Review* 2005-2006; 46(2): 377-418. Subject: 22.1

Itoh, Kenji; Andersen, Henning Boje; Madsen, Marlene Dyrlov; Østergaard, Doris; Ikeno, Masaaki. Patient views of adverse events: comparisons of self-reported healthcare staff attitudes with disclosure of accident information. *Applied Ergonomics* 2006 July; 37(4): 513-523. Subject: 9.8

Iversen, Knut Ivar; Høyer, Georg; Sexton, Harold C. Coercion and patient satisfaction on psychiatric acute wards. *International Journal of Law and Psychiatry* 2007 November-December; 30(6): 504-511. Subject: 9.5.3

Ives, J. Kant, curves and medical learning practice: a reply to Le Morvan and Stock. *Journal of Medical Ethics* 2007 February; 33(2): 119-122. Subject: 7.2

Iwanoswski, Piotr S. Informed consent procedure for clinical trials in emergency settings: the Polish perspective. *Science and Engineering Ethics* 2007 September; 13(3): 333-336. Subject: 18.3

J

Jackson, Chuck. Waste and whiteness: Zora Neale Hurston and the politics of eugenics. *African American Review* 2000 Winter; 34(4): 639-660. Subject: 15.5

Jacob, Marie-Andree. Frail connections: legal and psychiatric knowledge practices in U.S. adjudication over organ donations by children and incompetent adults. *In:* Freeman, Michael, ed. Children's Health and Children's Rights. Boston: Martinus Nijhoff Publishers, 2006: 219-252. Subject: 19.5

Jacobs, Lesley A. Justice in health care: can Dworkin justify universal access? *In:* Burley, Justine, ed. Dworkin and His Critics: With Replies by Dworkin. Malden, MA: Blackwell Pub., 2004: 134-149. Subject: 9.4

Jacobson, Nora; Gewurtz, Rebecca; Haydon, Emma. Ethical review of interpretive research: problems and solutions. *IRB: Ethics and Human Research* 2007 September-October; 29(5): 1-8. Subject: 18.4

Jacobson, Peter D.; Parmet, Wendy E. A new era of unapproved drugs: the case of Abigail Alliance v Von Eschenbach. *JAMA: The Journal of the American Medical Association* 2007 January 10; 297(2): 205-208. Subject: 9.7

Jaenisch, Rudolf. Nuclear cloning, embryonic stem cells, and gene transfer. *In:* Rasko, John E.J.; O'Sullivan, Gabrielle M.; Ankeny, Rachel A., eds. The Ethics of Inheritable Genetic Modification: A Dividing Line? Cambridge: Cambridge University Press, 2006: 35-55. Subject: 14.5

Jafarey, Aamir; Thomas, George; Ahmad, Aasim; Srinivasan, Sandhya. Asia's organ farms [editorial]. *Indian Journal of Medical Ethics* 2007 April-June; 4(2): 52-53. Subject: 19.5

Jagadeesh, N. Narco analysis leads to more questions than answers. *Indian Journal of Medical Ethics* 2007 January-March; 4(1): 9. Subject: 9.7

Jagsi, Reshma. Conflicts of interest and the physician-patient relationship in the era of direct-to-patient advertising. *Journal of Clinical Oncology* 2007 March 1; 25(7): 902-905. Subject: 8.1

Jaing, Tang-Her; Tsay, Pei-Kwei; Fang, En-Chen; Yang, Shu-Ho; Chen, Shih-Hsiang; Yang, Chao-Ping; Hung, Iou-Jih. "Do-not-resuscitate" orders in patients with cancer at a children's hospital in Taiwan. *Journal of Medical Ethics* 2007 April; 33(4): 194-196. Subject: 20.5.2

Jakusovaite, Irayda; Bankauskaite, Vaida. Teaching ethics in a masters program in public health in Lithuania. *Journal of Medical Ethics* 2007 July; 33(7): 423-432. Subject: 7.2

Jalali, Rakesh; Howie, Stephen. Conduct of clinical trials in developing countries [letters]. *Lancet* 2007 August 18-24; 370(9587): 562. Subject: 18.6

Subject = NRCBL Primary Classification Number; see inside front cover.

Jamieson, Suzanne. Genetic information and the Australian labour movement. *In:* Betta, Michela, ed. The Moral, Social, and Commercial Imperatives of Genetic Testing and Screening: the Australian Case. Dordrecht: Springer, 2006: 211-220. Subject: 15.3

Jansen-van der Weide, Marijke Catharina; Onwuteaka-Philipsen, Bregje Dorien; van der Wal, Gerrit. Quality of consultation and the project 'Support and Consultation on Euthanasia in the Netherlands' (SCEN). *Health Policy* 2007 January; 80(1): 97-106. Subject: 9.8

Janvier, Annie. How much emotion is enough? *Journal of Clinical Ethics* 2007 Winter; 18(4): 362-365. Subject: 20.5.2

Janvier, Annie; Bauer, Karen Lynn; Lantos, John D. Are newborns morally different from older children? *Theoretical Medicine and Bioethics* 2007; 28(5): 413-425. Subject: 4.4

Jasen, Patricia. Breast cancer and the politics of abortion in the United States. *Medical History* 2005 October; 49(4): 423-444. Subject: 9.5.1

Jauhar, Sandeep. Between comfort and care, a blurry line. *New York Times* 2007 September 18; p. F5. Subject: 20.4.1

Javitt, Gail H. In search of a coherent framework: options for FDA oversight of genetics tests. *Food and Drug Law Journal* 2007; 62(4): 617-652. Subject: 15.3

Jaworska, Agnieszka. Ethical dilemmas in neurodegenerative disease: respecting patients at the twilight of agency. *In:* Illes, Judy, ed. Neuroethics: Defining the Issues in Theory, Practice, and Policy. New York: Oxford University Press, 2006: 87-101. Subject: 17.1

Jayaraman, K.S. Database targets Parsi genes [news]. *Nature* 2007 March 29; 446(7135): 475. Subject: 15.1

Jayaraman, K.S. Indian scientists battle journal retraction [news]. *Nature* 2007 June 14; 447(7146): 764. Subject: 1.3.7

Jayasinghe, Saroj. Faith-based NGOs and healthcare in poor countries: a preliminary exploration of ethical issues. *Journal of Medical Ethics* 2007 November; 33(11): 623-626. Subject: 21.1

Jecker, Nancy S. Medical futility: a paradigm analysis. *HEC (Healthcare Ethics Committee) Forum* 2007 March; 19(1): 13-32. Subject: 20.5.1

Jefferson, Valeria. The ethical dilemma of genetically modified food. *Journal of Environmental Health* 2006 July-August; 69(1): 33-34. Subject: 15.1

Jenkins, John G. GMC guidance on confidentiality [letter]. *BMJ: British Medical Journal* 2007 December 15; 335(7632): 1226. Subject: 8.4

Jennings, Beth. The politics of end-of-life decision-making: computerised decision-support tools, physicians' jurisdiction and morality. *Sociology of Health and Illness* 2006 April; 28(3): 350-375. Subject: 20.5.1

Jennings, Bruce. Autonomy. *In:* Steinbock, Bonnie, ed. The Oxford Handbook of Bioethics. Oxford; New York: Oxford University Press, 2007: 72-89. Subject: 2.1

Jensen, Norman. Empathy and patient-physician conflicts: exploring respect. *JGIM: Journal of General Internal Medicine* 2007 October; 22(10): 1485. Subject: 8.1

Jersild, Paul. Theological and moral reflections on stem cell research. *Journal of Lutheran Ethics [electronic]* 2007 March; 7(3): 4 p. Accessed: http://www.elca.org/jle/article.asp?k=705 [2007 February 28]. Subject: 18.5.4

Jesani, Amar. Response: questions of science, law and ethics. *Indian Journal of Medical Ethics* 2007 January-March; 4(1): 10-11. Subject: 1.3.9

Jesani, Amar; Coutinho, Lester. AIDS vaccine trials in India: ethical benchmarks and unanswered questions [editorial]. *Indian Journal of Medical Ethics* 2007 January-March; 4(1): 2-3. Subject: 18.5.9

Jester, Penelope M.; Tilden, Samuel J.; Li, Yufeng; Whitley, Richard J.; Sullender, Wayne M. Regulatory challenges: lessons from recent West Nile virus trials in the United States. *Contemporary Clinical Trials* 2006 June; 27(3): 254-259. Subject: 18.2

Jizba, Laurel. Ethics in grant funded academia: issues and questions. *Journal of Information Ethics* 2007 Spring; 16(1): 42-52. Subject: 5.3

Jochemsen, Henk. Is cloning compatible with human rights and human dignity? *In:* Wagner, Teresa; Carbone, Leslie, eds. Fifty Years after the Declaration: the United Nations' Record on Human Rights. Lanham, MD: University Press of America, 2001: 33-43. Subject: 14.5

Joffe, Ari R. The ethics of donation and transplantation: are definitions of death being distorted for organ transplantation? *Philosophy, Ethics, and Humanities in Medicine [electronic]* 2007 November 25; 2(28): 7 p. Accessed: http://www.peh-med.com/ [2008 January 24]. Subject: 19.5

Joffe, Carole; Shields, Wayne C. Morality and the abortion provider. *Contraception* 2006 July; 74(1): 1-2. Subject: 12.3

Joffe, Steven; Fernandez, Conrad V.; Pentz, Rebecca D.; Ungar, David R.; Mathew, N. Ajoy; Turner, Curtis W.; Alessandri, Angela J.; Woodman, Catherine L.; Singer, Dale A.; Kodish, Eric. Involving children with cancer in decision-making about research participation. *Journal of Pediatrics* 2006 December; 149(6): 862-868. Subject: 18.5.2

John Paul II, Pope. Address To the Participants to the International Congress "Life-Sustaining Treatments and Vegetative State: Scientific Advances and Ethical Dilemmas" [official English version]. Vatican City: Magisterium 2004 March 20; 5 p. [Online]. Accessed: http://www.academiavita.org/template.jsp?sez=DocumentiMagistero

See SUBJECT HEADING KEY FOR SECTION II on inside back cover.

523

&pag=papi/gp_sv/gp_sv&lang=english [2004 March 29]. Subject: 20.5.1

John, Jill E. The child's right to participate in research: myth or misconception? *British Journal of Nursing* 2007 February 8-21; 16(3): 157-160. Subject: 18.5.2

John, S.D. How to take deontological concerns seriously in risk-cost-benefit analysis: a re-interpretation of the precautionary principle. *Journal of Medical Ethics* 2007 April; 33(4): 221-224. Subject: 9.1

Johnson, Claire. Repetitive, duplicate, and redundant publications: a review for authors and readers. *Journal of Manipulative and Physiological Therapies* 2006 September; 29(7): 505-509. Subject: 1.3.7

Johnson, Edward L., Jr.; Johnson, Larry W. Conversations in ethics. *JONA's Healthcare Law, Ethics, and Regulation* 2007 October-December; 9(4): 117-118. Subject: 7.4

Johnson, Jeannette L.; Vandermark, Nancy R. Ethics in prevention research with children. *In:* Kleinig, John; Einstein, Stanley, eds. Ethical Challenges for Intervening in Drug Use: Policy, Research and Treatment Issues. Huntsville, TX: Office of International Criminal Justice; Criminal Justice Center, Sam Houston State University, 2006: 259-279. Subject: 18.5.2

Johnson, Jonas T.; Niparko, John K.; Levine, Paul A.; Kennedy, David W.; Rudy, Susan F.; Weber, Pete; Weber, Randal S.; Benninger, Michael S.; Rosenfeld, Richard M.; Ruben, Robert J.; Smith, Richard J.H.; Sataloff, Robert Thayer; Weir, Neil. Standards for ethical publication. *American Journal of Otolaryngology* 2007 January-February; 28(1): 1-2. Subject: 1.3.7

Johnson, Jonas T.; Niparko, John K.; Levine, Paul A.; Kennedy, David W.; Rudy, Susan F.; Weber, Peter C.; Weber, Randal S.; Benniger, Michael S.; Rosenfeld, Richard J.; Ruben, Robert J.; Smith, Richard J.H.; Sataloff, Robert Thayer; Weir, Neil. Standards for ethical publication. *Ear, Nose, and Throat Journal* 2006 December; 85(12): 792, 795. Subject: 1.3.7

Johnson, Kevin A.; Kozel, F. Andrew; Laken, Steven J.; George, Mark S. The neuroscience of functional magnetic resonance imaging fMRI for deception detection. *American Journal of Bioethics* 2007 September; 7(9): 58-60. Subject: 17.1

Johnson, Kirk. Proposed Colorado measure on rights for human eggs. *New York Times* 2007 November 18; p. A23. Subject: 4.4

Johnson, Larry W. Practice pointers for the nurse leader: lessons in conducting an ethics consult. *JONA's Healthcare Law, Ethics, and Regulation* 2007 July-September; 9(3): 97-99. Subject: 9.6

Johnson, Luke. Embryonic stem cell research: a legitimate application of just-war theory? *Ethics and Medicine:*

An International Journal of Bioethics 2007 Spring; 23(1): 19-30. Subject: 18.5.4

Johnson, Martin H. Regulating the science and therapeutic application of human embryo research: managing the tension between biomedical creativity and public concern. *In:* Spencer, J.R.; du Bois-Pedain, Antje, eds. Freedom and Responsibility in Reproductive Choice. Portland: Hart Pub., 2006: 91-106. Subject: 18.5.4

Johnson, Martin; Haigh, Carol; Yates-Bolton, Natalie. Valuing of altruism and honesty in nursing students: a two-decade replication study. *Journal of Advanced Nursing* 2007 February; 57(4): 366-374. Subject: 4.1.3

Johnson, Sandra. New CDC guidelines for HIV screening: ethical implications for health care providers. *Health Care Ethics USA* 2007 Winter; 15(1): 5-7. Subject: 9.5.6

Johnson, Summer. A rebuttal to Dzur and Levin: Johnson on the legitimacy and authority of bioethics commissions. *Kennedy Institute of Ethics Journal* 2007 June; 17(2): 143-152. Subject: 2.4

Johnson, Summer. Making up is hard to do [review of After Harm: Medical Error and the Ethics of Forgiveness, by Nancy Berlinger]. *Hastings Center Report* 2007 March-April; 37(2): 45-46. Subject: 9.8

Johnson, Susan C. New reproductive technologies — discourse and dilemma. *In:* Garwood-Gowers, Austen; Tingle, John; Wheat, Kay, eds. Contemporary Issues in Healthcare Law and Ethics. Edinburgh; New York: Elsevier Butterworth-Heinemann, 2005: 115-131. Subject: 14.1

Johnson, W. Brad; Bacho, Roderick; Heim, Mark; Ralph, John. Multiple-role dilemmas for military mental health care providers. *Military Medicine* 2006 April; 171(4): 311-315. Subject: 17.2

Johnston, Carolyn; Haughton, Peter. Medical students' perceptions of their ethics teaching. *Journal of Medical Ethics* 2007 July; 33(7): 418-422. Subject: 7.2

Johnston, Carolyn; Liddle, Jane. The Mental Capacity Act 2005: a new framework for healthcare decision making. *Journal of Medical Ethics* 2007 February; 33(2): 94-97. Subject: 8.3.3

Johnston, Josephine. Tied up in nots over genetic parentage. *Hastings Center Report* 2007 July-August; 37(4): 28-31. Subject: 14.1

Johnston, Josephine; Wasunna, Angela A. Patents, biomedical research, and treatments: examining concerns, canvassing solutions. *Hastings Center Report* 2007 January-February; 37(1): S2-S35. Subject: 15.8

Johnston, Therese E. Issues surrounding protection and assent in pediatric research. *Pediatric Physical Therapy* 2006 Summer; 18(2): 133-140. Subject: 18.5.7

Joint Committee on Medical Genetics; Farndon, Peter A. Recording, using and sharing genetic information and test results: consent is the key in all medical specialties.

Clinical Medicine 2006 May-June; 6(3): 236-238. Subject: 15.3

Joly, Pierre-Benoit; Rip, Arie. A timely harvest. The public should be consulted on contentious research and development early enough for their opinions to influence the course of science and policy-making. *Nature* 2007 November 8; 450(7167): 174. Subject: 1.3.9

Jonas, Monique. The Baby MB case: medical decision making in the context of uncertain infant suffering. *Journal of Medical Ethics* 2007 September; 33(9): 541-544. Subject: 20.5.2

Jones, Christopher N. How much longer will patients trust us? [letter]. *BMJ: British Medical Journal* 2007 December 15; 335(7632): 1226. Subject: 8.4

Jones, Dan. The depths of disgust. Is there wisdom to be found in repugnance? Or is disgust 'the nastiest of all emotions', offering nothing but support to prejudice? *Nature* 2007 June 14; 447(7146): 768-771. Subject: 17.1

Jones, David S.; Perlis, Roy H. Pharmacogenetics, race, and psychiatry: prospects and challenges. *Harvard Review of Psychiatry* 2006 March-April; 14(2): 92-108. Subject: 15.1

Jones, D. Gareth. Anatomical investigations and their ethical dilemmas. *Clinical Anatomy* 2007 April; 20(3): 338-343. Subject: 20.1

Jones, D. Gareth. Neuroscience and the modification of human beings. *In:* Deane-Drummond, Celia; Scott, Peter Manley, eds. Future Perfect?: God, Medicine and Human Identity. New York: T and T Clark International, 2006: 87-99. Subject: 17.1

Jones, James W.; McCullough, Laurence B. Ethics of over-scheduling: when enough becomes too much. *Journal of Vascular Surgery* 2007 March; 45(3): 635-636. Subject: 9.8

Jones, James W.; McCullough, Laurence B. Ethics of unprofessional behavior that disrupts: crossing the line. *Journal of Vascular Surgery* 2007 February; 45(2): 433-435. Subject: 7.4

Jones, James W.; McCullough, Laurence B.; Richman, Bruce W. Ethics and professionalism: do we need yet another surgeons' charter? *Journal of Vascular Surgery* 2006 October; 44(4): 903-906. Subject: 4.1.2

Jones, James W.; McCullough, Laurence B.; Richman, Bruce W. My brother's keeper: uncompensated care for illegal immigrants. *Journal of Vascular Surgery* 2006 September; 44(3): 679-682. Subject: 9.5.10

Jones, James W.; McCullough, Laurence B.; Richman, Bruce W. Other people's money: ethics, finances, and bad outcomes. *Journal of Vascular Surgery* 2006 April; 43(4): 863-865. Subject: 9.5.1

Jones, James W.; McCullough, Laurence B.; Richman, Bruce W. Painted into a corner: unexpected complications in treating a Jehovah's Witness. *Journal of Vascular Surgery* 2006 August; 44(2): 425-428. Subject: 8.3.4

Jones, Leslie Sargent. The ethics of transcranial magnetic stimulation [letter and reply]. *Science* 2007 March 23; 315(5819): 1663-1664. Subject: 17.5

Jones, Marian Moser; Bayer, Ronald. Paternalism and its discontents: motorcycle helmet laws, libertarian values, and public health. *American Journal of Public Health* 2007 February; 97(2): 208-217. Subject: 9.1

Jones, Melinda. Adolescent gender identity and the courts. *In:* Freeman, Michael, ed. Children's Health and Children's Rights. Boston: Martinus Nijhoff Publishers, 2006: 121-148. Subject: 9.5.7

Jones, Nancy L. A code of ethics for the life sciences. *Science and Engineering Ethics* 2007 March; 13(1): 25-43. Subject: 1.3.9

Jones, Roland; Kingdon, David. Council of Europe recommendation on human rights and psychiatry: a major opportunity for mental health services [editorial]. *European Psychiatry* 2005 November; 20(7): 461-464. Subject: 17.1

Jones, Sian; Jones, Bridie. Advance directives and implications for emergency departments. *British Journal of Nursing* 2007 February 22-March 7; 16(4): 220-223. Subject: 20.5.4

Jonsen, Albert R. Guest editorial: a note on the notion of commercialism. *CQ: Cambridge Quarterly of Healthcare Ethics* 2007 Fall; 16(4): 368-373. Subject: 9.3.1

Jonsen, Albert R. How to appropriate appropriately: a comment on Baker and McCullough. *Kennedy Institute of Ethics Journal* 2007 March; 17(1): 43-54. Subject: 2.1

Jonsen, Albert R. The history of bioethics as a discipline. *In:* Khushf, George, ed. Handbook of Bioethics: Taking Stock of the Field From a Philosophical Perspective. Dordrecht; Boston: Kluwer Academic, 2004: 31-51. Subject: 2.2

Jørgensen, H.K.; Hartling, O.J. Anonymity in connection with sperm donation. *Medicine and Law: The World Association for Medical Law* 2007 March; 26(1): 137-143. Subject: 8.4

Jotkowitz, Alan B. Ethics consultation: whose ethics? *American Journal of Bioethics* 2007 February; 7(2): 41-42. Subject: 9.6

Jotkowitz, Alan; Porath, Avi; Volandes, Angelo E.; Paasche-Orlow, Michael K. Health literacy, access to care and outcomes of care. *American Journal of Bioethics* 2007 November; 7(11): 25-27; author reply W1-W2. Subject: 9.1

Jotkowitz, A.; Glick, Shimon; Gezundheit, B. Truth-telling in a culturally diverse world. *Cancer Investigation* 2006 December; 24(8): 786-789. Subject: 8.2

Joung, Phillan; Eggert, Marion. The cloning debate in South Korea. *In:* Roetz, Heiner, ed. Cross-Cultural Issues

See SUBJECT HEADING KEY FOR SECTION II on inside back cover.

525

in Bioethics: The Example of Human Cloning. New York: Rodopi, 2006: 155-178. Subject: 14.5

Joyce, Theodore; Kaestner, Robert. State reproductive policies and adolescent pregnancy resolution: the case of parental involvement laws. *Journal of Health Economics* 1996 October; 15(5): 579-607. Subject: 12.4.1

Juengst, Eric T. "Alter-ing" the human species? Misplaced essentialism in science policy. *In:* Rasko, John E.J.; O'Sullivan, Gabrielle M.; Ankeny, Rachel A., eds. The Ethics of Inheritable Genetic Modification: A Dividing Line? Cambridge: Cambridge University Press, 2006: 149-158. Subject: 15.1

Juengst, Eric T. Population genetic research and screening: conceptual and ethical issues. *In:* Steinbock, Bonnie, ed. The Oxford Handbook of Bioethics. Oxford; New York: Oxford University Press, 2007: 471-490. Subject: 15.11

Junker-Kenny, Maureen. Genetic perfection, or fulfilment of creation in Christ? *In:* Deane-Drummond, Celia; Scott, Peter Manley, eds. Future Perfect?: God, Medicine and Human Identity. New York: T and T Clark International, 2006: 155-167. Subject: 15.1

Justo, Luis; Erazun, Fabiana. Neuroethics and human rights. *American Journal of Bioethics* 2007 May; 7(5): 16-18. Subject: 17.1

Juthberg, Christina; Eriksson, Sture; Norberg, Astrid; Sundin, Karin. Perceptions of conscience in relation to stress of conscience. *Nursing Ethics* 2007 May; 14(3): 329-343. Subject: 4.1.3

K

Kaba, R.; Sooriakumaran, P. The evolution of the doctor-patient relationship. *International Journal of Surgery* 2007 February; 5(1): 57-65. Subject: 8.1

Kabasenche, William P.; Henry, Michael; Fishman, Jennifer R.; Youngner, Stuart J. Emotions, memory suppression, and identity. *American Journal of Bioethics* 2007 September; 7(9): 33-34; author reply W1-W3. Subject: 17.4

Kabir, M.; az-Zubair, Banu. Who is a parent? Parenthood in Islamic ethics. *Journal of Medical Ethics* 2007 October; 33(10): 605-609. Subject: 14.1

Kaczor, Christopher. Philosophy and theology: the authority of Pope John Paul II allocution; is ANH required for PVS patients?; papal allocution and Catholic tradition; human life as intrinsic good;. *National Catholic Bioethics Quarterly* 2007 Autumn; 7(3): 595-605. Subject: 20.5.1

Kaczor, Christopher Robert. All human beings are persons. *In his:* The Edge of Life: Human Dignity and Contemporary Bioethics. Dordrecht: Springer, 2005: 41-66. Subject: 4.4

Kaczor, Christopher Robert. An ethical assessment of Bush's guidelines for stem cell research. *In his:* The Edge

of Life: Human Dignity and Contemporary Bioethics. Dordrecht: Springer, 2005: 83-96. Subject: 18.5.4

Kaczor, Christopher Robert. Could artificial wombs end the abortion debate? *In his:* The Edge of Life: Human Dignity and Contemporary Bioethics. Dordrecht: Springer, 2005: 105-131. Subject: 14.1

Kaczor, Christopher Robert. How is the dignity of the person as agent recognized?: distinguishing intention from foresight. *In his:* The Edge of Life: Human Dignity and Contemporary Bioethics. Dordrecht: Springer, 2005: 67-81. Subject: 4.4

Kaczor, Christopher Robert. Moral absolutism and ectopic pregnancy. *In his:* The Edge of Life: Human Dignity and Contemporary Bioethics. Dordrecht: Springer, 2005: 97-103. Subject: 14.1

Kaebnick, Gregory. Putting concerns about nature in context: the case of agricultural biotechnology. *Perspectives in Biology and Medicine* 2007 Autumn; 50(4): 572-584. Subject: 15.1

Kaebnick, Gregory E. Secrets and open societies [editorial]. *Hastings Center Report* 2007 May-June; 37(3): 2. Subject: 21.3

Kaebnick, Gregory E. Small talk. *Hastings Center Report* 2007 January-February; 37(1): inside front cover. Subject: 5.1

Kaebnick, Gregory E. The problem with trust and sympathy [editorial]. *Hastings Center Report* 2007 March-April; 37(2): 2. Subject: 9.6

Kaebnick, Gregory E. What should HCR publish? [editorial]. *Hastings Center Report* 2007 November-December; 37(6): 2. Subject: 1.3.7

Kafarowski, Joanna. The woman/gender questions: best practices of conducting research with indigenous peoples in Canada. *NCEHR Communique CNERH* 2006 Spring; 14(1): 18-20. Subject: 18.5.1

Kahn, Jeffrey. Baseball, alcohol and public health. *American Journal of Bioethics* 2007 July; 7(7): 3. Subject: 9.5.1

Kahn, Jeffrey. What vaccination programs mean for research [editorial]. *American Journal of Bioethics* 2007 March; 7(3): 5-10. Subject: 9.7

Kahn, Jeffrey P. Organs and stem cells: policy lessons and cautionary tales. *Hastings Center Report* 2007 March-April; 37(2): 11-12. Subject: 18.5.4

Kahn, Jeffrey P. Why public health and politics don't mix [editorial]. *American Journal of Bioethics* 2007 November; 7(11): 3-4. Subject: 9.1

Kahn, Jeffrey; Mastroianni, Anna. The implications of public health for bioethics. *In:* Steinbock, Bonnie, ed. The Oxford Handbook of Bioethics. Oxford; New York: Oxford University Press, 2007: 671-695. Subject: 9.1

Kahnawake Schools Diabetes Prevention Project. Code of research ethics. Kahnawake Territory, Mohawk Nation via Quebec, Canada: Kahnawake Schools Diabetes Prevention Project, 2007; 31 p. [Online]. Accessed: http://www.ksdpp.org/i/ksdpp_code_of_research_ethics2007.pdf [2007 April 26]. Subject: 6

Kaimal, Girija; Steinberg, Annie G.; Ennis, Sara; Harasink, Sue Moyer; Ewing, Rachel; Li, Yuelin. Parental narratives about genetic testing for hearing loss: a one year follow up study. *Journal of Genetic Counseling* 2007 December; 16(6): 775-787. Subject: 15.2

Kaiser, Jocelyn. Attempt to patent artificial organism draws a protest [news]. *Science* 2007 June 15; 316(5831): 1557. Subject: 15.8

Kaiser, Jocelyn. Death prompts a review of gene therapy vector [news]. *Science* 2007 August 3; 317(5838): 580. Subject: 15.4

Kaiser, Jocelyn. Gene transfer an unlikely contributor to patient's death [news]. *Science* 2007 December 7; 318(5856): 1535. Subject: 15.4

Kaiser, Jocelyn. Issues with tissues [news brief]. *Science* 2007 June 29; 316(5833): 1829. Subject: 4.4

Kaiser, Jocelyn. Privacy policies take a toll on research, survey finds [news]. *Science* 2007 November 16; 318(5853): 1049. Subject: 1.3.9

Kaiser, Jocelyn. Questions remain on cause of death in arthritis trial [news]. *Science* 2007 September 21; 317(5845): 1665. Subject: 15.4

Kaiser, Jocelyn. Stung by controversy, biomedical groups urge consistent guidelines [news]. *Science* 2007 July 27; 317(5837): 441. Subject: 1.3.9

Kaiser, Matthias. Practical ethics in search of a toolbox: ethics of science and technology at the crossroads. *In:* Gunning, Jennifer; Holm, Søren, eds. Ethics, law, and society. Volume 2. Aldershot, Hants, England; Burlington, VT: Ashgate, 2006: 35-44. Subject: 15.1

Kakuk, Peter. The slippery slope of the middle ground: reconsidering euthanasia in Britain. *HEC (Healthcare Ethics Committee) Forum* 2007 June; 19(2):145-159. Subject: 20.5.1

Kaldjian, Lauris C.; Jones, Elizabeth W.; Rosenthal, Gary E. Facilitating and impeding factors for physicians' error disclosure: a structured literature review. *Joint Commission Journal on Quality and Patient Safety / Joint Commission Resources* 2006 April; 32(4): 188-198. Subject: 9.8

Kaldjian, Lauris C.; Jones, Elizabeth W.; Wu, Barry J.; Forman-Hoffman, Valerie L.; Levi, Benjamin H.; Rosenthal, Gary E. Disclosing medical errors to patients: attitudes and practices of physicians and trainees. *JGIM: Journal of General Internal Medicine* 2007 July; 22(7): 988-996 [see correction in JGIM: Journal of General Internal Medicine 2007 September; 22(9): 1384]. Subject: 9.8

Kaleebu, Pontiano. HIV vaccine trials in Uganda: personal experience as an investigator. *In:* AIDS Vaccine Handbook, 2nd edition: Global Perspectives. New York: AIDS Vaccine Advocacy Coalition, 2005: 145-151 [Online]. Accessed: http://www.avac.org/pdf/primer2/AVH_CH21.pdf [2006 March 8]. Subject: 9.1

Kalichman, Michael W. Responding to challenges in educating for the responsible conduct of research. *Academic Medicine* 2007 September; 82(9): 870-875. Subject: 18.2

Kalichman, Michael W.; Plemmons, Dena K. Reported goals for responsible conduct of research courses. *Academic Medicine* 2007 September; 82(9): 846-852. Subject: 18.2

Kaltiala-Heino, Riittakerttu; Fröjd, Sari. Severe mental disorder as a basic commitment criterion for minors. *International Journal of Law and Psychiatry* 2007 January-February; 30(1): 81-94. Subject: 17.7

Kamerow, Douglas. Great health care, guaranteed. *BMJ: British Medical Journal* 2007 May 26; 334(7603): 1086. Subject: 9.8

Kamerow, Douglas. Killing me softly. *BMJ: British Medical Journal* 2007 March 3; 334(7591): 454. Subject: 20.6

Kamm, F.M. Ending life. *In:* Rhodes, Rosamond; Francis, Leslie P.; Silvers, Anita, eds. The Blackwell Guide to Medical Ethics. Malden, MA: Blackwell Pub., 2007: 142-161. Subject: 20.7

Kamm, F.M. Ronald Dworkin's views on abortion and assisted suicide. *In:* Burley, Justine, ed. Dworkin and His Critics: With Replies by Dworkin. Malden, MA: Blackwell Pub., 2004: 218-240. Subject: 12.4.2

Kannan, Ramya. Code of ethics for doctors not being enforced, says Anbumani [news]. *Hindu* 2007 January 1 [Online]. Accessed: http://www.the hindu.com/2007/01/31/stories/2007013108640100.htm [2007 January 31]. Subject: 6

Kanter, James. Proposed ban on genetically modified corn in Europe [news]. *New York Times* 2007 November 23; p. C3. Subject: 15.1

Kanter, Steven L.; Wimmers, Paul F.; Levine, Arthur S. In-depth learning: one school's initiatives to foster integration of ethics, values, and the human dimensions of medicine. *Academic Medicine* 2007 April; 82(4): 405-409. Subject: 7.2

Kanungo, R. Ethics in research [editorial]. *Indian Journal of Medical Microbiology* 2006 January; 24(1): 5-6. Subject: 1.3.9

Kapiriri, Lydia; Norheim, Ole Frithjof; Martin, Douglas K. Priority setting at the micro-, meso- and macro-levels in Canada, Norway and Uganda. *Health Policy* 2007 June; 82(1): 78-94. Subject: 9.4

Kaplan, Kalman J. Zeno, Job and Terry Schiavo: the right to die versus the right to life. *Ethics and Medicine: An International Journal of Bioethics* 2007 Summer; 23(2): 95-102. Subject: 20.5.1

See SUBJECT HEADING KEY FOR SECTION II on inside back cover.

527

Kaposy, Chris. Can infants have interests in continued life? *Theoretical Medicine and Bioethics* 2007; 28(4): 301-330. Subject: 20.5.2

Kaposy, Chris; Sastre, María Teresa Muñoz; Peccarisi, Céline; Legrain, Elizabeth; Mullet, Etienne; Sorum, Paul. The real-life consequences of being denied access to an abortion. *American Journal of Bioethics* 2007 August; 7(8): 34-36; author reply W3. Subject: 12.5.1

Kapp, Marshall B. Pain control for dying patients: hastening death or ensuring comfort? *Journal of Opioid Management* 2006 May-June; 2(3): 128-129. Subject: 20.5.1

Kapp, Marshall B. Patient autonomy in the age of consumer-driven health care: informed consent and informed choice. *Journal of Legal Medicine* 2007 January-March; 28(1): 91-117. Subject: 9.1

Kapp, Marshall B. The US Supreme Court decision on assisted suicide and the prescription of pain medication: limit the celebration. *Journal of Opioid Management* 2006 March-April; 2(2): 73-74. Subject: 20.7

Kara, Mahmut Alpertunga. Applicability of the principle of respect for autonomy: the perspective of Turkey. *Journal of Medical Ethics* 2007 November; 33(11): 627-630. Subject: 8.1

Karbwang, Juntra; Crawley, Francis P. Need to strengthen ethics committees. SciDevNet: Science and Development Network 2007 November 12: 2 p. [Online]. Accessed: http://www.scidev.net/dossiers/index.cfm? fuseaction=dossierreaditem&dossier=5&type=3& itemid=687&language=1 [2008 February 21]. Subject: 18.2

Karenberg, Axel. Neurosciences and the Third Reich — Introduction. *Journal of the History of the Neurosciences* 2006 September; 15(3): 168-172. Subject: 20.5.1

Karlawish, Jason. Research on cognitively impaired adults. *In:* Steinbock, Bonnie, ed. The Oxford Handbook of Bioethics. Oxford; New York: Oxford University Press, 2007: 597-620. Subject: 18.3

Karlsson, Marit; Milberg, Anna; Strang, Peter. Dying with dignity according to Swedish medical students. *Supportive Care in Cancer* 2006 April; 14(4): 334-339. Subject: 20.3.2

Karpin, Isabel; Mykitiuk, Roxanne. Regulating inheritable genetic modification, or policing the fertile scientific imagination? A feminist legal response. *In:* Rasko, John E.J.; O'Sullivan, Gabrielle M.; Ankeny, Rachel A., eds. The Ethics of Inheritable Genetic Modification: A Dividing Line? Cambridge: Cambridge University Press, 2006: 193-222. Subject: 15.4

Karsjens, Kari L.; Whitney, Simon N.; McCullough, Laurence B. Exploring the nature of physician intent in "silent decisions". *American Journal of Bioethics* 2007 July; 7(7): 42-44; author reply W1-W3. Subject: 8.1

Karunaratne, A.S.; Myles P.S.; Ago M.J.; Komesaroff, P.A. Communication deficiencies in research and monitoring by ethics committees. *Internal Medicine Journal* 2006 February; 36(2): 86-91. Subject: 18.2

Kasachkoff, Tziporah. Drug addiction and responsibility for health care of drug addicts. *In:* Kleinig, John; Einstein, Stanley, eds. Ethical Challenges for Intervening in Drug Use: Policy, Research and Treatment Issues. Huntsville, TX: Office of International Criminal Justice; Criminal Justice Center, Sam Houston State University, 2006: 189-204. Subject: 9.5.9

Kasiske, Bertram L. Dialysis facility ownership and epoetin dosing in hemodialysis patients: a US physician perspective. *American Journal of Kidney Diseases* 2007 September; 50(3): 354-357. Subject: 19.3

Kass, Leon R. Science, religion, and the human future. *Commentary* 2007 April; 123(4): 36-48. Subject: 5.1

Kass, Leon R. The right to life and human dignity. *New Atlantis* 2007 Spring; (16): 23-40. Subject: 4.4

Kass, Nancy E.; Hyder, Adnan Ali; Ajuwon, Ademola; Appiah-Poku, John; Barsdorf, Nicola; Elsayed, Dya Eldin; Mokhachane, Mantoa; Mupenda, Bavon; Ndebele, Paul; Ndossi, Godwin; Sikateyo, Bornwell; Tangwa, Godfrey; Tindana, Paulline. The structure and function of research ethics committees in Africa: a case study. *PLoS Medicine* 2007 January; 4(1): e3: 0026-0031 [Online]. Accessed: http://www.plos.org/press/plme-04-01-Kass.pdf [2007 January 25]. Subject: 18.2

Kass, N.E.; Myers, R.; Fuchs, E.J.; Carson, K.A.; Flexner, C. Balancing justice and autonomy in clinical research with healthy volunteers. *Clinical Pharmacology and Therapeutics* 2007 August; 82(2): 219-227. Subject: 18.2

Kassam-Adams, Nancy; Newman, Elana. The reactions to research participation questionnaires for children and for parents (RRPQ-C and RRPQ-P). *General Hospital Psychiatry* 2002 September-October; 24(5): 336-342. Subject: 18.5.2

Kassirer, Jerome P. Assault on editorial independence: improprieties of the Canadian Medical Association [editorial]. *Journal of Medical Ethics* 2007 February; 33(2): 63-66. Subject: 1.3.7

Kassirer, Jerome P. Commercialism and medicine: an overview. *CQ: Cambridge Quarterly of Healthcare Ethics* 2007 Fall; 16(4): 377-386. Subject: 9.3.1

Kassirer, Jerome P. Pharmaceutical ethics? [review of Ethics and the Pharmaceutical Industry, edited by Michael A. Santoro and Thomas M. Gorrie]. *Open Medicine [electronic]* 2007; 1(1): 58-59. Accessed: http://www.openmedicine.ca [2007 April 19]. Subject: 9.7

Kassirer, Jerome P. Professional societies and industry supports: what is the quid pro quo? *Perspectives in Biology and Medicine* 2007 Winter; 50(1): 7-17. Subject: 7.4

Subject = NRCBL Primary Classification Number; see inside front cover.

Katavic, Vedran. Five-year report of Croatian Medical Journal's research integrity editor — policy, policing, or policing policy. *Croatian Medical Journal* 2006 April; 47(2): 220-227. Subject: 1.3.7

Katayama, Alyce C. U.S. ART practitioners soon to begin their forced march into a regulated future. *Journal of Assisted Reproduction and Genetics* 2003 July; 20(7): 265-270. Subject: 14.4

Katz, Leo. Choice, consent, and cycling: the hidden limitations of consent. *Michigan Law Review* 2006 February; 104(4): 627-670. Subject: 9.4

Kavanaugh, John F. In defense of human life [editorial]. *America* 2007 November 26; 197(17): 8. Subject: 12.3

Kaye, D.H. Bioethics, bench, and bar: selected arguments in Landry v. Attorney General. *Jurimetrics* 2000 Winter; 40(2): 193-216. Subject: 15.1

Kaye, D.H.; Smith, Michael E. DNA identification databases: legality, legitimacy, and the case for population-wide coverage. *Wisconsin Law Review* 2003; 2003(3): 413-459. Subject: 15.1

Kaye, Jane. Testing times: what is the legal situation when an adolescent wants a genetic test? *Clinical Ethics* 2007 December; 2(4): 176-180. Subject: 15.3

Kayser, Bengt; Mauron, Alexandre; Miah, Andy. Current anti-doping policy: a critical appraisal [debate]. *BMC Medical Ethics [electronic]* 2007; 8:2. 10 p. Subject: 9.5.9

Kearnes, Matthew; Wynne, Brian. On nanotechnology and ambivalence: the politics of enthusiasm. *NanoEthics* 2007 August; 1(2): 131-142. Subject: 5.4

Keim, Brandon. Gene therapy trials on trial: the unfortunate tale of Jolee Mohr. *GeneWatch* 2007 September-October; 20(5): 10-11. Subject: 15.4

Keim, Brandon. Tied up in red tape, European trials shut down [news]. *Nature Medicine* 2007 February; 13(2): 110. Subject: 18.2

Keiper, Adam. Nanoethics as a discipline? *New Atlantis* Spring 2007; (16): 55-67. Subject: 5.1

Kelk, Constantijn. Previously expressed wishes related to end of life decisions. *In:* Gevers, J.K.M.; Hondius, E.H.; Hubben, J.H., eds. Health Law, Human Rights and the Biomedicine Convention: Essays in Honour of Henriette Roscam Abbing. Leiden; Boston: Martinus Nijhoff Publishers, 2005: 131-145. Subject: 20.5.4

Keller, Johannes. In genes we trust: the biological component of psychological essentialism and its relationship to mechanisms of motivated social cognition. *Journal of Personality and Social Psychology* 2005 April; 88(4): 686-702. Subject: 15.1

Kelley, Amy S.; Gold, Heather T.; Roach, Keith W.; Fins, Joseph J. Differential medical and surgical house staff involvement in end-of-life decisions: a retrospective chart review. *Journal of Pain and Symptom Management* 2006 August; 32(2): 110-117. Subject: 20.5.1

Kelty, Miriam; Bates, Angela; Pinn, Vivian W. National Institutes of Health policy on the inclusion of women and minorities as subjects in clinical research. *In:* Gallin, John I.; Ognibene, Frederick P., eds. Principles and Practice of Clinical Research. 2nd edition. Oxford: Academic, 2007: 129-142. Subject: 18.5.1

Kemble, Sarah; King, Susanne L.; Fleck, Leonard M. The price of compromise: the Massachusetts health care reform [letters]. *Hastings Center Report* 2007 January-February; 37(1): 4-7. Subject: 9.3.1

Kendall, Marilyn; Harris, Fiona; Boyd, Kirsty; Sheikh, Aziz; Murray, Scott A.; Brown, Duncan; Mallinson, Ian; Kearney, Nora; Worth, Allison. Key challenges and ways forward in researching the "good death": qualitative in-depth interview and focus group study. *BMJ: British Medical Journal* 2007 March 10; 334(7592): 521-524. Subject: 18.5.7

Kendall, Sharon. Being asked not to tell: nurses' experiences of caring for cancer patients not told their diagnosis. *Journal of Clinical Nursing* 2006 September; 15(9): 1149-1157. Subject: 8.2

Kennedy, Donald. Turning the tables with Mary Jane [editorial]. *Science* 2007 May 4; 316(5825): 661. Subject: 9.5.7

Kennedy, Evelyn P.; MacPhee, Cyndee. Access to confidential sexual health services. *Canadian Nurse* 2006 September; 102(7): 29-31. Subject: 9.5.7

Kennedy, Holly P.; Renfrew, Mary J.; Madi, Banyana C.; Opoku, Dora; Thompson, Joyce B. The conduct of ethical research collaboration across international and culturally diverse communities. *Midwifery* 2006 June; 22(2): 100-107. Subject: 14.1

Kennett, Jeanette. Mental disorder, moral agency and the self. *In:* Steinbock, Bonnie, ed. The Oxford Handbook of Bioethics. Oxford; New York: Oxford University Press, 2007: 90-113. Subject: 4.3

Kennett, Jeannette; Matthews, Stephen. The moral goal of treatment in cases of dual diagnosis. *In:* Kleinig, John; Einstein, Stanley, eds. Ethical Challenges for Intervening in Drug Use: Policy, Research and Treatment Issues. Huntsville, TX: Office of International Criminal Justice; Criminal Justice Center, Sam Houston State University, 2006: 409-436. Subject: 9.5.9

Kenny, Belinda; Lincoln, Michelle; Balandin, Susan. A dynamic model of ethical reasoning in speech pathology. *Journal of Medical Ethics* 2007 September; 33(9): 508-513. Subject: 4.1.1

Kenny, Denis. Inheritable genetic modification as moral responsibility in a creative universe. *In:* Rasko, John E.J.; O'Sullivan, Gabrielle M.; Ankeny, Rachel A., eds. The Ethics of Inheritable Genetic Modification: A Dividing

See SUBJECT HEADING KEY FOR SECTION II on inside back cover.

529

Line? Cambridge: Cambridge University Press, 2006: 77-102. Subject: 15.1

Kenny, Michael G. A question of blood, race, and politics. *Journal of the History of Medicine and Allied Sciences* 2006 October; 61(4): 456-491. Subject: 19.4

Kenny, Nuala. Searching for doctor good: virtues for the twenty-first century. *In:* Kenny, Nuala; Shelton, Wayne, eds. Lost Virtue: Professional Character Development in Medical Education. Amsterdam; Oxford: Elsevier, 2006: 211-233. Subject: 4.1.2

Kenny, Nuala P. Codes and character: the pillars of professional ethics. *Journal of the American College of Dentists* 1998 Fall; 65(3): 5-8. Subject: 6

Kenny, Nuala P.; McMahon, Meghan; Flood, Colleen M.; Braun, Joshua A. Canadian media and health policy research: the limits of stories. *American Journal of Bioethics* 2007 August; 7(8): 19-21; author reply W1-W2. Subject: 1.3.7

Kenny, Nuala P.; Melnychuk, Ryan M.; Asada, Yukiko. The promise of public health: ethical reflections. *Canadian Journal of Public Health* 2006 September-October; 97(5): 402-404. Subject: 9.1

Kenny, Nuala; Chafe, Roger. Pushing right against the evidence: turbulent times for Canadian health care. *Hastings Center Report* 2007 September-October; 37(5): 24-26. Subject: 9.3.1

Kent, Alastair; Mintzes, Barbara. Should patient groups accept money from drug companies? [debate]. *BMJ: British Medial Journal* 2007 May 5; 334(7600): 934-935. Subject: 9.7

Kenyon, S.; Dixon-Woods, M.; Jackson, C.J.; Windridge, K.; Pitchforth, E. Participating in a trial in a critical situation: a qualitative study in pregnancy. *Quality and Safety in Health Care* 2006 April; 15(2): 98-101. Subject: 18.5.3

Keown, John. Defending the Council of Europe's opposition to euthanasia. *In:* McLean, Sheila A.M., ed. First Do No Harm: Law, Ethics, and Healthcare. Aldershot, England; Burlington, VT: Ashgate, 2006: 479-494. Subject: 20.5.1

Kerr, Cathel. Hoax raises awareness about organ shortages [news]. *CMAJ/JAMC: Canadian Medical Association Journal* 2007 July 17; 177(2): 135. Subject: 19.6

Kerridge, I.; Maguire, J.; Newby, D.; McNeill, P.M.; Henry, D.; Hill, S.; Day, R.; Macdonald, G.; Stokes, B.; Henderson, K. Cooperative partnerships or conflict-of-interest? A national survey of interaction between the pharmaceutical industry and medical organizations. *Internal Medicine Journal* 2005 April; 35(4): 206-210. Subject: 7.3

Kershaw, Sarah. U.S. rule limits emergency care for immigrants; shift in chemotherapy; New York used money under Medicaid for illegal residents. *New York Times* 2007 September 22; p. A1,A11. Subject: 9.5.10

Kesselheim, Aaron S.; Mello, Michelle M. Confidentiality laws and secrecy in medical research: improving public access to data on drug safety. Concealing clinical trial data from public scrutiny has implications for Americans' health. *Health Affairs* 2007 March-April; 26(2): 483-491. Subject: 18.2

Kessler, David A.; Levy, Douglas A. Direct-to-consumer advertising: is it too late to manage the risks? *Annals of Family Medicine* 2007 January/February; 5(1): 4-5. Subject: 20.3.2

Kettis-Lindblad, Åsa; Ring, Lena; Viberth, Eva; Hansson, Mats G. Genetic research and donation of tissue samples to biobanks. What do potential sample donors in the Swedish general public think? *European Journal of Public Health* 2006 August; 16(4): 433-440. Subject: 15.1

Kettner, Matthias. Die Herstellung einer öffentlichen Hegemonie. Humangenomforschung in der deutschen und der US-amerikanischen Presse [The making of a public hegemony. Human genome research in the German and US press, by Jürgen Gerhards and Mike Steffen Schäfer] [book review]. *Ethik in der Medizin* 2007 June; 19(2): 167-168. Subject: 15.1

Kettner, Matthias. Medizinethik in den Medien — Befunde und Aufgaben in Theorie und Praxis [Medical ethics in the media — findings and challenges in theory and practice]. *Ethik in der Medizin* 2006 December; 18(4): 353-358. Subject: 2.1

Keulartz, Jozef. Comments on Gert: Gert's common morality: old-fashioned or untimely? *In:* Korthals, Michiel; Bogers, Robert J., eds. Ethics for Life Scientists. Dordrecht, The Netherlands: Springer, 2004: 141-145. Subject: 2.1

Khalil, Susan S.; Silverman, Henry J.; Raafat, May; El-Kamary, Samer; El-Setouhy, Maged. Attitudes, understanding, and concerns regarding medical research amongst Egyptians: a qualitative pilot study. *BMC Medical Ethics [electronic]* 2007; 8(9): 12 p. Accessed: http://www.biomedcentral.com/content/pdf/1472-6939-8-9.pdf [2007 December 18]. Subject: 18.1

Khalili, Mohammed I. Organ trading in Jordan: bad news, good news. *Politics and the Life Sciences* 2007 March; 26(1): 12-14. Subject: 19.5

Kharaboyan, Linda; Knoppers, Bartha Maria; Avard, Denise; Nisker, Jeff. Understanding umbilical cord blood banking: what women need to know before deciding [editorial]. *Women's Health Issues* 2007 September-October; 17(5): 277-280. Subject: 19.4

Khoat, D.V.; Hong, L.D.; An, C.Q.; Ngu, D.; Reidpath, D.D. A situational analysis of HIV/AIDS-related discrimination in Hanoi, Vietnam. *AIDS Care* 2005 July; 17(Supplement 2): S181-S193. Subject: 9.5.6

Khong, T.Y.; Tanner, Alison R. Foetal and neonatal autopsy rates and use of tissue for research: the influence of 'organ retention' controversy and new consent process.

Journal of Paediatrics and Child Health 2006 June; 42(6): 366-369. Subject: 20.1

Khoo, Chong-Yew. Ethical issues in ophthalmology and vision research. *Annals of the Academy of Medicine, Singapore* 2006 July; 35(7): 512-516. Subject: 18.2

Khoury, Muin J.; Gwinn, Marta; Bowen, Scott J.; Westfall, John M.; Mold, James; Fagnan, Lyle; Jones, Loretta; Wells, Kenneth B. Genomics and public health research [letter and replies]. *JAMA: The Journal of the American Medical Association* 2007 June 6; 297(21): 2347-2348. Subject: 15.1

Khroutski, Konstantin S. BioCosmological approach in world bioethics. *Eubios Journal of Asian and International Bioethics* 2007 November; 17(6): 167-171. Subject: 2.1

Khushf, George. Introduction: taking stock of bioethics from a philosophical perspective. *In:* Khushf, George, ed. Handbook of Bioethics: Taking Stock of the Field From a Philosophical Perspective. Dordrecht; Boston: Kluwer Academic, 2004: 1-28. Subject: 2.1

Khushf, George. Open questions in the ethics of convergence. *Journal of Medicine and Philosophy* 2007 May-June; 32(3): 299-310. Subject: 5.1

Khushf, George. The ethics of NBIC convergence. *Journal of Medicine and Philosophy* 2007 May-June; 32(3): 185-196. Subject: 5.1

Kidd, J.; Finlayson, M. Navigating uncharted water: research ethics and emotional engagement in human inquiry. *Journal of Psychiatric and Mental Health Nursing* 2006 August; 13(4): 423-428. Subject: 4.1.3

Kieran, Shannon; Loescher, Lois J.; Lim, Kyung Hee. The role of financial factors in acceptance of clinical BRCA genetic testing. *Genetic Testing* 2007 Spring; 11(1): 101-110. Subject: 15.3

Kieve, Millie. Falling on deaf ears [comment]. *New Scientist* 2007 September 15-21; 195(2621): 24. Subject: 9.7

Kilmarx, Peter H.; Ramjee, Gita; Kitayaporn, Dwip; Kunasol, Prayura. Protection of human subjects' rights in HIV-preventive clinical trials in Africa and Asia: experiences and recommendations. *AIDS* 2001; 15 (suppl.5): S73-S79. Subject: 18.5.9

Kilpatrick, Dean G. The ethics of disaster research: a special section. *Journal of Traumatic Stress* 2004 October; 17(5): 361-362. Subject: 18.4

Kim, Jung-Ran; Fisher, Murray J.; Elliott, Doug. Undergraduate nursing students' knowledge and attitudes towards organ donation in Korea: implications for education. *Nurse Education Today* 2006 August; 26(6): 465-474. Subject: 7.2

Kim, Scott Y.; Kieburtz, Karl. Appointing a proxy for research consent after one develops dementia: the need for further study [editorial]. *Neurology* 2006 May 9; 66(9):

1298-1299 [Online]. Accessed: http://www.neurology. org/ [2007 April 16]. Subject: 18.3

Kim, Scott Y.H. Assessing and communicating the risks and benefits of gene transfer clinical trials. *Current Opinion in Molecular Therapeutics* 2006 October; 8(5): 384-389. Subject: 15.4

Kim, S. Joseph; Gordon, Elisa J.; Powe, Neil R. The economics and ethics of kidney transplantation: perspectives in 2006. *Current Opinion in Nephrology and Hypertension* 2006 November; 15(6): 593-598. Subject: 19.3

Kim, Tae-gyu. Korea mulls allowing research using cloned embryos [news]. *Korea Times* 2007 January 29: 2 p. [Online]. Accessed: http://times.hankooki.com/service/ print/Print.php?po=times.hankooki.com/1page/2 00701/ kt2007011917571310230.htm [2007 February 6]. Subject: 18.5.4

Kim, Yong-Soon; Park, Jin-Hee; Han, Sung-Suk. Differences in moral judgment between nursing students and qualified nurses. *Nursing Ethics* 2007 May; 14(3): 309-319. Subject: 4.1.3

Kimberly, Michael B.; Hoehn, K. Sarah; Feudtner, Chris; Nelson, Robert M.; Schreiner, Mark. Variation in standards of research compensation and child assent practices: a comparison of 69 institutional review board-approved informed permission and assent forms for 3 multi-c enter pediatric clinical trials. *Pediatrics* 2006 May; 117(5): 1706-1711. Subject: 18.3

Kimmelman, Jonathan. Clinical trials and SCID row: the ethics of phase 1 trials in the developing world. *Developing World Bioethics* 2007 December; 7(3): 128-135. Subject: 18.5.9

Kimmelman, Jonathan. Inventors as investigators: the ethics of patents in clinical trials. *Academic Medicine* 2007 January; 82(1): 24-31. Subject: 1.3.9

Kimmelman, Jonathan. Missing the forest: further thoughts on the ethics of bystander risk in medical research. *CQ: Cambridge Quarterly of Healthcare Ethics* 2007 Fall; 16(4): 483-490. Subject: 18.6

Kimmelman, Jonathan. The therapeutic misconception at 25: treatment, research, and confusion. *Hastings Center Report* 2007 November-December; 37(6): 36-42. Subject: 18.1

Kimmelman, Jonathan; Lott, Jason P.; Savulescu, Julian. Towards a global human embryonic stem cell bank: differential termination. *American Journal of Bioethics* 2007 August; 7(8): 52-53; author reply W4-W6. Subject: 18.5.4

Kimsma, Geritt K.; van Leeuwen; Evert. The role of family in euthanasia decision making. *HEC (Healthcare Ethics Committee) Forum* 2007 December; 19(4): 365-373. Subject: 20.5.1

King, Jean. Accepting tobacco industry money for research: has anything changed now that harm reduction is

See SUBJECT HEADING KEY FOR SECTION II on inside back cover.

531

on the agenda? *Addiction* 2006 August; 101(8): 1067-1069. Subject: 18.6

King, Lesley; Rowan, Andrew N. The mental health of laboratory animals. *In:* McMillan, Frank, ed. Mental Health and Well-Being in Animals. Ames, IA: Blackwell Pub., 2005: 259-276. Subject: 22.2

King, Nancy M.P. Genes and Tourette syndrome: scientific, ethical, and social implications. *Advances in Neurology* 2006; 99: 144-147. Subject: 15.1

King, Nancy M.P. The glass house: assessing bioethics. *In:* Eckenwiler, Lisa A.; Cohn, Felicia G., eds. The Ethics of Bioethics: Mapping the Moral Landscape. Baltimore, MD: Johns Hopkins University Press, 2007: 297-309. Subject: 2.1

King, Patricia A. Race, equity, health policy, and the African American community. *In:* Prograis, Lawrence J.; Pellegrino, Edmund D., eds. African American Bioethics: Culture, Race, and Identity. Washington, DC: Georgetown University Press, 2004: 67-92. Subject: 9.5.4

King, Patricia A.; Areen Judith; Gostin, Lawrence O. The human genome: pathways to health. *In their:* Law, Medicine and Ethics. New York: Foundation Press, 2006: 1-111. Subject: 15.1

King, Patricia A.; Areen, Judith; Gostin, Lawerence O. Reproduction and the new genetics. *In their:* Law, Medicine and Ethics. New York: Foundation Press, 2006: 509-654. Subject: 14.1

King, Patricia A.; Areen, Judith; Gostin, Lawrence O. Death and dying. *In their:* Law, Medicine and Ethics. New York: Foundation Press, 2006: 370-508. Subject: 20.5.1

King, Patricia A.; Areen, Judith; Gostin, Lawrence O. Private control of science and medicine. *In their:* Law, Medicine and Ethics. New York: Foundation Press, 2006: 112-207. Subject: 7.1

King, Patricia A.; Areen, Judith; Gostin, Lawrence O. The human body. *In their:* Law, Medicine and Ethics. New York: Foundation Press, 2006: 208-369. Subject: 18.1

Kinghorn, Warren A.; McEvoy, Matthew D.; Michel, Andrew; Balboni, Michael. Professionalism in modern medicine: does the emperor have any clothes? *Academic Medicine* 2007 January; 82(1): 40-45. Subject: 4.1.2

Kinlaw, Kathy. Prolonging living and dying. *In:* Eisen, Arri; Laderman, Gary, eds. Science, Religion, and Society: An Encyclopedia of History, Culture, and Controversy. Vol. 2. Armonk, NY: M.E. Sharpe, 2007: 731-738. Subject: 20.5.1

Kinoshita, Satomi. Respecting the wishes of patients in intensive care units. *Nursing Ethics* 2007 September; 14(5): 651-664. Subject: 20.5.1

Kintisch, Eli. New cell rules [news]. *Science* 2007 January 26; 315(5811): 449. Subject: 15.8

Kipnis, Kenneth. Harm and uncertainty in newborn intensive care. *Theoretical Medicine and Bioethics* 2007; 28(5): 393-412. Subject: 20.5.2

Kipnis, Kenneth. Medical confidentiality. *In:* Rhodes, Rosamond; Francis, Leslie P.; Silvers, Anita, eds. The Blackwell Guide to Medical Ethics. Malden, MA: Blackwell Pub., 2007: 104-127. Subject: 8.4

Kipnis, Kenneth. The expert ethics witness as teacher. *In:* Rasmussen, Lisa, ed. Ethics Expertise: History, Contemporary Perspectives, and Applications. Dordrecht: Springer, 2005: 269-279. Subject: 2.3

Kirby, David A. The devil in our DNA: a brief history of eugenics in science fiction films. *Literature and Medicine* 2007 Spring; 26(1): 83-108. Subject: 15.5

Kirby, Jeff; Simpson, Christy. An innovative, inclusive process for meso-level health policy development. *HEC (Healthcare Ethics Committee) Forum* 2007 June; 19(2):161-176. Subject: 9.6

Kirby, Neil. Treatment or crime? the status of stem cell therapies and research in South African law. *Medicine and Law: The World Association for Medical Law* 2007 March; 26(1): 95-115. Subject: 19.5

Kirk, Kenneth C. The Alaska Health Care Decisions Act, analyzed. *Alaska Law Review* 2005 December; 22(2): 213-253. Subject: 20.5.1

Kirkham, Sheryl R.; Browne, Annette J. Toward a critical theoretical interpretation of social justice discourses in nursing. *ANS: Advances in Nursing Science* 2006 October-December; 29(4): 324-339. Subject: 4.1.3

Kirklin, D. Framing, truth telling and the problem with non-directive counselling. *Journal of Medical Ethics* 2007 January; 33(1): 58-62. Subject: 8.1

Kirklin, D. Minding the gap between logic and intuition: an interpretative approach to ethical analysis. *Journal of Medical Ethics* 2007 July; 33(7): 386-389. Subject: 2.1

Kirklin, D. Truth telling, autonomy and the role of metaphor. *Journal of Medical Ethics* 2007 January; 33(1): 11-14. Subject: 8.2

Kirkpatrick, William J.; Reamer, Frederic G.; Sykulski, Marilyn. Social work ethics audits in health care settings: a case study. *Health and Social Work* 2006 August; 31(3): 225-228. Subject: 1.3.1

Kisely, Stephen; Smith, Mark; Lawrence, David; Cox, Martha; Campbell, Leslie Anne; Maaten, Sarah. Inequitable access for mentally ill patients to some medically necessary procedures. *CMAJ/JAMC: Canadian Medical Association Journal* 2007 March 13; 176(6): 779-784. Subject: 9.2

Kissoon, Niranjan. Bench-to-bedside review: humanism in pediatric critical care medicine - a leadership challenge. *Critical Care (London, England)* 2005 August; 9(4): 371-375. Subject: 8.1

Kitcher, Phillip. Scientific research — who should govern? *NanoEthics* 2007 December; 1(3): 177-184. Subject: 5.3

Kittay, Eva F. Beyond autonomy and paternalism: the caring transparent self. *In:* Nys, Thomas; Denier, Yvonne; Vandevelde, Toon, eds. Autonomy and Paternalism: Reflections on the Theory and Practice of Health Care. Leuven; Dudley, MA: Peeters, 2007: 23-70. Subject: 4.1.2

Kitzinger, Jenny; Williams, Clare. Forecasting the future: legitimizing hope and calming fears in the embryo stem-cell debate. *In:* Deane-Drummond, Celia; Scott, Peter Manley, eds. Future Perfect?: God, Medicine and Human Identity. New York: T and T Clark International, 2006: 129-142. Subject: 18.5.4

Kjølberg, Kamilla; Wickson, Fern. Social and ethical interactions with nano: mapping the early literature. *NanoEthics* 2007 August; 1(2): 89-104. Subject: 5.4

Klagsbrun, Francine. Multiple choice. Human reproductive technology enables many people to have children, but it poses dangers. *Moment* 1999 August; 24(4): 24-25. Subject: 14.4

Klahr, Saulo. One physician's exploration of the ethics in the practice of medicine. *Kidney International* 2006 August; 70(4): 613-614. Subject: 4.1.2

Klein, Roger D. Gene patents and genetic testing in the United States. As genetic testing moves into mainstream medicine, its restriction by gene patent holders will have far-reaching, detrimental effects on the healthcare system. *Nature Biotechnology* 2007 September; 25(9): 989-991. Subject: 15.8

Kleinig, John. Ethical issues on substance use intervention. *In:* Kleinig, John; Einstein, Stanley, eds. Ethical Challenges for Intervening in Drug Use: Policy, Research and Treatment Issues. Huntsville, TX: Office of International Criminal Justice; Criminal Justice Center, Sam Houston State University, 2006: 21-44. Subject: 9.5.9

Kleinig, John. Thinking ethically about needle and syringe programs. *In:* Kleinig, John; Einstein, Stanley, eds. Ethical Challenges for Intervening in Drug Use: Policy, Research and Treatment Issues. Huntsville, TX: Office of International Criminal Justice; Criminal Justice Center, Sam Houston State University, 2006: 121-132. Subject: 9.5.9

Kleinig, John. Thinking ethically about needle and syringe programs. *Substance Use and Misuse* 2006; 41(6-7): 815-825. Subject: 9.5.9

Kleinman, Daniel Lee; Kinchy, Abby J. Against the neoliberal steamroller? The Biosafety Protocol and the social regulation of agricultural biotechnologies. *Agriculture and Human Values* 2007 Summer; 24(2): 195-206. Subject: 15.7

Kline, A. David. Giftedness, humility and genetic enhancement. *Human Reproduction and Genetic Ethics: An International Journal* 2007; 13(2): 16-21. Subject: 15.1

Kline, Wendy. A new deal for the child: Ann Cooper Hewitt and sterilization in the 1930s. *In:* Currell, Susan; Cogdell, Christina, eds. Popular Eugenics: National Efficiency and American Mass Culture in the 1930s. Athens, OH: Ohio University Press, 2006: 17-43. Subject: 15.5

Klingberg-Allvin, Marie; Van Tam, Vu; Nga, Nguyen Tha; Ransjo-Arvidson, Anna-Berit; Johansson, Annika. Ethics of justice and ethics of care. Values and attitudes among midwifery students on adolescent sexuality and abortion in Vietnam and their implications for midwifery education: a survey by questionnaire and interview. *International Journal of Nursing Studies* 2007 January; 44(1): 37-46. Subject: 7.2

Klitzman, Robert. Additional implications of a national survey on ethics consultation in United States hospitals. *American Journal of Bioethics* 2007 February; 7(2): 47-48. Subject: 9.6

Klitzman, Robert. Clinicians, patients, and the brain. *In:* Illes, Judy, ed. Neuroethics: Defining the Issues in Theory, Practice, and Policy. New York: Oxford University Press, 2006: 229-241. Subject: 17.1

Klitzman, Robert. Pleasing doctors: when it gets in the way. *BMJ: British Medical Journal* 2007 September 8; 335(7618): 514. Subject: 8.1

Klitzman, Robert; Albala, Ilene; Siragusa, Joseph; Nelson, Kristen N.; Appelbaum, Paul S. The reporting of monetary compensation in research articles. *Journal of Empirical Research on Human Research Ethics* 2007 December; 2(4): 61-67. Subject: 18.1

Klitzman, Robert; Thorne, Deborah; Williamson, Jennifer; Chung, Wendy; Marder, Karen. Decision-making about reproductive choices among individuals at-risk for Huntington's disease. *Journal of Genetic Counseling* 2007 June; 16(3): 347-362. Subject: 15.2

Klitzman, Robert; Thorne, Deborah; Williamson, Jennifer; Marder, Karen. The roles of family members, health care workers, and others in decision-making processes about genetic testing among individuals at risk for Huntington disease. *Genetics in Medicine* 2007 June; 9(6): 358-371. Subject: 15.3

Kluge, Eike-Henner W. Sex selection: some ethical and policy considerations. *Health Care Analysis: An International Journal of Health Philosophy and Policy* 2007 June; 15(2): 73-89. Subject: 14.3

Klugman, Craig M.; Braun, Joshua A. Buying the fourth estate. *American Journal of Bioethics* 2007 August; 7(8): 16-18; author reply W1-W2. Subject: 1.3.7

Kmietowicz, Zosia. Doctors condemn use of drugs as weapon [news]. *BMJ: British Medical Journal* 2007 May 26; 334(7603): 1073. Subject: 21.3

See SUBJECT HEADING KEY FOR SECTION II on inside back cover.

533

Kmietowicz, Zosia. Doctors get advice on rights of children and young people [news]. *BMJ: British Medical Journal* 2007 September 29; 335(7621): 633. Subject: 9.5.7

Kmietowicz, Zosia. Doctors threaten to withdraw subscription to GMC [news]. *BMJ: British Medical Journal* 2007 July 7; 335(7609): 14. Subject: 9.7

Kmietowicz, Zosia. Dying patients are often not told of the closeness of death [news]. *BMJ: British Medical Journal* 2007 December 8; 335(7631): 1176. Subject: 20.4.1

Kmietowicz, Zosia. Make access to early abortions easier and quicker, say doctors. *BMJ: British Medical Journal* 2007 July 7; 335(7609): 14. Subject: 12.1

Kmietowicz, Zosia. Public support for hybrid embryos rises, poll shows [news]. *BMJ: British Medical Journal* 2007 September 8; 335(7618): 466-467. Subject: 15.1

Kmietowicz, Zosia. Regulator gives green light to using human-animal embryos [news]. *BMJ: British Medical Journal* 2007 September 15; 335(7619): 531. Subject: 15.1

Kmietowicz, Zosia. Repeal law that puts "FDA on the payroll of the industry" [news]. *BMJ: British Medical Journal* 2007 March 3; 334(7591): 447. Subject: 9.7

Knight, Andrew. The poor contribution of chimpanzee experiments to biomedical progress. *Journal of Applied Animal Welfare Science* 2007; 10(4): 281-308. Subject: 22.2

Knobel, Peter. An expanded approach to Jewish bioethics: a liberal/aggadic approach. *In:* Cutter, William, ed. Healing and the Jewish Imagination: Spiritual and Practical Perspectives on Judaism and Health. Woodstock, VT: Jewish Lights Pub., 2007: 171-183. Subject: 2.1

Knoppers, Bartha Maria; Joly, Yoly; Simard, Jacques; Durocher, Francine. The emergence of an ethical duty to disclose genetic research results: international perspectives. *European Journal of Human Genetics* 2006 November; 14(11): 1170-1178 [see correction in: European Journal of Human Genetics 2006 December; 14(12): 1322]. Subject: 15.1

Knoppers, Bartha Maria; Sallée, Clémentine. Ethical aspects of genome research and banking. *In:* Sensen, Chrisopher W., ed. Handbook of Genome Research: Genomics, Proteomics, Metabolomics, Bioinformatics, Ethical and Legal Issues. Volume 2. Weinheim: Wiley-VCH, 2005: 509-536. Subject: 15.1

Knowles, Lori P. The governance of reprogenetic technology: international models. *In:* Knowles, Lori P.; Kaebnick, Gregory E., eds. Reprogenetics: Law, Policy, and Ethical Issues. Baltimore: Johns Hopkins University Press, 2007: 127-143. Subject: 14.1

Knudsen, Hannah K.; Ducharme, Lori J.; Roman, Paul M. Racial and ethnic disparities in SSRI availability in substance abuse treatment. *Psychiatric Services* 2007 January; 58(1): 55-62. Subject: 9.5.4

Koeman, Jan H. Comments on Korthals: new public responsibilities for life scientists. *In:* Korthals, Michiel; Bogers, Robert J., eds. Ethics for Life Scientists. Dordrecht, The Netherlands: Springer, 2004: 171-174. Subject: 15.1

Kohane, Isaac C.; Mandl, Kenneth D.; Taylor, Patrick L.; Holm, Ingrid A.; Nigrin, Daniel J.; Kunkel, Louis M. Reestablishing the researcher-patient compact. *Science* 2007 May 11; 316(5826): 836-837. Subject: 18.2

Kohut, Kelly; Manno, Michael; Gallinger, Steven; Esplen, Mary Jane. Should healthcare providers have a duty to warn family members of individuals with an HNPCC-causing mutation? A survey of patients from the Ontario Familial Colon Cancer Registry. *Journal of Medical Genetics* 2007 June; 44(6): 404-407. Subject: 15.1

Kojima, Somei; Waikagul, Jitra; Rojekittikhun, Wichit; Keicho, Naoto. The current situation regarding the establishment of national ethical guidelines for biomedical research in Thailand and its neighboring countries. *Southeast Asian Journal of Tropical Medicine and Public Health* 2005 May; 36(3): 728-732. Subject: 18.2

Kok, Jeroen D. Is subfertility a medical condition? *Journal of Clinical Ethics* 2007 Spring; 18(1): 49-52. Subject: 14.4

Kolata, Gina. Researcher who helped start stem cell war may now end it [news]. *New York Times* 2007 November 22; p. A1, A28. Subject: 14.5

Kolata, Gina. Scientists bypass need for embryo to get stem cells; method using human skin is seen as defusing the debate over ethics [news]. *New York Times* 2007 November 21; p. A1, A23. Subject: 18.5.1

Kolber, Adam J. A limited defense of clinical placebo deception. *Yale Law and Policy Review* 2007; 26: 75-134 [Online]. Accessed:http://papers.ssrn.com/sol3/papers.cfm?abstract_id=967563 [2008 April 10]. Subject: 8.2

Kolber, Adam J. Pain detection and the privacy of subjective experience. *American Journal of Law and Medicine* 2007; 33(2-3): 433-456. Subject: 4.4

Kolber, Adam; Henry, Michael; Fishman, Jennifer R.; Youngner, Stuart J. Clarifying the debate over therapeutic forgetting. *American Journal of Bioethics* 2007 September; 7(9): 25-27; author reply W1-W3. Subject: 17.4

Kolenc, Antony B. Easing abortion's pain: can fetal pain legislation survive the new judicial scrutiny of legislative fact-finding? *Texas Review of Law and Politics* 2005 Fall; 10(1): 171-228. Subject: 12.4.2

Kollas, Chad D.; Boyer-Kollas, Beth. Closing the Schiavo case: an analysis of legal reasoning. *Journal of Palliative Medicine* 2006 October; 9(5): 1145-1163. Subject: 20.5.1

Kolmer, D.M. Beneken Genaamd; Tellings, A.; Garretsen, H.F.L.; Bongers, I.M.B. Communalization of health care: how to do it properly. *Medicine and Law: The World Association for Medical Law* 2007 March; 26(1): 53-68. Subject: 4.1.3

Komatsu, Yoshihiko. The age of a "revolutionized human body" and the right to die. *In:* LaFleur, William R.; Böhme, Gernot; Shimazono, Susumu, eds. Dark Medicine: Rationalizing Medical Research. Bloomington: Indiana University Press, 2007: 180-200. Subject: 20.5.1

Komesaroff, P. Ethical issues in the relationships involving medicine and industry: evolving problems require evolving. *Internal Medicine Journal* 2005 April; 35(4): 203-205. Subject: 7.3

Kompanje, Erwin J.O. 'No time to be lost!' Ethical considerations on consent for inclusion in emergency pharmacological research in severe traumatic brain injury in the European Union. *Science and Engineering Ethics* 2007 September; 13(3): 371-381. Subject: 18.3

Kon, Alexander A. Assent in pediatric research. *Pediatrics* 2006 May; 117(5): 1806-1810. Subject: 18.3.9

Kon, Alexander A. Neonatal euthanasia is unsupportable: the Groningen Protocol should be abandoned. *Theoretical Medicine and Bioethics* 2007; 28(5): 453-463. Subject: 20.5.2

Kon, Alexander A. Resident-generated versus instructor-generated cases in ethics and professionalism training. *Philosophy, Ethics and Humanities in Medicine [electronic]* 2006; 1(10): 6 p. Accessed: http://www.peh-med.com/content/1/1/10 [2006 August 15]. Subject: 7.2

Kon, Alexander A. The risky business of assessing research risk. *American Journal of Bioethics* 2007 March; 7(3): 21-22. Subject: 18.5.2

Kon, Alexander A.; Whitney, Simon N.; McCullough, Laurence B. Silent decisions or veiled paternalism? Physicians are not experts in judging character. *American Journal of Bioethics* 2007 July; 7(7): 40-42; author reply W1-W3. Subject: 8.1

Konda, Vani; Huo, Dezheng; Hermes, Gretchen; Liu, Michael; Patel, Roshan; Rubin, David T. Do patients with inflammatory bowel disease want genetic testing? *Inflammatory Bowel Disease* 2006 June; 12(6): 497-502. Subject: 15.3

Kondro, Wayne. American Medical Association boards implantable chip wagon [news]. *CMAJ/JAMC: Canadian Medical Association Journal* 2007 August 14; 177(4): 331-332. Subject: 1.3.12

Kondro, Wayne. Call for arm's-length national research integrity agency [news]. *CMAJ/JAMC: Canadian Medical Association Journal* 2007 March 13; 176(6): 749-750. Subject: 1.3.9

Kondro, Wayne; Hébert, Paul C. Research misconduct? What misconduct? = Inconduite scientifique? Quelle inconduite? [editorial]. *CMAJ/JAMC: Canadian Medical Association Journal* 2007 March 27; 176(7): 905, 907. Subject: 1.3.9

Koniaris, Leonidas G.; Sheldon, Jon P.; Zimmers, Teresa A. Can lethal injection for execution really be "fixed"? [commentary]. *Lancet* 2007 February 3-9; 369(9559): 352-353. Subject: 20.6

Konishi, Emiko; Davis, Anne J. Japan: the teaching of nursing ethics in Japan. *In:* Davis, Anne J.; Tschudin, Verena; de Raeve, Louise, eds. Essentials of Teaching and Learning in Nursing Ethics: Perspectives and Methods. New York: Churchill Livingstone Elsevier, 2006: 251-260. Subject: 4.1.3

Konotey-Ahulu, Felix I.D. Need for ethnic experts to tackle genetic public health [letter]. *Lancet* 2007 December 1-7; 370(9602): 1826-1827. Subject: 9.5.4

Koopmans, Joy; Hiraki, Susan; Ross, Laine Friedman. Attitudes and beliefs of pediatricians and genetic counselors regarding testing and screening for CF and G6PD: implications for policy. *American Journal of Medical Genetics. Part A* 2006 November 1; 140(21): 2305-2311. Subject: 15.3

Kopelman, Loretta M. Clinical trials for breast cancer and informed consent: how women helped make research a cooperative venture. *In:* Rawlinson, Mary C.; Lundeen, Shannon, eds. The Voice of Breast Cancer in Medicine and Bioethics. Dordrecht, Netherlands: Springer, 2006: 133-161. Subject: 18.5.3

Kopelman, Loretta M. Using the best interests standard to decide whether to test children for untreatable, late-onset genetic diseases. *Journal of Medicine and Philosophy* 2007 July-August; 32(4): 375-394. Subject: 15.3

Kopelman, Loretta M. When can children with conditions be in no-benefit, higher-hazard pediatric studies? *American Journal of Bioethics* 2007 March; 7(3): 15-17. Subject: 18.5.2

Kopelman, Loretta M.; Kopelman, Arthur E. Using a new analysis of the best interests standard to address cultural disputes: whose data, which values? *Theoretical Medicine and Bioethics* 2007; 28(5): 373-391. Subject: 8.3.2

Koper, Megan; Bubela, Tania; Caulfield, Timothy; Boon, Heather. Media portrayal of conflicts of interest in biomedical research. *Health Law Review* 2007; 15(3): 30-31. Subject: 1.3.9

Koplewicz, Harold S. Conflict of interest in the eyes of the beholder. *Journal of Child and Adolescent Psychopharmacology* 2006 October; 16(5): 511-512. Subject: 1.3.7

Korenromp, Marijke J.; Page-Christiaens, Godelieve C.M.L.; van den Bout, Jan; Mulder, Eduard J.H.; Visser, Gerard H.A. Maternal decision to terminate pregnancy in case of Down syndrome. *American Journal of*

See SUBJECT HEADING KEY FOR SECTION II on inside back cover.

535

Obstetrics and Gynecology 2007 February; 196(2): 149. Subject: 15.2

Körner, U.; Bondolfi, A.; Bühler, E.; Macfie, J.; Meguid, M.M.; Messing, B.; Oehmichen, F.; Valentini, L.; Allison, S.P. Ethical and legal aspects of enteral nutrition. *Clinical Nutrition* 2006 April; 25(2): 196-202. Subject: 20.5.1

Korthals, Michiel. New public responsibilities for life scientists. *In:* Korthals, Michiel; Bogers, Robert J., eds. Ethics for Life Scientists. Dordrecht, The Netherlands: Springer, 2004: 163-170. Subject: 15.1

Kory, Deborah. How psychologists aid torture. AlterNet (Independent Media Institute) 2007 June 27: 5 p. Accessed: http://www.alternet.org/story/55308 [2007 June 27]. Subject: 7.1

Koski, Edward Greg. Renegotiating the grand bargain: balancing prices, profits, people, and principles. *In:* Santoro, Michael A.; Gorrie, Thomas M., eds. Ethics and the Pharmaceutical Industry. Cambridge; New York: Cambridge University Press, 2005: 393-403. Subject: 9.7

Kostas-Polston, Elizabeth A.; Hayden, Susan J. Living ethics: contributing to knowledge building through qualitative inquiry. *Nursing Science Quarterly* 2006 October; 19(4): 304-310. Subject: 4.1.3

Kosunen, Tiina. Ethical implications of genetic testing and screening [abstract]. *In:* Østnor, Lars, ed. Bioetikk og teologi: Rapport fra Nordisk teologisk nettverk for bioetikks workshop i Stockholm 27.-29. September 1996. Oslo: Nord. teol. nettv. for bioetikk, 1996: 107-115. Subject: 15.3

Kothari, Sunil; Kirschner, Kristi L. Abandoning the golden rule: the problem with "putting ourselves in the patient's place". *Topics in Stroke Rehabilitation* 2006 Fall; 13(4): 68-73. Subject: 8.1

Kotsirilos, Vicki; Hassed, Craig S.; Arnold, Peter C.; Kerridge, Ian H.; McPhee, John R. Ethical and legal issues at the interface of complementary and conventional medicine [letters and reply]. *Medical Journal of Australia* 2004 November 15; 181(10): 581-582. Subject: 4.1.1

Kouri, Robert P. Achieving reproductive rights: access to emergency oral contraception and abortion in Quebec. *In:* Flood, Colleen M., ed. Just Medicare: What's In, What's Out, How We Decide. Buffalo, NY: University of Toronto Press, 2006: 168-190. Subject: 11.1

Kovac, Jeffrey. Moral rules, moral ideals, and use-inspired research. *Science and Engineering Ethics* 2007 June; 13(2): 159-169. Subject: 1.3.9

Kozanczyn, Christa; Collins, Katie; Fernandez, Conrad V. Offering results to research subjects: U.S. institutional review board policy. *Accountability in Research* 2007 October-December; 14(4): 255-267. Subject: 18.2

Krahn, Gloria L.; Hammond, Laura; Turner, Anne. A cascade of disparities: health and health care access for people with intellectual disabilities. *Mental Retardation and Developmental Disabilities Research Reviews* 2006; 12(1): 70-82. Subject: 9.5.3

Kreß, Hartmut. Ethik in der Medizin — Schlaglichter aus der Sicht protestantischer Ethik [Ethics in medicine — highlights from the point of view of Protestant ethics]. *Ethik in der Medizin* 2006 December; 18(4): 320-324. Subject: 2.1

Kressel, Laura M.; Chapman, Gretchen B.; Leventhal, Elaine. The influence of default options on the expression of end-of-life treatment preferences in advance directives. *JGIM: Journal of General Internal Medicine* 2007 July; 22(7): 1007-1010. Subject: 20.5.4

Krill, Edward J. What parents face with their child's life-threatening illness: comment on "How much emotion is enough?" and "Real life informs consent". *Journal of Clinical Ethics* 2007 Winter; 18(4): 369-372. Subject: 20.5.2

Krimsky, Sheldon. Publication bias, data ownership, and the funding effect in science: threats to the integrity of biomedical research. *In:* Wagner, Wendy; Steinzor, Rena, eds. Rescuing Science from Politics: Regulation and the Distortion of Scientific Research. Cambridge: Cambridge University Press, 2006: 61-85. Subject: 1.3.9

Krimsky, Sheldon. The ethical and legal foundations of scientific 'conflict of interest'. *In:* Lemmens, Trudo; Waring, Duff R., eds. Law and Ethics in Biomedical Research: Regulation, Conflict of Interest and Liability. Toronto; Buffalo: University of Toronto Press, 2006: 63-81. Subject: 1.3.9

Krimsky, Sheldon. The profit of scientific discovery and its normative implications. *Chicago-Kent Law Review* 1999; 75(1): 15-39. Subject: 15.8

Krimsky, Sheldon. When conflict-of-interest is a factor in scientific misconduct. *Medicine and Law: The World Association for Medical Law* 2007 September; 26(3): 447-463. Subject: 1.3.9

Krimsky, Sheldon; Simoncelli, Tania. Genetic privacy: new frontiers. *GeneWatch* 2007 September-October; 20(5): 3-10. Subject: 15.1

Krimsky, Sheldon; Simoncelli, Tania. Testing pesticides in humans: of mice and men divided by ten. *JAMA: The Journal of the American Medical Association* 2007 June 6; 297(21): 2405-2407. Subject: 18.2

Kring, Daria L. The patient self-determination act: has it reached the end of its life? *JONA's Healthcare Law, Ethics, and Regulation* 2007 October-December; 9(4): 125-133. Subject: 20.5.4

Krishna, Anurag. The ethics of research in children [editorial]. *Indian Pediatrics* 2005 May; 42(5): 419-423. Subject: 18.2

Krishna, Rajeev; Kelleher, Kelly; Stahlberg, Eric. Patient confidentiality in the research use of clinical medical databases. *American Journal of Public Health* 2007 April; 97(4): 654-658. Subject: 8.4

Kristinsson, Sigurdur. Autonomy and informed consent: a mistaken association? *Medicine, Health Care and Philosophy* 2007 September; 10(3): 253-264. Subject: 8.3.1

Krizova, Eva; Simek, Jiri. Theory and practice of informed consent in the Czech Republic. *Journal of Medical Ethics* 2007 May; 33(5): 273-277. Subject: 8.3.1

Krohmal, Benjamin J.; Emanuel, Ezekiel J. Access and ability to pay: the ethics of a tiered health care system. *Archives of Internal Medicine* 2007 March 12; 167(5): 433-437. Subject: 9.3.1

Krohmal, Benjamin J.; Emanuel, Ezekiel J. Tiers without tears: the ethics of a two-tier health care system. *In:* Steinbock, Bonnie, ed. The Oxford Handbook of Bioethics. Oxford; New York: Oxford University Press, 2007: 175-189. Subject: 9.1

Kröner, Hans-Peter. Eugenik: Zur Geschichte biomedizinischer Utopien. *In:* Neuer-Miebach, Therese; Wunder, Michael, eds. Bio-Ethik und die Zukunft der Medizin. Bonn: Psychiatrie-Verlag, 1998: 20-30. Subject: 15.5

Krones, Tanja; Schlüter, Elmar; Neuwohner, Elke; El Ansari, Susan; Wissner, Thomas; Richter, Gerd. What is the preimplantation embryo? *Social Science and Medicine* 2006 July; 63(1): 1-20. Subject: 15.2

Krosin, Michael T.; Klitzman, Robert; Levin, Bruce; Cheng, Jianfeng; Ranney, Megan L. Problems in comprehension of informed consent in rural and peri-urban Mali, West Africa. *Clinical Trials (London, England)* 2006; 3(3): 306-313. Subject: 18.3

Krousel-Wood, Marie; Muntner, Paul; Jannu, Ann; Hyre, Amanda; Breault, Joseph. Does waiver of written informed consent from the institutional review board affect response rate in a low-risk research study? *Journal of Investigative Medicine* 2006 May; 54(4): 174-179. Subject: 18.2

Krug, E.F., III. Law and ethics at the border of viability. *Journal of Perinatology* 2006 June; 26(6): 321-324. Subject: 20.5.2

Kruitenbrouwer, Frank. Private life: "frappez toujours". *In:* Gevers, J.K.M.; Hondius, E.H.; Hubben, J.H., eds. Health Law, Human Rights and the Biomedicine Convention: Essays in Honour of Henriette Roscam Abbing. Leiden; Boston: Martinus Nijhoff Publishers, 2005: 147-158. Subject: 8.4

Krumholz, Harlan. What have we learnt from Vioxx? *BMJ: British Medical Journal* 2007 January 20; 334(7585): 120-123. Subject: 1.3.9

Kubiak, Erik N.; Park, Samuel S.; Egol, Kenneth; Zuckerman, Joseph D.; Koval, Kenneth J. Increasingly conflicted: an analysis of conflicts of interest reported at the annual meetings of the Orthopaedic Trauma Associa-

tion. *Bulletin (Hospital for Joint Diseases (New York, N.Y.))* 2006; 63(3-4): 83-87. Subject: 1.3.9

Kuczewski, Mark G. Democratic ideals and bioethics commissions: the problem of expertise in an egalitarian society. *In:* Eckenwiler, Lisa A.; Cohn, Felicia G., eds. The Ethics of Bioethics: Mapping the Moral Landscape. Baltimore, MD: Johns Hopkins University Press, 2007: 83-94. Subject: 2.4

Kuczewski, Mark G. Empirical metaethical absolutism in contemporary Catholic bioethics [review of Contemporary Catholic Health Care Ethics, by David F. Kelly]. *Medical Humanities Review* 2005 Spring-Fall; 19(1-2): 76-80. Subject: 2.1

Kuczewski, Mark G. Ethics committees and case consultation: theory and practice. *In:* Khushf, George, ed. Handbook of Bioethics: Taking Stock of the Field From a Philosophical Perspective. Dordrecht; Boston: Kluwer Academic, 2004: 315-334. Subject: 9.6

Kuczewski, Mark G. Talking about spirituality in the clinical setting: can being professional require being personal? *American Journal of Bioethics* 2007 July; 7(7): 4-11. Subject: 8.1

Kuehn, Bridget M. Pediatrics group recommends public cord blood banking. *JAMA: The Journal of the American Medical Association* 2007 February 14; 297(6): 576. Subject: 19.4

Kukla, Rebecca. How do patients know? *Hastings Center Report* 2007 September-October; 37(5): 27-35. Subject: 8.3.1

Kukla, Rebecca. Resituating the principle of equipoise: justice and access to care in non-ideal conditions. *Kennedy Institute of Ethics Journal* 2007 September; 17(3): 171-202. Subject: 18.2

Kulakowski, Elliott C. Dealing with allegations of research misconduct: the other side of responsible conduct of research. *In:* Kulakowski, Elliott C.; Chronister, Lynne U., eds. Research Administration and Management. Sudbury, MA: Jones and Bartlett, 2006: 617-624. Subject: 1.3.9

Kullnat, Megan Wills. Boundaries. *JAMA: The Journal of the American Medical Association* 2007 January 24-31; 297(4): 343-344. Subject: 8.1

Kulynych, Jennifer. Intent to deceive: mental state and scienter in the new uniform federal definition of scientific misconduct. *Stanford Technology Law Review* 1998; 2: 35 p. [Online]. Accessed: http://stlr.stanford.edu/STLR/Articles/98_STLR_2/index.htm [2007 April 23]. Subject: 1.3.9

Kulynych, Jennifer J. Some thoughts about the evaluation of non-clinical functional magnetic resonance imaging. *American Journal of Bioethics* 2007 September; 7(9): 57-58. Subject: 17.1

Kulynych, Jennifer J. The regulation of MR neuroimaging research: disentangling the Gordian knot.

See SUBJECT HEADING KEY FOR SECTION II on inside back cover.

537

American Journal of Law and Medicine 2007; 33(2-3): 295-317. Subject: 18.2

Kumas, Gülsah; Öztunç, Gürsel; Alparslan, Z. Nazan. Intensive care unit nurses' opinions about euthanasia. *Nursing Ethics* 2007 September; 14(5): 637-650. Subject: 20.5.1

Kumra, Sanjiv; Ashtari, Manzar; Anderson, Britt; Cervellione, Kelly L.; Kan, Li. Ethical and practical considerations in the management of incidental findings in pediatric MRI studies. *Journal of the American Academy of Child and Adolescent Psychiatry* 2006 August; 45(8): 1000-1006. Subject: 9.5.7

Kuosmanen, Lauri; Hätönen, Heli; Malkavaara, Heikki; Kylmä, Jari; Välimäki, Maritta. Deprivation of liberty in psychiatry hospital care: the patient's perspective. *Nursing Ethics* 2007 September; 14(5): 597-607. Subject: 17.7

Kuppermann, Miriam; Learman, Lee A.; Gates, Elena; Gregorich, Steven E.; Nease, Robert F., Jr.; Lewis, James; Washington, A. Eugene. Beyond race or ethnicity and socioeconomic status: predictors of prenatal testing for Down syndrome. *Obstetrics and Gynecology* 2006 May; 107(5): 1087-1097. [see correction in: Obstetrics and Gynecology 2006 August; 108(2): 453]. Subject: 18.5.3

Kusmer, Ken. Indiana apologizes for role in eugenics. *ABC News* 2007 April 13: 2 p. [Online]. Accessed: http://abcnews.go.com/US/wireStory?id=3036919 [2008 February 21]. Subject: 15.5

Kusum, K. Sex selection in India. *In:* Dennerstein, Lorraine, ed. Women's Rights and Bioethics. Paris: UNESCO, 2000: 50-58. Subject: 14.3

Kutty, V. Raman. The study served no purpose. *Indian Journal of Medical Ethics* 2007 April-June; 4(2): 78. Subject: 18.5.2

Kuz, Kelly M. Young teenagers providing their own surgical consents: an ethical-legal dilemma for perioperative registered nurses. *Canadian Operating Room Nursing Journal* 2006 June; 24(2): 6-8, 10-11, 14-15. Subject: 9.5.7

Kvochak, Patricia A. Legal issues. *In:* Gallin, John I.; Ognibene, Frederick P., eds. Principles and Practice of Clinical Research. 2nd edition. Oxford: Academic, 2007: 109-120. Subject: 18.1

Kwak, Jung; Salmon, Jennifer R. Attitudes and preferences of Korean-American older adults and caregivers on end-of-life care. *Journal of the American Geriatrics Society* 2007 November; 55(11): 1867-1872. Subject: 20.4.1

Kwiecinski, Maureen. Limiting conflicts of interest arising from physician investment in specialty hospitals. *Specialty Law Digest: Health Care Law* 2006 January; (321). Subject: 7.3

Kwok, Timothy; Twinn, Sheila; Yan, Elsie. The attitudes of Chinese family caregivers of older people with demen-

tia towards life sustaining treatments. *Journal of Advanced Nursing* 2007 May; 58(3): 256-262. Subject: 20.5.1

Kyo-hun, Chin. Current debates on 'human cloning' in Korea. *In:* Roetz, Heiner, ed. Cross-Cultural Issues in Bioethics: The Example of Human Cloning. New York: Rodopi, 2006: 141-153. Subject: 14.5

L

Laabs, Carolyn Ann. Primary care nurse practitioners' integrity when faced with moral conflict. *Nursing Ethics* 2007 November; 14(6): 795-809. Subject: 4.1.3

Laaser, Ulrich; Donev, Donco; Bjegovic, Vesna; Sarolli, Ylli. Public health and peace [editorial]. *Croatian Medical Journal* 2002 April; 43(2):107-113. Subject: 9.1

Lacey, Debra. End-of-Life decision making for nursing home residents with dementia: a survey of nursing home social services staff. *Health and Social Work* 2006 August; 31(3): 189-199. Subject: 20.5.1

Ladas, Spiros D. Informed consent: still far from ideal? *Digestion* 2006; 73(2-3): 187-188. Subject: 8.3.1

Ladd, Paddy. Cochlear implantation, colonialism, and deaf rights. *In:* Komesaroff, Linda, ed. Surgical Consent: Bioethics and Cochlear Implantation. Washington, DC: Gallaudet University Press, 2007: 1-29. Subject: 9.5.1

Ladd, Rosalind Ekman. Rights of the autistic child. *In:* Freeman, Michael, ed. Children's Health and Children's Rights. Boston: Martinus Nijhoff Publishers, 2006: 87-98. Subject: 9.5.3

Ladd, Rosalind Ekman; Lawrence, Ryan E.; Curlin, Farr A. Some reflections on conscience. *American Journal of Bioethics* 2007 December; 7(12): 32-33; author reply W1-W2. Subject: 4.1.2

LaFleur, William R. Refusing utopia's bait: research, rationalizations, and Hans Jonas. *In:* LaFleur, William R.; Böhme, Gernot; Shimazono, Susumu, eds. Dark Medicine: Rationalizing Medical Research. Bloomington: Indiana University Press, 2007: 233-245. Subject: 18.1

LaFollette, Eva; LaFollette, Hugh. Private conscience, public acts [editorial]. *Journal of Medical Ethics* 2007 May; 33(5): 249-254. Subject: 2.1

LaFollette, Hugh; Lawrence, Ryan E.; Curlin, Farr A. The physician's conscience. *American Journal of Bioethics* 2007 December; 7(12): 15-17; author reply W1-W2. Subject: 4.1.2

Laine, Christine; Goodman, Steven N.; Griswold, Michael E.; Sox, Harold C. Reproducible research: moving toward research the public can really trust. *Annals of Internal Medicine* 2007 March 20; 146(6): 450-453. Subject: 1.3.9

Laine, Christine; Horton, Richard; DeAngelis, Catherine D.; Drazen, Jeffrey M.; Frizelle, Frank A.; Godlee, Fiona; Haug, Charlotte; Hébert, Paul C.; Kotzin,

---

Sheldon; Marusic, Ana; Sahni, Peush; Schroeder, Torben V.; Sox, Harold C.; Van Der Weyden, Martin B.; Verheugt, Freek W.A. Clinical trial registration: looking back and moving ahead [editorial]. *Annals of Internal Medicine* 2007 August 21; 147(4): 275-277. Subject: 18.6

Laine, Christine; Horton, Richard; DeAngelis, Catherine D.; Drazen, Jeffrey M.; Frizelle, Frank A.; Godlee, Fiona; Haug, Charlotte; Hébert, Paul C.; Kotzin, Sheldon; Marusic, Ana; Sahni, Peush; Schroeder, Torben V.; Sox, Harold C.; Van Der Weyden, Martin B.; Verheught, Freek W.A. Clinical trial registration: looking back and moving ahead [editorial]. *BMJ: British Medical Journal* 2007 June 9; 334(7605): 1177-1178. Subject: 18.6

Laine, Christine; Horton, Richard; DeAngelis, Catherine D.; Drazen, Jeffrey M.; Frizelle, Frank A.; Godlee, Fiona; Haug, Charlotte; Hébert, Paul C.; Kotzin, Sheldon; Marusic, Ana; Sahni, Peush; Schroeder, Torben V.; Sox, Harold C.; Van Der Weyden, Martin B.; Verheught, Freek W.A. Clinical trial registration: looking back and moving ahead [editorial]. *CMAJ/JAMC: Canadian Medical Association Journal* 2007 July 3; 177(1): 57-58. Subject: 18.6

Laine, Christine; Horton, Richard; DeAngelis, Catherine D.; Drazen, Jeffrey M.; Frizelle, Frank A.; Godlee, Fiona; Haug, Charlotte; Hébert, Paul C.; Kotzin, Sheldon; Marusic, Ana; Sahni, Peush; Schroeder, Torben V.; Sox, Harold C.; Van Der Weyden, Martin B.; Verheugt, Freek W.A. Clinical trial registration: looking back and moving ahead [editorial]. *JAMA: The Journal of the American Medical Association* 2007 July 4; 298(1): 93-94. Subject: 18.6

Laine, Christine; Horton, Richard; DeAngelis, Catherine D.; Drazen, Jeffrey M.; Frizelle, Frank A.; Godlee, Fiona; Haug, Charlotte; Hébert, Paul C.; Kotzin, Sheldon; Marusic, Ana; Sahni, Peush; Schroeder, Torben V.; Sox, Harold C.; Van Der Weyden, Martin B.; Verheugt, Freek W.A. Clinical trials registration: looking back and moving ahead [editorial]. *New England Journal of Medicine* 2007 June 28; 356(26): 2734-2736. Subject: 18.6

Laing, Jacqueline. The prohibition on eugenics and reproductive autonomy. *University of New South Wales Law Journal* 2006; 29(2): 261-265. Subject: 15.5

Lamey, Andy. Food fight!: Davis versus Regan on the ethics of eating beef. *Journal of Social Philosophy* 2007 Summer; 38(2): 331-348. Subject: 22.1

Lance, Mark Norris; Little, Margaret Olivia. Defending moral particularism. *In:* Dreier, James, ed. Contemporary Debates in Moral Theory. Oxford: Blackwell, 2006: 305-321. Subject: 2.1

Landau, Ruth. Assisted human reproduction: lessons of the Canadian experience. *Philosophy and Public Policy Quarterly* 2007 Winter-Spring; 27(1-2): 18-23. Subject: 14.1

Landeweer, Elleke; Berghmans, Ron; Abma, Tineke; Widdershoven, Guy. Coercive treatment in mental hospitals: legal regulations and experiences in the Netherlands. *EACME Newsletter* 2007 July; (17): 3 p. [Online]. Accessed: http://www.eacmeweb.com/newsletter/n17.htm [2007 August 15]. Subject: 18.5.1

Lane, Harlan. Ethnicity, ethics, and the deaf-world. *In:* Komesaroff, Linda, ed. Surgical Consent: Bioethics and Cochlear Implantation. Washington, DC: Gallaudet University Press, 2007: 42-69. Subject: 9.5.1

Lang, Slobodan; Kovacic, Luka; Šogoric, Selma; Brborovic, Ognjen. Challenges of goodness III: public health facing war. *Croatian Medical Journal* 2002 April; 43(2): 156-165. Subject: 9.1

Langan, John. Catholic perspectives on nutrition. *Ethics and Intellectual Disability* 2005 Summer; 9(1): 3-4. Subject: 20.5.1

Langer, Gary; Lyerly, Anne Drapkin; Faden, Ruth. Counting on embryos [letter and reply]. *Science* 2007 October 26; 318(5850): 566, 568. Subject: 18.5.4

Langlands, Nicola. Life after death: my life after a heart and lung transplant. *In:* Gunning, Jennifer; Holm, Søren, eds. Ethics, law, and society. Volume 2. Aldershot, Hants, England; Burlington, VT: Ashgate, 2006: 291-294. Subject: 19.2

Langlois, Natalie. Life-sustaining treatment law: a model for balancing a woman's reproductive rights with a pharmacist's conscientious objection. *Boston College Law Review* 2006 July; 47(4): 815-852. Subject: 11.1

Langone, Melissa. Promoting integrity among nursing students. *Journal of Nursing Education* 2007 January; 46(1): 45-47. Subject: 7.2

Langston, Anne L.; Johnston, Marie; Robertson, Clare; Campbell, Marion K.; Entwistle, Vikky A.; Marteau, Theresa M., McCallum, Marilyn; Ralston, Stuart H. Protocol for stage 1 of the GaP study (Genetic testing acceptability for Paget's disease of bone): an interview study about genetic testing and preventive treatment: would relatives of people with Paget's disease want testing and treatment if they were available? *BMC Health Services Research* 2006 June 8; 6: 71: 9 p. Subject: 15.3

Langston, Edward L.; Burns-Cox, C.J.; Halpin, David; Frost, C. Stephen; Hall, Peter. Ethical treatment of military detainees [letters]. *Lancet* 2007 December 15-21; 370(9604): 1999-2000. Subject: 21.4

Langworthy, Jennifer M.; le Fleming, Christine. Consent or submission? The practice of consent within UK chiropractic. *Journal of Manipulative and Physiological Therapeutics* 2005 January; 28(1): 15-24. Subject: 8.3.1

Lanoix, Monique. When cure entails care. *American Journal of Bioethics* 2007 March; 7(3): 34-36. Subject: 9.1

Länsimies-Antikainen, Helena; Pietilä, Anna-Maija; Laitinen, Tomi; Schwab, Ursula; Rauramaa, Rainer;

See SUBJECT HEADING KEY FOR SECTION II on inside back cover.

539

Länsimies, Esko. Evaluation of informed consent: a pilot study. *Journal of Advanced Nursing* 2007 July; 59(2): 146-154. Subject: 18.3

Lanter, Jennifer. Clinical research with cognitively impaired subjects: issues for nurses. *Dimensions of Critical Care Nursing* 2006 March-April; 25(2): 89-92. Subject: 18.5.6

Lantos, John D. At the Lok Nayak Hospital, Delhi. *Hastings Center Report* 2007 January-February; 37(1): 9. Subject: 9.5.7

Lantos, John D. Research in wonderland: does "minimal risk" mean whatever an institutional review board says it means? *American Journal of Bioethics* 2007 March; 7(3): 11-12. Subject: 18.5.2

Lantz, Cheryl M. Teaching spiritual care in a public institution: legal implications, standards of practice, and ethical obligations. *Journal of Nursing Education* 2007 January; 46(1): 33-38. Subject: 7.2

Lantz, Göran. Is health care ethics useful? *In:* Østnor, Lars, ed. Bioetikk og teologi: Rapport fra Nordisk teologisk nettverk for bioetikks workshop i Stockholm 27.-29. September 1996. Oslo: Nord. teol. nettv. for bioetikk, 1996: 101-106. Subject: 2.1

Lanza, Robert. Stem cell breakthrough: don't forget the ethics [letter]. *Science* 2007 December 21; 318(5858): 1865. Subject: 18.5.4

Lanzerath, Dirk. Die Eigenständigkeit der Bioethik und ihr Verhältnis zur Biopolitik [The autonomy of bioethics and its relationship to biopolitics]. *Ethik in der Medizin* 2006 December; 18(4): 364-368. Subject: 2.1

Larcher, Vic. Ethical issues in child protection. *Clinical Ethics* 2007 December; 2(4): 208-212. Subject: 9.5.7

Larcher, Vic; Hird, Michael F. Withholding and withdrawing neonatal intensive care. *Current Paediatrics* 2002 December; 12(6): 470-475. Subject: 20.5.2

Largent, Beverly A. Reaction to Universal Patient Acceptance: the perspective of a private practice dentist. *Journal of Dental Education* 2006 November; 70(11): 1202-1207. Subject: 9.2

Largent, Beverly A. When is it proper to refer a patient receiving public aid to another dentist? *Journal of the American Dental Association* 2006 March; 137(3): 395-396. Subject: 4.1.1

Lassen, Jesper; Gjerris, Mickey; Sandoe, Peter. After Dolly — ethical limits to the use of biotechnology on farm animals. *Theriogenology* 2006 March 15; 65(5): 992-1004. Subject: 22.3

Latham, Melanie. Cyberwoman and her surgeon in the twenty-first century. *In:* Garwood-Gowers, Austen; Tingle, John; Wheat, Kay, eds. Contemporary Issues in Healthcare Law and Ethics. Edinburgh; New York: Elsevier Butterworth-Heinemann, 2005: 233-249. Subject: 9.5.5

Latham, Stephen R. Justice and the financing of health care. *In:* Rhodes, Rosamond; Francis, Leslie P.; Silvers, Anita, eds. The Blackwell Guide to Medical Ethics. Malden, MA: Blackwell Pub., 2007: 341-353. Subject: 9.3.1

Latimer, Joanna. Becoming in-formed: genetic counselling, ambiguity and choice. *Health Care Analysis: An International Journal of Health Philosophy and Policy* 2007 March; 15(1): 13-23. Subject: 15.2

Lauridsen, S.M.R.; Norup, M.S.; Rossel, P.J.H. The secret art of managing healthcare expenses: investigating implicit rationing and autonomy in public healthcare systems. *Journal of Medical Ethics* 2007 December; 33(12): 704-707. Subject: 9.4

Laurie, Graeme. The autonomy of others: reflections on the rise and rise of patient choice in contemporary medical law. *In:* McLean, Sheila A.M., ed. First Do No Harm: Law, Ethics, and Healthcare. Aldershot, England; Burlington, VT: Ashgate, 2006: 131-149. Subject: 2.1

Lauritzen, Paul. Daniel Callahan and bioethics. Where the best arguments take him. *Commonweal* 2007 June 1; 134(11): 8-13. Subject: 2.2

Lavery, J.V.; Slutsky, A.S. Substitute decisions about genetic testing in critical care research: a glimpse behind the curtain. *Critical Care Medicine* 2006 April; 34(4): 1257-1259. Subject: 15.3

Lavieri, Robert R. The ethical mouse: be not like Icarus. *American Journal of Bioethics* 2007 May; 7(5): 57-58. Subject: 15.1

Lawlor, Rob. Moral theories in teaching applied ethics. *Journal of Medical Ethics* 2007 June; 33(6): 370-372. Subject: 2.3

Lawrence Livermore National Laboratory. Institutional Review Board. New investigator instructions: required reading for all new investigators. Livermore, CA: Lawrence Livermore National Laboratory, 2006 May 9; 4 p. [Online]. Accessed: http://www.llnl.gov/HumanSubjects/pi_instructions.html [2007 April 24]. Subject: 18.2

Lawrence, Ryan E.; Curlin, Farr A. Clash of definitions: controversies about conscience in medicine. *American Journal of Bioethics* 2007 December; 7(12): 10-14. Subject: 4.1.2

Lazarus, Jeffrey V.; Nielsen, Stine; Jakubcionyte, Rita; Kuliesyte, Esmeralda; Liljestrand, Jerker. Factors affecting attitudes towards medical abortion in Lithuania. *European Journal of Contraception and Reproductive Health Care* 2006 September; 11(3): 202-209. Subject: 12.5.2

Lazarus, J. Michael; Hakim, Raymond M. Dialysis facility ownership and epoetin dosing in hemodialysis patients: a dialysis provider's perspective. *American Journal* *of Kidney Diseases* 2007 September; 50(3): 366-370. Subject: 19.3

Le Coz, Pierre; Tassy, Sebastien. The philosophical moment of the medical decision: revisiting emotions felt, to improve ethics of future decisions. *Journal of Medical Ethics* 2007 August; 33(8): 470-472. Subject: 4.1.2

Leavine, Barbara Ann. Court ordered cesareans: can a pregnant woman refuse? *Houston Law Review* 1992 Spring; 29(1): 185-213. Subject: 9.5.5

Lebacqz, Karen. Choosing our children: the uneasy alliance of law and ethics in John Robertson's thought. *In:* Galston, Arthur W.; Peppard, Christiana Z., eds. Expanding Horizons in Bioethics. Dordrecht; Norwell, MA: Springer, 2005: 123-139. Subject: 15.1

Lebeer, Guy. Clinical ethics committees in Europe — assistance in medical decisions, fora for democratic debates, or bodies to monitor basic rights? *In:* Gunning, Jennifer; Holm, Søren, eds. Ethics, law, and society. Volume 1. Aldershot, Hants, England; Burlington, VT: Ashgate, 2005: 65-72. Subject: 9.6

Ledewitz, Bruce. Protecting posterity: economics, abortion politics, and the law. *Conservation Biology* 2006 August; 20(4): 940-941. Subject: 12.1

Ledford, Heidi. Death in gene therapy trial raises questions about private IRBs [news]. *Nature Biotechnology* 2007 October; 25(10): 1067. Subject: 15.4

Ledford, Heidi. Out of bounds. *Nature* 2007 January 11; 445(7124): 132-133. Subject: 15.1

Ledford, Heidi. Trial and error: the ethics committees that oversee research done in humans have been attacked from all sides. *Nature* 2007 August 2; 448(7153): 530-532. Subject: 18.2

Ledger, Sylvia Dianne. Euthanasia and assisted suicide: there is an alternative. *Ethics and Medicine: An International Journal of Bioethics* 2007 Summer; 23(2): 81-94. Subject: 20.7

Lee, A.; Wu, H.Y. Diagnosis disclosure in cancer patients — when the family says "no!". *Singapore Medical Journal* 2002 October; 43(10): 533-538. Subject: 8.2

Lee, Dong-Ik. The bioethics of 'quality of life' and 'sanctity of life'. *Dolentium Hominum* 2007; 22(2): 33-39. Subject: 4.4

Lee, Ellie. Young women, pregnancy, and abortion in Britain: a discussion of 'law in practice'. *International Journal of Law, Policy and the Family* 2004 December; 18(3): 283-304. Subject: 12.4.2

Lee, James E.; Kelly, D. Clay. Constitutional challenge to grave disability. *Journal of the American Academy of Psychiatry and the Law* 2007; 35(4): 534-535. Subject: 17.7

Lee, Joyce M.; Howell, Joel D. Tall girls: the social shaping of a medical therapy. *Archives of Pediatrics and Adolescent Medicine* 2006 October; 160(10):1035-1039. Subject: 9.5.7

Lee, K. Jane.; Havens, Peter L.; Sato, Thomas T.; Hoffman, George M.; Leuthner, Steven R. Assent for treatment: clinician knowledge, attitudes, and practice. *Pediatrics* 2006 August; 118(2): 723-730. Subject: 8.3.1

Lee, Robert. GM resistant: Europe and the WTO panel dispute on biotech products. *In:* Gunning, Jennifer; Holm, Søren, eds. Ethics, law, and society. Volume 1. Aldershot, Hants, England; Burlington, VT: Ashgate, 2005: 131-140. Subject: 15.1

Lee, S.S.-J. The ethical implications of stratifying by race in pharmacogenomics. *Clinical Pharmacology and Therapeutics* 2007 January; 81(1): 122-125. Subject: 15.11

Lee, Tricia K. Fujikawa. Emergency contraception in religious hospitals: the struggle between religious freedom and personal autonomy. *University of Hawai'i Law Review* 2004 Winter; 27(1): 65-109. Subject: 11.1

Legemaate, Johan. Psychiatry and human rights. *In:* Gevers, J.K.M.; Hondius, E.H.; Hubben, J.H., eds. Health Law, Human Rights and the Biomedicine Convention: Essays in Honour of Henriette Roscam Abbing. Leiden; Boston: Martinus Nijhoff Publishers, 2005: 119-130. Subject: 17.8

Legemaate; Johan; Verkerk, Marian; van Wijlick, Eric; de Graeff, Alexander. Palliative sedation in The Netherlands: starting-points and contents of a national guideline. *European Journal of Health Law* 2007 April; 14(1): 61-73. Subject: 20.4.1

Leget, Carlo. Retrieving the ars moriendi tradition. *Medicine, Health Care and Philosophy* 2007 September; 10(3): 313-319. Subject: 20.3.1

Leget, Carlo; Hoedemaekers, R. Teaching medical students about fair distribution of healthcare resources. *Journal of Medical Ethics* 2007 December; 33(12): 737-741. Subject: 9.4

Leget, Carlo; Olthuis, Gert. Compassion as a basis for ethics in medical education. *Journal of Medical Ethics* 2007 October; 33(10): 617-620. Subject: 2.3

Legge, M.; Fitzgerald, R.; Frank, N. A retrospective study of New Zealand case law involving assisted reproduction technology and the social recognition of 'new' family. *Human Reproduction* 2007 January; 22(1): 17-25. Subject: 14.2

Lehmann, Lisa Soleymani; Swartz, Katherine; Chin, Michael; Angell, Marcia; Daniels, Norman; Brock, Dan; Relman, Bud; Fein, Rashi. Harvard Medical School public forum: insuring the uninsured: does Massachusetts have the right model? 17 May 2007. *Journal of Clinical Ethics* 2007 Fall; 18(3): 270-293. Subject: 9.3.1

Lehrer, Jocelyn A.; Pantell, Robert; Tebb, Kathleen; Shafer, Mary-Ann. Forgone health care among U.S. adolescents: associations between risk characteristics and confidentiality concern. *Journal of Adolescent Health* 2007 March; 40(3): 218-226. Subject: 9.5.7

See SUBJECT HEADING KEY FOR SECTION II on inside back cover.

541

Leider, Robert. Quality of Life and Human Difference, edited by David Wasserman, Jerome Bickenbach, and Robert Wachbroit [book review]. *Ethics and Intellectual Disability* 2006 Spring; 9(2): 1, 4-5. Subject: 9.5.3

Leidinger, Friedrich. Müssen Demenzkranke ein "Sonderopfer für die Forschung" bringen? — Für eine neue Wissenschaft von der Demenz! *In:* Neuer-Miebach, Therese; Wunder, Michael, eds. Bio-Ethik und die Zukunft der Medizin. Bonn: Psychiatrie-Verlag, 1998: 106-119. Subject: 18.5.6

Leino-Kilpi, Helena. [Education in nursing ethics research] [editorial]. *Nursing Ethics* 2007 July; 14(4): 443-444. Subject: 7.2

Lemaire, François. Do all types of human research need ethics committee approval? *American Journal of Respiratory and Critical Care Medicine* 2006 August 15; 174(4): 363-364. Subject: 18.2

Lemiengre, Joke; de Casterlé, Bernadette Dierckx; Van Craen, Katleen; Schotsmans, Paul; Gastmans, Chris. Institutional ethics policies on medical end-of-life decisions: a literature review. *Health Policy* 2007 October; 83(2-3): 131-143. Subject: 20.5.1

Lemiengre, Joke; Dierckx de Casterlé, Bernadette; Verbeke, Geert; Guisson, Catherine; Schotsmans, Paul; Gastmans, Chris. Ethics policies on euthanasia in hospitals — a survey in Flanders (Belgium). *Health Policy* 2007 December; 84(2-3): 170-180. Subject: 20.5.1

Lemmens, Trudo. Commercialized medical research and the need for regulatory reform. *In:* Flood, Colleen M., ed. Just Medicare: What's In, What's Out, How We Decide. Buffalo, NY: University of Toronto Press, 2006: 396-426. Subject: 1.3.9

Lemmens, Trudo; Miller, Paul B. The human subjects trade: ethical, legal, and regulatory remedies to deal with recruitment incentives and to protect scientific integrity. *In:* Lemmens, Trudo; Waring, Duff R., eds. Law and Ethics in Biomedical Research: Regulation, Conflict of Interest and Liability. Toronto; Buffalo: University of Toronto Press, 2006: 132-179. Subject: 18.6

Leners, Debra Woodward; Roehrs, Carol; Piccone, Adam Vincent. Tracking the development of professional values in undergraduate nursing students. *Journal of Nursing Education* 2006 December; 45(12): 504-511. Subject: 4.1.3

Lenhard, Wolfgang; Breitenach, Erwin; Ebert, Harald; Schindelhauer-Deutscher, H. Joachim; Zang, Klaus D.; Henn, Wolfram. Attitudes of mothers towards their child with Down syndrome before and after the introduction of prenatal diagnosis. *Intellectual and Developmental Disabilities* 2007 April; 45(2): 98-102. Subject: 15.2

Lenk, Christian; Biller-Andorno, Nikola. Nanomedicine — emerging or re-emerging ethical issues? A discussion of four ethical themes. *Medicine, Health Care and Philosophy* 2007 June; 10(2): 173-184. Subject: 5.4

Lenzer, Jeanne. Advert for breast cancer gene test triggers inquiry [news]. *BMJ: British Medical Journal* 2007 September 22; 335(7620): 579. Subject: 15.3

Lenzer, Jeanne. Drug company tries to suppress internal memos [news]. *BMJ: British Medical Journal* 2007 January 13; 334(7584): 59. Subject: 9.7

Lenzer, Jeanne. Nigeria files criminal charges against Pfizer. *BMJ: British Medical Journal* 2007 June 9; 334(7605): 1181. Subject: 18.1

Lenzer, Jeanne. US Senate passes bill granting mandatory access to data [news]. *BMJ: British Medical Journal* 2007 November 3; 335(7626): 906. Subject: 1.3.9

Lerner, Barron H. Hero or victim? Barney Clark and the technological imperative. *In his:* When Illness Goes Public: Celebrity Patients and How We Look at Medicine. Baltimore: Johns Hopkins University Press, 2006: 180-200. Subject: 19.2

Lerner, Barron H. Subjects or objects? Prisoners and human experimentation. *New England Journal of Medicine* 2007 May 3; 356(18): 1806-1807. Subject: 18.5.5

Leroi, Armand Marie. The future of neo-eugenics: now that many people approve the elimination of certain genetically defective fetuses, is society closer to screening all fetuses for all known mutations? *EMBO Reports* 2006 December; 7(12): 1184-1187. Subject: 15.2

Lertsithichai, Panuwat. Health research, fair benefits and access to medicines. *Journal of the Medical Association of Thailand = Chotmaihet Thangphaet* 2006 April; 89(4): 558-564. Subject: 18.5.9

Lesser, Eugene A.; Starr, Jennifer; Kong, Xuan; Megerian, J. Thomas; Gozani, Shai N. Point-of-service nerve conduction studies: an example of industry-driven disruptive innovation in health care. *Perspectives in Biology and Medicine* 2007 Winter; 50(1): 40-53. Subject: 9.7

Letamo, Gobopamang. The discriminatory attitudes of health workers against people living with HIV. *PLoS Medicine* 2005 August; 2(8): e261 (0715-0716). Subject: 9.5.6

Létourneau, Lyne. The regulation of animal biotechnology: at the crossroads of law and ethics. *In:* Einsiedel, Edna; Timmermans, Frank, eds. Crossing Over: Genomics in the Public Arena. Calgary, Alberta, Canada: University of Calgary Press, 2005: 173-192. Subject: 15.1

Lett, Dan. Health Canada dithers while "fertility preservations" proceed [news]. *CMAJ/JAMC: Canadian Medical Association Journal* 2007 July 17; 177(2): 135-136. Subject: 14.6

Lett, Dan. Manitoba physicians consider DNR guidelines [news]. *CMAJ/JAMC: Canadian Medical Association Journal* 2007 January 30; 176(3): 310-311. Subject: 20.5.1

Levada, William. The magisterium's role in bioethics. *Origins* 2007 March 1; 36(37): 581-588. Subject: 2.1

Level of Care Study Investigators; Canadian Critical Care Trials Group; Cook, Deborah; Rocker, Graeme; Marshall, John; Griffith, Lauren; McDonald, Ellen; Guyatt, Gordon. Levels of care in the intensive care unit: a research program. *American Journal of Critical Care* 2006 May; 15(3): 269-379. Subject: 20.5.1

Levin, Leonard A.; Palmer, Julie Gage. Institutional review boards should require clinical trial registration. *Archives of Internal Medicine* 2007 August 13-27; 167(15): 1576-1580. Subject: 18.6

Levin, Philip D.; Sprung, Charles L. Intensive care triage — the hardest rationing decision of them all. *Critical Care Medicine* 2006 April; 34(4): 1250-1251. Subject: 9.4

Levin, Yuval. A middle ground for stem cells [op-ed]. *New York Times* 2007 January 19; p. A23. Subject: 18.5.4

Levine, Carol. Analyzing Pandora's box: the history of bioethics. *In:* Eckenwiler, Lisa A.; Cohn, Felicia G., eds. The Ethics of Bioethics: Mapping the Moral Landscape. Baltimore, MD: Johns Hopkins University Press, 2007: 3-23. Subject: 2.2

Levins, Susan. Teen leaves "his only hope" behind in U.S. After 20 months, 14-year-old with leukemia returns home, saying no more chemotherapy or bone marrow transplants. *Washington Post* 2007 January 11; p. B1, B5. Subject: 8.3.4

Levitt, Cheryl. Spirituality and family medicine. *In:* Meier, Augustine; O'Connor, Thomas St. James; VanKatwyk, Peter, eds. Spirituality and Health: Multidisciplinary Explorations. Waterloo, Ont.: Wilfrid Laurier University Press, 2005: 61-72. Subject: 9.1

Levitt, Mairi; Manson, Neil. My genes made me do it? The implications of behavioural genetics for responsibility and blame. *Health Care Analysis: An International Journal of Health Philosophy and Policy* 2007 March; 15(1): 33-40. Subject: 15.6

Levy, Barry S. Health and peace. *Croatian Medical Journal* 2002 April; 43(2):114-116. Subject: 9.1

Levy, Daniel R. The maternal-fetal conflict: the right of a woman to refuse a cesarean section versus the state's interest in saving the life of the fetus. *West Virginia Law Review* 2005 Fall; 108(1): 97-124. Subject: 9.5.5

Levy, Douglas E.; Youatt, Emily J; Shields, Alexandra E. Primary care physicians' concerns about offering a genetic test to tailor smoking cessation treatment. *Genetics in Medicine* 2007 December; 9(12): 842-849. Subject: 15.3

Levy, Neil. Rethinking neuroethics in the light of the extended mind thesis. *American Journal of Bioethics* 2007 September; 7(9): 3-11. Subject: 17.1

Levy, Neil. The social: a missing term in the debate over addiction and voluntary control. *American Journal of Bioethics* 2007 January; 7(1): 35-36. Subject: 17.1

Levy, Y. Practical aspects of the issue of patients refusing medical care. *Medicine and Law: The World Association for Medical Law* 2007 March; 26(1): 23-31. Subject: 8.3.4

Lewin, Tamar. Court says health coverage may bar birth-control pills. *New York Times* 2007 March 17; p. A11. Subject: 11.1

Lewis, Milton James. Medicine and euthanasia. *In his:* Medicine and Care of the Dying: A Modern History. Oxford; New York: Oxford University Press, 2007: 198-228. Subject: 20.5.1

Lewis, M. Jane; Peterson, Susan K. Perceptions of genetic testing for cancer predisposition among Ashkenazi Jewish women. *Community Genetics* 2007; 10(2): 72-81. Subject: 15.3

Lewis, Steven; Southern, Danielle A.; Maxwell, Colleen J.; Dunn, James R.; Noseworthy, Tom W.; Ghali, William A. What prosperous, highly educated Americans living in Canada think of the Canadian and US health care systems. *Open Medicine [electronic]* 2007; 1(2): E68-E74. Subject: 9.1

Lewis, William R.; Luebke, Donna L.; Johnson, Nancy J.; Harrington, Michael D.; Costantini, Ottorino; Aulisio, Mark P. Withdrawing implantable defibrillator shock therapy in terminally ill patients. *American Journal of Medicine* 2006 October; 119(10): 892-896. Subject: 20.4.1

Lexchin, Joel. The secret things belong unto the Lord our God: secrecy in the pharmaceutical arena. *Medicine and Law: The World Association for Medical Law* 2007 September; 26(3): 417-430. Subject: 1.3.9

Li, L.L.-M.; Cheong, K.Y.P.; Yaw, L.K.; Liu, E.H.C. The accuracy of surrogate decisions in intensive care scenarios. *Anaesthesia and Intensive Care* 2007 February; 35(1): 46-51. Subject: 8.3.3

Liang, Bryan A. Special doctor's docket. Lethal injection: policy considerations for medicine. *Journal of Clinical Anesthesia* 2006 September; 18(6): 466-470. Subject: 20.6

Liao, Lih Mei; Creighton, Sarah M. Requests for cosmetic genitoplasty: how should healthcare providers respond? *BMJ: British Medical Journal* 2007 May 26; 334(7603): 1090-1092. Subject: 9.5.1

Liao, S. Matthew; Goldschmidt-Clermont, Pascal J.; Sugarman, Jeremy. Ethical and policy issues relating to progenitor-cell-based strategies for prevention of atherosclerosis. *Journal of Medical Ethics* 2007 November; 33(11): 643-646. Subject: 18.1

Liao, S. Matthew; Savulescu, Julian; Sheehan, Mark. The Ashley treatment: best interests, convenience, and parental decision-making. *Hastings Center Report* 2007 March-April; 37(2): 16-20. Subject: 8.3.2

Liao, S. Matthew; Wasserman, David T.; Henry, Michael; Fishman, Jennifer R.; Youngner, Stuart J. Neuroethical concerns about moderating traumatic memo-

See SUBJECT HEADING KEY FOR SECTION II on inside back cover.

543

ries. *American Journal of Bioethics* 2007 September; 7(9): 38-40; author reply W1-W3. Subject: 17.4

Liaschenko, Joan. Teaching feminist ethics. *In:* Davis, Anne J.; Tschudin, Verena; de Raeve, Louise, eds. Essentials of Teaching and Learning in Nursing Ethics: Perspectives and Methods. New York: Churchill Livingstone Elsevier, 2006: 203-215. Subject: 2.3

Liaschenko, Joan; Peter, Elizabeth. Feminist ethics. *In:* Tschudin, Verena, ed. Approaches to Ethics: Nursing Beyond Boundaries. New York: Butterworth-Heinemann, 2003: 33-43. Subject: 4.1.3

Liaschenko, Joan; Peter, Elizabeth. Feminist ethics: a way of doing ethics. *In:* Davis, Anne J.; Tschudin, Verena; de Raeve, Louise, eds. Essentials of Teaching and Learning in Nursing Ethics: Perspectives and Methods. New York: Churchill Livingstone Elsevier, 2006: 181-190. Subject: 4.1.3

Lifson, Alan R.; Rybicki, Sarah L. Routine opt-out HIV testing [commentary]. *Lancet* 2007 February 17-23; 369(9561): 539-540. Subject: 9.5.6

Light, Donald W. Review of Richard Smith, The Trouble with Medical Journals [book review]. *American Journal of Bioethics* 2007 May; 7(5): 61-63. Subject: 1.3.7

Light, Terry R. Orthopaedic gifts: opportunities and obligations. *Journal of Bone and Joint Surgery. American volume* 2006 November; 88(11): 2521-2526. Subject: 9.7

Lilie, Hans. International legal limits to human cloning. *In:* Vöneky, Silja; Wolfrum, Rüdiger, eds. Human Dignity and Human Cloning. Leiden: Nijhoff, 2004: 125-132. Subject: 14.5

Lillquist, Erik; Sullivan, Charles A. The law and genetics of racial profiling in medicine. *Harvard Civil Rights-Civil Liberties Law Review* 2004 Summer; 39(2): 391-483. Subject: 15.11

Lim, E.-C.; Seet, R.C.S. Attitudes of medical students to placebo therapy. *Internal Medicine Journal* 2007 March; 37(3): 156-160. Subject: 7.2

Lin, Patrick. Nanotechnology bound: evaluating the case for more regulation. *NanoEthics* 2007 August; 1(2): 105-122. Subject: 5.4

Linacre Institute. Catholic medical decision-making on the concept of futility. *Linacre Quarterly* 2007 August; 74(3): 258-262. Subject: 20.5.1

Lind, Rebecca Ann; Lepper, Tammy Swenson. Sensitivity to research misconduct: a conceptual model. *Medicine and Law: The World Association for Medical Law* 2007 September; 26(3): 585-598. Subject: 1.3.9

Lindee, Susan. Experimental injury: wound ballistics and aviation medicine in mid-century America. *In:* LaFleur, William R.; Böhme, Gernot; Shimazono, Susumu, eds. Dark Medicine: Rationalizing Medical Research. Bloomington: Indiana University Press, 2007: 121-137. Subject: 18.5.1

Lindemann, Gesa. Die faktische Kraft des Normativen [The real power of normative ethics]. *Ethik in der Medizin* 2006 December; 18(4): 342-347. Subject: 2.1

Lindemann, Hilde. Breasts, wombs, and the body politic [review of Mass Hysteria: Medicine, Culture, and Mothers' Bodies, by Rebecca Kukla]. *Hastings Center Report* 2007 March-April; 37(2): 43-44. Subject: 9.5.5

Lindemann, Hilde. Obligations to fellow and future bioethicists: publication. *In:* Eckenwiler, Lisa A.; Cohn, Felicia G., eds. The Ethics of Bioethics: Mapping the Moral Landscape. Baltimore, MD: Johns Hopkins University Press, 2007: 270-277. Subject: 2.1

Linder, John F.; Meyers, Frederick J. Palliative care for prison inmates: "don't let me die in prison". *JAMA: The Journal of the American Medical Association* 2007 August 22-29; 298(8): 894-901. Subject: 20.4.1

Lindh, Inga-Britt; Severinsson, Elisabeth; Berg, Agneta. Moral responsibility: a relational way of being. *Nursing Ethics* 2007 March; 14(2): 129-140. Subject: 4.1.3

Lindsay, Robin; Graham, Helen. Relational narratives: solving an ethical dilemma concerning an individual's insurance policy. *In:* Tschudin, Verena, ed. Approaches to Ethics: Nursing Beyond Boundaries. New York: Butterworth-Heinemann, 2003: 53-60. Subject: 4.1.3

Linhorst, Donald. Individual rights, coercion, and empowerment. *In his:* Empowering People with Severe Mental Illness: A Practical Guide. New York: Oxford University Press, 2006: 40-64. Subject: 17.7

Lipman, Hannah I. Informed consent. *American Journal of Geriatric Cardiology* 2007 January-February; 16(1): 42-43. Subject: 9.5.2

Lippert-Rasmussen, Kasper. Why killing some people is more seriously wrong than killing others. *Ethics* 2007 July; 117(4): 716-738. Subject: 20.5.1

Lippman, Abby; Wertz, Dorothy C.; Fletcher, John C.; Nolan, Kathleen. Genetics. *In:* Baylis, Françoise; Downie, Jocelyn; Freedman, Benjamin; Hoffmaster, Barry; Sherwin, Susan, eds. Health Care Ethics in Canada. Toronto: Harcourt Brace Canada, 1995: 367-410. Subject: 15.3

Lipscomb, Martin; Snelling, Paul C. Moral content and assignment marking: an exploratory study. *Nurse Education Today* 2006 August; 26(6): 457-464. Subject: 7.2

Liptak, Adam. Florida panel urges steps for painless executions. *New York Times* 2007 March 2; p. A12. Subject: 20.6

Lipworth, W.; Ankeny, R.; Kerridge, I. Consent in crisis: the need to reconceptualize consent to tissue banking research. *Internal Medicine Journal* 2006 February; 36(2): 124-128. Subject: 15.1

Lisker, Rubén; Carnevale, Alessandra. Changing opinions of Mexican geneticists on ethical issues. *Archives of*

Medical Research 2006 August; 37(6): 794-803. Subject: 15.2

Liss, Howard. Publication bias in the pulmonary/allergy literature: effect of pharmaceutical company sponsorship. *The Israel Medical Association Journal* 2006 July; 8(7): 451-454. Subject: 9.7

Little, Melissa; Hall, Wayne; Orlandi, Amy. Delivering on the promise of human stem-cell research. What are the real barriers? *EMBO Reports* 2006 December; 7(12): 1188-1192. Subject: 18.5.4

Litton, Paul. "Nanoethics"? What's new? *Hastings Center Report* 2007 January-February; 37(1): 22-25. Subject: 5.1

Livingston, Ivor Lensworth; Carter, J. Jacques. Eliminating racial and ethnic disparities in health: a framework for action. *In:* Livingston, Ivor Lensworth, ed. Praeger Handbook of Black American Health: Policies and Issues Behind Disparities in Health, Vol. II. 2nd edition. Westport, CT: Praeger Publishers, 2004: 835-861. Subject: 9.5.4

Lizza, John P. Potentiality and human embryos. *Bioethics* 2007 September; 21(7): 379-385. Subject: 4.4

Lo, Bernard. Conflicts of interest. *In his:* Resolving Ethical Dilemmas: A Guide for Clinicians. 3rd edition. Philadelphia, PA: Lippincott Williams and Wilkins, 2005: 183-232. Subject: 7.3

Lo, Bernard. Decisions about life-sustaining interventions. *In his:* Resolving Ethical Dilemmas: A Guide for Clinicians. 3rd edition. Philadelphia, PA: Lippincott Williams and Wilkins, 2005: 103-152. Subject: 20.5.1

Lo, Bernard. Ethical issues in obstetrics and gynecology. *In his:* Resolving Ethical Dilemmas: A Guide for Clinicians. 3rd edition. Philadelphia, PA: Lippincott Williams and Wilkins, 2005: 249-255. Subject: 9.5.5

Lo, Bernard. Ethical issues in organ transplantation. *In his:* Resolving Ethical Dilemmas: A Guide for Clinicians. 3rd edition. Philadelphia, PA: Lippincott Williams and Wilkins, 2005: 264-271. Subject: 19.5

Lo, Bernard. Ethical issues in pediatrics. *In his:* Resolving Ethical Dilemmas: A Guide for Clinicians. 3rd edition. Philadelphia, PA: Lippincott Williams and Wilkins, 2005: 235-242. Subject: 8.3.2

Lo, Bernard. Ethical issues in psychiatry. *In his:* Resolving Ethical Dilemmas: A Guide for Clinicians. 3rd edition. Philadelphia, PA: Lippincott Williams and Wilkins, 2005: 256-263. Subject: 17.8

Lo, Bernard. Ethical issues in public health emergencies. *In his:* Resolving Ethical Dilemmas: A Guide for Clinicians. 3rd edition. Philadelphia, PA: Lippincott Williams and Wilkins, 2005: 280-284. Subject: 9.1

Lo, Bernard. Ethical issues in surgery. *In his:* Resolving Ethical Dilemmas: A Guide for Clinicians. 3rd edition.

Philadelphia, PA: Lippincott Williams and Wilkins, 2005: 243-248. Subject: 8.3.1

Lo, Bernard. Human papillomavirus vaccination programmes. *BMJ: British Medical Journal* 2007 August 25; 335(7616): 357-358. Subject: 9.5.1

Lo, Bernard. Principles in the ethical care of underserved patients. *In:* King, Talmadge E.; Wheeler, Margaret B., eds. Medical Management of Vulnerable and Underserved Patients: Principles, Practice, and Populations. New York: McGraw-Hill Medical Pub. Division, 2007: 47-55. Subject: 2.1

Lo, Bernard. Shared decision making. *In his:* Resolving Ethical Dilemmas: A Guide for Clinicians. 3rd edition. Philadelphia, PA: Lippincott Williams and Wilkins, 2005: 57-102. Subject: 8.3.3

Lo, Bernard. Testing for genetic conditions. *In his:* Resolving Ethical Dilemmas: A Guide for Clinicians. 3rd edition. Philadelphia, PA: Lippincott Williams and Wilkins, 2005: 272-279. Subject: 15.3

Lo, Bernard. The doctor-patient relationship. *In his:* Resolving Ethical Dilemmas: A Guide for Clinicians. 3rd edition. Philadelphia, PA: Lippincott Williams and Wilkins, 2005: 153-182. Subject: 8.1

Lock, Margaret. Inventing a new death and making it believable. *In:* van Dongen, Els; Fainzang, Sylvie, eds. Lying and Illness: Power and Performance. Amsterdam: Het Spinhuis; Piscataway, NJ: Transaction Publishers, 2005: 12-35. Subject: 20.2.1

Lock, Margaret. On making up the good-as-dead in a utilitarian world. *In:* Franklin, Sarah; Lock, Margaret, eds. Remaking Life and Death: Toward an Anthropology of the Biosciences. Santa Fe: School of American Research Press; Oxford: James Currey, 2003: 165-192. Subject: 20.2.1

Lock, Margaret. Utopias of health eugenics, and germline engineering. *In:* Nichter, Mark and Lock, Margaret, eds. New Horizons in Medical Anthropology: Essays in Honour of Charles Leslie. London: Routledge, 2002: 240-266. Subject: 15.1

Lockwood, Gillian M. Whose embryos are they anyway? *In:* Shenfield, Françoise; Sureau, Claude, eds. Contemporary Ethical Dilemmas in Assisted Reproduction. Abingdon: Informa Healthcare, 2006: 3-11. Subject: 14.6

Lockwood, Michael; Anscombe, G.E.M. Sins of omission? The non-treatment of controls in clinical trials. *Proceedings of the Aristotelian Society, Supplementary Volumes* 1983; 57: 207-227. Subject: 18.2

Loeben, Gregory; Stoehr, James D. Normative judgments, responsibility and executive function. *American Journal of Bioethics* 2007 January; 7(1): 27-29. Subject: 17.1

Loewy, Erich H.; Loewy, Roberta Springer. Framing issues in health care: do American ideals demand basic

See SUBJECT HEADING KEY FOR SECTION II on inside back cover.

545

health care and other social necessities for all? *Health Care Analysis: An International Journal of Health Philosophy and Policy* 2007 December; 15(4): 261-271. Subject: 9.2

Löfmark, Rurik; Nilstun, T.; Bolmsjö, I Ågren. From cure to palliation: concept, decision and acceptance. *Journal of Medical Ethics* 2007 December; 33(12): 685-688. Subject: 20.4.1

Logan, Lara. Switzerland's suicide tourists. *CBSNews.com* 2003 July 23: 2 p. [Online]. Accessed: http://www.cbsnews.com/stories/2003/02/12/60II/printable540332.shtml [2008 February 7]. Subject: 20.5.1

Loike, John D.; Tendler, Moshe D. Ethical dilemmas in stem cell research: human-animal chimeras. *Tradition* 2007 Winter; 40(4): 28-49. Subject: 18.5.4

Loike, John D.; Tendler, Moshe D. Molecular genetics, evolution, and Torah principles. *Torah u-Madda Journal* 2006-2007; 14: 173-192. Subject: 15.1

Lombardo, Paul A.; Dorr, Gregory M. Eugenics, medical education, and the public health service: another perspective on the Tuskegee syphilis experiment. *Bulletin of the History of Medicine* 2006 Summer; 80(2): 291-316. Subject: 15.5

London, Alex John. Clinical equipoise: foundational requirement or fundamental error? *In:* Steinbock, Bonnie, ed. The Oxford Handbook of Bioethics. Oxford; New York: Oxford University Press, 2007: 571-596. Subject: 2.1

London, Alex John. Two dogmas of research ethics and the integrative approach to human-subjects research. *Journal of Medicine and Philosophy* 2007 March-April; 32(2): 99-116. Subject: 18.1

London, Leslie. 'Issues of equity are also issues of rights': lessons from experiences in Southern Africa. *BMC Public Health* 2007 January 26; 7:14. Subject: 9.1

Long, Angela M. Why criminalizing sex selection techniques is unjust: an argument challenging conventional wisdom. *Health Law Journal* 2006; 14: 69-104. Subject: 14.3

Long, J. Michael. Student views of professional ethics. *Journal of the American College of Dentists* 1996 Spring; 63(1): 37-42. Subject: 4.1.1

Long, Robert Emmet. Dr. Kevorkian and assisted suicide. *In his:* Suicide. New York: H.W. Wilson, 1995: 76-114. Subject: 20.7

Long, Robert Emmet. Kevorkian's critics. *In his:* Suicide. New York: H.W. Wilson, 1995: 115-144. Subject: 20.7

Long, Robert Emmet. Nancy Cruzan and the "right to die". *In his:* Suicide. New York: H.W. Wilson, 1995: 54-75. Subject: 20.5.1

Longaker, Michael T.; Baker, Laurence C.; Greely, Henry T. Proposition 71 and CIRM — assessing the return

on investment. *Nature Biotechnology* 2007 May; 25(5): 513-521. Subject: 18.5.4

Lopus, Jane S.; Grimes, Paul W.; Becker, William E.;Pearson, Rodney A. Effects of human subjects requirements on classroom research: multidisciplinary evidence. *Journal of Empirical Research on Human Research Ethics* 2007 September; 2(3): 69-77. Subject: 18.6

Lorenz, Rolf J. Tübinger Initiative gegen die geplante Bio-Ethik-Konvention. *In:* Neuer-Miebach, Therese; Wunder, Michael, eds. Bio-Ethik und die Zukunft der Medizin. Bonn: Psychiatrie-Verlag, 1998: 159-162. Subject: 2.1

Lothen-Kline, Christine; Howard, Donna E.; Hamburger, Ellen K.; Worrell, Kevin D.; Boekeloo, Bradley O. Truth and consequences: ethics, confidentiality, and disclosure in adolescent longitudinal prevention research. *Journal of Adolescent Health* 2003 November; 33(5): 385-394. Subject: 8.4

Lötjönen, Salla. Research on human subjects. *In:* Gevers, J.K.M.; Hondius, E.H.; Hubben, J.H., eds. Health Law, Human Rights and the Biomedicine Convention: Essays in Honour of Henriette Roscam Abbing. Leiden; Boston: Martinus Nijhoff Publishers, 2005: 175-190. Subject: 18.2

Lott, Jason P.; Savulescu, Julian. Towards a global human embryonic stem cell bank. *American Journal of Bioethics* 2007 August; 7(8): 37-44. Subject: 18.5.4

Loue, Sana; Ioan, Beatrice. Legal and ethical issues in heroin diagnosis, treatment, and research. *Journal of Legal Medicine* 2007 April-June; 28(2): 193-221. Subject: 9.5.9

Loughlin, Michael. A platitude too far: 'evidence-based ethics'. Commentary on Borry (2006), Evidence-based medicine and its role in ethical decision-making. Journal of Evaluation in Clinical Practice 12, 306-311. *Journal of Evaluation in Clinical Practice* 2006 June; 12(3): 312-318 [see correction in: J Eval Clin Pract. 2006 August; 12(4): 471]. Subject: 9.1

Louhiala, P. How tall is too tall? On the ethics of oestrogen treatment for tall girls. *Journal of Medical Ethics* 2007 January; 33(1): 48-50. Subject: 9.5.7

Lowe, Jennifer; Pomerantz, Andrew M.; Pettibone, Jon C. The influence of payment method on psychologists' diagnostic decisions: expanding the range of presenting problems. *Ethics and Behavior* 2007; 17(1): 83-93. Subject: 17.2

Lown, Beth A.; Chou, Calvin L.; Clark, William D.; Haidet, Paul; White, Maysel Kemp; Krupat, Edward; Pelletier, Stephen; Weissmann, Peter; Anderson, M. Brownell. Caring attitudes in medical education: perceptions of deans and curriculum leaders. *JGIM: Journal of General Internal Medicine* 2007 November; 22(11): 1514-1522. Subject: 7.2

Lowrence, William W.; Collins, Francis S. Identifiability in genomic research. *Science* 2007 August 3; 317(5838): 600-602. Subject: 15.1

Luban, David. Torture and the professions [commentary]. *Criminal Justice Ethics* 2007 Summer-Fall; 26(2): 2, 58-66. Subject: 21.4

Lubowitz, James H. Randomize, then consent: a strategy for improving patient acceptance of participation in randomized controlled trials. *Arthroscopy* 2006 September; 22(9): 1007-1008. Subject: 18.2

Lucassen, Anneke; Clarke, Angus. Should families own genetic information [debate]. *BMJ: British Medical Journal* 2007 July 7; 335(7609): 22-23. Subject: 15.8

Luce, John M. Acknowledging our mistakes. *Critical Care Medicine* 2006 May; 34(5): 1575-1576. Subject: 9.8

Luce, John M. California's new law allowing surrogate consent for clinical research involving subjects with impaired decision-making capacity. *Intensive Care Medicine* 2003 June; 29(6): 1024-1025. Subject: 18.3

Ludwick, Ruth; Silva, Mary Cipriano. What would you do? Ethics and infection control. *Online Journal of Issues in Nursing [electronic]* 2007; 12(1): 8 p. Accessed: http://www.nursingworld.org/ojin/tocv12n1.htm [2007 March 15]. Subject: 9.1

Luft, Harold S.; Flood, Ann Barry; Escarce, José J. New policy on disclosures at Health Services Research. *Health Services Research* 2006 October; 41(5): 1721-1732. Subject: 1.3.7

Lumley, Judith; Daly, Jeanne. Authors' misconduct in the firing line. *Australian and New Zealand Journal of Public Health* 2006 October; 30(5): 403-404. Subject: 1.3.7

Luna, Florencia. AIDS, research, and acceptable codes. *In:* Luna, Florencia; Herissone-Kelly, Peter, ed. Bioethics and Vulnerability: a Latin American View. Amsterdam; New York: Rodopi, 2006: 87-109. Subject: 9.5.6

Luna, Florencia. Assisted reproduction and local experiences: women and context in Latin America. *In:* Luna, Florencia; Herissone-Kelly, Peter, ed. Bioethics and Vulnerability: a Latin American View. Amsterdam; New York: Rodopi, 2006: 63-71. Subject: 14.4

Luna, Florencia. Assumptions in the "standard of care" debate. *In:* Cohen, Jillian Clare; Illingworth, Patricia; Schüklenk, Udo, eds. The Power of Pills: Social, Ethical and Legal Issues in Drug Development, Marketing, and Pricing. London; Ann Arbor, MI: Pluto, 2006: 215-223. Subject: 9.7

Luna, Florencia. Internal reasons and abortion. *In:* Luna, Florencia; Herissone-Kelly, Peter, ed. Bioethics and Vulnerability: a Latin American View. Amsterdam; New York: Rodopi, 2006: 39-47. Subject: 12.3

Luna, Florencia. Research in developing countries. *In:* Steinbock, Bonnie, ed. The Oxford Handbook of Bioethics. Oxford; New York: Oxford University Press, 2007: 621-647. Subject: 18.2

Luna, Florencia. Social science research and respect for persons. *In:* Luna, Florencia; Herissone-Kelly, Peter, ed. Bioethics and Vulnerability: a Latin American View. Amsterdam; New York: Rodopi, 2006: 73-85. Subject: 18.2

Luna, Florencia. To procreate or not to procreate? AIDS and reproductive rights. *In:* Luna, Florencia; Herissone-Kelly, Peter, ed. Bioethics and Vulnerability: a Latin American View. Amsterdam; New York: Rodopi, 2006: 49-62. Subject: 9.5.6

Lundmark, Mikael. Vocation in theology-based nursing theories. *Nursing Ethics* 2007 November; 14(6): 767-780. Subject: 4.1.3

Lundqvist, Anita; Nilstun, Tore. Human dignity in paediatrics: the effects of health care. *Nursing Ethics* 2007 March; 14(2): 215-228. Subject: 9.5.7

Lunstroth, John; Goldman, Jan. Ethical intelligence from neuroscience: is it possible? *American Journal of Bioethics* 2007 May; 7(5): 18-20. Subject: 17.1

Lupton, M. Nanotechnology — salvation or damnation for humans? *Medicine and Law: The World Association for Medical Law* 2007 June; 26(2): 349-362. Subject: 5.4

Lustig, B. Andrew. Death, dying, euthanasia, and palliative care: perspectives from philosophy of medicine and ethics. *In:* Khushf, George, ed. Handbook of Bioethics: Taking Stock of the Field From a Philosophical Perspective. Dordrecht; Boston: Kluwer Academic, 2004: 441-471. Subject: 20.5.1

Lustig, B. Andrew. The church and the world: are there theological resources for a common conversation? *Christian Bioethics* 2007 May-August; (13)2: 225-244. Subject: 12.3

Lybecker, Kristina M. Social, ethical, and legal issues in drug development, marketing, and pricing policies: setting priorities: pharmaceuticals as private organizations and the duty to make money/maximize profits. *In:* Cohen, Jillian Clare; Illingworth, Patricia; Schüklenk, Udo, eds. The Power of Pills: Social, Ethical and Legal Issues in Drug Development, Marketing, and Pricing. London; Ann Arbor, MI: Pluto, 2006: 25-31. Subject: 9.7

Lyckholm, Laurie; Quillin, John. Equanimity abandoned? *American Journal of Bioethics* 2007 July; 7(7): 31-32. Subject: 8.1

Lyden, Martin. Capacity issues related to the health care proxy. *Mental Retardation* 2006 August; 44(4): 272-282. Subject: 8.3.1

Lyerly, Anne Drapkin; Faden, Ruth R. Willingness to donate frozen embryos for stem cell research. *Science* 2007 July 6; 317(5834): 46-47. Subject: 18.5.4

Lynch, Margaret A.; Glaser, Danya; Prior, Vivien; Inwood, Vivien. Following up children who have been abused: ethical considerations for research design. *Child*

See SUBJECT HEADING KEY FOR SECTION II on inside back cover.

547

Psychology and Psychiatry Review 1999 May; 4(2): 68-75. Subject: 18.5.2

Lynn, Joanne; Baily, Mary Ann; Bottrell, Melissa; Jennings, Bruce; Levine, Robert J.; Davidoff, Frank; Casarett, David; Corrigan, Janet; Fox, Ellen; Wynia, Matthew K.; Agich, George J.; O'Kane, Margaret; Speroff, Theodore; Schyve, Paul; Batalden, Paul; Tunis, Sean; Berlinger, Nancy; Cronenwett, Linda; Fitzmaurice, J. Michael; Neveloff Dubler, Nancy; James, Brent. The ethics of using quality improvement methods in health care. *Annals of Internal Medicine* 2007 May 1; 146(9): 666-673. Subject: 9.8

Lyren, Anne; Leonard, Ethan. Vaccine refusal: issues for the primary care physician. *Clinical Pediatrics* 2006 June; 45(5): 399-404. Subject: 9.5.1

Lysaught, M. Therese. Becoming one body: health care and cloning. *In:* Hauerwas, Stanley and Wells, Samuel, eds. The Blackwell Companion to Christian Ethics. Malden, MA : Blackwell, 2004: 263-275. Subject: 14.5

Lysaught, M. Therese. Making decisions about embryonic stem cell research. *In:* Hamel, Ronald, ed. Making Health Care Decisions: A Catholic Guide. Liguori, MO: Liguori Publications, 2006: 19-36. Subject: 18.5.4

M

Mabrouk, Patricia Ann. Introducing summer high school student-researchers to ethics in scientific research. *Journal of Chemical Education* 2007 June; 84(6): 952. Subject: 1.3.9

Macauley, Robert C. The role of substituted judgment in the aftermath of a suicide attempt. *Journal of Clinical Ethics* 2007 Summer; 18(2): 111-118. Subject: 9.5.5

MacDonald, Hannah. Relational ethics and advocacy in nursing: literature review. *Journal of Advanced Nursing* 2007 January; 57(2): 119-126. Subject: 8.1

Macdonald, Marilyn. Origins of difficulty in the nurse-patient encounter. *Nursing Ethics* 2007 July; 14(4): 510-521. Subject: 8.1

MacDougall, Iain C. Dialysis facility ownership and epoetin dosing in hemodialysis patients: a view from Europe. *American Journal of Kidney Diseases* 2007 September; 50(3): 358-361. Subject: 19.3

Macduff, Colin; McKie, Andrew; Martindale, Sheelagh; Rennie, Anne Marie; West, Bernice; Wilcock, Sylvia. A novel framework for reflecting on the functioning of research ethics review panels. *Nursing Ethics* 2007 January; 14(1): 99-116. Subject: 18.2

Macer, Darryl; Toledano, Sarah Jane; de Castro, Leonardo D.; Siruno, Lalaine H. Republication: in that case. *Journal of Bioethical Inquiry* 2007; 4(3): 239, 241-244. Subject: 9.5.7

Macer, D.R.J. Patent or perish? An ethical approach to patenting human genes and proteins. *Pharmacogenomics Journal* 2002; 2(6): 361-366. Subject: 15.8

Machado, Calixto; Kerein, Julius; Ferrer, Yazmina; Portela, Liana; de la C. García, Maria; Manero, José M. The concept of brain death did not evolve to benefit organ transplants. *Journal of Medical Ethics* 2007 April; 33(4): 197-200. Subject: 20.2.1

Machado, Calixto; Korein, J.; Ferrer, Y.; Portela, L.; de la C. García, M.; Chinchilla, M.; Machado, Y.; Machado, Y.; Manero, J.M. The Declaration of Sydney on human death. *Journal of Medical Ethics* 2007 December; 33(12): 699-703. Subject: 20.2.1

MacIntosh, Constance. Indigenous self-determination and research on human genetic material: a consideration of the relevance of debates on patents and informed consent, and the political demands on researchers. *Health Law Journal* 2005; 13: 213-251. Subject: 15.11

Macintosh, Constance. Jurisdictional roulette: constitutional and structural barriers to aboriginal access to health. *In:* Flood, Colleen M., ed. Just Medicare: What's In, What's Out, How We Decide. Buffalo, NY: University of Toronto Press, 2006: 193-215. Subject: 9.5.4

Mack, Jennifer W.; Wolfe, Joanne; Grier, Holocombe E.; Cleary, Paul D.; Weeks, Jane C. Communication about prognosis between parents and physicians of children with cancer: parent preferences and the impact of prognostic information. *Journal of Clinical Oncology* 2006 November 20; 24(33): 5265-5270. Subject: 8.2

MacKellar, Calum. Ethics and genetics of human behaviour [commentary]. *Ethics and Medicine: An International Journal of Bioethics* 2007 Spring; 23(1): 7-9. Subject: 15.6

MacKenzie, Catriona. Feminist bioethics and genetic termination. *Bioethics* 2007 November; 21(9): 515-516. Subject: 15.2

Mackenzie, Robin. Regulating reprogenetics: strategic sacralisation and semantic massage. *Health Care Analysis: An International Journal of Health Philosophy and Policy* 2007 December; 15(4): 305-319. Subject: 14.4

Mackie, Jocelyn E.; Taylor, Andrew D.; Finegold, David L.; Daar, Abdallah S.; Singer, Peter A. Lessons on ethical decision making from the bioscience industry. *PLoS Medicine* 2006 May; 3(5): e129. Subject: 15.1

Macklin, Ruth. Global health. *In:* Steinbock, Bonnie, ed. The Oxford Handbook of Bioethics. Oxford; New York: Oxford University Press, 2007: 696-720. Subject: 21.1

Macklin, Ruth. Reproductive rights and health in the developing world. *In:* Galston, Arthur W.; Peppard, Christiana Z., eds. Expanding Horizons in Bioethics. Dordrecht; Norwell, MA: Springer, 2005: 87-101. Subject: 14.1

Macklin, Ruth; Weir, Robert F.; Paris, John J.; Crone, Robert K.; Reardon, Frank; Jecker, Nancy S. Children and the elderly: who should decide? *In:* Baylis, Françoise;

Downie, Jocelyn; Freedman, Benjamin; Hoffmaster, Barry; Sherwin, Susan, eds. Health Care Ethics in Canada. Toronto: Harcourt Brace Canada, 1995: 277-318. Subject: 8.3.3

MacNeil, S. Danielle; Fernandez, Conrad V. Attitudes of research ethics board chairs toward disclosure of research results to participants: results of a national survey. *Journal of Medical Ethics* 2007 September; 33(9): 549-553. Subject: 18.2

Macpherson, Cheryl Cox. Global bioethics: did the Universal Declaration on Bioethics and Human Rights miss the boat? *Journal of Medical Ethics* 2007 October; 33(10): 588-590. Subject: 2.1

Macrina, Francis L. Scientific societies and promotion of the responsible conduct of research: codes, policies, and education. *Academic Medicine* 2007 September; 82(9): 865-869. Subject: 18.2

Madden, Deirdre. Assisted reproduction in the Republic of Ireland — a legal quagmire. *In:* Gunning, Jennifer; Holm, Søren, eds. Ethics, law, and society. Volume 2. Aldershot, Hants, England; Burlington, VT: Ashgate, 2006: 27-34. Subject: 14.4

Madoff, Ray D. Autonomy and end-of-life decision making: reflections of a lawyer and a daughter. *Buffalo Law Review* 2005 Summer; 53(3): 963-971. Subject: 20.5.4

Madsen, S.M.; Holm, S.; Riis, P. Attitudes towards clinical research among cancer trial participants and non-participants: an interview study using a grounded theory approach. *Journal of Medical Ethics* 2007 April; 33(4): 234-240. Subject: 18.1

Madueme, Hans. Addiction as an amoral condition? The case remains unproven. *American Journal of Bioethics* 2007 January; 7(1): 25-27. Subject: 17.1

Maehle, Andreas-Holger. Professional ethics and discipline: the Prussian Medical Courts of Honour, 1888-1920. *Medizinhistorisches Journal* 1999; 34(3-4): 309-338. Subject: 2.2

Maggon, Krishan. Regulatory reforms and GCP clinical trials with new drugs in India. *Clinical Trials (London, England)* 2004; 1(5): 461-467. Subject: 18.2

Magill, Gerard. Ethical and policy issues related to medical error and patient safety. *In:* McLean, Sheila A.M., ed. First Do No Harm: Law, Ethics, and Healthcare. Aldershot, England; Burlington, VT: Ashgate, 2006: 101-116. Subject: 9.8

Magnavita, Nicola. The unhealthy physician. *Journal of Medical Ethics* 2007 April; 33(4): 210-214. Subject: 7.4

Magnus, David. Playing it safe [editorial]. *American Journal of Bioethics* 2007 March; 7(3): 1-2. Subject: 18.5.2

Magnus, David; Tabor, Holly; Karkazis, Katrina. Transplants for developmentally delayed children. *Ethics*

and Intellectual Disability 2007 Winter; 10(1): 3-4. Subject: 19.6

Magrini, Nicola; Font, Maria. Direct to consumer advertising of drugs in Europe [editorial]. *BMJ: British Medical Journal* 2007 September 15; 335(7619): 526. Subject: 9.7

Maharaj, N.R.; Dhai, A.; Wiersma, R.; Moodley, J. Intersex conditions in children and adolescents: surgical, ethical, and legal considerations. *Journal of Pediatric and Adolescent Gynecology* 2005 December; 18(6): 399-402. Subject: 10

Maher, Brendan. His daughter's DNA. *Nature* 2007 October 18; 449(7164): 772-776. Subject: 15.1

Mahowald, Mary Briody. Preconception and prenatal decisions. *In her:* Bioethics and Women: Across the Life Span. Oxford; New York: Oxford University Press, 2006: 73-91, 250-251. Subject: 15.2

Mahowald, Mary Briody. Research issues. *In her:* Bioethics and Women: Across the Life Span. Oxford; New York: Oxford University Press, 2006: 214-229, 263-265. Subject: 18.5.3

Mahowald, Mary Briody; Sherwin, Susan; Overall, Christine. Assisted reproductive technologies. *In:* Baylis, Françoise; Downie, Jocelyn; Freedman, Benjamin; Hoffmaster, Barry; Sherwin, Susan, eds. Health Care Ethics in Canada. Toronto: Harcourt Brace Canada, 1995: 449-485. Subject: 14.4

Mahowald, Mary B. Prenatal testing for selection against disabilities. *CQ: Cambridge Quarterly of Healthcare Ethics* 2007 Fall; 16(4): 457-462. Subject: 15.2

Maitra, Robin T.; Harfst, Anja; Bjerre, Lise M.; Kochen, Michael M.; Becker, Annette. Do German general practitioners support euthanasia? Results of a nationwide questionnaire survey. *European Journal of General Practice* 2005 September-December; 11(3-4): 94-100. Subject: 20.7

Majumdar, Sisir K. History of evolution of the concept of medical ethics. *Bulletin of the Indian Institute of History of Medicine (Hyderabad)* 2003 January-June; 33(1): 17-31. Subject: 2.2

Makdisi, June Mary Zekan. The protection of embryonic life in the European Council's Convention on Biomedicine. *National Catholic Bioethics Quarterly* 2007 Spring; 7(1): 31-39. Subject: 15.3

Malakoff, Marion. Palliative care/physician-assisted dying: alternative or continuing care? *Care Management Journals* 2006 Spring; 7(1): 41-44. Subject: 20.4.1

Malek, Janet. Understanding risks and benefits in research on reproductive genetic technologies. *Journal of Medicine and Philosophy* 2007 July-August; 32(4): 339-358. Subject: 15.1

Malek, Janet; Kopelman, Loretta M. The well-being of subjects and other parties in genetic research and testing.

See SUBJECT HEADING KEY FOR SECTION II on inside back cover.

549

Journal of Medicine and Philosophy 2007 July-August; 32(4): 311-319. Subject: 15.1

Maloney, Dennis M. Another institution ordered to halt human subjects research projects. *Human Research Report* 2007 November; 22(11): 1-2. Subject: 18.6

Maloney, Dennis M. Case study: institutional review board (IRB) must review hundreds of protocols — again. *Human Research Report* 2007 November; 22(11): 6-7. Subject: 18.2

Maloney, Dennis M. Case study: university forms two new institutional review boards (IRBs) to rereview suspended studies. *Human Research Report* 2007 June; 22(6): 6-7. Subject: 18.2

Maloney, Dennis M. Case study: university is allowed to resume its human subjects research projects. *Human Research Report* 2007 December; 22(12): 6-7. Subject: 18.6

Maloney, Dennis M. Case study: university says it is doing what it can to earn right to resume research. *Human Research Report* 2007 July; 22(7): 6-7. Subject: 18.2

Maloney, Dennis M. Case study: university says it will modify many human subject protection procedures. *Human Research Report* 2007 August; 22(8): 6-7. Subject: 18.3

Maloney, Dennis M. Changes for expedited reviews by institutional review boards (IRBs). *Human Research Report* 2007 December; 22(12): 1-2. Subject: 18.2

Maloney, Dennis M. Court says actions of institutional review board (IRB) members were ethically wrong. *Human Research Report* 2007 April; 22(4): 8. Subject: 18.2

Maloney, Dennis M. Court says institutional review board (IRB) abdicated duty to protect children as subjects. *Human Research Report* 2007 January; 22(1): 8. Subject: 18.5.2

Maloney, Dennis M. Court says institutional review boards (IRBs) are not objective enough to protect human subjects. *Human Research Report* 2007 March; 22(3): 8. Subject: 18.2

Maloney, Dennis M. Federal office orders college to halt all federally-supported human research. *Human Research Report* 2007 September; 22(9): 1-2. Subject: 18.6

Maloney, Dennis M. Final guidance issued on protection of children as research subjects. *Human Research Report* 2007 February; 22(2): 1-2. Subject: 18.5.2

Maloney, Dennis M. Final guidance on adverse events for institutional review boards (IRBs). *Human Research Report* 2007 March; 22(3): 1-2. Subject: 18.2

Maloney, Dennis M. Impermissible research with prisoners and generally improper protocol reviews [case study]. *Human Research Report* 2007 April; 22(4): 6-7. Subject: 18.5.5

Maloney, Dennis M. In court: court says nontherapeutic research requires special informed consent measures. *Hu-*

man Research Report 2007 November; 22(11): 8. Subject: 18.3

Maloney, Dennis M. In court: former research subjects charge researchers and their institutions with negligence. *Human Research Report* 2007 August; 22(8): 8. Subject: 18.5.2

Maloney, Dennis M. In court: informed consent form lacked crucial information, says court. *Human Research Report* 2007 June; 22(6): 8. Subject: 18.5.2

Maloney, Dennis M. In court: legal principles for protecting human subjects. *Human Research Report* 2007 October; 22(10): 8. Subject: 18.2

Maloney, Dennis M. In court: researchers and a history of unethical behavior. *Human Research Report* 2007 December; 22(12): 8. Subject: 18.5.2

Maloney, Dennis M. In court: researchers didn't tell subject's mother of high lead levels in house until after blood tests. *Human Research Report* 2007 July; 22(7): 8. Subject: 18.3

Maloney, Dennis M. Informed consent issue dwarfed by ethics of research study itself. *Human Research Report* 2007 February; 22(2): 8. Subject: 18.5.2

Maloney, Dennis M. Institutional review board (IRB) accused of relying too much on subcommittee [case study]. *Human Research Report* 2007 February; 22(2): 6-7. Subject: 18.2

Maloney, Dennis M. Institutional review boards (IRBs), privacy rule, and subject recruitment. *Human Research Report* 2007 July; 22(7): 1-2. Subject: 18.2

Maloney, Dennis M. IRB members must receive continuing education on protection of human subjects. *Human Research Report* 2007 May; 22(5): 6-7. Subject: 18.2

Maloney, Dennis M. IRBs have some leeway on methods of informed consent. *Human Research Report* 2007 August; 22(8): 3. Subject: 18.3

Maloney, Dennis M. Research with adult subjects who have impaired decision-making capacity. *Human Research Report* 2007 October; 22(10): 1-2. Subject: 18.5.6

Maloney, Dennis M. Study's children lived in housing with various levels of possible lead exposure. *Human Research Report* 2007 May; 22(5): 8. Subject: 18.5.2

Maloney, Dennis M. University explains how it will strengthen its support for institutional review boards (IRBs) [case study]. *Human Research Report* 2007 October; 22(10): 6-7. Subject: 18.2

Maloney, Dennis M. University says its institutional review board (IRB) policies and procedures were just misunderstood. *Human Research Report* 2007 January; 22(1): 6-7. Subject: 18.2

Maloy, Richard H.W. Will new appointees to the Supreme Court be able to effect an overruling of Roe v.

Wade? *Western New England Law Review* 2005; 28(1): 29-55. Subject: 12.4.1

Malpas, Phillipa. Predictive genetic testing in children and respect for autonomy. *In:* Freeman, Michael, ed. Children's Health and Children's Rights. Boston: Martinus Nijhoff Publishers, 2006: 297-309. Subject: 15.3

Mameli, M. Reproductive cloning, genetic engineering and the autonomy of the child: the moral agent and the open future. *Journal of Medical Ethics* 2007 February; 33(2): 87-93. Subject: 14.5

Mamo, Laura. Negotiating conception: lesbians' hybrid-technological practices. *Science, Technology, and Human Values* 2007 May; 32(3): 369-393. Subject: 14.1

Manafa, Ogenna; Lindegger, Graham; Ijsselmuiden, Carel. Informed consent in an antiretroviral trial in Nigeria. *Indian Journal of Medical Ethics* 2007 January-March; 4(1): 26-30. Subject: 18.5.9

Manaouil, C.; Gignon, M.; Decourcelle, M.; Jardé, O. Law, ethics and medicine: a new legal frame for end of life in France. *Journal of Medical Ethics* 2007 May; 33(5): 278. Subject: 20.5.1

Mandell, M. Susan;; Zamudio, Stacy; Seem, Debbie; McGaw, Lin J.; Wood, Geri; Liehr, Patricia.; Ethier, Angela; D'Alessandro, Anthony M. National evaluation of healthcare provider attitudes toward organ donation after cardiac death. *Critical Care Medicine* 2006 December; 34(12): 2952-2958. Subject: 19.5

Manderson, Lenore; Kelaher, Margaret; Williams, Gail; Shannon, Cindy. The politics of community: negotiation and consultation in research on women's health. *Human Organization* 1998 Summer; 57(2): 222-229. Subject: 18.5.3

Mangalore, Roshni; Knapp, Martin. Equity in mental health. *Epidemiologia e Psichiatria Sociale* 2006 October-December; 15(4): 260-266. Subject: 9.2

Mangan, Katherine. Medical schools stop using dogs and pigs in teaching: training of future doctors now largely depends on new technologies rather than lab animals. *Chronicle of Higher Education* 2007 October 12; 54(7): A12. Subject: 7.2

Mangione, Lorraine; Forti, Rosalind; Iacuzzi, Catherine M. Ethics and endings in group psychotherapy: saying good-bye and saying it well. *International Journal of Group Psychotherapy* 2007 January; 57(1): 25-40; discussion 49-59, 61-66. Subject: 17.2

Mangon, R. The medical (ir)relevance of race and ethnicity in a multiethnic society. *Community Genetics* 2007; 10(3): 199. Subject: 15.11

Mann, Howard. Deception in the single-blind run-in phase of clinical trials. *IRB: Ethics and Human Research* 2007 March-April; 29(2): 14-17. Subject: 18.2

Mann, Howard. Evaluation of research design by research ethics committees: misleading reassurance and the need for substantive reforms. *American Journal of Bioethics* 2007 February; 7(2): 84-86. Subject: 18.2

Mann, Karen V. Learning and teaching in professional character development. *In:* Kenny, Nuala; Shelton, Wayne, eds. Lost Virtue: Professional Character Development in Medical Education. Amsterdam; Oxford: Elsevier, 2006: 145-183. Subject: 4.1.2

Mannaerts, Debbie; Mortier, Freddy. Minors and euthanasia. *In:* Freeman, Michael, ed. Children's Health and Children's Rights. Boston: Martinus Nijhoff Publishers, 2006: 255-277. Subject: 20.5.2

Manning, Christopher L. Institutional racism: article too strong? I think not. *BMJ: British Medical Journal* 2007 April 14; 334(7597): 761. Subject: 17.1

Mano, Max S.; Rosa, Daniela D.; Dal Lago, Lissandra. Multinational clinical trials in oncology and post-trial benefits for host countries: where do we stand? *European Journal of Cancer* 2006 November; 42(16): 2675-2677. Subject: 18.2

Mansi, James A.; Franco, Eduardo L.; de Pokomandy, Alexandra; Spence, Andrea R.; Burchell, Ann N.; T rottier, Helen; Mayrand, Marie-Hélène; Lau, Susie; Ferenczy, Alex; Brophy, James M.; Cassels, Alan K.; Nisker, Jeff; Lippman, Abby; Boscoe, Madeline; Shimmin, Carolyn. Vaccination against human papillomavirus. *CMAJ/JAMC: Canadian Medical Association Journal* 2007 December 4; 177(12): 1524-1528. Subject: 9.5.5

Manson, H. The role of the 'lifestyle' label and negative bias in the allocation of health resources for erectile dysfunction drugs: an ethics-based appraisal. *International Journal of Impotence Research* 2006 January-February; 18(1): 98-103.[see corrections in: International Journal of Impotence Research 2006 March-April; 18(2): 221]. Subject: 10

Mantese, Theresamarie; Pfeiffer, Christine; McClinton, Jacquelyn. Cosmetic surgery and informed consent: legal and ethical considerations. *Michigan Bar Journal* 2006 January; 85(1): 26-29 [Online]. Accessed: http://www.michbar.org/journal/pdf/pdf4article957.pdf [2007 May 11]. Subject: 8.3.1

Marco, Catherine A.; Moskop, John C.; Solomon, Robert C.; Geiderman, Joel M.; Larkin, Gregory L. Gifts to physicians from the pharmaceutical industry: an ethical analysis. *Annals of Emergency Medicine* 2006 November; 48(5): 513-521. Subject: 9.7

Marco, Catherine A.; Schears, Raquel M. Death, dying, and last wishes. *Emergency Medicine Clinics of North America* 2006 November; 24(4): 969-987. Subject: 20.4.1

Marcus, Robert; Firth, John. Should you tell patients about beneficial treatments that they cannot have? [debate]. *BMJ: British Medial Journal* 2007 April 21; 334(7598): 826-827. Subject: 9.4

See SUBJECT HEADING KEY FOR SECTION II on inside back cover.

551

Margetts, Barrie. Stopping the rot in nutrition science. *Public Health Nutrition* 2006 April; 9(2): 169-173. Subject: 1.3.9

Marik, Paul E. Should age limit admission to the intensive care unit? *American Journal of Hospice and Palliative Care* 2007 February-March; 24(1): 63-66. Subject: 9.5.2

Mariner, Wendy K. Medicine and public health: crossing legal boundaries. *Journal of Health Care Law and Policy* 2007; 10(1): 121-151. Subject: 9.1

Markel, Howard; Gostin, Lawrence O.; Fidler, David P. Extensively drug-resistant tuberculosis: an isolation order, public health powers, and a global crisis [commentary]. *JAMA: The Journal of the American Medical Association* 2007 July 4; 298(1): 83-86. Subject: 9.1

Marker, Rita L. Suicide by any other name. *Human Life Review* 2007 Winter; 33(1): 78-94. Subject: 20.7

Markowitz, Hal; Timmel, Gregory B. Animal well-being and research outcomes. *In:* McMillan, Frank, ed. Mental Health and Well-Being in Animals. Ames, IA: Blackwell Pub., 2005: 277-283. Subject: 22.2

Marks, Jonathan H. The bioethics of war [review of Oath Betrayed: Torture, Medical Complicity and the War on Terror, by Steven H. Miles; Bioethics and Armed Conflicts: Moral Dilemmas of Medicine and War, by Michael L. Gross]. *Hastings Center Report* 2007 March-April; 37(2): 41-42. Subject: 21.2

Marks, Stephen P. Tying Prometheus down: human rights issues of human genetic manipulation. *In:* Gruskin, Sofia; Grodin, Michael A.; Annas, George J.; Marks, Stephen P., eds. Perspectives on Health and Human Rights. New York: Routledge, 2005: 163-178. Subject: 15.1

Marquis, Don. Abortion revisited. *In:* Steinbock, Bonnie, ed. The Oxford Handbook of Bioethics. Oxford; New York: Oxford University Press, 2007: 395-415. Subject: 12.3

Marshall, Eliot. Sequencers of a famous genome confront privacy issues [news]. *Science* 2007 March 30; 315(5820): 1780. Subject: 15.10

Marshall, Ian E. Physicians and the pharmaceutical industry: a symbiotic relationship? *In:* Cohen, Jillian Clare; Illingworth, Patricia; Schüklenk, Udo, eds. The Power of Pills: Social, Ethical and Legal Issues in Drug Development, Marketing, and Pricing. London; Ann Arbor, MI: Pluto, 2006: 57-64. Subject: 9.7

Marshall, Jennifer; Martin, Toby; Downie, Jocelyn; Malisza, Krisztina. A comprehensive analysis of MRI research risks: in support of full disclosure. *Canadian Journal of Neurological Sciences* 2007 February; 34(1): 11-17. Subject: 18.3

Marshall, Jessica. Operating in whose interest? *New Scientist* 2007 January 13-19; 193(2586): 6-7. Subject: 9.5.3

Marshall, Mary Faith. ASBH and moral tolerance. *In:* Eckenwiler, Lisa A.; Cohn, Felicia G., eds. The Ethics of

Bioethics: Mapping the Moral Landscape. Baltimore, MD: Johns Hopkins University Press, 2007: 134-144. Subject: 2.1

Martin, Adrienne M. Tales publicly allowed: competence, capacity, and religious belief. *Hastings Center Report* 2007 January-February; 37(1): 33-40. Subject: 8.3.3

Martin, Emily. Pharmaceutical virtue. *Culture, Medicine and Psychiatry* 2006 June; 30(2): 157-174. Subject: 9.7

Martin, Rebecca A.; Robert, Jason Scott. Is risky pediatric research without prospect of direct benefit ever justified? *American Journal of Bioethics* 2007 March; 7(3): 12-15. Subject: 18.5.2

Martínez, Jaime Vidal. Biomedical research with human embryos: changes in the legislation on assisted reproduction in Spain. *Revista de Derecho y Genoma Humano = Law and the Human Genome Review* 2006 July-December; (25): 161-182. Subject: 18.5.4

Martínez-Alarcón, L.; Ríos, A.; Conesa, C.; Alcaraz, J.; González, M.J.; Ramírez, P.; Parrilla, P. Attitude of kidney patients on the transplant waiting list toward related-living donation. A reason for the scarce development of living donation in Spain. *Clinical Transplantation* 2006 November-December; 20(6): 719-724. Subject: 19.3

Marusic, Ana; Bates, Tamara; Anic, Ante; Marusic, Matko. How the structure of contribution disclosure statements affects validity of authorship: a randomized study in a general medical journal. *Current Medical Research and Opinion* 2006 June; 22(6): 1035-1044. Subject: 1.3.7

Marusic, Ana; Katavic, Vedran; Marusic, Matko. Role of editors and journals in detecting and preventing scientific misconduct: strengths, weakness, opportunities, and threats. *Medicine and Law: The World Association for Medical Law* 2007 September; 26(3): 545-566. Subject: 1.3.9

Maschke, Karen J. The federalist turn in bioethics? *Hastings Center Report* 2007 November-December; 37(6): 3. Subject: 2.1

Maschke, Karen J. The pressure to tolerate risk in human subjects research [review of Lesser Harms: The Morality of Risk in Medical Research by Sydney A. Halpern]. *Medical Humanities Review* 2005 Spring-Fall; 19(1-2): 39-44. Subject: 18.1

Masella, Richard S. Renewing professionalism in dental education: overcoming the market environment. *Journal of Dental Education* 2007 February; 71(2): 205-216. Subject: 4.1.1

Mason, J.K.; Laurie, G.T. Biomedical human research and experimentation. *In their:* Mason and McCall Smith's Law and Medical Ethics. Seventh ed. Oxford; New York: Oxford University Press, 2005: 648-684. Subject: 18.1

Mason, J.K.; Laurie, G.T. Consent to treatment. *In their:* Mason and McCall Smith's Law and Medical Ethics. Seventh ed. Oxford; New York: Oxford University Press, 2005: 348-411. Subject: 8.3.1

Mason, J.K.; Laurie, G.T. Euthanasia. *In their:* Mason and McCall Smith's Law and Medical Ethics. Seventh ed. Oxford; New York: Oxford University Press, 2005: 598-647. Subject: 20.5.1

Mason, J.K.; Laurie, G.T. Genetic information and the law. *In their:* Mason and McCall Smith's Law and Medical Ethics. Seventh ed. Oxford; New York: Oxford University Press, 2005: 206-252. Subject: 15.1

Mason, J.K.; Laurie, G.T. Health resources and dilemmas in treatment. *In their:* Mason and McCall Smith's Law and Medical Ethics. Seventh ed. Oxford; New York: Oxford University Press, 2005: 412-440. Subject: 9.4

Mason, J.K.; Laurie, G.T. Health rights and obligations in the European Union. *In their:* Mason and McCall Smith's Law and Medical Ethics. Seventh ed. Oxford; New York: Oxford University Press, 2005: 48-70. Subject: 9.2

Mason, J.K.; Laurie, G.T. Medical confidentiality. *In their:* Mason and McCall Smith's Law and Medical Ethics. Seventh ed. Oxford; New York: Oxford University Press, 2005: 253-294. Subject: 8.4

Mason, J.K.; Laurie, G.T. Medical futility. *In their:* Mason and McCall Smith's Law and Medical Ethics. Seventh ed. Oxford; New York: Oxford University Press, 2005: 539-597. Subject: 20.5.1

Mason, J.K.; Laurie, G.T. Mental health and human rights. *In their:* Mason and McCall Smith's Law and Medical Ethics. Seventh ed. Oxford; New York: Oxford University Press, 2005: 710-739. Subject: 17.1

Mason, J.K.; Laurie, G.T. Public health and the state/patient relationships. *In their:* Mason and McCall Smith's Law and Medical Ethics. Seventh ed. Oxford; New York: Oxford University Press, 2005: 29-47. Subject: 9.1

Mason, J.K.; Laurie, G.T. Research on children, fetuses and embryos. *In their:* Mason and McCall Smith's Law and Medical Ethics. Seventh ed. Oxford; New York: Oxford University Press, 2005: 685-709. Subject: 18.5.1

Mason, J.K.; Laurie, G.T. The body as property. *In their:* Mason and McCall Smith's Law and Medical Ethics. Seventh ed. Oxford; New York: Oxford University Press, 2005: 511-538. Subject: 4.4

Mason, J.K.; Laurie, G.T. The control of fertility. *In their:* Mason and McCall Smith's Law and Medical Ethics. Seventh ed. Oxford; New York: Oxford University Press, 2005: 120-167. Subject: 11.1

Mason, J.K.; Laurie, G.T. The diagnosis of death. *In their:* Mason and McCall Smith's Law and Medical Ethics. Seventh ed. Oxford; New York: Oxford University Press, 2005: 464-476. Subject: 20.2.1

Mason, J.K.; Laurie, G.T. The donation of organs and transplantation. *In their:* Mason and McCall Smith's Law

and Medical Ethics. Seventh ed. Oxford; New York: Oxford University Press, 2005: 477-510. Subject: 19.5

Mason, J.K.; Laurie, G.T. The management of infertility and childlessness. *In their:* Mason and McCall Smith's Law and Medical Ethics. Seventh ed. Oxford; New York: Oxford University Press, 2005: 71-119. Subject: 14.1

Mason, J.K.; Laurie, G.T. Treatment of the aged. *In their:* Mason and McCall Smith's Law and Medical Ethics. Seventh ed. Oxford; New York: Oxford University Press, 2005: 441-463. Subject: 9.5.2

Masood, Junaid; Hafeez, Azhar; Wiseman, Oliver; Hill, James T. Informed consent: are we deluding ourselves? A randomized controlled study. *BJU International* 2007 January; 99(1): 4-5. Subject: 8.3.1

Masson, Judith. Parenting by being; parenting by doing — in search of principles for founding families. *In:* Spencer, J.R.; du Bois-Pedain, Antje, eds. Freedom and Responsibility in Reproductive Choice. Portland: Hart Pub., 2006: 131-155. Subject: 14.1

Master, Zubin; McLeod, Marcus; Mendez, Ivar. Benefits, risks and ethical considerations in translation of stem cell research to clinical applications in Parkinson's disease. *Journal of Medical Ethics* 2007 March; 33(3): 169-173. Subject: 18.5.1

Master, Zubin; Williams-Jones, Bryn; Lott, Jason P.; Savulescu, Julian. The global HLA banking of embryonic stem cells requires further scientific justification. *American Journal of Bioethics* 2007 August; 7(8): 45-46; author reply W4-W6. Subject: 18.5.4

Masterton, Malin; Helgesson, Gert; Höglund, Anna T.; Hansson, Mats G. Queen Christina's moral claim on the living: justification of a tenacious moral intuition. *Medicine, Health Care and Philosophy* 2007 September; 10(3): 321-327. Subject: 20.1

Mastroianni, Anna; Kahn, Jeffrey. Swinging on the pendulum: shifting views of justice in human subjects research. *In:* Lemmens, Trudo; Waring, Duff R., eds. Law and Ethics in Biomedical Research: Regulation, Conflict of Interest and Liability. Toronto; Buffalo: University of Toronto Press, 2006: 47-60. Subject: 18.6

Mastroianni, George R. Kurt Gottschaldt's ambiguous relationship with National Socialism. *History of Psychology* 2006 February; 9(1): 38-54. Subject: 15.11

Matesanz, Mateu B. Advance statements: legal and ethical implications. *Nursing Standard* 2006 September 20-26; 21(2): 41-45. Subject: 20.5.4

Matsui, Kenji; Lie, Reidar K.; Kita, Yoshikuni. Two methods of obtaining informed consent in a genetic epidemiological study: effects on understanding. *Journal of Empirical Research on Human Research Ethics* 2007 September; 2(3): 39-48. Subject: 15.1

Matthews, Daryl; Wendler, Sheila. Ethical issues in the evaluation and treatment of death row inmates. *Current*

See SUBJECT HEADING KEY FOR SECTION II on inside back cover.

553

Opinion in Psychiatry 2006 September; 19(5): 518-521. Subject: 20.6

Matthews, Eric. Is autonomy relevant to psychiatric ethics? *In:* Nys, Thomas; Denier, Yvonne; Vandevelde, Toon, eds. Autonomy and Paternalism: Reflections on the Theory and Practice of Health Care. Leuven; Dudley, MA: Peeters, 2007: 129-146. Subject: 17.1

Matthews, Robert. Are you looking at me? Medical researchers keen to scour patients' data for insights into disease should get consent first or risk coming seriously unstuck. *New Scientist* 2007 August 4-10; 195(2615): 18. Subject: 18.3

Mauceri, Joseph M. Evolution and the embryo: the evidence for special creation. *Linacre Quarterly* 2007 February; 74(1): 30-49. Subject: 18.5.4

Mauler, Valerie. Improving public health: balancing ethics, culture, and technology. *Georgetown Journal of Legal Ethics* 2007 Summer; 20(3): 817-833. Subject: 2.1

Mavroforou, A.; Koumantakis, E.; Mavrophoros, D.; Michalodimitrakis, E. Medically assisted human reproduction: the Greek view. *Medicine and Law: The World Association for Medical Law* 2007 June; 26(2): 339-347. Subject: 14.2

May, Thomas; Craig, J.M.; Spellecy, Ryan. IRBs, hospital ethics committees, and the need for "translational informed consent". *Academic Medicine* 2007 July; 82(7): 670-674. Subject: 18.3

May, Thomas; Silverman, Ross D. Free-riding, fairness and the rights of minority groups in exemption from mandatory childhood vaccination. *Human Vaccines* 2005 January-February; 1(1): 12-15. Subject: 9.1

Mayer, Jane. The experiment. *The New Yorker* 2005 July 11 and 18; 61-71. Subject: 21.4

Mayer, Musa. Listen to all the voices: an advocate's perspective on early access to investigational therapies. *Clinical Trials* 2006; 3(2): 149-153. Subject: 18.2

Mayer, Susan. HFEA allows women to donate their eggs for research [news]. *BMJ: British Medical Journal* 2007 March 3; 334(7591): 445. Subject: 19.5

Mayers, Douglas L.; Chung, Jain; Kohlbrenner, Veronika M.; Hall, David B.; DeMasi, Ralph A.; Neubacher, Dietmar; Buss, Neil E.; Salgo, Miklos P. Seeking ethical designs for HIV clinical trials in treatment-experienced patients: an industry perspective. *AIDS Research and Human Retroviruses* 2006 November; 22(11): 1110-1112. Subject: 18.2

Mayman, Daniel M.; Guyer, Melvin. Defendants' constitutional rights against forced medication in Sell hearings. *Journal of the American Academy of Psychiatry and the Law* 2007; 35(1): 118-120. Subject: 17.8

Mayor, Susan. Fitting the drug to the patient. *BMJ: British Medical Journal* 2007 March 3; 334(7591): 452-453. Subject: 15.1

Mayor, Susan. Genome sequence of one person is published for first time [news]. *BMJ: British Medical Journal* 2007 September 15; 335(7619): 530-531. Subject: 15.10

Mayor, Susan. UK body wants consultation on human-animal hybrid research [news]. *BMJ: British Medical Journal* 2007 January 20; 334(7585): 112. Subject: 15.1

Mayor, Susan. UK study will reimburse part of cost of IVF to women who donate eggs for research [news]. *BMJ: British Medical Journal* 2007 September 22; 335(7620): 581. Subject: 14.4

Mazaris, Evangelos; Papalois, Vassilios E. Ethical issues in living donor kidney transplantation. *Experimental and Clinical Transplantation* 2006 December; 4(2): 485-497. Subject: 19.3

Maze, Claire D. Martino. Registered nurses' willingness to serve populations on the periphery of society. *Journal of Nursing Scholarship* 2006; 38(3): 301-306. Subject: 9.5.4

Mc Fleming, Jennifer. The governance of human genetic research databases in mental health research. *International Journal of Law and Psychiatry* 2007 May-June; 30(3): 182-190. Subject: 15.1

McAllister, Marc Chase. Human dignity and individual liberty in Germany and the United States as examined through each country's leading abortion cases. *Tulsa Journal of Comparative and International Law* 2004 Spring; 11(2): 491-520. Subject: 12.4.2

McAneeley, Lindsay N. Physician assisted suicide: expanding the laboratory to the state of Hawai'i. *University of Hawai'i Law Review* 2006 Winter; 29(1): 269-299. Subject: 20.7

McCabe, Helen. Nursing involvement in euthanasia: how sound is the philosophical support? *Nursing Philosophy* 2007 July; 8(3): 167-175. Subject: 20.5.1

McCabe, Helen. Nursing involvement in euthanasia:a 'nursing-as-healing-praxis' approach. *Nursing Philosophy* 2007 July; 8(3): 176-186. Subject: 20.5.1

McCallum, Jan M.; Arekere, Dhananjaya M.; Green, B. Lee; Katz, Ralph V.; Rivers, Brian M. Awareness and knowledge of the U.S. Public Health Service syphilis study at Tuskegee: implications for biomedical research. *Journal of Health Care the Poor and Underserved* 2006 November; 17(4): 716-733. Subject: 18.5.1

McCarron, Mary; McCallion, Philip. End-of-life care challenges for persons with intellectual disability and dementia: making decisions about tube feeding. *Intellectual and Developmental Disabilities* 2007 April; 45(2): 128-131. Subject: 20.5.1

McCarthy, Michael. US campaign tackles drug company influence over doctors. *Lancet* 2007 March 3-9; 369(9563): 730. Subject: 9.7

McCarthy, Patrick M.; Lamm, Richard D.; Sade, Robert M. Medical ethics collides with public policy: LVAD

---

for a patient with leukemia. *Annals of Thoracic Surgery* 2005 September; 80(3): 793-798. Subject: 9.5.1

McCartney, James J. Embryonic stem cell research and respect for human life: philosophical and legal reflections. *Albany Law Review* 2002; 65(3): 597-624. Subject: 18.5.4

McCarty, Catherine A.; Nair, Anuradha; Austin, Diane M.; Giampietro, Philip F. Informed consent and subject motivation to participate in a large, population-based genomics study: the Marshfield Clinic Personalized Medicine Research Project. *Community Genetics* 2007; 10(1): 2-9. Subject: 18.3

McClure, Katie B.; Delorio, Nicole M.; Schmidt, Terri A.; Chiodo, Gary; Gorman, Paul. A qualitative study of institutional review board members' experience reviewing research proposals using emergency exception from informed consent. *Journal of Medical Ethics* 2007 May; 33(5): 289-293. Subject: 18.2

McClusky, Joan. Tell public about brain death [letter]. *BMJ: British Medical Journal* 2007 June 9; 334(7605): 1179. Subject: 20.2

McConnell, Yarrow; Frager, Gerri; Levetown, Marcia. Decision making in pediatric palliative care. *In:* Carter, Brian S.; Levetown, Marcia, eds. Palliative Care for Infants, Children, and Adolescents: A Practical Handbook. Baltimore: Johns Hopkins University Press, 2004: 69-111. Subject: 20.4.2

McConville, Brad; Kelly, D. Clay. Cruel and unusual? Defining the conditions of confinement in the mentally ill. *Journal of the American Academy of Psychiatry and the Law* 2007; 35(4): 533-534. Subject: 9.5.3

McCubbin, Caroline N. Legal and ethico-legal issues in e-healthcare research projects in the UK. *Social Science and Medicine* 2006 June; 62(11): 2768-2773. Subject: 1.3.12

McCullough, Laurence B. Geroethics. *In:* Khushf, George, ed. Handbook of Bioethics: Taking Stock of the Field From a Philosophical Perspective. Dordrecht; Boston: Kluwer Academic, 2004: 507-523. Subject: 9.5.2

McCullough, Laurence B. John Gregory's medical ethics and the reform of medical practice in eighteenth-century Edinburgh. *Journal of the Royal College of Physicians of Edinburgh* 2006 March; 36(1): 86-92. Subject: 2.2

McCullough, Laurence B. The ethical concept of medicine as a profession: its origins in modern medical ethics and implications for physicians. *In:* Kenny, Nuala; Shelton, Wayne, eds. Lost Virtue: Professional Character Development in Medical Education. Amsterdam; Oxford: Elsevier, 2006: 17-27. Subject: 4.1.2

McCullough, Laurence B. Towards a professional ethics model of clinical ethics. *Journal of Medicine and Philosophy* 2007 January-February; 32(1): 1-6. Subject: 2.1

McCullough, Laurence B.; Coverdale, John H.; Chervenak, Frank A. Constructing a systematic review for argument-based clinical ethics literature: the example of concealed medications. *Journal of Medicine and Philosophy* 2007 January-February; 32(1): 65-76. Subject: 17.4

McCullough, Laurence B.; Coverdale, John H.; Chervenak, Frank A. Preventive ethics for including women of childbearing potential in clinical trials. *American Journal of Obstetrics and Gynecology* 2006 May; 194(5): 1221-1227. Subject: 18.5.3

McCullough, Laurence B.; McGuire, Amy L.; Whitney, Simon N.; Easton, Raul B.; Graber, Mark A.; Monnahan, Jay; Hughes, Jason. Consent: informed, simple, implied and presumed. *American Journal of Bioethics* 2007 December; 7(12): 49-50; author reply W3-W4. Subject: 8.3.1

McDaniel, Charlotte. Melding or meddling: compliance and ethics programs. *HEC (Healthcare Ethics Committee) Forum* 2007 June; 19(2): 97-107. Subject: 9.6

McDonald, Katherine; Hernandez, Brigida; Plemmons, Dena; Simmerling, Mary. Privacy in organizational research. *Journal of Empirical Research on Human Research Ethics* 2007 March; 2(1): 69-73. Subject: 18.2

McDougall, Rosalind. Parental virtue: a new way of thinking about the morality of reproductive actions. *Bioethics* 2007, May; 21(4): 181-190. Subject: 14.1

McGee, Ellen M.; Maguire, Gerald Q., Jr. Becoming borg to become immortal: regulating brain implant technologies. *CQ: Cambridge Quarterly of Healthcare Ethics* 2007 Summer; 16(3): 291-302. Subject: 5.1

McGee, Glenn; Bjarnadóttir, Dyrleif. Abuses of science in medical ethics. *In:* Rhodes, Rosamond; Francis, Leslie P.; Silvers, Anita, eds. The Blackwell Guide to Medical Ethics. Malden, MA: Blackwell Pub., 2007: 289-302. Subject: 2.1

McGee, Glenn; Johnson, Summer. Has the spread of HPV vaccine marketing conveyed immunity to common sense? [editorial]. *American Journal of Bioethics* 2007 July; 7(7): 1-2. Subject: 9.5.1

McGrath, Pam D.; Forrester, Kim. Ethico-legal issues in relation to end-of-life care and institutional mental health. *Australian Health Review* 2006 August; 30(3): 286-297. Subject: 20.5.1

McGrath, Pam; Henderson, David; Holewa, Hamish. Patient-centred care: qualitative findings on health professionals' understanding of ethics in acute medicine. *Journal of Bioethical Inquiry* 2006; 3(3): 149-160. Subject: 4.1.1

McGraw, Melanie P.; Perlman, Jeffrey; Chervenak, Frank A.; McCullough, Laurence B. Clinical concepts of futility and ethically justified limits on neonatal care: a case presentation of an infant with Apgar scores of 0 at 1, 5, and 10 minutes. *American Journal of Perinatology* 2006 April; 23(3): 159-162. Subject: 20.5.2

See SUBJECT HEADING KEY FOR SECTION II on inside back cover.

555

McGregor, J.; Dreifuss-Netter, F. France and the United States: the legal and ethical differences in assisted reproductive technology (ART). *Medicine and Law: The World Association for Medical Law* 2007 March; 26(1): 117-135. Subject: 14.2

McGuire, Amy L.; Cho, Mildred K.; McGuire, Sean E.; Caulfield, Timothy. The future of personal genomics. *Science* 2007 September 21; 317(5845): 1687. Subject: 15.1

McHale, Jean. Law reform, clinical research and adults without mental capacity — much needed clarification or a recipe for further uncertainty? *In:* McLean, Sheila A.M., ed. First Do No Harm: Law, Ethics, and Healthcare. Aldershot, England; Burlington, VT: Ashgate, 2006: 215-233. Subject: 18.5.6

McIntyre, Di; Whitehead, Margaret; Gilson, Lucy; Dahlgren, Göran; Tang, Shenglan. Equity impacts of neoliberal reforms: what should the policy responses be? *International Journal of Health Services* 2007; 37(4): 693-709. Subject: 9.1

McKay, Angela. Publicly accessible intuitions: "neutral reasons" and bioethics. *Christian Bioethics* 2007 May-August; (13)2: 183-197. Subject: 2.1

McKechnie, L.; Gill, A.B. Consent for neonatal research. *Archives of Disease in Childhood. Fetal and Neonatal Edition* 2006 September; 91(5): F374-F376. Subject: 18.5.4

McKee, M. Diane; O'Sullivan, Lucia F.; Weber, Catherine M. Perspectives on confidential care for adolescent girls. *Annals of Family Medicine* 2006 November-December; 4(6): 519-526. Subject: 9.5.7

McKenzie, Kwame; Bhui, Kamaldeep. Institutional racism in mental health care. *BMJ: British Medical Journal* 2007 March 31; 334(7595): 649-650. Subject: 17.1

McKerlie, Dennis. Justice and the elderly. *In:* Steinbock, Bonnie, ed. The Oxford Handbook of Bioethics. Oxford; New York: Oxford University Press, 2007: 190-208. Subject: 9.5.2

McKneally, Martin. Controversies in cardiothoracic surgery: should therapeutic cloning be supported to provide stem cells for cardiothoracic surgery research and treatment? [debate]. *Journal of Thoracic and Cardiovascular Surgery* 2006 May; 131(5): 937-940. Subject: 18.5.4

McKneally, Martin. Put my name on that paper: reflections on the ethics of authorship [editorial]. *Journal of Thoracic and Cardiovascular Surgery* 2006 March; 131(3): 517-519. Subject: 1.3.7

McKneally, Martin F. Beyond disclosure: managing conflicts of interest to strengthen trust in our profession. *Journal of Thoracic and Cardiovascular Surgery* 2007 February; 133(2): 300-302. Subject: 7.3

McKneally, Martin F. Managing expectations and fear: invited commentary on "Indecency in cardiac surgery: a memoir of my education at a Super-Esteemed Medical Place (SEMP)," by Dr. Edmund Erde. *Journal of Cardiac Surgery* 2007 January-February; 22(1): 49-50. Subject: 9.5.1

McLachlan, Hugh. Let's legalise cloning. *New Scientist* 2007 July 21-27; 195(2613): 20. Subject: 14.5

McLachlan, Hugh V.; Swales, J. Kim. Abortion. *In their:* From the Womb to the Tomb: Issues in Medical Ethics. Glasgow, Scotland: Humming Earth, 2007: 1-20. Subject: 12.1

McLachlan, Hugh V.; Swales, J. Kim. Altruism and blood donation. *In their:* From the Womb to the Tomb: Issues in Medical Ethics. Glasgow, Scotland: Humming Earth, 2007: 171-186. Subject: 19.4

McLachlan, Hugh V.; Swales, J. Kim. Embryology and human cloning. *In their:* From the Womb to the Tomb: Issues in Medical Ethics. Glasgow, Scotland: Humming Earth, 2007: 21-83. Subject: 4.4

McLachlan, Hugh V.; Swales, J. Kim. Health and health care — justice, rights and equality. *In their:* From the Womb to the Tomb: Issues in Medical Ethics. Glasgow, Scotland: Humming Earth, 2007: 187-249. Subject: 9.4

McLachlan, Hugh V.; Swales, J. Kim. Posthumous insemination and consent. *In their:* From the Womb to the Tomb: Issues in Medical Ethics. Glasgow, Scotland: Humming Earth, 2007: 251-267. Subject: 14.2

McLachlan, Hugh V.; Swales, J. Kim. Surrogate motherhood. *In their:* From the Womb to the Tomb: Issues in Medical Ethics. Glasgow, Scotland: Humming Earth, 2007: 85-170. Subject: 14.2

McLean, Margaret R. Religion, ethics, and the Human Genome Project. *In:* Eisen, Arri; Laderman, Gary, eds. Science, Religion, and Society: An Encyclopedia of History, Culture, and Controversy. Vol. 2. Armonk, NY: M.E. Sharpe, 2007: 787-794. Subject: 15.10

McLean, Michelle; Naidoo, Soornarain S. Medical students' views on the white coat: a South African perspective on ethical issues. *Ethics and Behavior* 2007 December; 17(4): 387-402. Subject: 7.1

McLean, Sheila A.M. From Bland to Burke: the law and politics of assisted nutrition and hydration. *In:* McLean, Sheila A.M., ed. First Do No Harm: Law, Ethics, and Healthcare. Aldershot, England; Burlington, VT: Ashgate, 2006: 431-446. Subject: 20.5.1

McLean, Sheila A.M. What and who are clinical ethics committees for? [editorial]. *Journal of Medical Ethics* 2007 September; 33(9): 497-500. Subject: 9.6

McLean, Sheila A.M.; Campbell, Alastair; Gutridge, Kerry; Harper, Helen. Human tissue legislation and medical practice: a benefit or a burden? *Medical Law International* 2007; 8(1): 1-21. Subject: 19.5

McLean, Sheila A.M.; Mason, J. Kenyon. Our inheritance, our future: their rights? *In:* Freeman, Michael, ed. Children's Health and Children's Rights. Boston: Martinus Nijhoff Publishers, 2006: 279-296. Subject: 15.3

McLean, Sheila; Williamson, Laura. The demise of UKXIRA and the regulation of solid-organ xenotransplantation in the UK [editorial]. *Journal of Medical Ethics* 2007 July; 33(7): 373-375. Subject: 19.1

McLeod, Carolyn. For dignity or money: feminists on the commodification of women's reproductive labour. *In:* Steinbock, Bonnie, ed. The Oxford Handbook of Bioethics. Oxford; New York: Oxford University Press, 2007: 258-281. Subject: 14.2

McLeod, Carolyn; Baylis, Françoise. Donating fresh versus frozen embryos to stem cell research: in whose interests. *Bioethics* 2007 November; 21(9): 465-477. Subject: 18.5.4

McManus, John; McClinton, Annette; Gerhardt, Robert; Morris, Michael. Performance of ethical military research is possible: on and off the battlefield. *Science and Engineering Ethics* 2007 September; 13(3): 297-303. Subject: 18.5.1

McMichael, Anthony J.; Bambrick, Hilary J. Greenhouse-gas costs of clinical trials. *Lancet* 2007 May 12–18; 369(9573): 1584-1585. Subject: 18.2

McMonagle, Ethan. Functional neuroimaging and the law: a Canadian perspective. *American Journal of Bioethics* 2007 September; 7(9): 69-70. Subject: 17.1

McMurray, David L., Jr. Genomics and ethnicity: using a tool in the U.S. Environmental Protection Agency's environmental justice toolkit. *Journal of Health Care Law and Policy* 2007; 10(1): 187-214. Subject: 15.11

McNamee, Michael John. Nursing Schadenfreude: the culpability of emotional construction. *Medicine, Health Care and Philosophy* 2007 September; 10(3): 289-299. Subject: 4.1.3

McNeil, Donald G., Jr. Drugs banned, world's poor suffer in pain. *New York Times* 2007 September 10; p. A1. A14, A15. Subject: 20.4.1

McNeil, Donald G., Jr. In India, a quest to ease the pain of the dying. *New York Times* 2007 September 11; p. F1, F5. Subject: 20.4.1

McNeil, Donald G., Jr. Japanese slowly shedding their misgivings about the use of painkilling drugs. *New York Times* 2007 September 10; p. A15. Subject: 20.4.1

McNeill, Paul M. Should bioethics play football? *Monash Bioethics Review* 2000 October; 19(4): 46-49. Subject: 2.1

McNeill, Paul M.; Kerridge, Ian H.; Arciuli, Catherine; Henry, David A.; Macdonald, Graham J.; Day, Richard O.; Hill, Suzanne R. Gifts, drug samples and other items given to medical specialists by pharmaceutical companies. *Journal of Bioethical Inquiry* 2006; 3(3): 139-148. Subject: 9.7

McNeill, P.M.; Kerridge, I.H.; Henry, D.A.; Stokes, B.; Hill, S.R.; Newby, D.; Macdonald, G.J; Day, R.O.; Maguire, J.; Henderson, K.M. Giving and receiving of gifts between pharmaceutical companies and medical spe-

cialists in Australia. *Internal Medicine Journal* 2006 September; 36(9): 571-578. Subject: 9.7

McPhate, Gordon. Ensoulment revised in response to genetics, neuroscience and out-of-body experiences. *In:* Deane-Drummond, Celia; Scott, Peter Manley, eds. Future Perfect?: God, Medicine and Human Identity. New York: T and T Clark International, 2006: 100-112. Subject: 15.6

McPhee, John; Stewart, Cameron. Recent developments. *Journal of Bioethical Inquiry* 2006; 3(3): 125-131. Subject: 2.1

McRae, Andrew D.; Weijer, Charles. Lessons from everyday lives: a moral justification for acute care research. *Critical Care Medicine* 2002 May; 30(5): 1146-1151. Subject: 18.2

Meckler, Laura. Kidney shortage inspires a radical idea: organ sales. *Wall Street Journal* 2007 November 13; p. A1 [Online]. Accessed: http://online.wsj.com/public/article_print/SB119490273908090431.html [2007 November 13]. Subject: 19.5

Meckler, Laura. More kidneys for transplants may go to young; policy to stress benefit to patient over length of time on wait list. *Wall Street Journal* 2007 March 10; p. A1, A7. Subject: 19.3

Meckler, Laura. What living organ donors need to know; even as transplants surge, data on long-term impact on givers remain scant. *Wall Street Journal* 2007 January 30; p. B1, B13. Subject: 19.5

Medical Research Council [MRC] (Great Britain). MRC guidance on open access to published research. London: Medical Research Council, 2006 October 1; 2 p. [Online]. Accessed: http://www.mrc.ac.uk/consumption/groups/public/documents/content/mrc002548.pdf [2007 April 4]. Subject: 1.3.7

Meghani, Zahra. Is personhood an illusion? *American Journal of Bioethics* 2007 January; 7(1): 62-63. Subject: 17.1

Mehta, Pritti. Promoting equality and diversity in UK biomedical and clinical research. *Nature Reviews Genetics* 2006 September; 7(9): 668. Subject: 9.5.4

Meier, Barry. Participants left uninformed in some halted medical trials. *New York Times* 2007 October 30; p. C1, C2. Subject: 18.6

Meier, Benjamin Mason; Mori, Larisa M. The highest attainable standard: advancing a collective human right to public health. *Columbia Human Rights Law Review* 2005 Fall; 37(1): 101-147. Subject: 9.1

Meilaender, Gilbert. Fitness fixation: why health is not a civic virtue. *Christian Century* 2007 October 16; 124(21): 8-9. Subject: 4.2

Meilaender, Gilbert. Genes as resources. *Hedgehog Review* 2002 Fall [Online]. Accessed: http://www.virginia.edu/iasc/hedgehog.html [2006 September 25]. Subject: 15.1

See SUBJECT HEADING KEY FOR SECTION II on inside back cover.

557

Meilaender, Gilbert. Human dignity and public bioethics. *New Atlantis* 2007 Summer; (17): 33-52. Subject: 2.1

Meininger, Herman P. Embedded in professional practice: ethics at the 12th World Congress of IASSID. *Ethics and Intellectual Disability* 2005 Winter; 8(2): 1-3. Subject: 9.5.3

Mello, Michelle M.; Joffe, Steven. Compact versus contract: industry sponsors' obligations to their research subjects. *New England Journal of Medicine* 2007 June 28; 356(26): 2737-2743. Subject: 18.3

Memarian, Robabeh; Salsali, Mahvash; Vanaki, Zohreh; Ahmadi, Fazlolah; Hajizadeh, Ebrahim. Professional ethics as an important factor in clinical competency in nursing. *Nursing Ethics* 2007 March; 14(2): 203-214. Subject: 4.1.3

Memis, Tekin. Debate on patentability of biotechnological studies in Turkey. *Revista de Derecho y Genoma Humano = Law and the Human Genome Review* 2007 January-June; (26): 121-135. Subject: 15.8

Memtsoudis, Stavros G.; Besculides, Melanie C.; Swamidoss, Cephas P. Do race, gender, and source of payment impact on anesthetic technique for inguinal hernia repair? *Journal of Clinical Anesthesia* 2006 August; 18(5): 328-333. Subject: 9.5.1

Mendelson, Danuta. Roman concept of mental capacity to make end-of-life decisions. *International Journal of Law and Psychiatry* 2007 May-June; 30(3): 201-212. Subject: 8.3.3

Menikoff, Jerry. Toward a general theory of research ethics. *Hastings Center Report* 2007 May-June; 37(3): 3. Subject: 18.2

Menon, Devidas; Stafinski, Tania; Martin, Douglas. Priority-setting for healthcare: who, how, and is it fair? *Health Policy* 2007 December; 84(2-3): 220-233. Subject: 9.4

Menzel, Paul. Allocation of scarce resources. *In:* Rhodes, Rosamond; Francis, Leslie P.; Silvers, Anita, eds. The Blackwell Guide to Medical Ethics. Malden, MA: Blackwell Pub., 2007: 305-322. Subject: 9.4

Mepham, Ben. Food ethics. *In:* Gunning, Jennifer; Holm, Søren, eds. Ethics, law, and society. Volume 1. Aldershot, Hants, England; Burlington, VT: Ashgate, 2005: 141-151. Subject: 15.1

Merati, T.; Supriyadi; Yuliana, F. The disjunction between policy and practice: HIV discrimination in health care and employment in Indonesia. *AIDS Care* 2005 July; 17(Supplement 2): S175-S179. Subject: 9.5.6

Mercurio, M.R. Parental authority, patient's best interest and refusal of resuscitation at borderline gestational age. *Journal of Perinatology* 2006 August; 26(8): 452-457. Subject: 20.5.2

Merlino, Joseph P. Psychoanalysis and ethics — relevant then, essential now. *Journal of the American Academy of*

Psychoanalysis and Dynamic Psychiatry 2006 Summer; 34(2): 231-247. Subject: 17.2

Merlo, D.F.; Knudsen, L.E.; Matusiewicz, K.; Niebrój, L.; Vähäkangas, K.H. Ethics in studies on children and environmental health. *Journal of Medical Ethics* 2007 July; 33(7): 408-413. Subject: 18.5.2

Merritt, Maria; Grady, Christine. Reciprocity and post-trial access for participants in antiretroviral therapy trials. *AIDS* 2006 September 11; 20(14): 1791-1794. Subject: 18.2

Mertes, Heidi.; Pennings, G. Oocyte donation for stem cell research. *Human Reproduction* 2007 March; 22(3): 629-634. Subject: 19.5

Mertl, Steve. B.C. seized three ailing sextuplets; blood transfusions carried out before parents could challenge move in court. *Toronto Star* 2007 February 1; 2 p. [Online]. Accessed: http://www.thestar.com/printArticle/177069 [2007 February 5]. Subject: 8.3.4

Mertz, Marcel. Complementary and alternative medicine: the challenges of ethical justification. *Medicine, Health Care and Philosophy* 2007 September; 10(3): 329-345. Subject: 4.1.1

Messer, Neil. Medicine, science and virtue. *In:* Deane-Drummond, Celia; Scott, Peter Manley, eds. Future Perfect?: God, Medicine and Human Identity. New York: T and T Clark International, 2006: 113-125. Subject: 18.2

Messikomer, Carla M. "Our options have changed . . . we will not call you back": communicating with my primary care physician. *Perspectives in Biology and Medicine* 2007 Summer; 50(3): 435-443. Subject: 8.1

Mettner, Jeanne. Code YOU: will genetic testing make personalized medicine a reality? *Minnesota Medicine* 2007 May; 90(5): 26-29. Subject: 15.3

Metzl, Jonathan M. If direct-to-consumer advertisements come to Europe: lessons from the USA. *Lancet* 2007 February 24-March 2; 369(9562): 704-706. Subject: 9.7

Metzl, Jonathan M.; Herzig, Rebecca M. Medicalisation in the 21st century: introduction. *Lancet* 2007 February 24-March 2; 369(9562):697-698. Subject: 4.2

Metzner, Jeffrey L.; Tardiff, Kenneth; Lion, John; Reid, William H.; Recupero, Patricia Ryan; Schetky, Diane H.; Edenfield, Bruce M.; Mattson, Marlin; Janofsky, Jeffrey S. Resource document on the use of restraint and seclusion in correctional mental health care. *Journal of the American Academy of Psychiatry and the Law* 2007; 35(4): 417-425. Subject: 9.5.3

Meyer, Erin K.G.; AuBuchon, James P. Conflicting duties: an ethical dilemma in transfusion medicine. *Medical Ethics Newsletter [Lahey Clinic]* 2007 Fall; 14(3): 3,7. Subject: 20.5.1

Meyers, Christopher. Clinical ethics consulting and conflict of interest structurally intertwined. *Hastings Center Report* 2007 March-April; 37(2): 32-40. Subject: 9.6

Meyers, Christopher. Personhood: empirical thing or rational concept? *American Journal of Bioethics* 2007 January; 7(1): 63-65. Subject: 17.1

Michaels, Mark H. Ethical considerations in writing psychological assessment reports. *Journal of Clinical Psychology* 2006 January; 62(1): 47-58. Subject: 17.1

Michalowski, Sabine. Advance refusals of life-sustaining medical treatment: the relativity of an absolute right. *Modern Law Review* 2005 November; 68(6): 958-982. Subject: 20.5.4

Michalsen, Andrej; Reinhart, Konrad. "Euthanasia": a confusing term, abused under the Nazi regime and misused in present end-of-life debate. *Intensive Care Medicine* 2006 September; 32(9): 1304-1310. Subject: 20.5.1

Mielke, Jens. Clinical ethics in the developing world: a case in point: in Zimbabwe. *Formosan Journal of Medical Humanities* 2007 July; 8(1-2): 15-20. Subject: 9.6

Mielke, J.; Ndebele, P. Making research ethics review work in Zimbabwe — the case for investment in local capacity. *Central African Journal of Medicine* 2004 November-December; 50(11-12): 115-119. Subject: 18.2

Milanovic, Fabien; Pontille, David; Cambon-Thomsen, Anne. Biobanking and data sharing: a plurality of exchange regimes. *Genomics, Society and Policy* 2007 April; 3(1): 17-30. Subject: 15.1

Miles, Steven H. Human genomic research ethics: changing the rules. *In:* Cohen, Jillian Clare; Illingworth, Patricia; Schüklenk, Udo, eds. The Power of Pills: Social, Ethical and Legal Issues in Drug Development, Marketing, and Pricing. London; Ann Arbor, MI: Pluto, 2006: 203-214. Subject: 15.1

Miller, Arthur G. What can the Milgram obedience experiments tell us about the Holocaust? Generalizing from the social psychology laboratory. *In:* Miller, Arthur G. ed. The Social Psychology of Good and Evil. New York: The Guilford Press, 2004: 193-239. Subject: 18.4

Miller, Franklin G.; Brody, Howard. Clinical equipoise and the incoherence of research ethics. *Journal of Medicine and Philosophy* 2007 March-April; 32(2): 151-165. Subject: 18.2

Miller, Franklin G.; Campbell, Eric G.; Vogeli, Christine; Weissman, Joel S. Financial relationships of institutional review board members [letter and reply]. *New England Journal of Medicine* 2007 March 1; 356(9): 965. Subject: 18.2

Miller, Franklin G.; Fins, Joseph J. Protecting human subjects in brain research: a pragmatic approach. *In:* Illes, Judy, ed. Neuroethics: Defining the Issues in Theory, Practice, and Policy. New York: Oxford University Press, 2006: 123-140. Subject: 18.2

Miller, Franklin G.; Wendler, David; Swartzman, Leora C. Deception in research on the placebo effect.

PLoS Medicine 2005 September; 2(9): e262 (0853-0859). Subject: 8.2

Miller, Franklin G.; Wertheimer, Alan. Facing up to paternalism in research ethics. *Hastings Center Report* 2007 May-June; 37(3): 24-34. Subject: 18.2

Miller, F.G.; Kaptchuk, T.J. Acupuncture trials and informed consent. *Journal of Medical Ethics* 2007 January; 33(1): 43-44. Subject: 18.3

Miller, Geoffrey. Ten days in Texas. *Hastings Center Report* 2007 July-August; 37(4): inside back cover. Subject: 20.5.1

Miller, Henry I. Two views of the emperor's new clones [letter]. *Nature Biotechnology* 2007 March; 25(3): 281. Subject: 14.5

Miller, Mark. Making decisions about advance health care directives. *In:* Hamel, Ronald, ed. Making Health Care Decisions: A Catholic Guide. Liguori, MO: Liguori Publications, 2006: 91-108. Subject: 20.5.4

Miller, Paul B.; Weijer, Charles. Equipoise and the duty of care in clinical research: a philosophical response to our critics. *Journal of Medicine and Philosophy* 2007 March-April; 32(2): 117-133. Subject: 18.2

Miller, Paul B.; Weijer, Charles. Revisiting equipoise; a response to Gifford. *Kennedy Institute of Ethics Journal* 2007 September; 17(3): 227-246. Subject: 18.2

Miller, Robin L.; Forte, Draco; Wilson, Bianca Della; Greene, George J. Protecting sexual minority youth from research risks: conflicting perspectives. *American Journal of Community Psychology* 2006 June; 37(3-4): 341-348. Subject: 18.3

Miller, Seumas. Privacy, confidentiality, and the treatment of drug addicts. *In:* Kleinig, John; Einstein, Stanley, eds. Ethical Challenges for Intervening in Drug Use: Policy, Research and Treatment Issues. Huntsville, TX: Office of International Criminal Justice; Criminal Justice Center, Sam Houston State University, 2006: 467-483. Subject: 8.4

Miller, Seumas; Selgelid, Michael. Ethical and philosophical consideration of the dual-use dilemma in the biological sciences. *Science and Engineering Ethics* 2007 December; 13(4): 523-580. Subject: 21.3

Miller, Suellen; Billings, Deborah L. Abortion and postabortion care: ethical, legal, and policy issues in developing countries. *Journal of Midwifery and Women's Health* 2005 July-August; 50(4): 341-343. Subject: 12.4.1

Miller, Tina; Bell, Linda. Consenting to what? Issues of access, gate-keeping and 'informed' consent. *In:* Mauthner, Melanie; Birch, Maxine; Jessop, Julie; Miller, Tina, eds. Ethics in Qualitative Research. London; Thousand Oaks, CA: Sage Publications Ltd., 2002: 53-69. Subject: 18.3

Miller, Vail M.; Volandes, Angelo E.; Paasche-Orlow, Michael K. Poor eHealth literacy and consumer-directed

See SUBJECT HEADING KEY FOR SECTION II on inside back cover.

559

health plans: a recipe for market failure. *American Journal of Bioethics* 2007 November; 7(11): 20-22; author reply W1-W2. Subject: 9.1

Mills, Ann; Werhane, Patricia; Gorman, Michael. The pharmaceutical industry and its obligations in the developing world. *In:* Cohen, Jillian Clare; Illingworth, Patricia; Schüklenk, Udo, eds. The Power of Pills: Social, Ethical and Legal Issues in Drug Development, Marketing, and Pricing. London; Ann Arbor, MI: Pluto, 2006: 32-40. Subject: 9.7

Mills, Eithne. Parents, children, and medical treatment: legal rights and responsibilities. *In:* Komesaroff, Linda, ed. Surgical Consent: Bioethics and Cochlear Implantation. Washington, DC: Gallaudet University Press, 2007: 70-87. Subject: 9.5.7

Mills, Peter. Recent issues in assisted reproduction: evolutions in science, law and ethics. *In:* Gunning, Jennifer; Holm, Søren, eds. Ethics, law, and society. Volume 1. Aldershot, Hants, England; Burlington, VT: Ashgate, 2005: 23-31. Subject: 18.6

Millum, Joseph; Emanuel, Ezekiel J. The ethics of international research with abandoned children. Research with abandoned children does not necessarily involve exploitation. *Science* 2007 December 21; 318(5858): 1874-1875. Subject: 18.5.2

Milwaukee Guild of the Catholic Medical Association. Checklist for Catholic hospitals. *Linacre Quarterly* 2007 May; 74(2): 159-163. Subject: 9.1

Miossec, Marie; Miossec, Pierre. New regulatory rules for clinical trials in the United States and the European Union: key points and comparisons. *Arthritis and Rheumatism* 2006 December; 54(12): 3735-3740. Subject: 18.2

Mirarchi, Ferdinando L.; Conti, Lucia. Living wills and DNR: is patient safety compromised? *Human Life Review* 2007 Fall; 33(4): 66-73. Subject: 20.5.4

Mishra, Pankaj Kumar; Ozalp, Faruk; Gardner,Roy S.; Arangannal, Arul; Murday, Andrew. Informed consent in cardiac surgery: is it truly informed? *Journal of Cardiovascular Medicine* 2006 September; 7(9): 675-681. Subject: 8.3.1

Mistry, Parul R. Donation after cardiac death: an overview. *Mortality* 2006 May; 11(2): 182-195. Subject: 20.2.1

Mitchell, Ellen M.H.; Trueman, Karen; Gabriel, Mosotho; Bock, Lindsey B. Bickers. Building alliances from ambivalence: evaluation of abortion values clarification workshops with stakeholders in South Africa. *African Journal of Reproductive Health* 2005 December; 9(3): 89-99. Subject: 12.5.1

Mitchell, Peter. Critics pan timid European response to TeGenero disaster. *Nature Biotechnology* 2007 May 25(5): 485-486. Subject: 18.2

Mitchell, Peter. EU cell therapy legislation. *Nature Biotechnology* 2007 June; 25(6): 614. Subject: 18.5.4

Mitchell, Susan L.; Teno, Joan M.; Intrator, Orna; Feng, Zhanlian; Mor, Vincent. Decisions to forgo hospitalization in advance dementia: a nationwide study. *Journal of the American Geriatrics Society* 2007 March; 55(3): 432-438. Subject: 20.5.1

Mitka, Mike. Stem cell legislation [news]. *JAMA: The Journal of the American Medical Association* 2007 February 14; 297(6): 581. Subject: 18.5.4

Mitton, Craig R.; McMahon, Meghan; Morgan, Steve; Gibson, Jennifer. Centralized drug review processes: are they fair? *Social Science and Medicine* 2006 July; 63(1): 200-211. Subject: 9.7

Mittra, James. Marginalising 'eugenic anxiety' through a rhetoric of 'liberal choice': a critique of the House of Commons Select Committee Report on reproductive technologies. *New Genetics and Society* 2007 August; 26(2): 159-179. Subject: 14.1

Miyazaki, Michiko. The history of abortion-related acts and current issues in Japan. *Medicine and Law: The World Association for Medical Law* 2007 December; 26(4): 791-799. Subject: 12.4.2

Mizani, Mehrdad A.; Baykal, N. A software platform to analyse the ethical issues of electronic patient privacy: the S3P example. *Journal of Medical Ethics* 2007 December; 33(12): 695-698. Subject: 8.4

Mody, Cyrus; McCray, Patrick; Roberts, Jody; Berne, Rosalyn; Lin, Patrick; Keiper, Adam. Debating nanoethics [letters and reply]. *New Atlantis* 2007 Summer; (17): 5-14. Subject: 5.4

Moerman, C.J.; Haafkens, J.A.; Söderström, M.; Rásky, É.; Maguire, P.; Maschewsky-Schneider, U.; Norstedt, M.; Hahn, D.; Reinerth, H.; McKevitt, M. Gender equality in the work of local research ethics committees in Europe: a study of practice of five countries. *Journal of Medical Ethics* 2007 February; 33(2): 107-112. Subject: 18.2

Moffatt, Barton; Elliott, Carl. Ghost marketing: pharmaceutical companies and ghostwritten journal articles. *Perspectives in Biology and Medicine* 2007 Winter; 50(1): 18-31. Subject: 1.3.7

Mohammadi, S. Mehrdad; Mohammadi, S. Farzad; Hedges, Jerris R. Conceptualizing a quality plan for healthcare: a philosophical reflection on the relevance of the health profession to society. *Health Care Analysis: An International Journal of Health Philosophy and Policy* 2007 December; 15(4): 337-361. Subject: 9.8

Mohan, Bannur Muthai. Misconceptions about narco analysis. *Indian Journal of Medical Ethics* 2007 January-March; 4(1): 7-8. Subject: 9.7

Mohindra, Raj. Obligations to treat, personal autonomy, and artificial nutrition and hydration. *Clinical Medicine* 2006 May-June; 6(3): 271-273. Subject: 20.5.1

Mohindra, R.K. Medical futility: a conceptual model. *Journal of Medical Ethics* 2007 February; 33(2): 71-75. Subject: 20.5.1

Molinelli, A.; Picchioni, D.M.; Celesti, R. Voluntary interruption of pregnancy in Europe: medico-legal issues and ethical approach to the regulation. *Minerva Ginecologica* 2005 April; 57(2): 217-223. Subject: 12.4.1

Möller, Hans-Jürgen. Ethical aspects of publishing [editorial]. *World Journal of Biological Psychiatry* 2006; 7(2): 66-69. Subject: 1.3.7

Molyneux, David. "And how is life going for you?" — an account of subjective welfare in medicine. *Journal of Medical Ethics* 2007 October; 33(10): 568-572. Subject: 4.4

Momen, Hooman; Gollogly, Laragh. Cross-cultural perspectives of scientific misconduct. *Medicine and Law: The World Association for Medical Law* 2007 September; 26(3): 409-416. Subject: 1.3.9

Monaghan, Peter. Panel warns psychological journals about corporate influence. *Chronicle of Higher Education* 2007 December 21; 54(17): A12. Subject: 1.3.7

Montagut, Jacques; Menezo, Yves. How to legislate in human reproduction: the French experience. *Journal of Assisted Reproduction and Genetics* 2003 July; 20(7): 287-289. Subject: 14.4

Montgomery, Charlotte. Research: keeping humans alive. *In her:* Blood Relations: Animals, Humans, and Politics. Toronto: Between the Lines, 2000: 80-127. Subject: 22.2

Montrose, J.L. Is negligence an ethical or a sociological concept? *Modern Law Review* 1958 May; 21(3): 259-264. Subject: 8.5

Moodley, Kaymanthri. Microbicide research in developing countries: have we given the ethical concerns due consideration? *BMC Medical Ethics* 2007; 8(10): 7 p. [electronic]. Accessed: http://www.biomedcentral. com/content/pdf/1472-6939-8-10.pdf [2007 December 18]. Subject: 9.5.6

Moodley, Kaymanthri. Teaching medical ethics to undergraduate students in post-apartheid South Africa, 2003-2006. *Journal of Medical Ethics* 2007 November; 33(11): 673-677. Subject: 2.3

Moodley, Kaymanthri; Myer, Landon. Health research ethics committees in South Africa 12 years into democracy. *BMC Medical Ethics [electronic]* 2007; 8(1): 8 p. Accessed: http://www.biomedcentral.com/1472-6939/8/1 [2007 February 21]. Subject: 18.2

Moore, Ilene N.; Snyder, Samuel Leason; Miller, Cynthia; An, Angel Qi; Blackford, Jennifer U.; Zhou, Chuan; Hickson, Gerald B. Confidentiality and privacy in health care from the patient's perspective: does HIPAA help? *Health Matrix: The Journal of Law-Medicine* 2007 Spring; 17(2): 215-272. Subject: 8.4

Moore, Mary E.; Berk, Stephen N.; Freedman, Benjamin; Salisbury, David A.; Schechter, Martin T. Research involving human subjects. *In:* Baylis, Françoise; Downie, Jocelyn; Freedman, Benjamin; Hoffmaster, Barry; Sherwin, Susan, eds. Health Care Ethics in Canada. Toronto: Harcourt Brace Canada, 1995: 319-364. Subject: 18.2

Moore, Molly; Gavard, Corinne. French plan to screen DNA of visa-seekers draws anger [news]. *Washington Post* 2007 September 21; p. A14. Subject: 15.1

Moore, Solomon. DNA exoneration leads to change in legal system; states pass new laws; new police procedures — prisoners gain evidence access [news]. *New York Times* 2007 October 1; p. A1, A22. Subject: 15.1

Moosa, M.R.; Kidd, M. The dangers of rationing dialysis treatment: the dilemma facing a developing country. *Kidney International* 2006 September; 70(6): 1107-1114. Subject: 19.3

Moran, Gordon. Rubber stamp-type decisions for funding of academic research: paradigms and conflicts of interest. *Journal of Information Ethics* 2007 Spring; 16(1): 53-58. Subject: 5.3

Moran, Maureen B. Ethical issues in research with human subjects. *Journal of the American Dietetic Association* 2006 September; 106(9): 1346, 1348. Subject: 18.2

Moran, Nuala. UK parses merits of value-based drug pricing. *Nature Biotechnology* 2007 April; 25(4): 369-370. Subject: 9.7

Moreno, Jonathan. Secret state experiments and medical ethics. *In:* Galston, Arthur W.; Peppard, Christiana Z., eds. Expanding Horizons in Bioethics. Dordrecht; Norwell, MA: Springer, 2005: 59-69. Subject: 18.5.1

Moreno, Jonathan D. Bioethics and bioterrorism. *In:* Steinbock, Bonnie, ed. The Oxford Handbook of Bioethics. Oxford; New York: Oxford University Press, 2007: 721-733. Subject: 21.3

Moreno, Jonathan D. Stumbling toward bioethics: human experiments policy and the early Cold War. *In:* LaFleur, William R.; Böhme, Gernot; Shimazono, Susumu, eds. Dark Medicine: Rationalizing Medical Research. Bloomington: Indiana University Press, 2007: 138-146. Subject: 18.2

Moreno, Jonathan D. The triumph of autonomy in bioethics and commercialism in American healthcare. *CQ: Cambridge Quarterly of Healthcare Ethics* 2007 Fall; 16(4): 415-419. Subject: 2.1

Moreno, Jonathan D.; Berger, Sam. Biotechnology and the new right: neoconservatism's red menace. *American Journal of Bioethics* 2007 October; 7(10): 7-13. Subject: 5.1

Moreno, Jonathan D.; Selgelid, Michael J. The dual-use dilemma [letter and reply]. *Hastings Center Report* 2007 September-October; 37(5): 6-7. Subject: 21.3

See SUBJECT HEADING KEY FOR SECTION II on inside back cover.

561

Moreno, Rui; Afonso, Susana. Ethical, legal and organizational issues in the ICU: prediction of outcome. *Current Opinion in Critical Care* 2006 December; 12(6): 619-623. Subject: 5.2

Morgan, Derek. Regulating the bio-economy: a preliminary assessment of biotechnology and law. *In:* Bennett, Belinda; Tomossy, George F., eds. Globalization and Health: Challenges for Health Law and Bioethics. Dordrecht: Springer, 2006: 59-69. Subject: 5.3

Morgan, Myfanwy; Hooper, Richard; Mayblin, Maya; Jones, Roger. Attitudes to kidney donation and registering as a donor among ethnic groups in the UK. *Journal of Public Health* 2006 September; 28(3): 226-234. Subject: 19.3

Morgan, Steven G. Direct-to-consumer advertising and expenditure on prescription drugs: a comparison of experiences in the United States and Canada. *Open Medicine [electronic]* 2007; 1(1): 37-45. Accessed: http://www.openmedicine.ca [2007 April 19]. Subject: 9.7

Morgenstern, L.; Laquer, M.; Treyzon, L. Ethical challenges of percutaneous endoscopic gastrostomy. *Surgical Endoscopy* 2005 March; 19(3): 398-400. Subject: 20.5.1

Morin, Karine: Green, Shane K. Professionalism in biomedical science. *American Journal of Bioethics* 2007 February; 7(2): 66-68. Subject: 14.5

Morioka, Masahiro. The ethics of human cloning and the sprout of human life. *In:* Roetz, Heiner, ed. Cross-Cultural Issues in Bioethics: The Example of Human Cloning. New York: Rodopi, 2006: 1-16. Subject: 14.5

Morreim, E. Haavi. Ties without tethers: bioethics corporate relations in the AbioCor artificial heart trial. *In:* Eckenwiler, Lisa A.; Cohn, Felicia G., eds. The Ethics of Bioethics: Mapping the Moral Landscape. Baltimore, MD: Johns Hopkins University Press, 2007: 181-190. Subject: 9.6

Morris, Albert W., Jr.; Gadson, Sandra L.; Burroughs, Valentine. "For the good of the patient," survey of the physicians of the National Medical Association regarding perceptions of DTC advertising, Part II, 2006. *Journal of the National Medical Association* 2007 March; 99(3): 287-293. Subject: 9.7

Morris, Jonathan. A brave new world of clones. *In his:* The Ethics of Biotechnology. Philadelphia: Chelsea Publishers, 2006: 30-55. Subject: 14.5

Morris, Jonathan. GM foods: what are they doing to our dinner? *In his:* The Ethics of Biotechnology. Philadelphia: Chelsea Publishers, 2006: 104-130. Subject: 15.1

Morris, Jonathan. Human cloning: should humans be cloned? *In his:* The Ethics of Biotechnology. Philadelphia: Chelsea Publishers, 2006: 56-82. Subject: 14.5

Morris, Jonathan. Stem cell research. *In his:* The Ethics of Biotechnology. Philadelphia: Chelsea Publishers, 2006: 83-104. Subject: 18.5.4

Morris, Kelly. Issues on female genital mutilation/cutting — progress and parallels. *Lancet* 2006 December; 368(special issue): S64-S66. Subject: 9.5.5

Morris, Marilyn C.; Fischbach, Ruth L.; Nelson, Robert M.; Schleien, Charles L. A paradigm for inpatient resuscitation research with an exception from informed consent. *Critical Care Medicine* 2006 October; 34(10): 2567-2575. Subject: 20.5.1

Morris, Stephen G. Canada's Assisted Human Reproduction Act: a chimera of religion and politics. *American Journal of Bioethics* 2007 February; 7(2): 69-70. Subject: 14.5

Morris, Stephen G. Neuroscience and the free will conundrum. *American Journal of Bioethics* 2007 May; 7(5): 20-22. Subject: 17.1

Morrison, Wynne. Thoughts on advance directives. *Journal of Palliative Medicine* 2006 April; 9(2): 483-484. Subject: 20.5.4

Morroni, Chelsea; Myer, Landon; Tibazarwa, Kemilembe. Knowledge of the abortion legislation among South African women: a cross-sectional study. *Reproductive Health [electronic]* 2006; 3:7, 5 p. Accessed: http://www.reproductive-health-journal.com/content/pdf/1742-4755-3-7.pdf [2007 April 19]. Subject: 12.5.2

Morse, Stephen J. Medicine and morals, craving and compulsion. *In:* Kleinig, John; Einstein, Stanley, eds. Ethical Challenges for Intervening in Drug Use: Policy, Research and Treatment Issues. Huntsville, TX: Office of International Criminal Justice; Criminal Justice Center, Sam Houston State University, 2006: 323-339. Subject: 9.5.9

Morse, Stephen J. Moral and legal responsibility and the new neuroscience. *In:* Illes, Judy, ed. Neuroethics: Defining the Issues in Theory, Practice, and Policy. New York: Oxford University Press, 2006: 33-50. Subject: 17.1

Morse, Stephen J. Voluntary control of behavior and responsibility. *American Journal of Bioethics* 2007 January; 7(1): 12-13. Subject: 17.1

Mor-Yosef, Shlomo; Weiss, Yuval; Birnbaum, Yair C. Problems and questions regarding the treatment of political leaders. *Israel Medical Association Journal* 2006 November; 8(11): 754-756. Subject: 21.4

Moscarillo, T.J.; Holt, H.; Perman, M.; Goldberg, S.; Cortellini, L.; Stoler, J.M.; DeJong, W.; Miles, B.J.; Albert, M.S.; Go, R.C.P.; Blacker, Deborah. Knowledge of and attitudes about Alzheimer disease genetics: report of a pilot survey and two focus groups. *Community Genetics* 2007; 10(2): 97-102. Subject: 15.1

Moselli, N.M.; Debernardi, F.; Piovano, F. Forgoing life sustaining treatments: differences and similarities between North America and Europe. *Acta Anaesthesiologica Scandinavica* 2006 November; 50(10): 1177-1186. Subject: 20.5.1

Subject = NRCBL Primary Classification Number; see inside front cover.

Moser, Albine; Houtepen, Rob; Widdershoven, Guy. Patient autonomy in nurse-led shared care: a review of theoretical and empirical literature. *Journal of Advanced Nursing* 2007 February; 57(4): 357-365. Subject: 8.1

Moses, Lyria Bennett. Understanding legal responses to technological change: the example of in vitro fertilization. *Minnesota Journal of Law, Science and Technology* 2004-2005; 6(2): 505-618. Subject: 14.4

Moses, Sarah. A just society for the elderly: the importance of justice as participation. *Notre Dame Journal of Law, Ethics and Public Policy* 2007; 21(2): 335-362. Subject: 9.5.2

Moskop, John C.; Easton, Raul B.; Graber, Mark A.; Monnahan, Jay; Hughes, Jason. Information disclosure and consent: patient preferences and provider responsibilities. *American Journal of Bioethics* 2007 December; 7(12): 47-49; author reply W3-W4. Subject: 8.3.1

Moskop, John C.; Geiderman, Joel M.; Hobgood, Cherri D.; Larkin, Gregory L. Emergency physicians and disclosure of medical errors. *Annals of Emergency Medicine* 2006 November; 48(5): 523-531. Subject: 9.8

Moskop, John C.; Iserson, Kenneth V. Triage in medicine, Part II: underlying values and principles. *Annals of Emergency Medicine* 2007 March; 49(3): 282-287. Subject: 9.4

Moss, Ralph W. Health checks, not shots: blanket vaccination against a sexually transmitted virus is the wrong way to protect women's health. *New Scientist* 2007 February 24-March 2; 193(2592): 20. Subject: 9.5.7

Mostert, Mark P. Cultures of death, old and new. *Human Life Review* 2007 Fall; 33(4): 54-65. Subject: 20.5.1

Moszynski, Peter. BMA backs police campaign against genital mutilation [news]. *BMJ: British Medical Journal* 2007 July 21; 335(7611): 116. Subject: 9.5.5

Motluk, Alison. Crisis of trust over sperm bank errors: a new register of DNA from donors and their offspring is exposing major gaps in US sperm bank regulation. *New Scientist* 2007 August 11-17; 195(2616): 6-7. Subject: 14.6

Mouradian, Wendy E. Band-Aid solutions to the dental access crisis: conceptually flawed — a response to Dr. David H. Smith. *Journal of Dental Education* 2006 November; 70(11): 1174-1179. Subject: 9.2

Moynihan, Ray. Attempt to undermine European ban on advertising drugs fails in France [news]. *BMJ: British Medical Journal* 2007 February 10; 334(7588): 279. Subject: 9.7

Moynihan, Ray. Direct to consumer advertising should not come to Europe. *BMJ: British Medical Journal* 2007 May 19; 334(7602): 1025. Subject: 9.7

Moynihan, Ray. EC report on drug advertising found to be "biased". *BMJ: British Medical Journal* 2007 June 23; 334(7607): 1290. Subject: 9.7

Moynihan, Ray. Healthcare giant advertises to children in Australia's classrooms [news]. *BMJ: British Medical Journal* 2007 September 29; 335(7621): 637. Subject: 9.7

Moynihan, Timothy J.; Schapira, Lidia. Preparing ourselves, our trainees, and our patients: a commentary on truthtelling. *Journal of Clinical Oncology* 2007 February 1; 25(4): 456-457. Subject: 8.2

Mueller, Paul S.; Hook, C.Christopher; Litin, Scott C. Physician preferences and attitudes regarding industry support of CME programs. *American Journal of Medicine* 2007 March; 120(3): 281-285. Subject: 9.7

Mueller, Paul S.; Montori, Victor M.; Bassler, Dirk; Koenig, Barbara A.; Guyatt, Gordon H. Ethical issues in stopping randomized trials early because of apparent benefit. *Annals of Internal Medicine* 2007 June 19; 146(12): 878-881. Subject: 18.2

Mujovic-Zornic, Hajrija. Legislation and patients' rights: some necessary remarks. *Medicine and Law: The World Association for Medical Law* 2007 December; 26(4): 709-719. Subject: 8.1

Mukherjee, Debjani; Levin, Rebecca L.; Heller, Wendy. The cognitive, emotional, and social sequelae of stroke: psychological and ethical concerns in post-stroke adaptation. *Topics in Stroke Rehabilitation* 2006 Fall; 13(4): 26-35. Subject: 9.5.1

Mukherjee, Raja; Eastman, Nigel; Turk, Jeremy; Hollins, Sheila. Fetal alcohol syndrome: law and ethics. *Lancet* 2007 April 7-13; 369(9568): 1149-1150. Subject: 9.5.9

Muller, David. GOMER. *Health Affairs* 2007 May-June; 26(3): 831-835. Subject: 20.4.1

Müller, Denis. The original risk: overtheologizing ethics and undertheologizing sin. *Christian Bioethics* 2007 January-April; (13)1: 7-23. Subject: 2.1

Müller, Fernando Suárez. On futuristic gerontology: a philosophical evaluation of Aubrey de Grey's SENS project. *International Journal of Applied Philosophy* 2007 Fall; 21(2): 225-239. Subject: 9.5.2

Müller-Hill, Benno. The silence of the scholars. *In:* LaFleur, William R.; Böhme, Gernot; Shimazono, Susumu, eds. Dark Medicine: Rationalizing Medical Research. Bloomington: Indiana University Press, 2007: 57-62. Subject: 1.3.9

Munson, Ronald. Organ transplantation. *In:* Steinbock, Bonnie, ed. The Oxford Handbook of Bioethics. Oxford; New York: Oxford University Press, 2007: 211-239. Subject: 19.5

Murdoch, C.J. Kant's antipode: Nietzsche's transvaluation of human dignity and its "implications" for biotechnology policy. *Health Law Review* 2007; 15(3): 24-26. Subject: 4.4

Murphy, Dominic; Dandeker, Christopher; Horn, Oded; Hotopf, Matthew; Hull, Lisa; Jones, Margaret;

See SUBJECT HEADING KEY FOR SECTION II on inside back cover.

563

Marteau, Theresa; Rona, Roberto; Wessely, Simon. UK armed forces responses to an informed consent policy for anthrax vaccination: a paradoxical effect? *Vaccine* 2006 April 12; 24(16): 3109-3114. Subject: 8.3.1

Murphy, Donald J.; Santilli, Sara. Elderly patients' preferences for long-term life support. *Archives of Family Medicine* 1998 September; 7(5): 484-488. Subject: 20.5.1

Murphy, Fiona; Byrne, Gobnait. Ethical issues regarding live kidney transplantation. *British Journal of Nursing* 2007 October 25-November 7; 16(19): 1224-1229. Subject: 19.3

Murphy, Julien S. Is pregnancy necessary? Feminist concerns about ectogenesis. *In:* Gelfand, Scott; Shook, John R., eds. Ectogenesis: Artificial Womb Technology and the Future of Human Reproduction. Amsterdam; New York: Editions Rodopi, B.V., 2006: 27-46. Subject: 14.1

Murphy, M. Dianne; Goldkind, Sara F. The regulatory and ethical challenges of pediatric research. *In:* Santoro, Michael A.; Gorrie, Thomas M., eds. Ethics and the Pharmaceutical Industry. Cambridge; New York: Cambridge University Press, 2005: 48-67. Subject: 9.7

Murphy, Timothy F. When 'emergency contraception' is neither. *American Journal of Bioethics* 2007 August; 7(8): W7. Subject: 11.1

Murray, Fiona. The stem cell market — patents and the pursuit of scientific progress. *New England Journal of Medicine* 2007 June 7; 356(23): 2341-2343. Subject: 18.5.4

Murray, Thomas H. Enhancement. *In:* Steinbock, Bonnie, ed. The Oxford Handbook of Bioethics. Oxford; New York: Oxford University Press, 2007: 491-515. Subject: 4.5

Murtagh, Ged M. Ethical reflection on the harm in reproductive decision-making. *Journal of Medical Ethics* 2007 December; 33(12): 717-720. Subject: 14.1

Musch, Timothy I.; Carroll, Robert G.; Lane, Pascale H.; Talman, William T. A broader view of animal research [letter]. *BMJ: British Medical Journal* 2007 February 10; 334(7588): 274. Subject: 22.2

Mushaben, Joyce; Giles, Geoffrey; Lennox, Sara. Women, men and unification: gender politics and the abortion struggle since 1989. *In:* Jarausch, K.H., ed. After Unity: Reconfiguring German Identities. Oxford: Berghahn Books, 1997: 137-172. Subject: 12.4.2

Musschenga, A.W.; van Luijn, H.E.M.; Keus, R.B.; Aaronson, N.K. Are risks and benefits of oncological research protocols both incommensurable and incompensable? *Accountability in Research* 2007 July-September; 14(3): 179-196. Subject: 18.2

Muula, Adamson. Malawi: ethical challenges of HIV and AIDS in Malawi, southern Africa. *In:* Davis, Anne J.; Tschudin, Verena; de Raeve, Louise, eds. Essentials of Teaching and Learning in Nursing Ethics: Perspectives and Methods. New York: Churchill Livingstone Elsevier, 2006: 291-300. Subject: 9.5.6

Muula, A.S. Ethical and practical consideration of women choosing cesarean section deliveries without "medical indication" in developing countries. *Croatian Medical Journal* 2007 February; 48(1): 94-102. Subject: 9.5.5

Muula, A.S.; Mfutso-Bengo, J.M. Responsibilities and obligations of using human research specimens transported across national boundaries. *Journal of Medical Ethics* 2007 January; 33(1): 35-38. Subject: 19.5

Myer, Landon; Moodley, Kaymanthri; Hendricks, Fahad; Cotton, Mark. Healthcare providers' perspectives on discussing HIV status with infected children. *Journal of Tropical Pediatrics* 2006 August; 52(4): 293-295. Subject: 9.5.6

Myers, Richard S. US law and conscientious objection in healthcare. *In:* Watt, Helen, ed. Cooperation, Complicity and Conscience: Problems in Healthcare, Science, Law and Public Policy. London: Linacre Centre, 2005: 296-315. Subject: 8.1

Mykitiuk, Roxanne; Nisker, Jeff; Bluhm, Robyn. The Canadian Assisted Human Reproduction Act: protecting women's health while potentially allowing human somatic cell nuclear transfer into non-human oocytes. *American Journal of Bioethics* 2007 February; 7(2): 71-73. Subject: 14.5

Myser, Catherine. The challenges of amnesia in assessing capacity, assigning a proxy, and deciding to forego life-prolonging medical treatment. *Journal of Clinical Ethics* 2007 Fall; 18(3): 262-269. Subject: 8.1

Myser, Catherine. White normativity in U.S. bioethics: a call and method for more pluralist and democratic standards and policies. *In:* Eckenwiler, Lisa A.; Cohn, Felicia G., eds. The Ethics of Bioethics: Mapping the Moral Landscape. Baltimore, MD: Johns Hopkins University Press, 2007: 241-259. Subject: 2.1

Myskja, Bjørn K. Lay expertise: why involve the public in biobank governance? *Genomics, Society and Policy* 2007 April; 3(1): 1-16. Subject: 15.1

Mysorekar, Uma. Spirituality in palliative care — a Hindu perspective. *In:* Puchalski, Christina M., ed. A Time for Listening and Caring: Spirituality and the Care of the Chronically Ill and Dying. Oxford; New York: Oxford University Press, 2006: 171-182. Subject: 20.4.1

N

Nagell, Hilde W. A penny for your thoughts — ethics in sponsored research. *In:* Gunning, Jennifer; Holm, Søren, eds. Ethics, law, and society. Volume 1. Aldershot, Hants, England; Burlington, VT: Ashgate, 2005: 45-60. Subject: 1.3.9

Najjar, Dany; Srinivasan, M.; Hammersmith, Kristin M.; Cohen, Elisabeth J.; Rapuano, Christopher J.;

Laibson, Peter R. Informed consent for Creutzfeldt-Jakob disease after corneal transplantation [letters and reply]. *Cornea* 2005 January; 24(1): 121-122. Subject: 8.3.1

Nakashima, David Y. Your body, your choice: how mandatory advance health-care directives are necessary to protect your fundamental right to accept or refuse medical treatment. *University of Hawai'i Law Review* 2004 Winter; 27(1): 201-231. Subject: 20.5.4

Naparstek, Yaakov. Ariel Sharon's illness: should we dedicate a medical journal issue to a single case study? *Israel Medical Association Journal* 2006 November; 8(11): 739-740. Subject: 1.3.7

Napier, Stephen. Human embryos as human subjects. *Ethics and Medics* 2007 September; 32(9): 3-4. Subject: 18.5.4

Nash, David A. "The profession of dentistry:" the University of Kentucky's curriculum in professional ethics. *Journal of the American College of Dentists* 1996 Spring; 63(1): 25-29. Subject: 7.2

Nash, Robert. Health-care workers in influenza pandemics [comment]. *Lancet* 2007 July 28-August 3; 370(9584): 300-301. Subject: 9.1

Natale, JoAnne E.; Joseph, Jill G.; Pretzlaff, Robert K.; Silber, Tomas J.; Guerguerian, Anne-Marie. Clinical trials in pediatric traumatic brain injury: unique challenges and potential responses. *Developmental Neuroscience* 2006; 28(4-5): 276-290. Subject: 18.5.7

Nathanson, Vivienne. Indian doctors are promised help to fight abuse of prisoners [news]. *BMJ: British Medical Journal* 2007 May 12; 334(7601): 972. Subject: 21.4

Nation, George A., III. Obscene contracts: the doctrine of unconscionability and hospital billing of the uninsured. *Kentucky Law Journal* 2005-2006; 94(1): 101-137. Subject: 9.3.1

National Catholic Bioethics Center. Brief comments on the CDF responses [from Statement of the NCBC On the CDF's "Responses to Certain Questions Concerning Artificial Nutrition and Hydration"]. *Ethics and Medics* 2007 November; 32(11): 3-4. Subject: 20.5.1

National Center for Ethics in Health Care (United States). IntegratedEthics [Integrated Ethics]: IntegratredEthics Toolkit — A Manual for the IntegratedEthics Program Officer; Ethics Consultation Toolkit — A Manual for the Ethics Consultation Coordinator; Ethical Leadership Toolkit — A Manual for the Ethical Leadership Coordinator; Preventive Ethics Toolkit — A Manual for the Preventive Ethics Coordinator. Washington, DC: National Center for Ethics in Health Care, Veterans Health Administration, 2007: multiple pages in 4 volumes. Subject: 9.6

National Health and Medical Research Council (Australia) [NHMRC]. Submission by the National Health and Medical Research Council to the review by the Federal Privacy Commissioner of the Private Sector Provisions of the Privacy Act 1988. Canberra: National Health and Medical Research Council, 2004 December 10; 39 p. [Online]. Accessed: http://www.nhmrc.gov.au/about/_files/psp.pdf [2007 April 18]. Subject: 8.4

National Health and Medical Research Council (Australia) [NHMRC]. The impact of privacy legislation on NHMRC stakeholders: comparative stakeholder analysis. Canberra: National Health and Medical Research Council, 2004 July; 51 p. [Online]. Accessed: http://www.nhmrc.gov.au/about/_files/st8.pdf [2007 April 18]. Subject: 8.4

National Institutes of Health [NIH] (United States). Stem Cell Information [government document]. Bethesda, MD: National Institutes of Health [NIH] (United States), 2007 May 4 and 2007 September 11: [34 p.] [Online]. Accessed: http://stemcells.nih.gov/staticresources/research/registry/PDFs/Registry.pdf [2007 October 2]. Subject: 18.5.4

National Institutes of Health [NIH] (United States); United States. Department of Health and Human Services. Office for Human Research Protections [OHRP]; Association of American Medical Colleges [AAMC]; American Society of Clinical Oncology [ASCO]. Alternative Models of IRB Review Workshop Summary Report, November 17-18, 2005. [Rockville, MD]: Office of Human Research Protections, [2007 April]: 7 p. [Online]. Accessed: http://www.hhs.gov/ohrp/sachrp/documents/AltModIRB.pdf [2007 April 30]. Subject: 18.2

National Institutes of Health [NIH] (United States). NIH Tracking/Inclusion Committee; Pinn, Vivian W.; Roth, Carl; Bates, Angela C.; Caban, Carlos; Jarema, Kim. Monitoring Adherence to the NIH Policy on the Inclusion of Women and Minorities as Subjects in Clinical Research. Comprehensive Report: Tracking of Human Subjects Research as Reported in Fiscal Year 2004 and Fiscal Year 2005. Bethesda, MD: National Institutes of Health [NIH], 2006: 147 p. Subject: 18.5.3

National Institutes of Health [NIH] (United States); United States. Department of Health and Human Services. Plan for implementation of Executive Order 13435: expanding approved stem cell lines in ethically responsible ways. Bethesda, MD: National Institutes of Health 2007 September 18; 25 p. [Online]. Accessed: http://stemcells.nih.gov/staticresources/policy/eo13435.pdf [2007 October 11]. Subject: 5.3

National Reference Center for Bioethics Literature. News from the National Reference Center for Bioethics Literature (NRCBL) and the National Information Resource on Ethics and Human Genetics (NIREHG). *Kennedy Institute of Ethics Journal* 2007 December; 17(4): 399-403. Subject: 2.1

Naudts, Kris; Ducatelle, Caroline; Kovacs, Jozsef; Laurens, Kristin; van den Eynde, Frederique; van Heeringen, Cornelis. Euthanasia: the role of the psychia-

See SUBJECT HEADING KEY FOR SECTION II on inside back cover.

565

trist. *British Journal of Psychiatry* 2006 May; 188: 405-409. Subject: 20.5.1

Neal, Joseph M.; Rathmell, James P. Scientific misconduct: no end in sight. *Regional Anesthesia and Pain Medicine* 2006 July-August; 31(4): 294-295. Subject: 1.3.9

Neal, Karama C.; Volandes, Angelo E.; Paasche-Orlow, Michael K. Health literacy: more than a one-way street. *American Journal of Bioethics* 2007 November; 7(11): 29-30; author reply W1-W2. Subject: 8.1

Neale, Ann. Who really wants health care justice? *Health Progress* 2007 January-February; 88(1): 40-43. Subject: 2.1

Needleman, Jacob. A philosopher's reflection on commercialism in medicine. *CQ: Cambridge Quarterly of Healthcare Ethics* 2007 Fall; 16(4): 433-438. Subject: 9.3.1

Neher, Jon O. You're fired. *Hastings Center Report* 2007 May-June; 37(3): 7-8. Subject: 9.5.9

Neil, David A.; Coady, C.A.J.; Thompson, J.; Kuhse, H. End-of-life decisions in medical practice: a survey of doctors in Victoria (Australia). *Journal of Medical Ethics* 2007 December; 33(12): 721-725. Subject: 20.5.1

Neill, Ushma S. Stop misbehaving! [editorial]. *Journal of Clinical Investigation* 2006 July; 116(7): 1740-1741. Subject: 1.3.9

Neitzke, Gerald. Confidentiality, secrecy, and privacy in ethics consultation. *HEC (Healthcare Ethics Committee) Forum* 2007 December; 19(4): 293-302. Subject: 9.6

Nelson, Erin L. Legal and ethical issues in ART "outcomes" research. *Health Law Journal* 2005; 13: 165-186. Subject: 14.4

Nelson, Hilde Lindemann. Four narrative approaches to bioethics. *In:* Khushf, George, ed. Handbook of Bioethics: Taking Stock of the Field From a Philosophical Perspective. Dordrecht; Boston: Kluwer Academic, 2004: 163-181. Subject: 2.1

Nelson, James Lindemann. Illusions about persons. *American Journal of Bioethics* 2007 January; 7(1): 65-66. Subject: 17.1

Nelson, James Lindemann. Synecdoche and stigma. *CQ: Cambridge Quarterly of Healthcare Ethics* 2007 Fall; 16(4): 475-478. Subject: 15.2

Nelson, James Lindemann. Testing, terminating, and discriminating. *CQ: Cambridge Quarterly of Healthcare Ethics* 2007 Fall; 16(4): 462-468. Subject: 15.2

Nelson, James Lindemann. Trusting bioethicists. *In:* Eckenwiler, Lisa A.; Cohn, Felicia G., eds. The Ethics of Bioethics: Mapping the Moral Landscape. Baltimore, MD: Johns Hopkins University Press, 2007: 47-55. Subject: 2.1

Nelson, James L.; Lindemann, Hilde. What families say about surrogacy: a response to "Autonomy and the family as (in)appropriate surrogates for DNR decisions". *Journal*

of Clinical Ethics 2007 Fall; 18(3): 219-226. Subject: 20.5.1

Nelson, Margaret K. Listening to Anna. *Health Affairs* 2007 May-June; 26(3): 836-840. Subject: 20.5.4

Nelson, Robert M. Including children in research: participation or exploitation? *In:* Santoro, Michael A.; Gorrie, Thomas M., eds. Ethics and the Pharmaceutical Industry. Cambridge; New York: Cambridge University Press, 2005: 68-79. Subject: 18.5.2

Nelson, William A.; Campfield, Justin. Ethical implications of transparency. Valid justification is required when withholding information. *Healthcare Executive* 2006 November-December; 21(6): 33-34. Subject: 8.2

Nelson, W.; Pomerantz, A.; Howard, K.; Bushy, A. A proposed rural healthcare ethics agenda. *Journal of Medical Ethics* 2007 March; 33(3): 136-139. Subject: 9.5.1

Neresini, Federico. Eve's sons. *New Genetics and Society* 2007 August; 26(2): 221-233. Subject: 14.5

Ness, Roberta B. Influence of the HIPAA Privacy Rule on health research. *JAMA: The Journal of the American Medical Association* 2007 November 14; 298(18): 2164-2170. Subject: 8.4

Netherlands. Ministry of Health, Welfare and Sport. Regional Euthanasia Review Committees. Annual report 2005. Arnhem: Regional Euthanasia Review Committees, 2006 April; 30 p. [Online]. Accessed: http://www.toetsingscommissieseuthanasie.nl/Images/Annual1%20Report%202005%20English_tcm12-2439.pdf [2007 April 17]. Subject: 20.5.1

Neuberger, James; Gimson, Alexander. Selfless adults and split donor livers [comment]. *Lancet* 2007 July 28-August 3; 370(9584): 299-300. Subject: 19.6

Neugebauer, Matthias. Der theologische Lebensbegriff Dietrich Bonhoeffers im Lichte aktueller Fragen um Euthanasie, Sterbehilfe und Zwangssterilisation. *In:* Gestrich, Christof; Neugebauer, Johannes, eds. Der Wert menschlichen Lebens: medizinische Ethik bei Karl Bonhoeffer und Dietrich Bonhoeffer. Berlin: Wichern-Verlag, 2006: 147-165. Subject: 20.5.1

Neumärker, Klaus-Jürgen. Karl Bonhoeffers Entscheidungen zur Zwangssterilisation und Euthanasie. Versuch einer ethischen Beurteilung unter Berücksichtigung D. Bonhoeffers. *In:* Gestrich, Christof; Neugebauer, Johannes, eds. Der Wert menschlichen Lebens: medizinische Ethik bei Karl Bonhoeffer und Dietrich Bonhoeffer. Berlin: Wichern-Verlag, 2006: 33-65. Subject: 20.5.1

Neutra, Raymond Richard; Cohen, Aaron; Fletcher, Tony; Michaels, David; Richter, Elihu D.; Soskolne, Colin L. Toward guidelines for the ethical reanalysis and reinterpretation of another's research. *Epidemiology* 2006 May; 17(3): 335-338. Subject: 1.3.9

Subject = NRCBL Primary Classification Number; see inside front cover.

Neutra, Raymond R. What to declare and why? *Epidemiology* 2006 May; 17(3): 244-245. Subject: 1.3.9

New York. Supreme Court. Queens County. Matter of Long Island Jewish Medical Center, Petitioner. Baby Doe, a Minor Patient, Respondent. [Date of Decision: 28 February 1996]. West's New York Supplement, 2d Series, 1996; 641: 989-992. Subject: 20.5.2

New Zealand. Parliament. Human Assisted Reproductive Technology Bill. Wellington, New Zealand: Government Printer, 2004. No.195-3 [Online]. Accessed: http://www.parliament.nz/NR/rdonlyres/65E86459-2539-4FC3-B6F5-03440792D310/46172/DBHOH_BILL_24 _49999999999997.pdf [2007 March 14]. Subject: 14.1

Newburger, Amy E.; Caplan, Arthur L. Taking ethics seriously in cosmetic dermatology. *Archives of Dermatology* 2006 December; 142(12): 1641-1642. Subject: 4.1.1

Newcombe, J.P.; Kerridge, I.H. Assessment by human research ethics committees of potential conflicts of interest arising from pharmaceutical sponsorship of clinical research. *Internal Medicine Journal* 2007 January; 37(1): 12-17. Subject: 18.2

Newdick, Christopher. The positive side of healthcare rights. *In:* McLean, Sheila A.M., ed. First Do No Harm: Law, Ethics, and Healthcare. Aldershot, England; Burlington, VT: Ashgate, 2006: 573-586. Subject: 9.2

Newell, Elizabeth R. Competency, consent, and electroconvulsive therapy: a mentally ill prisoner's right to refuse invasive medical treatment in Oregon's criminal justice system. *Lewis and Clark Law Review* 2005 Winter; 9(4): 1019-1045. Subject: 17.5

Newsom, Robert W. Seattle syndrome: comments on the reaction to Ashley X. *Nursing Philosophy* 2007 October; 8(4): 291-294. Subject: 9.5.3

Newton, Sam K.; Appiah-Poku, John. Opinions of researchers based in the UK on recruiting subjects from developing countries into randomized controlled trials. *Developing World Bioethics* 2007 December; 7(3): 149-156. Subject: 18.5.9

Newton, Sam K.; Appiah-Poku, John. The perspectives of researchers on obtaining informed consent in developing countries. *Developing World Bioethics* 2007 April; 7(1): 19-24. Subject: 18.3

Ng, Ernest Hung Yu; Liu, Athena; Chan, Celia H.Y.; Chan, Cecilia Lai Wan; Yeung, William Shu Biu; Ho, Pak Chung. Regulating reproductive technology in Hong Kong. *Journal of Assisted Reproduction and Genetics* 2003 July; 20(7): 281-286. Subject: 14.4

Nicholl, David J.; Jenkins, Trefor; Miles, Steven H.; Hopkins, William; Siddiqui, Adnan; Boulton, Frank, et. al. Biko to Guantanamo: 30 years of medical involvement in torture [letter]. *Lancet* 2007 September 8-14; 370(9590): 823. Subject: 21.4

Nichols, Len M. The moral case for covering children (and everyone else): policy analysis and evaluation cannot rest until there is real health care justice throughout the entire land. *Health Affairs* 2007 March-April; 26(2): 405-407. Subject: 9.3.1

Nicolasora, Nelson; Pannala, Rahul; Mountantonakis, Stavros; Shanmugam, Baia; DeGirolamo, Angela; Amoateng-Adjepong, Yaw.; Manthous, Constantine A. If asked, hospitalized patients will choose whether to receive life-sustaining therapies. *Journal of Hospital Medicine* 2006 May; 1(3): 161-167. Subject: 20.5.1

Nie, Jing-Bao; Campbell, Alastair V. Multiculturalism and Asian bioethics: cultural war or creative dialogue? *Journal of Bioethical Inquiry* 2007; 4(3): 163-167. Subject: 21.7

Nielsen, Louise Fuks; Møldrup, Claus. Lay perspective on pharmacogenetics and its application to future drug treatment: a Danish quantitative survey. *New Genetics and Society* 2007 December; 26(3): 309-324. Subject: 15.1

Niemann, Ulrich; Tag, Brigitte. Amputation bei einer Patientin mit einer Psychose in der Vorgeschichte? [Amputation in a patient with a history of psychosis]. *Ethik in der Medizin* 2007 June; 19(2): 128-138. Subject: 8.3.4

Nieto, Antonio; Mazon, Angel; Pamies, Rafael; Linana, Juan J.; Lanuza, Amparo; Jiménez, Fernando Oliver; Medina-Hernandez, Alejandra; Nieto, F. Javier. Adverse effects of inhaled corticosteroids in funded and nonfunded studies. *Archives of Internal Medicine* 2007 October 22; 167(19): 2047-2053. Subject: 1.3.9

Night, Susan S.; Lawrence, Ryan E.; Curlin, Farr A. Negotiating the tension between two integrities: a richer perspective on conscience. *American Journal of Bioethics* 2007 December; 7(12): 24-26; author reply W1-W2. Subject: 4.1.2

Niveau, Gérard. Relevance and limits of the principle of "equivalence of care" in prison medicine. *Journal of Medical Ethics* 2007 October; 33(10): 610-613. Subject: 9.2

Niveau, G.; Materi, J. Psychiatric commitment: over 50 years of case law from the European Court of Human Rights. *European Psychiatry* 2007 January; 22(1): 59-67. Subject: 17.7

Niveau, G.; Materi, J. Psychiatric commitment: over 50 years of case law from the European Court of Human Rights. *European Psychiatry* 2006 October; 21(7): 427-435. Subject: 17.7

Nixon, Ron. DNA tests find branches but few roots. *New York Times* 2007 November 25; p. A1, A7. Subject: 15.11

Noah, Barbara A. The role of religion in the Schiavo controversy. *Houston Journal of Health Law and Policy* 2006; 6(2): 319-346. Subject: 20.5.1

Noakes, J.; Pridham, G. The 'euthanasia' programme 1939-1945. *In their:* Nazism, 1919-1945, Volume 3: Foreign Policy, War and Racial Extermination: A Documen-

See SUBJECT HEADING KEY FOR SECTION II on inside back cover.

567

tary Reader. Exeter: University of Exeter Press, 1998-2001: 389-440. Subject: 20.5.1

Nobbs, Christopher. Probability potentiality. *CQ: Cambridge Quarterly of Healthcare Ethics* 2007 Spring; 16(2): 240-247. Subject: 9.5.7

Noble, John H., Jr. Declaration of Helsinki: dead [letter]. *BMJ: British Medical Journal* 2007 October 13; 335(7623): 736. Subject: 18.2

Nor, Siti Nurani Mohd. The ethics of human cloning: with reference to the Malaysian bioethical discourse. *In:* Roetz, Heiner, ed. Cross-Cultural Issues in Bioethics: The Example of Human Cloning. New York: Rodopi, 2006: 215-246. Subject: 14.5

Norcross, Alastair. Animal experimentation. *In:* Steinbock, Bonnie, ed. The Oxford Handbook of Bioethics. Oxford; New York: Oxford University Press, 2007: 648-667. Subject: 22.2

Nordby, Halvor. Meaning and normativity in nurse-patient interaction. *Nursing Philosophy* 2007 January; 8(1): 16-27. Subject: 8.1

Nordenfelt, Lennart. The logic of health concepts. *In:* Khushf, George, ed. Handbook of Bioethics: Taking Stock of the Field From a Philosophical Perspective. Dordrecht; Boston: Kluwer Academic, 2004: 205-222. Subject: 4.2

Nordmann, Alfred. Knots and strands: an argument for productive disillusionment. *Journal of Medicine and Philosophy* 2007 May-June; 32(3): 217-236. Subject: 5.1

Normile, Dennis. Japan Universities take action [news]. *Science* 2007 January 5; 315(5808): 26. Subject: 1.3.9

Normile, Dennis. Osaka University researchers reject demand to retract Science paper. *Science* 2007 June 22; 316(5832): 1681. Subject: 1.3.9

North Carolina Medical Board. Capital Punishment [position statement]. North Carolina Medical Board, adopted 2007 January; [Online]. Accessed: http://www. ncmedboard.org/Clients/NCBOM/Public/PublicMedia/ capitalpunishment.htm [2007 February 9]. Subject: 20.6

North, Carol S.; Pfefferbaum, Betty; Tucker, Phebe. Ethical and methodological issues in academic mental health research in populations affected by disasters: the Oklahoma City experience relevant to September 11, 2001. *CNS Spectrums* 2002 August; 7(8): 580-584. Subject: 18.4

Northcott, Michael S. In the waters of Babylon: the moral geography of the embryo. *In:* Deane-Drummond, Celia; Scott, Peter Manley, eds. Future Perfect?: God, Medicine and Human Identity. New York: T and T Clark International, 2006: 73-86. Subject: 18.5.4

Northern Ireland Targeting Social Need Renal Group; Kee, Frank; Reaney, Elizabeth; Savage, Gerard; O'Reilly, Dermot; Patterson, Chris; Maxwell, Peter; Fogarty, Damian. Are gatekeepers to renal services referring patients equitably? *Journal of Health Services Re-search and Policy* 2007 January; 12(1): 36-41. Subject: 19.3

Northoff, Georg. The influence of brain implants on personal identity and personality — a combined theoretical and empirical investigation in 'neuroethics'. *In:* Schramme, Thomas; Thome, Johannes, eds. Philosophy and Psychiatry. Berlin; New York: De Gruyter, 2004: 326-344. Subject: 17.1

Nota, L.; Ferrari, L.; Soresi, S.; Wehmeyer, M. Self-determination, social abilities and the quality of life of people with intellectual disability. *Journal of Intellectual Disability Research* 2007 November; 51(11): 850-865. Subject: 4.4

Novack, Gary D. Research ethics. *The Ocular Surface* 2006 April; 4(2): 103-104. Subject: 1.3.9

Novak, David. A Jewish argument for socialized medicine. *In his:* The Sanctity of Human Life. Washington, DC: Georgetown University Press, 2007: 91-110. Subject: 9.1

Novak, David. On the use of embryonic stem cells. *In his:* The Sanctity of Human Life. Washington, DC: Georgetown University Press, 2007: 1-89. Subject: 18.5.4

Novak, David. Physician-assisted suicide. *In his:* The Sanctity of Human Life. Washington, DC: Georgetown University Press, 2007: 111-171. Subject: 20.7

Novas, Carlos. What is the bioscience industry doing to address the ethical issues it faces? *PLoS Medicine* 2006 May; 3(5): e142. Subject: 15.1

Novotny, Thomas E.; Mordini, Emilio; Chadwick, Ruth; Pedersen, J. Martin; Fabbri, Fabrizio; Lie, Reidar; Thanachaiboot, Natapong; Mossialos, Elias; Permanand, Govin. Bioethical implications of globalization: an international consortium project of the European Commission. *PLoS Medicine* 2006 February; 3(2); e43: 0173-0176. Subject: 21.1

Nowak, Rachel. Egg-freezing: a reproductive revolution [news]. *New Scientist* 2007 March 24-30; 193(2596): 8-9. Subject: 14.6

Nowenstein, Graciela. Nemo censetur ignorare legem? Presumed consent to organ donation in France, from Parliament to hospitals. *In:* Garwood-Gowers, Austen; Tingle, John; Wheat, Kay, eds. Contemporary Issues in Healthcare Law and Ethics. Edinburgh; New York: Elsevier Butterworth-Heinemann, 2005: 173-188. Subject: 19.5

Nunes, Fred. Abortion: thinking clearly about controversial public policy. *African Journal of Reproductive Health* 2004 December; 8(3): 11-26. Subject: 12.5.2

Nussenblatt, Robert B.; Gottesman, Michael M. Rules to prevent conflict of interest for clinical investigators conducting human subjects research. *In:* Gallin, John I.; Ognibene, Frederick P., eds. Principles and Practice of Clinical Research. 2nd edition. Oxford: Academic, 2007: 121-127. Subject: 18.1

Nuzzo, Jennifer B. HHS proposes changes to federal quarantine regulations. *Biosecurity and bioterrorism: Biodefense Strategy, Practice, and Science* 2006; 4(1): 11-12. Subject: 9.1

Nuzzo, Jennifer B.; Henderson, Donald A.; O'Toole, Tara; Inglesby, Thomas V. Comments from the Center for Biosecurity of UPMC on proposed revisions to federal quarantine rules. *Biosecurity and Bioterrorism* 2006; 4(2): 204-206. Subject: 9.1

Nycum, Gillian; Reid, Lynette. The harm-benefit trade-off in "bad deal" trials. *Kennedy Institute of Ethics Journal* 2007 December; 17(4): 321-350. Subject: 15.4

Nyika, Aceme. Ethical and regulatory issues surrounding African traditional medicine in the context of HIV/AIDS. *Developing World Bioethics* 2007 April; 7(1): 25-34. Subject: 4.1.1

Nyrhinen, Tarja; Hietala, Marja; Puukka, Pauli; Leino-Kilpi, Helena. Consequences as ethical issues in diagnostic genetic testing — a comparison of the perceptions of patients/parents and personnel. *New Genetics and Society* 2007 April; 26(1): 47-63. Subject: 15.3

Nyrhinen, Tarja; Hietala, Marja; Puukka, Pauli; Leino-Kilpi, Helena. Privacy and equality in diagnostic genetic testing. *Nursing Ethics* 2007 May; 14(3): 295-308. Subject: 15.3

Nys, Herman. Organ transplantation. *In:* Gevers, J.K.M.; Hondius, E.H.; Hubben, J.H., eds. Health Law, Human Rights and the Biomedicine Convention: Essays in Honour of Henriette Roscam Abbing. Leiden; Boston: Martinus Nijhoff Publishers, 2005: 219-230. Subject: 19.5

Nys, Herman; Stultiëns, Loes; Borry, Pascal; Goffin, Tom; Dierickx, Kris. Patient rights in EU Member States after the ratification of the Convention on Human Rights and Biomedicine. *Health Policy* 2007 October; 83(2-3): 223-235. Subject: 9.2

Nys, Thomas. A bridge over troubled water: paternalism as the expression of autonomy. *In:* Nys, Thomas; Denier, Yvonne; Vandevelde, Toon, eds. Autonomy and Paternalism: Reflections on the Theory and Practice of Health Care. Leuven; Dudley, MA: Peeters, 2007: 147-165. Subject: 2.1

O

O'Beirne; Maeve; Stingl, Michael; Hayward, Sarah. Who reviews the projects of unaffiliated researchers for ethics? A case study from Alberta. *CQ: Cambridge Quarterly of Healthcare Ethics* 2007 Summer; 16(3): 346-355. Subject: 18.2

O'Brien, C.M.; Thorburn, T.G.; Sibbel-Linz, A.; McGregor, A.D. Consent for plastic surgical procedures. *Journal of Plastic, Reconstructive and Aesthetic Surgery* 2006; 59(9): 983-989. Subject: 8.3.1

O'Connell, Laurence J. Spirituality in palliative care: an ethical imperative. *In:* Puchalski, Christina M., ed. A Time for Listening and Caring: Spirituality and the Care of the Chronically Ill and Dying. Oxford; New York: Oxford University Press, 2006: 27-38. Subject: 20.4.1

O'Connor, Annette M.; Wennberg, John E.; Legare, France; Llewellyn-Thomas, Hilary A.; Moulton, Benjamin W.; Sepucha, Karen R.; Sodano, Andrea G.; King, Jaime S. Toward the 'tipping point': decision aids and informed patient choice. *Health Affairs* 2007 May-June; 26(3): 716-725. Subject: 8.3.1

O'Donnell, Charlie. Medical training: cooperation problems and solutions. *In:* Watt, Helen, ed. Cooperation, Complicity and Conscience: Problems in Healthcare, Science, Law and Public Policy. London: Linacre Centre, 2005: 118-127. Subject: 7.2

O'Dowd, Adrian. Doctors don't need second signature for abortion [news]. *BMJ: British Medical Journal* 2007 October 27; 335(7625): 844. Subject: 12.4.3

O'Dowd, Adrian. MPs back creation of hybrid embryos [news]. *BMJ: British Medical Journal* 2007 April 14; 334(7597): 764. Subject: 15.1

O'Dowd, Adrian. MPs want to drop second doctor's signature for abortion [news]. *BMJ: British Medical Journal* 2007 November 10; 335(7627): 960. Subject: 12.4.1

O'Dowd, Adrian. No evidence backs reduction in abortion time limit, minister says [news]. *BMJ: British Medical Journal* 2007 November 3; 335(7626): 903. Subject: 12.4.1

O'Dowd, Adrian. Report highlights abuse of older people's human rights. *BMJ: British Medical Journal* 2007 August 25; 335(7616): 367. Subject: 9.5.2

O'Dowd, Adrian. UK may allow creation of "cybrids" for stem cell research [news]. *BMJ: British Medical Journal* 2007 March 10; 334(7592): 495. Subject: 18.5.4

O'Gorman, Mary Lou. Spirituality in end-of-life care from a Catholic perspective: reflections of a hospital chaplain. *In:* Puchalski, Christina M., ed. A Time for Listening and Caring: Spirituality and the Care of the Chronically Ill and Dying. Oxford; New York: Oxford University Press, 2006: 139-154. Subject: 20.4.1

O'Grady, John C. Commentary: a British perspective on the use of restraint and seclusion in correctional mental health care. *Journal of the American Academy of Psychiatry and the Law* 2007; 35(4): 439-443. Subject: 9.5.3

O'Malley, Patricia. Pharmaceutical advertising and clinical nurse specialist practice. *Clinical Nurse Specialist CNS.* 2006 January-February; 20(1): 13-15. Subject: 1.3.7

O'Neil, Graeme. Australia ends stem-cell cloning ban [news brief]. *Nature Biotechnology* 2007 February; 25(2): 153. Subject: 18.5.4

See SUBJECT HEADING KEY FOR SECTION II on inside back cover.

569

O'Neil, Peter. China's doctors signal retreat on organ harvest. *CMAJ/JAMC: Canadian Medical Association Journal* 2007 November 20; 177(11): 1341. Subject: 19.5

O'Neill, Robert D. Xenotransplantation: the solution to the shortage of human organs for transplantation? *Mortality* 2006 May; 11(2): 211-231. Subject: 19.1

O'Reilly, Keviin B. [sic; Kevin]. Confronting eugenics: does the now discredited practice have relevance to today's technology? *AMnews* 2007 July 9: 6 p. [Online]. Accessed: http://www.ama-assn.org/amednews/2007/07/09/prsa0709.htm [2008 February 12]. Subject: 15.5

O'Rourke, Kevin. Ethical reflection continues post-Schiavo. *Health Care Ethics USA* 2007 Winter; 15(1): 13-14. Subject: 20.5.1

O'Sullivan, Gabrielle M. Ethics and welfare issues in animal genetic modification. *In:* Rasko, John E.J.; O'Sullivan, Gabrielle M.; Ankeny, Rachel A., eds. The Ethics of Inheritable Genetic Modification: A Dividing Line? Cambridge: Cambridge University Press, 2006: 103-129. Subject: 15.1

O'Toole, Brian. Four ways we approach ethics. *Journal of Dental Education* 2006 November; 70(11): 1152-1158. Subject: 4.1.1

O'Toole, Brian. Promoting access to oral health care: more than professional ethics is needed. *Journal of Dental Education* 2006 November; 70(11): 1217-1220. Subject: 9.2

O'Toole, Leslie C.; Sobel-Read, Kevin B. Pharmacist refusals: a new twist on the debate over individual autonomy. *Gender Medicine* 2006 March; 3(1): 13-17. Subject: 12.4.3

Obasogie, Osagie. Racial alchemy. It may not be long before genetic skin-lightening treatments are on sale, so it's time to stop pretending colour prejudice isn't a problem. *New Scientist* 2007 August 18-24; 195(2617): 17. Subject: 15.11

Oczak, Malgorzata; Niedzwienska; Agnieszka. Debriefing in deceptive research: a proposed new procedure. *Journal of Empirical Research on Human Research Ethics* 2007 September; 2(3): 49-59. Subject: 18.4

Odling-Smee, Lucy; Giles, Jim; Fuyuno, Ichiko; Cyranoski, David; Marris, Emma. Where are they now? [news]. *Nature* 2007 January 18; 445(7125): 244-245. Subject: 1.3.9

Oeming, Manfred. The Jewish perspective on cloning. *In:* Vöneky, Silja; Wolfrum, Rüdiger, eds. Human Dignity and Human Cloning. Leiden: Nijhoff, 2004: 35-45. Subject: 14.5

Offit, Kenneth; Kohut, Kelly; Clagett, Bartholt; Wadsworth, Eve A.; Lafaro, Kelly J.; Cummings, Shelly; White, Melody; Sagi, Michal; Bernstein, Donna; Davis, Jessica G. Cancer genetic testing and as-

sisted reproduction. *Journal of Clinical Oncology* 2006 October 10; 24(29): 4775-4782. Subject: 15.3

Ogbogu, Ubaka. Canada's approach to conflict-of-interest oversight [letter]. *CMAJ/JAMC: Canadian Medical Association Journal* 2007 August 14; 177(4): 375-376. Subject: 1.3.9

Ogilvie, Gina S.; Remple, Valencia P.; Marra, Fawziah; McNeil, Shelly A.; Naus, Monika; Pielak, Karen L.; Ehlen, Thomas G.; Dobson, Simon R.; Money, Deborah M.; Patrick, David M. Parental intention to have daughters receive the human papillomavirus vaccine [letters and reply]. *CMAJ/JAMC: Canadian Medical Association Journal* 2007 December 4; 177(12): 1506-1512. Subject: 9.5.7

Ogino, Miho. Eugenics, reproductive technologies, and the feminist dilemma in Japan. *In:* LaFleur, William R.; Böhme, Gernot; Shimazono, Susumu, eds. Dark Medicine: Rationalizing Medical Research. Bloomington: Indiana University Press, 2007: 223-232. Subject: 15.5

Oguamanam, Chidi. Biomedical orthodoxy and complementary and alternative medicine: ethical challenges of integrating medical cultures. *Journal of Alternative and Complementary Medicine* 2006 July-August; 12(6): 577-581. Subject: 4.1.1

Oguz, N. Yasemin; Kavas, M. Volkan; Aksu, Murat. Teaching thanatology: a qualitative and quantitative study. *Eubios Journal of Asian and International Bioethics* 2007 November; 17(6): 172-177. Subject: 2.3

Oh, Do-Youn; Kim, Jee-Hyun; Kim, Dong-Wan; Im, Seock-Ah; Kim, Tae-You; Heo, Dae Seog; Bang, Yung-Jue; Kim, Noe Kyeong. CPR or DNR? End-of-life decision in Korean cancer patients: a single center's experience. *Supportive Care in Cancer* 2006 February; 14(2): 103-108. Subject: 20.5.1

Ohlin, Jens David. Is the concept of the person necessary for human rights? *Columbia Law Review* 2005 January; 105(1): 209-249. Subject: 4.4

Øhrstrøm, Peter; Dyhrberg, Johan. Ethical problems inherent in psychological research based on Internet communication as stored information. *Theoretical Medicine and Bioethics* 2007; 28(3): 221-241. Subject: 18.4

Okarma, Tom. Adult stem cells won't do. *New Scientist* 2007 March 10-16; 193(2594): 20. Subject: 18.5.1

Okike, Kanu; Kocher, Mininder S.; Mehlman, Charles T.; Bhandari, Mohit. Conflict of interest in orthopaedic research. An association between findings and funding in scientific presentations. *Journal of Bone and Joint Surgery. American volume* 2007 March; 89(3): 608-613. Subject: 1.3.9

Öksüzoglu, Berna; Abali, Hüseyin; Bakar, Meral; Yildirim, Nuriye; Zengin, Nurullah. Disclosure of cancer diagnosis to patients and their relatives in Turkey: views of accompanying persons and influential factors in reaching those views. *Tumori* 2006 January-February; 92(1): 62-66. Subject: 8.2

Oldani, Michael. Can doctors take back the script? Understanding the total system of prescription generation. *Atrium* 2007 Summer; 4: 15-17, 28. Subject: 9.7

Oleson, Christopher; Rhonheimer, Martin; Cole, Basil. More on the contraceptive choice [letters and reply]. *National Catholic Bioethics Quarterly* 2007 Winter; 7(4): 649-654. Subject: 11.1

Oliffe, John; Thorne, Sally; Hislop, T.Gregory; Armstrong, Elizabeth-Anne. "Truth telling" and cultural assumptions in an era of informed consent. *Family and Community Health* 2007 January-March; 30(1): 5-15. Subject: 8.2

Oliver, D. A perspective on euthanasia. *British Journal of Cancer* 2006 October 23; 95(8): 953-954. Subject: 20.5.1

Olsen, Douglas. Editorial comment: nursing and other health care disciplines have a longstanding tradition of conscientious objection. *Nursing Ethics* 2007 May; 14(3): 277-279. Subject: 4.1.3

Olsen, Douglas P. Arranging live organ donation over the Internet: is it ethical? Each nurse must decide. *AJN: American Journal of Nursing* 2007 March; 107(3): 69-72. Subject: 19.5

Olsen, Douglas P. Unwanted treatment: what are the ethical implications? *AJN: American Journal of Nursing* 2007 September; 107(9): 51-53. Subject: 8.3.4

Olsen, Jørn. Kafka's truth-seeking dogs. *Epidemiology* 2006 May; 17(3): 242-243. Subject: 1.3.9

Olsson, I. Anna S.; Hansen, Axel K.; Sandøe, Peter. Ethics and refinement in animal research [letter]. *Science* 2007 September 21; 317(5845): 1680. Subject: 22.2

Olthuis, Gert; Leget, Carlo; Dekkers, Wim. Why hospice nurses need high self-esteem. *Nursing Ethics* 2007 January; 14(1): 62-71. Subject: 20.4.1

Omonzejele, Peter F. Obligation of non-maleficence: moral dilemma in physician-patient relationship. *Journal of Medicine and Biomedical Research* 2005 June; 4(1): 7 p. [Online]. Accessed: http://www.bioline.org.br/titles?id=jm&year=2005&vol=4&num=01&keys=V4N1 [2008 February 14]. Subject: 8.1

Onder, Robert. "People need a fairy tale": the embryonic stem cell and cloning debate in Missouri. *Missouri Medicine* 2006 March-April; 103(2): 106-111. Subject: 18.5.4

Orbinski, James; Beyrer, Chris; Singh, Sonal. Violations of human rights: health practitioners as witnesses. *Lancet* 2007 August 25-31; 370(9588): 698-704. Subject: 21.1

Oregon. Department of Human Services. Public Health Division. Death with Dignity Act history. Salem: Department of Human Services, n.d.; 1 p. [Online]. Accessed: http://egov.oregon.gov/DHS/ph/pas/docs/History.pdf [2007 April 12]. Subject: 20.5.1

Orentlicher, David. Bioethics and society: from the ivory tower to the state house. *In:* Eckenwiler, Lisa A.; Cohn, Felicia G., eds. The Ethics of Bioethics: Mapping the Moral Landscape. Baltimore, MD: Johns Hopkins University Press, 2007: 74-82. Subject: 21.1

Orlikoff, James E.; Totten, Mary K. Conflict of interest and governance. New approaches for a new healthcare environment. *Healthcare Executive* 2006 September-October; 21(5): 52, 54. Subject: 9.1

Ormond, K.E.; Iris, M.; Banuvar, S.; Minogue, J.; Annas, G.J.; Elias, S. What do patients prefer: informed consent models for genetic carrier testing. *Journal of Genetic Counseling* 2007 August; 16(4): 539-550. Subject: 15.3

Ormondroyd, E.; Moynihan, C.; Watson, M.; Foster, C.; Davolls, S.; Ardern-Jones, A.; Eeles, R. Disclosure of genetic research results after the death of the patient participant: a qualitative study of the impact on relatives. *Journal of Genetic Counseling* 2007 August; 16(4): 527-538. Subject: 15.1

Orr, Robert D. The role of Christian belief in public policy. *Christian Bioethics* 2007 May-August; (13)2: 199-209. Subject: 2.1

Orr, Robert D.; Lawrence, Ryan E.; Curlin, Farr A. The role of moral complicity in issues of conscience. *American Journal of Bioethics* 2007 December; 7(12): 23-24; author reply W1-W2. Subject: 4.1.2

Orr; Robert D.; Craig, Debra. Old enough [case study and commentaries]. *Hastings Center Report* 2007 November-December; 37(6): 15-16. Subject: 8.3.4

O'Rourke, Kevin D. Artificial nutrition and hydration and the Catholic tradition: the Terri Schiavo case had even members of Congress debating the issue. *Health Progress* 2007 May-June; 88(3): 50-54. Subject: 20.5.1

Ortega, N.L.; Bicaldo, B.F.; Sobritchea, C.; Tan, M.L. Exploring the realities of HIV/AIDS-related discrimination in Manila, Philippines. *AIDS Care* 2005 July; 17(Supplement 2): S153-S164. Subject: 9.5.6

Ortendahl, M. Risk in public health and clinical work [letter]. *Journal of Medical Ethics* 2007 April; 33(4): 246. Subject: 9.1

Osborn, Andrew. Three Russian doctors face trial for vaccine tests [news]. *BMJ: British Medial Journal* 2007 April 21; 334(7598): 817. Subject: 18.2

Ostrer, H.; Wilson, D.I.; Hanley, N.A. Human embryo and early fetus research. *Clinical Genetics* 2006 August; 70(2): 98-107. Subject: 18.5.4

Otlowski, Margaret. Donor perspectives on issues associated with donation of genetic samples and information: an Australian viewpoint. *Journal of Bioethical Inquiry* 2007; 4(2): 135-150. Subject: 19.5

See SUBJECT HEADING KEY FOR SECTION II on inside back cover.

571

Otto, Sheila. Memento . . . life imitates art: the request for an ethics consultation. *Journal of Clinical Ethics* 2007 Fall; 18(3): 247-251. Subject: 9.6

Ouellette, Alicia; Cohen, Beverly; Reider, Jacob. Practical, state, and federal limits on the scope of compelled disclosure of health records. *American Journal of Bioethics* 2007 March; 7(3): 46-48. Subject: 8.4

Outomuro, Delia; Lott, James P.; Savulescu, Julian. Moral dilemmas around a global human embryonic stem cell bank. *American Journal of Bioethics* 2007 August; 7(8): 47-48; author reply W4-W6. Subject: 18.5.4

Outomuro, Delia; Moreno, Jonathan D.; Berger, Sam. Critiques on biotechnology and the problem of pigeonholing philosophical thinking. *American Journal of Bioethics* 2007 October; 7(10): 25-27; author reply W1-W3. Subject: 5.1

Øverland, Gerhard. Survival lotteries reconsidered. *Bioethics* 2007 September; 21(7): 355-363. Subject: 19.5

Owen, Michael. Ethical review of social and behavioral science research. *In:* Kulakowski, Elliott C.; Chronister, Lynne U., eds. Research Administration and Management. Sudbury, MA: Jones and Bartlett, 2006: 543-556. Subject: 18.4

Owen, Paul. Government gives ground on NHS database. *Guardian Unlimited* 2006 December 18: 2 p. [Online]. Accessed: http://politics.guardian.co.uk/print/0,,329665933-110251,00.html [2006 December 18]. Subject: 1.3.12

Oyebode, Femi. The Mental Capacity Act 2005. *Clinical Medicine* 2006 March-April; 6(2): 130-131. Subject: 8.3.3

Ozakinci, Gozde; Humphris, Gerry; Steel, Michael. Provision of breast cancer risk information to women at the lower end of the familial risk spectrum. *Community Genetics* 2007; 10(1): 41-44. Subject: 15.2

Ozar, David T. Applying systems thinking to oral health care: commentary on Dr. Patricia H. Werhane's article. *Journal of Dental Education* 2006 November; 70(11): 1196-1197. Subject: 9.2

Ozar, David T. Basic oral health needs: a public priority. *Journal of Dental Education* 2006 November; 70(11): 1159-1165. Subject: 9.4

Ozar, David T. Conflicting values in oral health care. *Journal of the American College of Dentists* 1998 Fall; 65(3): 15-18. Subject: 4.1.1

Ozar, David T. Ethics, access, and care. *Journal of Dental Education* 2006 November; 70(11): 1139-1145. Subject: 9.2

Ozdogan, Mustafa; Samur, Mustafa; Artac, Mehmet; Yildiz, Mustafa; Savas, Burhan; Bozcuk, Hakan Sat. Factors related to truth-telling practice of physicians treating patients with cancer in Turkey. *Journal of Palliative Medicine* 2006 October; 9(5): 1114-1119. Subject: 18.5.1

Ozgen, C. Ethics in the use of new medical technologies for neurosurgery: "Islamic viewpoint". *Acta Neurochirurgica. Supplement* 2006; 98: 13-17. Subject: 17.1

P

Padela, Aasim I. Islamic medical ethics: a primer. *Bioethics* 2007 March; 21(3): 169-178. Subject: 2.1

Padgett, Barry L.; Haas, Thomas. An ethical wrinkle on the face of therapy claims. *Plastic Surgical Nursing* 2004 July-September; 24(3): 123-126. Subject: 9.5.1

Page, Cameron. Hope. *Hastings Center Report* 2007 November-December; 37(6): 12. Subject: 8.2

Page, Stacey A.; Mitchell, Ian. Patients' opinions on privacy, consent and the disclosure of health information for medical research. *Chronic Diseases in Canada* 2006; 27(2): 60-67. Subject: 8.4

Pagliari, Claudia; Detmer, Don; Singleton, Peter. Potential of electronic personal health records. *BMJ: British Medical Journal* 2007 August 18; 335(7615): 330-333. Subject: 8.4

Pakes, Francis. The legislation of euthanasia and assisted suicide: a tale of two scenarios. *International Journal of the Sociology of Law* 2005 June; 33(2): 71-84. Subject: 20.7

Palac, Diane M. Is it justified to breach confidentiality to protect a patient from abuse? *Medical Ethics Newsletter [Lahey Clinic]* 2007 Spring; 14(2): 3. Subject: 8.4

Paley, John. Caring as a slave morality: Nietzschean themes in nursing ethics. *Journal of Advanced Nursing* 2002 October; 40(1): 25-35. Subject: 4.1.1

Paley, John. Past caring: the limitations of one-to-one ethics. *In:* Davis, Anne J.; Tschudin, Verena; de Raeve, Louise, eds. Essentials of Teaching and Learning in Nursing Ethics: Perspectives and Methods. New York: Churchill Livingstone Elsevier, 2006: 149-164. Subject: 4.1.3

Palmboom, G.G.; Willems, D.L.; Janssen, N.B.A.T.; de Haes, J.C.J.M. Doctor's views on disclosing or withholding information on low risks of complication. *Journal of Medical Ethics* 2007 February; 33(2): 67-70. Subject: 8.3.1

Palmer, Julie Gage. Governmental regulation of genetic technology, and the lessons learned. *In:* Knowles, Lori P.; Kaebnick, Gregory E., eds. Reprogenetics: Law, Policy, and Ethical Issues. Baltimore: Johns Hopkins University Press, 2007: 20-63. Subject: 15.1

Palmer, Lyle J. UK Biobank: bank on it. *Lancet* 2007 June 16-22; 369(9578): 1980-1981. Subject: 15.1

Palmer, Robert Chi-Noodin; Palmer, Marianne Leslie. Ojibwe beliefs and rituals in end-of-life care. *In:* Puchalski, Christina M., ed. A Time for Listening and Caring: Spirituality and the Care of the Chronically Ill and Dying. Oxford; New York: Oxford University Press, 2006: 215-225. Subject: 20.4.1

Palombo, Enzo A.; Bhave, Mrinal. Genetically transformed healthcare: healthy children and parents. *In:* Betta, Michela, ed. The Moral, Social, and Commercial Imperatives of Genetic Testing and Screening: the Australian Case. Dordrecht: Springer, 2006: 185-199. Subject: 15.3

Panda, Mukta; Heath, Gregory W.; Desbiens, Norman A.; Moffitt, Benjamin. Research status of case reports for medical school institutional review boards [letter]. *JAMA: The Journal of the American Medical Association* 2007 September 19; 298(11): 1277-1278. Subject: 18.2

Pandya, Dipak P.; Dave, Jay. Protection of human subjects in clinical research: the pitfalls in clinical research. *Comprehensive Therapy* 2005 Spring; 31(1): 72-77. Subject: 18.2

Panforti, Maria Donata; Serio, Mario. Genetics and artificial procreation in Italy. *In:* Meulders-Klein, Marie-Thérèse; Deech, Ruth; Vlaardingerbroek, Paul, eds. Biomedicine, the Family and Human Rights. New York: Kluwer Law International, 2002: 253-257. Subject: 14.1

Pang, Ching Ling. A code of ethics for scientists. *Lancet* 2007 March 31-April 6; 369(9567): 1068. Subject: 1.3.9

Pang, Samantha Mei-che. The principle-based approach to nursing ethics: a critical analysis. *In:* Davis, Anne J.; Tschudin, Verena; de Raeve, Louise, eds. Essentials of Teaching and Learning in Nursing Ethics: Perspectives and Methods. New York: Churchill Livingstone Elsevier, 2006: 67-79. Subject: 4.1.3

Panicola, Michael R. Making decisions about medically administered nutrition and hydration. *In:* Hamel, Ronald, ed. Making Health Care Decisions: A Catholic Guide. Liguori, MO: Liguori Publications, 2006: 109-126. Subject: 20.5.1

Panikkar, Bindu; Brugge, Doug. The ethical issues in uranium mining research in the Navajo nation. *Accountability in Research* 2007 April-June; 14(2): 121-153. Subject: 18.5.1

Pantel, Johannes; Haberstroh, Julia. Psychopharmakaverordnung im Altenpflegeheim - Zwischen indikationsgeleiteter Therapie und „Chemical Restraint" = Psychotropic drug use in nursing homes - between adequate care and "chemical restraint". *Ethik in der Medizin* 2007 December; 19(4): 258-269. Subject: 17.4

Paonessa, Louis. Straightening out your heir: on the constitutionality of regulating the use of preimplantation technologies to select preembryos or modify the genetic profile thereof based on expected sexual orientation. *Rutgers Computer and Technology Law Journal* 2007; 33(2): 331-366. Subject: 15.2

Papadimitriou, John D.; Skiadas, Panayiotis; Mavrantonis, Constantino S.; Polimeropoulos, Vassilis; Papadimitriou, Dimitris J.; Papacostas, Kyriaki J. Euthanasia and suicide in antiquity: viewpoint of the dramatists and philosophers. *Journal of the Royal Society of Medicine* 2007 January; 100(1): 25-28. Subject: 20.5.1

Paradies, Yin C.; Montoya, Michael J.; Fullerton, Stephanie M. Racialized genetics and the study of complex diseases: the thrifty genotype revisited. *Perspectives in Biology and Medicine* 2007 Spring; 50(2): 203-227. Subject: 15.1

Paradise, Jordan; Janson, Christopher. Decoding the research exemption. *Nature Reviews Genetics* 2006 February; 7(2): 148-154. Subject: 15.8

Parens, Erik. Creativity, gratitude, and the enhancement debate. *In:* Illes, Judy, ed. Neuroethics: Defining the Issues in Theory, Practice, and Policy. New York: Oxford University Press, 2006: 75-86. Subject: 17.1

Parens, Erik; Knowles, Lori P. Reprogenetics and public policy: reflections and recommendations. *In:* Knowles, Lori P.; Kaebnick, Gregory E., eds. Reprogenetics: Law, Policy, and Ethical Issues. Baltimore: Johns Hopkins University Press, 2007: 253-294. Subject: 14.1

Paris, John. A life too burdensome. *Tablet* 2007 January 6; 261(8672): 10-11. Subject: 20.5.1

Paris, John J.; Moreland, Michael P.; Whitney, Simon N.; McCullough, Laurence B. Silence is not always golden in medical decision-making. *American Journal of Bioethics* 2007 July; 7(7): 39-40; author reply W1-W3. Subject: 8.1

Paris, John J.; Schreiber, Michael D.; Moreland, Michael P. Parental refusal of medical treatment for a newborn. *Theoretical Medicine and Bioethics* 2007; 28(5): 427-441. Subject: 8.3.4

Paris, J.J.; Billinngs, J.A.; Cummings, B.; Moreland, M.P. Howe v. MGH and Hudson v. Texas Children's Hospital: two approaches to resolving family-physician disputes in end-of-life care. *Journal of Perinatology* 2006 December; 26(12): 726-729. Subject: 20.5.1

Paris, J.J.; Graham, N.; Schreiber, M.D.; Goodwin, M. Approaches to end-of-life decision-making in the NICU: insights from Dostoevsky's The Grand Inquisitor. *Journal of Perinatology* 2006 July; 26(7): 389-391. Subject: 20.5.2

Parke, David W.; Durfee, David A.; Zacks, Charles M.; Orloff, Paul N., eds. Ethics in ophthalmology. *In their:* The Profession of Ophthalmology: Practice Management, Ethics, and Advocacy. San Francisco: American Academy of Ophthalmology, 2005: 167- 247. Subject: 4.1.2

Parker, C. Ethics for embryos. *Journal of Medical Ethics* 2007 October; 33(10): 614-616. Subject: 18.5.4

Parker, C. Perspectives on ethics. *Journal of Medical Ethics* 2007 January; 33(1): 21-23. Subject: 2.1

Parker, H. Stewart. Reply to "Poor trial design leaves gene therapy death a mystery" [letter]. *Nature Medicine* 2007 November; 13(11): 1276. Subject: 15.4

Parker, Lisa S. Bioethics as activism. *In:* Eckenwiler, Lisa A.; Cohn, Felicia G., eds. The Ethics of Bioethics: Mapping the Moral Landscape. Baltimore, MD: Johns Hopkins University Press, 2007: 144-157. Subject: 21.1

See SUBJECT HEADING KEY FOR SECTION II on inside back cover.

573

Parker, Lisa S. Ethical expertise, maternal thinking, and the work of clinical ethicists. *In:* Rasmussen, Lisa, ed. Ethics Expertise: History, Contemporary Perspectives, and Applications. Dordrecht: Springer, 2005: 165-207. Subject: 9.6

Parker, Lisa S.; Satkoske, Valerie B. Conflicts of interest: are informed consent an appropriate model and disclosure an appropriate remedy? *Journal of the American College of Dentists* 2007 Summer; 74(2): 19-26. Subject: 7.3

Parker, Malcolm. Patients as rational traders: response to Stewart and DeMarco. *Journal of Bioethical Inquiry* 2006; 3(3): 133-136. Subject: 8.1

Parker, Michael. Deliberation and moral courage: the UK Genethics Club as a case study. *Notizie di Politeia* 2006; 22(81): 78-83. Subject: 2.3

Parker, Michael. The best possible child. *Journal of Medical Ethics* 2007 May; 33(5): 279-283. Subject: 15.5

Parker, Richard G. Sexuality, health and human rights [editorial]. *American Journal of Public Health* 2007 June; 97(6): 972-973. Subject: 10

Parkes, Georgina; Hall, Ian. Gender dysphoria and cross-dressing in people with intellectual disability: a literature review. *Mental Retardation* 2006 August; 44(4): 260-271. Subject: 9.5.3

Parmet, Wendy E. Legal power and legal rights — isolation and quarantine in the case of drug-resistant tuberculosis. *New England Journal of Medicine* 2007 August 2; 357(5): 433-435. Subject: 9.1

Parmet, Wendy E. Pharmaceuticals, public health, and the law: a public health perspective. *In:* Cohen, Jillian Clare; Illingworth, Patricia; Schüklenk, Udo, eds. The Power of Pills: Social, Ethical and Legal Issues in Drug Development, Marketing, and Pricing. London; Ann Arbor, MI: Pluto, 2006: 77-87. Subject: 9.7

Parmet, Wendy E. Public health and constitutional law: recognizing the relationship. *Journal of Health Care Law and Policy* 2007; 10(1): 13-25. Subject: 9.1

Parry, Jane. A matter of life and death. China is moving towards changing its transplantation practices in a bid to gain wider international acceptance [news]. *BMJ: British Medical Journal* 2007 November 10; 335(7627): 961. Subject: 19.5

Parry, Sarah. (Re)constructing embryos in stem cell research: exploring the meaning of embryos for people involved in fertility treatments. *Social Science and Medicine* 2006 May; 62(10): 2349-2359. Subject: 18.5.4

Parsi, Kayhan; Braun, Joshua A. Media and health: are bioethicists just another interest group? *American Journal of Bioethics* 2007 August; 7(8): 18-19; author reply W1-W2. Subject: 1.3.7

Parsi, Kayhan; Kuczewski, Mark G. Failure to thrive: can education save the life of ethics consultation? *American Journal of Bioethics* 2007 February; 7(2): 37-39. Subject: 9.6

Parsons, Brian; Kennedy, Miriam. A review of recorded information given to patients starting to take clozapine and the development of guidelines on disclosure, a key component of informed consent. *Journal of Medical Ethics* 2007 October; 33(10): 564-567. Subject: 8.3.1

Parvizi, Javad; Tarity, T. David; Conner, Kyle; Smith, J. Bruce. Institutional review board approval: why it matters. *Journal of Bone and Joint Surgery. American volume* 2007 February; 89(2): 418-426. Subject: 18.2

Pascal, Chris B. Beyond the federal definition: other forms of misconduct. *In:* Kulakowski, Elliott C.; Chronister, Lynne U., eds. Research Administration and Management. Sudbury, MA: Jones and Bartlett, 2006: 523-530. Subject: 1.3.9

Pasternak, Ryan H.; Geller, Gail; Parrish, Catherine; Cheng, Tina L. Adolescent and parent perceptions on youth participation in risk behavior research. *Archives of Pediatrics and Adolescent Medicine* 2006 November; 160(11): 1159-1166. Subject: 18.5.2

Patel, Mitesh S.; Chernew, Michael E. The impact of the adoption of gag laws on trust in the patient-physician relationship. *Journal of Health Politics, Policy and Law* 2007 October; 32(5): 819-842. Subject: 9.3.2

Paterson, Brodie; Duxbury, Joy. Restraint and the question of validity. *Nursing Ethics* 2007 July; 14(4): 535-545. Subject: 17.3

Patfield, Martyn. The 'mentally disordered' provisions of the New South Wales Mental Health Act 1990: their ethical standing and effect on services. *Australasian Psychiatry* 2006 September; 14(3): 263-266. Subject: 9.5.3

Patni, Shalini; Wagstaff, John; Tofazzal, Nasima; Bonduelle, Myriam; Moselhi, Marsham; Kevelighan, Euan; Edwards, Steve. Metastatic unknown primary tumour presenting in pregnancy: a rarity posing an ethical dilemma. *Journal of Medical Ethics* 2007 August; 33(8): 442-443. Subject: 9.5.5

Patowary, S. Pharmacogenomics — therapeutic and ethical issues. *Kathmandu University Medical Journal (KUMJ)* 2005 October-December; 3(4): 428-430. Subject: 15.1

Patthoff, Donald. Defining the ethical organization in oral health care. *Journal of the American College of Dentists* 1998 Fall; 65(3): 24-26. Subject: 4.1.1

Patthoff, Donald E. How did we get here? Where are we going? Hopes and gaps in access to oral health care. *Journal of Dental Education* 2006 November; 70(11): 1125-1132. Subject: 4.1.1

Patthoff, Donald E. The need for dental ethicists and the promise of universal patient acceptance: response to Richard Masella's "Renewing professionalism in dental education". *Journal of Dental Education* 2007 February; 71(2): 222-226. Subject: 4.1.1

Pattinson, Shaun D. Designing donors. *In:* Gunning, Jennifer; Holm, Søren, eds. Ethics, law, and society. Volume 1. Aldershot, Hants, England; Burlington, VT: Ashgate, 2005: 251-256. Subject: 15.2

Patton, Cindy. Bullets, balance, or both: medicalisation in HIV treatment. *Lancet* 2007 February 24-March 2; 369(9562): 706-707. Subject: 9.5.6

Paul, Diane B. On drawing lessons from the history of eugenics. *In:* Knowles, Lori P.; Kaebnick, Gregory E., eds. Reprogenetics: Law, Policy, and Ethical Issues. Baltimore: Johns Hopkins University Press, 2007: 3-19. Subject: 15.5

Paul, Jobst. Das bio-ethische Netzwerk. *In:* Neuer-Miebach, Therese; Wunder, Michael, eds. Bio-Ethik und die Zukunft der Medizin. Bonn: Psychiatrie-Verlag, 1998: 60-71. Subject: 2.1

Paul, Norbert W.; Fangerau, Heiner. Why should we bother? Ethical and social issues in individualized medicine. *Current Drug Targets* 2006 December; 7(12): 1721-1727. Subject: 9.7

Pawelzik, Markus; Prinz, Aloys. The moral economics of psychotherapy. *In:* Schramme, Thomas; Thome, Johannes, eds. Philosophy and Psychiatry. Berlin; New York: De Gruyter, 2004: 370-386. Subject: 17.2

Paxson, Heather. Reproduction as spiritual kin work: orthodoxy, IVF, and the moral economy of motherhood in Greece. *Culture, Medicine and Psychiatry* 2006 December; 30(4): 481-505. Subject: 14.1

Paxton, Lynn A.; Hope, Tony; Jaffe, Harold W. Pre-exposure prophylaxis for HIV infection: what if it works? *Lancet* 2007 July 7-13; 370(9581): 89-93. Subject: 9.5.6

Pear, Robert. Medicare says it won't cover "preventable" hospital errors. *New York Times* 2007 August 19; p A1, A20. Subject: 9.8

Pearce, David. Economics and genetic diversity. *Futures* 1987; 19(6): 710-712. Subject: 15.1

Pearson, Helen. Cancer patients opt for unapproved drug [news]. *Nature* 2007 March 29; 446(7135): 474-475. Subject: 9.7

Pearson, Helen. Infertility researchers target uterus transplant [news]. *Nature* 2007 February 1; 445(7127): 466-467. Subject: 19.1

Pearson, Yvette. Never let me clone?: countering an ethical argument against the reproductive cloning of humans. *EMBO Reports* 2006 July; 7(7): 657-660. Subject: 14.5

Pearson, Yvette E. Storks, cabbage patches, and the right to procreate. *Journal of Bioethical Inquiry* 2007; 4(2): 105-115. Subject: 14.1

Pegoraro, Renzo; Putoto, Giovanni. Findings from a European survey on current bioethics training activities in hospitals. *Medicine, Health Care and Philosophy* 2007 March; 10(1): 91-96. Subject: 2.3

Peiffer, Jürgen. Phases in the postwar German reception of the "Euthanasia Program" (1939-1945) involving the killing of the mentally disabled and its exploitation by neuroscientists. *Journal of the History of the Neurosciences* 2006 September; 15(3): 210-244. Subject: 20.5.1

Pellegrino, Edmund D. Character formation and the making of good physicians [M575]. *In:* Kenny, Nuala and Shelton, Wayne, eds. Lost Virtue: Professional Character Development in Medical Education. Amsterdam; Oxford: Elsevier, 2006: 1-15. Subject: 4.1.2

Pellegrino, Edmund D. Culture and bioethics: where ethics and mores meet. *In:* Prograis, Lawrence J.; Pellegrino, Edmund D., eds. African American Bioethics: Culture, Race, and Identity. Washington, DC: Georgetown University Press, 2004: ix-xxi. Subject: 21.7

Pellegrino, Edmund D. Philosophy of medicine and medical ethics: a phenomenological perspective. *In:* Khushf, George, ed. Handbook of Bioethics: Taking Stock of the Field From a Philosophical Perspective. Dordrecht; Boston: Kluwer Academic, 2004: 183-202. Subject: 2.1

Pellegrino, Edmund D. Professing medicine, virtue based ethics, and the retrieval of professionalism. *In:* Walker, Rebecca L.; Ivanhoe, Philip J., eds. Working Virtue: Virtue Ethics and Contemporary Moral Problems. Oxford: Clarendon, 2007: 61-85. Subject: 4.1.2

Peltier, Bruce. Codes and colleagues: is there support for Universal Patient Acceptance. *Journal of Dental Education* 2006 November; 70(11): 1221-1225. Subject: 9.2

Peltier, Bruce. Response to unethical behavior in oral health care. *Journal of the American College of Dentists* 1998 Fall; 65(3): 19-23. Subject: 4.1.1

Penasa, Simone. The issue of constitutional law legitimacy on "human assisted reproduction" between reasonableness of the choices and effectiveness of the protection of all involved subjects. *Revista de Derecho y Genoma Humano = Law and the Human Genome Review* 2006 July-December; (25): 117-137. Subject: 15.2

Pence, Gregory. What's so good about natural motherhood? (In praise of unnatural gestation). *In:* Gelfand, Scott; Shook, John R., eds. Ectogenesis: Artificial Womb Technology and the Future of Human Reproduction. Amsterdam; New York: Editions Rodopi, B.V., 2006: 77-88. Subject: 14.1

Pence, Gregory E. Are genetic abortions eugenic? *In his:* The Elements of Bioethics. Boston: McGraw-Hill, 2007: 172-202. Subject: 15.5

Pence, Gregory E. Can research be just on people with schizophrenia? *In his:* The Elements of Bioethics. Boston: McGraw-Hill, 2007: 203-232. Subject: 18.5.6

See SUBJECT HEADING KEY FOR SECTION II on inside back cover.

575

Pence, Gregory E. Emotivism and banning some concep-tions. *In his:* The Elements of Bioethics. Boston: McGraw-Hill, 2007: 109-136. Subject: 14.4

Pence, Gregory E. Is there a duty to die? *In his:* The Ele-ments of Bioethics. Boston: McGraw-Hill, 2007: 233-262. Subject: 20.5.1

Pence, Gregory E. Kant on whether alcoholism is a dis-ease. *In his:* The Elements of Bioethics. Boston: McGraw-Hill, 2007: 21-51. Subject: 9.5.9

Pence, Gregory E. Kant's critique of adult organ dona-tion. *In his:* The Elements of Bioethics. Boston: McGraw-Hill, 2007: 52-80. Subject: 19.5

Pence, Gregory E. Lying to patients and ethical relativ-ism. *In his:* The Elements of Bioethics. Boston: McGraw-Hill, 2007: 1-20. Subject: 8.2

Pence, Gregory E. Terri Schiavo: when does personhood end? *In his:* The Elements of Bioethics. Boston: McGraw-Hill, 2007: 137-171. Subject: 20.5.1

Pence, Gregory E. Treating Jehovah's Witnesses profes-sionally. *In his:* The Elements of Bioethics. Boston: McGraw- Hill, 2007: 263-279. Subject: 8.3.4

Pence, Gregory E. Utilitarians vs. Kantians on stopping AIDS. *In his:* The Elements of Bioethics. Boston: McGraw-Hill, 2007: 81-108. Subject: 9.5.6

Peniston, Reginald L. Does an African American per-spective alter clinical ethical decision making at the bed-side? *In:* Prograis, Lawrence J.; Pellegrino, Edmund D., eds. African American Bioethics: Culture, Race, and Iden-tity. Washington, DC: Georgetown University Press, 2004: 127-136. Subject: 8.1

Pennicuik, Susan. The Australian law reform inquiry into genetic commission testing — a worker's perspective. *In:* Betta, Michela, ed. The Moral, Social, and Commercial Imperatives of Genetic Testing and Screening: the Austra-lian Case. Dordrecht: Springer, 2006: 201-210. Sub-ject: 15.3

Pennings, Guido. Directed organ donation: discrimina-tion or autonomy? *Journal of Applied Philosophy* 2007; 24(1): 41-49. Subject: 19.6

Pennings, Guido. International parenthood via procre-ative tourism. *In:* Shenfield, Françoise; Sureau, Claude, eds. Contemporary Ethical Dilemmas in Assisted Repro-duction. Abingdon: Informa Healthcare, 2006: 43-56. Subject: 14.1

Pennock, Robert T. Pre-existing conditions: genetic test-ing, causation, and the justice of medical insurance. *In:* Rhodes, Rosamond; Francis, Leslie P.; Silvers, Anita, eds. The Blackwell Guide to Medical Ethics. Malden, MA: Blackwell Pub., 2007: 407-424. Subject: 15.3

Pennsylvania Catholic Health Association. Draft princi-ples and guidelines for non-heart-beating organ donation. *National Catholic Bioethics Quarterly* 2007 Autumn; 7(3): 563-566. Subject: 20.2.1

Pentz, Rebecca D.; Billot, Laurent; Wendler, David. Research on stored biological samples: views of African American and White American cancer patients. *American Journal of Medical Genetics. Part A* 2006 April 1; 140(7): 733-739. Subject: 15.1

Pentz, Rebecca D.; Flamm, Anne L.; Sugarman, Jeremy; Cohen, Marlene Z.; Xu, Zhiheng; Herbst, Roy S.; Abbruzzese, James L. Who should go first in trials with scarce agents? The views of potential participants. *IRB: Ethics and Human Research* 2007 July-August; 29(4): 1-6. Subject: 18.2

Pentz, Rebecca D.; Joffe, Steven; Emanuel, Ezekiel J.; Schnipper, Lowell E.; Haskell, Charles M.; Tannock, Ian F. ASCO core values. *Journal of Clinical Oncology* 2006 December 20; 24(36): 5780-5782. Subject: 6

Peppin, Patricia. Directing consumption: direct-to-con-sumer advertising and global public health. *In:* Bennett, Belinda; Tomossy, George F., eds. Globalization and Health: Challenges for Health Law and Bioethics. Dor-drecht: Springer, 2006: 109-128. Subject: 9.7

Peppin, Patricia. The power of illusion and the illusion of power: direct-to-consumer advertising and Canadian health care. *In:* Flood, Colleen M., ed. Just Medicare: What's In, What's Out, How We Decide. Buffalo, NY: University of Toronto Press, 2006: 355-378. Subject: 9.7

Perel, Pablo; Roberts, Ian; Sena, Emily; Wheble, Philipa; Briscoe, Catherine; Sandercock, Peter; Macleod, Malcolm; Mignini, Luciano E.; Jayaram, Predeep; Khan, Khalid S. Comparison of treatment ef-fects between animal experiments and clinical trials: sys-tematic review. *BMJ: British Medical Journal* 2007 Jan-uary 27; 334(7586): 197-200. Subject: 22.2

Pérez-Cárceles, M.D.; Lorenzo, M.D.; Luna, A.; Osuna, E. Elderly patients also have rights. *Journal of Medical Ethics* 2007 December; 33(12): 712-716. Sub-ject: 9.5.2

Pergament, Eugene. Controversies and challenges of ar-ray comparative genomic hybridization in prenatal genetic diagnosis. *Genetics in Medicine* 2007 September; 9(9): 596-599. Subject: 15.2

Perkins, Alexis C.; Choi, Joanna Mimi; Kimball, Alexa B. Reporting of ethical review of clinical research submit-ted to the Journal of the American Academy of Dermatol-ogy. *Journal of the American Academy of Dermatology* 2007 February; 56(2): 279-284. Subject: 18.2

Perkins, Henry S. Controlling death: the false promise of advance directives. *Annals of Internal Medicine* 2007 July 3; 147(1): 51-57. Subject: 20.5.4

Perlis, Clifford S.; Harwood, Michael; Perlis, Roy H. Extent and impact of industry sponsorship conflicts of in-terest in dermatology research. *Journal of the American Academy of Dermatology* 2005 June; 52(6): 967-971. Sub-ject: 18.2

Perring, Christian. Against scientism, for personhood. *American Journal of Bioethics* 2007 January; 7(1): 67-68. Subject: 17.1

Perry, Joshua E. Biopolitics at the bedside: proxy wars and feeding tubes. *Journal of Legal Medicine* 2007 April-June; 28(2): 171-192. Subject: 20.5.1

Perry, Sandy; Woodall, Angela L.; Pressman, Eva K. Association of ultrasound findings with decision to continue Down syndrome pregnancies. *Community Genetics* 2007; 10(4): 227-230. Subject: 15.2

Perske, Robert. The "big bang" theory and Down syndrome. *Ethics and Intellectual Disability* 2007 Winter; 10(1): 1, 4-6. Subject: 9.5.3

Persson, Anders. Research ethics and the development of medical biotechnology. *Xenotransplantation* 2006 November; 13(6): 511-513. Subject: 15.1

Persson, Anders; Hemlin, Sven; Welin, Stellan. Profitable exchanges for scientists: the case of Swedish human embryonic stem cell research. *Health Care Analysis: An International Journal of Health Philosophy and Policy* 2007 December; 15(4): 291-304. Subject: 5.3

Pesce, Andrew. Abortion laws in Australia: time for consistency? *University of New South Wales Law Journal* 2006; 29(2): 224-226. Subject: 12.1

Pescosolido, Bernice A; Fettes, Danielle L.; Martin, Jack K.; Monahan, John; McLeod, Jane D. Perceived dangerousness of children with mental health problems and support for coerced treatment. *Psychiatric Services* 2007 May; 58(5): 619-625. Subject: 17.7

Peter, Elizabeth. Feminist ethics: a critique. *In:* Davis, Anne J.; Tschudin, Verena; de Raeve, Louise, eds. Essentials of Teaching and Learning in Nursing Ethics: Perspectives and Methods. New York: Churchill Livingstone Elsevier, 2006: 191-201. Subject: 4.1.3

Peters, Jeremy W. New Jersey requires HIV test in pregnancy. *New York Times* 2007 December 27; p. B3. Subject: 9.5.6

Peters, Matthew J. Should smokers be refused surgery? *BMJ: British Medical Journal* 2007 January 6; 334(7583): 20-21. Subject: 9.5.9

Peters, Philip G., Jr. The ambiguous meaning of human conception. *U.C. Davis Law Review* 2006 November; 40(1): 199-228. Subject: 4.4

Peters, Ted. Perfect humans or trans-humans? *In:* Deane-Drummond, Celia; Scott, Peter Manley, eds. Future Perfect?: God, Medicine and Human Identity. New York: T and T Clark International, 2006: 15-32. Subject: 15.1

Petersen, Alan. 'Biobanks' "engagements": engendering trust or engineering consent? *Genomics, Society and Policy* 2007 April; 3(1): 31-43. Subject: 15.1

Petersen, Alan. The genetic conception of health: is it as radical as claimed? *Health (London)* 2006 October; 10(4): 481-500. Subject: 15.1

Petersen, Alan; Anderson, Alison. A question of balance or blind faith?: scientists' and science policymakers' representations of the benefits and risks of nanotechnologies. *NanoEthics* 2007 December; 1(3): 243-256. Subject: 5.4

Petersen, Kerry. Classifying abortion as a health matter: the case for de-criminalising abortion laws in Australia. *In:* McLean, Sheila A.M., ed. First Do No Harm: Law, Ethics, and Healthcare. Aldershot, England; Burlington, VT: Ashgate, 2006: 353-368. Subject: 12.4.2

Petersen, Kerry. The rights of donor-conceived children to know the identity of their donor: the problem of the known unknowns and the unknown unknowns. *In:* Bennett, Belinda; Tomossy, George F., eds. Globalization and Health: Challenges for Health Law and Bioethics. Dordrecht: Springer, 2006: 151-167. Subject: 14.4

Peterson, Kathryn E. My father's eyes and my mother's heart: the due process rights of the next of kin in organ donation. *Valparaiso University Law Review* 2005 Fall; 40(1): 169. Subject: 19.5

Peterson, M. Should the precautionary principle guide our actions or our beliefs? *Journal of Medical Ethics* 2007 January; 33(1): 5-10. Subject: 5.2

Petri, J. Thomas. Altered nuclear transfer, gift, and mystery: an Aristotelian-Thomistic response to David L. Schindler. *National Catholic Bioethics Quarterly* 2007 Winter; 7(4): 729-747. Subject: 18.5.4

Petroni, Angelo Maria. Perspectives for freedom of choice in bioethics and health care in Europe. *In:* Engelhardt, H. Tristram, ed. Global Bioethics: The Collapse of Consensus. Salem, MA: M&M Scrivener Press, 2006: 238-270. Subject: 2.1

Petrova, Mila; Dale, Jeremy; Fulford, Bill (KWM). Values-based practice in primary care: easing the tensions between individual values, ethical principles and best evidence. *British Journal of General Practice* 2006 September; 56(530): 703-709. Subject: 4.1.2

Petsko, Gregory A. A matter of life and death. *Genome Biology* 2005; 6(5): 109.1-109.3. Subject: 20.5.1

Pharmaceutical Research and Manufacturers of America [PhRMA]. PhRMA guiding principles: direct to consumer advertisements about prescription medicines. Washington, DC: Pharmaceutical Research and Manufacturers of America, 2005 November; 10 p. [Online]. Accessed: http://www.phrma.org/files/DTCGuidingprinciples.pdf [2007 April 30]. Subject: 9.7

Phelps, Ceri; Wood, F.; Bennett, P.; Brain, K.; Gray, J. Knowledge and expectations of women undergoing cancer genetic risk assessment: a qualitative analysis of free-text questionnaire comments. *Journal of Genetic Counseling* 2007 August; 16(4): 505-514. Subject: 15.3

See SUBJECT HEADING KEY FOR SECTION II on inside back cover.

577

Phelps, Elizabeth A. The neuroscience of a person network. *American Journal of Bioethics* 2007 January; 7(1): 49-54. Subject: 17.1

Phillips, Helen. Impossible awakenings. *New Scientist* 2007 July 7-13; 195(2611): 40-43. Subject: 20.5.1

Phillips, Peter W.B.; Einsiedel, Edna. The future of genomics. *In:* Einsiedel, Edna; Timmermans, Frank, eds. Crossing Over: Genomics in the Public Arena. Calgary, Alberta, Canada: University of Calgary Press, 2005: 239-247. Subject: 15.1

Phillips, Susan. Ethical decision-making when caring for the noncompliant patient. *Journal of Infusion Nursing* 2006 September-October; 29(5): 266-271. Subject: 9.5.9

Phillips, Trisha B. Money, advertising and seduction in human subjects research. *American Journal of Bioethics* 2007 February; 7(2): 88-90. Subject: 18.1

Piccart, Martine; Goldhirsch, Aron. Keeping faith with trial volunteers: how best to serve patients' interests in large clinical trials? [commentary]. *Nature* 2007 March 8; 446(7132): 137-138. Subject: 18.1

Pigliucci, Massimo; Kaplan, Jonathan. On the concept of biological race and its applicability to humans. *Philosophy of Science* 2003 December; 70(5): 1161-1172. Subject: 15.11

Pijnenburg, Martien A.M.; Leget, Carlo. Who wants to live forever? Three arguments against extending the human lifespan. *Journal of Medical Ethics* 2007 October; 33(10): 585-587. Subject: 20.5.1

Pilarski, Linda M.; Mehta, Michael D.; Caulfield, Timothy; Kaler, Karan V.I.S.; Backhouse, Christopher J. Microsystems and nanoscience for biomedical applications: a view to the future. *In:* Hunt, Geoffrey; Mehta, Michael D., eds. Nanotechnology: Risk, Ethics and Law. London; Sterling, VA: Earthscan, 2006: 35-42. Subject: 5.4

Pimple, Kenneth D. Ethical issues in drug user treatment research. *In:* Kleinig, John; Einstein, Stanley, eds. Ethical Challenges for Intervening in Drug Use: Policy, Research and Treatment Issues. Huntsville, TX: Office of International Criminal Justice; Criminal Justice Center, Sam Houston State University, 2006: 205-216. Subject: 18.2

Pinto, Sharrel L.; Lipowski, Earlene; Segal, Richard; Kimberlin, Carole; Algina, James. Physicians' intent to comply with the American Medical Association's guidelines on gifts from the pharmaceutical industry. *Journal of Medical Ethics* 2007 June; 33(6): 313-319. Subject: 9.7

Pirakitikulr, Darlyn; Bursztajn, Harold J. Pride and prejudice: avoiding genetic gossip in the age of genetic testing. *Journal of Clinical Ethics* 2007 Summer; 18(2): 156-161. Subject: 15.3

Pirzadeh, Sara M.; McCarthy Veach, Patricia; Bartels, Dianne M.; Kao, Juihsien; LeRoy, Bonnie S. A national survey of genetic counselors' personal values. *Journal of*

Genetic Counseling 2007 December; 16(6): 763-773. Subject: 15.2

Pishchita, A. Presumed consent in the law of the Russian Federation on transplanting organs. *Medicine and Law: The World Association for Medical Law* 2007 March; 26(1): 179-188. Subject: 19.5

Pishchita, A.N. Elderly patients as a vulnerable category of the population requiring special legal protection with respect to the provision of medical care. *European Journal of Health Law* 2007 December; 14(4): 349-354. Subject: 9.5.2

Pitchers, M.; Stokes, A.; Lonsdale, R.; Premachandra, D.J.; Edwards, D.R. Research tissue banking in otolaryngology: organization, methods and uses, with reference to practical, ethical and legal issues. *Journal of Laryngology and Otology* 2006 June; 120(6): 433-438. Subject: 19.5

Pittman, Larry J. Embryonic stem cell research and religion: the ban on federal funding as a violation of the establishment clause. *University of Pittsburgh Law Review* 2006 Fall; 68(1): 131-190. Subject: 18.5.4

Pitts, Peter J. Settling for second best? [letter]. *Nature Biotechnology* 2007 July; 25(7): 715-716. Subject: 9.7

Pivetti, Monica. Natural and unnatural: activists' representations of animal biotechnology. *New Genetics and Society* 2007 August; 26(2): 137-157. Subject: 15.1

Plant; Margo; Knoppers, Bartha Maria. Umbilical cord blood banking in Canada: socio-ethical and legal issues. *Health Law Journal* 2005; 13: 187-212. Subject: 19.4

Plass, A.M.C. Informed consent for newborn screening? *Community Genetics* 2007; 10(4): 262-263. Subject: 15.3

Plemmons, Dena K.; Kalichman, Michael W. Reported goals for knowledge to be learned in responsible conduct of research courses. *Journal of Empirical Research on Human Research Ethics* 2007 June; 2(2): 57-66. Subject: 18.1

Ploem, Corrette. Freedom of research and its relation to the right to privacy. *In:* Gevers, J.K.M.; Hondius, E.H.; Hubben, J.H., eds. Health Law, Human Rights and the Biomedicine Convention: Essays in Honour of Henriette Roscam Abbing. Leiden; Boston: Martinus Nijhoff Publishers, 2005: 159-173. Subject: 18.6

Plotz, David. The ethics of enhancement: we can make ourselves stronger, fast, smarter. Should we? *Slate Magazine* 2003 March 12: 3 p. [Online]. Accessed: http://www.slate.com/id/2079310 [2007 April 12]. Subject: 4.5

Po, Alain Li Wan. Personalised medicine: who is an Asian? *Lancet* 2007 May 26 - June 1; 369(9575): 1770-1771. Subject: 15.11

Poff, Deborah. Community-based REBS: the experience of the British Columbia medical services foundation. *NCEHR Communique CNERH* 2006 Spring; 14(1): 24-25. Subject: 18.2

Pogge, Thomas. Montréal statement on the human right to essential medicines. *CQ: Cambridge Quarterly of Healthcare Ethics* 2007 Winter; 16(1): 97-108. Subject: 21.1

Poland, Susan. Court clarifies role of guardians in foregoing life-support. *Ethics and Intellectual Disability* 2005 Winter; 8(2): 3, 7. Subject: 20.5.2

Polansky, Samara. Overcoming the obstacles: a collaborative approach to informed consent in prenatal genetic screening. *Health Law Journal* 2006; 14: 21-43. Subject: 15.2

Polifroni, E. Carol. Ethical knowing and nursing education [editorial]. *Journal of Nursing Education* 2007 January; 46(1): 3. Subject: 7.2

Pollack, Andrew. A genetic test that very few need, marketed to the masses. *New York Times* 2007 September 11; p. C3. Subject: 15.3

Pollack, Andrew. Death in gene therapy treatment is still unexplained. *New York Times* 2007 September 18; p. A22. Subject: 15.4

Pollack, Andrew. Gene therapy study to resume after woman's death [news]. *New York Times* 2007 November 26; p. A16. Subject: 15.4

Pollack, Andrew. Round 2 [two] for biotech beets; after delay over safety fears, engineered crop will be planted [news]. *New York Times* 2007 November 27; p. C1, C2. Subject: 15.1

Pollack, Andrew. The dialysis business: fair treatment? *New York Times* 2007 September 16; p. BU1, BU10. Subject: 19.3

Pollack, Robert E. Von der religiösen Pflicht, die Natur und den Staat zu hinterfragen: D. Bonhoeffer über den Schutz und die Würde des menschlichen Lebens im Zusammenhang mit der modernen genetischen Medizin. *In:* Gestrich, Christof; Neugebauer, Johannes, eds. Der Wert menschlichen Lebens: medizinische Ethik bei Karl Bonhoeffer und Dietrich Bonhoeffer. Berlin: Wichern-Verlag, 2006: 66-97. Subject: 15.5

Pollard, Irina. Cloning technology. *In her:* Life, Love and Children: A Practical Introduction to Bioscience Ethics and Bioethics. Boston: Kluwer Academic Publishers, 2002: 145-155. Subject: 14.5

Pollard, Irina. Procreative technology: achievements and desired goals. *In her:* Life, Love and Children: A Practical Introduction to Bioscience Ethics and Bioethics. Boston: Kluwer Academic Publishers, 2002: 105-125. Subject: 14.1

Pollard, Irina. The recombinant DNA technologies. *In her:* Life, Love and Children: A Practical Introduction to Bioscience Ethics and Bioethics. Boston: Kluwer Academic Publishers, 2002: 127-143. Subject: 15.1

Pollard, Irina. The state of wellbeing: on the end of life care and euthanasia. *In her:* Life, Love and Children: A Practical Introduction to Bioscience Ethics and Bioethics.

Boston: Kluwer Academic Publishers, 2002: 97-103. Subject: 20.5.1

Porta, Nicolas; Frader, Joel. Withholding hydration and nutrition in newborns. *Theoretical Medicine and Bioethics* 2007; 28(5): 443-451. Subject: 20.5.2

Porter, Susan. Organ transplants. Part two: Questions and controversy. *Ohio State Medical Journal* 1984 January; 80(1): 33, 37, 39. Subject: 19.5

Post, Stephen G. The aging society and the expansion of senility: biotechnological and treatment goals. *In:* Steinbock, Bonnie, ed. The Oxford Handbook of Bioethics. Oxford; New York: Oxford University Press, 2007: 304-323. Subject: 9.5.2

Potter, Nancy Nyguist; Zanni, Guido R.; Stavis, Paul F. Querying the "community" in community mental health. *American Journal of Bioethics* 2007 November; 7(11): 42-43; author reply W3-W4. Subject: 17.1

Pottker-Fishel, Carrie Gene. Improper bedside manner: why state partner notification laws are ineffective in controlling the proliferation of HIV. *Health Matrix: The Journal of Law-Medicine* 2007 Winter; 17(1): 147-179. Subject: 9.5.6

Poulis, Ioannis. Bioethics and physiotherapy [editorial]. *Journal of Medical Ethics* 2007 August; 33(8): 435-436. Subject: 4.1.1

Powell, Sean T.; Allison, Matthew A.; Kalichman, Michael W. Effectiveness of a responsible conduct of research course: a preliminary study. *Science and Engineering Ethics* 2007 June; 13(2): 249-264. Subject: 1.3.9

Powell, Tia. Commentary: support for case-based analysis in decision making after a suicide attempt. *Journal of Clinical Ethics* 2007 Summer; 18(2): 119-121. Subject: 9.5.5

Powell, Tia. Cultural context in medical ethics: lessons from Japan. *Philosophy, Ethics, and Humanities in Medicine [electronic]* 2006; 1(4): 7 p. Accessed: http://www.peh-med.com/content/1/1/4 [2006 May 15]. Subject: 21.7

Powell, Tia. Wrestling satan and conquering dopamine: addiction and free will. *American Journal of Bioethics* 2007 January; 7(1): 14-15. Subject: 17.1

Power, Tara E.; Adams, Paul C.; Barton, James C.; Acton, Ronald T.; Howe, Edmund, III; Palla, Shana; Walker, Ann P.; Anderson, Roger; Harrison, Barbara. Psychosocial impact of genetic testing for hemochromatosis in the HEIRS study: a comparison of participants recruited in Canada and in the United States. *Genetic Testing* 2007 Spring; 11(1): 55-64. Subject: 15.3

Powers, Penny. Persuasion and coercion: a critical review of philosophical and empirical approaches. *HEC (Healthcare Ethics Committee) Forum* 2007 June; 19(2):125-143. Subject: 2.1

Powledge, Tabitha M. Looking at ART: is it time to scrutinize assisted reproduction? *Scientific American* 2002 April; 286(4): 20, 23. Subject: 14.4

See SUBJECT HEADING KEY FOR SECTION II on inside back cover.

579

Pozgar, George D. Acquired immunodeficiency syndrome. *In his:* Legal Aspects of Health Care Administration. 9th edition. Sudbury, MA: Jones and Bartlett Publishers, 2004: 375-385. Subject: 9.5.6

Pozgar, George D. Health care ethics. *In his:* Legal Aspects of Health Care Administration. 9th edition. Sudbury, MA: Jones and Bartlett Publishers, 2004: 387-422. Subject: 20.7

Pozgar, George D. Issues of procreation. *In his:* Legal Aspects of Health Care Administration. 9th edition. Sudbury, MA: Jones and Bartlett Publishers, 2004: 345-364. Subject: 12.4.2

Pozgar, George D. Patient consent. *In his:* Legal Aspects of Health Care Administration. 9th edition. Sudbury, MA: Jones and Bartlett Publishers, 2004: 313-334. Subject: 8.3.1

Pozgar, George D. Patient rights and responsibilities. *In his:* Legal Aspects of Health Care Administration. 9th edition. Sudbury, MA: Jones and Bartlett Publishers, 2004: 365-374. Subject: 8.1

Prainsack, Barbara. Research populations: biobanks in Israel. *New Genetics and Society* 2007 April; 26(1): 85-103. Subject: 15.1

Pray, W. Steven. Ethical, scientific, and educational concerns with unproven medications. *American Journal of Pharmaceutical Education* 2006 December 15; 70(6): 141. Subject: 9.7

Preminger, Beth A.; Fins, Joseph J. Face transplantation: an extraordinary case with lessons for ordinary practice. *Plastic and Reconstrive Surgery* 2006 September 15; 118(4): 1073-1074. Subject: 19.1

Prentice, David A. The whole truth about stem cells and relevant therapies. *Today's Christian Doctor* 2007 Summer; 38(2): 21-24. Subject: 18.5.1

Prentice, David A.; Tarne, Gene. Treating diseases with adult stem cells [letter]. *Science* 2007 January 19; 315(5810): 328. Subject: 18.5.1

Prescription Medicines Code of Practice Authority. Code of practice for the pharmaceutical industry 2006. London: Prescription Medicines Code of Practice Authority, 2006: 56 p. [Online]. Accessed: http://www.abpi.org.uk/links/assoc/PMCPA/Code06use.pdf [2007 April 17]. Subject: 6

Presenza, Louis J. Naltrexone as a "mandate" or as a choice: comments on "Judicially mandated naltrexone use by criminal offenders: a legal analysis". *Journal of Substance Abuse Treatment* 2006 September; 31(2): 129-130. Subject: 9.5.9

Preston, Julia. U.S. set to begin a vast expansion of DNA sampling; big effect on immigrants; law to cover most people detained or arrested by federal agents. *New York Times* 2007 February 5; p. A1, A15. Subject: 15.1

Price, Connie C.; Braun, Joshua A. Cinematic thinking: narratives and bioethics unbound. *American Journal of Bioethics* 2007 August; 7(8): 21-23; author reply W1-W2. Subject: 1.3.7

Prideaux, David; Rogers, Wendy. Audit or research: the ethics of publication. *Medical Education* 2006 June; 40(6): 497-499. Subject: 1.3.7

Pringle, Helen. Abortion and disability: reforming the law in South Australia. *University of New South Wales Law Journal* 2006; 29(2): 207-217. Subject: 12.4.2

Prior, Pauline M. Mentally disordered offenders and the European Court of Human Rights. *International Journal of Law and Psychiatry* 2007 November-December; 30(6): 546-557. Subject: 9.5.3

Probst, Janice C. Prisoners' dilemma: the importance of negative results. *Family Medicine* 2006 November-December; 38(10): 742-743. Subject: 1.3.7

Prograis, Lawrence J. An African American's internal perspective on biomedical ethics. *In:* Prograis, Lawrence J.; Pellegrino, Edmund D., eds. African American Bioethics: Culture, Race, and Identity. Washington, DC: Georgetown University Press, 2004: 153-158. Subject: 2.1

Proot, Ireen M.; ter Meulen, Ruud H.J.; Abu-Saad, Huda Huijer; Crebolder, Harry F.J.M. Supporting stroke patients' autonomy during rehabilitation. *Nursing Ethics* 2007 March; 14(2): 229-241. Subject: 9.5.2

Providence Center for Health Care Ethics. A primer for understanding the CDF's Responses regarding ANH for the PVS patient. *Health Care Ethics USA* 2007 Fall; 15(4): 15-9. Subject: 20.5.1

Prusak, Bernard G.; Malmqvist, Erik; Fenton, Elizabeth. Back to the future: Habermas's The Future of Human Nature [letters and reply]. *Hastings Center Report* 2007 March-April; 37(2): 4-6. Subject: 15.1

Pruss, Alexander R. Cooperation with past evil and use of cell-lines derived from aborted fetuses. *In:* Watt, Helen, ed. Cooperation, Complicity and Conscience: Problems in Healthcare, Science, Law and Public Policy. London: Linacre Centre, 2005: 89-104. Subject: 18.5.4

Pryor, Erica R.; Habermann, Barbara; Broome, Marion E. Scientific misconduct from the perspective of research coordinators: a national survey. *Journal of Medical Ethics* 2007 June; 33(6): 365-369. Subject: 1.3.9

Pugliese, Elizabeth. Organ trafficking and the TVPA: why one word makes a difference in international enforcement efforts. *Journal of Contemporary Health Law and Policy* 2007 Fall; 24(1): 181-208. Subject: 19.5

Puljak, Livia. Croatia founded a national body for ethics in science. *Science and Engineering Ethics* 2007 June; 13(2): 191-193. Subject: 2.4

Pullen, Sophie. Research on people: ethical considerations. *In:* Pullen, Sophie; Gray, Carol, eds. Ethics, Law and the Veterinary Nurse. New York: Elsevier Butterworth Heinemann, 2006: 117-128. Subject: 18.2

Pullman, D.; Hodgkinson, K. Genetic knowledge and moral responsibility: ambiguity at the interface of genetic research and clinical practice. *Clinical Genetics* 2006 March; 69(3): 199-203. Subject: 15.1

Pywell, Stephanie. Infant vaccination: a conflict of ethical imperatives? *In:* Garwood-Gowers, Austen; Tingle, John; Wheat, Kay, eds. Contemporary Issues in Healthcare Law and Ethics. Edinburgh; New York: Elsevier Butterworth-Heinemann, 2005: 213-232. Subject: 9.5.7

Q

Qiu, Jane. Chinese law aims to quell fear of failure [news]. *Nature* 2007 September 6; 449(7158): 12. Subject: 1.3.9

Qiu, Jane. To walk again. *New Scientist* 2007 November 10-16; 196(2629): 57-59. Subject: 18.1

Quest, Dale. Case vignette 1: a randomized double-blind double-dummy cross-over study of oral hexylinsulin monoconjugate 2 [PEGInsulin} versus insulin lispro for postprandial glycæmic control in adult patients with Type 2 diabetes mllitus. *NCEHR Communique CNERH* 2005 Spring; 13(1): 7-8. Subject: 18.2

Quest, Dale. Case vignette 2: a phase 2, randomized, double-blind, placebo-controlled study of DPE6591A in rheumatoid arthritis patients. *NCEHR Communique CNERH* 2005 Spring; 13(1): 9-10. Subject: 18.2

Quest, Dale. Case vignette 3: a multi-centered trial to compare TCT vs Clozapine for treatment-resistant schizophrenia. *NCEHR Communique CNERH* 2005 Spring; 13(1): 11-12. Subject: 18.2

Quigley, Muireann. A NICE fallacy. *Journal of Medical Ethics* 2007 August; 33(8): 465-466. Subject: 9.4

Quigley, Muireann. Non-human primates: the appropriate subjects of biomedical research? *Journal of Medical Ethics* 2007 November; 33(11): 655-658. Subject: 22.2

Quigley, Muireann. Property and the body: applying Honoré. *Journal of Medical Ethics* 2007 November; 33(11): 631-634. Subject: 19.5

Quill, Elizabeth. Congress considers higher fines for mistreating laboratory animals [news]. *Chronicle of Higher Education* 2007 October 26; 54(9): A20-A21. Subject: 22.2

Quill, Elizabeth. Researchers call for self-regulation in care of lab animals. *Chronicle of Higher Education* 2007 October 26; 54(9): A20, A22. Subject: 22.2

Quill, Timothy E. Legal regulation of physician-assisted death: the latest report cards. *New England Journal of Medicine* 2007 May 10; 356(19): 1911-1913. Subject: 20.7

Quill, Timothy E. Physician assisted death in vulnerable populations [editorial]. *BMJ: British Medical Journal* 2007 September 29; 335(7621): 625-626. Subject: 20.7

Quinlivan, Julie A.; Suriadi, Christine. Attitudes of new mothers towards genetics and newborn screening. *Journal of Psychosomatic Obstetrics and Gynaecology* 2006 March; 27(1): 67-72. Subject: 15.3

Quist, Norman. Hope, uncertainty, and lacking mechanisms. *Journal of Clinical Ethics* 2007 Winter; 18(4): 357-361. Subject: 9.6

Qureshi, K.; Gershon, R.R.M.; Sherman, M.F.; Straub, T.; Gebbie, E.; McCollum, M.; Erwin, M.J.; Morse, S.S. Health care workers' ability and willingness to report to duty during catastrophic disasters. *Journal of Urban Health: Bulletin of the New York Academy of Medicine* 2005 September; 82(3): 378-388. Subject: 7.1

R

Rabe, Marianne. Ethik in der Pflegeausbildung [Ethics in nursing education]. *Ethik in der Medizin* 2006 December; 18(4): 379-384. Subject: 2.3

Rabi, Suzanne M.; Patton, Lynn R.; Fjortoft, Nancy; Zgarrick, David P. Characteristics, prevalence, attitudes, and perceptions of academic dishonesty among pharmacy students. *American Journal of Pharmaceutical Education* 2006 August 15; 70(4): 73. Subject: 7.2

Rabin, Cheryl; Tabak, Nili. Healthy participants in phase I clinical trials: the quality of their decision to take part. *Journal of Clinical Nursing* 2006 August; 15(8): 971-979. Subject: 18.2

Rabin, Roni Caryn. As demand for donor eggs soars, high prices stir ethical concerns. *New York Times* 2007 May 15; p. F6. Subject: 14.4

Racher, Frances .E. The evolution of ethics for community practice. *Journal of Community Health Nursing* 2007 Spring; 24(1): 65-76. Subject: 18.6

Racine, Eric. HEC member perspectives on the case analysis process: a qualitative multi-site study. *HEC (Healthcare Ethics Committee) Forum* 2007 September; 19(3): 185-206. Subject: 9.6

Racine, Eric. Identifying challenges and conditions for the use of neuroscience in bioethics. *American Journal of Bioethics* 2007 January; 7(1): 74-75. Subject: 17.1

Racine, Eric; Illes, Judy. Emerging ethical challenges in advanced neuroimaging research: review, recommendations and research agenda. *Journal of Empirical Research on Human Research Ethics* 2007 June; 2(2): 1-10. Subject: 18.5.1

Racine, Eric; Illes, Judy. Neuroethical responsibilities. *Canadian Journal of Neurological Sciences* 2006 August; 33(3): 269-277, 260-268. Subject: 18.2

Racine, Eric; van der Loos, Hz Adriaan; Illes, Judy. Internet marketing of neuroproducts: new practices and

See SUBJECT HEADING KEY FOR SECTION II on inside back cover.

581

healthcare policy changes. *CQ: Cambridge Quarterly of Healthcare Ethics* 2007 Spring; 16(2): 181-194. Subject: 17.1

Racine, Eric; Waldman, Sarah; Palmour, Nicole; Risse, David; Illes, Judy. "Currents of hope": neurostimulation techniques in U.S. and U.K. print media. *CQ: Cambridge Quarterly of Healthcare Ethics* 2007 Summer; 16(3): 312-316. Subject: 17.5

Radden, Jennifer. Virtue ethics as professional ethics: the case of psychiatry. *In:* Walker, Rebecca L.; Ivanhoe, Philip J., eds. Working Virtue: Virtue Ethics and Contemporary Moral Problems. Oxford: Clarendon, 2007: 113-134. Subject: 17.1

Radulovic, Jelena; Stankovic, Bratislav. Genetic determinants of emotional behavior: legal lessons from genetic models. *DePaul Law Review* 2007 Spring; 56(3): 823-836. Subject: 15.6

Rady, Mohamed Y.; Verheijde, Joseph L.; McGregor, Joan. Organ donation after circulatory death: the forgotten donor? *Critical Care* 2006; 10(5): 166. Subject: 19.5

Rady, Mohamed Y.; Verheijde, Joseph L.; Spital, Aaron. Ethically increasing the supply of transplantable organs [letters]. *Annals of Internal Medicine* 2007 April 3; 146(7): 537-538. Subject: 19.5

Rafter, Nicole H. Claims-making and socio-cultural context in the first U.S. eugenics campaign. *Social Problems* 1992 February; 39(1): 17-34. Subject: 15.5

Raghavan, Ramesh. A question of faith. *JAMA: The Journal of the American Medical Association* 2007 April 4; 297(13): 1412. Subject: 12.3

Råholm, Maj-Britt. Caritative caring ethics: a description reflected through the Aristotelian terms phronesis, techne and episteme. *In:* Tschudin, Verena, ed. Approaches to Ethics: Nursing Beyond Boundaries. New York: Butterworth-Heinemann, 2003: 13-23. Subject: 4.1.3

Rai, Mohammad A.; Afzal, Omer. Organs in the bazaar: the end of the beginning? *Politics and the Life Sciences* 2007 March; 26(1): 10-11. Subject: 19.5

Raja, Kavitha. Patients' perspectives on medical information: results of a formal survey. *Indian Journal of Medical Ethics* 2007 January-March; 4(1): 16-17. Subject: 8.3.1

Rajczi, Alex. A critique of the innovation argument against a national health program. *Bioethics* 2007 July; 21(6): 316-323. Subject: 9.3.1

Rakhudu, M.A.; Mmelesi, A.M.M.; Myburgh, C.P.H.; Poggenpoel, M. Exploration of the views of traditional healers regarding the termination of pregnancy (TOP) law. *Curationis* 2006 August; 29(3): 56-60. Subject: 12.3

Rakowski, Eric. Reverence for life and the limits of state power. *In:* Burley, Justine, ed. Dworkin and His Critics: With Replies by Dworkin. Malden, MA: Blackwell Pub., 2004: 241-263. Subject: 12.4.2

Rakowski, Eric. Ronald Dworkin, reverence for life, and the limits of state power. *Utilitas* 2001 March; 13(1): 33-64. Subject: 12.3

Ralston, D. Christopher; Ho, Justin. Disability, humanity, and personhood: a survey of moral concepts. *Journal of Medicine and Philosophy* 2007 November-December; 32(6): 619-633. Subject: 4.4

Ramanthan, Mala; Jesani, Amar. Ethics in nutrition intervention research. *Indian Journal of Medical Ethics* 2007 April-June; 4(2): 76-77. Subject: 18.5.2

Rameix, Suzanne; Durand-Zaleski, Isabelle. Justice and the allocation of healthcare. *In:* Marinker, Marshall, ed. Constructive Conversations about Health: Policy and Values. Oxford; Seattle: Radcliffe Pub., 2006: 177-183. Subject: 9.4

Ramnarayan, Padmanabhan; Craig, Finella; Petros, Andy; Pierce, Christine. Characteristics of deaths occurring in hospitalised children: changing trends. *Journal of Medical Ethics* 2007 May; 33(5): 255-260. Subject: 20.4.2

Ranney, Megan L.; Gee, Erin M.; Merchant, Roland C. Nonprescription availability of emergency contraception in the United States: current status, controversies, and impact on emergency medicine practice. *Annals of Emergency Medicine* 2006 May; 47(5): 461-471. Subject: 11.1

Rao, Mahendra S. Are there morally acceptable alternatives to blastocyst derived ESC? *Journal of Cellular Biochemistry* 2006 August 1; 98(5): 1054-1061. Subject: 18.5.4

Rao, Mahendra S. Mired in the quagmire of uncertainty: The "catch-22" of embryonic stem cell research. *Stem Cells and Development* 2006 August; 15(4): 492-496. Subject: 18.5.4

Rapgay, Lobsang. A Buddhist approach to end-of-life care. *In:* Puchalski, Christina M., ed. A Time for Listening and Caring: Spirituality and the Care of the Chronically Ill and Dying. Oxford; New York: Oxford University Press, 2006: 131-137. Subject: 20.4.1

Raposo, Vera Lúcia; Osuna, Eduardo. Embryo dignity: the status and juridical protection of the in vitro embryo. *Medicine and Law: The World Association for Medical Law* 2007 December; 26(4): 737-746. Subject: 4.4

Rapp, Rayna. Cell life and death, child life and death: genomic horizons, genetic diseases, family stories. *In:* Franklin, Sarah; Lock, Margaret, eds. Remaking Life and Death: Toward an Anthropology of the Biosciences. Santa Fe: School of American Research Press; Oxford: James Currey, 2003: 129-164. Subject: 15.1

Rappaport, Z.H. Robotics and artificial intelligence: Jewish ethical perspectives. *Acta Neurochirurgica. Supplement* 2006; 98: 9-12. Subject: 17.1

Rapport, Frances L. Response to: 'Caring as a slave morality: Nietzschean themes in nursing ethics' by J. Paley (2002) Journal of Advanced Nursing 40, 25-35. *Journal of*

Advanced Nursing 2002 October; 40(1): 42-44. Subject: 4.1.1

Rapport, F.L.; Maggs, C.J. Titmuss and the gift relationship: altruism revisited. *Journal of Advanced Nursing* 2002 December; 40(5): 495-503. Subject: 19.4

Raskin, Joyce M.; Mazor, Nadav A. The artificial womb and human subject research. *In:* Gelfand, Scott; Shook, John R., eds. Ectogenesis: Artificial Womb Technology and the Future of Human Reproduction. Amsterdam; New York: Editions Rodopi, B.V., 2006: 159-182. Subject: 14.1

Rasko, John E.J.; O'Sullivan, Gabrielle M.; Ankeny, Rachel A. Is inheritable genetic modification the new dividing line? *In:* Rasko, John E.J.; O'Sullivan, Gabrielle M.; Ankeny, Rachel A., eds. The Ethics of Inheritable Genetic Modification: A Dividing Line? Cambridge: Cambridge University Press, 2006: 1-15. Subject: 15.4

Raspe, Heiner. Individuelle Gesundheitsleistungen in der vertragsärztlichen Versorgung - Eine medizinethische Diskussion = Individual health services within Germany's statutory health insurance system: ethical considerations. *Ethik in der Medizin* 2007 March; 19(1): 24-38. Subject: 9.1

Ratanakul, Pinit. Human cloning: Thai Buddhist perspectives. *In:* Roetz, Heiner, ed. Cross-Cultural Issues in Bioethics: The Example of Human Cloning. New York: Rodopi, 2006: 203-213. Subject: 14.5

Rathert, Cheryl; Fleming, David A. Ethical climates of HCOs and end-of-life moral conflict in care terms. *Organizational Ethics: Healthcare, Business, and Policy* 2006 Fall-Winter; 3(2): 101-111. Subject: 9.1

Ravindran, G.D. The study was unjustified and fallacious. *Indian Journal of Medical Ethics* 2007 April-June; 4(2): 81. Subject: 18.5.2

Ray, Ratna; Raju, Mohan. Attitude towards euthanasia in relation to death anxiety among a sample of 343 nurses in India. *Psychological Reports* 2006 August; 99(1): 20-26. Subject: 20.5.1

Rayasam, Renuka. Cloning around. An FDA ruling could spur growth at a Texas company that has the leg up on duplicating animals. *U.S. News and World Report* 2007 January 8; 142(1): 46-47. Subject: 22.3

Reach, G. Innovative therapies: some ethical considerations. *Diabetes and Metabolism* 2006 December; 32(5, Part 2): 527-531. Subject: 19.1

Reardon, Jenny. Decoding race and human difference in a genomic age. *Differences: A Journal of Feminist Cultural Studies* 2004 Fall; 15(3): 38-65. Subject: 15.11

Reches, Avinoam. Transparency with respect to the health of political leaders. *Israel Medical Association Journal* 2006 November; 8(11): 751-753. Subject: 8.4

Reckless, Ian. Patients and doctors: rights and responsibilities in the NHS. *Clinical Medicine* 2005 September-October; 5(5): 499-500. Subject: 9.1

Recupero, Patricia R.; Rainey, Samara E. Informed consent to e-therapy. *American Journal of Psychotherapy* 2005; 59(4): 319-331. Subject: 8.3.1

Redman, Barbara K. Responsibility for control; ethics of patient preparation for self-management of chronic disease. *Bioethics* 2007 June; 21(5): 243-250. Subject: 9.5.1

Rees, Charlotte E.; Knight, Lynn V. The trouble with assessing students' professionalism: theoretical insights from sociocognitive psychology. *Academic Medicine* 2007 January; 82(1): 46-50. Subject: 4.1.2

Rees, C.E.; Knight, L.V. "The stroke is eighty nine": understanding unprofessional behaviour through physician-authored prose. *Medical Humanities* 2007 June; 33(1): 38-43. Subject: 7.1

Reeves, Roy R.; Douglas, Sharon P.; Garner, Rosa T.; Reynolds, Marti D.; Silvers, Anita. The individual rights of the difficult patient [case study and commentaries]. *Hastings Center Report* 2007 March-April; 37(2): 13-15. Subject: 8.1

Reggiani, Andés H. "Drilling eugenics into people's minds": expertise, public opinion, and biopolitics in Alexis Carrel's Man, the Unknown. *In:* Currell, Susan; Cogdell, Christina, eds. Popular Eugenics: National Efficiency and American Mass Culture in the 1930s. Athens, OH: Ohio University Press, 2006: 70-90. Subject: 15.5

Rehmann-Sutter, Christoph. Controlling bodies and creating monsters: popular perceptions of genetic modifications. *In:* Rasko, John E.J.; O'Sullivan, Gabrielle M.; Ankeny, Rachel A., eds. The Ethics of Inheritable Genetic Modification: A Dividing Line? Cambridge: Cambridge University Press, 2006: 57-76. Subject: 15.1

Reich, Eugenie Samuel. Misconduct report kept under wraps. *New Scientist* 2007 November 24-30; 196(2631): 16. Subject: 1.3.9

Reid, Clare L.; Menon, David K. Researching incapacity: time to get our acts together [letter]. *BMJ: British Medical Journal* 2007 September 1; 335(7617): 415. Subject: 18.5.6

Reid, Lynette; Ram, Natalie; Brown, Blake. Compensation for gamete donation: the analogy with jury duty. *CQ: Cambridge Quarterly of Healthcare Ethics* 2007 Winter; 16(1): 35-43. Subject: 14.4

Reider, Alan E.; Dahlinghaus, Andrew B. The impact of new technology on informed consent. *Comprehensive Ophthalmology Update* 2006 November-December; 7(6): 299-302. Subject: 8.3.1

Reidpath, D.D.; Brijnath, B.; Chan, K.Y. An Asia Pacific six-country study on HIV-related discrimination: introduction. *AIDS Care* 2005 July; 17(Supplement 2): S117-S127. Subject: 9.5.6

Reidpath, D.D.; Chan, K.Y. HIV discrimination: integrating the results from a six-country situational analysis

See SUBJECT HEADING KEY FOR SECTION II on inside back cover.

583

in the Asia Pacific. *AIDS Care* 2005 July; 17(Supplement 2): S195-S204. Subject: 9.5.6

Reidpath, D.D.; Chan, K.Y. HIV/AIDS discrimination in the Asia Pacific [editorial]. *AIDS Care* 2005 July; 17(Supplement 2): S115-S116. Subject: 9.5.6

Reiheld, Alison; Easton, Raul B.; Graber, Mark A.; Monnahan, Jay; Hughes, Jason. Consent by survey: losing autonomy one percentage point at a time. *American Journal of Bioethics* 2007 December; 7(12): 53-54; author reply W3-W4,. Subject: 8.3.1

Reiling, Jennifer. Euthanasia as a romantic motive. *JAMA: The Journal of the American Medical Association* 2007 November 7; 298(17): 2076. Subject: 20.5.1

Reiman, Jeffrey. Being fair to future people: the non-identity problem in the original position. *Philosophy and Public Affairs* 2007 Winter; 35(1): 69-92. Subject: 4.4

Reiman, Jeffrey. The pro-life argument from substantial identity and the pro-choice argument from asymmetric value: a reply to Patrick Lee [see correction in Bioethics 2007 September; 21(7): 407]. *Bioethics* 2007 July; 21(6): 329-341. Subject: 12.3

Reinders, Hans S. Euthanasia and disability: comments on the Terry Schiavo case. *Ethics and Intellectual Disability* 2005 Summer; 9(1): 6-7. Subject: 20.5.1

Reinhardt, Uwe E. A social contract for twenty-first century health care: three-tier health care with bounty hunting. *In:* Engström, Timothy H.; Robison, Wade L., eds. Health Care Reform: Ethics and Politics. Rochester, NY: University of Rochester Press, 2006: 67-98. Subject: 9.3.2

Reis, Chen; Heisler, Michele; Amowitz, Lynn L.; Moreland, R. Scott; Mafeni, Jerome O.; Anyamele, Chukwuemeka; Iacopino, Vincent. Discriminatory attitudes and practices by health workers toward patients with HIV/AIDS in Nigeria. *PLoS Medicine* 2005 August; 2(8): e246 (0743-0752). Subject: 9.5.6

Reis, Elizabeth. Divergence or disorder: the politics of naming intersex. *Perspectives in Biology and Medicine* 2007 Autumn; 50(4): 535-543. Subject: 10

Reis, Linda M.; Baumiller, Robert; Scrivener, William; Yager, Geoffrey; Warren, Nancy Steinberg. Spiritual assessment in genetic counseling. *Journal of Genetic Counseling* 2007 February; 16(1): 41-52. Subject: 15.2

Reis, Shmuel; Biderman, Aya; Mitki, Revital; Borkan, Jeffrey M. Secrets in primary care: a qualitative exploration and conceptual model. *JGIM: Journal of General Internal Medicine* 2007 September; 22(9): 1246-1253. Subject: 8.4

Reisman, Anna B. Saving Sylvia Cleary. *Hastings Center Report* 2007 July-August; 37(4): 9-10. Subject: 9.5.9

Reiter-Theil, Stella; Mertz, Marcel; Meyer-Zehnder, Barbara. The complex roles of relatives in end-of-life decision-making: an ethical analysis. *HEC (Healthcare Ethics Committee) Forum* 2007 December; 19(4): 341-364. Subject: 20.5.1

Reitman, James. Shall we prolong life in order to give a patient time to decide about faith? *Today's Christian Doctor* 2007 Spring; 38(1): 28-29. Subject: 20.5.2

Relman, Arnold S. Medical professionalism in a commercialized health care market. *JAMA: The Journal of the American Medical Association* 2007 December 12; 298(22): 2668-2670. Subject: 9.3.1

Relman, Arnold S. The problem of commercialism in medicine. *CQ: Cambridge Quarterly of Healthcare Ethics* 2007 Fall; 16(4): 375-376. Subject: 9.3.1

Remennick, Larissa. The quest for the perfect baby: why do Israeli women seek prenatal genetic testing? *Sociology of Health and Illness* 2006 January; 28(1): 21-53. Subject: 15.3

Rennie, Stuart. Do the ravages of the HIV/AIDS epidemic ethically justify mandatory HIV testing? [letter]. *Developing World Bioethics* 2007 April; 7(1): 48-49. Subject: 9.5.6

Rennie, Stuart; Muula, Adamson S.; Westreich, Daniel. Male circumcision and HIV prevention: ethical, medical and public health tradeoffs in low-income countries: ethical challenges surrounding the implementation of male circumcision as an HIV prevention strategy. *Journal of Medical Ethics* 2007 June; 33(6): 357-361. Subject: 9.5.7

Rentmeester, Christy A. Should a good healthcare professional be (at least a little) callous? *Journal of Medicine and Philosophy* 2007 January-February; 32(1): 43-64. Subject: 8.1

Rentmeester, Christy A. "Why aren't you doing what we want?" Cultivating collegiality and communication between specialist and generalist physicians and residents. *Journal of Medical Ethics* 2007 May; 33(5): 308-310. Subject: 7.3

Renzong, Qiu. Cloning issues in China. *In:* Roetz, Heiner, ed. Cross-Cultural Issues in Bioethics: The Example of Human Cloning. New York: Rodopi, 2006: 51-75. Subject: 14.5

Replogle, Jill. Abortion debate heats up in Latin America. *Lancet* 2007 July 28-August 3; 370(9584): 305-306. Subject: 12.5.1

Resnick, Andrew S.; Mullen, James L.; Kaiser, Larry R.; Morris, Jon B. Patterns and predictions of resident misbehavior — a 10-year retrospective look. *Current Surgery* 2006 November-December; 63(6): 418-425. Subject: 7.2

Resnicoff, Steven H. Supplying human body parts: a Jewish law perspective. *DePaul Law Review* 2005-2006; 55: 851-874. Subject: 19.5

Resnik, David B. Access to medications and global justice. *In:* Cohen, Jillian Clare; Illingworth, Patricia; Schüklenk, Udo, eds. The Power of Pills: Social, Ethical

---

Subject = NRCBL Primary Classification Number; see inside front cover.

and Legal Issues in Drug Development, Marketing, and Pricing. London; Ann Arbor, MI: Pluto, 2006: 88-97. Subject: 9.7

Resnik, David B. Are the new EPA regulations concerning intentional exposure studies involving children overprotective? *IRB: Ethics and Human Research* 2007 September-October; 29(5): 15-19. Subject: 18.5.2

Resnik, David B. Embryonic stem cell patents and human dignity. *Health Care Analysis: An International Journal of Health Philosophy and Policy* 2007 September; 15(3): 211-222. Subject: 15.8

Resnik, David B. Intentional exposure studies of environmental agents on human subjects: assessing benefits and risks. *Accountability in Research* 2007 January-March; 14(1): 35-55. Subject: 18.1

Resnik, David B. Neuroethics, national security and secrecy. *American Journal of Bioethics* 2007 May; 7(5): 14-15. Subject: 17.1

Resnik, David B. Some recent challenges to openness and freedom in scientific publication. *In:* Korthals, Michiel; Bogers, Robert J., eds. Ethics for Life Scientists. Dordrecht, The Netherlands: Springer, 2004: 85-99. Subject: 1.3.9

Resnik, David B. The human genome: common resource but not common heritage. *In:* Korthals, Michiel; Bogers, Robert J., eds. Ethics for Life Scientists. Dordrecht, The Netherlands: Springer, 2004: 197-210. Subject: 15.8

Resnik, David B. The new EPA regulations for protecting human subjects: haste makes waste. *Hastings Center Report* 2007 January-February; 37(1): 17-21. Subject: 18.6

Resnik, David B.; Roman, Gerard. Health, justice, and the environment. *Bioethics* 2007, May; 21(4): 230-241. Subject: 16.1

Resnik, David B.; Vorhaus, Daniel B. Genetic modification and genetic determinism. *Philosophy, Ethics and Humanities in Medicine [electronic]* 2006; 1(9): 11 p. Accessed: http://www.peh-med.com/content/1/1/9 n.d. Subject: 15.2

Resnik, David B.; Wing, Steven. Lessons learned from the children's environmental exposure research study. *American Journal of Public Health* 2007 March; 97(3): 414-418. Subject: 18.5.2

Resnik, D.B. Responsibility for health: personal, social, and environmental. *Journal of Medical Ethics* 2007 August; 33(8): 444-445. Subject: 9.1

Rest, Kathleen M.; Halpern, Michael H. Politics and the erosion of federal scientific capacity: restoring scientific integrity to public health science. *American Journal of Public Health* 2007 November; 97(11): 1939-1944. Subject: 9.1

Resta, Robert G. Defining and redefining the scope and goals of genetic counseling. *American Journal of Medical*

Genetics. Part C, Seminars in Medical Genetics 2006 November 15; 142(4): 269-275. Subject: 15.2

Revheim, Nadine; Javitt, Daniel C. Reading disability goes beyond consent forms [letter]. *Psychiatric Services* 2007 April; 58(4): 566. Subject: 18.3

Revie, Linda. "More than just boots! The eugenic and commercial concerns behind A. R. Kaufman's birth controlling activities". *Canadian Bulletin of Medical History* 2006; 23(1): 119-143. Subject: 15.5

Revill, James; Dando, Malcolm R. A Hippocratic oath for life scientists. A Hippocratic-style oath in the life sciences could help to educate researchers about the dangers of dual-use research. *EMBO Reports* 2006 July; 7 Special No: S55-S60. Subject: 1.3.9

Rhodes, Rosamond. The professional responsibilities of medicine. *In:* Rhodes, Rosamond; Francis, Leslie P.; Silvers, Anita, eds. The Blackwell Guide to Medical Ethics. Malden, MA: Blackwell Pub., 2007: 71-87. Subject: 4.1.2

Rhodes, Rosamond; Smith, Lawrence G. Molding professional character. *In:* Kenny, Nuala; Shelton, Wayne, eds. Lost Virtue: Professional Character Development in Medical Education. Amsterdam; Oxford: Elsevier, 2006: 99-114. Subject: 4.1.2

Rich, Ben A. Causation and intent: persistent conundrums in end-of-life care. *CQ: Cambridge Quarterly of Healthcare Ethics* 2007 Winter; 16(1): 63-73. Subject: 20.5.1

Rich, Karen L. Using Buddhist Sangha as a model of communitarianism in nursing. *Nursing Ethics* 2007 July; 14(4): 466-477. Subject: 7.3

Richards, Edward P. Public health law as administrative law: example lessons. *Journal of Health Care Law and Policy* 2007; 10(1): 61-88. Subject: 9.1

Richards, F.H. Maturity of judgement in decision making for predictive testing for nontreatable adult-onset neurogenetic conditions: a case against predictive testing of minors. *Clinical Genetics* 2006 November; 70(5): 396-401. Subject: 15.3

Richards, Janet Radcliffe. Selling organs, gametes, and surrogacy services. *In:* Rhodes, Rosamond; Francis, Leslie P.; Silvers, Anita, eds. The Blackwell Guide to Medical Ethics. Malden, MA: Blackwell Pub., 2007: 254-268. Subject: 4.4

Richards, Martin. Genes, genealogies, and paternity: making babies in the twenty-first century. *In:* Spencer, J.R.; du Bois-Pedain, Antje, eds. Freedom and Responsibility in Reproductive Choice. Portland: Hart Pub., 2006: 53-72. Subject: 15.1

Richards, N. Life or death decisions in the NICU. *Journal of Perinatology* 2006 April; 26(4): 248-251. Subject: 20.5.2

Richards, R. Jason. How we got where we are: a look at informed consent in Colorado — past, present, and future.

See SUBJECT HEADING KEY FOR SECTION II on inside back cover.

585

Northern Illinois University Law Review 2005 Fall; 26(1): 69-99. Subject: 8.3.1

Richards, Tessa. My illness, my record. *BMJ: British Medical Journal* 2007 March 10; 334(7592): 510. Subject: 8.2

Richardson, A.; Sitton-Kent, L. Research ethics. *In:* Holland, Stephen, ed. Introducing Nursing Ethics: Themes in Theory and Practice. Salisbury: APS, 2004: 131-150. Subject: 18.1

Richardson, Genevra. Balancing autonomy and risk: a failure of nerve in England and Wales? *International Journal of Law and Psychiatry* 2007 January-February; 30(1): 71-80. Subject: 17.1

Richardson, Henry S. Gradations of researchers' obligation to provide ancillary care for HIV/AIDS in developing countries. *American Journal of Public Health* 2007 November; 97(11): 1956-1961. Subject: 9.5.6

Richman, Vincent V.; Richman, Alex. Enhancing research integrity [letter]. *CMAJ/JAMC: Canadian Medical Association Journal* 2007 August 14; 177(4): 375. Subject: 1.3.9

Richter, Gerd. Greater patient, family and surrogate involvement in clinical ethics consultation: the model of clinical ethics liaison service as a measure for preventive ethics. *HEC (Healthcare Ethics Committee) Forum* 2007 December; 19(4): 327-340. Subject: 9.6

Rie, Michael A.; Kofke, W. Andrew. Nontherapeutic quality improvement: the conflict of organizational ethics and societal rule of law. *Critical Care Medicine* 2007 February; 35(2, Supplement): S66-S84. Subject: 7.1

Riebschleger, Joanne; Nordquist, Gigi; Newman, Elana;. Mandated reporting. *Journal of Empirical Research on Human Research Ethics* 2007 March; 2(1): 61-64. Subject: 8.4

Ries, Nola M. Regulation of human stem cell research in Japan and Canada: a comparative analysis. *University of New Brunswick Law Journal = Revue de Droit de L'université du Nouveau Brunswick* 2005; 54: 62-74. Subject: 18.5.4

Rietjens, Judith A.C.; Bilsen, Johan; Fischer, Susanne; Van Der Heide, Agnes; Van Der Maas, Paul J.; Miccinessi, Guido; Norup, Michael; Onwuteaka-Philipsen, Bregje D.; Vrakking, Astrid M.; Van Der Wal, Gerrit. Using drugs to end life without an explicit request of the patient. *Death Studies* 2007 March; 31(3): 205-221. Subject: 20.5.1

Riggs, Billy J. Ethical considerations of integrating spiritual direction into psychotherapy. *Journal of Pastoral Care and Counseling* 2006 Winter; 60(4): 353-362. Subject: 17.2

Riley, Margaret Foster; Merrill, Richard A. Regulating reproductive genetics: a review of American bioethics commissions and comparison to the British Human Fertili-

sation and Embryology Authority. *Columbia Science and Technology Law Review* 2005; 6(1): 1-64. Subject: 2.4

Ringel, Steven P. Autonomy and ars moriendi. *Neurology* 2006 September 26; 67(6): 1101-1102. Subject: 19.2

Ríos, A.; Ramírez, P.; Martínez, L.; Montoya, M.J.; Lucas, D.; Alcaraz, J.; Rodríguez, M.M.; Rodríguez, J.M.; Parrilla, P. Are personnel in transplant hospitals in favor of cadaveric organ donation? Multivariate attitudinal study in a hospital with a solid organ transplant program. *Clinical Transplantation* 2006 November-December; 20(6): 743-754. Subject: 19.5

Ritchie, Karen; Freedman, Benjamin; Culver, Charles M.; Ferrell, Richard B.; Green, Ronald M. Competence and mental illness. *In:* Baylis, Françoise; Downie, Jocelyn; Freedman, Benjamin; Hoffmaster, Barry; Sherwin, Susan, eds. Health Care Ethics in Canada. Toronto: Harcourt Brace Canada, 1995: 231-273. Subject: 4.3

Ritter, Alison J.; Fry, Craig L.; Swan, Amy. The ethics of reimbursing drug users for public health research interviews: what price are we prepared to pay? [editorial]. *International Journal of Drug Policy* 2003 February; 14(1): 1-3. Subject: 18.5.1

Rivera, Roberto; Borasky, David; Rice, Robert; Carayon, Florence; Wong, Emelita. Informed consent: an international researchers' perspective. *American Journal of Public Health* 2007 January; 97(1): 25-30. Subject: 18.3

Robert, Jason Scott. Gene maps, brain scans, and psychiatric nosology. *CQ: Cambridge Quarterly of Healthcare Ethics* 2007 Spring; 16(2): 209-218. Subject: 17.1

Robert, Jason Scott. The science and ethics of making part-human animals in stem cell biology. *FASEB Journal: Official Publication of the Federation of American Societies for Experimental Biology* 2006 May; 20(7): 838-845. Subject: 15.1

Roberts, Laura Weiss; Coverdale, John; Louie, Alan K. Philanthropy, ethics, and leadership in academic psychiatry. *Academic Psychiatry* 2006 July-August; 30(4): 269-272. Subject: 17.1

Roberts, Laura Weiss; Johnson, Mark E.; Brems, Christiane; Warner, Teddy D. Preferences of Alaska and New Mexico psychiatrists regarding professionalism and ethics training. *Academic Psychiatry* 2006 May-June; 30(3): 200-204. Subject: 17.2

Roberts, Laura Weiss; Warner, Teddy D.; Dunn, Laura B.; Brody, Janet L.; Hammond, Katherine A. Green; Roberts, Brian B. Shaping medical students' attitudes toward ethically important aspects of clinical research: results of a randomized, controlled educational intervention. *Ethics and Behavior* 2007; 17(1): 19-50. Subject: 2.3

Robertson, Geoffrey. Health and human rights series [comment]. *Lancet* 2007 August 4-10; 370(9585): 368-369. Subject: 21.1

Subject = NRCBL Primary Classification Number; see inside front cover.

Robertson, John A. Compensation and egg donation for research. *Fertility and Sterility* 2006 December; 86(6): 1573-1575. Subject: 14.2

Robertson, John A. The virtues of muddling through. *Hastings Center Report* 2007 July-August; 37(4): 26-28. Subject: 14.1

Robertson, Michael. Part I: psychiatrists and social justice -- the concept of justice. *Journal of Ethics in Mental Health [electronic]* 2007 November; 2(2): 5 p. Accessed: http://www.jemh.ca [2008 January 24]. Subject: 9.5.3

Robertson, Michael. Part II: psychiatrists and social justice – when the social contract fails. *Journal of Ethics in Mental Health [electronic]* 2007 November; 2(2): 4 p. Accessed: http://www.jemh.ca [2008 January 24]. Subject: 9.5.3

Robertson, Michael; Walter, Garry. A critical reflection on utilitarianism as the basis for psychiatric ethics. Part I: Utilitarianism as ethical theory. *Journal of Ethics in Mental Health [electronic]* 2007 April; 2(1): 4 p. Accessed: http://www.jemh.ca [2007 July 31]. Subject: 17.1

Robertson, Michael; Walter, Gary. A critical reflection on utilitarianism as the basis for psychiatric ethics. Part II: Utilitarianism and psychiatry. *Journal of Ethics in Mental Health* 2007 April; 2(1): 4 p. Accessed: http://www.jemh.ca [2007 July 31]. Subject: 17.1

Robinson, David J.; O'Neill, Desmond. Access to health care records after death: balancing confidentiality with appropriate disclosure. *JAMA: The Journal of the American Medical Association* 2007 February 14; 297(6): 634-636. Subject: 8.4

Robinson, Ellen M.; Phipps, Marion; Purtilo, Ruth B.; Tsoumas, Angelica; Hamel-Nardozzi, Marguerite. Complexities in decision making for persons with disabilities nearing end of life. *Topics in Stroke Rehabilitation* 2006 Fall; 13(4): 54-67. Subject: 20.5.1

Robinson, Louise; Murdoch-Eaton, Deborah; Carter, Yvonne. NHS research ethics committees — still need more common sense and less bureaucracy [editorial]. *BMJ: British Medical Journal* 2007 July 7; 335(7609): 6. Subject: 18.2

Robinson, L.; Hutchings, D.; Corner, L.; Beyer, F.; Dickinson, H.; Vanoli, A.; Finch, T.; Hughes, J.; Ballard, C.; May, C.; Bond, J. A systematic literature review of the effectiveness of non-pharmacological interventions to prevent wandering in dementia and evaluations of the ethical implications and acceptability of their use. *Health Technology Assessment* 2006 August; 10(26): iii-108. Subject: 9.5.2

Robinson, Mary R.; Thiel, Mary Martha; Meyer, Elaine C. On being a spiritual care generalist. *American Journal of Bioethics* 2007 July; 7(7): 24-26. Subject: 8.1

Robison, Wade L. The moral crisis in health care. *In:* Engström, Timothy H.; Robison, Wade L., eds. Health Care Reform: Ethics and Politics. Rochester, NY: University of Rochester Press, 2006: 13-39. Subject: 9.1

Roche, Patricia A.; Annas, George J. New genetic privacy concerns: DNA donors may give up more than they realize. *GeneWatch* 2007 January-February; 20(1): 14-17. Subject: 15.1

Rocker, Graeme. Life-support limitation in the pre-hospital setting. *Intensive Care Medicine* 2006 October; 32(10): 1464-1466. Subject: 20.5.1

Rockwell, Lindsay E. Truthtelling. *Journal of Clinical Oncology* 2007 February 1; 25(4): 454-455. Subject: 8.2

Roco, Mihail C. Progress in governance of converging technologies integrated from the nanoscale. *Annals of the New York Academy of Sciences* 2006 December; 1093: 1-23. Subject: 4.5

Roden, Gregory J. Unborn persons, incrementalism and the silence of the lambs. *Human Life Review* 2007 Fall; 33(4): 22-32. Subject: 12.4.2

Rodgers, M.E. Human bodies, inhuman uses: public reactions and legislative responses to the scandals of bodysnatching. *In:* Garwood-Gowers, Austen; Tingle, John; Wheat, Kay, eds. Contemporary Issues in Healthcare Law and Ethics. Edinburgh; New York: Elsevier Butterworth-Heinemann, 2005: 151-172. Subject: 20.1

Rodgers, Sanda. Abortion denied: bearing the limits of law. *In:* Flood, Colleen M., ed. Just Medicare: What's In, What's Out, How We Decide. Buffalo, NY: University of Toronto Press, 2006: 107-136. Subject: 12.4.2

Rodriguez, Keri L.; Young, Amanda J. Patients' and healthcare providers' understandings of life-sustaining treatment: are perceptions of goals shared or divergent? *Social Science and Medicine* 2006 January; 62(1): 125-133. Subject: 20.5.4

Rodwin, Marc A. Medical commerce, physician entrepreneurialism, and conflicts of interest. *CQ: Cambridge Quarterly of Healthcare Ethics* 2007 Fall; 16(4): 387-397. Subject: 9.3.1

Roehr, Bob. More than 90% of US doctors receive drug company favours. *BMJ: British Medical Journal* 2007 April 28; 334(7599): 869. Subject: 9.7

Roelcke, Volker. Psychiatrie und Nervenheilkunde im Nationalsozialismus: Ärztliches Verhalten zwischen Bewährung und Versagen. *In:* Gestrich, Christof; Neugebauer, Johannes, eds. Der Wert menschlichen Lebens: medizinische Ethik bei Karl Bonhoeffer und Dietrich Bonhoeffer. Berlin: Wichern-Verlag, 2006: 13-32. Subject: 17.1

Roemer, Ruth; Roemer, Milton I. Comparative national public health legislation. *In:* Detels, Roger; McEwen, James; Beaglehole, Robert; Tanaka, Heizo, eds. Oxford Textbook of Public Health. Fourth edition. Oxford; New York: Oxford University Press, 2004: 337-357. Subject: 9.1

See SUBJECT HEADING KEY FOR SECTION II on inside back cover.

587

Roff, Sue Rabbitt. Self-interest, self-abnegation and self-esteem: towards a new moral economy of non-directed kidney donation. *Journal of Medical Ethics* 2007 August; 33(8): 437-441. Subject: 19.5

Rogers, Naomi. Race and the politics of polio: Warm Springs, Tuskegee, and the March of Dimes. *American Journal of Public Health* 2007 May; 97(5): 784-795. Subject: 9.5.4

Rogers, Wendy; Ballantyne, Angela; Draper, Heather. Is sex-selective abortion morally justified and should it be prohibited? *Bioethics* 2007 November; 21(9): 520-524. Subject: 14.3

Rogers-Hayden, Tee; Mohr, Alison; Pidgeon, Nick. Introduction: engaging with nanotechnology — engaging differently? *NanoEthics* 2007 August; 1(2): 123-130. Subject: 5.4

Rogowski, Wolf. Current impact of gene technology on healthcare: a map of economic assessments. *Health Policy* 2007 February; 80(2): 340-357. Subject: 15.3

Rohrich, Rod J.; Longaker, Michael T.; Cunningham, Bruce. On the ethics of composite tissue allotransplantation (facial transplantation) [editorial]. *Plastic and Reconstructive Surgery* 2006 May; 117(6): 2071-2073. Subject: 19.1

Rohter, Larry. In the Amazon, giving blood but getting nothing. *New York Times* 2007 June 20; p. A1, A4. Subject: 18.5.9

Rollin, Bernard. Ethics, biotechnology, and animals. *In:* Waldau, Paul; Patton, Kimberly, eds. A Communion of Subjects: Animals in Religion, Science, and Ethics. New York: Columbia University Press, 2006: 519-532. Subject: 15.1

Rollin, Bernard. The ethics of referral. *The Canadian Veterinary Journal.* 2006 July; 47(7): 717-718. Subject: 22.1

Rollin, Bernard E. An ethicist's commentary on characterizing of convenience euthanasia in ethical terms. *Canadian Veterinary Journal. La Revue Vétérinaire Canadienne* 2006 August; 47(8): 742. Subject: 22.1

Rollin, Bernard E. Animal mind: science, philosophy, and ethics. *Journal of Ethics* 2007; 11(3): 253-274. Subject: 22.1

Rollin, Bernard E. Of mice and men. *American Journal of Bioethics* 2007 May; 7(5): 55-57. Subject: 15.1

Romano, Gaetano. Perspectives and controversies in the field of stem cell research. *Drug News and Perspectives* 2006 September; 19(7): 433-439. Subject: 18.5.4

Romeo-Casabona, Carlos; Nicolas, Pilar. Research ethics committees in Spain. *In:* Beyleveld, D.; Townend, D.; Wright, J., eds. Research Ethics Committees, Data Protection and Medical Research in European Countries. Hants, England; Burlington, VT: Ashgate; 2005: 233-244. Subject: 18.2

Romeo-Malanda, Sergio; Nicol, Dianne. Protection of genetic data in medical genetics: a legal analysis in the European context. *Revista de Derecho y Genoma Humano = Law and the Human Genome Review* 2007 July-December; (27): 97-134. Subject: 15.3

Roongrerngsuke, Siriyupa; Phornprapha, Sarote. Perceptions of issues in biotechnology management in Thailand. *Eubios Journal of Asian and International Bioethics* 2007 November; 17(6): 185-190. Subject: 5.1

Roper, James E. Commentary on "Financial conflict and Vioxx: a public policy case study" by Tereskerz. *Organizational Ethics: Healthcare, Business, and Policy* 2006 Fall-Winter; 3(2): 120-124. Subject: 9.7

Ropert-Coudert, Yan. Recognition could support a science code of conduct [letter]. *Nature* 2007 May 17; 447(7142): 259. Subject: 1.3.9

Rorty, Mary V.; Mills, Ann E.; Werhane, Patricia H. Institutional practices, ethics, and the physician. *In:* Rhodes, Rosamond; Francis, Leslie P.; Silvers, Anita, eds. The Blackwell Guide to Medical Ethics. Malden, MA: Blackwell Pub., 2007: 180-197. Subject: 7.3

Roscam Abbing, Henriette. Human tissue research, individual rights and bio-banks. *In:* Gunning, Jennifer; Holm, Søren, eds. Ethics, law, and society. Volume 2. Aldershot, Hants, England; Burlington, VT: Ashgate, 2006: 7-15. Subject: 15.1

Roscam Abbing, Henriette D.C. Pharmacogenetics: a new challenge for health law. *Medicine and Law: The World Association for Medical Law* 2007 December; 26(4): 781-789. Subject: 15.1

Rose, Donald N. Respect for patient autonomy in forensic psychiatric nursing. *Journal of Forensic Nursing* 2005 Spring; 1(1): 23-27. Subject: 8.1

Rose, Nikolas. Beyond medicalisation. *Lancet* 2007 February 24-March 2; 369(9562): 700-702. Subject: 4.2

Rose, Nikolas S. At genetic risk. *In his:* Politics of Life Itself: Biomedicine, Power, and Subjectivity in the Twenty-first Century. Princeton: Princeton University Press, 2007: 106-130, 280-283. Subject: 15.3

Rose, Nikolas S. Race in the age of genomic medicine. *In his:* Politics of Life Itself: Biomedicine, Power, and Subjectivity in the Twenty-first Century. Princeton: Princeton University Press, 2007: 155-186, 287-291. Subject: 15.11

Rosen, Jeffrey. The brain on the stand: how neuroscience is transforming the legal system. *New York Times Magazine* 2007 March 11; p. 48-53, 70, 77, 82, 83. Subject: 4.3

Rosen, Michael R. Are stem cells drugs? The regulation of stem cell research and development. *Circulation* 2006 October 31; 114(18): 1992-2000. Subject: 18.5.4

Rosén, Per. Public dialogue on healthcare prioritisation. *Health Policy* 2006 November; 79(1): 107-116. Subject: 9.4

Subject = NRCBL Primary Classification Number; see inside front cover.

Rosen, Sydney; Sanne, Ian; Collier, Alizanne; Simon, Jonathon L. Rationing antiretroviral therapy for HIV/AIDS in Africa: choices and consequences. *PLoS Medicine* 2005 November; 2(11): e303: 1098-1104. Subject: 9.5.6

Rosenberg, Leah B.; Henry, Michael; Fishman, Jennifer R.; Youngner, Stuart J. Necessary forgetting: on the use of propranolol in post-traumatic stress disorder management. *American Journal of Bioethics* 2007 September; 7(9): 27-28; author reply W1-W3. Subject: 17.4

Rosenberg, Leah; Gehrie, Eric. Against the use of medical technologies for military or national security interests. *American Journal of Bioethics* 2007 May; 7(5): 22-24. Subject: 17.1

Rosenberg, Tina. Doctor or drug pusher? Pain is difficult to measure, and those who treat pain sufferers have to make highly subjective decisions about dosage levels of drugs that can be abused or even resold. When a doctor gets it wrong, is that bad medicine — or a drug felony? *New York Times Magazine* 2007 June 17; p. 48-55, 64, 68- 71. Subject: 9.5.9

Rosenthal, Meredith B.; Donohue, Julie M. Direct-to-consumer advertising of prescription drugs: a policy dilemma. *In:* Santoro, Michael A.; Gorrie, Thomas M., eds. Ethics and the Pharmaceutical Industry. Cambridge; New York: Cambridge University Press, 2005: 169-183. Subject: 9.7

Rosenthal, M. Sara. Patient misconceptions and ethical challenges in radioactive iodine scanning and therapy. *Journal of Nuclear Medicine Technology* 2006 September; 34(3): 143-150; quiz 151-152. Subject: 8.1

Roskies, Adina. A case study of neuroethics: the nature of moral judgment. *In:* Illes, Judy, ed. Neuroethics: Defining the Issues in Theory, Practice, and Policy. New York: Oxford University Press, 2006: 17-32. Subject: 17.1

Roskies, Adina L. The illusion of personhood. *American Journal of Bioethics* 2007 January; 7(1): 55-57. Subject: 17.1

Rosner, Fred. Commentary on "Jewish law and end-of-life decision making". *Journal of Clinical Ethics* 2007 Winter; 18(4): 396-398. Subject: 20.5.1

Rosner, Fred. Medical research in children: ethical issues. *Cancer Investigation* 2006 March; 24(2): 218-220. Subject: 18.5.2

Ross, Colin A. Ethics of CIA and military contracting by psychiatrists and psychologists. *Ethical Human Psychology and Psychiatry* 2007; 9(1): 25-34. Subject: 18.4

Ross, David B. The FDA and the case of ketek. *New England Journal of Medicine* 2007 April 19; 356(16): 1601-1604. Subject: 9.7

Ross, Joseph S.; Lackner, Josh E.; Lurie, Peter; Gross, Cary P.; Wolfe, Sidney; Krumholz, Harlan M. Pharmaceutical company payments to physicians: early experi-ences with disclosure laws in Vermont and Minnesota. *JAMA: The Journal of the American Medical Association* 2007 March 21; 297(11): 1216-1223. Subject: 9.7

Ross, Lainie Friedman. The moral status of the newborn and its implications for medical decision making. *Theoretical Medicine and Bioethics* 2007 October; 28(5): 349-355. Subject: 4.4

Ross, Lainie Friedman; Philipson, Louis H.; Voltarelli, Julio C.; Couri, Carlos E.B.; Stracieri, Ana B.P.L.; Oliveira, Maria C.; Moraes, Daniela A.; Fabiano, Pieroni; Coutinho, Marina; Malmegrim, Kelen C.R.; Foss-Freitas, Maria C.; Simões, Belinda P.; Foss, Milton C.; Squiers, Elizabeth; Burt, Richard K. Ethics of hematopoietic stem cell transplantation in type 1 diabetes mellitus [letter and reply]. *JAMA: The Journal of the American Medical Association* 2007 July 18; 298(3): 285-286. Subject: 18.5.2

Ross, Lainie Friedman; Siegler, Mark; Thistlethwaite, J. Richard, Jr. We need a registry of living kidney donors. *Hastings Center Report* 2007 November-December; 37(6): Inside back cover. Subject: 19.5

Rothman, Marc D.; Van Ness, Peter H.; O'Leary, John R.; Fried, Terri R. Refusal of medical and surgical interventions by older persons with advanced chronic disease. *JGIM: Journal of General Internal Medicine* 2007 July; 22(7): 982-987. Subject: 8.3.4

Rothman, S.M.; Rothman, D.J. The hidden cost of organ sale. *American Journal of Transplantation* 2006 July; 6(7): 1524-1528. Subject: 19.5

Rothstein, Mark A.; Talbott, Meghan K. Compelled authorizations for disclosure of health records: magnitude and implications. *American Journal of Bioethics* 2007 March; 7(3): 38-45. Subject: 8.4

Rothstein, Mark A.; Talbott, Meghan K. Response to open peer commentaries on "Compelled authorizations for disclosure of health records: magnitude and implications" [letter]. *American Journal of Bioethics* 2007 March; 7(3): W1-W3. Subject: 8.4

Rothstein, M.A.; Epps, P.G. Pharmacogenomics and the (ir)relevance of race. *Pharmacogenomics Journal* 2001; 1(2): 104-108. Subject: 15.11

Roucounas, Emmanuel. The Biomedicine Convention in Relation to other international instruments. *In:* Gevers, J.K.M.; Hondius, E.H.; Hubben, J.H., eds. Health Law, Human Rights and the Biomedicine Convention: Essays in Honour of Henriette Roscam Abbing. Leiden; Boston: Martinus Nijhoff Publishers, 2005: 23-34. Subject: 21.1

Rousseau, Paul. Allegations of euthanasia. *American Journal of Hospice and Palliative Care* 2006 October-November; 23(5): 422-423. Subject: 20.5.1

Rowley, Emma. On doing 'being ordinary': women's accounts of BRCA testing and maternal responsibility. *New*

See SUBJECT HEADING KEY FOR SECTION II on inside back cover.

589

Genetics and Society 2007 December; 26(3): 241-250. Subject: 15.3

Rowson, Richard. Nurses' difficulties with rights. *Nursing Ethics* 2007 November; 14(6): 838-840. Subject: 21.1

Roy, Elizabeth; Samuels, Sumerlee. Ethical debating: therapy for adolescents with eating disorders. *Journal of Pediatric Nursing* 2006 April; 21(2): 161-166. Subject: 9.5.7

Roy, Nobhojit; Madhiwalla, Neha; Pai, Sanjay A. Drug promotional practices in Mumbai: a qualitative study. *Indian Journal of Medical Ethics* 2007 April-June; 4(2): 57-61. Subject: 9.7

Royal College of Obstetricians and Gynaecologists (Great Britain). RCOG response to MRC consultation on the Code of Practice for the use of human stem cell lines. London: Royal College of Obstetricians and Gynaecologists, 2004 May 27; 3 p. [Online]. Accessed: http://www.rcog.org.uk/index.asp?PageID=1349 [2007 April 5]. Subject: 18.5.4

Royal College of Surgeons of England. Working party. Facial transplantation: working party report. London: Royal College of Surgeons of England, 2006 November; 48 p. [Online]. Accessed: http://www.rcseng.ac.uk/publications/docs/facial_transplant_report_2006.html/pdffile/downloadFile [2007 April 12]. Subject: 19.1

Rozsos, Elizabeth. Hungary: nursing and nursing ethics after the Communist era. *In:* Davis, Anne J.; Tschudin, Verena; de Raeve, Louise, eds. Essentials of Teaching and Learning in Nursing Ethics: Perspectives and Methods. New York: Churchill Livingstone Elsevier, 2006: 301-312. Subject: 4.1.3

Rózynska, Joanna; Czarkowski, Marek. Emergency research without consent under Polish Law. *Science and Engineering Ethics* 2007 September; 13(3): 337-350. Subject: 18.3

Rozzini, Renzo; Trabucchi, Marco. Advance directives and quality of end-of-life care: pros and cons in older people [letter]. *Journal of the American Geriatrics Society* 2007 September; 55(9): 1472. Subject: 20.5.4

Rubin, M.H. Is there a doctor in the house? *Journal of Medical Ethics* 2007 March; 33(3): 158-159. Subject: 7.1

Rubin, Susan B. If we think it's futile, can't we just say no? *HEC (Healthcare Ethics Committee) Forum* 2007 March; 19(1): 45-65. Subject: 20.5.1

Rudnick, Abraham. Other-consciousness and the use of animals as illustrated in medical experiments. *Journal of Applied Philosophy* 2007 May; 24(2): 202-208. Subject: 22.2

Rudnick, Abraham. Processes and pitfalls of dialogical bioethics. *Health Care Analysis: An International Journal of Health Philosophy and Policy* 2007 June; 15(2): 123-135. Subject: 2.1

Ruger, Jennifer Prah. Health, health care, and incompletely theorized agreements: a normative theory of health policy decision making. *Journal of Health Politics, Policy and Law* 2007 February; 32(1): 51-87. Subject: 9.3.1

Ruger, J.P. Ethics and governance of global health inequalities. *Journal of Epidemiology and Community Health* 2006 November; 60(11): 998-1003 [Online]. Accessed: http://jech.bmj.com/cgi/reprint/60/11/998 [2007 November 13]. Subject: 9.2

Ruger, J.P. Rethinking equal access: agency, quality, and norms. *Global Public Health* 2007 January; 2(1): 78-96. Subject: 9.2

Ruger, J.P. The moral foundations of health insurance. *Quarterly Journal of Medicine* 2007; 100(1): 5 p. [Online]. Accessed: http://papers.ssrn.com/sol3/papers.cfm?abstract_id=957971 [2007 November 12]. Subject: 9.3.1

Rushlow, Jenny. Rapid DNA database expansion and disparate minority impact. *GeneWatch* 2007 July-August; 20(4): 3-11. Subject: 15.1

Rushton, Cynda Hylton. Donation after cardiac death: ethical implications and implementation strategies. *AACN Advanced Critical Care* 2006 July-September; 17(3): 345-349. Subject: 19.5

Rusnak, A.J.; Chudley, A.E. Stem cell research: cloning, therapy and scientific fraud. *Clinical Genetics* 2006 October; 70(4): 302-305. Subject: 15.1

Russell, Barbara. The crucible of anorexia nervosa. *Journal of Ethics in Mental Health [electronic]* 2007 November; 2(2): 6 p. Accessed: http://www.jemh.ca [2008 January 24]. Subject: 8.3.4

Russell, Barbara J.; Pape, Deborah A. Ethics consultation: continuing its analysis. *Journal of Clinical Ethics* 2007 Fall; 18(3): 235-242. Subject: 9.6

Rust, Susanne; Spivak, Cary. Panel will investigate research firms' ethics: conflicts of interest exist, presidential candidate says. *Milwaukee Journal Sentinel: JS Online* 2007 May 5; 2 p. [Online]. Accessed: http://www.jsonline.com/ story/index.aspx?id=601378 [2007 May 7]. Subject: 1.3.9

Rutecki, Greg. Permissibility to accept refusal of potentially life-saving treatment. *Ethics and Medicine: An International Journal of Bioethics* 2007 Summer; 23(2): 77-80. Subject: 8.3.4

Rutherford, Alexandra. The social control of behavior control: behavior modification, individual rights, and research ethics in America, 1971-1979. *Journal of the History of the Behavioral Sciences* 2006 Summer; 42(3): 203-220. Subject: 18.4

Rutz, Berthold; Yeats, Siobhan. Patents: patenting of stem cell related inventions in Europe. *Biotechnology Journal* 2006 April; 1(4): 384-387. Subject: 15.8

Rx&D: Canada's Research-Based Pharmaceutical Companies. Rx&D code of conduct: 11 guiding principles. Ottawa: Rx&D: Canada's Research-Based Pharmaceutical Companies, 2006 Winter; 1 p. [Online]. Accessed: http://www.canadapharma.org/Industry_Publications/Code/GuidingPrinciples0512.pdf [2007 April 4]. Subject: 6

Rydell, Robert; Cogdell, Christina; Largent, Mark. The Nazi eugenics exhibit in the United States, 1934-43. *In:* Currell, Susan; Cogdell, Christina, eds. Popular Eugenics: National Efficiency and American Mass Culture in the 1930s. Athens, OH: Ohio University Press, 2006: 359-384. Subject: 15.5

S

Sabin, James E.; Cochran, David. Confronting trade-offs in health care: Harvard Pilgrim Health Care's organizational ethics program. *Health Affairs* 2007 July-August; 26(4): 1129-1134. Subject: 9.3.2

Sachedina, Abdulaziz. Brain death in Islamic jurisprudence. Charlottesville, VA: University of Virginia, n.d.: 7 p. [Online]. Accessed: http://people.virginia.edu/~aas/article/article6.htm [2007 April 2]. Subject: 20.2.1

Sachedina, Abdulaziz. The cultural and religious in Islamic biomedicine: the case of human cloning. *In:* Roetz, Heiner, ed. Cross-Cultural Issues in Bioethics: The Example of Human Cloning. New York: Rodopi, 2006: 263-290. Subject: 14.5

Sachs, Greg A.; Cassel, Christine K. Ethical aspects of dementia. *Neurologic Clinics* 1989 November; 7(4): 845-858. Subject: 17.1

Sade, Robert M. Reports of clinical trials: ethical aspects. *Journal of Thoracic and Cardiovascular Surgery* 2006 August; 132(2): 245-246. Subject: 18.2

Sade, Robert M.; Grande, David; Gorske, Arnold L.; Campbell, Eric G. A national survey of physician-industry relationships [letters and reply]. *New England Journal of Medicine* 2007 August 2; 357(5): 507-508. Subject: 9.7

Sade, Robert M.; Henry, Michael; Fishman, Jennifer R.; Youngner, Stuart J. On moralizing and hidden agendas: the pot and the kettle in political bioethics. *American Journal of Bioethics* 2007 September; 7(9): 42-43; author reply W1-W3. Subject: 17.4

Sade, Robert M.; Robinson, David J.; O'Neill, Desmond. Confidentiality of medical information after death [letter and reply]. *JAMA: The Journal of the American Medical Association* 2007 June 20; 297(23): 2585. Subject: 8.4

Sadeghi, Mahmoud. Islamic perspectives on human cloning. *Human Reproduction and Genetic Ethics: An International Journal* 2007; 13(2): 32-40. Subject: 14.5

Sadler, John Z. The rhetorician's craft, distinctions in science, and political morality. *Philosophy, Ethics, and Humanities in Medicine [electronic]* 2006; 1(7): 2 p.

Accessed: http://www.peh-med.com/content/1/1/7 [2006 August 15]. Subject: 17.1

SAEM Ethics Committee; Schears, R.M.; Watters, A.; Schmidt, T.A.; Marco, C.A.; Larkin, G.L.; Marshall, J.P.; Mason, J.D.; McBeth, B.D.; Mello, M.J. The Society for Academic Emergency Medicine position on ethical relationships with the biomedical industry. *Academic Emergency Medicine* 2007 February; 14(2): 179-181. Subject: 7.1

Säfken, Christian; Frewer, Andreas. The duty to warn and clinical ethics: legal and ethical aspects of confidentiality and HIV/AIDS. *HEC (Healthcare Ethics Committee) Forum* 2007 December; 19(4): 313-326. Subject: 9.6

Sagoff, Mark. A transcendental argument for the concept of personhood in neuroscience. *American Journal of Bioethics* 2007 January; 7(1): 72-73. Subject: 17.1

Sagoff, Mark. Further thoughts about the human neuron mouse. *American Journal of Bioethics* 2007 May; 7(5): 51-52. Subject: 15.1

Sahakian, Barbara; Morein-Zamir, Sharon. Professor's little helper. The use of cognitive-enhancing drugs by both ill and healthy individuals raises ethical questions that should not be ignored. *Nature* 2007 December 20-27; 450(7173): 1157-1159. Subject: 4.5

Sakaguchi, Misa; Maeda, Shoichi. Informed consent for anesthesia: survey of current practices in Japan. *Journal of Anesthesia* 2005; 19(4): 315-319. Subject: 8.3.1

Saks, Elyn R.; Dunn, Laura B.; Marshall, Barbara J.; Nayak, Gauri V.; Golshan, Shahrokh; Jeste, Dilip V. The California Scale of Appreciation: a new instrument to measure the appreciation component of capacity to consent to research. *American Journal of Geriatric Psychiatry* 2002 March-April; 10(2): 166-174. Subject: 18.3

Salek, Sam. Health economics and access to treatment. *In:* Gunning, Jennifer; Holm, Søren, eds. Ethics, law, and society. Volume 1. Aldershot, Hants, England; Burlington, VT: Ashgate, 2005: 59-64. Subject: 9.4

Sales, Becky; McKenzie, Nigel. Time to act on behalf of mentally disordered offenders. *BMJ: British Medical Journal* 2007 June 9; 334(7605): 1222. Subject: 9.5.3

Salladay, Susan A. Life and death disagreements [interview]. *Journal of Christian Nursing* 2007 January-March; 24(1): 38-40. Subject: 20.5.1

Salman, Rustam Al-Shahi, Stone, Jon; Warlow, Charles. What do patients think about appearing in neurology "grand rounds"? *Journal of Neurology, Neurosurgery, and Psychiatry* 2007 May; 78(5): 454-456. Subject: 17.1

Salomon, Fred; Ziegler, Andrea. Moral und Abhängigkeit - Ethische Entscheidungskonflikte im hierarchischen System Krankenhaus = Morals and dependency - ethical conflicts in the hierarchical system of a

See SUBJECT HEADING KEY FOR SECTION II on inside back cover.

591

hospital. *Ethik in der Medizin* 2007 September; 19(3): 174-186. Subject: 4.1.2

Salomon, F. Leben erhalten und Sterben ermöglichen.Entscheidungskonflikte in der Intensivmedizin = Saving life and permitting death. Decision conflicts in intensive medicine. *Der Anaesthesist* 2006 January; 55(1): 64-69. Subject: 20.5.1

Salter, Brian. Bioethics, politics and the moral economy of human embryonic stem cell science: the case of the European Union's Sixth Framework Programme. *New Genetics and Society* 2007 December; 26(3): 269-288. Subject: 18.5.4

Salter, Brian; Salter, Charlotte. Bioethics and the global moral economy: the cultural politics of human embryonic stem cell science. *Science, Technology and Human Values* 2007 September; 32(5): 554-581. Subject: 2.1

Salter, Frank K. On the ethics of defending genetic interests. *In his:* On Genetic Interests: Family, Ethnicity, and Humanity in an Age of Mass Migration. New Brunswick, NJ: Transaction Publishers, 2007: 283-323. Subject: 15.1

Salvi, Vinita; Damania, K. HIV, research, ethics and women. *Journal of Postgraduate Medicine* 2006 July-September; 52(3): 161-162. Subject: 18.5.9

Samanta, Ash; Samanta, Jo. Advance directives, best interests and clinical judgement: shifting sands at the end of life. *Clinical Medicine* 2006 May-June; 6(3): 274-278. Subject: 20.5.4

Sammet, Kai. Autonomy or protection from harm? Judgements of German courts on care for the elderly in nursing homes. *Journal of Medical Ethics* 2007 September; 33(9): 534-537. Subject: 9.5.2

Sammons, Helen M.; Atkinson, Maria; Choonara, Imti; Stephenson, Terence. What motivates British parents to consent for research? A questionnaire study. *BMC Pediatrics* 2007 March 9; 7: 12. Subject: 18.3

Samuels, Allison. Brutal case studies. A new book documents a true ethics horror story. *Newsweek* 2007 February 12; 149(7): 49. Subject: 18.5.1

Sandberg, David E. Growth attenuation in developmental disabilities. *Growth, Genetics and Hormones* 2007 March; 23(1): 12-13. Subject: 9.7

Sanders, Cheryl J. Religion and ethical decision making in the African American community: bioterrorism and the black postal workers. *In:* Prograis, Lawrence J.; Pellegrino, Edmund D., eds. African American Bioethics: Culture, Race, and Identity. Washington, DC: Georgetown University Press, 2004: 93-104. Subject: 21.3

Sanders, Karen; Fullbrook, Suzanne. Autonomy and care: respecting the wishes of the deceased patient. *British Journal of Nursing* 2007 March 22-April 11; 16(6): 360-361. Subject: 19.5

Sander-Staudt, Maureen. Of machine born? A feminist assessment of ectogenesis and artificial wombs. *In:*

Gelfand, Scott; Shook, John R., eds. Ectogenesis: Artificial Womb Technology and the Future of Human Reproduction. Amsterdam; New York: Editions Rodopi, B.V., 2006: 109-128. Subject: 14.1

Sanger, Carol. Regulating teenage abortion in the United States: politics and policy. *International Journal of Law, Policy and the Family* 2004 December; 18(3): 305-318. Subject: 12.4.2

Sangster, Catriona. 'Cooling corpses': Section 43 of the Human Tissue Act 2004 and organ donation. *Clinical Ethics* 2007 March; 2(1): 23-27. Subject: 19.5

Saniotis, Arthur. Changing ethics in medical practice: a Thai perspective. *Indian Journal of Medical Ethics* 2007 January-March; 4(1): 24-25. Subject: 9.3.1

Sanner, Margareta A. Giving and taking — to whom and from whom? People's attitudes toward transplantation of organs and tissues from different sources. *Clinical Transplantation* 1998 December; 12(6): 530-537. Subject: 19.5

Sanner, Margareta A. People's attitudes and reactions to organ donation. *Mortality* 2006 May; 11(2): 133-150. Subject: 19.5

Santosuosso, Amedeo; Sellaroli, Valentina; Fabio, Elisabetta. What constitutional protection for freedom of scientific research? *Journal of Medical Ethics* 2007 June; 33(6): 342-344. Subject: 1.3.9

Sanz-Ortiz, Jaime. Informed consent and sedation. *Clinical and Translational Oncology* 2006 February; 8(2): 94-97. Subject: 8.3.1

Sarig, Merav. Israeli surgeon is arrested for suspected organ trafficking [news]. *BMJ: British Medical Journal* 2007 May 12; 334(7601): 973. Subject: 19.5

Sass, Hans-Martin. Ethical risk in medical research. *In:* Shapiro, S.; Dinger, J.; Scriba, P., eds. Enabling Risk Assessment in Medicine: Farewell Symposium for Werner-Kari Raff. New Brunswick, NJ: Transaction Publishers, 2004: 83-93. Subject: 1.3.9

Sass, Hans-Martin. Fritz Jahr's 1927 concept of bioethics. *Kennedy Institute of Ethics Journal* 2007 December; 17(4): 279-295. Subject: 2.2

Sass, Hans-Martin. Let probands and patients decide about moral risk: stem cell research and medical treatment. *In:* Roetz, Heiner, ed. Cross-Cultural Issues in Bioethics: The Example of Human Cloning. New York: Rodopi, 2006: 425-444. Subject: 18.5.4

Sastre, María Teresa Muñoz; Peccarisi, Céline; Legrain, Elizabeth; Mullet, Etienne; Sorum, Paul. Acceptability in France of induced abortion for adolescents. *American Journal of Bioethics* 2007 August; 7(8): 26-32. Subject: 12.5.2

Satel, Sally. Desperately seeking a kidney. What you learn about people — and yourself — when you need them to donate an organ. *New York Times Magazine* 2007 December 16; p. 62-67. Subject: 19.5

Satia, Jessie A.; McRitchie, Susan; Kupper, Lawrence L.; Halbert, Chanita Hughes. Genetic testing for colon cancer among African-Americans in North Carolina. *Preventive Medicine* 2006 January; 42(1): 51-59. Subject: 15.3

Saugstad, Ola Didrik. Non-selective fetal reduction is malpractice. *Journal of Perinatal Medicine* 2006; 34(5): 355-358. Subject: 12.1

Saul, Stephanie. Doctors and drug makers: a move to end cozy ties. *New York Times* 2007 February 12; p. C10. Subject: 9.7

Saul, Stephanie. U.S. to review drug intended for one race. *New York Times* 2005 June 13: A1. Subject: 9.5.4

Saunders, John. More guidelines on research ethics? With its new research ethics guidelines, the UK Royal College of Physicians continues a useful tradition of providing guidance to medical researchers. [editorial]. *Journal of Medical Ethics* 2007 December; 33(12): 683-684. Subject: 18.2

Saunders, T.A.; Stein, D.J.; Dilger, J.P. Informed consent for labor epidurals: a survey of Society for Obstetric Anesthesia and Perinatology anesthesiologists from the United States. *International Journal of Obstetric Anesthesia* 2006 April; 15(2): 98-103. Subject: 9.5.5

Saunders, William L. Washington insider [summary]. *National Catholic Bioethics Quarterly* 2007 Winter; 7(4): 661-669. Subject: 2.1

Savage, Teresa A. Ethical issues in research with patients who have experienced stroke. *Topics in Stroke Rehabilitation* 2006 Fall; 13(4): 1-10. Subject: 18.5.6

Savell, Kristin. The legal significance of birth. *University of New South Wales Law Journal* 2006; 29(2): 200-206. Subject: 4.4

Saver, Cynthia. Legal and ethical aspects of publishing. *AORN Journal* 2006 October; 84(4): 571-5 [see correction in: AORN Journal 2006 December; 84(6): 950]. Subject: 1.3.7

Saver, Richard S. The costs of avoiding physician conflicts of interest: a cautionary tale of gainsharing regulation. *In:* Flood, Colleen M., ed. Just Medicare: What's In, What's Out, How We Decide. Buffalo, NY: University of Toronto Press, 2006: 281-306. Subject: 9.3.1

Savulescu, Julian. Autonomy, the good life, and controversial choices. *In:* Rhodes, Rosamond; Francis, Leslie P.; Silvers, Anita, eds. The Blackwell Guide to Medical Ethics. Malden, MA: Blackwell Pub., 2007: 17-37. Subject: 4.2

Savulescu, Julian. Genetic interventions and the ethics of enhancement of human beings. *In:* Steinbock, Bonnie, ed. The Oxford Handbook of Bioethics. Oxford; New York: Oxford University Press, 2007: 516-535. Subject: 15.1

Savulescu, Julian. In defence of procreative beneficence. *Journal of Medical Ethics* 2007 May; 33(5): 284-288. Subject: 15.5

Savulescu, Julian; Lawrence, Ryan E.; Curlin, Farr A. The proper place of values in the delivery of medicine. *American Journal of Bioethics* 2007 December; 7(12): 21-22; author reply W1-W2. Subject: 4.1.2

Sawyer, Robert J. Robot ethics [editorial]. *Science* 2007 November 16; 318(5853): 1037. Subject: 5.1

Sawyer, Susan M.; Cerritelli, Belinda; Carter, Lucy S.; Cooke, Mary; Glazner, Judith A.; Massie, John. Changing their minds with time: a comparison of hypothetical and actual reproductive behaviors in parents of children with cystic fibrosis. *Pediatrics* 2006 September; 118(3): e649-e656. Subject: 15.2

Sax, Joanna K. Reforming FDA policy for pediatric testing: challenges and changes in the wake of studies using antidepressant drugs. *Indiana Health Law Review* 2007; 4(1): 61-84. Subject: 18.5.2

Sayeed, Sadath A. Legal challenges at the limits of viability. *Medical Ethics Newsletter [Lahey Clinic]* 2007 Spring; 14(2): 6-8. Subject: 9.5.8

Sayers, Gwen M.; Gabe, Simon M. Restraint in order to feed: justifying a lawful policy for the UK. *European Journal of Health Law* 2007 April; 14(1): 3-20. Subject: 21.5

Sayers, G.M. Should research ethics committees be told how to think? *Journal of Medical Ethics* 2007 January; 33(1): 39-42. Subject: 18.2

Sayson, Ciriaco M. Asian bioethics: theoretical background. *Asian Biotechnology and Development Review* 2006 November; 9(1): 7-11 [Online}. Accessed: http://www.openj-gate.org/articlelist.asp?LatestYear=2007&JCode=104346&year=2006&vol=9&issue=1&ICode=583649 [2008 February 29]. Subject: 2.1

Schachter, Debbie; Kleinman, Irwin. Psychiatrists' documentation of informed consent: a representative survey. *Canadian Journal of Psychiatry* 2006 June; 51(7): 438-444. Subject: 8.3.1

Schaffer, Marjorie A. Ethical problems in end-of-life decisions for elderly Norwegians. *Nursing Ethics* 2007 March; 14(2): 242-257. Subject: 20.5.1

Schaller, Jean; Moser, Hugo; Begleiter, Michael L.; Edwards, Janice. Attitudes of families affected by adrenoleukodystrophy toward prenatal diagnosis, presymptomatic and carrier testing, and newborn screening. *Genetic Testing* 2007 Fall; 11(3): 296-302. Subject: 15.3

Schaub, Diana. Bioethics and "The Public Interest". *New Atlantis* 2007 Winter; 15: 135-140. Subject: 2.1

Schauenburg, H.; Hildebrandt, A. Public knowledge and attitudes on organ donation do not differ in Germany and Spain. *Transplantation Proceedings* 2006 June; 38(5): 1218-1220. Subject: 19.5

See SUBJECT HEADING KEY FOR SECTION II on inside back cover.

593

Scheetz, Mary D. The Teaching Scholars Program: a proposed approach for promoting research integrity. *Medicine and Law: The World Association for Medical Law* 2007 September; 26(3): 599-614. Subject: 1.3.9

Schefer, Markus. Geltung der Grundrechte vor der Geburt. *In:* Schreiber, Hans-Peter, ed. Biomedizin und Ethik: Praxis, Recht, Moral. Basel; Boston: Birkhäuser, 2004: 43-49, 96. Subject: 4.4

Scheirton, Linda S.; Mu, K.; Lohman, H.; Cochran, T.M. Error and patient safety: ethical analysis of cases in occupational and physical therapy practice. *Medicine, Health Care and Philosophy* 2007 September; 10(3): 301-311. Subject: 9.8

Schellings, Ron; Kessels, Alfons G.; ter Riet, Gerben; Knottnerus, J. André; Sturmans, Ferd. Randomized consent designs in randomized controlled trials: systematic literature search. *Contemporary Clinical Trials* 2006 August; 27(4): 320-332. Subject: 18.3

Schenkenberg, Thomas; Kochenour, Neil D.; Botkin, Jeffrey R. Ethical considerations in clinical care of the "VIP". *Journal of Clinical Ethics* 2007 Spring; 18(1): 56-63. Subject: 9.5.1

Schenker, Joseph G. Legal aspects of ART practice in Israel. *Journal of Assisted Reproduction and Genetics* 2003 July; 20(7): 250-259. Subject: 14.4

Schenker, Yael; Wang, Frances; Selig, Sarah Jane; Ng, Rita; Fernandez, Alicia. The impact of language barriers on documentation of informed consent at a hospital with on-site interpreter services. *JGIM: Journal of General Internal Medicine* 2007 November; 22(Supplement 2): 294-299. Subject: 9.5.4

Scheper-Hughes, Nancy. Rotten trade: millennial capitalism, human values and global justice in organs trafficking. *Journal of Human Rights* 2003 June; 2(2): 197-226. Subject: 19.5

Scheper-Hughes, Nancy. The ends of the body: commodity fetishism and the global traffic in organs. *SAIS Review* 2002 Winter-Spring; 22(1): 61-80 [Online]. Accessed: http://muse.jhu.edu/journals/sais_review/v022/22.1scheper.pdf [2007 February 21]. Subject: 19.5

Scherer, Yvonne.; Jezewski, Mary Ann; Graves, Brian; Wu, Yow-Wu; Bu, Xiaoyan. Advance directives and end-of-life decision making: survey of critical care nurses' knowledge, attitude, and experience. *Critical Care Nurse* 2006 August; 26(4): 30-40. Subject: 20.5.1

Schermer, M.H.N. Brave New World versus Island — utopian and dystopian views on psychopharmacology. *Medicine, Health Care and Philosophy* 2007 June; 10(2): 119-128. Subject: 17.4

Schexnayder, Stephen M.; Hester, D. Micah. A new perspective on community consultation in pediatric resuscitation research. *Critical Care Medicine* 2006 October; 34(10): 2684-2685. Subject: 20.5.2

Schicktanz, Silke. Why the way we consider the body matters — reflections on four bioethical perspectives on the human body. *Philosophy, Ethics, and Humanities in Medicine [electronic]* 2007 December 4; 2(30): 12 p. Accessed: http://www.peh-med.com/ [2008 January 24]. Subject: 4.4

Schiermeier, Quirin. Primate work faces German veto [news]. *Nature* 2007 April 26; 446(7139): 955. Subject: 22.2

Schildmann, Jan; Steger, Florian; Vollmann, Jochen. „Aufklärung im ärztlichen Alltag" - ein Lehrmodul zur integrierten Bearbeitung medizinethischer und -historischer Aspekte im neuen Querschnittsbereich GTE = "Informed consent" - an integrated teaching module on ethical and historical aspects for the new subject "history, theory, ethics of medicine". *Ethik in der Medizin* 2007 September; 19(3): 187-199. Subject: 7.2

Schillinger, Dean; Volandes, Angelo E.; Paasche-Orlow, Michael K. Literacy and health communication: reversing the 'inverse care law'. *American Journal of Bioethics* 2007 November; 7(11): 15-18; author reply W1-W2. Subject: 9.1

Schiltz, Elizabeth R. The disabled Jesus: a parent looks at the logic behind prenatal testing and stem cell research. *America* 2007 March 12; 196(9): 16-18. Subject: 15.3

Schleger, Heidi Albisser. „Alter" und „Kosten" - Faktoren bei Therapieentscheidungen am Lebensende? Eine Analyse informeller Wissensstrukturen bei Ärzten und Pflegenden = "Age" and "Costs" - factors in treatment decisions at the end-of-life? An analysis of informal knowledge structures of doctors and nurses. *Ethik in der Medizin* 2007 June; 19(2): 103-119. Subject: 20.5.1

Schlieter, Jens. Some aspects of the Buddhist assessment of human cloning. *In:* Vöneky, Silja; Wolfrum, Rüdiger, eds. Human Dignity and Human Cloning. Leiden: Nijhoff, 2004: 23-33. Subject: 14.5

Schlieter, Jens. Some observations on Buddhist thoughts on human cloning. *In:* Roetz, Heiner, ed. Cross-Cultural Issues in Bioethics: The Example of Human Cloning. New York: Rodopi, 2006: 179-202. Subject: 14.5

Schmaltz, Florian. Neurosciences and research on chemical weapons of mass destruction in Nazi Germany. *Journal of the History of the Neurosciences* 2006 September; 15(3): 186-209. Subject: 21.2

Schmelzer, Marilee. Institutional review boards: friend, not foe. *Gastroenterology Nursing: the Official Journal of the Society of Gastroenterology Nurses and Associates* 2006 January-February; 29(1): 80-81. Subject: 18.2

Schmidt, Charles. Putting the brakes on psychosis. *Science* 2007 May 18; 316(5827): 976-977. Subject: 17.1

Schmidt, Charlie. Negotiating the RNAi patent thicket. *Nature Biotechnology* 2007 March; 25(3): 273-275. Subject: 15.8

Schmidt, Eric B. Making someone child-sized forever: ethical considerations in inhibiting the growth of a developmentally disabled child. *Clinical Ethics* 2007 March; 2(1): 46-49. Subject: 9.5.3

Schmidt, Eric B. The parental obligation to expand a child's range of open futures when making genetic trait selections for their child. *Bioethics* 2007, May; 21(4): 191-197. Subject: 15.3

Schmidt, Harald. Patients' charters and health responsibilities. *BMJ: British Medical Journal* 2007 December 8; 335(7631): 1187-1189. Subject: 8.1

Schmidt, Harald. Whose dignity? Resolving ambiguities in the scope of "human dignity" in the Universal Declaration on Bioethics and Human Rights. *Journal of Medical Ethics* 2007 October; 33(10): 578-584. Subject: 2.1

Schmidt, Kurt W. Lost in translation — bridging gaps through procedural norms: comments on the papers of Capaldi and Tao. *In:* Engelhardt, H. Tristram, ed. Global Bioethics: The Collapse of Consensus. Salem, MA: M&M Scrivener Press, 2006: 180-206. Subject: 21.7

Schmidt, Kurt W.; Frewer, Andreas. Current problems of clinical ethics: confidentiality and end-of-life decisions — is silence always golden? Introduction. *HEC (Healthcare Ethics Committee) Forum* 2007 December; 19(4): 273-276. Subject: 9.6

Schmidt, Ulf. Turning the history of medical ethics from its head onto its feet: a critical commentary on Baker and McCullough. *Kennedy Institute of Ethics Journal* 2007 March; 17(1): 31-42. Subject: 2.1

Schmidt-Felzmann, Heike. Authority and influence in the psychotherapeutic relationship. *In:* Nys, Thomas; Denier, Yvonne; Vandevelde, Toon, eds. Autonomy and Paternalism: Reflections on the Theory and Practice of Health Care. Leuven; Dudley, MA: Peeters, 2007: 167-180. Subject: 17.2

Schneider, A. Patrick. Emergency contraception (EC) for victims of rape: ten myths. *Linacre Quarterly* 2007 August; 74(3): 181-203. Subject: 11.1

Schneider, Carl E. The cash nexus. *Hastings Center Report* 2007 July-August; 37(4): 11-12. Subject: 9.1

Schneider, Carl E. Void for vagueness. *Hastings Center Report* 2007 January-February; 37(1): 10-11. Subject: 8.3.1

Schneider, Ingrid. Oocyte donation for reproduction and research cloning — the perils of commodification and the need for European and international regulation. *Revista de Derecho y Genoma Humano = Law and the Human Genome Review* 2006 July-December; (25): 205-241. Subject: 14.1

Schneiderman, Lawrence J. Effect of ethics consultations in the intensive care unit. *Critical Care Medicine* 2006 November; 34(11, Supplement): S359-S363. Subject: 20.5.1

Schneiderman, Lawrence J. The media and the medical market. *CQ: Cambridge Quarterly of Healthcare Ethics* 2007 Fall; 16(4): 420-424. Subject: 9.3.1

Scholze, Simone. Setting standards for scientists. For almost ten years, COMEST has advised UNESCO on the formulation of ethical guidelines. *EMBO Reports* 2006 July; 7 Special No: S65-S67. Subject: 1.3.9

Schomerus, G.; Matschinger, H.; Angermeyer, M.C. Alcoholism: illness beliefs and resource allocation preferences of the public. *Drug and Alcohol Dependence* 2006 May 20; 82(3): 204-210. Subject: 9.5.9

Schonfeld, Toby L.; Romberger, Debra J.; Hester, D. Micah; Shannon, Sarah E. Resuscitating a bad patient. *Hastings Center Report* 2007 January-February; 37(1): 14-16. Subject: 20.5.1

Schonfeld, Toby; Gordon, Bruce; Amoura, Jean; Brown, Joseph Spencer. Money matters. *American Journal of Bioethics* 2007 February; 7(2): 86-88. Subject: 18.2

Schramme, Thomas. Coercive threats and offers in psychiatry. *In:* Schramme, Thomas; Thome, Johannes, eds. Philosophy and Psychiatry. Berlin; New York: De Gruyter, 2004: 357-369. Subject: 17.7

Schramme, Thomas. The significance of the concept of disease for justice in health care. *Theoretical Medicine and Bioethics* 2007; 28(2): 121-135. Subject: 9.4

Schreiber, Hans-Peter. Embryonen- und Stammzellforschung. *In:* Schreiber, Hans-Peter, ed. Biomedizin und Ethik: Praxis, Recht, Moral. Basel; Boston: Birkhäuser, 2004: 84-87, 96. Subject: 18.5.4

Schroeder, Doris. Editorial: rights and procreative liberty. *CQ: Cambridge Quarterly of Healthcare Ethics* 2007 Summer; 16(3): 325. Subject: 14.3

Schroeder, D. Benefit sharing: it's time for a definition. *Journal of Medical Ethics* 2007 April; 33(4): 205-209. Subject: 15.1

Schubert, David. Two views of the emperor's new clones [letter]. *Nature Biotechnology* 2007 March; 25(3): 282-283. Subject: 14.5

Schüklenk, Udo. More on publication ethics [editorial]. *Bioethics* 2007 March; 21(3): ii. Subject: 1.3.7

Schüklenk, Udo; Ashcroft, Richard. HIV vaccine trials: reconsidering the therapeutic misconception and the question of what constitutes trial related injuries [editorial]. *Developing World Bioethics* 2007 December; 7(3): ii-iv. Subject: 18.5.1

Schüklenk, Udo; Bello, Braimoh. Globalization and health: a developing world perspective on ethical and policy issues. *In:* Bennett, Belinda; Tomossy, George F., eds. Globalization and Health: Challenges for Health Law and Bioethics. Dordrecht: Springer, 2006: 13-25. Subject: 9.1

Schüklenk, Udo; Gartland, K.M.A. Confronting an influenza pandemic: ethical and scientific issues. *Biochemi-*

See SUBJECT HEADING KEY FOR SECTION II on inside back cover.

595

cal Society Transactions 2006 December; 34(Pt 6): 1151-1154. Subject: 9.1

Schüklenk, Udo; Kleinschmidt, Anita. Rethinking mandatory HIV testing during pregnancy in areas with high HIV prevalence rates: ethical and policy issues. *American Journal of Public Health* 2007 July; 97(7): 1179-1183. Subject: 9.5.6

Schulte, Peter F.J.; Stienen, Juan J.; Bogers, Jan; Cohen, Dan; van Dijk, Daniel; Lionarons, Wendell H.; Sanders, Sophia S.; Heck, Adolph H. Compulsory treatment with clozapine: a retrospective long-term cohort study. *International Journal of Law and Psychiatry* 2007 November-December; 30(6): 539-545. Subject: 17.4

Schultz, Heather Yarnall; Blalock, Elisabeth. Transparency is the key to the relationship between biomedical journals and medical writers. *Journal of Investigative Dermatology* 2007 April; 127(4): 735-737. Subject: 1.3.7

Schultz, Jane E. Corpus interruptus: biotech drugs, insurance providers and the treatment of breast cancer. *Journal of Bioethical Inquiry* 2007; 4(2): 93-102. Subject: 9.5.5

Schuppli, C.A.; Fraser, D. Factors influencing the effectiveness of research ethics committees. *Journal of Medical Ethics* 2007 May; 33(5): 294-301. Subject: 18.2

Schwab, Abraham. Getting rid of heroes. *Atrium* 2007 Summer; 4: 25-28. Subject: 1.3.1

Schwartz, Barry. The evolving relationship between specialists and general dentists: practical and ethical challenges. *Journal of the American College of Dentists* 2007 Spring; 74(1): 22-26. Subject: 7.3

Schwartz, Joan P. Integrity in research: individual and institutional responsibility. *In:* Gallin, John I.; Ognibene, Frederick P., eds. Principles and Practice of Clinical Research. 2nd edition. Oxford: Academic, 2007: 39-46. Subject: 18.2

Schwartz, John. DNA pioneer's genome blurs race lines. *New York Times* 2007 December 12; p. A22. Subject: 15.11

Schwartz, Linda; Woloshin, Steven. Participation in mammography screening: Women should be encouraged to decide what is right for them, rather than being told what to do. *BMJ: British Medical Journal* 2007 October 13; 335(7623): 731-732. Subject: 9.5.5

Schwartz, Peter. Stem cells: biopsy on frozen embryos [letter]. *Hastings Center Report* 2007 January-February; 37(1): 7-8. Subject: 18.5.4

Schwartz, Peter H.; Whitney, Simon N.; McCullough, Laurence B. Silence about screening. *American Journal of Bioethics* 2007 July; 7(7): 46-48; author reply W1-W3. Subject: 8.1

Schwenzer, Karen J.; Wang, Lijuan. Assessing moral distress in respiratory care practitioners. *Critical Care Medicine* 2006 December; 34(12): 2967-2973. Subject: 7.1

Sciolino, Elaine. Low turnout undercuts Portugal vote on abortion law. *New York Times* 2007 February 12; p. A3. Subject: 12.4.1

Sciolino, Elaine. Portugal to vote on putting end to abortion ban. *New York Times* 2007 February 11; p. A1, A14. Subject: 12.4.1

Scofield, Giles R. The war on error. *American Journal of Bioethics* 2007 February; 7(2): 44-45. Subject: 9.6

Scolding, Neil. Cooperation problems in science: use of embryonic/fetal material. *In:* Watt, Helen, ed. Cooperation, Complicity and Conscience: Problems in Healthcare, Science, Law and Public Policy. London: Linacre Centre, 2005: 105-117. Subject: 18.5.4

Scotland. Scottish Executive. Health Department. Chief Scientist Office. Scottish Ethics Advisory Group [SEAG]. Consultation Report: Review of NHS Research Ethics Committees. Edinburgh: Scottish Executive, 2006 January 12: 21 p. [Online]. Accessed: http://www.scotland.gov.uk/Publications/2006/01/12093352/0 [2007 May 7]. Subject: 18.2

Scotland. Scottish Executive. Health Department. Chief Scientist Office.; Nolan, Moira; Hinds, Alison. Review of the NHS Research Ethics Committee System [letter with Consultation attachments]. Edingurgh: Scottish Executive, 2005 June 30: 10 p. [Online]. Accessed: http://www.scotland.gov.uk/Publications/2005/07/0194908/49116 [2007 May 7]. Subject: 18.2

Scott, Charles L. Psychiatry and the death penalty. *Psychiatric Clinics of North America* 2006 September; 29(3): 791-804. Subject: 20.6

Scott, Christopher Thomas; Baker, Monya. Overhauling clinical trials. *Nature Biotechnology* 2007 March; 25(3): 287-292. Subject: 18.6

Scott, Joan A. Inheritable genetic modification: clinical applications and genetic counseling considerations. *In:* Rasko, John E.J.; O'Sullivan, Gabrielle M.; Ankeny, Rachel A., eds. The Ethics of Inheritable Genetic Modification: A Dividing Line? Cambridge: Cambridge University Press, 2006: 223-241. Subject: 15.2

Scott, Larry D. Living donor liver transplant — is the horse already out of the barn? *The American Journal of Gastroenterology* 2006 April; 101(4): 686-688. Subject: 19.5

Scott, P. Anne. Virtue, nursing and the moral domain of practice. *In:* Tschudin, Verena, ed. Approaches to Ethics: Nursing Beyond Boundaries. New York: Butterworth-Heinemann, 2003: 25-32. Subject: 4.1.3

Scully, Jackie Leach. Inheritable genetic modification and disability: normality and identity. *In:* Rasko, John E.J.; O'Sullivan, Gabrielle M.; Ankeny, Rachel A., eds. The Ethics of Inheritable Genetic Modification: A Dividing Line? Cambridge: Cambridge University Press, 2006: 175-192. Subject: 15.4

Scully, Jackie Leach; Banks, Sarah; Shakespeare, Tom W. Chance, choice and control: lay debate on prenatal social sex selection. *Social Science and Medicine* 2006 July; 63(1): 21-31. Subject: 14.3

Scully, Jackie Leach; Porz, Rouven; Rehmann-Sutter, Christoph. 'You don't make genetic test decisions from one day to the next' — using time to preserve moral space. *Bioethics* 2007, May; 21(4):208-217. Subject: 15.3

Seale, Clive. National survey of end-of-life decisions made by UK medical practitioners. *Palliative Medicine* 2006 January; 20(1): 3-10. Subject: 20.5.1

Seavilleklein, Victoria; Sherwin, Susan. The myth of the gendered chromosome: sex selection and the social interest. *CQ: Cambridge Quarterly of Healthcare Ethics* 2007 Winter; 16(1): 7-19. Subject: 14.3

Secretariat of the Convention on Biological Diversity. Bonn guidelines on access to genetic resources and fair and equitable sharing of the benefits arising out of their utilization. Montreal: Secretariat of the Convention on Biological Diversity, 2002: 20 p. [Online]. Accessed: http://www. biodiv.org/doc/publications/cbd-bonn-gdls-en.pdf [2007 April 16]. Subject: 15.1

Sedgh, Gilda; Henshaw, Stanley; Singh, Susheela; Åhman, Elizabeth; Shah, Iqbal H. Induced abortion: estimated rates and trends worldwide. *Lancet* 2007 October 13-19; 370(9595): 1338-1345. Subject: 12.5.2

Seedhouse, David. Ethics and health promotion. *In his:* Health Promotion: Philosophy, Prejudice, and Practice. 2nd edition. New York: J. Wiley, 2004: 197-213. Subject: 9.1

Segal, Judy Z. "Compliance" to "concordance": a critical view. *Journal of Medical Humanities* 2007 June; 28(2): 81-96. Subject: 8.1

Segal, Steven P.; Tauber, Alfred I.; Zanni, Guido R.; Stavis, Paul F. Revisiting Hume's law [comment and reply]. *American Journal of Bioethics* 2007 November; 7(11): 43-45; author reply W3-W4. Subject: 17.7

Selby, Patricia L. On whose conscience? Patient rights disappear under broad protective measures for conscientious objectors in health care. *University of Detroit Mercy Law Review* 2006 Summer; 83(4): 507-543. Subject: 9.2

Selgelid, Michael J. A tale of two studies: ethics, bioterrorism, and the censorship of science. *Hastings Center Report* 2007 May-June; 37(3): 35-43. Subject: 21.3

Selgelid, Michael J. Ethics and drug resistance. *Bioethics* 2007 May; 21(4): 218-229. Subject: 9.7

Selgelid, Michael J. Neugenics? *Monash Bioethics Review* 2000 October; 19(4): 9-33. Subject: 15.5

Sellaroli, Valentina; Cucca, Francesco; Santosuosso, Amedeo. Shared genetic data and the rights of involved people. *Revista de Derecho y Genoma Humano = Law and the Human Genome Review* 2007 January-June; (26): 193-231. Subject: 15.1

Sellman, Derek. Trusting patients, trusting nurses. *Nursing Philosophy* 2007 January; 8(1): 28-36. Subject: 8.1

Semin, Semih; Aras, Sahbal. Bioethics and Turkey: crossroads and challenges. *Politics and the Life Sciences* 2007 March; 26(1): 2-9. Subject: 9.1

Sen, Piyal; Gordon, Harvey; Adshead, Gwen; Irons, Ashley. Ethical dilemmas in forensic psychiatry: two illustrative cases. *Journal of Medical Ethics* 2007 June; 33(6): 337-341. Subject: 17.1

Seoul National University Investigation Committee. Summary of final report on Professor Hwang Woo-suk's research [press release]. Seoul, Korea: Seoul National University Investigation Committee, 2006 January 10; 4 p. [Online]. Accessed: http://www.useoul.edu/sk_board/ boards/sk_news_read.jsp?board=11940&p_tid=86373& p_rel=null&id=63459 [2007 April 26]. Subject: 1.3.9

Serebrovska, Zoya; Serebrovskaya, Tatiana; Di Pietro, Maria Luisa; Pyle, Rebecca. Fertility restoration by the cryopreservation of oocytes and ovarian tissue from the position of biomedical ethics: a review. *Fiziolohichnyi Zhurnal* 2006; 52(6): 101-108. Subject: 14.6

Sermon, K.D.; Michiels, A.; Harton, G.; Moutou, C.; Repping, S.; Scriven, P.N.; SenGupta, S.; Traeger-Synodinos, J.; Vesela, K.; Viville, S.; Wilton, L.; Harper, J.C. ESHRE PGD Consortium data collection VI: cycles from January to December 2003 with pregnancy follow-up to October 2004. *Human Reproduction* 2007 February; 22(2): 323-326. Subject: 15.2

Sernyak, Michael; Rosenheck, Robert. Experience of VA psychiatrists with pharmaceutical detailing of antipsychotic medications. *Psychiatric Services* 2007 October; 58(10): 1292-1296. Subject: 17.4

Serour, Gamal I. Religious perspectives on ethical issues in assisted reproductive technologies. *In:* Shenfield, Françoise, Sureau, Claude, eds. Contemporary Ethical Dilemmas in Assisted Reproduction. Abingdon: Informa Healthcare, 2006: 99-114. Subject: 14.4

Seymour, John. A question of autonomy? *In his:* Childbirth and the Law. Oxford; New York: Oxford University Press, 2000: 189-239. Subject: 9.5.5

Sforza, Teri. Thousands of human eggs may be mssing. Bankrupt egg-donor registry says fertility doctors may have transferred eggs without permission. *Orange County Register* 2007 July 23: 10 p. [Online]. Accessed: http:// www.ocregister.com/eggs-options-doctors-1782305-embryos-egg [2007 July 24]. Subject: 14.4

Shahzad, Qaiser. Playing God and the ethics of divine names: an Islamic paradigm for biomedical ethics. *Bioethics* 2007 October; 21(8): 413-418. Subject: 2.1

Shaibu, Sheila. Ethical and cultural considerations in informed consent in Botswana. *Nursing Ethics* 2007 July; 14(4): 503-509. Subject: 18.3

See SUBJECT HEADING KEY FOR SECTION II on inside back cover.

Shakespeare, Jocasta. A date with death [interview]. *Sunday Times Magazine (London)* 2006 April 16: 5 p. [Online]. Accessed: http://www.timesonline.co.uk/tol/life_and_style/article702621.ece?token=null&offset=24 [2008 February 7]. Subject: 20.5.1

Shakur, Haleema; Roberts, Ian; Barnetson, Lin; Coats, Tim. Clinical trials in emergency situations [editorial]. *BMJ: British Medical Journal* 2007 January 27; 334(7586): 165-166. Subject: 18.5.3

Sham, C.O.; Cheng, Y.W.; Ho, K.W.; Lai, P.H.; Lo, L.W.; Wan, H.L.; Wong, C.Y.; Yeung, Y.N.; Yuen, S.H.; Wong, A.Y. Do-not-resuscitate decision: the attitudes of medical and non-medical students. *Journal of Medical Ethics* 2007 May; 33(5): 261-265. Subject: 20.5.1

Shamoo, Adil E. Deregulating low-risk research. *Chronicle of Higher Education* 2007 August 3; 53(48): B16. Subject: 18.2

Shamoo, Adil E.; Schwartz, Jack. Universal and uniform protections of human subjects in research. *American Journal of Bioethics* 2007 December; 7(12): 7-9. Subject: 18.1

Shamoo, Adil; Woeckner, Elizabeth. Ethical flaws in the TeGenero trial. *American Journal of Bioethics* 2007 February; 7(2): 90-92. Subject: 18.1

Shanley, Mary L. The Baby Business: How Money, Science and Politics Drive the Commerce of Conception, by Debora L. Spar [book review]. *DePaul Journal of Health Care Law* 2007; 10(4): 557-566. Subject: 14.4

Shannon, Thomas A. Cloning, uniqueness, and individuality. *In:* Walter, James J.; Shannon, Thomas A., eds. Contemporary Issues in Bioethics: A Catholic Perspective. Lanham, MD: Rowman and Littlefield Publishers, 2005: 101-123. Subject: 14.5

Shannon, Thomas A. Reproductive technologies: ethical and religious issues. *In:* Walter, James J.; Shannon, Thomas A., eds. Contemporary Issues in Bioethics: A Catholic Perspective. Lanham, MD: Rowman and Littlefield Publishers, 2005: 125-143. Subject: 14.1

Shannon, Thomas A.; Walter, James J. Artificial nutrition and hydration: assessing the papal statement. *In their:* Contemporary Issues in Bioethics: A Catholic Perspective. Lanham, MD: Rowman and Littlefield Publishers, 2005: 257-261. Subject: 20.5.1

Shannon, Thomas A.; Walter, James J. Assisted nutrition and hydration and the Catholic tradition: the case of Terri Schiavo. *In their:* Contemporary Issues in Bioethics: A Catholic Perspective. Lanham, MD: Rowman and Littlefield Publishers, 2005: 269-280. Subject: 20.5.1

Shannon, Thomas A.; Walter, James J. Implications of the papal allocution on feeding tubes. *In their:* Contemporary Issues in Bioethics: A Catholic Perspective. Lanham, MD: Rowman and Littlefield Publishers, 2005: 263-268. Subject: 20.5.1

Shannon, Thomas A.; Walter, James J. The PVS patient and the forgoing/withdrawing of medical nutrition and hydration. *In their:* Contemporary Issues in Bioethics: A Catholic Perspective. Lanham, MD: Rowman and Littlefield Publishers, 2005: 231-256. Subject: 20.5.1

Shannon, Thomas A.; Wolter, Allan B. Reflections on the moral status of the pre-embryo. *In:* Walter, James J.; Shannon, Thomas A., eds. Contemporary Issues in Bioethics: A Catholic Perspective. Lanham, MD: Rowman and Littlefield Publishers, 2005: 67-90. Subject: 4.4

Shapiro, Dvorah S.; Friedmann, Reuven. To feed or not to feed the terminal demented patient — is there any question? *Israel Medical Association Journal* 2006 July; 8(7): 507-508. Subject: 20.5.1

Shapiro, Johanna; Rucker, Lloyd; Robitshek, Daniel. Teaching the art of doctoring: an innovative medical student elective. *Medical Teacher* 2006 February; 28(1): 30-35. Subject: 7.2

Shapiro, Kenneth. Animal experimentation. *In:* Waldau, Paul; Patton, Kimberly, eds. A Communion of Subjects: Animals in Religion, Science, and Ethics. New York: Columbia University Press, 2006: 533-543. Subject: 22.2

Shapiro, Samuel. Integrity of independent and industry-sponsored studies. *In:* Shapiro, S.; Dinger, J.; Scriba, P., eds. Enabling Risk Assessment in Medicine: Farewell Symposium for Werner-Kari Raff. New Brunswick, NJ: Transaction Publishers, 2004: 67-82. Subject: 1.3.9

Shapshay, Sandra; Pimple, Kenneth D. Participation in biomedical research is an imperfect moral duty: a response to John Harris. *Journal of Medical Ethics* 2007 July; 33(7): 414-417. Subject: 18.1

Sharma, Renuka M. The ethics of birth and death: gender infanticide in India. *Journal of Bioethical Inquiry* 2007; 4(3): 181-192. Subject: 20.5.2

Sharma, Shridhar. Human rights in psychiatric care: an Asian perspective. *Acta Psychiatrica Scandinavica* 2000 January; 101(399 Supplement): 97-101. Subject: 21.1

Sharma, S. Legalization of abortion in Nepal: the way forward. *Kathmandu University Medical Journal* 2004 July-September; 2(3): 177-178. Subject: 12.4.1

Sharp, Helen M. Ethical issues in the management of dysphagia after stroke. *Topics in Stroke Rehabilitation* 2006 Fall; 13(4): 18-25. Subject: 20.4.1

Sharp, Lesley A. Human, monkey, machine: the brave new world of human hybridity. *In her:* Bodies, Commodities, and Biotechnologies: Death, Mourning, and Scientific Desire in the Realm of Human Organ Transfer. New York: Columbia University Press, 2007: 77-115. Subject: 15.1

Sharp, Michael. The effect of genetic determinism and exceptionalism on law and policy. *Health Law Review* 2007; 15(3): 16-18. Subject: 15.1

Sharp, Richard R.; Foster, Morris W. Grappling with groups: protecting collective interests in biomedical research. *Journal of Medicine and Philosophy* 2007 July-August; 32(4): 321-337. Subject: 15.11

Sharpe, Virginia A. Strategic disclosure requirements and the ethics of bioethics. *In:* Eckenwiler, Lisa A.; Cohn, Felicia G., eds. The Ethics of Bioethics: Mapping the Moral Landscape. Baltimore, MD: Johns Hopkins University Press, 2007: 170-180. Subject: 2.1

Shatrugna, Veena. An extremely cynical study. *Indian Journal of Medical Ethics* 2007 April-June; 4(2): 79-80. Subject: 18.5.2

Shaw, Alison. The contingency of the 'genetic link' in the construction of kinship and inheritance — an anthropological perspective. *In:* Spencer, J.R.; du Bois-Pedain, Antje, eds. Freedom and Responsibility in Reproductive Choice. Portland: Hart Pub., 2006: 73-90. Subject: 15.1

Shaw, David. The body as unwarranted life support: a new perspective on euthanasia. *Journal of Medical Ethics* 2007 September; 33(9): 519-521. Subject: 20.5.1

Shaw, Jim. POLST: honoring wishes at the end of life. *Health Care Ethics USA* 2007 Winter; 15(1): 8-12. Subject: 20.5.4

Sheikh, Aziz; Esmail, Aneez. Should Muslims have faith based health services? *BMJ: British Medical Journal* 2007 January 13; 334(7584): 74-75. Subject: 9.5.4

Sheldon, Jane P.; Jayaratne, Toby Epstein; Feldbaum, Merle B.; DiNardo, Courtney D.; Petty, Elizabeth M. Applications and implications of advances in human genetics: perspectives from a group of Black Americans. *Community Genetics* 2007; 10(2): 82-92. Subject: 15.11

Sheldon, Sally. Reproductive technologies and the legal determination of fatherhood. *Feminist Legal Studies* 2005; 13(3): 349-362. Subject: 14.1

Sheldon, Tony. Cosmetic surgery gets under Dutch skin. *BMJ: British Medical Journal* 2007 September 15; 335(7619): 541. Subject: 4.5

Sheldon, Tony. Holland bans private stem cell therapy [news]. *BMJ: British Medical Journal* 2007 January 6; 334(7583): 12. Subject: 19.1

Sheldon, Tony. Incidence of euthanasia in the Netherlands falls [news]. *BMJ: British Medical Journal* 2007 May 26; 334(7603): 1075. Subject: 20.5.1

Sheldon, Tony. Undertakers offer cash incentive for organ donation in an attempt to encourage donors. *BMJ: British Medical Journal* 2007 June 2; 334(7604): 1131. Subject: 19.5

Shelton, B.L. Consent and consultation in genetic research on American Indians and Alaskan Natives. Nixon, NV: Indigenous Peoples Council on Biocolonialism, 2002; 3 p. [Online]. Accessed: http://www.ipcb.org/publications/briefing_papers/files/consent.html [2007 April 24]. Subject: 15.11

Shenfield, Françoise; Babinet, Charles; Teboul, Gérard. Human cloning: reproductive crime or therapeutic panacea — where are we now? *In:* Shenfield, Françoise, Sureau, Claude, eds. Contemporary Ethical Dilemmas in Assisted Reproduction. Abingdon: Informa Healthcare, 2006: 13-25. Subject: 14.5

Shenfield, Françoise; Sureau, Claude. The welfare of the child: whose responsibility? *In:* Shenfield, Françoise; Sureau, Claude, eds. Contemporary Ethical Dilemmas in Assisted Reproduction. Abingdon: Informa Healthcare, 2006: 73-83. Subject: 9.5.7

Sheperd, Lois. Terri Shiavo: unsettling the settled. *Loyola University Chicago Law Journal* 2006 Winter; 37(2): 297-341. Subject: 20.5.1

Shepherd, J.P.; Ho, M.; Shepherd, H.R.; Sivarajasingam, V. Confidential registration in health services: randomised controlled trial. *Emergency Medicine Journal* 2006 June; 23(6): 425-427. Subject: 8.4

Sheremeta, Lorraine. Nanotechnologies and the ethical conduct of research involving human subjects. *In:* Hunt, Geoffrey; Mehta, Michael D., eds. Nanotechnology: Risk, Ethics and Law. London; Sterling, VA: Earthscan, 2006: 247-258. Subject: 5.4

Sheremeta, Lorraine. Public meets private: challenges for informed consent and umbilical cord blood banking in Canada. *Health Law Review* 2007; 15(2): 23-29. Subject: 18.3

Sheridan, Kimberly; Zinchenko, Elena; Gardner, Howard. Neuroethics in education. *In:* Illes, Judy, ed. Neuroethics: Defining the Issues in Theory, Practice, and Policy. New York: Oxford University Press, 2006: 265-275. Subject: 17.1

Sherlock, Richard. Bioethics in liberal regimes: a review of the President's Council. *Ethics and Medicine: An International Journal of Bioethics* 2007 Fall; 23(3): 169-188. Subject: 2.4

Sherwin, C.M. Animal welfare: reporting details is good science [letter]. *Nature* 2007 July 19; 448(7151): 251. Subject: 22.2

Sherwin, Susan. Abortion through a feminist lens. *In:* Baylis, Françoise; Downie, Jocelyn; Freedman, Benjamin; Hoffmaster, Barry; Sherwin, Susan, eds. Health Care Ethics in Canada. Toronto: Harcourt Brace Canada, 1995: 441-447. Subject: 12.4.1

Sherwood, Mylaina L.; Buchinsky, Farrel J.; Quigley, Matthew R.; Donfack, Joseph; Choi, Sukgi S.; Conley, Stephen F.; Derkay, Craig S.; Myer, Charles M., III; Ehrlich, Garth D.; Post, J. Christopher. Unique challenges of obtaining regulatory approval for a multicenter protocol to study the genetics of RRP and suggested remedies. *Otolaryngology and Head and Neck Surgery* 2006 August; 135(2): 189-196. Subject: 15.1

See SUBJECT HEADING KEY FOR SECTION II on inside back cover.

599

Sheth, Hiten G.; Sheth, A.G. Frequent attenders to ophthalmic accident and emergency departments [letter]. *Journal of Medical Ethics* 2007 August; 33(8): 496. Subject: 17.1

Sheu, Shuh-Jen; Huang, Shu-He; Tang, Fu-In; Huang, Song-Lih. Ethical decision making on truth telling in terminal cancer: medical students' choices between patient autonomy and family paternalism. *Medical Education* 2006 June; 40(6): 590-598. Subject: 8.2

Shewmon, D. Alan; De Giorgio, Christopher M. Early prognosis in anoxic coma: reliability and rationale. *Neurologic Clinics* 1989 November; 7(4): 823-843. Subject: 9.5.1

Shiao, Judith Shu-Chu; Koh, David; Lo, Li-Hua; Lim, Meng-Kin; Guo, Yueliang Leon. Factors predicting nurses' consideration of leaving their job during the SARS outbreak. *Nursing Ethics* 2007 January; 14(1): 5-17. Subject: 9.5.1

Shickle, Darren. The Mental Capacity Act 2005. *Clinical medicine* 2006 March-April; 6(2): 169-173. Subject: 8.3.3

Shields, Wayne C.; Jordan, Beth. Adding value to reproductive health research: communicating about the moral dimensions of science. *Contraception* 2006 September; 74(3): 199-200. Subject: 14.1

Shiffrin, Seana Valentine. Autonomy, beneficence, and the permanently demented. *In:* Burley, Justine, ed. Dworkin and His Critics: With Replies by Dworkin. Malden, MA: Blackwell Pub., 2004: 195-217. Subject: 20.5.1

Shildrick, Margrit. Reconfiguring the bioethics of reproduction. *Philosophy in the Contemporary World* 2004 Spring-Summer; 11(1): 75-83. Subject: 14.1

Shim, Janet K.; Russ, Ann J.; Kaufman, Sharon R. Risk, life extension and the pursuit of medical possibility. *Sociology of Health and Illness* 2006 May; 28(4): 479-502. Subject: 9.5.2

Shimazono, Susumu. Why we must be prudent in research using human embryos: differing views of human dignity. *In:* LaFleur, William R.; Böhme, Gernot; Shimazono, Susumu, eds. Dark Medicine: Rationalizing Medical Research. Bloomington: Indiana University Press, 2007: 201-222. Subject: 18.5.4

Shirley, Jamie L. Limits of autonomy in nursing's moral discourse. *ANS: Advances in Nursing Science* 2007 January-March; 30(1): 14-25. Subject: 4.1.3

Shore, Nancy. Community-based participatory research and the ethics review process. *Journal of Empirical Research on Human Research Ethics* 2007 March; 2(1): 31-41. Subject: 18.6

Short, Robert. HFEA wants greater use of single embryo transfers in IVF [news]. *BMJ: British Medical Journal* 2007 April 14; 334(7597): 766. Subject: 14.4

Shorvon, Simon. The prosecution of research — experience from Singapore. *Lancet* 2007 June 2-8; 369(9576): 1835-1837. Subject: 1.3.9

Shostak, Sara; Ottman, Ruth. Ethical, legal, and social dimensions of epilepsy genetics. *Epilepsia* 2006 October; 47(10): 1595-1602. Subject: 15.3

Shrader-Frechette, Kristin. EPA's 2006 human-subjects rule for pesticide experiments. *Accountability in Research* 2007 October-December; 14(4): 211-254. Subject: 18.6

Shrader-Frechette, Kristin. Nanotoxicology and ethical conditions for informed consent. *NanoEthics* 2007 March; 1(1): 47-56. Subject: 5.4

Shragge, Jeremy E. A graduate student perspective on the accreditation of programs ensuring ethical research with humans. *NCEHR Communique CNERH* 2006 Spring; 14(1): 21-22. Subject: 18.2

Shroff, Sunil. Working towards ethical organ transplants. *Indian Journal of Medical Ethics* 2007 April-June; 4(2): 68-69. Subject: 19.5

Shuchman, Miriam. Commercializing clinical trials — risks and benefits of the CRO boom. *New England Journal of Medicine* 2007 October 4; 357(14): 1365-1368. Subject: 18.2

Shuchman, Miriam. Drug risks and free speech — can Congress ban consumer drug ads? *New England Journal of Medicine* 2007 May 31; 356(22): 2236-2239. Subject: 9.7

Shuchman, Miriam. FERPA, HIPAA, and the privacy of college students. *New England Journal of Medicine* 2007 July 12; 357(2): 109-110. Subject: 8.4

Shuster, Evelyne. Microarray genetic screening: a prenatal roadblock for life? *Lancet* 2007 February 10-16; 369(9560): 526-529. Subject: 15.2

Shute, Nancy. Unraveling your DNA's secrets. Do-it-yourself genetic tests promise to reveal your risk of coming down with a disease. But do they really deliver? *U.S. News and World Report* 2007 January 8; 142(1): 50-54, 57-58. Subject: 15.3

Shweder, Richard A. Protecting human subjects and preserving academic freedom: prospects at the University of Chicago. *American Ethnologists* 2006 November; 33(4): 507-518. Subject: 18.2

Sibbald, Robert; Downar, James; Hawryluck, Laura. Perceptions of "futile care" among caregivers in intensive care units. *CMAJ/JAMC: Canadian Medical Association Journal* 2007 November 6; 177(10): 1201-1208. Subject: 20.5.1

Sibley, Robert. When healers become killers: the doctor as terrorist. *CMAJ/JAMC: Canadian Medical Association Journal* 2007 September 11; 177(6): 688. Subject: 21.1

Siddiky, Abul. Should junior doctors be obtaining consent? *British Journal of Hospital Medicine* 2006 November; 67(11): M214. Subject: 8.3.1

Sidhu, Navdeep S.; Dunkley, Margaret E.; Egan, Melinda J. "Not-for-resuscitation" orders in Australian public hospitals: policies, standardised order forms and patient information leaflets. *Medical Journal of Australia* 2007 January 15; 186(2): 72-75. Subject: 20.5.1

Sieber, Joan E. Institutional introspection [editorial]. *Journal of Empirical Research on Human Research Ethics* 2007 December; 2(4): 1-2. Subject: 18.2

Sieber, Joan E. Respect for persons and informed consent: a moving target [editorial]. *Journal of Empirical Research on Human Research Ethics* 2007 September; 2(3): 1-2. Subject: 18.3

Siegel, Paul E.; Ellis, Norman R. Note on the recruitment of subjects for mental retardation research. *American Journal of Mental Deficiency* 1985 January; 89(4): 431-433. Subject: 18.5.6

Siegel-Itzkovich, Judy. Doctor's licence suspended after he admitted removing hundreds of ova without consent [news]. *BMJ: British Medical Journal* 2007 March 17; 334(7593): 557. Subject: 7.4

Siegel-Itzkovich, Judy. Israeli surgeons put their names on study they had not done. *BMJ: British Medical Journal* 2007 May 19; 334(7602): 1023. Subject: 1.3.9

Silberman, Jordan; Morrison, Wynne; Feudtner, Chris. Pride and prejudice: how might ethics consultation services minimize bias? *American Journal of Bioethics* 2007 February; 7(2): 32-34. Subject: 9.6

Silver, Ken; Sharp, Richard R. Ethical considerations in testing workers for the -Glu69 marker of genetic susceptibility to chronic beryllium disease. *Journal of Occupational and Environmental Medicine* 2006 April; 48(4): 434-443. Subject: 15.3

Silver, Lee M. Human-animal combinations. *In his:* Challenging Nature: The Clash of Science and Spirituality at the New Frontiers of Life. New York: Ecco, 2006: 172-187, 385-386. Subject: 15.1

Silver, Lee M. The battle for mother nature's genes. *In his:* Challenging Nature: The Clash of Science and Spirituality at the New Frontiers of Life. New York: Ecco, 2006: 278-293, 397-399. Subject: 15.1

Silver, Lee M. The embryonic soul. *In his:* Challenging Nature: The Clash of Science and Spirituality at the New Frontiers of Life. New York: Ecco, 2006: 98-124. Subject: 4.4

Silver, Lee M. The politics of cloning. *In his:* Challenging Nature: The Clash of Science and Spirituality at the New Frontiers of Life. New York: Ecco, 2006: 125-146, 376-380. Subject: 14.5

Silver, R.; Bernhardt, B.; Wilfond, B.; Geller, G. Genetic counselors' experiences of moral value conflicts with clients. *Journal of Genetic Counseling* 2007 December; 16(6): 690. Subject: 15.2

Silvers, Anita. Judgment and justice: evaluating health care for chronically ill and disabled patients. *In:* Rhodes, Rosamond; Francis, Leslie P.; Silvers, Anita, eds. The Blackwell Guide to Medical Ethics. Malden, MA: Blackwell Pub., 2007: 354-372. Subject: 9.4

Silversides, Ann. Slouching toward disclosure. *CMAJ/JAMC: Canadian Medical Association Journal* 2007 November 20; 177(11): 1342-1343. Subject: 9.8

Silversides, Ann. The wide gap between genetic research and clinical needs [news]. *CMAJ/JAMC: Canadian Medical Association Journal* 2007 January 30; 176(3): 315-316. Subject: 15.2

Simitis, Spiros. A convention on cloning — annotations to an almost unsolvable dilemma. *In:* Vöneky, Silja; Wolfrum, Rüdiger, eds. Human Dignity and Human Cloning. Leiden: Nijhoff, 2004: 167-182. Subject: 14.5

Simmons, Aaron. A critique of Mary Anne Warren's weak animal rights view. *Environmental Ethics* 2007 Fall; 29(3): 267-278. Subject: 22.1

Simon, Alfred. Editorial. *Ethik in der Medizin* 2007 June; 19(2): 89-90. Subject: 20.5.4

Simon, Jürgen. Biotechnology and law: biotechnology patents. Special considerations on the inventions with human material. *Revista de Derecho y Genoma Humano = Law and the Human Genome Review* 2006 July-December; (25): 139-159. Subject: 15.8

Simpson, Bob. Negotiating the therapeutic gap: prenatal diagnostics and termination of pregnancy in Sri Lanka. *Journal of Bioethical Inquiry* 2007; 4(3): 207-215. Subject: 15.2

Simpson, Bob. On parrots and thorns: Sri Lankan perspective on genetics, science and personhood. *Health Care Analysis: An International Journal of Health Philosophy and Policy* 2007 March; 15(1): 41-49. Subject: 15.1

Sinclair, Andrew H.; Schofield, Peter R. Human embryonic stem cell research: an Australian perspective. *Cell* 2007 January 26; 128(2): 221-223. Subject: 18.2

Singapore. Bioethics Advisory Committee. Personal Information in Biomedical Research: a report by the Bioethics Advisory Committee. Bioethics Advisory Committee 2007 May: 48 p.; Appendices 141 p. Subject: 18.1

Singeo, Lindsey. The patentability of the native Hawaiian genome. *American Journal of Law and Medicine* 2007; 33(1): 119-139. Subject: 15.8

Singer, Eleanor; Bossarte, Robert M. Incentives for survey participation when are they "coercive"? *American Journal of Preventive Medicine* 2006 November; 31(5): 411-418. Subject: 18.2

Singer, Lawrence E. Does mission matter? *Houston Journal of Health Law and Policy* 2006; 6(2): 347-377. Subject: 9.1

See SUBJECT HEADING KEY FOR SECTION II on inside back cover.

601

Singer, Peter. A convenient truth [op-ed]. *New York Times* 2007 January 26; p. A21. Subject: 9.5.3

Singer, Peter. Treating (or not) the tiniest babies. *Free Inquiry* 2007 June-July; 27(4): 20-21. Subject: 20.5.2

Singer, Peter; Wells, Deane. Ectogenesis. *In:* Gelfand, Scott; Shook, John R., eds. Ectogenesis: Artificial Womb Technology and the Future of Human Reproduction. Amsterdam; New York: Editions Rodopi, B.V., 2006: 9-25. Subject: 14.1

Singh, Jerome Amir; Abdool Karim, Salim S.; Abdool Karim, Quarrisha; Mlisana, Koleka; Williamson, Carolyn; Gray, Clive; Govender, Michelle; Gray, Andrew. Enrolling adolescents in research on HIV and other sensitive issues: lessons from South Africa. *PLoS Medicine* 2006 July; 3(7): e180 (0984-0988). Subject: 18.5.2

Singh, Jerome Amir; Govender, Michelle; Mills, Edward J. Do human rights matter to health? *Lancet* 2007 August 11-17; 370(9586): 521-527. Subject: 21.1

Siu, Lillian L. Clinical trials in the elderly — a concept comes of age. *New England Journal of Medicine* 2007 April 12; 356(15): 1575-1576. Subject: 18.5.7

Sizemore, Rebecca. Separating medical and ethical: helping families determine the best interests of loved ones. *Dimensions of Critical Care Nursing* 2006 September-October; 25(5): 216-220. Subject: 20.5.1

Skene, Loane. Legal rights of human bodies, body parts and tissue. *Journal of Bioethical Inquiry* 2007; 4(2): 129-133. Subject: 19.5

Skene, Loane. Life-prolonging treatment and patients' legal rights. *In:* McLean, Sheila A.M., ed. First Do No Harm: Law, Ethics, and Healthcare. Aldershot, England; Burlington, VT: Ashgate, 2006: 421-429. Subject: 20.5.1

Skene, Loane. Should the law limit genetic tests on embryos and foetuses? *University of New South Wales Law Journal* 2006; 29(2): 250-253. Subject: 15.2

Skene, Loane. Theft of DNA: do we need a new criminal offence? *In:* Gunning, Jennifer; Holm, Søren, eds. Ethics, law, and society. Volume 1. Aldershot, Hants, England; Burlington, VT: Ashgate, 2005: 85-94. Subject: 15.1

Skirton, Heather; Frazier, Lorraine Q.; Calvin, Amy O.; Cohen, Marlene Z. A legacy for the children — attitudes of older adults in the United Kingdom to genetic testing. *Journal of Clinical Nursing* 2006 May; 15(5): 565-573. Subject: 15.3

Slack, Catherine; Strode, Ann; Fleischer, Theodore; Gray, Glenda; Ranchod, Chitra. Enrolling adolescents in HIV vaccine trials: reflections on legal complexities from South Africa. *BMC Medical Ethics* 2007; 8:5; 8 p. [Online]. Accessed: http://www.biomedcentral.com/1472-6939/8/5 [2007 June 18]. Subject: 18.5.2

Slaughter, Louise M. Your genes and privacy [editorial]. *Science* 2007 May 11; 316(5826): 797. Subject: 15.3

Slaughter, Susan; Cole, Dixie; Jennings, Eileen; Reimer, Marlene A. Consent and assent to participate in research from people with dementia. *Nursing Ethics* 2007 January; 14(1): 27-40. Subject: 18.3

Slaven, Marcia Jacobson. First impressions: the experiences of a community member on a research ethics committee. *IRB: Ethics and Human Research* 2007 May-June; 29(3): 17-19. Subject: 18.2

Sleeboom-Faulkner, Margaret. Predictive genetic testing in Asia: social science perspectives on the bioethics of choice. *Journal of Bioethical Inquiry* 2007; 4(3): 193-195. Subject: 15.3

Sleeboom-Faulkner, Margaret. Social-science perspectives on bioethics: predictive genetic testing (PGT) in Asia. *Journal of Bioethical Inquiry* 2007; 4(3): 197-206. Subject: 15.3

Slieper, Chad F.; Hyle, Laurel R.; Rodriguez, Maria Alma. Difficult discharge: lessons from the oncology setting. *American Journal of Bioethics* 2007 March; 7(3): 31-32. Subject: 8.1

Slieper, Chad F.; Wasson, Katherine; Ramondetta, Lois M. From technician to professional: integrating spirituality into medical practice. *American Journal of Bioethics* 2007 July; 7(7): 26-27. Subject: 8.1

Sloan, Richard P. Ethical problems. *In his:* Blind Faith: The Unholy Alliance of Religion and Medicine. New York: St. Martin's Press, 2006: 181-206. Subject: 8.1

Sloane, Peter; DeRenzo, Evan G. The case of Mr. A.B. *Journal of Clinical Ethics* 2007 Winter; 18(4): 399-401. Subject: 9.5.2

Slocum, J. Michael. Legal issues in clinical trials. *In:* Kulakowski, Elliott C.; Chronister, Lynne U., eds. Research Administration and Management. Sudbury, MA: Jones and Bartlett, 2006: 189-206. Subject: 18.2

Slomka, Jacquelyn; McCurdy, Sheryl; Ratliff, Eric A.; Timpson, Sandra; Williams, Mark L. Perceptions of financial payment for research participation among African-American drug users in HIV studies. *JGIM: Journal of General Internal Medicine* 2007 October; 22(10): 1403-1409. Subject: 9.5.6

Slonina, Mary Irene. State v. physicians et al.: legal standards guiding the mature minor doctrine and the bioethical judgment of pediatricians in life-sustaining medical treatment. *Health Matrix: The Journal of Law-Medicine* 2007 Winter; 17(1): 181-214. Subject: 8.3.2

Slosar, John Paul. Medical futility in the post-modern context. *HEC (Healthcare Ethics Committee) Forum* 2007 March; 19(1): 67-82. Subject: 20.5.1

Sloth-Nielsen, Julia. Of newborns and nubiles: some critical challenges to children's rights in Africa in the era of HIV/AIDS. *In:* Freeman, Michael, ed. Children's Health and Children's Rights. Boston: Martinus Nijhoff Publishers, 2006: 73-85. Subject: 9.5.6

Subject = NRCBL Primary Classification Number; see inside front cover.

Slováčková, Birgita; Slováček, Ladislav. Moral judgement competence and moral attitudes of medical students. *Nursing Ethics* 2007 May; 14(3): 320-328. Subject: 7.2

Slowther, Anne-Marie. Determining best interests in patients who lack capacity to decide for themselves. *Clinical Ethics* 2007 March; 2(1): 19-21. Subject: 8.3.3

Slowther, Anne-Marie. The concept of autonomy and its interpretation in health care. *Clinical Ethics* 2007 December; 2(4): 173-175. Subject: 8.3.1

Slutsman, Julia; Buchanan, David; Grady, Christine. Ethical issues in cancer chemoprevention trials: considerations for IRBs and investigators. *IRB: Ethics and Human Research* 2007 March-April; 29(2): 1-6. Subject: 18.2

Smajdor, Anna. State-funded IVF will make us rich . . . or will it? *Journal of Medical Ethics* 2007 August; 33(8): 468-469. Subject: 14.4

Smajdor, Anna. The moral imperative for ectogenesis. *CQ: Cambridge Quarterly of Healthcare Ethics* 2007 Summer; 16(3): 336-345. Subject: 14.1

Smearman, Claire A. Drawing the line: the legal, ethical and public policy implications of refusal clauses for pharmacists. *Arizona Law Review* 2006 Fall; 48(3): 469-540. Subject: 11.1

Smith, Alexander K.; Davis, Roger B.; Krakauer, Eric L. Differences in the quality of the patient-physician relationship among terminally ill African-American and white patients: impact on advance care planning and treatment preferences. *JGIM: Journal of General Internal Medicine* 2007 November; 22(11): 1579-1582. Subject: 8.1

Smith, Charles B.; Battin, Margaret P.; Francis, Leslie P.; Jacobson, Jay A. Should rapid tests for HIV infection now be mandatory during pregnancy? Global differences in scarcity and a dilemma of technological advance. *Developing World Bioethics* 2007 August; 7(2): 86-103. Subject: 9.5.6

Smith, Craig S. France: doctors petition for euthanasia. *New York Times* 2007 March 10; p. A6. Subject: 20.5.1

Smith, David H. Band-Aid solutions to problems of access: their origins and limits. *Journal of Dental Education* 2006 November; 70(11): 1170-1173. Subject: 9.2

Smith, David H.; Davis, Dena S.; Cohen, Cynthia B.; Martin, Adrienne M. Taking religion seriously [letters and reply]. *Hastings Center Report* 2007 July-August; 37(4): 4-6. Subject: 8.3.3

Smith, George P. A compassionate death. *In his:* The Christian Religion and Biotechnology: A Search for Principled Decision-making. Dordrecht: Springer, 2005: 189-230. Subject: 20.5.1

Smith, George P. Freedom of scientific investigation. *In his:* The Christian Religion and Biotechnology: A Search for Principled Decision-making. Dordrecht: Springer, 2005: 85-148. Subject: 15.1

Smith, George P. Genetic enhancement. *In his:* The Christian Religion and Biotechnology: A Search for Principled Decision-making. Dordrecht: Springer, 2005: 149-188. Subject: 15.1

Smith, George P. Procreational autonomy or theological restraints. *In his:* The Christian Religion and Biotechnology: A Search for Principled Decision-making. Dordrecht: Springer, 2005: 61-84. Subject: 2.1

Smith, Ian A. A new defense of Quinn's Principle of Double Effect. *Journal of Social Philosophy* 2007 Summer; 38(2): 349-364. Subject: 2.1

Smith, Jayne L.; Cervero, Ronald M.; Valentine, Thomas. Impact of commercial support on continuing pharmacy education. *Journal of Continuing Education in the Health Professions* 2006 Fall; 26(4): 302-312. Subject: 4.1.1

Smith, Martin L.; Weise, Kathryn L. The goals of ethics consultation: rejecting the role of "ethics police". *American Journal of Bioethics* 2007 February; 7(2): 42-44. Subject: 9.6

Smith, Melissa. Patients and doctors: rights and responsibilities in the NHS (2). *Clinical Medicine* 2005 September-October; 5(5): 501-502. Subject: 9.1

Smith, Rachel A. Picking a frame for communicating about genetics: stigmas or challenges. *Journal of Genetic Counseling* 2007 June; 16(3): 289-298. Subject: 15.1

Smith, Richard. Curbing the influence of the drug industry: a British view. *PLoS Medicine* 2005 September; 2(9): e241 (0821-0823). Subject: 9.7

Smith, Richard. Lapses at the New England Journal of Medicine. *Journal of the Royal Society of Medicine* 2006 August; 99(8): 380-382. Subject: 1.3.7

Smith, Richard. The highly profitable but unethical business of publishing medical research. *Journal of Royal Society of Medicine* 2006 September; 99(9): 452-456. Subject: 1.3.7

Smith, Richard; Williams, Gareth. Should medical journals carry drug advertising? [debate]. *BMJ: British Medical Journal* 2007 July 14; 335(7610): 74-75. Subject: 1.3.7

Smith, Shane; Neaves, William; Teitelbaum, Steven; Prentice, David A.; Tarne, Gene. Adult versus embryonic stem cells: treatments [letter and reply]. *Science* 2007 June 8; 316(5830): 1422-1423. Subject: 18.5.1

Smith, Stephen W. Some realism about end of life: the current prohibition and the euthanasia underground. *American Journal of Law and Medicine* 2007; 33(1): 55-95. Subject: 20.7

Smith, Valerie. A patient's perspective on moral issues and universal oral health care. *Journal of the American College of Dentists* 2007 Fall; 74(3): 27-31. Subject: 9.5.1

Smith, Wesley J. "We never say no." The right-to-die movement abandons pretense. *Weekly Standard* 2006

See SUBJECT HEADING KEY FOR SECTION II on inside back cover.

April 27: 4 p. [Online]. Accessed: http://www. weeklystandard.com/Content/Public/Articles/000/000/ 012/124abkbr.asp [2008 February 7]. Subject: 20.5.1

Smyth, Tim. Bioethics, politics and policy development. *Monash Bioethics Review* 2000 October; 19(4): 38-45. Subject: 9.1

Snead, Carter. Neuroimaging, entrapment, and the predisposition to crime. *American Journal of Bioethics* 2007 September; 7(9): 60-61. Subject: 17.1

Snead, O. Carter. Assessing the Universal Declaration on Bioethics and Human Rights: implications for human dignity and the respect for human life. *National Catholic Bioethics Quarterly* 2007 Spring; 7(1): 53-71. Subject: 4.4

Snelling, Jeanne. Implications for providers and patients: a comment on the regulatory framework for preimplantation genetic diagnosis in New Zealand. *Medical Law International* 2007; 8(1): 23-49. Subject: 15.2

Snodgrass, Mary Ellen. Abortion. *In her:* Historical Encyclopedia of Nursing. Santa Barbara, CA: ABC-CLIO, 1999: 1-4. Subject: 12.1

Snodgrass, Mary Ellen. Nazi nurses. *In her:* Historical Encyclopedia of Nursing. Santa Barbara, CA: ABC-CLIO, 1999: 191-194. Subject: 20.5.1

Snowden, Claire; Elbourne, Diana; Garcia, Jo. Declining enrolment in a clinical trial and injurious misconceptions: is there a flipside to the therapeutic misconception? *Clinical Ethics* 2007 December; 2(4): 193-200. Subject: 18.3

Snowdon, Claire; Garcia, Jo; Elbourne, Diana. Making sense of randomization: responses of parents of critically ill babies to random allocation of treatment in a clinical trial. *Social Science and Medicine* 1997 November; 45(9): 1337-1355. Subject: 18.2

Snyder, Lois. Bioethics, assisted suicide, and the "right to die". *Annals of Clinical Psychiatry* 2001; 13(1): 13-18. Subject: 20.7

Snyder, Lois; Neubauer, Richard L. Pay-for-performance principles that promote patient-centered care: an ethics manifesto. *Annals of Internal Medicine* 2007 December 4; 147(11): 792-794. Subject: 9.3.1

Sobel, Richard. The HIPAA paradox: the privacy rule that's not. *Hastings Center Report* 2007 July-August; 37(4): 40-50. Subject: 8.4

Sobolski, Gregory K.; Flores, Leonardo; Emanuel, Ezekiel J. Institutional review board review of multicenter studies [letter]. *Annals of Internal Medicine* 2007 May 15; 146(10): 759. Subject: 18.2

Sobsey, Dick. Growth attenuation and indirect-benefit rationale. *Ethics and Intellectual Disability* 2007 Winter; 10(1): 1,2, 7-8. Subject: 8.3.2

Society for Adolescent Medicine. Guidelines for adolescent health research. 1995. *Journal of Adolescent Health* 2003 November; 33(5): 410-415. Subject: 18.2

Society for Adolescent Medicine; Santelli, John S.; Smith Rogers, Audrey; Rosenfeld, Walter D.; DuRant, Robert H.; Dubler, Nancy; Morreale, Madlyn; English, Abigail; Lyss, Sheryl; Wimberly, Yolanda; Schissel, Anna. Guidelines for adolescent health research. A position paper of the Society for Adolescent Medicine. *Journal of Adolescent Health* 2003 November; 33(5): 396-409. Subject: 18.2

Society for Assisted Reproductive Technology [SART]. Practice Committee; American Society for Reproductive Medicine [ASRM]. Practice Committee. Guidelines on number of embryos transferred. *Fertility and Sterility* 2006 November; 86(5, Supplement): S51-S52. Subject: 14.4

Society for Assisted Reproductive Technology [SART]. Practice Committee; American Society for Reproductive Medicine [ASRM]. Practice Committee. Revised mini- mum standards for practices offering assisted reproductive technologies. *Fertility and Sterility* 2006 November; 86(5, Supplement): S53-S56. Subject: 14.1

Society for Clinical Trials; Dickersin, Kay; Davis, Barry R.; Dixon, Dennis O.; George, Stephen L.; Hawkins, Barbara S.; Lachin, John; Peduzzi, Peter; Pocock, Stuart. The Society for Clinical Trials supports United States legislation mandating trials registration. Position paper. *Clinical Trials (London, England)* 2004; 1(5): 417-420. Subject: 18.2

Society of Obstetricians and Gynecologists of Canada; Canadian Fertility and Andrology Society; Min, Jason K.; Claman, Paul; Hughes, Ed. Guidelines for the number of embryos to transfer following in vitro fertilization. *Journal of Obstetrics and Gynaecology Canada* 2006 September; 28(9): 799-813. Subject: 14.4

Soini, Sirpa; Ibarreta, Dolores; Anastasiadou, Violetta; Aymé, Ségolène; Braga, Suzanne; Cornel, Martina; Coviello, Domenico A.; Evers-Kiebooms, Gerry; Geraedts, Joep; Gianaroli, Luca; Harper, Joyce; Kosztolanyi, György; Lundin, Kersti; Rodrigues-Cerezo, Emilio; Sermon, Karen; Sequeiros, Jorge; Tranebjaerg, Lisbeth; Kääriäinen, Helena. The interface between assisted reproductive technologies and genetics: technical, social, ethical, and legal issues. *European Journal of Human Genetics* 2006 May; 14(5): 588-645. Subject: 14.1

Soini, S. Preimplantation genetic diagnosis (PGD) in Europe: diversity of legislation a challenge to the community and its citizens. *Medicine and Law: The World Association for Medical Law* 2007 June; 26(2): 309-323. Subject: 15.2

Sokalska, Maria E. Implementation of the Convention in central Europe: the case of Poland. *In:* Gevers, J.K.M.; Hondius, E.H.; Hubben, J.H., eds. Health Law, Human

Rights and the Biomedicine Convention: Essays in Honour of Henriette Roscam Abbing. Leiden; Boston: Martinus Nijhoff Publishers, 2005: 255-268. Subject: 21.1

Sokol, Daniel K. Can deceiving patients be morally acceptable? *BMJ: British Medical Journal* 2007 May 12; 334(7601): 984-986. Subject: 8.2

Sokol, Daniel K. How the doctor's nose has shortened over time; a historical overview of the truth-telling debate in the doctor-patient relationship. *Journal of the Royal Society of Medicine* 2006 December; 99(12): 632-636. Subject: 8.2

Sokol, Daniel K. No patient is an island. *BMJ: British Medical Journal* 2007 September 15; 335(7619): 568. Subject: 9.6

Sokol, Daniel K. Virulent epidemics and scope of healthcare workers' duty of care. *Emerging Infectious Diseases* 2006 August; 12(8): 1238-1241. Subject: 7.1

Sokol, Daniel K. What would you do, doctor? *BMJ: British Medial Journal* 2007 April 21; 334(7598): 853. Subject: 8.1

Solomon, David. Domestic disarray and imperial ambition: contemporary applied ethics and the prospects for global bioethics. *In:* Engelhardt, H. Tristram, ed. Global Bioethics: The Collapse of Consensus. Salem, MA: M&M Scrivener Press, 2006: 335-361. Subject: 2.1

Solomon, Lewis D. Life extension: public policy aspects. *In his:* The Quest for Human Longevity: Science, Business, and Public Policy. New Brunswick, NJ: Transaction Publishers, 2006: 165-187. Subject: 20.5.1

Solomon, Margot R.; DeNatale, Mary Lou. Academic dishonesty and professional practice: a convocation. *Nurse Educator* 2000 November-December; 25(6): 270-271. Subject: 4.1.3

Solskone, Colin L.; Light, Andrew. Towards ethics guidelines for environmental epidemiologists. *Science of the Total Environment* 1996; 184(1-2): 137-147. Subject: 7.1

Somerville, Margaret A. The importance of empirical research in bioethics: the case of human embryo stem cell research = Importance de la recherche empirique en bioéthique: cas de la recherche sur les cellules souches embryonnaires humaines [editorial]. *JOGC: Journal of Obstetrics and Gynaecology Canada = JOGC: Journal d'Obstétrique et Gynécologie du Canada* 2005 October; 27(10): 929-932. Subject: 18.5.4

Song, John; Bartels, Dianne M.; Ratner, Edward R.; Alderton, Lucy; Hudson, Brenda; Ahluwalia, Jasjit S. Dying on the streets: homeless persons' concerns and desires about end of life care. *JGIM: Journal of General Internal Medicine* 2007 April; 22(4): 435-441. Subject: 20.4.1

Song, John; Ratner, Edward R.; Bartels, Dianne M.; Alderton, Lucy; Hudson, Brenda; Ahluwalia, Jasjit S. Experiences with and attitudes toward death and dying among homeless persons. *JGIM: Journal of General Internal Medicine* 2007 April; 22(4): 427-434. Subject: 20.3.1

Song, Sang-yong. The rise and fall of embryonic stem cell research in Korea. *Asian Biotechnology and Development Review* 2006 November; 9(1): 65-73 [Online]. Accessed: http://www.openj-gate.org/articlelist.asp?LatestYear=2007&JCode=104346&year=2006&vol=9&issue=1&ICode=583649 [2008 February 29]. Subject: 18.5.4

Soreth, Janice; Cox, Edward; Kweder, Sandra; Jenkins, John; Galson, Steven. Ketek — the FDA perspective. *New England Journal of Medicine* 2007 April 19; 356(16): 1675-1676. Subject: 9.7

Sox, Harold C. Medical professionalism and the parable of the craft guilds [editorial]. *Annals of Internal Medicine* 2007 December; 147(11): 809-810. Subject: 4.1.2

Spaemann, Robert. Christianity and western philosophy. *In:* Vöneky, Silja; Wolfrum, Rüdiger, eds. Human Dignity and Human Cloning. Leiden: Nijhoff, 2004: 47-51. Subject: 14.5

Spar, Debora. The egg trade — making sense of the market for human oocytes. *New England Journal of Medicine* 2007 March 29; 356(13): 1289-1291. Subject: 14.4

Spar, Debora; Harrington, Anna. Selling stem cell science: how markets drive law along the technological frontier. *American Journal of Law and Medicine* 2007; 33(4): 541-565. Subject: 18.5.4

Sparks, Richard C. Making decisions about end-of-life care. *In:* Hamel, Ronald, ed. Making Health Care Decisions: A Catholic Guide. Liguori, MO: Liguori Publications, 2006: 73-90. Subject: 20.5.1

Sparrow, Margaret June. Abortion politics and the impact on reproductive health care [editorial]. *Australian and New Zealand Journal of Obstetrics and Gynaecology* 2005 December; 45(6): 471-473. Subject: 11.1

Spece, Roy G.; Bernstein, Carol. Scientific misconduct and liability for the acts of others. *Medicine and Law: The World Association for Medical Law* 2007 September; 26(3): 477-491. Subject: 1.3.9

Spece, Roy G.; Bernstein, Carol. What is scientific misconduct, who has to (dis)prove it, and to what level of certainty? *Medicine and Law: The World Association for Medical Law* 2007 September; 26(3): 493-510. Subject: 1.3.9

Spellecy, Ryan; Zanni, Guido R.; Stavis, Paul F. Psychiatric outpatient commitment: one tool along a continuum [comment and reply]. *American Journal of Bioethics* 2007 November; 7(11): 45-47; author reply W3-W4. Subject: 17.7

Spence, Des. A time of change in abortion. *BMJ: British Medical Journal* 2007 December 15; 335(7632): 1266. Subject: 12.4.3

See SUBJECT HEADING KEY FOR SECTION II on inside back cover.

605

Spielman, Bethany. Faulty premise, premature conclusion: that money was extraneous to the research ethics of the TGN1412 study. *American Journal of Bioethics* 2007 February; 7(2): 93-94. Subject: 18.1

Spielthenner, Georg. Ordinary and extraordinary means of treatment. *Ethics and Medicine: An International Journal of Bioethics* 2007 Fall; 23(3): 145-158. Subject: 20.5.1

Spier, Raymond E. Some thoughts on the 2007 World Conference on Research Integrity. *Science and Engineering Ethics* 2007 December; 13(4): 383-386. Subject: 1.3.9

Spiers, Alexander S.D. Studies in animals should be more like those in humans [letter]. *BMJ: British Medical Journal* 2007 February 10; 334(7588): 274. Subject: 22.2

Spiess, Christian. Recognition and social justice: a Roman Catholic view of Christian bioethics of long-term care and community service. *Christian Bioethics* 2007 September-December;(13)3: 287-301. Subject: 2.1

Spike, Jeffrey P. Memory identity and capacity. *Journal of Clinical Ethics* 2007 Fall; 18(3): 252-255. Subject: 9.6

Spike, Jeffrey P. Who's guarding the henhouse? Ramifications of the Fox study. *American Journal of Bioethics* 2007 February; 7(2): 48-50. Subject: 9.6

Spinney, Laura. Therapy for autistic children causes outcry in France [comment]. *Lancet* 2007 August 25-31; 370(9588): 645-646. Subject: 9.5.7

Spital, Aaron; Snyder, Deborah J.; Miller, Franklin J.; Rosenstein, Donald L.; Domingo, Angela F.; Salvana, Edsel Maurice T.; Rady, Mohamed Y.; Verheijde, Joseph L.; McGregor, Joan; Hanto, Douglas W. Solicitation of deceased and living organ donors [letters and reply]. *New England Journal of Medicine* 2007 June 7; 356(23): 2427-2429. Subject: 19.5

Spital, Aaron; Steinberg, David. Living donor list exchanges disadvantage blood-group-O recipients [letter and reply]. *American Journal of Kidney Diseases* 2005 May; 45(5): 962. Subject: 19.5

Spitzer, Walter O. Minimizing bias and prejudice: special challenges for contractual research by academicians. *In:* Shapiro, S.; Dinger, J.; Scriba, P., eds. Enabling Risk Assessment in Medicine: Farewell Symposium for Werner-Kari Raff. New Brunswick, NJ: Transaction Publishers, 2004: 31-49. Subject: 1.3.9

Sporrong, Sofia Kälvemark; Arnetz, Bengt; Hansson, Mats G.; Westerholm, Peter; Höglund, Anna T. Developing ethical competence in health care organizations. *Nursing Ethics* 2007 November; 14(6): 825-837. Subject: 7.2

Sprague, Stuart. What part of spirituality don't you understand? *American Journal of Bioethics* 2007 July; 7(7): 28-29. Subject: 8.1

Spriggs, Merle. When "risk" and "benefit" are open to interpretation - as is generally the case. *American Journal of Bioethics* 2007 March; 7(3): 17-19. Subject: 18.5.2

Spurgeon, Brad. Doctors sign petition calling for euthanasia to be decriminalised [news]. *BMJ: British Medical Journal* 2007 March 17; 334(7593): 555. Subject: 20.5.1

Spurgeon, David. New York Times reveals payments to doctors by drug firms [news]. *BMJ: British Medical Journal* 2007 March 31; 334(7595): 655. Subject: 9.7

Sque, Magi; Payne, Sheila; Clark, Jill Macleod. Gift of life or sacrifice?: key discourses to understanding organ donor families' decision-making. *Mortality* 2006 May; 11(2): 117-132. Subject: 19.5

Srebnik, Debra S.; Russo, Joan. Consistency of psychiatric crisis care with advance directive instructions. *Psychiatric Services* 2007 September; 58(9): 1157-1163. Subject: 17.1

Sreenivasan, Gopal. Health care and equality of opportunity. *Hastings Center Report* 2007 March-April; 37(2): 21-31. Subject: 9.2

Sringernyuang, L.; Thaweesit, S.; Nakapiew, S. A situational analysis of HIV/AIDS-related discrimination in Bangkok, Thailand. *AIDS Care* 2005 July; 17(Supplement 2): S165-S174. Subject: 9.5.6

Srivastava, Ranjana. The art of letting go. *New England Journal of Medicine* 2007 July 5; 357(1): 3-5. Subject: 8.2

Stabile, Bonnie. Demographic profile of states with human cloning laws: morality policy meets political economy. *Politics and the Life Sciences* 2007 March; 26(1): 43-50. Subject: 14.5

Stafleu, F.R.; Tramper, R.; Vorstenbosch, J.; Joles, J.A. The ethical acceptability of animal experiments: a proposal for a system to support decisionmaking. *Lab Animal* 1999 July; 33(3): 295-303. Subject: 22.2

Stanberry, Benedict. Legal and ethical aspects of telemedicine. *Journal of Telemedicine and Telecare* 2006; 12(4): 166-175. Subject: 1.3.12

Standridge, John B. Of doctor conventions and drug companies. *Family Medicine* 2006 July-August; 38(7): 518-520. Subject: 7.3

Stange, Kurt C. In this issue: doctor-patient and drug company-patient communication. *Annals of Family Medicine* 2007 January/February; 5(1): 2-3. Subject: 9.7

Stansfield, Alison J.; Holland, A.J.; Clare, I.C.H. The sterilisation of people with intellectual disabilities in England and Wales during the period 1988 to 1999. *Journal of Intellectual Disability Research* 2007 August; 51(8): 569-570. Subject: 9.5.3

Starck, Christian. The human embryo is a person and not an object. *In:* Vöneky, Silja; Wolfrum, Rüdiger, eds. Human Dignity and Human Cloning. Leiden: Nijhoff, 2004: 63-67. Subject: 14.5

Stark, Patsy; Roberts, Chris; Newble, David; Bax, Nigel. Discovering professionalism through guided reflec-

---

tion. *Medical Teacher* 2006 February; 28(1): e25-e31. Subject: 4.1.2

Starks, Helene; Back, Anthony L.; Pearlman, Roberta A.; Koenig, Barbara A.; Hsu, Clarissa; Gordon, Judith R.; Bharucha, Ashok J. Family member involvement in hastened death. *Death Studies* 2007 February; 31(2): 105-130. Subject: 20.5.1

Statham, H.; Solomou, W.; Green, J. Late termination of pregnancy: law, policy and decision making in four English fetal medicine units. *BJOG: An International Journal of Obstetrics and Gynaecology* 2006 December; 113(12): 1402-1411. Subject: 12.4.2

Steckel, Cynthia M. Mandatory influenza immunization for health care workers — an ethical discussion. *AAOHN Journal* 2007 January; 55(1): 34-39. Subject: 9.1

Steers, Edward, Jr. Dr. Mudd and the "colored" witnesses. *Civil War History* 2000; 46(4): 324-336. Subject: 7.1

Steger, Florian. Neuropathological research at the "Deutsche Forschungsanstalt fuer Psychiatrie" (German Institute for Psychiatric Research) in Munich (Kaiser-Wilhelm-Institute). Scientific utilization of children's organs from the "Kinderfachabteilungen" (Children's Special Departments) at Bavarian State Hospitals. *Journal of the History of the Neurosciences* 2006 September; 15(3): 173-185. Subject: 20.5.2

Steigleder, Klaus. Medizinethik und Philosophie [Medical ethics and philosophy]. *Ethik in der Medizin* 2006 December; 18(4): 310-314. Subject: 2.1

Stein, Joel; Brady Wagner, Lynne C. Is informed consent a "yes or no" response? Enhancing the shared decision-making process for persons with aphasia. *Topics in Stroke Rehabilitation* 2006 Fall; 13(4): 42-46. Subject: 8.3.3

Stein, Rob. 'Embryo bank' stirs ethics fears; firm lets clients pick among fertilized eggs. *Washington Post* 2007 January 6; p. A1, A8. Subject: 14.4

Stein, Rob. First U.S. uterus transplant planned; some experts say risk isn't justified. *Washington Post* 2007 January 15; p. A1, A9. Subject: 19.1

Stein, Rob. New trend in organ donation raises questions; as alternative approach becomes more frequent, doctors worry that it puts donors at risk. *Washington Post* 2007 March 18;. Subject: 19.5

Stein, Rob. New zeal in organ procurement raises fears; donation groups say they walk a fine line, but critics see potential for abuses. *Washington Post* 2007 September 13; p. A1, A9. Subject: 19.5

Steinberg, Brian. New medical-device ads; old concerns: Can a knee implant be sold this way, and should it be? *Wall Street Journal* 2007 April 10; p. B6. Subject: 9.7

Steinberg, David. How much risk can medicine allow a willing altruist? *Journal of Clinical Ethics* 2007 Spring; 18(1): 12-17. Subject: 19.3

Steinberg, David. Reply to Valapour, "Living donor transplantation: the perfect balance of public oversight and medical responsibility". *Journal of Clinical Ethics* 2007 Spring; 18(1): 21-22. Subject: 19.5

Steinberg, Douglas. Consciousness is missing — and so is research. After the Schiavo controversy in the USA, obstacles still hinder the study of people with little or no awareness. *EMBO Reports* 2005 November; 6(11): 1009-1011. Subject: 4.4

Steinbock, Bonnie. Defining parenthood. *In:* Freeman, Michael, ed. Children's Health and Children's Rights. Boston: Martinus Nijhoff Publishers, 2006: 311-334. Subject: 14.4

Steinbock, Bonnie. Defining personhood. *In:* Spencer, J.R.; du Bois-Pedain, Antje, eds. Freedom and Responsibility in Reproductive Choice. Portland: Hart Pub., 2006: 107-128. Subject: 14.1

Steinbock, Bonnie. Moral status, moral value, and human embryos: implications for stem cell research. *In:* Steinbock, Bonnie, ed. The Oxford Handbook of Bioethics. Oxford; New York: Oxford University Press, 2007: 416-440. Subject: 4.4

Steinbrook, Robert. Guidance for guidelines. *New England Journal of Medicine* 2007 January 25; 356(4): 331-333. Subject: 5.2

Steinbrook, Robert. Organ donation after death. *New England Journal of Medicine* 2007 July 19; 357(3): 209-213. Subject: 19.5

Steiner, Hillel. The right to trade in human body parts. *In:* Seglow, Jonathan ed. The Ethics of Altruism. London: Frank Cass, 2004: 187-193. Subject: 19.5

Steinmetzer, Jan; Groß, Dominik; Duncker, Tobias Heinrich. Ethische Fragen im Umgang mit transidenten Personen - Limitierende Faktoren des gegenwärtigen Konzepts von "Transsexualität" = Ethical problems concerning transgender persons: limiting factors of present concepts of "transsexualism". *Ethik in der Medizin* 2007 March; 19(1): 39-54. Subject: 10

Steneck, Nicholas H.; Bulger, Ruth Ellen. The history, purpose, and future of instruction in the responsible conduct of research. *Academic Medicine* 2007 September; 82(9): 829-834. Subject: 18.2

Stephens, Carolyn; Porter, John; Nettleton, Clive; Willis, Ruth. UN Declaration on the Rights of Indigenous Peoples [letter]. *Lancet* 2007 November 24-30; 370(9601): 1756. Subject: 21.7

Stephenson, James A.; Staal, Mark A. An ethical decision-making model for operational psychology. *Ethics and Behavior* 2007; 17(1): 61-82. Subject: 17.1

Stephenson, Jeffrey. Assisted dying: a palliative care physician's view. *Clinical Medicine (London, England)* 2006 July-August; 6(4): 374-377. Subject: 20.7

See SUBJECT HEADING KEY FOR SECTION II on inside back cover.

607

Sternberg, Robert J. Not a case of black and white. *New Scientist* 2007 October 27-November 2; 196(2627): 24. Subject: 15.6

Steven, Megan S.; Pascual-Leone, Alvaro. Transcranial magnetic stimulation and the human brain: an ethical evaluation. *In:* Illes, Judy, ed. Neuroethics: Defining the Issues in Theory, Practice, and Policy. New York: Oxford University Press, 2006: 201-211. Subject: 17.1

Stevens, David. Medical martyrdom? *Today's Christian Doctor* 2007 Fall; 38(3): 18-21. Subject: 4.1.2

Stevens, M.L. Tina. Intellectual capital and voting booth bioethics: a contemporary historical critique. *In:* Eckenwiler, Lisa A.; Cohn, Felicia G., eds. The Ethics of Bioethics: Mapping the Moral Landscape. Baltimore, MD: Johns Hopkins University Press, 2007: 59-73. Subject: 2.1

Steward, Douglas O.; DeMarco, Joseph P. Rejoinder. *Journal of Bioethical Inquiry* 2006; 3(3): 137-138. Subject: 9.5.1

Stewart, Alexandra M. Mandating HPV vaccination: private rights, public good [letter]. *New England Journal of Medicine* 2007 May 10; 356(19): 1998-1999. Subject: 9.7

Stewart, Cameron. Introduction: the human body — The Land That Time Forgot. *Journal of Bioethical Inquiry* 2007; 4(2): 117-118. Subject: 19.1

Stewart, Cameron. Recent developments. *Journal of Bioethical Inquiry* 2007; 4(2): 81-84. Subject: 8.3.3

Stewart, Cameron. Recent developments. *Journal of Bioethical Inquiry* 2007; 4(3): 169-170. Subject: 14.5

Stillman, Robert J. A 47-year-old woman with fertility problems who desires a multiple pregnancy. *JAMA: The Journal of the American Medical Association* 2007 February 28; 297(8): 858-867. Subject: 14.4

Stirewalt, Karolyn. To release or not to release? When is it all right for physicians who treat injured workers to release medical information without their consent? *Minnesota Medicine* 2007 September; 90(9): 52-53. Subject: 8.4

Stobbart, L.; et al. "We saw human guinea pigs explode". *BMJ: British Medical Journal* 2007 March 17; 334(7593): 566-567. Subject: 1.3.7

Stocking, Carol B.; Houghham, Gavin W.; Danner, Deborah D.; Patterson, Marian B.; Whitehouse, Peter J.; Sachs, Greg A. Empirical assessment of a research advance directive for persons with dementia and their proxies. *Journal of the American Geriatrics Society* 2007 October; 55(10): 1609-1612. Subject: 17.1

Stocking, C.B.; Hougham, G.W.; Danner, D.D.; Patterson, M.B.; Whitehouse, P.J.; Sachs, G.A. Speaking of research advance directives: planning for future research participation. *Neurology* 2006 May 9; 66(9): 1361-1366. Subject: 20.5.4

Stolberg, Sheryl Gay. President calls for genetic privacy bill. *New York Times* 2007 January 18; p. A14. Subject: 8.4

Stollorz, Volker. Öffentlich und Industrie-geförderte Studien in der Perzeption der Medien — ein Dilemma. *In:* Shapiro, S.; Dinger, J.; Scriba, P., eds. Enabling Risk Assessment in Medicine: Farewell Symposium for Werner-Kari Raff. New Brunswick, NJ: Transaction Publishers, 2004: 125-142. Subject: 1.3.9

Stone, Judy. Ethical issues in human subjects research. *In her:* Conducting Clinical Research: A Practical Guide for Physicians, Nurses, Study Coordinators, and Investigators. Cumberland, MD: Mountainside MD Press, 2006: 145-172. Subject: 18.2

Stone, Judy. Society and politics. *In her:* Conducting Clinical Research: A Practical Guide for Physicians, Nurses, Study Coordinators, and Investigators. Cumberland, MD: Mountainside MD Press, 2006: 173-194. Subject: 18.6

Stone, Julie. Evaluating the ethical and legal content of professional codes of ethics. *In:* Allsop, Judith; Saks, Mike, eds. Regulating the Health Professions. London; Thousand Oaks, CA: Sage Publications, 2002: 62-75. Subject: 6

Stonington, Scott; Ratanakul, Pinit. Is there a global bioethics? End-of-life in Thailand and the case for local difference. California: Pacific Rim Research Program (University of California, Multi-Campus Research Unit), 2006: 7 p. [Online]. Accessed: http://repositories.cdlib.org/cgi/viewcontent.cgi?article=1021&context=pacrim [2007 October 1]. Subject: 20.5.1

Stoos, William Kevin. Who am I? Why am I? (The anguish of a clone). *Linacre Quarterly* 2007 May; 74(2): 171-173. Subject: 14.5

Storch, Janet. Building moral communities in health care [editorial]. *Nursing Ethics* 2007 September; 14(5): 569-570. Subject: 9.1

Storch, Janet L.; Kenny, Nuala. Shared moral work of nurses and physicians. *Nursing Ethics* 2007 July; 14(4): 478-491. Subject: 7.3

Storrow, Richard F. The bioethics of prospective parenthood: in pursuit of the proper standard for gatekeeping in infertility clinics. *Cardozo Law Review* 2007 April; 28(5):101-138. Subject: 14.1

Storz, Philipp; Kolpatzik, Kai; Perleth, Matthias; Klein, Silvia; Häussler, Bertram. Future relevance of genetic testing: a systematic horizon scanning analysis. *International Journal of Technology Assessment in Health Care* 2007 Fall; 23(4): 495-504. Subject: 15.3

Stossel, Thomas P. Regulation of financial conflicts of interest in medical practice and medical research: a damaging solution in search of a problem. *Perspectives in Biology and Medicine* 2007 Winter; 50(1): 54-71. Subject: 5.3

Stotland, Nada L.; Ross, Lainie F.; Clayton, Ellen W.; Mishtal, Joanna Z.; Chavkin, Wendy; Zarate, Victor; O'Connell, Patrick; Mistrot, Jacques; Parson, Ken-

neth C.; **Curlin, Farr A.; Lawrence, Ryan E.; Lantos, John D.** Religion, conscience, and controversial clinical practices [letters and reply]. *New England Journal of Medicine* 2007 May 3; 356(18): 1889-1892. Subject: 8.1

Straus, Joseph. Patentierung von Leben? *In:* Schreiber, Hans-Peter, ed. Biomedizin und Ethik: Praxis, Recht, Moral. Basel; Boston: Birkhäuser, 2004: 50-55. Subject: 15.8

Straus, Sharon; Stelfox, Tom. Whose life is it anyway? Capacity and consent in Canada. *CMAJ/JAMC: Canadian Medical Association Journal* 2007 November 20; 177(11): 1329. Subject: 8.3.3

Streeter, Oscar E.; Cuyjet, Aloysius B.; Norris, Keith; Hylton, Kevin. Issues surrounding the involvement of African Americans in clinical trials and other research. *In:* Livingston, Ivor Lensworth, ed. Praeger Handbook of Black American Health: Policies and Issues Behind Disparities in Health, Vol. II. 2nd edition. Westport, CT: Praeger Publishers, 2004: 808-834. Subject: 18.5.1

Stretton, Dean. Harriton v Stephens; Waller v James: wrongful life and the logic of non-existence. *Melbourne University Law Review* 2006 December; 30(3): 972-1001. Subject: 15.2

Strong, Carson. Case commentary: parental request for life-prolonging interventions. *HEC (Healthcare Ethics Committee) Forum* 2007 December; 19(4): 377-380. Subject: 20.5.2

Strong, Carson. Embryology, metaphysics, and common sense: a response to Gómez-Lobo [letter]. *Theoretical Medicine and Bioethics* 2007; 28(4): 337-340. Subject: 4.4

Strong, Carson; Feudtner, Chris; Carter, Brian S.; Rushton, Cynda H. Goals, values, and conflict resolution. *In:* Carter, Brian S.; Levetown, Marcia, eds. Palliative Care for Infants, Children, and Adolescents: A Practical Handbook. Baltimore: Johns Hopkins University Press, 2004: 23-43. Subject: 20.4.2

Strong, P.M.; Davis, A.G. Roles, role formats and medical encounters: a cross-cultural analysis of staff-client relationships in children's clinics. *Sociological Review* 1977 November; 25(4): 775-800. Subject: 8.1

Strous, Rael D. Hitler's psychiatrists: healers and researchers turned executioners and its relevance today. *Harvard Review of Psychiatry* 2006 January-February; 14(1): 30-37. Subject: 20.5.1

Strumolo, Adaline R. Prescription privacy. *American Journal of Law and Medicine* 2007; 33(4): 705-708. Subject: 9.7

Stultiëns, Loes; Goffin, Tom; Borry, Pascal; Dierickx, Kris; Nys, Herman. Minors and informed consent: a comparative approach. *European Journal of Health Law* 2007 April; 14(1): 21-46. Subject: 8.3.2

Suchman, Anthony L. Advancing humanism in medical education. *JGIM: Journal of General Internal Medicine* 2007 November; 22(11): 1630-1631. Subject: 7.2

Sugarman, Jeremy. Examining the provisions for research without consent in the emergency setting. *Hastings Center Report* 2007 January-February; 37(1): 12-13. Subject: 18.3

Sugarman, Jeremy. Roles of moral philosophy in appropriated bioethics: a response to Baker and McCullough. *Kennedy Institute of Ethics Journal* 2007 March; 17(1): 65-67. Subject: 2.1

Sugarman, Jeremy; Roter, Debra; Cain, Carole; Wallace, Roberta; Schmechel, Don; Welsh-Bohmer, Kathleen A. Proxies and consent discussions for dementia research. *Journal of the American Geriatrics Society* 2007 April; 55(4): 556-561. Subject: 18.3

Sugarman, Stephen D. Cases in vaccine court — legal battles over vaccines and autism. *New England Journal of Medicine* 2007 September 27; 357(13): 1275-1277. Subject: 9.7

Sui, Suli; Sleeboom-Faulkner, Margaret. Commercial genetic testing in mainland China: social, financial and ethical issues. *Journal of Bioethical Inquiry* 2007; 4(3): 229-237. Subject: 15.3

Sujdak Mackiewicz, Birgitta N. Artificial nutrition and hydration: advancing the conversation. *Health Care Ethics USA* 2007 Summer; 15(3): 9-11. Subject: 20.5.1

Sullivan, Scott M. The development and nature of the ordinary/extraordinary means distinction in the Roman Catholic tradition. *Bioethics* 2007 September; 21(7): 386-397. Subject: 20.5.1

Sulmasy, Dan. DCD policies: the devil is always in the details. *Health Care Ethics USA* 2007 Fall; 15(4): 22. Subject: 19.5

Sulmasy, Daniel P. 'Reinventing' the rule of double effect. *In:* Steinbock, Bonnie, ed. The Oxford Handbook of Bioethics. Oxford; New York: Oxford University Press, 2007: 114-149. Subject: 2.1

Sulmasy, Daniel P. Cancer care, money, and the value of life: whose justice? Which rationality? *Journal of Clinical Oncology* 2007 January 10; 25(2): 217-222. Subject: 9.5.1

Sulmasy, Daniel P. Who owns the human genome? *In:* Monsour, H. Daniel, ed. Ethics and the New Genetics: An Integrated Approach. Toronto: University of Toronto Press, 2007: 123-133. Subject: 15.8

Sun, Chiao-Yin.; Lee, Chin-Chan; Chang, Chiz-Tzung; Hung, Cheng-Chih; Wu, Mai-Szu. Commercial cadaveric renal transplant: an ethical rather than medical issue. *Clinical Transplantation* 2006 May-June; 20(3): 340-345. Subject: 19.5

Sunstein, Cass R. Slaughterhouse jive [review of Introduction to Animal Rights: Your Child or the Dog? by Gary

See SUBJECT HEADING KEY FOR SECTION II on inside back cover.

609

L. Francione]. *New Republic* 2001 January 29; 224(5): 40-45. Subject: 22.1

Sutcliffe, Alastair G.; Ludwig, Michael. Outcome of assisted reproduction. *Lancet* 2007 July 28-August 3; 370(9584): 351-359. Subject: 14.4

Sutherland, Elaine E. Is there a right not to procreate? *In:* McLean, Sheila A.M., ed. First Do No Harm: Law, Ethics, and Healthcare. Aldershot, England; Burlington, VT: Ashgate, 2006: 319-336. Subject: 14.1

Suzuki, Kohta; Hoshi, Kazuhiko; Minai, Junko; Yanaihara, Takumi; Takeda, Yasuhisa; Yamagata, Zentaro. Analysis of national representative opinion surveys concerning gestational surrogacy in Japan. *European Journal of Obstetrics, Gynecology, and Reproductive Biology* 2006 May 1; 126(1): 39-47. Subject: 14.2

Svenaeus, Fredrik. A Heideggerian defense of therapeutic cloning. *Theoretical Medicine and Bioethics* 2007; 28(1): 31-62. Subject: 14.5

Svenaeus, Fredrik. Do antidepressants affect the self? A phenomenological approach. *Medicine, Health Care and Philosophy* 2007 June; 10(2): 153-166. Subject: 17.4

Svenaeus, Fredrik. Psychopharmacology and the self: an introduction to the theme [editorial]. *Medicine, Health Care and Philosophy* 2007 June; 10(2): 115-117. Subject: 17.4

Svensson, Sara; Hansson, Sven Ove. Protecting people in research: a comparison between biomedical and traffic research. *Science and Engineering Ethics* 2007 March; 13(1): 99-115. Subject: 18.1

Swartz, Marvin S.; Swanson, Jeffrey W. Psychiatric advance directives and recovery-oriented care. *Psychiatric Services* 2007 September; 58(9): 1164. Subject: 17.1

Swazo, Norman K. Research integrity and rights of indigenous peoples: appropriating Foucault's critique of knowledge/power. *Studies in History and Philosophy of Biological and Biomedical Sciences* 2005 September; 36(3): 568-584. Subject: 18.2

Swede, Helen; Stone, Carol L.; Norwood, Alyssa R. National population-based biobanks for genetic research. *Genetics in Medicine* 2007 March; 9(3): 141-149. Subject: 15.1

Sweet, Victoria. Code pearl. *Health Affairs* 2008 January-February; 27(1): 216-220. Subject: 20.5.1

Sweet, Victoria. Thy will be done. *Health Affairs* 2007 May-June; 26(3): 825-830. Subject: 20.5.4

Swidler, Robert N.; Seastrum, Terese; Shelton, Wayne. Difficult hospital inpatient discharge decisions: ethical, legal and clinical practice issues. *American Journal of Bioethics* 2007 March; 7(3): 23-28. Subject: 9.5.2

Swift, Jennifer. Eggs don't come cheap. *New Scientist* 2007 December 8-14; 196(2633): 22. Subject: 14.5

Swindell, J.S. Facial allograft transplantation, personal identity and subjectivity. *Journal of Medical Ethics* 2007 August; 33(8): 449-453. Subject: 19.1

Swota, Alissa Hurwitz. Changing policy to reflect a concern for patients who sign out against medical advice. *American Journal of Bioethics* 2007 March; 7(3): 32-34. Subject: 8.1

Sylva, Douglas; Yoshihara, Susan. Rights by stealth: the role of UN human rights treaty bodies in the campaign for an international right to abortion. *National Catholic Bioethics Quarterly* 2007 Spring; 7(1): 97-128. Subject: 12.4.1

Synofzik, Matthis; Maetzler, Walter. Wie sollen wir Patienten mit Demenz behandeln? Die ethisch problematische Funktion der Antidementiva = How should we treat dementia patients? The ethically problematic function of antidementia drugs. *Ethik in der Medizin* 2007 December; 19(4): 270-280. Subject: 17.4

Szasz, Thomas. Honouring advance decisions: you don't in psychiatry [letter]. *BMJ: British Medical Journal* 2007 November 3; 335(7626): 900. Subject: 20.5.4

Szasz, Thomas. Secular humanism and "scientific psychiatry". *Philosophy, Ethics, and Humanities in Medicine [electronic]* 2006; 1(5): 5 p. Accessed: http://www.peh-med.com/content/1/1/5 [2006 July 19]. Subject: 17.1

Szumacher, Ewa. The feminist approach in the decision-making process for treatment of women with breast cancer. *Annals of the Academy of Medicine, Singapore* 2006 September; 35(9): 655-661. Subject: 9.5.5

T

Tabak, Nili. Israel: teaching nursing ethics in Israel: ancient values meet modern healthcare problems. *In:* Davis, Anne J.; Tschudin, Verena; de Raeve, Louise, eds. Essentials of Teaching and Learning in Nursing Ethics: Perspectives and Methods. New York: Churchill Livingstone Elsevier, 2006: 261-270. Subject: 4.1.3

Tabbara, Khalid; Al-Kawi, M. Zuheir. Ethics in medical research and publication. *Annals of Saudi Medicine* 2006 July-August; 26(4): 257-260. Subject: 1.3.9

Tabor, Holly K.; Cho, Mildred K. Ethical implications of array comparative genomic hybridization in complex phenotypes: points to consider in research. *Genetics in Medicine* 2007 September; 9(9): 626-631. Subject: 15.3

Tai, Michael Cheng-tek. The debate on establishing a biobank in Taiwan. *Asian Biotechnology and Development Review* 2006 November; 9(1): 31-36 [Online]. Accessed. http://www.openj-gate.org/articlelist.asp?LatestYear=2007&JCode=104346&year=2006&vol=9&issue=1&ICode=583649 [2008 February 29]. Subject: 15.1

Tai, Michael Cheng-tek; Hill, Donald. A Confucian perspective on bioethical principles in ethics consultation.

Subject = NRCBL Primary Classification Number; see inside front cover.

Clinical Ethics 2007 December; 2(4): 201-207. Subject: 20.5.2

Takala, Tuija. Concepts of "person" and "liberty," and their implications to our fading notions of autonomy. *Journal of Medical Ethics* 2007 April; 33(4): 225-228. Subject: 4.4

Takala, Tuija; Häyry, Matti. Benefiting from past wrongdoing, human embryonic stem cell lines, and the fragility of the German legal position. *Bioethics* 2007 March; 21(3): 150-159. Subject: 18.5.4

Takeshita, Naoki; Hanaoka, Kanako; Shibui, Yukihiro; Jinnai, Hikoyoshi; Abe, Yuji; Kubo, Harumi. Regulating assisted reproductive technologies in Japan. *Journal of Assisted Reproduction and Genetics* 2003 July; 20(7): 260-264. Subject: 14.4

Talbott, John A.; Mallott, David B. Professionalism, medical humanism, and clinical bioethics: the new wave — does psychiatry have a role? *Journal of Psychiatric Practice* 2006 November; 12(6): 384-390. Subject: 7.2

Talone, Patricia A. Making decisions about organ transplantation. *In:* Hamel, Ronald, ed. Making Health Care Decisions: A Catholic Guide. Liguori, MO: Liguori Publications, 2006: 55-72. Subject: 19.1

Talseth, Anne-Grethe; Gilje, Fredricka. Unburdening suffering: responses of psychiatrists to patients' suicide deaths. *Nursing Ethics* 2007 September; 14(5): 620-636. Subject: 4.4

Tan, Eng-King; Lee, Jennie; Hunter, Christine; Shinawi, Lina; Fook-Chong, S.; Jankovic, Joseph. Comparing knowledge and attitudes towards genetic testing in Parkinson's disease in an American and Asian population. *Journal of the Neurological Sciences* 2007 January 31; 252(2): 113-120. Subject: 15.3

Tang, Ping Fen; Johansson, Camilla; Wadensten, Barbro; Wenneberg, Stig; Ahlström, Gerd. Chinese nurses' ethical concerns in a neurological ward. *Nursing Ethics* 2007 November; 14(6): 810-824. Subject: 4.1.3

Tangwa, Godfrey B. How not to compare western scientific medicine with African traditional medicine. *Developing World Bioethics* 2007 April; 7(1): 41-44. Subject: 4.1.1

Tangwa, Godfrey B. Moral status of embryonic stem cells: perspective of an African villager. *Bioethics* 2007 October; 21(8): 449-457. Subject: 18.5.4

Tanne, Janice Hopkins. Drug advertisements in US paint a "black and white scenario" [news]. *BMJ: British Medical Journal* 2007 February 10; 334(7588): 279. Subject: 9.7

Tanne, Janice Hopkins. FDA places "black box" warnings on anaemia drugs amid reports of incentives to doctors. *BMJ: British Medical Journal* 2007 May 19; 334(7602): 1022. Subject: 9.7

Tanne, Janice Hopkins. Group asks US institutes to reveal industry ties [news]. *BMJ: British Medical Journal* 2007 January 20; 334(7585): 115. Subject: 5.3

Tanne, Janice Hopkins. Investigators to review conflicts of interest at NIH [news]. *BMJ: British Medical Journal* 2007 April 14; 334(7597): 767. Subject: 1.3.9

Tanne, Janice Hopkins. US campaign aims to end industry gifts and speaking fees [news]. *BMJ: British Medical Journal* 2007 February 24; 334(7590): 385. Subject: 9.7

Tanne, Janice Hopkins. US companies are fined for payments to surgeons [news]. *BMJ: British Medical Journal* 2007 November 24; 335(7629): 1065. Subject: 7.3

Tanne, Janice Hopkins. US Congress asked to suspend funding for Planned Parenthood [news]. *BMJ: British Medical Journal* 2007 November 3; 335(7626): 903. Subject: 12.4.2

Tanne, Janice Hopkins. US gene therapy trial is to restart, despite patient's death. *BMJ: British Medical Journal* 2007 December 8; 335(7631): 1172-1173. Subject: 15.4

Tanne, Janice Hopkins. US guidelines often influenced by industry [news]. *BMJ: British Medical Journal* 2007 January 27; 334(7586): 171. Subject: 9.7

Tanne, Janice Hopkins. US Supreme Court approves ban on "partial birth abortion". *BMJ: British Medical Journal* 2007 April 28; 334(7599): 866. Subject: 12.4.2

Tannert, Christof. The autonomy axiom and the cloning of humans. *Human Reproduction and Genetic Ethics: An International Journal* 2007; 13(1): 4-7. Subject: 14.5

Tao Lai-Po-wah, Julia. A Confucian approach to a "shared family decision model" in health care: reflections on moral pluralism. *In:* Engelhardt, H. Tristram, ed. Global Bioethics: The Collapse of Consensus. Salem, MA: M&M Scrivener Press, 2006: 154-179. Subject: 8.1

Tarantola, Daniel; Gruskin, Sofia. New guidance on recommended HIV testing and counselling [comment]. *Lancet* 2007 July 21-27; 370(9583): 202-203. Subject: 9.5.6

Tarzian, Anita J. Disability and slippery slopes [Perspective]. *Hastings Center Report* 2007 September-October; 37(5): inside back cover. Subject: 4.2

Task Force on Values, Ethics, and Rationing in Critical Care (VERICC); Truog, Robert D.; Brock, Dan W.; Cook, Deborah J.; Danis, Marion; Luce, John M.; Rubenfeld, Gordon D.; Levy, Mitchell M. Rationing in the intensive care unit. *Critical Care Medicine* 2006 April; 34(4): 958-963; quiz 971. Subject: 9.4

Tattersall, Martin H.N.; Kerridge, Ian H. Doctors behaving badly? [editorial]. *Medical Journal of Australia* 2006 September 18; 185(6): 299-300 [see correction in: Medical Journal of Australia 2006 November 20; 185(10): 576]. Subject: 7.3

Tauer, Carol A. Making decisions about genetic testing. *In:* Hamel, Ronald, ed. Making Health Care Decisions: A Catholic Guide. Liguori, MO: Liguori Publications, 2006: 37-53. Subject: 15.3

See SUBJECT HEADING KEY FOR SECTION II on inside back cover.

Taylor, Allyn L.; Bettcher, Douglas W.; Fluss, Sev S.; DeLand, Katherine; Yach, Derek. International health instruments: an overview. *In:* Detels, Roger; McEwen, James; Beaglehole, Robert; Tanaka, Heizo, eds. Oxford Textbook of Public Health. Fourth edition. Oxford; New York: Oxford University Press, 2004: 359-386. Subject: 21.1

Taylor, Holly A. Instead of revising half the story, why not rewrite the whole thing? *American Journal of Bioethics* 2007 March; 7(3): 19-21. Subject: 18.5.2

Taylor, Holly A. Moving beyond compliance: measuring ethical quality to enhance the oversight of human subjects research. *IRB: Ethics and Human Research* 2007 September-October; 29(5): 9-14. Subject: 18.6

Taylor, Joseph G. NICE, Alzheimer's and QALY. *Clinical Ethics* 2007 March; 2(1): 50-54. Subject: 4.4

Taylor, J.S. A "queen of hearts" trial of organ markets: why Scheper-Hughes's objections to markets in human organs fail. *Journal of Medical Ethics* 2007 April; 33(4): 201-204. Subject: 19.5

Taylor, Mark J. Data protection, shared (genetic) data and genetic discrimination. *Medical Law International* 2007; 8(1): 51-77. Subject: 15.1

Taylor, Patrick L. Research sharing, ethics and public benefit. *Nature Biotechnology* 2007 April; 25(4): 398-401. Subject: 18.5.4

Taylor, Patrick L. Rules of engagement. Is there an inherent conflict between public debate and free scientific inquiry? *Nature* 2007 November 8; 450(7167): 163-164. Subject: 1.3.9

Taylor, Robert M. Ethical aspects of medical economics. *Neurologic Clinics* 1989 November; 7(4): 883-900. Subject: 9.3.1

Taylor, Robert S. Self-ownership and transplantable human organs. *Public Affairs Quarterly* 2007 January; 21(1): 89-107. Subject: 19.1

te Braake, Trees. Research on human embryos. *In:* Gevers, J.K.M.; Hondius, E.H.; Hubben, J.H., eds. Health Law, Human Rights and the Biomedicine Convention: Essays in Honour of Henriette Roscam Abbing. Leiden; Boston: Martinus Nijhoff Publishers, 2005: 191-203. Subject: 18.5.4

Teagarden, J. Russell; Wynia, Matthew K. Ensuring fairness in coverage decisions: applying the American Medical Association Ethical Force Program's consensus report to managed care pharmacy. *American Journal of Health-System Pharmacy* 2006 September 15; 63(18): 1749-1754. Subject: 9.3.2

Tejeda, Heriberto A.; Green, Sylvan B.; Trimble, Edward L.; Ford, Leslie; High, Joseph L.; Ungersleider, Richard S.; Friedman, Michael A.; Brawley, Otis W. Representation of African Americans, Hispanics, and Whites in National Cancer Institute treatment trials. *Journal of National Cancer Institute* 1996 June 19; 88(12): 812-816. Subject: 18.5.1

Temkin, Elizabeth. Contraceptive equity: the birth control center of the International Workers Order. *American Journal of Public Health* 2007 October; 97(10): 1737-1745. Subject: 11.1

Temple, Robert; Stockbridge, Norman L. BiDil for heart failure in black patients: the U.S. Food and Drug Administration perspective. *Annals of Internal Medicine* 2007 January 2; 146(1): 52-62. Subject: 15.11

Templeton, Allan; Braude, Peter. Umbilical cord blood banking and the RCOG. *Lancet* 2007 March 31-April 6; 369(9567): 1077. Subject: 19.4

Templeton, Sarah-Kate. Doctors: let us kill disabled babies. *TimesOnline (London)* 2006 November 5: 2 p. [Online]. Accessed: http://www.timesonline.co.uk/tol/news/uk/article625477.ece [2007 December 12]. Subject: 20.5.2

Templeton, Sarah-Kate; Swinford, Steven. Haunted mother who backs mercy killing. *TimesOnline (London)* 2006 November 5: 2 p. [Online]. Accessed: http://www.timesonline.co.uk/tol/news/uk/article625550.ece [2007 December 12]. Subject: 20.5.2

ten Have, H.; Ang, T.W. UNESCO's Global Ethics Observatory. *Journal of Medical Ethics* 2007 January; 33(1): 15-16. Subject: 2.1

Ten, C.L. A child's right to a father. *Monash Bioethics Review* 2000 October; 19(4): 33-37. Subject: 14.4

Tenenbaum, Evelyn M.; Reese, Brian; Henry, Michael; Fishman, Jennifer R.; Youngner, Stuart J. Memory-altering drugs: shifting the paradigm of informed consent. *American Journal of Bioethics* 2007 September; 7(9): 40-42; author reply W1-W3. Subject: 17.4

Tengland, Per-Anders. A two-dimensional theory of health. *Theoretical Medicine and Bioethics* 2007; 28(4): 257-284. Subject: 4.2

Tengland, Per-Anders. Empowerment: a goal or a means for health promotion? *Medicine, Health Care and Philosophy* 2007 June; 10(2): 197-207. Subject: 9.1

ter Meulen, Ruud. Ethical issues of evidence-based medicine. *In:* Gunning, Jennifer; Holm, Søren, eds. Ethics, law, and society. Volume 1. Aldershot, Hants, England; Burlington, VT: Ashgate, 2005: 51-58. Subject: 18.2

Tercyak, Kenneth P.; Peshkin, Beth N.; Wine, Lauren A.; Walker, Leslie R. Interest of adolescents in genetic testing for nicotine addiction susceptibility. *Preventive Medicine* 2006 January; 42(1): 60-65. Subject: 15.3

Tereskerz, Patricia M. Financial conflict and Vioxx: a public policy case study. *Organizational Ethics: Healthcare, Business, and Policy* 2006 Fall-Winter; 3(2): 112-119. Subject: 9.7

Terrell White, Mary. A right to benefit from international research: a new approach to capacity building in less-developed countries. *Accountability in Research* 2007 April-June; 14(2): 73-92. Subject: 18.5.9

Terry, W.; Olson, L.G.; Ravenscroft, P.; Wilss, L.; Boulton-Lewis, G. Hospice patients' views on research in palliative care. *Internal Medicine Journal* 2006 July; 36(7): 406-413. Subject: 20.4.1

Tetali, Shailaja. The importance of patient privacy during a clinical examination. *Indian Journal of Medical Ethics* 2007 April-June; 4(2): 65. Subject: 8.4

Thachuk, Angela. The space in between: narratives of silence and genetic terminations. *Bioethics* 2007 November; 21(9): 511-514. Subject: 15.2

Tham, Joseph. Bioethics and anointing of the sick. *Linacre Quarterly* 2007 August; 74(3): 253-257. Subject: 2.1

Thamer, Mae; Zhang, Yi. Dialysis facility ownership and epoetin dosing in patients receiving hemodialysis: the authors respond. *American Journal of Kidney Diseases* 2007 October; 50(4): 538-541. Subject: 19.3

Theofrastous, Theodore C. Session 8: Canada and U.S. approaches to health care: how the Canadian and U.S. political, regulatory, and legal systems impact health care. *Canadian - United States Law Journal* 2005; 31: 269-280. Subject: 9.7

Thiele, Felix. Bioethics: its foundation and application in political decision making. *In:* Machamer, Peter; Wolters, Gereon, eds. Science, Values, and Objectivity. Pittsburgh, PA: University of Pittsburgh Press, 2004: 256-274. Subject: 2.1

Thomas, Charles R., Jr.; Pinto, Harlan A.; Roach, Mack, III; Vaughn, Clarence B. Participation in clinical trials: is it state-of the art treatment for African Americans and other people of color? *Journal of National Medical Association* 1994; 86(3): 177-182. Subject: 18.5.1

Thomas, Cordelia. Public dialogue and xenotransplantation. *Medicine and Law: The World Association for Medical Law* 2007 December; 26(4): 801-815. Subject: 2.4

Thomas, Florian P.; Beres, Alana; Shevell, Michael I. "A cold wind coming": Heinrich Gross and child euthanasia in Vienna. *Journal of Child Neurology* 2006 April; 21(4): 342-348. Subject: 20.5.2

Thomas, George. Response: such neat resolutions are not possible in India. *Indian Journal of Medical Ethics* 2007 January-March; 4(1): 34. Subject: 20.5.1

Thomas, N.; Murray, E.; Rogstad, K.E. Confidentiality is essential if young people are to access sexual health services. *International Journal of STD and AIDS* 2006 August; 17(8): 525-529. Subject: 8.4

Thomasma, David C. Virtue theory in philosophy of medicine. *In:* Khushf, George, ed. Handbook of Bioethics: Taking Stock of the Field From a Philosophical Perspective. Dordrecht; Boston: Kluwer Academic, 2004: 89-120. Subject: 2.1

Thompson, Clive. How to farm stem cells without losing your soul. *Wired Magazine* 2005 June; 13.06: 4 p. [Online]. Accessed: http://www.wired.com/wired/archive/13.06/stemcells.html [2006 October 12]. Subject: 18.5.4

Thompson, Paul B. Research ethics for animal biotechnology. *In:* Korthals, Michiel; Bogers, Robert J., eds. Ethics for Life Scientists. Dordrecht, The Netherlands: Springer, 2004: 105-120. Subject: 22.2

Thompson, Richard E. Look what's happened to medical ethics: broader horizons, updated ideas, fresh language. *Physician Executive* 2006 March-April; 32(2): 60-62. Subject: 2.1

Thomson, Mary M. Bringing research into therapy: liability anyone? *In:* Lemmens, Trudo; Waring, Duff R., eds. Law and Ethics in Biomedical Research: Regulation, Conflict of Interest and Liability. Toronto; Buffalo: University of Toronto Press, 2006: 183-205. Subject: 15.4

Thornton, Tim. Tacit knowledge as the unifying factor in evidence based medicine and clinical judgement. *Philosophy, Ethics, and Humanities in Medicine [electronic]* 2006; 1(2): 10 p. Accessed: http://www.peh-med.com/content/1/1/2 [2006 June 15]. Subject: 7.1

Thrush, Carol R.; Vander Putten, Jim; Rapp, Carla Gene; Pearson, L. Carolyn; Berry, Katherine Simms; O'Sullivan, Patricia S. Content validation of the Organizational Climate for Research Integrity (OCRI) Survey. *Journal of Empirical Research on Human Research Ethics* 2007 December; 2(4): 35-52. Subject: 1.3.9

Thurs, Daniel Patrick. No longer academic: models of commercialization and the construction of a nanotech industry. *Science as Culture* 2007 June; 16(2): 169-186. Subject: 5.4

Tice, Martha A. Patient safety: honoring advanced directives. *Home Healthcare Nurse* 2007 February; 25(2): 79-81. Subject: 20.5.4

Tierney, John. Are scientists playing God? It depends on your religion. *New York Times* 2007 November 20; p. F1, F2. Subject: 1.3.9

Tod, A.M.; Nicolson, P.; Allmark, P. Ethical review of health service research in the UK: implications for nursing. *Journal of Advanced Nursing* 2002 November; 40(4): 379-386. Subject: 18.2

Todres, Les; Galvin, Kathleen; Dahlberg, Karin. Lifeworld-led healthcare: revisiting a humanizing philosophy that integrates emerging trends. *Medicine, Health Care and Philosophy* 2007 March; 10(1): 53-63. Subject: 9.1

Toiviainen, Leila. 'The Globalisation of Nursing: Ethical, Legal and Political Issues' University of Surrey 10-11 July 2006: a summary of the deliberations of the concurrent working groups. *Nursing Ethics* 2007 March; 14(2): 258-263. Subject: 4.1.3

See SUBJECT HEADING KEY FOR SECTION II on inside back cover.

613

Toliušiene, Jolanta; Peicius, Eimantas. Changes in nursing ethics education in Lithuania. *Nursing Ethics* 2007 November; 14(6): 753-757. Subject: 7.2

Tollefsen, Christopher. Religious reasons and public healthcare deliberations. *Christian Bioethics* 2007 May-August; (13)2: 139-157. Subject: 2.1

Tollefsen, Christopher. Sic et non: some disputed questions in reproductive ethics. *In:* Khushf, George, ed. Handbook of Bioethics: Taking Stock of the Field From a Philosophical Perspective. Dordrecht; Boston: Kluwer Academic, 2004: 381-413. Subject: 14.1

Tomasini, Floris. Imagining human enhancement: whose future, which rationality? *Theoretical Medicine and Bioethics* 2007; 28(6): 497-507. Subject: 4.5

Tomes, Nancy. Patient empowerment and the dilemmas of late-modern medicalisation. *Lancet* 2007 February 24-March 2; 369(9562): 698-700. Subject: 4.2

Tomlinson, Thomas. Futility beyond CPR: the case of dialysis. *HEC (Healthcare Ethics Committee) Forum* 2007 March; 19(1): 33-43. Subject: 20.5.1

Tomossy, George F.; Ford, Jolyon. Globalization and clinical trials: compensating subjects in developing countries. *In:* Bennett, Belinda; Tomossy, George F., eds. Globalization and Health: Challenges for Health Law and Bioethics. Dordrecht: Springer, 2006: 27-45. Subject: 18.5.9

Tonelli, Mark R. What medical futility means to clinicians. *HEC (Healthcare Ethics Committee) Forum* 2007 March; 19(1): 83-93. Subject: 20.5.1

Tong, Rosemarie. Feminist approaches to bioethics. *In:* Khushf, George, ed. Handbook of Bioethics: Taking Stock of the Field From a Philosophical Perspective. Dordrecht; Boston: Kluwer Academic, 2004: 143-161. Subject: 2.1

Tong, Rosemarie. Gender-based disparities east/west: rethinking the burden of care in the United States and Taiwan. *Bioethics* 2007 November; 21(9): 488-499. Subject: 9.5.5

Tong, Rosemarie. Out of body gestation: in whose best interests? *In:* Gelfand, Scott; Shook, John R., eds. Ectogenesis: Artificial Womb Technology and the Future of Human Reproduction. Amsterdam; New York: Editions Rodopi, B.V., 2006: 59-76. Subject: 14.1

Tong, Rosemarie. Out-of-body gestation: in whose best interests? *Philosophy in the Contemporary World* 2004 Spring-Summer; 11(1): 65-74. Subject: 14.1

Tong, Rosemarie. Traditional and feminist bioethical perspectives on gene transfer: is inheritable genetic modification really the problem? *In:* Rasko, John E.J.; O'Sullivan, Gabrielle M.; Ankeny, Rachel A., eds. The Ethics of Inheritable Genetic Modification: A Dividing Line? Cambridge: Cambridge University Press, 2006: 159-173. Subject: 15.4

Tong, Rosemary. Stem-cell research and the affirmation of life. *Conscience* 2007 Autumn; 28(3): 19-23. Subject: 18.5.4

Tonks, Alison. Physician assisted deaths: no "slippery slope" in the Netherlands and Oregon. *BMJ: British Medical Journal* 2007 May 19; 334(7602): 1029. Subject: 20.7

Tonti-Filippini, Nicholas. Reproductive discrimination. *University of New South Wales Law Journal* 2006; 29(2): 254-260. Subject: 15.2

Tonti-Filippini, Nicholas. The need for ethics committees, and their role and function. *National Catholic Bioethics Quarterly* 2007 Winter; 7(4): 749-769. Subject: 18.2

Toop, Les; Mangin, Dee. Industry funded patient information and the slippery slope to New Zealand. *BMJ: British Medical Journal* 2007 October 6; 335(7622): 694-695. Subject: 9.7

Torjuul, Kirsti; Elstad, Ingunn; Sørlie, Venke. Compassion and responsibility in surgical care. *Nursing Ethics* 2007 July; 14(4): 522-534. Subject: 4.1.3

Torke, Alexia M.; Alexander, G. Caleb; Lantos, John; Siegler, Mark. The physician-surrogate relationship. *Archives of Internal Medicine* 2007 June 11; 167(11): 1117-1121. Subject: 8.3.3

Torrance, Rebecca J.; Lasome, Caterina E.M.; Agazio, Janice B. Ethics and computer-mediated communication: implications for practice and policy. *JONA: The Journal of Nursing Administration* 2002 June; 32(6): 346-353. Subject: 1.3.12

Torres, Mary Ann. The human right to health, national courts, and access to HIV/AIDS treatment: a case study from Venezuela. *In:* Gruskin Sofia; Grodin, Michael A.; Annas, George J.; Marks, Stephen P., eds. Perspectives on health and Human Rights. New York: Routledge, 2005: 507-516. Subject: 9.2

Toth-Fejel, Tihamer; Dodsworth, Chris; Lahl, Jennifer. Syntactic measures of bias (and a perspective on the essential issue of bioethics). *American Journal of Bioethics* 2007 October; 7(10): 40-42. Subject: 2.1

Touitou, Yvan.; Smolensky, Michael H.; Portaluppi, Francesco. Ethics, standards, and procedures of animal and human chronobiology research. *Chronobiology International* 2006; 23(6): 1083-1096. Subject: 22.1

Toumey, Chris. Privacy in the shadow of nanotechnology. *NanoEthics* 2007 December; 1(3): 211-222. Subject: 5.4

Tovino, Stacey A. Functional neuroimaging and the law: trends and directions for future scholarship. *American Journal of Bioethics* 2007 September; 7(9): 44-56. Subject: 17.4

Trachtman, Howard; Henry, Michael; Fishman, Jennifer R.; Youngner, Stuart J. Spinoza's passions. *American Journal of Bioethics* 2007 September; 7(9): 21-23; author reply W1-W3. Subject: 17.4

Trachtman, Howard; Volandes, Angelo E.; Paasche-Orlow, Michael K. Illiteracy ain't what it used to be. *American Journal of Bioethics* 2007 November; 7(11): 27-28; author reply W1-W2. Subject: 9.1

Travaline, John M. Medicine: Notes and abstracts. *National Catholic Bioethics Quarterly* 2007 Winter; 7(4): 793-808. Subject: 9.1

Travaline, John M. Understanding brain death diagnosis — II. *Ethics and Medics* 2007 April; 32(4): 3-4. Subject: 20.2.1

Treloar, Susan A.; Morley, Katherine I.; Taylor, Sandra D.; Hall, Wayne D. Why do they do it? A pilot study towards understanding participant motivation and experience in a large genetic epidemiological study of endometriosis. *Community Genetics* 2007; 10(2): 61-71. Subject: 15.1

Triggle, David J. Treating desires not diseases: a pill for every ill and an ill for every pill? *Drug Discovery Today* 2007 February; 12(3-4): 161-166. Subject: 9.7

Triner, Wayne; Jacoby, Liva; Shelton, Wayne; Burk, Mathew; Imarenakhue, Samual; Watt, James; Larkin, Gregory; McGee, Glenn. Exception from informed consent enrollment in emergency medical research: attitudes and awareness. *Academic Emergency Medicine* 2007 February; 14(2): 187-191. Subject: 8.3.1

Trivedi, Bijal. Researchers detour around stem-cell rules. *Chronicle of Higher Education* 2007 October 3; 54(6): A12-A15. Subject: 18.5.4

Trost, Bernd. Ethische Probleme der Pflegenden im Altenpflegeheim - Nachdenkliches zur Psychopharmakaverordnung aus Sicht eines Heimleitenden = Ethical questions in a long term care nursing home — Reflections of an administrator considering psychotropic prescriptions. *Ethik in der Medizin* 2007 December; 19(4): 281-288. Subject: 17.4

Trotter, Griffin. Editorial introduction: futility in the 21st century. *HEC (Healthcare Ethics Committee) Forum* 2007 March; 19(1): 1-12. Subject: 20.5.1

Trotter, Griffin. Left bias in academic bioethics. *In:* Eckenwiler, Lisa A.; Cohn, Felicia G., eds. The Ethics of Bioethics: Mapping the Moral Landscape. Baltimore, MD: Johns Hopkins University Press, 2007: 108-117. Subject: 2.1

Truog, Robert D. Tackling medical futility in Texas. *New England Journal of Medicine* 2007 July 5; 357(1): 1-3. Subject: 20.5.2

Tsaloglidou, Areti; Rammos, Kyriakos; Kiriklidis, Konstantinos; Zourladani, Athanasia; Matziari, Chrysoula. Nurses' ethical decision-making role in artificial nutritional support. *British Journal of Nursing* 2007 September 13-27; 16(16): 996-998. Subject: 20.5.1

Tschudin, Verena. Narrative ethics. *In her:* Approaches to Ethics: Nursing Beyond Boundaries. New York: Butterworth-Heinemann, 2003: 61-72. Subject: 4.1.3

Tsuneishi, Kei-chi. Unit 731 and the human skulls discovered in 1989: physicians carrying out organized crimes. *In:* LaFleur, William R.; Böhme, Gernot; Shimazono, Susumu, eds. Dark Medicine: Rationalizing Medical Research. Bloomington: Indiana University Press, 2007: 73-84. Subject: 18.5.1

Tucker, Jonathan B.; Hooper, Craig. Protein engineering: security implications; the increasing ability to manipulate protein toxins for hostile purposes has prompted calls for regulation. *EMBO Reports* 2006 July; 7 Special No: S14-S17. Subject: 5.3

Tucker, Kathryn L. Federalism in the context of assisted dying: time for the laboratory to extend beyond Oregon, to the neighboring state of California. *Willamette Law Review* 2005; 41(5): 863-880. Subject: 20.7

Tucker, Kathryn L. Privacy and dignity at the end of life: protecting the right of Montanans to choose aid in dying. *Montana Law Review* 2007 Summer; 68(2): 317-333. Subject: 20.5.1

Tuffrey-Wijne, Irene; Bernal, Jane; Butler, Gary; Hollins, Sheila; Curfs, Leopold. Using nominal group technique to investigate the views of people with intellectual disabilities on end-of-life care provision. *Journal of Advanced Nursing* 2007 April; 58(1): 80-89. Subject: 20.4.1

Tuffs, Annette. Doctors protest about fetal sex tests in early pregnancy. *BMJ: British Medial Journal* 2007 April 7; 334(7596): 712. Subject: 14.3

Tuffs, Annette. German council demands opt-out system for transplants [news]. *BMJ: British Medical Journal* 2007 May 12; 334(7601): 973. Subject: 19.5

Tuffs, Annette. German doctors may have to report patients who have piercings and beauty treatments [news]. *BMJ: British Medical Journal* 2007 November 3; 335(7626): 905. Subject: 4.5

Tuffs, Annette. German doctors: public enemy number one? *BMJ: British Medical Journal* 2007 May 26; 334(7603): 1087. Subject: 8.1

Tuffs, Annette. Media claim allocations of organs to Saudi patients was unfair [news]. *BMJ: British Medical Journal* 2007 September 29; 335(7621): 634. Subject: 19.6

Tuffs, Annette. Swiss hospitals admit to allowing assisted suicide on their wards under guidelines [news]. *BMJ: British Medical Journal* 2007 November 24; 335(7629): 1064-1065. Subject: 20.7

Tuohey, John F. A matrix for ethical decision making in a pandemic: the Oregon Tool for emergency preparedness. *Health Progress* 2007 November-December; 88(6): 20-25. Subject: 9.1

See SUBJECT HEADING KEY FOR SECTION II on inside back cover.

615

Tuohey, John F. Making access a priority: ethics has a vital role in fostering collaboration in health care. *Health Progress* 2007 March-April; 88(2): 67-72. Subject: 9.2

Tuohey, John F. Screening for aneuploidy: a complex ethical issue. *Health Care Ethics USA* 2007 Spring; 15(2): 4-8. Subject: 15.2

Tupasela, A. When legal worlds collide: from research to treatment in hereditary cancer prevention. *European Journal of Cancer Care* 2006 July; 15(3): 257-266. Subject: 9.5.1

Turale, Sue. Reflections on the ethics involved in international research. *Nursing and Health Sciences* 2006 September; 8(3): 131-132. Subject: 18.2

Turillazzi, E.; Fineschi, V. Female genital mutilation: the ethical impact of the new Italian law. *Journal of Medical Ethics* 2007 February; 33(2): 98-101. Subject: 9.5.5

Turner, Leigh. Global health inequalities and bioethics. *In:* Eckenwiler, Lisa A.; Cohn, Felicia G., eds. The Ethics of Bioethics: Mapping the Moral Landscape. Baltimore, MD: Johns Hopkins University Press, 2007: 229-240. Subject: 21.1

Turney, Jon. Inhuman, superhuman, or posthuman? Images of genetic futures. *In:* Einsiedel, Edna; Timmermans, Frank, eds. Crossing Over: Genomics in the Public Arena. Calgary, Alberta, Canada: University of Calgary Press, 2005: 225-235. Subject: 15.1

Turney, Lyn. Essentially whose? Genetic testing and the ownership of genetic information. *In:* Betta, Michela, ed. The Moral, Social, and Commercial Imperatives of Genetic Testing and Screening: the Australian Case. Dordrecht: Springer, 2006: 237-245. Subject: 15.3

Turone, Fabio. Court upholds demand for preimplantation genetic diagnosis [news]. *BMJ: British Medical Journal* 2007 October 6; 335(7622): 687. Subject: 15.2

Turone, Fabio. Doctor helps Italian patient die [news]. *BMJ: British Medical Journal* 2007 January 6; 334(7583): 9. Subject: 20.7

Turone, Fabio. New reproduction law reduces success rate [news]. *BMJ: British Medical Journal* 2007 July 14; 335(7610): 62. Subject: 14.4

Turpin, David L. Policies for biomedical journals address ethics, confidentiality, and corrections. *American Journal of Orthodontics and Dentofacial Orthopedics* 2006 December; 130(6): 693-695. Subject: 1.3.7

Turrens, Julio F. Teaching research integrity and bioethics to science undergraduates. *Cell Biology Education* 2005 Winter; 4(4): 330-334. Subject: 2.3

Turton, Frederick E.; Snyder, Lois. Physician-industry relations [letter]. *Annals of Internal Medicine* 2007 March 20; 146(6): 469. Subject: 7.3

Twine, Richard. Animal genomics and ambivalence: a sociology of animal bodies in agricultural biotechnology.

Genomics, Society and Policy 2007 August; 3(2): 99-117. Subject: 15.1

Twine, Richard. Thinking across species — a critical bioethics approach to enhancement. *Theoretical Medicine and Bioethics* 2007; 28(6): 509-523. Subject: 4.5

Twombly, Renee. Goal of maintaining public's trust brings research groups together on conflict-of-interest guidelines [news]. *Journal of the National Cancer Institute* 2005 November 2; 97(21): 1560-1561. Subject: 1.3.9

Tyshenko, Michael G. Management of natural and bioterrorism induced pandemics. *Bioethics* 2007 September; 21(7): 364-369. Subject: 21.3

U

Ubel, Peter A. Confessions of a bedside rationer: commentary on Hurst and Danis. *Kennedy Institute of Ethics Journal* 2007 September; 17(3): 267-269. Subject: 9.4

Udesky, Laurie. Push to mandate HPV vaccine triggers backlash in USA. *Lancet* 2007 March 24-30; 369(9566): 979-980. Subject: 9.5.7

Uhl, K.; Parekh, A.; Kweder, S. Females in clinical studies: where are we going? *Clinical Pharmacology and Therapeutics* 2007 April; 81(4): 600-602. Subject: 18.5.3

Uhrenfeldt, Lisbeth; Hall, Elisabeth O.C. Clinical wisdom among proficient nurses. *Nursing Ethics* 2007 May; 14(3): 387-398. Subject: 4.1.3

UK Biobank. UK Biobank ethics and governance framework [EGF]: Version 2.0. London: UK Biobank, 2006 July; 20 p. [Online]. Accessed: http://www.ukbiobank. ac.uk/docs/EGF_Version2_July%2006%20most%20 uptodate.pdf [2007 April 3]. Subject: 15.1

UK Cystic Fibrosis Database Steering Committee; Sims, Erika J.; Mugford, Miranda; Clark, Allan; Aitken, David; McCormick, Jonathan; Mehta, Gita; Mehta, Anil. Economic implications of newborn screening for cystic fibrosis: a cost of illness retrospective cohort study. *Lancet* 2007 April 7-13; 369(9568): 1187-1195. Subject: 15.3

Underwood, J.C.E. The impact on histopathology practice of new human tissue legislation in the UK. *Histopathology* 2006 September; 49(3): 221-228. Subject: 19.5

United Nations Educational, Scientific and Cultural Organisation [UNESCO]. Division of Ethics of Science and Technology. Guide No.2: bioethics committees at work: procedures and policies. Paris, France: United Nations Educational, Scientific and Cultural Organisation, Division of Ethics of Science and Technology, 2005; 72 p. [Online]. Accessed: http://unesdoc.unesco.org/images/ 0014/001473/147392e.pdf [2007 April 19]. Subject: 2.4

United Nations. Committee on Economic, Social, and Cultural Rights. General comment no. 14 (2000): the right to the highest attainable standard of health (Article 12 of the International Covenant on Economic, Social, and

---

Cultural Rights). *In:* Gruskin Sofia; Grodin, Michael A.; Annas, George J.; Marks, Stephen P., eds. Perspectives on health and Human Rights. New York: Routledge, 2005: 473-495. Subject: 9.2

United Network for Organ Sharing [UNOS]. Board of Directors. Directed Donation. Richmond, VA: United Network for Organ Sharing, 1996 June: 2 p. [Online]. Accessed: http://www.unos.org/Resources/bioethics.asp?index=10 [2007 May 8]. Subject: 19.5

United Network for Organ Sharing [UNOS]. Ethics Committee. Preferred status for organ donors: a report of the United Network for Organ Sharing Ethics Committee. Richmond, VA: United Network for Organ Sharing [UNOS], 1993 June 30: 4 p. [Online]. Accessed: http://www.unos.org/Resources/bioethics.asp?index=5 [2007 May 8]. Subject: 19.6

United Network for Organ Sharing [UNOS]. Ethics Committee, Organ Procurement and Transplantation Network [OPTN]. Allocation of Organs from Non-Directed Living Donors. Richmond, VA: United Network for Organ Sharing, 2002 June: 1 p. [Online]. Accessed: http://www.unos.org/Resources/bioethics.asp?index=9 [2007 May 8]. Subject: 19.6

United Network for Organ Sharing [UNOS]. Ethics Committee. Payment Subcommittee. Financial incentives for organ donation. A report of the payment subcommittee. Richmond, VA: United Network for Organ Sharing, 1993 June 30: 4 p. [Online]. Accessed: http://www.unos.org/Resources/bioethics.asp?index=3 [2007 May 8]. Subject: 19.5

United Network for Organ Sharing [UNOS]. Research to Practice Steering Committee; Metzger, Robert A.; Taylor, Gloria J.; McGaw, Lin J.; Weber, Phyllis G.; Delmonico, Francis L.; Prottas, Jeffrey M. Research to practice: a national consensus conference. *Progress in Transplantation* 2005 December; 15(4): 379-384. Subject: 19.5

United States. Congress. House. A bill to amend title 35, United States Code, to prohibit the patenting of human genetic material. Washington, DC: U.S. G.P.O., 2007. 15 p. [Online]. Accessed: http://frwebgate.access.gpo.gov/cgi-bin/useftp.cgi?IPaddress=162.140.64.21&filename=h977ih.pdf&directory=/diskb/wais/data/110_cong_bills [2007 April 4]. Subject: 15.8

United States. Congress. House. A bill to derive human pluripotent stem cell lines using techniques that do not harm human embryos. Washington, DC: U.S. G.P.O., 2007. 4 p. [Online]. Accessed: http://frwebgate.access.gpo.gov/cgi-bin/useftp.cgi?IPaddress=162.140.64.21&filename=h322ih.pdf&directory=/diskb/wais/data/110_cong_bills [17 January 2007]. Subject: 18.5.4

United States. Congress. House. A bill to prohibit discrimination on the basis of genetic information with respect to health insurance and employment. Washington,

DC: U.S. G.P.O., 2006. 81 p. [Online]. Accessed: http://frwebgate.access.gpo.gov/cgi-bin/useftp.cgi?IPaddress=162.140.64.21&filename=h493ih.pdf&directory=/diskb/wais/data/110_cong_bills [2007 January 20]. Subject: 15.3

United States. Congress. House. An act to amend the Public Health Service Act to provide for human embryonic stem cell research. Washington, DC: U.S. G.P.O., 2007. 4 p. [Online]. Accessed: http://frwebgate.access.gpo.gov/cgi-bin/useftp.cgi?IPaddress=162.140.64.21&filename=h3eh.pdf&directory=/diskb/wais/data/110_cong_bills [2007 January 17]. Subject: 18.5.4

United States. Congress. Senate. A bill to amend the Public Health Service Act to provide for human embryonic stem cell research. Washington, DC: U.S. G.P.O., 2007. 3 p. [Online]. Accessed: http://frwebgate.access.gpo.gov/cgi-bin/useftp.cgi?IPaddress=162.140.64.21&filename=s5pcs.pdf&directory=/diskb/wais/data/110_cong_bills [2007 April 10]. Subject: 18.5.4

United States. Congress. Senate. A bill to derive human pluripotent stem cell lines using techniques that do not knowingly harm embryos. Washington, DC: U.S. G.P.O., 2007. 5 p. [Online]. Accessed: http://frwebgate.access.gpo.gov/cgi-bin/useftp.cgi?IPaddress=162.140.64.21&filename=s5lis.pdf&directory=/diskb/wais/data/110_cong_bills [2007 January 17]. Subject: 18.5.4

United States. Congress. Senate. A bill to intensify research to derive human pluripotent stem cell lines. Washington, DC: U.S. G.P.O., 2007. 6 p. [Online]. Accessed: http://frwebgate.access.gpo.gov/cgi-bin/useftp.cgi?IPaddress=162.140.64.21&filename=s30hds.pdf&directory=/diskb/wais/data/110_cong_bills [2007 April 10]. Subject: 18.5.4

United States. Congress. Senate. A bill to prohibit discrimination on the basis of genetic information with respect to health insurance and employment. Washington, DC: U.S. G.P.O., 2007. 80 p. [Online]. Accessed: http://frwebgate.access.gpo.gov/cgi-bin/useftp.cgi?IPaddress=162.140.64.21&filename=s358is.pdf&directory=/diskb/wais/data/110_cong_bills [2007 February 20]. Subject: 15.3

United States. Congress. Senate. A bill to provide increased Federal funding for stem cell research, to expand the number of embryonic stem cell lines available for Federally funded research, to provide ethical guidelines for stem cell research, to derive human pluripotent stem cell lines using techniques that do not create an embryo or embryos for research or knowingly harm embryos, and for other purposes. Washington, DC: U.S. G.P.O., 2007. 15 p. [Online]. Accessed: http://frwebgate.access.gpo.gov/cgi-bin/useftp.cgi?IPaddress=162.140.64.21&filename=s363is.pdf&directory=/diskb/wais/data/110_cong_bills [2007 April 4]. Subject: 18.5.4

United States. Congress. Senate. A bill to secure the promise of personalized medicine for all Americans by ex-

See SUBJECT HEADING KEY FOR SECTION II on inside back cover.

617

panding and accelerating genomics research and initiatives to improve the accuracy of disease diagnosis, increase the safety of drugs, and identify novel treatments. Washington, DC: U.S. G.P.O., 2007. 32 p. [Online]. Accessed: http://frwebgate.access.gpo.gov/cgi-bin/useftp.cgi? IPaddress=162.140.64.21&filename=s976is.pdf& directory=/diskb/wais/data/110_cong_bills [2007 April 4]. Subject: 15.1

United States. Department of Defense [DOD]. Medical program support for detainee operations. Washington, DC: Department of Defense, [2310.08E], 2006 June 6; 10 p. [Online]. Accessed: http://www.dtic.mil/whs/directives/corres/pdf/ 231008p.pdf [2007 April 5]. Subject: 9.5.1

United States. Department of Education [DOE]. Protection of human subjects; proposed rule. *Federal Register* 1997 May 22; 62(99): 28155-28159 [Online]. Accessed: http:// frwebgate.access.gpo.gov/cgi-bin/multidb.cgi [2005 December 27]. Subject: 18.2

United States. Department of Energy [DOE]. Policy on research misconduct. *Federal Register* 2005 June 28; 70(123): 37010-37016 [Online]. Accessed: http://a257.g. akamaitech.net/7/257/2422/01jan20051800/edocket. access.gpo.gov/20 05/pdf/05-12645.pdf [2007 April 24]. Subject: 1.3.9

United States. Department of Health and Human Services [DHHS]. Protecting Personal Health Information in Research: Understanding the HIPAA Privacy Rule. Washington, DC: Department of Health and Human Services (HHS), 2003: 33 p. Subject: 18.2

United States. Department of Health and Human Services [DHHS]. Protection of human research subjects; notice of proposed rule making. *Federal Register* 2001 July 6; 66(130): 35576-35580 [Online]. Accessed: http:// frwebgate.access.gpo.gov/cgi-bin/multidb.cgi [2005 December 27]. Subject: 18.2

United States. Department of Health and Human Services [DHHS]; United States. Centers for Medicare and Medicaid Services. Medicare and Medicaid programs; hospital conditions of participation: patients' rights. Final rule. *Federal Register* 2006 December 8; 71(236): 71377-71428. Sub- ject: 9.3.1

United States. Department of Health and Human Services [DHHS]; United States. Centers for Medicare and Medicaid Services. Medicare and Medicaid programs; conditions for coverage for organ procurement organizations (OPOs). Final rule. *Federal Register* 2006 May 31; 71(104): 30981-31054. Subject: 19.5

United States. Department of Health and Human Services [DHHS]; United States. Centers for Medicare and Medicaid Services. Medicare program; physicians referrals to health care entities with which they have financial relationships; exceptions for certain electronic prescribing and electronic health records arrangements; final rule. *Federal Register* 2006 August 8; 71(152): 45139-45171. Subject: 9.3.1

United States. Department of Health and Human Services [DHHS]. Office of Research Integrity. Final rule on research misconduct: frequently asked questions and answers. Rockville, MD: Office of Research Integrity, 2005 June 13; 4 p. [Online]. Accessed: http://ori.dhhs.gov/policies/faq.shtml [2007 April 24]. Subject: 1.3.9

United States. Department of Health and Human Services [DHHS]. Office of the Secretary. Protection of human research subjects: delay of effective date. *Federal Register* 2001 March 19; 66(53): 15352 [Online]. Accessed: http:// frwebgate.access.gpo.gov/cgi-bin/multidb.cgi [2005 December 27]. Subject: 18.2

United States. Department of Health and Human Services [DHHS]. Office for Human Research Protections [OHRP]. Guidance on reporting and reviewing adverse events and unanticipated problems involving risks to subjects or others: Draft - October 11, 2005. Rockville, MD: Office for Human Research Protections, 2005 October 11: 25 p. [Online]. Accessed: http://www.hhs.gov/ohrp/requests/aerg.pdf [2007 March 10]. Subject: 18.2

United States. Department of Health and Human Services [DHHS]. Office for Human Research Protections [OHRP]. Guidance on reporting incidents to OHRP. Washington, DC: Office for Human Research Protections, 2005 May 27; 6 p. [Online]. Accessed: http://www.hhs. gov/ohrp/policy/incidreport_ohrp.html [2007 April 24]. Subject: 18.6

United States. Department of Health and Human Services [DHHS]. Office for Human Research Protections [OHRP]. Part 46 - Protection of human subjects [Revised 2005 June 23; Effective 2005 June 23]. Code of Federal Regulations: Title 45: Public Welfare 2005 October 1: 117-134 [Online]. Accessed: http://www.hhs.gov/ohrp/humansubjects/guidance/45cfr46.htm [2007 April 25]. Subject: 18.2

United States. Department of Health and Human Services [DHHS]. Office for Human Research Protections [OHRP]. Special protections for children as research subjects. Children involved as subjects in research: guidance on the HHS 45CFR 46.407 ("407") review process. Washington, DC: Office for Human Research Protections, 2005 May 26; 9 p. [Online]. Accessed: http://www.hhs.gov/ohrp/children/guidance_407process.html [2007 April 24]. Subject: 18.5.2

United States. Department of Health and Human Services [DHHS]. Office of Inspector General. Medicare and state health care programs: fraud and abuse; safe harbors for certain electronic prescribing and electronic health records arrangements under the anti-kickback statute; final rule. *Federal Register* 2006 August 8; 71(152): 45109-45137. Subject: 9.3.1

United States. Environmental Protection Agency [EPA]. Protections for subjects in human research; proposed rule. *Federal Register* 2005 September 12; 70(175): 53837-53866 [Online]. Accessed: http://www.frwebgate. access.gpo.gov/cgi-bin/multidb.cgi [2005 December 27]. Subject: 18.2

United States. Environmental Protection Agency [EPA]. Science Advisory Board. Comments on the use of Data from the Testing of Human Subjects: A Report by the Science Advisory Board and the FIFRA Scientific Advisory Panel. Washington, DC: Environmental Protection Agency, 2000 September 11: 40 p. [Online]. Accessed: http://www.epa.gov/sab/pdf/ec0017.pdf [2007 October 24]. Subject: 18.1

United States. Food and Drug Administration [FDA]. Expanded access to investigational drugs for treatment use. *Federal Register* 2006 December 14; 71(240): 75147-75168 [Online]. Accessed: http://www.fda.gov/ OHRMS/DOCKET/98fr/06-9684.pdf [2007 October 10]. Subject: 9.7

United States. Food and Drug Administration [FDA]. Human drugs and biologics; determination that informed consent is not feasible or is contrary to the best interests of recipients; revocation of 1990 interim final rule; establishment of new interim final rule. *Federal Register* 1999 October 5; 64(192): 54180-54189 [Online]. Accessed: http://www.fda.gov/oc/gcp/preambles/64fr54180.html [2005 December 27]. Subject: 18.5.1

United States. Food and Drug Administration [FDA]. Requirements on content and format of labeling for human prescription drug and biological products and draft guidances and two guidances for industry on the content and format of labeling for human prescription drug and biological products; final rule and notices [21 CFR Parts 201, 314, and 601]. *Federal Register* 2006 January 24; 71(15): 3922-3997. Subject: 9.7

United States. Food and Drug Administration [FDA]. Strengthening the regulation of clinical trials and bioresearch monitoring. *FDA Consumer* 2006 November-December; 40(6): 35. Subject: 18.2

United States. Food and Drug Administration [FDA]. Center for Drug Evaluation and Research [CDER], Office of the Commissioner; Center for Biologics Evaluation and Research [CBER]; Center for Devices and Radiological Health [CDRH];Good Clinical Practice Program [GCPP] (United States). Adverse Event Reporting — Improving Human Subject Protection. Guidance for Clinical Investigators, Sponsors and IRBs [draft guidance]. Rockville, MD: Food and Drug Administration 2007 April: 8 p. [Online]. Accessed: http://www. clinicalresearchresources.com/images/ fdaguidanceadvreport.pdf [2007 September 12]. Subject: 18.2

United States. Food and Drug Administration [FDA]. Center for Drug Evaluation and Research [CDER];

Center for Biologics Evaluation and Research [CBER]; Center for Devices and Radiological Health [CDRH] (United States). Protecting the Rights, Safety, and Welfare of Study Subjects — Supervisory Responsibilities of Investigators. Guidance for Industry [draft guidance]. Rockville, MD: Food and Drug Administration 2007 May: 16 p. [Online]. Accessed: http://www. clinicalresearchresources.com/images/ fdaguidancestudysub.pdf [2007 September 12]. Subject: 18.2

United States. Office of the President. Fact Sheet: Embryonic Stem Cell Research. *Washington, DC: The White House, Office of the Press Secretary,* 2001 August 9; 3 p. [Online]. Accessed: http://www.whitehouse.gov/news/releases/2001/ 08/text/20010809-1.html [2001 August 10]. Subject: 18.5.4

United States. Office of the President. Domestic Policy Council. Advancing stem cell science without destroying human life. Washington, DC: The White House, 2007 January 9: 64 p. [Online]. Accessed: http://www.whitehouse. gov/infocus/healthcare/stemcell_010907.pdf [2007 January 24]. Sub- ject: 18.5.4

United States. Supreme Court. U.S. Supreme Court partial-birth abortion decision. *Origins* 2007 May 3; 36(46): 749-753. Subject: 12.4.4

United States. Veterans Health Administration. National Ethics Committee. The ethics of palliative sedation as a therapy of last resort. *American Journal of Hospice and Palliative Care* 2006 December-2007 January; 23(6): 483-491. Sub- ject: 20.4.1

Upshur, Ross E.G.; Lavery, James V.; Tindana, Paulina O. Taking tissue seriously means taking communities seriously. *BMC Medical Ethics* 2007; 8(11): 1-6 [Online]. Accessed: http://www.biomedcentral.com/content/pdf/ 1472-6939-8-11.pdf [2008 January 24]. Subject: 19.5

Upshur, Ross; Buetow, Stephen; Loughlin,Michael; Miles, Andrew. Can academic and clinical journals be in financial conflict of interest situations? The case of evidence-based incorporated. *Journal of Evaluation in Clinical Practice* 2006 August; 12(4): 405-409. Subject: 1.3.7

Uscher-Pines, Lori; Duggan, Patrick S.; Garoon, Joshua P.; Karron, Ruth A.; Faden, Ruth R. Planning for an influenza pandemic: social justice and disadvantaged groups. *Hastings Center Report* 2007 July-August; 37(4): 32-39. Subject: 9.1

V

Vaithianathan, Rhema. Better the devil you know than the doctor you don't: is advertising drugs to doctors more harmful than advertising to patients? *Journal of Health Services Research and Policy* 2006 October; 11(4): 235-239. Subject: 1.3.7

Valaitis, J.A. Cultural diversity in health care: interpersonal and ethical considerations. *WMJ: Official Publica-*

See SUBJECT HEADING KEY FOR SECTION II on inside back cover.

619

tion of the State Medical Society of Wisconsin 2006 June; 105(4): 12-15. Subject: 21.7

Valapour, Maryam. Living donor transplantation: the perfect balance of public oversight and medical responsibility. *Journal of Clinical Ethics* 2007 Spring; 18(1): 18-20. Subject: 19.5

Vallgårda, Signild. When are health inequalities a political problem? *European Journal of Public Health* 2006 December; 16(6): 615-616. Subject: 9.2

van Aken, Jan. When risk outweighs benefit. Dual-use research needs a scientifically sound risk-benefit analysis and legally binding biosecurity measures. *EMBO Reports* 2006 July; 7 Special No: S10-S13. Subject: 5.3

Van Bogaert, Donna Knapp. Ethical considerations in African traditional medicine: a response to Nyika. *Developing World Bioethics* 2007 April; 7(1): 35-40. Subject: 4.1.1

van Bruchem-van de Scheur, G.G.; van der Arend, Arie J.G.; Spreeuwenberg, Cor; Abu-Saad, Huda Huijer; ter Meulen, Ruud H.J. Euthanasia and physician-assisted suicide in the Dutch homecare sector: the role of the district nurse. *Journal of Advanced Nursing* 2007 April; 58(1): 44-52. Subject: 20.5.1

Van Citters, Aricca D.; Naidoo, Umadevi; Foti, Mary Ellen. Using a hypothetical scenario to inform psychiatric advance directives. *Psychiatric Services* 2007 November; 58(11): 1467-1471. Subject: 17.1

van Delden, Johannes J.M. Terminal sedation: source of a restless ethical debate. *Journal of Medical Ethics* 2007 April; 33(4): 187-188. Subject: 20.5.1

van den Belt, Henk. Comments on Bulger: the responsible conduct of research, including responsible authorship and publication practices. *In:* Korthals, Michiel; Bogers, Robert J., eds. Ethics for Life Scientists. Dordrecht, The Netherlands: Springer, 2004: 63-66. Subject: 1.3.9

van den Brink-Muinen, Atie; van Dulmen, Sandra M.; de Haes, Hanneke C.J.M.; Visser, Adriaan Ph.; Schellevis, F.G.; Bensing, J.M. Has patients' involvement in the decision-making process changed over time? *Health Expectations* 2006 December; 9(4): 333-342. Subject: 8.3.1

van den Daele, Wolfgang. Moderne Tabus? — Zum Verbot des Klonens von Menschen. *In:* Schreiber, Hans-Peter, ed. Biomedizin und Ethik: Praxis, Recht, Moral. Basel; Boston: Birkhäuser, 2004: 77-83, 96. Subject: 14.5

van den Hoven, Jeroen; Vermaas, Pieter E. Nano-technology and privacy: on continuous surveillance outside the panopticon. *Journal of Medicine and Philosophy* 2007 May-June; 32(3): 283-297. Subject: 5.4

Van Denend, Toni; Finlayson, Marcia. Ethical decision making in clinical research: application of CELIBATE. *American Journal of Occupational Therapy* 2007 January-February; 61(1): 92-95. Subject: 18.2

van der Heide, Agnes; Onwuteaka-Philipsen, Bregje D.; Rurup, Mette L.; Buiting, Hilde M.; van Delden, Johannes J.M.; Hanssen-de Wolf, Johanna E.; Janssen, Anke G.J.M.; Pasman, H. Roeline W.; Rietjens, Judith A.C.; Prins, Cornelis J.M.; Deerenberg, Ingeborg M.; Gevers, Joseph K.M.; van der Maas, Paul J.; van der Wal, Gerrit. End-of-life practices in the Netherlands under the Euthanasia Act. *New England Journal of Medicine* 2007 May 10; 356(19): 1957-1965. Subject: 20.7

van der Loos, H.F. Machiel. Design and engineering ethics considerations for neurotechnologies. *CQ: Cambridge Quarterly of Healthcare Ethics* 2007 Summer; 16(3): 303-307. Subject: 17.1

van der Wal, Gerrit. Quality of care, patient safety, and the role of the patient. *In:* Gevers, J.K.M.; Hondius, E.H.; Hubben, J.H., eds. Health Law, Human Rights and the Biomedicine Convention: Essays in Honour of Henriette Roscam Abbing. Leiden; Boston: Martinus Nijhoff Publishers, 2005: 77-92. Subject: 9.8

Van Der Weyden, Martin B. Preventing and processing research misconduct: a new Australian code for responsible research: it all depends on compliance. *Medical Journal of Australia* 2006 May 1; 184(9): 430-431. Subject: 1.3.9

van der Zijpp, Akke. Comments on Düwell: research as a challenge for ethical reflection. *In:* Korthals, Michiel; Bogers, Robert J., eds. Ethics for Life Scientists. Dordrecht, The Netherlands: Springer, 2004: 157-159. Subject: 1.3.9

Van Dijk, Yehuda; Sonnenblick, Moshe. Enteral feeding in terminal dementia — a dilemma without a consensual solution. *Israel Medical Association Journal* 2006 July; 8(7): 503-504. Subject: 20.5.1

Van Dooren, Thom. Terminated seed: death, proprietary kinship and the production of (bio)wealth. *Science as Culture* 2007 March; 16(1): 71-93. Subject: 15.1

Van Groenou, Aneema A.; Bakes, Katherine Mary. Art, Chaos, Ethics, and Science (ACES): a doctoring curriculum for emergency medicine. *Annals of Emergency Medicine* 2006 November; 48(5): 532-537. Subject: 7.1

van Haselen, Robbert. Misconduct in CAM research: does it occur? *Complementary Therapies in Medicine* 2006 June; 14(2): 89-90. Subject: 1.3.9

van Hooft, Stan. Caring and ethics in nursing. *In:* Tschudin, Verena, ed. Approaches to Ethics: Nursing Beyond Boundaries. New York: Butterworth-Heinemann, 2003: 1-12. Subject: 4.1.3

van Hooft, Stan. Socratic dialogue: an example. *In:* Tschudin, Verena, ed. Approaches to Ethics: Nursing Beyond Boundaries. New York: Butterworth-Heinemann, 2003: 115-123. Subject: 7.2

Van Hoyweghen, Ine; Horstman, Klasien; Schepers, Rita. Genetic 'risk carriers' and lifestyle 'risk takers'.

Subject = NRCBL Primary Classification Number; see inside front cover.

Which risks deserve our legal protection in insurance? *Health Care Analysis: An International Journal of Health Philosophy and Policy* 2007 September; 15(3): 179-193. Subject: 15.3

van Leeuwen, Evert; Kimsma, Gerrit. Public policy and ending lives. *In:* Rhodes, Rosamond; Francis, Leslie P.; Silvers, Anita, eds. The Blackwell Guide to Medical Ethics. Malden, MA: Blackwell Pub., 2007: 220-237. Subject: 20.1

van Luijn, H.E.M.; Musschenga, A.W.; Keus, R.B.; Aaronson, N.K. Evaluating the risks and benefits of phase II and III clinical cancer trials: a look at institutional review board members in the Netherlands. *IRB: Ethics and Human Research* 2007 January-February; 29(1): 13-17. Subject: 18.2

Van Overwalle, Geertrui; van Zimmeren, Esther; Verbeure, Birgit; Matthijs, Gert. Models for facilitating access to patents on genetic inventions. *Nature Reviews Genetics* 2006 February; 7(2): 143-148. Subject: 15.8

Van Rosendaal, Guido M.A. Queue jumping: social justice and the doctor-patient relationship [editorial]. *Canadian Family Physician* 2006 December; 52(12): 1525-1526. Subject: 9.3.1

van Veen, E.-B.; Riegman, P.H.J.; Dinjens, W.N.M.; Lam, K.H.; Oomen, M.H.A.; Spatz, A.; Mager, R.; Ratcliffe, C.; Knox, K.; Kerr, D.; van Damme, B.; van de Vijver, M.; van Boven, H.; Morente, M.M.; Alonso, S.; Kerjaschki, D.; Pammer, J.; Lopez-Guerrero, J.A.; Llombart Bosch, A.; Carbone, A.; Gloghini, A.; Teodorovic, I.; Isabelle, M.; Passioukov, A.; Lejeune, S.; Therasse, P.; Oosterhuis, J.W. TuBaFrost 3: regulatory and ethical issues on the exchange of residual tissue for research across Europe. *European Journal of Cancer* 2006 November; 42(17): 2914-2923. Subject: 18.2

Van Vleet, Lee M. Between black and white. The gray area of ethics in EMS. *JEMS: A Journal of Emergency Medical Services* 2006 October; 31(10): 55-56, 58-63; quiz 64-65 [see correction in JEMS: A Journal of Emergency Medical Services 2006 December; 31(12): 16]. Subject: 9.5.1

Vanderpool, Harold Y. A revisionist look at the fall and rise of medical ethics [review of Disrupted Dialogue: Medical Ethics and the Collapse of Physician-Humanist Communication (1770-1980), by Robert M. Veatch]. *Medical Humanities Review* 2005 Spring-Fall; 19(1-2): 30-34. Subject: 2.2

Vanderwel, Marianne. Accreditation: the application of quality principles to the protection of human research subjects. *NCEHR Communique CNERH* 2006 Spring; 14(1): 9-11. Subject: 18.2

Vanlaere, Linus; Bouckaert, Filip; Gastmans, Chris. Care for suicidal older people: current clinical-ethical considerations. *Journal of Medical Ethics* 2007 July; 33(7): 376-381. Subject: 9.5.2

Vanlaere, Linus; Gastmans, Chris. Ethics in nursing education: learning to reflect on care practices. *Nursing Ethics* 2007 November; 14(6): 758-766. Subject: 7.2

Vannelli, Alberto; Battaglia, Luigi; Poiasina, Elia; Belli, Filiberto; Bonfanti, Giuliano; Gallino, Gianfrancesco; Vitellaro, Marco; De Dosso, Sara; Leo, Ermanno. The art of decision-making in surgery. To what extent does economics influence choice? *Chirurgia Italiana* 2006 November-December; 58(6): 717-722. Subject: 9.4

Vantsos, Miltiadis; Kiroudi, Marina. An Orthodox view of philanthropy and Church diaconia. *Christian Bioethics* 2007 September-December; (13)3: 251-268. Subject: 2.1

Varelius, Jukka. Execution by lethal injection, euthanasia, organ-donation and the proper goals of medicine. *Bioethics* 2007 March; 21(3): 140-149. Subject: 20.6

Varma, Sumeeta; Ratterman, Allison Griffin. Sharing data, DNA and tissue samples. *Journal of Empirical Research on Human Research Ethics* 2007 March; 2(1): 97-100. Subject: 19.5

Varma, Sumeeta; Wendler, David. Medical decision making for patients without surrogates. *Archives of Internal Medicine* 2007 September 10; 167(16): 1711-1715. Subject: 8.3.3

Vasgird, Daniel R. Prevention over cure: the administrative rationale for education in the responsible conduct of research. *Academic Medicine* 2007 September; 82(9): 835-837. Subject: 18.2

Vaslef, Steven N.; Cairns, Charles B.; Falletta, John M. Ethical and regulatory challenges associated with the exception from informed consent requirements for emergency research: from experimental design to institutional review board approval. *Archives of Surgery* 2006 October; 141(10): 1019-1023; discussion 1024. Subject: 18.3

Vastag, Brian. US aims to tighten rules on direct-to-consumer drug ads. *Nature Biotechnology* 2007 March; 25(3): 267. Subject: 9.7

Vathsala, A. Commercial renal transplantation — body parts for sale [editorial]. *Annals of the Academy of Medicine, Singapore* 2006 April; 35(4): 227-228. Subject: 19.5

Veatch, Robert M. Character formation in professional education: a word of caution. *In:* Kenny, Nuala; Shelton, Wayne, eds. Lost Virtue: Professional Character Development in Medical Education. Amsterdam; Oxford: Elsevier, 2006: 29-45. Subject: 7.2

Veatch, Robert M. Court authorizes withdrawing of ventilator and nutrition. *Ethics and Intellectual Disability* 2006 Spring; 9(2): 1, 3-4. Subject: 20.5.2

Veatch, Robert M. Is bioethics applied ethics? *Kennedy Institute of Ethics Journal* 2007 March; 17(1): 1-2. Subject: 2.1

Veatch, Robert M. The irrelevance of equipoise. *Journal of Medicine and Philosophy* 2007 March-April; 32(2): 167-183. Subject: 18.2

See SUBJECT HEADING KEY FOR SECTION II on inside back cover.

621

Veatch, Robert M. The roles of scientific and normative expertise in public policy formation: the anthrax vaccine case. *In:* Rasmussen, Lisa, ed. Ethics Expertise: History, Contemporary Perspectives, and Applications. Dordrecht: Springer, 2005: 211-225. Subject: 1.3.9

Veatch, Robert M.; Balint, John A.; Glannon, Walter; Cohen, Peter J.; Brudney, Daniel. Just deserts? [letters and reply]. *Hastings Center Report* 2007 May-June; 37(3): 4-6. Subject: 19.6

Veatch, Robert M.; Easton, Raul B.; Graber, Mark A.; Monnahan, Jay; Hughes, Jason. Implied, presumed, and waived-consent: the relative moral wrongs of under- and over-informing. *American Journal of Bioethics* 2007 December; 7(12): 39-41; author reply W3-W4. Subject: 8.3.1

Vedantam, Shankar. APA rules on interrogation abuse; psychologists' group bars member participation in certain techniques. *Washington Post* 2007 August 20; p. A3. Subject: 21.4

Veeman, Michele; Adamowicz, Wiktor; Hu, Wuyang; Hünnemeyer, Anne. Canadian attitudes to genetically modified food. *In:* Einsiedel, Edna; Timmermans, Frank, eds. Crossing Over: Genomics in the Public Arena. Calgary, Alberta, Canada: University of Calgary Press, 2005: 99-113. Subject: 15.1

Vendittelli, F.; Pons, J.C. Elective abortions for minors: impact of the new law in France. *European Journal of Obstetrics, Gynecology, and Reproductive Biology* 2007 January; 130(1): 107-113. Subject: 12.4.2

Ventura, Carla A. Arena; Mendes, Isabel Amelia Costa; Trevizan, Maria Auxiliadora. Psychiatric nursing care in Brazil: legal and ethical aspects. *Medicine and Law: The World Association for Medical Law* 2007 December; 26(4): 829-840. Subject: 21.1

Verheijde, Joseph L.; Rady, Mohamed Y.; McGregor, Joan L.; Easton, Raul B.; Graber, Mark A.; Monnahan, Jay; Hughes, Jason. Defining the scope of implied consent in the emergency department: shortchanging patient's right to self determination. *American Journal of Bioethics* 2007 December; 7(12): 51-52; author reply W3-W4. Subject: 8.3.1

Verhovek, Sam Howe. Parents halt growth of severely disabled girl. *Seattle Times* 2007 January 4; 3 p. [Online]. Accessed: http://seattletimes.nwsource.com [2007 January 8]. Subject: 4.4

Vernick, William J. How long to postpone an operation after a myocardial infarction? When perioperative consultants contradict the literature, leaving the anesthesiologist in the middle. *Journal of Clinical Anesthesia* 2006 August; 18(5): 325-327. Subject: 9.5.1

Vezeau, Toni M. Teaching professional values in a BSN program. *International Journal of Nursing Education Scholarship* 2006; 3: Article 25. Subject: 7.2

Vialettes, B.; Samuelian-Massat, C.; Valéro, R.; Béliard, S. The refusal of treatment in anorexia nervosa, an ethical conflict with three characters: "the girl, the family and the medical profession". Discussion in a French legislative context. *Diabetes and Metabolism* 2006 September; 32(4): 306-311. Subject: 8.3.4

Victoroff, Michael S. Guide to critical care ethics not ready for prime time [review of Critical Care Ethics: A Practice Guide from the ACCM Ethics Committee, edited by Dan R. Thompson and Heidi B. Kummer]. *Managed Care* 2006 July; 15(7): 14-16. Subject: 2.1

Viens, A.M. Addiction, responsibility and moral psychology. *American Journal of Bioethics* 2007 January; 7(1): 17-20. Subject: 17.1

Viens, A.M. Criminal law in the regulation of somatic cell nuclear transfer. *American Journal of Bioethics* 2007 February; 7(2): 73-75. Subject: 14.5

Viens, A.M. The use of functional neuroimaging technology in the assessment of loss and damages in tort law. *American Journal of Bioethics* 2007 September; 7(9): 63-65. Subject: 17.1

Vig, Elizabeth K.; Starks, Helene; Taylor, Janelle S.; Hopley, Elizabeth K.; Fryer-Edwards, Kelly. Surviving surrogate decision-making: what helps and hampers the experience of making medical decisions for others. *JGIM: Journal of General Internal Medicine* 2007 September; 22(9): 1274-1279. Subject: 8.3.3

Vila, J.J.; Jimenez, F.J.; Inarrairaegui, M.; Prieto, C.; Nantes, O.; Borda, F. Informed consent document in gastrointestinal endoscopy: understanding and acceptance by patients. *Revista Española de Enfermedades Digestivas: Organo Oficial de la Sociedad Española de Patología Digestiva* 2006 February; 98(2): 101-111. Subject: 9.5.1

Villanueva, Tiago. Portugal is ready to decriminalise abortion [news]. *BMJ: British Medical Journal* 2007 February 17; 334(7589): 332. Subject: 12.4.1

Vincent, Jean-Louis; Brun-Buisson, Christian; Niederman, Michael; Haenni, Christian; Harbarth, Stephan; Sprumont, Dominique; Valencia, Mauricio; Torres, Antoni. Ethics roundtable debate: a patient dies from an ICU-acquired infection related to methicillin-resistant Staphylococcus aureus — how do you defend your case and your team? *Critical Care* 2005 February; 9(1): 5-9. Subject: 20.5.1

Vitzthum, Wolfgang Graf. Back to Kant! An interjection in the debate on cloning and human dignity. *In:* Vöneky, Silja; Wolfrum, Rüdiger, eds. Human Dignity and Human Cloning. Leiden: Nijhoff, 2004: 87-106. Subject: 14.5

Vivaldelli, Joan. Therapeutic reciprocity: "A union through pain.". *AJN: American Journal of Nursing* 2007 July; 107(7): 74, 76. Subject: 4.1.3

Vlach, David L.; Daniel, Anasseril. Commentary: evolving toward equivalency in correctional mental health care

— a view from the maximum security trenches. *Journal of the American Academy of Psychiatry and the Law* 2007; 35(4): 436-438. Subject: 9.5.3

Vogel, Gretchen. Still waiting for cybrids. *Science* 2007 September 14; 317(5844): 1483. Subject: 18.5.4

Vogel, Lawrence. Natural-law Judaism? The genesis of bioethics in Hans Jonas, Leo Strauss, and Leon Kass. *In:* Schweiker, William; Johnson, Michael A.; Jung, Kevin, eds. Humanity Before God: Contemporary Faces of Jewish, Christian, and Islamic Ethics. Minneapolis, MN: Fortress Press, 2006: 209-237. Subject: 2.1

Volandes, Angelo E.; Abbo, Elmer D. Flipping the default: a novel approach to cardiopulmonary resuscitation in end-stage dementia. *Journal of Clinical Ethics* 2007 Summer; 18(2): 122-139. Subject: 20.5.4

Volandes, Angelo E.; Paasche-Orlow, Michael K. Health literacy, health inequality and a just healthcare system. *American Journal of Bioethics* 2007 November; 7(11): 5-10. Subject: 9.1

Vollmann, Jochen. Ethik in der klinischen Medizin — Bestandsaufnahme und Ausblick [Ethics in clinical medicine — taking stock and prospects]. *Ethik in der Medizin* 2006 December; 18(4): 348-352. Subject: 2.1

von Elm, Erik. Research integrity: collaboration and research needed. *Lancet* 2007 October 20-26; 370(9596): 1403-1404. Subject: 1.3.9

Voss Horrell, Sarah C.; MacLean, William E., Jr.; Conley, Virginia M. Patient and parent/guardian perspectives on the health care of adults with mental retardation. *Mental Retardation* 2006 August; 44(4): 239-248. Subject: 9.8

Vrakking, Astrid M.; van der Heide, Agnes; Onwuteaka-Philipsen, Bregje D.; van der Maas, Paul J.; van der Wal, Gerrit. Regulating physician-assisted dying for minors in the Netherlands: views of paediatricians and other physicians. *Acta Paediatrica* 2007 January; 96(1): 117-121. Subject: 20.5.2

W

Wade, Christopher H.; Wilfond, Benjamin S. Ethical and clinical practice considerations for genetic counselors related to direct-to-consumer marketing of genetic tests. *American Journal of Medical Genetics. Part C, Seminars in Medical Genetics* 2006 November 15; 142(4): 284-292; discussion 293. Subject: 15.2

Wade, Derick. Ethics of collecting and using healthcare data. *BMJ: British Medical Journal* 2007 June 30; 334(7608): 1330-1331. Subject: 9.8

Wade, Nicholas. Panel says data is flawed in a major stem cell report. *New York Times* 2007 February 28; p. A15. Subject: 1.3.9

Wadman, Meredith. Dolly: a decade on. *Nature* 2007 February 22; 445(7130): 800-801. Subject: 14.5

Wadman, Meredith. Poor trial design leaves gene therapy death a mystery [news]. *Nature Medicine* 2007 October; 13(10): 1124. Subject: 15.4

Wadman, Meredith. Stem-cell issue moves up the US agenda [news]. *Nature* 2007 April 19; 446(7138): 842. Subject: 18.5.4

Wadman, Meredith. US genetics bill blocked again [news]. *Nature* 2007 August 9; 448(7154): 631. Subject: 15.3

Wager, Elizabeth. What do journal editors do when they suspect research misconduct? *Medicine and Law: The World Association for Medical Law* 2007 September; 26(3): 535-544. Subject: 1.3.9

Wah, Julia Tao Lai Po. Dignity in long-term care for older persons: a Confucian perspective. *Journal of Medicine and Philosophy* 2007 September-October; 32(5): 465-481. Subject: 9.5.2

Wah, Julia Tao Lai Po; Chan, Ho Mun; Fan, Ruiping. Exploring the bioethics of long-term care. *Journal of Medicine and Philosophy* 2007 September-October; 32(5): 395-399. Subject: 9.5.2

Wainwright, Steven; Williams, Clare; Michael, Mike; Farsides, Bobbie; Cribb, Alan. Remaking the body? Scientists' genetic discourses and practices as examples of changing expectations on embryonic stem cell therapy for diabetes. *New Genetics and Society* 2007 December; 26(3): 251-268. Subject: 18.5.4

Waite, Michael. To tell the truth: the ethical and legal implications of disclosure of medical error. *Health Law Journal* 2005; 13: 1-33. Subject: 8.2

Wakefield, Claire E.; Kasparian, Nadine A.; Meiser, Bettina; Homewood, Judi; Kirk, Judy; Tucker, Kathy. Attitudes toward genetic testing for cancer risk after genetic counseling and decision support: a qualitative comparison between hereditary cancer types. *Genetic Testing* 2007 Winter; 11(4): 401-411. Subject: 15.3

Walker, Robert; Logan, T.K.; Clark, James J.; Leukefeld, Carl. Informed consent to undergo treatment for substance abuse: a recommended approach. *Journal of Substance Abuse Treatment* 2005 December; 29(4): 241-251. Subject: 9.5.9

Wall, Sarah. Organizational ethics, change, and stakeholder involvement: a survey of physicians. *HEC (Healthcare Ethics Committee) Forum* 2007 September; 19(3): 227-243. Subject: 9.1

Wallace, Barbara C. Ethical issues surrounding access to drug user counseling/treatment. *In:* Kleinig, John; Einstein, Stanley, eds. Ethical Challenges for Intervening in Drug Use: Policy, Research and Treatment Issues. Huntsville, TX: Office of International Criminal Justice; Criminal Justice Center, Sam Houston State University, 2006: 529-551. Subject: 9.5.9

See SUBJECT HEADING KEY FOR SECTION II on inside back cover.

623

Wallace, Helen. The UK National DNA Database. Balancing crime detection, human rights and privacy. *EMBO Reports* 2006 July; 7 Special No: S26-S30. Subject: 15.1

Walsh, Adrian; Lynch, Tony. Drug user counseling remuneration and ethics. *In:* Kleinig, John; Einstein, Stanley, eds. Ethical Challenges for Intervening in Drug Use: Policy, Research and Treatment Issues. Huntsville, TX: Office of International Criminal Justice; Criminal Justice Center, Sam Houston State University, 2006: 453-466. Subject: 9.5.9

Walter, James J. A Catholic reflection on embryonic stem cell research. *In:* Walter, James J.; Shannon, Thomas A., eds. Contemporary Issues in Bioethics: A Catholic Perspective. Lanham, MD: Rowman and Littlefield Publishers, 2005: 91-99. Subject: 18.5.4

Walter, James J. Perspectives on medical ethics: biotechnology and genetic medicine. *In:* Walter, James J.; Shannon, Thomas A., eds. Contemporary Issues in Bioethics: A Catholic Perspective. Lanham, MD: Rowman and Littlefield Publishers, 2005: 181-197. Subject: 15.1

Walter, James J. Terminal sedation: a Catholic perspective. *In:* Walter, James J.; Shannon, Thomas A., eds. Contemporary Issues in Bioethics: A Catholic Perspective. Lanham, MD: Rowman and Littlefield Publishers, 2005: 225-229. Subject: 20.4.1

Walter, James J. The meaning and validity of quality of life judgments in contemporary Roman Catholic medical ethics. *In:* Walter, James J.; Shannon, Thomas A., eds. Contemporary Issues in Bioethics: A Catholic Perspective. Lanham, MD: Rowman and Littlefield Publishers, 2005: 209-221. Subject: 4.4

Walter, James J. Theological parameters: Catholic doctrine on abortion in a pluralist society. *In:* Walter, James J.; Shannon, Thomas A., eds. Contemporary Issues in Bioethics: A Catholic Perspective. Lanham, MD: Rowman and Littlefield Publishers, 2005: 145-178. Subject: 12.3

Walter, James J. Theological perspectives on cancer genetics and gene therapy: the Roman Catholic tradition. *In:* Walter, James J.; Shannon, Thomas A., eds. Contemporary Issues in Bioethics: A Catholic Perspective. Lanham, MD: Rowman and Littlefield Publishers, 2005: 199-207. Subject: 15.4

Walters, LeRoy. Der Widerstand Paul Braunes und des Bonhoefferkreises gegen die "Euthanasie" - Programm der Nationalsozialisten. *In:* Gestrich, Christof; Neuge- bauer, Johannes, eds. Der Wert menschlichen Lebens: medizinische Ethik bei Karl Bonhoeffer und Dietrich Bonhoeffer. Berlin: Wichern-Verlag, 2006: 98-146. Subject: 20.5.1

Walther, Guy. Freiheitsentziehende Maßnahmen in Altenpflegeheimen - rechtliche Grundlagen und Alternativen der Pflege = Restraints in long-term care nursing homes for the elderly - legal aspects and alterna-

tives. *Ethik in der Medizin* 2007 December; 19(4): 289-300. Subject: 17.3

Waltz, Emily. Supreme Court boosts licensees in biotech patent battles [news]. *Nature Biotechnology* 2007 March; 25(3): 264-265. Subject: 5.3

Waltz, Emily. The body snatchers: rising demand has created a thriving market for human body parts — and not all of it above ground [news]. *Nature Medicine* 2006 May; 12(5): 487-488. Subject: 19.5

Waltz, Emily. Tracking down tissues. *Nature Biotechnology* 2007 November; 25(11): 1204-1206. Subject: 19.5

Wang, Grace; Watts, Carolyn. The role of genetics in the provision of essential public health services. *American Journal of Public Health* 2007 April; 97(4): 620-625. Subject: 15.1

Wang, Yanguang. The moral status of the human embryo in Chinese stem cell research. *Asian Biotechnology and Development Review* 2006 November; 9(1): 45-63 [Online]. Accessed: http://www.openj-gate.org/articlelist.asp?LatestYear=2007&JCode=104346&year=2006&vol=9&issue=1&ICode=583649 [2008 February 29]. Subject: 18.5.4

Warburg, Ronnie. Renal transplantation: living donors and markets for body parts — Halakha in concert with halakhic policy or public policy? *Tradition* 2007 Summer; 40(2): 14-48. Subject: 19.5

Ward, C.M. The breast-implant controversy: a medicomoral critique. *British Journal of Plastic Surgery* 2001; 54(4): 352-357. Subject: 9.5.5

Waring, Duff R.; Glass, Kathleen Cranley. Legal liability for harm to research participants: the case of placebo-controlled trials. *In:* Lemmens, Trudo; Waring, Duff R., eds. Law and Ethics in Biomedical Research: Regulation, Conflict of Interest and Liability. Toronto; Buffalo: University of Toronto Press, 2006: 206-227. Subject: 18.3

Warnick, Jason E.; Henry, Michael; Fishman, Jennifer R.; Youngner, Stuart J. Propranolol and its potential inhibition of positive post-traumatic growth. *American Journal of Bioethics* 2007 September; 7(9): 37-38; author reply W1-W3. Subject: 17.4

Warnock, Mary. The limits of rights-based discourse. *In:* Spencer, J.R.; du Bois-Pedain, Antje, eds. Freedom and Responsibility in Reproductive Choice. Portland: Hart Pub., 2006: 3-14. Subject: 15.1

Warwick, P. Health care policy. *In:* Holland, Stephen, ed. Introducing Nursing Ethics: Themes in Theory and Practice. Salisbury: APS, 2004: 189-208. Subject: 9.1

Wasserman, David. Addiction and disability: moral and policy issues. *In:* Kleinig, John; Einstein, Stanley, eds. Ethical Challenges for Intervening in Drug Use: Policy, Research and Treatment Issues. Huntsville, TX: Office of International Criminal Justice; Criminal Justice Center,

Sam Houston State University, 2006: 133-152. Subject: 9.5.9

Wasserman, David; Asch, Adrienne. Reply to Nelson. *CQ: Cambridge Quarterly of Healthcare Ethics* 2007 Fall; 16(4): 478-482. Subject: 15.2

Wasserman, J.; Flannery, M.A.; Clair, J.M. Raising the ivory tower: the production of knowledge and distrust of medicine among African Americans. *Journal of Medical Ethics* 2007 March; 33(3): 177-180. Subject: 9.5.4

Waterman, A.D.; Schenk, E.A.; Barrett, A.C.; Waterman, B.M.; Rodrigue, J.R.; Woodle, E.S.; Shenoy, S.; Jendrisak, M.; Schnitzler, M. Incompatible kidney donor candidates' willingness to participate in donor-exchange and non-directed donation. *American Journal of Transplantation* 2006 July; 6(7): 1631-1638. Subject: 19.3

Waters, Brent. Saving us from ourselves: Christology, anthropology and the seduction of posthuman medicine. *In:* Deane-Drummond, Celia; Scott, Peter Manley, eds. Future Perfect?: God, Medicine and Human Identity. New York: T and T Clark International, 2006: 183-195. Subject: 4.5

Watnick, Suzanne. Obesity: a problem of Darwinian proportions? *Advances in Chronic Kidney Disease* 2006 October; 13(4): 428-432. Subject: 9.5.1

Watson, Katie; Quill, Timothy. A conversation with Dr. Timothy Quill [interview]. *Atrium* 2007 Summer; 4: 18-20. Subject: 20.7

Watson, Rory. Developing countries need stronger research guidelines [news]. *BMJ: British Medical Journal* 2007 May 26; 334(7603): 1076. Subject: 18.6

Watson, Rory. Scientists welcome ruling on patent on breast cancer gene [news]. *BMJ: British Medical Journal* 2007 October 13; 335(7623): 740-741. Subject: 15.8

Watt, Helen. Cooperation problems in care of suicidal patients. *In her:* Cooperation, Complicity and Conscience: Problems in Healthcare, Science, Law and Public Policy. London: Linacre Centre, 2005: 139-147. Subject: 8.1

Watt, Helen. Embryos and pseudoembryos: parthenotes, reprogrammed oocytes and headless clones. *Journal of Medical Ethics* 2007 September; 33(9): 554-556. Subject: 4.4

Watts, Geoff. Animal testing: is it worth it? *BMJ: British Medical Journal* 2007 January 27; 334(7586): 182-184. Subject: 22.2

Watts, Geoff. Croatian academic is found guilty of plagiarism. *BMJ: British Medical Journal* 2007 May 26; 334(7603): 1077. Subject: 1.3.9

Watts, Geoff. Genes on ice. *BMJ: British Medical Journal* 2007 March 31; 334(7595): 662-663. Subject: 15.1

Watts, Geoff. Quite reasonably emotional. *Lancet* 2007 January 13-19; 369(9556): 90-91. Subject: 18.4

Watts, Geoff. The locked code. *BMJ: British Medical Journal* 2007 May 19; 334(7602): 1032-1033. Subject: 15.8

Watts, Geoff. UK Biobank gets 10% response rate as it starts recruiting volunteers in Manchester [news]. *BMJ: British Medical Journal* 2007 March 31; 334(7595): 659. Subject: 15.1

Watts, Jonathan. China introduces new rules to deter human organ trade. *Lancet* 2007 June 9-15; 369(9577): 1917-1918. Subject: 19.5

Wear, Stephen. Ethical expertise in the clinical setting. *In:* Rasmussen, Lisa, ed. Ethics Expertise: History, Contemporary Perspectives, and Applications. Dordrecht: Springer, 2005: 243-258. Subject: 9.6

Wear, Stephen. Informed consent. *In:* Khushf, George, ed. Handbook of Bioethics: Taking Stock of the Field From a Philosophical Perspective. Dordrecht; Boston: Kluwer Academic, 2004: 251-290. Subject: 8.3.1

Weatherall, David. Animal research: the debate continues. *Lancet* 2007 April 7-13; 369(9568): 1147-1148. Subject: 22.2

Weatherall, David; Munn, Helen. Moving the primate debate forward [editorial]. *Science* 2007 April 13; 316(5822): 173. Subject: 22.2

Weaver, Kathryn. Ethical sensitivity: state of knowledge and needs for further research. *Nursing Ethics* 2007 March; 14(2): 141-155. Subject: 1.3.1

Weaver, Kathryn; Morse, Janice M. Pragmatic utility: using analytical questions to explore the concept of ethical sensitivity. *Research and Theory for Nursing Practice* 2006 Fall; 20(3): 191-214. Subject: 4.1.1

Webster, Paul. Canadian soldiers and doctors face torture allegations. *Lancet* 2007 April 28 - May 4; 369(9571): 1419-1420. Subject: 7.4

Weed, Douglas L. Ethics and philosophy of public health. *In:* Khushf, George, ed. Handbook of Bioethics: Taking Stock of the Field From a Philosophical Perspective. Dordrecht; Boston: Kluwer Academic, 2004: 525-547. Subject: 9.1

Wehling, Martin. Probleme Industrie-geförderter klinischer Studien im öffentlichen Bereich. *In:* Shapiro, S.; Dinger, J.; Scriba, P., eds. Enabling Risk Assessment in Medicine: Farewell Symposium for Werner-Kari Raff. New Brunswick, NJ: Transaction Publishers, 2004: 21-30. Subject: 1.3.9

Wei, S.; Quigg, M.H.; Monaghan, Kristin G. Is cystic fibrosis carrier screening cost effective? *Community Genetics* 2007; 10(2): 103-109. Subject: 15.3

Weijer, Charles; Miller, P.B. Refuting the net risks test: a response to Wendler and Miller's "Assessing research risks systematically". *Journal of Medical Ethics* 2007 August; 33(8): 487-490. Subject: 18.2

See SUBJECT HEADING KEY FOR SECTION II on inside back cover.

625

Weil, Elizabeth. The needle and the damage done: lethal injections are often botched and sometimes painful. Doctors don't want to administer them. Is it time to kill this form of execution? *New York Times Magazine* 2007 February 11; p. 46-51. Subject: 20.6

Weimer, David L. Public and private regulation of organ transplantation: liver allocation and the final rule. *Journal of Health Politics, Policy and Law* 2007 February; 32(1): 9-49. Subject: 19.6

Weiner, Daniel E.; Levey, Andrew S. Dialysis facility ownership and epoetin dosing in hemodialysis patients: an overview. *American Journal of Kidney Diseases* 2007 September; 50(3): 349-353. Subject: 19.3

Weiner, Rory B. A cooperative beneficence approach to health care reform. *In:* Engström, Timothy H.; Robison, Wade L., eds. Health Care Reform: Ethics and Politics. Rochester, NY: University of Rochester Press, 2006: 209-239. Subject: 9.3.1

Weinfurt, Kevin P.; Allsbrook, Jennifer S.; Friedman, Joëlle Y.; Dinan, Michaela A.; Hall, Mark A.; Schulman, Kevin A.; Sugarman, Jeremy. Developing model language for disclosing financial interests to potential clinical research participants. *IRB: Ethics and Human Research* 2007 January-February; 29(1): 1-5. Subject: 18.2

Weiniger, C.F.; Elchalal, U.; Sprung, C.L.; Weissman, C.; Matot, I. Holy consent — a dilemma for medical staff when maternal consent is withheld for emergency caesarean section. *International Journal of Obstetric Anesthesia* 2006 April; 15(2): 145-148. Subject: 9.5.5

Weinstein, James N.; Clay, Kate; Morgan, Tamara S. Informed patient choice: patient-centered valuing of surgical risks and benefits. *Health Affairs* 2007 May-June; 26(3): 726-730. Subject: 8.3.1

Weisbrot, David. The imperative of the "new genetics": challenges for ethics, law, and social policy. *In:* Betta, Michela, ed. The Moral, Social, and Commercial Imperatives of Genetic Testing and Screening: the Australian Case. Dordrecht: Springer, 2006: 95-124. Subject: 15.1

Weisbrot, David; Opeskin, Brian. Insurance and genetics: regulating a private market in the public interest. *In:* Betta, Michela, ed. The Moral, Social, and Commercial Imperatives of Genetic Testing and Screening: the Australian Case. Dordrecht: Springer, 2006: 125-164. Subject: 15.3

Weisleder, Pedro. Inconsistency among American states on the age at which minors can consent to substance abuse treatment. *Journal for the American Academy of Psychiatry and the Law* 2007; 35(3): 317-322. Subject: 9.5.9

Weiss Roberts, Laura; Coverdale, John; Louie, Alan. Professionalism and the ethics-related roles of academic psychiatrists [editorial]. *Academic Psychiatry* 2005 November-December; 29(5): 413-415 [Online]. Accessed:

http://ap.psychiatryonline.org/ [2007 April 13]. Subject: 17.2

Weiss, Gail Garfinkel. What would you do? New issues in medical ethics. *Medical Economics* 2006 August 18; 83(16): 56-61, 63-64. Subject: 9.4

Weiss, Martin Meyer; Weiss, Peter D.; Weiss, Joseph B. Anthrax vaccine and public health policy. *American Journal of Public Health* 2007 November; 97(11): 1945-1951. Subject: 9.5.1

Weiss, Rick. Death points to risks in research; one woman's experience in gene therapy trial highlights weaknesses in the patient safety net. *Washington Post* 2007 August 6; p. A1, A7. Subject: 15.4

Weiss, Rick. House passes bill relaxing limits on stem cell research. *Washington Post* 2007 January 12; p. A4. Subject: 5.3

Weiss, Rick. Stem cells created with no harm to human embryos; but concerns are raised about the technique. *Washington Post* 2006 August 24; p. A3. Subject: 18.5.4

Weiss, Rick. Suspended gene therapy test had drawn early questions. *Washington Post* 2007 July 28; p. A9. Subject: 15.4

Weiss, Sheila Faith. Human genetics and politics as mutually beneficial resources: the case of the Kaiser Wilhelm Institute for Anthropology, Human Heredity and Eugenics during the Third Reich. *Journal of the History of Biology* 2006 Spring; 39(1): 41-88. Subject: 15.5

Weissbrodt, David; Pekin, Ferhat; Wilson, Amelia. Piercing the confidentiality veil: physician testimony in international criminal trials against perpetrators of torture. *Minnesota Journal of International Law* 2006 Winter; 15(1): 43-109. Subject: 8.4

Weissmann, Gerald. Science fraud: from patchwork mouse to patchwork data. *FASEB Journal: Official Publication of the Federation of American Societies for Experimental Biology* 2006 April; 20(6): 587-590. Subject: 1.3.9

Weitlauf, Julie C.; Ruzek, Josef I.; Westrup, Darrah A.; Lee, Tina; Keller, Jennifer. Empirically assessing participant perceptions of the research experience in a randomized clinical trial: the women's self-defense project as a case example. *Journal of Empirical Research on Human Research Ethics* 2007 June; 2(2): 11-24. Subject: 18.4

Welch, H. Gilbert; Woloshin, Steven; Schwartz, Lisa M. How two studies on cancer screening led to two results. *New York Times* 2007 March 13; p. F5, F8. Subject: 18.5.1

Wellcome Trust. Statement on the handling of allegations of research misconduct. London: Wellcome Trust, 2005 November; 6 p. [Online]. Accessed: http://www.wellcome.ac.uk/doc_WTD002756.html [2007 April 18]. Subject: 1.3.9

Wells, David A.; Ross, Joseph S.; Detsky, Allan S. What is different about the market for health care? *JAMA: The*

Journal of the American Medical Association 2007 December 19; 298(23): 2785-2787. Subject: 9.3.1

Wells, D.J.; Playle, L.C.; Enser, W.E.; Flecknell, P.A.; Gardiner, M.A.; Holland, J.; Howard, B.R.; Hubrecht, R.; Humphreys, K.R.; Jackson, I.J.; Lane, N.; Maconochie, M.; Mason, G.; Morton, D.B.; Raymond, R.; Robinson, V.; Smith, J.A.; Watt, N. Assessing the welfare of genetically altered mice. *Lab Animal* 2006 April; 40(2): 111-114. Subject: 15.1

Wells, Joseph K. Ethical dilemma and resolution: a case scenario. *Indian Journal of Medical Ethics* 2007 January-March; 4(1): 31-33. Subject: 9.5.2

Welsh, Ian; Plows, Alexandra; Evans, Robert. Human rights and genomics: science, genomics and social movements at the 2004 London Social Forum. *New Genetics and Society* 2007 August; 26(2): 123-135. Subject: 15.1

Wen, Chuck K.; Hudak, Pamela L.; Hwang, Stephan W. Homeless people's perceptions of welcomeness and unwelcomeness in healthcare encounters. *JGIM: Journal of General Internal Medicine* 2007 July; 22(7): 1011-1017. Subject: 9.5.10

Wendler, David; Kington, Raynard; Madans, Jennifer; Van Wye, Gretchen; Christ-Schmidt, Heidi; Pratt, Laura A.; Brawley, Otis W.; Gross, Cary P.; Emanuel, Ezekiel. Are racial and ethnic minorities less willing to participate in health research? *PLoS Medicine* 2006 February; 3(2): e19: 0201-0210. Subject: 18.5.1

Wendler, David; Varma, Sumeeta. Minimal risk in pediatric research. *Journal of Pediatrics* 2006 December; 149(6): 855-861. Subject: 18.5.2

Wendler, D.; Miller, F.G. Assessing research risks systematically: the net risks test. *Journal of Medical Ethics* 2007 August; 33(8): 481-486. Subject: 18.2

Werhane, Patricia H. Access, responsibility, and funding: a systems thinking approach to universal access to oral health. *Journal of Dental Education* 2006 November; 70(11): 1184-1195. Subject: 9.2

Werhane, Patricia H.; Gorman, Michael E. Intellectual property rights, access to life-enhancing drugs, and corporate moral responsibilities. *In:* Santoro, Michael A.; Gorrie, Thomas M., eds. Ethics and the Pharmaceutical Industry. Cambridge; New York: Cambridge University Press, 2005: 260-281. Subject: 9.7

Werner, Michael J.; Price, Elizabeth. Managing conflicts of interest: a survival guide for biotechs. *Nature Biotechnology* 2007 February; 25(2): 161-163. Subject: 18.2

Werner, Wolfgang Franz. "Euthanasie" und Widerstand in der Rheinprovinz. *In:* Faust, Anselm, ed. Verfolgung und Widerstand im Rheinland und in Westfalen 1933-1945. Köln: W. Kohlhammer, 1992: 224-233. Subject: 20.5.1

Werntoft, Elisbet; Hallberg, Ingalill R.; Edberg, Anna-Karin. Older people's reasoning about age-related priori-

tization in health care. *Nursing Ethics* 2007 May; 14(3): 399-412. Subject: 9.5.2

Wessely, Simon. When doctors become terrorists. *New England Journal of Medicine* 2007 August 16; 357(7): 635-637. Subject: 21.4

Wessler, Heinz Werner. The charm of biotechnology: human cloning and Hindu bioethics in perspective. *In:* Roetz, Heiner, ed. Cross-Cultural Issues in Bioethics: The Example of Human Cloning. New York: Rodopi, 2006: 247-262. Subject: 14.5

West, Robin L. Social justice, public health, and constitutional authority [review of Social Justice: The Moral Foundations of Public Health and Health Policy, by Madison Powers and Ruth Faden]. *DePaul Journal of Health Care Law* 2007; 10(4): 567-585. Subject: 9.1

Wexler, Barbara. Cloning. *In her:* Genetics and genetic engineering. 2005 ed. Detroit, MI: Thomson/Gale Group, 2006: 117-134. Subject: 14.5

Wexler, Barbara. Ethical issues and public opinion. *In her:* Genetics and genetic engineering. 2005 ed. Detroit, MI: Thomson/Gale Group, 2006: 163-173. Subject: 15.1

Wexler, Barbara. Genetic engineering and biotechnology. *In her:* Genetics and genetic engineering. 2005 ed. Detroit, MI: Thomson/Gale Group, 2006: 135-161. Subject: 15.1

Wexler, Barbara. Genetic testing. *In her:* Genetics and genetic engineering. 2005 ed. Detroit, MI: Thomson/Gale Group, 2006: 83-98. Subject: 15.3

Wexler, Barbara. The Human Genome Project. *In her:* Genetics and genetic engineering. 2005 ed. Detroit, MI: Thomson/Gale Group, 2006: 99-116. Subject: 15.10

Weyers, Heleen. Legal recognition of the right to die. *In:* Garwood-Gowers, Austen; Tingle, John; Wheat, Kay, eds. Contemporary Issues in Healthcare Law and Ethics. Edinburgh; New York: Elsevier Butterworth-Heinemann, 2005: 252-267. Subject: 20.5.1

Whetstine, Leslie; Streat, Stephen; Darwin, Mike; Crippen, David. Pro/con ethics debate: when is dead really dead? *Critical Care (London, England)* 2005; 9(6): 538-542. Subject: 20.2.1

Whiddett, Richard; Hunter, Inga; Engelbrecht, Judith; Handy, Jocelyn. Patients' attitudes towards sharing their health information. *International Journal of Medical Informatics* 2006 July; 75(7): 530-541. Subject: 8.4

White, Ben; Willmott, Lindy. Will you do as I ask? Compliance with instructions about health care in Queensland. *Queensland University of Technology Law and Justice Journal* 2004; 4(1): 77-87. Subject: 8.3.4

White, Caroline. Cancer expert attacks research paper [news]. *BMJ: British Medical Journal* 2007 September 8; 335(7618): 469. Subject: 1.3.9

See SUBJECT HEADING KEY FOR SECTION II on inside back cover.

627

White, Caroline. Doctors who give lethal injections should be punished, says Amnesty [news]. *BMJ: British Medical Journal* 2007 October 6; 335(7622): 690. Subject: 20.6

White, Douglas B.; Braddock, Clarence H., III; Bereknyei, Sylvia; Curtis, J. Randall. Toward shared decision making at the end of life in intensive care units: opportunities for improvement. *Archives of Internal Medicine* 2007 March 12; 167(5): 461-467. Subject: 20.5.1

White, Douglas B.; Curtis, J. Randall; Wolf, Leslie E.; Prendergast, Thomas J.; Taichman, Darren B.; Kuniyoshi, Gary; Acerra, Frank; Lo, Bernard; Luce, John M. Life support for patients without a surrogate decision maker: who decides? *Annals of Internal Medicine* 2007 July 3; 147(1): 34-40. Subject: 8.3.3

White, Douglas B.; Engelberg, Ruth A.; Wenrich, Marjorie D.; Lo, Bernard; Curtis, J. Randall. Prognostication during physician-family discussions about limiting life support in intensive care units. *Critical Care Medicine* 2007 February; 35(2): 442-448. Subject: 20.5.1

White, Gladys B. The development of reprogenetic policy and practice in the United States: looking to the United Kingdom. *In:* Knowles, Lori P.; Kaebnick, Gregory E., eds. Reprogenetics: Law, Policy, and Ethical Issues. Baltimore: Johns Hopkins University Press, 2007: 240-252. Subject: 14.1

White, Gladys B.; Moreno, Jonathan D.; Berger, Sam. The sky is falling . . . or maybe not: the moral necessity of technology assessment. *American Journal of Bioethics* 2007 October; 7(10): 20-21; author reply W1-W3. Subject: 5.2

White, Lawrence W. Corporatization of health care. *In:* Engström, Timothy H.; Robison, Wade L., eds. Health Care Reform: Ethics and Politics. Rochester, NY: University of Rochester Press, 2006: 99-115. Subject: 9.3.1

White, Mary Terrell. Bioethics without a map. *Medical Humanities Review* 2005 Spring-Fall; 19(1-2): 9-12. Subject: 21.1

White, Mary Terrell. Uncertainty and moral judgment: the limits of reason in genetic decision making. *Journal of Clinical Ethics* 2007 Summer; 18(2): 148-155. Subject: 15.3

Whiting, Demian. Inappropriate attitudes, fitness to practise and the challenges facing medical educators. *Journal of Medical Ethics* 2007 November; 33(11): 667-670. Subject: 7.2

Whitney, Simon N.; McCullough, Laurence B. Physicians' silent decisions: because patient autonomy does not always come first. *American Journal of Bioethics* 2007 July; 7(7): 33-38. Subject: 8.1

Whittaker, Peter. Stem cells, patents and ethics. *In:* Gunning, Jennifer; Holm, Søren, eds. Ethics, law, and society.

Volume 1. Aldershot, Hants, England; Burlington, VT: Ashgate, 2005: 17-21. Subject: 18.5.1

Whyte, J. Treatments to enhance recovery from the vegetative and minimally conscious states: ethical issues surrounding efficacy studies. *American Journal of Physical Medicine and Rehabilitation* 2007 February; 86(2): 86-92. Subject: 20.5.1

Wibulpolprasert, Suwit; Moosa, Sheena; Satyanarayana, K.; Samarage, Sarath; Tangcharoensathien, Viroj. WHO's web-based public hearings: hijacked by pharma? [letter]. *Lancet* 2007 November 24-30; 370(9601): 1754. Subject: 18.6

Wicclair, Mark R. Professionalism, religion and shared decision-making. *American Journal of Bioethics* 2007 July; 7(7): 29-31. Subject: 8.1

Wicclair, Mark R.; Lawrence, Ryan E.; Curlin, Farr A. The moral significance of claims of conscience in healthcare. *American Journal of Bioethics* 2007 December; 7(12): 30-31; author reply W1-W2. Subject: 4.1.2

Wichman, Alison. Institutional review boards. *In:* Gallin, John I.; Ognibene, Frederick P., eds. Principles and Practice of Clinical Research. 2nd edition. Oxford: Academic, 2007: 47-58. Subject: 18.2

Wicker, Sabine; Rabenau, Holger; Gottschalk, Rene; Spickhoff, Andreas. Nadelstichverletzung des behandelnden Arztes bei der Untersuchung einer nicht-einwilligungsfähigen Patientin - Darf ein HIV-Test durchgeführt werden? [Needlestick injuries incurred while examining a patient incapable of giving consent. May an HIV-test be performed?]. *Ethik in der Medizin* 2007 September; 19(3): 215-220. Subject: 9.5.6

Wickins, Jeremy. The ethics of biometrics: the risk of social exclusion from the widespread use of electronic identification. *Science and Engineering Ethics* 2007 March; 13(1): 45-54. Subject: 5.1

Wickson, Fern. From risk to uncertainty in the regulation of GMOs: social theory and Australian practice. *New Genetics and Society* 2007 December; 26(3): 325-339. Subject: 15.7

Widdershoven, Guy A.M. How to combine hermeneutics and Wide Reflective Equilibrium. *Medicine, Health Care and Philosophy* 2007 March; 10(1): 49-52. Subject: 2.1

Widdershoven, Guy; Berghmans, Ron. Coercion and pressure in psychiatry: lessons from Ulysses. *Journal of Medical Ethics* 2007 October; 33(10): 560-563. Subject: 17.1

Widdows, Heather. Conceptualising the self in the genetic era. *Health Care Analysis: An International Journal of Health Philosophy and Policy* 2007 March; 15(1): 5-12. Subject: 15.1

Widdows, Heather. Is global ethics moral neo-colonialism? An investigation of the issue in the context of bio-ethics. *Bioethics* 2007 July; 21(6): 305-315. Subject: 21.7

Subject = NRCBL Primary Classification Number; see inside front cover.

Wiechelt, Shelly A. Ethical issues surrounding access to treatment for substance misuse. *In:* Kleinig, John; Einstein, Stanley, eds. Ethical Challenges for Intervening in Drug Use: Policy, Research and Treatment Issues. Huntsville, TX: Office of International Criminal Justice; Criminal Justice Center, Sam Houston State University, 2006: 553-563. Subject: 9.5.9

Wiesemann, Claudia. Die Beziehung der Medizinethik zur Medizingeschichte und Medizinetheorie [The relationship of medical ethics to the history of medicine and medical theory]. *Ethik in der Medizin* 2006 December; 18(4): 337-341. Subject: 2.1

Wiesen, Jonathan; Kulak, David. "Male and female He created them:" revisiting gender assignment and treatment in intersex children. *Journal of Halacha and Contemporary Society* 2007 Fall; (54): 5-29. Subject: 10

Wiesing, Urban. Ethical aspects of limiting residents' work hours. *Bioethics* 2007 September; 21(7): 398-405. Subject: 7.2

Wiggins, Osborne P.; Schwartz, Michael Alan. Philosophical issues in psychiatry. *In:* Khushf, George, ed. Handbook of Bioethics: Taking Stock of the Field From a Philosophical Perspective. Dordrecht; Boston: Kluwer Academic, 2004: 473-488. Subject: 4.3

Wijdicks, Eelco F.M. The clinical criteria of brain death throughout the world: why has it come to this? = Les critères cliniques de mort encéphalique à travers le monde: pour quoi en arriver là [editorial]. *Canadian Journal of Anaesthesia* 2006 June; 53(6): 540-543. Subject: 20.2.1

Wikler, Daniel. Paternalism and the mildly retarded. *Philosophy and Public Affairs* 1979 Summer; 8(4): 377-392. Subject: 9.5.3

Wilan, Ken. Susan Wood. *Nature Biotechnology* 2007 May 25(5): 495. Subject: 11.1

Wilcockson, Michael. Abortion and infanticide. *In his:* Issues of Life and Death. London: Hodder and Stoughton Educational, 1999: 33-55. Subject: 12.3

Wilcockson, Michael. Euthanasia and doctors' ethics. *In his:* Issues of Life and Death. London: Hodder and Stoughton Educational, 1999: 57-71. Subject: 20.5.1

Wilcox, Allen J. On conflicts of interest. *Epidemiology* 2006 May; 17(3): 241. Subject: 1.3.9

Wild, Verina. Plädoyer für einen Einschluss schwangerer Frauen in Arzneimittelstudien = Towards the inclusion of pregnant women in drug trials. *Ethik in der Medizin* 2007 March; 19(1): 7-23. Subject: 18.5.3

Wildeman, Sheila. Access to treatment of serious mental illness: enabling choice or enabling treatment? *In:* Flood, Colleen M., ed. Just Medicare: What's In, What's Out, How We Decide. Buffalo, NY: University of Toronto Press, 2006: 231-257. Subject: 17.1

Wilder, Christine M.; Elbogen, Eric B.; Swartz, Marvin S.; Swanson, Jeffrey W.; Van Dorn, Richard A. Effect of patients' reasons for refusing treatment on implementing psychiatric advance directives. *Psychiatric Services* 2007 October; 58(10): 1348-1350. Subject: 17.1

Wildes, Kevin Wm. Global and particular bioethics. *In:* Engelhardt, H. Tristram, ed. Global Bioethics: The Collapse of Consensus. Salem, MA: M&M Scrivener Press, 2006: 362-379. Subject: 2.1

Wilfond, Benjamin S. The Ashley case: the public response and policy implications. *Hastings Center Report* 2007 September-October; 37(5): 12-13. Subject: 8.3.2

Wilkinson, Stephen. Eugenics and the criticism of bioethics [review of Genetic Politics: From Eugenics to Genome, by Ann Kerr and Tom Shakespeare]. *Ethical Theory and Moral Practice* 2007 August; 10(4): 409-418. Subject: 2.1

Wilkinson, Stephen. Separating conjoined twins: the case of Ladan and Lelah Bijani. *In:* Gunning, Jennifer; Holm, Søren, eds. Ethics, law, and society. Volume 1. Aldershot, Hants, England; Burlington, VT: Ashgate, 2005: 257-260. Subject: 4.4

Wilkinson, T.M. Individual and family decisions about organ donation. *Journal of Applied Philosophy* 2007; 24(1): 26-40. Subject: 19.5

Willett, Walter C.; Blot, William J.; Colditz, Graham A.; Folsom, Aaron R.; Henderson, Brian E.; Stampfer, Meir J. Merging and emerging cohorts: necessary but not sufficient. *Nature* 2007 January 18; 445(7125): 257-258. Subject: 15.1

Williams, Anne. The morning-after pill. *Human Reproduction and Genetic Ethics: An International Journal* 2007; 13(1): 8-36. Subject: 11.1

Williams, Charlotte J.; Shuster, John L.; Clay, Olivio J.; Burgio, Kathryn L. Interest in research participation among hospice patients, caregivers, and ambulatory senior citizens: practical barriers or ethical constraints? *Journal of Palliative Medicine* 2006 August; 9(4): 968-974. Subject: 18.5.7

Williams, Clare. Dilemmas in fetal medicine: premature application of technology or responding to women's choice? *Sociology of Health and Illness* 2006 January; 28(1): 1-20. Subject: 9.5.8

Williams, Hywel C.; Naldi, Luigi; Paul, Carle; Vahlquist, Anders; Schroter, Sara; Jobling, Ray. Conflicts of interest in dermatology. *Acta Dermato-Venereologica* 2006; 86(6): 485-497. Subject: 7.1

Williams, Kylie; Cocking, Dean. Counselor-client relationships and professional role morality. *In:* Kleinig, John; Einstein, Stanley, eds. Ethical Challenges for Intervening in Drug Use: Policy, Research and Treatment Issues. Huntsville, TX: Office of International Criminal Justice; Criminal Justice Center, Sam Houston State University, 2006: 341-365. Subject: 9.5.9

See SUBJECT HEADING KEY FOR SECTION II on inside back cover.

629

Williams, Monique M. Invisible, unequal, and forgotten: health disparities in the elderly. *Notre Dame Journal of Law, Ethics and Public Policy* 2007; 21(2): 441-478. Subject: 9.5.2

Williams, Robin; Johnson, Paul. Forensic DNA databasing: a European perspective. Interim report. Durham, Great Britain: University of Durham, 2005 June; 143 p. [Online]. Accessed: http://www.dur.ac.uk/resources/sass/Williams%20and%20Johnson%20Interim %20Report %202005-1.pdf [2007 April 25]. Subject: 15.1

Williamson, Graham R.; Prosser, Sue. Action research: politics, ethics and participation. *Journal of Advanced Nursing* 2002 December; 40(5): 587-593. Subject: 18.2

Williamson, Laura. Empirical assessments of clinical ethics services: implications for clinical ethics committees. *Clinical Ethics* 2007 December; 2(4): 187-192. Subject: 9.6

Willmott, Lindy. Surrogacy: ART's forgotten child. *University of New South Wales Law Journal* 2006; 29(2): 227-232. Subject: 14.2

Willmott, Lindy; White, Ben; Howard, Michelle. Refusing advance refusals: advance directives and life-sustaining medical treatment. *Melbourne University Law Review* 2006 April; 30(1): 211-243. Subject: 20.5.4

Willyard, Cassandra. Allegations of bias cloud conflicting reports on bisphenol A's effects [news]. *Nature Medicine* 2007 September; 13(9): 1002. Subject: 1.3.9

Willyard, Cassandra. Pfizer lawsuit spotlights ethics of developing world clinical trials [news]. *Nature Medicine* 2007 July; 13(7): 763. Subject: 18.5.9

Wilmot, Stephen. A fair range of choice: justifying maximum patient choice in the British National Health Service. *Health Care Analysis: An International Journal of Health Philosophy and Policy* 2007 June; 15(2): 59-72. Subject: 9.2

Wilson, James. Is respect for autonomy defensible? *Journal of Medical Ethics* 2007 June; 33(6): 353-356. Subject: 8.3.1

Wilson, James. Transhumanism and moral equality. *Bioethics* 2007 October; 21(8): 419-425. Subject: 4.5

Wilson, Kenneth; Schreier, Alan; Griffin, Angel; Resnik, David. Research records and the resolution of misconduct allegations at research universities. *Accountability in Research* 2007 January-March; 14(1): 57-71. Subject: 1.3.9

Wilson, Naomi. Professionals' experiences of addressing ethical issues in services for people with intellectual disabilities: a brief report. *Ethics and Intellectual Disability* 2005 Winter; 8(2): 5-7. Subject: 8.3.3

Wilson, Penelope; Sexton, Wendy; Singh, Andrea; Smith, Melissa; Durham, Stephanie; Cowie, Anne; Fritschi, Lin. Family experiences of tissue donation in Australia. *Progress in Transplantation* 2006 March; 16(1): 52-56. Subject: 19.5

Wilson, Roxanne M. Litigating on the new frontier: inroads on the duties of sponsors and investigators in clinical trials. *Federation of Defense and Corporate Counsel Quarterly* 2005 Fall; 56(1): 49-75. Subject: 18.2

Winau, Rolf. Experimentation on humans and informed consent: how we arrived where we are. *In:* LaFleur, William R.; Böhme, Gernot; Shimazono, Susumu, eds. Dark Medicine: Rationalizing Medical Research. Bloomington: Indiana University Press, 2007: 46-56. Subject: 18.3

Winfield, Marlene. For patients' sake, don't boycott e-health records. *BMJ: British Medical Journal* 2007 July 21; 335(7611): 158. Subject: 1.3.12

Wingfield, Joy. Researching the chemists: towards an integrated research agenda: conference report. *Clinical Ethics* 2007 March; 2(1): 42-44. Subject: 9.7

Wingfield, Joy. You pays your money and you takes your choice? *In:* Gunning, Jennifer; Holm, Søren, eds. Ethics, law, and society. Volume 2. Aldershot, Hants, England; Burlington, VT: Ashgate, 2006: 287-289. Subject: 9.7

Winickoff, David E. Governing stem cell research in California and the USA: towards a social infrastructure. *Trends in Biotechnology* 2006 September; 24(9): 390-394. Subject: 18.5.4

Winik, Lyric Wallwork. The 'good' Nazi doctor: according to the 'ways of Auschwitz'. *Moment* 1999 October; 24(5): 60-77. Subject: 21.4

Winslade, William J. Severe brain injury: recognizing the limits of treatment and exploring the frontiers of research. *CQ: Cambridge Quarterly of Healthcare Ethics* 2007 Spring; 16(2): 161-168. Subject: 17.1

Winslade, William J.; Carson, Ronald A. Foreword: the role of religion in health law and policy. *Houston Journal of Health Law and Policy* 2006; 6(2): 245-248. Subject: 2.1

Wirtz, Veronika; Cribb, Alan; Barber, Nick. Patient-doctor decision-making about treatment within the consultation — a critical analysis of models. *Social Science and Medicine* 2006 January; 62(1): 116-124. Subject: 8.1

Wirtz, Veronika; Cribb, Alan; Barber, Nick. The use of informed consent for medication treatment in hospital: a qualitative study of the views of doctors and nurses. *Clinical Ethics* 2007 March; 2(1): 36-41. Subject: 8.3.1

Wisely, J.; Lilleyman, J. Implementing the district hospital recommendations for the National Health Service Research Ethics Service in England. *Journal of Medical Ethics* 2007 March; 33(3): 168. Subject: 18.2

Witherspoon, Gerald S. What I learned from Schiavo. *Hastings Center Report* 2007 November-December; 37(6): 17-20. Subject: 20.5.4

Wnukiewicz-Kozlowka, Agata. The admissibility of research in emergency medicine. *Science and Engineering Ethics* 2007 September; 13(3): 315-332. Subject: 18.5.1

Wolf, Don P. An opinion on regulating the assisted reproductive technologies. *Journal of Assisted Reproduction and Genetics* 2003 July; 20(7): 290-292. Subject: 14.4

Wolf, J.H.; Foley, P. Hans Gerhard Creutzfeldt (1885-1964): a life in neuropathology. *Journal of Neural Transmission* 2005 August; 112(8): I-XCVII. Subject: 7.1

Wolf, Leslie E.; Zandecki, Jolanta. Conflicts of interest in research: how IRBs address their own conflicts. *IRB: Ethics and Human Research* 2007 January-February; 29(1): 6-12. Subject: 18.2

Wolfberg, Adam J. The patient as ally — learning the pelvic examination. *New England Journal of Medicine* 2007 March 1; 356(9): 889-890. Subject: 7.2

Wolff, Katharina; Brun, Wibecke; Kvale, Gerd; Nordin, Karin. Confidentiality versus duty to inform — an empirical study on attitudes towards the handling of genetic information. *American Journal of Medical Genetics. Part A* 2007 January 15; 143(2): 142-148. Subject: 8.4

Wolfrum, Rüdiger; Vöneky, Silja. Who is protected by human rights conventions? Protection of the embryo vs. scientific freedom and public health. *In:* Vöneky, Silja; Wolfrum, Rüdiger, eds. Human Dignity and Human Cloning. Leiden: Nijhoff, 2004: 133-143. Subject: 14.5

Wolinsky, Howard. The battle of Helsinki: two troublesome paragraphs in the Declaration of Helsinki are causing a furore over medical research ethics. *EMBO Reports* 2006 July; 7(7): 670-672. Subject: 18.2

Wolowelsky, Joel B.; Grazi, Richard V.; Brander, Kenneth; Freundel, Barry; Friedman, Michelle; Goldberg, Judah; Greenberger, Ben; Kaplan, Feige; Reichman, Edward; Zimmerman, Deena R. Sex selection and Halakhic ethics: a contemporary discussion. *Tradition* 2007 Spring; 40(1): 45-78. Subject: 14.3

Wolpe, Paul Root. Religious responses to neuroscientific questions. *In:* Illes, Judy, ed. Neuroethics: Defining the Issues in Theory, Practice, and Policy. New York: Oxford University Press, 2006: 289-296. Subject: 17.1

Wolpert, Lewis. Is cell science dangerous? *Journal of Medical Ethics* 2007 June; 33(6): 345-348. Subject: 15.1

Woltanski, A.; Cragun, R.; Myers, M.; Cragun, D. Views on religion and abortion: a comparison of genetic counselors and the general population. *Journal of Genetic Counseling* 2007 December; 16(6): 686. Subject: 15.2

Wolverson, M. Restrictive physical interventions. *In:* Holland, Stephen, ed. Introducing Nursing Ethics: Themes in Theory and Practice. Salisbury: APS, 2004: 111-130. Subject: 17.3

Wong, D.; Kyle, G. Some ethical considerations for the "off-label" use of drugs such as Avastin. *British Journal of*

Ophthalmology 2006 October; 90(10): 1218-1219. Subject: 9.7

Wong, J.G.; Clare, I.C.H.; Gunn, M.J.; Holland, A.J. Capacity to make health care decisions: its importance in clinical practice. *Psychological Medicine* 1999 March; 29(2): 437-446. Subject: 8.3.3

Wong, Sophia Isako. The moral personhood of individuals labeled "mentally retarded": a Rawlsian response to Nussbaum. *Social Theory and Practice* 2007 October; 33(4): 579-594. Subject: 9.5.3

Wong-Kim, Evaon; Song, Young; Vasgird, Daniel. Cultural competence of researchers. *Journal of Empirical Research on Human Research Ethics* 2007 March; 2(1): 83-85. Subject: 18.5.1

Wonkam, Ambroise; Hurst, Samia A. Acceptance of abortion by doctors and medical students in Cameroon [letter]. *Lancet* 2007 June 16-22; 369(9578): 1999. Subject: 12.4.3

Woodrow, Susannah R.; Jenkins, Anthony P. How thorough is the process of informed consent prior to outpatient gastroscopy? A study of practice in a United Kingdom District Hospital. *Digestion* 2006; 73(2-3): 189-197. Subject: 8.3.1

Woolfrey, Joan. Ectogenesis: liberation, technological tyranny, or just more of the same? *In:* Gelfand, Scott; Shook, John R., eds. Ectogenesis: Artificial Womb Technology and the Future of Human Reproduction. Amsterdam; New York: Editions Rodopi, B.V., 2006: 129-139. Subject: 14.1

Woolley, Douglas C.; Medvene, Louis J.; Kellerman, Rick D.; Base, Michelle; Mosack, Victoria. Do residents want automated external defibrillators in their retirement home? *Journal of the American Medical Directors Association* 2006 March; 7(3): 135-140. Subject: 9.5.2

Woolley, Karen L. Goodbye Ghostwriters!: How to work ethically and efficiently with professional medical writers. *Chest* 2006 September; 130(3): 921-923. Subject: 1.3.7

Workman, Stephen. Researching a good death [editorial]. *BMJ: British Medical Journal* 2007 March 10; 334(7592): 485-486. Subject: 18.5.7

World Federation of Neurology; Rosenberg, Roger N. World Federation of Neurology position paper on human stem cell research. *Journal of the Neurological Sciences* 2006 April 15; 243(1-2): 1-2. Subject: 18.5.4

World Health Organization [WHO]. Commission on Social Determinants of Health; Marmot, Michael. Achieving health equity: from root causes to fair outcomes. *Lancet* 2007 September 29-October 5; 370(9593): 1153-1163. Subject: 9.2

World Intellectual Property Organization [WIPO]. Study takes critical look at benefit sharing of genetic resources and traditional knowledge [press release]. Kuala Lumpur/Geneva/Nairobi: World Intellectual Property Or-

See SUBJECT HEADING KEY FOR SECTION II on inside back cover.

ganization, 2004 February 10; 4 p. [Online]. Accessed: http://www.wipo.int/edocs/prdocs/en/2004/wipo_pr_2004_373.html [2007 April 16]. Subject: 15.8

World Medical Association [WMA]. The World Medical Association statement on HIV/AIDS and the medical profession. *Indian Journal of Medical Ethics* 2007 April-June; 4(2): 84-86. Subject: 9.5.6

World Medical Association [WMA]. World Medical Association Declaration of Helsinki: Ethical principles for medical research involving human subjects. *Bulletin of the World Health Organization* 2001; 79(4): 373. Subject: 6

Wray, Matt. Three generations of imbeciles are enough: American eugenics and poor white trash. *In his:* Not Quite White: White Trash and the Boundaries of Whiteness. Durham: Duke University Press, 2006: 65-95. Subject: 15.5

Wright, Alexis A.; Katz, Ingrid T. Letting go of the rope — aggressive treatment, hospice care, and open access. *New England Journal of Medicine* 2007 July 26; 357(4): 324-327. Subject: 20.4.1

Wright, C.M.; Waterston, A.J.R. Relationships between paediatricians and infant formula milk companies. *Archives of Disease in Childhood* 2006 May; 91(5): 383-385. Subject: 9.5.7

Wright, Linda; MacRae, Susan; Gordon, Debra; Elliot, Esther; Dixon, David; Abbey, Susan; Richarson, Robert. Disclosure of misattributed paternity: issues involved in the discovery of unsought information. *Seminars in Dialysis* 2002 June; 15(3): 202-206. Subject: 8.2

Wright, Samuel L. Protecting the right to medical treatment from the war on terror: the status of Guantanamo Bay detainees. *Journal of Legal Medicine* 2007 January-March; 28(1): 135-149. Subject: 21.2

Wright, Wendy; Staible, Nancy. HPV mandates: parents trump politics. *Ethics and Medics* 2007 July; 32(7): 1-3. Subject: 8.3.2

Wrigley, A. Proxy consent: moral authority misconceived. *Journal of Medical Ethics* 2007 September; 33(9): 527-531. Subject: 8.3.3

Wu, Eugene B. The ethics of implantable devices. *Journal of Medical Ethics* 2007 September; 33(9): 532-533. Subject: 20.5.1

Wunder, Michael. Grafenecker Erklärung zur Bio-Ethik. *In:* Neuer-Miebach, Therese; Wunder, Michael, eds. Bio-Ethik und die Zukunft der Medizin. Bonn: Psychiatrie-Verlag, 1998: 182-195. Subject: 2.1

Würbel, Hanno. Publications should include an animal-welfare section [letter]. *Nature* 2007 March 15; 446(7133): 257. Subject: 22.2

Wüstner, Kerstin; Heinze, Ulrich. Attitudes towards pre-implantation genetic diagnosis — a German and Japanese comparison. *New Genetics and Society* 2007 April; 26(1): 1-27. Subject: 15.2

Wylie, Jean E.; Mineau, Geraldine P. Biomedical databases: protecting privacy and promoting research. *Trends in Biotechnology* 2003 March; 21(3): 113-116. Subject: 15.1

Wynia, Matthew K. Breaching confidentiality to protect the public: evolving standards of medical confidentiality for military detainees. *American Journal of Bioethics* 2007 August; 7(8): 1-5. Subject: 8.4

Wynia, Matthew K. Ethics and public health emergencies: restrictions on liberty [editorial]. *American Journal of Bioethics* 2007 February; 7(2): 1-5. Subject: 9.1

Wynia, Matthew K. Mandating vaccination: what counts as a "mandate" in public health and when should they be used? *American Journal of Bioethics* 2007 December; 7(12): 2-6. Subject: 9.5.1

X

Xenakis, B.G. Stephen. Military medical ethics under attack. *Journal of Ambulatory Care Management* 2006 October-December; 29(4): 342-344. Subject: 7.1

Y

Yacoub, Ahmed Abdel Aziz. Euthanasia. *In his:* The Fiqh of Medicine: Responses in Islamic Jurisprudence to Developments in Medical Science. London, UK: Ta-Ha Publishers Ltd., 2001: 159-201. Subject: 20.5.1

Yacoub, Ahmed Abdel Aziz. Reproduction and cloning. *In his:* The Fiqh of Medicine: Responses in Islamic Jurisprudence to Developments in Medical Science. London, UK: Ta-Ha Publishers Ltd., 2001: 233-253. Subject: 14.1

Yacoub, Ahmed Abdel Aziz. The prevention and termination of pregnancy. *In his:* The Fiqh of Medicine: Responses in Islamic Jurisprudence to Developments in Medical Science. London, UK: Ta-Ha Publishers Ltd., 2001: 202-232. Subject: 12.3

Yacoub, Ahmed Abdel Aziz. The rights and responsibilities of patients and those who treat them. *In his:* The Fiqh of Medicine: Responses in Islamic Jurisprudence to Developments in Medical Science. London, UK: Ta-Ha Publishers Ltd., 2001: 100-126. Subject: 8.5

Yacoub, Ahmed Abdel Aziz. Transplantation. *In his:* The Fiqh of Medicine: Responses in Islamic Jurisprudence to Developments in Medical Science. London, UK: Ta-Ha Publishers Ltd., 2001: 254-280. Subject: 19.5

Yalisove, Daniel. From the ivory tower to the trenches: teaching professional ethics to substance abuse counselors. *In:* Kleinig, John; Einstein, Stanley, eds. Ethical Challenges for Intervening in Drug Use: Policy, Research and Treatment Issues. Huntsville, TX: Office of International Criminal Justice; Criminal Justice Center, Sam Houston State University, 2006: 507-527. Subject: 2.3

Subject = NRCBL Primary Classification Number; see inside front cover.

Yamalik, Nermin. The responsibilities and rights of dental professionals 1. Introduction. *International Dental Journal* 2006 April; 56(2): 109-111. Subject: 4.1.1

Yamaori, Tetsuo. Strategies for survival versus accepting impermanence: rationalizing brain death and organ transplantation today. *In:* LaFleur, William R.; Böhme, Gernot; Shimazono, Susumu, eds. Dark Medicine: Rationalizing Medical Research. Bloomington: Indiana University Press, 2007: 165-179. Subject: 20.2.1

Yang, Julia A.; Kombarakaran, Francis A. A practitioner's response to the new health privacy regulations. *Health and Social Work* 2006 May; 31(2): 129-136. Subject: 8.3.1

Yang, Y.; Zhang, K.L.; Chan, K.Y.; Reidpath, D.D. Institutional and structural forms of HIV-related discrimination in health care: a study set in Beijing. *AIDS Care* 2005 July; 17(Supplement 2): S129-S140. Subject: 9.5.6

Yank, Veronia; Rennie, Drummond; Bero, Lisa A. Financial ties and concordance between results and conclusions in meta-analyses: retrospective cohort study. *BMJ: British Medical Journal* 2007 December 8; 335(7631): 1202-1205. Subject: 18.2

Yarborough, Mark; Sharp, Richard R. Bioethics consultation and patient advocacy organizations: expanding the dialogue about professional conflicts of interest. *CQ: Cambridge Quarterly of Healthcare Ethics* 2007 Winter; 16(1): 74-81. Subject: 9.6

Yarbrough, Susan; Klotz, Linda. Incorporating cultural issues in education for ethical practice. *Nursing Ethics* 2007 July; 14(4): 492-502. Subject: 7.2

Yates, Ferdinand D., Jr. Holding the hospital hostage. *American Journal of Bioethics* 2007 March; 7(3): 36-37. Subject: 9.5.2

Yates, Tom; Crane, Rosie; Burnett, Angela. Rights and the reality of healthcare charging in the United Kingdom. *Medicine, Conflict and Survival* 2007 October-December; 23(4): 297-304. Subject: 9.2

Yeates, Neil. Health Canada's new standards on conflict of interest. *CMAJ/JAMC: Canadian Medical Association Journal* 2007 October 9; 177(8): 900. Subject: 9.7

Yee, Samantha; Hitkari, Jason A.; Greenblatt, Ellen M. A follow-up study of women who donated oocytes to known recipient couples for altruistic reasons. *Human Reproduction* 2007 July; 22(7): 2040-2050. Subject: 14.2

Yoshikawa, Thomas T.; Ouslander, Joseph G. Integrity in publishing: update on policies and statements on authorship, duplicate publications and conflict of interest [editorial]. *Journal of the American Geriatrics Society* 2007 February; 55(2): 155-157. Subject: 1.3.7

Yoshimura, Yasunori. Bioethical aspects of regenerative and reproductive medicine. *Human Cell* 2006 May; 19(2): 83-86. Subject: 18.5.4

Young, Alison Harvison. Possible policy strategies for the United States: comparative lessons. *In:* Knowles, Lori P.; Kaebnick, Gregory E., eds. Reprogenetics: Law, Policy, and Ethical Issues. Baltimore: Johns Hopkins University Press, 2007: 226-239. Subject: 14.1

Young, Charles; Godlee, Fiona. Managing suspected research misconduct [editorial]. *BMJ: British Medical Journal* 2007 February 24; 334(7590): 378-379. Subject: 1.3.9

Young, John; Saleh, Fabien M. The prisoner's right to treatment. *Journal for the American Academy of Psychiatry and the Law* 2007; 35(3): 384-386. Subject: 17.8

Youngner, Stuart J. The definition of death. *In:* Steinbock, Bonnie, ed. The Oxford Handbook of Bioethics. Oxford; New York: Oxford University Press, 2007: 285-303. Subject: 20.2.1

Yu, Erika; Fan, Ruiping. A Confucian view of personhood and bioethics. *Journal of Bioethical Inquiry* 2007; 4(3): 171-179. Subject: 4.4

Yuen, R.K.N.; Lam, S.T.S.; Allison, D. Bioethics and prenatal diagnosis of foetal diseases. *Hong Kong Medical Journal* 2006 December; 12(6): 488. Subject: 15.2

Z

Zadunayski, Anna; Hicks, Matthew; Gibbard, Ben; Godlovitch, Glenys. Behind the screen: legal and ethical considerations in neonatal screening for prenatal exposure to alcohol. *Health Law Journal* 2006; 14: 105-127. Subject: 9.5.9

Zahedi, F.; Larijani, B.; Isikoglu, M.; Senol, Y.; Berkkanoglu, M.; Ozgur, K.; Donmez, L. Considerations of third-party reproduction in Iran [letter and reply]. *Human Reproduction* 2007 March; 22(3): 902-903. Subject: 14.2

Zali, Mohammad Reza; Shahraz, Saeed; Borzabadi, Shokoufeh. Bioethics in Iran: legislation as the main problem. *Archives of Iranian Medicine* 2002 July; 5(3): 7 p. [Online]. Accessed: http://www.ac.ir./AIM/0253/0253136.htm [2008 March 31]. Subject: 15.1

Zamperetti, N.; Proietti, R. End of life in the ICU: laws, rules and practices: the situation in Italy. *Intensive Care Medicine* 2006 October; 32(10): 1620-1622. Subject: 20.5.2

Zaner, Richard. Physicians and patients in relation: clinical interpretation and dialogues of trust. *In:* Khushf, George, ed. Handbook of Bioethics: Taking Stock of the Field From a Philosophical Perspective. Dordrecht; Boston: Kluwer Academic, 2004: 223-250. Subject: 8.1

Zaner, Richard M. A comment on community consultation. *American Journal of Bioethics* 2007 February; 7(2): 29-30. Subject: 9.6

Zanni, Guido R.; Stavis, Paul F. The effectiveness and ethical justification of psychiatric outpatient commitment. *American Journal of Bioethics* 2007 November; 7(11): 31-41. Subject: 17.7

See SUBJECT HEADING KEY FOR SECTION II on inside back cover.

633

Zarkowski, Pamela. Professional promises: summary and next steps. *Journal of Dental Education* 2006 November; 70(11): 1241-1245. Subject: 4.1.1

Zarocostas, John. UN calls for tougher rules to prevent sale of children's organs [news]. *BMJ: British Medical Journal* 2007 March 31; 334(7595): 656. Subject: 19.5

Zeiler, Kristin. Shared decision-making, gender and new technologies. *Medicine, Health Care and Philosophy* 2007 September; 10(3): 279-287. Subject: 15.2

Zeiler, Kristin. Who am I? When do "I" become another? An analytic exploration of identities, sameness and difference, genes and genomes. *Health Care Analysis: An International Journal of Health Philosophy and Policy* 2007 March; 15(1): 25-32. Subject: 15.1

Zetola, Nicola M. Association between rates of HIV testing and elimination of written consents in San Francisco [letter]. *JAMA: The Journal of the American Medical Association* 2007 March 14; 297(10): 1061. Subject: 9.5.6

Zettler, Patricia; Wolf, Leslie E.; Lo, Bernard. Establishing procedures for institutional oversight of stem cell research. *Academic Medicine* 2007 January; 82(1): 6-10. Subject: 18.6

Zezima, Katie. Not all are pleased at plan to offer birth control at Maine middle school. *New York Times* 2007 October 21; p. A22. Subject: 11.2

Zhai, Xiaomei; Qiu, Ren Zong. Perceptions of long-term care, autonomy, and dignity, by residents, family and caregivers: the Beijing experience. *Journal of Medicine and Philosophy* 2007 September-October; 32(5): 425-445. Subject: 9.5.2

Ziegler, Stephen J. Euthanasia and the administration of neuromuscular blockers without ventilation: should physicians fear prosecution? *Omega* 2006; 53(4): 295-310. Subject: 20.7

Ziegler, Stephen; Bosshard, Georg. Role of non-governmental organisations in physician assisted suicide. *BMJ: British Medical Journal* 2007 February 10; 334(7588): 295-298. Subject: 20.7

Zilberberg, Julie. Sex selection and restricting abortion and sex determination. *Bioethics* 2007 November; 21(9): 517-519. Subject: 14.3

Zimbardo, Philip. When good people do evil. *Yale Alumni Magazine* 2007 January-February; 70(3): 40-47. Subject: 18.4

Zimmerman, Richard K.; Tabbarah, Melissa; Nowalk, Mary P.; Raymund, Mahlon; Jewell, Ilene K.; Wilson, Stephen A; Ricci, Edmund M. Racial differences in beliefs about genetic screening among patients at inner-city neighborhood health centers. *Journal of the National Medical Association* 2006 March; 98(3): 370-377. Subject: 15.3

Zion, Deborah. Ethics, disease and obligation. *In:* Bennett, Belinda; Tomossy, George F., eds. Globalization and Health: Challenges for Health Law and Bioethics. Dordrecht: Springer, 2006: 47-57. Subject: 9.5.6

Zion, Deborah; Jureidini, Jon; Newman, Louise; Kyambi, Sarah; Zion, Deborah;. Republication: in that case [case study and commentaries]. *Journal of Bioethical Inquiry* 2006; 3(3): 193-202. Subject: 17.7

Zoloth, Laurie. Being in the world: neuroscience and the ethical agent. *In:* Illes, Judy, ed. Neuroethics: Defining the Issues in Theory, Practice, and Policy. New York: Oxford University Press, 2006: 61-73. Subject: 17.1

Zoloth, Laurie. I want you: notes toward a theory of hospitality. *In:* Eckenwiler, Lisa A.; Cohn, Felicia G., eds. The Ethics of Bioethics: Mapping the Moral Landscape. Baltimore, MD: Johns Hopkins University Press, 2007: 205-219. Subject: 2.1

Zucchi, Pierluigi; Honings, Bonifacio. Cancer pain, death and dying. *Dolentium Hominum* 2007; 22(2): 61-72. Subject: 4.4

Zucker, David J.; Taylor, Bonita E. Spirituality, suffering, and prayerful presence within Jewish tradition. *In:* Puchalski, Christina M., ed. A Time for Listening and Caring: Spirituality and the Care of the Chronically Ill and Dying. Oxford; New York: Oxford University Press, 2006: 193-214. Subject: 20.4.1

Zuckerberg, Joaquin. International human rights for mentally ill persons: the Ontario experience. *International Journal of Law and Psychiatry* 2007 November-December; 30(6): 512-529. Subject: 21.1

Zuckerman, Shachar; Lahad, Amnon; Shmueli, Amir; Zimran, Ari; Peleg, Leah; Orr-Urtreger, Avi; Levy-Lahad, Ephrat; Sagi, Michal. Carrier screening for Gaucher disease: lessons for low-penetrance, treatable diseases. *JAMA: The Journal of the American Medical Association* 2007 September 19; 298(11): 1281-1290. Subject: 15.3

Zuzelo, Patti Rager. Exploring the moral distress of registered nurses. *Nursing Ethics* 2007 May; 14(3): 344-359. Subject: 4.1.3

Zwart, Hub. Professional ethics and scholarly communication. *In:* Korthals, Michiel; Bogers, Robert J., eds. Ethics for Life Scientists. Dordrecht, The Netherlands: Springer, 2004: 67-80. Subject: 1.3.9

Zwart, Hub. Statements, declarations and the problems of ethical expertise [editorial]. *Genomics, Society and Policy* 2007 April; 3(1): ii-iv. Subject: 15.1

Zwart, Nijmegen Hub. Genomics and self-knowledge: implications for societal research and debate. *New Genetics and Society* 2007 August; 26(2): 181-202. Subject: 15.1

Zycinski, J. Bioethics, technology and human dignity: the Roman Catholic viewpoint. *Acta Neurochirurgica Supplement* 2006; 98: 1-7. Subject: 18.5.4

Subject = NRCBL Primary Classification Number; see inside front cover.

Zycinski, Joseph. Ethics in medical technologies: the Roman Catholic viewpoint. *Journal of Clinical Neuroscience* 2006 June; 13(5): 518-523. Subject: 18.5.4

Zypries, Brigitte. From procreation to generation? Constitutional and legal-political issues in bioethics. *In:* Vöneky, Silja; Wolfrum, Rüdiger, eds. Human Dignity and Human Cloning. Leiden: Nijhoff, 2004: 107-121. Subject: 14.1

Anonymous

Access to health care for undocumented migrants in Europe [editorial]. *Lancet* 2007 December 22-2008 January 4; 370(9605): 2070. Subject: 9.5.1

Access to investigational drugs in the USA [editorial]. *Lancet* 2007 August 18-24; 370(9587): 540. Subject: 9.2

Addicted to secrecy: sealed drug documents should be opened up [editorial]. *Nature* 2007 April 19; 446(7138): 832. Subject: 9.7

AMA unveils health care ethics program, toolkit. *Healthcare Benchmarks and Quality Improvement* 2006 January; 13(1): 9-11. Subject: 9.1

Amniotic fluid supplies 'repair kit' for later life. *New Scientist* 2007 January 13-19; 193(2586): 9. Subject: 18.5.4

Animal-human hybrid-embryo research [editorial]. *Lancet* 2007 September 15-21; 370(9591): 909. Subject: 15.1

Annex: Relevant international and national documents. *In:* Vöneky, Silja; Wolfrum, Rüdiger, eds. Human Dignity and Human Cloning. Leiden: Nijhoff, 2004: 185-319. Subject: 14.5

Approval of nonprescription sale of Plan B muddies ethical waters. *ED Management* 2006 October; 18(10): 109-112. Subject: 11.1

Ashley's treatment: unethical or compassionate? [editorial]. *Lancet* 2007 January 13-19; 369(9556): 80. Subject: 9.5.3

Avoiding a chimaera quagmire [editorial]. *Nature* 2007 January 4; 445(7123): 1. Subject: 15.1

Biotechnology. *In:* Marks, Stephen P., ed. Health and Human Rights: Basic International Documents. Cambridge, MA: Harvard University, Francois-Xavier Bagnoud Center for Health and Human Rights, 2004: 281-304. Subject: 15.1

Board games: the way research on human subjects is overseen in the United States requires reform [editorial]. *Nature* 2007 August 2; 448(7153): 511-512. Subject: 18.6

Brain death revisited. *Health Care Ethics USA* 2007 Winter; 15(1): 15. Subject: 20.2.1

Britain to let women donate eggs for research. *New York Times* 2007 February 22; p. A13. Subject: 18.5.4

. . . but not as we know it. Synthetic life is on the way, and we need to think about the consequences [editorial]. *New Scientist* 2007 October 20-26; 196(2626): 5. Subject: 15.1

California dreaming: universities should draw the line at certain types of support from the drug industry [editorial]. *Nature* 2007 July 26; 448(7152): 388. Subject: 5.3

California suicide bill 'implicitly anti-Catholic' [news brief]. *America* 2007 May 28; 196(19): 7. Subject: 20.7

California Supreme Court to hear case involving physicians who refused to perform IVF because of religious beliefs. *Kaiser Daily Women's Health Policy Reports* 2007 August 3: 2 p. [Online]. Accessed: http://kaisernetwork.org/daily_reports/rep_index.cfm?DR_ID=46645 [2007 August 3]. Subject: 14.4

Catholic hospitals will comply with flawed law [news]. *America* 2007 October 15; 197(11): 7. Subject: 11.1

Celebrities in the ED: managers often face both ethical and operational challenges. *ED Management* 2006 December; 18(12): 133-135. Subject: 9.4

Chimera research should be lightly regulated, not banned [editorial]. *Lancet* 2007 January 20-26; 369(9557): 164. Subject: 18.5.4

China: doctors agree not to take organs from prisoners. *New York Times* 2007 October 6; p. A9. Subject: 19.5

Connecticut Catholic hospitals will comply with Plan B law. *Origins* 2007 October 11; 37(18). Subject: 11.1

Criteria creep: the politically motivated extension of US stem-cell registry makes no scientific sense [editorial]. *Nature* 2007 October 18; 449(7164): 756. Subject: 18.5.4

Directive action required: Europe's handling of applications to grow genetically modified crops amounts to bad governance [editorial]. *Nature* 2007 December 13; 450(7172): 921. Subject: 15.1

Discriminating on genes: the United States is belatedly establishing necessary protections in law. Others, take note [editorial]. *Nature* 2007 July 5; 448(7149): 2. Subject: 15.3

Doctors who fail their patients [editorial]. *New York Times* 2007 February 13; p. A22. Subject: 8.2

Dolly's legacy: ten years on, mammalian cloning is moving forward with central societal issues remaining unresolved. Yet human reproductive cloning seems inevitable [editorial]. *Nature* 2007 February 22; 445(7130): 795. Subject: 14.5

Do-it-yourself science: how much involvement can patient advocates have in genetics? [editorial]. *Nature* 2007 October 18; 449(7164): 755-756. Subject: 15.1

Due process — right to medical access — Supreme Court of Canada holds that ban on private health insurance violates Quebec charter of human rights and freedoms — Chaoulli v. Quebec (Attorney General), 2005 S.C.C. 35, 29272, [2005] S.C.J. No. 33 QUICKLAW (June 9, 2005). *Harvard Law Review* 2005 December; 119(2): 677-684. Subject: 9.1

Dying in America, post Schiavo. *Health Care Ethics USA* 2007 Summer; 15(3): 20-21. Subject: 20.5.1

See SUBJECT HEADING KEY FOR SECTION II on inside back cover.

635

Educational advantage. *Journal of Empirical Research on Human Research Ethics* 2007 March; 2(1): 47-48. Subject: 7.2

Emergency contraception: what's happening? *Health Care Ethics USA* 2007 Summer; 15(3): 20. Subject: 11.1

Enhancing, not cheating. A broad debate about the use of drugs that improve cognition for both the healthy and the ill is needed [editorial]. *Nature* 2007 November 15; 450(7168): 320. Subject: 4.5

Enough talk already: governments should act on researchers' attempts to engage the public over nanotechnology [editorial]. *Nature* 2007 July 5; 448(7149): 1-2. Subject: 5.4

An erosion of conscientious objection? *Health Care Ethics USA* 2007 Spring; 15(2): 15. Subject: 8.1

Ethical ruling and matured eggs offer hope for fertility [news brief]. *Nature* 2007 July 5; 448(7149): 13. Subject: 14.4

Ethics briefings. *Journal of Medical Ethics* 2007 April; 33(4): 247-248. Subject: 2.1

Ethics Committee to review physician-drug industry ties [news]. *Minnesota Medicine* 2007 June; 90(6): 23. Subject: 9.7

The ethics of public health [editorial]. *Lancet* 2007 December 1-7; 370(9602): 1801. Subject: 9.1

Euthanasia in Belgium up by 10% [news]. *BMJ: British Medial Journal* 2007 April 7; 334(7596): 714. Subject: 20.5.1

Extraordinary measures. Perpetuating a vegetative, unresponsive life may not in every case protect human dignity. *Christian Century* 2007 October 16; 124(21): 5. Subject: 20.5.1

Federal funding for embryo research unlikely to rise after Bush. *BioEdge* 2007 August 22; 262: 4. Subject: 5.3

Flogging Gardasil. In its rush to market its human papillomavirus vaccine, Merck forgot to make a strong and compelling case for compulsory immunization [editorial]. *Nature Biotechnology* 2007 March; 25(3): 261. Subject: 9.7

Forever young: can stunting the growth of a disabled child ever be a good thing? [editorial]. *New Scientist* 2007 January 13-19; 193(2586): 3. Subject: 9.5.3

France: doctor convicted of euthanasia but avoids prison. *New York Times* 2007 March 16; p. A6. Subject: 20.5.1

The future of mind control. *Economist* 2002 May 25; 363(8274): 11. Subject: 17.3

Gene therapy trials for cystic fibrosis to begin in U.K. [news]. *GeneWatch* 2007 January-February; 20(1): 19. Subject: 15.4

Genome abuse [editorial]. *Nature* 2007 September 27; 449(7161): 377. Subject: 15.1

German stem-cell law should change, says ethics council [news]. *Nature* 2007 July 26; 448(7152): 399. Subject: 18.5.4

How to be good? Mentoring and training for ethical behaviour aren't all their cracked up to be [editorial]. *Nature* 2007 October 11; 449(7163): 638. Subject: 1.3.9

HPV vaccine (Gardasil). *Health Care Ethics USA* 2007 Winter; 15(1): 15-16. Subject: 9.5.1

Humane and compassionate elder care as a human right [editorial]. *Lancet* 2007 August 25-31; 370(9588): 629. Subject: 9.5.2

An inconvenient truth: research on human embryonic stem cells must go on [editorial]. *Nature* 2007 November 29; 450(7170): 585-586. Subject: 18.5.4

Infected patient's lawyer says risk wasn't disclosed. *New York Times* 2007 November 18; p. A35. Subject: 19.5

The Islamic Code of Medical Ethics. *World Medical Journal* 1982 September-October; 29(5): 78-80. Subject: 6

Italy: abortion blunder rekindles debate. *New York Times* 2007 August 28; p. A13. Subject: 12.4.2

Italy: Cardinal says patient should have the right to die. *New York Times* 2007 January 23; p. A9. Subject: 20.5.1

Italy: doctor cleared in right-to-die case. *New York Times* 2007 March 7; p. A9. Subject: 20.5.1

Italy: No disciplinary action for doctor in right-to-die case. *New York Times* 2007 February 2; p. A8. Subject: 20.5.1

Japanese universities fire researchers for misconduct [news brief]. *Nature* 2007 January 4; 445(7123): 12. Subject: 1.3.9

Leading by example [editorial]. *Nature* 2007 January 18; 445(7125): 229. Subject: 1.3.9

Legal and illegal organ donation [editorial]. *Lancet* 2007 June 9-15; 369(9577): 1901. Subject: 19.5

The long and winding road [editorial]. *Nature* 2007 September 27; 449(7161): 377. Subject: 18.5.4

Making abortion legal, safe, and rare [editorial]. *Lancet* 2007 July 28-August 3; 370(9584): 291. Subject: 12.5.1

Meanings of 'life' [editorial]. *Nature* 2007 June 28; 447(7148): 1031-1032. Subject: 15.1

Messing with home brews. Political moves to expand FDA oversight to home brews are a bad idea. [editorial]. *Nature Biotechnology* 2007 March; 25(3): 262. Subject: 15.3

Mohler would favor altering 'gay' fetus [news]. *Christian Century* 2007 April 3; 124(7): 15. Subject: 15.2

Myriad Genetic launches direct-to-consumer advertising of breast cancer gene test in Northeastern cities [news]. *Kaiser Daily Women's Health Policy Report* 2007 September 11: 2 p. [Online]. Accessed: http://kaisernetwork.org/daily_reports/print_report.cfm?DR_ID=47409&dr_cat=2 [2007 September 11]. Subject: 15.3

Subject = NRCBL Primary Classification Number; see inside front cover.

A necessary vaccine [editorial]. *New York Times* 2007 February 26; p. A20. Subject: 9.5.5

New Jersey Catholic hospitals support stem cell research, promote cord blood donation. *Health Care Ethics USA* 2007 Winter; 15(1): 15. Subject: 18.5.1

A new tool to promote research: the Law on Biomedical Research [editorial]. *Revista de Derecho y Genoma Humano = Law and the Human Genome Review* 2007 January-June; (26): 17-20. Subject: 15.1

NIH to make chimpanzee breeding moratoriam permanent [news]. *ATLA: Alternatives to Laboratory Animals* 2007 June; 35(3): 300. Subject: 22.2

Of animal eggs and human embryos [editorial]. *New York Times* 2007 September 24; p. A22. Subject: 15.1

Paid vs. unpaid donors. *Vox Sanguinis* 2006 January; 90(1): 63-70. Subject: 19.4

Pennsylvania: advance directive; DNR order. *Mental and Physical Disability Law Reporter* 2007 March-April; 31(2): 312-313. Subject: 20.5.4

The plagiarism policy of the American Journal of Nursing. *AJN: American Journal of Nursing* 2007 July; 107(7): 78-79. Subject: 1.3.7

Presumed consent. *ATLA:Alternatives to Laboratory Animals* 2007 August; 35(4): 379. Subject: 19.5

Probity gone nuts [editorial]. *Nature Biotechnology* 2007 May 25(5): 483. Subject: 9.7

Proyecto CHIMBRIDS: chimeras and hybrids in comparative European and international research-scientific, ethical, philosophical and legal aspects. *Revista de Derecho y Genoma Humano = Law and the Human Genome Review* 2007 July-December; (27): 227-243. Subject: 15.1

Pulling rank: why should US military personnel be singled out for genetic discrimination? [editorial]. *Nature* 2007 August 30; 448(7157): 969. Subject: 15.3

Replicator review: Nature has implemented a peer-reviewed policy for strong claims [editorial]. *Nature* 2007 November 22; 450(7169): 457-458. Subject: 14.5

Right to refuse treatment: prisoner's claim that conditioning eligibility for parole on taking potentially medically inappropriate medication violated his due process rights is not frivolous. *Journal of the American Academy of Psychiatry and the Law* 2007; 35(2): 260-262. Subject: 8.3.4

Risk, consent and IRB models. *Protecting Human Subjects* 2007 November (15): 1, 4-5. Subject: 18.6

Safeguarding clinical trials [editorial]. *Nature Medicine* 2007 February; 13(2): 107. Subject: 18.2

Science at WHO and UNICEF: the corrosion of trust [editorial]. *Lancet* 2007 September 22-28; 370(9592): 1007. Subject: 1.3.9

Should Ashkenazi Jews volunteer for gene studies? *Moment* 1999 April; 24(2): 32-33. Subject: 15.11

Spare the apes [news]. *New Scientist* 2007 September 15-21; 195(2621): 4. Subject: 22.2

Special report: Switzerland: an appointment with death. *Update (International Task Force on Euthanasia and Assisted Suicide)* 2007; 21(1): 4 p. [Online]. Accessed: http://www.internationaltaskforce.org/ina40.htm#2 [2008 February 4]. Subject: 20.5.1

Sterile thinking. *New Atlantis* 2007 Winter; 15: 143. Subject: 15.5

Timeout, not the final buzzer, in the stem cell debate [editorial]. *Lancet* 2007 December 1-7; 370(9602): 1802. Subject: 18.5.1

Treatments to keep disabled girl small stir debate. *Washington Post* 2007 January 5; p. A2. Subject: 9.5.3

Two sample policies: donation after cardiac death. *Health Care Ethics USA* 2007 Summer; 15(3): 12-18. Subject: 19.5

Uninformed consent? The US should revamp rules on informed consent to ensure that people have all the information and support they need before deciding to enroll in clinical trials [editorial]. *Nature Medicine* 2007 September; 13(9): 999. Subject: 18.3

An unwieldy hybrid [editorial]. *Nature* 2007 May 24; 447(7143): 353-354. Subject: 18.5.4

The use of human-animal hybrid embryos [news]. *ATLA: Alternatives to Laboratory Animals* 2007 March; 35(1): 8-9. Subject: 15.1

Vatican clarifies position on artificial nutrition [news]. *America* 2007 October 1; 197(9): 6. Subject: 20.5.1

When in doubt, disclose [editorial]. *Lancet* 2007 February 3-9; 369(9559): 344. Subject: 1.3.9

Who is accountable? [editorial]. *Nature* 2007 November 1; 450(7166): 1. Subject: 1.3.7

See SUBJECT HEADING KEY FOR SECTION II on inside back cover.

637

SECTION III:
MONOGRAPHS

SUBJECT ENTRIES

Section III: Monographs Contents

1 Ethics
    1.1 Philosophical Ethics 645
    1.2 Religious Ethics 645
    1.3 Applied and Professional Ethics
        1.3.1 General 647
        1.3.2 Business 647
        1.3.3 Education 647
        1.3.4 Engineering 648
        1.3.5 Government/Criminal Justice 648
        1.3.6 International Affairs 648
        1.3.7 Journalism/Mass Media 648
        1.3.9 Scientific Research 648
        1.3.10 Social Work 649
        1.3.11 Agriculture 649
        1.3.12 Information Technology 649

2 Bioethics
    2.1 General 649
    2.2 History of Health Ethics 653
    2.3 Education/Programs 653

3 Philosophy of Biology
    3.1 General 653
    3.2 Evolution and Creation 653

4 Philosophy of Medicine and Health
    4.1 Philosophy of Medicine, Nursing, & Other Health Professions
        4.1.1 General 653
        4.1.2 Philosophy of Medicine 654
        4.1.3 Philosophy of Nursing 654
    4.3 Concept of Mental Health 654
    4.4 Quality/Value of Life/Personhood 655

Only those classes with entries in this volume appear on this Contents list.

5 Science/Technology and Society

    5.1 General 655

    5.2 Technology Assessment 656

    5.3 Social Control of Science/Technology 656

    5.4 Nanotechnology 656

6 Codes of/Position Statements on Professional Ethics

    6 General 657

7 Sociology of Health Care

    7.1 General 657

    7.2 Education for Health Care Professionals 658

    7.3 Professional-Professional Relationship 659

    7.4 Professional Misconduct 659

8 Patient Relationships

    8.1 General 659

    8.2 Truth Disclosure 659

    8.3 Informed Consent

        8.3.1 General 659

        8.3.4 Right to Refuse Treatment 659

    8.4 Confidentiality 659

    8.5 Malpractice 660

9 Health Care

    9.1 General 660

    9.3 Health Care Economics

        9.3.1 General 662

        9.3.2 Managed Care 663

    9.4 Allocation of Health Care Resources 663

    9.5 Health Care for Specific Diseases or Groups

        9.5.1 General 663

        9.5.2 Health Care for the Aged 664

        9.5.3 Health Care for Mentally Disabled Persons 664

        9.5.4 Health Care for Minorities 664

        9.5.5 Health Care for Women 664

        9.5.6 Health Care for HIV Infection and AIDS 665

        9.5.7 Health Care for Minors 665

        9.5.8 Health Care for Embryos and Fetuses 666

        9.5.9 Health Care for Substance Abusers/Users of Controlled Substances . . 666

        9.5.10 Health Care for Indigents 666

    9.6 Ethics Committees/Consultation 666

    9.7 Drugs and Pharmaceutical Industry 666

    9.8 Quality of Health Care 667

10 Sexuality/Gender

    10 General 667

11 Contraception

    11.1 General 668

12 Abortion

    12.1 General 668

    12.3 Moral and Religious Aspects 668

    12.4 Legal Aspects

        12.4.1 General 668

    12.5 Social Aspects

        12.5.1 General 668

        12.5.2 Abortion: Demographic Surveys 669

13 Population

    13.1 General 669

    13.2 Population Growth 669

    13.3 Population Policy 669

14 Reproduction/Reproductive Technologies

    14.1 General 669

    14.2 Artificial Insemination and Surrogacy 670

    14.3 Sex Predetermination/Selection 670

    14.4 In Vitro Fertilization and Embryo Transfer 670

    14.5 Cloning 670

15 Genetics, Molecular Biology and Microbiology

    15.1 General 670

    15.2 Genetic Counseling/Prenatal Diagnosis 672

    15.3 Genetic Screening/Testing 672

    15.4 Gene Therapy/Transfer 673

    15.5 Eugenics 673

    15.6 Behavioral Genetics 673

    15.8 Genetic Patents 673

    15.9 Sociobiology 674

16 Environmental Quality

    16.1 General 674

17 The Neurosciences and Mental Health Therapies

    17.1 General 675

    17.2 Psychotherapy 677

    17.3 Behavior Modification 677

    17.4 Psychopharmacology 677

    17.5 Electrical Stimulation of the Brain 677

18 Human Experimentation

    18.1 General 677

    18.2 Policy Guidelines/Institutional Review Boards 678

    18.3 Informed Consent 678

    18.4 Behavior Research 678

    18.5 Research on Special Populations

        18.5.1 General 678

        18.5.2 Research on Newborns or Minors 678

        18.5.4 Research on Embryos or Fetuses 679

        18.5.5 Research on Prisoners. 679

        18.5.6 Research on Mentally Ill and Disabled Persons 679

    18.6 Social Control of Human Experimentation 679

19 Artificial and Transplanted Organs/Tissues

    19.1 General 679

    19.2 Artificial and Transplanted Hearts 679

    19.5 Donation/Procurement of Organs/Tissues 680

20 Death and Dying

    20.1 General 680

    20.2 Definition/Determination of Death

        20.2.1 General 680

    20.3 Attitudes Toward Death

        20.3.2 Health Providers 680

    20.4 Care of the Dying Patient

        20.4.1 General 680

        20.4.2 Care of the Dying Child. 681

    20.5 Prolongation of Life and Euthanasia

        20.5.1 General 681

        20.5.2 Allowing Minors to Die. 682

        20.5.4 Living Wills/Advance Directives 683

    20.6 Capital Punishment 683

    20.7 Suicide/Assisted Suicide 683

21 International/Political Dimensions of Biology and Medicine

    21.1 General 683

    21.3 Chemical and Biological Weapons 683

    21.4 Torture and Genocide. 684

    21.7 Cultural Pluralism. 686

22 Animal Welfare

    22.1 General 687

    22.2 Animal Experimentation 687

    22.3 Animal Production 687

New Journal Subscriptions. 687

New Publications from the Kennedy Institute of Ethics. 688

Title Index 691

National Reference Center for Bioethics Literature Classification Scheme:

                                    Inside Front Cover

SECTION III: MONOGRAPHS
SUBJECT ENTRIES

1.1 PHILOSOPHICAL ETHICS

Burley, Justine, ed. DWORKIN AND HIS CRITICS: WITH RE-PLIES BY DWORKIN. Malden, MA: Blackwell, 2004. 412 p. ISBN 0-631-19766-4. (Philosophers and Their Critics series; Vol. 11.) [HM665 .D86 2004] (1.1; 1.3.5; 4.4; 8.1; 9.4; 12.3; 14.5; 15.1; 20.5.1; 20.7)

Daleiden, Joseph L. THE SCIENCE OF MORALITY: THE INDI-VIDUAL, COMMUNITY AND FUTURE GENERATIONS. Amherst, NY: Prometheus Books, 1998. 534 p. ISBN 1-57392-225-0. [BJ57 .D35 1998] (1.1; 1.3.2; 9.5.9; 10; 12.3; 20.5.1)

Fan, Ruiping. SOCIAL JUSTICE IN HEALTH CARE: A CRITICAL APPRAISAL. Ann Arbor, MI: ProQuest Information & Learning/UMI, 1999. 219 p. (Publication Order No. AAT-9928532. Dissertation, (Ph.D. in Philosophy)—Rice University, 1999.) (1.1; 2.1; 9.1)

Geach, Mary and Gormally, Luke, eds. HUMAN LIFE, AC-TION, AND ETHICS: ESSAYS BY G.E.M. ANSCOMBE. Exeter/Charlottesville, VA: Imprint Academic, 2005. 298 p. ISBN 978-1-84540-061-3; ISBN 1-84540-061-5. (St. Andrews Studies in Philosophy and Public Affairs series.) [BJ1011 .A57 2005] (1.1; 1.2; 4.4; 18.3; 20.5.1; 20.5.2)

Held, Virginia. THE ETHICS OF CARE: PERSONAL, POLITI-CAL, AND GLOBAL. Oxford/New York: Oxford University Press, 2006. 211 p. ISBN 978-0-19-532590-4; ISBN 0-19-518099-2. [BJ1475 .H45 2006] (1.1; 10)

Kamm, F.M. INTRICATE ETHICS: RIGHTS, RESPONSIBILI-TIES, AND PERMISSIBLE HARM. Oxford/New York: Oxford University Press, 2007. 509 p. ISBN 978-0-19-518969-8; ISBN 0-19-518969-8. (Oxford Ethics Series.) [BJ1031 .K36 2007] (1.1)

Munro, Donald J. A CHINESE ETHICS FOR THE NEW CEN-TURY: THE CH'IEN MU LECTURES IN HISTORY AND CULTURE, AND OTHER ESSAYS ON SCIENCE AND CONFUCIAN ETHICS. Hong Kong: Chinese University Press, 2005. 158 p. ISBN 978-962-996-056-8; ISBN 962-996-056-7. (Ch'ien Mu Lectures series.) [BJ117 .M86 2005] (1.1; 5.1)

Nadler, Steven. SPINOZA'S ETHICS: AN INTRODUCTION. New York: Cambridge University Press, 2006. 281 p. ISBN 978-0-521-54479-5; ISBN 0-521-54479-3. (Cambridge Introductions to Key Philosophical Texts series. Gift of Max M. and Marjorie B. Kampelman.) [B3974 .N28 2006] (1.1; 1.2)

Teays, Wanda. SECOND THOUGHTS: CRITICAL THINKING FROM A MULTICULTURAL PERSPECTIVE. Mountain View, CA: Mayfield, 1996. 484 p. ISBN 1-55934-479-2. [BC177 .T4 1996] (1.1; 21.7)

Valls, Andrew, ed. RACE AND RACISM IN MODERN PHILOS-OPHY. Ithaca, NY: Cornell University Press, 2005. 293 p. ISBN 0-8014-7274-1. [HT1523 .R2517 2005] (1.1; 1.2; 1.3.5; 21.1)

Van Hooft, Stan. UNDERSTANDING VIRTUE ETHICS. Chesham, Bucks, UK: Acumen, 2006. 184 p. ISBN 1-84465-045-6. (Understanding Movements in Modern Thought series.) [BJ1531 .H66 2006] (1.1; 1.3.5; 12.3; 20.5.1)

Waal, Frans de. PRIMATES AND PHILOSOPHERS: HOW MO-RALITY EVOLVED. Princeton, NJ: Princeton University Press, 2006. 209 p. ISBN 978-0-691-12447-6; ISBN 0-691-12447-7. [BJ1311 .W14 2006] (1.1; 3.2; 22.1)

Walker, Rebecca L. and Ivanhoe, Philip J., eds. WORKING VIRTUE: VIRTUE ETHICS AND CONTEMPORARY MORAL PROBLEMS. Oxford: Clarendon Press/New York: Oxford University Press, 2007. 319 p. ISBN 978-0-19-927165-8. (Gift of the publisher.) [BJ1518 .W67 2007] (1.1; 1.3.2; 1.3.8; 4.1.2; 4.4; 16.1; 17.1; 22.1)

Wong, David B. NATURAL MORALITIES: A DEFENSE OF PLU-RALISTIC RELATIVISM. Oxford/New York: Oxford University Press, 2006. 293 p. ISBN 978-0-19-530539-5; ISBN 0-19-530539-6. [BJ37 .W66 2006] (1.1)

1.2 RELIGIOUS ETHICS

Abrahams, Israel and Fine, Lawrence. HEBREW ETHICAL WILLS. Philadelphia: Jewish Publication Society, 2006, c1926. 363 p. ISBN 978-0-8276-0827-6; ISBN 0-8276-0827-6. (Expanded facsimile edition. Foreword by Judah Goldin. Expanded and with a new introduction by Lawrence Fine. Originally published in 1926. "Edward E. Elson Classics." Gift of Max M. and Marjorie B. Kampelman.) [BJ1286 .W59 T69 2006] (1.2)

Berger, Yitzhak and Shatz, David, eds. JUDAISM, SCIENCE, AND MORAL RESPONSIBILITY. Lanham, MD: Rowman & Littlefield, 2006. 303 p. ISBN 978-0-7425-4596-0; ISBN 0-7425-4596-2. (The Orthodox Forum Series. Gift of Max M. and Marjorie B. Kampelman.) [BM538 .S3 O78 2002] (1.2; 5.1; 15.1; 17.1; 17.2)

Cohn-Sherbok, Dan. THE PARADOX OF ANTI-SEMITISM. London/New York: Continuum, 2006. 242 p. ISBN 978-0-8264-8896-1; ISBN 0-8264-8896-X. (Gift of Max M. and Marjorie B. Kampelman.) [DS145 .C5738 2006] (1.2; 21.4)

Csikszentmihalyi, Mark. MATERIAL VIRTUE: ETHICS AND THE BODY IN EARLY CHINA. Leiden: Brill, 2004. 402 p. ISBN 978-90-04-14196-4; ISBN 90-04-14196-0. (Sinica Leidensia series; Vol. 66. ISSN 0169-9563.) [BJ117 .C75 2004] (1.2; 4.4)

Cutter, William, ed. HEALING AND THE JEWISH IMAGINATION: SPIRITUAL AND PRACTICAL PERSPECTIVES ON JUDAISM AND HEALTH. Woodstock, VT: Jewish Lights, 2007. 219 p. ISBN 978-1-58023-314-9; ISBN 1-58023-314-7. (Gift of Max M. and Marjorie B. Kampelman.) [BM538 .H43 H43 2007] (1.2; 2.1; 9.1)

Dorff, Elliot N. THE UNFOLDING TRADITION: JEWISH LAW AFTER SINAI. New York: Aviv Press, 2005. 566 p. ISBN 978-0-916219-29-1; ISBN 0-916219-29-1. (Gift of Max M. and Marjorie B. Kampelman.) [BM521 .D67 2005] (1.2)

Ellenson, David. AFTER EMANCIPATION: JEWISH RELIGIOUS RESPONSES TO MODERNITY. Cincinnati, OH: Hebrew Union College Press, 2004. 547 p. ISBN 978-0-87820-223-2; ISBN 0-87820-223-4. (Gift of Max M. and Marjorie B. Kampelman.) [BM195 .E42 2004] (1.2)

Greenberg, Irving. FOR THE SAKE OF HEAVEN AND EARTH: THE NEW ENCOUNTER BETWEEN JUDAISM AND CHRISTIANITY. Philadelphia: Jewish Publication Society, 2004. 274 p. ISBN 978-0-8276-0807-8; ISBN 0-8276-0807-1. (Gift of Max M. and Marjorie B. Kampelman.) [BM535 .G73 2004] (1.2)

Hockenos, Matthew D. A CHURCH DIVIDED: GERMAN PROTESTANTS CONFRONT THE NAZI PAST. Bloomington: Indiana University Press, 2004. 269 p. ISBN 978-0-253-34448-9; ISBN 0-253-34448-4. (Gift of Max M. and Marjorie B. Kampelman.) [BR856 .H685 2004] (1.2; 1.3.5)

Katz, Steven T., ed. THE CAMBRIDGE HISTORY OF JUDAISM, VOLUME IV: THE LATE ROMAN-RABBINIC PERIOD. New York: Cambridge University Press, 2006. 1135 p. ISBN 978-0-521-77248-8; ISBN 0-521-77248-6. (Gift of Max M. and Marjorie B. Kampelman.) [BM155.2 .C35 v.4] (1.2)

Kellner, Menachem. MAIMONIDES' CONFRONTATION WITH MYSTICISM. Oxford, UK/Portland, OR: Littman Library of

Jewish Civilization, 2006. 343 p. ISBN 978-1-904113-29-4. (Littman Library of Jewish Civilization series. Gift of Max M. and Marjorie B. Kampelman.) [BM755 .M6 K45 2006] (1.2)

Kohn, Livia. COSMOS AND COMMUNITY: THE ETHICAL DIMENSION OF DAOISM. Cambridge, MA: Three Pines Press, 2004. 291 p. ISBN 1-931483-02-7. [BJ1290.8 .K65 2004] (1.2)

Malka, Salomon. EMMANUEL LEVINAS: HIS LIFE AND LEGACY. Pittsburgh, PA: Duquesne University Press, 2006. 330 p. ISBN 978-0-8207-0358-9; ISBN 0-8207-0358-3. (Foreword by Philippe Nemo. Translation from the French by Michael Kigel and Sonja Embree of: *La Vie et la trace*. Gift of Max M. and Marjorie B. Kampelman.) [B2430 .L484 M3413 2006] (1.2; Biography)

Na'im, 'Abd Allah Ahmad, ed. ISLAMIC FAMILY LAW IN A CHANGING WORLD: A GLOBAL RESOURCE BOOK. London/New York: Zed Books; Distributed in the US by: New York: Palgrave, 2002. 320 p. ISBN 1-84277-093-4. [KBP540.3 .I85 2002] (1.2; 7.1; 9.5.5; 9.5.7; 21.1; 21.7)

Peters, F.E. THE CHILDREN OF ABRAHAM: JUDAISM, CHRISTIANITY, ISLAM. Princeton, NJ: Princeton University Press, 2006. 237 p. ISBN 978-0-691-12769-9; ISBN 0-691-12769-7. ("A new edition." Princeton Classic Editions. Foreword by John L. Esposito. Gift of Max M. and Marjorie B. Kampelman.) [BM157 .P47 2006] (1.2)

Rizvi, Sayyid Muhammad. MARRIAGE & MORALS IN ISLAM. Qum: Ansariyan Publications, 2001. 122 p. ISBN 978-964-438-084-6; ISBN 964-438-084-3. (Fifth edition. Arabic title: *Akhlaq-i izdivaj dar Islam*.) (1.2; 10; 11.1; 12.1; 14.1)

Römelt, Josef. FREIHEIT, DIE MEHR IST ALS WILLKÜR: CHRISTLICHE ETHIK IN ZWISCHENMENSCHLICHER BEZIEHUNG, LEBENSGESTALTUNG, KRANKHEIT UND TOD. Regensburg: Pustet, 1997. 320 p. ISBN 3-7917-1538-0. (Handbuch der Moraltheologie series; Bd. 2.) [BJ1249 .R764 1997] (1.2; 2.1; 20.1)

Sacks, Jonathan. TO HEAL A FRACTURED WORLD: THE ETHICS OF RESPONSIBILITY. New York: Schocken Books, 2005. 280 p. ISBN 978-0-8052-1196-2. (Gift of Max M. and Marjorie B. Kampelman.) [BJ1451 .S23 2005] (1.2)

Schweiker, William; Johnson, Michael A.; and Jung, Kevin, eds. HUMANITY BEFORE GOD: CONTEMPORARY FACES OF JEWISH, CHRISTIAN, AND ISLAMIC ETHICS. Minneapolis, MN: Fortress Press, 2006. 326 p. ISBN 978-0-8006-3822-1; ISBN 0-8006-3822-0. [BJ1188 .H84 2006] (1.2; 1.3.5; 2.1; 4.4; 21.1; 21.2)

Seeskin, Kenneth. MAIMONIDES ON THE ORIGIN OF THE WORLD. New York: Cambridge University Press, 2005. 215 p. ISBN 978-0-521-69752-1; ISBN 0-521-69752-2. (Gift of Max M. and Marjorie B. Kampelman.) [B759 .M34 S44 2007] (1.2; 3.2)

Sorajjakool, Siroj. WHEN SICKNESS HEALS: THE PLACE OF RELIGIOUS BELIEF IN HEALTHCARE. Philadelphia:

Templeton Foundation Press, 2006. 149 p. ISBN 978-1-59947-090-0; ISBN 1-59947-090-X. [BL65 .M4 S67 2006] (1.2; 4.4; 9.1; 17.1)

Steinsaltz, Adin. THE TALMUD, THE STEINSALTZ EDITION, VOLUME 6: TRACTATE BAVA METZIA, PART VI. New York: Random House, 1993. 375 p. ISBN 978-0-679-41378-3; ISBN 0-674-41378-2. (Gift of Max M. and Marjorie B. Kampelman.) (1.2)

Steinsaltz, Adin. THE TALMUD, THE STEINSALTZ EDITION, VOLUME 19: TRACTATE SANHEDRIN, PART V. New York: Random House, 1999. 187 p. ISBN 978-0-375-50248-4; ISBN 0-375-50248-3. (Gift of Max M. and Marjorie B. Kampelman.) (1.2)

Telushkin, Joseph. A CODE OF JEWISH ETHICS, VOLUME 1: YOU SHALL BE HOLY. New York: Bell Tower, 2006. 559 p. ISBN 978-1-4000-4835-9; ISBN 1-4000-4835-4. (Gift of Max M. and Marjorie B. Kampelman.) [BJ1285.2 .T45 2006] (1.2; 1.1)

Tirosh-Samuelson, Hava, ed. WOMEN AND GENDER IN JEW-ISH PHILOSOPHY. Bloomington: Indiana University Press, 2004. 356 p. ISBN 0-253-21673-7. (Jewish Literature and Culture series. Proceedings of a conference held 25-26 February 2001, at Arizona State University. Gift of Max M. and Marjorie B. Kampelman.) [B5800 .W66 2004] (1.2; 1.1; 10)

Walzer, Michael, ed. LAW, POLITICS, AND MORALITY IN JU-DAISM. Princeton, NJ: Princeton University Press, 2006. 217 p. ISBN 978-0-691-12508-4; ISBN 0-691-12508-2. (Ethikon Series in Comparative Ethics. Gift of Max M. and Marjorie B. Kampelman.) [KBM524.14 .L39 2006] (1.2; 1.3.5; 21.2)

Wodzinski, Marcin. HASKALAH AND HASIDISM IN THE KINGDOM OF POLAND: A HISTORY OF CONFLICT. Oxford, UK/Portland, OR: Littman Library of Jewish Civilization, 2005. 335 p. ISBN 978-1-904113-08-9; ISBN 1-904113-08-7. (Littman Library of Jewish Civilization series. Translation from the Polish by Sarah Cozens and Agnieszka Mirowska of: *Oswiecenie zydowskie w Królestwie Polskim wobec chasydyzmu.* Gift of Max M. and Marjorie B. Kampelman.) [BM198.4 .P6 W6313 2005] (1.2)

Zohar, Zion, ed. SEPHARDIC AND MIZRAHI JEWRY: FROM THE GOLDEN AGE OF SPAIN TO MODERN TIMES. New York: New York University Press, 2005. 343 p. ISBN 978-0-8147-9706-8; ISBN 0-8147-9706-7. (Gift of Max M. and Marjorie B. Kampelman.) [DS135 .S7 S4525 2005] (1.2)

1.3.1 APPLIED AND PROFESSIONAL ETHICS: GENERAL

Bennett, Mark D. and Gibson, Joan McIver. A FIELD GUIDE TO GOOD DECISIONS: VALUES IN ACTION. Westport, CT: Praeger, 2006. 194 p. ISBN 978-0-275-98937-8; ISBN 0-275-98937-2. [HD30.23 .B462 2006] (1.3.1)

Boas, Franz. ANTHROPOLOGY & MODERN LIFE. New Bruns-wick, NJ: Transaction Publishers, 2004. 333 p. ISBN 978-0-7658-0535-5; ISBN 0-7658-0535-9. (Classics in Anthropology series. "With a new introduction and afterword by Herbert S. Lewis." Originally published as new and revised edition, New York: Norton, 1932.) [GN27 .B6 2004] (1.3.1; 15.5; 21.7)

Boxill, Jan, ed. SPORTS ETHICS: AN ANTHOLOGY. [Malden, MA]: Blackwell, 2003. 351 p. ISBN 978-0-631-21697-1; ISBN 0-631-21697-9. (Blackwell Philosophy Anthologies series.) [GV706.3 .S68 2003] (1.3.1; 1.3.3; 4.5; 9.5.9; 10; 15.1)

Peterson, Grethe B., ed. THE TANNER LECTURES ON HUMAN VALUES, VOLUME 26. Salt Lake City: University of Utah Press, 2006. 237 p. ISBN 978-0-87480-872-8; ISBN 0-87480-872-2. (Tanner Lectures on Human Values se-ries. ISSN 0275-7656.) (1.3.1; 1.3.5; 21.1; 21.2)

Pritchard, Michael S. PROFESSIONAL INTEGRITY: THINKING ETHICALLY. Lawrence: University Press of Kansas, 2006. 195 p. ISBN 0-7006-1446-X. [BJ1725 .P752 2006] (1.3.1; 1.3.2; 1.3.4; 4.1.1; 6)

1.3.2 APPLIED AND PROFESSIONAL ETHICS: BUSINESS

Brown, Marvin T. CORPORATE INTEGRITY: RETHINKING ORGANIZATIONAL ETHICS AND LEADERSHIP. Cam-bridge/New York: Cambridge University Press, 2005. 260 p. ISBN 978-0-521-60657-8; ISBN 0-521-60657-8. [HD60 .B766 2005] (1.3.2; 16.1; 21.7)

Hirschland, Matthew J. CORPORATE SOCIAL RESPONSIBIL-ITY AND THE SHAPING OF GLOBAL PUBLIC POLICY. New York: Palgrave Macmillan, 2006. 202 p. ISBN 978-1-4039-7453-2; ISBN 1-4039-7453-5. (Political Evo-lution and Institutional Change series. Gift of the pub-lisher.) [HD60 .H574 2006] (1.3.2; 1.3.5; 21.1)

1.3.3 APPLIED AND PROFESSIONAL ETHICS: EDUCATION

Campbell, Elizabeth. THE ETHICAL TEACHER. Maiden-head/Philadelphia: Open University Press, 2003. 178 p. ISBN 978-0-335-21218-7; ISBN 0-335-21218-2. (Profes-sional Learning series.) [LB1027 .C245 2003] (1.3.3)

Steinweis, Alan E. STUDYING THE JEW: SCHOLARLY ANTISEMITISM IN NAZI GERMANY. Cambridge, MA: Har-vard University Press, 2006. 203 p. ISBN 978-0-674-02205-8; ISBN 0-674-02205-X. [DS146 .G4 S73 2006] (1.3.3; 1.3.5; 21.4)

Strathern, Marilyn, ed. AUDIT CULTURES: ANTHROPOLOGI-CAL STUDIES IN ACCOUNTABILITY, ETHICS AND THE ACAD-EMY. London/New York: Routledge, 2000. 310 p. ISBN 978-0-415-23327-9; ISBN 0-415-23327-5. (European As-sociation of Social Anthropologists series.) [LB2324 .A87 2000] (1.3.3; 1.3.2; 1.3.5)

1.3.4 APPLIED AND PROFESSIONAL ETHICS: ENGINEERING

National Academy of Engineering (United States). EMERGING TECHNOLOGIES AND ETHICAL ISSUES IN ENGINEERING: PAPERS FROM A WORKSHOP, OCTOBER 14-15, 2003. Washington, DC: National Academies Press, 2004. 155 p. ISBN 0-309-09271-X. (Also available on the Web at: http://www.nap.edu.) [TA157 .E45 2004] (1.3.4; 1.3.9; 1.3.12; 5.3; 5.4; 17.1)

Spier, Raymond. ETHICS, TOOLS, AND THE ENGINEER. Boca Raton, FL: CRC Press, 2001. 306 p. ISBN 0-8493-3740-2. (Technology Management Series.) [BJ59 .S67 2001] (1.3.4; 1.1; 1.3.9; 5.1; 14.5; 15.1; 15.5; 18.1; 22.2)

1.3.5 APPLIED AND PROFESSIONAL ETHICS: GOVERNMENT/ CRIMINAL JUSTICE

Herf, Jeffrey. THE JEWISH ENEMY: NAZI PROPAGANDA DURING WORLD WAR II AND THE HOLOCAUST. Cambridge, MA: Belknap Press of Harvard University Press, 2006. 390 p. ISBN 978-0-674-02175-4; ISBN 0-674-02175-4. (Gift of Max M. and Marjorie B. Kampelman.) [D810 .P7 G337 2006] (1.3.5; 1.3.7; 15.5; 21.4)

Rawls, John. POLITICAL LIBERALISM. New York: Columbia University Press, 2005. 525 p. ISBN 978-0-231-13089-9; ISBN 0-231-13089-9. (Expanded edition. Columbia Classics in Philosophy series.) [JC578 .R37 2005] (1.3.5; 1.1)

Richardson, Henry S. DEMOCRATIC AUTONOMY: PUBLIC REASONING ABOUT THE ENDS OF POLICY. Oxford/New York: Oxford University Press, 2002. 316 p. ISBN 0-19-515091-0. (Oxford Political Theory series.) [JC423 .R485 2002] (1.3.5)

Smith, Steven B. READING LEO STRAUSS: POLITICS, PHILOSOPHY, JUDAISM. Chicago: University of Chicago Press, 2006. 256 p. ISBN 978-0-226-76402-3; ISBN 0-226-76402-8. (Gift of Max M. and Marjorie B. Kampelman.) [JA71 .S65 2006] (1.3.5; 1.2)

Zuckert, Catherine and Zuckert, Michael. THE TRUTH ABOUT LEO STRAUSS: POLITICAL PHILOSOPHY AND AMERICAN DEMOCRACY. Chicago: University of Chicago Press, 2006. 306 p. ISBN 978-0-226-99332-4; ISBN 0-226-99332-9. (Gift of Max M. and Marjorie B. Kampelman.) [RA395 .A3 M4184 2006] (1.3.5)

1.3.6 APPLIED AND PROFESSIONAL ETHICS: INTERNATIONAL AFFAIRS

Falk, Avner. FRATRICIDE IN THE HOLY LAND: A PSYCHOANALYTIC VIEW OF THE ARAB-ISRAELI CONFLICT. Madison: University of Wisconsin Press, Terrace Books, 2004. 271 p. ISBN 978-0-299-20250-7; ISBN 0-299-20250-X. (Gift of Max M. and Marjorie B. Kampelman.) [DS119.7 .F318 2004] (1.3.6; 1.3.5; 17.1; 21.2)

1.3.7 APPLIED AND PROFESSIONAL ETHICS: JOURNALISM/ MASS MEDIA

Kiernan, Vincent. EMBARGOED SCIENCE. Urbana: University of Illinois Press, 2006. 176 p. ISBN 978-0-252-03097-0; ISBN 0-252-03097-4. [Q225 .K48 2006] (1.3.7; 1.3.9; 5.3)

Lipson, Charles. DOING HONEST WORK IN COLLEGE: HOW TO PREPARE CITATIONS, AVOID PLAGIARISM, AND ACHIEVE REAL ACADEMIC SUCCESS. Chicago: University of Chicago Press, 2004. 189 p. ISBN 0-226-48473-4. (Chicago Guides to Writing, Editing, and Publishing series.) [PN171 .F56 L56 2004] (1.3.7; 1.3.3; 1.3.12)

1.3.9 APPLIED AND PROFESSIONAL ETHICS: SCIENTIFIC RESEARCH

Eaton, Margaret L. and Kennedy, Donald. INNOVATION IN MEDICAL TECHNOLOGY: ETHICAL ISSUES AND CHALLENGES. Baltimore, MD: Johns Hopkins University Press, 2007. 155 p. ISBN 978-0-8018-8526-6; ISBN 0-8018-8526-4. (Gift of the publisher.) [R855.3 .E28 2007] (1.3.9; 2.2; 5.1; 5.2; 9.7; 14.1; 18.1)

Goodman, Allegra. INTUITION: A NOVEL. New York: Dial Press/Random House, 2006. 344 p. ISBN 978-0-385-33612-3; ISBN 0-385-33612-8. [PS3557 .O5829 I58 2006] (1.3.9; 18.1; Fiction)

Israel, Mark and Hay, Iain. RESEARCH ETHICS FOR SOCIAL SCIENTISTS: BETWEEN ETHICAL CONDUCT AND REGULATORY COMPLIANCE. London/Thousand Oaks, CA: Sage, 2006. 193 p. ISBN 978-1-4129-0390-5; ISBN 1-4129-0390-4. [H62 .I785 2006] (1.3.9; 1.1; 5.2; 6; 8.4; 18.2; 18.3; 18.4)

Korthals, Michiel and Bogers, Robert J., eds. ETHICS FOR LIFE SCIENTISTS. Dordrecht/Nowell, MA: Springer, 2004. 219 p. ISBN 1-4020-3178-5. (Wageningen UR Frontis Series; Vol. 5.) [QH331 .E712 2004] (1.3.9; 1.3.2; 1.3.7; 3.1; 5.3; 15.1; 15.8; 22.2)

Kurtz, Paul and Koepsell, David, ed. SCIENCE AND ETHICS: CAN SCIENCE HELP US MAKE WISE MORAL JUDGMENTS? Amherst, NY: Prometheus Books, 2007. 359 p. ISBN 978-1-59102-537-5; ISBN 1-59102-537-0. (Gift of the publisher.) [Q175.35 .S346 2007] (1.3.9; 2.1; 5.1; 15.1)

Oliver, Paul. THE STUDENT'S GUIDE TO RESEARCH ETHICS. Maidenhead, UK/New York: Open University Press, 2003. 156 p. ISBN 0-335-21087-2. [Q180.55 .M67 O38 2003] (1.3.9; 1.3.7; 5.3; 8.4; 18.2; 18.3)

Wagner, Wendy and Steinzor, Rena, eds. RESCUING SCIENCE FROM POLITICS: REGULATION AND THE DISTORTION OF SCIENTIFIC RESEARCH. Cambridge/New York: Cambridge University Press, 2006. 304 p. ISBN 0-521-54009-7. (Gift of the publisher.) [Q125 .R4178 2006] (1.3.9; 1.3.5; 1.3.7; 5.3; 9.7; 16.1; 21.1)

1.3.10 APPLIED AND PROFESSIONAL ETHICS: SOCIAL WORK

Beckett, Chris and Maynard, Andrew. VALUES & ETHICS IN SOCIAL WORK: AN INTRODUCTION. London/Thousand Oaks, CA: Sage, 2005. 200 p. ISBN 978-0-4129-0140-6; ISBN 1-4129-0140-5. [HV10.5 .B354 2005] (1.3.10; 1.1; 1.2; 21.7)

Guttmann, David. ETHICS IN SOCIAL WORK: A CONTEXT OF CARING. Binghamton, NY: Haworth Press, 2006. 295 p. ISBN 978-0-7890-2853-2; ISBN 0-7890-2853-0. (Social Work Practice in Action series.) [HV40 .G98 2006] (1.3.10; 1.1; 6; 7.4; 8.4)

Reamer, Frederic G. ETHICAL STANDARDS IN SOCIAL WORK: A REVIEW OF THE NASW CODE OF ETHICS. Washington, DC: NASW Press, 2006. 303 p. ISBN 978-0-87101-371-1; ISBN 0-87101-371-1. (Second edition.) [HV40.8 .U6 R43 2006] (1.3.10; 6)

1.3.11 APPLIED AND PROFESSIONAL ETHICS: AGRICULTURE

Daniel, Pete. TOXIC DRIFT: PESTICIDES AND HEALTH IN THE POST-WORLD WAR II SOUTH. Baton Rouge: Louisiana State University Press in association with Smithsonian National Museum of American History, Washington, DC, 2005. 209 p. ISBN 0-8071-3098-2. (The Walter Lynwood Fleming Lectures in Southern History series.) [RA1270 .P4 D36 2005] (1.3.11; 1.3.5; 5.3; 7.1; 9.5.1; 16.1; 21.3)

1.3.12 APPLIED AND PROFESSIONAL ETHICS: INFORMATION TECHNOLOGY

Cavalier, Robert J. THE IMPACT OF THE INTERNET ON OUR MORAL LIVES. Albany: State University of New York Press, 2005. 249 p. ISBN 0-7914-6346-X. [TK5105.878 .C38 2005] (1.3.12; 1.3.3; 5.3; 10)

Grossman, Elizabeth. HIGH TECH TRASH: DIGITAL DEVICES, HIDDEN TOXINS, AND HUMAN HEALTH. Washington, DC: Island Press/Shearwater Books, 2006. 334 p. ISBN 978-1-55963-554-7; ISBN 1-55963-554-1. [TD799.85 G76 2006] (1.3.12; 9.1; 16.1; 21.1)

Otis, Laura. NETWORKING: COMMUNICATING WITH BODIES AND MACHINES IN THE NINETEENTH CENTURY. Ann Arbor: University of Michigan Press, 2001. 268 p. ISBN 978-0-472-11213-5; ISBN 0-472-11213-9. (Studies in Literature and Science series.) [TK5102.2 .O88 2001] (1.3.12; 5.1)

Pourciau, Lester J., ed. ETHICS AND ELECTRONIC INFORMATION IN THE TWENTY-FIRST CENTURY. West Lafayette, IN: Purdue University Press, 1999. 334 p. ISBN 1-55753-138-2. [T58.5 .E77 1999] (1.3.12; 1.1; 1.3.2; 5.3; 8.4; 15.10)

Smith, Ted; Sonnenfeld, David; and Pellow, David Naguib, eds. CHALLENGING THE CHIP: LABOR RIGHTS AND

ENVIRONMENTAL JUSTICE IN THE GLOBAL ELECTRONICS INDUSTRY. Philadelphia: Temple University Press, 2006. 357 p. ISBN 978-1-59213-330-4; ISBN 1-59213-330-4. (Foreword by Jim Hightower.) [HD9696 .A2 C425 2006] (1.3.12; 1.3.2; 16.1; 16.3; 21.1)

Spinello, Richard A. CYBERETHICS: MORALITY AND LAW IN CYBERSPACE. Sudbury, MA: Jones and Bartlett, 2006. 232 p. ISBN 0-7637-3783-6. (Third edition.) [TK5105.875 .I57 S68 2006] (1.3.12; 1.1; 5.3)

Vaagan, Robert, ed. [and] International Federation of Library Associations. THE ETHICS OF LIBRARIANSHIP: AN INTERNATIONAL SURVEY. München: Saur, 2002. 344 p. ISBN 3-598-21831-1. (IFLA Publications series; No. 101. Introduction for Alex Byrne.) [Z682.35 .P75 E84 2002] (1.3.12; 1.3.1; 6; 21.1)

Willinsky, John. THE ACCESS PRINCIPLE: THE CASE FOR OPEN ACCESS TO RESEARCH AND SCHOLARSHIP. Cambridge, MA: MIT Press, 2006. 287 p. ISBN 0-262-23242-1. (Digital Libraries and Electronic Publishing series.) [Z286 .O63 W55 2006] (1.3.12; 1.3.7; 1.3.9; 5.1)

2.1 BIOETHICS (GENERAL)

Abrams, Frederick R. DOCTORS ON THE EDGE: WILL YOUR DOCTOR BREAK THE RULES FOR YOU? Boulder, CO: Sentient Publications, 2006. 202 p. ISBN 978-1-59181-045-2; ISBN 1-59181-045-0. ("Twelve clinical cases described from caregiving physicians' viewpoints, with ethical discussions.") [R725.5 .A27 2006] (2.1; 1.3.1; 4.1.2)

Amarakone, Keith and Panesar, Sukhmeet S. ETHICS AND HUMAN SCIENCES. Edinburgh/New York: Elsevier Mosby, 2006. 211 p. ISBN 978-0-7234-3346-0; ISBN 0-7234-3346-1. [R724 .A5884 2006] (2.1)

Ashley, Benedict M.; DeBlois, Jean; and O'Rourke, Kevin D. HEALTH CARE ETHICS: A CATHOLIC THEOLOGICAL ANALYSIS. Washington, DC: Georgetown University Press, 2006. 328 p. ISBN 978-1-58901-116-8; ISBN 1-58901-116-3. (Fifth edition. Gift of the publisher.) [R724 .A74 2006] (2.1; 1.2)

Atighetchi, Dariusch. ISLAMIC BIOETHICS: PROBLEMS AND PERSPECTIVES. New York: Springer, 2007. 375 p. ISBN 978-1-4020-4961-3; ISBN 1-4020-4961-7. (International Library of Ethics, Law, and the New Medicine series; Vol. 31. Gift of the publisher.) [R725.59 .A884 2007] (2.1; 1.2)

Baylis, Françoise; Downie, Jocelyn; Freedman, Benjamin; Hoffmaster, Barry; and Sherwin, Susan. HEALTH CARE ETHICS IN CANADA. Toronto: Harcourt Brace Canada, 1995. 576 p. ISBN 0-7747-3288-1. (Gift of Françoise Baylis.) [R724 .H346 1995] (2.1; 21.7)

Benatar, David, ed. CUTTING TO THE CORE: EXPLORING THE ETHICS OF CONTESTED SURGERIES. Lanham, MD: Rowman & Littlefield, 2006. 236 p. ISBN 978-0-7425-5001-8; ISBN 0-7425-5001-X. (Gift of the publisher.) [RD27.7 .C87 2006] (2.1; 9.5.1; 9.5.7; 10; 18.2; 18.5.1; 19.1)

Bennett, Belinda and Tomossy, George F., eds. GLOBAL-IZATION AND HEALTH: CHALLENGES FOR HEALTH LAW AND BIOETHICS. Dordrecht: Springer, 2006. 218 p. ISBN 978-1-4020-4195-2; ISBN 1-4020-4195-0. (International Library of Ethics, Law, and the New Medicine series; Vol. 27. Gift of the publisher.) [RA441 .G5624 2006] (2.1; 9.1; 14.5; 15.1; 15.8; 18.6; 21.1)

Boitte, P.; Cadoré, B.; Jacquemin, D.; and Zorrilla, S. POUR UNE BIOÉTHIQUE CLINIQUE: MÉDICALISATION DE LA SOCIÉTÉ, QUESTIONNEMENT ÉTHIQUE ET PRATIQUES DE SOINS. Lille, France: Presses universitaires du Septentrion, 2002. 175 p. ISBN 2-85939-766-3. (Les Savoirs mieux series.) (2.1; 1.1; 8.1)

Brazier, Margaret and Cave, Emma. MEDICINE, PATIENTS AND THE LAW. London: Penguin Books, 2007. 541 p. ISBN 978-0-141-03020-3; ISBN 0-141-03020-8. (Fully revised fourth edition.) (2.1)

de Bouvet, Armelle and Cobbaut, Jean-Philippe, eds. UNE BIOÉTHIQUE POUR UN MONDE HABITABLE? LA BIOÉTHIQUE EN DISCUSSION. Villeneuve d'Ascq: Presses universitaires du Septentrion, 2006. 140 p. ISBN 978-2-859-39922-1; ISBN 2-859-39922-4. (Savoirs mieux series; No. 21. ISSN 1292-4385.) [R724 .B4855 2006] (2.1; 21.1)

De Vries, Raymond G.; Turner, Leigh; Orfali, Kristina; and Bosk, Charles, eds. THE VIEW FROM HERE: BIOETHICS AND THE SOCIAL SCIENCES. Malden, MA: Blackwell, 2007. 219 p. ISBN 978-1-4051-5269-3. (Sociology of Health and Illness Monograph series. "First published as Volume 28, Number 6 of *Sociology of Health and Illness*.") [R724 .V472 2007] (2.1; 7.1)

Eberl, Jason T. THOMISTIC PRINCIPLES AND BIOETHICS. London/New York: Routledge, 2006. 155 p. ISBN 0-415-77063-7. (Routledge Annals of Bioethics series; Vol. 2.) [QH332 .E24 2006] (2.1; 1.1; 4.4; 12.1; 14.5; 18.5.4; 19.5; 20.5.1)

Eckenwiler, Lisa A. and Cohn, Felicia, eds. THE ETHICS OF BIOETHICS: MAPPING THE MORAL LANDSCAPE. Baltimore, MD: Johns Hopkins University Press, 2007. 320 p. ISBN 978-0-8018-8612-0; ISBN 0-8018-8612-0. (Gift of the publisher.) [R724 .E8242 2007] (2.1; 1.3.1; 2.2; 2.3; 2.4; 21.1)

Engelhardt, H. Tristram, eds. GLOBAL BIOETHICS: THE COLLAPSE OF CONSENSUS. Salem, MA: M & M Scrivener Press, 2006. 396 p. ISBN 978-0-9764041-3-2; ISBN 0-9764041-3-3. (Conflicts and Trends: Studies in Values and Policies series.) [QH332 .G56 2006] (2.1; 1.1; 21.1; 21.7)

Feuillet-Le-Mintier, Brigitte. NORMATIVITÉ ET BIOMÉDECINE. Paris: Economica, 2003. 304 p. ISBN 2-7178-4723-5. (Collection Etudes juridiques series; No. 17. "Ouvrage publié avec le concours de la Mission recherche "Droit et justice".) [K3601 .N67 2003] (2.1; 1.1; 2.4; 9.6; 21.1)

Galston, Arthur W. and Peppard, Christiana Z., eds. EXPANDING HORIZONS IN BIOETHICS. New York: Springer-Verlag, 2005. 255 p. ISBN 1-4020-3061-4. [QH332 .E96 2005] (2.1; 5.1; 14.1; 15.3; 16.1; 18.1; 18.6; 20.4.1; 21.7)

Garwood-Gowers, Austen ; Tingle, John; Wheat, Kay, eds. CONTEMPORARY ISSUES IN HEALTHCARE LAW AND ETHICS. Edinburgh/New York: Elsevier Butterworth-Heinemann, 2005. 318 p. ISBN 0-7506-8832-7. [R724 .C664 2005] (2.1; 9.1; 14.1; 19.1; 20.5.1)

Grubb, Andrew and Laing, Judith, eds. PRINCIPLES OF MEDICAL LAW. Oxford: Oxford University Press, 2004. 1191 p. ISBN 0-19-926358-2. (Second edition.) [KD3395 .P75 2004] (2.1)

Gunning, Jennifer and Holm, Søren, eds. ETHICS, LAW AND SOCIETY, VOLUME I. Aldershot, England/Burlington, VT: Ashgate, 2005. 280 p. ISBN 0-7546-4583-5. [BJ1581.2 .E85 2005 v.1] (2.1; 1.3.2; 1.3.7; 1.3.9; 1.3.11; 1.3.12; 9.3.1; 9.6; 9.8; 14.1; 14.5; 15.8; 16.1; 17.1; 19.5; 20.1; 21.1)

Gunning, Jennifer and Holm, Søren, eds. ETHICS, LAW AND SOCIETY, VOLUME II. Aldershot, England/Burlington, VT: Ashgate, 2006. 326 p. ISBN 978-0-7546-4881-9; ISBN 0-7546-4881-8. (Gift of the publisher.) [BJ1581.2 .E85 2005 v.2] (2.1; 1.3.1; 1.3.2; 1.3.7; 1.3.9; 1.3.11; 5.3; 9.5.7; 9.7; 14.1; 15.1; 17.1; 18.5.1; 19.2; 21.1; 21.3; 21.4)

Hall, Mark A.; Bobinski, Mary Anne; and Orentlicher, David. BIOETHICS AND PUBLIC HEALTH LAW. New York: Aspen Publishers, 2005. 601 p. ISBN 0-7355-5205-3. [KF3775 .A7 H34 2005] (2.1; 8.1; 8.3.1; 9.1; 9.4; 14.1; 15.1; 19.6; 20.5.1)

Hamel, Ron, ed. MAKING HEALTH CARE DECISIONS: A CATHOLIC GUIDE. Liguori, MO: Liguori Publications, 2006. 138 p. ISBN 978-0-7648-1402-0; ISBN 0-7648-1402-8. (Co-published with the Catholic Health Association of the United States.) [R725.56 .M27 2006] (2.1; 1.2; 15.3; 19.1; 20.5.1; 20.5.4)

Hervé, Christian; Knoppers, Bartha Maria; Molinari, Patrick A.; and Moutel, Grégoire, eds. PLACE DE LA BIOÉTHIQUE EN RECHERCHE ET DANS LES SERVICES CLINIQUES. Paris: Dalloz, 2005. 215 p. ISBN 2-247-06003-X. (Thèmes & commentaires series. Actes series. Institut international de recherche en éthique biomédicale [IIREB], Séminaire d'experts, Université René-Descartes (Paris V), held 4-5 December 2003 in Paris.) (2.1; 8.1; 18.1; 18.2)

Hottois, G. QU'EST-CE QUE LA BIOÉTHIQUE? Paris: Librarie philosophique J. Vrin, 2004. 127 p. ISBN 2-7116-1667-8. (Chemins philosophiques series.) (2.1; 2.2)

Howard, Philip and Bogle, James. MEDICAL LAW AND ETHICS. Malden, MA: Blackwell, 2005. 204 p. ISBN 1-4051-1868-7. (Lecture Notes series.) [R724 .H69 2005] (2.1; 4.3; 7.3; 8.3.1; 8.4; 12.4.1; 14.2; 20.5.1)

Illingworth, Patricia and Parmet, Wendy E. ETHICAL HEALTH CARE. Upper Saddle River, NJ: Pearson Educational, 2006. 606 p. ISBN 0-13-045301-3. [R724 .I544

2006] (2.1; 1.1; 5.1; 6; 8.1; 8.4; 9.1; 9.3.1; 9.7; 9.8; 10; 14.5; 18.1; 21.1; 21.3; 21.7)

Jackson, Jennifer. ETHICS IN MEDICINE. Cambridge, UK/Malden, MA: Polity Press, 2006. 232 p. ISBN 978-0-7456-2569-0; ISBN 0-7456-2569-X. [R724 .J25 2006] (2.1; 2.3)

Jackson, Julia. ETHICS, LEGAL ISSUES, AND PROFESSIONALISM IN SURGICAL TECHNOLOGY. Clifton Park, NY: Delmar Learning, 2007. 368 p. ISBN 978-1-4018-5793-6; ISBN 1-4018-5793-0. (Gift of the publisher.) [RD32.3 .J33 2007] (2.1; 1.1; 1.3.1; 2.3; 4.1.2; 6; 9.1)

Jecker, Nancy S.; Jonsen, Albert R.; and Pearlman, Robert A., eds. BIOETHICS: AN INTRODUCTION TO THE HISTORY, METHODS, AND PRACTICE. Boston: Jones & Bartlett, 2007. 545 p. ISBN 978-0-7637-4314-7; ISBN 0-7637-4314-3. (Second edition.) [R724 .B4583 2007] (2.1; 2.2; 2.3)

Johnson, Alan G. and Johnson, Paul R.V. MAKING SENSE OF MEDICAL ETHICS: A HANDS-ON GUIDE. London: Hodder Arnold; Distributed in the USA: New York: Oxford University Press, 2007. 222 p. ISBN 978-0-340-92559-1; ISBN 0-340-92559-0. (Gift of the publisher.) [R724 .J64 2006] (2.1; 1.1)

Kaczor, Christopher. THE EDGE OF LIFE: HUMAN DIGNITY AND CONTEMPORARY BIOETHICS. Dordrecht: Springer, 2005. 155 p. ISBN 978-1-4020-3155-7; ISBN 1-4020-3155-6. (Philosophy of Medicine series; Vol. 85. Catholic Studies in Bioethics series; Vol. 4.) [QH332 .K325 2005] (2.1; 1.2; 4.4; 12.3; 14.1; 18.5.4; 20.5.1; 20.5.2; 20.6; 21.1)

Kaldjian, Lauris Christopher. PHYSICIAN INTEGRITY, RELIGIOUS BELIEF, AND THE ADEQUACY OF MEDICAL ETHICS. Ann Arbor, MI: ProQuest Information and Learning/UMI, 2004. 274 p. (Publication No. AAT-3152943. Dissertation, (Ph.D.)—Yale University, 2004.) (2.1; 1.2; 4.1.2)

King, Patricia A.; Areen, Judith; and Gostin, Lawrence O. LAW, MEDICINE AND ETHICS. New York: Foundation Press, 2006. 666 p. ISBN 978-1-58778-912-0; ISBN 1-58778-912-4. [KF3821 .A7 K56 2006] (2.1)

Kluge, Eike-Henner W., ed. READINGS IN BIOMEDICAL ETHICS: A CANADIAN FOCUS. Toronto: Pearson Prentice Hall, 2005. ISBN 0-13-120066-6. (Third edition.) [R724 .R435 2005] (2.1)

LaFrance, Arthur Birmingham. BIOETHICS: HEALTH CARE, HUMAN RIGHTS AND THE LAW. Newark, NJ: LexisNexis Matthew Bender, 2006. 1283 p. ISBN 978-1-4224-0587-1; ISBN 1-4224-0587-7. (Second edition.) [KF3821 .A7 L34 2006] (2.1; 21.1)

Laney, Dawn, ed. BIOMEDICAL ETHICS. San Diego, CA: Greenhaven Press, 2007. 183 p. ISBN 978-0-7377-2859-0; ISBN 0-7377-2859-0. (History of Issues series. Opposing Viewpoints series.) [R724 .B488 2007] (2.1; 4.5; 6; 8.1; 14.5; 15.3; 15.5; 18.1; 20.5.1; 20.7)

Lecaldano, Eugenio. DIZIONARIO DE BIOETICA. Roma-Bari: GLF Editori Laterza, 2002. 339 p. ISBN 88-

420-6730-X. (Manuali Laterza series; No. 173. Text in Italian; terminology in Italian, English, French, Spanish and German.) (2.1; Reference)

Levine, Carol, ed. TAKING SIDES: CLASHING VIEWS ON CONTROVERSIAL BIOETHICAL ISSUES. Dubuque, IA: McGraw Hill/Dushkin, 2006. 377 p. ISBN 978-0-07-312955-6; ISBN 0-07-312955-0. (Eleventh edition. Taking Sides series.) [QH332 .T34 2006] (2.1)

Lo, Bernard. RESOLVING ETHICAL DILEMMAS: A GUIDE FOR CLINICIANS. Philadelphia: Lippincott Williams & Wilkins, 2005. 309 p. ISBN 0-7817-5357-0. (Third edition.) [R724 .L59 2005] (2.1; 4.3; 7.4; 8.1; 8.2; 8.3.1; 8.3.3; 8.4; 9.1; 9.4; 9.6; 9.7; 15.3; 17.1; 19.1; 20.2.1; 20.5.1; 20.7)

Luna, Florencia. BIOETHICS AND VULNERABILITY: A LATIN AMERICAN VIEW. Amsterdam/New York: Rodopi, 2006. 177 p. ISBN 978-90-420-2073-3; ISBN 90-420- 2073-3. (Edited by Peter Herissone-Kelly. Value Inquiry Book Series; Vol. 180. Translation from the Spanish by Laura Pakter of: *Ensayos de bioética*.) [QH332 .L8613 2006] (2.1; 1.1; 7.1; 9.5.5; 9.5.6; 12.1; 14.1; 18.4)

Mahowald, Mary Briody. BIOETHICS AND WOMEN: ACROSS THE LIFE SPAN. Oxford/New York: Oxford University Press, 2006. 272 p. ISBN 978-0-19-517617-9; ISBN 0-19-517617-0. (Gift of the publisher.) [R725.5 .M34 2006] (2.1; 1.1; 8.1; 9.5.2; 9.5.5; 9.5.7; 14.1; 15.2; 15.3; 18.5.3; 20.4.1)

Mason, J.K. and Laurie, G.T. MASON AND McCALL SMITH'S LAW AND MEDICAL ETHICS. Oxford/New York: Oxford University Press, 2006. 774 p. ISBN 978-0-19-928239-5; ISBN 0-19-928239-0. (Seventh edition. Revised edition of: *Law and Medical Ethics*, by J.K. Mason, R.A. McCall Smith, and G.T. Laurie; 6th edition, 2002. Gift of the publisher.) [K3601 .M38 2005] (2.1; 1.3.5; 4.4; 8.3.1; 8.4; 9.1; 9.5.2; 14.1; 15.1; 18.1; 18.5.2; 18.5.4; 19.5; 20.2.1; 20.5.1; 21.1)

McLachlan, Hugh V. and Swales, J. Kim. FROM THE WOMB TO THE TOMB: ISSUES IN MEDICAL ETHICS. Glasgow, Scotland: Humming Earth, 2007. 279 p. ISBN 978-1-84622-011-1; ISBN 1-84622-011-4. (Gift of the author.) [R724 .M29226 2007] (2.1)

McLean, Sheila A.M., ed. FIRST DO NO HARM: LAW, ETHICS AND HEALTHCARE. Aldershot, Hampshire/Burlington, VT: Ashgate, 2006. 605 p. ISBN 978-0-7546-2614-5; ISBN 0-7546-2614-8. (Applied Legal Philosophy series. Gift of the publisher.) [K3601 .F57 2006] (2.1; 1.1; 8.3.1; 8.4; 14.1; 14.5; 15.1; 18.1; 20.5.1)

Meyers, Christopher. A PRACTICAL GUIDE TO CLINICAL ETHICS CONSULTING: EXPERTISE, ETHOS, AND POWER. Lanham, MD: Rowman & Littlefield, 2007. 115 p. ISBN 978-0-7425-4828-2; ISBN 0-7425-4828-7. [R724 .M438 2007] (2.1; 7.1; 9.6)

Neuer-Miebach, Therese and Wunder, Michael, eds. BIO-ETHIK UND DIE ZUKUNFT DER MEDIZIN. Bonn: Psychiatrie-Verlag, 1998. 200 p. ISBN 978-3-88414-227-1; ISBN 3-88414-227-5. [QH332 .B565 1998] (2.1)

Novak, David. THE SANCTITY OF HUMAN LIFE. Washington, DC: Georgetown University Press, 2007. 186 p. ISBN 978-1-58901-176-2; ISBN 1-58901-176-7. (Max M. and Marjorie B. Kampelman Collection on Jewish Ethics. Gift of the publisher.) [R725.57 .N68 2007] (2.1; 1.2; 4.4; 7.1; 8.1; 18.5.4; 20.5.1; 20.7)

Nys, Thomas; Denier, Yvonne; and Vandevelde, Toon, eds. AUTONOMY & PATERNALISM: REFLECTIONS ON THE THEORY AND PRACTICE OF HEALTH CARE. Leuven/Dudley, MA: Peeters, 2007. 183 p. ISBN 978-90-429-1880-1. (Ethical Perspectives Monographic Series; No. 5.) [R725.5 .A95 2007] (2.1; 1.1; 8.3.1; 17.2)

Østnor, Lars, ed. BIOETIKK OG TEOLOGI: RAPPORT FRA NORDISK TEOLOGISK NETTVERK FOR BIOETIKKS WORKSHOP I STOCKHOLM 27.-29. SEPTEMBER 1996. Oslo: Nordisk teologisk nettverk for bioetikk, 1996. 236 p. ISBN 82-993709-1-4. (Gift of Warren Reich.) [R725.56 .B46 1996] (2.1; 1.2)

Pence, Gregory E. THE ELEMENTS OF BIOETHICS. Boston: McGraw-Hill, 2007. 279 p. ISBN 978-0-07-313277-8; ISBN 0-07-313277-2. [R724 .P37 2007] (2.1; 1.1.; 2.3)

Pollard, Irina. LIFE, LOVE AND CHILDREN: A PRACTICAL INTRODUCTION TO BIOSCIENCE ETHICS AND BIOETHICS. Boston: Kluwer Academic, 2002. 269 p. ISBN 978-1-4020-7294-9; ISBN 1-4020-7294-5. [R725.5 .P655 2002] (2.1)

Prograis, Lawrence J. and Pellegrino, Edmund D., eds. AFRICAN AMERICAN BIOETHICS: CULTURE, RACE, AND IDENTITY. Washington, DC: Georgetown University Press, 2007. 169 p. ISBN 978-1-58901-164-9; ISBN 1-58901-164-3. ("Conference held on September 23-24, 2004 at Georgetown University entitled: Symposium on African American Perspectives in Bioethics and Second Annual Conference on Health Disparities.") [R724 .S937 2004] (2.1; 1.2; 8.1; 9.1; 9.5.4; 16.3; 21.3; 21.7)

Rasmussen, Lisa, ed. ETHICS EXPERTISE: HISTORY, CONTEMPORARY PERSPECTIVES, AND APPLICATIONS. Dordrecht: Springer, 2005. 278 p. ISBN 978-1-4020-3819-8; ISBN 1-4020-3819-4. (Philosophy and Medicine Series; Vol. 87.) [R724 .E84 2005] (2.1; 1.2; 4.1.2; 8.1; 9.5.1; 9.6; 9.7)

Rawlinson, Mary C. and Lundeen, Shannon, eds. THE VOICE OF BREAST CANCER IN MEDICINE AND BIOETHICS. Dordrecht: Springer, 2006. 207 p. ISBN 978-1-4020-4508-0; ISBN 1-4020-4508-5. (Philosophy and Medicine series; Vol. 88.) [RC280 .B8 V65 2006] (2.1; 4.1.2; 7.1; 7.2; 9.1; 9.5.1; 10; 18.3; 18.5.3; 21.1)

Requena Meana, Pablo. MODELOS DE BIOÉTICA CLÍNICA: PRESENTACIÓN CRÍTICA DEL PRINCIPIALISMO Y LA CASUÍSTICA. Roma: Edizioni Università della Santa Croce, 2005. 340 p. ISBN 88-8333-156-7. (Thesis, (Doctoratum in Theologia)—Pontificia Universitas Sanctae Crucis, 2005. Gift of the author.) (2.1; 1.1; 1.2)

Rhodes, Rosamond; Francis, Leslie P.; and Silvers, Anita, eds. THE BLACKWELL GUIDE TO MEDICAL ETHICS. Malden,

MA: Blackwell, 2007. 435 p. ISBN 978-1-4051-2584-0; ISBN 1-4051-2584-5. [R724 .B515 2007] (2.1; 8.1; 9.3.1; 9.4; 20.5.1)

Rogers, Wendy A. and Braunack-Mayer, Annette J. PRACTICAL ETHICS FOR GENERAL PRACTICE. Oxford: Oxford University Press, 2004. 210 p. ISBN 978-0-19-852504-2; ISBN 0-19-852504-4. [R724 .R615 2004] (2.1; 1.1; 4.1.2; 8.1)

Rose, Nikolas. THE POLITICS OF LIFE ITSELF: BIOMEDICINE, POWER, AND SUBJECTIVITY IN THE TWENTY-FIRST CENTURY. Princeton: Princeton University Press, 2006. 350 p. ISBN 978-0-691-12191-8; ISBN 0-691-12191-5. (Information series.) [R725.5 .R676 2007] (2.1; 15.1; 21.1)

Rothman, David J. and Rothman, Sheila M. TRUST IS NOT ENOUGH: BRINGING HUMAN RIGHTS TO MEDICINE. New York: New York Review Books, 2006. 213 p. ISBN 1-59017-140-3. (New York Review Books Collection series. Preface by Aryeh Neier.) [R724 .R626 2006] (2.1; 4.4; 9.1; 9.2; 9.5.6; 9.5.7; 18.3; 19.5; 21.1; 21.4)

Simón Vázquez, Carlos, ed. DICCIONARIO DE BIOÉTICA. Burgos: Editorial Monte Carmelo, 2006. 786 p. ISBN 978-84-8353-07-8; ISBN 84-8353-07-4. (Bioética; Serie Gran Formato. Gift of the editor.) (2.1; Reference)

Smith, George P. THE CHRISTIAN RELIGION AND BIOTECHNOLOGY: A SEARCH FOR PRINCIPLED DECISION-MAKING. Dordrecht/Norwell, MA: Springer, 2005. 251 p. ISBN 1-4020-3146-7. (International Library of Ethics, Law, and the New Medicine series; Vol. 25.) [TP248.185 .S63 2005] (2.1; 1.2; 4.5; 5.1; 14.1; 15.1; 20.5.1)

Spielman, Bethany J. BIOETHICS IN LAW. Totowa, NJ: Humana Press, 2007. 181 p. ISBN 978-1-58829-434-0; ISBN 1-58829-434-X. [KF3821 .S687 2007] (2.1; 1.3.5; 1.3.7; 1.3.8; 2.3; 2.4; 9.6; 18.2)

Spinsanti, Sandro. ETICA BIO-MEDICA. Cinisello Calsamo, Milano: Edizioni Paoline, 1987. 244 p. ISBN 88-215-1184-7. (Etica professionale e sociale series; No. 3. Gift of Warren Reich.) [R724 .S62 1987] (2.1)

Spinsanti, Sandro, ed. BIOETICA E GRANDI RELIGIONI. Milano: Edizioni Paoline, 1987. 155 p. ISBN 88-215-1362-9. (Etica professionalie e sociale series; No. 4. Includes translation of articles from the *Encyclopedia of Bioethics*. Gift of Warren Reich.) (2.1; 1.2)

Steinbock, Bonnie, ed. THE OXFORD HANDBOOK OF BIOETHICS. Oxford/New York: Oxford University Press, 2007. 747 p. ISBN 978-0-19-927335-5. (Gift of the publisher.) [QH332 .O94 2007] (2.1)

Vallero, Daniel A. BIOMEDICAL ETHICS FOR ENGINEERS: ETHICS AND DECISION MAKING IN BIOMEDICAL AND BIOSYSTEM ENGINEERING. Amsterdam/Boston: Elsevier/Academic Press, 2007. 408 p. ISBN 978-0-7506-8227-5; ISBN 0-7506-8227-2. (The Biomedical Engineering Series. Gift of the publisher.) [R856 .V35 2007] (2.1; 1.3.4; 2.3)

Numbers in ( ) = NRCBL Classification Numbers

Wallner, Jürgen. HEALTH CARE ZWISCHEN ETHIK UND RECHT. Wien, Austria: Facultas-WUV, 2007. 331 p. ISBN 978-3-7089-0048-3. (Gift of the author.) (2.1)

Walter, James J. and Shannon, Thomas A. CONTEMPO-RARY ISSUES IN BIOETHICS: A CATHOLIC PERSPECTIVE. Lanham, MD: Rowman & Littlefield, 2005. 293 p. ISBN 0-7425-5060-5. [R725.56 .W35 2005] (2.1; 1.1; 1.2; 4.4; 9.5.1; 12.3; 14.1; 14.5; 15.1; 18.5.4; 20.5.1)

Walters, LeRoy; Kahn, Tamar Joy; and Goldstein, Doris Mueller, eds. BIBLIOGRAPHY OF BIOETHICS, VOLUME 33. Washington, DC: Kennedy Institute of Ethics, Georgetown University, 2007. 848 p. ISBN 978-1-883913-14-4. (Bibliography of Bioethics series; Vol. 33. ISSN 0363-0161.) [Z6675 .E8 W34 v.33] (2.1; Reference)

Watt, Helen, ed. COOPERATION, COMPLICITY AND CON-SCIENCE: PROBLEMS IN HEALTHCARE, SCIENCE, LAW AND PUBLIC POLICY. London: Linacre Centre; Distributed in the U.S. by: Chicago University Press, 2005. 320 p. ISBN 0-906561-10-8. [QH332 .C685 2005] (2.1; 1.2; 7.2; 9.5.7; 12.4.1; 18.5.4; 20.7)

White, Stuart M. and Baldwin, Timothy J. LEGAL AND ETH-ICAL ASPECTS OF ANAESTHESIA: CRITICAL CARE AND PERIOPERATIVE MEDICINE. Cambridge/New York: Cam-bridge University Press, 2004. 216 p. ISBN 1-841102-09-1. [KD3395 .W48 2004] (2.1; 8.1)

Yacoub, Ahmed Abdel Aziz. THE FIQH OF MEDICINE: RE-SPONSES IN ISLAMIC JURISPRUDENCE TO DEVELOPMENTS IN MEDICAL SCIENCE. London: Ta-Ha Publishers, 2001. 349 p. ISBN 1-84200-025-X. (Publisher's address: 1 Wynne Road, postal code SW9 0BB.) [KBP3098 .Y33 2001] (2.1; 1.2; 7.4; 12.4.2; 14.1; 14.5; 19.1; 20.5.1; Reference)

2.2 BIOETHICS: HISTORY OF HEALTH ETHICS

Dossetor, John B. BEYOND THE HIPPOCRATIC OATH: A MEMOIR OF THE RISE OF MODERN MEDICAL ETHICS. Ed-monton: University of Alberta Press, 2005. 298 p. ISBN 0-88864-453-1. [R724 .D676 2005] (2.2; 2.1; 7.1; 7.2; 9.6; 18.2; 18.3; Biography)

2.3 BIOETHICS: EDUCATION/ PROGRAMS

Kenny, Nuala and Shelton, Wayne, eds. LOST VIRTUE: PRO-FESSIONAL CHARACTER DEVELOPMENT IN MEDICAL EDU-CATION. Amsterdam/Oxford: Elsevier, 2006. 233 p. ISBN 978-0-7623-1196-5; ISBN 0-7623-1196-7. (Advances in Bioethics series; Vol. 10.) [K3611 .E84 L67 2006] (2.3; 1.1; 1.3.1; 4.1.2; 7.2)

3.1 PHILOSOPHY OF BIOLOGY (GENERAL)

Rockwell, W. Teed. NEITHER BRAIN NOR GHOST: A NONDUALIST ALTERNATIVE TO THE MIND-BODY IDENTITY THEORY. Cambridge, MA: MIT Press, 2007. 231 p. ISBN 978-0-262-68167-4. ("A Bradford Book.") [B105 .M55 R63 2005] (3.1; 1.1; 17.1)

3.2 EVOLUTION AND CREATION

Brockman, John, ed. INTELLIGENT THOUGHT: SCIENCE VERSUS THE INTELLIGENT DESIGN MOVEMENT. New York: Vintage Books, 2006. 256 p. ISBN 978-0-307-27722-0; ISBN 0-307-27722-4. [BL262 .I58 2006] (3.2; 1.2)

Johnson, Anne Janette. THE SCOPES "MONKEY TRIAL". De-troit, MI: Omnigraphics, 2007. 245 p. ISBN 0-7808-0955-6. (Defining Moments series. Gift of the publisher.) [KF224 .S3 J64 2007] (3.2; 1.2; 1.3.3)

Mindell, David P. THE EVOLVING WORLD: EVOLUTION IN EVERYDAY LIFE. Cambridge, MA: Harvard University Press, 2006. 341 p. ISBN 0-674-02191-6. [QH371 .M54 2006] (3.2)

Wilson, Edward O. THE CREATION: AN APPEAL TO SAVE LIFE ON EARTH. New York: W.W. Norton, 2006. 175 p. ISBN 978-0-393-06217-5; ISBN 0-393-06217-1. [QH303 .W55 2006] (3.2; 3.1; 16.1)

4.1.1 PHILOSOPHY OF THE HEALTH PROFESSIONS (GENERAL)

Murray, Colin and Sanders, Peter. MEDICINE MURDER IN COLONIAL LESOTHO: THE ANATOMY OF A MORAL CRISIS. Edinburgh: Edinburgh University Press for the Interna-tional African Institute, 2005. 493 p. ISBN 0-7486-2284-5. (International African Library series; No. 31.) [BF1584 .L5 M87 2005] (4.1.1; 7.1)

Snyder, Lois, ed. COMPLEMENTARY AND ALTERNATIVE MEDICINE: ETHICS, THE PATIENT, AND THE PHYSICIAN. Totowa, NJ: Humana Press, 2007. 241 p. ISBN 978-1-58829-584-2; ISBN 1-58829-584-2. (Biomedical Ethics Reviews series; 2007.) [R724 .C662 2007] (4.1.1; 7.2; 8.1)

Thrangu, Khenchen Rinpoche. MEDICINE BUDDHA TEACH-INGS. Ithaca, NY: Snow Lion, 2004. 192 p. ISBN 978-1-55939-216-7; ISBN 1-55939-216-9. [BQ4690 .B5 T57 2004] (4.1.1; 1.2)

Wanjek, Christopher. BAD MEDICINE: MISCONCEPTIONS AND MISUSES REVEALED, FROM DISTANCE HEALING TO VI-TAMIN O. New York: John Wiley & Sons, 2003. 280 p. ISBN 978-0-471-43499-3; ISBN 0-471-43499-X. [R729.9 .W36 2003] (4.1.1)

White, Gillian. TALKING ABOUT SPIRITUALITY IN HEALTH CARE PRACTICE: A RESOURCE FOR THE MULTI-PROFES-SIONAL HEALTH CARE TEAM. London/Philadelphia: Jessica Kingsley, 2006. 175 p. ISBN 978-1-84310-305-9; ISBN 1-84310-305-2. [R725.55 .W53 2006] (4.1.1; 1.2; 9.1)

Yardley-Nohr, Terrie. ETHICS FOR MASSAGE THERAPISTS. Philadelphia: Lippincott Williams & Wilkins, 2007. 169 p. ISBN 978-0-7817-5339-5; ISBN 0-7817-5339-2. (LWW

Massage Therapy & Bodywork Educational Series.) [RM721 .Y37 2007] (4.1.1; 1.3.1; 6.; 8.1)

4.1.2 PHILOSOPHY OF MEDICINE

Egan, Erin A. and Surdyk, Patricia M. LIVING PROFESSION-ALISM: REFLECTIONS ON THE PRACTICE OF MEDICINE. Lanham, MD: Rowman & Littlefield, 2006. 115 p. ISBN 978-0-7425-4851-0; ISBN 0-7425-4851-1. (Practicing Bioethics series.) [R725.5 .L58 2006] (4.1.2; 2.1; 7.2; 8.1)

Kincaid, Harold and McKitrick, Jennifer, eds. ESTABLISH-ING MEDICAL REALITY: ESSAYS IN THE METAPHYSICS AND EPISTEMOLOGY OF BIOMEDICAL SCIENCE. Dordrecht: Springer, 2007. 236 p. ISBN 978-1-4020-5215-6; ISBN 1-4020-5215-4. (Philosophy and Medicine series; Vol. 90.) [R723 .E87 2007] (4.1.2; 1.1; 2.1; 4.2; 9.1; 15.11)

Stark, Andrew. THE LIMITS OF MEDICINE. Cambridge/New York: Cambridge University Press, 2006. 256 p. ISBN 978-0-521-67226-9; ISBN 0-521-67226-0. [R723 .S755 2006] (4.1.2; 7.1)

Stempsey, William E., ed. ELISHA BARTLETT'S PHILOSOPHY OF MEDICINE. Dordrecht: Springer, 2005. 239 p. ISBN 978-1-4020-3041-3; ISBN 1-4020-3041-X. (Philosophy and Medicine series; Vol. 83. Classics of Medical Ethics series; Vol. 2.) [R723 .E558 2005] (4.1.2; 2.2)

Sulmasy, Daniel P. A BALM FOR GILEAD: MEDITATIONS ON SPIRITUALITY AND THE HEALING ARTS. Washington, DC: Georgetown University Press, 2006. 154 p. ISBN 978-1-58901-122-9; ISBN 1-58901-122-8. (Gift of the publisher.) [R725.55 .S85 2006] (4.1.2; 1.2; 7.1; 8.1)

Szczeklik, Andrzej. CATHARSIS: ON THE ART OF MEDICINE. Chicago: University of Chicago Press, 2005. 161 p. ISBN 978-0-226-78869-2; ISBN 0-226-78869-5. (Translation from the Polish by Antonia Lloyd-Jones of: *Katharsis: O uzdrowicielskiej mocy natury i sztuki*. Foreword by Czeslaw Milosz.) [R723 .S9413 2005] (4.1.2; 7.1)

Toellner, Richard and Wiesing, Urban, eds. WISSEN, HANDELN, ETHIK: STRUKTUREN ÄRZTLICHEN HANDELNS UND IHRE ETHISCHE RELEVANZ. Stuttgart/New York: Gustav Fischer, 1995. 134 p. ISBN 3-437-11701-7. (Medizin Ethik [Jahrbuch des Arbeitskreises Medizinischer Ethik-Kommissionen in der Bundes-republik Deutschland] series; Bd. 6. ISSN 0936-9015.) (4.1.2; 8.3.3; 9.7; 18.3)

Van der Eijk, Philip J. MEDICINE AND PHILOSOPHY IN CLAS-SICAL ANTIQUITY: DOCTORS AND PHILOSOPHERS ON NA-TURE, SOUL, HEALTH AND DISEASE. Cambridge/New York: Cambridge University Press, 2005. 404 p. ISBN 978-0-521-81800-1; ISBN 0-521-81800-1. [R135 .E36 2005] (4.1.2; 4.2; 4.4; 6)

Wright, H.G. MEANS, ENDS AND MEDICAL CARE. Dordrecht: Springer, 2007. 179 p. ISBN 978-1-4020-5291-0; ISBN 1-4020-5291-X. (Philosophy and Medicine series; Vol. 92.) [R723 .W738 2007] (4.1.2; 2.1; 4.2)

4.1.3 PHILOSOPHY OF NURSING

Armstrong, Alan E. NURSING ETHICS: A VIRTUE-BASED AP-PROACH. Basingstoke, [England]/New York: Palgrave Macmillan, 2007. 240 p. ISBN 978-0-230-50688-6; ISBN 0-230-50688-7. [RT85 .A76 2007] (4.1.3; 1.1)

Davis, Anne J.; Tschudin, Verena; and De Raeve, Louise, [eds.]. ESSENTIALS OF TEACHING AND LEARNING IN NURS-ING ETHICS: PERSPECTIVES AND METHODS. Edinburgh/New York: Churchill Livingstone Elsevier, 2006. 370 p. ISBN 978-0-443-07480-6; ISBN 0-443-07480-1. (Forewords by Margretta Madden Styles and Mo Im Kim.) [RT85 .E72 2006] (4.1.3; 1.1; 2.1; 2.2; 2.3; 10; 21.1)

Dooley, Dolores and McCarthy, Joan. NURSING ETHICS: IRISH CASES AND CONCERNS. Dublin: Gill & Macmillan, 2005. 314 p. ISBN 978-0-7171-3576-9; ISBN 0-7171-3576-4. [RT85 .D66 2005] (4.1.3; 2.1; 8.1)

Falk, Ursula A. and Falk, Gerhard. DEVIANT NURSES AND IMPROPER PATIENT CARE: A STUDY OF FAILURE IN THE MEDI-CAL PROFESSION. Lewiston, NY: Edwin Mellen Press, 2006. 191 p. ISBN 978-0-7734-5967-0; ISBN 0-7734-5967-7. (Gift of the publisher.) [RT85.6 .F35 2006] (4.1.3; 7.1; 7.4; 8.1)

Gauwitz, Donna F. LEGAL & ETHICAL NURSING. Clifton Park, NY: Thomson Delmar Learning, 2007. 162 p. ISBN 978-1-4018-1183-9; ISBN 1-4018-1183-3. (Thomson Delmar Learning's Nursing Review series.) [RT85 .G34 2007] (4.1.3; 9.1; 21.7)

Sala, Roberta. ETICA E BIOETICA PER L'INFERMIERE. Rome: Carocci Faber, 2003. 187 p. ISBN 88-7466-062-6. (La professione infermieristica series; No. 40.) [RT85 .S35 2003] (4.1.3; 2.1)

Tschudin, Verena, ed. APPROACHES TO ETHICS: NURSING BEYOND BOUNDARIES. Edinburgh/New York: Butterworth-Heinemann, 2003. 138 p. ISBN 0-7506-5326-4. [RT85 .A676 2003] (4.1.3; 1.1; 7.1; 10; 21.1; 21.7)

van Hooft, Stan. CARING ABOUT HEALTH. Aldershot, Hants, England/Burlington, VT: Ashgate, 2006. 226 p. ISBN 978-0-7546-5358-5; ISBN 0-7546-5358-7. (Ashgate Studies in Applied Ethics series.) [RT85 .H67 2006] (4.1.3; 2.1; 2.3; 4.1.1; 4.4)

4.3 CONCEPT OF MENTAL HEALTH

Eriksen, Karen and Kress, Victoria E. BEYOND THE DSM STORY: ETHICAL QUANDARIES, CHALLENGES, AND BEST PRACTICES. Thousand Oaks, CA: Sage, 2005. 259 p. ISBN 0-7619-3032-9. [RC469 .E75 2005] (4.3; 8.4; 9.3.1; 10; 21.7)

Fulford, K.W.M. (Bill); Thornton, Tim; and Graham, George. OXFORD TEXTBOOK OF PHILOSOPHY AND PSYCHIA-TRY. New York: Oxford University Press, 2006. 872 p. ISBN 978-0-19-852695-7; ISBN 0-19-852695-4. (Inter-national Perspectives in Philosophy and Psychiatry series. "Includes CD containing 179 essential readings.") [RC437.5 .F85 2006] (4.3; 2.1; 8.3.1; 17.1)

Numbers in ( ) = NRCBL Classification Numbers

Jenkins, Richard, ed. QUESTIONS OF COMPETENCE: CULTURE, CLASSIFICATION AND INTELLECTUAL DISABILITY. Cambridge/New York: Cambridge University Press, 1998. 250 p. ISBN 0-521-62662-5. [HV3004 .Q47 1998] (4.3; 7.1; 9.5.3; 21.7)

Lewis, Bradley. MOVING BEYOND PROZAC, DSM, AND THE NEW PSYCHIATRY: THE BIRTH OF POSTPSYCHIATRY. Ann Arbor: University of Michigan Press, 2006. 198 p. ISBN 978-0-472-03117-7; ISBN 0-472-03117-1. (Corporealities: Discourses of Disability series.) [RC437.5 .L49 2006] (4.3; 7.1; 17.1; 17.4)

Madsen, Kristie and Leech, Peter. THE ETHICS OF LABELING IN MENTAL HEALTH. Jefferson, NC: McFarland, 2007. 200 p. ISBN 978-0-7864-2872-4. [RC469 .M346 2007] (4.3; 17.1)

Perlin, Michael L. THE HIDDEN PREJUDICE: MENTAL DISABILITY ON TRIAL. Washington, DC: American Psychological Association, 2000. 329 p. ISBN 1-55798-616-9. (The Law and Public Policy: Psychology and the Social Sciences series.) [KF480 .P474 2000] (4.3; 1.3.5; 7.1; 8.3.4; 9.2; 17.1; 17.7; 17.8)

4.4 QUALITY/ VALUE OF LIFE/PERSONHOOD

Aydede, Murat, ed. PAIN: NEW ESSAYS ON ITS NATURE AND THE METHODOLOGY OF ITS STUDY. Cambridge, MA: MIT Press, 2005. 423 p. ISBN 0-262-51188-6. ("A Bradford Book.") [RB127 .P332495 2005] (4.4)

Belshaw, Christopher. 10 GOOD QUESTIONS ABOUT LIFE AND DEATH. Malden, MA: Blackwell, 2005. 178 p. ISBN 978-1-4051-2604-5; ISBN 1-4051-2604-3. [BD431 .B387 2005] (4.4; 1.1; 1.2; 20.1)

Deane-Drummond, Celia and Scott, Peter Manley, eds. FUTURE PERFECT? GOD, MEDICINE AND HUMAN IDENTITY. London/New York: T&T Clark International, 2006. 219 p. ISBN 0-567-03079-2. (Gift of the publisher.) [QH438.7 .F88 2006] (4.4; 1.2; 2.1; 4.5; 5.1; 9.5.2; 15.1; 17.1; 18.5.4)

Johnston, Nancy E. and Scholler-Jaquish, Alwilda, eds. MEANING IN SUFFERING: CARING PRACTICES IN THE HEALTH PROFESSIONS. Madison: University of Wisconsin Press, 2007. 293 p. ISBN 0-299-22254-3. (Interpretive Studies in Healthcare and Human Sciences series; Vol. 6. Gift of the publisher.) [BF789 .S8 M43 2007] (4.4; 20.4.1; 20.4.2)

Perrotin, Catherine and Demaison, Michel, eds. LA DOULEUR ET LA SOUFFRANCE. Paris: Cerf, 2002. 219 p. ISBN 2-204-06876-4. [BJ1409 .D68 2002] (4.4; 1.2)

Shildrick, Margrit. EMBODYING THE MONSTER: ENCOUNTERS WITH THE VULNERABLE SELF. London/Thousand Oaks, CA: Sage, 2002. 153 p. ISBN 0-7619-7014-2. (Theory, Culture & Society series.) [BF697.5 .B63 S54 2002] (4.4; 10)

Shilling, Chris. THE BODY IN CULTURE, TECHNOLOGY AND SOCIETY. Thousand Oaks, CA: Sage, 2005. 247 p. ISBN 978-0-7619-7124-5; ISBN 0-7619-7124-6. (Theory, Culture & Society series.) [HM636 .S554 2005] (4.4)

West, Christopher and Pope John Paul. THEOLOGY OF THE BODY FOR BEGINNERS. West Chester, PA: Ascension Press, 2004. 151 p. ISBN 1-932645-34-9. [BX1795 .S48 W479 2004] (4.4; 1.2; 10)

5.1 SCIENCE, TECHNOLOGY AND SOCIETY (GENERAL)

Agazzi, Evandro. RIGHT, WRONG AND SCIENCE: THE ETHICAL DIMENSIONS OF THE TECHNO-SCIENTIFIC ENTERPRISE. Amsterdam [The Netherlands]/New York: Rodopi, 2004. 354 p. ISBN 90-420-0919-5. (Poznan Studies in the Philosophy of the Sciences and the Humanities series; Vol. 81. Monographs-in-Debate series. ISSN 0303-8157. Edited by Craig Dilworth.) [BJ57 .A3613 2004] (5.1; 5.3)

Alexander, Denis and White, Robert S. SCIENCE, FAITH, AND ETHICS: GRID OR GRIDLOCK? Peabody, MA: Hendrickson Publishers, 2006, 2005. 190 p. ISBN 978-1-59856-018-3; ISBN 1-59856-018-2. (Originally published as: *Beyond Belief*; Lion Hudson, Oxford, England, 2005.) [CB430 .A43 2006] (5.1; 1.2; 3.2; 14.5; 15.1; 16.1)

Altner, Günter. LEBEN IN DER HAND DES MENSCHEN: DIE BRISANZ DES BIOTECHNISCHEN FORTSCHRITTS. Darmstadt: Primus, 1998. 234 p. ISBN 3-89678-077-8. (5.1; 1.1; 1.2; 1.3.11; 2.1; 4.4; 5.3; 9.5.2; 14.5; 15.7; 15.8; 19.1; 20.5.1; 20.7; 22.2; 22.3)

Ball, Philip. THE DEVIL'S DOCTOR: PARACELSUS AND THE WORLD OF RENAISSANCE MAGIC AND SCIENCE. London: William Heineman, 2006. 435 p. ISBN 978-0-434-01134-6; ISBN 0-434-01134-7. [R147 .P2 B35 2006a] (5.1; Biography)

Carlson, Elof Axel. TIMES OF TRIUMPH, TIMES OF DOUBT: SCIENCE AND THE BATTLE FOR PUBLIC TRUST. Cold Spring Harbor, NY: Cold Spring Harbor Laboratory Press, 2006. 227 p. ISBN 0-87969-805-5. [Q175.35 .C37 2006] (5.1; 1.3.9; 2.1; 5.3; 9.7; 14.1; 15.5; 15.7; 16.1; 16.2; 18.5.4; 21.2)

Chase, Victor D. SHATTERED NERVES: HOW SCIENCE IS SOLVING MODERN MEDICINE'S MOST PERPLEXING PROBLEM. Baltimore, MD: Johns Hopkins University Press, 2006. 289 p. ISBN 0-8018-8514-0. (Gift of the publisher.) [RC350 .N48 C43 2006] (5.1; 4.4; 4.5)

Cranor, Carl F. TOXIC TORTS: SCIENCE, LAW, AND THE POSSIBILITY OF JUSTICE. Cambridge/New York: Cambridge University Press, 2006. 398 p. ISBN 978-0-521-86182-3; ISBN 0-521-86182-9. (Gift of the publisher.) [KF1299 .H39 C73 2006] (5.1; 1.3.5; 1.3.8; 7.1; 16.1)

Eisen, Arri and Laderman, Gary, eds. SCIENCE, RELIGION, AND SOCIETY: AN ENCYCLOPEDIA OF HISTORY, CULTURE, AND CONTROVERSY. Armonk, NY: M.E. Sharpe, 2007. 2 volumes [888 p.]. ISBN 978-0-7656-8064-8 [set]; ISBN 0-7656-8064-5 [set]. (Gift of the publisher.) [BL240.3

.S37 2007] (5.1; 1.2; 3.2; 4.1.1; 14.5; 15.1; 15.10; 16.1; 17.1; 20.4.1; 20.5.1)

Erickson, Mark. SCIENCE, CULTURE AND SOCIETY: UNDERSTANDING SCIENCE IN THE TWENTY-FIRST CENTURY. Cambridge, UK/Malden, MA: Polity, 2005. 241 p. ISBN 978-0-7456-2975-9; ISBN 0-7456-2975-X. [Q175.5 .E75 2005] (5.1; 5.3)

Frayling, Christopher. MAD, BAD AND DANGEROUS? THE SCIENTIST AND THE CINEMA. London: Reaktion Books, 2005. 239 p. ISBN 0-86189-255-1. [PN1995.9 .S267 F73 2005] (5.1)

Jordan, Diann. SISTERS IN SCIENCE: CONVERSATIONS WITH BLACK WOMEN SCIENTISTS ON RACE, GENDER, AND THEIR PASSION FOR SCIENCE. West Lafayette, IN: Purdue University Press, 2006. 240 p. ISBN 978-1-55753-386-9; ISBN 1-55753-386-5. [Q141 .S556 2006] (5.1; 10; Biography)

Kaplan, David M., ed. READINGS IN THE PHILOSOPHY OF TECHNOLOGY. Lanham, MD: Rowman & Littlefield, 2004. 512 p. ISBN 978-0-7425-1489-8; ISBN 0-7425-1489-7. [T14 .R39 2004] (5.1; 1.3.5; 1.3.9; 1.3.12; 4.4; 4.5; 5.3; 16.1)

Karafyllis, Nicole C., ed. BIOFAKTE: VERSUCH ÜBER DEN MENSCHEN ZWISCHEN ARTEFAKT UND LEBEWESEN. Paderborn: Mentis, 2003. 295 p. ISBN 978-3-89785-384-3; ISBN 3-89785-384-1. [BH301 .N3 B56 2003] (5.1; 1.1; 4.4; 15.1; 19.1)

Kennedy, Helena; Little, Miles; Pell, Cardinal George; Somerville, Margaret; Hilton, Douglas; and Jacobson, Lisa. ETHICALLY CHALLENGED: BIG QUESTIONS FOR SCIENCE. Carlton, Victoria, Australia: Miegunyah Press, 2007. 121 p. ISBN 978-0-522-85321-6; ISBN 0-522-85321-8. (Introduction by Sir Gustav Nossal. The Alfred Deakin Debate series; No. 2. Series edited by Jonathan Mills.) [Q175.35 .E825 2007] (5.1; 5.3)

Machamer, Peter and Wolters, Gereon, eds. SCIENCE, VALUES, AND OBJECTIVITY. Pittsburgh: University of Pittsburgh Press [and] Konstanz: Universitätsverlag Konstanz, 2004. 317 p. ISBN 978-0-8229-4237-5; ISBN 0-8229-4237-2. (Pittsburgh-Konstanz Series in the Philosophy and History of Science.) [Q175 .S3626 2004] (5.1; 1.1.; 1.3.9; 2.1; 5.3)

Shepherd-Barr, Kirsten. SCIENCE ON STAGE: FROM DOCTOR FAUSTUS TO COPENHAGEN. Princeton, NJ: Princeton University Press, 2006. 271 p. ISBN 978-0-691-12150-5; ISBN 0-691-12150-8. [PN1650 .S34 S54 2006] (5.1; 5.3; 7.1)

Silver, Lee M. CHALLENGING NATURE: THE CLASH OF SCIENCE AND SPIRITUALITY AT THE NEW FRONTIERS OF LIFE. New York: HarperCollins, 2006. 444 p. ISBN 978-0-06-058267-8; ISBN 0-06-058267-7. [Q175.35 .S55 2006] (5.1; 1.2; 5.3; 14.5; 15.1)

5.2 TECHNOLOGY ASSESSMENT

Schlich, Thomas and Tröhler, Ulrich, eds. THE RISKS OF MEDICAL INNOVATION: RISK PERCEPTION AND ASSESSMENT IN HISTORICAL CONTEXT. London/New York: Routledge, 2006. 291 p. ISBN 978-0-415-33481-5; ISBN 0-415-33481-0. (Routledge Studies in the Social History of Medicine series; Vol. 21.) [RA418.5 .M4 R55 2006] (5.2; 7.1; 20.2.1)

5.3 SOCIAL CONTROL OF SCIENCE/TECHNOLOGY

Gough, Michael, ed. POLITICIZING SCIENCE: THE ALCHEMY OF POLICYMAKING. Stanford, CA: Hoover Institution Press [and] Washington, DC: George C. Marshall Institute, 2003. 313 p. ISBN 0-8179-3932-6. (Hoover Institution Press Publication No. 517.) [Q175.5 .P65 2003] (5.3; 1.3.5; 1.3.9; 5.2; 9.5.1; 16.1; 16.2; 21.1; 21.3)

Greif, Karen F. and Merz, Jon F. CURRENT CONTROVERSIES IN THE BIOLOGICAL SCIENCES: CASE STUDIES OF POLICY CHALLENGES FROM NEW TECHNOLOGIES. Cambridge, MA: MIT Press, 2007. 385 p. ISBN 978-0-262-57239-2. (Basic Bioethics series. Gift of the publisher.) [R850 .G74 2007] (5.3; 1.3.9; 9.5.6; 14.1; 15.8; 15.10; 19.1)

Jaffe, Adam B. and Lerner, Josh. INNOVATION AND ITS DISCONTENTS: HOW OUR BROKEN PATENT SYSTEM IS ENDANGERING INNOVATION AND PROGRESS, AND WHAT TO DO ABOUT IT. Princeton, NJ: Princeton University Press, 2004. 236 p. ISBN 978-0-691-11725-6; ISBN 0-691-11725-X. [KF3120 .J34 2004] (5.3; 1.3.2; 9.3.1)

Kulakowski, Elliott C. and Chronister, Lynne U. RESEARCH ADMINISTRATION AND MANAGEMENT. Boston: Jones and Bartlett, 2006. 916 p. ISBN 0-7637-3277-X. (Gift of the publisher.) [Q180 .U5 R3816 2006] (5.3; 1.3.9; 1.3.12; 5.1; 5.2; 8.4; 18.2; 21.1; 22.2)

Mehta, Michael D., ed. BIOTECHNOLOGY UNGLUED: SCIENCE, SOCIETY, AND SOCIAL COHESION. Vancouver: UBC Press, 2005. 194 p. ISBN 0-7748-1133-1. [TP248.23 .B566 2005] (5.3; 1.3.11; 1.3.12; 15.1; 21.1)

Mowery, David C.; Nelson, Richard R.; Sampat, Bhaven N.; and Ziedonis, Arvids A. IVORY TOWER AND INDUSTRIAL INNOVATION: UNIVERSITY-INDUSTRY TECHNOLOGY TRANSFER BEFORE AND AFTER THE BAYH-DOLE ACT IN THE UNITED STATES. Stanford, CA: Stanford University Press, 2004. 241 p. ISBN 978-0-8047-4920-6; ISBN 0-8047-4920-5. (Innovation and Technology in the World Economy series.) [T174.3 .I96 2004] (5.3)

5.4 NANOTECHNOLOGY

Berne, Rosalyn W. NANOTALK: CONVERSATIONS WITH SCIENTISTS AND ENGINEERS ABOUT ETHICS, MEANING, AND BELIEF IN THE DEVELOPMENT OF NANOTECHNOLOGY. Mahwah, NJ: Lawrence Erlbaum Associates, 2006. 361 p. ISBN 978-0-8058-4810-6; ISBN 0-8058-4810-X. [T174.7 .B37 2006] (5.4; 5.3)

Numbers in ( ) = NRCBL Classification Numbers

Edwards, Steven A. THE NANOTECH PIONEERS: WHERE ARE THEY TAKING US? Weinheim: Wiley-VCH, 2006. 244 p. ISBN 978-3-527-31290-0; ISBN 3-527-31290-0. [T174.7 .E39 2006] (5.4; 5.3)

Hall, J. Storrs. NANOFUTURE: WHAT'S NEXT FOR NANOTECHNOLOGY. Amherst, NY: Prometheus Books, 2005. 333 p. ISBN 978-1-59102-287-9. (Foreword by K. Eric Drexler.) [T174.7 .H35 2005] (5.4; 14.1)

Hunt, Geoffrey and Mehta, Michael D., eds. NANOTECHNOLOGY: RISK, ETHICS AND LAW. London/Sterling, VA: Earthscan, 2006. 296 p. ISBN 978-1-84407-358-0; ISBN 1-84407-358-0. (Science and Society Series.) [T174.7 .N37525 2006] (5.4; 1.3.2; 1.3.9; 5.2; 5.3; 15.1; 16.1; 18.1; 21.1)

National Research Council (United States). Committee to Review of the National Nanotechnology Initiative. A MATTER OF SIZE: TRIENNIAL REVIEW OF THE NATIONAL NANOTECHNOLOGY INITIATIVE. Washington, DC: National Academies Press, 2006. 183 p. ISBN 978-0-309-10223-0; ISBN 0-309-10223-5. [T174.7 .M37 2006] (5.4; 5.3)

Shelley, Toby. NANOTECHNOLOGY: NEW PROMISES, NEW DANGERS. London: Zed Books; Black Point, Nova Scotia: Fernwood; Bangalore, India: Books for Change; Kuala Lumpur, Malaysia: SIRD; Cape Town, S.A.: David Philip; New York: Distributed in the USA exclusively by Palgrave Macmillan, 2006. 170 p. ISBN 978-1-84277-687-2 [Zed]; ISBN 1-84277-687-8 [Zed]; ISBN 983-2535-794 [SIRD]. [T174.7 .S54 2006] (5.4; 1.3.11; 5.2; 5.3; 9.1; 21.2)

6 CODES OF/ POSITION STATEMENTS ON PROFESSIONAL ETHICS

American Medical Association. Council on Ethical and Judicial Affairs; Southern Illinois University at Carbondale. School of Medicine; and Southern Illinois University of Carbondale. School of Law. CODE OF MEDICAL ETHICS: CURRENT OPINIONS WITH ANNOTATIONS. Chicago: AMA Press, 2006. 399 p. ISBN 978-1-57947-777-6; ISBN 1-57947-777-1. (2006-2007 edition.) [R725 .A55 2006-2007] (6; 1.3.2; 2.1)

Cooper, Melvin Wayne. THE CODE OF ETHICS OF THE AMERICAN MEDICAL ASSOCIATION: THE SOCIAL CONSTRUCTION OF MEDICAL MORALITY: THE ESTABLISHMENT OF A MASCULINE PROFESSION. Ann Arbor, MI: ProQuest Information and Learning/UMI, 2003. 220 p. (Publication No. AAT-3110970. Dissertation, (Ph.D. in the Humanities)—University of Texas at Dallas, 2003.) [LD5340.7 .C66865 2003] (6; 2.2; 4.1.2; 7.1; 8.1)

7.1 SOCIOLOGY OF HEALTH CARE (GENERAL)

Anderson, Warwick. COLONIAL PATHOLOGIES: AMERICAN TROPICAL MEDICINE, RACE, AND HYGIENE IN THE PHILIPPINES. Durham, NC: Duke University Press, 2006. 355 p. ISBN 978-0-8223-3843-7; ISBN 0-8223-3843-2. (Gift of

the publisher.) [RC962 .P6 A53 2006] (7.1; 9.1; 9.5.4; 9.7; 18.5.9)

Aronowitz, Robert A. UNNATURAL HISTORY: BREAST CANCER AND AMERICAN SOCIETY. Cambridge/New York: Cambridge University Press, 2007. 366 p. ISBN 978-0-521-82249-7; ISBN 0-521-82249-1. (Gift of the publisher.) [RC280 .B8 A783 2007] (7.1; 9.5.1)

Belli, Angela and Coulehan, Jack, eds. PRIMARY CARE: MORE POEMS BY PHYSICIANS. Iowa City: University of Iowa Press, 2006. 130 p. ISBN 978-1-58729-503-4; ISBN 1-58729-503-2. [PS591 .P48 P75 2006] (7.1)

Bliss, Michael. HARVEY CUSHING: A LIFE IN SURGERY. Oxford/New York: Oxford University Press, 2005. 591 p. ISBN 978-0-19-516989-8; ISBN 0-19-516989-1. [RD592.9 .C87 B55 2005] (7.1; 17.6; Biography)

Campo, Rafael. THE ENEMY. Durham, NC: Duke University Press, 2007. 99 p. ISBN 978-0-8223-3960-1. (Fifth collection of poetry. Gift of the publisher.) [PS3553 .A4883 E54 2007] (7.1)

Carter, Albert Howard III. OUR HUMAN HEARTS: A MEDICAL AND CULTURAL JOURNEY. Kent, OH: Kent State University Press, 2006. 206 p. ISBN 978-0-87338-863-4; ISBN 0-87338-863-1. (Literature and Medicine Series; No. 8.) [RC682 .C48 2006] (7.1)

Charon, Rita. NARRATIVE MEDICINE: HONORING THE STORIES OF ILLNESS. Oxford/New York: Oxford University Press, 2006. 266 p. ISBN 978-0-19-516675-0; ISBN 0-19-516675-2. [RC48 .C42 2006] (7.1; 4.4.; 8.1)

Crellin, John K. PUBLIC EXPECTATIONS AND PHYSICIANS' RESPONSIBILITIES: VOICES OF MEDICAL HUMANITIES. Oxford/Seattle: Radcliffe, 2005. 165 p. ISBN 1-85775-642-8. [R702 .C74 2005] (7.1; 1.3.9; 4.2; 5.3; 8.1)

Cribb, Alan. HEALTH AND THE GOOD SOCIETY: SETTING HEALTHCARE ETHICS IN SOCIAL CONTEXT. Oxford: Clarendon Press/New York: Oxford University Press, 2005. 236 p. ISBN 978-0-19-924273-3; ISBN 0-19-924273-9. (Issues in Biomedical Ethics series.) [R724 .C824 2005] (7.1; 7.2; 8.1; 9.1; 9.4)

DeShazer, Mary K. FRACTURED BORDERS: READING WOMEN'S CANCER LITERATURE. Ann Arbor: University of Michigan Press, 2005. 301 p. ISBN 978-0-472-0609-5; ISBN 0-472-06909-8. [PS169 .C35 D47 2005] (7.1; 9.5.1; 9.5.5; Biography)

Duffin, Jacalyn, ed. CLIO IN THE CLINIC: HISTORY IN MEDICAL PRACTICE. Oxford/New York: Oxford University Press, 2005. 334 p. ISBN 0-19-516128-9. [R133 .C56 2005] (7.1; 8.1; 9.5.6; 9.7; 17.1; 18.3)

Franklin, Sarah and Lock, Margaret, eds. REMAKING LIFE & DEATH: TOWARD AN ANTHROPOLOGY OF THE BIOSCIENCES. Santa Fe, NM: School of American Research Press [and] Oxford: John Currey, 2003. 372 p. ISBN 978-1-930618-20-6; ISBN 1-930618-20-4 [School of American Research Press]; ISBN 0-85255-932-1 [John Currey]. (School of

American Research Advanced Seminar Series.) [QP81 .R45 2003] (7.1; 14.5; 15.1; 18.5.4; 19.5; 20.3.1)

Henderson, Gail E.; Estroff, Sue E.; Churchill, Larry R.; King, Nancy M.P.; Oberlander, Jonathan; and Strauss, Ronald P., eds. THE SOCIAL MEDICINE READER, VOLUME 2: SOCIAL AND CULTURAL CONTRIBUTIONS TO HEALTH, DIFFERENCE, AND INEQUALITY. Durham, NC: Duke University Press, 2005. 320 p. ISBN 0-8223-3593-X. (Second edition. Gift of the publisher.) [RA418 .S6424 2005 v.2] (7.1; 8.1; 9.5.4; 9.5.5; 10; 15.5; 15.11; 18.1; 21.7)

Huth, Edward J. and Murray, T. Jock, eds. MEDICINE IN QUOTATIONS: VIEWS OF HEALTH AND DISEASE THROUGH THE AGES. Philadelphia: American College of Physicians, 2006. 581 p. ISBN 978-1-930513-67-9; ISBN 1-930513-67-4. (Second edition.) [R705 .M465 2006] (7.1)

Kelleher, David and Leavey, Gerard, eds. IDENTITY AND HEALTH. London/New York: Routledge, 2004. 207 p. ISBN 978-0-415-30792-5; ISBN 0-415-30792-9. [RA776.9 .I34 2004] (7.1; 1.2; 4.4; 9.5.2; 10; 17.1)

King, Nancy M.P.; Strauss, Ronald P.; Churchill, Larry R.; Estroff, Sue E.; Henderson, Gail E.; and Oberlander, Jonathan, eds. THE SOCIAL MEDICINE READER, VOLUME 1: PATIENTS, DOCTORS, AND ILLNESS. Durham, NC: Duke University Press, 2005. 294 p. ISBN 0-8223-3568-9. (Second edition. Gift of the publisher.) [RA418 .S6424 2005 v.1] (7.1; 2.2; 6; 7.2; 8.1; 8.2; 8.3.1; 8.3.4; 20.5.1)

Levy, Barry S. and Sidel, Victor W., eds. SOCIAL INJUSTICE AND PUBLIC HEALTH. Oxford/New York: Oxford University Press, 2006. 529 p. ISBN 978-0-19-517185-3; ISBN 0-19-517185-3. (Published in cooperation with the American Public Health Association.) [RA418 .S6423 2006] (7.1; 9.1; 9.5.1; 9.5.2; 9.5.4; 9.5.7; 9.5.10; 10; 16.3; 17.1; 21.1; 21.2)

Livingston, Julie. DEBILITY AND THE MORAL IMAGINATION IN BOTSWANA. Bloomington: Indiana University Press, 2005. 310 p. ISBN 0-253-21785-7. (African Systems of Thought series.) [HN806 .A8 L58 2005] (7.1; 4.2; 4.4; 9.1; 9.3.1)

Lucire, Yolande. CONSTRUCTING RSI: BELIEF AND DESIRE. Sydney: University of New South Wales Press, 2001. 216 p. ISBN 0-86840-778-X. [RD97.6 .L83 2003] (7.1; 4.2)

Mullan, Fitzhugh; Ficklen, Ellen; and Rubin, Kyna, eds. NARRATIVE MATTERS: THE POWER OF THE PERSONAL ESSAY IN HEALTH POLICY. Baltimore, MD: Johns Hopkins University Press, 2006. 293 p. ISBN 0-8018-8479-9. ("A Health Affairs Reader." Foreword by Abraham Verghese. Gift of the publisher.) [RA393 .N37 2006] (7.1; 9.1)

Neuberger, Julia. THE MORAL STATE WE'RE IN: A MANIFESTO FOR A 21ST CENTURY SOCIETY. London: HarperCollins Entertainment, 2005. 347 p. ISBN 978-0-00-718167-4; ISBN 0-00-718167-1. [HN400 .M26 N483 2005] (7.1; 1.3.5)

Oberlander, Jonathan; Churchill, Larry R.; Estroff, Sue E.; Hendersen, Gail E.; King, Nancy M.P.; and Strauss, Ronald P., eds. THE SOCIAL MEDICINE READER, VOLUME 3: HEALTH POLICY, MARKETS, AND MEDICINE. Durham, NC: Duke University Press, 2005. 288 p. ISBN 0-8223-3569-7. (Second edition. Gift of the publisher.) [RA418 .S6424 2005 v.3] (7.1; 9.1; 9.3.1; 9.3.2; 9.4; 9.5.2; 9.5.10; 14.4; 19.1; 21.1)

Porter, Roy, ed. THE CAMBRIDGE HISTORY OF MEDICINE. New York: Cambridge University Press, 2006. 408 p. ISBN 0-521-68289-4; ISBN 978-0-521-68289-3. [R131 .C229 2006] (7.1; 9.1; 9.7; Reference)

Siegrist, Johannes and Marmot, Michael, eds. SOCIAL INEQUALITIES IN HEALTH: NEW EVIDENCE AND POLICY IMPLICATIONS. Oxford/New York: Oxford University Press, 2006. 258 p. ISBN 978-0-19-856816-2; ISBN 0-19-856816-9. [RA418 .S64229 2006] (7.1; 9.1; 9.3.1)

Stone, John. THE SMELL OF MATCHES. Baton Rouge: Louisiana State University Press, 1972. 77 p. ISBN 0-8071-1477-4. (Gift of Warren Reich.) [PS3569 .T6413 S64 1988] (7.1)

van Dijck, José. THE TRANSPARENT BODY: A CULTURAL ANALYSIS OF MEDICAL IMAGING. Seattle: University of Washington Press, 2005. 193 p. ISBN 0-295-98490-2. (In Vivo: The Cultural Mediations of Biomedical Science series.) [RC78.7 .D53 D553 2005] (7.1; 1.3.7; 4.4; 5.1)

Woodward, John and Jütte, Robert, eds. COPING WITH SICKNESS: HISTORICAL ASPECTS OF HEALTH CARE IN A EUROPEAN PERSPECTIVE. Sheffield: European Association for the History of Medicine and Health Publications, 1995. 224 p. ISBN 0-9527045-0-1. (History of Medicine, Health and Disease Series.) (7.1; 4.1.1; 4.1.3; 9.3.1; 9.5.1; 9.5.6)

Wootton, David. BAD MEDICINE: DOCTORS DOING HARM SINCE HIPPOCRATES. Oxford/New York: Oxford University Press, 2006. 304 p. ISBN 978-0-19-280355-9; ISBN 0-19-280355-7. [R484 .W66 2006] (7.1; 9.1; 9.8)

7.2 EDUCATION FOR HEALTH CARE PROFESSIONALS

Gunderman, Richard B. ACHIEVING EXCELLENCE IN MEDICAL EDUCATION. New York: Springer, 2006. 179 p. ISBN 978-1-84628-296-6; ISBN 1-84628-296-9. (Gift of the publisher.) [R735 .G86 2006] (7.2; 4.1.2; 9.8)

Manning, Phil R. and DeBakey, L. MEDICINE: PRESERVING THE PASSION IN THE 21ST CENTURY. New York: Springer, 2004. 470 p. ISBN 0-387-00427-0. (Second edition. Revised edition of: *Medicine: Preserving the Passion*, 1987.) [R845 .M36 2004] (7.2; 1.3.1; 1.3.12; 4.1.2; 8.1; 9.8)

Pories, Susan; Jain, Sachin H. and Harper, Gordon, eds. THE SOUL OF A DOCTOR: HARVARD MEDICAL STUDENTS FACE LIFE AND DEATH. Chapel Hill, NC: Algonquin Books of Chapel Hill, 2006. 248 p. ISBN 978-1-56512-507-0; ISBN 1-56512-507-X. ("Selected essays from the Harvard Medical School Patient-Doctor course.") [R747 .H28 S68 2006] (7.2; 7.1; 8.1; Biography)

Rhodes, Teresa [Ann Reitsma]. IDENTIFICATION OF CURRICULAR AND EDUCATIONAL NEEDS FOR PRIMARY CARE PHYSICIANS IN DEALING WITH THE CLINICAL APPLICATION OF GENOMIC MEDICINE. Ann Arbor, MI: University Microfilms International [UMI], 2001. 97 p. (Publication No. AAT-3005661. Dissertation, (Ph.D.)—Medical University of South Carolina, College of Health Professionals, 2001.) (7.2; 7.1; 15.1)

7.3 PROFESSIONAL-PROFESSIONAL RELATIONSHIP

Bartholomew, Kathleen. SPEAK YOUR TRUTH: PROVEN STRATEGIES FOR EFFECTIVE NURSE-PHYSICIAN COMMUNICATION. Marblehead, MA: HCPro, 2005. 162 p. ISBN 978-1-57839-556-9; ISBN 1-57839-556-9. [RT86.4 .B37 2005] (7.3)

7.4 PROFESSIONAL MISCONDUCT

Celenza, Andrea. SEXUAL BOUNDARY VIOLATIONS: THERAPEUTIC, SUPERVISORY, AND ACADEMIC CONTEXTS. Lanham, MD: Jason Aronson, 2007. 267 p. ISBN 978-0-7657-0471-9; ISBN 0-7657-0471-4. [RC480.8 .C4 2007] (7.4; 7.1; 8.1; 10; 17.2)

8.1 PATIENT RELATIONSHIPS (GENERAL)

Bub, Barry. COMMUNICATION SKILLS THAT HEAL: A PRACTICAL APPROACH TO A NEW PROFESSIONALISM IN MEDICINE. Oxford/Seattle: Radcliffe, 2006. 165 p. ISBN 978-1-85775-664-7; ISBN 1-85775-664-9. (Gift of the publisher.) [R118 .B82 2006] (8.1; 4.4; 17.1; 17.2)

Dedicated Issue on Physician-Patient Communication. FAMILY MEDICINE 2002 May; 34(5): 310-393. (ISSN 0742-3225.) (8.1; 7.2; 9.1; 21.7)

Groopman, Jerome. HOW DOCTORS THINK. Boston: Houghton Mifflin, 2007. 307 p. ISBN 978-0-618-61003-7; ISBN 0-618-61003-0. [R723.5 .G75 2007] (8.1; 9.4)

Houser, Janet and Bokovoy, Joanna. CLINICAL RESEARCH IN PRACTICE: A GUIDE FOR THE BEDSIDE SCIENTIST. Sudbury, MA: Jones and Bartlett, 2006. 277 p. ISBN 978-0-7637-3875-4; ISBN 0-7637-3875-1. [R850 .H677 2006] (8.1; 4.1.3; 7.1)

Kirk, Timothy W. THE MORAL SIGNIFICANCE OF INTIMACY IN NURSE-PATIENT RELATIONSHIPS. Ann Arbor, MI: ProQuest Information and Learning/UMI, 2004. 149 p. (Publication No. AAT-3126350. Dissertation, (Ph.D. in Philosophy)—Villanova University, 2004.) (8.1; 4.1.3)

Launer, John. NARRATIVE-BASED PRIMARY CARE: A PRACTICAL GUIDE. Abingdon, Oxon, UK: Radcliffe Medical Press, 2002. 251 p. ISBN 978-1-85775-539-8; ISBN 1-85775-539-1. (Foreword by Trisha Greenhalgh.) (8.1; 7.2; 9.5.1)

Woods, David, ed. COMMUNICATION FOR DOCTORS: HOW TO IMPROVE PATIENT CARE AND MINIMIZE LEGAL RISKS. Oxford/San Francisco: Radcliffe Publishing, 2004. 125 p. ISBN 1-85775-895-1. (8.1; 7.1)

8.2 TRUTH DISCLOSURE

Dongen, Elis van and Fainzang, Sylvie, eds. LYING AND ILLNESS: POWER AND PERFORMANCE. Amsterdam: Het Spinhuis; Distributed in the USA, Canada, Latin-America and the UK by: New Brunswick, NJ: Transaction, 2005. 207 p. ISBN 90-5589-245-9. [BJ1421 .L95 2005] (8.2; 7.1; 8.1)

8.3.1 INFORMED CONSENT (GENERAL)

Beyleveld, Deryck and Brownsword, Roger. CONSENT IN THE LAW. Oxford/Portland, OR: Hart, 2007. 388 p. ISBN 978-1-84113-679-4; ISBN 1-84113-679-0. (Legal Theory Today series.) [K579 .I6 B49 2007] (8.3.1; 2.1; 4.3; 18.3; 21.1)

Brokken, David Allan. DISCLOSURE: THE HIDDEN WEAKNESS IN INFORMED CONSENT. Ann Arbor, MI: ProQuest Information and Learning/UMI, 2004. 471 p. (Publication No. AAT-3149399. Dissertation, (Ph.D. in Philosophy)—University of Minnesota, 2004.) (8.3.1; 8.2; 9.7; 18.3)

Pereira, André Gonçalo Dias. O CONSENTIMENTO INFORMADO NA RELAÇÃO MÉDICO-PACIENTE: ESTUDO DE DIREITO CIVIL. Coimbra, Portugal: Coimbra Editora, 2004. 699 p. ISBN 978-972-32-1247-1; ISBN 972-32-1247-1. (Faculdade de Direito da Universidade de Coimbra, Centro de Direito Biomédico series; No. 9.) [KKQ3119 .I54 P47 2004] (8.3.1)

Rodrigues, João Vaz. O CONSENTIMENTO INFORMADO PARA O ACTO MÉDICO NO ORDENAMENTO JURÍDICO PORTUGUÊS (ELEMENTOS PARA O ESTUDO DA MANIFESTAÇÃO DA VONTADE DO PACIENTE). Coimbra, Portugal: Coimbra Editora, 2001. 549 p. ISBN 978-972-32-1013-2; ISBN 972-32-1013-4. (Faculdade de Direito da Universidade de Coimbra, Centro de Direito Biomédico (series); 3.) [KKQ3119 .I54 R63 2001] (8.3.1; 2.1)

8.3.4 RIGHT TO REFUSE TREATMENT

Trotter, Griffin. THE ETHICS OF COERCION IN MASS CASUALTY MEDICINE. Baltimore, MD: Johns Hopkins University Press, 2007. 154 p. ISBN 978-0-8018-8551-8; ISBN 0-8018-8551-5. [RA645.5 .T76 2007] (8.3.4; 9.1)

8.4 CONFIDENTIALITY

Borten, Kate. A MARKETER'S GUIDE TO HIPAA: RESOURCES FOR CREATING EFFECTIVE AND COMPLIANT MARKETING. Marblehead, MA: HCPro, 2006. 126 p. ISBN 978-1-57839-875-1; ISBN 1-57839-875-4. (Foreword by Chris Houchens.) [RA979 .B67 2007] (8.4; 1.3.2; 9.3.1)

Herdman, Roger and Moses, Harold L. EFFECT OF THE HIPAA PRIVACY RULE ON HEALTH RESEARCH: PROCEED-

INGS OF A WORKSHOP PRESENTED TO THE NATIONAL CAN-
CER POLICY FORUM. Washington, DC: National Academies
Press, 2006. 93 p. ISBN 0-309-10291-X. [R864 .E34
2006] (8.4; 1.3.12; 8.2; 9.1)

8.5 MALPRACTICE

Baker, Tom. THE MEDICAL MALPRACTICE MYTH. Chicago:
University of Chicago Press, 2005. 214 p. ISBN
0-226-03648-0. [KF2905.3 .B35 2005] (8.5; 9.3.1; 9.8)

Trumpi, Pauline. DOCTORS WHO RAPE: MALPRACTICE AND
MISOGYNY. Rochester, VT: Schenkman Books, 1997.
324 p. ISBN 0-87047-108-2. (Publisher's address: PO Box
119, zip 05767.) [HV6561 .T79 1997] (8.5; 7.4; 8.1; 8.4;
10; 17.2)

9.1 HEALTH CARE (GENERAL)

Aday, Lu Ann, ed. REINVENTING PUBLIC HEALTH: POLICIES
AND PRACTICES FOR A HEALTHY NATION. San Francisco,
CA: Jossey-Bass/Wiley & Sons, 2005. 370 p. ISBN
0-7879-7561-3. (Foreword by Kenneth I. Shine.) [RA418
.R425 2005] (9.1; 7.1; 9.3.1)

Allsop, Judith and Saks, Mike, eds. REGULATING THE
HEALTH PROFESSIONS. London/Thousand Oaks, CA: Sage,
2002. 166 p. ISBN 978-0-7619-6740-8; ISBN
0-7619-6740-0. [RA395 .G6 R44 2002] (9.1; 4.1.1; 6;
21.1)

Almgren, Gunnar. HEALTH CARE POLITICS, POLICY, AND
SERVICES: A SOCIAL JUSTICE ANALYSIS. New York:
Springer Publishing Company, 2007. 347 p. ISBN
978-0-8261-0236-2; ISBN 0-8261-0236-0. [RA395 .A3
A4795 2007] (9.1; 7.1; 9.3.1; 9.5.2)

Balint, John; Philpott, Sean; Baker, Robert; and Strosberg,
Martin, eds. ETHICS AND EPIDEMICS. Amsterdam/Boston:
JAI Press/Elsevier, 2006. 253 p. ISBN 978-0-7623-
1311-2; ISBN 0-7623-1311-0. (Advances in Bioethics series; Vol. 9. ISSN 1479-3709.) (9.1; 2.2; 7.1; 8.1; 9.5.1;
9.5.6; 9.7; 18.5.9; 21.1; 21.3)

Bayer, Ronald; Gostin, Lawrence O.; Jennings, Bruce; and
Steinbock, Bonnie, eds. PUBLIC HEALTH ETHICS: THEORY,
POLICY, AND PRACTICE. Oxford/New York: Oxford Uni-
versity Press, 2007. 418 p. ISBN 978-0-19-518084-8;
ISBN 0-19-518085-2. (Gift of the publisher.) [RA427.25
.P82 2007] (9.1; 7.1; 8.3.1; 8.4; 9.5.7; 9.5.9; 9.7; 15.3;
16.1; 16.3; 21.3)

Behan, Pamela. SOLVING THE HEALTH CARE PROBLEM:
HOW OTHER NATIONS SUCCEEDED AND WHY THE UNITED
STATES HAS NOT. Albany: State University of New York
Press, 2006. 171 p. ISBN 978-0-7914-6837-1; ISBN
0-7914-6837-2. [RA395 .A3 B44 2006] (9.1; 9.2; 9.3.1)

Bhattacharya, Sanjoy; Harrison, Mark; and Worboys, Michael. FRACTURED STATES: SMALLPOX, PUBLIC HEALTH
AND VACCINATION POLICY IN BRITISH INDIA 1800-1947. New
Delhi: Orient Longman, 2005. 264 p. ISBN 978-81-
250-2866-6; ISBN 81-250-2866-8. (New Perspectives in

South Asian History series; No. 11.) [RA644 .S6 B53
2005] (9.1; 9.5.1; 9.7; 18.1)

Borowy, Iris and Gruner, Wolf D., eds. FACING ILLNESS IN
TROUBLED TIMES: HEALTH IN EUROPE IN THE INTERWAR
YEARS, 1918-1939. Frankfurt am Main/New York: Peter
Lang, 2005. 424 p. ISBN 3-631-51948-6 [Germany];
ISBN 0-8204-6542-9 [U.S.]. (Gift of the publisher.)
[RA418.3 .E85 F333 2005] (9.1; 7.1; 9.5.1; 15.5; 21.1)

Chard, Richard E. THE MEDIATING EFFECT OF PUBLIC OPIN-
ION ON PUBLIC POLICY: EXPLORING THE REALM OF HEALTH
CARE. Albany: State University of New York Press, 2004.
179 p. ISBN 978-0-7914-6053-5; ISBN 0-7914-6053-3.
(SUNY Series in Public Policy.) [RA395 .A3 C4795 2004]
(9.1; 1.3.5; 1.3.7; 7.1; 9.3.1; 21.1)

Davis-Floyd, Robbie and Johnson, Christine Barbara, eds.
MAINSTREAMING MIDWIVES: THE POLITICS OF CHANGE.
London/New York: Routledge, 2006. 559 p. ISBN 978-
0-415-93151-9; ISBN 0-415-93151-7. [RG950 .M34
2006] (9.1; 1.3.5; 7.1; 9.5.5; 9.5.7; 21.1)

Detels, Roger; McEwen, James; Beaglehole, Robert; and
Tanaka, Heizo, eds. OXFORD TEXTBOOK OF PUBLIC
HEALTH. New York: Oxford University Press, 2004.
1955 p. ISBN 978-0-19-850959-2; ISBN 0-19-850959-6.
(Fourth edition.) [RA425 .O9 2004] (9.1; 1.3.12; 7.1;
9.5.1; 21.1)

Dickinson, Taylor. MEDICINE, ITS MARKETPLACE, AND THE
AMERICAN DREAM. [s.l.]: Xlibris.com, 2003. 95 p. ISBN
1-4010-8846-5. [R855.3 .D53 2003] (9.1; 9.3.1; 9.5.1)

Engström, Timothy H. and Robison, Wade L., eds. HEALTH
CARE REFORM: ETHICS AND POLITICS. Rochester, NY: University of Rochester Press, 2006. 289 p. ISBN
1-58046-226-X. (Gift of the publisher.) [RA395 .A3 H432
2006] (9.1; 1.1; 9.3.1; 9.3.2; 21.1)

Flood, Colleen M., ed. JUST MEDICARE: WHAT'S IN, WHAT'S
OUT, HOW WE DECIDE. Toronto/Buffalo, NY: University of
Toronto Press, 2006. 458 p. ISBN 978-0-8020-8002-8;
ISBN 0-8020-8002-2. [KE3404 .J87 2006] (9.1; 1.3.5;
1.3.7; 9.3.1; 9.4; 9.5.4; 9.5.6; 11.1; 12.4.1; 15.1; 17.1)

Galgut, Damon. THE GOOD DOCTOR: A NOVEL. New York:
Grove Press, 2004, 2003. 215 p. ISBN 0-8021-4169-2.
[PR9369.3 .G28 G66 2004] (9.1; Fiction)

Hemenway, David. PRIVATE GUNS, PUBLIC HEALTH. Ann
Arbor: University of Michigan Press, 2004. 326 p. ISBN
0-472-11405-0. [RD96.3 .H45 2004] (9.1; 7.1; 20.1)

Institute of Medicine (United States). Board on Global
Health. Forum on Microbial Threats. THE THREAT OF PAN-
DEMIC INFLUENZA: ARE WE READY? WORKSHOP SUMMARY.
Washington, DC: National Academies Press, 2004. 411 p.
ISBN 0-309-09504-2. (Edited by Stacey L. Knobler,
Alison Mack, Adel Mahmoud and Stanley M. Lemon.
Also available on the Web at: http://www.nap.edu.)
[RA644 .I6 T48 2005] (9.1; 1.3.11; 7.1; 9.3.1; 9.5.1; 9.7;
22.3)

Israel, Barbara A.; Eng, Eugenia; Schulz, Amy J.; and Parker, Edith A., eds. METHODS IN COMMUNITY-BASED PARTICIPATORY RESEARCH FOR HEALTH. San Francisco: Jossey-Bass, 2005. 479 p. ISBN 978-0-7879-7562-3; ISBN 0-7879-7562-1. (Foreword by David Satcher.) [RA440.85 .M475 2005] (9.1; 7.1; 9.5.4; 16.1; 18.1)

Jamison, Dean T.; et al., eds. DISEASE CONTROL PRIORITIES IN DEVELOPING COUNTRIES. New York: Oxford University Press, 2006. 1401 p. ISBN 978-0-8213-6179-5; ISBN 0-8213-6179-1. (Second edition. World Bank. Disease Control Priorities Project.) [RA441.5 .D57 2006] (9.1; 9.3.1; 21.1)

Jamison, Dean T.; et al., eds. PRIORITIES IN HEALTH. Washington, DC: World Bank, 2006. 217 p. ISBN 978-0-8213-6260-0; ISBN 0-8213-6260-7. ("Disease Control Priorities in Developing Countries.") [RA441.5 .P75 2006] (9.1; 9.3.1; 21.1)

Jost, Timothy Stoltzfus. HEALTH CARE AT RISK: A CRITIQUE OF THE CONSUMER-DRIVEN MOVEMENT. Durham, NC: Duke University Press, 2007. 265 p. ISBN 978-0-8223-4124-6. (Gift of the publisher.) [RA394 .J67 2007] (9.1; 9.3.1; 9.8)

Klein, Rudolf. THE NEW POLITICS OF THE NHS: FROM CREATION TO REINVENTION. Oxford/Seattle: Radcliffe, 2006. 279 p. ISBN 978-1-84619-066-7; ISBN 1-84619-066-5. (Fifth edition.) [RA395 .G6 K64 2006] (9.1)

Le Fanu, James. THE RISE AND FALL OF MODERN MEDICINE. New York: Carroll & Graf, 2000, c1999. 426 p. ISBN 978-0-7867-0723-4; ISBN 0-7867-0723-1. (Originally published in Great Britain, 1999. Gift of Peter Inman.) [R149 .L45 2000] (9.1; 7.1; 15.1)

Marinker, Marshall, ed. CONSTRUCTIVE CONVERSATIONS ABOUT HEALTH: POLICY AND VALUES. Oxford/Seattle, WA: Radcliffe, 2006. 243 p. ISBN 978-1-84619-033-9; ISBN 1-84619-033-9. (Gift of the publisher.) [RA393 .C65 2006] (9.1; 1.1; 9.4)

Marmor, Theodore [and] Nuffield Trust for Research and Policy Studies in Health Services. FADS IN MEDICAL CARE MANAGEMENT AND POLICY. London: TSO [The Stationery Office], 2004. 72 p. ISBN 978-0-11-702863-0; ISBN 0-11-702863-0. (9.1; 9.3.1)

Mason, Diana J.; Leavitt, Judith K.; and Chaffee, Mary W. POLICY & POLITICS IN NURSING AND HEALTH CARE. St. Louis, MO: Saunders, 2002. 824 p. ISBN 978-0-7216-9534-1; ISBN 0-7216-9534-5. (Fourth edition.) [RT86.5 .P58 2002] (9.1; 4.1.3; 7.1; 7.3; 8.1; 9.3.1; 9.3.2; 9.4; 16.3; 21.1)

Mechanic, David. THE TRUTH ABOUT HEALTH CARE: WHY REFORM IS NOT WORKING IN AMERICA. New Brunswick, NJ: Rutgers University Press, 2006. 228 p. ISBN 978-0-8135-3887-7; ISBN 0-8135-3887-7. (Critical Issues in Health and Medicine series.) [JC251 .S8 Z83 2006] (9.1; 9.3.1; 9.4; 9.8; 17.1)

Meier, Augustine; O'Connor, Thomas St. James; and VanKatwyk, Peter, eds. SPIRITUALITY AND HEALTH: MULTIDISCIPLINARY EXPLORATIONS. Waterloo, Ontario: Wilfrid Laurier University Press, 2005. 316 p. ISBN 0-88920-477-2. [BL65 .M4 S675 2005] (9.1; 1.2; 4.4; 9.5.1; 17.1; 20.4.1)

Moerman, Daniel E. MEANING, MEDICINE, AND THE "PLACEBO EFFECT". Cambridge/New York: Cambridge University Press, 2002. 172 p. ISBN 978-0-521-00087-1; ISBN 0-521-00087-4. (Cambridge Studies in Medical Anthropology series; No. 9.) [R726.5 .M645 2002] (9.1; 4.4; 8.2; 9.7; 17.1; 21.7)

Morrison, Eileen E. ETHICS IN HEALTH ADMINISTRATION: A PRACTICAL APPROACH FOR DECISION MAKERS. Sudbury, MA: Jones and Bartlett, 2006. 354 p. ISBN 978-0-7637-2652-2; ISBN 0-7637-2652-4. (Gift of the publisher.) [RA394 .M67 2006] (9.1; 1.1; 6)

Palmer, Roy and Wetherill, Diana, eds. MEDICINE FOR LAWYERS. London: Royal Society of Medicine Press, 2005. 196 p. ISBN 978-1-85315-548-2; ISBN 1-85315-548-9. [KD3395 .A75 M44 2005] (9.1)

Pozgar, George D. LEGAL ASPECTS OF HEALTH CARE ADMINISTRATION. Boston: Jones and Barlett, 2004. 560 p. ISBN 0-7637-3182-X. (Ninth edition.) [KF3821 .P69 2004] (9.1; 1.3.5; 1.3.12; 2.1; 7.4; 8.3.1; 8.5; 9.3.2; 9.5.6; 9.8; 14.1)

Pozgar, George D. and Santucci, Nina M. STUDENT CASE LAW RESOURCE GUIDE TO ACCOMPANY LEGAL ASPECTS OF HEALTH CARE ADMINISTRATION. Boston: Jones and Bartlett, 2006. 64 p. ISBN 0-7637-4093-4. (Ninth edition.) [KF3821 .P69 2004 suppl.] (9.1; 1.3.5; 1.3.12; 2.1; 7.4; 8.3.1; 8.5; 9.3.2; 9.5.6; 9.8; 14.1)

Raffle, Angela E. and Gray, J.A. Muir. SCREENING: EVIDENCE AND PRACTICE. Oxford/New York: Oxford University Press, 2007. 317 p. ISBN 978-0-19-921449-5. [RA427.5 .R34 2007] (9.1; 9.8)

Rosner, David and Markowitz, Gerald. ARE WE READY? PUBLIC HEALTH SINCE 9/11. Berkeley: University of California Press [and] New York: Milbank Memorial Fund, 2006. 193 p. ISBN 978-0-520-25038-3; ISBN 0-520-25038-9. (California/Milbank Books on Health and the Public series; No. 15.) [RA395 .A3 R66 2006] (9.1; 1.3.5; 21.1; 21.3)

Saltman, Richard; Rico, Ana; and Boerma, Wienke, eds. PRIMARY CARE IN THE DRIVER'S SEAT: ORGANIZATIONAL REFORM IN EUROPEAN PRIMARY CARE. Maidenhead, England/New York: Open University Press, 2006. 251 p. ISBN 0-335-21365-0. (European Observatory on Health Systems and Policies Series.) [RA395 .E85 P75 2006] (9.1; 9.3.1; 9.4; 9.8; 21.1)

Seedhouse, David. HEALTH PROMOTION: PHILOSOPHY, PREJUDICE AND PRACTICE. Hoboken, NJ: John Wiley & Sons, 2004. 295 p. ISBN 978-0-470-84733-6; ISBN 0-470-84733-6. (Second edition.) [RA427.8 .S44 2004] (9.1)

Sheard, Sally and Donaldson, Liam. THE NATION'S DOCTOR: THE ROLE OF THE CHIEF MEDICAL OFFICER 1855-1998. Oxford/Seattle: Radcliffe Books, 2006. 238 p. ISBN 1-84619-001-0. (Foreword by Deni Pereira Gray. Published in association with The Nuffield Trust.) [RA395 .G6 S54 2006] (9.1)

Shore, David A., ed. THE TRUST PRESCRIPTION FOR HEALTHCARE: BUILDING YOUR REPUTATION WITH CONSUMERS. Chicago: Health Administration Press, 2005. 165 p. ISBN 978-1-56793-240-9; ISBN 1-56793-240-1. [RA395 .A3 S495 2005] (9.1; 7.1; 9.8)

Sloan, Richard P. BLIND FAITH: THE UNHOLY ALLIANCE OF RELIGION AND MEDICINE. New York: St. Martin's Press, 2006. 295 p. ISBN 978-0-312-34881-6; ISBN 0-312-34881-9. [BL65 .M4 S56 2006] (9.1; 1.2; 1.3.7)

Sneiderman, Barney; Irvine, John C.; and Osborne, Philip H. CANADIAN MEDICAL LAW: AN INTRODUCTION FOR PHYSICIANS, NURSES AND OTHER HEALTH CARE PROFESSIONALS. Toronto: Carswell, 2003. 766 p. ISBN 0-459-24074-9. (Third edition.) [KE3646 .S63 2003] (9.1; 2.1; 8.3.1; 8.5)

Stevens, Rosemary. THE PUBLIC-PRIVATE HEALTH CARE STATE: ESSAYS ON THE HISTORY OF AMERICAN HEALTH CARE POLICY. New Brunswick, NJ: Transaction Publishers, 2007. 362 p. ISBN 978-0-7658-0349-8; ISBN 0-7658-0349-6. [RA395 .A3 S817 2007] (9.1; 1.3.5; 9.3.1; 9.3.2; 9.5.10; 9.8)

9.3.1 HEALTH CARE ECONOMICS (GENERAL)

Atlas, Scott W., ed. POWER TO THE PATIENT: SELECTED HEALTH CARE ISSUES AND POLICY SOLUTIONS. Washington, DC: Hoover Institution Press, 2005. 62 p. ISBN 978-0-8179-4592-3; ISBN 0-8179-4592-X. (Hoover Institution Press Publication series; No. 532.) [RA410.53 .P685 2005] (9.3.1; 9.1)

Cohn, Jonathan. SICK: THE UNTOLD STORY OF AMERICA'S HEALTH CARE CRISIS—AND THE PEOPLE WHO PAY THE PRICE. New York: HarperCollins, 2007. 302 p. ISBN 978-0-06-058045-2; ISBN 0-06-058045-3. [RA395 .A3 C635 2007] (9.3.1; 9.1; 9.5.10)

Flood, Colleen M.; Roach, Kent; and Sossin, Lorne, eds. ACCESS TO CARE, ACCESS TO JUSTICE: THE LEGAL DEBATE OVER PRIVATE HEALTH INSURANCE IN CANADA. Toronto/Buffalo, NY: University of Toronto Press, 2005. 611 p. ISBN 978-0-8020-9420-9; ISBN 0-8020-9420-1. [KE3404 .A83 2005] (9.3.1; 1.3.8; 9.1; 9.2; 9.5.2; 21.1)

Gratzer, David. THE CURE: HOW CAPITALISM CAN SAVE AMERICAN HEALTH CARE. New York: Encounter Books, 2006. 233 p. ISBN 978-1-59403-153-3; ISBN 1-59403-153-3. [RA410.53 .G733 2006] (9.3.1; 9.1)

Halvorson, George C. HEALTH CARE REFORM NOW! A PRESCRIPTION FOR CHANGE. San Francisco: Jossey-Bass/John Wiley & Sons, 2007. 361 p. ISBN 978-0-7879-9752-6.

(Gift of the publisher.) [RA395 .A3 H3449 2007] (9.3.1; 9.1)

Hart, Julian Tudor. THE POLITICAL ECONOMY OF HEALTH CARE: A CLINICAL PERSPECTIVE. Bristol, UK: Polity Press, 2006. 320 p. ISBN 978-1-86134-808-1; ISBN 1-86134-808-8. (Health & Society Series.) [RA412.5 .G7 T83 2006] (9.3.1; 9.1)

Herzlinger, Regina. WHO KILLED HEALTH CARE? AMERICA'S $2 TRILLION MEDICAL PROBLEM—AND THE CONSUMER-DRIVEN CURE. New York: McGraw-Hill, 2007. 304 p. ISBN 978-0-07-148780-1; ISBN 0-07-148780-8. [RA395 .A3 H48 2007] (9.3.1; 1.3.2)

Hyman, David A. MEDICARE MEETS MEPHISTOPHELES. Washington, DC: Cato Institute, 2006. 138 p. ISBN 978-1-930865-92-1; ISBN 1-930865-92-9. [RA395 .A3 H96 2006] (9.3.1; 9.5.2)

Kotlikoff, Laurence J. THE HEALTHCARE FIX: UNIVERSAL INSURANCE FOR ALL AMERICANS. Cambridge, MA: MIT Press, 2007. 116 p. ISBN 978-0-262-11314-4. [RA412.2 .K68 2007] (9.3.1; 9.4; 9.5.2; 9.5.10)

Mahar, Maggie. MONEY-DRIVEN MEDICINE: THE REAL REASON HEALTH CARE COSTS SO MUCH. New York: Collins, 2006. 451 p. ISBN 978-0-06-076533-0; ISBN 0-06-076533-X. [RA410 .M34 2006] (9.3.1; 5.2; 9.3.2; 9.4; 9.7)

McDonough, Mary J. CAN A HEALTH CARE MARKET BE MORAL? A CATHOLIC VISION. Washington, DC: Georgetown University Press, 2007. 256 p. ISBN 978-1-58901-157-1; ISBN 1-58901-157-0. (Gift of the publisher.) [R725.56 .M334 2007] (9.3.1; 1.2; 1.3.2)

McLaughlin, Catherine G., ed. HEALTH POLICY AND THE UNINSURED. Washington, DC: Urban Institute Press, 2004. 311 p. ISBN 978-0-87766-719-3; ISBN 0-87766-719-5. [RA413.7 .U53 H434 2004] (9.3.1; 7.1)

Neumann, Peter J. USING COST-EFFECTIVENESS ANALYSIS TO IMPROVE HEALTH CARE: OPPORTUNITIES AND BARRIERS. Oxford/New York: Oxford University Press, 2005. 209 p. ISBN 978-0-19-517186-0; ISBN 0-19-517186-1. [RA410.5 .N48 2005] (9.3.1; 9.4)

Porter, Michael E. and Teisberg, Elizabeth Olmsted. REDEFINING HEALTH CARE: CREATING VALUE-BASED COMPETITION ON RESULTS. Boston: Harvard Business School Press, 2006. 506 p. ISBN 1-59139-778-2. [RA399 .A1 P67 2006] (9.3.1; 9.1; 9.8)

Ryder, Bruce, guest ed. *Symposium on Chaoulli*. OSGOODE HALL LAW JOURNAL 2006 Summer; 44(2): 249-375. (ISSN 0030-6185.) (9.3.1; 7.1; 9.2; 21.1)

Swartz, Katherine. REINSURING HEALTH: WHY MORE MIDDLE-CLASS PEOPLE ARE UNINSURED AND WHAT GOVERNMENT CAN DO. New York: Russell Sage Foundation, 2006. 203 p. ISBN 978-0-87154-787-3; ISBN 0-87154-787-2. [HG9396 .S93 2006] (9.3.1; 1.3.2; 1.3.5)

Wynia, Matthew K. and Schwab, Abraham P. ENSURING FAIRNESS IN HEALTH CARE COVERAGE: AN EMPLOYER'S GUIDE TO MAKING GOOD DECISIONS ON TOUGH ISSUES. New York: AMACOM, 2007. 225 p. ISBN 978-0-8144-7384-9; ISBN 0-8144-7384-9. (Gift of the publisher.) [HG9396 .W96 2007] (9.3.1; 1.3.2; 9.3.2)

9.3.2 MANAGED CARE

Verheijde, Joseph L. MANAGING CARE: A SHARED RESPONSIBILITY. Dordrecht: Springer, 2006. 219 p. ISBN 978-1-4020-4184-6; ISBN 1-4020-4184-5. (Issues in Business Ethics series; Vol. 22.) [RA395 .A3 V47 2006] (9.3.2; 9.3.1; 9.4)

9.4 ALLOCATION OF HEALTH CARE RESOURCES

Aaron, Henry J.; Schwartz, William B.; and Cox, Melissa. CAN WE SAY NO? THE CHALLENGE OF RATIONING HEALTH CARE. Washington, DC: Brookings Institution Press, 2005. 199 p. ISBN 978-0-8157-0121-7; ISBN 0-8157-0121-7. [RA410.5 .A23 2005] (9.4; 9.3.1; 21.1)

Newdick, Christopher. WHO SHOULD WE TREAT? RIGHTS, RATIONING, AND RESOURCES IN THE NHS. Oxford/New York: Oxford University Press, 2005. 278 p. ISBN 0-19-926418-X. (Second edition.) [KD3210 .N49 2005] (9.4; 8.3.1; 9.1; 9.2; 9.3.1)

9.5.1 HEALTH CARE FOR SPECIFIC DISEASES/ GROUPS (GENERAL)

Beamish, Rob and Ritchie, Ian. FASTEST, HIGHEST, STRONGEST: A CRITIQUE OF HIGH-PERFORMANCE SPORT. New York: Routledge, 2006. 194 p. ISBN 978-0-415-77043-9; ISBN 0-415-77043-2. (Routledge Critical Studies in Sport series. Gift of the publisher.) [RC1230 .B45 2006] (9.5.1; 9.5.9)

Carroll-Johnson, Rose Mary; Gorman, Linda M.; and Bush, Nancy Jo, eds. PSYCHOSOCIAL NURSING CARE ALONG THE CANCER CONTINUUM. Pittsburgh, PA: Oncology Nursing Society, 2006. 670 p. ISBN 978-1-890504-57-1; ISBN 1-890504-57-2. (Second edition.) [RC266 .P79 2006] (9.5.1; 1.1; 1.2; 4.4; 7.1; 8.1; 9.5.9; 15.3; 17.1; 19.1; 20.7; 21.7)

Clark, Rachel. A LONG WALK HOME. Abingdon: Radcliffe Medical Press, 2002. 142 p. ISBN 978-1-85775-906-8; ISBN 1-85775-906-0. (With Naomi Jefferies, John Hasler and David Pendleton.) (9.5.1; Biography)

Cueto, Marcos. COLD WAR, DEADLY FEVERS: MALARIA ERADICATION IN MEXICO, 1955-1975. Washington, DC: Woodrow Wilson Center Press/Baltimore, MD: Johns Hopkins University Press, 2007. 264 p. ISBN 978-0-8018-8645-4; ISBN 0-8018-8645-7. (Gift of the publisher.) [RC162 .M6 C84 2007] (9.5.1; 7.1; 21.1)

Denniston, George C.; Gallo, Pia Grassivaro; Hodges, Frederick M.; Milos, Marilyn Fayre; and Viviani, Franco, eds. BODILY INTEGRITY AND THE POLITICS OF CIRCUMCI-

SION: CULTURE, CONTROVERSY, AND CHANGE. New York: Springer, 2006. 269 p. ISBN 978-1-4020-4915-6; ISBN 1-4020-4915-3. [GN484 .B63 2006] (9.5.1; 9.5.5; 9.5.7; 10; 17.1; 21.1; 21.7)

Gaudet, Marcia. CARVILLE: REMEMBERING LEPROSY IN AMERICA. Jackson: University Press of Mississippi, 2004. 221 p. ISBN 978-1-57806-693-3; ISBN 1-57806-693-X. (Foreword by James Carville.) [RA644 .L3 G38 2004] (9.5.1; 7.1)

Herzog, Albert A., Jr., ed. DISABILITY ADVOCACY AMONG RELIGIOUS ORGANIZATIONS: HISTORIES AND REFLECTIONS. Binghamton, NY: Haworth Pastoral Press, 2006. 233 p. ISBN 978-0-7890-3290-4; ISBN 0-7890-3290-2. (Co-published simultaneously as *Journal of Religion, Disability & Health* 2006; 10(1-2).) [BV4460 .D57 2006] (9.5.1; 1.2; 9.5.3)

Institute of Medicine (United States). Board on Population Health and Public Health Practice. Committee on Gulf War and Health: A Review of the Medical Literature Relative to the Gulf War Veterans' Health. GULF WAR AND HEALTH, VOLUME 4: HEALTH EFFECTS OF SERVING IN THE GULF WAR. Washington, DC: National Academies Press, 2006. 275 p. ISBN 978-0-309-10176-9; ISBN 0-309-10176-X. [DS79.744 .M44 G83 2000 v.4] (9.5.1; 16.3; 21.2; 21.3)

Johnson, Harriet McBryde. TOO LATE TO DIE YOUNG: NEARLY TRUE TALES FROM A LIFE. New York: Henry Holt, 2005. 261 p. ISBN 978-0-8050-7594-6; ISBN 0-8050-7594-1. [HV3013 .J65 A3 2005] (9.5.1; Biography)

Komesaroff, Linda, ed. SURGICAL CONSENT: BIOETHICS AND COCHLEAR IMPLANTATION. Washington, DC: Gallaudet University Press, 2007. 203 p. ISBN 978-1-56368-349-7. [RF305 .S85 2007] (9.5.1; 7.1; 8.3.2; 9.5.7; 17.1)

Kuczynski, Alex. BEAUTY JUNKIES: INSIDE OUR $15 BILLION OBSESSION WITH COSMETIC SURGERY. New York: Doubleday, 2006. 290 p. ISBN 978-0-385-50853-7; ISBN 0-385-50853-0. [RD118 .K83 2006] (9.5.1; 4.4; 9.5.2; 9.5.5)

Lerner, Barron H. WHEN ILLNESS GOES PUBLIC: CELEBRITY PATIENTS AND HOW WE LOOK AT MEDICINE. Baltimore, MD: Johns Hopkins University Press, 2006. 334 p. ISBN 0-8018-8462-4. [R703 .L47 2006] (9.5.1; 1.3.7; 7.1; 7.2; 8.4; 9.5.6; 17.1; 18.5.1; 19.2; Biography)

Mullen, Thomas. THE LAST TOWN ON EARTH. London: Fourth Estate, 2006. 394 p. ISBN 0-00-723499-6. ("Deep in the forests of the Pacific Northwest, a mill town called Commonwealth votes to quarantine itself in the wake of the 1918 flu pandemic, setting up guards to prevent anyone from coming in or out, but a violent confrontation with a tired, hungry, and cold soldier will have devastating repercussions for the entire town.") [PS3613 .U447 L37 2006] (9.5.1; 1.1; Fiction)

Parke, David W., ed. THE PROFESSION OF OPHTHALMOLOGY: PRACTICE MANAGEMENT, ETHICS, AND ADVOCACY.

San Francisco: American Academy of Ophthalmology, 2005. 352 p. ISBN 1-56055-494-0. (Publisher's address: 655 Beach Street, Suite 300, Box 7424, zip 94109-1336.) [RE72 .P76 2005] (9.5.1; 2.1; 4.1.2; 6; 9.3.1)

Tayman, John. THE COLONY: THE HARROWING TRUE STORY OF THE EXILES OF MOLOKAI. New York: Scribner, 2006. 421 p. ISBN 978-0-7432-3300-2; ISBN 0-7432-3300-X. (Subtitle on cover: *The Harrowing True Story of the Exiles of Molokai.* "A Lisa Drew Book.") [RA644 .L3 T39 2006] (9.5.1; 7.1; 9.1; 18.5.5)

9.5.2 HEALTH CARE FOR THE AGED

Gillick, Muriel R. THE DENIAL OF AGING: PERPETUAL YOUTH, ETERNAL LIFE, AND OTHER DANGEROUS FANTASIES. Cambridge, MA: Harvard University Press, 2006. 341 p. ISBN 978-0-674-02148-8; ISBN 0-674-02148-7. [HQ1064 .U5 G45 2006] (9.5.2; 9.3.1; 20.5.1)

Margolies, Luisa. MY MOTHER'S HIP: LESSONS FROM THE WORLD OF ELDERCARE. Philadelphia: Temple University Press, 2004. 339 p. ISBN 1-59213-238-3. (Foreword by Walter M. Bortz II.) [RC954.4 .M37 2004] (9.5.2; 8.1; 9.5.5; Biography)

Pruchno, Rachel A. and Smyer, Michael A., eds. CHALLENGES OF AN AGING SOCIETY: ETHICAL DILEMMAS, POLITICAL ISSUES. Baltimore, MD: Johns Hopkins University Press, 2007. 448 p. ISBN 978-0-8018-8648-5; ISBN 0-8018-8648-1. (Gift of the publisher.) [RA408 .A3 C47 2007] (9.5.2; 7.1; 9.1)

Thane, Pat, ed. A HISTORY OF OLD AGE. Los Angeles, CA: J. Paul Getty Museum, 2005. 320 p. ISBN 978-0-89236-834-1; ISBN 0-89236-834-9. (Getty Trust Publications series. First published in the United Kingdom by: Thames & Hudson, 2005.) [HQ1061 .H54 2005] (9.5.2; 7.1)

9.5.3 HEALTH CARE FOR MENTALLY DISABLED PERSONS

Johnson, Kelley and Traustadóttir, Rannveig, eds. DEINSTITUTIONALIZATION AND PEOPLE WITH INTELLECTUAL DISABILITIES: IN AND OUT OF INSTITUTIONS. London/Philadelphia: Jessica Kingsley, 2005. 293 p. ISBN 978-1-84310-101-7; ISBN 1-84310-101-7. [HV3004 .D42 2005] (9.5.3; 2.2; 9.3.1; 9.5.5; 17.7)

Switzky, Harvey N. and Greenspan, Stephen, eds. WHAT IS MENTAL RETARDATION? IDEAS FOR AN EVOLVING DISABILITY IN THE 21ST CENTURY. Washington, DC: American Association on Mental Retardation, 2006. 358 p. ISBN 978-0-940898-94-6; ISBN 0-940898-93-4. (Revised and updated edition. AAMR Books and Research Monographs series.) [RC570 .W527 2006] (9.5.3; 1.2; 4.3; 9.5.7; 15.6)

9.5.4 HEALTH CARE FOR MINORITIES

Gary, Lisa Ché. UNDERSTANDING RACIAL HEALTH CARE DISPARITIES: THE ROLE OF CONSUMER EMPOWERMENT, CONSUMER EXPECTATIONS AND NEGATIVE HEALTH CARE EXPERIENCES. Ann Arbor, MI: ProQuest Information and

Learning/UMI, 2005. 166 p. (Publication No. AAT-3194655. Dissertation, (Ph.D.)—Yale University, 2005.) (9.5.4; 7.1; 8.1; 9.4; 9.8)

Livingston, Ivor Lensworth, ed. PRAEGER HANDBOOK OF BLACK AMERICAN HEALTH: POLICIES AND ISSUES BEHIND DISPARITIES IN HEALTH. Westport, CT: Praeger, 2004. 2 volumes. [911 p.]. ISBN 0-313-32477-8 (set). (Foreword by David Satcher.) [RA448.5 .N4 H364 2004] (9.5.4; 7.1; 9.1; 9.5.6; 9.5.9; 10; 15.11; 17.1)

Molina, Natalia. FIT TO BE CITIZENS? PUBLIC HEALTH AND RACE IN LOS ANGELES, 1879-1939. Berkeley: University of California Press, 2006. 279 p. ISBN 0-520-24648-9. (American Crossroads series; No. 20.) [RA448.4 .M65 2006] (9.5.4; 7.1; 9.1)

Waxler-Morrison, Nancy; Anderson, Joan M.; Richardson, Elizabeth; and Chambers, Natalie A., eds. CROSS-CULTURAL CARING: A HANDBOOK FOR HEALTH PROFESSIONALS. Vancouver: UBC Press, 2005. 365 p. ISBN 978-0-7748-1255-9; ISBN 0-7748-1255-9. (Second edition.) [RA563 .M56 C76 2005] (9.5.4; 7.1; 8.1; 21.7)

9.5.5 HEALTH CARE FOR WOMEN

Allen, Ann Taylor. FEMINISM AND MOTHERHOOD IN WESTERN EUROPE, 1890- 1970: THE MATERNAL DILEMMA. New York: Palgrave Macmillan, 2005. 354 p. ISBN 1-4039-6236-7. (Rose Fitzgerald Kennedy Collection on Women, Infants and Children.) [HQ759 .A45 2005] (9.5.5; 7.1; 9.5.7; 10; 11.1; 15.5; 21.1)

Apple, Rima D. PERFECT MOTHERHOOD: SCIENCE AND CHILDREARING IN AMERICA. New Brunswick, NJ: Rutgers University Press, 2006. 209 p. ISBN 978-0-8135-3843-3; ISBN 0-8135-3843-2. (Rose Fitzgerald Kennedy Collection on Women, Infants and Children. Gift of the publisher.) [HQ759 .A58 2006] (9.5.5; 7.1; 8.1)

Billison, Janet Mancini and Fluehr-Lobban, Carolyn, eds. FEMALE WELL-BEING: TOWARD A GLOBAL THEORY OF SOCIAL CHANGE. London/New York: Zed Books; Distributed in the USA exclusively by: Palgrave Macmillan, 2005. 432 p. ISBN 1-84277-009-8. (Rose Fitzgerald Kennedy Collection on Women, Infants and Children.) [HQ1154 .F4415 2005] (9.5.5; 7.1; 9.5.7; 21.1; 21.7)

Dennerstein, Lorraine, ed. WOMEN'S RIGHTS AND BIOETHICS. Paris: UNESCO, 2000. 215 p. ISBN 978-92-3-103765-8; ISBN 92-3-103765-X. (Rose Fitzgerald Kennedy Collection on Women, Infants and Children.) [R725.5 .W665 2000] (9.5.5; 2.1; 9.5.7; 10; 14.1; 16.3; 21.1)

Langer, Oded, ed. THE DIABETES IN PREGNANCY DILEMMA: LEADING CHANGE WITH PROVEN SOLUTIONS. Lanham, MD: University Press of America, 2006. 720 p. ISBN 978-0-7618-3270-6; ISBN 0-7618-3270-X. (Rose Fitzgerald Kennedy Collection on Women, Infants and Children.) [RG580 .D5 D522 2006] (9.5.5; 1.1.; 4.4; 9.5.1; 9.5.8; 15.1; 17.4)

Rivkin-Fish, Michele. WOMEN'S HEALTH IN POST-SOVIET RUSSIA: THE POLITICS OF INTERVENTION. Bloomington: Indiana University Press, 2005. 253 p. ISBN 0-253-21767-9. (New Anthropologies of Europe series.) (Rose Fitzgerald Kennedy Collection on Women, Infants and Children.) [RA564.85 .R56 2005] (9.5.5; 1.3.5; 7.1; 9.1; 10; 14.1; 21.1)

Schwartz, Marie Jenkins. BIRTHING A SLAVE: MOTHERHOOD AND MEDICINE IN THE ANTEBELLUM SOUTH. Cambridge, MA: Harvard University Press, 2006. 401 p. ISBN 978-0-674-02202-7; ISBN 0-674-02202-5. (Rose Fitzgerald Kennedy Collection on Women, Infants and Children.) [RG518 .U5 S34 2006] (9.5.5; 7.1; 9.5.4; 14.1)

Seymour, John. CHILDBIRTH AND THE LAW. Oxford/New York: Oxford University Press, 2000. 391 p. ISBN 978-0-19-826468-2; ISBN 0-19-826468-2. (Rose Fitzgerald Kennedy Collection on Women, Infants and Children.) [K4366 .S47 2000] (9.5.5; 8.3.1; 8.5; 9.5.8; 9.5.9; 11.4)

Solinger, Rickie. PREGNANCY AND POWER: A SHORT HISTORY OF REPRODUCTIVE POLITICS IN AMERICA. New York: New York University Press, 2005. 303 p. ISBN 978-0-8147-9827-0; ISBN 0-8147-9827-6. (Rose Fitzgerald Kennedy Collection on Women, Infants and Children.) [HQ766.5 .U5 S67 2005] (9.5.5; 2.2; 7.1; 9.5.4; 10; 11.1; 12.4.1; 14.1; 21.1)

Wagner, Marsden. BORN IN THE USA: HOW A BROKEN MATERNITY SYSTEM MUST BE FIXED TO PUT MOTHERS AND INFANTS FIRST. Berkeley: University of California Press, 2006. 295 p. ISBN 978-0-520-24596-9; ISBN 0-520-24596-2. (Rose Fitzgerald Kennedy Collection on Women, Infants and Children.) [RG518 .U5 W34 2006] (9.5.5; 4.1.3; 8.5; 9.5.7)

9.5.6 HIV INFECTION AND AIDS

Abdool Karim, S.S. and Abdool Karim, Q., eds. HIV/AIDS IN SOUTH AFRICA. Cambridge/Cape Town: Cambridge University Press, 2005. 592 p. ISBN 978-0-521-61629-4; ISBN 0-521-61629-8. [RA643.86 .S6 H585 2005] (9.5.6; 7.1; 9.3.1; 9.5.5; 9.5.7; 9.5.9; 10; 19.4; 21.1)

Beck, Eduard J.; Mays, Nicholas; Whiteside, Alan W.; and Zuniga, José M., eds. THE HIV PANDEMIC: LOCAL AND GLOBAL IMPLICATIONS. Oxford/New York: Oxford University Press, 2006. 799 p. ISBN 978-0-19-852843-2; ISBN 0-19-852843-4. [RA643.8 .H582 2006] (9.5.6; 7.1; 9.1; 9.3.1; 9.7; 21.1; 21.7)

Culshaw, Rebecca. SCIENCE SOLD OUT: DOES HIV REALLY CAUSE AIDS? Berkeley, CA: North Atlantic Books, 2007. 96 p. ISBN 978-1-55643-642-0; ISBN 1-55643-642-4. (The Terra Nova Series. Foreword by Harvey Bialy.) [RA643.8 .S35 2007] (9.5.6; 1.3.9; 7.1)

de Waal, Alex. AIDS AND POWER: WHY THERE IS NO POLITICAL CRISIS—YET. London/New York: Zed Books; Cape Town: David Philip in Association with the International African Institute and the Royal African Society; Distributed in the USA exclusively by: New York: Palgrave Macmillan, 2006. 147 p. ISBN 978-1-84277-707-7; ISBN

1-84277-707-6. (African Arguments series.) [RA643.86 .S6 D49 2006] (9.5.6; 1.3.5; 7.1; 21.1)

Foster, Geoff; Levine, Carol; and Williamson, John, eds. A GENERATION AT RISK: THE GLOBAL IMPACT OF HIV/AIDS ON ORPHANS AND VULNERABLE CHILDREN. Cambridge/New York: Cambridge University Press, 2005. 312 p. ISBN 978-0-521-65264-3; ISBN 0-521-65264-2. (Rose Fitzgerald Kennedy Collection on Women, Infants and Children.) [RA643.8 .G46 2005] (9.5.6; 1.2; 7.1; 9.5.7; 21.1)

Frasca, Tim. AIDS IN LATIN AMERICA. New York: Palgrave Macmillan, 2005. 261 p. ISBN 1-4039-6944-2. [RA643.86 .L29 F73 2005] (9.5.6; 1.2; 7.1; 10; 21.1)

Iliffe, John. THE AFRICAN AIDS EPIDEMIC: A HISTORY. Athens: Ohio University Press, 2006. 214 p. ISBN 0-8214-1688-X. [RA643.86 .A35 I43 2006] (9.5.6; 7.1; 21.1)

Kontaratos, Nikolas. DISSECTING A DISCOVERY: THE REAL STORY OF HOW THE RACE TO UNCOVER THE CAUSE OF AIDS TURNED SCIENTISTS AGAINST DISEASE, POLITICS AGAINST SCIENCE, NATION AGAINST NATION. [s.l.]: Xlibris, 2006. 395 p. ISBN 978-1-4257-0627-2; ISBN 1-4257-0627-4. (9.5.6; 1.3.9; 7.1; 9.1; 19.4; 21.1)

Kramer, Larry. THE NORMAL HEART. New York: New American Library, 1985. 123 p. ISBN 0-452-25798-0. (Introduction by Andrew Holleran and foreword by Joseph Papp. "A Plume Book." Gift of Warren Reich.) [PS3561 .R252 N6 1985] (9.5.6; 7.1)

Patterson, Amy S. THE POLITICS OF AIDS IN AFRICA. Boulder, CO: Lynne Rienner, 2006. 226 p. ISBN 978-1-58826-452-7; ISBN 1-58826-452-1. (Challenge and Change in African Politics series.) [RA643.86 A35 P378 2006] (9.5.6; 1.3.5; 21.1)

9.5.7 HEALTH CARE FOR MINORS

Freeman, Michael, ed. CHILDREN'S HEALTH AND CHILDREN'S RIGHTS. Leiden/Boston: Martinus Nijhoff, 2006. 337 p. ISBN 978-90-04-14894-9; ISBN 90-04-14894-9. (Rose Fitzgerald Kennedy Collection on Women, Infants and Children.) [HQ789 .C453 2006] (9.5.7; 1.1; 8.3.2; 8.3.3; 9.5.1; 9.5.3; 9.5.6; 10; 15.3; 17.1; 18.5.2; 19.5; 20.5.2; 21.1)

Gallego, Ruben. WHITE ON BLACK. Orlando, FL: Harcourt, 2004. 168 p. ISBN 0-15-101227-X. (Translation from the Russian by Marian Schwartz of: *Beloe na chernom*; St. Petersburg: Limbus Press, 2004. Rose Fitzgerald Kennedy Collection on Women, Infants and Children.) [PG3491.96 .N78 B4513 2006] (9.5.7; Biography)

Richardson, Jeanita W. THE COST OF BEING POOR: POVERTY, LEAD POISONING, AND POLICY IMPLEMENTATION. Westport, CT: Praeger, 2005. 204 p. ISBN 0-275-96912-6. [RA1231 .L4 R535 2005] (9.5.7; 9.1; 9.5.10)

9.5.8 HEALTH CARE FOR EMBRYOS AND FETUSES

Blizzard, Deborah. LOOKING WITHIN: A SOCIOCULTURAL EXAMINATION OF FETOSCOPY. Cambridge, MA: MIT Press, 2007. 253 p. ISBN 978-0-262-02616-1; ISBN 0-262- 02616-3. (Basic Bioethics series. Gift of the publisher.) [RG628.3 .F47 B57 2007] (9.5.8; 7.1; 15.2; 21.7)

Casey, Gerald. BORN ALIVE: THE LEGAL STATUS OF THE UNBORN CHILD IN ENGLAND AND THE U.S.A.. Chichester, UK: Barry Rose Law, 2005. 250 p. ISBN 1-902681-46-0. (Rose Fitzgerald Kennedy Collection on Women, Infants and Children.) [KD744 .U53 C37 2005] (9.5.8; 2.2; 4.4; 9.5.5)

9.5.9 HEALTH CARE FOR SUBSTANCE ABUSERS/ USERS OF CONTROLLED SUBSTANCES

Brandt, Allan M. THE CIGARETTE CENTURY: THE RISE, FALL, AND DEADLY PERSISTENCE OF THE PRODUCT THAT DEFINED AMERICA. New York: Basic Books, 2007. 600 p. ISBN 978-0-465-07047-3; ISBN 0-465-07047-7. [HD9130.8 .U5 B72 2007] (9.5.9; 7.1)

Cook, Christopher C.H. ALCOHOL, ADDICTION AND CHRISTIAN ETHICS. Cambridge/New York: Cambridge University Press, 2006. 221 p. ISBN 978-0-521-85182-4; ISBN 0-521-85182-3. (New Studies in Christian Ethics series; Vol. 27.) [BV4596 .A48 C66 2006] (9.5.9; 1.2)

Council of Europe. DRUG ADDICTION. Strasbourg: Council of Europe, 2005. 173 p. ISBN 978-92-871-5639-6; ISBN 92-871-5639-5. (Ethical Eye series. Also published in French as: *Regard éthique—la toxicomanie*; ISBN 92-871-5638-7.) [RC564 .D77 2005] (9.5.9; 9.5.5; 9.5.8; 21.1)

Fish, Jefferson M., ed. DRUGS AND SOCIETY: U.S. PUBLIC POLICY. Lanham, MD: Rowman & Littlefield, 2006. 228 p. ISBN 0-7425-4245-9. [HV5825 .D813 2006] (9.5.9; 1.3.5; 9.1)

Kleinig, John and Einstein, Stanley, eds. ETHICAL CHALLENGES FOR INTERVENING IN DRUG USE: POLICY, RESEARCH AND TREATMENT ISSUES. Huntsville, TX: Office of International Criminal Justice [OIJC], Sam Houston State University, Criminal Justice Center, 2006. 770 p. ISBN 978-0-942511-65-9; ISBN 0-942511-65-4. (The Uncertainty Series; Vol. 5.) [HV4998 .E74 2006] (9.5.9; 1.3.9; 2.3; 4.3; 6; 8.1; 8.3.1; 8.4; 10; 17.2; 18.5.1; 18.5.2)

9.5.10 HEALTH CARE FOR INDIGENTS

Engel, Jonathan. POOR PEOPLE'S MEDICINE: MEDICAID AND AMERICAN CHARITY CARE SINCE 1965. Durham, NC: Duke University Press, 2006. 318 p. ISBN 0-8223-3695-2. [RA412.4 .E54 2006] (9.5.10; 7.1; 9.1; 9.3.1; 9.3.2; 9.5.2; 9.5.4; 9.5.6)

King, Jr., Talmadge E. and Wheeler, Margaret B. MEDICAL MANAGEMENT OF VULNERABLE AND UNDERSERVED PA-

TIENTS: PRINCIPLES, PRACTICE, AND POPULATIONS. New York: McGraw-Hill Medical, 2007. 454 p. ISBN 978-0-07-144331-9; ISBN 0-07-144331-2. [RA418.5 .P6 M437 2007] (9.5.10; 7.1; 9.1; 9.3.1; 9.5.4; 9.5.5; 9.5.6; 9.5.7; 16.3; 17.1)

9.6 ETHICS COMMITTEES/ CONSULTATION

Post, Linda Farber; Blustein, Jeffrey; and Dubler, Nancy Neveloff. HANDBOOK FOR HEALTH CARE ETHICS COMMITTEES. Baltimore, MD: Johns Hopkins University Press, 2007. 327 p. ISBN 0-8018-8448-9. (Gift of the publisher.) [R725.3 .P67 2007] (9.6; 2.1; 4.3; 8.2; 8.3.1; 9.2; 9.4; 9.5.7; 20.5.1)

9.7 DRUGS AND PHARMACEUTICAL INDUSTRY

Allen, Arthur. VACCINE: THE CONTROVERSIAL STORY OF MEDICINE'S GREATEST LIFESAVER. New York: W.W. Norton, 2007. 523 p. ISBN 978-0-393-05911-3; ISBN 0-393-05911-1. [RA638 .A45 2007] (9.7; 9.5.1)

Blech, Jörg. DIE KRANKHEITSERFINDER: WIE WIR ZU PATIENTEN GEMACHT WERDEN. Frankfurt am Main: S. Fischer, 2004. 255 p. ISBN 3-10-004410-X. (Ninth edition.) [R724 .B55 2004] (9.7; 8.1; 9.3.1; 9.5.9)

Brody, Howard. HOOKED: ETHICS, THE MEDICAL PROFESSION, AND THE PHARMACEUTICAL INDUSTRY. Lanham, MD: Rowman & Littlefield, 2007. 367 p. ISBN 978-0-7425-5218-0; ISBN 0-7425-5218-7. (Explorations in Bioethics and the Medical Humanities series.) [R724 .B763 2007] (9.7; 1.3.2; 9.3.1)

Brookes, Tim. A WARNING SHOT: INFLUENZA AND THE 2004 FLU VACCINE SHORTAGE. Washington, DC: American Public Health Association, 2005. 92 p. ISBN 978-0-87553-049-9; ISBN 0-87553-049-4. [RA644 .I6 B76 2005] (9.7; 9.1; 9.4; 9.5.1; 21.1)

Cohen, Jillian Clare; Illingworth, Patricia; and Schüklenk, Udo, eds. THE POWER OF PILLS: SOCIAL, ETHICAL AND LEGAL ISSUES IN DRUG DEVELOPMENT, MARKETING, AND PRICING. London/Ann Arbor, MI: Pluto Press, 2006. 297 p. ISBN 978-0-7453-2402-9; ISBN 0-7453-2402-9. [RA401 .A1 P69 2006] (9.7; 1.3.2; 5.3; 9.1; 9.3.1; 15.1; 21.1)

Greene, Jeremy A. PRESCRIBING BY NUMBERS: DRUGS AND THE DEFINITION OF DISEASE. Baltimore, MD: Johns Hopkins University Press, 2007. 318 p. ISBN 978-0-8018-8477-1; ISBN 0-8018-8477-2. [RM262 .G74 2007] (9.7; 4.2; 7.1; 9.5.1)

Law, Jacky. BIG PHARMA: EXPOSING THE GLOBAL HEALTHCARE AGENDA. New York: Carroll & Graf, 2006. 266 p. ISBN 978-0-7867-1783-5; ISBN 0-7867-1783-1. (Gift of the publisher.) [HD9665.5 .L385 2006] (9.7; 1.3.2; 5.3; 9.1; 15.1; 21.1)

Madden, Michael. THE SEESAW SYNDROME. Dallas: Durban House, 2003. 236 p. ISBN 978-1-930754-42-3;

ISBN 1-930754-42-6. ("*The Seesaw Syndrome* by Michael Madden is an original and superbly crafted novel of greed, corruption, and ill-gotten gains as drug executives and medical researchers exploit the power they have for tremendous profit. A new drug is debuted, that promotes weight recovery among patients ravaged by cancer—yet when a young woman and her mentor learns that it promotes tumor growth as well, they are marked as targets to be silenced...") [PS3553 .O5545 C6 2003] (9.7; 9.5.1; Fiction)

Santoro, Michael A. and Gorrie, Thomas M., eds. ETHICS AND THE PHARMACEUTICAL INDUSTRY. Cambridge/New York: Cambridge University Press, 2005. 492 p. ISBN 978-0-521-85496-2; ISBN 0-521-85496-2. [HD9665.5 .E85 2005] (9.7; 1.3.2; 1.3.5; 5.3; 8.3.1; 9.2; 9.3.1; 9.4; 9.5.6; 9.5.7; 18.2; 18.5.2; 18.5.4; 21.1)

Schacter, Bernice. THE NEW MEDICINES: HOW DRUGS ARE CREATED, APPROVED, MARKETED, AND SOLD. Westport, CT: Praeger, 2006. 267 p. ISBN 0-275-98141-X. [RM301.25 .S34 2006] (9.7; 1.3.5; 1.3.7; 9.3.1; 18.2; 22.2)

Strandberg, Kenneth M. ESSENTIALS OF LAW AND ETHICS FOR PHARMACY TECHNICIANS. Boca Raton, FL: CRC Press/Taylor & Francis, 2007. 196 p. ISBN 978-1-4200-4556-7; ISBN 1-4200-4556-3. (Second edition. CRC Press Pharmacy Education Series. "A CRC title.") [KF2915 .P452 S77 2007] (9.7; 1.3.5; 4.1.1)

Zellmer, William A. THE CONSCIENCE OF A PHARMACIST: ESSAYS ON VISION AND LEADERSHIP FOR A PROFESSION. Bethesda, MD: American Society of Health-System Pharmacists, 2002. 196 p. ISBN 978-1-58528-030-8; ISBN 1-58528-030-5. [RS100 .Z45 2002] (9.7; 4.1.1; 9.3.1; 9.8)

9.8 QUALITY OF HEALTH CARE

Coulter, Angela; Entwistle, Vikki; and Gilbert David. INFORMING PATIENTS: AN ASSESSMENT OF THE QUALITY OF PATIENT INFORMATION MATERIALS. London: King's Fund, 1998. 219 p. ISBN 978-1-85717-214-0; ISBN 1-85717-214-0. (9.8; 8.2)

Grol, Richard; Wensing, Michel; and Eccles, Martin. IMPROVING PATIENT CARE: THE IMPLEMENTATION OF CHANGE IN CLINICAL PRACTICE. Edinburgh/New York: Elsevier Butterworth Heinemann, 2005. 290 p. ISBN 0-7506-8819-X. [RA441 .I47 2005] (9.8; 9.1)

Gutkind, Lee, ed. RAGE AND RECONCILIATION: INSPIRING A HEALTH CARE REVOLUTION. Dallas, TX: Southern Methodist University Press, 2005. 214 p. ISBN 0-87074-503-4. (Medical Humanities Series. Includes an 80-minute audio CD with three essays read by actors and panel discussion.) [R729.8 .R34 2005] (9.8; 7.3; 9.3.2)

Institute of Medicine (United States). Board on Health Care Services. Committee on Identifying Priority Areas for Quality Improvement. PRIORITY AREAS FOR NATIONAL ACTION: TRANSFORMING HEALTH CARE QUALITY. Washington, DC: National Academies Press, 2003. 143 p. ISBN 0-309-08543-8. (Quality Chasm Series. Edited by Karen

Adams and Janet M. Corrigan.) [RA399 .A3 I562 2003] (9.8; 9.1)

Meulen, Ruud ter; Biller-Andomo, Nikola; Lenk, Christian; and Lie, Reidar, K., eds. EVIDENCE-BASED PRACTICE IN MEDICINE AND HEALTH CARE: A DISCUSSION OF THE ETHICAL ISSUES. Berlin/New York: Springer, 2005. 184 p. ISBN 3-540-22239-1. [R723.7 .E96 2005] (9.8; 1.3.5; 1.3.9; 7.1; 7.2; 8.1; 9.1; 9.3.2; 9.5.1; 9.5.2; 9.5.4; 17.1; 21.1)

Mulcahy, Linda. DISPUTING DOCTORS: THE SOCIO-LEGAL DYNAMICS OF COMPLAINTS ABOUT MEDICAL CARE. Maidenhead, UK/Philadelphia: Open University Press, 2003. 173 p. ISBN 978-0-335-21245-6; ISBN 0-335-21245-1. [RA412.5 .G7 M85 2003] (9.8; 8.1; 9.1)

Peraino, Robert A. THE CONSUMER'S GUIDE TO MEDICAL MISTAKES: INFORMATION YOU NEED BEFORE BECOMING A PATIENT. New York: Vantage Press, 2005. 87 p. ISBN 978-0-533-15128-8; ISBN 0-533-15128-7. (9.8; 8.2; 8.3.1; 9.5.2)

Runciman, Bill; Merry, Alan; and Walton, Merrilyn. SAFETY AND ETHICS IN HEALTHCARE: A GUIDE TO GETTING IT RIGHT. Aldershot, England/Burlington, VT: Ashgate, 2007. 334 p. ISBN 978-0-7546-4437-8. [R725 .R86 2007] (9.8; 5.2; 9.4)

10 SEXUALITY/ GENDER

Adams, Vincanne and Pigg, Stacy Leigh, eds. SEX IN DEVELOPMENT: SCIENCE, SEXUALITY, AND MORALITY IN GLOBAL PERSPECTIVE. Durham, NC: Duke University Press, 2005. 342 p. ISBN 0-8223-3491-7. [HQ18 .D44 S49 2005] (10; 1.3.5; 9.5.6; 11.1; 21.1)

Andrew, Barbara S.; Keller, Jean; and Schwartzman, Lisa H., eds. FEMINIST INTERVENTIONS IN ETHICS AND POLITICS: FEMINIST ETHICS AND SOCIAL THEORY. Lanham, MD: Rowman & Littlefield, 2005. 245 p. ISBN 978-0-7425-4269-3; ISBN 0-7425-4269-6. (Feminist Constructions series.) [BJ1395 .F4495 2005] (10; 1.1; 1.3.5)

Brown, Sarah. TREATING SEX OFFENDERS: AN INTRODUCTION TO SEX OFFENDER TREATMENT PROGRAMMES. Cullompton, UK/Portland, OR: Willan Publishing, 2005. 282 p. ISBN 978-1-84392-122-6; ISBN 1-84392-122-7. [RC560 .S47 B76 2005] (10; 9.5.1; 17.1; 21.1)

Drescher, Jack and Zucker, Kenneth J., eds. EX-GAY RESEARCH: ANALYZING THE SPITZER STUDY AND ITS RELATION TO SCIENCE, RELIGION, POLITICS, AND CULTURE. New York: Harrington Park Press, 2006. 352 p. ISBN 978-1-56023-557-6; ISBN 1-56023-557-8. [RC558 .E98 2006] (10; 7.1; 17.2; 18.4)

Ferrell, Robyn. COPULA: SEXUAL TECHNOLOGIES, REPRODUCTIVE POWERS. Albany: State University of New York Press, 2006. 175 p. ISBN 0-7914-6754-6. (SUNY Series in Gender Theory.) [HQ1075 .F474 2006] (10; 1.1; 4.4; 5.1; 14.1)

Green, Ronald M., ed. RELIGION AND SEXUAL HEALTH: ETHICAL, THEOLOGICAL, AND CLINICAL PERSPECTIVES. Dordrecht/Boston: Kluwer Academic, 1992. 232 p. ISBN 978-0-7923-1752-4; ISBN 0-7923-1752-1. (Theology and Medicine series; Vol. 1.) [BL65 .S4 R45 1992] (10; 1.2)

Hird, Myra J. SEX, GENDER, AND SCIENCE. New York: Palgrave Macmillan, 2004. 197 p. ISBN 1-4039-2177-6. [HQ1190 .H567 2004] (10; 7.1; 22.1)

Lloyd, Elisabeth A. THE CASE OF THE FEMALE ORGASM: BIAS IN THE SCIENCE OF EVOLUTION. Cambridge, MA: Harvard University Press, 2005. 311 p. ISBN 978-0-674-02246-1; ISBN 0-674-02246-7. [QP251 .L56 2005] (10; 3.2)

McLaren, Angus. IMPOTENCE: A CULTURAL HISTORY. Chicago: University of Chicago Press, 2007. 332 p. ISBN 978-0-226-50076-8; ISBN 0-226-50076-4. [RC889 .M345 2007] (10; 7.1; 9.7; 15.5)

Mui, Constance L. and Murphy, Julien S., eds. GENDER STRUGGLES: PRACTICAL APPROACHES TO CONTEMPORARY FEMINISM. Lanham, MD: Rowman & Littlefield, 2002. 369 p. ISBN 0-7425-1255-X. (Feminist Constructions series.) [HQ1190 .G473 2002] (10; 9.5.3; 9.5.5; 9.5.7; 9.5.8)

Overall, Christine. THINKING LIKE A WOMAN: PERSONAL LIFE AND POLITICAL IDEAS. Toronto: Sumach Press, 2001. 271 p. ISBN 1-894549-05-8. [HQ1190 .O963 2001] (10; 2.1; 9.5.5)

Rosenfeld, Dana and Faircloth, Christopher A., eds. MEDICALIZED MASCULINITIES. Philadelphia: Temple University Press, 2006. 263 p. ISBN 1-59213-098-4. [RA564.83 .M43 2006] (10; 7.1; 9.5.1; 9.7; 17.4)

Russo, Giovanni. EVANGELIUM AMORIS: CORSO DI MORALE FAMILIARE E SESSUALE. Messina: Editrice Coop. S.Tom. a r.l. [and] Torino: Editrice Elledici, 2007. 270 p. ISBN 978-88-01-03746-3. (Manuali di Panteno, Teologia series; No. 3. Gift of the author.) (10; 1.2)

Tuchman, Arleen Marcia. SCIENCE HAS NO SEX: THE LIFE OF MARIE ZAKRZEWSKA, M.D.. Chapel Hill: University of North Carolina Press, 2006. 336 p. ISBN 978-0-8078-3020-8; ISBN 0-8078-3020-8. (Studies in Social Medicine series.) [R692 .T83 2006] (10; 7.1; Biography)

11.1 CONTRACEPTION (GENERAL)

Cline, David P. CREATING CHOICE: A COMMUNITY RESPONDS TO THE NEED FOR ABORTION AND BIRTH CONTROL, 1961-1973. New York: Palgrave Macmillan, 2006. 290 p. ISBN 978-1-4039-6814-2; ISBN 1-4039-6814-4. (Palgrave Studies in Oral History series.) [HQ766.5 .U5 C46 2006] (11.1; 10; 12.3; 12.5.1; 12.5.2)

Johnson, John W. GRISWOLD V. CONNECTICUT: BIRTH CONTROL AND CONSTITUTIONAL RIGHT TO PRIVACY. Lawrence: University Press of Kansas, 2005. 266 p. ISBN 0-7006-1378-1. (Landmark Law Cases & American Society series.) [KF224 .G75 J64 2005] (11.1; 1.3.5; 8.4)

12.1 ABORTION (GENERAL)

Press, Eyal. ABSOLUTE CONVICTIONS: MY FATHER, A CITY, AND THE CONFLICT THAT DIVIDED AMERICA. New York: Henry Holt, 2006. 292 p. ISBN 978-0-8050-7731-5; ISBN 0-8050-7731-6. [HQ767.5 .U5 P735 2006] (12.1; 4.4; 12.3; 12.4.1; 12.5.1)

12.3 ABORTION: MORAL AND RELIGIOUS ASPECTS

Coope, Christopher Miles. WORTH AND WELFARE IN THE CONTROVERSY OVER ABORTION. New York: Palgrave/Macmillan, 2006. 350 p. ISBN 978-0-333-76018-5; ISBN 0-333-76018-2. [HQ767.15 .C68 2006] (12.3; 4.4)

Derr, Mary Krane; MacNair, Rachel; and Naranjo-Huebl, Linda, eds. PROLIFE FEMINISM: YESTERDAY AND TODAY. [Kansas City, MO]: Feminism and Nonviolence Studies Association, 2005. 474 p. ISBN 978-1-4134-9576-8; ISBN 1-4134-9576-1. (Expanded second edition. Original edition published by Sulzburger & Graham, 1995. "To order additional copies of this book, contact: Xlibris Corporation, www.Xlibris.com.") [HQ767.5 .U5 P77 2005] (12.3; 9.5.5; 10; 14.1; 15.5)

Page, Cristina. HOW THE PRO-CHOICE MOVEMENT SAVED AMERICA: FREEDOM, POLITICS, AND THE WAR ON SEX. New York: Basic Books, 2006. 236 p. ISBN 978-0-465-05489-3; ISBN 0-465-05489-7. [HQ767.5 .U5 P334 2006] (12.3; 10; 11.1; 21.1)

12.4.1 ABORTION: LEGAL ASPECTS (GENERAL)

Dellapenna, Joseph W. DISPELLING THE MYTHS OF ABORTION HISTORY. Durham, NC: Carolina Academic Press, 2006. 1283 p. ISBN 0-89089-509-0. [KF3771 .D45 2005] (12.4.1; 4.4; 7.1; 12.3; 12.4.2; 12.5.1; 15.5; 20.5.2)

Eser, Albin and Koch, Hans-Georg. ABORTION AND THE LAW: FROM INTERNATIONAL COMPARISON TO LEGAL POLICY. The Hague: TMC Asser Press, Distributed in North America by: West Nyack, NY: Cambridge University Press, 2005. 325 p. ISBN 978-90-6704-197-3; ISBN 90-6704-197-1. [K5181 .E84 2005] (12.4.1; 7.1; 12.3; 12.5.2)

12.5.1 ABORTION: SOCIAL ASPECTS (GENERAL)

Boltanski, Luc. LA CONDITION FOETALE: UNE SOCIOLOGIE DE L'ENGENDREMENT ET DE L'AVORTEMENT. Paris: Gallimard, 2004. 420 p. ISBN 978-2-07-076702-1; ISBN 2-07-076702-7. (NRF essais series.) [HQ767.5 .F7 B64 2004] (12.5.1)

Halkias, Alexandra. THE EMPTY CRADLE OF DEMOCRACY: SEX, ABORTION, AND NATIONALISM IN MODERN GREECE. Durham, NC: Duke University Press, 2004. 413 p. ISBN 0-8223-3323-6. [HQ767.5 .G8 H35 2004] (12.5.1; 1.3.5; 4.4; 10; 11.1; 14.1; 21.1)

Haussman, Melissa. ABORTION POLITICS IN NORTH AMER-ICA. Boulder, CO: Lynne Rienner, 2005. 211 p. ISBN 1-58826-336-3. [HV767.5 .U5 H38 2005] (12.5.1; 9.1; 9.2; 12.4.2; 14.1; 21.1)

12.5.2 ABORTION: SOCIAL ASPECTS—DEMOGRAPHIC SURVEYS

Wainer, Jo. LOST: ILLEGAL ABORTION STORIES. Melbourne: Melbourne University Press, 2006. 214 p. ISBN 978-0-522-85231-8; ISBN 0-522-85231-9. (Foreword by Helen Garner.) [HQ767.5 .A9 L67 2006] (12.5.2; 7.1; 12.4.2; 12.4.3)

13.1 POPULATION (GENERAL)

Al-Kawthari, Muhammad Ibn Adam. BIRTH CONTROL AND ABORTION IN ISLAM. Santa Barbara, CA: White Thread Press, 2006. 80 p. ISBN 978-1-933764-00-9; ISBN 1-933764-00-7. [KBP3124 .A45 2006] (13.1; 1.2; 12.3)

13.2 POPULATION GROWTH

Russell, Claire and Russell, W.M.S. POPULATION CRISES AND POPULATION CYCLES. London: Galton Institute, 1999. 124 p. ISBN 0-9504066-5-1. (Publisher's address: 19 Northfields Prospect, Northfields, London SW18 1PE.) [HB871 .R86 1999] (13.2; 13.1)

13.3 POPULATION POLICY

Accampo, Elinor. BLESSED MOTHERHOOD, BITTER FRUIT: NELLY ROUSSEL AND THE POLITICS OF FEMALE PAIN IN THIRD REPUBLIC FRANCE. Baltimore, MD: Johns Hopkins University Press, 2006. 312 p. ISBN 978-0-8018-8404-7; ISBN 0-8018-8404-7. [HQ764 .R68 A33 2006] (13.3; 7.1; 10; 21.1; Biography)

Greenhalgh, Susan and Winckler, Edwin A. GOVERNING CHINA'S POPULATION: FROM LENINIST TO NEOLIBERAL BIOPOLITICS. Stanford, CA: Stanford University Press, 2005. 394 p. ISBN 0-8047-4879-9. [HQ767.5 .C6 G74 2005] (13.3; 7.1)

14.1 REPRODUCTION/ REPRODUCTIVE TECHNOLOGIES (GENERAL)

Bleiklie, Ivar; Goggin, Malcolm L.; and Rothmayr, Christine, eds. COMPARATIVE BIOMEDICAL POLICY: GOVERNING ASSISTED REPRODUCTIVE TECHNOLOGIES. London, UK/New York: Routledge, 2004. 284 p. ISBN 978-0-415-32547-9; ISBN 0-415-32547-1. (Routledge/ECPR Studies in European Political Science series; Vol. 32.) [RG133.5 .C668 2004] (14.1; 1.3.5; 4.4; 14.5; 18.5.4; 21.1; 21.7)

Brosens, Ivo, ed. THE CHALLENGE OF REPRODUCTIVE MEDI-CINE AT CATHOLIC UNIVERSITIES: TIME TO LEAVE THE CAT-ACOMBS. LeuvenDudley, MA: Peeters, 2006. 263 p. ISBN 978-90-429-1762-0; ISBN 90-429-1762-8. [RG133 .C43 2006] (14.1; 1.2; 2.1; 2.3; 5.2; 9.5.5; 9.5.8; 11.1; 12.3)

Chneiweiss, Hervé and Nau, Jean-Yves. BIOÉTHIQUE, AVIS DE TEMPÊTES: LES NOUVEAUX ENJEUX DE LA MAÎTRISE DU VIVANT. Paris: Alvik, 2003. 207 p. ISBN 2-914833-11-3. [RG133.5 .C49 2003] (14.1; 1.1.; 2.1; 7.1; 14.4)

Cobb, Matthew. THE EGG & SPERM RACE: THE SEVEN-TEENTH-CENTURY SCIENTISTS WHO UNRAVELLED THE SE-CRETS OF SEX, LIFE, AND GROWTH. London/New York: Pocket Books, 2007. 333 p. ISBN 978-1-4165-2600-1; ISBN 0-4165-2600-5. (Originally published: Free Press, Great Britain, 2006.) [QP251 .C615 2007] (14.1; 7.1; 10)

Gelfand, Scott and Shook, John R., eds. ECTOGENESIS: AR-TIFICIAL WOMB TECHNOLOGY AND THE FUTURE OF HUMAN REPRODUCTION. Amsterdam/New York: Editions Rodopi, B.V., 2006. 197 p. ISBN 978-90-420-2081-1; ISBN 90-420-2081-4. (Value Inquiry Book Series; Vol. 184.) [RG155 .E38 2006] (14.1; 1.1; 9.5.5; 10; 18.5.3; 18.5.4)

Hodges, Sarah, ed. REPRODUCTIVE HEALTH IN INDIA: HIS-TORY, POLITICS, CONTROVERSIES. Hyderabad, India: Orient Longman, in association with the Wellcome Trust Centre for the History of Medicine, 2006. 264 p. ISBN 978-81-250-2939-7; ISBN 81-250-2939-7. (New Perspectives in South Asian History series; Vol. 13. Publisher's address: 3-6-752 Himayatnagar, Hyderabad 500 029 (A.P.), India.) [HQ766.5 .I5 R473 2006] (14.1; 1.2; 7.1; 9.5.5; 10; 13.3; 15.5; 21.1)

Keller, Eve. GENERATING BODIES AND GENDERED SELVES: THE RHETORIC OF REPRODUCTION IN EARLY MODERN ENG-LAND. Seattle: University of Washington Press, 2007. 248 p. ISBN 978-0-295-98641-8; ISBN 0-295-98641-7. (In Vivo: The Cultural Mediations of Biomedical Science series. "A Samuel and Althea Stroum Book.") [RG518 .G7 K45 2007] (14.1; 4.4; 7.1; 10)

Mundy, Liza. EVERYTHING CONCEIVABLE: HOW ASSISTED REPRODUCTION IS CHANGING MEN, WOMEN, AND THE WORLD. New York: Alfred A. Knopf, 2007. 406 p. ISBN 978-1-4000-4428-3. [RG133.5 .M86 2007] (14.1; 7.1; 14.2; 14.4; 14.6)

Shenfield, Françoise and Sureau, Claude, eds. CONTEMPO-RARY ETHICAL DILEMMAS IN ASSISTED REPRODUCTION. Abingdon, Oxon, UK: Informa Healthcare; Distributed in North and South America by: Boca Raton, FL: Taylor & Francis, 2006. 120 p. ISBN 978-0-415-37131-5; ISBN 0-415-37131-7. (Gift of the publisher.) [RG133.5 .C6685 2006] (14.1; 1.2; 4.4; 9.5.7; 14.5; 15.2; 21.1)

Shepard, Bonnie. RUNNING THE OBSTACLE COURSE TO SEX-UAL AND REPRODUCTIVE HEALTH: LESSONS FROM LATIN AMERICA. Westport, CT: Praeger, 2006. 215 p. ISBN 0-275-97066-3. [HQ1236.5 .L37 S54 2006] (14.1; 9.5.5; 10; 21.1)

Spencer, J.R. and Du Bois-Pedain, Antje, eds. FREEDOM AND RESPONSIBILITY IN REPRODUCTIVE CHOICE. Oxford/Portland, OR: Hart Publishing, 2006. 201 p. ISBN 978-1-84113-582-3; ISBN 1-84113-582-8. [K3611 .A77 F73 2006] (14.1; 1.1; 4.5; 7.1; 14.4; 15.1; 18.5.4)

See inside front cover for NRCBL Classification Scheme

Tittle, Peg, ed. SHOULD PARENTS BE LICENSED? DEBATING THE ISSUES. Amherst, NY: Prometheus Books, 2004. 364 p. ISBN 1-59102-094-8. (Contemporary Issues Series.) [HQ755.8 .S5326 2004] (14.1; 1.3.5; 4.4; 8.1; 9.5.3; 9.5.5; 9.5.7; 11.1; 11.3; 15.2; 15.4; 15.5)

14.2 ARTIFICIAL INSEMINATION AND SURROGACY

Institute of Medicine (United States) [IOM]. Committee on Assessing the Medical Risks of Human Oocyte Donation for Stem Cell Research; Institute of Medicine (United States). Board on Health Sciences Policy; and National Research Council (United States). Board on Life Sciences. ASSESSING THE MEDICAL RISKS OF HUMAN OOCYTE DONATION FOR STEM CELL RESEARCH: WORKSHOP REPORT. Washington, DC: National Academies Press, 2007. 95 p. ISBN 978-0-309-10355-8; ISBN 0-309-10355-X. (Edited by Linda Giudice, Eileen Santa, and Robert Pool. Public Workshop on Assessing the Medical Risks of Human Oocyte Donation for Stem Cell Research held 28 September 2006 in Burlingame, California.) [RG134 .I55 2007] (14.2; 5.2; 17.1; 18.5.1)

McWhinnie, Alexina M. FAMILIES FOLLOWING ASSISTED CONCEPTION: WHAT DO WE TELL OUR CHILD? Dundee: University of Dundee, Department of Social Work, 1996. 50 p. ISBN 1-873153-23-6. (14.2; 7.1; 9.5.7; 14.4; 17.1)

McWhinnie, Alexina M. WHO AM I? EXPERIENCES OF DONOR CONCEPTION. Warwickshire: Idreos Education Trust, 2006. 66 p. ISBN 978-0-9554031-0-1; ISBN 0-9554031-0-3. (Publisher's address: 15 Wathen Road, Leaminton Spa, Warwickshire, postal code CV32 5UX.) (14.2; 4.4; 8.1; 17.1)

Raposo, Vera Lúcia. DE MÃE PARA MÃE: QUESTÕES LEGAIS E ÉTICAS SUSCITADAS PELA MATERNIDADE DE SUBSTITUIÇÃO. Coimbra, Portugal: Coimbra Editora, 2005. 198 p. ISBN 972-32-1345-1. (Faculdade de Direito da Universidade de Coimbra, Centro de Direito Biomédico (Series); 10.) [KKQ619 .R37 2005] (14.2; 14.1; 14.4)

14.3 SEX PREDETERMINATION/ SELECTION

Aravamudan, Gita. DISAPPEARING DAUGHTERS: THE TRAGEDY OF FEMALE FOETICIDE. New Delhi/New York: Penguin Books, 2007. 188 p. ISBN 978-0-14-310170-3; ISBN 0-14-310170-6. [HQ767.5 .I5 A74 2007] (14.3; 12.3; 12.4.1; 12.5.1; 15.2)

14.4 IN VITRO FERTILIZATION AND EMBRYO TRANSFER

Creating Life? Examining the Legal, Ethical and Medical Issues of Assisted Reproductive Technologies [Part 1]. JOURNAL OF GENDER, RACE AND JUSTICE 2005 Fall; 9(1): 1-136. (ISSN 1550-7815.) (14.4; 1.2; 4.4; 8.4; 9.5.5; 10; 14.2; 14.6; 15.3; 18.5.4)

Denham, Melinda M. EXPERIENCES OF IN VITRO FERTILIZATION DONOR EGG RECIPIENTS: THE IMPACT OF TECHNOLOGY ON REPRODUCTION. Ann Arbor, MI: ProQuest Information and Learning/UMI, 2005. 342 p. (Publication No. AAT-3196109. Dissertation, (Ph.D. in Anthropology)—State University of New York at Albany, 2005.) (14.4; 5.1; 7.1; 14.1)

Poilpot, Marie-Paule, ed. ÉTHIQUE ET BIOÉTHIQUE: L'ASSISTANCE MÉDICALE À LA PROCRÉATION. Ramonville Saint-Agne, France: Editions Erès, 1999. 114 p. ISBN 2-86586-696-3. (Collection Fondation pour l'enfance series.) (14.4; 2.1; 15.3)

Throsby, Karen. WHEN IVF FAILS: FEMINISM, INFERTILITY, AND THE NEGOTIATION OF NORMALITY. Basingstoke, Hampshire/New York: Palgrave Macmillan, 2004. 223 p. ISBN 978-1-4039-3554-0; ISBN 1-4039-3554-8. [RG135 .T495 2004] (14.4; 10)

14.5 CLONING

Gerdes, Louise, ed. CLONING. Farmington Hills, MI: Greenhaven Press/Thomson Gale, 2006. 141 p. ISBN 0-7377-3220-2. (Introducing Issues with Opposing Viewpoints series. Gift of the publisher.) [QH442.2 .C5645 2006] (14.5; 18.5.4; 22.2)

Roetz, Heiner, ed. CROSS-CULTURAL ISSUES IN BIOETHICS: THE EXAMPLE OF HUMAN CLONING. Amsterdam/New York: Rodopi, 2006. 470 p. ISBN 90-420-1609-4. (At the Interface / Probing the Boundaries series; Vol. 27. "A volume in the *Making Sense Of:* project 'Health, Illness and Disease'.") [QH442.2 .C768 2006] (14.5; 1.2; 2.1; 4.4; 5.3; 15.1; 18.5.4; 21.7)

Vöneky, Silja and Wolfrum, Rüdiger, eds. HUMAN DIGNITY AND HUMAN CLONING. Leiden/Boston: Martinus Nijhoff, 2004. 319 p. ISBN 978-90-04-14233-6; ISBN 90-04-14233-9. [QH442.2 .H876 2004] (14.5; 1.2; 4.4; 14.1; 15.1; 18.5.4)

Woodward, John, ed. THE ETHICS OF HUMAN CLONING. Detroit, MI: Thomson/Gale; Farmington Hills, MI: Greenhaven Press, 2005. 111 p. ISBN 0-7377-2187-1. (At Issue Series.) [QH438.7 .E844 2005] (14.5)

15.1 GENETICS, MOLECULAR BIOLOGY AND MICROBIOLOGY (GENERAL)

Carroll, Sean B. THE MAKING OF THE FITTEST: DNA AND THE ULTIMATE FORENSIC RECORD OF EVOLUTION. New York: W.W. Norton, 2006. 301 p. ISBN 978-0-393-06163-5; ISBN 0-393-06163-9. (Illustrations by Jamie W. Carroll and Leanne M. Olds.) [QP624 .C37 2006] (15.1; 3.2)

Castle, David; Cline, Cheryl; Daar, Abdallah S.; Tsamis, Charoula; and Singer, Peter. SCIENCE, SOCIETY, AND THE SUPERMARKET: THE OPPORTUNITIES AND CHALLENGES OF NUTRIGENOMICS. Hoboken, NJ: Wiley-Interscience, 2007. 163 p. ISBN 978-0-471-77000-8; ISBN 0-471-77000-0. [QP144 .G45 S35 2007] (15.1; 1.3.11; 9.7)

Crichton, Michael. NEXT: A NOVEL. New York: HarperCollins, 2006. 431 p. ISBN 978-0-06-087298-4; ISBN 0-06-087298-5. ("We live in a time of momentous scientific leaps; a time when it's possible to sell our eggs and sperm online for thousands of dollars or test our spouses for genetic maladies. We live in a time when one fifth of all our genes are owned by someone else, and an unsuspecting person and his family can be pursued cross-country because they happen to have certain valuable genes within their chromosomes.") [PS3553 .R48 N48 2006] (15.1; Fiction)

Deane-Drummond, Celia. GENETICS AND CHRISTIAN ETHICS. Cambridge/New York: Cambridge University Press, 2006. 281 p. ISBN 978-0-521-53637-0; ISBN 0-521-53637-5. (Gift of the publisher.) [QH438.7 .D43 2006] (15.1; 1.1; 1.2; 9.5.5; 10; 15.2; 15.3; 15.4; 15.5; 15.7; 15.8)

Einsiedel, Edna F. and Timmermans, Frank, eds. CROSSING OVER: GENOMICS IN THE PUBLIC ARENA. Calgary: University of Calgary Press, 2005. 261 p. ISBN 978-1-55238-191-5; ISBN 1-55238-191-9. (Essays from the conference held April 25-27, 2003, Kananaskis, Alberta.) [TP248.23 .C765 2005] (15.1; 1.3.2; 1.3.7; 1.3.11; 1.3.12; 5.3; 5.4; 18.5.4; 22.2)

Entine, Jon, ed. LET THEM EAT PRECAUTION: HOW POLITICS IS UNDERMINING THE GENETIC REVOLUTION IN AGRICULTURE. Washington, DC: AEI Press, 2006. 203 p. ISBN 0-8447-4200-7. [SB106 .B56 L48 2006] (15.1; 1.3.11; 5.3; 9.7; 15.7)

Flaman, Paul. GENETIC ENGINEERING, CHRISTIAN VALUES AND CATHOLIC TEACHING. New York: Paulist Press, 2002. 138 p. ISBN 0-8091-4089-6. [QH442 .F525 2002] (15.1; 1.2; 1.3.11; 2.3; 4.5; 8.4; 14.5; 15.2; 15.3; 15.5; 18.5.4; 19.1; 21.3; 22.2)

Fox, Charles W. and Wolf, Jason B., eds. EVOLUTIONARY GENETICS: CONCEPTS AND CASE STUDIES. New York: Oxford University Press, 2006. 592 p. ISBN 978-0-19-516818-1; ISBN 0-19-516818-6. [QH390 .E94 2006] (15.1; 3.1; 3.2; 15.11)

Gibbon, Sahra. BREAST CANCER GENES AND THE GENDERING OF KNOWLEDGE: SCIENCE AND CITIZENSHIP IN THE CULTURAL CONTEXT OF THE "NEW" GENETICS. Basingstoke, Hampshire/New York: Palgrave Macmillan, 2007. 221 p. ISBN 978-1-4039-9901-6; ISBN 1-4039-9901-5. [RC280 .B8 G478 2007] (15.1; 7.1; 9.5.5; 10)

Gunn, Moira A. WELCOME TO BIOTECH NATION: MY UNEXPECTED ODYSSEY INTO THE LAND OF SMALL MOLECULES, LEAN GENES, AND BIG IDEAS. New York: AMACOM, 2007. 258 p. ISBN 978-0-8144-0923-7; ISBN 0-8144-0923-7. (Gift of the publisher.) [TP248.215 .G86 2007] (15.1; 1.3.11; 14.5)

Harris, John. ENHANCING EVOLUTION: THE ETHICAL CASE FOR MAKING BETTER PEOPLE. Princeton, NJ: Princeton University Press, 2007. 242 p. ISBN 978-0-691-12844-3. [QH442 .H37 2007] (15.1; 4.5; 14.5; 18.5.4)

Hedgecoe, Adam. THE POLITICS OF PERSONALISED MEDICINE: PHARMACOGENETICS IN THE CLINIC. Cambridge/New York: Cambridge University Press, 2004. 208 p. ISBN 0-521-60265-3. (Cambridge Studies in Society and the Life Sciences series.) [RM301.3 .G45 H435 2004] (15.1; 5.2; 5.3; 7.1; 9.3.1; 9.5.2; 9.7; 15.3)

Holmes, Frederic Lawrence, ed. RECONCEIVING THE GENE: SEYMOUR BENZER'S ADVENTURES IN PHAGE GENETICS. New Haven, CT: Yale University Press, 2006. 334 p. ISBN 978-0-300-11078-4; ISBN 0-300-11078-2. [QH429.2 .S49 2006] (15.1; Biography)

Jean, Patrice and Régent, Julie. Université de Nouvelle-Calédonie. Département de droit. ÉTHIQUE ET GÉNÉTIQUE: ACTES DU COLLOQUE DE NOUMÉA, 25 JUILLET 1997. Paris: Harmattan, 2000. 257 p. ISBN 2-7384-9512-5. [QH438.7 .E854 2000] (15.1; 1.1; 1.2; 1.3.5; 2.1; 15.5; 15.6)

Kate, Kerry ten and Laird, Sarah A., eds. THE COMMERCIAL USE OF BIODIVERSITY: ACCESS TO GENETIC RESOURCES AND BENEFIT-SHARING. London: Earthscan, 2000, 1999. 398 p. ISBN 978-1-85383-941-2; ISBN 1-85383-941-8. (Second edition.) [QH75 .K384 2000] (15.1; 1.3.2; 1.3.5; 1.3.11; 9.7)

Knowles, Lori P. and Kaebnick, Gregory E., eds. REPROGENETICS: LAW, POLICY, AND ETHICAL ISSUES. Baltimore, MD: Johns Hopkins University Press, 2007. 302 p. ISBN 978-0-8018-8524-2; ISBN 0-8018-8524-8. (Gift of the publisher.) [QH442 .R472 2007] (15.1; 2.4; 5.3; 9.1; 14.1; 14.5; 15.5; 18.5.4; 21.1; 21.7)

Kuszler, Patricia; Battuello, Kathryn; and O'Connor, Sean. GENETIC TECHNOLOGIES AND THE LAW. Durham, NC: Carolina Academic Press, 2007. 1007 p. ISBN 978-0-89089-621-1; ISBN 0-89089-621-6. (Carolina Academic Press Law Casebook Series.) [KF3827 .G4 K87 2007] (15.1; 1.3.5; 1.3.11; 8.2; 9.1; 15.3; 18.2)

Lawler, Peter Augustine. STUCK WITH VIRTUE: THE AMERICAN INDIVIDUAL AND OUR BIOTECHNOLOGICAL FUTURE. Wilmington, DE: ISI Books, 2005. 262 p. ISBN 978-1-932236-84-2; ISBN 1-932236-84-8. (Religion and Contemporary Culture series. "Each chapter was originally prepared as a lecture of presentation.") [JC311 .L39 2005] (15.1; 5.1; 15.5; 15.9)

McManis, Charles R., ed. BIODIVERSITY AND THE LAW: INTELLECTUAL PROPERTY, BIOTECHNOLOGY AND TRADITIONAL KNOWLEDGE. London/Sterling, VA: Earthscan, 2007. 484 p. ISBN 978-1-84407-349-8; ISBN 1-84407-349-1. (Gift of the publisher.) [K3488 .B566 2007] (15.1; 1.3.11; 5.3; 15.8; 16.1; 21.1; 21.7)

Melo-Martín, Immaculada de. TAKING BIOLOGY SERIOUSLY: WHAT BIOLOGY CAN AND CANNOT TELL US ABOUT MORAL AND PUBLIC POLICY ISSUES. Lanham, MD: Rowman & Littlefield, 2005. 161 p. ISBN 0-7425-4920-8. [QH438.7 .M45 2005] (15.1; 3.1; 7.1; 14.5)

Mitchell, C. Ben; Pellegrino, Edmund D.; Elshtain, Jean Bethke; Kilner, John F.; Rae, Scott B. BIOTECHNOLOGY

AND THE HUMAN GOOD. Washington, DC: Georgetown University Press, 2007. 210 p. ISBN 978-1-58901-138-0; ISBN 1-58901-138-4. (Gift of the publisher.) [TP248.23 .B567 2007] (15.1; 1.2; 2.1; 4.4; 4.5; 5.4)

Monsour, Daniel, ed. ETHICS AND THE NEW GENETICS: AN INTEGRATED APPROACH. Toronto/Buffalo, NY: University of Toronto Press, 2007. 198 p. ISBN 978-0-8020-9273-1. ("Originated as papers presented at a think tank sponsored in 2002 by the Canadian Catholic Bioethics Institute.") [QH438.7 .E838 2007] (15.1; 1.2; 2.1)

Morris, Jonathan. THE ETHICS OF BIOTECHNOLOGY. New York: Chelsea House, 2006. 158 p. ISBN 0-7910-8520-1. (Biotechnology in the 21st Century series. Gift of the publisher.) [TP248.23 .M67 2006] (15.1; 1.3.11; 14.5; 18.5.4; 22.2; Reference)

National Academies (United States). Keck Futures Initiative. THE GENOMIC REVOLUTION: IMPLICATIONS FOR TREATMENT AND CONTROL OF INFECTIOUS DISEASE: WORKING GROUP SUMMARIES. Washington, DC: National Academies Press, 2006. 118 p. ISBN 978-0-309-10109-7; ISBN 0-309-10109-3. (Also available on the Web at: http://www.nap.edu.) [RA643 .N38 2005] (15.1; 9.5.1; 9.7)

Nationaler Ethikrat. BIOBANKEN: CHANCE FÜR DEN WISSENSCHAFTLICHEN FORTSCHRITT ODER AUSVERKAUF DER "RESSOURCE" MENSCH? Berlin: Nationaler Ethikrat, 2003. 113 p. ("Vorträge der Jahrestagung des Nationalen Ethikrates 2002." Gift of DRZE.) (15.1; 1.3.12; 2.4; 9.7; 15.10; 15.11)

Rasko, John E.J.; O'Sullivan, Gabrielle M.; and Ankeny, Rachel A., eds. THE ETHICS OF INHERITABLE GENETIC MODIFICATION: A DIVIDING LINE? Cambridge/New York: Cambridge University Press, 2006. 315 p. ISBN 0-521-52973-5. (Gift of the publisher.) [RG133.5 .E8475 2006] (15.1; 2.1; 4.5; 5.3; 10; 14.5; 15.2; 15.4; 18.5.4; 22.2)

Reilly, Philip R. THE STRONGEST BOY IN THE WORLD: HOW GENETIC INFORMATION IS RESHAPING OUR LIVES. Cold Spring Harbor, NY: Cold Spring Harbor Laboratory Press, 2006. 278 p. ISBN 978-0-87969-801-0; ISBN 0-87969-801-2. [QH431 .R383 2006] (15.1; 1.3.11; 15.2; 18.5.4; 22.1)

Roof, Judith. THE POETICS OF DNA. Minneapolis: University of Minnesota Press, 2007. 243 p. ISBN 978-0-8166-4998-3; ISBN 0-8166-4998-7. (Posthumanities series; No. 2.) [QP624 .R66 2007] (15.1; 15.6; 1.1)

Sandel, Michael J. THE CASE AGAINST PERFECTION: ETHICS IN THE AGE OF GENETIC ENGINEERING. Cambridge, MA: Belknap Press of Harvard University Press, 2007. 162 p. ISBN 978-0-674-01927-0; ISBN 0-674-01927-X. (Gift of the publisher.) [QH438.7 .S2634 2007] (15.1; 1.2; 4.5; 15.5)

Schreiber, Hans-Peter, ed. BIOMEDIZIN UND ETHIK: PRAXIS – RECHT – MORAL. Basel: Birkhäuser, 2004. 142 p. ISBN 3-7643-7065-3. [R724 .B576 2004] (15.1; 1.3.12; 4.4; 14.5; 15.2; 15.5; 15.8; 18.5.4)

Sensen, Christoper, ed. HANDBOOK OF GENOME RESEARCH: GENOMICS, PROTEOMICS, METABOLOMICS, BIOINFORMATICS, ETHICAL AND LEGAL ISSUES. New York: Wiley-VCH, 2005. 2 volumes. [618 p.]. ISBN 978-3-527-31348-8; ISBN 3-527-31348-6. [QH447 .H35 2005] (15.1; 1.3.2; 1.3.12; 4.4; 5.3; 9.7)

Stephenson, Frank H. DNA: HOW THE BIOTECH REVOLUTION IS CHANGING THE WAY WE FIGHT DISEASE. Amherst, NY: Prometheus Books, 2007. 333 p. (Gift of the publisher.) [TP248.215 .S74 2007] (15.1; 9.5.1; 9.5.2; 9.5.6; 14.5; 15.4; 15.10; 18.5.4; 21.3)

Wexler, Barbara. GENETICS AND GENETIC ENGINEERING. Detroit: Thomson/Gale, 2006. 186 p. ISBN 1-4144-0415-8. (2005 edition. Information Plus Reference Series. ISSN 1546-6426. Gift of the publisher.) [QH430 .W49 2006] (15.1; 3.2; 14.5; 15.3; 15.10; 16.1)

Zimmern, Ron and Cook, Christopher. GENETICS AND HEALTH: POLICY ISSUES FOR GENETIC SCIENCE AND THEIR IMPLICATIONS FOR HEALTH AND HEALTH SERVICES. London: Nuffield Trust for Research and Policy Studies in Health Services/ The Stationery Office [and] Cambridge: Public Health Genetics, 2000. 82 p. ISBN 978-0-11-702675-9; ISBN 0-11-702675-1. (Foreword by John Wyn Owen. "Nuffield Trust Genetics Scenario Project".) (15.1; 5.1; 9.1)

15.2 GENETIC COUNSELING/ PRENATAL DIAGNOSIS

Resta, Robert G., ed. PSYCHE AND HELIX: PSYCHOLOGICAL ASPECTS OF GENETIC COUNSELING: ESSAYS BY SEYMOUR KESSLER, PH.D.. New York: Wiley-Liss, 2000. 180 p. ISBN 978-0-471-35055-2; ISBN 0-471-35055-9. [RB155.7 .P78 2000] (15.2; 17.1)

Schöne-Seifert, Bettina; Krüger, Lorenz; and Toellner, Richard, eds. HUMANGENETIK: ETHISCHE PROBLEME DER BERATUNG, DIAGNOSTIK UND FORSCHUNG: [TAGUNG VOM 21. BIS 23. NOVEMBER 1991 IN GÖTTINGEN]. Stuttgart: Gustav Fischer, 1993. 322 p. ISBN 3-437-11471-9. (Medizin-Ethik series; Bd. 4. Dokumentation der Jahresversammlung des Arbeitskreises Medizinischer Ethik-Kommissionen in der Bundesrepublik Deutschland, Köln, 1991.) (15.2; 1.1; 2.4; 8.4; 9.7; 15.3; 15.10)

Weil, Jon. PSYCHOSOCIAL GENETIC COUNSELING. Oxford/New York: Oxford University Press, 2000. 297 p. ISBN 978-0-19-512066-0; ISBN 0-19-512066-3. (Oxford Monographs on Medical Genetics series; No. 41.) [RB155.7 .W45 2000] (15.2; 7.1; 17.1)

15.3 GENETIC SCREENING/ TESTING

Betta, Michael, ed. THE MORAL, SOCIAL, AND COMMERCIAL IMPERATIVES OF GENETIC TESTING AND SCREENING: THE AUSTRALIAN CASE. Dordrecht/New York: Springer, 2006. 268 p. ISBN 978-1-4020-4618-6; ISBN 1-4020-4618-9. (International Library of Ethics, Law, and the New Medicine series; Vol. 30.) [RB155.65 .M67 2006] (15.3; 1.3.12; 7.1; 8.4; 9.3.1; 15.5)

Numbers in ( ) = NRCBL Classification Numbers

Fuchs, Michael; Lanzerath, Dirk; and Schmidt, Matthias C., eds. PRÄDIKTIVE GENETISCHE TESTS: "HEALTH PURPOSES" UND INDIKATIONSSTELLUNG ALS KRITERIEN DER ANWENDUNG. Bonn: Institut für Wissenschaft und Ethik, 2004. 158 p. ISBN 3-936020-01-9. (Ethik in Biowissenschaften und Medizin Forschungsbeiträge, Reihe A series; Bd. 2. ISSN 1617-8742. Gift of DRZE.) (15.3; 1.1; 7.1; 8.3.1; Reference)

Guttmacher, Alan E.; Collins, Francis S.; and Drazen, Jeffrey M., eds. GENOMIC MEDICINE: ARTICLES FROM THE NEW ENGLAND JOURNAL OF MEDICINE. Baltimore, MD: Johns Hopkins University Press/Boston: New England Journal of Medicine, 2004. 179 p. ISBN 978-0-8018-7979-1; ISBN 0-8018-7979-5. (Foreword by Elias Zerhouni.) [RB155 .G463 2004] (15.3; 9.7; 15.1; 15.4)

Hashiloni-Dolev, Yael. A LIFE (UN)WORTHY OF LIVING: REPRODUCTIVE GENETICS IN ISRAEL AND GERMANY. Dordrecht: Springer, 2007. 195 p. ISBN 978-1-4020-5217-0. (International Library of Ethics, Law, and the New Medicine series; Vol. 34.) [RG133.5 .H382 2007] (15.3; 4.4; 7.1; 12.1; 14.1; 15.2)

Parthasarathy, Shobita. BUILDING GENETIC MEDICINE: BREAST CANCER, TECHNOLOGY, AND THE COMPARATIVE POLITICS OF HEALTH CARE. Cambridge, MA: MIT Press, 2007. 271 p. ISBN 978-0-262-16242-5. (Inside Technology series.) [RC268.44 .B73 P37 2007] (15.3; 5.1; 7.1; 21.7)

Pupecki, Sandra R. GENETIC SCREENING: NEW RESEARCH. New York: Nova Science Publishers, 2006. 214 p. ISBN 978-1-60021-006-8; ISBN 1-60021-006-6. (Nova Biomedical series.) [RB155.65 .G46 2006] (15.3; 15.2)

15.4 GENE THERAPY/ TRANSFER

Naff, Clay Farris, ed. GENE THERAPY. Detroit: Thomson Gale; Farmington Hills, MI: Greenhaven Press, 2005. 224 p. ISBN 978-0-7377-1968-3; ISBN 0-7377-1968-2. (Exploring Science and Medical Discoveries series.) [RB155.8 .G46172 2005] (15.4; 1.2; 2.2; 18.1)

15.5 EUGENICS

Barth, Gernot and Henseler, Joachim, eds. KINDHEIT, EUGENIK, SOZIALPÄDAGOGIK: FESTSCHRIFT ZUM 60. GEBURTSTAG VON JÜRGAN REYER. Baltmannsweiler: Schneider Hohengehren, 2004. 193 p. ISBN 978-3-89676-814-8; ISBN 3-89676-814-X. [LA724 .K563 2004] (15.5; 1.3.3)

Currell, Susan and Cogdell, Christina, eds. POPULAR EUGENICS: NATIONAL EFFICIENCY AND AMERICAN MASS CULTURE IN THE 1930S. Athens: Ohio University Press, 2006. 406 p. ISBN 978-0-8214-1691-4; ISBN 0-8214-1691-X. (Gift of the publisher.) [HQ755.5 .U5 P66 2006] (15.5; 1.3.7; 2.2; 7.1; 10; 11.3; 21.4)

Glad, John. FUTURE HUMAN EVOLUTION: EUGENICS IN THE TWENTY-FIRST CENTURY. Schuylkill Haven, PA: Hermitage Publishers, 2006. 136 p. ISBN 1-55779-154-6. (Preface by Seymour W. Itzkoff. "Exerpts from this book have appeared in *Mankind Quarterly* and *Jewish Press*." Also available on the Web at: http://www.whatwemaybe.org. Publisher's address: PO Box 578, postal code 17972-0578. Gift of the publisher.) [HQ751 .G52 2006] (15.5; 2.2; 11.3; 13.1; 15.6; 21.4)

Tankard Reist, Melinda. DEFIANT BIRTH: WOMEN WHO RESIST MEDICAL EUGENICS. North Melbourne, Victoria: Spinifex Press, 2006. 338 p. ISBN 978-1-876756-59-8; ISBN 1-876756-59-4. [HQ751 .D43 2006] (15.5; 9.5.5; 9.5.7; 9.5.8)

Turda, Marius and Weindling, Paul, eds. BLOOD AND HOMELAND: EUGENICS AND RACIAL NATIONALISM IN CENTRAL AND SOUTHEAST EUROPE, 1900-1940. Budapest/New York: Central European University Press, 2007. 467 p. ISBN 978-963-7326-81-3; ISBN 963-7326-81-2. [HQ755.5 .E8 B56 2007] (15.5; 1.2; 1.3.5; 2.2; 9.1; 15.11)

Winfield, Ann Gibson. EUGENICS AND EDUCATION IN AMERICA: INSTITUTIONALIZED RACISM AND THE IMPLICATIONS OF HISTORY, IDEOLOGY, AND MEMORY. New York: Peter Lang, 2007. 195 p. ISBN 978-0-8204-8146-3. (Complicated Conversation: A Book Series of Curriculum Studies; Vol. 18. ISSN 1534-2816.) [LC212.2 .W56 2007] (15.5; 1.3.3; 2.2)

Wray, Matt. NOT QUITE WHITE: WHITE TRASH AND THE BOUNDARIES OF WHITENESS. Durham, NC: Duke University Press, 2006. 213 p. ISBN 978-0-8223-3873-4; ISBN 0-8223-3873-4. [E184 .A1 W83 2006] (15.5; 2.2; 7.1)

15.6 BEHAVIORAL GENETICS

Institute of Medicine (United States). Board on Health Sciences Policy. Committee on Assessing Interactions Among Social, Behavioral, and Genetic Factors in Health. GENES, BEHAVIOR, AND THE SOCIAL ENVIRONMENT: MOVING BEYOND THE NATURE/NURTURE DEBATE. Washington, DC: National Academies Press, 2006. 368 p. ISBN 978-0-309-10196-7; ISBN 0-309-10196-4. (Edited by Lyla M. Hernandez and Dan G. Blazer. Also available on the Web at: http://www.nap.edu.) [QH457 .G458 2006] (15.6; 7.1; 10; 15.1; 15.9; 22.2)

Joseph, Jay. THE MISSING GENE: PSYCHIATRY, HEREDITY, AND THE FRUITLESS SEARCH FOR GENES. New York: Algora, 2006. 324 p. ISBN 978-0-87586-410-5; ISBN 0-87586-410-4. [RC455.4 .G4 J675 2006] (15.6; 11.3; 15.1; 15.5; 17.1)

15.8 GENETIC PATENTS

Dutfield, Graham. INTELLECTUAL PROPERTY, BIOGENETIC RESOURCES AND TRADITIONAL KNOWLEDGE. London/Sterling, VA: Earthscan, 2004. 258 p. ISBN 978-1-84407-048-0; ISBN 1-84407-048-4. [SB123.3 .D879 2004] (15.8; 1.3.11; 15.7; 21.1)

Jaenichen, Hans-Rainer; McDonell, Leslie A.; Haley, James F.; and Hosoda, Yoshinori. FROM CLONES TO CLAIMS: THE EUROPEAN PATENT OFFICE'S CASE LAW ON THE PATENTABILITY OF BIOTECHNOLOGY INVENTIONS IN COMPARISON TO THE UNITED STATES AND JAPANESE PRAC-

TICE. Cologne: Carl Heymanns, 2006. 881 p. ISBN 978-3-452-24738-4; ISBN 3-452-24738-4. (Fourth edition. Heymanns Intellectual Property series. Previously published as: *The European Patent Office's Case Law on the Patentability of Biotechnology Inventions in Comparison to the United States Practice in Case Law.*) [KJC2751 .B56 J34 2006] (15.8; 1.3.11; 9.7; 15.1; 21.1; 22.2; 22.3)

Nationaler Ethikrat [Germany]. THE PATENTING OF BIO-TECHNOLOGICAL INVENTIONS INVOLVING THE USE OF BIO-LOGICAL MATERIAL OF HUMAN ORIGIN: OPINION. Berlin: German National Ethics Council, 2005. 40 p. (Translation from the German of: *Zur Patentierung biotechnologischer Erfindungen unter Verwendung biologischen Materials menschlichen Ursprungs: Stellungnahme.* Gift of DRZE.) (15.8; 2.4)

Remédio Marques, J.P.; de Ética, Conselho Dinamarquês. PATENTES DE GENES HUMANOS? Coimbra, Portugal: Coimbra Editora, 2001. 154 p. ISBN 978-972-32-1022-4; ISBN 972-32-1022-3. (Faculdade de Direito da Universidade de Coimbra, Centro de Direito Biomédico series; No. 4.) [KKQ1210 .B56 R46 2001] (15.8)

15.9 SOCIOBIOLOGY

Dawkins, Richard. THE SELFISH GENE. Oxford/New York: Oxford University Press, 2006. 360 p. ISBN 978-0-19-929115-1. (30th anniversary edition, with a new introduction by the author.) [QH437 .D38 2006] (15.9; 3.2; 15.6)

Salter, Frank. ON GENETIC INTERESTS: FAMILY, ETHNICITY, AND HUMANITY IN AN AGE OF MASS MIGRATION. New Brunswick, NJ: Transaction, 2007. 388 p. ISBN 978-1-4128-0596-1; ISBN 1-4128-0596-1. (With a new introduction by the author. Originally published: Frankfurt am Main/New York: Peter Lang, 2003.) [QH431 .S274 2007] (15.9)

16.1 ENVIRONMENTAL QUALITY (GENERAL)

Baber, Walter F. and Bartlett, Robert V. DELIBERATIVE EN-VIRONMENTAL POLITICS: DEMOCRACY AND ECOLOGICAL RATIONALITY. Cambridge, MA: MIT Press, 2005. 276 p. ISBN 0-262-52444-9. [GE170 .B33 2005] (16.1; 1.3.5)

Bratspies, Rebecca M. and Miller, Russell A., eds. TRANSBOUNDARY HARM IN INTERNATIONAL LAW: LESSONS FROM THE TRAIL SMELTER ARBITRATION. Cambridge/New York: Cambridge University Press, 2006. 347 p. ISBN 978-0-521-85643-0; ISBN 0-521-85643-4. (Gift of the publisher.) [KZ6148 .T73 2006] (16.1; 1.3.6)

British Medical Association. HEALTH & ENVIRONMENTAL IMPACT ASSESSMENT: AN INTEGRATED APPROACH. London: Earthscan, 1998. 243 p. ISBN 978-1-85383-541-4; ISBN 1-85383-541-2. (Environment and Health Series.) (16.1; 9.1; 9.3.1; 16.3)

Brown, Lester R. PLAN B 2.0: RESCUING A PLANET UNDER STRESS AND A CIVILIZATION IN TROUBLE. New York: W.W. Norton, 2006. 365 p. ISBN 978-0-393-32831-8; ISBN

0-393-32831-7. (Updated and expanded edition of: *Plan B: Rescuing a Planet Under Stress and a Civilization in Trouble*; 2003.) [HC79 .E5 B7595 2006] (16.1; 1.3.2)

Checker, Melissa. POLLUTED PROMISES: ENVIRONMENTAL RACISM AND THE SEARCH FOR JUSTICE IN A SOUTHERN TOWN. New York: New York University Press, 2005. 275 p. ISBN 978-0-8147-1658-8; ISBN 0-8147-1658-X. [GE235 .G4 C46 2005] (16.1; 9.5.4)

Corburn, Jason. STREET SCIENCE: COMMUNITY KNOWL-EDGE AND ENVIRONMENTAL HEALTH JUSTICE. Cambridge, MA: MIT Press, 2005. 271 p. ISBN 0-262-53272-7. (Urban and Industrial Environments series.) [RA565 .C67 2005] (16.1; 5.2; 5.3; 9.1; 9.5.4)

Garte, Seymour. WHERE WE STAND: A SURPRISING LOOK AT THE REAL STATE OF OUR PLANET. New York: AMACOM, 2008. 290 p. ISBN 978-0-8144-0910-7; ISBN 0-8144-0910-5. [GF75 .G36 2008] (16.1)

Girardet, Herbert, ed. SURVIVING THE CENTURY: FACING CLIMATE CHAOS AND OTHER GLOBAL CHALLENGES. London/Sterling, VA: Earthscan, 2007. 210 p. ISBN 978-1-84407-458-7. (Published in cooperation with the World Future Council. Gift of the publisher.) [HC79 .E5 S86453 2007] (16.1; 1.3.11)

Keiter, Robert B. KEEPING FAITH WITH NATURE: ECOSYS-TEMS, DEMOCRACY, AND AMERICA'S PUBLIC LANDS. New Haven, CT: Yale University Press, 2003. 434 p. ISBN 978-0-300-09273-8; ISBN 0-300-09273-3. [GE180 .K45 2003] (16.1)

Louka, Elli. INTERNATIONAL ENVIRONMENTAL LAW: FAIR-NESS, EFFECTIVENESS, AND WORLD ORDER. Cambridge/New York: Cambridge University Press, 2006. 518 p. ISBN 978-0-521-68759-1; ISBN 0-521-68759-4. (Gift of the publisher.) [K3585 .L68 2006] (16.1; 21.1; 22.1)

Magee, Mike. HEALTHY WATERS: WHAT EVERY HEALTH PROFESSIONAL SHOULD KNOW ABOUT WATER. Bronxville, NY: Spencer Books, 2005. 142 p. ISBN 978-1-889793-16-0; ISBN 1-889793-16-7. (Gift of the author.) [RA591 .M218 2005] (16.1; 1.3.11)

Ott, Konrad. ÖKOLOGIE UND ETHIK: EIN VERSUCH PRAKTISCHER PHILOSOPHIE. Tübingen: Attempto Verlag, 1994. 188 p. ISBN 978-3-89308-162-2; ISBN 3-89308-162-3. (Second edition. Ethik in den Wissenschaften series; Bd. 4.) [QH540.5 .O88 1994] (16.1; 1.1)

Pellow, David Naguib and Park, Lisa Sun-Hee. THE SILI-CON VALLEY OF DREAMS: ENVIRONMENTAL INJUSTICE, IM-MIGRANT WORKERS, AND THE HIGH-TECH GLOBAL ECONOMY. New York: New York University Press, 2002. 303 p. ISBN 978-0-8147-6710-8; ISBN 0-8147-6710-9. (Critical America series.) [HC107 .C22 S376 2002] (16.1; 1.3.11; 1.3.12; 9.5.4)

Sandler, Ronald L. CHARACTER AND ENVIRONMENT: A VIR-TUE-ORIENTED APPROACH TO ENVIRONMENTAL ETHICS. New York: Columbia University Press, 2007. 201 p. ISBN

978-0-231-14106-2. (Gift of the publisher.) [GE42 .S26 2007] (16.1; 1.1; 1.3.11; 15.1)

Shrader-Frechette, Kristin. TAKING ACTION, SAVING LIVES: OUR DUTIES TO PROTECT ENVIRONMENTAL AND PUBLIC HEALTH. Oxford/New York: Oxford University Press, 2007. 299 p. ISBN 978-0-19-532546-1. (Gift of the publisher.) [RA566 .S37 2007] (16.1; 1.3.5; 5.3; 9.1)

Zwaan, B. van der and Petersen, Arthur, eds. SHARING THE PLANET: POPULATION, CONSUMPTION, SPECIES: SCIENCE AND ETHICS FOR A SUSTAINABLE AND EQUITABLE WORLD. Delft [Netherlands]: Eburon Publishers, 2003. 264 p. ISBN 978-90-5166-986-2; ISBN 90-5166-986-0. (Papers presented at the symposium "Sharing the Planet: Population, Consumption, Species," held 12-14 June 2002 in Groningen, The Netherlands.) [HC79 .E5 S426 2003] (16.1; 13.1; 21.1; 22.1)

17.1 THE NEUROSCIENCES AND MENTAL HEALTH THERAPIES (GENERAL)

American Psychiatric Nurses Association; International Society of Psychiatric-Mental Health Nurses; and American Nurses Association. PSYCHIATRIC-MENTAL HEALTH NURSING: SCOPE AND STANDARDS OF PRACTICE. Silver Spring, MD: American Nurses Association, 2007. 151 p. ISBN 978-1-55810-250-7; ISBN 1-55810-250-7. [RC440 .P7427 2007] (17.1; 4.1.3)

Andreasen, Nancy C. THE CREATING BRAIN: THE NEURO-SCIENCE OF GENIUS. New York: Dana Press, 2005. 197 p. ISBN 978-1-932594-07-2; ISBN 1-932594-07-8. [QP393 .A53 2005] (17.1; 15.6)

Andrews, Jonathan and Digby, Anne, eds. SEX AND SECLU-SION, CLASS AND CUSTODY: PERSPECTIVES ON GENDER AND CLASS IN THE HISTORY OF BRITISH AND IRISH PSYCHIATRY. Amsterdam/New York: Rodopi, 2004. 338 p. ISBN 90-420-1176-9. (The Wellcome Series in the History of Medicine, Clio Medica; No. 73. ISSN 0045-7183.) [RC455.4 .S45 S49 2004] (17.1; 4.3; 7.1; 9.5.5; 10; 15.5; 17.7)

Bailey, Jon S. and Burch, Mary R. ETHICS FOR BEHAVIOR ANALYSTS: A PRACTICAL GUIDE TO THE BEHAVIOR ANA-LYST CERTIFICATION BOARD GUIDELINES FOR RESPONSI-BLE CONDUCT. Mahwah, NJ: Lawrence Erlbaum Associates, 2005. 296 p. ISBN 978-0-8058-5118-2; ISBN 0-8058-5118-6. [RC473 .B43 B355 2005] (17.1; 2.3; 4.1.1; 4.3)

Bersoff, Donald N. ETHICAL CONFLICTS IN PSYCHOLOGY. Washington, DC: American Psychological Association, 2003. 573 p. ISBN 1-59147-050-1. (Third edition.) [BF76.4 .E814 2003] (17.1; 1.3.12; 2.3; 4.3; 6. 7.3; 7.4; 8.1; 8.4; 9.3.2; 9.5.6; 9.5.7; 10; 17.2; 18.2; 18.3; 18.4; 22.2)

Bloom, Floyd E., ed. BEST OF THE BRAIN FROM SCIENTIFIC AMERICAN. New York: Dana Press, 2007. 270 p. ISBN 978-1-932594-22-5. (Gift of the publisher.) [QP360.5 .B52 2007] (17.1)

Brown, Thomas E. ATTENTION DEFICIT DISORDER: THE UNFOCUSED MIND IN CHILDREN AND ADULTS. New Haven, CT: Yale University Press, 2005. 360 p. ISBN 978-0-300-10641-1; ISBN 0-300-10641-6. (Yale University Press Health & Wellness series. Anonymous gift.) [RJ506 .H9 B765 2005] (17.1; 9.5.1; 9.5.7; 17.4)

Bush, Shane S. ETHICAL DECISION MAKING IN CLINICAL NEUROPSYCHOLOGY. Oxford/New York: Oxford University Press, 2007. 157 p. ISBN 978-0-19-532822-6. (Oxford Workshop Series: American Academy of Clinical Neuropsychology. Gift of the publisher.) [RC386.6 .N48 B87 2007] (17.1; 2.1; 7.4; 8.1)

Corrigan, Patrick W., ed. ON THE STIGMA OF MENTAL ILL-NESS: PRACTICAL STRATEGIES FOR RESEARCH AND SOCIAL CHANGE. Washington, DC: American Psychological Association, 2005. 343 p. ISBN 1-59147-189-3. [RC455.2 .P85 O5 2005] (17.1; 1.3.5; 1.3.7; 4.3; 7.1)

Druss, Richard G. THE PSYCHOLOGY OF ILLNESS: IN SICK-NESS AND IN HEALTH. Washington, DC: American Psychiatric Press, 1995. 114 p. ISBN 978-0-88048-661-3; ISBN 0-88048-661-9. [R726.5 .D78 1995] (17.1; 8.1; 17.2; 17.4)

Farber, Seth. UNHOLY MADNESS: THE CHURCH'S SURREN-DER TO PSYCHIATRY. Downers Grove, IL: InterVarsity Press, 1999. 162 p. ISBN 0-8308-1939-8. [BT732.4 .F37 1999] (17.1; 1.2)

Ford, Gary G. ETHICAL REASONING FOR MENTAL HEALTH PROFESSIONALS. Thousand Oaks, CA: SAGE Publications, 2006. 393 p. ISBN 0-7619-3094-9. [RC455.2 .E8 F67 2006] (17.1; 1.1; 1.3.1; 1.3.2; 1.3.12; 4.3; 6; 7.2; 8.1; 9.6; 18.5.6)

Foucault, Michel. PSYCHIATRIC POWER: LECTURES AT THE COLLEGE DE FRANCE, 1973-74. Basingstoke, Hampshire [England]/New York: Palgrave Macmillan, 2006. 382 p. ISBN 978-1-4039-6922-1; ISBN 1-4039-6922-1. (Translation from the French by Graham Burchell of *Le pouvoir psychiatrique*, Seuil/Gallimard, 2003, in the series: Hautes études. Edited by Jacques Lagrange. François Ewald and Alessandro Fontana, general editors. Gift of the publisher.) [RC437.5 .F6813 2006] (17.1; 4.1.2; 7.1; 17.7)

Frank, Richard G. and Glied, Sherry A. BETTER BUT NOT WELL: MENTAL HEALTH POLICY IN THE UNITED STATES SINCE 1950. Baltimore, MD: Johns Hopkins University Press, 2006. 183 p. ISBN 0-8018-8443-8. (Foreword by Rosalynn Carter.) [RA790.6 .F723 2006] (17.1; 7.1; 9.1; 9.3.1)

Gale, Colin and Howard, Robert. PRESUMED CURABLE: AN ILLUSTRATED CASEBOOK OF VICTORIAN PSYCHIATRIC PA-TIENTS IN BETHLEM HOSPITAL. Petersfield, UK/Philadelphia: Wrightson Biomedical, 2003. 128 p. ISBN 978-1-871816-48-8; ISBN 1-871816-48-3. [RC450 .G7 G35 2003] (17.1; 7.1)

Garland, Brent, ed. NEUROSCIENCE AND THE LAW: BRAIN, MIND, AND THE SCALES OF JUSTICE: A REPORT ON AN INVI-TATIONAL MEETING CONVENED BY THE AMERICAN ASSOCI-ATION FOR THE ADVANCEMENT OF SCIENCE AND THE DANA

FOUNDATION. New York: Dana Press [and] Washington, DC: AAAS, 2004. 229 p. ISBN 1-932594-04-3. (Papers presented at a workshop held in September 2003 in Washington, DC, with four additional commissioned papers.) [KF2910 .N45 N48 2004] (17.1; 1.1; 7.1)

Glannon, Walter. BIOETHICS AND THE BRAIN. Oxford/New York: Oxford University Press, 2007. 235 p. ISBN 978-0-19-530778-8; ISBN 0-19-530778-X. [RC343 .G53 2007] (17.1; 2.1; 3.1; 9.7; 17.6; 18.1; 20.2.1)

Green, Stephen A. and Bloch, Sidney, eds. AN ANTHOLOGY OF PSYCHIATRIC ETHICS. Oxford/New York: Oxford University Press, 2006. 498 p. ISBN 978-0-19-856488-1; ISBN 0-19-856488-0. (Gift of the publisher.) [RC455.2 .E8 A58 2006] (17.1; 1.1; 1.3.5; 2.1; 4.3; 7.3; 8.1; 8.3.1; 8.4; 9.4; 17.4; 18.3; 18.5.6; 20.6)

Guimón, José. INEQUITY AND MADNESS: PSYCHOSOCIAL AND HUMAN RIGHTS ISSUES. New York: Kluwer Academic/Plenum Publishers, 2001. 225 p. ISBN 0-306-46674-0. [RC455.2 .E8 G85 2001] (17.1; 4.3; 7.1; 17.7; 21.1)

Herlihy, Barbara and Corey, Gerald, [eds.]. ACA ETHICAL STANDARDS CASEBOOK. Alexandria, VA: American Counseling Association, 2006. 264 p. ISBN 978-1-55620-255-1; ISBN 1-55620-255-5. (Sixth edition. Publication Order No. 72839. Publisher's address: 5999 Stevenson Avenue, zip 22304; Website: http://www.counseling.org .Gift of the publisher.) [BF637 .C6 A37 2006] (17.1; 1.3.1; 1.3.3; 6; 8.1; 8.3.1; 8.4; 9.5.4; 9.5.7; 20.7; 21.7)

Houser, Rick; Wilczenski, Felicia L.; and Ham, MaryAnna. CULTURALLY RELEVANT ETHICAL DECISION-MAKING IN COUNSELING. Thousand Oaks, CA: Sage, 2006. 334 p. ISBN 978-1-4129-0587-9; ISBN 1-4129-0587-7. [BF637 .C6 H676 2006] (17.1; 1.1; 1.2; 8.3.2; 8.4; 9.5.7; 20.5.1; 20.7; 21.7)

Illes, Judy, ed. NEUROETHICS: DEFINING THE ISSUES IN THEORY, PRACTICE, AND POLICY. Oxford/New York: Oxford University Press, 2006. 329 p. ISBN 978-0-19-856721-9; ISBN 0-19-856721-9. (Foreword by Arthur Caplan. Gift of the publisher.) [RC343 .N44 2006] (17.1; 1.2; 1.3.3; 4.5; 7.1; 15.1; 17.5; 18.2)

Jones, Caroline (contributing editor); Shillito-Clarke, Carol; Syme, Gabrielle; Hill, Derek; Casemore, Roger; and Murdin, Lesley. QUESTIONS OF ETHICS IN COUNSELLING AND THERAPY. Buckingham, UK/Philadelphia: Open University Press, 2000. 190 p. ISBN 978-0-335-20610-0; ISBN 0-335-20610-7. [RC455.2 .E8 Q475 2000] (17.1; 4.1.1; 8.1)

Jones, Petre, ed. DOCTORS AS PATIENTS. Oxford/Seattle: Radcliffe, 2005. 200 p. ISBN 978-1-85775-887-0; ISBN 1-85775-887-0. (Foreword by Michael Shooter.) [R727.9 .D63 2005] (17.1; 7.1; 7.4)

Kandel, Eric R. IN SEARCH OF MEMORY: THE EMERGENCE OF A NEW SCIENCE OF MIND. New York: W.W. Norton, 2006. 510 p. ISBN 978-0-393-05863-5; ISBN 0-393-

05863-8. (Gift of Max M. and Marjorie B. Kampelman.) [RC339.52 .K362 A3 2006] (17.1; Biography)

Kendler, Howard H. AMORAL THOUGHTS ABOUT MORALITY: THE INTERSECTION OF SCIENCE, PSYCHOLOGY, AND ETHICS. Springfield, IL: Charles C. Thomas, 2000. 198 p. ISBN 978-0-398-07029-8; ISBN 0-398-07029-6. [BF76.4 .K47 2000] (17.1; 15.6; 21.1; 21.7)

Leff, Julian and Warner, Richard. SOCIAL INCLUSION OF PEOPLE WITH MENTAL ILLNESS. Cambridge/New York: Cambridge University Press, 2006. 192 p. ISBN 978-0-521-61536-5; ISBN 0-521-61536-4. [RC454 .L374 2006] (17.1; 7.1)

Linhorst, Donald M. EMPOWERING PEOPLE WITH SEVERE MENTAL ILLNESS: A PRACTICAL GUIDE. Oxford/New York: Oxford University Press, 2006. 353 p. ISBN 978-0-19-517187-7; ISBN 0-19-517187-X. ("Presents a model of empowerment and then applies it to seven areas that have potential to empower people with severe mental illness, including treatment planning, housing, employment, and others. Provides practitioners, administrators, and policymakers with specific guidelines and actions to promote empowerment.") [HV3006 .A4 L56 2006] (17.1; 7.1; 9.1; 18.5.6)

Marland, Hilary. DANGEROUS MOTHERHOOD: INSANITY AND CHILDBIRTH IN VICTORIAN BRITAIN. Houndmills, Basingstoke, Hampshire/New York: Palgrave Macmillan, 2004. 303 p. ISBN 978-1-4039-2038-6; ISBN 1-4039-2038-9. [RG851 .M37 2004] (17.1; 4.3; 8.1; 9.5.5; 20.5.2)

McGrath, Joanna Collicutt. ETHICAL PRACTICE IN BRAIN INJURY REHABILITATION. Oxford/New York: Oxford University Press, 2007. 182 p. ISBN 978-0-19-856899-5; ISBN 0-19-856889-1. [RD594 .M34 2007] (17.1; 2.1; 4.4)

Merkel, R.; Boer, G.; Fegert, J.; Galert, T.; Hartmann, D.; Nuttin, B.; and Rosahl, S. INTERVENING IN THE BRAIN: CHANGING PSYCHE AND SOCIETY. Berlin/New York: Springer, 2007. 533 p. ISBN 978-3-540-46476-1. (Ethics of Science and Technology Assessment series; Vol. 29 = Wissenschaftsethik und Technikfolgenbeurteilung series; Bd. 29. ISSN 1860-4803.) [RC350 .N48 I58 2007] (17.1; 4.4; 4.5; 15.4)

Mitchell, Robert W. DOCUMENTATION IN COUNSELING RECORDS: AN OVERVIEW OF ETHICAL, LEGAL, AND CLINICAL ISSUES. Alexandria, VA: American Counseling Association, 2007. 109 p. ISBN 978-1-55620-273-5; ISBN 1-55620-273-3. (Third edition.) [RC466 .M57 2006] (17.1; 1.3.12; 8.1; 9.3.1; 10)

Nunn, Chris. DE LA METTRIE'S GHOST: THE STORY OF DECISIONS. New York: Macmillan, 2005. 228 p. ISBN 978-1-4039-9495-0; ISBN 1-4039-9495-1. [B808.9 .N88 2005] (17.1; 1.1; 3.1)

Orr, Jackie. PANIC DIARIES: A GENEALOGY OF PANIC DISORDER. Durham, NC: Duke University Press, 2006. 362 p. ISBN 0-8223-3623-5. (Gift of the publisher.) [RC535 .O77 2006] (17.1; 5.3; 7.1; 17.4; 21.2)

Osborne, Lawrence. AMERICAN NORMAL: THE HIDDEN WORLD OF ASPERGER SYNDROME. New York: Copernicus Books/Springer-Verlag, 2002. 224 p. ISBN 0-387-95307-8. [RC553 .A88 O83 2002] (17.1; 4.3; 9.5.7; 17.4)

Sahmland, Irmtraut, ed. "HALTESTATION PHILIPPSHOSPITAL": EIN PSYCHIATRISCHES ZENTRUM - KONTINUITÄT UND WANDEL 1535-1904-2004: EINE FESTSCHRIFT ZUM 500. GEBURTSTAG PHILIPPS VON HESSEN. Marburg: Jonas-Verlag, 2004. 494 p. ISBN 978-3-89445-341-1; ISBN 3-89445-341-9. (Historische Schriftenreihe des Landeswohlfahrtsverbandes Hessen; Quellen und Studien series; Bd. 10.) [RC450 .G32 R543 2004] (17.1; 1.3.5; 7.1; 15.5; 20.5.1; 21.4)

Schramme, Thomas and Thome, Johannes, eds. PHILOSO-PHY AND PSYCHIATRY. Berlin/New York: Walter De Gruyter, 2004. 391 p. ISBN 978-3-11-017800-5; ISBN 3-11-017800-1. [RC437.5 .P45 2004] (17.1; 1.1; 4.3; 7.1; 17.2; 17.7)

Weinberg, Darin. OF OTHERS INSIDE: INSANITY, ADDIC-TION, AND BELONGING IN AMERICA. Philadelphia: Temple University Press, 2005. 226 p. ISBN 978-1-59213-404-5; ISBN 1-59213-404-1. (Foreword by Bryan S. Turner.) [HV3009 .W45 2005] (17.1; 7.1; 9.5.9; 9.5.10)

17.2 PSYCHOTHERAPY

Corey, Gerald; Corey, Marianne Schneider; and Callanan, Patrick. ISSUES AND ETHICS IN THE HELPING PROFESSIONS. Belmont, CA: Brooks/Cole/Thomson Learning, 2007. 552 p. ISBN 978-0-534-61443-0; ISBN 0-534-61443-4. (Seventh edition.) [RC455.2 .E8 C66 2007] (17.2; 1.3.1; 4.1.1; 8.1; 8.4; 21.7)

Danto, Elizabeth Ann. FREUD'S FREE CLINICS: PSYCHO-ANALYSIS & SOCIAL JUSTICE, 1918-1938. New York: Colum-bia University Press, 2005. 348 p. ISBN 978-0-231-13180-3; ISBN 0-231-13180-1. [BF173 .D365 2005] (17.2; 7.1; 9.5.10; 10)

Goldberg, Arnold. MORAL STEALTH: HOW "CORRECT BE-HAVIOR" INSINUATES ITSELF INTO PSYCHOTHERAPEUTIC PRACTICE. Chicago: University of Chicago Press, 2007. 150 p. ISBN 978-0-226-30120-4; ISBN 0-226-30120-6. [RC480.8 .G65 2007] (17.2; 1.3.1; 4.1.1; 8.1)

Pope, Kenneth S., Sonne, Janet L. and Greene, Beverly. WHAT THERAPISTS DON'T TALK ABOUT AND WHY: UNDER-STANDING TABOOS THAT HURT US AND OUR CLIENTS. Washington, DC: American Psychological Association, 2006. 199 p. ISBN 1-59147-401-9. (Forewords by Melba J.T. Vasquez and Gerald p. Koocher.) [RC480.5 .P636 2006] (17.2; 7.4; 8.1; 10)

Rogers, Carl R. ON BECOMING A PERSON: A THERAPIST'S VIEW OF PSYCHOTHERAPY. Boston: Houghton Mifflin, 1995. 420 p. ISBN 0-395-75531-X. (Reprint of earlier edi-tion.) [RC480.5 .R62 1995] (17.2; 1.1; 4.4; 8.1)

Wilcoxon, S. Allen; Remley, Jr., Theodore P.; Gladding, Samuel T.; and Huber, Charles H. ETHICAL, LEGAL, AND PROFESSIONAL ISSUES IN THE PRACTICE OF MARRIAGE AND FAMILY THERAPY. Upper Saddle River, NJ: Pearson Merrill Prentice Hall, 2007. 428 p. ISBN 978-0-13-112034-1; ISBN 0-13-112034-4. (Fourth edition.) [RC488.5 .H8 2007] (17.2; 1.3.1)

17.3 BEHAVIOR MODIFICATION

Lemov, Rebecca. WORLD AS LABORATORY: EXPERIMENTS WITH MICE, MAZES, AND MEN. New York: Hill & Wang, 2005. 291 p. ISBN 978-0-8090-7464-8; ISBN 0-8090-7464-8. [HM668 .L46 2005] (17.3; 1.3.5; 1.3.9; 5.1; 7.2; 17.2; 18.4)

17.4 PSYCHOPHARMACOLOGY

Dubovsky, Steven L. and Dubovsky, Amelia N. PSYCHOTROPIC DRUG PRESCRIBER'S SURVIVAL GUIDE: ETH-ICAL MENTAL HEALTH TREATMENT IN THE AGE OF BIG PHARMA. New York: W.W. Norton, 2007. 187 p. ISBN 978-0-393-70510-2; ISBN 0-393-70510-2. [RM315 .D878 2007] (17.4; 9.3.1; 9.7; 18.6)

Lakoff, Andrew. PHARMACEUTICAL REASON: KNOWLEDGE AND VALUE IN GLOBAL PSYCHIATRY. Cambridge/New York: Cambridge University Press, 2005. 206 p. ISBN 978-0-521-54666-9; ISBN 0-521-54666-4. (Cambridge Studies in Society and the Life Sciences series.) [RM315 .L29 2005] (17.4; 7.1; 15.6)

Leventhal, Allan M. and Martell, Christopher R. THE MYTH OF DEPRESSION AS DISEASE: LIMITATIONS AND AL-TERNATIVES TO DRUG TREATMENT. Westport, CT: Praeger, 2006. 178 p. ISBN 978-0-275-98976-7; ISBN 0-275-98976-3. (Contemporary Psychology series. ISSN 1546-668X. Foreword by Marsha Linehan.) [RC537 .L477 2006] (17.4; 17.2; 17.3)

Vuckovich, Paula K. JUSTIFYING COERCION: NURSES' EX-PERIENCES MEDICATING INVOLUNTARY PSYCHIATRIC PA-TIENTS. Ann Arbor, MI: ProQuest Information and Learning/UMI, 2003. 184 p. (Publication No. AAT-3088677. Dissertation (Ph.D. in Nursing)—University of San Diego, 2003.) [RC440 .V8 2003] (17.4; 8.3.4; 17.3; 17.7; 17.8)

17.5 ELECTRICAL STIMULATION OF THE BRAIN

Dukakis, Kitty and Tye, Larry. SHOCK: THE HEALING POWER OF ELECTROCONVULSIVE THERAPY. New York: Avery/Penguin, 2006. 289 p. ISBN 978-1-58333-265-8; ISBN 1-58333-265-0. [RC485 .D85 2006] (17.5)

18.1 HUMAN EXPERIMENTATION (GENERAL)

Evans, Imogen; Thornton, Hazel; and Chalmers, Iain. TESTING TREATMENTS: BETTER RESEARCH FOR BETTER HEALTHCARE. London: British Library, 2006. 116 p. ISBN 0-7123-4909-X. [R850 .E98 2006] (18.1; 9.8; 18.2)

Gallin, John I. and Ognibene, Frederick P., eds. PRINCI-PLES AND PRACTICE OF CLINICAL RESEARCH. Amsterdam/Boston: Elsevier/Academic Press, 2007. 430 p. ISBN 978-0-12-369440-9; ISBN 0-12-369440-X. (Second edition.) [R850 .G35 2007] (18.1; 5.3; 18.2; 18.5.1)

Lafleur, William R.; Böhme, Gernot; and Shimazono, Susumu, eds. DARK MEDICINE: RATIONALIZING UNETHICAL MEDICAL RESEARCH. Bloomington, IN: Indiana University Press, 2007. 259 p. ISBN 978-0-253-34872-2. (Bioethics and the Humanities series. Gift of the publisher.) [R853 .H8 D37 2007] (18.1; 2.2; 18.3; 18.5.1; 18.5.8; 18.6; 21.3)

Lemmens, Trudo and Waring, Duff R., eds. LAW AND ETHICS IN BIOMEDICAL RESEARCH: REGULATION, CONFLICT OF INTEREST, AND LIABILITY. Toronto/Buffalo, NY: University of Toronto Press, 2006. 267 p. ISBN 978-0-8020-8643-2; ISBN 0-8020-8643-8. (Gift of the publisher.) [KE3663 .M38 L39 2006] (18.1; 1.3.5; 1.3.9; 5.3; 9.7; 18.2; 18.3; 18.5.1)

18.2 HUMAN EXPERIMENTATION: POLICY GUIDELINES/ IRB

Amdur, Robert J. and Bankert, Elizabeth A. INSTITUTIONAL REVIEW BOARD MEMBER HANDBOOK. Sudbury, MA: Jones and Bartlett, 2007. 230 p. ISBN 978-0-7637-4122-8; ISBN 0-7637-4122-1. (Second edition. Gift of the publisher.) [R852.5 .A463 2007] (18.2; 18.3; 18.5.1; Reference)

Fayers, Peter and Hays, Ron, eds. ASSESSING QUALITY OF LIFE IN CLINICAL TRIALS: METHODS AND PRACTICE. Oxford/New York: Oxford University Press, 2005. 467 p. ISBN 978-0-19-852769-5; ISBN 0-19-852769-1. (Second edition. Revised edition of: *Quality of Life Assessment in Clinical Trials*, edited by Maurice J. Staquet, Ron D. Hays, and Peter M. Fayers, 1998.) [R853 .C55 A88 2005] (18.2; 4.4)

Mazur, Dennis J. EVALUATING THE SCIENCE AND ETHICS OF RESEARCH ON HUMANS: A GUIDE FOR IRB MEMBERS. Baltimore, MD: Johns Hopkins University Press, 2007. 252 p. ISBN 978-0-8018-8502-0; ISBN 0-8018-8502-7. (Gift of the publisher.) [R853 .H8 M39 2007] (18.2; 8.4; 18.3)

Stone, Judy. CONDUCTING CLINICAL RESEARCH: A PRACTICAL GUIDE FOR PHYSICIANS, NURSES, STUDY COORDINATORS, AND INVESTIGATORS. Cumberland, MD: Mountainside MD Press, 2006. 427 p. ISBN 978-0-974917-80-1; ISBN 0-974917-80-X. (Publisher's address: 725 Park Street, Suite 400, zip 21502; tel. 301-777-8089; fax 301-777-8699. Gift of the publisher.) [R853 .C55 S76 2006] (18.2; 9.3.1; 18.3; 18.6; Reference)

18.3 HUMAN EXPERIMENTATION: INFORMED CONSENT

Menikoff, Jerry and Richards, Edward P. WHAT THE DOCTOR DIDN'T SAY: THE HIDDEN TRUTH ABOUT MEDICAL RESEARCH. Oxford/New York: Oxford University Press, 2006. 321 p. ISBN 978-0-19-514797-1; ISBN 0-19-514797-9. [R853 .C55 M46 2006] (18.3; 1.3.9; 8.2; 18.1; 18.5.2; 18.5.6)

18.4 BEHAVIORAL RESEARCH

Mauthner, Melanie; Birch, Maxine; Jessop, Julie; and Miller, Tina, eds. ETHICS IN QUALITATIVE RESEARCH. London/Thousand Oaks, CA: SAGE, 2002. 172 p. ISBN 978-0-7619-7309-6; ISBN 0-7619-7309-5. [H62 .E777 2002] (18.4; 7.1; 10; 18.2; 18.3)

Wolcott, Harry F. SNEAKY KID AND ITS AFTERMATH: ETHICS AND INTIMACY IN FIELDWORK. Walnut Creek, CA: AltaMira Press, 2002. 222 p. ISBN 0-7591-0312-7. [LB45 .W65 2002] (18.4; 1.3.3; 1.3.9; 7.1; 17.1)

18.5.1 RESEARCH ON SPECIAL POPULATIONS (GENERAL)

De La Rosa, Mario R.; Segal, Bernard; and Lopez, Richard, eds. CONDUCTING DRUG ABUSE RESEARCH WITH MINORITY POPULATIONS: ADVANCES AND ISSUES. New York: Haworth Press, 1999. 297 p. ISBN 978-0-7890-0530-4; ISBN 0-7890-0530-1. (Co-published simultaneously as: *Drugs & Society*, 1999; 14(1-2).) [HV5824 .E85 C66 1999] (18.5.1; 7.1; 9.5.4; 9.5.9; 18.3; 18.5.2; 21.7)

Hornblum, Allen M. SENTENCED TO SCIENCE: ONE BLACK MAN'S STORY OF IMPRISONMENT IN AMERICA. University Park: Pennsylvania State University Press, 2007. 207 p. ISBN 978-0-271-03336-5. [R853 .H8 H673 2007] (18.5.1; 18.6)

Tucker, Todd. THE GREAT STARVATION EXPERIMENT: THE HEROIC MEN WHO STARVED SO THAT MILLIONS COULD LIVE. New York: Free Press, 2006. 270 p. ISBN 978-0-7432-7030-4; ISBN 0-7432-7030-4. [RC627 .S7 T83 2006] (18.5.1; 6; 7.1; 18.2; 18.3; 21.2)

Washington, Harriet A. MEDICAL APARTHEID: THE DARK HISTORY OF MEDICAL EXPERIMENTATION ON BLACK AMERICANS FROM COLONIAL TIMES TO THE PRESENT. New York: Doubleday, 2006. 501 p. ISBN 978-0-385-50993-0; ISBN 0-385-50993-6. [R853 .H8 W37 2006] (18.5.1; 1.3.9; 2.2; 9.5.4; 10; 15.5; 16.2; 18.3; 18.5.2; 18.6)

18.5.2 RESEARCH ON NEWBORNS AND MINORS

Alderson, Priscilla and Morrow, Virginia. ETHICS, SOCIAL RESEARCH AND CONSULTING WITH CHILDREN AND YOUNG PEOPLE. Ilford: Barnado's, 2004. 171 p. ISBN 1-904659-07-1. (Revised and updated from: *Listening to Children: Children, Ethics and Social Research*; Barnardo's, 1995. Rose Fitzgerald Kennedy Collection on Women, Infants and Children.) (18.5.2; 8.4; 9.3.1; 18.2; 18.3)

Fraser, Sandy; Lewis, Vicky; Ding, Sharon; Kellett, Mary; and Robinson, Chris, eds. DOING RESEARCH WITH CHILDREN AND YOUNG PEOPLE. London/Thousand Oaks, CA: SAGE, in association with the Open University, 2004. 294 p. ISBN 978-0-7619-4381-5; ISBN 0-7619-4381-1. (Rose Fitzgerald Kennedy Collection on Women, Infants and Children.) [HQ767.85 .D65 2004] (18.5.2; 7.1)

Leadbeater, Bonnie; Banister, Elizabeth; Benoit, Cecilia; Jansson, Mikael; Marshall, Anne; Riecken, Ted, eds. ETHICAL ISSUES IN COMMUNITY-BASED RESEARCH WITH CHILDREN AND YOUTH. Toronto/Buffalo, NY: University of Toronto Press, 2006. 266 p. ISBN 978-0-8020-4882-0; ISBN 0-8020-4882-X. (Rose Fitzgerald Kennedy Collection on Women, Infants and Children.) [HV715 .E85 2006] (18.5.2; 1.3.3; 8.4; 9.5.4; 9.5.7; 18.2; 18.3; 18.4; 20.7)

Lewis, Vicky; Kellett, Mary; Robinson, Chris; Fraser, Sandy and Ding, Sharon, eds. THE REALITY OF RESEARCH WITH CHILDREN AND YOUNG PEOPLE. London: Sage in association with the Open University, 2004. 306 p. ISBN 978-0-7619-4379-2. [HQ767.85 .R435 2004] (18.5.2; 18.2; 21.1)

18.5.4 RESEARCH ON EMBRYOS AND FETUSES

Cohen, Cynthia B. RENEWING THE STUFF OF LIFE: STEM CELLS, ETHICS, AND PUBLIC POLICY. Oxford/New York: Oxford University Press, 2007. 311 p. ISBN 978-0-19-530524-1. [QH588 .S83 C46 2007] (18.5.4; 1.2; 4.4; 9.3.1; 14.5; 15.1; 18.1; 18.6; 21.1)

Fox, Cynthia. CELL OF CELLS: THE GLOBAL RACE TO CAPTURE AND CONTROL THE STEM CELL. New York: W.W. Norton, 2007. 546 p. ISBN 978-0-393-05877-2; ISBN 0-393-05877-8. [QH588 .S83 F69 2007] (18.5.4; 9.5.1; 14.5; 18.5.1; 21.1; 15.1)

German National Ethics Council = Nationaler Ethikrat. THE IMPORT OF HUMAN EMBRYONIC STEM CELLS: OPINION. Berlin: German National Ethics Council, 2002. 55 p. (December 2001. Translation from the German by Philip Slotkin. Gift of DRZE.) (18.5.4; 2.4; 4.4)

Herold, Eve. STEM CELL WARS: INSIDE STORIES FROM THE FRONTLINES. New York: Palgrave Macmillan, 2006. 238 p. ISBN 978-1-4039-7499-0; ISBN 1-4039-7499-3. [QH588 .S83 H47 2006] (18.5.4; 1.3.9; 12.3; 15.1; 21.1)

Jochemsen, Henk, ed. HUMAN STEM CELLS: SOURCE OF HOPE AND OF CONTROVERSY: A STUDY OF THE ETHICS OF HUMAN STEM CELL RESEARCH AND THE PATENTING OF RELATED INVENTIONS. Chicago: The Bioethics Press, 2005. 162 p. ISBN 978-0-9711599-5-2; ISBN 0-9711599-5-5. ("This report is produced by the Prof. dr. G.A. Lindeboom Institute, Centre for Medical Ethics, Ede, The Netherlands and the Business Ethics Center of Jerusalem, Israel. It is endorsed by the Council for Biotechnology Policy of Wilberforce Forum, Washington, DC, the Center for Bioethics and Human Dignity, Bannockburn, IL, and the Christian Legal Society, Washington, DC.") (18.5.4; 1.3.2; 4.4; 9.3.1; 15.1; 15.8)

Kelly, Evelyn B. STEM CELLS. Westport, CT: Greenwood Press, 2007. 203 p. ISBN 978-0-313-33763-5; ISBN 0-313-33763-2. (Health and Medical Issues Today series. ISSN 1558-7592.) [QH588 .S83 K45 2007] (18.5.4; 15.1; 18.5.1)

Solo, Pam and Pressburg, Gail. THE PROMISE AND POLITICS OF STEM CELL RESEARCH. Westport, CT: Praeger, 2007. 171 p. ISBN 978-0-275-99038-1; ISBN 0-275-99038-9. (Foreword by Mary Tyler Moore. Published in cooperation with the Civil Society Institute.) [QH588 .S83 S65 2007] (18.5.4; 1.2; 1.3.5; 5.3; 15.1; 21.1)

18.5.5 RESEARCH ON PRISONERS

Institute of Medicine (United States). Board on Health Sciences Policy. Committee on Ethical Considerations for Revisions to DHHS Regulations for Protection of Prisoners Involved in Research. ETHICAL CONSIDERATIONS FOR RESEARCH INVOLVING PRISONERS. Washington, DC: National Academies Press, 2007. 265 p. ISBN 978-0-309-10119-6; ISBN 0-309-10119-0. (Edited by Lawrence O. Gostin, Cori Vanchieri and Andrew Pope. Also available on the Web: http://www.nap.edu.) [R853 .H8 E8253 2007] (18.5.5; 1.3.5; 18.2; 18.3; Reference)

18.5.6 RESEARCH ON MENTALLY ILL AND DISABLED PERSONS

Everitt, Brian S. and Wessely, Simon. CLINICAL TRIALS IN PSYCHIATRY. London/New York: Oxford University Press, 2004. 189 p. ISBN 0-19-852642-3. [RC454 .E866 2004] (18.5.6; 7.1; 17.1; 18.2)

18.6 SOCIAL CONTROL OF HUMAN EXPERIMENTATION

Epstein, Steven. INCLUSION: THE POLITICS OF DIFFERENCE IN MEDICAL RESEARCH. Chicago: University of Chicago Press, 2007. 413 p. ISBN 978-0-226-21309-5; ISBN 0-226-21309-9. (Chicago Studies in Practices of Meaning series.) [R853 .S64 E67 2007] (18.6; 10; 18.2; 18.5.1; 18.5.3)

19.1 ARTIFICIAL AND TRANSPLANTED ORGANS OR TISSUES (GENERAL)

Seder, Robert. TO THE MARROW. Fort Lee, NJ: CavanKerry Press, 2005. 177 p. ISBN 978-0-9723045-6-6; ISBN 0-9723045-6-8. [RC280 .L9 S43 2005] (19.1; 9.5.1; Biography)

19.2 ARTIFICIAL AND TRANSPLANTED HEARTS

Wailoo, Keith; Livingston, Julie; and Guarnaccia, Peter, eds. A DEATH RETOLD: JESICA SANTILLAN, THE BUNGLED TRANSPLANT, AND PARADOXES OF MEDICAL CITIZENSHIP. Chapel Hill: University of North Carolina Press, 2006. 378 p. ISBN 978-0-8078-5773-1; ISBN 0-8078-5773-4. (Studies in Social Medicine series.) [RD598.35 .T7 D43 2006] (19.2; 8.3.1; 9.5.4; 19.4; 19.6; 21.1)

19.5 DONATION/ PROCUREMENT OF ORGANS/TISSUES

Healy, Kieran. LAST BEST GIFTS: ALTRUISM AND THE MARKET FOR HUMAN BLOOD AND ORGANS. Chicago: University of Chicago Press, 2006. 193 p. ISBN 978-0-226-32237-7; ISBN 0-226-32237-8. [RD129.5 .H43 2006] (19.5; 9.3.1)

20.1 DEATH AND DYING (GENERAL)

Carol, Anne. LES MÉDECINS ET LA MORT: XIXe-XXe SIÈCLE. Paris: Aubier, 2004. 335 p. ISBN 2-70-072331-7. (Collection historique series.) [R505 .C35 2004] (20.1; 2.2; 4.4; 7.1; 9.1; 20.2.1; 20.5.1)

Crawford, Robert. CAN WE EVER KILL? London: Darton, Longman & Todd, 2000. 225 p. ISBN 978-0-232-52358-4; ISBN 0-232-52358-4. (Previous edition published by: Fount, London, 1991.) [BJ1409.5 .C73 2000] (20.1; 1.2; 4.4; 12.3; 20.5.1; 20.6; 20.7; 21.2)

Dickinson, George E. and Leming, Michael R., eds. ANNUAL EDITIONS: DYING, DEATH, AND BEREAVEMENT 2005/2006. New York: McGraw-Hill, 2005. 221 p. ISBN 0-07-310204-0. (Eighth edition. Dying, Death, and Bereavement series. ISSN 1096-4223.) (20.1; 1.2; 8.1; 20.3.1; 20.4.1; 20.5.1; 20.7)

Garces-Foley, Kathleen, ed. DEATH AND RELIGION IN A CHANGING WORLD. Armonk, NY: M.E. Sharpe, 2006. 322 p. ISBN 978-0-7656-1221-2; ISBN 0-7656-1221-6. [BL504 .D363 2006] (20.1; 1.2; 1.3.7; 20.3.1)

Jewish Theological Library. FROM THIS WORLD TO THE NEXT: JEWISH APPROACHES TO ILLNESS, DEATH & THE AFTERLIFE. New York: Library, Jewish Theological Seminary of America, 1999. 120 p. ("This publication is issued in conjunction with the exhibition *From This World to the Next: Jewish Approaches to Illness, Death and the Afterlife*, held at the Library of The Jewish Theological Seminary of America from December 12, 1999 to April 12, 2000." Gift of Max M. and Marjorie B. Kampelman.) [BM712 .F78 1999] (20.1; 1.2)

Kellehear, Allan. A SOCIAL HISTORY OF DYING. Cambridge/New York: Cambridge University Press, 2007. 297 p. ISBN 978-0-521-69429-2; ISBN 0-521-69429-9. [HQ1073 .K44 2007] (20.1; 7.1)

MacDonald, Helen. HUMAN REMAINS: DISSECTION AND ITS HISTORIES. New Haven, CT: Yale University Press, 2006. 220 p. ISBN 0-300-11699-3. (First published in 2005 by Melbourne University Press, Australia. Gift of the publisher.) [QM33.4 .M33 2006] (20.1; 7.1; 7.2; 9.5.10; 18.5.5; 20.3.2)

Park, Katharine. SECRETS OF WOMEN: GENDER, GENERATION, AND THE ORIGINS OF HUMAN DISSECTION. New York: Zone Books, 2006. 419 p. ISBN 978-1-890951-67-2; ISBN 1-890951-67-6. [QM33.4 .P37 2006] (20.1; 4.4; 7.1; 9.5.5; 10)

Shannon, Joyce Brennfleck, ed. DEATH AND DYING SOURCEBOOK: BASIC CONSUMER HEALTH INFORMATION ABOUT END-OF-LIFE CARE AND RELATED PERSPECTIVES AND ETHICAL ISSUES, INCLUDING END-OF-LIFE SYMPTOMS AND TREATMENTS, PAIN MANAGEMENT, QUALITY-OF-LIFE CONCERNS, THE USE OF LIFE SUPPORT, PATIENTS' RIGHTS AND PRIVACY ISSUES, ADVANCE DIRECTIVES, PHYSICIAN-ASSISTED SUICIDE, CAREGIVING, ORGAN AND TISSUE DONATION, AUTOPSIES, FUNERAL ARRANGEMENTS, AND GRIEF, ALONG WITH STATISTICAL DATA, INFORMATION ABOUT THE LEADING CAUSES OF DEATH, A GLOSSARY, AND DIRECTORIES OF SUPPORT GROUPS AND OTHER RESOURCES. Detroit, MI: Omnigraphics, 2006. 653 p. ISBN 0-7808-0871-1. (Second edition. Health Reference Series. Gift of the publisher.) [R726.8 .D3785 2006] (20.1; 4.4; 8.1; 9.4; 9.5.1; 19.5; 20.4.1; 20.4.2; 20.5.1; 20.5.4; 20.7; Reference)

20.2.1 DEFINITION OR DETERMINATION OF DEATH (GENERAL)

Wiedebach, Hartwig. HIRNTOD ALS WERTVERHALT: MEDIZINETHISCHE BAUSTEINE AUS JONAS COHNS WERTWISSENSCHAFT UND MAIMONIDES' THEOLOGIE. Münster: Lit, 2003. 84 p. ISBN 3-8258-7098-7. (Naturwissenschaft, Philosophie, Geschichte series; Bd. 20. Gift of Max M. and Marjorie B. Kampelman.) (20.2.1; 2.1; 4.4)

20.3.2 ATTITUDES TOWARD DEATH: HEALTH PERSONNEL

Chen, Pauline W. FINAL EXAM: A SURGEON'S REFLECTIONS ON MORTALITY. New York: Alfred A. Knopf, 2007. 267 p. ISBN 978-0-307-26353-7; ISBN 0-307-26353-3. [RD27.35 .C47 A3 2007] (20.3.2; 4.1.2; 20.4.1; Biography)

Weisse, Allen B. LESSONS IN MORTALITY: DOCTORS AND PATIENTS STRUGGLING TOGETHER. Columbia: University of Missouri Press, 2006. 182 p. ISBN 978-0-8262-1666-3; ISBN 0-8262-1666-8. [R726.5 .W454 2006] (20.3.2; 7.1; 8.1)

20.4.1 CARE OF THE DYING PATIENT (GENERAL)

Brenner, Daniel S.; Blanchard, Tsvi; Fins, Joseph J.; and Hirschfield, Bradley. EMBRACING LIFE & FACING DEATH: A JEWISH GUIDE TO PALLIATIVE CARE. New York: CLAL—National Jewish Center for Learning and Leadership, 2002. 96 p. ISBN 978-0-9633329-0-5; ISBN 0-9633329-0-2. (Preface by Senator Joseph Lieberman. Gift of Max M. and Marjorie B. Kampelman.) (20.4.1; 1.2; 4.4; 20.5.1; 20.5.4)

Carlson, Melissa Diane Aldridge. THE IMPACT OF FOR-PROFIT OWNERSHIP ON HOSPICE CARE. Ann Arbor, MI: ProQuest Information and Learning/UMI, 2005. 171 p. (Publication No. AAT-3194630. Dissertation, (Ph.D.)—Yale University, 2005.) (20.4.1; 7.1; 9.3.1; 9.4)

Craig, Janet B. RACIAL AND SOCIOECONOMIC DIFFERENCES IN FAMILY PERCEIVED BARRIERS TO SATISFACTION WITH

Numbers in ( ) = NRCBL Classification Numbers

END-OF-LIFE CARE: AN EXPLORATORY STUDY. Ann Arbor, MI: ProQuest Information and Learning/UMI, 2003. 141 p. (Publication No. AAT-3073662. Dissertation, (Ph.D. of Health Administration)—Medical University of South Carolina, College of Health Professions, 2003.) (20.4.1; 7.1)

Hurwitz, Peter Joel; Picard, Jacques; and Steinberg, Avraham, eds. JEWISH ETHICS AND THE CARE OF END-OF-LIFE PATIENTS: A COLLECTION OF RABBINICAL, BIOETHICS, PHILOSOPHICAL, AND JURISTIC OPINIONS. Jersey City, NJ: KTAV, in association with the Institute for Jewish Studies, University of Basel, Basel, Switzerland, 2006. 254 p. ISBN 0-88125-921-7. (Gift of Max M. and Marjorie B. Kampelman.) [BM635.4 .J384 2006] (20.4.1; 1.2; 4.4; 8.1; 20.3.1; 20.4.2; 20.5.1; 20.7)

Lewis, Milton J. MEDICINE AND CARE OF THE DYING: A MODERN HISTORY. Oxford/New York: Oxford University Press, 2007. 277 p. ISBN 978-0-19-517548-6; ISBN 0-19-517548-4. [R726.8 .L484 2007] (20.4.1; 1.2; 2.2; 4.4; 20.5.1)

Pharaoh, Gill. CARING FOR THE DYING AT HOME: A PRACTICAL GUIDE. London: Free Association Books, 2004. 168 p. ISBN 1-85343-739-5. (Foreword by Tony Benn.) [R726.8 .P43 2004] (20.4.1; 20.4.2; 20.5.1; 20.7)

Puchalski, Christina M. A TIME FOR LISTENING AND CARING: SPIRITUALITY AND THE CARE OF THE CHRONICALLY ILL AND DYING. Oxford/New York: Oxford University Press, 2006. 458 p. ISBN 978-0-19-514682-0; ISBN 0-19-514682-4. [R726.8 .T56 2006] (20.4.1; 1.2; 17.2; 20.3.1; 20.4.2; 20.5.4)

Rowe, Charles. THE DIOCESAN BISHOP'S PASTORAL RESPONSIBILITY FOR THE ETHICALLY CORRECT CARE OF CRITICALLY ILL PATIENTS AT CATHOLIC HEALTH CARE FACILITIES IN THE UNITED STATES. Romae: [s.n.], 2006. 375 p. (Dissertation, [Doctoratum in Theologie Morali]—Pontificia Universitas Lateranensis Academia Alfonsiana, Institutum Superius Theologiae Moralis, 2006. Gift of the author.) [RC86.95 .R69 2006] (20.4.1; 1.2; 1.3.2; 2.2; 4.1.2; 8.1; 9.1; 20.5.1)

Smith, David H. PARTNERSHIP WITH THE DYING: WHERE MEDICINE AND MINISTRY SHOULD MEET. Lanham, MD: Rowman & Littlefield, 2005. 135 p. ISBN 0-7425-4466-4. [BV4460.6 .S65 2005] (20.4.1; 1.2; 20.3.1)

Woods, Simon. DEATH'S DOMINION: ETHICS AT THE END OF LIFE. Maidenhead, England/New York: Open University Press, 2007. 178 p. ISBN 978-0-335-21160-9; ISBN 0-335-21160-7. (Facing Death series.) [R726 .W657 2007] (20.4.1; 1.1; 2.1; 2.2; 4.4; 20.5.1; 20.7)

20.4.2 CARE OF THE DYING CHILD

Bauman, Renea A. THE LIVED EXPERIENCE OF NURSES PROVIDING FUTILE CARE TO CRITICALLY ILL NEWBORNS AND INFANTS. Ann Arbor, MI: ProQuest Learning and Information/UMI, 2003. 71 p. (Publication No. AAT-1412197. Thesis, (M.S.)—Southern Connecticut State University,

2003. Rose Fitzgerald Kennedy Collection on Women, Infants and Children.) (20.4.2; 4.1.3; 7.1; 17.1; 20.5.2)

Carter, Brian S. and Levetown, Marcia, eds. PALLIATIVE CARE FOR INFANTS, CHILDREN, AND ADOLESCENTS: A PRACTICAL HANDBOOK. Baltimore, MD: Johns Hopkins University Press, 2004. 399 p. ISBN 978-0-8018-8005-X; ISBN 0-8018-8005-X. (Foreword by Kathleen M. Foley. Rose Fitzgerald Kennedy Collection on Women, Infants and Children.) [RJ249 .P356 2004] (20.4.2; 1.2; 7.1; 9.5.6; 15.1; 17.1; 20.5.2; Reference)

Goldman, Ann; Hain, Richard; and Liben, Stephen, eds. OXFORD TEXTBOOK OF PALLIATIVE CARE FOR CHILDREN. Oxford/New York: Oxford University Press, 2006. 661 p. ISBN 978-0-19-852653-7; ISBN 0-19-852653-9. (Oxford Textbook Series.) [RJ249 .O94 2006] (20.4.2; 1.2; 4.1.1; 4.4; 8.1; 9.5.6; 9.5.7; 9.7; 9.8; 17.1; 20.3.2; 20.3.3; 21.1)

McKelvey, Robert S. WHEN A CHILD DIES: HOW PEDIATRIC PHYSICIANS AND NURSES COPE. Seattle: University of Washington Press, 2006. 319 p. ISBN 978-0-295-98653-1; ISBN 0-295-98653-0. [R726.8 .M355 2006] (20.4.2; 8.1; 17.1; 20.3.2)

20.5.1 PROLONGATION OF LIFE AND EUTHANASIA (GENERAL)

Blasius, Dirk. EUTHANASIE IN HADAMAR: DIE NATIONALSOZIALISTISCHE VERNICHTUNGSPOLITIK IN HESSISCHEN ANSTALTEN: BEGLEITBAND: EINE AUSSTELLUNG DES LANDESWOHLFAHRTSVERBANDES HESSEN. Kassel: Landeswohlfahrtsverbandes Hessen, 1991. 259 p. ISBN 3-89203-015-4. (Historische Schriftenreihe des Landeswohlfahrtsverbandes Hessen, Kataloge series; Bd. 1. Max M. and Marjorie B. Kampelman Collection on Jewish Ethics.) [R726 .E7918 1991] (20.5.1; 1.3.5; 15.5; 17.1; 21.4)

Bryant, Michael S. CONFRONTING THE "GOOD DEATH": NAZI EUTHANASIA ON TRIAL, 1945-1953. Boulder: University Press of Colorado, 2005. 269 p. ISBN 0-87081-809-0. [D804.5 .H35 B79 2005] (20.5.1; 1.3.5; 2.2; 4.1.2; 17.1; 21.4)

Cassell, Joan. LIFE AND DEATH IN INTENSIVE CARE. Philadelphia: Temple University Press, 2005. 233 p. ISBN 1-59213-336-3. [RD49 .C37 2005] (20.5.1; 4.4; 8.1; 20.3.2; 20.4.1; 21.1)

Csikai, Ellen L. and Jones, Barbara, ed. TEACHING RESOURCES FOR END OF LIFE AND PALLIATIVE CARE COURSES. Chicago: Lyceum Books, 2007. 324 p. ISBN 978-1-933478-10-4. (Gift of the publisher.) [HV687 .T43 2007] (20.5.1; 1.2; 1.3.10; 2.3; 7.2; 9.5.6; 17.1; 20.1; 20.4.2; 21.7; Reference)

Gestrich, Christof and Neugebauer, Johannes, eds. DER WERT MENSCHLICHEN LEBENS: MEDIZINISCHE ETHIK BEI KARL BONHOEFFER UND DIETRICH BONHOEFFER. Berlin: Wichern-Verlag, 2006. 165 p. ISBN 978-3-88981-207-0; ISBN 3-88981-207-4. ("Papers from the Tenth Dietrich Bonhoeffer Vorlesung, held April 29-30, 2005, in Berlin, organized by the Theologische Fakultät of

Humboldt-Universität zu Berlin and the Union Theological Seminary in New York." Gift of LeRoy Walters.) [R726 .D524 2005] (20.5.1; 1.2; 1.3.5; 2.2; 11.3; 15.5; 21.4)

Greve, Michael. DIE ORGANISIERTE VERNICHTUNG "LEBENSUNWERTEN LEBENS" IM RAHMEN DER "AKTION T4": DARGESTELLT AM BEISPIEL DES WIRKENS UND DER STRAFRECHTLICHEN VERFOLGUNG AUSGEWÄHLTER NS-TÖTUNGSÄRZTE. Herbolzheim: Centaurus Verlag, 2006. 145 p. ISBN 978-3-8255-0123-5; ISBN 3-8255-0123-X. (Second edition. Reihe Geschichtswissenschaft series; Bd. 43. ISSN 0177-2767.) [R726 .G75 1998] (20.5.1; 1.3.5; 2.2; 4.1.2; 21.2; 21.4; Biography)

Hamel, Ronald P. and Walter, James J., eds. ARTIFICIAL NUTRITION AND HYDRATION AND THE PERMANENTLY UNCONSCIOUS PATIENT: THE CATHOLIC DEBATE. Washington, DC: Georgetown University Press, 2007. 294 p. ISBN 978-1-58901-178-6; ISBN 1-58901-178-3. (Gift of the publisher.) [RB150 .C6 A78 2007] (20.5.1; 1.2)

Hübener, Kristina, ed. and Heinze, Martin. BRANDENBURGISCHE HEIL- UND PFLEGEANSTALTEN IN DER NS-ZEIT. Berlin: Be.bra Wissenschaft, 2002. 480 p. ISBN 978-3-89809-301-9; ISBN 3-89809-301-8. (Schriftenreihe zur Medizin-Geschichte des Landes Brandenburg series; Bd. 3. ISSN 1611-8456. Max M. and Marjorie B. Kampelman Collection on Jewish Ethics.) [R726 .B748 2002] (20.5.1; 1.3.5; 2.2; 11.3; 15.5; 17.1)

Israël, Lucien and Lévy, Elisabeth. LES DANGERS DE L'EUTHANASIE. Paris: Syrtes, 2002. 153 p. ISBN 2-84545-051-6. (Preface by d'Alain Besançon de l'Institut.) [R726 .I848 2002] (20.5.1; 7.1; 9.3.1)

Kaul, Friedrich Karl. DIE PSYCHIATRIE IM STRUDEL DER "EUTHANASIE": EIN BERICHT ÜBER DIE ERSTE INDUSTRIEMÄßIG DURCHGEFÜHRTE MORDAKTION DES NAZIREGIMES. Köln/Frankfurt am Main: Europäische Verlagsanstalt, 1979. 234 p. ISBN 978-3-4342-5107-1; ISBN 3-4342-5107-3. [D804 .G4 K37 1979] (20.5.1; 1.3.5; 2.2; 15.5; 21.4)

Legros, Bérengère. L'EUTHANASIE ET LE DROIT: ÉTAT DES LIEUX SUR UN SUJET MÉDIATISÉ. Bordeaux: Les Etudes hospitalières, 2006. 201 p. ISBN 978-2-84874-067-6; ISBN 2-84874-067-1. (Second edition. Essentiel series. ISSN 1631-9702.) (20.5.1; 2.2)

Schulze, Dietmar. DIE LANDESANSTALT NEURUPPIN IN DER NS-ZEIT. Berlin-Brandenburg: Be.bra Wissenschaft, 2004. 222 p. ISBN 978-3-937233-12-3; ISBN 3-937233-12-1. (Schriftenreihe zur Medizin-Geschichte des Landes Brandenburg series; Bd. 8. ISSN 1611-8456.) [RC450 .G32 N48 2004] (20.5.1; 1.3.5; 2.2; 15.5; 17.1; 21.4)

Shannon, Thomas A. and Faso, Charles N. LET THEM GO FREE: A GUIDE FOR WITHDRAWING LIFE SUPPORT. Washington, DC: Georgetown University Press, 2007. 61 p. ISBN 978-1-58901-140-3; ISBN 1-58901-140-6. (On front cover: *With a Family Prayer Service.*) [RC86.7 .S53 2007] (20.5.1; 1.2; 20.4.1)

Thiele, Felix, ed. AKTIVE UND PASSIVE STERBEHILFE: MEDIZINISCHE, RECHTSWISSENSCHAFTLICHE UND PHILOSOPHISCHE ASPEKTE. München: Wilhelm Fink, 2005. 285 p. ISBN 978-3-7705-3838-6; ISBN 3-7705-3838-2. (Neuzeit & Gegenwart: Philosophie in Wissenschaft und Gesellschaft series.) (20.5.1; 4.4; 20.4.1; 21.1)

Tolmein, Oliver. SELBSTBESTIMMUNGSRECHT UND EINWILLIGUNGSFÄHIGKEIT: DER ABBRUCH DER KÜNSTLICHEN ERNÄHRUNG BEI PATIENTEN IM VEGETATIVE STATE IN RECHTSVERGLEICHENDER SICHT: DER KEMPTENER FALL UND DIE VERFAHREN CRUZAN UND BLAND. Frankfurt am Main: Mabuse-Verlag, 2004. 311 p. ISBN 3-935964-73-0. (Mabuse-Verlag Wissenschaft series; Bd. 81.) [K3611 .E95 T65 2004] (20.5.1; 1.3.5; 8.3.1; 8.3.3; 20.2.1; 20.4.1)

Tuchel, Johannes. "KEIN RECHT AUF LEBEN": BEITRÄGE UND DOKUMENTE ZUR ENTRECHTUNG UND VERNICHTUNG "LEBENSUNWERTEN LEBENS" IM NATIONALSOZIALISMUS. Berlin: Wissenschaftlicher Autoren-Verlag, 1984. 122 p. ISBN 978-3-88840-221-0; ISBN 3-88840-221-2. [KK8364 .K45 1984] (20.5.1; 1.3.5; 2.2; 15.5; 21.4)

Tulloch, Gail. EUTHANASIA—CHOICE AND DEATH. Edinburgh: Edinburgh University Press, 2005. 158 p. ISBN 0-7486-2247-0. (Contemporary Ethical Debates series.) [R726 .T85 2005] (20.5.1; 9.4; 20.2.1; 20.5.2; 21.7)

20.5.2 ALLOWING MINORS TO DIE

Beddies, Thomas and Hübener, Kristina, eds. KINDER IN DER NS-PSYCHIATRIE. Berlin-Brandenburg: Be.bra Wissenschaft, 2004. 205 p. ISBN 3-937233-14-8. (Schriftenreihe zur Medizin-Geschichte des Landes Brandenburg series; Bd. 10. ISSN 1611-8456.) [R726 .K497 2004] (20.5.2; 1.3.5; 2.2; 9.5.7; 15.5; 17.1; 21.4)

Bhatnagar, Rashmi Dube; Dube, Renu; and Dube, Reena. FEMALE INFANTICIDE IN INDIA: A FEMINIST CULTURAL HISTORY. Albany: State University New York Press, 2005. 320 p. ISBN 0-7914-6328-1. [HV6541 .I5 B53 2005] (20.5.2; 7.1; 9.5.5; 9.5.7; 10; 21.7)

Kramar, Kirsten Johnson. UNWILLING MOTHERS, UNWANTED BABIES: INFANTICIDE IN CANADA. Vancouver: UBC Press, 2005. 227 p. ISBN 0-7748-1176-5. (Law and Society Series.) [KE8910 .K73 2005] (20.5.2; 1.3.5; 7.1; 9.5.5; 9.5.7)

Lantos, John D. and Meadow, William L. NEONATAL BIOETHICS: THE MORAL CHALLENGES OF MEDICAL INNOVATION. Baltimore, MD: Johns Hopkins University Press, 2006. 177 p. ISBN 0-8018-8344-X. (Gift of the publisher.) [RJ253.5 .L37 2006] (20.5.2; 2.2; 5.3; 9.1; 9.3.1; 9.5.7; 20.4.2)

Miller, Geoffrey. EXTREME PREMATURITY: PRACTICES, BIOETHICS, AND THE LAW. Cambridge/New York: Cambridge University Press, 2007. 232 p. ISBN 978-0-521-68053-0; ISBN 0-521-68053-0. [RJ250 .M55 2007] (20.5.2; 1.1; 2.1; 4.4; 20.4.2)

Numbers in ( ) = NRCBL Classification Numbers

Winner, Brenda. TEN PERFECT FINGERS. Fayetteville, NC: Old Mountain Press, 1999. 131 p. ISBN 978-1-884778-64-3; ISBN 1-884778-64-X. ("Brenda Winner tells the story of her anencephalic daughter, Jarren, and how Brenda helped establish the first medical protocol in America to utilize anencephalic infants as organ donors as a tribute to her daughter's life and to help other parents.") [RG629 .B73 W55 1999] (20.5.2; 19.5; Biography)

20.5.4 LIVING WILLS/ ADVANCE DIRECTIVES

Mirarchi, Ferdinando L. WHAT'S THE PATIENT'S CODE STATUS? Erie, PA: Patient Advanced Educational Resource (P.A.E.R.), 2000. 51 p. ISBN 978-1-889014-58-6; ISBN 1-889014-58-3. ("A Patient and Family Reference Guide to Components of Advanced Directives and Code of Status Designation." Publisher's address: http://www.paer.org.) [R726.2 .M57 2000] (20.5.4)

20.6 CAPITAL PUNISHMENT

Henderson, Harry. CAPITAL PUNISHMENT. New York: Facts on File, 2006. 316 p. ISBN 978-0-8160-5708-5; ISBN 0-8160-5708-7. (Third edition. Library in a Book series.) [KF9227 .C2 F53 2006] (20.6; 1.3.5; Reference)

20.7 SUICIDE/ ASSISTED SUICIDE

Lewis, Penney. ASSISTED DYING AND LEGAL CHANGE. Oxford/New York: Oxford University Press, 2007. 217 p. ISBN 978-0-19-921287-3. [K5178 .L49 2007] (20.7; 1.1; 8.1; 20.5.1)

Long, Robert Emmet, ed. SUICIDE. New York: H.W. Wilson, 1995. 162 p. ISBN 0-8242-0869-2. (The Reference Shelf series; Vol. 67, No. 2.) [HV6545 .S815 1995] (20.7; 8.1; 9.5.7; 10; 20.5.1)

Mitchell, John B. UNDERSTANDING ASSISTED SUICIDE: NINE ISSUES TO CONSIDER. Ann Arbor: University of Michigan Press, 2007. 221 p. ISBN 978-0-472-06996-5; ISBN 0-472-06996-9. [R726 .M565 2007] (20.7; 1.1; 1.2; 4.4; 20.5.1)

21.1 INTERNATIONAL AND POLITICAL DIMENSIONS OF BIOLOGY AND MEDICINE (GENERAL)

Gevers, J.K.M.; Hondius, E.H.; Hubben, J.H., eds. HEALTH LAW, HUMAN RIGHTS AND THE BIOMEDICINE CONVENTION: ESSAYS IN HONOUR OF HENRIETTE ROSCAM ABBING. Leiden/Boston: Martinus Nijhoff Publishers, 2005. 271 p. ISBN 90-04-14822-1. (International Studies in Human Rights series; Vol. 85.) [KJC6227 .H43 2005] (21.1; 5.3; 8.4; 9.2; 9.3.1; 9.4; 9.8; 15.1; 17.1; 18.1; 18.5.4; 19.1; 20.5.1)

Gruskin, Sofia; Grodin, Michael A.; Annas, George J.; and Marks, Stephen P., eds. PERSPECTIVES ON HEALTH AND HUMAN RIGHTS. New York: Routledge, 2005. 649 p. ISBN 0-415-94807-X. ("Follow-up/companion volume to:

Health and Human Rights, 1999.") [RA427.25 .P47 2005] (21.1; 7.2; 9.1; 9.2; 9.5.1; 9.5.5; 9.5.6; 9.5.7; 9.7; 10; 14.1; 14.5; 15.1; 20.6)

Marks, Stephen P., ed. HEALTH AND HUMAN RIGHTS: BASIC INTERNATIONAL DOCUMENTS. Cambridge, MA: Harvard University, François-Xavier Bagnoud Center for Health and Human Rights; Distributed by: Harvard University Press, 2004. 318 p. ISBN 978-0-674-01809-9; ISBN 0-674-01809-5. (Harvard Series on Health and Human Rights.) [RA418 .H3872 2004] (21.1; 4.4; 6; 7.1; 9.1; 9.2; 9.5.5; 9.5.6; 9.5.7; 14.1; 15.1; 18.2; 21.4)

O'Neil, Jr., Edward. AWAKENING HIPPOCRATES: A PRIMER ON HEALTH, POVERTY, AND GLOBAL SERVICE. Chicago: American Medical Association, 2006. 502 p. ISBN 978-1-57947-772-1; ISBN 1-57947-772-0. ("A comprehensive overview of the current state of world poverty and health, directed to the health care provider interested in volunteering abroad.") [HC79 .P6 O55 2006] (21.1; 1.2; 1.3.6; 9.1; 9.5.6; 9.5.10; Biography)

Ponnuru, Ramesh. THE PARTY OF DEATH: THE DEMOCRATS, THE MEDIA, THE COURTS, AND THE DISREGARD FOR HUMAN LIFE. Washington, DC: Regnery Publishing; Distributed to the trade by: Lanham, MD: National Book Network, 2006. 303 p. ISBN 1-59698-004-4. [HQ767.5 .U5 P66 2006] (21.1; 1.3.5; 1.3.7; 4.4; 12.3; 15.1; 18.5.4; 20.5.1; 20.5.2)

Wagner, Teresa and Carbone, Leslie, eds. FIFTY YEARS AFTER THE DECLARATION: THE UNITED NATIONS' RECORD ON HUMAN RIGHTS. Lanham, MD: University Press of America [and] Washington, DC: Family Research Council, 2001. 162 p. ISBN 0-7618-1842-1. [K3240 .F54 2001] (21.1; 1.2; 1.3.5; 9.2; 9.5.7; 9.7; 10; 13.3; 14.5; 20.5.1)

21.3 CHEMICAL AND BIOLOGICAL WEAPONS

Geißler, Erhard. BIOLOGISCHE WAFFEN – NICHT IN HITLERS ARSENALEN: BIOLOGISCHE UND TOXIN-KAMPFMITTEL IN DEUTSCHLAND VON 1915 BIS 1945. Münster, Germany: Lit-Verlag, 1999. 905 p. ISBN 3-8258-2955-1; ISBN 3-8258-2955-3. (Second edition. Studien zur Friedensforschung series; Bd. 13.) [UG447.8 .G447 1999] (21.3; 1.3.5; 5.3; 21.2)

Khardori, Nancy, ed. BIOTERRORISM PREPAREDNESS: MEDICINE, PUBLIC HEALTH, POLICY. Weinheim: Wiley-VCH, 2006. 261 p. ISBN 978-3-527-31235-1; ISBN 3-527-31235-8. (Gift of the publisher.) (21.3; 1.3.6; 9.1; 9.5.7; 21.1)

National Research Council (United States). Committee on Advances in Technology and the Prevention of Their Application to Next Generation Biowarfare Threats [and] Institute of Medicine (United States). Board on Global Health. GLOBALIZATION, BIOSECURITY, AND THE FUTURE OF THE LIFE SCIENCES. Washington, DC: National Academies Press, 2006. 299 p. ISBN 978-0-309-10032-8; ISBN 0-309-10032-1. (Also available on the Web at: http://www.nap.edu.) [HV6433.3 .G56 2006] (21.3; 15.1; 21.2)

Tucker, Jonathan B. WAR OF NERVES: CHEMICAL WARFARE FROM WORLD WAR I TO AL-QAEDA. New York: Pantheon, 2006. 479 p. ISBN 0-375-42229-3. [UG447 .T83 2006] (21.3; 21.2)

Vilensky, Joel A. DEW OF DEATH: THE STORY OF LEWISITE, AMERICA'S WORLD WAR I WEAPON OF MASS DESTRUCTION. Bloomington: Indiana University Press, 2005. 213 p. ISBN 978-0-253-34612-4; ISBN 0-253-34612-6. [UG447.5 .L48 V35 2005] (21.3; 16.1; 21.2)

21.4 TORTURE AND GENOCIDE

Bagaric, Mirko and Clarke, Julie. TORTURE: WHEN THE UN-THINKABLE IS MORALLY PERMISSIBLE. Albany: State University of New York Press, 2007. 114 p. ISBN 978-0-7914-7154-8. [HV8593 .B35 2007] (21.4; 1.3.5)

Bloxham, Donald and Kushner, Tony. THE HOLOCAUST: CRITICAL HISTORICAL APPROACHES. Manchester/New York: Manchester University Press; Distributed in the USA by: New York: Palgrave, 2005. 238 p. ISBN 978-0-7190-3779-5; ISBN 0-7190-3779-4. (Gift of Max M. and Marjorie B. Kampelman.) [D804.348 .B56 2005] (21.4)

Brown-Fleming, Suzanne. THE HOLOCAUST AND CATHO-LIC CONSCIENCE: CARDINAL ALOISIUS MUENCH AND THE GUILT QUESTION IN GERMANY. Notre Dame, IN: University of Notre Dame Press, published in association with the United States Holocaust Memorial Museum, 2006. 240 p. ISBN 978-0-268-02187-0; ISBN 0-268-02187-2. (Gift of Max M. and Marjorie B. Kampelman.) [BX4705 .M755 B76 2006] (21.4; 1.2; 21.2)

Cesarani, David. BECOMING EICHMANN: RETHINKING THE LIFE, CRIMES, AND TRIAL OF A "DESK MURDERER". Cambridge, MA: Da Capo Press, 2006. 458 p. ISBN 978-0-306-81476-1; ISBN 0-306-81476-5. (First published in Great Britain as: *Eichmann: His Life and Crimes*; William Heinemann, 2006. Gift of Max M. and Marjorie B. Kampelman.) [DD247 .E5 C47 2006] (21.4; 1.3.5; 15.5; Biography)

Deák, István. ESSAYS ON HITLER'S EUROPE. Lincoln: University of Nebraska Press, 2001. 222 p. ISBN 978-0-8032-6630-8; ISBN 0-8032-6630-8. (Gift of Max M. and Marjorie B. Kampelman.) [DS135 .E83 D43 2001] (21.4; 1.3.5; 15.5)

Dean, Martin. COLLABORATION IN THE HOLOCAUST: CRIMES OF THE LOCAL POLICE IN BELORUSSIA AND UKRAINE, 1941-1944. New York: St. Martin's Press, published in association with the United States Holocaust Memorial Museum, 2000. 241 p. ISBN 978-0-312-22056-3; ISBN 0-312-22056-1. (Gift of Max M. and Marjorie B. Kampelman.) [DS135 .B38 D43 2000] (21.4; 1.3.5)

Dickinson, John K. GERMAN AND JEW: THE LIFE AND DEATH OF SIGMUND STEIN. Chicago: Ivan R. Dee, published in association with the United States Holocaust Memorial Museum, 2001. 345 p. ISBN 978-1-56663-404-5; ISBN 1-56663-404-0. (Introduction by Raul Hilberg. Originally published by: Quadrangle Books, Chicago, 1967. Gift of

Max M. and Marjorie B. Kampelman.) [DS135 .G33 D45 2001] (21.4; Biography)

Eisenstein, Bernice. I WAS A CHILD OF HOLOCAUST SURVI-VORS. New York: Riverhead Books, 2006. 187 p. ISBN 978-1-59448-918-1; ISBN 1-59448-918-1. (Gift of Max M. and Marjorie B. Kampelman.) [F1059.5 .T689 J53 2006] (21.4; Biography)

Faust, Anselm, ed. and Zimmermann, Michael. VERFOLGUNG UND WIDERSTAND IM RHEINLAND UND IN WESTFALEN, 1933-1945. Köln: W. Kohlhammer, 1992. 254 p. ISBN 978-3-17-011586-6; ISBN 3-17-011586-3. (Schriften zur politischen Landeskunde Nordrhein-Westfalens series; Bd. 7.) [DD801 .P48 V47 1992] (21.4; 1.3.5; 15.5)

Friedländer, Saul. THE YEARS OF EXTERMINATION: NAZI GERMANY AND THE JEWS, 1939-1945. New York: HarperCollins, 2007. 870 p. ISBN 978-0-06-019043-9; ISBN 0-06-019043-4. (Gift of Max M. and Marjorie B. Kampelman.) [D804.3 .F753 2007] (21.4; 1.3.5; 15.5)

Gallo, Patrick J., ed. PIUS XII, THE HOLOCAUST AND THE RE-VISIONISTS. Jefferson, NC: McFarland, 2006. 218 p. ISBN 978-0-7864-2374-9; ISBN 0-7864-2374-3. [BX1378 .P57 2006] (21.4; 1.2; 1.3.5)

Glassner, Martin Ira and Krell, Robert, eds. AND LIFE IS CHANGED FOREVER: HOLOCAUST CHILDHOODS REMEM-BERED. Detroit: Wayne State University Press, 2006. 356 p. ISBN 978-0-8143-3173-6; ISBN 0-8143-3173-4. (Landscapes of Childhood series. Gift of Max M. and Marjorie B. Kampelman.) [D804.48 .A53 2006] (21.4; Biography)

Goldensohn, Leon. THE NUREMBERG INTERVIEWS: AN AMERICAN PSYCHIATRIST'S CONVERSATIONS WITH THE DE-FENDANTS AND WITNESSES. New York: Vintage Books/Random House, 2005. 490 p. ISBN 978-1-4000-3043-9; ISBN 1-4000-3043-9. (Edited and with an introduction by Robert Gellately. Gift of Max M. and Marjorie B. Kampelman.) [KZ1176 .G65 2004] (21.4; 1.3.5; 21.2)

Good, Michael. THE SEARCH FOR MAJOR PLAGGE: THE NAZI WHO SAVED JEWS. Bronx, NY: Fordham University Press, 2006. 271 p. ISBN 978-0-8232-2441-8; ISBN 0-8232-2441-4. (Expanded, second edition. Gift of Max M. and Marjorie B. Kampelman.) [DS135 .L5 G662 2006] (21.4; 1.3.5; 15.5; Biography)

Greenberg, Karen J., ed. THE TORTURE DEBATE IN AMER-ICA. Cambridge/New York: Cambridge University Press, 2006. 414 p. ISBN 978-0-521-67461-4; ISBN 0-521-67461-1. [JC599 .U5 T665 2006] (21.4; 1.3.5; 1.3.8; 21.1; 21.2)

Gross, Jan T. FEAR: ANTI-SEMITISM IN POLAND AFTER AUSCHWITZ. New York: Random House, 2006. 303 p. ISBN 978-0-375-50924-7; ISBN 0-375-50924-0. (Gift of Max M. and Marjorie B. Kampelman.) [DS146 .P6 G76 2006] (21.4)

Numbers in ( ) = NRCBL Classification Numbers

Katz, Steven T., ed. THE IMPACT OF THE HOLOCAUST ON JEWISH THEOLOGY. New York: New York University Press, 2005. 310 p. ISBN 0-8147-4784-1. ("Contains papers presented at two conferences entitled *Jewish Thought after the Holocaust,* held in Ashkelon, Israel, in 1999 and 2001." Gift of Max M. and Marjorie B. Kampelman.) [BM645 .H6 I47 2005] (21.4; 1.2; 1.3.5)

Kaufman, Debra; Herman, Gerlad; Ross, James; and Phillips, David, eds. FROM THE PROTOCOLS OF THE ELDERS OF ZION TO HOLOCAUST DENIAL TRIALS: CHALLENGING THE MEDIA, THE LAW AND THE ACADEMY. London/Portland, OR: Vallentine Mitchell, 2007. 131 p. ISBN 978-0-85303-642-5. (Gift of Max M. and Marjorie B. Kampelman.) [D804.355 .F76 2007] (21.4; 1.3.5; 1.3.7)

Klempner, Mark. THE HEART HAS REASONS: HOLOCAUST RESCUERS AND THEIR STORIES OF COURAGE. Cleveland, OH: Pilgrim Press, 2006. 235 p. ISBN 978-0-8298-1699-0; ISBN 0-8298-1699-2. (Gift of Max M. and Marjorie B. Kampelman.) [D804.65 .K55 2006] (21.4; 21.2; Biography)

Kugelmass, Jack; Boyarin, Jonathan; and Baker, Zachary M., eds. FROM A RUINED GARDEN: THE MEMORIAL BOOKS OF POLISH JEWRY. Bloomington, IN: Indiana University Press, published in association with the United States Holocaust Memorial Museum, 1998. 353 p. ISBN 978-0-253-21187-3; ISBN 0-253-21187-5. (Second, expanded edition. Translation from the Yiddish by Jack Kugelmass and Jonathan Boyarin; geographical index and bibliography by Zachary M. Baker. Gift of Max M. and Marjorie B. Kampelman.) [DS135 .P6 F77 1998] (21.4)

Langer, Lawrence L. USING AND ABUSING THE HOLOCAUST. Bloomington: Indiana University Press, 2006. 165 p. ISBN 978-0-253-34745-9; ISBN 0-253-34745-9. (Jewish Literature and Culture series. Gift of Max M. and Marjorie B. Kampelman.) [D804.195 .L357 2006] (21.4)

Levi, Primo and De Benedetti, Leonardo. AUSCHWITZ REPORT. London/New York: Verso, 2006. 97 p. ISBN 978-1-84467-092-5; ISBN 1-84467-092-9. (Translation from the Italian by Judith Woolf of: *Rapporto sull'organizzazione igienico-sanitaria del campo di concentramento per ebrei di Monowitz (Auschwitz, Alta Silesia),* Minerva Medica, 1946. Edited by Robert S.C. Gordon. Gift of Max M. and Marjorie B. Kampelman.) [D805.5 .A96 L4713 2006] (21.4; 1.3.5; 9.5.1)

Majer, Diemut. "NON-GERMANS" UNDER THE THIRD REICH: THE NAZI JUDICIAL AND ADMINISTRATIVE SYSTEM IN GERMANY AND OCCUPIED EASTERN EUROPE, WITH SPECIAL REGARD TO OCCUPIED POLAND, 1939-1945. Baltimore, MD: Johns Hopkins University Press, published in association with the United States Holocaust Memorial Museum, 2003. 1033 p. ISBN 978-0-8018-6493-3; ISBN 0-8018-6493-3. (Translation from the German by Peter Thomas Hill, Edward Vance Humphrey, and Brian Levin of: *"Fremdvölkische" im Dritten Reich*; Harald Boldt Verlag, Boppard am Rhein, 1981; 2nd edition: Oldenbourg Verlag, München, 1993. Gift of Max M. and Marjorie B. Kampelman.) [KK6050 .M3413 2003] (21.4; 1.3.5; 1.3.8; 15.5)

Mendelsohn, Daniel. THE LOST: A SEARCH FOR SIX OF SIX MILLION. New York: HarperCollins, 2006. 512 p. ISBN 978-0-06-054297-9; ISBN 0-06-054297-7. (Gift of Max M. and Marjorie B. Kampelman.) [E184.37 .M48 L67 2006] (21.4; Biography)

Miles, Steven. OATH BETRAYED: TORTURE, MEDICAL COMPLICITY, AND THE WAR ON TERROR. New York: Random House, 2006. 220 p. ISBN 978-1-4000-6578-3; ISBN 1-4000-6578-X. [R725.5 .M55 2006] (21.4; 1.3.5; 2.1; 4.1.2; 21.2)

Moyn, Samuel. A HOLOCAUST CONTROVERSY: THE TREBLINKA AFFAIR IN POSTWAR FRANCE. Waltham, MA: Brandeis University Press, published by Hanover, NH: University Press of New England, 2005. 220 p. ISBN 978-1-58465-509-1; ISBN 1-58465-509-7. (Tauber Institute for the Study of European Jewry Series. Gift of Max M. and Marjorie B. Kampelman.) [D804.348 .M69 2005] (21.4; 1.3.5)

Nayman, Shira. AWAKE IN THE DARK: STORIES. New York: Scribner, 2006. 290 p. ISBN 978-0-7432-9268-9; ISBN 0-7432-9268-5. (Gift of Max M. and Marjorie B. Kampelman.) [PS3614 .A96 A97 2006] (21.4; Fiction)

Némirovsky, Irène. SUITE FRANÇAISE. New York: Alfred A. Knopf, 2007. 395 p. ISBN 978-1-4000-4473-3; ISBN 1-4000-4473-1. (Originally published in France; Éditions Denoël, Paris, 2004. Translation from the French by Sandra Smith; originally published in Great Britain by Chatto & Windus, London, 2006. Gift of Max M. and Marjorie B. Kampelman.) [PQ2627 .E4 S8518 2007] (21.4; Fiction)

Noakes, J. and Pridham, G., eds. NAZISM, 1919-1945, VOLUME 3: FOREIGN POLICY, WAR AND RACIAL EXTERMINATION: A DOCUMENTARY READER. Exeter, UK: University of Exeter Press, 2001. 678 p. ISBN 0-85989-602-1. (Exeter Studies in History series.) [DD256.5 .N59 1998 v.3] (21.4; 1.3.5; 15.5; 20.5.1; 21.2)

Ogilvie, Sarah A. and Miller, Scott. REFUGE DENIED: THE ST. LOUIS PASSENGERS AND THE HOLOCAUST. Madison: University of Wisconsin Press, 2006. 203 p. ISBN 978-0-299-21980-2; ISBN 0-299-21980-1. (Gift of Max M. and Marjorie B. Kampelman.) [DS135 .G5 A156 2006] (21.4; Biography)

Poznanski, Renée. JEWS IN FRANCE DURING WORLD WAR II. Hanover, NH: Brandeis University Press/University Press of New England in association with the United States Holocaust Memorial Museum, 2001. 601 p. ISBN 978-1-58465-144-4; ISBN 1-58465-144-X. (Translation from the French by Nathan Bracher of: *Juifs en France pendant la Seconde Guerre mondiale*; Hachette, 1997. Gift of Max M. and Marjorie B. Kampelman.) [DS135 .F83 P7913 2001] (21.4; 1.3.5; 15.5)

Rayski, Adam. THE CHOICE OF THE JEWS UNDER VICHY: BETWEEN SUBMISSION AND RESISTANCE. Notre Dame, IN: University of Notre Dame Press, published in association with the United States Holocaust Memorial Museum, 2005. 388 p. ISBN 978-0-268-04021-5; ISBN 0-268-04021-4. (Translation from the French by Will

Sayers of: *Le choix des Juifs sous Vichy: Entre soumission et résistance*; Éditions La Découverte, Paris, 1992. Foreword by François Bédarida. Gift of Max M. and Marjorie B. Kampelman.) [DS135 .F83 R3913 2005] (21.4; 1.3.5; 15.5)

Rittner, Carol; Roth, John K.; and Whitworth, Wendy, eds. GENOCIDE IN RWANDA: COMPLICITY OF THE CHURCHES? St. Paul, MN: Paragon House, 2004. 319 p. ISBN 1-55778-837-5. [DT450.435 .G474 2004] (21.4; 1.2; 21.1; 21.2)

Rosenfarb, Chava. THE TREE OF LIFE: A TRILOGY OF LIFE IN THE LODZ GHETTO, BOOK ONE: ON THE BRINK OF THE PRECIPICE, 1939. Madison: University of Wisconsin Press, c1985, 2004. 314 p. ISBN 978-0-299-20454-9; ISBN 0-299-20454-5. (Library of World Fiction series. Reprint. Originally published: Scribe, Melbourne, Australia, c1985. Translation from the Yiddish by the author in collaboration with Goldie Morgentaler of: *Boym fun lebn*. Gift of Max M. and Marjorie B. Kampelman.) [PJ5129 .R597 B613 2004 v.1] (21.4; Fiction)

Rosenfarb, Chava. THE TREE OF LIFE: A TRILOGY OF LIFE IN THE LODZ GHETTO: BOOK TWO: FROM THE DEPTHS I CALL YOU, 1940-1942. Madison, WI: University of Wisconsin Press/Terrace Books, c1985, [2005]. 398 p. ISBN 978-0-299-20454-7; ISBN 0-299-20454-5. (Library of World Fiction series. Reprint. Originally published: Scribe, Melbourne, Australia, 1985. Translation from the Yiddish by the author in collaboration with Goldie Morgentaler of: *Boym fun lebn*, 1972. Gift of Max M. and Marjorie B. Kampelman.) [PJ5129 .R597 B613 2004 v.2] (21.4; Fiction)

Rosenfarb, Chava. THE TREE OF LIFE: A TRILOGY OF LIFE IN THE LODZ GHETTO: BOOK THREE: THE CATTLE CARS ARE WAITING, 1942-1944. Madison, WI: University of Wisconsin Press/Terrace Books, c1985, [2005]. 376 p. ISBN 978-0-299-22124-9; ISBN 0-299-22124-5. (Library of World Fiction series. Reprint. Originally published: Scribe, Melbourne, Australia, c1985. Translation from the Yiddish by the author in collaboration with Goldie Morgentaler of: *Boym fun lebn*, 1972. Gift of Max M. and Marjorie B. Kampelman.) [PJ5129 .R597 B613 2004 v.3] (21.4; Fiction)

Satloff, Robert. AMONG THE RIGHTEOUS: LOST STORIES FROM THE HOLOCAUST'S LONG REACH INTO ARAB LANDS. New York: PublicAffairs, 2006. 251 p. ISBN 978-1-58648-399-9; ISBN 1-58648-399-4. (Gift of Max M. and Marjorie B. Kampelman.) [DS135 .A68 S28 2006] (21.4; 1.3.5; 15.5)

Sebastian, Mihail. JOURNAL, 1935-1944: THE FASCIST YEARS. Chicago, IL: Ivan R. Dee, published in association with the United States Holocaust Memorial Museum, 2000. 641 p. ISBN 978-1-56663-326-0; ISBN 1-56663-326-5. (Translation from the Romanian by Patrick Camiller, with an introduction and notes by Radu Ioanid. Gift of Max M. and Marjorie B. Kampelman.) [DS135 .R73 S38713 2000] (21.4; Biography)

Simmons, Cynthia and Perlina, Nina. WRITING THE SIEGE OF LENINGRAD: WOMEN'S DIARIES, MEMOIRS AND DOCU-MENTARY PROSE. Pittsburgh: University of Pittsburgh Press, c2002, 2005. 242 p. ISBN 978-0-8229-5869-7; ISBN 0-8229-5869-4. (Pitt Series in Russian and East European Studies. Foreword by Richard Bidlack.) [D764.3 .L4 S56 2002] (21.4; 10; 21.2; Biography)

Trunk, Isaiah. LÓDZ GHETTO: A HISTORY. Bloomington: Indiana University Press, 2006. 493 p. ISBN 978-0-253-34755-8; ISBN 0-253-34755-6. ("Published in association with the United States Holocaust Memorial Museum." Introduction by Israel Gutman. Translation from the Yiddish by Robert Moses Shapiro of: *Lodzher geto*, 1962. Gift of Max M. and Marjorie B. Kampelman.) [DS135 .P62 T7813 2006] (21.4; 1.3.5; 15.5)

Wiesel, Elie. NIGHT. New York: Hill and Wang, 2006. 120 p. ISBN 978-0-374-50001-6; ISBN 0-374-50001-0. (With a new preface by the author. Translation from the French by Marion Wiesel, Éditions de Minuit, 1958. Distributed in Canada by: Douglas & McIntyre. Gift of Max M. and Marjorie B. Kampelman.) [D810 .J4 W513 2006] (21.4; Biography)

Zabarko, Boris, ed. HOLOCAUST IN THE UKRAINE. London/Portland, OR: Vallentine Mitchell, 2005. 394 p. ISBN 978-0-85303-524-4; ISBN 0-85303-524-5. (The Library of Holocaust Testimonies series. Translation by Marina Guba. Gift of Max M. and Marjorie B. Kampelman.) [DS135 .U43 A154 2005] (21.4)

21.7 CULTURAL PLURALISM

Achenbach, Thomas M. and Rescorla, Leslie A. MULTICULTURAL UNDERSTANDING OF CHILD AND ADOLESCENT PSYCHOPATHOLOGY: IMPLICATIONS FOR MENTAL HEALTH ASSESSMENT. New York: Guilford Press, 2007. 322 p. ISBN 978-1-59385-348-8; ISBN 1-59385-348-3. [RJ499 .M85 2006] (21.7; 4.3; 7.1; 9.5.7; 18.5.2)

Fouad, Nadya A. and Arredondo, Patricia. BECOMING CULTURALLY ORIENTED: PRACTICAL ADVICE FOR PSYCHOLOGISTS AND EDUCATORS. Washington, DC: American Psychological Association, 2007. 162 p. ISBN 978-1-59147-424-1; ISBN 1-59147-424-8. (Gift of the publisher.) [BF637 .C6 F576 2007] (21.7; 8.1; 17.1)

Li, Xiaorong. ETHICS, HUMAN RIGHTS AND CULTURE: BEYOND RELATIVISM AND UNIVERSALISM. Basingstoke [England]/New York: Palgrave Macmillan, 2006. 274 p. ISBN 978-1-4039-8548-4; ISBN 1-4039-8548-0. [JC571 .L52828 2006] (21.7; 21.1)

Lipson, Juliene G. and Dibble, Suzanne L., eds. CULTURE & CLINICAL CARE. San Francisco: UCSF Nursing Press, 2005. 487 p. ISBN 978-0-943671-22-2; ISBN 0-943671-22-1. [RT86.54 .C852 2005] (21.7; 7.1)

Loue, Sana. ASSESSING RACE, ETHNICITY, AND GENDER IN HEALTH. New York: Springer Sciences-Business Media, 2006. 158 p. ISBN 978-0-387-32461-6; ISBN 0-387-32461-5. [RA448.4 .L679 2006] (21.7; 9.1; 9.5.4; 10)

Mace, Ruth; Holden, Clare J.; and Shennan, Stephen, eds. THE EVOLUTION OF CULTURAL DIVERSITY: A PHYLOGEN-

ETIC APPROACH. London: UCL Press, 2005. 291 p. ISBN 978-1-84472-065-1; ISBN 1-84472-065-9. (UCL Series.) [GN360 .E885 2005] (21.7; 15.1)

Tseng, Wen-Shing. HANDBOOK OF CULTURAL PSYCHIATRY. San Diego, CA: Academic Press, 2001. 855 p. ISBN 0-12-701632-5. [RC455.4 .E8 T763 2001] (21.7; 17.1; 17.2)

22.1 ANIMAL WELFARE (GENERAL)

Armstrong, Susan J. and Botzler, Richard G., eds. THE ANIMAL ETHICS READER. London/New York: Routledge, 2003. 588 p. ISBN 0-415-27589-X. (Gift of Peter Inman.) [HV4708 .A548 2003] (22.1; 1.3.12; 14.5; 15.1; 19.5; 22.2; 22.3; Reference)

Bekoff, Marc. ANIMAL PASSIONS AND BEASTLY VIRTUES: REFLECTIONS ON REDECORATING NATURE. Philadelphia: Temple University Press, 2006. 303 p. ISBN 1-59213-348-7. (Animals, Culture and Society series. Foreword by Jane Goodall.) [QL785 .B36 2006] (22.1)

Dunayer, Joan. SPECIESISM. Derwood, MD: Ryce Publishing; Distributed by: New York: Lantern Books, 2004. 204 p. ISBN 978-0-9706475-6-6; ISBN 0-9706475-6-5. [HV4708 .D87 2004] (22.1)

Fetzer, James J. THE EVOLUTION OF INTELLIGENCE: ARE HUMANS THE ONLY ANIMALS WITH MINDS? Chicago: Open Court, 2005. 272 p. ISBN 978-0-8126-9459-8; ISBN 0-8126-9459-7. [BF431 .F415 2005] (22.1; 3.1; 15.6; 15.9)

Marie, M.; Edwards, S.; Gandini, G.; Reiss, M.; and von Borell, E. ANIMAL BIOETHICS: PRINCIPLES AND TEACHING METHODS. Wageningen, [The Netherlands]: Wageningen Academic Publishers, 2005. 360 p. ISBN 90-76998-58-2. [HV4712 .A54 2005] (22.1; 1.2; 1.3.11; 2.3; 4.1.1; 22.2; 22.3)

McMillan, Franklin D., ed. MENTAL HEALTH AND WELL-BEING IN ANIMALS. Ames, IA: Blackwell, 2005. 301 p. ISBN 978-0-8138-0489-7; ISBN 0-8138-0489-2. [SF745 .M46 2005] (22.1; 1.3.11; 3.1; 4.4; 22.2; 22.3)

Montgomery, Charlotte. BLOOD RELATIONS: ANIMALS, HUMANS, AND POLITICS. Toronto: Between the Lines, 2000. 337 p. ISBN 1-896357-39-3. [HV4768 .M66 2000] (22.1; 1.3.11; 22.2; 22.3)

Palmeri, Frank, ed. HUMANS AND OTHER ANIMALS IN EIGHTEENTH-CENTURY BRITISH CULTURE: REPRESENTATION, HYBRIDITY, ETHICS. Aldershot, Hants, England/ Burlington, VT: Ashgate, 2006. 217 p. ISBN 978-0-7546-5475-9; ISBN 0-7546-5475-3. [PR448 .A55 H86 2006] (22.1)

Pullen, Sophie and Gray, Carol, eds. ETHICS, LAW, AND THE VETERINARY NURSE. Edinburgh/New York: Butterworth Heinemann Elsevier, 2006. 174 p. ISBN 978-0-7506-8844-4; ISBN 0-7506-8844-0. [SF774.5 .E84 2006] (22.1; 1.2; 1.3.5; 4.1.1; 22.2)

Rhoades, Rebecca H. THE HUMANE SOCIETY OF THE UNITED STATES EUTHANASIA TRAINING MANUAL. Washington, DC: Humane Society of the United States, 2002. 174 p. ISBN 0-9658942-6-6. (Publisher's address: 2100 L Street, NW, zip 20037.) [SF756.394 .R46 2002] (22.1; 9.7; 20.5.1; Reference)

Rollin, Bernard E. AN INTRODUCTION TO VETERINARY MEDICAL ETHICS: THEORY AND CASES. Ames, IA: Blackwell, 2006. 331 p. ISBN 978-0-8138-0399-9; ISBN 0-8138-0399-3. (Second edition.) [SF756.39 .R65 2006] (22.1; 1.3.11; 22.2; 22.3)

Waldau, Paul and Patton, Kimberley, eds. A COMMUNION OF SUBJECTS: ANIMALS IN RELIGION, SCIENCE, AND ETHICS. New York: Columbia University Press, 2006. 686 p. ISBN 978-0-231-13642-6; ISBN 0-231-13642-0. [BL439 .C66 2006] (22.1; 1.2; 1.3.11; 4.4; 16.1; 22.2; 22.3)

22.2 ANIMAL EXPERIMENTATION

Institute for Laboratory Animal Research (United States). International Workshop on the Development of Science-Based Guidelines for Laboratory Animal Care Program Committee, National Research Council. THE DEVELOPMENT OF SCIENCE-BASED GUIDELINES FOR LABORATORY ANIMAL CARE: PROCEEDINGS OF THE NOVEMBER 2003 INTERNATIONAL WORKSHOP. Washington, DC: National Academies Press, 2004. 248 p. ISBN 0-309-09302-3. (Also available on the Web at: http://www.nap.edu.) [SF406 .D48 2004] (22.2; 1.3.11; 16.1; 21.1)

Mason, Peter. THE BROWN DOG AFFAIR: THE STORY OF A MONUMENT THAT DIVIDED A NATION. London: Two Sevens Publishing, 2005. 118 p. ISBN 978-0-9529854-0-2; ISBN 0-9529854-0-3. (Second edition.) [HV4943 .G7 M37 1997] (22.2)

Silverman, Jerald; Suckow, Mark A.; Murthy, Sreekant, eds. THE IACUC HANDBOOK. Boca Raton, FL: CRC Press, 2007. 652 p. ISBN 978-0-8493-4010-9; ISBN 0-8493-4010-1. (Second edition.) [HV4708 .I23 2007] (22.2; 1.3.5)

22.3 ANIMAL PRODUCTION

Ferrières, Madeleine. SACRED COW, MAD COW: A HISTORY OF FOOD FEARS. Ithaca, NY: Columbia University Press, 2006. 399 p. ISBN 978-0-231-13192-6; ISBN 0-231-13192-5. (Arts and Traditions of the Table: Perspectives on Culinary History series. Translation from the French by Jody Gladding of: *Histoire des peurs alimentaires: Du Moyen âge à l'aube du xxe Siècle*; Édition du Seuil, 2002.) [RC622 .F47613 2006] (22.3; 1.3.11)

Greger, Michael. BIRD FLU: A VIRUS OF OUR OWN HATCHING. New York: Lantern Books, 2006. 465 p. ISBN 978-1-59056-098-3; ISBN 1-59056-098-1. (Foreword by Kennedy Shortridge.) [RA644 .I6 G74 2006] (22.3; 1.3.11)

NEW JOURNAL SUBSCRIPTIONS

NANOETHICS 2007 March; 1(1). Quarterly. ISSN 1871-4757. Publisher's address: Springer Netherlands,

Van Godewijckstraat 30, Dordrecht 3311 GX Netherlands; Website: http://www.springeronline.com.) (5.4)

NEW PUBLICATIONS FROM THE KENNEDY INSTITUTE OF ETHICS

Walters, LeRoy; Kahn, Tamar Joy; and Goldstein, Doris Mueller, eds. BIBLIOGRAPHY OF BIOETHICS, VOLUME 33. Washington, DC: Kennedy Institute of Ethics, Georgetown University, 2007. 848 p. ISBN 978-1-883913-14-4. (ISSN 0363-0161. Pricing and availability from: Mara Snyder, Kennedy Institute of Ethics, Georgetown University, Box 571212; tel: 1-888-BIO-ETHX [U.S. and Canada only] [or] 202-687-6689; fax: 202-687-6770; email: mrm37@georgetown.edu.) [Z6675 .E8 W34 v.33] (2.1; Reference)

BIOETHICS SEARCHER'S GUIDE TO ONLINE INFORMATION RESOURCES. Washington, DC: National Reference Center for Bioethics Literature [and] National Information Resource on Ethics and Human Genetics, Kennedy Institute of Ethics, Georgetown University, 2008. 173 p. (May 2008. Pricing and availability from: Mara Snyder, Kennedy Institute of Ethics, Georgetown University, Box 571212; tel: 1-888-BIO-ETHX [U.S. and Canada only] [or] 202-687-6689; fax: 202-687-6770; email: mrm37@georgetown.edu.) (2.1; Reference)

SECTION IV:
MONOGRAPHS

TITLE INDEX

SECTION IV: MONOGRAPHS TITLE INDEX

A

Abortion and the Law: From International Comparison to Legal Policy. *Section III:* 12.4.1 —Eser, Albin and Koch, Hans-Georg.

Abortion Politics in North America. *Section III:* 12.5.1 —Haussman, Melissa.

Absolute Convictions: My Father, a City, and the Conflict That Divided America. *Section III:* 12.1 —Press, Eyal.

ACA Ethical Standards Casebook. *Section III:* 17.1 —Herlihy, Barbara and Corey, Gerald, [eds.].

Access Principle: The Case for Open Access to Research and Scholarship, The. *Section III:* 1.3.12 —Willinsky, John.

Access to Care, Access to Justice: The Legal Debate Over Private Health Insurance in Canada. *Section III:* 9.3.1 —Flood, Colleen M.; Roach, Kent; and Sossin, Lorne, eds.

Achieving Excellence in Medical Education. *Section III:* 7.2 —Gunderman, Richard B.

African AIDS Epidemic: A History, The. *Section III:* 9.5.6 —Iliffe, John.

African American Bioethics: Culture, Race, and Identity. *Section III:* 2.1 —Prograis, Lawrence J. and Pellegrino, Edmund D., eds.

After Emancipation: Jewish Religious Responses to Modernity. *Section III:* 1.2 —Ellenson, David.

AIDS and Power: Why There Is No Political Crisis—Yet. *Section III:* 9.5.6 —de Waal, Alex.

AIDS in Latin America. *Section III:* 9.5.6 —Frasca, Tim.

Aktive und passive Sterbehilfe: medizinische rechtswissenschaftliche und philosophische Aspekte. *Section III:* 20.5.1 —Thiele, Felix, ed.

Alcohol, Addiction and Christian Ethics. *Section III:* 9.5.9 —Cook, Christopher C.H.

American Normal: The Hidden World of Asperger Syndrome. *Section III:* 17.1 —Osborne, Lawrence.

Among the Righteous: Lost Stories from the Holocaust's Long Reach into Arab Lands. *Section III:* 21.4 —Satloff, Robert.

Amoral Thoughts About Morality: The Intersection of Science, Psychology, and Ethics. *Section III:* 17.1 —Kendler, Howard H.

And Life Is Changed Forever: Holocaust Childhoods Remembered. *Section III:* 21.4 —Glassner, Martin Ira and Krell, Robert, eds.

Animal Bioethics: Principles and Teaching Methods. *Section III:* 22.1 —Marie, M.; Edwards, S.; Gandini, G.; Reiss, M.; and von Borell, E.

Animal Ethics Reader, The. *Section III:* 22.1 —Armstrong, Susan J. and Botzler, Richard G., eds.

Animal Passions and Beastly Virtues: Reflections on Redecorating Nature. *Section III:* 22.1 —Bekoff, Marc.

Annual Editions: Dying, Death, and Bereavement 2005/2006. *Section III:* 20.1 —Dickinson, George E. and Leming, Michael R., eds.

Anthology of Psychiatric Ethics, An. *Section III:* 17.1 —Green, Stephen A. and Bloch, Sidney, eds.

Anthropology & Modern Life. *Section III:* 1.3.1 —Boas, Franz.

Approaches to Ethics: Nursing Beyond Boundaries. *Section III:* 4.1.3 —Tschudin, Verena, ed.

Are We Ready? Public Health Since 9/11. *Section III:* 9.1 —Rosner, David and Markowitz, Gerald.

Artificial Nutrition and Hydration and the Permanently Unconscious Patient: The Catholic Debate. *Section III:* 20.5.1 —Hamel, Ronald P. and Walter, James J., eds.

Assessing Quality of Life in Clinical Trials: Methods and Practice. *Section III:* 18.2 —Fayers, Peter and Hays, Ron, eds.

Assessing Race, Ethnicity, and Gender in Health. *Section III:* 21.7 —Loue, Sana.

Assessing the Medical Risks of Human Oocyte Donation for Stem Cell Research: Workshop Report. *Section III:* 14.2 —Giudice, Linda; Santa, Eileen; and Pool, Robert, eds.; Institute of Medicine (United States) [IOM]. Committee on Assessing the Medical Risks of Human Oocyte Donation for Stem Cell Research; Institute of Medicine (United States). Board on Health Sciences Policy; and National Research Council (United States). Board on Life Sciences.

Assisted Dying and Legal Change. *Section III:* 20.7 —Lewis, Penney.

Attention Deficit Disorder: The Unfocused Mind in Children and Adults. *Section III:* 17.1 —Brown, Thomas E.

Audit Cultures: Anthropological Studies in Accountability, Ethics and the Academy. *Section III:* 1.3.3 —Strathern, Marilyn, ed.

Auschwitz Report. *Section III:* 21.4 —Levi, Primo and De Benedetti, Leonardo.

Autonomy & Paternalism: Reflections on the Theory and Practice of Health Care. *Section III:* 2.1 —Nys, Thomas; Denier, Yvonne; and Vandevelde, Toon, eds.

Awake in the Dark: Stories. *Section III:* 21.4 —Nayman, Shira.

Awakening Hippocrates: A Primer on Health, Poverty, and Global Service. *Section III:* 21.1 —O'Neil, Jr., Edward.

B

Bad Medicine: Doctors Doing Harm Since Hippocrates. *Section III:* 7.1 —Wootton, David.

Bad Medicine: Misconceptions and Misuses Revealed, from Distance Healing to Vitamin O. *Section III:* 4.1.1 —Wanjek, Christopher.

Balm for Gilead: Meditations on Spirituality and the Healing Arts, A. *Section III:* 4.1.2 —Sulmasy, Daniel P.

Beauty Junkies: Inside Our $15 Billion Obsession with Cosmetic Surgery. *Section III:* 9.5.1 —Kuczynski, Alex.

Becoming Culturally Oriented: Practical Advice for Psychologists and Educators. *Section III:* 21.7 —Fouad, Nadya A. and Arredondo, Patricia.

Becoming Eichmann: Rethinking the Life, Crimes, and Trial of a "Desk Murderer". *Section III:* 21.4 —Cesarani, David.

Best of the Brain from Scientific American. *Section III:* 17.1 —Bloom, Floyd E., ed.

Better But Not Well: Mental Health Policy in the United States Since 1950. *Section III:* 17.1 —Frank, Richard G. and Glied, Sherry A.

Beyond the DSM Story: Ethical Quandaries, Challenges, and Best Practices. *Section III:* 4.3 —Eriksen, Karen and Kress, Victoria E.

Beyond the Hippocratic Oath: A Memoir of the Rise of Modern Medical Ethics. *Section III:* 2.2 —Dossetor, John B.

Bibliography of Bioethics, Volume 33. *Section III:* 2.1 —Walters, LeRoy; Kahn, Tamar Joy; and Goldstein, Doris Mueller, eds.

Big Pharma: Exposing the Global Healthcare Agenda. *Section III:* 9.7 —Law, Jacky.

Biobanken: Chance für den wissenschaftlichen Fortschritt oder Ausverkauf der "Ressource" Mensch? *Section III:* 15.1 —Nationaler Ethikrat.

Biodiversity and the Law: Intellectual Property, Biotechnology and Traditional Knowledge. *Section III:* 15.1 —McManis, Charles R., ed.

Bioethics: An Introduction to the History, Methods, and Practice. *Section III:* 2.1 —Jecker, Nancy S.; Jonsen, Albert R.; and Pearlman, Robert A., eds.

Bioethics and Public Health Law. *Section III:* 2.1 —Hall, Mark A.; Bobinski, Mary Anne; and Orentlicher, David.

Bioethics and the Brain. *Section III:* 17.1 —Glannon, Walter.

Bioethics and Vulnerability: A Latin American View. *Section III:* 2.1 —Luna, Florencia.

Bioethics and Women: Across the Life Span. *Section III:* 2.1 —Mahowald, Mary Briody.

Bioethics: Health Care, Human Rights and the Law. *Section III:* 2.1 —LaFrance, Arthur Birningham.

Bioethics in Law. *Section III:* 2.1 —Spielman, Bethany J.

Bio-Ethik und die Zukunft der Medizin. *Section III:* 2.1 —Neuer-Miebach, Therese and Wunder, Michael, eds.

Bioéthique, avis de tempêtes: les nouveaux enjeux de la maîtrise du vivant. *Section III:* 14.1 —Chneiweiss, Hervé and Nau, Jean-Yves.

Bioéthique pour un monde habitable? La bioéthique en discussion, Une. *Section III:* 2.1 —de Bouvet, Armelle and Cobbaut, Jean-Philippe, eds.

Bioetica e grandi religioni. *Section III:* 2.1 —Spinsanti, Sandro, ed.

Bioetikk og teologi: Rapport fra Nordisk teologisk nettverk for bioetikks workshop i Stockholm 27.-29. september 1996. *Section III:* 2.1 —Østnor, Lars, ed.; Nordisk teologisk nettverk for bioetikk.

Biofakte: Versuch über den Menschen zwischen Artefakt und Lebewesen. *Section III:* 5.1 —Karafyllis, Nicole C., ed.

Biologische Waffen — nicht in Hitlers Arsenalen: Biologische und Toxin-Kampfmittel in Deutschland von 1915 bis 1945. *Section III:* 21.3 —Geißler, Erhard.

Biomedical Ethics. *Section III:* 2.1 —Laney, Dawn, ed.

Biomedical Ethics for Engineers: Ethics and Decision Making in Biomedical and Biosystem Engineering. *Section III:* 2.1 —Vallero, Daniel A.

Biomedizin und Ethik: Praxis – Recht – Moral. *Section III:* 15.1 —Schreiber, Hans-Peter, ed.

Biotechnology and the Human Good. *Section III:* 15.1 —Mitchell, C. Ben; Pellegrino, Edmund D.; Elshtain, Jean Bethke; Kilner, John F.; Rae, Scott B.

Biotechnology Unglued: Science, Society, and Social Cohesion. *Section III:* 5.3 —Mehta, Michael D., ed.

Bioterrorism Preparedness: Medicine, Public Health, Policy. *Section III:* 21.3 —Khardori, Nancy, ed.

Bird Flu: A Virus of Our Own Hatching. *Section III:* 22.3 —Greger, Michael.

Birth Control and Abortion in Islam. *Section III:* 13.1 —Al-Kawthari, Muhammad Ibn Adam.

Birthing a Slave: Motherhood and Medicine in the Antebellum South. *Section III:* 9.5.5 —Schwartz, Marie Jenkins.

Blackwell Guide to Medical Ethics, The. *Section III:* 2.1 —Rhodes, Rosamond; Francis, Leslie P.; and Silvers, Anita, eds.

Blessed Motherhood, Bitter Fruit: Nelly Roussel and the Politics of Female Pain in Third Republic France. *Section III:* 13.3 —Accampo, Elinor.

Blind Faith: The Unholy Alliance of Religion and Medicine. *Section III:* 9.1 —Sloan, Richard P.

Blood and Homeland: Eugenics and Racial Nationalism in Central and Southeast Europe, 1900-1940. *Section III:* 15.5 —Turda, Marius and Weindling, Paul, eds.

Blood Relations: Animals, Humans, and Politics. *Section III:* 22.1 —Montgomery, Charlotte.

Bodily Integrity and the Politics of Circumcision: Culture, Controversy, and Change. *Section III:* 9.5.1 —Denniston, George C.; Gallo, Pia Grassivaro; Hodges, Frederick M.; Milos, Marilyn Fayre; and Viviani, Franco, eds.

Body in Culture, Technology and Society, The. *Section III:* 4.4 —Shilling, Chris.

Born Alive: The Legal Status of the Unborn Child in England and the U.S.A. *Section III:* 9.5.8 —Casey, Gerald.

Born in the USA: How a Broken Maternity System Must Be Fixed to Put Mothers and Infants First. *Section III:* 9.5.5 —Wagner, Marsden.

Brandenburgische Heil- und Pflegeanstalten in der NS-Zeit. *Section III:* 20.5.1 —Hübener, Kristina, ed. and Heinze, Martin.

Breast Cancer Genes and the Gendering of Knowledge: Science and Citizenship in the Cultural Context of the "New" Genetics. *Section III:* 15.1 —Gibbon, Sahra.

Brown Dog Affair: The Story of a Monument That Divided a Nation, The. *Section III:* 22.2 —Mason, Peter.

Building Genetic Medicine: Breast Cancer, Technology, and the Comparative Politics of Health Care. *Section III:* 15.3 —Parthasarathy, Shobita.

C

Cambridge History of Judaism, Volume IV: The Late Roman-Rabbinic Period, The. *Section III:* 1.2 —Katz, Steven T., ed.

Cambridge History of Medicine, The. *Section III:* 7.1 —Porter, Roy, ed.

Can a Health Care Market Be Moral? A Catholic Vision. *Section III:* 9.3.1 —McDonough, Mary J.

Can We Ever Kill? *Section III:* 20.1 —Crawford, Robert.

Can We Say No? The Challenge of Rationing Health Care. *Section III:* 9.4 —Aaron, Henry J.; Schwartz, William B.; and Cox, Melissa.

Canadian Medical Law: An Introduction for Physicians, Nurses and Other Health Care Professionals. *Section III:* 9.1 —Sneiderman, Barney; Irvine, John C.; and Osborne, Philip H.

Capital Punishment. *Section III:* 20.6 —Henderson, Harry.

Caring About Health. *Section III:* 4.1.3 —van Hooft, Stan.

Caring for the Dying at Home: A Practical Guide. *Section III:* 20.4.1 —Pharaoh, Gill.

Carville: Remembering Leprosy in America. *Section III:* 9.5.1 —Gaudet, Marcia.

Case Against Perfection: Ethics in the Age of Genetic Engineering, The. *Section III:* 15.1 —Sandel, Michael J.

Case of the Female Orgasm: Bias in the Science of Evolution, The. *Section III:* 10 —Lloyd, Elisabeth A.

Catharsis: On the Art of Medicine. *Section III:* 4.1.2 —Szczeklik, Andrzej.

Cell of Cells: The Global Race to Capture and Control the Stem Cell. *Section III:* 18.5.4 —Fox, Cynthia.

Challenge of Reproductive Medicine at Catholic Universities: Time to Leave the Catacombs, The. *Section III:* 14.1 —Brosens, Ivo, ed.

Challenges of an Aging Society: Ethical Dilemmas, Political Issues. *Section III:* 9.5.2 —Pruchno, Rachel A. and Smyer, Michael A., eds.

Challenging Nature: The Clash of Science and Spirituality at the New Frontiers of Life. *Section III:* 5.1 —Silver, Lee M.

See inside front cover for NRCBL Classification Scheme

Challenging the Chip: Labor Rights and Environmental Justice in the Global Electronics Industry. *Section III:* 1.3.12 —Smith, Ted; Sonnenfeld, David; and Pellow, David Naguib, eds.

Character and Environment: A Virtue-Oriented Approach to Environmental Ethics. *Section III:* 16.1 —Sandler, Ronald L.

Childbirth and the Law. *Section III:* 9.5.5 —Seymour, John.

Children of Abraham: Judaism, Christianity, Islam, The. *Section III:* 1.2 —Peters, F.E.

Children's Health and Children's Rights. *Section III:* 9.5.7 —Freeman, Michael, ed.

Chinese Ethics for the New Century: The Ch'ien Mu Lectures in History and Culture, and Other Essays on Science and Confucian Ethics, A. *Section III:* 1.1 —Munro, Donald J.

Choice of the Jews Under Vichy: Between Submission and Resistance, The. *Section III:* 21.4 —Rayski, Adam.

Christian Religion and Biotechnology: A Search for Principled Decision-making, The. *Section III:* 2.1 —Smith, George P.

Church Divided: German Protestants Confront the Nazis, A. *Section III:* 1.2 —Hockenos, Matthew D.

Cigarette Century: The Rise, Fall, and Deadly Persistence of the Product That Defined America, The. *Section III:* 9.5.9 —Brandt, Allan M.

Clinical Research in Practice: A Guide for the Bedside Scientist. *Section III:* 8.1 —Houser, Janet and Bokovoy, Joanna.

Clinical Trials in Psychiatry. *Section III:* 18.5.6 —Everitt, Brian S. and Wessely, Simon.

Clio in the Clinic: History in Medical Practice. *Section III:* 7.1 —Duffin, Jacalyn, ed.

Cloning. *Section III:* 14.5 —Gerdes, Louise, ed.

Code of Ethics of the American Medical Association: The Social Construction of Medical Morality: The Establishment of a Masculine Profession, The. *Section III:* 6 —Cooper, Melvin Wayne.

Code of Jewish Ethics, Volume 1: You Shall Be Holy, A. *Section III:* 1.2 —Telushkin, Joseph.

Code of Medical Ethics: Current Opinions with Annotations. *Section III:* 6 —American Medical Association. Council on Ethical and Judicial Affairs; Southern Illinois University at Carbondale. School of Medicine; and Southern Illinois University of Carbondale. School of Law.

Cold War, Deadly Fevers: Malaria Eradication in Mexico, 1955-1975. *Section III:* 9.5.1 —Cueto, Marcos.

Collaboration in the Holocaust: Crimes of the Local Police in Belorussia and Ukraine, 1941-1944. *Section III:* 21.4 —Dean, Martin.

Colonial Pathologies: American Tropical Medicine, Race, and Hygiene in the Philippines. *Section III:* 7.1 —Anderson, Warwick.

Colony: The Harrowing True Story of the Exiles of Molokai, The. *Section III:* 9.5.1 —Tayman, John.

Commercial Use of Biodiversity: Access to Genetic Resources and Benefit-Sharing, The. *Section III:* 15.1 —Kate, Kerry ten and Laird, Sarah A., eds.

Communication for Doctors: How to Improve Patient Care and Minimize Legal Risks. *Section III:* 8.1 —Woods, David, ed.

Communication Skills That Heal: A Practical Approach to a New Professionalism in Medicine. *Section III:* 8.1 —Bub, Barry.

Communion of Subjects: Animals in Religion, Science, and Ethics, A. *Section III:* 22.1 —Waldau, Paul and Patton, Kimberley, eds.

Comparative Biomedical Policy: Governing Assisted Reproductive Technologies. *Section III:* 14.1 —Bleiklie, Ivar; Goggin, Malcolm L.; and Rothmayr, Christine, eds.

Complementary and Alternative Medicine: Ethics, the Patient, and the Physician. *Section III:* 4.1.1 —Snyder, Lois, ed.

Condition foetale: Une sociologie de l'engendrement et de l'avortement, La. *Section III:* 12.5.1 —Boltanski, Luc.

Conducting Clinical Research: A Practical Guide for Physicians, Nurses, Study Coordinators, and Investigators. *Section III:* 18.2 —Stone, Judy.

Conducting Drug Abuse Research with Minority Populations: Advances and Issues. *Section III:* 18.5.1 —De La Rosa, Mario R.; Segal, Bernard; and Lopez, Richard, eds.

Confronting the "Good Death": Nazi Euthanasia on Trial, 1945-1953. *Section III:* 20.5.1 —Bryant, Michael S.

Conscience of a Pharmacist: Essays on Vision and Leadership for a Profession, The. *Section III:* 9.7 —Zellmer, William A.; American Society of Health-System Pharmacists.

Consent in the Law. *Section III:* 8.3.1 —Beyleveld, Deryck and Brownsword, Roger.

Consentimento Informado na Relação Médico-Paciente: Estudo de Direito Civil, O. *Section III:* 8.3.1 —Pereira, André Gonçalo Dias.

Consentimento Informado para o Acto Médico no Ordenamento Jurídico Português (Elementos para o estudo da manifestação da vontade do paciente), O. *Section III:* 8.3.1 —Rodrigues, João Vaz.

Constructing RSI: Belief and Desire. *Section III:* 7.1 —Lucire, Yolande.

Constructive Conversations About Health: Policy and Values. *Section III:* 9.1 —Marinker, Marshall, ed.

Consumer's Guide to Medical Mistakes: Information You Need Before Becoming a Patient, The. *Section III:* 9.8 —Peraino, Robert A.

Contemporary Ethical Dilemmas in Assisted Reproduction. *Section III:* 14.1 —Shenfield, Françoise and Sureau, Claude, eds.

Contemporary Issues in Bioethics: A Catholic Perspective. *Section III:* 2.1 —Walter, James J. and Shannon, Thomas A.

Contemporary Issues in Healthcare Law and Ethics. *Section III:* 2.1 —Garwood-Gowers, Austen ; Tingle, John; Wheat, Kay, eds.

Cooperation, Complicity and Conscience: Problems in Healthcare, Science, Law and Public Policy. *Section III:* 2.1 —Watt, Helen, ed.

Coping with Sickness: Historical Aspects of Health Care in a European Perspective. *Section III:* 7.1 —Woodward, John and Jütte, Robert, eds.

Copula: Sexual Technologies, Reproductive Powers. *Section III:* 10 —Ferrell, Robyn.

Corporate Integrity: Rethinking Organizational Ethics and Leadership. *Section III:* 1.3.2 —Brown, Marvin T.

Corporate Social Responsibility and the Shaping of Global Public Policy. *Section III:* 1.3.2 —Hirschland, Matthew J.

Cosmos and Community: The Ethical Dimension of Daoism. *Section III:* 1.2 —Kohn, Livia.

Cost of Being Poor: Poverty, Lead Poisoning, and Policy Implementation, The. *Section III:* 9.5.7 —Richardson, Jeanita W.

Creating Brain: The Neuroscience of Genius, The. *Section III:* 17.1 —Andreasen, Nancy C.

Creating Choice: A Community Responds to the Need for Abortion and Birth Control, 1961-1973. *Section III:* 11.1 —Cline, David P.

Creation: An Appeal to Save Life on Earth, The. *Section III:* 3.2 —Wilson, Edward O.

Cross-Cultural Caring: A Handbook for Health Professionals. *Section III:* 9.5.4 —Waxler-Morrison, Nancy; Anderson, Joan M.; Richardson, Elizabeth; and Chambers, Natalie A., eds.

Cross-Cultural Issues in Bioethics: The Example of Human Cloning. *Section III:* 14.5 —Roetz, Heiner, ed.

Crossing Over: Genomics in the Public Arena. *Section III:* 15.1 —Einsiedel, Edna F. and Timmermans, Frank, eds.

Culturally Relevant Ethical Decision-Making in Counseling. *Section III:* 17.1 —Houser, Rick; Wilczenski, Felicia L.; and Ham, MaryAnna.

Culture & Clinical Care. *Section III:* 21.7 —Lipson, Juliene G. and Dibble, Suzanne L., eds.

Cure: How Capitalism Can Save American Health Care, The. *Section III:* 9.3.1 —Gratzer, David.

Current Controversies in the Biological Sciences: Case Studies of Policy Challenges from New Technologies. *Section III:* 5.3 —Greif, Karen F. and Merz, Jon F.

Cutting to the Core: Exploring the Ethics of Contested Surgeries. *Section III:* 2.1 —Benatar, David, ed.

Cyberethics: Morality and Law in Cyberspace. *Section III:* 1.3.12 —Spinello, Richard A.

D

Dangerous Motherhood: Insanity and Childbirth in Victorian Britain. *Section III:* 17.1 —Marland, Hilary.

Dangers de l'euthanasie, Les. *Section III:* 20.5.1 —Israël, Lucien and Lévy, Elisabeth.

Dark Medicine: Rationalizing Unethical Medical Research. *Section III:* 18.1 —Lafleur, William R.; Böhme, Gernot; and Shimazono, Susumu, eds.

De La Mettrie's Ghost: The Story of Decisions. *Section III:* 17.1 —Nunn, Chris.

Death and Dying Sourcebook: Basic Consumer Health Information About End-of-Life Care and Related Perspectives and Ethical Issues, Including End-of-life Symptoms and Treatments, Pain Management, Quality-of-Life Concerns, the Use of Life Support, Patients' Rights and Privacy Issues, Advance Directives, Physician-Assisted Suicide, Caregiving, Organ and Tissue Donation, Autopsies, Funeral Arrangements, and Grief, along with Statistical Data, Information About the Leading Causes of Death, a Glossary, and Directories of Support Groups and Other Resources. *Section III:* 20.1 —Shannon, Joyce Brennfleck, ed.

Death and Religion in a Changing World. *Section III:* 20.1 —Garces-Foley, Kathleen, ed.

Death Retold: Jesica Santillan, the Bungled Transplant, and Paradoxes of Medical Citizenship, A. *Section III:* 19.2 —Wailoo, Keith; Livingston, Julie; and Guarnaccia, Peter, eds.

Death's Dominion: Ethics at the End of Life. *Section III:* 20.4.1 —Woods, Simon.

Debility and the Moral Imagination in Botswana. *Section III:* 7.1 —Livingston, Julie.

Defiant Birth: Women Who Resist Medical Eugenics. *Section III:* 15.5 —Tankard Reist, Melinda.

Deinstitutionalization and People with Intellectual Disabilities: In and Out of Institutions. *Section III:* 9.5.3 —Johnson, Kelley and Traustadóttir, Rannveig, eds.

Deliberative Environmental Politics: Democracy and Ecological Rationality. *Section III:* 16.1 —Baber, Walter F. and Bartlett, Robert V.

Democratic Autonomy: Public Reasoning About the Ends of Policy. *Section III:* 1.3.5 —Richardson, Henry S.

Denial of Aging: Perpetual Youth, Eternal Life, and Other Dangerous Fantasies, The. *Section III:* 9.5.2 —Gillick, Muriel R.

Development of Science-Based Guidelines for Laboratory Animal Care: Proceedings of the November 2003 International Workshop, The. *Section III:* 22.2 —Institute for Laboratory Animal Research (United States). International Workshop on the Development of Science-Based Guidelines for Laboratory Animal Care Program Committee, National Research Council.

Deviant Nurses and Improper Patient Care: A Study of Failure in the Medical Profession. *Section III:* 4.1.3 —Falk, Ursula A. and Falk, Gerhard.

Devil's Doctor: Paracelsus and the World of Renaissance Magic and Science, The. *Section III:* 5.1 —Ball, Philip.

Dew of Death: The Story of Lewisite, America's World War I Weapon of Mass Destruction. *Section III:* 21.3 —Vilensky, Joel A.

Diabetes in Pregnancy Dilemma: Leading Change with Proven Solutions, The. *Section III:* 9.5.5 —Langer, Oded, ed.

Diccionario de Bioética. *Section III:* 2.1 —Simón Vázquez, Carlos, ed.

Die Psychiatrie im Strudel der "Euthanasie": Ein Bericht über die erste industriemäßig durchgeführte Mordaktion des Naziregimes. *Section III:* 20.5.1 —Kaul, Friedrich Karl.

Diocesan Bishop's Pastoral Responsibility for the Ethically Correct Care of Critically Ill Patients at Catholic Health Care Facilities in the United States. *Section III:* 20.4.1 —Rowe, Charles.

Disability Advocacy Among Religious Organizations: Histories and Reflections. *Section III:* 9.5.1 —Herzog, Albert A., Jr., ed.

Disappearing Daughters: The Tragedy of Female Foeticide. *Section III:* 14.3 —Aravamudan, Gita.

Disclosure: The Hidden Weakness in Informed Consent. *Section III:* 8.3.1 —Brokken, David Allan.

Disease Control Priorities in Developing Countries. *Section III:* 9.1 —Jamison, Dean T.; et al., eds.

Dispelling the Myths of Abortion History. *Section III:* 12.4.1 —Dellapenna, Joseph W.

Disputing Doctors: The Socio-Legal Dynamics of Complaints About Medical Care. *Section III:* 9.8 —Mulcahy, Linda.

Dissecting a Discovery: The Real Story of How the Race to Uncover the Cause of AIDS Turned Scientists Against Disease, Politics Against Science, Nation Against Nation. *Section III:* 9.5.6 —Kontaratos, Nikolas.

Dizionario de bioetica. *Section III:* 2.1 —Lecaldano, Eugenio.

DNA: How the Biotech Revolution Is Changing the Way We Fight Disease. *Section III:* 15.1 —Stephenson, Frank H.

Doctors as Patients. *Section III:* 17.1 —Jones, Petre, ed.

Doctors on the Edge: Will Your Doctor Break the Rules for You? *Section III:* 2.1 —Abrams, Frederick R.

Doctors Who Rape: Malpractice and Misogyny. *Section III:* 8.5 —Trumpi, Pauline.

Documentation in Counseling Records: An Overview of Ethical, Legal, and Clinical Issues. *Section III:* 17.1 —Mitchell, Robert W.

Doing Honest Work in College: How to Prepare Citations, Avoid Plagiarism, and Achieve Real Academic Success. *Section III:* 1.3.7 —Lipson, Charles.

Doing Research with Children and Young People. *Section III:* 18.5.2 —Fraser, Sandy; Lewis, Vicky; Ding, Sharon; Kellett, Mary; and Robinson, Chris, eds.

Douleur et la souffrance. *Section III:* 4.4 —Perrotin, Catherine and Demaison, Michel, eds.

Drug Addiction. *Section III:* 9.5.9 —Council of Europe.

Drugs and Society: U.S. Public Policy. *Section III:* 9.5.9 —Fish, Jefferson M., ed.

Dworkin and His Critics: with Replies by Dworkin. *Section III:* 1.1 —Burley, Justine, ed.

E

Ectogenesis: Artificial Womb Technology and the Future of Human Reproduction. *Section III:* 14.1 —Gelfand, Scott and Shook, John R., eds.

Edge of Life: Human Dignity and Contemporary Bioethics, The. *Section III:* 2.1 —Kaczor, Christopher.

Effect of the HIPAA Privacy Rule on Health Research: Proceedings of a Workshop Presented to the National Cancer Policy Forum. *Section III:* 8.4 —Herdman, Roger and Moses, Harold L.

Egg & Sperm Race: The Seventeenth-Century Scientists Who Unravelled the Secrets of Sex, Life, and Growth. *Section III:* 14.1 —Cobb, Matthew.

Elements of Bioethics, The. *Section III:* 2.1 —Pence, Gregory E.

Elisha Bartlett's Philosophy of Medicine. *Section III:* 4.1.2 —Stempsey, William E., ed.

Embargoed Science. *Section III:* 1.3.7 —Kiernan, Vincent.

Embodying the Monster: Encounters with the Vulnerable Self. *Section III:* 4.4 —Shildrick, Margrit.

Embracing Life & Facing Death: A Jewish Guide to Palliative Care. *Section III:* 20.4.1 —Brenner, Daniel S.; Blanchard, Tsvi; Fins, Joseph J.; and Hirschfield, Bradley.

Emerging Technologies and Ethical Issues in Engineering: Papers from a Workshop, October 14-15, 2003. *Section III:* 1.3.4 —National Academy of Engineering (United States).

Emmanuel Levinas: His Life and Legacy. *Section III:* 1.2 —Malka, Salomon.

Empowering People with Severe Mental Illness: A Practical Guide. *Section III:* 17.1 —Linhorst, Donald M.

Empty Cradle of Democracy: Sex, Abortion, and Nationalism in Modern Greece, The. *Section III:* 12.5.1 —Halkias, Alexandra.

Enemy, The. *Section III:* 7.1 —Campo, Rafael.

Enhancing Evolution: The Ethical Case for Making Better People. *Section III:* 15.1 —Harris, John.

Ensuring Fairness in Health Care Coverage: An Employer's Guide to Making Good Decisions on Tough Issues. *Section III:* 9.31 —Wynia, Matthew K. and Schwab, Abraham P.

Essays on Hitler's Europe. *Section III:* 21.4 —Deák, István.

Essentials of Law and Ethics for Pharmacy Technicians. *Section III:* 9.7 —Strandberg, Kenneth M.

Essentials of Teaching and Learning in Nursing Ethics: Perspectives and Methods. *Section III:* 4.1.3 —Davis, Anne J.; Tschudin, Verena; and De Raeve, Louise, [eds.].

Establishing Medical Reality: Essays in the Metaphysics and Epistemology of Biomedical Science. *Section III:* 4.1.2 —Kincaid, Harold and McKitrick, Jennifer, eds.

Ethical Challenges for Intervening in Drug Use: Policy, Research and Treatment Issues. *Section III:* 9.5.9 —Kleinig, John and Einstein, Stanley, eds.

Ethical Conflicts in Psychology. *Section III:* 17.1 —Bersoff, Donald N.

Ethical Considerations for Research Involving Prisoners. *Section III:* 18.5.5 —Institute of Medicine (United States). Board on Health Sciences Policy. Committee on Ethical Considerations for Revisions to DHHS Regulations for Protection of Prisoners Involved in Research.

Ethical Decision Making in Clinical Neuropsychology. *Section III:* 17.1 —Bush, Shane S.

Ethical Health Care. *Section III:* 2.1 —Illingworth, Patricia and Parmet, Wendy E.

Ethical Issues in Community-Based Research with Children and Youth. *Section III:* 18.5.2 —Leadbeater, Bonnie; Banister, Elizabeth; Benoit, Cecilia; Jansson, Mikael; Marshall, Anne; Riecken, Ted, eds.

Ethical, Legal, and Professional Issues in the Practice of Marriage and Family Therapy. *Section III:* 17.2 —Wilcoxon, S. Allen; Remley, Jr., Theodore P.; Gladding, Samuel T.; and Huber, Charles H.

Ethical Practice in Brain Injury Rehabilitation. *Section III:* 17.1 —McGrath, Joanna Collicutt.

Ethical Reasoning for Mental Health Professionals. *Section III:* 17.1 —Ford, Gary G.

Ethical Standards in Social Work: A Review of the NASW Code of Ethics. *Section III:* 1.3.10 —Reamer, Frederic G.

Ethical Teacher, The. *Section III:* 1.3.3 —Campbell, Elizabeth.

Ethically Challenged: Big Questions for Science. *Section III:* 5.1 —Kennedy, Helena; Little, Miles; Pell, Cardinal George; Somerville, Margaret; Hilton, Douglas; and Jacobson, Lisa.

Ethics and Electronic Information in the Twenty-First Century. *Section III:* 1.3.12 —Pourciau, Lester J., ed.

Ethics and Epidemics. *Section III:* 9.1 —Balint, John; Philpott, Sean; Baker, Robert; and Strosberg, Martin, eds.

Ethics and Human Sciences. *Section III:* 2.1 —Amarakone, Keith and Panesar, Sukhmeet S.

Ethics and the New Genetics: An Integrated Approach. *Section III:* 15.1 —Monsour, Daniel, ed..

Ethics and the Pharmaceutical Industry. *Section III:* 9.7 —Santoro, Michael A. and Gorrie, Thomas M., eds.

Ethics Expertise: History, Contemporary Perspectives, and Applications. *Section III:* 2.1 —Rasmussen, Lisa, ed.

Ethics for Behavior Analysts: A Practical Guide to the Behavior Analyst Certification Board Guidelines for Responsible Conduct. *Section III:* 17.1 —Bailey, Jon S. and Burch, Mary R.

Ethics for Life Scientists. *Section III:* 1.3.9 —Korthals, Michiel and Bogers, Robert J., eds.

Ethics for Massage Therapists. *Section III:* 4.1.1 —Yardley-Nohr, Terrie.

Ethics, Human Rights and Culture: Beyond Relativism and Universalism. *Section III:* 21.7 —Li, Xiaorong.

Ethics in Health Administration: A Practical Approach for Decision Makers. *Section III:* 9.1 —Morrison, Eileen E.

Ethics in Medicine. *Section III:* 2.1 —Jackson, Jennifer.

Ethics in Qualitative Research. *Section III:* 18.4 —Mauthner, Melanie; Birch, Maxine; Jessop, Julie; and Miller, Tina, eds.

Ethics in Social Work: A Context of Caring. *Section III:* 1.3.10 —Guttmann, David.

Ethics, Law and Society, Volume I. *Section III:* 2.1 —Gunning, Jennifer and Holm, Søren, eds.

Ethics, Law and Society, Volume II. *Section III:* 2.1 —Gunning, Jennifer and Holm, Søren, eds.

Ethics, Law, and the Veterinary Nurse. *Section III:* 22.1 —Pullen, Sophie and Gray, Carol, eds.

Ethics, Legal Issues, and Professionalism in Surgical Technology. *Section III:* 2.1 —Jackson, Julia.

Ethics of Bioethics: Mapping the Moral Landscape, The. *Section III:* 2.1 —Eckenwiler, Lisa A. and Cohn, Felicia, eds.

Ethics of Biotechnology, The. *Section III:* 15.1 —Morris, Jonathan.

Ethics of Care: Personal, Political, and Global, The. *Section III:* 1.1 —Held, Virginia.

Ethics of Coercion in Mass Casualty Medicine, The. *Section III:* 8.3.4 —Trotter, Griffin.

Ethics of Human Cloning, The. *Section III:* 14.5 —Woodward, John, ed.

Ethics of Inheritable Genetic Modification: A Dividing Line?, The. *Section III:* 15.1 —Rasko, John E.J.; O'Sullivan, Gabrielle M.; and Ankeny, Rachel A., eds.

Ethics of Labeling in Mental Health, The. *Section III:* 4.3 —Madsen, Kristie and Leech, Peter.

Ethics of Librarianship: An International Survey, The. *Section III:* 1.3.12 —Vaagan, Robert, ed.; International Federation of Library Associations.

Ethics, Social Research and Consulting with Children and Young People. *Section III:* 18.5.2 —Alderson, Priscilla and Morrow, Virginia.

Ethics, Tools, and the Engineer. *Section III:* 1.3.4 —Spier, Raymond.

Éthique et bioéthique: L'assistance médicale à la procréation. *Section III:* 14.4 —Poilpot, Marie-Paule, ed.

Éthique et génétique: actes du colloque de Mouméa, 25 juillet 1997. *Section III:* 15.1 —Jean, Patrice and Régent, Julie; Université de Nouvelle-Calédonie. Département de droit.

Etica bio-medica. *Section III:* 2.1 —Spinsanti, Sandro.

Etica e bioetica per l'infermiere. *Section III:* 4.1.3 —Sala, Roberta.

Eugenics and Education in America: Institutionalized Racism and the Implications of History, Ideology, and Memory. *Section III:* 15.5 —Winfield, Ann Gibson.

Euthanasia—Choice and Death. *Section III:* 20.5.1 —Tulloch, Gail.

Euthanasie in Hadamar: Die nationalsozialistische Vernichtungspolitik in hessischen Anstalten: Begleitband: eine Ausstellung des Landeswohlfahrtsverbandes Hessen. *Section III:* 20.5.1 —Blasius, Dirk.

Evaluating the Science and Ethics of Research on Humans: A Guide for IRB Members. *Section III:* 18.2 —Mazur, Dennis J.

Evangelium amoris: corso di morale familiare e sessuale. *Section III:* 10 —Russo, Giovanni.

Everything Conceivable: How Assisted Reproduction Is Changing Men, Women, and the World. *Section III:* 14.1 —Mundy, Liza.

Evidence-Based Practice in Medicine and Health Care: A Discussion of the Ethical Issues. *Section III:* 9.8 —Meulen, Ruud ter; Biller-Andomo, Nikola; Lenk, Christian; and Lie, Reidar, K., eds.

Evolution of Cultural Diversity: A Phylogenetic Approach, The. *Section III:* 21.7 —Mace, Ruth; Holden, Clare J.; and Shennan, Stephen, eds.

Evolution of Intelligence: Are Humans the Only Animals with Minds?, The. *Section III:* 22.1 —Fetzer, James J.

Evolutionary Genetics: Concepts and Case Studies. *Section III:* 15.1 —Fox, Charles W. and Wolf, Jason B., eds.

Evolving World: Evolution in Everyday Life, The. *Section III:* 3.2 —Mindell, David P.

Ex-Gay Research: Analyzing the Spitzer Study and Its Relation to Science, Religion, Politics, and Culture. *Section III:* 10 —Drescher, Jack and Zucker, Kenneth J., eds.

Expanding Horizons in Bioethics. *Section III:* 2.1 —Galston, Arthur W. and Peppard, Christiana Z., eds.

Experiences of In Vitro Fertilization Donor Egg Recipients: The Impact of Technology on Reproduction. *Section III:* 14.4 —Denham, Melinda M.

Extreme Prematurity: Practices, Bioethics, and the Law. *Section III:* 20.5.2 —Miller, Geoffrey.

F

Facing Illness in Troubled Times: Health in Europe in the Interwar Years. *Section III:* 9.1 —Borowy, Iris and Gruner, Wolf D., eds.

Fads in Medical Care Management and Policy. *Section III:* 9.1 —Marmor, Theodore.

Families Following Assisted Conception: What Do We Tell Our Child? *Section III:* 14.2 —McWhinnie, Alexina M.

Family Medicine 2002 May; 34(5). *Section III:* 8.1 —Dedicated Issue on Physician-Patient Communication.

Fastest, Highest, Strongest: A Critique of High-Performance Sport. *Section III:* 9.5.1 —Beamish, Rob and Ritchie, Ian.

Fear: Anti-Semitism in Poland after Auschwitz. *Section III:* 21.4 —Gross, Jan T.

Section III arranged by NRCBL Classification Number, then Author or Title

Female Infanticide in India: A Feminist Cultural History. *Section III:* 20.5.2 —Bhatnagar, Rashmi Dube; Dube, Renu; and Dube, Reena.

Female Well-Being: Toward a Global Theory of Social Change. *Section III:* 9.5.5 —Billison, Janet Mancini and Fluehr-Lobban, Carolyn, eds.

Feminism and Motherhood in Western Europe, 1890-1970: The Maternal Dilemma. *Section III:* 9.5.5 —Allen, Ann Taylor.

Feminist Interventions in Ethics and Politics: Feminist Ethics and Social Theory. *Section III:* 10 —Andrew, Barbara S.; Keller, Jean; and Schwartzman, Lisa H., eds.

Field Guide to Good Decisions: Values in Action, A. *Section III:* 1.3.1 —Bennett, Mark D. and Gibson, Joan McIver.

Fifty Years after the Declaration: The United Nations' Record on Human Rights. *Section III:* 21.1 —Wagner, Teresa and Carbone, Leslie, eds.

Final Exam: A Surgeon's Reflections on Mortality. *Section III:* 20.3.2 —Chen, Pauline W.

Fiqh of Medicine: Responses in Islamic Jurisprudence to Developments in Medical Science, The. *Section III:* 2.1 —Yacoub, Ahmed Abdel Aziz.

First Do No Harm: Law, Ethics and Healthcare. *Section III:* 2.1 —McLean, Sheila A.M., ed.

Fit to Be Citizens? Public Health and Race in Los Angeles, 1879-1939. *Section III:* 9.5.4 —Molina, Natalia.

For the Sake of Heaven and Earth: The New Encounter Between Judaism and Christianity. *Section III:* 1.2 —Greenberg, Irving.

Fractured Borders: Reading Women's Cancer Literature. *Section III:* 7.1 —DeShazer, Mary K.

Fractured States: Smallpox, Public Health and Vaccination Policy in British India, 1800-1947. *Section III:* 9.1 —Bhattacharya, Sanjoy; Harrison, Mark; and Worboys, Michael.

Fratricide in the Holy Land: A Psychoanalytic View of the Arab-Israeli Conflict. *Section III:* 1.3.6 —Falk, Avner.

Freedom and Responsibility in Reproductive Choice. *Section III:* 14.1 —Spencer, J.R. and Du Bois-Pedain, Antje, eds.

Freiheit, die mehr ist als Willkür: christliche Ethik in zwischenmenschlicher Beziehung, Lebensgestaltung, Krankheit und Tod. *Section III:* 1.2 —Römelt, Josef.

Freud's Free Clinics: Psychoanalysis & Social Justice, 1918-1938. *Section III:* 17.2 —Danto, Elizabeth Ann.

From a Ruined Garden: The Memorial Books of Polish Jewry. *Section III:* 21.4 —Kugelmass, Jack; Boyarin, Jonathan; and Baker, Zachary M., eds.

From Clones to Claims: The European Patent Office's Case Law on the Patentability of Biotechnology Inventions in Comparison to the United States and Japanese Practice. *Section III:* 15.8 —Jaenichen, Hans-Rainer; McDonell, Leslie A.; Haley, James F.; and Hosoda, Yoshinori.

From the Protocols of the Elders of Zion to Holocaust Denial Trials: Challenging the Media, the Law and the Academy. *Section III:* 21.4 —Kaufman, Debra; Herman, Gerlad; Ross, James; and Phillips, David, eds.

From the Womb to the Tomb: Issues in Medical Ethics. *Section III:* 2.1 —McLachlan, Hugh V. and Swales, J. Kim.

From This World to the Next: Jewish Approaches to Illness, Death & the Afterlife. *Section III:* 20.1 —Jewish Theological Library.

Future Human Evolution: Eugenics in the Twenty-First Century. *Section III:* 15.5 —Glad, John.

Future Perfect? God, Medicine and Human Identity. *Section III:* 4.4 —Deane-Drummond, Celia and Scott, Peter Manley, eds.

G

Gender Struggles: Practical Approaches to Contemporary Feminism. *Section III:* 10 —Mui, Constance L. and Murphy, Julien S., eds.

Gene Therapy. *Section III:* 15.4 —Naff, Clay Farris, ed.

Generating Bodies and Gendered Selves: The Rhetoric of Reproduction in Early Modern England. *Section III:* 14.1 —Keller, Eve.

Generation at Risk: The Global Impact of HIV/AIDS on Orphans and Vulnerable Children, A. *Section III:* 9.5.6 —Foster, Geoff; Levine, Carol; and Williamson, John, eds.

Genes, Behavior, and the Social Environment: Moving Beyond the Nature/Nurture Debate. *Section III:* 15.6 —Institute of Medicine (United States). Board on Health Sciences Policy. Committee on Assessing Interactions Among Social, Behavioral, and Genetic Factors in Health.

Genetic Engineering, Christian Values and Catholic Teaching. *Section III:* 15.1 —Flaman, Paul.

Genetic Screening: New Research. *Section III:* 15.3 —Pupecki, Sandra R.

Genetic Technologies and the Law. *Section III:* 15.1 —Kuszler, Patricia; Battuello, Kathryn; and O'Connor, Sean.

Genetics and Christian Ethics. *Section III:* 15.1 —Deane-Drummond, Celia.

Genetics and Genetic Engineering. *Section III:* 15.1 —Wexler, Barbara.

Genetics and Health: Policy Issues for Genetic Science and Their Implications for Health and Health Services. *Section III:* 15.1 —Zimmern, Ron and Cook, Christopher.

Genocide in Rwanda: Complicity of the Churches? *Section III:* 21.4 —Rittner, Carol; Roth, John K.; and Whitworth, Wendy, eds.

Genomic Medicine: Articles from the New England Journal of Medicine. *Section III:* 15.3 —Guttmacher, Alan E.; Collins, Francis S.; and Drazen, Jeffrey M., eds.

Genomic Revolution: Implications for Treatment and Control of Infectious Disease: Working Group Summaries, The. *Section III:* 15.1 —National Academies (United States). Keck Futures Initiative.

German and Jew: The Life and Death of Sigmund Stein. *Section III:* 21.4 —Dickinson, John K.

Global Bioethics: The Collapse of Consensus. *Section III:* 2.1 —Engelhardt, H. Tristram, eds.

Globalization and Health: Challenges for Health Law and Bioethics. *Section III:* 2.1 —Bennett, Belinda and Tomossy, George F., eds.

Globalization, Biosecurity, and the Future of the Life Sciences. *Section III:* 21.3 —National Research Council (United States). Committee on Advances in Technology and the Prevention of Their Application to Next Generation Biowarfare Threats; Institute of Medicine (United States). Board on Global Health.

Good Doctor: A Novel, The. *Section III:* 9.1 —Galgut, Damon.

Governing China's Population: From Leninist to Neoliberal Biopolitics. *Section III:* 13.3 —Greenhalgh, Susan and Winckler, Edwin A.

Great Starvation Experiment: The Heroic Men Who Starved So That Millions Could Live, The. *Section III:* 18.5.1 —Tucker, Todd.

Griswold v. Connecticut: Birth Control and the Constitutional Right to Privacy. *Section III:* 11.1 —Johnson, John W.

Gulf War and Health, Volume 4: Health Effects of Serving in the Gulf War. *Section III:* 9.5.1 —Institute of Medicine (United States). Board on Population Health and Public Health Practice. Committee on Gulf War and Health: A Review of the Medical Literature Relative to the Gulf War Veterans' Health.

H

"Haltestation Philippshospital": ein psychiatrisches Zentrum - Kontinuität und Wandel 1535-1904-2004; eine Festschrift zum 500. Geburtstag Philipps von Hessen. *Section III:* 17.1 —Sahmland, Irmtraut, ed.

Handbook for Health Care Ethics Committees. *Section III:* 9.6 —Post, Linda Farber; Blustein, Jeffrey; and Dubler, Nancy Neveloff.

Handbook of Cultural Psychiatry. *Section III:* 21.7 —Tseng, Wen-Shing.

Handbook of Genome Research: Genomics, Proteomics, Metabolomics, Bioinformatics, Ethical and Legal Issues. *Section III:* 15.1 —Sensen, Christopher, ed.

Harvey Cushing: A Life in Surgery. *Section III:* 7.1 —Bliss, Michael.

Haskalah and Hasidism in the Kingdom of Poland: A History of Conflict. *Section III:* 1.2 —Wodzinski, Marcin.

Healing and the Jewish Imagination: Spiritual and Practical Perspectives on Judaism and Health. *Section III:* 1.2 —Cutter, William, ed.

Health & Environmental Impact Assessment: An Integrated Approach. *Section III:* 16.1 —British Medical Association.

Health and Human Rights: Basic International Documents. *Section III:* 21.1 —Marks, Stephen P., ed.

Health and the Good Society: Setting Healthcare Ethics in Social Context. *Section III:* 7.1 —Cribb, Alan.

Health Care at Risk: A Critique of the Consumer-Driven Movement. *Section III:* 9.1 —Jost, Timothy Stoltzfus.

Health Care Ethics: A Catholic Theological Analysis. *Section III:* 2.1 —Ashley, Benedict M.; DeBlois, Jean; and O'Rourke, Kevin D.

Health Care Ethics in Canada. *Section III:* 2.1 —Baylis, Françoise; Downie, Jocelyn; Freedman, Benjamin; Hoffmaster, Barry; and Sherwin, Susan.

Health Care Politics , Policy, and Services: A Social Justice Analysis. *Section III:* 9.1 —Almgren, Gunnar.

Health Care Reform: Ethics and Politics. *Section III:* 9.1 —Engström, Timothy H. and Robison, Wade L., eds.

Health Care Reform Now! A Prescription for Change. *Section III:* 9.3.1 —Halvorson, George C.

Health Care zwischen Ethik und Recht. *Section III:* 2.1 —Wallner, Jürgen.

Health Law, Human Rights and the Biomedicine Convention: Essays in Honour of Henriette Roscam Abbing. *Section III:* 21.1 —Gevers, J.K.M.; Hondius, E.H.; Hubben, J.H., eds.

Health Policy and the Uninsured. *Section III:* 9.3.1 —McLaughlin, Catherine G., ed.

Health Promotion: Philosophy, Prejudice and Practice. *Section III:* 9.1 —Seedhouse, David.

Healthcare Fix: Universal Insurance for All Americans, The. *Section III:* 9.3.1 —Kotlikoff, Laurence J.

Healthy Waters: What Every Health Professional Should Know About Water. *Section III:* 16.1 —Magee, Mike.

Heart Has Reasons: Holocaust Rescuers and Their Stories of Courage, The. *Section III:* 21.4 —Klempner, Mark.

Hebrew Ethical Wills. *Section III:* 1.2 —Abrahams, Israel and Fine, Lawrence.

Hidden Prejudice: Mental Disability on Trial, The. *Section III:* 4.3 —Perlin, Michael L.

High Tech Trash: Digital Devices, Hidden Toxins, and Human Health. *Section III:* 1.3.12 —Grossman, Elizabeth.

Hirntod als Wertverhalt: Medizinethische Bausteine aus Jonas Cohns Wertwissenschaft und Maimonides' Theologie. *Section III:* 20.2.1 —Wiedebach, Hartwig.

History of Old Age, A. *Section III:* 9.5.2 —Thane, Pat, ed.

HIV Pandemic: Local and Global Implications, The. *Section III:* 9.5.6 —Beck, Eduard J., Mays, Nicholas; Whiteside, Alan W.; and Zuniga, José M., eds.

HIV/AIDS in South Africa. *Section III:* 9.5.6 —Abdool Karim, S.S. and Abdool Karim, Q., eds.

Holocaust and Catholic Conscience: Cardinal Aloisius Muench and the Guilt Question in Germany, The. *Section III:* 21.4 —Brown-Fleming, Suzanne.

Holocaust Controversy: The Treblinka Affair in Postwar France, A. *Section III:* 21.4 —Moyn, Samuel.

Holocaust: Critical Historical Approaches, The. *Section III:* 21.4 —Bloxham, Donald and Kushner, Tony.

Holocaust in the Ukraine. *Section III:* 21.4 —Zabarko, Boris, ed.

Hooked: Ethics, the Medical Profession, and the Pharmaceutical Industry. *Section III:* 9.7 —Brody, Howard.

How Doctors Think. *Section III:* 8.1 —Groopman, Jerome.

How the Pro-Choice Movement Saved America: Freedom, Politics, and the War on Sex. *Section III:* 12.3 —Page, Cristina.

Human Dignity and Human Cloning. *Section III:* 14.5 —Vöneky, Silja and Wolfrum, Rüdiger, eds.

Human Life, Action, and Ethics: Essays by G.E.M. Anscombe. *Section III:* 1.1 —Geach, Mary and Gormally, Luke, eds.

Human Remains: Dissection and Its Histories. *Section III:* 20.1 —MacDonald, Helen.

Human Stem Cells: Source of Hope and of Controversy: A Study of the Ethics of Human Stem Cell Research and the Patenting of Related Inventions. *Section III:* 18.5.4 —Jochemsen, Henk, ed.

Humane Society of the United States Euthanasia Training Manual, The. *Section III:* 22.1 —Rhoades, Rebecca H.

Humangenetik: ethische Probleme der Beratung, Diagnostik und Forschung: [Tagung vom 21. bis 23. November 1991 in Göttingen]. *Section III:* 15.2 —Schöne-Seifert, Bettina; Krüger, Lorenz; and Toellner, Richard, eds.

Humanity Before God: Contemporary Faces of Jewish, Christian, and Islamic Ethics. *Section III:* 1.2 —Schweiker, William; Johnson, Michael A.; and Jung, Kevin, eds.

Humans and Other Animals in Eighteenth-Century British Culture: Representation, Hybridity, Ethics. *Section III:* 22.1 —Palmeri, Frank, ed.

I

I Was a Child of Holocaust Survivors. *Section III:* 21.4 —Eisenstein, Bernice.

IACUC Handbook, The. *Section III:* 22.2 —Silverman, Jerald; Suckow, Mark A.; Murthy, Sreekant, eds.

Identification of Curricular and Educational Needs for Primary Care Physicians in Dealing with the Clinical Application of Genomic Medicine. *Section III:* 7.2 —Rhodes, Teresa [Ann Reitsma].

Identity and Health. *Section III:* 7.1 —Kelleher, David and Leavey, Gerard, eds.

Impact of For-Profit Ownership on Hospice Care, The. *Section III:* 20.4.1 —Carlson, Melissa Diane Aldridge.

Impact of the Holocaust on Jewish Theology, The. *Section III:* 21.4 —Katz, Steven T., ed.

Impact of the Internet on Our Moral Lives, The. *Section III:* 1.3.12 —Cavalier, Robert J.

Import of Human Embryonic Stem Cells: Opinion, The. *Section III:* 18.5.4 —German National Ethics Council = Nationaler Ethikrat.

Impotence: A Cultural History. *Section III:* 10 —McLaren, Angus.

Improving Patient Care: The Implementation of Change in Clinical Practice. *Section III:* 9.8 —Grol, Richard; Wensing, Michel; and Eccles, Martin.

In Search of Memory: The Emergence of a New Science of Mind. *Section III:* 17.1 —Kandel, Eric R.

Inclusion: The Politics of Difference in Medical Research. *Section III:* 18.6 —Epstein, Steven.

Inequity and Madness: Psychosocial and Human Rights Issues. *Section III:* 17.1 —Guimón, José.

Informing Patients: An Assessment of the Quality of Patient Information Materials. *Section III:* 9.8 —Coulter, Angela; Entwistle, Vikki; and Gilbert David.

Innovation and Its Discontents: How Our Broken Patent System Is Endangering Innovation and Progress, and What to Do About It. *Section III:* 5.3 —Jaffe, Adam B. and Lerner, Josh.

See inside front cover for NRCBL Classification Scheme

Innovation in Medical Technology: Ethical Issues and Challenges. *Section III:* 1.3.9 —Eaton, Margaret L. and Kennedy, Donald.

Institutional Review Board Member Handbook. *Section III:* 18.2 —Amdur, Robert J. and Bankert, Elizabeth A.

Intellectual Property, Biogenetic Resources and Traditional Knowledge. *Section III:* 15.8 —Dutfield, Graham.

Intelligent Thought: Science Versus the Intelligent Design Movement. *Section III:* 3.2 —Brockman, John, ed.

International Environmental Law: Fairness, Effectiveness, and World Order. *Section III:* 16.1 —Louka, Elli.

Intervening in the Brain: Changing Psyche and Society. *Section III:* 17.1 —Merkel, R.; Boer, G.; Fegert, J.; Galert, T.; Hartmann, D.; Nuttin, B.; and Rosahl, S.

Intricate Ethics: Rights, Responsibilities, and Permissible Harm. *Section III:* 1.1 —Kamm, F.M.

Introduction to Veterinary Medical Ethics: Theory and Cases, An. *Section III:* 22.1 —Rollin, Bernard E.

Intuition: A Novel. *Section III:* 1.3.9 —Goodman, Allegra.

Islamic Bioethics: Problems and Perspectives. *Section III:* 2.1 —Atighetchi, Dariusch.

Islamic Family Law in a Changing World: A Global Resource Book. *Section III:* 1.2 —Na'im, 'Abd Allah Ahmad, ed.

Issues and Ethics in the Helping Professions. *Section III:* 17.2 —Corey, Gerald; Corey, Marianne Schneider; and Callanan, Patrick.

Ivory Tower and Industrial Innovation: University-Industry Technology Transfer Before and After the Bayh-Dole Act in the United States. *Section III:* 5.3 —Mowery, David C.; Nelson, Richard R.; Sampat, Bhaven N.; and Ziedonis, Arvids A.

J

Jewish Enemy: Nazi Propaganda During World War II and the Holocaust, The. *Section III:* 1.3.5 —Herf, Jeffrey.

Jewish Ethics and the Care of End-of-Life Patients: A Collection of Rabbinical, Bioethical, Philosophical, and Juristic Opinions. *Section III:* 20.4.1 —Hurwitz, Peter Joel; Picard, Jacques; and Steinberg, Avraham, eds.

Jews in France during World War II. *Section III:* 21.4 —Poznanski, Renée.

Journal of Gender, Race and Justice 2005 Fall; 9(1). *Section III:* 14.4 —Creating Life? Examining the Legal, Ethical and Medical Issues of Assisted Reproductive Technologies [Part 1].

Journal, 1935-1944: The Fascist Years. *Section III:* 21.4 —Sebastian, Mihail.

Judaism, Science, and Moral Responsibility. *Section III:* 1.2 —Berger, Yitzhak and Shatz, David, eds.

Just Medicare: What's In, What's Out, How We Decide. *Section III:* 9.1 —Flood, Colleen M., ed.

Justifying Coercion: Nurses' Experiences Medicating Involuntary Psychiatric Patients. *Section III:* 17.4 —Vuckovich, Paula K.

K

Keeping Faith with Nature: Ecosystems, Democracy, and America's Public Lands. *Section III:* 16.1 —Keiter, Robert B.

"Kein Recht auf Leben": Beiträge und Dokumente zur Entrechtung und Vernichtung "lebensunwerten Lebens" im Nationalsozialismus. *Section III:* 20.5.1 —Tuchel, Johannes.

Kinder in der NS-Psychiatrie. *Section III:* 20.5.2 —Beddies, Thomas and Hübener, Kristina, eds.

Kindheit, Eugenik, Sozialpädagogik: Festschrift zum 60. Geburtstag von Jürgen Reyer. *Section III:* 15.5 —Barth, Gernot and Henseler, Joachim, eds.

Krankheitserfinder: Wie wir zu Patienten gemacht werden, Die. *Section III:* 9.7 —Blech, Jörg.

L

L'euthanasie et le droit: état des lieux sur un sujet médiatisé. *Section III:* 20.5.1 —Legros, Bérengère.

Landesanstalt Neuruppin in der NS-Zeit, Die. *Section III:* 20.5.1 —Schulze, Dietmar.

Last Best Gifts: Altruism and the Market for Human Blood and Organs. *Section III:* 19.5 —Healy, Kieran.

Last Town on Earth, The. *Section III:* 9.5.1 —Mullen, Thomas.

Law and Ethics in Biomedical Research: Regulation, Conflict of Interest, and Liability. *Section III:* 18.1 —Lemmens, Trudo and Waring, Duff R., eds.

Law, Medicine and Ethics. *Section III:* 2.1 —King, Patricia A.; Areen, Judith; and Gostin, Lawrence O.

Law, Politics, and Morality in Judaism. *Section III:* 1.2 —Walzer, Michael, ed.

Leben in der Hand des Menschen: die Brisanz des biotechnischen Fortschritts. *Section III:* 5.1 —Altner, Günter.

Legal & Ethical Nursing. *Section III:* 4.1.3 —Gauwitz, Donna F.

Legal and Ethical Aspects of Anaesthesia: Critical Care and Perioperative Medicine. *Section III:* 2.1 —White, Stuart M. and Baldwin, Timothy J.

Legal Aspects of Health Care Administration. *Section III:* 9.1 —Pozgar, George D.

Lessons in Mortality: Doctors and Patients Struggling Together. *Section III:* 20.3.2 —Weisse, Allen B.

Let Them Eat Precaution: How Politics Is Undermining the Genetic Revolution in Agriculture. *Section III:* 15.1 —Entine, Jon, ed.

Let Them Go Free: A Guide for Withdrawing Life Support. *Section III:* 20.5.1 —Shannon, Thomas A. and Faso, Charles N.

Life and Death in Intensive Care. *Section III:* 20.5.1 —Cassell, Joan.

Life, Love and Children: A Practical Introduction to Bioscience Ethics and Bioethics. *Section III:* 2.1 —Pollard, Irina.

Life (Un)Worthy of Living: Reproductive Genetics in Israel and Germany, A. *Section III:* 15.3 —Hashiloni-Dolev, Yael.

Limits of Medicine, The. *Section III:* 4.1.2 —Stark, Andrew.

Lived Experience of Nurses Providing Futile Care to Critically Ill Newborns and Infants, The. *Section III:* 20.4.2 —Bauman, Renea A.

Living Professionalism: Reflections on the Practice of Medicine. *Section III:* 4.1.2 —Egan, Erin A. and Surdyk, Patricia M.

Lódz Ghetto: A History. *Section III:* 21.4 —Trunk, Isaiah.

Long Walk Home, A. *Section III:* 9.5.1 —Clark, Rachel.

Looking Within: A Sociocultural Examination of Fetoscopy. *Section III:* 9.5.8 —Blizzard, Deborah.

Lost: A Search for Six of Six Million, The. *Section III:* 21.4 —Mendelsohn, Daniel.

Lost: Illegal Abortion Stories. *Section III:* 12.5.2 —Wainer, Jo.

Lost Virtue: Professional Character Development in Medical Education. *Section III:* 2.3 —Kenny, Nuala and Shelton, Wayne, eds.

Lying and Illness: Power and Performance. *Section III:* 8.2 —Dongen, Elis van and Fainzang, Sylvie, eds.

M

Mad, Bad and Dangerous? The Scientist and the Cinema. *Section III:* 5.1 —Frayling, Christopher.

Mãe para Mãe: Questões Legais e Éticas Suscitadas pela Maternidade de Substituição, De. *Section III:* 14.2 —Raposo, Vera Lúcia.

Maimonides' Confrontation with Mysticism. *Section III:* 1.2 —Kellner, Menachem.

Maimonides on the Origin of the World. *Section III:* 1.2 —Seeskin, Kenneth.

Mainstreaming Midwives: The Politics of Change. *Section III:* 9.1 —Davis-Floyd, Robbie and Johnson, Christine Barbara, eds.

Making Health Care Decisions: A Catholic Guide. *Section III:* 2.1 —Hamel, Ron, ed.

Making of the Fittest: DNA and the Ultimate Forensic Record of Evolution, The. *Section III:* 15.1 —Carroll, Sean B.

Making Sense of Medical Ethics: A Hands-On Guide. *Section III:* 2.1 —Johnson, Alan G. and Johnson, Paul R.V.

Managing Care: A Shared Responsibility. *Section III:* 9.3.2 —Verheijde, Joseph L.

Marketer's Guide to HIPAA: Resources for Creating Effective and Compliant Marketing, A. *Section III:* 8.4 —Borten, Kate.

Marriage & Morals in Islam. *Section III:* 1.2 —Rizvi, Sayyid Muhammad.

Mason and McCall Smith's Law and Medical Ethics. *Section III:* 2.1 —Mason, J.K. and Laurie, G.T.

Material Virtue: Ethics and the Body in Early China. *Section III:* 1.2 —Csikszentmihalyi, Mark.

Matter of Size: Triennial Review of the National Nanotechnology Initiative, A. *Section III:* 5.4 —National Research Council (United States). Committee to Review of the National Nanotechnology Initiative.

Meaning in Suffering: Caring Practices in the Health Professions. *Section III:* 4.4 —Johnston, Nancy E. and Scholler-Jaquish, Alwilda, eds.

Meaning, Medicine, and the "Placebo Effect". *Section III:* 9.1 —Moerman, Daniel E.

Means, Ends and Medical Care. *Section III:* 4.1.2 —Wright, H.G.

Médecins et la mort: XIXe-XXe siècle, Les. *Section III:* 20.1 —Carol, Anne.

Mediating Effect of Public Opinion on Public Policy: Exploring the Realm of Health Care, The. *Section III:* 9.1 —Chard, Richard E.

Medical Apartheid: The Dark History of Medical Experimentation on Black Americans from Colonial Times to the Present. *Section III:* 18.5.1 —Washington, Harriet A.

Medical Law and Ethics. *Section III:* 2.1 —Howard, Philip and Bogle, James.

Medical Malpractice Myth, The. *Section III:* 8.5 —Baker, Tom.

Medical Management of Vulnerable and Underserved Patients: Principles, Practice, and Populations. *Section III:* 9.5.10 —King, Jr., Talmadge E. and Wheeler, Margaret B.

Medicalized Masculinities. *Section III:* 10 —Rosenfeld, Dana and Faircloth, Christopher A., eds.

Medicare Meets Mephistopheles. *Section III:* 9.3.1 —Hyman, David A.

Medicine and Care of the Dying: A Modern History. *Section III:* 20.4.1 —Lewis, Milton J.

Medicine and Philosophy in Classical Antiquity: Doctors and Philosophers on Nature, Soul, Health and Disease. *Section III:* 4.1.2 —Van der Eijk, Philip J.

Medicine Buddha Teachings. *Section III:* 4.1.1 —Thrangu, Khenchen Rinpoche.

Medicine for Lawyers. *Section III:* 9.1 —Palmer, Roy and Wetherill, Diana, eds.

Medicine in Quotations: Views on Health and Disease Through the Ages. *Section III:* 7.1 —Huth, Edward J. and Murray, T. Jock, eds.

Medicine, Its Marketplace, and the American Dream. *Section III:* 9.1 —Dickinson, Taylor.

Medicine Murder in Colonial Lesotho: The Anatomy of a Moral Crisis. *Section III:* 4.1.1 —Murray, Colin and Sanders, Peter.

Medicine, Patients and the Law. *Section III:* 2.1 —Brazier, Margaret and Cave, Emma.

Medicine: Preserving the Passion in the 21st Century. *Section III:* 7.2 —Manning, Phil R. and DeBakey, L.

Mental Health and Well-Being in Animals. *Section III:* 22.1 —McMillan, Franklin D., ed.

Methods in Community-Based Participatory Research for Health. *Section III:* 9.1 —Israel, Barbara A.; Eng, Eugenia; Schulz, Amy J.; and Parker, Edith A., eds.

Missing Gene: Psychiatry, Heredity, and the Fruitless Search for Genes, The. *Section III:* 15.6 —Joseph, Jay.

Modelos de bioética clínica: Presentación crítica del principialismo y la casuística. *Section III:* 2.1 —Requena Meana, Pablo.

Money-Driven Medicine: The Real Reason Health Care Costs So Much. *Section III:* 9.3.1 —Mahar, Maggie.

Moral Significance of Intimacy in Nurse-Patient Relationships, The. *Section III:* 8.1 —Kirk, Timothy W.

Moral, Social, and Commercial Imperatives of Genetic Testing and Screening: The Australian Case, The. *Section III:* 15.3 —Betta, Michael, ed.

Moral State We're In: A Manifesto for a 21st Century Society, The. *Section III:* 7.1 —Neuberger, Julia.

Moral Stealth: How "Correct Behavior" Insinuates Itself into Psychotherapeutic Practice. *Section III:* 17.2 —Goldberg, Arnold.

Moving Beyond Prozac, DSM, and the New Psychiatry: The Birth of Postpsychiatry. *Section III:* 4.3 —Lewis, Bradley.

Multicultural Understanding of Child and Adolescent Psychopathology: Implications for Mental Health Assessment. *Section III:* 21.7 —Achenbach, Thomas M. and Rescorla, Leslie A.

My Mother's Hip: Lessons from the World of Eldercare. *Section III:* 9.5.2 —Margolies, Luisa.

Myth of Depression as Disease: Limitations and Alternatives to Drug Treatment, The. *Section III:* 17.4 —Leventhal, Allan M. and Martell, Christopher R.

N

Nanofuture: What's Next for Nanotechnology. *Section III:* 5.4 —Hall, J. Storrs.

Nanotalk: Conversations with Scientists and Engineers About Ethics, Meaning, and Belief in the Development of Nanotechnology. *Section III:* 5.4 —Berne, Rosalyn W.

Nanotech Pioneers: Where Are They Taking Us?, The. *Section III:* 5.4 —Edwards, Steven A.

Nanotechnology: New Promises, New Dangers. *Section III:* 5.4 —Shelley, Toby.

Nanotechnology: Risk, Ethics and Law. *Section III:* 5.4 —Hunt, Geoffrey and Mehta, Michael D., eds.

Narrative Matters: The Power of the Personal Essay in Health Policy. *Section III:* 7.1 —Mullan, Fitzhugh; Ficklen, Ellen; and Rubin, Kyna, eds.

Narrative Medicine: Honoring the Stories of Illness. *Section III:* 7.1 —Charon, Rita.

Narrative-Based Primary Care: A Practical Guide. *Section III:* 8.1 —Launer, John.

Nation's Doctor: The Role of the Chief Medical Officer, 1855-1998, The. *Section III:* 9.1 —Sheard, Sally and Donaldson, Liam.

Natural Moralities: A Defense of Pluralistic Relativism. *Section III:* 1.1 —Wong, David B.

Nazism, 1919-1945, Volume 3: Foreign Policy, War and Racial Extermination: A Documentary Reader. *Section III:* 21.4 —Noakes, J. and Pridham, G., eds.

Neither Brain nor Ghost: A Nondualist Alternative to the Mind-Body Identity Theory. *Section III:* 3.1 —Rockwell, W. Teed.

Neonatal Bioethics: The Moral Challenges of Medical Innovation. *Section III:* 20.5.2 —Lantos, John D. and Meadow, William L.

Networking: Communicating with Bodies and Machines in the Nineteenth Century. *Section III:* 1.3.12 —Otis, Laura.

Neuroethics: Defining the Issues in Theory, Practice, and Policy. *Section III:* 17.1 —Illes, Judy, ed.

Neuroscience and the Law: Brain, Mind, and the Scales of Justice: A Report on an Invitational Meeting Convened by the American Association for the Advancement of Science and the Dana Foundation. *Section III:* 17.1—Garland, Brent, ed.

New Medicines: How Drugs Are Created, Approved, Marketed, and Sold, The. *Section III:* 9.7 —Schacter, Bernice.

New Politics of the NHS: From Creation to Reinvention, The. *Section III:* 9.1 —Klein, Rudolf.

Next: A Novel. *Section III:* 15.1 —Crichton, Michael.

Night. *Section III:* 21.4 —Wiesel, Elie.

"Non-Germans" Under the Third Reich: The Nazi Judicial and Administrative System in Germany and Occupied Eastern Europe, with Special Regard to Occupied Poland, 1939-1945. *Section III:* 21.4 —Majer, Diemut.

Normal Heart, The. *Section III:* 9.5.6 —Kramer, Larry.

Normativité et biomédecine. *Section III:* 2.1 —Feuillet-Le-Mintier, Brigitte.

Not Quite White: White Trash and the Boundaries of Whiteness. *Section III:* 15.5 —Wray, Matt.

Nuremberg Interviews: An American Psychiatrist's Conversations with the Defendants and Witnesses, The. *Section III:* 21.4 —Goldensohn, Leon.

Nursing Ethics: A Virtue-Based Approach. *Section III:* 4.1.3 —Armstrong, Alan E.

Nursing Ethics: Irish Cases and Concerns. *Section III:* 4.1.3 —Dooley, Dolores and McCarthy, Joan.

O

Oath Betrayed: Torture, Medical Complicity, and the War on Terror. *Section III:* 21.4 —Miles, Steven.

Of Others Inside: Insanity, Addiction, and Belonging in America. *Section III:* 17.1 —Weinberg, Darin.

Ökologie und Ethik: ein Versuch praktischer Philosophie. *Section III:* 16.1 —Ott, Konrad.

On Becoming a Person: A Therapist's View of Psychotherapy. *Section III:* 17.2 —Rogers, Carl R.

On Genetic Interests: Family, Ethnicity, and Humanity in an Age of Mass Migration. *Section III:* 15.9 —Salter, Frank.

On the Stigma of Mental Illness: Practical Strategies for Research and Social Change. *Section III:* 17.1 —Corrigan, Patrick W., ed.

Organisierte Vernichtung "lebensunwerten Lebens" im Rahmen der "Aktion T4" : Dargestellt am Beispiel des Wirkens und der strafrechtlichen Verfolgung ausgewählter NS-Tötungsärzte, Die. *Section III:* 20.5.1 —Greve, Michael.

Osgoode Hall Law Journal 2006 Summer; 44(2). *Section III:* 9.3.1 . Ryder, Bruce, guest ed.

Our Human Hearts: A Medical and Cultural Journey. *Section III:* 7.1 —Carter, Albert Howard III.

Oxford Handbook of Bioethics, The. *Section III:* 2.1 —Steinbock, Bonnie, ed.

Oxford Textbook of Palliative Care for Children. *Section III:* 20.4.2 —Goldman, Ann; Hain, Richard; and Liben, Stephen, eds.

Oxford Textbook of Philosophy and Psychiatry. *Section III:* 4.3 —Fulford, K.W.M. (Bill); Thornton, Tim; and Graham, George.

Oxford Textbook of Public Health. *Section III:* 9.1 —Detels, Roger; McEwen, James; Beaglehole, Robert; and Tanaka, Heizo, eds.

P

Pain: New Essays on Its Nature and the Methodology of Its Study. *Section III:* 4.4 —Aydede, Murat, ed.

Palliative Care for Infants, Children, and Adolescents: A Practical Handbook. *Section III:* 20.4.2 —Carter, Brian S. and Levetown, Marcia, eds.

Panic Diaries: A Genealogy of Panic Disorders. *Section III:* 17.1 —Orr, Jackie.

Paradox of Anti-Semitism, The. *Section III:* 1.2 —Cohn-Sherbok, Dan.

Partnership with the Dying: Where Medicine and Ministry Should Meet. *Section III:* 20.4.1 —Smith, David H.

Party of Death: The Democrats, the Media, the Courts, and the Disregard for Human Life, The. *Section III:* 21.1 —Ponnuru, Ramesh.

Patentes de Genes Humanos? *Section III:* 15.8 —Remédio Marques, J.P. and de Ética, Conselho Dinamarquês.

Patenting of Biotechnological Inventions Involving the Use of Biological Material of Human Origin: Opinion, The. *Section III:* 15.8 —Nationaler Ethikrat.

Perfect Motherhood: Science and Childbearing in America. *Section III:* 9.5.5 —Apple, Rima D.

Perspectives on Health and Human Rights. *Section III:* 21.1 —Gruskin, Sofia; Grodin, Michael A.; Annas, George J.; and Marks, Stephen P., eds.

Pharmaceutical Reason: Knowledge and Value in Global Psychiatry. *Section III:* 17.4 —Lakoff, Andrew.

Philosophy and Psychiatry. *Section III:* 17.1 —Schramme, Thomas and Thome, Johannes, eds.

Physician Integrity, Religious Belief, and the Adequacy of Medical Ethics. *Section III:* 2.1 —Kaldjian, Lauris Christopher.

Pius XII, the Holocaust and the Revisionists. *Section III:* 21.4 —Gallo, Patrick J., ed.

Place de la bioéthique en recherche et dans les services cliniques. *Section III:* 2.1 —Hervé, Christian; Knoppers, Bartha Maria; Molinari, Patrick A.; and Moutel, Grégoire, eds.

Plan B 2.0: Rescuing a Planet Under Stress and a Civilization in Trouble. *Section III:* 16.1 —Brown, Lester R.

Poetics of DNA, The. *Section III:* 15.1 —Roof, Judith.

Policy & Politics in Nursing and Health Care. *Section III:* 9.1 —Mason, Diana J.; Leavitt, Judith K.; and Chaffee, Mary W.

Political Economy of Health Care: A Clinical Perspective, The. *Section III:* 9.3.1 —Hart, Julian Tudor.

Political Liberalism. *Section III:* 1.3.5 —Rawls, John.

Politicizing Science: The Alchemy of Policymaking. *Section III:* 5.3 —Gough, Michael, ed.

Politics of AIDS in Africa, The. *Section III:* 9.5.6 —Patterson, Amy S.

Politics of Life Itself: Biomedicine, Power, and Subjectivity in the Twenty-First Century, The. *Section III:* 2.1 —Rose, Nikolas.

Politics of Personalised Medicine: Pharmacogenetics in the Clinic, The. *Section III:* 15.1 —Hedgecoe, Adam.

Polluted Promises: Environmental Racism and the Search for Justice in a Southern Town. *Section III:* 16.1 —Checker, Melissa.

Poor People's Medicine: Medicaid and American Charity Care Since 1965. *Section III:* 9.5.10 —Engel, Jonathan.

Popular Eugenics: National Efficiency and American Mass Culture in the 1930s. *Section III:* 15.5 —Currell, Susan and Cogdell, Christina, eds.

Population Crises and Population Cycles. *Section III:* 13.2 —Russell, Claire and Russell, W.M.S.

Pour une bioéthique clinique: Médicalisation de la société, questionnement éthique et pratiques de soins. *Section III:* 2.1 —Boitte, P.; Cadoré, B.; Jacquemin, D.; and Zorrilla, S.

Power of Pills: Social, Ethical and Legal Issues in Drug Development, Marketing, and Pricing, The. *Section III:* 9.7 —Cohen, Jillian Clare; Illingworth, Patricia; and Schüklenk, Udo, eds.

Power to the Patient: Selected Health Care Issues and Policy Solutions. *Section III:* 9.3.1 —Atlas, Scott W., ed.

Practical Ethics for General Practice. *Section III:* 2.1 —Rogers, Wendy A. and Braunack-Mayer, Annette J.

Practical Guide to Clinical Ethics Consulting: Expertise, Ethos, and Power, A. *Section III:* 2.1 —Meyers, Christopher.

Prädiktive Gentests: "Health Purposes" und Indikationsstellung als Kriterien der Anwendung. *Section III:* 15.3 —Fuchs, Michael; Lanzerath, Dirk; and Schmidt, Matthias C., eds.

Praeger Handbook of Black American Health: Policies and Issues Behind Disparities in Health. *Section III:* 9.5.4 —Livingston, Ivor Lensworth, ed.

Pregnancy and Power: A Short History of Reproductive Politics in America. *Section III:* 9.5.5 —Solinger, Rickie.

Prescribing By Numbers: Drugs and the Definition of Disease. *Section III:* 9.7 —Greene, Jeremy A.

Presumed Curable: An Illustrated Casebook of Victorian Psychiatric Patients in Bethlem Hospital. *Section III:* 17.1 —Gale, Colin and Howard, Robert.

Primary Care in the Driver's Seat: Organizational Reform in European Primary Care. *Section III:* 9.1 —Saltman, Richard; Rico, Ana; and Boerma, Wienke, eds.

Primary Care: More Poems by Physicians. *Section III:* 7.1 —Belli, Angela and Coulehan, Jack, eds.

Primates and Philosophers: How Morality Evolved. *Section III:* 1.1 —Waal, Frans de.

Principles and Practice of Clinical Research. *Section III:* 18.1 —Gallin, John I. and Ognibene, Frederick P., eds.

Principles of Medical Law. *Section III:* 2.1 —Grubb, Andrew and Laing, Judith, eds.

Priorities in Health. *Section III:* 9.1 —Jamison, Dean T.; et al., eds.

Priority Areas for National Action: Transforming Health Care Quality. *Section III:* 9.8 —Institute of Medicine (United States). Board on Health Care Services. Committee on Identifying Priority Areas for Quality Improvement.

Private Guns, Public Health. *Section III:* 9.1 —Hemenway, David.

Profession of Ophthalmology: Practice Management, Ethics, and Advocacy, The. *Section III:* 9.5.1 —Parke, David W., ed.

Professional Integrity: Thinking Ethically. *Section III:* 1.3.1 —Pritchard, Michael S.

ProLife Feminism: Yesterday and Today. *Section III:* 12.3 —Derr, Mary Krane; MacNair, Rachel; and Naranjo-Huebl, Linda, eds.

Promise and Politics of Stem Cell Research, The. *Section III:* 18.5.4 —Solo, Pam and Pressburg, Gail.

Psyche and Helix: Psychological Aspects of Genetic Counseling: Essays by Seymour Kessler, Ph.D. *Section III:* 15.2 —Resta, Robert G., ed.

Psychiatric Power: Lectures at the College de France, 1973-74. *Section III:* 17.1 —Foucault, Michel.

Psychiatric-Mental Health Nursing: Scope and Standards of Practice. *Section III:* 17.1 —American Psychiatric Nurses Association; International Society of Psychiatric-Mental Health Nurses; and American Nurses Association.

Psychology of Illness: In Sickness and In Health, The. *Section III:* 17.1 —Druss, Richard G.

Psychosocial Genetic Counseling. *Section III:* 15.2 —Weil, Jon.

Psychosocial Nursing Care Along the Cancer Continuum. *Section III:* 9.5.1 —Carroll-Johnson, Rose Mary; Gorman, Linda M.; and Bush, Nancy Jo, eds.

Psychotropic Drug Prescriber's Survival Guide: Ethical Mental Health Treatment in the Age of Big Pharma. *Section III:* 17.4 —Dubovsky, Steven L. and Dubovsky, Amelia N.

Public Expectations and Physicians' Responsibilities: Voices of Medical Humanities. *Section III:* 7.1 —Crellin, John K.

Public Health Ethics: Theory, Policy, and Practice. *Section III:* 9.1 —Bayer, Ronald; Gostin, Lawrence O.; Jennings, Bruce; and Steinbock, Bonnie, eds.

Public-Private Health Care State: Essays on the History of American Health Care Policy, The. *Section III:* 9.1 —Stevens, Rosemary.

Q

Qu'est-ce que la bioéthique? *Section III:* 2.1 —Hottois, G.

Questions of Competence: Culture, Classification and Intellectual Disability. *Section III:* 4.3 —Jenkins, Richard, ed.

Questions of Ethics in Counselling and Therapy. *Section III:* 17.1 —Jones, Caroline (contributing editor); Shillito-Clarke, Carol; Syme, Gabrielle; Hill, Derek; Casemore, Roger; and Murdin, Lesley.

R

Race and Racism in Modern Philosophy. *Section III:* 1.1 —Valls, Andrew, ed.

Racial and Socioeconomic Differences in Family Perceived Barriers to Satisfaction with End-of-Life Care: An Exploratory Study. *Section III:* 20.4.1 —Craig, Janet B.

Rage and Reconciliation: Inspiring a Health Care Revolution. *Section III:* 9.8 —Gutkind, Lee, ed.

Reading Leo Strauss: Politics, Philosophy, Judaism. *Section III:* 1.3.5 —Smith, Steven B.

Readings in Biomedical Ethics: A Canadian Focus. *Section III:* 2.1 —Kluge, Eike-Henner W., ed.

Readings in the Philosophy of Technology. *Section III:* 5.1 —Kaplan, David M., ed.

Reality of Research with Children and Young People, The. *Section III:* 18.5.2 —Lewis, Vicky; Kellett, Mary; Robinson, Chris; Fraser, Sandy; and Ding, Sharon, eds.

Reconceiving the Gene: Seymour Benzer's Adventures in Phage Genetics. *Section III:* 15.1 —Holmes, Frederic Lawrence, ed.

Redefining Health Care: Creating Value-Based Competition on Results. *Section III:* 9.3.1 —Porter, Michael E. and Teisberg, Elizabeth Olmsted.

Refuge Denied: The St. Louis Passengers and the Holocaust. *Section III:* 21.4 —Ogilvie, Sarah A. and Miller, Scott.

Regulating the Health Professions. *Section III:* 9.1 —Allsop, Judith and Saks, Mike, eds.

Reinsuring Health: Why More Middle-Class People Are Uninsured and What Government Can Do. *Section III:* 9.3.1 —Swartz, Katherine.

Reinventing Public Health: Policies and Practices for a Healthy Nation. *Section III:* 9.1 —Aday, Lu Ann, ed.

Religion and Sexual Health: Ethical, Theological, and Clinical Perspectives. *Section III:* 10 —Green, Ronald M., ed.

Remaking Life & Death: Toward an Anthropology of the Biosciences. *Section III:* 7.1 —Franklin, Sarah and Lock, Margaret, eds.

Renewing the Stuff of Life: Stem Cells, Ethics, and Public Policy. *Section III:* 18.5.4 —Cohen, Cynthia B.

Reproductive Health in India: History, Politics, Controversies. *Section III:* 14.1 —Hodges, Sarah, ed.

Reprogenetics: Law, Policy, and Ethical Issues. *Section III:* 15.1 —Knowles, Lori P. and Kaebnick, Gregory E., eds.

Rescuing Science from Politics: Regulation and the Distortion of Scientific Research. *Section III:* 1.3.9 —Wagner, Wendy and Steinzor, Rena, eds.

Research Administration and Management. *Section III:* 5.3 —Kulakowski, Elliott C. and Chronister, Lynne U.

Research Ethics for Social Scientists: Between Ethical Conduct and Regulatory Compliance. *Section III:* 1.3.9 —Israel, Mark and Hay, Iain.

Resolving Ethical Dilemmas: A Guide for Clinicians. *Section III:* 2.1 —Lo, Bernard.

Right, Wrong and Science: The Ethical Dimensions of the Techno-Scientific Enterprise. *Section III:* 5.1 —Agazzi, Evandro.

Rise and Fall of Modern Medicine, The. *Section III:* 9.1 —Le Fanu, James.

Risks of Medical Innovation: Risk Perception and Assessment in Historical Context, The. *Section III:* 5.2 —Schlich, Thomas and Tröhler, Ulrich, eds.

Running the Obstacle Course to Sexual and Reproductive Health: Lessons from Latin America. *Section III:* 14.1 —Shepard, Bonnie.

S

Sacred Cow, Mad Cow: A History of Food Fears. *Section III:* 22.3 —Ferrières, Madeleine.

Safety and Ethics in Healthcare: A Guide to Getting It Right. *Section III:* 9.8 —Runciman, Bill; Merry, Alan; and Walton, Merrilyn.

Sanctity of Human Life, The. *Section III:* 2.1 —Novak, David.

Science and Ethics: Can Science Help Us Make Wise Moral Judgments? *Section III:* 1.3.9 —Kurtz, Paul and Koepsell, David, ed.

Science, Culture and Society: Understanding Science in the Twenty-First Century. *Section III:* 5.1 —Erickson, Mark.

Science, Faith, and Ethics: Grid or Gridlock? *Section III:* 5.1 —Alexander, Denis and White, Robert S.

Science Has No Sex: The Life of Marie Zakrzewska, M.D. *Section III:* 10 —Tuchman, Arleen Marcia.

Science of Morality: The Individual, Community and Future Generations, The. *Section III:* 1.1 —Daleiden, Joseph L.

Science on Stage: From Doctor Faustus to Copenhagen. *Section III:* 5.1 —Shepherd-Barr, Kirsten.

Science, Religion, and Society: An Encyclopedia of History, Culture, and Controversy. *Section III:* 5.1 —Eisen, Arri and Laderman, Gary, eds.

Science, Society, and the Supermarket: The Opportunities and Challenges of Nutrigenomics. *Section III:* 15.1 —Castle, David; Cline, Cheryl; Daar, Abdallah S.; Tsamis, Charoula; and Singer, Peter.

Science Sold Out: Does HIV Really Cause AIDS? *Section III:* 9.5.6 —Culshaw, Rebecca.

Science, Values, and Objectivity. *Section III:* 5.1 —Machamer, Peter and Wolters, Gereon, eds.

Scopes "Monkey Trial", The. *Section III:* 3.2 —Johnson, Anne Janette.

Screening: Evidence and Practice. *Section III:* 9.1 —Raffle, Angela E. and Gray, J.A. Muir.

Search for Major Plagge: The Nazi Who Saved Jews, The. *Section III:* 21.4 —Good, Michael.

Second Thoughts: Critical Thinking from a Multicultural Perspective. *Section III:* 1.1 —Teays, Wanda.

Secrets of Women: Gender, Generation, and the Origins of Human Dissection. *Section III:* 20.1 —Park, Katharine.

Seesaw Syndrome, The. *Section III:* 9.7 —Madden, Michael.

Selbstbestimmungsrecht und Einwilligungsfähigkeit: der Abbruch der künstlichen Ernährung bei Patienten im vegetative state in rechtsvergleichender Sicht: der Kemptener Fall und die Verfahren Cruzan und Bland. *Section III:* 20.5.1 —Tolmein, Oliver.

Selfish Gene, The. *Section III:* 15.9 —Dawkins, Richard.

Sentenced to Science: One Black Man's Story of Imprisonment in America. *Section III:* 18.5.1 —Hornblum, Allen M.

Sephardic and Mizrahi Jewry: From the Golden Age of Spain to Modern Times. *Section III:* 1.2 —Zohar, Zion, ed.

Sex and Seclusion, Class and Custody: Perspectives on Gender and Class in the History of British and Irish Psychiatry. *Section III:* 17.1 —Andrews, Jonathan and Digby, Anne, eds.

Sex, Gender, and Science. *Section III:* 10 —Hird, Myra J.

Sex in Development: Science, Sexuality, and Morality in Global Perspective. *Section III:* 10 —Adams, Vincanne and Pigg, Stacy Leigh, eds.

Sexual Boundary Violations: Therapeutic, Supervisory, and Academic Contexts. *Section III:* 7.4 —Celenza, Andrea.

Sharing the Planet: Population, Consumption, Species: Science and Ethics for a Sustainable and Equitable World. *Section III:* 16.1 —Zwaan, B. van der and Petersen, Arthur, eds.

Shattered Nerves: How Science Is Solving Modern Medicine's Most Perplexing Problem. *Section III:* 5.1 —Chase, Victor D.

Shock: The Healing Power of Electroconvulsive Therapy. *Section III:* 17.5 —Dukakis, Kitty and Tye, Larry.

Should Parents Be Licensed? Debating the Issues. *Section III:* 14.1 —Tittle, Peg, ed.

Sick: The Untold Story of America's Health Care Crisis—and the People Who Pay the Price. *Section III:* 9.3.1 —Cohn, Jonathan.

Silicon Valley of Dreams: Environmental Injustice, Immigrant Workers, and the High-Tech Global Economy, The. *Section III:* 16.1 —Pellow, David Naguib and Park, Lisa Sun-Hee.

Sisters in Science: Conversations with Black Women Scientists on Race, Gender, and Their Passion for Science. *Section III:* 5.1 —Jordan, Diann.

Smell of Matches, The. *Section III:* 7.1 —Stone, John.

Sneaky Kid and Its Aftermath: Ethics and Intimacy in Fieldwork. *Section III:* 18.4 —Wolcott, Harry F.

Social History of Dying, A. *Section III:* 20.1 —Kellehear, Allan.

Social Inclusion of People with Mental Illness. *Section III:* 17.1 —Leff, Julian and Warner, Richard.

Social Inequalities in Health: New Evidence and Policy Implications. *Section III:* 7.1 —Siegrist, Johannes and Marmot, Michael, eds.

Social Injustice and Public Health. *Section III:* 7.1 —Levy, Barry S. and Sidel, Victor W., eds.

Social Justice in Health Care: A Critical Appraisal. *Section III:* 1.1 —Fan, Ruiping.

Social Medicine Reader, Volume 1: Patients, Doctors, and Illness, The. *Section III:* 7.1 —King, Nancy M.P.; Strauss, Ronald P.; Churchill, Larry R.; Estroff, Sue E.; Henderson, Gail E.; and Oberlander, Jonathan, eds.

Social Medicine Reader, Volume 2: Social and Cultural Contributions to Health, Difference, and Inequality, The. *Section III:* 7.1 —Henderson, Gail E.; Estroff, Sue E.; Churchill, Larry R.; King, Nancy M.P.; Oberlander, Jonathan; and Strauss, Ronald P., eds.

Social Medicine Reader, Volume 3: Health Policy, Markets, and Medicine, The. *Section III:* 7.1 —Oberlander, Jonathan; Churchill, Larry R.; Estroff, Sue E.; Hendersen, Gail E.; King, Nancy M.P.; and Strauss, Ronald P., eds.

Solving the Health Care Problem: How Other Nations Succeeded and Why the United States Has Not. *Section III:* 9.1 —Behan, Pamela.

Soul of a Doctor: Harvard Medical Students Face Life and Death, The. *Section III:* 7.2 —Pories, Susan; Jain, Sachin H.; and Harper, Gordon, eds.

Speak Your Truth: Proven Strategies for Effective Nurse-Physician Communication. *Section III:* 7.3 —Bartholomew, Kathleen.

Speciesism. *Section III:* 22.1 —Dunayer, Joan.

Spinoza's Ethics: An Introduction. *Section III:* 1.1 —Nadler, Steven.

Spirituality and Health: Multidisciplinary Explorations. *Section III:* 9.1 —Meier, Augustine; O'Connor, Thomas St. James; and VanKatwyk, Peter, eds.

Sports Ethics: An Anthology. *Section III:* 1.3.1 —Boxill, Jan, ed.

Stem Cell Wars: Inside Stories from the Frontlines. *Section III:* 18.5.4 —Herold, Eve.

Stem Cells. *Section III:* 18.5.4 —Kelly, Evelyn B.

Street Science: Community Knowledge and Environmental Health Justice. *Section III:* 16.1 —Corburn, Jason.

Strongest Boy in the World: How Genetic Information Is Reshaping Our Lives, The. *Section III:* 15.1 —Reilly, Philip R.

Stuck with Virtue: The American Individual and Our Biotechnological Future. *Section III:* 15.1 —Lawler, Peter Augustine.

Student Case Law Resource Guide to Accompany Legal Aspects of Health Care Administration. *Section III:* 9.1 —Pozgar, George D. and Santucci, Nina M.

Student's Guide to Research Ethics, The. *Section III:* 1.3.9 —Oliver, Paul.

Studying the Jew: Scholarly Antisemitism in Nazi Germany. *Section III:* 1.3.3 —Steinweis, Alan E.

Suicide. *Section III:* 20.7 —Long, Robert Emmet, ed.

Suite Française. *Section III:* 21.4 —Némirovsky, Irène.

Surgical Consent: Bioethics and Cochlear Implantation. *Section III:* 9.5.1 —Komesaroff, Linda, ed.

Surviving the Century: Facing Climate Chaos and Other Global Challenges. *Section III:* 16.1 —Girardet, Herbert, ed.

T

Taking Action, Saving Lives: Our Duties to Protect Environmental and Public Health. *Section III:* 16.1 —Shrader-Frechette, Kristin.

Taking Biology Seriously: What Biology Can and Cannot Tell Us About Moral and Public Policy Issues. *Section III:* 15.1 —Melo-Martín, Immaculada de.

Taking Sides: Clashing Views on Controversial Bioethical Issues. *Section III:* 2.1 —Levine, Carol, ed.

Talking About Spirituality in Health Care Practice: A Resource for the Multi-Professional Health Care Team. *Section III:* 4.1.1 —White, Gillian.

Talmud, the Steinsaltz Edition, Volume 6: Tractate Bava Metzia, Part VI, The. *Section III:* 1.2 —Steinsaltz, Adin.

Talmud, the Steinsaltz Edition, Volume 19: Tractate Sanhedrin, Part V, The. *Section III:* 1.2 —Steinsaltz, Adin.

Tanner Lectures on Human Values, Volume 26. *Section III:* 1.3.1 —Peterson, Grethe B., ed.

Teaching Resources for End of Life and Palliative Care Courses. *Section III:* 20.5.1 —Csikai, Ellen L. and Jones, Barbara, ed.

10 Good Questions About Life and Death. *Section III:* 4.4 —Belshaw, Christopher.

Ten Perfect Fingers. *Section III:* 20.5.2 —Winner, Brenda.

Testing Treatments: Better Research for Better Healthcare. *Section III:* 18.1 —Evans, Imogen; Thornton, Hazel; and Chalmers, Iain.

Theology of the Body for Beginners. *Section III:* 4.4 —West, Christopher and Pope John Paul.

Thinking Like a Woman: Personal Life and Political Ideas. *Section III:* 10 —Overall, Christine.

Thomistic Principles and Bioethics. *Section III:* 2.1 —Eberl, Jason T.

Threat of Pandemic Influenza: Are We Ready? Workshop Summary. *Section III:* 9.1 —Institute of Medicine (United States). Board on Global Health. Forum on Microbial Threats.

Time for Listening and Caring: Spirituality and the Care of the Chronically Ill and Dying, A. *Section III:* 20.4.1 —Puchalski, Christina M.

Times of Triumph, Times of Doubt: Science and the Battle for Public Trust. *Section III:* 5.1 —Carlson, Elof Axel.

To Heal a Fractured World: The Ethics of Responsibility. *Section III:* 1.2 —Sacks, Jonathan.

To the Marrow. *Section III:* 19.1 —Seder, Robert.

Too Late to Die Young: Nearly True Tales from a Life. *Section III:* 9.5.1 —Johnson, Harriet McBryde.

Torture Debate in America, The. *Section III:* 21.4 —Greenberg, Karen J., ed.

Torture: When the Unthinkable Is Morally Permissible. *Section III:* 21.4 —Bagaric, Mirko and Clarke, Julie.

Toxic Drift: Pesticides and Health in the Post-World War II South. *Section III:* 1.3.11 —Daniel, Pete.

Toxic Torts: Science, Law, and the Possibility of Justice. *Section III:* 5.1 —Cranor, Carl F.

Transboundary Harm in International Law: Lessons from the Trail Smelter Arbitration. *Section III:* 16.1 —Bratspies, Rebecca M. and Miller, Russell A., eds.

Transparent Body: A Cultural Analysis of Medical Imaging, The. *Section III:* 7.1 —van Dijck, José.

Treating Sex Offenders: An Introduction to Sex Offender Treatment Programmes. *Section III:* 10 —Brown, Sarah.

Tree of Life: A Trilogy of Life in the Lodz Ghetto, Book One: On the Brink of the Precipice, 1939, The. *Section III:* 21.4 —Rosenfarb, Chava.

Tree of Life: A Trilogy of Life in the Lodz Ghetto: Book Two: From the Depths I Call You, 1940-1942, The. *Section III:* 21.4 —Rosenfarb, Chava.

Tree of Life: A Trilogy of Life in the Lodz Ghetto: Book Three: The Cattle Cars Are Waiting, 1942-1944, The. *Section III:* 21.4 —Rosenfarb, Chava.

Trust Is Not Enough: Bringing Human Rights to Medicine. *Section III:* 2.1 —Rothman, David J. and Rothman, Sheila M.

Trust Prescription for Healthcare: Building Your Reputation with Consumers, The. *Section III:* 9.1 —Shore, David A., ed.

Truth About Health Care: Why Reform Is Not Working in America, The. *Section III:* 9.1 —Mechanic, David.

Truth About Leo Strauss: Political Philosophy and American Democracy, The. *Section III:* 1.3.5 —Zuckert, Catherine and Zuckert, Michael.

U

Understanding Assisted Suicide: Nine Issues to Consider. *Section III:* 20.7 —Mitchell, John B.

Understanding Racial Health Care Disparities: The Role of Consumer Empowerment, Consumer Expectations and Negative Heatlh Care Experiences. *Section III:* 9.5.4 —Gary, Lisa Ché.

Understanding Virtue Ethics. *Section III:* 1.1 —Van Hooft, Stan.

Unfolding Tradition: Jewish Law After Sinai, The. *Section III:* 1.2 —Dorff, Elliot N.

Unholy Madness: The Church's Surrender to Psychiatry. *Section III:* 17.1 —Farber, Seth.

Unnatural History: Breast Cancer and American Society. *Section III:* 7.1 —Aronowitz, Robert A.

Unwilling Mothers, Unwanted Babies: Infanticide in Canada. *Section III:* 20.5.2 —Kramar, Kirsten Johnson.

Using and Abusing the Holocaust. *Section III:* 21.4 —Langer, Lawrence L.

Using Cost-Effectiveness Analysis to Improve Health Care: Opportunities and Barriers. *Section III:* 9.3.1 —Neumann, Peter J.

V

Vaccine: The Controversial Story of Medicine's Greatest Lifesaver. *Section III:* 9.7 —Allen, Arthur.

Values & Ethics in Social Work: An Introduction. *Section III:* 1.3.10 —Beckett, Chris and Maynard, Andrew.

Verfolgung und Widerstand im Rheinland und in Westfalen, 1933-1945. *Section III:* 21.4 —Faust, Anselm, ed. and Zimmermann, Michael.

View from Here: Bioethics and the Social Sciences, The. *Section III:* 2.1 —De Vries, Raymond G.; Turner, Leigh; Orfali, Kristina; and Bosk, Charles, eds.

Voice of Breast Cancer in Medicine and Bioethics, The. *Section III:* 2.1 —Rawlinson, Mary C. and Lundeen, Shannon, eds.

W

War of Nerves: Chemical Warfare from World War I to Al-Qaeda. *Section III:* 21.3 —Tucker, Jonathan B.

Warning Shot: Influenza and the 2004 Flu Vaccine Shortage, A. *Section III:* 9.7 —Brookes, Tim.

Welcome to Biotech Nation: My Unexpected Odyssey into the Land of Small Molecules, Lean Genes, and Big Ideas. *Section III:* 15.1 —Gunn, Moira A.

Wert menschlichen Lebens: Medizinische Ethik bei Dietrich Bonhoeffer und Karl Bonhoeffer, Der. *Section III:* 20.5.1 —Gestrich, Christof and Neugebauer, Johannes, eds.

What is Mental Retardation? Ideas for an Evolving Disability in the 21st Century. *Section III:* 9.5.3 —Switzky, Harvey N. and Greenspan, Stephen, eds.

What the Doctor Didn't Say: The Hidden Truth About Medical Research. *Section III:* 18.3 —Menikoff, Jerry and Richards, Edward P.

What Therapists Don't Talk About and Why: Understanding Taboos That Hurt Us and Our Clients. *Section III:* 17.2 —Pope, Kenneth S.; Sonne, Janet L.; and Greene, Beverly.

What's the Patient's Code Status? *Section III:* 20.5.4 —Mirarchi, Ferdinando L.

When a Child Dies: How Pediatric Physicians and Nurses Cope. *Section III:* 20.4.2 —McKelvey, Robert S.

When Illness Goes Public: Celebrity Patients and How We Look at Medicine. *Section III:* 9.5.1 —Lerner, Barron H.

When IVF Fails: Feminism, Infertility, and the Negotiation of Normality. *Section III:* 14.4 —Throsby, Karen.

When Sickness Heals: The Place of Religious Belief in Healthcare. *Section III:* 1.2 —Sorajjakool, Siroj.

Where We Stand: A Surprising Look at the Real State of Our Planet. *Section III:* 16.1 —Garte, Seymour.

White on Black. *Section III:* 9.5.7 —Gallego, Ruben.

Who Am I? Experiences of Donor Conception. *Section III:* 14.2 —McWhinnie, Alexina M.

Who Killed Health Care? America's $2 Trillion Medical Problem—and the Consumer-Driven Cure. *Section III:* 9.3.1 —Herzlinger, Regina.

Who Should We Treat? Rights, Rationing, and Resources in the NHS. *Section III:* 9.4 —Newdick, Christopher.

Wissen, Handeln, Ethik: Strukturen ärztlichen Handelns und ihre ethische Relevanz. *Section III:* 4.1.2 —Toellner, Richard and Wiesing, Urban, eds.

Women and Gender in Jewish Philosophy. *Section III:* 1.2 —Tirosh-Samuelson, Hava, ed.

Women's Health in Post-Soviet Russia: The Politics of Intervention. *Section III:* 9.5.5 —Rivkin-Fish, Michele.

Women's Rights and Bioethics. *Section III:* 9.5.5 —Dennerstein, Lorraine, ed.

Working Virtue: Virtue Ethics and Contemporary Moral Problems. *Section III:* 1.1 —Walker, Rebecca L. and Ivanhoe, Philip J., eds.

World as Laboratory: Experiments with Mice, Mazes, and Men. *Section III:* 17.3 —Lemov, Rebecca.

Worth and Welfare in the Controversy over Abortion. *Section III:* 12.3 —Coope, Christopher Miles.

Writing the Siege of Leningrad: Women's Diaries, Memoirs and Documentary Prose. *Section III:* 21.4 —Simmons, Cynthia and Perlina, Nina.

Y

Years of Extermination: Nazi Germany and the Jews, 1939-1945, The. *Section III:* 21.4 —Friedländer, Saul.

BIBLIOGRAPHY OF BIOETHICS
SUBJECT HEADING KEY FOR SECTION II

Section II lists the primary classification number for each document. To find the full citation in Section I, or to find documents on the same subject, refer to the list below, which indicates all possible subject headings for each primary classification number appearing in Section II.

1.3.1	Professional Ethics
1.3.7	Journalism and Publishing
1.3.9	Biomedical Research/ Research Ethics and Scientific Misconduct
1.3.12	Telemedicine and Informatics
2.1	Bioethics and Medical Ethics, *or its subdivisions:* /Legal Aspects, /Philosophical Aspects, *or* /Religious Aspects
2.2	Bioethics and Medical Ethics/ History
2.3	Bioethics and Medical Ethics/ Education
2.4	Bioethics and Medical Ethics/ Commissions
4.1.1	Professional Ethics
4.1.2	Philosophy of Medicine
4.1.3	Nursing Ethics and Philosophy
4.2	Health, Concept of
4.3	Mental Health, Concept of
4.4	Quality and Value of Life
4.5	Enhancement
5.1, 5.2	Biomedical Research
5.3	Biomedical Research/ Social Control of Science and Technology
5.4	Nanotechnology
6	Codes of Ethics
7.1	Sociology of Medicine
7.2	Medical Education
7.3	Professional Professional Relationship
7.4	Malpractice and Professional Misconduct
8.1	Patient Relationships
8.2	Truth Disclosure
8.3.1	Informed Consent
8.3.2	Informed Consent/ Minors
8.3.3	Informed Consent/ Incompetents
8.3.4	Treatment Refusal
8.4	Confidentiality
8.5	Malpractice and Professional Misconduct
9.1	Health Care *or* Public Health
9.2	Right to Health Care
9.3.1	Health Care/ Health Care Economics
9.3.2	Health Care/ Health Care Economics/ Managed Care Programs
9.4	Resource Allocation
9.5.1	Care for Specific Groups
9.5.2	Care for Specific Groups/ Aged
9.5.3	Care for Specific Groups/ Mentally Disabled
9.5.4	Care for Specific Groups/ Minorities
9.5.5	Care for Specific Groups/ Women
9.5.6	AIDS *or its subdivisions:* /Confidentiality, /Human Experimentation, *or* /Legal Aspects
9.5.7	Care for Specific Groups/ Minors
9.5.8	Care for Specific Groups/ Fetuses
9.5.9	Care for Specific Groups/ Substance Abusers
9.5.10	Care for Specific Groups/ Indigents
9.6	Ethicists and Ethics Committees
9.7	Drug Industry
9.8	Health Care/ Health Care Quality
10	Sexuality
11.1, 11.2, 11.4	Contraception
12.1	Abortion
12.3	Abortion/ Moral and Religious Aspects
12.4.1, 12.4.2, 12.4.3, 12.4.4	Abortion/ Legal Aspects
12.5.1, 12.5.2, 12.5.3	Abortion/ Social Aspects
14.1	Reproductive Technologies
14.2	Artificial Insemination and Surrogate Mothers
14.3	Sex Determination
14.4	In Vitro Fertilization
14.5	Cloning *or its subdivision:* Cloning /Legal Aspects
14.6	Cryobanking of Sperm, Ova, and Embryos